Strength of Materials

Materials

(Mechanics of Solids)

in
SI Units

[A Textbook for Students of B.E./B.Tech./B.Arch., AMIE,
UPSC (Engineering Services) and Other Engineering Examinations]

Er. R.K. Rajput

*M.E. (Hons.) Gold Medalist; Grad.-Mech. Engg. & Elect. Engg.; MIE (India);
MSESI; MISTE; CE (India)*

Recipient of :
"Best Teacher (Academic) Award"
"Distinguished Author Award"
"Jawahar Lal Nehru Memorial Gold Medal"
for an Outstanding Research Paper
(Institution of Engineers—India)
Principal (Formerly)
Punjab College of Information Technology
PATIALA

S. CHAND & COMPANY LTD.

(AN ISO 9001 : 2000 COMPANY)

RAM NAGAR, NEW DELHI - 110 055

S. CHAND & COMPANY LTD.
(An ISO 9001 : 2000 Company)

Head Office: 7361, RAM NAGAR, NEW DELHI - 110 055
Phone: 23672080-81-82, 9899107446, 9911310888 Fax 91-11-23677446
Shop at: **schandgroup.com**; e-mail: **info@schandgroup.com**

Branches :

Ahmedabad : 1st Floor, Heritage, Near Gujarat Vidhyapeeth, Ashram Rod, **Ahmedabad** - 380 014, Ph: 27541965, 27542369, ahmedabad@schandgroup.com

Bengaluru : No. 6, Ahuja Chambers, 1st Cross, Kumara Krupa Road, **Bengaluru** - 560 001, Ph: 22268048, 22354008, bangalore@schandgroup.com

Bhopal : Bajaj Tower, Plot No. 243, Lala Lajpat Rai Colony, Raiser Road, **Bhopal** - 462 011, Ph: 4274723. bhopal@schandgroup.com

Chandigarh : S.C.O. 2419-20, First Floor, Sector - 22-C (Near Aroma Hotel), **Chandigarh** -160 022, Ph: 2725443, 2725446, chandigarh@schandgroup.com

Chennai : 152, Anna Salai, **Chennai** - 600 002, Ph: 28460026, 2846027, chennai@schandgroup.com

Coimbatore : No. 5, 30 Feet Road, Krishnasamy Nagar, Ramanathapuram, **Coimbatore** -641045, Ph: 0422-2323620 coimbatore@schandgroup.com **(Markting Office)**

Cuttack : 1st Floor, Bhartia Tower, Badambadi, **Cuttack** -753 009, Ph: 233258; 2332581, cuttack@schandgroup.com

Dehradun : 1st Floor, 20, New Road, Near Dwarka Store, **Dehradun** 248 001, Ph: 2711101, 2710861, dehradun@schandgroup.com

Guwahati : Pan Bazar, **Guwahati** - 781 001, Ph: 2738811, 2735640 gwahati@schandgroup.com

Hyderabad : Padma Plaza, H.No. 3-4-630, Opp. Ratna College, Narayaaguda, **Hyderabad** - 500029, Ph: 24651135, 24744815, hyderabad@schandgroup.com

Jaipur : A-14, Janta Store Shopping Complex, University Marg, Bpu Nagar, **Jaipur** - 302 05, Ph: 2719126, jaipur@schandgroup.com

Jalandhar : Mai Hiran Gate, **Jalandhar** - 144 008, Ph: 2401630, 500060, jalandhar@schandgroup.com

Jammu : 67/B, B-Block, Gandhi Nagar, **Jammu** - 180 004, (M) 09886651464 **(Marketing Ofce)**

Kochi : Kachapilly Square, Mullassery Canal Road, Ernakulam, Kochi - 682 011, Ph: 237807, cochin@schandgroup.com

Kolkata : 285/J, Bipin Bihari Ganguli Street, **Kolkata** - 700 012, Ph: 2367459, 22373914, kolkata@schandgroup.com

Lucknow : Mahabeer Market, 25 Gwynne Road, Aminabad, **Lucknow** -226 018, Ph: 2626801,2284815, lucknow@schandgroup.com

Mumbai : Blackie House, 103/5, Walchand Hirachand Marg, Opp. GP.O., **Mumbai** - 400 001, Ph: 22690881, 22610885, mumbai@schandgroup.com

Nagpur : Karnal Bag, Model Mill Chowk, Umrer Road, **Nagpur** - 440032, Ph: 2723901, 2777666, nagpur@schandgroup.com

Patna : 104, Citicentre Ashok, Govind Mitra Road, **Patna** - 800 004, Ph: 2300489, 2302100, patna@schandgroup.com

Pune : 291/1, Ganesh Gayatri Complex, 1st Floor, Somwarpeth, Near Jain Mandir, **Pune** - 411 011, Ph: 64017298, pune@schandgroup.com **(Marketing Office)**

Raipur : Kailash Residency, Plot No. 4B, Bottle House Road, Shankar Nagar, **Raipur** - 492007, Ph: 09981200834, raipur@schandgroup.com **(Marketing Office)**

Ranchi : Flat No. 104, Sri Draupadi Smriti Apartments, East of Jaipal Singh Stadium, Neel Ratan Street, Upper Bazar, **Ranchi** - 834 001, Ph: 2208761, ranchi@schandgroup.com **(Marketing Office)**

siliguri : 122, Raja Ram Mohan Roy Road, East Vivekanandapally, P.O., **Siliguri**-734001 Dist., Jalpaiguri, (W.B.) Ph: 0353-2520750 **(Marketing Office)**

Visakhapatnam : Plot No. 7, 1st Floor, Allipuram Extension, Opp. Radhakrishna Towers, Seethanmadhara North Extn., **Visakhapatnam** - 530 013, (M) 09347580841, visakhapatnam@schandgroup.com **(Marketing Office)**

First Edition 1998
Subsequent Editions and Reprints 1999, 2001, 2003 (Twice), 2004
First Multicolour Illustrative Revised Edition 2006
Reprints 2007 (Twice), 2008, 2009
Fifth Revised Edition 2010; Reprint 2011
Reprint with Correction 2012

ISBN : 81-219-2594-0   **Code** : 10A 310

PRINTED IN INDIA

By Rajendra Ravindra Printers Pvt. Ltd., 7361, Ram Nagar, New Delhi -110 055 and published by S. Chand & Company Ltd., 7361, Ram Nagar, New Delhi -110 055.

PREFACE TO THE FIFTH EDITION

I am pleased to present the **fifth edition** of this book. The warm reception, which the previous editions and reprints of this book have enjoyed all over India and abroad, has been a matter of great satisfaction to me.

Besides revising the whole book thoroughly a new chapter **(Chapter No. 23) : "UNIVERSITIES' QUESTIONS (*Recent*) – with SOLUTIONS"** has been added to make the book still more a useful unit.

Any suggestions for improvement of the book will be warmly received and incorporated in the next edition.

Er. R. K. Rajput
(Author)

PREFACE TO THE FIRST EDITION

The primary purpose of writing this book is to make available to the student community, a book which deals with the various topics in the subject of *Strength of Materials* exhaustively. I have taken special care to present the subject-matter in a lucid, direct and easily understandable style. A large number of worked out simple, moderate and difficult problems are arranged in a systematic manner to enable the students to grasp the subject effectively, from examination point of view.

The book comprises of 21 chapters (including advance topics) covering the syllabi in the subject of "Strength of Materials" of all the Indian Universities and Competitive Examinations as well. The other important feature of the book is that it contains *Experiments* at the end of the chapters to enable the students to have an access to the practical aspects of the subject.

Besides the above features, the book contains *Typical Examples* (useful for students appearing in competitive examinations in particular and other students in general), *Highlights*, *Objective Type Questions* and a large number of *Unsolved Examples*. All these features go to make this treatise a complete and comprehensive book on the subject. It can be candidly said that this book will prove a boon to the students appearing for Engineering Undergraduate Examinations as well as for Competitive Examinations.

The author's thanks are due to his wife Ramesh Rajput without whose cooperation and encouragement this book would have never been materialised.

In the end the author wishes to express his gratitude to Shri Ravindra Kumar Gupta, Managing Director, S. Chand & Company Ltd., New Delhi for taking pains in bringing out this book in a short span of time.

Although every care has been taken to make the book free of errors, yet the author shall be obliged, if errors present are brought to his notice.

Constructive criticism will be warmly received.

(Author)

Dedicated
to
My Wife
Ramesh Rajput

NOMENCLATURE

A = Area of cross-section, m^2

a = Rankine constant

B, b = Width, m

C = Modulus of rigidity, GN/m^2

D, d = Depth, diameter, m

E = Modulus of elasticity, GN/m^2

e = Linear strain, eccentricity (m)

e_v = Volumetric strain

F_c = Ventrifugal force, N

G = Centroid of area/lamina

g = Acceleration due to gravity, 9.81 m/s^2

h = Height, m

I = Second moment of area, m^4

I_p = Polar moment of inertia, m^4

I_{xy} = Product moment of area, m^4

K = Bulk modulus of elasticity, GN/m^2; radius of gyration, m

k = Stiffness of spring (N/mm)

l = Length, m

M = Bending moment, Nm

m = Mass, kg

$\dfrac{1}{m}$ = Poisson's ratio

N = Number of leaves; speed, r.p.m.

n = Number of leaves; speed, r.p.m.

P = Force, N

p = Pressure, N/mm^2 (or MN/m^2)

R, r = Radius, m

T = Torque, Nm

V = Volume, m^3

W = Load, N

w = Load per unit length, N/m

x, y = Cartesian co-ordinates

y = Distance, deflection, m

Z = Section modulus

r, θ = Polar co-ordinates

α = Co-efficient of linear expansion, m/m°C

α, θ, β = Angle, rad.

ρ = Density, kg/m^3

ϕ = Shear strain

δ = Deflection, shrinkage allowance, m

ω = Angular velocity, rad/s

μ = Co-efficient of friction

σ = Normal stress, N/mm^2 (MN/m^2)

τ = Shear stress, N/mm^2 (MN/m^2)

σ_c = Circumferential (or hoop) stress, N/mm^2 (MN/m^2)

σ_l = Longitudinal stress, N/mm^2 (MN/m^2)

σ_r = Radical stress, N/mm^2 (MN/m^2)

$\sigma_1, \sigma_2, \sigma_3$ = Principal streses, N/mm^2 (MN/m^2)

CONTENTS

16. Columns and Struts

17. Analysis of Framed Structures

18. Theories of Failure

19. Stresses Due to Rotation

20. Bending of Curved Bars

21. Unsymmetrical Bending and Shear Centre

22. Competitive Examinations (*UPSC, GATE etc.*) Questions with Solutions

23. Universities Questions (*Recent*) with Solutions

Additional Objective Type Questions (Including Questions for ESE, CSE, GATE etc., exams.)

Material Testing–Experiments

Index

Simple Stresses and Strains

1.1. CLASSIFICATION OF LOADS

A load may be defined as the combined effect of external forces acting on a body. The loads may be classified as : (*i*) dead loads, (*ii*) live or fluctuating loads, (*iii*) inertia loads or forces, and (*iv*) centrifugal loads or forces.

The other way of classification is (*i*) tensile loads, (*ii*) compressive loads, (*iii*) torsional or twisting loads, (*iv*) bending loads, and (*v*) shearing loads.

The load may also be a 'point' (or concentrated) or 'distributed.'

Point load. A point load or concentrated load is one which is *considered to act at a point*. In actual practice, the load has to be distributed over a small area, because, such small knife-edge contacts are generally neither possible, nor desirable.

Distributed load. A distributed load is one which is *distributed or spread in some manner over the length of the beam*. If the spread is uniform, (*i.e.*, at the uniform rate, say *w* kN or N/metre run) it is said to be uniformly distributed load and is abbreviated as u.d.l. If the spread is not at uniform rate, it is said to be non-uniformly distributed load. *Triangulary* and *trapezoidally* distributed loads fall under this category.

1.2. STRESS

When a body is acted upon by some load or external force, it undergoes deformation (*i.e., change in shape or dimensions*) which increases gradually. During deformation, the material of the body resists the tendency of the load to deform the body, and when the load influence is taken over by the internal resistance of the material of the body, it becomes stable. This *internal resistance which the body offers to meet with the load is called* **stress.**

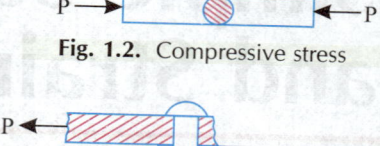

Fig. 1.1. Tensile stress

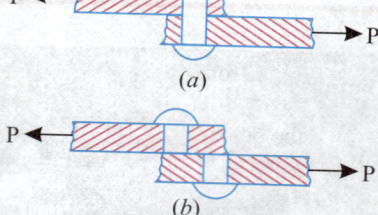

Fig. 1.2. Compressive stress

Stress can be considered either as total stress or unit stress. Total stress represents the total resistance to an external effect and is expressed in N, kN or MN. Unit stress represents the resistance developed by a unit area of cross-section, and is expressed in kN/m^2 or MN/m^2 or N/mm^2. For the remainder of this text, the word stress will be used to signify unit stress.

The various types of stresses may be classified as :

1. Simple or direct stress

    (*i*) Tension   (*ii*) Compression   (*iii*) Shear.

2. Indirect stress

    (*i*) Bending   (*ii*) Torsion.

3. Combined stress. Any possible combination of types 1 and 2.

This chapter deals with simple stresses only.

Fig. 1.3. (*a*) Rivet resisting shear.
(*b*) Rivet failure due to shear.

1.3. SIMPLE STRESS

Simple stress is often called *direct stress* because it develops under direct loading conditions. That is, simple tension and simple compression occur when the applied force, called load, is in line with the axis of the member (axial loading) (Figs. 1.1 and 1.2), and simple shear occurs, when equal,

The chain link at the bottom of the bulldozer spreads the weight and provides a strong grip.

parallel, and opposite forces tend to cause a surface to slide relative to the adjacent surface (Fig. 1.3).

In certain loading situations, the stresses that develop are not simple stresses. For example, referring to Fig. 1.4, the member is subjected to a load which is perpendicular to the axis of the member (transverse loading) (Fig. 1.5). This will cause the member to bend, resulting in deformation of the material and stresses being developed internally to resist the deformation. All three types of stresses—tension, compression and shear—will develop, but they will not be simple stresses, since they were not caused by direct loading.

Bronze nut.

When any type of simple stress σ (sigma) develops, we can calculate the magnitude of the stress by,

$$\sigma = \frac{P}{A} \qquad \qquad ... (1.1)$$

where,   σ = Stress, kN/m² or N/mm²,

P = Load $\begin{bmatrix} \text{external force causing} \\ \text{stress to develop} \end{bmatrix}$, kN or N, and

A = Area over which stress develops, m² or mm².

It may be noted that in cases of either simple tension or simple compression, the areas which resist the load are perpendicular to the direction of forces. When a member is subjected to simple shear, the resisting area is parallel to the direction of the force. Common situations causing shear stresses are shown in Figs. 1.3 and 1.4.

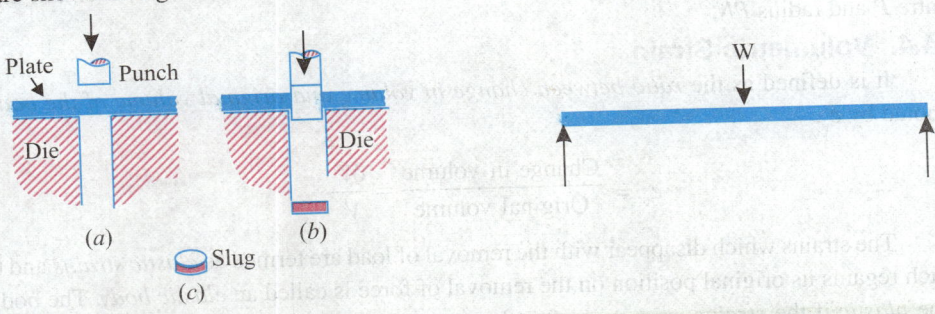

Fig. 1.4. (a) Punch approaching plate;
     (b) Punch shearing plate;
     (c) Slug showing sheared area.

Fig. 1.5. Simply supported beam.
           (Transverse loading)

1.4. STRAIN

Any element in a material subjected to stress is said to be strained. *The **strain** (e) is the deformation produced by stress.* The various types of strains are explained below :

1.4.1. Tensile Strain

A piece of material, with uniform cross-section, subjected to a uniform axial tensile stress, will increase its length from l to $(l + \delta l)$, (Fig. 1.6) and the increment of length δl is the actual deformation of the material. The fractional deformation or the tensile strain is given by

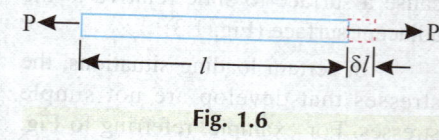

Fig. 1.6

$$e_t = \frac{\delta l}{l} \qquad ...(1.2)$$

1.4.2. Compressive Strain

Under compressive forces, a similar piece of material would be reduced in length (Fig. 1.7) from l to $(l - \delta l)$.

The fractional deformation again gives the strain e_c,

where,

$$e_c = \frac{\delta l}{l} \qquad ...(1.2a)$$

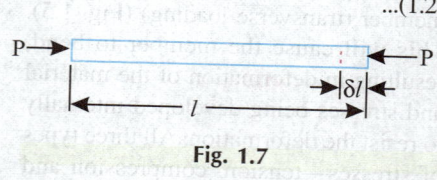

Fig. 1.7

1.4.3. Shear Strain

In case of a shearing load, a shear strain will be produced which is *measured by the angle through which the body distorts.*

In Fig. 1.8 is shown a rectangular block $LMNP$ fixed at one face and subjected to force F. After application of force, it distorts through an angle ϕ and occupies new position $LM'N'P$. The shear strain (e_s) is given by

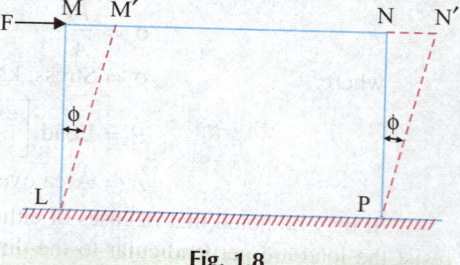

$$e_s = \frac{NN'}{NP} = \tan \phi$$

$$= \phi \text{ (radians) .....since}$$

ϕ is very small.

Fig. 1.8

The above result has been obtained by assuming NN' equal to arc (as NN' is small) drawn with centre P and radius PN.

1.4.4. Volumetric Strain

It is defined as the *ratio between change in volume and original volume of the body,* and is denoted by e_v.

$$\therefore \qquad e_v = \frac{\text{Change in volume}}{\text{Original volume}} = \frac{\delta V}{V} \qquad ...(1.3)$$

The strains which disappear with the removal of load are termed as *elastic strains* and the body which regains its original position on the removal of force is called an *elastic body.* The body is said to be *plastic* if the *strains exist even after the removal of external force.* There is always a limiting value of load up to which the strain totally disappears on the removal of load—the stress corresponding to this load is called *elastic limit.*

Robert Hooke discovered experimentally that within elastic limit, stress varies *directly* as strain

i.e., $\qquad$ Stress $\propto$ Strain

or, $\qquad \dfrac{\text{Stress}}{\text{Strain}} = $ a constant

This constant is termed as *Modulus of elasticity.*

(i) Young's modulus:

It is the *ratio between tensile stress and tensile strain or compressive stress and compressive strain*. It is denoted by *E*. It is the same as modulus of elasticity.

or, $E = \dfrac{\sigma}{e} \left[= \dfrac{\sigma_t}{e_t} \text{ or } \dfrac{\sigma_c}{e_c} \right]$

$$...(1.4)$$

(ii) Modulus of rigidity:

It is defined as the *ratio of shear stress* τ *(tau) to shear strain* and is denoted by *C*, *N* or *G*. It is also called *shear modulus of elasticity*.

Heavy duty punching machine.

or, $\dfrac{\tau}{e_s} = C, N \text{ or } G$ $...(1.5)$

(iii) Bulk or volume modulus of elasticity:

It may be defined as the *ratio of normal stress* (on each face of a solid cube) *to volumetric strain* and is denoted by the letter *K*.

or, $\dfrac{\sigma_n}{e_v} = K$ $...(1.6)$

Example 1.1. *A square steel rod 20 mm × 20 mm in section is to carry an axial load (compressive) of 100 kN. Calculate the shortening in a length of 50 mm. E = 2.14 × 10⁸ kN/m².*

Solution.

Area, $A = 0.02 \times 0.02 = 0.0004 \text{ m}^2$

Length, $l = 50 \text{ mm or } 0.05 \text{ m}$

Load, $P = 100 \text{ kN}$

$E = 2.14 \times 10^8 \text{ kN/m}^2$

Shortening of the rod δl:

Stress, $\sigma = \dfrac{P}{A}$

∴ $\sigma = \dfrac{100}{0.0004} = 250000 \text{ kN/m}^2$

Also, $E = \dfrac{\text{Stress}}{\text{Strain}}$

or, $\text{Strain} = \dfrac{\text{Stress}}{E} = \dfrac{250000}{2.14 \times 10^8}$

or, $\dfrac{\delta l}{l} = \dfrac{250000}{2.14 \times 10^8}$

∴ $\delta l = \dfrac{250000}{2.14 \times 10^8} \times l = \dfrac{250000}{2.14 \times 10^8} \times 0.05$

$= 0.0000584 \text{ m or } 0.0584 \text{ mm}$

Hence, the shortening of the rod = **0.0584 mm** **(Ans.)**

Example 1.2. *A hollow cast-iron cylinder 4 m long, 300 mm outer diameter, and thickness of metal 50 mm is subjected to a central load on the top when standing straight. The stress produced is 75000 kN/m². Assume Young's modulus for cast-iron as 1.5×10^8 kN/m² and find (i) magnitude of the load, (ii) longitudinal strain produced, and (iii) total decrease in length.*

Solution. Outer diameter, $\qquad D = 300\ \text{mm} = 0.3\ \text{m}$

Thickness, $\qquad\qquad\qquad t = 50\ \text{mm} = 0.05\ \text{m}$

Length, $\qquad\qquad\qquad l = 4\ \text{m}$

Stress produced, $\qquad\qquad \sigma = 75000\ \text{kN/m}^2$

$\qquad\qquad\qquad\qquad\qquad E = 1.5 \times 10^8\ \text{kN/m}^2$

Inner diameter of the cylinder, $d = D - 2t = 0.3 - 2 \times 0.05 = 0.02\ \text{m}$

(i) Magnitude of the load P :

Using the relation, $\qquad\qquad \sigma = \dfrac{P}{A}$

or, $\qquad P = \sigma \times A = 75000 \times \dfrac{\pi}{4}\,(D^2 - d^2) = 75000 \times \dfrac{\pi}{4}\,(0.3^2 - 0.2^2)$

or, $\qquad\qquad \mathbf{P = 2945.2\ kN}$ **(Ans.)**

(ii) Longitudinal strain produced, e :

Using the relation,

Strain, $\qquad\qquad\qquad e = \dfrac{\text{Stress}}{E} = \dfrac{75000}{1.5 \times 10^8}$

$\qquad\qquad\qquad\qquad = \mathbf{0.0005}$ **(Ans.)**

(iii) Total decrease in length, δl :

Using the relation,

$\qquad\qquad \text{Strain} = \dfrac{\text{Change in length}}{\text{Original length}} = \dfrac{\delta l}{l}$

$\qquad\qquad 0.0005 = \dfrac{\delta l}{4}$

$\qquad\qquad\quad \delta l = 0.0005 \times 4\ \text{m} = 0.002\ \text{m} = 2\ \text{mm}$

Hence, *decrease in length* = **2 mm** **(Ans.)**

Depending on the purpose and function, machine parts are made of different alloys and materials.

Example 1.3. *The following observations were made during a tensile test on a mild steel specimen 40 mm in diameter and 200 mm long.*

Elongation with 40 kN load (within limit of proportionality),

$$\delta l = 0.0304 \ mm$$

Yield load $= 161 \ kN$

Maximum load $= 242 \ kN$

Length of specimen at fracture

$$= 249 \ mm$$

Determine :

(*i*) *Young's modulus of elasticity,.*   (*ii*) *Yield point stress,*

(*iii*) *Ultimate stress, and*   (*iv*) *Percentage elongation.*

Solution.

(*i*) **Young's modulus of elasticity** E **:**

Stress, $$\sigma = \frac{P}{A} = \frac{40}{\frac{\pi}{4} \times (0.04)^2} = 3.18 \times 10^4 \ kN/m^2$$

Strain, $$e = \frac{\delta l}{l} = \frac{0.0304}{200} = 0.000152$$

∴ $$E = \frac{Stress}{Strain} = \frac{3.18 \times 10^4}{0.000152}$$

$$= \mathbf{2.09 \times 10^8 \ kN/m^2} \ \ \mathbf{(Ans.)}$$

(*ii*) **Yield point stress :**

Yield point stress $$= \frac{\text{Yield point load}}{\text{Area}}$$

To bear high compressive and variable stresses, automobile wheels are made of alloys.

$$= \frac{161}{\frac{\pi}{4} \times (0.04)^2} = \mathbf{12.8 \times 10^4 \ kN/m^2} \quad \textbf{(Ans.)}$$

(iii) Ultimate stress :

$$\text{Ultimate stress} = \frac{\text{Maximum load}}{\text{Area}}$$

$$= \frac{242}{\frac{\pi}{4} \times (0.04)^2} = \mathbf{19.2 \times 10^4 \ kN/m^2} \quad \textbf{(Ans.)}$$

(iv) Percentage elongation :

$$\text{Percentage elongation} = \frac{\text{Length of specimen at fracture} - \text{original length}}{\text{Original length}}$$

$$= \frac{249 - 200}{200} = 0.245 = \mathbf{24.5\%} \quad \textbf{(Ans.)}$$

Example 1.4. *A steel wire 2 m long and 3 mm in diameter is extended by 0.75 mm when a weight W is suspended from the wire. If the same weight is suspended from a brass wire, 2.5 m long and 2 mm in diameter, it is elongated by 4.64 mm. Determine the modulus of elasticity of brass if that of steel be 2.0×10^5 N/mm^2.*

(AMIE Summer, 2000)

Solution. *Given :* $l_s = 2$ m, $d_s = 3$ mm, $\delta l_s = 0.75$ mm, $E_s = 2.0 \times 10^5$ N/mm^2; $l_b = 2.5$ m, $d_b = 2$ mm, $\delta l_b = 4.64$ mm.

Modulus of elasticity of brass, E_b :

From Hooke's law, we know that

$$\delta l = \frac{Pl}{AE}$$

where, δl = Extension, l = Length, A = Cross-sectional area, and

E = Modulus of elasticity.

Parts made from sand casting using grey cast iron, ductile iron and copper alloys.

Case I : *For steel wire :*

$$\delta l_s = \frac{P l_s}{A_s E_s}$$

or,

$$0.75 = \frac{P \times (2 \times 1000)}{\left(\frac{\pi}{4} \times 3^2\right) \times 2.0 \times 10^5}$$

or,

$$P = 0.75 \times \left(\frac{\pi}{4} \times 3^2\right) \times 2.0 \times 10^5 \times \frac{1}{2000} \qquad \ldots(i)$$

Case II : *For brass wire:*

$$\delta l_b = \frac{P l_b}{A_b E_b}$$

$$4.64 = \frac{P \times (2.5 \times 1000)}{\left(\frac{\pi}{4} \times 2^2\right) \times E_b}$$

or,

$$P = 4.64 \times \left(\frac{\pi}{4} \times 2^2\right) \times E_b \times \frac{1}{2500} \qquad \ldots(ii)$$

Equating eqns. (*i*) and (*ii*), we get

$$0.75 \times \left(\frac{\pi}{4} \times 3^2\right) \times 2.0 \times 10^5 \times \frac{1}{2000} = 4.64 \times \left(\frac{\pi}{4} \times 2^2\right) \times E_b \times \frac{1}{2500}$$

or,

$$\boxed{E_b = 0.909 \quad 10^5 \text{ N/mm}^2 \text{ (Ans.)}}$$

Example 1.5. *A steel bar is 900 mm long; its two ends are 40 mm and 30 mm in diameter and the length of each rod is 200 mm. The middle portion of the bar is 15 mm in diameter and 500 mm long. If the bar is subjected to an axial tensile load of 15 kN, find its total extension.*

Take E = 200 GN/m² (G stands for giga and 1G = 10⁹).

Solution. Refer to Fig. 1.9.

Load, $P = 15$ kN

Area, $A_1 = \dfrac{\pi}{4}$ $40^2 = 1256.6$ mm² $= 0.001256$ m²

Area, $A_2 = \dfrac{\pi}{4}$ $15^2 = 176.7$ mm² $= 0.0001767$ m²

Area, $A_3 = \dfrac{\pi}{4}$ $30^2 = 706.8$ mm² $= 0.0007068$ m²

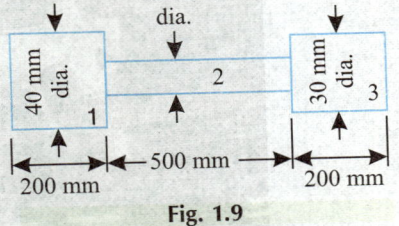

Fig. 1.9

Lengths : $l_1 = 200$ mm $= 0.2$ m, $l_2 = 500$ mm $= 0.5$ m and $l_3 = 200$ mm $= 0.2$ m

Total extension of the bar :

Let δl_1, δl_2 and δl_3 be the extensions in the parts 1, 2 and 3 of the steel bar respectively.

Then, $\delta l_1 = \dfrac{P l_1}{A_1 E}$, $\delta l_2 = \dfrac{P l_2}{A_2 E}$, $\delta l_3 = \dfrac{P l_3}{A_3 E}$ $\left[\because E = \dfrac{\sigma}{e} = \dfrac{P/A}{\delta l / l} = \dfrac{P \cdot l}{A \cdot \delta l} \text{ or } \delta l = \dfrac{Pl}{AE} \right]$

Total extension of the bar,

$$\delta l = \delta l_1 + \delta l_2 + \delta l_3$$

$$= \frac{P l_1}{A_1 E} + \frac{P l_2}{A_2 E} + \frac{P l_3}{A_3 E} = \frac{P}{E} \left[\frac{l_1}{A_1} + \frac{l_2}{A_2} + \frac{l_3}{A_3} \right]$$

$$= \frac{15 \times 10^3}{200 \times 10^9} \left[\frac{0.20}{0.001256} + \frac{0.50}{0.0001767} + \frac{0.20}{0.0007068} \right]$$

$$= 0.0002454 \text{ m} = 0.2454 \text{ mm}$$

Hence, total extension of the steel bar **= 0.2454 mm (Ans.)**

Example 1.6. *The bar shown in Fig. 1.10 is subjected to a tensile load of 50 kN. Find the diameter of the middle portion if the stress is limited to 130 MN/m². Find also the length of the middle portion if the total elongation of the bar is 0.15 mm. Take E = 200 GN/m².*

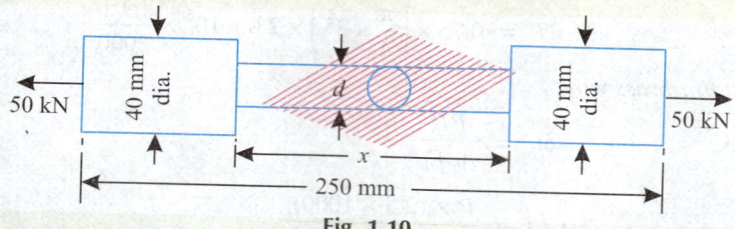

Fig. 1.10

Solution. Magnitude of tensile load, $P = 50$ kN

Stress in the middle portion, $\sigma = 130$ MN/m²

Total elongation of the bar, $\delta l = 0.15$ mm $= 0.15 \times 10^{-3}$ m

Modulus of elasticity, $E = 200$ GN/m²

Diameter of the middle portion, d :

Now, stress in the middle portion, $\sigma = \dfrac{P}{A} = \dfrac{50 \times 1000}{(\pi/4)\, d^2} = 130 \times 10^6$

∴ $d = \left[\dfrac{50 \times 1000}{(\pi/4) \times 130 \times 10^6} \right]^{1/2}$

$= 0.0221$ m or 22.1 mm

Hence, diameter of the middle portion **= 22.1 mm (Ans.)**

Marine crane under testing.

Length of the middle portion :

Let the length of the middle portion $= x$ metre

Stress in the end portions,
$$\sigma' = \frac{50 \times 1000}{\pi/4 \times \left(\dfrac{40}{1000}\right)^2}$$

$$= 39.79 \times 10^6 \text{ N/m}^2$$

$\therefore$ Elongation of the end portion $= \sigma' \times \dfrac{(0.25 - x)}{E}$

Also, elongation of the end portions + extension of the middle portion $= 0.15 \times 10^{-3}$

$$\frac{39.79 \times 10^6 \times (0.25 - x)}{200 \times 10^9} + \frac{130 \times 10^6 \times x}{200 \times 10^9} = 0.15 \times 10^{-3}$$

$$39.79 \times 10^6 \times (0.25 - x) + 130 \times 10^6 \times x = 200 \times 10^9 \times 0.15 \times 10^{-3}$$

Dividing both sides by 39.79×10^6, we get

$$0.25 - x + 3.267\, x = 0.754$$

$\therefore$
$$x = 0.222 \text{ m or } 222 \text{ mm}$$

Hence, length of the middle portion = **222 mm** **(Ans.)**

Example 1.7. *A steel tie rod 50 mm in diameter and 2.5 m long is subjected to a pull of 100 kN. To what length the rod should be bored centrally so that the total extension will increase by 15 per cent under the same pull, the bore being 25 mm diameter? Take E = 200 GN / m².*

Solution. Refer to Fig. 1.11 (a, b).

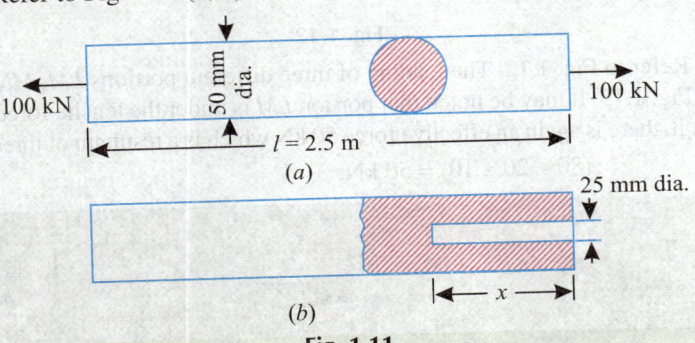

Fig. 1.11

Diameter of the steel tie rod $= 50 \text{ mm} = 0.05 \text{ m}$

Length of the steel rod, $l = 2.5 \text{ m}$

Magnitude of the pull, $P = 100 \text{ kN}$

Diameter of the bore $= 25 \text{ mm} = 0.025 \text{ m}$

Modulus of elasticity, $E = 200 \times 10^9 \text{ N/m}^2$

Length of the bore, x :

Stress in the solid rod, $\sigma = \dfrac{P}{A} = \dfrac{100 \times 1000}{(\pi/4) \times (0.05)^2} = 50.92 \times 10^6 \text{ N/m}^2$

Elongation of the solid rod, $\delta l = \dfrac{\sigma l}{E} = \dfrac{50.92 \times 10^6 \times 2.5}{200 \times 10^9}$

$$= 0.000636 \text{ m or } 0.636 \text{ mm}$$

Elongation after the rod is bored $= 1.15 \times 0.636 = 0.731 \text{ mm}$

Area at the reduced section $= \dfrac{\pi}{4}(0.05^2 - 0.025^2) = 0.001472 \text{ m}^2$

Stress in the reduced section, $\sigma' = \dfrac{100 \times 1000}{0.001472} = 67.9 \times 10^6 \text{ N/m}^2$

∴ Elongation of the rod $= \dfrac{\sigma(2.5-x)}{E} + \dfrac{\sigma'.x}{E} = 0.731 \times 10^{-3}$

or, $= \dfrac{50.92 \times 10^6 (2.5-x)}{200 \times 10^9} + \dfrac{67.9 \times 10^6 x}{200 \times 10^9} = 0.731 \times 10^{-3}$

or, $= 50.92 \times 10^6 (2.5-x) + 67.9 \times 10^6 x$

$= 200 \times 10^9 \times 0.731 \times 10^{-3}$

or, $= (2.5-x) + 1.33\, x = 2.87$

or, $x = 1.12 \text{ m}$

Hence, length of the bore $= 1.12 \text{ m}$ **(Ans.)**

Example 1.8. *A brass bar having cross-sectional area of 1000 mm² is subjected to axial forces shown in the Fig. 1.12. Find the total elongation of the bar. Modulus of elasticity of brass = 100 GN/m².*

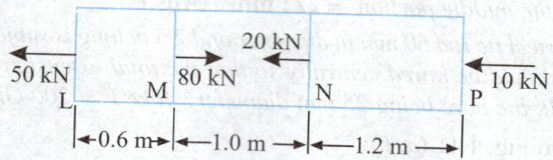

Fig. 1.12

Solution. Refer to Fig. 1.12. The loading of three different portions *LM*, *MN* and *NP* is shown separately in the Fig. 1.13. It may be noted that portion *LM* is under the tensile force 50 kN to the left, and to the right of it, there is again an effective force 50 kN which is a resultant of three forces to its right

i.e. $(80 - 20 - 10) = 50$ kN.

A metal fabrication unit.

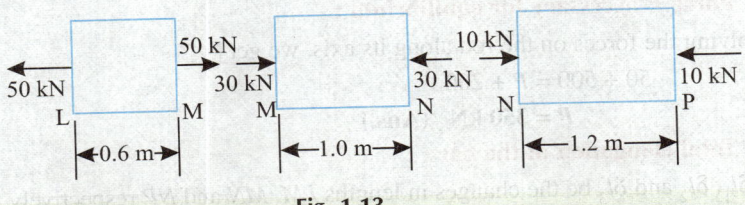

Fig. 1.13

Similarly, in portion *MN*, the compressive force on the left is 30 kN (*i.e.*, 80 – 50) and 30 kN on the right (*i.e.*, 20 + 10). In *NP*, the compressive load is 10 kN, (*i.e.*, 80 – 50 – 20) and on the right, there is already a compressive load of 10 kN. So, we observe that the bar is in equilibrium under the action of these forces.

Total elongation of the bar :

Let δl_1, δl_2 and δl_3 be the changes in length *LM*, *MN* and *NP* respectively.

Then, $\qquad \delta l_1 = \dfrac{P_1 l_1}{AE}$ .... increase (+)

$\qquad\qquad \delta l_2 = \dfrac{P_2 l_2}{AE}$ .... decrease (–)

$\qquad\qquad \delta l_3 = \dfrac{P_3 l_3}{AE}$ .... decrease (–)

∴ Net change in length,

$$\delta l = \delta l_1 - \delta l_2 - \delta l_3$$

$$= \frac{P_1 l_1}{AE} - \frac{P_2 l_2}{AE} - \frac{P_3 l_3}{AE} = \frac{1}{AE}(P_1 l_1 - P_2 l_2 - P_3 l_3)$$

$$= \frac{10^3}{1000 \times 10^{-6} \times 100 \times 10^9}(50 \times 0.6 - 30 \times 1 - 10 \times 1.2)$$

$$= \frac{1}{10^5}(30 - 30 - 12) = \frac{1}{10^5} \times (-12)$$

$$= -0.00012 \text{ m} = -0.12 \text{ mm}$$

Negative sign indicates that the bar is **shortened by 0.12 mm** **(Ans.)**

Example 1.9. *A member LMNP is subjected to point loads as shown in Fig. 1.14.*

Calculate :

(i) Force P necessary for equilibrium.

(ii) Total elongation of the bar.

Take E = 210 GN/m²

Solution. Refer to Fig. 1.14

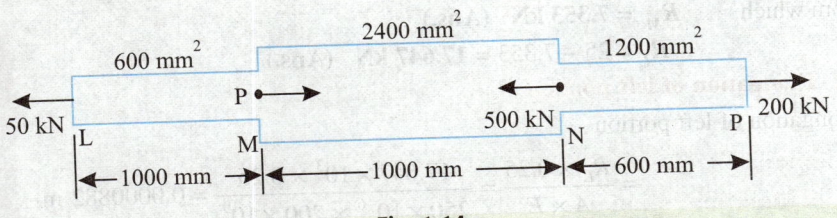

Fig. 1.14

(i) Force P necessary for equilibrium :

Resolving the forces on the rod along its axis, we get

$$50 + 500 = P + 200$$

$$P = 350 \text{ kN} \quad \textbf{(Ans.)}$$

(ii) Total elongation of the bar :

Let δl_1, δl_2 and δl_3 be the changes in lengths LM, MN and NP respectively.

Then,
$$\delta l_1 = \frac{P_1 l_1}{A_1 E} = \frac{50 \times 1000 \times 1}{600 \times 10^{-6} \times 210 \times 10^9} = 3.97 \times 10^{-4} \text{ m} \quad ....\text{increase } (+)$$

$$\delta l_2 = \frac{P_2 l_2}{A_2 E} = \frac{300 \times 1000 \times 1}{2400 \times 10^{-6} \times 210 \times 10^9} = 5.95 \times 10^{-4} \text{ m} \quad ....\text{decrease } (-)$$

$$\delta l_3 = \frac{P_3 l_3}{A_3 E} = \frac{200 \times 1000 \times 0.6}{1200 \times 10^{-6} \times 210 \times 10^9} = 4.76 \times 10^{-4} \text{ m} \quad ....\text{increase } (+)$$

$\therefore$ Total elongation, $\delta l = \delta l_1 - \delta l_2 + \delta l_3 = 3.97 \times 10^{-4} - 5.95 \times 10^{-4} + 4.76 \times 10^{-4}$

$$= 10^{-4} (3.97 - 5.95 + 4.76) = 2.78 \times 10^{-4} \text{ m, or, } 0.278 \text{ mm}$$

Hence, total elongation of the bar = **0.278 mm** **(Ans.)**

Example 1.10. In Fig. 1.15 is shown a steel bar of cross-sectional area 250 mm² held firmly by the end supports and loaded by an axial force of 25 kN.

Determine :

(i) Reactions at L and M.

(ii) Extension of the left portion.

$$E = 200 \text{ GN/m}^2.$$

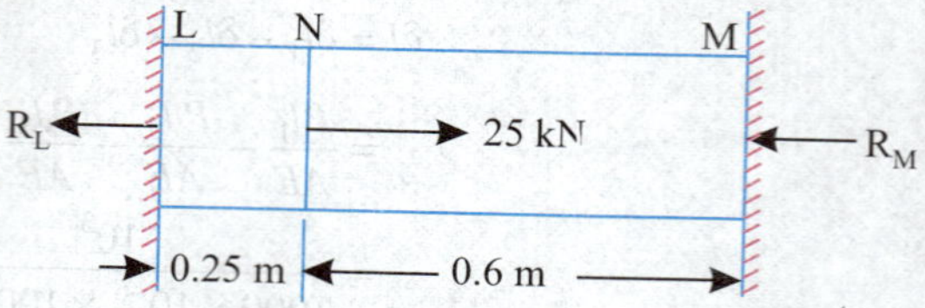

Solution. Refer to Fig. 1.15

(i) Reactions at L and M :

As the bar is in equilibrium.

$\therefore$
$$R_L + R_M = 25 \text{ kN} \quad ...(i)$$

Also, since total length of the bar remains unchanged,

$\therefore$ Extension in LN = contraction in MN

$$\frac{R_L \times 0.25}{A \times E} = \frac{R_M \times 0.6}{A \times E}$$

$$R_L \times 0.25 = R_M \times 0.6$$

$$R_L = \frac{R_M \times 0.6}{0.25} = 2.4 \, R_M$$

Substituting the value of R_L in (i), we get

$$2.4 \, R_M + R_M = 25$$

From which
$$R_M = 7.353 \text{ kN} \quad \textbf{(Ans.)}$$

$\therefore$
$$R_L = 25 - 7.353 = 17.647 \text{ kN} \quad \textbf{(Ans.)}$$

(ii) Elongation of left portion :

Elongation of left portion

$$= \frac{R_L \times 0.25}{A \times E} = \frac{17.647 \times 10^3 \times 0.25}{250 \times 10^{-6} \times 200 \times 10^9} = 0.0000882 \text{ m}$$

$$= \textbf{0.0882 mm} \quad \textbf{(Ans.)}$$

Fig. 1.15

Example 1.11. *A straight uniform bar AD is clamped at both ends and loaded as shown in Fig. 1.16. Initially the bar is stress free. Determine the stresses in all the three parts (AB, BC, CD) of the bar if the cross-sectional area of bar is 1000 mm².*

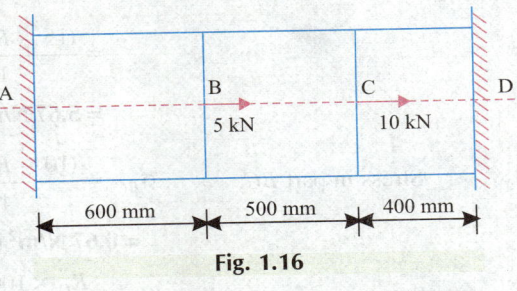

Fig. 1.16

Solution. Refer to Figs. 1.16 and 1.17

Let R_A and R_D be the reactions at the supports A and D respectively. For equilibrium of the bar AD, these reactions must act towards the left.

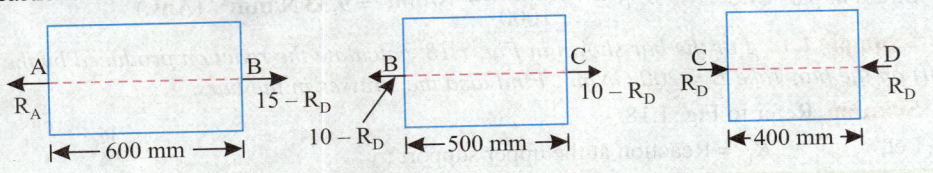

Fig. 1.17

Hence $$R_A + R_D = 5 + 10 = 15 \text{ kN} \qquad \qquad ...(i)$$

Fig. 1.17 shows the free-body diagram for the parts AB, BC and CD.

Let δl_{AB}, δl_{BC} and δl_{CD} be the extensions in the parts AB, BC and CD respectively.

Then $$\delta l_{AB} = \frac{(15 - R_D) \times l_{AB}}{A_{AB} \times E} \qquad \qquad ...... \text{(extension)}$$

$$\delta l_{BC} = \frac{(10 - R_D) \times l_{BC}}{A_{BC} \times E} \qquad \qquad ...... \text{(extension)}$$

$$\delta l_{CD} = \frac{R_D \times l_{CD}}{A_{CD} \times E} \qquad \qquad ...... \text{(compression)}$$

Since the *supports are rigid*, therefore elongation of AB and BC shall be *equal to* the compressions of CD.

Hence, $$\delta l_{AB} + \delta l_{BC} = \delta l_{CD}$$

Substituting the values, we have

$$\frac{(15 - R_D) \times 600}{1000 \times E} + \frac{(10 - R_D) \times 500}{1000 \times E} = \frac{R_D \times 400}{1000 \times E}$$

$$(\because A_{AB} = A_{BC} = A_{CD} = 1000 \text{ mm}^2 \text{ .... given})$$

or, $6 (15 - R_D) + 5 (10 - R_D) = 4 R_D$

or, $90 - 6 R_D + 50 - 5 R_D = 4 R_D$

or, $15 R_D = 140$

∴ $$R_D = \frac{140}{15} = 9.33 \text{ kN}$$

From eqn. (*ii*), we have

$$R_A + 9.33 = 15$$

or, $$R_A = 15 - 9.33 = 5.67 \text{ kN}$$

Stress in part AB, $$\sigma_{AB} = \frac{\text{Load in part } AB}{\text{Cross-sectional area of part } AB}$$

$$= \frac{(15 - R_D) \times 1000}{1000} \text{ N/mm}^2 = \frac{15 - 9.33}{1000} \times 1000$$

$$= \textbf{5.67 N/mm}^2 \textbf{ (tensile) (Ans.)}$$

Stress in part BC, $\sigma_{BC} = \dfrac{(10 - R_D) \times 1000}{1000} \text{ N/mm}^2 = \dfrac{10 - 9.33}{1000} \times 1000$

$$= \textbf{0.67 N/m}^2 \textbf{ (tensile) (Ans.)}$$

Stress in part CD, $\sigma_{CD} = \dfrac{R_D \times 1000}{1000} \text{ N/mm}^2 = \textbf{9.33 N/mm}^2 \textbf{ (Ans.)}$

Example 1.12. *For the bar shown in Fig. 1.18, calculate the reaction produced by the lower support on the bar. Take E = 200 GN/m². Find also the stresses in the bars.*

Solution. Refer to Fig. 1.18.

Let, R_1 = Reaction at the upper support ;

R_2 = Reaction at the lower support when the

bar touches it.

If the bar MN finally rests on the lower support, we have

$$R_1 + R_2 = 55 \text{ kN} = 55000 \text{ N}$$

For bar LM, the total force

$$= R_1 = 55000 - R_2 \text{ (tensile)}$$

For bar MN, the total force $= R_2$ (compressive)

∴ δl_1 = Extension of LM

$$= \frac{(55000 - R_2) \times 1.2}{(110 \times 10^{-6}) \times 200 \times 10^{9}} \text{ m}$$

and, δl_2 = Contraction of MN

$$= \frac{R_2 \times 2.4}{220 \times 10^{-6} \times 200 \times 10^{9}}$$

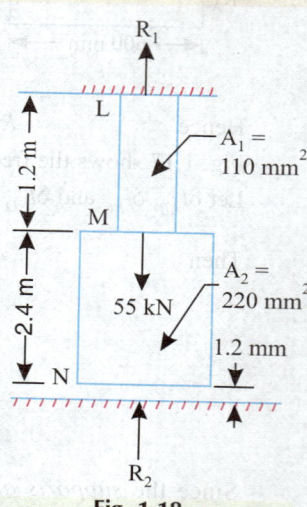

Fig. 1.18

In order that N rests on the lower support, we have from compatibility equation,

$$\delta l_1 - \delta l_2 = 1.2/1000 = 0.0012 \text{ m}$$

or, $\dfrac{(55000 - R_2) \times 1.2}{(110 \times 10^{-6}) \times 200 \times 10^{9}} - \dfrac{R_2 \times 2.4}{220 \times 10^{-6} \times 200 \times 10^{9}} = 0.0012$

or, $2 \,(55000 - R_2) \times 1.2 - 2.4 \, R_2 = (220 \times 10^{-6}) \times 200 \times 10^{9} \times 0.0012$

or, $132000 - 2.4 \, R_2 - 2.4 \, R_2 = 52800$

or, $4.8 \, R_2 = 79200$

∴ $R_2 = 16500 \text{ N}$ or **16.5 kN (Ans.)**

and, $R_1 = 55 - 16.5 = \textbf{38.5 kN}$ **(Ans.)**

Stress in LM

$$= \frac{R_1}{A_1} = \frac{38.5}{110 \times 10^{-6}} = 0.350 \times 10^{6} \text{ kN/m}^2$$

$$= \textbf{350 MN/m}^2 \textbf{ (Ans.)}$$

Tractors are especially built to run on rough and bumpy terrains. The engine provides high torque at low speeds.

Stress in MN

$$= \frac{R_2}{A_2} = \frac{16.5}{220 \times 10^{-6}} = 0.075 \times 10^6 \text{ kN/m}^2$$

$$= 75 \text{ MN/m}^2 \quad \textbf{(Ans.)}$$

Example 1.13. *A 700 mm length of aluminium alloy bar is suspended from the ceiling so as to provide a clearance of 0.3 mm between it and a 250 mm length of steel bar as shown in Fig. 1.19. $A_{al} = 1250 \text{ mm}^2$, $E_{al} = 70 \text{ GN/m}^2$, $A_s = 2500 \text{ mm}^2$, $E_s = 210 \text{ GN/m}^2$. Determine the stress in the aluminium and in the steel due to a 300 kN load applied 500 mm from the ceiling.*

Solution. Refer to Fig. 1.19. On application of load of 300 kN at Q, the portion LQ will move forward and come in contact with N so that QM and NP will both be under compression. LQ will elongate, while QM and NP will contract and the net elongation will be equal to gap of 0.3 mm between M and N.

Let, $\qquad \sigma_1$ = Tensile stress in LQ,

$\qquad\qquad \sigma_2$ = Compressive stress in QM, and

$\qquad\qquad \sigma_3$ = Compressive stress in NP.

Elongation of $\quad LQ = \dfrac{\sigma_1 \times 0.5}{70 \times 10^9}$ m (+)

Contraction of $\quad QM = \dfrac{\sigma_2 \times 0.2}{70 \times 10^9}$ m (−)

Contraction of $\quad NP = \dfrac{\sigma_3 \times 0.25}{210 \times 10^9}$ m (−)

But, force in $\quad QM$ = force in NP

$\therefore \quad \sigma_2 \times 1250 \times 10^{-6} = \sigma_3 \times 2500 \times 10^{-6}$

$\therefore \qquad\qquad \sigma_3 = \dfrac{\sigma_2}{2}$

$\therefore$ Contraction of $NP = \dfrac{\sigma_2 \times 0.25}{2 \times 210 \times 10^9}$ m

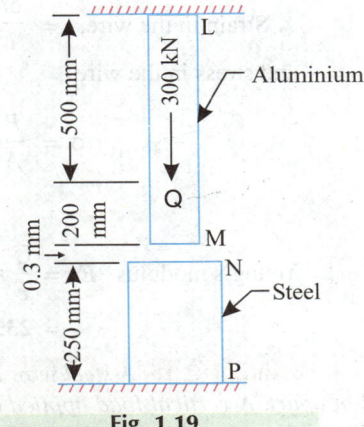

Fig. 1.19

$\therefore$ Net elongation $= \dfrac{\sigma_1 \times 0.5}{70 \times 10^9} - \dfrac{\sigma_2 \times 0.2}{70 \times 10^9} - \dfrac{\sigma_2 \times 0.25}{2 \times 210 \times 10^9}$

This must be equal to 0.0003 m (given).

$\therefore \quad \dfrac{\sigma_1 \times 0.5}{70 \times 10^9} - \dfrac{\sigma_2 \times 0.2}{70 \times 10^9} - \dfrac{\sigma_2 \times 0.25}{2 \times 210 \times 10^9} = 0.0003$

$3\,\sigma_1 - 1.2\,\sigma_2 - 0.25\,\sigma_2 = 2 \times 210 \times 10^9 \times 0.0003$

$3\,\sigma_1 - 1.45\,\sigma_2 = 2 \times 210 \times 10^9 \times 0.0003$ ...(i)

Tensile force in *LQ* + compressive force in *QM* = 300000

$1250 \times 10^{-6} \times \sigma_1 + 1250 \times 10^{-6} \times \sigma_2 = 300000$

$\sigma_1 + \sigma_2 = 2.4 \times 10^8 \text{ N/m}^2$ ...(ii)

Solving (*i*) and (*ii*), we get

$\sigma_1 = 1.065 \times 10^8 \text{ N/m}^2$

$= \textbf{106.5 MN/m}^2$ (*tensile*) **(Ans.)**

$\sigma_2 = 1.335 \times 10^8 \text{ N/m}^2$

$= \textbf{133.5 MN/m}^2$ (*compressive*) **(Ans.)**

$\sigma_3 = \dfrac{\sigma_2}{2} = 0.667 \times 10^8 \text{ N/m}^2$

$= \textbf{66.7 MN/m}^2$ (*compressive*) **(Ans.)**

Example 1.14. *Two parallel steel wires 6 m long, 10 mm diameter are hung vertically 70 mm apart and support a horizontal bar at their lower ends. When a load of 9 kN is attached to one of the wires, it is observed that the bar is 2.4° to the horizontal. Find 'E' for wire.*

Solution. Refer to Fig. 1.20.

Two wires *LM* and *ST* made of steel, each 6 m long and 10 mm diameter are fixed at the supports and a load of 9 kN is applied on wire *ST*. Let the inclination of the bar after the application of the load be θ.

The extension in the length of steel wire *ST*,

$\delta l = 70 \tan \theta = 70 \times \tan 2.4°$

$= 70 \times 0.0419 = 2.933 \text{ mm} = 0.00293 \text{ m}$

$\therefore$ Strain in the wire, $e = \dfrac{\delta l}{l} = \dfrac{0.00293}{6} = 0.000488$

and stress in the wire

$\sigma = \dfrac{P}{A} = \dfrac{9000}{\dfrac{\pi}{4} \times \left(\dfrac{10}{1000}\right)^2} = 11.46 \times 10^7 \text{ N/m}^2$

Fig. 1.20

Young's modulus $E = \dfrac{\sigma}{e} = \dfrac{11.46 \times 10^7}{0.000488} = 235 \times 10^9 \text{ N/m}^2$

$= \textbf{235 GN/m}^2$ **(Ans.)**

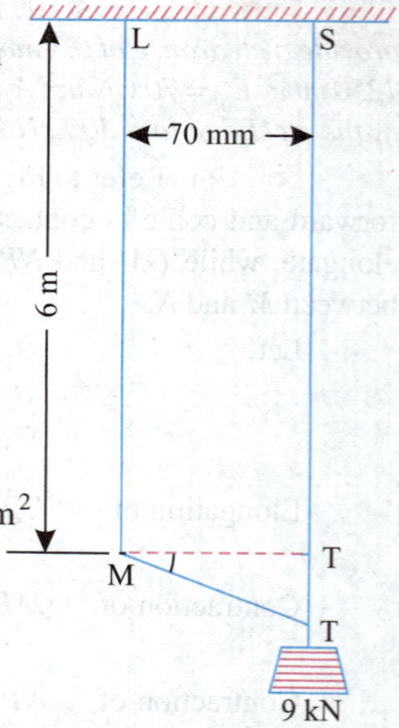

Example 1.15. *A steel wire 1 mm diameter is stretched horizontally between two fixed points 2 m apart. A vertical load applied at the mid-span of the wire causes a vertical displacement of 45 mm of the point of application of the load applied. What will be the stress induced in the wire and the load applied ? Neglect the weight of the wire. Take E for the wire material as 200 GN/m².*

Solution. Refer to Fig. 1.21 (*a*).

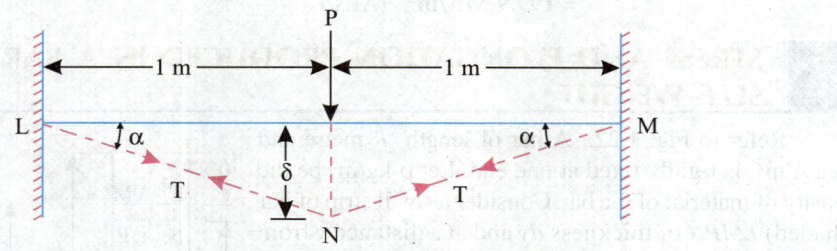

Fig. 1.21 (a)

Steel wire diameter, $d = 1$ mm $= 0.001$ m

Vertical displacement or deflection,

$$\delta = 45 \text{ mm} = 0.045 \text{ m}$$

$E = 200$ GN/m^2, Length of span $= 2$ m

Load, P :, Stress in the wire :

Fig. 1.21 (*b*) has been drawn by taking line *AB* (representing load *P*) of any length and then from the points *A* and *B*, the lines *AC* and *BC* have been drawn *parallel* to *LN* and *MN* to represent the tension *T* produced in each half portion of the wire.

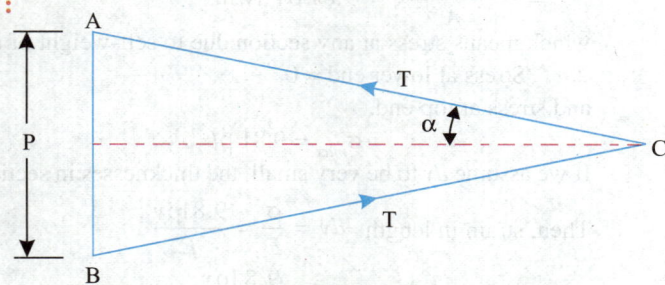

Now from ∆ABC,

Fig. 1.21 (b)

$$P = 2T \sin \alpha, \ T = \frac{P}{2 \sin \alpha} = \frac{P}{2 \tan \alpha} \begin{bmatrix} \text{since when } \alpha \text{ is very} \\ \text{small, } \sin \alpha = \tan \alpha \end{bmatrix}$$

∴

$$T = \frac{P}{2 \times \dfrac{0.045}{1}} = 11.11 \ P$$

∴ Stress in the wire

$$= \frac{11.11 \ P}{\pi / 4 \times (0.001)^2} = 14.14 \times 10^6 \ P \ \text{N/m}^2$$

And change in length (δl) in each half portion of the wire

$$= \sqrt{(0.045)^2 + 1^2} - 1 = 1.001 - 1 = 0.001 \text{ m}$$

And, strain

$$= \frac{\delta l}{l} = \frac{0.001}{1} = 0.001$$

But,

$$\frac{\text{Stress}}{\text{Strain}} = E$$

∴

$$\frac{14.14 \times 10^6 \ P}{0.001} = 200 \times 10^9$$

∴

$$P = \frac{200 \times 10^9 \times 0.001}{14.14 \times 10^6} = \textbf{14.14 N} \ \ \textbf{(Ans.)}$$

Stress in the wire $= 14.14 \times 10^6 \times 14.14$ N/m²

$$= \textbf{199.9 MN/m}^2 \quad \textbf{(Ans.)}$$

1.5. STRESS AND ELONGATION PRODUCED IN A BAR DUE TO ITS SELF-WEIGHT

Refer to Fig. 1.22. A bar of length 'l' metre and area A m², is rigidly fixed at one end. Let ρ kg/m³ be the density of material of the bar. Consider a small strip of bar (shaded) *LMPN* of thickness dy and at a distance y from the free end. Now the force acting down at NP = weight of bar $NPTS = Ay\rho$.

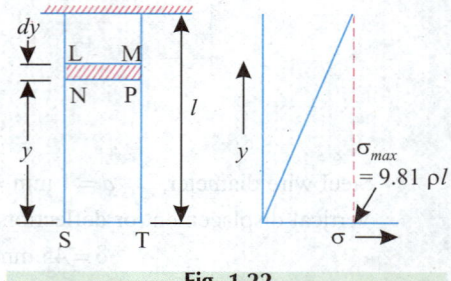

Fig. 1.22

Also, stress at section NP is given by

$$\sigma = \frac{\text{Force at } NP}{\text{Area of cross-section of the bar}}$$

$$= \frac{Ay\rho \times 9.81}{A} = 9.81 \, \rho y \text{ N/m}^2$$

which means stress at any section due to self-weight of the bar is directly proportional to y,

∴ Stress at lower end = 0

and stress at top end,

$$\sigma_{max} = 9.81 \, \rho l$$

If we assume dy to be very small, the thicknesses in sections *LM* and *NP* are *practically equal*.

Then, strain in length $dy = \dfrac{\sigma}{E} = \dfrac{9.81 \rho y}{E}$

Extension in length $dy = \dfrac{9.81 \rho y}{E} \, dy$

∴ Total elongation of the bar,

$$\delta l = \int_0^l \frac{9.81 \, \rho y}{E} \, dy = \frac{9.81 \, \rho}{E} \int_0^l y \, dy = \frac{9.81 \, \rho}{E} \left| \frac{y^2}{2} \right|_0^l$$

$$\delta l = \frac{9.81 \, \rho l^2}{2E} \qquad \qquad \qquad ...(1.7)$$

1.6. TIE BAR OF UNIFORM STRENGTH

Fig. 1.23 shows a tie bar of uniform strength carrying a load P newtons. Let the stress everywhere be σ N/m².

Let,

$\rho =$ Density of material in kg/m³,

$A =$ Area of cross-section at QQ, and

$A + dA =$ Area of cross-section at NN.

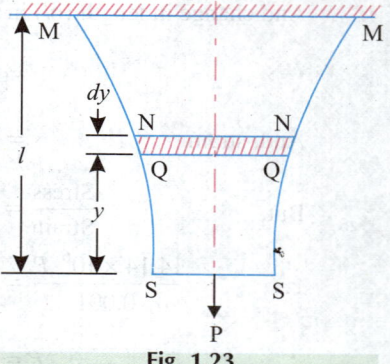

Fig. 1.23

If A varies from A_1 to A_2 for the sections *SS* to *MM*, we have

For section $\qquad SS : A_1 \, \sigma = P$

or, $\qquad \qquad \sigma = \dfrac{P}{A_1} \qquad \qquad ...(1.8)$

For section QQ : $A.\sigma = P + 9.81 \times$ mass of *SQ* in kg $\qquad$ ...(1.9)

For section NN :

$$(A + dA)\,\sigma = P + 9.81 \times \text{mass of portion } SQ \text{ in kg} + 9.81 \times \text{mass of portion}$$
$$QN \text{ in kg}$$
$$= A\sigma + 9.81 \times \text{mass of portion } QN \text{ in kg} \quad \text{...from eqn. (1.9)}$$

i.e., $\qquad A.\sigma + dA.\sigma = A.\sigma + 9.81 \times (A.dy)\,\rho$

or, $\qquad \dfrac{dA}{A} = \dfrac{9.81\,\rho}{\sigma}\,dy$ $\qquad$ ...(1.10)

Integrating $\quad \left[ln\,A\right]_{A_1}^{A} = \left[\dfrac{P}{\sigma}\,y\right]_0^{y}$

or, $\qquad ln\,(A/A_1) = \dfrac{\rho y}{\sigma}$ or $A = A_1 e^{9.81\rho y/\sigma}$

At, $\qquad\qquad y = l,\ A = A_2$

$\therefore \qquad\qquad A_2 = A_1.e^{9.81\rho l/\sigma}$ $\qquad$ ...(1.11)

Example 1.16. *A vertical tie of uniform strength is 20 metres long. If the area of the bar at the lower end is 600 mm², find the area at the upper end when the tie is to carry a load of 800 kN. The material of the tie weighs 80 kN/m³.*

Solution. Area of the tie at the bottom, $A_1 = 600 \text{ mm}^2 = 600 \times 10^{-6} \text{ m}^2$

Intensity of stress, $\sigma = \dfrac{800 \times 1000}{600 \times 10^{-6}} = 1333.3 \text{ MN/m}^2$

Density of the material,

$$\rho = \dfrac{80 \times 1000}{9.81} \text{ kg/m}^3 = 8155 \text{ kg/m}^3$$

Area at the upper end (A_2) is given by,

$$A_2 = A_1.\,e^{9.81\,\rho l/\sigma} \qquad\qquad \text{[Eqn. (1.11)]}$$
$$= 600 \times 10^{-6} \times e^{(9.81 \times 8155 \times 20/1333.3 \times 10^6)}$$
$$= 600.036 \times 10^{-6} \text{ m}^2 \text{ or } 600.036 \text{ mm}^2$$

Hence, area of the tie at the upper end = **600.036 mm²** **(Ans.)**

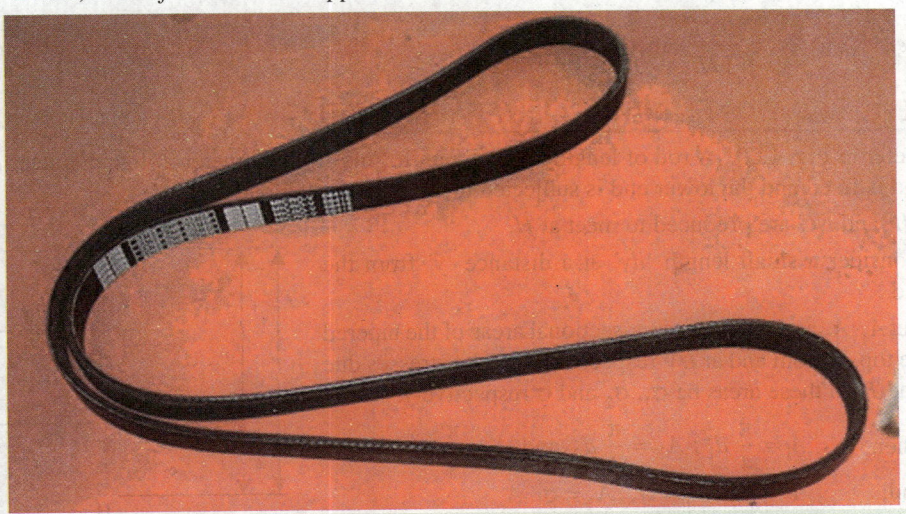

Although they look simple, drive belts of pulleys should be designed with care. If the belts fail while running, machines can be damaged due to sudden shift of loads.

1.7. STRESS IN A BAR DUE TO ROTATION

A bar of length l revolving about Y-axis at an angular speed of ω rad/sec is shown in Fig. 1.24.

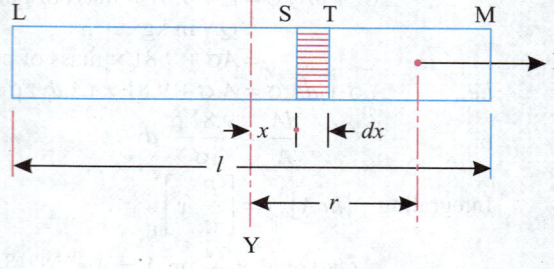

Tensile force on the element ST

$\qquad$ = Centrifugal force P
$\qquad$ on the part TM

$\qquad$ = $\omega^2 rM$

(where, M = mass of the part TM)

Thus, if

$\qquad A$ = Area of cross-section of the bar (in m²), and

$\qquad \rho$ = Density of the material (in kg/m³),

Fig. 1.24

then, $\qquad P = \rho A \left(\dfrac{l}{2} - x\right) \omega^2 \left[x + \dfrac{1}{2}\left(\dfrac{l}{2} - x\right)\right]$

or, $\qquad P = 1/2\ \rho A \omega^2\ (l^2/4 - x^2)$ newton

i.e. Stress, $\sigma_c = 1/2\ \rho \omega^2\ (l^2/4 - x^2)$ N/m² $\qquad\qquad$ ...(1.12)

Maximum stress will occur at

$\qquad x = 0 \qquad i.e., \qquad \sigma_{max} = \dfrac{1}{8}\ \rho\omega^2 l^2$ $\qquad\qquad$ ...(1.13)

Extension of the element

$\qquad ST = \dfrac{\sigma_c}{E}\ dx$

Thus, substituting the value of σ_c from eqn. (1.12) and integrating, the total extension in the bar is given by

$$\delta l = 2\int_0^{l/2} \frac{1}{2E}\ \rho\omega^2\ (l^2/4 - x^2)\ dx = \frac{\rho\omega^2}{E}\left[\frac{l^2}{4}x - \frac{x^3}{3}\right]_0^{l/2}$$

i.e., $\qquad \delta l = \dfrac{\rho\omega^2 l^3}{12E}$ metre $\qquad\qquad$ ...(1.14)

1.8. ELONGATION IN CASE OF A TAPER ROD

Refer to Fig. 1.25. A rod of length 'l' tapers uniformly from diameter d_1 to a diameter d_2. Its wider end is fixed and the lower end is subjected to an axial tensile load P.

MF and NG are produced to meet at H.

Consider a small length 'dy' at a distance 'y' from the lower end.

Let A_1, A_2 and A be the cross-sectional areas of the tapered rod at the top, bottom and at UV respectively and let stresses due to the load P on these areas be σ_1, σ_2 and σ respectively.

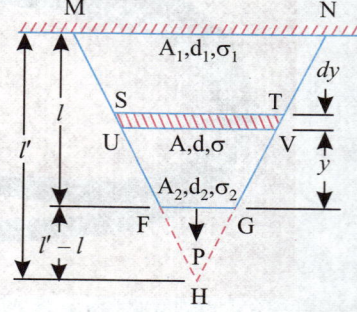

Then, $\qquad A_1 = \dfrac{\pi}{4}\ d_1^2;\ A_2 = \dfrac{\pi}{4}\ d_2^2$ and $A = \dfrac{\pi}{4}\ d^2$

and, $\qquad P = \sigma_1 . A_1 = \sigma_2 . A_2 = \sigma . A$

or, $\qquad \sigma = \dfrac{\sigma_2 A_2}{A}$

Fig. 1.25

Elongation of the small strip,

$$STVU = \frac{\sigma}{E} \cdot dy$$

$$= \frac{\sigma_2 A_2}{AE} \cdot dy = \frac{\sigma_2}{E} \cdot \frac{d_2^2}{d^2} \cdot dy \qquad \qquad ...(i)$$

From similar triangles FGH and UVH

$$\frac{d_2}{d} = \frac{l' - l}{l' - l + y}$$

Putting the value of $\frac{d_2}{d}$ in equation (i), we get

Elongation in the small strip $STVU$

$$= \frac{\sigma_2}{E} \cdot \frac{(l' - l)^2}{(l' - l + y)^2}$$

∴ Total elongation or extension in the taper rod,

$$\delta l = \int_0^l \frac{\sigma_2}{E} \cdot \frac{(l' - l)^2}{(l' - l + y)^2} \cdot dy = \frac{\sigma_2 (l' - l)^2}{E} \int_0^l \frac{1}{(l' - l + y)^2} \cdot dy$$

$$= \frac{\sigma_2 (l' - l)^2}{E} \left| \frac{-1}{l' - l + y} \right|_0^l = \frac{\sigma_2 (l' - l)^2}{E} \cdot \left[\frac{-1}{l' - l + l} - \frac{-1}{l' - l + 0} \right]$$

$$= \frac{\sigma_2 (l' - l)^2}{E} \times \frac{l}{l' (l' - l)} = \frac{\sigma_2 (l' - l)}{l' E} l \qquad \qquad ...(ii)$$

Also from similar triangles FGH and MNH,

$$\frac{l' - l}{l'} = \frac{d_2}{d_1}$$

Substituting this value in equation (ii), we get

$$\delta l = \frac{\sigma_2 l}{E} \cdot \frac{d_2}{d_1}$$

or, $$\delta l = \frac{Pl}{A_2 E} \cdot \frac{d_2}{d_1} = \frac{Pl}{(\pi/4) d_2^2 .E} \cdot \frac{d_2}{d_1}$$

i.e., $$\delta l = \frac{4Pl}{\pi E d_1 d_2} \qquad \qquad ...(1.15)$$

Example 1.17. *If a tension test bar is found to taper uniformly from $(D - a)$ to $(D + a)$, prove that the error involved in using the mean diameter to calculate Young's modulus is* $\left(\frac{10 a}{D}\right)^2$ *per cent.*

Solution. We know that,

Extension in a taper rod $= \dfrac{4 Pl}{\pi E d_1 d_2}$

but, $$d_1 = (D + a) \ (given) \ and, \ d_2 = (D - a) \ (given)$$

∴ Extension in rod, $$\delta l = \frac{4 Pl}{\pi E (D + a) (D - a)} = \frac{4 Pl}{\pi E (D^2 - a^2)}$$

∴ $$E = \frac{4 Pl}{(D^2 - a^2).\delta l}$$

This gives the actual value of Young's modulus, if the diameters are measured correctly at every section.

Let the value of Young's modulus, as calculated with mean diameter D be E'.

Then,

$$E' = \frac{4\,Pl}{\pi D^2 .\delta l}$$

∴ %age error

$$= \frac{E - E'}{E} \times 100$$

$$= \left[\frac{\dfrac{4\,Pl}{\pi\,(D^2 - a^2).\delta l} - \dfrac{4\,Pl}{\pi D^2 .\delta l}}{\dfrac{4\,Pl}{\pi\,(D^2 - a^2).\delta l}}\right] \times 100 = \left[\frac{\dfrac{1}{D^2 - a^2} - \dfrac{1}{D^2}}{\dfrac{1}{D^2 - a^2}}\right] \times 100$$

$$= (D^2 - a^2)\left[\frac{a^2}{(D^2 - a^2)\,D^2}\right] \times 100 = \frac{100\,a^2}{D^2} = \left(\frac{10\,a}{D}\right)^2$$

∴ %age error $= \left(\dfrac{\mathbf{10a}}{\mathbf{D}}\right)^2$ **(Ans.)**

Example 1.18. *A rod is hung vertically and is firmly fixed at the top position as shown in Fig. 1.26. If its length is l and tapers from a diameter d_1 to d_2 uniformly, determine its extension due to its own weight. Take density of rod material as ρ.*

Solution. Refer to Fig. 1.26.

Let H be the point where the lines SU and TV meet, after being produced. Now, the extension in the length of the rod due to its self-weight = extension of conical rod SHT due to its self-weight – extension in UHV due to its self-weight – extension in rod length l due to the weight of UHV.

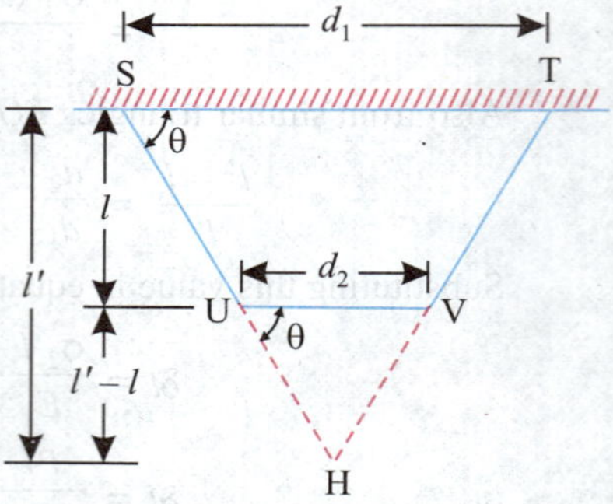

But, extension in conical SHT of length l' due to its self-weight

$$= \frac{9.81\,\rho\,l'^2}{6\,E}\left[\begin{array}{l}\text{where } \rho \text{ is in kg/m}^3 \text{ and length}\\ \text{and diameters are in metres.}\end{array}\right]$$

and, extension in UHV of length $(l' - l)$ due to self-weight

$$= \frac{9.81\,\rho\,(l' - l)^2}{6\,E}$$

and, weight of UHV

$$= \frac{\pi\,(UV)^2}{4} \times \left(\frac{l' - l}{3}\right) \times 9.81\,\rho$$

$$= \frac{9.81\,\rho\,\pi}{12}\,(l' - l)\,d_2^2 = P \text{ (say)}$$

Also, extension of tapering rod $SUVT$ due to P

$$= \frac{4\,Pl}{\pi E d_1 d_2} = \frac{4\,(9.81\,\rho)\,\pi}{12}\,(l' - l)\,d_2^2 \times \frac{l}{\pi\,E d_1 d_2}$$

$$= \frac{9.81\,\rho\,(l' - l)\,l}{3\,E}\cdot\frac{d_2}{d_1}$$

Fig. 1.26

∴ Extension in the rod due to its own weight,

$$\delta l = \frac{9.81 \, \rho \, l'^2}{6E} - \frac{9.81 \, \rho \, (l' - l)^2}{6E} - \frac{9.81 \, \rho \, (l' - l) \, l}{3E} \times \frac{d_2}{d_1}$$

Also,
$$\left. \begin{array}{l} \dfrac{\dfrac{d_2}{2}}{l' - l} = \cot \theta \\[2mm] \text{and,} \quad \dfrac{d_1}{2l'} = \cot \theta \end{array} \right] \quad \text{From Fig 1.26}$$

∴
$$\frac{\dfrac{d_2}{2}}{l' - l} = \frac{d_1}{2l'} \quad \text{or} \quad \frac{d_2}{l' - l} = \frac{d_1}{l'}$$

or,
$$l' = \frac{d_1 \, l}{d_1 - d_2} \quad \text{or} \quad l' - l = \frac{d_1 \, l}{d_1 - d_2} - l$$

$$= \left[\frac{d_1 - d_1 + d_2}{d_1 - d_2} \right] l = \frac{d_2 \, l}{d_1 - d_2}$$

∴
$$\delta l = \frac{9.81 \, \rho \left[\dfrac{d_1 \, l}{d_1 - d_2} \right]^2}{6E} - \frac{9.81 \, \rho \left[\dfrac{d_2 \, l}{d_1 - d_2} \right]^2}{6E} - \frac{9.81 \, \rho}{3E} \left(\frac{d_2 \, l}{d_1 - d_2} \right) l \times \frac{d_2}{d_1}$$

$$= \frac{9.81 \, \rho \, l^2}{3E} \left[\frac{d_1^2}{2 \, (d_1 - d_2)^2} - \frac{d_2^2}{2 \, (d_1 - d_2)^2} - \frac{d_2^2}{(d_1 - d_2) \, d_1} \right]$$

∴
$$\delta l = \frac{9.81 \, \rho \, l^2}{6E} \left[\frac{d_1^3 + 2d_2^3 - 3d_1 \, d_2^2}{d_1 \, (d_1 - d_2)^2} \right] \quad \textbf{(Ans.)}$$

A twin- cylinder motorcycle engine.

Example 1.19. *A vertical rod tapers uniformly from a diameter of 65 mm at the top to 35 mm at the bottom. It is rigidly fixed at the upper end and is subjected to an axial load of 25 kN. If its length is 2.5 m determine the total extension in the bar.*

Take density of material $= 2 \times 10^5$ kg/m^3 and Young's modulus $= 210$ GN/m^2.

Solution. Extension in the rod due to its self-weight

$$= \frac{9.81 \, \rho \, l^2}{6E} \left[\frac{d_1^3 + 2d_2^3 - 3d_1 \, d_2^2}{d_1 \, (d_1 - d_2)^2} \right]$$

$$= \frac{9.81 \times 2 \times 10^5 \times 2.5^2}{6 \times 210 \times 10^9} \left[\frac{(0.065)^3 + 2 \times (0.035)^3 - 3 \times 0.065 \times (0.035)^2}{0.065 \, (0.065 - 0.035)^2} \right]$$

$$= 2.02 \times 10^{-5} \text{ m}$$

Also extension due to external load

$$= \frac{4 \, P l}{\pi E d_1 \, d_2}$$

$$= \frac{4 \times 25000 \times 2.5}{\pi \times 210 \times 10^9 \times 0.065 \times 0.075} = 0.000166 \text{ m}$$

∴ **Total extension in the bar**

$$= 2.02 \times 10^{-5} + 0.000166 = 0.0001862 \text{ m} = \textbf{0.1862 mm} \quad \textbf{(Ans.)}$$

1.9. ELONGATION OF A CONICAL BAR DUE TO ITS SELF-WEIGHT

Refer to Fig. 1.27. *MNH* is a conical bar of length *l* with its wider end of diameter '*d*' fixed rigidly at *MN*.

Let ρ kg/m^3 be the density of the bar material. Consider a small strip of thickness *dy* at a distance *y* from the lower end.

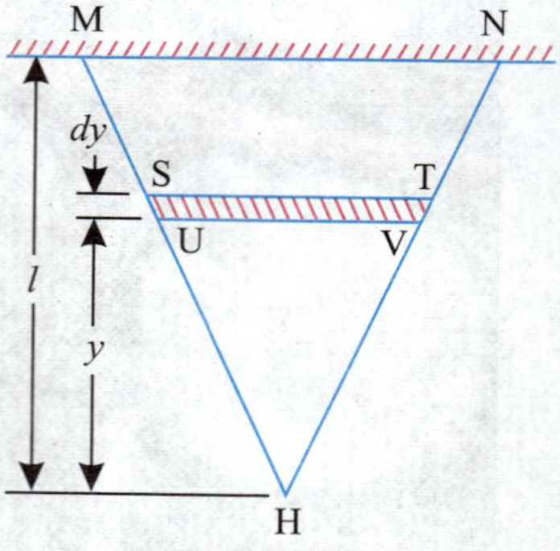

Weight of portion $UVH = 9.81 \, \rho \, \frac{\pi \, d_s^2}{4} \cdot \frac{y}{3}$

where, d_s is the diameter of the strip.

From similar triangles *MNH* and *UVH*,

$$\frac{MN}{UV} = \frac{l}{y} \quad \text{or} \quad UV = \frac{MNy}{l} = \frac{d \times y}{l}$$

∴ Weight of *UVH* $= 9.81 \, \rho . \frac{\pi}{4} \left[\frac{d \times y}{l} \right]^2 . \frac{y}{3}$

$$= 9.81 \times \frac{\pi}{12} . \rho \, \frac{d^2}{l^2} . y^3$$

Fig. 1.27

∴ Stress at section *UV*

$$= \frac{\text{Force at } UV}{\text{Cross-sectional area at } UV}$$

$$= \frac{\text{Weight of } UVH}{\dfrac{\pi d_s^2}{4}} = \frac{9.81 \times \dfrac{\pi}{12} . \rho \, \dfrac{d^2}{l^2} . y^3}{\dfrac{\pi}{4} . \left[\dfrac{d \times y}{l} \right]^2} = \frac{9.81 \, \rho y}{3}$$

$\therefore$ Extension in $dy = \dfrac{9.81\,\rho y}{3E}.dy$

and total extension in the bar,

$$\delta l = \int_0^l \frac{9.81\,\rho y}{3E}.dy = \frac{9.81\,\rho}{3E}\int_0^l y\,dy$$

$$= \frac{9.81\,\rho}{3E}\left|\frac{y^2}{2}\right|_0^l = \frac{9.81\,\rho l^2}{6E}$$

$\therefore$ $\qquad \delta l = \dfrac{9.81\,\rho l^2}{6E}$ $\qquad$ ...(1.16)

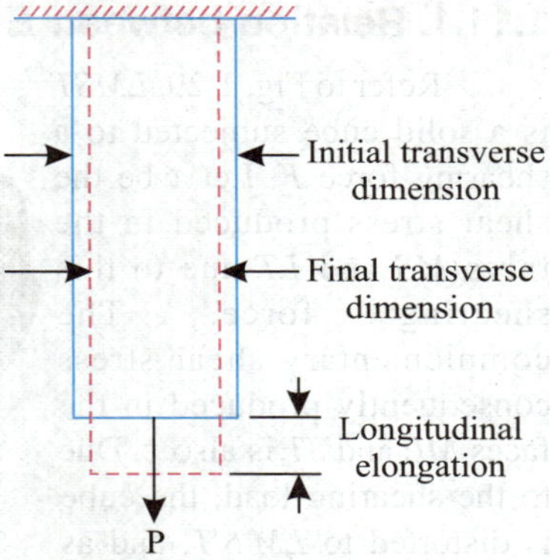

Initial transverse dimension

Final transverse dimension

Longitudinal elongation

P

Fig. 1.28

1.10. POISSON'S RATIO

If a body is subjected to a load, its length changes; ratio of this change in length to the original length is known as *linear or primary strain*. Due to this load, the dimensions of the body change; in all directions at right angles to its line of application the strains thus produced are called *lateral or secondary or transverse strains* and are of *nature opposite to that of primary strains*. For example, if the load is tensile, there will be an increase in length and a corresponding decrease in cross-sectional area of the body (Fig. 1.28). In this case, linear or primary strain will be tensile and secondary or lateral or transverse strain compressive.

The ratio of lateral strain to linear strain is known as **Poisson's ratio**.

i.e., Poisson's ratio,

$$\mu = \frac{\text{Lateral strain or transverse strain}}{\text{Linear or primary strain}} = \frac{1}{m}$$

where m is a constant and its value varies between 3 and 4 for different materials.

Table 1.1 gives the average values of Poisson's ratio for common materials.

TABLE 1.1. Poisson's Ratio for Some of the Common Materials		
S. No.	*Material*	*Poisson's ratio*
1.	Aluminium	0.330
2.	Brass	0.340
3.	Bronze	0.350
4.	Cast iron	0.270
5.	Concrete	0.200
6.	Copper	0.355
7.	Steel	0.288
8.	Stainless steel	0.305
9.	Wrought iron	0.278

1.11. RELATIONS BETWEEN THE ELASTIC MODULII

Relations exist between the elastic constants for any specific material and these relations hold good for all materials within the elastic range. The relations result from the fact that the application of any particular type of stress necessarily produces other types of stress at other places in the material. Further, each of the stresses produces its corresponding strain and all the strains produced must be consistent.

1.11.1. Relation between *E* and *C*

Refer to Fig. 1.29. *LMST* is a solid cube subjected to a shearing force *F*. Let τ be the shear stress produced in the faces *MS* and *LT* due to this shearing force. The complementary shear stress consequently produced in the faces *ML* and *ST* is also τ. Due to the shearing load, the cube is distorted to *LM'S'T*, and as such, the edge *M* moves to *M'*, *S* to *S'* and the diagonal *LS* to *L'S*.

Modern requirements of high quality and fuel efficiency, demand high reliability and performance of the machine parts.

Shear strain $= \phi = \dfrac{SS'}{ST}$

Also, shear strain $= \dfrac{\tau}{C}$

∴ $\dfrac{SS'}{ST} = \dfrac{\tau}{C}$ ...(i)

On the diagonal *LS'*, draw a perpendicular *SN* from *S*.

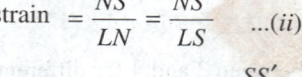

Now diagonal strain $= \dfrac{NS'}{LN} = \dfrac{NS'}{LS}$ ...(ii)

$$NS' = SS' \cos 45° = \dfrac{SS'}{\sqrt{2}}$$

Fig. 1.29

[∠ *LS' T* is assumed to be equal to ∠*LST* since *SS'* is very small.]

and, $\quad LS = ST \times \sqrt{2}$

Putting the value of *LS* in (ii), we get

Diagonal strain $= \dfrac{SS'}{\sqrt{2}\, ST \times \sqrt{2}} = \dfrac{SS'}{2\, ST}$

But, $\dfrac{SS'}{ST} = \dfrac{\tau}{C}$

∴ Diagonal strain $= \dfrac{\tau}{2C} = \dfrac{\sigma_n}{2C}$ ...(iii)

where σ_n is the normal stress due to shear stress τ. The net strain in the direction of diagonal *LS*

$$= \dfrac{\sigma_n}{E} + \dfrac{\sigma_n}{mE}$$

[Since the diagonals *LS* and *MT* have normal tensile and compressive stress σ_n, respectively.]

$$= \dfrac{\sigma_n}{E}\left[1 + \dfrac{1}{m}\right]$$ ...(iv)

Comparing equations (iii) and (iv), we get

$$\dfrac{\sigma_n}{2C} = \dfrac{\sigma_n}{E}\left[1 + \dfrac{1}{m}\right]$$

i.e.,
$$E = 2C \left[1 + \frac{1}{m} \right] \qquad \ldots(1.17)$$

1.11.2. Relation between *E* and *K*

If the solid cube in question is subjected to σ_n (normal compressive stress) on all the faces, the

direct strain in each axis $= \dfrac{\sigma_n}{E}$ (compressive) and lateral strain in other axis $= \dfrac{\sigma_n}{mE}$ (tensile).

$\therefore$ Net compressive strain in each axis

$$= \frac{\sigma_n}{E} - \frac{\sigma_n}{mE} - \frac{\sigma_n}{mE} = \frac{\sigma_n}{E} \left[1 - \frac{2}{m} \right]$$

Volumetric strain (e_v) in this case will be,

$$e_v = 3 \times \text{linear strain} = 3 \times \frac{\sigma_n}{E} \left[1 - \frac{2}{m} \right]$$

But,
$$e_v = \frac{\sigma_n}{K}$$

$\therefore$
$$\frac{\sigma_n}{K} = \frac{3\sigma_n}{E} \left[1 - \frac{2}{m} \right] \qquad \text{or,} \qquad E = 3K \left[1 - \frac{2}{m} \right] \qquad \ldots(1.18)$$

The relation between *E*, *C* and *K* can be established by eliminating *m* from the equations (1.17) and (1.18) as follows :

From equation (1.17),

$$m = \frac{2C}{E - 2C}$$

$$E = 3K \left[1 - \frac{2}{2C / (E - 2C)} \right]$$

or,
$$E = 3K \left[1 - \frac{E - 2C}{C} \right]$$

or,
$$\frac{E}{3K} = \frac{C - E + 2C}{C} = \frac{3C - E}{C}$$

or,
$$\frac{E}{3K} + \frac{E}{C} = 3$$

or,
$$EC + 3KE = 9KC$$

or,
$$E(3K + C) = 9KC$$

or,
$$E = \frac{9KC}{3K + C} \qquad \ldots(1.19)$$

> **Note.** When a square or rectangular block subjected to a shear load is in equilibrium, the shear stress in one plane is always associated with a complementary shear stress (of equal value) in the other plane at right angles to it.

Example 1.20. *A concrete cylinder of diameter 150 mm and length 300 mm when subjected to an axial compressive load of 240 kN resulted in an increase of diameter by 0.127 mm and a decrease in length of 0.28 mm. Compute the value of Poisson's ratio $\mu \left(= \dfrac{l}{m} \right)$ and modulus of elasticity E.*

Solution. Diameter of the cylinder, $d = 150$ mm

Length of the cylinder, $l = 300$ mm

Increase in diameter, $\delta d = 0.127$ mm (+)

Decrease in length $l = 0.28$ mm (−)

Axial compressive load, $P = 240$ kN

Poisson's ratio, μ :

We know that,

Linear strain $= \dfrac{\delta l}{l} = \dfrac{0.28}{300} = 0.000933$

and, lateral strain $= \dfrac{\delta d}{d} = \dfrac{0.127}{150} = 0.000846$

∴ Poisson's ratio, $\mu = \dfrac{\text{Lateral strain}}{\text{Linear strain}} = \dfrac{0.000846}{0.000933} = 0.907$

Modulus of elasticity, E :

Using the relation, $E = \dfrac{\text{Stress}}{\text{Strain (linear)}} = \dfrac{P/A}{\delta l / l}$

$$E = \dfrac{240/\left(\dfrac{\pi}{4} \times 0.15^2\right)}{(0.00028/0.3)} = \dfrac{240 \times 4 \times 0.3}{\pi \times 0.15^2 \times 0.00028}$$

$= 14.55 \times 10^6$ kN/m^2 = 14.55 GN/m^2

∴ **Young's modulus,** $E = \textbf{14.55 GN/m}^2$ **(Ans.)**

Example 1.21. *For a given material, Young's modulus is 110 GN/m^2 and shear modulus is 42 GN/m^2. Find the Bulk modulus and lateral contraction of a round bar of 37.5 mm diameter and 2.4 m length when stretched 2.5 mm.*

Solution. Young's modulus,

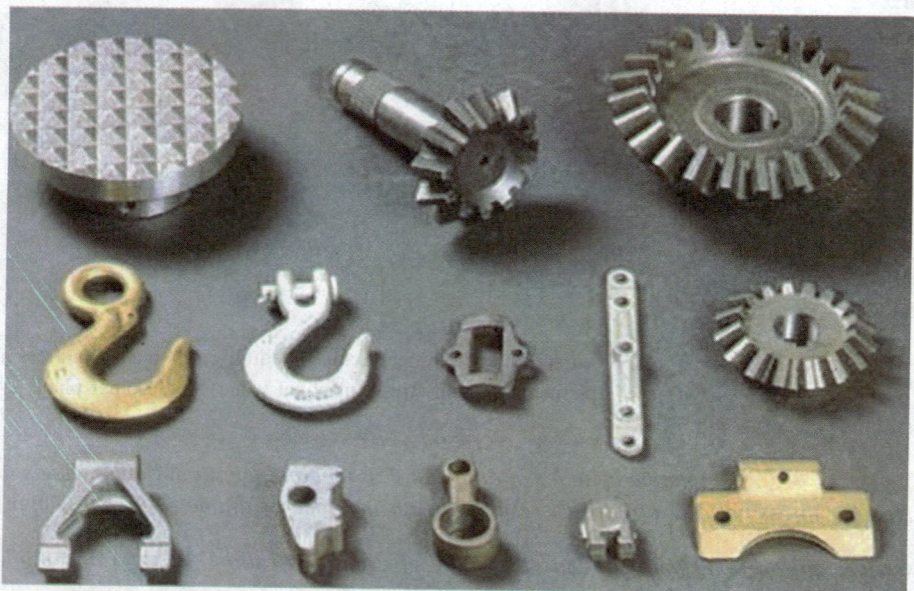

Parts made from forging. Applicable materials: Carbon steel, stainless steel, steel alloy, copper alloy, etc.

$$E = 110 \text{ GN/m}^2$$

Shear modulus,

$$C = 42 \text{ GN/m}^2$$

Diameter of round bar,

$$d = 37.5 \text{ mm} = 0.0375 \text{ m}$$

Length of round bar,

$$l = 2.4 \text{ m}$$

Extension of bar,

$$\delta l = 2.5 \text{ mm} = 0.0025 \text{ m}$$

Bulk modulus, K :

We know that,

$$E = 2C\left(1 + \frac{1}{m}\right)$$

$$110 \times 10^9 = 2 \times 42 \times 10^9 \left(1 + \frac{1}{m}\right)$$

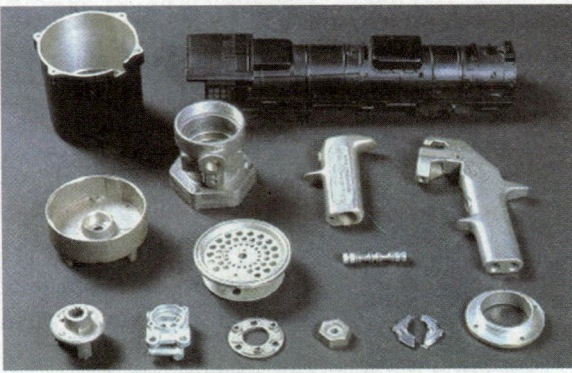

Parts made from die-casting. Applicable materials : aluminium alloy and zinc alloy.

$$\frac{1}{m} = 1.31 - 1 = 0.31 \quad \text{or} \quad m = \frac{1}{0.31} = 3.22$$

Substituting this value of m in the equation

$$K = \frac{mE}{3(m-2)} \quad K = \frac{3.22 \times 110 \times 10^9}{3(3.22-2)}$$

$$= \textbf{96.77 GN/m}^2 \quad \textbf{(Ans.)}$$

Lateral contraction, δd :

Longitudinal strain,

$$\frac{\delta l}{l} = \frac{0.0025}{2.4} = 0.00104$$

and, lateral strain

$$= 0.00104 \times \frac{1}{m} = 0.00104 \times \frac{1}{3.22} = 0.000323$$

∴ Lateral contraction,

$$\delta d = 0.000323 \, d$$
$$= 0.000323 \times 37.5 = \textbf{0.0121 mm} \quad \textbf{(Ans.)}$$

Example 1.22. *The following data relate to a bar subjected to a tensile test :*

Diameter of the bar,	*d = 30 mm (= 0.03 m)*
Tensile load,	*P = 54 kN*
Gauge length,	*l = 300 mm (= 0.3 m)*
Extension of the bar,	*δl = 0.112 mm*
Change in diameter,	*δd = 0.00366 mm*

Calculate : (i) Poisson's ratio; (ii) The values of three modulii.

Solution.

(i) Poisson's ratio $\dfrac{1}{m}$ or (μ) :

Stress,

$$\sigma = \frac{P}{A} = \frac{54}{\frac{\pi}{4}d^2} = \frac{54}{\frac{\pi}{4} \times (0.03)^2}$$

$$= 76394 \text{ kN/m}^2 = 76.4 \text{ MN/m}^2$$

Linear strain

$$= \frac{\delta l}{l} = \frac{0.112}{30} = 3.73 \times 10^{-4}$$

Lateral strain

$$= \frac{\delta d}{d} = \frac{0.00366}{30} = 1.22 \times 10^{-4}$$

∴ Poisson's ratio,

$$\mu = \frac{1}{m} = \frac{\text{Lateral strain}}{\text{Linear strain}}$$

$$= \frac{1.22 \times 10^{-4}}{3.73 \times 10^{-4}} = 0.327 \quad \textbf{(Ans.)}$$

(ii) The values of three modulii, *E*, *C* and *K* :

We know that, $E = \dfrac{\text{Stress}}{\text{Strain}} = \dfrac{76.4}{3.73 \times 10^{-4}}$

$$= \textbf{2.05} \times \textbf{10}^{\textbf{5}} \textbf{ MN/m}^{\textbf{2}} \quad \textbf{(Ans.)}$$

Also, $E = 2C\left[1 + \dfrac{1}{m}\right]$ ...(1.17)

or, $C = \dfrac{E}{2\left[1 + \dfrac{1}{m}\right]} = \dfrac{2.05 \times 10^5}{2\,(1 + 0.327)}$

$$= \textbf{0.77} \times \textbf{10}^{\textbf{5}} \textbf{ MN/m}^{\textbf{2}} \quad \textbf{(Ans.)}$$

Again, $E = 3K\left[1 - \dfrac{2}{m}\right]$

∴ $K = \dfrac{E}{3\left[1 - \dfrac{2}{m}\right]} = \dfrac{2.05 \times 10^5}{3\,(1 - 2 \times 0.327)}$

$$= \textbf{1.97} \times \textbf{10}^{\textbf{5}} \textbf{ MN/m}^{\textbf{2}} \quad \textbf{(Ans.)}$$

Example 1.23. *A C.I. flat, 300 mm long and of 30 mm × 50 mm uniform section, is acted upon by the following forces uniformly distributed over the respective cross-section; 25 kN in the direction of length (tensile); 350 kN in the direction of the width (compressive); and 200 kN in the direction of thickness (tensile). Determine the change in volume of the flat.*

Take E = 140 GN/m², and m = 4.

Solution. Refer to Fig. 1.30.

The stresses in the direction of the axes (*X, Y, Z*) are :

$$\sigma_x = \frac{350000}{0.03 \times 0.3}$$

$$= 38.8 \times 10^6 \text{ N/m}^2 \text{ (compressive)}$$

$$\sigma_y = \frac{25000}{0.05 \times 0.03}$$

$$= 16.67 \times 10^6 \text{ N/m}^2 \quad \text{(tensile)}$$

$$\sigma_z = \frac{200000}{0.3 \times 0.05} = 13.33 \times 10^6 \text{ N/m}^2$$

The strains along the three principal directions are :

$$e_x = -\frac{38.8 \times 10^6}{E} - \frac{16.67 \times 10^6}{mE} - \frac{13.33 \times 10^6}{mE}$$

$$= -\frac{10^6}{E}\left[38.8 + \frac{16.67}{m} + \frac{13.33}{m}\right]$$

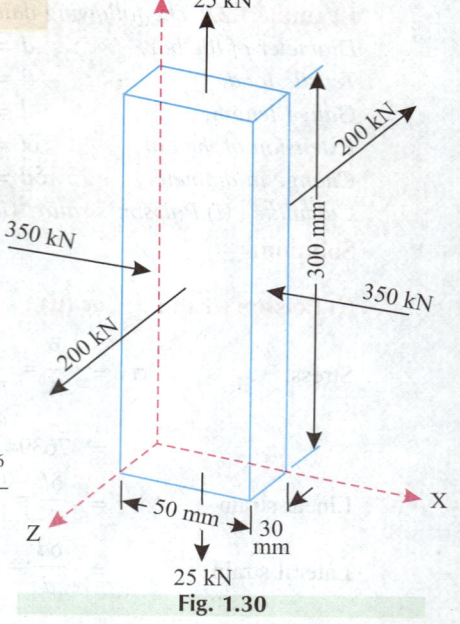

Fig. 1.30

$$= -\frac{10^6}{E}\left[38.8 + \frac{30}{m}\right]$$

$$e_y = \frac{16.67 \times 10^6}{E} + \frac{38.8 \times 10^6}{mE} - \frac{13.33 \times 10^6}{mE}$$

$$= \frac{10^6}{E}\left[16.67 + \frac{38.8}{m} - \frac{13.33}{m}\right]$$

$$= \frac{10^6}{E}\left[16.67 + \frac{25.47}{m}\right]$$

$$e_z = \frac{13.3 \times 10^6}{E} - \frac{16.67 \times 10^6}{mE} + \frac{38.8 \times 10^6}{mE}$$

$$= \frac{10^6}{E}\left[13.33 - \frac{16.67}{m} + \frac{38.8}{m}\right]$$

$$= \frac{10^6}{E}\left[13.33 + \frac{22.13}{m}\right]$$

Volumetric strain $e_v = e_x + e_y + e_z$

∴
$$e_v = -\frac{10^6}{E}\left[38.8 + \frac{30}{m}\right] + \frac{10^6}{E}\left[16.67 + \frac{25.47}{m}\right] + \frac{10^6}{E}\left[13.33 + \frac{22.13}{m}\right]$$

$$= \frac{10^6}{E}\left[-38.8 - \frac{30}{m} + 16.67 + \frac{25.47}{m} + 13.33 + \frac{22.13}{m}\right]$$

$$= \frac{10^6}{E}\left[-8.8 + \frac{17.6}{m}\right] = \frac{10^6}{E}\left[-8.8 + \frac{17.6}{m}\right] = \frac{10^6}{E}(-4.4)$$

$$= \frac{10^6}{140 \times 10^9}(-4.4) = -0.0000314 \text{ (Comp.)}$$

Change in volume (*decrease* in volume)

$$\delta V = e_v \times V \qquad\qquad \left[\because e_v = \frac{\delta V}{V}\right]$$

$$= 0.0000314 \times 0.3 \times 0.03 \times 0.05 \text{ m}^3$$

$$= 0.0000314 \times 0.3 \times 0.03 \times 0.05 \times 10^9 \text{ mm}^3 = \mathbf{14.13 \text{ mm}^3} \quad \textbf{(Ans.)}$$

Example 1.24. *A bar of steel is 60 mm × 60 mm in section and 180 mm long. It is subjected to a tensile load of 300 kN along the longitudinal axis and tensile loads of 750 kN and 600 kN on the lateral faces. Find the change in the dimensions of the bar and the change in volume.*

Take: *E= 200 GN/m², and*

$$\frac{1}{m} = 0.3.$$

Solution. Fig. 1.31 shows the bar subjected to the given load system.

Machine parts.

Change in dimensions of the bar :

Let the reference axes X, Y and Z be chosen as shown in the Fig. 1.31.

The stresses along these axes are :

$$\sigma_x = \frac{600 \times 1000}{60 \times 180 \times 10^{-6}} = 55.55 \times 10^6 \text{ N/m}^2 \text{ (tensile)}$$

$$\sigma_y = \frac{750 \times 1000}{60 \times 180 \times 10^{-6}}$$

$$= 69.44 \times 10^6 \text{ N/m}^2 \text{ (tensile)}$$

$$\sigma_z = \frac{300 \times 1000}{60 \times 60 \times 10^{-6}}$$

$$= 83.33 \times 10^6 \text{ N/m}^2 \text{ (tensile)}$$

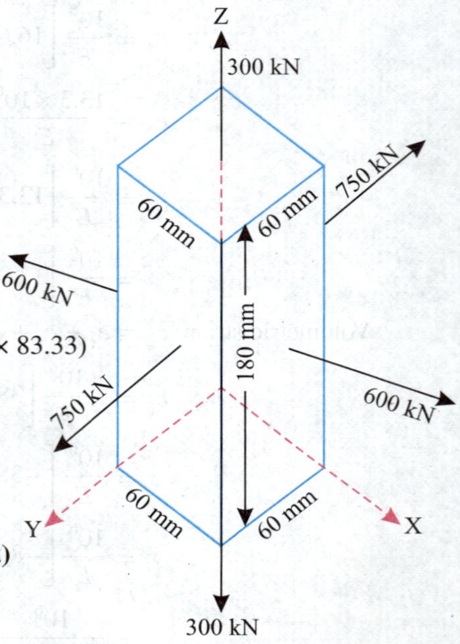

Strain along the X-axis,

$$e_x = \frac{\sigma_x}{E} - \frac{\sigma_y}{mE} - \frac{\sigma_z}{mE}$$

$$= \frac{1 \times 10^6}{200 \times 10^9} (55.55 - 0.3 \times 69.44 - 0.3 \times 83.33)$$

$$= \frac{1}{200 \times 10^3} (55.55 - 20.83 - 25)$$

$$= 4.85 \times 10^{-5}$$

∴ *Increase in dimension parallel to X-axis*

$$= 4.85 \times 10^{-5} \times 60 = \mathbf{0.00291 \text{ mm}} \quad \textbf{(Ans.)}$$

Strain along the Y-axis,

$$e_y = \frac{\sigma_y}{E} - \frac{\sigma_x}{mE} - \frac{\sigma_z}{mE}$$

$$= \frac{1 \times 10^6}{200 \times 10^9} (69.44 - 0.3 \times 55.55 - 0.3 \times 83.33)$$

$$= \frac{1}{200 \times 10^3} (69.44 - 16.66 - 25) = 1.389 \times 10^{-4}$$

Increase in dimension parallel to Y-axis

$$= 1.389 \times 10^{-4} \times 60 = \mathbf{0.00833 \text{ mm}} \quad \textbf{(Ans.)}$$

Strain along the Z-axis,

$$e_z = \frac{\sigma_z}{E} - \frac{\sigma_x}{mE} - \frac{\sigma_y}{mE}$$

$$= \frac{1 \times 10^6}{200 \times 10^9} (83.33 - 0.3 \times 55.55 - 0.3 \times 69.44)$$

$$= \frac{1}{200 \times 10^3} (83.33 - 16.66 - 20.83) = 2.292 \times 10^{-4}$$

Increase in dimension parallel to Z-axis

$$= 2.291 \times 10^{-4} \times 180 = \mathbf{0.0412 \text{ mm}} \quad \textbf{(Ans.)}$$

Change in volume, δV :

Volumetric strain,

Fig. 1.31

$$e_v = \frac{\delta V}{V}$$

But, $e_v = e_x + e_y + e_z$

$$= 4.85 \times 10^{-5} + 1.389 \times 10^{-4} + 2.292 \times 10^{-4}$$

$$= 10^{-4} (0.485 + 1.389 + 2.292) = 4.166 \times 10^{-4}$$

∴ $\delta V = e_v \times V = 4.166 \times 10^{-4} \times (60 \times 60 \times 180)$ mm^3

$$= \mathbf{269.9 \ mm^3} \ \textbf{(Ans.)}$$

1.12. STRESSES INDUCED IN COMPOUND TIES OR STRUTS

Frequently ties consist of two materials, fastened together to prevent uneven straining of the two materials. In these cases, it is interesting to calculate the distribution of the load between the materials. It will be assumed that the two materials also are symmetrically distributed about the axis of the bar, as with a cylindrical rod encased in a tube (Fig. 1.32).

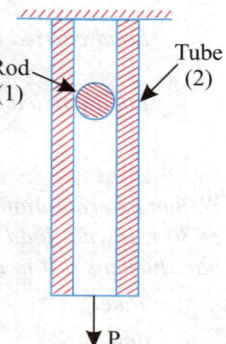

If an axial load P is applied to the bar,

$$P = \sigma_1 A_1 + \sigma_2 A_2 \qquad ...(1.20)$$

where, σ_1 and σ_2 are the stresses induced and A_1 and A_2 are the cross-sectional areas of the materials.

The strains produced, e_1 and e_2 are equal.

∴ $e_1 = e_2$

Fig. 1.32

∴ $\dfrac{\sigma_1}{E_1} = \dfrac{\sigma_2}{E_2} \Rightarrow \dfrac{\sigma_1}{\sigma_2} = \dfrac{E_1}{E_2}$ $\qquad ...[1.20\ (a)]$

Example 1.25. *A concrete column of cross-sectional area 400 mm × 400 mm is reinforced by four longitudinal 50 mm diameter round steel bars placed at each corner. If the column carries a compressive load of 300 kN, determine :*

(i) Loads carried;

(ii) The compressive stress produced in the concrete and steel bars.

Young's modulus of elasticity of steel is 15 times that of concrete.

Solution. Refer to Fig. 1.33.

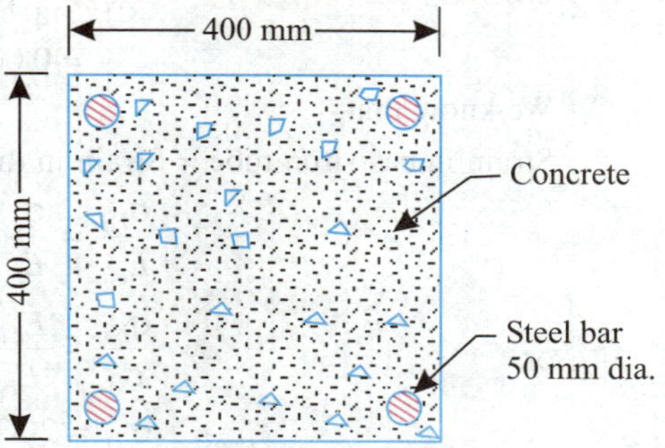

Cross-sectional area of the column

$$= 0.4 \times 0.4 = 0.16 \ m^2$$

Area of steel bars,

$$A_s = 4 \times \frac{\pi}{4} \times (0.05)^2$$

$$= 0.00785 \ m^2$$

∴ Area of concrete,

$$A_c = 0.16 - 0.00785 = 0.1521 \ m^2$$

Since the steel bars and concrete shorten by the *same amount* under the compressive load,

∴ Strain in steel bars = Strain in concrete

or, $e_s = e_c$

$$\frac{\sigma_s}{E_s} = \frac{\sigma_c}{E_c} \quad \text{or,} \quad \sigma_s = \sigma_c \cdot \frac{E_s}{E_c} = 15 \ \sigma_c \quad (\because \ E_s = 15 \ E_c)$$

Fig. 1.33

Also, load shared by steel bars + load shared by concrete = 300000 N

or, $P_s + P_c$ = 300000 N

or, $\sigma_s \times A_s + \sigma_c \times A_c$ = 300000

15 $\sigma_c \times$ 0.00785 + $\sigma_c \times$ 0.1521 = 300000

σ_c (15 × 0.00785 + 0.1521) = 300000

or, compressive stress in concrete,

$$\sigma_c = 1.11 \times 10^6 \text{ N/m}^2 = \textbf{1.11 MN/m}^2 \textbf{ (Ans.)}$$

and, compressive stress in steel bars,

$$\sigma_s = 15 \ \sigma_c = 15 \times 1.11 \times 10^6$$
$$= 16.65 \times 10^6 \text{ N/m}^2 = \textbf{16.65 MN/m}^2 \textbf{ (Ans.)}$$

Load carried by steel bars, $P_s = \sigma_s \times A_s = 16.65 \times 10^6 \times 0.00785$
$$= 130.7 \text{ kN} \approx \textbf{131 kN (Ans.)}$$

Load carried by concrete, $P_c = 1.11 \times 10^6 \times 0.1521$
$$= 168.9 \approx \textbf{169 kN (Ans.)}$$

Example 1.26. *A copper rod of 40 mm diameter is surrounded tightly by a cast-iron tube of 80 mm external diameter, the ends being firmly fastened together. When put to a compressive load of 30 kN, what load will be shared by each ? Also determine the amount by which the compound bar shortens if it is 2 m long.*

Take: $E_{c.i}$ = 175 GN/m^2,

and E_c = 75 GN/m^2.

Solution. Refer to Fig. 1.34.

Diameter of the copper rod   = 0.04 m

∴   Area of copper rod,

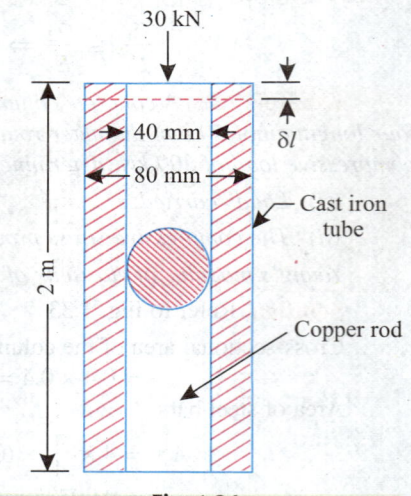

$$A_c = \frac{\pi}{4} \times 0.04^2$$
$$= 0.0004 \ \pi \text{m}^2$$

External diameter of cast iron tube = 0.08 m

∴   Area of cast iron tube,

$$A_{ci} = \frac{\pi}{4} (0.08^2 - 0.04^2)$$
$$= 0.0012 \ \pi \text{ m}^2$$

We know that,

Strain in cast-iron tube = Strain in the copper rod

$$\frac{\sigma_{c.i}}{E_{c.i}} = \frac{\sigma_c}{E_c}$$

or, $\dfrac{\sigma_{c.i}}{\sigma_c} = \dfrac{E_{c.i}}{E_c} = \dfrac{175 \times 10^9}{75 \times 10^9} = 2.33$

Fig. 1.34

∴ $\sigma_{c.i.} = 2.33 \ \sigma_c$ ...(*i*)

Also, total load = Load shared by cast-iron tube + load shared by copper rod.

or, $P = P_{c.i.} + P_c$

$$30 = \sigma_{c.i.} . A_{c.i} + \sigma_c . A_c$$

or, $30 = \sigma_{c.i.} \times 0.0012\pi + \sigma_c \times 0.0004 \ \pi$ ...(*ii*)

Substituting the value of $\sigma_{c.i.}$ from (*i*) in (*ii*), we get

$$30 = 2.33 \ \sigma_c \times 0.0012 \ \pi + \sigma_c \times 0.0004 \ \pi$$

$$= \sigma_c (0.008785 + 0.001257)$$
$$= 0.010042 \, \sigma_c$$
$$\sigma_c = \frac{30}{0.010042} = 2987.5 \text{ kN/m}^2$$

And, from equation (i),

$$\sigma_{c.i.} = 2.33 \, \sigma_c = 2.33 \times 2987.5$$
$$= 6960.8 \text{ kN/m}^2$$

Load shared by the copper rod,

$$P_c = \sigma_c A_c = 2987.5 \times 0.0004 \, \pi$$
$$= \textbf{3.75 kN} \quad \textbf{(Ans.)}$$

Load shared by cast iron tube,

$$P_{ci} = 30 - 3.75 = \textbf{26.25 kN} \quad \textbf{(Ans.)}$$

Strain $= \dfrac{\sigma_c}{E_c}$

or, $\quad \dfrac{\sigma_c}{E_c} = \dfrac{\text{Decrease in length}}{\text{Original length}} = \dfrac{\delta l}{l}$

∴ $\quad \delta l = \dfrac{\sigma_c}{E_c} \times l = \dfrac{2987.5 \times 10^3}{75 \times 10^9} \times 2$

$$= 0.0000796 \text{ m} \quad \text{or} \quad 0.0796 \text{ mm}$$

Hence, decrease in length = **0.0796 mm** **(Ans.)**

Example 1.27. *A solid steel cylinder 500 mm long and 70 mm diameter is placed inside an aluminium cylinder having 75 mm inside diameter and 100 mm outside diameter. The aluminium cylinder is 0.16 mm longer than the steel cylinder. An axial load of 500 kN is applied to the bar and cylinder through rigid cover plates as shown in Fig. 1.35. Find the stress developed in the steel cylinder and aluminium tube. Assume for steel, E = 220 GN/m² and for aluminium, E = 70 GN/m².*

Solution. Refer to Fig. 1.35.

Since the aluminium cylinder is 0.16 mm longer than the steel cylinder, the load required to compress this cylinder by 0.16 mm will be found as follows :

$$E = \frac{\text{Stress}}{\text{Strain}} = \frac{P/A}{\delta l/l} = \frac{Pl}{A.\delta l}$$

or, $\quad P = \dfrac{E.A.\delta l}{l}$

$$= \frac{70 \times 10^9 \times \dfrac{\pi}{4}(0.1^2 - 0.075^2) \times 0.00016}{0.50016} = 76944 \text{ N}$$

Bicycle brakes are the forerunners of the modern automobile brakes.

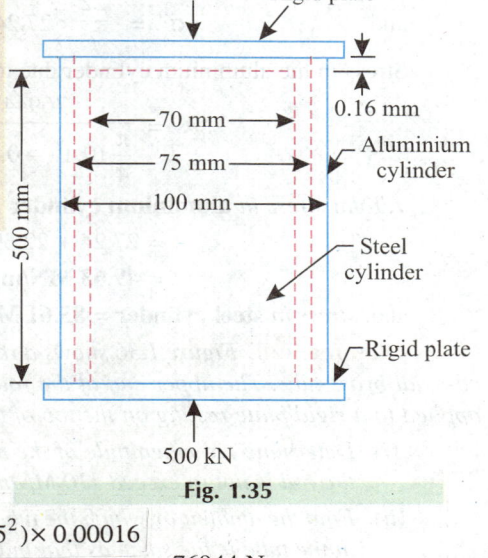

Fig. 1.35

When the aluminium cylinder is compressed by its extra length 0.16 mm, the load then shared by both aluminium as well as steel cylinder will be,

$$500000 - 76944 = 423056 \text{ N}$$

Let,
$$e_s = \text{Strain in steel cylinder,}$$
$$e_a = \text{Strain in aluminium cylinder,}$$
$$\sigma_s = \text{Stress produced in steel cylinder, and}$$
$$\sigma_a = \text{Stress produced in aluminium cylinder.}$$
$$E_s = 220 \text{ GN/m}^2$$
$$E_a = 70 \text{ GN/m}^2$$

As both the cylinders are of the same length and are compressed by the same amount,

$$\therefore \qquad e_s = e_a$$

or,
$$\frac{\sigma_s}{E_s} = \frac{\sigma_a}{E_a}$$

or,
$$\sigma_s = \frac{E_s}{E_a} \cdot \sigma_a$$

$$= \frac{220 \times 10^9}{70 \times 10^9} \times \sigma_a = \frac{22}{7} \sigma_a$$

Also,
$$P_s + P_a = P$$

or,
$$\sigma_s . A_s + \sigma_a . A_a = 423056$$

$$\frac{22}{7} \sigma_a \times A_s + \sigma_a \times A_a = 423056$$

$$\therefore \qquad \sigma_a = \frac{423056}{\dfrac{22}{7} A_s + A_a} = \frac{423056}{\dfrac{22}{7} \times \dfrac{\pi}{4} \times 0.07^2 + \dfrac{\pi}{4}(0.1^2 - 0.075^2)}$$

$$= \frac{423056}{0.012095 + 0.003436} = 27.24 \times 10^6 \text{ N/m}^2 = 27.24 \text{ MN/m}^2$$

and,
$$\sigma_s = \frac{22}{7} \times 27.24 = 85.61 \text{ MN/m}^2$$

Stress in the aluminium cylinder due to load 76944 N

$$= \frac{76944}{\dfrac{\pi}{4}(0.1^2 - 0.075^2)} = 22.39 \times 10^6 \text{ N/m}^2 = 22.39 \text{ MN/m}^2$$

∴ Total stress in **aluminium cylinder**

$$= 27.24 + 22.39$$
$$= \textbf{49.63 MN/m}^2 \quad \textbf{(Ans.)}$$

and, *stress in* **steel cylinder** = **85.61 MN/m²** **(Ans.)**

Example 1.28. *Figure 1.36 shows a round steel rod supported in a recess and surrounded by co-axial brass tube. The upper end of the rod is 0.1mm below that of the tube and an axial load is applied to a rigid plate resting on the top of the tube.*

(i) *Determine the magnitude of the maximum permissible load if the compressive stress in the rod is not to exceed 110 MN/m² and that in the tube is not to exceed 80 MN/m².*

(ii) *Find the amount by which the tube will be shortened by the load if the compressive stress in the tube is the same as that in the rod.*

Take:
$$E_s = 200 \text{ GN/m}^2;$$
$$E_b = 100 \text{ GN/m}^2.$$

(Panjab University)

Tower bridge, London. In bridges different members bear different magnitudes of tension, compression and shear stresses.

Solution. Refer to Fig. 1.36.

(*i*) **Maximum permissible load, *P* :**

Area of steel rod,

$$A_s = \frac{\pi}{4} \times \left(\frac{30}{1000}\right)^2 = 706.86 \times 10^{-6} \text{ m}^2$$

Area of brass tube,

$$A_b = \frac{\pi}{4} \times \left[\left(\frac{50}{1000}\right)^2 - \left(\frac{45}{1000}\right)^2\right] = 373.06 \times 10^{-6} \text{ m}^2$$

Let σ_s and σ_b MN/m^2 be the stresses induced in the steel rod and brass tube respectively.
Since the sum of the loads carried by the steel and brass is equal to the total load,

∴ Total load, $P = \sigma_s A_s + \sigma_b A_b = 706.86 \times 10^{-6} \sigma_s + 373.06 \times 10^{-6} \sigma_b$ ...(*i*)

When the top plate makes contact with steel and compresses it, the compression of the brass will exceed that of steel by 0.1 mm.

i.e., $\delta l_b = \delta l_s + 0.0001$ m

or, $\dfrac{\sigma_b \, l_b}{E_b} = \dfrac{\sigma_s \, l_s}{E_s} + 0.0001$

or, $\dfrac{\sigma_b \times 0.3}{100 \times 10^3} = \dfrac{\sigma_s \times 0.4}{200 \times 10^3} + 0.0001$

or, $\sigma_b = 0.667 \, \sigma_s + 33.3$ ...(*ii*)

The maximum stress of 110 MN/m^2 in the rod and 80 MN/m^2 in the tube will not occur simultaneously, rather the magnitudes of induced stresses in the two materials will be related to each other by eqn. (*ii*). Thus if it is assumed that the stress in steel is the limiting case, then from eqn. (*ii*) for $\sigma_s = 110$ MN/m^2,

$\sigma_b = 0.667 \times 110 + 33.3$

$= 106.7$ MN/m^2

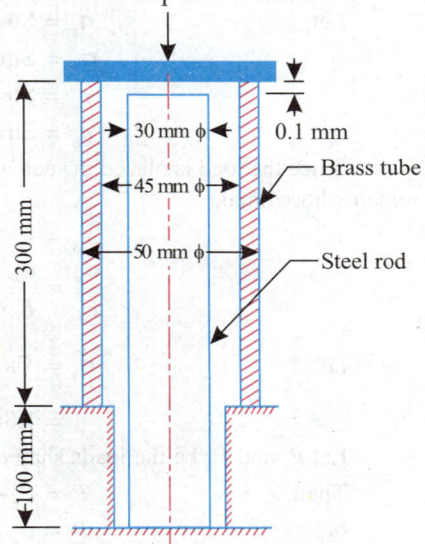

Fig. 1.36

which is more than the allowable stress in brass leading to the conclusion that the assumption is wrong. Then again assuming the stress in brass to be the limiting case, from (ii) we have, for $\sigma_b = 80 \text{ MN/m}^2$,

$$80 = 0.667\,\sigma_s + 33.3$$

∴
$$\sigma_s = \frac{80 - 33.3}{0.667} = 70 \text{ MN/m}^2$$

which is safe, being less than the maximum allowable stress in steel. Thus substituting the values of σ_b and σ_s in eqn. (i), we get

$$P = 706.86 \times 10^{-6} \times 70 + 373.06 \times 10^{-6} \times 80$$

$$= 0.079325 \text{ MN or } 79.325 \text{ kN}$$

Hence, maximum permissible load, $P = \textbf{79.325 kN}$ (**Ans.**)

(**ii**) **The amount by which the tube will be shortened, δl_b :**

In this case
$$\sigma_s = \sigma_b = 110 \text{ MN/m}^2 \quad (Given)$$

$$\delta l_b = \frac{\sigma_b l_b}{E_b} = \frac{110 \times 0.3}{100 \times 10^3} = 0.00033 \text{ m or } 0.33 \text{ mm}$$

Hence,
$$\delta l_b = \textbf{0.33 mm} \quad (\textbf{Ans.})$$

Example 1.29. *Two vertical rods, one of steel and the other of bronze, are rigidly fastened at upper ends at a horizontal distance of 760 mm apart. Each rod is 3 m long and 25 mm in diameter. A horizontal cross-piece connects the lower ends of the bars. Where should a load of 4.5 kN be placed on the cross-piece so that it remains horizontal after being loaded?*

Determine the stresses in each rod.

Given:
$$E_s = 210 \text{ GN/m}^2;$$
$$E_b = 112.5 \text{ GN/m}^2.$$

Solution. Refer to Fig. 1.37.

Length of each rod   = 3 m

Diameter of each rod   = 25 mm = 0.025 m

Load, $P = 4.5$ kN

Let, σ_s = Stress in steel rod,

σ_b = Stress in brass rod,

e_s = Strain in steel rod, and

e_b = Strain in bronze rod.

Since the load is placed in such a manner that the cross-piece remains horizontal,

∴ $e_s = e_b$

or, $\dfrac{\sigma_s}{E_s} = \dfrac{\sigma_b}{E_b}$

i.e., $\sigma_s = \sigma_b \cdot \dfrac{E_s}{E_b} = \sigma_b \cdot \dfrac{210 \times 10^9}{112.5 \times 10^9}$

$= 1.867\,\sigma_b$ ...(i)

Let P_s and P_b be the loads shared by the steel and bronze rods respectively.

Then, $P = P_s + P_b$

or, $P = \sigma_s \cdot A_s + \sigma_b \cdot A_b$ ... (ii)

where, A_s = Cross-sectional area of steel rod, and

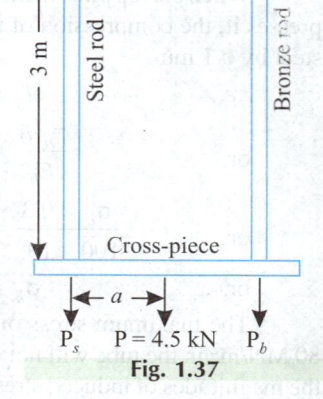

Fig. 1.37

A_b = Cross-sectional area of bronze rod.

$\therefore$ $$P = 1.867\, \sigma_b \times \frac{\pi}{4} \times (0.025)^2 + \sigma_b \times \frac{\pi}{4} \times (0.025)^2$$

$$4500 = 0.000916\, \sigma_b + 0.000491\, \sigma_b$$

$$\sigma_b = 3.19 \times 10^6 \text{ N/m}^2 = 3.19 \text{ MN/m}^2$$

and, $$\sigma_s = 1.867 \times 3.19 = 5.956 \text{ MN/m}^2$$

$\therefore$ $$P_b = \sigma_b A_b = 3.19 \times 10^6 \times \frac{\pi}{4} \times (0.025)^2 = 1566 \text{ N}$$

Let 'a' be the distance from the steel rod where the load P should be placed so that the cross - piece remains horizontal after being loaded.

Then, $$P_b \times 760 = P \times a$$

$\therefore$ $$a = \frac{P_b \times 760}{P} = \frac{1566 \times 760}{4500}$$

$$= 265 \text{ mm} \quad \textbf{(Ans.)}$$

Example 1.30. *A beam weighing 450 N is held in a horizontal position by three vertical wires, one attached to each end of the beam, one to the middle of its length. The outer wires are of brass of diameter 1.25 mm and the central wire is of steel of diameter 0.625 mm. If the beam is rigid and wires of the same length and unstressed before the beam is attached, estimate the stresses induced in the wires. Take Young's modulus for brass as 86 GN/m² and for steel 210 GN/m².*

Solution. Refer to Fig. 1.38.

Let, P_b = Load taken by the brass wire, and

P_s = Load taken by the steel wire.

Then, $2P_b + P_s = P$ ...(i)

Since the beam is *horizontal*, all wires will extend by the *same amount.*

i.e., $$e_b = e_s$$

($\because$ Length of each wire is same.)

where, e_b = Strain in brass wire, and

e_s = Strain in steel wire.

$$\frac{\sigma_b}{E_b} = \frac{\sigma_s}{E_s}$$

$$\frac{P_b}{A_b \cdot E_b} = \frac{P_s}{A_s \cdot E_s}$$

or, $$P_s = \frac{P_b A_s E_s}{A_b E_b}$$

Bronze flanges.

$$= \frac{P_b \times \dfrac{\pi}{4} \times (0.625 \times 10^{-3})^2 \times 210 \times 10^9}{\dfrac{\pi}{4} \times (1.25 \times 10^{-3})^2 \times 86 \times 10^9}$$

or, $$P_s = 0.61\, P_b$$ ...(ii)

Substituting the value of P_s in eqn. (i), we get

$$2\, P_b + 0.61\, P_b = P$$

$$2.61\, P_b = 450$$

$$P_b = 172.4 \text{ N}$$

and, $\quad P_s = 0.61 \times 172.4 = 105.2 \text{ N}$

Now, *stress, induced in the brass wire,*

$$\sigma_b = \frac{P_b}{A_b} = \frac{172.4}{\dfrac{\pi}{4} \times (1.25 \times 10^{-3})^2} = 1.40 \times 10^8 \text{ N/m}^2$$

$$= \mathbf{140 \ MN/m^2} \quad \textbf{(Ans.)}$$

and, *stress induced in a steel wire,*

$$\sigma_s = \frac{105.2}{\dfrac{\pi}{4} \times (0.625 \times 10^{-3})^2} = 3.429 \times 10^8 \text{ N/m}^2$$

$$= \mathbf{342.9 \ MN/m^2} \quad \textbf{(Ans.)}$$

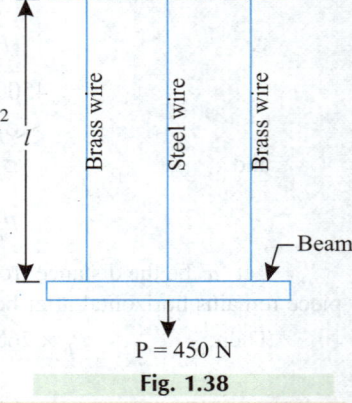

Fig. 1.38

Example 1.31. *A rigid bar is supported by three rods in the same vertical plane and equidistant. The outer rods are of brass and of length 600 mm and diameter 30 mm. The central rod is of steel of 900 mm length and of 37.5 mm diameter. Calculate the forces in the bars due to an applied force P, if the bar remains horizontal after the load has been applied. Take* $\dfrac{E_s}{E_b} = 2.$

Solution. Refer to Fig. 1.39. Let suffix 1 be used for brass and 2 for steel.

$$A_1 = \frac{\pi}{4} \times (30)^2 = 225 \ \pi \ \text{mm}^2$$

$$A_2 = \frac{\pi}{4} \times (37.5)^2 = 351.56 \ \pi \ \text{mm}^2$$

From statical equilibrium, we have

$$2P_1 + P_2 = P \qquad ...(i)$$

From compatibility equation, we have

$$\delta l_1 = \delta l_2$$

$$\frac{P_1 \, l_1}{A_1 \, E_1} = \frac{P_2 \, l_2}{A_2 \, E_2}$$

or, $\quad P_2 = \dfrac{P_1 \, A_2 \, E_2 \, l_1}{A_1 \, E_1 \, l_2}$

$$= P_1 \times \frac{351.56 \ \pi}{225 \ \pi} \times \frac{2}{1} \times \frac{600}{900}$$

$$= 2.08 \ P_1 \qquad ...(ii)$$

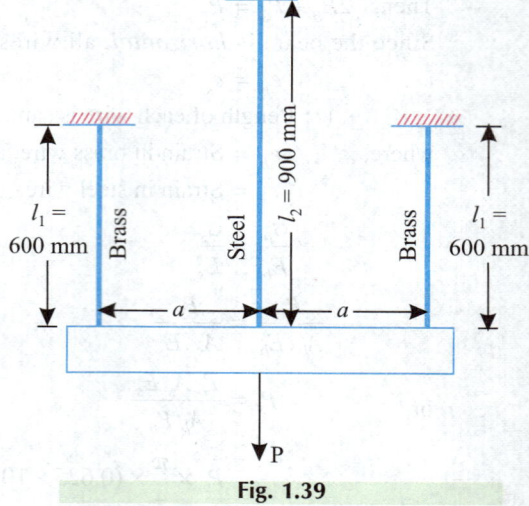

Substituting in (*i*), we get

$$2P_1 + 2.08 \ P_1 = P$$

or, $\quad P_1 = \mathbf{0.245 \ P} \quad \textbf{(Ans.)}$

and, $\quad P_2 = 2.08 \times 0.245 \ P = \mathbf{0.51 \ P} \quad \textbf{(Ans.)}$

Fig. 1.39

Example 1.32. *Two copper rods and one steel rod together support a load as shown in Fig. 1.40. If the stresses in copper and steel are not to exceed* $60 \times 10^6 \text{ N/m}^2$ *and* $120 \times 10^6 \text{ N/m}^2$*, respectively, find the safe load that can be supported. Young's modulus for steel is twice that of copper.*

Solution. Refer to Fig. 1.40.

Each rod will be compressed to the same extent. Let each rod contract by δ metre, then $\delta = e_s \times l_s$ $= e_c \times l_c$ (e_s and e_c being strains in steel and copper respectively)

$$\therefore \qquad \frac{e_s}{e_c} = \frac{l_c}{l_s} = \frac{0.15}{0.20} = 0.75$$

Fig. 1.40

Let the stresses in steel and copper be σ_s and σ_c respectively.

Then, $\qquad \sigma_s = e_s \times E_s \quad$ and, $\sigma_c = e_c \times E_c$

$$\therefore \qquad \frac{\sigma_s}{\sigma_c} = \frac{e_s}{e_c} \times \frac{E_s}{E_c} = 0.75 \times 2 = 1.5$$

$$\therefore \qquad \sigma_s = 1.5\, \sigma_c$$

If the steel is permitted to reach its safe stress of 120×10^6 N / m^2, the corresponding stress in copper will be $\dfrac{120 \times 10^6}{1.5} = 80 \times 10^6$ N/m² for copper which exceeds the safe stress of 60×10^6 N/m². Therefore, let copper be allowed to reach its safe stress of 60×10^6 N/m², the corresponding stress in steel will be $60 \times 10^6 \times 1.5 = 90 \times 10^6$ N/m².

$\therefore$ *Total load,P* $\quad$ = Load on steel + load on copper

$$= P_s + P_c = \sigma_s \times A_s + \sigma_c \times A_c$$
$$= 90 \times 10^6 \times 0.03 \times 0.03 + 2 \times 60 \times 10^6 \times 0.025 \times 0.025$$

$$= 156000 \text{ N} \quad \text{or } \textbf{156 kN} \quad \textbf{(Ans.)}$$

Bushings.

Example 1.33. *A steel bolt and sleeve assembly is shown in Fig. 1.41. The nut is tightened up on the tube through the rigid end blocks until the tensile force in the bolt is 40 kN. If an external load 30 kN is then applied to the end blocks, tending to pull them apart, estimate the resulting force in the bolt and sleeve.* **(AMIE Winter, 2001)**

Solution. Area of steel bolt, $\quad A_b = \dfrac{\pi}{4} \times \left(\dfrac{25}{1000}\right)^2 = 4.908 \times 10^{-4}$ m²

Area of steel sleeve, $\qquad A_s = \dfrac{\pi}{4} \times \left[\left(\dfrac{62.5}{1000}\right)^2 - \left(\dfrac{50}{1000}\right)^2\right] = 1.104 \times 10^{-3}$ m²

Forces in the bolt and sleeve:

(i) Stresses due to tightening the nut:

Let, $\qquad\qquad \sigma_b$ = Stress developed in steel bolt due to tightening the nut, and

$\qquad\qquad \sigma_s$ = Stress developed in steel sleeve due to tightening the nut.

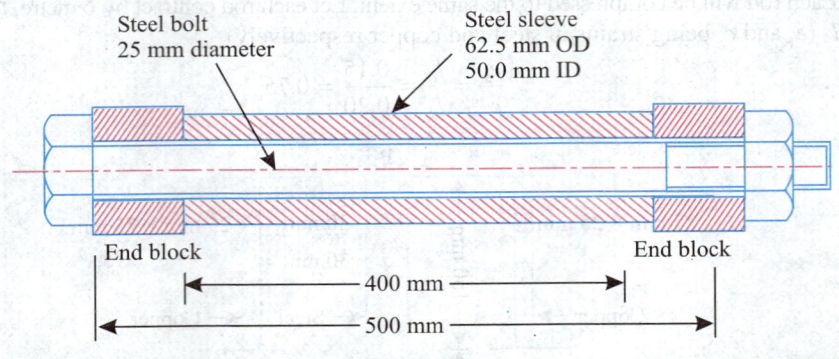

Fig. 1.41

Tensile force in the steel bolt $= 40$ kN $= 0.04$ MN (given)

Hence, $\qquad \sigma_b \times A_b = 0.04$

or, $\qquad \sigma_b \times 4.908 \times 10^{-4} = 0.04$

$\therefore \qquad \sigma_b = \dfrac{0.04}{4.908 \times 10^{-4}} = 81.5$ MN/m^2 (tensile)

Compressive force in steel sleeve $= 0.04$ MN

$$\sigma_s \times A_s = 0.04$$

or, $\qquad \sigma_s \times 1.104 \times 10^{-3} = 0.04$

$\therefore \qquad \sigma_b = \dfrac{0.04}{1.104 \times 10^{-3}} = 36.23$ MN/m^2 (compressive)

(ii) *Stresses due to tensile force:*

Let the stresses developed due to tensile force of 30 kN ($= 0.03$ MN) in steel bolt and sleeve be σ'_b and σ'_s respectively.

Then, $\qquad \sigma'_b \times A_b + \sigma'_s \times A_s = 0.03$

$$\sigma'_b \times 4.908 \times 10^{-4} + \sigma'_s \times 1.104 \times 10^{-3} = 0.03 \qquad \qquad ...(1)$$

In a compound system with an external tensile load, elongation caused in each will be the *same*.

$$\delta l_b = \frac{\sigma'_b}{E_b} \times l_b$$

or, $\qquad \delta l_b = \dfrac{\sigma'_b}{E_b} \times 0.5 \qquad \qquad (\because\ l_b = 500 \text{ mm} = 0.5 \text{ m})$

and, $\qquad \delta l_s = \dfrac{\sigma'_s}{E_s} \times 0.4 \qquad \qquad (\because\ l_s = 400 \text{ mm} = 0.4 \text{ m})$

But, $\qquad \delta l_b = \delta l_s$

$\therefore \qquad \dfrac{\sigma'_b}{E_b} \times 0.5 = \dfrac{\sigma'_s}{E_s} \times 0.4$

or, $\qquad \sigma'_b = 0.8\, \sigma'_s \qquad \qquad (\because\ E_b = E_s)\ ...(2)$

Substituting this value in eqn. (1), we get

$$0.8\, \sigma'_s \times 4.908 \times 10^{-4} + \sigma'_s \times 1.104 \times 10^{-3} = 0.03$$

$$\sigma'_s (0.8 \times 4.908 \times 10^{-4} + 1.104 \times 10^{-3}) = 0.03$$

$\therefore \qquad \sigma'_s = 20$ MN/m^2 (tensile)

and, $\qquad \sigma'_b = 0.8 \times 20 = 16$ MN/m^2 (tensile)

Resulting stress in steel bolt,

$$(\sigma_b)_r = \sigma_b + \sigma'_b = 81.5 + 16 = 97.5 \text{ MN/m}^2$$

Resulting stress in steel sleeve,

$$(\sigma_s)_r = \sigma_s + \sigma'_s = 36.23 - 20 = 16.23 \text{ MN/m}^2 \text{ (compressive)}$$

Resulting force in steel bolt $= (\sigma_b)_r \times A_b$

$$= 97.5 \times 4.908 \times 10^{-4} = \textbf{0.0478} \text{ (\textit{tensile}) \textbf{MN} (Ans.)}$$

Resulting force in steel sleeve $= (\sigma_s)_r \times A_s$

$$= 16.23 \times 1.104 \times 10^{-3} = \textbf{0.0179 MN} \text{ (\textit{compressive}) (Ans.)}$$

Example 1.34. *A steel rod 20 mm diameter passes centrally through a steel tube 25 mm internal diameter and 40 mm external diameter. The tube is 750 mm long and is closed by rigid washers of negligible thickness which are fastened by nuts threaded on the rod. The nuts are tightened until the compressive load on the tube is 20 kN. Calculate the stresses in the tube and the rod.*

Find the increase in these stresses when one nut is tightened by one quarter of a turn relative to the other. There are 0.4 threads per mm length. Take E = 200 GN / m².

Solution. Refer to Fig. 1.42.

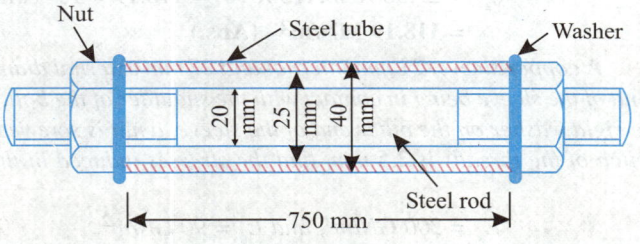

Nut   Steel tube   Washer

20 mm   25 mm   40 mm

Steel rod

750 mm

Fig. 1.42

Area of steel rod, $\quad A_{sr} = \dfrac{\pi}{4} \times (0.02)^2 = 0.0003142 \text{ m}^2$

Area of steel tube, $\quad A_{st} = \dfrac{\pi}{4} \times [\, (0.04)^2 - (0.025)^2 \,] = 0.000766 \text{ m}^2$

Compressive load on the steel tube = Tensile load in the steel rod

i.e., $\quad \sigma_{st_1} \times A_{st} = \sigma_{sr_1} \times A_{sr} = 20000$

where, σ_{st_1} and σ_{sr_1} are the compressive and tensile stresses in steel tube and steel rod respectively in N/m².

or, $\qquad 0.000766 \, \sigma_{st_1} = 0.003142 \, \sigma_{sr_1} = 20000$

or, $\qquad \sigma_{st_1} = \dfrac{20000}{0.000766} = 26.1 \times 10^6 \text{ N/m}^2$

$$= \textbf{26.1 MN/m}^2 \text{ (comp.) (Ans.)}$$

and, $\qquad \sigma_{sr_1} = \dfrac{20000}{0.0003142} = 63.6 \times 10^6 \text{ N/m}^2 \text{ (tensile)}$

$$= \textbf{63.6 MN/m}^2 \text{ (tensile) (Ans.)}$$

When nut is tightened by one quarter, let σ_{sr_2} and σ_{st_2} be the additional stresses produced in the rod and the tube respectively.

Distance travelled by the nut in one quarter of a revolution

$$= \dfrac{1}{4} \times \dfrac{0.001}{0.4} = 0.000625 \text{ m}$$

Also, $\quad\quad\quad$ 0.000625 = Extension in rod + contraction in tube

$$= \frac{\sigma_{sr_2}}{E} \times l + \frac{\sigma_{st_2}}{E} \times l = \frac{l}{E}(\sigma_{sr_2} + \sigma_{st_2}) \quad\quad ...(i)$$

Again, $\quad\quad\quad \sigma_{sr_2} \times A_{sr} = \sigma_{st_2} \times A_{st}$

or, $\quad\quad\quad \sigma_{sr_2} = \frac{\sigma_{st_2} \times A_{st}}{A_{sr}} = \frac{\sigma_{st_2} \times 0.000766}{0.0003142} = 2.438\,\sigma_{st_2} \quad\quad ...(ii)$

Substituting the value of σ_{sr_2} in equation (i), we get

$$0.000625 = \frac{0.75}{200}(2.438\,\sigma_{st_2} + \sigma_{st_2})$$

or, $\quad\quad\quad \sigma_{st_2} = \frac{0.000625 \times 200 \times 10^9}{0.75 \times 3.438}$

or, $\quad\quad\quad \sigma_{st_2} = 48.478 \times 10^6 \text{ N/m}^2$

$$= \textbf{48.478 MN/m}^2 \quad \textbf{(Ans.)}$$

and, $\quad\quad\quad \sigma_{sr_2} = 2.438 \times 48.478 \times 10^6 = 118.19 \times 10^6 \text{ N/m}^2$

$$= \textbf{118.19 MN/m}^2 \quad \textbf{(Ans.)}$$

Example 1.35. *A copper sleeve, 21 mm internal and 27 mm external diameter, surrounds a 20 mm steel bolt, one end of the sleeve being in contact with the shoulder of the bolt. The sleeve is 60 mm long. After putting a rigid washer on the other end of the sleeve, a nut is screwed on the bolt through 10 degrees. If the pitch of the threads is 2.5 mm, find the stresses induced in the copper sleeve and steel bolt.*

Take: $\quad\quad\quad E_s = 200 \text{ GN/m}^2,$ *and* $E_c = 90 \text{ GN/m}^2$ $\quad\quad$ **(Panjab University)**

Solution. Area of copper sleeve,

$$A_c = \frac{\pi}{4}\left[\left(\frac{27}{1000}\right)^2 - \left(\frac{21}{1000}\right)^2\right] = 226.19 \times 10^{-6} \text{ m}^2$$

Area of steel bolt, $\quad\quad A_s = \frac{\pi}{4}\left(\frac{20}{1000}\right)^2 = 314.16 \times 10^{-6} \text{ m}^2$

Length of the sleeve $\quad\quad = 60 \text{ mm or } 0.06 \text{ m}$

Pitch of the threads, $\quad\quad p = 2.5 \text{ mm}$

Number of degrees through which nut is screwed on the bolt = 10 degrees

Stress induced in steel, σ_s :

Stress induced in copper, σ_c :

Due to rotation the nut will move along the axis of the bolt through a distance δ (say), such that

$$= \frac{10}{360} \times \text{pitch} = \frac{10}{360} \times 2.5 = 0.0694 \text{ mm or } 0.0694 \times 10^{-3} \text{ m}$$

Further, the axial movement of the nut = Compression in the copper sleeve + stretch in the steel bolt

i.e. $\quad\quad\quad \delta = e_c l_c + e_s l_s$

$$0.0694 \times 10^{-3} = 0.06\,(e_c + e_s) \quad\quad\quad (\because\ l_c = l_s = 0.06 \text{ m})$$

or, $\quad\quad\quad \dfrac{0.0694 \times 10^{-3}}{0.06} = \dfrac{\sigma_c}{E_c} + \dfrac{\sigma_s}{E_s} \quad\quad\quad ...(i)$

Also the tube and the bolt, in equilibrium, will carry the same load.

i.e.,
$$P_c = P_s$$
or
$$\sigma_c A_c = \sigma_s A_s$$

or
$$\sigma_c = \frac{\sigma_s A_s}{A_c} = \frac{\sigma_s \times 314.16 \times 10^{-6}}{226.19 \times 10^{-6}} = 1.39\,\sigma_s \qquad \qquad ...(ii)$$

From (*i*) and (*ii*), we get

$$\frac{0.0694 \times 10^{-3}}{0.06} = \frac{1.39\,\sigma_s}{90 \times 10^9} + \frac{\sigma_s}{200 \times 10^9}$$

$$1.156 \times 10^{-3} = \frac{\sigma_s}{10^{-9}}\left(\frac{1.39}{90} + \frac{1}{200}\right) = 0.0204\,\sigma_s \times 10^{-9}$$

∴
$$\sigma_s = \frac{1.156 \times 10^{-3} \times 10^9}{0.0204} = \mathbf{56.66\ MN/m^2}\ (tensile)\ \textbf{(Ans.)}$$

and,
$$\sigma_c = 1.39 \times 56.66 = \mathbf{78.75\ MN/m^2}\ (compressive)\ \textbf{(Ans.)}$$

Example 1.36. *A copper sleeve, 20 mm internal and 26 mm external diameter, surrounds a 18 mm steel bolt, one end of the sleeve being in contact with the shoulder of the bolt. After putting a rigid washer on the other end of the tube, a nut is screwed on the bolt until the compressive stress in sleeve is 75 MN/m². Find the range of the external axial load that can be applied to this assembly if the stress in the bolt is never to be compressive and in the sleeve never to be zero.*

Take: $E_s = 200\ GN/m^2,$
and $E_c = 90\ GN/m^2$

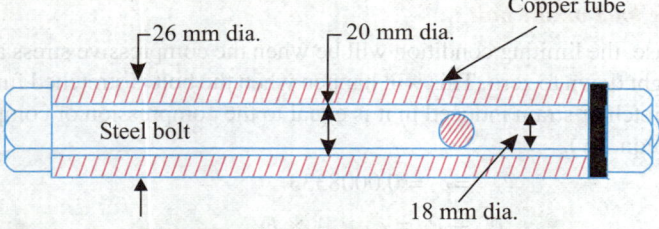

Fig. 1.43

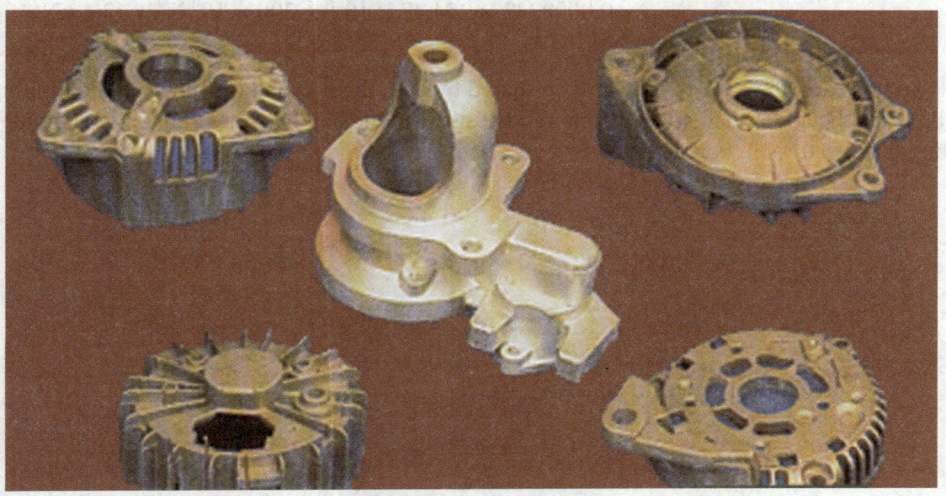

Aluminium brackets.

Solution. Area of steel bolt,

$$A_s = \frac{\pi}{4} \times \left(\frac{18}{1000}\right)^2 = 254.47 \times 10^{-6} \text{ m}^2$$

Area of copper sleeve, $A_c = \frac{\pi}{4} \times \left[\left(\frac{26}{1000}\right)^2 - \left(\frac{20}{1000}\right)^2\right] = 216.77 \times 10^{-6} \text{ m}^2$

When the tube is in *compression* due to the tightening of the nut, the bolt will be under *tension* and

$$P_c = P_s \quad \text{or,} \quad \sigma_c A_c = \sigma_s A_s$$

(where, σ_c and σ_s are the stresses in copper tube and steel bolt respectively.)

or $$\sigma_s = \frac{\sigma_c A_c}{A_s} = \frac{75 \times 216.77}{254.47} = 63.89 \text{ MN/m}^2 \qquad ...(i)$$

∴ $$P_s \ (= P_c) = \sigma_s A_s = 63.89 \times 254.47 \times 10^{-6}$$
$$= 0.01626 \text{ MN or } 16.26 \text{ kN} \qquad ...(ii)$$

Strain in steel, $$e_s = \frac{\sigma_s}{E_s} = \frac{63.89 \times 10^6}{200 \times 10^9} = 0.0003194 \text{ (tensile)} \qquad ...(iii)$$

Strain in copper, $$e_c = \frac{\sigma_c}{E_c} = \frac{75 \times 10^6}{90 \times 10^9} = 0.0008333 \text{ (compressive)} \qquad ...(iv)$$

The following two cases can arise depending on whether the load, compressive or tensile, is applied on the bolt or on the tube.

1st Case: **Steel bolt**

(i) Tensile load on the bolt :

In this case, the limiting condition will be when the compressive stress already existing in the tube is just brought down to zero. This will happen when the bolt is stretched further to the extent that the additional stretch or strain induced in it is equal to the compression or compressive strain in tube as calculated in eqn. (*iv*).

$$e_{s_2} = e_c = 0.0008333$$
$$P_2 = P_{s_2} = e_{s_2} \times A_s \times E_s$$
$$= 0.0008333 \times 254.47 \times 10^{-6} \times 200 \times 10^9 \text{ N or } \textbf{42.4 kN} \qquad ...(v)$$

(ii) Compressive load on the bolt :

In this case, the applied load will be shared by both the bolt and the tube and the limiting case will be when the additional induced compressive stress in the bolt becomes equal in magnitude to the existing tensile stress in bolt as given by eqn. (*i*).

$$\sigma_{s_3} = \sigma_s \qquad ...(vi)$$
$$P_3 = P_{s_3} + P_{c_3} \qquad ...(vii)$$

and, $$e_{s_3} = e_{c_3}$$

or, $$\frac{\sigma_s}{E_s} = \frac{P_{c_3}}{A_c E_c} \quad \text{or} \quad P_{c_3} = \frac{\sigma_s A_c E_c}{E_s} \qquad ...(viii)$$

Now, substituting values in eqn. (*vii*), we get

$$P_3 = \sigma_s A_s + \frac{\sigma_s A_c E_c}{E_s} = \sigma_s \left(A_s + \frac{A_c E_c}{E_s}\right)$$

$$= 63.89 \times 10^6 \left(254.47 \times 10^{-6} + \frac{216.77 \times 10^{-6} \times 90 \times 10^9}{200 \times 10^9} \right) N$$

$$= \mathbf{22.49 \ kN} \ (compressive) \qquad \qquad ...(ix)$$

2nd Case : **Copper tube**

(i) *Tensile load on tube*:

In this case again the load will be shared by both the bolt and the tube and the stress in the tube will be the limiting case and the additional stress induced in the tube will become equal to 75 MN/m²

i.e., $\qquad \qquad \sigma_{c4} = 75 \ MN/m^2 \qquad \qquad ...(x)$

$$P_4 = P_{s_4} + P_{c_4} \qquad \qquad ...(xi)$$

$$e_{s_4} = e_{c_4}$$

or, $\qquad \qquad \dfrac{\sigma_{s_4}}{E_s} = \dfrac{\sigma_{c_4}}{E_c} \left[or \ \dfrac{\sigma_{s_4}}{\sigma_{c_4}} = \dfrac{E_s}{E_c} \right]$

Thus, $\qquad \qquad P_4 = \sigma_{s_4} A_s + \sigma_{c_4} A_c = \sigma_{c_4} \left(\dfrac{\sigma_{s_4}}{\sigma_{c_4}} A_s + A_c \right) = \sigma_{c_4} \left(\dfrac{E_s}{E_c} A_s + A_c \right)$

$$= 75 \times 10^6 \left[\frac{200 \times 10^9}{90 \times 10^9} \times 254.47 \times 10^6 + 216.77 \times 10^{-6} \right] N$$

$$= \mathbf{58.67 \ kN} \ (tensile) \qquad \qquad ...(xii)$$

(ii) *Compressive load applied on tube* :

Here this load will be borne only by tube and the limiting condition will be when the stress in bolt reduces to zero. This will happen when the compressive strain induced in the tube due to external load is just equal in magnitude to the strain e_s in the bolt as given in eqn. (*iii*).

Then, $\qquad \qquad e_{c_5} = e_s \qquad \qquad ...(xiii)$

$$P_s = A_c \, e_{c5} \, E_c = 216.77 \times 10^{-6} \times 0.0003194 \times 90 \times 10^9 \ N \ or \ \mathbf{6.23 \ kN} \ (compressive) \quad ...(xiv)$$

From eqns. (*v*), (*ix*) (*xii*) and (*xiv*) it is evident that for the given condition to be fulfilled, the allowable range of the external load is from **6.23 kN** (*compressive*) **to 42.4 kN** (*tensile*) **(Ans.)**

Example 1.37. *A compound bar consists of a central steel strip 40 mm wide and 5 mm thick placed between two strips of brass each 40 mm wide and x mm thick. The strips are firmly fixed together to form a compound bar of rectangular section 40 mm wide and (2x + 5 mm) thick. Determine :*

(*i*) *The thickness of the brass strips which will make the apparent modulus of elasticity of the compound bar equal to $160 \times 10^3 \ MN/m^2$.*

(*ii*) *The maximum axial pull the bar can then carry if the stress is not to exceed 160 MN/m², in either the brass or the steel.*

Take: $\qquad \qquad E_s = 207 \ GN/m^2 \ and \ E_b = 114 \ GN/m^2.$

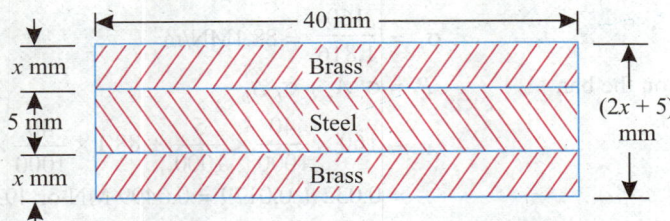

Fig. 1.44

Solution. Let,
$$P = \text{Load on the compound bar (in } MN),$$
$$\sigma_s = \text{Stress in steel strip (in MN/m}^2), \text{ and}$$
$$\sigma_b = \text{Stress in brass strip (in MN/m}^2)$$

(i) Thickness of the brass strip, x :

Strain in steel = Strain in brass

$$\therefore \quad \frac{\sigma_s}{E_s} = \frac{\sigma_b}{E_b} \quad \text{or} \quad \frac{\sigma_s}{\sigma_b} = \frac{E_s}{E_b} = \frac{207}{114}$$

$$\therefore \quad \sigma_s = 1.816 \, \sigma_b$$

Also, total load = Load on steel + load on brass

$$P = \sigma_s A_s + \sigma_b A_b$$

$$= 1.816 \, \sigma_b \left(\frac{40}{1000} \times \frac{5}{1000} \right) + \sigma_b \times \left(\frac{40}{1000} \times \frac{2x}{1000} \right)$$

$$= \sigma_b (363.2 + 80 x) \times 10^{-6}$$

Area of the composite section,

$$A = \frac{40}{1000} \times \frac{5}{1000} + \frac{40}{1000} \times \frac{2x}{1000} = (200 + 80 x) \times 10^{-6} \text{ m}^2$$

Apparent Young's modulus,

$$E = 160 \times 10^3 \text{ MN/m}^2$$

$$\therefore \quad \text{Strain, } e = \frac{P}{AE} = \frac{\sigma_b (363.2 + 80 x) \times 10^{-6}}{(200 + 80 x) \times 10^{-6} \times 160 \times 10^3}$$

This must be equal to the strain of brass or steel

$$= \frac{\sigma_b}{E_b} = \frac{\sigma_b}{114 \times 10^3}$$

$$\therefore \quad \frac{\sigma_b (363.2 + 80 x) \times 10^{-6}}{(200 + 80 x) \times 10^{-6} \times 160 \times 10^3} = \frac{\sigma_b}{114 \times 10^3}$$

or, $\quad 114 (363.2 + 80 \, x) = (200 + 80x) \times 160$

or, $\quad 363.2 + 80 \, x = 280.7 + 112.28 \, x$

or, $\quad 32.28 \, x = 82.5$

$$\therefore \quad x = 2.55 \text{ mm}$$

Hence, thickness of the brass strip = **2.55 mm (Ans.)**

(ii) Maximum axial pull in the bar :

Since $\sigma_s = 1.816 \, \sigma_b$ and since the stress in either brass or steel should not exceed 160 MN/m²,

Let, $\quad \sigma_s = 160 \text{ MN/m}^2,$

then, $\quad \sigma_b = \dfrac{160}{1.816} = 88.1 \text{MN/m}^2$

$\therefore$ Load on the bar, $\quad P = \sigma_s A_s + \sigma_b A_b$

$$= 160 \times \frac{40}{1000} \times \frac{5}{1000} + 88.1 \times \frac{40}{1000} \times \frac{2 \times 2.55}{1000}$$

$$= 0.032 + 0.0179 = 0.0499 \text{ MN or } 49.9 \text{ kN}$$

Hence, maximum pull in the bar = **49.9 kN (Ans.)**

1.13. THERMAL STRESSES AND STRAINS

If the temperature of a body is lowered or raised its dimensions will decrease or increase correspondingly. If these changes, however, are checked the stresses thus developed in the body are called *temperature stresses* and corresponding strains are called *temperature strains*.

Let, l = Length of a bar of uniform cross-section,

t_1 = Initial temperature of the bar,

t_2 = Final temperature of the bar, and

α = Co-efficient of linear expansion.

The extension in the bar due to rise in temperature will be

$$= \alpha\,(t_2 - t_1)l.$$

If this elongation in the bar is prevented by some external force or by fixing the bar ends, the temperature strain thus produced will be given by,

$$\text{Temperature strain} = \frac{\alpha\,(t_2 - t_1)\,l}{l} = \alpha\,(t_2 - t_1) \ \text{(compressive)} \qquad \text{...(1.21)}$$

∴ Temperature stress developed $= \alpha\,(t_2 - t_1)\,E$ (compressive)   ...(1.22)

If, however, the *temperature of the bar is lowered, the temperature strain and stress will be tensile in nature.*

Example 1.38. *A steel rod 15 m long is at a temperature of 15°C. Find the free expansion of the length when the temperature is raised to 65°C. Find the temperature stress produced when :*

(i)   *The expansion of the rod is prevented;*

(ii) *The rod is permitted to expand by 6 mm.*

Take: $\alpha = 12 \times 10^{-6}$ per°C,

and $E = 200$ GN/m².

Solution. Free expansion of the rod

$$= l\,\alpha\,(t_2 - t_1) = 15 \times 12 \times 10^{-6}\,(65 - 15)$$
$$= 0.009 \text{ m} = 9 \text{ mm}$$

(i)   When the expansion is fully prevented :

Temperature stress, $= \alpha\,(t_2 - t_1)\,E = 12 \times 10^{-6}\,(65 - 15) \times 200 \times 10^9$
$$= 120 \times 10^6 \text{ N/m}^2$$
$$= \textbf{120 MN/m}^2 \ \textbf{(Ans.)}$$

(ii)   When the rod is permitted to expand by 6 mm :

In this case, the amount of expansion prevented

$$= 9 - 6 = 3 \text{ mm}$$

∴ Strain, $e = \dfrac{\text{Expansion prevented}}{\text{Original length}} = \dfrac{3}{15 \times 1000} = 0.0002$

∴   Temperature stress,

$$= e \times E = 0.0002 \times 200 \times 10^9 = 40 \times 10^6 \text{ N/m}^2$$
$$= \textbf{40 MN/m}^2 \ \textbf{(Ans.)}$$

Example 1.39. *Calculate the values of the stress and strain in portions AC and CB of the steel bar shown in Fig. 1.45. A close fit exists at both of the rigid supports at room temperature and the temperature is raised by 75°C. Take E = 200 GPa and α = 12 × 10⁻⁶ /°C for steel. Area of cross-sections of AC is 400 mm² and of BC is 800 mm².*

(AMIE Winter, 20

Solution. *Given*: Rise in temperature, $t = 75°C$

$$E = 200 \text{ GPa} = 200 \times 10^9 \text{ N/m}^2$$
$$\alpha = 12 \times 10^{-6}/°C$$
$$A_{AC} = 400 \text{ mm}^2 = 400 \times 10^{-6} \text{ m}^2$$
$$A_{BC} = 800 \text{ mm}^2 = 800 \times 10^{-6} \text{ m}^2$$
$$l_{AC} = 0.3 \text{ m}; \, l_{BC} = 0.3 \text{ m}$$

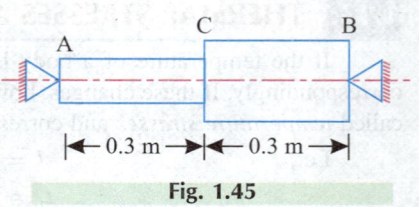

Fig. 1.45

Values of stress and strain in portions *AC* and *CD* :

Elongation in portion *AC*, $\delta l_{AC} = l_{AC} \, \alpha \, t$

$$= 0.3 \times 12 \times 10^{-6} \times 75 = 270 \times 10^{-6} \text{ m}$$

Strain in portion $AC = \dfrac{\delta l_{AC}}{l_{AC}} = \dfrac{270 \times 10^{-6}}{0.3} = \textbf{900} \times \textbf{10}^{-6}$ **(Ans.)**

Compressive stress in portion AC

$$= E \times \text{strain in portion } AC$$
$$= 200 \times 10^9 \times 900 \times 10^{-6} = 180 \times 10^6 \text{ N/m}^2 \text{ or } \textbf{180 MN/m}^2 \text{ (Ans.)}$$

Strain in portion $CB = \dfrac{\delta l_{CB}}{l_{CB}} = \dfrac{0.3 \times 12 \times 10^{-6} \times 75}{0.3} = \textbf{900} \times \textbf{10}^{-6}$ **(Ans.)**

Compressive stress in portion CB

$$= E \times \text{strain in portion } CB$$
$$= 200 \times 10^9 \times 900 \times 10^{-6} = \textbf{180 MN/m}^2 \text{ (Ans.)}$$

Example 1.40. *A copper flat measuring 60 mm × 30 mm is brazed to another 60 mm × 60 mm mild steel flat as shown in Fig. 1.46. If the combination is heated through 120°C, determine :*

(i) *The stress produced in each of the bar,*

(ii) *Shear force which tends to rupture the brazing, and*

(iii) *Shear stress.*

Take: $\alpha_c = 18.5 \times 10^{-6} \text{ per°C};$

$\alpha_s = 12 \times 10^{-6} \text{ per°C};$

$E_c = 110 \text{ GN/m}^2;$

$E_s = 220 \text{ GN/m}^2$

Length of each flat = 400 mm.

Rollers on the above frames drive the conveyor belts meant for material handling.

Solution. Refer to Fig. 1.46.

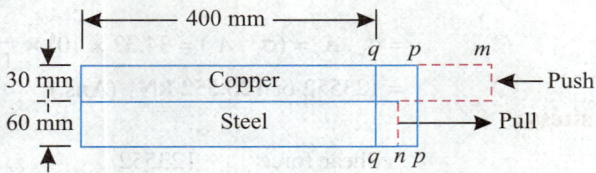

Fig. 1.46

(i)   Stress produced in each bar :

As α_c is greater than α_s, therefore extension of copper flat will be more than that of steel flat, but since they are brazed together, *the former will try to pull the latter and the latter to push the former; finally, however, they will become stable at certain position pp after compromise.*

Increase in length of copper flat,

$$qm = \alpha_c (t_2 - t_1) \, l$$

Increase in length of steel flat, qn

$$= \alpha_s (t_2 - t_1) \, l$$

Now compressive strain in copper flat,

$$e_c = \frac{pm}{l} = \frac{qm - qp}{l} \qquad , \text{or,} \qquad e_c = \frac{qm}{l} - \frac{qp}{l}$$

$$= \alpha_c (t_2 - t_1) - e \qquad\qquad ...(i)$$

where,

$$\frac{qp}{l} = e = \text{common strain}$$

Tensile strain in steel,

$$e_s = \frac{np}{l} = \frac{qp - qn}{l} = e - \alpha_s (t_2 - t_1) \qquad\qquad ...(ii)$$

Adding (i) and (ii), we get

$$e_c + e_s = \alpha_c (t_2 - t_1) - \alpha_s (t_2 - t_1)$$
$$= (\alpha_c - \alpha_s)(t_2 - t_1)$$

But,

$$e_c = \frac{\sigma_c}{E_c}, e_s = \frac{\sigma_s}{E_s} \quad \text{or} \quad \frac{\sigma_c}{E_c} + \frac{\sigma_s}{E_s} = (\alpha_c - \alpha_s)(t_2 - t_1)$$

or,

$$\frac{\sigma_c}{110 \times 10^9} + \frac{\sigma_s}{220 \times 10^9} = (18.5 \times 10^{-6} - 12 \times 10^{-6}) \times 120$$

or,

$$2\,\sigma_c + \sigma_s = 220 \times 10^9 \times 120\,(18.5 \times 10^{-6} - 12 \times 10^{-6})$$

or,

$$2\,\sigma_c + \sigma_s = 171.6 \times 10^6 \qquad\qquad ...(iii)$$

At position pp,

Pull on steel = Push on copper

$$\sigma_s . A_s = \sigma_c . A_c$$

$$\sigma_s \times \frac{60}{1000} \times \frac{60}{1000} = \sigma_c \times \frac{60}{1000} \times \frac{30}{1000}$$

$$\sigma_s = \frac{\sigma_c}{2} = 0.5\,\sigma_c$$

Putting this value of σ_s in (iii), we get

$$2\,\sigma_c + 0.5\,\sigma_c = 171.6 \times 10^6$$

or,

$$\sigma_c = 68.64 \times 10^6 \text{ N/m}^2$$

$$= \textbf{68.64 MN/m}^2 \text{ (Ans.)}$$

and, $\sigma_s = 0.5 \sigma_c = 34.32 \text{ MN/m}^2$ **(Ans.)**

(ii) Shear force :

Shear force $= \sigma_s . A_s = (\sigma_c . A_c) = 34.32 \times 10^6 \times \dfrac{60}{1000} \times \dfrac{60}{1000}$

$= 123552$ or **123.552 kN (Ans.)**

(iii) Shear stress :

Shear stress $= \dfrac{\text{Shear force}}{\text{Shear area}} = \dfrac{123552}{\dfrac{400}{1000} \times \dfrac{60}{1000}}$

$= 5.148 \times 10^6 \text{ N/m}^2$

$= \textbf{5.148 MN/m}^2$ **(Ans.)**

Example 1.41. *A steel bar is placed between two copper bars, each having the same area and length as steel bar at 20°C. At this stage, they are rigidly connected together at both the ends. When the temperature is raised to 320°C, the length of the bars increases by 1.5 mm. Determine the original length and final stresses in the bars.*

Take: $E_s = 220 \text{ GN/m}^2; E_c = 110 \text{ GN/m}^2;$
$\alpha_s = 0.000012 \text{ per°C};$
$\alpha_c = 0.0000175 \text{ per°C}.$

Solution. Refer to Fig. 1.47.

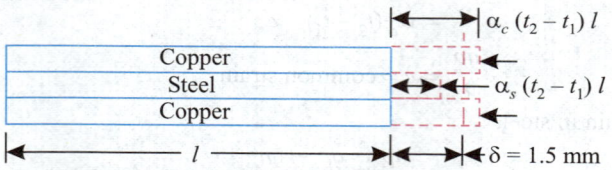

Fig. 1.47

Let the cross-sectional area of steel bar $= A$

∴ Cross-sectional area of two copper bars $= 2A$

Free expansion of steel bar $= \alpha_s (t_2 - t_1) l$

Free expansion of copper bar $= \sigma_c (t_2 - t_1) l$

Let δ be the actual expansion of each bar such that $\alpha_s (t_2 - t_1) l < \delta < \sigma_c (t_2 - t_1) l$

∵ *Steel is in tension and copper is in compression.*

Let σ_s and σ_c be the stresses in steel and copper respectively.

Also, $e_c + e_s = (\alpha_c - \alpha_s) (t_2 - t_1)$

$= (0.0000175 - 0.000012) (320 - 20)$

$= 0.00165$ ...(i)

When the system is in equilibrium,

Pull in steel = Push in copper

$\sigma_s . A_s = \sigma_c . A_c$

$\sigma_s \times A = \sigma_c \times 2A$

∴ $\sigma_s = 2 \sigma_c$ ...(ii)

∴ $e_c = \dfrac{\sigma_c}{E_c}$ and $e_s = \dfrac{\sigma_s}{E_s}$

∴ Eqn. (i) can be written as

$$\frac{\sigma_c}{E_c} + \frac{\sigma_s}{E_s} = 0.00165$$

$$\frac{\sigma_c}{110 \times 10^9} + \frac{2\sigma_c}{220 \times 10^9} = 0.00165$$

$$2\sigma_c + 2\sigma_c = 220 \times 10^9 \times 0.00165$$

$$\sigma_c = 90.75 \times 10^6 \text{ N/m}^2 \text{ or } \textbf{90.75 MN/m}^2 \quad \textbf{(Ans.)}$$

and,

$$\sigma_s = 2 \times 90.75 = \textbf{181.5 MN/m}^2 \quad \textbf{(Ans.)}$$

Also,

$$e_c = \alpha_c (t_2 - t_1) - e$$

∴

$$e = \alpha_c (t_2 - t_1) - e_c$$

$$= 0.0000175 (320 - 20) - \frac{\sigma_c}{E_c}$$

$$= 0.0000175 \times 300 - \frac{90.75 \times 10^6}{110 \times 10^9}$$

$$= 0.00525 - 0.000825 = 0.004425$$

∴

$$\delta = e \times l = 1.5 \text{ mm or } 0.0015 \text{ m (given)}$$

∴

$$l = \frac{0.0015}{0.004425} = 0.339 \text{ m}$$

Hence, original length $l = \textbf{0.339 m} \quad \textbf{(Ans.)}$

Example 1.42. *A composite bar madeup of aluminium and steel is held between two supports as shown in Fig. 1.48. The bars are stress-free at a temperature of 40°C. What will be the stresses in the two bars when the temperature is 20°C if (i) the supports are non-yielding, and (ii) the supports come nearer to each other by 0.1 mm. It can be assumed that the change of temperature is uniform all along the length of the bar.*

Take:
$E_s = 210 \text{ GN/m}^2;$
$E_a = 74 \text{ GN/m}^2;$
$\alpha_s = 11.7 \times 10^{-6} \text{ per } °C;$
$\alpha_a = 23.4 \times 10^{-6} \text{ per } °C.$

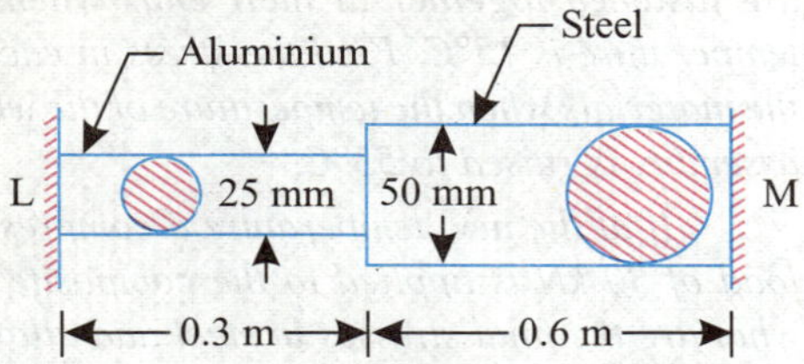

Fig. 1.48

Solution. Refer to Fig. 1.48.

Let us assume that the support at *L* is removed so that the contraction of the bar is allowed freely.

Fall of temperature   = 40 – 20 = 20°C

Contraction of steel bar $= \alpha_s (t_2 - t_1) l_s$

$= 11.7 \times 10^{-6} \times 20 \times 0.6 = 0.0001404 \text{ m}$

Contraction of aluminium bar $= \alpha_a (t_2 - t_1) l_a = 23.4 \times 10^{-6} \times 20 \times 0.3 = 0.0001404$

Total contraction $= 0.0001404 + 0.0001404 = 0.0002808 \text{ m}$ ....(*i*)

Now, let a force *P* be applied to the end *L* till this end is brought in contact with the left hand support. Let this force cause a stress 'σ' in the steel bar. The stress in the aluminium bar will be 4 σ, since the area of the aluminium bar is one-fourth that of steel bar.

Extension of steel due to *P* $= \frac{\sigma}{210 \times 10^9} \times 0.6 \text{ m}$

Extension of aluminium bar due to *P*

$= \frac{4\sigma}{74 \times 10^9} \times 0.3 \text{ m}$

∴ Total extension $= \dfrac{0.6\sigma}{210 \times 10^9} + \dfrac{1.2\sigma}{74 \times 10^9}$ m ...(ii)

(i) When the supports do not yield :

$$\frac{0.6\sigma}{210 \times 10^9} + \frac{1.2\sigma}{74 \times 10^9} = 0.0002808$$

$$0.002857\ \sigma + 0.0162\ \sigma = 0.0002808 \times 10^9 = 280800$$

$$\sigma = 14.73 \times 10^6 \text{ N/m}^2 \text{ or } 14.73 \text{ MN/m}^2$$

∴ *Stress in steel bar* = **14.73 MN/m²**

and, *stress in aluminium bar* $= 4 \times 14.73 =$ **58.92 MN/m²** **(Ans.)**

(ii) When the supports yield by 0.1 mm :

$$\frac{0.6\sigma}{210 \times 10^9} + \frac{1.2\sigma}{74 \times 10^9} = 0.0002808 - \frac{0.1}{1000}$$

$$0.002857\ \sigma + 0.0162\ \sigma = 10^9\ (0.0002808 - 0.0001) = 180800$$

∴ $\sigma = 9.49 \times 10^6 \text{ N/m}^2 \text{ or } 9.49 \text{ MN/m}^2$

∴ Stress in the *steel bar* = **9.49 MN/m²** **(Ans.)**

and, stress in the *aluminium bar* $= 4 \times 9.49 =$ **37.96 MN/m²** **(Ans.)**

Example 1.43. *A flat bar of aluminium alloy 25 mm wide and 5 mm thick is placed between two steel bars each 25 mm wide and 10 mm thick to form a composite bar 25 mm × 25 mm as shown in Fig. 1.49. The three bars are fastened together at their ends when the temperature is 15°C. Find the stress in each of the materials when the temperature of the whole assembly is raised to 55°C.*

If at the new temperature a compressive load of 30 kN is applied to the composite bar what are the final stresses in steel and alloy ?

Take: $E_s = 200 \text{ GN/m}^2;$

$E_{al} = \dfrac{200}{3} \text{ GN/m}^2;$

$\alpha_s = 1.2 \times 10^{-5} \text{ per}°C;$

$\alpha_{al} = 2.3 \times 10^{-5} \text{ per}°C.$

Automobile engine parts.

Solution. Refer to Fig. 1.49.

Area of aluminium, $A_{al} = 25 \times 5 = 125 \text{ mm}^2$ or $125 \times 10^{-6} \text{ m}^2$

Area of steel, $A_s = 2 \times 25 \times 10 = 500 \text{ mm}^2$ or $500 \times 10^{-6} \text{ m}^2$

(i) Stresses due to rise of temperature :

If the two members had been free to expand, then

Free expansion of steel $= \alpha_s\, t l_s$

Free expansion of aluminium $= \alpha_{al}\, t l_{al}$

But since the members are fastened to each other at the ends, final expansion of each member would be the same. Let this expansion be δ. The free expansion of aluminium is greater than δ while the free expansion of steel is less than δ. Hence the *steel* is subjected to *tensile stress* while aluminium is subjected to *compressive stress*. Let σ_s and σ_{al} be the stresses in steel and aluminium respectively.

The whole system will be in equilibrium when,

Total tension (pull) in steel = Total compression (push) in aluminium

$$\sigma_s A_s = \sigma_{al} A_{al}$$

or, $\qquad \sigma_s \times 500 \times 10^{-6} = \sigma_{al} \times 125 \times 10^{-6}$

$\therefore \qquad\qquad \sigma_s = \dfrac{\sigma_{al}}{4} = 0.25\,\sigma_{al}$

Final increase in length of steel = Final increase in length of aluminium

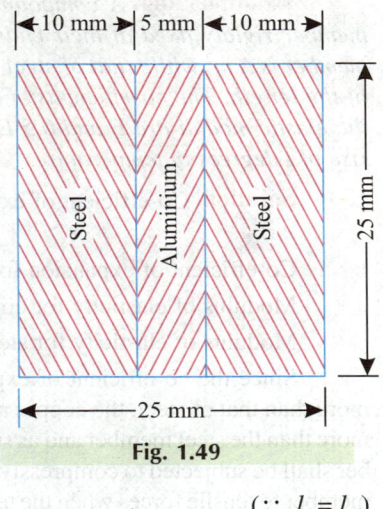

Fig. 1.49

$\therefore \qquad \alpha_s\, t l_s + \dfrac{\sigma_s}{E_s}\, l_s = \alpha_{al}\, t l_{al} - \dfrac{\sigma_{al}}{E_{al}}\, l_{al}$

or, $\qquad \alpha_s\, t + \dfrac{\sigma_s}{E_s} = \alpha_{al}\, t - \dfrac{\sigma_{al}}{E_{al}} \qquad\qquad (\because\ l_s = l_{al})$

But, $\qquad\qquad t = 55 - 15 = 40°C$

$\therefore\ 1.2 \times 10^{-5} \times 40 + \dfrac{0.25\,\sigma_{al}}{200 \times 10^9} = 2.3 \times 10^{-5} \times 40 - \dfrac{\sigma_{al}}{\dfrac{200}{3} \times 10^9}$

or, $\quad 1.2 \times 10^{-5} \times 40 \times 200 \times 10^9 + 0.25\,\sigma_{al} = 2.3 \times 10^{-5} \times 40 \times 200 \times 10^9 - 3\,\sigma_{al}$

$\qquad\qquad 96 \times 10^6 + 0.25\,\sigma_{al} = 184 \times 10^6 - 3\,\sigma_{al}$

or, $\qquad\qquad\qquad \boldsymbol{\sigma_{al} = 27.07\ MN/m^2}$ (*compressive*) **(Ans.)**

and, $\qquad\qquad \boldsymbol{\sigma_s} = 0.25\,\sigma_{al} = 0.25 \times 27.7 = \boldsymbol{6.76\ MN/m^2}$ (*tensile*) **(Ans.)**

(ii) Stresses due to external compressive load 30 kN :

Let $\sigma_s{}'$ and $\sigma_{al}{}'$ be the stresses due to external loading in steel and aluminium respectively.

Strain in steel, $\qquad\qquad e_s = $ Strain in aluminium, e_{al}

i.e., $\qquad\qquad\qquad \dfrac{\sigma_s'}{E_s} = \dfrac{\sigma_{al}'}{E_{al}}$

$\therefore \qquad\qquad \sigma_s' = \dfrac{E_s}{E_{al}}\,\sigma_{al}' = \dfrac{200}{200/3}\,\sigma_{al}' = 3\,\sigma_{al}'$

But, load on steel + load on aluminium = Total load

$$\sigma_s' A_s + \sigma_{al}' A_{al} = 30 \times 1000$$

or, $\quad 3\,\sigma_{al}' \times 500 \times 10^{-6} + \sigma_{al}' \times 125 \times 10^{-6} = 30000$

$\therefore \qquad\qquad\qquad \sigma_{al}' = 18.46\ MN/m^2$ (compressive)

and, $\qquad\qquad\qquad \sigma_s' = 3\,\sigma_{al}'$

$\qquad\qquad\qquad\qquad = 3 \times 18.46 = 55.38\ MN/m^2$ (compressive)

Final stress :

Stress in aluminium $\qquad = \sigma_{al} + \sigma_{al}'$

$\qquad\qquad\qquad\qquad = 27.07 + 18.46 = \boldsymbol{45.53\ MN/m^2}$ (*compressive*) **(Ans.)**

Stress in steel $\qquad\qquad = \sigma_s + \sigma_s' = -6.76 + 55.38$

$\qquad\qquad\qquad\qquad = \boldsymbol{48.62\ MN/m^2}$ (*compressive*) **(Ans.)**

Example 1.44. *A compound bar is made-up by connecting a steel member and a copper member rigidly fixed at their ends as shown in Fig. 1.50. The cross-sectional area of the copper member is A m² while that of steel member is 2A m² for half the length and A m² for the other half of the length. The co-efficients of expansion of steel and copper are α and 1.3 α respectively while the elastic moduli are E and 0.5 E respectively. Estimate the stress induced in the members due to rise of t degrees in temperature.* **(Jodhpur University)**

Solution. Co-efficient of expansion for copper,

$$\alpha_c = 1.3\,\alpha$$

Co-efficient of expansion for steel, $\alpha_s = \alpha$

Modulus of elasticity for copper, $E_c = 0.5\,E$

Modulus of elasticity for steel, $E_s = E$

Since the co-efficient of expansion of copper is more than that of steel, the copper member shall expand more than the steel member and as such the copper member shall be subjected to compressive forces and the steel member to tensile forces when the temperature rises. The compound bar shall stabilise at an intermediate expansion between that of copper and steel.

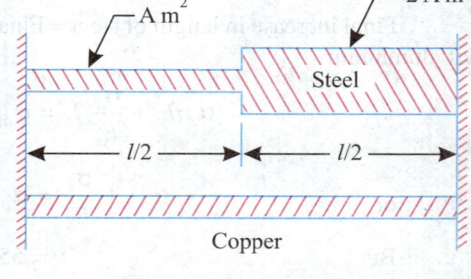

Fig. 1.50

Due to restraint at the ends,

Compressive stress set-up in copper bar = σ_c

Tensile stress set up in steel bar of cross-sectional area

$$A = \sigma_s$$

Tensile stress set up in steel bar of cross-sectional area

$$2A = \sigma'_s$$

Also, total tension (pull) in steel bar = Compression (push) in copper bar

∴ $\qquad \sigma_s \times A = \sigma'_s \times 2A = \sigma_c \times A$ or $\sigma_c = \sigma_s = 2\,\sigma'_s$

Inside a machine shop.

If l_a is the actual length of the compound bar due to rise in temperature, then

For copper rod :

$$l\,(1 + \alpha_c\, t) - l_a = \frac{\sigma_c}{E_c} \times l \qquad \dots(i)$$

For steel rod :

$$l_a - l\,(1 + \alpha_s\, t) = \frac{\sigma_s}{E_s} \times \frac{l}{2} + \frac{\sigma_s'}{E_s} \times \frac{l}{2} \qquad \dots(ii)$$

Adding (*i*) and (*ii*), we get

$$(\alpha_c - \alpha_s)\, t = \frac{\sigma_c}{E_c} + \frac{\sigma_s}{2\,E_s} + \frac{\sigma_s'}{2\,E_s}$$

or ,

$$(1.3\alpha - \alpha)\, t = \frac{\sigma_c}{0.5\,E} + \frac{\sigma_c}{2\,E} + \frac{\sigma_c}{4\,E} = \frac{11\sigma_c}{4\,E}$$

or,

$$0.3\,\alpha\, t = \frac{11\sigma_c}{4\,E}$$

or,

$$\sigma_c = \sigma_s = 0.109\,E\,\alpha\, t \quad \textbf{(Ans.)}$$

∴

$$\sigma'_s = \frac{\sigma_s}{2} = 0.0545\,E\,\alpha\, t \quad \textbf{(Ans.)}$$

Example 1.45. *Two steel rods, one of 75 mm diameter and the other of 55 mm diameter, are joined end to end by means of a turn buckle. The other end of each rod is rigidly fixed and there is initially a small tension in the rods. If the effective length of each rod is 4.5 m, find the increase in this tension when the turn buckle is turned by one-quarter of a turn. On one end of the bigger diameter rod there are 0.15 threads per mm length, while there are 0.2 threads per mm length on the other rod.*

Neglect the extension of the turn buckle. Find also what rise in temperature would nullify the increase in tension.

Take: $E = 200\ GN/m^2$, *and* $\alpha = 12 \times 10^{-6}\ per\,°C$.

Solution. Refer to Fig. 1.51.

75 mm dia   2   1   55 mm dia.

Fig. 1.51

Cross-sectional area of rod '1' $A_1 = \dfrac{\pi}{4} \times \left(\dfrac{55}{1000}\right)^2 = 0.002376\ m^2$

Cross-sectional area of rod '2' $A_2 = \dfrac{\pi}{4} \times \left(\dfrac{75}{1000}\right)^2 = 0.004418\ m^2$

When the turn buckle is turned by one-quarter of a turn,

Extension of rod '1' $= \dfrac{1}{4} \times \dfrac{1}{0.2 \times 1000} = 0.00125\ m$

Extension of rod '2' $= \dfrac{1}{4} \times \dfrac{1}{0.15 \times 1000} = 0.001666\ m$

∴ Total extension of the two rods $= 0.00125 + 0.001666 = 0.002916\ m$

Let the tension in each rod be '*P*'.

∴ Total extension in the two bars $= \dfrac{Pl_1}{A_1 E} + \dfrac{Pl_2}{A_2 E} = \dfrac{P \times 4.5}{200 \times 10^9} \left(\dfrac{1}{A_1} + \dfrac{1}{A_2} \right) (\because l_1 = l_2 = 4.5 \text{ m})$

$$= \dfrac{4.5 \, P}{200 \times 10^9} \left(\dfrac{1}{0.002376} + \dfrac{1}{0.004418} \right) \text{m} = 0.002916 \text{ m}$$

Solving, we get $P = 200241$ N or 200.241 kN

In order to nullify this tension by rise of temperature, the total expansion of the two rods must be equal to 0.002916 m.

Let the rise of temperature $(t_2 - t_1)$ be $t°C$

∴ $\alpha \cdot t \cdot l = 0.002916$

$12 \times 10^{-6} \times t \times 4.5 = 0.002916$

∴ $t = \dfrac{0.002916}{12 \times 10^{-6} \times 4.5} = \mathbf{54°C}$ **(Ans.)**

Example 1.46. *A steel rod 40 mm in diameter is fixed concentrically in a brass tube having outside and inside diameters of 60 mm and 50 mm respectively. Both the rod and the tube are 0.5 m long and their ends are level. The compound rod is held between two stops which are exactly 0.5 m apart and the temperature of the bar is then raised by 50°C.*

(i) Find the stresses in the rod and the tube if the distance between the stops (a) remains constant (b) is increased by 0.2 mm.

(ii) Find the increase in the distance between the stops if the force exerted between them is 60 kN.

Take: E_s = 200 GN/m²; E_b = 90 GN/m²;
α_s = 1.2 × 10⁻⁵ per °C; α_b = 2.1 × 10⁻⁵ per °C.

Closeup view of lapping machine.

Solution.

(i) (a) When the distance between the stops remains constant :

Stress in steel, $\sigma_s = \alpha_s \cdot t \cdot E_s = 1.2 \times 10^{-5} \times 50 \times 200 \times 10^9$

$= \mathbf{120 \ MN/m^2}$ **(Ans.)**

Stress in brass,

$\sigma_b = \alpha_b \cdot t \cdot E_b = 2.1 \times 10^{-5} \times 50 \times 90 \times 10^9$

$= \mathbf{94.5 \ MN/m^2}$ **(Ans.)**

(b) When the distance between the stops is increased by 0.2 mm :

Strain in steel, $e_s = \dfrac{\text{Expansion prevented}}{\text{Original length}} = \dfrac{\alpha_s \, t \, l - (0.2/1000)}{0.5}$

$$= \dfrac{1.2 \times 10^{-5} \times 50 \times 0.5 - 0.0002}{0.5} = 0.0002$$

∴ *Stress in steel,* $\sigma_s = e_s E_s = 0.0002 \times 200 \times 10^9$

$$= 40 \times 10^6 \text{ N/m}^2 = \textbf{40 MN/m}^2 \quad \text{(compressive) (Ans.)}$$

Strain in brass, $\qquad e_b = \dfrac{\text{Expansion prevented}}{\text{Original length}}$

$$= \frac{\alpha_b\, tl - 0.0002}{0.5} = \frac{2.1 \times 10^{-5} \times 50 \times 0.5 - 0.0002}{0.5}$$

$$= 0.00065$$

∴ *Stress in brass,* $\qquad \sigma_b = e_b\, E_b = 0.00065 \times 90 \times 10^9$

$$= 58.5 \times 10^6 \text{ N/m}^2 = \textbf{58.5 MN/m}^2 \text{ (compressive) (Ans.)}$$

(ii) When the force exerted between the stops is 60 kN:

Let δ metre be the expansion of the composite member.

Strain in steel, $\qquad e_s = \dfrac{\alpha_s\, tl - \delta}{l} = \dfrac{1.2 \times 10^{-5} \times 50 \times 0.5 - \delta}{0.5}$

$$= 0.0006 - 2\,\delta$$

∴ Stress in steel, $\qquad \sigma_s = e_s\, E_s = (0.0006 - 2\,\delta) \times 200 \times 10^9$

Similarly, strain in brass, e_b

$$= \frac{\alpha_b\, tl - \delta}{l} = \frac{2.1 \times 10^{-5} \times 50 \times 0.5 - \delta}{0.5}$$

$$= (0.00105 - 2\,\delta)$$

∴ Stress in brass, $\qquad \sigma_b = e_b\, E_b = (0.00105 - 2\,\delta) \times 90 \times 10^9$

Area of steel rod, $\qquad A_s = \dfrac{\pi}{4} \times \left(\dfrac{40}{1000}\right)^2 = 1256.6 \times 10^{-6} \text{ m}^2$

Area of brass tube, $\qquad A_b = \dfrac{\pi}{4} \times \left[\left(\dfrac{60}{1000}\right)^2 - \left(\dfrac{50}{1000}\right)^2\right] = 863.9 \times 10^{-6} \text{ m}^2$

Load on steel + load on brass = Total load between the stops

$$\sigma_s A_s + \sigma_b A_b = P$$

or, $\quad (0.0006 - 2\,\delta) \times 200 \times 10^9 \times 1256.6 \times 10^{-6} + (0.00105 - 2\,\delta) \times$

$$90 \times 10^9 \times 863.9 \times 10^{-6} = 60000$$

or, $\quad (0.0006 - 2\,\delta) \times 2.513 \times 10^8 + (0.00105 - 2\,\delta) \times 0.7775 \times 10^8 = 60000$

or, $\quad 15.078 \times 10^4 + 8.164 \times 10^4 - \delta\, [5.026 + 1.555] \times 10^8 = 60000$

or, $\quad 23.242 \times 10^4 - 6.581 \times 10^8\, \delta = 60000 = 6 \times 10^4$

∴ $\qquad \delta = \dfrac{(23.242 - 6) \times 10^4}{6.581 \times 10^8} \times 10^3 \text{ mm} = 0.262 \text{ mm}$

*Hence, the increase in distance between the stops = **0.262 mm** (Ans.)*

Example 1.47. *A steel tube 2.4 cm external diameter and 1.8 cm internal diameter encloses a copper rod 1.5 cm diameter to which it is rigidly connected at the two ends. If at a temperature of 10°C there is no longitudinal stress, calculate the stresses in the rod and the tube when the temperature is raised to 200°C.*

$\qquad$ *Given:* $\qquad\quad E_s = 2.1 \times 10^5 \text{ N/mm}^2; \qquad E_c = 1.0 \times 10^5 \text{ N/mm}^2;$

$\qquad\qquad\qquad\qquad \alpha_s = 11 \times 10^{-6}/°C; \qquad\qquad \alpha_c = 18 \times 10^{-6}/°C$

$\hfill$ **(AMIE Summer, 2002)**

Solution. *Given:*

External diameter of steel tube,

$$D_s = 2.4 \text{ cm} = 24 \text{ mm}$$

Internal diameter of steel tube,
$$d_s = 1.8 \text{ cm} = 18 \text{ mm}$$
Diameter of copper rod, $D_c = 1.5 \text{ cm} = 15 \text{ mm}$
Initial temperature, $t_1 = 10°C$
Final temperature, $t_2 = 200°C$
Then, rise in temperature, $t = t_2 - t_1 = 200 - 10 = 190°C$
$$E_s = 2.1 \times 10^5 \text{ N/mm}^2, \alpha_s = 11 \times 10^{-6}/°C;$$
$$E_c = 1.0 \times 10^5 \text{ N/mm}^2, \alpha_c = 18 \times 10^{-6}/°C.$$

Stresses in the rod and the tube, σ_c, σ_s :

Refer to Fig. 1.52. Since $\alpha_c > \alpha_s$, elongation of copper will naturally be more than that of steel for the same rise of temperature but since they are rigidly jointed at each end, the copper rod will venture to pull the steel tube along with it; whereas the steel tube will struggle to bring the copper rod back. Ultimately, they will compromise and become stable at certain common position.

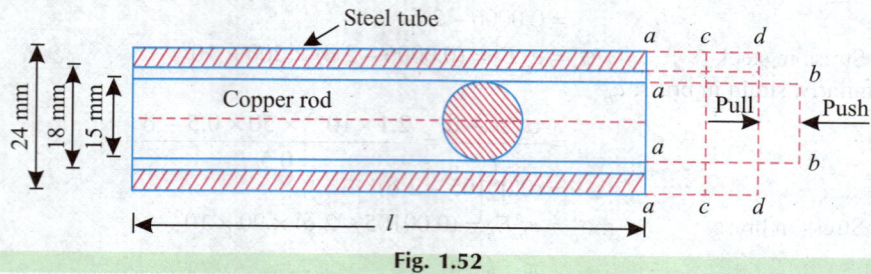

Fig. 1.52

Extension of copper rod when free to expand $= ab = l\,\alpha_c\,.\,t$
Extension of steel rod when free to expand $= a_c = l\,\alpha_s\,t$

Being connected together, suppose they compromise at the position dd; which means that steel tube will be pulled from c to d, and the copper rod pushed back from b to d. In this way *steel is under tension* and the *copper is under compression*.

∴ Compressive strain in copper rod,

$$e_c = \frac{bd}{l} = \frac{ab - ad}{l} = \frac{ab}{l} - \frac{ad}{l}$$

$$= \alpha_c.t - \frac{ad}{l} \qquad\qquad (\because ab = l\,.\,\alpha_c\,.\,t)$$

or, $\qquad\qquad e_c = \alpha_c\,.\,t - e \qquad$ (where, $\dfrac{ad}{l}$ = common strain = e) ...(i)

Tensile strain in steel tube, $e_s = \dfrac{cd}{l} = \dfrac{ad - ac}{l} = \dfrac{ad}{l} - \dfrac{ac}{l}$

$$= \frac{ad}{l} - \alpha_s\,.\,t \qquad\qquad (\because ac = l.\alpha_s\,.\,t)$$

or, $\qquad\qquad e_s = e - \alpha_s\,.\,t \qquad\qquad$ ...(ii)

Adding eqns. (i) and (ii), we have

$$e_c + e_s = \alpha_c\,.\,t - \alpha_s\,.\,t = t\,(\alpha_c - \alpha_s)$$

But, $\qquad\qquad e_c = \dfrac{\sigma_c}{E_c}$ and, $e_s = \dfrac{\sigma_s}{E_s}$

where, $\qquad\qquad \sigma_c$ = Temperature stress in copper, and

$\qquad\qquad\qquad \sigma_s$ = Temperature stress in steel.

$$\therefore \qquad \frac{\sigma_c}{E_c} + \frac{\sigma_s}{E_s} = t\,(\alpha_c - \alpha_s)$$

or, $\qquad \dfrac{\sigma_c}{1.0 \times 10^5} + \dfrac{\sigma_s}{2.1 \times 10^5} = 190\,(18 \times 10^{-6} - 11 \times 10^{-6})$

or, $\qquad \dfrac{\sigma_c}{1.0 \times 10^5} + \dfrac{\sigma_s}{2.1 \times 10^5} = 190 \times 7 \times 10^{-6}$

or, $\qquad 2.1\,\sigma_c + \sigma_s = 2.1 \times 10^5 \times 190 \times 7 \times 10^{-6}$

or, $\qquad 2.1\,\sigma_c + \sigma_s = 279.3 \qquad \qquad …(iii)$

But at the stabilized or common position dd,

Push on copper rod = Pull on steel tube

i.e., $\qquad \qquad \sigma_c . A_c = \sigma_s . A_s$

or, $\qquad \sigma_c \times \dfrac{\pi}{4} \times 15^2 = \sigma_s \times \dfrac{\pi}{4} \times (24^2 - 18^2)$

or, $\qquad \qquad \sigma_c \times 225 = \sigma_s\,(576 - 324)$

or, $\qquad \sigma_s = \sigma_c \times \dfrac{225}{(576 - 324)} = 0.89\,\sigma_c \qquad …(iv)$

Substituting for σ_s in eqn. (iii), we have

$$2.1\,\sigma_c + 0.89\,\sigma_c = 279.3$$

or, $\qquad \qquad 2.99\,\sigma_c = 279.3$

$\therefore \qquad \qquad \sigma_c = \dfrac{279.3}{2.99} = $ **93.41 N/mm²** (**Ans.**)

A lapping machine is used to fine-finish the gear teeth.

From eqn. (iv), we get $\qquad \sigma_s = 0.89 \times 93.41 = $ **83.13 N/mm²** (**Ans.**)

Example 1.48. *A copper bar 50 mm in diameter is placed within a steel tube 75 mm external diameter and 50 mm internal diameter of exactly the same length. The two pieces are rigidly fixed together by two pins 18 mm in diameter, one at each end passing through the bar and tube. Calculate the stresses induced in the copper bar, steel tube and pins if the temperature of the combination is raised by 50°C.*

Take: $\qquad E_s = 210\ GN/m^2;$
$\qquad \qquad E_c = 105\ GN/m^2;$
$\qquad \qquad \alpha_s = 11.5 \times 10^{-6}\ per\ °C;$
$\qquad \qquad \alpha_c = 17 \times 10^{-6}\ per\ °C.$

(Engineering Services)

Pin (18 mm diameter)

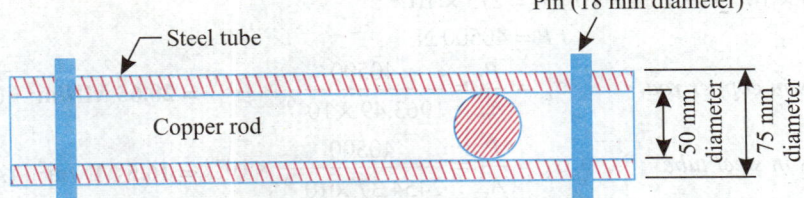

Steel tube

Copper rod

50 mm diameter

75 mm diameter

Fig. 1.53

Solution. Diameter of the copper bar = 50 mm

External diameter of the steel tube = 75 mm

Internal diameter of the steel tube = 50 mm

Diameter of each pin = 18 mm

Rise in temperature, $t = 50\ °C$

Area of copper bar, $\qquad A_c = \dfrac{\pi}{4} \times \left(\dfrac{50}{1000}\right)^2 = 1963.49 \times 10^{-6}\ m^2$

Area of steel tube, $\qquad A_s = \dfrac{\pi}{4} \times \left[\left(\dfrac{75}{1000}\right)^2 - \left(\dfrac{50}{1000}\right)^2\right] = 2454.37 \times 10^{-6}\ m^2$

$$\sigma_c = ?, \ \sigma_s = ?, \ \tau_{pin} = ?$$

Free expansion of copper bar $= \alpha_c\, t\, l$

Free expansion of steel tube $= \alpha_s\, t\, l$

Difference in free expansion $= (\alpha_c - \alpha_s)\, tl = (17 \times 10^{-6} - 11.5 \times 10^{-6}) \times 50 \times l$

$$= 275 \times 10^{-6}\, l \qquad\qquad\qquad ...(i)$$

A compressive force P (newton) exerted by the steel tube on the copper rod opposes the extra expansion of the copper rod, and in return, the copper rod exerts an equal tensile force P to pull the steel tube so that the consequent combined effect of reduction in the length of copper rod and the increase in length of steel tube equalises the difference in free expansions of the two.

Reduction in the length of copper rod due to force P

Another closeup view of a lapping machine.

$$= \frac{Pl}{A_c\, E_c} = \frac{Pl}{1963.49 \times 10^{-6} \times 105 \times 10^9}\ \text{metre}$$

Increase in length of steel tube due to force P

$$= \frac{Pl}{A_s\, E_s} = \frac{Pl}{2454.37 \times 10^{-6} \times 210 \times 10^9}\ \text{metre}$$

But, $\qquad \dfrac{Pl}{1963.49 \times 10^{-6} \times 105 \times 10^9} + \dfrac{Pl}{2454.37 \times 10^{-6} \times 210 \times 10^9} = 275 \times 10^{-6}\, l$

$$4.85 \times 10^{-9}\, P + 1.94 \times 10^{-9}\, P = 275 \times 10^{-6}$$

$\therefore \qquad\qquad\qquad\qquad P = 40500\ N$

Stress in copper rod, $\qquad \sigma_c = \dfrac{P}{A_c} = \dfrac{40500}{1963.49 \times 10^{-6}} \times 10^{-6} = \mathbf{20.63\ MN/m^2}$ **(Ans.)**

Stress in steel tube $\qquad \sigma_s = \dfrac{P}{A_s} = \dfrac{40500}{2454.37 \times 10^{-6}} \times 10^{-6} = \mathbf{16.5\ MN/m^2}$ **(Ans.)**

Shear stress in pins, $\qquad \tau_{pin} = \dfrac{P}{2 \times A_{pin}} \qquad\qquad$ [since each of the pin is in *double shear*]

$$= \frac{40500}{2 \times \pi/4 \times \left(\dfrac{18}{1000}\right)^2} \times 10^{-6} = \mathbf{79.57\ MN/m^2}\ \textbf{(Ans.)}$$

Example 1.49. *A steel rod of 10 mm diameter passes centrally through a copper tube of external diameter 40 mm and internal diameter 30 mm and of length 2m. The tube is closed at each end by 20 mm thick steel plates which are screwed by the nuts. The nuts are tightened until the copper tube is reduced to length 1.9996 m. Find the stresses in the rod and the tube.*

If the whole assembly is heated through 60°C, what are then the stresses in the rod and the tube, assuming that the thickness of the plates remains unchanged ?

Take:

$E_s = 210 \ GN/m^2$;
$E_c = 100 \ GN/m^2$;
$\alpha_s = 12 \times 10^{-6} \ per \ °C$;
$\alpha_c = 17.5 \times 10^{-6} \ per \ °C$.

Solution. Refer to Fig. 1.54.

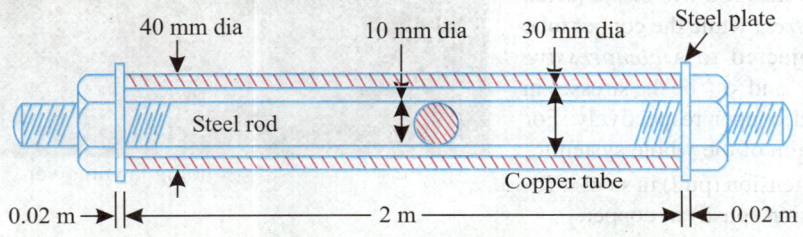

Fig. 1.54

Area of the steel rod, $\quad A_s = \dfrac{\pi}{4} \times \left(\dfrac{10}{1000}\right)^2 = 78.54 \times 10^{-6} \ m^2$

Area of the copper tube, $\quad A_c = \dfrac{\pi}{4} \times \left[\left(\dfrac{40}{1000}\right)^2 - \left(\dfrac{30}{1000}\right)^2\right] = 549.78 \times 10^{-6} \ m^2$

(i) Stresses due to tightening of the nuts :

When the nuts are tightened, the steel rod will be subjected to tensile stress and the copper tube will be subjected to compressive stress.

Let, $\qquad\qquad \sigma_s$ = Tensile stress in steel rod, and

σ_c = Compressive stress in copper tube.

As there is no external load,

Push on the copper tube = Pull in the steel rod

$$\sigma_c \times \dfrac{\pi}{4}\left[(0.04^2 - 0.03^2)\right] = \sigma_s \times \dfrac{\pi}{4} \times 0.01^2 \quad or \quad \sigma_s = 7\sigma_c \qquad ...(i)$$

Strain in the copper tube, $\quad \sigma_c = \dfrac{2 - 1.9996}{2} = 0.0002$

∴ *Stress in the copper tube,*

$$\sigma_c = e_c \times E_c = 0.0002 \times 100 \times 10^9 \times 10^{-6}$$
$$= 20 \ MN/m^2 \ (compressive) \ \textbf{(Ans.)}$$

and, *stress in the steel rod,* $\quad \boldsymbol{\sigma_s = 7\sigma_c = 7 \times 20}$
$$\boldsymbol{= 140 \ MN/m^2} \ (tensile) \ \textbf{(Ans.)}$$

(ii) Stresses due to rise of temperature :

If the two members had been free to expand, then

Free expansion of steel $\quad = \alpha_s \, t \, l_s$

Free expansion of copper = $\alpha_c \, t \, l_c$

Since α_c is greater than α_s, the free expansion of copper is greater than the free expansion of steel. But since the ends are provided with washers and nuts, the members are not free to expand fully. Final expansion of each of the members will be the same. Let this final expansion be δ. The free expansion of copper is greater than δ, while the free expansion of steel is less than δ. Hence the steel rod will be subjected to a *tensile stress* while the copper tube will be subjected to a *compressive stress*. Let $\sigma_s{}'$ and $\sigma_c{}'$ be the stresses in the steel and copper respectively. For the equilibrium of the whole system,

A riding lawnmower.

Total tension (pull) in steel = Total compression (push) in copper.

i.e.,
$$\sigma_s' A_s = \sigma_c' A_c$$

or,
$$\sigma_s' = \sigma_c' \frac{A_c}{A_s} = 7\sigma_c'$$

Final expansion of steel = Final expansion of copper

$$\alpha_s \, t \, l_s + \frac{\sigma_s'}{E_s}.l_s = \alpha_c \, t \, l_c - \frac{\sigma_c'}{E_c} \, l_c$$

But
$$t = 60°C$$
$$l_s = 2 + 2 \times 0.02 = 2.04 \text{ m}$$
and
$$l_c = 2\text{m}$$

$\therefore$
$$12 \times 10^{-6} \times 60 \times 2.04 + \frac{7\sigma_c'}{210 \times 10^9} \times 2.04$$

$$= 17.5 \times 10^{-6} \times 60 \times 2 - \frac{\sigma_c'}{100 \times 10^9} \times 2$$

or,
$$12 \times 10^{-6} \times 60 \times 2.04 \times 210 \times 10^9 + 7\sigma'_c \times 2.04$$
$$= 17.5 \times 10^{-6} \times 60 \times 2 \times 210 \times 10^9 - 2.1 \times 2 \, \sigma'_c$$

$$3.084 \times 10^8 + 14.28 \, \sigma'_c = 4.41 \times 10^8 - 4.2 \, \sigma'_c$$

$\therefore$
$$\sigma'_c = 0.0717 \times 10^8 \text{ N/m}^2$$

or,
$$= 7.17 \text{ } MN/m^2 \text{ (compressive)}$$

and,
$$\sigma'_s = 7 \, \sigma'_c = 7 \times 7.17 = 50.19 \text{ MN/m}^2 \text{ (tensile)}$$

Therefore, the final stresses due to tightening the nuts and rise of temperature will be as follows:

Stress in copper
$$= \sigma_c + \sigma'_c = 20 + 7.17$$
$$= \mathbf{27.17 \text{ MN/m}^2} \text{ } (\textit{compressive}) \text{ } (\textbf{Ans.})$$

Stress in steel
$$= \sigma_s + \sigma'_s = 140 + 50.19$$
$$= \mathbf{190.19 \text{ MN/m}^2} \text{ } (\textit{tensile}) \text{ } (\textbf{Ans.})$$

Example 1.50. *A brass rod of 6 mm diameter and 1 m long is joined at one end to a rod of steel 6 mm diameter and 1.3 m long. The compound rod is placed in a vertical position with the steel rod at the top and connected top and bottom are connected to rigid fixing in such a way that it is carrying a tensile load of 3.5 kN.*

An attachment is fixed at the junction of the two rods and to this a vertical axial load of 1.3 kN is applied downwards as shown in Fig. 1.55. Calculate the stresses in steel and brass. The temperature is then raised through 30°C. What are the final stresses in steel and brass ?

Take: $E_s = 200 \ GN/m^2; \ E_b = 85 \ GN/m^2;$
  $\alpha_s = 12 \times 10^{-6} \ per°C; \ \alpha_b = 19 \times 10^{-6} \ per°C.$ **(Panjab University)**

Solution. Refer to Fig. 1.55.

$$A_s = A_b = \frac{\pi}{4} \times \left(\frac{6}{1000}\right)^2$$

$$= 28.27 \times 10^{-6} \ m^2$$

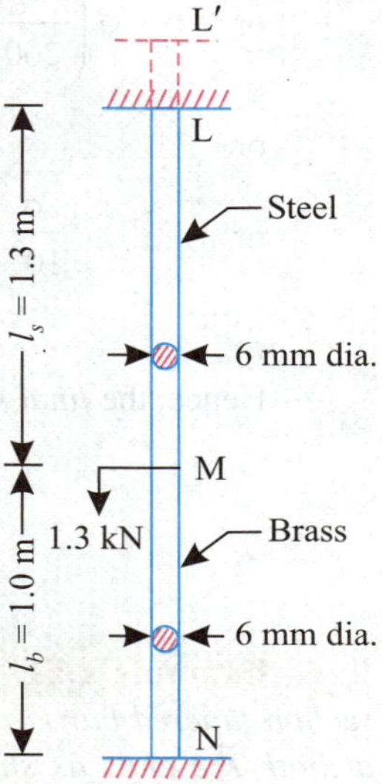

— In the *first case*, when the composite bar *LMN* is held in fixtures, both steel and brass portions of the rod carry the tensile load of 3.5 kN. Thus, the initial tensile stresses in each of the metals will be equal, as their areas of cross-section are equal.

i.e., $\sigma_{s_1} = \sigma_{b_1} = \dfrac{3.5 \times 1000}{28.27 \times 10^{-6}} \times 10^{-6}$

$$= 123.8 \ MN/m^2 \quad \text{(tensile)}$$

— When the load of 1.3 kN is applied at junction *M*, it will induce tensile stress in steel and compressive stress in brass. The magnitudes of the loads P_s and P_b due to the action of 1.3 kN load at *M*, will be such that the stretch in steel will be *equal* in magnitude to the compression in brass, because total *length LN remains constant* :

i.e., $P_s + P_b = 1.3 \times 1000 = 1300 \ N$

and, $\delta l_s = \delta l_b$

or, $\dfrac{P_s l_s}{A_s E_s} = \dfrac{P_b l_b}{A_b E_b}$

or, $P_s = \dfrac{E_s}{E_b} \cdot \dfrac{l_b}{l_s} \cdot P_b$ $(\because A_s = A_b)$

$$= \frac{200 \times 10^9}{85 \times 10^9} \times \frac{1}{1.3} \times P_b = 1.81 \ P_b$$

∴ $1.81 \ P_b + P_b = 1300$

or, $P_b = 462.6 \ N$

and, $P_s = 1300 - 462.6 = 837.4 \ N$

Thus, the corresponding additional stresses in steel and brass, due to the application of 1300 N load are

$$\sigma_{s_2} = \frac{P_s}{A_s} = \frac{837.4}{28.27 \times 10^{-6}} \times 10^{-6} = 29.62 \ MN/m^2 \ \text{(tensile)}$$

$$\sigma_{b_2} = \frac{P_b}{A_b} = \frac{462.6}{28.27 \times 10^{-6}} \times 10^{-6} = 16.36 \ MN/m^2 \ \text{(compressive)}$$

Fig. 1.55

Then the *resultant stresses* are :

$$\sigma_{br} = \sigma_{b_1} - \sigma_{b_2} = 123.8 - 16.36 = \textbf{107.44 MN / m}^2 \text{ (tensile)} \quad \textbf{(Ans.)}$$

$$\sigma_{sr} = \sigma_{s_1} + \sigma_{s_2} = 123.8 + 29.62 = \textbf{153.42 MN/m}^2 \text{ (tensile)} \quad \textbf{(Ans.)}$$

— In the absence of fixtures, the rod *LN* would have expanded freely due to temperature rise to *NL'*. But the pressure of fixtures effectively compresses the rod *LN* by an amount *LL'* which is equal to $(\alpha_s \, l_s \, t + \alpha_b \, l_b \, t)$. Thus if σ is the compressive stress in steel and brass (it will be same in both as the cross-sectional areas are the same) due to temperature rise, then we have

$$\frac{\sigma}{E_s} \, l_s + \frac{\sigma}{E_b} \, l_b = \alpha_s \, l_s \, t + \alpha_b \, l_b \, t$$

or, $$\sigma \left(\frac{1.3}{200 \times 10^9} + \frac{1}{85 \times 10^9} \right) = 30 \, (12 \times 10^{-6} \times 1.3 + 19 \times 10^{-6} \times 1)$$

or, $$\frac{\sigma}{10^9} \left(\frac{1.3}{200} + \frac{1}{85} \right) = 30 \times 10^{-6} \, (12 \times 1.3 + 19)$$

or, $$\frac{\sigma}{10^9} \, (0.0065 + 0.0117) = 1038 \times 10^{-6}$$

∴ $$\sigma = \frac{1038 \times 10^{-6} \times 10^9}{0.0065 + 0.0117} \times 10^{-6} \text{ MN/m}^2 \text{ (compressive)}$$

Hence, the *final stresses* in steel and brass are :

$$\sigma_{sf} = \sigma_s - \sigma = 153.42 - 57.03$$

$$= \textbf{96.39 MN/m}^2 \text{ (tensile)} \quad \textbf{(Ans.)}$$

$$\sigma_{bf} = \sigma_b - \sigma = 107.44 - 57.03$$

$$= \textbf{50.41 MN/m}^2 \text{ (tensile)} \quad \textbf{(Ans.)}$$

Example 1.51. *A circular section tapered bar is rigidly fixed at both the ends as shown in the Fig.1.56. If the temperature is raised by 40°C, calculate the stress in the bar. Take E = 200 GN / m^2 and α = 12 × 10^{-6} per°C.*

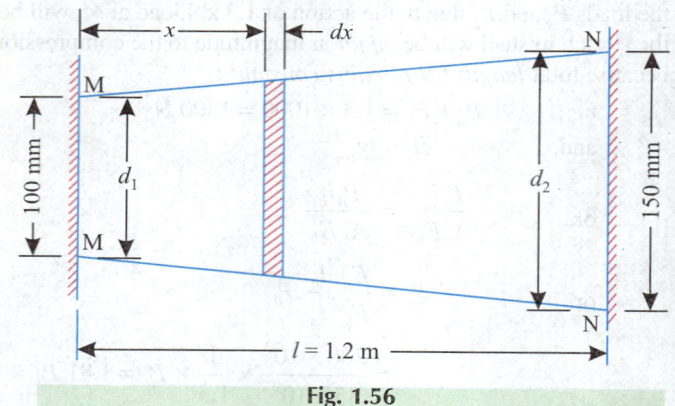

Fig. 1.56

Solution. Refer to Fig. 1.56.

When the temperature is raised by *t*°C, a compressive force *P* is induced. Since this force is same for all cross-sections, maximum stress will be at section *MM*. If the bar were free to expand its expansion would be :

$$\delta l = \alpha \, l \, (t_2 - t_1)$$

$$= 12 \times 10^{-6} \times 1.2 \times 40 = 0.000576 \text{ m} \qquad \qquad \text{...(i)}$$

The force induced will be a force *P* which is required to prevent a free expansion of 0.000576 m.

For an element *dx*, the expansion is

$$\delta x = \frac{P dx}{A_x E}$$

∴ Total extension $$\delta l = \int_0^l \frac{P dx}{A_x E}$$

Earthmoving equipment.

But,

$$A_x = (\pi/4) \left(d_1 + \frac{d_2 - d_1}{l} . x \right)^2 = (\pi/4) \ (d_1 + kx)^2$$

where,

$$k = \frac{d_2 - d_1}{l}$$

∴

$$\delta l = \frac{4P}{\pi E} \int_0^l \frac{dx}{(d_1 + kx)^2} = -\frac{AP}{\pi kE} \left[\frac{1}{d_1 + kx} \right]_0^l$$

$$= -\frac{4 Pl}{E \ (d_2 - d_1)} \left[\frac{1}{d_1 + d_2 - d_1} - \frac{1}{d_1} \right] = \frac{4 \ Pl}{\pi E d_1 \ d_2} \qquad ...(ii)$$

Equating (i) and (ii), we get

$$\alpha l \ (t_2 - t_1) = \frac{4 \ Pl}{\pi E d_1 \ d_2}$$

From which,

$$P = \frac{\pi E \alpha \ (t_2 - t_1) \ d_1 \ d_2}{4}$$

Maximum stress induced is given by

$$\sigma = [\ (\pi/4) \ E \alpha \ (t_2 - t_1) \ d_1 \ d_2] \div \{\pi/4 \ d_1^2\}$$

or,

$$\sigma = E \alpha \ (t_2 - t_1) \frac{d_2}{d_1}$$

Substituting the numerical values, we get

$$\sigma = 200 \times 10^9 \times 12 \times 10^{-6} \times 40 \times \frac{0.150}{0.100}$$

$$= 144 \times 10^6 \ \text{N/m}^2 \ = \textbf{144 MN/m}^2 \ \textbf{(Ans.)}$$

1.14. HOOP STRESS

If a thin tyre of steel or any other metal is to be shrunk on to a wheel, the diameter of the tyre is to be slightly smaller than that of the wheel so that it does not come out easily. Let D be the diameter of the wheel and d that of the tyre. The temperature of the tyre is to be increased by $t°$ so that it increases in diameter from d to D. When tyre is slipped on to the wheel and the temperature falls, the steel tyre will try to come to its original diameter and, in doing so, the hoop stress (tensile) will be set up.

Temperature strain, $e = \dfrac{\text{Contraction prevented}}{\text{Original length}} = \dfrac{\pi D - \pi d}{\pi d}$

∴ $\qquad e = \dfrac{D - d}{d}$

∴ Circumferential stress or *hoop stress* developed due to fall of temperature $= e\,E$

$\qquad = \dfrac{D - d}{d}\,E$ $\qquad\qquad\qquad$ ...(1.23)

Example 1.52. *A rigid wheel 1.2 m in diameter is to be provided with a thin steel tyre. If the stress in the steel tyre is not to exceed 120 MN/m², find (i) The minimum diameter of the tyre. (ii) The minimum temperature to which tyre is to be raised so that it can be fitted over the wheel.*

Take: $\qquad\qquad E = 200\ GN/m^2$,

and, $\qquad\qquad \alpha = 12 \times 10^{-6}$ *per* °C.

Solution. Let D and d be the diameter of the rigid wheel and least diameter of the steel tyre respectively.

(*i*) Minimum diameter of the tyre :

Strain in the tyre $\qquad = \dfrac{D - d}{d}$

Stress in the tyre, $\qquad \sigma = \left(\dfrac{D - d}{d}\right)E$

or, $\qquad\qquad \left(\dfrac{D}{d} - 1\right) = \dfrac{\sigma}{E} = \dfrac{120 \times 10^6}{200 \times 10^9} = 0.0006$

∴ $\qquad\qquad \dfrac{D}{d} = 1.0006$ or $\dfrac{d}{D} = 0.9994$

∴ $\qquad\qquad d = 0.9994 \times 1.20 = \mathbf{1.1993\ m}$ **(Ans.)**

(*ii*) Minimum temperature :

Let the steel be subjected to a temperature rise of t°C *i.e.*, $(t_2 - t_1)$

∴ $\qquad\qquad \pi D = \pi d\,(1 + \alpha\,t)$

$\qquad\qquad 1 + \alpha\,t = \dfrac{D}{d} = 1.0006$

$\qquad\qquad \alpha\,t = 0.0006$

∴ $\qquad\qquad t = \dfrac{0.0006}{12 \times 10^{-6}} = 50°C$

Hence, rise in temperature = **50°C** **(Ans.)**

TYPICAL EXAMPLES (For Competitive Examinations)

Example 1.53. *A cylindrical block of concrete is encompassed by a close-fitting thin steel tube of radius r and wall thickness t as shown in Fig. 1.57. If the concrete block is subjected to a uniform axial compressive stress σ_x, what should be the ratio $\dfrac{r}{t}$ so that the axial strain in the tube becomes equal to the axial strain in the concrete block? The moduli and Poisson's ratios for steel and concrete are 200 GPa, 2.29 and 13.35 GPa, 0.25 respectively.*

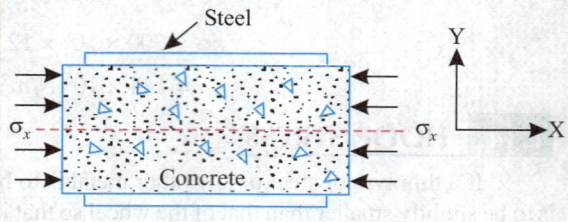

Fig. 1.57

Solution. *Given*:

$E_s = 200 \ GPa = 200 \times 10^9 \ \text{N/m}^2, \ \dfrac{1}{m_s} = 0.29$;

$E_c = 13.35 \ GPa = 13.35 \times 10^9 \ \text{N/m}^2, \ \dfrac{1}{m_c} = 0.25$

Ratio $\dfrac{r}{t}$:

Axial strain in the steel tube $= \dfrac{\sigma_s}{E_s} - \dfrac{\sigma_s}{m_s \ E_s}$

Axial strain in the concrete block $= \dfrac{\sigma_c}{E_c} - \dfrac{\sigma_c}{m_c \ E_c}$

But, axial strain in the steel tube = axial strain in the concrete block   ...(Given)

Hydraulic roller lifters of a crane.

$$\dfrac{\sigma_s}{E_s} - \dfrac{\sigma_s}{m_s E_s} = \dfrac{\sigma_c}{E_c} - \dfrac{\sigma_c}{m_c E_c}$$

$$\dfrac{\sigma_s}{200 \times 10^9} - \dfrac{0.29 \ \sigma_s}{200 \times 10^9} = \dfrac{\sigma_c}{13.35 \times 10^9} - \dfrac{0.25 \ \sigma_c}{13.35 \times 10^9}$$

$$\dfrac{\sigma_s}{200}(1 - 0.29) = \dfrac{\sigma_c}{13.35}(1 - 0.25)$$

or,
$$\dfrac{0.71 \ \sigma_s}{200} = \dfrac{0.75 \ \sigma_c}{13.35}$$

or,
$$\sigma_s = 15.82 \ \sigma_c$$

Cross-sectional area of steel tube,

$$\begin{aligned} A_s &= \pi \ (r+t)^2 - \pi \ r^2 \\ &= \pi \ [(r+t)^2 - r^2] \\ &= \pi \ [r^2 + 2rt + t^2 - r^2] \\ &= \pi \ (2rt + t^2) \simeq 2 \ \pi rt \end{aligned}$$

(neglecting t^2 as it is very small compared to r).

Cross-sectional area of concrete,

$$A_c = \pi r^2$$

Since there is no external load, it follows that,

Total compressive force in concrete block = Total tensile force in steel tube

$$\sigma_c A_c = \sigma_s A_s$$
$$\sigma_c \times \pi \ r^2 = 15.82 \ \sigma_c \times 2 \ \pi rt$$

or,
$$\dfrac{r}{t} = \textbf{31.64} \quad \textbf{(Ans.)}$$

Example 1.54. *A pile of uniform section is embedded in soil by a depth h. The pile supports a structural load P at its top which is transferred to the soil entirely by friction, shown in Fig. 1.58. The variation of friction (f) along the depth of the pile is given by $f_y = ky^2$, where y is the elevation above the bottom of the pile.*

Determine the total shortening of the pile.

Solution. Refer to Fig. 1.59 (*i*), (*ii*) and (*iii*).

Fig. 1.59 (*ii*) shows the variation of f with y. Let us determine the value of constant k by equating total resistance for the full height h and equating it to the external load P, neglecting the self weight of the pile.

∴ Total frictional resistance = P

$$\int_0^h ky^2 \, . \, dy = P$$

or

$$\frac{kh^3}{3} = P$$

i.e.,

$$k = \frac{3P}{h^3} \qquad \qquad ...(1)$$

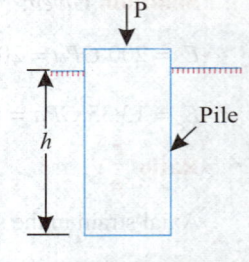

Fig. 1.58

Now, the total compressive force, at any height y above the bottom of the pile, will be equal to the frictional resistance of the clay for the height y.

∴ Total compressive force, $P_y = \int_0^y ky^2 \, dy = \frac{ky^3}{3}$

$$= \frac{3P}{h^3} \, . \, \frac{y^3}{3} = \frac{Py^3}{h^3} \qquad \text{(substituting for } k) \qquad ...(2)$$

The variation of P_y is shown in Fig. 1.59 (*iii*)

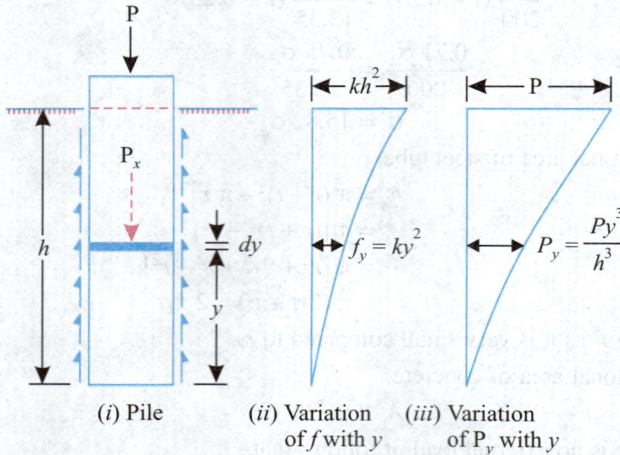

(*i*) Pile   (*ii*) Variation   (*iii*) Variation
of *f* with *y*   of P$_y$ with *y*

Fig. 1.59

Assuming that the compressive force is constant for a length dy, the shortening of small length dy of the pile is

$$= \frac{P_y \, . \, dy}{AE}$$

∴ Total shortening of pile $= \displaystyle\sum_{y=0}^{y=h} \frac{P_y \, . \, dy}{AE}$

$$= \int_0^h \frac{Py^3}{h^3} \, . \, \frac{dy}{AE} = \frac{P}{AEh^3} \int_0^h y^3 \, dy \qquad \left[\because P_y = \frac{Py^3}{h^3} \right]$$

$$= \frac{P}{AEh^3} \, . \, \frac{h^4}{4} = \frac{Ph}{4 \, AE} \qquad \textbf{(Ans.)}$$

Example 1.55. *A flat steel plate of trapezoidal form of uniform thickness of 20 mm tapers uniformly from a width 100 mm to 200 mm in a length of 800 mm. If an axial tensile force of 100 kN is applied at each end, find the elongation of the plate.*

Take, E = 205 GN/m².

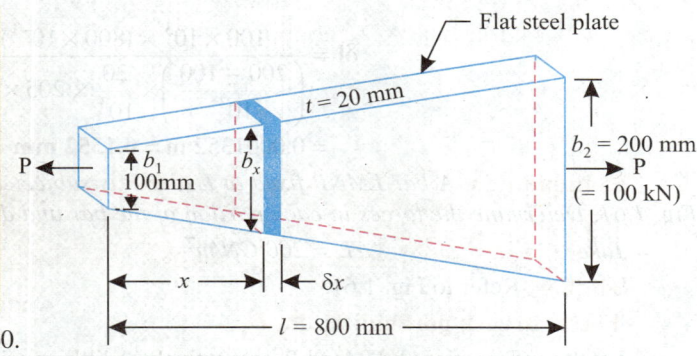

Fig. 1.60

Solution. Refer to Fig. 1.60.

Width, $b_1 = 100$ mm, $b_2 = 200$ mm

Thickness, $t = 20$ mm

Length, $l = 800$ mm

Axial force (tensile), $P = 100$ kN

Elongation of the plate, δl :

Consider a small section of length δx at a distance x from the width b_1.

The width at the section, $b_x = b_1 + \dfrac{b_2 - b_1}{l} x = b_1 + kx,$

where, $k = \dfrac{b_2 - b_1}{l}$

Area $= (b_1 + k x) t$

Now, extension of a short length δx

$= \dfrac{P \, \delta x}{(b_1 + kx) \, tE}$

∴ Total extension δl of the bar is given by,

$$\delta l = \int_0^l \frac{P \, dx}{(b_1 + kx) \, t \, E}$$

$$= \frac{P}{tE} \cdot \frac{1}{k} \Big[\log_e (b_1 + kx) \Big]_0^l$$

$$= \frac{P}{kt \, E} \left(\log_e \frac{b_1 + kl}{b_1} \right)$$

$$= \frac{P}{kt \, E} \log_e \frac{b_2}{b_1}$$

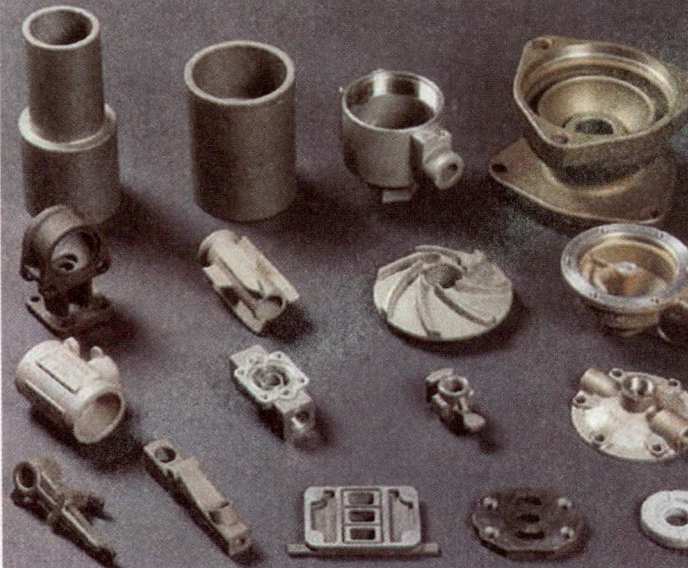

Parts made of investment casting. Applicable materials: Aluminium alloy, copper alloy, carbon steel, stainless steel, steel alloy, etc.

Putting $k = \dfrac{b_2 - b_1}{l}$, we get

$$\delta l = \frac{Pl}{(b_2 - b_1) \, tE} \log_e \frac{b_2}{b_1}$$

Substituting the numerical values, we get

$$\delta l = \frac{100 \times 10^3 \times (800 \times 10^{-3})}{\left(\dfrac{200 - 100}{10^3}\right) \times \dfrac{20}{10^3} \times 205 \times 10^9} \log_e \frac{200}{100}$$

$$= 0.0001352 \text{ m} = \mathbf{0.1352 \text{ mm}} \quad \textbf{(Ans.)}$$

Example 1.56. *A bar LMNP fixed at L and P is subjected to axial forces as shown in the Fig. 1.61. Determine the forces in each portion of the bar and displacement of points M and N.*

Take $E = 200 \text{ GN/m}^2$

Solution. Refer to Fig. 1.61.

Forces in each portion : P_1, P_2, P_3 **:**

Let the end reactions be R_1 and R_2 respectively. A little consideration will show that the bar *LM* will be in *tension* while bar *NP* will be in *compression*. Hence the directions of R_1 and R_2 will be as marked.

The free body diagrams of the portions *LM*, *MN* and *NP* are shown in Fig. 1.61 (*b*)

For portion LM :

Force $P_1 = R_1$; $l_1 = 0.5$ m ; $A_1 = 1000$ mm$^2 = 1000 \times 10^{-6}$ m^2

For portion NP :

Force $P_3 = R_2$; $l_3 = 1$ m; $A_3 = 2000$ mm$^2 = 2000 \times 10^{-6}$ m^2

For portion MN :

Let the axial force be P_2 (compressive). Considering forces on the left ; $P_2 = 50 - R_1$. Similarly, considering forces on the right, $P_2 = R_2 - 100$

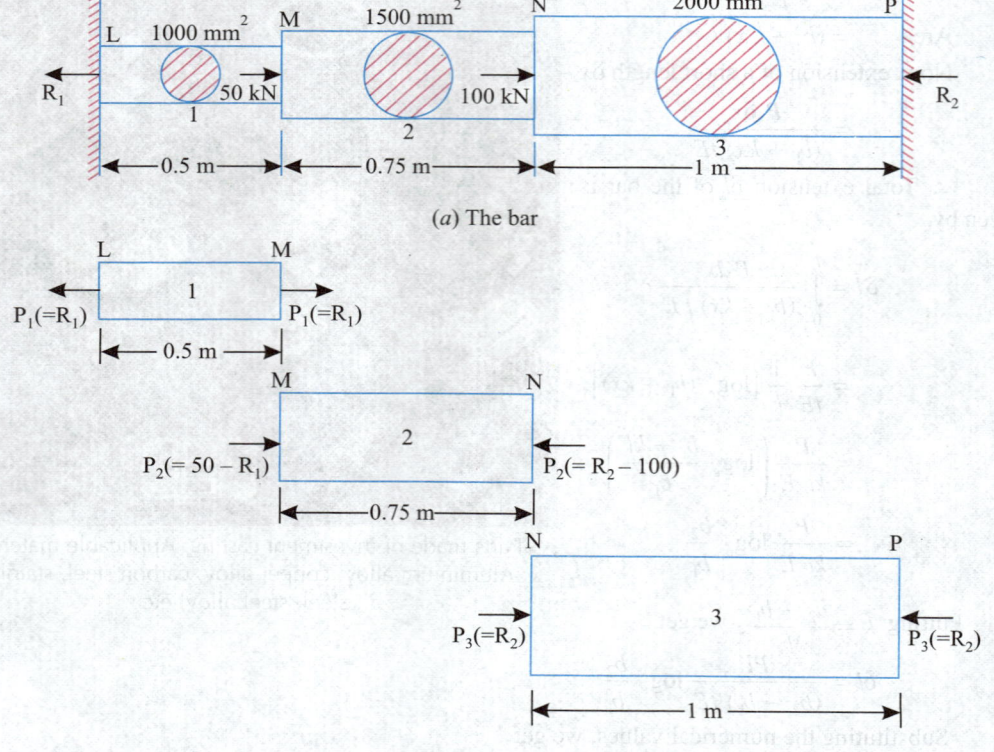

(*a*) The bar

(*a*) Free body diagram

Fig. 1.61

Thus, $50 - R_1 = R_2 - 100$

or, $R_1 + R_2 = 100 + 50 = 150$ kN   ...(i)

(The above equation can also be obtained by considering the static equilibrium of the bar)

Now,   Extension of LM = Compression of MN and NP

∴ $\delta l_1 = \delta l_2 + \delta l_3$

$$\frac{P_1 l_1}{A_1 E} = \frac{P_2 l_2}{A_2 E} + \frac{P_3 l_3}{A_3 E}$$

Substituting the values, we get

$$\frac{(R_1 \times 10^3) \times 0.5}{1000 \times 10^{-6} \times 200 \times 10^9} = \frac{(50 - R_1) \times 10^3 \times 0.75}{1500 \times 10^{-6} \times 200 \times 10^9} + \frac{R_2 \times 10^3 \times 1}{2000 \times 10^{-6} \times 200 \times 10^9}$$

or, $R_1 \times 0.5 = \dfrac{(50 - R_1) \times 0.75}{1.5} + \dfrac{R_2}{2}$

or, $1.5\, R_1 = 75 - 1.5\, R_1 + 1.5\, R_2$

or, $3\, R_1 - 1.5\, R_2 = 75$

or, $R_1 - 0.5\, R_2 = 25$ kN   ...(ii)

From (i) and (ii), we get $R_2 = \dfrac{125}{1.5} = 83.33$ kN

and, $R_1 = 66.67$ kN

Hence, $P_1 = R_1 = \mathbf{66.67}$ **kN** (*tensile*) **(Ans.)**

$P_2 = 50 - R_1 = 50 - 66.67 = -16.67$ kN

$\qquad = \mathbf{16.67}$ **kN** (*tensile*) **(Ans.)**

$P_3 = R_2 = \mathbf{83.33}$ **kN** (*compressive*) **(Ans.)**

Displacement of point M, δl_1 :

$$\delta l_1 = \frac{P_1 l_1}{A_1 E} = \frac{66.67 \times 10^3 \times 0.5}{1000 \times 10^{-6} \times 200 \times 10^9}$$

$$= 0.1666 \times 10^{-3} \text{ m or } \mathbf{0.1666 \text{ mm}} \text{ (Ans.)}$$

A twin-cylinder IC engine.

Displacement of point, N :

Displacement of point $\qquad N = \delta l_1 + \delta l_2$

where, $\qquad \delta l_2 = \dfrac{P_2 \, l_2}{A_2 \, E} = \dfrac{16.67 \times 10^3 \times 0.75}{1000 \times 10^{-6} \times 200 \times 10^9}$

$\qquad\qquad\qquad\qquad = 4.17 \times 10^{-5}$ m or 0.0417 mm

$\therefore$ *Displacement of point* $\qquad N = 0.1666 + 0.0417 =$ **0.2083 mm**

$$\left[\text{Displacement of point can also be calculated as follows:} \right.$$
$$\left. \delta l_3 = \dfrac{P_3 \, l_3}{A_3 \, E} = \dfrac{83.33 \times 10^3 \times 1}{2000 \times 10^{-6} \times 200 \times 10^9} = 2.083 \times 10^{-3} \text{ m or } 0.2083 \text{ mm} \right]$$

Example 1.57. *Fig. 1.62 shows three bars made of copper, zinc and aluminium and of equal length rigidly connected at their ends. They have cross-sectional areas of 250 mm², 375 mm² and 500 mm² respectively. If the compound member is subjected to a longitudinal pull of 125 kN, estimate the proportion of load carried on each rod, and the induced stresses.*

Take: $\qquad\qquad\qquad E_{cu} = 130 \text{ GN/m}^2,$
$\qquad\qquad\qquad\qquad E_{zn} = 100 \text{ GN/m}^2,$
and, $\qquad\qquad\qquad E_{al} = 80 \text{ GN/m}^2.$

Solution. Refer to Fig. 1.62.

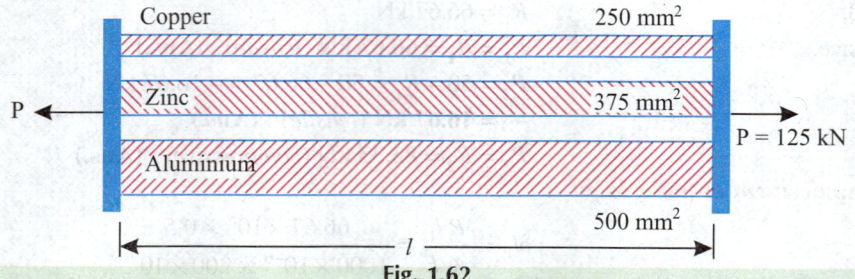

Fig. 1.62

Area of copper bar, $\qquad A_{cu} = 250 \text{ mm}^2 = 250 \times 10^{-6} \text{ m}^2$
Area of zinc bar, $\qquad A_{zn} = 375 \text{ mm}^2 = 375 \times 10^{-6} \text{ m}^2$
Area of aluminium $\qquad A_{al} = 500 \text{ mm}^2 = 500 \times 10^{-6} \text{ m}^2$
Longitudinal pull, $\qquad P = 125 \text{ kN}$
$\qquad\qquad E_{cu} = 130 \text{ GN/m}^2; \ E_{zn} = 100 \text{ GN/m}^2, \ E_{al} = 80 \text{ GN/m}^2$

Loads carried by each bar, P_{cu}, P_{zn}, P_{al} :

Considering equilibrium of the bar, we have

$$P_{cu} + P_{zn} + P_{al} = P = 125 \text{ kN} \qquad\qquad\qquad\qquad ...(i)$$

Since all the bars are rigidly connected at their ends, their deformations will be *equal*.

$\therefore \qquad\qquad \dfrac{P_{cu} \, l}{A_{cu} \, E_{cu}} = \dfrac{P_{zn} \, l}{A_{zn} \, E_{zn}} = \dfrac{P_{al} \, l}{A_{al} \, E_{al}}$

Hence, $\qquad P_{zn} = P_{cu} \times \dfrac{A_{zn}}{A_{cu}} \times \dfrac{E_{zn}}{E_{cu}} = \dfrac{375 \times 10^{-6}}{250 \times 10^{-6}} \times \dfrac{100}{130} \times P_{cu} = \dfrac{15}{13} P_{cu}$

and, $\qquad P_{al} = P_{cu} \times \dfrac{A_{al}}{A_{cu}} \times \dfrac{E_{al}}{E_{cu}} = \dfrac{500 \times 10^{-6}}{250 \times 10^{-6}} \times \dfrac{80}{130} = \dfrac{16}{13} \times P_{cu} = \dfrac{16}{13} P_{cu}$

Substituting the values of P_{zn} and P_{al} in eqn. (*i*), we get,

$$P_{cu} + \frac{15}{13} P_{cu} + \frac{16}{13} P_{cu} = 125 \quad \text{or} \quad \frac{44}{13} P_{cu} = 125$$

$\therefore$

$$P_{cu} = \frac{125 \times 13}{44} = \textbf{36.93 kN} \quad \textbf{(Ans.)}$$

and,

$$P_{zn} = \frac{15}{13} \times 36.93 = \textbf{42.61 kN} \quad \textbf{(Ans.)}$$

and,

$$P_{al} = \frac{16}{13} \times 36.93 = \textbf{45.45 kN} \quad \textbf{(Ans.)}$$

Stresses induced in the bar σ_{cu}, σ_{zn}, σ_{al} :

$$\sigma_{cu} = \frac{P_{cu}}{A_{cu}} = \frac{36.93 \times 10^3}{250 \times 10^{-6}} = \textbf{147.72 MN/m}^2 \quad \textbf{(Ans.)}$$

$$\sigma_{zn} = \frac{P_{zn}}{A_{zn}} = \frac{42.61 \times 10^3}{375 \times 10^{-6}} = \textbf{113.63 MN/m}^2 \quad \textbf{(Ans.)}$$

$$\sigma_{al} = \frac{P_{al}}{A_{al}} = \frac{45.45 \times 10^3}{500 \times 10^{-6}} = \textbf{90.9 MN/m}^2 \quad \textbf{(Ans.)}$$

Example 1.58. *A rigid bar LMNS pinned at M and connected to two vertical rods, is shown in Fig. 1.63. Assuming that the bar was initially horizontal and the rods stress free, determine the stress in each rod after the load P = 500 N is applied.*

Take: $\qquad E_{steel}$ = 200 GN/m²,

and, $\qquad E_{al}$ = 100 GN/m².

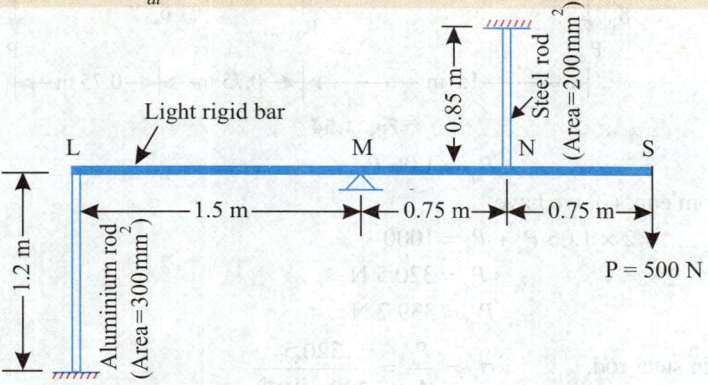

Fig. 1.63

Solution. *Steel rod:*

Length, $\qquad l_s$ = 0.85 m

Area, $\qquad A_s$ = 200 mm² = 200 × 10⁻⁶ m²

Aluminium rod :

Length, $\qquad l_{al}$ = 1.2 m

Area, $\qquad A_{al}$ = 300 mm² = 300 × 10⁻⁶ m²

Stress in each rod, σ_s, σ_{al} :

Refer to Fig. 1.64

Let, P_s = Force in the steel rod, and

P_{al} = Force in the aluminium rod.

At the equilibrium condition, we have

$$P_{al} \times 1.5 + P_s \times 0.75 = P \times 1.5$$

$$2\,P_{al} + P_s = 2P = 1000 \quad ...(i)$$

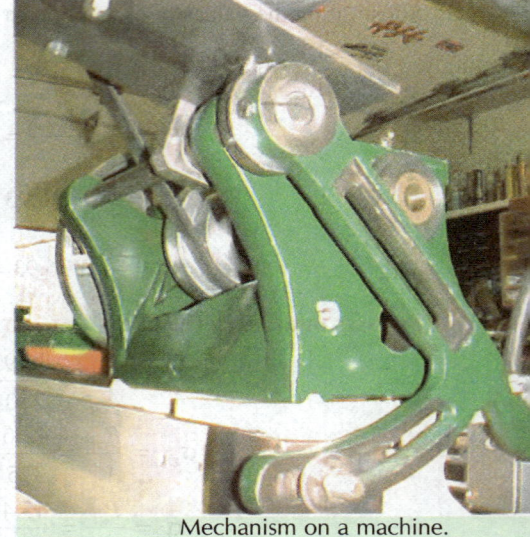

Also, $\dfrac{\delta l_{al}}{1.5} = \dfrac{\delta l_s}{0.75}$

$\therefore$ $\delta l_{al} = 2\,\delta l_s$

or, $\dfrac{P_{al} \times l_{al}}{A_{al} \times E_{al}} = 2 \times \dfrac{P_s \times l_s}{A_s \times E_s}$

or, $\dfrac{P_{al} \times 1.2}{300 \times 10^{-6} \times 100 \times 10^9}$

$$= 2 \times \dfrac{P_s \times 0.8}{200 \times 10^{-6} \times 200 \times 10^9}$$

Mechanism on a machine.

or, $P_{al} = 2 \times P_s \times \dfrac{0.85}{1.2} \times \dfrac{300 \times 10^{-6} \times 100 \times 10^9}{200 \times 10^{-6} \times 200 \times 10^9} = 1.06\,P_s$

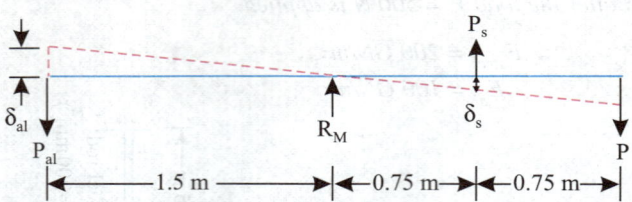

Fig. 1.64

i.e., $P_{al} = 1.06\,P_s$ ...(ii)

Hence, from eqn. (*i*), we have

$$2 \times 1.06\,P_s + P_s = 1000$$

or, $P_s = 320.5$ N

and, $P_{al} = 339.7$ N

$\therefore$ Stress in steel rod, $\sigma_s = \dfrac{P_s}{A_s} = \dfrac{320.5}{200 \times 10^{-6}}$

$$= 1.6 \times 10^{-6} \text{ N/m}^2$$

$$= 1.6 \times 10^6 \times 10^{-6} \text{ MN/m}^2$$

i.e., $\sigma_s = \mathbf{1.6\ MN/m^2}$ **(Ans.)**

and, stress in aluminium rod, $\sigma_{al} = \dfrac{P_{al}}{A_{al}} = \dfrac{339.7}{300 \times 10^{-6}} = 1.13 \times 10^6 \text{ N/m}^2$

$$= 1.13 \times 10^6 \times 10^{-6} \text{ MN/m}^2$$

i.e., $\sigma_{al} = \mathbf{1.13\ MN/m^2}$ **(Ans.)**

Example 1.59. *Determine the reduction in length in a circular tapered steel bar with a cylindrical hole under a compressive force of 60 kN as shown in Fig. 1.65.*

Take: $E = 2.1 \times 10^5 \text{ MN/m}^2 \text{ (or } 2.1 \times 10^5 \text{ N/mm}^2)$

Solution. Refer to Fig. 1.65.

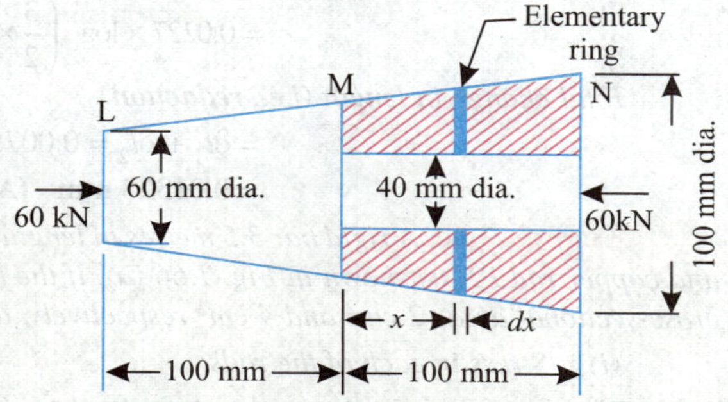

Portion LM :

Fig. 1.65

$d_2 = 60$ mm

Diameter at M, $d_1 = \dfrac{100 + 60}{2}$

$\qquad = 80$ mm

Length $\qquad LM = 100$ mm

Compressive force, $\quad P = 60$ kN

Contraction in length LM is given by

$$\delta l_1 = \frac{4Pl}{\pi E d_1 d_2} \qquad \text{...[Eqn. (1.8)]}$$

$$= \frac{4 \times (60 \times 1000) \times 100}{\pi \times 2.1 \times 10^5 \times 80 \times 60} = 0.00758 \text{ mm}$$

Portion MN :

Consider an elementary ring of length dx at a distance of x from M.

Diameter at x, $\qquad dx = 80 + \left(\dfrac{100 - 80}{100}\right) \times x = 80 + 0.2\,x$

Area of cross-section, $\quad A_x = \dfrac{\pi}{4}\left[(80 + 0.2x)^2 - 40^2\right]$

$$= (\pi/4)\,(80 + 0.2x + 40)\,(80 + 0.2x - 40)$$

$$= (\pi/4)\,(120 + 0.2x)\,(40 + 0.2x)$$

$$= (\pi/100)\,(600 + x)\,(200 + x) \quad \begin{bmatrix} \text{... multiplying denominator} \\ \text{and numerator by 25} \end{bmatrix}$$

Stress, $\qquad \sigma_x = \dfrac{P}{A_x} = \dfrac{60 \times 1000 \times 100}{\pi\,(600 + x)\,(200 + x)}$

Strain, $\qquad e_x = \dfrac{\sigma_x}{E}$

Change in length over $dx = e_x.dx$

Total change in length for MN,

$$\delta l_2 = \int\limits_0^{100} e_x \cdot dx$$

$$= \int\limits_0^{100} \frac{6 \times 10^6}{\pi E\,(600 + x)\,(200 + x)}\,dx$$

$$= \frac{6 \times 10^6}{\pi \times 2.1 \times 10^5} \int\limits_0^{100} \frac{1}{400}\left[\frac{1}{200 + x} - \frac{1}{600 + x}\right] dx$$

$$= \frac{6 \times 10^6}{\pi \times 2.1 \times 10^5} \times \frac{1}{400}\left| \log_e (200 + x) - \log_e (600 + x) \right|_0^{100}$$

$$= 0.0227\left(\log_e \frac{300}{200} - \log_e \frac{700}{600}\right)$$

$$= 0.0227\left(\log_e \frac{3}{2} - \log_e \frac{7}{6}\right)$$

$$= 0.0227 \times \log_e \left(\frac{3}{2} \times \frac{6}{7} \right) = 0.005705 \text{ mm}$$

Total change in length (*i.e.*, **reduction**)

$$= \delta l_1 + \delta l_2 = 0.00758 + 0.005705$$

$$= \textbf{0.01328 mm} \quad \textbf{(Ans.)}$$

Example 1.60. *A rigid bar 3.5 metres in length is hinged at L and is supported by steel rod SM and copper rod PN as shown in Fig. 1.66 (a). If the lengths of SM and PN are 1 m and 0.75 m and cross-sectional areas 2 cm² and 4 cm² respectively, determine :*

(*i*) *Stress in each of the rods;*

(*ii*) *Elongation of the steel rod if a load of 100 kN is applied on the bar at a distance of 2.5 metres from the hinge.*

Take: $\qquad\qquad E_c = 1.2 \times 10^8 \text{ kN/m}^2,$

and $\qquad\qquad E_s = 2 \times 10^8 \text{ kN/m}^2$

The bar is horizontal prior to the application of the load.

Solution. Refer to Fig. 1.66.

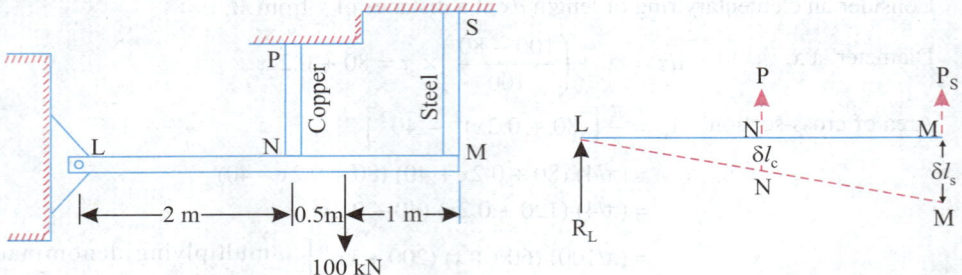

Fig. 1.66

A diesel generator.

Let, R_L = Reaction at the hinge,

P_c = Load taken by copper rod,

P_s = Load taken by steel rod,

δl_c = Extension in the copper rod, and

δl_s = Extension in the steel rod.

(i) Stress in each rod :

When the bar is in equilibrium,

$$\Sigma V = 0, \ i.e., \ P_c + P_s + R_L = 100$$

and, $\Sigma M = 0, \ i.e., \ P_c \times 2 + P_s \times 3.5$

$$= 100 \times 2.5 \qquad \qquad ...(i) \ \text{(moments about } L)$$

From similar Δs LNN' and LMM'

$$\frac{NN'}{MM'} = \frac{\delta l_c}{\delta l_s} = \frac{LN}{LM} = \frac{2}{3.5} = 0.572$$

or, $\delta l_c = 0.572 \ \delta l_s$

Also, $\delta l_c = \dfrac{P_c \, l_c}{A_c \, E_c}$ and $\delta l_s = \dfrac{P_s \, l_s}{A_s \, E_s}$

or, $\dfrac{P_c \times 0.75}{4 \times 10^{-4} \times 1.2 \times 10^8} = \dfrac{0.572 \, P_s \times 1.0}{2 \times 10^{-4} \times 2 \times 10^8}$

or, $P_c = 0.9152 \, P_s$ $...(ii)$

Putting value of P_c in (i), we get

$$0.9152 \, P_s \times 2 + P_s \times 3.5 = 100 \times 2.5$$

$$1.8304 \, P_s + 3.5 \, P_s = 250$$

∴ $P_s = 46.9 \ \text{kN}$

and, $P_c = 0.9152 \times 46.9 = 42.9 \ \text{kN}$

∴   Stress in steel rod, $\sigma_s = \dfrac{P_s}{A_s} = \dfrac{46.9}{2 \times 10^{-4}} = 2.345 \times 10^5 \ \text{kN/m}^2$

i.e., $\boldsymbol{\sigma_s = 2.345 \times 10^5 \ \text{kN/m}^2 \ \text{(Ans.)}}$

Stress in copper rod, $\sigma_c = \dfrac{P_c}{A_c} = \dfrac{42.9}{4 \times 10^{-4}} = 1.072 \times 10^5 \ \text{kN/m}^2$

i.e., $\boldsymbol{\sigma_c = 1.072 \times 10^5 \ \text{kN/m}^2 \ \text{(Ans.)}}$

(ii) Elongation of the steel rod, δl_s :

$$\delta l_s = \frac{P_s \, l_s}{A_s \, E_s} = \frac{46.9 \times 1.0}{2 \times 10^{-4} \times 2 \times 10^8} = 0.00117 \ \text{m or } 1.17 \ \text{mm}$$

i.e., $\boldsymbol{\delta l_s = 1.17 \ \text{mm} \ \text{(Ans.)}}$

HIGHLIGHTS

1. A load may be defined as the combined effect of external forces acting on a body.

    The loads may be classified as :

    (i) Dead loads, (ii) Live or fluctuating loads, (iii) Inertia loads or forces, and (iv) Centrifugal loads or forces; Or (i) Tensile loads, (ii) Compressive loads, (iii) Torsional or twisting loads, (iv) Bending loads, and (v) Shearing loads; Or (i) Point loads, (ii) distributed loads.

2. The internal resistance which the body offers to meet the load is called *stress*. When any type of *simple* stress develops, we can calculate the magnitude of the stress by

    $$\sigma = \frac{P}{A}$$

    where, σ = Stress, P = Load,

    and, A = Area over which stress develops.

3. The strain (e) is the deformation produced by the stress.

    Tensile strain, $e_t = \dfrac{\delta l}{l} = \dfrac{\text{Increase in length}}{\text{Original length}}$

    Compressive strain, $e_c = \dfrac{\delta l}{l} = \dfrac{\text{Decrease in length}}{\text{Original length}}$

    Shear strain, $e_s = \tan \phi = \phi$ radian. (Since ϕ is very small.)

    Volumetric strain, $e_v = \dfrac{\delta V}{V}$

4. Hooke's law states that within elastic limit, stress varies directly as strain.

    i.e., stress $\propto$ strain or $\dfrac{\text{Stress}}{\text{Strain}} = $ constant

    Modulus of elasticity $E = \dfrac{\sigma}{e}$

    Modulus of rigidity, C (or N, or G) $= \dfrac{\tau}{e_s}$

    Bulk modulus of elasticity $K = \dfrac{\sigma_n}{e_v}$.

5. Stress (σ) and elongation (δl) produced in a bar due to its own weight :

    $$\sigma = 9.81 \rho \, y \text{ N/m}^2, \, \delta l = \frac{9.81 \, \rho l^2}{2E}$$

    (where, ρ = density of material, kg/m^3)

6. Elongation in case of a tapered rod,

    $$\delta l = \frac{4Pl}{\pi E d_1 d_2}$$

7. Elongation of a conical bar due to its self-weight,

    $$\delta l = \frac{9.81 \, \rho l^2}{6E}$$

8. In case of a bar of uniform strength,

    $$A_2 = A_1 \, e^{9.81 \rho l / \sigma}$$

9. Maximum stress (σ_{max}) and the total extension (δl) in a bar due to rotation :

(a) 45   (b) 60

(c) 90   (d) 180.

25. Which of the following statements is *incorrect* ?

 (a) Stress is directly proportional to strain within elastic limit.

 (b) The stress is force per unit area.

 (c) Hooke's law holds good up to the breaking point.

 (d) The ratio of linear stress to linear strain is called Young's modulus.

26. When a bar is subjected to a change of temperature and its deformation is prevented, which of the following stresses is induced ?

 (a) Thermal stress (b) Shear stress

 (c) Tensile stress   (d) Compressive stress.

27. Temperature stress developed in a bar depends upon which of the following ?

 (a) Co-efficient of linear expansion

 (b) Change of temperature

 (c) Young's modulus

 (d) All of the above.

28. The temperature strain in a bar is .... proportional to the change in temperature.

 (a) directly

 (b) indirectly

 (c) either (a) or (b)

 (d) none of the above.

29. .... strain is the deformation of the bar per unit length in the direction of the force.

 (a) Volumetric   (b) Shear

 (c) Lateral   (d) Linear.

30. Strain in a direction at right angles to the direction of applied force is known as

 (a) lateral strain

 (b) shear strain

 (c) volumetric strain

 (d) none of the above.

31. A localised compressive stress at the area of contact between two members is known as .... stress.

 (a) shear   (b) crushing

 (c) bending   (d) tensile.

32. Maximum stress (σ_{max}) induced in a bar of length l, rotating at an angular velocity w, is given by

 (a) $\dfrac{1}{2}\rho\,\omega^2 l^2$ (b) $\dfrac{1}{8}\rho\,\omega^2 l^2$

 (c) $\rho\,\omega^2 l^2$ (d) $\rho\,\omega l^2$.

B. Fill in the blank/say 'Yes' or 'No' :

33. The combined effect of external forces acting on a body is called a ....

34. A .... load is one which is considered to act at a point.

35. The internal resistance which the body offers to meet the load is called ....

36. Simple stress is often called direct stress.

37. The strain is the deformation produced by stress.

38. The ratio between change in volume and original volume of a body is called volumetric strain.

39. The ratio between tensile stress and tensile strain or compressive stress and compressive strain is called shear modulus of elasticity.

40. The ratio of normal stress to volumetric strain is called .... modulus of elasticity.

41. Elongation produced in a bar due to its self-weight $= \dfrac{9.81\,\rho l^2}{2E}$.

42. Elongation of conical bar due to its self-weight = ...

43. The ratio of lateral strain to linear strain is known as ....

44. When a square or rectangular block subjected to a shear load is in equilibrium, the shear stress in one plane is always associated with a complementary shear stress (of equal value) in the other plane at .... angles to it.

A truck wheel is being machined.

ANSWERS

1. (*c*)	**2.** (*c*)	**3.** (*a*)	**4.** (*d*)	**5.** (*a*)	**6.** (*d*)	**7.** (*a*)
8. (*a*)	**9.** (*a*)	**10.** (*d*)	**11.** (*a*)	**12.** (*c*)	**13.** (*a*)	**14.** (*a*)
15. (*b*)	**16.** (*b*)	**17.** (*a*)	**18.** (*a*)	**19.** (*c*)	**20.** (*c*)	**21.** (*b*)
22. (*c*)	**23.** (*a*)	**24.** (*c*)	**25.** (*c*)	**26.** (*a*)	**27.** (*d*)	**28.** (*a*)
29. (*d*)	**30.** (*a*)	**31.** (*b*)	**32.** (*b*)	**33.** load	**34.** point	**35.** stress
36. Yes	**37.** Yes	**38.** Yes	**39.** No	**40.** bulk	**41.** Yes	**42.** $\dfrac{9.81\rho l^2}{6E}$
43. Poisson's ratio,	**44.** right.					

UNSOLVED EXAMPLES

Stress, Strain and Young's Modulus :

1. A bar 0.3 m long is 50 mm square in section for 120 mm of its length, 25 mm diameter for 80 mm and of 40 mm diameter for the remaining length. If a tensile force of 100 kN is applied to the bar, calculate the maximum and minimum stresses produced in it, and the total elongation. Take $E = 200$ GN/m^2 and assume uniform distribution of load over the cross-sections.

[**Ans.** 204 MN/m^2, 40 MN/m^2, 0.1453 mm]

2. A brass rod 20 mm diameter was subjected to a tensile load of 40 kN. The extension of the rod was found to be 254 divisions in the 200 mm extension meter. If each division is equal to 0.001 mm, find the elastic modulus of brass.

[**Ans.** 100.25 GN/m^2]

3. Find the total extension of the bar, shown in Fig. 1.67 for an axial pull of 40 kN.

[**Ans.** 0.225 mm]

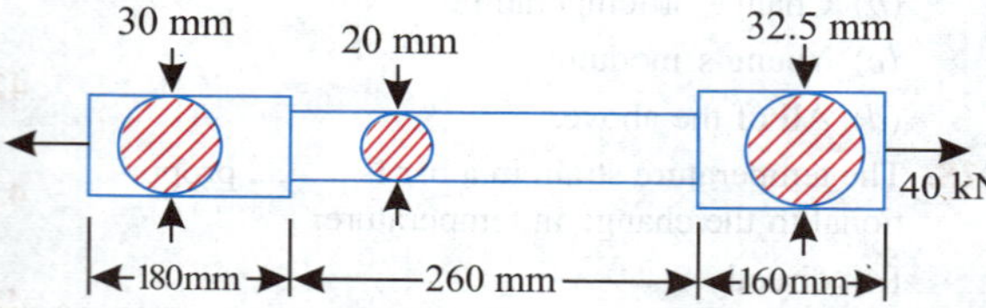

Fig. 167

4. A brass bar having a cross-sectional area of 1000 mm^2 is subjected to axial forces as shown in Fig. 1.68. Find the total change in length of the bar. Take $E_s = 105$ GN/m^2.

[**Ans.** 0.1143 m (decrease)]

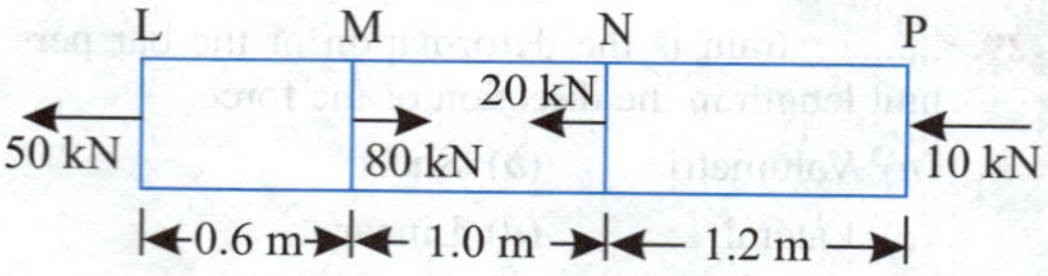

Fig. 1.68

5. A tie bar has enlarged ends of square section 60 mm × 60 mm as shown in Fig. 1.69. If the middle portion of the bar is also of square section, find the size and length of the middle portion, if the stress there is 140 MN/m^2, the total extension of the bar is 0.14 mm.

Take $E = 200$ GN/m^2 .

[**Ans.** 25 mm × 25 mm, 1.79 m]

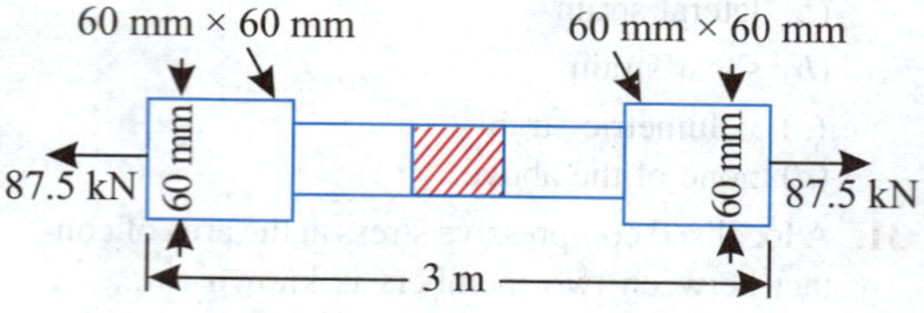

Fig. 1.69

6. A circular rod 0.2 m long, tapers from 20 mm diameter at one end to 10 mm diameter at the other. On applying an axial pull of 6 kN, it was found to extend by 0.068 mm. Find the Young's modulus of the material of the rod.

[**Ans.** 112.3 GN/m^2]

7. A vertical rod 3 m long is rigidly fixed at upper end and carries an axial tensile load of 50 kN force. The rod tapers uniformly from a diameter of 50 mm at the top to 30 mm at the bottom. Calculate the total extension of the bar. Take density of material = 1 × 10^5 kg/m^3 and $E = 210$ GN/m^2.

[**Ans.** 0.6217 mm]

Composite Sections :

8. A solid steel bar, 40 mm diameter, 2 m long passes centrally through a copper tube of internal diameter 40 mm, thickness of metal 5 mm and length 2 m. The ends of the bar and tube are brazed together and a tensile load of 150 kN is applied axially to the compound bar. Find the stresses in the steel and copper, and extension of the compound bar. Assume $E_c = 100$ GN/m^2 and $E_s = 200$ GN/m^2.

    [**Ans.** 93.1 MN/m^2, 46.55 MN/m^2, 0.931 mm]

9. A weight of 100 kN is suspended from a roof by two steel rods each 5 m long, joined to form an included angle of 100°. If the working stress is 56 MN/m^2, find the diameter of the rods and the amount each is stretched by the weight. Neglect weight of rods and take $E = 210$ GN/m^2.

    [**Ans.** 42.1 mm, 1.33 mm].

10. Three parallel wires in the same vertical plane jointly support a load of 15 kN. The middle wire is of steel and is 1 m long, while the outer ones of brass, the length of each being 1.05 m. The area of cross-section of each wire is 200 mm^2. After the wires have been so adjusted as to each carry one-third of the load, a further load of 35 kN is added. Find the stress in each wire and fraction of the whole load carried by the steel wire.   Take $E_s = 200$ GN/m^2 and $E_b = 80$ GN/m^2.

    [**Ans.** $\sigma_b = 62.8$ MN/m^2, $\sigma_s = 124.32$ MN/m^2, 49.7 %]

11. A reinforced concrete column is 300 mm × 300 mm in section. The column is provided with 8 bars each of 20 mm diameter. The column carries a load of 360 kN. Find the stresses in concrete and the steel bars. Take $E_s = 210$ GN/m^2 and $E_c = 14$ GN/m^2.

    [**Ans.** 2.87 MN/m^2, 43.05 MN/m^2]

12. A reinforced concrete column is 300 mm in diameter and has 4 steel bars each of 12 mm diameter embedded in it. If the allowable stresses in steel and concrete are 65.0 MN/m^2 and 4.0 MN/m^2 respectively, calculate the safe axial load which the column can carry. Take $E_s = 15\ E_c$.

    [**Ans.** 308.078 kN]

13. A compound tube consists of a steel tube 170 mm external diameter and 10 mm thickness and an outer brass tube 190 mm external diameter and 10 mm thickness. The two tubes are of the same length. The compound tube carries an axial load of 1 MN. Find the stresses and the load carried by each tube and the amount by which it shortens. Length of each tube is 0.15 m. Take $E_s = 200$ GN/m^2 and $E_b = 100$ GN/m^2.

    [**Ans.** $\sigma_s = 127.34$ MN/m^2, $\sigma_b = 63.67$ MN/m^2 $P_s = 0.64$ MN, $P_b = 0.36$ MN; 0.096 mm]

14. Two steel plates each 25 mm thick, are held together with the help of 20 mm diameter and 50 mm long steel bolts. If the pitch of the threads is 2.5 mm, find the increase in stress in the shank of the bolt

    when the nut is turned through $\dfrac{1}{200}$ of a turn with respect to the head of the bolt, assuming that the

    plates do not deform.. Take $E_s = 205$ GN/m^2.   [**Ans.** 51.25 MN/m^2]

    [**Hint.** Stretch in 50 mm length of the shank,

    $$\delta l = 2.5 \times \frac{1}{200} = 0.0125 \text{ mm}$$

    $$\therefore \quad \text{strain, } e = \frac{0.0125}{50}$$]

15. A rigid cross bar is supported horizontally by two vertical bars, L and M, of equal length and hanging from their tops. The bars L and M are 0.6 m apart. The cross bar stays horizontal even after a vertical force of 6 kN is applied to it at a point 0.4 m from M. If the stress in L is 200 MN/m^2, find the stress in M and the areas of cross-section of the two rods.

    Take $E_L = 200$ GN/m^2 and $E_M = 130$ GN/m^2 [**Ans.** 130 MN/m^2, 20.0 mm^2, 15.385 mm^2]

16. A 28 mm diameter steel bar, 400 mm long is placed centrally within a brass tube having an inside diameter of 30 mm and outside diameter of 40 mm. The bar is shorter in length than the tube by 0.12 mm. While the bar and tube are held vertically on a rigid horizontal platform, a compressive force of

60 kN is applied at the top of the tube through a rigid plate. Determine the stresses induced in both the bar and the tube. Take E_s = 200 GN/m^2 and E_b =100 GN/m^2.

[**Ans.** 48.85 MN/m^2 (comp.); 54.42 MN/m^2 (comp.)]

17. A solid steel bar of 70 mm diameter and 0.5 m long is placed inside an aluminium tube having 75 mm inside diameter and 100 mm outside diameter. The aluminium cylinder is 0.15 mm longer than the steel bar. An axial compressive load of 600 kN is applied to the bar and the cylinder through the rigid cover plates. Find the stresses developed in the steel bar and the aluminium cylinder. Take E_s = 200 GN/m^2 and E_c = 70 GN/m^2.

[**Ans.** σ_s = 106.8 MN/m^2 ; σ_{al} = 54.99 MN/m^2]

18. Fig. 1.70 shows a load supported by two copper rods and one steel rod. If the stresses in copper and steel are not to exceed 60 MN/m^2 and 120 MN/m^2, find the safe load that can be supported.

Take E_s = 2E_c.

[**Ans.** 223.2 kN]

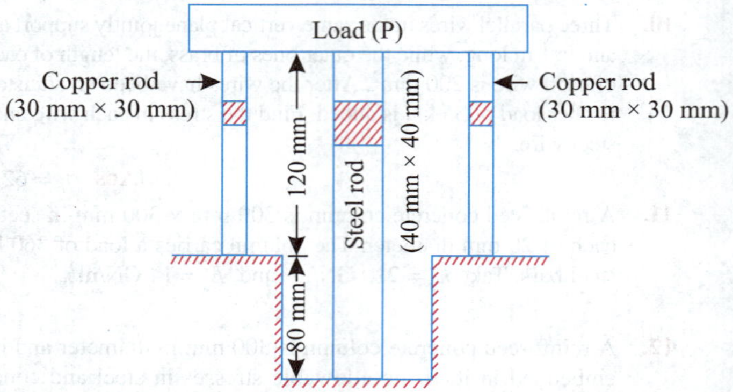

19. A vertical tie of uniform strength is 18 m long. If the area of the bar at the lower end is 500 mm^2, find the area at the upper end when the tie is to carry a load of 700 kN. The material of the tie weighs 8 × 10^5 N/mm^3.

[**Ans.** 500.5 mm^2]

Fig. 1.70

20. A 25 mm diameter rod of steel passes centrally through a copper tube 63 mm external diameter, 40 mm internal diameter, and 1.30 m long. The tube is closed by rigid washers of negligible thickness and nuts threaded on the rod. Find the stresses in each when the nuts are tightened until the tube is reduced in length by 0.13 mm. Find the increase in the stresses if one nut is tightened by 1/2 of a turn relative to the other, there being 0.4 threads per mm. Take E_s = 200 GN/m^2 and E_c = 100 GN/m^2.

[**Ans.** σ_c = 10 MN/m^2 (comp.), σ_s = 37.9 MN/m^2 (tensile), Increase in σ_c = 33.2 MN/m^2 (comp.), Increase in σ_s = 125.8 MN/m^2 (tensile)]

21. A steel rod 18 mm in diameter passes centrally through a steel tube 30 mm in external diameter and 2.5 mm thickness. The tube is 0.75 m long and is closed by rigid washers of negligible thickness which are fastened by nuts threaded on the rod. The nuts are tightened until the compressive load on the tube is 20 kN. Calculate the stresses in the tube and the rod.

[**Ans.** σ_{st} = 90.15 MN/m^2 (comp.); σ_{sr} = 76.52 MN/m^2 (tensile)]

22. A compound bar consists of a central steel strip 25 mm wide and 6.4 mm thick placed between two strips of brass each 25 mm wide and x mm thick. The strips are firmly fixed together to form a compound bar of rectangular section 25 mm wide and (2x + 6.4) mm thick. Determine :

   (i) The thickness of the brass strips which will make the apparent modulus of elasticity of compound bar 157 GN/m^2.

   (ii) The maximum axial pull the bar can then carry if the stress is not to exceed 157 MN/m^2, in either the brass or the steel.   Take E_s = 207 GN/m^2 and E_b = 114 GN/m^2.

[**Ans.** (i) x = 3.726 mm, (ii) 41.226 kN]

23. A weight of 50 kN is hanging from three wires of equal length, middle one is of steel and the other two are of copper. If 300 mm^2 is the cross-sectional area of each wire, then find out the load shared by each. Take E_s = 210 GN/m^2, and E_b = 120 GN/m^2.

[**Ans.** σ_c = 13.6 kN, σ_s = 22.8 kN]

Poisson's Ratio :

24. A steel bar of rectangular cross-section 20 mm × 10 mm is subjected to a pull of 20 kN in the direction of its length. Taking $E = 204$ GN/m^2 and m = 10/3, find the percentage decrease of cross-section.
[**Ans.** 0.0294%]

25. A vertical circular bar 20 mm diameter, 3 m long carries a tensile load of 150 kN. Calculate :

(*i*) Elongation ;

(*ii*) Decrease in diameter ; and

(*iii*) Volumetric strain. [**Ans.** (*i*) 15.9 mm, (*ii*) 0.0265 mm, (*iii*) 0.00265]

26. A steel bar 300 mm long, 50 mm wide and 12 mm thick is subjected to an axial pull of 84 kN. Find the change in length, width, thickness and volume of the bar. Take $E = 200$ GN/m^2 and Poisson's ratio = 0.32. [**Ans.** $\delta l = 0.21$ mm (increase), $\delta b = 0.0112$ mm (decrease), $\delta t = 0.0027$ mm (decrease), $\delta V = 45.36$ mm^3 (increase)]

27. A rectangular block 250 mm × 100 mm × 80 mm is subjected to axial loads as follows :

480 kN (tensile) in the direction of its length.

900 kN (tensile) on the 250 mm × 80 mm faces.

1000 kN (compressive) on the 250 mm × 100 mm faces.

Taking $E = 200$ GN/m^2 and Poisson's ratio as 0.25 find the following :

(*i*) Change in volume of the block.

(*ii*) Values of the modulus of rigidity and bulk modulus for material of the block.

[**Ans.** (*i*) 325 mm^3 (increase), (*ii*) $C = 80$ GN/m^2, $K = 133$ GN/m^2]

28. A bar of 30 mm diameter is subjected to a pull of 60 kN. The measured extension on gauge length of 200 mm is 0.09 mm and the change in diameter is 0.0039 mm. Calculate the Poisson's ratio and the values of the three modulii.

[**Ans.** $E = 188.62$ GN/m^2 ; $C = 73.17$ GN/m^2 ; $K = 148.9$ GN/m^2]

Thermal Stresses and Strains :

29. A railway line is laid so that there is no stress in the rails at 60°C. Calculate the stress in the rails at 20°C, if

(*i*) no allowance is made for contraction ;

(*ii*) there is an allowance of 5 mm for contraction per rail

The rails are 30 mm long. $E = 210$ GN/m^2, $\alpha = 0.000012$ per °C.

[**Ans.** (*i*) 100.8 MN/m^2, (*ii*) 65.8 MN/m^2]

30. A steel rod 30 mm diameter and 300 mm long is subjected to tensile force P acting axially. The temperature of the rod is then raised through 80°C and the total extension measured as 0.35 mm. Calculate the value of P. Take $E_s = 200$ GN/m^2 and $\alpha_s = 12 \times 10^{-6}$ per °C.

[**Ans.** 6.29 kN]

Vertical honing machine. Honing is abrasive process employing bonded abrasives for the improvement of surface finish.

31. A steel bar is placed between two copper bars each having the same area and length as the steel bar at 15°C. At this stage, they are rigidly connected together at both the ends. When the temperature is raised to 315°C, the length of the bars increases by 1.5 mm. Determine the original length and final stresses in the bar. Take $E_s = 210$ GN/m^2; $E_c = 110$ GN/m^2;

    $\alpha_s = 0.000012$ per °C; $\alpha_c = 0.0000175$ per °C.

    [**Ans.** 0.34 m, $\sigma_c = 84.5$ MN/m^2 (comp.), $\sigma_s = 169$ MN/m^2 (tensile)]

32. A gun metal rod 22 mm diameter screwed at the ends passes through a steel tube having 30 mm and 25 mm external and internal diameters respectively. The temperature of the whole assembly is raised to 126°C and the nuts on the rods are then screwed tightly on the ends of the tube. Find the intensity of stress in the rod and the tube when the common temperature has fallen to 16°C. Take $E_s = 210$ GN/m^2, $E_g = 94$ GN/m^2, $\alpha_s = 12 \times 10^{-6}$ per°C, and $\alpha_g = 20 \times 10^{-6}$ per °C.

    [**Ans.** $\sigma_g = 46.27$ MN/m^2 (tensile), $\sigma_s = 81.43$ MN/m^2 (comp.)]

33. A steel rod of diameter 20 mm passes centrally through a tight-fitting copper tube of external diameter 40 mm. The tube is closed by rigid washers of negligible thickness and nuts threaded on the rod. The nuts are tightened till compressive load on the tube is 50 kN. Find the stresses in the rod and the tube when the temperature of the assembly falls 50°C below room temperature. Take, $E_s = 210$ GN/m^2, $E_c = 100$ GN/m^2, $\alpha_s = 0.000012$ per °C, and $\alpha_c = 0.000018$ per °C.

    [**Ans.** Net stress in rod = 123.2 MN/m^2, Net stress in tube = 41.1 MN/m^2)]

34. Two steel plates, each 25 mm thick, are riveted together forming a lap joint by using 25 mm diameter steel rivets. Before placing the rivets in the drilled holes, they are heated to 600 °C and then the heads formed. Find the gripping force exerted by each rivet on the plates when the assembly has cooled down to room temperature of 20°C. Take $E_s = 200$ GN/m^2, and $\alpha_s = 11.6 \times 10^{-6}$ per °C. Assume that the plates do not deform. [**Ans.** 660.5 kN]

35. A brass bar of 25 mm diameter is enclosed in a steel tube of 25 mm internal diameter and 50 mm external diameter. Both of them are 1 m long at room temperature and fastened rigidly to each other at the ends.

    (*i*) If the room temperature is 20°C, find to what temperature the assembly should be heated so as to generate a compressive stress of 48.7 MN/m^2 in brass. What is then stress in steel ?

    (*ii*) If the composite bar is then pulled with a force of 50 kN, find the resulting stresses in both the materials.

    Take $E_s = 200$ GN/m^2, and $E_b = 100$ GN/m^2;

    $\alpha_s = 11.6 \times 10^{-6}$ per °C, and $\alpha_b = 18.7 \times 10^{-6}$ per °C.

    [**Ans.** (*i*) 80 °C, 16.23 MN/m^2 (tensile), (*ii*) 34.17 MN/m^2 (compressive), 45.3 MN/m^2 (tensile)]

36. A steel rod, 12 mm in diameter, passes centrally through a copper tube 2.5 m long and having 36 mm and 48 mm as internal and external diameters respectively. The tube is closed at each end by 24 mm thick steel plates which are secured by nuts. The nuts are tightened until the copper tube is reduced in length by 0.50 mm. The whole assembly is then raised in temperature by 60 °C. Calculate the stresses in copper and steel before and after the rise of temperature, assuming that the thickness of the plates remains unchanged.

    Take $E_s = 210$ GN/m^2, $E_c = 105$ GN/m^2, $\alpha_s = 1.2 \times 10^{-5}$ per °C, and $\alpha_c = 1.75 \times 10^{-5}$ per °C

    [**Ans.** (*i*) $\sigma_c = 21.34$ MN/m^2 (comp.), $\sigma_s = 149.38$ MN/m^2 (tensile)

    (*ii*) $\sigma_c = 28.61$ MN/m^2 (comp.), $\sigma_s = 200.27$ MN/m^2 (tensile)]

37. Three short pillars, each 500 mm^2 in section, support a weight of 200 kN. The central pillar is of steel and the outer ones are of copper. The pillars are so adjusted that at a temperature of 15°C each carries equal load. The temperature is then raised to 115 °C. Estimate the stress in each pillar at 115 °C.

    Take $E_s = 210$ GN/m^2 and $E_c = 80$ GN/m^2;

    $\alpha_s = 1.2 \times 10^{-5}$/°C ;

    $\alpha_c = 1.85 \times 10^{-5}$/°C ; [**Ans.** $\sigma_s = 75.55$ MN/mm^2 (comp.), $\sigma_c = 162.22$ MN/m^2 (comp.)]

Principal Stresses and Strains

Chapter 2

Brake calipers and wheels of a racing car

2.1. STRESSES IN A TENSILE MEMBER

Refer Fig. 2.1. When a bar of uniform sectional area A is subjected to an axial load P, then the stress acting on a cross-section given by LL normal to the axis is P/A. Considering another section given by the plane MM inclined at θ to LL, the area cut by the plane is $\dfrac{A}{\cos\theta}$. Let the normal stress across MM be σ_n.

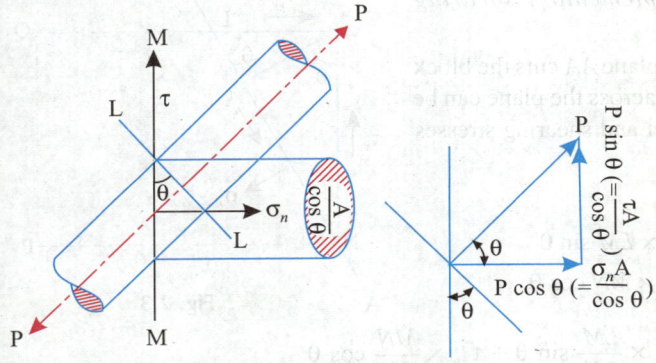

Fig. 2.1

Resolving perpendicular to *MM*

$$\sigma_n \cdot \frac{A}{\cos \theta} = P \cos \theta$$

∴ $$\sigma_n = \frac{P}{A} \cos^2 \theta \qquad \qquad ...(2.1)$$

Further there will be a shearing stress of τ acting parallel to *MM* and resolving in this direction

$$\tau \cdot \frac{A}{\cos \theta} = P \sin \theta$$

$$\tau = \frac{P}{A} \sin \theta \cos \theta \qquad \qquad ...(2.2)$$

This means that *when a rod is subjected to pure tension, both tensile and shearing stresses are produced. In a material under direct compression. the corresponding stresses would be compressive and shearing*. It is possible that the shearing stress produced may be more important than the applied stress. The greatest shearing stress may be calculated as follows :

Since, $$\tau = \frac{P}{A} \sin \theta \cos \theta = \frac{P}{A} \frac{\sin 2\theta}{2} \qquad \qquad ...(2.2a)$$

The greatest value is when $\sin 2\theta = 1$ or $\theta = 45°$

∴ The greatest shearing stress produced

$$\tau_{max} = \frac{P}{2A} \qquad \qquad ...(2.2\ b)$$

Note. The failure of concrete in compression normally occurs across the shear planes at 45° to the applied load.

2.2. STRESSES DUE TO PURE SHEARING

Fig. 2.2 shows a rectangular block *LMNP*, of unit depth perpendicular to the paper. Let shearing stresses τ_{xy} act along the faces *ML* and *PN*. For equilibrium there may be neither a resultant force nor a resultant couple. Since $\tau_{xy} \times ML = \tau_{xy} \times PN$ there is no resultant horizontal force, but the couple due to these is $\tau_{xy} \times ML \times MN$. If shearing stresses τ'_{xy} are introduced on the surfaces *PL* and *MN* to balance the outstanding couple for equilibrium, then

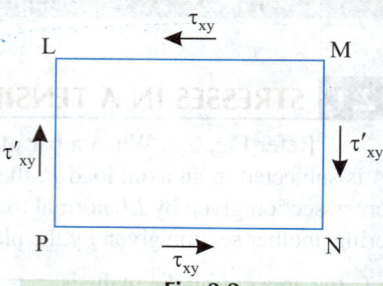

Fig. 2.2

$$\tau'_{xy} \times MN \times ML = \tau_{xy} \times ML \times MN \qquad ..(2.3)$$

or, $$\tau_{xy} = \tau'_{xy}$$

Hence, for *equilibrium, complementary shearing stresses* τ'_{xy} *must be introduced.*

Refer to Fig. 2.3. If an arbitrary plane *AA* cuts the block at an angle θ to *LM*, the stresses acting across the plane can be determined by resolution. Let the direct and shearing stresses across *AA* be σ_n and τ respectively.

Resolving perpendicular to *AA* :

$$\sigma_n \times LN = \tau_{xy} \times LM \sin \theta$$
$$+ \tau_{xy} \times MN \cos \theta$$

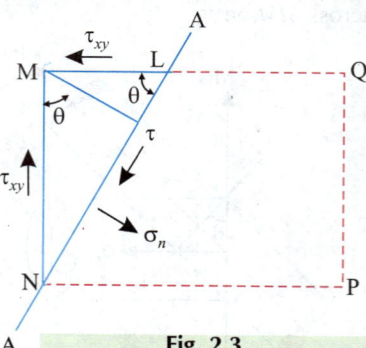

Fig. 2.3

∴ $$\sigma_n = \tau_{xy} \times \frac{LM}{LN} \sin \theta + \tau_{xy} \times \frac{MN}{LN} \cos \theta$$

$$= \tau_{xy} \cos \theta \sin \theta + \tau_{xy} \sin \theta \cos \theta$$

$$= 2 \tau_{xy} \sin \theta \cos \theta$$

$$= \tau_{xy} \sin 2\theta \qquad \qquad ...(2.4)$$

The maximum value will occur when $2\theta = 90°$ or $\sigma_n = \tau_{xy}$ when $\theta = 45°$. And the minimum when $2\theta = -90°$ or $\sigma_n = -\tau_{xy}$ (compressive). In other words a *case of a pure shear is equivalent to a direct tensile stress and a direct compressive stress acting perpendicular to each other.*

Resolving parallel to *AA* :

$$\tau \times LN = \tau_{xy} \times MN \sin \theta - \tau_{xy} \times LM \cos \theta$$

$$\tau = \tau_{xy} \frac{MN}{LN} \sin \theta - \tau_{xy} \times \frac{LM}{LN} \cos \theta$$

$$= \tau_{xy} \times \sin \theta \sin \theta - \tau_{xy} \cos \theta \cos \theta = \tau_{xy} (\sin^2 \theta - \cos^2 \theta)$$

$$= (-) \tau_{xy} \cos 2\theta \qquad \qquad ...(2.5)$$

∴ τ will be zero when $\theta = \pm 45°$

2.3. TWO MUTUALLY PERPENDICULAR DIRECT STRESSES

Refer to Fig. 2.4. At any point in a material where stress is acting, it is possible to assume that the point consists of a very small triangular block, such that the stresses act across the faces of the block. Consider that direct stresses σ_x and σ_y act across the faces *LM* and *MN* and that the block has unit depth perpendicular to *LMN*. Let the stresses τ and σ_n act on the same plane at an angle θ to *LM*.

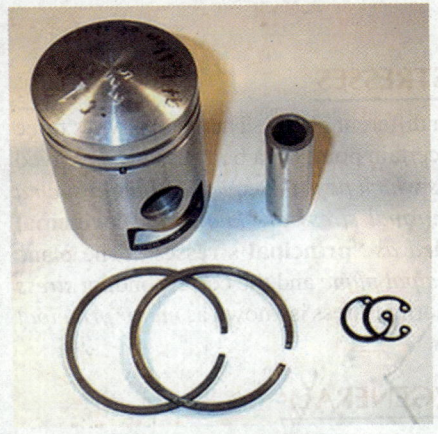

Automobile piston and piston rings.

Fig. 2.4

Resolving *normal* to *LN*:

$$\sigma_n \times LN = \sigma_x \times LM \cos \theta + \sigma_y \ MN \sin \theta$$

$$\sigma_n = \sigma_x \times \frac{LM}{LN} \cos \theta + \sigma_y \frac{MN}{LN} \sin \theta$$

$$= \sigma_x \cos^2 \theta + \sigma_y \sin^2 \theta = \frac{\sigma_x}{2} \times 2 \cos^2 \theta + \frac{\sigma_y}{2} \times 2 \sin^2 \theta$$

$$= \frac{\sigma_x}{2} (1 - \sin^2 \theta + \cos^2 \theta) + \frac{\sigma_y}{2} (1 - \cos^2 \theta + \sin^2 \theta)$$

$$= \frac{\sigma_x + \sigma_y}{2} + \sigma_x \left[\frac{\cos^2 \theta - \sin^2 \theta}{2} \right] - \sigma_y \left[\frac{\cos^2 \theta - \sin^2 \theta}{2} \right]$$

$$= \frac{\sigma_x + \sigma_y}{2} + \frac{\sigma_x - \sigma_y}{2} \cos 2\theta \qquad \qquad ...(2.6)$$

When $\theta = 0$, then $\quad \sigma_n = \dfrac{\sigma_x + \sigma_y}{2} + \dfrac{\sigma_x - \sigma_y}{2} = \sigma_x \qquad \qquad ...(2.6a)$

and, when $\theta = \pi/2$, then $\sigma_n = \dfrac{\sigma_x + \sigma_y}{2} - \dfrac{\sigma_x - \sigma_y}{2} = \sigma_y \qquad \qquad ...(2.6b)$

Resolving *parallel* to *LN*:

$$\tau \times LN = \sigma_x \times LM \, \sin\theta - \sigma_y \times MN \, \cos\theta$$

$$\tau = \sigma_x \times \frac{LM}{LN} \sin\theta - \sigma_y \times \frac{MN}{LN} \cos\theta$$

$$= \sigma_x \cos\theta \sin\theta - \sigma_y \sin\theta \cos\theta = (\sigma_x - \sigma_y) \sin\theta \cos\theta$$

$$= \frac{\sigma_x - \sigma_y}{2} \sin 2\theta \qquad \qquad ...(2.7)$$

The maximum value of τ occurs when $2\theta = \pi/2$ or $\theta = \pi/4$ and then

$$\tau_{max} = \frac{\sigma_x - \sigma_y}{2} \qquad \qquad ...(2.7a)$$

The *resultant stress*,

$$\sigma_r = \sqrt{\sigma_n^2 + \tau^2}$$

$\tan\phi = \dfrac{\tau}{\sigma_n}$ where ϕ is the angle which the resultant stress makes with the normal to the plane and is called **obliquity**.

2.4. PRINCIPAL PLANES AND PRINCIPAL STRESSES

A body may be subjected to stresses in one plane or in different planes. There are always three mutually perpendicular planes along which the stresses at a certain point (in a body) can be resolved completely into stresses normal to these planes. *These planes which pass through the point in such a manner that the resultant stress across them is totally a normal stress are known as* **"principal planes"** *and normal stresses across these planes are termed as* **"principal stresses"**. The plane carrying the maximum normal stress is called the *major principal plane* and the corresponding *stress the major principal stress*. The plane carrying the minimum normal stress is known as *minor principal plane* and the corresponding *stress as minor principal stress*.

2.5. TWO-DIMENSIONAL STRESS SYSTEM (GENERAL)

When the stresses at some point are considered to be acting on a small triangular block at a point as in article 2.3, then a general stress system will consist of direct and shearing stresses acting across the faces of the block. Consider (Fig. 2.5) some plane *LN* at angle θ to the plane of the stress.

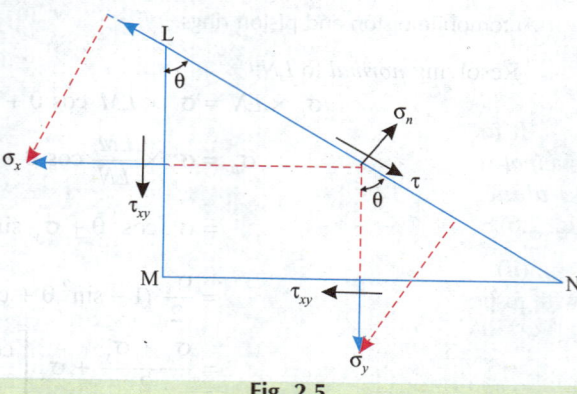

Resolving *normal* to *LN* :

$$\sigma_n \times LN = \tau_{xy} \, LM \, \sin\theta + \sigma_x \, LM \, \cos\theta$$

$$+ \tau_{xy} \, MN \, \cos\theta + \sigma_y \, MN \, \sin\theta$$

Fig. 2.5

$$\therefore \qquad \sigma_n = \tau_{xy}\,\frac{LM}{LN}\sin\theta + \sigma_x\,\frac{LM}{LN}\cos\theta + \tau_{xy}\,\frac{MN}{LN}\cos\theta + \sigma_y\,\frac{MN}{LN}\sin\theta$$

$$= \tau_{xy}\cos\theta\sin\theta + \sigma_x\cos^2\theta + \tau_{xy}\sin\theta\cos\theta + \sigma_y\sin^2\theta$$

$$= 2\,\tau_{xy}\sin\theta\cos\theta + \sigma_x\cos^2\theta + \sigma_y\sin^2\theta$$

$$= \tau_{xy}\sin 2\theta + \frac{\sigma_x+\sigma_y}{2} + \frac{\sigma_x-\sigma_y}{2}\cos 2\theta \qquad\qquad ...(2.8)$$

Resolving *along LN*:

$$\tau\times LN = \tau_{xy}\times MN\sin\theta + \sigma_x\times LM\sin\theta - \tau_{xy}\times LM\cos\theta - \sigma_y\times MN\cos\theta$$

$$\therefore \qquad \tau = \tau_{xy}\,\frac{MN}{LN}\sin\theta + \sigma_x\,\frac{LM}{LN}\sin\theta - \tau_{xy}\,\frac{LM}{LN}\cos\theta - \sigma_y\,\frac{MN}{LN}\cos\theta$$

$$= \tau_{xy}\sin\theta\sin\theta + \sigma_x\cos\theta\sin\theta - \tau_{xy}\cos\theta\cos\theta - \sigma_y\sin\theta\cos\theta$$

$$= \tau_{xy}\sin^2\theta + \sigma_x\sin\theta\cos\theta - \tau_{xy}\cos^2\theta - \sigma_y\sin\theta\cos\theta$$

$$= \tau_{xy}\,(\sin^2\theta - \cos^2\theta) + \left(\frac{\sigma_x-\sigma_y}{2}\right)\sin 2\theta$$

$$= \left[\frac{\sigma_x-\sigma_y}{2}\right]\sin 2\theta - \tau_{xy}\cos 2\theta \qquad\qquad ...(2.9)$$

(*i*) In order to find out the *principal stresses*, the *maximum* and *minimum* values of σ_n must be obtained.

Differentiating σ_n w.r.t. θ in equation (2.8), we get

$$\frac{d(\sigma_n)}{d\theta} = 2\,\tau_{xy}\cos 2\theta - \frac{2(\sigma_x-\sigma_y)}{2}\sin 2\theta$$

Equating this to zero for maximum σ_x, we get

$$0 = \left(\frac{\sigma_x-\sigma_y}{2}\right)\sin 2\theta - \tau_{xy}\cos 2\theta$$

Comparing with equation (2.9), we have $\tau = 0$

Hence *for a principal plane there may be no shear stress acting.*

Also, $\qquad \left(\dfrac{\sigma_x-\sigma_y}{2}\right)\sin 2\theta = \tau_{xy}\cos 2\theta$

or, $\qquad \tan 2\theta = \dfrac{2\,\tau_{xy}}{\sigma_x-\sigma_y} \qquad ...(2.10\,a)$

It follows that for any particular system, the *principal stresses may be calculated by considering the planes which carry zero shearing stress* (Fig. 2.6).

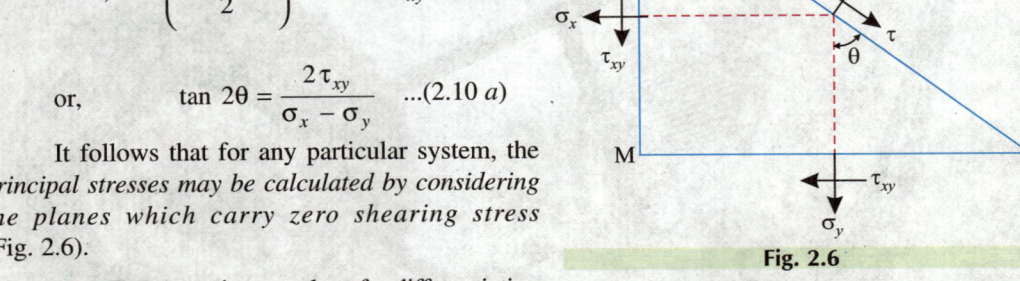

Fig. 2.6

(*ii*) To get maximum value of τ, differentiating τ with respect to θ in eqn. (2.9) and equating to zero, we get

$$\frac{d(\tau)}{d\theta} = \left(\frac{\sigma_x-\sigma_y}{2}\right)2\cos 2\theta + \tau_{xy}\sin 2\theta\times 2 = 0$$

or, $\qquad (\sigma_x-\sigma_y)\cos 2\theta = -2\,\tau_{xy}\sin 2\theta$

or, $$\tan 2\theta = \frac{-(\sigma_x - \sigma_y)}{2\,\tau_{xy}}$$

Also, $\qquad \cot(180° - 2\theta_1) = -\cot 2\theta_1$

and, $\qquad \cot(360° - 2\theta_2) = -\cot 2\theta_2$

where θ_1 and θ_2 are the inclinations of *maximum shear stress with the plane of tensile stress* σ_x.

$\therefore \qquad \cot(180° - 2\theta_1) = \dfrac{2\,\tau_{xy}}{\sigma_x - \sigma_y}$ $\qquad\qquad$ ...(2.10b)

and, $\qquad \cot(360° - 2\theta_2) = \dfrac{2\,\tau_{xy}}{\sigma_x - \sigma_y}$

Resolving *normal* to *LM*:

$$\sigma_x \times LM + \tau_{xy} \times MN = \sigma_n \times LN \cos\theta$$

Dividing both sides by *LN*, we get

$$\sigma_x \cos\theta + \tau_{xy} \sin\theta = \sigma_n \cos\theta \qquad \left[\because \frac{LM}{LN} = \cos\theta, \ \frac{MN}{LN} = \sin\theta \right]$$

$$\sigma_x + \tau_{xy} \tan\theta = \sigma_n \qquad\qquad\qquad ...(2.11)$$

Resolving *parallel* to *LM*:

$$\sigma_y \times MN + \tau_{xy} \times LM = \sigma_n \times LN \sin\theta$$

Dividing both sides by *LN*, we get

$$\sigma_y \sin\theta + \tau_{xy} \cos\theta = \sigma_n \sin\theta$$

$$\sigma_y + \tau_{xy} \cot\theta = \sigma_n \qquad\qquad\qquad ...(2.11\,a)$$

Hence, $\qquad \tau_{xy} \tan\theta = \sigma_n - \sigma_x \qquad\qquad\qquad (i)$

and, $\qquad \tau_{xy} \cot\theta = \sigma_n - \sigma_y \qquad\qquad\qquad (ii)$

$$\tau_{xy}^2 = (\sigma_n - \sigma_x)(\sigma_n - \sigma_y)$$

or, $\qquad \tau_{xy}^2 = \sigma_n^2 - (\sigma_x + \sigma_y)\,\sigma_n + \sigma_x \sigma_y$

A twin-cylinder motorbike is being assembled.

Solving,

$$\sigma_n \ (= \sigma \ \text{say}) = \frac{\sigma_x + \sigma_y}{2} \pm \frac{1}{2} \sqrt{(\sigma_x + \sigma_y)^2 - 4\sigma_x \sigma_y + 4\tau_{xy}^2}$$

i.e.

$$\sigma = \frac{\sigma_x + \sigma_y}{2} \pm \frac{1}{2} \sqrt{(\sigma_x - \sigma_y)^2 + 4\tau_{xy}^2} \qquad \qquad ...(2.12)$$

∴ Major principal stress,

$$\sigma_1 = \frac{\sigma_x + \sigma_y}{2} + \sqrt{\left(\frac{\sigma_x - \sigma_y}{2}\right)^2 + \tau_{xy}^2} \qquad \qquad ...(2.12\ a)$$

Minor principal stress,

$$\sigma_2 = \frac{\sigma_x + \sigma_y}{2} - \sqrt{\left(\frac{\sigma_x - \sigma_y}{2}\right)^2 + \tau_{xy}^2} \qquad \qquad ...(2.12\ b)$$

Also,

$$\tau_{max} = \frac{\sigma_1 - \sigma_2}{2} \quad \text{or} \quad \tan 2\theta = \frac{2\tau_{xy}}{\sigma_x - \sigma_y}$$

or,

$$\tan 2\theta = \frac{2\tau_{xy}}{\sigma_x - \sigma_y}$$

(Also $\sigma_1 + \sigma_2 = \sigma_x + \sigma_y$)

2.6. GRAPHICAL METHODS

2.6.1. Mohr's Circle

A German scientist Otto Mohr devised a graphical method for finding out the normal and shear stresses on any interface of an element when it is subjected to two perpendicular stresses. This method is explained as follows:

2.6.1.1. Mohr's circle construction for "like stresses"

Refer to Fig. 2.7. *Steps of construction*:

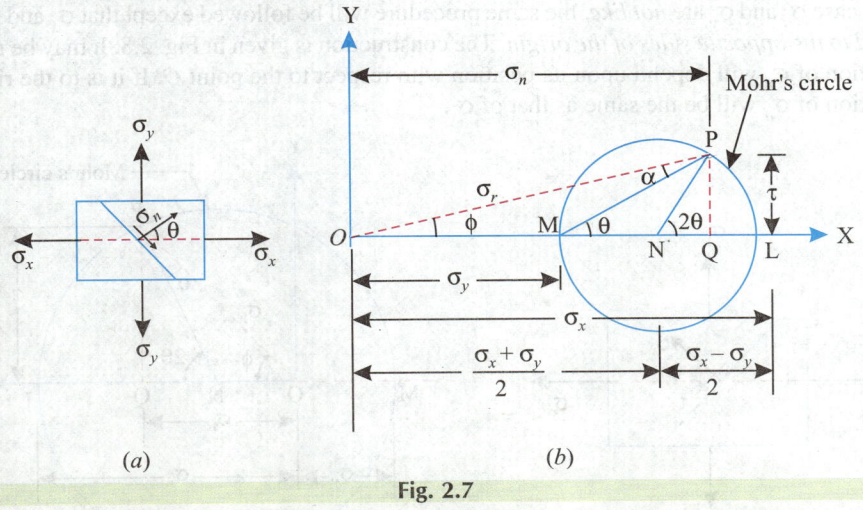

(a)	(b)

Fig. 2.7

1. Using some suitable scale, measure OL and OM equal to σ_x and σ_y respectively on the axis OX.

2. Bisect LM at N.

3. With N as centre and NL or NM radius, draw a circle.

4. At the centre N draw a line NP at an angle 2θ, in the same direction as the normal to the plane makes with the direction of σ_x. In Fig. 2.7 (*a*) which represents the stress system, the normal to the plane makes an angle θ with the direction of σ_x in the anticlockwise direction. The line NP therefore, is drawn in the anticlockwise direction.

5. From P, drop a perpendicular PQ on the axis OX. PQ will represent τ and OQ σ_n.

Now, from stress diagram

$$NP = NL = \frac{\sigma_x - \sigma_y}{2}$$

$$PQ = NP \sin 2\theta = \frac{\sigma_x - \sigma_y}{2} \sin 2\theta = \tau \qquad \text{(Eqn. 2.7)}$$

Similarly,

$$OQ = ON + NQ = \frac{\sigma_x + \sigma_y}{2} + \frac{\sigma_x - \sigma_y}{2} \cos 2\theta = \sigma_n \qquad \text{(Eqn. 2.6)}$$

Also, from stress circle, τ is maximum when

$$2\theta = 90°, \text{ or } \theta = 45°$$

and,

$$\tau_{max} = \frac{\sigma_x - \sigma_y}{2} \qquad \text{(Eqn. 2.7 }a\text{)}$$

Sign conventions used:

(*i*) In order to mark τ in stress system, we will take the *clockwise shear as positive* and *anticlockwise shear as negative.*

(*ii*) *Positive* values of τ will be *above* the axis and *negative* values *below* the axis.

(*iii*) If θ is in the anticlockwise direction, the radius vector will be above the axis and θ will be reckoned positive. If θ is in the clockwise direction, it will be negative and the radius vector will be below the axis.

(*iv*) *Tensile stress will be reckoned positive and will be plotted to the right of the origin O. Compressive stress will be reckoned negative and will be plotted to the left of the origin O.*

2.6.1.2. Mohr's circle construction for "unlike stresses"

In case σ_x and σ_y are *not like*, the same procedure will be followed except that σ_x and σ_y *will be measured to the opposite sides of the origin.* The construction is given in Fig. 2.8. It may be noted that the direction of σ_n will depend upon its position with respect to the point O. If it is to the right of O, the direction of σ_n will be the same as that of σ_x.

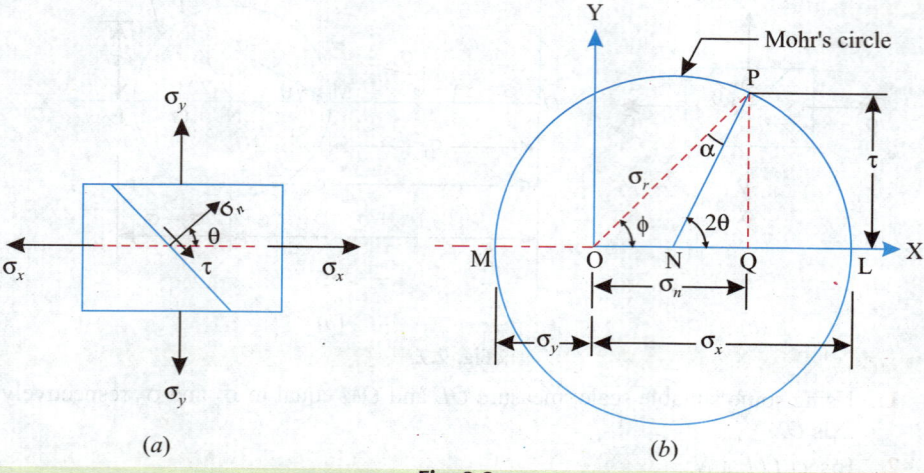

(*a*)   (*b*)

Fig. 2.8

2.6.1.3. Mohr's circle construction for two perpendicular direct stresses with state of simple shear

Refer to Fig. 2.9. Following steps of construction are followed if the material is subjected to direct stresses σ_x and σ_y along with a state of simple shear:

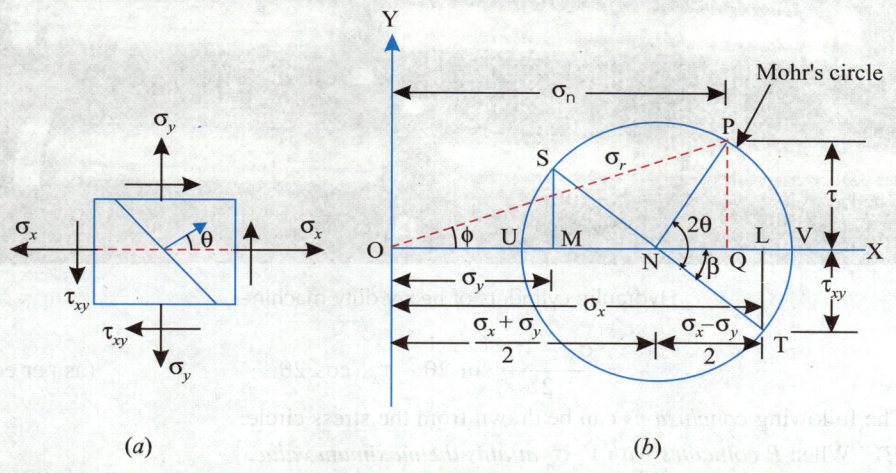

(a)   (b)

Fig. 2.9

1. Using some suitable scale, measure $OL = \sigma_x$ and $OM = \sigma_y$ along the axis OX.
2. At L draw LT perpendicular to OX and equal to τ_{xy}. LT has been drawn downward (as per sign conventions adopted) because τ_{xy} is acting up with respect to the plane across which σ_x is acting, tending to rotate it in the *anticlockwise direction* and is *negative*.
3. Similarly, make MS perpendicular to OX and equal to τ_{xy}, but above OX.
4. Join ST to cut the axis in N.
5. With N as centre and NS or NT as radius, draw a circle.
6. At N make NP at angle 2θ with NT in the anticlockwise direction.
7. Draw PQ perpendicular to the axis. PQ will give τ while OQ will give σ_n and OP will give σ_r.

Proof. Let the radius of the stress circle be R.

Then,
$$R = \sqrt{NL^2 + LT^2} = \sqrt{\left[\frac{\sigma_x - \sigma_y}{2}\right]^2 + \tau_{xy}^2}$$

Also,
$$R \cos \beta = NL = \frac{\sigma_x - \sigma_y}{2}$$

$$R \sin \beta = LT = \tau_{xy}$$

Now,
$$OQ = ON + NQ = ON + R \cos (2\theta - \beta)$$
$$= ON + R \cos 2\theta \cos \beta + R \sin 2\theta \sin \beta$$
$$= \frac{\sigma_x + \sigma_y}{2} + \frac{\sigma_x - \sigma_y}{2} \cos 2\theta + \tau_{xy} \sin 2\theta \qquad \text{(as per eqn. 2.8)}$$
$$= \sigma_n$$

Similarly,
$$PQ = R \sin (2\theta - \beta) = R \sin 2\theta \cos \beta - R \cos 2\theta \sin \beta$$

Hydraulic cylinders of heavy duty machines.

$$= \frac{\sigma_x - \sigma_y}{2} \sin 2\theta - \tau_{xy} \cos 2\theta \qquad \text{(as per eqn. 2.9)}$$

The following *conclusions* can be drawn from the stress circle:

(*i*)   When *P coincides with V*, σ_n *attains the maximum value.*

$$\sigma_{n(max)} = OV = ON + NV$$

$$= \frac{\sigma_x + \sigma_y}{2} + \sqrt{\left[\frac{\sigma_x - \sigma_y}{2}\right]^2 + \tau_{xy}^2} \qquad \left[\begin{array}{l} \because NV = NT \text{ and,} \\ NT^2 = NL^2 + LT^2 \end{array}\right]$$

$\sigma_{n\,(max)}$ (or σ_1) is known as *major principal stress.*

$$\tau = 0; \ \sigma_{r\,(max)} = \sigma_{n\,(max)}$$

$$\tan 2\theta = \tan \beta = \frac{\tau_{xy}}{\left[\dfrac{\sigma_x - \sigma_y}{2}\right]} = \frac{2\tau_{xy}}{\sigma_x - \sigma_y} \qquad \text{(as per eqn. 2.10)}$$

(*ii*)   When *P coincides with U*, σ_n *attains the minimum value*,

$$\sigma_{n\,(min)} = OU = ON - NU$$

$$= \frac{\sigma_x + \sigma_y}{2} - \sqrt{\left[\frac{\sigma_x - \sigma_y}{2}\right]^2 + \tau_{xy}^2} \qquad \left[\begin{array}{l} \because NU = NS = NT \text{ and} \\ NU^2 = NT^2 = NL^2 + LT^2 \end{array}\right]$$

$\sigma_{n(min)}$ (or σ_2) is known as *minor principal stress.*

$$\tau = 0; \ \sigma_{r\,(min)} = \sigma_{n\,(min)} \ ; \ \theta = \ 90° + \beta/2$$

(*iii*)   When $2\theta = \beta + 90°$, τ *attains the maximum value*,

$$\tau_{max} = \sqrt{\left[\frac{\sigma_x - \sigma_y}{2}\right]^2 + \tau_{xy}^2} = \frac{\sigma_1 - \sigma_2}{2}$$

When, $\qquad \qquad 2\theta = \beta + 270°$

$$\tau_{max} = -\sqrt{\left[\frac{\sigma_x - \sigma_y}{2}\right]^2 + \tau_{xy}^2}$$

2.6.1.4. Mohr's circle construction for principal stresses

Refer to Fig. 2.10. The following are the *steps* of construction :

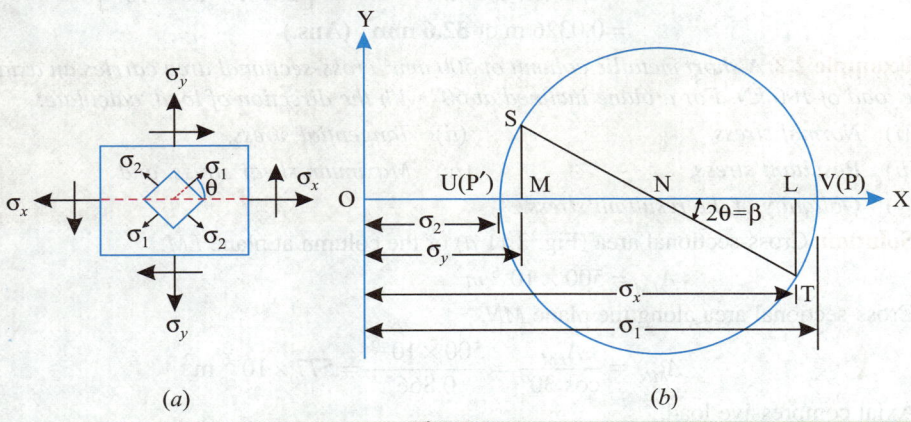

(a)   (b)

Fig. 2.10

1. Mark OL and OM proportional to σ_x and σ_y.
2. At L and M, erect perpendiculars $LT = MS$ proportional to τ_{xy} in appropriate directions.
3. Join ST, intersecting the axis in N.

Since $\tau = 0$, NV represents the major principal plane, P coinciding with V. Similarly NP' represents minor principal plane, P' coinciding with U.

$$OV = ON + NV = \frac{\sigma_x + \sigma_y}{2} + R, \quad \text{where } R \text{ is the radius of the circle.}$$

$$= \frac{\sigma_x + \sigma_y}{2} + \sqrt{\left(\frac{\sigma_x - \sigma_y}{2}\right)^2 + \tau_{xy}^2} = \sigma_1$$

Similarly,

$$OU = ON - NU$$

$$= \frac{\sigma_x + \sigma_y}{2} - \sqrt{\left(\frac{\sigma_x - \sigma_y}{2}\right)^2 + \tau_{xy}^2} = \sigma_2$$

$$\tan \beta = \frac{LT}{LN} = \frac{\tau_{xy}}{\dfrac{\sigma_x - \sigma_y}{2}}$$

$$= \frac{2\tau_{xy}}{\sigma_x - \sigma_y} = \tan 2\theta \qquad \qquad \text{(where, } \beta = 2\theta)$$

Example 2.1. *A circular bar is subjected to an axial pull of 100 kN. If the maximum intensity of shear stress on any oblique plane is not to exceed 60 MN/m², determine the diameter of the bar.*

Solution. Axial pull, $P = 100$ kN

Maximum intensity of shear stress,

$$\tau_{max} = 60 \text{ MN/m}^2$$

Diameter of the bar, d:

Cross-sectional area of the bar,

$$A = (\pi/4) \times d^2$$

But the maximum shear stress produced,

$$\tau_{max} = \frac{P}{2A} \qquad \qquad \text{(Eqn. 2.2 } b)$$

$$60 \times 10^6 = \frac{100 \times 1000}{2 \times (\pi/4) \times d^2} \quad \text{or} \quad d = \left[\frac{100 \times 1000 \times 4}{2 \times \pi \times 60 \times 10^6} \right]^{1/2}$$

$$= 0.0326 \text{ m or } \textbf{32.6 mm} \quad \textbf{(Ans.)}$$

Example 2.2. *A short metallic column of 500 mm² cross-sectional area carries an axial compressive load of 100 kN. For a plane inclined at 60° with the direction of load, calculate:*

   (i)   Normal stress,   *(ii)   Tangential stress,*

 (iii)   Resultant stress,   *(iv)   Maximum shear stress, and*

  (v)   Obliquity of the resultant stress.

Solution. Cross-sectional area (Fig. 2.11 *a*) of the column at plane *LM*,

$$A_{LM} = 500 \times 10^{-6} \ m^2$$

Cross-sectional area along the plane *MN*,

$$A_{MN} = \frac{A_{LM}}{\cos 30°} = \frac{500 \times 10^{-6}}{0.866} = 577 \times 10^{-6} \ m^2$$

Axial compressive load,

$$P = 100 \text{ kN}$$

Let, P_n = Normal force on the plane *MN*, and

                P_t = Tangential force on the plane *MN*.

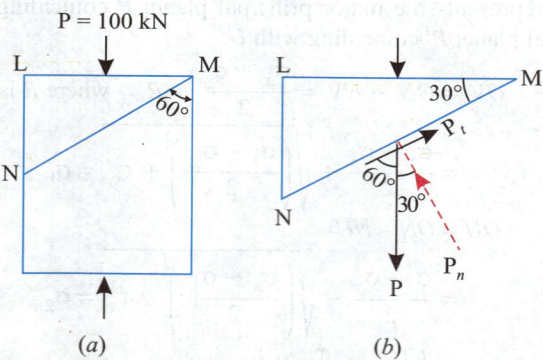

(a)   *(b)*

Fig. 2.11

For equilibrium of wedge *LMN* (Fig. 2.11*b*) resolve the foces in *perpendicular* and *parallel* directions to the plane *MN*.

Then, $P_n = P \cos 30° = 100 \times 0.866 = 86.6 \text{ kN}$

and, $P_t = P \sin 30° = 100 \times 0.5 = 50 \text{ kN}$

 (i)   **Normal stress, σ_n:**

$$\sigma_n = \frac{P_n}{A_{MN}} = \frac{86.6 \times 1000}{577 \times 10^{-6}}$$

$$= \textbf{150.1 MN / m}^2 \ (\textit{compressive}) \ \textbf{(Ans.)}$$

(ii)   **Tangential stress, τ :**

$$\tau = \frac{P_t}{A_{MN}} = \frac{50 \times 1000}{577 \times 10^{-6}} = \textbf{86.6 MN/m}^2 \ (\textit{shear}) \ \textbf{(Ans.)}$$

(iii)   **Resultant stress, σ_r:**

$$\sigma_r = \sqrt{\sigma_n^2 + \tau^2} = \sqrt{(150.1)^2 + (86.6)^2}$$

= 173.29 MN/m² (compressive) (Ans.)

(iv) **Maximum shear stress, τ_{max}:**

$$\tau_{max} = \frac{P}{2A_{LM}} = \frac{100 \times 1000}{2 \times 500 \times 10^{-6}}$$

= **100 MN/m² (Ans.)**

(v) **The obliquity of the resultant stress, ϕ:**

$$\tan \phi = \frac{\tau}{\sigma_n} = \frac{86.6}{150.1} = 0.5769$$

$$\phi = \tan^{-1} 0.5769 = 30° \quad \textbf{(Ans.)}$$

Example 2.3. *A prismatic bar carrying an axial tensile stress σ_x is cut by an oblique section LM as shown in Fig. 2.12. If the normal and shear stresses on this section are 90 MN/m² and 30 MN/m² respectively, find the value of σ_x and the angle θ defining the aspect of section LM.*

(Aligarh University)

Fig. 2.12

Solution. Let the thickness or depth of the bar normal to the plane of the paper be *unity*. Resolving forces at right angles and along the plane LM, we get

$$\sigma_n \times LM = \sigma_x \times \cos(90 - \theta) \times LM$$

or, $\sigma_n = \sigma_x \sin \theta$ ...(i)

and, $\tau \times LM = \sigma_x \cos \theta \times LM$

or, $\tau = \sigma_x \cos \theta$ ...(ii)

since, $\sigma_n = 90$ MN/m²   (given)

$\tau = 30$ MN/m²   (given)

From (*i*) and (*ii*), we get

$$90 = \sigma_x \sin \theta, \ 30 = \sigma_x \cos \theta$$

thus, $\sigma_x = \sqrt{90^2 + 30^2}$

= **94.87 MN/m² (Ans.)**

and, $\dfrac{\sigma_x \sin \theta}{\sigma_x \cos \theta} = \dfrac{90}{30} = 3$

or, $\tan \theta = 3$

or,   **$\theta = 71°33'$ (Ans.)**

Example 2.4. *The principal stresses in the wall of a container (Fig. 2.13) are 40 MN/m² and 80 MN/m². Determine the normal, shear and resultant stresses in magnitude and direction in a plane, the normal of which makes an angle of 30° with the direction of maximum principal stress.*

Solution. *Given*: $\sigma_x = 80$ MN/m² (tensile);

$\sigma_y = 40$ MN/m² (tensile);

$\theta = 30°$.

CNC machined parts.

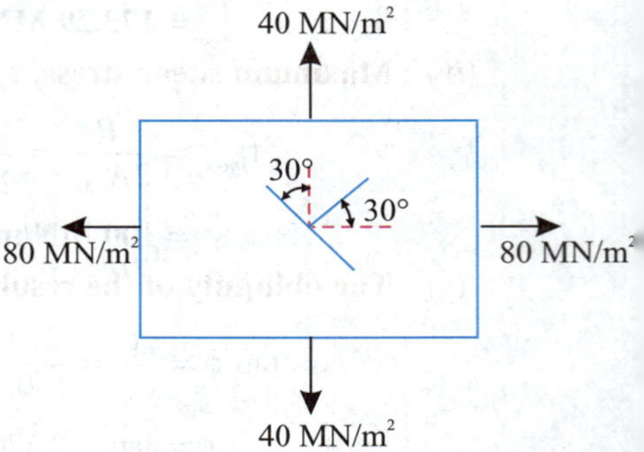

(i) **Normal stress, σ_n:**

$$\sigma_n = \frac{\sigma_x + \sigma_y}{2} + \frac{\sigma_x - \sigma_y}{2} \cos 2\theta \qquad \text{(Eqn. 2.6)}$$

$$= \frac{80 + 40}{2} + \frac{80 - 40}{2} \cos 60°$$

$$= 60 + 10 = \mathbf{70 \ MN/m^2} \quad \textbf{(Ans.)}$$

(ii) **Shear stress, τ :**

$$\tau = \frac{\sigma_x - \sigma_y}{2} \sin 2\theta = \frac{80 - 40}{2} \sin 60°$$

$$...\text{(Eqn. 2.7)}$$

$$= \mathbf{17.32 \ MN/m^2} \quad \textbf{(Ans.)}$$

(iii) **Resultant stress, σ_r, ϕ:**

$$\sigma_r = \sqrt{\sigma_n^2 + \tau^2} = \sqrt{70^2 + 17.32^2}$$

i.e., $\sigma_r = \mathbf{72.11 \ MN/m^2}$ **(Ans.)**

If ϕ is the angle that the resultant makes with the normal to the plane, then

$$\tan \phi = \frac{\tau}{\sigma_n} = \frac{17.32}{70} = 0.2474$$

$$\phi = \mathbf{13° \ 54'} \quad \textbf{(Ans.)}$$

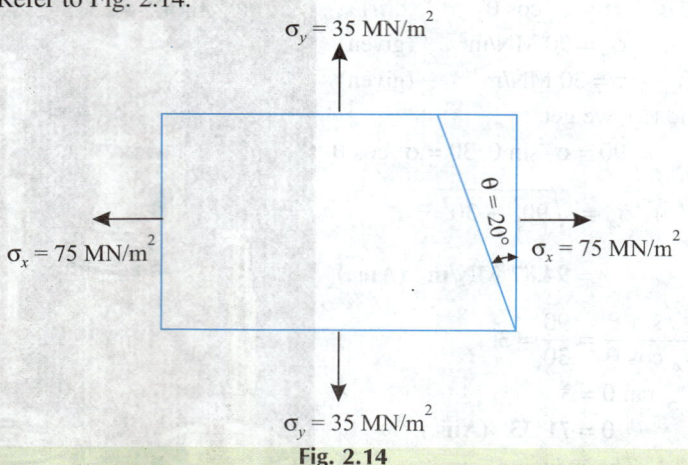

Fig. 2.13

Example 2.5. *The principal stresses at a point across two perpendicular planes are 75 MN/m² (tensile) and 35 MN/m² (tensile). Find the normal, tangential stresses and the resultant stress and its obliquity on a plane at 20° with the major principal plane.*

Solution. Refer to Fig. 2.14.

Fig. 2.14

Analytical method: *Given*; $\sigma_x = 75 \ MN/m^2$ (tensile);

$$\sigma_y = 35 \ MN/m^2 \text{ (tensile)}; \ \theta = 20°.$$

Normal stress, $\quad \sigma_n = \dfrac{\sigma_x + \sigma_y}{2} + \dfrac{\sigma_x - \sigma_y}{2} \cos 2\theta = \dfrac{75 + 35}{2} + \dfrac{75 - 35}{2} \cos (2 \times 20°)$

$$= 55 + 20 \cos 40° = 70.32 \ MN/m^2 \text{ (tensile)}$$

$$\sigma_n = \mathbf{70.32 \ MN/m^2} \text{ (tensile)} \quad \textbf{(Ans.)}$$

Tangential stress, $\quad \tau = \dfrac{\sigma_x - \sigma_y}{2} \sin 2\theta = \dfrac{75 - 35}{2} \sin 40° = 20 \sin 40° = 12.85 \ MN/m^2$

Hence, $\tau = 12.85$ **MN/m² (Ans.)**

Resultant stress, $\sigma_r = \sqrt{\sigma_n^2 + \tau^2} = \sqrt{(70.32)^2 + (12.85)^2} = 71.48$ MN/m²

Hence, $\sigma_r = 71.48$ **MN/m² (Ans.)**

Obliquity φ: $\tan \phi = \dfrac{\tau}{\sigma_n} = \dfrac{12.85}{70.32} = 0.1827$

∴ $\phi = 10°21'$ **(Ans.)**

Graphical method:

Refer to Fig. 2.15.

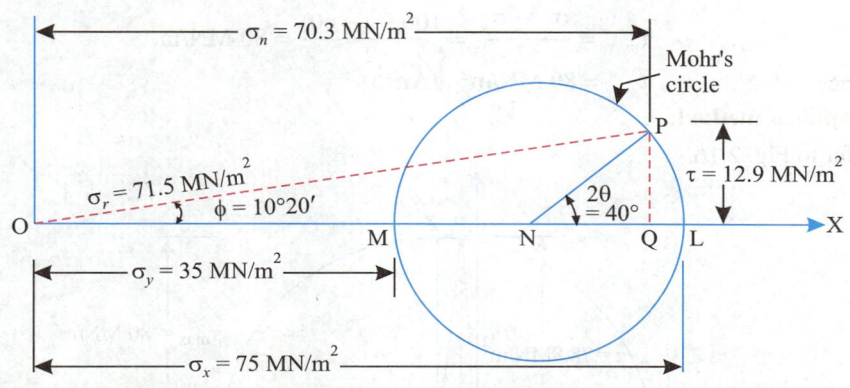

Fig. 2.15

— Plot $OL = \sigma_x = 75$ MN/m² and $OM = \sigma_y = 35$ MN/m².

— Bisect ML to get N and draw the circle with radius NL as shown.

— Make NP at angle 40° (*i.e.* 2θ) to NL.

— Drop perpendicular PQ.

From Mohr's circle, we have

$\sigma_n = OQ = 70.3$ **MN/m² (Ans.)**

$\tau = PQ = 12.9$ **MN/m² (Ans.)**

$\sigma_r = OP = 71.5$ **MN/m² (Ans.)**

$\phi = 10° 20'$ **(Ans.)**

Example 2.6. *At a point in a stressed body the principal stresses are 100 MN/m² (tensile) and 60 MN/m² (compressive). Determine the normal stress and the shear stress on a plane inclined at 50° to the axis of major principal stress. Also calculate the maximum shear stress at the point.*

Solution. *Given:* $\sigma_x = + 100$ MN/m²;

$\sigma_y = - 60$ MN/m²;

(– sign is used as the stress is *compressive*)

$\theta = 50°$

Analytical method:

Normal stress,

Bearing and splined shaft.

$$\sigma_n = \frac{\sigma_x + \sigma_y}{2} + \frac{\sigma_x - \sigma_y}{2} \cos 2\theta$$

$$= \frac{100 + (-60)}{2} + \frac{100 - (-60)}{2} \cos 100°$$

$$= 20 + 80 \cos 100° = 6.1 \ MN/m^2 \ (tensile)$$

Hence, $\sigma_n = 6.1 \ MN/m^2 \ (tensile) \ (Ans.)$

Shear stress/tangential stress,

$$\tau = \frac{\sigma_x - \sigma_y}{2} \sin 2\theta = \frac{100 - (-60)}{2} \sin 100° = 78.78 \ MN/m^2$$

Hence, $\tau = 78.78 \ MN/m^2 \ (shear) \ (Ans.)$

Maximum shear stress,

$$\tau_{max} = \frac{\sigma_x - \sigma_y}{2} = \frac{100 - (-60)}{2} = 80 \ MN/m^2$$

Hence, $\tau_{max} = 80 \ MN/m^2 \ (Ans.)$

Graphical method:

Refer to Fig. 2.16.

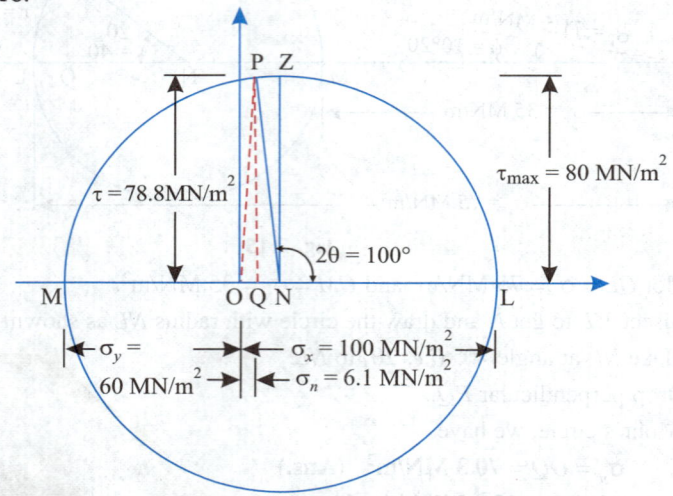

Fig. 2.16

— Plot $OL = \sigma_x = 100 \ MN/m^2$ (tensile)

   $OM = \sigma_y = 60 \ MN/m^2$ (compressive).

— Bisect ML to get N and draw the circle with radius NL as shown.

— Make NP at angle $100°$ (i.e. 2θ) to NL

— Drop perpendicular PQ. Join OP

From Mohr's circle, we have

$$\sigma_n = OQ = 6.1 \ MN/m^2 \ (tensile) \ (Ans.)$$
$$\tau = PQ = 78.8 \ MN/m^2 \ (shear) \ (Ans.)$$
$$\tau_{max} = NZ = 80 \ MN/m^2 \ (shear) \ (Ans.)$$

Example 2.7. *A point is subjected to perpendicular stresses of 50 MN/m² and 30 MN/m², both tensile. Calculate the normal, tangential stresses and resultant stress and its obliquity on a plane making an angle of 30° with the axis of second stress.*

Solution. $\sigma_x = 50 \ MN/m^2$ (tensile), $\sigma_y = 30 \ MN/m^2$ (tensile), $\theta = 90° - 30° = 60°$

Analytical method:

Normal stress,
$$\sigma_n = \frac{\sigma_x + \sigma_y}{2} + \frac{\sigma_x - \sigma_y}{2} \cos 2\theta$$
$$= \frac{50 + 30}{2} + \frac{50 - 30}{2} \cos 120° = 40 + 10 \cos 120°$$
$$= \textbf{35 MN/m}^2 \textbf{ (Ans.)}$$

Tangential stress,
$$\tau = \frac{\sigma_x - \sigma_y}{2} \sin 2\theta = \frac{50 - 30}{2} \sin 120°$$
$$= \textbf{8.66 MN/m}^2 \textbf{ (Ans.)}$$

Resultant stress,
$$\sigma_r = \sqrt{\sigma_n^2 + \tau^2} = \sqrt{35^2 + 8.66^2}$$
$$= \textbf{36.05 MN/m}^2 \textbf{ (Ans.)}$$

$$\tan \phi = \frac{\tau}{\sigma_n} = \frac{8.66}{35} = 0.2474$$

$$\boldsymbol{\phi} \textbf{ (obliquity)} = \textbf{13°54'} \textbf{ (Ans.)}$$

Graphical method:

Refer to Fig. 2.17.

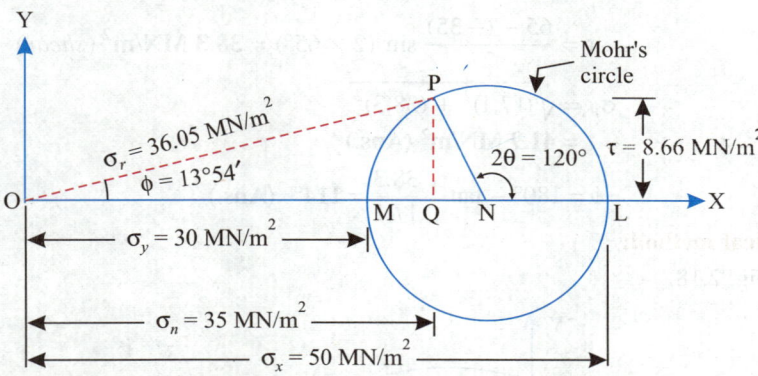

Fig. 2.17

— Plot $OL = \sigma_x = 50$ MN/m^2 (tensile) and $OM = \sigma_y = 30$ MN/m^2 (tensile).

— Bisect ML to get N and draw the circle with radius NL as shown.

— Make NP at an angle 120° (i.e. 2θ) to NL.

— Drop perpendicular PQ. Join OP.

From Mohr's circle, we have
$$\sigma_n = OQ = \textbf{35 MN/m}^2 \textit{ (tensile)} \textbf{ (Ans.)}$$
$$\tau = PQ = \textbf{8.66 MN/m}^2 \textit{ (shear)} \textbf{ (Ans.)}$$
$$\sigma_r = OP = \textbf{36.05 MN/m}^2 \textbf{ (Ans.)}$$
$$\text{Obliquity } \phi = \textbf{13°54'} \textbf{ (Ans.)}$$

Example 2.8. *Draw the Mohr's stress circle for direct stresses of 65 MN/m^2 (tensile) and 35 MN/m^2 (compressive) and estimate the magnitude and direction of the resultant stresses on planes making angles of 20° and 65° with the plane of the first principal stress. Find also the normal and tangential stresses on these planes.*

Solution. *Given:* $\sigma_x = + 65$ MN/m^2,

$\sigma_y = -35$ MN/m^2, θ = 20° and 65°

Analytical method:

Case 1:
$$\sigma_n = \frac{\sigma_x + \sigma_y}{2} + \frac{\sigma_x - \sigma_y}{2} \cos 2\theta$$

$$= \frac{65 + (-35)}{2} + \frac{65 - (-35)}{2} \cos (2 \times 20°)$$

$$= \textbf{53.3 MN/m}^2 \textit{ (tensile)} \textbf{ (Ans.)}$$

$$\tau = \frac{\sigma_x - \sigma_y}{2} \sin 2\theta = \frac{65 - (-35)}{2} \sin (2 \times 20°)$$

$$= \textbf{32.1 MN/m}^2 \textit{ (shear)} \textbf{ (Ans.)}$$

$$\sigma_r = \sqrt{\sigma_n^2 + \tau^2} = \sqrt{(53.3)^2 + (32.1)^2}$$

$$= \textbf{62.2 MN/m}^2 \textbf{ (Ans.)}$$

$$\tan \phi = \frac{\tau}{\sigma_n} = \frac{32.1}{53.3}$$

$$\phi = \textbf{31°} \textbf{ (Ans.)}$$

Case 2:
$$\sigma_n = \frac{65 + (-35)}{2} + \frac{65 - (-35)}{2} \cos (2 \times 65°) = -17.1 \text{ MN}/\text{m}^2$$

$$= \textbf{17.1 MN/m}^2 \textit{ (compressive)} \textbf{ (Ans.)}$$

$$\tau = \frac{65 - (-35)}{2} \sin (2 \times 65°) = \textbf{38.3 MN/m}^2 \textit{ (shear)} \textbf{ (Ans.)}$$

$$\sigma_r = \sqrt{(17.1)^2 + (38.3)^2}$$

$$= \textbf{41.9 MN/m}^2 \textbf{ (Ans.)}$$

$$\phi = 180° - \tan^{-1} \frac{38.3}{17.1} = \textbf{114°} \textbf{ (Ans.)}$$

Graphical method:

Refer Fig. 2.18.

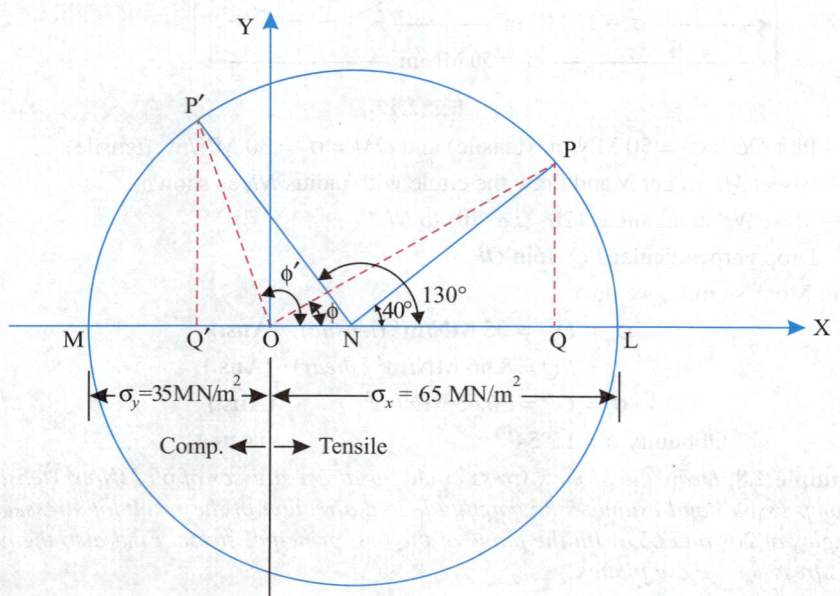

Fig. 2.18

— Plot $OL = 65$ MN/m^2 and $OM = 35$ MN/m^2.

— Bisect LM at N and draw the Mohr circle with NL as radius.

— Draw NP at 40° (*i.e.* 2θ) to NL and NP' at 130° (*i.e.* 2θ) to NL.

— Drop PQ and $P'Q'$ perpendicular to LM.

From Mohr's circle:

First plane (Case 1)

$\sigma_n = OQ = 53.3$ MN/m^2 (*tensile*) **(Ans.)**

$\tau = PQ = 32.1$ MN/m^2 (*shear*) **(Ans.)**

$\sigma_r = OP = 62.2$ MN/m^2 **(Ans.)**

$\phi = 31°$ **(Ans.)**

Second plane (Case 2)

$\sigma_n = OQ' = 17.1$ MN/m^2 **(Ans.)**

$\tau = P'Q' = 38.3$ MN/m^2 **(Ans.)**

$\sigma_r = OP' = 41.9$ MN/m^2 **(Ans.)**

$\phi = 114°$ **(Ans.)**

Belt drive, differential pulley, shaft and motor.

Example 2.9. *The principal stresses acting on an element subjected to plane stresses are 12 and 4.5 MN/m^2 (both tensile) respectively (See Fig. 2.19). Find out the position of the plane AA when the resultant stress makes the maximum angle with the normal to the plane.*

Solution. Refer Fig. 2.19.

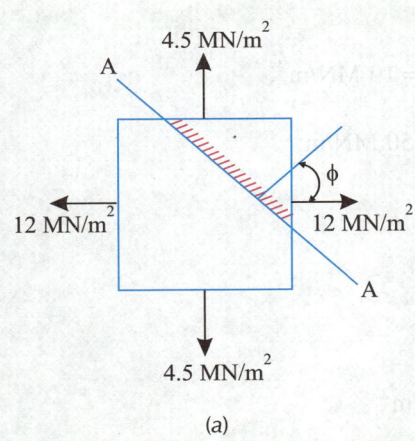

(a)

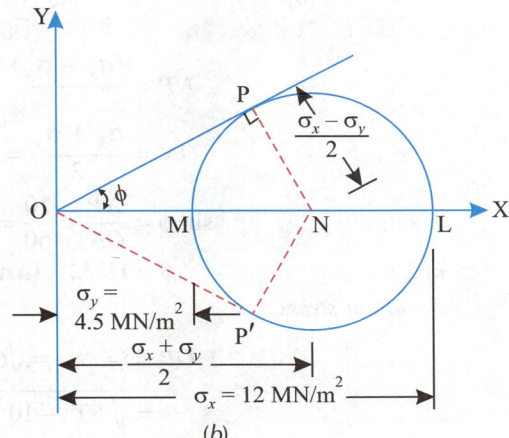

(b)

Fig. 2.19

The problem requires that the angle φ should be a *maximum*. Fig 2.19 (*b*) shows Mohr's circle for the state of stress in Fig. 2.19 (*a*). *The angle φ will be maximum when the line OP is tangential to Mohr's circle.* Hence in Fig. 2.19 (*b*), *OP* is drawn tangential to the circle and the angle φ is measured as 27°. The line *OP'* could also be drawn on the other side of the circle as shown in dotted line. The plane *AA* is defined by the angle φ in Fig. 2.19 (*a*).

The analytical solution of the problem is somewhat lengthy. Mohr's circle however enables one to obtain the following result (which is very useful in *soil mechanics*).

$$\sin \phi = \frac{NP}{ON} = \frac{\frac{1}{2}(\sigma_x - \sigma_y)}{\frac{1}{2}(\sigma_x + \sigma_y)}$$

$$= \frac{\frac{1}{2}(12 - 4.5)}{\frac{1}{2}(12 + 4.5)} = 0.4545$$

∴ $\phi = 27°$ (Ans.)

Example 2.10. *A piece of material is subjected to two compressive stresses at right angles, their values being 40 MN/m² and 60 MN/m². Find the position of the plane across which the resultant stress is most inclined to the normal, and determine the value of this resultant stress.*

Solution. In Fig. 2.20 (a) the angle θ is inclined to the plane of the 40 MN/m² compression. In Fig. 2.20 (b), $OL = 60$ MN/m², $OM = 40$ MN/m². The maximum angle ϕ is obtained when OP is a tangent to the stress circle.

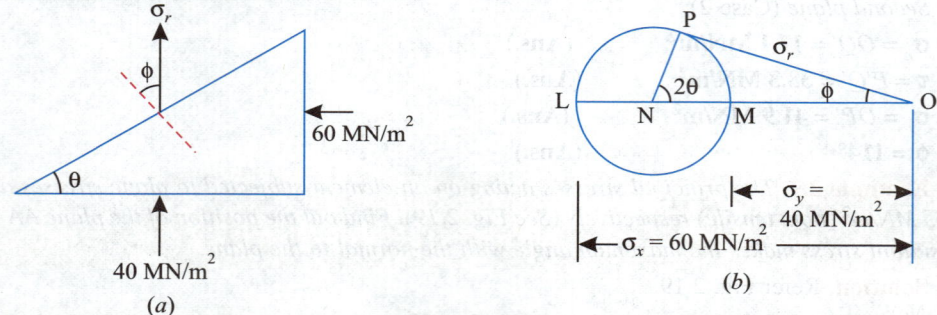

Fig. 2.20

$$NP = \frac{(\sigma_x - \sigma_y)}{2} = \frac{60 - 40}{2} = 10 \text{ MN/m}^2,$$

$$ON = \frac{\sigma_x + \sigma_y}{2} = \frac{60 + 40}{2} = 50 \text{ MN/m}^2$$

Then, $\quad \sin \phi = \dfrac{NP}{ON} = \dfrac{10}{50} = 0.2$

i.e., $\quad \phi = 11°32'$ **(Ans.)**

Resultant stress,

$$OP = (= \sigma_r) = \sqrt{ON^2 - NP^2}$$

$$= \sqrt{50^2 - 10^2} = 49 \text{ MN/m}^2$$

Hence, $\quad \sigma_r = 49$ **MN/m² (Ans.)**

Also, $\quad 2\theta = 90 - \phi$

$$\theta = \frac{90 - \phi}{2} = \frac{90 - 11° 32'}{2}$$

$$= 39°14'$$

which gives the position of the plane required. **(Ans.)**

Example 2.11. *The principal tensile stresses at a point across two perpendicular planes are 120 MN/m² and 60 MN/m². Find:*

(i) *The normal and tangential stresss and the resultant stress and its obliquity on a plane at 20° with the major principal plane.*

(ii) *The intensity of stress which acting alone can produce the same maximum strain. Take Poisson's ratio = 1/4.*

Solution. Refer to Fig. 2.21.

Given: $\sigma_x = 120$ MN/m² (tensile),

$\sigma_y = 60$ MN/m² (tensile), and

$\theta = 20°$

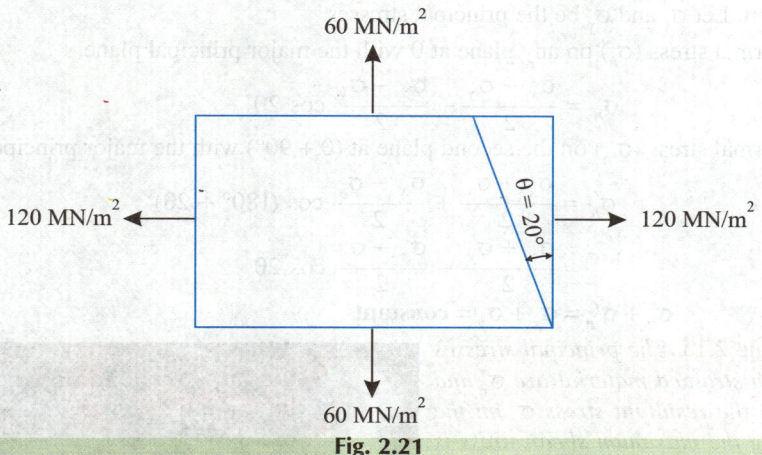

Fig. 2.21

(i) σ_n, τ, σ_r :

Normal stress,

$$\sigma_n = \frac{\sigma_x + \sigma_y}{2} + \frac{\sigma_x - \sigma_y}{2} \cos 2\theta$$

$$= \frac{120 + 60}{2} + \frac{120 - 60}{2} \cos 40°$$

$$= 90 + 30 \cos 40°$$

$$= \textbf{112.98 MN/m}^2 \ (\textit{tensile}) \ \textbf{(Ans.)}$$

Tangential stress,

$$\tau = \frac{\sigma_x - \sigma_y}{2} \sin 2\theta = \frac{120 - 60}{2} \sin 40° = 30 \sin 40°$$

$$= \textbf{19.28 MN/m}^2 \ (\textit{shear}) \ \textbf{(Ans.)}$$

Resultant stress,

$$\sigma_r = \sqrt{\sigma_n^2 + \tau^2} = \sqrt{112.98^2 + 19.28^2}$$

$$= \textbf{114.6 MN/m}^2 \ \textbf{(Ans.)}$$

Obliquity,

$$\phi = \tan^{-1} \frac{\tau}{\sigma_n} = \tan^{-1} \frac{19.28}{112.98}$$

$$= \textbf{9°41'} \ \textbf{(Ans.)}$$

(ii) **Intensity of stress, σ:**

Maximum strain

$$= \frac{\sigma_x}{E} - \frac{\sigma_y}{mE} = \frac{1}{E}\left(\sigma_x - \frac{\sigma_y}{m}\right)$$

$$= \frac{1}{E}\left(120 - 60 \times \frac{1}{4}\right) = \frac{105}{E}$$

Let σ be the stress which acting alone can produce the same maximum strain.

$$\frac{\sigma}{E} = \frac{105}{E}$$

i.e. $\sigma = 105$ MN/m² **(Ans.)**

Example 2.12. *Show that in a strained material subjected to two-dimensional stress, the sum of the normal components of stresses on any two mutually perpendicular planes is constant.*

Solution. Let σ_x and σ_y be the principal stresses.

The normal stress (σ_n) on any plane at θ with the major principal plane,

$$\sigma_n = \frac{\sigma_x + \sigma_y}{2} + \frac{\sigma_x - \sigma_y}{2}\cos 2\theta \qquad \qquad ...(i)$$

The normal stress (σ'_n) on the second plane at $(\theta + 90°)$ with the major principal plane,

$$\sigma'_n = \frac{\sigma_x + \sigma_y}{2} + \frac{\sigma_x - \sigma_y}{2}\cos(180° + 2\theta)$$

$$= \frac{\sigma_x + \sigma_y}{2} - \frac{\sigma_x - \sigma_y}{2}\cos 2\theta \qquad \qquad ...(ii)$$

$$\sigma_n + \sigma'_n = \sigma_x + \sigma_y = \text{constant} \qquad \qquad ......\text{ Proved.}$$

Example 2.13. *The principal stresses at a point in a strained material are σ_x and σ_y. Show that the resultant stress σ_r on the plane carrying the maximum shear stress is given by:*

$$\sigma_r = \left[\frac{\sigma_x^2 + \sigma_y^2}{2}\right]^{1/2}$$

Solution. The plane carrying the maximum shear stress is at $\theta = 45°$ with the major principal plane. On this plane the resultant stress,

$$\sigma_r = \sqrt{\sigma_x^2 \cos^2 \theta + \sigma_y^2 \sin^2 \theta}$$

$$= \sqrt{\sigma_x^2 \cos^2 45° + \sigma_y^2 \sin^2 45°}$$

$$= \left[\frac{\sigma_x^2 + \sigma_y^2}{2}\right]^{1/2}$$

Automobile wheel and suspension spring in assembly

$$......\text{ Proved.}$$

Example 2.14. *At a point in a bracket the stresses on two mutually perpendicular planes are 35 MN/m² (tensile) and 15 MN/m² (tensile). The shear stress across these planes is 9 MN/m². Find the magnitude and direction of the resultant stress on a plane making an angle of 40° with the plane of first stress. Find also the normal and tangential stresses on the planes.*

Solution.

Analytical method:

Given: $\sigma_x = 35$ MN/m² (tensile),

$\sigma_y = 15$ MN/m² (tensile),

$\tau_{xy} = 9$ MN/m^2 shear, and $\theta = 40°$

Normal stress:

We know that, $\sigma_n = \dfrac{\sigma_x + \sigma_y}{2} + \dfrac{\sigma_x - \sigma_y}{2} \cos 2\theta + \tau_{xy} \sin 2\theta$

$\qquad\qquad = \dfrac{35 + 15}{2} + \dfrac{35 - 15}{2} \cos (2 \times 40°) + 9 \sin (2 \times 40°)$

$\qquad\qquad = 25 + 1.74 + 8.86$

$\qquad\qquad = \mathbf{35.6}$ **MN/m^2** **(Ans.)**

Tangential stress:

$\tau = \dfrac{\sigma_x - \sigma_y}{2} \sin 2\theta - \tau_{xy} \cos 2\theta$

$\qquad = \dfrac{35 - 15}{2} \sin (2 \times 40°) - 9 \cos (2 \times 40°)$

$\qquad = \mathbf{8.29}$ **MN/m^2 (Ans.)**

Resultant stress:

$\sigma_r = \sqrt{(35.6)^2 + (8.29)^2} = \mathbf{36.55}$ **MN/m^2** **(Ans.)**

$\tan \phi = \dfrac{\tau}{\sigma_n} = \dfrac{8.29}{35.6}$

i.e. $\qquad\qquad \phi = \mathbf{13°}$ **(Ans.)**

Graphical method:

Refer Fig. 2.22

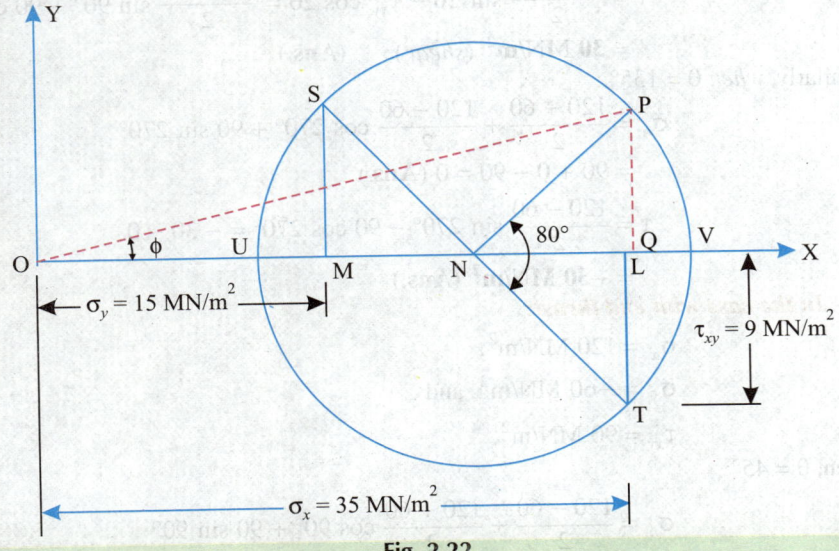

Fig. 2.22

— Plot $OL = 35$ MN/m^2 and $OM = 15$ MN/m^2.

— Drop perpendicular LT and MS, each 9 MN/m^2 as shown in Fig. 2.22.

— Join ST to get N and draw the Mohr's circle to pass through S and T.

— Draw NP at 80° to NT.

— Draw perpendicular to OQ.

From Mohr's circle, we have

$$\sigma_n = OQ = 35.6 \text{ MN/m}^2 \text{ (tensile)} \quad \text{(Ans.)}$$
$$\tau = PQ = 8.29 \text{ MN/m}^2 \text{ (shear)} \quad \text{(Ans.)}$$
$$\sigma_r = OP = 36.55 \text{ MN/m}^2 \quad\quad\quad \text{(Ans.)}$$
$$\phi = 13° \quad\quad\quad\quad\quad\quad\quad \text{(Ans.)}$$

Example 2.15. *When a certain thin-walled tube is subjected to internal pressure and torque the stresses in the tube wall are (a) 120 MN/m² (tensile) (b) 60 MN/m² (tensile) in a direction at right angles to (a), (c) complementary shear stress of 90 MN/m² in the directions of (a) and (b).*

(i) *Calculate the normal and tangential stresses on the two planes which are equally inclined to (a) and (b).*

(ii) *What are the results if due to an end thrust; (b) is compressive, (a) and (c) being unchanged.*

Solution. *Given :* $\quad \sigma_x = 120 \text{ MN/m}^2 \text{ (tensile)},$
$$\sigma_y = 60 \text{ MN/m}^2 \text{ (tensile), and}$$
$$\tau_{xy} = 90 \text{ MN/m}^2 \text{ (shear)}.$$

(i) $\sigma_n = ?, \tau = ?,$

When, $\theta = 45°$:

$$\sigma_n = \frac{\sigma_x + \sigma_y}{2} + \frac{\sigma_x - \sigma_y}{2} \cos 2\theta + \tau_{xy} \sin 2\theta$$

$$= \frac{120 + 60}{2} + \frac{120 - 60}{2} \cos 90° + 90 \sin 90° = 90 + 0 + 90$$

$$= \textbf{180 MN/m}^2 \textit{ (tensile)} \quad \textbf{(Ans.)}$$

$$\tau = \frac{\sigma_x - \sigma_y}{2} \sin 2\theta - \tau_{xy} \cos 2\theta = \frac{120 - 60}{2} \sin 90° + 90 \cos 90°$$

$$= \textbf{30 MN/m}^2 \textit{ (shear)} \quad \textbf{(Ans.)}$$

Similarly, *when* $\theta = 135°$:

$$\sigma_n = \frac{120 + 60}{2} + \frac{120 - 60}{2} \cos 270° + 90 \sin 270°$$

$$= 90 + 0 - 90 = 0 \textbf{ (Ans.)}$$

$$\tau = \frac{120 - 60}{2} \sin 270° - 90 \cos 270° = -30 - 0$$

$$= \textbf{–30 MN/m}^2 \textbf{ (Ans.)}$$

(ii) *In the case with end thrust:*

$$\sigma_x = 120 \text{ MN/m}^2,$$

$$\sigma_y = -60 \text{ MN/m}^2, \text{ and}$$

$$\tau_{xy} = 90 \text{ MN/m}^2.$$

when, $\theta = 45°$:

$$\sigma_n = \frac{120 - 60}{2} + \frac{120 + 60}{2} \cos 90° + 90 \sin 90°$$

$$= 30 + 0 + 90 = \textbf{120 MN/m}^2 \textbf{ (Ans.)}$$

$$\tau = \frac{120 - (-60)}{2} \sin 90° - 90 \cos 90°$$

$$= 90 - 0 = \textbf{90 MN/m}^2 \quad \textbf{(Ans.)}$$

when, $\theta = 135°$:

$$\sigma_n = \frac{120 - 60}{2} + \frac{120 + 60}{2} \cos 270° + 90 \sin 270°$$

$$= 30 + 0 - 90 = -\textbf{60 MN/m}^2 \textbf{ (Ans.)}$$

$$\tau = \frac{120 - (-60)}{2} \sin 270° - 90 \cos 270°$$

$$= -90 - 0 = -\textbf{90 MN/m}^2 \textbf{ (Ans.)}$$

Example 2.16. *At a point in a piece of material there is a tensile stress of 90 MN/m² upon the horizontal plane, and compressive stress of 45 MN/m² upon the vertical plane. There is also a shear stress of 45 MN/m² on each of these planes. Determine the planes of maximum shear stress and value of maximum shear stress at the point from the first principles.*

Solution. We know that:

$$\tau = \frac{\sigma_x - \sigma_y}{2} \sin 2\theta - \tau_{xy} \cos 2\theta \text{ (For derivation refer article 2.5)}$$

Here, $\sigma_x = 90$ MN/m² (tensile),

$\sigma_y = 45$ MN/m² (comp.), and

$\tau_{xy} = 45$ MN/m² (shear).

Also, $\cot (180 - 2\theta_1) = \dfrac{2\tau_{xy}}{\sigma_x - \sigma_y}$ (eqn. 2.10 b)

(where, θ_1 = inclination of maximum shear with the plane of tensile stress σ_x)

$$= \frac{2 \times 45}{90 - (-45)} = \frac{90}{135} = \frac{2}{3}$$

or, $180 - 2\theta_1 = 56° \ 18'$

$2\theta_1 = 180° - 56° \ 18' = 123° \ 42'$

or, $\boldsymbol{\theta_1 = 61° \ 51'}$ **(Ans.)**

∴ $\tau_{max} = \dfrac{90 - (-45)}{2} \sin (2 \times 61°51') - 45 \cos (2 \times 61°51')$

$$= \textbf{81.1 MN/m}^2 \textbf{ (Ans.)}$$

Example 2.17. *An element in a stressed material has tensile stress of 500 MN/m² and a compressive stress of 350 MN/m² acting on two mutually perpendicular planes and equal shear stresses of 100 MN/m² on these planes. Find principal stresses and position of the principal planes. Find also maximum shearing stress.*

Solution. *Analytical method:*

Given $\sigma_x = 500$ MN/m² (tensile),

$\sigma_y = 350$ MN/m² (comp.), and

$\tau_{xy} = 100$ MN/m² (shear).

Principal stresses; $\boldsymbol{\sigma_1, \sigma_2}$:

We know that,

$$\sigma = \frac{\sigma_x + \sigma_y}{2} \pm \sqrt{\left[\frac{\sigma_x - \sigma_x}{2}\right]^2 + \tau_{xy}^2}$$

$$= \frac{500 + (-350)}{2} \pm \sqrt{\left[\frac{500 - (-350)}{2}\right]^2 + (100)^2}$$

$$= 75 \pm 436.6$$
$$\sigma_1 = 511.6 \text{ MN/m}^2$$
$$\sigma_2 = -361.6 \text{ MN/m}^2$$

Hence, *Maximum principal stress,*

$$\sigma_1 = \textbf{511.6 MN/m}^2 \text{ (tensile)} \quad \textbf{(Ans.)}$$

and, *Minimum principal stress,*

$$\sigma_2 = \textbf{361.6 MN/m}^2 \text{ (compressive)} \quad \textbf{(Ans.)}$$

Position of principal planes θ_1, θ_2:

Also,
$$\tan 2\theta = \frac{2\tau_{xy}}{\sigma_x - \sigma_y} = \frac{2 \times 100}{500 - (-350)} = 0.2353$$

from which,
$$2\theta = 13° \, 14' \text{ or } 193° \, 14'$$
$$\boldsymbol{\theta_1 = 6°37' \text{ and } \theta_2 = 96° \, 37' \textbf{ (Ans.)}}$$

Maximum shear stress, τ_{max}:

$$\tau_{max} = \frac{\sigma_1 - \sigma_2}{2} = \frac{511.6 - (-361.6)}{2}$$
$$= \textbf{436.6 MN/m}^2 \textbf{ (Ans.)}$$

Graphical method:

Refer to Fig. 2.23.

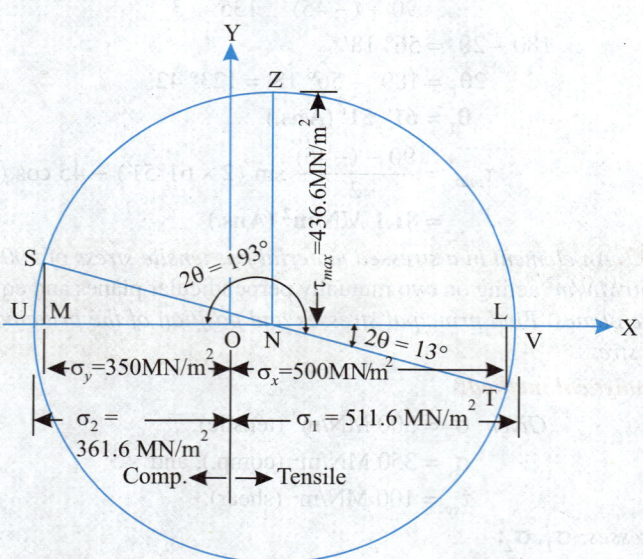

Fig. 2.23

— Plot $OL = 500 \text{ MN/m}^2$ (tensile), and
$OM = 350 \text{ MN/m}^2$ (compressive).

— Bisect LM at N and from L and M draw perpendiculars LT and MS each 100 MN/m^2 (shear stress) to LM.

— With N as centre and radius NT draw the circle.

From Mohr's circle:

$$\sigma_1 = OV = \textbf{511.6 MN/m}^2 \ \textit{(tensile)} \ \textbf{(Ans.)}$$

$$\sigma_2 = OU = \textbf{361.6 MN/m}^2 \ \textit{(comp.)} \ \textbf{(Ans.)}$$

$$\tau_{max} = NZ = \textbf{436.6 MN/m}^2 \ \ \textit{(Shear)} \ \textbf{(Ans.)}$$

$$2\theta = 13° \ \text{or} \ 193°$$

$$\theta_1 = \textbf{6° 30'}$$

and,

$$\theta_2 = \textbf{96° 30'} \ \ \textbf{(Ans.)}$$

Example 2.18. *At a point in a bracket the stresses on two mutually perpendicular planes are 400 MN/m² tensile and 300 MN/m² tensile. The shear stress across these planes is 200 MN/m².*

Determine graphically or otherwise the magnitude and directions of principal stresses and maximum shear stress.

Solution. *Analytical method*:

Given: $\sigma_x = 400$ MN/m² (tensile),

$\sigma_y = 300$ MN/m² (tensile), and

$\tau_{xy} = 200$ MN/m² (shear).

Principal stresses:

We know that,

$$\sigma = \frac{\sigma_x + \sigma_y}{2} \pm \sqrt{\left[\frac{\sigma_x - \sigma_y}{2}\right]^2 + \tau_{xy}^2}$$

$$= \frac{400 + 300}{2} \pm \sqrt{\left[\frac{400 - 300}{2}\right]^2 + 200^2}$$

$$= 350 \pm \sqrt{2500 + 40000} = 350 \pm 206$$

Maximum principal stress,

$$\sigma_1 = 350 + 206 = \textbf{556 MN/m}^2 \ \textit{(tensile)} \ \textbf{(Ans.)}$$

Minimum principal stress,

$$\sigma^2 = 350 - 206 = \textbf{144 MN/m}^2 \ \textit{(tensile)} \ \ \textbf{(Ans.)}$$

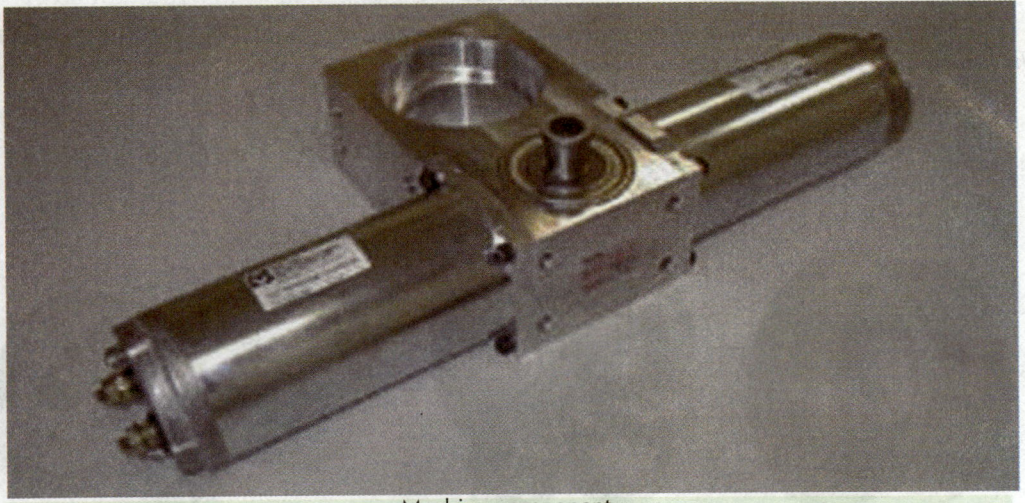

Machine component.

Directions of principal stresses, θ_1, θ_2 :

Also,
$$\tan 2\theta = \frac{2\tau_{xy}}{\sigma_x - \sigma_y} = \frac{2 \times 200}{400 - 300} = \frac{400}{100} = 4$$

or,
$$2\theta = 76° \text{ or } 256°$$

∴
$$\theta_1 = 38°, \theta_2 = 128°$$

Maximum shearing stress, τ_{max}:

$$\tau_{max} = \frac{\sigma_1 - \sigma_2}{2} = \frac{556 - 144}{2}$$
$$= \textbf{206 MN/m}^2 \textbf{ (Ans.)}$$

The directions of maximum shear stress (206 MN/m^2) with plane of $\sigma_x = 45° + 38° = $ **83° (Ans.)**

Graphical solution:

Fig. 2.24 shows the graphical solution, which is self-explanatory. The graphical result is shown in the figure itself.

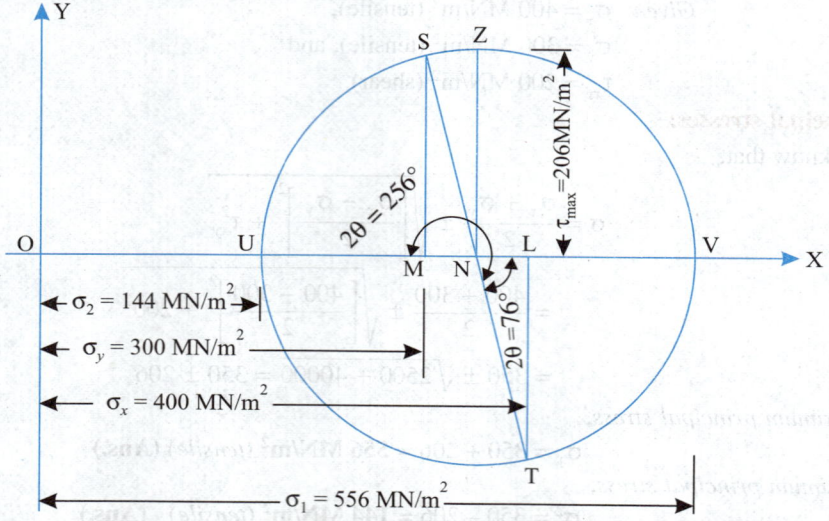

Fig. 2.24

Example 2.19. *The state of stress at a point of a machine is shown in Fig. 2.25 (a). Determine the principal stresses and maximum shear stress and its inclination.*

Solution. Refer Fig. 2.25 (a)

Analytical method:

Given: $\sigma_x = 50$ MN/m^2 (tensile),
$\sigma_y = 30$ MN/m^2 (comp.), and

Pulleys with grooves for belt drive.

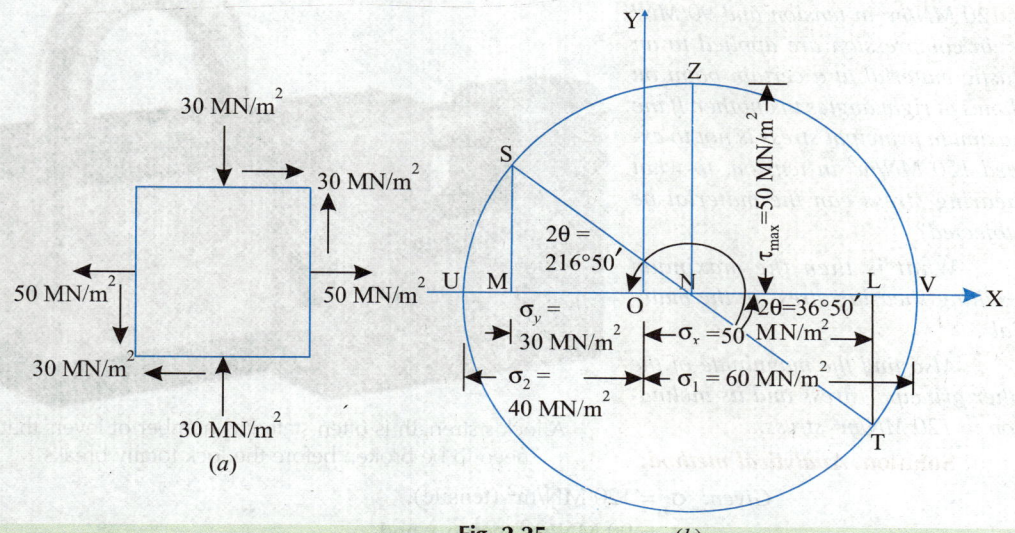

Fig. 2.25

Principal stresses σ_1, σ_2:

Using the relation:

$$\sigma = \left[\frac{\sigma_x + \sigma_y}{2}\right] \pm \sqrt{\left[\frac{\sigma_x - \sigma_y}{2}\right]^2 + \tau_{xy}^2}$$

$$= \frac{50 + (-30)}{2} \pm \sqrt{\left[\frac{50 - (-30)}{2}\right]^2 + 30^2}$$

$$= 10 \pm \sqrt{40^2 + 30^2} = 10 \pm 50$$

or, $\sigma_1 = 60$ MN/m².
        $\sigma_2 = -40$ MN/m².

Hence, *Maximum* principal stress,

$\sigma_1 = 60$ **MN/m² (tensile)   (Ans.)**

Minimum principal stress,

$\sigma_2 = 40$ **MN/m² (comp.)   (Ans.)**

Directions of principal stresses, θ_1, θ_2:

Using the relation:

$$\tan 2\theta = \frac{2\tau_{xy}}{\sigma_x - \sigma_y} = \frac{2 \times 30}{50 - (-30)} = \frac{60}{80} = \frac{3}{4}$$

or, $2\theta = 36° 50'$ or $216° 50'$

$\theta_1 = 18° 25'$ **and** $\theta_2 = 108° 25'$ **(Ans.)**

Maximum shearing stress, τ_{max}:

$$\tau_{max} = \frac{\sigma_1 - \sigma_2}{2} = \frac{60 - (-40)}{2} = 50 \text{ MN/m}^2 \text{ (Ans.)}$$

The directions of maximum shear stress 50 MN/m² with plane of σ_x

$$= 45° + 18°25' = 63°25' \text{ (Ans.)}$$

Graphical solution:

Fig. 2.25(*b*) shows the graphical solution which is self-explanatory. The graphical result is shown in the figure itself.

Example 2.20. *Direct stresses of 120 MN/m² in tension and 90 MN/m² in compression are applied to an elastic material at a certain point on planes at right angles to another. If the maximum principal stress is not to exceed 150 MN/m² in tension, to what shearing stress can the material be subjected?*

What is then the maximum resulting shearing stress in the material ?

Also find the magnitude of the other principal stress and its inclination to 120 MN/m² stress.

A lock's strength is often stated as number of levers that need to be broken before the lock totally breaks.

Solution. *Analytical method*:

$$\text{Given: } \sigma_x = 120 \text{ MN/m}^2 \text{ (tensile)},$$
$$\sigma_y = 90 \text{ MN/m}^2 \text{ (comp.), and}$$
$$\sigma_1 = 150 \text{ MN/m}^2 \text{ (tensile).}$$

Shearing stress τ_{xy}:

Using the relation:

$$\sigma_1 = \frac{\sigma_x + \sigma_y}{2} + \sqrt{\left[\frac{\sigma_x - \sigma_y}{2}\right]^2 + \tau_{xy}^2}$$

$$150 = \frac{120 + (-90)}{2} + \sqrt{\left[\frac{120 - (-90)}{2}\right]^2 + \tau_{xy}^2}$$

$$150 = 15 + \sqrt{105^2 + \tau_{xy}^2}$$

or, $$135 = \sqrt{105^2 + \tau_{xy}^2}$$

or, $$(135)^2 = 105^2 + \tau_{xy}^2$$

or, $$\tau_{xy} = \sqrt{(135)^2 - (105)^2}$$

$$= \textbf{84.8 MN/m}^2 \textbf{ (shear)} \quad \textbf{(Ans.)}$$

Now, *minimum principal stress*,

$$\sigma_2 = \frac{120 + (-90)}{2} - \sqrt{\left[\frac{120 - (-90)}{2}\right]^2 + (84.8)^2}$$

$$= 15 - 135 = -120 \text{ MN/m}^2$$

$$\sigma_2 = \textbf{120 MN/m}^2 \textbf{ (comp.)} \quad \textbf{(Ans.)}$$

Maximum shear stress τ_{max}:

Using the relation:

$$\tau_{max} = \frac{\sigma_1 - \sigma_2}{2} = \frac{150 - (-120)}{2} = \textbf{135 MN/m}^2 \textbf{ (Ans.)}$$

Directions of principal stresses θ_1, θ_2:

We know that,

$$\tan 2\theta = \frac{2\tau_{xy}}{\sigma_x - \sigma_y} = \frac{2 \times 84.8}{120 - (-90)}$$

or, $\quad 2\theta = 39°$ or $219°$

$\quad\quad \theta_1 = 19° \ 30'$

and, $\quad \theta_2 = 109° \ 30'$

The inclination of the plane of σ_2 (120 MN/m²) is 109°30' with the plane of 120 MN/m² (tensile) stress. **(Ans.)**

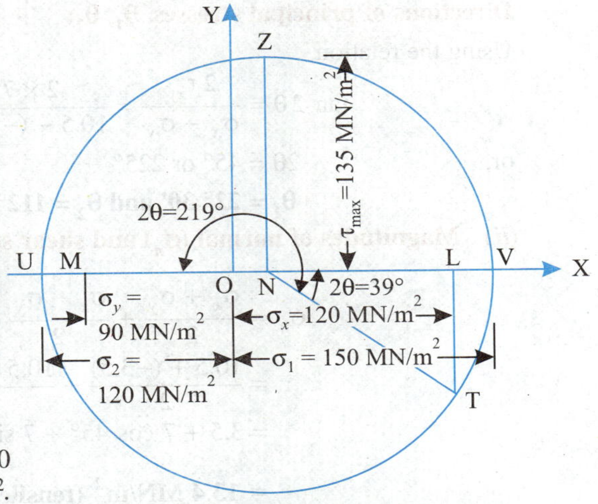

Graphical solution. :

Refer to Fig. 2.26.

— Plot $OL = 120$ MN/m², $OM = 90$ MN/m² and $OV = \sigma_1 = 150$ MN/m².

— Bisect LM at N.

— With N as centre and NV as radius draw a circle.

— Drop LT perpendicular to OV meeting the circle at T.

From *Mohr's circle:* $\quad \tau_{xy} = LT = \textbf{84.8 MN/m}^2$ **(Ans.)**

$\quad\quad\quad \tau_{max} = NZ = \textbf{135 MN/m}^2$ **(Ans.)**

$\quad\quad\quad \sigma_2 = OU = \textbf{120 MN/m}^2$ *(comp.)* **(Ans.)**

$\quad\quad\quad \theta_2 = \textbf{109° 30'}$ **(Ans.)**

Fig. 2.26

Example 2.21. *Two mutually perpendicular planes of an element of material are subjected to direct stresses of 10.5 MN/m² (tensile) and 3.5 MN/m² (comp.) and shear stress of 7 MN/m².*

Find graphically or otherwise:

(i) The magnitude and direction of principal stresss, and

(ii) Magnitude of the normal and shear stresses on a plane on which the shear stress is maximum.

Solution. *Analytical method*:

$$\text{Given: } \sigma_x = 10.5 \text{ MN/m}^2 \text{ (tensile)},$$
$$\sigma_y = 3.5 \text{ MN/m}^2 \text{ (comp.)}, \text{ and}$$
$$\tau_{xy} = 7 \text{ MN/m}^2.$$

(i) Principal stresses σ_1, σ_2:

We know that;

$$\sigma = \left(\frac{\sigma_x + \sigma_y}{2}\right) \pm \sqrt{\left[\frac{\sigma_x - \sigma_y}{2}\right]^2 + \tau_{xy}^2}$$

$$= \frac{10.5 + (-3.5)}{2} \pm \sqrt{\left[\frac{10.5 - (-3.5)}{2}\right]^2 + 7^2}$$

$$= 3.5 \pm \sqrt{7^2 + 7^2} = 3.5 \pm 9.9$$

or, $\quad \sigma_1 = 13.4$ MN/m², $\sigma_2 = -6.4$ MN/m²

Hence, *Maximum principal stress,*

$$\sigma_1 = \textbf{13.4 MN/m}^2 \text{ (tensile)} \quad \textbf{(Ans.)}$$

Minimum principal stress,

$$\sigma_2 = \textbf{6.4 MN/m}^2 \text{ (comp.)} \quad \textbf{(Ans.)}$$

Directions of principal stresses, θ_1, θ_2:

Using the relation,

$$\tan 2\theta = \frac{2\tau_{xy}}{\sigma_x - \sigma_y} = \frac{2 \times 7}{10.5 - (-3.5)}$$

or,

$$2\theta = 45° \text{ or } 225°$$

$$\boldsymbol{\theta_1 = 22° \ 30' \text{ and } \theta_2 = 112° \ 30' \text{ (Ans.)}}$$

(ii) Magnitudes of normal (σ_n) and shear stresses (τ) :

$$\sigma_n = \frac{\sigma_x + \sigma_y}{2} + \frac{\sigma_x - \sigma_y}{2} \cos 2\theta + \tau_{xy} \sin 2\theta$$

$$= \frac{10.5 + (-3.5)}{2} + \frac{10.5 - (-3.5)}{2} \cos 45° + 7 \sin 45°$$

$$= 3.5 + 7 \cos 45° + 7 \sin 45°$$

$$= \textbf{13.4 MN/m}^2 \textbf{ (tensile)   (Ans.)}$$

$$\tau = \frac{\sigma_x - \sigma_y}{2} \sin 2\theta - \tau_{xy} \cos 2\theta$$

$$= \frac{10.5 - (-3.5)}{2} \sin 45° - 7 \cos 45° = \textbf{0 (Ans.)}$$

Graphical solution :

Refer to Fig. 2.27.

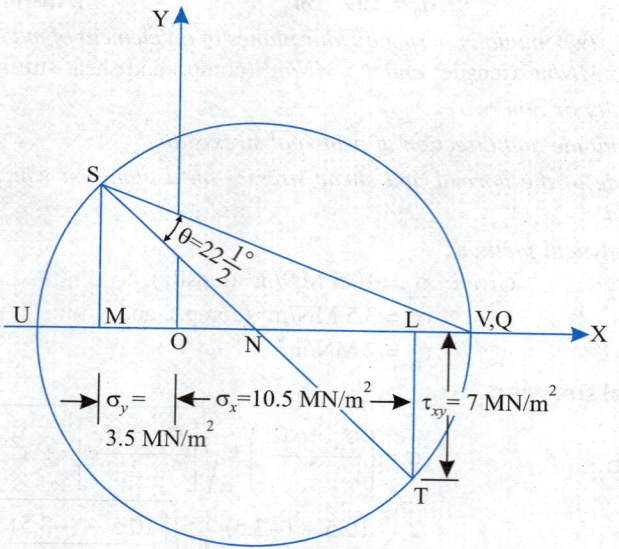

Fig. 2.27

— Draw the circle as per usual procedure.

— At S make $NSQ = 22\frac{1}{2}°$. This meets V *i.e.* Q and V coincide.

Measure OV, VQ

$$\sigma_n = OV = 13.4 \text{ MN/m}^2 \text{ (tensile)}$$

$$\tau = VQ = \text{Zero}$$

Note. Had the point Q not coincided with the point V, the procedure to find σ_n and τ would have been as follows:

— From Q draw perpendicular to UV, meeting UV at say A.

— Join OQ and NQ

Then, $\sigma_n = OA$, and

$\tau = QA$; OQ however gives the resultant (σ_r) of these (σ_n and τ) stresses.

Example 2.22. *At a point in a material under stress, the intensity of the resultant stress on a certain plane is 50 MN/m² (tensile) inclined at 30° to the normal of that plane. The stress on a plane at right angles to this has a normal tensile component of intensity of 30 MN/m². Find:*

(i)   The resultant stress on the second plane.

(ii) The principal planes and stresses.

(iii) The plane of maximum shear and its intensity.

Solution. Refer to Fig. 2.28 (a, b)

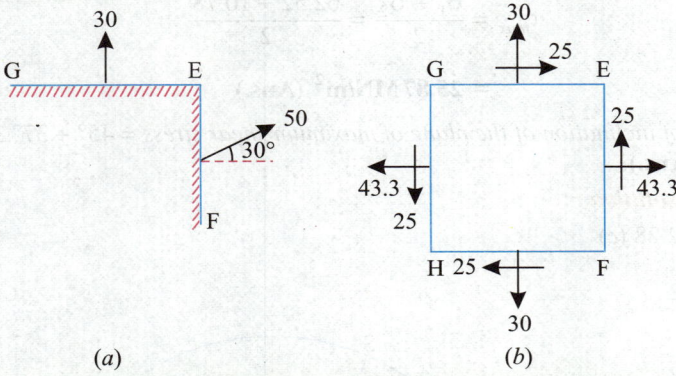

Fig. 2.28

Analytical method:

  (i) The resultant stress on the second plane :

The tangential stress on the first plane is

$$\tau_{xy} = 50 \sin 30° = 25 \text{ MN/m}^2$$

Hence on the second plane the complementary shear stress is 25 MN/m².

Resultant stress on the second plane is

$$\sigma_r = \sqrt{(30)^2 + (25)^2}$$

$$= \textbf{39 MN/m}^2 \textbf{ (Ans.)}$$

(ii) Principal planes and stresses :

The intensity of stress normal to the first plane is

$$= 50 \cos 30° = 43.3 \text{ MN/m}^2$$

Fig. 2.28 (b) shows the final stress system from which the principal stresses are:

$$\sigma = \frac{\sigma_x + \sigma_y}{2} \pm \sqrt{\left[\frac{\sigma_x - \sigma_y}{2}\right]^2 + \tau_{xy}^2}$$

$$= \frac{43.3 + 30}{2} \pm \sqrt{\left[\frac{43.3 - 30}{2}\right]^2 + 25^2}$$

$$= 36.65 \pm \sqrt{(6.65)^2 + 25^2} = 36.65 \pm 25.87$$

Maximum principal stress,

$$\sigma_1 = 62.52 \text{ MN/m}^2 \text{ (tensile)} \quad \textbf{(Ans.)}$$

Minimum principal stress,

$$\sigma_2 = 10.78 \text{ MN/m}^2 \text{ (tensile)} \quad \textbf{(Ans.)}$$

Direction of the principal stresses (θ_1, θ_2) *are:*

$$\tan 2\theta = \frac{2\tau_{xy}}{\sigma_x - \sigma_y} = \frac{2 \times 25}{43.3 - 30} = 3.76$$

or,

$$2\theta = 75° \text{ or } 255°$$

$$\boldsymbol{\theta_1 = 37° \ 30' \text{ and } \theta_2 = 127° \ 30'} \quad \textbf{(Ans.)}$$

(iii) Maximum shear stress, τ_{max} :

$$\tau_{max} = \frac{\sigma_1 - \sigma_2}{2} = \frac{62.52 - 10.78}{2}$$

$$= 25.87 \text{ MN/m}^2 \quad \textbf{(Ans.)}$$

The angle of inclination of the plane of maximum shear stress = 45° + 37° 30' = **82° 30' with the plane EF.** **(Ans.)**

Graphical solution :

Refer Fig. 2.28 (c)

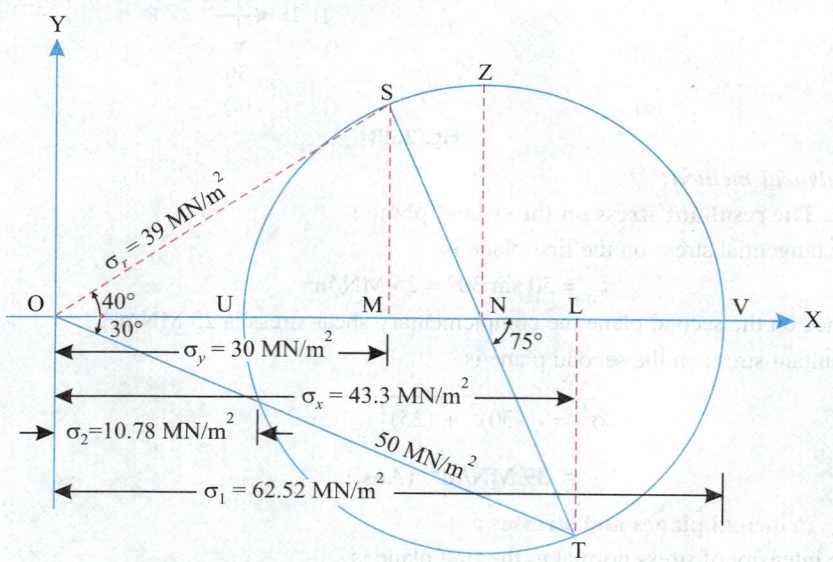

Fig. 2.28 (c)

— On the plane *EF*, σ_n is positive (being tensile) and τ is negative (being in anticlockwise direction). Hence draw *OT* at an inclination of 30° to *OX* in the clockwise direction and make *OT* = 50 MN/m².

— Draw *TL* perpendicular to the axis *OX*.

— Set off *OM* = 30 MN/m² to the right of *O* since the stress on *EG* is tensile and positive.

— Bisect *LM* to get *N*.

— With N as centre and NT as the radius, draw Mohr's circle.

From *Mohr's circle*:

$$\sigma_1 = OV = \textbf{62.52 MN/m}^2 \text{ (tensile)} \quad \textbf{(Ans.)}$$

$$\sigma_2 = OU = \textbf{10.78 MN/m}^2 \text{ (tensile)} \quad \textbf{(Ans.)}$$

$$\angle LNT = \textbf{75}° \; (\theta_1 = 37° \, 30') \quad \textbf{(Ans.)}$$

Resultant stress on the second plane, $OS = \textbf{39 MN/m}^2$ **(inclination 40°).**

Maximum shear stress $NZ = \textbf{25.86 MN/m}^2$. The angle of inclination of the plane of maximum shear stress $\frac{1}{2}\angle TNZ = $ **82° 30′ with the line NT** (*i.e.* with plane EF) **(Ans.)**

Example 2.23. *A thin cylindrical tube 80 mm internal diameter and 5 mm thick, is closed at the ends and is subjected to an internal pressure of 6 MN / m². A torque of 2009.6 Nm is also applied to the tube. Find the hoop stress, longitudinal stress, maximum and minimum principal stresses and maximum shear stress.*

Solution. Internal Pressure, $p = 6$ MN/m²

Internal diameter of the tube, $d = 80$ mm or 0.08 m

Thickness of the tube, $\qquad t = 5$ mm or 0.005 m

Torque applied to the tube, $\quad T = 2009.6$ Nm

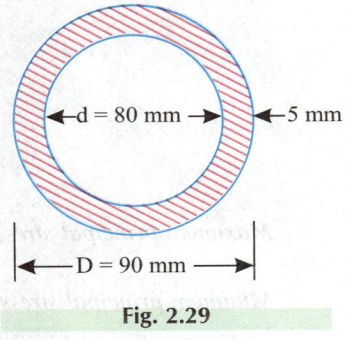

Hoop (or circumferential stress), σ_c:

$$\sigma_c = \frac{pd}{2t} = \frac{6 \times 10^6 \times 0.08}{2 \times 0.005}$$

$$= \textbf{48 MN/m}^2 \quad \textbf{(Ans.)}$$

Fig. 2.29

Longitudinal stress, σ_l:

$$\sigma_l = \frac{pd}{4t} = \frac{6 \times 10^6 \times 0.08}{4 \times 0.005}$$

$$= \textbf{24 MN/m}^2 \quad \textbf{(Ans.)}$$

Clutch gears.

Torsional shear stress, τ:

$$T = \tau \times \frac{\pi}{16}\left(\frac{D^4 - d^4}{D}\right)$$

$$2009.6 = \tau \times \frac{\pi}{16}\left[\frac{0.09^4 - 0.8^4}{0.09}\right]$$

$$2009.6 = \tau \times 5.37 \times 10^{-5}$$

$$\therefore \qquad \tau = \frac{2009.6}{5.377 \times 10^{-5}}$$

$$= 37.37 \text{ MN/m}^2 \quad \textbf{(Ans.)}$$

Principal stresses, σ_1, σ_2, and τ_{max}:

Now, $\sigma_x = 48 \text{ MN/m}^2$ (tensile), $\sigma_y = 24 \text{ MN/m}^2$ (tensile), and $\tau_{xy} = 37.37 \text{ MN/m}^2$ (shear). The principal stresses are given by:

$$\sigma = \frac{\sigma_x + \sigma_y}{2} \pm \sqrt{\left[\frac{\sigma_x - \sigma_y}{2}\right]^2 + \tau_{xy}^2}$$

$$= \frac{48 + 24}{2} \pm \sqrt{\left[\frac{48 - 24}{2}\right]^2 + (37.37)^2}$$

$$= 36 \pm 39.25$$

Maximum principal stress,

$$\sigma_1 = \textbf{75.25 MN/m}^2 \; (tensile) \quad \textbf{(Ans.)}$$

Minimum principal stress,

$$\sigma_2 = \textbf{36} - \textbf{39.25} = - \textbf{3.25 MN/m}^2$$
$$= \textbf{3.25 MN/m}^2 \; (comp.) \quad \textbf{(Ans.)}$$

Maximum shear stress,

$$\tau_{max} = \frac{\sigma_1 - \sigma_2}{2} = \frac{75.25 - (-3.25)}{2}$$

$$= \textbf{39.25 MN/m}^2 \quad \textbf{(Ans.)}$$

2.6.2. Ellipse of Stress

Refer to Fig. 2.30. To draw an ellipse of stress, following are the *steps* of construction:

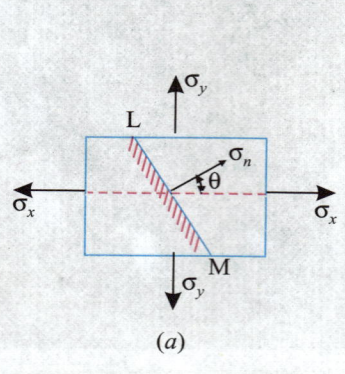

(a)

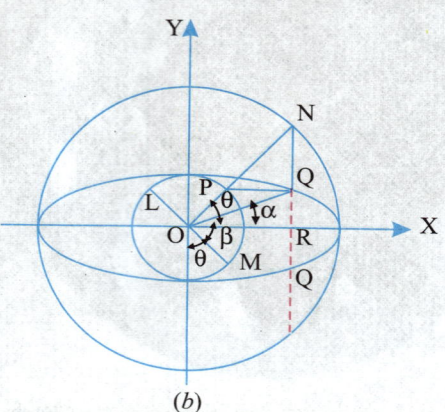

(b)

Fig. 2.30

1. Taking O as centre draw two circles with radii proportional to σ_x and σ_y respectively.
2. Through O draw ON perpendicular to the interface LM as shown, to cut the inner circle in P and outer circles in N.
3. Through N, draw a line NR perpendicular to OX and through P, draw line PQ perpendicular to OY to cut the line NR in Q. Join OQ.

From the diagram,

$$OR = \sigma_x \cos \theta$$
$$QR = \sigma_y \sin \theta$$
$$OQ = \sqrt{OR^2 + QR^2}$$
$$= \sqrt{\sigma_x^2 \cos^2 \theta + \sigma_y^2 \sin^2 \theta} = \sigma_r$$

Hence OQ gives the resultant σ_r for that plane,

and

$$\tan \alpha = \frac{\sigma_y \sin \theta}{\sigma_x \cos \theta} = \frac{\sigma_y}{\sigma_x} \tan \theta.$$

The point Q may be established for different values of θ. The locus of Q will give an *ellipse*.

Example 2.24. *At a point the principal stresses are 100 MN/m² and 60 MN/m² both tensile. Find by the ellipse of stress the resultant stress on a plane inclined at 35° to the major principal stress.*

Solution. The given plane is at 35° to the major principal stress or at 55° to the major principal plane. Refer to Fig. 2.31.

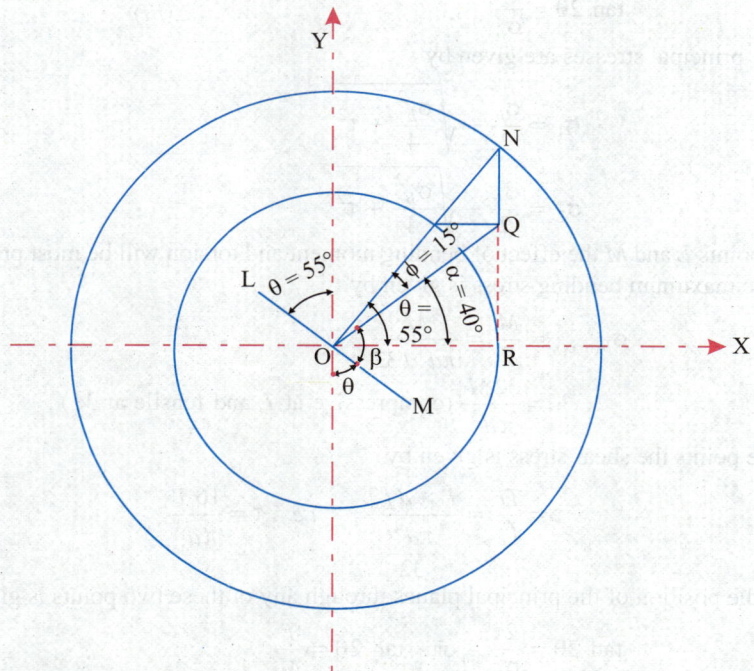

Fig. 2.31

— Take O as centre and draw two circles with radii proportional to 100 MN/m² and 60 MN/m² respectively.

— Through O draw ON perpendicular to the interface LM as shown, to cut the inner circle in P and outer circle in N.

— Through N draw line NR perpendicular to OX and through P, draw line PQ perpendicular to OY to cut the line NR in Q. Join OQ.

Then, resultant stress $\quad \sigma_r = OQ = 75.5 \text{ MN/m}^2$,

$$\alpha = \angle ROQ = 40°$$

and, obliquity $\quad\quad\quad \phi = \angle POQ = 15°. \quad \text{(Ans.)}$

2.7. COMBINED BENDING AND TORSION

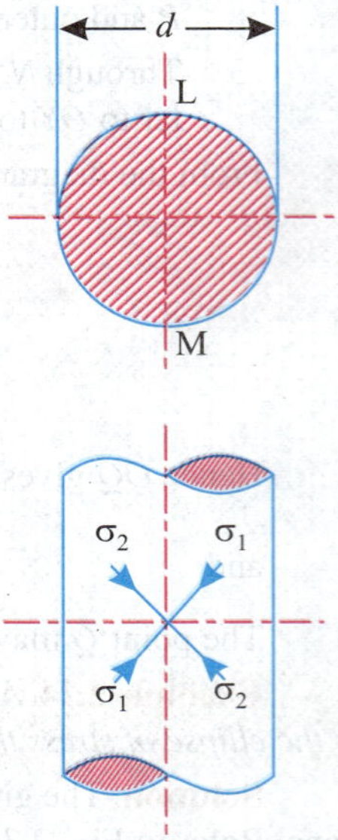

Consider a shaft of diameter d subject to a bending moment M and a twisting moment T at a section.

The *bending stress* (σ_b) at any point in the section at a radius r and at a distance y from the neutral axis is given by

$$\sigma_b = \frac{M}{I} y \quad\quad ...(i)$$

where, I = Moment of inertia of the section about the neutral axis.

The *shear stress* (τ) at the point is given by

$$\tau = \frac{T}{I_P} r \quad\quad ...(ii)$$

where, $\quad\quad I_p$ = Polar moment of inertia.

The location of the principal planes through the point is given by

$$\tan 2\theta = \frac{2\tau}{\sigma_b} \quad\quad ...(iii)$$

and, the principal stresses are given by

Fig. 2.32

$$\sigma_1 = \frac{\sigma_b}{2} + \sqrt{\frac{\sigma_b^2}{4} + \tau^2} \quad\quad ...(2.13)$$

and,

$$\sigma_2 = \frac{\sigma_b}{2} - \sqrt{\frac{\sigma_b^2}{4} + \tau^2} \quad\quad ...(2.14)$$

At the points L and M the effect of bending moment and torsion will be most predominant. At these points the maximum bending stress is given by

$$\sigma_{b(max)} = \frac{M}{Z} = \frac{M}{\pi d^3 / 32}$$

$$= \frac{32 M}{\pi d^3} \quad \text{(compressive at } L \text{ and tensile at } M\text{)} \quad\quad ...(iv)$$

At these points the shear stress is given by

$$\tau = \frac{Tr}{I_p} = \frac{T \times d/2}{\dfrac{\pi d^4}{32}} \quad i.e. \quad \tau = \frac{16 T}{\pi d^3} \quad\quad ...(v)$$

Hence the position of the principal planes through any of these two points is given by,

$$\tan 2\theta = \frac{2\tau}{\sigma_b} \quad \text{or} \quad \tan 2\theta = \frac{T}{M}$$

The principal stresses are given by

$$\sigma_1 = \frac{\sigma_b}{2} + \sqrt{\frac{\sigma_b^2}{4} + \tau^2} = \frac{16}{\pi d^3}\left\{M + \sqrt{M^2 + T^2}\right\} \quad\quad ...(2.15)$$

and,

$$\sigma_2 = \frac{\sigma_b}{2} - \sqrt{\frac{\sigma_b^2}{4} + \tau^2} = \frac{16}{\pi d^3}\left\{M - \sqrt{M^2 + T^2}\right\} \quad\quad ...(2.16)$$

The maximum shear stress is given by,

$$\tau_{max} = \frac{\sigma_1 - \sigma_2}{2} \qquad\qquad ...(2.17)$$

$$\tau_{max} = \frac{16}{\pi d^3} \sqrt{M^2 + T^2} \qquad\qquad ...(2.18)$$

Example 2.25. *A cylinder (500 mm internal diameter and 20 mm wall thickness) with closed ends is subjected simultaneously to an internal pressure of 0.60 MPa, bending moment 64000 Nm and torque 16000 Nm. Determine the maximum tensile stress and shearing stress in the wall.*

(AMIE Winter, 2002)

Solution. *Given:* $d = 500$ mm $= 0.5$ m; $t = 20$ mm $= 0.02$ m;
$p = 0.60$ MPa $= 0.6$ MN/m^2; $M = 64000$ Nm $= 0.064$ MNm;
$T = 16000$ Nm $= 0.016$ MNm.

Maximum tensile stress:

First let us determine the principle stresses σ_1 and σ_2 assuming this as a *thin cylinder.*

We know, $\sigma_1 = \dfrac{pd}{2t} = \dfrac{0.6 \times 0.5}{2 \times 0.02} = 7.5$ MN/m^2

and, $\sigma_2 = \dfrac{pd}{4t} = \dfrac{0.6 \times 0.5}{4 \times 0.02} = 3.75$ MN/m^2

Next consider effect of combined bending moment and torque on the walls of the cylinder.

Then the principal stresses σ'_1 and σ'_2 are given by,

$$\sigma'_1 = \frac{16}{\pi d^3}\left[M + \sqrt{M^2 + T^2} \right] \qquad \text{[Eqn. (2.15)]}$$

and, $$\sigma'_2 = \frac{16}{\pi d^3}\left[M - \sqrt{M^2 + T^2} \right] \qquad \text{[Eqn. (2.16)]}$$

Crane engine.

$$\therefore \qquad \sigma_1' = \frac{16}{\pi \times (0.5)^3} \left[0.064 + \sqrt{0.064^2 + 0.016^2} \right] = 5.29 \text{ MN/m}^2$$

and,

$$\sigma_2' = \frac{16}{\pi \times (0.5)^3} \left[0.064 - \sqrt{0.064^2 + 0.016^2} \right] = -0.08 \text{ MN/m}^2$$

Maximum tensile stress $\sigma_I = \sigma_1 + \sigma' = 7.5 + 5.29 = \textbf{12.79 MN/m}^2$ **(Ans.)**

Maximum shearing stress, τ_{max} :

We know, $\qquad \tau_{max} = \dfrac{\sigma_I - \sigma_{II}}{2}$

$$\sigma_{II} = \sigma_2 + \sigma_2' = 3.75 - 0.08 = 3.67 \text{ MN/m}^2 \text{ (tensile)}$$

$$\therefore \qquad \tau_{max} = \frac{12.79 - 3.67}{2} = \quad \textbf{4.56 MN/m}^2 \text{ (Ans.)}$$

Example 2.26. *A shaft section 100 mm in diameter is subjected to a bending moment of 4000 Nm and a torque of 6000 Nm. Find:*

(i) *The maximum direct stress induced on the section, and specify the position of the plane on which it acts.*

(ii) *What stress acting alone can produce the same maximum strain?*

Take Poisson's ratio = 0.3.

Solution. Diameter of the shaft section, $d = 100$ mm $= 0.1$ m

Bending moment, $\qquad M = 4000$ Nm

Torque, $\qquad T = 6000$ Nm

Poisson's ratio, $\qquad \dfrac{1}{m} = 0.3$

(i) $\sigma_1 = ?$ $\sigma_2 = ?$

The principal stresses are given by

$$\sigma_1 = \frac{16}{\pi d^3} (M + \sqrt{M^2 + T^2}) \qquad \text{...(Eqn. 2.15)}$$

and, $\qquad \sigma_2 = \dfrac{16}{\pi d^3} (M - \sqrt{M^2 + T^2}) \qquad \text{...(Eqn. 2.16)}$

$$\therefore \qquad \sigma_1 = \frac{16}{\pi \times (0.1)^3} (4000 + \sqrt{4000^2 + 6000^2})$$

$$= \frac{16}{\pi \times (0.1)^3} (4000 + 7211) = 0.57 \times 10^8 \text{ N/m}^2$$

$$= \textbf{57 MN/m}^2 \text{ (Ans.)}$$

$$\sigma_2 = \frac{16}{\pi (0.1)^3} (4000 - \sqrt{4000^2 + 6000^2})$$

$$= \frac{16}{\pi \times (0.1)^3} (4000 - 7211) = -0.1635 \times 10^8 \text{ N/m}^2$$

$$= \textbf{-16.35 MN/m}^2 \text{ (Ans.)}$$

Position of the major principal planes:

$$\tan 2\theta = \frac{T}{M} = \frac{6000}{4000} = 1.5$$

$$2\theta = \tan^{-1} 1.5 = 56.3°$$

$$\boldsymbol{\theta = 28.15°} \text{ (Ans.)}$$

Maximum strain:

$$e_1 = \frac{\sigma_1}{E} - \frac{\sigma_2}{mE} = \frac{1}{E}\left(\sigma_1 - \frac{\sigma_2}{m}\right) = \frac{10^6}{E}\left[57 - \{0.3 \times (-16.35)\}\right]$$

$$= \frac{10^6}{E}(57 + 0.3 \times 16.35) = \frac{61.9 \times 10^6}{E}$$

(ii) Stress which can produce the same maximum strain, σ:

Now, maximum strain $= \dfrac{61.9 \times 10^6}{E}$ ...(as above)

But, $\dfrac{61.9 \times 10^6}{E} = \dfrac{\sigma}{E}$

∴ $\sigma = \mathbf{61.9 \ MN/m^2}$ **(Ans.)**

2.8. ANALYSIS OF STRAIN

2.8.1. Direct Strain on an Oblique Plane Due to a Direct Pull on a Plane

Refer Fig. 2.33. Let a rectangular body *LMNS* be extended to *LTUS* under the action of direct stress (pull) σ.

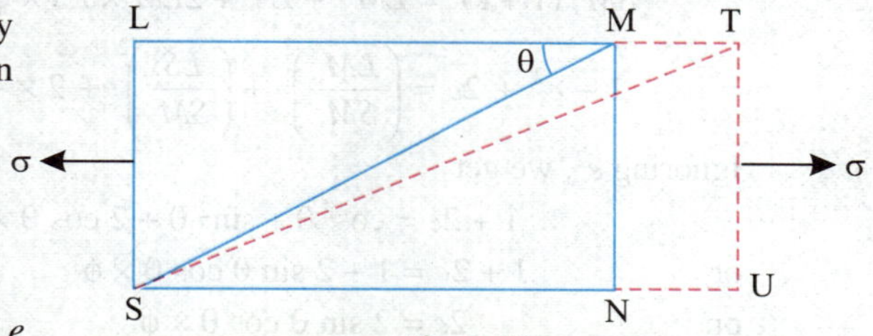

Extension in length $\quad LM = MT$

Longitudinal strain, $\quad e_x = \dfrac{MT}{LM}$

∴ $\quad MT = LM \times e_x$

$LT = LM + MT$

Fig. 2.33

$$= LM + LM \times e_x = LM(1 + e_x)$$

Let *e* be the strain in the diagonal *SM*

$$ST = SM + SM \times e = SM(1 + e)$$

$$ST^2 = LT^2 + LS^2$$

$$= [LM(1 + e_x)]^2 + LS^2$$

$$= LM^2(1 + 2e_x) + LS^2 \text{ (Ignoring } e_x^2 \text{ as } e_x \text{ is a very small fraction)}$$

$$= LM^2 + LS^2 + 2LM^2 \times e_x$$

$$= SM^2 + 2LM^2 \times e_x \qquad (\because LM^2 + LS^2 = SM^2)$$

But, $\quad ST = SM(1 + e)$

∴ $\quad ST^2 = SM^2(1 + 2e)$ (ignoring e^2)

∴ $\quad SM^2(1 + 2e) = SM^2 + 2LM^2 \times e_x$

or, $\quad 1 + 2e = 1 + 2\left(\dfrac{LM}{SM}\right)^2 \times e_x \quad \text{or} \quad e = \left(\dfrac{LM}{SM}\right)^2 \times e_x$

or, $\quad e = e_x \cos^2 \theta$ \hfill (2.19)

2.8.2. Direct Strain on an Oblique Plane Due to Shear Stress τ

Let the body LMNS be deformed to *L'M'NS* under the action of shear stress τ (Fig. 2.34)

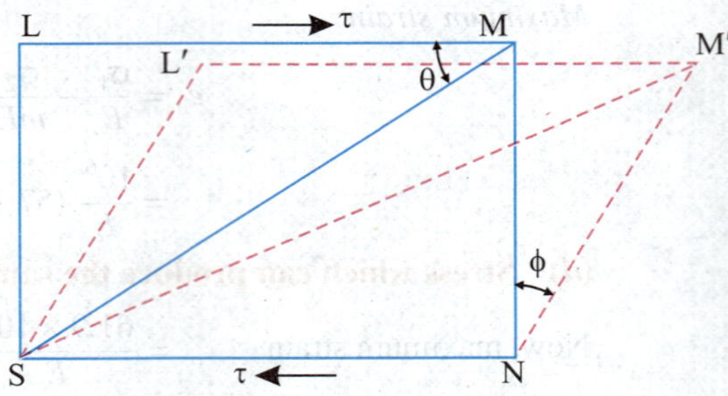

Shear Strain $= \dfrac{MM'}{MN} = \tan \phi = \phi$

or, $MM' = MN \times \phi$

Let, $e = $ Strain in SM

$SM' = SM + SM \times e$

$= SM\,(1 + e)$

$(SM')^2 = (LM')^2 + (LS)^2$

$= (LM + MM')^2 + LS^2$

$= (LM + MN \times \phi)^2 + LS^2$

$= (LM + LS \times \phi)^2 + LS^2$ $\qquad (\because MN = LS)$

$= LM^2 + LS^2 \times \phi^2 + 2\,LM \times LS \times \phi + LS^2$

Ignoring $LS^2 \times \phi^2$, we get

$(SM')^2 = LM^2 + LS^2 + 2LM \times LS \times \phi$

$SM^2\,(1 + e)^2 = LM^2 + LS^2 + 2LM \times LS \times \phi$

$1 + e^2 + 2e = \left(\dfrac{LM}{SM}\right)^2 + \left(\dfrac{LS}{SM}\right)^2 + 2 \times \dfrac{LM}{SM} \times \dfrac{LS}{SM} \times \phi$

Ignoring e^2, we get

$1 + 2e = \cos^2 \theta + \sin^2 \theta + 2 \cos \theta \times \sin \theta \times \phi$

or, $1 + 2e = 1 + 2 \sin \theta \cos \theta \times \phi$

or, $2e = 2 \sin \theta \cos \theta \times \phi$

$\therefore \qquad e = \sin \theta \cos \theta \times \phi$

or, $e = \phi \times \dfrac{\sin 2\theta}{2}$ $\qquad\qquad$ ...(2.20)

i.e. the linear strain on the plane SM inclined at an angle θ to the direction of the shear stress τ is equal to the shear strain due to action of τ multiplied by $\dfrac{\sin 2\theta}{2}$.

2.8.3. Direct Strain on an Oblique Plane Due to Two Normal Pulls and a Shear Force

Let the body $LMNS$ be deformed to $L_1 M_2 N_1 S$ under the action of the direct normal stresses σ_x and σ_y and the shear stress τ_{xy} (Fig. 2.35)

Fig. 2.34

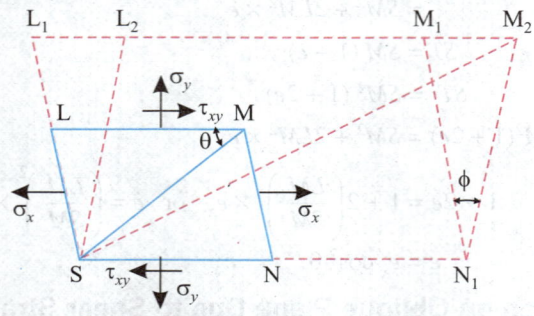

Fig. 2.35

Let, $e_x = $ Longitudinal strain in the direction of σ_x,

$e_y = $ Longitudinal strain in the direction of σ_y,

e = Longitudinal strain in the plane SM, and

ϕ = Shear strain in the block.

Now, $e_x = \dfrac{SN_1 - SN}{SN}$

$\therefore \quad SN_1 = SN\,(1 + e_x)$

and, $\quad e_y = \dfrac{SL_1 - SL}{SL}$

$\therefore \quad SL_1 = SL\,(1 + e_y)$ and, $e = \dfrac{SM_2 - SM}{SM}$

$\therefore \quad SM_2 = SM\,(1 + e)$

Also, $\quad \phi = \dfrac{M_1 M_2}{M_1 N_1}$

$\therefore \quad M_1 M_2 = M_1 N_1 \times \phi$

Fitting piston rings.

Now, $SM_2{}^2 = SL_1{}^2 + L_1 M_2{}^2 = SL_1{}^2 + (L_1 M_1 + M_1 M_2)^2$

$= SL_1{}^2 + (L_1 M_1 + M_1 N_1 \times \phi)^2$

$= SL_1{}^2 + (L_1 M_1 + SL_1 \times \phi)^2 \qquad (\because M_1 N_1 = SL_1)$

$= SL_1{}^2 + L_1 M_1{}^2 + 2 L_1 M_1 \times SL_1 \times \phi \qquad$ ignoring term having ϕ^2

$\therefore \quad SL_1{}^2 + L_1 M_1{}^2 + 2 L_1 M_1 \times SL_1 \times \phi = SM^2 (1 + e)^2$

i.e. $\qquad SM^2 (1 + 2e) = SL_1{}^2 + SN_1{}^2 + 2 \times SN_1 \times SL_1 \times \phi \quad (\because L_1 M_1 = SN_1)$

ignoring e^2

$SM^2 (1 + 2e) = SL^2 (1 + e_y)^2 + SN^2 (1 + e_x)^2$

$+ 2 \times SN (1 + e_x)\, SL (1 + e_y) \times \phi$

$SM^2 (1 + 2e) = SL^2 (1 + 2e_y) + SN^2 (1 + 2e_x) + 2 \times SN \times SL \times \phi$

[Ignoring squares of strain and products of the strains *i.e.* $e_x^2, e_y^2, e_x \times \phi, e_y \times \phi$]

$1 + 2e = \left(\dfrac{SN}{SM}\right)^2 (1 + 2e_x) + \left(\dfrac{SL}{SM}\right)^2 (1 + 2e_y)\; 2 \times \phi \times \dfrac{SN}{SM} \times \dfrac{SL}{SM}$

$= (1 + 2e_x)\cos^2 \theta + (1 + 2e_y) \sin^2 \theta + 2\phi \times \cos\theta \sin\theta$

$= \cos^2 \theta + 2 e_x \cos^2 \theta + \sin^2 \theta + 2 e_y \sin^2 \theta + \phi \times \sin 2\theta$

$= 1 + 2 e_x \cos^2 \theta + 2 e_y \sin^2 \theta + \phi \times \sin 2\theta$

$\therefore \qquad e = e_x \cos^2 \theta + e_y \sin^2 \theta + \phi \times \dfrac{\sin 2\theta}{2}$

Thus the direct strain e in a plane SM (inclined at an angle θ to the axis of σ_x) due to direct normal stresses σ_x and σ_y and the shear stress τ_{xy} is given by

$$e = e_x \cos^2 \theta + e_y \sin^2 \theta + \phi \times \dfrac{\sin 2\theta}{2} \qquad \qquad ...(2.21)$$

Value of θ for the maximum and minimum value of e:

To find the value of θ for the maximum and minimum value of e we put $\dfrac{de}{d\theta} = 0$

$$e = e_x \cos^2 \theta + e_y \sin^2 \theta + \phi \times \frac{\sin 2\theta}{2}$$

$$= e_x \left(\frac{1 + \cos 2\theta}{2} \right) + e_y \left(\frac{1 - \cos 2\theta}{2} \right) + \phi \times \frac{\sin 2\theta}{2}$$

$$= \frac{e_x + e_y}{2} + \frac{e_x \cos 2\theta - e_y \cos 2\theta}{2} + \phi \times \frac{\sin 2\theta}{2}$$

$$= \frac{e_x + e_y}{2} + \frac{(e_x - e_y) \cos 2\theta}{2} + \phi \times \frac{\sin 2\theta}{2} \qquad ...(i)$$

$$\frac{de}{d\theta} = -(e_x - e_y) \sin 2\theta + \phi \times \cos 2\theta = 0$$

$$\therefore \qquad \frac{\sin 2\theta}{\cos 2\theta} = \frac{\phi}{e_x - e_y} \quad \text{or} \quad \tan 2\theta = \frac{\phi}{e_x - e_y}$$

i.e. $\qquad 2\theta = \tan^{-1} \left(\frac{\phi}{e_x - e_y} \right) \quad \text{or} \quad 2\theta = \tan^{-1} \left(\frac{\phi}{e_x - e_y} \right) + 180°$

To find the maximum and minimum value of e:

$$\cos 2\theta = \frac{1}{\sec 2\theta} = \pm \frac{1}{\sqrt{1 + \tan^2 2\theta}}$$

$$= \pm \frac{1}{\sqrt{1 + \left(\dfrac{\phi}{e_x - e_y} \right)^2}} = \frac{1}{\sqrt{1 + \dfrac{\phi^2}{(e_x - e_y)^2}}} = \sqrt{\frac{(e_x - e_y)^2}{\phi^2 + (e_x - e_y)^2}}$$

Now substituting the values of $\cos 2\theta$ and $\tan 2\theta$ in eqn. (*i*), we get

$$e = \frac{e_x + e_y}{2} + \frac{e_x - e_y}{2} \cos 2\theta + \phi \times \frac{\sin 2\theta}{2}$$

$$= \frac{e_x + e_y}{2} + \frac{\cos 2\theta}{2} [(e_x - e_y) + \phi \tan 2\theta]$$

$$= \frac{e_x + e_y}{2} \pm \frac{1}{2} \sqrt{\frac{(e_x - e_y)^2}{\phi^2 - (e_x - e_y)^2}} \left[(e_x - e_y) + \frac{\phi^2}{e_x - e_y} \right]$$

$$= \frac{e_x + e_y}{2} \pm \frac{1}{2} \sqrt{\frac{(e_x - e_y)^2}{\phi^2 + (e_x - e_y)^2}} \left[\frac{(e_x - e_y)^2 + \phi^2}{e_x - e_y} \right] = \frac{e_x + e_y}{2} \pm \frac{1}{2} \sqrt{(e_x - e_y)^2 + \phi^2}$$

We get *two values* of wholly direct strains *i.e.* the principal strains

i.e. $\qquad e_{(max)} = \dfrac{e_x + e_y}{2} + \dfrac{1}{2} \sqrt{(e_x - e_y)^2 + \phi^2} \qquad ...(2.22)$

$$\theta = \frac{1}{2} \tan^{-1} \frac{\phi}{e_x - e_y} \qquad ...[2.22\ (a)]$$

and, $\qquad e_{(min)} = \dfrac{e_x + e_y}{2} - \dfrac{1}{2} \sqrt{(e_x - e_y)^2 + \phi^2} \qquad ...(2.23)$

$$\theta = \frac{1}{2} \left(\tan^{-1} \frac{\phi}{e_x - e_y} + 180° \right) = \frac{1}{2} \tan^{-1} \frac{\phi}{e_x - e_y} + 90° \qquad ...[2.23\ (a)]$$

i.e., these two planes of principal strains are at right angles to each other.

$$\tan 2\theta = \frac{\phi}{e_x - e_y} = \frac{\tau_{xy}/C}{\left(\dfrac{\sigma_x}{E} - \dfrac{\sigma_y}{mE}\right) - \left(\dfrac{\sigma_y}{E} - \dfrac{\sigma_x}{mE}\right)} = \frac{\tau_{xy}/C}{\dfrac{\sigma_x}{E} + \dfrac{\sigma_x}{mE} - \dfrac{\sigma_y}{E} - \dfrac{\sigma_y}{mE}}$$

$$= \frac{\tau_{xy}/C}{\dfrac{\sigma_x}{E}\left(1 + \dfrac{1}{m}\right) - \dfrac{\sigma_y}{E}\left(1 + \dfrac{1}{m}\right)} = \frac{\tau_{xy}/C}{\dfrac{1}{E}\left(1 + \dfrac{1}{m}\right)(\sigma_x - \sigma_y)}$$

$$= \frac{\tau_{xy}}{\sigma_x - \sigma_y}\frac{E}{C\left(\dfrac{m+1}{m}\right)}$$

We know, $E = 2C\left(\dfrac{m+1}{m}\right)$

∴ $\tan 2\theta = \dfrac{2\,\tau_{xy}}{\sigma_x - \sigma_y}$

This shows that maximum and minimum principal strains and stresses act in the same directions.

2.8.4. Strains on an Inclined Section Due to Two Perpendicular Normal Strains

Let e_1 and e_2 be the principal strains on rectangular block *LMNS*. Let *L'M'N'S* be its shape after it is strained (Fig. 2.36).

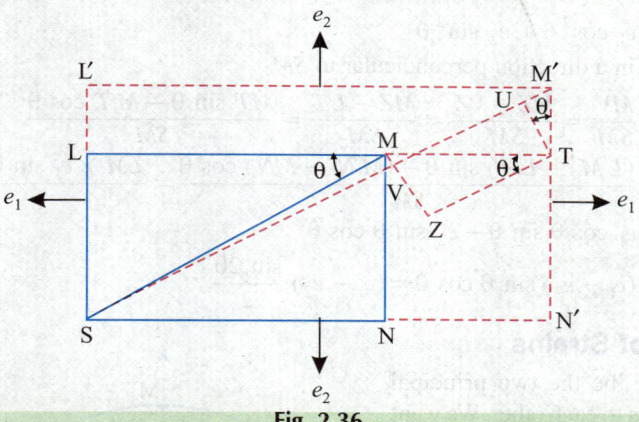

Fig. 2.36

$$e_1 = \frac{L'M' - LM}{LM}\ ;\ e_2 = \frac{M'N' - MN}{MN}$$

— From M draw MV perpendicular on *M'S*.

— Produce *LM* to meet *M'N'* in *T*.

— From *T* draw *TU* perpendicular on *M'S*.

— From *T* draw a line parallel to *SM'* to meet *MV* produced at *Z*.

Strain in *SM* (inclined at an angle θ to *LM*) in direction *SM*

$$= \frac{SM' - SM}{SM} = \frac{VM'}{SM} = \frac{VU + UM'}{SM}$$

$$= \frac{MT \cos \theta + M'T \sin \theta}{SM} = \frac{(L'M' - LM) \cos \theta + (M'N' - MN) \sin \theta}{SM}$$

$$= \frac{LM \times e_1 \cos \theta + MN \times e_2 \sin \theta}{SM}$$

Tyres of automobile need different considerations from the metallic wheels over which they fit.

$$= e_1 \cos^2 \theta + e_2 \sin^2 \theta \qquad \qquad ...(2.24)$$

Strain in SM in a direction perpendicular to SM

$$= \frac{MV}{SM} = \frac{MZ - VZ}{SM} = \frac{MZ - UT}{SM} = \frac{MT \sin \theta - M'T \cos \theta}{SM}$$

$$= \frac{(L'M' - LM) \sin \theta - (M'N' - MN) \cos \theta}{SM} = \frac{LM \times e_1 \sin \theta - MN \times e_2 \cos \theta}{SM}$$

$$= e_1 \cos \theta \sin \theta - e_2 \sin \theta \cos \theta$$

$$= (e_1 - e_2) \sin \theta \cos \theta = (e_1 - e_2) \frac{\sin 2\theta}{2} \qquad \qquad ...(2.25)$$

2.8.5. Ellipse of Strains

Let e_1 and e_2 be the two principal strains at right angles to each other. We want to find strains in the direction making an angle θ with the direction of e_1.

Refer to Fig. 2.37.

— Draw two concentric circles with centre O and radii equal to e_1 and e_2.

— Draw ON making an angle θ with the direction of e_1.

— From N draw NR perpendicular to OL.

— From P draw PQ perpendicular on NR.

— Join OQ.

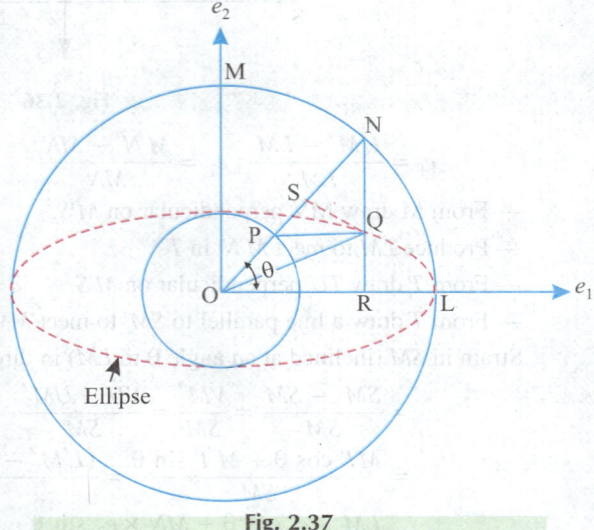

Fig. 2.37

Then OQ gives the resultant strain in a direction at an angle θ to the direction of e_1.

Resolve OQ into two components (OS and SQ), one along ON and the other at right angles to it. Then,

$$OS = e_1 \cos^2 \theta + e_2 \sin^2 \theta \text{, and}$$

$$SQ = \frac{e_1 - e_2}{2} \sin 2\theta$$

For values of θ varying from $0°$ to $360°$, the locus of the point Q will be an ellipse (shown dotted) with $2e_1$ as the major axis and $2e_2$ as the minor axis.

2.8.6. Mohr's Circle of Strains

From article 2.8.5 we have e, the strain along OS due to e_1 and e_2 principal strains is given by

$$e = e_1 \cos^2 \theta + e_2 \sin^2 \theta$$

$$= e_1 \left(\frac{\cos^2 \theta}{2} + \frac{\cos^2 \theta}{2} + \frac{\sin^2 \theta}{2} - \frac{\sin^2 \theta}{2} \right) + e_2 \left(\frac{\sin^2 \theta}{2} + \frac{\sin^2 \theta}{2} + \frac{\cos^2 \theta}{2} - \frac{\cos^2 \theta}{2} \right)$$

$$= \frac{e_1}{2} + \frac{e_1}{2} (\cos^2 \theta - \sin^2 \theta) + \frac{e_2}{2} + \frac{e_2}{2} (\sin^2 \theta - \cos^2 \theta) = \frac{e_1 + e_2}{2} + \frac{e_1 - e_2}{2} \cos 2\theta$$

The strain along SQ,

$$e' = \frac{e_1 - e_2}{2} \sin 2\theta$$

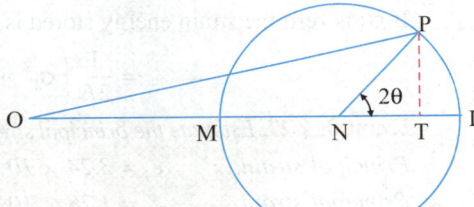

Refer to Fig. 2.38.

— Take $OL = e_1$ and $OM = e_2$

— Bisect LM at N

— With N as centre and NL as radius draw the circle.

— Take $\angle LNP = 2\theta$

Fig. 2.38

— From P drop PT perpendicular on OL.

$$OT = OM + MN + NT$$

$$= e_2 + \frac{e_1 - e_2}{2} + NP \cos 2\theta$$

$$= \frac{e_1 + e_2}{2} + \frac{e_1 - e_2}{2} \cos 2\theta = e$$

$$PT = NP \sin 2\theta$$

$$= \frac{e_1 - e_2}{2} \sin 2\theta = e'$$

— Join OP; then OP gives the *resultant strain*.

2.8.7. Principal Strains and Strain Energy due to Principal Stresses

2.8.7.1. Principal strains due to principal stresses

Steel pulleys.

If σ_1, σ_2 and σ_3 be the three principal stresses acting on three mutually perpendicular planes, $\frac{1}{m}$ be the Poisson's ratio, E the modulus of elasticity for the material of the body then the strains e_1,

e_2 and e_3 in the directions of respective stresses are:

$$e_1 = \frac{\sigma_1}{E} - \frac{\sigma_2}{mE} - \frac{\sigma_3}{mE} \qquad \qquad ...(i)$$

$$e_2 = \frac{\sigma_2}{E} - \frac{\sigma_3}{mE} - \frac{\sigma_1}{mE} \qquad \qquad ...(ii)$$

$$e_3 = \frac{\sigma_3}{E} - \frac{\sigma_1}{mE} - \frac{\sigma_2}{mE} \qquad \qquad ...(iii)$$

2.8.7.2. Strain energy due to principal stresses

Let e_1, e_2 and e_3 be the respective strains in the directions of the three principal stresses σ_1, σ_2 and σ_3. The strain energy stored is

$$= \frac{1}{2} \sigma_1 e_1 + \frac{1}{2} \sigma_2 e_2 + \frac{1}{2} \sigma_3 e_3$$

$$= \frac{1}{2} \sigma_1 \left(\frac{\sigma_1}{E} - \frac{\sigma_2}{mE} - \frac{\sigma_3}{mE} \right) + \frac{1}{2} \sigma_2 \left(\frac{\sigma_2}{E} - \frac{\sigma_3}{mE} - \frac{\sigma_1}{mE} \right)$$

$$+ \frac{1}{2} \sigma_3 \left(\frac{\sigma_3}{E} - \frac{\sigma_1}{mE} - \frac{\sigma_2}{mE} \right)$$

$$= \frac{1}{2E} \left(\sigma_1^2 + \sigma_2^2 + \sigma_3^2 \right) - \frac{1}{mE} \left(\sigma_1 \sigma_2 + \sigma_2 \sigma_3 + \sigma_3 \sigma_1 \right) \qquad ...(2.26)$$

If σ_3 is zero the strain energy stored is

$$= \frac{1}{2E} \left(\sigma_1^{\ 2} + \sigma_2^{\ 2} - \frac{2}{m} \sigma_1 \sigma_2 \right) \qquad \qquad ...(2.27)$$

Example 2.27. *Estimate the principal stresses acting on steel plate which gave the following results:*

Principal strain, $\qquad e_1 = 3.24 \times 10^{-4}$

Principal strain, $\qquad e_2 = 1.28 \times 10^{-4}$

Modulus of elasticity, $\quad E = 200 \ GN/m^2$

Poisson's ratio, $\qquad \dfrac{1}{m} = 0.25$

Solution. Let σ_1 and σ_2 be the principal stresses.

Now, $\qquad\qquad\qquad e_1 = \dfrac{\sigma_1}{E} - \dfrac{\sigma_2}{mE}$

∴ $\qquad 3.24 \times 10^{-4} = \dfrac{\sigma_1}{200 \times 10^9} - 0.25 \times \dfrac{\sigma_2}{200 \times 10^9}$

or, $\qquad \sigma_1 - 0.25 \ \sigma_2 = 3.24 \times 10^{-4} \times 200 \times 10^9 = 64.8 \times 10^6 \qquad \qquad ...(i)$

Similarly, $\qquad\qquad e_2 = \dfrac{\sigma_2}{E} - \dfrac{\sigma_1}{mE}$

$\qquad 1.28 \times 10^{-4} = \dfrac{\sigma_2}{200 \times 10^9} - \dfrac{\sigma_1}{200 \times 10^9} \times 0.25$

∴ $\qquad \sigma_2 - 0.25 \ \sigma_1 = 1.28 \times 10^{-4} \times 200 \times 10^9 = 25.6 \times 10^6 \qquad \qquad ...(ii)$

Solving (*i*) and (*ii*), we get

$$\sigma_1 = 75.95 \times 10^6 \ N/m^2 \ \text{ or } \ \textbf{75.95 MN/m}^2 \ \ \textbf{(Ans.)}$$

and, $\qquad\qquad\qquad \boldsymbol{\sigma_2 = 44.59 \ MN/m^2} \ \ \textbf{(Ans.)}$

Example 2.28. *A flat bar 80 mm × 10 mm is subjected to an axial pull of 160 kN. One side of the bar is polished and lines are ruled on it to form a square of 50 mm side; one diagonal of the square being along the middle line of the polished side. If E = 205 GN/m² and Poisson's ratio is 0.25, calculate the alteration in the angles and sides of the square.*

Solution. Cross-sectional area of the bar,

$$A = 80 \times 10 = 800 \text{ mm}^2 \text{ or } 800 \times 10^{-6} \text{ m}^2$$

∴ Longitudinal stress,

$$\sigma = \frac{160 \times 10^3}{800 \times 10^{-6}} = 200 \text{ MN/m}^2$$

e_x strain in the diagonal along the axis of the bar

$$= \frac{\sigma}{E} = \frac{200 \times 10^6}{205 \times 10^9} = 0.000975 \text{ (tensile)}$$

e_y strain in the other diagonal

$$= e_x \times \frac{1}{m} = 0.000975 \times 0.25 = 0.000244 \text{ (compressive)}$$

Strain on each side of the square

$$= e_x \cos^2 45° - e_y \sin^2 45° = 0.000975 \times \frac{1}{2} - 0.000244 \times \frac{1}{2} = 0.0003655$$

∴ *Change in length of the sides*

$$= 0.0003655 \times 50 = \textbf{0.018275 mm} \text{ (Ans.)}$$

Change in the angle of the square $= \phi$

$$= \text{shear strain} = e_x - e_y$$

$$= 0.000975 - (- 0.000244) = \textbf{0.001219 radian (Ans.)}$$

The angles in line with the axis of the bar will be decreased and the other two angles increased by this amount.

Example 2.29. *A material is subjected to two mutually perpendicular linear strains together with a shear strain. One of the linear strains is 0.00025 tensile. Determine the magnitudes of the other linear strain and the shear strain if the principal strains are 0.0001 compressive and 0.0003 tensile.* **(AMIE Summer, 2001)**

Solution. *Given:* $e_x = 0.00025$ (tensile); $e_1 = + 0.0003$ (tensile); $e_2 = - 0.0001$ (compressive)
e_y; e_x **(shear strain):**

Graphical method:

Refer to Fig. 2.39.

— Plot OL ($= e_x$) = 0.00025 (tensile)

— Plot OV ($= e_1$) = 0.0003 (tensile)

— Plot OU ($= e_2$) = 0.0001 (compressive)

— Bisect UV and N.

— With N as centre and NV as radius draw a circle.

— Draw LT perpendicular to OV meeting the circle at T.

— Join TN and produce it to meet at point S in the circle.

Draw SM perpendicular to UV. From Mohr's circle, we have

$$e_y = OM = \textbf{0.005 (compressive)}$$
(Ans.)

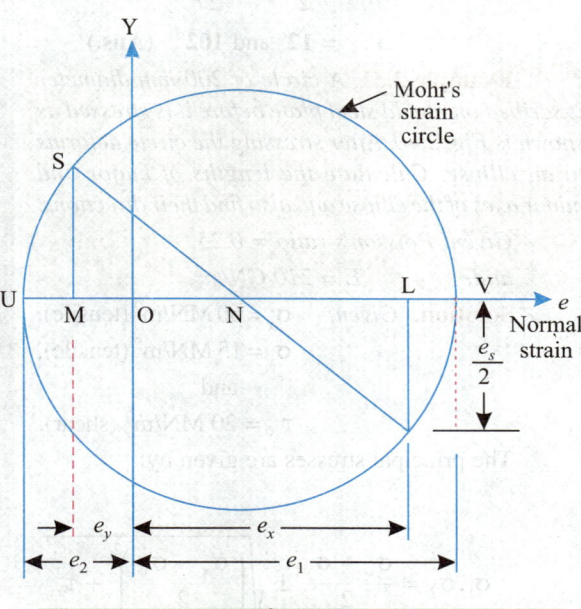

Fig. 2.39

$$\frac{e_s}{2} = LT = 0.000132$$

or, $e_s = \mathbf{0.000264}$ **(Ans.)**

Example 2.30. *A block of material is subjected to a tensile strain of 12×10^{-6} and a compressive strain of 15×10^{-6} on planes at right angles to each other. There is also a shear strain of 12×10^{-6} and there is no strain on planes at right angles to the above planes. Calculate the principal strian in magnitude and direction.*

Solution. $e_x = 12 \times 10^{-6}$;

$e_y = -15 \times 10^{-6}$; $\phi = 12 \times 10^{-6}$.

Principal strain,

$$e = \frac{1}{2}(e_x + e_y) \pm \frac{1}{2}\sqrt{(e_x - e_y)^2 + \phi^2}$$

∴

$$e = \frac{12 \times 10^{-6} - 15 \times 10^{-6}}{2} \pm \frac{1}{2}\sqrt{[12 \times 10^{-6} - (-15 \times 10^{-6})]^2 + (12 \times 10^{-6})^2}$$

$$= -1.5 \times 10^{-6} \pm \frac{1}{2} \times 10^{-6}\sqrt{729 + 144}$$

$$= -1.5 \times 10^{-6} \pm \frac{1}{2} \times 10^{-6} \times 29.55$$

$$= -16.275 \times 10^{-6} \quad \text{or} \quad +13.275 \times 10^{-6}$$

∴ *Principal strains are:*

13.275×10^{-6} *(tensile)* **(Ans.)**

16.275×10^{-6} *(compressive)* **(Ans.)**

$$\theta = \frac{1}{2}\tan^{-1}\frac{\phi}{e_x - e_y} = \frac{1}{2}\tan^{-1}\frac{12 \times 10^{-6}}{12 \times 10^{-6} - (-15 \times 10^{-6})}$$

$$= \frac{1}{2}\tan^{-1}\frac{12}{27}$$

$$= \mathbf{12°} \text{ and } \mathbf{102°} \quad \textbf{(Ans.)}$$

Example 2.31. *A circle of 200 mm diameter is scribed on a mild steel plate before it is stressed as shown is Fig. 2.40. After stressing the circle deforms to an ellipse. Calculate the lengths of major and minor axes of the ellipse and also find their directions.*

Given: Poisson's ratio = 0.25,

and, $E = 210 \text{ GN/m}^2$.

Solution. *Given:* $\sigma_x = 60 \text{ MN/m}^2$ (tensile),

$\sigma_y = 15 \text{ MN/m}^2$ (tensile),

and

$\tau_{xy} = 20 \text{ MN/m}^2$ (shear).

The principal stresses are given by:

Pulleys.

$$\sigma_1, \sigma_2 = \frac{\sigma_x + \sigma_y}{2} \pm \sqrt{\left(\frac{\sigma_x - \sigma_y}{2}\right)^2 + \tau_{xy}^2}$$

$$= \frac{60 + 15}{2} \pm \sqrt{\left[\frac{60 - 15}{2}\right]^2 + 20^2}$$

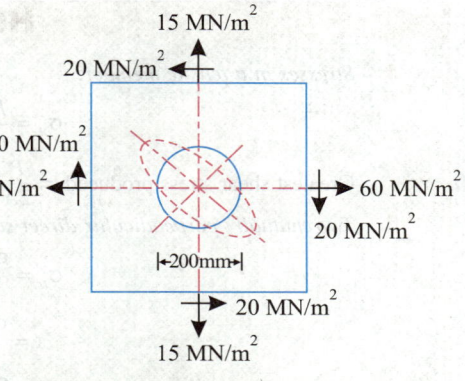

$$= 37.5 \pm \sqrt{506.25 + 400} = 37.5 \pm 30.1$$

∴ *Maximum principal stress,*

$\sigma_1 = 67.6$ MN/m^2 (tensile),

and, *minimum principal stress,*

$\sigma_2 = 7.4$ MN/m^2 (tensile).

Directions of principal stresses are calculated as follows:

$$\tan 2\theta = \frac{2\tau_{xy}}{\sigma_x - \sigma_y} = \frac{2 \times 20}{60 - 15} = \frac{40}{45} = 0.888$$

or, $2\theta = 41°36'$ or $221°36'$

∴ $\theta_1 = 20° 48'$ and $\theta_2 = 110° 48'$ **(Ans.)**

Fig. 2.40

The major principal strain

$$e_1 = \frac{\sigma_1}{E} - \frac{\sigma_2}{mE} = \frac{1}{E}\left(\sigma_1 - \frac{\sigma_2}{m}\right)$$

or, $e_1 = (67.6 - 7.4 \times 0.25) \dfrac{10^6}{210 \times 10^9} = 0.000313$ (increase)

∴ Increase in diameter $= e_1 \times d = 0.000313 \times 0.2 = 0.0000626$ m

The minor principal strain

$$e_2 = \frac{\sigma_2}{E} - \frac{\sigma_1}{mE} = \frac{1}{E}\left(\sigma_2 - \frac{\sigma_1}{m}\right)$$

or, $e_2 = (7.4 - 67.6 \times 0.25) = \dfrac{10^6}{210 \times 10^9}$

$$= -0.00004524 \text{ (decrease)}$$

∴ Decrease in diameter

$$= e_2 \times d = 0.00004524 \times 0.2 = 0.000009048 \text{ m}$$

Hence, *magnitude of diameter along the direction* σ_1

$$= d + e_1 \times d = 0.2 + 0.0000626$$

$$= \mathbf{0.2000626 \ m \ (Ans.)}$$

and, *magnitude of diameter along the direction of* σ_2

$$= d - e_2 \times d = 0.2 - 0.000009048$$

$$= \mathbf{0.199999952 \ m \ (Ans.)}$$

Hence circle will be converted into an ellipse having major axis = 0.2000626 m at 20°48′ and the minor axis = 0.199999952 m at 110°48' with the vertical centre line (measured in the anticlockwise direction).

HIGHLIGHTS

1. *Stresses in a tensile member:*

$$\sigma_n = \frac{P}{A} \cos^2 \theta, \, \tau = \frac{P}{A} \sin \theta \cos \theta$$

Greatest shear stress produced $= \dfrac{P}{2A}$

2. *Two mutually perpendicular direct stresses:*

$$\sigma_n = \frac{\sigma_x + \sigma_y}{2} + \frac{\sigma_x - \sigma_y}{2} \cos 2\theta \qquad \ldots(i)$$

$$\tau = \frac{\sigma_x - \sigma_y}{2} \sin 2\theta \qquad \ldots(ii)$$

$$\tau_{max} = \frac{\sigma_x - \sigma_y}{2} \qquad \ldots(iii)$$

$$\sigma_r = \sqrt{\sigma_n^2 + \tau^2} \qquad \ldots(iv)$$

$$\tan \phi = \frac{\tau}{\sigma_n} \qquad \ldots(v)$$

3. The planes which pass through the points in such a manner that the resultant stress across them is totally a normal stress are known as *"Principal planes"* and normal stresses across these planes are termed as *"Principal stresses"*. The plane carrying the maximum normal stress is called the *'major principal plane'* and the stress is called the *'major principal stress'*. The plane carrying the minimum normal stress is known as *'minor principal plane'* and the stress is known as *'minor principal stress'*.

4. *Two dimensional stress (General):*

$$\sigma_n = \frac{\sigma_x + \sigma_y}{2} + \frac{\sigma_x - \sigma_y}{2} \cos 2\theta + \tau_{xy} \sin 2\theta \qquad \ldots(i)$$

$$\tau = \frac{\sigma_x - \sigma_y}{2} \sin 2\theta - \tau_{xy} \cos 2\theta \qquad \ldots(ii)$$

$$\sigma_r = \sqrt{\sigma_n^2 + \tau^2} \qquad \ldots(iii)$$

$$\tan \phi = \frac{\tau}{\sigma_n} \qquad \ldots(iv)$$

Principal stresses, maximum shear stress, angles of inclination:

$$\sigma = \frac{\sigma_x + \sigma_y}{2} \pm \sqrt{\left[\frac{\sigma_x - \sigma_y}{2}\right]^2 + \tau_{xy}^2}$$

Major principal stress, $\quad \sigma_1 = \dfrac{\sigma_x + \sigma_y}{2} + \sqrt{\left[\dfrac{\sigma_x - \sigma_y}{2}\right]^2 + \tau_{xy}^2} \qquad \ldots(i)$

Minor principal stress, $\quad \sigma_2 = \dfrac{\sigma_x + \sigma_y}{2} - \sqrt{\left[\dfrac{\sigma_x - \sigma_y}{2}\right]^2 + \tau_{xy}^2} \qquad \ldots(ii)$

Maximum shear stress, $\quad \tau_{max} = \dfrac{\sigma_1 - \sigma_2}{2} \qquad \ldots(iii)$

$$\tan 2\theta = \frac{2\tau_{xy}}{\sigma_x - \sigma_y} \qquad \ldots(iv)$$

5. *Combined bending and torsion:*

$$\sigma_1 = \frac{\sigma_b}{2} + \sqrt{\frac{\sigma_b^2}{4} + \tau^2} = \frac{16}{\pi d^3}\left(M + \sqrt{M^2 + T^2}\right)$$

$$\sigma_2 = \frac{\sigma_b}{2} - \sqrt{\frac{\sigma_b^2}{4} + \tau^2} = \frac{16}{\pi d^3}(M - \sqrt{M^2 + T^2})$$

$$\tau_{max} = \frac{\sigma_1 - \sigma_2}{2} = \frac{16}{\pi d^3}\sqrt{M^2 + T^2}$$

6. If e_x and e_y be the strains in mutually perpendicular direction then strain in any direction is

$$e = e_x \cos^2\theta + e_y \sin^2\theta$$

7. If ϕ be the shear strain then direct strain in any direction is

$$e = \phi \times \frac{\sin 2\theta}{2}$$

8. Principal strains due to direct and shear strains are

$$e = \frac{e_x + e_y}{2} \pm \sqrt{(e_x - e_y)^2 + \phi^2}$$

Principal strains e_1, e_2 and e_3 are:

$$e_1 = \frac{\sigma_1}{E} - \frac{\sigma_2}{mE} - \frac{\sigma_3}{mE}$$

$$e_2 = \frac{\sigma_2}{E} - \frac{\sigma_3}{mE} - \frac{\sigma_1}{mE}$$

$$e_3 = \frac{\sigma_3}{E} - \frac{\sigma_1}{mE} - \frac{\sigma_2}{mE}$$

Strain energy due to principal stresses:

$$= \frac{1}{2E}(\sigma_1^2 + \sigma_2^2 + \sigma_3^2) - \frac{1}{mE}(\sigma_1\sigma_2 + \sigma_2\sigma_3 + \sigma_3\sigma_1)$$

UNSOLVED EXAMPLES

1. If at a point within the material, the minimum and maximum principal stresses are 30 MN/m² and 90 MN/m² respectively both tensile, determine the normal and shearing stress on a plane passing through the point and making an angle of tan⁻¹ 0.25 with the plane on which the maximum principal stress acts.   [**Ans.** σ_n = 86.4 MN/m², τ = 13.5 MN/m²]

2. The principal tensile stresses at a point across two perpendicular planes are 80 MN/m² and 40 MN/m². Find the normal, tangential, and resultant stresses and its obliquity on a plane at 20° with the major principal plane.

    [**Ans.** σ_n = 75 MN/m² (tensile), τ = 12.86 MN/m², σ_r = 76.4 MN/m², ϕ = 9°40′]

3. Determine the normal, shear and resultant stress in magnitude and direction in a plane, the normal of which makes an angle of 30° with the direction of 30 MN/m² stress (tensile). The value of other tensile stress is 15 MN/m².

    [**Ans.** σ_n = 26.25 MN/m² (tensile), τ = 6.5 MN/m² (shear), σ_r = 27.03 MN/m²]

4. A piece of steel plate is subjected to perpendicular stresses 6 MN / m² (tensile) and 4 MN/m² (compressive). Calculate the normal and tangential stresses and magnitude and direction of the resultant stress on the interface whose normal makes an angle of 30° with the axis of the second stress.

    [**Ans.** 0.5 MN/m², 4.33 MN/m², 4.36 MN/m² 83.4°]

5. At a point in a strained material, the principal stresses are 100 MN/m² (tensile) and 40 MN/m² (compressive). Determine the resultant stress in magnitude and direction on a plane inclined at 60° to the axis of major principal stress. What is the resultant stress and maximum intensity of shear stress in the material at the point?

    [**Ans.** σ_n = 5 MN/m² (compressive), τ = 60.62 MN/m², σ_r = 60.8 MN/m², τ_{max} = 70 MN/m²]

6. At a certain point in a strained material the principal stresses are 100 MN/m^2 and 40 MN/m^2, both tensile. Find the normal, tangential and resultant stresses across a plane through the point at 48° to the major principal plane, using Mohr's circle of stress.

[**Ans.** $\sigma_n = 66.9$ MN/m^2, $\tau = 29.8$ MN/m^2, $\sigma_r = 73.2$ MN/m^2, $\phi = 24°$]

7. When a certain thin walled tube is subjected to internal pressure and torque the stresses in the tube wall are (*i*) 60 MN/m^2 (tensile), (*ii*) 30 MN/m^2 (tensile) in a direction at right angles to (*i*), (*iii*) complementary shear stress of 45 MN/m^2 in the directions of (*i*) and (*ii*). Calculate the normal and tangential stresses on the two planes which are equally inclined to (*i*) and (*ii*).

[**Ans.** $\sigma_n = 90$ MN/m^2, $\tau = 15$ MN/m^2 (when $\theta = 45°$); $\sigma_n = $ zero, $\tau = -15$ MN/m^2 (when $\theta = 135°$)]

8. Mutually perpendicular faces of a square element of a thin plate are subjected to normal and shear stresses of 63 MN/m^2 (tensile), 47.2 MN/m^2 (compressive) and 39.4 MN/m^2 (shear). Determine graphically or otherwise the magnitudes and directions of the principal stresses and the greatest shearing stress.

[**Ans.** $\sigma_1 = 75.6$ MN/m^2 (tensile), $\sigma_2 = 59.8$ MN/m^2 (compressive), $\theta_1 = 17° 46'$, $\theta_2 = 107° 46'$, $\tau_{max} = 7.9$ MN/m^2]

9. At a point in an elastic material under strain, there are normal stresses of 50 MN/m^2 and 30 MN/m^2 respectively at right angles to each other with a shearing stress of 25 MN/m^2. Find the principal stresses and position of principal planes if (*i*) 50 MN/m^2 is tensile and 30 MN/m^2 is also tensile (*ii*) 50 MN/m^2 is tensile and 30 MN/m^2 is compressive. Find also the maximum shear stress and its plane in both the cases.

[**Ans.** (*i*) $\sigma_1 = 66.9$ MN/m^2 (tensile), $\sigma_2 = 13$ MN/m^2 (tensile), $\theta = 34° 6'$ and $\theta = 124° 6'$, $\tau_{max} = 26.9$ MN/m^2 at 76° 6' (*ii*) $\sigma_1 = 57.1$ MN/m^2, $\sigma_2 = 37$ MN/m^2 (comp.), $\theta_1 = 16°$, $\theta_2 = 106°$, $\tau_{max} = 47.1$ MN/m^2 at 61°.]

10. Fig. 2.41 shows the state of stress of a point in a two dimensional stressed body. Determine the magnitudes and directions of the principal stresses.

[**Ans.** 163.25 MN/m^2 (comp.), 36.75 MN/m^2 (comp.), $\theta_1 = 54°30'$ and $\theta_2 = 144° 30'$]

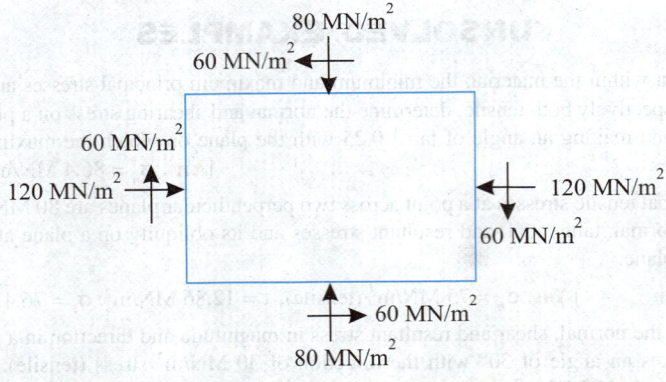

Fig. 2.41

11. A rectangular block of material is subjected to a tensile stress of 100 MN/m^2 on one plane and a tensile stress of 50 MN/m^2 on a plane at right angles, together with the shear stresses of 60 MN/m^2 on the same planes. Find:

   (*i*) The magnitude of the principal stresses.   (*ii*) The directions of the principal planes.

   (*iii*) The magnitude of the greatest shear stress.

   [**Ans.** (*i*) $\sigma_1 = 140$ MN/m^2 (tensile), $\sigma_2 = 10$ MN/m^2 (tensile), $\theta_1 = 33° 41'$, $\theta_2 = 123° 41'$, $\tau_{max} = 65$ MN/m^2 will occur on planes at 33° 41' + 45° = 78°41' and 78° 41' + 90° = 168°41' with the plane carrying the normal stress of 10 MN/m^2]

12. A circle of 100 mm diameter is drawn on a mild steel plate before it is stressed as shown in Fig. 2.42. Find the lengths of the major and minor axes of the ellipse formed as a result of the deformation of the circle marked.

$E = 2 \times 10^5$ MN/m^2

and $\dfrac{1}{m} = \dfrac{1}{4}$

[**Ans.** Major axis = 100.05 mm, minor axis = 99.9875 mm]

13. Refer to Fig. 2.43. Determine the magnitude of principal stresses and maximum shearing stress and direction of planes carrying these stresses.

[**Ans.** 184.36 MN/m^2, 17°33',
− 128.48 MN/m^2, 107°33',
156.42 MN/m^2, 62°33']

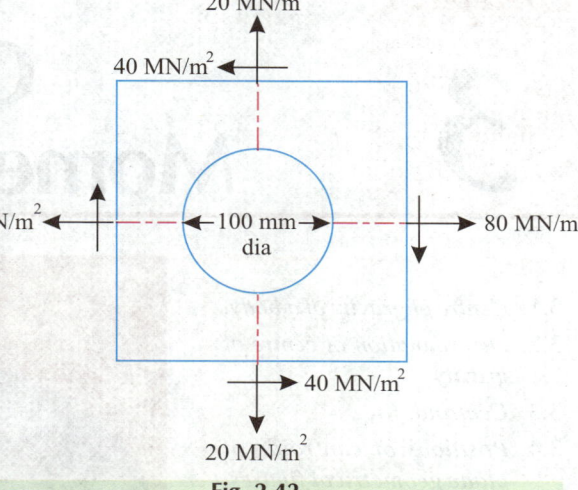

Fig. 2.42

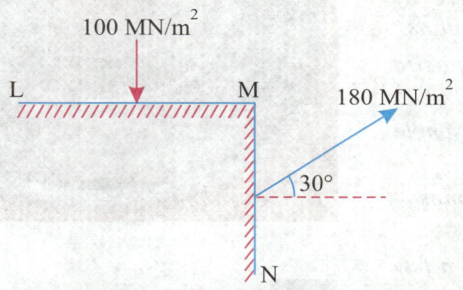

Fig. 2.43

Riding lawnmower.

Chapter 3

Centroid and Moment of Inertia

3.1. CENTRE OF GRAVITY OF A BODY

A body comprises of several parts and its every part possesses weight. *Weight is the force of attraction between a body and the earth and is proportional to mass of the body.* The weights of all parts of a body can be considered as parallel forces directed towards the centre of the earth. Therefore, they may be combined into a resultant force whose magnitude is equal to their algebraic sum. If a supporting force, equal and opposite to the resultant, is applied to the body along the line of action of the resultant, the body will be in equilibrium. This line of action will pass through the centre of gravity of the body. Thus, **centre of gravity of the body** may be defined *as the point through which the whole weight of a body may be assumed to act.* The centre of gravity of a body or an object is usually denoted by c.g. or simply by *G*. The position of c.g. depends upon shape of the body and this may or may not necessarily be within the boundary of the body.

3.2. DETERMINATION OF CENTRE OF GRAVITY

The centre of gravity of some objects may be found by balancing the object on a point. Take a thin plate of thickness *t*, shown in Fig. 3.1. Draw the diagonals of the upper and lower faces to intersect at *J* and *K* respectively. If the plate is placed on point at *K*, the plate will not fall. That is, it is balanced. If suspended from *J*, the plate will hang horizontally. The centre of gravity of the plate is at the centre of the line *JK*.

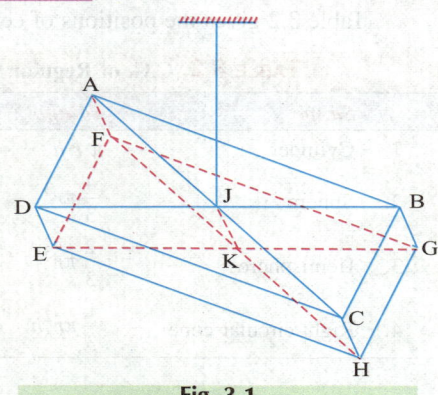

Fig. 3.1

Again, if we suspend a uniform rod by a string (Fig. 3.2.) and move the position of the string until the rod hangs vertically, we can determine that the centre of gravity of the rod lies at its centre. Through the use of similar procedures it can be established that a body which has an axis, or line, of symmetry has its centre of gravity located on that line, or axis. Of course, if a body has more than one axis of symmetry, the centre of gravity must lie at the intersection of the axes.

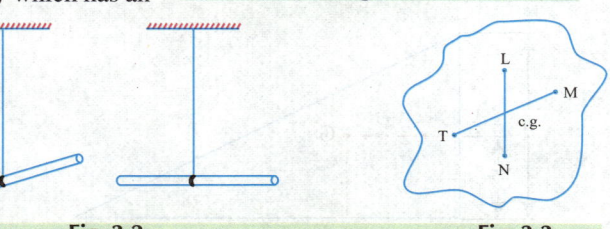

Fig. 3.2   **Fig. 3.3**

Another method for determining the centre of gravity is by suspension. Take an object, the section of which is shown in Fig. 3.3. Suspend it from point *L*. The body will not come to rest until its resultant weight is vertically downward from *L*. Through *L*, draw a vertical line *LN*. Then suspend the body from a point *M*, and let it come to rest. Through *M* draw a vertical line *MT*. The point of intersection *LN* and *MT* is the position of the centre of gravity.

3.3. CENTROID

The **centroid or centre of area** is defined *as the point where the whole area of the figure is assumed to be concentrated.* Thus centroid can be taken as quite analogous to centre of gravity when bodies have area only and not weight.

3.4. POSITIONS OF CENTROIDS OF PLANE GEOMETRICAL FIGURES

Table 3.1 gives the positions of centroids of some plane geometrical figures.

	Shape	Area	$\bar{x}$	$\bar{y}$	
		TABLE 3.1. Centroids of Plane Geometrical Figures			
1.	Rectangle	bh	$b/2$	$h/2$	Fig. 3.4
2.	Triangle	$\dfrac{bh}{2}$	$b/3$	$h/3$	Fig. 3.5
3.	Circle	$\dfrac{\pi}{4}d^2$	$\dfrac{d}{2}$	$\dfrac{d}{2}$	Fig. 3.6
4.	Semicircle	$\dfrac{\pi}{8}d^2$	$\dfrac{d}{2}$	$\dfrac{4r}{3\pi}(=0.424\,r)$	Fig. 3.7
5.	Quadrant	$\dfrac{\pi}{16}d^2$	$\dfrac{4r}{3\pi}(=0.424\,r)$	$\dfrac{4r}{3\pi}(=0.424\,r)$	Fig. 3.8
6.	Trapezium	$(a+b)\dfrac{h}{2}$	$\dfrac{a^2+b^2+ab}{3\,(a+b)}$	$\dfrac{(2a+b)}{(a+b)}\times\dfrac{h}{3}$	Fig. 3.9

3.5. POSITIONS OF CENTRE OF GRAVITY OF REGULAR SOLIDS

Table 3.2 gives the positions of centre of gravity of regular solids.

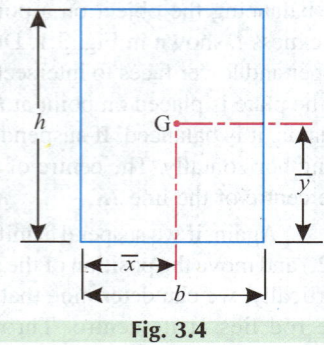

Fig. 3.4

Shape	Volume	c.g.
	TABLE 3.2. C.G. of Regular Solids	
1. Cylinder	$\pi r^2 h$	Fig. 3.10
2. Sphere	$\dfrac{4}{3}\pi r^3$	Fig. 3.11
3. Hemisphere	$\dfrac{2}{3}\pi r^3$	Fig. 3.12
4. Right circular cone	$\dfrac{1}{3}\pi r^2 h$	Fig. 3.13

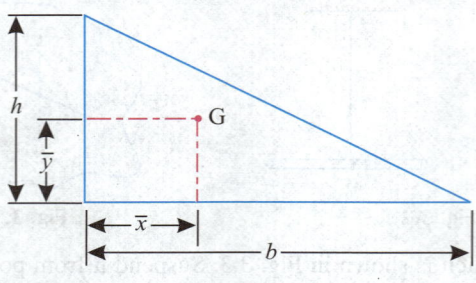

Fig. 3.5

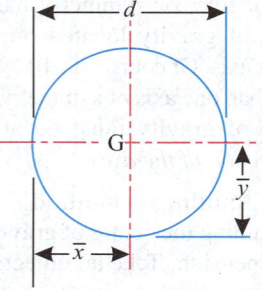

Fig. 3.6

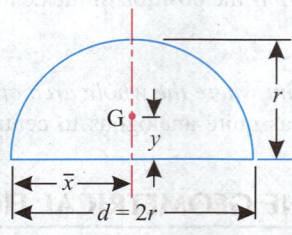

Fig. 3.7

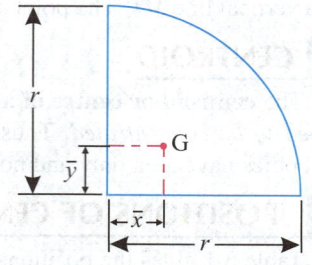

Fig. 3.8

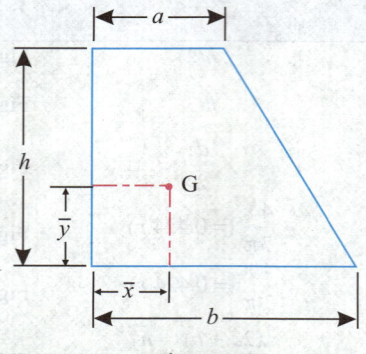

Fig. 3.9

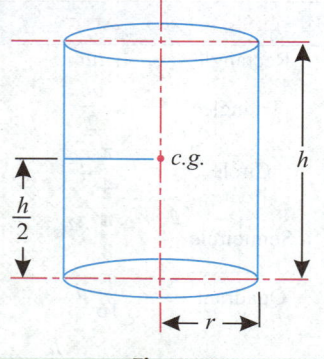

Fig. 3.10

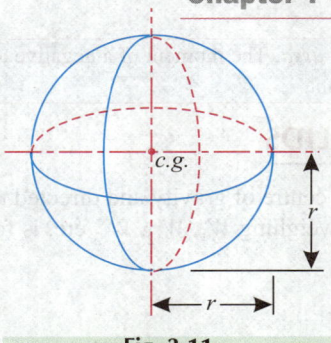

Fig. 3.11

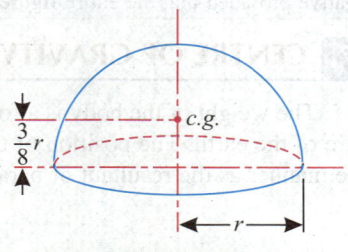

Fig. 3.12

3.6. CENTROIDS OF COMPOSITE AREAS

The location of the centroid of a plane figure can be thought of as the average distance of the area to an axis. Usually the axes involved will be the X and Y axes. In determining the location of the centroid it is found advantageous to place the X-axis through the lowest point and the Y-axis through the left edge of the figure. This places the plane area entirely within the first quadrant where X and Y distances are positive (Fig. 3.14.). Then divide the area into simple areas such as rectangles, triangles, etc. (Fig. 3.15.). Take the moment of each single area about the Y-axis. Sum up the moments about the Y-axis. Since the centroid of the composite figure is the point at which the entire area is assumed to be concentrated, the moment of the entire area about the Y axis must be equal to the moments of its component parts about the Y-axis.

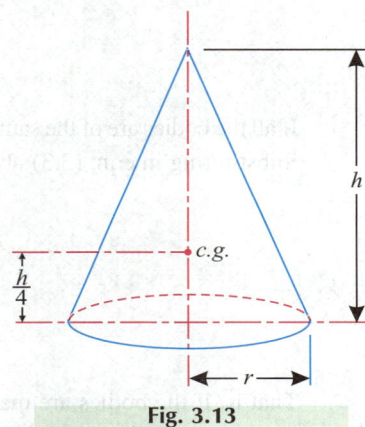

Fig. 3.13

That is, $(a_1 + a_2 ...... + a_n)\ \bar{x} = a_1 x_1 + a_2 x_2 + ... + a_n x_n$

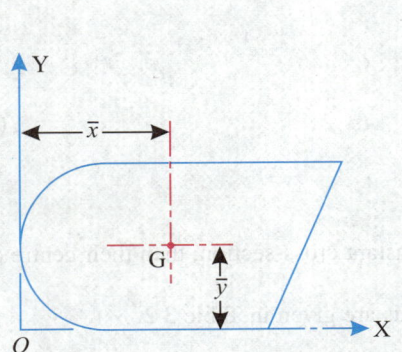

Fig. 3.14. Centroid of a composite area.

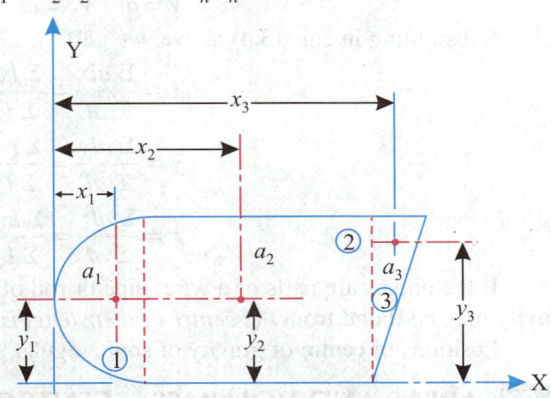

Fig. 315. Composite area divided into simple areas.

$$\bar{x} = \frac{a_1 x_1 + a_2 x_2 + ... + a_n x_n}{a_1 + a_2 + ... + a_n} \text{ or } \bar{x} = \frac{\Sigma\ ax}{\Sigma\ a} \qquad ...(3.1)$$

Following the same procedure for moments about the X-axis:

$$\bar{y} = \frac{a_1 y_1 + a_2 y_2 + ... + a_n y_n}{a_1 + a_2 + ... + a_n} \text{ or } \bar{y} = \frac{\Sigma ay}{\Sigma a} \qquad ...(3.2)$$

Note. If a *hole* exists in the plane figure, treat it as a *negative area*. The moment of a negative area will be negative provided that the entire figure lies in the first quadrant.

3.7. CENTRE OF GRAVITY OF SIMPLE SOLIDS

The weight of the body is a force acting at its own centre of gravity and directed towards the centre of the earth. The position of the centre of bodies weighing W_1, W_2, W_3 etc. is found in the same manner as the resultant of parallel forces.

$$\left. \begin{aligned} \bar{x} &= \frac{\Sigma\,Wx}{\Sigma\,W} \\[2mm] \bar{y} &= \frac{\Sigma\,Wy}{\Sigma\,W} \\[2mm] \bar{z} &= \frac{\Sigma\,Wz}{\Sigma\,W} \end{aligned} \right\} \qquad \ldots(3.3)$$

If all the bodies are of the same material having density ρ, then $W = \rho\,V_1$, $W_2 = \rho\,V_2$, $W_3 = \rho\,V_3$ etc. Substituting in eqn. (3.3) above, we have

$$\left. \begin{aligned} \bar{x} &= \frac{\Sigma\,\rho\,Vx}{\Sigma\,\rho\,V} = \frac{\Sigma\,Vx}{\Sigma\,V} \\[2mm] \bar{y} &= \frac{\Sigma\,\rho\,Vy}{\Sigma\,\rho\,V} = \frac{\Sigma\,Vy}{\Sigma\,V} \\[2mm] \bar{z} &= \frac{\Sigma\,\rho\,Vz}{\Sigma\,\rho\,V} = \frac{\Sigma\,Vz}{\Sigma\,V} \end{aligned} \right\} \qquad \ldots(3.4)$$

That is, if the bodies are made of the same material and are of the same density throughout, the centre of gravity of the bodies is their centre of volume. If the bodies are of the same cross-section but perhaps of different lengths,

$$V = al_1, \ V_2 = al_2, \ V_3 = al_3 \text{ etc.}$$

Substituting in eqn. (3.4) above, we get

$$\bar{x} = \frac{\Sigma\,alx}{\Sigma\,al} = \frac{\Sigma\,lx}{\Sigma\,l}$$

$$\bar{y} = \frac{\Sigma\,aly}{\Sigma\,al} = \frac{\Sigma\,ly}{\Sigma\,l} \qquad \ldots(3.5)$$

$$\bar{z} = \frac{\Sigma\,alz}{\Sigma\,al} = \frac{\Sigma\,lz}{\Sigma\,l}$$

If the bodies are parts of a wire, pipe or rod of constant cross-section, then their centre of gravity may be found from the *centre of their lengths*.

Positions of centre of gravity of some regular solids are given in Table 3.2.

3.8. AREAS AND VOLUMES – CENTROID METHOD

Since the centre of gravity of an area or a body is the point at which the area or mass of the body may be assumed to be concentrated, it can be said that the distance through which an area or a body moves is the same as the distance described by its centre of gravity. This relation is used in finding areas and volumes. Thus, a line moving parallel to its original position is said to generate an area that is equal to the *length of the line multiplied by the distance through which its centroid moves*. That is, area is equal to length times width. Also, an area moving parallel to its original position is said to develop the same volume of prism that is equal to *the area multiplied by the distance through which the centroid moves.* That is, *a volume is equal to the area of the base times the altitudes.*

Similarly, a line rotating about one end will develop the area of a circle. A right triangle rotating about either leg will develop the volume of a cone. In each case line *or area moves through a distance equal to length of a path described by the centroid of either the line or the area.* Many determinations of areas or volumes are simplified by the use of this method.

3.9. CENTRE OF GRAVITY IN A FEW SIMPLE CASES

1. C.G. of solid right circular cone:

Refer to Fig. 3.16. Let *ABC* be the cone and *AD* its axis. Consider an elementary circular plate *PQ* cut-off by two planes parallel to the base *BC* at distances y and $y + dy$ from A, and having its centre at M.

Let, $\qquad AD = h, BD = r, PM = r'$

Triangles *APM*, *ABD* are similar.

$$\frac{AM}{MP} = \frac{AD}{BD}$$

$\therefore \qquad \dfrac{y}{r'} = \dfrac{h}{r}, \textit{ i.e. } r' = \dfrac{yr}{h}$

If w be the density of the material, weight of *PQ*

$$= \pi r'^2 \, dy.w = \frac{\pi r^2 \, y^2}{h^2} \cdot dy.w$$

The c.g. of *PQ* is at *M*. Hence the distance of c.g. of the cone *from A*

Machine part with bearing.

$$= \frac{\displaystyle\sum_{y=0}^{y=h} \frac{\pi \, r^2 y^2}{h^2} \, dy.w.y}{\displaystyle\sum_{y=0}^{y=h} \frac{\pi \, r^2 y^2}{h^2} \cdot dy.w} = \frac{\displaystyle\int_0^h y^3 \, dy}{\displaystyle\int_0^h y^2 \, dy} = \frac{\dfrac{h^4}{4}}{\dfrac{h^3}{3}} = \frac{3h}{4}$$

$$= \left[h - \frac{3h}{4}\right] = \frac{h}{4} \textit{ from the base} \qquad ...(3.6)$$

Hence the c.g. of a solid cone lies on the axis at a height one-fourth of the total height from the base.

2. C.G. of a thin hollow right circular cone:

Refer to Fig. 3.17. Let *ABC* be the cone and *AD* its axis. Consider a circular ring cut off by planes of *PQ* and *P'Q'* parallel to the base *BC* at distances y and $y + dy$ from A.

Let the radius of *PQ* $= r'$

$\qquad BD = r, AD = h$

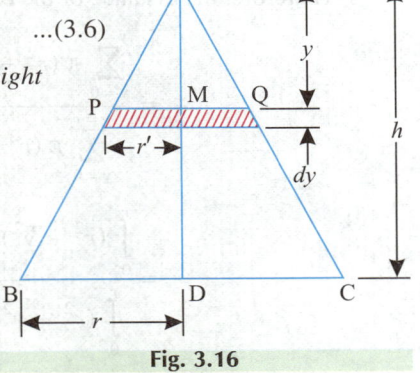

Fig. 3.16

Semi-vertical angle of the cone

$$= \angle BAD = \alpha$$

Clearly, $PP' = dy \sec \alpha$

As in (1) above, $r' = \dfrac{ry}{h}$

Area of elementary ring

$$= 2\pi \, r' \, PP' = 2\pi \cdot \frac{ry}{h} \, dy \sec \alpha$$

If w be the weight per unit area of the material, weight of the ring.

$$= 2\pi \cdot \frac{ry}{h} \cdot dy \sec \alpha \cdot w$$

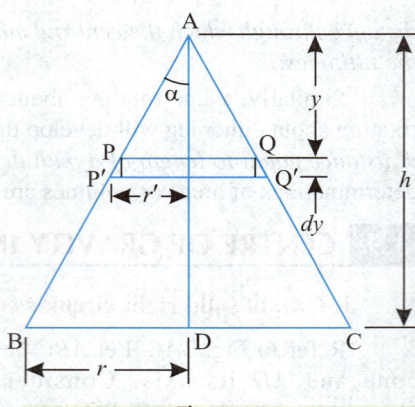

Fig. 3.17

The c.g. of the ring lies on AD at distance y from A. Hence the distance of the c.g. of the cone from A

$$= \frac{\displaystyle\sum_{y=0}^{y=h} 2\pi \frac{ry}{h} \, dy \sec \alpha \, w.y}{\displaystyle\sum_{y=0}^{y=h} 2\pi \frac{ry}{h} \, dy \sec \alpha \cdot w} = \frac{\displaystyle\int_0^h y^2 \, dy}{\displaystyle\int_0^h y \, dy} = \frac{\dfrac{h^3}{3}}{\dfrac{h^2}{2}} = 2/3 \, h$$

$$= (h - 2/3h) = h/3 \text{ from the base} \qquad \qquad ...(3.7)$$

Hence the c.g. of a thin hollow cone lies on the axis at a height one-third of the total height above the base.

3. C.G. of a solid hemisphere:

Refer Fig. 3.18. Let ACB be the hemisphere of radius r, and OC its central radius. Consider an elementary circular plate PQ cut off by planes parallel to AB at distances y and $y + dy$ from AB.

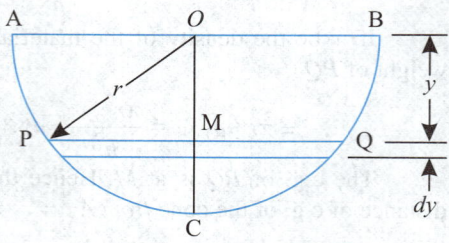

$$PM^2 = OP^2 - OM^2 = r^2 - y^2$$

Weight of $\quad PQ = \pi (r^2 - y^2) \times dy \cdot w$

Fig. 3.18

where w is the weight of unit volume of the material.

The c.g. of PQ lies on OC at distance y from O.

Therefore, the distance of the c.g. of the hemisphere from O.

$$= \frac{\displaystyle\sum_{y=0}^{y=r} \pi (r^2 - y^2) \, dy \cdot w \cdot y}{\displaystyle\sum_{y=0}^{y=r} \pi (r^2 - y^2) \, dy \cdot w}$$

$$= \frac{\displaystyle\int_0^r (r^2 - y^2) \, y \, dy}{\displaystyle\int_0^r (r^2 - y^2) \, dy} = \frac{\left| \dfrac{r^2 y^2}{2} - \dfrac{y^4}{4} \right|_0^r}{\left| r^2 y - \dfrac{y^3}{3} \right|_0^r}$$

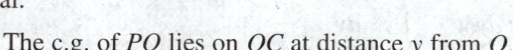

$$= \frac{\dfrac{r^4}{4}}{\dfrac{2r^3}{3}} = \frac{3r}{8} \qquad \qquad \text{...(3.8)}$$

Hence c.g. of a solid hemisphere lies on the central radius at distance $\dfrac{3r}{8}$ from the plane base, where r is the radius of the hemisphere.

4. C.G. of a thin hollow hemisphere:

In Fig. 3.18, if the hemisphere is hollow of negligible thickness, then PQ is a ring whose area = $2\pi r\, dy$, by mensuration.

Weight of $PQ = 2\pi r \cdot dy \cdot w$

where w is the weight per unit area of the material.

∴ Distance of the c.g. of the hemisphere from $O = \bar{y}$ (say)

$$= \frac{\displaystyle\int_0^r 2\pi r\, dy\, w \cdot y}{\displaystyle\int_0^r 2\pi r\, dy \cdot w} = \frac{\displaystyle\int_0^r y\, dy}{\displaystyle\int_0^r dy} = \frac{\dfrac{r^2}{2}}{r} = r/2. \qquad \text{...(3.9)}$$

Hence the c.g. of a hollow hemisphere bisects the central radius.

5. C.G. of a semi-circular lamina:

Let Fig. 3.18 represent a semi-circular plate of radius r.

The length of the elementary strip $PQ = 2 \times PM = 2\sqrt{(r^2 - y^2)}$

Area of $PQ = 2\sqrt{r^2 - y^2}\, dy$

If w be the weight per unit area of the material, weight of $PQ = 2\sqrt{r^2 - y^2}\, dy \cdot w$

The c.g. of PQ lies on OC at distance y from O.

$$\bar{y} = \frac{\displaystyle\int_0^r 2\sqrt{r^2 - y^2}\, dy \cdot wy}{\displaystyle\int_0^r 2\sqrt{r^2 - y^2}\, dy \cdot w} = \frac{\displaystyle\int_0^r y\sqrt{r^2 - y^2}\, dy}{\displaystyle\int_0^r \sqrt{r^2 - y^2}\, dy}$$

Now,

$$\int_0^r y\sqrt{r^2 - y^2}\, dy = -\frac{1}{2}\int_0^r (r^2 - y^2)^{1/2}\, (-2y)\, dy$$

$$= -\frac{1}{2}\left|\frac{2}{3}(r^2 - y^2)^{3/2}\right|_0^r = \frac{r^3}{3} \quad \text{and} \quad \int_0^r \sqrt{r^2 - y^2}\, dy$$

$$= \left| y\frac{(r^2 - y^2)}{2} + \frac{r^2}{2}\sin^{-1}\frac{y}{2}\right|_0^r = \frac{r^2}{2} \times \frac{\pi}{2} = \frac{\pi r^2}{4}$$

∴

$$\bar{y} = \frac{\dfrac{r^3}{3}}{\dfrac{\pi r^2}{4}} = \frac{4r}{3\pi} \qquad \qquad \text{...(3.10)}$$

Hence the c.g. of a semi-circular lamina lies on the central radius at distance $\dfrac{4r}{3\pi}$ from the bounding diameter, where r is the radius of the plate.

6. **C.G. (or centroid) of semi-circular arc:**

Refer to Fig. 3.19. Let OC be the central radius and P an element of arc subtending angle $d\theta$ at O.

By symmetry, the c.g. of the arc lies on OC.

Length of elemental arc $P = r\, d\theta$,

where r is the radius of the arc.

Distance of P from $AB = r \cos \theta$

∴ The distance of c.g. of the whole arc from AB

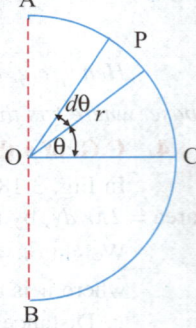

$$= \frac{\displaystyle\int_{-\pi/2}^{\pi/2} r\, d\theta \cdot r \cos \theta}{\displaystyle\int_{-\pi/2}^{\pi/2} r\, d\theta}$$

$$= \frac{r \left|\sin \theta\right|_{-\pi/2}^{\pi/2}}{\left|\theta\right|_{-\pi/2}^{\pi/2}} = \frac{2r}{\pi} \qquad ...(3.11)$$

Fig. 3.19

WORKED EXAMPLES
Centroid—Areas/Laminas

Example 3.1. *Find out the position of the centroid of L section as shown in Fig. 3.20.*

Solution. Refer to Fig. 3.20.

Divide the composite figure into *two* simple areas:

(*i*) Rectangle (160 mm × 40 mm) ...(1)

(*ii*) Rectangle (80 mm × 40 mm) ...(2)

Tracked bearing housings.

To determine the location of the centroid of the plane figure we have the following table:

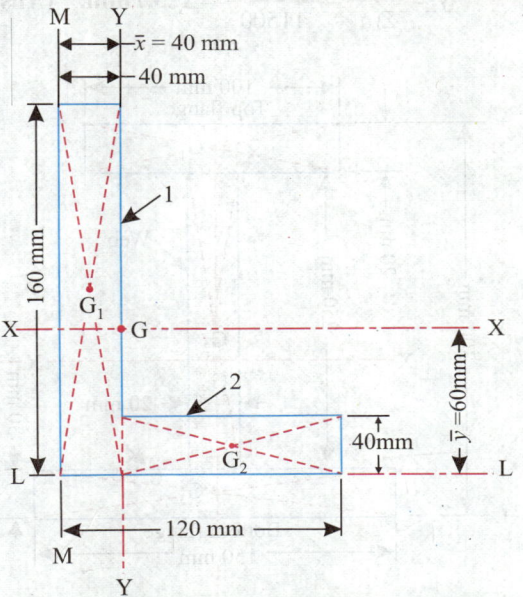

Fig. 3.20

Components	Area a (mm²)	Centroidal distance 'x' from MM (mm)	Centroidal distance 'y' from LL (mm)	ax (mm³)	ay (mm³)
Rectangle (1)	160 × 40 = 6400	20	80	128000	512000
Rectangle (2)	80 × 40 = 3200	80	20	256000	64000
	9600 (Σ a)	–	–	384000 (Σ ax)	576000 (Σ ay)

$$\bar{x} = \frac{\Sigma\, ax}{\Sigma a} = \frac{384000}{9600} = \textbf{40 mm.} \ \ \textbf{(Ans.)} \qquad \bar{y} = \frac{\Sigma ay}{\Sigma a} = \frac{576000}{9600} = \textbf{60 mm.} \ \ \textbf{(Ans.)}$$

Example 3.2. *Determine the position of the centroid of I-section as shown in Fig. 3 .21.*

Solution. Refer to Fig. 3.21

Divide the composite figure into the simple areas:

   (*i*) Rectangle (100 mm × 20 mm) — top flange   ....(1)

   (*ii*) Rectangle (250 mm × 20 mm) — web   ....(2)

   (*iii*) Rectangle (150 mm × 30 mm) — bottom flange   ....(3)

   Since the composite area is symmetrical about Y-axis, so the centroid will lie on Y-axis and we shall, therefore, find $\bar{y}$ only.

To determine the location of the centroid of the plane figure we have the following table:

Components	Area 'a' (mm²)	Centroidal distance 'y' from LL (mm)	ay (mm³)
Rectangle (1)	100 × 20 = 2000	290	580000
Rectangle (2)	250 × 20 = 5000	155	775000
Rectangle (3)	150 × 20 = 4500	15	67500
	11500 (Σa)	–	1422500 (Σ ay)

$$\bar{y} = \frac{\Sigma \, ay}{\Sigma \, a} = \frac{1422500}{11500} = \textbf{123.7 mm.} \quad \textbf{(Ans.)}$$

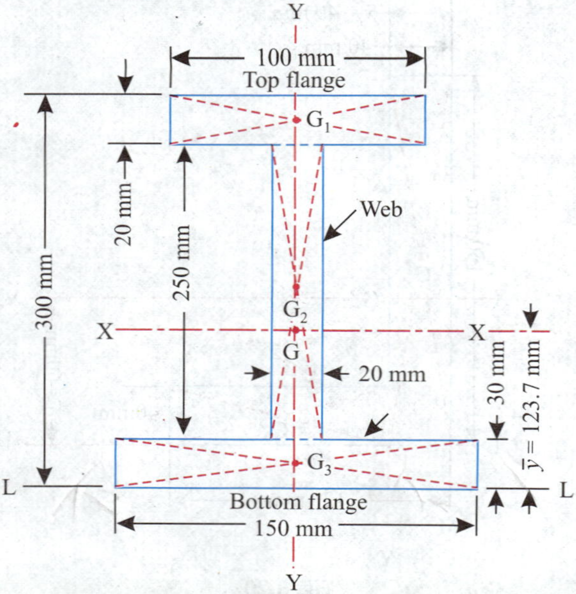

Fig. 3.21

Example 3.3. *Determine the centroid of the shaded area shown in Fig. 3.22.*

Solution. Refer to Fig. 3.23

Divide the composite figure into four simple areas:

   (*i*)   Triangle – marked (1)

  (*ii*)   Rectangle – (200 mm × 187.5 mm) – marked (2)

 (*iii*)   Triangle – marked (3)

 (*iv*)   Semicircle (radius = 150 mm) – marked (4)

Triangle (1), rectangle (2) and triangle (3) are the *positive areas* and the semicircle is a *negative area.*

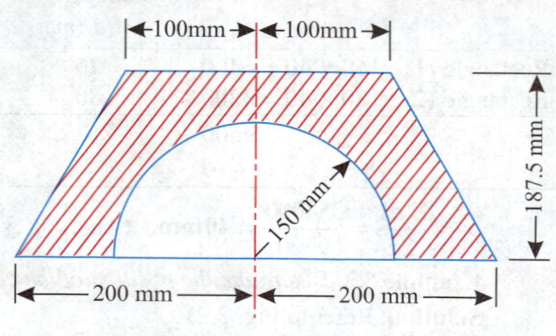

Fig. 3.22

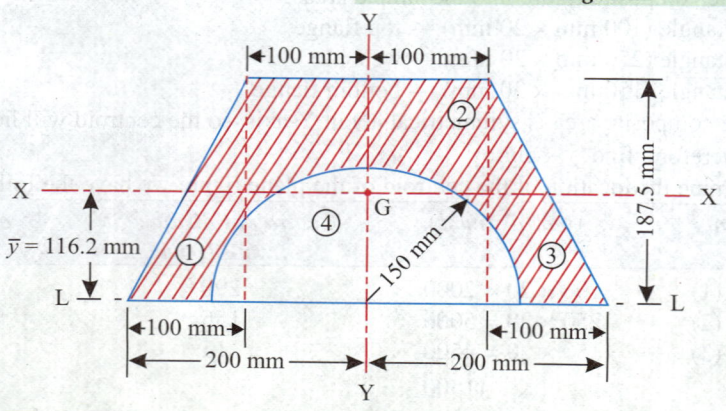

Fig. 3.23

Since the composite area is symmetrical about Y-axis, so the centroid will lie on Y-axis and we shall, therefore, find out $\bar{y}$ only.

To determine $\bar{y}$ we have the following table:

Components	Area 'a' (mm^2)	Centroidal distance 'y' from LL (mm)	ay (mm^3)
Triangle (1)	$\dfrac{1}{2} \times 100 \times 187.5 = 9375$ (+)	$\dfrac{187.5}{3} = 62.5$	585937 (+)
Rectangle (2)	$200 \times 187.5 = 37500$ (+)	$\dfrac{187.5}{2} = 93.75$	3515625 (+)
Triangle (3)	$\dfrac{1}{2} \times 100 \times 187.5 = 9375$ (+)	$\dfrac{187.5}{3} = 62.5$	585937 (+)
Semi-circle (4)	$\dfrac{\pi r^2}{2} = \dfrac{4 \times 150^2}{2} = 35343$ (−)	$\dfrac{4r}{3\pi} = \dfrac{4 \times 150}{3\pi} = 63.7$	2251349 (−)
	$a = 20907$	−	$\Sigma\, ay = 2436150$

∴

$$\bar{y} = \frac{\Sigma\, ay}{\Sigma a} = \frac{2436150}{20907} = \textbf{116.2 mm}$$

Hence the centroid of the composite figure lies at a distance **116.2 mm from LL,** the axis of reference.   **(Ans.)**

Centre of Gravity–Solids

Example 3.4. *A hemisphere of diameter 60 mm is placed on the top of a cylinder, whose diameter is also 60 mm. The height of the cylinder is 75 mm. Find the common c.g. of the composite body.*

Solution. Refer Fig. 3.24.

Since the composite solid is symmetrical about Y-axis, so the c.g. will lie on Y-axis and we shall, therefore, find out $\bar{y}$ only.

Volume of hemisphere

$$V_1 = \frac{2}{3}\pi r^3 = \frac{2}{3} \times \pi \times 30^3 = 18000\,\pi \text{ mm}^3$$

Volume of cylinder

$$V_2 = \pi r^2 h = \pi \times 30^2 \times 75$$
$$= 67500\,\pi \text{ mm}^3.$$

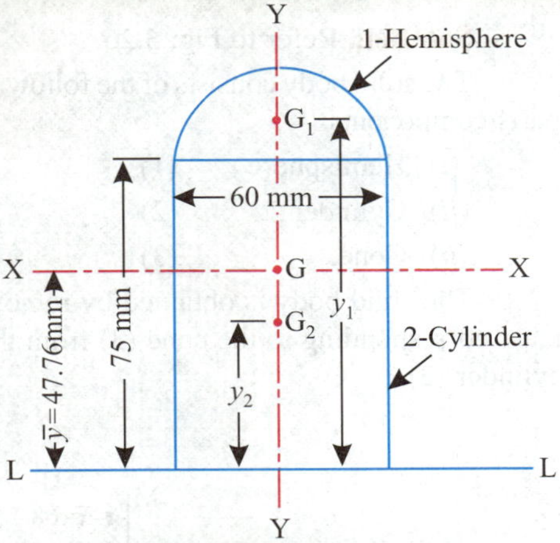

Fig. 3.24

To determine $\bar{y}$ of composite solid, we have the following table:

Components	Volume (mm^3) V	Centroidal distance 'y' from LL (mm)	Vy (mm^4)
Hemisphere (1)	$18000\,\pi$	$y_1 = 75 + \dfrac{3r}{8}$ $= 75 + \dfrac{3 \times 30}{8}$ $= 86.25$	$1552500\,\pi$
Cylinder (2)	$67500\,\pi$	$y_2 = \dfrac{75}{2} = 37.5$	$2531250\,\pi$
Total	$\Sigma V = 85500\,\pi$	−	$\Sigma Vy = 4083750\,\pi$

$$\overline{y} = \frac{\sum Vy}{\sum V} = \frac{4083750\,\pi}{85500\,\pi}$$

$$= 47.76 \text{ mm.}$$

Hence the c.g. of the composite body lies at a distance of **47.76 mm from LL**, the axis of reference. **(Ans.)**

> **Note.** In this case since the individual components (hemisphere and cylinder) are made of same material so we have considered volumes only (instead of weights) but in case the components are made of different materials then we should take weights instead of volumes (because densities are different).

IC engine.

Example 3.5. *Determine the centre of gravity of a homogeneous solid body of revolution as shown in Fig. 3.25.*

Solution. Refer to Fig. 3.26.

The solid body consists of the following three parts/components:

(*i*) Hemisphere   (1)

(*ii*) Cylinder   (2)

(*iii*) Cone.   (3)

The solid body is obtained by removing the mass corresponding to the cone (3) from the composite body consisting of hemisphere (1) and the cylinder (2).

Fig. 3.25

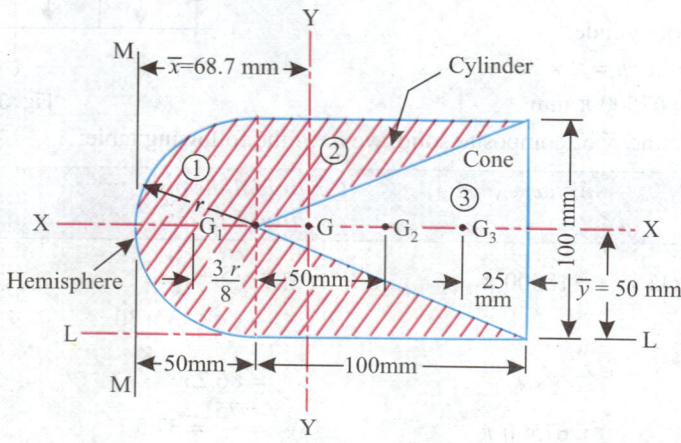

Fig. 3.26

Since the body is symmetrical about X-axis, therefore we shall find out $\bar{x}$ only, with the following table:

Components	Volume V (mm^3)	Centroidal distance 'x' from MM (mm)	Vx (mm^4)
Hemisphere (1)	$\frac{2}{3}\pi r^3 = \frac{2}{3}\pi \times 50^3$ $= 83333\ \pi\ (+)$	$50 - \frac{3r}{8} = 50 - \frac{3 \times 50}{8}$ $= 31.25$	$2604156\ \pi\ (+)$
Cylinder (2)	$\pi r^2 h = \pi \times 50^2 \times 100$ $= 250000\ \pi\ (+)$	$50 + \frac{100}{2} = 100$	$25000000\ \pi\ (+)$
Cone (3)	$\frac{1}{3}\pi r^2 h = \frac{1}{3} \times 50^2 \times 100$ $= 83333\ \pi\ (-)$	$50 + 100 - \frac{100}{4} = 125$	$10416625\ \pi\ (-)$
	$\Sigma\ V = 250000\ \pi$	–	$\Sigma\ Vx = 17187531\ \pi$

$$\bar{x} = \frac{\Sigma\ Vx}{\Sigma\ V} \cdot \frac{17187531\ \pi}{250000\ \pi} = \textbf{68.7 mm} \quad \textbf{(Ans.)} \quad \text{and} \quad \bar{y} = \frac{100}{2} = \textbf{50 mm} \quad \textbf{(Ans.)}$$

Example 3.6. *A frustrum of a solid right circular cone having base circle diameter 100 mm, height 100 mm and top diameter 50 mm has an axial hole of diameter 25 mm. Find the c.g. of the solid.*

Solution. Refer to Fig. 3.27. The frustrum of cone can be obtained by removing cone (2) of base diameter 50 mm and height 100 mm from another cone (1) of base diameter 100 mm and height 200 mm. The given solid will be obtained by removing a cylinder (3) of diameter 25 mm and height 100 mm from this frustrum.

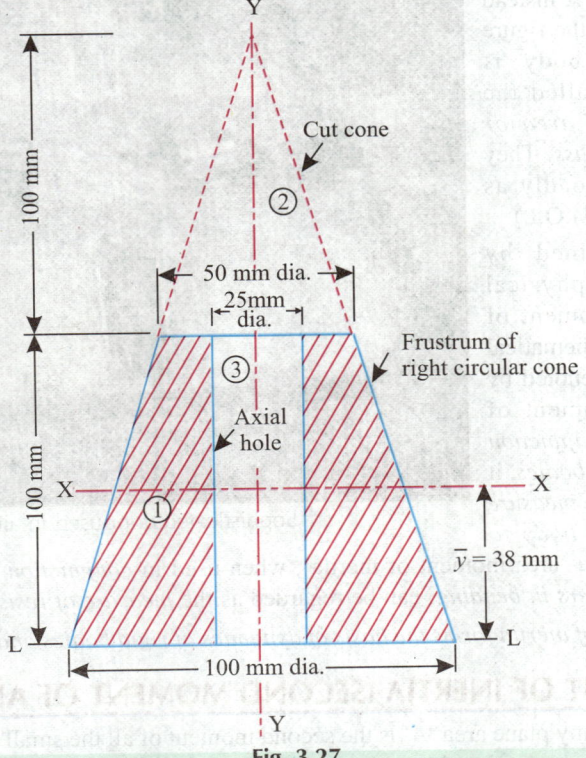

Fig. 3.27

Since the solid is symmetrical about Y-axis, we shall find out $\bar{y}$ only with the following table:

Components	Volume $(V)(mm^3)$	Centroidal distance 'y' from LL (mm)	Vy (mm^4)
Cone (1)	$\dfrac{1}{3} \times \pi \times 50^2 \times 200 = 166666\ \pi\ (+)$	$\dfrac{200}{4} = 50$	$8333300\ \pi\ (+)$
Cone (2)	$\dfrac{1}{3} \times \pi \times 25^2 \times 100 = 20833\ \pi\ (-)$	$\dfrac{100}{4} + 100 = 125$	$2604125\ \pi\ (-)$
Cylinder (3)	$\pi \times (12.5)^2 \times 100 = 15625\ \pi\ (-)$	$\dfrac{100}{2} = 50$	$781250\ \pi\ (-)$
	$\Sigma\ V = 130208\ \pi$	—	$\Sigma\ Vy = 4947925\ \pi$

$$\therefore \qquad \bar{y} = \frac{\Sigma\ Vy}{\Sigma\ V} = \frac{4947925\ \pi}{130208\ \pi} = 38\ \text{mm} \quad \text{(Ans.)}$$

3.10. MOMENT OF INERTIA (M.O.I) — INTRODUCTION

The moment of a force (also called the *first moment of force*) about any point is the product of the force and the perpendicular distance between them. If this first moment is again multiplied by the perpendicular distance between them, the product so obtained is called the *second moment of force or moment of moment of the force*. If instead of force, the area of the figure or mass of the body is considered, it is called the *second moment of area or second moment of mass*. They are also termed broadly as *moments of inertia* (M.O.I.)

When examined by itself, there is no physical significance for moment of inertia. It is just a mathematical expression usually denoted by *I*. When 'mass moment of inertia' is used in *conjunction with rotation of rigid bodies*, it can be regarded as the *measure of resistance of the body to*

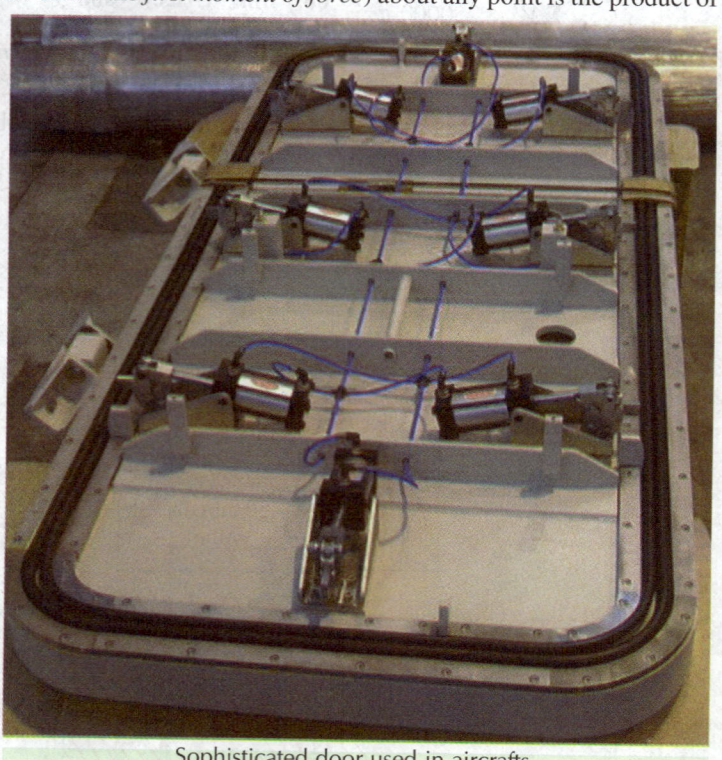

Sophisticated door used in aircrafts.

rotation. Similarly the 'area moment of inertia', when used in *conjunction with the deflection or deformation of members in bending*, can be regarded as the *measure of resistance to bending*.

• *The moment of inertia forms the basis of dynamics of rigid bodies and strength of materials.*

3.11. MOMENT OF INERTIA (SECOND MOMENT OF AN AREA)

Moment of inertia of any plane area '*A*' is the second moment of all the small areas '*dA*' comprising the area '*A*' about any axis in the plane of area *A*. Referring to Fig. 3.28,

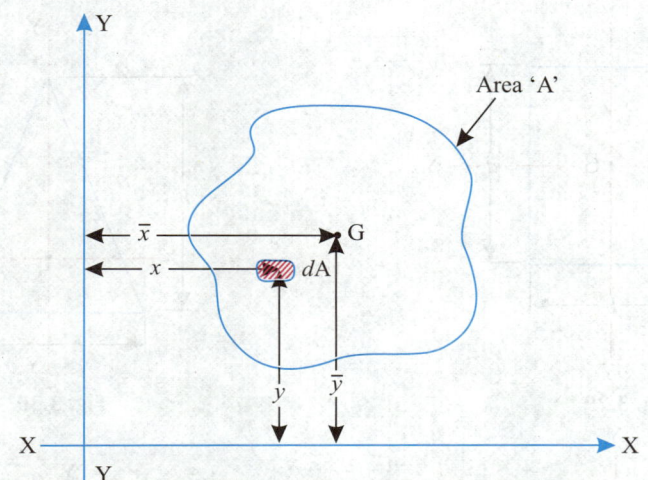

Fig. 3.28

I_{yy} = Moment of inertia about yy = $\Sigma\,(dA \cdot x)\,x$

where $(dA \cdot x)$ is the first moment of area dA about YY and $(dA \cdot x)\,x$ is the moment of the first moment (called second moment) of the area dA about the same axis YY.

$\therefore$
$$I_{yy} = \Sigma\,dA \cdot x^2 = \int_A dAx^2 \qquad\qquad ...(3.12)$$

Similarly,
$$I_{xx} = \Sigma\,dAy^2 = \int_A dAy^2 \qquad\qquad ...(3.13)$$

Table 3.3 gives the moment of inertia of simple areas.

TABLE 3.3. **Moment of Inertia for Simple Areas**

	Shape	Figure	Moment of inertia
1.	Rectangle	Fig. 3.29	$I_{xx} = \dfrac{bd^3}{12}$
			$I_{yy} = \dfrac{db^3}{12}$
2.	Triangle	Fig. 3.30	$I_{xx} = \dfrac{bh^3}{36}$
			$\left(= \dfrac{bh^3}{12}\ \text{about the base}\right)$
3.	Circle	Fig. 3.31	$I_{xx} = I_{yy} = \dfrac{\pi r^4}{4}$
4.	Semicircle	Fig. 3.32	$I_{xx} = 0.11\ r^4$
			$I_{yy} = \dfrac{\pi r^4}{8}$
5.	Quarter Circle	Fig. 3.33	$I_{xx} = I_{yy} = 0.055\ r^4$
6.	Ellipse	Fig. 3.34	$I_{xx} = \dfrac{\pi\,ab^3}{4}$
			$I_{yy} = \dfrac{\pi ba^3}{4}$

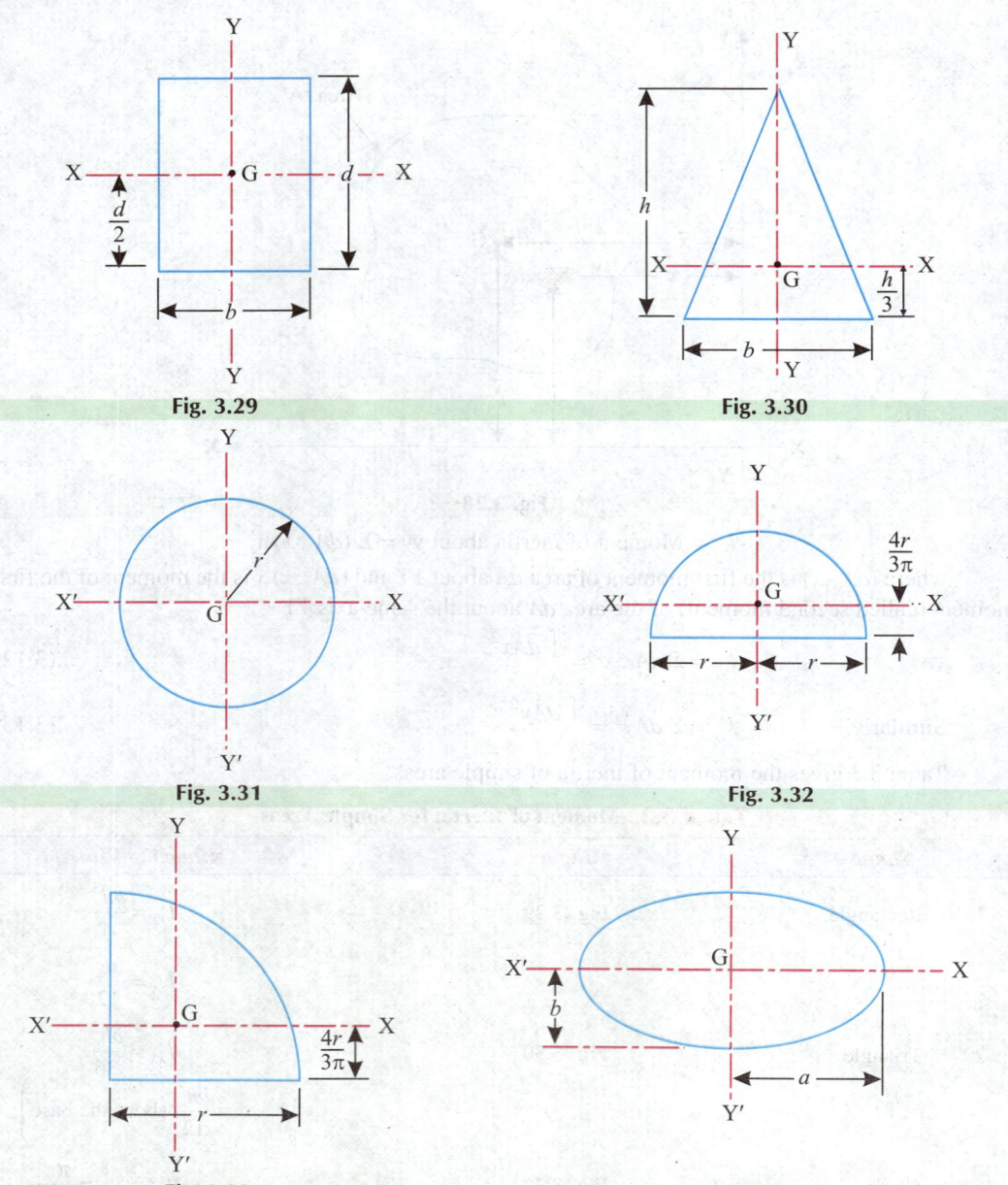

Fig. 3.29

Fig. 3.30

Fig. 3.31

Fig. 3.32

Fig. 3.33

Fig. 3.34

3.12. THEOREM OF PARALLEL AXES (OR TRANSFER FORMULA)

The theorem of parallel axes states: *"The moment of inertia of a lamina about any axis in the plane of the lamina equals the sum of moment of inertia about a parallel centroidal axis in the plane of lamina and the product of the area of the lamina and square of the distance between the two axes."*

In Fig. 3.35 is shown a lamina of area *A*. Let *LM* be the axis in the plane of lamina about which the moment of inertia of the lamina is required to be found out. Let *XX* be the centroidal axis in the plane of the lamina and parallel to the axis *LM*. Let '*h*' be the distance between the two axes *XX* and *LM*.

It may be assumed that the lamina consists of an infinite number of small elemental components parallel to the axis *XX*. Consider one such elemental component at a distance y from the axis *XX*. The distance of the elemental component from the axis *LM* will be $(h \pm y)$ accordingly as the elemental component and the axis *LM* are on the opposite sides of *XX* or on the same side of *XX*.

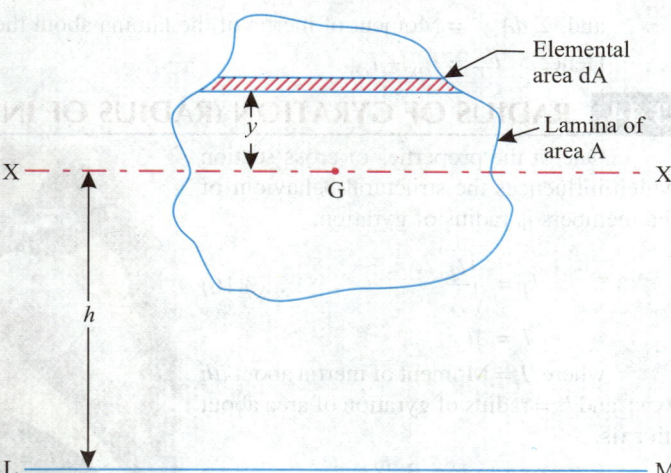

Fig. 3.35

Moment of inertia of the elemental component about the axis

$$LM = dA (h \pm y)^2$$

∴ Moment of inertia of the whole lamina about the axis *LM*

$$= I_{LM} = \Sigma\, dA\, (h \pm y)^2 = \Sigma\, dA h^2 + \Sigma dA\, y^2 \pm 2\, \Sigma\, dAh\, y$$
$$= h^2\, \Sigma\, dA + \Sigma dA\, y^2 \pm 2\, h\, \Sigma\, dA\, y$$

But, $\quad \Sigma dA = A,\ h^2\, \Sigma dA = Ah^2$

$\Sigma dA\, y^2$ = Moment of inertia of the lamina about the axis *XX*, (*i.e.*, I_{XX} or I_G) and

$\Sigma\, dA\, y = 0$ since *XX* is a centroidal axis.

∴ $\quad\quad I_{LM} = I_{XX}$ (or I_G) $+ Ah^2$ $\quad\quad\quad$ ...(3.14)

<h2>3.13.   THEOREM OF PERPENDICULAR AXES</h2>

The theorem of perpendicular axes states : *"If I_{OX} and I_{OY} be the moments of inertia of a lamina about mutually perpendicular axes OX and OY in the plane of the lamina and I_{OZ} be the moment of inertia of the lamina about an axis (OZ) normal to the lamina and passing through the point of intersection of the axes OX and OY then, $I_{OZ} = I_{OX} + I_{OY}$."*

Refer Fig. 3.36. Let *OX* and *OY* be the two mutually perpendicular axes lying in the plane of the lamina. Let *OZ* be axis normal to the lamina and passing through *O*.

Consider an elemental component of area *dA* of the lamina. Let the distance of this elemental component from the axis *OZ*, i.e. from *O* be *r*.

Then, moment of inertia of the elemental component about $OZ, = dA \times r^2$.

If the co-ordinates of the elemental components be (x, y) referred to the axes *OX* and *OY*, we have

$$r^2 = x^2 + y^2$$

Moment of inertia of the elemental component about the axis *OZ*

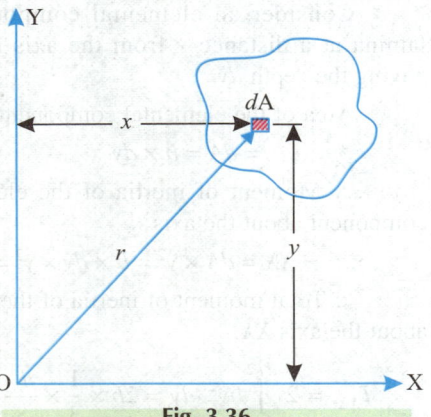

Fig. 3.36

$$= dA\, (x^2 + y^2) = dA\, x^2 + dA\, y^2$$

∴ Total moment of inertia of the lamina about the axis *OZ*

$$= I_{OZ} = \Sigma\, (dA\, x^2 + dA\, y^2)$$
$$= \Sigma\, dA\, x^2 + \Sigma dA\, y^2$$

But, $\Sigma\, dA\, x^2$ = Moment of inertia of the lamina about the axis $OY = I_{OY}$

and, $\Sigma\, dA\, y^2$ = Moment of inertia of the lamina about the axis $OX = I_{OX}$

Hence, $I_{OZ} = I_{OX} + I_{OY}$ ...(3.15)

3.14. RADIUS OF GYRATION (RADIUS OF INERTIA)

One of the properties of cross-section which influences the structural behaviour of the members is radius of gyration.

$$k_i = \sqrt{\frac{I_i}{A}} \qquad ...(3.16)$$

$(\because\ I_i = A k_i^2)$

where, I_i = Moment of inertia about ith axis; and k_i = radius of gyration of area about ith axis.

Referring to Fig. 3.28,

$$k_{XX} = \sqrt{\frac{I_{XX}}{A}}, \qquad k_{YY} = \sqrt{\frac{I_{YY}}{A}}$$

Members when subjected to axial forces tend to buckle. The load at which members will buckle is proportional to the square of the radius of the gyration. The radius

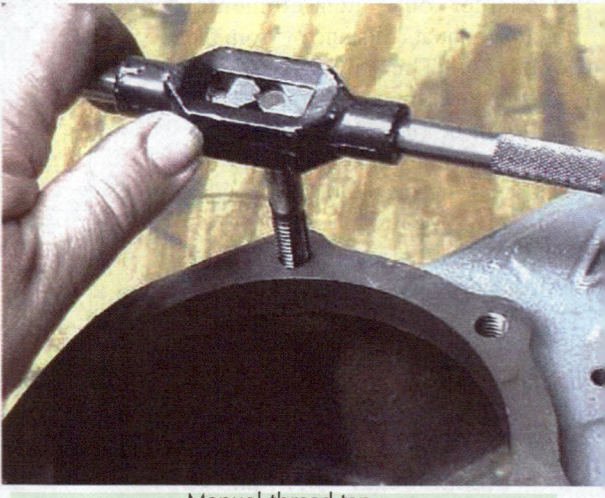

Manual thread tap.

of gyration is usually referred to with respect to centroidal axes system of the section.

A dimensionless quantity called unit radius of inertia is defined as ratio of M.O.I. to square of area.

3.15. MOMENT OF INERTIA OF LAMINAE OF DIFFERENT SHAPES

1. Rectangular lamina:

Refer to Fig. 3.37. Let a rectangular lamina be 'b' units wide and 'd' units deep.

(*a*) *Moment of inertia about centroidal axis XX parallel to the width*:

Consider an elemental component of lamina at a distance y from the axis XX and having the depth dy.

Area of the elemental component

$$= dA = b \times dy$$

∴ Moment of inertia of the elemental component about the axis

$$XX = dA \times y^2 = b \times dy \times y^2 = by^2\, dy$$

∴ Total moment of inertia of the lamina about the axis XX,

$$I_{XX} = 2 \int_0^{d/2} by^2\, dy = 2b \times \frac{1}{3} \times \frac{d^3}{8} = \frac{bd^3}{12}$$

...(3.17)

Fig. 3.37

Similarly, moment of inertia about the centroidal axis YY parallel to the depth,

$$I_{YY} = \frac{db^3}{12} \qquad \qquad ...(3.18)$$

(b) *Moment of inertia about an axis LL passing through the bottom edge or top edge*:

By the theorem of parallel axis, the moment of inertia about the axis *LL* is given by

$$I_{LL} = I_{XX} + Ah^2$$

In this case $A = bd$

$$h = \frac{d}{2}$$

$$I_{XX} = \frac{bd^3}{12}$$

$$I_{LL} = \frac{bd^3}{12} + bd \times \left(\frac{d}{2}\right)^2 = \frac{bd^3}{12} + \frac{bd^3}{4} = \frac{bd^3}{3} \qquad ...(3.19)$$

Similarly, the moment of inertia about the axis *MM* is given by,

$$I_{MM} = \frac{db^3}{3} \qquad \qquad ...(3.20)$$

If *G* be the centroid of the lamina, the axis through the centroid and normal to the plane of the lamina is called the *polar axis*. Let I_p be the moment of inertia about the polar axis, I_p is called the *polar moment of inertia*.

By the theorem of perpendicular axes,

$$I_p = I_{XX} + I_{YY} = \frac{bd^3}{12} + \frac{db^3}{12} \qquad ...(3.21)$$

2. **Rectangular lamina with a centrally situated rectangular hole:**

Refer to Fig. 3 .38. Let in a rectangular lamina $B \times D$, a rectangular hole $b \times d$ be made centrally.

Moment of inertia of the lamina about any axis = Moment of inertia of bigger rectangle– moment of inertia of the smaller rectangle.

For example,

$$I_{XX} = \frac{BD^3}{12} - \frac{bd^3}{12} \qquad ...(3.22)$$

3. **Moment of inertia of a triangular lamina :**

Refer to Fig. 3.39. Let *ABC* be a triangular lamina of base *b* and altitude *h*.

(a) *Moment of inertia of a triangle about an axis LL through the vertex and parallel to the axis*:

The triangle may be considered to consist of a number of infinitely small elemental components parallel to the base. Consider one such elemental component at a distance *y* from the vertex and of thickness *dy*. Width of elemental component

$$= b' = \frac{b}{h} y$$

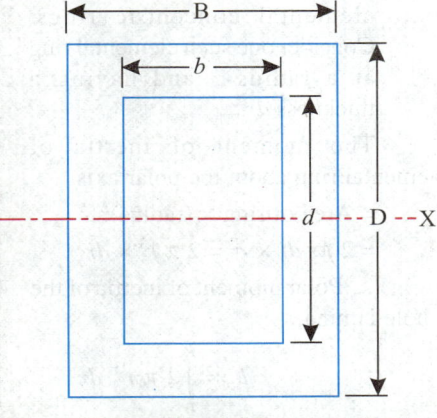

Fig. 3.38

∴   Area of elemental component

$$= b' \, dy = \frac{b}{h} \, y \, dy$$

∴   Moment of inertia of the elemental component about the axis LL

$$= \frac{b}{h} \, y \, dy \, y^2 = \frac{b}{h} \, y^3 \, dy$$

∴   Moment of inertia of the lamina about the axis LL,

$$I_{LL} = \frac{b}{h} \int_0^h y^3 \, dy = \frac{bh^3}{4} \qquad \qquad \qquad ...(3.23)$$

Fig. 3.39

(b) *Moment of inertia of a triangle about the centroidal axis parallel to the base:*

Let XX be the centroidal axis. This axis is at distance of $2/3h$ from the vertex.

Applying the theorem of parallel axes, we have

$$I_{LL} = I_{XX} + A \left(\frac{2}{3} h \right)^2 \qquad ∴ \quad \frac{bh^3}{4} = I_{XX} + \frac{bh}{2} \times \frac{4}{9} h^2$$

$$I_{XX} = \frac{bh^3}{4} - \frac{2}{9} bh^3 \qquad \text{or} \qquad I_{XX} = \frac{bh^3}{36} \qquad \qquad ...(3.24)$$

(c) *Moment of inertia of a triangle about the base:*

Applying the theorem of parallel axes again, we have

$$I_{BC} = I_{XX} + A \left(\frac{h}{3} \right)^2 = \frac{bh^3}{36} + \frac{bh}{2} \times \frac{h^2}{9}$$

$$= \frac{bh^3}{36} + \frac{bh^3}{18} = \frac{bh^3}{12} \qquad \qquad \qquad ...(3.25)$$

4. Moment of inertia of a circular lamina:

Fig. 3.40 shows a circular lamina of radius R. The lamina may be considered as consisting of elemental concentric rings. Consider one such elemental ring at a radius r and having a thickness dr.

The moment of inertia of elemental ring about the polar axis

= Area of ring × (radius)^2

= $2 \pi r \, dr \times r^2 = 2 \pi r^3 \times dr$

∴ Polar moment of inertia of the whole lamina,

$$I_p = \int_0^R 2 \pi r^3 \, dr$$

Splined hub.

$$∴ \qquad I_p = \frac{2 \pi R^4}{4} = \frac{\pi R^4}{2} \qquad .... (3.26)$$

If D be the diameter of the lamina,

$$D = 2R$$

$$\therefore \quad I_p = \frac{\pi}{2}\left(\frac{D}{2}\right)^4$$

$$= \frac{\pi}{32}D^4$$

But, $\quad I_{XX} = I_{YY}$ and $I_{XX} + I_{YY} = I_p = \dfrac{\pi D^4}{32}$

$$\therefore \quad I_{XX} = I_{YY} = \frac{\pi D^4}{32} \qquad ...(3.27)$$

5. Moment of inertia of a circular lamina with a centrally situated circular hole: Refer to Fig. 3.41. Let D and d be the external and internal diameters of the lamina respectively.

Polar moment of inertia = Polar moment of inertia of the bigger circle–polar moment of inertia of the smaller circle.

$$\therefore \quad I_p = \frac{\pi D^4}{32} - \frac{\pi d^4}{32}$$

$$I_{XX} = I_{YY} = \frac{I_p}{2} = \frac{\pi D^4}{64} - \frac{\pi d^4}{64}$$

$$= \frac{\pi}{64}(D^4 - d^4) \qquad ...(3.27\ a)$$

6. Moment of inertia of a semi- circular lamina: In Fig. 3.42 is shown a semi-circle of radius R. Let LM be the base of the semi-circle.

The moment of inertia of a circular lamina about

a diameter LM $= \dfrac{\pi R^4}{4} = \dfrac{\pi D^4}{64}$

$\therefore$ Moment of inertia of the semi-circle about LM

$$I_{LM} = \frac{\pi R^4}{8} = \frac{\pi D^4}{128} \qquad ...(3.28)$$

Let XX be centroidal axis parallel to the base LM. Let h be the distance between the axis XX and LM. We have,

$$h = \frac{4R}{3\pi} = \frac{4}{3\pi} \times \frac{D}{2} = \frac{2D}{3\pi}$$

By theorem of parallel axes, we have

$$I_{LM} = I_{XX} + Ah^2$$

$$\therefore \quad \frac{\pi R^4}{8} = I_{XX} + \frac{\pi R^2}{2} \times \left(\frac{4R}{3\pi}\right)^2$$

From which, $I_{XX} = \dfrac{\pi R^4}{8} - \dfrac{8\pi R^4}{9\pi^2}$

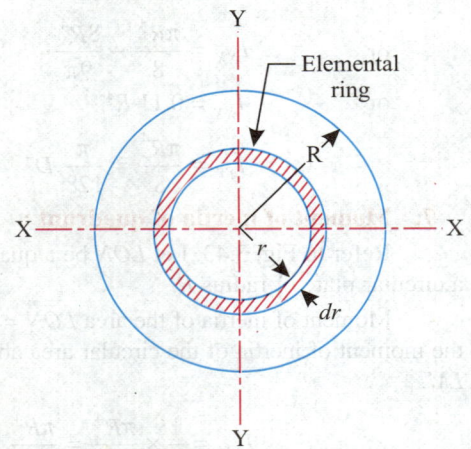

Fig. 3.40

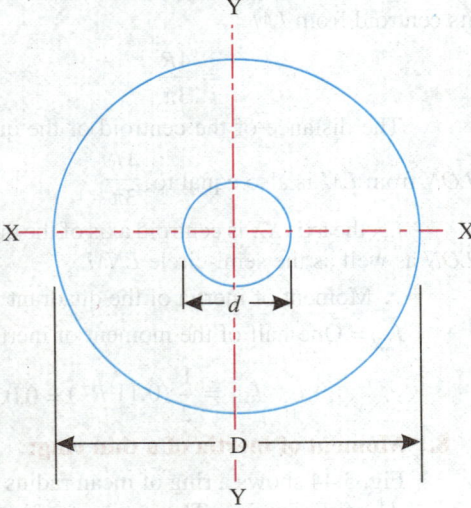

Fig. 3.41

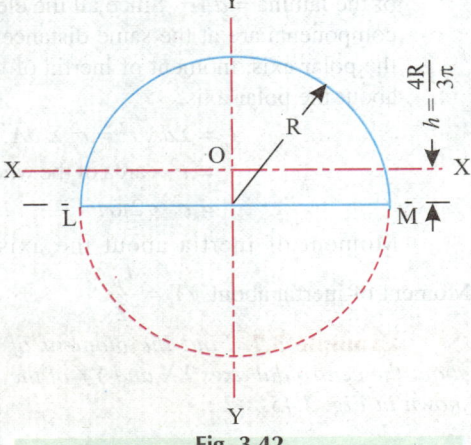

Fig. 3.42

or $\qquad I_{XX} = \dfrac{\pi R^4}{8} - \dfrac{8R^4}{9\pi}$

or $\qquad I_{XX} = 0.11\ R^4$ $\qquad\qquad$ ...(3.29)

$$I_{YY} = \dfrac{\pi R^4}{8} = \dfrac{\pi}{128}\ D^4 \qquad\qquad ...(3.30)$$

7. Moment of inertia of quadrant :

Refer to Fig. 3.43. Let LON be a quadrant of a circular plate of radius R.

Moment of inertia of the area $LON = 1/4$th of the moment of inertia of the circular area about axis LM.

$\therefore \qquad I_{LM} = \dfrac{1}{4} \times \dfrac{\pi R^4}{4} = \dfrac{\pi R^4}{16}$

Consider the semi-circle LNM. Distance of its centroid from LM.

$$= \dfrac{4R}{3\pi}$$

The distance of the centroid of the quadrant LON from LM is also equal to $\dfrac{4R}{3\pi}$

i.e. the axis XX is centroid axis of the quadrant LON as well as the semi-circle LNM.

$\therefore$ Moment of inertia of the quadrant about axis XX,

I_{XX} = One-half of the moment of inertia of the semi-circle about XX.

$\therefore \qquad I_{XX} = \dfrac{1}{2}\ (0.11\ R^4) = 0.055\ R^4$ $\qquad\qquad$ ...(3.31)

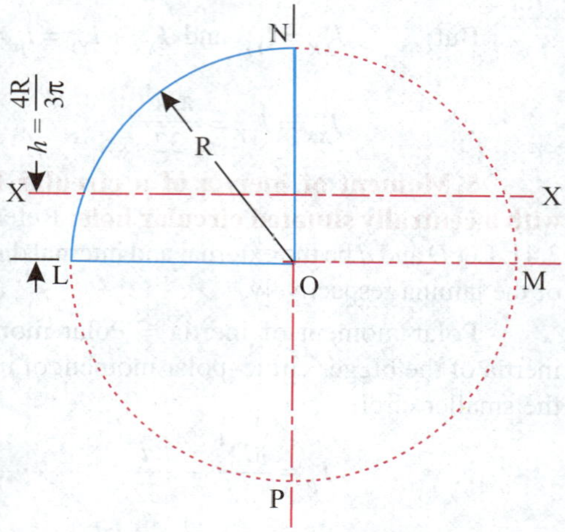

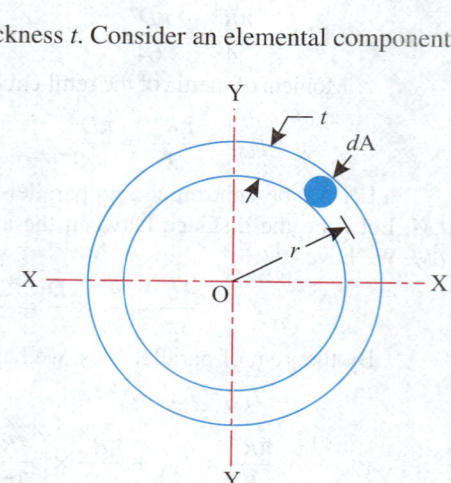

Fig. 3.43

8. Moment of inertia of a thin ring:

Fig. 3.44 shows a ring of mean radius r and of thickness t. Consider an elemental component dA of the lamina. The moment of inertia of this elemental component about the polar axis of the lamina $= dA r^2$. Since all the elemental components are at the same distance r from the polar axis, moment of inertia of the ring about the polar axis,

$I_p = \Sigma dA\ r^2 = r^2\ \Sigma\ dA$

$\qquad = r^2 \times$ area of the whole ring

$\qquad = r^2 \times 2\pi rt \qquad$ ...(3.32)

Moment of inertia about the axis XX =

Moment of inertia about $YY = \dfrac{I_p}{2}$

Fig. 3.44

Example 3.7. *Find the moment of inertia about the centroidal axes XX and YY of the section shown in Fig. 3.45.*

Solution. Refer to Fig. 3.46.

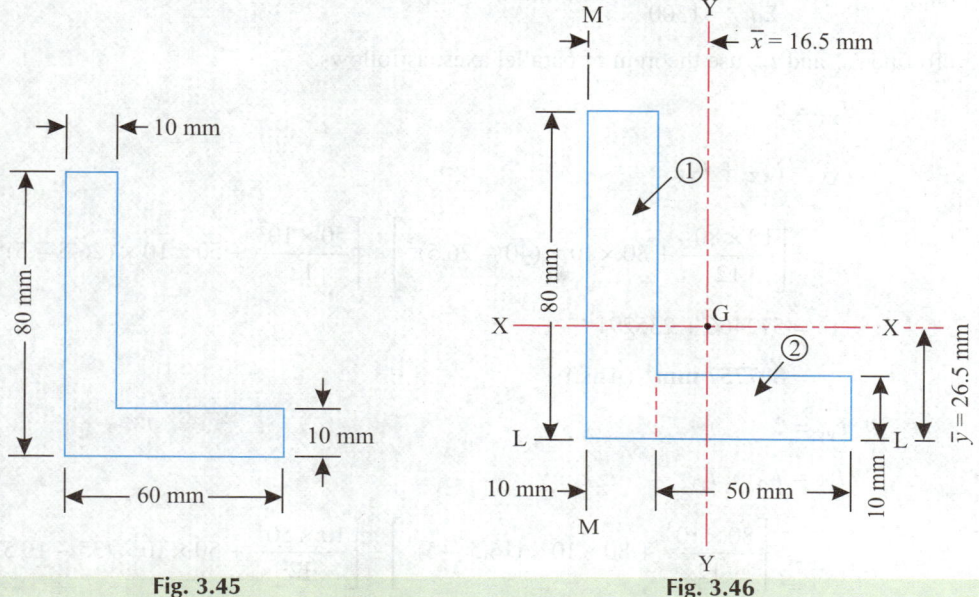

Fig. 3.45	Fig. 3.46

To determine the location of centroid of the section we have the following table:

Components	Area a (mm²)	Centroidal distance 'x' from MM (mm)	Centroidal distance 'y' from LL (mm)	ax (mm³)	ay (mm³)
Rectangle (1)	80 × 10 = 800	5	40	4000	32000
Rectangle (2)	50 × 10 = 500	35	5	17500	2500
Total	Σa = 1300	–	–	Σax = 21500	Σay = 34500

$$\therefore \quad \bar{x} = \frac{\Sigma ax}{\Sigma a} = \frac{21500}{1300} = 16.5 \text{ mm}$$

Racing car wheels.

and, $\bar{y} = \dfrac{\Sigma ay}{\Sigma a} = \dfrac{34500}{1300} = 26.5$ mm

To find I_{XX} and I_{YY} use theorem of parallel axes, as follows:

$I_{XX} = ?$

$I_{XX} = I_{XX_1} + I_{XX_2}$

$$= \left[\frac{10 \times 80^3}{12} + 80 \times 10 \times (40 - 26.5)^2\right] + \left[\frac{50 \times 10^3}{12} + 50 \times 10 \times (26.5 - 5)^2\right]$$

$$= 572466 + 235291$$

$$= \mathbf{807757} \ \textbf{mm}^4 \ \textbf{(Ans.)}$$

$I_{YY} = ?$

$I_{YY} = I_{YY_1} + I_{YY_2}$

$$= \left[\frac{80 \times 10^3}{12} + 80 \times 10 \times (16.5 - 5)^2\right] + \left[\frac{10 \times 50^3}{12} + 50 \times 10 \times (35 - 16.5)^2\right]$$

$$= 112466 + 275291 = \mathbf{387757} \ \textbf{mm}^4 \ \ \textbf{(Ans.)}$$

Example 3.8. *Calculate the moment of inertia about horizontal and vertical gravity axes* (I_{XX} *and* I_{YY}) *of the section shown in Fig. 3.47.*

Solution. Refer Fig. 3.48.

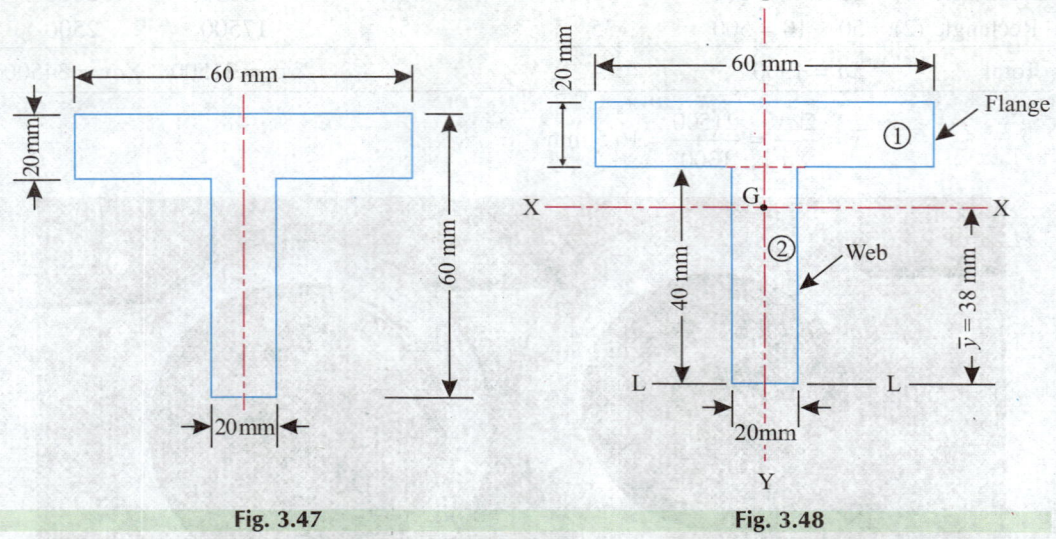

Fig. 3.47	**Fig. 3.48**

Since the section is symmetrical about Y-axis, therefore we shall find out $\bar{y}$ only, for which we have the following table:

Components	Area 'a' (mm²)	Centroidal distance 'y' from LL (mm)	ay (mm³)
Rectangle (1)	$60 \times 20 = 1200$	$40 + \dfrac{20}{2} = 50$	60000
Rectangle (2)	$40 \times 20 = 800$	$\dfrac{40}{2} = 20$	16000
	$\Sigma a = 2000$	–	$\Sigma ay = 76000$

$$\therefore \qquad \bar{y} = \frac{\Sigma ay}{\Sigma a} = \frac{76000}{2000} = 38 \text{ mm}$$

To find I_{XX} and I_{YY} use the theorem of parallel axes as follows:

$$I_{XX} = ?, \quad I_{YY} = ?$$

$$I_{XX} = I_{XX_1} + I_{XX_2}$$

$$I_{XX} = \left[\frac{60 \times 20^3}{12} + 60 \times 20 \times (50 - 38)^2\right] + \left[\frac{20 \times 40^3}{12} + 40 \times 20 \times (38 - 20)^2\right]$$

$$= 212800 + 365866$$

$$= \mathbf{578666 \text{ mm}^4} \quad \textbf{(Ans.)}$$

$$I_{YY} = \left[\frac{20 \times 60^3}{12}\right] + \left[\frac{40 \times 20^3}{12}\right] = 360000 + 26666$$

$$= \mathbf{386666 \text{ mm}^4} \quad \textbf{(Ans.)}$$

Example 3.9. *A T-beam is made up of two plates and two angles as shown in Fig. 3.49. Determine the moment of inertia of T-section about an axis passing through the centroid of the section and parallel to the top plate.*

Solution. Refer Fig. 3.50

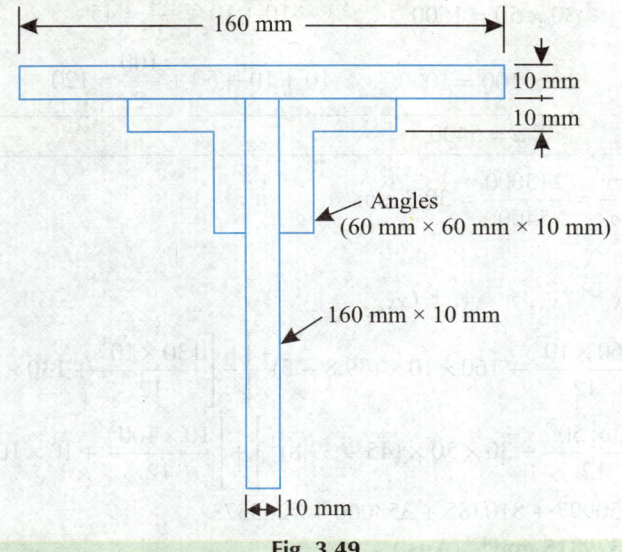

160 mm

10 mm

10 mm

Angles (60 mm × 60 mm × 10 mm)

160 mm × 10 mm

10 mm

Fig. 3.49

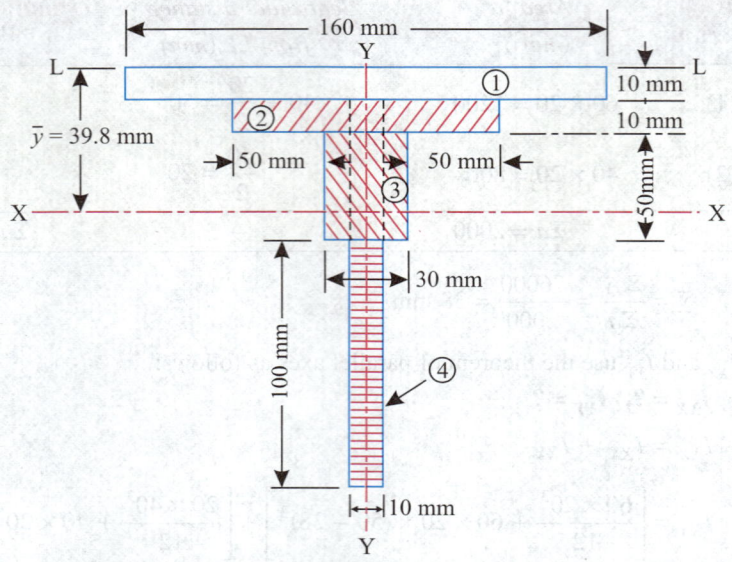

Fig. 3.50

The given section is divided into four components as shown. Since the section is symmetrical about Y-axis we shall find out $\bar{y}$ for which we have the following table:

Components	Area 'a' (mm^2)	Centroidal distance 'y' from LL (mm)	ay (mm^3)
Rectangle (1)	$160 \times 10 = 1600$	$\dfrac{10}{2} = 5$	8000
Rectangle (2)	$130 \times 10 = 1300$	$10 + \dfrac{10}{2} = 15$	19500
Rectangle (3)	$30 \times 50 = 1500$	$10 + 10 + \dfrac{50}{2} = 45$	67500
Rectangle (4)	$10 \times 100 = 1000$	$10 + 10 + 50 + \dfrac{100}{2} = 120$	120000
	$\Sigma\,a = 5400$		$\Sigma\,ay = 215000$

$$\therefore \quad \bar{y} = \frac{\Sigma ay}{\Sigma a} = \frac{215000}{5400} = 39.8 \text{ mm}$$

$I_{XX} = ?$

$$I_{XX} = I_{XX_1} + I_{XX_2} + I_{XX_3} + I_{XX_4}$$

$$I_{XX} = \left[\frac{160 \times 10^3}{12} + 160 \times 10 \times (39.8 - 5)^2\right] + \left[\frac{130 \times 10^3}{12} + 130 \times 10 \times (39.8 - 15)^2\right]$$

$$+ \left[\frac{30 \times 50^3}{12} + 30 \times 50 \times (45 - 39.8)^2\right] + \left[\frac{10 \times 100^3}{12} + 10 \times 100 \times (120 - 39.8)^2\right]$$

$$= 1950997 + 810385 + 353060 + 7265373$$

$$= \mathbf{10379815 \ mm^4} \quad \textbf{(Ans.)}$$

Example 3.10. *A compound section is formed by riveting 20 cm × 1 cm flat plates, one on each flange of (ISL 300) 30 cm × 15 cm I-section girder. Find for the compound section:*

(i) I_{XX}, *(ii)* I_{YY}, *and (iii) The least radius of gyration.*

The properties of I-section are as follows:

$I_{XX} = 7332.9$ *cm*^{4}, $I_{YY} = 376.2$ *cm*^{4}, *area* $= 48.08$ *cm*^{2}

Solution. Refer to Fig. 3.51.

Since the section is symmetrical about both XX and YY axes, therefore the c.g. of the whole section will lie at G as shown in Fig. 3.51.

(i) **Moment of inertia about X-axis , I_{XX}:**

$I_{XX} = I_{XX}$ for I-section + I_{XX} for two plates

$$= 7332.9 + 2 \left[\frac{20 \times 1^3}{12} + 20 \times 1 \times \left(15 + \frac{1}{2} \right)^2 \right] = 7332.9 + 2\,(1.67 + 4805)$$

$$= \textbf{16946.24 cm}^4 \quad \textbf{(Ans.)}$$

(ii) **Moment of inertia about Y-axis I_{YY} :**

$I_{YY} = I_{YY}$ for I-section + I_{YY} for two plates $= 376.2 + 2 \left[\dfrac{1 \times 20^3}{12} \right]$

$$= \textbf{1709.53 cm}^4 \quad \textbf{(Ans.)}$$

(iii) **Least value of radius of gyration, k_{least} :**

$$k_{least} = \sqrt{\frac{\text{Least value of moment of inertia}}{\text{Total area of the section}}} = \sqrt{\frac{1709.53}{48.08 + 2 \times 20 \times 1}} = \sqrt{\frac{1709.53}{88.08}}$$

$$= \textbf{4.4 cm. (Ans.)}$$

Marine hydraulic unit.

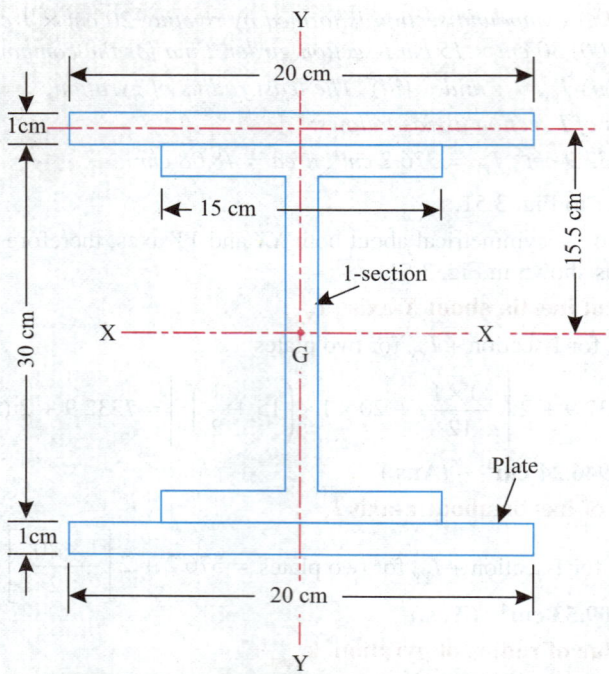

Fig. 3.51

Example 3.11. *Find the centroidal moment of inertia of the shaded area shown in Fig. 3.52.*

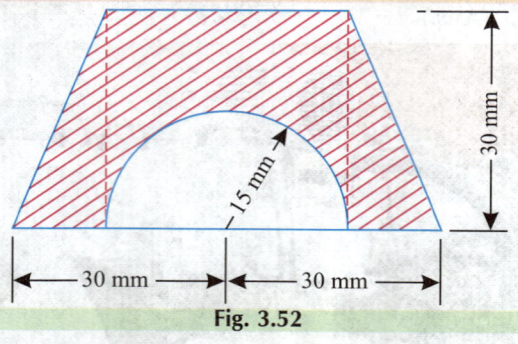

Fig. 3.52

Solution. Refer to Fig. 3.53.

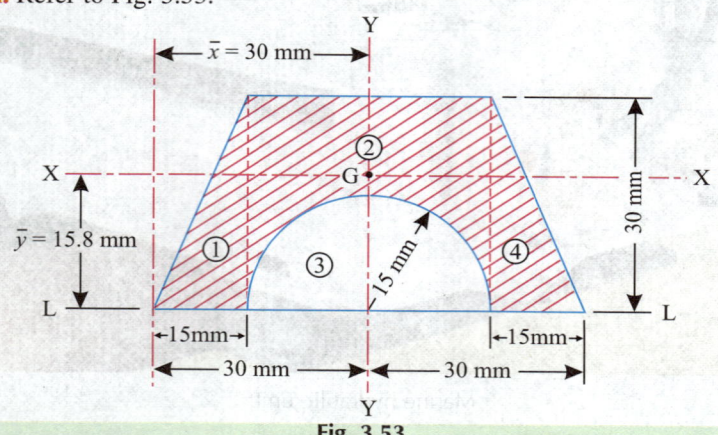

Fig. 3.53

Since the figure is symmetrical about *YY*-axis, therefore we shall find $\bar{y}$ for which we have the following table:

Component	Area 'a' (mm^2)	Centroidal distance 'y' from LL (mm)	ay (mm^3)
Triangle (1)	$\dfrac{1}{2} \times 15 \times 30 = 225 \ (+)$	$\dfrac{30}{3} = 10$	2250 (+)
Rectangle (2)	$30 \times 30 = 900 \ (+)$	$\dfrac{30}{2} = 15$	13500 (+)
Semicircle (3)	$\dfrac{\pi \times 15^2}{2} = 353.4 \ (-)$	$\dfrac{4 \times 15}{3\pi} = 6.37$	2251.2 (−)
Triangle (4)	$\dfrac{1}{2} \times 15 \times 30 = 225 \ (+)$	$\dfrac{30}{3} = 10$	2250 (+)
	$\Sigma a = 996.6$		$\Sigma ay = 15748.8$

$$\therefore \quad \bar{y} = \frac{\Sigma ay}{\Sigma a} = \frac{15748.8}{996.6} = 15.8 \text{ mm}$$

$I_{XX} = ?$

$$I_{XX} = I_{XX_1} + I_{XX_2} - I_{XX_3} + I_{XX_4}$$

$$I_{XX} = \left[\frac{15 \times (30)^3}{36} + \frac{1}{2} \times 15 \times 30 \left(15.8 - \frac{30}{3} \right)^2 \right] + \left[\frac{30 \times 30^3}{12} + 30 \times 30 \times (15.8 - 15)^2 \right]$$

$$- \left[0.11 \times 15^4 + \frac{\pi \times 15^2}{2} \times \left(15.8 - \frac{4 \times 15}{3\pi} \right)^2 \right]$$

$$+ \left[\frac{15 \times 30^3}{36} + \frac{1}{2} \times 15 \times 30 \left(15.8 - \frac{30}{3} \right)^2 \right]$$

Tractors being transported on load carriers.

$$= 18819 + 68076 - 37023 + 18819$$
$$= 68691 \text{ mm}^4$$

i.e. $I_{XX} = \mathbf{68691 \text{ mm}^4}$ **(Ans.)**

$I_{YY} = ?$

$I_{YY} = I_{YY_1} + I_{YY_2} - I_{YY_3} + I_{YY_4}$

$$I_{YY} = \left[\frac{30 \times 15^3}{36} + \frac{1}{2} \times 30 \times 15 \times \left(30 - \frac{2}{3} \times 15\right)^2 + \left[\frac{30 \times 30^3}{12}\right] - \left[\frac{\pi \times 30^4}{64 \times 2}\right]\right]$$

$$+ \left[\frac{30 \times 15^3}{36} + \frac{1}{2} \times 30 \times 15 \times \left(30 - \frac{2}{3} \times 15\right)^2\right]$$

$$= 92812.5 + 67500 - 19880.4 + 92812.5$$
$$= 233244.6 \text{ mm}^4$$

i.e. $I_{YY} = \mathbf{233244.6 \text{ mm}^4}$ **(Ans.)**

Example 3.12. *For the shaded area shown in Fig. 3.54 find the following:*

(*i*)   *The position of the centroid.*

(*ii*) *The second moment of area about the base.*

(*iii*) *The radius of gyration about the base.*

Solution.

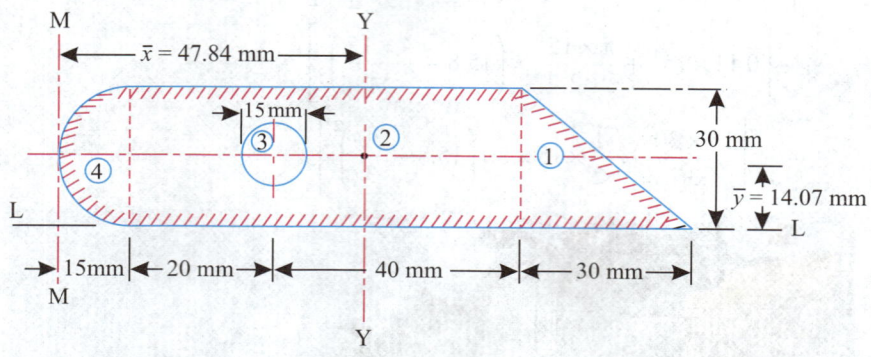

Fig. 3.54

(*i*) **The position of the centroid:**

To determine the location of the centroid of the shaded area we have the following table:

Component	Area 'a' (mm^2)	Centroidal distance 'x' from MM (mm)	Centroidal distance 'y' from LL (mm)	ax (mm^3)	ay (mm^3)
Triangle (1)	$\frac{1}{2} \times 30 \times 30 = 450$	$15 + 20 + 40 + \frac{30}{3}$ $= 85$	$\frac{30}{3} = 10$	38250	4500
Rectangle (2)	$60 \times 30 = 1800$	$15 + \frac{60}{2} = 45$	$\frac{30}{2} = 15$	81000	27000

Circle (3)	$\pi \times 7.5^2 = 176.71(-)$	$15 + 20 = 35$	$\dfrac{30}{2} = 15$	$6184.8\ (-)$ $\quad$ $2650.6\ (-)$
Semicircle (4)	$\dfrac{\pi \times 15^2}{2} = 353.43$	$15 - \dfrac{4 \times 15}{3\pi}$ $= 8.63$	$\dfrac{30}{2} = 15$	3050.5 $\quad\quad$ 5301.4

			$\Sigma\, ax$	$\Sigma\, ay$
$\Sigma a = 2426.72$	$-$	$-$	$= 116115.2$	$= 34150.8$

$$\bar{x} = \frac{\Sigma\, ax}{\Sigma a} = \frac{116115.2}{2426.72} = 47.84 \text{ mm} \quad \textbf{(Ans.)}$$

$$\bar{y} = \frac{\Sigma\, ay}{a} = \frac{34150.8}{2426.72} = 14.07 \text{ mm} \quad \textbf{(Ans.)}$$

(ii) The second moment of area (M.O.I) about the base, I_{LL} :

Applying parallel axes theorem,

$$I_{LL} = I_G + Ah^2$$

where, I_G = M.O.I. about C.G,

$\quad\quad\quad A$ = Area of the lamina, and

$\quad\quad\quad h$ = Distance between the C.G. and the reference line (i.e., LL in this case).

∴ M.O.I. about base LL,

$\quad\quad I_{LL}$ = M.O.I. of triangle (1) + M.O.I of rectangle (2) –

$$M.O.I. \text{ of circle (3)} + M.O.I. \text{ of semicircle (4)}$$

$$= \left[\frac{30 \times 30^3}{12}\right] + \left[\frac{60 \times 30^3}{12} + 60 \times 30 \times 15^2\right] - \left[\frac{\pi \times 15^4}{64} + \frac{\pi \times 15^2}{4} \times 15^2\right]$$

$$+ \left[\frac{\pi \times 30^4}{64 \times 2} + \frac{\pi \times 30^2}{8} \times 15^2\right]$$

$$= 67500 + 540000 - 42245.8 + 99402$$

$$= 664656 \text{ mm}^4 = \textbf{664656 mm}^4 \quad \textbf{(Ans.)}$$

(iii) The radius of gyration about the base, k_{LL}:

$$I_{LL} = Ak_{LL}^2$$

$$\therefore \quad k_{LL} = \sqrt{\frac{I_{LL}}{A}} = \sqrt{\frac{664656}{2426.72}} = 16.55 \text{ mm}$$

$$\textbf{(Ans.)}$$

Example 3.13. *Determine the moment of inertia of the shaded area about the edge LM as shown in Fig. 3.55.*

Solution.

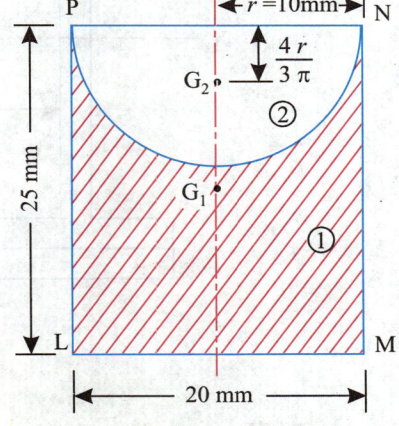

Fig. 3.55

$$I_{LM} = ?$$

$\quad I_{LM}$ = M.O.I. of rectangle (1) – M.O.I. of semicircle (2)

$$= \left[\frac{20 \times 25^3}{12} + 20 \times 25 \times \left(\frac{25}{2}\right)^2\right]$$

$$-\left[0.11 \times 10^4 + \frac{\pi \times 10^2}{2} \times \left(25 - \frac{4 \times 10}{3\pi}\right)^2\right]$$

$$= 104166 - 68771$$

$$= 35395 \text{ mm}^4 \quad \text{(Ans.)}$$

Example 3.14. *A column is made up of two rolled steel joists of I-section 16 cm × 8 cm × 1 cm with plate 20 cm × 1 cm riveted with flanges one each on the top and at the bottom. The edges of the plates flush with outside edges of joists flanges. Find for the compound section.*

(i) I_{XX};   (ii) I_{YY};   (iii) *Least radius of gyration.*

Solution. Refer to Fig. 3.56.

Area of cross-section of the column,

$$A = 2[8 \times 1 \times 2 + 14 \times 1] + 2 \times 20 \times 1$$
$$= 2 \times 30 + 2 \times 20$$
$$= 100 \text{ cm}^2$$

As the section is symmetrical, the c.g. will lie at the point of intersection of two axes of symmetry as shown in Fig. 3.56.

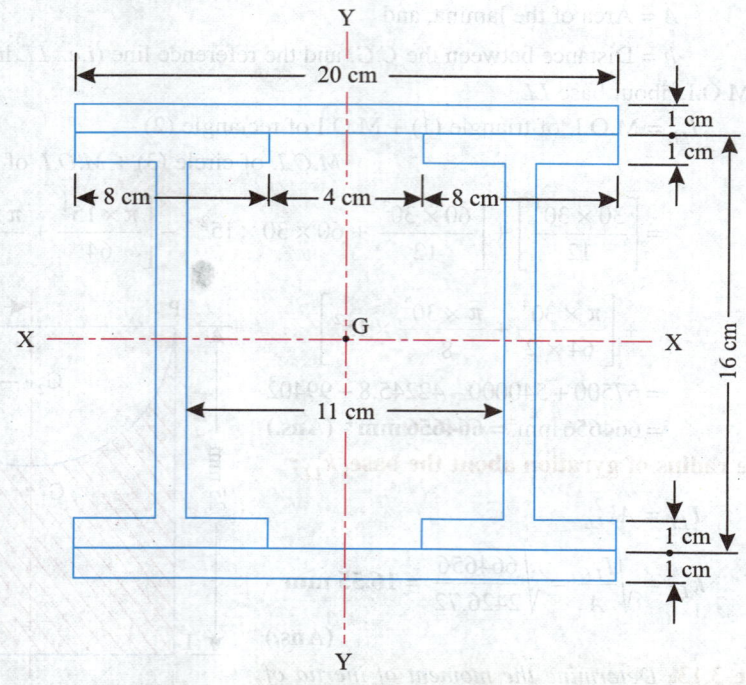

Fig. 3.56

(i) Moment of inertia about X-axis, I_{XX} :

$$I_{XX} = 4\left[\frac{8 \times 1^3}{12} + 8 \times 1 \times (8 - 0.5)^2\right] + 2\left[\frac{1 \times 14^3}{12}\right]$$

$$+ 2\left[\frac{20 \times 1^3}{12} + 20 \times 1 \times (8 + 0.5)^2\right]$$

$$= 1802.6 + 457.3 + 2893.3$$

$$= \textbf{5153.2 cm}^4 \quad \textbf{(Ans.)}$$

(ii) **Moment of inertia about Y-axis, I_{YY} :**

I_{YY} for one I-section only

$$= 2 \left[\frac{1 \times 8^3}{12} \right] + \frac{14 \times 1^3}{12} = 85.33 + 1.16 = 86.49 \text{ cm}^4$$

I_{YY} for whole section

$$= 2 \left[86.49 + 30 \times 6^2 \right] + 2 \times \frac{1 \times 20^3}{12}$$

$$= 2332.98 + 1333.33 = \textbf{3666.31 cm}^4 \quad \textbf{(Ans.)}$$

(iii) **Least radius of gyration, k_{least} :**

$$k_{least} = \sqrt{\frac{\text{Least value of moment of inertia}}{\text{Total area of the section}}} = \sqrt{\frac{3666.31}{100}}$$

$$= \textbf{6.05 cm} \quad \textbf{(Ans.)}$$

Example 3.15. *A fabricated girder system consists of two plates 30 cm × 1.5 cm and two ISLC 300 channels 30 cm × 10 cm set 10 cm apart as shown in Fig. 3.57. The properties of a 30 cm × 10 cm channel section are as follows:*

$I_{XX} = 6047.9 \text{ cm}^4$, $I_{YY} = 34.60$ *cm*^{4}, *area = 42.11 cm*^{2}, *axis yy from back of channel = 2.55 cm.*

Calculate the value of moment of inertia and radii of gyration about the horizontal and vertical axes through the centroid of the combined section. Neglect the effect of rivets etc.

Pulley.

Solution. Refer to Fig. 3.57.

Since the section is symmetrical the centroid will lie on the intersection of two axes of symmetry, *LL* and *MM* as shown,

$$I_{LL} = I_{LL} \text{ or } I_{XX} \text{ for channels} + I_{LL} \text{ for plates}$$

$$= 2 \times 6047.9 + 2 \left[\frac{30 \times (1.5)^3}{12} + 30 \times 1.5 \times \left(15 + \frac{1.5}{2} \right)^2 \right] = 12095.8 + 22342.5$$

$$= \textbf{34438.3 cm}^4. \quad \textbf{(Ans.)}$$

$$I_{MM} = I_{MM} \text{ for channels} + I_{MM} \text{ for plates}$$

$$= 2 \left(I_{YY} \text{ for channels} + Ah^2 \right) + I_{MM} \text{ for plates}$$

$$= 2 \left[346 + 42.11 \left(\frac{10}{2} + 2.55 \right)^2 \right] + 2 \times \frac{1.5 \times 30^3}{12} = 5492.75 + 6750$$

$$= \textbf{12242.75 cm}^4 \quad \textbf{(Ans.)}$$

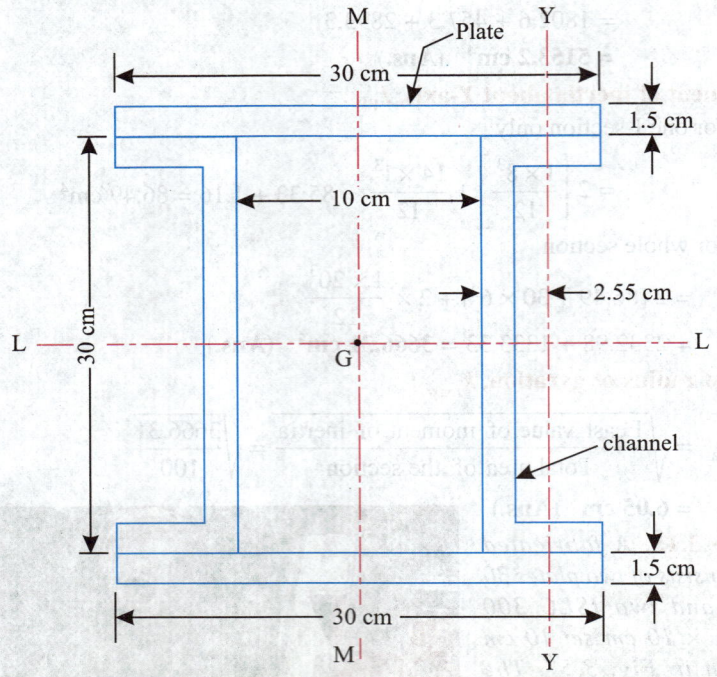

Fig. 3.57

Radius of gyration about *LL* axis,

$$k_{LL} = \sqrt{\frac{I_{LL}}{\text{Area}}} = \sqrt{\frac{34438.3}{2 \times 42.11 + 2 \times 30 \times 1.5}}$$

$$= \textbf{14.06 cm.} \quad \textbf{(Ans.)}$$

Radius of gyration about *MM* axis,

$$k_{MM} = \sqrt{\frac{I_{MM}}{\text{Area}}} = \sqrt{\frac{12242.75}{2 \times 42.11 + 2 \times 30 \times 1.5}}$$

$$= \textbf{8.38 cm.} \quad \textbf{(Ans.)}$$

Example 3.16. *A stanchion is built up of three ISMB 225 as in Fig. 3.58. Calculate second moment of area of the built-up-section about the principal axes LL and MM, and the least radius of gyration.*

Solution. Refer to Fig. 3.58

From steel tables for one *ISMB* 225
Area, $a = 39.72$ cm^2

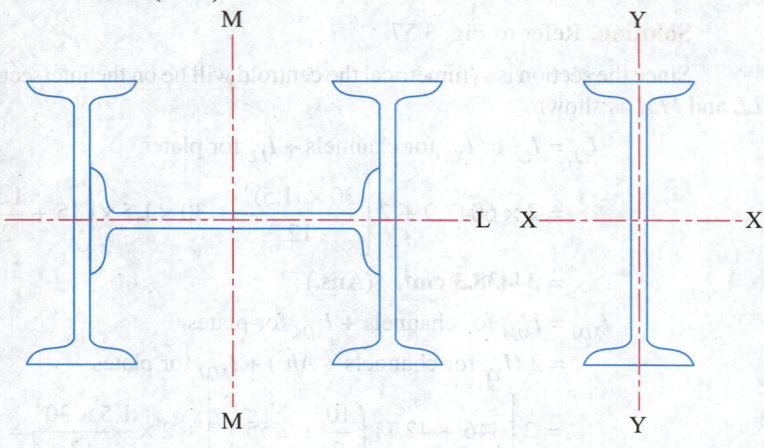

Fig. 3.58

Web thickness $= 6.5$ mm

$$I_{XX} = 3441.8 \text{ cm}^4$$
$$I_{YY} = 218.3 \text{ cm}^4$$

Since axis LL passes through the centroids of all the three beams,

$$I_{LL} = I_{XX} \text{ for two extreme beams} + I_{YY} \text{ for central beam}$$
$$= 2 \times 3441.8 + 218.3$$
$$= 7101.9 \text{ cm}^4. \quad \textbf{(Ans.)}$$

Axis MM passes through the centroids of central beam but is parallel to axis YY of two extreme beams.

$$\therefore \ I_{MM} = I_{XX} \text{ for central beam} + 2 [I_{YY} + \text{area} \times \text{(distance between axis } MM \text{ and } YY \text{ of extreme beam)}^2]$$

$$= 3441.8 + \left[218.3 + 39.72 \left(\frac{22.5}{2} + \frac{0.65}{2} \right)^2 \right]$$

$$= 14521.82 \text{ cm}^4. \quad \textbf{(Ans.)}$$

Least radius of gyration $= \sqrt{\dfrac{\text{Least of } I_{LL} \text{ or } I_{MM}}{\text{Area of compound beam}}} = \sqrt{\dfrac{7101.9}{3 \times 39.72}}$

$$= 7.72 \text{ cm}. \quad \textbf{(Ans.)}$$

3.16. PRODUCT OF INERTIA

Consider a plane area or section (area 'A') as shown in the Fig. 3.59. Further consider an elemental area 'dA' at a distance x and y from the YY-axis, and XX-axis respectively (as shown in Fig. 3.59). Then $\Sigma\, xy\, dA$ is *defined as the product of inertia of the cross-section.*

Mathematically, $\qquad I_{XY} = \Sigma xy\, dA = \displaystyle\int_A xy\, dA$ \hfill ...(3.33)

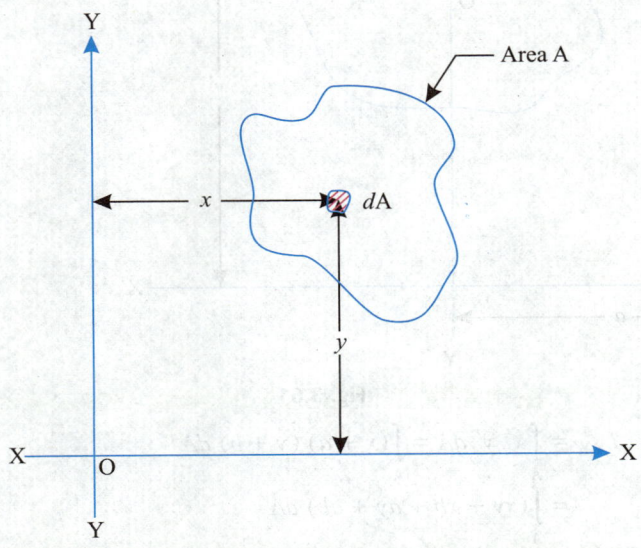

Fig. 3.59

Sign of product of inertia:

The sign of product of inertia depends upon the co-ordinates of the various small areas of the plane figure with reference to the co-ordinate axes *XX* and *YY* about which the product of inertia is to be found out as shown in Fig. 3.60.

Example. *In quadrant IV the sign of X co-ordinate is positive and that of Y co-ordinate is negative, giving the sign for the product of inertia as negative (+ve) × (−ve) = (−ve).*

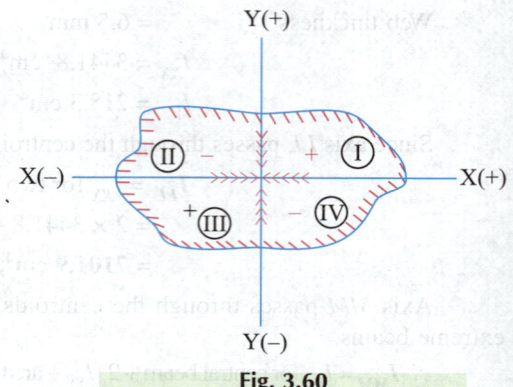

Fig. 3.60

3.16.1. Parallel Axes Theorem for Product of Inertia

When the product of inertia of a plane area or figure is known about its centroidal axes *XX* and *YY*, it is possible to transform the moments of inertia about any other two axes *X'X'* and *Y'Y'* parallel to the axes *XX* and *YY*.

Consider an area '*A*' with axes as shown in Fig. 3.61, *dA* being small area (hatched). Refering to Fig. 3.61,

Pin wrench socket.

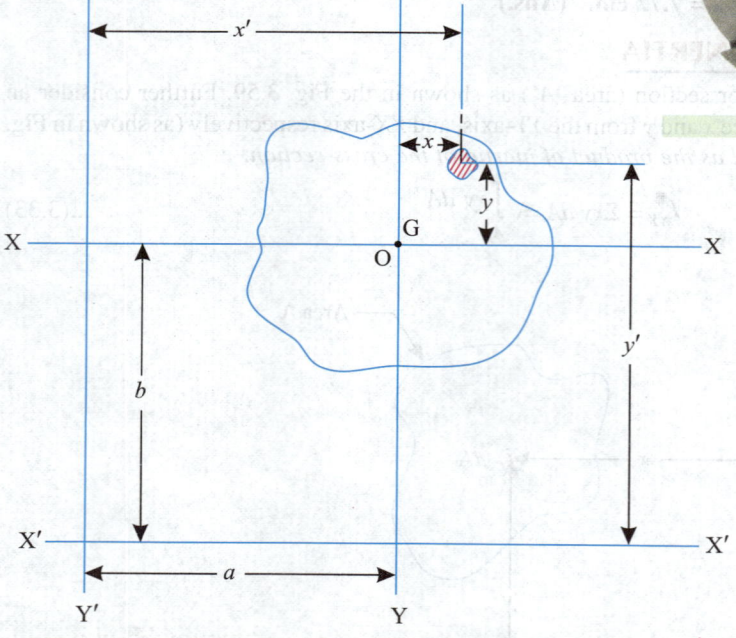

Fig. 3.61

$$I_{XY'} = \int_A x' \, y' \, dA = \int_A (x + a) \, (y + b) \, dA$$

$$= \int_A (xy + xb + ay + ab) \, dA$$

$$= \int_A xy \, dA + b \int_A x \, dA + a \int_A y \, dA + ab \int_A dA$$

$$= I_{xy} + 0 + 0 + abA$$

Hence, $\qquad I_{X'Y'} = I_{xy} + A\,ab$ $\qquad\qquad\qquad$ ...(3.34)

This is the *transfer formula for the product of inertia.*

3.16.2. Principal Axes and Principal Moments of Inertia

If *the two axes about which the product of inertia is found, are such, that the product of inertia becomes zero, the two axes are then called the* **principal axes.** *The moment of inertia about a principal axis is called the* **principal moment of inertia.**

3.16.2.1. Determination of principal moments of inertia and directions of principal axes

Refer to Fig. 3.62. *XX* and *YY* are the axes through the centroid '*O*' of the section and *X′ X′* and *Y′ Y′* are the two more axes inclined at angle θ to the axes *XX* and *YY*. *dA* is the small area with co-ordinates (*x*, *y*) with respect to axes *XX* and *YY* and co-ordinates (*x′*, *y′*) with respect of axes *X′X′* and *Y′Y′* . ∠*XOX′* = θ.

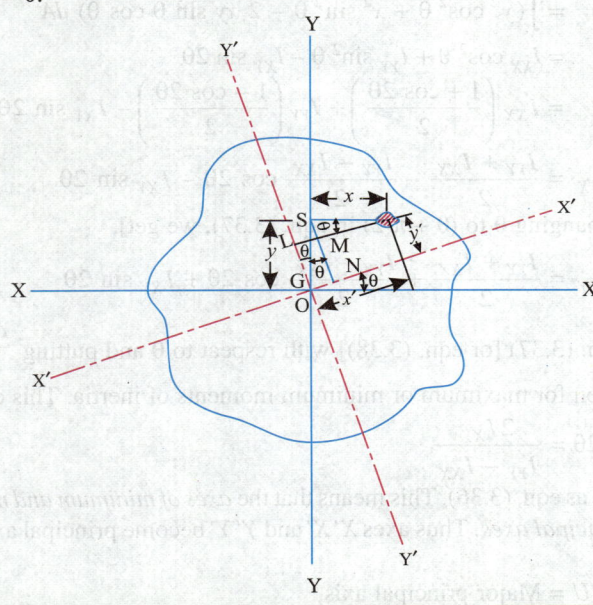

Fig. 3.62

Now, $\qquad x' = LM + x\cos\theta$

$$= ON + x\cos\theta$$

$$= y\sin\theta + x\cos\theta$$

$$y' = OL = MN = SN - SM$$

$$= y\cos\theta - x\sin\theta$$

Now, $I_{X'Y'} = \displaystyle\int x'\,y'\,dA$

$$= \int (x\cos\theta + y\sin\theta)(y\cos\theta - x\sin\theta)\,dA$$

$$= \int (xy\cos^2\theta - x^2\sin\theta\cos\theta + y^2\sin\theta\cos\theta - xy\sin^2\theta)\,dA$$

$$= \int xy(\cos^2\theta - \sin^2\theta)\,dA - \int x^2\frac{\sin 2\theta}{2}\,dA + \int y^2\frac{\sin 2\theta}{2}\,dA$$

$$= \cos 2\theta \int xy\,dA - \frac{\sin 2\theta}{2}\left[\int x^2\,dA - \int y^2\,dA\right]$$

$$= I_{XY} \cos 2\theta - \frac{\sin 2\theta}{2} (I_{YY} - I_{XX})$$

$$\therefore \qquad I_{X'Y'} = I_{XY} \cos 2\theta - \frac{I_{YY} - I_{XX}}{2} \sin 2\theta \qquad\qquad ...(3.35)$$

This is the *general equation for finding the product of inertia about any two perpendicular axes if the moments of inertia and product of inertia about two other perpendicular axes are known.*

When $I_{X'Y'} = 0$, the axes $X'X'$ and $Y'Y'$ will become the principal axes. Under this condition, we get the following relation from eqn. (3.35).

$$\tan 2\theta = \frac{2 I_{XY}}{I_{YY} - I_{XX}} \qquad\qquad ...(3.36)$$

This equation (3.36) locates the direction of the principal axes by giving two values of θ differing by 90°.

Now, $\qquad I_{XX'} = \int y'^2 \, dA = \int (y \cos \theta - x \sin \theta)^2 \, dA$

$$= \int (y^2 \cos^2 \theta + x^2 \sin^2 \theta - 2\, xy \sin \theta \cos \theta) \, dA$$

$$= I_{XX} \cos^2 \theta + I_{YY} \sin^2 \theta - I_{XY} \sin 2\theta$$

$$= I_{XX} \left(\frac{1 + \cos 2\theta}{2} \right) + I_{YY} \left(\frac{1 - \cos 2\theta}{2} \right) - I_{XY} \sin 2\theta$$

$$\therefore \qquad I_{XX'} = \frac{I_{YY} + I_{XX}}{2} - \frac{I_{YY} - I_{XX}}{2} \cos 2\theta - I_{XY} \sin 2\theta \qquad\qquad ...(3.37)$$

Similarly [or by changing θ to $(\theta + \pi/2)$ in eqn. (3.37), we get],

$$I_{Y'Y'} = \frac{I_{YY} + I_{XX}}{2} + \frac{I_{YY} - I_{XX}}{2} \cos 2\theta + I_{XY} \sin 2\theta \qquad\qquad ...(3.38)$$

Differentiating eqn. (3.37) [or eqn. (3.38)] with respect to θ and putting $\dfrac{d \, I_{XX'}}{d\theta} = 0$,

we get the condition for maximum or minimum moments of inertia. This condition is:

$$\tan 2\theta = \frac{2 I_{XY}}{I_{YY} - I_{XX}} \qquad\qquad ...(3.39)$$

Eqn. (3.39) is same as eqn. (3.36). This means that the *axes of minimum and maximum moments of inertia are also the principal axes.* Thus axes $X' X'$ and $Y' Y'$ become principal axes if θ is given by eqn. (3.36) or (3.39).

Let, $\qquad\qquad UU$ = Major principal axis,

$\qquad\qquad\qquad VV$ = Minor principal axis,

$\qquad\qquad\qquad I_{UU}$ = Maximum moment of inertia, and

$\qquad\qquad\qquad I_{VV}$ = Minimum moment of inertia.

The values of I_{UU} and I_{VV} are given by :

$$I_{UU} = \left[\frac{I_{YY} + I_{XX}}{2} \right] + \sqrt{ \left[\frac{I_{YY} - I_{XX}}{2} \right]^2 + (I_{XY})^2 } \qquad\qquad ...(3.40)$$

and, $\qquad\qquad I_{VV} = \left[\frac{I_{YY} + I_{XX}}{2} \right] - \sqrt{ \left[\frac{I_{YY} - I_{XX}}{2} \right]^2 + (I_{XY})^2 } \qquad\qquad ...(3.41)$

Eqns. (3.40) and (3.41) can be used to find the values of I_{UU} and I_{VV} while the eqn. (3.36) will give the inclination of the principal axes.

Adding eqns. (3.40) and (3.41), we get

$$I_{UU} + I_{VV} = I_{XX} + I_{YY} \qquad\qquad ...(3.42)$$

Adding eqns. (3.37) and (3.3 8), we get

$$I_{X'Y'} + I_{Y'Y'} = I_{XX} + I_{YY}$$ ...(3.43)

Hence from eqns. (3.41) and (3.42) it is clear that the *sum of moments of inertia about any two perpendicular axes remains constant.*

3.16.2.2. Mohr's circle for principal moments of inertia

Mohr's circle can be drawn as follows:

Refer to Fig. 3.63.

1. Take origin O and from it measure $OL = I_{YY}$ and $OM = I_{XX}$ along the horizontal I-axis.

2. At L draw upward perpendicular LW equal to I_{XY} if I_{XY} is positive (and a downward perpendicular if I_{XY} is negative).

3. Bisect LM at N.

4. With N as the centre and NW as radius, draw a circle cutting OL and OM extended at S and T respectively.

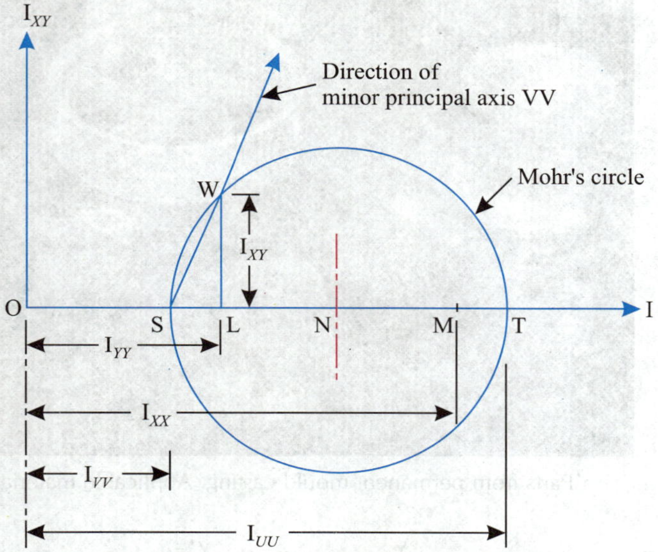

Then OS gives 1_{VV} and OT gives I_{UU}

Fig. 3.63

The direction SW gives the direction of minor principal axis and a direction perpendicular to it, is the direction of major principal axis.

Note. I_{XX} and I_{YY} are always *positive*.

Example 3.17. *Find the product of inertia for the section shown in the Fig. 3.64 about axes XX and YY and hence find the product of inertia about parallel axes X' X' and Y' Y' through centroid.*

Solution. Refer Fig. 3.64.

The entire area is divided into two rectangles (1) and (2) separated by a dotted line. The position of c.g. of each area (G_1 and G_2) and total area (G) is shown in the Fig. 3.64.

$$I_{XY} = ?$$

The product of inertia of rectangle (1) about XX and YY

= Product of inertia of rectangle (1) about its c.g. (G_1) +

$A\, ab$ (parallel axes theorem)

$$= 0 + (200 \times 20)\,(+ 100)\,(- 10)$$

$$= -400 \times 10^4 \text{ mm}^4 \quad \left[\begin{array}{l} \because \quad a = +100 \\ \text{and, } b = -10 \end{array} \right]$$

The product of inertia of rectangle (2) about XX and YY

= Product of inertia of rectangle (2) about its c.g. (G_2) + Aab (parallel axes theorem)

$$= 0 + (20 \times 100)\,(+ 10)\,(- 70) = -140 \times 10^4 \text{ mm}^4$$

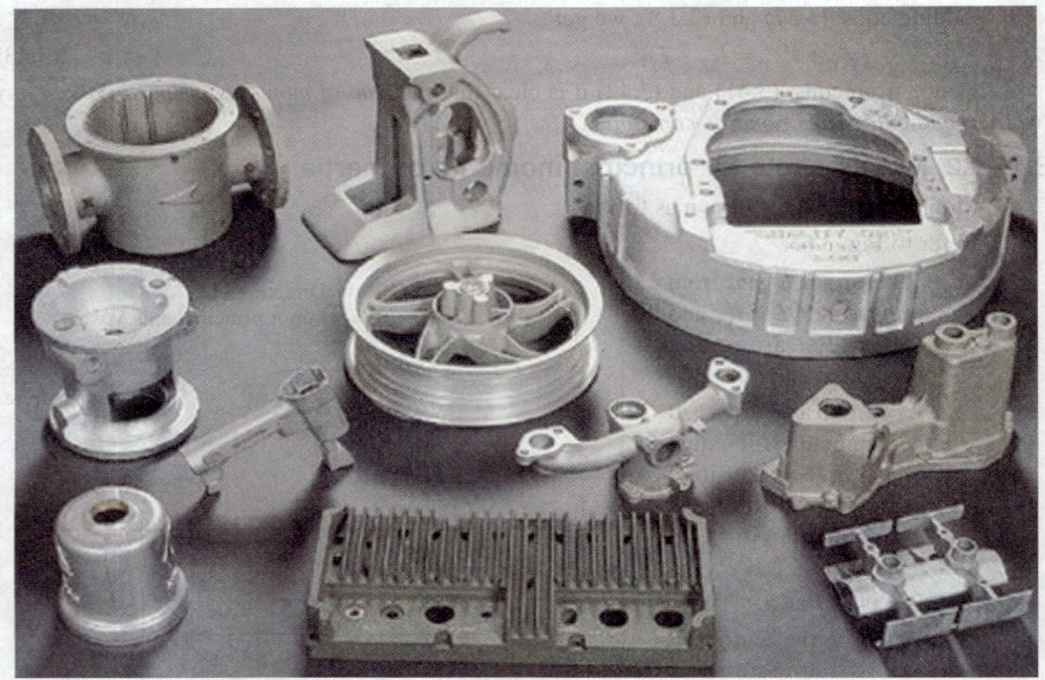

Parts from permanent mould casting. Applicable materials: Aluminium alloy, zinc alloy.

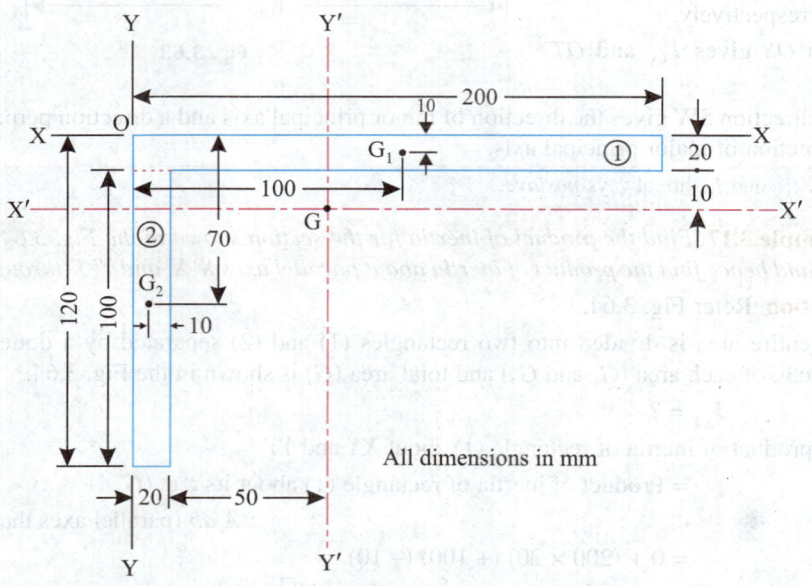

All dimensions in mm

Fig. 3.64

∴ I_{XY} = Total product of inertia = $-400 \times 10^4 - 140 \times 10^4$

= -540×10^4 mm^4 (Ans.)

$I_{X'Y'}$ = ?

$I_{X'Y'}$ = Product of inertia about $X'X'$ and $Y'Y'$ (parallel axes theorem)

$$= I_{XY} - A\,ab$$

$$= -540 \times 10^4 - (20 \times 200 + 20 \times 100)\,(+70)\,(-30) = +720 \times 10^4 \text{ mm}^4$$

Hence, $\quad I_{X'\,Y'} = +720 \times 10^4 \text{ mm}^4$ **(Ans.)**

Example 3.18. *Find the product of inertia for the plane hatched area about the axes XX and YY shown in Fig. 3.65.*

Solution. Fig. 3.65.

The whole area is divided into the following three areas:

(*i*) Rectangle (1);

(*ii*) Triangle (2);

(*iii*) Semi-circle (3).

Let, $\ I_{\bar X \bar Y}$ = Product of inertia for each area about its centroidal axes parallel to the given axes

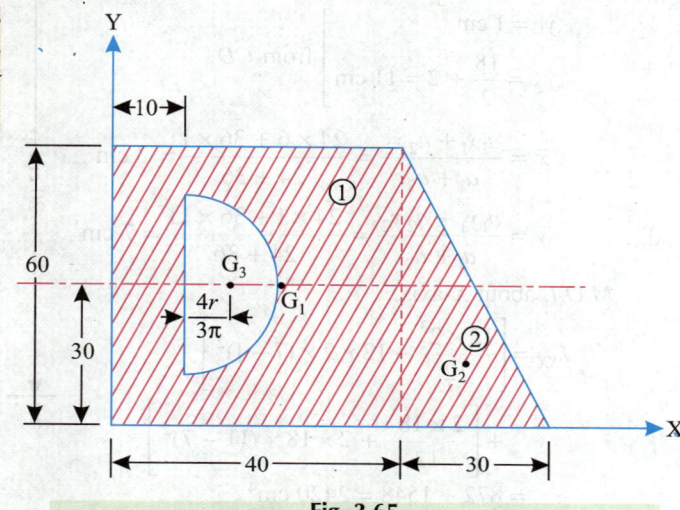

Fig. 3.65

For area (1) i.e. Rectangle (40×60) :

$$I_{XY} = I_{\bar X \bar Y} + A\,ab = 0 + 40 \times 60 \times (+20)\,(+30) = +144 \times 10^4 \text{ mm}^4 \ (+)$$

For area (2) i.e. Triangle:

$$I_{XY} = I_{\bar X \bar Y} + A\,ab$$

$$= -\frac{(30)^2\,(60)^2}{72} + \frac{30 \times 60}{2}\,(+50)\,(+20) \qquad \left[\because \text{For a triangle} \quad I_{\bar x \bar y} = -\frac{b^2\,h^2}{72} \right]$$

$$= -45000 + 900000 = 855000 \ \text{ or } \ 85.5 \times 10^4 \text{ mm}^4$$

For area (3) i.e. semi-circle:

$$I_{XY} = I_{\bar X \bar Y} + A\,ab$$

$$= 0 + \frac{\pi\,r^2}{2}\left(10 + \frac{4r}{3\pi}\right)(+30) = \frac{\pi \times 10^2}{2}\left(10 + \frac{4 \times 10}{3\pi}\right)(+30)$$

$$= 50\,\pi\left(300 + \frac{400}{\pi}\right) = 6.71 \times 10^4 \text{ mm}^4 \ (-)$$

Total I_{XY} for whole hatched area,

$$= +144 \times 10^4 + 85.5 \times 10^4 - 6.71 \times 10^4 = 222.79 \times 10^4 \text{ mm}^4$$

Hence, I_{xy} **(total)** $= 222.79 \times 10^4 \text{ mm}^4$ **(Ans.)**

Example 3.19. *Find the principal moments of inertia and directions of principal axes for the angle section shown in Fig. 3.66 by (i) Calculations (ii) Mohr's circle construction.*

Solution. Refer to Fig. 3.67.

(*i*) *By calculations:*

Area $a_1 = 12 \times 2 = 24 \text{ cm}^2$

Area $a_2 = 18 \times 2 = 36$ cm^2

$$\left.\begin{array}{l} x_1 = 6 \text{ cm} \\ x_2 = 1 \text{ cm} \end{array}\right\} \text{from } AB$$

$$\left.\begin{array}{l} y_1 = 1 \text{ cm} \\ y_2 = \dfrac{18}{2} + 2 = 11 \text{ cm} \end{array}\right\} \text{from } CD$$

$$\therefore \quad \bar{x} = \frac{a_1 x_1 + a_2 x_2}{a_1 + a_2} = \frac{24 \times 6 + 36 \times 1}{24 + 36} = 3 \text{ cm}$$

and,

$$\bar{y} = \frac{a_1 y_1 + a_2 y_2}{a_1 + a_2} = \frac{24 \times 1 + 36 \times 11}{24 + 36} = 7 \text{ cm}$$

M.O.I. about X-axis,

$$I_{XX} = \left[\frac{12 \times 2^3}{12} + 12 \times 2 \times (7-1)^2\right]$$

$$+ \left[\frac{2 \times 18^3}{12} + 2 \times 18 \times (11-7)^2\right]$$

$$= 872 + 1548 = 2420 \text{ cm}^4$$

M.O.I. about Y-axis,

$$I_{YY} = \left[\frac{2 \times 12^3}{12} + 2 \times 12 \times (6-3)^2\right] + \left[\frac{18 \times 2^3}{12} + 18 \times 2 \times (3-1)^2\right]$$

$$= 504 + 156 = 660 \text{ cm}^4$$

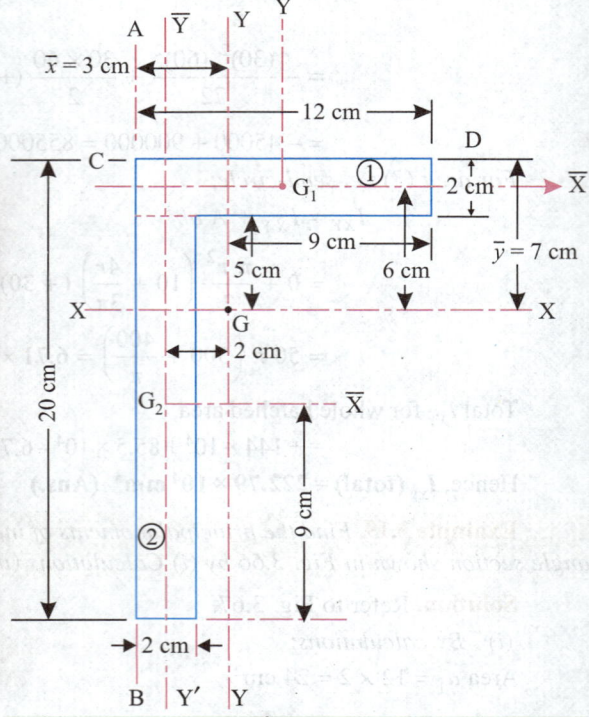

Fig. 3.66

Hub of a tractor wheel.

Fig. 3.67

Product of inertia I_{XY} :

I_{XY} for rectangle (1)

$$= I_{\bar{X}\bar{Y}} + A\, ab$$
$$= 0 + (12 \times 2) \times 3 \times 6 = 432 \text{ cm}^4$$

I_{XY} for rectangle (2)

$$= I_{\bar{X}\bar{Y}} + A\, ab$$
$$= 0 + (18 \times 2)(-2)(-4) = 288 \text{ cm}^4$$

∴ Total $\quad I_{XY} = I_{XY}$ for rectangle (1) $+ I_{XY}$ for rectangle (2)

$$= 432 + 288 = 720 \text{ cm}^4$$

Directions of principal axes :

$$\tan 2\theta = \frac{2\, I_{XY}}{I_{YY} - I_{XX}} = \frac{2 \times 720}{660 - 2420} = -0.818$$

or $\quad\quad 2\theta = -39.3°\quad$ or $\quad (180 - 39.3°)$

∴ $\quad\quad \theta = -19.7°\quad$ or $\quad\quad 70.3°$

Principal moments of inertia :

Major principal moment of inertia,

$$I_{UU} = \left[\frac{I_{YY} + I_{XX}}{2}\right] + \sqrt{\left[\frac{I_{YY} - I_{XX}}{2}\right]^2 + I_{XY}{}^2}$$

$$= \left[\frac{660 + 2420}{2}\right] + \sqrt{\left[\frac{660 - 2420}{2}\right]^2 + (720)^2}$$

$$= 1540 + 1137 = 2677 \text{ cm}^4$$

i.e. $\quad\quad \mathbf{I_{UU} = 2677 \text{ cm}^4}$ **(Ans.)**

Minor principal moment of inertia,

$$I_{VV} = \left[\frac{I_{YY} + I_{XX}}{2}\right] - \sqrt{\left[\frac{I_{YY} - I_{XX}}{2}\right]^2 + I_{XY}{}^2}$$

$$= \left[\frac{660 + 2420}{2}\right] - \sqrt{\left[\frac{660 - 2420}{2}\right]^2 + (720)^2}$$

$$= 1540 - 1137 = 403 \text{ cm}^4$$

i.e. $\quad\quad \mathbf{I_{VV} = 403 \text{ cm}^4}$

To know the inclination of *UU* or *VV* axis, put $\theta = -19.7°$ in the general equation for product of inertia (Eqn. 3.37).

$$I_{XX'} = \frac{I_{YY} + I_{XX}}{2} - \frac{I_{YY} - I_{XX}}{2} \cos 2\theta - I_{XY} \sin 2\theta$$

$$= \left(\frac{660 + 2420}{2}\right) - \left(\frac{660 - 2420}{2}\right) \cos(2 \times -19.7°) - 720 \sin(2 \times -19.7°)$$

$$= 1540 + 680 + 457 = 2677 \text{ cm}^4$$

$$= \text{major principal moment of inertia.}$$

∴ *The axis UU is inclined at – 19.7° to X-axis and axis VV is inclined at 70.3° to X-axis.*

Refer to Fig. 3.68.

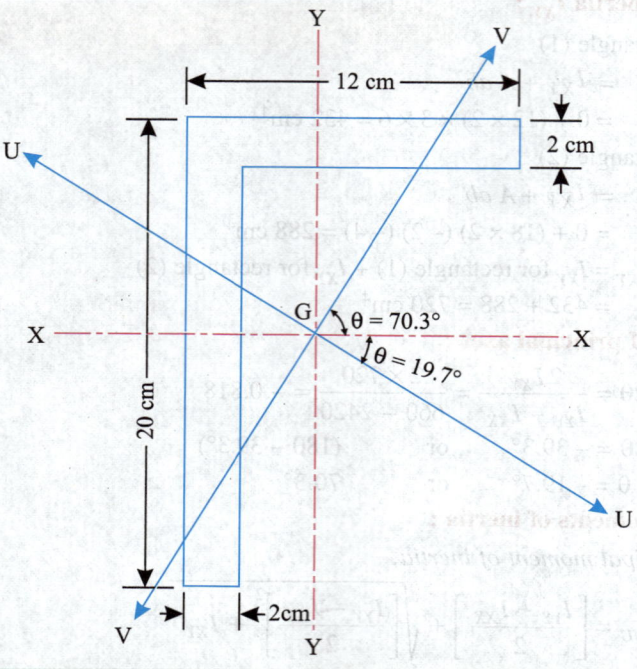

Fig. 3.68

(*ii*) *By Mohr's circle construction:*

The various steps of construction are as follows: Refer to Fig. 3.69.

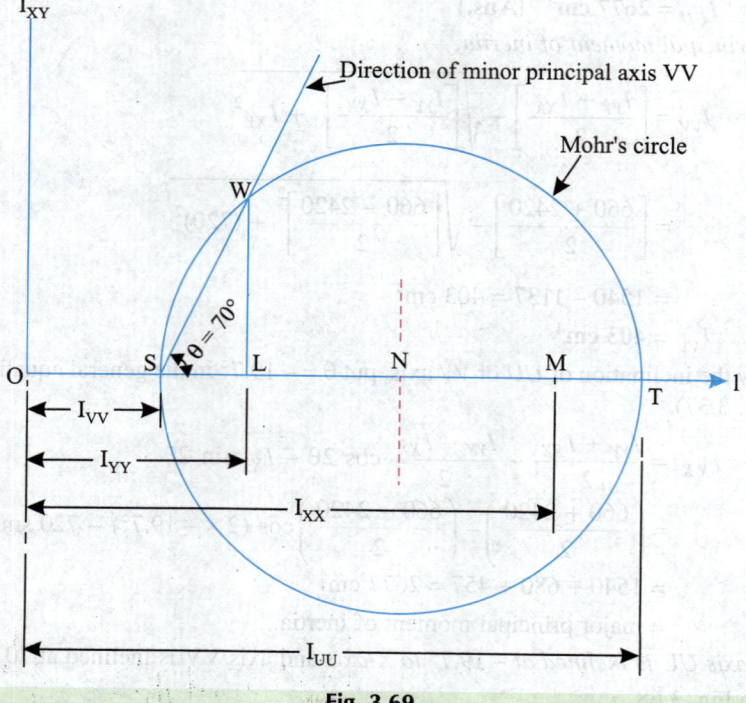

Fig. 3.69

1. Take origin 0 and from it measure $OL = I_{YY} = 660\ cm^4$ (according to some scale) and $OM = I_{XX} = 2420\ cm^4$ along the horizontal I-axis.

2. At L draw upward perpendicular LW equal to $I_{XY} = 720\ cm^4$ (since I_{XY} is $+ ve$).

3. Bisect LM at N.

4. With N as the centre and NW as radius, draw a circle cutting OL and OM extended at S and T respectively.

Ignition switch of a petrol engine.

Then, $OT = I_{UU} = 2677\ cm^4$, and $OS = I_{VV} = 403\ cm^4$

Minor principal axis is inclined at 70°.

TYPICAL EXAMPLES (For Competitive Examinations)

Example 3.20.

(*i*) *With reference to the shaded area of Fig. 3.70 locate the centroid with respect to OL and OM.*

(*ii*) *Determine the moment of inertia of the area about the horizontal axis passing through the centroid.*

Solution.

(*i*) **Centroid of the area:**

For location of the centroid of the shaded area we have the following table:

Components	Area 'a' (mm^2)	Centroidal distance 'x' from OM (mm)	Centroidal distance 'y' from OL (mm)	ax (mm^3)	ay (mm^3)
Quarter circle (1)	$\dfrac{\pi \times 30^2}{4}$ $= 706.85\ (-)$	$\dfrac{4 \times 30}{3\pi} = 12.73$	$30 - \dfrac{4 \times 30}{3\pi}$ $= 17.27$	8998.2 (–)	12207.3 (–)
Square (2)	$30 \times 30 = 900\ (+)$	$\dfrac{30}{2} = 15$	$\dfrac{30}{2} = 15$	13500	13500 (+)
Quarter circle (3)	$\dfrac{\pi \times 30^2}{4}$ $= 706.85\ (+)$	$30 + \dfrac{4 \times 30}{3\pi}$ $= 42.73$	$\dfrac{4 \times 30}{3\pi} = 12.73$	30203.7	8998.2 (+)
	$\Sigma\, a = 900$	–	–	$\Sigma\, ax = 34705.5$	$\Sigma\, ay = 10290.9$

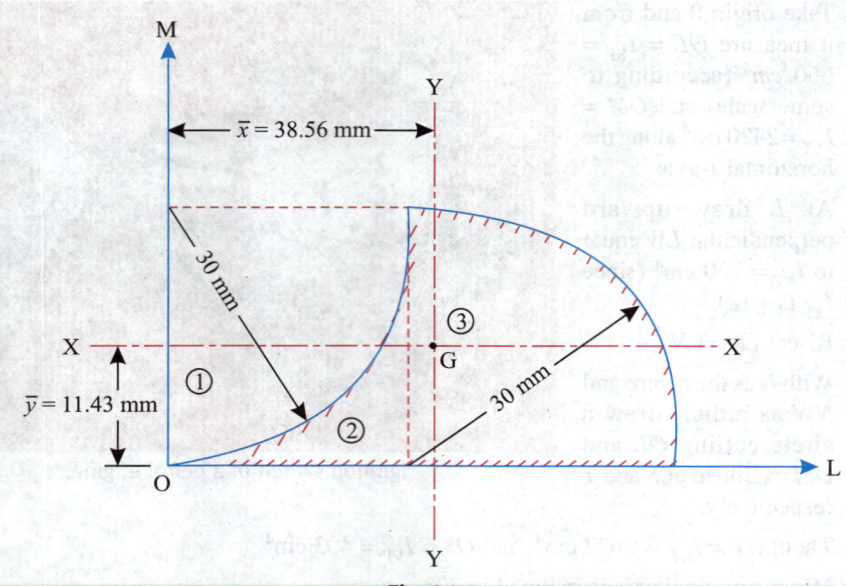

Fig. 3.70

$$\bar{x} = \frac{\Sigma \, ax}{\Sigma \, a} = \frac{34705.5}{900} = \mathbf{38.56 \ mm} \quad \textbf{(Ans.)}$$

$$\bar{y} = \frac{\Sigma \, ay}{\Sigma \, a} = \frac{10290.9}{900} = \mathbf{11.43 \ mm} \quad \textbf{(Ans.)}$$

(ii) **M.O.I. about *XX* axis:**

$$I_{XX} = -I_{XX_1} + I_{XX_2} + I_{XX_3}$$

$$= -\left[0.055 \times 30^4 + \frac{\pi \times 30^2}{4} \times \left(18.57 - \frac{4 \times 30}{3\pi} \right)^2 \right]$$

$$+ \left[\frac{30 \times (30)^3}{12} + 30 \times 30 \times (15 - 11.43)^2 \right]$$

$$+ \left[0.055 \times 30^4 + \frac{\pi \times 30^2}{4} \times \left(\frac{4 \times 30}{3\pi} - 11.43 \right)^2 \right]$$

$$- [44550 + 24091] + [67500 + 11470] + [44550 + 1198]$$

$$= -68641 + 78970 + 45748$$

Hence, $I_{XX} = \mathbf{56077 \ mm^4}$ **(Ans.)**

Example 3.21. *The plane polygonal lamina has one circular cut out of 1.6 m diameter and one rectangular cut out of 4 m × 3 m as shown in Fig. 3.71. Locate the centroid of the remainder and find the second moment of its area about the centroidal axis parallel to the base.*

Solution. Refer to Fig. 3.71.

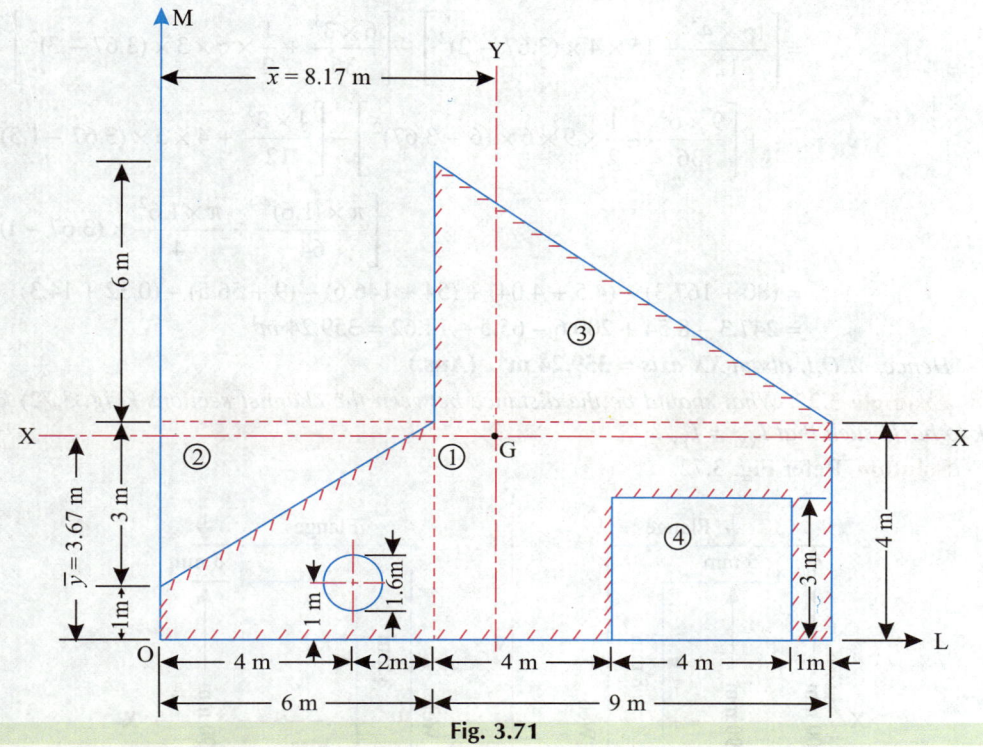

Fig. 3.71

(i) Centroid of the polygonal lamina:

For the location of the centroid of the polygonal lamina we have the following table:

Components	Area 'a' (mm^2)	Centroidal distance 'x' from OM (mm)	Centroidal distance 'y' from OL (mm)	ax (mm^3)	ay (mm^3)
Rectangle (1)	$15 \times 4 = 60$	$\dfrac{15}{2} = 7.5$	$\dfrac{4}{2} = 2$	450	120
Triangle (2)	$\dfrac{1}{2} \times 6 \times 3 = 9\ (-)$	$\dfrac{6}{3} = 2$	$1 + \dfrac{2}{3} \times 3 = 3$	$18\ (-)$	$27\ (-)$
Triangle (3)	$\dfrac{1}{2} \times 9 \times 6 = 27$	$6 + \dfrac{9}{3} = 9$	$4 + \dfrac{6}{3} = 6$	243	162
Rectangle (4)	$4 \times 3 = 12\ (-)$	$10 + \dfrac{4}{2} = 12$	$\dfrac{3}{2} = 1.5$	$144\ (-)$	$18\ (-)$
Circle (5)	$\dfrac{\pi \times 1.6^2}{4} = 2.01\ (-)$	4	1	$8.04\ (-)$	$2.01\ (-1)$
	$\Sigma a = 63.99$	–	–	$\Sigma ax = 522.96$	$\Sigma ay = 234.99$

$$\therefore \quad \bar{x} = \frac{\Sigma ax}{\Sigma a} = \frac{522.96}{63.99} = 8.17 \text{ m} \quad \textbf{(Ans.)}$$

$$\bar{y} = \frac{\Sigma ay}{\Sigma a} = \frac{234.99}{63.99} = 3.67 \text{ m} \quad \textbf{(Ans.)}$$

(ii) M.O.I. about XX axis, I_{XX}:

$$I_{XX} = I_{XX_1} - I_{XX_2} + I_{XX_3} - I_{XX_4} - I_{XX_5}$$

$$= \left[\frac{15 \times 4^3}{12} + 15 \times 4 \times (3.67 - 2)^2\right] - \left[\frac{6 \times 3^3}{36} + \frac{1}{2} \times 6 \times 3 \times (3.67 - 3)^2\right]$$

$$+ \left[\frac{9 \times 6^3}{36} + \frac{1}{2} \times 9 \times 6 \times (6 - 3.67)^2\right] - \left[\frac{4 \times 3^3}{12} + 4 \times 3 \times (3.67 - 1.5)^2\right]$$

$$- \left[\frac{\pi \times (1.6)^4}{64} + \frac{\pi \times 1.6^2}{4} \times (3.67 - 1)^2\right]$$

$$= (80 + 167.3) - (4.5 + 4.04) + (54 + 146.6) - (9 + 56.5) - (0.32 + 14.3)$$

$$= 247.3 - 8.54 + 200.6 - 65.5 - 14.62 = 359.24 \ m^4$$

Hence, M.O.I. about XX axis = **359.24 m⁴**   **(Ans.)**

Example 3.22. *What should be the distance between the channel sections (Fig. 3.72) kept back to back, such that $I_{XX} = I_{YY}$?*

Solution. Refer Fig. 3.72.

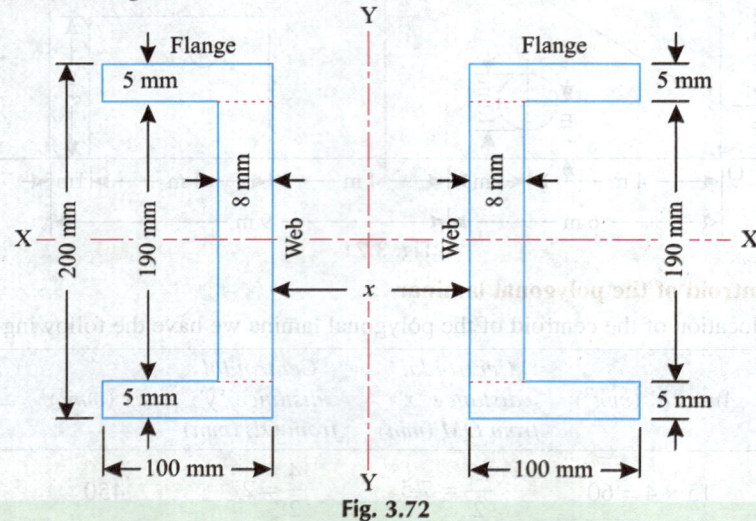

Fig. 3.72

Let, x = Clear distance between the channel sections.

$I_{XX} = I_{YY}$ (given)

Now let us find out I_{XX} and I_{YY} separately and equate them.

I_{YY} = M.O.I. of flanges (4 Nos.) + M.O.I. of webs (2 Nos.)

$$= 4\left[\frac{100 \times 5^3}{12} + 100 \times 5 \times (100 - 2.5)^2\right] + 2\left[\frac{8 \times 190^3}{12}\right]$$

$$= 4 (1041.6 + 4753125) + 2 (4572666) = 28.16 \times 10^6 \ mm^4$$

I_{YY} = M.O.I. of flanges (4 Nos.) + M.O.I. of webs (2 Nos.)

$$= 4\left[\frac{5 \times 100^3}{12} + 5 \times 100\left(50 + \frac{x}{2}\right)^2\right] + 2\left[\frac{190 \times 8^3}{12} + 190 \times 8\left(\frac{8}{2} + \frac{x}{2}\right)^2\right]$$

$$= 4\left[416666 + 500\left(2500 + \frac{x^2}{4} + 50x\right)\right] + 2\left[8106.7 + 1520\left(16 + \frac{x^2}{4} + 4x\right)\right]$$

$$= 1666664 + 5000000 + 500x^2 + 100000x + 16213 + 48640 + 760x^2 + 12160x$$

$$= 1260\, x^2 + 112160x + 6731517$$

But, $\qquad I_{XX} = I_{YY}$ (given)

∴ $\qquad 28.16 \times 10^6 = 1260\, x^2 + 112160\, x + 6731517$

or $x^2 + 89\, x - 17006 = 0$, $\quad x = \dfrac{-89 + \sqrt{89^2 + 4 \times 17006}}{2} = \dfrac{-89 + 275.6}{2} = 93.3$ mm

Hence, distance between the two channel sections = **93.3 mm** (Ans.)

HIGHLIGHTS

Centroid:

1. *Weight* is the force of attraction between a body and the earth and is proportional to mass of the body.

2. *Centre of gravity of the body* may be defined as the point through which the whole weight of the body may be assumed to act. The centre of gravity of some objects may be found by balancing the object on a point, another method is by suspension.

3. The *centroid or centre of area* is defined as the point where the whole area of the figure is assumed to be concentrated.

Inside view of an aircraft.

4. *Positions of centroids* $(\bar{x}, \bar{y})$:

Rectangle	:	(b/2, h/2)
Triangle	:	(b/3, h/3)
Circle	:	(d/2, d/2)
Semi-circle	:	(r, 0.424r)
Quadrant	:	(0.424r, 0.424r)

where b, h, r, d have the usual meanings.

5. *Positions of C.G. of regular solids:*

Shape	Volume	C.G.
Cylinder	$\pi r^2 h$	h/2 from the base
Sphere	$\dfrac{4}{3}\pi r^3$	r (from the lowest point)
Hemisphere	$\dfrac{2}{3}\pi r^3$	3/8 r from the base
Right-circular cone	$\dfrac{1}{3}\pi r^2 h$	h/4 from the base

6. *Centroid of composite areas:*

$$\bar{x} = \frac{a_1 x_1 + a_2 x_2 + ... + a_n\, x_n}{a_1 + a_2 + ... + a_n} = \frac{\Sigma\, ax}{\Sigma\, a}$$

$$\bar{y} = \frac{a_1 y_1 + a_2 y_2 + ... + a_n y_n}{a_1 + a_2 + ... + a_n} = \frac{\Sigma \, ay}{\Sigma \, a}$$

If a hole exists in plane figure, treat it as a negative area. The moment of negative area will be negative provided that the entire figure lies in the first quadrant.

7. **Centre of gravity of simple solids:**

$$\bar{x} = \frac{\Sigma \, Wx}{\Sigma \, W}; \quad \bar{y} = \frac{\Sigma \, Wy}{\Sigma \, W}; \quad \bar{z} = \frac{\Sigma \, Wz}{\Sigma \, W}.$$

If all the bodies are made of same material,

$$\bar{x} = \frac{\Sigma \, V x}{\Sigma \, V}; \quad \bar{y} = \frac{\Sigma \, V y}{\Sigma \, V}; \quad \bar{z} = \frac{\Sigma \, V z}{\Sigma \, V}.$$

If the bodies are part of a wire, pipe or rod of constant cross-section, then their centre of gravity may be found from the *centre of their lengths*.

$$\bar{x} = \frac{\Sigma \, lx}{\Sigma \, l}; \quad \bar{y} = \frac{\Sigma \, ly}{\Sigma \, l}; \quad \bar{z} = \frac{\Sigma \, lz}{\Sigma \, l}.$$

Moment of Inertia:

8. *Moment of inertia for simple areas:*

Rectangle $\quad\quad\quad\quad I_{XX} = \dfrac{bd^3}{12}, I_{YY} = \dfrac{db^3}{12}$

Triangle $\quad\quad\quad\quad I_{XX} = \dfrac{bh^3}{36}$

$\quad\quad\quad$ *About the base* $= \dfrac{bh^3}{12}$

Circle $\quad\quad\quad\quad\quad I_{XX} = \dfrac{\pi d^4}{64}$

Semi-circle $\quad\quad\quad I_{XX} = 0.11 \, r^4, \ I_{YY} = \dfrac{\pi d^4}{128} \left(= \dfrac{\pi r^4}{8} \right)$

Quadrant $\quad\quad\quad\quad I_{XX} = 0.055 \, r^4$

9. *The moment of inertia forms the basis of dynamics of rigid bodies and strength of materials.*

10. *The theorem of parallel axes states : "The moment of inertia of a lamina about any axis in the plane of lamina equals the sum of moment of inertia about a parallel centroidal axis in the plane of lamina and the product of the area of lamina and square of the distance between the two axes."*

i.e. $\quad\quad\quad\quad\quad I_{LM} = I_G + A \, h^2$

where, $\quad\quad\quad\quad$ LM = Axis about which M.O.I. is required,

$\quad\quad\quad\quad\quad\quad\quad$ A = Area of lamina, and

$\quad\quad\quad\quad\quad\quad\quad$ h = Distance between centroidal and given axes.

11. *The theorem of perpendicular axes states : "If I_{OX} and I_{OY} be the moments of inertia of a lamina about mutually perpendicular axes OX and OY in the plane of the lamina and I_{OZ} be the moment of inertia of the lamina about an axis (OZ) normal to the lamina and passing through the point of intersection of the axes OX and OY, then,*

$$I_{OZ} = I_{OX} + I_{OY}"$$

12. The radius of gyration of the area about its *i*th axis, $k_i = \sqrt{\dfrac{I_i}{A}}$

where, $\qquad I_i$ = M.O.I. about ith axis; and A = area.

13. Product of inertia, $I_{XY} = \int_A xy \, dA$

Transfer formula for the product of inertia is given by:

$$I_{X'Y'} = I_{XY} + A \, ab.$$

Also $\qquad I_{XY'} = I_{XY} \cos 2\theta - \dfrac{I_{YY} - I_{XX}}{2} \sin 2\theta$

$$I_{X'X'} = \dfrac{I_{YY} + I_{XX}}{2} - \dfrac{Y_{YY} - I_{XX}}{2} \cos 2\theta - I_{XY} \sin 2\theta$$

$$I_{Y'Y'} = \dfrac{I_{YY} + I_{XX}}{2} + \dfrac{I_{YY} - I_{XX}}{2} \cos 2\theta + I_{XY} \sin 2\theta$$

$$\tan 2\theta = \dfrac{2 I_{XY}}{I_{YY} - I_{XX}}$$

$\qquad I_{UU}$ = M.O.I. about minor principal axis

$$= \dfrac{I_{YY} + I_{XX}}{2} + \sqrt{\left[\dfrac{I_{YY} - I_{XX}}{2}\right]^2 + (I_{XY})^2}$$

and, $\qquad I_{VV}$ = M.O.I. about minor principal axis

$$= \dfrac{I_{YY} + I_{XX}}{2} - \sqrt{\left[\dfrac{I_{YY} - I_{XX}}{2}\right]^2 + (I_{XY})^2}$$

also, $\qquad I_{UU} + I_{VV} = I_{XX} + I_{YY}$ and $I_{X'X'} + I_{Y'Y'} = I_{XX} + I_{YY}$

OBJECTIVE TYPE QUESTIONS

Choose the Correct Answer:

1. Which of the following forms the basis of rigid bodies and strength of materials?

   (a) Centroid
   (b) Centre of gravity
   (c) Moment of inertia
   (d) Any of the above.

2. "The moment of inertia of lamina about any axis in the plane of the lamina equals the sum of moment of inertia about a parallel centroidal axis in the plane of lamina and the product of the area of the lamina and square of the distance between the two axes."

   The above theorem is known as

   (a) Theorem of parallel axes
   (b) Theorem of perpendicular axes
   (c) Three moments theorem
   (d) None of the above.

3. "If I_{OX} and I_{OY} be moments of inertia of a lamina about mutually perpendicular axes OX and OY in the plane of the lamina and I_{OZ} be the moment of inertia of the lamina about an axis (OZ) normal to the lamina and passing through the point of intersection of the axes OX and OY then $I_{OZ} = I_{OX} + I_{OY}$."

   The above theorem is known as

   (a) Theorem of perpendicular axes
   (b) Theorem of parallel axes
   (c) Three moments theorem
   (d) None of the above.

4. Moment of inertia of a rectangle about its XX-axis is given by

   (a) $\dfrac{bd^3}{12}$ (b) $\dfrac{db^3}{12}$

   (c) $\dfrac{bd^2}{6}$ (d) $\dfrac{bd^3}{6}$.

5. Moment of inertia of a semi-circle about its XX-axis is given by

   (a) $0.22 \, r^3$ (b) $0.11 \, r^4$
   (c) $0.14 \, r^4$ (d) $0.2 \, r^4$.

6. Moment of inertia of a quadrant about its XX-axis is given by
   (a) 0.055 r^A (b) 0.04 r^A=
   (c) 0.06 r^A (d) r^A.

7. The moment of inertia about a principal axis is called
   (a) mass moment of inertia
   (b) second moment of area
   (c) principal moment of inertia

   (d) any of the above.

8. If the two axes about which the product of inertia is found, are such that the product of inertia becomes zero, the two axes are called
   (a) centroidal axes
   (b) major and minor axes
   (c) principal axes
   (d) any of the above.

ANSWERS

1. (c)   2. (a)   3. (a)   4. (a)   5. (b)   6. (a)   7. (c)   8. (c).

THEORETICAL QUESTIONS

1. Define the terms 'Centroid' and 'Centre of gravity'.
2. Give positions of c.g. of the following solids :
   Right circular cone, cylinder, sphere and hemisphere.
3. State and explain 'Theorem of parallel axes'.
4. Enunciate 'Theorem of perpendicular axes'.
5. Derive expressions for M.O.I. of the following:
   (i) Rectangular lamina;   (ii) Circular lamina;
   (iii) Semi-circular lamina;   (iv) Thin ring.
6. What is 'Radius of gyration'?
7. What is "Product of inertia"?
8. Explain the "Parallel axes theorem for product of inertia".

UNSOLVED EXAMPLES–CENTROID/C.G.

1. Locate the centroid of the piece of sheet metal shown in Fig. 3.73.
   (**Ans.** $\bar{x}$ = 45.9 mm; $\bar{y}$ = 31.9 mm)

2. Find the centroid of the lamina shown in Fig. 3.74.
   (**Ans.** $\bar{x}$ = 25 mm; $\bar{y}$ = 35 mm)

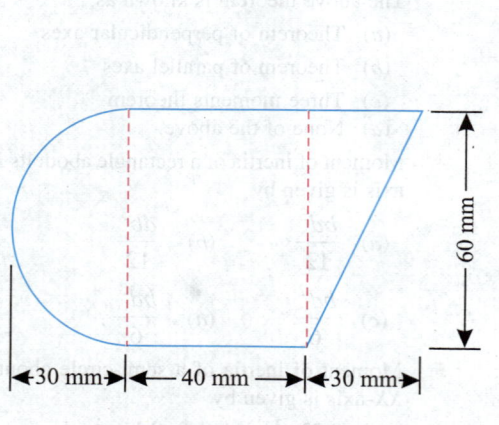

Fig. 3.73

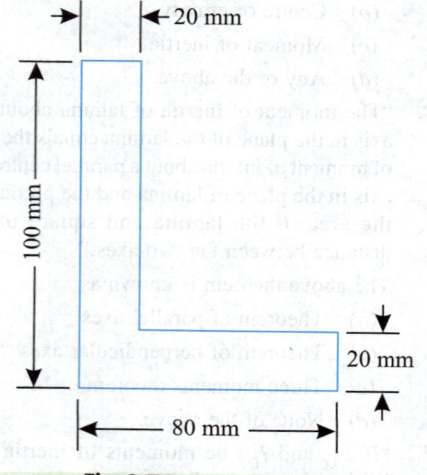

Fig. 3.74

3. Find the centroid of the lamina shown in Fig. 3.75.

(**Ans.** $\bar{x}$ = 18.64 mm from the bottom)

4. Fig. 3.76 shows an unsymmetrical *I*-section. The upper flange is 60 mm × 7.5 mm, the lower flange is 120 mm × 10 mm, the overall depth is 160 mm and the thickness of the metal at the web 5 mm. Find the distance of c.g. from the bottom of the lower flange.

(**Ans.** 56.8 mm)

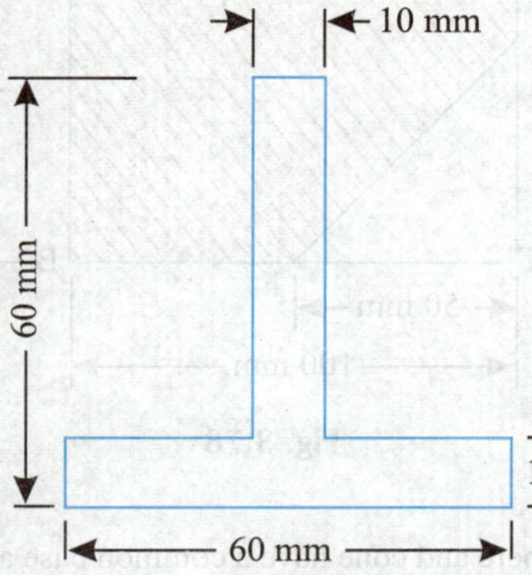

Fig. 3.75 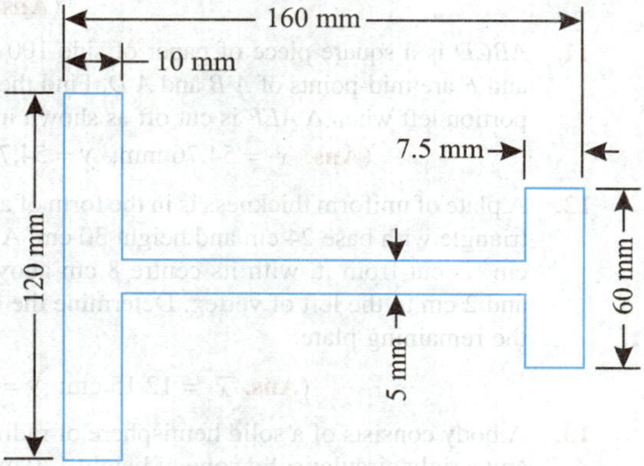   **Fig. 3.76**

5. A square sheet of metal has a square of one quarter of the original area cut from the corner as shown in Fig. 3.77. Calculate the position of the c.g. of the remaining portion of the sheet.

(**Ans.** Distance of c.g. from *A B* or *A D* = $\frac{5}{12}$ *A B*)

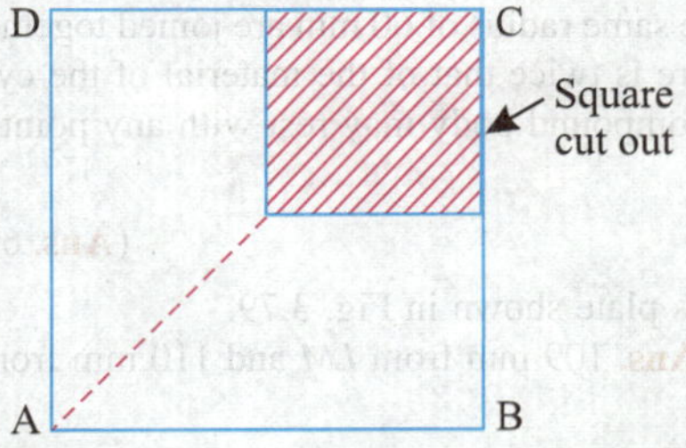

Fig. 3.77

6. From a circular plate of diameter 60 mm is cut out a circle whose diameter is a radius of the plate. Find the c.g. of the remainder.

(**Ans.** 5 mm from the centre.)

7. A hollow cast iron column has 300 mm external diameter and 225 mm inside diameter. In casting, the bore got eccentric such that the thickness varies from 25 mm on one side and 50 mm on the other side. Find the position of c.g. of the section.

(**Ans.** 166.7 mm)

8. From a rectangular sheet of metal *ABCD*, in which *AB* = 40 cm and *BC* = 60 cm a triangular piece *ABX* is removed, such that *AX* = *BX* = 25 cm. Calculate the distance of centre of gravity of the remainder.

(**Ans.** 33.57 cm from the base.)

A manual cutting machine with flywheel.

9. Find the c.g. of a semi-circular section having the outer and inner radii 200 mm and 160 mm respectively. **(Ans.** $\bar{y}$ = 112 mm)

10. Where must a hole of radius 2.5 mm be punched out in a circular disc of radius 1 m, if c.g. of the remainder is 50 mm from centre of disc.
   (Ans. 0.625 m)

11. *ABCD* is a square piece of paper of side 100 mm and *E* and *F* are mid-points of *A B* and *A D*. Find the c.g. of the portion left when Δ *AEF* is cut off as shown in Fig. 3.78.
   (Ans. $\bar{x}$ = 54.76 mm; $\bar{y}$ = 54.76 mm)

12. A plate of uniform thickness is in the form of an isosceles triangle with base 24 cm and height 30 cm. A hole of 25 cm^2 is cut from it, with its centre 8 cm above the base and 2 cm to the left of vertex. Determine the centroid of the remaining plate.

   (Ans. $\bar{x}$ = 12.15 cm; $\bar{y}$ = 10.15 cm)

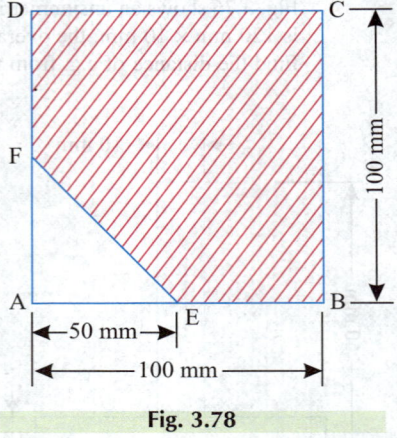

Fig. 3.78

13. A body consists of a solid hemisphere of radius 100 mm and a right circular solid cone of height 120 mm. The hemisphere and cone have a common base and are made of the same material. Find the position of the c.g. of the compound body.

   (Ans. 132.2 mm)

14. A solid right circular cone has its base scooped out, so that the hollow is right cone on the same base. Find the height upto which the cone should be scooped out so that the c.g. of the remainder coincides with the vertex of the hollow.

15. A right circular cylinder and a hemisphere, having the same radius of 60 mm are joined together face to face. The density of the material of the hemisphere is twice that of the material of the cylinder. Find the greatest height of the cylinder so that the compound body may rest with any point of the hemisphere in contact with a horizontal plane.

   (Ans. 60 mm)

16. Determine the centre of gravity of thin homogeneous plate shown in Fig. 3.79.
   (Ans. 109 mm from *LM* and 110 mm from *MN*)

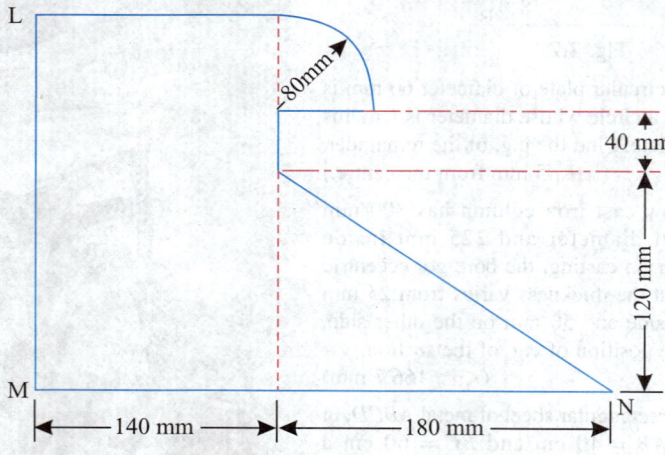

Fig. 3.79

17. Find the centroid of the area shown in Fig. 3.80. (**Ans.** $\bar{x} = 34.8$ mm; $\bar{y} = 89.8$ mm)
18. Determine the position of centroid of an area enclosed by two semi-circles and their common diameter as shown in Fig. 3.81. Take $r_1 = 180$ mm and $r_2 = 135$ mm.

(**Ans.** $\bar{x} = 0$; $\bar{y} = 103$ mm)

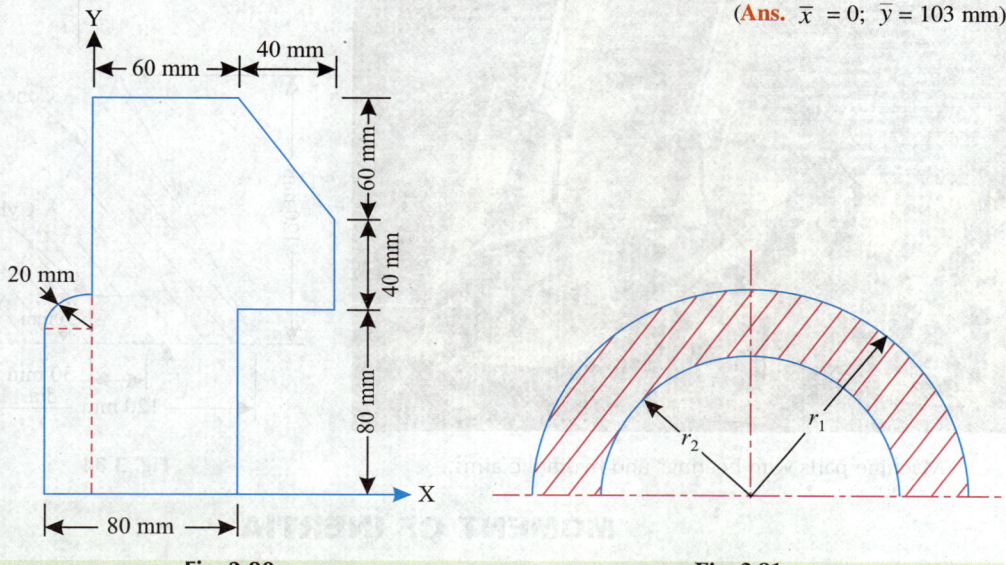

Fig. 3.80   **Fig. 3.81**

19. A semi-circle of diameter 100 mm is drawn on a paper sheet. A triangle with diameter of the semi-circle as base is cut out of the semi-circle as shown in Fig. 3.82. Determine the position of centroid/ c.g. (**Ans.** $\bar{x} = 0$; $\bar{y} = 29.2$ mm)
20. A solid right circular cylinder has its base scooped out (Fig. 3.83) so that the hollow is a right circular cone on the same base and having the same height as the cylinder. Find the c.g. of the remainder. (**Ans.** 5/8 h)

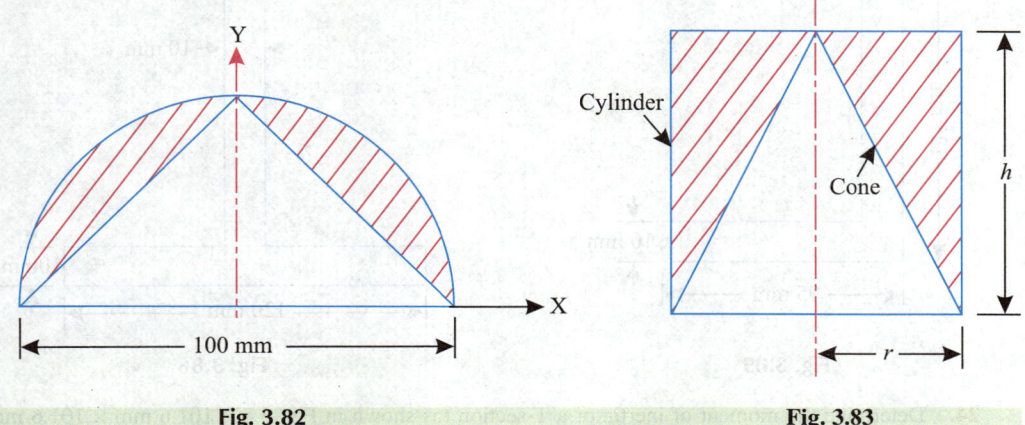

Fig. 3.82   **Fig. 3.83**

21. A solid steel cone is made lighter by removing a part of the material from it in the form of a cylinder as shown in Fig. 3.84. Determine the position of the c.g. of the remaining metal.

(**Ans.** 37.6 mm)

Machine parts with bearings and hydraulic arms.

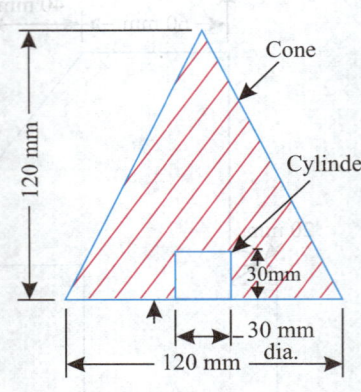

Fig. 3.84

MOMENT OF INERTIA

22. Find I_{XX} and I_{YY} for the unequal angle section 125 × 95 × 10 mm shown in Fig. 3.85.

(**Ans.** $I_{XX} = 330.8 \times 10^4$ mm⁴; $I_{YY} = 101.3 \times 10^4$ mm⁴)

23. Find the moment of inertia about the horizontal axis through the c.g. of the section shown in Fig. 3.86.

(**Ans.** $I_{XX} = 590.4 \times 10^4$ mm⁴)

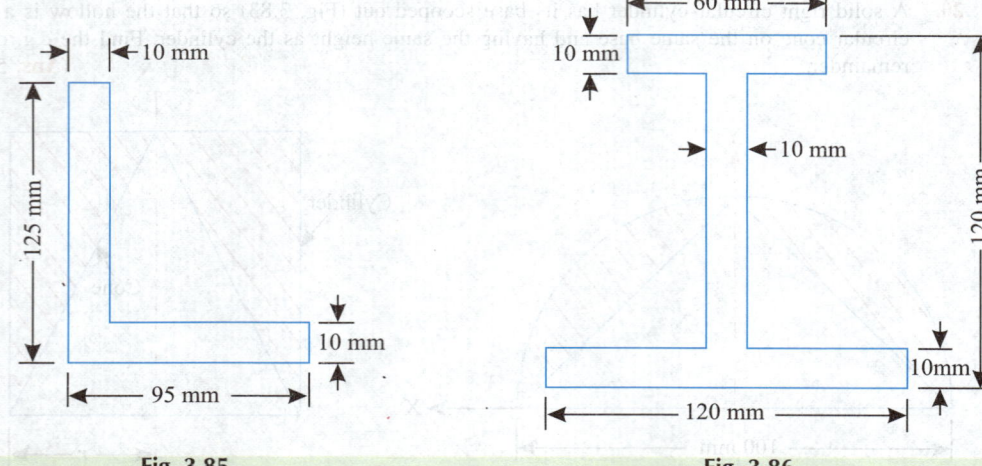

Fig. 3.85   **Fig. 3.86**

24. Determine the moment of inertia of a T-section (as shown in Fig. 3.87) 101.6 mm × 101.6 mm × 12.7 mm about an axis passing through the centre of the section and perpendicular to the stem or vertical leg.

(**Ans.** $I_{XX} = 231.4 \times 10^4$ mm⁴)

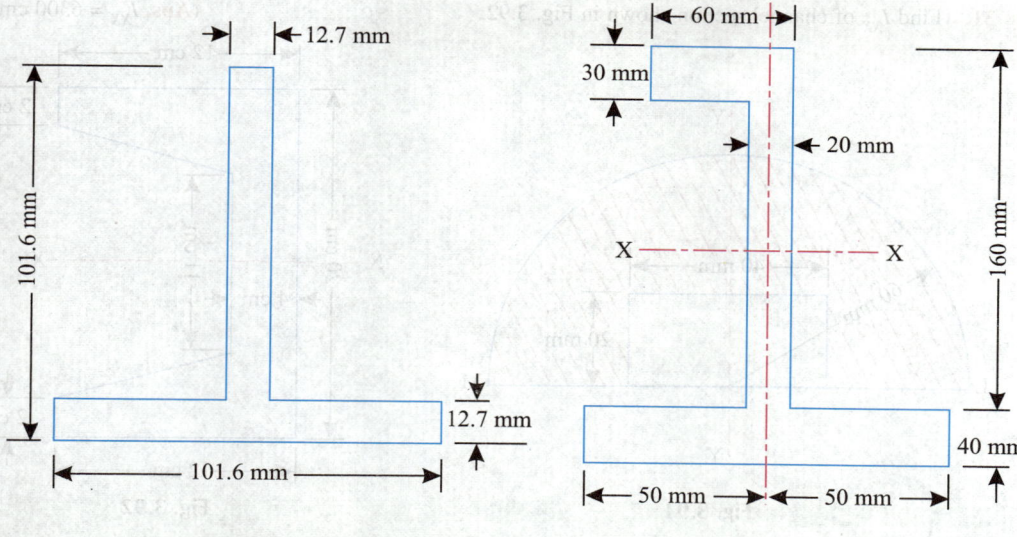

Fig. 3.87   **Fig. 3.88**

25. Find the moment of inertia about the centroidal axis *XX* for the lamina shown in Fig. 3.88.

(**Ans.** $I_{XX} = 4015.38 \times 10^4$ mm^4)

26. Find the centroidal axis of the lamina shown in Fig. 3.89 parallel to the base. Find also the moment of inertia about this centroidal axis.
(**Ans.** 40.8 mm; 385.53×10^4 mm^4)

27. A channel section $16 \times 8 \times 2$ cm stands with flanges horizontal. Determine moment of inertia about *XX* and *YY* axes passing through the centroid of the section.
(**Ans.** $I_{XX} = 1863.7$ cm^4; $I_{YY} = 302$ cm^4)

28. A short column is built by riveting flanges of T-sections of size 30 mm × 60 mm × 5 mm to form a cross of 60 mm × 60 mm. Determine I_{XX} and I_{YY} and the least radius of gyration square. Neglect the effect of rivets.

(**Ans.** $I_{XX} = 94580$ mm^4; $I_{YY} = 180520$ mm^4; $k^2_{\text{least}} = 111.3$ mm^2)

29. Find the moment of inertia of section shown hatched about *LM* for Fig. 3.90.

(**Ans.** 35327 cm^4)

30. For the section shown in Fig. 3.91 find the position of c.g. and calculate the moment of inertia about the horizontal centroidal axis.

(**Ans.** c.g. is along *YY* at 28 mm from *LM*, and I_{XX} = 169.77×10^4 mm^4)

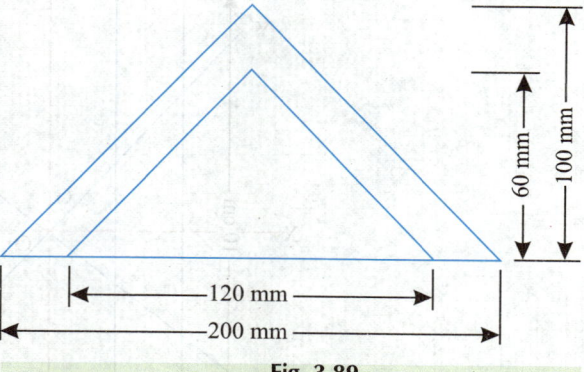

Fig. 3.89

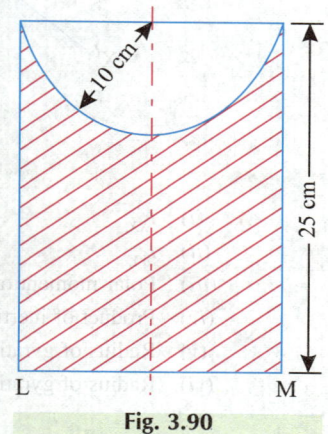

Fig. 3.90

31. Find I_{XX} of channel section shown in Fig. 3.92. (**Ans.** $I_{XX} = 6300$ cm⁴)

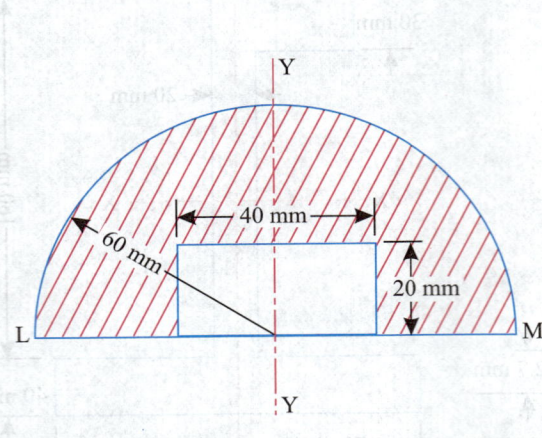

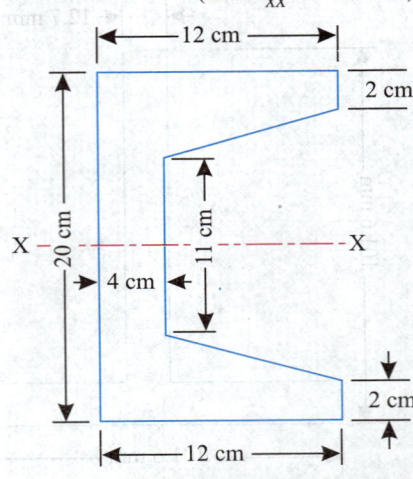

Fig. 3.91   Fig. 3.92

32. For the angle shown in Fig. 3.93, find :

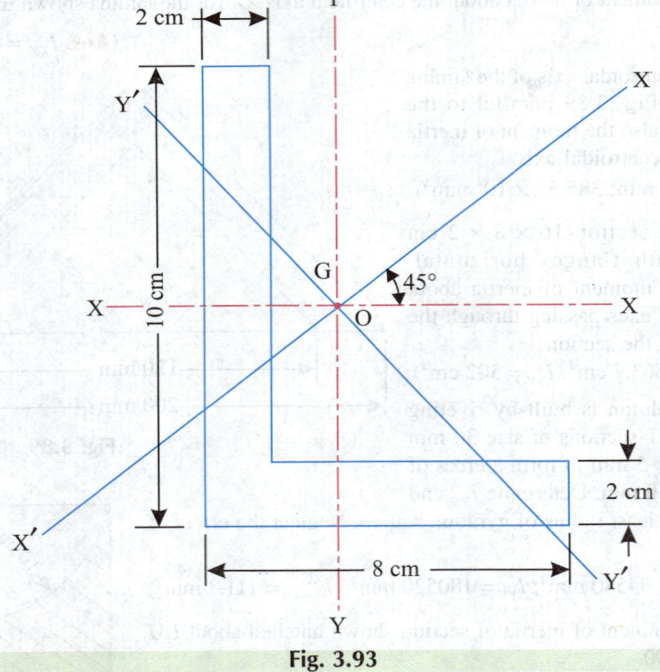

Fig. 3.93

A. (*i*) $I_{XX,}$
   (*ii*) $I_{YY,}$
   (*iii*) Polar moment of inertia about an axis through *G*,
   (*iv*) Product of inertia about *XX* and *YY* axes,
   (*v*) Radius of gyration about *XX* axis, and
   (*vi*) Radius of gyration about *YY* axis.

B. The inclination of the principal axes and principal moments of inertia (*a*) by calculations and (*b*) by Mohr's circle construction.

C. Moments of inertia and product of inertia (*i.e.* $I_{X'X'}$, $I_{Y'Y'}$ and $I_{X'Y'}$) if the axis $X'X'$ is inclined at 45° to XX axis as shown in Fig. 3.93.

Ans. A.   (*i*) 290.67 cm⁴   (*ii*) 162.677 cm⁴   (*iii*) 453.334 cm⁴
   (*iv*) – 120 cm⁴   (*v*) 3.014 cm   (*vi*) 3.555 cm.

B. 362.667 cm⁴; 90.667 cm⁴; θ = 30.9° and 120.9°.

C. $I_{Y'Y'}$ = 106.667 cm⁴; $I_{X'X'}$ = 346.667 cm⁴ and $I_{X'Y'}$ = 64 cm⁴.

Pulley with drive dog.

Bending Moments and Shearing Forces

4.1. INTRODUCTION

When any structure is loaded, stresses are induced in the various parts of the structure, and in order to calculate the stresses, where the structure is supported at a number of points, the *bending moments and shearing forces* acting must also be calculated. In general, a structure may be considered to consist of a series of beams, linked together in some way, and further, the complete structure may be treated as a beam with an elaborate cross-section. Calculations can be made progressively first on the structure as a whole and then on the individual parts.

4.2. SOME BASIC DEFINITIONS

Beam. Beam is a *structural member which is acted upon by a system of extenal loads at right angles to the axis.*

Bending. Bending implies deformation of a bar produced by loads perpendicular to its axis as well as force-couples acting in a plane passing through the axis of the bar.

Plane bending. If the plane of loading passes through one of the principal centroidal axes of the cross-section of the beam, the bending is said to be *plane* (or *direct*).

Oblique bending. If the plane of loading does *not* pass through one of the principal centroidal axes of the cross-section of the beam, the bending is said to be *oblique*.

Point load. A *point load or concentrated load is one which is considered to act at a point*. In actual practice, the load has to be distributed over a small area because such small knife-edge contacts are generally neither possible nor desirable.

Distributed load. A *distributed load is one which is distributed or spread in some manner over the length of the beam*. If the spread is uniform (*i.e.* at the uniform rate, say w kN/metre run) it is said to be uniformly distributed load and is abbreviated as U.D.L. If the spread is not at uniform rate, it is said to be non-uniformly distributed load. *Triangular* and *Trapezium* distributed loads fall under this category.

4.3. CLASSIFICATION OF BEAMS

Depending upon the type of supports beams are classified as follows:

1. **Cantilever.** A cantilever is a beam whose one end is fixed and the other end free. Fig. 4.1 shows a cantilever with end A rigidly fixed into its support and the other end B free. The length between A and B is known as the length of cantilever.

2. **Simply (or freely) supported beam.** A simply supported beam is one whose ends freely rest on walls or columns or knife edges (Fig. 4.2). In all such cases, the reactions are *always upwards*.

3. **Overhanging beam.** An overhanging beam is one in which the supports are not situated at the ends *i.e.* one or both the ends project *beyond the supports*. In Fig. 4.3, C and D are two supports and both the ends A and B of the beam are overhanging beyond the supports C and D respectively.

4. **Fixed beam.** A fixed beam is one whose both ends are rigidly fixed or built in into its supporting walls or columns (Fig. 4.4).

5. **Continuous beam.** A continuous beam is one which has *more than two supports* (Fig. 4.5). The supports at the extreme left and right are called the *end supports* and all the other supports, except the extreme, are called *intermediate supports*.

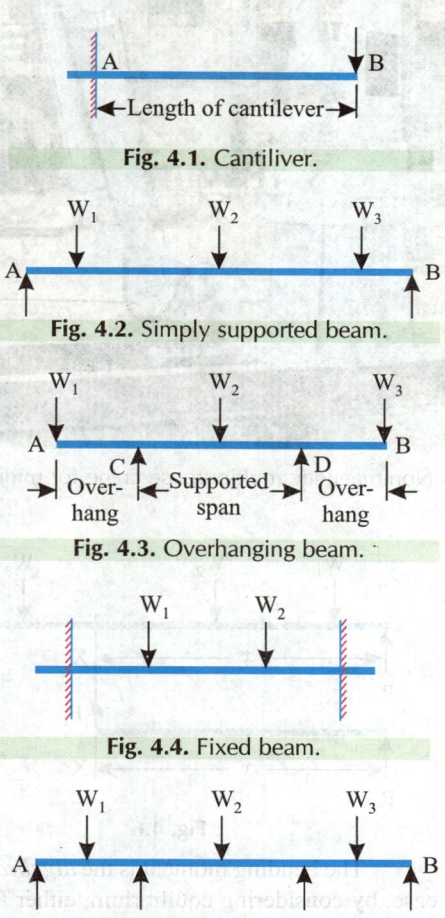

Fig. 4.1. Cantiliver.

Fig. 4.2. Simply supported beam.

Fig. 4.3. Overhanging beam.

Fig. 4.4. Fixed beam.

Fig. 4.5. Continuous beam

It may be noted that the first three types of beams (*i.e.*, cantilevers, simply supported beams and overhanging beams) are known as *Statically Determinate Beams* as the *reactions of these beams at their supports can be determined by the use of equations of static equilibrium and the reactions are independent of the deformation of beams*. The last two types of beams (*i.e.* fixed beams and continuous beams) are known as *Statically Indeterminate Beams* as their reactions at supports cannot be determined by the use of equations of static equilibrium.

Non-magnetic multipurpose crane for minehunters.

4.4. SHEARING FORCE (S.F.) AND BENDING MOMENT (B.M.)

When a beam, which is in equilibrium under a series of forces, is cut in some section X, and the beam to the left to the section remains in equilibrium (Fig. 4.6), then some force must act at the section. Prior to cutting, this force would be provided by the adjacent material, and would act tangentially to the section. Hence there will be a *shearing force* at the section. Numerically this shearing force will be given by the algebraic sum of the forces to the left or to the right of the section. As a convention, an upwad force to the left of a section is counted as producing *negative* shearing force. Similarly an upward force to the right of the section will produce *positive* shearing force.

Considering further the equilibrium of the material to the left of the section X (Fig. 4.6), it follows that there can be no resultant moment to the left of the section. Hence, any moment produced by the forces acting on the beam must be balanced by an equal and opposite moment produced by the internal forces acting in the beam at the section. This is the *bending moment* at the section.

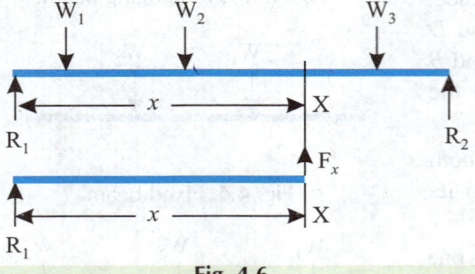

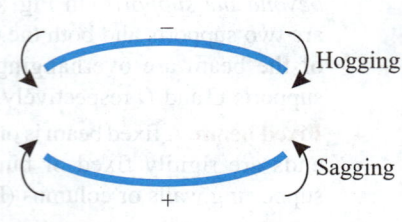

Fig. 4.6   **Fig. 4.7**

The bending moment is the *algebraic sum of moments to the left or right of the section*. In each case, by considering equilibrium, either for forces or moments, the resultant, caused by the applied forces to one side of the section is balanced by the bending moment and shearing force acting at the section. The "*sign conventions*" for bending moments is that a beam in "**hogging**" condition is subject to *negative* bending moment, and one in a "**sagging**" condition to *positive* bending moment (Fig. 4.7).

4.5. SIGN CONVENTIONS

For writing the general expressions for bending moment, and shearing force we shall be adopting the following sign conventions:

Shearing Force. A shearing force having an *upward direction* to the *right* hand side of a section or *downwards to the left of the section* will be taken '*positive*'. Similarly, a '*negative*' shearing force will be one that has a *downward direction to the right of the section or upward direction to the left of the section.*

Bending Moment. A bending moment causing *concavity upwards* will be taken as '*positive*' and called as *sagging bending moment,* a bending moment causing *convexity upwards* will be taken as '*negative*' and called *hogging bending moment.*

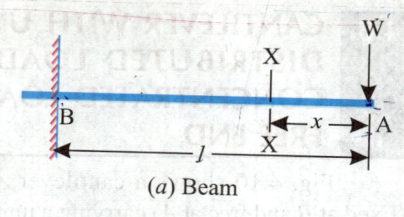

(a) Beam

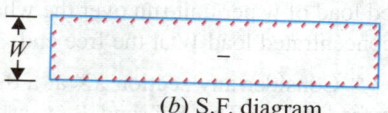

(b) S.F. diagram

4.6. CANTILEVER WITH AN END LOAD

Refer to Fig. 4.8. A cantilever is supported at one end only, being built-in at its support, giving it a fixed slope at that point.

Consider a section *XX* at a distance *x* from the free end *A*.

S.F. at $\quad X = S_x = -W$

B.M. at $\quad X = M_x = -Wx$

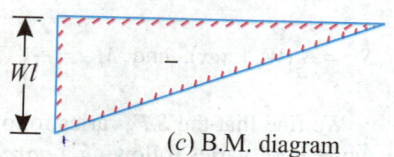

(c) B.M. diagram

Fig. 4.8

Thus we find that *S.F.* is constant at all sections of the member between *A* and *B*. But *B.M.* at any section is proportional to the distance of the section from the free end.

At $\qquad x = 0,$

i.e at *A,* $\qquad$ B.M. = 0

At $\qquad x = l,$

i.e. at *B,* $\qquad$ B.M. = *Wl*

S.F. and B.M. diagrams are shown in Fig. 4.8.

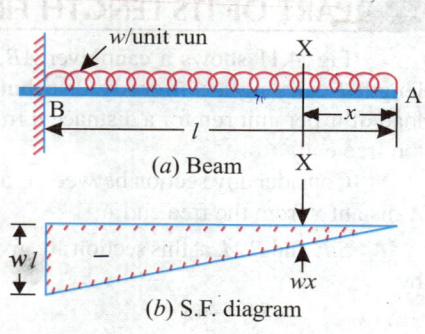

(a) Beam

(b) S.F. diagram

4.7. CANTILEVER WITH UNIFORMLY DISTRIBUTED LOAD

Let the load be distributed over the whole length of the beam, the loading being *w* per unit run (Fig. 4.9).

Consider the section *XX* at a distance *x* from the free end *A*.

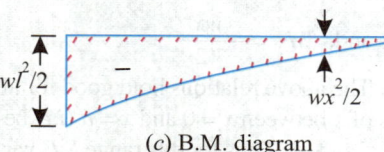

(c) B.M. diagram

Fig. 4.9

S.F. at $X = S_x = -wx$

B.M. at $X = M_x = -w.x.\dfrac{x}{2}$

$$= -\frac{wx^2}{2}$$

Thus we find that the variation of the shear-force is according to a *linear law*, while the variation of bending moment is according to *parabolic law*.

At $\qquad x = 0, S_x = 0$ and $M_x = 0$

At $\qquad x = l, S_x = -wl$

and, $\qquad M_x = -\dfrac{wl^2}{2}$

4.8. CANTILEVER WITH UNIFORMLY DISTRIBUTED LOAD AND A CONCENTRATED LOAD AT THE FREE END

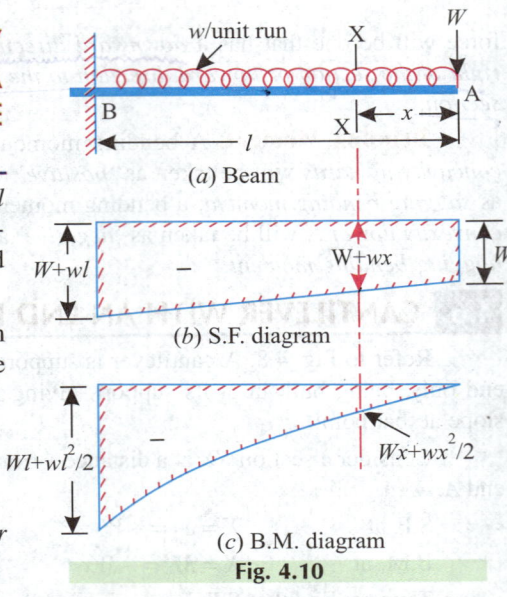

(*a*) Beam

Fig. 4.10 shows a cantilever *AB* of length *l* (fixed at *B* and free at *A*) carrying a uniformly distributed load of *w* per unit run over the whole length and a concentrated load *W* at the free end.

Consider any section *XX* at a distance *x* from the free end *A*. The *S.F.* and *B.M.* at section *X*, are respectively given by

$$S_x = -(W + wx), \text{ and } M_x = -\left(Wx + \frac{wx^2}{2}\right)$$

We find that the *S.F.* varies following a *linear law* while *B.M.* varies following a *parabolic law*.

(*b*) S.F. diagram

(*c*) B.M. diagram

Fig. 4.10

4.9. CANTILEVER CARRYING UNIFORMLY DISTRIBUTED LOAD FOR A PART OF ITS LENGTH FROM THE FREE END

Fig. 4.11 shows a cantilever *AB* of length *l* carrying a uniformly distributed load of *w* per unit run for a distnace *a* from the free end.

Consider any section between *C* and *A* distant *x* from the free end *A*.

S.F. and *B.M.* at this section are given by

$$S_x = -wx$$

and, $$M_x = -\frac{wx^2}{2}$$

The above relations hold good for all values of *x* between *x* = 0 and *x* = *a* (*i.e.* between *C* and *A*). Thus for this range *S.F.* varies following a *linear law* while the *B.M.* varies following a *parabolic law*.

At *x* = 0 $S_x = 0$
and, $M_x = 0$
At *x* = *a* $S_x = -wa$

and, $$M_x = -\frac{wa^2}{2}$$

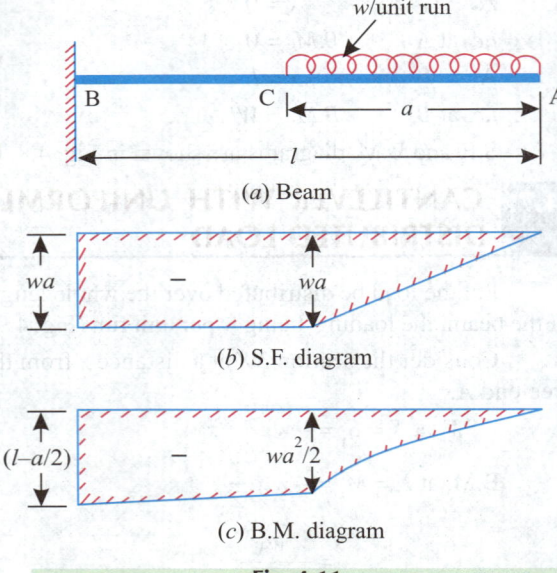

(*a*) Beam

(*b*) S.F. diagram

(*c*) B.M. diagram

Fig. 4.11

Now consider any section between *B* and *C* at a distance *x* from the end *A*.

The *S.F.* and *B.M.* at the section are given by

$$S_x = -wx$$

and, $$M_x = -wa\left(x - \frac{a}{2}\right)$$

Thus between B and C, S.F. is constant at $-wa$ but *B.M.* varies according to *linear law*.

At $x = a$,

$$M_x = -wa\,(a - a/2) = -\frac{wa^2}{2}$$

At $x = l$,

$$M_x = -wa\,(l - a/2)$$

Example 4.1. *Draw the S.F. and B.M. diagrams for cantilever loaded as shown in Fig. 4.12 (a).*

Solution. S.F. calculations:

S.F. between A and C

i.e $\qquad S_{A-C} = -1$ kN

S.F. at D *i.e.* $\qquad S_D = -1 - 2 \times 2 = -5$ kN

S.F. will change uniformly from -1 kN to -5 kN

S.F. at E *i.e.* $\qquad S_E = -5 - 1 = -6$ kN

S.F. between E and B

i.e. $\qquad S_{E-B} = -6$ kN

S.F. diagram is shown in Fig. 4.12 (b)

B.M. calculations:

B.M. at A *i.e.* $\qquad M_A = 0$

B.M. at C *i.e.* $\qquad M_C = -1 \times 1 = -1$ kNm

B.M. at D *i.e.* $\qquad M_D = -1 \times (1 + 2) - 2 \times 2$
$$\times\, 2/2 = -3 - 4$$
$$= -7 \text{ kNm}$$

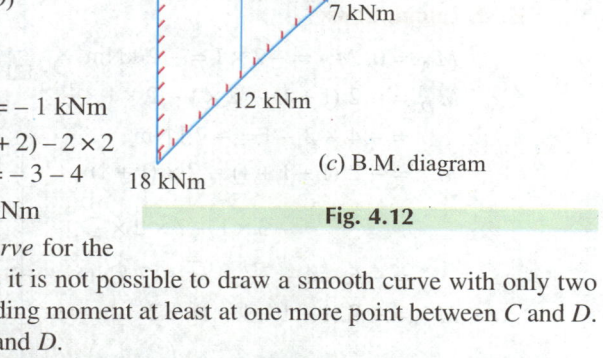

(a) Beam

(b) S.F. diagram

(c) B.M. diagram

Fig. 4.12

The *B.M.* diagram is a *parabolic curve* for the portion *CD* on which there is U.D.L. Since it is not possible to draw a smooth curve with only two points, so it is necessary to find out the bending moment at least at one more point between C and D. Let us consider point F in the *middle* of C and D.

Thames Bridge, London. Shear, bending and compressive stresses are important considerations while building bridges.

B.M. at F *i.e.* $M_F = -1 \times (1+1) - 2 \times 1 \times \dfrac{1}{2} = -2 - 1 = -3$ kNm

B.M. at E *i.e.* $M_E = -1 \times 4 - 2 \times 2 \left(\dfrac{2}{2} + 1\right) = -4 - 8 = -12$ kNm

B.M. at B *i.e.* $M_B = -1 \times 5 - 2 \times 2 \left(\dfrac{2}{2} + 1 + 1\right) - 1 \times 1 = -5 - 12 - 1 = -18$ kNm

B.M. diagram is shown in Fig. 4.12 (c).

Example 4.2. *Draw the S.F. and B.M. diagrams for a cantilever loaded as shown in Fig. 4.13 (a).*

Solution. S.F. calculations:

$S_A = -2$ kN

$S_C = -2 - 2 = -4$ kN

From C to D S.F. will change uniformly from -4 kN to -6 kN and total value of S.F. at D *i.e.* $S_D = -6 - 3 = -9$ kN. From D to E S.F. will change uniformly from -9 kN to -11 kN.

$S_F = -11 - 1 = -12$ kN

$S_B = -12$ kN

S.F. diagram is shown in Fig. 4.13 (b)

B.M. calculations:

$M_A = 0, \ M_C = -2 \times 1 = -2$ kNm

$M_D = -2(1+1) - 2 \times 1 - 2 \times 1 \times \dfrac{1}{2}$

$\quad = -4 - 2 - 1 = -7$ kNm

$M_E = -2(1+1+1) - 2 \times (1+1)$

$\quad\quad - 3 \times 1 - 2 \times 2 \times \dfrac{2}{2}$

$\quad = -6 - 4 - 3 - 4 = -17$ kNm

$M_F = -2(1+1+1+1) - 2$

$\times (1+1+1) - 3(1+1) - 2 \times 2 \left(\dfrac{2}{2} + 1\right)$

$\quad = -8 - 6 - 6 - 8 = -28$ kNm

$M_B = -2(1+1+1+1+1)$

$\quad -2(1+1+1+1) - 3(1+1+1)$

$\quad -2 \times 2 \left(\dfrac{2}{2} + 1 + 1\right) - 1 \times 1$

$\quad = -10 - 8 - 9 - 12 - 1 = -40$ kNm

B.M. diagram is shown in Fig. 4.13 (c)

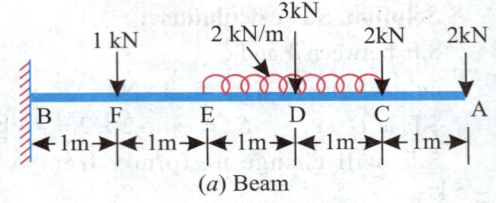

(a) Beam

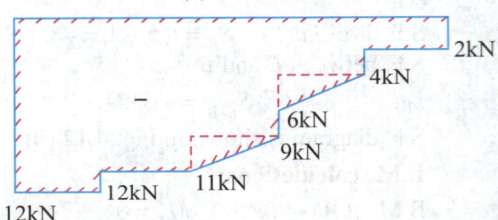

(b) S.F. diagram

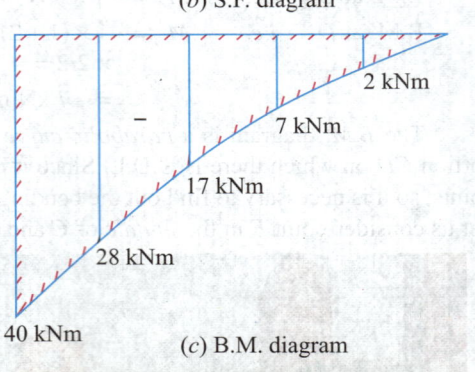

(c) B.M. diagram

Fig. 4.13

4.10. CANTILEVER CARRYING LOAD WHOSE INTENSITY VARIES UNIFORMLY FROM ZERO AT THE FREE END TO *w* PER UNIT RUN AT THE FIXED END

Fig. 4.14 shows a cantilever AB of length l carrying a load whose intensity varies uniformly from zero at the free end to w per unit run at the fixed end. Let the intensity of loading at XX, at a distance x from the free end A be w_x per unit run.

$\therefore \quad w_x = \dfrac{w}{l} \cdot x$ since the intensity of load increases uniformly form zero at the free end to w at the fixed.

$\therefore$ Load acting for an elemental distance dx from $x = w_x.dx$, thus the total load acting for any distance between $x = a$ and $x = b$

$$= \sum_{x=a}^{x=b} w_x \cdot dx$$

= Area of load diagram between $x = a$ and $x = b$.

Hence we arrive at an important conclusion that the *total distributed load acting on any segment equals the area of the load diagram on that segment.*

S.F. and B.M. at distance x from the end A are given by,

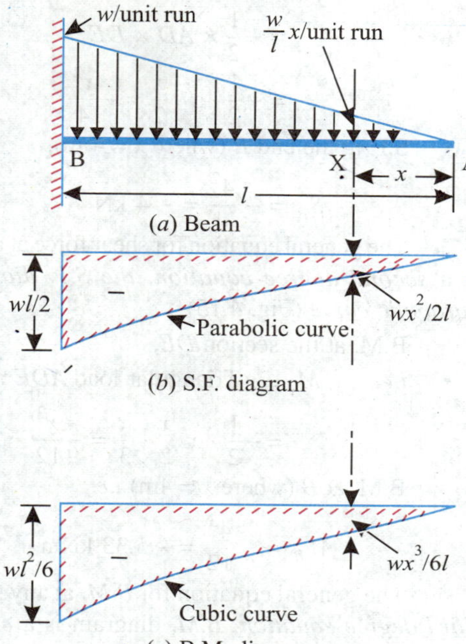

S_x = Area of the load diagram between X and A

$$= -\frac{1}{2} \cdot x.w_x = -\frac{1}{2} \cdot x \cdot \frac{w}{l} \cdot x$$

$$= -\frac{wx^2}{2l}$$

and, $\qquad M_x$ = Moment of the load acting on XA about X

= Area of load diagram between X and A ×

distance of centroid of this diagram from X

$$= -\frac{wx^2}{2l} \cdot \frac{x}{3} = -\frac{wx^3}{6l}$$

At $x = 0$, $\qquad S_x = 0$

and, $\qquad M_x = 0$

At $x = l$, $\qquad S_x = -\dfrac{wl}{2}$

and, $\qquad M_x = -\dfrac{wl^2}{6}$

The S.F. changes following a *parabolic law* while the B.M. changes following a *cubic law*.

Example 4.3. *Draw the B.M. and S.F. diagrams for the cantilever loaded as shown in Fig. 4.15 (a).*

Solution. Consider a section at a distance x from A. Rate of loading DE at the section is obtained by similarity of $\Delta s\ ABC$ and ADE

$$\frac{DE}{AD} = \frac{BC}{AB} \quad \text{or} \quad \frac{DE}{x} = \frac{2}{4}$$

or, $\qquad DE = \dfrac{x}{2}$ kN/m

S.F. at the section,

i.e. $\qquad S_x = -$ Triangular load ADE

w/unit run $\qquad \dfrac{w}{l}x$/unit run

(a) Beam

(b) S.F. diagram

Parabolic curve

$wl/2$ $\qquad wx^2/2l$

(c) B.M. diagram

Cubic curve

$wl^2/6$ $\qquad wx^3/6l$

Fig. 4.14

$= \frac{1}{2}bh$

$= \frac{1}{2} \times l \times \frac{w}{l}(x)xt$

$= \frac{1}{2} \times wl \times \frac{x}{3}(a)$

$= \frac{wl^2}{6}$

$$= -\frac{1}{2} \times AD \times DE$$

$$= -\frac{1}{2} \cdot x \cdot \frac{x}{2} = -\frac{x^2}{4}$$

S.F. at the end B (where $x = 4$ m),

$$S_B = -\frac{4^2}{4} = -4 \text{ kN}$$

The general equation for shear force at any section is a *second degree equation, the S.F. diagram is a parabolic curve* (Fig. 4.15)

B.M. at the section DE,

i.e. $M_x = -$ Triangular load $ADE \times AD/3$

$$= -\frac{1}{2} \cdot x \cdot \frac{x}{2} \cdot \frac{x}{3} = -\frac{x^3}{12}$$

B.M. at B (where $x = 4$m) *i.e.*

$$M_B = -\frac{4^3}{12} = -5.33 \text{ kNm}$$

The general equation for *B.M.* at any section is a *third degree equation. B.M.* diagram is a *cubic parabolic* (Fig. 4.15 (c)).

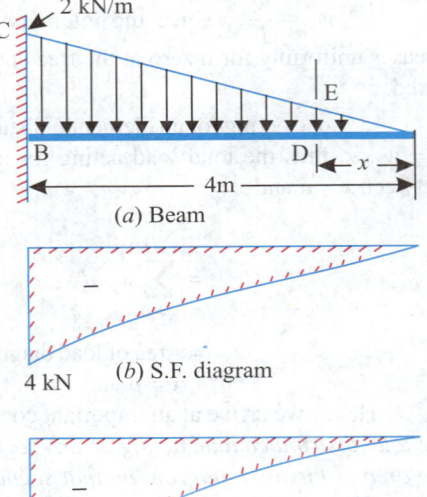

(a) Beam

(b) S.F. diagram

(c) B.M. diagram

Fig. 4.15

| 4.11. | CANTILEVER CARRYING LOAD WHOSE INTENSITY VARIES UNIFORMLY FROM ZERO AT THE FIXED POINT END TO *w* PER UNIT RUN AT THE FREE END |

Fig. 4.16 shows a cantilever AB of length l (fixed at B and free at A) carrying a load whose intensity varies uniformly from zero at the fixed end to w per unit run at the free end. It is convenient to find *S.F.* and *B.M.* at any section by considering the left part of the section. Consider any section XX at the distance x from the fixed end B.

Mechanisms are very sensitive to bending stresses.

Let M_B be the reacting moment or fixing moment at B.

S.F. at X = Algebraic sum of forces on BX

$\therefore \qquad S_x = -\dfrac{wl}{2} + \dfrac{x}{2} \cdot \dfrac{wx}{l} = -\dfrac{wl}{2} + \dfrac{wx^2}{2l}$

B.M. at X = Algebraic sum of moment of forces and reactions on BX about X.

$\therefore \qquad M_x = \dfrac{wl}{2} x - \dfrac{wx^2}{2l} \cdot \dfrac{x}{3} - M_B$ $\quad\left[\begin{array}{l}\text{where, } M_B = \text{moment of total load about } B \\ \qquad\qquad = \dfrac{wl}{2} \cdot \dfrac{2l}{3} = \dfrac{wl^2}{3}\end{array}\right]$

$\qquad = \dfrac{wl}{2} x - \dfrac{wx^3}{6l} - \dfrac{wl^2}{3}$

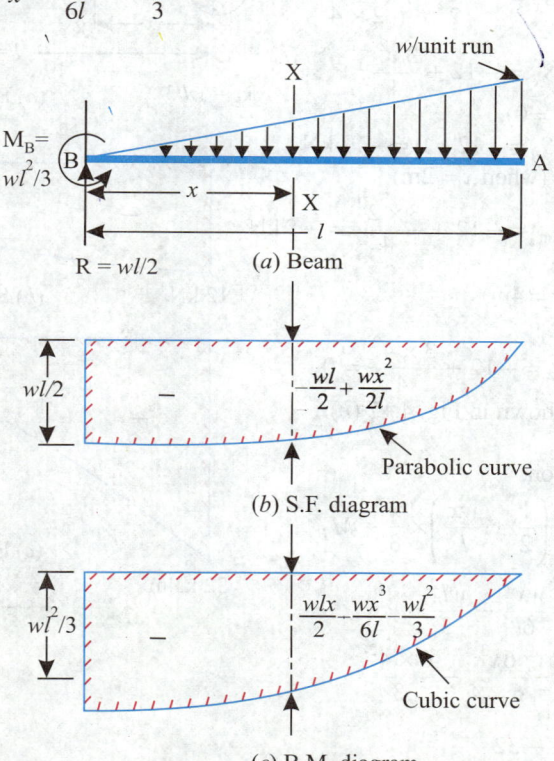

(a) Beam

(b) S.F. diagram

(c) B.M. diagram

Fig. 4.16

At $x = 0$ *i.e.* at B, $S_B = -\dfrac{wl}{2}$ and $M_B = -\dfrac{wl^2}{3}$

At $x = l$ *i.e.* at A, $S_A = -\dfrac{wl}{2} + \dfrac{wl^2}{2l} = 0$

and, $\qquad M_A = \dfrac{wl}{2} \cdot l - \dfrac{wl^3}{6l} - \dfrac{wl^2}{3} = \dfrac{wl^2}{2} - \dfrac{wl^2}{6} - \dfrac{wl^2}{3} = 0$

Example 4.4. *Draw the shear force and bending moment diagrams for a cantilever loaded as shown in Fig. 4.17 (a).*

Solution. It is convenient to find the *S.F.* and *B.M.* at any section by considering the left part of the section.

Let M_B be the reacting or fixing moment at B

$\therefore$ $\qquad M_B$ = Moment of total load about $B = \left(\dfrac{1}{2} \times l \times w\right) \times \left(\dfrac{2}{3} l\right) = \dfrac{wl^2}{3}$

Consider any section XX distant x from the fixed end B.

$$S_x = -\frac{wl}{2} + \left(\frac{1}{2} \cdot x \cdot \frac{wx}{l}\right)$$

or, $\qquad S_x = -\dfrac{wl}{2} + \dfrac{wx^2}{2l}$

or, $\qquad S_x = -\dfrac{6 \times 4}{2} + \dfrac{6 \times x^2}{2 \times 4}$

or, $\qquad S_x = -12 + \dfrac{3x^2}{4}$.

S.F at B (when $x = 0$),

$$S_B = -12 + 0 = -12 \text{ kN}$$

S.F. at mid-point (when $x = 2$m)

$$S_{\text{mid-point}} = -12 + \frac{3 \times 2^2}{4} = -9 \text{ kN}$$

S.F. at A (when $x = 4$m)

$$S_A = -12 + \frac{3 \times 4^2}{4} = 0$$

S.F. diagram is shown in Fig. 4.17 (b)

B.M. at the section,

$$M_x = R_B \times x - \left(\frac{1}{2} \cdot x \cdot \frac{wx}{l}\right) \times \frac{x}{3} - M_B$$

$$= \frac{wl}{2} \times x - \frac{wx^3}{6l} - \frac{wl^2}{3}$$

$$= \frac{6 \times 4 \times x}{2} - \frac{6x^3}{6 \times 4} - \frac{6 \times 4^2}{3}$$

$$= 12x - \frac{x^3}{4} - 32$$

B.M. at B (when $x = 0$) $\qquad M_B = -32$ kNm

B.M. at mid span (when $x = 2$m)

$$M_{\text{mid span}} = 12 \times 2 - \frac{2^3}{4} - 32 = 24 - 2 - 32 = -10 \text{ kNm}$$

B.M. at A (when $x = 4$m) $\quad M_A = 12 \times 4 - \dfrac{4^3}{4} - 32 = 48 - 16 - 32 = 0$

B.M. diagram is shown in Fig. 4.17 (c).

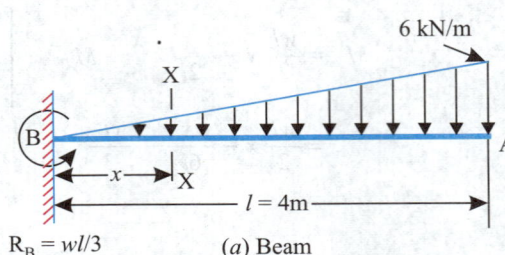

(a) Beam

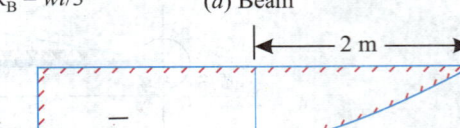

12 kN $\qquad$ (b) S.F. diagram

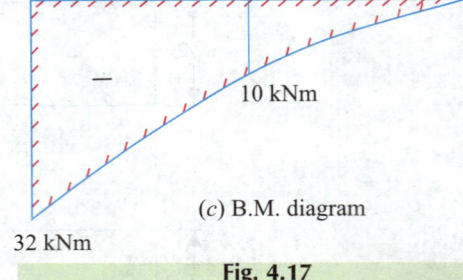

32 kNm $\qquad$ (c) B.M. diagram

Fig. 4.17

CANTILEVERS WITH COUPLES

Example 4.5. *Find the reaction at the fixed end of the cantilever loaded as shown in Fig.4.18 (a). Draw also the S.F. and B.M. diagrams.*

Solution. Total vertical load on the cantilever = 20 + 30 = 50 kN

∴ Vertical reaction at B = 50 kN (upwards)

Taking moments about B:

Sum of moments

= – 20 × 8 – 30 (couple at C) – 30 × 4 + 20 (couple at E)

= – 160 – 30 – 120 + 20 = – 290 kNm

(*i.e. clockwise*)

Hence at B the support will provide a balancing or reacting moment of 290 kNm *anticlockwise*.

Hence the reaction at B will consist of an upward reacting force 50 kN and an anticlockwise reacting moment of 290 kNm.

S.F. calculations:

S_{A-C-D} = – 20 kN

S_{D-E-B} = – 20 – 30 = – 50 kN

S.F. diagram is shown in Fig. 4.18 (*b*)

B.M. calculations:

$M_A = 0$

B.M. just on the right hand side of C

= – 20 × 2 = – 40 kNm

B.M. just on the left hand side of C

= – 40 – 30 = – 70 kNm

M_D = – 20 × 4 – 30 = – 110 kNm

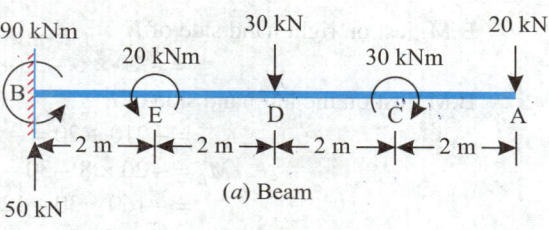

(*a*) Beam

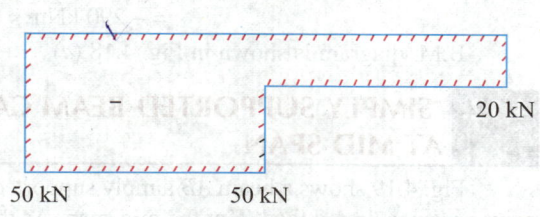

(*b*) S.F. diagram

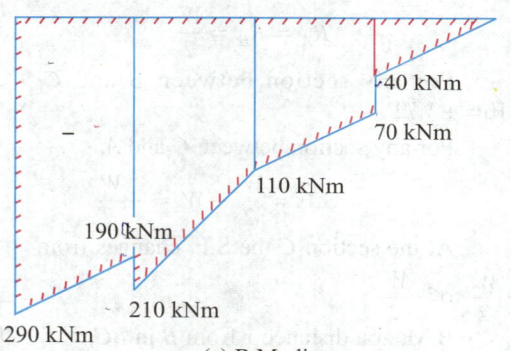

(*c*) B.M. diagram

Fig. 4.18

A machine part is being processed on a lathe.

B.M. just on right hand side of E
$$= -20 \times 6 - 30 - 30 \times 2 = -210 \text{ kNm}$$

B.M. just on the left hand side of E
$$= -210 + 20 = -190 \text{ kNm}$$
$$M_B = -20 \times 8 - 30 - 30 \times 4 + 20$$
$$= -160 - 30 - 120 + 20$$
$$= -290 \text{ kNm} = \text{reacting moment at } B$$

B.M. diagram is shown in Fig. 4.18 (c).

4.12. SIMPLY SUPPORTED BEAM CARRYING A CONCENTRATED LOAD AT MID SPAN

Fig. 4.19 shows a beam AB simply supported at the ends A and B. Let its span be l and carry a point or concentrated load W at the mid span. As the load is symmetrically placed on the span, reaction at each support is $W/2$.

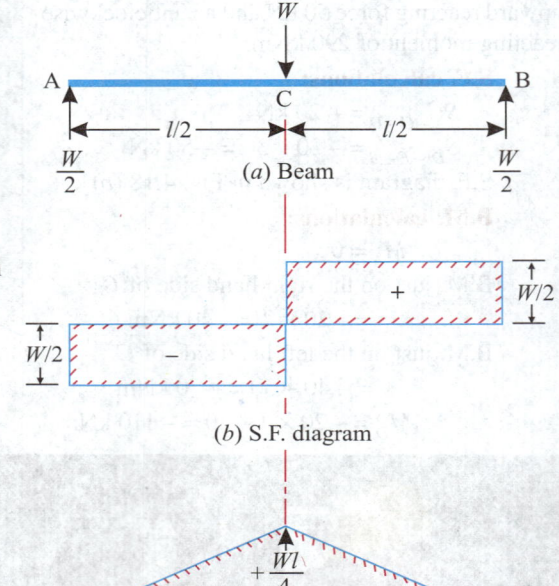

$$\therefore \qquad R_A = R_B = \frac{W}{2}$$

For any section between B and C,
S.F. $= + W/2$

For any section between C and A,
$$\text{S.F.} = \frac{W}{2} - W = -\frac{W}{2}$$

At the section C the S.F. changes from $+\dfrac{W}{2}$ to $-\dfrac{W}{2}$.

B.M. at a distance x from B in BC,
$$M_x = +\frac{Wx}{2}$$

B.M. at B (where $x = 0$),
$$M_B = 0$$

(a) Beam

(b) S.F. diagram

B.M. at C (where $x = l/2$),
$$M_C = \frac{Wl}{4}$$

B.M. at a distance x from B in CA,
$$M_x = +\frac{Wx}{4} - W(x - l/2)$$

(c) B.M. diagram

Fig. 4.19

B.M. at C where $x = l/2$, $\qquad M_C = +\dfrac{Wl}{4}$

B.M. at A where $x = l$, $\qquad M_A = \dfrac{Wl}{2} - \dfrac{Wl}{2} = 0$

Note. B.M. at supports in case of simply supported beams is always zero.

4.13. SIMPLY SUPPORTED BEAM CARRYING A CONCENTRATED LOAD 'NOT' AT MID SPAN

Let the beam AB of span l (Fig. 4.20) carry a concentrated load W at a distance a from A and a distance b from B.

Taking moments about A to determine the support reactions, we have:

$$R_B \times l = W \times a$$

$$\therefore \quad R_B = \frac{Wa}{l}$$

But, $\quad R_A + R_B = W$

$$\therefore \quad R_A = W - \frac{Wa}{l} = \frac{W(l-a)}{l} = \frac{Wb}{l}$$

S.F. just to the left of B is $+\dfrac{Wa}{l}$.

It remains constant upto C.

S.F. just to the left of C is

$$\left(+\frac{Wa}{l} - W\right) = \left[\frac{-W(l-a)}{l}\right] = -\frac{Wb}{l}$$

It remains constant upto A.

B.M. at the supports A and $B = 0$

B.M. at $\quad C, M_C = \dfrac{Wa}{l} \times b = \dfrac{Wab}{l}$

Thus B.M. is *maximum* at C where S.F. *changes* sign.

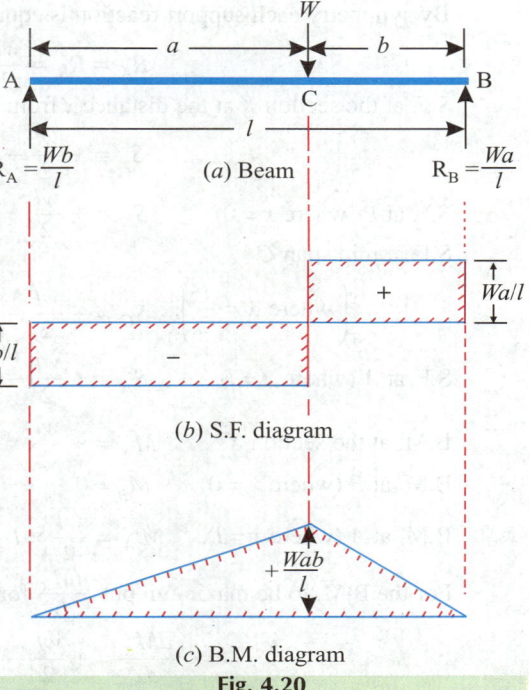

(a) Beam

(b) S.F. diagram

(c) B.M. diagram

Fig. 4.20

4.14. SIMPLY SUPPORTED BEAM CARRYING A UNIFORMLY DISTRIBUTED LOAD OF *w* PER UNIT RUN OVER THE WHOLE SPAN

Fig. 4·21 shows a simply supported beam AB carrying a uniformly distributed load (U.D.L.) of w per unit over the whole span.

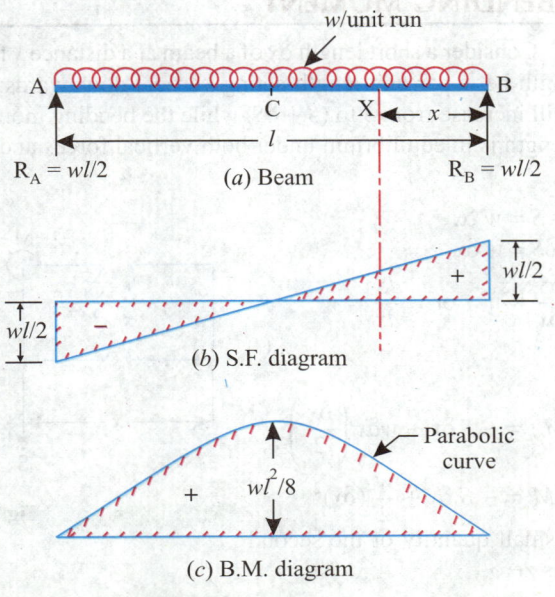

(a) Beam

(b) S.F. diagram

(c) B.M. diagram

Fig. 4.21

By symmetry each support reaction is equal, *i.e.,*

$$R_A = R_B = \frac{wl}{2}$$

S.F. at the section X at the distance x from B,

$$S_x = +\frac{wl}{2} - wx$$

S.F. at B (where $x = 0$) $S_B = +\frac{wl}{2}$

S.F. at mid span C

$$\left(\text{where } x = \frac{l}{2}\right), \quad S_C = +\frac{wl}{2} - \frac{wl}{2} = 0$$

S.F. at A (where $x = l$), $S_A = +\frac{wl}{2} - wl = -\frac{wl}{2}$

B.M. at the section X. $M_x = +\frac{wl}{2}x - \frac{wx^2}{2}$ ...(*i*)

B.M. at B (where $x = 0$), $M_B = 0$

B.M. at A (where $x = l$), $M_A = \frac{wl}{2} \times l - \frac{wl^2}{2} = 0$

For the B.M. to be maximum put $\dfrac{dM_x}{dx}$ for eqn. (*i*) equal to zero.

$$\therefore \qquad \frac{dM_x}{dx} = +\frac{wl}{2} - wx = 0$$

i.e, $x = \dfrac{l}{2}$ [for the B.M. to be *maximum*, it may be noted that, S.F. *at this point is zero*]

$$\therefore \qquad M_{\max} = +\frac{wl}{2} \times \frac{l}{2} - \frac{w}{2} \times \left[\frac{l}{2}\right]^2 = +\frac{wl^2}{8}$$

4.15. GENERAL RELATION BETWEEN THE LOAD, THE SHEARING FORCE AND THE BENDING MOMENT

Refer to Fig. 4·22. Consider a short length δx of a beam at a distance x from some origin. Let the load over this short length be w per unit length acting vertical downwards; then the shearing force over this short length will increase from S to $(S + \delta S)$ while the bending moment increases from M to $(M + \delta M)$. This short length is in equilibrium under both vertical forces and couples.

Vertical forces:

$$(S + \delta S) - S = w\,\delta x$$
$$\therefore \qquad \delta S = w.\delta x$$

or, $\qquad \dfrac{\delta S}{\delta x} = w \quad$ or $\quad \dfrac{dS}{dx} = w$

...(*i*)

Couples:

$$M - (M + \delta M) = -S\,\delta x + w\delta x\left(\frac{\delta x}{2}\right)$$

or, $\qquad -\delta M = -S.\delta x + \dfrac{w}{2}(\delta x)^2$

Since $(\delta x)^2$ is a small quantity of the second order, it may be taken as zero.

$$\therefore \qquad -\delta M = -S\,\delta x \quad \text{or} \quad S = \frac{\delta M}{\delta x}$$

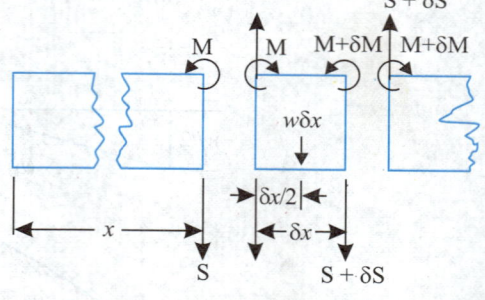

Fig. 4.22

Unlike land automobile, helicopter flies in the air. Failure of popeller blades or shaft can be fatal.

and in the $\underset{\delta x \to 0}{\text{Lt.}} \ S = \dfrac{dM}{dx}$ ...(ii)

Putting these relations into integral form we have,

$$S = \int w \, dx \qquad ...(iii)$$

$$M = \int s \, dx \qquad ...(iv)$$

From equation (i) it is concluded that *the rate of change of S.F. at any section represents the rate of loading at the section.*

From equation (ii) it is concluded that the *rate of change of B.M. at any section represents the S.F. at that section.*

The *B.M. at M shall be maximum or minimum when* $\dfrac{dM}{dx} = 0$ *i.e., S = 0.* Thus at the sections *where S.F. is zero or changes sign (because then it passes through zero) the B.M. is either maximum or minimum.*

Example 4.6. *Draw the S.F. and B.M. diagrams for simply supported beam loaded as shown in Fig. 4.23 (a).*

Solution. To determine the support reactions taking moments about *A*, we get

$$R_B \times 4 = 2 \times 1 + 4 \,(1 + 1) + 2 \,(1 + 1 + 1)$$
$$= 2 + 8 + 6 = 16$$

$$R_B = \frac{16}{4} = 4 \text{ kN}$$

But, $\qquad R_A + R_B = 2 + 4 + 2 = 8 \text{ kN}$

$\therefore \qquad R_A = 8 - R_B = 8 - 4 = 4 \text{ kN}$

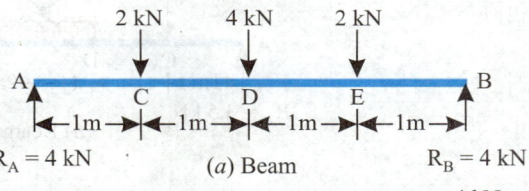

(a) Beam

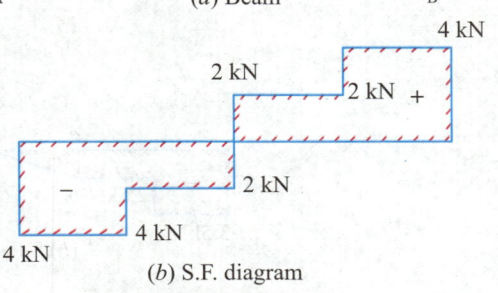

(b) S.F. diagram

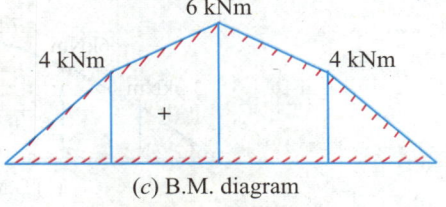

(c) B.M. diagram

Fig. 4.23

S.F. calculations:

$$S_{B-E} = + 4 \text{ kN}$$
$$S_{E-D} = 4 - 2 = 2 \text{ kN}$$
$$S_{D-C} = 2 - 4 = - 2 \text{ kN}$$
$$S_{C-A} = - 2 - 2 = - 4 \text{ kN}$$

S.F. diagram is shown in Fig. 4.23 (*b*)

B.M. calculations:

$$M_B = 0$$
$$M_E = 4 \times 1 = 4 \text{ kNm}$$
$$M_D = 4 (1 + 1) - 2 \times 1 = 8 - 2 = 6 \text{ kNm}$$
$$M_C = 4 (1 + 1 + 1) - 2 (1 + 1) - 4 \times 1$$
$$= 12 - 4 - 4 = 4 \text{ kNm}$$
$$M_A = 4 (1 + 1 + 1 + 1) - 2 (1 + 1 + 1) - 4 (1 + 1) - 2 \times 1$$
$$= 16 - 6 - 8 - 2 = 0$$

B.M. diagram is shown in Fig. 4.23 (*c*).

Example 4.7. *Draw the shear force and bending diagrams for the beam shown loaded in Fig. 4.24 (a). Clearly mark the position of the maximum bending moment and determine its value.*

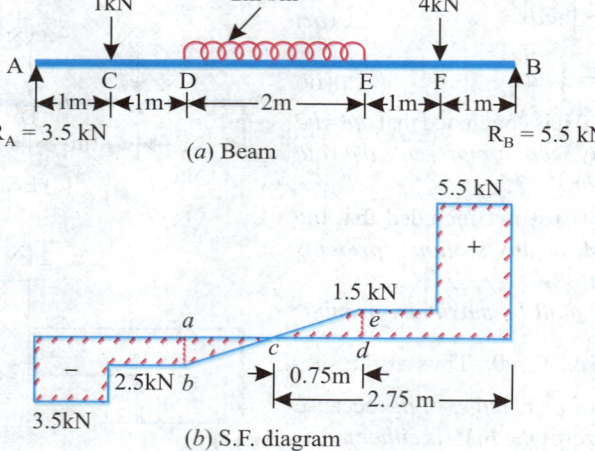

(*a*) Beam

(*b*) S.F. diagram

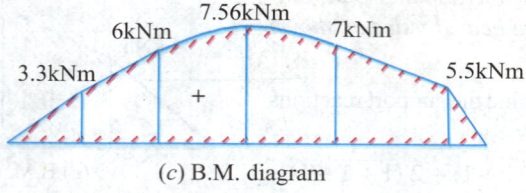

(*c*) B.M. diagram

Fig. 4.24

Solution. Let R_A and R_B be the respective support reactions at A and B. Taking moments about A, we have:

$$R_B \times 6 = 1 \times 1 + 2 \times 2 \left(\frac{2}{2} + 1 + 1 \right) + 4 \times 5 = 1 + 12 + 20 = 33$$

$$\therefore \qquad R_B = \frac{33}{6} = 5.5 \text{ kN}$$

But, $\qquad R_A + R_B = 1 + 2 \times 2 + 4 = 9$

$$\therefore \qquad R_A = 9 - R_B = 9 - 5.5$$
$$= 3.5 \text{ kN}$$

S.F. calculations:

$$S_{B-F} = 5.5 \text{ kN}$$
$$S_{F-E} = 5.5 - 4 = 1.5 \text{ kN}$$

From E to D S.F. changes from $+ 1.5$ kN to $- 2.5$ kN

$$S_{D-C} = -2.5 \text{ kN}$$
$$S_{C-A} = -2.5 - 1$$
$$= -3.5 \text{ kN}$$

S.F. diagram is shown in Fig. 4.24 (b)

To locate the point between D and E where S.F. is zero we have from the two similar Δs abc and cde:

Industrial gas turbine.

$$\frac{ab}{ac} = \frac{de}{cd} ; \frac{2.5}{2 - cd} = \frac{1.5}{cd}$$

or, $\qquad 2.5 \, cd = 1.5 \, (2 - cd)$

$$2.5 \, cd = 3 - 1.5 \, cd$$

$\therefore \qquad cd = 0.75 \text{ m}$

B.M. calculations:

$$M_B = 0$$
$$M_F = 5.5 \times 1 = 5.5 \text{ kNm}$$
$$M_E = 5.5 \times 2 - 4 \times 1 = 7 \text{ kNm}$$

B.M. at the point where S.F. is zero is the maximum and is

$$M_{\max} = 5.5 \times 2.75 - 4 \times (2.75 - 1) - 2 \times 0.75 \times \frac{0.75}{2}$$
$$= 15.125 - 7.0 - 0.5625 = 7.56 \text{ kNm}$$

$$M_D = 5.5 \times 4 - 4 \times 3 - 2 \times 2 \times \frac{2}{2} = 22 - 12 - 4 = 6 \text{ kNm}$$

$$M_C = 5.5 \times 5 - 4 \times 4 - 2 \times 2 \left(\frac{2}{2} + 1 \right) = 27.5 - 16 - 8 = 3.5 \text{ kNm}$$

$$M_A = 0$$

B.M. diagram for the portion DE which carries U.D.L. shall be a *parabolic curve*.

B.M. diagram is shown in Fig. 4.24 (c)

Example 4.8. *Draw the B.M.and S.F. diagrams for the beam shown in Fig. 4.25 (a)*

Solution. To find reactions R_A and R_B, taking moments above A, we get

$$R_B \times 2 = 1 \times 0.5 \times 0.5/2 + 2 \times 0.5 + 5 \times 1.5 + 1 \times 1 \times \left(\frac{1}{2} + 0.5 + 0.5 \right)$$

$$= 0.125 + 1 + 7.5 + 1.5 = 10.125$$

or, $\qquad R_B = 5.06 \text{ say } 5.1 \text{ kN}$

But,

$$R_A + R_B = 1 \times 0.5 + 2 + 5 + 1 \times 1$$
$$= 0.5 + 2 + 5 + 1 = 8.5 \text{ kN}$$
$$\therefore \quad R_A = 3.4 \text{ kN}$$

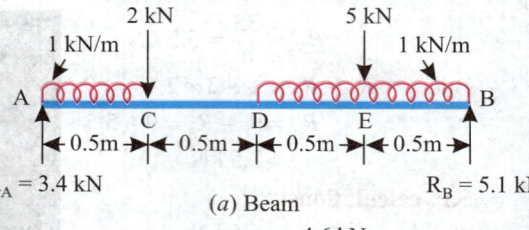

(a) Beam

S.F. calculations:

$$S_B = 5.1 \text{ kN}$$
$$S_E = 5.1 - 1 \times 0.5 - 5 = -0.4 \text{ kN}$$
$$S_{D-C} = -0.4 - 1 \times 0.5 = -0.9 \text{ kN}$$
$$S_C = -0.9 - 2 = -2.9 \text{ kN}$$
$$S_A = -2.9 - 1 \times 0.5 = -3.4 \text{ kN}$$

S.F. diagram is shown in Fig. 4.25 (b).

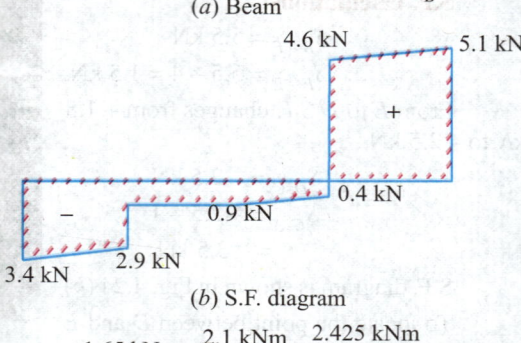

(b) S.F. diagram

B.M. calculations:

$$M_B = 0$$
$$M_E = 5.1 \times 0.5 - 1 \times 0.5 \times 0.5/2$$
$$= 2.425 \text{ kNm}$$

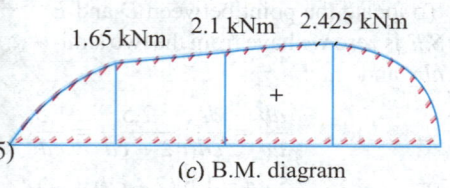

(c) B.M. diagram

$$M_D = 5.1 \times 1 - 5 \times 0.5 - 1 \times 1 \times \frac{1}{2}$$
$$= 5.1 - 2.5 - 0.5 = 2.1 \text{ kNm}$$
$$M_C = 5.1 \times 1.5 - 5 \times 1 - 1 \times 1 \,(0.5 + 0.5)$$
$$= 7.65 - 5 - 1 = 1.65 \text{ kNm}$$

Maximum S.F. is 5.1 kN at *B* and the

Fig. 4.25

maximum *B.M.* is 2.425 kNm at *E*.

B.M. diagram is shown in Figure 4.25 (c).

4.16. SIMPLY SUPPORTED BEAM CARRYING A LOAD WHOSE INTENSITY VARIES UNIFORMLY FROM ZERO AT EACH END TO *w* PER UNIT RUN AT THE MID SPAN

For 4·26 shows a simply supported beam *AB* of span *l* carrying the load mentioned above.

Total load on beam $\qquad = w \times \dfrac{l}{2}$

$$\therefore \qquad R_A = R_B = \frac{wl}{4}$$

Rate of loading at a section at a distance *x* from *B*

$$\frac{EF}{BE} = \frac{CD}{BC} \qquad \text{(By similarity of } \Delta s \; BCD \text{ and } BEF)$$

$$\therefore \qquad EF = \frac{CD \times BE}{BC} = \frac{w \times x}{l/2} = \frac{2\,wx}{l}$$

∴ S.F. at the section is

$$S_x = +\frac{wl}{4} - \frac{1}{2} \times x \times \frac{2\,wx}{l} = \frac{wl}{4} - \frac{wx^2}{l}$$

S.F. at *B* (where $x = 0$), $S_B = +\dfrac{wl}{4}$

S.F. at *C* (where $x = l/2$)

$$S_C = +\frac{wl}{4} - \frac{w}{l}\,(l/2)^2 = 0$$

S.F. at A,

$$S_A = + \frac{wl}{4} - w \times \frac{l}{2} = - \frac{wl}{4}$$

The general equation from $S.F.$ is a second degree equation representing a parabola, so the $S.F.$ diagram is *parabolic*.

$B.M.$ at the section under consideration is

$$M_x = \frac{wl}{4} x - \left(\frac{1}{2} x \times \frac{2wx}{l} \right)$$

$$\times x/3 = \frac{wl}{4} x - \frac{wx^3}{3l}$$

$B.M.$ is zero at both the supports A and B since the beam is simply supported.

For the B.M. to be *maximum* we have

$$\frac{dM_x}{dx} = \frac{wl}{4} - \frac{wx^2}{l} = 0$$

i.e., $x = l/2$ (here the S.F. is zero)

The maximum $B.M.$ is thus at the mid span and is,

$$M_{max} = \frac{wl}{4} \times \frac{l}{2} - \frac{w}{3l} \times \left[\frac{l}{2} \right]^3 = \frac{wl^2}{8} - \frac{wl^2}{24} = + \frac{wl^2}{12}$$

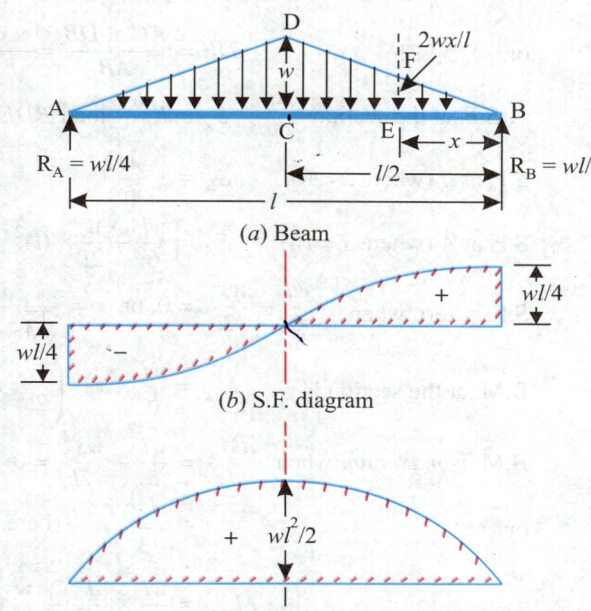

(a) Beam

(b) S.F. diagram

(c) B.M. diagram

Fig. 4.26

4.17. SIMPLY SUPPORTED BEAM CARRYING A LOAD WHOSE INTENSITY VARIES UNIFORMLY FROM ZERO AT ONE END TO w PER UNIT RUN AT THE OTHER END

Fig. 4·27 shows a simply supported beam AB of span l carrying a load whose intensity varies uniformly from zero at the right end B to w per unit run at the left end A.

Let R_A and R_B be the vertical reactions at A and B. For the equilibrium of the beam taking moments about A, we have:

$$R_B \times l = \left(\frac{1}{2} \times l \times w \right) \times \frac{l}{3}$$

$$\therefore \qquad R_B = \frac{wl}{6}$$

But, $R_A + R_B = \frac{wl}{2}$

$$\therefore \qquad R_A = \frac{wl}{2} - \frac{wl}{6} = \frac{wl}{3}$$

Rate of loading at a section at a distance of x from B:

$$\frac{DE}{DB} = \frac{AC}{AB}$$

(from similar Δs BAC and BDE)

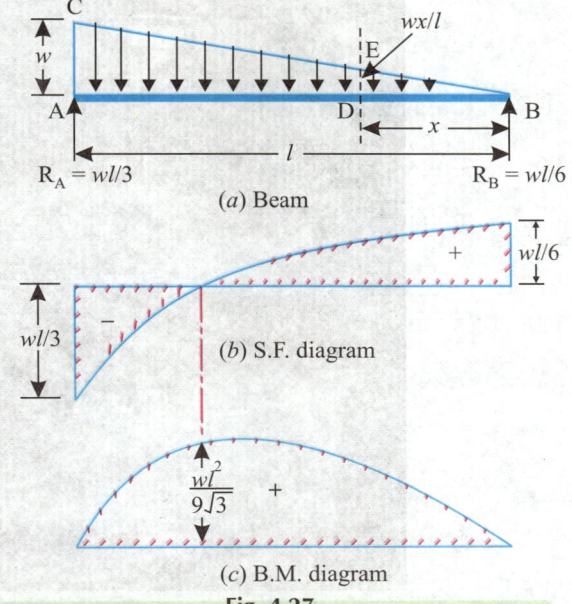

(a) Beam

(b) S.F. diagram

(c) B.M. diagram

Fig. 4.27

or,
$$DE = \frac{AC \times DB}{AB} = \frac{wx}{l}$$

∴ S.F. at the section,
$$S_x = R_B - \text{load } BDE = \frac{wl}{6} - \frac{1}{2} \times x \times \frac{wx}{l} = \frac{wl}{6} - \frac{wx^2}{2l}$$

S.F. at B (where $x = 0$),
$$S_B = + \frac{wl}{6}$$

S.F. at A (where $x = l$),
$$S_A = \frac{wl}{6} - \frac{w}{2l} \times (l)^2 = \frac{wl}{6} - \frac{wl}{2} = -\frac{wl}{3}$$

S.F. is zero when
$$\frac{wl}{6} - \frac{wx^2}{2l} = 0 \text{ or } x = \frac{l}{\sqrt{3}}$$

B.M. at the section is
$$M_x = \frac{wl}{6} \times x - \left(\frac{1}{2} \times x \times \frac{wx}{l}\right) \times \frac{x}{3} = \frac{wlx}{6} - \frac{wx^3}{6l}$$

B.M. is *maximum* when
$$\frac{dM_x}{dx} = \frac{wl}{6} - \frac{wx^2}{2l} = 0$$

or,
$$x = \frac{l}{\sqrt{3}} \qquad \text{(here } S.F. \text{ is zero)}$$

Thus,
$$M_{max} = \frac{wl}{6} \times \frac{l}{\sqrt{3}} - \frac{w}{6l}\left[\frac{l}{\sqrt{3}}\right]^3 = \frac{wl^2}{6\sqrt{3}} - \frac{wl^3}{18\sqrt{3}l} = \frac{wl^2}{9\sqrt{3}}$$

B.M. at both the supports A and B is *zero* since the beam is *simply supported*.

A horizontal boring machine.

Example 4.9. *Calculate the values of maximum and minimum bending moments and shearing forces for the simply supported beam loaded as shown in Fig. 4.28 (a).Draw the bending moment and shearing force diagrams to scale indicating the significant values along the beam.*

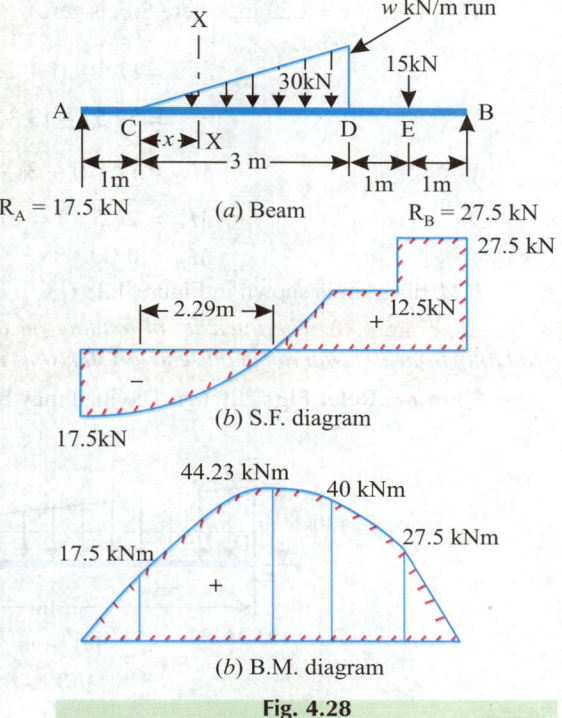

(a) Beam

(b) S.F. diagram

(b) B.M. diagram

Fig. 4.28

Solution. To find reactions R_A and R_B taking moments above A, we get

$$R_B \times 6 = 30\left(3 \times \frac{2}{3} + 1\right) + 15 \times 5$$

$$= 90 + 75 = 165$$

$$\therefore \qquad R_B = \frac{165}{6} = 27.5 \text{ kN}$$

But, $R_A + R_B = 30 + 15 = 45$

$$\therefore \qquad R_A = 45 - 27.5 = 17.5 \text{ kN}$$

If the rate of loading at D is w, then

$$\frac{1}{2} \times 3 \times w = 30$$

$$\therefore \qquad w = \frac{30 \times 2}{3} = 20 \text{ kN/m}$$

S.F. calculations:

Starting from left side:

$$S_A = -17.5 \text{ kN}$$

$$S_x = -17.5 + \frac{1}{2} \times x \times (\text{rate of loading at } XX)$$

i.e.,

$$S_x = -17.5 + \frac{1}{2} \times x \times \frac{20x}{3}$$

or,

$$S_x = -17.5 + \frac{10x^2}{3} \qquad \qquad \qquad \dots (i)$$

When,

$$S_x = 0$$

i.e.,

$$-17.5 + \frac{10x^2}{3} = 0$$

or,

$$10x^2 = 52.5 \qquad \text{or} \qquad x = 2.29 \text{ m}$$

$$S_{D-E} \text{ (putting } x = 3 \text{ m)} = -17.5 + \frac{10 \times 3^2}{3} = 12.5 \text{ kN}$$

(Eqn. (*i*) giving S.F. for *CD* is *second degree equation representing a parabola.*)

$$S_{E-B} = 12.5 + 15 = 27.5 \text{ kN}$$

S.F. diagram is shown in Figure 4.28 (*b*).

B.M. calculations:

$$M_A = 0$$

$$M_C = 17.5 \times 1 = 17.5 \text{ kNm}$$

$$M_x = 17.5\,(1 + x) - \left(\frac{1}{2} \times x \times \frac{20x}{3}\right) \times \frac{x}{3} = 17.5\,(1 + x) - \frac{10x^3}{9}$$

M_{max} (when $x = 2.29$ m, where S.F. is zero)

$$= 17.5 \times (1 + 2.29) - \frac{10 \times (2.29)^3}{9}$$

$$= 57.57 - 13.34 = 44.23 \text{ kNm}$$

$$M_D = 17.5 (1 + 3) - \frac{10 \times 3^3}{9} = 70 - 30 = 40 \text{ kNm}$$

$$M_E = 27.5 \times 1 = 27.5 \text{ kNm}$$

$$M_B = 0$$

B.M. diagram is shown in Figure 4.28 (c).

Example 4.10. *The intensity of loading on a simply suported beam of 4 m span increases gradually from 30 kN/m run at one end to 130 kN/m run at the other. Draw the S.F. and B.M. diagrams.*

Solution. Refer Fig. 4.29 (a). The load may be divided into:

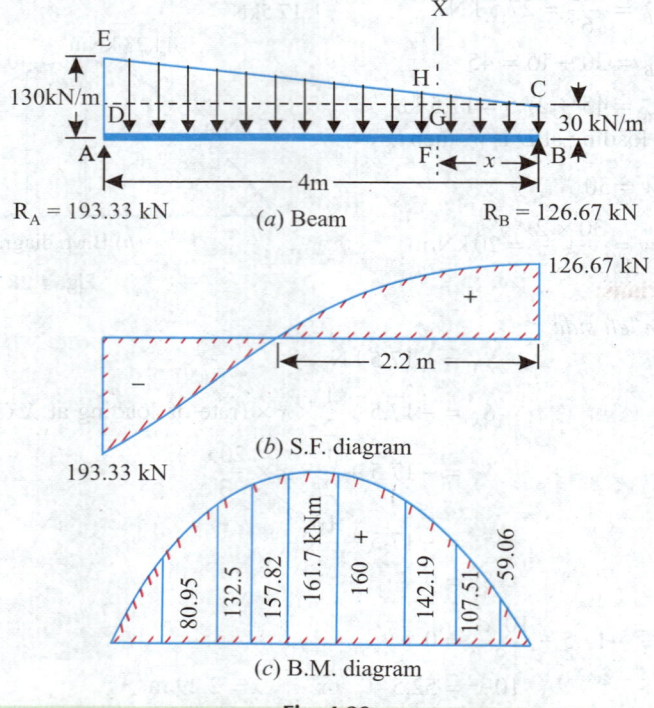

(a) Beam

(b) S.F. diagram

(c) B.M. diagram

Fig. 4.29

 (*i*) A uniformly distributed load (U.D.L.) of 30 kN/m run throughout the span.

 (*ii*) A uniformly increasing triangular load which is zero at the right end *B* and increases unifornly to 100 kN/m run at the left end *A*.

To find reactions R_A and R_B taking moments about *A*, we get

$$R_B \times 4 = 30 \times 4 \times \frac{4}{2} + \frac{1}{2} \times 4 \times 100 \times \frac{4}{3} = 240 + 266.67$$

∴ $\qquad R_B = 126.67$ kN

But, $\qquad R_A + R_B = 30 \times 4 + \frac{1}{2} \times 4 \times 100 = 120 + 200 = 320$ kN

∴ $\qquad R_A = 320 - 126.67 = 193.33 \text{ } kN$

S.F. calculations:

Consider any section XX at a distance x from B. Rate of loading at the section is indicated by the line $FH = (FG + GH)$.

From the two similar Δs CDE and CGH, we have

$$\frac{GH}{GC} = \frac{DE}{DC}$$

or, $$GH = \frac{DE \times GC}{DC} = \frac{100 \times x}{4} = 25\,x$$

$\therefore$ Rate of loading at $x = 30 + 25\,x$

S.F. at the section,

$$S_x = 126.67 - (\text{load } BCGF + \text{load } CGH)$$

$$= 126.67 - (30\,x + \frac{1}{2} \times x \times 25\,x)$$

$$= 126.67 - 30\,x - 12.5\,x^2$$

				TABLE 4.1					
x metre	0	0.5	1	1.5	2.0	2.5	3.0	3.5	4.0
S_x (kN)	126.67	108.54	84.17	53.54	16.67	− 26.45	− 75.83	− 131.45	− 193.33
M_x (kNm)	0	59.06	107.51	142.19	160	157.82	132.5	80.95	0

B.M. calculations:

B.M. at the section, $$M_x = 126.67\,x - 30\,x \times \frac{x}{2} - 12.5\,x^2 \times \frac{x}{3}$$

$$= 126.67x - 15\,x^2 - \frac{12.5}{3}\,x^3$$

For maximum B.M. $S_x = 0$

i.e., $126.67 - 30x - 12.5\,x^2 = 0$

$$x^2 + 2.4\,x - 10.13 = 0$$

$\therefore$ $$x = \frac{-2.4 + \sqrt{(2.4)^2 + 4 \times 10.13}}{2} = 2.2 \text{ m (neglecting } - \text{ve value)}$$

$$M_{max} = 126.67 \times 2.2 - 15 \times 2.2^2 - \frac{12.5}{3} \times 2.2^3$$

$$= 278.67 - 72.6 - 44.37 = 161.7 \text{ kNm}$$

In order to plot the S.F. and B.M. diagrams their values have been calculated and tabulated at 0.5 m intervals (Table 4.1).

S.F. and B.M. diagrams are shown in Fig. 4.29 (b) and (c) respectively.

Example 4.11. *Fig. 4.30 (a) shows a beam AB of length 4 m acted upon by the forces and moments. Draw the B.M. and S.F. diagrams.*

Solution. To determine reactions R_A and R_B taking moments about A, we get

$$R_B \times 4 = 20 \times 1.5 \times 1.5/2 + 30 + 20 \times 3 = 22.5 + 30 + 60$$

$\therefore$ $$R_B = 28.12 \text{ kN}$$

But, $$R_A + R_B = 20 \times 1.5 + 20 = 50 \text{ kN}$$

$\therefore$ $$R_A = 50 - 28.12 = 21.88 \text{ kN}$$

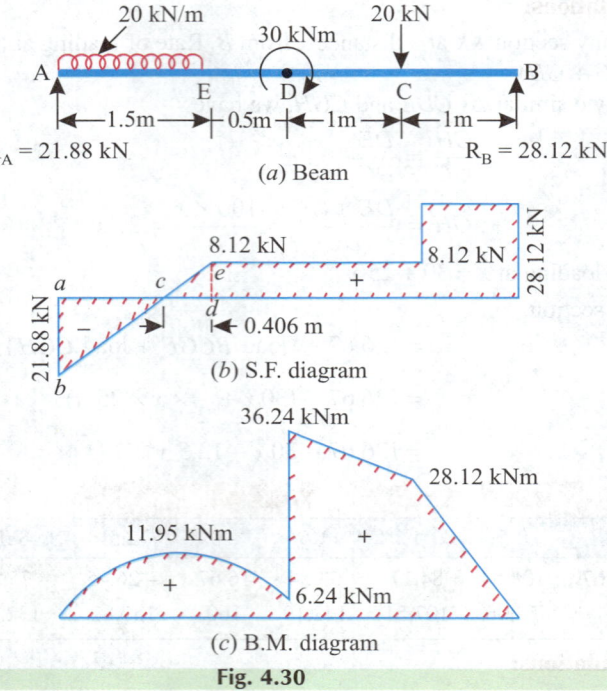

(a) Beam

(b) S.F. diagram

(c) B.M. diagram

Fig. 4.30

S.F. calculations:

$$S_{B-C} = 28.12 \text{ kN}$$

$$S_{C-D-E} = 28.12 - 20 = 8.12 \text{ kN}$$

$$S_A = 8.12 - 20 \times 1.5 = -21.88 \text{ kN}$$

Model of a dome shaped roof of a building. Beams in the roof should bear bending as well as shear stresses.

To locate the point between A and E where the S.F. is zero we have from similar Δs abc and cde.

$$\frac{ab}{ac} = \frac{de}{cd} \quad \text{or} \quad \frac{21.88}{(1.5 - cd)} = \frac{8.12}{cd}$$

or, $21.88 \, cd = 8.12 \, (1.5 - cd)$

or, $21.88 \, cd = 12.18 - 8.12 \, cd$

or, $30 \, cd = 12.18$

$\therefore$ $cd = 0.406 \, m$ (Fig. 4.30 (b))

B.M. calculations:

$$M_B = 0$$
$$M_C = 28.12 \times 1 = 28.12 \text{ kNm}$$
$$M_D \text{ (right)} = 28.12 \times 2 - 20 \times 1 = 36.24 \text{ kNm}$$
$$M_D \text{ (left)} = 36.24 - 30 = 6.24 \text{ kNm}$$
$$M_{max} \text{ (0.406 m from } E) = 28.12 \, (1 + 1 + 0.5 + 0.406)$$
$$- 20 \, (1 + 0.5 + 0.406) - 30 - 20 \times 0.406 \times 0.406/2$$
$$= 81.72 - 38.12 - 30 - 1.65 = 11.95 \text{ kNm}$$

Example 4.12. *A beam of 20 m span, hinged at its both ends is loaded as shown in Fig. 4.31 (a). Determine the reactions at the ends and draw the B.M. and S.F. diagrams.*

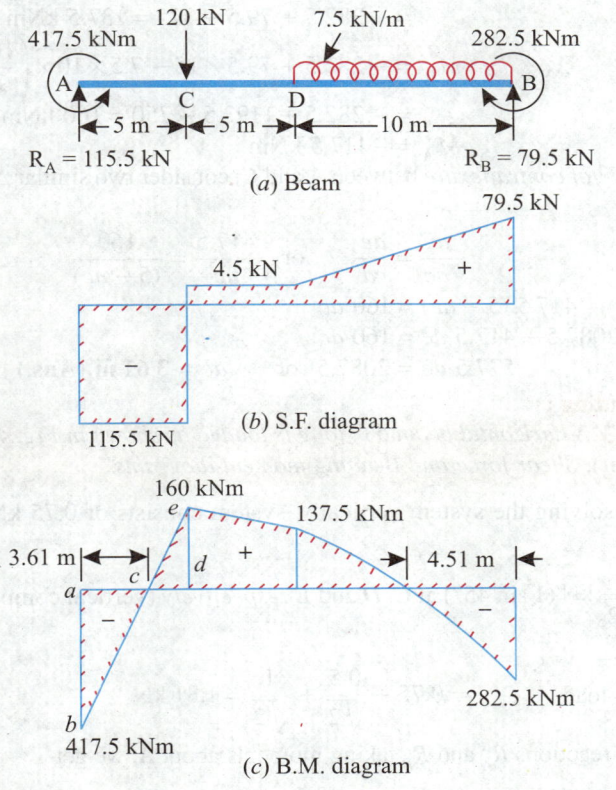

(a) Beam

(b) S.F. diagram

(c) B.M. diagram

Fig. 4.31

Solution. To determine reactions R_A and R_B taking moment about A, we get

$$R_B \times 20 + 417.5 = 120 \times 5 + 7.5 \times 10 \times \left(\frac{10}{2} + 5 + 5\right) + 282.5$$

or, $R_B \times 20 + 417.5 = 600 + 1125 + 282.5$

or , $R_B = \textbf{79.5 kN}$ **(Ans.)**

But, $R_A + R_B = 120 + 7.5 \times 10 = 195$ kN

                $R_A = 195 - 79.5 = \textbf{115.5 kN}$ **(Ans).**

S.F. calculations:

$$S_B = 79.5 \text{ kN}$$
$$S_{D-C} = 79.5 - 7.5 \times 10 = 4.5 \text{ kN}$$
$$S_{C-A} = 4.5 - 120 = -115.5 \text{ kN}$$

S.F. diagram is shown in Fig. 4.31 (*b*)

B.M. calculations:

$$M_B = -282.5 \text{ kNm}$$

For *BD*, at distance x from B,

$$M_x = -282.5 + 79.5\,x - 7.5 \times x \times \frac{x}{2}$$
$$= -282.5 + 79.5x - 3.75\,x^2 \qquad \text{...(Parabolic)}$$

This is *zero* at $x = 4.51$ m (*point of contraflexure*)

$$M_D = -282.5 + 79.5 \times 10 - 7.5 \times 10 \times \frac{10}{2}$$
$$= -282.5 + 79.5 - 375 = 137.5 \text{ kNm}$$

$$M_C = -282.5 + 79.5 \times 15 - 7.5 \times 10 \times \left(\frac{10}{2} + 5\right)$$
$$= -282.5 + 1192.5 - 750 = 160 \text{ kNm}$$
$$M_A = -417.5 \text{ kNm}$$

To locate *point of contraflexure* between *A* and *C*, consider two similar Δs *abc* and *cde* (Figure 4.31 (*c*)):

$$\frac{ab}{ac} = \frac{de}{cd}, \quad \text{or,} \quad \frac{417.5}{ac} = \frac{160}{(5 - ac)}$$

or, $417.5\,(5 - ac) = 160\,ac$

            $2087.5 - 417.5\,ac = 160\,ac$

            $577.5\,ac = 2087.5$ or $ac = \textbf{3.61 m (Ans.)}$

Inclined Loading :

Example 4.13. *A horizontal beam 6 m long is loaded as shown in Fig. 4.32 (a). Construct the Axial force (or thrust), Shear force and Bending moment diagrams.*

Solution. Resolving the system vertically, system consists of 0.75 kN (1.5 sin 30°), $\dfrac{0.5}{\sqrt{2}}$ (0.5 sin 45°) and $\dfrac{1}{\sqrt{2}}$ kN (1 sin 45°) at *C, D* and *E respectively* (vertical components shown dotted) in Fig. 4.32 (*a*).

Total vertical load $= 0.75 + \dfrac{0.5}{\sqrt{2}} + \dfrac{1}{\sqrt{2}} = 1.81$ kN

To determine reactions R_A and R_B taking moments about *A*, we get

$$R_B \times 6 = \frac{1}{\sqrt{2}} \times 1 + \frac{0.5}{\sqrt{2}} \times 2.5 + 0.75 \times 4 = 4.59$$

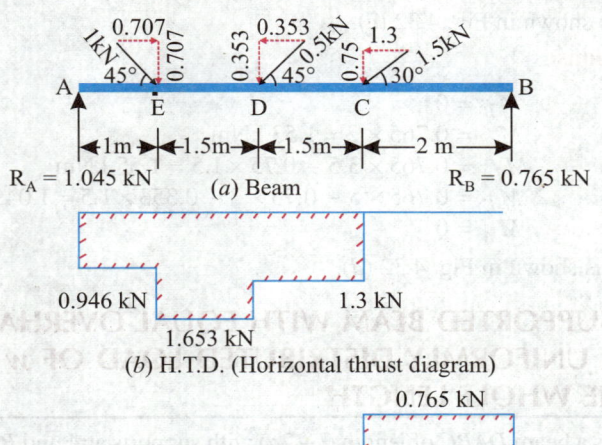

(a) Beam

$R_A = 1.045$ kN $R_B = 0.765$ kN

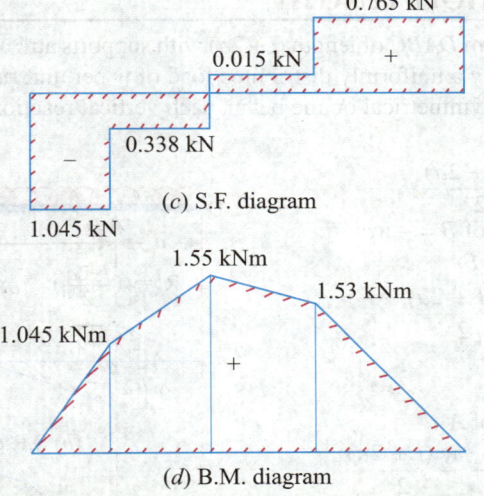

0.946 kN   1.3 kN

1.653 kN

(b) H.T.D. (Horizontal thrust diagram)

0.765 kN

0.015 kN   +

0.338 kN

−

(c) S.F. diagram

1.045 kN

1.55 kNm

1.53 kNm

1.045 kNm

+

(d) B.M. diagram

Fig. 4.32

∴ $R_B = 0.765$ kN

But, $R_A + R_B$ = Total vertical load = 1.81

∴ $R_A = 1.81 - 0.765 = 1.045$ kN

H.T.D. (horizontal thrust diagram) calculations:

Horizontal system consists of 1.3 kN (1.5 cos 30°), 0.353 kN (0.5 cos 45°), 0.707 kN (1 cos 45°) at C, D and E respectively (horizontal components shown dotted) in Fig. 4.32 (a).

The force at E is opposite in direction to those of C and D.

Horizontal reaction at hinge A

$$= 1.3 + 0.353 - 0.707 = 0.946 \text{ kN}$$

H.T.D. is shown in Fig. 4.32 (b).

S.F. calculations:

$$S_{B-C} = 0.765 \text{ kN}$$
$$S_{C-D} = 0.765 - 0.75 = 0.015 \text{ kN}$$
$$S_{D-E} = 0.015 - 0.353 = -0.338 \text{ kN}$$
$$S_{E-A} = -0.338 - 0.707 = -1.045 \text{ kN}$$

S.F. diagram is shown in Fig. 4.32 (c).

B.M. calculations:

$$M_B = 0$$
$$M_C = 0.765 \times 2 = 1.53 \text{ kNm}$$
$$M_D = 0.765 \times 3.5 - 0.75 \times 1.5 = 1.55 \text{ kNm}$$
$$M_E = 0.765 \times 5 - 0.75 \times 3 - 0.353 \times 1.5 = 1.045 \text{ kNm}$$
$$M_A = 0$$

B.M. diagram is shown in Fig. 4.32 (d).

4.18. SIMPLY SUPPORTED BEAM WITH EQUAL OVERHANGS AND CARRYING A UNIFORMLY DISTRIBUTED LOAD OF w PER UNIT RUN OVER THE WHOLE LENGTH

Fig. 4.33 shows a beam $DABC$ of length $(l + 2a)$ with supports at A and B so that $AB = l$ and $DA = BC = a$. Let the beam carry a uniformly distributed load of w per unit run over the entire length.

Since the loading is symmetrical on the beam, each vertical reaction equals half the total load on the beam,

$$\therefore \quad R_A = R_B = \frac{w\,(l + 2a)}{2}$$

S. F. just to the right of $B = -wa$

S.F. just to the left of B

$$= -wa + \frac{w\,(l + 2a)}{2}$$

$$= +\frac{wl}{2}$$

S.F. just to the right of A

$$= -w(l + a) + \frac{w\,(l + 2a)}{2}$$

$$= -\frac{wl}{2}$$

S.F. just to the left of A

$$= -w\,(l + a) + 2 \times \frac{w\,(l + 2a)}{2}$$

$$= +wa$$

The S.F. is $+\dfrac{wl}{2}$ at B and $-\dfrac{wl}{2}$ at A as such that it will be zero at the mid point between A and B.

B.M. at

$$B = -wa \times \frac{a}{2} = -\frac{wa^2}{2}$$

B.M. at the point of zero shear *i.e.* at a distance $(a + l/2)$ from C is the maximum, and is,

$$M_{max} = \left[-\frac{w}{2}\,(a + l/2)^2 + \frac{w(l + 2a)}{2} \times l/2 \right]$$

$$= \frac{w}{2}\left[\frac{l^2}{4} - a^2 \right]$$

If $a < l/2$ then M_{max} will be +ve and B.M. diagram shall be as shown in Fig. 4.33 (a).

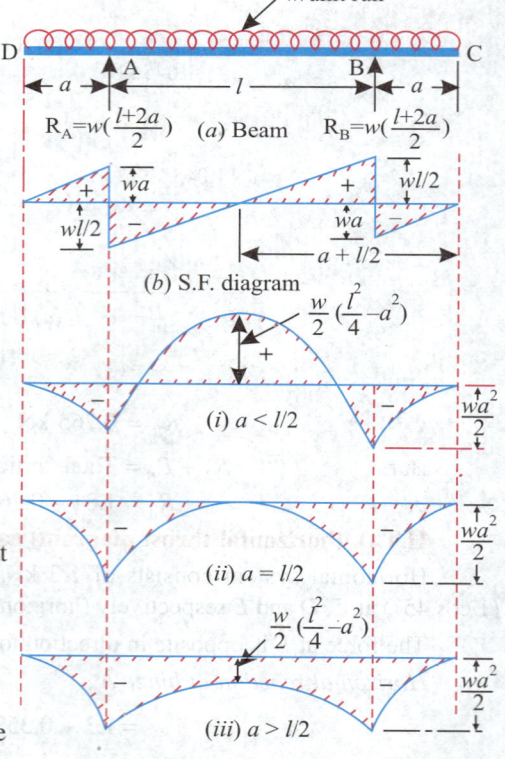

$$R_A = w\left(\frac{l+2a}{2}\right) \quad \text{(a) Beam} \quad R_B = w\left(\frac{l+2a}{2}\right)$$

(b) S.F. diagram

$$\frac{w}{2}\left(\frac{l^2}{4} - a^2\right)$$

(i) $a < l/2$

(ii) $a = l/2$

$$\frac{w}{2}\left(\frac{l^2}{4} - a^2\right)$$

(iii) $a > l/2$

(c) B.M. diagram

Fig. 4.33

If $a = l/2$ then M_{max} shall be zero and *B.M.* diagram shall be as shown in Fig. 4.33 (*b*).
If $a > l/2$ then M_{max} shall be negative and *B.M.* diagram shall be as shown in Fig. 4.33 (*c*).
If *B.M.* is zero at a distance *x* from either end then

$$-\frac{wx^2}{2} + \frac{w(l + 2a)}{2} (x - a) = 0$$

or, $x = \dfrac{(l + 2a) \pm \sqrt{l^2 - 4a^2}}{2}$

4.19. THE POINTS OF CONTRAFLEXURE

The bending moments of opposite nature always produce curvatures of beams in opposite directions. *In a beam if the bending moment changes sign at a point, the point itself having zero bending moment, the beam changes curvature at this point of zero bending moment and this point is called the* **point of contraflexure.** *So at a point of contraflexure the beam flexes in opposite direction. The point of contraflexure is called the point of inflexion or a virtual hinge. The point of contraflexure can be found by setting the bending moment equation in terms of x equal to zero for part of a span where bending moment is likely to change sign.*

Overhanging Beams:

Example 4.14. *Draw S.F. and B.M. diagrams for the loaded beam shown in Fig. 4.34 (a).*

Solution. To determine reactions R_A and R_B taking moments about *A*, we get

$R_B \times 5 = 5.5 \times 2 + 2 \times 5 \times \dfrac{5}{2} + 2 \times 7 = 11 + 25 + 14 = 50$

∴ $R_B = 10$ kN
But, $R_A + R_B = 5.5 + 2 \times 5 + 2 = 17.5$ kN
∴ $R_A = 175 - 10 = 7.5$ kN

S.F. calculations:

$S_{D-B} = -2$ kN
$S_B = -2 + 10 = 8$ kN
$S_C = 8 - 2 \times 3 - 5.5 = -3.5$ kN $R_A = 7.5$ kN
$S_A = -3.5 - 2 \times 2 = -7.5$ kN

S.F. diagram is shown in Fig. 4.34 (*b*).

B.M. calculations:

$M_D = 0$
$M_B = -2 \times 2 = -4$ kNm
B.M. mid way between *B* and *C*,

$M_P = -2 \times 3.5 + 10 \times 1.5 - 2$

        $\times 1.5 \times \dfrac{1.5}{2}$

     $= -7 + 15 - 2.25 = 5.75$ kNm
$M_C = -2 \times 5 + 10 \times 3 - 2 \times 3 \times 3/2$
     $= -10 + 30 - 9 = 11$ kNm

B.M. mid way between *A* and *C*,

$M_Q = -2 \times 6 + 10 \times 4 - 2 \times 4 \times 4/2 - 5.5 \times 1$
     $= -12 + 40 - 16 - 5.5 = 6.5$ kNm

or, $\left[\begin{array}{l} M_Q = +7.5 \times 1 - 2 \times 1 \times \dfrac{1}{2} \\ = 6.5 \text{ kNm ... From left side} \end{array}\right]$

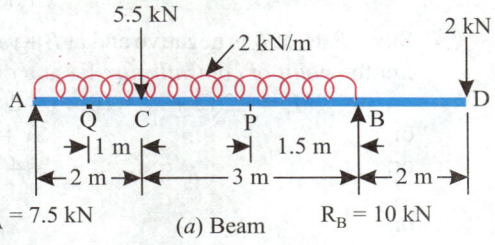

(*a*) Beam

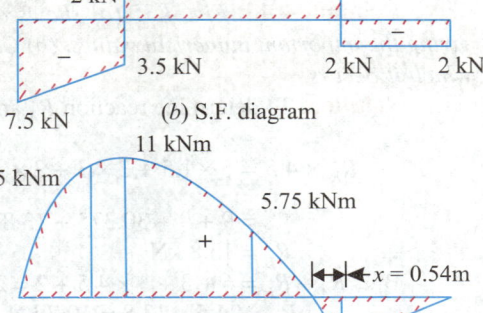

(*b*) S.F. diagram

(*c*) B.M. diagram

Fig. 4.34

Grinding machine.

Since *B.M.* at *B* is negative and at *P* + *ve*, therefore, the *B.M.* will cross zero line between them. Let the *point of contraflexure* lie at a distance *x* from *B*.

$$M_x = -2(2+x) + 10 \times x - 2 \times x \times x/2 = 0$$

or, $-4 - 2x + 10x - x^2 = 0$

or, $-4 + 8x - x^2 = 0$ or $x^2 - 8x + 4 = 0$

or, $x = \dfrac{8 \pm \sqrt{64 - 16}}{2} = \dfrac{8 \pm 6.93}{2}$

from which $x = 0.54$ m, the other value of x being inadmissible.
B.M. diagram is shown in Fig. 4.34 (*c*).

Example 4.15. *Fig. 4.35 (a) shows a loaded beam. (a) Sketch the B.M. and S.F. diagrams giving the important numerical values. (b) Calculate the maximum bending moment and the point at which it occurs.*

Solution. To determine reaction R_A and R_B taking moments about *A*, we get

$$R_B \times 4.5 = 6 \times 1.5 + 3 \times 3 + 3 \times 4.5 \times \frac{4.5}{2} + 2 \times (4.5 + 2.4)$$

$$= 9 + 9 + 30.375 + 13.8$$

∴ $R_B = 13.8$ kN

But, $R_A + R_B = 6 + 3 + 3 \times 4.5 + 2 = 24.5$ kN

∴ $R_A = 24.5 - 13.8 = 10.7$ kN

S.F. calculations:

$$S_{E-B} = -2 \text{ kN}$$
$$S_B = -2 + 13.8 = 11.8 \text{ kN}$$
$$S_D = 11.8 - 3 \times 1.5 - 3 = 4.3 \text{ kN}$$
$$S_C = 4.3 - 3 \times 1.5 - 6 = -6.2 \text{ kN}$$
$$S_A = -6.2 - 3 \times 1.5 = -10.7 \text{ kN}$$

To locate the point between C and D where S.F. is zero we have from similar triangles abc and cde.

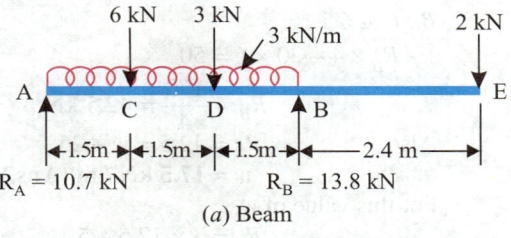

$$\frac{ab}{ac} = \frac{de}{cd}$$

$$\frac{0.2}{(1.5-cd)} = \frac{4.3}{cd}$$

or, $\quad 0.2\,cd = 4.3\,(1.5-cd)$

or, $\quad 0.2\,cd = 6.45 - 4.3\,cd$

or, $\quad cd = 1.43$ m

At this point *B.M.* will be *maximum*.

B.M. calculations:

$$M_E = 0$$
$$M_B = -2 \times 2.4 = -4.8 \text{ kNm}$$
$$M_D = -2 \times (2.4 + 1.5) + 13.8$$
$$\times 1.5 - 3 \times 1.5 \times \frac{1.5}{2}$$
$$= -7.8 + 20.7 - 3.375$$
$$= 9.52 \text{ kNm}$$
$$M_{max} = -2 \times (2.4 + 1.5 + 1.43) + 13.8$$
$$\times (1.5 + 1.43) - 3 \times 2.93$$
$$\times \frac{2.93}{2} - 3 \times 1.43$$
$$= -10.66 + 40.43 - 12.87 - 4.29$$
$$= 12.61 \text{ kNm}$$

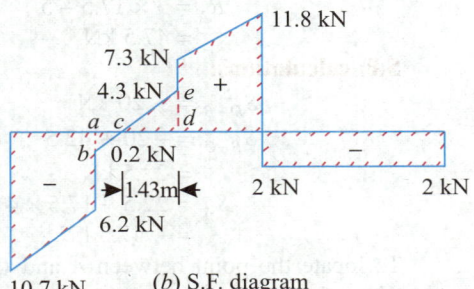

(a) Beam

(b) S.F. diagram

(c) B.M. diagram

Fig. 4.35

$$M_C = -2 \times 5.4 + 13.8 \times 3 - 3 \times 1.5 - 3 \times 3 \times \frac{3}{2}$$
$$= -10.8 + 41.4 - 4.5 - 13.5 = 12.6 \text{ kNm.}$$
$$M_A = 0$$

Since *B.M.* at B is negative and at D is positive, therefore, the *B.M.* will cross zero line between them. Let the point of contraflexure lie at a distance x from B.

$$M_x = -2 \times (2.4 + x) + 13.8 \times x - 3 \times x \times \frac{x}{2} = 0$$

or, $\qquad\qquad -4.8 - 2x + 13.8x - 1.5x^2 = 0$

or, $\qquad\qquad 1.5x^2 - 11.8x + 4.8 = 0$

from which $x = 0.43$ m, the other value of x being inadmissible.

B.M. diagram is shown in Fig. 4.35 (c).

Example 4.16. *For the beam loaded as shown in Fig. 4.36 (a) calculate the value of U.D.L. w so that B.M. at C is 50 kNm. Draw the S.F. and B.M. diagrams for this beam for the calculated value of w. Locate the point of contraflexure, if any.*

Solution. Taking moments about A, we get

$$R_B \times 8 = w \times 4 \times 4/2 + 20 \times 10 = 8w + 200$$

or $\qquad\qquad R_B = (w + 25) \text{ kN}$

But, $R_A + R_B = w \times 4 + 20$

$\therefore \qquad\qquad R_A = (4w + 20) - (w + 25)$

$$= (3w - 5) \text{ kN}$$

B.M. at *C* is:

$$R_B \times 4 - 20 \times 6 = 50$$

$$\therefore \qquad R_B = \frac{170}{4} = 42.5 \text{ kN}$$

Also, $\qquad R_B = 42.5 = w + 25$

$$\therefore \qquad \boldsymbol{w = 17.5 \text{ kN / m (Ans.)}}$$

For this value of *w*,

$$R_A = 3 \times 17.5 - 5$$
$$= 47.5 \text{ kN}$$

S.F. calculations:

$$S_{D-B} = -20 \text{ kN}$$
$$S_{B-C} = -20 + 42.5$$
$$= 22.5 \text{ kN}$$
$$S_A = 22.5 - 17.5 \times 4$$
$$= -47.5 \text{ kN}$$

To locate the point between *A* and *C* where the *S.F.* is zero we have from the similar Δs *abc* and *cde*,

$$\frac{ab}{ac} = \frac{de}{cd}$$

$$\frac{47.5}{4 - cd} = \frac{22.5}{cd}$$

or, $\qquad 47.5 \, cd = 22.5 \, (4 - cd)$

$$47.5 \, cd = 90 - 22.5 \, cd$$
$$cd = 1.28 \text{ m} \quad \text{Fig. 4.36 } (b)$$

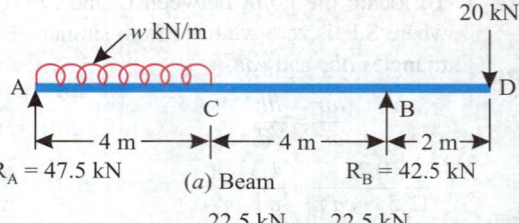

(a) Beam

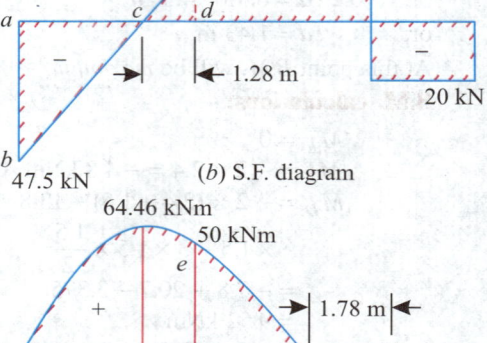

(b) S.F. diagram

(c) B.M. diagram

Fig. 4.36

B.M. calculations:

$$M_D = 0$$
$$M_B = -20 \times 2 = -40 \text{ kNm}$$
$$M_C = -20 \times 6 + 42.5 \times 4$$
$$= -120 + 170 = 50 \text{ kNm}$$

Maximum *B.M.* will occur at the point where *S.F.* is *zero*. In this case, this point is at a distance of 1.28 m from *C*.

$$\therefore \qquad M_{max} = -20 \times (2 + 4 + 1.28) + 42.5 \, (4 + 1.28) - 17.5 \times 1.28 \times \frac{1.28}{2}$$
$$= -145.6 + 224.4 - 14.34$$
$$= 64.46 \text{ kNm}$$

B.M. diagram between *C* and *A* will be *parabolic* as the load on *AC* is U.D.L.

$$M_A = 0$$

To find point of contraflexure, consider two similar Δs *abc* and *cde* Fig. 4.36 (*c*),

such that, $\qquad \dfrac{ab}{ac} = \dfrac{de}{cd} \qquad \dfrac{40}{ac} = \dfrac{50}{(4 - ac)}$

or, $\qquad 40 \, (4 - ac) = 50 \, ac$

$$160 - 40 \, ac = 50 \, ac$$
$$ac = 1.78 \text{ m}$$

i.e. **at 1.78 m from** *C* **(Ans.)**

Example 4.17. *A beam AB 10 metres long carries a uniformly distributed load of 20 kN/m over its entire length together with concentrated load of 50 kN at the left end A and 80 kN at the end B [Fig. 4.37 (a)]. The beam is to be supported at two props at the same level, 6 metres apart, so that the reaction is the same at each. Determine the positions of the supports and draw S.F. and B.M. diagrams. Find the value of maximum B.M. Locate the points of contraflexure (or inflexion), if any.*

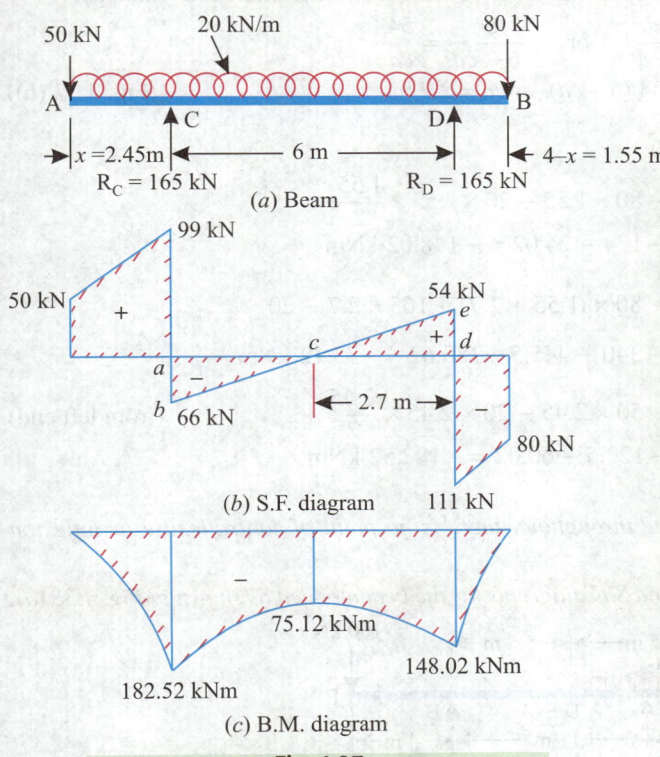

(a) Beam

(b) S.F. diagram

(c) B.M. diagram

Fig. 4.37

Heavy duty boring machine.

Solution. Refer Fig. 4.37 (a). Since two support reactions are equal,

$$\therefore \qquad R_C = R_D$$

But, $\qquad R_C + R_D = 50 + 20 \times 10 + 80 = 330$ kN

$$\therefore \qquad R_C = R_D = 165 \text{ kN}$$

Assume the overhang on the left side to be x metres, the overhang then, on the right side is

$$(10 - 6 - x) = (4 - x) \text{ m}$$

To determine the value of x taking moments about A, we get

$$R_C \times x + R_D \, (6 + x) = 20 \times 10 \times \frac{10}{2} + 80 \times 10$$

or $\qquad 165x + 165 \, (6 + x) = 1000 + 800$

or $\qquad 165x + 990 + 165x = 1800$

or $\qquad x = 2.45 \text{ m}$

$\therefore$ Overhang $\qquad AC = 2.45$ m, and

Overhang $\qquad BD = 4 - 2.45 = 1.55$ m

S.F. calculations:

$$S_B = -80 \text{ kN}$$

$$S_D = -80 - 20 \times 1.55 + 165 = 54 \text{ kN}$$
$$S_C = 54 - 20 \times 6 + 165 = +99 \text{ kN}$$

(S.F. changes from – 66 kN to + 99 kN)

$$S_A = 50 \text{ kN } (i.e.\ 90 - 20 \times 2.45)$$

For the point of zero S.F. in the span CD we have from the two similar Δs abc and cde.

$$\frac{ab}{ac} = \frac{de}{cd} \quad \text{or} \quad \frac{66}{6 - cd} = \frac{54}{cd}$$

or $\qquad 66\ cd = 54\ (6 - cd)$ or $cd = 2.7$ m $\qquad$ (Fig. 4.37 (b))

B.M. calculations:

$$M_B = 0$$
$$M_D = -80 \times 1.55 - 20 \times 1.55 \times \frac{1.55}{2}$$
$$= -124 - 24.02 = -148.02 \text{ kNm}$$

B.M. at the point of zero S.F. is

$$= -80 \times (1.55 + 2.7) + 165 \times 2.7 - 20$$
$$-(1.55 + 2.7)\frac{(1.55 + 2.7)}{2} = -340 + 445.5 - 180.62 = -75.12 \text{ kNm}$$

$$M_C = -50 \times 2.45 - 20 \times 2.45 \times \frac{2.45}{2} \qquad \text{(from left end)}$$
$$= -122.5 - 60.02 = -182.52 \text{ kNm}$$

B.M. is maximum at support C.

Since the B.M. remains negative throughout, there is no point of contraflexure or inflexion [Fig. 4.37 (c)].

Example 4.18. *Draw the B.M. and S.F. diagrams for the beam loaded as shown in Fig. 4.38 (a).*

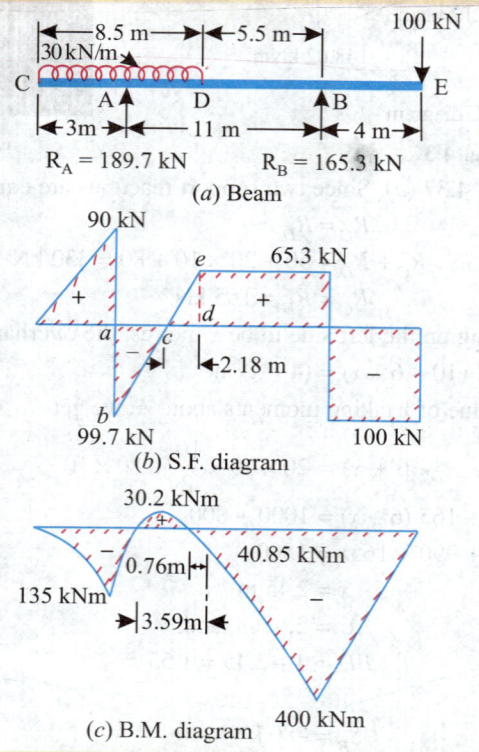

(a) Beam

(b) S.F. diagram

(c) B.M. diagram

Fig. 4.38

Machines in a machine shop.

Solution. To determine reactions R_A and R_B taking moments about A, we get

$$R_B \times 11 + 30 \times 3 \times \frac{3}{2} = 30 \times 5.5 \times \frac{5.5}{2} + 100 \times 15$$

or $R_B \times 11 + 135 = 453.75 + 1500$

or $R_B = 165.3 \text{ kN}$

But, $R_A + R_B = 30 \times 8.5 + 100 = 255 + 100 = 355 \text{ kN}$

∴ $R_A = 355 - 165.3 = 189.7 \text{ kN}$

S.F. calculations:

$$S_{E-B} = -100 \text{ kN}$$

$$S_{B-D} = -100 + 165.3 = 65.3 \text{ kN}$$

$$S_A = 65.3 - 30 \times 5.5 + 189.7 = 90 \text{ kN}$$

(S.F. changes from -99.7 kN to 90 kN)

$$S_C = 90 - 30 \times 3 = 0$$

To locate the point where the S.F. is *zero* consider two similar triangles *abc* and *cde*.

$$\frac{ab}{ac} = \frac{de}{cd} ; \quad \frac{99.7}{5.5 - cd} = \frac{65.3}{cd}$$

or $99.7 \, cd = 65.3 \, (5.5 - cd)$

or $cd = 2.18 \, m$ [(Fig. 4.38 (*b*)]

B.M. calculations:

$$M_E = 0$$

$$M_B = -100 \times 4 = -400 \text{ kNm}$$

$$M_D = -100 \times 9{\cdot}5 + 165{\cdot}3 \times 5{\cdot}5 = -950 + 909{\cdot}15$$
$$= -40{\cdot}85 \text{ kNm}$$

Maximum B.M. between A and D is at 2.18 m from D. Its value is:

$$M_{max} = -100 \times (4 + 5{\cdot}5 + 2{\cdot}18) + 165{\cdot}3 (5{\cdot}5 + 2{\cdot}18) - 30 \times 2{\cdot}18 \times \frac{2{\cdot}18}{2}$$

$$= -1168 + 1269{\cdot}5 - 71{\cdot}28 = 30{\cdot}2 \text{ kNm}$$

$$M_A = -30 \times 3 \times \frac{3}{2} = -135 \text{ kNm} \qquad \text{...(From left end)}$$

To *locate point of inflexion* take B.M. at a distance x from D (in AD) and equate it to *zero*.

$$-100 \times (4 + 5{\cdot}5 + x) + 165{\cdot}3 \times (5{\cdot}5 + x) - 30 \times x \times \frac{x}{2} = 0$$

$$= -950 - 100x + 909{\cdot}15 + 165{\cdot}3x - 15x^2 = 0$$

or $15x^2 - 65{\cdot}3x + 40{\cdot}85 = 0$

or $x^2 - 4{\cdot}35x + 2{\cdot}72 = 0$

$\therefore$

$$x = \frac{4{\cdot}35 \pm \sqrt{4{\cdot}35^2 - 4 \times 2{\cdot}72}}{2} = \frac{4{\cdot}35 \pm 2{\cdot}83}{2}$$

or $x = 3{\cdot}59 \text{ m or } 0{\cdot}76 \text{ m}$ (Fig. 4.38 (c))

Overhanging Beam with Couple

Example 4·19. *Draw the shearing force and bending moment diagrams for a beam shown in Fig. 4·39.*

Solution. The horizontal force of 5 kN acting on the top of a 1m lever at C causes a clockwise moment of 5 kNm at C and a horizontal thrust of 5 kN in the beam. Since the horizontal thrust in beam effects neither the shear force in the beam nor the B.M. in it, for purpose of analysing the shear and the bending moment in the beam the given force of 5 kN can be replaced by a clockwise movement of 5 kNm [Fig. 4·39 (b)] acting at C.

To determine reactions R_A and R_B taking moments about A, we get

$$R_B \times 6 = 2 \times 4 \times \frac{4}{2} + 5 +$$
$$10 (4 + 2 + 2) = 16 + 5 + 80$$

$\therefore$ $R_B = \dfrac{101}{6} = 16{\cdot}83 \text{ kN}$

But, $R_A + R_B = 2 \times 4 + 10 = 18$

$\therefore$ $R_A = 18 - R_B = 18 - 16{\cdot}83$
$$= 1{\cdot}17 \text{ kN}$$

S.F. calculations:

$$S_{D-B} = -10 \text{ kN}$$

$$S_{B-C} = -10 + 16{\cdot}83 = 6{\cdot}83 \text{ kN}$$

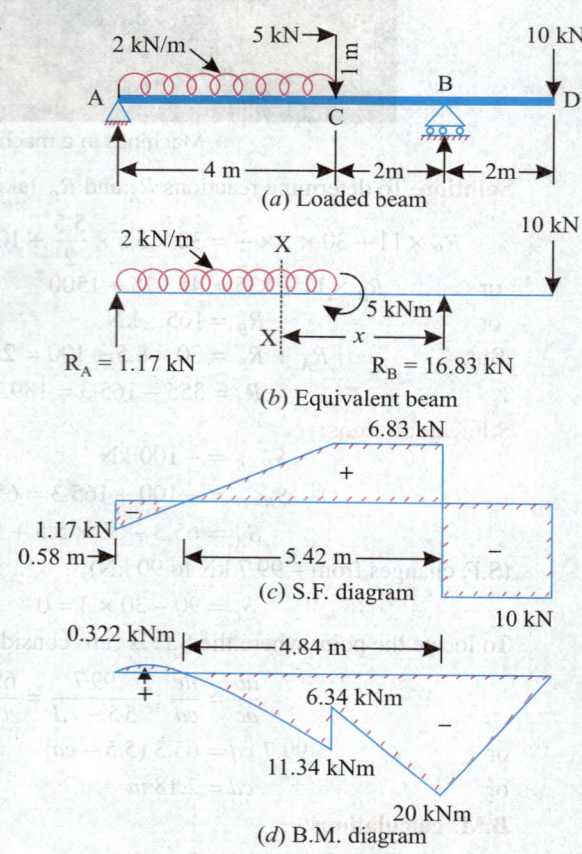

(a) Loaded beam

(b) Equivalent beam

(c) S.F. diagram

(d) B.M. diagram

Fig. 4.39

$$S_A = 6.83 - 2 \times 4 = -1.17 \text{ kN}$$

S.F. at a distance x from B in AC is

$$S_x = -10 + 16.83 - 2(x - 2)$$

To locate the point of zero shear we have

$$S_x = 0$$

i.e. $-10 + 16.83 - 2(x-2) = 0$

or, $+6.83 - 2x + 4 = 0$

or, $x = 5.415$ m say 5.42 m

or, $(6 - 5.42) = 0.58$ m from A.

S.F. diagram is shown in Fig. 4.39 (c).

B.M. calculations:

$$M_D = 0$$
$$M_B = -10 \times 2 = -20 \text{ kNm}$$
$$M_C = -10 \times 4 + 16.83 \times 2 - 5 = -11.34 \text{ kNm}$$
$$M_{max} \ (5.42 \text{ m from } B) = -10 \times (2 + 5.42) + 16.83 \times 5.42 - 5$$
$$- \frac{2 \times (5.42 - 2)^2}{2}$$
$$= -74.2 + 91.218 - 5 - 11.696 = 0.322 \text{ kNm}$$
$$M_A = 0$$

To locate the *point of contraflexure* between A and C, find out the bending moment at the section XX at a distance x from B in AC and equate it to *zero* as given below:

$$M_x = -10(2 + x) + 16.83 \times x - 5 - \frac{2 \times (x - 2)^2}{2} = 0$$

or, $-20 - 10x + 16.83x - 5 - (x - 2)^2 = 0$

or, $-25 + 6.83x - (x^2 - 4x + 4) = 0$

or, $-25 + 6.83x - x^2 + 4x - 4 = 0$

or, $x^2 - 10.83x + 29 = 0$

$$\therefore \quad x = \frac{+10.83 \pm \sqrt{(10.83)^2 - 4 \times 29}}{2} = \frac{+10.83 \pm 1.135}{2}$$

$$= 5.98 \text{ m} \quad \text{or} \quad 4.84 \text{ m}$$

The value of $x = 5.98$ m is discarded being impossible since the beam is only 6 m long to the left of B.

The B.M. is thus zero at 4.84 m from B (Fig. 4.39 (d)).

Example 4.20. *Calculate the reactions at A and D for the beam shown in Fig. 4.40. Draw the bending moment and shear force diagrams showing all important values.*

(AMIE, Winter-2001)

Solution. Refer Fig. 4.40 and 4.41.

Fig. 4.40 shows an overhanging beam ABCDF supported by a roller support at A and a hinged support at D. In the figure, a load of 4 kN is applied through a

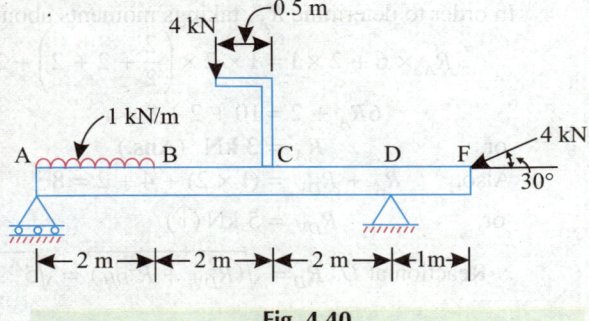

Fig. 4.40

Iron ceiling of a building.

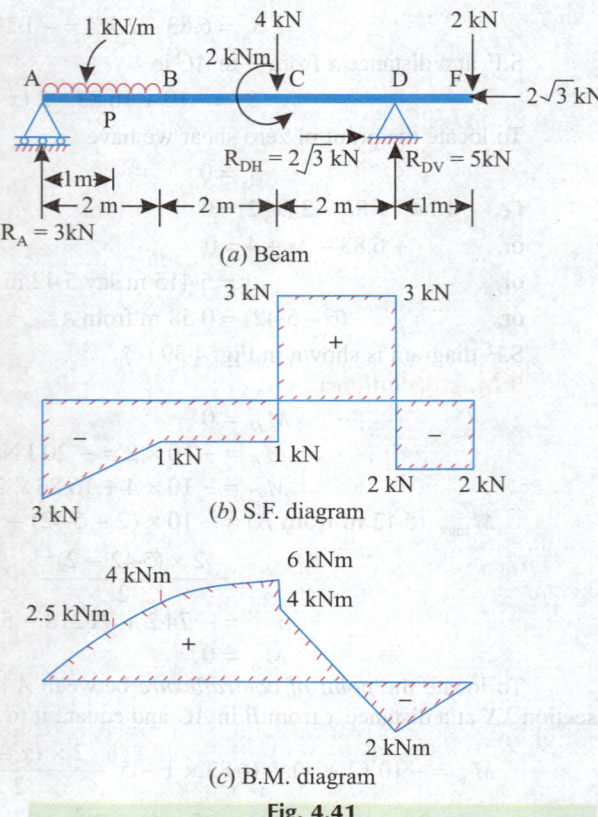

(a) Beam

(b) S.F. diagram

(c) B.M. diagram

Fig. 4.41

bracket 0·5 m away from the point C. Now apply equal and opposite load of 4 kN at C. This will be equivalent to a anticlockwise couple of the value of (4 × 0·5) = 2 kNm acting at C together with a vertical downward load of 4 kN at C. This is shown in Fig. 4·41. Complete Fig. 4·41 (a) and show U.D.L. (1 kN/m) over the part AB, a point load of 2 kN vertically downward at F, and a horizontal load of $2\sqrt{3}$ kN as shown.

Reactions at A and D:

Let, R_A = Reaction at roller A,

        R_{DV} = Vertically component of the reaction at the hinged support D, and

        R_{DH} = Horizontal component of the reaction at the hinged support D.

Obviously $R_{DH} = 2\sqrt{3}$ kN ($\rightarrow$)

In order to determine R_A, takings moments about D, we get

$$R_A \times 6 + 2 \times 1 = 1 \times 2 \times \left(\frac{2}{2} + 2 + 2\right) + 2 + 4 \times 2$$

$$6R_A + 2 = 10 + 2 + 8$$

or, $R_A = \textbf{3 kN}$ **(Ans.)**

Also, $R_A + R_{DV} = (1 \times 2) + 4 + 2 = 8$

or, $R_{DV} = 5$ kN ($\uparrow$)

∴ Reaction at D, $R_D = \sqrt{(R_{DV}^2 + R_{DH}^2)} = \sqrt{5^2 + (2\sqrt{3})^2} = \textbf{6.08 kN}$ **(Ans.)**

Inclination with horizontal $= \theta$

$$= \tan^{-1} \frac{5}{2\sqrt{3}} = 55 \cdot 3° \text{ (Ans.)}$$

S.F. calculations:

$$S_F = -2 \text{ kN}$$
$$S_D = -2 + 5 = 3 \text{ kN}$$
$$S_C = 3 - 4 = -1 \text{ kN}$$
$$S_B = -1 \text{ kN}$$
$$S_A = -1 - (1 \times 2) = -3 \text{ kN}$$

S.F. diagram is shown in Fig. 4·41 (*b*)

B.M. calculations:

$$M_F = 0$$
$$M_D = -2 \times 1 = -2 \text{ kNm}$$
$$M_C = [-2(1+2) + 5 \times 2] + 2 = 6 \text{ kNm}$$

[The bending moment increases from 4 kNm in (*i.e.*, $-2(1+2) + 5 \times 2$) to 6 kNm as shown in Fig. 4·41 (*c*)].

$$M_B = -2(1+2+2) + 5(2+2) - 4 \times 2 + 2$$
$$= -10 + 20 - 8 + 2 = 4 \text{ kNm}$$
$$M_P = -2\left(1 + 2 + 2 + \frac{2}{2}\right) + 5(2+2+1) - 4(2+1) + 2 - 1 \times 1 \times \frac{1}{2}$$
$$= -12 + 25 - 12 + 2 - 0 \cdot 5 = 2 \cdot 5 \text{ kNm}$$
$$M_A = 0$$

B.M. diagram is shown in Fig. 4·41 (*c*).

4·20. LOADING AND B.M. DIAGRAMS FROM S.F. DIAGRAMS

If the S.F. diagram for a beam is given, (or the *B.M.* diagram is given) the loading diagrams for the beam can very easily be drawn, if the following points are remembered:

1. The S.F. diagram will consist of *rectangle* (or series of rectangles), if the beam is loaded with *point loads*.
2. The S.F. diagram will consist of *inclined lines* for the portion on which U.D.L. is acting.
3. The S.F. diagram will consist of *parabolic lines* for the portion over which *triangular or trapezium load* distribution is acting.
4. The S.F. diagram will consist of 'cubics' for the portion over which *parabolic load* distribution is acting.
5. The B.M. diagram will consist of *inclined lines* if the beam is loaded with *free point loads*.
6. The B.M. diagram will consist of *parabolic lines* for portion over which *U.D.L. is acting*.
7. The B.M. diagram will consist of '*cubic*' or *third degree polynomials if the load distribution is a triangle*.
8. The B.M. diagram will consist of *fourth degree polynomial if the load distribution is parabolic*.

Example 4·21. *Fig. 4·42 (a) shows a S.F. diagram for a beam with rests on two supports, one being at the left hand end. Deduce from S.F. diagram the loading on the beam. Draw the B.M. diagram giving principal values.*

Solution.

— By inspection of the diagram, the other support will be at *B* where the *S.F.* changes sign.
— Since the *S.F.* is constant from *C* to *B*, there is a point load of 1 kN at *C*.
— The value of $R_B = 4 + 1 = 5$ kN

— Since the *S.F. decreases* from 4 kN at *B* to 1 kN at *D linearly*, there is a U.D.L. from *B* to *D* of an intensity $\frac{4-1}{3} = 1$ kN/m.

— Since there is a drop of *S.F.* there is a point load at *D* = 1 + 1·75 = 2·75 kN

— Since the S.F. varies from 1·75 kN at *D* to 3·75 kN at *A linearly*, there is a U.D.L. on *DA* of intensity $\frac{3·75 - 1·75}{2} = 1$ kN/m

— The reaction at *A* is evidently 3·75 kN. The loading is shown in Fig. 4·42 (*b*).

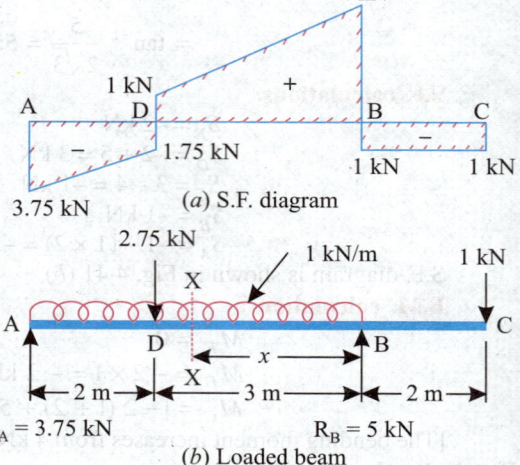

(*a*) S.F. diagram

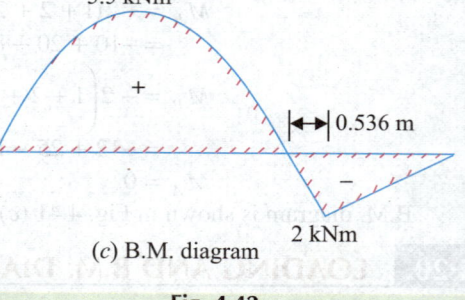

(*b*) Loaded beam

B.M. calculations:

$M_C = 0$

$M_B = -1 \times 2 = -2$ kNm

B.M. will be maximum at *D* (*i.e.* 5 m from *C*) where *S.F.* changes sign.

∴ $M_{max} = M_D$

$$= -1 \times 5 + 5 \times 3 - 1 \times 3 \times \frac{3}{2}$$

$$= 5·5 \text{ kNm}$$

(*c*) B.M. diagram

Fig. 4.42

To locate *point of inflexion* (or contraflexure) find out the bending moment at section *XX* at a distance *x* from *B* and equate to *zero* as follows:

$$M_x = -1(2 + x) + 5 \times x - \frac{1 \times x \times x}{2} = 0$$

$$-2 - x + 5x - \frac{x^2}{2} = 0$$

$$x^2 - 8x + 4 = 0$$

$$x = \frac{8 \pm \sqrt{(64 - 16)}}{2} = \frac{8 \pm 6·928}{2}$$

$$= 7·46 \text{ m} \quad \text{or} \quad 0·536 \text{ m}$$

The value of *x* = 7·46 m is discarded being impossible, since the beam is only 5 m long to the left of *B*.

Hence the point of contraflexure lies at a distance of *0·536 m from B* [Fig. 4·42 (*c*)].

Hinges in Beams

Example 4·22. *Draw the B.M. and S.F. diagrams for a beam loaded as shown in Fig. 4·43 (a).*

Solution. As the hinge passes on the reaction of one side to the other and the *B.M.* at the hinge is zero, the given beam can be assumed to be as shown in Fig. 4·43 (*b*).

Reaction $R_C = R_D = \frac{10 \times 8}{2} = 40$ kN

Beam *ABC* thus is simply supported at *A* and *B* and has an overhang *BC* carrying a point load of $R_C = 40$ kN (in addition to U.D.L. of 10 kN/m).

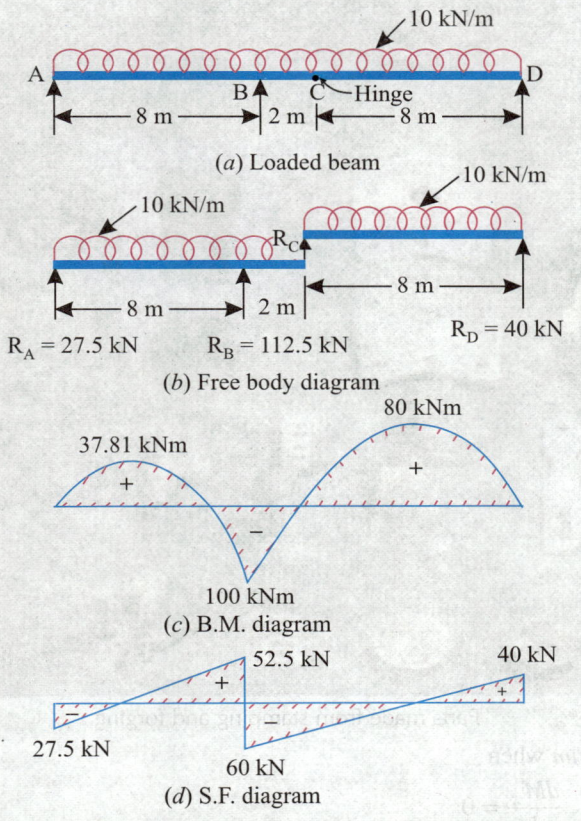

(a) Loaded beam

(b) Free body diagram

(c) B.M. diagram

(d) S.F. diagram

Fig. 4.43

For *part ABC*, to determine support reactions R_A and R_B take moments about A as follows:

$$R_B \times 8 = R_C \times 10 + 10 \times 10 \times \frac{10}{2} = 40 \times 10 + 500$$

$$R_B = \frac{900}{8} = 112.5 \text{ kN}$$

But, $R_A + R_B = 10 \times 10 + 40 = 140 \text{ kN}$

∴ $R_A = 140 - R_B = 140 - 112 \cdot 5 = 27 \cdot 5 \text{ kN}$

B.M. calculations:

Part CD of the beam can be treated as simply supported and the maximum bending moment will be

$$= \frac{wl^2}{8} = \frac{10 \times 8^2}{8} = 80 \text{ kNm}$$

$$M_C = 0; \; M_D = 0$$

$$M_B = -40 \times 2 - \frac{10 \times 2 \times 2}{2} = -100 \text{ kNm}$$

B.M. at a distance x from A is

$$M_x = 27 \cdot 5x - \frac{10 \times x \times x}{2} = 27 \cdot 5x - 5x^2$$

It shall be zero at $x = \frac{27 \cdot 5}{5} = 5 \cdot 5 \text{ m from A}$

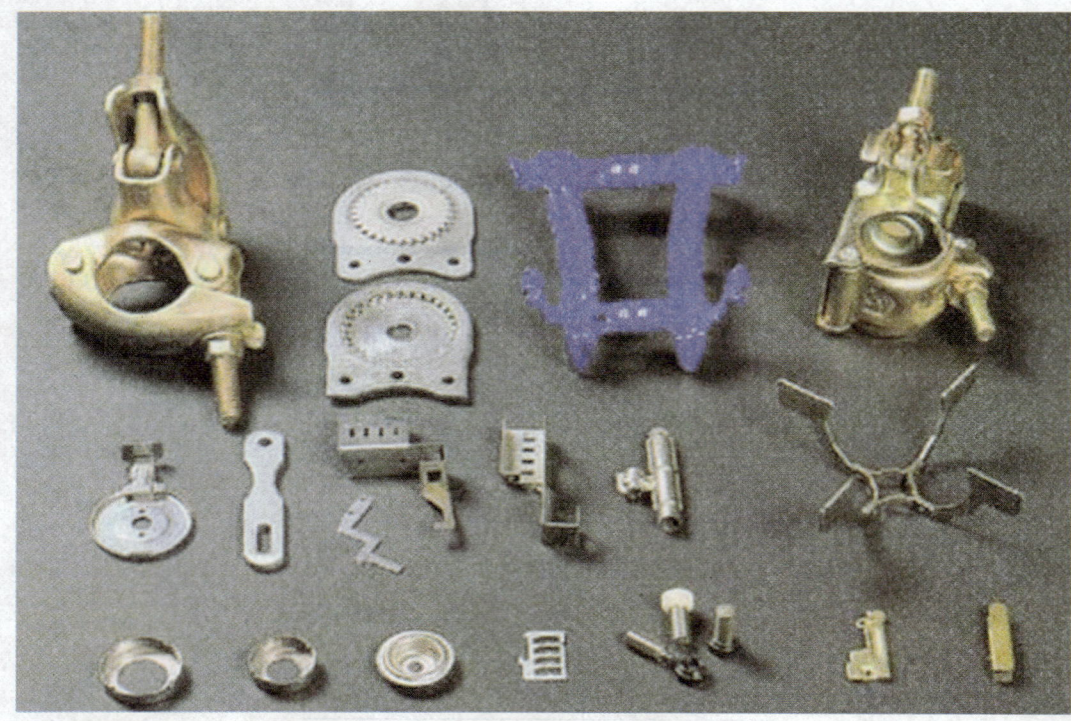

Parts made from stamping and forging.

B.M. is *maximum* when

$$\frac{dM_x}{dx} = 0$$

i.e. When $27 \cdot 5 - 10x = 0$

or,   when $x = 2 \cdot 75$ m

∴ $M_{max} = 27 \cdot 5 \times 2 \cdot 75 - 5 \times 2 \cdot 75^2 = 75 \cdot 625 - 37 \cdot 8125 = 37 \cdot 81$ kNm

Fig. 4·43 (*c*) shows the B.M. diagram.

S.F. calculations:

S_D is 40 kN at *D* and gradually changes to $(40 - 10 \times 8) = -40$ kN at *C*. Just to the right of *B* the S.F. is $(-40 - 10 \times 2) = -60$ kN and at *B* it changes to $(-60 + 112 \cdot 5) = 52 \cdot 5$ kN. From *B* it gradually changes to $(52 \cdot 5 - 10 \times 8) = -27 \cdot 5$ kN at *A*.

Fig. 4·43 (*d*) shows the S.F. diagram.

TYPICAL EXAMPLES (For Competitive Examinations)

Example 4·23. *A beam ABC is loaded and supported as shown in Fig. 4·44 (a). Find the magnitude of the clockwise moment M to be applied at C so that the reaction at B will be 30 kN upward and then draw the shear force and bending moment diagrams for the beam.*

Solution. Refer Fig. 4·44

Reaction at *B* = 30 kN...(Given)

Let us first find out reaction at *C* and moment *M*.

Taking moment about *C* (Refer Fig. 4·44 (*a*)), we have

$$30 \times 8 + 40 \times 4 + M = 20 \times 3 \left(\frac{3}{2} + 6 \right) + 10 \times 2$$

or, $240 + 160 + M = 450 + 20 = 470$

M = 70 kNm (Ans.)

Also, $R_B + R_C + 40 = 20 \times 3 + 10$

or $30 + R_C + 40 = 70$

∴ $R_C = 0$

S.F. calculations: Refer to Fig. 4·44 (b)

$S_C = 0; S_F = -10$ kN
$S_E = -10 + 40 = 30$ kN, $S_D = 30$ kN
$S_B = 30 - 20 \times 2 + 30 = +20$ kN

(changes from -10 kN to $+20$ kN)

$S_A = 0$

B.M. calculations: (Refer to Fig. 4·44 (c))

$M_C = -70$ kNm; $M_F = -70$ kNm
$M_E = -70 - 10 \times 2 = -90$ kNm
$M_D = -70 - 10 \times 4 + 40 \times 2 = -30$ kNm

S.F. changes sign at G. Let us calculate x
from similar Δs abc and cde.

$$\frac{ab}{ac} = \frac{de}{ec} \quad \text{or} \quad \frac{10}{2-x} = \frac{30}{x}$$

or, $10x = 60 - 30x$ or $x = 1.5$ m

∴ $M_G = -70 - 10 \times (2 + 2 + 1.5) + 40(2 + 1.5) - 20 \times 1.5 \times \frac{1.5}{2}$

$= -70 - 55 + 140 - 22.5 = -7.5$ kNm

$M_B = -70 - 10 \times 6 + 40 \times 4 - 20 \times 2 \times 2/2$

$= -70 - 60 + 160 - 40 = -10$ kNm; $M_A = 0$

S.F. and B.M. diagrams for the beam are shown in Fig. 4·44 (b) and (c).

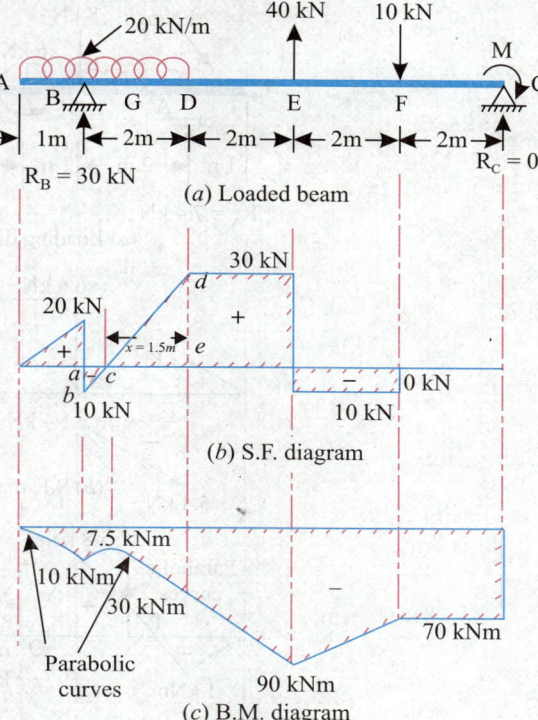

(a) Loaded beam

(b) S.F. diagram

(c) B.M. diagram

Fig. 4.44

Example 4.24. *Sketch the shear force and bending moment diagrams showing salient values for the beam loaded as shown in the Fig. 4·45 (a). Indicate the point of contraflexure.*

Solution. Refer to Fig. 4·45 (a)

Let us calculate reactions at A and B.

Taking moments about A, we get

$$R_B \times 5 + 2 \times 1 \times \frac{1}{2} = 8 \times 2 + 6 + 4 \times 7 + 2 \times 2 \times 2/2$$

or, $R_B \times 5 + 1 = 16 + 6 + 28 + 4 = 54$

∴ $R_B = 10.6$ kN

Also, $R_A + R_B = 2 \times 3 + 8 + 4 = 18$ kN

∴ $R_A = 18 - 10.66 = 7.4$ kN

S.F. calculations:

$S_F = -4$ kN
$S_B = -4 + 10.6 = 6.6$ kN
$S_E = +6.6$ kN
$S_D = 6.6 - 8.0 = -1.4$ kN

(changes from 6·6 kN to -1.4 kN)

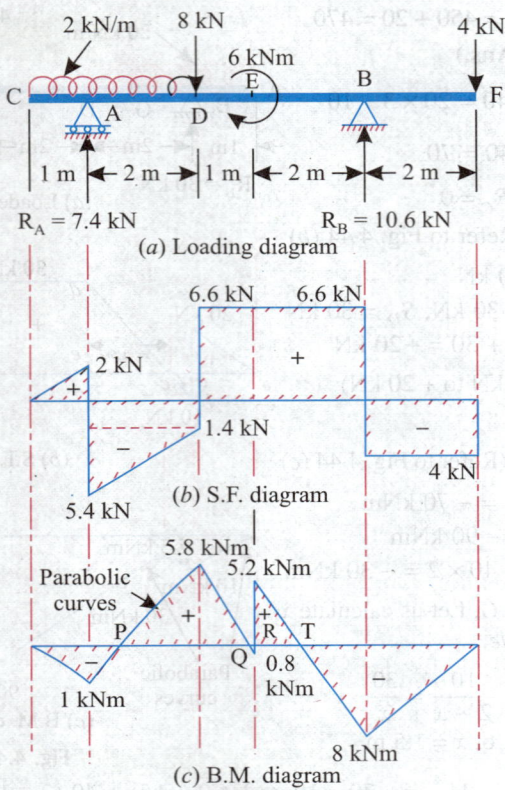

(a) Loading diagram

(b) S.F. diagram

(c) B.M. diagram

Fig. 4.45

$$S_A = -1 \cdot 4 - 2 \times 2 + 7 \cdot 4 = 2 \text{ kN}$$

(changes from $-5 \cdot 4$ kN to $+2$ kN)

$$S_C = 0$$

Note. In S.F. calculations applied moment 6 kNm should not be considered since it is a moment not a force.

B.M. calculations:

$$M_F = 0$$
$$M_B = -4 \times 2 = -8 \text{ kNm}$$
$$M_E = (-4 \times 4 + 10 \cdot 6 \times 2) - 6 = -0 \cdot 8 \text{ kNm}$$

(changes from $+5 \cdot 2$ kNm to $-0 \cdot 8$ kNm)

$$M_D = -4 \times 5 + 10 \cdot 6 \times 3 - 6 = 5 \cdot 8 \text{ kNm}$$
$$M_A = -4 \times 7 + 10 \cdot 6 \times 5 - 6 - 8 \times 2 - 2 \times 2 \times 2/2$$
$$= -28 + 53 - 6 - 16 - 4 = -1 \text{ kNm}$$
$$M_C = 0$$

P, Q, R and T are the points of contraflexures.

Example 4·25. *Draw shear force and bending moment diagrams for the beam shown in Fig. 4·46 (a).*

Solution. There is an inclined load 14·14 kN acting at A at 45°, the vertical component ($V_A = 14 \cdot 14 \sin 45° = 10$ kN) only takes part in calculations of S.F. and B.M. for the beam. Horizontal component ($H_A = 14.14 \cos 45° = 10$ kN) is absorbed by the hinged support at E (i.e. $H_A = R_{EH} = 10$ kN)

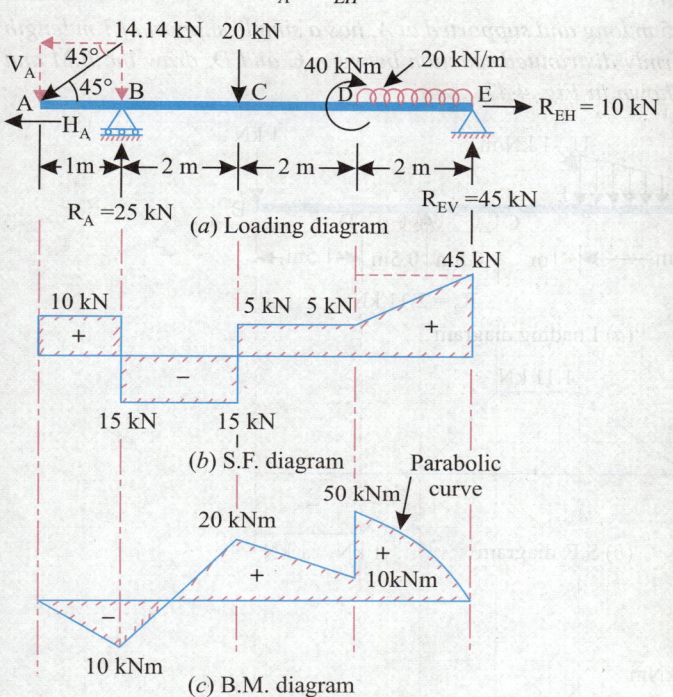

(a) Loading diagram

(b) S.F. diagram

(c) B.M. diagram

Fig. 4.46

Depending on the working temperature, metal sheets can be cold rolled or hot rolled.

Let us calculate the reactions at B and E. Taking moments about E (hinged support), we get

$$R_B \times 6 + 40 = V_A \times 7 + 20 \times 4 + 20 \times 2 \times 2/2$$
$$R_B \times 6 + 40 = 10 \times 7 + 80 + 40$$
$$\therefore \qquad R_B = 25 \text{ kN}$$

Also, $\qquad R_{EV} + R_B = 10 + 20 + 20 \times 2 = 70$ kN

$$\therefore \qquad R_{EV} = 70 - 25 = 45 \text{ kN}$$

S.F. calculations:

$$S_E = 45 \text{ kN}$$
$$S_D = 45 - 20 \times 2 = 5 \text{ kN}$$
$$S_C = 5 - 20 = -15 \text{ kN}$$

(changes from + 5 kN to – 15 kN)

$$S_B = -15 + 25 = +10 \text{ kN}$$

(changes from – 15 kN to + 10 kN)

$$S_A = 10 \text{ kN}$$

B.M. calculations:

$$M_E = 0$$
$$M_D = (+45 \times 2 - 20 \times 2 \times 2/2) - 40 = 10 \text{ kNm}$$

(changes from +50 kNm to 10 kNm)

$$M_C = 45 \times 4 - 20 \times 2 \times (2/2 + 2) - 40$$
$$= 180 - 120 - 40 = 20 \text{ kNm}$$
$$M_B = 45 \times 6 - 20 \times 2 \times (2/2 + 2 + 2) - 40 - 20 \times 2$$
$$= 270 - 200 - 40 - 40 = -10 \text{ kNm}$$

(or considering L.H.S.: $-10 \times 1 = -10$ kNm)

$$M_A = 0$$

Example 4·26. *A beam AB, 6·5 m long and supported at A, has a simple support of 1 m length between C and D. Assuming a uniformly distributed reaction between C and D, draw the B.M and S.F. diagrams for the loaded beam shown in Fig. 4·47.*

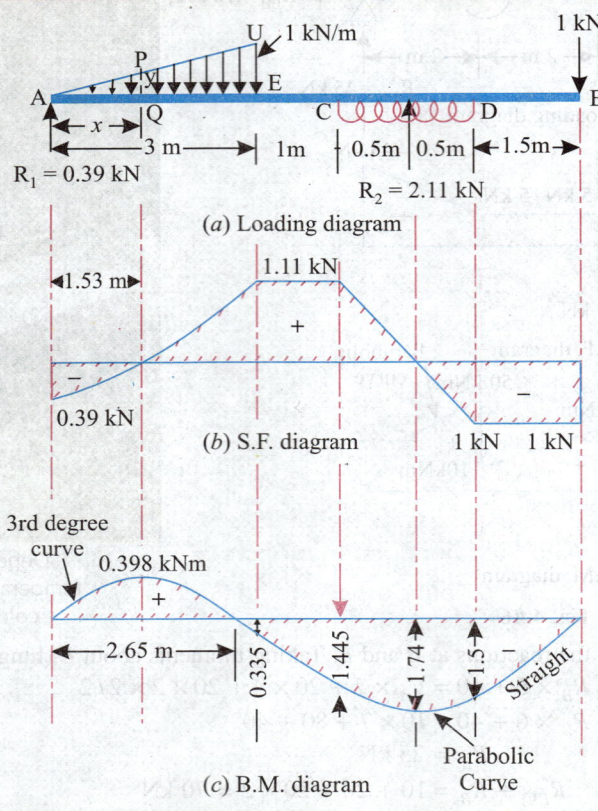

Fig. 4.47

Solution. Reactions. Refer to Fig. 4·47 (*a*). Since the reaction at *CD* is uniformly distributed over a length of the 1m, it can be replaced by its resultant R_2, acting at the centre of *CD*. Taking moments about *A*, we get

$$R_2 \times 4·5 = \left(\frac{1}{2} \times 3 \times 1\right) \times (2/3 \times 3) + 1 \times 6·5 \quad \text{or} \quad R_2 = 2·112 \text{ kN say } 2·11 \text{ kN}$$

Taking moments about R_2, we get

$$R_1 \times 4·5 + 1 \times 2 = \left(\frac{1}{2} \times 3 \times 1\right)(1/3 \times 3 + 1·5) \quad \text{or} \quad R_1 = 0·388 \text{ kN say } 0·39 \text{ kN}$$

(Check: $R_1 + R_2 = 2·11 + 0·39 = 2·5$ kN = total load)

∴ U.D.L. between *C* and *D*

$$r = \frac{2·11}{1} = 2·11 \text{ kN/m}$$

S.F. calculations:

$$S_B = -1 \text{ kN}; \ S_D = -1 \text{kN}$$
$$S_C = -1 + (1 \times r) = -1 + 2·11 = 1·11 \text{ kN}$$

The variation of S.F. between D and C is linear. The S.F. is zero at a distance of $\dfrac{1}{2 \cdot 11} = 0 \cdot 474$ m from D.

$$S_E = 1 \cdot 11 \text{ kN}$$

(S.F. is constant between C and E).

Since the loading between E to A is triangular, the variation of S.F. between E to A will be parabolic. At any distance x from A, we have

$$S_x = -R_1 + \frac{1}{2} x \cdot y = -R_1 + 1/2 \times x \times \frac{x}{3} = -R_1 + \frac{x^2}{6} \qquad \ldots(i)$$

$$\left[\text{In similar } \Delta s \; APQ \text{ and } AUE \; ; \; \frac{y}{x} = \frac{1}{3} \text{ or } y = \frac{x}{3} \right]$$

At $\qquad x = 0, S_A = -R_1 = -0.39 \text{ kN}$

At $\qquad x = 3, S_E = -0.39 + \dfrac{3^2}{6} = 1 \cdot 11 \text{ kN}$

To get the value of x for zero S.F. equate (i) to zero.

i.e. $\qquad -R_1 + \dfrac{x^2}{6} = 0$

or $\qquad x^2 = 6 R_1 = 6 \times 0 \cdot 39 = 2 \cdot 34$

$\therefore \qquad x = 1 \cdot 53 \text{ m}$

B.M. calculations:

$$M_B = 0$$
$$M_D = -1 \times 1 \cdot 5 - 1 \cdot 5 \text{ kNm}$$

Since the reaction between D and C is uniformly distributed, the B.M. will vary parabolically. Maximum negative bending moment will occur at a distance of $0 \cdot 47$ m from D (where S.F. is zero).

Water turbines.

$$\therefore \qquad M_{max} = -1 \times (1{\cdot}5 + 0{\cdot}474) + 2{\cdot}11 \times 0{\cdot}474 \times \frac{0{\cdot}474}{2}$$

$$= -1{\cdot}74 \text{ kNm}$$

$$M_C = -1 \times 2{\cdot}5 + 2{\cdot}11 \times 0{\cdot}5 = -1{\cdot}445 \text{ kNm}$$

$$M_E = -1 \times 3{\cdot}5 + 2{\cdot}11 \times 1{\cdot}5 = -0{\cdot}335 \text{ kNm}$$

The maximum positive bending moment will be at distance 1·53 m from A, where S.F. is zero.

$$\therefore \qquad M_{max} = R_1 \times x - \frac{1}{2} \times x \times y \times \left(\frac{x}{3}\right) \quad \text{(where } y = x/3) \qquad \text{...(proved earlier)}$$

i.e.

$$M_{max\,(x=1{\cdot}53\,m)} = 0{\cdot}39 \times 1{\cdot}53 - \frac{1}{2} \times 1{\cdot}53 \times \frac{1{\cdot}53}{3} \times \frac{1{\cdot}53}{3}$$

$$= 0{\cdot}398 \text{ kNm}$$

The B.M. will be zero at a distance, say x' from A, so that

$$M_{x'} = 0{\cdot}39\,x' - \frac{1}{2}\,x'\left(1 \times \frac{x'}{3}\right)\left(\frac{x'}{3}\right) = 0{\cdot}39x' - \frac{x'^3}{18} = 0$$

which gives $\qquad x' = 0$ or 2·65 m

Thus the B.M. is zero at a distance of 2·65 m from A, or 3·85 m from B.

Example 4·27. *Fig. 4.48 (a) is the shearing force diagram for a beam which rests on two supports, one being at the left hand end. Deduce from S.F.D. the loading on the beam. Draw the B.M.D. with principal values.*

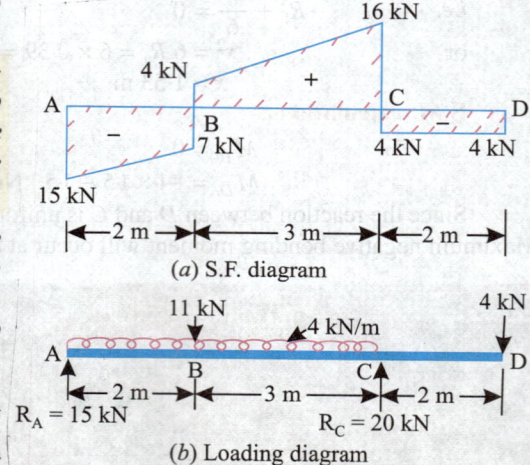

(a) S.F. diagram

Solution. By the inspection of the diagram, the other support will be at C, where the S.F. changes sign.

— Since the S.F. is constant from D to C, there is a point load of 4 kN at D. The value of $R_C = 16 + 4 = 20$ kN. Since the *S.F.* decreases from 16 kN at C to 4 kN at B linearly, there is a U.D.L. from C to B of intensity $\frac{16-4}{3} = 4$ kN / m.

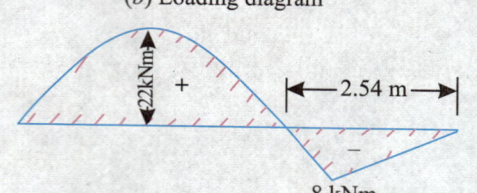

(b) Loading diagram

— Since there is a drop of *S.F.*, there is a point load at B = 4 + 7 = 11 kN.

— Again, since the *S.F.* varies from 7 kN at B to 15 kN at A linearly, there is U.D.L. on AB of intensity $\frac{15-7}{2} = 4$ kN / m.

(c) B.M. diagram

Fig. 4.48

The reaction at A is evidently 15 kN. The loading has been drawn on the beam (See Fig. 4·48 (b)).

B.M. calculations:

$$M_D = 0$$
$$M_C = -4 \times 2 = -8 \text{ kNm}$$

Between CB,

$$M_x = -4x + 20\,(x-2) - 4\,(x-2)\left(\frac{x-2}{2}\right)$$

$$= -4x + 20\,(x-2) - 2\,(x-2)^2$$

B.M. will be maximum at $x = 5$m where S.F. changes sign.

$$\therefore \qquad M_{max} = -4 \times 5 + 20 \,(5 - 2) - 2 \,(5 - 2)^2$$
$$= -20 + 60 - 18 = +22 \text{ kNm}$$

For B.M. to be zero,

$$M_x = -4x + 20 \,(x - 2) - 2 \,(x - 2)^2 = 0$$

i.e. $\qquad -4x + 20x - 40 - 2 \,(x^2 - 4x + 4) = 0$

or $\qquad 16x - 40 - 2x^2 + 8x - 8 = 0$

or $\qquad -2x^2 + 24x - 48 = 0$

or $\qquad x^2 - 12x + 24 = 0 \text{ or } x = 2 \cdot 54 \text{ m}$

Fig. 4·48 (c) shows the B.M.D.

HIGHLIGHTS

1. Beam is a structural member which is acted upon by a system of loads at right angles to the axis.
2. Depending upon the type of supports, beams are *classified* as follows:
   (i) Cantilever;   (ii) Simply or freely supported beams;
   (iii) Overhanging beam;   (iv) Fixed beam;
   (v) Continuous beam.
3. Numerically the shearing force is given by the algebraic sum of the forces to the left or to the right of the section.
4. The bending moment at a section is algebraic sum of moment to the left or right of the section.
5. In case of simply supported beams, the first step before analysing it for shear force and bending moment is to determine support reactions.
6. (i) For 'point loads' on a beam/cantilever the S.F. diagram shows horizontal straight lines and the B.M. diagram shows inclined straight lines.
   (ii) For U.D.L. on a beam/cantilever the S.F. diagram shows inclined straight lines and the B.M. shows parabolic curve.
7. In case of inclined loads resolve them into vertical and horizontal components. Horizontal components cause only thrust in the beam. S.F. and B.M. are caused by the vertical components.
8. At a hinge in a beam there is nil B.M. but there is S.F.
9. B.M. is maximum at a point where the S.F. is zero or where it changes directions from +ve to −ve or vice versa.
10. The point of contraflexure or point of inflexion is the point where the B.M. is zero.

OBJECTIVE TYPE QUESTIONS

Choose the Correction Answer:

1. If the plane of loading passes through one of the principal centroidal axes of the cross-section of the beam, the bending is said to be
   (a) plane
   (b) oblique
   (c) either of the above
   (d) none of the above.

2. A _____ load is one which is considered to act a point.
   (a) point
   (b) uniformly distributed
   (c) trapezoidal
   (d) none of the above.

3. If the plane of loading does not pass through one of the principal centroidal axes of the cross-section of the beam, the beam is said to be
   (a) plane
   (b) oblique
   (c) either of the above
   (d) none of the above.

4. A cantilever is a beam whose
   (a) one end is fixed and the other end free
   (b) both ends are fixed
   (c) both ends are simply supported
   (d) none of the above

5. In which of the following beams, the supports are *not* situated at the ends?
   (a) Cantilever beam
   (b) Simply supported beam
   (c) Overhanging beam
   (d) None of the above.

6. A continuous beam is one which has
   (a) less than two supports
   (b) two supports only
   (c) more than two supports
   (d) none of the above.

7. Which of the following are the statically determinate beams?
   (a) Cantilevers
   (b) Simply supported beams
   (c) Overhanging beams
   (d) All of the above.

8. Which of the following are the statically indeterminate beams?
   (a) Fixed beams
   (b) Continuous beams
   (c) Both (a) and (b)
   (d) none of the above.

9. In a cantilever with uniformly distributed load the shearing force varies following a
   (a) linear law
   (b) parabolic law
   (c) either of the above
   (d) none of the above.

10. In a cantilever, carrying a load whose intensity varies uniformly from zero at the free end to w per unit run at the fixed end, the *S.F.* changes following a
    (a) linear law   (b) parabolic law
    (c) cubic law   (d) none of theabove.

11. In a cantilever, carrying a load whose intensity varies uniformly from zero at the free end to w per unit run at the fixed end, the *B.M.* changes following a
    (a) cubic law   (b) parabolic law
    (c) linear law   (d) none of the above.

12. In a cantilever of length l carrying a load whose intensity varies uniformly from zero at the free end to w per unit run at the fixed end, the maximum bending moment is
    (a) $\dfrac{wl}{3}$ (b) $\dfrac{wl^2}{3}$
    (c) $\dfrac{wl^2}{6}$ (d) $\dfrac{wl^2}{24}$.

13. Bending moment at supports in case of simply supported beams is always
    (a) less than unity   (b) more than unity
    (c) zero   (d) none of the above.

14. In a simply supported beam carrying a uniformly distributed load of w per unit run over the whole span the maximum B.M. is equal to
    (a) $\dfrac{wl^2}{4}$ (b) $\dfrac{wl^3}{6}$
    (c) $\dfrac{wl^2}{8}$ (d) $\dfrac{wl^3}{8}$.

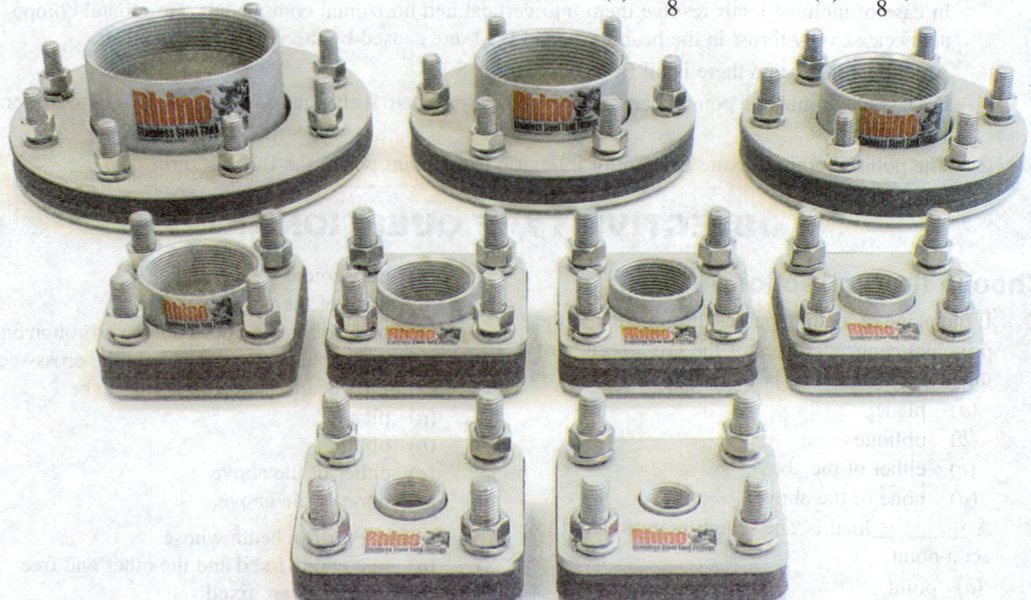

Couplings used in tanks and pipes.

15. In a simply supported beam carrying a load whose intensity varies uniformly from zero at one end to w per unit run at the mid span the maximum bending moment is equal to

    (a) $\dfrac{wl^2}{4}$ (b) $\dfrac{wl^2}{8}$

    (c) $\dfrac{wl^2}{12}$ (d) $\dfrac{wl^2}{24}$.

16. In a simply supported beam carrying a load whose intensity varies uniformly from zero at

    one end to w per unit run at the other end the maximum B.M. is equal to

    (a) $\dfrac{wl^2}{8}$ (b) $\dfrac{wl^2}{12}$

    (c) $\dfrac{wl^2}{24}$ (d) $\dfrac{wl^2}{9\sqrt{3}}$.

17. The point of contraflexure is also called
    (a) the point of inflexion
    (b) a virtual hinge
    (c) either of the above
    (d) none of the above.

ANSWERS

1. (a)	2. (a)	3. (b)	4. (a)	5. (c)	6. (c)
7. (d)	8. (c)	9. (a)	10. (b)	11. (a)	12. (c)
13. (c)	14. (c)	15. (c)	16. (d)	17. (c).	

UNSOLVED EXAMPLES

1. A horizontal cantilever 5 metres long carries a point load of 1 kN at the free end and a U.D.L. of 0.5 kN/m over a length of 3m from the free end. Draw the S.F. and B.M. diagrams.
   [**Ans.** $S_{max} = -2.5$ kN, $M_{max} = -10.25$ kNm]

2. A 6m long cantilever carries loads of 2 kN and 3 kN at 2m and 5m respectively from fixed end and a U.D.L. of 10 kN over its entire length. Draw S.F. and B.M. diagrams.
   [**Ans.** $S_{max} = -15$ kN, $M_{max} = -49$ kNm]

3. A cantilever $PQRS$, 7m long is fixed at P such that $PQ = QR = 2m$ and $RS = 3m$. It carries loads of 5, 3 and 2 kN at Q, R and S respectively in addition to U.D.L. of 1 kN/m run between P and Q and 2 kN/m run between R and S. Draw shearing force and bending moment diagrams.
   [**Ans.** $S_{max} = -18$ kN, $M_{max} = -71$ kNm]

4. A simply supported beam of 16 m efective span carries the concentrated loads of 4 kN, 5 kN and 3 kN at distances 3, 7 and 11 m respectively from the left support. Calculate maximum shearing force and bending moment. Draw S.F. and B.M. diagrams.
   [**Ans.** $S_{max} = -7$ kN, $M_{max} = -33.30$ kNm]

5. Draw the S.F. and B.M. diagrams for the beam shown in Fig. 4·49. Also find the position and magnitude of the maximum bending moment.
   [**Ans.** Position of maximum B.M. = 2·675 m from right end; $M_{max} = 3·155$ kNm]

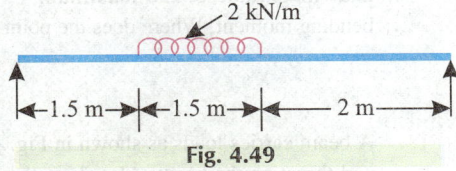

Fig. 4.49

6. A beam AB 10 m long has supports at its ends A and B. It carries a point load of 2·5 kN at 3m from A and a point load of 2·5 kN at 7m from A and a uniformly distributed load of 0·5 kN/m between the point loads. Draw the shearing force and bending moment diagrams for the beam.
   [**Ans.** $M_{max} = 11·5$ kNm]

7. The intensity of loading in a simply supported beam of 8m span varies gradually from 2 kN/m at one end to 6 kN/m at the other end. Draw the shearing force and bending moment diagrams.
   [**Ans.** $M_{max} = 32·22$ kNm]

8. A beam $ABCD$ 10m long is supported at B and C. The overhangs AB and CD are 2m and 3m respectively. The overhang AB carries U.D.L. of 1 kN/m and CD carries U.D.L. of 0·5 kN/m. In addition, there are point loads of 1 kN, 2 kN and 1 kN, at distances of 1·5m, 3m and 8m from A respectively. Find the reactions and draw B.M. and S.F. diagrams.
   [**Ans.** $R_B = 4.45$ kN; $R_C = 3.05$ kN; $M_B = 2.5$ kNm; $M_C = 3.25$ kNm]

9. A beam *ABCD*, 10m long, is simply supported at *B* and *C* which are 4m apart, and overhangs the support *B* by 3m. The overhanging part *AB* carries U.D.L.of 1 kN/m and the part *CD* carries U.D.L. of 0·5 kN/m. Calculate the position and magnitude of the least value of the bending moment between the supports. Draw the S.F. and B.M. diagrams. **[Ans.** 0·617 kNm at 1·1 m from *C*]

10. A horizontal beam *AB* 8m long is supported at *A* and *C*, 6m from *A*. The beam supports a U.D.L. of 1·5 kN/m over its entire length and also concentrated loads of 3 kN and 1·5 kN at *D* and *B* respectively, *D* being 2m from *A*. Draw the S.F. and B.M. diagrams for the beam. Where does the maximum B.M. occur and what is its value? **[Ans.** 8 kNm at 2m from *A*].

11. A beam *ABCD* is simply supported at *B* and *C*, 5m apart, and the overhanging parts *AB* and *CD* are 1·5 m long respectively. The beam carries a U.D.L. of 3 kN between *A* and *C* and there is concentrated load of 2 kN at *D*. Draw the S.F. and B.M. diagrams. Calculate the position of maximum B.M. between *B* and *C*. **[Ans.** M_{max} = 4.275 kNm at 3.76m from *A*]

12. Draw the S.F. and B.M. diagrams of the beam shown in Fig. 4·50. Indicate on the diagrams values of the shearing forces and bending moments at significant points. Also show the location and magnitude of maximum bending moment.

    [Ans. M_{max} = 20 kNm at 2m to the left of 4 kN load]

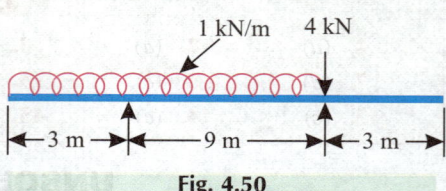

Fig. 4.50

13. A horizontal beam *AB* of length 8m is simply supported at *A* and *B*. It carries U.D.L. of 3 kN/m over the entire span and a clockwise moment of 12 kNm is applied in the plane of the beam at a point *C*, 5m from *A*. Draw the shearing force and bending moment diagrams and determine the position and magnitude of maximum bending moment.

    [Ans. 27 kNm at *C*].

14. A beam *ABCD* shown in Fig. 4·51 is simply supported at *B* and *C*, It carries a point load at the free end *A*, a U.D.L. between *B* and *C* and an anticlockwise moment, in the plane of the beam, applied at the free end *D*. Draw the S.F. and B.M. diagrams and determine the position and magnitude of the maximum bending moment. Where does the point of contraflexure lie?

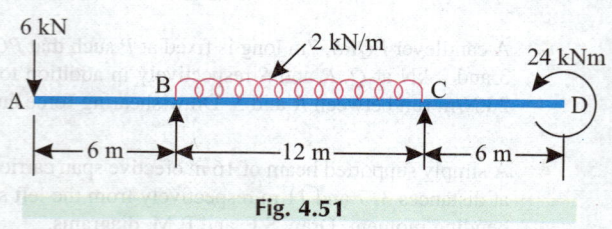

Fig. 4.51

    [Ans. M_{max} = 36·25 kNm at a distance of 3·5m from support *C*;
    point of contraflexure = 9·52m from support *C*].

15. A beam carries loads as shown in Fig. 4·52. Estimate the maximum bending moment, shearing force and thrust on the beam. Also draw the corresponding diagrams.

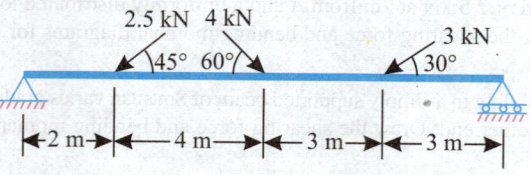

Fig. 4.52

    [Ans. M_{max} = 14·42 kNm; *S* = 3·58 kN, *T* = 2·36 kN].

16. S.F. diagram for a loaded beam is shown in Fig. 4·53. Determine the loading and the nature of the beam and sketch it neatly.

Hence determine B.M. diagram indicating important ordinates and the points of contraflexure if any.

[Ans. U.D.L. of 1 kN/m between A and C, load of 9 kN at C; supports B and D; $M_A = M_D = 0$; $M_B = -2$ kNm; 28 kNm; B.M. is zero at 0·26 m from B in BC].

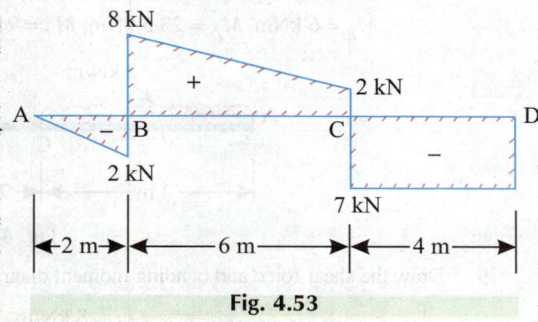

Fig. 4.53

17. Draw the shear force and bending moment diagrams for the beam loaded as shown in Fig. 4·54.

[Ans. $R_A = 8·75$ kN, $R_D = 46·25$ kN; $M_D = -5$ kNm, $M_B = -2·5$ kNm, $M_{max (AB)} = 3·86$ kNm]

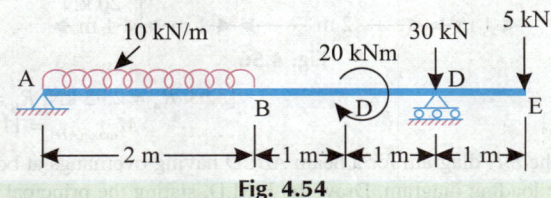

Fig. 4.54

18. An overhanging beam ABC is simply supported at A and B over a span of 6m and BC overhangs by 3m. If the supported span AB carries central concentrated load of 8 kN and overhanging span BC carries 2 kN/m completely draw S.F. and B.M. diagrams indicating salient points.

[Ans. $R_A = 2·5$ kN. $R_B = 11·5$ kN, $M_B = -9$ kNm, $M_B = +7·5$ kNm]

19. Draw shear force and bending moment diagrams for the beam shown in Fig. 4·55. Show all important values.

[Ans. $R_A = 9·25$ kN, $R_B = 8·75$ kN; $M_E = 6$ kNm

Petrol water pump with flywheel.

$M_B = 6$ kNm, $M_D = 23 \cdot 5$ kNm, $M_C = 21$ kNm]

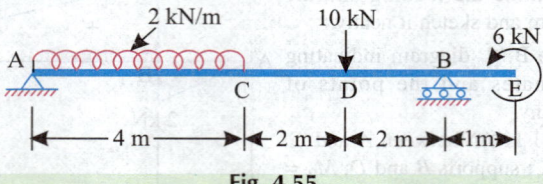

Fig. 4.55

20. Draw the shear force and bending moment diagrams for the beam loaded as shown in Fig. 4·56.

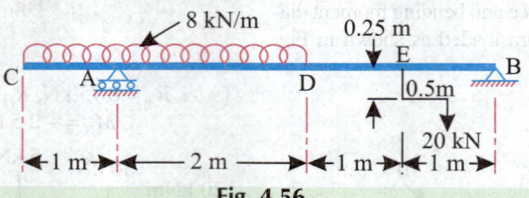

Fig. 4.56

[**Ans.** $R_A = 23 \cdot 5$ kN, $R_B = 20 \cdot 5$ kN; $M_D = 11$ kNm,
$M_{\text{max (AD)}} = 11 \cdot 02$ kNm; $M_A = -4$ kNm]

21. Fig. 4·57 shows the *S.F.* diagram for a beam *ABCD* having overhangs at both the supports. From the S.F.D. deduce the loading diagram. Draw the B.M.D. stating the principal values.

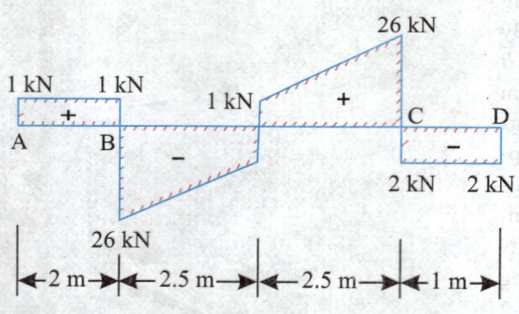

Fig. 4.57

[**Ans.** Loading: 1 kN at *A*; UDL of 10 kN/m from *B* and *C*; 2 kN at *E*;
2 kN at *D*; $M_B = -2$ kNm; $M_C = -2$ kNm; $M_E = +31 \cdot 75$ kNm]

Bending Stresses in Beams

5.1. THEORY OF SIMPLE BENDING (BENDING EQUATION)

When a beam is loaded it is bent and subjected to bending moments. Consequently, longitudinal or bending stresses are induced in cross-section. In order to determine the practical utility of any beam, it is very necessary to establish a relationship between the radius of curvature to which the beam bends, the bending moment, the bending stress and its cross-sectional dimensions. The equation which connects these quantities is known as the "*bending equation*".

Assumptions in 'Theory of bending':

1. The material of the beam is perfectly homogeneous throughout.

2. The stress induced is proportional to the strain and at no place the stress exceeds the elastic limit.

3. The value of modulus of elasticity (E) is same, for the fibres of the beam under compression or under tension.

4. The transverse section of the beam, which is plane before bending, remains plane after bending.

5. There is no resultant pull or push on the cross-section of the beam.

6. The loads are applied in the plane of bending.

7. The transverse section of the beam is symmetrical about a line passing through the centre of gravity in the plane of bending.

8. The radius of curvature of the beam before bending is very large in comparison to its transverse dimensions.

As a result of a bending moment or couple, a length of beam will take up a curved shape, and a very short length may be treated as a part of the arc of a circle. It follows that at the outer radii the material will be in tension and at the inner radii in compression, and at some radius there will be no stress. This layer of the meterial is the *neutral layer* or *neutral axis.*

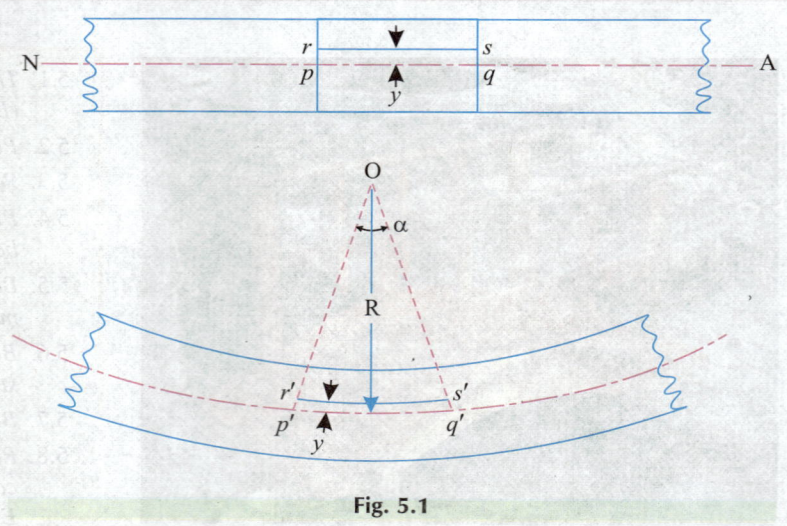

Fig. 5.1 shows a longitudinal section of a beam; the neutral layer (axis) N.A. being bent to form an arc of a circle of radius R. The neutral layer is then, before bending, the length pq, which after bending becomes $p'q'$.

Fig. 5.1

Consider some layer rs at a distance y from pq, which after bending becomes $r's'$. Let $p'q'$ subtend an angle α at the centre of curvature.

∴ $p'q' = R\alpha$ and $r's' = (R - y)\alpha$

Initially the parallel layers would have equal lengths, so that $pq = rs$ and since there is no stress at the neutral layer, then there is no strain.

∴ $p'q' = pq$

Now the strain in rs $= \dfrac{rs - r's'}{rs}$ but $rs = pq = p'q'$

∴ Strain $= \dfrac{p'q' - r's'}{rs}$

But, $p'q' = R\alpha$

and, $r's' = (R - y)\alpha$; strain $= \dfrac{R\alpha - (R - y)\alpha}{R\alpha} = \dfrac{y}{R}$

Now if the stress in $rs = \sigma$ and Young's modulus is E, then

Strain $= \dfrac{\sigma}{E}$

$\dfrac{\sigma}{E} = \dfrac{y}{R}$ or $\dfrac{\sigma}{y} = \dfrac{E}{R}$ ...(5.1)

If a transverse section of the beam is now considered (Fig. 5.2), let a strip of area δa, lie at a distance y from the neutral axis.

Then, the normal force on this area $(\delta a) = \dfrac{E}{R} y \delta a$

Now the moment of this force about the neutral axis is

$$= \frac{E}{R} y \, \delta a \times y \text{ or } \frac{E}{R} y^2 \, \delta a$$

This is the resisting moment of the material caused by the stress produced, and the total resisting moment is

$$= \Sigma \frac{E}{R} y^2 \, \delta a \text{ or } \frac{E}{R} \Sigma y^2 \, \delta a$$

And $\Sigma y^2 \, \delta a$ is the second moment of area about the neutral axis, I_{NA}.

$\therefore$ Resisting moment $M = \dfrac{E}{R} \times I$

But since the resisting moment balances the applied bending moment,

$\therefore$ $\qquad M = \dfrac{E}{R} \times I \text{ or } \dfrac{M}{I} = \dfrac{E}{R}$

But, $\qquad \dfrac{E}{R} = \dfrac{\sigma}{y}$

$\therefore$ $\qquad \dfrac{M}{I} = \dfrac{\sigma}{y} = \dfrac{E}{R}$ ...(5.2)

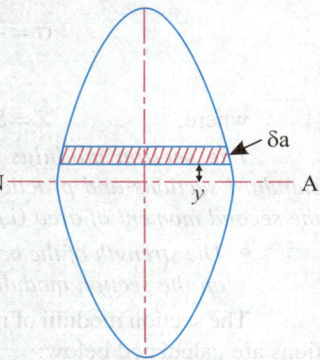

Fig. 5.2

where,
$\qquad M =$ Moment of resistance,
$\qquad I =$ Moment of inertia of the section about neutral axis (N.A.),
$\qquad E =$ Young's modulus of elasticity,
$\qquad R =$ Radius of curvature of N.A., and
$\qquad \sigma =$ Bending stress.

The above equation is known as the **'Bending equation'**.

5.2. POSITION OF NEUTRAL AXIS

Consider the cross-section of a beam (Fig. 5.2), there will be no resultant force on the section for condition of equilibrium.

The force acting on a small area δa at a distance 'y' from the neutral axis is given by:

$$\delta F = \sigma \cdot \delta a = \frac{E}{R} y \cdot \delta a$$

or, the total force normal to the section,

$$F = \frac{E}{R} \Sigma y \cdot \delta a$$

$\therefore$ For zero resultant force $\Sigma y \cdot \delta a = 0$.

Now $\Sigma y \cdot \delta a$ is the moment of the sectional area about the neutral axis, and since this moment is zero, the axis must pass through the centre of area.

Hence the *neutral axis or neutral layer, passes through the centre of area.*

5.3. SECTION MODULUS

Referring to the bending equation, $\dfrac{M}{I} = \dfrac{\sigma}{y}$, we have

$$\sigma = \frac{My}{I} = \frac{M}{I/y} \quad \text{or} \quad \sigma = \frac{M}{Z}$$
$$...(5.3)$$

where, $\qquad Z = $ Section modulus $= I/y$

The section modulus is usually quoted for all standard sections and practically is of greater use than the second moment of area (i.e., M.O.I.).

- *The strength of the beam section depends mainly on the section modulus.*

The section modulii of rectangular and circular sections are calculated below:

Fig. 5.3

(i) Rectangular section:

Fig. 5.3 shows a rectangular section of width b and depth d. Let the horizontal centroidal axis be neutral axis.

Section modulus, $Z = \dfrac{\text{Moment of inertia about the neutral axis}}{\text{Distance of the most distant point of the section from the neutral axis}}$

$$= \frac{I}{y_{max}}$$

But, $\qquad\qquad I = \dfrac{bd^3}{12}$ and, $y_{max} = \dfrac{d}{2}$

$\therefore \qquad\qquad Z = \dfrac{bd^3/12}{d/2} = \dfrac{bd^2}{6}$ $\qquad\qquad ...(5.4)$

(Moment of resistance, $M = \sigma Z = \sigma \times \dfrac{1}{6} bd^2$)

Colorado river bridge.

(ii) Hollow rectangular section:

Refer to Fig. 5.4.

Moment of inertia about the neutral axis,

$$I = \frac{BD^3}{12} - \frac{bd^3}{12} = \frac{1}{12}(BD^3 - bd^3)$$

$$y_{max} = D/2$$

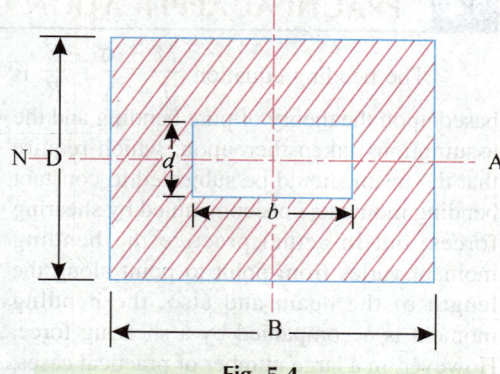

∴ Section modulus $Z = \dfrac{I}{y_{max}}$

$$= \frac{(BD^3 - bd^3)/12}{D/2} = \left(\frac{BD^3 - bd^3}{6D}\right)$$

...(5.5)

Fig. 5.4

$$\left[\begin{array}{c} \text{Moment of resistance} \\[2mm] M = \sigma\,Z = \sigma \times \left(\dfrac{BD^3 - bd^3}{6D}\right) \end{array}\right]$$

(iii) Circular section:

Refer to Fig. 5.5.

Moment of inertia of the section about the neutral axis,

$$I = \frac{\pi d^4}{64}$$

$$y_{max} = d/2$$

∴ Section modulus Z

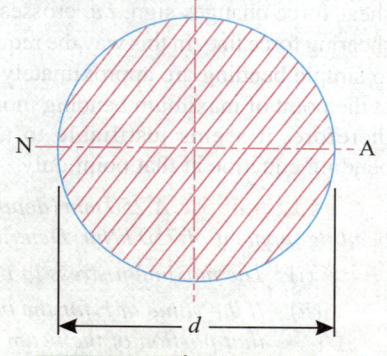

$$= \frac{I}{y_{max}} = \frac{\pi d^4/64}{d/2} = \frac{\pi d^3}{32}$$

...(5.6)

Fig. 5.5

$$\left(\text{Moment of resistance} = \sigma \cdot Z = \sigma \times \frac{\pi d^3}{32}\right)$$

(iv) Hollow circular section:

Refer to Fig. 5.6.

Moment of inertia of the section about the neutral axis,

$$I = \frac{\pi}{64}(D^4 - d^4) \text{ and } y_{max} = \frac{D}{2}$$

∴ Section modulus,

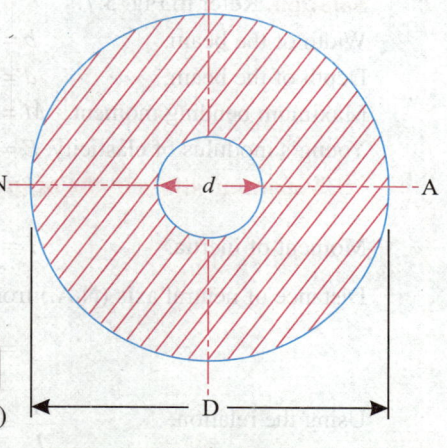

$$Z = \frac{I}{y_{max}} = \frac{\pi(D^4 - d^4)}{64} \times \frac{2}{D} = \frac{\pi}{32}\left(\frac{D^4 - d^4}{D}\right)$$

...(5.7)

Fig. 5.6

[Moment of resistance,

$$M = \sigma \times Z = \sigma \times \frac{\pi}{32}\frac{(D^4 - d^4)}{D}]$$

5.4 PRACTICAL APPLICATION OF BENDING EQUATION

The bending equation $\dfrac{M}{I} = \dfrac{\sigma}{y} = \dfrac{E}{R}$ is based upon the theory of pure bending and the assumptions taken thereupon, which require that the beam should be subjected to constant bending moments unaccompanied by shearing forces, but in actual practice the bending moment varies from point to point along the length of the beam and also, the bending moment is accompanied by a shearing force. However, in a large number of practical cases, the bending moment is maximum when the shear force changes sign, *i.e.* crosses the zero shearing force line. In this way the requirements to simple bending are approximately satisfied at the point of maximum bending moment and therefore, it seems justifiable to apply the bending equation at that point only.

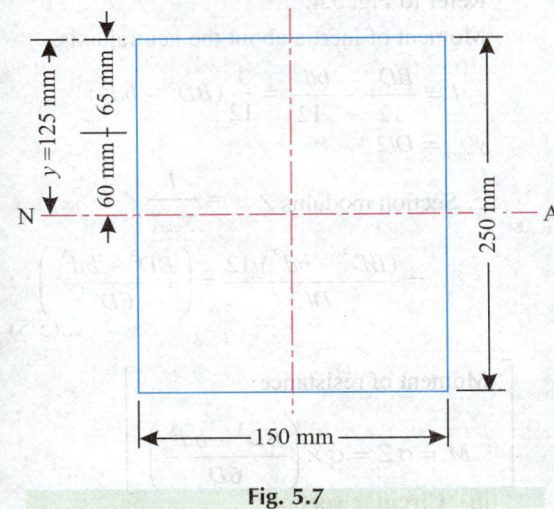

Fig. 5.7

Example 5.1. *A 250 mm (depth) × 150 mm (width) rectangular beam is subjected to maximum bending moment of 750 kNm. Determine:*

   (i) *The maximum stress in the beam.*

   (ii) *If the value of E for the beam material is 200 GN/m², find out the radius of curvature for that portion of the beam where the bending is maximum.*

   (iii) *The value of the longitudinal stress at a distance of 65 mm from the top surface of the beam.*

Solution. Refer to Fig. 5.7.

Width of the beam, $b = 150 \text{ mm} = 0.15 \text{ m}$

Depth of the beam, $d = 250 \text{ mm} = 0.25 \text{ m}$

Maximum bending moment, $M = 750 \text{ kNm}$

Young's modulus of elasticity, $E = 200 \text{ GN/m}^2$.

(i) Maximum stress in the beam:

Moment of inertia, $I = \dfrac{bd^3}{12} = \dfrac{0.15 \times 0.25^3}{12} = 0.0001953 \text{ m}^4$

Distance of neutral axis (N.A.) from the top surface of the beam,

$$y = \frac{d}{2} = \frac{0.25}{2} = 0.125 \text{ m}$$

Using the relation, $\dfrac{M}{I} = \dfrac{\sigma}{y}$, we get

$$\sigma = \frac{M \cdot y}{I} = \frac{750 \times 10^3 \times 0.125}{0.0001953}$$

$$= 4.8 \times 10^8 \text{ N/m}^2 \text{ or } 480 \text{ MN/m}^2$$

Hence, the maximum stress in the beam = **480 MN/m²**   **(Ans.)**

(ii) Radius of curvature, R:

Using the relation, $\dfrac{M}{I} = \dfrac{E}{R}$, we get

$$R = \dfrac{EI}{M} = \dfrac{200 \times 10^9 \times 0.0001953}{750 \times 10^3} = 52.08 \text{ m} \quad \textbf{(Ans.)}$$

(iii) Longitudinal stress at a distance of 65 mm from top surface of the beam, σ_1:

Using the relation, $\dfrac{M}{I} = \dfrac{\sigma}{y} = \dfrac{\sigma_1}{y_1}$, we get

$$\sigma_1 = \dfrac{M \cdot y_1}{I}$$

$$= \dfrac{750 \times 10^3 \times (60 \times 10^{-3})}{0.0001953} \times 10^{-6} \text{ MN/m}^2 \; (\because \; y_1 = 125 - 65 = 60 \text{ mm})$$

$$= 230.4 \text{ MN/m}^2 \quad \textbf{(Ans.)}$$

Example 5.2. *A symmetrical section 200 mm deep has a moment of inertia of 2.26×10^{-5} m^4 about its neutral axis. Determine the longest span over which, when simply supported, the beam would carry a uniformly distributed load of 4 kN/m run without the stress due to bending exceeding 125 MN/m^2.*

Solution.

Depth of the symmetrical section, $d = 200$ mm

Moment of inertia about neutral axis, $I = 2.26 \times 10^{-5}$ m^4

Uniformly distributed load, $w = 4$ kN/m run

Maximum bending stress, $\sigma = 125$ MN/m^2.

Longest span, l :

Using the relation

$$\dfrac{M}{I} = \dfrac{\sigma}{y}, \text{ we have}$$

$$M = \dfrac{\sigma I}{y} = \dfrac{125 \times 10^6 \times 2.26 \times 10^{-5}}{0.2/2}$$

$$= 28.25 \times 10^3 \text{ Nm} = 28.25 \text{ kNm}$$

In tall buildings such as WTC towers steel beams are used to hold concrete structure.

Also the maximum bending moment due to uniformly distributed load is

$$= \frac{wl^2}{8} = \frac{4 \times l^2}{8} = 0.5 l^2 \text{ kNm}; \ 0.5 l^2 = 28.25$$

From which, $\qquad l = 7.516 \text{ m}$ **(Ans.)**

Example 5.3. *Determine the dimensions of joist of a timber for span 8 m to carry a brick wall 200 mm thick and 5 m high, if the density of brick work is 1850 kg/m³ and the maximum permissible stress is limited to 7.5 MN/m². Given that the depth of joist is twice the width.*

Solution.

Length of span, $\qquad l = 8 \text{ m}$

Thickness of brick wall $\quad = 200 \text{ mm or } 0.2 \text{ m}$

Height of the wall $\qquad = 5 \text{ m}$

Maximum permissible stress $\ = 7.5 \text{ MN/m}^2$

Density of brick work $\qquad = 1850 \text{ kg/m}^3$

Width of the joist, b:

Depth of the joist, d:

Total weight of the wall

$$W = \text{Length of span} \times \text{thickness of wall} \times \text{height of wall}$$
$$\times \text{ density of brick work per m}^3$$
$$= 8 \times 0.2 \times 5 \times (1850 \times 9.81)$$
$$= 145188 \text{ N or } 0.145 \text{ MN (say)}$$

Maximum B.M. $\qquad = \dfrac{Wl}{8} = \dfrac{0.145 \times 8}{8} = 0.145 \text{ MNm}$

Moment of resistance $\quad M = \dfrac{\sigma I}{y} = \dfrac{\sigma \times bd^3/12}{d/2}$ $\qquad \left(\because \dfrac{M}{I} = \dfrac{\sigma}{y} \text{ or } M = \dfrac{\sigma I}{y} \right)$

$$= \sigma \times \frac{bd^2}{6} = \frac{7.5 \times bd^2}{6} \text{ MNm.}$$

where, b and d are in metres.

Equating moment of resistance to B.M., we get

$$\frac{7.5 \times bd^2}{6} = 0.145$$

or, $\qquad\qquad bd^2 = \dfrac{0.145 \times 6}{7.5} = 0.116$

or, $\qquad\qquad b \times (2b)^2 = 0.116 \qquad\qquad\qquad (\because d = 2b)$

From which, $\qquad b = \left(\dfrac{0.116}{4} \right)^{\frac{1}{3}}$

$$b = 0.307 \text{ m or } \textbf{307 mm} \quad \textbf{(Ans.)}$$

and, $\qquad\qquad d = 2b = 2 \times 307 = \textbf{614 mm} \quad \textbf{(Ans.)}$

Example 5.4. *A beam consists of a symmetrical rolled steel joist. The beam is simply supported at its ends and carries a point load at the centre of the span. If the maximum stress due to bending is 140 MPa, find the ratio of the depth of the beam section to span in order that the central deflection may not exceed $\dfrac{1}{480}$ of the span.*

Take, $E = 200$ GPa. $\qquad\qquad\qquad\qquad\qquad\qquad\qquad$ **(AMIE Summer, 2001)**

Solution. *Given:* $\sigma = 140$ MPa; $\delta l = \dfrac{1}{480}$ of the span (l); $E = 200$ GPa.

$$\frac{d}{l} = ?$$

Let, $\quad\quad W$ = Point load at the centre of the span, and

$\quad\quad\quad\quad \delta l$ = Central deflection due to the load W.

Then, $\quad\quad \delta l = \dfrac{Wl^3}{48\,EI} = \dfrac{1}{480} \times l$

$\therefore \quad\quad EI = \dfrac{Wl^3 \times 480}{48\,l} = 10 Wl^2$

or, $\quad\quad 200 \times 10^9 \times I = 10\,Wl^2$

or, $\quad\quad I = \dfrac{10 Wl^2}{200 \times 10^9}$

From the bending equation, we can write

$$\frac{M}{I} = \frac{\sigma}{y} \quad\quad\quad \left[\text{where, } M = \frac{Wl}{4}\right]$$

$$\frac{\dfrac{Wl}{4}}{\dfrac{10Wl^2}{200 \times 10^9}} = \frac{140 \times 10^6}{\left(\dfrac{d}{2}\right)}$$

$$\frac{Wl}{4} \times \frac{200 \times 10^9}{10Wl^2} = \frac{2 \times 140 \times 10^6}{d}$$

or, $\quad\quad \dfrac{200 \times 10^9}{4 \times 10l} = \dfrac{2 \times 140 \times 10^6}{d}$

or, $\quad\quad \dfrac{d}{l} = \dfrac{2 \times 140 \times 10^6 \times 4 \times 10}{200 \times 10^9} = \mathbf{0.056}$ **(Ans.)**

Example 5.5. *A floor has to carry a load of 12 kN per square metre. The floor is supported on rectangular joists each 300 mm × 100 mm and 5 m long. Calculate the distance apart (from the centre to centre) at which joists should be placed so that the maximum stress in the joists should not exceed 8 MN/m².*

Solution. Moment of inertia of each joist,

$$I = \frac{bd^3}{12} = \frac{0.1(0.3)^3}{12} = 2.25 \times 10^{-4}\ \text{m}^4$$

Maximum B.M. (M) to which each joist is subjected,

$$M = \frac{\sigma I}{y} = \frac{8 \times 2.25 \times 10^{-4}}{0.30/2} = 0.012\,\text{MNm} = 12\,\text{kNm} \quad\quad ...(i)$$

Let x be the spacing, in metres, of the joists from centre to centre. Each joist will share half of the load of the floor between two floor joists on either side. The length of each joist is 5 m. Therefore the total load on each joist (if the spacing between two joists is x from centre to centre) will be

$$= 12 \times x \times 5 = 60\,x\ \text{kN}$$

$\therefore$ Maximum B.M. $\quad = \dfrac{Wl}{8} = \dfrac{60x \times 5}{8}\ \text{kNm} \quad\quad ...(ii)$

From eqns. (*i*) and (*ii*), we get

$$12 = \frac{60x \times 5}{8}$$

or,

$$x = \frac{12 \times 8}{60 \times 5} = 0.32\,\text{m or } \mathbf{320\,mm} \quad \textbf{(Ans.)}$$

Example 5.6. *A cast iron water main 12 metres long, of 500 mm inside diameter and 25 mm wall thickness runs full of water and is supported at its ends. Calculate the maximum stress in the metal if density of cast iron is 7200 kg/m³ and that of water is 1000 kg/m³.*

Solution. Refer Fig. 5.8

Inside diameter of C.I. water main, $d = 500\,\text{mm} = 0.5\,\text{m}$

Wall thickness, $t = 25\,\text{mm} = 0.025\,\text{m}$

∴ Outside diameter of water main,

$$D = d + 2t = 500 + 2 \times 25 = 550 = 0.55\,\text{m}$$

Now cross-sectional area of main $= \pi/4\,(0.55^2 - 0.5^2) = 0.04123\,\text{m}^2$

Weight of water main per metre length

$$= 0.04123 \times 1 \times 7200 \times 9.81\,\text{N} = 2912.16\,\text{N}$$

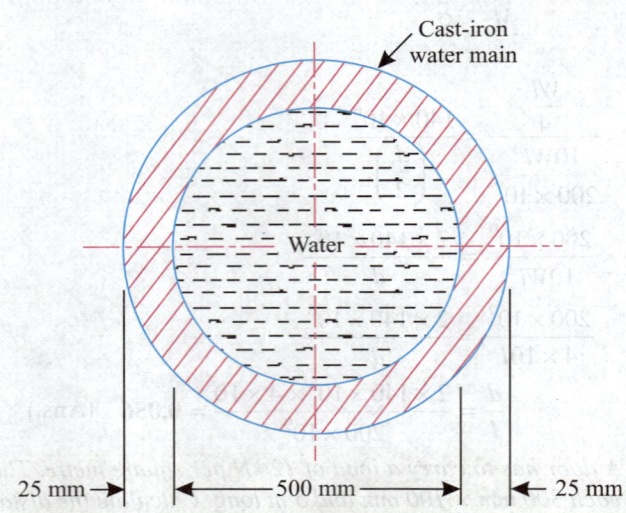

Cast-iron water main

Water

25 mm   500 mm   25 mm

Fig. 5.8

Weight of water in one metre long main

$$= (\pi/4) \times (0.5)^2 \times 1 \times 1000 \times 9.81\,\text{N} = 1926.19\,\text{N}$$

Total weight of pipe (per metre length) when full of water

$$= 2912.16 + 1926.19 = 4838.35\,\text{N}$$

Bending moment, $M = \dfrac{wl^2}{8} = \dfrac{4838.35 \times 12^2}{8}\,\text{Nm} = 87090.3\,\text{Nm}$

Moment of inertia, $I = (\pi/64)\,[(0.55)^4 - (0.5)^4] = 1.42384 \times 10^{-3}\,\text{m}^4$

$$y = \frac{D}{2} = \frac{0.55}{2} = 0.275\,\text{m}$$

Using the relation, $\dfrac{M}{I} = \dfrac{\sigma}{y}$, we get

$$\sigma = \frac{M \cdot y}{I} = \frac{87090.3 \times 0.275}{1.42384 \times 10^{-3}} \times 10^{-6} = 16.82 \, \text{MN/m}^2 \quad \text{(Ans.)}$$

Example 5.7. *A hollow circular bar having outside diameter twice the inside diameter is used as a beam. From the bending moment diagram of the beam, it is found that the bar is subjected to a bending moment of 40 kNm. If the allowable bending stress in the beam is to be limited to 100 MN/m², find the inside diameter of the bar.*

Solution.

Let, $\qquad\qquad\qquad d$ = Inside diameter of hollow circular bar.

Then outside diameter $\qquad D = 2d$

Bending moment to which the bar is subjected to

$$= M = 40 \text{ kNm}$$

Allowable bending stress, $\sigma = 100 \text{ MN/m}^2$

Inside diameter of the bar, d :

Moment of inertia, $\qquad I = \pi/64 \, (D^4 - d^4) = \pi/64 \, [(2d)^4 - d^4] = \dfrac{15}{64} \pi d^4$

$$y = \frac{D}{2} = \frac{2d}{2} = d$$

Now using the relation, $\dfrac{M}{I} = \dfrac{\sigma}{y}$, we get,

$$\sigma = \frac{My}{I}$$

or, $\qquad\qquad 100 \times 10^6 = \dfrac{40 \times 1000 \times d}{\dfrac{15}{64} \pi d^4}$ $\qquad\qquad (\because 1 \text{ MN/m}^2 = 10^6 \text{ N/m}^2)$

Vertical boring machine.

or,
$$d^3 = \frac{40 \times 1000 \times 64}{15\pi \times 100 \times 10^6} = 0.0005432$$

or,
$$d = 0.0816 \text{ m or } \textbf{81.6 mm} \quad \textbf{(Ans.)}$$

Example 5.8. *Two wooden planks 150 mm × 50 mm each are connected to form a T-section of a beam. If a moment of 3.4 kNm is applied around the horizontal neutral axis, inducing tension below the neutral axis, find the stresses at the extreme fibres of the cross-section. Also calculate the total tensile force on the cross-section.*

Solution. *Given:* Moment = 3.4 kNm = 3.4×10^3 Nm

Divide the section into rectangles as shown in the Fig. 5.9.

Component	Area 'a' (mm²)	Centroidal distance from the bottom face, y (mm)	ay (mm³)
Rectangle (1)	150 × 50 = 7500	175	1312500
Rectangle (2)	150 × 50 = 7500	75	562500
	Σ a = 15000		Σ ay = 1875000

∴ Distance of the neutral axis *XX* from the bottom face,
$$\bar{y} = \frac{\Sigma ay}{\Sigma a} = \frac{1875000}{15000} = 125 \text{ mm}$$

$$I_{XX} = \left[\frac{150 \times 50^3}{12} + 150 \times 50 \times (175 - 125)^2\right]$$
$$+ \left[\frac{50 \times 150^3}{12} + 150 \times 50(125 - 75)^2\right]$$
$$= (1562500 + 18750000) + (14062500 + 18750000)$$
$$= 5312.5 \times 10^4 \text{ mm}^4 \text{ or } 5312.5 \times 10^{-8} \text{ m}^4$$

Extreme fibre stresses: Refer to fig. 5.10.

Distance of *c. g.* from the upper extreme fibre,
$$y_c = 200 - 125 = 75 \text{ mm, and } y_t = 125 \text{ mm}$$
Now using the relation,

$$\frac{M}{I} = \frac{\sigma}{y} \text{ with usual notations, we get}$$

$$\sigma = \frac{M \times y}{I}$$

∴ Tensile stress,
$$\sigma_t = \frac{M \times y_t}{I} = \frac{3.4 \times 10^3 \times (125 \times 10^{-3})}{5312.5 \times 10^{-8}}$$
$$= 8 \times 10^6 \text{ N/m}^2 \text{ or } \textbf{8 MN/m}^2 \quad \textbf{(Ans.)}$$

Compression stress,
$$\sigma_c = \frac{M \times y_c}{I} = \frac{3.4 \times 10^3 \times (75 \times 10^{-3})}{5312.5 \times 10^{-8}}$$
$$= 4.8 \times 10^6 \text{ N/m}^2 \text{ or } \textbf{4.8 MN/m}^2 \quad \textbf{(Ans.)}$$

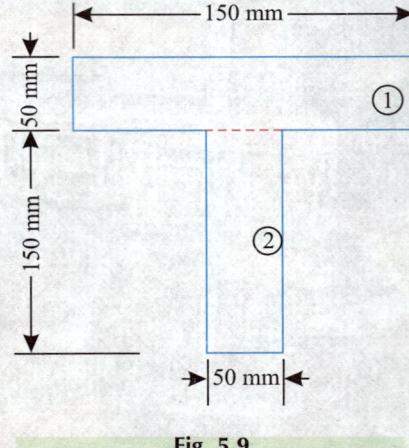

Fig. 5.9

Total tensile force on the cross-section :

We know that the section is subjected to compressive stresses above the *xx*-axis *i.e.*, neutral axis, and tensile stresses below it as shown in Fig. 5.10

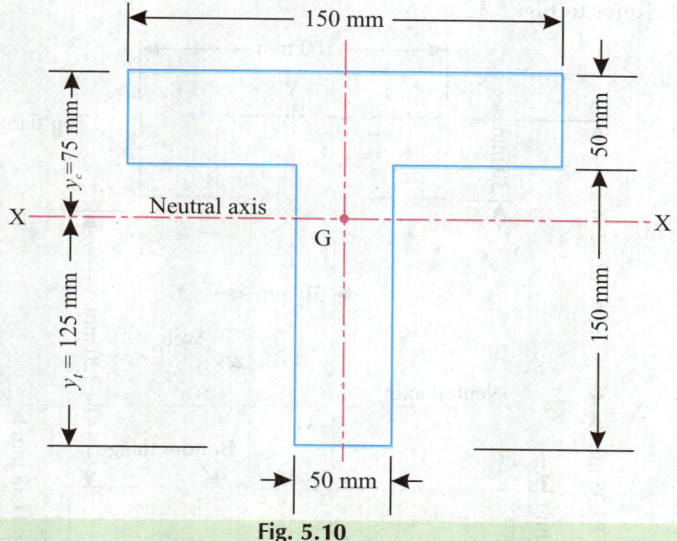

Fig. 5.10

Area in tension	$= 125 \times 50 \times 10^{-6} = 6.25 \times 10^{-3}$ m^2
∴ *Total tensile force*	= Average tensile stress × area in tension

$$= \left(\frac{8 \times 10^6 + 0}{2} \right) \times 6.25 \times 10^{-3}$$

$$= 25000 \text{ N or } \textbf{25 kN} \quad \textbf{(Ans.)}$$

Example 5.9. *A beam simply supported at ends and having cross-section as shown in Fig.5.11 is loaded with a U.D.L., over whole of its span. If the beam is 8 m long, find the U.D.L. if maximum permissible bending stress in tension is limited to 30 MN/m^2 and in compression to 45 MN/m^2. What are the actual maximum bending stresses set up in the section?*

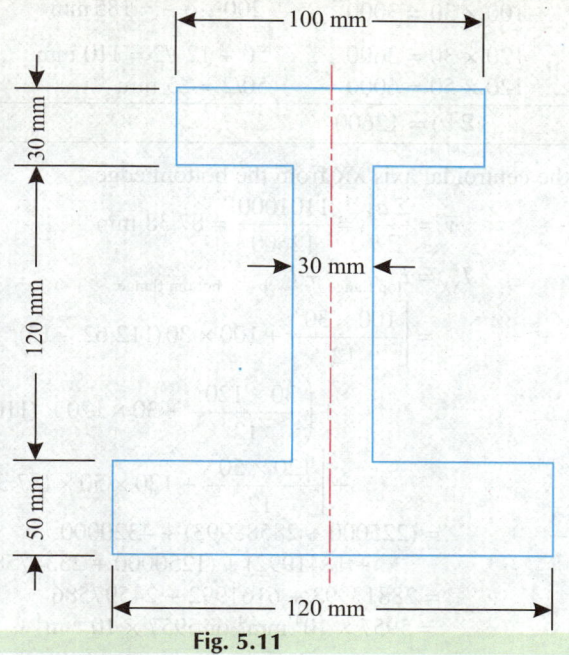

Fig. 5.11

Solution. Refer to Fig. 5.12.

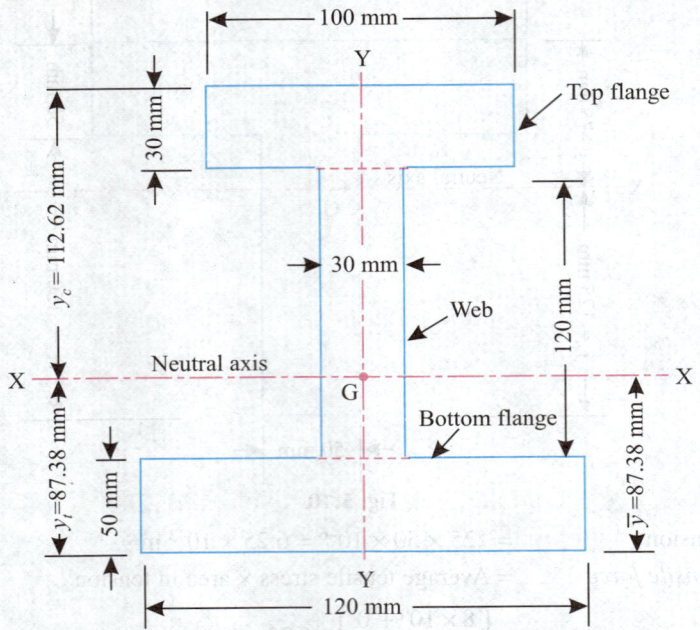

Fig. 5.12

The following table gives the calculations for location of horizontal neutral axis:

Component	Area 'a' (mm^2)	Centroidal distance from bottom edge, 'y' (mm)	ay (mm^3)
Top flange	100 × 30 = 3000	$200 - \dfrac{30}{2} = 185$ mm	555000
Web	120 × 30 = 3600	50 + 120/2 = 110 mm	396000
Bottom flange	120 × 50 = 6000	50/2 = 25 mm	150000
Total	(Σ a) = 12600		(Σ ay) = 1101000

Distance of the centroidal axis *XX* from the bottom edge,

$$\bar{y} = \frac{\Sigma ay}{\Sigma a} = \frac{1101000}{12600} = 87.38 \text{ mm}$$

$$I_{XX} = I_{\text{top flange}} + I_{\text{web}} + I_{\text{bottom flange}}$$

$$= \left[\frac{100 \times 30^3}{12} + 100 \times 30(112.62 - 15)^2\right]$$

$$+ \left(\frac{30 \times 120^3}{12} + 30 \times 120 \times (110 - 87.38)^2\right)$$

$$+ \left[\frac{120 \times 50^3}{12} + 120 \times 50 \times (87.38 - 25)^2\right]$$

$$= (225000 + 28588993) + 4320000$$

$$+ 1841992) + (1250000 + 23347586)$$

$$= 28813993 + 6161992 + 24597586$$

$$= 5957 \times 10^4 \text{ mm}^4 \text{ or } 5957 \times 10^{-8} \text{ m}^4$$

18th Century boring machine.

Maximum bending moment,

$$M = \frac{wl^2}{8} = \frac{w \times 8^2}{8} = 8w$$

(where, w = uniformly distributed load)

For *tension side* of the I-section :

$$\frac{M}{I} = \frac{\sigma_t}{y_t}$$

$$M = I \times \frac{\sigma_t}{y_t} = \frac{5957 \times 10^{-8} \times 30 \times 10^6}{87.38 \times 10^{-3}}$$

$$= 20452 \text{ Nm} = 20.452 \text{ kNm}$$

For *compression side* of the I-section :

$$M = \frac{I \times \sigma_c}{y_c} = \frac{5957 \times 10^{-8} \times 45 \times 10^6}{112.62 \times 10^{-3}}$$

$$= 23802.6 \text{ Nm} = 23.8026 \text{ kNm}$$

∴ Moment of resistance $M = 20.452$ kNm (*i.e.* minimum of the above two values)

∴ $\qquad\qquad 8w = 20.452$

or $\qquad\qquad w = \dfrac{20.452}{8} = \textbf{2.556 kN/m}$ (Ans.)

Actual maximum stress in the top-most fibres of the beam

$$= \frac{M}{I} y_c = \frac{20.452 \times 10^3 \times 112.62 \times 10^{-3}}{5957 \times 10^{-8}}$$

$$= 38.6 \times 10^6 \text{ N/m}^2 = \textbf{38.6 MN/m}^2 \text{ (\textit{compressive})} \qquad \textbf{(Ans.)}$$

Actual maximum stress in the bottom-most fibres of the beam

$$= 30 \text{ MN/m}^2 \text{ (tensile)} \quad \text{(Ans.)}$$

Example 5.10. *Fig. 5.13 shows a cost iron bracket of cross-section of I-form. Find:*

(*i*) *Position of the neutral axis and the moment of inertia of the section about the neutral axis.*

(*ii*) *Determine the maximum bending moment that should be imposed on this section if the tensile stress in the top flange is not to exceed 40 MN/m². What is then the value of the compressive stress in the bottom flange?*

Solution. Refer Fig. 5.13 and 5.14.

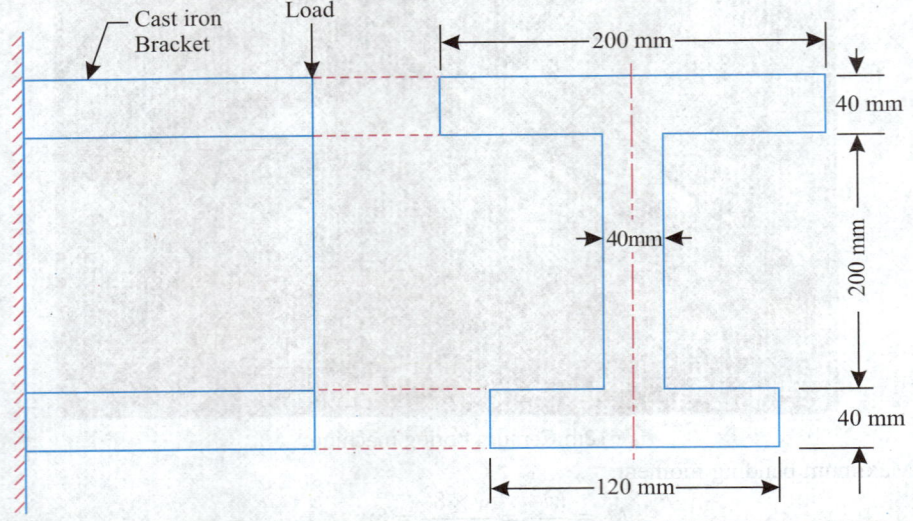

Fig. 5.13

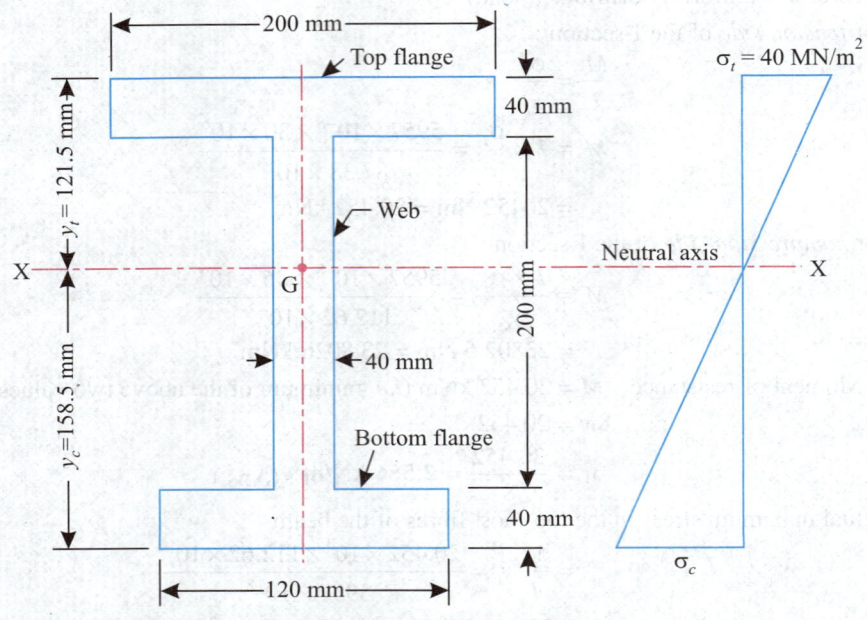

Fig. 5.14

(i) Position of the neutral axis:

The section may be split into three rectangular components. The table of calculations is given below:

Component	Area 'a' (mm^2)	Centroidal distance from bottom edge, 'y' (mm)	ay (mm^3)
Top flange	$200 \times 40 = 8000$	260 mm	2080000
Web	$200 \times 40 = 8000$	140 mm	1120000
Bottom flange	$120 \times 40 = 4800$	20 mm	96000
Total	$(\Sigma a) = 20800$		$(\Sigma ay) = 3296000$

∴ Distance of the neutral axis from the bottom edge,

$$\bar{y} = \frac{\Sigma ay}{\Sigma a} = \frac{3296000}{20800} = \textbf{158.5 mm} \quad \textbf{(Ans.)}$$

Hence, $y_c = 158.5$ mm

and, $y_t = (40 + 200 + 40) - 158.5 = 121.5$ mm

M.O.I. $I_{XX} = ?$

$I_{XX} = I_{\text{top flange}} + I_{\text{web}} + I_{\text{bottom flange}}$

$$= \left[\frac{200 \times 40^3}{12} + 200 \times 40 \times (121.5 - 20)^2\right]$$

$$+ \left[\frac{40 \times 200^3}{12} + 40 \times 200 \times (158.5 - 140)^2\right]$$

$$+ \left[\frac{120 \times 40^3}{12} + 120 \times 40 \times (158.5 - 20)^2\right]$$

$$= 83484667 + 29404667 + 92714800$$

$$= 2.056 \times 10^8 \text{ mm}^4 = 2.056 \times 10^{-4} \text{ m}^4$$

(ii) Maximum bending moment, M:

Using the relation,

$$\frac{M}{I} = \frac{\sigma}{y}, \text{ we get}$$

$$\frac{M}{I} = \frac{\sigma_t}{y_t}$$

or, $$M = \frac{\sigma_t}{y_t} \times I = \frac{40 \times 10^6 \times 2.056 \times 10^{-4}}{121.5 \times 10^{-3}}$$

$$= 67687.2 \text{ Nm} = \textbf{67.7 kNm (say)} \quad \textbf{(Ans.)}$$

Compressive stress in the bottom flange, σ_c:

Using the relation:

$$\frac{M}{I} = \frac{\sigma_c}{y_c} \quad \text{or} \quad \sigma_c = \frac{M \cdot y_c}{I}$$

or, $$\sigma_c = \frac{(67.7 \times 10^3) \times (158.5 \times 10^{-3})}{2.056 \times 10^{-4}}$$

$$= 52.19 \times 10^6 \text{ N/m}^2 = \textbf{52.19 MN/m}^2 \quad \textbf{(Ans.)}$$

Example 5.11. *The horizontal beam of section shown in Fig. 5.15 is 4 m long and is simply supported at the ends. Calculate the maximum uniformly distributed load it can carry if the tensile and compressive stresses must not exceed 25 MN/m² and 45 MN/m² respectively.*

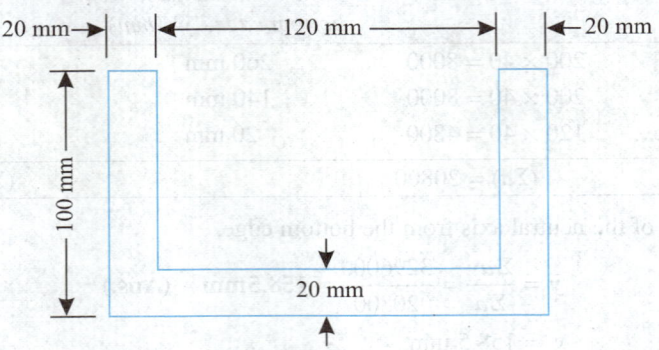

Fig. 5.15

Solution. Divide the section into three rectangles (components) as shown in the Fig. 5.16.

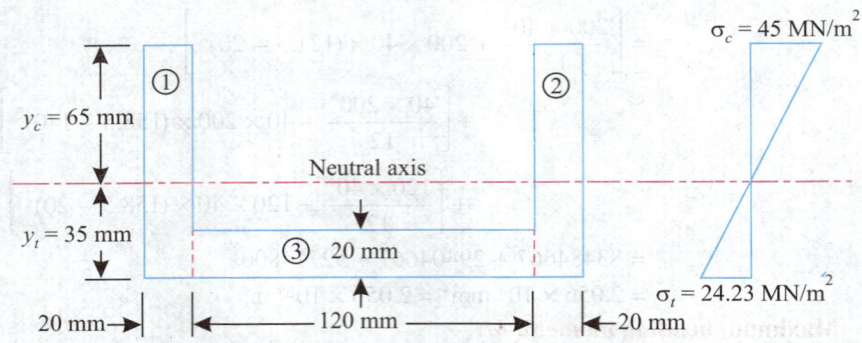

Fig. 5.16

The areas of the individual components, their centroidal distances from the bottom edge and their moments about the bottom edge are tabulated below :

Component	Area 'a' (mm²)	Centroidal distance from bottom edge, 'y' (mm)	ay (mm³)
Rectangle (1)	100 × 20 = 2000	50	100000
Rectangle (2)	100 × 20 = 2000	50	100000
Rectangle (3)	120 × 20 = 2400	10	24000
Total	(Σ a) = 6400		(Σ ay) = 224000

Distance of the neutral axis from the bottom edge,

$$\bar{y} = \frac{\Sigma ay}{\Sigma a} = \frac{224000}{6400} = 35 \, \text{mm}$$

$$y_t = 35 \, \text{mm and } y_c = 100 - 35 = 65 \, \text{mm}$$

$$I_{XX} = I_{\text{rect.}-1} + I_{\text{rect.}-2} + I_{\text{rect.}-3}$$

$$= \left[\frac{20 \times 100^3}{12} + 20 \times 100 \times (50 - 35)^2 \right]$$

$$+ \left[\frac{20 \times 100^3}{12} + 20 \times 100 \times (50 - 35)^2 \right]$$

$$+ \left[\frac{120 \times 20^3}{12} + 120 \times 20 \times (35 - 10)^2 \right]$$

$$= 2 \times 2116666 + 1580000$$

$$= 5813332 \text{ mm}^4 = 581.33 \times 10^{-8} \text{ m}^4$$

We know that, $\dfrac{M}{I} = \dfrac{\sigma_t}{y_t} = \dfrac{\sigma_c}{y_c}$

or, $\dfrac{\sigma_c}{\sigma_t} = \dfrac{y_c}{y_t} = \dfrac{65}{35} = 1.857$

So, the ratio of compressive and tensile stresses is 1.857.

If the beam is loaded such that compressive stress (σ_c) should not exceed 45 MN/m^2, the permissible limit, the corresponding tensile stresses (σ_t) according to the above ratio, will be:

$$45 \times \frac{1}{1.857} = 24.23 \text{ MN/m}^2$$

Augur (component) used in soil boring machines.

which is within the permissible limit; but in case, the beam is loaded so as to produce tensile stress of 25 MN/m^2, the corresponding compressive stress would be $25 \times 1.857 = 46.42$ MN /m^2 which is more than the permissible limit given in the question.

Hence the actual stress produced in the beam should be 45 MN/m^2 compressive and 24.23 MN/m^2 tensile and the U.D.L. (uniformly distributed load) can be determined from these values.

Using the relation: $\dfrac{M}{I} = \dfrac{\sigma}{y} = \dfrac{\sigma_t}{y_t} = \dfrac{\sigma_c}{y_c}$, we get

$$M = \frac{\sigma_t}{y_t} \cdot I = \frac{\sigma_c}{y_c} \cdot I$$

Substituting, $\sigma_t = 24.23$ MN/m^2 (and not the given value)

$$\sigma_c = 45 \text{ MN/m}^2, \ y_t = 35 \text{ mm},$$

$$y_c = 65 \text{ mm}.$$

$$I = I_{XX} = 581.33 \times 10^{-8} \text{ m}^4.$$

$$M = \frac{\sigma_t}{y_t} \cdot I$$

$$= 24.23 \times 10^6 \times \frac{(581.33 \times 10^{-8})}{35 \times 10^{-3}} \text{ Nm or } 4.0245 \text{ kNm}$$

Let w be the uniformly distributed load in kN/m run

$$M_{max} = \frac{wl^2}{8} = \frac{w \times 4^2}{8}$$

Equating the maximum bending moment and the resisting moment M, we have

$$\frac{w \times 4^2}{8} = 4.0245$$

or,
$$w = \frac{4.0245 \times 8}{4^2}$$

i.e. $w = \textbf{2.012 kN/m run}$ **(Ans.)**

Example 5.12. *Fig. 5.17 shows the cross-section of a cast-iron beam. When this beam is subjected to a bending moment the tensile stress at the bottom edge is 30 MN/m². Calculate:*

   (*i*)   *The value of the bending moment.*

   (*ii*) *Stress induced at the top edge.*

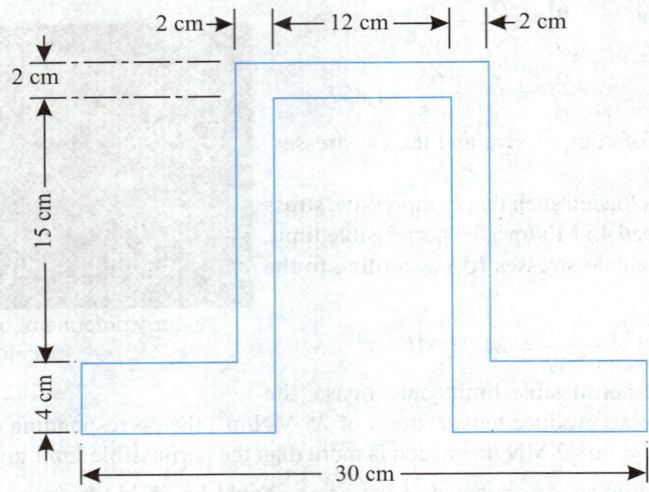

Fig. 5.17

Solution. In order to locate the position of neutral axis, divide the section into five rectangles as shown in Fig. 5.18.

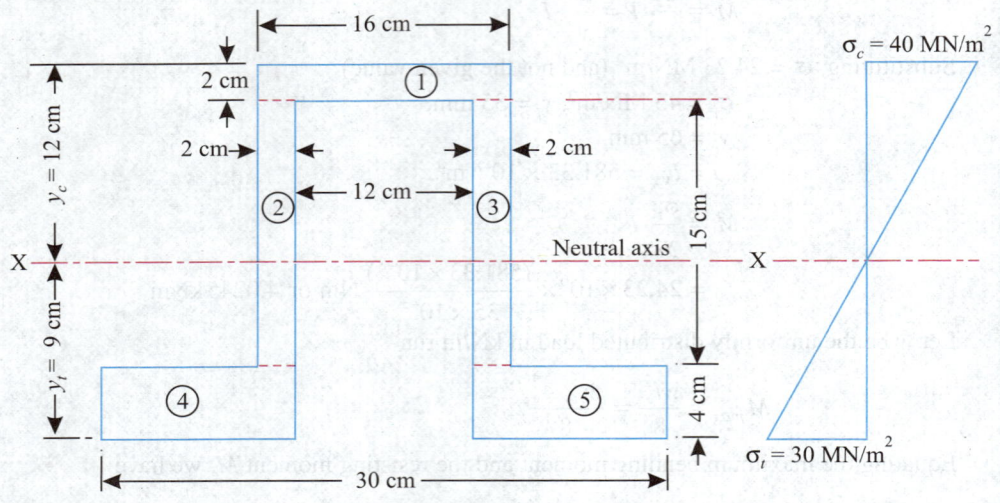

Fig. 5.18

The areas of the individual components, their centroidal distances from the bottom edge and their moments about the bottom edge are tabulated below:

Component	Area 'a' (cm²)	Centroidal distance from bottom edge, 'y' (cm)	ay (cm³)
Rectangle (1)	16 × 2 = 32	20	640
Rectangle (2) and (3)	2 × 15 × 2 = 60	11.5	690
Rectangle (4) and (5)	2 × 9 × 4 = 72	2	144
Total	Σ a = 164	–	Σ ay = 1474

Distance of the neutral axis from the bottom edge,

$$\bar{y} = \frac{\Sigma ay}{\Sigma a} = \frac{1474}{164} = 8.90 \text{ say 9 cm}$$

∴ $y_t = 9$ cm

and, $y_c = 12$ cm

$$I_{XX} = I_{rect. -1} + I_{rect. - (2 \& 3)} + I_{rect. - (4 \& 5)}$$

$$= \left[\frac{16 \times 2^3}{12} + 16 \times 2 \times (12 - 1)^2\right] + 2 \times \left[\frac{2 \times 15^3}{12} + 2 \times 15 \times (11.5 - 9)^2\right]$$

$$+ 2\left(\frac{9 \times 4^3}{12} + 9 \times 4 \times (9 - 2)^2\right)$$

$$= 3883 + 2 \times 750 + 2 \times 1812$$

$$= 9007 \text{ cm}^4 \text{ or } 9007 \times 10^{-8} \text{ m}^4$$

(i) Bending moment M:

Using the relation,

$$\frac{M}{I} = \frac{\sigma}{y} = \frac{\sigma_t}{y_t}, \text{ we get}$$

$$M = \frac{\sigma_t}{y_t} \times I = \frac{30 \times 10^6 \times 9007 \times 10^{-8}}{9 \times 10^{-2}}$$

$$= 30023 \text{ Nm or } \textbf{30 kNm (say) (Ans.)}$$

(ii) Stress at the top edge σ_c:

Using the relation, $\dfrac{M}{I} = \dfrac{\sigma}{y} = \dfrac{\sigma_c}{y_c}$, we get

$$\sigma_c = \frac{M}{I} \times y_c$$

$$= \frac{(30 \times 10^3) \times 12 \times 10^{-2}}{9007.6 \times 10^{-8}}$$

$$= 40 \times 10^6 \text{ N/m}^2 \text{ or } \textbf{40 MN/m}^2 \text{ (Ans.)}$$

Automobile carburettor.

Example 5.13. *A steel stanchion is built of a rolled steel joist of I section 45 cm × 20 cm united by 1.5 cm thick and 30 cm wide plates fastened on each flange. The length of the stanchion is 5 m and is freely supported at both ends. For the I section:* $I_{XX} = 35060$ cm⁴ *Find:*

(i) *Moment of inertia of the enlarged section about XX-axis.*

(ii) *Greatest central point load the beam will carry if the bending stress is not to exceed 120 MN/m².*

(iii) *Minimum length of the 30 cm × 1.5 cm plates.*

Solution. Fig. 5.19 shows the stanchion.

(i) M.O.I. of enlarged section:

Moment of inertia of the enlarged   =   Moment of inertia of I-section about *XX* axis +
section about *XX*-axis   moment of inertia of the plates about *XX* axis.

$$\therefore \quad I_{XX} = 35060 + 2\left[\frac{30 \times 1.5^3}{12} + 30 \times 1.5 \times \left(\frac{45}{2} + \frac{1.5}{2}\right)^2\right]$$

$$= 35060 + 2\,(8.437 + 24325.313)$$

$$= \mathbf{83727.5\ cm^4} \quad \textbf{(Ans.)}$$

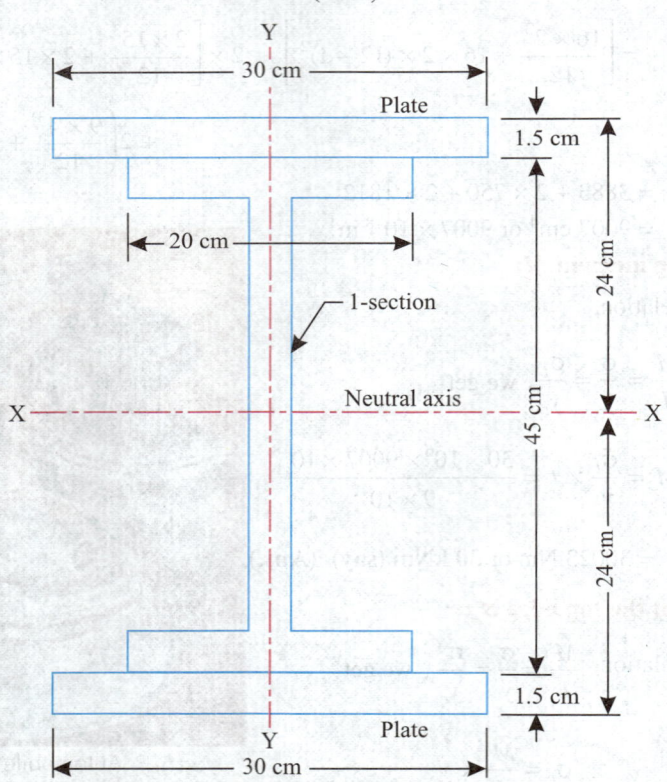

Fig. 5.19

(ii) Greatest central point load W:

We know that,

Maximum B.M. $= \dfrac{Wl}{4} = \dfrac{W \times 5}{4} = 1.25\,W$

But moment of resistance,

$$M = \frac{\sigma}{y} \times I = \frac{120 \times 10^6}{24 \times 10^{-2}} \times 83727.5 \times 10^{-8} = 418637.5 \text{ Nm}$$

∴ $1.25\ W = 418637.5$

or, $W = \dfrac{418637.5}{1.25} = 334910\ \text{N} = \textbf{334.91 kN}$ **(Ans.)**

(iii) Minimum length of the plates:

Suppose the cover plates are absent for a distance of x metres from each support. Then at these points the bending moment *must not exceed moment of resistance of I section* alone *i.e.*

$$\sigma \times \frac{I}{y} = 120 \times 10^6 \times \frac{35060 \times 10^{-8}}{24 \times 10^{-2}} = 175300 \text{ Nm}$$

∴ Bending moment at x metres from each support

$$= \frac{W}{2} \times x = 175300 \text{ or } \frac{334910}{2} \cdot x = 175300$$

from which $x = 1.05$ m.

Hence leaving 1.05 metre from each support, for the middle $5 - 2.1 = 2.9$ metres, the cover plates should be provided.   **(Ans.)**

Example 5.14. *A simply supported beam and its cross-section are as shown in Fig. 5.20. The beam carries a load W = 20 kN as shown. Its self-weight is 7 kN/m. Calculate the maximum normal stress at 1–1.*

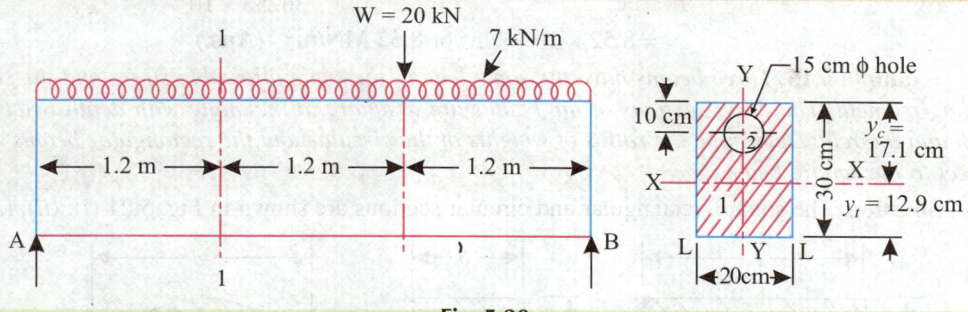

Fig. 5.20

Solution. To determine reaction at the supports taking moments about A, we get

$$R_B \times 3.6 = 20 \times 2.4 + 7 \times 3.6 \times 3.6/2 = 93.36$$

∴ $R_B = \dfrac{93.36}{3.6} = 25.93\,\text{kN}$

But, $R_A + R_B = 20 + 7 \times 3.6 = 45.2$ kN

∴ $R_A = 19.27$ kN

∴ Bending moment at 1–1,

$$M = 25.93 \times 2.4 - 20 \times 1.2 - 7 \times 2.4 \times 2.4/2 = 18.072 \text{ kNm}$$

Refer to cross-section of the beam. The areas of the individual components, their centroidal distances from the bottom edge (LL) and their moments about the bottoms edge are tabulated below:

Component	Area 'a' (cm^2)	Centroidal distance from the bottom edge, 'y' (cm)	ay (cm^3)
Rectangle (1)	$30 \times 20 = 600$	15	9000
Circular hole (2)	$(\pi/4) \times 15^2 = 176.7\ (-)$	20	3534 (–)
Total	$\Sigma a = 423.3$	–	$\Sigma ay = 5466$

Distance of the neutral axis (XX axis) from the bottom edge,

$$\bar{y} = \frac{\Sigma ay}{\Sigma a} = \frac{5466}{423.3} = 12.9 \text{ cm}$$

$$y_t = 12.9 \text{ cm and } y_c = 30 - 12.9 = 17.1 \text{ cm}$$

$$I_{XX} = I_{\text{rect.}} - I_{\text{hole}}$$

$$= \left[\frac{20 \times 30^3}{12} + 20 \times 30 \times (15 - 12.9)^2 \right]$$

$$-[(\pi/64) \times 15^4 + (\pi/4) \times 15^2 \times (17.1 - 10)^2]$$

$$= 47646 - 11393 = 36253 \text{ cm}^4 = 36253 \times 10^{-8} \text{ m}^4$$

Maximum normal stress at 1–1:

Using the relation : $\dfrac{M}{I} = \dfrac{\sigma}{y}$ with usual notations, we get

[We shall take the value of $y = y_c = 17.1$ cm (i.e. greater of the two values between y_t and y_c)]

$$\frac{M}{I} = \frac{\sigma_c}{y_c} \quad \text{or} \quad \sigma_c = \frac{M\,y_c}{I} = \frac{(18.072 \times 10^3) \times (17.1 \times 10^{-2})}{36253 \times 10^{-8}}$$

$$= 8.52 \times 10^6 \text{ N/m}^2 \text{ or } \textbf{8.52 MN/m}^2 \quad \textbf{(Ans.)}$$

Example 5.15. *Three beams have the same length, the same allowable stress and the same bending moment. The cross-sections of the beams are a square, a rectangle with depth twice the width and a circle. Determine the ratios of weights of the circular and the rectangular beams with respect to the square beam.*

Solution. The square, rectangular and circular sections are shown in Fig. 5.21 (i), (ii), (iii).

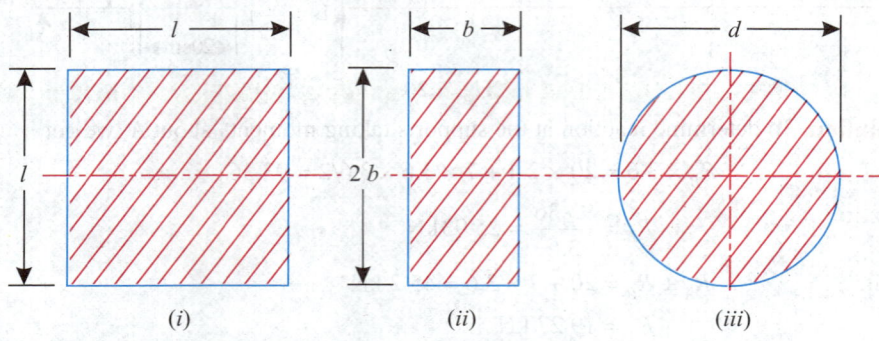

(i) (ii) (iii)

Fig. 5.21

Let, l = Side of the square beam,

b = Width of a rectangular beam,

$2b$ = Depth of the rectangular beam, and

d = Diameter of a circular section.

As all the three beams have the same allowable stress (σ) and bending moment (M), therefore the modulus of section (Z) of the three beams must be equal.

Modulus of section for a *square beam*,

$$Z_1 = \frac{I}{y} = \frac{l \times l^3/12}{l/2} = \frac{l^4/12}{l/2} = \frac{l^3}{6} \qquad ...(i)$$

Modulus of section for a *rectangular beam*,

$$Z_2 = \frac{I}{y} = \frac{b \times (2b)^3/12}{2b/2} = \frac{2b^3}{3} \qquad ...(ii)$$

and modulus of section for a *circular beam*,

$$Z_3 = \frac{I}{y} = \frac{\pi/64 \times d^4}{d/2} = \frac{\pi}{32} \times d^3 \qquad ...(iii)$$

Equating (*i*) and (*ii*), we get

$$\frac{l^3}{6} = \frac{2b^3}{3}$$

$$\therefore \quad l^3 = 4b^3 \qquad \text{or} \qquad b = 0.63\, l \qquad ...(iv)$$

Now equating (*i*) and (*iii*), we get

$$\frac{l^3}{6} = \frac{\pi}{32} d^3$$

$$\therefore \qquad l^3 = \frac{3\pi}{16} \times d^3$$

or, $$d^3 = \frac{16}{3\pi} l^3 \quad \text{or,} \quad d = 1.193\, l$$

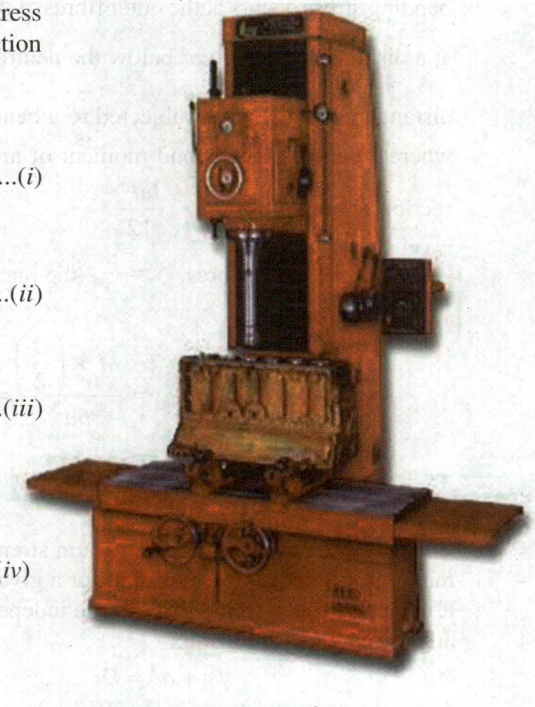

Boring machine.

Since the weights of the beams are proportional to the cross-sectional areas of the respective beams therefore,

$$\frac{\text{Weight of square beam}}{\text{Weight of rectangular beam}} = \frac{\text{Cross-sectional area of square beam}}{\text{Cross-sectional area of rectangular beam}}$$

$$= \frac{l^2}{2b^2} = \frac{l^2}{2 \times (0.63 l)^2} = \textbf{1.259} \quad \textbf{(Ans.)}$$

and, $$\frac{\text{Weight of square beam}}{\text{Weight of circular beam}} = \frac{\text{Cross-sectional area of square beam}}{\text{Cross-sectional area of circular beam}}$$

$$= \frac{l^2}{(\pi/4) \times d^2} = \frac{l^2}{(\pi/4)(1.193 l)^2} = \textbf{0.895} \quad \textbf{(Ans.)}$$

Example 5.16. *If the beam cross-section is rectangular having a width of 75 mm, determine the required depth such that maximum bending stress induced in the beam does not exceed 40 MN/m².*

(AMIE Summer, 2000)

Solution. *Given:* $\quad b = 75$ mm $= 0.075$ m;

$$\sigma_{max} = 40 \text{ MN/m}^2.$$

Depth of the beam, *d*:

Fig. 5.22 shows a rectangular section of width b ($= 0.075$ m) and depth d metres. The bending is considered to take place about the horizontal neutral axis N.A. shown in the figure. The maximum bending stress occurs at the outer fibres of the rectangular section at a distance $\dfrac{d}{2}$ above or below the neutral axis. Any fibre at a distance y from N.A. is subjected to a bending stress $\sigma = \dfrac{My}{I}$, where I denotes the second moment of area of the rectangular section about the N.A. *i.e.* $\dfrac{bd^3}{12}$.

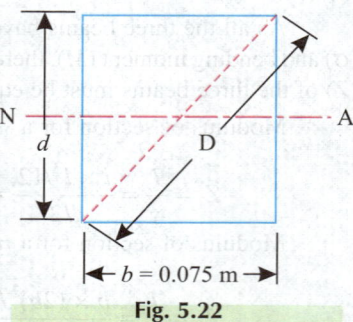

Fig. 5.22

At the outer fibres, $y = \dfrac{d}{2}$, the maximum bending stress there becomes

$$\sigma_{max} = \frac{M \times \left(\dfrac{d}{2}\right)}{\dfrac{bd^3}{12}} = \frac{M}{\dfrac{bd^2}{6}} \qquad \text{...}(i)$$

or, $$M = \sigma_{max} \cdot \frac{bd^2}{6} \qquad \text{...}(ii)$$

For the condition of maximum strength *i.e.* maximum moment M, the product bd^2 must be a maximum, since σ_{max} is constant for a given material. To maximize the quantity bd^2 we realise that it must be expressed in terms of one independent variable, say, b, and we may do this from the right angle triangle relationship.

$$b^2 + d^2 = D^2$$

or, $$d^2 = D^2 - b^2$$

Multiplying both sides by b, we get

$$bd^2 = bD^2 - b^3$$

To maximize bd^2 we take the first derivative of expression with respect to b and set it equal to zero, as follows:

$$\frac{d}{db}(bd^2) = \frac{d}{db}(bD^2 - b^3)$$

$$= D^2 - 3b^2 = b^2 + d^2 - 3b^2$$

$$= d^2 - 2b^2 = 0 \qquad \text{...}(iii)$$

Solving, we have depth, $d = \sqrt{2}\,b$

This is the desired ratio in order that the beam will carry a maximum moment M.

It is to be noted that the expression appearing in the denominator of the right side of eqn. (*i*) *i.e.* $\dfrac{bd^2}{6}$, is the section modulus (Z) of a rectangular bar. Thus, it follows, the section modulus is actually the quantity to be maximized for greatest strength of the beam.

Using the relation (*iii*), we have

$$d = \sqrt{2} \times 0.075 = 0.106 \text{ m}$$

Now, $$M = \sigma_{max} \times Z = \sigma_{max} \times \frac{bd^2}{6}$$

Substituting the values, we get

$$M = 40 \times \frac{0.075 \times (0.106)^2}{6} = 0.005618 \text{ MNm}$$

$$\sigma_{max} = \frac{M}{Z} = \frac{0.005618}{[0.075 \times (0.106)^2 / 6]} = 40 \text{ MN/m}^2$$

Hence, the required depth d = 0.106 m = **106 mm**   **(Ans.)**

Example 5.17.

(i) *Calculate the dimensions of the strongest section that can be cut out of a circular log of wood 25 cm in diameter.*

(ii) *Also specify the safe maximum span for the beam of rectangular section when it is to carry a U.D.L. of 2.5 kN per metre and the bending stresses are limited to 10 MN/m².*

(iii) *What will be the longest span if the circular log itself is used as a beam ?*

Solution. Refer Fig. 5.23.

(i) Dimensions of the strongest section:

A *strongest beam is one which offers maximum resistance* (which depends upon *section modulus*).

We know that,

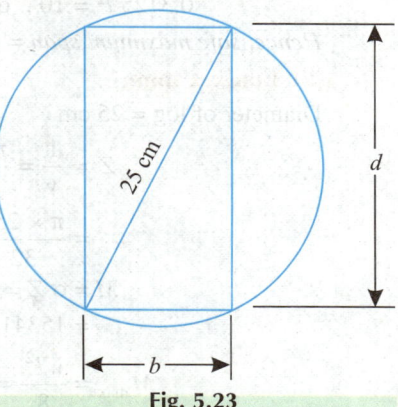

$$\frac{M}{I} = \frac{\sigma}{y}$$

or,

$$M = \sigma \times \frac{I}{y}$$

or,

$$M = \sigma \times Z \qquad \left(\because \frac{I}{y} = Z \right)$$

Fig. 5.23

The value of *M* will be maximum if the value of *Z* is maximum as the value of stress depends upon the material.

For making maximum use of the log of the wood, the *corners of the section to be selected must lie on the circumference.*

Let, b = Width of the section, and

d = Depth of the section.

The diagonal of the section = Diameter of circular log = 25 cm

∴ $b^2 + d^2 = 25^2$ or $d^2 = 25^2 - b^2$

Now, $Z = \dfrac{I}{y} = \dfrac{bd^3/12}{d/2} = \dfrac{bd^2}{6}$

$$= \frac{b(25^2 - b^2)}{6} = \frac{b(625 - b^2)}{6}$$

or, $Z = \dfrac{625b - b^3}{6}$

For *Z* to be maximum, $\dfrac{dZ}{db}$ should be equal to zero.

∴ $\dfrac{dZ}{db} = \dfrac{d}{db}\left(\dfrac{625b - b^3}{6} \right)$

or, $b^2 = \dfrac{625}{3}$ or **b = 14.43 cm** **(Ans.)**

Now, $d^2 = 25^2 - 14.43^2 = 416.78$

∴   **d = 20.41 cm** **(Ans.)**

(ii) Safe maximum span :

Now, $Z = \dfrac{bd^2}{6} = \dfrac{14.43 \times (20.41)^2}{6} = 1001.85 \text{ cm}^3 = 1001.85 \times 10^{-6} \text{ m}^3$

$$M = \sigma Z = (10 \times 10^6) \times 1001.85 \times 10^{-6} \text{ Nm}$$
$$= 10018.5 \text{ Nm} = 10 \text{ kNm (say)} \qquad ...(i)$$

We know that, in case of beam (length l) carrying a U.D.L. of 2.5 kN/m run,

$$M_{max} = \frac{wl^2}{8} = \frac{2.5l^2}{8} \text{ kNm}$$

or, $\qquad M_{max} = 0.3125\ l^2 \qquad\qquad ...(ii)$

Equating (i) and (ii), we get

$$0.3125\ l^2 = 10 \quad \text{or} \quad l = 5.66 \text{ m}$$

Hence, safe maximum span = **5.66 m** **(Ans.)**

(iii) Longest span:

Diameter of log = 25 cm

$\therefore \qquad Z = \dfrac{1}{y} = \dfrac{\pi d^4/64}{d/2} = \dfrac{\pi d^3}{32}$

$$= \frac{\pi \times 25^3}{32} = 1534.18 \text{ cm}^3 = 1534.18 \times 10^{-6} \text{ m}^3$$

$\therefore \qquad M = \sigma Z = (10 \times 10^6) \times 1534.18 \times 10^{-6}$

$$= 15341.8 \text{ Nm} = 15.34 \text{ kNm} \qquad ...(iii)$$

$$M_{max} = \frac{wl^2}{8} = \frac{2.5 \times l^2}{8} = 0.3125 l^2 \qquad ...(iv)$$

Equating (iii) and (iv), we get

$$0.3125\ l^2 = 15.34 \qquad \text{or} \qquad l = 7 \text{ m}.$$

Hence, longest span for circular log = **7 m** **(Ans.)**

Old fashioned bridge.

5.5 BEAM OF HETEROGENEOUS MATERIALS (FLITCHED BEAM)

It is common to reinforce beams to withstand loads greater than normally allowable, and in some cases different materials may be used to provide the reinforcement symmetrically disposed about the neutral axis.

The following expressions hold in all cases:

1. *The total resisting moment at any section in the sum of resisting moments caused by the individual material making up the section.*

Thus, if M_1 and M_2 are the resisting moments then the total resisting moment (M),

$$M = M_1 + M_2 \qquad \qquad ...(5.8)$$

2. *For each material, the same radius of curvature, that of the neutral axis, will apply.*

Thus the bending equation becomes for:

Material 1:
$$\frac{M_1}{I_1} = \frac{\sigma_1}{y_1} = \frac{E_1}{R_1}$$

and, Material 2:
$$\frac{M_2}{I_2} = \frac{\sigma_2}{y_2} = \frac{E_2}{R_2}$$

or, equating the radii of curvature

$$\frac{M_1}{E_1 I_1} = \frac{M_2}{E_2 I_2} \qquad \qquad ...(5.9)$$

Further, we get
$$\frac{\sigma_1}{y_1 E_1} = \frac{\sigma_2}{y_2 E_2}$$

Hence the dimensions of the section control the stress ratio at the outer fibres of each material, and therefore control the values of the resisting moments.

By the use of these equations, the *maximum allowable bending moment may be found out.*

FLITCHED BEAMS

Example 5.18. *A flitched timber beam consists of two joists 100 mm wide and 300 mm deep with a steel plate 200 mm deep and 15 mm thick placed symmetrically in between and clamped to them. Calculate the total moment of resistance of the section if the allowable stress in joint is 9 MN/m^2.*

Given : $E_s = 20 E_w$

Solution. Fig. 5.24 shows the arrangement and also the stress distribution in timber and steel.

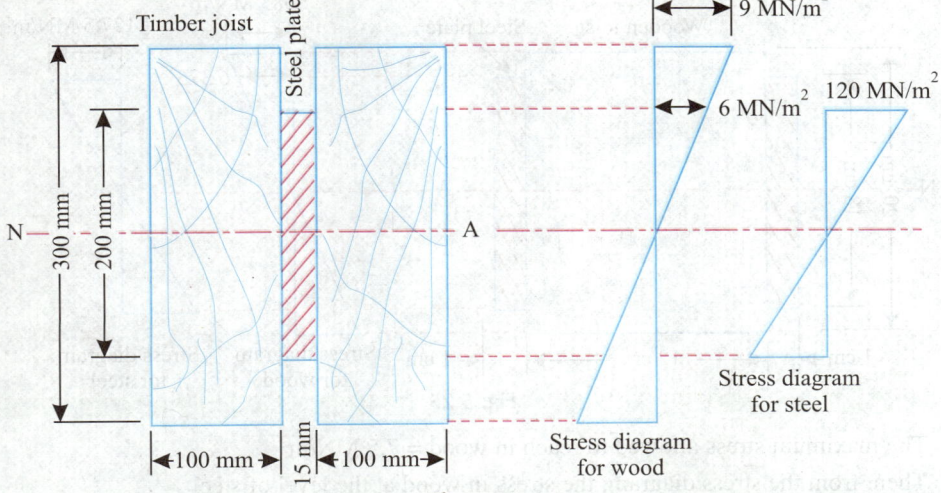

Fig. 5.24

From the stress diagram, the stress in wood, at the level of steel is

$$= 9 \times \frac{100}{150} = 6 \, \text{MN/m}^2$$

Since the strain in steel = strain in wood, we have maximum stress in steel,

$$\sigma_s = 6 \times \frac{E_s}{E_w} = 6 \times 20 = 120 \, \text{MN/m}^2$$

$$\left[\begin{array}{l} \because \dfrac{\sigma_s}{E_s} = \dfrac{\sigma_w}{E_w} \\[2mm] \therefore \quad \sigma_s = \sigma_w \times \dfrac{E_s}{E_w} \\[2mm] \text{where } \dfrac{E_s}{E_w} = \text{modular ratio} = m_r \end{array} \right]$$

Now, section modulus for wood

$$Z_w = \frac{b d^2}{6} = \frac{200 \times 300^2}{6} = 3 \times 10^6 \, \text{mm}^3 = 3 \times 10^{-3} \, \text{m}^3$$

and section modulus for steel,

$$Z_s = \frac{b d^2}{6} = \frac{15 \times 200^2}{6} = 10^5 \, \text{mm}^3 = 10^{-4} \, \text{m}^3$$

Moment of resistance of timber section,

$$M_w = \sigma_w \, Z_w = 9 \times 3 \times 10^{-3} = 0.027 \, \text{MNm}$$

Moment of resistance of steel section

$$M_s = \sigma_s \, Z_s = 120 \times 10^{-4} = 0.012 \, \text{MNm}$$

Total moment of resistance offered by the composite reaction,

$$M = M_w + M_s = 0.027 + 0.012$$
$$= 0.039 \, \text{MNm or } \textbf{39 kNm} \quad \textbf{(Ans.)}$$

Example 5.19. *A flitched beam consists of a wooden joist 12 cm wide and 20 cm deep strengthened by a steel plate 1 cm thick and 18 cm deep, one on either side of the joist. If the stresses in wood and steel are not to exceed 7.5 MN/m² and 127.5 MN/m², find the moment of resistance of the section of the beam. Take $E_s = 20 \, E_w$.*

Solution. Fig. 5.25 shows the arrangement and also the stress distribution in timber and steel.

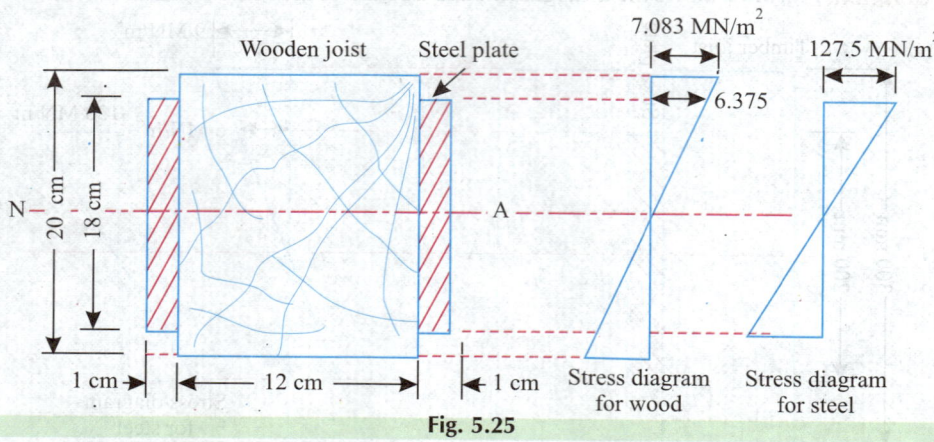

Fig. 5.25

The maximum stress allowed to reach in wood = 7.5 MN/m²

Then, from the stress diagram, the stress in wood at the level of steel

$$= 7.5 \times \frac{9}{10} = 6.75 \, \text{MN/m}^2$$

Since the strain in wood = strain in steel, we have maximum stress in steel $= 6.75 \times \dfrac{E_s}{E_w} =$ $67.5 \times 20 = 135$ MN/m². But the permissible stress in steel is only 127.5 MN/m².

Hence the maximum stress in wood should not be allowed to reach 7.5 MN/m².

Let the maximum stress in steel be allowed to reach 127.5 MN/m².

∴ Stress in steel at 9 cm from the N.A. (neutral axis) = 127.5 MN/m²

∴ Stress in wood at 9 cm from N.A.

$$= \frac{127.5}{20} = 6.375 \text{ MN/m}^2$$

∴ Stress in wood at 10 cm from the N.A.

$$= \frac{10}{9} \times 6.375 = 7.083 \text{ MN/m}^2$$

This stress is less than the safe stress of 7.5 MN/m².

Section modulus of wood,

$$Z_w = \frac{bd^2}{6} = \frac{12 \times 20^2}{6} = 800 \text{ cm}^3 = 800 \times 10^{-6} \text{ m}^3$$

Section modulus of steel,

$$Z_s = \frac{bd^2}{6} = \frac{2 \times 18^2}{6} = 108 \text{ cm}^3 = 108 \times 10^{-6} \text{ m}^3$$

Moment of resistance of *timber section*,

$$M_w = \sigma_w Z_w = (7.083 \times 10^6) \times 800 \times 10^{-6} = 5666 \text{ Nm}$$

Moment of resistance of *steel section*,

$$M_s = \sigma_s Z_s = (127.5 \times 10^6) \times 108 \times 10^{-6} = 13770 \text{ Nm}$$

Total moment of resistance offered by the *composite section*,

$$M = M_w + M_s = 5666 + 13770$$
$$= \textbf{19436 Nm (Ans.)}$$

Example 5.20. *A flitched timber beam made up of steel and timber has a section as shown in Fig. 5.25. Determine the moment of resistance of the beam. Take* $\sigma_s = 100$ *MN/m² and* $\sigma_w = 5$ *MN/m².*

Solution. Refer to Fig.5.26.

Section modulus of wood,

$$Z_w = \frac{bd^2}{6} = \frac{100 \times 200^2}{6}$$
$$= 66.6 \times 10^4 \text{ mm}^3 = 66.6 \times 10^{-5} \text{ m}^4$$

Section modulus of steel,

$$Z_s = \frac{bd^2}{6} = \frac{10 \times 200^2}{6}$$
$$= 66.6 \times 10^3 \text{ mm}^3 = 66.6 \times 10^{-6} \text{ m}^3$$

Moment of resistance of *timber section*,

$$M_w = \sigma_w Z_w = (5 \times 10^6) \times 66.6 \times 10^{-5}$$
$$= 3330 \text{ Nm}$$

Moment of resistance of *steel section*,

$$M_s = \sigma_s Z_s = (100 \times 10^6) \times 66.6 \times 10^{-6}$$
$$= 6660 \text{ Nm}$$

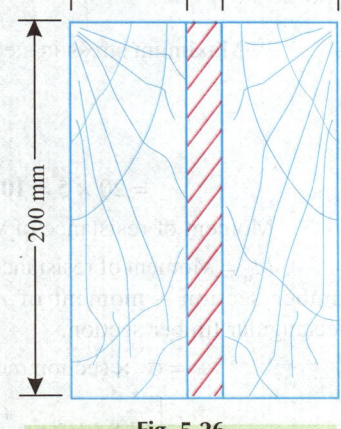

Fig. 5.26

Total moment of resistance offered by the *composite section*,

$$M = M_w + M_s = 3330 + 6660 = \textbf{9990 Nm} \quad \textbf{(Ans.)}$$

Example 5.21. *A flitched beam is made up of two timber joists 10 cm wide and 24 cm deep with a 2 cm thick steel plate 16 cm deep placed symmetrically between them and firmly attached to both. The plate is recessed into groove cut in the inner faces of the joists so that the overall dimensions of the built-up section may be taken as 20 cm × 24 cm.*

Calculate the moment of resistance of the combined section when the maximum bending stress in timber is 7.5 MN/m². What is then the maximum stress in steel?

Take: $E_s = 20\ E_w$, where E_s and E_w are the Young's modulii for steel and timber.

Solution. Refer Fig. 5.27.

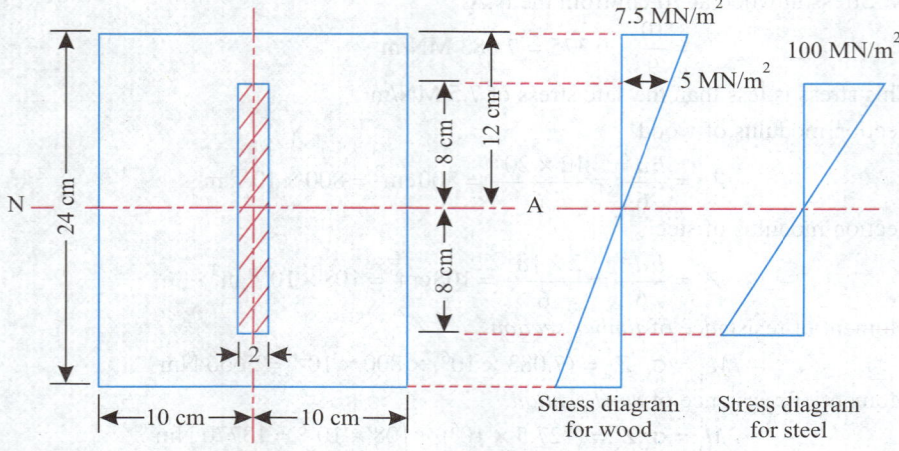

Fig. 5.27

Let the maximum stress in timber (σ_w) be 7.5 MN/m²

Hence stress in timber at 12 cm from N.A. (neutral axis),

$\sigma_w = 7.5$ MN/m²

∴ Stress in wood at 8 cm from the neutral axis,

$\sigma_w' = 7.5 \times 8/12 = 5$ MN/m²

∴ Maximum stress in steel $= m_r \times 5$

$$\left(\text{where, } m_r = \frac{E_s}{E_w} = 20\right)$$

$= 20 \times 5 = \textbf{100 MN/m}^2 \textbf{ (Ans.)}$

Moment of resistance of wood,

M_w = Moment of resistance of 20 cm × 24 cm rectangular timber section – moment of resistance of 2 cm × 16 cm rectangular timber section.

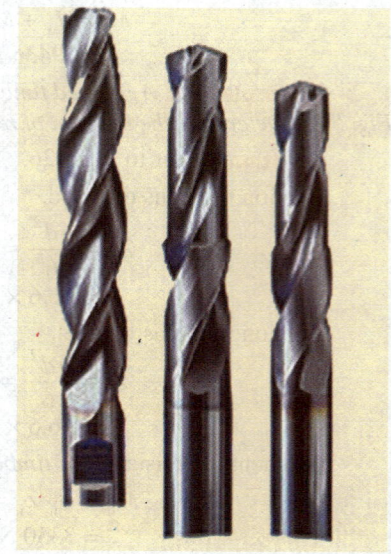

Drilling machine tools.

$= \sigma_w \times$ section modulus $- \sigma_w' \times$ section modulus

$$= \left[(7.5 \times 10^6) \times \frac{0.2 \times 0.24^2}{6}\right] - \left[(5 \times 10^6) \times \frac{0.02 \times 0.16^2}{6}\right]$$

$= 14400 - 426.6 = 13973.4$ Nm

Moment of resistance of steel,

$$M_s = \sigma_s \times \text{section modulus} = (100 \times 10^6) \times \frac{0.02 \times 0.16^2}{6}$$
$$= 8533.3 \text{ Nm}$$

$\therefore$ Total moment of resistance,

$$M = M_w + M_s = 13973.4 + 8533.3 = \textbf{22506.7 Nm} \quad \textbf{(Ans.)}$$

Example 5.22. *A timber beam 16 cm wide and 20 cm deep is to be reinforced by bolting on two steel flitches each 16 cm × 1 cm in section. Find the moment of resistance when:*

(i) The flitches are attached symmetrically at the top and bottom and

(ii) The flitches are attached symmetrically at the sides.

Allowable stress in timber is 6 MN/m². What is the maximum stress in steel in each case?

Take: $E_s = 20 E_w$, where E_s and E_w are Young's moduli for steel and timber/wood.

Solution.

(*i*) When the flitches are attached symmetrically at top and bottom:

Refer to Fig. 5.28.

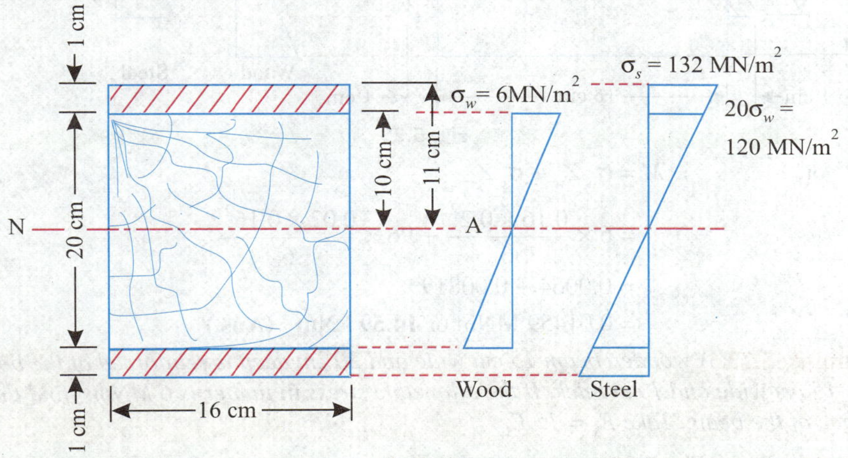

Fig. 5.28

Let, σ_w = Extreme fibre stress for wood, and

σ_s = Extreme fibre stress for steel.

$$\sigma_s = \frac{11}{10} \times (20\sigma_w) = 22\sigma_w$$

$\therefore$ When, $\sigma_w = 6$ MN/m²,

$$\sigma_s = 22 \times 6 = 132 \text{ MN/m}^2$$

Moment of resistance of the section = Moment of resistance of timber section + moment of resistance of steel section

$$= \sigma_s Z_w + \sigma_s Z_s$$
$$= 6 \times \frac{0.16 \times 0.2^2}{6} + \left[132 \times \frac{0.16 \times 0.22^2}{6} - 120 \times \frac{0.16 \times 0.2^2}{6}\right]$$
$$= 0.0064 + (0.1704 - 0.128) = 0.0488 \text{ MNm or } \textbf{48.8 kNm} \quad \textbf{(Ans.)}$$

(ii) When the flitches are attached symmetrically at the sides:

Refer to Fig. 5.29.

In this case when the maximum stress in wood is $\sigma_w = 6$ MN/m^2

The maximum stress in steel is $\sigma_s = \left(\dfrac{8}{10} \times 6\right) \times 20 = 96$ MN/m^2

Moment of resistance of the section,

$\qquad$ M = Moment of resistance of the timber section + moment of resistance of steel section

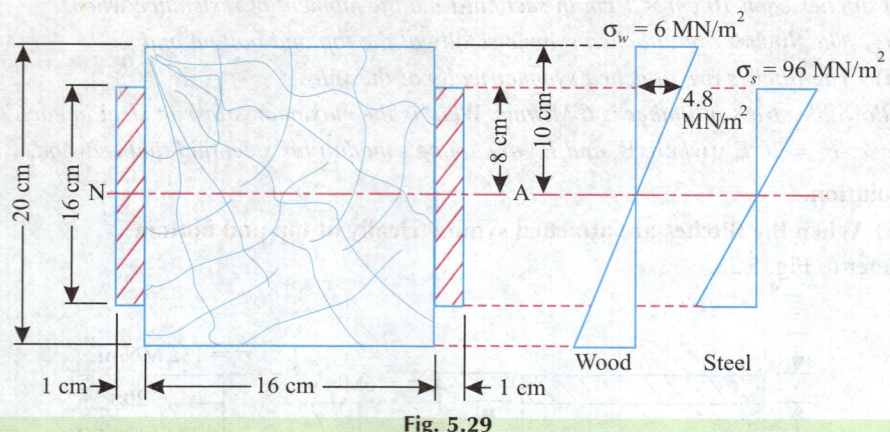

Fig. 5.29

or,$\qquad\qquad$ $M = \sigma_w Z_w + \sigma_s Z_s$

$$= 6 \times \frac{0.16 \times 0.2^2}{6} + 96 \times \frac{0.02 \times 0.16^2}{6}$$

$$= 0.0064 + 0.00819$$

$$= 0.01459 \text{ MNm or } \textbf{14.59 kNm} \quad \textbf{(Ans.)}$$

Example 5.23. *A wooden beam 15 cm wide and 20 cm deep is reinforced at the bottom by a steel plate 15 cm wide and 1 cm thick. If the allowable stress in timber is 6 MN/m^2 find the moment of resistance of the beam. Take $E_s = 15\ E_w$.*

Solution. Refer Fig. 5.30.

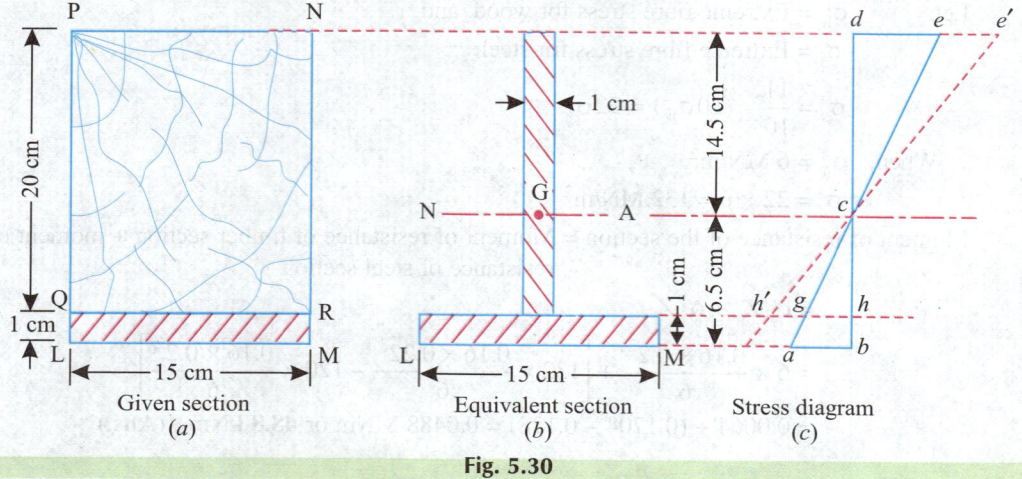

Given section	Equivalent section	Stress diagram
(a)	(b)	(c)

Fig. 5.30

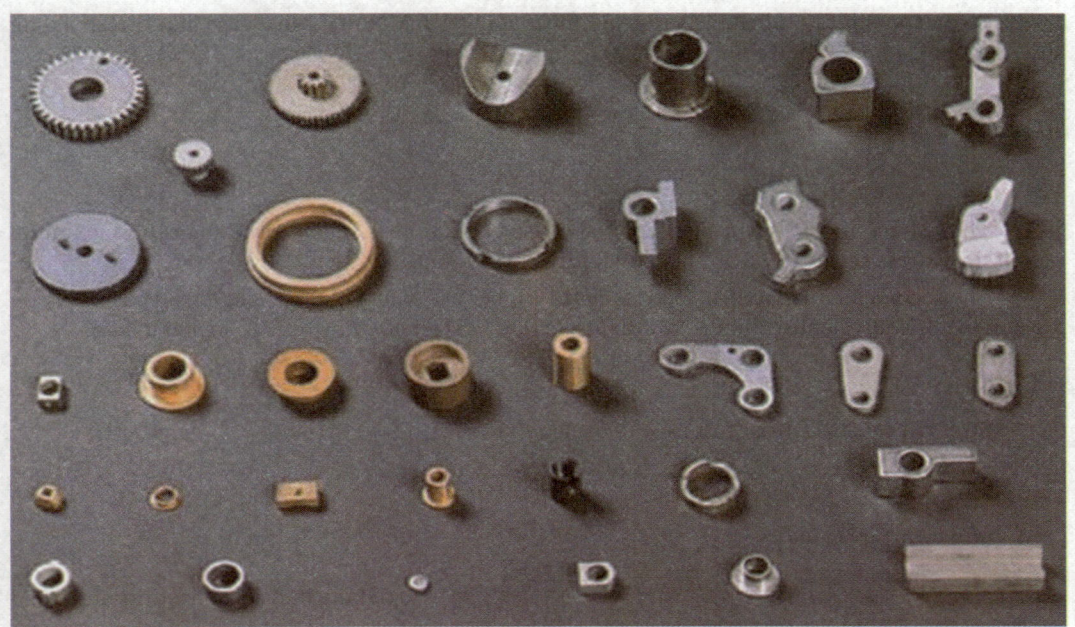

Components made from powder metallurgy. Applicable materials: Iron alloy and copper alloy.

Steel section equivalent to the given wooden section shall be 20 cm deep and:

$$\frac{15}{m_r} = \frac{15}{15} = 1 \text{ cm wide} \qquad \left(\because m_r = \frac{E_s}{E_w} = 15 \right)$$

Calculations for centroid for the equivalent section are tabulated on next page:

Component	Area 'a' of the component (cm²)	Distance 'y' of the centroid from LM (cm)	ay (cm³)
Horizontal rectangle	$15 \times 1 = 15$	$1/2 = 0.5$	$15 \times 0.5 = 7.5$
Vertical rectangle	$20 \times 1 = 20$	$\dfrac{20}{2} + 1 = 11$	$20 \times 11 = 220$
Total	$\Sigma a = 35$	—	$\Sigma ay = 227.5$

Distance of centroid of equivalent section from *LM*,

$$\bar{y} = \frac{\Sigma ay}{\Sigma a} = \frac{227.5}{35} = 6.5 \text{ cm}$$

$$I_{XX} = \left[\frac{15 \times 1^3}{12} + 15 \times 1 \times (6.5 - 0.5)^2 \right] + \left[\frac{1 \times 20^3}{12} + 1 \times 20 \times (14.5 - 10)^2 \right]$$

$$= 541.25 + 1071.66 = 1612.9 \text{ cm}^4 = 1612.9 \times 10^{-8} \text{ m}^4$$

Maximum permissible stress in timber

$$\sigma_w = 6 \text{ MN/m}^2$$

∴ Maximum permissible stress in steel (at the corresponding section in the equivalent beam),

$$\sigma_s = m_r \times 6 = 15 \times 6 = 90 \text{ MN/m}^2 \qquad \left(\text{where, } m_r = \frac{E_s}{E_w} = 15 \right)$$

So one possibility is that the stress in timber at QR is 6 MN/m^2 in that case the corresponding stress in steel is $hh' = 90$ MN/m^2.

Since, $ch = 5.5$ cm and $cd = 14.5$ cm,

$$de' = \frac{14.5}{5.5} \times 90 = 237.3 \text{ MN/m}^2$$

Thus stress in top surface of equivalent steel section is much more than the maximum permissible stress of 90 MN/m^2. This possibility is ruled out.

The second alternative is that stress in the timber at PN is 6 MN/m^2 and thus the stress de in top surface of equivalent section is 90 MN/m^2, stress in steel at QR, (from Δs cgh and cde).

$$gh = \frac{de}{cd} \times ch = \frac{90}{14.5} \times 5.5 = 34.14 \text{ MN/m}^2$$

Stress in steel at LM is (from Δs. abc and cde)

$$ab = \frac{de}{cd} \times bc = \frac{90}{14.5} \times 6.5 = 40.34 \text{ MN/m}^2$$

Moment of resistance,

$$M = 90 \times \frac{I}{y} = 90 \times \frac{1612.9 \times 10^{-8}}{(0.21 - 0.065)} \quad \text{(taking into account stress at } NP \text{ and corresponding value of } y)$$

$$= 0.01 \text{ MNm or } \textbf{10 kNm} \quad \textbf{(Ans.)}$$

$$\left[\text{(or, } 40.34 \times \frac{1612.9 \times 10^{-8}}{0.065} \quad \text{(taking into account stress at } LM \text{ and corresponding value of } y) \right.$$

$$\left. = 0.01 \text{ MNm or } 10 \text{ kNm} \right]$$

5.6 BEAMS OF UNIFORM STRENGTH

Normally the beams which are called upon to carry the loads have the uniform section throughout their length. But where they are loaded the bending moment varies from section to section of beams along their length. The stresses which are in the extreme fibres of the beams above and below the neutral axis also vary from section to section along their length. In this way all extreme fibres which can be loaded upto maximum permissible stress (say σ_{max}) are not loaded to their capacity. Hence in beams of uniform section there is a considerable wastage of material. *If a beam is suitably designed so that every extreme fibre along its length is loaded to its maximum permissible stress by varying the section, it is known as a* **beam of uniform strength**.

If at every section the extreme fibre stress is equal to σ_{max} then *section modulus (Z) of the beam at any section should be proportional to the bending moment at that section.*

For getting the beams of uniform strength the sections of the beam may be **varied by:**

(i) *Keeping the width constant throughout and varying the depth.*

(ii) *Keeping the depth constant throughout the length and varying the width.*

(iii) *By varying both, i.e. width and depth in a suitable manner.*

(Circular beams of uniform strength can be made by varying diameter in such a way that M/Z is constant.)

Examples of beams of uniform strength: *Box girders, plate girders, latticed girders, carriage springs, electric poles of varying diameters,* etc.

Note. The most common way of keeping the beam of uniform strength is by keeping the *width uniform and varying the depth.*

A. Cantilever

Case I: *Cantilever carrying a concentrated load at the free end:*

Consider a cantilever of length *l*, of rectangular section and carrying a concentrated load at the free end.

Bending moment at any section, $M_x = Wx$ (numerically)

Section modulus, $Z = \dfrac{bd^2}{6}$

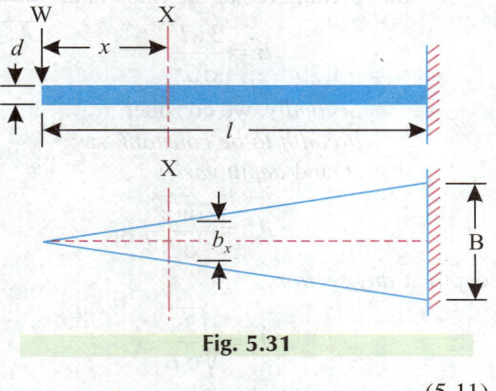

and, $\dfrac{M}{Z} = \dfrac{Wx}{bd^2/6} = \dfrac{6Wx}{bd^2}$

Fig. 5.31

...(5.11)

In this case either *b* is varied and *d* is kept constant throughout or *b* is kept constant and *d* is varied along the length.

Now, $\qquad \sigma = \dfrac{M}{Z} = \dfrac{6W}{d^2}\left(\dfrac{x}{b}\right)$

— To achieve the uniform strength, *b is uniformly increased* from zero at one end to maximum at the fixed end as shown in Fig. 5.31

where, $\qquad b_x = \dfrac{6Wx}{\sigma d^2}$

and, breadth of the fixed end,

$$B = \dfrac{6Wl}{\sigma d^2}$$

— *Secondly,* the width of the cantilever section is constant, say *B*, throughout and *depth varies.*

Then, $\qquad \sigma = \dfrac{6Wx}{Bd^2}$

or, $\qquad d_x = \sqrt{\left(\dfrac{6Wx}{\sigma B}\right)}$

as shown in the Fig. 5.32 and at the fixed end, depth of the section

$$= \sqrt{\left(\dfrac{6Wl}{\sigma B}\right)}$$

Case II. *Cantilever carrying a uniformly distributed load w per unit run:*

Consider a cantilever carrying a U.D.L. *w* per unit run. Bending moment at any section,

$$M_x = \dfrac{wx^2}{2} \text{ (numerically)}$$

— Let the depth of the rectangular section be kept constant as *d* and breadth varies for the uniform strength σ. Then breadth at any section (b_x),

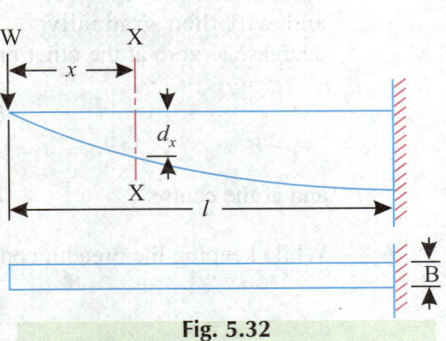

Fig. 5.32

$$b_x = \frac{3wx^2}{\sigma d^2}$$

and, breadth (B) at the fixed end,

$$B = \frac{3wl^2}{\sigma d^2}$$

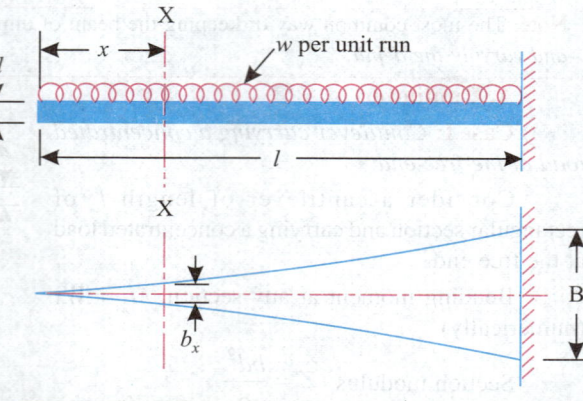

— *Secondly*, we consider *breadth to be constant*, say B, and *depth varies*

i.e. $\qquad d_x^2 = \frac{3wx^2}{\sigma B}$ or

depth at *any section*,

$$d_x = \sqrt{\frac{3w}{\sigma B}} \cdot x$$

and at the fixed end,

$$D = \sqrt{\frac{3w}{\sigma B}} \cdot l$$

(See Fig. 5.34)

Fig. 5.33

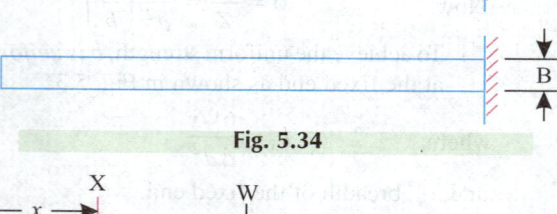

Fig. 5.34

B. Simply supported beam

Case I: *A beam simply supported at the ends and carrying a point load at the mid span:*

Consider a beam of length l, simply supported at the ends and carrying a concentrated load W at the mid span.

Bending moment at any

section $M_x = \dfrac{Wx}{2}$

— If the depth of the section is kept constant (See Fig. 5.35), the breadth of the beam will increase linearly upto the centre of the span and will then gradually

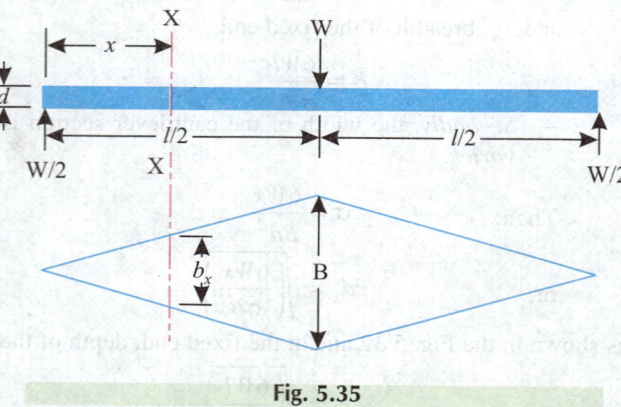

Fig. 5.35

decrease to zero at the other end. Breadth at any section,

$$b_x = \frac{3Wx}{\sigma d^2},$$

and at the centre, $\qquad B = \frac{3W}{\sigma d^2} \times \frac{l}{2} = \frac{3Wl}{2\sigma d^2}$

— While keeping the breadth constant (See Fig. 5.36) as B, depth at any section,

$$d_x = \sqrt{\frac{3Wx}{\sigma B}},$$

and at the centre,

$$D = \sqrt{\frac{3Wl}{2\sigma B}}$$

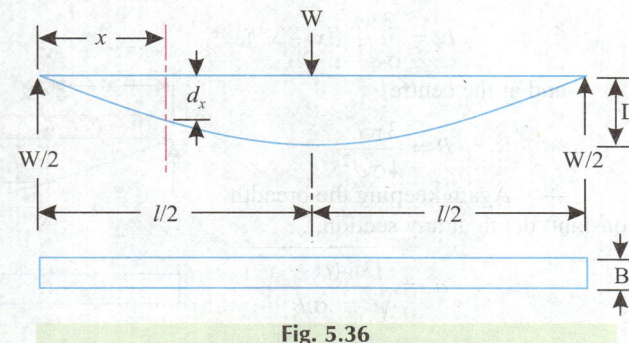

Fig. 5.36

Case II: *A Beam simply supperted at the ends and carrying U.D.L.:*

Consider a beam of length *l*, simply supported at the ends and carrying a U.D.L. of *w* per unit run.

Bending moment at any section, $M_x = \dfrac{wl}{2}x - \dfrac{wx^2}{2} = \dfrac{w}{2}(lx - x^2)$

— Keeping the depth constant (See Fig. 5.37), breadth at any section,

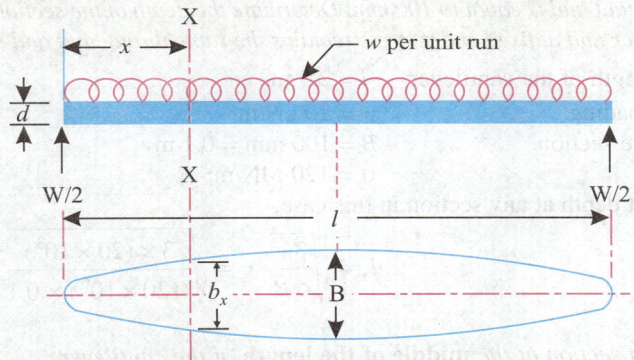

Fig. 5.37

Bridges are constructed with beams.

$$b_x = \frac{3w}{\sigma d^2}(lx - x^2),$$

and at the centre,

$$B = \frac{3wl^2}{4\sigma d^2}$$

— Again keeping the breadth constant, depth at any section,

$$d_x = \sqrt{\frac{3w(lx - x^2)}{\sigma B}}$$

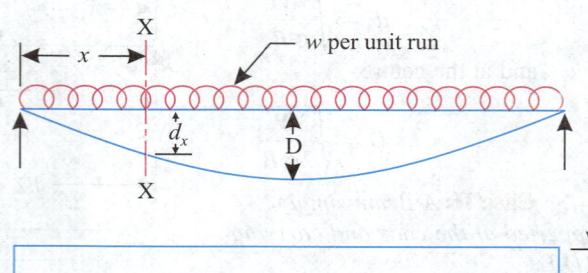

Fig. 5.38

and at the centre,

$$D = \sqrt{\frac{3w}{\sigma B} \cdot \frac{l}{2}}$$

Example 5.24. *A cantilever 2.5 m long carries a U.D.L. of 20 kN/m run. The breadth of the section remains constant and is equal to 100 mm. Determine the depth of the section at the middle of the length of the cantilever and at fixed end if stress remains the same throughout and equal to 120 MN/m².*

Solution. Length of the cantilever, $l = 2.5$ m.

Rate of the loading, $w = 20$ kN/m

Breadth of the section, $B = 100$ mm $= 0.1$ m

Stress, $\sigma = 120$ MN/m²

We know that depth at any section in this case,

$$d_x = \sqrt{\frac{3w}{\sigma B}} \cdot x = \sqrt{\frac{3 \times (20 \times 10^3)}{(120 \times 10^6) \times 0.1}} \cdot x = 0.0707\,x$$

Hence :

(i) *Depth of section at the* **middle of the length** *of the cantilever*

$$= 0.0707 \times \frac{l}{2} = 0.0707 \times \frac{2.5}{2}$$

$$= 0.088 \text{ m or } \textbf{88 mm} \quad \textbf{(Ans.)}$$

(ii) *Depth at the* **fixed end**

$$= 0.0707 \times l = 0.0707 \times 2.5$$

$$= 0.176 \text{ m or } \textbf{176 mm} \quad \textbf{(Ans.)}$$

Example 5.25. *A beam of uniform strength and varying rectangular section is simply supported over a span of 4 metres. It carries a uniformly distributed load of 15 kN/m run. The uniform strength is 100 MN/m². Determine:*

(i) *The depth at a distance of 1.5 m from one end if the breadth is same throughout and equal to 200 mm.*

(ii) *The breadth at the centre of the span if the depth is constant throughout the length of the beam and is equal to 150 mm.*

Solution.

Span, $l = 4$ m

Rate of loading, $w = 15$ kN/m

Uniform strength, $\sigma = 100$ MN/m²

(i) Depth at a distance of 1.5 m from one end, *d*:

Breadth is constant: $B = 200$ mm $= 0.2$ m.   ...(Given)

We know that, in this case, depth at any section is given by,

$$d_x = \sqrt{\frac{3w}{\sigma B}(lx - x^2)}$$

Now when, $x = 1.5$ m

$$d = \sqrt{\frac{3 \times (15 \times 1000)}{(100 \times 10^6) \times 0.2}(4 \times 1.5 - 1.5^2)}$$

$$= 0.092 \text{ m or } \textbf{92 mm} \quad \textbf{(Ans.)}$$

(ii) Breadth at the centre of the span, B:

Depth is constant: $d = 150$ mm $= 0.15$ m   ...(Given)

We know that, in this case, breadth at any section is given by,

$$b_x = \frac{3w}{\sigma d^2}(lx - x^2)$$

Now, when $x = 4/2 = 2$m,

$$B = \frac{3 \times (15 \times 1000)}{(100 \times 10^6) \times 0.15^2}(4 \times 2 - 2^2)$$

$$= 0.08 \text{ m or } \textbf{80 mm} \quad \textbf{(Ans.)}$$

5.7. BIMETALLIC STRIP

If two metal strips, having different co-efficients of thermal expansion and brazed together are heated, the assembly will bend due to change in temperature.

Fig. 5.39 shows a composite bar of rectangular strips of metal A and metal B permanently joined together.

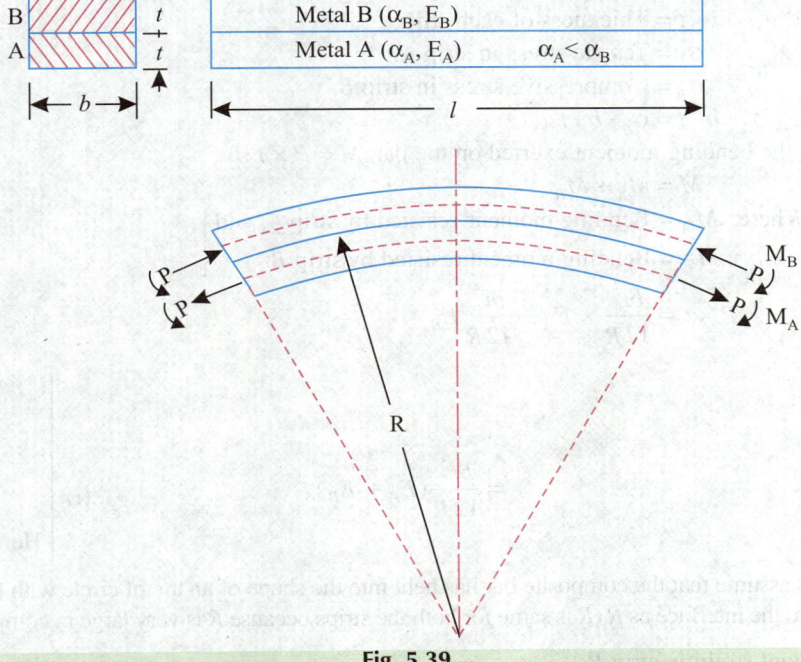

Fig. 5.39

Let, α_A = Co-efficient of linear expansion of metal A,

E_A = Young's modulus of elasticity of metal A,

α_B = Coefficient of linear expansion of metal B,

E_B = Young's modulus of elasticity of metal B,

$T°$ = Rise of temperature,

and, $\alpha_A < \alpha_B$ (or $\alpha_B > \alpha_A$).

When the composite bar is heated through $T°$ it will bend because $\alpha_B > \alpha_A$ and both the strips will deform together introducing compressive stress in metal B and tensile stress in metal A. Because $\alpha_A < \alpha_B$, metal A will exert compressive force on metal B and the interface *reducing* its free expansion of α_b lT and metal B will exert tensile force on metal A and further *increasing* its free expansion α_A lT.

Now, for equilibrium of the composite bar, compressive force on the strip of metal B = Tensile force on the strip of metal A.

A wind turbine annulus gear being machined on CNC boring machine.

If, b = Breadth of each strip,

t = Thickness of each strip,

σ_A = Tensile stress in strip A,

σ_B = Compressive stress in strip B,

Then, $\sigma_A \cdot b \cdot t = \sigma_B \cdot b \cdot t$

Also, the bending moment exerted on the bar, $M = P \times t$

But, $M = M_A + M_B$

$$\left[\begin{array}{l} \text{Where, } M_A = \text{Bending moment resisted by strip } A, \text{ and} \\ \quad M_B = \text{Bending moment resisted by strip } B. \end{array} \right]$$

$$P \times t = \frac{bt^3}{12R} \times E_A + \frac{bt^3}{12R} E_B$$

$$= \frac{bt^3}{12R}(E_A + E_B) \qquad ...(i)$$

$$\left[\begin{array}{c} \dfrac{M}{I} = \dfrac{E}{R} \\[2mm] M = \dfrac{I}{R} E \\[2mm] \text{Here } I = \dfrac{bt^3}{12} \end{array} \right]$$

Let us assume that the composite bar has bent into the shape of an arc of circle with the radius of curvature upto the interface as R (R is same for both the strips because R is very large in comparison to t).

Resultant strain in strip B, $e_B = \dfrac{t}{2R}$

$$\left[\because \frac{M}{I} = \frac{\sigma}{y} = \frac{E}{R} \text{ or } \frac{\sigma}{E} = \frac{y}{R} \text{ or } e = \frac{y}{R} = \frac{t}{2R}\right]$$

where $y = t/2$ in this case.

Resultant strain in strip A, $e_A = -t/2R$

Moreover resultant strain in strip B,

$$e_B = -\frac{P}{btE_B} + \alpha_B T \qquad \left[\because e = \frac{\sigma}{E} = \frac{P/A}{E} = \frac{P}{btE}\right]$$

and, resultant strain in strip A, $e_A = +\dfrac{P}{bt\,E_A} + \alpha_A T$

Difference of strains, $e_B - e_A = \left[\dfrac{t}{2R} - \left(-\dfrac{t}{2R}\right)\right] = \left(-\dfrac{P}{btE_B} + \alpha_B T\right) - \left(\dfrac{P}{bt\,E_A} + \alpha_A T\right)$

or, $\dfrac{t}{R} = (\alpha_B - \alpha_A)T - \dfrac{P}{bt}\left(\dfrac{1}{E_A} + \dfrac{1}{E_B}\right)$

or, $(\alpha_B - \alpha_A)T = \dfrac{t}{R} + \dfrac{P}{bt}\left(\dfrac{1}{E_A} + \dfrac{1}{E_B}\right)$ ...(ii)

Substituting the value of P from (i) in (ii), we get :

$$(\alpha_B - \alpha_A)T = \frac{t}{R} + \frac{bt^2}{12R}(E_A + E_B) \times \frac{1}{bt}\left(\frac{1}{E_A} + \frac{1}{E_B}\right)$$

$$= \frac{t}{R} + \frac{t}{12R}\frac{(E_A + E_B)^2}{E_A E_B} = \frac{t}{R}\left(1 + \frac{E_A^2 + E_B^2 + 2E_A E_B}{12 E_A E_B}\right)$$

$$= \frac{t}{12R E_A E_B}\left(E_A^2 + E_B^2 + 14 E_A E_B\right)$$

∴ Radius of curvature, $R = \dfrac{E_A^2 + E_B^2 + 14 E_A E_B}{12 E_A E_B (\alpha_B - \alpha_A)} \times \dfrac{t}{T}$ ...(5.12)

Example 5.26. *The following data relate to a bimetallic strip made of brass and steel strips:*

Width of each strip,	*b = 12 mm.*
Thickness of each strip,	*t = 2.4 mm.*
Temperature rise,	*T = 120 °C*

$\alpha_B = 19 \times 10^{-6} / °C, \ E_B = 0.9 \times 10^5 \ N/mm^2$

$\alpha_s = 11 \times 10^{-6} / °C, \ E_s = 2 \times 10^5 \ N/mm^2$

The composite strip is initially straight.

Find the radius of the bend.

Solution. Radius of the bend, R:

We know that, $R = \dfrac{E_s^2 + E_B^2 + 14 E_s E_B}{12 E_s E_B (\alpha_B - \alpha_s)} \times \dfrac{t}{T}$

Let, $\dfrac{E_s}{E_B} = m_r = \dfrac{2 \times 10^5}{0.9 \times 10^5} = 2.22$

Then, $R = \dfrac{\left(m_r + \dfrac{1}{m_r}\right) + 14}{12(\alpha_B - \alpha_s)} \times \dfrac{t}{T}$

(Dividing numerator and denominator by $E_s E_B$)

$$= \frac{\left(2.22 + \dfrac{1}{2.22}\right) + 14}{12(19 \times 10^{-6} - 11 \times 10^{-6})} \times \frac{2.4}{120} = \frac{16.67}{12 \times 10^{-6}\,(19 - 11)} \times \frac{2.4}{120}$$

$$= 3473 \text{ mm or } \mathbf{3.473 \ m} \quad (\mathbf{Ans.})$$

5.8. REINFORCED CEMENT CONCRETE (R.C.C.)

Since concrete is extremely weak in tension (though strong in compression) it has to be reinforced where it is subjected to tensile stresses. The reinforcing material for concrete should possess the following *characteristics*:

1. It should have a high tensile strength, high modulus of elasticity.
2. It should have same temperature of co-efficient of expansion and contraction as concrete so that thermal stresses do not develop.
3. It should be cheap and easily available.

Almost all the above requirements are fulfilled by **steel** and thus it is the commonly used as a reinforcing material. The steel employed for reinforcement may be mild steel, medium tensile, tensile steel, hard drawn steel, high tensile steel etc. which have different chemical composition and molecular structures. The commonly used form of reinforcement is round bars, but other sections can also be used. Adhesion of concrete with flats or angle irons is less than on round or square bars. Flat bars are more useful in case of tanks and pipes etc.

Concrete and steel make a very good composite because *cement concrete contracts during setting and firmly grips the steel reinforcement*. Most common mix of concrete is 1 : 2 : 4 (*i.e.* one part of cement, *two* parts of sand and *four* parts of aggregate by volume). The stresses developed in

Inside view of a machine shop.

R.C.C. due to temperature changes are negligible since co-efficient of thermal expansion of steel and concrete is more or less the same.

The following **assumptions** are made in R.C.C. for the design purposes:

1. A transverse section which is plane before bending remains plane after bending.
2. All tensile stresses are taken up by steel reinforcement only and the resistance of concrete to tension is nil.
3. Concrete is homogeneous and elastic. It is also of uniform quality and strength. The stress, therefore, is proportional to strain. It also follows that the stress diagram on the compression side will be triangle, the resultant compressive stress acting at the centre of the triangle.

R.C.C. beam–Rectangular section:

Refer to Fig. 5.40.

Let,
b = Breadth of the section,
d = Depth of the reinforcement,
h = Distance of neutral axis from the compression face,
σ_s = Maximum stress (tensile) developed in steel,
σ_c = Maximum stress (compressive) developed in concrete,
e_t = Tensile strain (developed in steel), and
e_c = Compressive strain (developed in concrete).

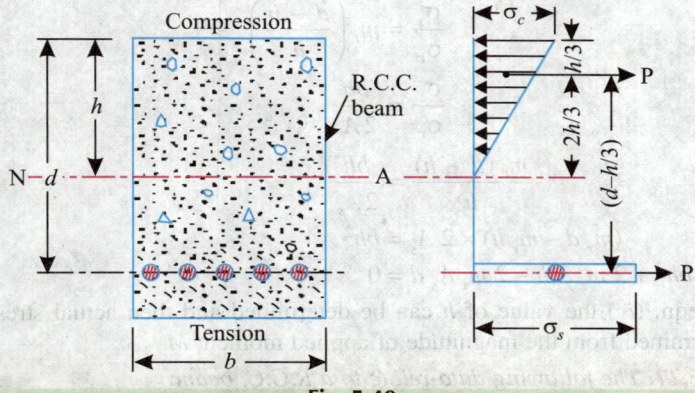

Fig. 5.40

Since the strain in any layer is proportional to its distance from the neutral axis, therefore, $e_c \propto h$ and $e_t \propto (d-h)$

or,
$$\frac{e_c}{e_t} = \frac{h}{d-h} \quad \text{or} \quad \frac{\sigma_c}{E_c} \times \frac{E_s}{\sigma_s} = \frac{h}{d-h}$$

or,
$$\frac{\sigma_s}{\sigma_c} = \frac{E_s}{E_c} \times \frac{d-h}{h}$$

$$\left[\begin{array}{l} \text{where, } E_s = \text{Young's modulus of steel, and} \\ \qquad\quad E_c = \text{Young's modulus of concrete.} \end{array} \right]$$

$$= m_r \times \frac{d-h}{h}$$

$$\left(\text{where, } m_r = \frac{E_s}{E_c} = \text{modular ratio}\right)$$

$\therefore$
$$\frac{\sigma_s}{\sigma_c} = m_r \left(\frac{d-h}{h}\right) \qquad \qquad ...(i)$$

When the beam is under the action of pure bending, then the resultant force P in steel is the same as the resultant force P in concrete, i.e.

$$P = \frac{\sigma_c \, A_c}{2} = \sigma_s \, A_s \left[\begin{array}{l} \text{where,} \quad A_c = b \times h \text{ (for concrete),} \\ \qquad A_s = \text{Area of cross-section of steel reinforcement} \end{array} \right]$$

Since the stress in concrete varies linearly along the depth h, therefore, the mean stress in concrete is taken as $\dfrac{\sigma_c}{2}$

$$\therefore \qquad \frac{\sigma_c}{2}(b \cdot h) = \sigma_s \cdot A_s \qquad \qquad \qquad ...(ii)$$

The resultant compressive force P in concrete and tensile force P in steel form a couple resisting the applied bending moment such that

$$M = P(d - h/3) = \frac{\sigma_c}{2}(b \cdot h)(d - h/3)$$

$$= \sigma_s \cdot A_s \, (d - h/3) \qquad \qquad \qquad ...(iii)$$

— In case allowable stresses (maximum) in steel are given, then we can find σ_s / σ_c and hence determine the value of h for given dimensions of a beam by using eqn. (i).

— Then, by using eqn. (ii) the area of steel reinforcement is determined.

— Finally the moment of resistance can be found by using eqn. (iii). This is known as **"Economic section"** in which the allowable values of stresses in steel and concrete have been realised.

— When the dimensions of the beam i.e. b and d, area of steel reinforcement A_s are given then from eqns. (i) and (ii) the value of h can be determined as follows:

$$\frac{\sigma_s}{\sigma_c} = m_r \left(\frac{d - h}{h} \right) \qquad \qquad \qquad ...(i)$$

or,

$$\frac{\sigma_s}{\sigma_c} = \frac{bh}{2 A_s} \qquad \qquad \qquad ...(ii)$$

or,

$$\frac{m_r \, (d - h)}{h} = \frac{bh}{2 A_s} \qquad \qquad \qquad ...(iii)$$

or,

$$(m_r \, d - m_r \, h) \times 2 \, A_s = bh^2$$

or,

$$bh^2 + 2 \, m_r \, A_s h - 2 \, m_r \, A_s \, d = 0 \qquad \qquad \qquad ...(iv)$$

From the eqn. (iv) the value of h can be determined and then actual stresses in steel and concrete are determined from the magnitude of applied moment M.

Example 5.27. *The following data relate to a R.C.C. beam:*

Width of the beam	*= 200 mm*
Depth of the beam	*= 400 mm*
Maximum allowable stress in steel	*= 144 MN/m²*
Maximum allowable stress in concrete	
	= 90 MN/m²
Modular ratio	*= 16*

Neglecting the weight of steel reinforcement and assuming the density of concrete as 23 kN/m³, find :

   (i)   *The area of steel reinforcement required, if both the stresses are developed and steel reinforcement is 60 mm above the tension face.*

  (ii) *The uniformly distributed load that can be carried over a span of 6 metres.*

Solution.

Width of the beam, $b = 200$ mm

Depth of the beam   = 400 mm

Cover for steel   = 60 mm

Distance of steel reinforcement from the compression face,

$$d = 400 - 60$$
$$= 340 \text{ mm}$$

Allowable stress in steel,

$$\sigma_s = 144 \text{ MN/m}^2$$

Allowable stress in concrete,

$$\sigma_c = 9 \text{ MN/m}^2$$

Modules ratio, $m_r = 16$

(i) Area of steel reinforcement, A_s:

As both the allowable stresses are to be realised, we shall find out the *economic section* of the beam.

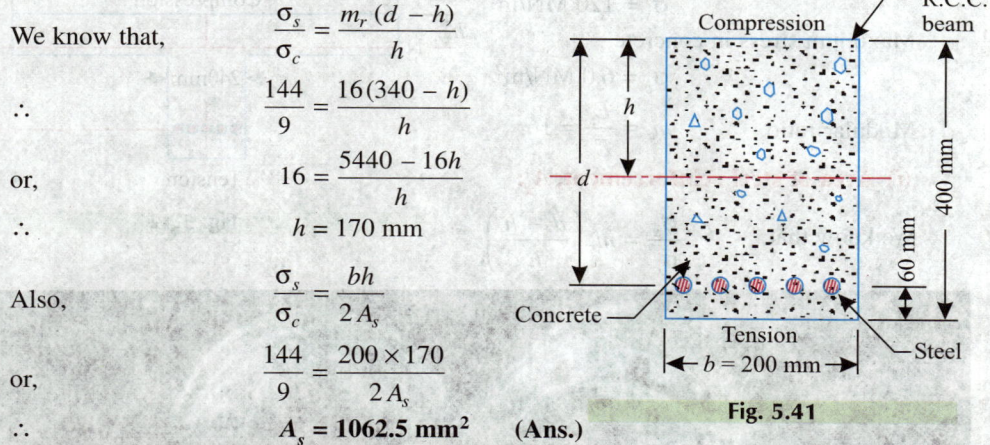

Fig. 5.41

We know that,

$$\frac{\sigma_s}{\sigma_c} = \frac{m_r\,(d - h)}{h}$$

∴

$$\frac{144}{9} = \frac{16(340 - h)}{h}$$

or,

$$16 = \frac{5440 - 16h}{h}$$

∴

$$h = 170 \text{ mm}$$

Also,

$$\frac{\sigma_s}{\sigma_c} = \frac{bh}{2\,A_s}$$

or,

$$\frac{144}{9} = \frac{200 \times 170}{2\,A_s}$$

∴

$$A_s = 1062.5 \text{ mm}^2 \quad \textbf{(Ans.)}$$

(ii) Uniformly distributed load carried by the beam:

Moment of resistance is given by,

$$M = \frac{\sigma_c}{2} bh \left(d - \frac{h}{3} \right) = \left(\frac{9 \times 10^3}{2} \right) \left[(200 \times 10^{-3}) \times (170 \times 10^{-3}) \right.$$

$$\left. \left\{ 340 \times 10^{-3} - \frac{170 \times 10^{-3}}{3} \right\} \right] \text{kNm}$$

$$= 4500 \times 10^{-9} \left[(200 \times 170) \times \left(340 - \frac{170}{3} \right) \right] \text{kNm} = 43.35 \text{ kNm}$$

Maximum bending moment

$$= \frac{wl^2}{8} = \frac{w \times 6^2}{8} = 4.5\,w \text{ kNm} = 43.35$$

$$\left[\begin{array}{l} \text{where, } w = \text{uniformly distributed load in kN/m, and} \\ l = \text{span length} = 6 \text{ m} \end{array} \right]$$

∴

$$w = 9.633 \text{ kN/m}$$

Weight of beam per metre run

$$= (200 \times 10^{-3}) \times (400 \times 10^{-3}) \times 1 \times 23 = 1.84 \text{ kN/m}$$

Hence, uniformly distributed load carried by the beam

$$= 9.633 - 1.84 = \textbf{7.793 kN/m run (Ans.)}$$

Example 5.28. *Fig. 5.42 shows a R.C.C. beam of T-section which has maximum stresses of 6.0 MN/m² and 120 MN/m² in concrete and steel respectively. If the modular ratio of steel and concrete is 16 and the neutral axis lies within the full width of the section, find:*

(i) *Area of steel reinforcement;*

(ii) *Moment of resistance.*

Solution. Let, h = Distance of the neutral axis from the compression face.

Breadth of concrete section in compression,

$$b = 720 \text{ mm}$$

Depth of steel reinforcement from the compression face,

$$d = 360 \text{ mm}$$

Maximum stress in steel,

$$\sigma_s = 120 \text{ MN/m}^2$$

Maximum stress in concrete,

$$\sigma_c = 6.0 \text{ MN/m}^2$$

Modular ratio,

$$m_r = \frac{E_s}{E_c} = 16$$

(*i*) **Area of steel reinforcement, A_s:**

We know that, $\dfrac{\sigma_s}{\sigma_c} = m_r \left(\dfrac{d - h}{h} \right)$

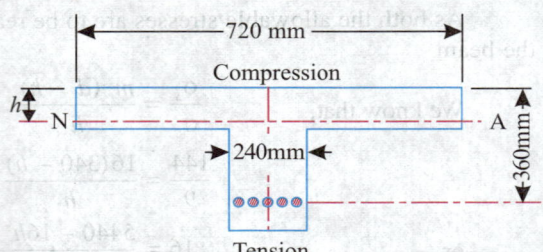

Fig. 5.42

Auto wheel parts.

or, $\dfrac{120}{6} = 16\left(\dfrac{360-h}{h}\right)$

or, $20\,h = 16\,(360-h)$ or $h = 160$ mm

Now, $\dfrac{\sigma_c}{2}\cdot bh = \sigma_s \cdot A_s$

or, $A_s = \dfrac{\sigma_c}{\sigma_s} \times \dfrac{bh}{2} = \left(\dfrac{6}{120}\right) \times \dfrac{720 \times 160}{2} = 2880\,\text{mm}^2$

(*ii*) **Moment of resistance, *M* :**

$M = \sigma_s \cdot A_s\,(d - h/3)$

$= 120 \times 10^3 \times 2880 \times 10^{-6}\left[\left(360 - \dfrac{160}{3}\right) \times 10^{-3}\right]\text{kNm}$

$= 105.98$ kNm **(Ans.)**

TYPICAL EXAMPLES (For Competitive Examinations)

Example 5.29. (*a*) *What are the assumptions made in deriving the flexure formula for symmetrical bending? Which of them is violated in unsymmetrical bending?*

(*b*) *A cantilever beam (Fig.5.43) of length l has a right triangular cross-section and is loaded by a concentrated load P at the end. Calculate the stress at A.*

(AMIE Summer, 2000)

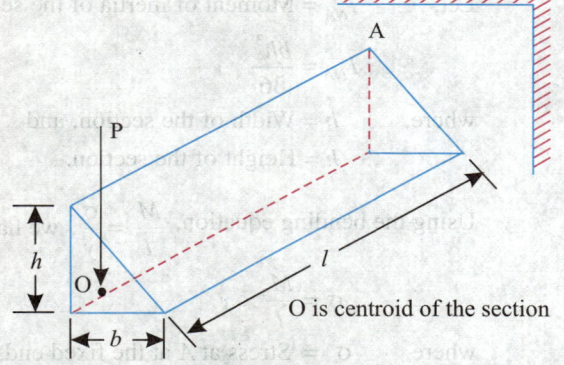

O is centroid of the section

Fig. 5.43

Solution.

(*a*) *Assumptions* made in deriving the flexure formula for symmetrical bending are:

1. The value of the Young's Modulus is the same for the beam material in tension as well as compression.

2. A transverse section of the beam, which is plane before bending will remain plane after bending.

3. The material of the beam is homogeneous and isotropic (Isotropic means having the same elastic properties in all the directions.)

4. The elastic limit is not exceeded.

5. The resultant pull or thrust on a transverse section of the beam is zero.

Violation of the assumption in unsymmetrical bending:

The assumption (2) mentioned above will not be true in unsymmetrical bending, because there exists a shearing force. As such the plane sections in such a case will, therefore, be distorted.

(*b*) Refer Fig. 5.43 and Fig. 5.44 (*i*), (*ii*), (*iii*).

Fig. 5.43 shows the given centilever beam of length *l* and loaded by a concentrated load *P* at the free end.

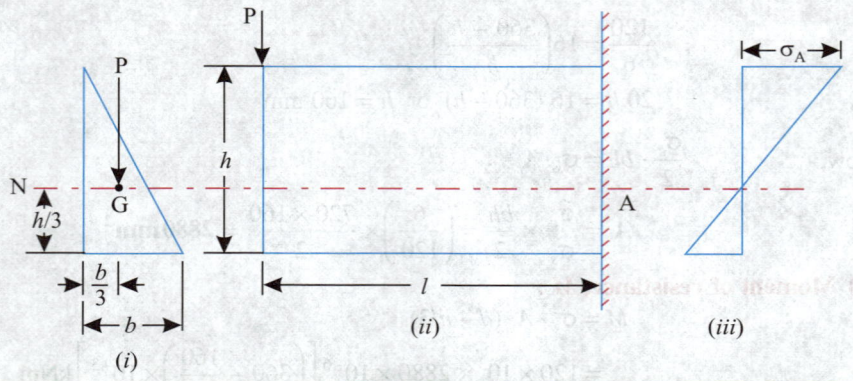

Fig. 5.44

Fig. 5.44 (*i*) shows the right triangular section of the beam. The load *P* acts at the centroid. Fig. 5.44 (*ii*) shows the same beam with its right hand end as fixed and the load *P* acts at its free end.

N.A. is the neutral axis. Fig. 5.44 (*iii*) shows the stress distribution.

Let, I_{NA} = Moment of inertia of the section about N.A.

$$I_{NA} = \frac{bh^3}{36}$$

where,   *b* = Width of the section, and

        h = Height of the section.

Using the bending equation, $\dfrac{M}{I} = \dfrac{\sigma}{y}$, we have

$$\sigma = \frac{M}{I} \times y$$

where,   σ = Stress at *A* at the fixed end,

        M = Bending moment at the fixed end = *P* × *l*, and

        y = Depth of the point *A* from N.A. = $\dfrac{2}{3}h$.

Substituting the values, we get

$$\sigma = \frac{P \times l}{\dfrac{bh^3}{36}} \times \frac{2}{3}h$$

or, $\sigma = \dfrac{P \times l \times 36}{bh^3} \times \dfrac{2}{3}h = \dfrac{\mathbf{24\,Pl}}{\mathbf{bh^2}}$ **(Ans.)**

Example 5.30. *A simply supported beam 100 mm wide and 120 mm deep is 5 m long and carries a load of 8 kN at the mid span. The load is not vertical, but inclined at an angle of 30° to the vertical, the line of action passing through the centroid of the section. Find the locations and magnitudes of maximum tensile and compressive stresses set up due to bending.*

Solution. *Given*: Width of the beam, *b* = 100 mm = 0.1 m

    Depth of the beam, *d* = 120 mm = 0.12 m

    Length of the beam, *l* = 5 m.

Locations and magnitudes of maximum tensile and compressive stresses, $(\sigma_t)_{max}$, $(\sigma_c)_{max}$:

Refer Fig. 5.45. Resolve the force 8 kN into two components-one normal to the beam and the other parallel to the beam.

The component normal to the beam = 8 cos 30° = 6.93 kN.

This component causes the bending of the beam.

Using the relation:

$$\frac{M}{I} = \frac{\sigma}{y} \text{ or } \sigma = \frac{My}{I}$$

where

$$M = \frac{Wl}{4} = \frac{6.93 \times 5}{4} = 8.66 \text{ kNm, and}$$

$$I = \frac{bd^3}{12} = \frac{0.1 \times 0.12^3}{12} = 1.44 \times 10^{-5} \text{ m}^4$$

$$y_t = y_c = \frac{0.12}{2} = 0.06 \text{ m}$$

Substituting the values, we get

$$\sigma_t = \sigma_c = \frac{8.66 \times 0.06}{1.44 \times 10^{-5}} \times 10^{-3} \text{ MN/m}^2 = 36.08 \text{ MN/m}^2$$

Hence, $(\sigma_t)_{max} = (\sigma_c)_{max} = \textbf{36.08 MN/m}^2$ **(Ans.)**

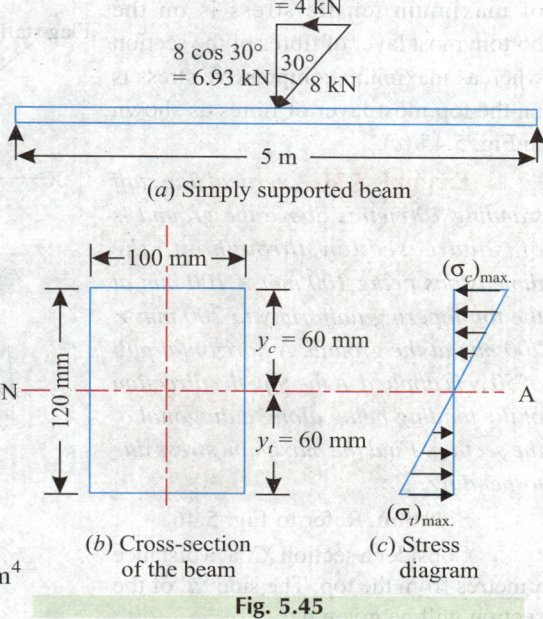

8 sin 30° = 4 kN

8 cos 30° = 6.93 kN 30° 8 kN

5 m

(a) Simply supported beam

100 mm

120 mm

$y_c = 60$ mm

N — A

$y_t = 60$ mm

$(\sigma_c)_{max.}$

$(\sigma_t)_{max.}$

(b) Cross-section of the beam

(c) Stress diagram

Fig. 5.45

Cold-rolled channels being transported on a truck.

It may be noted that the location of maximum tensile stress is on the bottom most layer of fibres of the section whereas maximum compressive stress is on the top most layer of fibres as shown in Fig. 5.45 (c).

Example 5.31. *A vertical flag staff standing 10 metres above the ground is of square section throughout, the dimensions being 100 mm × 100 mm at the top tapering uniformly to 200 mm × 200 mm at the ground. A horizontal pull of 50 N is applied at the top, the direction of the loading being along a diagonal of the section. Find the maximum stress due to bending.*

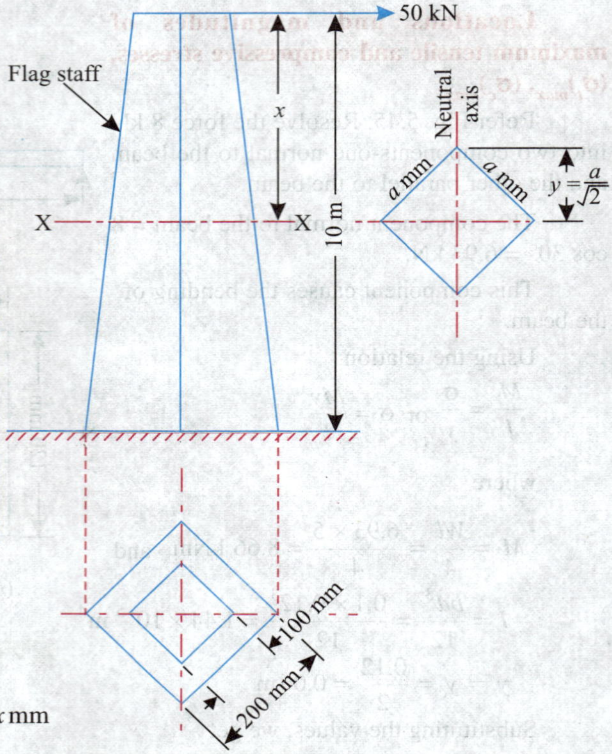

Solution. Refer to Fig. 5.46.

Consider a section *XX* at a distance *x* metres from the top. The side '*a*' of the section will be given by:

$$a = \left(100 + \frac{x}{10} \times 100\right) = 100 + 10x \, \text{mm}$$

If '*d*' is the diagonal, then

Fig. 5.46

$$d = \sqrt{2}\, a = \sqrt{2}\,(100 + 10x) \, \text{mm}$$

B.M. at

$$XX = 50x \, \text{Nm} = 50 \times 10^3 x \, \text{N.mm}$$

$$I_{XX} = \frac{a^4}{12}$$

$$y = \frac{d}{2} = \frac{\sqrt{2}\, a}{2} = \frac{a}{\sqrt{2}} = (100 + 10x)/\sqrt{2}$$

Now using the bending eqn., $\dfrac{M}{I} = \dfrac{\sigma}{y}$, we get

$$\therefore \qquad \sigma = \frac{My}{I} = \frac{(50 \times 10^3 x) \times (100 + 10x) \times 12}{\sqrt{2}\,(100 + 10x)^4} = \frac{60 \times 10^4 x}{\sqrt{2}\,(100 + 10x)^3}$$

The location of the section having maximum bending stress will be found by setting

$$\frac{d\sigma}{dx} = 0$$

$$\therefore \qquad \frac{d\sigma}{dx} = \frac{60 \times 10^4}{\sqrt{2}} \left[\frac{(100 + 10x)^3 - x \times 3\,(100 + 10x)^2 \times 10}{(100 + 10x)^6}\right] = 0$$

i.e., $\qquad (100 + 10x)^3 - 30x\,(100 + 10x)^2 = 0$

or, $\qquad x = 5 \, \text{m}$

$$\therefore \qquad \sigma_{max} = \frac{60 \times 10^4 \times 5}{\sqrt{2}\,(100 + 50)^3} = 0.628 \, \text{N/mm}^2 \; (\text{MN/m}^2) \quad (\textbf{Ans.})$$

Example 5.32. *Prove that the moment of resistance of a beam of square section with its diagonal in the plane of bending is increased by flattening the top and bottom corners as shown in Fig. 5.47 and that the moment of resistance is maximum when y = 8/9 y₁.*

Solution. Refer to Fig. 5.47.

Consider an elementary strip of width *b* and thickness δd at a distance *d* from neutral axis (*XX* axis).

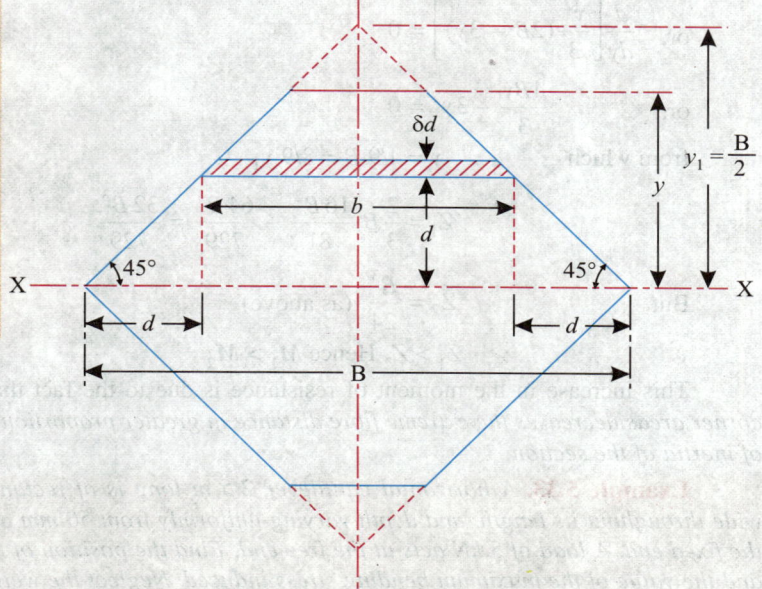

Fig. 5.47

Moment of inertia of elemental strip,

$$\delta I_{XX} = b \cdot \delta d \cdot d^2$$
$$= (B - 2d) \cdot \delta d \cdot d^2$$

$$= (B d^2 - 2d^3)\, \delta d$$

$$\left[\begin{array}{l} \because \ b + 2d = B \\ \text{or } b = B - 2d \\ \ ...\text{Fig.5.47} \end{array} \right]$$

∴ Moment of inertia of the whole section,

$$I_{XX} = 2\int_0^y (Bd^2 - 2d^3)\, \delta d = \frac{y^3}{3}(2B - 3y) = I_1 \ \text{(say)} \qquad ...(i)$$

When $\quad y = \dfrac{B}{2}, \ I_{XX} = \dfrac{B^3}{24}\left(2B - \dfrac{3B}{2}\right) = \dfrac{B^4}{48} = I_2 \qquad \text{(say)} \qquad ...(ii)$

Now $\qquad M_1 = \sigma_1 Z_1$

and $\qquad M_2 = \sigma_2 Z_2$

∴ $M_1 > M_2$ if $Z_1 > Z_2$ for the same value of σ.

$$Z_1 = \frac{I_1}{y} = \frac{I}{y} \times \frac{y^3}{3}(2B - 3y)$$

$$= \frac{y^2}{3}(2B - 3y) = \frac{2}{3}By^2 - y^3$$

$$Z_2 = \frac{I_2}{B/2} = \frac{B^4/48}{B/2} = \frac{B^3}{24}$$

For Z_1 to be *maximum*,

$$\frac{dZ_1}{dy} = 0$$

or, $\dfrac{d}{dy}\left[\dfrac{y^2}{3}(2B - 3y)\right] = 0$

or, $\dfrac{4By}{3} - 3y^2 = 0$

from which $y = 4/9\ B = 8/9\ y_1$ $(\because y_1 = B/2)$

$$Z_1 = \frac{2}{3}B \cdot \frac{16B^2}{81} - \frac{64\,B^3}{729} = \frac{32\,B^3}{729}$$

But, $Z_2 = \dfrac{B^3}{24}$ (as above)

$\therefore$ $Z_1 > Z_2$ Hence $M_1 > M_2$

This increase in the moment of resistance is due to the fact that the *removal of the small corner areas decreases the extreme fibre distance in greater proportion than it reduces the moment of inertia of the section.*

Example 5.33. *A horizontal cantilever 2.5 m long is of rectangular cross-section 50 mm wide throughout its length, and depth varying uniformly from 50 mm at the free end to 150 mm at the fixed end. A load of 3 kN acts at the free end. Find the position of the highest stressed section, and the value of the maximum bending stress induced. Neglect the weight of the cantilever itself.*

Solution. Refer Fig. 5.48

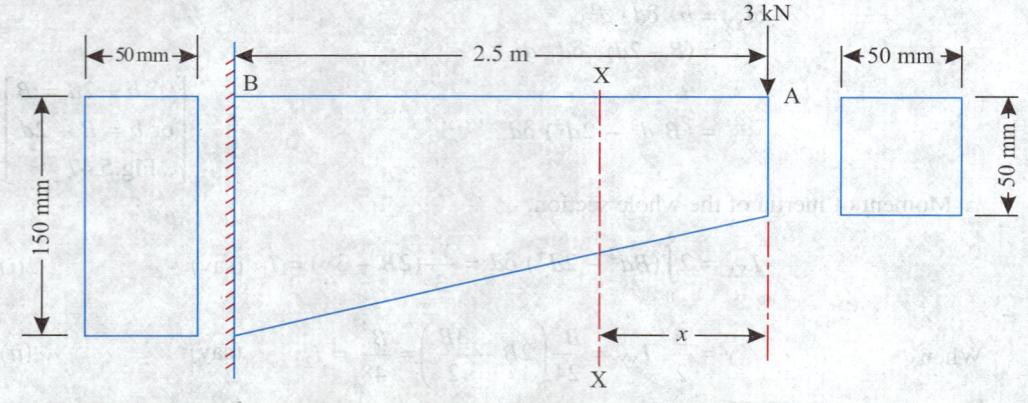

Fig. 5.48

Length of the cantilever, $l = 2.5$ m.

Load at A, $W = 3$ kN $= 3000$ N.

Position of highest stressed section:

Let, $x =$ Distance in metres of the section from A, which is highest stressed.

Now, bending moment at XX,

$$M_x = 3000\ x \text{ Nm or } 3 \times 10^6\ x \text{ N.mm}$$

and, depth of the cantilever at XX,

$$d = 50 + \frac{150 - 50}{2.5}x = 50 + 40\,x \text{ mm}$$

Using the relation: $\dfrac{M}{I} = \dfrac{\sigma}{y}$, we get

$$\sigma = \frac{M \times y}{I} = \frac{3 \times 10^6 \, x \times \left(\dfrac{50 + 40x}{2}\right)}{50 \times (50 + 40x)^3 /12}$$

$$= \frac{3 \times 10^6 \, x \times 6}{50 (50 + 40x)^2} \, \text{N/mm}^2 = \frac{360000x}{(50 + 40x)^2}$$

... (i)

Now for σ to be maximum, differentiate the above equation and equate it zero, i.e.

$$\frac{d\sigma}{dx} = \frac{d}{dx}\left[\frac{360000 \, x}{(50 + 40x)^2}\right]$$

$$= \frac{(50 + 40x)^2 \times 1 - x \times 2(50 + 40x) \times 40}{(50 + 40x)^4} = 0$$

or, $(50 + 40x)^2 = 80x (50 + 40x)$

or, $50 + 40x = 80x$ or **$x = 1.25$ m** (Ans.)

Maximum bendings stress induced:

Now, $$\sigma = \frac{360000 \, x}{(50 + 40x)^2}$$

[Eqn. (i)]

Trusses in a structure.

Inserting $x = 1.25$ m, we get

$$\sigma_{max} = \frac{360000 \times 1.25}{(50 + 40 \times 1.25)^2} = \textbf{45 N/mm}^2 \, \textbf{(MN/m}^2\textbf{)} \textbf{(Ans.)}$$

Example 5.34. *Fig. 5.49 (a) shows the cross-section of a compound bar which has been formed by brazing a brass strip 25 mm × 12 mm to a steel strip 25 mm × 6 mm. The compound bar is bent to a circular shape with neutral axis parallel to ST. Find :*

(i)   The position of the neutral axis and the ratio of the maximum stress in steel to that in the brass.

(ii) The radius of curvature of the neutral surface and resisting moment of the section when the maximum stress in steel becomes 72.6 MN/m².

Take E_s = 200 GN/m² and E_b = 80 GN/m².

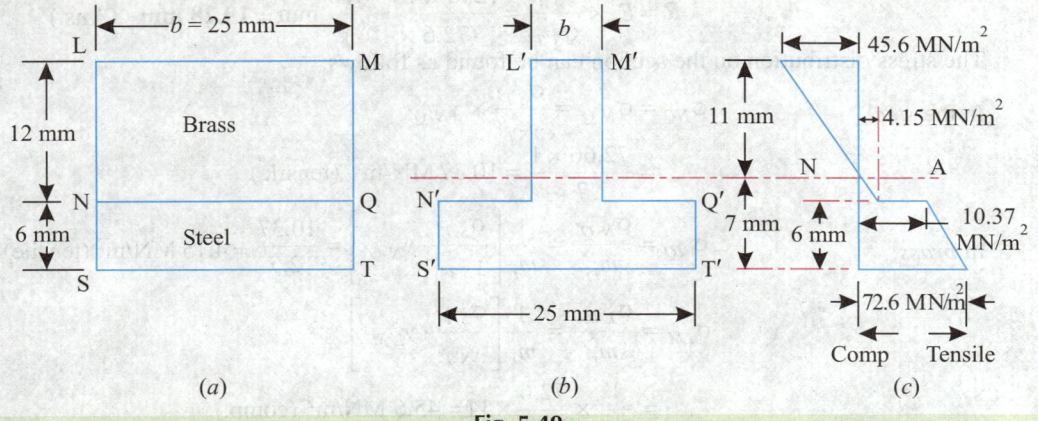

(a)   (b)   (c)

Fig. 5.49

Solution.

Dimensions of brass strip	$= 25 \text{ mm} \times 12 \text{ mm}$
Dimensions of steel strip	$= 25 \text{ mm} \times 6 \text{ mm}$
Modulus of elasticity of steel,	$E_s = 200 \text{ GN/m}^2$
Modulus of elasticity of brass,	$E_b = 80 \text{ GN/m}^2$

Ratio of modulii,

$$m_r = \frac{E_s}{E_b} = \frac{200}{80} = 2.5$$

∴

$$b' = \frac{b}{m_r} = \frac{25}{2.5} = 10 \text{ mm}$$

(i) $\bar{y} = ? \dfrac{\sigma_s}{\sigma_b} = ?$

Taking moments about $S'T'$ (See Fig. 5.49b).

$$25 \times 6 \times 6/2 + 12 \times 10 \times \left(\frac{12}{2} + 6 \right) = (25 \times 6 + 12 \times 10) \bar{y}$$

$$450 + 1440 = 270 \bar{y}$$

∴

$$\bar{y} = 7 \text{ mm (Ans.)}$$

Now from bending eqn.

$$\frac{\sigma}{y} = \frac{E}{R}, \text{ we have}$$

$$\frac{\sigma_{S'T'}}{\sigma_{L'M'}} = \frac{y_{S'T'}}{y_{L'M'}} = \frac{7}{11}$$

But,

$$\sigma_{ST} = \sigma_{S'T'} \text{ and } \sigma_{LM} = \sigma_{L'M'} / m_r$$

∴

$$\frac{\sigma_{ST}}{\sigma_{LM}} \left(i.e. \frac{\sigma_s}{\sigma_b} \right) = \frac{7}{11} \times m_r = \frac{7}{11} \times 2.5 = \mathbf{1.59 \text{ (Ans.)}}$$

It must be noted here that if σ_{ST} were tensile in nature then σ_{LM} would be compressive in nature.

(ii) Radius of curvature R; Resisting moment M:

For evaluating radius of curvature, using the eqn.,

$$\frac{\sigma}{y} = \frac{E}{R}, \text{ we get}$$

$$R = \frac{Ey}{\sigma}$$

and, since the maximum stress in steel occurs at ST or $S'T'$

$$R = E_s \times \frac{y_{S'T'}}{\sigma_{S'T'}} = \frac{(200 \times 10^9) \times 7}{(72.6 \times 10^6)} \text{ mm} = \mathbf{19.28 \text{ mm} \quad (Ans.)}$$

The stress distribution on the section can be found as follows:

In the *steel*,

$$\sigma_{NQ} = \sigma_{N'Q'} = \frac{\sigma_{S'T'}}{y_{S'T'}} \times y_{N'Q'}$$

$$= \frac{72.66 \times 1}{7} = 10.37 \text{ MN/m}^2 \text{ (tensile)}$$

In *brass*,

$$\sigma_{NQ} = \frac{\sigma_{N'Q'}}{m_r} = \frac{1}{m_r} \left[\frac{\sigma_{S'T'}}{y_{S'T']}} \cdot y_{N'Q'} \right] = \frac{10.37}{2.5} = 4.15 \text{ MN/m}^2 \text{ (tensile)}$$

$$\sigma_{LM} = \frac{\sigma_{L'M'}}{m_r} = \frac{1}{m_r} \left[\frac{\sigma_{S'T'}}{y_{S'T'}} \cdot y_{L'M'} \right]$$

$$= \frac{1}{2.5} \times \frac{72.6}{7} \times 11 = 45.6 \text{ MN/m}^2 \text{ (comp.)}$$

The stress distribution diagram is shown in Fig. 5.45 (c)

The resisting moment of the section (M) is given by:

$$\frac{M}{I_{NA}} = \frac{E}{R}$$

[where, I_{NA} = Moment of inertia about the neutral axis (N.A.)]

or,

$$M = \frac{E}{R} \cdot I_{NA}$$

Now,

$$I_{NA} = \left[\frac{25 \times 6^3}{12} + 25 \times 6(7 - 6/2)^2 \right]$$

$$+ \left[\frac{10 \times 12^3}{12} + 10 \times 12 \times (11 - 6)^2 \right]$$

$$= (450 + 2400) + (1440 + 3000)$$

$$= 7290 \text{ mm}^4 = 7290 \times 10^{-12} \text{ m}^4$$

∴

$$M = \frac{200 \times 10^9}{19.28 \times 10^{-3}} \times 7290 \times 10^{-12}$$

$$= 75622 \text{ Nm or } \textbf{75.622 kNm} \textbf{ (Ans.)}$$

Example 5.35. *A metallic tube of internal diameter 25 mm and 5 mm thickness is simply supported on a span of 1 m and the load at the mid span just sufficient to bring the stress to elastic limit level, is found to be 900 N. Four such tubes are firmly clamped to each other to form a single beam, the centres of the tubes forming a square of 35 mm side with two sides of this square horizontal. Calculate the maximum central load which this composite beam can carry if the stress is not to exceed the elastic limit.* **(Panjab University)**

Solution. Refer to Fig. 5.50.

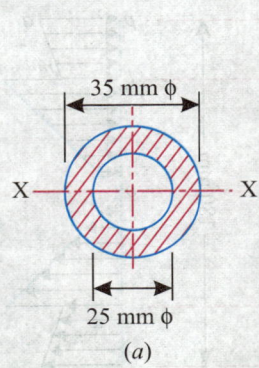

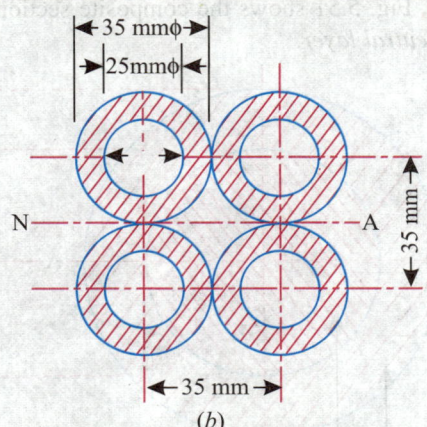

(a)   (b)

Fig. 5.50

Internal diameter of the tube, $d = 25$ mm

Thickness of the tube, $t = 5$ mm

∴ Outside diameter of the tube:

$$D = d + 2t = 25 + 2 \times 5 = 35 \text{ mm}$$

For the single tube with load (P) at mid span,

$$M_{max} = \frac{P \times 1}{4} \text{ Nm} = \frac{900 \times 1}{4} = 225 \text{ Nm}$$

$$\sigma_{max} = \frac{M_{max}}{I_{XX}} \cdot \frac{D}{2}$$

Now, $$I_{XX} = \frac{\pi}{64}(D^4 - d^4) = \frac{\pi}{64}(35^4 - 25^4) = 54487 \text{ mm}^4$$

$$\therefore \qquad \sigma_{max} = \frac{225 \times 1000}{54487} \times \frac{35}{2} = 72.3 \text{ N/mm}^2$$

For the *composite section*, $$I_{NA} = 4\left[\frac{\pi}{64}(35^4 - 25^4) + \frac{\pi}{4}(35^2 - 25^2) \times \left(\frac{35}{2}\right)^2\right]$$

$$= 4(54487 + 144317) = 795216 \text{ mm}^4$$

Thus if P' is the load which can be carried by the composite tube, then using the bending eqn. we have:

$$\frac{M_{max}}{I_{NA}} = \frac{\sigma_{max}}{y}$$

or, $$\sigma_{max} = \frac{M_{max} \times y}{I_{NA}} = \frac{(P' \times 1000/4) \times 35}{795216}$$

or, $$72.3 = \frac{(P' \times 1000/4)}{795216} \times 35 = \frac{P' \times 1000 \times 35}{4 \times 795216}$$

$$\therefore \qquad P' = \frac{72.3 \times 4 \times 795216}{1000 \times 35} = 6570 \text{ N or } \textbf{6.57 kN} \quad \textbf{(Ans.)}$$

Example 5.36. *A steel bar 120 mm in diameter is completely encased in an aluminium tube of 180 mm outer diameter and 120 mm inner diameter so as to make a composite beam. The composite beam is subjected to a bending moment of 15 kNm. Determine the maximum stress due to bending in each material. Take $E_s = 3 E_{al}$.*

Solution. Fig. 5.51 shows the composite section. *Strain in any layer is proportional to its distance from neutral layer.*

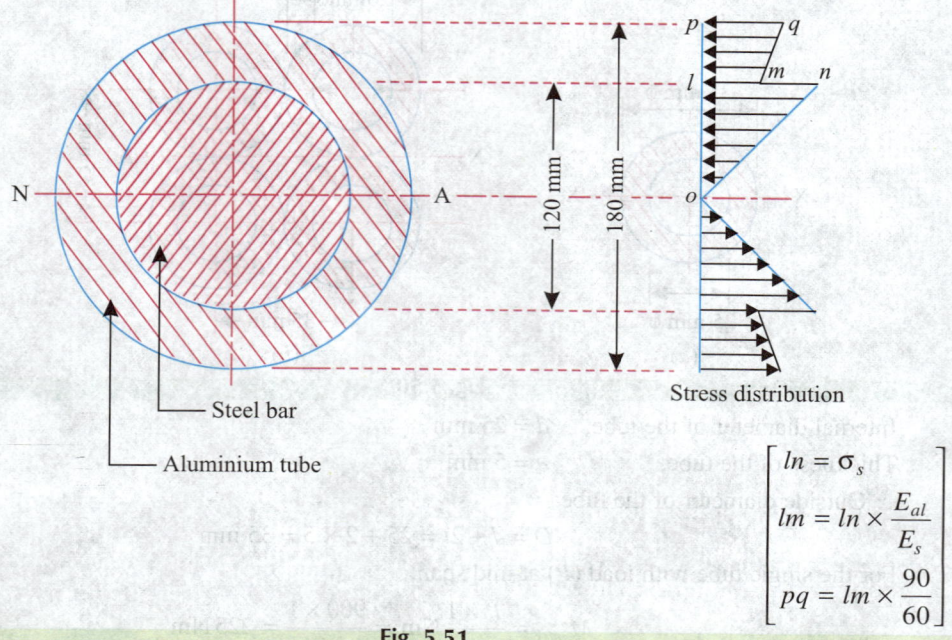

Steel bar

Aluminium tube

Stress distribution

$$\begin{bmatrix} ln = \sigma_s \\ lm = ln \times \dfrac{E_{al}}{E_s} \\ pq = lm \times \dfrac{90}{60} \end{bmatrix}$$

Fig. 5.51

Let, σ_s =   Maximum stress developed in steel,

    σ_{al} =   Maximum stress developed in aluminium, and

    E_s, E_{al} =   Modulii of elasticity for steel and aluminium respectively.

Now, strain in steel at a distance of 60 mm, from N.A.

$$= \frac{\sigma_s}{E_s}$$

and, strain in aluminium layer at a distance of 60 mm from N.A.

$$= \frac{\sigma_s}{E_s}$$

∴ Strain in aluminium layer at a distance of 90 mm from N.A.

$$= \frac{90}{60} \times \frac{\sigma_s}{E_s} = 1.5 \times \frac{\sigma_s}{E_s}$$

Stress in aluminium layer at a distance of 90 mm from N.A. (or the maximum stress)

Power generation flow sleeve of an aircraft.

$$\sigma_{al} = \frac{1.5\sigma_s}{E_s} \times E_{al} = \frac{1.5\sigma_s}{3E_{al}} \times E_{al} = 0.5\sigma_s \quad [\because E_s = 3\, E_{al} \text{ ... given}]$$

Bending moment, $M = M_s + M_{al}$

$$= \sigma_s\, Z_s + \sigma_{al} \times Z_{al} \qquad ...(i)$$

where, M_s =   Moment of resistance of steel bar,

    M_{al} =   Moment of resistance of aluminium tube,

    Z_s =   Section modulus of steel section, and

    Z_{al} =   Section modulus of aluminium section.

Now,
$$Z_s = \frac{\pi \times 120^3}{32} \text{ mm}^3$$
$$= 169646 \text{ mm}^3$$

$$\left[\because Z = \frac{I}{y} = \frac{(\pi/64)\,d^4}{d/2} = \frac{\pi d^3}{32}\right]$$

$$Z_{al} = \frac{\pi(180^4 - 120^4)}{32 \times 180} = 459458 \text{ mm}^4$$

$$\left[\because Z = \frac{I}{y} = \frac{(\pi/64)(D^4 - d^4)}{D/2} = \frac{\pi(D^4 - d^4)}{32 D}\right]$$

Substituting the values in eqn. (i), we get

$$15 \times 10^3 = \sigma_s \times 169646 \times 10^{-9} + \sigma_{al} \times 459458 \times 10^{-9}$$

or, $15 \times 10^3 = \sigma_s \times 169646 \times 10^{-9} + 0.5\,\sigma_s \times 459458 \times 10^{-9}$ [$\because M = 15$ kNm ... given]

$$= \sigma_s \times 10^{-9}\,(169646 + 229729) = \sigma_s \times 399375 \times 10^{-9}$$

∴
$$\sigma_s = \frac{15 \times 10^3}{399375 \times 10^{-9}} \text{ N/m}^2 = \mathbf{37.56\, MN/m^2} \quad \textbf{(Ans.)}$$

and, $\sigma_{al} = 0.5\,\sigma_s = 0.5 \times 37.56 = \textbf{18.78 MN/m}^2$ **(Ans.)**

Example 5.37. *Fig. 5.52 shows a T-beam (120 mm deep) used as a beam with simply supported ends, so that the flange comes under tension. The material of the beam can be subjected to 75 MN/m² in compression and 25 MN/m² in tension. It is desired to achieve a balanced design so that the largest possible bending stresses are reached simultaneously. Find:*

    (i)   The width of the flange.

    (ii) The magnitude of concentrated load that can be applied to the beam at its centre if the span length is 6 metres.

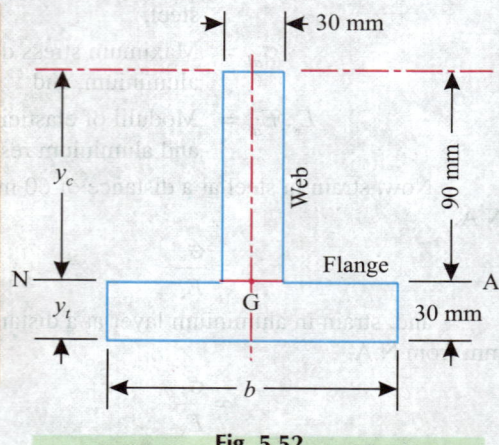

Fig. 5.52

Solution. Tensile stress in the flange,

$$\sigma_t = 25 \text{ MN/m}^2 \ (Given)$$

Compressive stress in the web,

$$\sigma_c = 75 \text{ MN/m}^2 \ (Given)$$

(i) The width of the flange, b:

Since due to bending in a layer stress is *proportional to its distance from the neutral layer*

∴ $\sigma_c \propto y_c$ or $75 \propto y_c$ and $\sigma_t \propto y_t$ or $25 \propto y_t$

But, $y_c + y_t = 120$ mm

∴ $y_c = 90$ mm and $y_t = 30$ mm

This means that neutral axis is passing at the *inter-section of flange and the web.*

Area of web $= 90 \times 30 = 2700 \text{ mm}^2$

Area of flange $= 30 \times b = 30\,b \text{ mm}^2$

Taking moments about the top, we get,

$$y_c = \frac{2700 \times 45 + 30b\,(90+15)}{2700 + 30b}$$

or, $90 = \dfrac{121500 + 3150b}{2700 + 30b}$

or, $90\,(2700 + 30\,b) = 121500 + 3150\,b$

$$243000 + 2700\,b = 121500 + 3150\,b$$

$$450\,b = 121500$$

∴ $b = \textbf{270 mm} \ \ \textbf{(Ans.)}$

(ii) Central load, W:

Moment of inertia,

$$I_{XX} = \left[\frac{30 \times 90^3}{12} + 30 \times 90 \times 45^2\right] + \left[\frac{270 \times 30^3}{12} + 270 \times 30 \times 15^2\right]$$

$$= (1822500 + 5467500) + (607500 + 1822500)$$

$$= 9720000 = 9.72 \times 10^{-6} \text{ m}^4$$

Section modulus, $Z_c = \dfrac{I_{XX}}{y_c} = \dfrac{9.72 \times 10^{-6}}{90 \times 10^{-3}} = 1.08 \times 10^{-4} \text{ m}^3$

Section modulus, $Z_t = \dfrac{I_{XX}}{y_t} = \dfrac{9.72 \times 10^{-6}}{30 \times 10^{-3}} = 3.24 \times 10^{-4} \text{ m}^3$

Bending moment (permissible),

$$M = \sigma_c Z_c = \sigma_t Z_t = 75 \times 1.08 \times 10^{-4} = 0.0081 \text{ MNm} = 8.1 \text{ kNm}$$

Maximum bending moment (at the centre of the beam), $M_{max} = \dfrac{Wl}{4}$

$$\begin{bmatrix} \text{where, } l = \text{span length} = 6\,\text{m, and} \\ W = \text{central load.} \end{bmatrix}$$

or, $8.1 = \dfrac{W \times 6}{4}$ $\qquad \therefore$ **$W = 5.4$ kN (Ans.)**

Example 5.38. *An I-section girder (Fig. 5.53) simply supported over a span of 8 metres carries a uniformly distributed load of 100 kN/m over the entire span. The beam is strengthened by the addition of 15 mm thick flanges, wherever necessary. Determine the length and width of the flange plates such that maximum stress due to bending does not exceed 125 MN/m².*

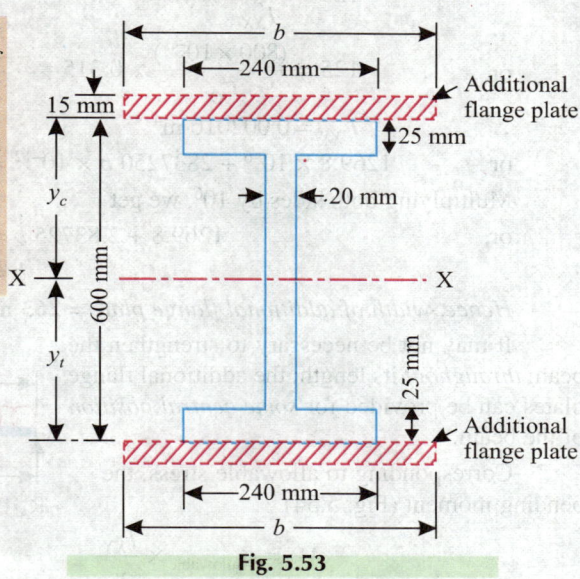

Fig. 5.53

Solution. Span of the beam = 8 m

U.D.L., w = 100 kN/m

Allowable stress, σ = 100 MN/m²

Length and width of flange plates :

As the section is symmetrical, therefore,

$$y_c = y_t$$

Moment of inertia about XX axis,

$$I_{XX} = \left[\frac{240 \times 600^3}{12} - \frac{220 \times 550^3}{12} \right]$$

$$= 10^6 \left[\frac{240}{12} \times 6^3 - \frac{220}{12} \times 5.5^3 \right]$$

$$= 1269.8 \times 10^6 \text{ mm}^4 = 1269.8 \times 10^{-6} \text{ m}^4$$

Maximum bending moment (at the centre),

$$M_{max} = \frac{wl^2}{8} = \frac{108 \times 8^2}{8} = 800\,\text{kNm}$$

Maximum stress developed,

$$\sigma_{max} = \frac{M_{max}}{I_{XX}} \times y_t$$

$$= \frac{800}{1269.8 \times 10^{-6}} \times (300 \times 10^{-3}) \times 10^{-3} \text{ MN/m}^2 = 189\,\text{MN/m}^2$$

Since σ_{max} (= 189 MN/m²) is more than the allowable stress of 125 MN/m², therefore, it is necessary to *add* flange plates to strengthen the section.

Let the width of the flange plates = b

The thickness of the flange plates = 15 mm

Moment of inertia of the section with additional flange plates

$$I'_{XX} = I_{XX} + 2\left[\frac{b \times 15^3}{12} + b \times 15 \times (300 + 7.5)^2\right]$$

$$= I_{XX} + 2\,[281.25\,b + 1418343.8\,b]$$

$$= I_{XX} + 2837250\,b$$

$$= (1269.8 \times 10^{-6} + 2837250\,b \times 10^{-12})\ \text{m}^4$$

$$y'_t = 300 + 15 = 315\ \text{mm} = 0.315\ \text{m}$$

Allowable stress,

$$125 = \frac{M_{max}}{I'_{XX}} \times y'_t$$

or,

$$125 = \frac{(800 \times 10^{-3})}{I'_{XX}} \times 0.315$$

∴

$$I'_{XX} = 0.002016\ \text{m}^4$$

or,

$$1269.8 \times 10^{-6} + 2837250\,b \times 10^{-12} = 0.002016$$

Multiplying both sides by 10^6, we get

or,

$$1269.8 + 2.83725\,b = 0.002016 \times 10^6$$

∴

$$b = 263\ \text{mm}$$

*Hence, width of additional flange plate = **263 mm** (Ans.)*

It may not be necessary to strengthen the beam *throughout* its length; the additional flange plates can be provided for *some central position* of the beam.

Corresponding to allowable stress, the bending moment (Fig. 5.54)

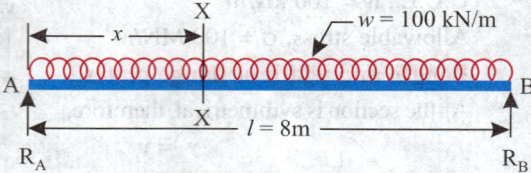

Fig. 5.54

$$M_x = \sigma_{allowable} \times \frac{I_{XX}}{y_t}$$

$$= 125 \times 10^6 \times \frac{1269.8 \times 10^{-6}}{0.3}$$

$$= 0.529 \times 10^6\ \text{Nm or } 529\ \text{kNm}$$

Refer to Fig. 5.54.

Reactions,

$$R_A = R_B = \frac{100 \times 8}{2} = 400\,\text{kN}$$

$$M_x = R_A \times x - \frac{wx^2}{2} = 400 \times x - \frac{100 \times x^2}{2}$$

$$= 400x - 50x^2\ \text{kNm} = 529\ \text{(as above)}$$

i.e.

$$400x - 50x^2 = 529\ \text{ or } x^2 - 8x + 10.58 = 0$$

$$x = \frac{8 \pm \sqrt{64 - 4 \times 10.58}}{2} = \frac{8 \pm 4.656}{2} = 6.328\ \text{m or } 1.672\ \text{m}$$

This shows that with a length of 1.672 m from both the sides, there is no necessity of additional plates because $M_x < 529$ kNm

∴ Length of additional flange plates

$$= 8 - 1.672 - 1.672 = \textbf{4.656 m}\ \ \textbf{(Ans.)}$$

Bimetallic strip

Example 5.39. *A bimetallic strip is formed by using strips of copper and steel. Each strip is 60 mm wide and 12 mm thick. Both the strips are fastened together so that no relative movement can take place between them. The bimetallic strip is now heated through 120°C. Assuming that both the strips bend by the same radius and stresses are transmitted only through the end connections, find :*

(i) *Radius of the bend.*

(ii) *Maximum tensile and compressive stresses in both.*

Given: $E_s = 200 \ GN/m^2;$ $E_c = 100 \ GN/m^2;$

       $\alpha_s = 11 \times 10^{-6}/°C;$ $\alpha_c = 18 \times 10^{-6}/°C.$

Solution. Thickness of strip, $t = 12$ mm

Width of strip, $b = 60$ mm

Temperature rise, $T = 120°C$

(*i*) Radius of the bend, *R*:

We know that the radius of the bend is given by:

$$R = \frac{E_s^2 + E_c^2 + 14 E_s E_c}{12 E_s E_c (\alpha_c - \alpha_s)} \times \frac{t}{T} \qquad \text{...[Eqn. (5.12)]}$$

where, $\dfrac{E_s}{E_c} = m_r = \dfrac{200}{100} = 2;$

then, $R = \dfrac{m_r + 1/m_r + 14}{12(\alpha_c - \alpha_s)} \times \dfrac{t}{T}$

(Dividing numerator and denominator by $E_s E_c$)

$$= \frac{2 + \dfrac{1}{2} + 14}{12(18 \times 10^{-6} - 11 \times 10^{-6})} \times \frac{12}{120}$$

$$= 19643 \text{ mm or } \textbf{19.643 m (Ans.)}$$

(*ii*) Maximum tensile and compressive stresses in the strips :

Because $\alpha_c > \alpha_s$ *tensile stress will be developed in steel and compressive stress in copper.*

Compressive force in copper = Tensile force in steel (due to temperature rise) *i.e.*

i.e. $P_c = \dfrac{bt^2}{12 R}(E_s + E_c)$

19th century pistol.

Direct stresses:

Direct compressive stress in copper

$$= \frac{P_c}{bt} = \frac{t}{12R}(E_s + E_c)$$

$$= \frac{12}{12 \times 19643}(200 \times 10^9 + 100 \times 10^9) \times 10^{-6} = 15.27 \, \text{MN/m}^2$$

i.e. Direct stress in copper,

$$(\sigma_c)_d = 15.27 \, \text{MN/m}^2 \, (\text{compressive})$$

∴ Direct stress in steel

$$(\sigma_s)_d = 15.27 \, \text{MN/m}^2 \, (\text{tensile})$$

Stresses due to bending:

Bending moment shared by copper,

$$M_c = \frac{E_c}{R} \times I_c = (\sigma_c)_b \times \frac{bt^2}{2}$$

Stress due to bending,

$$(\sigma_c)_b = \pm \frac{E_C}{R} \times \frac{bt^3}{12} \times \frac{6}{bt^2} \qquad \left[\because I_c = \frac{bt^3}{12} \right]$$

$$= \pm \frac{E_c \, t}{2R} = \pm \frac{100 \times 10^9 \times 12}{2 \times 19643} \times 10^{-6} = \pm 30.54 \, \text{MN/m}^2$$

Maximum stress developed in copper strip

$$= (\sigma_c)_b + (\sigma_c)_d = 30.54 + 15.27$$

$$= \textbf{45.81 MN/m}^2 \, (\textit{compressive}) \quad \textbf{(Ans.)}$$

Minimum stress developed in copper strip

$$= (\sigma_c)_b - (\sigma_c)_d = 30.54 - 15.27 = \textbf{15.27 MN/m}^2 \, (\textit{tensile}) \quad \textbf{(Ans.)}$$

Similarly, bending stress in steel,

$$(\sigma_s)_b = \pm \frac{E_s \cdot t}{2R} = \frac{200 \times 10^9 \times 12}{2 \times 19643} \times 10^{-6} = \pm 61.09 \, \text{MN/m}^2$$

Maximum stress in steel strip

$$= (\sigma_s)_b + (\sigma_s)_d = 61.09 + 15.27$$

$$= \textbf{76.36 MN/m}^2 \, (\textit{tensile}) \quad \textbf{(Ans.)}$$

Minimum stress in steel strip

$$= (\sigma_s)_b - (\sigma_s)_d = 61.09 - 15.27$$

$$= \textbf{45.82 MN/m}^2 \, (\textit{compressive}) \quad \textbf{(Ans.)}$$

HIGHLIGHTS

1. Bending equation is given as follows : $\dfrac{M}{I} = \dfrac{\sigma}{y} = \dfrac{E}{R}$

    where, $M =$ Moment of resistance,

    $I =$ Moment of inertia of the section about neutral axis (N.A.),

    $E =$ Young's modulus of elasticity,

    $R =$ Radius of curvature of neutral axis, and

    $\sigma =$ Bending stress.

2. Section modulus (Z) is given by the relation, $Z = \dfrac{I}{y}$

3. In case of a flitched beam:

$$M = M_1 + M_2 \qquad \qquad \text{...(i)}$$

$$\frac{M_1}{E_1 I_1} = \frac{M_2}{E_2 I_2} \qquad \qquad \text{...(ii)}$$

$$\frac{\sigma_1}{y_1 E_1} = \frac{\sigma_2}{y_2 E_2} \qquad \qquad \text{...(iii)}$$

4. For getting the beam of uniform strength the sections of the beam may be varied by:
   - (i) Keeping the width constant throughout and varying the depth.
   - (ii) Keeping the depth constant throughout the length and varying the width.
   - (iii) By varying both, *i.e.* width and depth in a suitable manner.

   The most common way of keeping the beam of uniform strength is by *keeping the width uniform and varying the depth.*

5. *Bimetallic Strip*:

   Radius of curvature, $R = \dfrac{E_A^2 + E_B^2 + 14 E_A E_B}{12 E_A E_B (\alpha_B - \alpha_A)} \times \dfrac{t}{T}$

OBJECTIVE TYPE QUESTIONS

Choose the Correct Answer:

1. The .... establishes a relationship between the radius of curvature to which the beam bends, the bending moment, the bending stresses and its (beam) cross-sectional dimensions.
   - (a) bending equation
   - (b) torsion equation
   - (c) either (a) or (b)
   - (d) none of the above.

2. In theory of bending which of the following assumptions is made?
   - (a) The material of the beam is perfectly homogeneous throughout.
   - (b) The stress induced is proportional to the strain and at no place the stress exceeds the elastic limit.
   - (c) The value of modulus of elasticity (E) is same, for the fibres of the beam under compression or under tension.
   - (d) The loads are applied in the plane of bending
   - (e) All of the above.

3. The bending equation is written as
   - (a) $\dfrac{I}{M} = \dfrac{\sigma}{y} = \dfrac{E}{R}$
   - (b) $\dfrac{M}{I} = \dfrac{\sigma^2}{y} = \dfrac{E^2}{R^2}$
   - (c) $\dfrac{M}{I} = \dfrac{\sigma}{y} = \dfrac{E}{R}$
   - (d) $\dfrac{M^2}{I} = \dfrac{\sigma^2}{y} = \dfrac{E^2}{R}$.

4. The strength of the beam mainly depends on
   - (a) bending moment
   - (b) c.g. of the section
   - (c) section modulus
   - (d) its weight.

5. In case of a circular section the section modulus is given as
   - (a) $\dfrac{\pi d^2}{16}$
   - (b) $\dfrac{\pi d^3}{16}$
   - (c) $\dfrac{\pi d^3}{32}$
   - (d) $\dfrac{\pi d^4}{64}$

6. For getting the beams of uniform strength the sections of the beam may be varied by
   - (a) keeping the width constant throughout and varying the depth
   - (b) keeping the depth constant throughout the length and varying the width
   - (c) varying both, *i.e.* width and depth in a suitable manner
   - (d) All of the above.

7. Circular beams of uniform strength can be made by varying diameter in such a way that
   - (a) $\dfrac{M}{Z}$ is constant (b) $\dfrac{\sigma}{y}$ is constant
   - (c) $\dfrac{E}{R}$ is constant (d) $\dfrac{M}{R}$ is constant.

8. The most common way of keeping the beam of uniform strength is by
   - (a) keeping the width uniform and varying the depth
   - (b) keeping the depth uniform and varying the width
   - (c) varying both width and depth
   - (d) none of the above.

UNSOLVED EXAMPLES

1. A 10 m long cast iron pipe of 400 mm inside diameter and 25 mm wall thickness runs full of water and is supported at its ends. Determine maximum stress intensity in metal if density of cast-iron is 7200 kg/m^3 and that of water is 1000 kg/m^3. [**Ans.** 13.61 MN/m^2]

2. A water main of 1.2 m internal diameter and 12 mm thick is running full. If the bending stress is not to exceed 56 MN/m^2, find the greatest span on which the pipe may be freely supported. Take densities of steel and water as 7680 kg/m^3 and 1000 kg/m^3 respectively. [**Ans.** 20.33 m]

3. A rectangular beam 20 cm deep by 10 cm wide is subjected to maximum bending moment of 500 kNm. Determine the maximum stress in the beam. If the value of E for the material is 200 GN/m^2, find out the radius of curvature for that portion of the beam where the bending moment is maximum. [**Ans.** 750 MN/m^2, 26.67 m]

4. An unequal angle bar 150 mm × 75 mm, thickness of metal 8 mm, is used as joist freely supported over a span of 3 metres with its longer leg placed vertically. Find the safe uniformly distributed load that it can carry if the maximum permissible bending stress is 120 MN/m^2. Also calculate the maximum tensile stress under the load. [**Ans.** 4.535 kN/m, 65.7 MN/m^2]

5. The moment of inertia of a symmetrical section of a beam about its neutral axis is 2640 cm^4 and its depth is 20 cm. Determine the longest span over which, when simply supported, the beam would carry a uniformly distributed load of 6 kN/m run without the stress due to bending exceeding 120 MN/m^2. [**Ans.** 5.93 m]

6. Find the dimensions of a timber joist of span 10 m to carry a brick wall 20 cm thick and 4 metre high, if the density of brickwork is 1900 kg/m^3 and the maximum permissible stress is limited to 8 MN/m^2. The depth of joist is to be twice its width. [**Ans.** b = 330 mm d = 660 mm]

7. A floor has to carry a load of 15 kN per square metre. The floor is supported on the floor joists 30 cm × 10 cm and 5 m long. Determine the spacing of the joists from centre to centre such that bending stress does not exceed 8.5 MN/m^2. [**Ans.** 272 mm]

8. A beam of *I*-section shown in Fig. 5.55 is subjected to a bending moment of 10 kNm at its neutral axis. Find the maximum stress induced in the beam. [**Ans.** 61.64 MN/m^2]

9. A *T*-section beam having flange 2 cm × 10 cm and web 10 cm × 2 cm is simply supported over a span of 6 m. It carries a U.D.L. of 3 kN/m run including its own weight over its entire span; together with a load of 2.5 kN at mid span. Find the maximum tensile and compressive stresses occurring in the beam section. [**Ans.** 12.94 MN/m^2 (comp.); 25.87 MN/m^2 (tensile)]

10. Find the dimensions of the strongest rectangular beam that can be cut out of log of a wood 180 mm diameter. [**Ans.** b = 103.9 mm, d = 147 mm]

11. An *I* section girder 12 cm deep has the following cross-sectional dimensions: *Top flange* 6 cm wide by 1 cm thick, *bottom flange* 12 cm wide by 1 cm thick, *web* 1 cm thick. The girder is 5 m long simply supported over a span of 3 m, overhanging both supports by the same amount and it carries a concentrated load of 2 kN each end. Find the maximum stress in the material due to bending. [**Ans.** 24.32 MN/m^2]

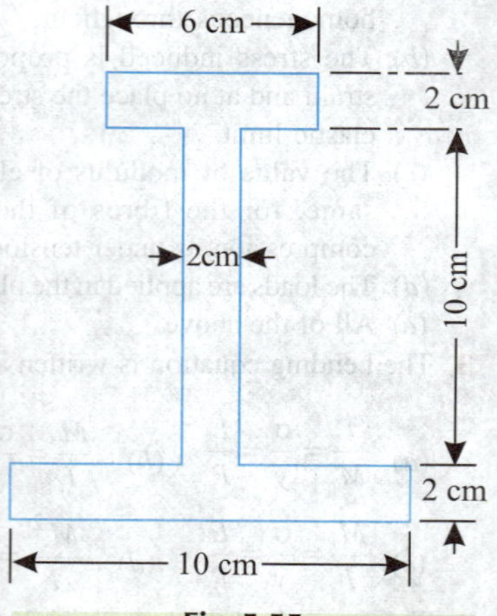

12. A beam of a *T*-section is used as a cantilever with the flange uppermost. The flange is 12 cm wide and 2 cm deep and the web is 1.5 cm wide and 12 cm deep whilst the cantilever is 2 m long.

Fig. 5.55

Determine the maximum permissible load which may be suspended from the end of the cantilever if the limiting stresses in tension and compression are 90 and 150 MN/m² respectively.

[**Ans.** 5.46 kN]

13. A cast iron beam section is of *I*-section with a top flange 8 cm × 2 cm thick, bottom flange 16 cm × 4 cm thick and the web 20 cm deep and 2 cm thick. The beam is freely supported on a span of 5 metres. If the tensile stress is not to exceed 20 MN/m², find the safe uniformly distributed load which the beam can carry. Find also the maximum compressive stress.

[**Ans.** 6.82 kN/m, σ_c = 37.34 MN/m²]

14. A horizontal beam of section shown in Fig. 5.56 is 3 m long and is simply supported at the ends. Find the maximum U.D.L. it can carry if the compressive and tensile stresses must not exceed 56 MN/m² and 30 MN/m² respectively.

[**Ans.** 6.55 kN/m]

15. A beam is made up of two wooden joists each 15 cm × 24 cm deep with a steel plate 24 cm × 1 cm thick sandwiched between them symmetrically and firmly united to them. Calculate the moment of resistance of the beam taking safe bending stresses for wood and steel as 10 MN/m² and 140 MN/m² respectively; modular ratio of steel and wood being 15.

[**Ans.** 26.88 kNm]

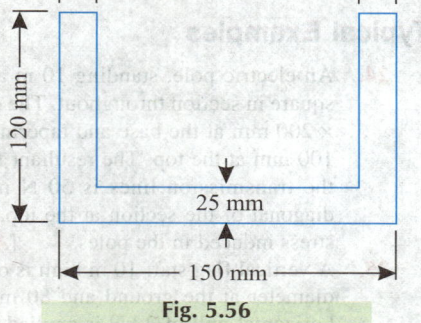

Fig. 5.56

16. A timber beam 30 cm × 20 cm is to be flitched by two steel plates 30 cm × 1 cm to be firmly fixed one on each side of the vertical faces of the beam. Calculate the moment of resistance of the beam taking safe working stresses for timber and steel as 15 MN/m² and 300 MN/m² ; modular ratio 20.

[**Ans.** 135 kNm]

17. A wooden beam 40 cm × 30 cm is to be flitched by two steel flitches of 30 cm × 2 cm to be firmly fixed one on each side of the vertical faces of the beam. Calculate the ratio of moment of resistance of timber and steel. Given E_s/E_w = 15.

[**Ans.** 1.185]

18. A flitched beam is made of two timber joists each 8 cm wide by 24 cm deep and a steel plate 1.5 cm thick and 16 cm deep, placed symmetrically between them and firmly fixed in place. Taking modulus of elasticity of steel as 20 times that of timber, find the maximum intensity of stress in steel if it is not to exceed 6 MN/m² in timber. Also determine the total amount of resistance of the composite beam.

[**Ans.** 80 MN/m², 14.336 kNm]

19. A timber beam 15 cm wide and 20 cm deep is to be reinforced by bolting on two steel flitches each 15 cm by 1.25 cm in section. Find the moment of resistance when (*i*) the flitches are attached symmetrically at top and bottom; and (*ii*) the flitches are attached symmetrically at the sides. Allowable stress in timber is 6 MN/m². What is the maximum stress in steel in each case? Take $E_s = 20\ E_w$.

[**Ans.** (*i*) σ_s = 135 MN/m², *M* = 56.86 kNm (*ii*) σ_s = 90 MN/m², *M* = 14.43 kNm]

20. A flitched beam consists of a wooden joist 12 cm wide and 20 cm deep strengthened by a steel plate 1 cm thick and 18 cm deep, one on either side of the joist. If the stresses in wood and steel are not to exceed 7 MN/m² and 120 MN/m², find the moment of resistance of the section of the beam. Take $E_s = 20\ E_w$.

[**Ans.** 18.29 kNm]

Beams of uniform strength

21. A cantilever 2.5 m long carries a load of 20 kN at the free end. The cantilever is of rectangular section with constant breadth of 50 mm but of variable depth. So as to have a cantilever of uniform strength determine the depth at intervals of 0.5 m. from the free end if the uniform strength is 100 N/mm² (100 MN/m²).

[**Ans.** 0, 109.5 mm, 154.9 mm, 189.7 mm, 219 mm, 244.9 mm]

22. A beam of 4 m length and simply supported at the ends carries a point load of 40 kN at the centre. The beam is of rectangular section with uniform breadth 100 mm throughout. If the beam has uniform strength throughout and is equal to 120 MN/m², determine the depth of the section at quarter spans from the ends and at mid point.

[**Ans.** 100 mm, 141.4 mm]

Bimetallic strips

23. The following data pertain to a bimetallic strip made of copper and steel strips:
    Width of each strip, $b = 50$ mm
    Thickness of each strip, $t = 15$ mm
    Temperature rise, $T = 100°C$,
    $\alpha_c = 18 \times 10^{-6}/°C$, $E_c = 1 \times 10^5$ N/mm^2
    $\alpha_s = 11 \times 10^{-6}/°C$, $E_s = 2 \times 10^5$ N/mm^2
    The composite strip is initially straight. Find the radius of
the bend. [**Ans.** 29.46 m]

Typical Examples

24. An electric pole, standing 10 m above the ground level, is square in section throughout. The dimensions being 200 mm × 200 mm at the base and tapering uniformly to 100 mm × 100 mm at the top. The resultant force due to the tension in the transmission lines is 50 N in the direction along the diagonal of the section at the top. Determine the maximum stress induced in the poles. [**Ans.** $\sigma_{max} = 0.628$ MN/m^2]

25. A vertical flag staff 10 m high is of circular section 200 mm diameter at the ground and 80 mm diameter at the top. A horizontal pull of 2 kN is applied at the top as shown in the Fig. 5.57. Calculate the maximum stress due to bending.
    [**Ans.** 39.3 MN/m^2]

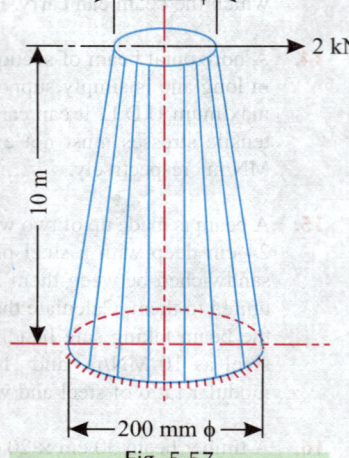

Fig. 5.57

26. A 25 mm diameter bronze rod is enclosed tightly in a steel tube of outside diameter of 50 mm. Find resisting moment of the composite section assuming that the allowable stresses in steel and bronze do not exceed 150 MN/m^2 and 120 MN/m^2 respectively. Take $E_s = 200$ GN/m^2, $E_b = 100$ GN/m^2
    [**Ans.** 1.7725 Nm]

[**Hint:** The two metals bend to the same radius of curvature, therefore, from the bending equation:

$$\frac{1}{R} = \frac{\sigma_s}{E_s \, y_s} = \frac{\sigma_b}{E_b \, y_b}, \text{ from which } \sigma_s = 4\sigma_b \Big]$$

27. A T-beam of depth 120 mm is used as a beam with simply supported ends, so that the flange comes under tension. The material of the beam can be subjected to 30 MN/m^2 in tension and 90 MN/m^2 in compression. It is desired to achieve a balanced design so that the largest possible bending stresses are reached simultaneously. Find:
    (*i*)   The width of the flange.
    (*ii*) The magnitude of a concentrated load that can be applied to the beam at the centre if the length of the span is 4 metres.
    [**Ans.** 270 mm; 9.72 kN]

28. Fig. 5.58 shows a beam of built up section simply supported at its ends with a span of 5 m. It carries a concentrated load of 12 kN at a distance of 2 metres from the left hand support. Calculate the maximum bending stress induced in the beam.
    [**Ans.** 100.44 MN/m^2]

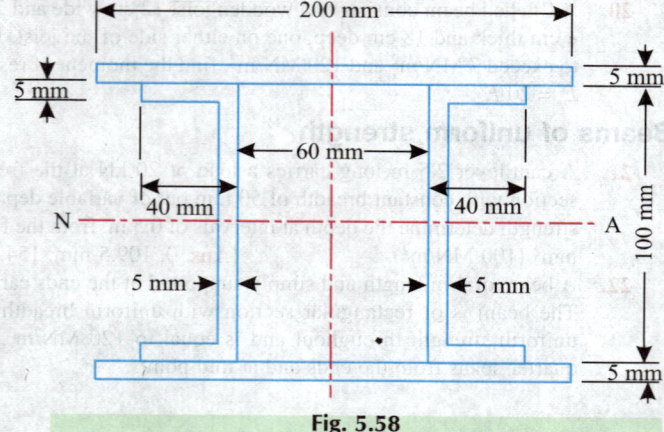

Fig. 5.58

R.C.C.

29. A reinforced concrete beam is 200 mm wide and 400 mm deep. The maximum allowable stresses in steel and concrete are 120 MN/m^2 and 7.5 MN/m^2 respectively. If modular ratio is 16 and the density of the concrete is 2360 kg/m^3, find:

    (*i*) The area of steel reinforcement required if both the stresses are developed and steel reinforcement is 60 mm above the tension face.

    (*ii*) The uniformly distributed that can be carried over a span of 5 metres.

    Neglect the weight of steel reinforcement.   [**Ans.** (*i*) 1062.5 mm^2; (*ii*) 9.672 kN/m run]

30. A R.C.C. beam is of rectangular section 250 mm wide and 550 mm deep. Steel reinforcement of 1200 mm^2 is placed 50 mm above the tension face. The maximum stress in concrete is 5 MN/m^2. If the modular ratio is 15, calculate:

    (*i*) The stress in steel;

    (*ii*) Moment of resistance.   [**Ans.** (*i*) 107.198 MN/m^2; (*ii*) 55.5 kNm]

Packing machine.

Combined Direct and Bending Stresses

Metal key cutting machine

6.1. INTRODUCTION

Invariably we come across cases where a member is called upon to take not only a bending moment but also a direct load, *e.g.* a tall chimney subjected to a wind pressure or earthquake shocks, a retaining wall resisting earth pressure or dam resisting water pressure, a hook, innumerable machine parts, gauge parts, crane jibs, etc. The member is under two stresses (*i*) *due to bending* moment and (*ii*) *due to the direct load*. In this chapter we shall consider how the two stresses have to be considered and how they vary across a cross-section.

6.2. LOAD ACTING ECCENTRICALLY TO ONE AXIS

Consider a short-column subjected to a direct load W the line of section of which is parallel to the axis of the column and intersects an axis of symmetry (*i.e.* geometric axis) at a distance e (known as eccentricity) from the centroid of the section, shown in Fig. 6.1.

Assume two equal and opposite loads W each, applied at the centroid C of the section. Application of two equal and opposite loads at C does not alter the loading pattern of the column as the two assumed loads cancel out each other.

The combined effect of the given load W at D and the assumed upward load W at C is a clockwise couple $M = W \times e$, leaving an axial load W to cause stresses directly.

Thus an eccentric load W is equivalent to an axial load W and a moment equal to $W \times e$.

Let, A = Area of cross-section of the column,

σ_d = Direct stress due to load W applied axially,

σ_b = Bending stress at a distance y from the neutral axis (N.A.), and

σ = Resultant of direct and bending stresses.

Then,
$$\sigma_b = \frac{M \times y}{I} = \frac{(W \times e)\, y}{I}$$

The bending stress is tensile if y is measured to the left of the N.A. (towards the face away from load) and is compressive if y is measured to the right of N.A. (towards the face nearer the load).

Resultant stress is given by,

$$\sigma_{max} = \sigma_d + \sigma_b = \frac{W}{A} + \frac{(W \times e)\, y}{I}$$

(when σ_b is compressive)   ...[6.1(a)]

$$\sigma_{min} = \sigma_d - \sigma_b = \frac{W}{A} - \frac{(W \times e)\, y}{I}$$

(When σ_b is tensile)   ...[6.1(b)]

If $\sigma_d > \sigma_b$, the stress throughout the section will be of the same sign. If, however $\sigma_d < \sigma_b$, the stress will change sign, being partly tensile and partly compressive across the section. Thus there can be three possible stress distributions as shown in Fig. 6.2.

When, $\sigma_d = \sigma_b$, $\sigma_{max} = 2\,\sigma_d$, and $\sigma_{min} = 0$.

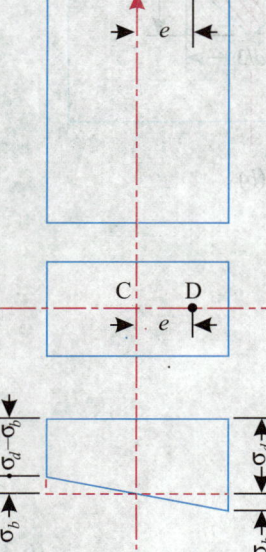

Fig. 6.1

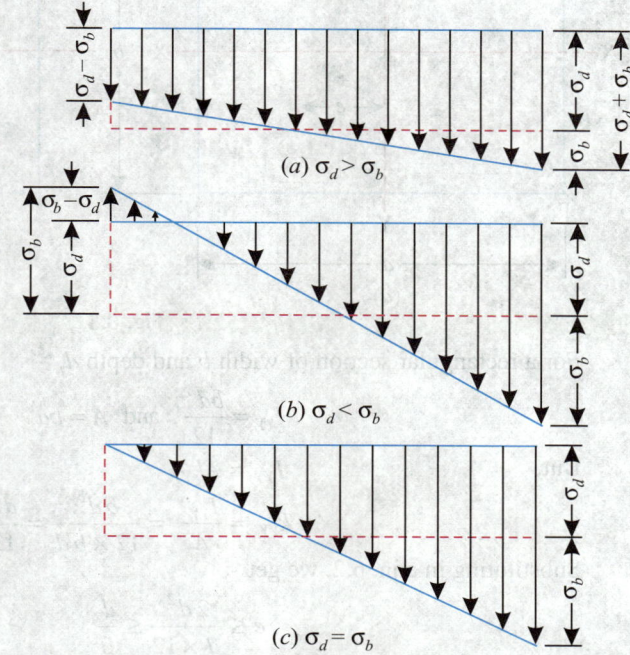

(a) $\sigma_d > \sigma_b$

(b) $\sigma_d < \sigma_b$

(c) $\sigma_d = \sigma_b$

Fig. 6.2

6.3. CONDITION FOR NO TENSION IN THE SECTION

Fig. 6.2 (*b*) indicates that $\sigma_d < \sigma_b$ and, therefore, stress changes sign, being partly tensile and partly compressive across the section. Since masonry is not capable of taking tension, we have to ensure, that no face of a masonry structure develops tension to avoid failure due to cracking. This limits the eccentricity *e* to a certain value which now we shall investigate for different sections.

For *no reverse stress*,

$$\sigma_d \geq \sigma_b \geq \frac{M}{Z}$$

$$\frac{W}{A} \geq \frac{W \times e \times d}{2I} \geq \frac{W \times e \times d}{2Ak^2} \quad \left[\begin{array}{l} \text{For symmetrical section} \\ y = y_t = y_c = d/2 \end{array} \right]$$

or,

$$e \leq \frac{2k^2}{d} \qquad \qquad \text{...(6.2)}$$

where,

d = Depth of the section, and

k = Radius of gyration.

Thus, *for no tension in the section, the eccentricity must not exceed* $\dfrac{2k^2}{d}$.

Case I. Rectangular section:

Let a rectangular section LMNP Fig. 6.3 (*a*) be loaded at a point distant '*e*' along *XX*-axis and off the *YY*-axis as shown. It is about the latter axis (*YY*) about which the bending will take place.

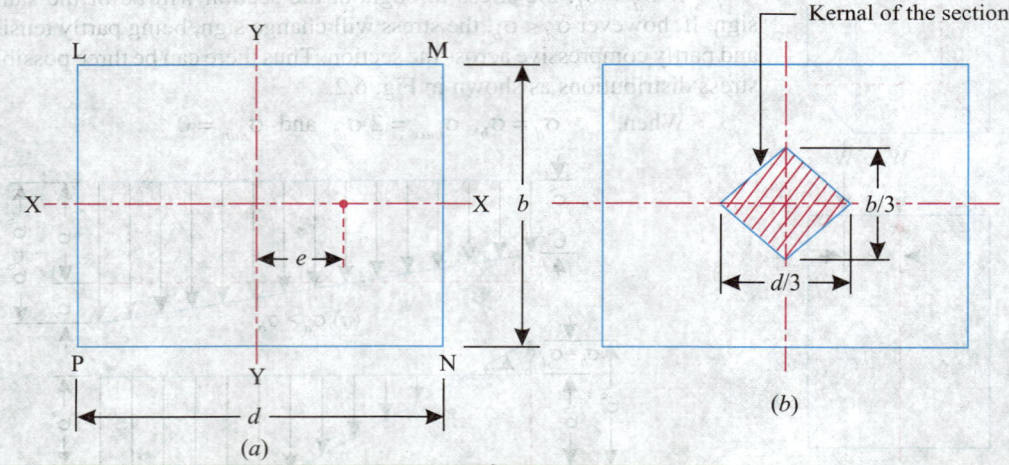

Fig. 6.3

For a rectangular section of width *b* and depth *d*,

$$I_{YY} = \frac{bd^3}{12}, \quad \text{and} \quad A = bd$$

But,

$$I_{YY} = Ak_{YY}^2$$

∴

$$k_{YY}^2 = \frac{I_{YY}}{A} = \frac{bd^3}{12 \times bd} = \frac{d^2}{12}$$

Substituting in eqn. 6.2, we get

$$e \leq \frac{2d^2}{d \times 12} \leq \frac{d}{6} \qquad \qquad \text{...(6.3)}$$

Therefore, to see that no reverse stress occurs, the load should not be placed at a distance more than *d*/6 on either side of the centroid on *XX*-axis. Hence the limit of eccentricity (*e*)

$$= d/6 + d/6 = d/3$$

Thus the stress will be of the same sign throughout the section if the load line is within the middle third of the section.

Similarly if the load is placed on *YY*-axis, off the *XX*-axis, middle third on *YY* axis *i.e.* "*b/3*" is safe zone. If the four points of middle third distances on *XX* and *YY* axis are joined, a rhombus of diamond shape is obtained as shown in Fig. 6.3 (*b*) which is known as the ***core or kernel of the section***. *If the load is placed anywhere inside the rhombus, the reverse stress will not occur in any part of the entire rectangular section.*

Case II. Hollow rectangular section:

Refer to Fig. 6.4.

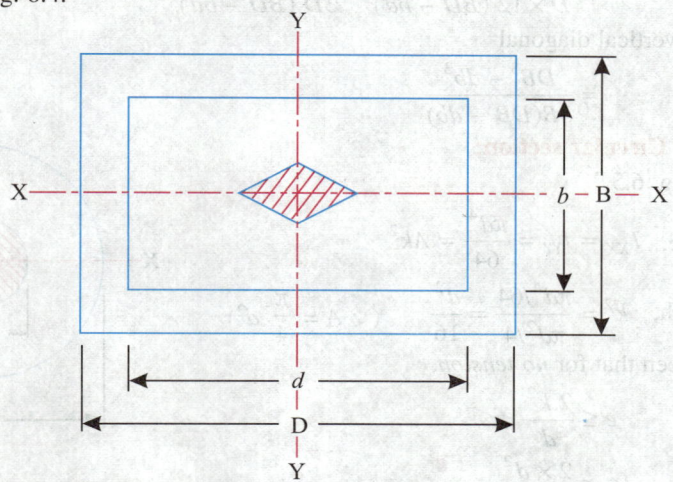

Fig. 6.4

Bending moment increases with the distance from the support.

Moment of inertia about YY-axis,

$$I_{YY} = \frac{BD^3}{12} - \frac{bd^3}{12} = \frac{BD^3 - bd^3}{12}$$

But,

$$I_{YY} = A.k_{YY}^2$$

$$\therefore \quad k_{YY}^2 = \frac{I_{YY}}{A} = \frac{(BD^3 - bd^3)}{12(BD - bd)} \qquad [\because \text{Area } A = (BD - bd)]$$

Horizontal diagonal of rhombus (*i.e.* $2e$)

$$= \frac{2 \times 2 (BD^3 - bd^3)}{D \times 12 (BD - bd)} = \frac{BD^3 - bd^3}{3D (BD - bd)} \qquad ...(6.4)$$

Similarly, vertical diagonal

$$= \frac{DB^3 - db^3}{3B(DB - db)} \qquad ...(6.4a)$$

Case III. Circular section:

Refer to Fig. 6.5.

In this case, $I_{XX} = I_{YY} = \dfrac{\pi d^4}{64} = Ak^2$

From which, $k^2 = \dfrac{\pi d^4/64}{\pi d^2/4} = \dfrac{d^2}{16} \qquad \left(\because A = \dfrac{\pi}{4} d^2\right)$

We have seen that for *no tension*,

$$e \le \frac{2 k^2}{d}$$

$$\therefore \qquad e \le \frac{2 \times d^2}{d \times 16}$$

or $\qquad e \le d/8 = d/8$

Hence diameter of *kernel*

$$= 2 e = 2 \times d/8 = d/4. \qquad ...(6.5)$$

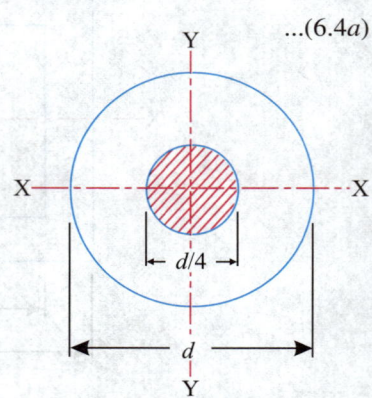

Fig. 6.5

Thus, in order that tension is not developed, the load must fall within *middle fourth of the section*.

Case IV. Hollow circular section:

Refer to Fig. 6.6.

$$I_{xx} = I_{YY} = \pi/64 \,(D^4 - d^4) = Ak^2$$

From which, $k^2 = \dfrac{\pi/64 \,(D^4 - d^4)}{\pi/4 \,(D^2 - d^2)}$

$$\left[\because A = \frac{\pi}{4} \,(D^2 - d^2)\right]$$

$$= \frac{D^2 + d^2}{16}$$

For *no tension*,

$$e \le \frac{2 k^2}{d}$$

$$\le \frac{2}{D} \left(\frac{D^2 + d^2}{16}\right) \le \frac{D^2 + d^2}{8 D}$$

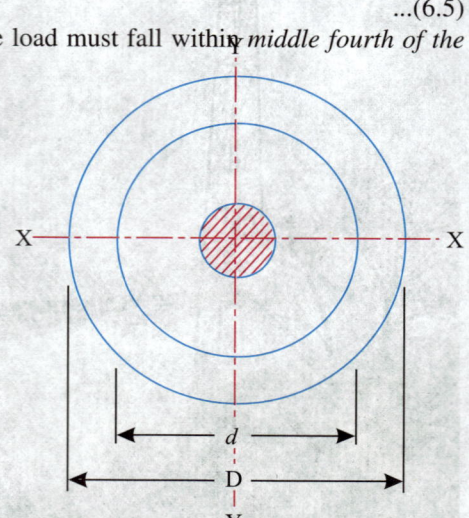

Fig. 6.6

∴ Diameter of kernel

$$= 2e = \frac{D^2 + d^2}{4\,D} \qquad ...(6.6)$$

Example 6.1. *A rectangular strut is 20 cm wide and 15 cm thick. It carries a load of 60 kN at an eccentricity of 2 cm in a plane bisecting the thickness. Find the maximum and minimum intensities of stress in the section.*

Solution. Refer to Fig. 6.7.

Width of cross-section,

$$b = 20 \text{ cm} = 0.2 \text{ m}$$

Depth of cross-section,

$$d = 15 \text{ cm} = 0.15 \text{ m}$$

Eccentricity,

$$e = 2 \text{ cm} = 0.02 \text{ m}$$

Load, $\qquad W = 60 \text{ kN}$

Maximum and minimum intensities of stresses,

$\sigma_{max}; \sigma_{min}:$

Direct stress,

$$\sigma_d = \frac{W}{A} = \frac{60}{0.2 \times 0.15}$$

$$= 2000 \text{ kN/m}^2 \text{ or } 2 \text{ MN/m}^2$$

Bending stress,

$$\sigma_b = \frac{M}{Z} = \frac{W \times e}{Z}$$

Now, $\qquad Z = \dfrac{I}{y} = \dfrac{d\,b^3/12}{b/2} = \dfrac{d\,b^2}{6}$

$$= \frac{0.15 \times (0.2)^2}{6} = 0.001 \text{ m}^3$$

∴ $\qquad \sigma_b = \dfrac{60 \times 0.02}{0.001} = 1200 \text{ kN/m}^2 \text{ or } 1.2 \text{ MN/m}^2$

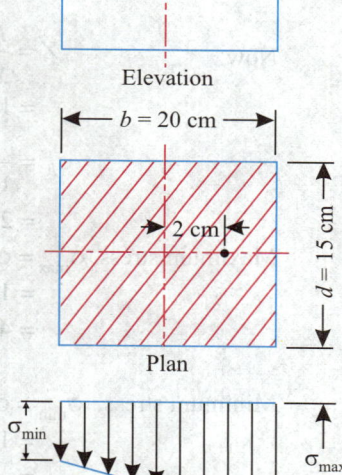

60 kN

Elevation

$b = 20$ cm

Plan

Fig. 6.7

Hence maximum stress,

$$\sigma_{max} = \sigma_d + \sigma_b$$

$$= 2 + 1.2 = \textbf{3.2 MN/m}^2 \textit{ (compressive)} \textbf{ (Ans.)}$$

and, minimum stress,

$$\sigma_{min} = \sigma_d - \sigma_b = 2 - 1.2$$

$$= \textbf{0.8 MN/m}^2 \textit{ (compressive)} \textbf{ (Ans.)}$$

Example 6.2. *A short column of hollow cylindrical section 25 cm outside diameter and 15 cm inside diameter carries a vertical load of 400 kN along one of the diameter planes 10 cm away from the axis of the column. Find the extreme intensities of stresses and state their nature.*

Solution. Refer to Fig. 6.8.

External diameter, $D = 25$ cm

Internal diameter, $\quad d = 15$ cm

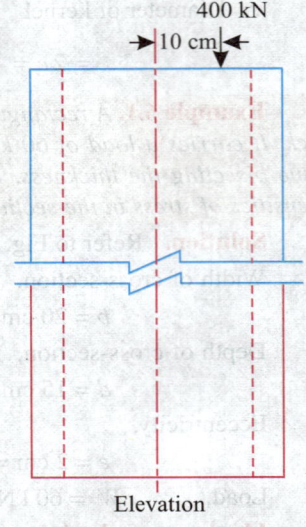

$\therefore$ Area

$$A = \frac{\pi}{4}(D^2 - d^2) = \frac{\pi}{4}(25^2 - 15^2)$$
$$= 314.2 \ cm^2 = 314.2 \times 10^{-4} \ m^2$$

Load, $\qquad W = 400 \ kN$

Eccentricity, $\qquad e = 10 \ cm = 0.1 \ m$

Maximum and minimum stresses,

σ_{max} ; σ_{min} :

Direct stress, $\qquad \sigma_d = \dfrac{W}{A}$

$$= \frac{400}{314.2 \times 10^{-4}} \times 10^{-3} \ MN/m^2$$
$$= 12.73 \ MN/m^2$$

Bending stress, $\qquad \sigma_d = \dfrac{W \times e}{Z}$

Now, $\qquad Z = \dfrac{I}{y} = \dfrac{(\pi/64)(25^4 - 15^4)}{25/2}$

$$= 1335 \ cm^3 = 1335 \times 10^{-6} \ m^3$$

$\therefore \qquad \sigma_b = \dfrac{400 \times 0.1}{1335 \times 10^{-6}} \times 10^{-3} \ MN/m^2$

$$= 29.96 \ MN/m^2$$

Maximum stress, $\sigma_{max} = \sigma_d + \sigma_b$

$$= 12.73 + 29.96$$
$$= \mathbf{42.69 \ MN/m^2}$$
$$\textit{(compressive)} \quad \textbf{(Ans.)}$$

Minimum stress, $\sigma_{min} = \sigma_d - \sigma_b$

$$= 12.73 - 29.96$$
$$= -17.23 \ MN/m^2$$
$$= \mathbf{17.23 \ MN/m^2} \ \textit{(tensile)} \quad \textbf{(Ans.)}$$

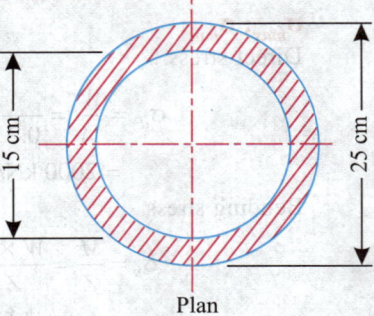

Elevation

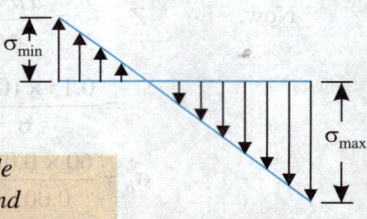

Plan

Fig. 6.8

Example 6.3. *A load of 75 kN is carried by a column made of cast-iron. The external and internal diameters are 200 mm and 180 mm respectively. If the eccentricity of the load is 35 mm, find:*

(i) *The maximum and minimum stress intensities.*

(ii) *Upto what eccentricity there is no tensile stress in the column?*

Solution. *Given:* Outside diameter,

$$D = 200 \ mm = 0.2 \ m$$

Inside diameter, $\qquad d = 180 \ mm = 0.18 \ mm$

Eccentricity, $\qquad e = 35 \ mm = 0.035 \ m$

Load, $\qquad W = 75 \ kN$

(i) Maximum and minimum stress intensities, σ_{max} ; σ_{min}:

Direct stress, $\qquad \sigma_d = \dfrac{W}{A} = \dfrac{75}{(\pi/4)(0.2^2 - 0.18^2)} \times 10^{-3} \ MN/m^2$

$$= 12.56 \ MN/m^2$$

Bending stress,

$$\sigma_b = \frac{W \times e}{Z}$$

But,

$$Z = \frac{I}{y} = \frac{(\pi/64)\,(D^4 - d^4)}{D/2}$$

$$= \frac{\pi}{64}\left[\frac{0.2^4 - 0.18^4}{0.2/2}\right]$$

$$= 2.7 \times 10^{-4} \text{ m}^3$$

$$\therefore \qquad \sigma_b = \frac{75 \times 0.035}{2.7 \times 10^{-4}} \times 10^{-3} \text{ MN/m}^2$$

$$= 9.72 \text{ MN/m}^2$$

$$\therefore \qquad \sigma_{max} = \sigma_d + \sigma_b$$

$$= 12.56 + 9.72$$

$$= \textbf{22.28 MN/m}^2 \,(comp.) \quad \textbf{(Ans.)}$$

and, $\sigma_{min} = \sigma_d - \sigma_b$

$$= 12.56 - 9.72$$

$$= \textbf{2.84 MN/m}^2 \,(comp.) \quad \textbf{(Ans.)}$$

(ii) Eccentricity for no tensile stress, e:

To have no tensile stress,

$$\frac{W \times e}{Z} = \frac{W}{A}$$

or, $\qquad e = \frac{Z}{A} = \frac{2.7 \times 10^{-4}}{(\pi/4)\,(0.2^2 - 0.18^2)}$

$$= 0.0452 \text{ m} = \textbf{45.2 mm} \quad \textbf{(Ans.)}$$

19th Century boring machine.

Example 6.4. *A short column of 20 cm external diameter and 15 cm internal diameter, when subjected to a load the stress measurements indicate that the stress varies from 150 MN/m^2 compressive at one end to 25 MN/m^2 tensile on the other end. Estimate the load and distance of the line of action from the axis of the column.*

Solution. *Given*: External diameter,

$$D = 20 \text{ cm}$$

Internal diameter,

$$d = 15 \text{ cm}$$

Area, $\quad A = \frac{\pi}{4}\,(20^2 - 15^2) = 137.5 \text{ cm}^2 = 137.5 \times 10^{-4} \text{ m}^2$

$\sigma_{max} = 150 \text{ MN/m}^2 \text{ (comp.)}$

$\sigma_{min} = 25 \text{ MN/m}^2 \text{ (tensile)}$

Load W ; Eccentricity e:

$$Z = \frac{I}{y} = \frac{(\pi/64)\,(D^4 - d^4)}{D/2} = \frac{(\pi/64)\,(20^4 - 15^4)}{20/2}$$

$$= 536.9 \text{ cm}^3 = 536.9 \times 10^{-6} \text{ m}^3$$

We know that,

Maximum stress, σ_{max} (comp.), $\quad 150 = \dfrac{W}{A} + \dfrac{W \times e}{Z}$ ...(i)

Minimum stress, σ_{min} (tensile), $\quad 25 = \dfrac{W \times e}{Z} - \dfrac{W}{A}$ ...(ii)

Subtracting (ii) from (i), we get $\quad 125 = \dfrac{2W}{A}$

From which, $\qquad W = \dfrac{125\,A}{2} = \dfrac{125 \times 137.5 \times 10^{-4}}{2}$

$$= 0.859 \text{ MN or } \mathbf{859\ kN} \quad \textbf{(Ans.)}$$

Now adding (i) and (ii), we get $\quad 175 = \dfrac{2W \times e}{Z}$

From which, $\qquad e = \dfrac{175\,Z}{2W} = \dfrac{175 \times 536.9 \times 10^{-6}}{2 \times 0.859} \text{ m}$

$$= 0.05469 \text{ m} = \mathbf{54.69\ mm} \quad \textbf{(Ans.)}$$

Example 6.5. *A hollow rectangular masonry pier is 1.2 m × 0.8 m, overall, the wall thickness being 0.15 m. A vertical load of 100 kN is transmitted in the vertical plane bisecting 1.2 m side at an eccentricity of 0.1 m from the geometric axis of the section. Calculate the maximum and minimum stress intensities in the section.*

Solution. Refer to Fig. 6.9.

Given: Outside dimensions of the section

$\qquad = 1.2 \text{ m} \times 0.8 \text{ m}$

Thickness $\qquad = 0.15 \text{ m}$

Inside dimensions of the section

$\qquad = (1.2 - 2 \times 0.15)$

$\qquad \times (0.8 - 2 \times 0.15)$

$\qquad = 0.9 \text{ m} \times 0.5 \text{ m}$

Area $\qquad = (1.2 \times 0.8)$

$\qquad - (0.9 \times 0.5) = 0.51 \text{ m}^2$

Eccentricity, $\ e = 0.1 \text{ m}$

Load, $\qquad W = 100 \text{ kN}$

Maximum and Minimum stress intensities, σ_{max} ; σ_{min}:

Direct stress,

$$\sigma_d = \dfrac{W}{A} = \dfrac{100}{0.51} = 196 \text{ kN/m}^2$$

Bending stress,

$$\sigma_b = \dfrac{M}{Z} = \dfrac{W \times e}{Z}$$

$$I = \dfrac{1.20 \times 0.8^3}{12} - \dfrac{0.9 \times 0.5^3}{12}$$

$$= 0.041825 \text{ m}^4$$

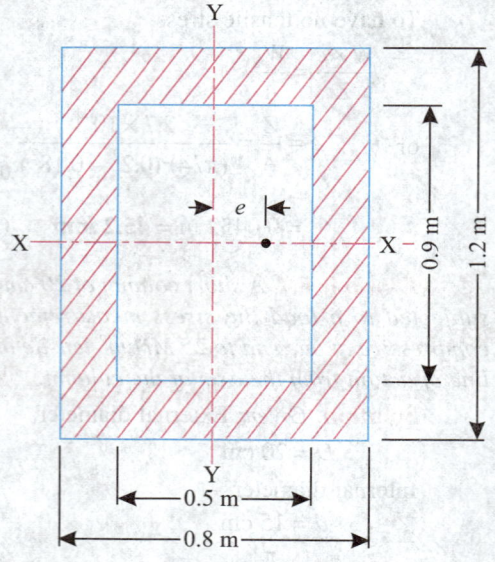

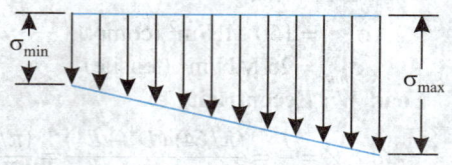

Fig. 6.9

$$y = \frac{0.8}{2} = 0.4 \text{ m}$$

$$Z = \frac{I}{y} = \frac{0.041825}{0.4} = 0.10456 \text{ m}^3$$

Hence,

$$\sigma_b = \frac{100 \times 0.1}{0.10456} = 95.6 \text{ kN/m}^2$$

∴ *Maximum stress,*

$$\sigma_{max} = \sigma_d + \sigma_b$$
$$= 196 + 95.6$$
$$= \textbf{291.6 kN/m}^2 \text{ (comp.)} \quad \textbf{(Ans.)}$$

Minimum stress, $\sigma_{min} = \sigma_d - \sigma_b$
$$= 196 - 95.6$$
$$= \textbf{100.4 kN/m}^2 \text{ (comp.)} \quad \textbf{(Ans.)}$$

Example 6.6. *A short column of I section 25 cm × 20 cm has a cross-sectional area 52 cm² and maximum radius of gyration 10.7 cm. A vertical load W kN acts through the centroid of the section together with a parallel load of W/4 kN acting through a point on the centre line of the web distant 6 cm from the centroid. Calculate the greatest allowable value of W if the maximum stress is not to exceed 65 MN/m². What is then the minimum stress?*

Solution. Refer to Fig. 6.10.

Given: Cross-sectional area,

$$A = 52 \text{ cm}^2 = 52 \times 10^{-4} \text{ m}^2$$

Maximum radius of gyration,

$$k = 10.7 \text{ cm} = 0.107 \text{ m}$$

Eccentricity, $\quad e = 6 \text{ cm} = 0.06 \text{ m}$

Maximum stress, $\sigma_{max} = 65 \text{ MN/m}^2$

Total load on the section

$$= W + \frac{W}{4} = 1.25 \, W \text{ kN}$$

Greatest value of W (kN):

Moment of inertia about N.A.

$$I = A \, k^2 = 52 \times (10.7)^2$$
$$= 5953.48 \text{ cm}^4$$
$$= 5953.48 \times 10^{-8} \text{ m}^4$$

Direct stress, $\quad \sigma_d = \frac{1.25 W}{52 \times 10^{-4}} \times 10^{-3} = 0.24 \, W \text{ MN/m}^2$

Bending stress, $\quad \sigma_b = \frac{M}{Z} = \frac{(W/4) \times 0.06}{Z} \times 10^{-3} \text{ MN/m}^2$

$$= \frac{0.015 \, W}{Z} \times 10^{-3} \text{ MN/m}^2 = \frac{0.015 \, W}{(I/y)} \times 10^{-3} \text{ MN/m}^2$$

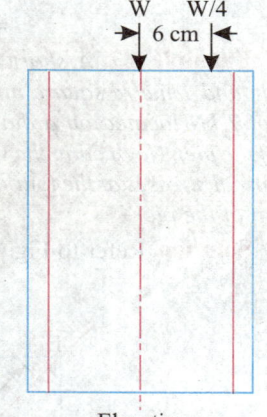

Elevation

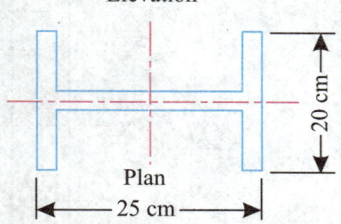

Plan

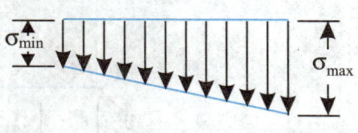

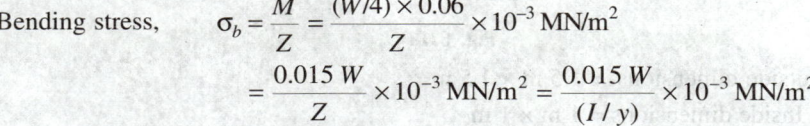

Fig. 6.10

$$= \frac{0.015\,W}{(5953.48 \times 10^{-8}/0.125)} \times 10^{-3}\ \text{MN/m}^2 \qquad \left[\because\ y = \frac{0.25}{2} = 0.125\ \text{m}\right]$$

$$= 0.0315\ W\ \text{MN/m}^2$$

But, $\sigma_d + \sigma_b = 65$ ...(Given)

$\therefore\ 0.24\,W + 0.0315\,W = 65$

From which, $W = \dfrac{65}{0.24 + 0.0315} = \mathbf{239.4\ kN}$ **(Ans.)**

Minimum stress, σ_{min} :

$$\sigma_{min} = \sigma_d - \sigma_b = 0.24\,W - 0.0315\,W$$
$$= W\,(0.24 - 0.0315)$$
$$= 239.4\,(0.24 - 0.0315) \times 10^{-3}$$
$$= \mathbf{49.9\ MN/m^2}\ \ \mathbf{(Ans.)}$$

Example 6.7. *A short hollow pier 1.5 metre square outside and 1 metre square inside, supports a vertical point load of 7 kN located on a diagonal 0.8 m from the vertical axis of the pier. Neglecting the self-weight of the pier, calculate the normal stresses at the four outside corners on a horizontal section of the pier.*

Solution. Refer to Fig. 6.11.

Marine machine.

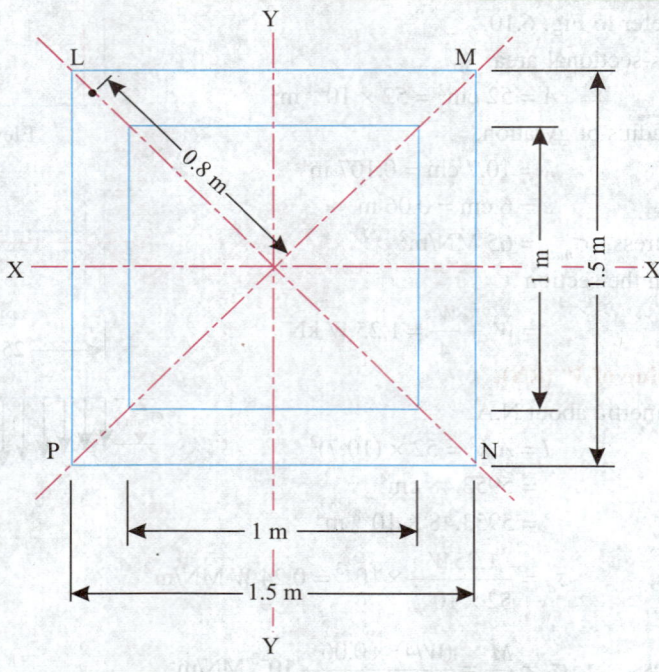

Fig. 6.11

Given: Outside dimensions $= 1.5\ \text{m} \times 1.5\ \text{m}$

Inside dimensions $= 1\ \text{m} \times 1\ \text{m}$

Load, $W = 7\ \text{kN}$

Eccentricity, $e = 0.8\ \text{m}$

Normal stresses at the four outside corners:

Direct stress,
$$\sigma_d = \frac{W}{A} = \frac{7}{(1.5^2 - 1^2)}$$
$$= 5.6 \text{ kN/m}^2 \text{ (comp.)}$$

Moment of inertia about the diagonal
$$I = \frac{1}{12}(1.5^4 - 1^4) = 0.3385 \text{ m}^4$$
$$Z = \frac{I}{y_{max}} = \frac{0.3385}{1.5/\sqrt{2}} = 0.319 \text{ m}^3$$

∴ Stress due to moment at the corners L and N,
$$\sigma_b = \frac{M}{Z} = \frac{W \times e}{Z} = \frac{7 \times 0.8}{0.319} = 17.55 \text{ kN/m}^2$$

At the corners M and P there will be no bending stress. Hence the stresses at the various corners are as follows:

Stress at corner L = 5.6 + 17.55 = **23.15 kN/m² (comp.) (Ans.)**

Stress at corner M or P = **5.6 kN/m² (comp.) (Ans.)**

Stress at corner N = 5.6 − 17.55

= **11.95 kN/m² (tensile) (Ans.)**

Example 6.8. *A masonry pier of 3 m × 4 m supports a load of 40 kN as shown in the Fig. 6.12.*

  (i) *Find the stresses developed at each corner of the pier.*

  (ii) *What additional load should be placed at the centre of the pier, so that there is no tension anywhere in the pier section?*

  (iii) *What are the stresses at the corners with the additional load in the centre?*

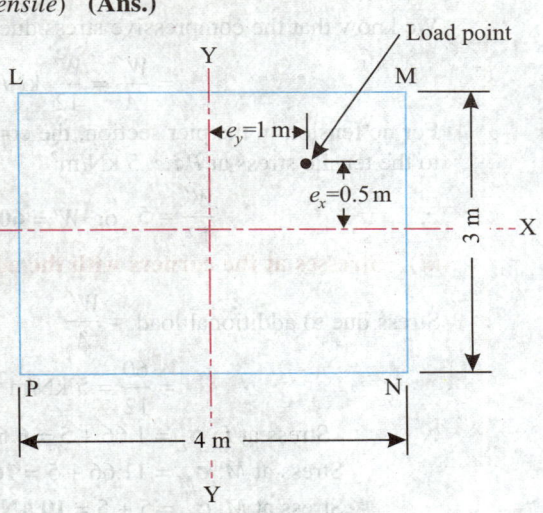

Fig. 6.12

Solution. *Given*: Width, $b = 4$ m

Thickness, $\qquad d = 3$ m

∴ Area, $\qquad A = 4 \times 3 = 12 \text{ m}^2$

Load, $\qquad W = 40$ kN

M.O.I. about XX-axis, $I_{XX} = \dfrac{4 \times 3^3}{12} = 9 \text{ m}^4$

and, $\qquad I_{YY} = \dfrac{3 \times 4^3}{12} = 16 \text{ m}^4$

Eccentricity about XX-axis,
$$e_x = 0.5 \text{ m}$$
and, $\qquad e_y = 1$ m

∴ Moment, $\qquad M_x = W \times e_x = 40 \times 0.5 = 20$ kNm

Similarly, $\qquad M_y = W \times e_y = 40 \times 1.0 = 40$ kNm

Distance between YY-axis and the corners L as well as M, $x = 2$ m

Distance between XX-axis and the corners L as well as P,
$$y = 1.5 \text{ m}$$

(i) Stresses developed at each corner:

From the geometry of the figure, we find that the stress at L,

$$\sigma_L = \frac{W}{A} + \frac{M_x \times y}{I_{XX}} - \frac{M_y \times x}{I_{YY}} = \frac{40}{12} + \frac{20 \times 1.5}{9} - \frac{40 \times 2}{16}$$

$$= 3.33 + 3.33 - 5.0 = \textbf{1.66 kN/m}^2 \quad \textbf{(Ans.)}$$

Similarly,

$$\sigma_M = \frac{W}{A} + \frac{M_x \times y}{I_{XX}} + \frac{M_y \times x}{I_{YY}} = \frac{40}{12} + \frac{20 \times 1.5}{9} + \frac{40 \times 2}{16}$$

$$= 3.33 + 3.33 + 5 = \textbf{11.66 kN/m}^2 \quad \textbf{(Ans.)}$$

$$\sigma_N = \frac{W}{A} - \frac{M_x \times y}{I_{XX}} + \frac{M_y \times x}{I_{YY}} = \frac{40}{12} - \frac{20 \times 1.5}{9} + \frac{40 \times 2}{16}$$

$$= 3.33 - 3.33 + 5 = \textbf{5 kN/m}^2 \quad \textbf{(Ans.)}$$

$$\sigma_P = \frac{W}{A} - \frac{M_x \times y}{I_{XX}} - \frac{M_y \times x}{I_{YY}} = \frac{40}{12} - \frac{20 \times 1.5}{9} - \frac{40 \times 2}{16}$$

$$= 3.33 - 3.33 - 5 = -5.0 \text{ kN/m}^2$$

$$= \textbf{5.0 kN/m}^2 \textbf{ (tension)} \quad \textbf{(Ans.)}$$

(ii) Additional load at the centre for no tension in the pier section, W' (kN):

We know that the compressive stress due to the load W'

$$= \frac{W'}{A} = \frac{W'}{12} \text{ kN/m}^2$$

For no tension in the pier section, the compressive stress due to the load W' should be equal to the tensile stress at P *i.e.* 5 kN/m²

∴ $$\frac{W'}{12} = 5 \quad \text{or} \quad W' = \textbf{60 kN} \quad \textbf{(Ans.)}$$

(iii) Stresses at the corners with the additional load in the centre:

Stress due to additional load $= \dfrac{W'}{A}$

$$= \frac{60}{12} = 5 \text{ kN/m}^2$$

∴   Stress at L, $\sigma_L = 1.66 + 5 = \textbf{6.66 kN/m}^2$ **(Ans.)**

Stress at M, $\sigma_M = 11.66 + 5 = \textbf{16.66 kN/m}^2$ **(Ans.)**

Stress at M, $\sigma_N = 5 + 5 = \textbf{10 kN/m}^2$ **(Ans.)**

Stress at N, $\sigma_P = -5 + 5 = \textbf{0}$ **(Ans.)**

6.4.   WIND PRESSURE ON CHIMNEYS

Chimneys are tall structures subjected to horizontal wind pressure. Due to horizontal wind force the base of the chimney shaft is subjected to bending moment. Thus at the base of such tall chimneys, there will be *bending stress due to wind force and direct* (or axial) *stress due to self-weight.*

The direct stress, $\sigma_d = \dfrac{W}{A}$ where W is the weight of chimney and A is the cross-sectional area. The

bending stress, $\sigma_b = \dfrac{M}{Z}$ where M is the bending moment due to horizontal axial pressure and Z is modulus of section.

The pressure due to wind is usually specified as a horizontal force on a unit area of a vertical plane, on which it acts normally. The shape of the object exposed to wind will, however, affect the magnitude of the force exerted by the wind on it. If the area exposed to the wind pressure is curved (as in the case of a cylindrical chimney shaft) the magnitude of the force will be less than when the area is a flat surface. The reduction factor K, depending upon the *shape of the area exposed to wind* is called the **co-efficient of wind resistance.** Its *value varies* from 0.5 to 0.75. For *cylindrical shafts, its value will be 2/3, unless stated otherwise.*

The cross-section of chimneys may be hollow, square or hollow circle. The cross-section may be uniform throughout the height or be tapering towards the top.

The total horizontal wind force is calculated by using the following relation:

Chimneys due to their height are susceptible to high bending stresses due to wind force.

$$P = K p A_p \qquad \qquad \dots(6.7)$$

where,

P = Total horizontal wind force,

K = Co-efficient of wind resistance,

p = Horizontal intensity of wind pressure, and

A_p = Projected area of the surface exposed to wind.

For rectangular and square chimneys of uniform cross-section:

Value of K is taken as unity *i.e.* $K = 1$

Projected area, $A = b \times h$

where, b = Width of size exposed to wind, and

h = Height of chimney.

P will act at h/2.

For uniform circular chimney shafts:

Value of K may be taken as 2/3 unless stated otherwise.

Project area $A = D \times h$

where, D = External diameter of chimney shaft, and

h = Height of chimney.

Example 6.9. *A masonry chimney 24 metres high, of uniform circular section, 3.5 metres external diameter and 2 metres internal diameter is subjected to a horizontal wind pressure of 1 kN/m² of projected area. Find the maximum and minimum stress intensities at the base, if the specific weight of masonry is 22 kN/m³.*

Solution. Refer to Fig. 6.13.

Given: External diameter, $D = 3.5$ m

Internal diameter, $D = 2$ m

$\therefore$ Area, $A = \dfrac{\pi}{4}(3.5^2 - 2^2) = 6.48 \text{ m}^2$

Wind pressure, $p = 1 \text{ kN/m}^2$

Specific weight of masonry, $w = 22 \text{ kN/m}^3$

Height of the chimney, $h = 24 \text{ m}$

Maximum and minimum stress intensities,

σ_{max} ; σ_{min} :

Direct stress,

$$\sigma_d = \frac{W}{A}$$

where, $W = A \times h \times w$, W being weight of the chimney

$$= 6.48 \times 24 \times 22 = 3421.44 \text{ kN}$$

$\therefore \qquad \sigma_d = \dfrac{3421.44}{6.48} = 528 \text{ kN/m}^2$

Moment of inertia,

$$I = \frac{\pi}{64}(3.5^4 - 2^4) = 6.58 \text{ m}^4$$

$$y = 3.5/2 = 1.75 \text{ m}$$

$\therefore$ Section modulus

$$Z = \frac{I}{y} = \frac{6.58}{1.75} \text{ m}^3 = 3.76 \text{ m}^3$$

The intensity of wind pressure on the projected area is given $i.e.,$

$$p = 1 \text{ kN/m}^2$$

Horizontal wind force,

$$P = Kp \times A_p, \text{ where } A_p \text{ is the projected area}$$

$$= \frac{2}{3} \times 1 \times 24 \times 3.5 = 56 \text{ kN}$$

$\left(\text{where, } K = \text{co-efficient of wind resistance} = \dfrac{2}{3} \right)$

Bending moment at the base,

$$M = P \times h/2 = 56 \times \frac{24}{2} = 672 \text{ kNm}$$

$\therefore$ Bending stress,

$$\sigma_b = \frac{M}{Z} = \frac{672}{3.76} = 178.7 \text{ kN/m}^2$$

$$\sigma_{max} = \sigma_d + \sigma_b = 528 + 178.7 = \textbf{706.7 kN/m}^2 \ (comp.) \ \textbf{(Ans.)}$$

$$\sigma_{min} = \sigma_d - \sigma_b = 528 - 178.7 = \textbf{349.3 kN/m}^2 \ (comp.) \ \textbf{(Ans.)}$$

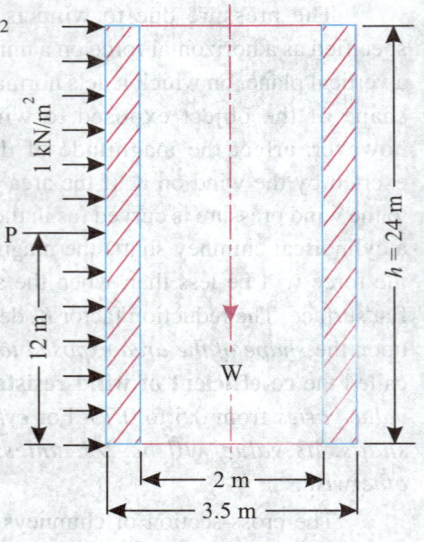
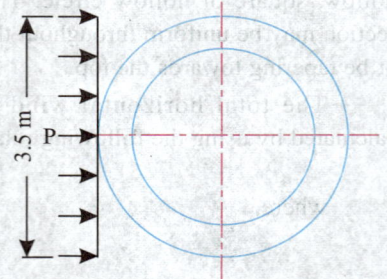
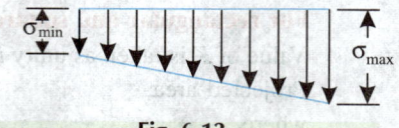

Fig. 6.13

Example 6.10. *A cylindrical chimney, 22 m high, of uniform circular section has 4 metres external diameter and 2 metres internal diameter. The intensity of horizontal wind pressure is 1.2 kN/m². Find the maximum and minimum normal stress intensities at the section, if the specific weight of masonry is 22 kN/m³.*

Solution. *Given:* External diameter, $D = 4 \text{ m}$

Internal diameter, $d = 2 \text{ m}$

Area, $\qquad A = \dfrac{\pi}{4}(4^2 - 2^2) = 9.426 \text{ m}^2$

Horizontal wind pressure, $\qquad p = 1.2 \text{ kN/m}^2$

Height of chimney, $\qquad h = 22 \text{ m}$

Specific weight of masonry, $\qquad w = 22 \text{ kN/m}^3$

The section is circular (being cylindrical shaft). The horizontal intensity of wind pressure, p is given. The *given intensity of wind pressure is not on the projected area*. Hence it is to be multiplied by the co-efficient of wind pressure, K. Since the value of K is not given, take $K = 2/3$ (usually the value of K for circular surface is assumed as 0.6).

Maximum and minimum stress intensities σ_{max} ; σ_{min} :

Direct stress, $\qquad\qquad \sigma_d = W/A$

Where, W (self weight of chimney)

$\qquad = A \times h \times w = 9.426 \times 22 \times 22 = 4562.2 \text{ kN}$

$\therefore \qquad \sigma_d = \dfrac{4562.2}{9.426} = 484 \text{ kN/m}^2$

Moment of inertia,

$\qquad I = \dfrac{\pi}{64}(D^4 - d^4)$

$\qquad = \pi/64\,(4^4 - 2^4)$

$\qquad = 11.78 \text{ m}^4; \; y = 4/2 = 2 \text{ m}$

$\therefore$ Section modulus,

$\qquad Z = \dfrac{I}{y} = \dfrac{11.78}{2} = 5.89 \text{ m}^3$

Horizontal wind pressure,

$\qquad P = K\,p\,A_p$, where A_p is the projected area

$\qquad = 2/3 \times 1.2 \times 22 \times 4 = 70.4 \text{ kN}$

Bending moment at the base,

$\qquad M = P \times h/2 = 70.4 \times \dfrac{22}{2} = 774.4 \text{ kNm}$

$\therefore$ Bending stress,

$\qquad \sigma_b = \dfrac{M}{Z} = \dfrac{774.4}{5.89} = 131.5 \text{ kN/m}^2$

$\qquad \sigma_{max} = \sigma_d + \sigma_b = 484 + 131.5$

$\qquad\qquad = \mathbf{615.5 \text{ kN/m}^2}$ (*comp.*) $\quad$ **(Ans.)**

$\qquad \sigma_{min} = \sigma_d - \sigma_b = 484 - 131.5$

$\qquad\qquad = \mathbf{352.5 \text{ kN/m}^2}$ (*comp.*) $\quad$ **(Ans.)**

Example 6.11. *A square chimney 24 m high, has an opening of 1.25 m × 1.25 m inside. The external dimensions are 2.5 m × 2.5 m. The horizontal intensity of wind pressure is 1.3 kN/m² and the specific weight of masonry is 22 kN/m³. Calculate the maximum and minimum stress intensities at the base of the chimney.*

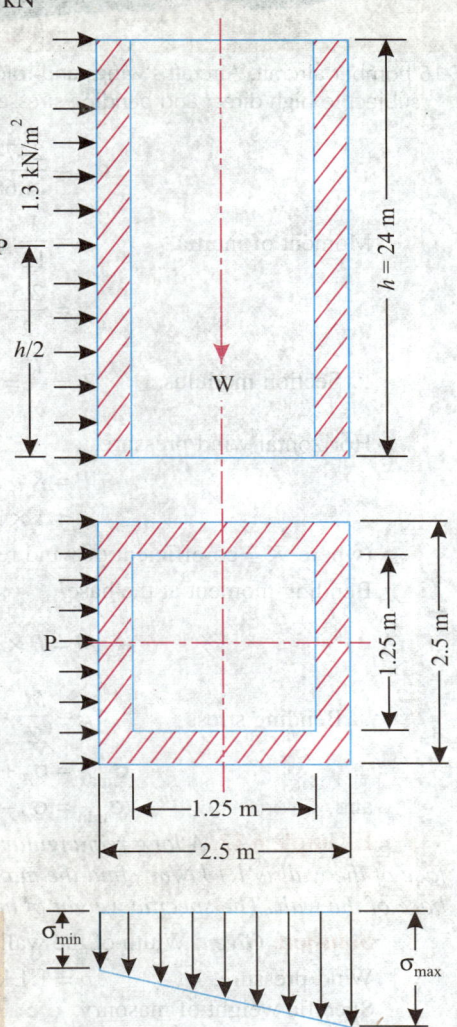

Fig. 6.14

Solution. Refer to Fig. 6.14.

F-16 bomber aircraft. Aircraft's wings and rotors are subject to high direct and bending stresses.

Outside dimensions
$$= 2.5 \text{ m} \times 2.5 \text{ m}$$

Inside dimensions
$$= 1.25 \text{ m} \times 1.25 \text{ m}$$

Area, $A = (2.5 \times 2.5) - (1.25 \times 1.25) = 4.687 \text{ m}^2$

Height of the chimney $\qquad h = 24 \text{ m}$

Horizontal intensity of pressure,
$$p = 1.3 \text{ kN/m}^2$$

Specific weight, $\qquad w = 22 \text{ kN/m}^3$

Maximum and minimum stresses, σ_{max} ; σ_{min}:

Direct stress, $\qquad \sigma_d = \dfrac{W}{A}$

where, W (self-weight of the chimney),
$$= A \times h \times w = 4.687 \times 24 \times 22 = 2474.7 \text{ kN}$$

$\therefore \qquad \sigma_d = \dfrac{2474.7}{4.687} = 528 \text{ kN/m}^2 \text{ (comp.)}$

Moment of inertia, $\qquad I = \dfrac{1}{12}(2.5 \times 2.5^3 - 1.25 \times 1.25^2) = 3.051 \text{ m}^4$

$$y = \dfrac{2.5}{2} = 1.25 \text{ m}$$

$\therefore$ Section modulus, $\qquad Z = \dfrac{I}{y} = \dfrac{3.051}{1.25} = 2.44 \text{ m}^3$

Horizontal wind pressure,
$$P = K p A_p \quad (K = 1)$$
$$= 1 \times 1.3 \times 24 \times 2.5 = 78 \text{ kN}$$

(where, K = co-efficient of wind resistance)

Bending moment at the base,
$$M = P \times h/2 = 78 \times \dfrac{24}{2} = 936 \text{ kNm}$$

$\therefore$ Bending stress, $\qquad \sigma_b = \dfrac{M}{Z} = \dfrac{936}{2.44} = 383.6 \text{ kN/m}^2$

$\therefore \qquad \sigma_{max} = \sigma_d + \sigma_b = 528 + 383.6 = \textbf{911.6 kN/m}^2$ *(comp.)* **(Ans.)**

and, $\qquad \sigma_{min} = \sigma_d - \sigma_b = 528 - 383.6 = \textbf{144.4 kN/m}^2$ *(comp.)* **(Ans.)**

Example 6.12. *A long rectangular wall is 2.5 m wide. If the maximum wind pressure on the face of the wall is 1.1 kN/m^2, find the maximum height of the wall so that there is no tension in the base of the wall. The specific weight of masonry is 22 kN/m^3.*

Solution. *Given*: Width of the wall = 2.5 m

Wind pressure, $\qquad p = 1.1 \text{ kN/m}^2$

Specific weight of masonry,
$$w = 22 \text{ kN/m}^3$$

Height of the wall, h:

Consider 1 m length of the wall.

Weight per metre length of the wall,

$$W = 2.5 \times 1 \times h \times 22 = 55\,h$$

∴ Direct stress, $\qquad \sigma_d = \dfrac{W}{A} = \dfrac{55\,h}{2.5 \times 1} = 22\,h$ (comp.)

Wind pressure on 1 m length of wall,

$$P = h \times 1 \times 1.1 = 1.1\,h \text{ kN}$$

Bending moment at the base,

$$M = P \times h/2 = 1.1\,h \times h/2 = 0.55\,h^2 \text{ kNm}$$

Moment of inertia, $\qquad I = \dfrac{1 \times 2.5^3}{12} = 1.302 \text{ m}^4$

$$y = 2.5/2 = 1.25 \text{ m}$$

Section modulus, $\qquad Z = \dfrac{I}{y} = \dfrac{1.302}{1.25} = 1.0416 \text{ m}^3$

∴ Bending stress, $\qquad \sigma_b = \dfrac{M}{Z} = \dfrac{0.55\,h^2}{1.0416} = 0.528\,h^2 \text{ kN/m}^2$

For no tension to develop at the base of the wall, $\sigma_a - \sigma_b = 0$

or, $\qquad 22\,h - 0.528\,h^2 = 0$

or, $\qquad h = \dfrac{22}{0.528} = \mathbf{41.67 \ m}$ **(Ans.)**

Example 6.13. *A chimney has external and internal dimensions of 2 m × 2 m and × 1 m × 1 m respectively. The height of the chimney is 14 m. Find the maximum and minimum stress intensities at the base when it is subjected to horizontal wind pressure of 1.4 kN/m² acting in the direction of one of the diagonals of the chimney. Specific weight of masonry is 22 kN/m³.*

Solution. Refer to Fig. 6.15.

Given: External dimensions $\quad = 2 \text{ m} \times 2 \text{ m}$

Internal dimensions $\qquad\qquad = 1 \text{ m} \times 1 \text{ m}$

∴ Area, $\quad A = (2 \times 2) - (1 \times 1) = 3 \text{ m}^2$

Height of the chimney, $\qquad h = 14 \text{ m}$

Horizontal wind pressure, $\qquad p = 1.4 \text{ kN/m}^2$

Specific weight of masonry, $\quad w = 22 \text{ kN/m}^3$

Maximum and minimum stress intensities,

σ_{max} ; σ_{min}:

Weight of chimney,

$$W = A \times h \times w = 3 \times 14 \times 22 = 942 \text{ kN}$$

Direct stress,

$$\sigma_d = \dfrac{W}{A} = \dfrac{924}{3} = 308 \text{ kN/m}^2$$

Bending takes place about the diagonal *LM*.

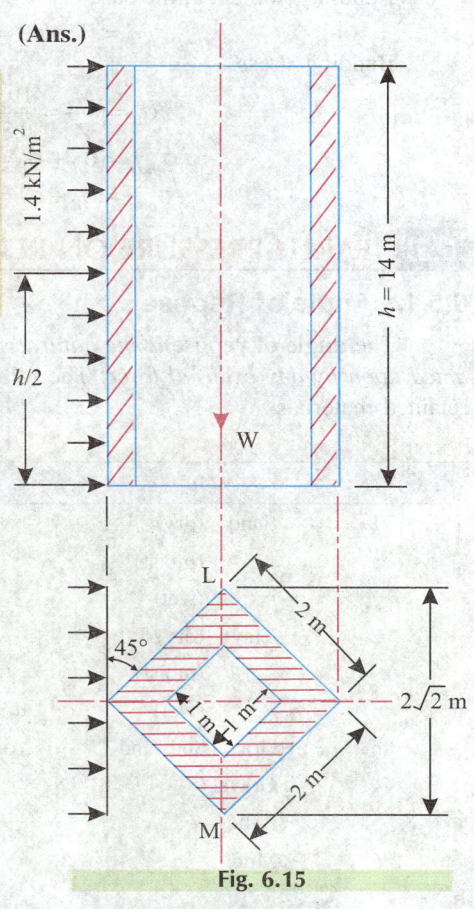

Fig. 6.15

$$I_{LM} = \frac{1}{12}(BH^3 - bh^3) \times 2 \text{ with usual notations}$$

$$= \frac{1}{12}\left[2\sqrt{2} \times (\sqrt{2})^3 - \sqrt{2} \times \left(\frac{1}{\sqrt{2}}\right)^3\right] \times 2$$

$$= \frac{1}{12}\left(2 \times 4 - \frac{1}{2}\right) \times 2 = 1.25 \text{ m}^4$$

$$\left[\text{or } I_{LM} = \frac{1}{12}(2^4 - 1^4) = 1.25 \text{ m}^4\right]$$

$$Z_{LM} = \frac{I_{LM}}{y} = \frac{1.25}{\sqrt{2}} = 0.884 \text{ m}^3$$

Horizontal wind force,

$$P = K . p . A_p \text{ where, } A_p \text{ is the projected area.}$$
$$K = \sin^2\theta = \sin^2 45° = 0.5$$
(where K = co-efficient of wind resistance)
$$A_p = h \times \text{diagonal} = 14 \times 2\sqrt{2} = 39.592 \text{ m}^2$$
$$\therefore \qquad P = 0.5 \times 1.4 \times 39.592 = 27.7 \text{ kN}$$

Bending moment at the base,

$$M = P \times h/2 = 27.7 \times 14/2 = 193.9 \text{ kN m}$$

Bending stress,

$$\sigma_b = \frac{M}{Z} = \frac{193.9}{0.884} = 219.3 \text{ kN/m}^2$$

$$\therefore \qquad \sigma_{max} = \sigma_d + \sigma_b = 308 + 219.3 = \mathbf{527.3 \text{ kN/m}^2} \quad \textbf{(Ans.)}$$
$$\sigma_{min} = \sigma_d - \sigma_b = 308 - 219.3 = \mathbf{88.7 \text{ kN/m}^2} \quad \textbf{(Ans.)}$$

6.5. EARTH PRESSURE ON RETAINING WALLS

6.5.1. Angle of Repose

The **angle of repose** *is the natural slope of the materials which they tend to take up if not acted upon by any external force*. The following table gives the characteristics of most common retained materials.

TABLE 6.1

S.No.	Material		Weight, kN/m³	Angle of repose (ϕ)
1.	Sand	(dry)	14.4 to 16	30°
		(moist)	15 to 17.6	35°
		(wet)	17.6 to 20	25°
2.	Clay	(dry)	19.2 to 22.4	30°
		(moist)	12.2 to 25.6	45°
		(wet)	19.2 to 25.6	15°
3.	Gravel and sand		16 to 17.6	25° to 30°
4.	Gravel		14.4	40°
5.	Ashes		6.4	40°
6.	Mud		16.8 to 25.6	0°

6.5.2. Earth Pressure

Retaining walls in case of construction of roads in hilly areas retain earth at their back. In this case, the intensity of pressure varies linearly from zero at the top to $K_a w_e h$ at the bottom,

where,

K_a = Co-efficient of active pressure,

w_e = Density of earth, and

h = Height of earth retained.

The value of K_a is determined by the following relationship:

$$K_a = \frac{1 - \sin \phi}{1 + \sin \phi}, \text{ where } \phi = \text{angle of repose for the earth.}$$

∴ For 1 m length of retaining wall,

$$P = \frac{w_e h^2}{2} \left(\frac{1 - \sin \phi}{1 + \sin \phi} \right) \qquad \qquad ...(6.8)$$

This formula for finding the value of P is known as **Rankine's formula.**

The details of the above formula may be given as follows:

Refer to Fig. 6.16. Consider a most common trapezoidal cross-section of a retaining wall 1 m long, retaining earth on its vertical side (top of earth surface is horizontal).

The maximum lateral earth pressure at any depth y below the horizontal top is given by

$$p_y = w_e y \left(\frac{1 - \sin \phi}{1 + \sin \phi} \right)$$

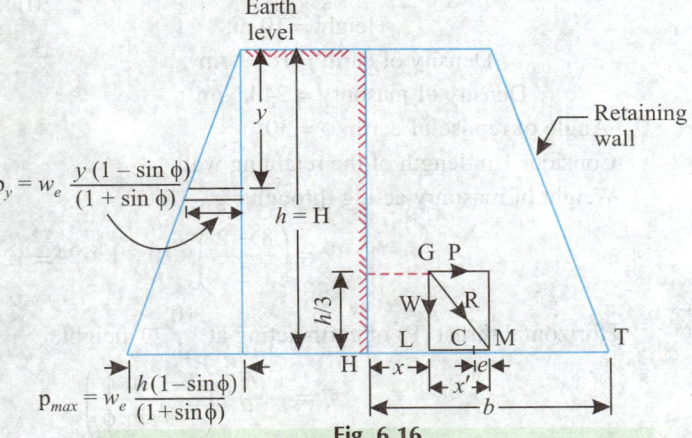

Fig. 6.16

$p_y = w_e \dfrac{y (1 - \sin \phi)}{(1 + \sin \phi)}$

$p_{max} = w_e \dfrac{h(1 - \sin\phi)}{(1 + \sin\phi)}$

Since ϕ, the angle of repose, for the particular earth is constant, so the multiplying factor $\dfrac{1 - \sin \phi}{1 + \sin \phi}$ also remains constant at all depths. The lateral earth pressure is directly proportional to the earth density and the depth and thus earth pressure diagram will be triangular as shown in the Fig. 6.16. The maximum earth pressure occurs at the bottom of the retaining wall and is equal to : $w_e h \left(\dfrac{1 - \sin \phi}{1 + \sin \phi} \right)$. The c.g. of the earth pressure diagram (triangle) lies at $h/3$ from the bottom and so the total horizontal or lateral pressure acts at this point.

Cockpit of F-16 aircraft.

Average pressure on the ratining wall

$$= \frac{w_e h}{2}\left(\frac{1 - \sin \phi}{1 + \sin \phi}\right).$$

Example 6.14. *A masonry retaining wall of trapezoidal section is 10 m high and retains earth which is level upto the top. The width at the top is 2 m and at the bottom 8 m and the exposed face is vertical. Find the maximum and minimum intensities of normal stress at the base.*

Take: Density of earth = 16 kN/m³
Density of masonry = 24 kN/m³
Angle of repose of earth = 30°

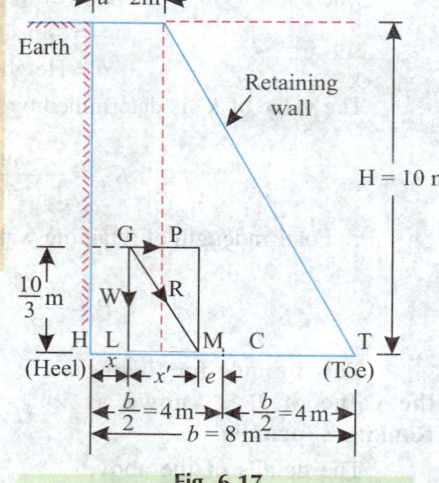

Fig. 6.17

Solution. Refer to Fig. 6.17.

Top width = 2 m;
Base width = 8 m;
Height = 10 m;
Density of earth = 16 kN/m³;
Density of masonry = 24 kN/m³;

Angle of repose of earth, $\phi = 30°$.

Consider 1 m length of the retaining wall.

Weight of masonry acting through c.g.,

$$W = \left(\frac{8 + 2}{2}\right) \times 10 \times 1 \times 24 = 1200 \text{ kN}$$

Horizontal thrust (P) of earth acting at $\frac{10}{3}$ m height,

$$P = \frac{w_e H^2}{2}\left(\frac{1 - \sin \phi}{1 + \sin \phi}\right)$$

$$= \frac{16 \times 10^2}{2}\left(\frac{1 - \sin 30°}{1 + \sin 30°}\right) = 266.67 \text{ kN}$$

Let x metre be distance of c.g. from the vertical base. Dividing the trapezium into a rectangle and a triangle and taking moments about the vertical face, we get

$$10 \times 2 \times \frac{2}{2} + \frac{1}{2} \times 6 \times 10 \times \left(\frac{6}{3} + 2\right) = \left(10 \times 2 + \frac{1}{2} \times 6 \times 10\right) x$$

$$20 + 120 = 50\, x$$

∴ $x = 2.8$ m.

From the Fig. 6.17,

$$\tan \alpha = \frac{P}{W} = \frac{266.67}{1200}$$

Also, $$\tan \alpha = \frac{x'}{10/3}$$

∴ $$\frac{x'}{10/3} = \frac{266.67}{1200}$$

or $\qquad x' = \dfrac{266.67 \times 10}{3 \times 1200} = 0.74$ m.

Now, $\qquad x + x' = 2.8 + 0.74 = 3.54$ m

which is *well within two-third base width*.

Eccentricity of resultant thrust,

$$e = 4 - 3.54 = 0.46 \text{ m}$$

B.M. due to eccentricity e,

$$M = W \times e = 1200 \times 0.46 = 552 \text{ kN m}$$

Stress due to B.M.:

Using the relation: $\qquad \dfrac{M}{I} = \dfrac{\sigma}{y}$

∴ Bending stress, $\qquad \sigma_b = \dfrac{My}{I} = \dfrac{(W \times e) \times \dfrac{8}{2}}{\dfrac{1 \times 8^3}{12}} = \dfrac{552 \times 4 \times 12}{8^3} = \pm 51.75 \text{ kN/m}^2$

Direct stress, $\qquad \sigma_d = \dfrac{\text{Weight of masonry}}{\text{Area of base}} = \dfrac{1200}{8 \times 1} = 150 \text{ kN/m}^2$

$$\sigma_{max} = \sigma_d + \sigma_b = 150 + 51.75$$
$$= \mathbf{201.75 \text{ kN/m}^2} \textit{ (comp.)} \textbf{ (Ans.)}$$

$$\sigma_{min} = \sigma_d - \sigma_b = 150 - 51.75$$
$$= \mathbf{98.25 \text{ kN/m}^2} \textit{ (comp.)} \textbf{ (Ans.)}$$

Example 6.15. *A retaining wall is 3 m wide at the top and 8 m wide at the bottom and is 18 m high. It is subjected to earth pressure on the back. If the weight of masonry is 25 kN/m³ and that of earth 16 kN/m³ and the angle of repose of earth be 30° and top of the earth is horizontal and level with the top of the wall, find the maximum and minimum intensities of pressure on the base. Examine the stability of the wall if μ = 0.62.*

Solution. Refer to Fig. 6.18.

Top width of the wall, $a = 3$ m

Base width of the wall,

$$b = 8 \text{ m}$$

Height of the wall $\qquad = 18$ m

Angle of repose of earth,

$$\phi = 30°$$

Considering 1 m length of wall, we have :

Weight of masonry,

$$W = \left(\dfrac{a+b}{2}\right) \times 25 \times 18$$

$$= \dfrac{3+8}{2} \times 25 \times 18$$

$$= 2475 \text{ kN}$$

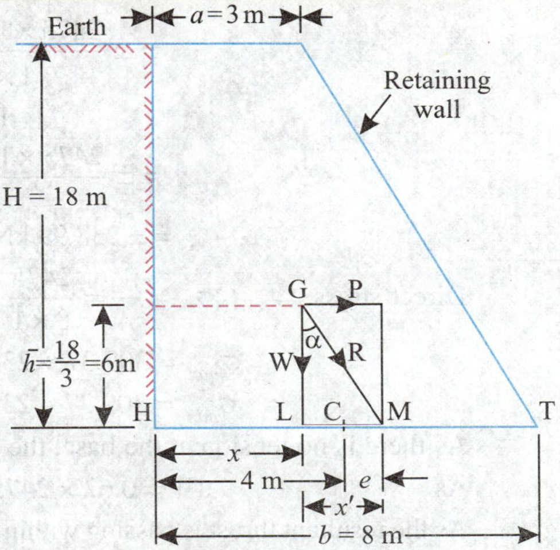

Earth pressure, $\qquad P = \dfrac{w_e H^2}{2}\left[\dfrac{1-\sin\phi}{1+\sin\phi}\right]$

$$= \dfrac{16 \times 18^2}{2}\left(\dfrac{1-\sin 30°}{1+\sin 30°}\right) = 864 \text{ kN}$$

Fig. 6.18

Distance of *c.g.* from the vertical face,

$$x = \frac{a^2 + ab + b^2}{3(a + b)}$$

$$= \frac{3^2 + 3 \times 8 + 8^2}{3(3 + 8)}$$

$$= \frac{9 + 24 + 64}{33} = 2.94 \text{ m}$$

(The value of *x* may also be
found out by taking moments)

From Fig. 6.18,

$$\tan \alpha = \frac{P}{W} = \frac{864}{2475}$$

Also,

$$\tan \alpha = \frac{x'}{18/3} = \frac{x'}{6}$$

∴

$$\frac{x'}{6} = \frac{864}{2475}$$

∴

$$x' = 2.09 \text{ m}$$

∴ The resultant thrust cuts the base at $(x + x') = (2.94 + 2.09) = 5.03$ m from H (Heel).
Eccentricity of the resultant thrust,

$$e = (x + x') - \frac{8}{2}$$

$$= 5.03 - 4 = 1.03 \text{ m}$$

Bending stress,

$$\sigma_b = \pm \frac{M}{I} \cdot y = \pm \frac{(W \times e) \times y}{I}$$

$$= \pm \frac{2475 \times 1.03 \times 8/2}{\frac{1 \times 8^3}{12}}$$

$$= \pm \frac{2475 \times 1.03 \times 4 \times 12}{8^3}$$

$$= \pm 238.99 \text{ kN/m}^2$$

Direct stress,

$$\sigma_d = \frac{W}{A} = \frac{2475}{8 \times 1} = 309.37 \text{ kN/m}^2$$

$$\sigma_{max.} = 309.37 + 238.99 = \textbf{548.36 kN/m}^2 \textbf{ (comp.)} \qquad \textbf{(Ans.)}$$

$$\sigma_{min.} = 309.37 - 238.99 = \textbf{70.38 kN/m}^2 \textbf{ (comp.)} \qquad \textbf{(Ans.)}$$

As there is no tension at the base, the *failure cannot occur due to tension.*

As $\qquad \mu W = 0.62 \times 2475 = 1534.5 \text{ kN} > P$, the *wall cannot fail by sliding.*

As the resultant thrust is passing within the base, therefore, the *wall is safe against overturning.*

Hence the wall is safe in every way.

Example 6.16. *A masonry retaining wall, trapezoidal in cross-section is 12 m high and has one face vertical and other battered 1 in 6. It retains earth at its vertical face, level with its top. Calculate its base width for no tension at base. Earth weighs 16 kN/m³ and masonry weighs 24/m³, angle of repose of earth = 30°.*

Solution. Refer to Fig. 6.19.

Let, a = width at the top

                 b = width at the base

Then, $a = b - 12 \times 1/6 = b - 2$

Consider 1 m length of the wall.

Weight of masonry,

$$W = \left(\frac{a+b}{2}\right) \times H \times 1 \times 24$$

$$= \frac{(b-2)+b}{2} \times 12 \times 24$$

$$= 144(2b - 2) \text{ kN}$$

Total earth pressure,

$$P = \frac{w_e H^2}{2}\left(\frac{1 - \sin\phi}{1 + \sin\phi}\right)$$

$$= \frac{16 \times 12^2}{2}\left(\frac{1 - \sin 30°}{1 + \sin 30°}\right) = 384 \text{ kN}$$

Also, $x = \dfrac{a^2 + ab + b^2}{3(a+b)}$

$$= \frac{(b-2)^2 + (b-2)b + b^2}{3(b-2+b)} = \frac{b^2 - 4b + 4 + b^2 - 2b + b^2}{6(b-1)}$$

$$= \frac{3b^2 - 6b + 4}{6(b-1)}$$

Again, $HM = x + x' = \dfrac{3b^2 - 6b + 4}{6(b-1)} + \dfrac{P}{W} \cdot \dfrac{h}{3}$

$$= \frac{3b^2 - 6b + 4}{6(b-1)} + \frac{384}{144(2b-2)} \times \frac{12}{3}$$

$$= \frac{3b^2 - 6b + 4}{6(b-1)} + \frac{5.33}{(b-1)}$$

But for tension,

$$HM = x + x' = \frac{2b}{3}$$

∴ $\dfrac{3b^2 - 6b + 4}{6(b-1)} + \dfrac{5.33}{(b-1)} = \dfrac{2b}{3}$

or, $3b^2 - 6b + 4 + 6 \times 5.33 = 6(b-1) \times 0.666\,b$

or, $3b^2 - 6b + 35.98 = 4(b-1)b$

or, $3b^2 - 6b + 35.98 = 4b^2 - 4b$

or, $b^2 + 2b - 35.98 = 0$

or, $b = \dfrac{-2 \pm \sqrt{4 + 4 \times 35.98}}{2}$

$$= \frac{-2 \pm 12.16}{2}$$

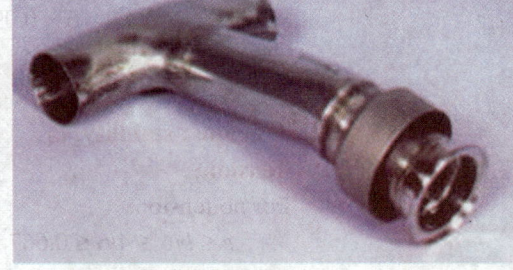

Air duct Tee of an aircraft.

= 5.08 m (neglecting – ve sign). **(Ans.)**

∴   Top width = $b - 2$

= 5.08 – 2 = **3.08 m**   **(Ans.)**

Example 6.17. *A masonry retaining wall is 1 m wide at the top and 4 m at the bottom and retains water level at its top. The wall is 5 m high. Test the stability of wall against:*

(i) *Tension,*   (ii) *Crushing.*

(iii) *Sliding, and*   (iv) *Overturning.*

    Given: Weight of masonry = 24 kN/m³; bearing capacity of soil = 240 kN/m³; co-efficient of friction, μ *= 0.6.*

Solution. Refer to Fig. 6.20.

$a = 1$ m, $b = 4$ m, $h = 5$ m

Consider 1 m length of the wall.

$$W = \left(\frac{1+4}{2}\right) \times 5 \times 1 \times 24 = 300 \text{ kN}$$

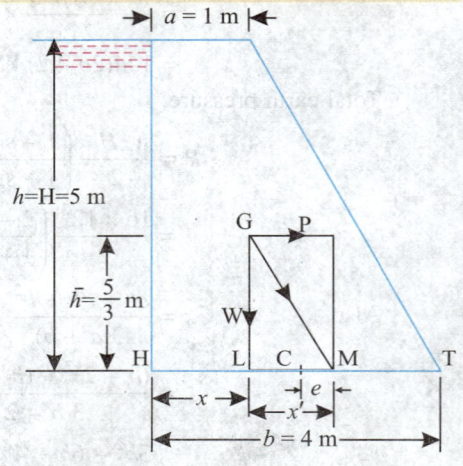

$$P = \frac{wh^2}{2} = \frac{10 \times 5^2}{2} = 125 \text{ kN}$$

$$x = \frac{a^2 + ab + b^2}{3\,(a+b)}$$

$$= \frac{1^2 + 1 \times 4 + 4^2}{3\,(1+4)} = \frac{1+4+16}{15} = \frac{21}{15}$$

= 1.4 m from the vertical face.

$HM = x + x'$

Fig. 6.20

$$= 1.4 + \frac{P}{W} \cdot \frac{h}{3} = 1.4 + \frac{125}{300} \times \frac{5}{3} = 2.094 \text{ m}$$

∴ $$e = (x + x') - \frac{b}{2}$$

$$= 2.094 - \frac{4}{2} = 0.094 \text{ m}$$

Again, $$\sigma_{max} = \frac{W}{b}\left(1 + \frac{6e}{b}\right)$$

$$= \frac{300}{4}\left(1 + \frac{6 \times 0.094}{4}\right)$$

$$= 85.57 \text{ kN/m}^2$$

Check against stability:

(i) Tension:

    For no tension:

        $e \le b/6 \le 4/6 \le 0.667$ m

    but 'e' as found out is 0.094 m, hence the wall is *safe against tension.*

(ii) Crushing:

    σ_{max} should be less than the bearing power of the soil

      $\sigma_{max} = 85.57 \text{ kN/m}^2 < 240 \text{ kN/m}^2$

    Hence the wall is *safe against crushing.*

(iii) **Sliding:**

The horizontal component of resultant R (should be less than frictional resistance),

$$R_H = P = 125 \text{ kN}$$

Frictional resistance $\quad = \mu W = 0.5 \times 300 = 180 \text{ kN}$

∴ Factor of safety against sliding

$$= \frac{\mu W}{P}$$

$$= \frac{180}{125} = 1.44$$

Since it is greater than one and hence the retaining wall is *safe against sliding.*

(iv) **Overturning:**

Overturning moment $\quad = P \times h/3$

$$= 125 \times 5/3 = 208.33 \text{ kN m}$$

Restoring moment $\quad = W (b - x)$

$$= 300 (4 - 1.4)$$

$$= 780 \text{ kN m}$$

∴ Factor of safety against overturning

$$= \frac{\text{Restoring moment}}{\text{Overturning moment}}$$

$$= \frac{780}{208.33} = 3.74$$

Hence *safe.*

TYPICAL EXAMPLES (For Competitive Examinations)

Example 6.18. *Fig. 6.21 shows a rectangular plate with a hole drilled in it. Determine the greatest and the least intensities of stress at the critical cross-section of the plate when subjected to an axial pull of 90 kN.*

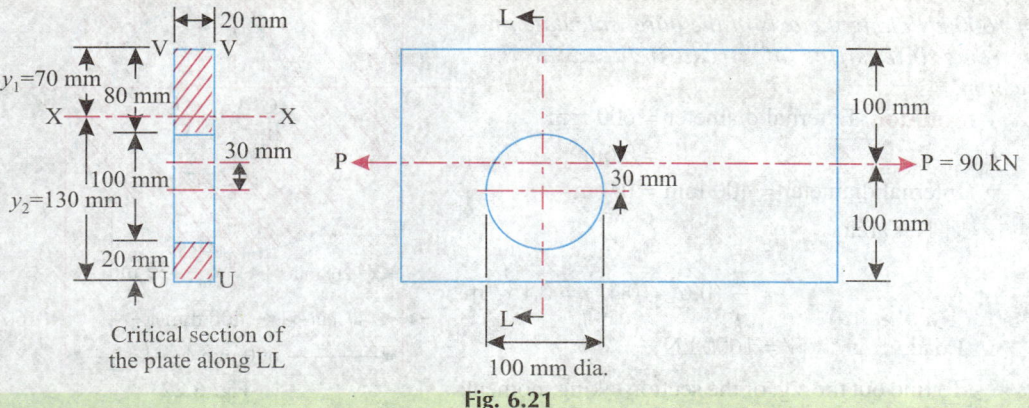

Critical section of the plate along LL

100 mm dia.

Fig. 6.21

Solution. Area at the critical/weakest section $= 80 \times 20 + 20 \times 20 = 2000 \text{ mm}^2 = 2 \times 10^{-3} \text{ m}^2$

Axial load, $P = 90 \text{ kN}$

For locating centroidal axis, taking moments about the edge UU, we have

$$y_2 = \frac{20 \times 20 \times 10 + 80 \times 20 \times 160}{(20 \times 20) + (80 \times 20)} = 130 \text{ mm} = 0.13 \text{ m}$$

$$y_1 = 200 - 130 = 70 \text{ mm} = 0.07 \text{ m}$$

Moment of inertia about the *XX*-axis,

$$I_{XX} = \left[\frac{20 \times 80^3}{12} + 20 \times 80 \times (70 - 40)^2 \right] + \left[\frac{20 \times 20^3}{12} + 20 \times 20 \times (130 - 10)^2 \right]$$

$$= 2293333 + 5773333 = 8066666 \text{ mm}^4 = 8.066 \times 10^{-6} \text{ m}^4$$

Eccentricity, $e = 100 - y_1 = 100 - 70 = 30 \text{ mm}$

[or $e = y_2 - 100 = 130 - 100 = 30 \text{ mm}$]

Direct stress, $\sigma_d = \dfrac{P}{A} = \dfrac{90}{2 \times 10^{-3}} \times 10^{-3} \text{ MN/m}^2 = 45 \text{ MN/m}^2$ (*i.e.* tensile)

Bending stress, $\sigma_b = \pm \dfrac{My}{I} = \pm \dfrac{(P \times e) \times y}{I}$

∴ *Maximum* stress along the edge *UU*,

$$\sigma_{UU} = -45 - \frac{(P \times e)\, y_2}{I_{XX}} = -45 - \frac{90 \times (30 \times 10^{-3}) \times 0.13}{8.066 \times 10^{-6}} \times 10^{-3}$$

$$= -88.5 \text{ kN/m}^2$$

∴ $\boldsymbol{\sigma_{UU} = 88.5 \text{ MN/m}^2}$ (*tensile*) **(Ans.)**

Minimum stress along the edge *VV*,

$$\sigma_{VV} = -45 + \frac{(P \times e) \times y_1}{I_{XX}} = -45 + \frac{90 \times (30 \times 10^{-3}) \times 0.07}{8.066 \times 10^{-6}} \times 10^{-3}$$

$$= -21.57 \text{ MN/m}^2$$

∴ $\boldsymbol{\sigma_{VV} = 21.57 \text{ MN/m}^2}$ (*tensile*) **(Ans.)**

Example 6.19. *A short hollow cast iron column having an external diameter of 600 mm and inside diameter 400 mm was cast in a factory. On inspection it was found that the bore is eccentric as shown in the Fig. 6.22. If the column carries a load of 1600 kN along the axis of the bore, calculate the extreme intensities of stress induced in the section.*

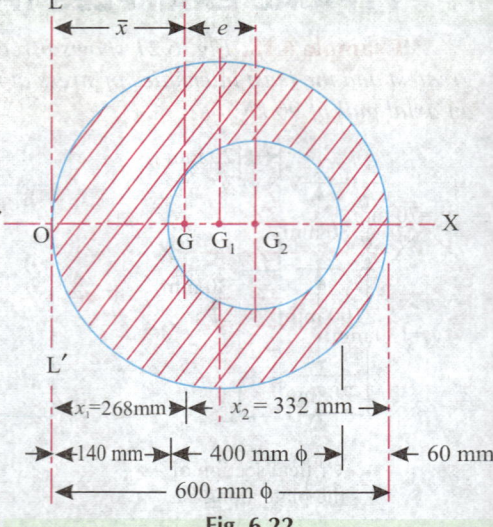

Fig. 6.22

Solution. External diameter = 600 mm

= 0.6 m

Internal diameter = 400 mm = 0.4 m

∴ Net area

$$= \frac{\pi}{4} (0.6^2 - 0.4^4) = 0.157 \text{ m}^2$$

Load, $W = 1600 \text{ kN}$

To find out the c.g. of the section taking moments about *LL'*, we get

$$OG = \bar{x} = \frac{(\pi/4) \times 0.6^2 \times OG_1 - (\pi/4) \times 0.4^2 \times OG_2}{(\pi/4) \times 0.6^2 - (\pi/4) \times 0.4^2}$$

(where $OG_1 = 0.3$ m, $OG_2 = 0.34$ m)

$$= \frac{(\pi/4) \times 0.6^2 \times 0.3 - (\pi/4) \times 0.4^2 \times 0.34}{(\pi/4) \times 0.6^2 - (\pi/4) \times 0.4^2} = 0.268 \text{ m} = 268 \text{ mm}$$

So,
$$GG_1 = 0.3 - 0.268 = 0.032 \text{ m}$$
$$GG_2 = 0.34 - 0.268 = 0.072 \text{ m}$$

∴ Eccentricity $= OG_2 - \bar{x} = 0.34 - 0.268 = 0.072 \text{ m} = 72 \text{ mm}$

Moment, $M = 1600 \times 0.072 = 115.2 \text{ kNm}$

$$I_{YY} = \left[\frac{\pi}{64} \times (0.6)^4 + \frac{\pi}{4} \times 0.6^2 \times (GG_1)^2\right] - \left[\frac{\pi}{64} \times 0.4^4 + \frac{\pi}{4} \times 0.4^2 \times (GG_2)^2\right]$$

$$= \left[\frac{\pi}{64} \times (0.6)^4 + \frac{\pi}{4} \times 0.6^2 \times 0.032^2\right] - \left[\frac{\pi}{64} \times 0.4^4 + \frac{\pi}{4} \times 0.4^2 \times 0.072^2\right]$$

$$= \frac{\pi}{64}(0.6^4 - 0.4^4) + \frac{\pi}{4}[0.6^2 \times 0.032^2 - 0.4^2 \times 0.072^2]$$

$$= 5.105 \times 10^{-3} - 3.619 \times 10^{-4} = 4.743 \times 10^{-3} \text{ m}^4$$

Bending stress (*tensile*)
$$\sigma_{bt} = \frac{Mx_1}{I_{YY}} = \frac{115.2 \times 0.268}{4.743 \times 10^{-3}} \times 10^{-3} = 6.51 \text{ MN/m}^2$$

Bending stress (compressive)
$$\sigma_{bc} = \frac{Mx_2}{I_{YY}} = \frac{115.2 \times 0.332}{4.743 \times 10^{-3}} \times 10^{-3} = 8.06 \text{ MN/m}^2$$

Direct stress, $\sigma_d = \dfrac{W}{A} = \dfrac{1600}{0.157} \times 10^{-3} = 10.19 \text{ MN/m}^2 \text{ (comp.)}$

∴ *Maximum compressive stress,*
$$\sigma_{max} = \sigma_d + \sigma_{bc} = 10.19 + 8.06 = \textbf{18.25 MN/m}^2 \quad \textbf{(Ans.)}$$

Minimum compressive stress,
$$\sigma_{min} = \sigma_d - \sigma_{bt} = 10.19 - 6.51 = \textbf{3.68 MN/m}^2 \quad \textbf{(Ans.)}$$

Example 6.20. *A steel rod 40 mm diameter passes through a copper tube 60 mm internal diameter and 80 mm external diameter. Rigid cover plates are provided at each end of the tube and steel rod passes through these cover plates also. Nuts are screwed on the projecting ends of the rod so that the cover plates put pressure on the ends of the tube. If the centre of the rod is 10 mm out of the centre of the tube, determine the maximum stress in the copper tube when one of the nuts is tightened to produce a linear strain of 0.002 in the rod.*

Take: $E_{\text{steel}} = 210 \text{ GN/m}^2$, and $E_{\text{copper}} = 105 \text{ GN/m}^2$.

Solution. Diameter of the steel rod, $d_s = 40$ mm

External diameter of the copper tube, $D_c = 80$ mm

Internal diameter of the copper tube, $d_c = 60$ mm

∴ Area of cross-section of the steel rod,

Gears made of powder metallurgy.

$$A_s = \frac{\pi}{4} \times 40^2 = 1256.6 \text{ mm}^2 = 1256.6 \times 10^{-6} \text{ m}^2$$

and area of cross-section of the copper tube,

$$A_c = \frac{\pi}{4}(80^2 - 60^2) = 2199.1 \text{ mm}^2 = 2199.1 \times 10^{-6} \text{ m}^2$$

Strain in the steel rod, $e_s = 0.002$

For equilibrium,

Pull in the steel rod (P_s) = Push in copper tube (P_c)

$$\sigma_s \times A_s = \sigma_c \times A_c$$

$$(e_s \times E_s) \times A_s = \sigma_c \times A_c$$

$$0.002 \times 210 \times 10^9 \times 1256.6 \times 10^{-6} = \sigma_c \times 2199.1 \times 10^{-6}$$

or, $$527.77 \times 10^3 = \sigma_c \times 2199.1 \times 10^{-6}$$

∴ $$\sigma_c = 240 \text{ MN/m}^2 \text{ (compressive)} \quad \text{...(direct stress)}$$

Refer to Fig. 6.23.

Eccentricity, $$e = 10 \text{ mm} = 0.01 \text{ m}$$

Moment, $$M = P \times e \text{ (when } P = P_s = P_c) = 527.77 \times 0.01 \text{kNm}$$

Maximum bending stress in the tube, $$\sigma_b = \sigma_c' = \pm \frac{My}{I_{YY}}$$

where, $$y = \frac{80}{2} = 40 \text{ mm} = 0.04 \text{ m}$$

and, $$I_{YY} = \frac{\pi}{64} (80^4 - 60^4) \text{ mm}^4 = 1.374 \times 10^{-6} \text{ m}^4$$

∴ $$\sigma_b = \pm \frac{(527.77 \times 0.01) \times 0.04}{1.374 \times 10^{-6}} \times 10^{-3} = 153.6 \text{ MN/m}^2$$

∴ Maximum stress in the copper tube

$$(\sigma_c)_{max} = \sigma_c + \sigma_c' = 240 + 153.6 = \textbf{393.6 MN/m}^2 \quad \textbf{(Ans.)}$$

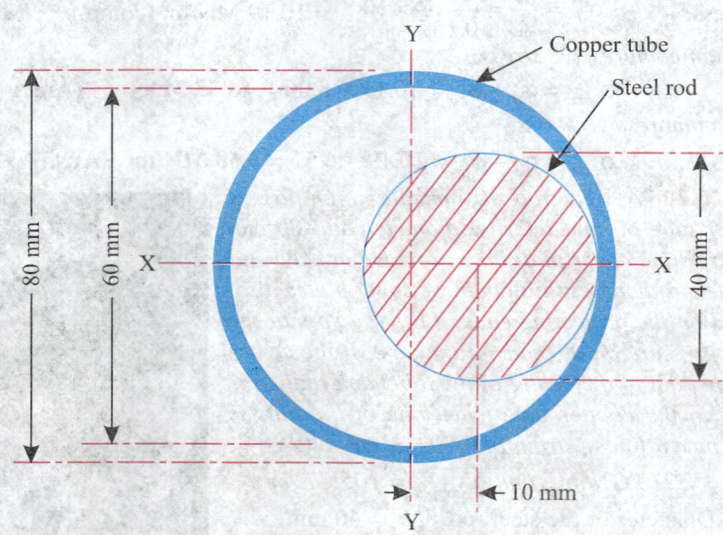

Y — Copper tube

Steel rod

80 mm   60 mm   X —— X   40 mm

10 mm

Y

Fig. 6.23

Example 6.21. *A tapering chimney of hollow circular section is 30 metres high. Its external diameter at the base is 2.4 m and at the top it is 1.6 m. It is subjected to the wind pressure of 2.2 kN/m² of the projected area. If the weight of the chimney is 4000 kN and internal diameter at the base is 0.8 m, determine the maximum and minimum stress intensities at the base.*

Solution. External diameter (at the base) = 2.4 m

Internal diameter (at the base) = 0.8 m

∴ Area of cross-section (at the base),

$$A = \frac{\pi}{4} (2.4^2 - 0.8^2) = 4.02 \text{ m}^2$$

Wind pressure, $p = 2.2$ kN/m^2

Weight of chimney, $W = 4000$ kN

Direct stress at the base due to weight of the chimney,

$$\sigma_d = \frac{W}{A} = \frac{400}{4.02} = 995 \text{ MN/m}^2$$

(compressive)

Projected area of the exposed surface of the chimney,

$$A_p = \left(\frac{2.4 + 1.6}{2}\right) \times 30 = 60 \text{ m}^2$$

Total force due to wind,

$$P = p \times A_p = 2.2 \times 60 = 132 \text{ kN}$$

Distance of the centroid of trapezoid *pqrs* from the base

$$= \frac{1.6 \times 30 \times \dfrac{30}{2} + 2\left[\left(\dfrac{2.4 - 1.6}{2 \times 2}\right) \times 30 \times \dfrac{30}{3}\right]}{1.6 \times 30 + 2 \times \dfrac{1}{2}\left(\dfrac{2.4 - 1.6}{2}\right) \times 30}$$

$$= \frac{720 + 120}{48 + 12} = 14 \text{ m}$$

Bending moment,

$$M = P \times 14 = 132 \times 14 = 1848 \text{ kNm}$$

Bending stress,

$$\sigma_b = \pm \frac{M \times y}{I}$$

where,

$$y = \frac{2.4}{2} = 1.2 \text{ m}$$

$$I = \frac{\pi}{64}(2.4^4 - 0.8^4) = 1.608 \text{ m}^4$$

$$\therefore \sigma_b = \pm \frac{1848 \times 1.2}{1.608} = 1379.1 \text{ MN/m}^2$$

Maximum stress at the base,

$$\sigma_{max} = \sigma_d + \sigma_b = 995 + 1379.1$$

$$= \textbf{2374.1 MN/m}^2 \textit{ (compressive)}$$

(Ans.)

Minimum stress at the base,

$$\sigma_{min} = \sigma_d - \sigma_b = 995 - 1379.1$$

$$= \textbf{– 384.1 MN/m}^2$$

$$= \textbf{384.1 MN/m}^2 \textit{ (tensile)} \textbf{ (Ans.)}$$

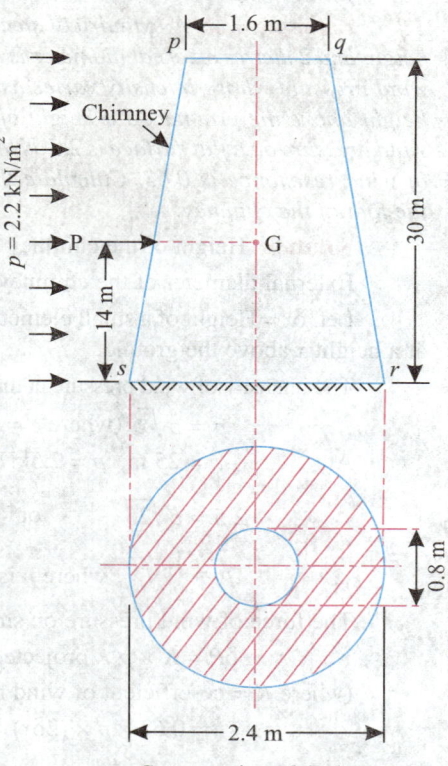

Cross-section of the base

Fig. 6.24

Machine assembled from casted components.

Example 6.22. *A cylindrical steel chimney of 45 metres height and 2 metres external diameter is exposed to a horizontal wind pressure whose intensity varies as the square root of the height above the ground. At a height of 25 m, the intensity of wind pressure on a flat surface is 2.5 kN/m², and the co-efficient of wind resistance is 0.62. Calculate the bending moment at the foot of the chimney.*

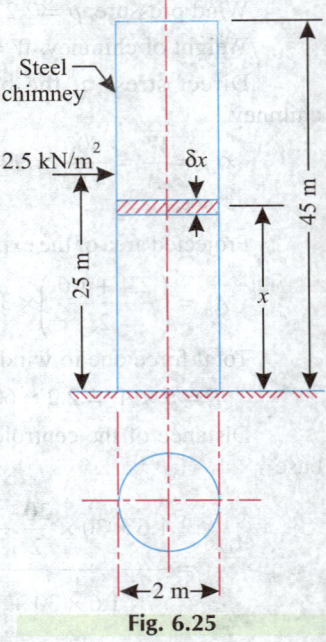

Fig. 6.25

Solution. Height of the chimney = 45 m

External diameter of the chimney = 2 m

Let, δx = Height of a small element of the exposed surface at a height x above the ground.

The intensity of wind pressure at that height is given by,

$$p = c\sqrt{x} \quad \text{(where } c = \text{constant)}$$

At, $\qquad x = 25$ m, $p = 2.5\text{kN/m}^2 \qquad \qquad \text{...(given)}$

$\therefore \qquad 2.5 = c\sqrt{25} \qquad \text{or} \qquad c = \dfrac{1}{2}$

$\therefore \qquad p = \dfrac{1}{2}\sqrt{x}$, where p is in kN/m² and x in metres.

The force of wind pressure on small element δx is

$$\delta P = K \times p \times \text{projected area}$$

(where K = co-efficient of wind resistance)

$$= 0.62 \times p \times (2\delta x)$$

$$= 0.62 \times \frac{1}{2}\sqrt{x} \times (2\delta x) = 0.62\sqrt{x}\,\delta x$$

Moment of δP about the base $= \delta P . x$

or, $\qquad \delta M = 0.62\sqrt{x} . x\,\delta x = 0.62\,x^{3/2}\,\delta x$

$\therefore$ Total moment at the base,

$$M = \Sigma x . \delta P = \int_0^{45} 0.62\,x^{3/2}\,dx = 0.62\left[\frac{2}{5}x^{5/2}\right]_0^{45}$$

$$= \frac{0.62 \times 2}{5}(45)^{5/2} = 3358.8 \text{ kNm}$$

i.e., $\qquad M = 3368.8$ **kNm (Ans.)**

Example 6.23. *Fig. 6.26 shows a forged crank with a section along the line LL. The thrust P of the connecting rod acts on the crank at an angle θ as shown. If the stress at the section LL were not to exceed 30 N/mm², find the maximum value of the thrust P.*

Assume the ZZ axis passes through the centroid of the section LL.

Solution. Allowable stress at the section *LL*,

$$\sigma_{max} = 30 \text{ N/mm}^2$$

Area of cross-section at the section *LL*,

$$A = 150 \times 75 - 30 \times 100 = 8250 \text{ mm}^2$$

Moment of inertia about the neutral axis,

$$I_{NA} = \frac{75 \times 150^3}{12} - \frac{30 \times (100)^3}{12} = 18.59 \times 10^6 \text{ mm}^4$$

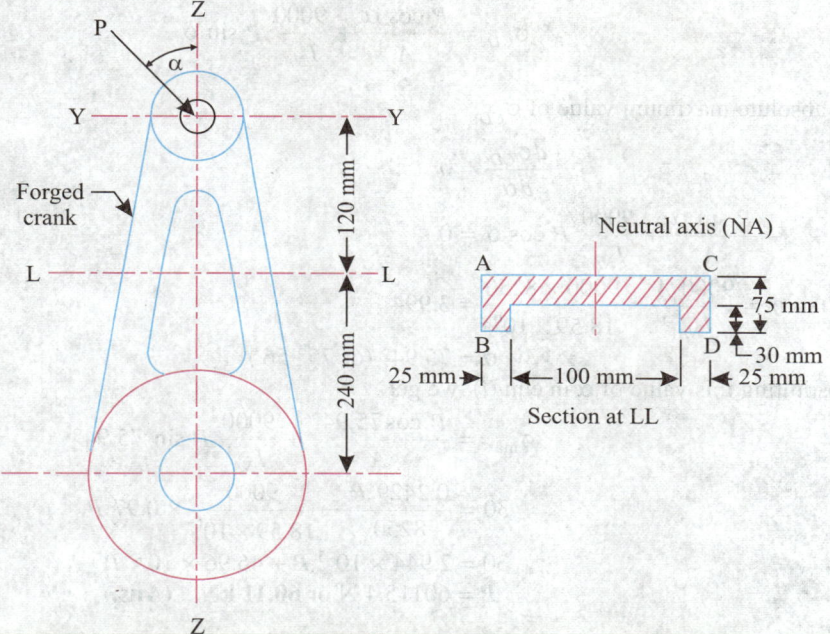

Fig. 6.26

Thrust P:

The thrust P can be resolved into the following two components:

(i) $P_z = P \cos \alpha$ (This component gives direct stress on the section)

(ii) $P_y = P \sin \alpha$

[This component gives rise to a bending moment of magnitude $(P_y \times 120)$ over the section *LL*, due to which bending stress will be induced.]

The resultant compressive stress along the line *CD*, for any value of α, will be greatest and its magnitude is given by

$$\sigma_{CD} = \frac{P_z}{A} + \frac{120 \, P_y}{I_{NA}} \times \left(\frac{150}{2}\right)$$

In machine boats and ships, the propeller shaft and blades, located at the rear, below water, bear high bending and direct stresses.

or,
$$\sigma_{CD} = \frac{P \cos \alpha}{A} + \frac{9000}{I_{NA}} P \sin \alpha \qquad ...(i)$$

For absolute maximum value of σ_{CD},

$$\frac{d\sigma_{CD}}{d\alpha} = 0$$

$\therefore \qquad -\dfrac{P \sin \alpha}{A} + \dfrac{9000}{I_{NA}} P \cos \alpha = 0$

or, $\tan \alpha = \dfrac{9000 \, A}{I_{NA}} = \dfrac{9000 \times 8250}{18.59 \times 10^6} = 3.994$

$\therefore \qquad\qquad\qquad\qquad \alpha = 75.94° \text{ (or } 75° \, 56')$

Substituting this value of α in eqn (i), we get

$$\sigma_{max} = \frac{P \cos 75.94°}{A} + \frac{9000}{I_{NA}} P \sin 75.94°$$

$$30 = \frac{0.2429 \, P}{8250} + \frac{9000}{18.59 \times 10^6} \times 0.97 \, P$$

or, $\qquad\qquad 30 = 2.944 \times 10^{-5} P + 46.96 \times 10^{-5} P$

$\therefore \qquad\qquad P = 60115.4 \text{ N or } \textbf{60.11 kN} \quad \textbf{(Ans.)}$

HIGHLIGHTS

1. For *no tension* in the section, the eccentricity must not exceed $\dfrac{2k^2}{d}$ *i.e.*, $e \le \dfrac{2k^2}{d}$

   where, $\qquad\qquad\qquad\qquad d =$ Depth of the section, and

   $\qquad\qquad\qquad\qquad\qquad k =$ Radius of gyration.

2. In case of *rectangular section*, the stress will be of the same sign throughout the section if the load line lies within the *middle third* of *the section*.

3. In case of a *circular section* the diameter of the kernel is $d/4$, where d is the diameter of the section.

4. In case of a hollow circular section the diameter of the kernel is $\dfrac{D^2 + d^2}{4D}$, where D and d are the external and internal diameters respectively of the section.

5. The total horizontal wind force (P) is calculated by using the following relation:

   $$P = K p A_p$$

   where, $\qquad\qquad\qquad\qquad K =$ Co-efficient of wind resistance,

   $\qquad\qquad\qquad\qquad\qquad p =$ Horizontal intensity of wind pressure, and

   $\qquad\qquad\qquad\qquad A_p =$ Projected area of the surface exposed to wind.

OBJECTIVE TYPE QUESTIONS

1. An eccentric load W, with eccentricity e, is equivalent to

   (*a*) an axial load W

   (*b*) a moment equal to $W \times e$

   (*c*) both (*a*) and (*b*)

   (*d*) none of the above.

2. For no tension in the section, the eccentricity must not exceed

   (*a*) $\dfrac{k^2}{d}$ $\qquad\qquad$ (*b*) $\dfrac{2k^2}{d}$

   (*c*) $\dfrac{4k^2}{d}$ $\qquad\qquad$ (*d*) $\dfrac{k}{\sqrt{d}}$.

(where, d = depth of the section, k = radius of gyration)

3. _____ loading induces, direct and bending stress at the section.
   (a) Uniformly distributed
   (b) Eccentric
   (c) Either of the above
   (d) None of the above.

4. The diameter of kernel of a circular section is
   (a) $\dfrac{d}{2}$
   (b) $\dfrac{d}{3}$
   (c) $\dfrac{d}{4}$
   (d) $\dfrac{d}{\sqrt{2}}$.

5. The diameter of kernel of a hollow circular section is
   (a) $\dfrac{D+d}{D}$
   (b) $\dfrac{D^2+d^2}{D}$
   (c) $\dfrac{D^2+d^2}{2D}$
   (d) $\dfrac{D^2+d^2}{4D}$.

6. In a rectangular section the stress will be of the same sign throughout the section if the load lies within the_____of the section.
   (a) middle third
   (b) middle half
   (c) either of the above
   (d) none of the above.

7. The total horizontal wind force = Co-efficient of wind resistance × horizontal intensity of wind pressure × _____.
   (a) cross-sectional area
   (b) projected area
   (c) either of the above
   (d) none of the above.

8. The brick chimney is stable if the resultant thrust lies within the middle _____
   (a) third
   (b) half
   (c) either of the above
   (d) none of the above.

ANSWERS

1. (c) 2. (b) 3. (b) 4. (c) 5. (d) 6. (a) 7. (b) 8. (a)

THEORETICAL QUESTIONS

1. What are the effects of eccentric loads on a short column?
2. Draw the stress distribution diagram in case of an eccentrically loaded column.
3. What is the middle third rule?
4. Give the limits of eccentricity in the following cases:
   (i) Rectangular section (ii) Circular section (iii) Hollow circular section.

UNSOLVED PROBLEMS

1. A short hollow cylindrical column has 300 mm external diameter and 25 mm metal thickness. It carries a vertical load of 1000 kN which is off the geometric axis by 20 mm. Calculate the maximum and minimum normal stress intensities induced in the section.
   [**Ans.** σ_{max} = 60.87 MN/m^2 (comp.); σ_{min} = 31.73 MN/m^2 (comp.)]

2. A cast iron column of hollow circular section has a projecting bracket carrying a load of 100 kN. The load line is off the axis of the column by 300 mm. The external diameter of the column is 300 mm and thickness of metal 25 mm. Find the maximum and minimum stress intensities.
   [**Ans.** σ_{max} = 26.49 MN/m^2 (comp.); σ_{min} = 17.23 MN/m^2 (tensile)]

3. A short hollow cast iron column having 25 cm external diameter and 20 cm internal diameter carries an axial load of 250 kN. It also carries a load of 50 kN on a bracket, such that the load line is 50 cm from the axis of the column. Find the maximum and minimum stress intensities on the section.
   [**Ans.** σ_{max} = 44.6 MN/m^2 (comp.); σ_{min} = 10.6 MN/m^2 (comp.)]

4. A hollow cast iron column of rectangular section 60 cm × 40 cm overall and 50 cm × 20 cm internally, carries a load of 1800 kN which is off the geometric axis by 10 cm in the vertical plane bisecting the thickness. Calculate the extreme intensities of stress induced in the section.
   [**Ans.** σ_{max} = 23.4 MN/m^2 (comp.); σ_{min} = 2.31 MN/m^2 (comp.)]

5. A short column of 100 mm diameter carries an eccentric load of 100 kN at an eccentricity of 5 mm from its axis. Determine the maximum compressive stress developed in the material. To what maximum extent can the load be increased without fear of exceeding the safe compressive strength of 35 MN/m² ? [**Ans.** σ_{max} = 17.82 MN/m², 196.36 kN]

6. A hollow rectangular masonry pier is 120 cm × 80 cm wide and 15 cm thick. A vertical load of 200 kN is transmitted in the vertical plane bisecting 120 cm side and at an eccentricity of 10 cm from the geometric axis of the section. Calculate the maximum and minimum stress intensities in the section. [**Ans.** σ_{max} = 0.61 MN/m² (comp.); σ_{min} = 0.17 MN/m² (comp.)]

7. A hollow rectangular column is having external and internal dimensions as 2.4 m × 1.8 m and 1.2 m × 1.2 m respectively. Calculate the safe load, that can be placed at an eccentricity of 0.5 m on a plane bisecting the longer side, if the maximum compressive stress is not to exceed 5 MN/m². [**Ans.** 720 kN]

8. A horizontal cross-section of a hollow rectangular pier is 2 m long and 1.5 m wide, overall, the wall thickness being 0.25 m. A vertical load of 350 kN is eccentric from the geometric centre of the section by 'e' cm measured along the centre line parallel to the longer sides. Calculate 'e' so that the maximum compressive stress on the section is twice the minimum. [**Ans.** e = 16 cm]

9. A hollow cast iron column of rectangular section 600 mm deep × 300 mm wide overall, thickness of metal 50 mm carries a load W in the vertical plane bisecting the width at an eccentricity e. If the extreme intensities of stress induced in the section are 8 MN/m² at one end to 50 MN/m² at the other, both compressive, evaluate W and e. [**Ans.** W = 232 MN; e = 100 mm]

10. A rectangular pier is subjected to a compressive load of 450 kN as shown in the Fig. 6.27.

    Find the stress intensities on all the four corners of the pier.

    [**Ans.** σ_L = 0.45 MN/m² (tensile); σ_M = 0.15 MN/m² (comp.); σ_N = 1.05 MN/m²; σ_P = 0.45 MN/m²]

11. A masonry chimney, 25 metres high, of uniform circular section, 4 metres external diameter and 2 metres internal diameter is subjected to a horizontal wind pressure of 1.2 kN/m² of projected area. Find the maximum and minimum stress intensities at the base, if the specific weight of masonry is 22 kN/m².

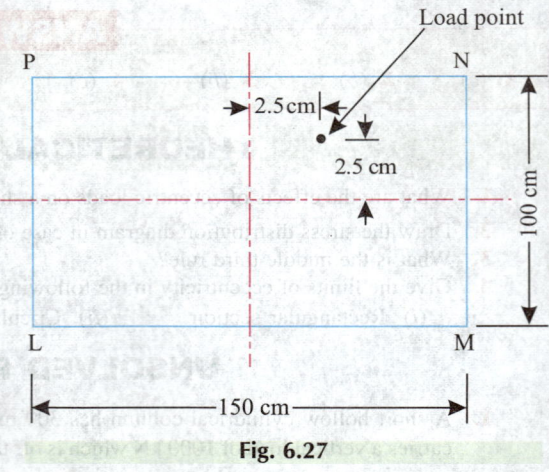

Fig. 6.27

[**Ans.** σ_{max} = 804.7 kN/m² (comp.); σ_{min} = 295.3 kN/m² (comp.)]

12. A square chimney 20 metres high, has an opening of 1.2 m × 1.2 m inside. The external dimensions are 2.4 m × 2.4 m. The horizontal intensity of wind pressure is 1.4 kN/m² and the specific weight of masonry is 22 kN/m³. Calculate the maximum and minimum stress intensities at the base of the chimney. [**Ans.** σ_{max} = 751 kN/m² (cop.); σ_{min} = 128 kN/m² (comp.)]

13. A cylindrical chimney 20 m high, of uniform circular section, is of 5 m external diameter and 3 m internal diameter. The intensity of horizontal wind pressure is 1.5 kN/m². Find maximum and minimum normal stress intensities at the base section, if the specific weight of masonry is 22 kN/m³. (**Hint:** Take K = 2/3) [**Ans.** σ_{max} = 533 kN/m² (comp.); σ_{min} = 346 kN/m² comp.)]

14. A long rectangular wall is 2 m wide. If the maximum wind pressure on the face wall is 1.2 kN/m², find the maximum height of wall so that there is no tension in the base of the wall. The wall masonry weighs 22 kN/m³. [**Ans.** h = 24.44 m]

15. A short cast-iron column has an external diameter of 200 mm and internal diameter of 160 mm, the distance between the centres of the outer and inner circles due to the displacement of the core during casting is 6 mm. A load of 400 kN acts through a vertical centre line passing through the centre of outer circle. Calculate the values of the greatest and least compressive stresses in a horizontal cross-section of the column.

[**Ans.** 46 MN/m^2, 26.7 MN/m^2]

16. A tapering chimney of hollow circular section is 45 m high. Its external diameter at the base is 3.6 m and at the top it is 2.4 m. It is subjected to the wind pressure of 2.2 kN / m^2 of the projected area. If the weight of the chimney is 6000 kN and the internal diameter at the base is 1.2 m, determine the maximum and minimum stress intensities at the base.

Sand and ceramic casting workshop.

[**Ans.** 2041.7 MN/m^2 (comp.); 715.5 MN/m^2 (tensile)]

17. A cylindrical chimney shaft 20 m high is of hollow circular section, 2.4 m external and 1 m internal diameters. The intensity of the horizontal wind pressure varies as $x^{2/3}$, where x is the height above the ground. If density of structure is 22.4 kN/m^3, co-efficient of wind resistance is 0.6 and wind pressure at a height of 27 m is 1.8 kN/m^2 determine the maximum and minimum intensities of stress at the base.

[**Ans.** 0.06897 MN/m^2; 0.2062 MN/m^2]

7

Shearing Stresses

7.1.   INTRODUCTION

The variation of bending stress from point to point in a beam has already been considered (in chapter 5) ignoring the effect of the shearing forces. Now since at every point it is assumed that the particles of materials are in equilibrium, the effect of the shearing forces and bending moments must be consistent.

There is involvement of two factors. The applied shearing force will be distributed as a shearing stress across transverse sections of the beam. But at each point on a section the transverse shearing stress will produce a complementary "*horizontal*" shearing stress; *i.e.* there will be shearing stresses acting between successive longitudinal layers of the beam, tending to resist sliding between these layers. *These longitudinal shearing stresses will balance the variation of bending stresses along the beam.* It should be noticed that if the *bending moment is consistent; there is no shearing force*, and hence no shearing stress. This is consistent, since without a variation of bending stress between successive transverse sections there can be no longitudinal shearing stresses.

7.2. SHEARING STRESS VARIATION

Let us consider a short slice of a beam (Fig. 7.1) of length δx, with a variation of bending moment over its length from M to $(M + \delta M)$. Take a layer rs at a distance y from the neutral axis; let the width of this layer be b and hence its area $(\delta x \times b)$. Between the successive faces of the slice there will be an excess force, since the stress at the face pr will be less than that on the faces qs. This excess force will be balanced by the shearing stress acting along the layer rs.

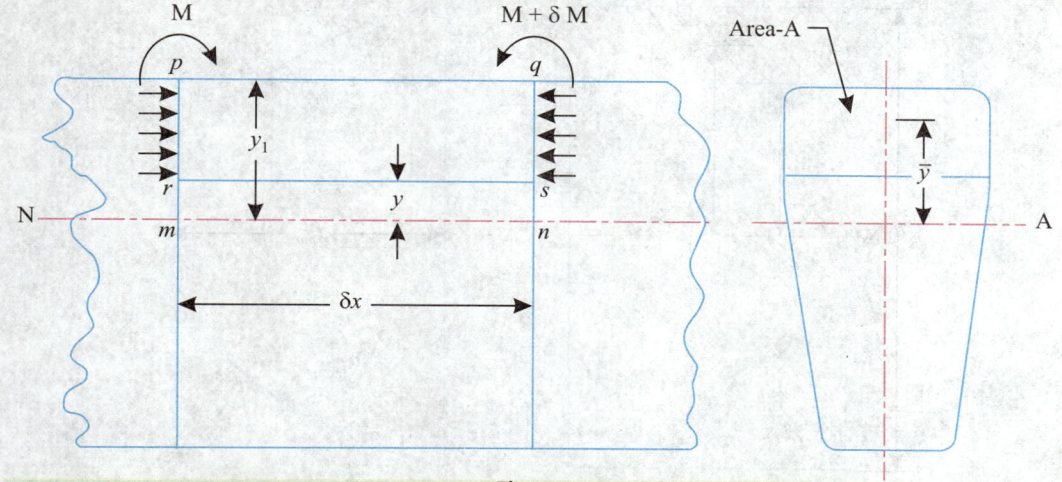

Fig. 7.1

Then the total force on the face pr,

$$= \sum_{y}^{y_1} \frac{My}{I} (\delta y \times b) = \int_{y}^{y_1} \frac{M}{I} (yb)\, dy$$

and total force on sq

$$= \int_{y}^{y_1} \frac{(M + \delta M)}{I} (y \cdot b)\, dy$$

or, the excess force between sq and rp

$$= \int_{y}^{y_1} \frac{\delta M}{I} (y \cdot b)\, dy = \frac{\delta M}{I} \int_{y}^{y_1} (y \cdot b)\, dy$$

But, $\displaystyle\int_{y}^{y_1} y.b\, dy$ is the moment of area of the face pr about the neutral axis.

or, $$\int_{y}^{y_1} y.b\, dy = A\bar{y}$$

This excess force is balanced by a shearing stress τ acting along rs. The force due to this stress is $\tau \times (b \cdot \delta x)$

$\therefore$ $$\tau \times (b.\delta x) = \frac{\delta M}{I} A\bar{y} \quad \text{or} \quad \tau = \frac{\delta M}{\delta x} \times \frac{1}{Ib} \times A\bar{y}$$

In the limit $\delta x \to 0$

$$\frac{\delta M}{\delta x} = S, \text{ the shearing stress } \tau = \frac{S A \bar{y}}{I b} \qquad \text{...(7.1)}$$

where b is the width of the beam at the point where the shearing force is considered, and $A\bar{y}$ is the moment of the sectional area above that point about the neutral axis.

7.3. VARIATION OF SHEAR STRESS IN BEAM CROSS-SECTION

1. Rectangular section:

Let b be the width and d the depth of rectangular section (Fig. 7.2). Let τ be the shear stress in a layer at distance y from N.A. where a particular section is subjected to shear force S.

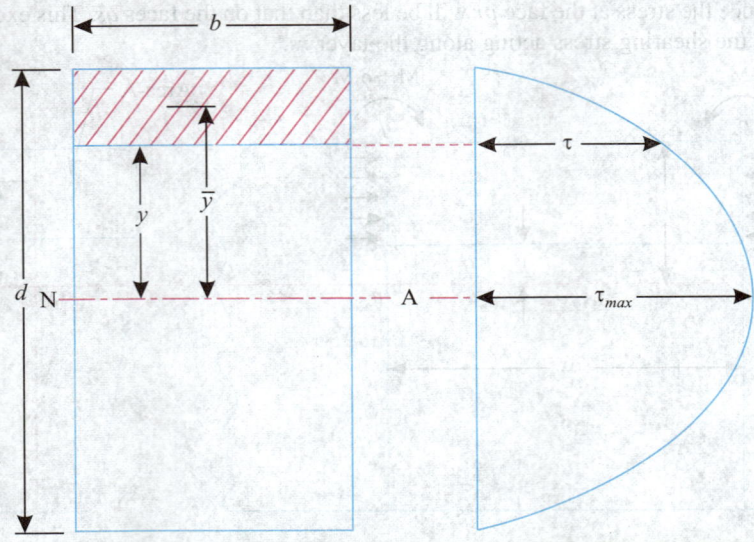

Fig. 7.2

Shaded area,
$$A = b\left(\frac{d}{2} - y\right), \quad \bar{y} = \frac{1}{2}(d/2 + y), \quad I = \frac{b\,d^3}{12}$$

$\therefore$
$$A\bar{y} = \frac{b}{2}\left(\frac{d}{2} - y\right)\left(\frac{d}{2} + y\right) = \frac{b}{2}\left(\frac{d^2}{4} - y^2\right)$$

Now,
$$\tau = \frac{S\,A\,\bar{y}}{I\,.\,b} = \frac{S \times \dfrac{b}{2}\left(\dfrac{d^2}{4} - y^2\right)}{\dfrac{bd^3}{12} \times b}$$

i.e.
$$\tau = \frac{12S \times b}{b \times bd^3 \times 2}\left(\frac{d^2}{4} - y^2\right) \qquad \qquad ...(7.2)$$

The maximum value of τ is at the neutral axis when $y = 0$

or
$$\tau_{max} = \frac{12S}{bd^3} \times \frac{d^2}{8} = \frac{3}{2} \times \frac{S}{bd}$$

But $\dfrac{S}{bd}$ is the mean shear stress (τ_{mean}) at the section considered.

Hence,
$$\tau_{max} = \frac{3}{2}\,\tau_{mean} \qquad \qquad ...[7.2(a)]$$

2. Solid circular section:

Fig. 7.3 shows a solid circular section of radius R. Consider an elementary strip of thickness δy at a distance y from N.A. Let the width of the strip be b, then
$$b = 2\sqrt{R^2 - y^2}$$

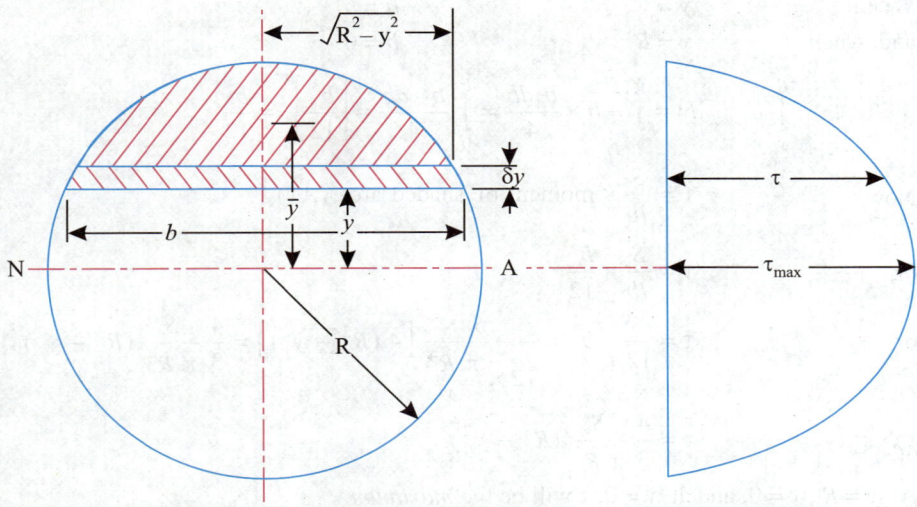

Fig. 7.3

Area of the elementary strip, $\delta a = b \times \delta y = 2 \sqrt{R^2 - y^2} \times \delta y$

Moment of elemental area about N.A. $= \delta a \times y$

$$= 2 \sqrt{(R^2 - y^2)} \times y \, \delta y$$

For shaded area, total moment $= \int\limits_{y}^{R} 2y \sqrt{(R^2 - y^2)} \, dy = \int\limits_{y}^{R} b . y . dy$

Now, $b = 2 \sqrt{R^2 - y^2}$ $\qquad \therefore \quad b^2 = 4 (R^2 - y^2)$

Differentiating both sides, we get

$$2b . db = -4 \times 2y \, dy \qquad \text{or} \qquad y \, dy = -\frac{b db}{4}$$

In a bridge compression, bending and shear are the major forces.

When, $y = y$ $b = b$

and, when $y = R$ $b = 0$

$$\therefore \qquad \int_{y}^{R} b.y.dy = \int_{b}^{0} -b \times \frac{b.db}{4} = \int_{0}^{b} \frac{b^2.db}{4} = \left[\frac{b^3}{12}\right]_{0}^{b} = \frac{b^3}{12}$$

Now, $\qquad \tau = \dfrac{S}{Ib} \times$ moment of shaded area $(A\overline{y})$

or, $\qquad \tau = \dfrac{S}{Ib} \times \dfrac{b^3}{12}$

or, $\qquad \tau = \dfrac{S}{12\,I} b^2 = \dfrac{S}{12 \times \dfrac{\pi\,R^4}{4}} \left[4\,(R^2 - y^2)\right] = \dfrac{4}{3}\dfrac{S}{\pi\,R^4}(R^2 - y^2)$

i.e. $\qquad \tau = \dfrac{4}{3}\dfrac{S}{\pi\,R^4}(R^2 - y^2)$ $\qquad\qquad$ ...(7.3)

At $y = R$, $\tau = 0$, and at $y = 0$, τ will be the *maximum*

Hence, $\qquad \tau_{\text{max}} = \dfrac{4}{3}\dfrac{S}{\pi\,R^2}$

Since, $\qquad \tau_{\text{mean}} = \dfrac{S}{\pi\,R^2}$

$\therefore \qquad \tau_{\text{max}} = \dfrac{4}{3}\,\tau_{\text{mean}}$ $\qquad\qquad$ ...[7.3(a)]

3. Symmetrical I-section:

Fig. 7.4 shows a symmetrical I-section.

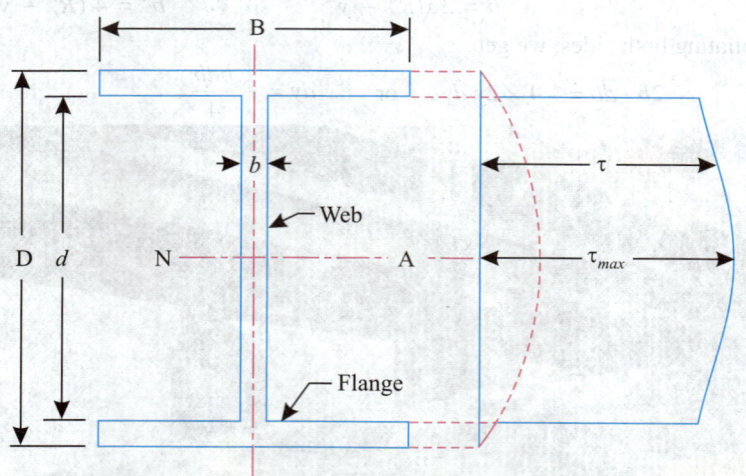

Fig. 7.4

Shearing stress in flanges:

$$\tau = \frac{S}{IB} \times A\overline{y}$$

$$A\overline{y} = B\left(\frac{D}{2} - y\right) \times \frac{1}{2} \times (D/2 + y) = \frac{B}{2}\left(\frac{D^2}{4} - y^2\right)$$

$$\therefore \qquad \tau = \frac{S}{IB} \times \frac{B}{2}\left(\frac{D^2}{4} - y^2\right) = \frac{S}{2I}\left(\frac{D^2}{4} - y^2\right) \qquad \qquad ...(7.4)$$

Shearing stress in web:

$$\tau = \frac{S}{I\,b}\,A\bar{y}$$

$$A\bar{y} = B.\frac{(D-d)}{2} \times \left(\frac{D+d}{4}\right) + \left(\frac{d}{2} - y\right)b \times \frac{1}{2}\left(\frac{d}{2} + y\right)$$

$$= \frac{B}{8}\,(D^2 - d^2) + \frac{b}{2}\left(\frac{d^2}{4} - y^2\right)$$

$$\therefore \qquad \tau = \frac{S}{8\,Ib}\left[B\,(D^2 - d^2) + b\,(d^2 - 4y^2)\right] \qquad \qquad ...(7.5)$$

The distribution of shearing stress caused by these separate effects is shown in Fig. 7.4. The value of τ will be maximum at $y = 0$

$$\therefore \qquad \tau_{max} = \frac{S}{8\,Ib}\left[B\,(D^2 - d^2) + bd^2\right] \qquad \qquad ...(7.6)$$

It may noted that at the inner surface of the flange, owing to the change in width of the section, the numerical value of the shearing stress changes from

$$\frac{S}{8\,I}\,(D^2 - d^2) \text{ to } \frac{S}{8\,I} \times \frac{B}{b} \times (D^2 - d^2)$$

Note. The idea of horizontal shearing stress becomes very important in those problems *where beams are built up of numerous parts riveted together.* Where outer reinforcements are riveted to the flanges of beams, it is necessary to calculate the shearing force acting at the junction, and hence to estimate the required number of rivets.

7.4. SHEAR STRESS DISTRIBUTION FOR TYPICAL SECTIONS

Fig. 7.5 shows shear stress distribution for some typical sections.

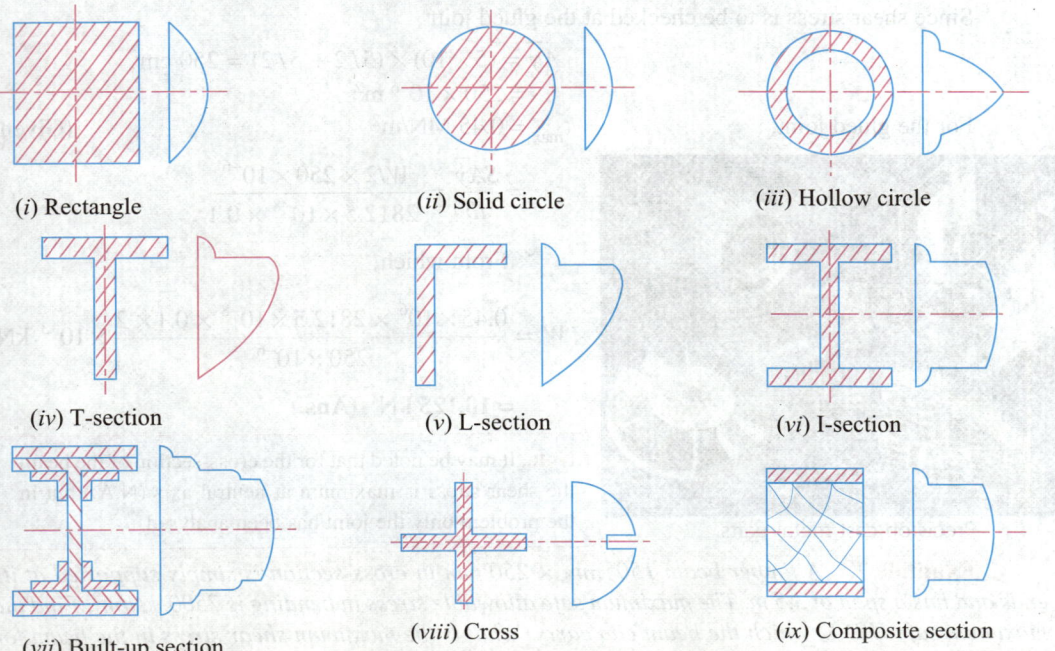

(*i*) Rectangle   (*ii*) Solid circle   (*iii*) Hollow circle

(*iv*) T-section   (*v*) L-section   (*vi*) I-section

(*vii*) Built-up section   (*viii*) Cross   (*ix*) Composite section

Fig. 7.5. Shear stress distribution for some typical sections.

Example 7.1. *A laminated wooden beam 10 cm wide and 15 cm deep is made up of three 5 × 10 cm wide planks glued together (Fig. 7.6) to resist longitudinal shear. The beam is simply supported over a span of 2 metres. If the allowable shearing stress in the glued joint is 0.45 MN/m² find the safe concentrated load that the beam may carry at its centre.*

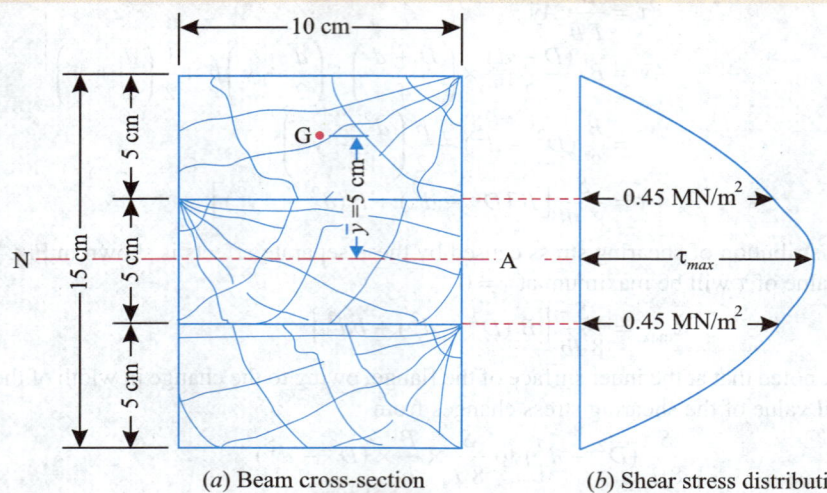

(*a*) Beam cross-section   (*b*) Shear stress distribution

Fig. 7.6

Solution. Let, W = Maximum safe concentrated load on the beam.

Then, maximum shear force for the beam $= \dfrac{W}{2}$

Moment of inertia for the beam, $\quad I = \dfrac{10 \times 15^3}{12} = 2812.5 \text{ cm}^4 = 2812.5 \times 10^{-8} \text{ m}^4$

Since shear stress is to be checked at the glued joint,

∴ $\qquad A\bar{y} = (5 \times 10) \times (5/2 + 5/2) = 250 \text{ cm}^3$

$\qquad\qquad\qquad = 250 \times 10^{-6} \text{ m}^2$

For the glued joint, $\qquad\qquad \tau_{\text{max}} = 0.45 \text{ MN/m}^2 \qquad\qquad$ (Given)

$$= \frac{SA\bar{y}}{Ib} = \frac{W/2 \times 250 \times 10^{-6}}{2812.5 \times 10^{-8} \times 0.1}$$

From which,

$$W = \frac{0.45 \times 10^6 \times 2812.5 \times 10^{-8} \times 0.1 \times 2}{250 \times 10^{-6}} \times 10^{-3} \text{ kN}$$

$$= \textbf{10.125 kN} \quad \textbf{(Ans.)}$$

Precision cast metal parts.

Note. It may be noted that for the cross-section of the beam the shear stress is maximum at neutral axis (N.A.) but in the problem only the joint has been analysed.

Example 7.2. *A timber beam 150 mm × 250 mm in cross-section is simply supported at its ends and has a span of 3.5 m. The maximum safe allowable stress in bending is 7500 kN/m². Find the maximum safe U.D.L. which the beam can carry. What is the maximum shear stress in the beam for the U.D.L. calculated?*

Solution. Refer to Fig. 7.7.

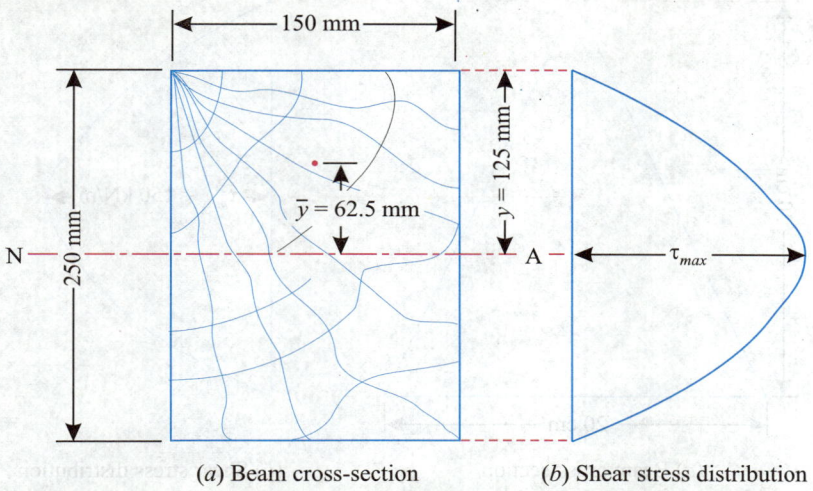

(*a*) Beam cross-section   (*b*) Shear stress distribution

Fig. 7.7

Let w be the U.D.L. in kN/m.

Then maximum bending moment at the centre of beam of length l,

$$M = \frac{wl^2}{8} \text{ kNm} = \frac{w \times 3.5^2}{8} = 1.53\, w \qquad ...(i)$$

Moment of inertia of the section, $\quad I = \dfrac{0.15 \times 0.25^3}{12} = 1.953 \times 10^{-4}\ \text{m}^4$

Distance of the topmost fibre from N.A.,

$$y = 0.25 / 2 = 0.125 \text{ m}$$

Using the relation,

$$\frac{M}{I} = \frac{\sigma}{y}, \quad \text{we have}$$

$$M = \frac{\sigma I}{y} = \frac{7500 \times 1.953 \times 10^{-4}}{0.125} = 11.718 \text{ kNm} \qquad ...(ii)$$

Equating (*i*) and (*ii*), we get $1.53\, w = 11.718$ or $w = $ **7.66 kN/m (Ans.)**

Due to this U.D.L. the maximum shearing force will be same as the reactions.

$$S_{max} = \frac{wl}{2} = \frac{7.66 \times 3.5}{2} = 13.4 \text{ kN}$$

∴ Maximum shear stress,

$$\tau_{max} = \frac{S\,A\bar{y}}{Ib} = \frac{13.4 \times 0.15 \times 0.125 \times 0.0625}{1.953 \times 10^{-4} \times 0.15}$$

$$= \textbf{536 kN/m}^2 \text{ (Ans.)} \qquad \left[\begin{array}{l} \because \quad A = 0.15 \times 0.125 \text{ m}^2 \\ \bar{y} = \dfrac{0.125}{2} = 0.0625 \text{ m} \end{array}\right]$$

Example 7.3. *A beam of triangular section having base width 20 cm and height of 30 cm is subjected to a shear force of 3 kN. Find the value of maximum shear stress, and sketch the shear stress distribution alongwith the depth of the beam.*

Solution. Refer to Fig. 7.8.

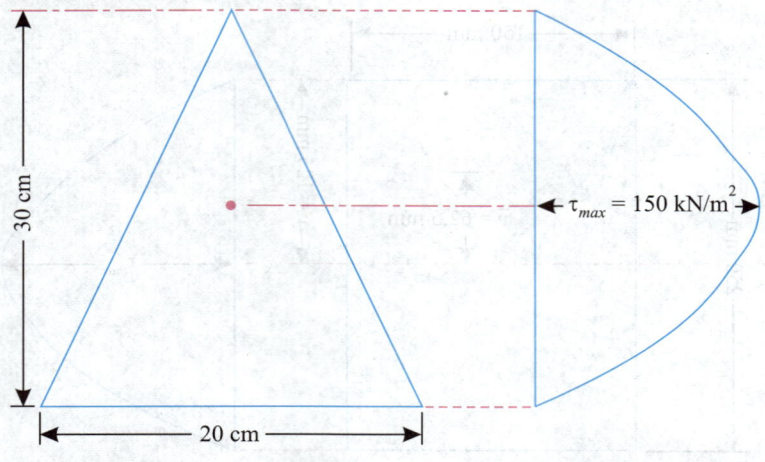

(*a*) Beam cross-section   (*b*) Shear stress distribution

Fig. 7.8

Given: Base width, $b = 20$ cm $= 0.2$ m
Height, $h = 30$ cm $= 0.3$ m

$\therefore$ Area $A = \dfrac{1}{2}\,bh = \dfrac{1}{2} \times 0.2 \times 0.3 = 0.03$ m^2

Shear foce $S = 3$ kN

Maximum shear stress τ_{max}:

We know that mean stress,

$$\tau_{mean} = \frac{S}{A} = \frac{3}{0.03} = 100 \text{ kN/m}^2$$

Now, using the relation,

$$\tau_{max} = \frac{3}{2}\,\tau_{mean} = 3/2 \times 100 = \mathbf{150 \text{ kN/m}^2} \quad \textbf{(Ans.)}$$

Example 7.4. *A circular beam 150 mm diameter is subjected to a shear force of 7 kN. Calculate the value of maximum shear stress, and sketch variation of shear stress along the depth of the beam.*

Solution. Refer to Fig. 7.9.

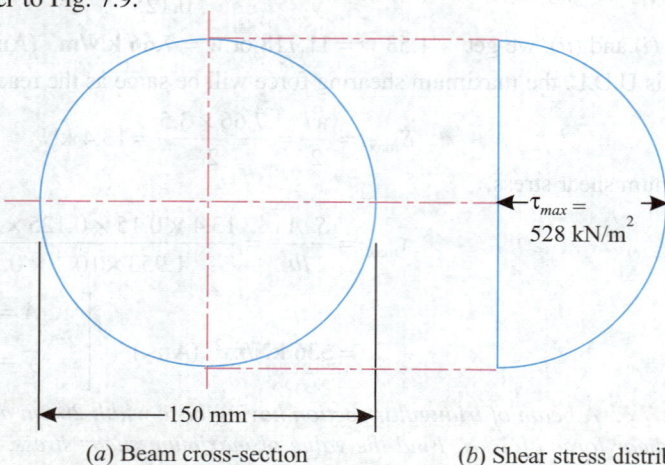

(*a*) Beam cross-section   (*b*) Shear stress distribution

Fig. 7.9

Given: Diameter of the beam $= 150 \text{ mm} = 0.15 \text{ m}$

$\therefore$ Area, $A = \dfrac{\pi}{4} \times 0.15^2 = 0.01767 \text{ m}^2$

Shear force, $S = 7 \text{ kN}$

Maximum shear stress, τ_{max}:

We know that mean stress, $\tau_{mean} = \dfrac{S}{A} = \dfrac{7}{0.01767} = 396 \text{ kN/m}^2$

Now, using the relation, $\tau_{max} = \dfrac{4}{3} \tau_{mean}$ .....[7.3 (a)]

$= \dfrac{4}{3} \times 396 = \mathbf{528 \text{ kN/m}^2 \text{ (Ans.)}}$

Example 7.5. *An I-section, with rectangular ends, has the following dimensions:*

> *Flanges:* $15 \text{ cm} \times 2 \text{ cm}$
>
> *Web*   : $30 \text{ cm} \times 1 \text{ cm}$

Find the maximum shearing stress developed in the beam for a shearing force of 10 kN.

Solution. Refer to Fig. 7.10.

Overall width, $B = 15 \text{ cm} = 0.15 \text{ m}$

Overall depth, $D = 34 \text{ cm} = 0.34 \text{ m}$

Flange thickness, $t_f = 2 \text{ cm} = 0.02 \text{ m}$

Depth of web, $d = 30 \text{ cm} = 0.3 \text{ m}$

Width of web, $b = 1 \text{ cm} = 0.01 \text{ m}$

Shearing force, $S = 10 \text{ kN}$

Maximum shear stress developed τ_{max} :

Moment of inertia of the section about the neutral axis,

$$I = \left[\dfrac{0.15 \times (0.34)^3}{12} - \dfrac{0.14 \times (0.30)^3}{12} \right]$$

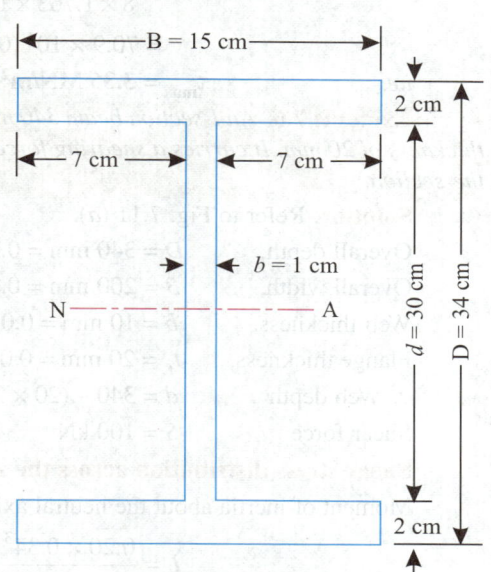

Fig. 7.10

Bridge with more metallic members and little concrete.

$$= [4.913 \times 10^{-4} - 3.15 \times 10^{-4}] = 1.763 \times 10^{-4} \text{ m}^4$$

$$\left[\text{Also } I = 2 \left(\frac{25 \times 2^3}{12} + 15 \times 2 \times (15 + 1)^2 \right) + \frac{1 \times 30^3}{12} \right.$$

$$\left. = 2 \times 7690 + 2250 = 17630 \text{ cm}^4 = 1.763 \times 10^{-4} \text{ m}^4 \right]$$

We know that maximum shear stress occurs at the neutral axis;

Using the relation,

$$\tau_{max} = \frac{S}{8 \; Ib} \left[B \, (D^2 - d^2) + bd^2 \right] \quad \text{with usual notations, we have}$$

$$\tau_{max} = \frac{10}{8 \times 1.763 \times 10^{-4} \times 0.01} \left[0.15 \, (0.34^2 - 0.30^2) + 0.01 \times (0.3)^2 \right]$$

$$= 70.9 \times 10^4 \, (0.00384 + 0.0009) = 3360 \text{ kN/m}^2 \text{ or } 3.36 \text{ MN/m}^2$$

i.e. $\qquad \tau_{max} = \textbf{3.36 MN/m}^2$ **(Ans.)**

Example 7.6. *An I-section beam 340 mm × 200 mm has a web thickness of 10 mm and flange thickness of 20 mm. It carries a shearing force of 100 kN. Sketch the shear stress distribution across the section.*

Solution. Refer to Fig. 7.11 (*a*).

Overall depth,	$D = 340$ mm $= 0.34$ m
Overall width,	$B = 200$ mm $= 0.2$ m
Web thickness,	$b = 10$ mm $= 0.01$ m
Flange thickness,	$t_f = 20$ mm $= 0.02$ m
∴ Web depth	$d = 340 - (20 \times 2) = 300$ mm $= 0.3$ m
Shear force	$S = 100$ kN

Shear stress distribution across the section:

Moment of inertia about the neutral axis,

$$I = \frac{0.20 \times 0.34^3}{12} - \frac{0.19 \times 0.30^3}{12} = 2.276 \times 10^{-4} \text{ m}^4$$

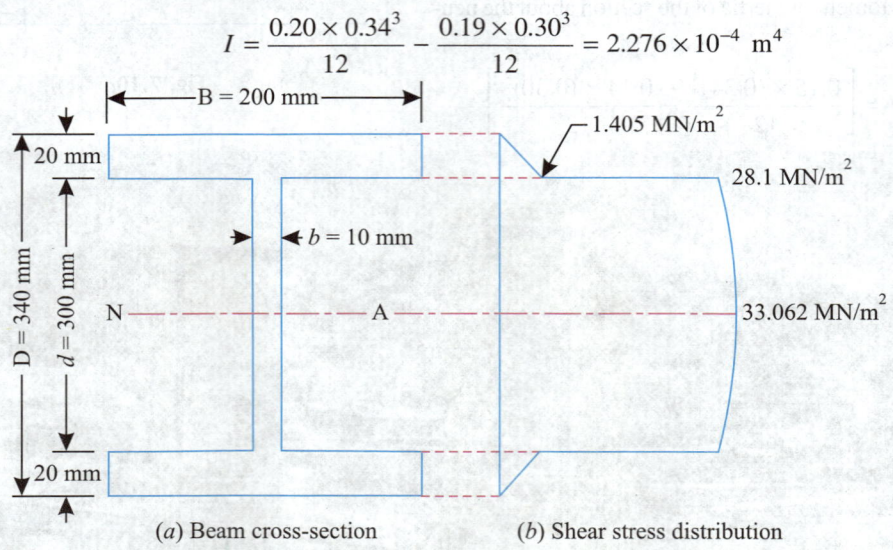

(*a*) Beam cross-section   (*b*) Shear stress distribution

Fig. 7.11

We know that the shear stress at the upper edge of the flange is zero, and the shear stress at the joint

$$= \frac{S}{8I} (D^2 - d^2) = \frac{100}{8 \times 2.267 \times 10^{-4}} (0.34^2 - 0.30^2) = 1405 \text{ kN/m}^2$$

$$= 1.405 \text{ MN/m}^2$$

The shear stress at the junction suddenly increases from

$$1.405 \text{ MN/m}^2 \text{ to } 1.405 \times \frac{B}{b} = 1.405 \times \frac{0.20}{0.01} = 28.1 \text{ MN/m}^2$$

Now using the relation,

$$\tau_{max} = \frac{S}{8\ Ib}\left[B(D^2 - d^2) + bd^2\right], \text{ we get}$$

$$\tau_{max} = \frac{100}{8 \times 2.276 \times 10^{-4} \times 0.01}\left[0.2\,(0.34^2 - 0.30^2) + 0.01 \times (0.3)^2\right]$$

$$= 549.21 \times 10^4\,(0.00512 + 0.0009) = 33062 \text{ kN/m}^2 = 33.062 \text{ MN/m}^2$$

Complete shear stress distribution is shown in Fig. 7.11 (b).

Example 7.7. *An I-section with rectangular ends has the following dimensions:*

Flanges :   *10 cm × 1 cm*

Web :   *12 cm × 1 cm*

If this section is subjected to a bending moment of 5 kNm and a shearing force of 5 kN, find the maximum tensile and shear stresses induced in it.

Solution. Refer to Fig. 7.12.

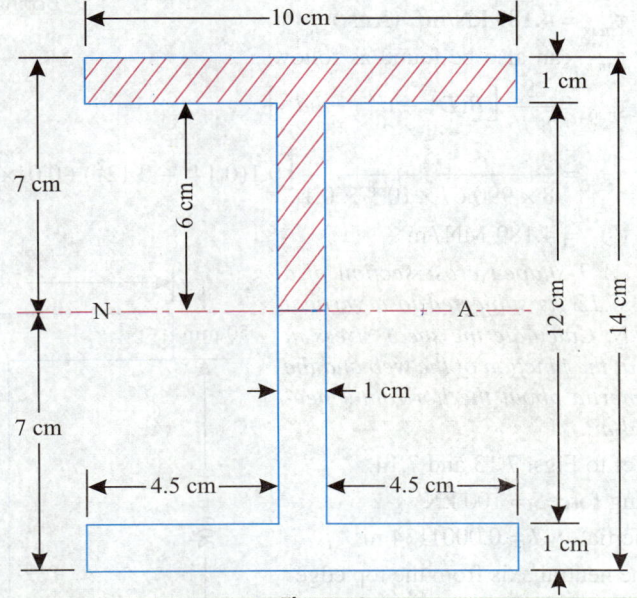

Fig. 7.12

Moment of inertia of the section about neutral axis (N.A.),

$$I = \frac{10 \times 14^3}{12} - \frac{9 \times 12^3}{12} = 2286.67 - 1296$$

$$= 990.67 \text{ cm}^4 = 990.67 \times 10^{-8} \text{ m}^4$$

Bending moment to which the section is subjected to, $M = 5$ kNm

If σ_t is the maximum tensile stress developed at outer face of the flange then from the relation,

$$\frac{M}{I} = \frac{\sigma_t}{y_t} \quad \text{or,} \quad \sigma_t = \frac{M}{I} \times y_t = \frac{5 \times 0.07}{990.67 \times 10^{-8}} \times 10^{-3} \text{ MN/m}^2$$

or $\sigma_t = 35.33$ **MN/m² (Ans.)**

Shear stress is obviously maximum in the web at N.A.

We know that,

$$\tau_{max} = \frac{S\ \overline{Ay}}{Ib}\ \text{with usual notations}$$

$$S = 5\ \text{kN} \qquad (Given)$$

$$I = 990.67 \times 10^{-8}\ \text{m}^4\ (Calculated\ above)$$

$$b = 0.01\ \text{m} \qquad (Given)$$

To find $\overline{Ay}$ split the shaded area above N.A. into two rectangles 10 cm × 1 cm and 1 cm × 6 cm

$$\overline{Ay} = (10 \times 1) \times (6 + 1/2) + (1 \times 6) \times 3$$

$$= 83\ \text{cm}^3 = 83 \times 10^{-6}\ \text{m}^3$$

$$\therefore \qquad \tau_{max} = \frac{5 \times 83 \times 10^{-6}}{990.67 \times 10^{-8} \times 0.01} \times 10^{-3}\ \text{MN/m}^2$$

(substituting values)

or $\tau_{max} = 4.189$ **kN/m² (Ans.)**

$$\left[\begin{array}{l}
\tau_{max}\ \text{can also be found as follows:} \\[4pt]
\tau_{max} = \frac{S}{8\ Ib}\left[B(D^2 - d^2) + bd^2 \right] \\[8pt]
\quad = \frac{5}{8 \times 990.67 \times 10^{-8} \times 0.1}\left[0.1(0.14^2 - 0.12^2) + 0.01 \times 0.12^2\right] \times 10^{-3} \\[8pt]
\quad = 4.189\ \text{MN/m}^2
\end{array}\right]$$

Boring machine chuck.

Example 7.8. *A T-shaped cross-section of a beam shown in Fig. 7.13 is subjected to a vertical shear force of 100 kN. Calculate the shear stress at the neutral axis and at the junction of the web and the flange. Moment of inertia about the horizontal neutral axis is 0.0001134 m⁴.*

Solution. Refer to Figs. 7.13 and 7.14.

Given: Shearing force $S = 100$ kN

Moment of inertia $I = 0.0001134$ m⁴

Distance of the neutral axis from the top edge

$$= \frac{(0.20 \times 0.05) \times 0.025 + (0.2 \times 0.05)\left(\dfrac{0.2}{2} + 0.05\right)}{(0.2 \times 0.05) + (0.2 \times 0.05)}$$

$$= \frac{0.00025 + 0.0015}{0.01 + 0.01} = 0.0875\ \text{m} = 87.5\ \text{mm}$$

Fig. 7.13

We know that the shear stress at the top edge of the flange, and bottom of the web is zero.

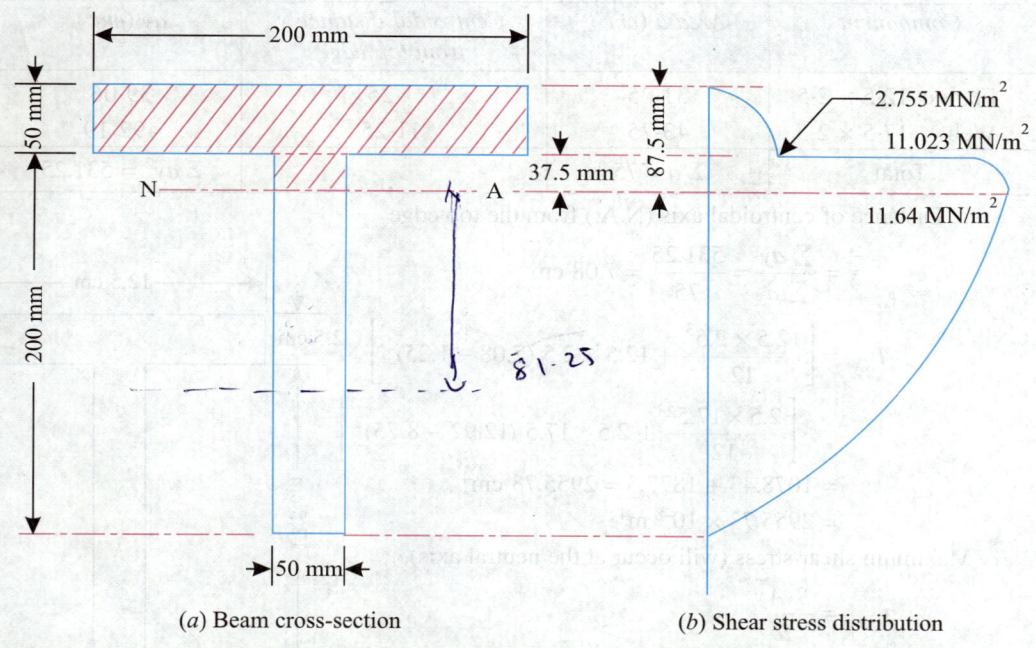

(a) Beam cross-section   (b) Shear stress distribution

Fig. 7.14

Shear stress at the neutral axis,

$$\tau_{NA} = \frac{S\,A\overline{y}}{Ib}$$

where, $A\overline{y} = (0.2 \times 0.05) \times (0.0875 - 0.025) + \left(0.0375 \times 0.05 \times \frac{0.0375}{2}\right)$

$$= 0.000625 + 0.00003516 = 0.00066 \text{ m}^3$$

∴ $\tau_{NA} = \dfrac{100 \times 0.00066}{0.0001134 \times 0.05}$ (substituting the values)

$$= 11640 \text{ kN/m}^2 = \textbf{11.64 MN/m}^2 \text{ (Ans.)}$$

Shear stress in the web just at the junction of web and flange

$$= \frac{S\,A\overline{y}}{Ib} = \frac{100 \times (0.2 \times 0.05) \times (0.025 + 0.0375)}{0.0001134 \times 0.05}$$

$$= 11023 \text{ kN/m}^2 = \textbf{11.023 MN/m}^2 \text{ (Ans.)}$$

Shear stress in the flange just at the junction of the flange and web

$$= \frac{100 \times (0.2 \times 0.05) \times (0.025 + 0.0375)}{0.0001134 \times 0.2}$$

$$= 2755 \text{ kN/m}^2 = \textbf{2.755 MN/m}^2 \text{ (or} = 0.05/0.2 \times 11.023 = 2.755 \text{ MN/m}^2) \textbf{ (Ans.)}$$

Example 7.9. *A simply supported beam carries a U.D.L. of intensity 2.5 kN/metre over entire span of 5 metres. The cross-section of the beam is a T-section having the dimensions as shown in Fig. 7.15. Calculate the maximum shear stress for the section of the beam.*

Solution. Refer to Figs. 7.15 and 7.16.

Let us first calculate the position of the neutral axis and moment of inertia of the section about the neutral axis. The relevant calculations are given on next page:

Component	Area a (cm^2)	Centroidal distance y from top edge	ay (cm^2)
Flange 12·5 × 2·5	31·25	1·25	39·06
Web 17·5 × 2·5	43·75	11·25	492·19
Total	$\Sigma a = 75$		$\Sigma ay^2 = 531.25$

∴ Distance of centroidal axis (N.A.) from the top edge

$$\bar{y} = \frac{\Sigma\, ay}{\Sigma\, a} = \frac{531.25}{75} = 7.08 \text{ cm}$$

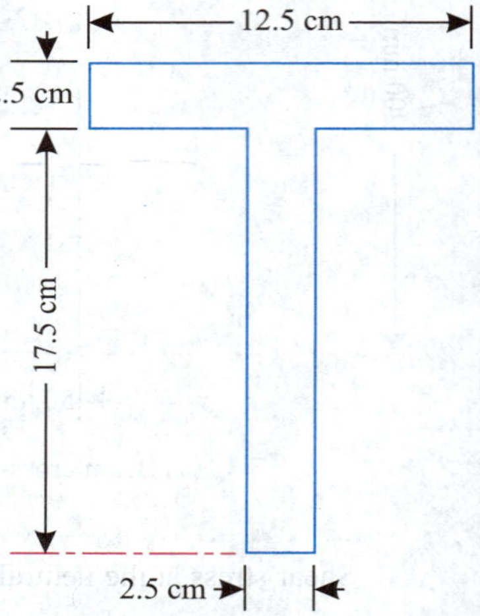

$$I_{XX} = \left[\frac{12.5 \times 2.5^3}{12} + 12.5 \times 2.5\,(7.08 - 1.25)^2\right]$$

$$+ \left[\frac{2.5 \times 17.5^3}{12} + 2.5 \times 17.5\,(12.92 - 8.75)^2\right]$$

$$= 1078.43 + 1877.3 = 2955.73 \text{ cm}^4$$

$$= 2955.73 \times 10^{-8} \text{ m}^4$$

Maximum shear stress (will occur at the neutral axis),

$$\tau_{max} = \frac{S\,A\bar{y}}{Ib}$$

Maximum shear force,

$$S = \frac{wl}{2} = \frac{2.5 \times 5}{2} = 6.25 \text{ kN}$$

$$A = 2.5 \times 12.92 = 32.3 \text{ cm}^2 = 32.3 \times 10^{-4} \text{ m}^2$$

$$\bar{y} = \frac{12.92}{2} = 6.46 \text{ cm} = 0.0646 \text{ m}$$

$$\therefore \quad \tau_{max} = \frac{6.25 \times (32.3 \times 10^{-4}) \times 0.0646}{2955.73 \times 10^{-8} \times (2.5 \times 10^{-2})} \times 10^{-3} \text{ MN/m}^2 = 1.765 \text{ MN/m}^2$$

Hence, t_{max} = **1.765 MN/m^2** **(Ans.)**

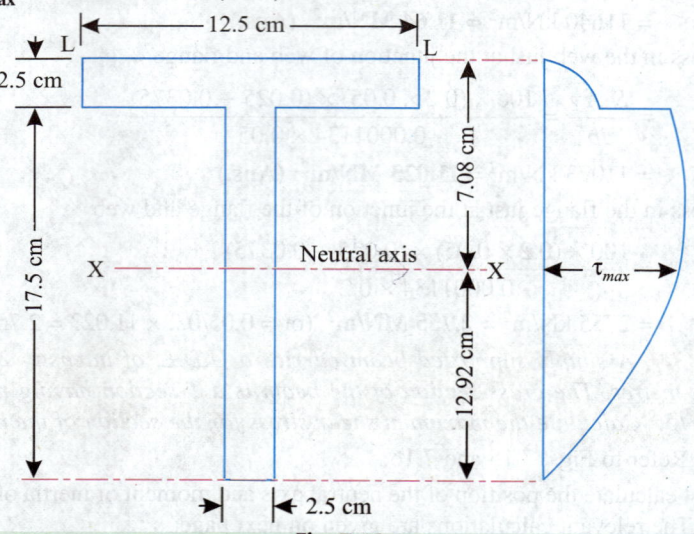

Fig. 7.15

Fig. 7.16

Example 7.10. *A beam of channel section 120 mm × 60 mm has uniform thickness of 15 mm. Draw diagram showing the distribution of shear stress for a vertical section where shearing force is 50 kN. Find the ratio between maximum and mean shear stresses.*

Solution. Refer to Fig. 7.17.

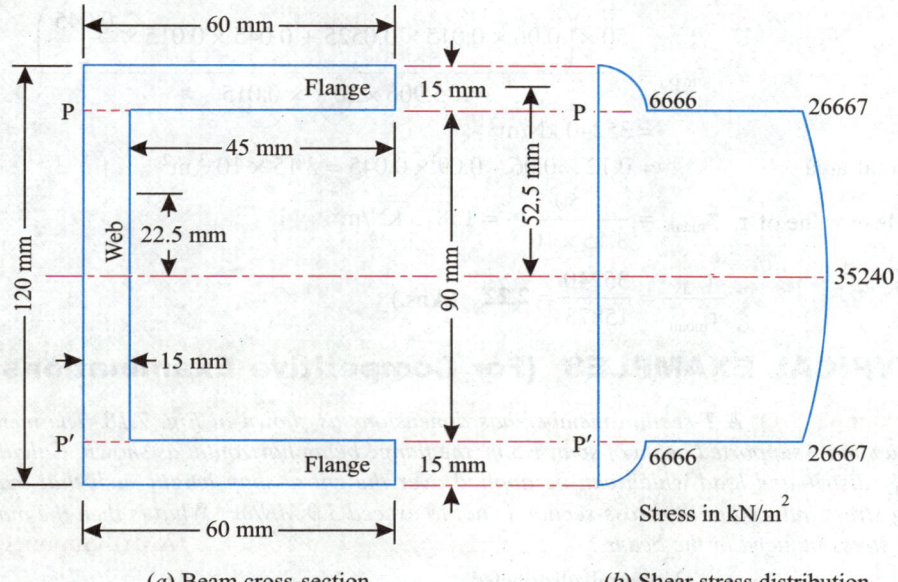

(*a*) Beam cross-section (*b*) Shear stress distribution

Fig. 7.17

Moment of inertia of the section passing through the neutral axis,

$$I = \frac{0.06 \times 0.12^3}{12} - \frac{0.045 \times 0.09^3}{12} = 5.906 \times 10^{-6} \text{ m}^4$$

Intensity of shear stress (τ) at the top and the bottom is zero.

The value of τ at sections P and P' in *flange* is given by,

$$\tau_P = \frac{S \, A\bar{y}}{Ib} = \frac{50 \times 0.06 \times 0.015 \times \left(\dfrac{0.09}{2} + \dfrac{0.015}{2}\right)}{5.906 \times 10^{-6} \times 0.06} = 6666 \text{ kN/m}^2$$

Bronze metal components.

The value of τ at sections P and P' in *web* is given by

$$\tau_P = \frac{S\,A\bar{y}}{Ib} = \frac{50 \times 0.06 \times 0.015 \times 0.0525}{5.906 \times 10^{-6} \times 0.015} = 26667 \text{ kN/m}^2$$

The value of τ at web at *neutral axis* (N.A.)

$$\tau_{max} = \frac{50 \times \left(0.06 \times 0.015 \times 0.0525 + 0.045 \times 0.015 \times \dfrac{0.045}{2}\right)}{5.906 \times 10^{-6} \times 0.015}$$

$$= 35240 \text{ kN/m}^2$$

Total area $= 0.12 \times 0.06 - 0.09 \times 0.045 = 3.15 \times 10^{-3} \text{ m}^2$

Mean value of τ, $\tau_{mean} = \dfrac{50}{3.15 \times 10^{-3}} = 15873 \text{ kN/m}^2$

$\therefore \qquad \dfrac{\tau_{max}}{\tau_{mean}} = \dfrac{35240}{15873} = \mathbf{2.22}$ **(Ans.)**

TYPICAL EXAMPLES (For Competitive Examinations)

Example 7.11. *A T-section member has dimensions as shown in Fig. 7.18. The member is used as a simply supported beam of span 1.5 m, the flange being horizontal as shown. Calculate the uniformly distributed load which can be applied over the entire span length such that maximum shearing stress induced in the cross-section is not to exceed 3.0 MN/m². What is then the maximum bending stress induced in the beam ?* **(AMIE Summer, 2001)**

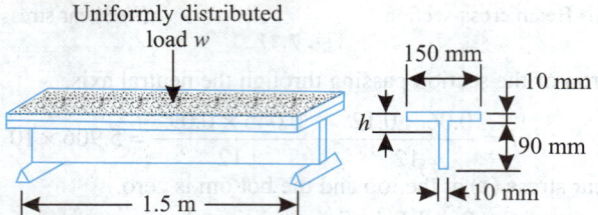

Fig. 7.18

Solution. *Given:* $\quad l = 1.5$ m; $\qquad \tau_{max} = 3.0$ MN/m²

Uniformly distributed load, *w* :

Distance of the neutral axis from the top edge,

$$h = \frac{(150 \times 10) \times 5 + (90 \times 10) \times (45 + 10)}{(150 \times 10) + (90 \times 10)} \quad 23.75 \text{ mm}$$

Next, let us calculate the moment of inertia of the section about the neutral axis N.A.

$$I_{NA} = \frac{150 \times 10^3}{12} + (150 \times 10) \times (23.75 - 5)^2 + \frac{10 \times 90^2}{12} + 90 \times 10 \times (76.25 - 45)^2$$

$$= 12500 + 527343.75 + 607500 + 878906.25 = 2026250 \text{ mm}^4$$

$$= 2026250 \times 10^{-12} \text{ m}^4$$

We know that the shear stress is maximum at N.A.

Let τ_{max} be the maximum shear stresss in N/m²

Then $\qquad \tau_{NA} = \tau_{max} = \dfrac{S A \bar{y}}{Ib}$ ...(i) where, S = Maximum shear force,

$A\bar{y}$ = Moment of area above N.A.,

I = Moment of inertia of the whole section about the N.A., and

b = Breadth of the web.

In this case, $A\bar{y} = 150 \times 10 \, (23.75 - 5) + 10 \times (23.75 - 10) \times \left(\dfrac{23.75 - 10}{2}\right)$

$$= 28125 + 945.3 = 29070.3 \text{ mm}^3 = 29070.3 \times 10^{-9} \text{ m}^3$$

Substituting the value in eqn. (i), we get

$$3.0 \times 10^6 = \frac{S \times 29070.3 \times 10^{-9}}{2026250 \times 10^{-12} \times (10 \times 10^{-3})}$$

∴ $\quad S = \dfrac{3.0 \times 10^6 \times 2026250 \times 10^{-12} \times (10 \times 10^{-3})}{29070.3 \times 10^{-9}} = 2091 \text{ N}$

We know that, for a simply supported beam of span l metres and loaded with a uniformly distributed load wN per metre run, the maximum shearing force will be at the support reactions.

$$\frac{wl}{2} = S \quad \text{or} \quad \frac{w \times 1.5}{2} = 2091$$

∴ $\quad w = 2788 \text{ N/m} \text{ or } \textbf{2.788 kN/m} \textbf{ (Ans.)}$

Maximum bending stress, σ_{max} :

Using the relation,

$$\frac{M}{I} = \frac{\sigma}{y}$$

where, $M = \dfrac{wl^2}{8} = \dfrac{2788 \times 1.5^2}{8} = 784.125 \text{ Nm}$

∴ $\quad \sigma_t = \dfrac{M \, y_t}{I} = \dfrac{784.125 \times (23.75 \times 10^{-3})}{2026250 \times 10^{-12}} \times 10^{-6}$ $\qquad (\because h = y_t)$

$$= 9.19 \text{ MN/m}^2 \text{ (tensile)}$$

and, $\quad \sigma_c = \dfrac{M \times y_c}{I} = \dfrac{784.125 \times \left[(100 - 23.75) \times 10^{-3}\right]}{2026250 \times 10^{-12}} \times 10^{-6}$ $\quad [(\because y_c = (100 - h)]$

$$= 29.51 \text{ MN/m}^2 \text{ (comp.)}$$

∴ Maximum bendings stress induced,

$$\sigma_{max} = \textbf{29.51 MN/m}^2 \textbf{ (comp.)} \textbf{ (Ans.)}$$

Example 7.12. *Fig. 7.19 shows the vertical cross-section of a beam which at this section is subjected to a vertical shearing force of 12 kN. Sketch the distribution of shearing stress and find the ratio of the maximum shear stress to the mean shear stress.*

Solution. Refer to Fig. 7.20.

We shall first determine the position of neutral axis and the M.O.I. of the section about the neutral axis (N.A.); the relevant calculations are given on next page.

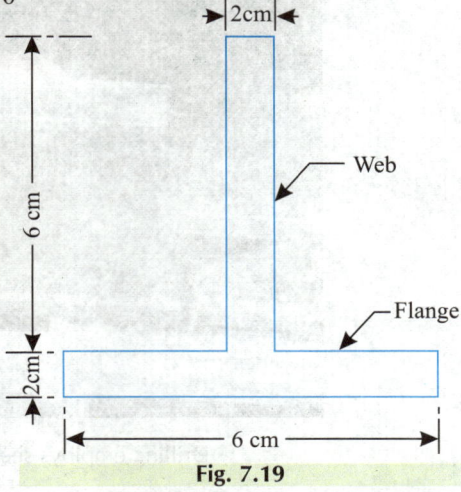

Fig. 7.19

Component	Area a (cm^2)	Centroidal distance y from LL. (cm)	ay (cm^3)
Web	$6 \times 2 = 12$	$6/2 + 2 = 5$	$12 \times 5 = 60$
Flange	$6 \times 2 = 12$	$2/2 = 1$	$12 \times 1 = 12$
Total	$(\Sigma a) = 24$		$(\Sigma ay) = 72$

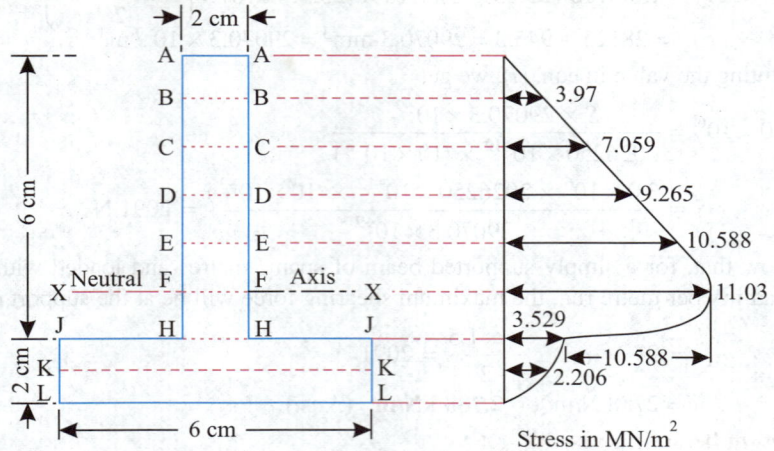

(a) Beam cross-section   (b) Shear stress distribution

Fig. 7.20

Distance of the centroidal axis (N.A.) from bottom edge,

$$\bar{y} = \frac{\Sigma ay}{\Sigma a} = \frac{72}{24} = 3 \text{ cm}$$

$$I_{XX} = \left[\frac{6 \times 2^3}{12} + 6 \times 2 \, (3-1)^2\right] + \left[\frac{2 \times 6^3}{12} + 6 \times 2 \times (5-3)^2\right]$$

$$= 52 + 84 = 136 \text{ cm}^4 = 136 \times 10^{-8} \text{ m}^4$$

Drilling employs shearing operation on machine parts.

For sketching the shear distribution let us determine shear stress at every 1 cm from the top as shown in Fig. 7.20 (*a*). Calculations for stresses at each of the sections are given in the table below:

Section	Area 'A'(cm^2)	$\bar{y}$ (cm)	b (cm)	$\dfrac{S\,A\bar{y}}{Ib}$ (MN/m^2)	Remarks
1	2	3	4	5	6
AA	0	—	0	0	$I = 136\ cm^4$
BB	$2 \times 1 = 2\ cm^2$ $= 2 \times 10^{-4}\ m^2$	$5 - 1/2 = 4.5\ cm$ $= 0.045\ m$	2 cm $= 0.02\ m$	$\dfrac{12 \times 2 \times 10^{-4} \times 0.045}{136 \times 10^{-8} \times 0.02}$ $\times 10^{-3} = 3.97$	$= 136 \times 10^{-8}\ m^4$ $S = 12\ kN.$
CC	$2 \times 2 = 4\ cm^2$ $= 4 \times 10^{-4}\ m^2$	$5 - 1 = 4\ cm$ $= 0.04\ m$	2 cm $= 0.02\ m$	$\dfrac{12 \times 4 \times 10^{-4} \times 0.04}{136 \times 10^{-8} \times 0.02}$ $\times 10^{-3} = 7.059$	Area 'A' is between the section and the top and $\bar{y}$ the distance of its centroid from N.A.
DD	$2 \times 3 = 6\ cm^2$ $= 6 \times 10^{-4}\ m^2$	$5 - 3/2 = 3.5\ cm$ $= 0.035\ m$	2 cm $= 0.02\ m$	$\dfrac{12 \times 6 \times 10^{-4} \times 0.035}{136 \times 10^{-8} \times 0.02}$ $\times 10^{-3} = 9.265$	
EE	$2 \times 4 = 8\ cm^2$ $= 8 \times 10^{-4}\ m^2$	$5 - 4/2 = 3\ cm$ $= 0.03\ m$	2 cm $= 0.02\ m$	$\dfrac{12 \times 8 \times 10^{-4} \times 0.03}{136 \times 10^{-8} \times 0.02}$ $\times 10^{-3} = 10.588$	
FF	$2 \times 5 = 10\ cm^2$ $= 10 \times 10^{-4}\ m^2$	$5 - 5/2 = 2.5\ cm$ $= 0.025\ m$	2 cm $= 0.02\ m$	$\dfrac{12 \times 10 \times 10^{-4} \times 0.025}{136 \times 10^{-8} \times 0.02}$ $\times 10^{-3} = 11.03$	
HH when b = 2 cm	$6 \times 2 = 12\ cm^2$ $= 12 \times 10^{-4}\ m^2$	$2\ cm = 0.02\ m$	2 cm $= 0.02\ m$	$\dfrac{12 \times 12 \times 10^{-4} \times 0.02}{136 \times 10^{-8} \times 0.02}$ $\times 10^{3} = 10.588$	
JJ when b = 6 cm	$6 \times 2 = 12\ cm^2$ $= 12 \times 10^{-4}\ m^2$	$2\ cm = 0.02\ m$	6 cm $= 0.06\ m$	$\dfrac{12 \times 12 \times 10^{-4} \times 0.02}{136 \times 10^{-8} \times 0.06}$ $\times 10^{-3} = 3.529$	
Section	Area 'A'(cm^2)	$\bar{y}$ (cm)	b (cm)	$\dfrac{S\,A\bar{y}}{Ib}$ (MN/m^2)	
KK	$6 \times 1 = 6\ cm^2$ $= 6 \times 10^{-4}\ m^2$	2.5 cm $= 0.025\ m$	6 cm $= 0.06\ m$	$\dfrac{12 \times 6 \times 10^{-4} \times 0.025}{136 \times 10^{-8} \times 0.06}$ $\times 10^{-3} = 2.206$	Area 'A' is between the section and the bottom and $\bar{y}$ the distance of its centroid from the neutral axis (N.A.)
LL	0	—	6 cm $= 0.06\ m$	0	

Fig. 7.20 (*b*) shows the shear stress distribution across the cross-section.

τ_{max} (at neutral axis) = 11.03 MN/m^2

$$\tau_{mean} = \frac{S}{\text{Area of cross-section}} = \frac{12}{24 \times 10^{-4}} \times 10^{-3}\ MN/m^2 = 5\ MN/m^2$$

$$\frac{\tau_{max}}{\tau_{mean}} = \frac{11.03}{5} = 2.206 \quad \text{(Ans.)}$$

Example 7.13. *A steel section shown in Fig. 7.21 is subjected to a shear force 200 kN. Determine the shear stress at the important points and sketch the shear distribution diagram.*

Solution. Since the section is symmetrical about XX and YY axes therefore c.g. of the section will lie on the geometrical centroid of the section. For the purpose of moment of inertia and shear stress, the two semi-circular grooves may be assumed to be together and considered as one circular hole of 200 mm diameter. Therefore M.O.I. of the section about its c.g.

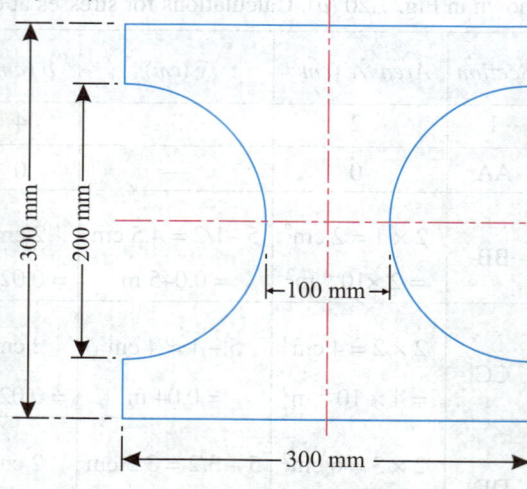

Fig. 7.21

$$I = \frac{0.3 \times 0.3^3}{12} - \frac{\pi}{64} \times 0.2^4$$
$$= 5.964 \times 10^{-4} \, \text{m}^4$$

We know that the shear stress at the extreme edges L and Q (Fig. 7.22) is zero.

Let us first find out shear stress at M.

Area, $\qquad A = 0.3 \times 0.05 = 0.015 \, \text{m}^2$

$$\bar{y} = 0.1 + \frac{0.05}{2} = 0.125 \, \text{m}$$

$$b = 0.3 \, \text{m}$$

Using the relation, $\qquad \tau = \dfrac{S\,A\bar{y}}{Ib}$, we get

$$\tau_M = \frac{200 \times 0.015 \times 0.125}{5.964 \times 10^{-4} \times 0.3} = \textbf{2096 kN/m}^2 \quad \textbf{(Ans.)}$$

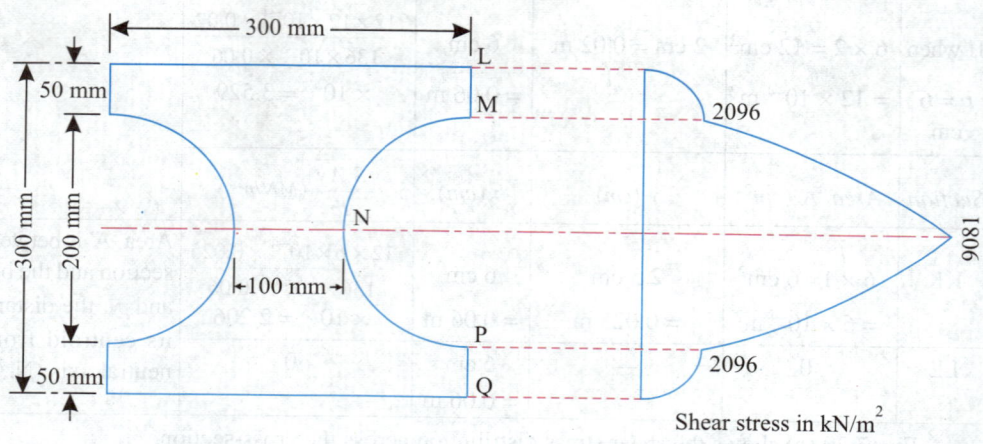

(a) Steel section

(b) Shear stress distribution

Fig. 7.22

The *shear stress at P* will also be equal to **2096 kN/m²**. Now let us find out the shear stress at N.

$$A\bar{y} = 0.3 \times 0.15 \times 0.075 - \left(\frac{\pi \times 0.1^2}{2} \times \frac{4 \times 0.1}{3\pi}\right)$$

$$= 0.003375 - 0.0006667 = 0.002708 \text{ m}^3$$

$$b = 0.1 \text{ m}$$

Again using the relation, $\tau = \dfrac{S\,A\bar{y}}{Ib}$, we get,

$$\tau_N = \frac{200 \times 0.002708}{5.964 \times 10^{-4} \times 0.1}$$

$$= \textbf{9081 kN/m}^2$$

The shear stress distribution diagram is shown in Fig. 7. 22 (*b*).

Example 7.14. *A cast-iron bracket, subjected to bending, has a cross-section of I-shape with unequal flanges as shown in Fig. 7.23. If the maximum compressive stress is not to exceed 20 MN/m², what is the bending moment, the section can take? If the section is subjected to 80 kN, draw the shear stress distribution over the depth of the section.*

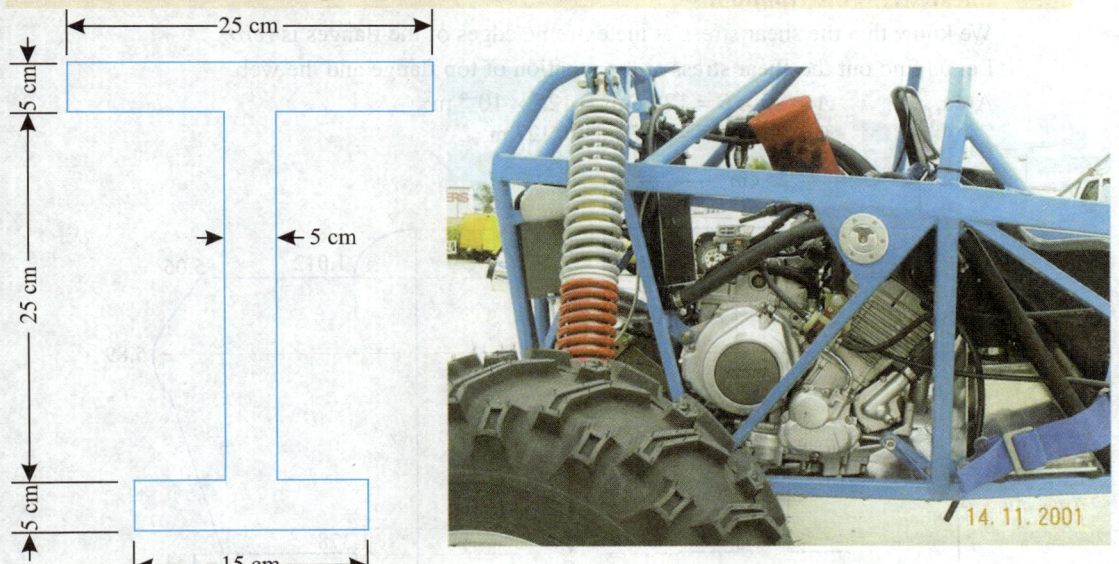

Fig. 7.23   Motor car with shock absorbers.

Solution. *Given.* Compressive stress in top flange,

$$\sigma_c = 20 \text{ MN/m}^2$$

Shear force, $S = 80$ kN

Bending moment the section can take, *M*:

We shall first determine the position of neutral axis and M.O.I. of the section about the N.A. The relevant calculations are given below:

Component	Area, *a* (cm²)	Centroidal distance y from the bottom edge (cm)	*ay* (cm³)
Top flange	25 × 5 = 125	5/2 + 25 + 5 = 32.5	4062.5
Web	25 × 5 = 125	25/2 + 5 = 17.5	2187.5
Bottom flange	15 × 5 = 75	5/2 = 2.5	187.5
Total	Σa = 325	–	Σay = 6437.5

Distance of the centroidal axis from the bottom edge,

$$\bar{y} = \frac{\sum ay}{\sum a} = \frac{6437.5}{325} = 19.8 \text{ cm (or 15.2 cm from the top edge)}$$

$$I_{XX} = \left[\frac{25 \times 5^3}{12} + 25 \times 5 \times (15.2 - 2.5)^2\right] + \left[\frac{5 \times 25^3}{12} + 5 \times 25 \,(19.8 - 17.5)^2\right]$$

$$+ \left[\frac{15 \times 5^3}{12} + 15 \times 5 \,(19.8 - 2.5)^2\right]$$

$$= 20421.66 + 7171.66 + 22603 = 50196.3 \text{ cm}^4 = 5.0196 \times 10^{-4} \text{ m}^4$$

Using the relation,

$$\frac{M}{I} = \frac{\sigma}{y} \text{ with usual notations, we have}$$

$$M = \frac{\sigma}{y} \times I = \frac{20}{0.152} \times 5.0196 \times 10^{-4} \text{ MNm} = \textbf{0.066 MNm}$$

Shear stress distribution:

We know that the shear stress at the extreme edges of the flanges is *zero*.

Let us find out the shear stress at the junction of top flange and the web.

Area, $A = 25 \times 5 = 125 \text{ cm}^2 = 125 \times 10^{-4} \text{ m}^2$

$\bar{y} = 15.2 - 2.5 = 12.7 \text{ cm} = 0.127 \text{ m}$

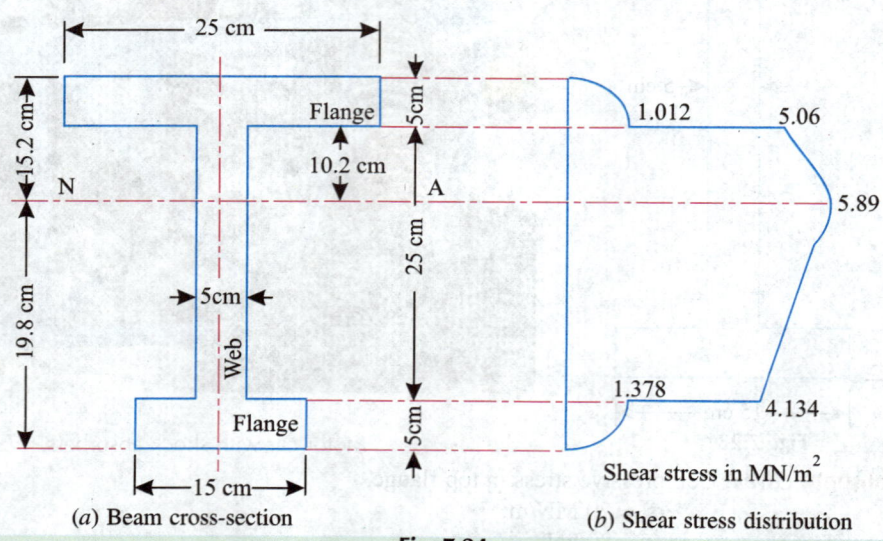

(*a*) Beam cross-section (*b*) Shear stress distribution

Fig. 7.24

Using the relation,

$$\tau = \frac{SA\bar{y}}{Ib}, \text{ we get}$$

$$\tau = \frac{80 \times 125 \times 10^{-4} \times 0.127}{5.0196 \times 10^{-4} \times 0.25} \times 10^{-3} \text{ MN/m}^2 = 1.012 \text{ MN/m}^2$$

The shear stress then changes from

1.012 MN/m^2 to $1.012 \times \dfrac{0.25}{0.05} = 5.06 \text{ MN/m}^2$

We know that the *shear stress is maximum at the neutral axis.*

$\therefore$
$$A\,\bar{y} = [25 \times 5 \times (15.2 - 2.5)] + \left[10.2 \times 5 \times \frac{10.2}{2}\right]$$
$$= 1578.5 + 260.1 = 1847.6 \text{ cm}^3 = 1847.6 \times 10^{-6} \text{ m}^3$$

Again using the relation, $\tau = \dfrac{SA\bar{y}}{Ib}$ with usual notations, we get

$$\tau = \frac{80 \times 1847.6 \times 10^{-6}}{5.0196 \times 10^{-4} \times 0.05} \times 10^{-3} = 5.89 \text{ MN/m}^2$$

Now let us find out the stress at the junction of the bottom flange and the web.

Area,
$$A = 15 \times 5 = 75 \text{ cm}^2 = 75 \times 10^{-4} \text{ m}^2$$
$$\bar{y} = 19.8 - 5/2 = 17.3 \text{ cm} = 0.173 \text{ m}$$

Using the relation :
$$\tau = \frac{SA\bar{y}}{Ib} \text{ with usual rotations, we have :}$$

$$\tau = \frac{80 \times 75 \times 10^{-4} \times 0.173}{5.0196 \times 10^{-4} \times 0.15} \times 10^{-3} = 1.378 \text{ MN/m}^2$$

The shear stress suddenly changes from 1.378 to $1.378 \times \dfrac{0.15}{0.05} = 4.134 \text{ MN/m}^2$

The shear stress distribution diagram is shown in Fig. 7.24 (b).

Example 7.15. *A beam has triangular cross-section with base b and height h, and is used with the base horizontal. Calculate the intensity of maximum shear stress and plot the variation of shear stress intensity along the section.*

Solution. Refer to Fig. 7.25.

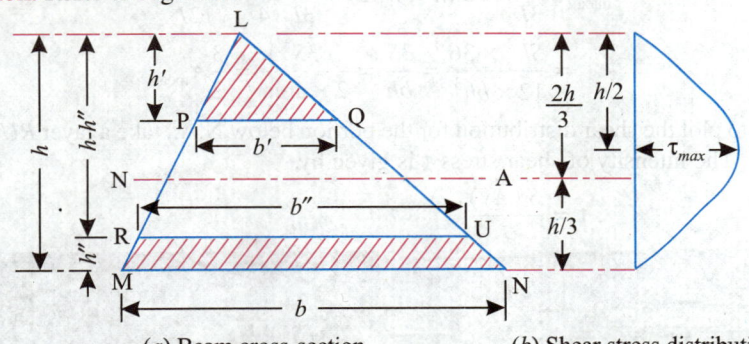

(a) Beam cross-section   (b) Shear stress distribution

Fig. 7.25

Moment of inertia about neutral axis,

$$I = \frac{bh^3}{36}$$

At any section from the apex L,

$$\tau = \frac{S\,(A\bar{y})}{Ib'}$$

where,
$$b' = \frac{bh'}{h}$$

$$(A\bar{y}) = \left(\frac{1}{2}\,b'\,h'\right)\left(\frac{2h}{3} - \frac{2h'}{3}\right)$$
$$= \frac{1}{3}\,b'\,h'\,(h - h')$$

$$\left[\begin{array}{l} \text{From similar } \Delta s \\ LPQ \text{ and } LMN \\ \dfrac{h'}{h} = \dfrac{b'}{b} \\ \text{or } b' = \dfrac{bh'}{h} \end{array}\right]$$

$$\therefore \qquad \tau = \frac{S}{Ib'} \frac{1}{3} b' h' (h - h') = \frac{S}{3\,I} h' (h - h') \qquad \qquad ...(i)$$

The variation is therefore, parabolic.

At $\qquad\qquad h' = 0, \tau = 0$

At N.A., $\qquad\qquad \tau = \dfrac{S}{3\,I} (2/3\ h)\ (1/3\ h)$

$$= \frac{2}{27} \frac{S\,h^2}{I} = \frac{2}{27} \times \frac{36\,S}{b\,h^3}\,h^2 = \frac{72}{27}\frac{S}{bh} = \frac{8}{3}\frac{S}{bh} \qquad\qquad ...(ii)$$

$$= \frac{4}{3}\frac{S}{\dfrac{1}{2}\,bh} = \frac{4}{3}\frac{S}{\text{area}} = \frac{4}{3}\,\tau_{\text{mean}}$$

Thus the intensity at N.A. is 4/3 of mean.

For maximum intensity above N.A.

$$\frac{d\tau}{dh'} = 0$$

i.e., $\qquad \dfrac{d}{dh'}\left[\dfrac{S}{3I}\,h'\,(h - h')\right] = h - 2h' = 0 \ \text{ or } \ h' = h/2$

Hence, maximum stress occurs at half the height.

Substituting $h' = h/2$ in eqn. (*i*), we get :

$$\tau_{\text{max}} = \frac{S}{3I}\,h/2\,(h - h/2) = \frac{S}{3I}\frac{h^2}{4} = \frac{Sh^2}{12I}$$

$$= \frac{Sh^2 \times 36}{12 \times bh^3} = \frac{3S}{bh} = \frac{3S}{2 \times \text{area}} = \frac{3}{2}\,\tau_{\text{mean}} \qquad\qquad ...(iii)$$

Similarly to plot the shear distribution for the portion below N.A., take a layer *RU* at a distance h'' from the base. The intensity of shear stress τ is given by,

$$\tau = \frac{S\,(A\bar{y})}{I}$$

A double-decker electric locomotive.

where, $b'' = RU = b/h\ (h - h').....$ From similar Δs LRU & LMN

$$(A\bar{y}) = \frac{1}{2}\ (b + b'')\ h'' \left[\frac{h}{3} - \frac{h''}{3} \left(\frac{b + 2b''}{b + b'}\right)\right]$$

$$= \frac{h''}{6} \left[b + \frac{b}{h}\ (h - h'')\right] \left[h - h'' \left\{\frac{b + \dfrac{2b}{h}\ (h - h'')}{b + \dfrac{b}{h}\ (h - h'')}\right\}\right]$$

$$= \frac{bh''}{6\ h} \left\{(2h - h'')\right\} \left[h - h'' \left\{\frac{3h - 2h''}{2h - h''}\right\}\right]$$

$$= \frac{bh''}{6h} \left\{2h^2 - 4h''\ h + 2h''^2\right\} = \frac{bh''}{3\ h}\ (h - h'')^2$$

$$\tau = \frac{S}{I\ \dfrac{b}{h}\ (h - h'')} \cdot \frac{bh''}{3\ h}\ (h - h'')^2 = \frac{S\ h''}{3I}\ (h - h'') \qquad\qquad ...(iv)$$

Thus the law of variation is *parabolic* and is similar to that in eqn. (*i*)

At $h'' = 0,\ \tau = 0$

At $h'' = h/3,\ \tau_{N.A} = \dfrac{S}{3\ I} \times \dfrac{h}{3}\ (h - h/3) = \dfrac{2}{27}\ \dfrac{Sh^2}{I}$

which is the same as found earlier.

The shear stress distribution is shown in Fig. 7.25.

Example 7.16. *A beam of square section subject to a shear force S is so placed that one of its diagonals is horizontal. Sketch shear stress distribution for the section.*

Solution. Refer to Fig. 7.26.

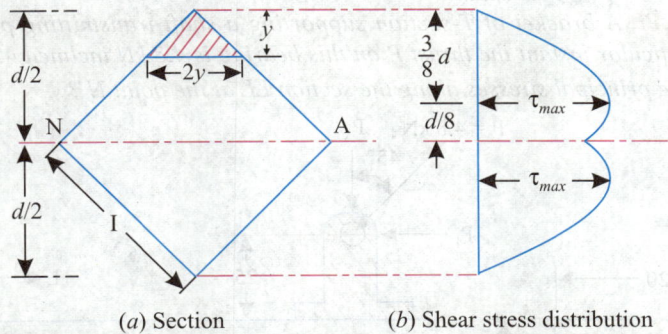

(*a*) Section   (*b*) Shear stress distribution

Fig. 7.26

Let, l = Length of the side of the square.

Then diagonal, $d = \sqrt{l^2 + l^2} = l\ \sqrt{2}$

Moment of inertia of the section about neutral axis (N.A.),

$$I = 2 \times \frac{d \times (d/2)^3}{12} = \frac{d^4}{48}$$

The value of shear stress at a distance y from the top edge,

$$\tau = \frac{S\ A\bar{y}}{I\ b}, \text{ with usual notations}$$

$$= \frac{S}{d^4/48 \times 2y} \times \left(\frac{1}{2} \times 2y \times y\right) \times \left(\frac{d}{2} - \frac{2y}{3}\right)$$

$$= \frac{4\,Sy}{d^4}\,(3d - 4y) \qquad\qquad\qquad\qquad ...(i)$$

For shear stress to be maximum ,

$$\frac{d\tau}{dy} = 0$$

or, $\qquad \dfrac{d}{dy}\left[\dfrac{4\,Sy}{d^4}\,(3d - 4y)\right] = 0$

$\therefore \qquad\qquad\qquad 3d - 8y = 0 \quad$ or $\qquad y = 3d/8$

The shear stress is maximum at

$$\left(\frac{d}{2} - \frac{3d}{8}\right) = d\,/\,8 \text{ from N.A.}$$

From eqn. (i), $\qquad \tau_{max} = \dfrac{4S \times 3d}{8d^4} \times \left(3d - 4 \times \dfrac{3}{8}d\right)$

$$= \frac{3S}{2d^3}\left(3d - \frac{3d}{2}\right) = \frac{9S}{4d^2} = \frac{9S}{4 \times (l\sqrt{2})^2} = \frac{9S}{8l^2}$$

$$\tau_{N.A} = \frac{4S}{d^4} \times (d\,/\,2) \times (3d - 4 \times d/2) = \frac{2S}{d^3} \times d = \frac{2S}{d^2} = \frac{2S}{(l\sqrt{2})^2}$$

$$= \frac{S}{l^2} = \frac{S}{\text{Area of cross-section}} = \tau_{\text{mean}}$$

Thus the shear stress at the neutral axis is the mean stress.

$$\frac{\tau_{\text{max}}}{\tau_{\text{mean}}} = \frac{9S\,/\,8l^2}{S\,/\,l^2} = \mathbf{9/8}$$

Example 7.17. *A bracket of T-section supporting a shaft transmitting power is shown in Fig. 7.27. At a particular instant the thrust P on this bearing is 4.8 kN inclined 45° to the vertical. What are the principal stresses along the section LL at the point N ?*

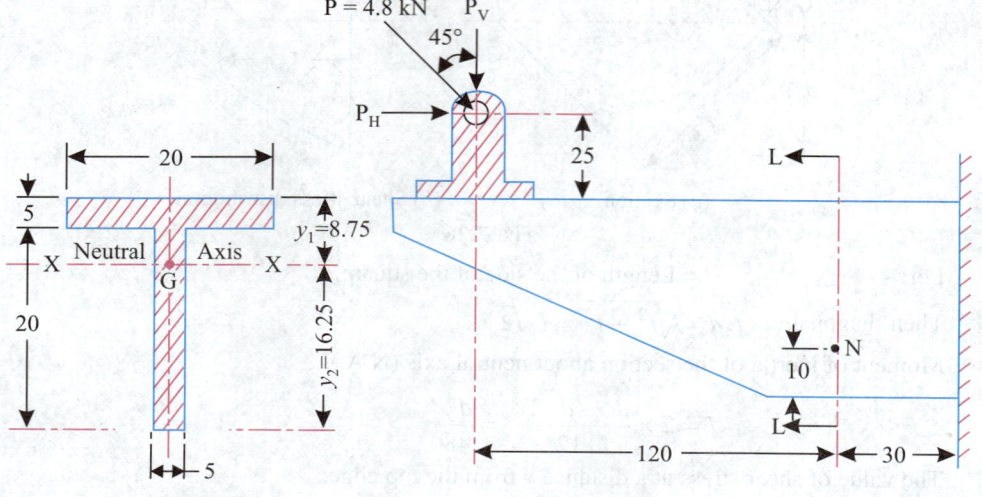

Section at LL   All dimensions in mm

Fig. 7.27

Solution. Thrust on the bearing,

$$P = 4.8 \text{ kN}$$

Vertical component of P,

$$P_V = 4.8 \cos 45° = 3.394 \text{ kN}$$

Horizontal component of P,

$$P_H = 4.8 \sin 45° = 3.394 \text{ kN}$$

Due to vertical component P_V there will be
(i) a bending moment, and (ii) a shearing force on the section.
Bending moment at the section LL,

$$M = P_V \times 120 = 3.394 \times (120 \times 10^{-3}) = 0.4073 \text{ kNm}$$

Shear force at the section $LL = P_V = 3.394 \text{ kN}$

T-section. Since the T-section is symmetrical about YY-axis, therefore, c.g.. lies on the YY axis.
The distance of the c.g. from the top edge,

$$y_1 = \frac{20 \times 5 \times 2.5 + 20 \times 5 \times 15}{20 \times 5 + 20 \times 5} = \frac{250 + 1500}{100 + 100} = 8.75 \text{ mm}$$

then, $\qquad y_2 = 25 - 8.75 = 16.25 \text{ mm}$

Now, I_{NA} or $I_{XX} = \left[\dfrac{20 \times 5^3}{12} + 20 \times 5 \, (8.75 - 2.5)^2 \right] + \left[\dfrac{5 \times 20^3}{12} + 5 \times 20 \times (16 \cdot 25 - 10)^2 \right]$

$$= (4114.58 + 7239.58) = 11354 \text{ mm}^4 = 11.354 \times 10^{-9} \text{ m}^4$$

Due to the bending moment M, the compressive stress at N,

$$\sigma' = \frac{M}{I_{XX}} \times y$$

where, $\qquad y = y_2 - 10 = 16.25 - 10 = 6.25 \text{ mm} = 0.00625 \text{ m}$

$$\therefore \qquad \sigma' = \frac{0.4073}{11.354 \times 10^{-9}} \times 0.00625 \times 10^{-3} \text{ MN/m}^2 = 224.2 \text{ MN/m}^2 \text{ (comp.)}$$

This bulldozer's bucket here has semi-circular motion powered by hydraulic arms.

Shear stress at $y = 6.25$ mm from the neutral axis

$$\tau = \frac{S\ A\bar{y}}{I_{XX}\ b} = \frac{3.394 \times (10 \times 5 \times 10^{-6}) \times (5 \times 10^{-3})}{11.354 \times 10^{-9} \times (5 \times 10^{-3})} \times 10^{-3} \ \text{MN/m}^2$$

$$= 14.95 \ \text{MN/m}^2$$

Now, eccentricity of horizontal component P_H,

$$e = 25 + 8.75 = 33.75 \ \text{mm} = 0.03375 \ \text{m}$$

∴ Bending moment due to P_H

$$M' = P_H \times e = 3.394 \times 0.03375 = 0.1145 \ \text{kNm}$$

Compressive stress on the section due to P_H,

$$\sigma_d = \frac{P_H}{A} = \frac{3.394}{(20 \times 5 + 5 \times 20) \times 10^{-6}}$$

$$= 16970 \ \text{kN/m}^2 \ \text{or} \ 16.97 \ \text{MN/m}^2 \ \text{(comp.)}$$

Tensile stress on the point N due to M',

$$\sigma'' = \frac{M'\ y}{I_{XX}} = \frac{0.1145 \times [(16.25 - 10) \times 10^{-3}]}{11.345 \times 10^{-9}} \times 10^{-3} \ \text{MN/m}^2$$

$$= 63.03 \ \text{MN/m}^2 \ \text{(tensile)}$$

Net direct stress at the point N

$$= -224.2 - 16.97 + 63.03 = -178.14 \ \text{MN/m}^2$$

Shear stress at point N, $\tau = 14.95 \ \text{MN/m}^2$

Principal stress at point N:

Principal stresses (σ_1, σ_2) are calculated as follows:

$$\sigma_1, \sigma_2 = -\frac{178.14}{2} \pm \sqrt{\left(\frac{178.14}{2}\right)^2 + (14.95)^2}$$

$$= -89.07 \pm \sqrt{7933.46 + 223.5} = -89.07 \pm 90.3 \ \text{or} \ -179.37, 1.23$$

∴ $\sigma_1 = -179.37 \ \text{MN/m}^2$

or , $\sigma_1 = \textbf{179.37 MN/m}^2 \ (compressive) \ \textbf{(Ans.)}$

and, $\sigma_2 = \textbf{+1.23 MN/m}^2 \ (tensile) \ \ \textbf{(Ans.)}$

Example 7.18. *Fig. 7.28 (a) shows R.S.J 30 cm × 15 cm. At a certain section it has to resist a bending moment of 150 kNm and a shear force of 300 kN. Find the principal stresses:*

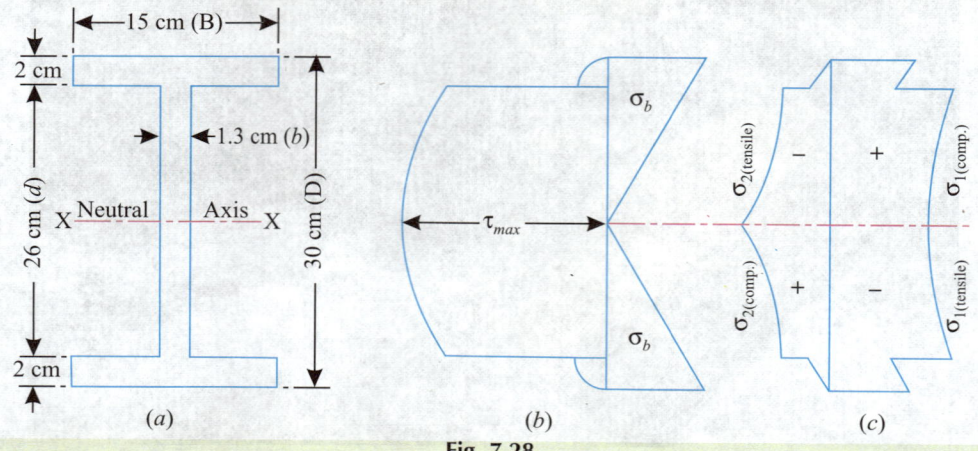

Fig. 7.28

(i)   *At the top;*

(ii) *In the flange at 13 cm from neutral axis;*

(iii) *In the web at 13 cm from neutral axis;*

(iv) *At the neutral axis.*

Show the variation of principal stresses along the section.

Solution. Bending moment to be resisted = 150 kNm

Shear force to be resisted = 300 kN

Moment of inertia of R.S.J about the neutral axis,

$$I_{XX} = \frac{15 \times 30^3}{12} - \frac{(15 - 1.3)\,26^3}{12}$$

$$= 33750 - 20066 = 13684 \text{ cm}^4 = 1.3684 \times 10^{-4} \text{ m}^4$$

Principal stresses, σ_1 ; σ_2:

(i)   *At the top :*

Bending stress,

$$\sigma_b = \frac{My}{I} = \frac{150 \times (15 \times 10^{-2})}{1.3684 \times 10^{-4}} \times 10^{-3} \text{ MN/m}^2 = 164.4 \text{ MN/m}^2 \text{ (comp.)}$$

Shear stress, $\tau = 0$

$$\sigma_1 = \textbf{164.4 MN/m}^2 \textit{ (comp)} \text{ ; } \sigma_2 = \textbf{0} \text{ (Ans.)}$$

(ii) *In the flange at 13 cm from the neutral axis:*

Bending stress,

$$\sigma_b = 164.4 \times \left(\frac{13}{15}\right) = 142.5 \text{ MN/m}^2$$

Shear stress, $\tau = \dfrac{S}{8I}(D^2 - d^2) = \dfrac{300}{8 \times 1.3684 \times 10^{-4}}$

$$\times [(0.3)^2 - (0.26)^2] \times 10^{-3} \text{ MN/m}^2$$

$$= 27.4 \times 10^4 \times (0.09 - 0.0676) \times 10^{-3} = 6.14 \text{ MN/m}^2$$

$$\sigma_1, \sigma_2 = \frac{142.5}{2} \pm \sqrt{\left(\frac{142.5}{2}\right)^2 + (6.14)^2} = 71.25 \pm 71.51$$

$$\sigma_1 = \textbf{142.76 MN/m}^2 \textit{ (comp.)}, \textbf{and}$$

$$\sigma_2 = \textbf{0.26 MN/m}^2 \textit{(tensile)} \textbf{ (Ans.)}$$

(iii) *In the web at 13 cm from the neutral axis:*

Bending stress $\sigma_b = 142.5 \text{ MN/m}^2$

Shear stress, $\tau = \dfrac{S}{8I}(D^2 - d^2) \times \dfrac{B}{b} = 6.14 \times \dfrac{15}{1.3} = 70.85 \text{ MN/m}^2$

$$\therefore \qquad \sigma_1, \sigma_2 = \frac{142.5}{2} \pm \sqrt{\left(\frac{142.5}{2}\right)^2 + (70.85)^2} = 71.25 \pm 100.48$$

$$\sigma_1 = \textbf{171.73 MN/m}^2 \textit{(comp)}, \textbf{and}$$

$$\sigma_2 = \textbf{29.23 MN/m}^2 \textit{(tensile)} \textbf{ (Ans.)}$$

[**Note.** We find from above that the intensity of principal stress (171.73 MN/m²) at the *junction* may exceed that at the extreme outside layers of the section when it is subjected to heavy bending moment and shearing force simultaneously.]

(iv) *At the neutral axis:*

Bending stress, $\sigma_b = 0$

Shearing stress, $\tau = \dfrac{S}{8I}(D^2 - d^2) \times \dfrac{B}{b} + \dfrac{Sd^2}{8I}$

$$= 70.85 + \dfrac{300 \times 10^{-3} \times 0.26^2}{8 \times 1.3684 \times 10^{-4}} = 87.38 \text{ MN/m}^2$$

∴ $\sigma_1 = \sigma_2 = 89.38 \text{ MN/m}^2$ **(Ans.)**

Fig. 7.28 (c) shows the variation of principal stresses along the section.

HIGHLIGHTS

1. The longitudinal shearing stresses balance the variation of bending stresses along the beam.

2. If the bending moment is consistent there will be no shearing stresses.

3. In case of *rectangular section*, $\tau_{max} = 3/2\ \tau_{mean}$

4. In case of *circular section*, $\tau_{max} = 4/3\ \tau_{mean}$.

5. In case of a symmetrical *I*-section,

$$\tau_{max} = \dfrac{S}{8\,Ib}[B\,(D^2 - d^2) + bd^2]$$

where, S = Shearing force,

               I = Moment of inertia about N.A.,

               B = Width of the flange (or section),

               b = Width or thickness of web,

               D = Overall depth of the section, and

               d = Depth/height of the web.

OBJECTIVE TYPE QUESTIONS

Choose the Correct Answer:

1. If the bending moment is consistent there will be no .... stresses.
   - (a) tensile
   - (b) compressive
   - (c) shearing
   - (d) none of the above.

2. In case of a rectangular section
   - (a) $\tau_{max} = \dfrac{1}{2}\tau_{mean}$
   - (b) $\tau_{max} = \tau_{mean}$
   - (c) $\tau_{max} = 3/2\ \tau_{mean}$
   - (d) $\tau_{max} = 5/2\ \tau_{mean}$.

3. In case of a circular section
   - (a) $\tau_{max} = 3/2\ \tau_{mean}$
   - (b) $\tau_{max} = \tau_{mean}$
   - (c) $\tau_{max} = 2/3\ \tau_{mean}$
   - (d) $\tau_{max} = 4/3\ \tau_{mean}$.

4. In case of a circular section the maximum shear stress is ...... percent more than the mean shear stress.
   - (a) 10
   - (b) 20
   - (c) 33.33
   - (d) 66.66.

5. In the case of an *I*-section beam maximum shear stress is at

   - (a) the junction of the top flange and web
   - (b) middle of the web
   - (c) either (a) or (b)
   - (d) none of the above.

6. A square section with side x of a beam is subjected to a shear force S, the magnitude of shear stress at the top edge of the square is
   - (a) $\dfrac{1.5\ S}{x^2}$
   - (b) $\dfrac{S}{x^2}$
   - (c) $\dfrac{0.55}{x^2}$
   - (d) Zero.

7. In *I*-section of a beam subjected to transverse shear force S the maximum shear stress is developed
   - (a) at the centre of the web
   - (b) at the top of edge of the top flange
   - (c) at the bottom edge of the top flange
   - (d) none of the above.

ANSWERS

1. (c)   2. (c)   3. (d)   4. (c)   5. (b)   6. (d)   7. (a)

THEORETICAL QUESTIONS

1. Derive an expression for the shear stress τ at a point in a transverse section subjected to shear force S.

2. Obtain an expression for shearing stress developed in a beam. State the assumptions on which the above is based. Show that for rectangular section distribution of shearing stress is parabolic.

3. Show the shear stress variations in the following sections:

    (*i*) Rectangle   (*ii*) Solid circle   (*iii*) Hollow circle   (*iv*) I-section

    (*v*) T-section   (*vi*) L-section   (*vii*) Cross   (*viii*) Built up section.

UNSOLVED EXAMPLES

1. A wooden beam 15 cm wide, 30 cm deep and 3 m long is carrying a U.D.L of 15 kN/m. Determine the maximum shear stress and sketch the variation of shear stress along the depth of the beam. [**Ans.** 0.75 MN/m^2]

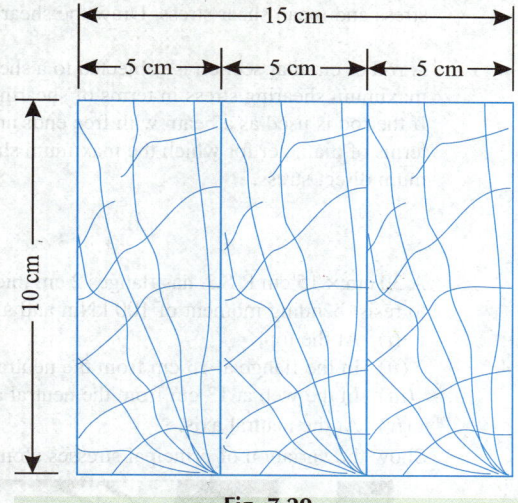

2. A circular beam of 100.5 mm diameter is subjected to a shear force of 10 kN. Calculate the value of maximum shear stress, and sketch the variation of the shear stress along the depth of the beam. [**Ans.** 1538 kN/m^2]

3. A laminated wooden beam 10 cm wide and 15 cm deep is made of three 5 cm × 10 cm wide planks glued together, as shown in Fig. 7.29 to resist longitudinal shear. The beam is simply supported over a span of 2.5 m. If allowable shearing stress in the glued joint is 0.55 MN/m^2, then, find the safe point load that the beam may carry at its centre.

    [**Ans.** 12.375 kN]

Fig. 7.29

4. An *I*-section with rectangular ends has the following dimensions:

Flanges   :   150 mm × 20 mm

Web   :   300 mm × 10 mm

Total Depth   :   340 mm

Determine the maximum shearing stress developed in the beam for the shearing force of 25 kN.

    [**Ans.** 8.4 MN/m^2]

5. An *I* beam has flanges 10 cm wide and 1 cm thick and web 12 cm high and 1 cm thick. If this section is subjected to a bending moment of 10 kNm and a shearing force of 10 kN, find the maximum tensile and shear stresses induced in it. [**Ans.** 70.66 MN/m^2, 8.378 MN/m^2]

6. An *I*-section having flanges 200 mm × 20 mm and web 400 mm × 15 mm is used as a beam. If at a section, it is subjected to a shear force of 150 kN, find the greatest intensity of shear stress in the beam and show also the variation of shear stress across the section. [**Ans.** 26.323 MN/m^2]

7. A beam of span *l* metres simply supported at the ends, carries a central load *W*. The beam section has an overall depth of 29 cm, with horizontal flanges each 15 cm × 2 cm and a vertical web 25 cm × 1 cm. If the maximum shear stress is to be 45 MN/m^2 when maximum bending stress is 150 MN/m^2, calculate the value of the centrally applied point load *W* and the span *l*.[**Ans.** *W* = 228.4 kN; *l* = 2.221 m]

8. A beam of channel section 120 mm × 60 mm has uniform thickness of 15 mm. Draw diagram showing the distribution of shear stress for a vertical section where shearing force is 150 kN. Find the ratio between maximum and mean shearing stresses.

$$[\text{Ans. } 2.22, \tau_{max} = 105.72 \text{ MN/m}^2, \tau_{mean} = 47.61 \text{ MN/m}^2]$$

9. Find the maximum shear stress induced by a load of 4 kN in the vertical section of a hollow beam of a square section if the outside width is 10 cm and the thickness of material is 2 cm.

$$[\text{Ans. } 1.35 \text{ MN/m}^2]$$

10. Calculate the ratio of maximum to mean shear stress in an *I*-beam, 200 mm wide × 350 mm deep, having the flanges 25 mm thick and web 12.5 mm thick. Also find the percentage of the total shearing force carried by the web. [**Ans.** 3.57, 92.9%]

11. A beam of *T*-section, symmetrical about the vertical axis, has the flange 120 mm × 10 mm and the web 100 mm × 10 mm. What is the percentage of the shearing force carried by the web? [**Ans.** 93.5%]

12. A tube of hollow square section, 50 mm square outside and 6 mm uniform thickness, is subjected to a shearing force of 50 kN acting in the direction of a diagonal. Find the maximum shearing stress produced. [**Ans.** 105.9 MN/m²]

13. A beam of square section of side 20 mm is placed so that the plane of bending is parallel to the diagonal. The shearing force at a section is 20 kN. Calculate the values of the maximum shearing stress and mean shear stress. Draw the shear stress distribution diagram for the section.

$$[\text{Ans. } 168.75 \text{ MN/m}^2; 50 \text{ MN/m}^2]$$

14. A rod of circular section is subjected to a shearing force on a plane perpendicular to its axis. Find the maximum shearing stress in terms of shearing force and rod diameter.

If the rod is used as a beam with free ends and a central concentrated load, express the free length in terms of diameter for which the maximum shearing stress, due to the shearing force, is half the maximum direct stress.

$$[\text{Ans. } l = \frac{2}{3} d \,]$$

15. A 30 cm × 15 cm R.S.J. has flanges 2 cm thick and web 1.3 cm thick. At a certain cross-section it has to resist bending moment of 100 kNm and a shear force of 200 kN. Find the principal stresses :
   (*i*) At the top,
   (*ii*) In the flange at 13 cm from the neutral axis,
   (*iii*) In the web at 13 cm from the neutral axis, and
   (*iv*) At the neutral axis.
Show the variation of principal stresses along the section.

A bridge.

Deflection of Beams

8.1. INTRODUCTION

It is observed that when a beam or a cantilever is subjected to some type of loading it deflects from its initial/original position. The amount of deflection depends upon its cross-section and bending moment. These days *strength* and *stiffness* are the two main design criteria for a beam or a cantilever.

According to *strength criterion* of the beam design, the beam should be adequately strong *to resist shear force and bending moment*. In other words the beam should be able to resist shear stresses and bending stresses. But according to *stiffness* (being mathematically calculated as $\frac{W}{\delta}$ where W is the applied load and δ is the maximum deflection or sag) *criterion* of the beam design, which is equally important, the beam should be adequately stiff to resist deflection. In other words, the *beam should be stiff enough not to deflect more than the permissible limit*.

8.2. BEAM DEFLECTION

When a load is placed on a beam the beam tends to

defect or sag as shown in Fig. 8.1. Deflection plays a significant role in the design of structures and machines, If floor beams, or joists deflect too far, the plaster on the ceiling under them may crack. Although no damage to the structure may result, the apppearance on the ceiling may be ruined. Also, a floor supported by such beams may be so out of level that its usefulness for machinery may be impaired.

Under load the neutral axis becomes a curved line and is called the **elastic curve.** *The deflection 'y' is vertical distance between a point on the elastic curve and the unloaded neutral axis.*

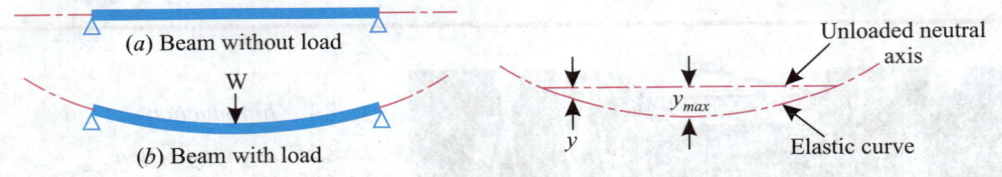

(a) Beam without load

(b) Beam with load

y_{max}

Unloaded neutral axis

Elastic curve

Fig. 8.1   **Fig. 8.2**

8.3. RELATION BETWEEN SLOPE, DEFLECTION AND RADIUS CURVATURE

Fig. 8.3 shows a small portion *AB* of a beam bent into an arc.

Let, ds = Length of beam *AB*,

C = Centre of the arc (into which the beam has been bent),

α = Angle which the tangent at *A* makes with *XX*-axis, and

$\alpha + d\alpha$ = Angle which the tangent at *B* makes with *XX*-axis.

We find, from the geometry of the figure, that

$$\angle ACB = d\alpha \text{ and } ds = R \, d\alpha.$$

∴ $R = \dfrac{ds}{d\alpha} = \dfrac{dx}{d\alpha}$ (assuming $ds = dx$)

or, $\dfrac{1}{R} = \dfrac{d\alpha}{dx}$ ...(i)

If the co-ordinates of point *A* are *x* and *y*, then

$$\tan \alpha = \dfrac{dy}{dx} \text{ or } \alpha = \dfrac{dy}{dx}$$

(taking tan $\alpha = \alpha$ since α is very small)

Differentiating the above equation w.r.t. *x*, we get

$$\dfrac{d\alpha}{dx} = \dfrac{d^2 y}{dx^2}$$

$$\dfrac{1}{R} = \dfrac{d^2 y}{dx^2}$$

$$\left[\because \ \dfrac{1}{R} = \dfrac{d\alpha}{dx} \text{ as at } (i) \right]$$

Also we know that,

Connecting rod and crankshaft of an IC engine.

$$\frac{M}{I} = \frac{E}{R} \quad \text{or} \quad M = E \times \frac{I}{R}$$

Now substituting $\frac{1}{R} = \frac{d^2y}{dx^2}$ in the above equation we get,

$$M = EI \frac{d^2y}{dx^2} \qquad \qquad \ldots(8.1)$$

The above equation is based *only on bending moment* (effect of shear force, being very small, has been *neglected*)

8.4. SIGN CONVENTIONS

To find out the slope and deflection of a centre line of a beam at any point proper sign conventions will have to be taken into account, the following **sign conventions** will be used:

1. x is *positive* when measured towards *right.*
2. y is *negative* when measured *downwards.*
3. M (bending moment) is *negative* when *hogging.*
4. Slope is *negative* when the rotation is *clockwise.*

8.5. SLOPE AND DEFLECTION AT A SECTION

The important methods used for finding out the slope and deflection at a section in a loaded beam are discussed as follows :

1. Double integration method.
2. Moment area method.
3. Macaulay's method.

● The first two methods are suitable for a *single load,* whereas the last one is suitable for *several loads.*

● *"Moment area method" is more useful as compared to double integral method because many problems which do not have a simple mathematical solution can be simplified by the bending moment area method.*

8.6. DOUBLE INTEGRATION METHOD

Cantilevers:

Case I. Cantilever beam– concentrated load at free end:

Fig. 8.4 shows cantilever with a concentrated load W acting at free end.

Let the moment of inertia of the section of the cantilever about the neutral axis be I. Consider a section XX at a distance x from the fixed end $A.$

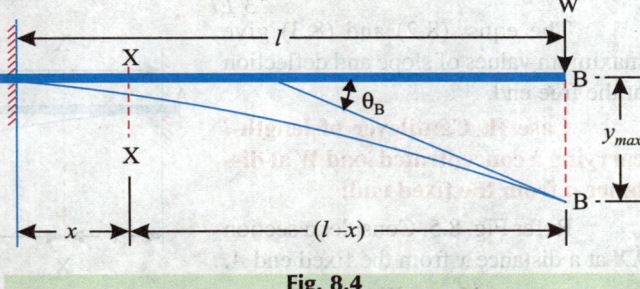

Fig. 8.4

$$M_x = -W(1-x)$$

$$\therefore \qquad EI \frac{d^2y}{dx^2} = -W(l-x)$$

Integrating, we get

$$EI \frac{dy}{dx} = -W\left(lx - \frac{x^2}{2}\right) + C_1$$

(where, C_1 = constant of integration)

At A (fixed end):

$$x = 0 \text{ and } \frac{dy}{dx} = 0$$

$$\therefore \qquad C_1 = 0$$

$$\therefore \qquad EI \frac{dy}{dx} = -W\left(lx - \frac{x^2}{2}\right)$$

...(i) *Slope equation*

Slope at B: Putting $x = l$, we have

$$\theta_B = \frac{dy}{dx} = -\frac{1}{EI} W\left(l \times l - \frac{l^2}{2}\right)$$

$$= -\frac{Wl^2}{2 \, EI}$$

i.e. $\qquad \theta_B = -\dfrac{Wl^2}{2 \, EI}$ ...(8.2)

Open view of an IC engine crankcase.

To get deflection, integrating eqn. (i) above, we get

$$EI \, y = -W\left(l \cdot \frac{x^2}{2} - \frac{x^3}{6}\right) + C_2$$ (where, C_2 = constant of integration)

At A (fixed end):

$$x = 0 \text{ and } y = 0$$

$$\therefore \qquad C_2 = 0$$

$$EI \, y = -W\left(l \cdot \frac{x^2}{2} - \frac{x^3}{6}\right)$$

...(ii) *Deflection equation*

Deflection at B: Putting $x = l$, we get

$$y_B = -\frac{W}{EI}\left(l \cdot \frac{l^2}{2} - \frac{l^3}{6}\right) = -\frac{Wl^3}{3 \, EI}$$

Downward deflection of $B = \dfrac{Wl^3}{3 \, EI}$ ...(8.3)

The eqns. (8.2) and (8.3) give maximum values of slope and deflection at the free end.

Case II. Cantilever of length l carrying a concentrated load W at distance a from the fixed end:

Refer Fig. 8.5. Consider a section XX at a distance x from the fixed end A.

$$M_x = -W(a - x)$$

$$\therefore \quad EI \frac{d^2y}{dx^2} = -W(a - x)$$

Fig. 8.5

Integrating, we get $\quad EI \dfrac{dy}{dx} = -W\left(ax - \dfrac{x^2}{2}\right) + C_1$ (where, C_1 = constant of integration)

At fixed end A: $x = 0$, $\dfrac{dy}{dx} = 0$

$\therefore$ $\qquad\qquad C_1 = 0$

$\therefore$ $\qquad EI\dfrac{dy}{dx} = -W\left(ax - \dfrac{x^2}{2}\right)$ ...(i) *Slope equation*

Slope at C: Putting $x = a$, we get

$$\theta_C = \dfrac{dy}{dx} = -\dfrac{W}{EI}\left(a \cdot a - \dfrac{a^2}{2}\right) = -\dfrac{Wa^2}{2EI}$$

i.e., $\qquad\qquad \theta_C = -\dfrac{Wa^2}{2EI}$

As there is no load on the portion BC there will be no B.M. in that portion and the portion will not bend; it shall be straight.

$$\theta_B = \theta_C = -\dfrac{Wa^2}{2EI} \qquad ...(8.4)$$

To get deflection, integrating eqn. (i), we get

$$EI\,y = -W\left(a \cdot \dfrac{x^2}{2} - \dfrac{x^3}{6}\right) + C_2$$

(where, C_2 = constant of integration)

At fixed end A: $\quad x = 0$, $y = 0$

$\therefore$ $\qquad\qquad C_2 = 0$

$\therefore$ $\qquad EI\,y = -W\left(a \cdot \dfrac{x^2}{2} - \dfrac{x^3}{6}\right)$

Levers

Deflection at C: Putting $x = a$, we get

$$y_C = -\dfrac{W}{EI}\left(a \cdot \dfrac{a^2}{2} - \dfrac{a^3}{6}\right) = -\dfrac{Wa^3}{3EI}$$

Downward deflection at $C = \dfrac{Wa^3}{3EI}$ ...(8.5)

But, $y_c = BD$ (Fig. 8.5) and $B'D = DC' \tan\theta_C = BC \times \theta_C$

[Since θ_C is small, therefore $\tan\theta_C = \theta_C$]

$$B'D = (l - a) \times \left(-\dfrac{Wa^2}{2EI}\right)$$

But, $\qquad y_B = BB' = BD + BD' = -\dfrac{Wa^3}{3EI} - \dfrac{Wa^2}{2EI}(l - a)$

Downward deflection of $B = \dfrac{Wa^3}{3EI} + \dfrac{Wa^2}{2EI}(l - a)$ ...(8.6)

Case III. Cantilever of length *l* carrying uniformly distributed load *w* per unit run over whole length :

Refer Fig. 8.6. Consider a section XX at a distance x from fixed end A.

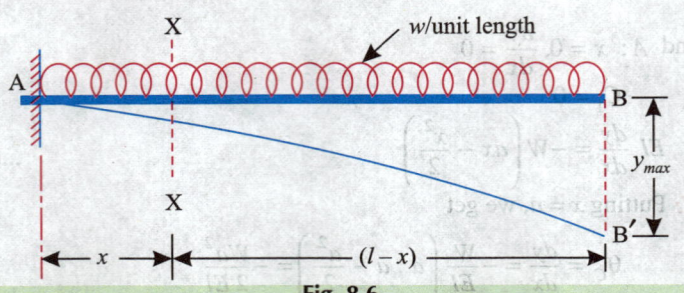

Fig. 8.6

$$M_x = -\frac{w(l-x)^2}{2}$$

or $$EI\frac{d^2y}{dx^2} = -\frac{w}{2}(l-x)^2$$

Integrating, we get $$EI\frac{dy}{dx} = +\frac{w}{6}(l-x)^3 + C_1 \qquad \text{(where, } C_1 = \text{constant of integration)}$$

At fixed end A: $$x = 0, \frac{dy}{dx} = 0$$

$$\therefore \qquad C_1 = -\frac{wl^3}{6}$$

$$\therefore \qquad EI\frac{dy}{dx} = +\frac{w}{6}(l-x)^3 - \frac{wl^3}{6} \qquad \qquad ...(i)\ Slope\ equation$$

Slope at B: Putting $x = l$, we have

$$EI\cdot\theta_B = \frac{dy}{dx} = \frac{w}{6}(l-l)^3 - \frac{wl^3}{6} = -\frac{wl^3}{6}$$

i.e., $$\theta_B = -\frac{wl^3}{6EI}\left(=-\frac{Wl^2}{6EI}\right) \qquad \text{(where } W = w.l) \ ...(8.7)$$

To get deflection, integrating eqn. (i), we get

$$EI\,y = -\frac{w}{24}(l-x)^4 - \frac{wl^3}{6}x + C_2$$

$$\text{(where, } C_2 = \text{constant of integration)}$$

At the fixed end A: $$x = 0, y = 0$$

$$0 = -\frac{wl^4}{24} + C_2$$

$$\therefore \qquad C_2 = \frac{wl^4}{24}$$

Hence, $$EI\,y = -\frac{w}{24}(l-x)^4 - \frac{wl^3}{6}x + \frac{wl^4}{24} \qquad ...(ii)\ Deflection\ equation$$

Deflection at B: Putting $x = l$, we get

$$EI\,y_B = -\frac{w}{24}(l-l)^4 - \frac{wl^3}{6}\times l + \frac{wl^4}{24}$$

$$= -\frac{wl^4}{6} + \frac{wl^4}{24} = -\frac{wl^4}{8}$$

$$\therefore \quad y_B = -\frac{wl^4}{8\,EI}\left(=-\frac{Wl^3}{8\,EI}\right) \qquad\qquad \text{(where, } W = wl\text{)}$$

Downward deflection of B

$$= \frac{wl^4}{8\,EI}\left(=\frac{Wl^3}{8\,EI}\right) \quad ...(8.8)$$

The eqns. (8.7 and 8.8) give slope and deflection at B which are *maximum*.

Case IV. Cantilever of length *l* carrying uniformly distributed load of *w* per unit run for a distance *a* from the fixed end.

Refer Fig. (8.7). Consider a section XX at a distance x from fixed end A.

$$M_x = -\frac{w(a-x)^2}{2}$$

$$\therefore \quad EI\frac{d^2y}{dx^2} = -\frac{w(a-x)^2}{2}$$

Fig. 8.7

Integrating for slope, we have

$$EI\frac{dy}{dx} = +\frac{w}{2}\left[\frac{(a-x)^3}{3}\right] + C_1$$

(where, C_1 = constant of integration)

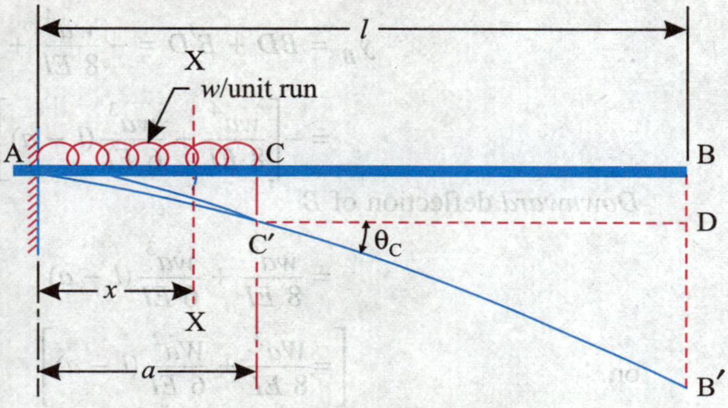

When, $\qquad x = 0, \dfrac{dy}{dx} = 0$

$$\therefore \qquad C_1 = -\frac{wa^3}{6}$$

Hence, $\qquad EI\cdot\dfrac{dy}{dx} = +\dfrac{w}{2}\times\dfrac{(a-x)^3}{3} - \dfrac{wa^3}{6} \qquad ...(i)\ Slope\ equation$

Slope at C: Putting $x = a$, we get

$$\theta_C = \frac{dy}{dx} = -\frac{wa^3}{6\,EI}$$

Since portion BC is not loaded it does not bend and remains straight, therefore,

$$\theta_B = \theta_C = -\frac{wa^3}{6\,EI}\left(=-\frac{Wa^2}{6\,EI}\right) \qquad \text{(where, } W = wa\text{) }...(8.9)$$

To get deflection integrating eqn. (i), we get

$$EI\,y = -\frac{w}{2}\frac{(a-x)^4}{12} - \frac{wa^3}{6}x + C_2 \text{ (where, } C_2 = \text{constant of integration)}$$

When, $\qquad x = 0, y = 0$

$$\therefore \qquad C_2 = +\frac{wa^4}{24}$$

Hence, $\qquad EI\,y = -\dfrac{w}{2}\dfrac{(a-x)^4}{12} - \dfrac{wa^3}{6}x + \dfrac{wa^4}{24} \qquad ...(ii)\ Deflection\ equation$

Deflection at C: Putting $x = a$, we get

$$EI\,y_c = -\frac{wa^4}{6} + \frac{wa^4}{24} = -\frac{wa^4}{8}$$

$$y_C = -\frac{wa^4}{8\,EI}\left(=-\frac{Wa^3}{8\,EI}\right) \qquad \text{(where, } W = w\cdot a\text{)} \quad ...(8.10)$$

$$CC' = BD = -\frac{wa^4}{8\,EI}$$

But, $\qquad B'D = C'D \tan\theta_C = BC \times \theta_C \qquad (\because \tan\theta_C = \theta_C \text{ when } \theta_C \text{ is small})$

$$= (l - a) \times \left(-\frac{wa^3}{6\,EI}\right) \qquad \left(\because \theta_C = -\frac{wa^3}{6\,EI}\right)$$

$\therefore \qquad y_B = BD + B'D = -\frac{wa^4}{8\,EI} + (l - a) \times \left(-\frac{wa^3}{6\,EI}\right)$

$$= -\left[\frac{wa^4}{8\,EI} + \frac{wa^3}{6\,EI}(l - a)\right]$$

Downward deflection of B

$$= \frac{wa^4}{8\,EI} + \frac{wa^3}{6\,EI}(l - a)$$

or, $\qquad \left[= \frac{Wa^3}{8\,EI} + \frac{Wa^2}{6\,EI}(l - a)\right] \qquad\qquad ...(8.11)$

Case V. Cantilever of length l carrying a uniformly distributed load of w per unit run on a part of span from the free end.:

It is obvious from the Fig. 8.8 (*a, b, c*) that to get result in case (*a*) take the differences of result in case (*b*) and case (*c*), thus:

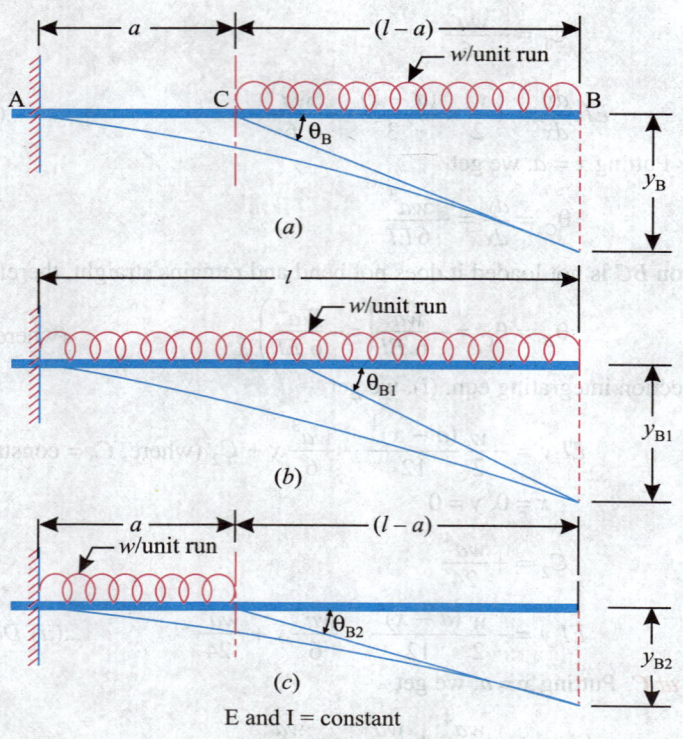

E and I = constant

Fig. 8.8

$$\theta_B = \theta_{B_1} - \theta_{B_2} \text{ and } y_B = y_{B_1} - y_{B_2}$$

But from the previous articles we have,

$$\theta_{B_1} = \frac{wl^3}{6\ EI}; \ y_{B_1} = -\frac{wl^4}{8\ El}$$

$$\theta_{B_2} = \frac{wa^3}{6\ EI}; \ y_{B_2} = -\left[\frac{wa^4}{8\ EI} + \frac{wa^3}{6\ EI}\ (l-a)\right]$$

Slope at B:

$$\theta_B = \theta_{B_1} - \theta_{B_2} = -\frac{wl^3}{6\ EI} - \left(-\frac{wa^3}{6\ EI}\right)$$

$$= -\frac{wl^3}{6\ EI} + \frac{wa^3}{6\ EI} = -\frac{w}{6\ EI}(l^3 - a^3)$$

i.e. $\quad \theta_B = -\dfrac{w}{6\ EI}\ (l^3 - a^3)$

Deflection at B:

$$y_B = y_{B_1} - y_{B_2}$$

$$= -\frac{wl^4}{8\ EI} - \left[-\left\{\frac{wa^4}{8\ EI} + \frac{wa^3}{6\ EI}(l-a)\right\}\right]$$

$$= -\frac{wl^4}{8\ EI} + \frac{wa^4}{8\ EI} + \frac{wa^3}{6\ EI}(l-a)$$

$$= -\frac{wl^4}{8\ EI} + \frac{wa^4}{8\ EI} + \frac{wa^3 l}{6\ EI} - \frac{wa^4}{6\ EI}$$

$$= -\frac{w}{8\ EI}(l^4 - a^4) - \frac{wa^3}{6\ EI}(l-a)$$

$$= -\frac{w}{24\ EI}\left[3(l^4 - a^4) - 4a^3(l-a)\right]$$

$$= -\frac{w}{24\ EI}(3l^4 - 3a^4 - 4a^3 l + 4a^4)$$

$$= -\frac{w}{24\ EI}(3l^4 - 4la^3 + a^4)$$

i.e. Downward deflection of B $= \dfrac{w}{24\ EI}(3l^4 - 4la^3 + a^4)$ ...(8.13)

A view of the declination shaft, gear and mounting platform in a mechanical automation.

Case VI. Cantilever of length *l* with a moment applied at the free end:

Refer Fig. 8.9. Consider a section *XX* at a distance *x* from the fixed end.

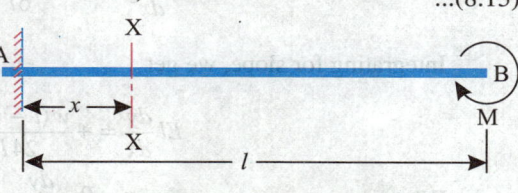

Fig. 8.9

$$M_x = -M; \quad EI\ \frac{d^2y}{dx^2} = -M$$

Integrating for slope, we have

$$EI\ \frac{dy}{dx} = -M.x + C_1$$

(where, C_1 = constant of integration)

When, $\quad x = 0, \dfrac{dy}{dx} = 0$

∴ $\qquad C_1 = 0$

Hence, $\qquad EI\dfrac{dy}{dx} = -M \cdot x$ $\qquad ..(i)$ *Slope equation*

Slope at B : Putting $x = l$, we get

$$\theta_B = \dfrac{dy}{dx} = -\dfrac{Ml}{EI} \qquad\qquad ...(8.14)$$

$$El\, y = -M \cdot \dfrac{x^2}{2} + C_2 \qquad (\text{where, } C_2 = \text{constant of integration})$$

When, $\qquad x = 0, y = 0$

∴ $\qquad C_2 = 0$

Hence, $\qquad El\, y = -M \cdot \dfrac{x^2}{2}$ $\qquad ...(ii)$ *Deflection equation*

Deflection at B: Putting $x = l$, we get

$$El\, y_B = -\dfrac{Ml^2}{2} \quad \therefore \quad y_B = -\dfrac{Ml^2}{2\,El}$$

i.e. Downward deflection of $\,B = \dfrac{Ml^2}{2\,EI}$ $\qquad ...(8.15)$

Case VII. Cantilever of length *l* carrying a distributed load when intensity varies uniformly from zero at the free end to *w* per unit run at the fixed ends.

Refer Fig. 8.10. Consider a section *XX* at a distance *x* from the fixed end.

Intensity of loading at the section *XX* $= \dfrac{w}{l}(l - x)$ per unit run

The B.M. at the section *XX*,

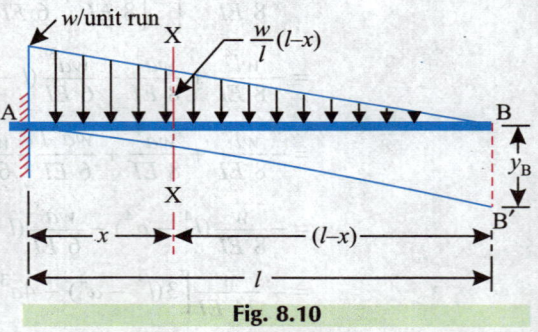

Fig. 8.10

$$M_x = -\dfrac{1}{2}(l - x) \times \dfrac{w}{l}(l - x) \times \dfrac{(l - x)}{3} = -\dfrac{w(l - x)^3}{6l}$$

∴ $$El\dfrac{d^2y}{dx^2} = -\dfrac{w(l - x)^3}{6l}$$

Integrating for slope, we get

$$El\dfrac{dy}{dx} = +\dfrac{w(l - x)^4}{24\,l} + C_1 \qquad (\text{where, } C_1 = \text{constant of integration})$$

When, $\qquad x = 0, \dfrac{dy}{dx} = 0 \quad \therefore \quad C_1 = -\dfrac{wl^3}{24}$

Hence, $\qquad El\dfrac{dy}{dx} = +\dfrac{w(l - x)^4}{24\,l} - \dfrac{wl^3}{24}$ $\qquad ...(i)$ *Slope equation*

Slope at B: Putting $x = l$, we get

$$\theta_B = \dfrac{dy}{dx} = -\dfrac{wl^3}{24\,El} \qquad\qquad ...(8.16)$$

To get deflection, integrating again, we have

$$El \ y = -\frac{w(l-x)^5}{120l} = -\frac{wl^3}{24}x + C_2 \qquad (\text{where, } C_2 = \text{constant of integration})$$

When $x = 0$, $y = 0$ $\qquad \therefore \ 0 = -\frac{wl^4}{120} + C_2$ or $C_2 = \frac{wl^4}{120}$

Hence, $\quad EI \ y = -\frac{w(l-x)^5}{120l} - \frac{wl^3}{24}x + \frac{wl^4}{120}$ $\qquad$ ...(ii) *Deflection equation*

Deflection at B: Putting $x = l$, we get

$$EI \ y_B = -\frac{wl^4}{24} + \frac{wl^4}{120} = -\frac{wl^4}{30}$$

or, $\qquad y_B = -\frac{wl^4}{30 \, EI}$

i.e. *Downward* deflection of B

$$= \frac{wl^4}{30 \, EI}$$

Case VIII. Cantilever of length 1 carrying a distributed load whose intensity varies uniformly from zero at the fixed end to w per unit run at the free end:

Refer Fig. 8.11. It is obvious that the deflection at

$$B = \begin{bmatrix} \text{Deflection at } B \text{ due to uniform-} \\ \text{ly distributed load of } w \text{ per unit} \\ \text{run over the whole.} \end{bmatrix} - \begin{bmatrix} \text{deflection at } B \text{ due to a distributed load where} \\ \text{intensity varies uniformly from zero at free end} \\ \text{end to } w \text{ per unit run at the fixed end.} \end{bmatrix}$$

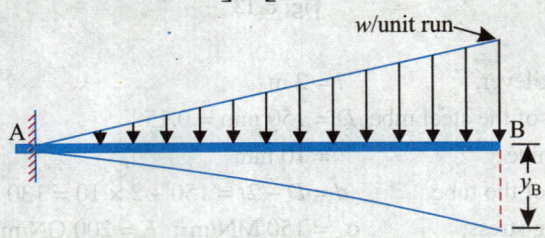

Fig. 8.11

Deflection *downward* at B, $\qquad y_B = \frac{wl^4}{8 \, EI} - \frac{wl^4}{30 \, EI} = \frac{11}{120} \frac{wl^4}{EI}$

WORKED EXAMPLES (Cantilevers)

Example 8.1. *A cantilever 1.5 m long carries a uniformly distributed load over the entire length. Find the deflection at the free end if the slope at the free end is 1.5°.*

Solution. Length of the cantilever,

$\qquad l = 1.5$ m

Slope at the free end,

$$= 1.5° = \frac{\pi}{180} \times 1.5 \text{ radian}$$

Deflection at B, y_B :

Slope at the free end

$$= \frac{wl^3}{6 \, EI} = \frac{\pi}{180} \times 1.5$$

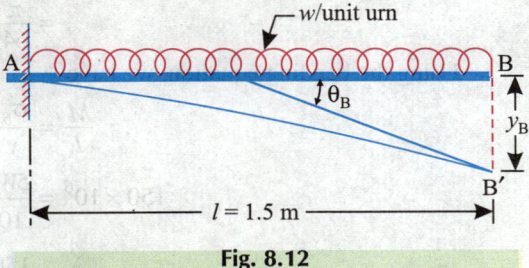

Fig. 8.12

∴ $$\frac{wl^3}{EI} = \frac{\pi \times 1.5 \times 6}{180} = \frac{\pi}{20}$$

Deflection at the free end, $$y_B = \frac{wl^4}{8EI} = \frac{wl^3}{EI} \times \frac{l}{8} = \frac{\pi}{20} \times \frac{1.5}{8}$$

$$= 0.02945 \text{ m or } 29.45 \text{ mm}$$

i.e. $$y_B = \textbf{29.45 mm} \quad \textbf{(Ans.)}$$

Example 8.2. *A 2 meters long cantilever made of steel tube of section 150 mm external diameter and 10 mm thick is loaded as shown in the fig. 8.13 (a). If E = 200 GN/m² calculate:*

(i) *The value of W so that the maximum bending stress is 150 MN/m².*

(ii) *The maximum deflection for the loading.*

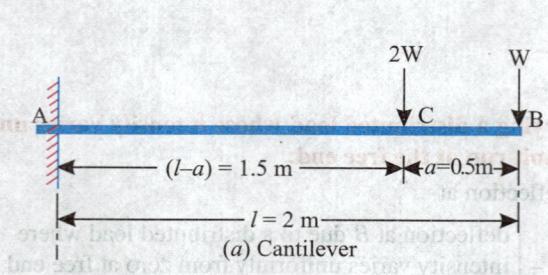

(a) Cantilever

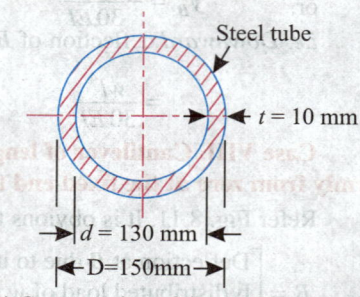

Steel tube

(b) Cross-section of the cantilever

Fig. 8.13

Solution.

Length of the cantilever, $l = 2$ m

External diameter of the steel tube, $D = 150$ mm $= 0.15$

Thickness of the tube $= 10$ mm

Internal diameter of the tube, $d = D - 2t = 150 - 2 \times 10 = 130$ mm $= 0.13$ m

Maximum bending stress, $\sigma_b = 150$ MN/m²; $E = 200$ GN/m²

(i) Load, W (N):

Maximum B.M, $M = Wl + 2W(l - a) = W(3l - 2a)$

$$= W(3 \times 2 - 2 \times 0.5) = 5\,W \text{ Nm}$$

Moment of inertia, $I = \frac{\pi}{64}(D^4 - d^4)$

$$= \frac{\pi}{64}\left[(0.15)^4 - (0.13)^4\right] = 10.8 \times 10^{-6} \text{ m}^4$$

Using the bending equation, we have

$$\frac{M}{I} = \frac{\sigma_b}{y} \quad \text{or} \quad \sigma_b = \frac{M \times y}{I}$$

$$150 \times 10^6 = \frac{5W \times (D/2)}{10.8 \times 10^{-6}} = \frac{5W \times 0.075}{10.8 \times 10^{-6}}$$

∴ $$W = \frac{150 \times 10^6 \times 10.8 \times 10^{-6}}{5 \times 0.075} = 4320 \text{ N}$$

Hence, $$W = \textbf{4320 N} \quad \textbf{(Ans.)}$$

(ii) The maximum deflection, δ:

Total deflection at the free end = Deflection at the free end due to the load W alone + deflection at the free end due to the load $2W$,

$$\delta = \frac{Wl^3}{3EI} + \left[\frac{2W(l-a)^3}{3EI} + \frac{2W(l-a)^2}{2EI}.a\right]$$

$$= \frac{Wl^3}{3EI} + \frac{2W(l-a)^3}{3EI} + \frac{W(l-a)^2}{EI}.a$$

$$= \frac{W}{3EI}\left[l^3 + 2(l-a)^3 + 3(l-a)^2.a\right]$$

$$= \frac{W}{3EI}\left[l^3 + 2l^3 - 6l^2a + 6la^2 - 2a^3 + 3l^2a - 6la^2 + 3a^3\right]$$

$$= \frac{W}{3EI}\left[3l^3 - 3l^2a + a^3\right] \quad \text{or} \quad \delta = \frac{W}{3EI}\left[3l^2(l-a) + a^3\right]$$

$$= \frac{4320}{3 \times 200 \times 10^9 \times 10.8 \times 10^{-6}}\left[3 \times 2^2(2-0.5) + 0.5^3\right] \times 10^3 \text{ mm}$$

$$= 12.08 \text{ mm}$$

Hence, *maximum deflection* = **12.08 mm** (**Ans.**)

Example 8.3. *A cantilever of 3 metres length and of uniform rectangular cross section 150 mm wide and 300 mm deep is loaded with a 30 kN load at its free end. In addition to this it carries a uniformly distributed load of 20 kN per metre run over its entire length, Calculate:*

(i) The maximum slope and maximum deflection.

(ii) The slope and deflection at 2 metres for the fixed end.

Take, E = 210 GN/m²

(Bombay University)

Steel bars used in deep hole drilling like petroleum wells.

Solution. Length of the cantilever, l = 3 m

Cross section: width, b = 150 mm = 0.15 m; depth, d = 300 mm = 0.3 m

$$\therefore \quad I = \frac{bd^3}{12} = \frac{0.15 \times 0.3^3}{12} = 337.5 \times 10^{-6} \text{ m}^4; \ E = 210 \text{ GN/m}^2$$

(i) Maximum slope (θ_{max}), maximum deflection (y_{max}):

Consider a section XX at a distance x from the fixed end (Fig. 8.14)

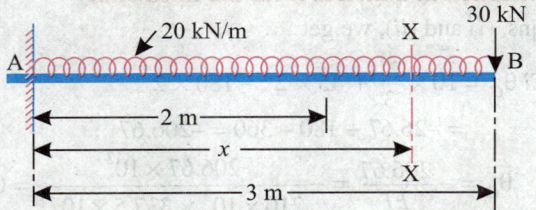

Fig. 8.14

$$M_x = -30(3-x) - \frac{20(3-x)^2}{2}$$

$$= -30(3 - x) - 10(9 + x^2 - 6x)$$
$$= -90 + 30x - 90 - 10x^2 + 60x$$
$$= -10x^2 + 9x - 180$$

$\therefore$ $\qquad EI \dfrac{d^2y}{dx^2} = M_x = -10x^2 + 90x - 180$

Integrating, we get

$$EI \dfrac{dy}{dx} = -\dfrac{10x^3}{3} + 45x^2 - 180x + C_1 \quad \text{(where, } C_1 = \text{constant of integration)}$$

When, $\qquad\qquad x = 0, \dfrac{dy}{dx} = 0$

$\therefore$ $\qquad\qquad\qquad C_1 = 0$

Hence, $\qquad EI \dfrac{dy}{dx} = -\dfrac{10x^3}{3} + 45x^2 - 180x$ $\qquad$ *(i) Slope equation*

Maximum slope is obviously at the free end B.

$\therefore$ Putting $\qquad\qquad x = 3.0$, we get

$$EI\,\theta_{max} = -\dfrac{10 \times 3^3}{3} + 45 \times 3^2 - 180 \times 3$$
$$= -90 + 405 - 540 = -225$$

$\therefore$ $\qquad \theta_{max} = -\dfrac{225}{EI} = \dfrac{225 \times 10^3}{210 \times 10^9 \times 337.5 \times 10^{-6}} = -0.003175$

Hence, $\qquad \boldsymbol{\theta_{max} = -0.003175}$ **radian (Ans.)**

Integrating eqn. (*i*) we get

$$EI\,y = -10 \times \dfrac{x^4}{12} + 45 \times \dfrac{x^3}{3} - 180 \times \dfrac{x^2}{2} + C_2$$
$$= -\dfrac{5x^4}{6} + 15x^3 - 90x^2 + C_2$$

When, $\qquad\qquad x = 0, \; y = 0 \qquad \therefore C_2 = 0$

Hence, $\qquad EI\,y = -\dfrac{5x^4}{6} + 15x^3 - 90x^2$ $\qquad$ ...(*ii*) *Deflection equation*

Maximum deflection is at the free end B.

$\therefore$ Putting $x = 3$, we get

$$EI\,y_{max} = -\dfrac{5 \times 3^4}{6} + 15 \times 3^3 - 90 \times 3^2 = -67.5 + 405 - 810 = -472.5$$

$$y_{max} = -\dfrac{472.5}{EI} = -\dfrac{472.5 \times 10^3}{210 \times 10^9 \times 337.5 \times 10^{-6}} \times 10^3 \text{ mm} = -6.67 \text{ mm}$$

Hence, $\qquad \boldsymbol{y_{max} = 6.67}$ **mm** *(downward)* **(Ans.)**

(*ii*) Slope and deflection at 2 metres from the fixed end:

Putting $x = 2$ in eqns. (*i*) and (*ii*), we get

$$EI\,\theta_C = 10 \times \dfrac{2^3}{3} + 45 \times 2^2 - 180 \times 2$$
$$= -26.67 + 180 - 360 = -206.67$$

$\therefore$ $\qquad \theta_C = -\dfrac{206.67}{EI} = -\dfrac{206.67 \times 10^3}{210 \times 10^9 \times 337.5 \times 10^{-6}} = 0.00292$

Hence, $\qquad \boldsymbol{\theta_C = -0.00292}$ **radian (or $-0.167°$) (Ans.)**

and, $\qquad EI y_C = -\dfrac{5 \times 2^4}{6} + 15 \times 2^3 - 90 \times 2^2 = -13.33 + 120 - 360 = -253.33$

∴
$$y_C = -\frac{253.3}{EI} = -\frac{253.3 \times 10^3}{210 \times 10^9 \times 337.5 \times 10^{-6}} \times 10^3 \ mm = -3.57 \ mm$$

i.e., $\qquad y_C = \textbf{3.57 mm} \ (downward) \ \textbf{(Ans.)}$

Example 8.4. *A 250 mm long cantilever of rectangular section 40 mm wide and 30 mm deep carries a uniformly distributed load. Calculate the value of w if the maximum deflection in the cantilever is not to exceed 0.5 mm.*

Take, $\quad E = 70 \ GN/m^2$.

Solution. Length of the cantilever, $l = 250 \ mm = 0.25 \ m$

Width, $\qquad\qquad b = 40 \ mm = 0.04 \ m$

Depth, $\qquad\qquad d = 30 \ mm = 0.03 \ m$

Moment of inertia, $\qquad I = \dfrac{bd^3}{12} = \dfrac{0.04 \times 0.03^3}{12} = 9 \times 10^{-8} \ m^4$

Young's Modulus, $\qquad E = 70 \ GN/m^2$

Maximum deflection, $\quad y_{max} = 0.5 \ mm = 0.0005 \ m$

Value of w (N/m run):

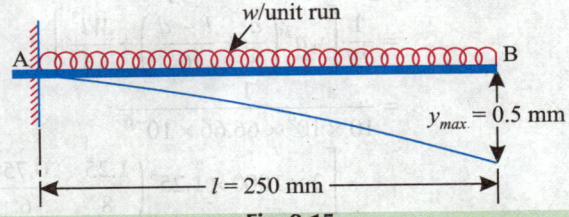

Fig. 8.15

Maximum deflection, $\qquad y_{max} = \dfrac{wl^4}{8 \ EI} = 0.0005$

or, $\qquad \dfrac{w \times (0.25)^4}{8 \times 70 \times 10^9 \times 9 \times 10^{-8}} = 0.0005$

$$w = \frac{0.0005 \times 8 \times 70 \times 10^9 \times 9 \times 10^{-8}}{(0.25)^4}$$

$$= 6451 \ N/m \ or \ 6.451 \ kN/m$$

Hence, $\qquad\qquad\qquad\qquad w = \textbf{6.451 kN/m} \ \textbf{(Ans).}$

Example 8.5. *A cantilever 2 metres long is of rectangular section 100 mm wide and 200 mm deep. It carries a uniformly distributed load of 2 kN per unit metre length for a length of 1.25 metres from the fixed end a point load of 0.8 kN at the free end. Find the deflection at the free end.*

Take, $E = 10 \ GN/m^2$.

Solution. Refer to Fig. 8.16.

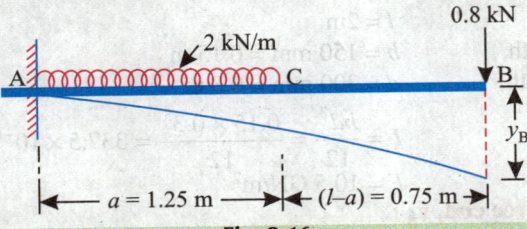

Fig. 8.16

Length of cantilever, $\qquad l = 2$ m

Cross section: width, $\qquad b = 100$ mm $= 0.1$ m

depth, $\qquad d = 200$ mm $= 0.2$ m

Moment of inertia, $\qquad I = \dfrac{bd^3}{12} = \dfrac{0.1 \times 0.2^3}{12} = 66.66 \times 10^{-6}$ m^4

$$w = 2 \text{ kN/m}; \quad E = 10 \text{ GN/m}^2$$

Deflection at the free end, y_B:

Deflection at the free end B, y_B (*downward*)

$= $ Deflection at B due to uniformly distributed load

$\qquad\qquad\qquad\qquad$ + deflection at B due to point load at B

$$= \left[\frac{wa^4}{8\,EI} + \frac{wa^3}{6\,EI}(l-a) \right] + \frac{Wl^3}{3\,EI}$$

$$= \left[\frac{wa^3}{EI} \left(\frac{a}{8} + \frac{l-a}{6} \right) \right] + \frac{Wl^3}{3EI}$$

$$= \frac{1}{EI} \left[wa^3 \left(\frac{a}{8} + \frac{l-a}{6} \right) + \frac{Wl^3}{3} \right]$$

$$= \frac{1}{10 \times 10^9 \times 66.66 \times 10^{-6}}$$

$$\left[2 \times 1000 \times 1.25^3 \left(\frac{1.25}{8} + \frac{0.75}{6} \right) + \frac{0.8 + 1000 \times 2^3}{3} \right]$$

$$= \frac{1}{666.6 \times 10^3} [1098.63 + 2133.33]$$

$$= 4.848 \times 10^{-3} \text{m} = 4.848 \text{ mm}$$

Hence, $\qquad\qquad\qquad y_B = \textbf{4.848 mm (Ans.)}$

Example 8.6. *A 2 metres long cantilever of rectangular section 150 mm wide and 300 mm deep is loaded as shown in the fig. 8.17. Calculate the deflection at the free end.*

Take, E = 10.5 GN/m^2.

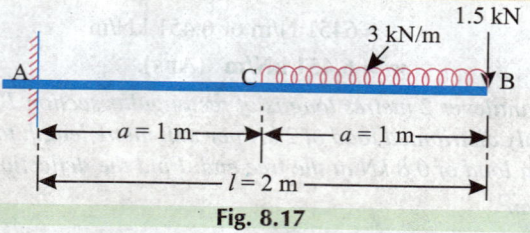

Fig. 8.17

Solution. Length of the cantilever,

$$l = 2\text{m}$$

Cross section: width, $\qquad b = 150$ mm $= 0.15$ m

depth, $\qquad d = 300$ mm $= 0.3$ m

Moment of inertia, $\qquad I = \dfrac{bd^3}{12} = \dfrac{0.15 \times 0.3^3}{12} = 337.5 \times 10^{-6}$ m^4

Young's Modulus, $\qquad E = 10.5$ GN/m^2

Deflection at the free end, y_B:

Deflection due to uniformly distributed load at the free end

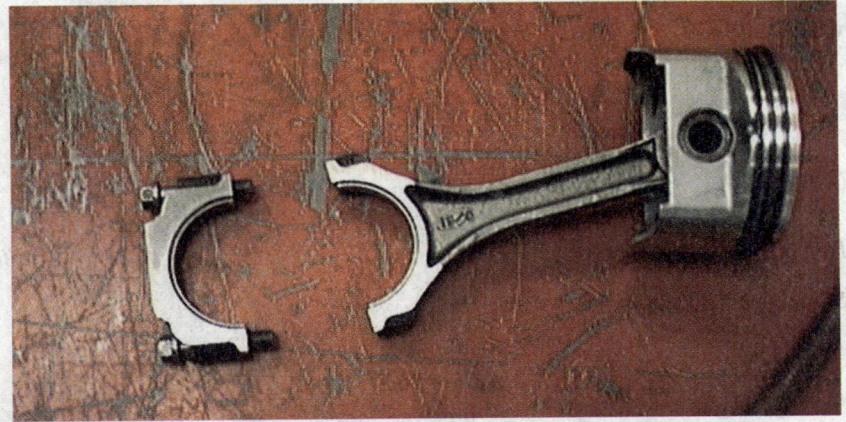

Piston, piston rings and connecting rod.

$$= \frac{w}{24\,EI}(3l^4 - 4l\,a^3 + a^4) \qquad\qquad \text{...(Eqn. 8.13)}$$

$$= \frac{3 \times 1000}{24 \times 10.5 \times 10^9 \times 337.5 \times 10^{-6}}\ [3 \times 2^4 - 4 \times 2 \times 1^3 + 1^3] \times 10^3 \text{ mm}$$

$$= 35.2 \times 10^{-6} \times 41 \times 1000 = 1.443 \text{ mm}$$

Deflection at the free end due to point load 1.5 kN

$$= \frac{Wl^3}{3\,EI} = \frac{1.5 \times 1000 \times 2^3}{3 \times 10.5 \times 10^9 \times 337.5 \times 10^{-6}} \times 10^3 \text{ mm} = 1.128 \text{ mm}$$

Total deflection (*downward*) at the free end, y_B

     = Deflection due to uniformly distributed load

                + deflection due to the point load

     = 1.443 + 1.128 = 2.57 mm

Hence, y_B **= 2.57 mm (Ans.)**

SIMPLY SUPPORTED BEAMS

Case I. Simply supported beam of span l carrying a point load at mid span:

Fig. 8.18 shows a simply supported beam AB of span l carrying a point load W at the mid span C.
Since the load is symmetrically applied the
maximum deflection (y_{max}) will occur at mid span.

Each vertical reaction equals $\dfrac{W}{2}$.

Consider the left half AC of the span.

The B.M. at any section XX in AC distant
x from A is given by,

$$EI \frac{d^2y}{dx^2} = +\frac{W}{2} x$$

Fig. 8.18

Integrating, we get

$$EI \frac{dy}{dx} = \frac{Wx^2}{4} + C_1$$

(where, C_1 = constant of integration)

when,
$$x = \frac{l}{2}, \frac{dy}{dx} = 0$$

$$0 = \frac{W}{4}(l/2)^2 + C_1$$

$\therefore \qquad C_1 = \frac{Wl^2}{16}$

Hence, $\qquad EI\frac{dy}{dx} = \frac{Wx^2}{4} - \frac{Wl^2}{16}$ ...(i) Slope equation

Slope at A : Putting $x = 0$, we get

$$\theta_A = \frac{dy}{dx} = -\frac{Wl^2}{16\,EI}$$

i.e., $\qquad \theta_A = -\frac{Wl^2}{16\,EI}$ ...(8.19)

Integrating the slope equation, we get

$$EIy = \frac{Wx^3}{12} - \frac{Wl^2}{16}x + C_2$$

When, $\qquad x = 0, y = 0 \qquad \therefore C_2 = 0$

Hence, $\qquad EI\,y = \frac{Wx^3}{12} - \frac{Wl^2}{16}\cdot x$ ...(ii) Deflection equation

Deflection at C: Putting $x = l/2$, we get

$$EI\,y_C = \frac{W \times (l/2)^3}{12} - \frac{Wl^2}{16}(l/2)$$

$$= \frac{Wl^3}{96} - \frac{Wl^3}{32} = -\frac{Wl^3}{48}$$

$\therefore \qquad y_C = -\frac{Wl^3}{48\,EI}$

Hence, (downward) deflection of C

$$= \frac{Wl^3}{48\,EI}$$ ...(8.20)

Case II. Simply supported beam of span l carrying a uniformly distributed load of w per unit run over the whole span:

Fig. 8.19 shows a simply supported beam AB of span l carrying a uniformly distributed load w per unit run over the whole span. Each vertical reaction equals $\frac{wl}{2}$.

Consider a section XX at a distance x from the end A.

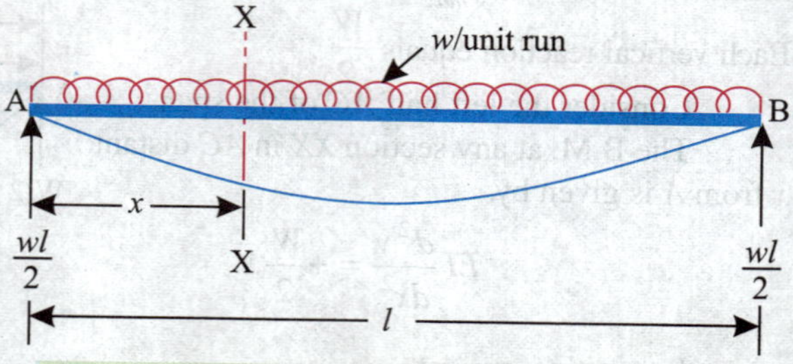

Fig. 8.19

$$M_x = \frac{wl}{2}\cdot x - \frac{wx^2}{2}$$

$$EI \frac{d^2 y}{dx^2} = \frac{wl}{2} x - \frac{wx^2}{2}$$

Integrating, we get

$$EI \frac{dy}{dx} = \frac{wl}{4} x^2 - \frac{wx^3}{6} + C_1 \qquad \text{(where, } C_1 = \text{constant of integration)}$$

The loading being symmetrical, the maximum deflection will occur at mid span and hence the slope at mid span equals zero.

i.e.,
$$x = \frac{l}{2}, \quad \frac{dy}{dx} = 0$$

$$0 = \frac{wl}{4} \left(\frac{l}{2}\right)^2 - \frac{w}{6} \left(\frac{l}{2}\right)^3 + C_1 = \frac{wl^3}{16} - \frac{wl^3}{48} + C_1$$

or,
$$C_1 = -\frac{wl^3}{24}$$

Hence,
$$EI \frac{dy}{dx} = \frac{wl}{4} x^2 - \frac{wx^3}{6} - \frac{wl^3}{24} \qquad \text{(i) slope equation}$$

Slope at A : Putting $x = 0$, we get

$$EI\theta_A = -\frac{wl^3}{24}$$

$$\therefore \qquad \theta_A = -\frac{wl^3}{24 EI} \qquad \qquad \qquad ...(8.21)$$

Integrating the slope equation, we get

$$EIy = \frac{wl x^3}{12} - \frac{wx^4}{24} - \frac{wl^3}{24} x + C_2 \qquad \text{(where, } C_2 = \text{constant of integration)}$$

When,
$$x = 0, y = 0 \qquad \therefore C_2 = 0$$

Hence,
$$EIy = +\frac{wl x^3}{12} - \frac{wx^4}{24} - \frac{wl^3}{24} \cdot x \qquad \text{(ii) Deflection equation}$$

Deflection at mid span, y_{max} : Putting $x = l/2$, we get

$$EIy_{max} = +\frac{wl}{12} (l/2)^3 - \frac{w}{24} (l/2)^4 - \frac{wl^3}{24} \cdot l/2$$

$$= +\frac{wl^4}{96} - \frac{wl^4}{384} - \frac{wl^4}{48} = -\frac{5wl^4}{384}$$

$$y_{max} = -\frac{5 wl^4}{384 EI}$$

Hence, maximum *downward* deflection

$$= \frac{5wl^4}{384 EI} \qquad \qquad \qquad ...(8.22)$$

WORKED EXAMPLES (Simply Supported Beams)

Example 8.7. *A girder of uniform section and constant depth is freely supported over a span of 3 metres. If the point load at the mid span is 30 kN and $I_{xx} = 15.614 \times 10^{-6}$ m^4, calculate:*

(i) *The central deflection.*

(ii) *The slopes at the ends of the beam.*

Take : $E = 200$ GN/m^2

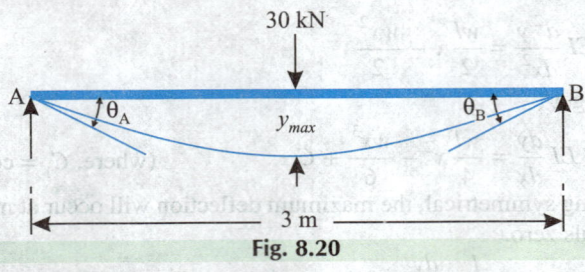

Fig. 8.20

Solution. Length of the span, $l = 3$ m

Load at the mid span, $\quad W = 30$ kN

$$I_{XX} = 15.614 \times 10^{-6} \text{ m}^4, E = 200 \text{ GN/m}^2$$

(i) Central deflection, y_{max} :

$$y_{max} = \frac{Wl^3}{48\,EI} = \frac{30 \times 10^3 \times 3^3}{48 \times 200 \times 10^9 \times 15.614 \times 10^{-6}} \times 10^3 \text{ mm}$$

$$= 5.4 \text{ mm } i.e. \ y_{max} = \mathbf{5.4 \text{ mm}} \quad \textbf{(Ans.)}$$

(ii) Slopes at the ends of the beam :

Slope at end A, $\qquad \theta_A = -\dfrac{Wl^2}{16\,EI} = -\dfrac{30 \times 1000 \times 3^2}{16 \times 200 \times 10^9 \times 15.614 \times 10^{-6}}$

$$= -0.0054 \text{ radian} = -0.0054 \times \frac{180}{\pi} = \mathbf{-0.309°} \quad \textbf{(Ans.)}$$

Slope at the end B, $\qquad \theta_B = +0.0054$ radian or $\mathbf{0.309°}$ *(downward)* **(Ans.)**

Example 8.8. *A steel girder of 6m length acting as a beam carries a uniformly distributed load w N/m run throughout its length. If $I = 30 \times 10^{-6}$ m^4 and depth 270 mm, calculate:*

(i) The magnitude of w so that the maximum stress developed in the beam section does not exceed 72 MN/m^2.

(ii) The slope and deflection (under this load) in the beam at a distance of 1.8 m from one end.

Take : E = 200 GN/m^2

Gear teeth are subject to high bending and shear stresses.

Solution. Length of the beam, $l = 6$m

Moment of inertia, $\qquad I = 30 \times 10^{-6}$ m^4

Maximum stress developed, $\sigma_b = 72$ MN/m^2

Young's modulus, $\qquad E = 200$ GN/m^2

Depth of section, $\qquad d = 270$ mm $= 0.27$ m

(i) Magnitude of w :

Maximum bending occurs at the centre of the beam.

$$M_{max} = \frac{wl^2}{8} = \frac{w \times 6^2}{8} = 4.5\,w$$

Maximum stress will occur at the extreme layers at a distance of $\pm d/2$ from the neutral axis (where d is the depth of the section).

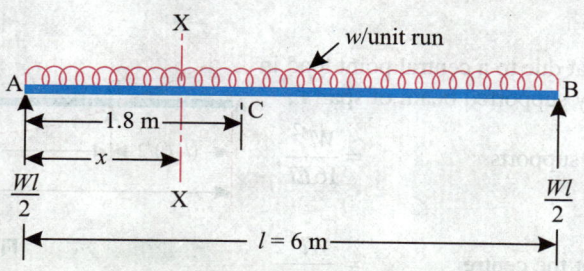

Fig. 8.21

Now,
$$\frac{M}{I} = \frac{\sigma_b}{y}$$

or,
$$\sigma_b = \frac{M \times y}{I} = \frac{4.5\,w \times (0.27/2)}{30 \times 10^{-6}} \quad \text{(where, } y = d/2\text{)}$$

or,
$$72 \times 10^6 = \frac{4.5\,w \times 0.135}{30 \times 10^{-6}}$$

∴
$$w = \frac{72 \times 10^6 \times 30 \times 10^{-6}}{4.5 \times 0.135} = 3555\,\text{N/m} \quad \textbf{(Ans.)}$$

(ii) Slope and deflection at $x = 1.8$ m:

Considering a section XX at a distance x from the end A, we have

$$EI \frac{dy}{dx} = \frac{wl\,x^2}{4} - \frac{wx^3}{6} - \frac{wl^3}{24} \qquad \text{...(Slope equation)}$$

$$EIy = \frac{wl\,x^3}{12} - \frac{w\,x^4}{24} - \frac{wl^3 x}{24} \qquad \text{...(Deflection equation)}$$

Slope at point C, at $x = 1.8$ m:

$$EI\,\theta_C = \frac{3555 \times 6 \times 1.8^2}{4} - \frac{3555 \times 1.8^3}{6} - \frac{3555 \times 6^3}{24}$$

$$= 3555\,(4.86 - 0.972 - 9) = -18173.2$$

∴
$$\theta_C = -\frac{18173.2}{200 \times 10^9 \times 30 \times 10^{-6}} = -0.00303\,\text{radian} \quad \text{or} \quad -0.173°$$

Hence,
$$\theta_C = -\mathbf{0.173°} \quad \textbf{(Ans.)}$$

Deflection at point C, at $x = 1.8$ m:

$$EIy_C = \frac{3555 \times 6 \times 1.8^3}{12} - \frac{3555 \times 1.8^4}{24} - \frac{3555 \times 6^3 \times 1.8}{24}$$

$$3555\,(2.916 - 0.4374 - 16.2) = -48779.6$$

$$y_c = -\frac{48779.6}{200 \times 10^9 \times 30 \times 10^{-6}} = -0.00813\,\text{m} = -8.13\,\text{mm}$$

Downward deflection of point C

$$= 8.13\,\text{mm} \quad \textbf{(Ans.)}$$

Example 8.9. *A uniform beam of length l is simply and symmetrically supported on a span l'. Find the ratio $\dfrac{l}{l'}$ so that the upward deflection at each end equals the downward deflection at mid span, due to a central point load.*

Solution. Length of the beam = l; length of the span = l'

Ratio l/l' :

We know that due to a central point load in the case of a simply supported beam of span l',

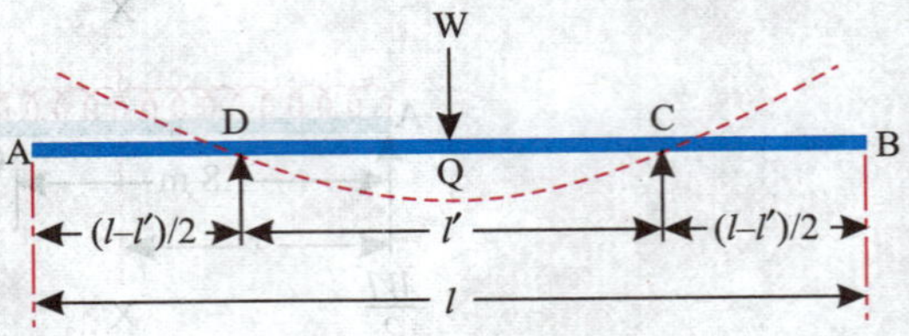

Slope at the supports $\quad = \dfrac{Wl'^2}{16\,EI},$

and

Deflection at the centre $\quad = \dfrac{Wl'^3}{48\,EI}$

Hence, $\qquad \theta_D = \dfrac{Wl'^2}{16\,EI}$

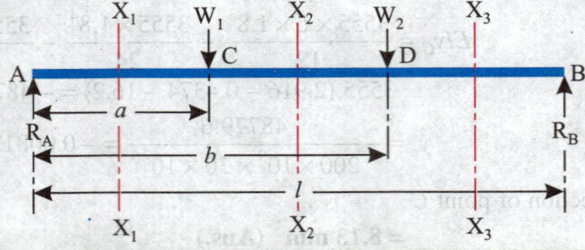

Fig. 8.22

Since the angle is very small $\quad y_A = \theta_D \times AD = \dfrac{Wl'^2}{16\,EI} \times \left(\dfrac{l - l'}{2}\right)$ \hfill ...(i)

Also deflection at the centre, $y_Q = \dfrac{Wl'^3}{48\,EI}$ \hfill ...(ii)

Equation (i) and (ii), we get.

$$\frac{Wl'^2}{16\,EI} \times \left(\frac{l - l'}{2}\right) = \frac{Wl'^3}{48\,EI}$$

or, $\qquad\qquad (l - l') = \dfrac{2}{3}l'$

or, $\qquad\qquad\qquad \dfrac{l}{l'} = \dfrac{5}{3}$ **(Ans.)**

8.7. MACAULAY'S METHOD

— In Macaulay's method a single equation is formed for all loadings on a beam, *the equation is constructed in such a way that the constants of integration apply to all portions of the beam.* This method is also called **method of singularity functions.**

— This is a convenient method for determining the deflection of a beam subjected to *point loads* or *in general discontinuous loads.*

This method is explained as follows :

Fig. 8.23 shows a beam of span l simply supported at A and B and carrying the point loads W_1 and W_2 at distances a and b from the end A. Let R_A and R_B be the reactions at A and B respectively.

Fig. 8.23

— Consider a section X_1X_1 between A and C distant x from A. The bending moment is given by:

$$M_x = R_A \cdot x \qquad\qquad ...(i)$$

This expression (for the bending moment) holds good for all values of x between $x = 0$ and $x = a$.

— Consider a section $X_2 X_2$ between C and D and distant x from end A. The bending moment is given by :

$$M_x = R_A \cdot x - W_1 (x - a) \qquad ...(ii)$$

This expression holds good for all values of x between $x = a$ and $x = b$.

— Consider a section $X_3 X_3$ between D and B distant x from A. The bending moment is given by:

$$M_x = R_A \cdot x - W_1 (x - a) - W_2 (x - b) \qquad ...(ii)$$

This expression holds good for all values of x between $x = b$ and $x = l$.

— *At any section*, in general, the bending moment is given by:

$$M_x = EI \frac{d^2 y}{dx^2} = R_A \times x \left| -W_1 (x - a) \right| - W_2 (x - b) \qquad ...(8.23)$$

In the above equation it may be noted that *as the magnitude of x goes on increasing so that the law of loading changes, additional expressions appear.*

For value of x between:

(i) $x = 0$ and $x = a$ ..... only the *first term* of the above equation should be considered.

(ii) $x = a$ and $x = b$ ..... only the *first two terms* of the above equation should be considered.

(iii) $x = b$ and $x = l$ .... *all the terms* of the above equation should be considered.

Integration eqn. (8.23), we get the general expression for slope as follows:

$$EI \frac{dy}{dx} = R_A \cdot \frac{x^2}{2} + C_1 \left| -\frac{W_1 (x - a)^2}{2} \right| - \frac{W_2 (x - b)^2}{2} \qquad ...Slope \ equation \ (8.24)$$

The following points are worth noting :

(i)   The constant of integration C_1 should be written after the first term of the above equation.

(ii) The quantity $(x - a)$ should be integrated as $\dfrac{(x - a)^2}{2}$ and not as $\dfrac{x^2}{2} - ax$.

Similarly the quantity $(x - b)$ should be integrated as a whole *i.e.* as $\dfrac{(x - b)^2}{2}$

(iii) The constant C_1 is valid for all value of x.

Integrating eqn. (8.24) we get the *deflection* equation,

$$EIy = R_A \times \frac{x^3}{6} + C_1 x + C_2 \left| -\frac{W_1 (x - a)^3}{6} \right| - \frac{W_2 (x - b)^3}{6} \qquad ...Deflection \ equation \ (8.25)$$

It may be further noted that :

(i) $(x - a)^2$ has been integrated to $\dfrac{(x - a)^3}{3}$ and $(x - b)^2$ has been integrated to $\dfrac{(x - b)^3}{3}$

(ii) Constant C_2 is written after $C_1 x$. The constant C_2 is *valid for all values of x.*

(iii) If the end conditions are known, the constants C_1 and C_2 can be evaluated.

Example: If the beam is simply supported the deflection is zero at A and B, *i.e.* when $x = 0$ and $x = l$, $y = 0$

Putting $x = 0$ and $y = 0$ in deflection equation, we get $C_2 = 0$

Putting $x = l$ and $y = 0$ in the deflection equation the constant C_1 can be calculated.

When the constants C_1 and C_2 are known the slope and deflection at any section at any section can be determined.

Example 8.10. *A beam AB of length l simply supported at the ends carries a point load W at a distance a from the left end. Find :*

(i) The deflection under the load.

(ii) The maximum deflection.

Solution. Fig. 8.24 shows a beam *AB* of span *l* carrying a point load *W* at *C*.

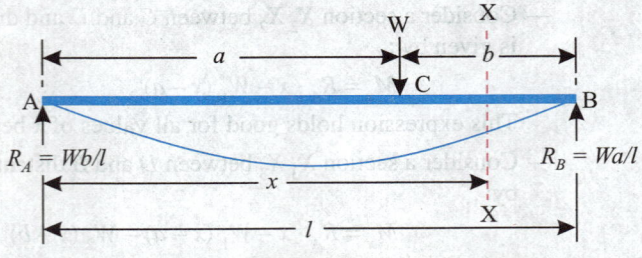

Fig. 8.24

An IC engine intake manifold.

Let, $AC = a$, $CB = b$, and $a > b$

To find reactions taking moments about *A*, we get

$$R_B \times l = W \times a$$

∴ $$R_B = \frac{Wa}{l}$$

Also, $$R_A + R_B = W$$

∴ $$R_A = W - \frac{Wa}{l} = W\left(1 - \frac{a}{l}\right)$$

$$= W\left(1 - \frac{a}{l}\right) = W\left(\frac{l-a}{l}\right) = \frac{Wb}{l} \qquad (\because l - a = b)$$

Hence, $$R_A = \frac{Wb}{l} \text{ and } R_B = \frac{Wa}{l}$$

The bending moment at any section *XX* at a distance *x* from the end *A*, following Macaulay's method, is given by:

$$M_x = EI \frac{d^2y}{dx^2} = \frac{Wb}{l} \cdot x \left| -W(x-a) \right. \qquad ...(i)$$

Integrating for slope, we get

$$EI\frac{dy}{dx} = \frac{Wb}{l} \cdot \frac{x^2}{2} + C_1 \left| -\frac{W(x-a)^2}{2} \right. \qquad \text{...(ii)}$$

Integrating again for deflection, we get

$$EIy = \frac{Wb}{l} \cdot \frac{x^3}{6} + C_1 x + C_2 \left| -\frac{W(x-a)^3}{6} \right. \qquad \text{...(iii)}$$

At A, the deflection is zero, *i.e.* when $x = 0$, $y = 0$

$$\therefore \qquad C_2 = 0$$

At B, the deflection is zero, *i.e.* when $x = l$, $y = 0$

$$\therefore \qquad 0 = \frac{Wb}{l} \cdot \frac{l^3}{6} + C_1 l - \frac{W(l-a)^3}{6}$$

$$\therefore \qquad C_1 l = \frac{W(l-a)^3}{6} - \frac{Wbl^2}{6} = \frac{Wb^3}{6} - \frac{Wbl^2}{6} \qquad (\because \; l - a = b)$$

$$= -\frac{Wb}{6}(l^2 - b^2) \quad \text{or} \quad C_1 = -\frac{Wb}{6l}(l^2 - b^2)$$

Hence the *slope* and *deflection* at any section are given by

$$EI\frac{dy}{dx} = \frac{Wbx^2}{2l} - \frac{Wb}{6l}(l^2 - b^2) - \frac{W(x-a)^2}{2} \qquad \text{...Slope equation}$$

$$EIy = \frac{Wbx^3}{6l} - \frac{Wb}{6l}(l^2 - b^2)x \left| -\frac{W(x-a)^3}{6} \right. \qquad \text{....Deflection equation}$$

(i) Deflection under the load, y_C :

To find y_C, putting $x = a$ in the deflection equation, we get

$$EI\,y_C = \frac{Wba^3}{6l} - \frac{Wb}{6l}(l^2 - b^2)a = -\frac{Wba}{6l}(l^2 - b^2 - a^2)$$

But, $\qquad l = a + b$

$$\therefore \qquad EIy_C = -\frac{Wba}{6l}(a^2 + b^2 + 2ab - b^2 - a^2) = -\frac{Wba}{6l}(2ab) = -\frac{Wa^2b^2}{3l}$$

$$\therefore \qquad y_C = -\frac{Wa^2b^2}{3EIl} \quad \textbf{(Ans.)}$$

(ii) The maximum deflection, y_{max} :

The maximum deflection will occur, on the larger segment AC. Moreover the *slope is zero at the point of maximum deflection*. Therefore, equating the slope at a section in AC to zero we get,

$$0 = \frac{Wbx^2}{2l} - \frac{Wb}{6l}(l^2 - b^2)$$

$$\therefore \qquad x^2 = \frac{l^2 - b^2}{3}$$

or $\qquad x = \sqrt{\frac{l^2 - b^2}{3}} \quad \text{or} \quad \sqrt{\frac{a^2 + 2ab}{3}}$

The maximum deflection can be obtained by putting this value of x in deflection equation.

Thus, $\qquad EIy_{max} = \frac{Wb}{6l}\left(\frac{l^2 - b^2}{3}\right)^{3/2} - \frac{Wb}{6l}(l^2 - b^2)\left(\frac{l^2 - b^2}{3}\right)^{1/2}$

$$= -\frac{Wb}{6l}(l^2 - b^2)^{3/2}\left[\frac{1}{\sqrt{3}} - \frac{1}{(3)^{3/2}}\right] = -\frac{Wb}{6l}(l^2 - b^2)^{3/2}\left[\frac{1}{\sqrt{3}} - \frac{1}{3\sqrt{3}}\right]$$

$$= -\frac{Wb}{6l}(l^2 - b^2)^{3/2} \times \frac{2\sqrt{3}}{9} = -\frac{Wb(l^2 - b^2)^{3/2}}{9\sqrt{3}\,l}$$

∴ $$y_{max} = -\frac{Wb(l^2 - b^2)^{3/2}}{9\sqrt{3}\,EIl} \quad \textbf{(Ans.)} \qquad \left[\begin{array}{l}\text{Putting } l = (a + b), \text{ we get} \\[2mm] y_{max} = -\dfrac{Wb(a^2 + 2ab)^{3/2}}{9\sqrt{3}\,EIl} \quad \textbf{Ans.}\end{array}\right]$$

Example 8.11. *A beam with a span of 4.5 metres carries a point load of 30 kN at 3 metres from the left support. If for the section,* $I_{XX} = 54.97 \times 10^{-6}\ m^4$ *and* $E = 200\ GN/m^2$, *find :*

(i) *The deflection under the load.*

(ii) *The position and amount of maximum deflection.*

Solution. Refer Fig. 8.25. The span of the beam, $l = 4.5$ m

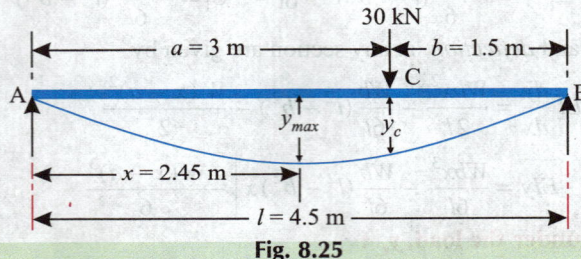

Fig. 8.25

Point load, $W = 30$ kN; $a = 3$m; $b = 1.5$ m

Moment of inertia,

$$I = 54.97 \times 10^{-6}\ \text{m}^4$$

$$E = 200\ \text{GN/m}^2.$$

(i) The deflection under the load, y_c :

$$y_c = -\frac{Wa^2 b^2}{3EIl}$$

$$= -\frac{30 \times 1000 \times 3^2 \times 1.5^2}{3 \times 200 \times 10^9 \times 54.97 \times 10^{-6} \times 4.5}$$

$$= -0.00409\ \text{m} = -4.09\ \text{mm}$$

Downward deflection under the load = **4.09 mm** **(Ans.)**

(ii) The position (x) and amount of maximum deflection (y_{max}) :

Position, $$x = \sqrt{\frac{l^2 - b^2}{3}} = \sqrt{\frac{4.5^2 - 1.5^2}{3}} = \textbf{2.45 m } \textit{from the left end.} \textbf{ (Ans.)}$$

Maximum deflection,

$$y_{max} = -\frac{Wb(l^2 - b^2)^{3/2}}{9\sqrt{3}\,EIl}$$

$$= -\frac{30 \times 1000 \times 1.5 \times (4.5^2 - 1.5^2)^{3/2}}{9\sqrt{3} \times 200 \times 10^9 \times 54.97 \times 10^{-6} \times 4.5} \times 10^3\ \text{mm} = -4.456\ \text{mm}$$

Hence, downward deflection = **4.456 mm** **(Ans.)**

Example 8.12. *A steel girder of uniform section, 14 metres long is simply supported at its ends. It carries concentrated loads of 90 kN and 60 kN at two points 3 metres and 4.5 metres from the two ends respectively. Calculate :*

(i)   *The deflection of the girder at the points under the two loads.*

(ii)   *The maximum deflection.*

Take : $I = 64 \times 10^{-4}\ m^4$ *and* $E = 210 \times 10^{-6}\ kN/m^2$.

Solution. Span of the steel girder,

$$l = 14.0\ \text{m}$$

Moment of inertia, $I = 64 \times 10^{-4}\ \text{m}^4$

Young's modulus $E = 210 \times 10^6\ \text{kN/m}^2$

Let R_A and R_B be the reactions at the support A and B respectively.

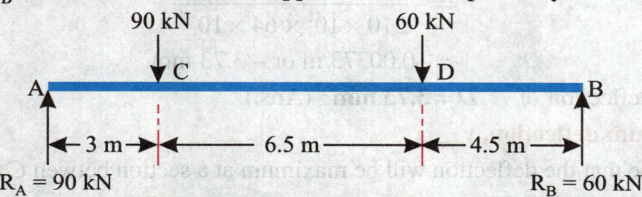

Fig. 8.26

Taking moments about A, we get

$$R_B \times 14 = 90 \times 3 + 60 \times 9.5 = 840$$

∴ $R_B = 60\ \text{kN}$

Also, $R_A + R_B = 90 + 60 = 150$

∴ $R_A = 150 - 60 = 90\ \text{kN}$

Consider any section XX at a distance x from the end A, following Macaulay's method, the bending moment is given by :

$$M_x = EI\frac{d^2 y}{dx^2} = 90x \left| -90(x-3) \right| - 60(x-9.5) \qquad \text{...(i)}$$

Integrating, we get

$$EI\frac{dy}{dx} = 45x^2 + C_1 \left| -45(x-3)^2 - 30(x-9.5)^2 \right. \qquad \text{...(ii)}$$

Integrating again, we get

$$EIy = 15x^3 + C_1 x + C_2 \left| -15(x-3)^3 - 10(x-9.5)^3 \right. \qquad \text{...(iii)}$$

When, $x = 0, y = 0$

∴ $C_2 = 0;$

When, $x = 14\ \text{m}, y = 0$

$$0 = 15 \times (14)^3 + C_1 \times 14 - 15(14-3)^3 - 10(14-9.5)^3$$

$$= 41160 + 14\,C_1 - 19965 - 911.25 = 14\,C_1 + 20283.75$$

∴ $C_1 = -1448.84$

Hence the deflection at any section is given by:

$$EIy = 15x^3 - 1448.84x \left| -15(x-3)^3 - 10(x-9.5)^3 \right.$$

...*Deflection equation.*

(i) y_C and y_D :

Deflectioin at C, y_C

          Putting $x = 3\ \text{m}$ in the deflection equation, we get

$$EIy_C = 15 \times 3^3 - 1448.84 \times 3 = 405 - 4346.52 = -3941.52$$

$$\therefore \qquad y_C = -\frac{3941.52}{EI} = -\frac{3941.52}{210 \times 10^6 \times 64 \times 10^{-4}} = -0.00293 \text{ m}$$

or, $\qquad = -2.93 \text{ mm}$

Downward deflection of C

$$= \textbf{2.93 mm} \quad \textbf{(Ans.)}$$

Deflection at D, y_D : Putting x = 9·5 mm in the deflection equation, we get

$$EIy_D = 15 \times 9.5^3 - 1448.84 \times 9.5 - 15 (9.5 - 3)^3$$

$$= 12860.6 - 13764 - 4119.4 = -5022.8$$

$$\therefore \qquad y_D = -\frac{5022.8}{EI}$$

$$= -\frac{5022.8}{210 \times 10^6 \times 64 \times 10^{-4}}$$

$$= -0.00373 \text{ m or} -3.73 \text{ mm}$$

Downward deflection of $\quad D = \textbf{3.73 mm} \quad \textbf{(Ans.)}$

(ii) Maximum deflection, y_{max} :

Let us assume that the deflection will be maximum at a section betwen C and D. *Equating the slope at the section to zero*, we get

$$EI\frac{dy}{dx} = 45 x^2 - 1448.84 - 45 (x - 3)^2 = 0$$

or, $\qquad 45 x^2 - 1448.84 - 45 (x^2 - 6x + 9) = 0$

or, $\qquad 45 x^2 - 1448.84 - 45 x^2 + 270x - 405 = 0$

or, $\qquad 270x = +1853.84$

$\therefore \qquad x = 6.87 \text{ m}$

Putting this value of x in the deflection equation, we get

$$EI \, y_{max} = 15 \times 6.87^3 - 1448.84 \times 6.87 - 15 (6.87 - 3)^3$$

$$= 4863.6 - 9953.5 - 869.4 = -5959.3$$

$$\therefore \qquad y_{max} = -\frac{5959.3}{EI} = -\frac{5959.3}{210 \times 10^6 \times 64 \times 10^{-4}}$$

$$= 0.0043 \text{ m or} -4.43 \text{ mm}$$

Downward deflection, $\quad y_{max} = \textbf{4.43 mm} \quad \textbf{(Ans.)}$

Example 8.13. *A beam AB of 4 metres span is simply supported at the ends and is loaded as shown in Fig 8.27. Determine:*

(i) *Deflection at C,*

(ii) *Maximum deflection, and*

(iii) *Slope at the end A.*

Given : E = 200 × 10⁶ kN/m², and I = 20 × 10⁻⁶ m⁴.

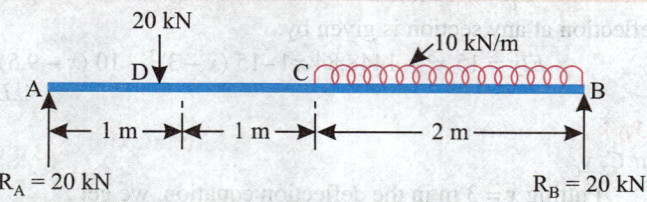

Fig. 8.27

Solution. Span of the beam, $l = 4\,\text{m}$, $E = 200 \times 10^6\,\text{kN/m}^2$, $I = 20 \times 10^{-6}\,\text{m}^4$

To calculate reaction at B taking moments about A, we get

$$R_B \times 4 = 20 \times 1 + 10 \times 2\left(\frac{2}{2} + 1 + 1\right) = 80$$

∴ $R_B = 20\,\text{kN}$

Also, $R_A + R_B = 20 + 10 \times 2 = 40$

∴ $R_A = 20\,\text{kN}$

Using Macaulay's method consider any section XX at a distance x from the end A; the bending moment at the section XX,

$$M_x = EI\frac{d^2y}{dx^2} = 20x \,\Big|\, -20(x-1) \,\Big|\, -\frac{10(x-2)^2}{2} \qquad \ldots(i)$$

Integrating, we get

$$EI\frac{dy}{dx} = 10x^2 + C_1 \,\Big|\, -10(x-1)^2 \,\Big|\, -\frac{5}{3}(x-2)^3 \qquad \ldots(ii)$$

Integrating agian, we get

$$EIy = \frac{10}{3}x^3 + C_1 x + C_2 \,\Big|\, -\frac{10}{3}(x-1)^3 \,\Big|\, -\frac{5}{12}(x-2)^4 \qquad \ldots(iii)$$

When, $x = 0$, $y = 0$

∴ $C_2 = 0$;

When, $x = 4\,\text{m}$, $y = 0$

∴ $$0 = \frac{10}{3} \times 4^3 + 4C_1 - \frac{10}{3}(4-1)^3 - \frac{5}{12}(4-2)^4$$

$$= 213.33 + 4C_1 - 90 - 6.67$$

∴ $C_1 = -29.16$

Hence, the slope and deflection equations are

$$EI\frac{dy}{dx} = 10x^2 - 29.16 \,\Big|\, -10(x-1)^2 - 5/3(x-2)^3 \qquad \ldots\textit{Slope equation}$$

and, $$EIy = \frac{10}{3}x^3 - 29.16x \,\Big|\, -\frac{10}{3}(x-1)^3 \,\Big|\, -\frac{5}{12}(x-2)^4 \qquad \ldots\textit{Deflection equation}$$

(i) **Deflection at C, y_C :**

Putting $x = 2\,\text{m}$ in the deflection equation, we get

$$EI\,y_C = \frac{10}{3} \times 2^3 - 29.16 \times 2 - \frac{10}{3}(2-1)^3$$

$$= 26.67 - 58.32 - 3.33 = -34.98$$

∴ $$y_C = -\frac{34.98}{EI} = -\frac{34.98}{200 \times 10^6 \times 20 \times 10^{-6}} \times 10^3\,\text{mm} = -8.74\,\text{mm}$$

Hence, $y_C = \textbf{8.74 mm}$ *(downward)* **(Ans.)**

(ii) **Maximum deflection, y_{max} :**

The maximum deflection will be very near to the mid-point C. Let us assume that it occurs in the section between D and C.

For maximum deflection equating the slope at the section to zero, we get

$$EI\frac{dy}{dx} = 10x^2 - 29.16 - 10(x-1^2) = 0$$

or, $$10x^2 - 29.16 - 10(x^2 - 2x + 1) = 0$$

or, $$10x^2 - 29.16 - 10x^2 + 20x - 10 = 0$$

or, $$x = \frac{39.16}{20} = 1.958\,\text{m}$$

Putting the value of x in the deflection equation, we get

$$EI\,y_{max} = \frac{10}{3} \times (1.958)^3 - 29.16 \times 1.958 - \frac{10}{3}(1.958 - 1)^3$$

$$= 25.02 - 57.09 - 2.93 = -35$$

$$\therefore \qquad y_{max} = -\frac{35}{EI} = -\frac{35}{200 \times 10^6 \times 20 \times 10^{-6}} \times 10^3\,\text{mm} = -8.75\,\text{mm}$$

i.e. $\qquad y_{max} = \textbf{8.75 mm}$ (*downward*) (**Ans.**)

(iii) Slope at the end A, θ_A :

Putting $x = 0$ in the slope equation, we get

$$EI\frac{dy}{dx} = -29.16$$

$$\therefore \qquad \theta_A = \frac{dy}{dx} = -\frac{29.16}{EI} = -\frac{29.16}{200 \times 10^6 \times 20 \times 10^{-6}} = -0.00729\,\text{radian}$$

$$= -0.00729 \times \frac{180}{\pi} = -0.417°$$

Hence $\qquad \theta_A = \textbf{-0.417°}$ (**Ans.**)

Example 8.14. *A beam AB of span 8 metres is simply supported at the ends. It carries a uniformly distributed of 30 kN/m over its entire length and a concentrated load of 60 kN at 3 metres from the support A. Determine the maximum deflection in the beam and the location where the deflection occurs.*

Take: $\qquad E = 200 \times 10^6\,kN/m^2, I$
$\qquad\qquad = 80 \times 10^{-4}\,m^4$

Solution. Length of span of the beam,

$$l = 8\,\text{m}$$

Moment of inertia, $\qquad I = 80 \times 10^{-4}\,\text{m}^4$

Young's modulus, $\qquad E = 200 \times 10^6\,\text{kN/m}^2$

To calculate reaction at B taking moments about A, we get

Shock-absorber and axle-shaft of an autom

$$R_B \times 8 = 60 \times 3 + 30 \times 8 \times 8/2 = 1140$$

$$\therefore \qquad R_B = 142.5\,\text{kN}$$

Also, $\qquad R_A + R_B = 60 + 30 \times 8 = 300\,\text{kN}$

$$\therefore \qquad R_A = 300 - 142.5 = 157.5\,\text{kN}$$

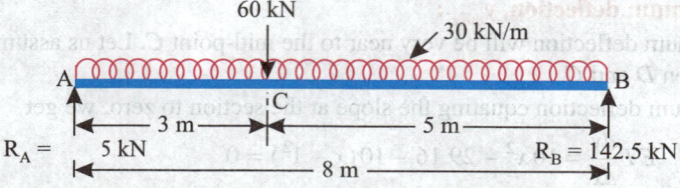

Fig. 8.28

Using Macaulay's method consider any section XX at a distance x from the end A; the bending moment at the section XX,

$$M_x = EI\frac{d^2y}{dx^2} = 157.5x - \frac{30x^2}{2} \Bigg| -60(x-3) \qquad ...(i)$$

On successively integrating the above equation, we have

$$EI\frac{dy}{dx} = \frac{157.5x^2}{2} - 5x^3 + C_1 \Bigg| -30(x-3)^2 \qquad ...(ii)$$

$$EIy = \frac{157.5x^3}{6} - \frac{5x^4}{4} + C_1x + C_2 \Bigg| -10(x-3)^3 \qquad ...(iii)$$

When, $x = 0$, $y = 0$

∴ $C_2 = 0$

When, $x = 8\text{m}$, $y = 0$

∴ $0 = \frac{157.5 \times 8^3}{6} - \frac{5 \times 8^4}{4} + 8C_1 - 10(8-3)^3$

$$= 13440 - 5120 + 8C_1 - 1250 = 7070 + 8C_1$$

∴ $C_1 = -883.75$

Hence, the slope and deflection equations are

$$EI\frac{dy}{dx} = \frac{157.5x^2}{2} - 5x^3 - 883.75 \Bigg| -30(x-3)^2 \qquad ...\textit{Slope equation}$$

$$EIy = \frac{157.5x^3}{6} - \frac{5x^3}{4} - 883.75x \Bigg| -10(x-3)^3 \qquad ...\textit{Deflection equation}$$

(i)   The maximum deflection and its location :

For maximum deflection, equating the slope at the section to zero, we get

$$EI\frac{dy}{dx} = \frac{157.5x^2}{2} - 5x^3 - 883.75 - 30(x-3)^2 = 0$$

By trial and error, $x = 3.92$ m satisfies it.

∴ Deflection is **maximum** at a distance of **3.92 metres from A** (Ans.)

To get maximum deflection (y_{max}) putting this value of x in the deflection equation, we have

$$EIy_{max} = \frac{157.5 \times 3.92^3}{6} - \frac{5 \times 3.92^4}{4} - 883.75 \times 3.92 - 10 \times (3.92 - 3)^3$$

$$= 1581.2 - 295.16 - 3464.3 - 7.78 = -2186$$

∴ $y_{max} = -\frac{2186}{EI} = -\frac{2186}{200 \times 10^6 \times 80 \times 10^{-4}} \times 10^3 \text{ mm} = -1.366 \text{ mm}$

Hence, $y_{max} = \textbf{1.366 mm}$ (*downward*) (Ans.)

Example 8.15. *A beam AB of span 8m is simply supported at the ends A and B and is loaded as shown in Fig. 8.29. If E = 200×10^6 kN/m² and I = $120 \times 10^{-6} m^4$ determine:*

(i)   *Deflection at the mid-span.*

(ii)   *Maximum deflection.*

(iii)   *Slope at the end A.*

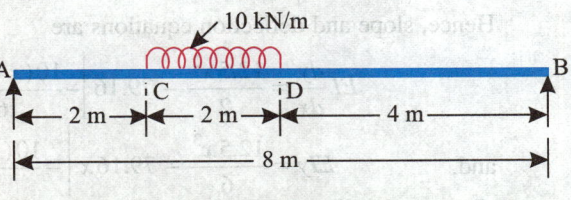

Fig. 8.29

Solution. Length of span of the beam = 8 m

Young's modulus,
$$E = 200 \times 10^6 \text{ kN/m}^2$$

Moment of inertia,
$$I = 120 \times 10^{-6} \text{ m}^4$$

To get reaction at B taking moments about A, we get
$$R_B \times 8 = 10 \times 2 \times \left(\frac{2}{2} + 2\right) = 60$$

∴
$$R_B = 7.5 \text{ kN}$$

Also,
$$R_A + R_B = 10 \times 2 = 20 \text{ kN}$$

∴
$$R_A = 20 - 7.5 = 12.5 \text{ kN}$$

In order that the general expression for the bending moment at any section may be expressed in the form suitable for application of Macaulay's method the loading on the beam is arranged as shown in Fig. 8.30.

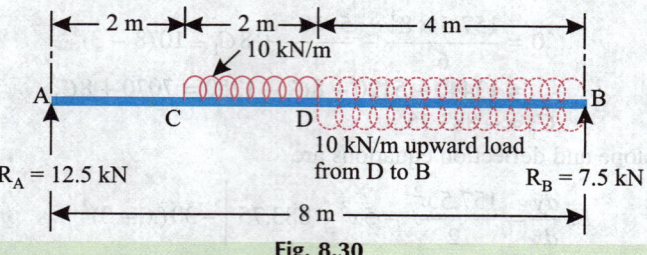

Fig. 8.30

Using Macaulay's method consider any section XX at a distance x from the end A; the bending moment at the section XX,

$$M_x = EI \frac{d^2y}{dx^2} = 12.5x \left| -\frac{10(x-2)^2}{2} \right| + \frac{10(x-4)^2}{2} \qquad ...(i)$$

On successively integrating the above equation, we get

$$EI \frac{dy}{dx} = \frac{12.5x^2}{2} + C_1 \left| -\frac{10(x-2)^3}{6} \right| + \frac{10(x-4)^3}{6} \qquad ....(ii)$$

$$EIy = \frac{12.5x^3}{6} + C_1 x + C_2 \left| -\frac{10(x-2)^4}{24} \right| + \frac{10(x-4)^4}{24} \qquad ...(iii)$$

When, $x = 0, y = 0, \quad C_2 = 0$

When, $x = 8\text{m}, y = 0$

∴
$$0 = \frac{12.5 \times 8^3}{6} + 8C_1 - \frac{10(8-2)^4}{24} + \frac{10(8-4)^4}{24}$$

$$= 1066.67 + 8C_1 - 540 + 106.67 = 8C_1 + 633.34$$

∴
$$C_1 = -79.16$$

Hence, slope and deflection equations are

$$EI \frac{dy}{dx} = \frac{12.5x^2}{2} - 79.16 \left| -\frac{10(x-2)^3}{6} \right| + \frac{10(x-4)^3}{6} \qquad ...Slope\ equation$$

and,
$$EIy = \frac{12.5x^3}{6} - 79.16x \left| -\frac{10(x-2)^4}{24} \right| + \frac{10(x-4)^4}{24} \qquad ...Deflection\ equation$$

(i) Deflection at the mid span, y_D :

For deflection at the mid span putting x = 4m in deflection equation, we get

$$Ely_D = \frac{12.5 \times 4^3}{6} - 79.16 \times 4 - \frac{10(4-2)^4}{24}$$

$$= 133.33 - 316.64 - 6.67 = -189.98$$

$$\therefore \quad y_D = \frac{189.98}{200 \times 10^6 \times 120 \times 10^{-6}} \times 10^3 \, mm = -7.195 \, mm$$

Hence, deflection at the mid span = **7.915 mm** (*downward*) **(Ans.)**

(ii) Maximum deflection, y_{max}:

Position of maximum deflection :

Let us assume that the deflection will be maximum between C and D. Equating the slope to zero, we have

$$EI\frac{dy}{dx} = \frac{12.5 x^2}{2} - 79.16 - \frac{10(x-2)^3}{6} = 0$$

or, $$6.25 x^2 - 79.16 - \frac{10(x-2)^3}{6} = 0$$

Solving the above equation by trial and error, we get $x = 3.75$ m

Putting $x = 3.75$ m in the deflection equation, we have

$$Ely_{max} = \frac{12.5 \times 3.75^3}{6} - 79.16 \times 3.75 - \frac{10(3.75-2)^4}{24}$$

$$= 109.86 - 296.85 - 3.91 = -190.9$$

$$y_{max} = -\frac{190.9}{200 \times 10^6 \times 120 \times 10^{-6}} \times 10^3 \, mm = -7.954 \, mm$$

Hence, *maximum deflection* = **7.954 mm** (*downward*) **(Ans.)**

(iii) Slope at the end A, θ_A :

For slope at A, putting $x = 0$ in the slope equation, we have

$$EI\frac{dy}{dx} = -79.16 \text{ or } EI\theta_A = -79.16$$

$$\therefore \quad \theta_A = -\frac{79.16}{200 \times 10^6 \times 120 \times 10^{-6}} = -0.00329 \, radian$$

$$= -0.00329 \times \frac{180}{\pi} = -0.188°$$

Hence, $\qquad \theta_A = -\textbf{0.188°}$ **(Ans.)**

Example 8.16. *A horizontal beam of uniform section and length l rests on supports at its ends. It carries a uniformly distributed load of w per unit run for a distance a from the right end. Calculate the value of a for which the maximum deflection will occur at the left end of uniformly distributed load.*

Solution. Fig. 8.31 shows the beam with the loading. To find reaction at A taking moments about B, we get

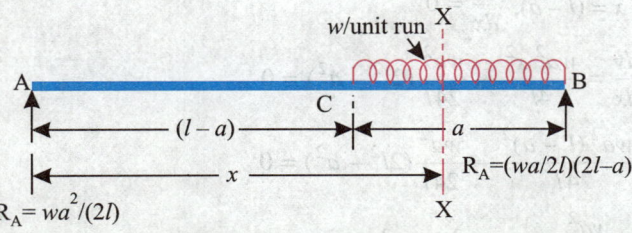

Fig. 8.31

$$R_A \times l = \frac{wa^2}{2}$$

$$\therefore \qquad R_A = \frac{wa^2}{2l}$$

Also $\quad R_A + R_B = wa$

$$\therefore \qquad R_B = wa - \frac{wa^2}{2l} = \frac{wa}{2l}(2l - a)$$

Using Macaulay's method consider any section XX at a distance x from the end A; the bending moment at the section,

Another view of shock-absorber and axle rod of an automobile.

$$M_x = EI\frac{d^2y}{dx^2} = \frac{wa^2}{2l}x \left| -\frac{w}{2}[x - (l - a)]^2 \right. \qquad ...(i)$$

On successively integrating the above equation, we have

$$EI\frac{dy}{dx} = \frac{wa^2}{4l}x^2 + C_1 \left| -\frac{w}{6}[x - (l - a)]^3 \right. \qquad ...(ii)$$

and, $\qquad EIy = \frac{wa^2}{12l} \cdot x^3 + C_1 x + C_2 \left| -\frac{w}{24}[x - (l - a)]^4 \right. \qquad ...(iii)$

When, $\qquad x = 0, y = 0$

$\therefore \qquad C_2 = 0$

When, $\qquad x = l, y = 0$

$$\therefore \qquad 0 = \frac{wa^2 l^3}{12l} + C_1 l - \frac{wa^4}{24}$$

or, $\qquad C_1 l = \frac{wa^4}{24} - \frac{wa^2 l^2}{12} = -\frac{wa^2}{24}(2l^2 - a^2)$

$\therefore \qquad C_1 = -\frac{wa^2}{24l}(2l^2 - a^2)$

Hence, the *slope* and *deflection equations* are :

$$EI\frac{dy}{dx} = \frac{wa^2 x^2}{4l} - \frac{wa^2}{24l}(2l^2 - a^2) \left| -\frac{w}{6}[x - (l - a)]^3 \right. \qquad ...Slope\ equation$$

$$EIy = \frac{wa^2 x^3}{12l} - \frac{wa^2}{24l}(2l^2 - a^2)x \left| -\frac{w}{24}[x - (l - a)]^4 \right. \qquad ...Deflection\ equation$$

For the condition that the maximum deflection should occur at C, we have :

When, $\qquad x = (l - a), \dfrac{dy}{dx} = 0$

$$\therefore \qquad EI\frac{dy}{dx} = \frac{wa^2 x^2}{4l} - \frac{wa^2}{24l}(2l^2 - a^2) = 0$$

or, $\qquad \dfrac{wa^2 (l - a)^2}{4l} - \dfrac{wa^2}{24l}(2l^2 - a^2) = 0$

or, $\qquad \dfrac{wa^2}{24l}[6(l - a)^2 - (2l^2 - a^2)] = 0$

or, $\dfrac{wa^2}{24l}[6(l^2 - 2la + a^2) - (2l^2 - a^2)] = 0$

or, $6l^2 - 12l\,a + 6\,a^2 - 2l^2 + a^2 = 0$

or, $7\,a^2 - 12l\,a + 4\,l^2 = 0$

or, $a = \dfrac{12l \pm \sqrt{144l^2 - 4 \times 7 \times 4l^2}}{14} = \dfrac{12l \pm \sqrt{144l^2 - 112l^2}}{14}$

$= \dfrac{12l \pm \sqrt{32l^2}}{14} + \dfrac{12l \pm 5.65l}{14} = 0.453l$ (neglecting +ve sign)

i.e., $a = 0.453\,l$

Putting this value of a in the value of C_1, we get

$$C_1 = -\dfrac{w(0.453l)^2}{24l}[2l^2 - (0.453l)^2]$$

$$= -0.00855\,wl^3\,(2 - 0.453^2) = -0.015345\,wl^3$$

Putting this value of C_1 in deflection equation, we get

$$EIy_{max} = \dfrac{w(0.453l)^2}{12l}(l - 0.453l)^3 - 0.015345\,wl^3\,(l - 0.453l)$$

$$= 0.0171\,wl \times (0.547l)^3 - 0.015345\,wl^3 \times 0.547l$$

$$= wl^4\,(0.0171 \times 0.547^3 - 0.015345 \times 0.547)$$

$$= wl^4\,(0.0027987 - 0.0083937)$$

$$= -0.005595\,wl^4 \text{ or } y_{max} = -\dfrac{0.005595\,wl^4}{EI}$$

Hence, *maximum deflection* $= \dfrac{\mathbf{0.005595}\,wl^4}{EI}$ *(downward)* **(Ans.)**

Example 8.17. *A simply supported beam carries the triangularly distributed symmetrical load as shown in Fig. 8.32. Find :*

  (i) *The slope at end A.*

  (ii) *The deflection at the centre.*

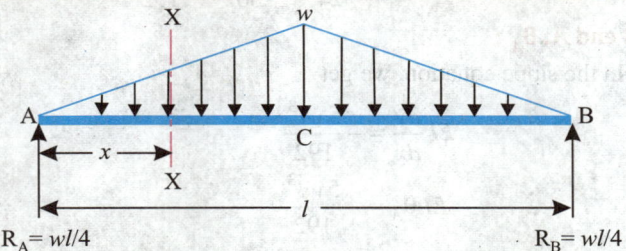

Fig. 8.32

Solution. The support reactions at A and B, by symmetry, are equal and each is one-half of the total load on the beam AB.

∴ $$R_A = R_B = \dfrac{1}{2}\left(\dfrac{1}{2}wl\right) = \dfrac{wl}{4}$$

Consider a section XX at a distance x from the end A.

Load intensity at $XX = \dfrac{w}{l/2} \cdot x = \dfrac{2wx}{l}$

The bending moment at the section is given by,

$$M_x = EI \frac{d^2 y}{dx^2} = \frac{wl}{4} \cdot x - \left(\frac{1}{2} \cdot x \cdot \frac{2wx}{l} \right) \times \frac{x}{3} \qquad ...(i)$$

$$= \frac{wlx}{4} - \frac{wx^3}{3l}$$

On successively integrating the above equation, we have

$$EI \frac{dy}{dx} = \frac{wlx^2}{8} - \frac{wx^4}{12l} + C_1 \qquad ...(ii)$$

$$EIy = \frac{wlx^3}{24} - \frac{wx^5}{60l} + C_1 x + C_2 \qquad ...(iii)$$

When, $\qquad x = 0, y = 0$

∴ $\qquad C_2 = 0;$

When, $\qquad x = l/2, \dfrac{dy}{dx} = 0$

∴ $\qquad 0 = \dfrac{wl}{8} \times (l/2)^2 - \dfrac{w}{12l} \times (l/2)^4 + C_1$

or, $\qquad C_1 + \dfrac{wl^3}{32} - \dfrac{wl^3}{192} = 0$

∴ $\qquad C_1 = -\dfrac{5wl^3}{192}$

Hence slope and deflection equations are

$$EI \frac{dy}{dx} = \frac{wl\,x^2}{8} - \frac{wx^4}{12l} - \frac{5wl^3}{192} \qquad ...Slope\ equation$$

$$EIy = \frac{wl\,x^3}{24} - \frac{wx^5}{60l} - \frac{5wl^3}{192} x \qquad ...Deflection\ equation$$

(i) Slope at end A, θ_A :

Putting $x = 0$ in the slope equation, we get

$$EI \frac{dy}{dx} = -\frac{5wl^3}{192}$$

or, $\qquad EI\,\theta_A = -\dfrac{5wl^3}{192} \qquad \left(\because \theta_A = \dfrac{dy}{dx} \right)$

∴ $\qquad \theta_A = -\dfrac{5wl^3}{192\,EI} \qquad$ **(Ans.)**

(ii) The deflection at the centre, y_C :

Putting $x = l/2$ in the deflection equation, we get

$$EIy = \frac{wl}{24} \times (l/2)^3 - \frac{w}{60l} \times (l/2)^5 - \frac{5wl^3}{192} \times l/2$$

$$= \frac{wl^4}{192} - \frac{wl^4}{1920} - \frac{5wl^4}{384} = \frac{wl^4}{1920}(10 - 1 - 25)$$

$$= -\frac{16\,wl^4}{1920} = -\frac{wl^4}{120}$$

$$y_C = -\frac{wl^4}{120\,EI}$$

Hence, deflection at the centre $= \dfrac{Wl^4}{120\,EI}$ *(downward)* **(Ans.)**

Example 8.18. *A beam 6m long is subjected to a 450 kNm clockwise couple as shown in the Fig. 8.33. If the uniform flexural rigidly (EI) of the beam is 8×10^4 kNm², determine :*

(i) *The deflection at the point of application of the couple.*

(ii) *The maximum deflection.*

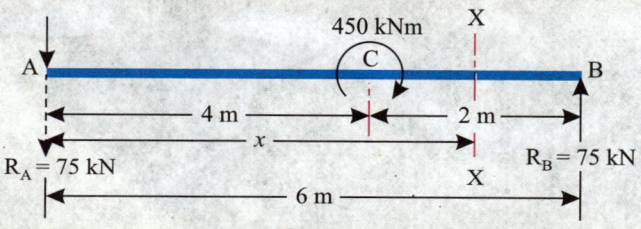

Fig. 8.33

Solution. Span of the beam $AB = 6$m

Flexural rigidity, $EI = 8 \times 10^4$ kNm²

To find reaction at B taking moments about A, we get

$$R_B \times 6 = 450$$

∴ $R_B = 75$ kN↑ and $R_A = 75$ kN ↓

Consider any section XX at a distance x from the end A. The bending moment at the section, following Macaulay's method, is given by:

$$M_x = EI\,\frac{d^2y}{dx^2} = -75\,x\,\big| +450 = -75\,x\,\big| + 450(x-4)^0 \qquad ...(i)$$

On successively differentiating the above equation, we get

$$EI\,\frac{dy}{dx} = -\frac{75\,x^2}{2} + C_1\,\bigg| + 450(x-4)^1 \qquad ...(ii)$$

and, $$EIy = -\frac{75\,x^3}{6} + C_1 x + C_2\,\bigg| + 225(x-4)^2 \qquad ...(iii)$$

When, $x = 0, y = 0$ ∴ $C_2 = 0$

When, $x = 6$ m, $y = 0$

∴ $$0 = -\frac{75 \times 6^3}{6} + 6C_1 + 225(6-4)^2$$

or, $6C_1 - 2700 + 900 = 0$ ∴ $C_1 = 300$

Hence, *the slope and deflection equations are :*

$$EI\,\frac{dy}{dx} = -\frac{75\,x^2}{2} + 300 + 450(x-4) \qquad ...Slope\ equation$$

and, $$EIy = -\frac{75\,x^3}{6} + 300\,x + 225(x-4)^2 \qquad ....Deflection\ equation$$

(i) The deflection at the point of application of the couple, y_C :

Putting $x = 4$ m in the deflection equation, we get

$$EIy_C = -\frac{75 \times 4^3}{6} + 300 \times 4 = 400$$

∴

$$y_C = \frac{400}{EI} = \frac{400}{8 \times 10^4} \times 10^3 \text{ mm} = 5 \text{ mm}$$

i.e.

$$y_C = 5 \text{ mm } (upward) \text{ (Ans.)}$$

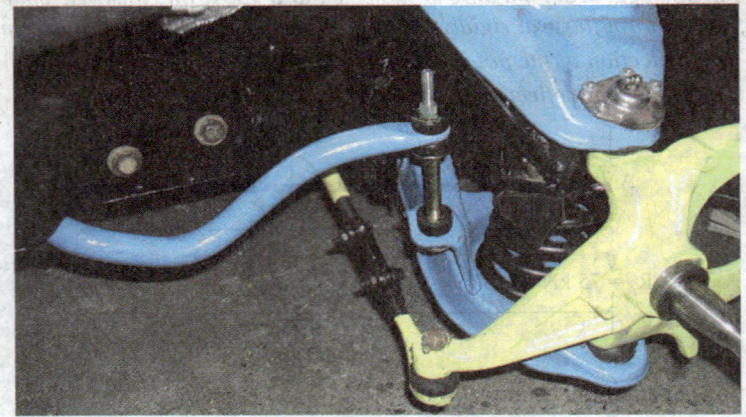

Another view of shock-obsorber, axle rod and part of steering mechanism of an automobile.

(ii) The maximum deflection, y_{max} :

The maximum deflection will occur in the larger segment. Equating the slope to zero, we get

$$EI\frac{dy}{dx} = -\frac{75x^2}{2} + 300 = 0 \text{ or } x = 2.828 \text{ m}$$

Putting $x = 2.828$ m in the deflection equation, we get

$$EIy_{max} = -\frac{75 \times (2.828)^3}{6} + 300 \times 2.828 = -282.71 + 848.4 = 565.69$$

∴

$$y_{max} = \frac{565.69}{EI} = \frac{565.69}{8 \times 10^4} \times 10^3 = 7.07 \text{ mm}$$

Hence, *maximum deflection* = **7.07 mm (upward)**   **(Ans.)**

Example 8.19. *A beam 6 metres long is loaded as shown in the Fig. 8.34. If the flexural rigidity (EI) of the beam is 8×10^4 kNm^2 find the deflection at point C.*

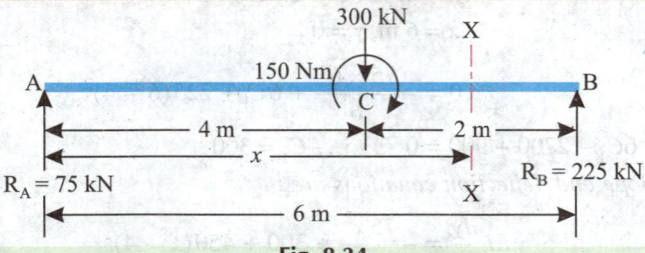

Fig. 8.34

Solution. Length of the beam span = 6 m

Flexural rigidity, $EI = 8 \times 10^4$ kNm^2

Deflection at point C :

Taking moments about A, we get

$$R_B \times 6 = 300 \times 4 + 150$$

∴ $$R_B = 225 \text{ kN}$$

Also, $$R_A + R_B = 300 \text{ kN}$$

∴ $$R_A = 300 - 225 = 75 \text{ kN}$$

Consider any section XX at a distance x from the end A. Using Macaulay's method, the bending moment at the section is given by :

$$M_x = EI \frac{d^2 y}{dx^2} = 75 x \, \bigg| -300 (x - 4) + 150$$

The above equation is rearranged as follows:

$$EI \frac{dy^2}{dx^2} = 75 x \, \bigg| -300 (x - 4) + 150 (x - 4)^0 \qquad \text{...(i)}$$

On successively integrating the above equation, we get

$$EI \frac{dy}{dx} = \frac{75 x^2}{2} + C_1 \, \bigg| -150 (x - 4)^2 + 150 (x - 4) \qquad \text{...(ii)}$$

and, $$EIy = \frac{25 x^3}{2} + C_1 x + C_2 \, \bigg| -50 (x - 4)^3 + 75 (x - 4)^2 \qquad \text{...(iii)}$$

When, $$x = 0, y = 0 \quad \therefore C_2 = 0$$

When, $$x = 6 \text{m}, y = 0$$

∴

$$0 = \frac{25 \times 6^3}{2} + 6 C_1 - 50 (6 - 4)^3 + 75 (6 - 4)^2$$

$$= 2700 + 6 C_1 - 400 + 300 \quad \therefore C_1 = -433.33$$

∴ Deflection equation is written as

$$EIy = \frac{25 x^3}{2} - 433.33 x - 50 (x - 4)^3 + 75 (x - 4)^2 \qquad \text{...Deflection equation}$$

To get deflection at the point C, putting $x = 4$m in the above equation, we get

$$EIy_C = \frac{25 \times 4^3}{2} - 433.33 \times 4 = -933.32$$

∴ $$y_C = -\frac{933.32}{EI} = -\frac{933.32}{8 \times 10^4} \times 10^3 \text{ mm} = -11.66 \text{ mm}$$

Hence, $$y_C = \textbf{11.66 mm} \ (downward) \ \ \textbf{(Ans.)}$$

OVERHANGING BEAMS

Example 8.20. *A 6 metres long beam carries point loads as shown in the Fig. 8.35. Determine maximum deflection and state where it occurs.*

Given: $E = 200 \times 10^6 \text{ kN/m}^2$ and $I = 120 \times 10^{-6} \text{ m}^4$

Solution. Refer to Fig. 8.35.

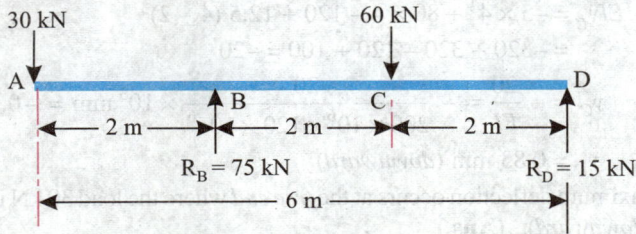

Fig. 8.35

Length of the beam = 6m

Young's modulus,
$$E = 200 \times 10^6 \, \text{kN/m}^2$$

Moment of inertia,
$$I = 120 \times 10^{-6} \, \text{m}^4$$

To find the support reactions taking moments about B, we get
$$R_D \times 4 + 30 \times 2 = 60 \times 2;$$
$$\therefore \qquad R_D = 15 \, \text{kN}$$

Also, $\qquad R_B + R_D = 30 + 60 = 90 \, \text{kN}$
$$\therefore \qquad R_B = 90 - 15 = 75 \, \text{kN}$$

Using Macaulay's method, consider any section XX at a distance x from the end A; the bending moment at the section XX,

$$M_x = EI \frac{d^2 y}{dx^2} = -30 \, x \mid +75 \, (x-2) \mid - 60(x-4)$$

Integrating, we get

$$EI \frac{dy}{dx} = -15 \, x^2 + C_1 \mid +37.5 \, (x-2)^2 \mid - 30 \, (x-4)^2$$

Integrating again, we get
$$EIy = -5x^3 + C_1 x + C_2 \mid + 12.5 \, (x-2)^3 \mid - 10 \, (x-4)^3$$

When, $\qquad x = 2\text{m}, y = 0 \qquad \therefore \; 0 = -5 \times 2^3 + C_1 \times 2 + C_2$

$\therefore \qquad 2C_1 + C_2 = 40$

When, $\qquad x = 6\text{m}, y = 0$

$\therefore$
$$0 = -5 \times 6^3 + 6C_1 + C_2 + 12.5 \, (6-2)^3 - 10 \, (6-4)^3$$
$$= -1080 + 6C_1 + C_2 + 800 - 80$$

$\therefore \qquad 6C_1 + C_2 = 360$

Solving (*iv*) and (*v*), we get
$$C_1 = 80 \text{ and } C_2 = -120$$

Equation for deflection will now be
$$EIy = -5x^3 + 80x - 120 \mid + 12.5 \, (x-2)^3 \mid - 10 \, (x-4)^3 \qquad \qquad ...Deflection \; equation$$

At $\qquad x = 0, y = y_A$

$\therefore \qquad EIy_A = -120$

or, $\qquad y_A = \dfrac{-120}{EI} = -\dfrac{120}{200 \times 10^6 \times 120 \times 10^{-6}}$
$$= -0.005 \, \text{m or } -5 \, \text{mm}$$

Hence, $\qquad y_A = 5 \, \text{mm} \; (downward)$

At $\qquad x = 4\text{m}, y = y_C$, deflection at the point C.

$\therefore \qquad EIy_C = -5 \times 4^3 + 80 \times 4 - 120 + 12.5 \, (4-2)^3$
$$= -320 + 320 - 120 + 100 = -20$$

$\therefore \qquad y_C = -\dfrac{20}{EI} = -\dfrac{20}{200 \times 10^6 \times 120 \times 10^{-6}} \times 10^3 \, \text{mm} = -0.83 \, \text{mm}$

Hence, $\qquad y_C = 0.83 \, \text{mm} \; (downward)$

In this case maximum deflection occurs at the *free end* where the load 30 kN is applied; *maximum deflection* = **5mm** (*downward*) **(Ans.)**

Example 8.21. *An overhanging beam ABC is loaded as shown in the Fig. 8.36. Determine*

(i) *Deflection at C.*

(ii) *Maximum deflection between A and B.*

Take: $E = 200 \times 10^6 kN/m^2$ and $I = 24 \times 10^{-6} m^4$

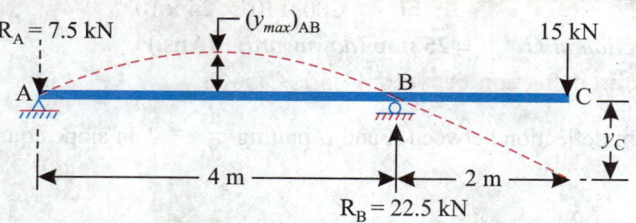

Fig. 8.36

Solution. Young's modulus,

$$E = 200 \times 10^6 \text{ kN/m}^2$$

Moment of inertia,

$$I = 24 \times 10^{-6} \text{ m}^4$$

Taking moments about A, we get

$$R_B \times 4 = 15 \times 6$$

$\therefore \qquad R_B = \dfrac{90}{4} = 22.5 \text{ kN}$

Also, $\quad R_A + R_B = 15 \text{ kN}$

$\therefore \qquad R_A = 15 - 22.5 = -7.5 \text{ kN}$

Consider any section XX at a distance x from the end A. Using Macaulay's method the B.M. at the section is given

$$M_x = EI \frac{d^2 y}{dx^2} = -7.5\,x \,\Big|\, +22.5\,(x-4)$$

Gear cone is used to transmit power to multiple gears from a single drive shaft. The shaft driving the cone, undergoes high bending and torsional stresses.

$$\qquad \qquad \dots(i)$$

On successively integrating the above equation, we get

$$EI \frac{dy}{dx} = -\frac{7.5\,x^2}{2} + C_1 \,\Big|\, + \frac{22.5}{2}(x-4)^2 \qquad \dots(ii)$$

and, $\quad EIy = \dfrac{2.5}{2}x^3 + C_1 x + C_2 \,\Big|\, + \dfrac{22.5}{6}(x-4)^3 \qquad \dots(iii)$

When, $\qquad x = 0,\ y = 0 \qquad\qquad \therefore\ C_2 = 0$

When, $\qquad x = 4 \text{ m},\ y = 0$

$\therefore \qquad 0 = -\dfrac{2.5}{2} \times (4)^3 + 4C_1 \qquad \therefore\ C_1 = 20$

Hence slope and deflection equation are:

$$EI \frac{dy}{dx} = -\frac{7.5\,x^2}{2} + 20 + \frac{22.5}{2}(x-4)^2 \qquad \dots \textit{Slope equation}$$

$$EIy = -\frac{2.5}{2}x^3 + 20x + \frac{22.5}{6}(x-4)^3 \qquad \dots \textit{Deflection equation}$$

(i) Deflection at y_c :

Putting $x = 6$ m in the deflection equation, we get

$$EIy_c = -\frac{2.5}{2} \times (6)^3 + 20 \times 6 + \frac{22.5}{6}(6-4)^3$$

$$= -270 + 120 + 30 = -120$$

$$\therefore \qquad y_C = -\frac{120}{EI} = -\frac{120}{200 \times 10^6 \times 24 \times 10^{-6}} \times 10^3 \text{ mm} = -25 \text{ mm}$$

Hence, *deflection at C* = **25 mm** *(downward)* **(Ans.)**

(ii) Maximum deflection between A and B, $(y_{max})_{AB}$:

For maximum deflection between A and B putting $\dfrac{dy}{dx} = 0$ in slope equation upto B, we get

$$-\frac{7.5 x^2}{2} + 20 = 0 \text{ or } 7.5 \, x^2 = 40$$

or

$$x = \sqrt{\frac{40}{7.5}} = 2.31 \text{ m}$$

Putting the value of $x = 2.31$ m in the deflection equation, we get

$$EI (y_{max})_{AB} = -\frac{2.5}{2} x^3 + 20 x$$

$$= -\frac{2.5}{2} \times (2.31)^3 + 20 \times 2.31 = 30.79$$

$$\therefore \qquad (y_{max})_{AB} = \frac{30.79}{EI} = \frac{30.79}{200 \times 10^6 \times 24 \times 10^{-6}} \times 10^3 \text{ mm} = 6.414 \text{ mm}$$

Hence, $(y_{max})_{AB} = \textbf{6.414 mm}$ *(upward)* **(Ans.)**

Example 8.22. *A beam ABC is loaded as shown in the Fig. 8.37. If $E = 200 \times 10^6$ kN/m^2 and $I = 9 \times 10^{-5}$ m^4 determine:*

  (i)   Slope at the end A.

  (ii) Deflection at the free end C.

  (iii) Maximum deflection .

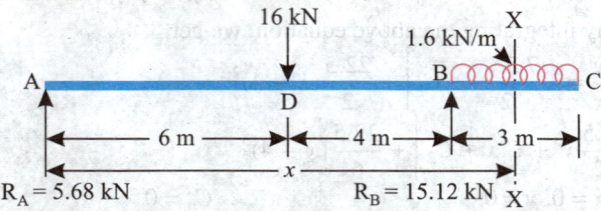

Fig. 8.37

Solution. Young's modulus,

$$E = 200 \times 10^6 \text{ kN/m}^2$$

Moment of inertia, $\qquad I = 9 \times 10^{-5} \text{ m}^4$

Taking moments about A, we get

$$R_B \times 10 = 16 \times 6 + 1.6 \times 3\left(\frac{3}{2} + 4 + 6\right) = 96 + 55.2 = 151.2$$

$$\therefore \qquad R_B = 15.12 \text{ kN}$$

Also, $\qquad R_A + R_B = 16 + 1.6 \times 3 = 20.8 \text{ kN}$

$$\therefore \qquad R_A = 20.8 - 15.12 = 5.68 \text{ kN}$$

Consider any section XX at a distance x from the end A. Using Macaulay's method, the B.M. at the section is given by:

$$M_x = EI\frac{d^2y}{dx^2} = 5.68x \left|\; -16(x-6) \;\right| + 15.12(x-10) - \frac{1.6(x-10)^2}{2}$$

On successively differentiating the above equation, we get

$$EI\frac{dy}{dx} = \frac{5.68}{2}x^2 + C_1 \left|\; -8(x-6)^2 + \frac{15.12}{2}(x-10)^2 - \frac{1.6}{6}(x-10)^3 \right. \qquad ...(i)$$

$$EIy = \frac{5.68}{6}x^3 + C_1x + C_2 \left|\; -\frac{8}{3}(x-6)^3 + \frac{15.12}{6}(x-10)^3 - \frac{1.6}{24}(x-10)^4 \right. \qquad ...(ii)$$

When, $x = 0, y = 0$

∴ $C_2 = 0$

When, $x = 10$ m, $y = 0$

Hence, from (ii) we have

$$0 = \frac{5.68}{6}\times 10^3 + 10C_1 - \frac{8}{3}(10-6)^3$$

or, $10\,C_1 + 946.67 - 170.67 = 0$

∴ $C_1 = -77.6$

Gear cone is being tested in the laboratory.

Hence, the slope and deflection equations are :

$$EI\frac{dy}{dx} = \frac{5.68}{2}x^2 - 77.6 \left|\; -8(x-6)^2 + \frac{15.12}{2}(x-10)^2 - \frac{1.6}{6}(x-10)^3 \right.$$

...Slope equation

and, $$EIy = \frac{5.68}{6}x^3 - 77.6x \left|\; -\frac{8}{3}(x-6)^3 + \frac{15.12}{6}(x-10)^3 - \frac{1.6}{24}(x-10)^4 \right.$$

...Deflection equation

(i) Slope at the end A, θ_A :

To get slope at the end A, putting $x = 0$ in slope equation, we have

$$EI\theta_A = -77.6 \qquad \left(\because \frac{dy}{dx} = \theta_A\right)$$

∴
$$\theta_A = -\frac{77.6}{EI} = -\frac{77.6}{200 \times 10^6 \times 9 \times 10^{-5}}$$

$$= 0.00431 \, \text{radian} = 0.00431 \times \frac{180}{\pi} = 0.247°$$

Hence,
$$\theta_A = \mathbf{0.247°} \quad (\textbf{Ans.})$$

(ii) Deflection at the free end C, y_C :

To get deflection at C, putting $x = 13$ m in the deflection equation, we get

$$EI \, y_C = \frac{5.68}{6} \times 13^3 - 77.6 \times 13 \; \Big| -\frac{8}{3}(13-6)^3 + \frac{15.12}{6}(13-10)^3 - \frac{1.6}{24}(13-10)^4$$

$$= 2079.8 - 1008.8 - 914.6 + 68 - 5.4 = 219$$

∴
$$y_C = \frac{219}{EI} = \frac{219}{200 \times 10^6 \times 9 \times 10^{-5}} \times 10^3 \, \text{mm} = 12.17 \, \text{mm (upward)}$$

Hence, *deflection at the free end C = **12.17 mm** (upward)* **(Ans.)**

(iii) Maximum deflection, y_{max} :

The maximum deflection between A and B will obviously occur near the mid-span, in the portion AD. Putting $\dfrac{dy}{dx} = 0$ in the slope equation (including the terms applicable for point D), we get

$$0 = \frac{5.68}{2}x^2 - 77.6$$

∴
$$x = \sqrt{\frac{77.6 \times 2}{5.68}} = 5.23 \, \text{m}$$

Now, putting the value of $x = 5.23$ m in the deflection equation, we get

$$EI \, y_{max} = \frac{5.68}{6} \times (5.23)^3 - 77.6 \times 5.23 = -270.42$$

∴
$$y_{max} = -\frac{270.42}{EI}$$

$$= -\frac{270.42}{200 \times 10^6 \times 9 \times 10^{-5}} \times 10^3 \, \text{mm} = -15.02 \, \text{mm}$$

Hence, *maximum deflection = **15.02 mm** (downward)* **(Ans.)**

Example 8.23. *Determine the following for an overhanging beam ABC supported at A and B and loaded as shown in Fig. 8.38 :*

(i) Deflection at the free end C;   *(ii) Maximum deflection between A and B.*

Take :   *$E = 200 \times 10^6 \, kN/m^2$ and $I = 13.5 \times 10^{-6} \, m^4$*

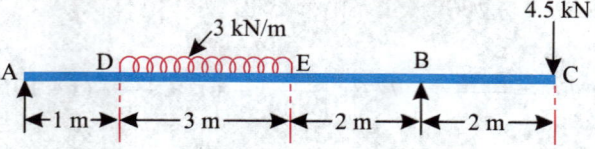

Fig. 8.38

Solution. Young's modulus,
$$E = 200 \times 10^6 \text{ kN/m}^2$$
Moment of inertia,
$$I = 13.5 \times 10^{-6} \text{ m}^4$$
Taking moments about A, we get
$$R_B \times 6 = 3 \times 3 \left(\frac{3}{2} + 1 \right) + 4.5 \times 8 = 22.5 + 36$$
∴ $$R_B = 9.75 \text{ kN}$$
Also, $$R_A + R_B = 3 \times 3 + 4.5 = 13.5 \text{ kN}$$
∴ $$R_A = 13.5 - 9.75 = 3.75 \text{ kN}$$
For continuity, extend U.D.L. (uniformly distributed load) from E to C, and apply U.D.L. *upwards* from E to C as shown in fig. 8.39.

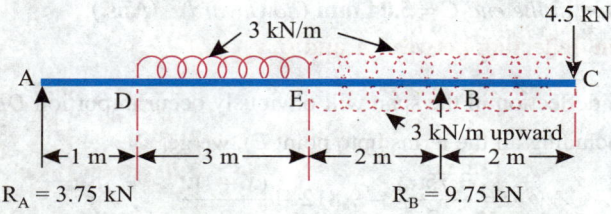

Fig. 8.39

Consider any section XX at distance x from the end A. Using Macaulay's method, the bending moment at the section is given by:

$$M_x = EI \frac{d^2 y}{dx^2} = 3.75 x \left| -\frac{3(x-1)^2}{2} \right| + \frac{3(x-4)^2}{2} + \left| 9.75 (x-6) \right| \quad \dots(i)$$

On successively integrating the above equation, we get

$$EI \frac{dy}{dx} = \frac{3.75}{2} x^2 + C_1 \left| -\frac{(x-1)^3}{2} \right| + \frac{(x-4)^3}{2} \left| + \frac{9.75}{2} (x-6)^2 \right. \quad \dots(ii)$$

and, $$EIy = \frac{3.75}{6} x^3 + C_1 x + C_2 \left| -\frac{(x-4)^4}{8} \right| + \frac{(x-4)^4}{8} + \frac{9.75 (x-6)^3}{6} \quad \dots(iii)$$

When, $$x = 0, y = 0$$
∴ $$C_2 = 0;$$
When, $$x = 6\text{m}, y = 0$$
∴ $$0 = \frac{3.75}{6} \times 6^3 + 6 C_1 - \frac{(6-1)^4}{8} + \frac{(6-4)^4}{8}$$
$$= 135 + 6 C_1 - 78.125 + 2$$
∴ $$C_1 = -9.812$$

Hence, the *slope and deflection equations* are:

$$EI \frac{dy}{dx} = \frac{3.75}{2} x^2 - 9.812 \left| -\frac{(x-1)^3}{2} \right| + \frac{(x-4)^3}{2} \left| + \frac{9.75}{2} (x-6)^2 \right.$$

...Slope equation

$$EIy = \frac{3.75}{6}x^3 - 9.812x \left| -\frac{(x-1)^4}{8} \right| + \frac{(x-4)^4}{8} + \frac{9.75(x-6)^3}{6}$$

$$...Deflection\ equation$$

(i) Deflection at the end C, y_C :

To get the deflection at the end C, putting $x = 8$ m in the deflection equation, we get

$$EIy_C = \frac{3.75}{6} \times 8^3 - 9.812 \times 8 - \frac{(8-1)^4}{8} + \frac{(8-4)^4}{8} + \frac{9.75(8-6)^3}{6}$$

$$= 320 - 78.49 - 300.12 + 32 + 13 = -13.61$$

$$\therefore \qquad y_C = -\frac{13.61}{EI} = -\frac{13.61}{200 \times 10^6 \times 13.5 \times 10^{-6}} \times 10^3 \text{ mm} = -5.04 \text{ mm}$$

Hence, *deflection at the end C = **5.04 mm*** (*downward*) **(Ans.)**

(ii) Maximum deflection between A and B, y_{max} :

The maximum deflection in the span will obviously occur in portion DE. Putting $\frac{dy}{dx} = 0$ in the slope equation (including all the terms upto point E), we get

$$0 = \frac{3.75x^2}{2} - 9,812 - \frac{(x-1)^3}{2}$$

or, $$(x-1)^3 - 3.75 x^2 + 19.624 = 0$$

Solving this by trial and error, we get $x = 2.46$ m. Putting this value of $x = 2.46$ m in the deflection equation, we get

$$EI\ y_{max} = \frac{3.75}{6} \times 2.46^3 - 9.812 \times 2.46 - \frac{(2.46-1)^4}{8}$$

$$= 9.3 - 24.14 - 0.57 = -15.41$$

$$\therefore \qquad y_{max} = -\frac{15.41}{EI}$$

$$= -\frac{15.41}{200 \times 10^6 \times 13.5 \times 10^{-6}} \times 10^3 \text{ mm} = -5.7 \text{ mm}$$

Hence, *maximum deflection = **5.7 mm*** (*downward*) **(Ans.)**

8.8. MOMENT AREA METHOD

The moment area method is partially convenient is case of beams acted upon with *point loads* in which case *bending moment area consists of triangles and reactangles.* In the case of distributed load the determination of the position of centroid itself involves integration and as such it no longer remains simpler than Macaulay's method. However, this method may be conveniently used in certain standard cases of distributed load where the position of the centroid of the bending moment area is known.

Consider a beam AB Fig. 8.40 (*a*) carrying such a load that it has bending diagram as shown in Fig. 8.40 (*b*). Let the beam bend into $AC'D'B$ as shown in Fig. 8.40 (*c*).

Now consider an element of small length CD of the beam at a distance x from B as shown in Fig. 8.40 (*a, b*).

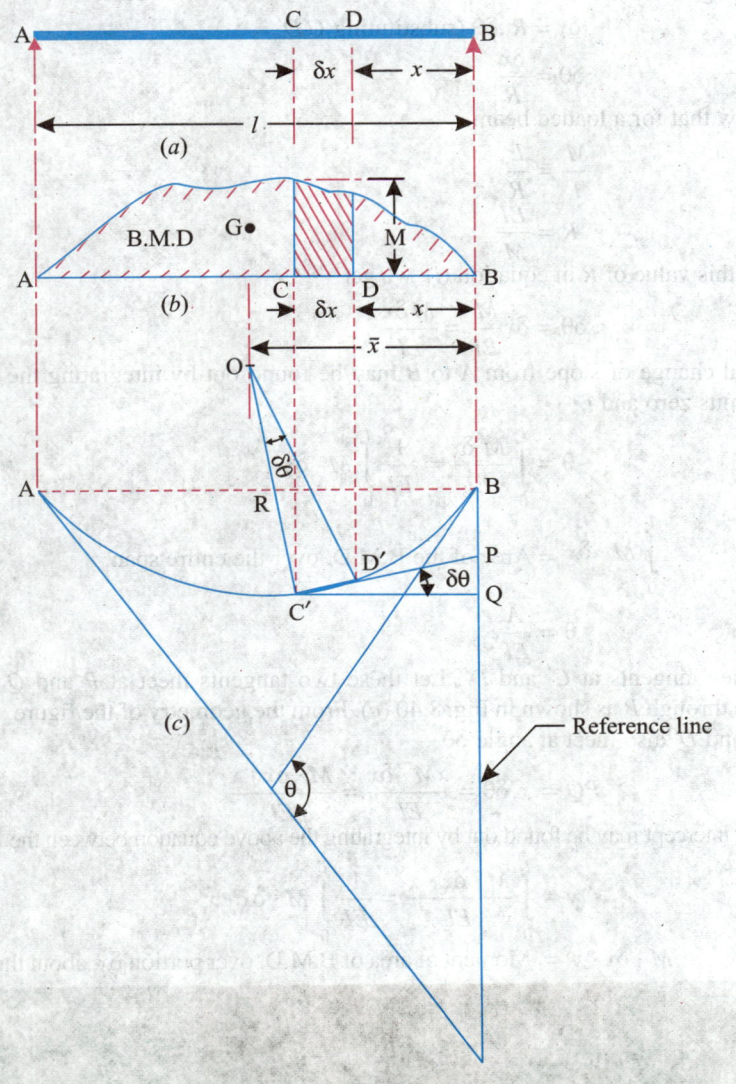

Fig. 8.40

Let, M = Bending moment between C and D,

δx = Length of the CD,

R = Radius of the bent up beam,

$\delta\theta$ = **Angle included between the tangents at C' and D', facing the reference line or it is the change of slope over the elementary portion δx.**

= also angle between the two normals,

A = Area of bending moment diagram (B.M.D.) over entire span,

$\bar{x}$ = Horizontal distance of centre of gravity G of the entire B.M.D area from the reference line, and

θ = The angle in radians, included between the tangents drawn at the extremities of the beam, *i.e.* at A and B and facing the reference line.

From the geometry of the bent up beam, we find that $C'D' = R\,\delta\theta$

or, $\qquad\qquad \delta x = R\cdot\delta\theta$ (substituting $C'D' = \delta x$)

$\therefore \qquad\qquad \delta\theta = \dfrac{\delta x}{R}$ $\qquad\qquad\qquad\qquad\qquad$ ...(i)

We know that for a loaded beam,

$$\frac{M}{I} = \frac{E}{R}$$

or, $\qquad\qquad R = \dfrac{EI}{M}$

Putting this value of R in equation (i), we get

$$\delta\theta = \delta x\,\frac{M}{EI} = \frac{M\,\delta x}{EI} \qquad\qquad ...(ii)$$

The total change of slope from A to B may be found out by integrating the above equation between the limits zero and l.

$\therefore \qquad\qquad \theta = \displaystyle\int_0^l \frac{M\,\delta x}{EI} = \frac{1}{EI}\int_0^l M\cdot\delta x$

But, $\qquad \displaystyle\int_0^l M\cdot\delta x = $ Area of the B.M.D. over the entire span.

$\therefore \qquad\qquad \theta = \dfrac{A}{EI}$ $\qquad\qquad\qquad\qquad\qquad$ ...(8.26)

Now draw tangents at C' and D'. Let these two tangents meet at P and Q on the vertical (reference) line through B as shown in Fig. 8.40 (c). From the geometry of the figure, we find that the tangents at C' and D' also meet at angle $\delta\theta$,

and, $\qquad\qquad PQ = x\cdot\delta\theta = \dfrac{xM\cdot\delta x}{EI} = \dfrac{M\cdot\delta x\cdot x}{EI} \qquad\qquad ...(iii)$

The total intercept may be found out by integrating the above equation between the limits zero and l.

$\therefore \qquad\qquad y = \displaystyle\int_0^l \frac{M\cdot\delta x\cdot x}{EI} = \frac{1}{EI}\int_0^l M\cdot\delta x\cdot x$

But, $\qquad\qquad M\cdot\delta x\cdot x = $ Moment of area of B.M.D. over portion δx, about the reference line.

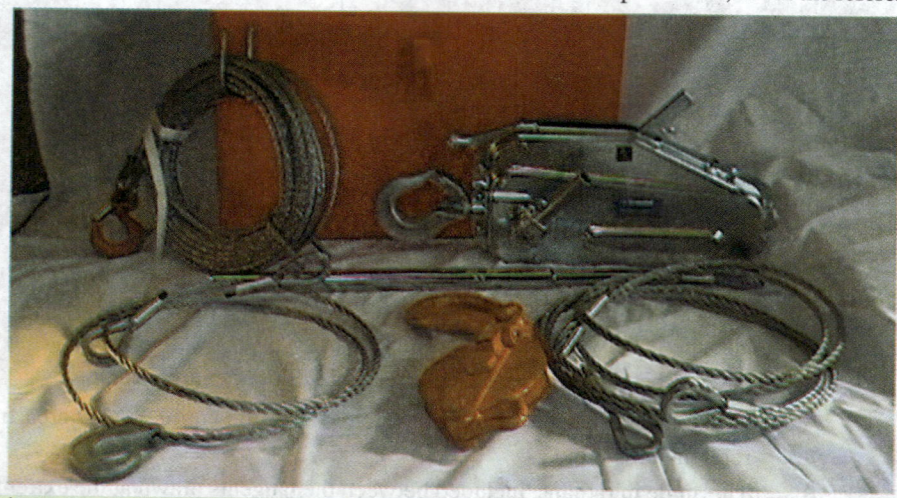

This kit is used in hoists for lifting weights, using metal ropes. The kit is called 'hoist rescue kit.'

∴ $\int_{0}^{l} M \cdot \delta x \cdot x$ = Moment of area of B.M.D. over the entire span δx, about the reference line

$$= A\bar{x} \qquad \therefore \qquad y = \frac{A\bar{x}}{EI} \qquad ...(8.27)$$

The product EI is known as *flexural rigidity of the beam*. The results given in equations (8.26) and (8.27) are known as **Mohr's theorems**.

Note. The area of the bending moment diagram (B.M.D.) above the zero line will be taken as positive and vice versa.

Expressions for A (area of B.M. diagram) and $\bar{x}$ (horizontal distance of c.g. of B.M. diagram) for some familiar B.M. diagrams:

1. *Rectangle :*

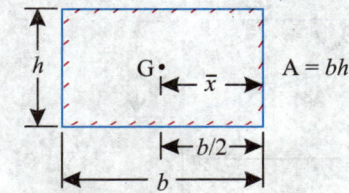

$A = bh$

2. *Triangle :*

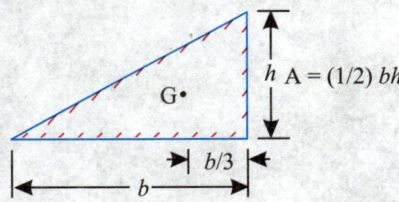

$A = (1/2) bh$

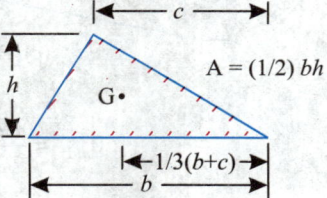

$A = (1/2) bh$

3. *Curves :*
  (i) $y = kx^2$

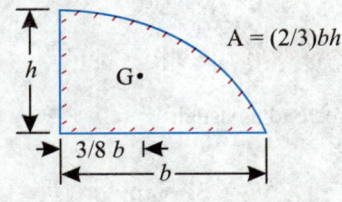

$A = (2/3)bh$

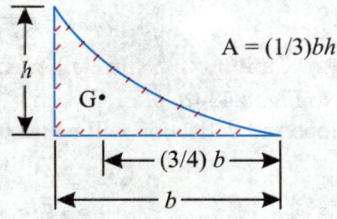

$A = (1/3)bh$

(ii) $y = kx^n$

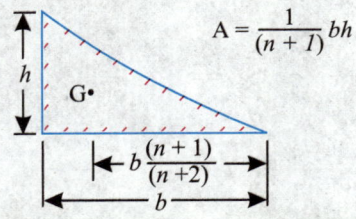

$A = \frac{1}{(n+1)} bh$

(iii) Sine curve

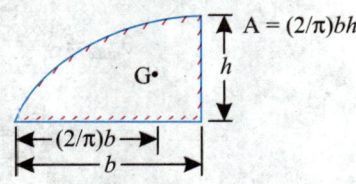

$A = (2/\pi)bh$

Fig. 8.41

8.8.1. Determination of Maximum Slope and Deflection in Important Cases.

Case I. Cantilever beam with a concentrated load at the free end:

Fig. 8.42 (*a, b, c*) shows a cantilever with a concentrated load W acting at free end, the elastic curve and B.M. diagram respectively.

The slope and deflection will be maximum at the free end. We known that,

$$\theta_{max} = \frac{A}{EI}$$

From Fig. 8.42 (*c*) $A = \frac{1}{2}(l)(Wl) = \frac{Wl^2}{2}$

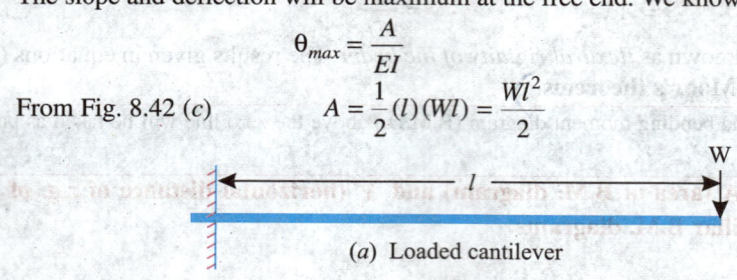

(*a*) Loaded cantilever

(*b*) Elastic curve

(*c*) B.M. diagram

Fig. 8.42

∴ $$\theta_{max} = \frac{Wl^2}{2EI} \qquad\qquad ...(8.28)$$

Also, $y_{max} = \dfrac{A\bar{x}}{EI} = \dfrac{\dfrac{Wl^2}{2} \times \dfrac{2l}{3}}{EI} = \dfrac{Wl^3}{3EI}$ $\left[\because \bar{x} = \dfrac{2}{3}l \right]$

Thus, $$y_{max} = \frac{Wl^3}{3EI} \qquad\qquad ...(8.29)$$

Case II. Cantilever beam with a concentrated load at any point:

Refer to Fig. 8.43 (*a, b, c*)

The slope and deflection will be maximum at the free end as usual.

$$\theta_{max} = \frac{A}{EI}$$

But, $A = \frac{1}{2}(a)(Wa) = \frac{Wa^2}{2}$

∴ $$\theta_{max} = \frac{Wa^2}{2EI} \qquad\qquad ...(8.30)$$

and, $y_{max} = \dfrac{A\bar{x}}{EI} = \dfrac{Wa^2}{2EI}\left(l - a + \dfrac{2}{3}a \right)$

or,
$$y_{max} = \frac{Wa^2}{2EI}\left(l - \frac{a}{3}\right) \qquad \ldots(8.31)$$

Also,
$$y_a = \frac{A\bar{x}}{EI} = \frac{Wa^2}{2EI}\left(\frac{2}{3}a\right)$$

or,
$$y_a = \frac{Wa^3}{3EI} \qquad \ldots(8.32)$$

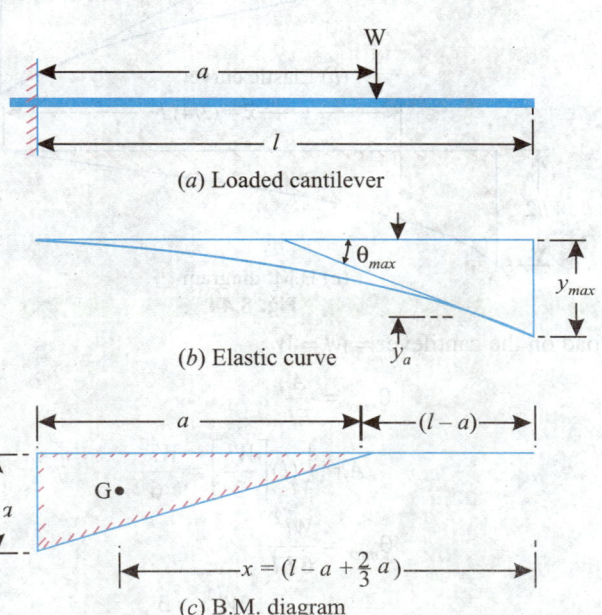

(a) Loaded cantilever

(b) Elastic curve

(c) B.M. diagram

Fig. 8.43

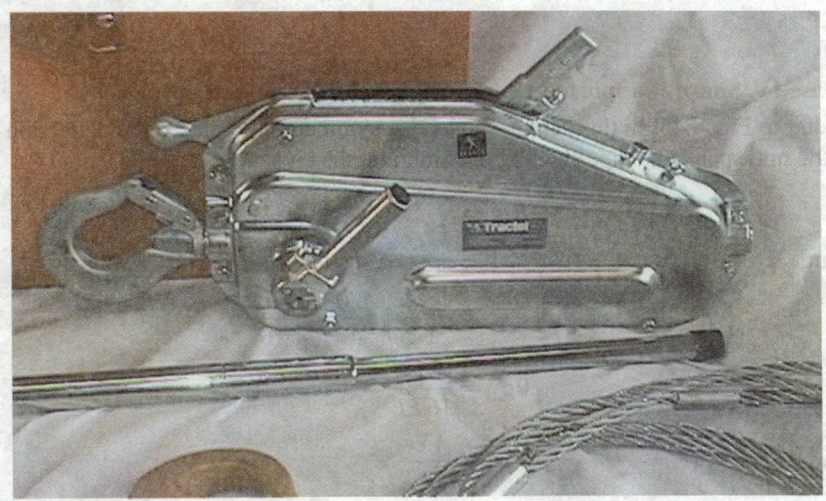

The above part of hoist rescue equipment holds and locks the metal rope as and when required. This is also used in heavy cranes and pile drivers.

Case III. Cantilever beam with uniformly distributed load:

Refer to Fig. 8.44

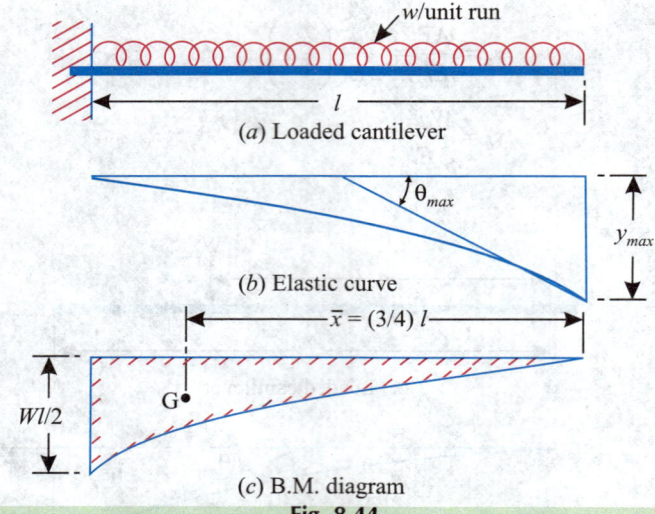

(a) Loaded cantilever

(b) Elastic curve

(c) B.M. diagram

Fig. 8.44

The total load on the cantilever = $wl = W$

Now, $$\theta_{max} = \frac{A}{EI}$$

But, $$A = \frac{1}{3}(l)\left[\frac{Wl}{2}\right] = \frac{Wl^2}{6}$$

∴ $$\theta_{max} = \frac{Wl^2}{6EI} \qquad \qquad ...(8.33)$$

and, $$y_{max} = \frac{A\bar{x}}{EI} = \frac{\dfrac{Wl^2}{6}\times\dfrac{3}{4}l}{EI} \qquad \qquad \left(\because \bar{x} = \frac{3}{4}l\right)$$

or, $$y_{max} = \frac{Wl^3}{8EI} \qquad \qquad ...(8.34)$$

Case IV. Simply supported beam with concentrated load at the centre:

Refer to Fig. 8.45. In this case as the load is symmetrically placed the deflection will be maximum at the mid span and slope shall be maximum at the ends.

$$\theta_{max} = \frac{A}{EI}$$

But, $$A = \text{area of shaded triangle [Fig. 8.45 }(c)]$$

$$= \frac{1}{2}\left(\frac{1}{2}\right)\left(\frac{Wl}{4}\right) = \frac{Wl^2}{16}$$

∴ $$\theta_{max} = \frac{Wl^2}{16EI} \qquad \qquad ...(8.35)$$

and, $$y_{max} = \frac{A\bar{x}}{EI}$$

$$= \frac{Wl^2}{16EI}\times\frac{2}{3}\left(\frac{l}{2}\right) \qquad \qquad \left[\because \bar{x} = \frac{2}{3}\left(\frac{l}{2}\right)\right]$$

$$= \frac{Wl^3}{48\,EI};$$

Hence, $y_{max} = \dfrac{Wl^3}{48\,EI}$ ...(8.36)

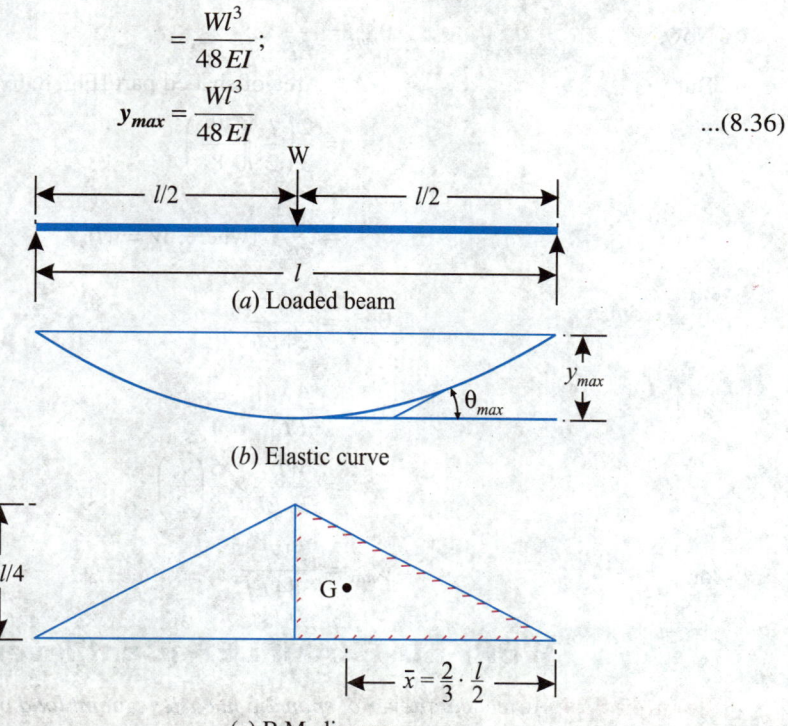

(a) Loaded beam

(b) Elastic curve

(c) B.M. diagram

Fig. 8.45

Case V. Simply supported beam with uniformly distributed load:

Refer to Fig. 8.46. In this case also the deflection will be maximum at the mid span and slope will be maximum at the ends.

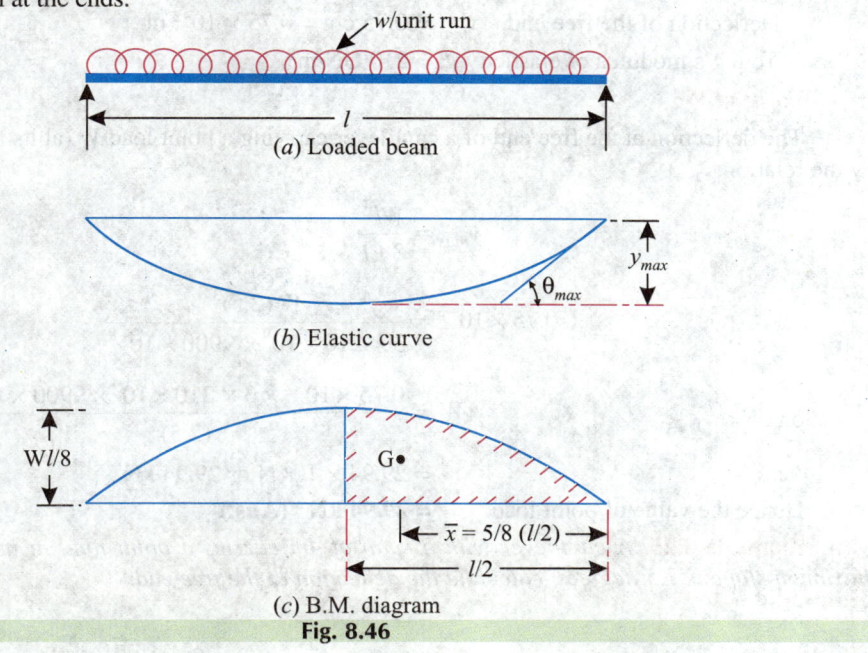

(a) Loaded beam

(b) Elastic curve

(c) B.M. diagram

Fig. 8.46

Now, $$\theta_{max} = \frac{A}{EI}$$

But, $$A = \text{Area of shaded part [Fig. 8.46 (c)]}$$

$$= \frac{2}{3}\left(\frac{l}{2}\right)\left(\frac{Wl}{8}\right)$$

$$= \frac{Wl^2}{24} \quad (\text{where, } W = wl)$$

∴ $$\theta_{max} = \frac{Wl^2}{24\,EI} \qquad \qquad ...(8.37)$$

and, $$y_{max} = \frac{A\bar{x}}{EI}$$

$$= \frac{Wl^2}{24\,EI} \times \frac{5}{8}\left(\frac{l}{2}\right) \qquad \left[\because \bar{x} = \frac{5}{8}\left(\frac{l}{2}\right)\right]$$

or, $$y_{max} = \frac{5Wl^3}{384\,EI} \qquad \qquad ...(8.38)$$

WORKED EXAMPLES (Cantilevers)

Example 8.24. *A steel cantilever of span 2.5 m carries a point load of W kN at its free end. The moment of inertia of the section of the cantilever is 9900 cm⁴. If the deflection at the free end is not to exceed 0.75 cm, what must be the value of W? Take E = 210 GN/m².*

Solution. Refer to Fig. 8.42

   Given: Span of cantilever, $l = 2.5$ m

   M.O.I. of the section, $I = 9900$ cm⁴ $= 9900 \times 10^{-8}$ m⁴

   Deflection of the free end, $y_{max} = 0.75$ cm $= 0.75 \times 10^{-2}$ m

   Young's modulus of elasticity, $E = 210$ GN/m²

Value of W:

The deflection at the free end of a cantilever carrying a point load W (at its free end) is given by the relation:

$$y_{max} = \frac{Wl^3}{3\,EI} \qquad \qquad \text{(Eqn. 8.29)}$$

$$0.75 \times 10^{-2} = \frac{W(2.5)^3}{3 \times 210 \times 10^9 \times 9900 \times 10^{-8}}$$

or, $$W = \frac{0.75 \times 10^{-2} \times 3 \times 210 \times 10^9 \times 9900 \times 10^{-8}}{(2.5)^3}$$

$$= 29.94 \times 10^3 \text{ N or } 29.94 \text{ kN}$$

Hence the value of point load $W = \mathbf{29.94}$ **kN** **(Ans.)**

Example 8.25. *A cantilever with a span of 4m carries a point load at its free end. If the maximum slope is 1.5 degrees, calculate the deflection at the free end.*

Solution. Refer to Fig. 8.42.

Given: Span of the cantilever, $l = 4$ m

Maximum slope, $\theta_{max} = 1.5°$

Deflection at the free end y_{max} :

We know that, in this case,

Maximum slope, $$\theta_{max} = \frac{Wl^2}{2 EI}$$

...(i)

and, maximum deflection, $$y_{max} = \frac{Wl^3}{3 EI}$$

...(ii)

The picture shows splined shaft, gears and bearing of a machine part.

But, $$\frac{Wl^2}{2 EI} = 1.5° = 1.5 \times \frac{\pi}{180} = 0.02618 \text{ radian (given)}$$

or, $$\frac{Wl^2}{EI} = 2 \times 0.02618 = 0.05236$$

Substituting this value of $\dfrac{Wl^2}{EI}$ in equation (ii), we get

$$y_{max} = \frac{Wl^2}{EI} \times \frac{l}{3} = 0.05236 \times \frac{4}{3} = 0.06981 \text{ m}$$

Hence, *deflection at the free end = **69.81 mm** (**Ans.**)*

Example 8.26. *A cantilever beam with a span of 3m carries a point load 30 kN at a distance of 2m from the fixed end. Determine the slope and deflection at the free end and at the point where load is applied.*

Take: *M.O.I. of the section = 11924 cm⁴, and E = 200 GN/m².*

Solution. Refer to Fig. 8.47.

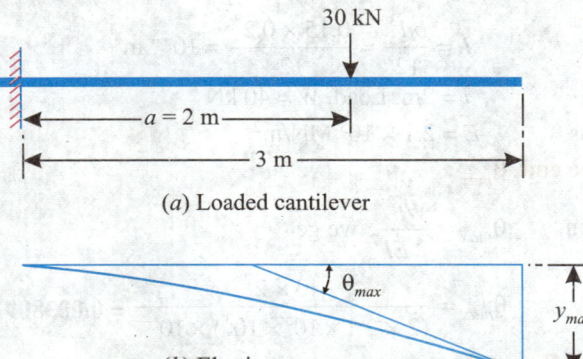

(a) Loaded cantilever

(b) Elastic curve

Fig. 8.47

Given: Span of cantilever beam,

$l = 3$ m

Distance of point load from the fixed end,

$a = 2$ m

Magnitude of the point load,
$$W = 30 \text{ kN}$$

M.O.I., $\qquad I = 11924 \text{ cm}^4 = 11924 \times 10^{-8} \text{ m}^4$

We know that in this case: Slope,

$$\theta_{max} = \frac{Wa^2}{2EI}$$

$$= \frac{(30 \times 1000) \times 2^2}{2 \times 200 \times 10^9 \times 11924 \times 10^{-8}} = \textbf{0.002516 radian.} \quad \textbf{(Ans.)}$$

Deflection at the free end,

$$y_{max} = \frac{wa^2}{2EI}\left[l - \frac{a}{3}\right] = \frac{(30 \times 1000) \times 2^2}{2 \times 200 \times 10^9 \times 11924 \times 10^{-8}}\left[3 - \frac{2}{3}\right]$$

$$= 0.00587 \text{ m} = \textbf{5.87 mm} \quad \textbf{(Ans.)}$$

Deflection at a distance $\quad a = 2\text{m}$

i.e. where the load acts, $\quad y_a = \dfrac{Wa^3}{3EI}$

$$= \frac{(30 \times 1000) \times (2)^3}{3 \times 200 \times 10^9 \times 11924 \times 10^{-8}}$$

$$= 0.00335 \text{ m} = \textbf{3.35 mm} \quad \textbf{(Ans.)}$$

Example 8.27. *A cantilever 150 mm wide and 200 mm deep projects 2m out of a wall, and is carrying a point load of 40 kN at the free end. Determine the slope and deflection of the cantilever at the free end.*

Take : $\qquad\qquad E = 2.1 \times 10^5 \text{ MN/m}^2.$

Solution. *Given*: Width of the section,

$$b = 150 \text{ mm} = 0.15 \text{ m}$$

Depth of the section, $\qquad d = 200 \text{ mm} = 0.2 \text{ m}$

∴ M.O.I, $\qquad I = \dfrac{bd^3}{12} = \dfrac{0.15 \times 0.2^3}{12} = 10^{-4} \text{ m}^4$

Length $\qquad\qquad l = 2\text{m}; \text{ Load}, W = 40 \text{ kN}$

Young's modulus, $\qquad E = 2.1 \times 10^5 \text{ MN/m}^2$

Slope at the free end, θ_{max} :

Using the relation, $\qquad \theta_{max} = \dfrac{Wl^2}{2EI},$ we get

$$\theta_{max} = \frac{40 \times 2^2}{(2 \times 2.1 \times 10^5 \times 10^3) \times 10^{-4}} = \textbf{0.003809 radian} \quad \textbf{(Ans.)}$$

Deflection at the free end, y_{max} :

Using the relation : $\qquad y_{max} = \dfrac{Wl^3}{3EI},$ we get

$$y_{max} = \frac{40 \times 2^3}{3 \times (2.1 \times 10^5 \times 10^3) \times 10^{-4}}$$

$$= 0.005079 \text{ m} = \textbf{5.079 mm} \quad \textbf{(Ans.)}$$

Example 8.28. *A steel cantilever projecting 3 metres from a wall is loaded with a uniformly distributed load of 20 kN/m run. Find the slope and deflection of the beam if the moment of inertia of the beam section is 7550 cm⁴.*

Take : E = 210 GN/m².

Solution. *Given*: Length, $\qquad l = 3$ metres

$\qquad\qquad\qquad$ U.D.L., $w = 20$ kN/m

∴ $\qquad\qquad$ Total load $= wl = 20 \times 3 = 60$ kN

Moment of inertia, $\qquad I = 7550$ cm⁴ $= 7550 \times 10^{-8}$ m⁴

Slope, θ_{max} :

Using the relation: $\qquad \theta_{max} = \dfrac{Wl^2}{6EI} = \dfrac{(60 \times 1000) \times (3)^2}{6 \times 210 \times 10^9 \times 7550 \times 10^{-8}}$

$\qquad\qquad\qquad\qquad = \mathbf{0.005676\ radian}$ **(Ans.)**

Maximum deflection, y_{max} :

Using the relation: $\qquad y_{max} = \dfrac{Wl^3}{8EI} = \dfrac{(60 \times 1000) \times (3)^3}{8 \times 210 \times 10^9 \times 7550 \times 10^{-8}}$

$\qquad\qquad\qquad\qquad = 0.01277$ m $= \mathbf{12.77\ mm}$ **(Ans.)**

Example 8.29. *A cantilever 3m long is loaded with a uniformly distributed load of 15 kN/m over a length of 2m from the fixed end. Determine the slope and deflection at the free end of the cantilever.*

Take : $\qquad\qquad\qquad E = 2.1 \times 10^8$ kN/m², and $I = 0.000095$ m⁴.

Solution. Refer to Fig. 8.48

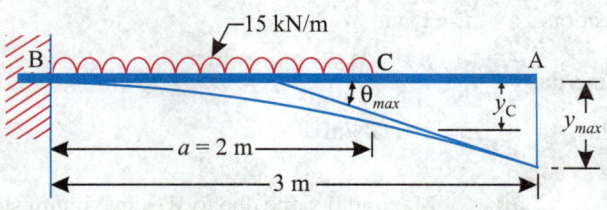

Fig. 8.48

Given:

Length, $\qquad\qquad\qquad l = 3$ m

U.D.L., $\qquad\qquad\qquad w = 15$ kN/m

Loaded length, $\qquad\qquad a = 2$m

Total load, $\qquad\qquad\qquad W = 15a = 15 \times 2 = 30$ kN

Moment of inertia $\qquad\quad I = 0.000095$ m⁴

Young's modulus $\qquad\quad E = 2.1 \times 10^8$ kN/m²

Slope at the free end, θ_{max} :

Using the relation, $\qquad \theta_{max} = \dfrac{Wa^2}{6EI}$, we get

$\qquad\qquad\qquad \theta_{max} = \dfrac{30 \times 2^2}{6 \times 2.1 \times 10^8 \times 0.000095} = \mathbf{0.001\ radian}$ **(Ans.)**

Deflection at the free end, y_{max} :

Using the relation: $\qquad y_{max} = \dfrac{Wa^2}{8EI} + \dfrac{Wa^2}{6EI}(l - a),$ with usual notations, we have

$$y_{max} = \frac{30 \times 2^2}{8 \times 2.1 \times 10^8 \times 0.000095} + \frac{30 \times 2^2}{6 \times 2.1 \times 10^8 \times 0.000095} \quad (3-2)$$

$$= 0.0015 + 0.001 = 0.0025 \text{ m} = \textbf{2.5 mm} \quad \textbf{(Ans.)}$$

Example 8.30. *A cantilever beam of 3m span is 15 cm wide and 25 cm deep. It carries a uniformly distributed load of 20 kN/m over its whole span and 25 kN load at the free end. Calculate the maximum slope and deflection.*

Take E = 210 GN/m²

Solution. Refer Fig. 8.49

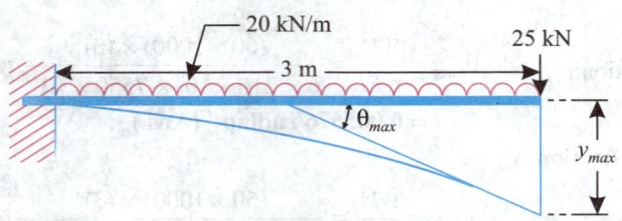

Fig. 8.49

Given: Length, $l = 3$m

U.D.L., $w = 20$ kN/m

Total U.D.L., $W = 20 \times 3 = 60$ kN

Point Load, $W_1 = 25$ kN

Width of the section, $b = 15$ cm

Depth of the section, $d = 25$ cm

∴ Moment of inertia, $I = \dfrac{bd^3}{12} = \dfrac{15 \times 25^3}{12} = 19531.25 \text{ cm}^4 = 19531.25 \times 10^{-8} \text{ m}^4$

Young's modulus, $E = 210$ GN/m²

Maximum slope, θ_{max} :

θ_{max} = Maximum slope due to W + maximum slope due to W_1

$$= \frac{Wl^2}{6EI} + \frac{W_1 l^2}{2EI} = \frac{l^2}{2EI}\left[\frac{W}{3} + W_1\right]$$

Drive and transmission shaft of an automobile.

$$= \frac{3^2}{2 \times 210 \times 10^9 \times 19531.25 \times 10^{-8}} \left(\frac{60000}{3} + 25 \times 1000 \right)$$

or, $\theta_{max} = 0.004937$ radian **(Ans.)**

Maximum deflection y_{max} :

Using the relation, $y_{max} = \frac{Wl^3}{8EI} + \frac{W_1 l^3}{3EI}$, we get

$$y_{max} = \frac{l^3}{EI} \left[\frac{W}{8} + \frac{W_1}{3} \right]$$

$$= \frac{3^3}{210 \times 10^9 \times 19531.25 \times 10^{-8}} \times \left[\frac{60000}{8} + \frac{25000}{3} \right]$$

$$= 0.01042 \text{ m or } \mathbf{10.42 \text{ mm}} \text{ (Ans.)}$$

Example 8.31. *A cantilever 2.5 m long is carrying a load of 25 kN at free end and 35 kN at a distance of 1.3 m from the end. Find the slope and deflection at the free end. Take $E = 2.0 \times 10^8$ kN/m² and $I = 1.5 \times 10^{-4}$ m⁴.*

Solution. Refer to Fig. 8.50

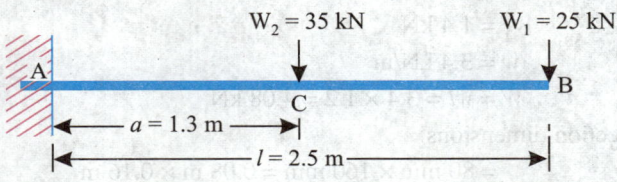

Fig. 8.50

Given: Length, $l = 2.5$ m

Load, $W_1 = 25$ kN

Load, $W_2 = 35$ kN

Length *AC*, $a = 1.3$ m

Young's modulus, $E = 2.0 \times 10^8$ kN/m²

Moment of inertia $I = 1.5 \times 10^{-4}$ m⁴

Slope at the free end, θ_{max}:

Using the relation, $\theta_{max} = \frac{W_1 l^2}{2EI} + \frac{W_2 a^2}{2EI} = \frac{1}{2EI}(W_1 l^2 + W_2 a^2)$

$$= \frac{1}{2 \times 2 \times 10^8 \times 1.5 \times 10^{-4}} \times (25 \times 2.5^2 + 35 \times 1.3^2)$$

or, $\theta_{max} = 0.00359$ radian **(Ans.)**

Deflection at the free end, y_{max}:

Using the relation, $y_{max} = \left[\frac{W_1 l^3}{3EI} \right] + \left[\frac{W_2 a^3}{3EI} + \frac{W_2 a^2}{2EI} (l - a) \right]$, we get

$$y_{max} = \left[\frac{25 \times 2.5^3}{3 \times 2 \times 10^8 \times 1.5 \times 10^{-4}} \right] + \left[\left\{ \frac{35 \times 1.3^3}{3 \times 2 \times 10^8 \times 1.5 \times 10^{-4}} \right. \right.$$

$$+ \left\{ \frac{35 \times 1.3^2}{2 \times 2 \times 10^8 \times 1.5 \times 10^{-4}} (2.5 - 1.3) \right\} \Bigg]$$

$$= 0.00434 + (0.000854 + 0.001183)$$

$$= 0.006377 \text{ m} = \textbf{6.377 mm} \quad \textbf{(Ans.)}$$

Example 8.32. *A cantilever 2m long is loaded with a point load of 1.4 kN at free end and distributed load of 3.4 kN per metre run over 1.2 metre from the fixed end. If the section is rectangular 80 mm × 160 mm, calculate the deflection at the free end.*

Take $\qquad E = I \times 10^7 \text{ kN/m}^2.$

Solution. Refer to Fig. 8.51

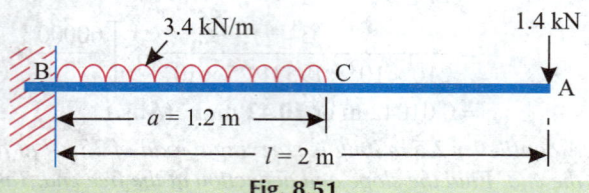

Fig. 8.51

Given: Length, $\qquad l = 2\text{m}$

Distance, $\qquad a = 1.2 \text{ m}$

Point load, $\qquad W_1 = 1.4 \text{ kN}$

U.D.L., $\qquad w = 3.4 \text{ kN/m}$

Total U.D.L., $\qquad W = wl = 3.4 \times 1.2 = 4.08 \text{ kN}$

Rectangular section dimensions

$$= 80 \text{ mm} \times 160 \text{ mm} = 0.08 \text{ m} \times 0.16 \text{ m}$$

∴ Moment of inertia,

$$I = \frac{0.08 \times 0.16^3}{12} = 2.73 \times 10^{-5} \text{ m}^4$$

Deflection at the free end, y_{max} :

$$y_{max} = \text{Deflection due to load } W + \text{deflection due to load } W_1$$

$$= \left[\frac{Wa^3}{8\,EI} + \frac{Wa^2}{6\,EI} (l - a) \right] + \frac{W_1 l^3}{3\,EI}$$

$$= \left[\frac{4.08 \times 1.2^3}{8 \times 10^7 \times 2.73 \times 10^{-5}} + \frac{4.08 \times 1.2^2 \, (2.0 - 1.2)}{6 \times 10^7 \times 2.73 \times 10^{-5}} \right]$$

$$+ \frac{1.4 \times 2^3}{3 \times 10^7 \times 2.73 \times 10^{-5}}$$

$$= 0.003228 + 0.002869 + 0.013675$$

or, $\qquad y_{max} = 0.01977 \text{ m} = 19.77 \text{ mm}$

Hence, *total deflection at the free end =* **19.77 mm** **(Ans.)**

Example 8.33. *A cantilever 100 mm wide and 200 mm deep is loaded as shown in Fig. 8.52. Find the slope and deflection at the free end A.*

Take : $\qquad E = 2.1 \times 10^8 \text{ kN/m}^2.$

Solution. Refer to Fig. 8.52.

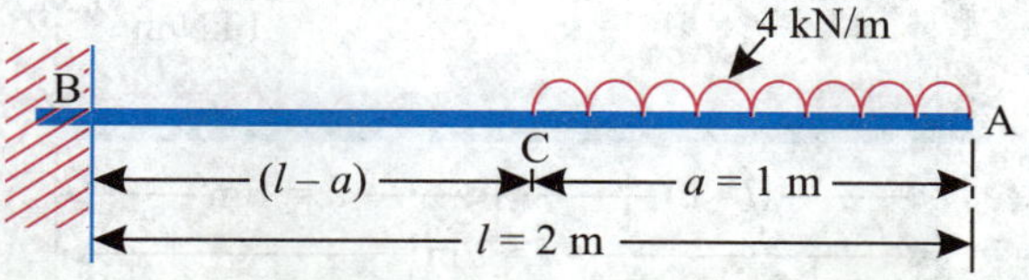

Fig. 8.52

Given: Length, $l = 2$ m
Loaded length, $a = 1$ m
Width, $b = 100$ mm $= 0.1$ m
Depth, $d = 200$ mm $= 0.2$ m
∴ Moment of inertia,

$$I = \frac{bd^3}{12} = \frac{0.1 \times 0.2^3}{12} = 6.666 \times 10^{-5} \text{ m}^4$$

U.D.L., $w = 4$ kN/m.
Young's modulus,

$$E = 2.1 \times 10^8 \text{ kN/m}^2$$

Slope at the free end A, θ_A :

Using the relation,

$$\theta_A = \left[\frac{(wl)\,l^2}{6\,EI}\right] - \left[\frac{\{w(l-a)\}\,(l-a)^2}{6\,EI}\right], \text{ we get}$$

$$\theta_A = \frac{1}{6\,EI}\left[(wl) \times l^2 - \{w(l-a)\}(l-a)^2\right]$$

$$= \frac{1}{6 \times 2.1 \times 10^8 \times 6.666 \times 10^{-5}} \times [(4 \times 2) \times 2^2 - 4(2-1) \times (2-1)^2]$$

$$= \frac{1}{6 \times 2.1 \times 10^8 \times 6.666 \times 10^{-5}}\,(32 - 4)$$

or $\theta_A = \mathbf{0.000333\ radian}$ **(Ans.)**

Deflection at the free end B, y_B :

Using the relation:

$$y_B = \left[\frac{(wl)\,l^3}{8\,EI}\right] - \left[\frac{\{w(l-a)\}\,(l-a)^3}{8\,EI} + \frac{\{w(l-a)\}\,(l-a)^2 \times a}{6\,EI}\right]$$

$$= \frac{(4 \times 2) \times 2^3}{8 \times 2.1 \times 10^8 \times 6.666 \times 10^{-5}} - \left[\frac{\{4(2-1)\}\,(2-1)^3}{8 \times 2.1 \times 10^8 \times 6.666 \times 10^{-5}}\right.$$

$$\left. + \frac{\{4(2-1)\}(2-1)^2 \times 1}{6 \times 2.1 \times 10^8 \times 6.666 \times 10^{-5}}\right]$$

$$= 0.0005714 - (0.00003571 + 0.00004762)$$

or, $y_B = 0.000488$ m $= \mathbf{0.488\ mm}$ **(Ans.)**

Example 8.34. *A cantilever 2 metres long is loaded a shown in Fig. 8.53. Calculate the deflection at the free end if the section is rectangular, 100 mm × 200 mm.*

Take : $E = 1.05 \times 10^7$ kN / m²

Solution. Refer to Fig. 8.53.

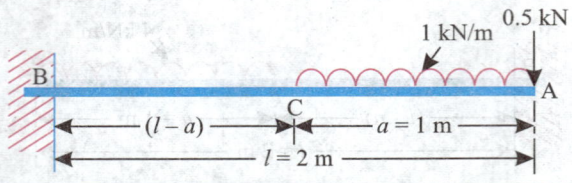

Fig. 8.53

Given: Length, $l = 2$ m

Length AC, $a = 1$m

Width, $b = 100$ mm $= 0.1$ m

Depth, $d = 200$ mm $= 0.2$ m

∴. Moment of inertia,

$$I = \frac{bd^3}{12} = \frac{0.1 \times 0.2^3}{12} = 6.666 \times 10^{-5} \text{ m}^4$$

U.D.L., $w = 1$ kN/m

Point load, $W_1 = 0.5$ kN

Young's modulus, $E = 1.05 \times 10^7$ kN/m²

Deflection at the free end A, y_A :

y_A = Deflection due to U.D.L. + deflection due to point load

$$= \left[\frac{(wl)\, l^3}{8\, EI}\right] - \left[\frac{(w(l-a)(l-a)^3}{8\, EI} + \frac{\{w(l-a)\}(l-a)^2 \times a}{6\, EI}\right] + \frac{W_1\, l^3}{3\, EI}$$

$$= \left[\frac{(1 \times 2) \times 2^3}{8 \times 1.05 \times 10^7 \times 6.666 \times 10^{-5}}\right] - \left[\frac{1(2-1) \times (2-1)^3}{8 \times 1.05 \times 10^7 \times 6.666 \times 10^{-5}}\right.$$

$$\left. + \frac{1(2-1)(2-1)^2 \times 1}{6 \times 1.05 \times 10^7 \times 6.666 \times 10^{-5}}\right] + \frac{0.5 \times 2^3}{3 \times 1.05 \times 10^7 \times 6.666 \times 10^{-5}}$$

$$= 0.002857 - (0.0001785 + 0.0002381) + 0.0019049$$

or, $y_A = 0.004345$ m $= \textbf{4.345 mm}$ **(Ans.)**

Output shaft and gears.

Example 8.35. *A cantilever beam of 4m span carries a U.D.L. of 3 kN/m over its entire span and a point load of 3 kN at free end. If the same beam is simply supported at two ends, what point load at the centre should it carry to have same deflection as the cantilever ?*

Solution. *Given:*

Length, $l = 4\text{m}$

U.D.L., $w = 3\text{ kN/m}$

Total U.D.L., $W = w \times l = 3 \times 4 = 12\text{ kN}$

Point load, $W_1 = 3\text{ kN}$

1st case: *Cantilever* :

Maximum deflection at the free end,

$$y_{max} = \text{Deflection due to } W + \text{deflection due to } W_1$$

$$= \frac{Wl^3}{8\,EI} + \frac{W_1 l^3}{3\,EI} = \frac{l^3}{EI}\left[\frac{W}{8} + \frac{W_1}{3}\right]$$

$$= \frac{l^3}{EI}\left[\frac{12}{8} + \frac{3}{3}\right] = \frac{2.5 l^3}{EI} \qquad \text{...}(i)$$

2nd Case : *Simply supported beam*:

Let W' be the central point load on the simply supported beam which produces same deflection as in the case of cantilever.

Then, $\qquad y'_{max} = \dfrac{W' l^3}{48\,EI}$ \qquad \text{...}(ii)

Equating (*i*) and (*ii*), we get

$$y_{max} = y'\text{max} \quad \text{(given)}$$

or, $\qquad \dfrac{2.5 l^3}{EI} = \dfrac{W' l^3}{48\,EI}$

From which $\qquad W' = 2.5 \times 48 = 120\text{ kN}$

Thus point load at the centre $W' = $ **120 kN** (**Ans.**)

Example 8.36. *A steel tube cantilever 4m long has outside diameter 120 mm and thickness 10 mm. It carries a uniformly distributed load w kN/m for 3m from the fixed end. Find w and deflection at the free end if the maximum stress due to bending is 7×10^4 kN / m².*

Take : $E = 2.1 \times 10^8$ kN/m².

Solution. Refer Fig. 8.54.

Given: Length, $l = 4\text{ m}$

Loaded length, $a = 3\text{m}$

Outside diameter, $D = 120\text{ mm} = 0.12\text{ m}$

Thickness, $t = 10\text{ mm} = 0.01\text{ m}$

∴ Inside diameter $d = D - 2t = 120 - 2 \times 10 = 100\text{ mm} = 0.1\text{ m}$

Moment of inertia, $= \dfrac{\pi}{64}(D^4 - d^4) = \dfrac{\pi}{64}[(0.12)^4 - (0.1)^4] = 5.27 \times 10^{-6}\text{ m}^4$

U.D.L. $= w\text{ kN/m}$

Total U.D.L., $W = w \times 3 = 3\,w\text{kN}$

Bending stress, $\sigma_b = 7 \times 10^4\text{ kN/m}^2$

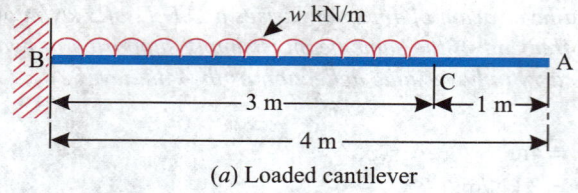

(a) Loaded cantilever

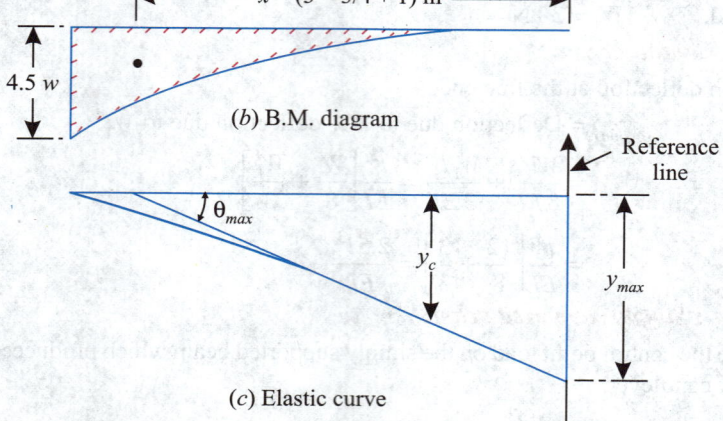

(b) B.M. diagram

(c) Elastic curve

Fig. 8.54

U.D.L., w:

Bending moment at the fixed end

$$M = \frac{W \times 3}{2} = \frac{3w \times 3}{2} = 4.5 \, w \, \text{kNm}$$

Using equation, $\dfrac{M}{I} = \dfrac{\sigma_b}{y}$, We get,

$$M = \frac{\sigma_b}{y} \cdot I$$

$$= \frac{7 \times 10^4 \times 5.27 \times 10^{-6}}{0.06} = 6.148 \, \text{kNm} \qquad \left(\because y = \frac{0.12}{2} = 0.06 \, \text{m} \right)$$

Equating (i) and (ii), we get,

$$4.5 \, w = 6.148$$

From which, $\qquad w = \dfrac{6.148}{4.5}$ or **w = 1.36 kN/m** **(Ans.)**

SIMPLY SUPPORTED BEAMS

Example 8.37. *A wooden beam 150 mm wide and 250 mm deep has a span of 4 metres. Determine the load, that can be placed at its centre to cause the beam a deflection of 12 mm. Take E = 6 × 10⁶ kN/m² .*

Also find the maximum slope.

Solution. *Given*: Span, $l = 4 \text{m}$

Deflection, $\qquad y_{max} = 12 \text{ mm} = 0.012 \text{ m}$

Width of the section, $\quad b = 150 \text{ mm} = 0.15 \text{ m}$

Depth of the section, $d = 250$ mm $= 0.25$ m

Moment of inertia, $I = \dfrac{bd^3}{12} = \dfrac{0.15 \times 0.25^3}{12} = 1.953 \times 10^{-4}$ m^4

Central load, W :

Using the relation, $y_{max} = \dfrac{Wl^3}{48\,EI}$, with usual notations, we get,

$$0.012 = \dfrac{W \times 4^3}{48 \times 6 \times 10^6 \times 1.953 \times 10^{-4}}$$

or, $\qquad W = \dfrac{0.012 \times 48 \times 6 \times 10^6 \times 1.953 \times 10^{-4}}{4^3}$

or, $\qquad W = \mathbf{10.55\ kN}$ **(Ans.)**

Maximum slope, θ_{max} :

Using the relation :

$$\theta_{max} = \dfrac{Wl^2}{16\,EI} = \dfrac{10.55 \times 4^2}{16 \times 6 \times 10^6 \times 1.953 \times 10^{-4}}$$

or, $\qquad \boldsymbol{\theta_{max} = 0.009}$ **radian (Ans.)**

Example 8.38. *A timber beam of rectangular section 10 cm wide and 25 cm deep is simply supported over a span of 4m. What uniformly distributed load in kN/m should the beam carry to produce a central deflection of 0.6 cm? Calculate the slope also.*

Take : E = 11 GN/m^2.

Solution. *Given:* Length of span,

$$l = 4 \text{ m}$$

Central deflection, $y_{max} = 0.6$ cm $= 0.6 \times 10^{-2}$ m

Width of section, $\qquad b = 10$ cm

Depth of the section, $\quad d = 25$ cm

Moment of inertia, $\qquad I = \dfrac{bd^3}{12} = \dfrac{10 \times 25^3}{12}$

$$= 13020.8 \text{ cm}^4 = 13020.8 \times 10^{-8} \text{ m}^4$$

Young's modulus, $\qquad E = 11$ GN/m^2

U.D.L., w in kN/m:

Using the relation : $\quad y_{max} = \dfrac{5Wl^3}{384\,EI}$, we get,

[where, $W = wl\ (= w \times 4)$ kN]

$$0.6 \times 10^{-2} = \dfrac{5(w \times 4 \times 1000) \times (4)^3}{384 \times 11 \times 10^9 \times 13020.8 \times 10^{-8}}$$

or, $\qquad w = \dfrac{0.6 \times 10^{-2} \times 384 \times 11 \times 10^9 \times 13020.8 \times 10^{-8}}{5 \times 4 \times 1000 \times (4)^3}$

or, $\qquad w = \mathbf{2.578\ kN/m}$ **(Ans.)**

Maximum slope, θ_{max} :

Using the relation : $\theta_{max} = \dfrac{Wl^2}{24\,EI}$, we get,

$$\theta_{max} = \frac{(2.578 \times 1000 \times 4) \times 4^2}{24 \times 11 \times 10^9 \times 13020.8 \times 10^{-8}}$$

or, $\theta_{max} = 0.004799$ **radian   (Ans.)**

Example 8.39. *A simply supported beam of I-section, 4m long, carries a total uniform load of 40 kN and a concentrated load of 70 kN at mid span.*

(i) *Find the maximum deflection of the beam.*

(ii) *If permissible deflection is limited to $\dfrac{1}{360}$ of the span, is this beam acceptable based on deflection ?*

(iii) *Find slope at the ends.*

Take : $E = 2.1 \times 10^8 \ kN/m^2$;
$I = 8.98 \times 10^{-5} \ m^4.$

Solution. Refer Fig. 8.55.

Given. Length of span,

$$l = 4 \text{ m}$$

Total uniform load,

$$W = 40 \text{ kN}$$

Mid span load, $W_1 = 70 \text{ kN}$

Young's modulus, $E = 2.1 \times 10^8 \text{ kN/m}^2$

Moment of inertia, $I = 8.98 \times 10^{-5} \text{ m}^4$

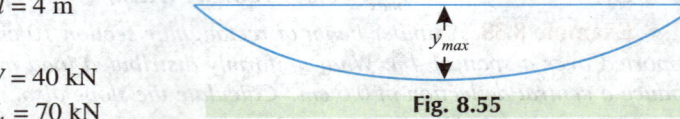

Fig. 8.55

(i) **Maximum deflection of the beam, y_{max}:**

We know that, y_{max} = Deflection due to load W + deflection due to W_1

$$= \frac{5W\,l^3}{384\,EI} + \frac{W_1\,l^3}{48\,EI} = \frac{l^3}{EI}\left(\frac{5}{384}W + \frac{W_1}{48}\right)$$

Shaft as shown above behaves like a beam with different loads acting at different pulleys.

$$= \frac{4^3}{2.1 \times 10^8 \times 8.98 \times 10^{-5}} \times \left(\frac{5}{384} \times 40 + \frac{70}{48} \right)$$

$$y_{max} = 0.006717 \text{ m} = \textbf{6.717 mm} \quad \textbf{(Ans.)}$$

(ii) Is the beam acceptable?

The maximum permissible deflection is

$$y = \frac{l}{360} = \frac{4}{360} = 0.01111 \text{m} = 11.11 \text{mm}$$

Since actual y_{max} = 6.717 mm, *the beam is acceptable, based on deflection.* **(Ans.)**

(iii) Slope at the end, θ_{max} :

We know that $\qquad \theta_{max}$ = Slope due to W + slope due to W_1

$$= \frac{Wl^2}{24 EI} + \frac{W_1 l^2}{16 EI} = \frac{l^2}{EI} \left(\frac{W}{24} + \frac{W_1}{16} \right)$$

$$= \frac{4^2}{2.1 \times 10^8 \times 8.98 \times 10^{-5}} \times \left(\frac{40}{24} + \frac{70}{16} \right)$$

or, $\qquad \theta_{max} = \textbf{0.005126 radian} \quad \textbf{(Ans.)}$

Example 8.40. *A beam of 4m span is carrying a point load of 40 kN at a distance of 3m from the left end. Calculate the slope at the two supports and deflection under the load. Also calculate maximum deflection.*

Take : $\qquad\qquad EI = 2.6 \times 10^7 \text{ N-m}^2.$

Solution. Refer to Fig. 8·56.

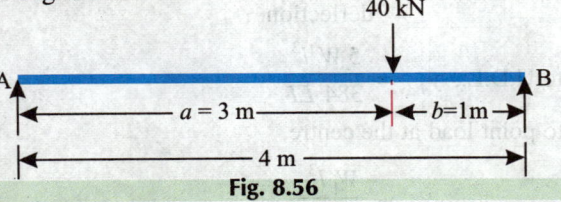

Fig. 8.56

Given: Length of span, $\qquad l = 4$ m

Magnitude of point load, $\quad W = 40$ kN

Distance between load and left end,

$$a = 3 \text{ m}$$

∴ Distance $\qquad b = 4 - 3 = 1$ m

Value of $EI = 2·6 \times 10^7$ N-m²

Slope at A, θ_A:

Using the relation: $\qquad \theta_A = \frac{Wb}{6 EI l} (l^2 - b^2)$, with usual notations, we get,

$$\theta_A = \frac{40 \times 1000 \times 1}{6 \times 2·6 \times 10^7 \times 4} (4^2 - 1^2)$$

or, $\qquad \theta_A = \textbf{0·0009615 radian} \quad \textbf{(Ans.)}$

Slope at B, θ_B :

Using the relation $\qquad \theta_B = \frac{Wa}{6 EI l} (l^2 - a^2)$, we get,

$$\theta_B = \frac{(40 \times 1000) \times 3}{6 \times 2·6 \times 10^7 \times 4} (4^2 - 3^2)$$

or, $\qquad\qquad\qquad\qquad \theta_B = 0\cdot001346$ radian   (Ans.)

Deflection under the load, y_C :

Using the relation, $\qquad\qquad y_C = \dfrac{Wab}{6\ EI\ l}\ (l^2 - a^2 - b^2)$, we get

$$y_C = \frac{(40 \times 1000) \times 3 \times 1}{6 \times 2\cdot6 \times 10^7 \times 4} \times (4^2 - 3^2 - 1^2)$$

or, $\qquad\qquad\qquad\qquad y_C = 0\cdot001154$ m $= 1\cdot154$ mm   (Ans.)

Maximum deflection, y_{max} :

Using the relation, $\qquad\qquad y_{max} = \dfrac{Wa}{9\sqrt{3}\ EI\ l}\ (l^2 - a^2)^{3/2}$, we get,

$$y_{max} = \frac{(40 \times 1000) \times 3}{9\sqrt{3} \times 2\cdot6 \times 10^7 \times 4} \times (4^2 - 3^2)^{3/2}$$

or, $\qquad\qquad\qquad\qquad y_{max} = 0\cdot00137$ m $= 1\cdot37$ mm   (Ans.)

Example 8·41. *A rectangular beam 150 mm (wide) × 240 mm (deep) is simply supported at the ends on a span of 4 m and carries a uniformly distributed load of 4 kN / m on whole span. What point load at the centre should it carry so that maximum deflection is doubled ?*

Solution. *Given:* Length, $\qquad l = 4$ m

U.D.L., $\qquad\qquad\qquad\qquad w = 4$ kN / m

Total U.D.L., $\qquad\qquad\qquad W = w \times l = 4 \times 4 = 16$ kN

Cross-section of the beam $\qquad = 150$ mm $\times 250$ mm $= 0\cdot15$ m $\times 0\cdot25$ m

Let, $\qquad\qquad\qquad\qquad W_1 =$ Point load at the centre which will produce double the deflection.

Deflection due to U.D.L., $\ y_{max} = \dfrac{5\ W\ l^3}{384\ EI}$

Deflection due to point load at the centre,

$$y'_{max} = \frac{W_1\ l^3}{48\ EI}$$

But, $\qquad\qquad\qquad\qquad y'_{max} = 2\ y_{max}$ $\qquad\qquad\qquad\qquad\qquad$ ...(Given)

$$\frac{W_1\ l^3}{48\ EI} = 2 \times \frac{5\ W\ l^3}{384\ EI}$$

∴ $\qquad\qquad\qquad \dfrac{W_1\ l^3}{48\ EI} = 2 \times \dfrac{5 \times 16 \times l^3}{384\ EI}$

$$W_1 = 2 \times \frac{5 \times 16 \times 48}{384} = 20 \text{ kN}$$

*Hence, the point load at the centre = **20 kN**   (Ans.)*

Example 8·42. *A simply supported beam of 5 m span has rectangular section 150 mm wide and 250 mm deep. The beam carries a point load of 10 kN at its centre and the deflection at the centre is 12·5 mm. Neglecting the self-weight of the beam calculate:*

(i) Young's modulus;

(ii) Slope at supports.

Solution. *Given:* Length, $\qquad l = 5$ m

Width, $\qquad$ $b = 150$ mm $= 0.15$ m

Depth, $\qquad$ $d = 250$ mm $= 0.25$ m

$\therefore$ Moment of inerti $I = \dfrac{bd^3}{12}$

$$= \dfrac{0.15 \times 0.25^3}{12} = 1.953 \times 10^{-4} \text{ m}^4$$

Point load, $\qquad$ $W_1 = 10$ kN

Deflection at the centre,

$\qquad$ $y_{max} = 12.5$ mm $= 0.0125$ m

(i) Young's modulus, E :

Using the relatio $y_{max} = \dfrac{W \, l^3}{48 \, EI}$, we get

$$0.0125 = \dfrac{10 \times 5^3}{48 \times E \times 1.953 \times 10^{-4}}$$

or, $\qquad$ $E = \dfrac{10 \times 5^3}{0.0125 \times 48 \times 1.953 \times 10^{-4}}$

or, $\qquad$ $E = 1.06 \times 10^8 \text{ kN/m}^2$ **(Ans.)**

(ii) Slope at each support, θ :

Using the relation,

$$\theta = \dfrac{Wl^2}{16 \, EI}, \text{ we get}$$

$$\theta = \dfrac{10 \times 5^2}{16 \times 1.06 \times 10^8 \times 1.953 \times 10^{-4}}$$

or, $\qquad$ $\theta = 0.000754$ radian **(Ans.)**

Example 8·43. *A simply supported beam 5 m long carries concentrated loads of 10 kN each at points 1 m from the ends.*

Calculate:

(i) Maximum slope and deflection of the beam, and

(ii) Slope and deflection under each load.

Take : EI = 1·2 × 10⁴ kNm².

Solution. Refer to Fig. 8·57.

Given: Length, $\qquad$ $l = 5$ m

Two concentrated loads: 10 kN each

$\qquad$ $EI = 1.2 \times 10^4 \text{ kNm}^2$

(i) Maximum slope, θ_{max} :

Using the relation,

$$\theta_{max} = \dfrac{A}{EI}, \text{ we get} \qquad (\text{where, } A = \text{area of B.M.D. between } E \text{ and } B)$$

$$\theta_{max} = \dfrac{1}{EI} (\text{area } klm + \text{area } lpnm) = \dfrac{1}{1.2 \times 10^4} \left[\dfrac{1}{2} \times 1 \times 10 + 10 \times 1.5 \right]$$

or, $\qquad$ $\theta_{max} = 0.001666$ radian **(Ans.)**

Maximum deflection, y_{max} :

Using the relation :

$$y_{max} = \frac{A\bar{x}}{EI} = \frac{1}{EI} \text{ (Moment of area } klm + \text{ moment of area } lpnm)$$

or,

$$y_{max} = \frac{1}{1 \cdot 2 \times 10^4} \times \left[\left\{ \frac{1}{2} \times 1 \times 10 \right\} \times \frac{2}{3} + (10 \times 1 \cdot 5) \left\{ 1 + \frac{3}{4} \right\} \right]$$

$$= \frac{1}{1 \cdot 2 \times 10^4} [3 \cdot 33 + 26 \cdot 25] = 0 \cdot 002465 \text{ m}$$

or,

$$y_{max} = 2 \cdot 465 \text{ mm} \quad \text{(Ans.)}$$

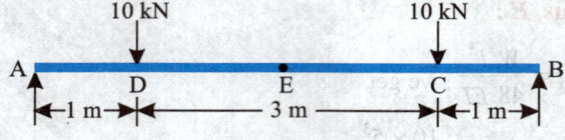

(a) Loaded beam

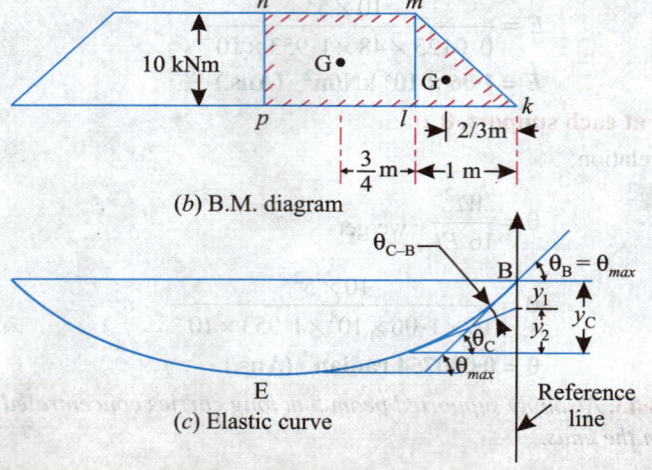

(b) B.M. diagram

(c) Elastic curve

Fig. 8.57

(ii) Slope under each load θ_C, θ_D :

As shown in Fig. 8·57 (c)

θ_C = Slope at C.

θ_{C-B} = The angle included between the tangents drawn at C and B facing reference line

$$= \frac{A}{EI}$$

(where, A = area of B.M.D. between C and B) = $\dfrac{1}{1 \cdot 2 \times 10^4} \left(\dfrac{1}{2} \times 1 \times 10 \right)$

or,

$$\theta_{C-B} = 0 \cdot 0004166 \text{ rad.}$$

$$\theta_C = \theta_B - \theta_{C-B}$$

But,

$$\theta_B = \theta_{max} = 0 \cdot 001666$$

∴ $\theta_C = 0.001666 - 0.0004166$

or $\theta_C = 0.001249$ **radian** **(Ans.)** $(\theta_D = \theta_C \; ...$ by symmetry$)$

Deflection under each load y_C, y_D :

As is clear from Fig. 8·57 (c),

y_C = The deflection under the load at C

    $= y_1 + y_2$

where, y_1 = The distance intercepted on the reference line between the tangents drawn at C and B.

$$= \frac{1}{EI} \times (\text{Moment of area of B.M.D. between } C \text{ and } B \text{ about the reference line})$$

$$= \frac{1}{1 \cdot 2 \times 10^4} \left(\frac{1}{2} \times 1 \times 10 \times 2/3 \right) = 0.000277 \text{ m} = 0.277 \text{ mm}$$

$y_2 = \theta_C \times 1$ $\left[\begin{array}{l} \theta_c \text{ being very small, the distance } y_2 \text{ is taken} \\ \text{as an arc of radius 1 m with } C \text{ as its centre.} \end{array} \right]$

     $= 0.001249 \times 1 = 0.001249 \text{ m} = 1.249 \text{ mm}$

∴ $y_C = 0.277 + 1.249$

or, $y_C = 1.526$ **mm** **(Ans.)** $(y_D = y_C \; ....$ by symmetry$)$

8·9. CONJUGATE BEAM METHOD

— *Conjugate method (or Funicular analogy or method of elastic weights) is a special case of the moment area method.*

— The methods like moment area method, Macaulay's method, discussed earlier are convenient for cases where the *beam is of uniform flexural rigidity*. In cases where the flexural rigidity is not uniform these methods become laborious and the *conjugate beam method* presents a very easy approach for solution of the problems.

— When the deflection equation is differentiated successively, we get the relations as given below:

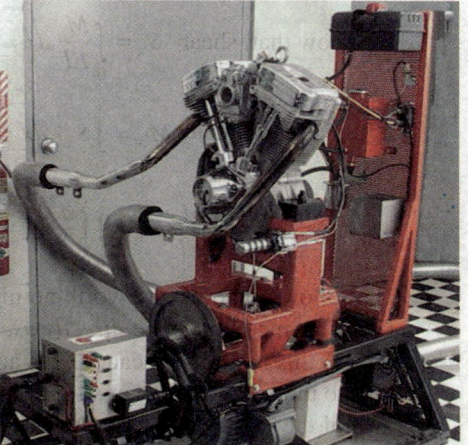

A twin-cylinder motorcycle engine is being tested on a dynamometer. Dynamometers measure the power of engines.

$$EI.y = Deflection \qquad \qquad ...(i)$$

$$EI \cdot \frac{dy}{dx} = Slope \qquad \qquad ...(ii)$$

$$EI \frac{d^2 y}{dx^2} = Moment = M \qquad \qquad ...(iii)$$

$$EI \cdot \frac{d^3 y}{dx^3} = Shear = S = \frac{dM}{dx} \qquad \qquad ...(iv)$$

$$EI \cdot \frac{d^4 y}{dx^4} = Load = \frac{dS}{dx} = \frac{d^2 M}{dx^2} \qquad \qquad ...(v)$$

From above we find that a similarity of relation exists among load, shear and moment [eqns. (v), (iv) and (iii)] and among moment, slope and deflection [eqns. (iii), (ii) and (i)]. This gives an idea of a method by which slope and deflection can be calculated by using the usual methods of calculating shear and bending moments in a beam. In this method, the beam (known as *conjugate beam*) is *loaded with not the actual loads, but with the elastic weight* $\dfrac{M}{EI}$ *corresponding to the actual load.*

Conjugate Beam Theorem I :

"The slope at any section of a loaded beam relative to the original axis of the beam, is equal to the shear in the conjugate beam at the corresponding section."

We know that, $\qquad$ load $= w = \dfrac{M}{EI}$

∴ $\qquad$ Shear $= S_x = \displaystyle\int_0^x w \cdot dx = \int_0^x \dfrac{M}{EI}\, dx$

But, $\qquad \displaystyle\int_0^x \dfrac{M}{EI}\, dx = \int_0^x \dfrac{d^2 y}{dx^2} = \dfrac{dy}{dx} = slope$

Conjugate Beam Theorem II :

"The deflection at any given section of a loaded beam, relative to the original position, is equal to the bending moment at the corresponding section of the conjugate beam."

We know that, shear $\quad S_x = \displaystyle\int_0^x \dfrac{M}{EI}\, dx$

∴ Bending moment, $M_x = \displaystyle\int_0^x S_x \cdot dx = \int_0^x \int_0^x \dfrac{M}{EI}\, dx$

But, $\qquad \displaystyle\int_0^x \int_0^x \dfrac{M}{EI}\, dx = \int_0^x \int_0^x \dfrac{d^2 y}{dx^2} = \int_0^x \dfrac{dy}{dx} = y = deflection$ $\qquad$ ...Proved

The following points are worth noting for the **conjugate beam method:**

(*i*) This method can be *directly used only for simply supported beams.*

(*ii*) In this method for cantilevers and fixed beams, artificial constraints need to be applied to the conjugate beam so that it is supported in a manner consistent with the constraints of the real beam.

The conjugate beam method is illustrated in the following examples.

SIMPLY SUPPORTED BEAMS

Example 8·44. *A simply supported beam is carrying a load W at the centre. Calculate the slopes at its ends and the central deflection, using conjugate beam method.*

Solution. Fig. 8·58 (*a*) shows the real beam; B.M. diagram for the real beam is shown in the Fig. 8·58 (*b*)

The $\dfrac{M}{EI}$ diagram of the real beam becomes the elastic weight or loading for the conjugate beam. The conjugate beam $A'C'B'$ (corresponding to the real beam ACB) with the loading is shown in Fig. 8·58 (*c*).

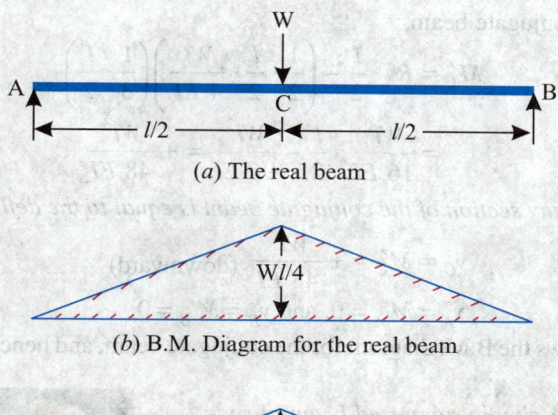

(a) The real beam

(b) B.M. Diagram for the real beam

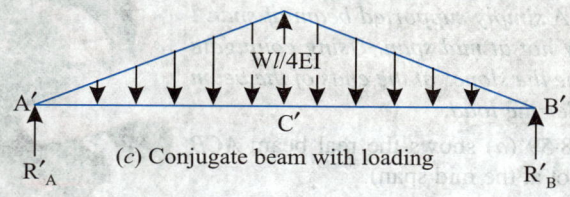

(c) Conjugate beam with loading

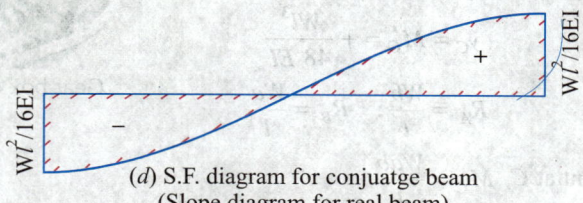

(d) S.F. diagram for conjuatge beam
(Slope diagram for real beam)

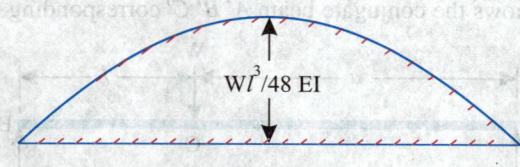

(e) B.M. diagram for conjugate beam
(Deflection diagram for real beam)

Fig. 8.58

For the conjugate beam:

$$R'_A = \frac{1}{2}\left(\frac{1}{2}l \cdot \frac{Wl}{4EI}\right) = \frac{Wl^2}{16\ EI} \qquad \therefore\ S'_A = -R'_A = -\frac{Wl^2}{16\ EI}$$

But *shear at any section of the conjugate beam is equal to the slope of the real beam.*

Hence,
$$\theta_A = \text{Slope at the end } A \text{ of real beam} = -\frac{Wl^2}{16\ EI}$$

Similarly,
$$\theta_B = S'_B = R'_B = +\frac{Wl^2}{16\ EI}$$

Fig. 8·58 *(d)* shows the S.F. diagram for the conjugate beam and hence the slope diagram for the real diagram. Since the S.F. is zero at C the slope θ_C at the centre of the real beam is zero.

Again, for the conjugate beam,

$$M'_C = R'_A \cdot \frac{l}{2} - \left(\frac{1}{2} \cdot \frac{l}{2} \cdot \frac{Wl}{4\,EI} \right) \left(\frac{1}{3} \cdot \frac{l}{2} \right)$$

$$= \frac{Wl^2}{16\,EI} \cdot \frac{l}{2} - \frac{Wl^3}{96\,EI} = + \frac{Wl^3}{48\,EI}$$

But the *B.M. at any section of the conjugate beam is equal to the deflection of the real beam.*

Hence, $\qquad y_C = M'_C = + \dfrac{Wl^3}{48\,EI}$ (downward)

Also, $\qquad y_A = M'_A = 0$, and $y_B = M'_B = 0$

Fig. 8·57 (*e*) shows the B.M. diagram for the conjugate beam, and hence the deflection diagram for the real beam.

Example 8·45. *A simply supported beam of span l carries a point load W not at mid span. Using conjugate beam method determine the slopes at the ends of the beam and the deflection under the load.*

Gear box of a motorcycle.

Solution. Fig. 8·59 (*a*) shows the real beam *ACB* with a point load *W* (not at the mid span).

Support reactions at *A* and *B* are :

$$y_C = M'_C = + \frac{Wl^3}{48\,EI}$$

$$R_A = \frac{Wb}{l}; \qquad R_B = \frac{Wa}{l}$$

Bending moment at *C*, $M_C = \dfrac{Wab}{l}$

— B.M. diagram for the real beam is shown is Fig. 8·59 (*b*)

— Fig. 8·59 (*c*) shows the conjugate beam *A′ B′ C′* corresponding to real beam *ACB*.

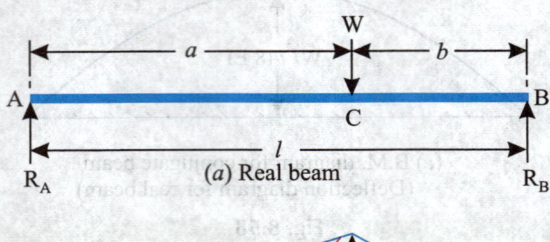

(*a*) Real beam

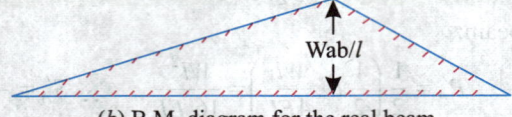

(*b*) B.M. diagram for the real beam

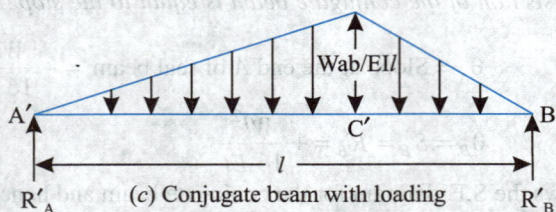

(*c*) Conjugate beam with loading

Fig. 8.59

To find reaction R'_B for the conjugate beam taking moments about A', we have

$$R'_B \times l = \frac{1}{2} \times a \times \frac{Wab}{EIl} \times \frac{2}{3} a + \frac{1}{2} \times b \times \frac{Wab}{EIl} \left(a + \frac{b}{3} \right)$$

$$= \frac{Wa^3 b}{3\,EIl} + \frac{Wab^2}{6\,EIl} (3a + b)$$

$$\therefore \quad R'_B = \frac{Wa^3 b}{3\,EIl^2} + \frac{Wab^2}{6\,EIl^2} (3a + b)$$

$$= \frac{Wab}{6\,EIl^2} (2a^2 + 3ab + b^2) \qquad \qquad \text{...(i)}$$

$$= \frac{Wab}{6\,EIl^2} [2a^2 + 3a\,(l - a) + (l - a)^2]$$

$$= \frac{Wab}{6\,EIl^2} [2a^2 + 3al - 3\,a^2 + l^2 + a^2 - 2al]$$

$$= \frac{Wab}{6\,EIl^2} (l^2 + al) = \frac{Wab}{6\,EIl} (l + a)$$

i.e., $$R'_B = \frac{Wab}{6\,EIl} (l + a)$$

Also, $$R'_A + R'_B = \frac{1}{2} \times l \times \frac{Wab}{EIl} = \frac{Wab}{2\,EI}$$

$$\therefore \quad R'_A = \frac{Wab}{2\,EI} - \frac{Wab}{6\,EIl} (l + a) = \frac{Wab}{6\,EIl} [3l - (l + a)]$$

$$= \frac{Wab}{6\,EIl} [2l - a] = \frac{Wab}{6\,EIl} (l + l - a)$$

$$= \frac{Wab}{6\,EIl} (l + b) \qquad \qquad (\because \; l = a + b)$$

Hence slope at A, $$\theta_A = \frac{Wab}{6\,EIl} (l + b) \text{ (Ans.)}$$

and, slope at B, $$\theta_B = \frac{Wab}{6\,EIl} (l + a) \text{ (Ans.)}$$

Deflection at C, $y_C = $ B.M. at C' for the conjugate beam

$$= R'_B \times b - \frac{1}{2} \times b \times \frac{Wab}{EIl} \times \frac{b}{3} = \frac{Wab}{6\,EIl} (l + a) \times b - \frac{Wab^3}{6\,EIl}$$

$$= \frac{Wab^2}{6\,EIl} (l + a - b) = \frac{Wab^2}{6\,EIl} [l + a - (l - a)]$$

$$= \frac{Wab^2}{6\,EIl} \times 2a = \frac{Wa^2 b^2}{3\,EIl}$$

Hence, $$y_C = \frac{Wa^2 b^2}{3\,EIl} \text{ (Ans.)}$$

Example 8·46. *For the beam shown in the Fig. 8·60 (a) determine the following:*

(i)   *Slope at end A;*

(ii) *Deflection at the mid span;*

(iii) *Maximum deflection.*

Given : $I = 8 \times 10^{-5}$ m⁴ *and* $E = 200 \times 10^{6}$ kN/m².

Solution. *Given:* $I = 8 \times 10^{-5}$ m⁴, and $E = 200 \times 10^{6}$ kN/m²

Refer to Fig. 8·60 (a).

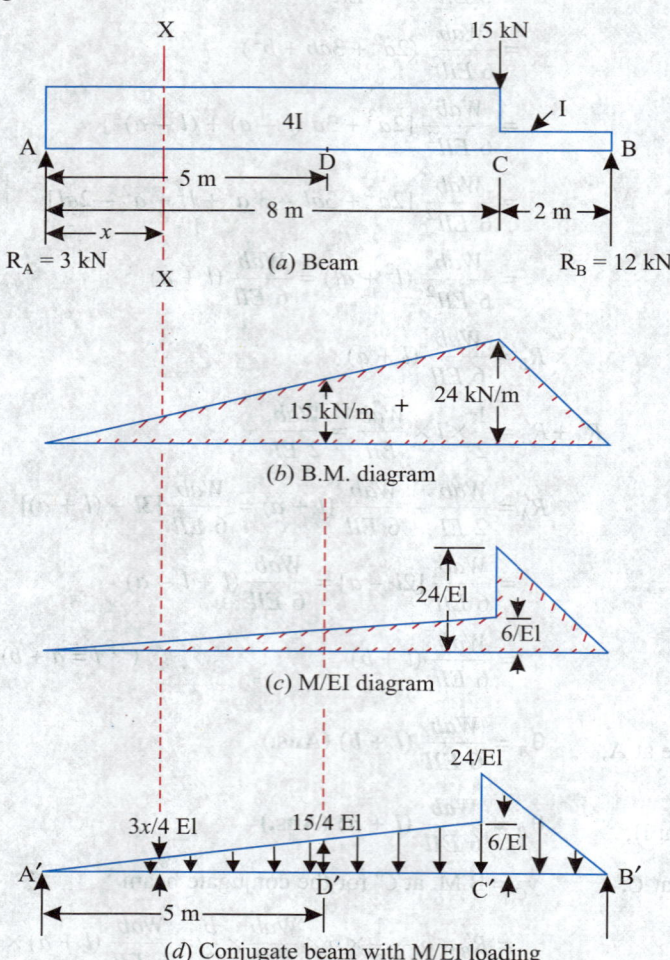

(a) Beam

(b) B.M. diagram

(c) M/EI diagram

(d) Conjugate beam with M/EI loading

Fig. 8.60

To find reaction at *B* taking moments about *A*, we get

$$R_B \times 10 = 15 \times 8 = 120 \qquad \therefore \qquad R_B = 12 \text{ kN}$$

Also, $\qquad\qquad R_A + R_B = 15 \text{ kN} \qquad \therefore \qquad R_A = 15 - 12 = 3 \text{ kN}$

∴ B.M. at *C*, $\qquad M_C = 12 \times 2 = 24 \text{ kNm}$

B.M. diagram for the given beam is shown in the Fig. 8·60 (b)

Fig. 8·60 (c) shows the conjugate beam *A'B'C'* with loading.

Mechanism of an oil supply system.

To find the reaction R'_A taking moments about the point B', we get

$$R'_A \times 10 = \left[\left(\frac{1}{2} \times 8 \times \frac{6}{EI} \right) \times \left(\frac{8}{3} + 2 \right) \right] + \left[\left(\frac{1}{2} \times 2 \times \frac{24}{EI} \right) \times \frac{2}{3} \times 2 \right]$$

$$= \left(\frac{24}{EI} \times \frac{14}{3} \right) + \left(\frac{24}{EI} \times \frac{4}{3} \right) = \frac{112}{EI} + \frac{32}{EI} = \frac{144}{EI}$$

$\therefore$ $\qquad\qquad R'_A = \dfrac{72}{5\,EI}$

Also, $\qquad R'_A + R'_B = \dfrac{1}{2} \times 8 \times \dfrac{6}{EI} + \dfrac{1}{2} \times 2 \times \dfrac{24}{EI} = \dfrac{24}{EI} + \dfrac{24}{EI} = \dfrac{48}{EI}$

$\therefore$ $\qquad\qquad R'_B = \dfrac{48}{EI} - \dfrac{72}{5\,EI} = \dfrac{168}{5\,EI}$

(i) Slope at the end A, θ_A :

$$\theta_A = \text{S.F. at } A' = -R'_A$$

$$= -\frac{72}{5\,EI} = -\frac{72}{5 \times 200 \times 10^6 \times 8 \times 10^{-5}} = -0.0009 \text{ radian}$$

Hence, $\qquad\qquad \theta_A = -\textbf{0.0009 radian}$ **(Ans.)**

(ii) Deflection at the mid span, y_D :

B.M. at the point D of the real beam,

$$M_D = 3 \times 5 = 15 \text{ kNm} \qquad \therefore \quad \left(\frac{M}{EI} \right)_D = \frac{15}{(4I)E} = \frac{15}{4\,EI}$$

Hence, $\qquad\qquad y_D = M'_D$ for the conjugate beam

$$= R'_A \times 5 - \frac{1}{2} \times 5 \times \frac{15}{4\,EI} \times \frac{5}{3} = \frac{72}{5\,EI} \times 5 - \frac{125}{8\,EI}$$

$$= \frac{72}{EI} - \frac{125}{8\,EI} = \frac{56.37}{EI}$$

$$= \frac{56 \cdot 37}{200 \times 10^6 \times 8 \times 10^{-5}} \times 10^3 \text{ mm} = 3 \cdot 52 \text{ mm}$$

Hence, $\quad y_D = 3 \cdot 52 \text{ mm}$ (*downward*) **(Ans.)**

(*iii*) Maximum deflection, y_{max} :

To find the maximum deflection, consider a section XX at a distance x from the end A;

$$M_x = R_A \cdot x = 3x \qquad \therefore \left(\frac{M}{EI}\right)_x = \frac{3x}{E(4I)} = \frac{3x}{4 EI}$$

Hence, $\dfrac{dy}{dx}$ at x of real beam = S.F. of conjugate beam = S'_x

$$= R'_A - \frac{1}{2} \times x \times \frac{3x}{4 EI}$$

Equating this to zero at the point of maximum deflection, we get

$$R'_A - \frac{1}{2} x \times \frac{3x}{4 EI} = 0$$

or, $\qquad \dfrac{72}{5 EI} - \dfrac{3x^2}{8 EI} = 0 \qquad \therefore x = 6 \cdot 196 \text{ m}$

Now, $\qquad y_{max} = M'_x$ of conjugate beam at $x = 6 \cdot 196 \text{ m}$

$$= R'_A \times x - \frac{1}{2} \times x \times \frac{3x}{4 EI} \times \frac{6 \cdot 196}{3}$$

$$= \frac{72}{5 EI} \times 6 \cdot 196 - \frac{1}{2} \times 6 \cdot 196 \times \frac{3 \times 6 \cdot 196}{4 EI} \times \frac{6 \cdot 196}{3}$$

$$= \frac{89 \cdot 22}{EI} - \frac{29 \cdot 73}{EI} = \frac{59 \cdot 49}{EI} = \frac{59 \cdot 49}{200 \times 10^6 \times 8 \times 10^{-5}} \times 1000$$

$$= 3 \cdot 72 \text{ mm}$$

Hence, $\qquad y_{max} = 3 \cdot 72 \text{ mm}$ **(Ans.)**

Example 8·47. *Using conjugate beam method, for the beam shown in Fig. 8·61 find the slopes and deflections at A, B, C, and D. Given: E = 200 × 10⁶ kN/m² and I = 300 × 10⁻⁴ m⁴. Neglect the weight of the beam.*

Solution. *Given:* $\qquad E = 200 \times 10^6 \text{ kN/m}^2; I = 300 \times 10^{-4} \text{ m}^4$

Refer to Fig. 8·61 (*a*).

To find the reaction at B, taking moments about A, we get

$$R_B \times 30 = 75 \times 10 + 150 \times 20 = 3750$$

$\therefore \qquad R_B = 125 \text{ kN}$

Also, $\qquad R_A + R_B = 75 + 150 = 225 \text{ kN}$

$\therefore \qquad R_A = 225 - 125 = 100 \text{ kN}$

B.M. at C, $\qquad M_C = 100 \times 10 = 1000 \text{ kNm};$

B.M. at D, $\qquad M_D = 125 \times 10 = 1250 \text{ kNm}$

Fig. 8·61 (*b*) shows the B.M. diagram for the given beam.

Fig. 8·61 (*c*) shows the conjugate beam with loading.

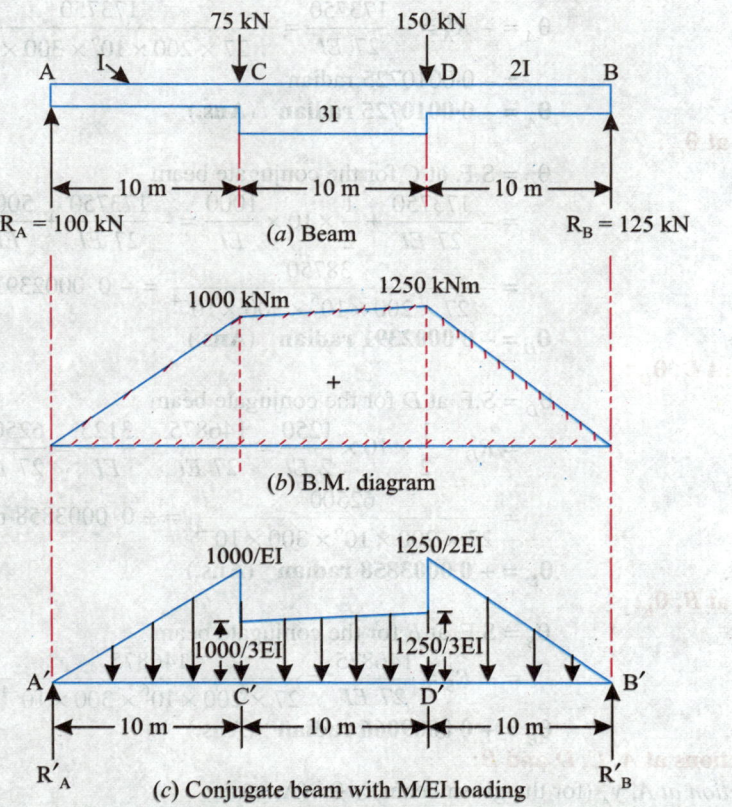

(a) Beam

(b) B.M. diagram

(c) Conjugate beam with M/EI loading

Fig. 8.61

To find reaction at B' taking moments about A', we get

$$R'_B \times 30 = \frac{1}{2} \times 10 \times \frac{1000}{EI} \times \frac{20}{3} + \frac{1000}{3\,EI} \times 10 \times 15$$

$$+ \frac{1}{2} \times 10 \times \frac{250}{3\,EI} \times \left(\frac{20}{3} + 10\right) + \frac{1}{2} \times 10 \times \frac{1250}{2\,EI} \times \left(20 + \frac{10}{3}\right)$$

$$= \frac{100000}{3\,EI} + \frac{150000}{3\,EI} + \frac{62500}{9\,EI} + \frac{218750}{3\,EI} = \frac{1468750}{9\,EI}$$

$$\therefore \qquad R'_B = \frac{146875}{27\,EI}$$

Also, $\qquad R'_A + R'_B = \frac{1}{2} \times 10 \times \frac{1000}{EI} + \frac{1000}{3\,EI} \times 10 + \frac{1}{2} \times 10 \times \frac{250}{3\,EI} + \frac{1}{2} \times 10 \times \frac{1250}{2\,EI}$

$$= \frac{5000}{EI} + \frac{10000}{3\,EI} + \frac{1250}{3\,EI} + \frac{3125}{EI} = \frac{35625}{3\,EI}$$

$$\therefore \qquad R'_A = \frac{35625}{3\,EI} - \frac{146875}{27\,EI} = \frac{173750}{27\,EI}$$

Slopes at A, C, D and B :

Slope at A, θ_A :

Slope at A for the given beam = S.F. at A for the conjugate beam

i.e.
$$\theta_A = -R'_A = -\frac{173750}{27\ EI} = -\frac{173750}{27 \times 200 \times 10^6 \times 300 \times 10^{-4}}$$
$$= -0.0010725 \text{ radian}$$

Hence,
$$\boldsymbol{\theta_A = -0.0010725 \text{ radian} \quad (Ans.)}$$

Slope at θ_C :

$$\theta_C = \text{S.F. at } C \text{ for the conjugate beam}$$
$$= -\frac{173750}{27\ EI} + \frac{1}{2} \times 10 \times \frac{1000}{EI} = -\frac{173750}{27\ EI} + \frac{5000}{EI} = -\frac{38750}{27\ EI}$$
$$= -\frac{38750}{27 \times 200 \times 10^6 \times 300 \times 10^{-4}} = -0.0002391 \text{ radian}$$

Hence,
$$\boldsymbol{\theta_D = -0.0002391 \text{ radian} \quad (Ans.)}$$

Slope at D, θ_D :

$$\theta_D = \text{S.F. at } D \text{ for the conjugate beam}$$
$$= R'_D - \frac{1}{2} \times 10 \times \frac{1250}{2\ EI} = \frac{146875}{27\ EI} - \frac{3125}{EI} = \frac{62500}{27\ EI}$$
$$= \frac{62500}{27 \times 200 \times 10^6 \times 300 \times 10^{-4}} = +0.0003858 \text{ radian}$$

Hence,
$$\boldsymbol{\theta_D = +0.0003858 \text{ radian} \quad (Ans.)}$$

Slope at B, θ_B:

$$\theta_B = \text{S.F. at } B \text{ for the conjugate beam}$$
$$= +R'_B = \frac{146875}{27\ EI} = \frac{146875}{27 \times 200 \times 10^6 \times 300 \times 10^{-4}} = +0.0009066$$

Hence,
$$\boldsymbol{\theta_B = +0.0009066 \text{ radian} \quad (Ans.)}$$

Deflections at A, C, D and B:

Deflection at A, y_A (for the given beam) $= 0$ **(Ans.)**

Deflection at C, y_C :

$$y_C = \text{B.M. at } C \text{ for the conjugate beam } (= M'_C)$$
$$= R'_A \times 10 - \frac{1}{2} \times 10 \times \frac{1000}{EI} \times \frac{10}{3}$$

The picture shows a crack in the inner race of a bearing. Such cracks can appear if the bearing is designed for below normal loads, defective materials, using the bearing for loads higher than normal or using for prolonged durations.

$$= \frac{173750}{27\ EI} \times 10 - \frac{50000}{3\ EI} = \frac{1287500}{27\ EI}$$

$$= \frac{1287500}{27 \times 200 \times 10^6 \times 300 \times 10^{-4}} \times 10^3\ \text{mm} = 7.94\ \text{mm}$$

Hence, $y_C = \textbf{7·94 mm}$ (*downward*) **(Ans.)**

Deflection at D, y_D:

$\qquad y_D$ = B.M. at D' for the conjugate beam (= M'_D)

$$= R'_B \times 10 - \frac{1}{2} \times 10 \times \frac{1250}{2\ EI} \times \frac{10}{3}$$

$$= \frac{146875}{27\ EI} \times 10 - \frac{31250}{3\ EI} = \frac{1187500}{27\ EI} = \frac{1187500}{27 \times 200 \times 10^6 \times 300 \times 10^{-4}} \times 10^3$$

$$= 7.33\ \text{mm}$$

Hence, $y_D = \textbf{7·33 mm}$ (*downward*) **(Ans.)**

Deflection at B, y_B:

$\qquad y_B = \textbf{0}$ **(Ans.)**

Example 8·48. *Using conjugate beam method find the mid span deflection of the beam shown in the Fig. 8·62 (a).*

$\qquad$ *Take:* $\qquad\qquad E = 200 \times 10^6\ kN/m^2\ and\ I = 200 \times 10^{-4}\ m^4$

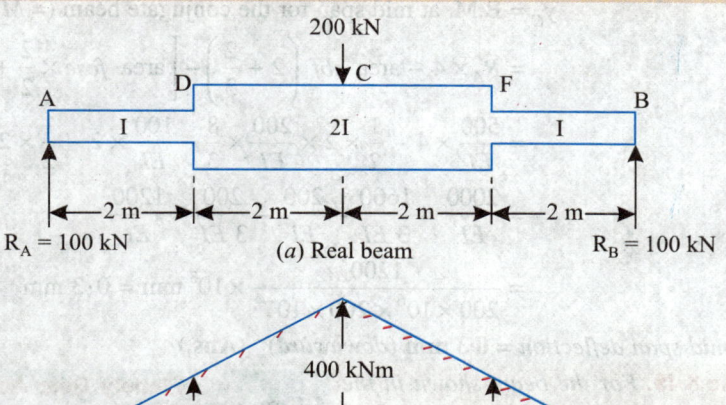

(a) Real beam

(b) B.M. diagram

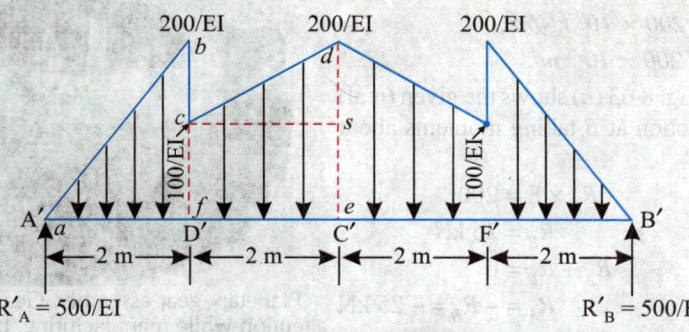

(c) Conjugate beam with M/EI loading

Fig. 8.62

Solution. *Given:* $E = 200 \times 10^6$ kN/m^2, $I = 200 \times 10^{-4}$ m^4

Fig. 8·62 (a) shows the given (real) beam. The reactions at the supports A and B are:

$$R_A = R_B = \frac{200}{2} = 100 \text{ kN}$$

Now,

$$M_F = 100 \times 2 = 200 \text{ kNm}$$
$$M_C = 100 \times 4 = 400 \text{ kNm},$$
$$M_D = 100 \times 2 = 200 \text{ kNm}$$

Fig. 8·62 (b) shows the B.M. diagram

Fig. 8·62 (c) shows the conjugate beam with loading.

Since the load on the conjugate beam is symmetrical the two support reactions are equal, each being equal to area *abcdea*.

∴
$$R'_A (= R'_B) = \text{Area of triangle } abf + \text{area of trapezium } fcde$$

$$= \frac{1}{2} \times 2 \times \frac{200}{EI} + \frac{1}{2}\left(\frac{100}{EI} + \frac{200}{EI}\right) \times 2$$

$$= \frac{200}{EI} + \frac{300}{EI} = \frac{500}{EI}$$

Mid span deflection for the beam,

$$y_C = \text{B.M. at mid span for the conjugate beam } (= M'_C)$$

$$= R'_A \times 4 - \text{area } abf\left(2 + \frac{2}{3}\right) - \left[\text{area } fcse \times \frac{2}{2} + \text{area } cds \times \frac{2}{3}\right]$$

$$= \frac{500}{EI} \times 4 - \frac{1}{2} \times 2 \times \frac{200}{EI} \times \frac{8}{3} - \frac{100}{EI} \times 2 - \frac{1}{2} \times 2 \times \frac{100}{EI} \times \frac{2}{3}$$

$$= \frac{2000}{EI} - \frac{1600}{3\,EI} - \frac{200}{EI} - \frac{200}{3\,EI} = \frac{1200}{EI}$$

$$= \frac{1200}{200 \times 10^6 \times 200 \times 10^{-4}} \times 10^3 \text{ mm} = 0.3 \text{ mm}$$

*Hence, mid span deflection = **0·3 mm** (downward)* **(Ans.)**

Example 8·49. *For the beam shown in the Fig. 8·63 (a) using conjugate beam method determine:*

   (i) Slope at the ends;

   (ii) Deflection at the centre.

Take: $E = 200 \times 10^6$ kN/m^2;
     $I = 200 \times 10^{-5}$ m^4.

Solution. Fig. 8·63 (a) shows the given (real) beam. To find reaction at B taking moments about A, we get

$$R_B \times 8 = 200$$

∴
$$R_B = 25 \text{ kN}$$

Also,
$$R_A + R_B = 0$$

∴
$$R_A = -R_B = -25 \text{ kN}$$

Planety Gear Assembly

1. Case
2. Outer pinion
3. Outer pinion shaft.
4. Inner pinion
5. Inner pinion shaft

Planetary gear assembly. Gears need special attention while manufacturing, because they bear high stresses and their failure can lead to stoppage of machines.

Bending moment about C,
$$M_C = 25 \times 4 = 100 \text{ kNm}.$$
Fig. 8·63 (b) shows the B.M. diagram for the real beam.
Fig. 8·63 (c) shows the conjugate beam with the loading.

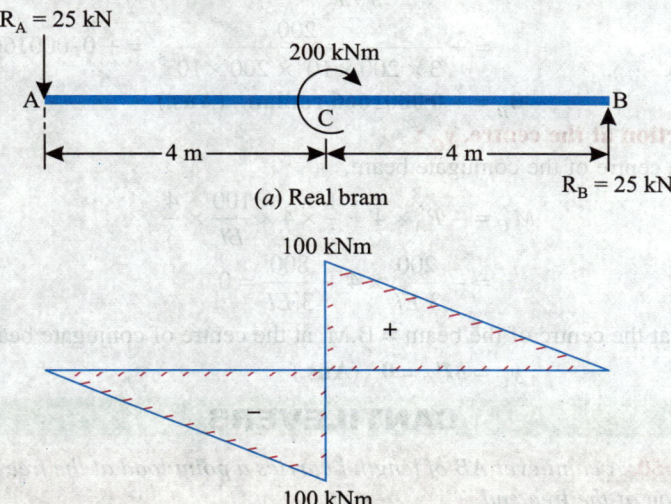

(a) Real bram

(b) B.M. Diagram

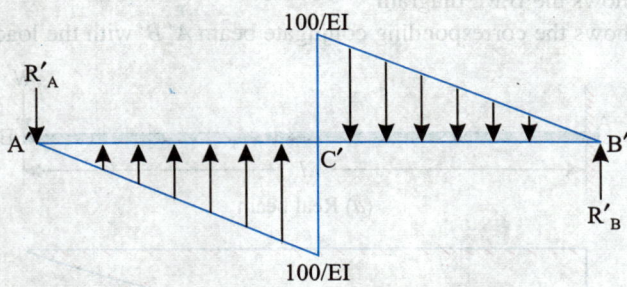

(c) Conjugate beam with M/EI loading

Fig. 8.63

Taking moments about B' of the conjugate beam, we get

$$R'_A \times 8 + \frac{1}{2} \times 4 \times \frac{100}{EI} \times \left(\frac{4}{3} + 4\right) = \frac{1}{2} \times 4 \times \frac{100}{EI} \times (2/3 \times 4)$$

$$R'_A \times 8 + \frac{3200}{3\,EI} = \frac{1600}{3\,EI}$$

or,
$$R'_A = -\frac{200}{3\,EI} \left(= \frac{200}{3\,EI} \downarrow\right)$$

∴
$$R'_B = +\frac{200}{3\,EI} \ (\uparrow)$$

(i) Slope at the ends :

Slope at the end A of real beam (θ_A)

= Shear at end A' of conjugate beam

∴
$$\theta_A = +\frac{200}{3\,EI} = \frac{200}{3 \times 200 \times 10^6 \times 200 \times 10^{-5}}$$

$$= + 0.0001666 \text{ radian}$$

Hence, $\qquad \theta_A = 0 \cdot 0001666 \text{ radian} \quad \text{(Ans.)}$

Similarly, $\qquad \theta_B = R'_B = + \dfrac{200}{3 \, EI}$

$$= + \frac{200}{3 \times 200 \times 10^6 \times 200 \times 10^{-5}} = + 0 \cdot 0001666 \text{ radian}$$

Hence, $\qquad \theta_B = + 0 \cdot 0001666 \text{ radian} \quad \text{(Ans.)}$

(ii) Deflection at the centre, y_C :

B.M. at the centre of the conjugate beam,

$$M'_C = - R'_A \times 4 + \frac{1}{2} \times 4 \times \frac{100}{EI} \times \frac{4}{3}$$

$$= - \frac{200}{3 \, EI} \times 4 + \frac{800}{3 \, EI} = 0$$

Deflection at the centre of the beam = B.M. at the centre of conjugate beam

∴ $\qquad y_C = M'_C = 0 \quad \text{(Ans.)}$

CANTILEVERS

Example 8·50. *A cantilever AB of length l carries a point load at the free end. Calculate the slope and deflection at the free end.*

Solution. Fig. 8·64 (*a*) shows the given beam *AB* with a point load at the free end.

Fig. 8·64 (*b*) shows the B.M. diagram.

Fig. 8·64 (*c*) shows the corresponding conjugate beam *A' B'* with the loading.

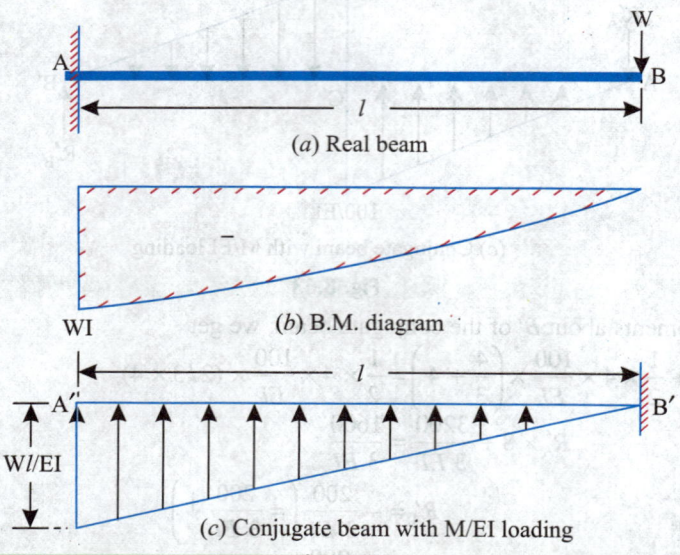

(*a*) Real beam

(*b*) B.M. diagram

(*c*) Conjugate beam with M/EI loading

Fig. 8.64

Slope at B, θ_B:

Slope at *B* for the given beam

$$= \text{S.F. at } B' \text{ for the conjugate beam}$$

$$= \frac{1}{2} \, l \, \frac{Wl}{EI} = \frac{Wl^2}{2 \, EI}$$

$$\therefore \qquad \theta_B = -\frac{Wl^2}{2\,EI} \quad \textbf{(Ans.)}$$

Deflection at B, y_B:

Deflection at B for the given beam = B.M. at B' for the conjugate beam

$$= \frac{1}{2} \times l \times \frac{Wl}{EI} \times \frac{2}{3}\,l = \frac{Wl^3}{3\,EI}$$

$$\therefore \qquad y_B = -\frac{Wl^3}{3\,EI} \quad \textbf{(Ans.)}$$

Example 8·51. *A cantilever AB is loaded as shown in the Fig. 8·65 (a). Draw the slope and deflection diagrams for the cantilever.*

Solution. Fig. 8·65 (*a*) and (*b*) shows the given (real) beam and its B.M. diagram respectively. Fig. 8·65 (*c*) shows the conjugate beam *A'C'B'*.

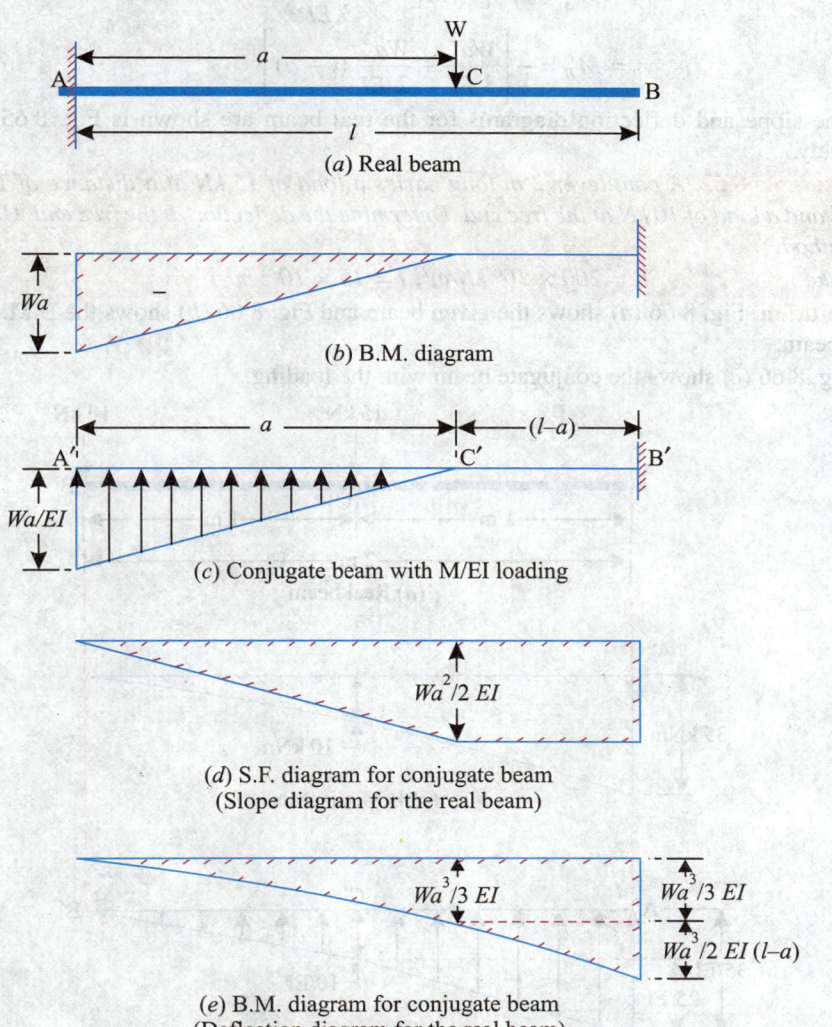

(*a*) Real beam

(*b*) B.M. diagram

(*c*) Conjugate beam with M/EI loading

(*d*) S.F. diagram for conjugate beam
(Slope diagram for the real beam)

(*e*) B.M. diagram for conjugate beam
(Deflection diagram for the real beam)

Fig. 8.65

$$S'_A = 0; \quad M'_A = 0; \quad S'_C = -\frac{1}{2} \times a \times \frac{Wa}{EI} = -\frac{Wa^2}{2\,EI}$$

$$M'_C = -\frac{1}{2} \times a \times \frac{Wa}{EI} \times \frac{2}{3} a = -\frac{Wa^3}{3\,EI}, \quad S'_B = S'_C = -\frac{Wa^2}{2\,EI}$$

$$M'_B = -\frac{1}{2} \times a \times \frac{Wa}{EI} \left\{ \frac{2}{3} a + (l-a) \right\} = -\left[\frac{Wa^3}{3\,EI} + \frac{Wa^2}{2\,EI} (l-a) \right]$$

Hence, for the real beam,

$$\theta_A = S'_A = 0; \quad \theta_C = S'_C = -\frac{Wa^2}{2\,EI}$$

$$\theta_B = S'_B = \theta_C = -\frac{Wa^2}{2\,EI}$$

$$y_A = M'_A = 0; \quad y_C = M'_C = -\frac{Wa^3}{3\,EI}$$

$$y_B = M'_B = -\left[\frac{Wa^3}{3\,EI} + \frac{Wa^2}{2\,EI} (l-a) \right]$$

The slope and deflection diagrams for the real beam are shown is Fig. 8·65 (*d*) and (*e*) respectively.

Example 8·52. *A cantilever 2 m long caries a load of 15 kN at a distance of 1 m from the fixed end and a load of 10 kN at the free end. Determine the deflection at the free end. Use conjugate beam method.*

Take : $\qquad E = 200 \times 10^6 \text{ kN/m}^2, \ I = 15 \times 10^{-6} \text{ m}^4.$

Solution. Fig. 8·66 (*a*) shows the given beam and Fig. 8·66 (*b*) shows the B.M. diagram for the real beam.

Fig. 8·66 (*c*) shows the conjugate beam with the loading.

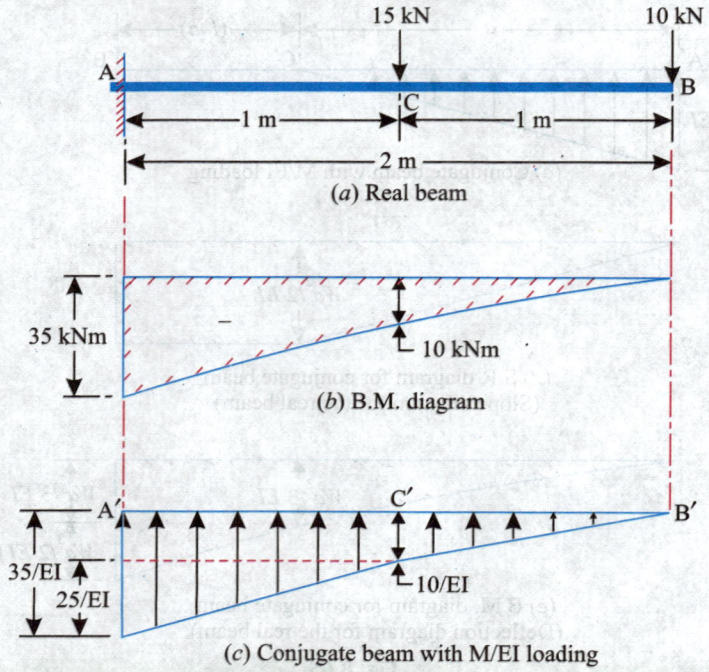

(*a*) Real beam

(*b*) B.M. diagram

(*c*) Conjugate beam with M/EI loading

Fig. 8.66

The deflection at the free end B, y_B = The B.M. at B' (conjugate beam)

$\therefore$

$$y_B = -\left[\frac{10}{EI} \times 1 \times \left(\frac{1}{2} + 1\right) + \frac{1}{2} \times 1 \times \frac{25}{EI} \times \left(\frac{2}{3} \times 1 + 1\right) + \frac{1}{2} \times 1 \times \frac{10}{EI} \times \frac{2}{3}\right]$$

$$= -\frac{1}{EI}\left(15 + \frac{125}{6} + \frac{10}{3}\right) = -\frac{235}{6EI}$$

$$= -\frac{235}{6 \times 200 \times 10^6 \times 15 \times 10^{-6}} \times 10^3 = -13\cdot055 \text{ mm}$$

Hence, *deflection at the free end* = **13·055 mm** (*downward*)　　(**Ans.**)

Example 8·53. *A cantilever of length 6 metres carries a uniformly distributed load of 10 kN/m over the whole length. If $E = 200 \times 10^6$ kN/m^2 and $I = 30 \times 10^{-5}$ m^4 determine the following, using conjugate beam method*:

　(*i*)　*Slope at the free end;*

　(*ii*)　*Deflection at the free end.*

Solution. $E = 200 \times 10^6$ kN/m^2, $I = 30 \times 10^{-5}$ m^4,

$$M_A = 10 \times 6 \times \frac{6}{2} = 180 \text{ kNm}$$

Fig. 8·67 (*a*) shows the real beam and the B.M. diagram is shown in Fig. 8·67 (*b*)

Fig. 8·67 (*c*) shows the corresponding conjugate beam $A'B'$ with $\dfrac{M}{EI}$ loading.

Shaft fitted with gears and pulleys for belt drive.

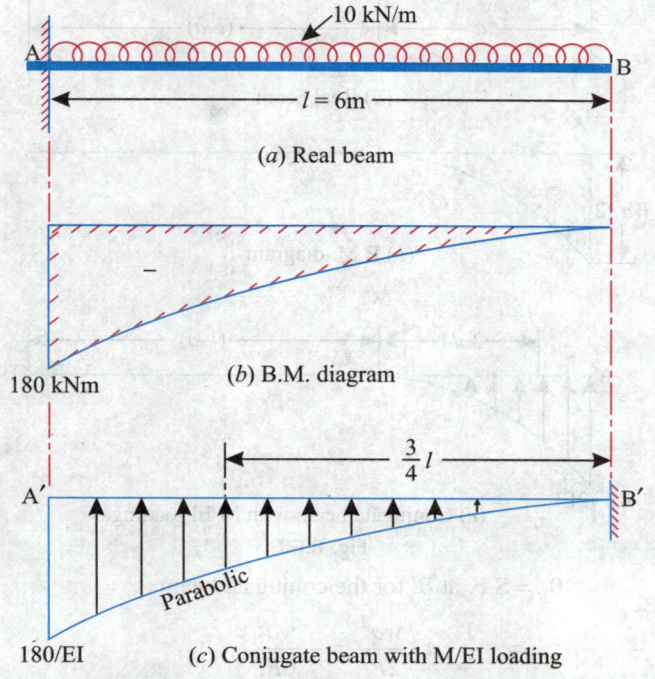

(*a*) Real beam

(*b*) B.M. diagram

180 kNm

180/EI　　(*c*) Conjugate beam with M/EI loading

Fig. 8.67

(*i*) Slope at the free·end, θ_B :

Slope at the free end, θ_B

$$= \text{S.F. at } B' \text{ for the conjugate beam}$$

$$= -\frac{1}{3} \times l \times \frac{180}{EI} = -\frac{60\,l}{EI} = -\frac{60 \times 6}{200 \times 10^6 \times 30 \times 10^{-5}}$$

$$= 0.006 \text{ radian}$$

Hence, $\qquad \theta_B = \textbf{0·006 radian} \quad \textbf{(Ans.)}$

(*ii*) Deflection at the free end, y_B:

Deflection at B for the given beam = B.M. at B' for conjugate beam

$$\therefore \qquad y_B = -\frac{1}{3} \cdot l \cdot \frac{180}{EI} \times \frac{3}{4} l = -\frac{45\,l^2}{EI} = -\frac{45 \times 6^2}{EI} = -\frac{1620}{EI}$$

$$= -\frac{1620}{200 \times 10^6 \times 30 \times 10^{-5}} \times 1000 = -27 \text{ mm}$$

Hence, $\qquad y_B = \textbf{27 mm} \; (downward) \quad \textbf{(Ans.)}$

Example 8·54. *A cantilever of span l is carrying uniformly distributed load of w per unit run on a length a from the fixed end. Determine the slope and deflection at the free end. Use conjugate beam method.* **(Mysore University)**

Solution. Fig. 8·68 (*a*) shows the real beam.

Fig. 8·68 (*b*) shows the B.M. diagram for the real beam.

Fig. 8·68 (*c*) shows the conjugate beam with the loading.

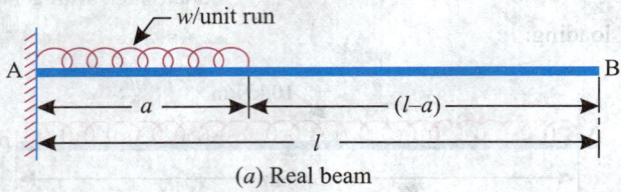

(a) Real beam

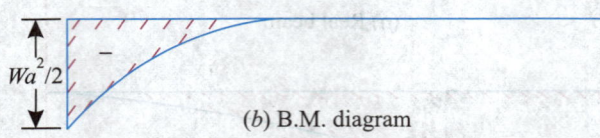

(b) B.M. diagram

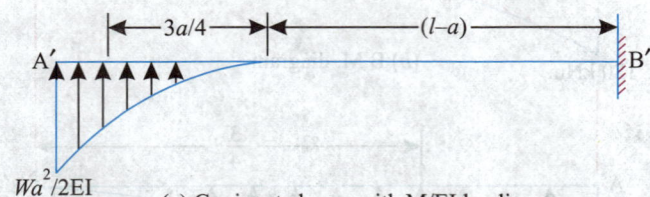

(c) Conjugate beam with M/EI loading

Fig. 8.68

Slope at B, $\qquad \theta_B = \text{S.F. at } B' \text{ for the conjugate beam}$

$$= -\frac{1}{3} \cdot a \cdot \frac{wa^2}{2EI} = -\frac{wa^3}{6EI}$$

Hence, $\theta_B = -\dfrac{wa^3}{6EI}$ (Ans.)

Deflection at B, y_B = B.M. at B' for the conjugate beam

$$= -\frac{1}{3} \cdot a \cdot \frac{wa^2}{2EI} \left[(l - a) + \frac{3a}{4} \right] = -\left[\frac{wa^3}{6EI} (l - a) + \frac{wa^4}{8EI} \right]$$

Hence, $y_B = \dfrac{wa^3}{6EI} (l - a) + \dfrac{wa^4}{8EI}$ (*downward*) (Ans.)

Conjugate Beam Method for Overhanging Beams

(*a*) *Beams with an overhang on one side:*

Fig. 8·69 shows an overhanging beam *ACBD* with an overhang on the right side. The deflection at *A* and *B* will be zero irrespective of the nature of loading on the beam. The conjugate beam, corresponding to the real beam should be such that the B.M. at *A* and *B* are zero. The conjugate beam *A′C′B′D′* fixed at *D′* and having a hinge at *B′* satisfies the requirements [Fig. 8·69 (*b*)]. Free end *D* of the real beam shall have slope as well as deflection, therefore, the corresponding point *D′* is fixed so that it has shearing force (S.F.) and bending moment (B.M.).

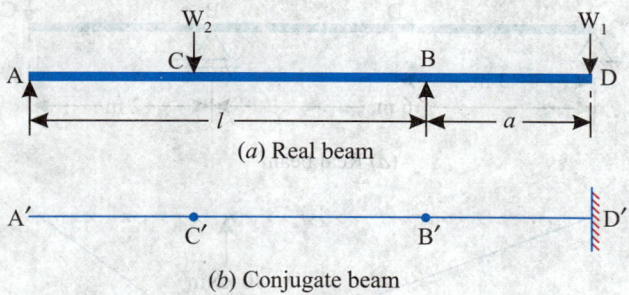

(*a*) Real beam

(*b*) Conjugate beam

Fig. 8.69

(*b*) *Beams with overhangs on both sides:*

Fig. 8·70 (*a*) shows a beam *CADBE* with overhangs on both the sides. It will have slope and deflection at *C* and *E* but no deflection at *A* and *B*. The conjugate beam should be such as to satisfy the following conditions: (*i*) It should have both S.F. and B.M. at *C* and *E*; (*ii*) It should have no B.M. at *A* and *B* and (*iii*) It should have S.F. at both *A* and *B*. All these conditions are satisfied by a conjugate beam *C′A′B′E′* having ends *C′* and *E′* fixed (so that there are both S.F. and B.M. at both the fixed ends) and having hinges at *A′* and *B′* (so that there is no B.M. at these two points) as shown in Fig. 8·70 (*b*).

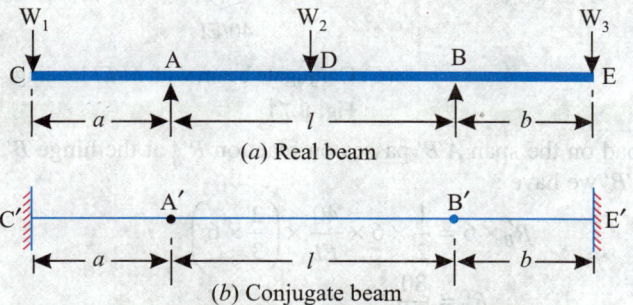

(*a*) Real beam

(*b*) Conjugate beam

Fig. 8.70

Example 8·55. *Using conjugate beam method determine slope at C and A and deflection of the central point D of the beam of uniform cross-section shown in Fig. 8·71 (a).*

    Take: $E = 200 \times 10^6 \text{ kN/m}^2 \; ; I = 120 \times 10^{-6} \text{ m}^4$

Solution. $E = 200 \times 10^6 \text{ kN/m}^2$, $I = 120 \times 10^{-6} \text{ m}^4$

Fig. 8·71 (a) shows the real beam *ADBC* with the loading. The B.M. diagram for this beam is shown in Fig. 8·71 (b).

Shaft with groove for belt drive.

— Since there is no deflection at *A* and *B* there will be slope at both the points, the conjugate beam *A'B'C'* should have hinges both at *A'* and *B'* so that for the conjugate beam there are no bending moments but there are shear forces at *A'* and *B'*.

— At *C* there are both slope and deflection so conjugate beam should have fixity at *C'* so that there are both the S.F. and B.M. at *C'*.

— Fig. 8·71 (c) shows the conjugate beam corres ponding to the given (real) beam.

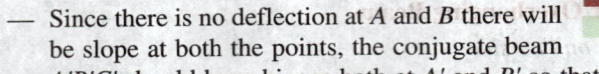

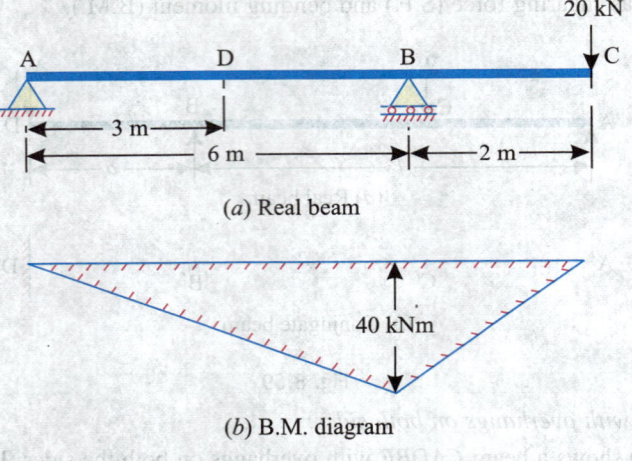

(a) Real beam

(b) B.M. diagram

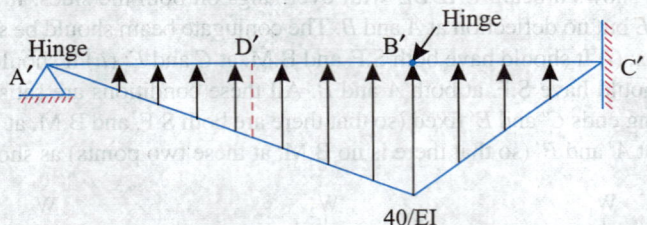

(C) Conjugate beam with *M/EI* loading

Fig. 8.71

    Triangular load on the span *A'B'* passes on reaction R'_B at the hinge *B'*. By taking moments about *A'* for span *A'B'* we have :

$$R'_B \times 6 = \frac{1}{2} \times 6 \times \frac{40}{EI} \times \left(\frac{2}{3} \times 6 \right)$$

∴
$$R'_B = \frac{80}{EI}$$

Slope at C, θ_C :

Slope at C = S.F. at C' for the conjugate beam

i.e.,
$$\theta_C = R'_B + \text{load on } B'C'$$
$$= \frac{80}{EI} + \frac{1}{2} \times 2 \times \frac{40}{EI} = \frac{120}{EI} = \frac{120}{200 \times 10^6 \times 120 \times 10^{-6}}$$
$$= 0.005 \text{ radian}$$

Hence,
$$\theta_C = 0.005 \text{ radian} \quad (Ans.)$$

Slope at A, θ_A :

Reaction
$$R'_A = \text{Load on } A'B' - R'_B = \frac{1}{2} \times 6 \times \frac{40}{EI} - \frac{80}{EI} = \frac{40}{EI}$$

∴ Slope at A,
$$\theta_A = R'_A = \frac{40}{EI}$$
$$= \frac{40}{200 \times 10^6 \times 120 \times 10^{-6}} = 0.00166 \text{ radian}$$

Hence,
$$\theta_A = 0.00166 \text{ radian} \quad (Ans.)$$

Deflection at D, y_D :

$$\frac{\text{Rate of loading at } D'}{\frac{40}{EI}} = \frac{3}{6} \qquad \qquad \text{(From similarity of } \Delta s)$$

∴ Rate of loading
$$= \frac{40}{EI} \times \frac{3}{6} = \frac{20}{EI}$$

$$y_D = \text{B.M. at } D' \text{ for the conjugate beam}$$
$$= R'_A \times 3 - \left(\frac{1}{2} \times 3 \times \frac{20}{EI} \right) \times \left(\frac{1}{3} \times 3 \right)$$
$$= \frac{40}{EI} \times 3 - \frac{30}{EI} = \frac{90}{EI} = \frac{90}{200 \times 10^6 \times 120 \times 10^{-6}} \times 10^3 = 3.75 \text{ mm}$$

Hence,
$$y_D = 3.75 \text{ mm} \quad (Ans.)$$

Example 8·56. *For the beam ABC shown in Fig. 8·72 (a) find the value of P in terms of w if the deflection at the end C has to be zero. Use conjugate beam method.*

(Baroda University)

Solution. Fig. 8·72 (a) shows the given beam with the loading.

— As there is no deflection at A or B but there is change of slope at both these points, therefore, the conjugate beam shall have hinges at A' and B' [Fig. 8·72 (b)].

— Also there is both deflection and change of slope at C, therefore, the conjugate beam shall have fixity at C' [Fig. 8·72 (b)].

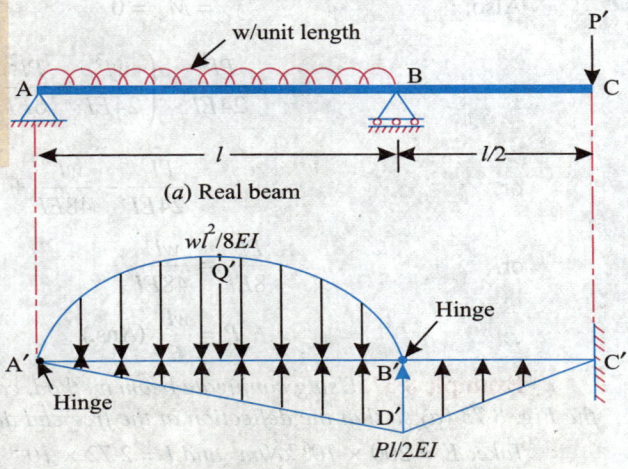

(a) Real beam

(b) Conjugate beam with M/EI loading

Fig. 8.72

$- \dfrac{M}{EI}$ diagram for the beam due to load P only is $\triangle A'C'D'$. Since P causes only $-ve$ moment throughout the full length of the beam therefore it is shown only on one side of the conjugate beam.

The uniformly distributed load (U.D.L.) on span AB causes $+ve$ moment, therefore, $\dfrac{M}{EI}$ diagram $A'Q'B'$ due to U.D.L. is shown as a load on the conjugate beam on the other side of the beam.

Reaction R_1 at B' due to U.D.L.

$$= \frac{1}{2} \times \left(\frac{2}{3} \times \frac{wl^2}{8EI} \times l \right) = \frac{wl^3}{24EI}$$

Reaction R_2 at B' due to triangular load $A'B'D'$ (by taking moments of area about A') is

$$R_2 \times l = \frac{1}{2} l \times \frac{Pl}{2EI} \times \frac{2}{3} l = \frac{Pl^3}{6EI}$$

$$\therefore \qquad R_2 = \frac{Pl^3}{6EI}$$

Shaft gear reducer.

∴ Net reaction on the hinge B' due to loads on beam $A'B'$

$$R'_B = R_1 - R_2 = \frac{wl^3}{24EI} - \frac{Pl^3}{6EI}$$

B.M. at C',
$$M'_C = \left(\frac{1}{2} \times \frac{l}{2} \times \frac{Pl}{2EI} \right) \times \left(\frac{2}{3} \times \frac{l}{2} \right) - R'_B \times \frac{l}{2}$$

$$= \frac{Pl^3}{24EI} - \left(\frac{wl^3}{24EI} - \frac{Pl^2}{6EI} \right) \times \frac{l}{2}$$

Also, $\qquad y_C = M'_c = 0$ ...(Given)

$$\therefore \qquad \frac{Pl^3}{24EI} - \left(\frac{wl^3}{24EI} - \frac{Pl^2}{6EI} \right) \times \frac{l}{2} = 0$$

or, $\qquad \dfrac{Pl^3}{24EI} - \dfrac{wl^4}{48EI} + \dfrac{Pl^3}{12EI} = 0$

or, $\qquad \dfrac{Pl^3}{8EI} = \dfrac{wl^4}{48EI}$

$$\therefore \qquad P = \frac{wl}{6} \quad \textbf{(Ans.)}$$

Example 8·57. *Using conjugate beam method, calculate the U.D.L. w over a beam shown in the Fig. 8·73 (a) so that the deflection at the free end does not exceed 10 mm.*

Take: $E = 200 \times 10^6$ kN/m² and $I = 2·72 \times 10^{-4}$ m⁴.

Solution. Fig. 8·73 (a) shows the real beam ABC with the loading. The conjugate beam $A'B'C'$ with loading is shown in Fig. 8·73 (b).

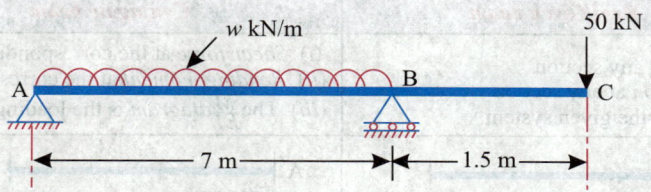

(a) Real beam

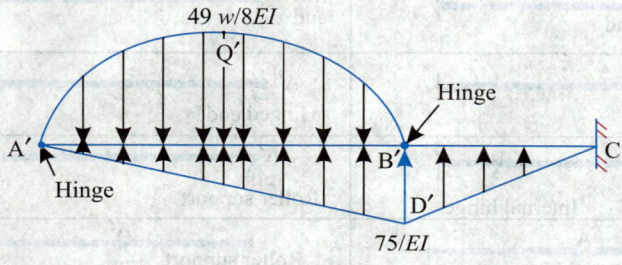

(b) Conjugate beam with M/EI loading

Fig. 8.73

Loads due to $A'Q'B'$ and $A'B'D'$ shall pass reactions at the hinge B'.

Reaction R_1 due to load $A'Q'B'$ is given by:

$$R_1 = \frac{1}{2}\left(\frac{2}{3} \times \frac{49w}{8EI} \times 7\right) = \frac{14\cdot29w}{EI}$$

Reaction R_2 due to load $A'B'D'$ is given by:

$$R_2 \times 7 = \left(\frac{1}{2} \times 7 \times \frac{75}{EI}\right) \times \left(\frac{2}{3} \times 7\right)$$

∴ $$R_2 = \frac{175}{EI}$$

Deflection at C, $\quad y_C = $ B.M. at C'

$$= (R_1 - R_2) \times 1\cdot5 - \frac{1}{2} \times 1\cdot5 \times \frac{75}{EI} \times \left(\frac{2}{3} \times 1\cdot5\right)$$

$$= \left(\frac{14\cdot29w}{EI} - \frac{175}{EI}\right) \times 1\cdot5 - \frac{56\cdot25}{EI}$$

$$= \frac{21\cdot43w}{EI} - \frac{262\cdot5}{EI} - \frac{56\cdot25}{EI} = \frac{1}{EI}(21\cdot43w - 318\cdot75)$$

But, $\qquad y_c = 10$ mm $= 0\cdot01$ m. ...(Given)

∴ $\quad \dfrac{1}{EI}(21\cdot43w - 318\cdot75) = 0\cdot01$

or, $\qquad 21\cdot43w - 318\cdot75 = 0\cdot01 \; EI$

or, $\qquad 21\cdot43w - 318\cdot75 = 0\cdot01 \times 200 \times 10^6 \times 2\cdot72 \times 10^{-4} = 544$

$$w = \textbf{40·26 kN/m} \quad \textbf{(Ans.)}$$

TABLE 8.1. Relation between the given beam and conjugate beam

Real/Given beam	Conjugate beam
1. (i) *Slope* at any section (ii) *Deflection* at any section (iii) *Loading*-the given system	(i) *Shear force* at the corresponding section (ii) *Bending moment* at the corresponding section (iii) The $\frac{M}{EI}$ *diagram* is the loading system
2. Fixed end $\overset{A}{\vert}$ ———————————⏐	$A'\vert$ ———————————⏐ Free end
3. A ———————————⏐ Free end	Fixed end $\overset{A'}{\vert}$ ———————————⏐
4. A⌂———————————⏐ Hinged end	A'⌂———————————⏐ Hinged end
5. A△———————————⏐ Roller support Internal hinge	A' △———————————⏐ Roller support
6. ⏐———————————A○———————————⏐ Internal hinge	Roller support △ ———————————A'⏐ (or internal pin)
7. ⏐———————————A○ Link ———————————⏐	⏐———————————A' ○———————————⏐ Roller
8. ⏐———————————A △———————————⏐ Roller	⏐———————————A' ○ Link ———————————⏐

Fig. 8.74

8·10.   PROPPED CANTILEVERS AND BEAMS

In a cantilever or a beam, to avoid excessive deflection or to reduce the values of bending moments, some support other than the existing ones may have to be provided; this *additional support* is known as a **Prop.** The prop renders the structure *indeterminate i.e.* the beam or the cantilever that has been propped cannot be analysed by simple equations of statics; to analyse the structure additional equations from considerations of slope or deflection are utilised.

Sheet coil processing equipment.

Example 8·58. *A cantilever AB of span 6 m is fixed at the end A and propped at the end B. It carries a point load of 50 kN at the mid span. Level of the prop is the same as that of the fixed end.*

(i)   *Determine reaction at the prop;*

(ii)   *Draw the S.F. and B.M. diagrams.*

Solution. Fig. 8·75 (*a*) shows the cantilever *AB* propped at the end *B* and carrying a point load of 50 kN at the mid span.

Fig. 8·75 (*b*) shows the B.M. diagram due to only 50 kN acting at *C*.

Fig. 8·75 (c) shows the B.M. diagram if the load 50 kN were not acting at C and only the prop reaction R_B were acting at B.

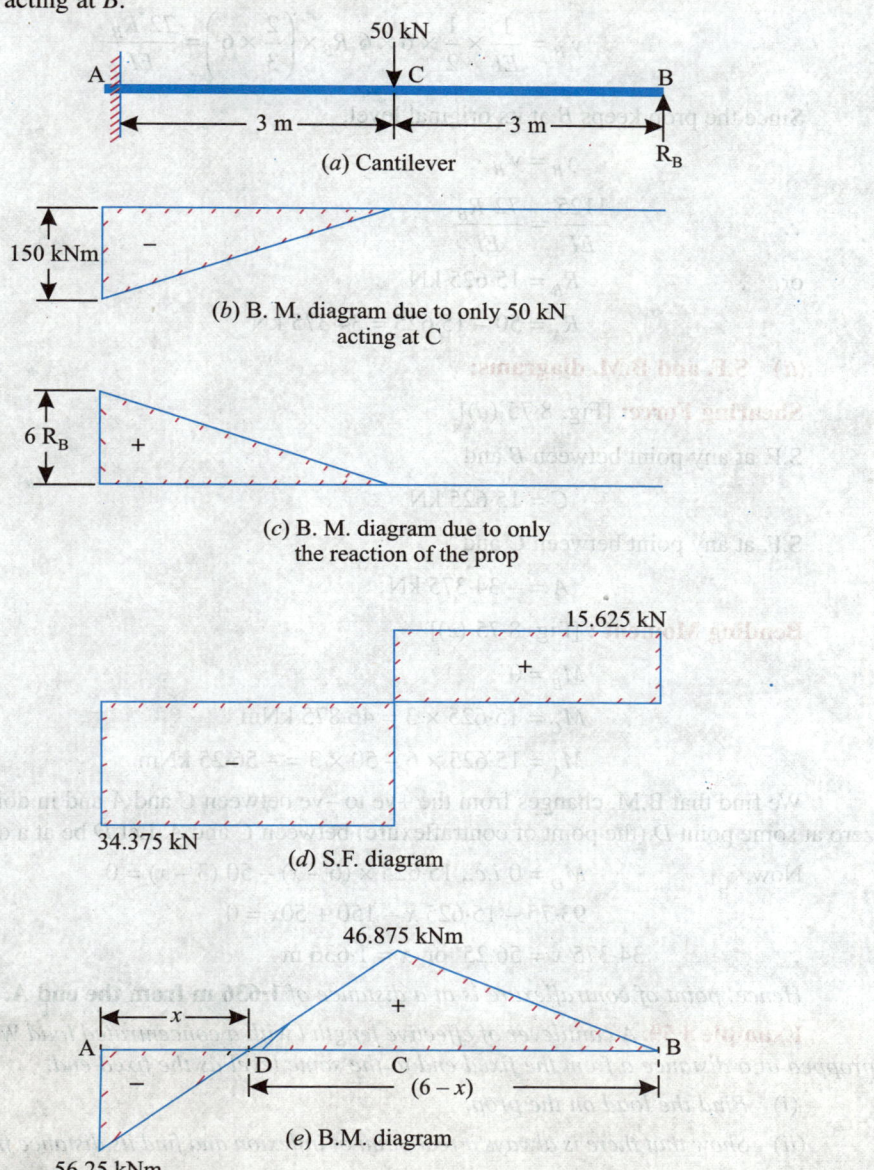

(a) Cantilever

150 kNm

(b) B. M. diagram due to only 50 kN acting at C

$6 R_B$

(c) B. M. diagram due to only the reaction of the prop

15.625 kN

34.375 kN

(d) S.F. diagram

46.875 kNm

56.25 kNm

(e) B.M. diagram

Fig. 8.75

(i) Reaction at the prop, R_B :

Using moment area method the downward deflection of B (y_B) due to only load 50 kN at C is given by:

$$y_B = \text{Moment of } \frac{M}{EI} \text{ diagram between } A \text{ and } B \text{ about } B \text{ [Fig. 8·73 (b)]}$$

$$= \frac{1}{EI} \times \frac{1}{2} \times 3 \times 150 \times \left(\frac{2}{3} \times 3 + 3 \right)$$

$$= \frac{1125}{EI}$$

Upward deflection (y'_B) of B due to only the prop reaction R_B at B is given by:

$$y'_B = \frac{1}{EI} \times \frac{1}{2} \times 6 \times 6\ R_B \times \left(\frac{2}{3} \times 6\right) = \frac{72\ R_B}{EI}$$

Since the prop keeps B at its original level,

$$y_B = y'_B$$

∴ $$\frac{1125}{EI} = \frac{72\ R_B}{EI}$$

or, $$R_B = 15 \cdot 625 \text{ kN}$$

$$R_A = 50 - 15 \cdot 625 = 34 \cdot 375 \text{ kN}$$

(ii) S.F. and B.M. diagrams:

Shearing Force: [Fig. 8·75 (*d*)]

S.F. at any point between B and

$$C = 15 \cdot 625 \text{ kN}$$

S.F. at any point between C and

$$A = -34 \cdot 375 \text{ kN}$$

Bending Moment : [Fig. 8·75 (*e*)]

$$M_B = 0$$

$$M_C = 15 \cdot 625 \times 3 = 46 \cdot 875 \text{ kNm}$$

$$M_A = 15 \cdot 625 \times 6 - 50 \times 3 = -56 \cdot 25 \text{ kNm}$$

We find that B.M. changes from the +ve to –ve between C and A and in doing so it becomes zero at some point D (the point of contraflexure) between C and A. Let D be at a distance x from A.

Now, $$M_D = 0 \text{ i.e., } 15 \cdot 625 \times (6 - x) - 50\ (3 - x) = 0$$

$$93 \cdot 75 - 15 \cdot 625\ x - 150 + 50x = 0$$

∴ $$34 \cdot 375\ x = 56 \cdot 25 \text{ or } x = 1 \cdot 636 \text{ m}$$

Hence, point of contraflexure is at a distance of **1·636 m from the end A. (Ans.)**

Example 8·59. *A cantilever of effective length l with a concentrated load W at the free end is propped at a distance a from the fixed end to the same level as the fixed end.*

(i) Find the load on the prop;

(ii) Show that there is always a real point of inflexion and find its distance from the fixed end.

(Kurukshetra University)

Solution. Fig. 8·76 (*a*) shows the cantilever AB propped at C and carrying a point load W at B.

B.M. at the prop $$C = -W\ (l - a)$$

B.M. at A due to only load

$$W = -Wl$$

B.M. at A due to only prop reaction $P = Pa$

Fig. 8·76 (*b*) shows the conjugate beam with $\dfrac{M}{EI}$ loading.

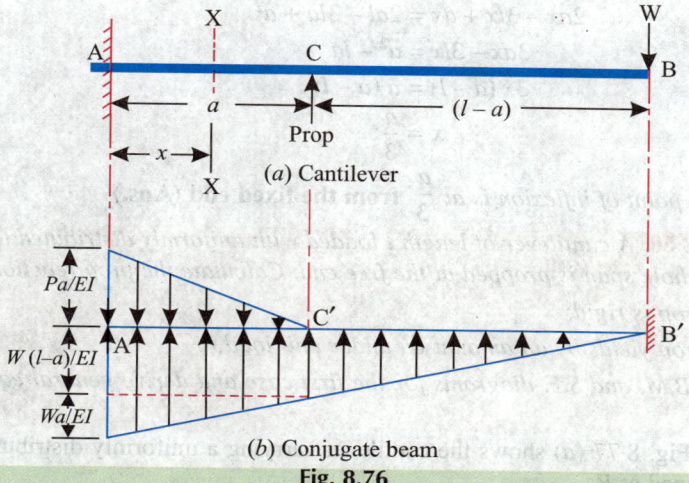

(a) Cantilever

(b) Conjugate beam
Fig. 8.76

(i) Load on the prop :

Deflection at C, y_C = B.M. at C' for the conjugate beam.

Now since prop level at C is the same as that of the fixed end, deflection of C is zero and hence the B.M. at C' for the conjugate beam is *zero*.

$$\left[\frac{1}{2}\, a \times \frac{Pa}{EI} \times \frac{2}{3}\, a\right] - \left[\left\{\left(\frac{W\,(l-a)}{EI} \times a\right) \times \frac{a}{2}\right\} - \left\{\left(\frac{1}{2}\, a \times \frac{Wa}{EI}\right) \times \frac{2}{3}\, a\right\}\right] = 0$$

$$\frac{Pa^3}{3EI} - \frac{Wa^2\,(l-a)}{2EI} - \frac{Wa^3}{3EI} = 0 \qquad\qquad ...(i)$$

∴ $$P = \frac{W\,(3l-a)}{2a} \quad \textbf{(Ans.)}$$

(ii) Real point of inflexion :

The B.M. is always negative between B and C. Between C and A the B.M. is negative due to load W at B and positive due to prop reaction P at C. So, if at all, the point of inflexion can lie between C and A this is possible only if $Pa > Wl$.

But, $$Pa = \frac{W\,(3l-a)}{2} = W\left(\frac{3l}{2} - \frac{a}{2}\right) \qquad ...\text{[From eqn. }(i)]$$

Since a can vary from zero to l, Pa varies from $\frac{3}{2}\,Wl$ to Wl. Thus Pa is *never less* than Wl and hence *there is always a real point of inflexion.*

Consider a section XX at a distance x $(x < a)$ from the end A.

B.M. at the section, $\quad M_x = -\,W\,(l-x) + P\,(a-x)$

For the point of inflexion, put $\quad M = 0$

∴ $\quad -\,W\,(l-x) + P\,(a-x) = 0$

Substituting for P from equation (i), we have

$$-\,W\,(l-x) + \frac{W\,(3l-a)}{2a}\,(a-x) = 0$$

$$2a\,(l-x) = (3l-a)\,(a-x)$$

$$2al - 2ax = 3la - 3lx - a^2 + ax$$

$$2ax - 3lx + ax = 2al - 3la + a^2$$
$$3ax - 3lx = a^2 - la$$
$$3x (a - l) = a (a - l)$$

or,
$$x = \frac{a}{3}$$

Hence, *the point of inflexion is at* $\dfrac{a}{3}$ *from the fixed end (Ans.)*

Example 8·60. *A cantilever of length l loaded with uniformly distributed load of w per unit length over the whole span is propped at the free end. Calculate the prop reaction if*

(i) *the prop is rigid;*

(ii) *the prop yields by an amount δu under unit load.*

Draw the B.M. and S.F. diagrams for the first case and derive general equations for slope and deflection.

Solution. Fig. 8·77 (*a*) shows the cantilever carrying a uniformly distributed load of *w* per unit run and propped at *B*.

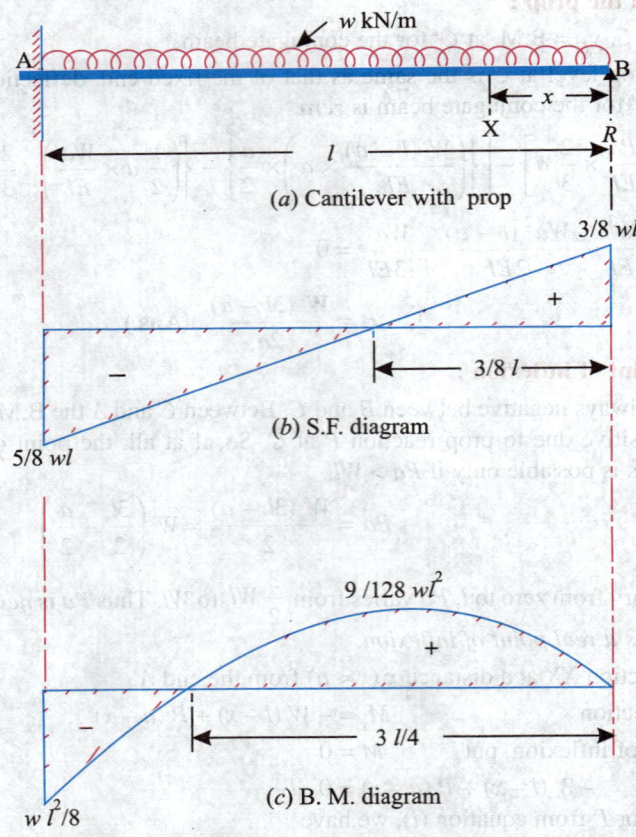

(*a*) Cantilever with prop

(*b*) S.F. diagram

(*c*) B. M. diagram

Fig. 8.77

When the prop is rigid :

The *downward* deflection of *B* due to the U.D.L. (in the absence of the prop),

$$y_B = \frac{wl^4}{8EI} \qquad \qquad ...(\text{Eqn. 8·8})$$

The *upward* deflection of B due to the prop. reaction R (in the absence of load),

$$y'_B = \frac{Rl^3}{3EI}$$ ..(Fig. 8·3)

Equating $y_B = y'_B$, we get, $\dfrac{wl^4}{8EI} = \dfrac{Rl^3}{3EI}$

$\therefore$ $R = \dfrac{3}{8} wl$ **(Ans.)**

Consider any section XX at a distance x from B.
The equation for S.F. is written as

$$S_x = R - wx = \frac{3}{8} wl - wx$$

When, $x = 0$, $S_B = \dfrac{3}{8} wl$

When, $x = l$, $S_B = \dfrac{3}{8} wl - wl = -\dfrac{5}{8} wl$

Now equating $S_x = 0$,

we get, $\dfrac{3}{8} wl - wx = 0$ $\therefore x = \dfrac{3}{8} l$

Fig. 8·77 (b) shows the S.F. diagram.
The B.M. at the section is given by:

Machinery to manufacture high precision equipment.

$$M_x = EI \frac{dy^2}{dx^2} = R \cdot x - \frac{wx^2}{2} = \frac{3}{8} wl\, x - \frac{wx^2}{2}$$ ...(ii)

When, $x = 0$, $M_B = 0$

When, $x = l$, $M_A = \dfrac{3}{8} wl^2 - \dfrac{wl^2}{2} = -\dfrac{wl^2}{8}$

When, $x = \dfrac{3}{8} l,$

$$M = \frac{3}{8} wl \times \frac{3}{8} l - \frac{w}{2} \times \left(\frac{3}{8} l\right)^2 = \frac{9}{128} wl^2$$

To find the point of contraflexure, equate (ii) to zero.

$\dfrac{3}{8} wlx - \dfrac{wx^2}{2} = 0$ or $x = \dfrac{3}{4} l$

The B.M. diagram will be parabolic as shown in the Fig. 8·77 (c)

Integrating eqn. (ii), we get

$$EI \frac{dy}{dx} = \frac{3}{8} wl \cdot \frac{x^2}{2} - \frac{wx^3}{6} + C_1$$

When, $x = l$, $\dfrac{dy}{dx} = 0$

$\therefore$ $0 = \dfrac{3}{8} wl \times \dfrac{l^2}{2} - \dfrac{wl^3}{6} + C_1$

or, $C_1 = \dfrac{wl^3}{6} - \dfrac{3wl^3}{16} = -\dfrac{wl^3}{48}$

$\therefore$ $EI \dfrac{dy}{dx} = \dfrac{3}{16} wlx^2 - \dfrac{wx^3}{6} - \dfrac{wl^3}{48}$ ...(1)

Integrating the above equation, we get

$$EIy = \frac{3}{16} \cdot wl \frac{x^3}{3} - \frac{wx^4}{24} - \frac{wl^3 x}{48} + C_2$$

When, $\qquad x = 0, y = 0 \qquad \therefore C_2 = 0$

$\therefore \qquad EIy = \dfrac{wlx^3}{16} - \dfrac{wx^4}{24} - \dfrac{wl^3x}{48}$ ...(2)

The eqns. (1) and (2) are equations for *slope and deflection at any point respectively.*

(ii) When the prop yields under the load:

When the prop sinks under the load, the sinking of the prop will be given by, $\delta = R \cdot \delta u$

Hence, we have $\qquad y_B = y'_B + \delta$

or $\qquad \dfrac{wl^4}{8EI} = \dfrac{Rl^3}{3EI} + R \cdot \delta u = R\left(\dfrac{l^3}{3EI} + \delta u\right) = R\left(\dfrac{l^3 + 3EI\,\delta u}{3EI}\right)$

or $\qquad R = \dfrac{3}{8} \times \dfrac{wl^4}{l^3 + 3EI\,\delta u}$ **(Ans.)**

Example 8·61. *For the propped cantilever shown in Fig. 8·78 (a) find the support reactions and draw the bending moment diagram.*

Solution. Fig. 8·78 (a) shows the cantilever with loading and propped at B.

Fig. 8·78 (b) shows the B.M. diagram due to U.D.L. only.

Fig. 8·78 (c) shows the B.M. diagram due to prop-reaction R_B only.

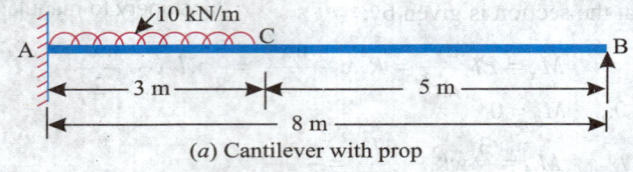

(a) Cantilever with prop

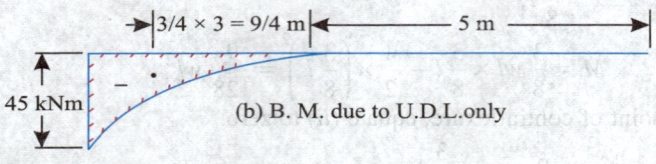

(b) B. M. due to U.D.L. only

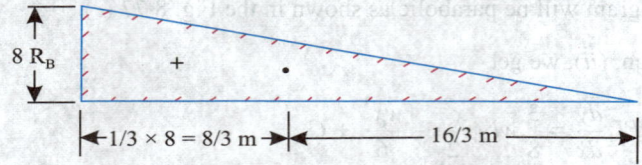

(c) B. M. due to prop-reaction only

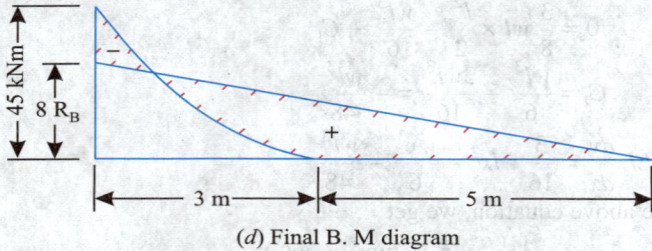

(d) Final B. M diagram

Fig. 8.78

Since the end A is fixed, the tangent to the elastic curve at A shall be horizontal. Deflection of B, which is zero in the present case, with respect to the tangent to elastic curve at A shall be equal to $\dfrac{1}{EI}$ times the algebraic sum of the moments of the two bending moment diagrams about B.

i.e. $\qquad \dfrac{1}{EI}\left[\left\{-\dfrac{1}{3}\times 3\times 45\times\left(\dfrac{9}{4}+5\right)\right\}+\left\{\dfrac{1}{2}\times 8\times 8\,R_B\times\dfrac{16}{3}\right\}\right]=0$

$\qquad -326{\cdot}25+170{\cdot}67\,R_B=0$

$\therefore \qquad\qquad\qquad R_B = \mathbf{1{\cdot}91\ kN}\ \ \textbf{(Ans.)}$

Also, $\qquad\qquad R_A + R_B = 10\times 3 = 30$

$\therefore \qquad\qquad R_A = 30 - 1{\cdot}91 = \mathbf{28{\cdot}09\ kN}\ \ \textbf{(Ans.)}$

Fig. 8·78 (d) shows the final B.M. diagram (obtained by superimposing the two bending moment diagrams [Fig. 8·76 (b) and Fig. 8·76 (c)].

Example 8·62. *A cantilever of effective length 8 metres carries a uniformly distributed load of 20 kN/m run over the whole length. If the cantilever is propped on a point 2 metres from the free end and the level of the prop adjusted so that there is no deflection at the free end, what is the reaction at the prop and the deflection of the beam at that point ?*

Take: $E = 200\times 10^6$ kN/m^2 and $I = 120\times 10^{-4}$ m^4.

Draw the B.M. and S.F. diagrams for the cantilever.

Solution. Fig. 8·79 (a) shows the given cantilever with the prop.

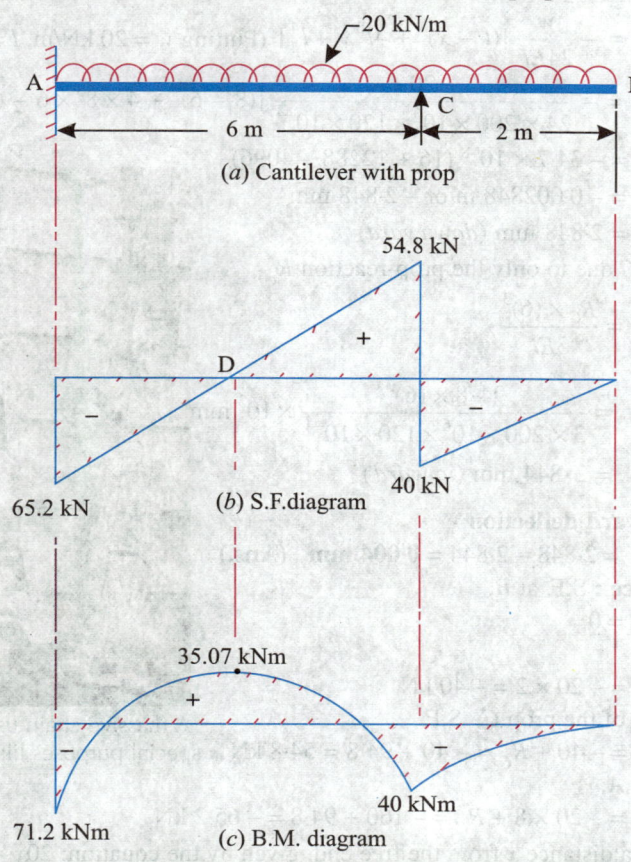

(a) Cantilever with prop

(b) S.F.diagram

(c) B.M. diagram

Fig. 8.79

Downward deflection at the free end B of the cantilever AB without prop

$$= \frac{wl^4}{8EI} = \frac{20 \times 8^4}{8EI} = \frac{10240}{EI}$$

Upward deflection at end B of the cantilever under the effect of only the prop-reaction at C

$$= \frac{R_C \times (6)^3}{3EI} + \frac{R_C \times (6)^2}{2EI} \times (8-6) \qquad \text{...(Eqn. 8·6)}$$

$$= \frac{72\,R_C}{EI} + \frac{36\,R_C}{EI} = \frac{108\,R_C}{EI}$$

For having no deflection of the free end B, equate the above two deflections.

$$\therefore \qquad \frac{10240}{EI} = \frac{108\,R_C}{EI}$$

or, $\qquad R_c = \textbf{94·8 kN} \ \textbf{(Ans.)}$

Downward deflection due to only the U.D.L. at a distance x from A is given by

$$EI\,y = -\frac{w}{24}(l-x)^4 - \frac{wl^3}{6}x + \frac{wl^4}{24} \qquad \text{...(case III Art. 8·6)}$$

$\therefore$ Deflection at C due to only the U.D.L. is

$$EI\,y_C = -\frac{w}{24}[(l-x)^4 + 4l^3x - l^4]$$

$$\therefore \qquad y_C = -\frac{w}{24EI}[(l-x)^4 + 4l^3x - l^4] \ \text{(Putting } w = 20 \text{ kN/m, } l = 8 \text{ m and } x = 6\text{m)}$$

$$= -\frac{20}{24 \times 200 \times 10^6 \times 120 \times 10^{-4}}[(8-6)^4 + 4 \times 8^3 \times 6 - (8)^4]$$

$$= -34·7 \times 10^{-8}(16 + 12288 - 4096)$$

$$= -0·002848 \text{ m or} -2·848 \text{ mm}$$

$$= 2·848 \text{ mm } (downward)$$

Deflection at C due to only the prop-reaction R_c

$$= \frac{R_C \times (6)^3}{3EI}$$

$$= \frac{94·8 \times (6)^3}{3 \times 200 \times 10^6 \times 120 \times 10^{-4}} \times 10^3 \text{ mm}$$

$$= 2·844 \text{ mm } (upward)$$

$\therefore$ **Net downward deflection**

$$= 2·848 - 2·844 = \textbf{0·004 mm} \ \textbf{(Ans.)}$$

Shearing Force : S.F. at B,

$$S_B = 0$$

Just before C,

$$\text{S.F.} = -20 \times 2 = -40 \text{ kN}$$

Just to the left of the prop C, S.F.

$$= -40 + R_C = -40 + 94·8 = 54·8 \text{ kN}$$

A massive chain used in industries for special purposes like material handling.

At the support A,

$$\text{S.F.} = -20 \times 8 + R_C = -160 + 94·8 = -65·2 \text{ kN}$$

S.F. is zero at a distance x from the free end, given by the equation: $20x - 94·8 = 0$

$$\therefore \qquad x = 4·74 \text{ m}$$

Bending moment: $M_C = -\dfrac{20 \times 2^2}{2} = -40$ kNm

$$M_A = -\dfrac{20 \times 8^2}{2} + 94\cdot8 \times 6 = -71\cdot2 \text{ kNm}$$

In span AC the B.M. is maximum at D [Fig. 8·79 (b)] where the S.F. is zero

i.e., at $\qquad\qquad x = 4\cdot74$ m

$\therefore \qquad\qquad M_D = -\dfrac{20 \times 4\cdot74^2}{2} + 94\cdot8\,(4\cdot74 - 2) = 35\cdot07$ kNm

S.F. and B.M. diagrams are shown in Fig. 8·79 (b) and 8·79 (c) respectively.

Example 8·63. *For the propped beam shown in Fig. 8·80 (a) determine the support reaction R_B and draw the S.F. and B.M. diagrams.*

Solution. Fig. 8·80 (a) shows a cantilever AB with loading and a prop at B.
Downward deflection of the free end due to only the varying load on AB

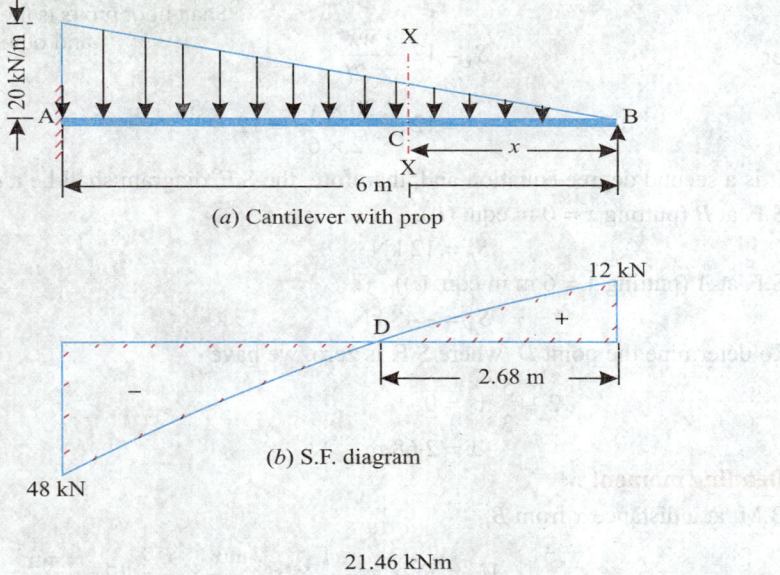

(a) Cantilever with prop

(b) S.F. diagram

(c) B. M. diagram

Fig. 8.80

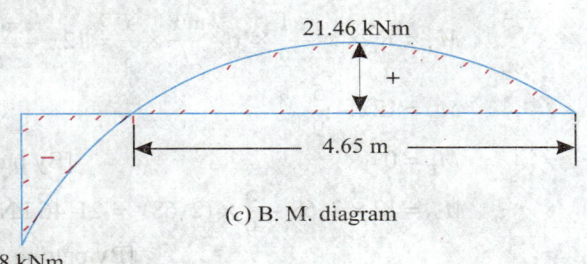

$$= \dfrac{wl^4}{30\ EI} \qquad\qquad\qquad \text{...[Eqn. 8·17]}$$

(Where, $w = 20$ kN/m and $l = 6$ m)

$$= \dfrac{20 \times 6^4}{30\ EI} = \dfrac{864}{EI}$$

Upward deflection of free end B caused by the prop-reaction R_B

$$= \frac{R_B \times l^3}{3EI} = \frac{R_B \times (6)^3}{3EI} = \frac{72\, R_B}{EI} \qquad \ldots[\text{Eqn. 8·3}]$$

Since the prop is level with fixed end and there is no deflection at B hence both the above deflections are *equal*.

$$\therefore \qquad \frac{72\, R_B}{EI} = \frac{864}{EI}$$

or $\qquad R_B = 12\ \text{kN} \quad \textbf{(Ans.)}$

Consider a section XX at a distance x from B.

Rate of loading at the section $= \dfrac{wx}{l}$

Shearing force:

S.F. at C (*i.e.*, at the section XX),

$$S_x = 12 - \frac{1}{2} \times x \times \frac{wx}{l}$$

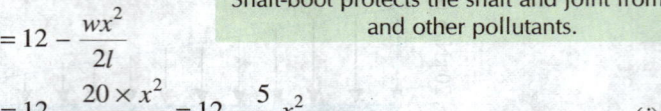

Shaft-boot protects the shaft and joint from dust and other pollutants.

or $\qquad S_x = 12 - \dfrac{wx^2}{2l}$

$$= 12 - \frac{20 \times x^2}{2 \times 6} = 12 - \frac{5}{3}\, x^2 \qquad \ldots(i)$$

It is a second degree equation and, therefore, the S.F. diagram shall be a *curve*.

S.F. at B (putting $x = 0$ in eqn. (i)),

$$S_B = 12\ \text{kN}$$

S.F. at A (putting $x = 6$ m in eqn. (i))

$$S_A = -48\ \text{kN}$$

To determine the point D where S.F. is zero, we have

$$12 - \frac{5}{3}\, x^2 = 0$$

$$\therefore \qquad x = 2\cdot68\ \text{m}$$

Bending moment :

B.M. at a distance x from B,

$$M_x = R_B \times x - \frac{1}{2} \times x \times \frac{wx}{l} \times \frac{x}{3} = 12x - \frac{wx^3}{6l}$$

or $\qquad M_x = 12x - \dfrac{5}{9}\, x^3 \qquad \ldots(ii)$

$\therefore \qquad M_B = 0 \qquad\qquad$ [By putting $x = 0$ in eqn. (ii)]

$$M_D = 12 \times 2\cdot68 - \frac{5}{9} \times (2\cdot68)^3 = 21\cdot46\ \text{kNm}$$

[By putting $x = 2\cdot68$ m in eqn. (ii)]

Since S.F. is zero at D, the B.M. here is *maximum*.

$$M_A = 12 \times 6 - \frac{5}{9} \times (6)^3 = -48\ \text{kNm}$$

To get point of contraflexure, put the eqn. (ii) equal to zero.

$$\therefore \qquad 12x - \frac{5}{9}\, x^3 = 0, \qquad x = 4\cdot65\ \text{m}$$

Fig. 8·80 (b) shows the S.F. diagram.

Fig. 8·80 (c) shows the B.M. diagram.

Example 8·64. *For the cantilever shown in Fig. 8·81 (a) determine:*

(i) *The end fixing moment;*

(ii) *Reaction at the freely supported end.*

Draw the B.M. diagram also.

Solution. Fig. 8·81 (*a*) shows the given cantilever

Fig. 8·81 (*b*) shows the B.M. Diagram due to moment 50 kNm only

Fig. 8·81 (*c*) shows the B.M. diagram due to only prop-reaction R_B.

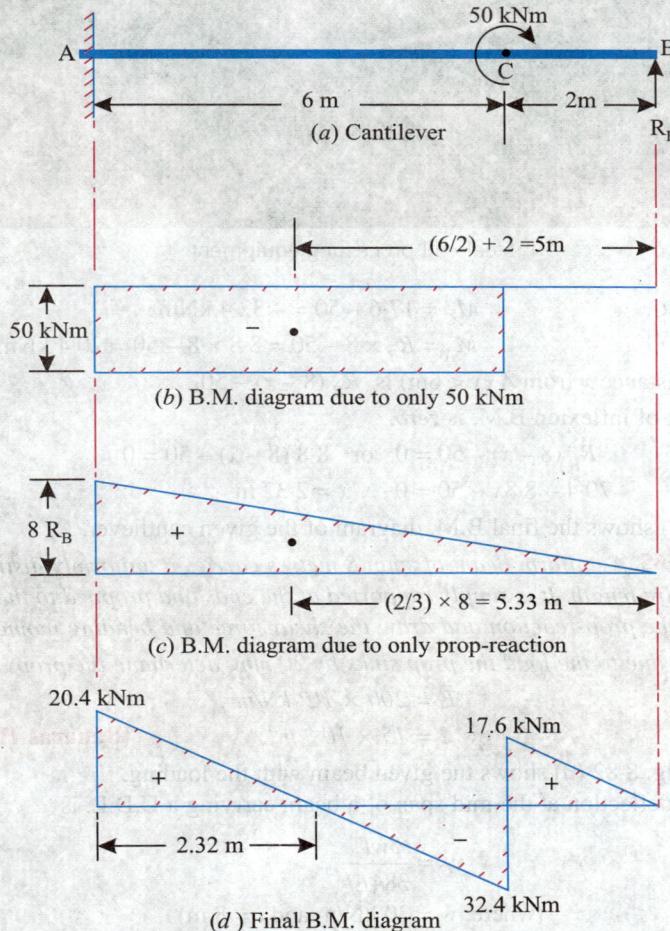

(*a*) Cantilever

(*b*) B.M. diagram due to only 50 kNm

(*c*) B.M. diagram due to only prop-reaction

(*d*) Final B.M. diagram

Fig. 8.81

Level of prop at *B* is the same as that of the fixed end *A*, therefore deflection at *B* is zero. As such the algebraic sum of moments of the two B.M. diagrams about vertical axis through *B* is zero.

$$\therefore \qquad -50 \times 6 \times 5 + \frac{1}{2} \times 8 \times 8 \, R_B \times 5 \cdot 33 = 0$$

$$\therefore \qquad R_B = 8 \cdot 8 \text{ kN}$$

For span BC: $\qquad M_C = R_B \times 2 = 8 \cdot 8 \times 2 = 17 \cdot 6 \text{ kNm}$

Steel coil processing equipment.

For Span AC: $M_C = 17.6 - 50 = -32.4$ kNm

$M_A = R_B \times 8 - 50 = 8.8 \times 8 - 50 = 20.4$ kNm

B.M. at a distance x from A ($x < 6$m) is $R_B (8 - x) - 50$.

At the point of inflexion B.M. is *zero*.

∴ $R_B (8 - x) - 50 = 0$ or $8.8 (8 - x) - 50 = 0$

or, $70.4 - 8.8x - 50 = 0$ ∴ $x = 2.32$ m

Fig. 8.81 (*d*) shows the final B.M. diagram of the given cantilever.

Example 8.65. *A uniform beam of span 8 metres carries a uniformly distributed load of 30 kN / m over its entire length. It is simply supported at the ends and propped to the same level at the centre. Calculate the prop-reaction and draw the shear force and bending moment diagrams.*

If, however, due to the load the prop sinks by 20 mm, determine the prop and end reactions.

Take: $E = 200 \times 10^6$ kN/m²;

$I = 15 \times 10^{-6}$ m⁴. **(Banaras Hindu University)**

Solution. Fig. 8.82 (*a*) shows the given beam with the loading.

Downward deflection at the mid span of a beam carrying a U.D.L. is

$$y_C = \frac{5wl^4}{384EI}$$ ...(Eqn. 8.22)

(where, $w = 30$ kN/m and $l = 8$ m)

Similarly upward deflection due to only the prop-reaction R_C is

$$y'_C = \frac{R_C l^3}{48EI}$$ ...(8.20)

Since the prop is at the same level end supports,

∴ $y_C = y'_C$

or, $$\frac{5wl^4}{384EI} = \frac{R_C l^3}{48EI}$$

or,
$$R_C = \frac{5wl}{8} = \frac{5 \times 30 \times 8}{8} = 150 \text{ kN}$$

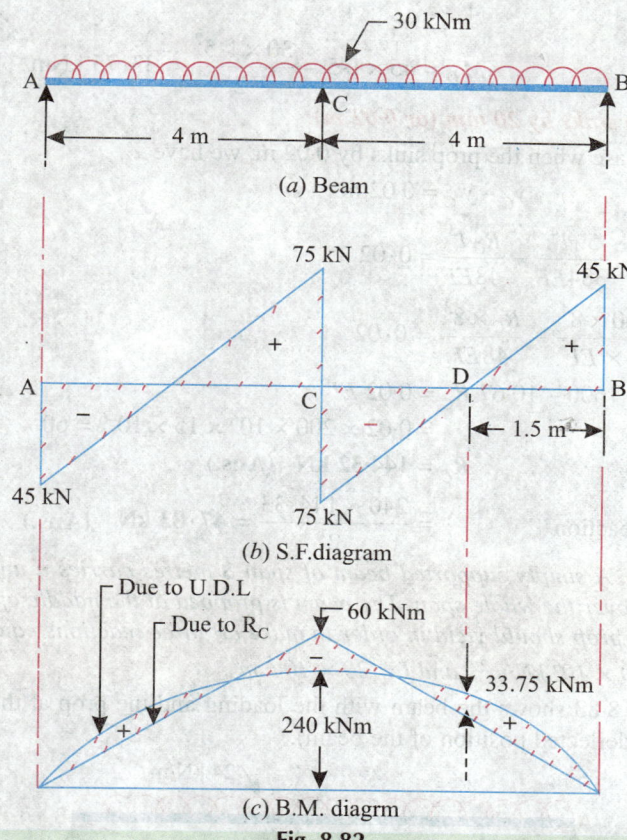

(a) Beam

(b) S.F.diagram

(c) B.M. diagrm

Fig. 8.82

By symmetry, reactions at A and B are equal and each is

$$\frac{30 \times 8 - 150}{2} = 45 \text{ kN}$$

S.F. diagram [Fig. 8·82 (b)] :

$$S_B = + 45 \text{ kN}$$
$$S_C = 45 - 120 = - 75 \text{ kN} \qquad \text{...(span } BC)$$
$$S_C = 45 - 120 + 150 = + 75 \text{ kN} \qquad \text{...(span } AC)$$
$$S_A = - 45 \text{ kN}$$

To find the point D, where S.F. is zero, we have from similarity of Δs

$$\frac{75}{45} = \frac{CD}{BD} = \frac{4 - BD}{BD}$$

∴ $$BD = 1·5 \text{ m}$$

B.M. diagram [Fig. 8·82 (c)] :

B.M. at mid span due to only U.D.L.

$$= \frac{wl^2}{8} = \frac{30 \times 8^2}{8} = 240 \text{ kN m}$$

B.M. at mid span due to only the prop-reaction (R_C = 150 kN) is

$$-\frac{R_C\, l}{4} = -\frac{150 \times 8}{4} = -300 \text{ kNm}$$

$$M_D = 45 \times 1 \cdot 5 - \frac{30 \times 1 \cdot 5^2}{2} = 33 \cdot 75 \text{ kNm}$$

When the prop sinks by 20 mm (or 0·02 m):

In the second case when the prop sinks by 0·02 m, we have

$$y_C - y'_C = 0 \cdot 02 \text{ m}$$

∴ $$\frac{5wl^4}{384EI} - \frac{R_C\, l^3}{48EI} = 0 \cdot 02$$

or, $$\frac{5 \times 30 \times 8^4}{384 \times EI} - \frac{R_C \times 8^3}{48EI} = 0 \cdot 02$$

$$1600 - 10 \cdot 67\, R_C = 0 \cdot 02\, EI$$
$$= 0 \cdot 02 \times 200 \times 10^6 \times 15 \times 10^{-6} = 60$$

∴ $$R_C = \textbf{144·32 kN} \quad \textbf{(Ans.)}$$

∴ Each end reaction $$= \frac{240 - 144 \cdot 33}{2} = \textbf{47·83 kN} \quad \textbf{(Ans.)}$$

Example 8·66. *A simply supported beam of span 8 metres carries a uniformly distributed load of 24 kN/m run over the whole span. The beam is propped at the middle of the span. Find the amount by which the prop should yield in order to make all three reactions equal.*

Take : E = 200 × 10⁶ kN/m², and I = 20 × 10⁻⁵ m⁴.

Solution. Fig. 8·83 shows the beam with the loading and the prop at the mid span (dotted curve represents the deflected position of the beam).

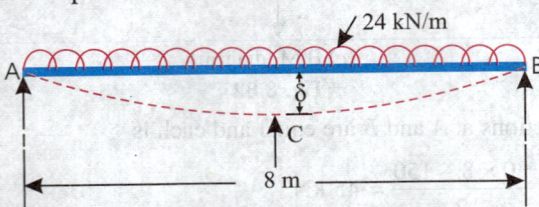

Fig. 8.83

Let the prop at C sink by δ so that each support carries equal load, then

$$R_A = R_B = R_C = \frac{24 \times 8}{3} = 64 \text{ kN}$$

Downward deflection of beam due to U.D.L. alone,

$$y_C = \frac{5\, wl^4}{384\, EI} = \frac{5 \times 24 \times 8^4}{384\, EI} = \frac{1280}{EI}$$

Upward deflection of beam due to prop-reaction R_C at C,

$$y'_C = \frac{R_C\, l^3}{48\, EI} = \frac{64 \times 8^3}{48\, EI} = \frac{682 \cdot 67}{EI}$$

Also, $$y_C - y'_C = \delta$$

∴ $$\frac{1280}{EI} - \frac{682 \cdot 67}{EI} = \delta$$

or
$$\delta = \frac{597 \cdot 33}{EI} = \frac{597 \cdot 33}{200 \times 10^6 \times 20 \times 10^{-5}} \times 10^3 = 14 \cdot 93 \text{ mm}$$

Hence,
$$\delta = 14 \cdot 93 \text{ mm} \quad \text{(Ans.)}$$

TYPICAL EXAMPLES (For Competitive Examinations)

Example 8·67. *For the beam shown in the Fig. 8·84 determine:*

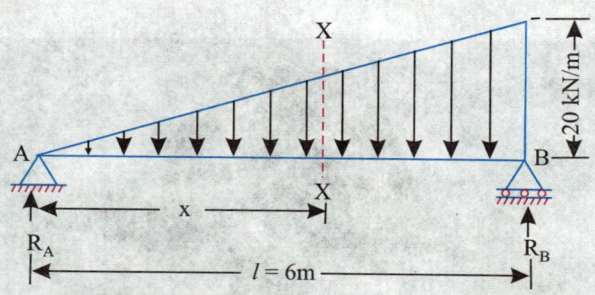

Fig. 8.84

  (*i*) *Maximum deflection;*   (*ii*) *Mid span deflection.*

Take: $E = 200 \times 10^6 \text{ kN/m}^2; I = 15 \times 10^{-5} \text{ m}^4.$

Solution. To find the reaction R_B taking moments about A, we have

$$R_B \times 6 = \frac{1}{2} \times 6 \times 20 \times \left(\frac{2}{3} \times 6\right) = 240$$

∴
$$R_B = 40 \text{ kN}$$

and,
$$R_A = \frac{1}{2} \times 6 \times 20 - 40 = 20 \text{ kN}$$

Rate of loading at the section XX

$$= \frac{wx}{l} = \frac{20x}{6} = \frac{10x}{3}$$

∴
$$M_x = 20 \times x - \frac{1}{2} \times x \times \frac{10x}{3} \times \frac{x}{3}$$

$$= 20x - \frac{5x^3}{9}$$

But,
$$M_x = EI \frac{d^2 y}{dx^2} = 20x - \frac{5x^3}{9}$$

Integrating both sides, we get

$$EI \frac{dy}{dx} = 10x^2 - \frac{5x^4}{36} + C_1 \qquad \qquad ...(i)$$

(where, C_1 = constant of integration)

Integrating eqn. (*i*), we get

$$EIy = \frac{10x^3}{3} - \frac{x^5}{36} + C_1 x + C_2 \qquad \qquad ...(ii)$$

(where, C_2 = constant of integration)

When, $x = 0, y = 0$ ∴ $C_2 = 0$

When, $x = 6m,\ y = 0$ $\therefore\quad 0 = \dfrac{10}{3} \times 6^3 - \dfrac{6^5}{36} + 6C_1$

or, $= 720 - 216 + 6C_1$ $\therefore\quad C_1 = -84$

Hence, $EI \dfrac{dy}{dx} = 10x^2 - \dfrac{5x^4}{36} - 84$ ...(1) *Slope equation*

and, $EIy = \dfrac{10x^3}{3} - \dfrac{x^5}{36} - 84x$ ...(2) *Deflection equation*

Prototype of a petrol engine in a laboratory.

For *maximum deflection* put

$$\dfrac{dy}{dx} = 0.$$

$\therefore$ $0 = 10x^2 - \dfrac{5x^4}{36} - 84$

$0 = 360x^2 - 5x^4 - 3024$ or $x^4 - 72\,x^2 + 604\cdot 8 = 0$

$\therefore$ $x^2 = \dfrac{72 \pm \sqrt{72^2 - 4 \times 604\cdot 8}}{2} = \dfrac{72 - 52\cdot 58}{2} = 9\cdot 71$

(Neglecting +ve sign since x cannot be more than 6 m)

or, $x = 3\cdot 12$ m

Thus deflection is maximum at 3·12 m distance from left hand support.

(i) Maximum deflection y_{max}:

For getting y_{max},   put $x = 3\cdot 12$ m in the deflection equation (2).

$$EIy_{max} = \dfrac{10 \times (3\cdot 12)^3}{3} - \dfrac{(3\cdot 12)^5}{36} - 84 \times 3\cdot 12$$

$$= 101.24 - 8.21 - 262.08 = -169.05$$

$\therefore$ $y_{max} = -\dfrac{169\cdot 05}{EI} = -\dfrac{169\cdot 05}{200 \times 10^6 \times 15 \times 10^{-5}} \times 10^3\ \text{mm} = -5\cdot 635\ \text{mm}$

Hence, $y_{max} = 5.635$ mm (downward) (Ans.)

(ii) **Deflection at the mid span:**

For mid span deflection, put $x = 3$ m in the deflection equation (2).

$$EIy = \frac{10 \times 3^3}{3} - \frac{(3)^5}{36} - 84 \times 3 = 90 - 6.75 - 252 = -168.75$$

$$\therefore \qquad y = -\frac{168.75}{EI} = -\frac{168.75}{200 \times 10^6 \times 15 \times 10^{-5}} \times 10^3 = -5.625 \text{ mm}$$

Hence, mid span deflection = 5.625 mm (downward) **(Ans.)**

Example 8·68. *Determine by Macaulay's method or otherwise the deflection at C and D in the beam shown in Fig. 8·85 (a).*

Take: $E = 200 \times 10^6$ kN/m^2 and $I = 20 \times 10^{-5}$ m^4 **(Rajasthan University)**

Solution. Fig. 8·85 (a) shows the beam with the loading.

To find reaction R_B taking moments about A, we have

$$R_B \times 6 = 20 \times 3 \times \frac{3}{2} + 20 \times 4 = 90 + 80 = 170 \qquad \therefore R_B = 28.33 \text{ kN}$$

and, $R_A = (20 \times 3 + 20) - 28.33 = 51.67$ kN

On part CB of the given beam assume a U.D.L. @ 20 kN/m applied both from above and below so that both these added loads neutralise each other and the net effect remains unchanged [Fig. 8·85 (b)].

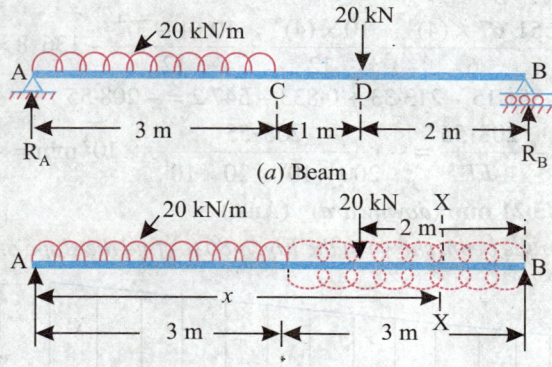

(a) Beam

(b) Beam with modified loading

Fig. 8.85

Consider a section XX at a distance x from the support A.

$$M_x = EI \frac{d^2y}{dx^2} = 51.67x - 10x^2 - 20(x-4) + 10(x-3)^2 \qquad ...(i)$$

By integrating the above equation successively, we get

$$EI \frac{dy}{dx} = \frac{51.67x^2}{2} - \frac{10x^3}{3} - 10(x-4)^2 + \frac{10(x-3)^3}{3} + C_1 \qquad ...(ii)$$

and, $$EIy = \frac{51.67x^3}{6} - \frac{10x^4}{12} - \frac{10(x-4)^3}{3} + \frac{10(x-3)^4}{12} + C_1 x + C_2 \qquad ...(iii)$$

(where, C_1 and C_2 are the constants of integration)

When, $x = 0, y = 0$ $\therefore$ $C_2 = 0$

When, $x = 6$ m, $y = 0$

$$\therefore \quad 0 = \frac{51 \cdot 67 \times 6^3}{6} - \frac{10 \times 6^4}{12} - \frac{10(6-4)^3}{3} + \frac{10(6-3)^4}{12} + 6C_1$$

$$= 1860 - 1080 - 26 \cdot 67 + 67 \cdot 5 + 6C_1$$

$$\therefore \quad C_1 = -136 \cdot 8$$

Hence the slope and deflection equations are:

$$EI\frac{dy}{dx} = \frac{51 \cdot 67x^2}{2} - \frac{10x^3}{3} - 10(x-4)^2 + \frac{10(x-3)^3}{3} - 136 \cdot 8 \quad \text{...Slope equation}$$

$$EIy = \frac{51 \cdot 67x^3}{6} - \frac{10x^4}{12} - \left[\frac{10(x-4)^3}{3}\right] + \left[\frac{10(x-3)^4}{12}\right] - 136 \cdot 8x \quad \text{...Deflection equation}$$

Deflection at C, y_C :

Deflection y_C at C, where $x = 3$ m (neglecting square brackets in which the terms become –ve) is:

$$EIy_C = \frac{51 \cdot 67 \times 3^3}{6} - \frac{10 \times 3^4}{12} - 136 \cdot 8 \times 3 = 232 \cdot 5 - 67 \cdot 5 - 410 \cdot 4 = -245 \cdot 4$$

$$\therefore \quad y_C = -\frac{245 \cdot 4}{EI} = -\frac{245 \cdot 4}{200 \times 10^6 \times 20 \times 10^{-5}} \times 10^3 \text{ mm} = -6 \cdot 135 \text{ mm}$$

i.e. $\quad y_C = \mathbf{6 \cdot 135}$ **mm** *(downward)* **(Ans.)**

Deflection at D, y_D :

To obtain deflection y_D at D put $x = 4$ m in the deflection equation.

$$EIy_D = \frac{51 \cdot 67 \times (4)^3}{6} - \frac{10 \times (4)^4}{12} + \frac{10(4-3)^4}{12} - 136 \cdot 8 \times 4$$

$$= 551 \cdot 15 - 213 \cdot 33 + 0 \cdot 833 - 547 \cdot 2 = -208 \cdot 55$$

$$\therefore \quad y_D = -\frac{208 \cdot 55}{EI} = -\frac{208 \cdot 55}{200 \times 10^6 \times 20 \times 10^{-5}} \times 10^3 \text{ mm} = -5 \cdot 21 \text{ mm}$$

Hence, $\quad y_D = \mathbf{5 \cdot 21}$ **mm** *(downward)* **(Ans.)**

Example 8·69. *For the beam shown in the Fig. 8·86 (a) find the slope and deflection equations.*

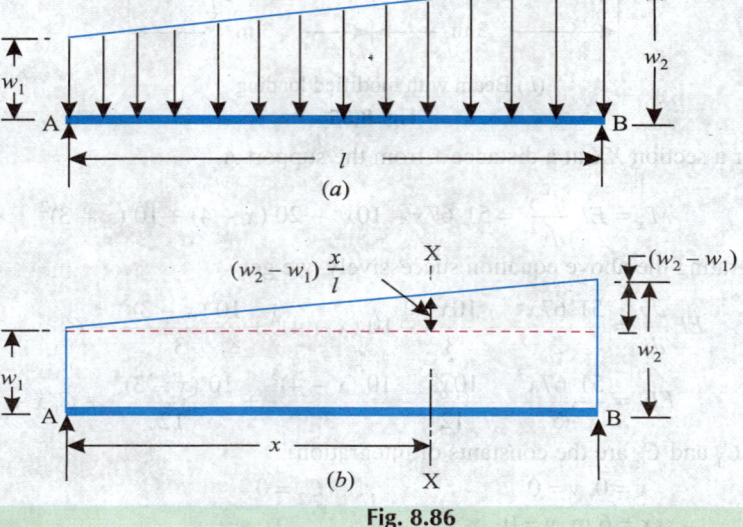

Fig. 8.86

Solution. To find reaction R_A taking moments about B, we get

$$R_A \cdot l = (w_1 l) \times \frac{l}{2} + \frac{1}{2} \times l \times (w_2 - w_1) \times l/3$$

$$= \frac{w_1 l^2}{2} + (w_2 - w_1) \times \frac{l^2}{6} = \frac{w_1 l^2}{2} + \frac{w_2 l^2}{6} - \frac{w_1 l^2}{6}$$

$$= \frac{w_1 l^2}{3} + \frac{w_2 l^2}{6}$$

$$\therefore \qquad R_A = \frac{w_1 l}{3} + \frac{w_2 l}{6}$$

Consider any section XX at a distance x from the support A.

$$M_x = EI \frac{d^2 y}{dx^2} = \left(\frac{w_1 l}{3} + \frac{w_2 l}{6} \right) x - \frac{w_1 x^2}{2} - \left[\frac{1}{2} x \times (w_2 - w_1) \frac{x}{l} \right] \times \frac{x}{3}$$

or $\qquad EI \frac{d^2 y}{dx^2} = \frac{l}{6} (2w_1 + w_2) x - \frac{w_1 x^2}{2} - \frac{1}{6} (w_2 - w_1) \frac{x^3}{l}$ ...(i)

Integrating successively the eqn. (i), we get

$$EI \frac{dy}{dx} = \frac{l}{12} (2w_1 + w_2) x^2 - \frac{w_1 x^3}{6} - \frac{(w_2 - w_1)}{24} \cdot \frac{x^4}{l} + C_1$$ ...(ii)

and, $\qquad EIy = \frac{l}{36} (2w_1 + w_2) x^3 - \frac{w_1}{24} x^4 - \frac{w_2 - w_1}{24} \cdot \frac{x^5}{5l} + C_1 x + C_2$ ...(iii)

(where, C_1 and C_2 are the constants of integration)

When $\qquad x = 0, y = 0$

$\therefore \qquad C_2 = 0$ [from eqn. (iii)]

When $\qquad x = l, y = 0$

$\therefore$

$$0 = \frac{l}{36} (2w_1 + w_2) l^3 - \frac{w_1}{24} l^4 - \frac{(w_2 - w_1)}{120} \cdot l^4 + C_1 l$$ [from eqn.(iii)]

$$= \frac{l^4}{360} (20 w_1 + 10 w_2 - 15 w_1 - 3 w_2 + 3 w_1) + C_1 l$$

$$= \frac{l^4}{360} (8 w_1 + 7 w_2) + C_1 l$$

$\therefore \qquad C_1 = - \frac{l^3}{360} (8 w_1 + 7 w_2)$

Thus substituting the values of C_1 and C_2 in eqns. (ii) and (iii), the slope and deflection equations are :

$$EI \frac{dy}{dx} = \frac{l}{12} (2 w_1 + w_2) x^2 - \frac{w_1 x^3}{6} - \frac{(w_2 - w_1)}{24} \cdot \frac{x^4}{l} - \frac{l^3}{360} (8 w_1 + 7 w_2)$$

...Slope equation **(Ans.)**

$$EIy = \frac{l}{36} (2 w_1 + w_2) x^3 - \frac{w_1}{24} x^4 - \frac{w_2 - w_1}{120} \cdot \frac{x^5}{l} - \frac{l^3}{360} (8 w_1 + 7 w_2) x$$

...Deflection equation **(Ans.)**

Example 8·70. *A long flat strip 50 mm wide and 3·2 mm thick is lying on a flat horizontal plane. One end of the strip is now lifted 25 mm from the plane by a vertical force applied at the end. The strip is so long that other end remains undisturbed. Calculate the maximum stress in the steel.*

Take : Weight of strip = 7800 kg/m³, and E = 205 GN/m². **(Panjab University)**

Solution. Width of the strip, $b = 50$ mm

Thickness of the strip, $t = 3.2$ mm

Weight of steel $= 7800$ kg/m^3

Young's modulus, $E = 205$ GN/m^2

Refer to Fig. 8·87.

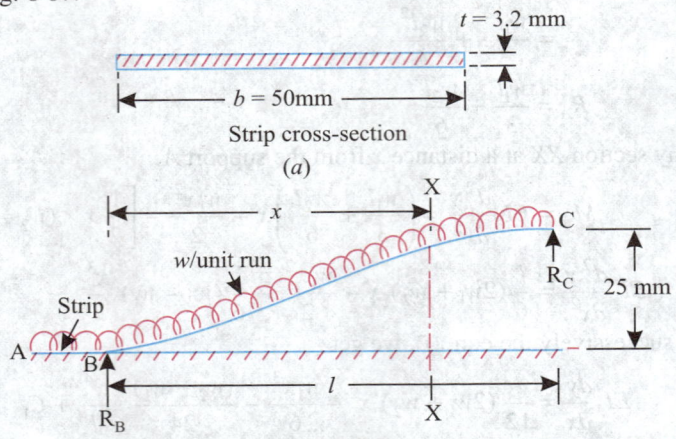

Strip cross-section

(a)

Fig. 8.87

As far as portion *AB* is concerned it does not bend since the ground reaction just balances the weight of the strip, but at the point *B* where the strip leaves the plane there will be the reaction R_B and the conditions are as though the length *AB* were cut.

At *B* slope is *zero*.

Taking the origin at *B* and assuming the end *C* to be simply supported, the bending moment equation at the section *XX* at a distance *x* from *B* is written as :

Tiny IC engine cylinders.

$$M_x = EI \frac{d^2 y}{dx^2} = R_B \times x - \frac{wx^2}{2} \qquad ...(i)$$

Now, at $x = l, M_x = 0$, whence $R_B = \dfrac{wl}{2}$

Integrating successively the eqn. (*i*), we get

$$EI \frac{dy}{dx} = R_B \times \frac{x^2}{2} - \frac{wx^3}{6} + C_1 \qquad ...(ii)$$

and, $$EIy = R_B \times \frac{x^3}{6} - \frac{wx^4}{24} + C_1 x + C_2 \qquad ...(iii)$$

(where, C_1 and C_2 are the constants of integration)

When, $x = 0, \dfrac{dy}{dx} = 0$ $\quad \therefore C_1 = 0$ [From eqn. (*ii*)]

When, $x = 0, y = 0$ $\quad \therefore C_2 = 0$

Thus the slope and deflection equations are :

$$EI \frac{dy}{dx} = \frac{wl}{2} \cdot \frac{x^2}{2} - \frac{wx^3}{6} \qquad \text{...Slope equation}$$

and,

$$EIy = \frac{wl}{2} \cdot \frac{x^3}{6} - \frac{wx^4}{24} \qquad \text{...Deflection equation}$$

Now, when $x = l, y = 25$ mm $= 0.025$ m, we have

$$0.025 \ EI = \frac{wl^4}{12} - \frac{wl^4}{24} = \frac{wl^4}{24}$$

or,

$$\frac{wl^4}{24 \ EI} = 0.025 \quad \text{or} \quad \frac{wl^4}{EI} = 0.6$$

or,

$$l^4 = \left(\frac{0.6 \ EI}{w} \right) m^4$$

Now, moment of inertia of strip, $I = \dfrac{bt^3}{12} = \dfrac{50 \times (3 \cdot 2)^3}{12}$ mm^4

$$= 136 \cdot 53 \text{ mm}^4 = 136 \cdot 53 \times 10^{-12} \text{ m}^4.$$

Weight of the strip per metre length,

$$w = (b \times t \times 1 \times 7800) \times 9 \cdot 81, \text{ where } b \text{ and } t \text{ are in metres.}$$

$$= 50 \times 10^{-3} \times 3 \cdot 2 \times 10^{-3} \times 7800 \times 9 \cdot 81 = 12 \cdot 24 \text{ N/m}$$

$\therefore$

$$l^4 = \frac{0 \cdot 6 \times 205 \times 10^9 \times 136 \cdot 53 \times 10^{-12}}{12 \cdot 24} = 1 \cdot 37 \text{ m}^4$$

or,

$$l = 1 \cdot 08 \text{ m}$$

Then,

$$R_B = \frac{wl}{2} = \frac{12 \cdot 24 \times 1 \cdot 08}{2} = 6 \cdot 61 \text{ N}$$

For maximum bending moment

$$\frac{dM_x}{dx} = 0$$

$\therefore$

$$\frac{d}{dx} \left(R_B \times x - \frac{wx^2}{2} \right) = 0$$

or

$$R_B - wx = 0$$

$\therefore$

$$x = \frac{R_B}{w} = \frac{wl/2}{w} = \frac{l}{2}$$

$\therefore$

$$M_{max} = + R_B \cdot \frac{l}{2} - \frac{w \cdot (l/2)^2}{2}$$

$$= + \frac{wl}{2} \cdot \frac{l}{2} - \frac{wl^2}{8} = \frac{wl^2}{8}$$

$$= \frac{12 \cdot 24 \times 1 \cdot 08^2}{8} = + 1 \cdot 785 \text{ Nm}$$

+ve sign indicates that the moment is *sagging i.e.* the *upper fibres are in compression.*

$\therefore$ Maximum bending stress (σ_b) is given by:

$$\frac{M}{I} = \frac{\sigma_b}{y}$$

or,

$$\sigma_b = \frac{My}{I} = \frac{1 \cdot 785 \times \left(\dfrac{3 \cdot 2}{2} \times 10^{-3} \right)}{136 \cdot 53 \times 10^{-12}} = 20 \cdot 92 \times 10^6 \text{ N/m}^2$$

= 20·92 MN/m²

Hence, $\sigma_b = $ **20·92 MN/m²** **(Ans.)**

Example 8·71. *A 1·5 m long horizontal cantilever, tapers in section from 200 mm deep by 75 mm wide at the fixed end to 75 mm square at the extreme end. It carries a load of 2·7 kN. Calculate the deflection of the free end.*

Take: $E = 14 \times 10^6 \ kN/m^2.$ **(Panjab University)**

Solution. Refer to Fig. 8·88

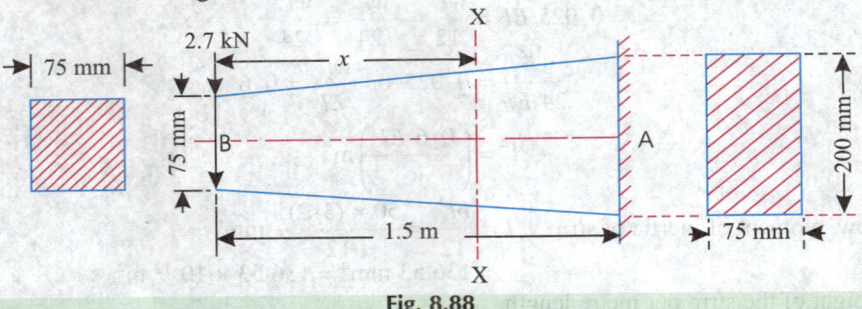

Fig. 8.88

Load, $W = 2\cdot7$ kN

Young's modulus,

$E = 14 \times 10^6 \ kN/m^2$

$I_A = \dfrac{75 \times 200^3}{12} = 50 \times 10^6 \ mm^4 = 50 \times 10^{-6} \ m^4$

$I_B = \dfrac{75^4}{12} = 2\cdot64 \times 10^6 \ mm^4 = 2\cdot64 \times 10^{-6} \ m^4$

$I_{XX} = \left[2\cdot64 + (50 - 2\cdot64)\dfrac{x}{l}\right] \times 10^{-6} \ m^4$

$= \left(2\cdot64 + 47\cdot36\dfrac{x}{l}\right) \times 10^{-6} \ m^4$

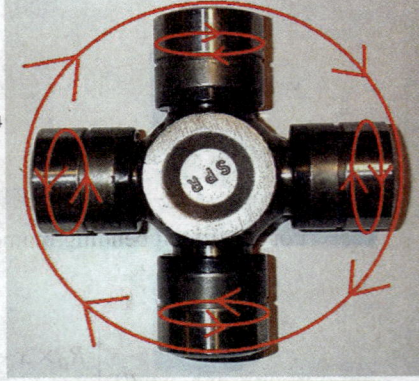

Schematic view of a universal joint.

Consider a section *XX* at a distance *x* from the end *B*.

$$M_x = EI_{XX}\dfrac{d^2 y}{dx^2} = -2\cdot7x \qquad\qquad\qquad ...(i)$$

or, $E\dfrac{d^2 y}{dx^2} = -\dfrac{2\cdot7 \ xl}{(2\cdot64 \ l + 47\cdot36x) \times 10^{-6}}$

$= -\dfrac{(2\cdot7 \times 1\cdot5)\ 10^6 \ x}{(2\cdot64 \times 1\cdot5 + 47\cdot36x)} = -\dfrac{4\cdot05x \times 10^6}{(47\cdot36x + 3\cdot96)}$

or, $E \cdot \dfrac{d^2 y}{dx^2} = -8\cdot56 \times 10^4 + \dfrac{3\cdot39 \times 10^5}{47\cdot36x + 3\cdot96}$ \qquad\qquad ...(ii)

Integrating eqn (*ii*), we get

$$E \cdot \dfrac{dy}{dx} = 8\cdot56 \times 10^4 \ x + \dfrac{3\cdot39 \times 10^5}{47\cdot36} \cdot \log_e (47\cdot36x + 3\cdot96) + C_1$$

(where, C_1 = constant of integration)

When, $x = l = 1\cdot5$ m, $\dfrac{dy}{dx} = 0$

∴ $$0 = -8.56 \times 10^4 \times 1.5 + \frac{3.39 \times 10^5}{47.36} \log_e (47.36 \times 1.5 + 3.96) + C_1$$

or, $$C_1 = 12.84 \times 10^4 - 7.16 \times 10^3 \log_e 75$$
$$= 12.84 \times 10^4 - 7.16 \times 10^3 \times 4.32 = +9.75 \times 10^4$$

∴ $$E \frac{dy}{dx} = -8.56 \times 10^4 x + \frac{3.39 \times 10^5}{47.36} \log_e (47.36x + 3.96) + 9.75 \times 10^4$$

...(iii) (Slope equation)

Integrating eqn (iii), we get

$$Ey = -8.56 \times 10^4 \times \frac{x^2}{2} + \frac{7.16 \times 10^3 (47.36x + 3.96)}{47.36}$$
$$\times [\log_e (47.36x + 3.96) - 1] + 9.75 \times 10^4 x + C_2$$
$$= -4.28 \times 10^4 x^2 + 1.512 \times 10^3 (47.36x + 3.96)$$
$$\times [\log_e (47.36x + 3.96) - 1] + 9.75 \times 10^4 x + C_2$$

Now, when $x = 1.5m, y = 0$

∴ $$0 = -4.28 \times 10^4 \times 1.5^2 + 1.512 \times 10^2 (47.36 \times 1.5 + 3.96)$$
$$[\log_e (47.36 \times 1.5 + 3.96) - 1] + 9.75 \times 10^4 \times 1.5 + C_2$$
$$= -9.63 \times 10^4 + 11340 [\log_e (75) - 1] + 14.625 \times 10^4 + C_2$$
$$= +4.995 \times 10^4 + 11340 (4.32 - 1) + C_2$$

∴ $$C_2 = -4.995 \times 10^4 - 11340 \times 3.32 = -8.76 \times 10^4$$

Hence, $$EIy = 4.28 \times 10^4 x^2 + 1.512 \times 10^2 (47.36x + 3.96) [\log_e (47.36x + 3.96) - 1]$$
$$+ 9.75 \times 10^4 x - 8.76 \times 10^4 \qquad ...Deflection\ equation$$

Thus, when $x = 0$, the deflection at the free end B is

$$Ey_B = 1.512 \times 10^2 \times 3.96 (\log_e 3.96 - 1) - 8.76 \times 10^4$$
$$= 56.89 - 8.76 \times 10^4 \simeq -8.76 \times 10^4$$

∴ $$y_B = -\frac{8.76 \times 10^4}{14 \times 10^6} \times 10^3 \ mm = -6.26 \ mm$$

Hence, $$y_B = 6.26 \ mm \ (downward) \ \ (Ans.)$$

Example 8·72. *A cantilever of uniform strength is to be turned from a mid steel bar 50 mm in diameter. A load of 12 kN is to be supported from the free end. If the maximum permissible stress is limited to 84 N/mm² determine:*

(i) The maximum permissible length of the cantilever;

(ii) The deflection at the free end.

Take: $\qquad E = 2 \times 10^5 \ N/mm^2.$

Solution. Refer to Fig. 8·89.

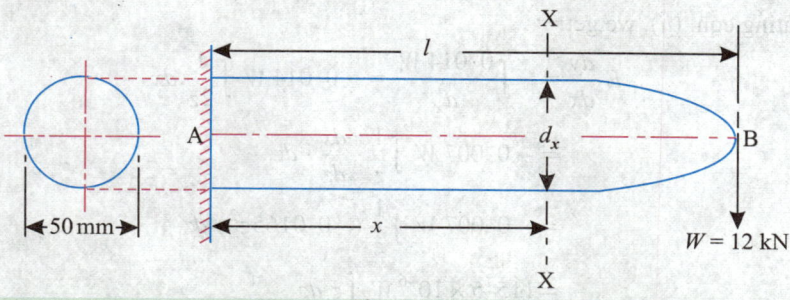

Fig. 8.89

Diameter at the end A, $d_A = 50$ mm $= 0.05$ m

Load at the free end, $W = 12$ kN $= 12000$ N

Maximum permissible stress,

$$\sigma = 84 \text{ N/mm}^2$$

Young's modulus, $E = 2 \times 10^5$ N/mm^2

(i) The maximum permissible length of the cantilever, l:

Consider any section XX at a distance x from the fixed end A. The bending moment at the section,

$$M_x = -W(l - x) = W(l - x) \text{ in magnitude}$$

Also, $$W(l - x) = \sigma \times Z = \sigma \times \frac{\pi}{32} d_x^3$$

$$\left[\begin{array}{l} \text{where, } d_x = \text{diameter of the cantilever at the section, and} \\ Z = \text{section modulus.} \end{array} \right]$$

$$\therefore \qquad d_x^3 = \frac{32\,W\,(l-x)}{\pi\,\sigma} = \frac{32 \times 12000\,(l-x)}{\pi \times 84} = 1455\,(l-x) \qquad ...(i)$$

When, $x = 0, d_A = 50$ mm

$$\therefore \qquad (50)^3 = 1455\,l$$

or, $$l = 85.9 \text{ mm} \quad \textbf{(Ans.)}$$

(ii) The deflection at the free end, y_B:

$$M_x = EI_{XX} \frac{d^2 y}{dx^2} = -W(l - x)$$

or, $$E \frac{d^2 y}{dx^2} = -\frac{W(l-x)}{\dfrac{\pi}{6} \times d_x^4} = -\frac{64W(l-x)}{\pi\,d_x^4}$$

$$= -\frac{64W(l-x)}{\pi \times 1455\,(l-x)\,d_x}$$

or, $$E \frac{d^2 y}{dx^2} = -\frac{0.014\,W}{d_x}$$

Let, $$d_x = 2z \qquad ...(ii)$$

Now from eqn. (i), we have

$$l - x = \frac{d_x^3}{1455} = \frac{8z^3}{1455} \quad \text{or} \quad x = l - \frac{8z^3}{1455}$$

$$\frac{dx}{dz} = -\frac{24z^2}{1455} = -0.0165\,z^2 \qquad ...(iii)$$

Integrating eqn. (ii), we get

$$E \frac{dy}{dx} = -\int \frac{0.014\,W}{d_x} = -0.014\,W \int \frac{1}{2z}\,dx$$

$$= -0.007\,W \int \frac{1}{z} \cdot \frac{dx}{dz} \cdot dz$$

$$= -0.007\,W \int \frac{1}{z}\,(-0.0165z^2)\,dz$$

$$= 115.5 \times 10^{-6}\,W \int z\,dz$$

$$= 115 \cdot 5 \times 10^{-6} \, W \left(\frac{z^2}{2} \right) + C_1 \text{ (where } C_1 = \text{constant of integration)}$$

At the fixed end A: when, $z = \dfrac{d_x}{2} = \dfrac{50}{2} = 25$ mm, $\dfrac{dy}{dx} = 0$

∴ $$0 = 115 \cdot 5 \times 10^{-6} \, W \left(\frac{25^2}{2} \right) + C_1$$

∴ $$C_1 = -0 \cdot 036 \, W$$

Hence, $$E \frac{dy}{dx} = 57 \cdot 75 \times 10^{-6} \, W \, z^2 - 0 \cdot 036 \, W$$

Integrating further, we get

$$Ey = 57 \cdot 75 \times 10^{-6} \, W \int (z^2 - 623 \cdot 4) \, \frac{dx}{dz} \cdot dz$$

$$= 57 \cdot 75 \times 10^{-6} \, W \int (z^2 - 623 \cdot 4)(-0 \cdot 0165 \, z^2) \, dz$$

$$= -0 \cdot 953 \times 10^{-6} \, W \int (z^4 - 623 \cdot 4 \, z^2) \, dz$$

$$= -0 \cdot 953 \times 10^{-6} \, W \left(\frac{z^5}{5} - \frac{623 \cdot 4 \, z^3}{3} \right) + C_2$$

(where, C_2 = constant of integration)

At A: $$z = \frac{d_x}{2} = 25, \; y = 0$$

∴ $$0 = -0 \cdot 953 \times 10^{-6} \, W \left[\frac{(25)^5}{5} - \frac{623 \cdot 4}{3} (25)^3 \right] + C_2$$

$$= 1 \cdot 23 \, W + C_2$$

∴ $$C_2 = -1 \cdot 23 \, W$$

Hence, $$Ey = -0 \cdot 953 \times 10^{-6} \, W \left(\frac{z^5}{5} - \frac{623 \cdot 4 \, z^3}{3} \right) - 1 \cdot 23 \, W$$

At B: $$z = 0 \text{ and } y = y_B$$

∴ $$Ey_B = -1 \cdot 23 \, W$$

or, $$y_B = -\frac{1 \cdot 23 \times 12000}{2 \times 10^5} = -0 \cdot 0738 \text{ mm}$$

Hence, $$y_B = 0 \cdot 0738 \text{ mm (downward) \quad (Ans.)}$$

Example 8·73. *A cantilever of a circular section tapers uniformly from a diameter d at the free end to 2d at the fixed end. It carries a concentrated load W at the free end. Calculate the diameter of a cantilever of uniform cross-section, which would have the same deflection.*

Solution. Fig. 8·90 shows the given cantilever with a point load W at the end. The upper and lower faces of the cantilever have been extended to meet at the point C. Obviously $AB = BC = l$

Consider any section XX at a distance x from the point C.

The diameter at the section,

$$d_x = \frac{x}{l} \cdot d$$

∴ $$I_{XX} = \frac{\pi}{64} \cdot \left(\frac{xd}{l} \right)^4 = \frac{\pi}{64} \cdot \frac{x^4 d^4}{l^4}$$

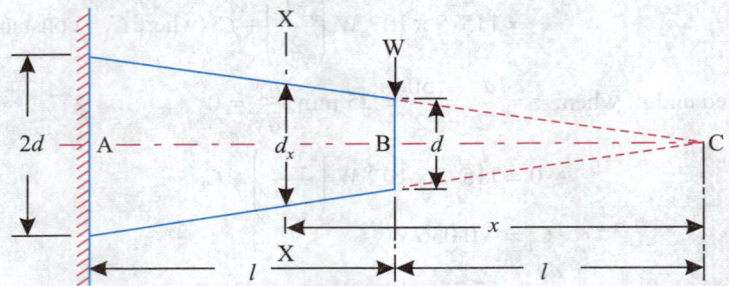

Fig. 8.90

The bending moment at the section is

$$M_x = EI_{XX} \frac{d^2y}{dx^2} = -W(x-l)$$

∴

$$E \frac{d^2y}{dx^2} = -\frac{W(x-l)}{\dfrac{\pi}{64} \cdot \dfrac{x^4 d^4}{l^4}}$$

$$= -\frac{64}{\pi d^4} Wl^4 \left(\frac{x-l}{x^4} \right)$$

or,

$$E \frac{d^2y}{dx^2} = -z(x^{-3} - lx^{-4}),$$

where,

$$z = \frac{64 \, Wl^4}{\pi d^4} \qquad \qquad ...(i)$$

Integrating both sides, we get

$$E \cdot \frac{dy}{dx} = -z \left[\frac{x^{-2}}{-2} - l \cdot \frac{x^{-3}}{-3} \right] + C_1$$

(where, C_1 = constant of integration)

When,

$$x = 2l, \frac{dy}{dx} = 0$$

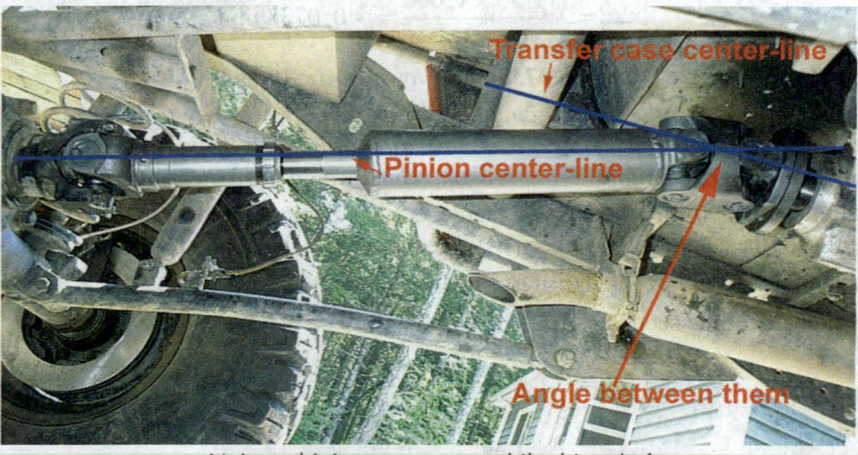

Universal joint on an automobile drive shaft.

$$\therefore \qquad 0 = -z \left[-\frac{1}{2} \times \frac{1}{4l^2} + \frac{l}{3} \times \frac{1}{8l^3} \right] + C_1$$

$$= -z \left[-\frac{1}{8l^2} + \frac{1}{24l^2} \right] + C_1$$

$$= z \times \frac{1}{12l^2} + C_1$$

$$\therefore \qquad C_1 = -\frac{z}{12l^2}$$

$$\therefore \qquad E \cdot \frac{dy}{dx} = -z \left(-\frac{x^{-2}}{2} + \frac{lx^{-3}}{3} \right) - \frac{z}{12l^2} \qquad \qquad ...(ii)$$

Integrating further, we get

$$Ey = -z \left(\frac{x^{-1}}{2} - \frac{lx^{-2}}{6} + \frac{x}{12\,l^2} \right) + C_2$$

When, $\qquad x = 2l, y = 0$

$$\therefore \qquad 0 = -z \left[\frac{1}{4l} - \frac{1}{24l} + \frac{1}{6l} \right] + C_2$$

$$= -z \times \frac{3}{8l} + C_2$$

$$\therefore \qquad C_2 = \frac{3z}{8l}$$

$$\therefore \qquad Ey = -z \left[\frac{1}{2x} - \frac{l}{6x^2} + \frac{x}{12l^2} - \frac{3}{8l} \right]$$

For the free end, $\qquad x = l$

$$\therefore \qquad y_B = -\frac{z}{E} \left(\frac{1}{2l} - \frac{l}{6l^2} + \frac{l}{12l^2} - \frac{3}{8l} \right)$$

$$= -\frac{z}{E} \left(\frac{1}{2l} - \frac{l}{6l} + \frac{l}{12l} - \frac{3}{8l} \right) = -\frac{z}{24El}$$

Substituting the value of z, we get

$$y_B = -\frac{64Wl^4}{\pi \, d^4} \times \frac{1}{24 \, El} = \frac{64}{24 \, \pi} \cdot \frac{Wl^3}{Ed^4} \text{ (downward)} \qquad ...(1)$$

Let, $\qquad D = $ diameter of a cantilever of uniform cross-section

Then, $\qquad I = \frac{\pi}{64} D^4$

$$y_B = \frac{Wl^3}{3 \, EI} = \frac{Wl^3}{3E \left(\frac{\pi}{64} D^4 \right)} = \frac{64}{3\pi} \cdot \frac{Wl^3}{ED^4} \qquad ...(2)$$

Equating (1) and (2), we get

$$\frac{64}{3\pi} \cdot \frac{Wl^3}{ED^4} = \frac{64}{24 \, \pi} \cdot \frac{Wl^3}{Ed^4}$$

or, $\qquad D^4 = 8d^4$

$$\therefore \qquad \mathbf{D = 1 \cdot 682d} \text{ (Ans.)}$$

TABLE 8·2. Slopes and Deflections for Different Loadings on Cantilevers and Beams

S. No.	Type of loading	Maximum bending moment	Slope	Maximum deflection
1	2	3	4	5
1.		$M_A = -Wl$	$\theta_B = -\dfrac{Wl^2}{2\,EI}$	$y_{max} = -\dfrac{Wl^3}{3\,EI}$
2.		$M_A = -Wa$	$\theta_B = -\dfrac{Wa^2}{2\,EI}$	$y_{max} = -\dfrac{Wa^2}{6\,EI}\,(3l-a)$
3.		$M_A = -\dfrac{wl^2}{2}\left(=-\dfrac{Wl}{2}\right)$ where, $W = wl$ (total load on the cantilever)	$\theta_B = -\dfrac{wl^3}{6\,EI}\left(=\dfrac{Wl^2}{6\,EI}\right)$	$y_{max} = -\dfrac{wl^4}{8\,EI}\left(=-\dfrac{Wl^3}{8\,EI}\right)$
4.		$M_A = -\dfrac{wa^2}{2}$	$\theta_B = \theta_C = -\dfrac{wa^3}{6\,EI}$ $\left(=-\dfrac{Wa^2}{6\,EI}\right)$ where, $W = wa$	$y_{max} = -\left[\dfrac{wa^4}{8\,EI} + \dfrac{wa^3}{6\,EI}(l-a)\right]$ $\left(=-\left[\dfrac{Wa^3}{8\,EI} + \dfrac{Wa^2}{6\,EI}(l-a)\right]\right)$

1	2	3	4	5
5.		$M_A = -w(l-a)\left[\dfrac{l-a}{2}+a\right]$	$\theta_B = -\dfrac{w}{6EI}(l^3-a^3)$	$y_{max} = -\left[\dfrac{w}{24EI}\right.$ $\left.(3l^4-4la^3+a^4)\right]$
6.		$M_A = -M$	$\theta_B = -\dfrac{Ml}{EI}$	$y_{max} = -\dfrac{Ml^2}{2EI}$
7.		$M_A = -\dfrac{wl^2}{6}$	$\theta_B = -\dfrac{wl^3}{24EI}$	$y_{max} = -\dfrac{wl^4}{30EI}$
8.		$M_C = \dfrac{Wl}{4}$	$\theta_A = -\dfrac{Wl^2}{16EI}$ $\theta_B = +\dfrac{Wl^2}{16EI}$	$y_{max}\,(=y_C) = -\dfrac{Wl^3}{48EI}$
9.		$M_C = \dfrac{Wab}{l}$	$\theta_A = -\dfrac{Wb(l^2-b^2)}{6EIl}$ $\theta_B = +\dfrac{Wa(l^2-a^2)}{6EIl}$	$y_{max} = -\dfrac{Wb(l^2-b^2)^{3/2}}{9\sqrt{3}\,EIl}$ $\left(at\ x = \sqrt{\dfrac{l^2-b^2}{3}}\right)$ $y_C = -\dfrac{Wa^2b^2}{3EIl}$

1	2	3	4	5
10.	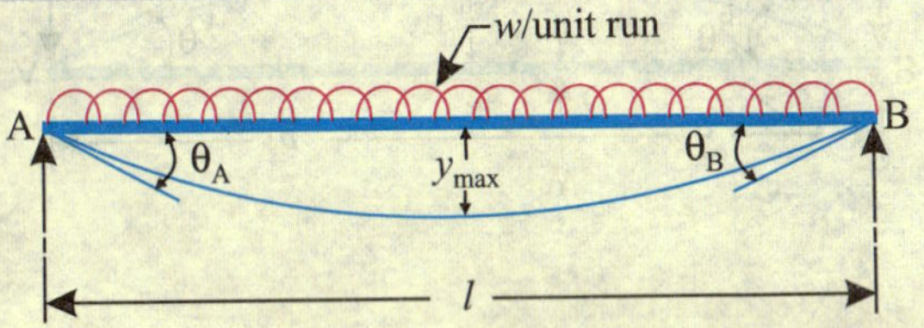	$M_{max} = \dfrac{wl^2}{8} = \dfrac{Wl}{8}$ where, W $= wl$ (total load on the beam).	$\theta_A = -\dfrac{wl^3}{24\,EI}\left(=-\dfrac{Wl^2}{24\,EI}\right)$ $\theta_B = +\dfrac{wl^3}{24\,EI}\left(=+\dfrac{Wl^2}{24\,EI}\right)$	$y_{max} = -\dfrac{5\,wl^4}{384\,EI}$ $\left[=-\dfrac{5\,Wl^3}{384\,EI}\right]$
11.	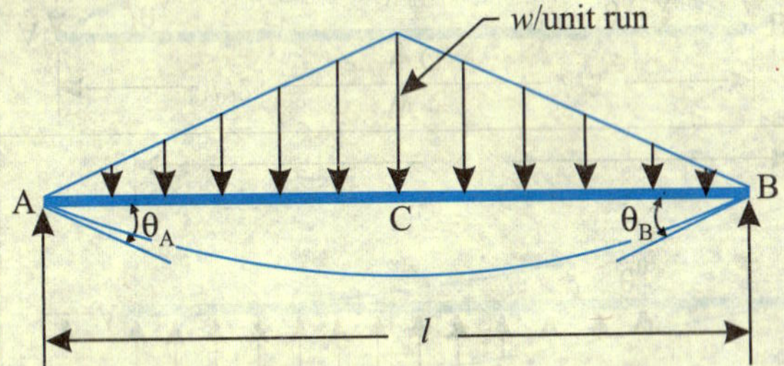	$M_{max} = \dfrac{wl^2}{9\sqrt{3}}$	$\theta_A = -\dfrac{7\,wl^3}{360\,EI}$ $\theta_B = +\dfrac{wl^3}{45\,EI}$	$y_{max} = -\dfrac{2\cdot5\,wl^4}{384\,EI}$ (at $x = 0\cdot519\,l$ from A)
12.	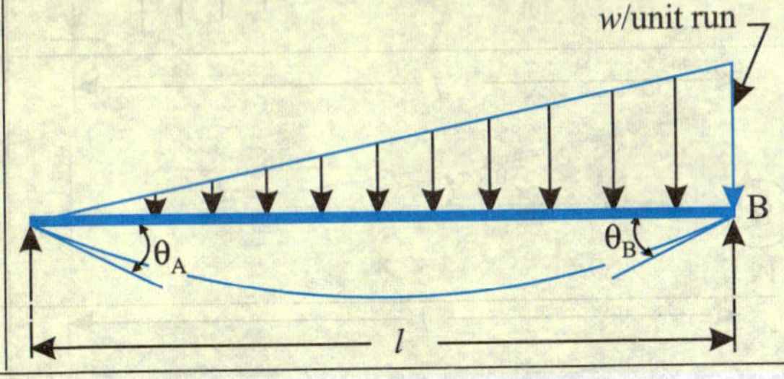	$M_{max} = \dfrac{wl^2}{12}$	$\theta_A = -\dfrac{5\,wl^3}{192\,EI}$ $\theta_B = +\dfrac{5\,wl^3}{192\,EI}$	$y_{max}\,(=y_C) = -\dfrac{wl^4}{120\,EI}$

Sign conventions used :

Slope: Clockwise — —
Counter-clockwise + +
Deflection: upward + +
Downward — —

Fig. 8.91

HIGHLIGHTS

1. The important methods used for finding out the slope and deflection at a section in a loaded beam are :

    (i) Double integration method;   (ii) Moment area method;   (iii) Macaulay's method.

    The first two methods are suitable for a *single load*, whereas the last one is suitable for *several loads*.

2. *Moment area method* is more useful as compared to double integral method because many problems which do not have a simple mathematical solution can be simplified by the bending moment area method.

3. **Conjugate beam method :**

    Theorem I.   "The slope at any section of a loaded beam relative to the original axis of the beam is equal to the shear in the conjugate beam at the corresponding section."

    Theorem II.   "The deflection at any given section of a loaded beam,, relative to the original position, is equal to the bending moment at the corresponding section of the conjugate beam."

OBJECTIVE TYPE QUESTIONS

Choose the Correct Answer:

1. The amount of deflection of a beam subjected to some type of loading depends upon

    (a) cross-section

    (b) bending moment

    (c) either (a) and (b)

    (d) both (a) and (b).

2. The slope and deflection at a section in a loaded beam can be found out by which of the following methods?

    (a) Double integration method

    (b) Moment area method

    (c) Macaulay's method

    (d) Any of the above.

3. The deflection at the free end of a cantilever of length l carrying a point load W at its free end is given as:

    (a) $\dfrac{Wl}{2EI}$ (b) $\dfrac{Wl^2}{2EI}$

    (c) $\dfrac{Wl^3}{2EI}$ (d) $\dfrac{Wl^3}{3EI}$.

4. A cantilever of length l is carrying a uniformly distributed load of w per unit run over the whole span. The deflection at the free end is given as:

    (a) $\dfrac{wl^3}{4EI}$ (b) $\dfrac{wl^2}{4EI}$

    (c) $\dfrac{wl^4}{8EI}$ (d) $\dfrac{wl^4}{16EI}$.

5. A cantilever of length l is carrying a uniformly distributed load of w per unit run for a distance a from fixed end. The slope at the free end is given as:

    (a) $\dfrac{wa^3}{6EI}$ (b) $\dfrac{wa^3}{8EI}$

    (c) $\dfrac{wa^3}{12EI}$ (d) $\dfrac{wa^3}{24EI}$.

6. A cantilever AB of length l has a moment M applied at free end. The deflection at the free end B is given as :

    (a) $\dfrac{M^2l}{EI}$ (b) $\dfrac{Ml^2}{2EI}$

    (c) $\dfrac{Ml}{2EI}$ (d) $\dfrac{Ml^3}{2EI}$.

7. A cantilever AB of length l is carrying a distributed load whose intensity varies uniformly from zero at the free end to w per unit at the fixed end. The deflection at the free end is given as:

    (a) $\dfrac{wl^2}{30EI}$ (b) $\dfrac{wl^3}{30EI}$

    (c) $\dfrac{wl^4}{30EI}$ (d) $\dfrac{wl^5}{30EI}$.

8. A cantilever AB of length l is carrying a distributed load whose intensity varies uniformly from zero at the fixed end to w per unit run at the free end. The deflection at free end is given as:

(a) $\dfrac{wl^3}{48EI}$ (b) $\dfrac{wl^4}{30EI}$

(c) $\dfrac{6}{120}\cdot\dfrac{wl^4}{EI}$ (d) $\dfrac{11}{120}\cdot\dfrac{wl^4}{EI}$.

9. A simply supported beam of span l is carrying a uniformly distributed load of w per unit run over the whole span. The maximum deflection in this case is given as:

(a) $\dfrac{wl^4}{48EI}$ (b) $\dfrac{wl^3}{30EI}$

(c) $\dfrac{5wl^4}{384EI}$ (d) $\dfrac{wl^4}{384EI}$.

Crane engine has many cylinders to supply the high power needed.

10. A simply supported beam of span l is carrying point load W at the mid span. What is the deflection at the centre of the beam?

(a) $\dfrac{Wl^2}{48EI}$ (b) $\dfrac{Wl^3}{48EI}$

(c) $\dfrac{5Wl^3}{348EI}$ (d) $\dfrac{11}{120}\cdot\dfrac{Wl^3}{EI}$.

ANSWERS

1. (d)   2. (d)   3. (d)   4. (c)   5. (a)   6. (b)   7. (c)

8. (d)   9. (c)   10. (b).

UNSOLVED EXAMPLES

1. A 3 metres long cantilever carries a uniformly distributed load over the entire length. If the slope at the free end is one degree, what is the deflection at the free end?
   [Ans. 39·27 mm]

2. A 250 mm long cantilever of rectangular section 48 mm wide and 36 mm deep carries a uniformly distributed load. Calculate the value of load w if the maximum deflection in the cantileveris not to exceed 1 mm. Take $E = 70 \times 10^9$ GN/m².   **[Ans. 26·75 kN/m]**

3. A cantilever of uniform cross-section of length l carries two point loads, W at the free end and $2W$ at a distance a from the free end. Find the maximum deflection due to this loading. If the cantilever is a steel tube of circular section 100 mm external diameter and 6 mm thick and $l = 1·5$ m and $a = 0·6$ metres determine the value of W so that the maximum bending stress is 160 MN/m² and calculate the maximum deflection for the loading. Take $E = 200$ GN/m². **[Ans. $W = 1905$ N; $\delta = 10·16$ mm]**

4. A 2 metres long cantilever is loaded with a point load of 500 N at the free end. If the section is rectangular 80 mm (wide) × 160 mm (deep), and $E = 10$ GN/m², calculate slope and deflection,
   (i) at the free end of the cantilever;
   (ii) at a distance of 0·6 m from the free end.
   [Ans. (i) 0·21°; 4·88 mm (downward) (ii) 0·19°; 2·75 mm (downward)]

5. A cantilever 2 m long is of rectangular section 120 mm wide and 240 mm deep. It carries a uniformly distributed load of 2·5 kN per metre length for a length of 1·25 metres from the fixed end and a point load of 1 kN at the free end. Find the deflection at the free end. Take $E = 10$ GN/m² **[Ans. 2·922 mm]**

6. A 2 metres long cantilever of rectangular section 100 mm wide and 200 mm deep carries a uniformly distributed load of 2 kN per metre run for a length of 1 m from the free end and a point load of 1 kN at the free end. Calculate the deflection at the free end. Take $E = 10·5$ GN/m².   **[Ans. 8·69 mm]**

7. A girder of uniform section and constant depth is freely supported over a span of 2 metres. Calculate the central deflection and slopes at the ends of the beam under a central load of 20 kN.

    Given: $I_{XX} = 7.807 \times 10^{-6}$ m^4, and $E = 200$ GN/m^2 [**Ans.** 2.13 mm; $-0.183°$; $+0.183°$]

8. A steel girder of 5 m length acting as a beam carries a uniformly distributed load of 20 N/m run throughout its length. If $I = 25.02 \times 10^{-6}$ m^4 and depth = 225 mm,

    (i) Calculate the magnitude of w so that the maximum stress developed in the beam section does not exceed 60 MN/m^2;

    (ii) Determine slope and deflection (under this load) in the beam at a distance of 1.5 m from one end. Take $E = 200$ GN/m^2.

         [**Ans.** (i) 4270 N/m; (ii) $\pm 0.145°$, -5.646 mm]

9. A simply supported 6 metres long rolled steel joist carries a uniformly distributed load of 8 kN/metre length. Determine slope and deflection at a distance of 2 metres from one end of the beam.

         [**Ans.** $\pm 0.276°$, -16.3 mm]

Macaulay's Method:

10. A beam with a span of 6 metres carries a point load of 40 kN at 4 metres from the left support. If, for the section $I_{XX} = 73.3 \times 10^{-6}$ m^4 and $E = 200$ GN/m^2, find :

    (i) The deflection under the load;

    (ii) The position and amount of maximum deflection.

         [**Ans.** (i) 9.7 mm; (ii) 3.27 m from the left end; 10.56 mm]

11. A steel girder of uniform section, 14 metres long, is simply supported at its ends. It carries concentrated loads of 120 kN and 80 kN at two points 3 metres and 4.5 metres from the two ends respectively.

    (i) Calculate the deflection of the girder at the two points under the two loads;

    (ii) The maximum deflection.

    Take: $I = 16 \times 10^{-4}$ m^4, and $E = 210 \times 10^6$ kN/m^2

         [**Ans.** (i) 15.64 mm, 19.93 mm (downward) (ii) 23.6 mm (downward)]

12. A beam of uniform section, 10 metres long, is simply supported at the ends. It carries point loads of 100 kN and 60 kN at distances of 2 m and 5 m respectively from the left end. Calculate :

    Cylinder of a car engine.

    (i) The deflection under each load;

    (ii) The maximum deflection.

    Given : $E = 200 \times 10^6$ N/m^2, and $I = 118 \times 10^{-4}$ m^4

         [**Ans.** (i) 4.34 mm, 6.79 mm (downward (ii) 6.76 mm (downward)]

13. A beam *ABCD*, 6 m long carries a point load of 20 kN at end *A* and 40 kN at point *C*, at a distance of 4 metres from *A*. The beam is supported over a span of 4 metres at points *B* and *D*. If $E = 200 \times 10^6$ kN/m^2 and $I = 80 \times 10^{-6}$ m^4 determine the maximum deflection and state where it occurs.

         [**Ans.** 5 mm at *A*]

14. A beam *AB* of 8 m span is simply supported at the ends. It carries a point load of 10 kN at a distance of 1 m from the end *A* and a uniformly distributed load of 5 kN / m for a length of 2 m from the end *B*. If $I = 10 \times 10^{-6}$ m^4, determine :

    (i) Deflection at the mid-span;   (ii) Maximum deflection;   (iii) Slope at the end *A*.

         [**Ans.** (i) 48.747 mm (downward) (ii) 8.75 mm (downward) (iii) $-0.417°$]

15. A beam AB of span 8 m is simply supported at ends A and B and is loaded as shown in the Fig. 8·92. Taking $E = 200 \times 10^6$ kN/m^2 and $I = 8.6 \times 10^8$ mm^4 find the position and magnitude of the maximum deflection.

    [**Ans.** 3·82 m from A, $y_{max} = -16.33$ mm]

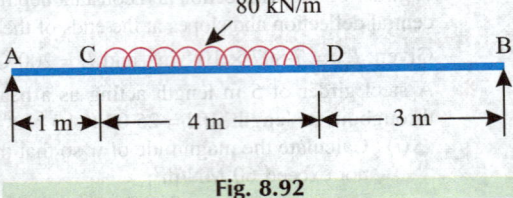

Fig. 8.92

16. A beam AB of span 6 metres and of flexural rigidity $EI = 8 \times 10^4$ kNm^2 is subjected to a clockwise couple of 600 kNm at a distance of 4 m from the left end. Find :

    (*i*) The deflection at the point of application of the couple;

    (*ii*) The maximum deflection. [**Ans.** (*i*) 6·67 mm (upward), (*ii*) 9·43 mm (upward)]

17. An overhanging beam ABC 6 m long is supported at A and B such that $AB = 4$ m. It is loaded with point load of 10 kN at the end C. If $E = 200 \times 10^6$ kN/m^2 and $I = 12 \times 10^{-6}$ m^4 determine:

    (*i*) Deflection at the point C;   (*ii*) Maximum deflection between A and B.

    [**Ans.** (*i*) 33·3 mm (downward) (*ii*) 8·55 mm (upward)]

18. A beam ABC 13 m long is supported at A and B, such that $AB = 10$ m and overhang $BC = 3$ m. It carries a point load of 4 kN from the end A and a uniformly distributed load of 0·4 kN/m over the entire overhang. Determine :

    (*i*) Slope at the end A;   (*ii*) Deflection at the free end C;   (*iii*) Maximum deflection

    Take: $E = 200 \times 10^6$ kN/m^2, and $I = 3 \times 10^{-5}$ m^4.

    [**Ans.** (*i*) 0·178°; (*ii*) 9·14 mm (upward); (*iii*) 11·27 mm (downward)]

19. An overhanging beam ABC 8 m long is supported at A and B such that $AB = 6$ m and the overhang $BC = 2$ m. It has a point load of 3 kN at the end C and a uniformly distributed load of 2 kN/m run for a length of 3 m at a distance of 1 m from the end A. If $E = 200 \times 10^6$ kN/m^2 and $I = 4.5 \times 10^{-6}$ m^4 determine :

    (*i*) Deflection at the free end C;   (*ii*) Maximum deflection between A and B.

    [**Ans.** (*i*) 10·08 mm (downward), (*ii*) 11·4 mm (downward)]

Moment Area Method:

20. A cantilever 15 cm wide and 20 cm. deep projects 1·5 m out of a wall, and is carrying a point load of 20 kN at the free end. Find the slope and deflection of the cantilever at the free end. Take $E = 210$ GN/m^2.   [**Ans.** $\theta_{max} = 0.00108$ rad.; $y_{max} = 1.08$ mm]

21. A steel cantilever of 2·5 m effective length carries a load of 25 kN at its free end. If the deflection at the free end is not to exceed 0·5 cm what must be the I value of the section of the cantilever? Take $E = 210$ GN/m^2 [**Ans.** $I = 12400$ cm^4]

22. Calculate the deflection at free end of a steel cantilever 7 m long when it carries uniformly distributed load of 50 kN.

    Take: $E = 2 \times 10^8$ kN/m^2, and $I = 0.0009$ m^4.   [**Ans.** 11·9 mm]

23. A cantilever of uniform section is loaded with 20 kN at its free end. In addition to this, a U.D.L. of 10 kN / m run is provided over entire span. Calculate maximum deflection and slope. The cantilever is 3 m long, 10 cm wide and 30 cm deep. $E = 210$ GN/m^2.

    [**Ans.** 0·594 cm, 0·002857 radian]

24. A cantilever projecting 2·5 metres from a wall is loaded with a U.D.L. of 120 kN. Determine the moment of inertia of the beam section, if the deflection of beam at the free end be 10 mm.

    Take $E = 2.05 \times 10^8$ kN/m^2.   [**Ans.** 0·0001143 m^4]

25. A cantilever 2·5 m long is loaded with a U.D.L. of 10 kN per metre run over a length of 1·5 m from the fixed end. Determine the slope and deflection at the free end of the cantilever. Take: $I = 9500$ cm^4, and $E = 210$ GN/m^2.   [**Ans.** 0·00028 rad.; 0·6 mm]

26. A hollow steel tube of outside diameter 240 mm and thickness of metal 25 mm is used as a cantilever beam 2·5 m long. It carries a U.D.L. of 15 kN/m run for 1·75 m from the fixed end. Calculate maximum slope and deflection of the beam.

[**Ans.** 0·000677 radian; 1·396 mm]

27. A cantilever 3 m long carries two point loads, 60 kN each, at distances of 0·75 m and 1·75 m respectively from the fixed end. Determine the deflection at the free end. Take $E = 200$ GN/m^2 and $I = 12689400$ cm^4.

[**Ans.** 0·00151 mm]

28. A beam 3 metres long, simply supported at its ends, is carrying a point load (W) at its centre. If the slope at the ends of the beam is not to exceed 1°, find the deflection at the centre of the beam.

[**Ans.** 17·45 mm]

29. A beam of 5 metres span is carrying a point load of 30 kN at a distance 3·75 metres from the left end. Calculate the slopes at the two supports and deflection under the load. Take $EI = 2·6 \times 10^7$ Nm^2.

[**Ans.** 0·00113 rad., 0·00158 rad., 1·69 mm]

30. A simply supported beam of circular cross-section is 5 m long and is of 150 mm diameter. What will be the maximum value of the central load if the deflection of the beam does not exceed 12·45 mm. Also calculate the slope at the supports. Take $E = 2 \times 10^8$ kN/m^2.

[**Ans.** 23·77 kN, 0·00747 rad.]

31. A timber beam is 10 cm wide by 20 cm deep and carries a load of 1·5 kN per m run over a span of 4 m. Find the deflection at the centre. Take $E = 9·1$ GN/m^2.

[**Ans.** 8·33 mm]

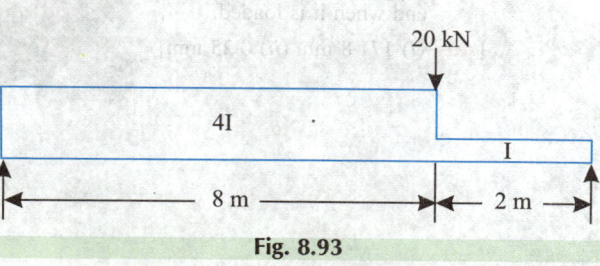

Cylinder studs of a car engine. The ribs are meant to dissipate the heat and cool the engine cylinder.

32. A beam of uniform section, length 1·92 metres is freely supported at the ends. It carries concentrated load of 120 kN at 0·72 metre from the right support. If moment of inertia for beam section is 0·0000212 m^4 and E for material = $1·35 \times 10^8$ kN/m^2 find the deflection of beam under the load.

[**Ans.** 5·42 mm]

33. A simply supported beam of 4 m span carries a U.D.L. of 20 kN/m on the whole span and in addition carries a point load of 40 kN at the centre of span. Calculate the slope at the ends and maximum deflection of the beam. Take $E = 200$ GN/m^2 and $I = 5000$ cm^4.

[**Ans.** $\theta_{max} = 0·00932$ rad.; $y_{max} = 14·21$ mm]

34. A beam 4 metres long is freely supported at the ends. It carries concentrated loads of 20 kN each at points 1 metre from the ends. Calculate the maximum slope and deflection of the beam and slope and deflection under each load. $EI = 13000$ kNm^2.

[**Ans.** 0·0023 rad., 2·82 mm, 0·00153 rad., 2·043 mm]

Conjugate Beam Method :

35. For the beam shown in the Fig. 8·93, using conjugate beam method, determine the following:

    (i) Slope at the end A;

    (ii) Deflection at its end A;

    (iii) Maximum deflection.

    Take: $E = 200 \times 10^6$ kN/m^2, and $I = 8 \times 10^{-5}$ m^4.

    [**Ans.** (i) 0·0012 rad., (ii) 4·7 mm,

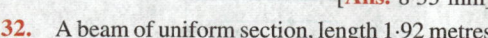

Fig. 8.93

(iii) 4·96 mm]

36. Using conjugate beam method find out the slope at the support A and the vertical deflection at the point B in terms of EI for the section AC shown in the Fig. 8·94. Take $I = I_{BC}$ and $I_{AB} = 2I_{BC}$

$$\left[\textbf{Ans.}\ \frac{21{\cdot}5}{10EI},\ \frac{42}{10EI}\right]$$

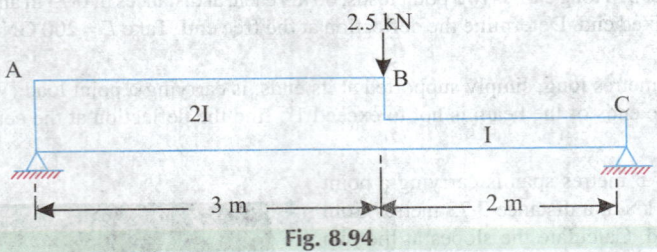

Fig. 8.94

37. A beam $A\,BCD$ is simply supported at its ends A and D over a span of 30 metres. It is made up of three portions $A\,B$, BC and CD each 10 metres in length. The moments of inertia of sections of these portions are I, $3\,I$ and $2\,I$ respectively, where $I = 300 \times 10^{-4}$ m^4. The beam carries a point load of 225 kN at B and a point load of 450 kN at C. If $E = 200 \times 10^6$ kN/m^2, determine:

    (*i*) Slopes at A, B, C, and D;   (*ii*) Deflection at A, B, C and D.

Neglect the weight of the beam.

    [**Ans.** (*i*) – 0·003218 rad.; – 0·0007176 rad.; + 0·001157 rad.; + 0·00272 rad., (*ii*) zero, 23·84 mm, 21·99 mm, zero]

38. Using Conjugate beam method find the mid span deflection of the beam shown in the Fig. 8·95. Take: $E = 200 \times 10^6$ kN/m^2, and $I = 200 \times 10^{-4}$ m^4.   [**Ans.** 0·45 mm]

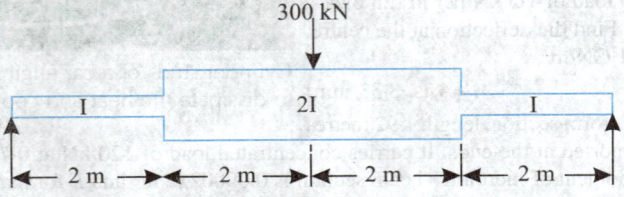

Fig. 8.95

39. A cantilever of uniform strength is to be turned from a mid steel bar 50 mm in diameter. A load of 5 kN is to be supported from the free end, and the maximum stress is to be limited to 70 N/mm^2. If $E = 2 \times 10^5$ N/mm^2, determine:

    (*i*) The maximum permissible length of the cantilever;

    (*ii*) The deflection of the free end when it is loaded.

    [**Ans.** (*i*) 171·8 mm (*ii*) 0·25 mm]

Cylinder stud layout of a car engine.

Fixed and Continuous Beams

Chapter

9

9.1. INTRODUCTION

Fixed Beams

A **fixed beam** (also called *built-in or encaster beam*) *is a beam the ends of which are constrained or built-in to remain in horizontal position* (Fig. 9.1).

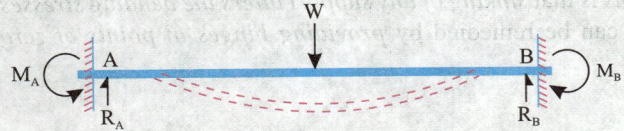

Fig. 9.1. Fixed beam.

Following points are worth noting:

1. Due to the fixidity, the slope of the beam is zero at each end, and a couple or moment will be induced at each end to satisfy this condition.

2. The induced moments M_A and M_B will be in the opposite direction to that of moments due to external loading.

3. *End moments* in case of fixed beams tend to bend the beam with *convexity upwards* whereas the *normal downward loads* tend to bend the beam

with *concavity upwards*. The condition of greatest strength will be realised when the greatest hogging moments are equal in the magnitude to the greatest sagging moments.

4. In case of a fixed beam there are four unknowns: R_A, R_B, M_A and M_B. Thus, the two statics equations must be supplemented by two additional equations arising from deformations.

For the same spans and loads, the *fixed beams* claim the following **advantages** *over simply supported beams*:

1. These have lesser values for maximum bending moments.
2. These have lesser values for maximum deflection.

Fixed beams (as compared to simply supported beams) entail the following **demerits**.

1. Large stresses are set up by temperature changes.
2. Special care has to be taken in aligning supports accurately at the same level.
3. Large stresses are set if a little sinking of one support takes place.
4. Frequent fluctuations in loading (specially in case of moving loads) render the degree of fixity at the ends very uncertain.

Several of the drawbacks of the fixed beams can be obviated by *having two cantilevers* at the ends and bridging the gap by hinging a beam to their free ends.

Continuous Beams

A **continuous beam** *is one which is supported on more than two supports* (Fig. 9.2). For usual loadings on the beam hogging moments causing convexity upwards occur at the supports and sagging moments causing concavity upwards occur at mid spans.

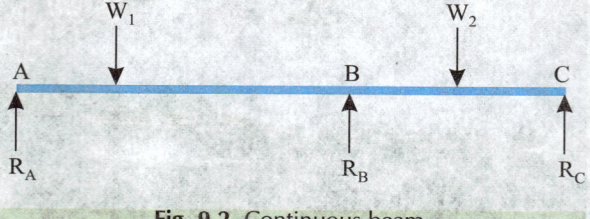

Fig. 9.2. Continuous beam.

Continuous beams claim the following **advantages** over simply supported beams:—

1. The maximum bending moment in case of a continuous beam is much less than in case of a simply supported beam of same span carrying same loads.
2. In case of a continuous beam, the average bending moment is lesser and hence lighter materials of construction can be used to resist the bending moment.
3. In case of a continuous beam, the bending moment is more over the supports than at mid spans and hence the weight of the beam does not affect the stresses materially.

The *drawback of continuous beams* is that *sinking of any support alters the bending stresses appreciably*. This drawback, however, can be remedied by *providing hinges at points of zero bending moments*.

9.2. FIXED BEAMS

9.2.1. Analysis of a Fixed Beam

Fig. 9.3 shows a fixed beam AB carrying point loads W_1 and W_2.

Let, R_{fA} = Reaction at the support A,

R_{fB} = Reaction at the support B, and

M_A, M_B = Fixed end moments.

The analysis of the beam may be divided in the following *stages*:—

1. In the *first stage* [Fig. 9.3 (a)] the beam is treated as simply supported beam carrying the points loads W_1 and W_2 [Fig. 9.3 (b)] shows the B.M. diagram for this condition. The bending moment at any section is a *sagging moment*. Let R_A and R_B be the reactions at A and B for this condition.

2. In the *second stage* [Fig. 9.3 (c)] the simply supported beam is considered *under the action of only the end couples* M_A and M_B without the given loading.

Let, R = Reaction at each end under this condition, and $M_B > M_A$.

Then, $R = \dfrac{M_B - M_A}{l}$

— If $M_B > M_A$, the reaction R is upwards at B and downwards at A.

— The bending moment M'_x, at any section, is a *hogging moment*. Fig. 9.3 (d) shows the B.M. diagram for this condition. The resultant (final) diagram is drawn by combining the above two B.M. diagrams as shown in Fig. 9.3 (f).

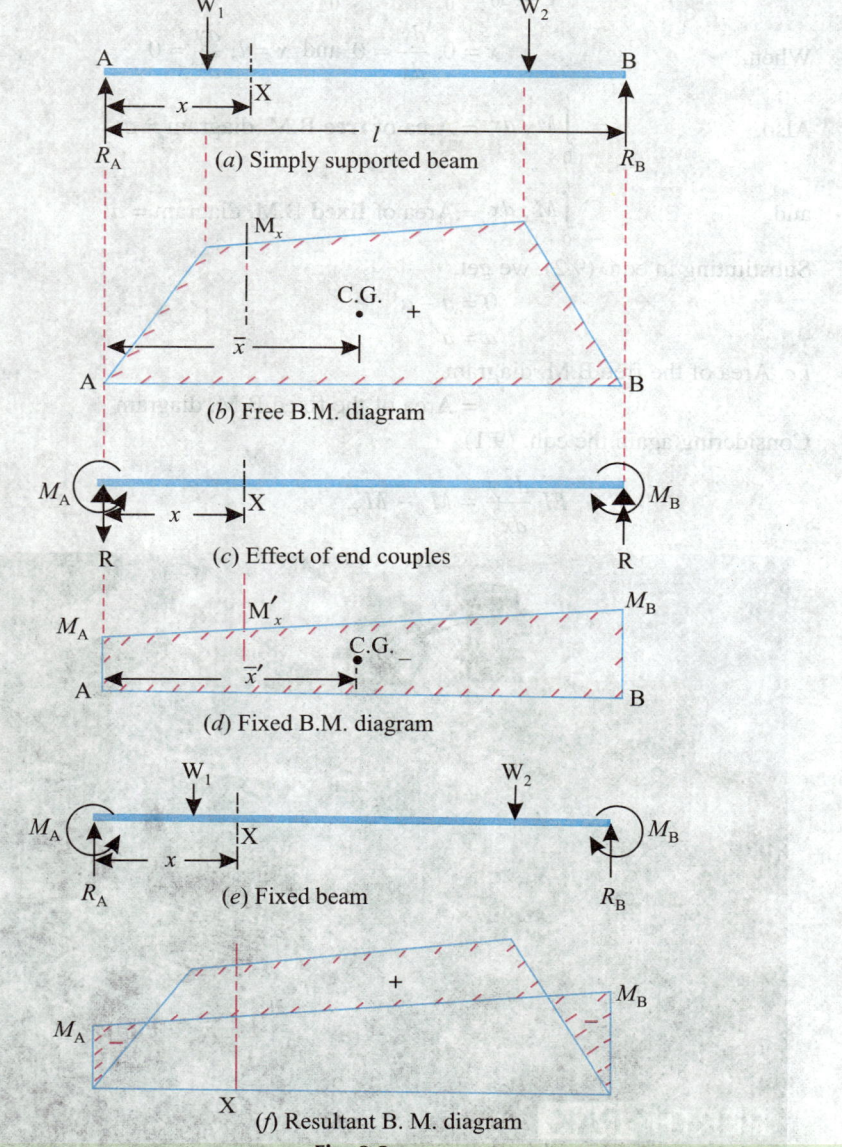

(a) Simply supported beam

(b) Free B.M. diagram

(c) Effect of end couples

(d) Fixed B.M. diagram

(e) Fixed beam

(f) Resultant B. M. diagram

Fig. 9.3

The final reactions are given as follows :

$$R_{fA} = R_A - R$$
$$R_{fB} = R_B + R$$

Now, the actual bending moment at any section X, distant x from the end A is given by

$$EI \frac{d^2 y}{dx^2} = M_x - M_x' \qquad \qquad ...(9.1)$$

Integrating both sides, we get,

$$EI \left[\frac{dy}{dx} \right]_0^l = \int_0^l M_x \, dx - \int_0^l M_x' \, dx \qquad \qquad ...(9.2)$$

When, $\qquad \qquad x = 0, \dfrac{dy}{dx} = 0$ and $x = l, \dfrac{dy}{dx} = 0$

Also, $\qquad \qquad \displaystyle\int_0^l M_x \, dx$ = Area of **free** B.M. diagram = a

and, $\qquad \qquad \displaystyle\int_0^l M_x' \, dx$ = Area of **fixed** B.M. diagram = a'

Substituting in eqn. (9.2), we get,

$$0 = a - a'$$

$\therefore \qquad \qquad a = a' \qquad \qquad ...(9.3)$

i.e. Area of the free B.M. diagram

$\qquad \qquad \qquad$ = Area of the fixed B.M. diagram

Considering again the eqn. (9.1)

$$EI \frac{d^2 y}{dx^2} = M_x - M_x'$$

Vegetable packing machine. In machines with many components, some parts act as beams.

Multiplying both sides by x, we get,

$$EIx \frac{d^2y}{dx^2} = M_x \cdot x - M'_x \cdot x$$

Integrating both sides, we have

$$\int_0^l EIx \frac{d^2y}{dx^2} \, dx = \int_0^l M_x \cdot x \, dx - \int_0^l M'_x \cdot x \, dx$$

$$\therefore \qquad EI \left[x \frac{dy}{dx} - y \right]_0^l = a\bar{x} - a'\bar{x}' \qquad \qquad ....(9.4)$$

where, $\bar{x}$ = Distance of the centroid of the free B.M. diagram from A, and

$\bar{x}'$ = Distance of the centroid of the fixed B.M. diagram from A.

When, $x = 0, y = 0$

$$\frac{dy}{dx} = 0$$

and $x = l, y = 0 \qquad \frac{dy}{dx} = 0$

Substituting these values in eqn. (9.4), we get,

$$0 = a\bar{x} - a'\bar{x}'$$

or, $a\bar{x} = a'\bar{x}'$

or, $\bar{x} = \bar{x}' \qquad \qquad (\because a = a')$

There by using the conditions,

$$a = a'$$

and, $\bar{x} = \bar{x}'$

The values of unknown M_A and M_B can be found out.

Some standard cases of a fixed beam are discussed below:—

Case I. Fixed beam carrying a point load at midspan:

Fig. 9.4 shows a fixed beam AB of span l carrying a point (or concentrated) load W at the midspan and the free and fixed B.M. diagrams (end moments M_A and M_B being equal due to symmetry).

By equating the areas of free and fixed B.M. diagrams, we have

$$a = a', \quad \frac{1}{2} \times l \times \frac{Wl}{4} = M_A \times l$$

$$\therefore \qquad M_A = \frac{Wl}{8} \quad \text{and} \quad M_B = M_A = \frac{Wl}{8}$$

B.M. at midspan $= M_x - M'_x = \frac{Wl}{4} - \frac{Wl}{8} = +\frac{Wl}{8}$

Due to symmetry the reactions R_A and R_B are equal.

$$\therefore \qquad R_A = R_B = \frac{W}{2}$$

Now, for the beam S.F. and B.M. diagrams can be easily drawn as shown in Fig. 9·4. Evidently, two points of contraflexure occur at $\frac{l}{4}$ from the ends (Fig. 9·4).

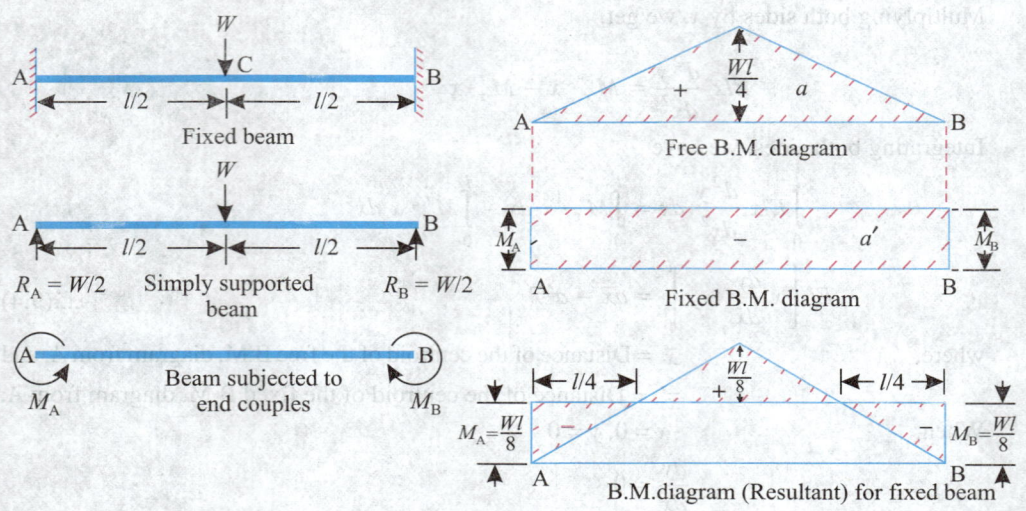

Fig. 9.4

Slope and deflection:

The B.M. at any section in AC, distant x from the end A, is given by

$$EI\frac{d^2y}{dx^2} = M_x - M'_x = \frac{W}{2}x - \frac{Wl}{8}$$

Integrating both sides, we get

$$EI\frac{dy}{dx} = \frac{Wx^2}{4} - \frac{Wl}{8}x + C_1$$

...Slope equation

When, $x = 0$, $\dfrac{dy}{dx} = 0$ ∴ $C_1 = 0$

Integrating again, we get

$$EIy = \frac{Wx^3}{12} - \frac{Wl}{8} \times \frac{x^2}{2} + C_2$$

...Deflection equation

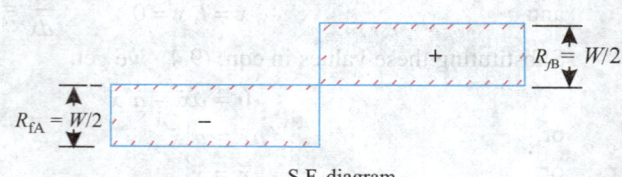

Blades used to cut metallic and non-metallic sheets: Thin kerf saw blades.

When, $x = 0$, $y = 0$ ∴ $C_2 = 0$

Maximum deflection occurs at midspan *i.e.* at $x = l/2$

$$\therefore \quad EIy_C = \frac{W}{12}\left(\frac{l}{2}\right)^3 - \frac{Wl}{16} \times \left(\frac{l}{2}\right)^2 = \frac{Wl^3}{96} - \frac{Wl^3}{64} = -\frac{Wl^3}{192}$$

$$\therefore \qquad y_C = -\frac{Wl^3}{192\,EI} \qquad\qquad ...(9.5)$$

(= 1/4 of the deflection for a simply supported beam)

Case II. Fixed beam carrying a concentrated load eccentrically placed on the beam:

Fig. 9·5 shows a beam AB of span l carrying a concentrated load W at C eccentrically on the span such that $AC = a$ and $BC = b$.

Evidently the free B.M. diagram is a triangle whose altitude is $\dfrac{Wab}{l}$.

Let M_A and M_B be the fixing moments at the ends A and B respectively.

$\because$ Area of the free B.M. diagram = Area of fixed B.M. diagram

$$\therefore \qquad \frac{1}{2} \times l \times \frac{Wab}{l} = \left(\frac{M_A + M_B}{2}\right) \times l$$

$$\therefore \qquad M_A + M_B = \frac{Wab}{l} \qquad\qquad ...(i)$$

Also, $\bar{x}$ (centroidal distance of free B.M. diagram from the end A) = $\bar{x}'$ (centroidal distance of fixed B.M. diagram from the end A)

But, $$\bar{x} = \frac{l + a}{3} \text{ from } A$$

and, $$\bar{x}' = \frac{M_A + 2M_B}{M_A + M_B} \times \frac{l}{3}$$

$$\therefore \qquad \frac{l + a}{3} = \frac{M_A + 2M_B}{M_A + M_B} \times \frac{l}{3}$$

or, $$M_A + 2M_B = \left(\frac{M_A + M_B}{l}\right)(l + a)$$

But, $$M_A + M_B = \frac{Wab}{l} \qquad\qquad \text{[from eqn } (i)]$$

$$\therefore \qquad M_A + 2M_B = \frac{Wab}{l^2}(l + a) \qquad\qquad ...(ii)$$

Subtracting eqn. (i) from eqn. (ii), we get

$$M_B = \frac{Wab}{l^2}(l + a) - \frac{Wab}{l} = \frac{Wab}{l^2}(l + a - l) = \frac{Wa^2b}{l^2}$$

Putting this value of M_B in eqn. (i), we get

$$M_A + \frac{Wa^2b}{l^2} = \frac{Wab}{l}$$

$$\therefore \qquad M_A = \frac{Wab}{l} - \frac{Wa^2b}{l^2} = \frac{Wab}{l^2}(l - a) = \frac{Wab^2}{l^2} \qquad [\because l - a = b]$$

Thus the fixing moments at A and B are :

$$M_A = \frac{Wab^2}{l^2} \qquad\qquad ...(9.6)$$

and, $$M_B = \frac{Wa^2b}{l^2} \qquad\qquad ...(9.7)$$

It may be noted that : if $a > b$, then $M_B > M_A$.

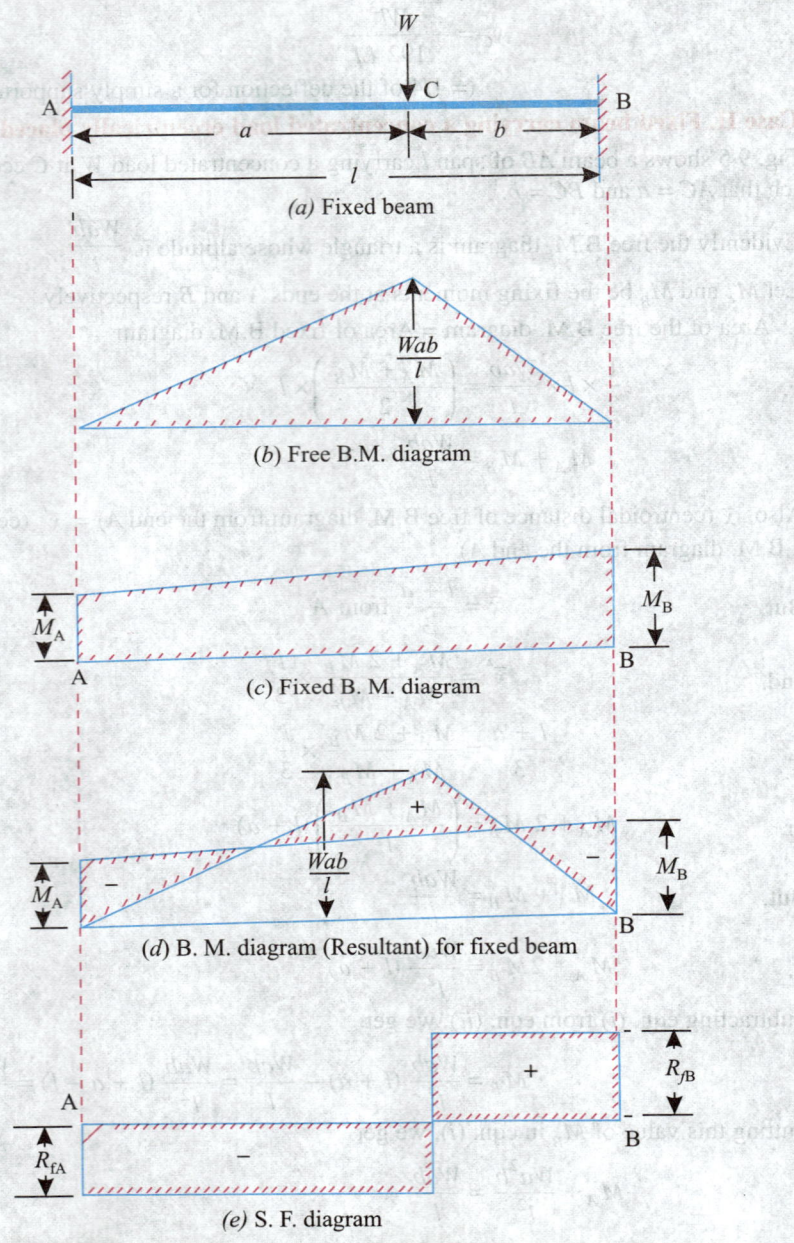

(a) Fixed beam

(b) Free B.M. diagram

(c) Fixed B. M. diagram

(d) B. M. diagram (Resultant) for fixed beam

(e) S. F. diagram

Fig. 9.5

Slope and deflection:

The bending moment at any section, distant x from the end A, is given by

$$EI \frac{d^2 y}{dx^2} = M_x - M'_x$$

$$= \frac{Wb}{l} x - \left[M_A + \frac{M_B - M_A}{l} \cdot x \right] - W(x - a)$$

But $M_A + \dfrac{M_B - M_A}{l} \cdot x = \dfrac{Wab^2}{l^2} + \dfrac{\dfrac{Wa^2b}{l^2} - \dfrac{Wab^2}{l^2}}{l} \cdot x$

$$= \dfrac{Wab^2}{l^2} + \dfrac{Wab}{l^3}(a - b) \cdot x$$

$\therefore \qquad EI \dfrac{d^2y}{dx^2} = \dfrac{Wb}{l} \cdot x - \dfrac{Wab^2}{l^2} - \dfrac{Wab}{l^3}(a - b)\,x \ \Big| - W(x - a)$

or, $\qquad EI \dfrac{d^2y}{dx^2} = \dfrac{Wb}{l}(l^2 - a^2 + ab)\,x - \dfrac{Wab^2}{l^2} \ \Big| - W(x - a)$

$$= \dfrac{Wb}{l^3}\Big[(a + b)^2 - (a^2 - ab)\Big]x - \dfrac{Wab^2}{l^2} \ \Big| - W(x - a)$$

$$= \dfrac{Wb}{l^3}(a^2 + b^2 + 2ab - a^2 + ab)\,x - \dfrac{Wab^2}{l^2} \ \Big| - W(x - a)$$

$$= \dfrac{Wb}{l^3}(3ab + b^2)\,x - \dfrac{Wab^2}{l^2} \ \Big| - W(x - a)$$

or, $\qquad EI \dfrac{d^2y}{dx^2} = \dfrac{Wb^2}{l^3}(3a + b)\,x - \dfrac{Wab^2}{l^2} \ \Big| - W(x - a)$

Integrating both sides, we get

$$EI \dfrac{dy}{dx} = \dfrac{Wb^2(3a + b)\,x^2}{2l^3} - \dfrac{Wab^2}{l^2}x + C_1 \ \Big| - W\dfrac{(x - a)^2}{2}$$

...Slope equation

When, $\qquad x = 0, \dfrac{dy}{dx} = 0 \qquad \therefore C_1 = 0$

Integrating again, we get

$$EIy = \dfrac{Wb^2(3a + b)\,x^3}{6l^3} - \dfrac{Wab^2 x^2}{2l^2} + C_2 \ \Big| - \dfrac{W(x - a)^3}{6}$$

..Deflection equation

When, $\qquad x = 0, y = 0 \qquad \therefore C_2 = 0$

Deflection under the load:

Substituting $x = a$ in the deflection equation, we get

$$EIy_C = \dfrac{Wb^2(3a + b)\,a^3}{6l^3} - \dfrac{Wab^2 \times a^2}{2l^2}$$

$$= -\dfrac{Wa^3 b^2}{6l^3}(-3a - b + 3l)$$

$$= -\dfrac{Wa^3 b^2}{6l^3}[-3a - b + 3(a + b)]$$

$$[\because l = (a + b)]$$

$$= -\dfrac{Wa^3 b^2}{6l^3}(-3a - b + 3a + 3b)$$

$$= -\dfrac{Wa^3 b^2}{6l^3} \times 2b = -\dfrac{Wa^3 b^3}{3l^3}$$

$\therefore \qquad y_C = -\dfrac{Wa^3 b^3}{3l^3\,EI}$ ...(9.8)

Blades used to cut metallic and non-metallic sheets: Wide kerf saw blades.

Maximum deflection:

Let $a > b$.

Maximum deflection will occur between A and C. To achieve this condition let us equate the slope to zero.

$$\frac{Wb^2 (3a + b) x^2}{2l^3} - \frac{Wab^2}{l^2} x = 0$$

On simplification, we get $x = \dfrac{2al}{3a + b}$

Putting this value in deflection equation, we get

$$EIy_{max} = \frac{Wb^2 (3a + b)}{6l^3} - \left(\frac{2al}{3a + b}\right)^3 - \frac{Wab^2}{2l^2} \left(\frac{2al}{3a + b}\right)^2$$

$$= -\frac{Wb^2}{6l^3} \left(\frac{2al}{3a + b}\right)^2 \left[3al - \frac{(3a + b)(2al)}{3a + b}\right]$$

$$= -\frac{Wb^2}{6l^3} \times \frac{4a^2 l^2}{(3a + b)^2} \times (al) = -\frac{2}{3} \times \frac{Wa^3 b^2}{(3a + b)^2}$$

$$\therefore \qquad y_{max} = -\frac{2}{3} \frac{Wa^3 b^2}{(3a + b)^2 \, EI} \qquad\qquad\qquad ...(9.9)$$

Points of contraflexure:

— For the point of contraflexure in AC,

$$M = \frac{Wb^2}{l^3} (3a + b) x - \frac{Wab^2}{l^2} = 0$$

or $\qquad \dfrac{Wb^2}{l^3} (3a + b) x = \dfrac{Wab^2}{l^2}$

or $\qquad x = \dfrac{al}{3a + b}$, distance of the point of contraflexure *from end A.*

— For the point of contraflexure *in BC*,

$$M = \frac{Wb^2}{l^3} (3a + b) x - \frac{Wab^2}{l^2} - W (x - a) = 0$$

or $\qquad \dfrac{Wb^2}{l^3} (3a + b) x = \dfrac{Wab^2}{l^2} + W (x - a)$

or $\qquad \dfrac{b^3 (3a + b) x}{l^3} = \dfrac{ab^2}{l^2} + (x - a)$ or $x \left[\dfrac{b^2 (3a + b)}{l^3} - 1\right] = \dfrac{ab^2}{l^2} - a$

or $\qquad x \left[\dfrac{3ab^2 + b^3 - l^3)}{l^3}\right] = \dfrac{ab^2 - al^2}{l^2}$ or $x (3ab^2 + b^3 - l^3) = ab^2 l - al^3$

or $\qquad x [3ab^2 + b^3 - (a + b)^3] = al (b^2 - l^2)$ $\qquad\qquad$ [$\because l = (a + b)$]

or $\qquad x [3ab^2 + b^3 - a^3 - b^3 - 3a^2 b - 3ab^2] = al [b^2 - (a + b)^2]$

or $\qquad x (- a^3 - 3a^2 b) = al (b^2 - a^2 - b^2 - 2ab)$

or $\qquad - a^2 x (a + 3b) = - a^2 l (a + 2b)$

or $\qquad x = \dfrac{l (a + 2b)}{(a + 3b)}$, distance of *point of contraflexure* from end A.

Example 9·1. *A fixed beam of 6 m span is loaded with point loads of 150 kN at distance 2 m from each support. Draw the B.M. and S.F. diagrams. Find also the maximum deflection.*

Take: $E = 2 \times 10^8$ kN/m², and $I = 8 \times 10^8$ mm⁴.

Solution. Fig. 9·6 (*a*) shows the fixed beam *AB* carrying point loads of 150 kN each. The fixed moments M_A and M_B are equal (due to symmetry). Free and fixed B.M. diagrams are also shown in the Fig. 9·6. (*b, c*).

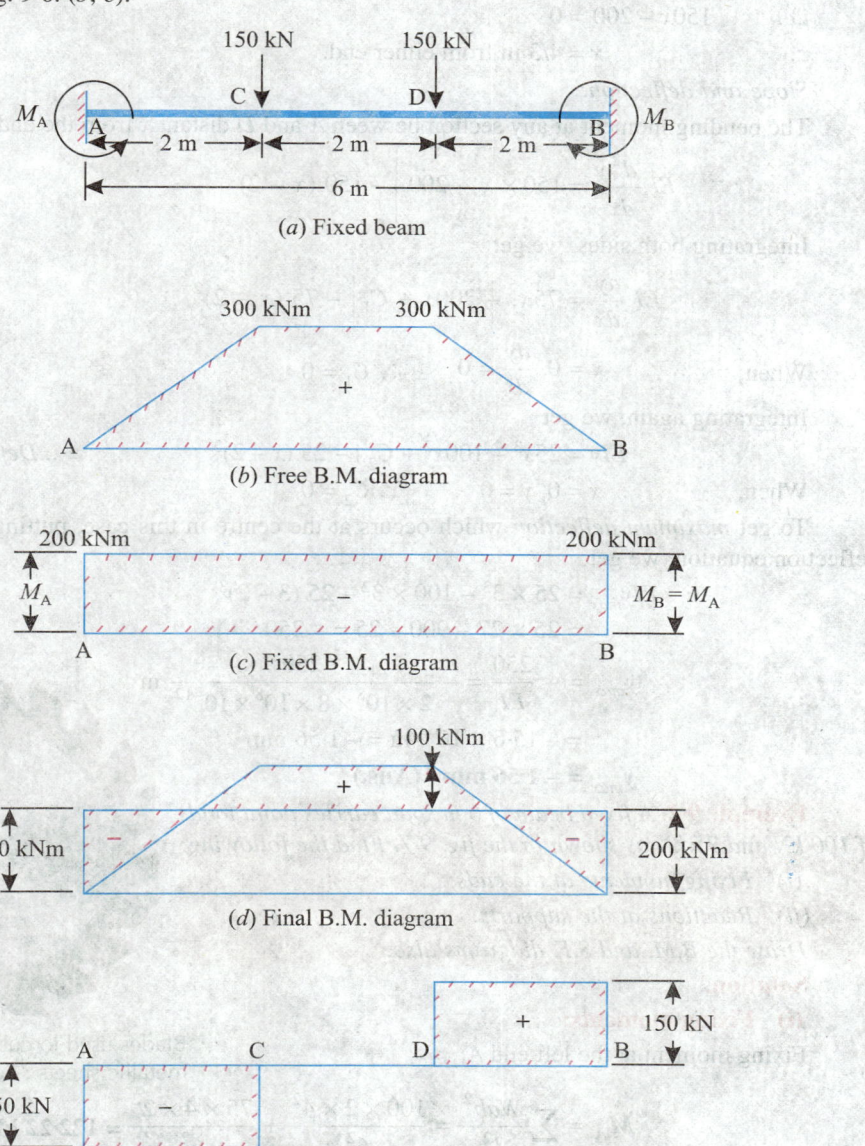

(*a*) Fixed beam

(*b*) Free B.M. diagram

(*c*) Fixed B.M. diagram

(*d*) Final B.M. diagram

(*e*) S.F. diagram

Fig. 9.6

By equating the areas of free and fixed B.M. diagrams, we have,

$$M_A \times 6 = 1/2 \,(6 + 2) \times 300$$

or $$M_A = 200 \text{ kNm}$$

∴ B.M. at the centre = 300 – 200 = 100 kNm

Points of contraflexure:

B.M. (actual) at any section in AC at a distant x from A is given by

$$M = \text{Free moment} - \text{fixed moment} = 150 \times x - 200$$

To get point of contraflexure equate $M = 0$.

i.e. $150x - 200 = 0$

∴ $x = 4/3$ m from either end.

Slope and deflection:

The bending moment at any section between A and D distant x from the end A is given by

$$EI \frac{d^2y}{dx^2} = 150 \times x - 200 \,\Big|\, -150 (x-2)$$

Integrating both sides, we get

$$EI \frac{dy}{dx} = 75x^2 - 200x + C_1 \,\Big|\, -75 (x-2)^2 \qquad \text{...Slope equation}$$

When, $x = 0,\ \dfrac{dy}{dx} = 0$ ∴ $C_1 = 0$

Integrating again, we get

$$EIy = 25x^3 - 100x^2 + C_2 \,|\, -25 (x-2)^3 \qquad \text{...Deflection equation}$$

When, $x = 0,\ y = 0$ ∴ $C_2 = 0$

To get *maximum deflection* which occurs at the centre in this case, putting $x = 3$ m in the deflection equation, we get

$$EIy_{max} = 25 \times 3^3 - 100 \times 3^2 - 25 (3-2)^3$$

$$= 25 \times 27 - 900 - 25 = -250$$

∴ $y_{max} = -\dfrac{250}{EI} = -\dfrac{250}{2 \times 10^8 \times 8 \times 10^8 \times 10^{-12}}$ m

$$= -15.6 \times 10^{-4} \text{ m} = -1.56 \text{ mm}$$

$$y_{max} = -1.56 \text{ mm} \quad \textbf{(Ans.)}$$

Example 9·2. *A fixed beam of 6 m span carries point loads of 100 kN and 75 kN as shown in the fig. 9·7. Find the following:*

  (i)   *Fixing moments at the ends;*

 (ii)   *Reactions at the supports.*

Draw the B.M. and S.F. diagrams also.

Solution.

(i) **Fixing moments:**

Fixing moment at the left end A,

$$M_A = \sum \frac{Wab^2}{l^2} = \frac{100 \times 2 \times 4^2}{6^2} + \frac{75 \times 4 \times 2^2}{6^2} = \textbf{122·22 kNm} \quad \textbf{(Ans.)}$$

Fixing moment at the right end B,

$$M_B = \sum \frac{Wa^2b}{l^2} = \frac{100 \times 2^2 \times 4}{6^2} + \frac{75 \times 4^2 \times 2}{6^2} = \textbf{111·11 kNm} \quad \textbf{(Ans.)}$$

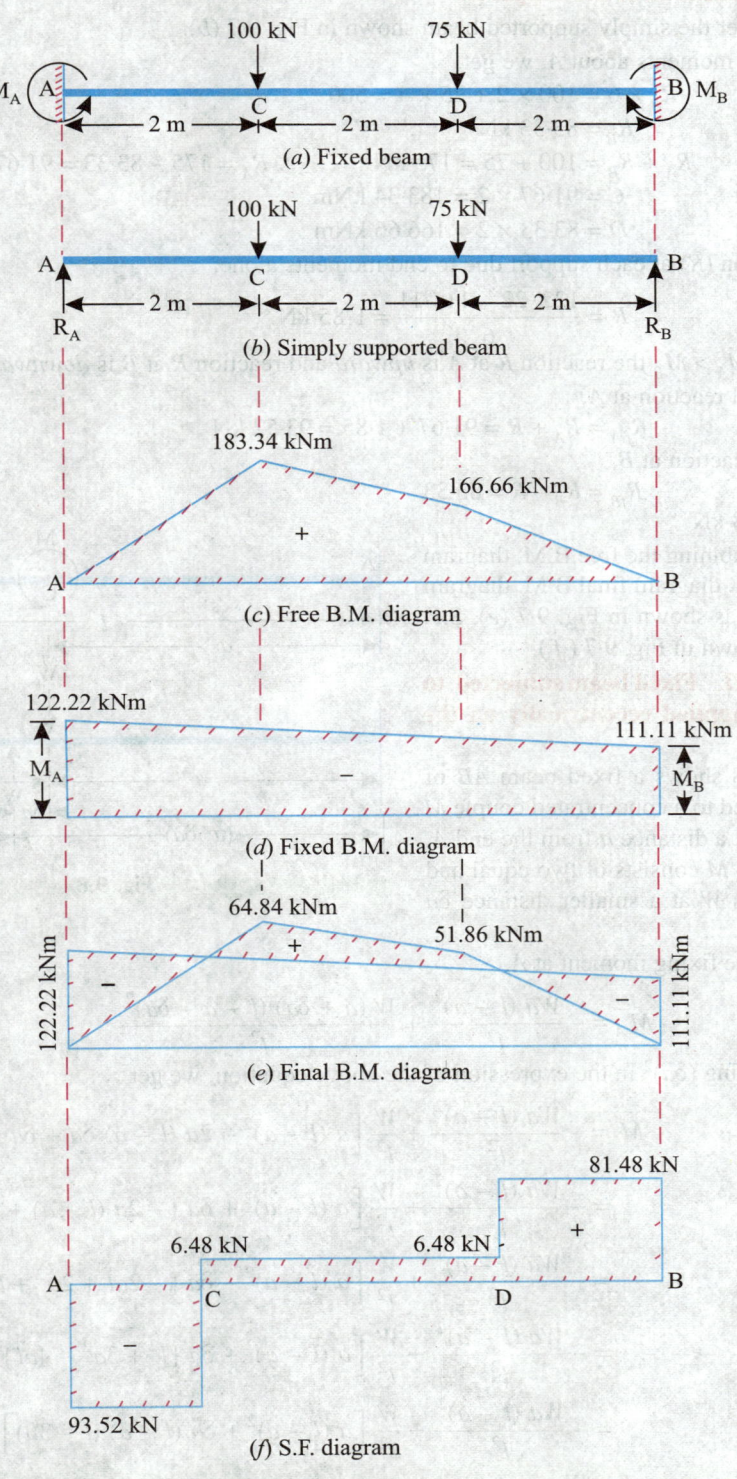

100 kN   75 kN

M_A A   C   D   B M_B

← 2 m → ← 2 m → ← 2 m →

(a) Fixed beam

100 kN   75 kN

A   C   D   B

← 2 m → ← 2 m → ← 2 m →

R_A R_B

(b) Simply supported beam

183.34 kNm

166.66 kNm

+

A   B

(c) Free B.M. diagram

122.22 kNm

111.11 kNm

M_A M_B

−

(d) Fixed B.M. diagram

64.84 kNm

51.86 kNm

+

122.22 kNm

−   −

111.11 kNm

(e) Final B.M. diagram

81.48 kN

6.48 kN   6.48 kN

+

A   C   D   B

93.52 kN

−

(f) S.F. diagram

Fig. 9.7

(ii) **Reactions at the supports:**

Consider the simply supported beam shown in Fig. 9·7 (*b*).

Taking moments about *A*, we get,

$$R_B \times 6 = 100 \times 2 + 75 \times 4 = 500$$
$$R_B = 83\text{·}33 \text{ kN}$$

Also, $R_A + R_B = 100 + 75 = 175$ kN $\therefore R_A = 175 - 83\text{·}33 = 91\text{·}67$ kN

B.M. at $C = 91\text{·}67 \times 2 = 183\text{·}34$ kNm

B.M. at $D = 83\text{·}33 \times 2 = 166\text{·}66$ kNm

Reaction (*R*) at each support due to end moments alone,

$$R = \frac{122\text{·}22 - 111\text{·}11}{6} = 1\text{·}85 \text{ kN}$$

Since $M_A > M_B$ the reaction R at A is *upward* and reaction R at B is *downward*.

$\therefore$ Final reaction at *A*,

$$R_{fA} = R_A + R = 91\text{·}67 + 1\text{·}85 = 93\text{·}52 \text{ kN}$$

Final reaction at *B*,

$$R_{fB} = R_B - R = 83\text{·}33 - 1\text{·}85 = 81\text{·}48 \text{ kN}$$

By combining the free B.M. diagram and fixed B.M. diagram final B.M. diagram can be drawn as shown in Fig. 9·7 (*e*). S.F. diagram is shown in Fig. 9·7 (*f*).

Case III. Fixed beam subjected to a couple M applied eccentrically on the span:

Fig. 9·8 shows a fixed beam *AB* of span *l* subjected to a concentrated couple *M* applied at *C* at a distance *a* from the end *A*. Let the couple *M* consists of two equal and opposite loads *W* at a smaller distance δ*a* apart.

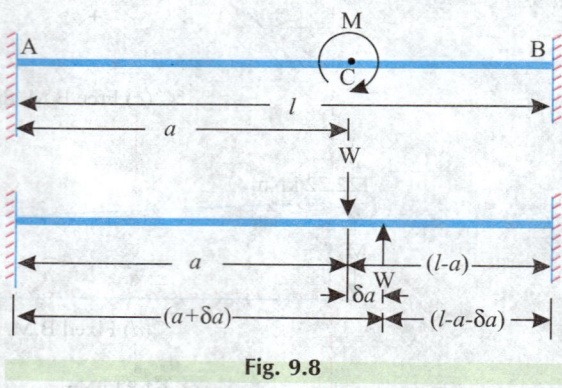

Fig. 9.8

Now, the fixing moment at *A*,

$$M_A = -\frac{Wa(l-a)^2}{l^2} + \frac{W(a+\delta a)(l-a-\delta a)^2}{l^2} \qquad \ldots(9.10)$$

Neglecting $(\delta a)^2$ in the expression of the above equation, we get

$$M = -\frac{Wa(l-a)^2}{l^2} + \frac{W}{l^2}\left[a(l-a)^2 - 2a(l-a)\,\delta a + \delta a\,(l-a)^2\right]$$

$$= -\frac{Wa(l-a)^2}{l^2} + \frac{W}{l^2}\left[a(l-a)^2 + \delta a\,\{-2a(l-a) + (l-a)^2\}\right]$$

$$= -\frac{Wa(l-a)^2}{l^2} + \frac{W}{l^2}\left[a(l-a)^2 - \delta a\,\{-2al + 2a^2 + l^2 + a^2 - 2al\}\right]$$

$$= -\frac{Wa(l-a)^2}{l^2} + \frac{W}{l^2}\left[a(l-a)^2 + \delta a\,\{l^2 + 3a^2 - 4al\}\right]$$

$$= -\frac{Wa(l-a)^2}{l^2} + \frac{W}{l^2}\left[a(l-a)^2 + \delta a\,(l-a)(l-3a)\right]$$

$$= \frac{W}{l^2}\left[-a(l-a)^2 + a(l-a)^2 + \delta a\,(l-a)(l-3a)\right]$$

$$= \frac{W\delta a}{l^2}(l-a)(l-3a).$$

When δa is small, M is equal to the couple $W\cdot\delta a$.

$\therefore \qquad M_A = \dfrac{M}{l^2}(l-a)(l-3a)$ \hfill ...(9·11)

Similarly, it can be proved that

$\qquad M_B = \dfrac{M}{l^2} a(2l-3a)$ \hfill ...(9·12)

Example 9·3. *A fixed beam of 6 m span is subjected to a concentrated couple of 150 kNm applied at a section 4 m from the left end. Find the end moments from the first principles.*

Draw B.M. and S.F. diagrams also.

Solution. Fig. 9·9 (a) shows the fixed beam. Assuming A as the origin, the unknowns are the reaction R_A and the end moment M_A ; let their directions be as shown in Fig. 9·9 (b). (After calculations if these unknowns work out to be negative then it means that they act in directions opposite to the assumed).

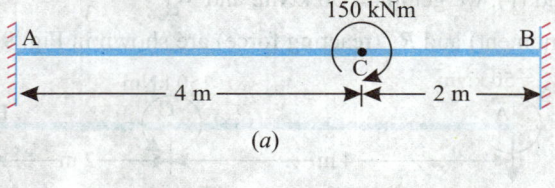

(a)

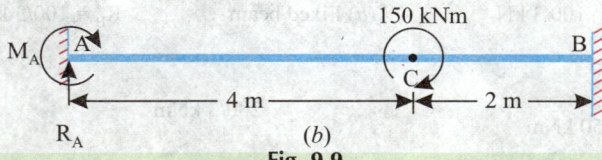

(b)

Fig. 9.9

The bending at any section, at a distance x from the end A, is given by

$$EI\frac{d^2y}{dx^2} = R_A x + M_A \;\bigg|\; + 150$$

The above expression, to facilitate application of Macaulay's Method, is arranged as follows:

$$EI\frac{d^2y}{dx^2} = R_A x + M_A \;\bigg|\; + 150(x-4)^0 \qquad ...(i)$$

Integrating both sides, we get

$$EI\frac{dy}{dx} = \frac{R_A \cdot x^2}{2} + M_A \cdot x + C_1 \;\bigg|\; + 150(x-4)$$

$$...(ii)$$
$$...\text{(Slope equation)}$$

When, $\quad x = 0, \dfrac{dy}{dx} = 0 \qquad \therefore C_1 = 0$

Integrating again, we get

$$EIy = \frac{R_A \cdot x^3}{6} + \frac{M_A \cdot x^2}{2} + C_2 \;\bigg|\; + 75(x-4)^2$$

$$...(iii)$$
$$...\text{(Deflection equation)}$$

When, $\quad x = 0, y = 0, \qquad \therefore C_2 = 0$

Blades used to cut metallic and non-metallic sheets: Split scoring saw blades.

When, $\qquad x = 6m, \qquad dy/dx = 0 \qquad \left(\because \text{At } B, \dfrac{dy}{dx} = 0 \right)$

Substituting these values in eqn. (ii), we get

$$0 = \frac{R_A \times 6^2}{2} + M_A \times 6 + 150 \,(6 - 4)$$

or, $\qquad 18R_A + 6M_A + 300 = 0$

or, $\qquad 3R_A + M_A = -50 \qquad\qquad\qquad ...(iv)$

At B, the deflection is zero.

∴ When $\qquad\qquad\qquad x = 6m, y = 0.$

Substituting these values in eqn. (iii) we get

$$0 = \frac{R_A \times 6^3}{6} + \frac{M_A \times 6^2}{2} + 75\,(6 - 4)^2 = 36R_A + 18M_A + 300$$

or, $\qquad 6R_A + 3M_A = -50 \qquad\qquad\qquad ...(v)$

Solving (iv) and (v), we get $M_A = 50$ kNm, and $R_A = -\dfrac{100}{3}$ kN

M_A (reacting moment) and R_A (reacting force) are shown in Fig. 9·10 (a).

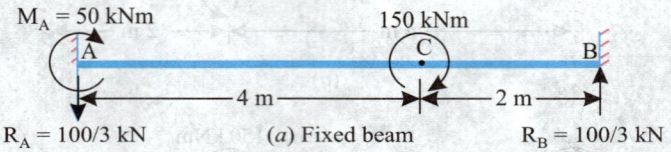

(a) Fixed beam

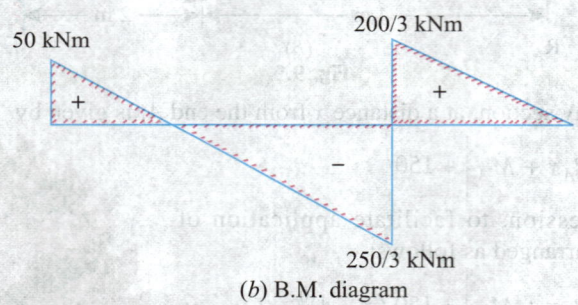

(b) B.M. diagram

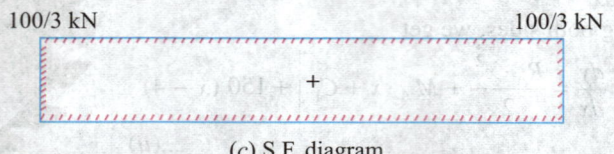

(c) S.F. diagram

Fig. 9.10

B.M. calculations:

$$M_A = +50 \text{ kNm} \quad (sagging)$$

$$M_C^A = 50 - \frac{100}{3} \times 4 = \frac{-250}{3} \text{ kNm} \quad (hogging)$$

$$M_C^B = -\frac{250}{3} + 150 = +\frac{200}{3} \text{ kNm} \quad (sagging)$$

$$M_B = 50 - \frac{100}{3} \times 6 + 150 = 0$$

B.M. Diagram for the beam is shown in Fig. 9·10 (b).

S.F. diagram:

The shear force at any section of the beam = 100/3 kN. S.F. diagram for the beam is shown in Fig. 9·10 (c).

Case IV.   Fixed beam with uniformly distributed load:

Fig. 9·11 (a) shows a fixed beam AB of length l carrying uniformly distributed load w per unit length throughout its length.

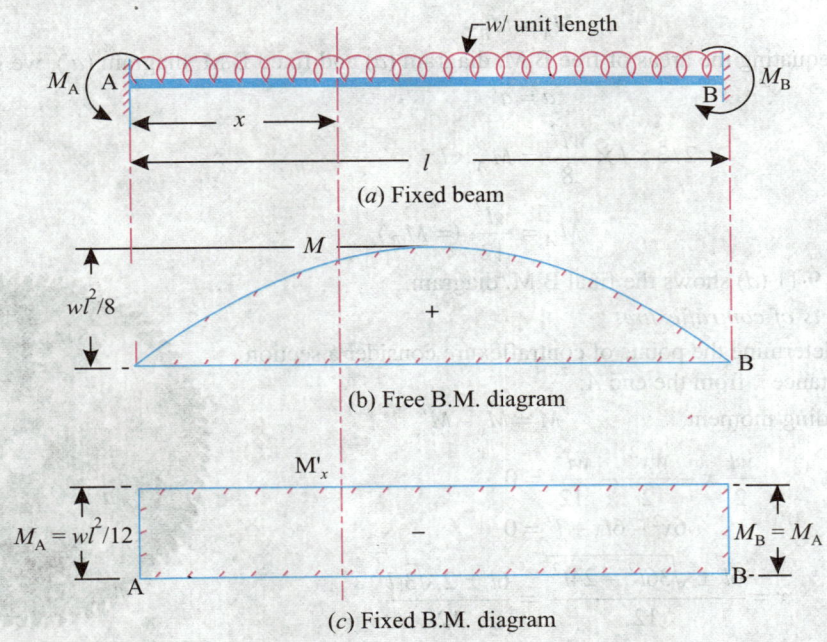

(a) Fixed beam

(b) Free B.M. diagram

(c) Fixed B.M. diagram

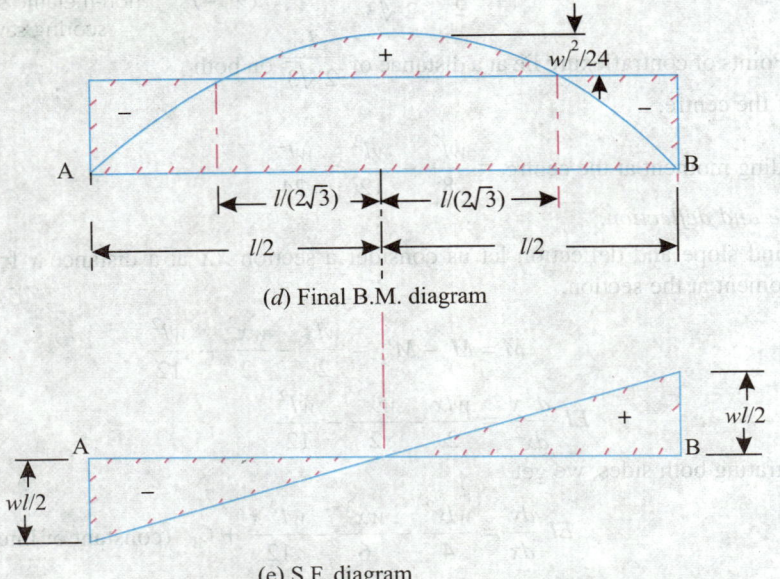

(d) Final B.M. diagram

(e) S.F. diagram

Fig. 9.11

If the beam is simply supported the maximum bending moment occurs at the centre and

$M_{max} = \dfrac{wl^2}{8}$; free B.M. diagram is shown in Fig. 9·11. (b).

B.M. diagram due to fixing couples M_A and M_B is shown in Fig. 9·11 (c).

Since the beam is *symmetrically loaded about its centre.*

∴ Reactions, $R_A = R_B = \dfrac{wl}{2}$ and fixing couples,

$$M_A = M_B$$

By equating the areas of free B.M. diagram (a) and fixed B.M. diagram (a'), we get

$$a = a'$$

i.e. $2/3 \times l \times \dfrac{wl^2}{8} = M_A \times l$

∴ $M_A = \dfrac{wl^2}{12} \ (= M_B)$ ...(9·13)

Fig. 9·11 (d) shows the final B.M. diagram.

Points of contraflexure:

To determine the points of contraflexure consider a section XX at a distance x from the end A.

Bending moment, $M = M_x - M'_x$

or, $\dfrac{wl}{2} x - \dfrac{wx^2}{2} - \dfrac{wl^2}{12} = 0$

or, $6x^2 - 6lx + l^2 = 0$

∴ $x = \dfrac{6l \pm \sqrt{36l^2 - 24l^2}}{12} = \dfrac{6l \pm 2\sqrt{3}\, l}{12}$

or, $x = \dfrac{l}{2} \pm \dfrac{l}{2\sqrt{3}}$ ...(9·14)

i.e. Points of contraflexure lie at a distance of $\dfrac{l}{2\sqrt{3}}$ on both the sides of the centre.

Bending moment at the centre $= \dfrac{wl^2}{8} - \dfrac{wl^2}{12} = \dfrac{wl^2}{24}$ ...(9·15)

Slope and deflection:

To find slope and deflection let us consider a section XX at a distance x from the end A. Bending moment at the section,

$$M = M_x - M'_x = \dfrac{wlx}{2} - \dfrac{wx^2}{2} - \dfrac{wl^2}{12}$$

or, $EI \dfrac{d^2 y}{dx^2} = \dfrac{wlx}{2} - \dfrac{wx^2}{2} - \dfrac{wl^2}{12}$ ...(i)

Integrating both sides, we get

$$EI \dfrac{dy}{dx} = \dfrac{wlx^2}{4} - \dfrac{wx^3}{6} - \dfrac{wl^2 x}{12} + C_1 \ \text{(constant of integration)}$$

When, $\qquad x = 0, \dfrac{dy}{dx} = 0$, (at the fixed end)

$\therefore \qquad C_1 = 0$

$$EI \dfrac{dy}{dx} = \dfrac{wlx^2}{4} - \dfrac{wx^3}{6} - \dfrac{wl^2 x}{12} \qquad \qquad ...(ii)$$

Integrating the eqn. (ii), we get

$$EIy = \dfrac{wlx^3}{12} - \dfrac{wx^4}{24} - \dfrac{wl^2 x^2}{24} + C_2 \quad \text{(constant of integration)}$$

When, $\qquad x = 0, y = 0$ (fixed end)

$\therefore \qquad C_2 = 0$

$$EIy = \dfrac{wlx^3}{12} - \dfrac{wx^4}{24} - \dfrac{wl^2 x^2}{24}$$

Maximum deflection takes place at the centre,

$x = l/2, y = y_{max}$ (Since the beam is symmetrically loaded about the centre)

$$\therefore \qquad EIy_{max} = \dfrac{wl \times (l/2)^3}{12} - \dfrac{w(l/2)^4}{24} - \dfrac{wl^2 (l/2)^2}{24}$$

$$= \dfrac{wl^4}{96} - \dfrac{wl^4}{384} - \dfrac{wl^4}{96}$$

$$y_{max} = - \dfrac{wl^4}{384 \, EI} \qquad \qquad ...(9.16)$$

(–ve sign indicates *downward* deflection)

This is the 1/5 of that for a freely supported beam with uniformly distributed load.

Example 9·4. *A fixed beam of 6 m span carries a uniformly distributed load of 2 kN / m run. If E = 2 × 10⁸ kN / m² and I = 0·48 × 10⁻⁴ m⁴, find:*

(i) *Bending moment at the centre;*

(ii) *Maximum deflection.*

Solution. Length of the span, $\quad l = 6m$

Rate of loading, $\qquad \qquad w = 2 kN/m$

$$E = 2 \times 10^8 \text{ kN/m}^2$$

$$I = 0.48 \times 10^{-4} \text{ m}^4$$

(i) B.M. at the centre :

We know that B.M. (final) at the centre of the beam

$$= \dfrac{wl^2}{24} = \dfrac{5 \times 6^2}{24} = \textbf{7·5 kNm} \quad \textbf{(Ans.)}$$

(ii) Maximum deflection, y_{max} :

$$y_{max} = - \dfrac{wl^4}{384 \, EI} = - \dfrac{2 \times 6^4}{384 \times 2 \times 10^8 \times 0.48 \times 10^{-4}} \times 10^3 \text{ mm}$$

$$= - \textbf{0·703 mm} \quad \textbf{(Ans.)}$$

Example 9·5. *A fixed beam of 8m span carries a uniformly distributed load of 40 kN/m run over 4 m length starting from left hand end and a concentrated load of 80 kN at a distance of 6 m from the left hand end. Find:*

(i) *Moments at the supports;*

(ii) *Deflection at centre of the beam.*

Take : EI = 15000 kNm².

Solution. Fig. 9·12 shows the fixed beam AB of 8 m span and carrying uniformly distributed load and a concentrated load.

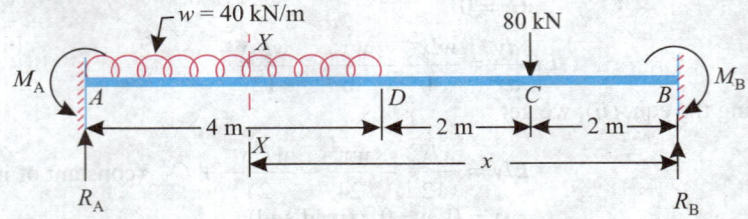

Fig. 9.12

Let,
R_A = Reaction at the support A,
R_B = Reaction at the support B,
M_A = Moment at the support A, and
M_B = Moment at the support B.

(i) Moments at the supports, M_A and M_B :

Taking the origin at B and x positive towards left, consider a section XX at a distance x from the end B in the position DA.

Let the directions of R_B and M_B be as shown (Negative results in any case would mean the actual direction is opposite to the assumed one).

Bending moment at the section,

$$M = -M_B + R_B \times x - 80\,(x-2) - \frac{w\,(x-4)^2}{2}$$

$$= -M_B + R_B \times x - 80\,(x-2) - 20\,(x-4)^2 \qquad (\because w = 40 \text{ kN / m})$$

or
$$EI\,\frac{d^2y}{dx^2} = -M_B + R_B \times x - 80\,(x-2) - 20\,(x-4)^2 \qquad \qquad ...(i)$$

A medium sized tunnel-boring machine. Cutting tips of this machine are made of titanium carbide.

Integrating both sides, we get

$$EI \frac{dy}{dx} = - M_B \cdot x + R \times \frac{x^2}{2} - 40 (x - 2)^2 - \frac{20 (x - 4)^3}{3} + C_1$$

(where, C_1 = constant of integration)

When, $\qquad x = 0, \dfrac{dy}{dx} = 0,$ at fixed end B

∴ $\qquad C_1 = 0$

∴ $\qquad EI \dfrac{dy}{dx} = - M_B \cdot x + R_B \times \dfrac{x^2}{2} - 40 (x - 2)^2 - \dfrac{20 (x - 4)^3}{3}$ ...(ii)

Integrating eqn. (ii), we get

$$EIy = - M_B \cdot \frac{x^2}{2} + R_B \cdot \frac{x^3}{6} - \frac{40 (x - 2)^3}{3} - \frac{20 (x - 4)^4}{12} + C_2$$

(where, C_2 = constant of integration)

When, $\qquad x = 0, y = 0,$ at fixed end D

∴ $\qquad C_2 = 0$

$$EIy = - M_B \cdot \frac{x^2}{2} + R \times \frac{x^3}{6} - \frac{40 (x - 2)^3}{3} - \frac{20 (x - 4)^4}{12}$$ ...(iii)

When, $\qquad x = 8m$ (at the end A), $\dfrac{dy}{dx} = 0,\ y = 0$

Substituting these value in eqns. (ii) and (iii), we get

$$0 = - 8M_B + 32 R_B - 1440 - \frac{1280}{3}$$

or, $\qquad 0 = M_B - 4R_B + 180 + \dfrac{160}{3}$

or, $\quad M_B - 4R_B + \dfrac{700}{3} = 0$ ...(iv)

and, $\qquad 0 = - 32 M_B + \dfrac{512}{6} R_B - 2880 - \dfrac{1280}{3}$

or, $\qquad 0 = M_B - 8/3\ R_B + 90 + 40/3$

or, $M_B - \dfrac{8}{3} R_B + \dfrac{310}{3} = 0$ ...(v)

Solving eqns. (iv) and (v), we get

$\qquad R_B = 97 \cdot 5$ kN

and, $\qquad$ **$M_B = 156.6$ kNm (–) (Ans.)**

Also, $\qquad R_A + R_B = 40 \times 4 + 80 = 240$ kN

∴ $\qquad R_A = 240 - 97 \cdot 5 = 142 \cdot 5$ kN

Equation for B.M.,

$$M = - M_B + R_B \cdot x - 80 (x - 2) - 20 (x - 4)^2$$
$$= - 156 \cdot 6 + 97 \cdot 5 \times x - 80 (x - 2) - 20 (x - 4)^2$$

When, $\qquad x = 8m, M = M_A$

∴ $\qquad M_A = -156.6 + 97.5 \times 8 - 80\,(8-2) - 20\,(8-4)^2$

$\qquad\qquad\qquad = -156.6 + 780 - 480 - 320$

i.e. $\qquad M_A = 176.6 \text{ kNm } (-) \quad \textbf{(Ans.)}$

(ii) Deflection at the centre, y_D :

From eqn. (*iii*), putting $x = 4$m, we get

$$EIy_D = -M_B \times \frac{4^2}{2} + R_B \times \frac{4^3}{6} - \frac{40\,(4-2)^3}{3} - 0$$

$$= -156.6 \times 8 + 97.5 \times \frac{64}{6} - \frac{320}{3}$$

$$= -1252.8 + 1040 - 106.66 = -319.46$$

∴ $\qquad\qquad y_D = -\dfrac{319.46}{EI}$

$$= \frac{319.46}{15000} = -0.02129 \text{ m} = -21.29 \text{ mm}$$

Hence, deflection at the centre = **21.29 mm** *(downward)* **(Ans.)**

Example 9·6. *A fixed beam of 6 m span carries uniformly distributed load of 10 kN/m for a distance 4m from the left hand end.*

Find the fixing moments at the supports.

Solution. Fig. 9·13 shows a fixed beam *AB* of span *l* (= 6m) carrying a uniformly distributed load of *w*/unit run (= 10 kN/m) for a distance *a* (= 4m) from the end *A*.

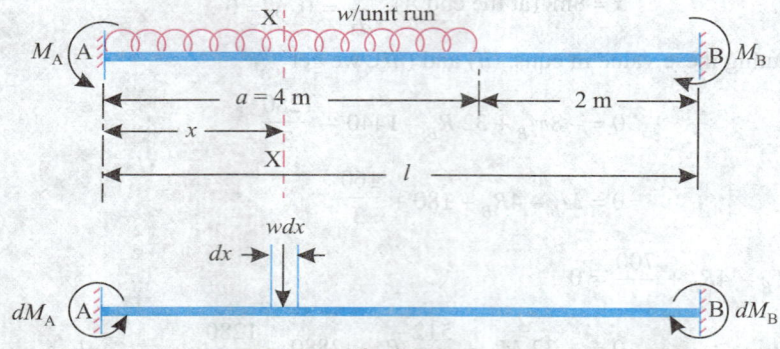

Fig. 9.13

Let us consider any section *XX* at a distance *x* from the end *A*.

Load acting for an elemental distance $dx = wdx$

The fixed end moments due to the elemental load (wdx) will be as follows:

$$dM_A = (wdx)\,\frac{x\,(l-x)^2}{l^2}, \quad dM_B = (wdx)\,\frac{x^2\,(l-x)}{l^2}$$

∴ Total fixing moment at *A*,

$$M_A = \int_0^a \frac{wx\,(l-x)^2}{l^2}\,dx = \int_0^a \frac{w}{l^2}\left[x\,(l^2 + x^2 - 2lx)\right]dx$$

$$= \int_0^a \frac{w}{l^2}\left(xl^2 + x^3 - 2lx^2\right)dx = \frac{w}{l^2}\left[l^2 \times \frac{x^2}{2} + \frac{x^4}{4} - 2l\cdot\frac{x^3}{3}\right]_0^a$$

$$= \frac{w}{l^2}\left[\frac{l^2 a^2}{2} + \frac{a^4}{4} - \frac{2la^3}{3}\right]$$

i.e.
$$M_A = \frac{wa^2}{12l^2}(6l^2 - 8la + 3a^2) \qquad \text{...(i)}$$

Similarly total fixing moment at B,

$$M_B = \int_0^a \frac{wx^2(l-x)}{l^2}dx = \frac{w}{l^2}\int_0^a (x^2l - x^3)\,dx$$

$$= \frac{w}{l^2}\left[\frac{x^3}{3}\cdot l - \frac{x^4}{4}\right]_0^a = \frac{w}{l^2}\left[\frac{a^3 \cdot l}{3} - \frac{a^4}{4}\right] = \frac{wa^3}{12l^2}(4l - 3a)$$

i.e.
$$M_B = \frac{wa^3}{12l^2}(4l - 3a) \qquad \text{...(ii)}$$

> **Note.** In the case the load covers the whole span, putting $a = l$, in the expressions for M_A and M_B, we have
>
> $$M_A = \frac{wl^2}{12\,l^2}(6l^2 - 8l^2 + 3l^2) = \frac{wl^2}{12}, \text{ and } M_B = \frac{wl^3}{12\,l^2}(4l - 3l) = \frac{wl^2}{12}$$

Now, to find the moments at the supports, in the problem, putting $w = 10$ kN/m, $a = 4$m and $l = 6$m, we get

$$M_A = \frac{wa^2}{12l^2}(6l^2 - 8la + 3a^2) = \frac{10 \times 4^2}{12 \times 6^2}(6 \times 6^2 - 8 \times 6 \times 4 + 3 \times 4^2)$$

$$= \frac{160}{432}(216 - 192 + 48) = 26.66 \text{ kNm}$$

i.e. **$M_A = 26.66$ kNm (Ans.)**

and,
$$M_B = \frac{wa^3}{12l^2}(4l - 3a) = \frac{10 \times 4^3}{12 \times 6^2}(4 \times 6 - 3 \times 4)$$

$$= \frac{640}{432}(24 - 12) = 17.77 \text{ kNm}$$

i.e. **$M_B = 17.77$ kNm (Ans.)**

Case V. Fixed beam carrying a triangular load whose intensity varies from zero at one end to w per unit run at the other end:

Fig. 9·14 shows a fixed beam AB of span l carrying a *triangular load* whose intensity varies from zero at the end A to w per unit run at the other end B.

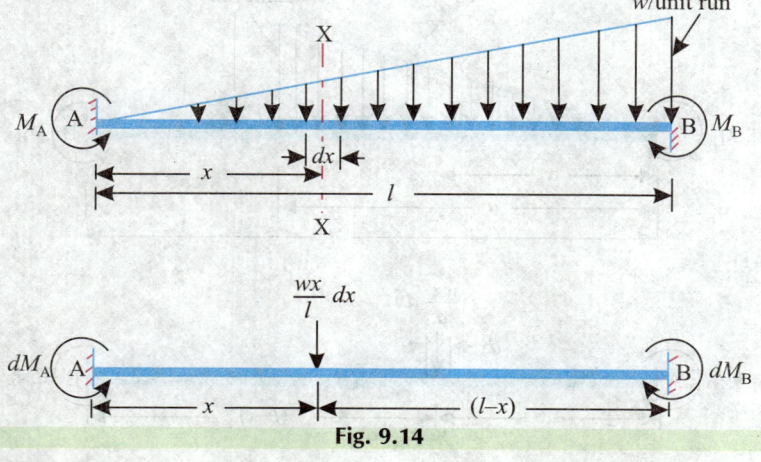

Fig. 9.14

Consider any section XX at a distance x from the end A. Intensity of loading at $XX = \dfrac{wx}{l}$

∴ Load acting for an elemental distance $dx = \dfrac{wx}{l}\, dx$

The fixed moments due to this elemental load are :

$$dM_A = \left(\frac{wx}{l}\, dx\right)\frac{x\,(l-x)^2}{l^2} = \frac{wx^2\,(l-x)^2\, dx}{l^3}$$

and, $$dM_B = \left(\frac{wx}{l}\, dx\right)\frac{x^2\,(l-x)}{l^2} = \frac{wx^3\,(l-x)\, dx}{l^3}$$

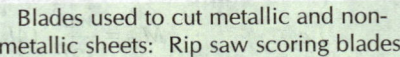

Blades used to cut metallic and non-metallic sheets: Rip saw scoring blades.

∴ Total fixing moment at A,

$$M_A = \int_0^l \frac{wx^2\,(l-x)^2}{l^2}\, dx = \frac{w}{l^2}\int_0^l x^2\,(l^2 + x^2 - 2lx)\, dx$$

$$= \frac{w}{l^3}\int_0^l \left(x^2 l^2 + x^4 - 2lx^3\right) dx = \frac{w}{l^3}\left[l^2 \times \frac{x^3}{3} + \frac{x^5}{5} - 2l\cdot\frac{x^4}{4}\right]_0^l$$

$$= \frac{w}{l^3}\left[l^2 \times \frac{l^3}{3} + \frac{l^5}{5} - 2l \times \frac{l^4}{4}\right] = \frac{w}{l^3}\left[\frac{l^5}{3} + \frac{l^5}{5} - \frac{l^5}{2}\right] = \frac{w}{l^3} \times \frac{l^5}{30} = \frac{wl^2}{30}$$

i.e. $\qquad M_A = \dfrac{wl^2}{30}$ $\qquad\qquad\qquad\qquad\qquad\qquad\qquad\qquad$ ...(9·17)

Similarly, $M_B = \displaystyle\int_0^l \frac{wx^3\,(l-x)\, dx}{l^3} = \frac{w}{l^3}\int_0^l x^3\,(l-x)\, dx$

$$= \frac{w}{l^3}\int_0^l (x^3 l - x^4)\, dx = \frac{w}{l^3}\left[\frac{x^4}{4}\cdot l - \frac{x^5}{5}\right]_0^l$$

$$= \frac{w}{l^3}\left(\frac{l^4}{4}\cdot l - \frac{l^5}{5}\right) = \frac{w}{l^3} \times \frac{l^5}{20} = \frac{wl^2}{20}$$

i.e. $\qquad M_B = \dfrac{wl^2}{20}$ $\qquad\qquad\qquad\qquad\qquad\qquad\qquad\qquad$ ...(9·18)

Case VI. Fixed beam carrying a triangular load for a given distance from one end:

Fig. 9·15 shows a fixed beam AB of span l carrying a triangular loading covering a distance a from the end A. Let the intensity of load varies from zero to w per unit run.

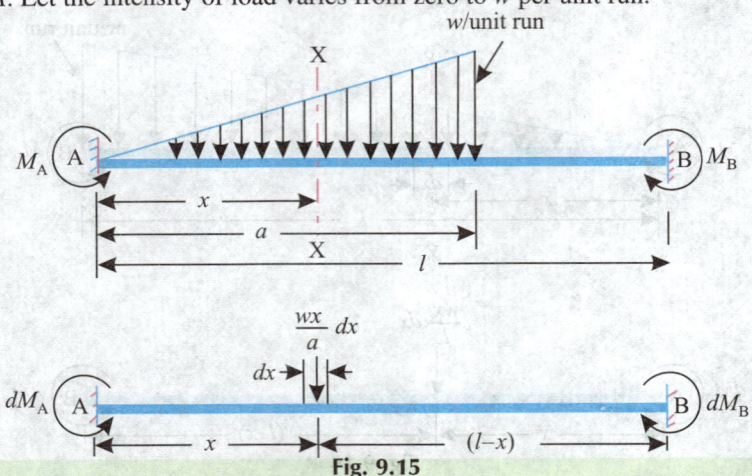

Fig. 9.15

Consider any section XX at a distance x from the end A in the loading range.

Intensity of loading at $XX = \dfrac{wx}{a}$

$\therefore$ Load acting for an elemental distance $dx = \dfrac{wx}{a}\, dx$

The fixed moments due to this elemental load are:

$$dM_A = \left(\frac{wx}{a}\, dx\right) \frac{x\,(l-x)^2}{l^2} = \frac{wx^2\,(l-x)^2}{al^2}\, dx$$

and,

$$dM_B = \left(\frac{wx}{a}\, dx\right) \frac{x^2\,(l-x)}{l^2} = \frac{wx^3\,(l-x)}{al^2}\, dx$$

Total fixing moment at A,

$$M_A = \int_0^a \frac{wx^2\,(l-x)^2}{al^2}\, dx = \frac{w}{al^2} \int_0^a x^2\,(l^2 + x^2 - 2lx)\, dx$$

$$= \frac{w}{al^2} \int_0^a (x^2 l^2 + x^4 - 2lx^3)\, dx = \frac{w}{al^2}\left[\frac{x^3}{3}.l^2 + \frac{x^5}{5} - \frac{2lx^4}{4}\right]_0^a$$

$$= \frac{w}{al^2}\left[\frac{a^3.l^2}{3} + \frac{a^5}{5} - \frac{2la^4}{4}\right] = \frac{w}{al^2}.\frac{a^3}{30}(10l^2 + 6a^2 - 15la)$$

or

$$M_A = \frac{wa^2}{30l^2}(10l^2 + 6a^2 - 15la) \qquad \ldots(9\cdot19)$$

Similarly the total fixing moment at B,

$$M_B = \int_0^a \frac{wx^3\,(l-x)}{al^2}\, dx = \frac{w}{al^2} \int_0^a x^3\,(l-x)\, dx$$

$$= \frac{w}{al^2} \int_0^a (x^3 l - x^4)\, dx = \frac{w}{al^2}\left[\frac{x^4}{4}l - \frac{x^5}{5}\right]_0^a$$

$$= \frac{w}{al^2}\left(\frac{a^4}{4}.l - \frac{a^5}{5}\right) = \frac{w}{al^2}.\frac{a^4}{20}(5l - 4a) = \frac{w}{al^2}.\frac{a^4}{20}(5l - 4a)$$

i.e.

$$M_B = \frac{wa^3}{20l^2}(5l - 4a) \qquad \ldots(9\cdot20)$$

Note. In the case when the triangular loading covers the whole span, putting $a = l$ in expressions for M_A and M_B, we have

$$M_A = \frac{w}{l \times l^2}.\frac{l^3}{30}\left(10l^2 + 6l^2 - 15l^2\right) = \frac{wl^2}{30}$$

and,

$$M_B = \frac{w}{l^2 \times l^2} \times \frac{l^4}{20}(5l - 4l) = \frac{wl^2}{20}$$

9·2·2. Fixed beam with ends at different levels (Effect of sinking of supports)

Fig. 9·16 shows a fixed beam AB of span l whose ends A and B are fixed at different levels; δ being the difference of level between the ends. Let the end A be at a higher level than the end B.

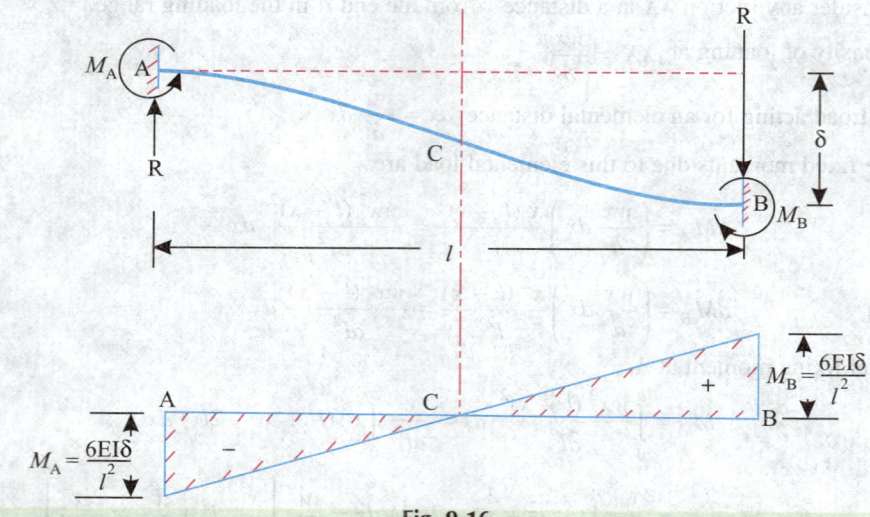

Fig. 9.16

Let, $\qquad M_A$ = Fixing moment at the end A, and

$\qquad M_B$ = Fixing moment at the end B.

[Evidently, for this case, M_A is negative (hogging) and M_B is positive (M_A and M_B being equal numerically).]

$\qquad\qquad\qquad\qquad$ R = Reaction at each support.

Consider any section at a distance x from the end A. Since the rate of loading is zero , we have

$$EI\,\frac{d^4y}{dx^4} = 0 \qquad\qquad\qquad\qquad \text{...(i)}$$

Integrating, we get,

Shear force $\qquad\qquad = EI\,\dfrac{d^3y}{dx^3} = C_1 \qquad$ (where, C_1 = constant of integration)

When, $\qquad\qquad\qquad x = 0$, S.F. = $+R \qquad \therefore\ C_1 = R$

$\therefore \qquad\qquad EI\,\dfrac{d^3y}{dx^3} = R \qquad\qquad\qquad\qquad \text{...(ii)}$

Integrating again, we get

B.M. at any section $\qquad = EI\,\dfrac{d^2y}{dx^2} = Rx + C_2 \qquad$ (where, C_2 = constant of integration)

When, $\qquad\qquad\qquad x = 0$, B.M. = $-M_A \qquad \therefore\ C_2 = -M_A$

$\therefore \qquad\qquad EI\,\dfrac{d^2y}{dx^2} = R.x - M_A \qquad\qquad\qquad \text{...(iii)}$

Integrating again, we have

$$EI\,\frac{dy}{dx} = R\cdot\frac{x^2}{2} - M_A.x + C_3 \qquad\qquad \text{...(Slope equation)}$$

(where, C_3 = constant of integration)

When, $\qquad\qquad\qquad x = 0,\ \dfrac{dy}{dx} = 0 \qquad\qquad \therefore\ C_3 = 0$

$\therefore \qquad\qquad EI\,\dfrac{dy}{dx} = \dfrac{R}{2}\,x^2 - M_A x \qquad\qquad\qquad \text{...(iv)}$

Integrating again, we have

$$Ely = \frac{R}{2} \cdot \frac{x^3}{3} - M_A \frac{x^2}{2} + C_4$$

...(*Deflection equation*)

(where, C_4 = constant of integration)

When, $\quad\quad x = 0, y = 0 \quad \therefore C_4 = 0$

$\therefore \quad\quad Ely = \frac{R}{2} \frac{x^3}{3} - M_A \cdot \frac{x^2}{2}$ ...(*v*)

When, $\quad\quad x = l, y = -\delta$

$$- EI\delta = \frac{Rl^3}{6} - \frac{M_A l^2}{2}$$ ...(*vi*)

We also know that at B, $x = l$, and $dy/dx = 0$.
Substituting these values in eqn. (*iv*) we get

$$0 = \frac{R}{2} \cdot l^2 - M_A l$$

$\therefore \quad\quad R = \frac{2M_A}{l}$ ...(*vii*)

Substituting this value of R in eqn. (*vi*), we get

$$- EI\delta = \frac{2M_A}{l} \cdot \frac{l^3}{6} - \frac{M_A l^2}{2} = \frac{M_A l^2}{3} - \frac{M_A l^2}{2}$$

$$EI\delta = \frac{M_A l^2}{6} \quad\quad \therefore M_A = \frac{6 EI\delta}{l^2}$$ ...(9·21)

and, $\quad\quad R_A = \frac{2M_A}{l} = \frac{12 EI\delta}{l^3}$ ...(9·22)

Hence the bending moment at any section distant x from A is given by,

$$M = EI \frac{d^2 y}{dx^2} = R \cdot x - M_A$$ ...[Eqn. (*iii*)]

or, $\quad\quad M = \frac{2M_A}{l} \cdot x - \frac{6EI\delta}{l^2}$ ...(9·23)

$\therefore$ Bending moment at B, [Putting $x = l$ in eqn. (9·23)]

$$M_B = \frac{2M_A}{l} \cdot l - \frac{6EI\delta}{l^2} = \frac{2 \times 6EI\delta}{l^2} - \frac{6EI\delta}{l^2} = \frac{6EI\delta}{l^2}$$

i.e. $\quad\quad M_B = \frac{6EI\delta}{l^2}$ ...(9·24)

Thus it may be concluded that when the ends of a fixed beam are at different levels, the fixing moment at each end is numerically equal to $\frac{6EI\delta}{l^2}$. At the *higher end* the moment is a hogging moment and at the *lower end* this moment is a *sagging* one.

Example 9·7. *A fixed beam of 5 m span carries a point load of 150 kN at 3 m from the left end. If the right end sinks by 1·5 mm find the fixing moments as well as reactions at the supports. Take: $E = 2 \times 10^8$ kN/m^2 and $I = 10000$ cm^4.*

Solution. Fig. 9·17 (*a*) shows the fixed beam with loading. Due to loading the fixing moment at each support will be a hogging moment but due to sinking of supports the moment at each end will be $\frac{6EI\delta}{l^2}$, the nature of moments being *hogging* at the higher end and *sagging* at the lower end.

Blades used to cut metallic and non-metallic sheets: Panel saw blades.

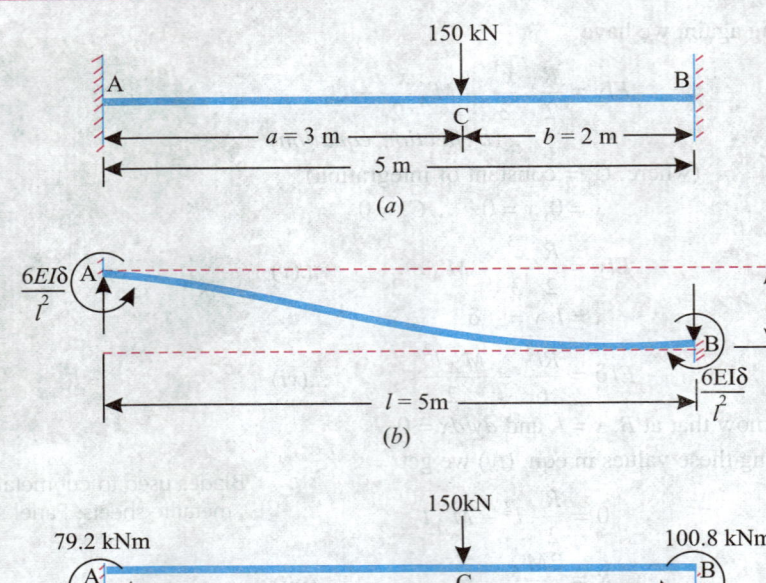

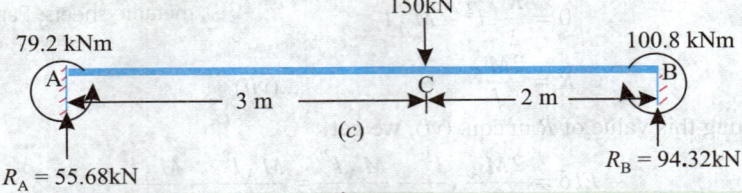

Fig. 9.17

Fixing moment at A:

$$M_A = -\frac{Wab^2}{l^2} - \frac{6EI\delta}{l^2}$$

$$= -\left[\frac{150 \times 3 \times 2^2}{5^2} + \frac{6 \times 2 \times 10^8 \times 10000 \times 10^{-8} \times 1 \cdot 5 \times 10^{-3}}{5^2}\right]$$

$$= -(72 + 7 \cdot 2) = -79 \cdot 2 \text{ kNm}$$

i.e. $M_A = \textbf{79·2 kNm (hogging)}$ **(Ans.)**

Fixing moment at B:

$$M_B = -\frac{Wa^2b}{l^2} + \frac{6EI\delta}{l^2}$$

$$= -\frac{150 \times 3^2 \times 2}{5^2} + \frac{6 \times 2 \times 10^8 \times 10000 \times 10^{-8} \times 1 \cdot 5 \times 10^{-3}}{5^2}$$

$$= -108 + 7 \cdot 2 = -100 \cdot 8 \text{ kNm}$$

i.e. $M_B = \textbf{100·8 kNm (hogging)}$ **(Ans.)**

Reactions at A and B:

R_A = Reaction due to load with simply supported condition

          – reaction due to end moments

$$= \frac{150 \times 2}{5} - \frac{(100 \cdot 8 - 79 \cdot 2)}{5} = 60 - 4 \cdot 32 = 55 \cdot 68 \text{ kN}$$

i.e. $R_A = \textbf{55·68 kN}$ **(Ans.)**

and, R_B = Reaction due to simply supported condition

          + reaction due to end moments

$$= \frac{150 \times 3}{5} + \frac{(100 \cdot 8 - 79 \cdot 2)}{5} = 90 + 4 \cdot 32 = 94 \cdot 32 \text{ kN}$$

i.e. $R_B = \textbf{94·32 kN}$ **(Ans.)**

Example 9·8. *A fixed beam of 6 m span carries a uniformly distributed load of 10 kN/m run on the entire beam. The level of right hand support sinks by 5 mm below that of the left hand end. If E = 2·08 × 10⁸ kN/m² and I = 4·52 × 10⁻⁵ m⁴ find:*

(i) *Support moments;*   (ii) *Support reactions;*   (iii) *Deflection at the centre.*

Solution. Fig. 9·18 shows fixed beam of 6 m span with uniformly distributed load of 10 kN/m over its entire span.

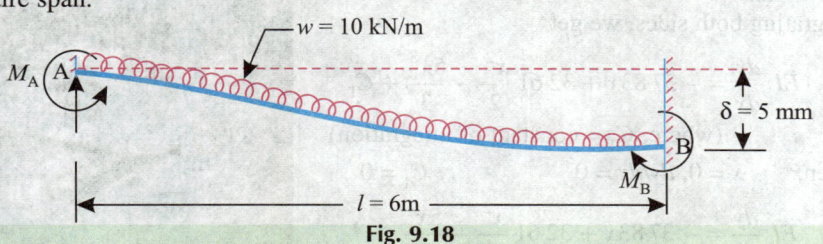

Fig. 9.18

Rate of loading, $w = 10$ kN/m

Amount by which the support B sinks,

$$\delta = 5\text{mm} = 5 \times 10^{-3} \text{ m}, \quad E = 2\cdot08 \times 10^8 \text{ kN/m}^2, \quad I = 4\cdot52 \times 10^{-5} \text{ m}^4$$

(i) Supports moments, M_A and M_B :

If the beam support does not sink, then

$$M'_A = M'_B = -\frac{Wl^2}{12}$$

$$R'_A = R'_B = \frac{wl}{2}$$

After the sinking of the support B, support moments will be given as follows:

$$M_A = -\frac{wl^2}{12} - \frac{6EI\delta}{l^2}$$

$$= -\frac{10 \times 6^2}{12} - \frac{6 \times 2\cdot08 \times 10^8 \times 4\cdot52 \times 10^{-5} \times 5 \times 10^{-3}}{6^2}$$

$$= -30 - 7\cdot83 = -37\cdot83 \text{ kNm}$$

i.e. $M_A = -\textbf{37·83 kNm}$ **(Ans.)**

$$M_B = -\frac{wl^2}{12} + \frac{6EI\delta}{l^2} = -30 + 7\cdot83 = -22\cdot17 \text{ kNm}$$

i.e. $M_B = -\textbf{22·17 kNm}$ **(Ans.)**

(ii) Support reactions, R_A and R_B :

$$R_A = \frac{wl}{2} + \frac{12EI\delta}{l^3} = \frac{10 \times 6}{2} + \frac{12 \times 2\cdot08 \times 10^8 \times 4\cdot52 \times 10^{-5} \times 5 \times 10^{-3}}{6^3}$$

$$= 30 + 2\cdot61 = \textbf{32·61 kN} \textbf{(Ans.)}$$

$$R_B = \frac{wl}{2} - \frac{12EI\delta}{l^3}$$

$$= 30 - 2\cdot61 = \textbf{27·39 kN} \textbf{(Ans.)}$$

(iii) Deflection at the centre, y_C:

Consider a section XX at a distance x from the end A.

B.M. at the section,

$$M = M_A + R \cdot x - \frac{wx^2}{2}$$

or, $\quad EI \dfrac{d^2y}{dx^2} = -37.83 + 32.61x - \dfrac{10x^2}{2}$

Integrating both sides, we get

$$EI \frac{dy}{dx} = -37.83x + 32.61 \frac{x^2}{2} - \frac{5x^3}{3} + C_1$$

(where, C_1 = constant of integration)

When, $\quad x = 0$, $dy/dx = 0$ $\qquad \therefore\ C_1 = 0$

$\therefore \quad EI \dfrac{dy}{dx} = -37.83x + 32.61 \dfrac{x^2}{2} - \dfrac{5x^3}{3}$

Integrating again, we get

$$EI\, y = -37.83 \frac{x^2}{2} + 32.61 \times \frac{x^3}{6} - \frac{5x^4}{12} + C_2$$

(where, C_2 = constant of integration)

Automated boring machine.

When, $\quad x = 0$, $y = 0$ $\qquad \therefore\ C_2 = 0$

$\therefore \qquad EI\, y = -37.83 \times \dfrac{x^2}{2} + 32.61 \times \dfrac{x^3}{6} - \dfrac{5x^4}{12}$

When, $\quad x = 3\text{m}$

$$EI\, y_C = -37.83 \times \frac{3^2}{2} + 32.61 \times \frac{3^3}{6} - \frac{5 \times 3^4}{12}$$

$$= -170.23 + 146.74 - 33.75 = -57.24$$

$\therefore$ Deflection at the centre,

$$y_C = -\frac{57.24}{EI}$$

$$= -\frac{57.24}{2.08 \times 10^8 \times 4.52 \times 10^{-5}} \times 10^3 \text{ mm} = -6.088 \text{ mm}$$

Hence, $\quad y_C = \textbf{6.088 mm}$ *(downward)* **(Ans.)**

Example 9.9. *A fixed beam AB, of 4 m span, carries a uniformly distributed load of 20 kN/m over its entire length. Support B, which was initially at the level of support A sinks by 10 mm. Draw the B.M. diagram for the beam.*

Take: $E = 2 \times 10^8$ kN/m^2; $I = 6 \times 10^{-5}$ m^4.

Solution. Rate of loading, $w = 20$ kN/m

Length of span, $\qquad l = 4$ m

$$E = 2 \times 10^8 \text{ kN/m}^2$$

$$I = 6 \times 10^{-5} \text{ m}^4$$

Fixed end moments due to uniformly distributed load,

$$M_A' = M_B' = -\frac{wl^2}{12}$$

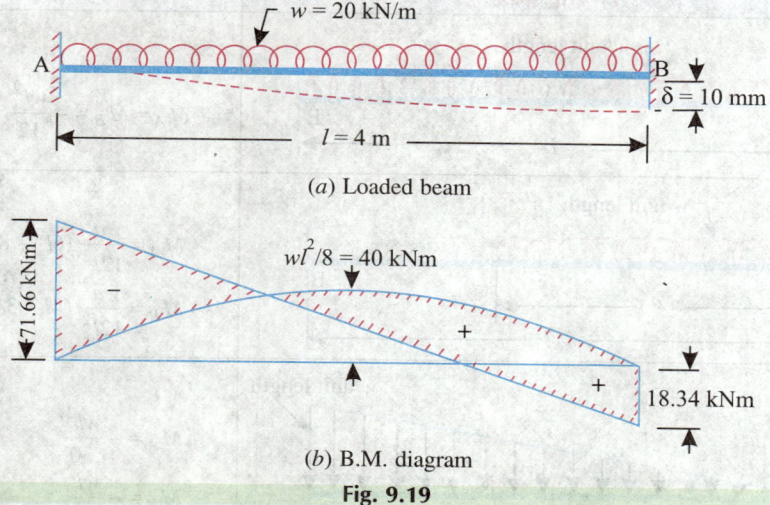

(a) Loaded beam

(b) B.M. diagram

Fig. 9.19

After the sinking of support B, support moments will be as follows:

$$M_A = -\frac{wl^2}{12} - \frac{6EI\delta}{l^2}$$

$$= -\frac{20 \times 4^2}{12} - \frac{6 \times 2 \times 10^8 \times 6 \times 10^{-5} \times 10 \times 10^{-3}}{4^2}$$

$$= -26 \cdot 66 - 45 = -71 \cdot 66 \text{ kNm}$$

$$M_B = -\frac{wl^2}{12} + \frac{6EI\delta}{l^2} = -26 \cdot 66 + 45 = +18 \cdot 34 \text{ kNm}$$

Fig 9.19 (b) shows the B.M. diagram for the beam.

TABLE 9·1. Fixed Beam Loadings

S.No.	Type of loading	Fixed end moments
1.	A, B beam with load W at centre, $l/2$ and $l/2$, total l	$M_A = M_B = -\dfrac{Wl}{8}$
2.	A, B beam with load W at distance a from A and b from B, total l	$M_A = -\dfrac{Wab^2}{l^2}$ $M_B = -\dfrac{Wa^2b}{l^2}$
3.	A, B beam with moment M at distance a, total l	$M_A = \dfrac{M}{l^2}(l-a)(l-3a)$ $M_B = \dfrac{M}{l^2}a(2l-3a)$

S.No.	Type of loading	Fixed end moments
4.	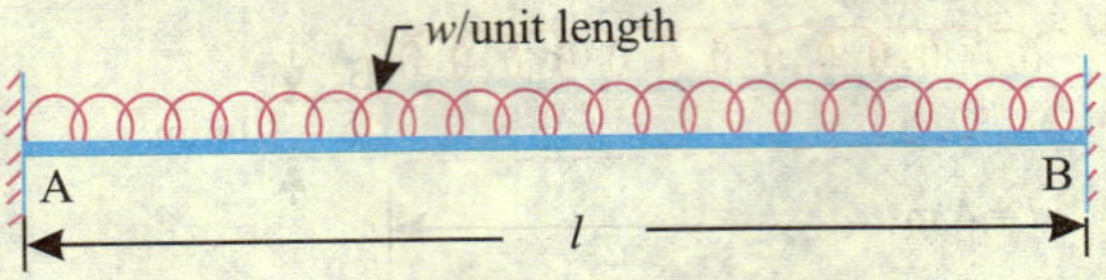 w/unit length, A ⟷ l ⟷ B	$M_A = M_B = -\dfrac{wl^2}{12}$
5.	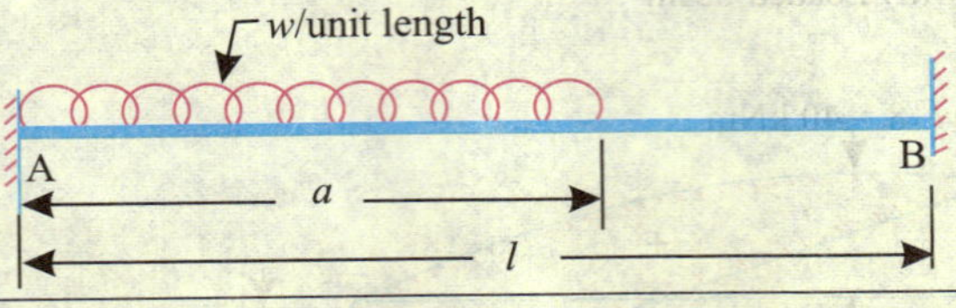 w/unit length, A ⟷ a ⟷ B, l	$M_A = \dfrac{wa^2}{12l^2}(6l^2 - 8la + 3a^2)$ $M_B = \dfrac{wa^3}{12l^2}(4l - 3a)$
6.	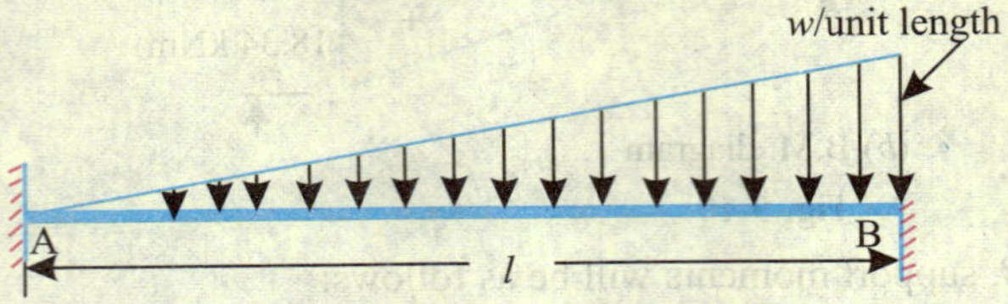 w/unit length, A ⟷ l ⟷ B	$M_A = -\dfrac{wl^2}{30}$ $M_B = -\dfrac{wl^2}{20}$
7.	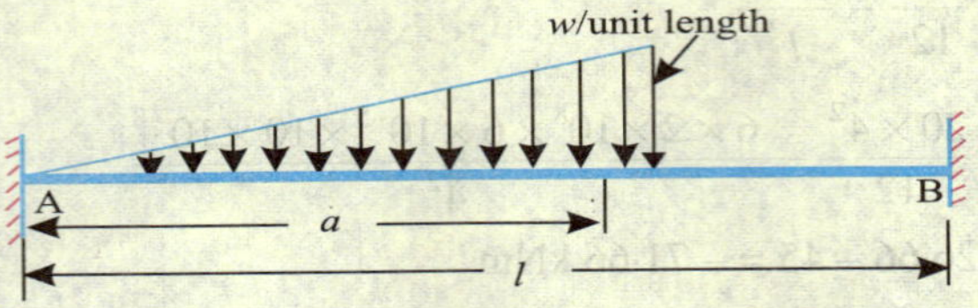 w/unit length, A ⟷ a ⟷ B, l	$M_A = -\dfrac{wa^2}{30l^2}(10l^2 + 6a^2 - 15la)$ $M_B = -\dfrac{wa^3}{20l^2}(5l - 4a)$
8.	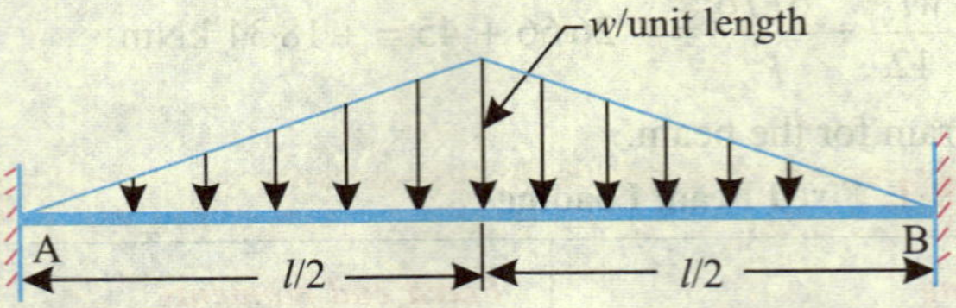 w/unit length, A ⟷ l/2 ⟷ l/2 ⟷ B	$M_A = M_B = -\dfrac{5wl^2}{96}$
9.	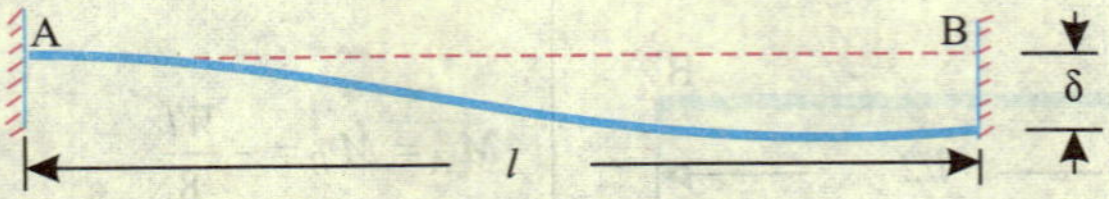 A ⟷ l ⟷ B, δ	$M_A = -\dfrac{6EI\delta}{l^2}$ $M_B = \dfrac{6EI\delta}{l^2}$

Fig. 9.20

9·3. CONTINUOUS BEAMS

9·3·1. Introduction

A **continuous beam** *is one which is supported on more than two supports.* The continuous beam is statically indeterminate and can be analysed by various methods; here only one method *viz;* the *theorem of three moments* or the *Clapeyron's theorem method* will be discussed.

9·3·2. Clapeyron's Theorem of Three Moments

Fig. 9·21 (*a*) shows a continuous beam, with its only two spans *AB* and *BC* considered. They may have any loading pattern.

A massive crane part is being shipped on a trailer.

Fig. 9·21 (b) shows the B.M. diagram for each of the two spans treating both beams *AB* and *BC* to be freely supported and independent of each other.

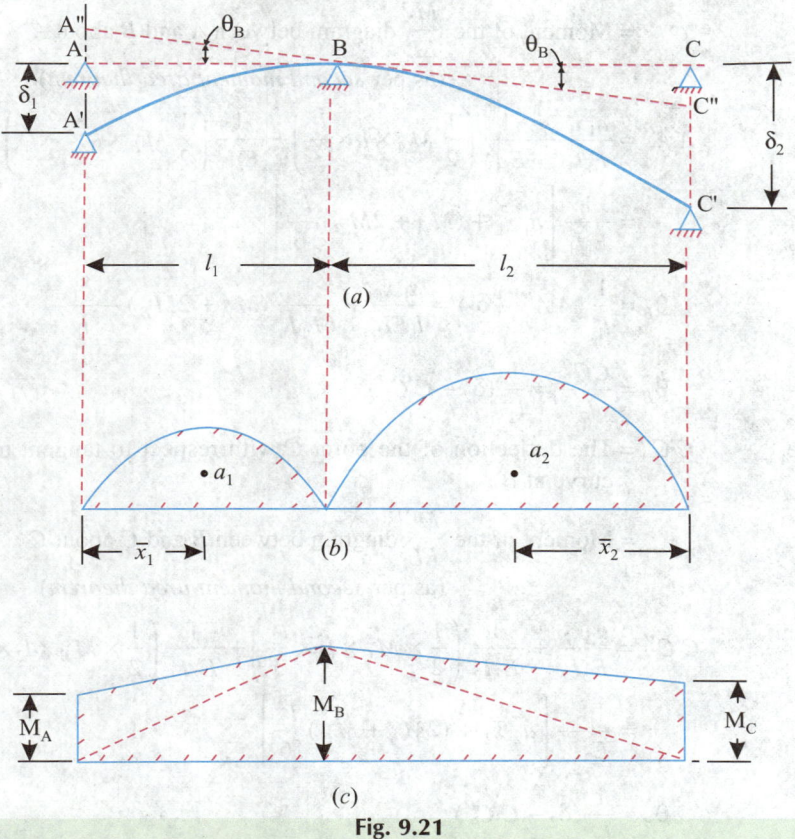

Fig. 9.21

Under the effect of the applied loads let the beam *ABC* takes the shape of *A'B'C'* (The supports *A* and *C* sinking by different amounts with respect to support *B*). Draw *A"BC"* a tangent to the elastic curve at *B* and let *AA"* and *CC"* be the intercepts made by this tangent on the verticals through *A* and *C*.

Let,
l_1 = Length of the span AB,

l_2 = Length of the span BC,

a_1 = Area of B.M. diagram for span AB (simply supported),

a_2 = Area of the B.M. diagram for span BC (simply supported),

$\bar{x}_1$ = Distance of the centroid of a_1 from A,

$\bar{x}_2$ = Distance of the centroid of a_2 from C,

$E_1 I_1$ = Flexural rigidity for the span AB,

$E_2 I_2$ = Flexural rigidity for the span BC,

δ_1 = Sinking of support A with respect to support B,

δ_2 = Sinking of support C with respect to support B, and

θ_B = Angle made by the tangent to elastic curve at B with the horizontal.

Now,
$$\theta_B = \frac{AA''}{l_1} = \frac{1}{l_1}(A'A'' - \delta_1)$$

where,
$A'A''$ = The deflection of the point A with respect to the tangent to the elastic curve at B.

$$= \text{Moment of the } \frac{M}{EI} \text{ diagram between } A \text{ and } B \text{ about } A$$

(as per *second moment area theorem*)

∴
$$A'A'' = \frac{a_1 \bar{x}_1}{E_1 I_1} + \frac{1}{E_1 I_1}\left(\frac{1}{2} M_A \times l_1 \times \frac{l_1}{3}\right) + \frac{1}{E_1 I_1}\left(\frac{1}{2} M_B \times l_1 \times \frac{2}{3} l_1\right)$$

$$= \frac{1}{E_1 I_1}\left[a_1 \bar{x}_1 + (M_A + 2M_B)\frac{l^2}{6}\right]$$

∴
$$\theta_B = \frac{1}{l_1}(A'A'' - \delta_1) = \frac{a_1 \bar{x}_1}{l_1 E_1 I_1} + \frac{1}{6E_1 I_1}(M_A + 2M_B) - \frac{\delta_1}{l_1} \qquad ....(i)$$

Also,
$$\theta_B = \frac{CC''}{l_2} = \frac{1}{l_2}(\delta_2 - C'C'')$$

where,
$C'C''$ = The deflection of the point C with respect to tangent to the elastic curve at B

$$= \text{Moment of the } \frac{M}{EI} \text{ diagram between } B \text{ and } C \text{ about } C$$

(as per *second moment area theorem*)

∴
$$C'C'' = \frac{a_2 \bar{x}_2}{E_2 I_2} + \frac{1}{E_2 I_2}\left(\frac{1}{2} \times M_C \times l_2 \times \frac{l_2}{3}\right) + \frac{1}{E_2 I_2}\left(\frac{1}{2} \times M_B \times l_2 \times \frac{2}{3} l_2\right)$$

$$= \frac{1}{E_2 I_2}\left[a_2 \bar{x}_2 + (2M_B + M_C)\frac{l^2}{6}\right]$$

∴
$$\theta_B = \frac{1}{l_2}(\delta_2 - C'C'')$$

$$= \frac{\delta_2}{l_2} - \frac{a_2 \bar{x}_2}{l_2 E_2 I_2} - \frac{l_2}{6E_2 I_2}(2M_B + M_C) \qquad ...(ii)$$

Equating equations (*i*) and (*ii*), we get

$$\frac{a_1 \bar{x}_1}{l_1 E_1 I_1} + \frac{l_1}{6 E_1 I_1}(M_A + 2M_B) - \frac{\delta_1}{l_1} = \frac{\delta_2}{l_2} - \frac{a_2 \bar{x}_2}{l_2 E_2 I_2} - \frac{l_2}{6 E_2 I_2}(2M_B + M_C)$$

or

$$\frac{M_A l_1}{6 E_1 I_1} + \frac{2 M_B}{6}\left(\frac{l_1}{E_1 I_1} + \frac{l_2}{E_2 I_2}\right) + \frac{M_C l_2}{6 E_2 I_2} + \frac{a_1 \bar{x}_1}{l_1 E_1 I_1} + \frac{a_2 \bar{x}_2}{l_2 E_2 I_2} = \frac{\delta_1}{l_1} + \frac{\delta_2}{l_2}$$

or

$$\frac{M_A l_1}{E_1 I_1} + 2 M_B\left(\frac{l_1}{E_1 I_1} + \frac{l_2}{E_2 I_2}\right) + \frac{M_C l_2}{E_2 I_2} + \frac{6 a_1 \bar{x}_1}{l_1 E_1 I_1} + \frac{6 a_2 \bar{x}_2}{l_2 E_2 I_2} = \frac{6\delta_1}{l_1} + \frac{6\delta_2}{l_2} \qquad ...(9.25)$$

Eqn. (9.25) is the ***most general form of the theorem of three moments.***

Following points are worth noting:

(*i*)　All *sagging* moments are taken as + *ve* and all *hogging* moments as – *ve*.

(*ii*)　δ_1 and δ_2 are taken as positive if after sinking supports A and C lie *below* the support of reference B.

(*iii*)　For any two consecutive spans if the end supports lie below the central support the value of δ for the particular support is taken as + *ve* otherwise – *ve*.

The following are the *special cases of theorem of three moments*:

Case I. When all supports remain at the same level:

In case all supports remain at the same level then δ_1 and δ_2 are zero and the eqn. (9.25) reduces to

$$\frac{M_A l_1}{E_1 I_1} + 2 M_B\left(\frac{l_1}{E_1 I_1} + \frac{l_2}{E_2 I_2}\right) + \frac{M_C l_2}{E_2 I_2} + \frac{6 a_1 \bar{x}_1}{l_1 E_1 I_1} + \frac{6 a_2 \bar{x}_2}{l_2 E_2 I_2} = 0 \qquad ...(9.26)$$

Case II. When flexural rigidity for spans is same (and supports are at the same level):

In case flexural rigidity for span is same *i.e.* if $E_1 I_1 = E_2 I_2$ and the supports are at the same level the eqn. (9.25) reduces to

$$M_A l_1 + 2 M_B (l_1 + l_2) + M_C l_2 + \frac{6 a_1 \bar{x}_1}{l_1} + \frac{6 a_2 \bar{x}_2}{l_2} = 0 \qquad ...(9.27)$$

The eqn. (9.27) gets further simplified if there are only two spans and the beam is freely supported at A and C, then $M_A = M_C = 0$.

Case III. When continuous beam has a fixed end:

The theorem of three moments, for a fixed end of the beam, can be modified by imagining a span of length l_0 and moment of inertia ∝ beyond the support and applying the theorem of three moments as usual (Refer Fig. 9.22).

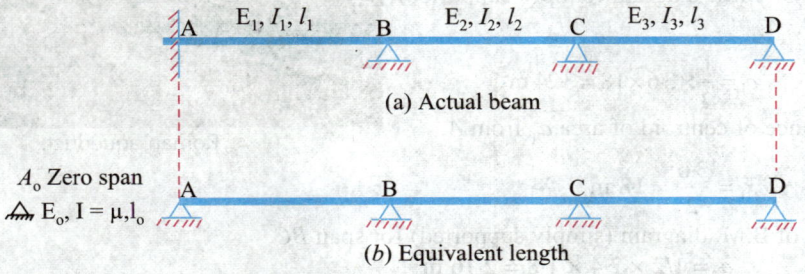

(a) Actual beam

(b) Equivalent length

Fig. 9.22

Example 9·10. *Fig. 9·23 shows a beam simply supported at the supports A and C and is continuous over the support B. Assuming EI is constant draw the bending moment and shear force diagrams.*

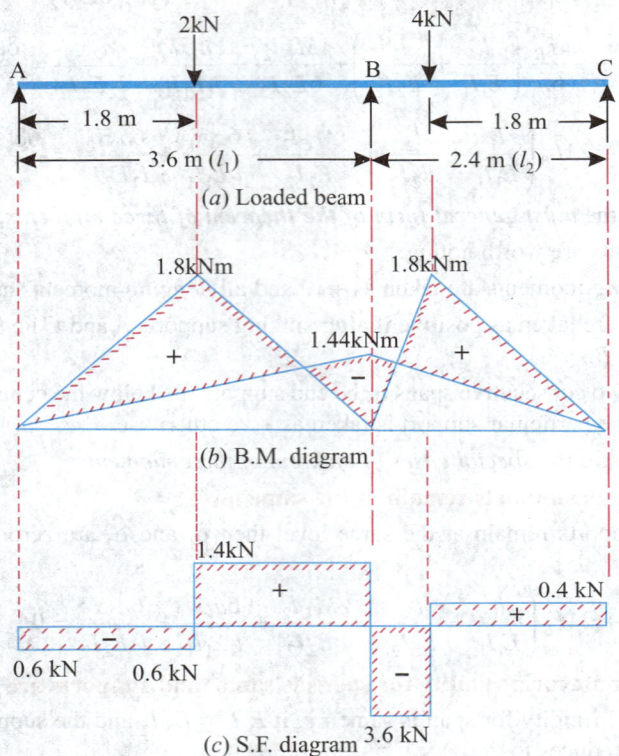

(a) Loaded beam

(b) B.M. diagram

(c) S.F. diagram

Fig. 9.23

Solution.

Let, M_A = Moment at the support A,

M_B = Moment at the support B, and

M_C = Moment at the support C.

The bending moments under the loads 2 kN and 4 kN (treating the span AB and BC as simply supported) are 1·8 kNm each.

Area of B.M. diagram (simply supported) for span AB,

$$a_1 = \frac{1}{2} \times 3·6 \times 1·8 = 3·24 \text{ m}^2$$

Distance of centroid of area a_1 from A,

$$\bar{x}_1 = \frac{3·6}{2} = 1·8 \text{ m}$$

Area of B.M. diagram (simply supported) for span BC,

$$a_2 = 1/2 \times 2·4 \times 1·8 = 2·16 \text{ m}^2$$

Distance of centroid of area a_2 from C,

$$\bar{x}_2 = \frac{2·4 + 1·8}{3} = 1·4 \text{ m}$$

Roman aqueduct.

Note. In a triangle (Fig. 9·24) the distances of the centroid are given as follows:

$$x_1 = \frac{l+a}{3} \qquad\qquad\qquad \text{(from } A\text{)}$$

$$x_2 = \frac{l+b}{3} \qquad\qquad\qquad \text{(from } B\text{)}$$

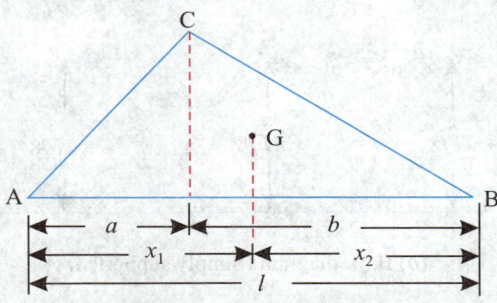

Fig. 9.24

Since the beam is freely supported at A and C, the support moments $M_A = M_C = 0$

Using the relation when there is no sinking of supports and EI is constant, we have

$$M_A l_1 + 2M_B (l_1 + l_2) + M_C l_2 + \frac{6a_1 \bar{x}_1}{l_1} + \frac{6a_2 \bar{x}_2}{l_2} = 0$$

or $\qquad 0 + 2M_B (3\cdot6 + 2\cdot4)\ 0 + \dfrac{6 \times 3\cdot24 \times 1\cdot8}{3\cdot6} + \dfrac{6 \times 2\cdot16 \times 1\cdot4}{2\cdot4} = 0$

or $\qquad 12M_B + 9\cdot72 + 7\cdot56 = 0$

∴ $\qquad\qquad M_B = -1\cdot44$ kN m

Support reactions R_A, R_B and R_C :

For span BC, taking moments about B, we get

$$R_C \times 2\cdot4 - 4 \times 0\cdot6 + 1\cdot44 = 0$$

∴ $\qquad\qquad R_C = \textbf{0·4 kN}$

For span AB, taking moment about B, we get

$$R_A \times 3\cdot6 - 2 \times 1\cdot8 + 1\cdot44 = 0$$

∴ $\qquad\qquad R_A = \textbf{0·6 kN}$

Also, $\qquad\qquad R_A + R_B + R_C = 2 + 4 = 6$ kN

∴ $\qquad\qquad R_B = 6 - 0\cdot4 - 0\cdot6 = \textbf{5 kN}$

B.M. and S.F. diagrams are shown in Fig. 9·23 (b) and (c) respectively.

Example 9·11. *Fig. 9·25 shows a continuous beam ABCD having three equal spans of length l each. It carries a uniformly distributed load w/unit length over its entire length. It is freely supported on all supports, which are at the same level.*

Draw the B.M. and S.F. diagrams for this beam.

Solution. Fig. 9·25 (a) shows the loaded beam.

Support moments: By symmetry we have

$$M_B = M_C$$

and, $\qquad\qquad M_A = M_D = 0 \qquad\qquad (\because \text{ Beam is freely supported})$

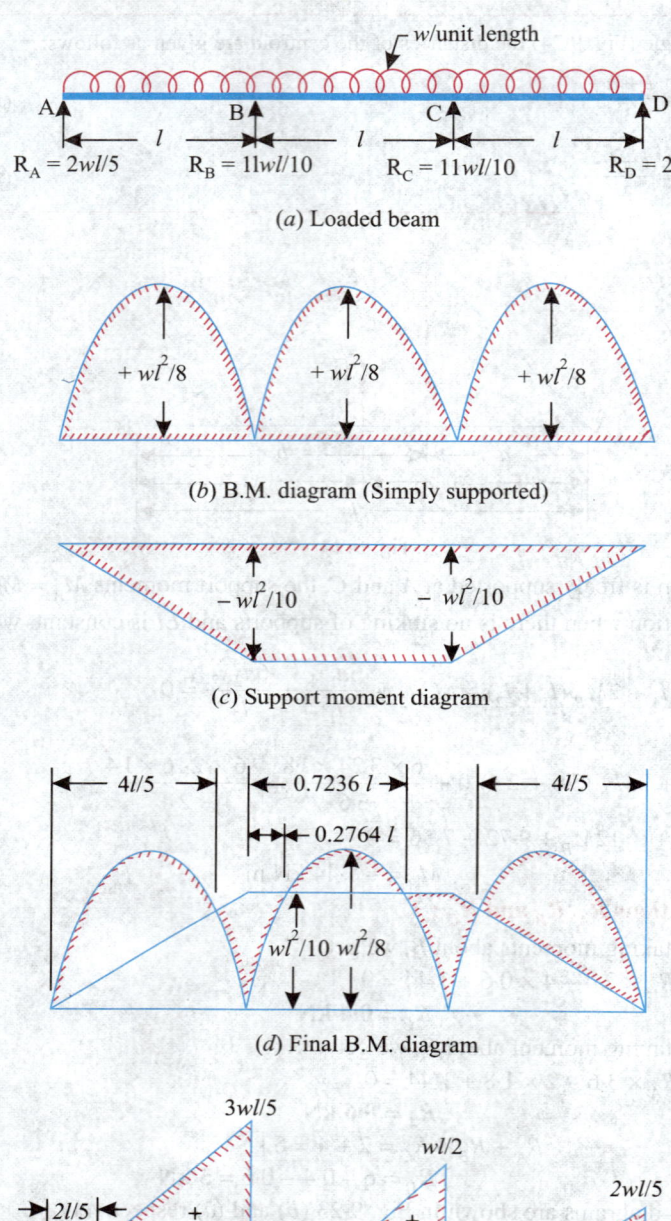

w/unit length

A B C D

$\xleftarrow{\quad l \quad}$ $\xleftarrow{\quad l \quad}$ $\xleftarrow{\quad l \quad}$

$R_A = 2wl/5$ $R_B = 11wl/10$ $R_C = 11wl/10$ $R_D = 2wl/5$

(a) Loaded beam

$+ wl^2/8$ $+ wl^2/8$ $+ wl^2/8$

(b) B.M. diagram (Simply supported)

$- wl^2/10$ $- wl^2/10$

(c) Support moment diagram

$\xleftarrow{\ 4l/5\ }$ $\xleftarrow{\ 0.7236\ l\ }$ $\xleftarrow{\ 4l/5\ }$

$\xleftarrow{}0.2764\ l$

$wl^2/10\ wl^2/8$

(d) Final B.M. diagram

$3wl/5$

$wl/2$

$2wl/5$

$\xrightarrow{}|2l/5|\xleftarrow{}$ $+$ $+$ $+$ $\xrightarrow{}|2l/5|\xleftarrow{}$

$2wl/5$ $-$ $-$

$wl/2$

$\xleftarrow{\ l/2\ }$ $3wl/5$

(e) S. F. diagram

Fig. 9.25

B.M. diagram for all spans (treating each to be independent and simply supported) is shown in Fig. 9·25 (b).

Area of B.M. diagram for each of three spans

$$a_1 = a_2 = a_3 = \frac{2}{3} \times \frac{wl^2}{8} \times l = \frac{wl^3}{12}$$

Similarly, distance of the centroid of the B.M. diagram for each span from either support,

$$\bar{x}_1 = \bar{x}_2 = \bar{x}_3 = \frac{l}{2}$$

Now applying *theorem of three moments* to spans AB and BC, we get

$$M_A l + 2M_B (l + l) + M_C l + \frac{6 \times \frac{wl^3}{12} \times l/2}{l} + \frac{6 \times \frac{wl^3}{12} \times \frac{l}{2}}{l} = 0$$

∴ $$0 + 4M_B l + M_C l + \frac{wl^3}{2} = 0 \qquad (\because M_A = 0)$$

or $$5M_B l + \frac{wl^3}{2} = 0 \qquad (\because M_B = M_C)$$

∴ $$M_B = -\frac{wl^2}{10} = M_C$$

Support moment diagram is shown in Fig. 9·25 (*c*) and the final B.M. diagram of the beam is shown in Fig. 9·25 (*d*).

Support reactions:

Again by symmetry, reactions $R_A = R_D$, and $R_B = R_C$
For the span CD, taking moments about C, we get

$$R_D \times l + M_C - wl \times \frac{l}{2} = 0 \qquad \therefore R_D \times l + \frac{wl^2}{10} - \frac{wl^2}{2} = 0$$

∴ $$R_D = \frac{2wl}{5} = R_A$$

Modern constructions have multiple number of beams forming integral part of them.

Also, $R_A + R_B + R_C + R_D = w \times 3l$

But, $R_A = R_D$ and $R_B = R_C$ ∴ $2(R_A + R_B) = 3wl$

or, $2\left(\dfrac{2wl}{5} + R_B\right) = 3wl$ ∴ $R_B = \dfrac{11\,wl}{10}$

S.F. diagram is shown in Fig. 9·25 (e).

In the S.F. diagram, by the similarity of triangles, we find that:

(i) S.F. is zero at 0·4 l from A in span AB,

(ii) S.F. is zero at 0·4 l from D in span CD, and

(iii) S.F. is zero at middle point of span BC.

Obviously the + ve B.M. is *maximum* at points of *zero shear*.

— Thus maximum + ve B.M. in span AB or CD is at $\dfrac{2}{5}\,l$ from A or D respectively and is

$$= R_A \times \frac{2}{5}\,l - \left(w \times \frac{2}{5}\,l\right) \times \left(\frac{1}{2} \times \frac{2}{5}\,l\right)$$

$$= \frac{2wl}{5} \times \frac{2}{5}\,l - \frac{2wl^2}{25} = \frac{2}{25}\,wl^2$$

— In span BC, maximum + ve B.M. is at mid span of BC and is

$$= \frac{wl^2}{8} - \frac{wl^2}{10} = \frac{wl^2}{40}$$

Point of contraflexure.

For spans AB and CD:

Let the B.M. be zero at a distance x from A, then

$$R_A \times x - \frac{wx^2}{2} = 0 \qquad \therefore \; x = \frac{2R_A}{w} = \frac{2}{w} \times \frac{2}{5}\,wl = \frac{4l}{5}$$

Thus the point of contraflexure in span AB is $\dfrac{4l}{5}$ from A. Similarly in span CD it is at $\dfrac{4l}{5}$ from D.

For span BC:

Let the B.M. be zero at a distance x from B, then

$$\frac{wl}{2}\,x - \frac{wl^2}{10} - \frac{wx^2}{2} = 0 \quad \text{or} \quad x^2 - lx + \frac{l^2}{5} = 0$$

∴ $$x = \frac{l \pm \sqrt{l^2 - \dfrac{4l^2}{5}}}{2} = \frac{l}{2}\left(1 \pm \frac{1}{\sqrt{5}}\right)$$

$$= \frac{l}{2}\,(1 \pm 0{\cdot}4472) = 0{\cdot}7236\,l,\; 0{\cdot}2764\,l$$

i.e. $x = 0{\cdot}7236\,l$ and $0{\cdot}2764\,l$

Example 9·12. *A continuous beam ABCD of uniform cross-section is loaded as shown in Fig. 9·26. Find:*

(i) Bending moments at the supports B and C; (ii) Reactions at the supports.

Draw B.M. and S.F. diagrams also.

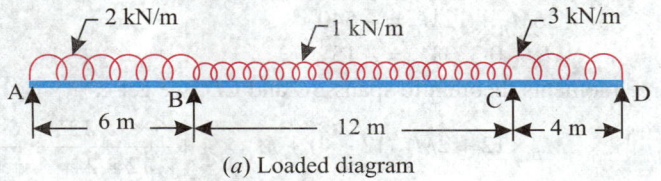

(a) Loaded diagram

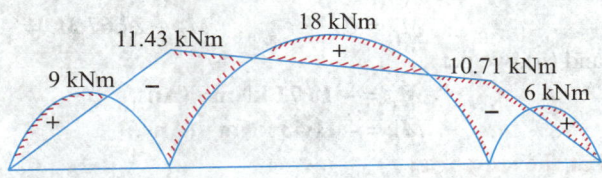

(b) B.M. diagram

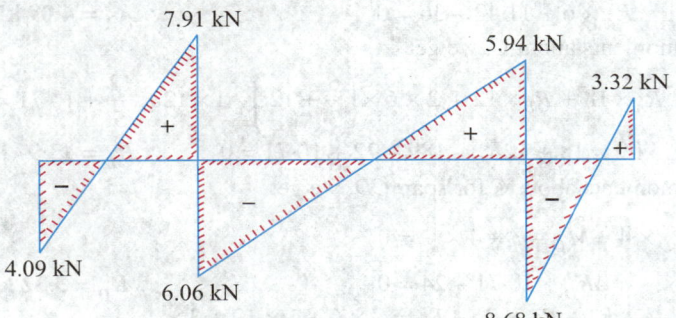

(c) S.F. diagram

Fig. 9.26

Solution. *For span AB:* Max B.M. $= \dfrac{wl^2}{8} = \dfrac{2 \times 6^2}{8} = 9$ kNm

For span BC: Max B.M. $= \dfrac{1 \times 12^2}{8} = 18$ kNm

For span CD: Max B.M. $= \dfrac{3 \times 4^2}{8} = 6$ kNm

Applying *three moments theorem* for span *AB* and *BC*, we get

$$M_A \times 6 + 2M_B (6 + 12) + 12M_C + \frac{6a_1 \bar{x}_1}{l_1} + \frac{6a_2 \bar{x}_2}{l_2} = 0$$

$$\left[\begin{array}{l} \text{where} \qquad a_1 \bar{x}_1 = \dfrac{w_1 l_1^3}{12} \times \dfrac{l_1}{2} = \dfrac{w_1 l_1^4}{24}, \\[3mm] \text{and} \qquad a_2 \bar{x}_2 = \dfrac{w_2 l_2^3}{12} \times \dfrac{l_2}{2} = \dfrac{w_2 l_2^4}{24} \end{array} \right]$$

or, $6M_A + 36M_B + 12M_C + \dfrac{6 \times 2 \times 6^4}{6 \times 24} + \dfrac{6 \times 1 \times 12^4}{12 \times 24} = 0$

or, $6M_A + 36M_B + 12M_C + 108 + 432 = 0$

But, $M_A = 0$

$$\therefore \qquad 36M_B + 12M_C = -540$$

or, $\qquad M_B + 0.333M_C = -15 \qquad \qquad \qquad \qquad ...(i)$

Applying three moments theorem to spans BC and CD, we get

$$M_B \times 12 + 2M_C(12+4) + M_D \times 4 + \frac{6 \times 1 \times 12^4}{12 \times 24} + \frac{6 \times 3 \times 4^4}{4 \times 24} = 0$$

$$12M_B + 32M_C + 4M_D + 432 + 48 = 0$$

But, $\qquad \qquad M_D = 0 \qquad \qquad \therefore \ M_B + 2.667\,M_C = -40 \qquad \qquad ...(ii)$

From eqns. (i) and (ii), we get

$$M_C = -\mathbf{10.71\ kNm} \quad \textbf{(Ans.)}$$

and, $\qquad \qquad \qquad M_B = -\mathbf{11.43\ kNm} \quad \textbf{(Ans.)}$

(ii) Reactions at the supports:

— Taking moments about B, we get

$$R_A + 6 + M_B - 2 \times 6 \times \frac{6}{2} = 0$$

$$R_A \times 6 + 11.43 - 36 = 0 \qquad \qquad \therefore R_A = \mathbf{4.09\ kN} \quad \textbf{(Ans.)}$$

— Taking moments about C, we get

$$R_A \times 18 + R_B \times 12 - 2 \times 6 \times \left(\frac{6}{3} + 12\right) - 1 \times 12 \times \frac{12}{5} + 10.71 = 0$$

$$4.09 \times 18 + 12R_B - 180 - 72 + 10.71 = 0 \qquad \therefore R_B = \mathbf{13.97\ kN} \quad \textbf{(Ans)}$$

— Taking moments about C for span CD, we get

$$R_D \times 4 + M_C - 3 \times 4 \times \frac{4}{2} = 0$$

$$4R_D + 10.71 - 24 = 0 \qquad \qquad \therefore R_D = \mathbf{3.32\ kN} \quad \textbf{(Ans.)}$$

Also, $\qquad \qquad R_A + R_B + R_C + R_D = 2 \times 6 + 1 \times 12 + 3 \times 4$

or, $\qquad 4.09 + 13.97 + R_C + 3.32 = 12 + 12 + 12 \qquad \therefore R_C = \mathbf{14.62\ kN} \quad \textbf{(Ans.)}$

The B.M. and S.F. diagrams are shown in Fig. 9·26.

An airplane wings, body, etc. partly treated like beams, for design considerations.

Example 9·13. *For the continuous loaded beam ABCD shown in Fig. 9·27, find:*
(i) Moments at the supports;   (ii) Reactions at the supports.
Draw the B.M. and S.F. diagrams also.

Solution. Refer to Fig. 9·27 (a).

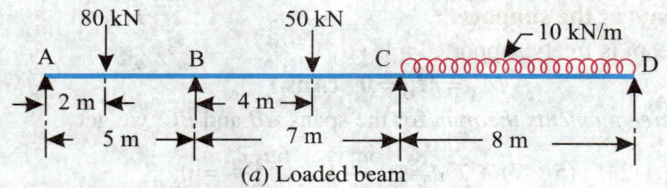

(a) Loaded beam

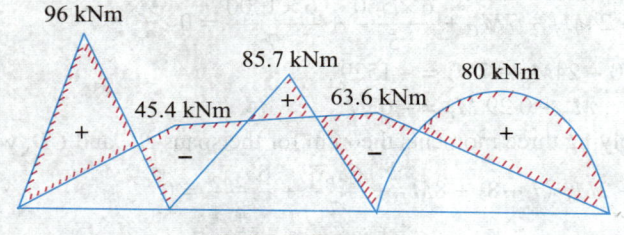

(b) B.M. diagram

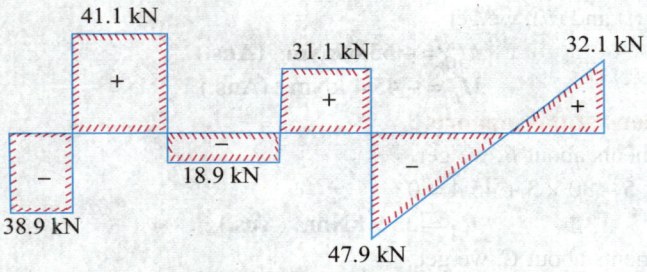

(c) S.F. diagram

Fig. 9.27

For span AB:

$$\text{Max B.M.} = \frac{wab}{l} = \frac{80 \times 2 \times 3}{5} = 96 \text{ kNm}$$

$$a = \frac{1}{2} \times 5 \times 96 = 240, \ \bar{x} = \frac{1}{3}(5+2) = 2.33 \text{ m}$$

∴

$$a\bar{x} = 240 \times 2.333 = 560$$

For span BC:

$$\text{Max B.M.} = \frac{50 \times 4 \times 3}{7} = 85.7 \text{ kNm,}$$

$$a = \frac{1}{2} \times 7 \times 85.7 = 300$$

With *C* as origin,

$$a\bar{x} = 300 \times \frac{1}{3}(7+3) = 1000$$

With *B* as origin,

$$a\bar{x} = 300 \times \frac{1}{3}(7+4) = 1100$$

For span CD:

$$\text{Max B.M.} = \frac{wl^2}{8} = \frac{10 \times 8^2}{2} = 80 \text{ kNm}$$

$$a = \frac{2}{3} \times 80 \times 8 = 426.67, \ \bar{x} = \frac{8}{2} = 4 \text{ m}$$

∴

$$a\bar{x} = 426.67 \times 4 = 1706.7 \text{ with } D \text{ as origin}$$

(i) Moments at the supports :

Since the beam is freely supported

∴ $\qquad M_A = M_D = 0$ **(Ans.)**

Applying *three moments theorem* for the spans AB and BC, we get

$$5M_A + 2M_B(5+7) + 7M_C + \frac{6a_1\bar{x}_1}{5} + \frac{6a_2\bar{x}_2}{7} = 0$$

$$5M_A + 24M_B + 7M_C + \frac{6 \times 560}{5} + \frac{6 \times 1000}{7} = 0$$

$$5 \times 0 + 24M_B + 7M_C = -1529$$

$$M_B + 0.29\,M_C = -63.7 \qquad\qquad ...(i)$$

Similarly, applying three moments theorem for the spans BC and CD, we have

$$7M_B + 2M_C(7+8) + 8M_D + \frac{6a_1\bar{x}_1}{7} + \frac{6a_2\bar{x}_2}{8} = 0$$

$$7M_B + 30M_C + 8 \times 0 + \frac{6 \times 1100}{7} + \frac{6 \times 1706.7}{8} = 0$$

or, $\qquad 7M_B + 30\,M_C = -2222.88 \qquad \therefore\ M_B + 4.28M_C = -317.6 \qquad ....(ii)$

From eqns. (*i*) and (*ii*), we get

$$M_C = -63.6\ \text{kNm} \quad \text{(Ans.)}$$

$$M_B = -45.4\ \text{kNm} \quad \text{(Ans.)}$$

(ii) Reactions at the supports :

Taking moments about B, we get

$$R_A \times 5 - 80 \times 3 + 45.4 = 0$$

$$R_A \simeq 38.9\ \text{kNm} \quad \text{(Ans.)}$$

Taking moments about C, we get

$$R_A(5+7) - 80(3+7) + R_B \times 7 - 50 \times 3 + 63.6 = 0$$

$$38.9 \times 12 - 800 + 7R_B - 150 + 63.6 = 0$$

∴ $\qquad\qquad R_B = 60\ \text{kN} \quad \text{(Ans)}$

Taking moments about C for the span CD, we get

$$R_D \times 8 - \frac{10 \times 8^2}{2} + 63.6 = 0 \qquad\qquad \therefore\ R_D = 32.1\ \text{kN} \quad \text{(Ans.)}$$

Also, $\qquad R_A + R_B + R_C + R_D = 80 + 50 + 10 \times 8 = 210$

$$38.9 + 60 + R_C + 32.1 = 210 \qquad\qquad \therefore\ R_C = 79\ \text{kN} \quad \text{(Ans.)}$$

B.M. and S.F. diagrams are shown in Fig. [(9.27 (*b*), (*c*)].

9·3·3. Beams with Overhangs

When a continuous beam has overhangs on one side or on both sides then the moment for the support that the beam overhangs is determined by treating the overhanging portion as a cantilever. The rest of the analysis is done by applying the theorem of three moments as per usual procedure. The complete procedure is explained in the following example.

Example 9·14. *A continuous beam ABCDE carrying a uniformly distributed load of w/unit length rests on three supports B, C and D, all at the same level. It has two equal overhangs of length l_0 on either side. Assuming EI constant, find the ratio of $\dfrac{l_0}{l}$ for the three support reactions to be equal.*

Solution. Fig. 9·28 (*a*) shows the loaded beam. Fig. 9·28 (*b*) shows the B.M. diagram in which each span has been treated as simply supported and independent of other spans.

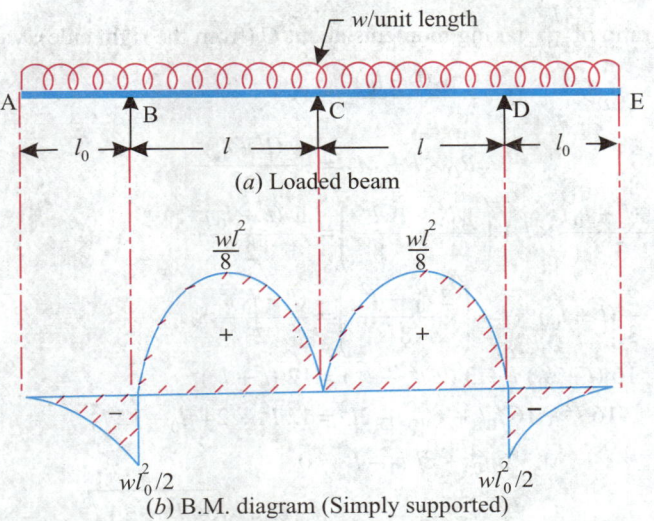

(a) Loaded beam

(b) B.M. diagram (Simply supported)

Fig. 9.28

$$M_B = M_D = -\frac{wl_0^2}{2} \qquad \qquad ...(i)$$

Total load on the beam

$$= 2w(l + l_0)$$

If the support reactions are equal, then $\quad R_B = R_C = R_D = \dfrac{2w(l + l_0)}{3} \qquad \qquad ...(ii)$

Applying the *"Theorem of three moments"* to the beam, we have

$$M_B \times l + 2M_C(l + l) + M_D \times l + \frac{6 \times \left(\dfrac{2}{3} \times \dfrac{wl^2}{8} \times l\right) \times \dfrac{l}{2}}{l} + \frac{6 \times \left(\dfrac{2}{3} \times \dfrac{wl^2}{8} \times l\right) \times \dfrac{l}{2}}{l} = 0$$

or, $\quad -\dfrac{wl_0^2}{2} \times l + 4M_C \times l - \dfrac{wl_0^2}{2} \times l + \dfrac{wl^3}{2} = 0$

or, $$M_C = \left(\frac{wl_0^2}{4} - \frac{wl^2}{8}\right)$$

Chimneys of this thermal power plant are built taking wind forces and aerodynamics into consideration.

To find the ratio of $\dfrac{l_0}{l}$ taking moments about C (from the right side), we get

$$R_D \times l + M_C - \frac{w(l + l_0)^2}{2} = 0$$

or,

$$R_D \times l + M_C = \frac{w(l + l_0)^2}{2}$$

or,

$$\frac{2w(l + l_0)}{3} \times l + \left(\frac{wl_0^2}{4} - \frac{wl^2}{8}\right) = \frac{w(l + l_0)^2}{2}$$

or,

$$\frac{2}{3}(l + l_0) \times l + \left(\frac{2l_0^2 - l^2}{8}\right) = \frac{(l + l_0)^2}{2}$$

or,

$$16(l + l_0)l + 3(2l_0^2 - l^2) = 12(l + l_0)^2$$

or,

$$16l^2 + 16l_0l + 6l_0^2 - 3l^2 = 12l^2 + 24ll_0 + 12l_0^2$$

or,

$$6l_0^2 + 8ll_0 - l^2 = 0$$

∴

$$l_0 = \frac{-8l + \sqrt{64l^2 + 24l^2}}{12} = \frac{-8l + 9.38l}{12} = 0.115\,l$$

i.e.

$$\frac{l_0}{l} = 0.115 \quad \text{(Ans.)}$$

Hence, when the ratio $l_0/l = 0.115$ the three support reactions will be equal.

9.3.4. Sinking of Supports

The bending moment and shearing force along a continuous beam are amply modified when sinking of support is taken into account. While applying the general equation (9.25) great care must be taken while adopting the sign convention. The procedure is explained in the following example.

Example 9.15. *Fig. 9.29 shows a continuous beam ABCD (with loads) of uniform cross section. If the support B sinks by 10 mm, find the following:*

(i)   Moments at the supports;

(ii) Reactions at the supports;

(iii) Points of contraflexure.

Take: E = 2 × 10⁸ kN/m²,

and   I = 8.5 × 10⁻⁵ m⁴.   **(Delhi University)**

Solution.

(i)   Moments at the supports:

Refer to Fig. 9.29 (*a*). Since the beam is freely supported, therefore

$$M_A = M_D = 0 \quad \text{(Ans.)}$$

The B.M. diagram shown in Fig. 9.29 (*b*) has been drawn by treating each span to be freely supported and independent of other spans.

Applying the *three moments theorem* to spans *AB* and *BC*, we get

$$M_A \times 6 + 2M_B(6 + 7) + M_C \times 7 + \frac{6 \times \left(\frac{1}{2} \times 6 \times 60\right) \times 3}{6} + \frac{6 \times \left(\frac{1}{2} \times 7 \times 85.7\right) \times \left(\frac{7 + 4}{3}\right)}{7}$$

$$= 6\,EI\left(-\frac{10}{1000} \times \frac{1}{6} - \frac{10}{1000} \times \frac{1}{7}\right)$$

(After sinking *A* and *C* will be above the middle support *B* hence δ_A and δ_C are -*ve*)

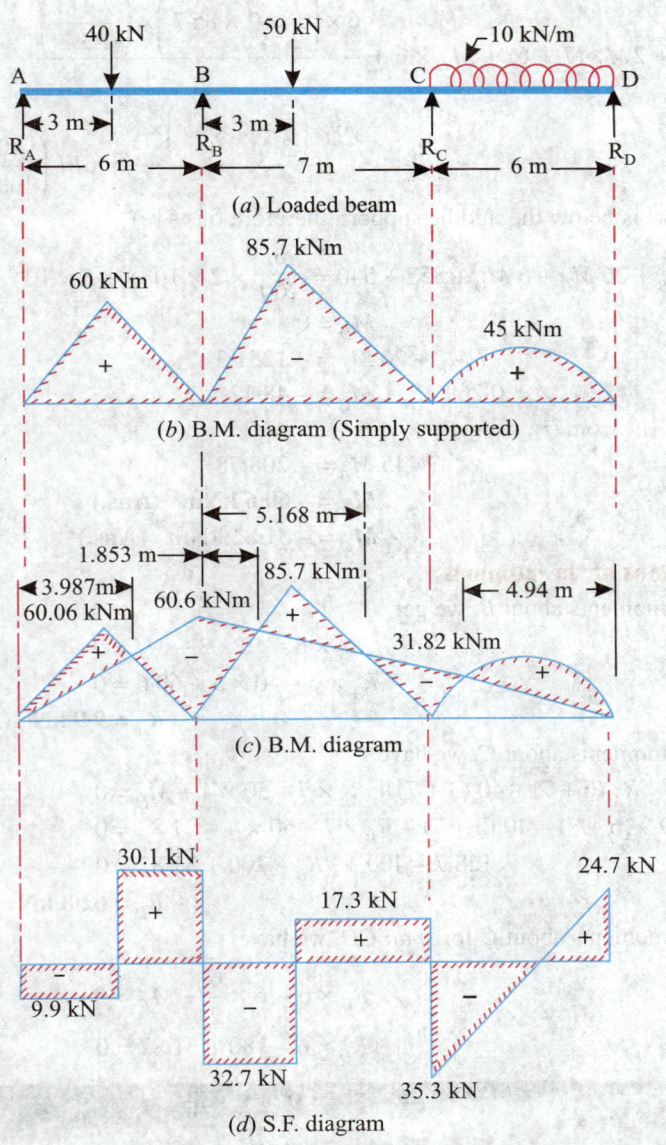

(a) Loaded beam

(b) B.M. diagram (Simply supported)

(c) B.M. diagram

(d) S.F. diagram

Fig. 9.29

Since $M_A = 0$

$\therefore \qquad 26 M_B + 7 M_C + 540 + 942 \cdot 7 = -\dfrac{6 EI}{100} \times \dfrac{13}{42}$

or, $\qquad 26 M_B + 7 M_C = -540 - 942 \cdot 7 - \dfrac{6 \times 2 \times 10^8 \times 8 \cdot 5 \times 10^{-5}}{100} \times \dfrac{13}{42}$

or, $\qquad 26\, M_B + 7\, M_C = -1798 \cdot 4$

or, $\qquad 3 \cdot 714\, M_B + M_C = -256 \cdot 9 \qquad\qquad\qquad …(i)$

Now applying the *three moments theorem* to spans *BC* and *CD*, we get

$$M_B \times 7 + 2M_C(7+6) + M_D \times 6 + \frac{6 \times \left(\frac{1}{2} \times 7 \times 85.7\right) \times \left(\frac{7+3}{3}\right)}{7}$$

$$+ \frac{6 \times \left(\frac{2}{3} \times 6 \times 45\right) \times 3}{6} = 6EI \left(\frac{10}{1000} \times \frac{1}{7} + 0\right)$$

(End support is below the middle support, therefore δ is +ve)

∴ $\quad 7 M_B + 26 M_C + 6 M_D + 857 + 540 = \frac{6}{700} \times 2 \times 10^8 \times 8.5 \times 10^{-5} = 145.7$

But, $\qquad\qquad\qquad\qquad\qquad M_D = 0$

∴ $\qquad\qquad\qquad\qquad 7 M_B + 26 M_C = -1251.3$

or, $\qquad\qquad\qquad\qquad 0.269 M_B + M_C = -48.12$ $\qquad\qquad$ ...(ii)

Subtracting (ii) from (i), we get

$$3.445 M_B = -208.78$$

$$M_B = -60.6 \text{ kNm} \quad (\textbf{Ans.})$$

and, $\qquad\qquad\qquad\qquad M_C = -31.82 \text{ kNm} \quad (\textbf{Ans.})$

(ii) Reactions at the supports:

— Taking moments about B, we get

$$R_A \times 6 - 40 \times 3 + M_B = 0$$

or, $\qquad\qquad\qquad\qquad R_A \times 6 - 40 \times 3 + 60.6 = 0$

∴ $\qquad\qquad\qquad\qquad\qquad\qquad R_A = 9.9 \text{ kN} \quad (\textbf{Ans.})$

— Taking moments about C, we have

$$R_A(6+7) - 40(3+7) + R_B \times 7 - 50 \times 4 + M_C = 0$$

or, $\quad 9.9(6+7) - 40(3+7) + R_B \times 7 - 50 \times 4 + 31.82 = 0$

or, $\qquad\qquad\qquad 128.7 - 400 + 7R_B - 200 + 31.82 = 0$

∴ $\qquad\qquad\qquad\qquad\qquad\qquad R_B = 62.8 \text{ kN} \quad (\textbf{Ans.})$

— Taking moments about C for *span CD*, we have

$$R_D \times 6 - 6 \times \frac{6}{2} + M_C = 0$$

or, $\qquad\qquad\qquad\qquad R_D \times 6 - 180 + 31.82 = 0$

This Roman aqueduct, built in the Ist century A.D., is considered as a engineering marvel of the ancient times for its design before the modern technological revolution.

∴ $\qquad\qquad\qquad\qquad\qquad\qquad\qquad\qquad R_D = 24\cdot7$ kN **(Ans.)**

— Taking moments about B for *span BD*, we have

$$R_D(6+7) - 10 \times 6\left(\frac{6}{2} + 7\right) + R_C \times 7 - 50 \times 3 + M_B = 0$$

$$24\cdot7 \times 13 - 600 + 7 R_C - 150 + 60\cdot6 = 0$$

∴ $\qquad\qquad\qquad\qquad\qquad\qquad\qquad\qquad R_C = 52\cdot6$ kN **(Ans.)**

$$\Bigg[\text{Check} \ : \ R_A + R_B + R_C + R_D = 9\cdot9 + 62\cdot8 + 24\cdot7 + 52\cdot6 = 150 \text{ kN}$$
$$= \text{load on the beam} = 40 + 50 + 10 \times 6 = 150 \text{ kN} \Bigg]$$

(*iii*) Points of contraflexure:

Span AB:

Let B.M. be zero at a distance x from A, then

$$R_A \cdot x - 40(x-3) = 0$$
$$9\cdot9x - 40(x-3) = 0$$
$$9\cdot9x - 40x + 120 = 0$$

∴ $\qquad\qquad\qquad\qquad\qquad\qquad\qquad x = 3\cdot987$ m from A **(Ans.)**

Span BC:

In the span *BC*, we find that there are two points of contraflexure: one between B and 50 kN load and the other between 50 kN load and C.

Let B.M. be zero at a distance x from B, then

$$R_A(6+x) - 40(3+x) + R_B \cdot x \, | - 50(x-30) = 0$$

— To get the *first* point of contraflexure, we neglect the term $50(x-3)$ we have:

$$R_A(6+x) - 40(3+x) + R_B \cdot x = 0$$
$$9\cdot9(6+x) - 40(3+x) + 62\cdot8x = 0$$

or, $\qquad\qquad\qquad 59\cdot4 + 9\cdot9x - 120 - 40x - 62\cdot8x = 0$

or, $\qquad\qquad\qquad\qquad\qquad\qquad 32\cdot7x = 60\cdot6$

∴ $\qquad\qquad\qquad\qquad\qquad\qquad\qquad x = 1\cdot853$ m from B **(Ans.)**

— To get the *second* point of contraflexure, considering the term $50(x-3)$ also, we have

$$9\cdot9(6+x) - 40(3+x) + 62\cdot8x - 50(x-3) = 0$$
$$59\cdot4 + 9\cdot9x - 120 - 40x + 62\cdot8x - 50x + 150 = 0$$
$$17\cdot3x = 89\cdot4$$

∴ $\qquad\qquad\qquad\qquad\qquad\qquad\qquad x = 5\cdot168$ m from B **(Ans.)**

Span CD:

Let the B.M. be zero at a distance x from D, then

$$R_D \cdot x - 10 \times x \times \frac{x}{2} = 0$$
$$24\cdot7x - 5x^2 = 0$$

∴ $\qquad\qquad\qquad\qquad\qquad\qquad\qquad x = 4\cdot94$ m from D **(Ans.)**

9·3·5. Continuous Beams with Fixed Ends

To analyse a continuous beam with one or both of its ends fixed, procedure given in Art. 9·3·2 (Case III) may be used. This is clarified in the following example.

Example 9·16. *Fig. 9·30 shows a beam fixed at one end. Support B settles by 10 mm on loading. Draw the B.M. and S.F. diagrams; given that:*

$$E = 2 \times 10^8 \ kN/m^2,$$

and,
$$I = 100 \times 10^{-6} \ m^4.$$

Solution.
— Fig. 9·30 (a) shows a loaded beam.
— Assume a span C_0. [Fig. 9·30 (b)] of length l, and of moment of inertia ∞.
— Fig. 9·30 (c) shows the B.M. diagram if each span were freely supported.

Support moments:
Now,
$$M_A = -20 \times 2 = -40 \ kNm$$

For span BC:

$$M_{max} = \frac{wl^2}{8} = \frac{8 \times 8^2}{8} = 64 \ kNm$$

Applying *three moments theorem* to spans *AB* and *BC*, we get

$$M_A \times \frac{6}{EI} + 2M_B\left(\frac{6}{EI} + \frac{8}{EI}\right) + M_C \times \frac{8}{EI} + \frac{6 \times \left(\frac{2+6}{2} \times 100\right) \times 3}{6\,EI} + \frac{6 \times \left(\frac{2}{3} \times 64 \times 8\right) \times 4}{8\,EI}$$

$$= 6\left(-\frac{10}{1000} \times \frac{1}{6} - \frac{10}{100} \times \frac{1}{8}\right)$$

[*Because after sinking the middle support is lower than the side supports; δ_A and δ_C are -ve.*]

$$\therefore \quad 6 \times (-40) + 28\,M_B + 8\,M_C + 1200 + 1024 = -\frac{6\,EI}{100} \times \frac{7}{24}$$

or,
$$-240 + 28\,M_B + 8\,M_C + 2224 = -\frac{6 \times 2 \times 10^8 \times 100 \times 10^{-6}}{100} \times \frac{7}{24}$$

or,
$$28\,M_B + 8\,M_C = -2334$$

or,
$$3 \cdot 5\,M_B + M_C = -291 \cdot 75 \qquad \qquad ...(i)$$

Applying the *three moments theorem* to span *BC* and the assumed span CC_0, we get

$$M_B \times \frac{8}{EI} + 2M_C\left(\frac{8}{EI} + 0\right) + M'_C \times \frac{l_0}{\infty} + \frac{6 \times \left(\frac{2}{3} \times 64 \times 8\right) \times 4}{EI \times 8} + 0 = 6\left(\frac{10}{1000} \times \frac{1}{8} + 0\right)$$

When machines have to perform in extreme temperatures, temperature stresses should be considered in addition to the other stresses, while designing.

[Because the end support for the span BCC_0 is lower than the middle support C, δ_B is +ve.]

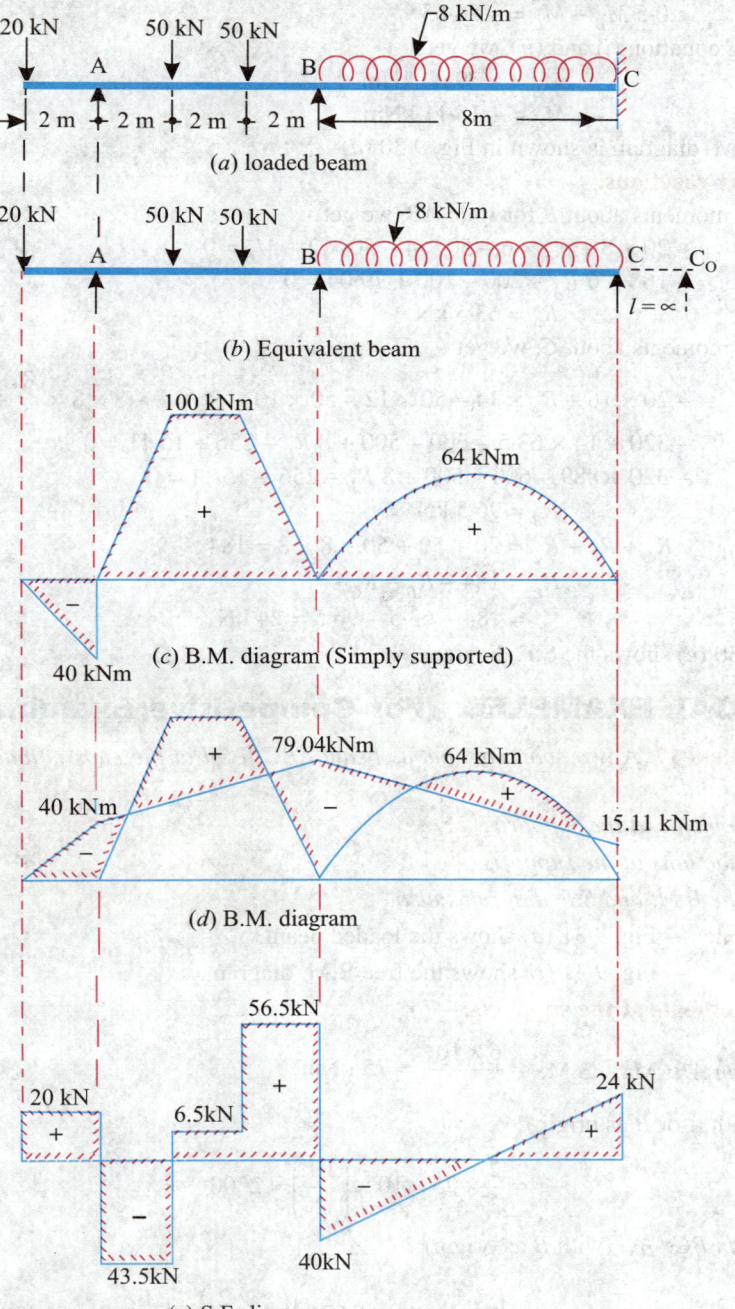

(a) loaded beam

(b) Equivalent beam

(c) B.M. diagram (Simply supported)

(d) B.M. diagram

(e) S.F. diagram

Fig. 9.30

∴ $$8\,M_B + 16\,M_C + 1024 = \frac{6}{800}\,EI$$

or, $$8\,M_B + 16\,M_C + 1024 = \frac{6}{800} \times 2 \times 10^8 \times 100 \times 10^{-6}$$

or, $\qquad\qquad 8\,M_B + 16\,M_C = -874$

or, $\qquad 0.5\,M_B + M_C = -54.63$ $\qquad\qquad\qquad\qquad\qquad$ ...(ii)

Solving equation (i) and (ii), we get

$$M_B = -79.04 \text{ kNm}$$

and, $\qquad\qquad M_C = -15.11 \text{ kNm}$

The B.M. diagram is shown in Fig. 9·30 (d).

Support reactions:

Taking moments about B for *span AB*, we get

$$-20 \times 8 + R_A \times 6 - 50 \times 4 - 50 \times 2 + M_B = 0$$

$$-160 + 6\,R_A - 200 - 100 + 79.04 = 0$$

$\therefore \qquad\qquad\qquad R_A = 63.5 \text{ kN}$

Taking moments about C, we get

$$-20 \times 16 + R_A \times 14 - 50 \times 12 - 50 \times 10 + R_B \times 8 - 8 \times 8 \times \frac{8}{2} + M_C = 0$$

$$-320 + 14 \times 63.5 - 600 - 500 + 8\,R_B - 256 + 15.11 = 0$$

$$-320 + 889 - 600 - 500 + 8\,R_B - 256 + 15.11 = 0$$

$\therefore \qquad\qquad\qquad R_B = 96.5 \text{ kN}$

Also, $\qquad R_A + R_B + R_C = 20 + 50 + 50 + 8 \times 8 = 184$

$\therefore \qquad\qquad\qquad R_C = 184 - R_B - R_C$

$$= 184 - 63.5 - 96.5 = 24 \text{ kN}$$

Fig. 9·30 (e) shows the S.F. diagram.

TYPICAL EXAMPLES (For Competitive Examinations)

Example 9·17. *A two span continuous beam ABC fixed at the ends is loaded as shown in Fig. 9·31 (a). Find:*

(i) *Moments at the supports;*

(ii) *Reactions at the supports.*

Draw the B.M. and S.F. diagrams also.

Solution. — Fig. 9·31 (a) shows the loaded beam,

$\qquad\qquad$ — Fig. 9·31 (c) shows the free B.M. diagram.

(i) **Moments at the supports:**

For span AB: Max. B.M. $= \dfrac{6 \times 10^2}{8} = 75 \text{ kNm}$

$a\bar{x}$ (with A or B as origin)

$$= \frac{2}{3} \times 75 \times 10 \times \left(\frac{10}{2}\right) = 2500$$

For span BC: $a\bar{x}$ (with B as origin)

$$= \left(-\frac{1}{2} \times 4 \times 68.571\right)\left(4 \times \frac{2}{3}\right) + \left(\frac{1}{2} \times 3 \times 51.429\right)\left(4 + \frac{3}{3}\right) = 20$$

$a\bar{x}$ (with C as origin) $\quad = \left(-\frac{1}{2} \times 4 \times 68.571\right)\left(3 \times \frac{4}{3}\right)$

$$+ \left(\frac{1}{2} \times 3 \times 51.429\right)\left(3 + \frac{2}{3}\right) = -440$$

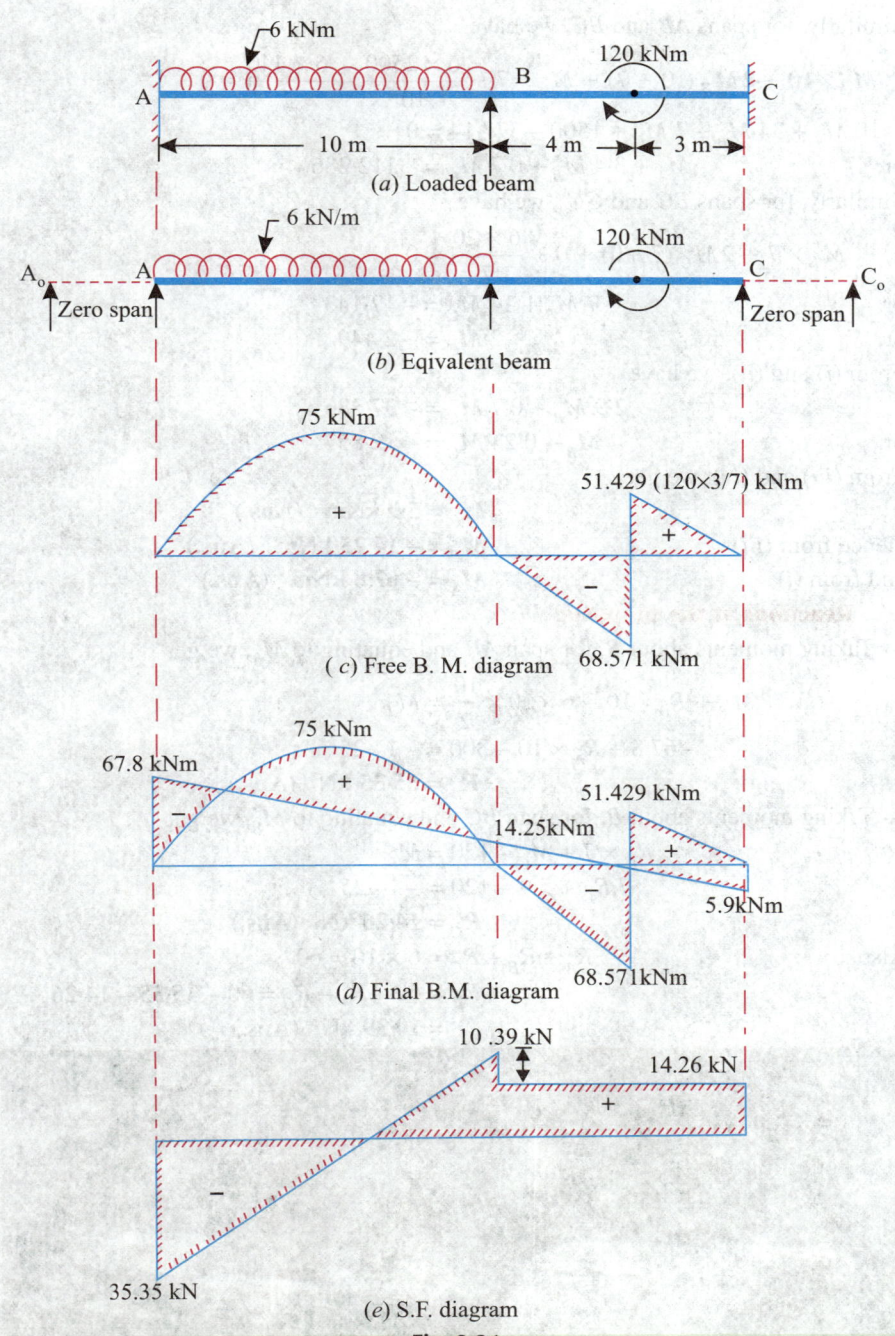

(a) Loaded beam

(b) Eqivalent beam

(c) Free B. M. diagram

(d) Final B.M. diagram

(e) S.F. diagram

Fig. 9.31

Applying *three moments theorem* for the spans A_0A and AB, we get

$$0 + 2 M_A (0 + 10) + M_B (10) + 0 + \frac{6 \times 2500}{10} = 0$$

or, $20 M_A + 10 M_B = -1500$

or, $M_A + 0\cdot5 M_B = -75$ ...(i)

Similarly, for spans AB and BC, we have

$$M_A \times 10 + 2M_B(10+7) + M_C \times 7 + \frac{6 \times 2500}{10} - \frac{6 \times 440}{7} = 0$$

$$10\,M_A + 34\,M_B + 7\,M_C + 1500 - 377 \cdot 14 = 0$$

or, $\qquad M_A + 3 \cdot 4\,M_B + 0 \cdot 7\,M_C = -112 \cdot 286 \qquad\qquad ...(ii)$

Similarly, for spans BC and CC_0, we have

$$M_B \times 7 + 2M_C(7+0) + 0 + \frac{6 \times 20}{7} = 0$$

or, $\qquad\qquad 7\,M_B + 14\,M_C = -17 \cdot 143$

or, $\qquad\qquad M_B + 2\,M_C = -2 \cdot 449 \qquad\qquad ...(iii)$

From (i) and (ii), we have

$$2 \cdot 9\,M_B + 0 \cdot 7\,M_C = -37 \cdot 286$$

or, $\qquad\qquad M_B + 0 \cdot 24\,M_C = -12 \cdot 857 \qquad\qquad ...(iv)$

From (iii) and (iv), we get

$$M_C = 5 \cdot 9 \text{ kNm} \quad \textbf{(Ans.)}$$

Hence from (iii) $\qquad\qquad M_B = -14 \cdot 25 \text{ kNm} \quad \textbf{(Ans.)}$

and from (i) $\qquad\qquad M_A = -67 \cdot 8 \text{ kNm} \quad \textbf{(Ans.)}$

(ii) Reactions at the supports:

— Taking moments about B, for span AB and equating to M_B, we get

$$M_A + R_A \times 10 - 6 \times 10 \times \frac{10}{2} = M_B$$

$$-67 \cdot 8 + R_A \times 10 - 300 = -14 \cdot 25$$

$\therefore \qquad\qquad R_A = 35 \cdot 35 \text{ kN} \quad \textbf{(Ans.)}$

— Taking moments about B, for span BC and equating to M_B, we get

$$R_C \times 7 + M_C - 120 = M_B$$

$$7\,R_C + 5 \cdot 9 - 120 = -14 \cdot 25$$

$\therefore \qquad\qquad R_C = 14 \cdot 26 \text{ kN} \quad \textbf{(Ans.)}$

Also, $\qquad\qquad R_A + R_B + R_C = 6 \times 10 = 60$

$\therefore \qquad\qquad R_B = 60 - R_A - R_C = 60 - 35 \cdot 35 - 14 \cdot 26$

$$= 10 \cdot 39 \text{ kN} \quad \textbf{(Ans.)}$$

Tunnel boring machine that was used to dig the under-sea tunnel between France and England.

HIGHLIGHTS

1. A fixed beam (also called built-in or encaster beam) is a beam the ends of which are constrained or built-in to remain in horizontal position.

2. If a beam is fixed at both ends, then slope and deflection at the both ends are zero.

3. For a fixed beam AB of length l:

$$a\bar{x} + a'\bar{x}' = 0$$

where $a\bar{x}$ = first moment of area a of the B.M. diagram about the point A (considering the beam to be simply supported at the ends);

$a'\bar{x}'$ = first moment of area a' of the B.M. diagram due to support moment about the point A.

4. For a beam AB of length l, fixed at both the ends carrying a concentrated load at its centre:

Fixing moments/couples, $\quad M_A = M_B = -\dfrac{Wl}{8}$

B.M. at the centre of the beam $= +\dfrac{Wl}{8}$

Deflection, $\quad y_{max} = +\dfrac{Wl^3}{192\,EI} \quad$ (at the centre)

5. For a beam AB of length l, fixed at both the ends, carrying a load W at a distance of a from end A, ($a > l/2$):

$$M_A = \frac{W\,ab^2}{l}, \; M_B = \frac{W\,a^2 b}{l^2}$$

$$y_{max} = -2/3\,\frac{W\,a^3 b^2}{(3a+b)^2\,EI}$$

Point of contraflexure:

$$x = \frac{al}{3a+b}$$

$\qquad\qquad\qquad\qquad\qquad\qquad\qquad\qquad$ ... from the end A

(for point of contraflexure in AC)

$$x = \frac{l\,(a+2b)}{a+3b}$$

$\qquad\qquad\qquad\qquad\qquad\qquad\qquad\qquad$ ... from the end A

(for point of contraflexure in BC)

6. For a beam AB of length l, fixed at both the ends carrying a uniformly distributed load w throughout its length:

Fixing moments/couples, $\quad M_A = M_B = -\dfrac{wl^2}{12}$

B.M. at the centre of the beam $= +\dfrac{wl^2}{24}$

Deflection, $\quad y_{max} = -\dfrac{wl^4}{384\,EI} \quad$ (at the centre)

7. In a fixed beam, if one support sinks by δ, then fixing moment at the ends due to sinking of support is $\dfrac{6\,EI\delta}{l^2}$ where l is the length between the supports.

8. A *continuous beam* is one which is supported on more than two supports.

9. The most general form of the "*Theorem of three moments*" is given as follows:

$$\frac{M_A l_1}{E_1 I_1} + 2M_B\left(\frac{l_1}{E_1 I_1} + \frac{l_2}{E_2 I_2}\right) + \frac{M_C l_2}{E_2 I_2} + \frac{6a_1 \bar{x}_1}{l_1 E_1 I_1} + \frac{6a_2 \bar{x}_2}{l_2 E_2 I_2} = \frac{6\delta_1}{l_1} + \frac{6\delta_2}{l_2}$$

where,

l_1 = Length of span AB,

l_2 = Length of span BC,

$a_1 \bar{x}_1$ = First moment of B.M. diagram for span AB, considering the origin at A,

$a_2 \bar{x}_2$ = First moment of B.M. diagram for span BC, considering the origin at C,

$E_1 I_1$ = Flexural rigidity for the span AB,

$E_2 I_2$ = Flexural rigidity for the span BC,

δ_1 = Sinking of the support A with respect to support B, and

δ_2 = Sinking of the support C with respect to support B.

OBJECTIVE TYPE QUESTIONS

Choose the Correct Answer:

1. A beam of length l, fixed at both ends, carries a point load W at its centre. If EI is the flexural rigidity of the beam, the maximum deflection in the beam is

    (a) $Wl^3/48\ EI$ (b) $Wl^3/192\ EI$

    (c) $Wl^3/96\ EI$ (d) $Wl^3/24\ EI$.

2. A beam of length l, fixed at both ends carries a uniformly distributed load of w per unit length. If EI is the flexural rigidity, then maximum deflection in the beam is

    (a) $wl^4/192\ EI$ (b) $wl^4/24\ EI$

    (c) $wl^4/384\ EI$ (d) $wl^4/12\ EI$.

3. A beam of length l fixed at both the ends carries a uniformly distributed load w per unit length throughout the span.

    The bending moment at the ends is

    (a) $wl^2/8$ (b) $wl^2/12$

    (c) $wl^3/12$ (d) $wl^2/24$.

4. A beam of length 6 metres carries a point load 120 kN at its centre. The beam is fixed at both ends. The fixing moment at the ends is

    (a) 40 kNm   (b) 90 kNm

    (c) 120 kNm   (d) 150 kNm.

5. A fixed beam of length l, sinks at one end by an amount d, if EI is the flexural rigidity of the beam, the fixing couple at the ends is

    (a) $2\ EI\ d/l^2$ (b) $4\ EI\ d/l^2$

    (c) $6\ EI\ d/l^2$ (d) $8\ EI\ d/l^2$.

6. A beam fixed at both the ends carries a uniformly distributed load of 20 kN/m over entire span of 6 m. The bending moment at the centre of the beam is

    (a) 10 kNm   (b) 30 kNm

    (c) 60 kNm   (d) 90 kNm.

7. A beam of length 4 m, fixed at both ends carries a point load of 120 kN at the centre. If EI for the beam is 20000 kNm^2, deflection at the centre of beam is:

    (a) 1·0 mm   (b) 2·0 mm

    (c) 5·0 mm   (d) 10·0 mm.

8. A continuous beam 12 m long, supported over two spans 6 m each, carries a concentrated load of 40 kN each at the centre of each span. The bending at the centre of the support is

    (a) 30 kNm   (b) 45 kNm

    (c) 90 kNm   (d) 150 kNm.

Side view of a vertical boring machine.

ANSWERS

1. (b)	**2.** (c)	**3.** (b)	**4.** (b)	**5.** (c)	**6.** (b)	**7.** (b)	**8.** (b)

UNSOLVED EXAMPLES

Fixed Beams:

1. A fixed beam of 6 m span carries two point loads of 600 kN each at 2 metres from each end. Draw the B.M. and S.F. diagrams and find the maximum deflection.
   Take: $E = 2 \times 10^8$ kN/m^2, and $I = 9 \times 10^8$ mm^4. **[Ans. 3·12 mm]**

2. A fixed beam of 6 m span carries two point loads of 200 kN and 150 kN at distances 2 m and 4 m from the left end.
   Find: (i) Fixing moments at the ends; (ii) Reactions at the supports.
   Draw the B.M. and S.F. diagrams also.
   [Ans. 244·45 kNm, 222·22 kNm; 187·03 kN, 162·97 kN]

3. A fixed beam *AB* of 5 m span is subjected to a concentrated couple of 50 kNm applied at a point *C* distant 3 m from the end *A*. The couple acts in the vertical longitudinal plane of symmetry of the beam in clockwise direction.
   (i) Find the end moments from first principles using Macaulay's method.
   (ii) Draw the B.M. and S.F. diagrams.
   Assume *EI* constant throughout. **[Ans. M_A = 16 kNm (sagging), M_B = 6 kNm (hogging).]**

4. A fixed beam of 6 m span carries a uniformly distributed load of 5 kN/m run. Find the maximum deflection of the beam.
   Take: $E = 2 \times 10^8$ kN/m^2, and $I = 0·48 \times 10^{-4}$ m^4. **[Ans. – 1·76 mm]**

5. A fixed beam of 8 m span carries uniformly distributed load of *w* kN/m run. Determine the value of *w* for the following cases:
   (i) When maximum bending moment is not to exceed 36 kNm;
   (ii) When maximum deflection is not to exceed $\dfrac{1}{2000}$ of the length. Take $EI = 7200$ kNm^2
   [Ans. (i) 6·75 kN/m run, (ii) 2·7 kN/m run]

6. A fixed beam of 6 m span carries concentrated loads of 80 kN each at a distance of 2 m from each end. Determine;
   (i) Reactions at the supports;
   (ii) Moments at the supports;
   (iii) Deflection at the centre of the beam.
   Take: $E = 2·05 \times 10^8$ kN/m^2 and $I = 3200 \times 10^{-8}$ m^4.
   [Ans. 80 kN each; – 106·66 kNm; 21·32 mm]

7. A fixed beam of 8 m span carries a uniformly distributed load of 20 kN/m run over 4 m length starting from left hand end and a concentrated load of 40 kN at a distance of 6 m from the left hand end.
   If $E = 15000$ kNm^2, find:
   (i) Moments at the supports; (ii) Deflection at the centre of the beam.
   [Ans. – 88·33 kNm, – 78·33 kNm; (ii) – 10·6 mm]

8. A fixed beam *AB* of 5 m span carries a point load of 200 kN at 3 m from the left end. If the right end sinks by 1 mm find the fixing moments as well as reactions at the supports.

   Take: $E = 2 \times 10^8$ kN/m^2 and $I = 10000$ cm^4.
   $$\left[\text{Ans. } \begin{array}{l} M_A = 100\text{·}8 \text{ kNm (hogging),} \\ M_B = 139\text{·}2 \text{ kNm (hogging)} \\ R_A = 72\text{·}32 \text{ kN, } R_B = 127\text{·}68 \text{ kN} \end{array} \right]$$

9. A fixed beam AB of 7 m span carries a uniformly distributed load of 20 kN/m run on the entire beam. The level of right support sinks by 10 mm below that of the left hand end. If $E = 2.08 \times 10^8$ kN/m^2 and $I = 4.52 \times 10^{-5}$ m^4 find:

    (*i*) Moments at the supports;

    (*ii*) Reactions at the supports;

    (*iii*) Deflection at the centre.

$$\left[\begin{array}{l} \textbf{Ans.} \ (i) \ M_A = 93.178 \text{ kNm}, \ M_B = -70.154 \text{ kNm} \\ (ii) \ \ R_A = 73.29 \text{ kN}, \ R_B = 66.71 \text{ kN} \\ (iii) \ \ y_C = -18.3 \text{ mm} \end{array}\right]$$

10. A fixed beam of 4 m span carries a concentrated load of 60 kN at the middle of the beam. The level of the right hand support is 6 mm below the level of left hand support. If $E = 2 \times 10^8$ kN/m^2, and $I = 6 \times 10^{-5}$ m^4 find:

    (*i*) Support reactions;   (*ii*) Support moments;   (*iii*) Deflection at the centre.

$$\left[\begin{array}{l} \textbf{Ans.} (i) \ 43.5 \text{ kN}, 1.5 \text{ kN} \\ (ii) \ -57 \text{ kNm}, -3 \text{ kNm} \\ (iii) \ -4.67 \text{ mm.} \end{array}\right]$$

Continuous Beams:

11. A continuous beam, 12 m long supported over spans $AB = BC = CD = 4$ m, carries a uniformly distributed load of 3 kN/m run over span AB, a concentrated load of 4 kN at a distance of 1 m from point B on support BC and a load of 3 kN at the centre of the span CD, find:

    (*i*) Support moments;   (*ii*) Support reactions.

Draw the B.M. diagram for the continuous beam.

$$\left[\begin{array}{l} \textbf{Ans.} M_A = 0; \ M_B = -4.05 \text{ kNm}; \ M_C = -1.05 \text{ kNm} \\ R_A = 4.9875 \text{ kN}; \ R_B = 10.7625 \text{ kN}; \\ R_C = 2.0125 \text{ kN}, \ R_D = 1.2375 \text{ kN} \end{array}\right]$$

12. A beam $ABCD$, 16 m long is continuous over three spans; $AB = 6$ m; $BC = 5$ m and $CD = 5$ m, the supports being at the same level. There is a uniformly distributed load of 20 kN/m over BC. On AB, there is a point load of 80 kN at 2 m from A and on CD, there is a point load of 60 kN at 3 m from D.

Calculate: (*i*) Support moments;   (*ii*) Support reactions.

$$\left[\begin{array}{l} \textbf{Ans.} (i) \ M_B = 56.8 \text{ kNm}; \ M_C = 45.8 \text{ kNm} \\ (ii) \ \ R_A = 43.9 \text{ kN}; \ \ \ \ R_B = 86.3 \text{ kN}; \\ R_C = 93 \text{ kN}, \ \ \ \ \ R_D = 14.8 \text{ kN} \end{array}\right]$$

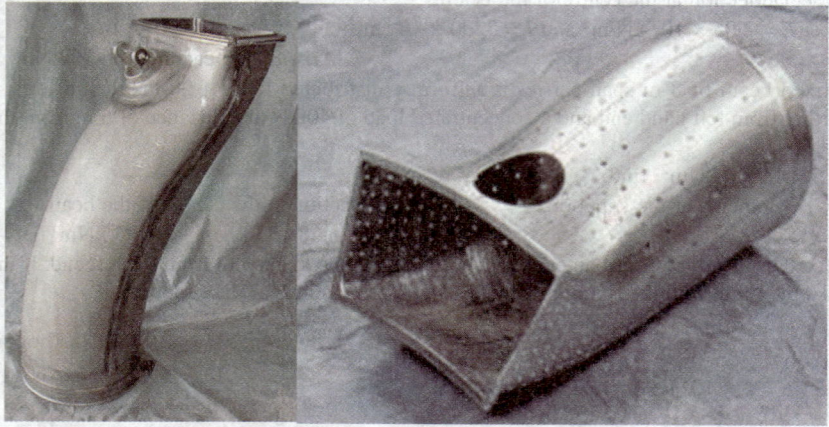

Ducts that direct air at high pressures and temperatures, in an aircraft.

13. A continuous beam ABC 8 metres long consists of two spans $AB = 3$ m and $BC = 5$ m. The span AB carries load of 50 kN/m while the span BC carries a load of 10 kN/m. Find:

    (*i*) Support moment at B;   (*ii*) Reactions at the supports.

        [**Ans.** (*i*) $M_B = 179 \cdot 7$ kNm (*ii*) $R_A = 48 \cdot 4$ kN; $R_B = 192 \cdot 5$ kN, $R_C = 59 \cdot 1$ kN]

14. A continuous beam $ABCDE$ 12 metres long consists of four spans of 3 metres each. The beam carries a uniformly distributed load of 40 kN/m over the whole span. Find:

    (*i*) Moments at the supports;   (*ii*) Reactions at the supports.

$$\left[\begin{array}{l} \textbf{Ans.} \quad (i) \;\; M_A = M_E = 0; \; M_B = M_D = \dfrac{270}{7} \text{ kNm}, \; M_C = \dfrac{180}{7} \text{ kNm} \\[4mm] \quad\;\; (ii) \;\; R_A = R_E = \dfrac{330}{7} \text{ kN}, \; R_B = R_D = \dfrac{960}{7} \text{ kN}, \; R_C = \dfrac{780}{7} \text{ kN} \end{array}\right]$$

15. A continuous beam ABC consists of two spans AB of length 4 m, and BC of length 3 m. The span AB carries a point load of 100 kN at its middle point. The span BC carries a point load of 120 kN at 1 m from C. Find:

    (*i*) Moments at the supports;   (*ii*) Reactions at the supports.

$$\left[\begin{array}{l} \textbf{Ans.} \, (i) \; M_A = M_C = 0; \; M_B = \dfrac{460}{7} \text{ kNm} \\[4mm] \quad\;\; (ii) \; R_A = \dfrac{235}{7} \text{ kN}; \; R_B = \dfrac{1000}{7} \text{ kN}, \; R_C = \dfrac{305}{7} \text{ kN} \end{array}\right]$$

16. The right hand span of a continuous beam carries a uniform load of intensity w over half its length as shown in Fig. 9·32. Determine the B.M. at the two intermediate supports.

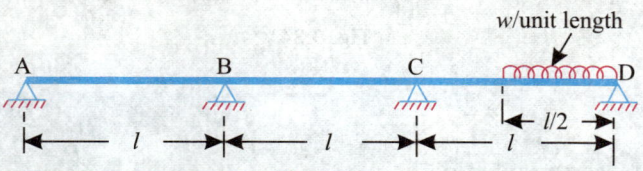

Fig. 9.32

$$\left[\textbf{Ans.} \;\; M_B = -\dfrac{7wl^2}{960}; \; M_C = -\dfrac{28\,wl^2}{960}\right]$$

17. A straight elastic beam of uniform section rests on four similar elastic supports which are placed l meters apart. The supports are such that they compress by d for each unit of load upon them. Show that when a uniformly distributed load of amount W comes on the beam, the reactions at the centre points are each

$$\dfrac{W\left(\dfrac{11}{6} + \dfrac{3\,EId}{l^3}\right)}{\left(5 + \dfrac{12\,EId}{l^3}\right)}.$$

18. A continuous beam $ABCD$, 20 m long is loaded as shown in the Fig. 9·33. If the support B sinks by 10 mm below A and C find the support moments.

Take: $E = 2 \cdot 1 \times 10^8$ kN/m²; $I = 85 \times 10^6$ mm⁴.

        [**Ans.** $M_B = M_D = 0$, $M_A = -27.4$ kNm; $M_C = -73$ kNm]

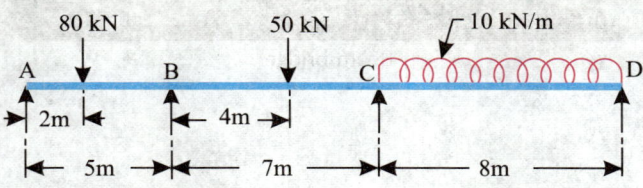

Fig. 9.33

19. A continuous beam *ABC* consists of two consecutive spans *AB* and *BC* 4 metres each and carries a distributed load of 60 kN per metre run. The end *A* is fixed and the end *C* is simply supported. Find:

(*i*) Support moments;   (*ii*) Support reactions.

$$\left[\begin{array}{l}\textbf{Ans. } (i)\ M_A = 68\text{·}57 \text{ kNm (hogging)};\ M_B = 102\text{·}85 \text{ kNm (hogging)}\\ \qquad (ii)\ R_A = 111\text{·}43 \text{ kN};\ R_B = 274\text{·}28 \text{ kNm};\ R_C = 94\text{·}29 \text{ kN}\end{array}\right]$$

20. A continuous beam consists of three successive spans of 8 metres, 10 metres and 6 metres and carries loads of 60 kN/m run, 40 kN/m run and 80 kN/m run respectively on the spans. Find:

(*i*) Bending moments at the supports;   (*ii*) Reactions at the supports.

$$\left[\begin{array}{l}\textbf{Ans. } (i)\ M_A = M_D = 0,\ M_B = 401\text{·}6 \text{ kNm (hogging)},\ M_C = 322 \text{ kNm (hogging)}\\ \qquad (ii)\ R_A = 189\text{·}8 \text{ kN};\ R_B = 498\text{·}2 \text{ kN};\ R_C = 485\text{·}7 \text{ kN},\ R_D = 186\text{·}3 \text{ kN}\end{array}\right]$$

21. For the beam shown in the Fig. 9·34 find:

(*i*) Support moments;   (*ii*) Support reactions.

$$\left[\begin{array}{l}\textbf{Ans.} (i)\ M_A = 30 \text{ kNm (hogging)},\ M_B = 22 \text{ kNm},\ M_C = 39\text{·}5 \text{ kNm},\ M_D = 0\\ \qquad (ii)\ R_A = 82\text{·}67 \text{ kN};\ R_B = 96\text{·}49 \text{ kN};\ R_C = 139\text{·}01\text{kN};\ R_D = 61\text{·}83\text{kN}\end{array}\right]$$

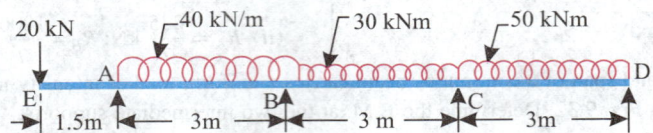

Fig. 9.34

Beam bridge.

Thin Shells

10.1. INTRODUCTION

In order to meet with several requirements, the fluids are stored under pressure in pressure vessels or shells and transmitted from one place to the other through pipes. Pressure vessels are made of cast iron, sheet steel and nonferrous alloys. Special material is used for chemical vessels. Vessels of spherical and cylindrical form are used for storing fluids under pressure *e.g.* steam boilers, air compressors, tanks and water tanks. Spheres are used for containing gas under pressure. Liquids and gases causing internal pressure in a closed vessel are referred as fluids. *When it is a gas, the pressure is constant in all parts of vessel. In case of liquid, the pressure is lowest at the top and increases with depth.* When the vessels are empty, they are subjected to an atmospheric pressure both internally and externally and hence the *resultant effect* of atmospheric pressure is *nil.*

10.2. THIN CYLINDRICAL SHELLS

A cylindrical vessel or shell may be *thin* or *thick* depending upon the thickness of the plate in relation to the *in-*

ternal diameter of the cylinder. The ratio of $\dfrac{t}{d} = \dfrac{1}{20}$ can be considered suitable line of demarcation between *thin* and *thick* cylinders.

In thin cylinders, the stress may be assumed uniformly distributed over the wall thickness. *Boilers, tanks, steam pipes, water pipes etc. are usually considered as thin cylinders. Thin cylinders are frequently required to operate under pressures upto30 MN/m² or more, for high pressures such as 250 MN/m² or more, thick cylinders are used.*

When these cylinders are subjected to internal fluid pressures the following two types of stresses are developed:

1. *Hoop or circumferential stresses.* These act in a tangential direction to the circumference of the shell.
2. *Longitudinal stresses.* These act parallel to the longitudinal axis of the shell.
3. *Radial stresses.* These act radially and are *too small and can be neglected.*

These three stresses are *mutually perpendicular and are principal stresses.*

10.2.1. Circumferential or Hoop Stresses

Let, d = Internal diameter of the cylinder,

t = Thickness of the cylinder,

p = Internal pressure (gauge) in the cylinder, and

σ_c = Circumferential or hoop stress.

Bursting force (pressure) = Resisting strength

$$pdl = 2\,lt\,\sigma_c$$

or, $$\sigma_c = \frac{pd}{2t} \qquad \text{(Fig. 10·1 } a,b,c\text{)} \qquad \qquad ...(10\cdot1)$$

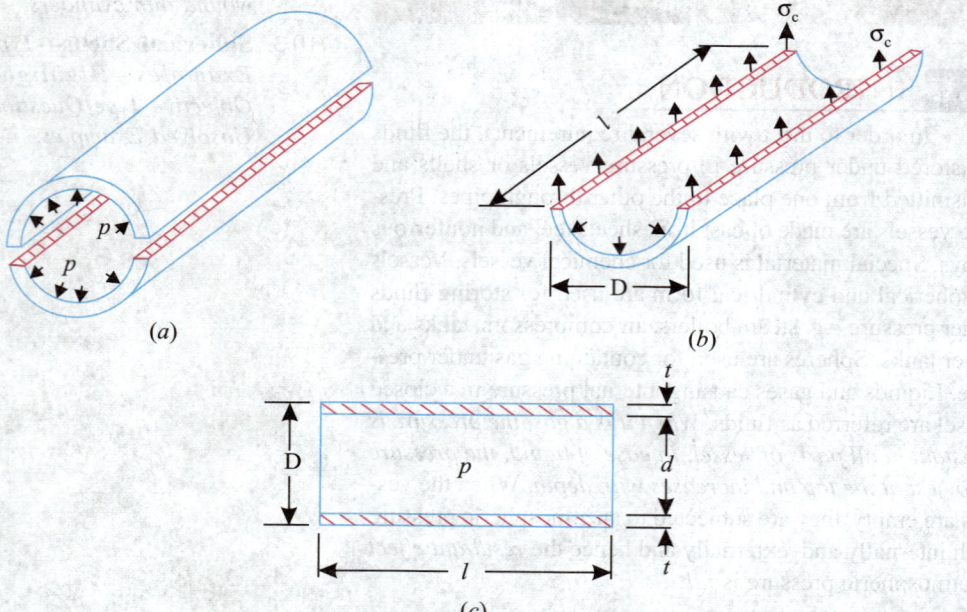

Fig. 10.1

10.2.2. Longitudinal Stresses

Now suppose the cylinder under consideration has its two ends covered with two end plates connected to them as shown in Fig. 10·2 (a, b).

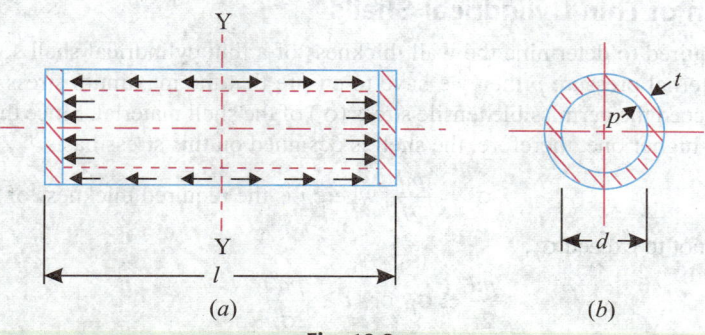

Fig. 10.2

Let, σ_l = Longitudinal stress produced in the shell.

Now, Pressure on the ends = Resisting force

$$p \times \frac{\pi}{4}d^2 = \pi dt.\sigma_l \text{ from which, } \sigma_l = \frac{pd}{4t} \qquad ...(10\cdot2)$$

Therefore, *the circumferential stress or hoop (σ_c) is twice as great as the longitudinal stress (σ_l) and in no case should the hoop stress be greater than the permissible stress in the material of the cylinder.*

Oil tanks on a goods train.

10.2.3. Maximum Shear Stress

In a cylindrical shell, at any point on its circumference, there is a set of two mutually perpendicular stresses σ_c and σ_l which are principal stresses and as such the planes in which these act are the principal planes. The maximum shear stress is found as follows:

Maximum shear stress,
$$\tau_{max} = \frac{\sigma_c - \sigma_l}{2}$$

$$= \frac{\dfrac{pd}{2t} - \dfrac{pd}{4t}}{2} = \frac{pd}{8t}$$

i.e.
$$\tau_{max} = \frac{pd}{8t} \qquad ...(10\cdot3)$$

10.2.4. Design of Thin Cylindrical Shells

If it is required to determine the wall thickness of a thin cylindrical shell so that it can with-stand a given internal pressure p then we have to ensure that the maximum stress developed in the shell does not exceed the permissible tensile stress (σ_t) of the shell material. Since the circumferential or hoop stress is higher one, therefore, the shell is designed on this stress basis.

Now, $\qquad\qquad \sigma_c = \dfrac{pd}{2t}$, where t is the required thickness of the shell.

But σ_c is not to exceed σ_t,

$\therefore \qquad\qquad \dfrac{pd}{2t} \leq \sigma_t \ \text{ or } \ t \geq \dfrac{pd}{2\sigma_t} \qquad\qquad\qquad …(10\cdot4)$

10.2.5. Cylindrical Shell with Hemispherical Ends

Fig. 10·3 shows a thin cylindrical shell with hemispherical ends.

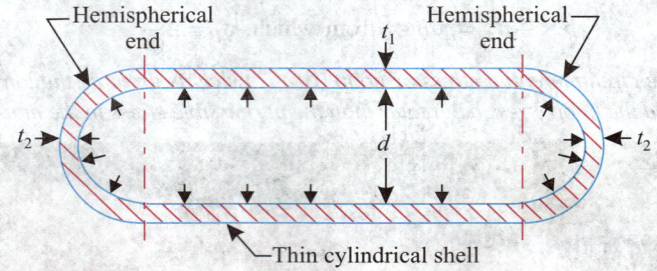

Fig. 10.3

Let, $\qquad\qquad\qquad d$ = Internal diameter of the cylinder,

$\qquad\qquad\qquad t_1$ = Wall thickness of cylindrical portion, and

$\qquad\qquad\qquad t_2$ = Wall thickness of hemispherical portion.

Circumferential stress developed in cylindrical portion

$$\sigma_{c_1} = \frac{pd}{2t_1}$$

Longitudinal stress in cylindrical portion,

$$\sigma_{l_1} = \frac{pd}{4t_1}$$

Circumferential strain in cylindrical portion,

$$e_{c_1} = \frac{\sigma_{c_1}}{E} - \frac{\sigma_{l_1}}{mE} = \frac{2\sigma_{l_1}}{E} - \frac{\sigma_{l_1}}{mE}$$

i.e. $\qquad\qquad\qquad = \dfrac{pd}{4t_1 E}\left(2 - \dfrac{1}{m}\right) \qquad\qquad\qquad …(i)$

Circumferential stress developed in hemispherical portion,

$$\sigma_{c_2} = \frac{pd}{4t_2}$$

Circumferential strain in hemispherical portion,

$$e_{c_2} = \frac{\sigma_{c_2}}{E} - \frac{\sigma_{c_2}}{mE}$$

$$e_{c_2} = \frac{pd}{4t_2 E}\left(1 - \frac{1}{m}\right) \qquad \qquad ...(ii)$$

In order that there is *no distortion* at the junction of cylindrical and hemispherical portions the circumferential strains in the two have to be *equal*.

i.e. $$e_{c_1} = e_{c_2}$$

$$\frac{pd}{4t_1 E}\left(2 - \frac{1}{m}\right) = \frac{pd}{4t_2 E}\left(1 - \frac{1}{m}\right)$$

∴ $$\frac{t_2}{t_1} = \frac{\left(1 - \frac{1}{m}\right)}{\left(2 - \frac{1}{m}\right)} = \frac{m-1}{2m-1} \qquad \qquad ...(10\cdot5)$$

Since $(m-1)$ is always less than $(2m-1)$, whatever be the value of m, thus t_2 shall always be less than t_1, i.e. the *hemispherical end is always thinner than the cylindrical portion*.

So that the maximum stress may be *same* in both cylindrical and hemispherical portions, we have

$$\sigma_{c_1} = \sigma_{c_2}$$

$$\frac{pd}{2t_1} = \frac{pd}{4t_2} \quad \text{or} \quad \frac{t_2}{t_1} = 0\cdot5 \qquad \qquad ...(10\cdot6)$$

Example 10·1. *A thin cylindrical shell of diameter 300 mm and wall thickness 6 mm has hemispherical ends. If there is no distortion of the junction under pressure determine the thickness of hemispherical ends.*

Take: E = 208 GN/m², and Poisson's ratio = 0·3

Solution. Diameter of the thin cylindrical shell,

$$d = 300 \text{ mm}$$

Thickness of cylindrical portion, $t_1 = 6$ mm

Poisson's ratio, $$\frac{1}{m} = 0\cdot3 \quad \text{or} \quad m = 3\cdot333$$

Modulus of elasticity, $E = 208$ GN/m²

Thickness of hemispherical ends, t_2:

For no distortion at junction of cylindrical and hemispherical portions,

$$\frac{t_2}{t_1} = \frac{m-1}{2m-1} = \frac{3\cdot333-1}{2\times3\cdot333-1} = 0\cdot4117$$

∴ $$t_2 = 0\cdot4117 \times t_1 = 0\cdot4117 \times 6 = 2\cdot47 \text{ mm}$$

i.e. $$t_2 = 2\cdot47 \text{ mm} \quad \textbf{(Ans.)}$$

10·2·6. Built-up Cylindrical Shells

The formulae derived for σ_c and σ_l have been obtained with the presumption that the cylindrical shell is *seamless i.e. solid drawn*. But in actual practice, cylindrical shells of large diameters, such as boiler shells, etc. are not seamless (without joints), but instead are built-up by longitudinal and circumferential joints. The longitudinal joints reduce the resisting strength of the shell plate against bursting and circumferential joints reduce the resisting strength of the plate against tearing due to pressure on the end plates.

Let,
$$\eta_l = \text{Efficiency of the longitudinal joint, and}$$
$$\eta_c = \text{Efficiency of the circumferential joint.}$$

Then, Bursting force = Resisting strength

i.e. $\qquad pdl = 2lt\,\sigma_c\,\eta_l \quad \text{or} \quad \sigma_c = \dfrac{pd}{2t\eta_l}$ ...(10·7)

Similarly, Pressure on the ends = Resisting force

i.e. $\qquad p \times \dfrac{\pi}{4}d^2 = \pi d.t.\sigma_l.\eta_c \quad \text{or} \quad \sigma_l = \dfrac{pd}{4t\eta_c}$ ...(10·8)

10·2·7. Change in Dimensions of a Thin Cylindrical Shell due to an Internal Pressure

A cylindrical shell, due to circumferential and longitudinal stresses, will increase in length as well as undergo a change in dimensions resulting in change of its volume.

Let, $\qquad l$ = Length of the shell,

$\qquad\qquad d$ = Diameter of the shell,

$\qquad\qquad t$ = Thickness of the shell,

$\qquad\qquad p$ = Intensity of pressure, and

$\qquad 1/m$ = Poisson's ratio

$$\left[= \dfrac{\text{Lateral strain}}{\text{Longitudinal strain}} \right]$$

Direct strain due to $\sigma_c = \dfrac{\sigma_c}{E}$

Direct strain due to $\sigma_l = \dfrac{\sigma_l}{E}$

Net circumferential strain,

Stainless steel T-duct used to direct gase and liquids.

$\qquad e_c$ = Direct strain – lateral strain due to direct strain $\dfrac{\sigma_l}{E}$

i.e. $\qquad e_c = \dfrac{\delta d}{d} = \dfrac{\sigma_c}{E} - \dfrac{1}{m}\dfrac{\sigma_l}{E} = \dfrac{\sigma_c}{E} - \dfrac{\sigma_c}{2mE}$, since $\sigma_l = \dfrac{\sigma_c}{2}$,

$\qquad\qquad = \dfrac{\sigma_c}{E}\left(1 - \dfrac{1}{2m}\right) = \dfrac{pd}{2tE}\left(1 - \dfrac{1}{2m}\right)$ ...(10·9)

and, net longitudinal strain,

$\qquad e_l = \dfrac{\delta l}{l} = \dfrac{\sigma_l}{E} - \dfrac{1}{m}\dfrac{\sigma_c}{E} = \dfrac{\sigma_l}{E} - \dfrac{2\sigma_l}{mE}$ $\qquad (\because \sigma_c = 2\sigma_l)$

$\qquad\qquad = \dfrac{pd}{4tE}\left(1 - \dfrac{2}{m}\right)$ ...(10·10)

The volumetric strain,

$\qquad e_v$ = Algebraic sum of net strains in all axes

$\qquad\qquad$ = net longitudinal strain + 2 × net circumferential strain

$\qquad\qquad = e_l + 2e_c$

The net circumferential strain has been taken twice because it has the effect of changing the diameter of the shell, necessarily both in XX and YY axes.

Also, $\qquad e_v = \dfrac{\delta V}{V} = \dfrac{\text{Change in volume}}{\text{Original volume}}$

$\therefore \qquad \delta V = e_v \times V = (e_l + 2e_c)\,V$

$\qquad\qquad = \left[\dfrac{pd}{4tE}\left(1 - \dfrac{2}{m}\right) + 2 \times \dfrac{pd}{2tE}\left(1 - \dfrac{1}{2m}\right)\right]V$

$$= \frac{pd}{2tE}\left[\left(\frac{1}{2} - \frac{1}{m}\right) + \left(2 - \frac{1}{m}\right)\right] V$$

$$= \frac{pdV}{2tE}\left(\frac{5}{2} - \frac{2}{m}\right) \qquad \qquad ...(10\cdot11)$$

where, $\qquad\qquad V = \dfrac{\pi}{4}\,d^2 l$

Changes in length (δl) and diameter (δd) may be found from equations (10·9) and (10·10) as follows:

$$\delta d = e_1 \times d = \frac{pd}{2tE}\left(1 - \frac{1}{2m}\right) \times d$$

$$= \frac{pd^2}{2tE}\left(1 - \frac{1}{2m}\right) \qquad \qquad ...(10\cdot12)$$

$$\delta l = e_2 \times l = \frac{pd}{4tE}\left(1 - \frac{2}{m}\right) l = \frac{pdl}{4tE}\left(1 - \frac{2}{m}\right) \qquad \qquad ...(10\cdot13)$$

> **Note.** The above formulae hold good only in case of *seamless shells*.

Example 10·2. *Calculate the bursting pressure for a cold drawn seamless steel tubing of 60 mm inside diameter with 2 mm wall thickness. The ultimate strength of steel is 380 MN/m².*

Solution. Diameter of the tube,
$$d = 60 \text{ mm (or 0·06 m)}.$$
Thickness of the tube, $\quad t = 2$ mm (or 0·002 m)
Ultimate stress, $\qquad \sigma_c = 380$ MN/m²
Bursting pressure p:

Using the relation, $\qquad \sigma_c = \dfrac{pd}{2t}$, we have

$$380 \times 10^6 = \frac{p \times 0\cdot06}{2 \times 0\cdot002}$$

or, $\qquad\qquad p = \dfrac{380 \times 10^6 \times 2 \times 0\cdot002}{0\cdot06} \times 10^{-6} \text{ MN/m}^2$.

$$= \mathbf{25\cdot33 \ MN/m^2} \ \textbf{(Ans.)}$$

Example 10·3. *Calculate the thickness of the metal required for a cast-iron main 800 mm in diameter for water at a pressure head of 100 m if the maximum permissible tensile stress is 20 MN/m² and weight of water is 10 kN/m³.*

Solution. Diameter of cast-iron main,
$$d = 800 \text{ mm} = 0\cdot8 \text{ m}$$
Maximum permissible stress,
$$\sigma_c = 20 \text{ MN/m}^2$$
Weight of water, $\qquad w = 10$ kN/m³
Thickness of metal, t:
We know that,

$$p = wh, \text{ where '}w\text{' is the specific weight of water and '}h\text{' is the head}$$
$$\text{of water in the main.}$$

$\therefore \qquad\qquad p = 10 \times 10^3 \times 100 = 10^6 \text{ N/m}^2 \text{ or 1 MN/m}^2$

Using the relation, $\qquad \sigma_c = \dfrac{pd}{2t}$

$$t = \frac{pd}{2\sigma_c} = \frac{1 \times 10^6 \times 0.8}{2 \times 20 \times 10^6}$$

$$= 0.02 \text{ m or } \textbf{20 mm} \quad \textbf{(Ans.).}$$

Example 10·4. *A cylindrical water tank of height 25 m, inside diameter 2·2 m, having vertical axis is open at the top. The tank is made of steel having yield stress of 210 MN/m². Determine the thickness of steel used when the tank is full of water.*

Given: Efficiency of the longitudinal joint = 70 %; Factor of safety = 3.

Solution. Factor of safety $= \dfrac{\text{Yield stress}}{\text{Working stress}} = \dfrac{210}{3} = 70 \text{ MN/m}^2$

The pressure will be *maximum at the base*. This pressure must be resisted by steel sheet and therefore the thickness must be determined by taking the maximum pressure into consideration.

Pressure at the base $\qquad p = wh$

$$= 10 \times 10^3 \times 25 \times 10^{-6} \text{ MN/m}^2 \quad \text{or} \quad (\because w = 10 \text{ kN/m}^3)$$

$$= 0.25 \text{ MN/m}^2$$

Since the tank is open at the top, there will only be hoop (or circumferential) stress which, under any circumstances, should not exceed the working stress.

We know, $\qquad\qquad \sigma_c = \dfrac{pd}{2t\eta_l}$

$$70 \times 10^6 = \frac{0.25 \times 10^6 \times 2.2}{2 \times t \times 0.7}$$

$$\therefore \qquad\qquad t = \frac{0.25 \times 10^6 \times 2.2}{70 \times 10^6 \times 2 \times 0.7}$$

$$= 0.0056 \text{ m or } 5.6 \text{ mm say } \textbf{6 mm} \quad \textbf{(Ans.)}$$

Example 10·5. *A boiler shell is to be made of 15 mm thick plate having tensile stress of 120 MN/m². If the efficiencies of the longitudinal and circumferential joints are 70% and 30% respectively, determine:*

(i) Maximum permissible diameter of the shell for an internal pressure of 2 MN/m².

(ii) Permissible intensity of internal pressure when the shell diameter is 1·5 m.

Solution. (*i*) Let, $\qquad d$ = Maximum permissible diameter.

Consider circumferential stress as 120 MN m²:

$$\sigma_c = \frac{pd}{2\eta_l}$$

$$120 \times 10^6 = \frac{2 \times 10^6 \times d}{2 \times 15 \times 10^{-3} \times 0.7}$$

$$d = \frac{120 \times 10^6 \times 2 \times 15 \times 10^{-3} \times 0.7}{2 \times 10^6} = 1.26 \text{ m}$$

Consider longitudinal stress as 120 MN m²:

$$\sigma_l = \frac{pd}{4\eta_c}$$

$$120 \times 10^6 = \frac{2 \times 10^6 \times d}{4 \times 15 \times 10^{-3} \times 0.3}$$

$$\therefore \qquad\qquad d = \frac{120 \times 10^6 \times 4 \times 15 \times 10^{-3} \times 0.3}{2 \times 10^6} = 1.08 \text{ m}$$

In order to satisfy both the conditions, **maximum diameter, d = 1·08 m** (*minimum* of the above two values) **(Ans)**.

> **Note.** If we provide bigger diameter (*i.e.* 1·26 m) then the longitudinal stress
>
> $$\left(\sigma_l = \frac{pd}{4t\eta_c} = \frac{2 \times 10^6 \times 1\cdot26}{4 \times 15 \times 10^{-3} \times 0\cdot3} = 140 \text{ MN/m}^2 \right)$$
>
> will be more than the permissible stress (120 MN / m²).

(*ii*) Again $\sigma_c = \dfrac{pd}{2t\eta_l}$

$$120 \times 10^6 = \frac{p \times 1\cdot5}{2 \times 15 \times 10^{-3} \times 0\cdot7}$$

$$p = \frac{120 \times 10^6 \times 2 \times 15 \times 10^{-3} \times 0\cdot7}{1\cdot5} = 1\cdot68 \text{ MN/m}^2$$

and, $\sigma_l = \dfrac{pd}{4t\eta_c}$

$$120 \times 10^6 = \frac{p \times 1\cdot5}{4 \times 15 \times 10^{-3} \times 0\cdot3}$$

∴ $$p = \frac{120 \times 10^6 \times 4 \times 15 \times 10^{-3} \times 0\cdot3}{1\cdot5} = 1\cdot44 \text{ MN/m}^2$$

Hence, *the permissible intensity of pressure,*

$$p = 1\cdot44 \text{ MN/m}^2 \text{ (minimum of two values) (Ans.)}$$

Example 10·6. *A cylindrical air drum is 2·25 m in diameter with plates 1·2 cm thick. The efficiencies of the longitudinal and circumferential joints are respectively 75% and 40%. If the tensile stress in the plating is to be limited to 120 MN/m² find the maximum safe air pressure.*

Solution. Let the maximum safe air pressure = p MN/m²

Consider circumferential stress as 120 MN/m²:

$$\sigma_c = \frac{pd}{2t\eta_l}$$

$$120 \times 10^6 = \frac{p \times 10^6 \times 2\cdot25}{2 \times 1\cdot2 \times 10^{-2} \times 0\cdot75}$$

∴ $$p = \frac{120 \times 10^6 \times 2 \times 1\cdot2 \times 10^{-2} \times 0\cdot75}{10^6 \times 2\cdot25} = 0\cdot96 \text{ MN/m}^2$$

Considering longitudinal stress as 120 MN/m²:

$$\sigma_l = \frac{pd}{4t\eta_c}$$

$$120 \times 10^6 = \frac{p \times 10^6 \times 2\cdot25}{4 \times 1\cdot2 \times 10^{-2} \times 0\cdot40}$$

∴ $$p = \frac{120 \times 10^6 \times 4 \times 1\cdot2 \times 10^{-2} \times 0\cdot40}{10^6 \times 2\cdot25} = 1\cdot024 \text{ MN/m}^2$$

Hence, *the maximum safe air pressure* = **0·96 MN/m²** (*i.e. minimum of the above two values* **(Ans.)**)

Note: If we provide larger pressure (*i.e.* 1·024 MN/m²) then the circumferential stress

$$\left(\sigma_c = \frac{pd}{2t\eta_l} = \frac{1 \cdot 024 \times 10^6 \times 2 \cdot 25}{2 \times 1 \cdot 20 \times 10^{-2} \times 0 \cdot 75} = 128 \text{ MN/m}^2 \right)$$

will be more thant he permissible stress (120 MN/m²).

Oil tank.

Example 10·7. *A cylindrical vessel whose ends are closed by means of rigid flange plates is made of steel plate 3 mm thick. The internal length and diameter of vessel are 50 cm and 25 cm respectively. Determine the longitudinal and circumferential stresses in the cylindrical shell due to an internal fluid pressure of 3 MN/m². Also calculate increase in length, diameter and volume of the vessel.*

Take: $E = 200 \text{ GN/m}^2$, *and* $\dfrac{1}{m} = 0 \cdot 3$.

Solution. Length of cylindrical vessel,

$$l = 50 \text{ cm}$$

Internal diameter, $d = 25 \text{ cm}$

Thickness, $t = 0 \cdot 3 \text{ cm}$

Fluid pressure, $p = 3 \text{ MN/m}^2$

Young's modulus, $E = 200 \text{ GN/m}^2$

Poisson's ratio, $\dfrac{1}{m} = 0 \cdot 3$

Circumferential or hoop stress σ_c:

$$\sigma_c = \frac{pd}{2t} = \frac{3 \times 10^6 \times (25 \times 10^{-2})}{2 \times (0 \cdot 3 \times 10^{-2})}$$

$$= 125 \times 10^6 \text{ N/m}^2 \text{ or } 125 \text{ MN/m}^2$$

i.e. $\sigma_c = \mathbf{125 \text{ MN/m}^2}$ **(Ans.)**

Longitudinal stress, σ_l :

$$\sigma_l = \frac{pd}{4t} = \frac{3 \times 10^6 \times (25 \times 10^{-2})}{4 \times (0 \cdot 3 \times 10^{-2})}$$

$$= 62 \cdot 5 \times 10^6 \text{ N/m}^2 \text{ or } 62 \cdot 5 \text{ MN/m}^2$$

i.e. $\sigma_l = \mathbf{62 \cdot 5 \text{ MN/m}^2}$ **(Ans.)**

Change in dimensions:

Let,
$$\delta d = \text{Change in diameter,}$$
$$\delta l = \text{Change in length,}$$
$$e_c = \text{Circumferential strain, and}$$
$$e_l = \text{Longitudinal strain.}$$

Then,
$$e_c = \frac{\sigma_c}{E} - \frac{\sigma_l}{mE} = \frac{125 \times 10^6}{E} - \frac{62 \cdot 5 \times 10^6 \times 0 \cdot 3}{E}$$

$$= \frac{10^6}{200 \times 10^9}(125 - 62 \cdot 5 \times 0 \cdot 3) = 0 \cdot 0005312$$

Change in diameter,
$$\delta d = e_c \times d = 0 \cdot 0005312 \times 25$$
$$= \mathbf{0 \cdot 01328 \ cm \ (increase) \ (Ans.)}$$

Similarly,
$$e_l = \frac{\sigma_l}{E} - \frac{\sigma_c}{mE} = \frac{62 \cdot 5 \times 10^6}{E} - \frac{125 \times 10^6 \times 0 \cdot 3}{E}$$

$$= \frac{10^6}{200 \times 10^9}(62 \cdot 5 - 125 \times 0 \cdot 3) = 0 \cdot 000125$$

∴ Change in length,
$$\delta l = e_l \times l = 0 \cdot 000125 \times 50$$
$$= \mathbf{0 \cdot 00625 \ cm \ (increase) \ (Ans.)}$$

Volumetric strain,
$$e_v = e_l + 2e_c = 0 \cdot 000125 + 2 \times 0 \cdot 0005312 = 0 \cdot 0011874$$

∴ Change in volume,
$$\delta V = e_v \times V = 0 \cdot 0011874 \times \frac{\pi}{4} \times 25^2 \times 50$$
$$= \mathbf{29 \cdot 15 \ cm^3 \ (increase) \ (Ans.)}$$

Example 10·8. *A cylindrical shell 3 m long which is closed at the ends has an internal diameter of 1 m and a wall thickness of 15 mm. Calculate the circumferential and longitudinal stresses induced and also change in the dimensions of the shell if it is subjected to an internal pressure of 1·5 MN/m².*

Take: $E = 200 \ GN/m^2$, *and* $\dfrac{1}{m} = 0 \cdot 3$

Solution. Length of the shell, $\quad l = 3 \ \text{m}$

Internal diameter, $\qquad\qquad d = 1 \ \text{m}$

Wall thickness, $\qquad\qquad t = 15 \ \text{mm (or } 15 \times 10^{-3} \ \text{m)}$

Internal pressure, $\qquad\qquad p = 1 \cdot 5 \ \text{MN/m}^2$

Young's modulus, $\qquad\qquad E = 200 \ \text{GN/m}^2$

Poisson's ratio, $\qquad\qquad \dfrac{1}{m} = 0 \cdot 3$

Circumferential stress, σ_c:

Using the relation:

$$\sigma_c = \frac{pd}{2t} = \frac{1 \cdot 5 \times 10^6 \times 1}{2 \times 15 \times 10^{-3}} = 50 \times 10^6 \ \text{N/m}^2 \quad \text{or} \quad 50 \ \text{MN/m}^2$$

i.e. $\qquad\qquad \boldsymbol{\sigma_c = 50 \ \text{MN/m}^2} \ \textbf{(Ans.)}$

Longitudinal stress, σ_l :

Using the relation:

$$\sigma_l = \frac{pd}{4t} = \frac{1.5 \times 10^6 \times 1}{4 \times 15 \times 10^{-3}} = 25 \times 10^6 \text{ N/m}^2 \text{ or } 25 \text{ MN/m}^2$$

i.e. $\qquad \sigma_l = \textbf{25 MN/m}^2$ **(Ans.)**

Change in diameter, δd:

We know that, $\qquad e_c = \dfrac{\sigma_c}{E}\left(1 - \dfrac{1}{2m}\right) = \dfrac{50 \times 10^6}{200 \times 10^9}\left(1 - \dfrac{1}{2} \times 0.3\right) = 0.0002125$

But, $\qquad e_c = \dfrac{\delta d}{d}$

$\therefore \qquad \delta d = e_c \times d = 0.0002125 \times 1 \times 10^3 \text{ mm} = 0.2125 \text{ mm}$

i.e. $\qquad \delta d = \textbf{0.2125 mm (increase)} \textbf{ (Ans.)}$

Change in length, δl:

Again, $\qquad e_l = \dfrac{\sigma_l}{E}\left(1 - \dfrac{2}{m}\right) = \dfrac{25 \times 10^6}{200 \times 10^9}(1 - 2 \times 0.3) = 0.00005$

But, $\qquad e_l = \dfrac{\delta l}{l}$

$\therefore \qquad \delta l = e_l \times l = 0.00005 \times 3 \times 10^3 \text{ mm} = 0.15 \text{ mm}$

i.e. $\qquad \delta l = \textbf{0.15 mm (increase)} \textbf{ (Ans.)}$

Change in volume, δV:

$$e_v = e_l + 2e_c = 0.00005 + 2 \times 0.0002125 = 0.000475$$

But, $\qquad e_v = \dfrac{\delta V}{V}$

$\therefore \qquad \delta V = e_v \times V = 0.000475 \times \dfrac{\pi}{4}d^2 l$

$$= 0.000475 \times \frac{\pi}{4} \times 1^2 \times 3 \text{ m}^3 = 0.0011193 \text{ m}^3$$

or, $\qquad \delta V = \textbf{1119.3} \times \textbf{10}^3 \text{ mm}^3$ **(Ans.)**

Example 10·9. *A built up cylindrical shell of 300 mm diameter, 3 m long and 6 mm thick is subjected to an internal pressue of 2 MN/m². Calculate the change in length, diameter and volume of the cylinder under that pressure if the efficiencies of the longitudinal and circumferential joints are 80% and 50% respectively. E = 200 GN/m²; m = 3.5.*

Solution. Diameter of the shell $d = 300 \text{ mm} = 0.3 \text{ m}$

Length of the shell, $\qquad l = 3 \text{m}$

Thickness of the shell, $\qquad t = 6 \text{ mm} = 0.006 \text{ m}$

Internal pressure, $\qquad p = 2 \text{ MN/m}^2$

Poisson's ratio, $\qquad \dfrac{1}{m} = \dfrac{1}{3.5}$

Young's modulus, $\qquad E = 200 \text{ GN/m}^2$

Efficiency of longitudinal joint,

$$\eta_l = 80\%$$

Efficiency of circumferential joint,

$$\eta_c = 50\%$$

We know that net circumferential strain,

$$e_c = \frac{\sigma_c}{E} - \frac{\sigma_l}{mE} \qquad \qquad ...(i)$$

and, net longitudinal strain,

$$e_l = \frac{\sigma_l}{E} - \frac{\sigma_c}{mE} \qquad \qquad \qquad ...(ii)$$

Also,
$$\sigma_c = \frac{pd}{2t\eta_l} = \frac{2 \times 10^6 \times 0.3}{2 \times 0.006 \times 0.8}$$
$$= 62.5 \times 10^6 \text{ N/m}^2 \text{ or } 62.5 \text{ MN/m}^2$$

and,
$$\sigma_l = \frac{pd}{4t\eta_c} = \frac{2 \times 10^6 \times 0.3}{4 \times 0.006 \times 0.5}$$
$$= 50 \times 10^6 \text{ N/m}^2 \text{ or } 50 \text{ MN/m}^2$$

Now substituting these values in (i) and (ii), we get

$$e_c = \frac{62.5 \times 10^6}{E} - \frac{50 \times 10^6}{3.5 \times E}$$

$$= \frac{10^6}{200 \times 10^9}\left(62.5 - \frac{50}{3.5}\right) = 0.000241$$

and,
$$e_l = \frac{50 \times 10^6}{E} - \frac{62.5 \times 10^6}{3.5 \times E}$$

$$= \frac{10^6}{200 \times 10^9}\left(50 - \frac{62.5}{3.5}\right) = 0.000161$$

Change in diameter, δd:

Using the relation: $e_c = \dfrac{\delta d}{d}$ or $\delta d = e_c \times d$

∴ $\delta d = 0.000241 \times 0.3 = 0.0000723$ m

δd = 0.0723 mm (increase) (Ans.)

Change in length, δl:

Using the relation:

$$e_l = \frac{\delta l}{l} \quad \text{or} \quad \delta l = e \times l$$

∴ $\delta l = 0.000161 \times 3 = 0.000483$

or, **δl = 0.483 mm (increase) (Ans.)**

Change in volume δV:

$$e_v = \frac{\delta V}{V} \quad \text{or} \quad \delta V = e_v \times V$$

∴ $\delta V = (e_l + 2e_c) \times V$

$$= (0.000161 + 2 \times 0.000241) \times \frac{\pi}{4} d^2 l$$

$$= (0.000161 + 2 \times 0.000241) \times \frac{\pi}{4} \times (0.3)^2 \times 3 = 0.0001363 \text{ m}^3$$

= 136300 mm³ (increase) (Ans.)

Caution. Do not attempt to solve this problem by using the relations $e_c = \dfrac{\sigma_c}{E}\left(1 - \dfrac{1}{2m}\right)$ and

$e_l = \dfrac{\sigma_l}{E}\left(1 - \dfrac{2}{m}\right)$ because the shell is not seamless and has different values of η_l and η_c.

Pipeline work in progress.

Example 10·10. *A copper cylinder, 90 cm long, 40 cm external diameter and wall thickness of 6 mm has its both ends closed by rigid blank flanges. It is initially full of oil at atmospheric pressure. Calculate the additional volume of oil which must be pumped into it in order to raise the oil pressure 5 MN/m² above atmospheric pressure. For copper, assume E = 100 GN/m² and Poisson's ratio = $\frac{1}{3}$.*

Take bulk modulus of oil as 2·6 GN/m².

Solution. Initial volume, $V = \dfrac{\pi}{4} d^2 l = \dfrac{\pi}{4} \times (40 - 2 \times 0.6)^2 \times 90 = 106427 \text{ cm}^3$

$$E = 100 \text{ GN/m}^2, \quad \frac{1}{m} = \frac{1}{3}$$

Bulk modulus of oil, $\quad K_{oil} = 2.6 \text{ GN/m}^2$

We know, $\qquad \dfrac{\delta V_1}{V} = \dfrac{pd}{2tE}\left[\dfrac{5}{2} - \dfrac{2}{m}\right]$

$$= \frac{5 \times 10^6 (40 - 2 \times 0.6) \times 10^{-2}}{2 \times 6 \times 10^{-3} \times 100 \times 10^9}\left[\frac{5}{2} - 2 \times \frac{1}{3}\right] = 0.002964$$

∴ Increase in volume, $\delta V_1 = 0.002964 \times 106427 = 315.45 \text{ cm}^3$.

In addition to this, the liquid will shrink in volume as follows:

$$\frac{5 \times 10^6}{\dfrac{\delta V_2}{V}} = 2.6 \times 10^9 \qquad\qquad \left[\frac{\sigma_n}{e_v} = \frac{p}{e_v} = K\right]$$

or, $\qquad\qquad \delta V_2 = \dfrac{5 \times 10^6 \times 106427}{2.6 \times 10^9} = 204.67 \text{ cm}^3$

∴ Net addition of oil which must be pumped into the shell,

$$\delta V = \delta V_1 + \delta V_2 = 315.45 + 204.67$$

or, $\qquad\qquad \boldsymbol{\delta V = 520.12 \text{ cm}^3}$ **(Ans.)**

Example 10·11. *A cylindrical shell 90 cm long and 20 cm internal diameter having thickness of metal as 8 mm is filled with fluid at atmospheric pressure. If an additional 20 cm³ of fluid is pumped into the cylinder, find (i) the pressure exerted by the fluid on the cylinder and (ii) the hoop stress induced.*

Solution. Pressure exerted by the fluid on the cylinder, p:

Let p, e_c and e_l be the pressure exerted by the fluid on the cylinder, circumferential and longitudinal strains respectively. *Assume* $\dfrac{1}{m}$ (Poisson's value) = 0·3 and $E = 200$ GN/m².

Using the relation:

$$e_c = \frac{\sigma_c}{E}\left(1 - \frac{1}{2m}\right) = \frac{pd}{2tE}\left(1 - \frac{1}{2m}\right)$$

$$= \frac{p \times 20 \times 10^{-2}}{2 \times 8 \times 10^{-3} \times 200 \times 10^9}\left(1 - \frac{1}{2} \times 0.3\right)$$

$$= 0.05312\, p \times 10^{-9} \text{ where } p \text{ is in N/m}^2$$

and,

$$e_l = \frac{\sigma_l}{E}\left(1 - \frac{2}{m}\right) = \frac{pd}{4tE}\,(1 - 2 \times 0.3)$$

$$= \frac{p \times 20 \times 10^{-2}}{4 \times 8 \times 10^{-3} \times 200 \times 10^9}\,(1 - 2 \times 0.3) = 0.0125p \times 10^{-9}$$

Now, using the relation:

$$\delta V = V\,(e_l + 2e_c)$$

$$20 \times 10^{-6} = \frac{\pi}{4} \times (20 \times 10^{-2})^2 \times 90 \times 10^{-2}$$

$$(0.0125p \times 10^{-9} + 2 \times 0.05312p \times 10^{-9})$$

$$\left[\because\ e_v = \frac{\delta V}{V} = e_l + 2e_c \atop \therefore\ \delta V = V\,(e_l + 2e_c)\right]$$

$$20 \times 10^{-6} = 0.02828 \times 10^{-9}\,(0.0125p + 0.1062p)$$

or,

$$20 \times 10^{-6} = 0.02828 \times 10^{-9} \times 0.1187\,p$$

∴

$$p = \frac{20 \times 10^{-6}}{0.02828 \times 10^{-9} \times 0.1187} \times 10^{-6} \text{ MN/m}^2$$

or,

$$p = \mathbf{5{\cdot}958 \text{ MN/m}^2} \quad \textbf{(Ans.)}$$

(ii) Hoop stress induced, σ_c:

Let, σ_c = Hoop (or circumferential) stress induced.

Then,

$$\sigma_c = \frac{pd}{2t} = \frac{5.958 \times 10^6 \times 20 \times 10^{-2}}{2 \times 8 \times 10^{-3}} \times 10^{-6} \text{ MN/m}^2$$

$$= \mathbf{74{\cdot}475 \text{ MN/m}^2}\ \textbf{(Ans.)}$$

Example 10·12. *A boiler drum consists of a cylindrical portion 4 m long, 1·5 m in diameter and 2·25 cm thick. It is closed by hemispherical ends. In a hydraulic test to 6 MN/m², how much additional water will be pumped in after initial filling at atmospheric pressure? The circumferential strain at the junction of the cylinder and hemisphere may be assumed as same for both.*

$$E = 200\ GN/m^2, K\ (for\ water) = 2{\cdot}13GN/m^2,\ and\ \frac{1}{m} = 0{\cdot}3$$

Solution. For cylindrical portion:

Circumferential/hoop stress,

$$\sigma_c = \frac{pd}{2t} = \frac{6 \times 10^6 \times 1.5}{2 \times 2.25 \times 10^{-2}} = 200 \times 10^6 \text{ N/m}^2 \text{ or } 200 \text{ MN/m}^2$$

Longitudinal stress,

$$\sigma_l = \frac{pd}{4t} = \frac{6 \times 10^6 \times 1.5}{2 \times 2.25 \times 10^{-2}} = 100 \times 10^6 \text{ N/m}^2 \text{ or } 100 \text{ MN/m}^2$$

Circumferential (or hoop strain),

$$e_c = \frac{\sigma_c}{E}\left(1 - \frac{1}{2m}\right) = \frac{200 \times 10^6}{200 \times 10^9}\left(1 - \frac{1}{2} \times 0.3\right) = 0.00085$$

and, longitudinal strain, $\quad e_l = \frac{\sigma_l}{E}\left(1 - \frac{2}{m}\right) = \frac{100 \times 10^6}{200 \times 10^9}(1 - 2 \times 0.3) = 0.0002$

Volumetric strain $\quad\quad e_v = e_l + 2e_c = 0.0002 + 2 \times 0.00085 = 0.0019$

But, $\quad\quad\quad\quad\quad e_v = \dfrac{\delta V}{V}$

$\therefore$ Increase in volume, $\quad \delta V = e_v \times V = 0.0019 \times \dfrac{\pi}{4} \times d^2 \times l = 0.0019 \times \dfrac{\pi}{4} \times 1.5^2 \times 4$

$$= 0.013432 \text{ m}^3 \quad\quad\quad\quad ...(i)$$

For hemispherical ends:

$$e = e_c \text{ (Given)} \quad\quad\quad\quad \therefore \; e_v = 3e_c$$

and, change in volume (increase)

$$= e_v \times V = 3e_c \times V = 3 \times 0.00085 \times \frac{4}{3}\pi r^3$$

$$= 3 \times 0.00085 \times \frac{4}{3}\pi\left(\frac{1.5}{2}\right)^3$$

$$= 0.0045068 \text{ m}^3 \quad\quad\quad\quad ...(ii)$$

Decrease in the volume of water due to water pressure can be found from the relation,

$$K = \frac{\sigma_n}{e_v}$$

$$\left[\begin{array}{l} \text{where, } K = \text{Bulk modulus,} \\ \sigma_n = \text{Normal stress } (= p), \text{ and} \\ e_v = \text{Volumetric strain,} \end{array}\right]$$

or, $\quad\quad\quad\quad e_v = \dfrac{\sigma_n}{K} = \dfrac{\delta V}{V} \quad \text{or} \quad \delta V = \dfrac{\sigma_n}{K} \times V$

Pipe network in a factory.

$$= \frac{6 \times 10^6}{2 \cdot 13 \times 10^9} \left[\frac{\pi}{4} \times 1 \cdot 5^2 \times 4 + \frac{4}{3} \pi \times (0 \cdot 75)^3 \right]$$

$$= \frac{6 \times 10^6}{2 \cdot 13 \times 10^9} (7 \cdot 0695 + 1 \cdot 7673) = 0 \cdot 02489 \text{ m}^3 \qquad \ldots(iii)$$

∴ **Total additional volume of water**

$$= (i) + (ii) + (iii) = 0 \cdot 013432 + 0 \cdot 0045068 + 0 \cdot 02489$$

$$= 0 \cdot 0428288 \text{ m}^3 = \textbf{42828·8 cm}^3 \quad \textbf{(Ans.)}$$

10·2·8. Wire Wound Cylinders

When thin cylindrical shell is subjected to internal fluid pressure tensile circumferential stress is developed, which is twice the longitudinal stress. Thus, the chances of bursting the cylinder longitudinally are more than those for circumferential failure of the cylinder. Therefore, it is necessary to strengthen the cylinder longitudinally for the following purposes:

(i) To increase the pressure-carrying capacity of the cylinder;

(ii) To reduce the chances of longitudinal burst.

So that the above objectives are achieved, the cylinder is *wound with layers of wire kept under tension*. A wire tightly wound around the cylinder being itself in tension gives rise to compressive stresses in the cylinder which to a great extent neutralises the tensile stresses in the cylindrical shell. Fluid pressure inside the shell does increase the initial tensile stresses in the wire around the cylinder.

The resultant circumferential stress in the cylinder is the *sum of the initial compressive stress due to wire winding and further tensile stress due to internal pressure*. The resultant circumferential stress in the *wire* is the *sum of two tensile stresses developed due (i) wire winding under tension and (ii) internal pressure in the cylinder*. Thus the pressure-carrying capacity of the cylindrical shell is increased.

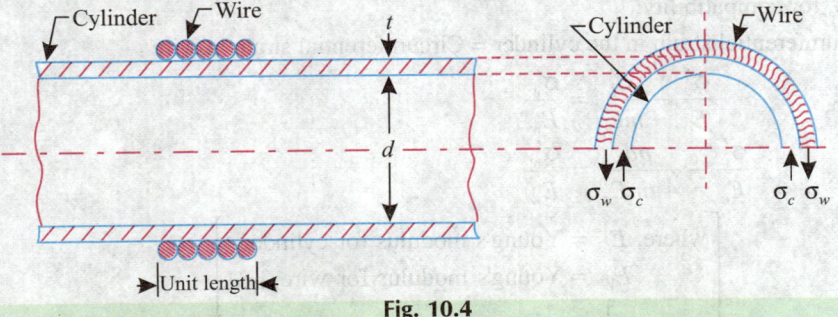

Fig. 10.4

Fig. 10·4 shows a cylinder around which wire is closely wound under a tensile stress.

Let, d = Diameter of the cylinder,

          t = Wall thickness of the cylinder,

      d_w = Diameter of the wire,

        n = Number of turns of wire per unit length $\left(= \frac{1}{d_w} \right)$,

      σ_w = Initial tension with which wire is wound,

      σ_c = Compressive circumferential stress developed in the cylinder,

      σ_c' = Circumferential stress developed in the cylinder ⎤ When the cylinder is

      σ_w' = Stress developed in the wire, and   ⎥ subjected to internal

       σ_l' = Longitudinal stress developed in the cylinder. ⎦ pressure p.

Before admitting fluid into the cylinder :

Tensile force exerted by wire per unit length

$$= 2 \times \frac{\pi}{4} d_w^2 \times \sigma_w \times n$$

Compressive force developed in the cylinder

$$= 2 \times t \times (1) \times \sigma_c = 2\sigma_c . t$$

For equilibrium,

$$2\sigma_c . t = 2 \times \frac{\pi}{4} d_w^2 \times \sigma_w \times n$$

But,

$$n = \frac{1}{d_w}$$

∴

$$2\sigma_c . t = 2 \times \frac{\pi}{4} d_w^2 \times \sigma_w \times \frac{1}{d_w}$$

∴

$$\sigma_c = \frac{\pi d_w}{4t} . \sigma_w \qquad \qquad \dots (10.14)$$

After admitting fluid into the cylinder:

When the wire wound cylinder is subjected to internal pressure, we have:

Longitudinal bursting force $= p \times \dfrac{\pi}{4} d^2 = \sigma'_l \times \pi dt$ (for equilibrium)

or,

$$\sigma'_l = \frac{pd}{4t} \qquad \qquad \dots (10.15)$$

Diametral bursting force per unit length $= p.d.l$

$$= \sigma'_c \times 2t \times 1 + \sigma'_w \times 2 \times \frac{\pi}{4} d_w^2 \times \frac{1}{d_w} \qquad \text{(for equilibrium)}$$

or,

$$pd = \sigma'_c \times 2t + \sigma'_w \times \frac{\pi d_w}{2} \qquad \qquad \dots (10.16)$$

Also, for compatibility:

Circumferential strain in the cylinder = Circumferential strain in wire

Then,

$$\frac{\sigma'_c}{E_c} - \frac{\sigma'_l}{mE_c} = \frac{\sigma'_w}{E_w}$$

$$\frac{\sigma'_c}{E_c} - \frac{pd}{4 \, tmE_c} = \frac{\sigma'_w}{E_w} \qquad \qquad \dots (10.17)$$

$$\left[\begin{array}{l} \text{where, } E_C = \text{Young's modulus for cylinder,} \\ \qquad E_W = \text{Young's modulus for wire, and} \\ \qquad \dfrac{1}{m} = \text{Poisson's ratio for cylinder.} \end{array} \right]$$

The stresses σ'_c and σ'_w can be calculated from equations (10.16) and (10.17).

Resultant stress in the wire $= \sigma_w + \sigma_w'$ [(10.18 (a))]

Resultant circumferential stress in the cylinder

$$= \sigma'_c - \sigma_c \qquad \qquad [(10.18 \, (b))]$$

Example 10.13. *A cast iron cylinder of 200 mm inner diameter and 12.5 mm thick is closely wound with a layer of 4 mm diameter steel wire under a tensile stress of 55 MN/m². Determine the stresses set up in the cylinder and steel wire if water under a pressure of 3 MN/m² is admitted in the cylinder. Take: $E_{c.i} = 100$ GN/m², $E_s = 200$ GN/m², and Poisson's ratio = 0.25.*

Solution. Inner diameter of the cast iron cylinder,

$$d = 200 \text{ mm} = 0.2 \text{ m}$$

Wall thickness of the cast iron cylinder,
$$t = 12 \cdot 5 \text{ mm} = 0 \cdot 0125 \text{ m}$$

Diameter of the steel wire, $\quad d_w = 4 \text{ mm} = 0 \cdot 004 \text{ m}$

Initial tension with which wire is wound,
$$\sigma_w = 55 \text{ MN/m}^2$$

Pressure with which water is admitted in the cylinder,
$$p = 3 \text{ MN/m}^2$$

Young's modulus for cast iron cylinder,
$$E_{c.i} (= E_c) = 100 \text{ GN/m}^2$$

Young's modulus for wire, $\quad E_s (= E_w) = 200 \text{ GN / m}^2$

Poisson's ratio, $\quad \dfrac{1}{m} = 0 \cdot 25$

Stresses set up in the cylinder and steel wire:

Before admitting water into the cylinder:

Tensile force exerted by wire per unit length = Compressive force developed in the cylinder

$$2 \times \frac{\pi}{4} d_w^2 \times \sigma_w \times n = 2 \times t \times 1 \times \sigma_c$$

∴ Initial circumferential compression in cylinder,

$$\sigma_c = \frac{\pi d_w}{4t} \times \sigma_w = \frac{\pi \times 0 \cdot 004 \times 55}{4 \times 0 \cdot 0125} = 13 \cdot 82 \text{ MN/m}^2$$

After admitting water into the cylinder:

Due to internal pressure, the longitudinal stress developed in the cylinder,

$$\sigma_l' = \frac{pd}{4t} = \frac{3 \times 0 \cdot 2}{4 \times 0 \cdot 0125} = 12 \text{ MN/m}^2$$

Now, total bursting force = Total resisting force..............per unit length

$$p \times d \times 1 = \sigma_c' \times 2t \times 1 + \sigma_w' \times 2 \times \frac{\pi}{4} d_w^2 \times n$$

or $\qquad p.d = \sigma_c'.2t + \sigma_w' \times \dfrac{\pi d_w}{2} \qquad \left(\because n = \dfrac{1}{d_w} \right)$

or $\qquad 3 \times 0 \cdot 2 = \sigma_c' \times 2 \times 0 \cdot 0125 + \sigma_w' \times \dfrac{\pi \times 0 \cdot 004}{2} \qquad \text{...}(i)$

and, Circumferential strain in cylinder = Circumferential strain in wire

$$\frac{\sigma_c'}{E_c} - \frac{\sigma_l'}{mE_c} = \frac{\sigma_w'}{E_w}$$

where, σ_c' = Circumferential stress in cylinder, and

σ_w' = Circumferential stress in wire.

$$\therefore \quad \frac{\sigma_c' \times 10^6}{100 \times 10^9} - \frac{12 \times 10^6}{100 \times 10^9} \times 0 \cdot 25 = \frac{\sigma_w' \times 10^6}{200 \times 10^9}$$

$$2 \sigma_c' - 2 \times 12 \times 0 \cdot 25 = \sigma_w'$$

or $\qquad 2 \sigma_c' - 6 = \sigma_w' \qquad \text{...}(ii)$

Substituting the value of σ_w' in eqn. (i), we get

$$3 \times 0 \cdot 2 = \sigma_c' \times 2 \times 0 \cdot 0125 + (2\sigma_c' - 6) \times \frac{\pi \times 0 \cdot 004}{2}$$

or, $0.6 = 0.025\ \sigma'_c + 0.0125\ \sigma'_c - 0.0377$

$\sigma'_c = 17\ MN/m^2$

∴ $\sigma'_w = 2 \times 17 - 6 = 28\ MN/m^2$

Resultant stress in the cylinder

$$= \sigma'_c - \sigma_c$$

$$= 17 - 13.82$$

$$= \textbf{3·18 MN/m}^2 \textbf{ (tensile)}\ \textbf{(Ans.)}$$

Resultant stress in the wire $= \sigma_w + \sigma'_w$

$$= 55 + 28 = \textbf{83 MN/m}^2 \textbf{ (tensile)}\ \textbf{(Ans.)}$$

Example 10·14. *A gun metal tube of 100 mm bore, wall thickness 2·5 mm is closely wound externally by a steel wire 1 mm diameter. Determine the tension under which the wire must be wound on the tube, if an internal radial pressure of 3 MN/m² is required before the tube is subjected to the tensile stress in the circumferential direction.*

Take: For gun metal: E = 102 GN/m²; $\dfrac{1}{m} = 0.35$

For steel: E = 210 GN/m².

Water tank.

Solution. Inside diameter of the tube,

$$d = 100\ mm = 0.1\ m$$

Wall thickness of the tube, $t = 2.5\ mm = 0.0025\ m$

Diameter of steel wire, $d_w = 1mm = 0.001\ m$

Internal radial pressure, $p = 3\ MN/m^2$

Young's modulus for tube, $E_{tube} = 102\ GN/m^2$

Young's modulus for wire, $E_{wire} = 210\ GN/m^2$

Poisson's ratio, $\dfrac{1}{m} = 0.35$

Tension under which wire must be wound, σ_w:

Before subjecting the tube to internal pressure:

Tensile force exerted by wire per unit length = Compressive force developed in the cylinder

$$2 \times \frac{\pi}{4}\ d_w^2 \times \sigma_w \times n = 2 \times t \times 1 \times \sigma_c$$

∴ $\sigma_c = \dfrac{\pi d_w}{4t} \times \sigma_w = \dfrac{\pi \times 0.001}{4 \times 0.0025}\ \sigma_w = 0.3142\ \sigma_w$ ...(i)

(where, σ_c = Initial circumferential compression in the cylinder)

After subjecting the tube to internal pressure:

Due to internal pressure, the longitudinal stress developed in the cylinder,

$$\sigma'_l = \frac{pd}{4t} = \frac{3 \times 0.1}{4 \times 0.0025} = 30\ MN/m^2$$

Total bursting force = Total resisting force..........per unit length (for equilibrium)

$$p \times d \times 1 = \sigma'_c \times 2t \times 1 + \sigma'_w \times 2 \times \frac{\pi}{4}\ d_w^2 \times n$$

or, $p.d = \sigma'_c \times 2t + \sigma'_w \times \dfrac{\pi d_w}{2}$

$$3 \times 0.1 = \sigma'_c \times 2 \times 0.0025 + \sigma'_w \times \frac{\pi \times 0.001}{2}$$

or, $\qquad 0.3 = 0.005\,\sigma'_c + 0.00157\,\sigma'_w \qquad \qquad ...(i)$

Also, Circumferential strain in cylinder = Circumferential strain in wire

$$\frac{\sigma'_c}{E_{tube}} - \frac{\sigma'_l}{mE_{tube}} = \frac{\sigma'_w}{E_{wire}}$$

$$\frac{\sigma'_c \times 10^6}{102 \times 10^9} - \frac{30 \times 10^6}{102 \times 10^9} \times 0.35 = \frac{\sigma'_w \times 10^6}{210 \times 10^9}$$

$$2.059\,\sigma'_c - 21.62 = \sigma'_w \qquad \qquad ...(ii)$$

Substituting the value of the σ'_w in eqn. (i), we get

$$0.3 = 0.005\,\sigma'_c + 0.00157\,(2.059\,\sigma'_c - 21.62)$$
$$0.3 = 0.005\,\sigma'_c + 0.00323\,\sigma'_c - 0.0339$$

∴ $\qquad \qquad \sigma'_c = 40.57 \text{ MN}/\text{m}^2$

and, $\qquad \sigma'_w = 2.059 \times 40.57 - 21.62 = 61.9 \text{ MN}/\text{m}^2$

But, $\qquad \qquad \sigma'_c - \sigma_c = 0$

∴ $\qquad \qquad \sigma'_c = \sigma_c = 40.57 \text{ MN}/\text{m}^2$ (compressive)

Also, $\qquad \qquad \sigma_c = 0.3142\,\sigma_w$ (from eqn. (i))

∴ $\qquad \qquad 40.57 = 0.3142\,\sigma_w$

$$\boldsymbol{\sigma_w = 129.12 \text{ MN/m}^2 \quad \text{(Ans.)}}$$

10.3. SPHERICAL SHELLS

Refer to Fig. 10.5.

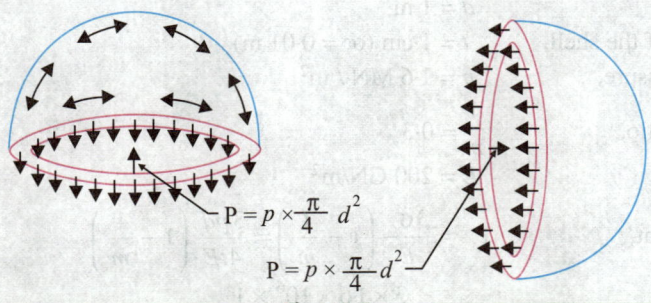

$$P = p \times \frac{\pi}{4}\,d^2$$
$$P = p \times \frac{\pi}{4}\,d^2$$

Fig. 10.5

Let, $\qquad \qquad d$ = Diameter of the spherical shell,

$\qquad \qquad t$ = Thickness of the shell,

$\qquad \qquad p$ = Intensity of pressure, and

$\qquad \qquad \sigma_c$ = Circumferential (or hoop) stress produced.

Now, Bursting force = Resisting force

$$p \times \frac{\pi}{4}\,d^2 = \pi d t \cdot \sigma_c$$

or, $\qquad \qquad \sigma_c = \frac{pd}{4t} \qquad \qquad ...(10.19)$

The above equation for stress is true only when the shell is *seamless*, but in case it is *built-up*, then

$$\sigma_c = \frac{pd}{4t\eta} \qquad \qquad ...(10.20)$$

where, η is the joint efficiency.

Change in dimensions:

Strain in any diametrical plane axis

$$e = \frac{\delta d}{d} = \frac{\sigma_c}{E} - \frac{\sigma_c}{mE}$$

$$= \frac{\sigma_c}{E}\left(1 - \frac{1}{m}\right) = \frac{pd}{4tE}\left(1 - \frac{1}{m}\right) \qquad \qquad ...(10\cdot21)$$

e_v = Algebraic sum of strains in all the three axis.

$$= e + e + e = 3e$$

Also, $$e_v = \frac{\delta V}{V}$$

or, $$\delta V = e_v \times V = 3e \times V$$

$$= \frac{3\,pd}{4tE}\left(1 - \frac{1}{m}\right) \times V$$

where, $$V = \frac{4}{3}\pi r^3 \quad \text{or} \quad \frac{\pi d^3}{6}$$

Change in diameter may be found from equation (10·21) as follows:

$$\delta d = \frac{pd}{4tE}\left(1 - \frac{1}{m}\right) \times d = \frac{pd^2}{4tE}\left(1 - \frac{1}{m}\right) \qquad \qquad ...(10\cdot23)$$

Example 10·15. *Calculate the increase in volume of a spherical shell 1 m in diameter and 1 cm thick when it is subjected to an internal pressure of 1·6 MN/m². Take E = 200 GN/m², and $\frac{1}{m} = 0.3$.*

Solution. Diameter of the shell,

$$d = 1 \text{ m}$$

Thickness of the shell, $$t = 1 \text{ cm (or} = 0\cdot01 \text{ m)}$$

Internal pressure, $$p = 1\cdot6 \text{ MN / m}^2$$

Poisson's ratio, $$\frac{1}{m} = 0\cdot3$$

$$E = 200 \text{ GN/m}^2$$

We know that, $$e_v = \frac{3\sigma_c}{E}\left(1 - \frac{1}{m}\right) = \frac{3pd}{4tE}\left(1 - \frac{1}{m}\right)$$

$$= \frac{3 \times 1\cdot6 \times 10^6 \times 1}{4 \times 0\cdot01 \times 200 \times 10^9}(1 - 0\cdot3) = 0\cdot00042$$

Change in volume, $$\delta V = e_v \times V$$

$$= 0\cdot00042 \times \frac{4}{3}\pi r^3 = 0\cdot00042 \times \frac{4}{3}\pi \times (0\cdot5)^3$$

$$= 0\cdot0002199 \text{ m}^3 \text{ or } 219\cdot9 \text{ cm}^3$$

i.e. $$\delta V = 219\cdot9 \text{ cm}^3 \text{ (Ans.)}$$

Example 10·16. *A thin spherical shell 1 m in diameter with its wall of 1·2 cm thickness is filled with a fluid at atmospheric pressure. What intensity of pressure will be developed in it if 175 cm³ more of fluid is pumped into it? Also, calculate the circumferential stress at that pressure and the increase in diameter.*

Take: E = 200 GN/m², and $\frac{1}{m} = 0\cdot3$

Solution. Diameter of the spherical shell,

$$d = 1 \text{ m}$$

Thickness of the wall,

$$t = 1.2 \text{ cm (or} = 0.012 \text{ m)}$$

Volume of fluid pumped,

$$\delta V = 175 \text{ cm}^3 = 175 \times 10^{-6} \text{ m}^3$$

Poisson's ratio, $\quad \dfrac{1}{m} = 0.3$

Young's modulus, $E = 200 \text{ GN/m}^2$

Increase in diameter δd:

Volumetric strain,

$$e_v = \frac{\delta V}{V} = \frac{175 \times 10^{-6}}{\dfrac{4}{3}\,\pi \times (0.5)^3} = 0.000334$$

But, $\qquad e_v = 3e$

or, $\qquad e = \dfrac{e_v}{3} = \dfrac{0.000334}{3} = 0.000111$

Also, $\qquad e = \dfrac{\delta d}{d}$

∴ $\qquad \delta d = e \times d = 0.000111 \times 1 = 0.000111 \text{ m or } 0.111 \text{ mm}$

Water tank.

Hence, increase in diameter = **0·111 mm (Ans.)**

Circumferential stress, σ_c:

Using the relation:

$$e = \frac{\sigma_c}{E}\left(1 - \frac{1}{m}\right), \text{ we get}$$

$$0.000111 = \frac{\sigma_c}{200 \times 10^9}\,(1 - 0.3)$$

∴ $\qquad \sigma_c = \dfrac{0.000111 \times 200 \times 10^9}{0.7} = 31.714 \times 10^6 \text{ N/m}^2 \text{ or}$

$$= 31.714 \text{ MN/m}^2$$

Pressure p:

Using the relation:

$$\sigma_c = \frac{pd}{4t}, \text{ we get}$$

$$p = \frac{\sigma_c \times 4t}{d} = \frac{31.714 \times 10^6 \times 4 \times 0.012}{1} = 1.522 \times 10^6 \text{ N/m}^2$$

or, $\qquad p = \textbf{1·522 MN/m}^2 \textbf{ (Ans.)}$

TYPICAL EXAMPLES (For Competitive Examinations)

Example 10·17. *A steel tube having a bore of 150 mm, wall thickness 2·25 mm is plugged at each end to form a closed cylinder with external length of 450 mm. If the tube is filled with oil and is subjected to a compressive load of 90 kN determine:*

(i) The pressure produced on oil;

(ii) The resulting circumferential stress in the tube wall.

Given: For oil: K (bulk modulus) = 2800 MN/m²;

   For steel: *E = 210 GN/m², and Poisson's ratio = 0·28.*

Solution. Inside diameter of the steel tube,

$$d = 150 \text{ mm} = 0.15 \text{ m}$$

Wall thickness of the steel tube,

$$t = 2.25 \text{ mm} = 0.00225 \text{ m}$$

Compressive load, $\quad P = 90 \text{ kN}$

Bulk modulus for oil, $\quad K = 2800 \text{ MN/m}^2$

Young's modulus for steel, $E = 210 \text{ GN/m}^2$

Poisson's ratio, $\quad \dfrac{1}{m} = 0.28$

(i) The pressure produced on oil, p:

Circumferential stress developed in cylinder,

$$\sigma_c = \frac{pd}{2t} = \frac{p \times 0.15}{2 \times 0.00225} = 33.33\, p$$

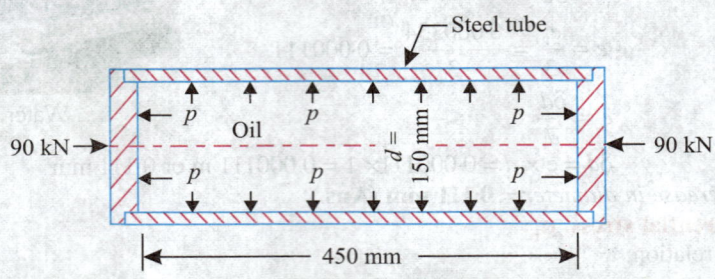

Fig. 10.6

Let, $\qquad \sigma_l = $ Longitudinal stress developed (MN/m^2).

For equilibrium we have,

$$p \times \frac{\pi}{4} d^2 + \sigma_l \times \pi dt = 90 \times 10^{-3}$$

$$p \times \frac{\pi}{4} \times 0.15^2 + \sigma_l \times \pi \times 0.15 \times 0.00225 = 90 \times 10^{-3}$$

$$0.01767\, p + 0.00106\, \sigma_l = 0.09$$

$\therefore \qquad\qquad \sigma_l = \dfrac{0.09 - 0.01767\, p}{0.00106} = 84.9 - 16.67\, p \text{ (compressive)}$

Longitudinal strain, $\quad e_l = \dfrac{\sigma_l}{E} - \dfrac{\sigma_c}{mE}$

Circumferential strain, $\quad e_c = \dfrac{\sigma_c}{E} - \dfrac{\sigma_l}{mE}$

Volumetric strain, $\qquad e_v = 2e_c + e_l$

$$= \left(\frac{2\sigma_c}{E} - \frac{2\sigma_l}{mE} \right) + \left(\frac{\sigma_l}{E} - \frac{\sigma_c}{mE} \right)$$

$$= \frac{1}{E} [2\sigma_c - 2 \times 0.28\sigma_l + \sigma_l - 0.28\sigma_c]$$

i.e. $\qquad\qquad e_v = \dfrac{1}{E}(1.72\, \sigma_c + 0.44\, \sigma_l)$ $\qquad\qquad$ ...(*i*)

Now bulk modulus for oil,

$$K = \frac{\sigma_n}{e_v} = \frac{p}{e_v} \quad \text{or} \quad e_v = \frac{p}{K}$$

Since e_v is negative in this case,

$$\therefore \qquad e_v = -\frac{p}{K} \qquad\qquad\qquad ...(ii)$$

From eqns. (*i*) and (*ii*), we get

$$-\frac{p}{K} = \frac{1}{E}(1.72\sigma_c + 0.44\,\sigma_l)$$

or, $\qquad -\dfrac{p\times10^6}{2800\times10^6} = \dfrac{1}{210\times10^9}(1.72\,\sigma_c + 0.44\,\sigma_l)\times10^6$

or, $\qquad -p\times\dfrac{210\times10^3}{2800} = 1.72\times33.33p - 0.44(84.9 - 16.67p)$ $\quad(\because \sigma_l$ is negative$)$

or, $\qquad\qquad -75p = 57.32p - 37.36 + 7.33p$

or, $\qquad\qquad 139.65p = 37.36$

$\therefore \qquad\qquad\qquad$ **$p = 0.267$ MN / m^2 (Ans.)**

(*ii*) Circumferential stress, σ_c:

The resulting circumferential stress,

$$\sigma_c = 33.33\,p$$
$$= 33.33\times0.267 = \textbf{8.9 MN/m}^2 \quad\textbf{(Ans.)}$$

Example 10·18. *The ends of a thin cylinder, 180 mm internal diameter and wall thickness 3·0 mm are closed by rigid plates and it is then filled with a liquid. When an axial compressive force of 33·6 kN is applied to the cylinder, the pressure of the liquid rises by 72 kN/m^2. If E = 200 GN/m^2 and Poisson's ratio = 0·3, find the bulk modulus of the liquid.*

Solution. Internal diameter of the cylinder,

$$d = 180 \text{ mm} = 0.18 \text{ m}$$

Wall thickness of the cylinder, $\quad t = 3.0 \text{ mm} = 0.003 \text{ m}$

Axial compressive force, $\qquad p = 33.6 \text{ kN}$

Pressure rise of the liquid, $\qquad p = 72 \text{ kN/m}^2$

Young's modulus, $\qquad\qquad E = 200 \text{ GN/m}^2$

Poisson's ratio, $\qquad\qquad \dfrac{1}{m} = 0.3$

Bulk modulus of the liquid, K (GN/m^2):

As no liquid is *pumped in or is allowed to escape*, therefore, volumetric strain of the liquid = compressive volumetric strain of the shell

Thus if P be the axial force on the shell which raises the pressure of liquid to p, then

Longitudinal stress, $\qquad\qquad \sigma_l = \dfrac{p}{\pi dt} - \dfrac{pd}{4t}$

$$= \frac{33.6\times10^{-3}}{\pi\times0.18\times0.003} - \frac{72\times10^{-3}\times0.18}{4\times0.003} \text{ (MN/m}^2)$$

$$= 19.80 - 1.08 = 18.72 \text{ MN/m}^2 \text{ (compressive)}$$

Circumferential stress, $\qquad\quad \sigma_c = \dfrac{pd}{2t}$

$$= \frac{72\times10^{-3}\times0.18}{2\times0.003} = 2.16 \text{ MN/m}^2 \text{ (tensile)}$$

Longitudinal strain, $\quad e_l = \dfrac{\sigma_l}{E} - \dfrac{\sigma_c}{mE}$

$$= -\frac{18 \cdot 72 \times 10^6}{200 \times 10^9} - \frac{2 \cdot 16 \times 10^6}{200 \times 10^9} \times 0 \cdot 3 \qquad (\because \sigma_l \text{ is negative})$$

$$= -\frac{10^6}{200 \times 10^9} (18 \cdot 72 + 2 \cdot 16 \times 0 \cdot 3)$$

$$= 96 \cdot 8 \times 10^{-6} \text{ (compressive)}$$

Circumferential strain, $\quad e_l = \dfrac{\sigma_c}{E} - \dfrac{\sigma_l}{mE}$

$$= \frac{2 \cdot 16 \times 10^6}{200 \times 10^9} + \frac{18 \cdot 72 \times 10^6}{200 \times 10^9} \times 0 \cdot 3 \quad (\because \sigma_l \text{ is negative})$$

$$= \frac{10^6}{200 \times 10^9} (2 \cdot 16 + 18 \cdot 72 \times 0 \cdot 3) = 38 \cdot 8 \times 10^{-6} \text{ (tensile)}$$

$\therefore$ Volumetric strain of the shell,

$$e_v = 2e_c + e_l = 2 \times 38 \cdot 8 \times 10^{-6} - 96 \cdot 8 \times 10^{-6}$$
$$= 19 \cdot 2 \times 10^{-6} \text{ (compressive)} \qquad \qquad ...(i)$$

But the compressive volumetric strain of the liquid

$$e_v = \frac{p}{K} = \frac{72 \times 10^{-3}}{K} \qquad \qquad ...(ii)$$

(where K is in GN/m^2)

Thus equating (i) and (ii), we get

$$19 \cdot 2 \times 10^{-6} = \frac{72 \times 10^3}{K \times 10^9}$$

$\therefore \qquad\qquad K = \dfrac{72 \times 10^3}{19 \cdot 2 \times 10^{-6} \times 10^9} = 3 \cdot 75 \text{ GN/m}^2$

i.e. $\qquad\qquad$ **K = 3·75 GN/m^2 (Ans.)**

Example 10·19. *A solid cylindrical test piece, 75 mm long and 50 mm diameter, is enclosed within a hollow pressure vessel. With the test piece in the vessel 20×10^3 mm^3 of oil is required just to fill the pressure vessel. Measurement shows that a further 50 mm^3 of oil has to be pumped into the vessel to raise the oil pressure to 7 MN/m^2.*

The experiment is repeated using the same pressure vessel and oil, but without the test piece inside the vessel. This time, after initially filling the pressure vessel, a further 364 mm^3 of oil are needed to raise the pressure to 7 MN/m^2.

The test piece is made of aluminium, for which Young's modulus and Poisson's ratio are 70 GN/m^2 and 0·3 respectively.

Find the bulk modulus of oil. **(Panjab University)**

Solution. Diameter of the cylindrical test piece,
$$d = 50 \text{ mm} = 0 \cdot 05 \text{ m}$$

Length of the test piece, $\quad = 75 \text{ mm} = 0 \cdot 075 \text{ m}$

Plastic containers are gaining popularity because of their rust free functioning, low cost and portability.

Oil pressure, $\qquad p = 7 \text{ MN/m}^2$

Young's modulus, $\qquad E = 70 \text{ GN/m}^2$

Poisson's ratio, $\qquad \dfrac{1}{m} = 0.3$

Bulk modulus of oil, K_{oil}:

Vessel with aluminium test piece:

$$\text{Volume of vessel} = \text{Volume of aluminium test piece} + \text{additional volume of oil}$$
$$\text{to fill the vessel}$$
$$= \frac{\pi}{4} \times 0.05^2 \times 0.075 + 20 \times 10^3 \times 10^{-9}$$
$$= 0.0001472 + 0.0\,0002 = 167.2 \times 10^{-6} \text{ m}^3$$

Let, $\qquad \delta V_1 = $ *Increase* in volume of vessel (due to pressure in the vessel)

Decrease in volume (due to compression) of the aluminium piece,

$$\delta V_2 = \frac{p}{K_{aluminium}} \times \text{volume of aluminium piece.}$$

But, $\qquad K_{aluminium} = \dfrac{E}{3\left(1 - \dfrac{2}{m}\right)} = \dfrac{70}{3\,(1 - 2 \times 0.3)} = 58.33 \text{ GN/m}^2$

$\therefore \qquad \delta V_2 = \dfrac{7 \times 10^6}{58.33 \times 10^9} \times \left(\dfrac{\pi}{4} \times 0.05^2 \times 0.075\right)$

$$= 0.0176 \times 10^{-6} \text{ m}^3$$

Decrease in the volume of oil (due to compression)

$$\delta V_3 = \frac{p}{K_{oil}} \times \text{volume of oil}$$
$$= \frac{7 \times 10^6}{K_{oil}} \times 20 \times 10^3 \times 10^{-9} = \frac{140}{K_{oil}}$$

Also, $\qquad \delta V_1 + \delta V_2 + \delta V_3 = 50 \times 10^{-9} \qquad\qquad\qquad ...(\text{Given})$

or, $\quad \delta V_1 + 0.0176 \times 10^{-6} + \dfrac{140}{K_{oil}} = 50 \times 10^{-9}$

$$\delta V_1 + \frac{140}{K_{oil}} = 50 \times 10^{-9} - 0.0176 \times 10^{-6}$$

Vessel without aluminium test piece:

Increase in volume of vessel $\qquad = \delta V_1$

Decrease in volume of oil $\qquad = \dfrac{p}{K_{oil}} \times \text{volume of vessel} = \dfrac{7 \times 10^6}{K_{oil}} \times 167.2 \times 10^{-6}$

or, $\delta V_1 + \dfrac{7 \times 10^6 \times 167.2 \times 10^{-6}}{K_{oil}} = 364 \times 10^{-9} \qquad\qquad\qquad ...(ii)$

Subtracting equation (*ii*) from (*i*), we have

$$\frac{140}{K_{oil}} - \frac{7 \times 10^6 \times 167.2 \times 10^{-6}}{K_{oil}} = 50 \times 10^{-9} - 0.0176 \times 10^{-6} - 364 \times 10^{-9}$$

$$\frac{1}{K_{oil}}\,(140 - 1170.4) = 10^{-9}\,(50 - 17.6 - 364) - \frac{1030.4}{K_{oil}} = -10^{-9} \times 331.6$$

$$\therefore \qquad \qquad K_{oil} = \frac{1030 \cdot 4}{331 \cdot 6} \times 10^9 = 3 \cdot 1 \text{ GN/m}^2 \text{ (Ans.)}$$

Example 10·20. *A cylinder made of bronze 180 mm outside diameter and 15 mm thick is strengthened by a single layer of steel wire 2·25 mm diameter wound over it under a constant stress of 75 MN/m². The cylinder is subjected to an internal pressure of 27 MN/m², with rise in temperature of the cylinder by 120°C. Assuming the cylinder to be a thin shell with closed ends, determine the final values of:*

(*i*) *The stress in the wire;*

(*ii*) *The circumferential stress in the cylinder wall;*

(*iii*) *The radial pressure between the wire and the cylinder.*

Take: For steel: E = 208 GN/m², α = 11·18 × 10⁻⁶ per °C

For brass: E = 90 GN/m², α = 18·6 × 10⁻⁶ per °C,

Poisson's ratio = 0·32.

Solution. Outside diameter of the cylinder,

$$D = 180 \text{ mm}$$

Wall thickness, $\qquad \qquad t = 15 \text{ mm}$

Inside diameter, $\qquad \qquad d = D - 2t = 180 - 2 \times 15 = 150 \text{ mm or } 0 \cdot 15 \text{ m}$

Diameter of the steel wire, $\qquad d_w = 2 \cdot 25 \text{ mm} = 0 \cdot 00225 \text{ m}$

Initial tension in the wire, $\qquad \sigma_w = 75 \text{ MN/m}^2$

Internal pressure, $\qquad \qquad p = 27 \text{ MN/m}^2$

Rise in temperature of cylinder $\qquad = 120°C$

Young's modulus for cylinder, $\quad E_c = 90 \text{ GN/m}^2$

Young's modulus for wire, $\qquad E_w = 208 \text{ GN/m}^2$

$$\alpha_{brass} = 18 \cdot 6 \times 10^{-6} \text{ per } °C$$

Poisson's ratio, $\qquad \qquad \dfrac{1}{m} = 0 \cdot 32$

$$\alpha_{steel} = 11 \cdot 8 \times 10^{-6} \text{ per } °C.$$

(*i*)   **Stress in the wire:**

Before subjecting the cylinder to internal pressure:

Tensile force exerted by wire per unit length = Compressive force developed in the cylinder

$$2 \times \frac{\pi}{4} d_w^2 \times \sigma_w \times n = 2t \times 1 \times \sigma_c$$

or, $\qquad \qquad 2 \times \dfrac{\pi}{4} d_w^2 \times \sigma_w \times \dfrac{1}{d_w} = 2.t.\sigma_c$

$$\therefore \qquad \qquad \sigma_c = \frac{\pi d_w}{4t} \sigma_w$$

(where, σ_c = Initial circumferential compression in the cylinder)

$$= \frac{\pi \times 0 \cdot 00225}{4 \times 0 \cdot 015} \times 75 = 8 \cdot 84 \text{ MN/m}^2 \text{ (compressive)}$$

After subjecting the cylinder to internal pressure:

Due to internal pressure, the longitudinal stress developed in the cylinder,

$$\sigma_t' = \frac{pd}{4t} = \frac{27 \times 0 \cdot 15}{4 \times 0 \cdot 015} = 67 \cdot 5 \text{ MN/m}^2$$

Total bursting force = Total resisting force (per unit length) ... *for equilibrium*

$$p \times d \times 1 = \sigma_c' \times 2t \times 1 + \sigma_w' \times 2 \times \frac{\pi}{4} d_w^2 \times n$$

$$p.d = \sigma'_c \times 2t + \sigma'_w \times \frac{\pi d_w}{2} \qquad \left(\because n = \frac{1}{d_w} \right)$$

$$27 \times 0.15 = \sigma'_c \times 2 \times 0.015 + \sigma'_w \times \frac{\pi \times 0.00225}{2}$$

$$4.05 = 0.03 \, \sigma'_c + 0.0035 \, \sigma'_w$$

$$\sigma'_c + 0.116 \, \sigma'_w = 135 \qquad \qquad \ldots(i)$$

Equating the strains in cylinder and wire, we have

$$\frac{\sigma'_c}{E_c} - \frac{\sigma'_l}{mE_c} + \alpha_{brass} \times 120 = \frac{\sigma'_w}{E_w} + \alpha_{steel} \times 120$$

$$\frac{\sigma'_c \times 10^6}{90 \times 10^9} - \frac{67.5 \times 10^6}{90 \times 10^9} \times 0.32 + 18.6 \times 10^{-6} \times 120 = \frac{\sigma'_w \times 10^6}{208 \times 10^9} + 11.8 \times 10^{-6} \times 120$$

Multiplying both sides by 90×10^9, we get

$$\sigma'_c \times 10^6 - 21.6 \times 10^6 + 18.6 \times 10^{-6} \times 120 \times 90 \times 10^9 = 0.432 \, \sigma'_w \times 10^6 + 11.8$$
$$\times 10^{-6} + 120 \times 90 \times 10^9$$

$$\sigma'_c \times 10^6 - 21.6 \times 10^6 + 200.88 \times 10^6 = 0.432 \, \sigma'_w \times 10^6 + 127.44 \times 10^6$$

$$\sigma'_c = 0.432 \, \sigma'_w - 51.84 \qquad \qquad \ldots(ii)$$

Substituting the value of σ'_c in eqn (i), we get

$$0.432 \, \sigma'_w - 51.84 + 0.116 \, \sigma'_w = 135$$

or, $$0.548 \, \sigma'_w = 186.84$$

or, $$\sigma'_w = 340.9 \text{ MN/m}^2 \text{ (tensile)}$$

and, $$\sigma'_c = 135 - 0.116 \times 340.9 = 95.46 \text{ MN/m}^2 \text{ (tensile) (from equation } (i)\text{)}$$

Final stress in the wire = $\sigma_w + \sigma'_w = 75 + 340.9 = \textbf{415·MN/m}^2$ **(tensile) (Ans.)**

IC engine cylinders are similar to pressure vessels.

(*ii*) **The circumferential stress in the cylinder wall:**

Final circumferential stress in the cylinder wall

$$= \sigma_c + \sigma_c'$$

$$= -8.84 + 95.46$$

$$= \mathbf{86.62\,MN/m^2}\ (\textit{tensile})\ \textbf{(Ans.)}$$

(*iii*) **The radial pressure between the wire and the cylinder, p_r:**

The radial pressure is due to wire winding and temperature rise.

Final circumferential stress

$$= \frac{pd}{2t} - \frac{p_r D}{2t}$$

or $86.62 = \dfrac{27 \times 0.15}{2 \times 0.015} - \dfrac{p_r \times 0.18}{2 \times 0.015}$

or $86.62 = 135 - 6\,p_r$

∴ $\mathbf{p_r = 8.06\,MN/m^2}$ **(Ans.)**

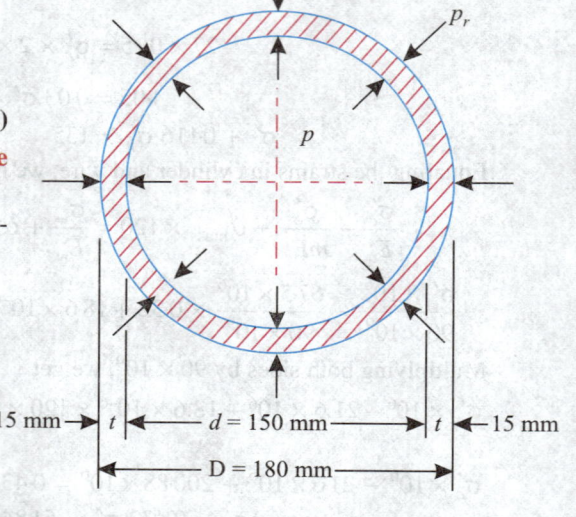

15 mm → t ← d = 150 mm → t ← 15 mm

D = 180 mm

Fig. 10.7

HIGHLIGHTS

1. A cylindrical vessel or shell may be thin or thick depending upon the thickness of the plate in relation to internal diameter of the cylinder. The ratio of $\dfrac{t}{d} = \dfrac{1}{20}$ can be considered as a suitable line of demarcation between thin and thick cylinders. In thin cylinders, the stresses may be assumed uniformly distributed over the wall thickness.

2. *Thin cylinders* are frequently required to operate under pressure upto 30 MN/m² or more; for high pressures such as 250 MN/m² or more, *thick cylinders* are used.

3. *Hoop or circumferential stress:*

$$\sigma_c = \frac{pd}{2t} \qquad \qquad \text{...for a \textit{seamless} shell}$$

and,

$$\sigma_c = \frac{pd}{2t\eta_l} \qquad \qquad \text{...for a \textit{built up} shell}$$

4. *Longitudinal stress:*

$$\sigma_l = \frac{pd}{4t} \qquad \qquad \text{...for a \textit{seamless} shell}$$

and,

$$\sigma_l = \frac{pd}{4t\eta_c} \qquad \qquad \text{...for a \textit{built up} shell}$$

where,

η_l = Efficiency of the longitudinal joint, and

η_c = Efficiency of the circumferential joint.

5. *Maximum shear stress (τ_{max}):* $\tau_{max} = \dfrac{pd}{8t}$

6. *Cylindrical shell with hemispherical ends:*

$$\frac{t_2}{t_1} = \frac{1 - \dfrac{1}{m}}{2 - \dfrac{1}{m}} \quad \text{...for \textit{no distortion} at the junction of cylindrical and hemispherical portions.}$$

$\dfrac{t_2}{t_1} = 0.5$ ...for maximum stress to be *same* in both cylindrical and hemispherical portion

where,

t_1 = Wall thickness of cylindrical portion, and

t_2 = Wall thickness of hemispherical portion.

7. *Change in dimensions of a thin cylindrical shell due to an internal pressure:*

$$e_c = \frac{\delta d}{d} = \frac{\sigma_c}{E} - \frac{1}{m}\frac{\sigma_l}{E} \qquad \text{...}(i)$$

$$e_l = \frac{\delta l}{l} = \frac{\sigma_l}{E} - \frac{1}{m}\frac{\sigma_c}{E}$$

$$e_l = e_l + 2\,e_c \qquad \text{...}(iii)$$

$$e_v = \frac{\delta V}{V}$$

In case of a seamless shell,

$$\delta V = \frac{p d V}{2tE}\left(\frac{5}{2} - \frac{2}{m}\right) \text{ where, } V \text{ is the volume of the shell.}$$

$$\delta d = \frac{p d^2}{2tE}\left(1 - \frac{1}{2m}\right) \qquad \text{...}(ii)$$

$$\delta l = \frac{p d l}{4tE}\left(1 - \frac{2}{m}\right) \qquad \text{...}(iii)$$

8. *Wire wound thin cylinders:*

Important relations/formulae are:

$$\sigma_c = \frac{\pi d_w}{4t}.\sigma_w \qquad \text{...}(i)$$

$$\sigma_l' = \frac{p d}{4t} \qquad \text{...}(ii)$$

$$p.d = \sigma_c' \times 2t + \sigma_w' \times \frac{\pi d_w}{2} \qquad \text{...}(iii)$$

$$\frac{\sigma_c'}{E_c} - \frac{p d}{4 t m E_c} = \frac{\sigma_w'}{E_w} \qquad \text{...}(iv)$$

The stresses σ_c' and σ_w' can be calculated from eqns. (*iii*) and (*iv*).

Resultant stress in wire $= \sigma_w + \sigma_w'$

Resultant stress in the cylinder

$= \sigma_c' - \sigma_c$

where,

d = Diameter,

t = Wall thickness of the cylinder,

d_w = Diameter of the wire,

σ_w = Initial tension with which wire is wound,

σ_c = Compressive circumferential stress developed in the cylinder,

σ_c' = Circumferential stress developed in the cylinder,

σ_w' = Stress developed in the wire,

σ_l' = Longitudinal stress developed in the cylinder,

When the cylinder is subjected to internal pressure p.

E_c = Young's modulus for cylinder,

E_w = Young's modulus for wire, and

$\dfrac{1}{m}$ = Poisson's ratio for cylinder.

9. Hoop or circumferential stress developed in a spherical shell is given as,

$$\sigma_c = \frac{pd}{4t} \qquad\qquad\qquad \text{...for a } \textit{seamless } \text{shell}$$

$$\sigma_c = \frac{pd}{4t\eta} \qquad\qquad\qquad \text{...for a } \textit{built-up } \text{shell}$$

where η is the joint efficiency.

10. *Change in dimentions of a spherical shell:*

$$e = \frac{\delta d}{d} = \frac{\sigma_c}{E} - \frac{1}{m}\frac{\sigma_c}{E} = \frac{\sigma_c}{E}\left(1 - \frac{1}{m}\right) = \frac{pd}{4tE}\left(1 - \frac{1}{m}\right) \qquad \text{...(i)}$$

$$\left[\text{or } \delta d = \frac{pd^2}{4tE}\left(1 - \frac{1}{m}\right)\right]$$

$$e_v = e + e + e = 3e \qquad\qquad\qquad\qquad \text{...(ii)}$$

$$e_v = \frac{\delta V}{V} \qquad\qquad\qquad\qquad \text{...(iii)}$$

$$\delta V = e_v \times V = \frac{3pd}{4tE}\left(1 - \frac{1}{m}\right) \times V \qquad\qquad \text{...(iv)}$$

where,

$$V = \frac{4}{3}\pi r^3 \text{ or } \frac{\pi d^3}{6}$$

OBJECTIVE TYPE QUESTIONS

Choose the Correct Answers:

1. Pressure vessels are made of
   - (a) non-ferrous materials
   - (b) sheet steel
   - (c) cast iron
   - (d) any of the above.

2. When a thin cylindrical shell is subjected to internal fluid pressure, which of the following stress is developed in its wall?
   - (a) Circumferential stress
   - (b) Longitudinal stress
   - (c) both (a) and (b)
   - (d) none of the above.

3. Chemical vessels are made of which of the following materials?
   - (a) Non-ferrous materials
   - (b) Sheet metal
   - (c) Cast iron
   - (d) Special material.

4. Vessels used for storing fluid under pressure are called...
   - (a) cylinders
   - (b) spheres
   - (c) shells
   - (d) none of the above.

5. A shell with wall thickness small compared to internal diameter $\left(\dfrac{d}{t} \geq 20\right)$ is called...
   - (a) thin shell
   - (b) thick shell
   - (c) either of the above
   - (d) none of the above.

6. Which of the following are usually considered as thin cylinders?
   - (a) Boilers
   - (b) Tanks
   - (c) Steam pipes
   - (d) Water pipes
   - (e) All of the above.

7. Thin cylinders are frequently required to operate under pressures upto

(a) 5 MN/m² (b) 15 MN/m²

(c) 30 MN/m² (d) 250 MN/m².

8. Longitudinal stresses act... to the longitudinal axis of the shell.

(a) parallel

(b) perpendicular

(c) either of the above

(d) none of the above.

9. In a thin shell circumferential stress (s_c) is given by

(a) $\sigma_c = \dfrac{pd}{2m_l}$ (b) $\sigma_c = \dfrac{pd}{2m_c}$

(c) $\sigma_c = \dfrac{pd}{tm_l}$ (d) $\sigma_c = \dfrac{pd^2}{e\eta_c}$

10. In case of seamless shell the change in volume (dV) is given by

(a) $\delta V = \dfrac{pdV}{2tE}\left(\dfrac{5}{2} - \dfrac{2}{m}\right)$

(b) $\delta V = \dfrac{pdV}{4tE}\left(\dfrac{5}{2} - \dfrac{2}{m}\right)$

(c) $\delta V = \dfrac{p^2 dV}{2tE}\left(\dfrac{5}{2} - \dfrac{2}{m}\right)$

(d) $\delta V = \dfrac{pd^2V}{2tE}\left(\dfrac{7}{2} - \dfrac{2}{m}\right)$

ANSWERS

1. (d)	**2.** (c)	**3.** (d)	**4.** (c)	**5.** (a)	**6.** (e)	**7.** (c)
8. (a)	**9.** (a)	**10.** (a).				

UNSOLVED EXAMPLES

Cylindrical Shells:

1. The diameter of a cylindrical shell made of steel is 3 m. The sheet is subjected to an internal pressure of 1 MN/m² gauge. Find out the thickness of the shell plate if ultimate tensile stress of the mild steel is 480 MN/m². Longitudinal joint efficiency of the shell is 80% and factor of safety is 6.

 [**Ans.** $t = 24$ mm]

2. A vertical cylindrical gasoline storage tank, made of 20 mm thick mild steel plate, has to withstand maximum pressure of 1 MN/m². Calculate the diameter of the tank if stress is 240 MN/m², factor of safety 2·5 and joint efficiency 85%.

 [**Ans.** $d = 0.327$m]

3. A gas cylinder of internal diameter 1·5 metres is 30 mm thick. Find the allowable pressure of the gas inside the cylinder if the tensile stress in the material is not to exceed 100 MN/m².

 [**Ans.** $p = 4$ MN/m²]

4. What thickness of metal would be required for cast-iron water pipe 80 cm in diameter under a head of 100 m? Assume the permissible tensile stress for cast-iron as 20 MN/m².

 [**Ans.** $t = 2$ cm]

Pneumatic cylinder.

5. A boiler shell is to be made of 15 mm thick plate having a limiting tensile stress of 120 MN/m². If the longitudinal and circumferential efficiencies are 70% and 30% respectively, determine what maximum diameter of the shell would be allowed for a maximum pressure of 2 MN/m². [**Ans.** $d = 1.08$ m]

6. The gauge pressure in a boiler of 1·5 m diameter and 12·5 mm thickness is 2 MN/m²; find the longitudinal and circumferential stresses in the boiler plate and circumferential, longitudinal and volumetric strains. Take E = 200 GN/m² and Poisson's ratio = 0·25.

   [**Ans.** σ_c = 120 MN/m²; σ_l = 60 MN/m², 5·25 × 10⁻⁴; 1·5 × 10⁻⁴; 12 × 10⁻⁴]

7. In question 6, if the boiler was built up and the longitudinal and circumferential efficiencies of the joints were 80% and 50% respectively, what would have been the values of the stresses then?

   [**Ans.** σ_c = 150 MN/m²; σ_l = 120 MN/m²]

8. A cylindrical shell 3m long and 50 cm in diameter and 1.25 cm thick is at atmospheric pressure. What would be its dimensions when it is subjected to an internal pressure of 2 MN/m²? E = 200 GN/m² and m = 4. [**Ans.** t = 300.015 cm; d = 50.00875 cm]

9. A cylindrical tank open at top and having vertical axis, is of 2·5 m inside diameter and 22 m high. The tank is filled with water and is made of structural steel with a yield point of 220 MN/m². Determine the thickness of the tank if (*i*) longitudinal joint is 100% efficient and (*ii*) longitudinal joint is 75% efficient. Assume factor of safety as 3. [**Ans.** 3·75 mm; 5 mm]

10. A vertical steam boiler is of 2 metres internal diameter and 4 metres high. It is constructed with 20 mm thick plates for a working pressure of 1 MN/m². The end plates are flat and are not stayed. Calculate:

    (*i*) The stress in the circumferential plates due to resisting the bursting effect.

    (*ii*) The stress in the circumferential plate due to the pressure on the end plates.

    (*iii*) The increase in length, diameter and volume.

    Assume the Poisson's ratio as 0·3 and E = 200 GN/m².

    [**Ans.** σ_c = 50 MN/m²; σ_l = 250 MN/m²; δd = 0.042 cm; δl = 0.02 cm and δV = 5970 cm³]

11. The air vessel of torpedo is of 50 cm external diameter and 1 cm thick, the length being 1·8 m. Find the change in external diameter and length when charged to 10 MN/m² internal pressure. E = 200 GN/m² and Poisson's ratio = 0·3. [**Ans.** 0·0531 cm, 0·045 cm]

12. A thin cylindrical steel shell of diameter 225 mm and wall thickness 4·5 mm has spherical ends. Determine the thickness of hemispherical ends, if there is no distortion of the junction under pressure.

    [**Ans.** 1·85 mm]

13. A 200 mm diameter cast iron pipe has thickness of 12 mm and is closely wound with a layer of 5 mm diameter steel wire under a tensile stress of 60 MN/m². If now water under a pressure of 3·5 MN/m² is admitted into the pipe, find the stresses induced in the pipe and steel wire.

    Take: $E_{c.i}$ = 100 GN/m²; E_s = 200 GN/m²; Poisson's ratio = 0·3.

    [**Ans.** 0·27 MN/m² (comp.)... Pipe, 89·97 MN/m² (tensile)...Wire]

14. A gun metal tube of 50 mm bore, wall thickness 1·25 mm is closely wound externally by a steel wire 0·5 mm diameter. Determine the tension under which the wire must be wound on the tube, if an internal radial pressure of 1·5 MN/m² is required before the tube is subjected to the tensile stress in the circumferential direction.

    Take: For tube: E = 102 GN/m²;

    For wire: E = 210 GN/m².

    Poisson's ratio = 0·35. [**Ans.** 64·56 MN/m²]

15. A copper tube 30 mm bore and 3 mm thick is plugged at its ends. It is just filled with water at atmospheric pressure. If an axial compressive load of 8 kN is applied to the plugs, find by how much the water pressure will increase? The plugs are assumed to be rigid and fixed to the tube.

    Take: E = 100 GN/m²; Bulk modulus = 2·2 GN/m²; Poisson's ratio = 0·33.

    [**Ans.** 0·176 kN/m²]

16. A cylinder made of bronze 120 mm outside diameter and 10 mm thick is strengthened by a single layer of steel wire, 1·5 mm diameter, wound over it under a constant stress of 50 MN/m². The cylinder is subjected to an internal pressure of 18MN/m² with rise in temperature of cylinder by 80°C. Assuming the cylinder to be a thin shell with closed ends, determine the final values of:

    (*i*) The stress in the wire;

    (*ii*) The circumferential stress in the cylinder wall;

    (*iii*) The radial pressure between the wire and the cylinder.

Take: For steel: $E = 208$ GN/m^2; $\alpha = 11.8 \times 10^{-6}$ per °C.

For brass: $E = 90$ GN/m^2; $\alpha = 18.6 \times 10^{-6}$ per °C.

Poisson's ratio = 0.32.   **[Ans.** (i) 277.27 MN/m^2, (ii) 57.75 MN/m^2, (iii) 5.37 MN/m^2**]**

Spherical Shells:

17. A spherical shell of 1.5 m diameter has 1 cm thick wall. Determine the pressure that can increase its volume by 100 cm^3.

Take: $E = 200$ GN/m^2; $\dfrac{1}{m} = 0.3$.   **[Ans.** 0.144 MN/m^2**]**

18. A bronze spherical shell is made of 1.5 cm thick plate. It is subjected to an internal pressure of 1 MN/m^2. If the permissible stress in the bronze is 55 MN/m^2, calculate the diameter of the spherical shell taking the efficiency as 80%.   **[Ans.** 2.64 m**]**

19. A vessel in the shape of a thin spherical shell 40 cm radius and 1 cm shell thickness is completely filled with a fluid at atmospheric pressure. Additional fluid is then pumped till the pressure increases by 6 MN/m^2. Find the volume of this additional fluid, given that $\dfrac{1}{m} = 0.26$ and $E = 100$ GN/m^2 for the shell material.   **[Ans.** 603.2 cm^3**]**

20. A thin spherical shell 1.5 m in diameter, with its wall of 1.25 cm thickness is filled with the fluid at atmospheric pressure. What intensity of pressure will be developed in it if 160 cm^3 more of fluid is pumped into it? Also calculate the hoop stress at that pressure and increase in diameter.

Take: $E = 200$ GN/m^2; $m = \dfrac{10}{3}$.

[Ans. $\sigma_c = 8.62$ MN/m^2; $p = 0.287$ MN/m^2; $\delta d = 0.00452$ cm**]**

Thick Shells

11·1. THICK CYLINDERS

11·1·1. Introduction

Thick cylinders are the cylindrical vessels, contain-*ing fluid under pressure and whose wall thickness is not small* ($t \geq d/20$).

- Unlike thin shells the *radial stress* in the wall thickness is *not negligible*, rather it varies from the inner surface where it is equal to the magnitude of the fluid pressure to the outer surface where usually it is equal to zero if exposed to the atmosphere.

- *Circumferential stress also varies along the thickness.*

- The variation in the radial as well as circumferential stresses across the thickness are obtained with the help of **Lame's theory.**

11·1·2. Lame's Theory

The *assumptions* made in *Lame's theory* are as follows:

1. The material is homogeneous and isotropic.

2. Plane sections perpendicular to the longitudinal axis of the cylinder remain plane after the application of internal pressure.

3. The material is stressed within the elastic limit.

4. All the fibres of the material are free to expand or contract independently without being constrained by the adjacent fibres.

A thick cylinder subjected to internal and external radial stress (pressure) is shown in Fig. 11·1. Consider an elemental ring of internal radius r and thickness dr.

Let, r_1 = Internal radius of the thick cylinder,

          r_2 = External radius of the thick cylinder,

          l = Length of the cylinder,

          p_1 = Pressure on the inner surface of the cylinder,

          p_2 = Pressure on the outer surface of the cylinder,

          σ_r = Internal radial stress (pressure) on the elemental ring,

   $(\sigma_r + d\sigma_r)$ = External radial stress (pressure) on the elemental ring, and

          σ_c = Circumferential stress on the elemental ring.

The conditions for equilibrium on one half of the elemental ring are (similar to those in the case of thin cylinder) are as follows:

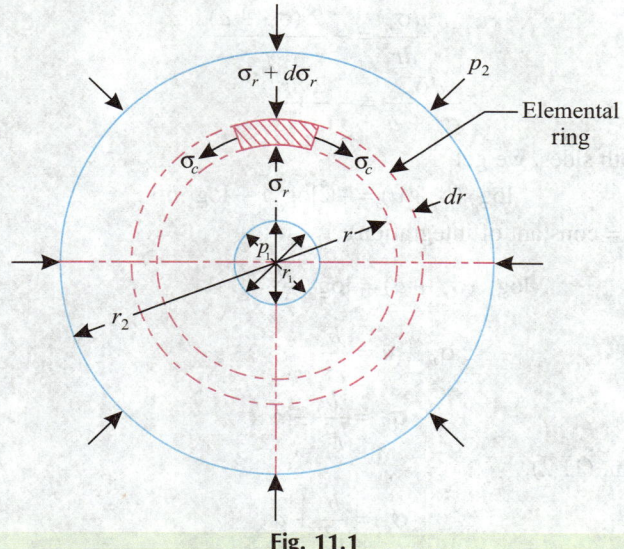

Fig. 11.1

Bursting force:
$$(\sigma_r \times 2rl) - [(\sigma + d\sigma_r) \times 2(r + dr)l] = 2l[-\sigma_r\, dr - rd\sigma_r - dr \cdot d\sigma_r]$$
$$= -2l(\sigma_r dr + r\, d\sigma_r)$$

(neglecting the product of small quantities)

Resisting force $= 2\sigma_c\, l\, dr$

Equating the resisting force to bursting force (for equilibrium), we get
$$2\,\sigma_c . l . dr = -2l\,(\sigma_r dr + rd\sigma_r)$$

or, $\sigma_c = -\sigma_r - r \cdot \dfrac{d\sigma_r}{dr}$ ...(11·1)

Now let us obtain another relation between the radial stress (pressure) and circumferential (or hoop) stress by using the condition that the longitudinal strain (e_l) at any point in the section is same.

The longitudinal stress, $\qquad \sigma_l = \dfrac{p_1 \times \pi \, r_1^2}{\pi \, (r_2^2 - r_1^2)} = \dfrac{p_1 \, r_1^2}{r_2^2 - r_1^2}$

Hence at any point in the section of the elemental ring considered above, the following three principal stresses exist:

   (i)   The radial stress (pressure), σ_r

   (ii)   The circumferential stress, σ_c

  (iii)   The longitudinal tensile stress, σ_l.

Since the longitudinal strain (e_l) is constant, we have

$$e_l = \frac{\sigma_l}{E} - \frac{\sigma_c}{mE} + \frac{\sigma_r}{mE} = \text{constant}$$

$$\left(\text{where, } \frac{1}{m} = \text{Poisson's ratio}\right)$$

But, since σ_l, m and E are constant

$\therefore \qquad\qquad\qquad\qquad \sigma_c - \sigma_r = \text{constant}$

Let, $\qquad\qquad\qquad\qquad \sigma_c - \sigma_r = 2a \qquad\qquad\qquad\qquad\qquad\qquad ...(11 \cdot 2)$

Putting $\sigma_c = (\sigma_r + 2a)$ in eqn. (11·1), we get

$$(\sigma_r + 2a) = -\sigma_r - r \cdot \frac{d\sigma_r}{dr}$$

$\therefore \qquad\qquad\qquad\qquad \dfrac{d\sigma_r}{dr} = -\dfrac{2\,(\sigma_r + a)}{r}$

or, $\qquad\qquad\qquad\qquad \dfrac{d\sigma_r}{\sigma_r + a} = -\dfrac{2\,dr}{r} \qquad\qquad\qquad\qquad\qquad ...(11.3)$

Integrating both sides, we get

$$\log_e (\sigma_r + a) = -2\log_e r + \log_e b$$

(where $\log_e b = $ constant of integration)

$\therefore \qquad\qquad\qquad\qquad \log_e (\sigma_r + a) = \log_e \dfrac{b}{r^2}$

or, $\qquad\qquad\qquad\qquad \sigma_r + a = \dfrac{b}{r^2}$

or, $\qquad\qquad\qquad\qquad \sigma_r = \dfrac{b}{r^2} - a \qquad\qquad\qquad\qquad\qquad\qquad ...(11 \cdot 4)$

Also, from eqn. (11·2),

$$\sigma_c = \frac{b}{r^2} + a \qquad\qquad\qquad\qquad\qquad\qquad ...(11 \cdot 5)$$

The equations (11·4) and (11·5) are called **Lame's equations**.

The constants a and b can be evaluated from the known internal and external radial pressure and radius.

It may be noted that in the above equations σ_r is *compressive* and σ_c is tensile.

Sign conventions:

Circumferential (or hoop stress) will be taken +ve if tensile and negative when compressive, while compressive radial stress will be taken +ve and tensile negative.

11·1·2·1. Special Cases

Case I. External pressure $= p_2$; internal pressure $= p_1$:

    At, $\qquad\qquad\qquad\qquad r = r_1, \sigma_r = p_1$, and

at $\qquad r = r_2, \sigma_r = p_2$

Substituting in eqn. (11·4), we have

$$p_1 = \frac{b}{r_1^2} - a \qquad\qquad \text{...(i)}$$

and, $\qquad\qquad p_2 = \frac{b}{r_2^2} - a \qquad\qquad \text{...(ii)}$

Steel tanks in a wine plant.

Now, from (i) and (ii), we get

$$b = \frac{r_1^2 r_2^2}{(r_2^2 - r_1^2)} (p_1 - p_2) \qquad\qquad \text{...(11·6(a))}$$

and, $\qquad\qquad a = \frac{p_1 r_1^2 - p_2 r_2^2}{r_2^2 - r_1^2} \qquad\qquad \text{...(11·6(b))}$

Substituting eqn. (11·6) in eqns. (11·4) and (11·5), we get

$$\sigma_r = \frac{1}{(r_2^2 - r_1^2)} \left[(p_2 r_2^2 - p_1 r_1^2 + \frac{r_1^2 r_2^2}{r^2} (p_1 - p_2) \right] \qquad\qquad \text{...(11·7)}$$

and, $\qquad\qquad \sigma_c = \frac{1}{r_2^2 - r_1^2} \left[(p_1 r_1^2 - p_2 r_2^2 + \frac{r_1^2 r_2^2}{r^2} (p_1 - p_2) \right] \qquad\qquad \text{...(11·8)}$

Case II: Internal pressure = p_1; external pressure = zero:

When there is only internal pressure and outer surface of the cylinder is exposed to atmospheric pressure, then

At, $\qquad\qquad r = r_1, \sigma_r = p_1,$ and

at, $\qquad\qquad r = r_2, \sigma_r = p_2 = 0$

Substituting in Eqn (11·4), we get

$$p_1 = \frac{b}{r_1^2} - a \qquad\qquad \text{...(i)}$$

$$p_2 = \frac{b}{r_2^2} - a = 0 \qquad \qquad ...(ii)$$

From (i) and (ii), we get

$$b = p_1 \left[\frac{r_1^2 r_2^2}{r_2^2 - r_1^2} \right] \qquad \qquad ...(11.9(a))$$

and,

$$a = p_1 \left[\frac{r_1^2}{r_2^2 - r_1^2} \right] \qquad \qquad ...(11.9(b))$$

Substituting eqn. (11·9) in eqns. (11·4) and (11·5), we get

$$\sigma_r = \frac{p_1 r_1^2}{r_2^2 - r_1^2} \left[\frac{r_2^2}{r^2} - 1 \right] \qquad \qquad ...(11 \cdot 10)$$

and,

$$\sigma_c = \frac{p_1 r_1^2}{r_2^2 - r_1^2} \left[\frac{r_2^2}{r^2} + 1 \right] \qquad \qquad ...(11 \cdot 11(a))$$

At,

$$r = r_1 : \ (\sigma_c)_{r_1} = p_1 \cdot \frac{r_2^2 + r_1^2}{r_2^2 - r_1^2}$$

At,

$$r = r_2 : \ (\sigma_c)_{r_2} = p_1 \cdot \frac{2 \, r_1^2}{r_2^2 - r_1^2}$$

Fig. 11·2 shows the graph between *stress vs. radius*. It is evident from the graph that the maximum values of both σ_r and σ_c occur at the *inner surface*.

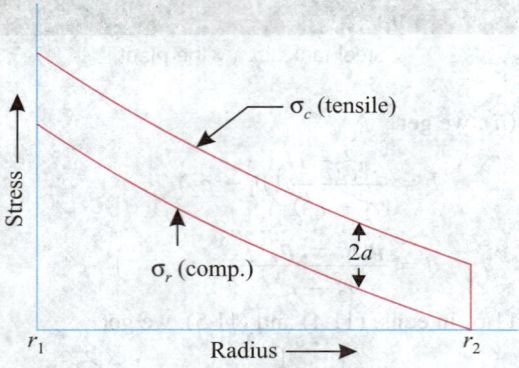

Fig. 11.2

Further for the closed ends cylinder, the longitudinal stress is

$$\sigma_l = \frac{p_1 \, \pi r_1^2}{\pi \, (r_2^2 - r_1^2)} = \frac{p_1 \, r_2^2}{r_2^2 - r_1^2} = a \qquad \qquad ...(11 \cdot 11(b))$$

Case III: External pressure = p_2; internal pressure = zero:

At, $\qquad \qquad r = r_1, \sigma_r = 0 \ $ and $\ $ at $r = r_2, \sigma_r = p_2$

Substituting in eqn (11·4), we get

$$0 = \frac{b}{r_1^2} - a \qquad \qquad ...(i)$$

and, $\qquad p_2 = \dfrac{b}{r_2^2} - a$ $\hfill$ ...(ii)

From (i) and (ii), we have

$$b = -\frac{p_2 r_1^2 r_2^2}{(r_2^2 - r_1^2)} \hfill ...(11\cdot12(a))$$

and, $\qquad a = -\dfrac{p_2 r_2^2}{r_2^2 - r_1^2}$ $\hfill$ ...(11·12(b))

∴ The maximum circumferential (or hoop) stress at radius r_1,

$$\sigma_c = -p_2 \left(\frac{2 r_2^2}{r_2^2 - r_1^2} \right) \text{ from eqn. (11·5)} \hfill ...(11\cdot13)$$

Case IV: Solid circular shaft subjected to external pressure p_2:

We know that, $\qquad \sigma_r = \dfrac{b}{r^2} - a$

$$\sigma_c = \frac{b}{r^2} + a$$

Since σ_r and σ_c are not infinite at $r = 0$, it follows that 'b' must be *zero* so that

$$\sigma_r = -a \hfill ...(11\cdot14(a))$$

and, $\qquad \sigma_c = a$ $\hfill$ ...(11·14(b))

11·1·2.2. Longitudinal and Shear Stresses

In case of a cylinder with closed ends, the tensile longitudinal stress (if it is assumed constant over the cross-section), is given by

$$\sigma_l = \frac{\text{Force acting on the end cover due to internal pressure}}{\text{Area of cross-section of the cylinder}}$$

$$= \frac{\pi\, r_1^2 \cdot p}{\pi\, (r_2^2 - r_1^2)} = \left[\frac{r_1^2}{r_2^2 - r_1^2} \right] p \hfill ...(11\cdot15)$$

Further, if no torque acts on the cylinder, then σ_c, σ_r and σ_l will all be the principal stresses; σ_r being *compressive* and σ_c as well as σ_l *tensile*. Moreover, $\sigma_c > \sigma_l$. Thus the maximum shear stress is given by

Pressure vessel.

$$\tau_{max} = \left[\frac{\sigma_c - (-\sigma_r)}{2} \right]$$

$$= \frac{1}{2} \left[\left(\frac{b}{r^2} + a \right) - \left(a - \frac{b}{r^2} \right) \right] \qquad [\because \sigma_r \text{ is compressive}]$$

i.e. $\qquad\qquad \tau_{max} = \dfrac{b}{r^2}$ $\qquad\qquad\qquad\qquad\qquad\qquad$ ...(11·16)

Thus the absolute maximum value of τ_{max} will be on the inner surface, but it will act on the plane which is inclined to the longitudinal axis at 45°.

Example 11·1. *A pipe of 200 mm internal diameter and 50 mm thickness carries a fluid at a pressure of 10 MN/m². Calculate the maximum and minimum intensities of circumferential stresses across the section.*

Also sketch the radial stress (pressure) distribution and circumferential stress distribution across the section.

Solution. Internal radius of the pipe,

$$r_1 = 200/2 = 100 \text{ mm} = 0 \cdot 1 \text{ m}$$

External radius of the pipe, $r_2 = (200 + 2 \times 50)/2 = 150 \text{ mm} = 0 \cdot 15 \text{ m}$

Pressure of the fluid, $\qquad p_1 = 10 \text{ MN/m}^2$

Circumferential stresses (max. and min.):

The *Lame's equations* are:

$$\sigma_r = \frac{b}{r^2} - a \qquad\qquad\qquad\qquad ...(i)$$

and, $\qquad\qquad\qquad\qquad \sigma_c = \dfrac{b}{r^2} + a \qquad\qquad\qquad\qquad ...(ii)$

At, $\qquad\qquad\qquad\qquad r = 0 \cdot 1 \text{ m}, \sigma_r = 10 \text{ MN/m}^2$

and, $\qquad\qquad$ when, $r = 0 \cdot 15 \text{ m}, \sigma_r = 0$

Substituting in (*i*), we get

$$10 = \frac{b}{0 \cdot 1^2} - a$$

$$0 = \frac{b}{0 \cdot 15^2} - a$$

Solving the above equations, we get

$$b = 0 \cdot 18 \text{ and } a = 8$$

At, $\qquad\qquad\qquad\qquad r = 0 \cdot 1 \text{ m}, \ \sigma_r = \dfrac{0 \cdot 18}{0 \cdot 1^2} - 8 = 10 \text{ MN/m}^2$

and, $\qquad\qquad\qquad\qquad \sigma_c = \dfrac{0 \cdot 18}{0 \cdot 1^2} + 8 = \textbf{26 MN/m}^2 \textbf{ (Ans.)}$

At, $\qquad\qquad\qquad\qquad r = 0 \cdot 15 \text{ m}, \ \sigma_r = \dfrac{0 \cdot 18}{0 \cdot 15^2} - 8 = 0$

and, $\qquad\qquad\qquad\qquad \sigma_c = \dfrac{0 \cdot 18}{0 \cdot 15^2} + 8 = \textbf{16 MN/m}^2 \textbf{ (Ans.)}$

Fig. 11·3(*a*, *b*), shows the distribution of radial stress (pressure) and circumferential stress across the section.

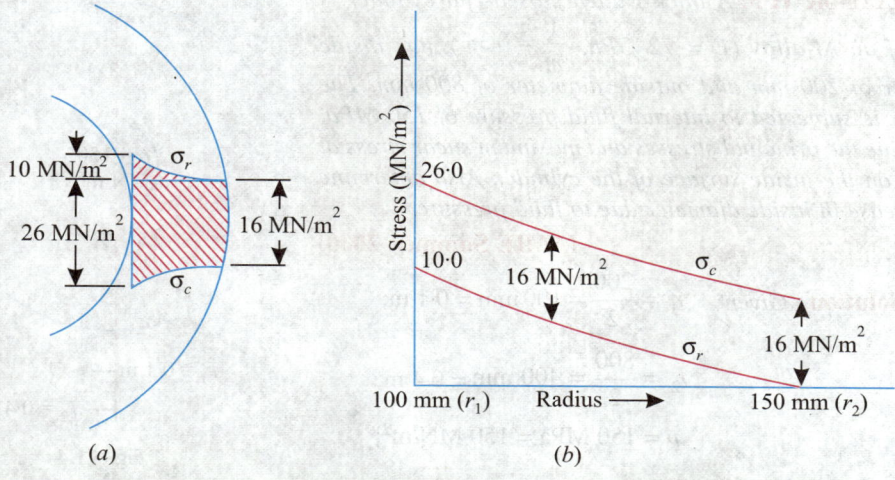

10 MN/m^2

σ_r

26 MN/m^2

16 MN/m^2

σ_c

Stress (MN/m^2)

26·0

10·0

16 MN/m^2

σ_c

σ_r

16 MN/m^2

100 mm (r_1)   Radius ⟶   150 mm (r_2)

(a)   (b)

Fig. 11.3

Example 11·2. *Calculate the thickness of metal necessary for a cylindrical shell of internal diameter 160 mm to withstand an internal pressure of 25 MN/m^2, if maximum permissible tensile stress is 125 MN/m^2.*

Solution. Internal radius of the shell, $r_1 = \dfrac{160}{2} = \text{mm} = 0{\cdot}08$ m

Internal pressure, $p_1 = 25$ MN/m^2

Maximum permissible tensile stress, $\sigma_c = 125$ MN/m^2

Thickness of the metal:

The *Lame's equations* are $\sigma_r = \dfrac{b}{r^2} - a$ ...(i)

and, $\sigma_c = \dfrac{b}{r^2} + a$ ...(ii)

At, $r = 0{\cdot}08$ m, $\sigma_r = 25$ MN/m^2

and, $\sigma_{c(max)} = 125$ MN/m^2

Substituting in (i) and (ii), we get $25 = \dfrac{b}{(0{\cdot}08)^2} - a$ ...(iii)

or, $25 = 156{\cdot}25\,b - a$ ...(iii)

and, $125 = \dfrac{b}{(0{\cdot}08)^2} + a$

or, $125 = 156{\cdot}25\,b + a$ ...(iv)

From (iii) and (iv), we have $b = 0{\cdot}48$ and $a = 50$

At, $r = r_2, \sigma_r = 0$

∴ $0 = \dfrac{0{\cdot}48}{r_2^2} - 50$

or, $r_2 = 0{\cdot}098$ m or 98 mm

∴ **Thickness of metal** $= r_2 - r_1 = 98 - 80 = $ **18 mm** **(Ans.)**

Example 11·3. *A thick walled closed-end cylinder is made of an Al-alloy ($E = 72$ GPa, $\dfrac{1}{m} = 0.33$), has inside diameter of 200 mm and outside diameter of 800 mm. The cylinder is subjected to internal fluid pressure of 150 MPa. Determine the principal stresses and maximum shear stress at a point on the inside surface of the cylinder. Also determine the increase in inside diameter due to fluid pressure.*

(AMIE Summer, 2000)

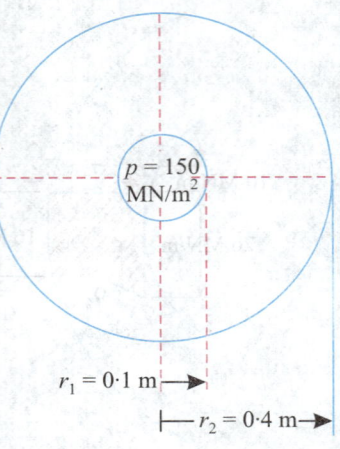

$p = 150$ MN/m²

$r_1 = 0.1$ m

$r_2 = 0.4$ m

Fig. 11.4

Solution. *Given:* $r_1 = \dfrac{200}{2} = 100 \text{ mm} = 0.1 \text{ m};$

$$r_2 = \dfrac{800}{2} = 400 \text{ mm} = 0.4 \text{ m};$$

$$p = 150 \text{ MPa} = 150 \text{ MN/m}^2;$$

$$E = 72 \text{ GPa} = 72 \times 10^9 \text{ N/m}^2;$$

$$\dfrac{1}{m} = 0.33$$

Principal stresses and maximum shear stress:

Using the conditions in *Lame's equation*:

$$\sigma_r = \dfrac{b}{r^2} - a$$

At, $r = 0.1 \text{ m}, \quad \sigma_r = +p = 150 \text{ MN/m}^2$

$r = 0.4 \text{ m}, \quad \sigma_r = 0$

Substituting the values in the above equation, we have

$$150 = \dfrac{b}{(0.1)^2} - a \qquad \qquad \text{...(i)}$$

$$0 = \dfrac{b}{(0.4)^2} - a \qquad \qquad \text{...(ii)}$$

Diesel and petrol tanks are a common sight in industries.

Subtracting (*ii*) from (*i*), we get

$$150 = b\left(\frac{1}{0\cdot1^2} - \frac{1}{0\cdot4^2}\right) = 93\cdot75\,b$$

$$\therefore \qquad b = 1\cdot6$$

Substituting for *b* in eqn. (*i*), we have

$$150 = \frac{1\cdot6}{0\cdot1^2} - a$$

$$\therefore \qquad a = 10$$

The circumferential (or hoop) stress by *Lame's equation*, is given by

$$\sigma_c = \frac{b}{r^2} + a$$

$$\therefore \qquad (\sigma_c)_{max}, \text{ at } r\,(=r_1) = 0\cdot1 \text{ m} = \frac{1\cdot6}{0\cdot1^2} + 10 = 170 \text{ MN/m}^2 \text{ (tensile), and}$$

$$(\sigma_c)_{min}, \text{ at } r\,(=r_2) = 0\cdot4 \text{ m} = \frac{1.6}{0.4^2} + 10 = 20 \text{ MN/m}^2 \text{ (tensile)}$$

∴ *Principal stresses are* **170 MN/m² and 20 MN/m²** **(Ans.)**

Maximum shear stress, $\quad \tau_{max} = \dfrac{(\sigma_c)_{max} - (\sigma_c)_{min}}{2}$

$$= \frac{170 - 20}{2} = \textbf{75 MN/m}^2 \text{ (Ans.)}$$

Increase in inside diameter, δd_1 :

We know, longitudinal (or axial) stress,

$$\sigma_l = \frac{p r_1^2}{r_2^2 - r_1^2} = \frac{150 \times (0\cdot1)^2}{(0\cdot4)^2 - (0\cdot1)^2} = 10 \text{ MN/m}^2$$

Circumferential (or hoop) strain at the inner radius, is given by:

$$(e_c)_{r_1} = \frac{1}{E}\left[\sigma_c + \frac{1}{m}(\sigma_r - \sigma_l)\right]$$

$$= \frac{1}{72 \times 10^9}\,[170 \times 10^6 + 0\cdot33\,(150 - 10) \times 10^6] = 0\cdot003$$

Also, $\qquad (e_c)_{r_1} = \dfrac{\delta d_1}{d_1}$

or, $\qquad 0\cdot003 = \dfrac{\delta d_1}{0\cdot2}$

$$\therefore \qquad \delta d_1 = 0\cdot003 \times 0\cdot2 = 0\cdot0006 \text{ m or } \textbf{0\cdot6 mm} \text{ (Ans.)}$$

Example 11·4. *Find the ratio of thickness to internal diameter for a tube subjected to internal pressure when the pressure is 5/8 of the value of the maximum permissible circumferential stress.*

Find the increase in internal diameter of such a tube 100 mm internal diameter when the internal pressure is 80 MN/m².

Also find the change in wall thickness.

Take: $\qquad E = 205 \text{ GN/m}^2, \text{ and } \dfrac{1}{m} = 0\cdot29$

Solution. Internal radius of the tube,

$$r_1 = \frac{100}{2} = 50 \text{ mm} = 0\cdot05 \text{ m}$$

Internal pressure, $\qquad p_1 = \sigma_r = 80 \text{ MN/m}^2$

$$E = 205 \text{ GN/m}^2$$

$$\frac{1}{m} = 0.29$$

Thickness (t) :
Internal diameter (d)

From eqn. (11.11a), at $r = r_1$)

$$(\sigma_c)_{max} = p_1 \left[\frac{r_2^2 + r_1^2}{r_2^2 - r_1^2} \right] \quad \text{or} \quad \frac{(\sigma_c)_{max}}{p_1} = \frac{r_2^2 + r_1^2}{r_2^2 - r_1^2}$$

$$\frac{8}{5} = \frac{(r_2 / r_1)^2 + 1}{(r_2 / r_1)^2 - 1}$$

From equation (11.11(a)), at $r = r_1$

From which, $\qquad \dfrac{r_2}{r_1} = 2.08 \text{ or } r_2 = 2.08 \ r_1$

$\therefore \qquad t = r_2 - r_1 = 2.08 \ r_1 - r_1 = 1.08 \ r_1$

so that, $\qquad \dfrac{t}{d} = \dfrac{1.08 \ r_1}{2 \ r_1} = 0.54$

Hence, $\qquad \dfrac{t}{d} = \mathbf{0.54}$ **(Ans.)**

Increase in internal diameter, δd:

At, $\qquad r = 0.05 \text{ m}, \sigma_r = p_1 = 80 \text{ MN/m}^2 \text{ (comp.)}$

and, $\qquad \sigma_c = 8/5 \times 80 = 128 \text{ MN/m}^2 \text{ (tensile)} \qquad \left[\because \dfrac{\sigma_r}{\sigma_c} = 5/8 \right]$

Further longitudinal stress, $\sigma_l = \dfrac{p_1 \, r_1^2}{r_2^2 - r_1^2}$ (assuming closed ends)

$$= \frac{80 \times 0.05^2}{(0.05 \times 2.08)^2 - (0.05)^2} = 24 \text{ MN/m}^2 \text{ (tensile)}$$

The circumferential strain at $r = 0.05$ m, is then given by,

$$e_c = \frac{\sigma_c}{E} + \frac{1}{m} \frac{\sigma_r}{E} - \frac{1}{m} \frac{\sigma_l}{E}$$

$$= \frac{1}{205 \times 10^9} [128 + 0.29 (80 - 24)] \times 10^6 = 0.7036 \times 10^{-3}$$

$\therefore \qquad e_c = \dfrac{\delta d}{d} = 0.7036 \times 10^{-3}$

or, $\qquad \delta d = 0.7036 \times 10^{-3} \times 100 = 0.07036 \text{ mm}$

Hence, $\qquad \delta d = \mathbf{0.07306 \text{ mm}}$ **(Ans.)**

Change in wall thickness:

At, $\qquad r = r_2 = 2.08 \times 0.05 = 0.104 \text{ m}$

$$\sigma_r = 0$$

and, $\qquad \sigma_c - \sigma_r = 2a = (\sigma_c)_{r_1} - (\sigma_r)_{r_1}$

(Since $\sigma_c - \sigma_r = $ constant from eqn. (11.2))

or, $\sigma_c - 0 = 128 - 80$.

i.e. $\sigma_c = 48$ MN/m² (tensile)

∴

$$e_c = \frac{1}{E}\left[\sigma_c - \frac{1}{m}\sigma_l + \frac{1}{m}\sigma_r\right]$$

$$= \frac{1}{205 \times 10^9}[48 - 0.29 \times 24 + 0.29 \times 0] \times 10^6 = 0.2 \times 10^{-3}$$

∴ Increase in external diameter,

$$(\delta d)_{r_2} = 2r_2\,e_c$$

$$= 2 \times 0.104 \times 0.2 \times 10^{-3}\text{ m} = 0.0416\text{ mm}$$

$$\left(\because e_c = \frac{(\delta d)_{r_2}}{2r_2}\right)$$

∴ *Decrease in wall thickness*

$$= \frac{1}{2}(0.07036 - 0.0416) = \mathbf{0.01438\ mm}\quad\textbf{(Ans.)}$$

Example 11·5. *A steel cylinder of 1000 mm inside diameter is to be designed for an internal pressure of 4·8 MN/m². Calculate:*

(i) The thickness if the maximum shearing stress is not to exceed 21 MN/m².

(ii) The increase in volume, due to working pressure, if the cylinder is 7 m long with closed ends.

Neglect any constraints due to ends.

Take: E = 200 GN/m²; Poisson's ratio = 1/3.

Solution. Inside radius of the cylinder,

$$\frac{1000}{2} = 500\text{ mm} = 0.5\text{m}$$

Internal pressure,

$$p_1 = 4.8\text{ MN/m}^2$$

Maximum shearing stress,

$$\tau_{max} = 21\text{ MN/m}^2$$

Length of the cylinder,

$$l = 7\text{ m}$$

$$E = 200\text{ GN/m}^2$$

Poisson's ratio,

$$\frac{1}{m} = 1/3$$

Submarine's body is designed treating it as a pressure vessel.

(i) Wall thickness:

The *Lame's equations* are:

$$\sigma_r = \frac{b}{r^2} - a\ \text{(comp.)}$$

$$\sigma_c = \frac{b}{r^2} + a\ \text{(tensile)}$$

∴

$$\tau_{max} = \frac{1}{2}[\sigma_c - (-\sigma_r)] = \frac{b}{r^2}$$

Absolute maximum shearing stress will act on the internal surface.

i.e. $$21 = \frac{b}{r_1^2} = \frac{b}{(0.5)^2} \qquad \therefore \qquad b = 5.25$$

Also at the inner surface, $$\sigma_r = 4.8 = \frac{b}{r_1^2} - a$$

or, $$4.8 = \frac{5.25}{0.5^2} - a$$

$\therefore$ $$a = 16.2$$

Further at, $$r = r_2$$

$$\sigma_r = 0 = \frac{b}{r_2^2} - a \quad \text{or} \quad \frac{b}{r_2^2} = a$$

i.e. $$r_2 = \sqrt{\frac{b}{a}} = \sqrt{\frac{5.25}{16.2}} = 0.57 \text{ m}$$

$\therefore$ Wall thickness $= r_2 - r_1 = 0.57 - 0.5 = 0.07$ m or 70 mm

Hence, *wall thickness* = **70 mm** **(Ans.)**

(ii) Increase in volume, δV:

At, $$r = r_1, \ \sigma_c = \frac{b}{r_1^2} + a = 21 + 16.2 = 37.2 \ \text{MN/m}^2 \ \text{(tensile)}$$

Longitudinal stress, $$\sigma_l = \frac{p_1 \, r_1^2}{r_2^2 - r_1^2} = \frac{4.8 \times 0.5^2}{0.57^2 - 0.5^2} = 16.02 \ \text{MN/m}^2$$

$\therefore$ $$(e_c)_{r_1} = \frac{1}{E}\left[\sigma_c + \frac{1}{m}(\sigma_r - \sigma_l)\right]$$

$$= \frac{1}{200 \times 10^9}\left[37.2 + \frac{1}{3}(4.8 - 16.02)\right] \times 10^6$$

$$= 0.1673 \times 10^{-3}$$

Also, $$(e_l)_{r_1} = \frac{1}{E}\left[\sigma_l + \frac{1}{m}(\sigma_r - \sigma_c)\right]$$

$$= \frac{1}{200 \times 10^9}\left[16.02 + \frac{1}{3}(4.8 - 37.2)\right] \times 10^6$$

$$= 0.0261 \times 10^{-3}$$

$\therefore$ Volumetric strain (e_v) $$= \frac{\delta V}{V} = 2(e_c)_{r_1} + (e_l)_{r_1}$$

$$= 2 \times 0.1673 \times 10^{-3} + 0.0261 \times 10^{-3} = 0.3607 \times 10^{-3}$$

or, $$\delta V = 0.3607 \times 10^{-3} \times \text{(internal volume)}$$

$$= 0.3607 \times 10^{-3} \times \frac{\pi}{4} \times 1^2 \times 7$$

or, $$\delta V = \mathbf{0.001983 \ m^3} \ \textbf{(increase)} \ \textbf{(Ans.)}$$

Example 11·6. *To measure the longitudinal and circumferential strains, strain gauges were fixed on the outer surface of a closed thick cylinder of diameter ratio 2·5 . At an internal pressure of 276 MN/m² these strains were recorded as 11·016 × 10⁻⁵ and 44·28 × 10⁻⁵ respectively. Determine:*

(i) Young's modulus; (ii) Modulus of rigidity.

Solution. Radius (or diameter) ratio,

$$\frac{r_2}{r_1} = 2.5$$

Internal pressure, $\qquad p_1(\sigma_r) = 276 \text{ MN/m}^2$

Longitudinal strain, $\qquad e_l = 11{\cdot}016 \times 10^{-5}$

Circumferential strain, $\qquad e_c = 44{\cdot}28 \times 10^{-5}$

(i) Young's modulus, E:

The *Lame's equations* are: $\qquad \sigma_r = \dfrac{b}{r^2} - a$ $\hfill ...(i)$

$$\sigma_c = \dfrac{b}{r^2} + a \hspace{3cm} ...(ii)$$

At, $\qquad r = r_2$

$$\sigma_r = 0 = \dfrac{b}{r_2^2} - a$$

i.e., $\qquad a = \dfrac{b}{r_2^2}$ $\hfill ...(iii)$

$$(or \;\; b = a r_2^2)$$

At, $\qquad r = r_1$

$$\sigma_r = 276 = \dfrac{b}{r_1^2} - a$$

or, $\qquad 276 = \dfrac{a r_2^2}{r_1^2} - a \;\; (\because b = a r_2^2) = a \left[\dfrac{r_2}{r_1}\right]^2 - a$

$$= a\,(2{\cdot}5^2 - 1) = 5{\cdot}25\,a$$

$\therefore \qquad a = \dfrac{276}{5{\cdot}25} = 52{\cdot}57 \text{ MN/m}^2$

Again, $\qquad$ at $r = r_2$

$$\sigma_c = \dfrac{b}{r_2^2} + a = a + a = 2a = 2 \times 52{\cdot}57 = 105{\cdot}14 \text{ MN/m}^2$$

Part of a submarine being transported.

$$\sigma_l = \frac{p_1 \, r_1^2}{r_2^2 - r_1^2} = \frac{p_1}{\left(\dfrac{r_2}{r_1}\right)^2 - 1} = \frac{276}{2 \cdot 5^2 - 1} = 52 \cdot 57 \text{ MN/m}^2 \text{ (tensile)}$$

$$\sigma_r = 0 \text{ (comp.)}$$

∴ Circumferential strain, $e_c = \dfrac{1}{E}\left[\sigma_c - \dfrac{1}{m}(\sigma_l - \sigma_r)\right] \times 10^6$

$$44 \cdot 28 \times 10^{-5} = \frac{1}{E}\left[105 \cdot 14 - \frac{1}{m}(52 \cdot 57 + 0)\right] \times 10^6$$

or, $$E = \frac{10^{11}}{44 \cdot 28}\left[105 \cdot 14 - \frac{52 \cdot 57}{m}\right] \qquad \qquad ...(iv)$$

Similarly, $e_l = \dfrac{1}{E}\left[\sigma_l - \dfrac{1}{m}(\sigma_c + \sigma_r)\right] \times 10^6$

$$11 \cdot 016 \times 10^{-5} = \frac{1}{E}\left[52 \cdot 57 - \frac{1}{m} \times 105 \cdot 14)\right]$$

or, $$E = \frac{10^{11}}{11 \cdot 016}\left[52 \cdot 57 - \frac{105 \cdot 14}{m}\right] \qquad \qquad ...(v)$$

Equating (iv) and (v), we get

$$= \frac{10^{11}}{44 \cdot 28}\left(105 \cdot 14 - \frac{52 \cdot 57}{m}\right) = \frac{10^{11}}{11 \cdot 016}\left(52 \cdot 57 - \frac{105 \cdot 14}{m}\right)$$

$$= 11 \cdot 016\left(105 \cdot 14 - \frac{52 \cdot 57}{m}\right) = 44 \cdot 28\left(52 \cdot 57 - \frac{105 \cdot 14}{m}\right)$$

or, $$1158 \cdot 22 - \frac{579 \cdot 11}{m} = 2327 \cdot 79 - \frac{4655 \cdot 59}{m}$$

or, $$\frac{4076 \cdot 48}{m} = 1169 \cdot 57$$

or, $$\frac{1}{m} = 0 \cdot 287$$

Substituting the value of $\dfrac{1}{m}$ in (iv), we get

$$E = \frac{10^{11}}{44 \cdot 28}(105 \cdot 14 - 52 \cdot 57 \times 0 \cdot 287)$$

$$= 203 \times 10^9 \text{ N/m}^2 \text{ or } 203 \text{ GN/m}^2$$

i.e. $\quad E = \mathbf{203 \ GN/m^2}$ **(Ans.)**

(ii) Modulus of rigidity, C:

Using the relation, $\quad C = \dfrac{E}{2\left(1 + \dfrac{1}{m}\right)} = \dfrac{203}{2(1 + 0 \cdot 287)} = 78 \cdot 86 \text{ GN/m}^2$

i.e. $\quad C = \mathbf{78 \cdot 86 \ GN/m^2}$ **(Ans.)**

Example 11·7. *A thick cylinder of 150 mm outside radius and 100 mm inside radius is subjected to an external pressure of 30 MN/m² and internal pressure of 60 MN/m². Calculate the maximum shear stress in the material of the cylinder at the inner radius* **(AMIE Winter, 2001)**

Sol. *Given:* $r_1 = 100$ mm $= 0.1$ m; $r_2 = 150$ mm $= 0.15$ m, External pressure, $p_2 = 30$ MN/m²; Internal pressure, $p_1 = 60$ MN/m².

Maximum shear stress, $(\tau_{max})_{r_1}$

Hoop stress (when p_1 and p_2 are known) is given by the relation:

$$\sigma_c = \frac{1}{r_2^2 - r_1^2}\left[p_1 r_1^2 - p_2 r_2^2 + \frac{r_1^2 r_2^2}{r^2}(p_1 - p_2) \right] \qquad \text{...[Eqn. (11.8)]}$$

$\therefore \quad (\sigma_c)_{r_1} = \dfrac{1}{r_2^2 - r_1^2}\left[p_1 r_1^2 - p_2 r_2^2 + r_2^2 (p_1 - p_2) \right] \qquad \text{(substituting } r = r_1\text{)}$

$$= \frac{1}{0.15^2 - 0.1^2}\left[60 \times 0.1^2 - 30 \times 0.15^2 + 0.15^2 (60 - 30) \right]$$

$$= \frac{1}{0.0125}(0.6 - 0.675 + 0.675) = 48 \text{ MN/m}^2 \text{ (tensile)}$$

Radial stress, in this case, is given by the relation:

$$\sigma_r = \frac{1}{r_2^2 - r_1^2}\left[p_2 r_2^2 - p_1 r_1^2 + \frac{r_1^2 r_2^2}{r^2}(p_1 - p_2) \right]$$

$\therefore \quad (\sigma_r)_{r_1} = \dfrac{1}{r_2^2 - r_1^2}\left[p_2 r_2^2 - p_1 r_1^2 + \dfrac{r_1^2 r_2^2}{r_1^2}(p_1 - p_2) \right]$

$$= \frac{1}{r_2^2 - r_1^2}\left[p_2 r_2^2 - p_1 r_1^2 + r_2^2 (p_1 - p_2) \right]$$

$$= \frac{1}{0.15^2 - 0.1^2}\left[30 \times 0.15^2 - 60 \times 0.1^2 + 0.15^2 (60 - 30) \right]$$

$$= \frac{1}{0.0125}\left[0.675 - 0.6 + 0.675 \right] = 60 \text{ MN/m}^2 \text{ (compressive)}$$

Maximum shear stress,

$$(\tau_{max})_{r_1} = \frac{(\sigma_c)_{r_1} - (\sigma_r)_{r_1}}{2} = \frac{48 - (-60)}{2} = \mathbf{54 \text{ MN/m}^2} \quad \textbf{(Ans.)}$$

$$[\because (\sigma_r)_{r_1} \text{ is compressive}]$$
$$\text{(assuming cylinder having closed ends)}$$

Controls inside a submarine.

Example 11·8. *An external pressure of 10 MN/m² is applied to a thick cylinder of internal diameter 150 mm and external diameter 300 mm. If the maximum hoop stress permitted on the inside wall is 35 MN/m², calculate:*

   (i)   *The maximum internal pressure that can be applied;*

   (ii) *The change in outside diameter if cylinder has the closed ends.*

Take: $E = 210$ *GN/m², and* $\dfrac{1}{m} = 0{\cdot}3$ **(AMIE Summer, 2002)**

Solution. *Given:* External pressure = 10 MN/m²;

$$r_1 = \frac{150}{2} = 75 \text{ mm} = 0{\cdot}075 \text{ m}; \quad r_2 = \frac{300}{2} = 150 \text{ mm};$$

$$(\sigma_c)_{r_1} = 35 \text{ MN/m}^2; \quad E = 210 \text{ GN/m}^2; \quad \frac{1}{m} = 0{\cdot}3.$$

Fig. 11·5. shows a cylindrical section subjected to external and internal radial pressures.

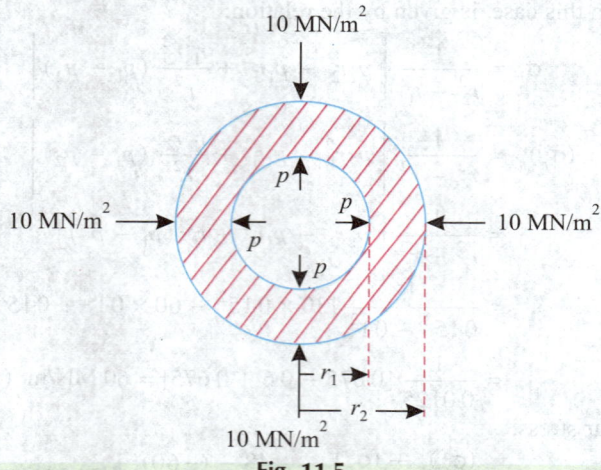

Fig. 11.5

(i) Maximum internal pressure, p:

Boundary conditions:

At radius $r_1 = 0{\cdot}075$ m, $p_{r_1} = p$ MN/m²

At radius $r_2' = 0{\cdot}15$ m, $p_{r_2} = 10$ MN/m²

Taking a and b as Lame's constants,

$$p = \frac{b}{0{\cdot}075^2} - a \qquad\qquad\qquad ...(i)$$

and, $$10 = \frac{b}{0{\cdot}15^2} - a \qquad\qquad\qquad ...(ii)$$

Subtracting eqn. (ii) and (i), we have

$$(p - 10) = b\left[\frac{1}{0{\cdot}075^2} - \frac{1}{0{\cdot}15^2}\right] = b\,[177{\cdot}78 - 44{\cdot}44]$$

or, $p - 10 = 133{\cdot}34\, b$ $\therefore$ $b = \dfrac{p - 10}{133{\cdot}34}$

From eqn. (ii), we have

$$a = \frac{b}{0{\cdot}15^2} - 10 = \frac{p - 10}{133{\cdot}34 \times 0{\cdot}15^2} - 10$$

or, $$a = \frac{p - 40}{3}$$

We know, $$\sigma_c = \frac{b}{r^2} + a \quad \text{or} \quad (\sigma_c)_{r_1} = \frac{b}{r_1^2} + a$$

or, $$35 = \frac{p - 10}{133 \cdot 34} \times \frac{1}{0 \cdot 075^2} + \frac{p - 40}{3}$$

or, $$35 = \frac{p - 10}{0 \cdot 75} + \frac{p - 40}{3} = 1 \cdot 333p - 13 \cdot 333 + 0 \cdot 333p - 13 \cdot 333$$

or, $$35 = 1 \cdot 666p - 26 \cdot 666$$

∴ $$p = \textbf{37 MN/m}^2 \quad \textbf{(Ans.)}$$

Alternatively

Using the eqn. (11·8), we have

$$\sigma_c = \frac{1}{r_2^2 - r_1^2} \left[p_1 r_1^2 - p_2 r_2^2 + \frac{r_1^2 r_2^2}{r^2}(p_1 - p_2) \right]$$

or, $$(\sigma_c)_{r_1} = \frac{1}{r_2^2 - r_1^2} \left[p_1 r_1^2 - p_2 r_2^2 + r_2^2 (p_1 - p_2) \right] \qquad \text{(Putting } r = r_1)$$

or, $$35 = \frac{1}{0 \cdot 15^2 - 0 \cdot 075^2} \left[p_1 \times 0 \cdot 075^2 - 10 \times 0 \cdot 15^2 + 0 \cdot 15^2 (p_1 - 10) \right]$$

$$= \frac{1}{0 \cdot 016875} \left[0 \cdot 005625\, p_1 - 0 \cdot 225 + 0 \cdot 0225\, p_1 - 0 \cdot 225) \right]$$

$$35 \times 0 \cdot 016875 = (0 \cdot 02815\, p_1 - 0 \cdot 45)$$

∴ $$p_1 = \frac{(35 \times 0 \cdot 016875 + 0 \cdot 45)}{0 \cdot 02815} = 37 \text{ MN/m}^2$$

(ii) Change in outside diameter:

Hoop stress at r_1,

$$(\sigma_c)_{r_1} = 37 \times \left[\frac{r_2^2 + r_1^2}{r_2^2 - r_1^2} \right]$$

Tank heads

$$= 37 \times \left[\frac{0.15^2 + 0.075^2}{0.15^2 - 0.075^2} \right] = 61.67 \text{ MN/m}^2$$

Hoop stress at r_2,

$$(\sigma_c)_{r_2} = 37 \times \left[\frac{2r_1^2}{r_2^2 - r_1^2} \right]$$

$$= 37 \times \left[\frac{2 \times 0.075^2}{0.15^2 - 0.075^2} \right] = 24.67 \text{ MN/m}^2$$

Longitudinal stress, $\qquad \sigma_l = p \times \left[\frac{r_1^2}{(r_2^2 - r_1^2)} \right] \qquad$ ...for a cylinder with closed ends

$$= 37 \times \left[\frac{0.075^2}{0.15^2 - 0.075^2} \right] = 12.33 \text{ MN/m}^2$$

Circumferential or hoop strain at outer radius,

$$(\sigma_c)_{r_2} = \left[\frac{24.67}{E} - \frac{12.33}{mE} - \frac{10}{mE} \right] \times 10^6$$

$$= \frac{10^6}{E} [24.67 - 0.3 \times 12.33 + 0.3 \times 10]$$

$$= \frac{10^6}{E} \times 23.97$$

∴ *Change in diameter* $\qquad = (\sigma_c)_{r_2} \times d_2 = \frac{10^6}{E} \times 23.97 \times d_2$

$$= \frac{10^6}{210 \times 10^9} \times 23.97 \times 300 \text{ mm} = \mathbf{0.03424 \text{ mm}} \quad \textbf{(Ans.)}$$

11·1·3. Design of Thick Cylindrical Shell

We know that the maximum values of σ_c and σ_r occur at the inner radius r_1. Thus at any radius r, the three principal stresses are as given below:

(*i*) Radial stress, σ_r (comp.)

(*ii*) Circumferential stress σ_c (tensile),

$$\sigma_c = \frac{p_1 r_1^2}{r_2^2 - r_1^2} \left[\frac{r_2^2}{r_1^2} + 1 \right]$$

$$= p_1 \left[\frac{r_2^2 + r_1^2}{r_2^2 - r_1^2} \right] = P_1 \frac{K^2 + 1}{K^2 - 1} \qquad \left(\text{where } K = \frac{r_2}{r_1} \right)$$

(*iii*) Longitudinal stress, $\quad \sigma_l = \frac{p_1 r_1^2}{r_2^2 - r_1^2} = \frac{p_1}{K^2 - 1}$ (tensile).

The *circumferential stress* (σ_c) is the *largest* out of the above mentioned three principal stresses.

The safe ratio of thickness of wall to the bore of the tube for a given pressure, can be found on the basis of any of the five theories of elastic failure.

1. **Maximum principal stress criteria :**

If σ is the simple stress for elastic failure, then

$$\sigma_r \left(\frac{K^2 + 1}{K^2 - 1} \right) \leq \sigma \qquad \qquad ...(11 \cdot 17)$$

2. **Maximum principal strain criteria:**

Maximum strain,
$$e_c = \frac{\sigma_c}{E} + \frac{\sigma_r}{mE} - \frac{\sigma_l}{mE}$$

∴
$$\sigma = e_c E \geq \left(\sigma_c + \frac{\sigma_r}{m} - \frac{\sigma_l}{m} \right)$$

or
$$\sigma \geq \sigma_r \left[\frac{K^2 + 1}{K^2 - 1} + \frac{1}{m} - \frac{1}{m} \cdot \frac{1}{K^2 - 1} \right]$$

or
$$\sigma \geq \sigma_r \left[\frac{K^2 + 1}{K^2 - 1} + \frac{1}{m} \cdot \frac{K^2 - 2}{K^2 - 1} \right] \qquad \qquad ...(11 \cdot 18)$$

If there is no longitudinal stress in the shell, we have

$$\sigma = e_c E \geq \sigma_r \left[\frac{K^2 + 1}{K^2 - 1} + \frac{1}{m} \right] \qquad \qquad ...(11 \cdot 19)$$

3. **Maximum shear stress criteria:**

The maximum shear intensity $= \dfrac{\sigma_c + \sigma_r}{2} = \dfrac{\sigma_r}{2} \left[\dfrac{K^2 + 1}{K^2 - 1} + 1 \right] = \dfrac{\sigma_r}{2} \cdot \dfrac{2K^2}{K^2 - 1}$

Hence,
$$\frac{\sigma}{2} \geq \frac{\sigma_r}{2} \cdot \frac{2K^2}{K^2 - 1} \text{ or } \sigma \geq \sigma_r \cdot \frac{2K^2}{K^2 - 1} \qquad \qquad ...(11 \cdot 20)$$

4. **Maximum strain energy:**

We know that the strain energy (U) per unit volume under the three principal stresses is given by,

$$U = \frac{1}{2E} \left[\sigma_1^2 + \sigma_2^2 + \sigma_3^2 - \frac{2}{m} (\sigma_1 \sigma_2 + \sigma_2 \sigma_3 + \sigma_3 \sigma_1) \right]$$

Substituting $\sigma_1 = \sigma_c = \sigma_r \left[\dfrac{K^2 + 1}{K^2 - 1} \right]$, $\sigma_2 = \sigma_r$

and,
$$\sigma_3 = \sigma_l = \sigma_r \cdot \frac{1}{K^2 - 1}, \text{ we get}$$

$$U = \frac{\sigma_r^2}{2E} \frac{\left[2 \left(1 + \dfrac{1}{m} \right) K^4 + 3 \left(1 - \dfrac{2}{m} \right) \right]}{(K^2 - 1)^2}$$

The energy for a single tensile stress,

$$\sigma = \frac{\sigma^2}{2E}$$

Hence we get the criteria:
$$\sigma \geq \sigma_r \cdot \frac{\sqrt{2 \left(1 + \dfrac{1}{m} \right) K^4 + 3 \left(1 - \dfrac{1}{m} \right)}}{K^2 - 1} \qquad \qquad ...(11 \cdot 21)$$

Similarly, it can be shown that if σ_l is neglected, we have

$$\sigma \geq \dfrac{\sigma_r \sqrt{2\left(1 + \dfrac{1}{m}\right) K^4 + 2\left(1 + \dfrac{1}{m}\right)}}{(K^2 - 1)} \qquad \text{...(11·22)}$$

5. Maximum shear strain energy:

It can be shown that the shear strain energy U per unit volume is $\dfrac{\sigma_r^2}{2E} \cdot \dfrac{3K^4}{(K-1)^2}$

Hence, $\qquad\qquad\qquad \sigma \geq \sigma_r \dfrac{\sqrt{3K^4}}{(K^2 - 1)} \qquad\qquad\qquad \text{...(11·23)}$

If σ_l is neglected, we have $\qquad \sigma \geq \sigma_r \dfrac{\sqrt{3K^4 + 1}}{(K^2 - 1)} \qquad\qquad \text{...(11·24)}$

Example 11·9. *A thick cylindrical shell is of 300 mm internal diameter and has to withstand an internal pressure of 30 MN/m². If the permissible tensile stress of the shell is 150 MN/m² and $\dfrac{1}{m} = 0.3$ calculate the thickness of metal necessary for the cylinder on the basis of the following criteria:*

(i)   Maximum principal stress;
(ii) Maximum principal strain;
(iii) Maximum shear;
(iv) Maximum strain energy.

Neglect the longitudinal direct stress.

These fury type tank washers shoot cleaning liquid as high-speed jets to clean the inside of tanks.

Solution. Internal radius of the shell, $r_1 = \dfrac{300}{2} = 150$ mm

Internal pressure $\qquad\qquad\qquad p(\sigma_r) = 30$ MN/m²

Permissible tensile stress, $\qquad\qquad \sigma = 150$ MN/m²

Poisson's ratio, $\qquad\qquad\qquad \dfrac{1}{m} = 0.3$

Metal thickness:

(i) Maximum principal stress criteria:

We know that, $\qquad\qquad \sigma_r \left(\dfrac{K^2 + 1}{K^2 - 1}\right) \leq \sigma \qquad\qquad$ ...from eqn. (11·17)

$$\left(\text{where, } K = \dfrac{r_2}{r_1}\right)$$

or, $\qquad\qquad\qquad\qquad \dfrac{K^2 + 1}{K^2 - 1} \leq \dfrac{150}{30} \leq 5$

From which, $\qquad\qquad\qquad K^2 \leq 1.5$

or, $\qquad\qquad\qquad\qquad K = 1.225$

i.e. $\qquad\qquad\qquad\qquad \dfrac{r_2}{r_1} = 1.225$

or, $\qquad\qquad\qquad\qquad r_2 = 1.225 \times 150 = 183.75$ mm

$\therefore$ Metal thickness $= r_2 - r_1 = 183 \cdot 75 - 150 = \textbf{33·75 mm}$ (Ans.)

(ii) Maximum principal strain criteria:

We know that, $\sigma_r \left[\dfrac{K^2+1}{K^2-1} + \dfrac{1}{m} \right] \le \sigma$

from eqn. (11·19)

or $\dfrac{K^2+1}{K^2-1} + 0.3 \le \dfrac{150}{30} \le 5$

or $\dfrac{K^2+1}{K^2-1} = 4.7$

$\therefore$ $K^2 = 1 \cdot 54$ or $K = 1 \cdot 24$

or, $r_2 = K r_1 = 1 \cdot 24 \times 150 = 186$ mm

$\therefore$ Metal thickness $= r_2 - r_1$

$= 186 - 150 = \textbf{36 mm}$ (Ans.)

Stainless steel tank.

(iii) Maximum shear stress criteria:

We know that, $\dfrac{2\,\sigma_r\,K^2}{K^2-1} \le \sigma$

or, $\dfrac{K^2}{K^2-1} \le \dfrac{\sigma}{2\,\sigma_r} \le \dfrac{150}{2 \times 30} = 2 \cdot 5$

or, $K^2 = 1 \cdot 67$

or, $K = 1 \cdot 29$

$\therefore$ $r_2 = K r_1 = 1 \cdot 29 \times 150 = 193 \cdot 5$ mm

$\therefore$ Metal thickness $= r_2 - r_1 = 193 \cdot 5 - 150 = \textbf{43·5 mm}$ (Ans.)

(iv) Maximum strain energy criteria:

We know that,

$$\sigma^2 \ge \frac{2\,\sigma_r^2 \left[K^4 \left(1 + \dfrac{1}{m} \right) + \left(1 - \dfrac{1}{m} \right) \right]}{(K^2-1)^2}$$

or, $150^2 (K^2-1)^2 = 2 \times 30^2 (1 \cdot 3 K^4 + 0 \cdot 7)$

or, $12 \cdot 5 (K^4 - 2K^2 + 1) = 1 \cdot 3 K^4 + 0 \cdot 7$

or, $11 \cdot 2 K^4 - 25 K^2 + 11 \cdot 8 = 0$

or, $K^4 - 2 \cdot 23 K^2 + 1 \cdot 054 = 0$

or, $K^2 = \dfrac{2 \cdot 23 + \sqrt{2 \cdot 23^2 - 4 \times 1 \cdot 054}}{2} = 1 \cdot 49$

or, $K = 1 \cdot 22$

$\therefore$ $r_2 = K r_1 = 1 \cdot 22 \times 150 = 183$ mm

$\therefore$ Metal thickness $= r_2 - r_1 = 183 - 150 = \textbf{33 mm}$ (Ans.)

11·1·4. Compound or Shrunk Cylinders

It has been seen in case of thin cylinders that when a thin cylinder is wound with wire under tension, the whole shell will be in a state of initial compression; but when fluid is admitted the circumferential (or hoop) stress is reduced. This method can be also used in case of thick cylinders to reduce the maximum circumferential stress.

Another method which is usually used for the design of gun tube is to shrink one tube on to another. By keeping proper initial difference in common radii, the required shrinkage pressure at the junction can be developed.

— The initial shrinkage pressure p' will set *initial compressive circumferential stresses in inner tube and initial tensile circumferential stresses in the outer tube.*

— On the admittance of fluid, the internal pressure will be jointly resisted by the internal and external tubes. Due to the fluid pressure, tensile *circumferential stresses will be set in the inner and outer tubes.*

— The final circumferential (or hoop) stresses in the tubes will be the *algebraic sum of initial stresses and stresses due to fluid pressure.*

— The distribution of stresses in a compound cylinder is shown in Fig. 11·6.

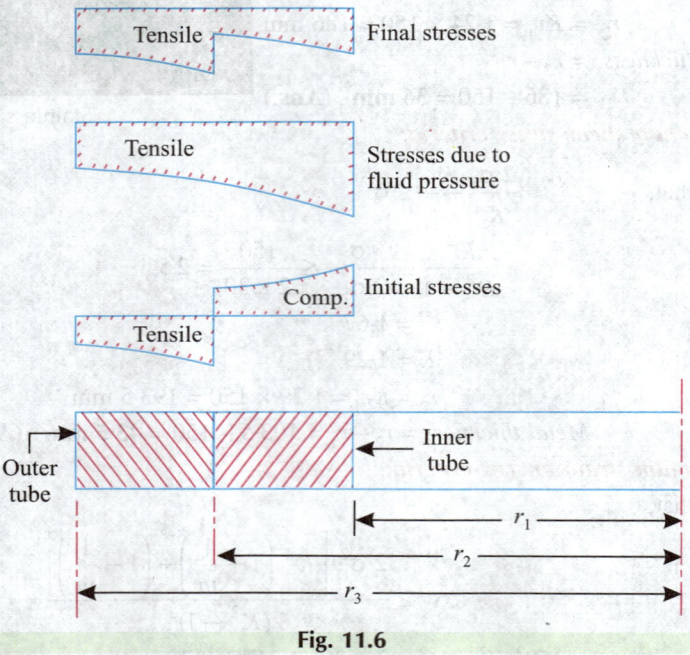

Fig. 11.6

Initial stresses:

Let,

$\qquad r_2$ = Radius at the common surface, and

$\qquad p$ = The shrinkage pressure.

Inner tube:

$\qquad$ Inner radius = r_1; outer radius = r_2

The *Lame's equations* are:

$$\sigma_r = \frac{b}{r^2} - a \qquad \qquad \text{...(1)}$$

and,

$$\sigma_c = \frac{b}{r^2} + a \qquad \qquad \text{...(2)}$$

At,

$\qquad r = r_1, \sigma_r = 0;$

$\qquad r = r_2, \sigma_r = p$

Hence,

$$\frac{b}{r^2} - a = 0 \qquad \qquad \text{...(i)}$$

and,
$$\frac{b}{r_2^2} - a = p \qquad \qquad ...(ii)$$

Outer tube:

Inner radius = r_2; outer radius = r_3

$$\sigma_r = \frac{b'}{r^2} - a' \qquad \qquad ...(3)$$

and,
$$\sigma_c = \frac{b'}{r^2} + a' \qquad \qquad ...(4)$$

At,
$$r = r_3, \sigma_r = 0$$
$$r = r_2, \sigma_r = p$$

Hence,
$$\frac{b'}{r_3^2} - a' = 0 \qquad \qquad ...(iii)$$

and,
$$\frac{b'}{r_2^2} - a' = p \qquad \qquad (iv)$$

The four constants a, b, a' and b' can be found out from equations (i) to (iv); knowing the four constants, the initial circumferential stresses can be found from eqns. (2) and (4).

Stresses due to fluid pressure:

In the compound cylinder, when the *fluid is admitted*, the intensity of pressure is resisted *jointly by the section* (of the compound cylinder). By applying the *Lame's equations* to the compound cylinder (under internal pressure), we have

$$\sigma_r = \frac{b}{r^2} - a \qquad \qquad ...(i)$$

and,
$$\sigma_c = \frac{b}{r^2} + a \qquad \qquad ...(ii)$$

At,
$$r = r_1, \ \sigma_r = p_1; \qquad \qquad ...(iii)$$
$$r = r_3, \ \sigma_r = 0 \qquad \qquad ...(iv)$$

The two constants a and b can be calculated from equations (iii) and (iv). Thus, the intensity of circumferential stress due to fluid pressure can be calculated.

Final stresses:

Final stresses (in the compound cylinder) = Initial stresses + stresses due to fluid pressure.

11·1·5. Necessary Difference of Radii for Shrinkage

In order to find out the necessary difference of radii for shrinkage let us proceed as follows:

Let, r_2 = The common radius of the tubes at junction after the shrinakge on,

$\delta r'$ = The difference between the outer radius of inner tube and r_2, and

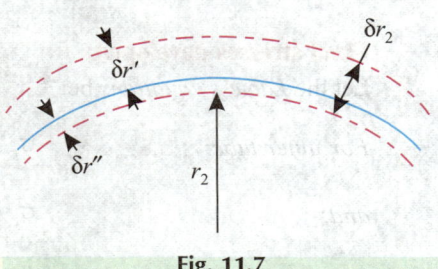

Fig. 11.7

$\delta r''$ = The difference between the inner radius of the outer tube and r_2.

If δr_2 is the difference in the radii before shrinking on, we have

$$\delta r_2 = \delta r' + \delta r'' \qquad \qquad ...(i)$$

For the *inner tube*; the circumferential strain at the common radius r_2 is given by

$$\frac{\delta r'}{r_2} = \frac{1}{E}\left[\left(\frac{b}{r_2^2}+a\right)+\frac{p}{m}\right] \text{comp.} \quad \text{...(ii)}$$

Similarly, for the *outer tube*, the circumferential strain at the common radius r_2 is given by

$$\frac{\delta r''}{r_2} = \frac{1}{E}\left[\left(\frac{b'}{r_2^2}+a'\right)+\frac{p}{m}\right] \text{tensile}$$

$$\text{...(iii)}$$

From eqns. (*ii*) and (*iii*), we have

$$\frac{\delta r' + \delta r''}{r_2} = \frac{\delta r_2}{r_2} = \frac{1}{E}\left[\left(\frac{b'}{r_2^2}+a'\right)-\left(\frac{b}{r_2^2}+a\right)\right]$$

$$\text{...(11·25)}$$

Vertical stainless steel tanks.

Thus, the original difference of radii at the junction radius r_2, before shrinking on, divided by r_2 is equal to the algebraic difference of the circumferential stresses for the two tubes at the junction divided by E.

Example 11·10. *A compound cylinder, formed by shrinking one tube to another is subjected to an internal pressure of 90 MN/m². Before the fluid is admitted, the internal and external diameters of the compound cylinder are 180 mm and 300 mm respectively and the diameter at the junction is 240 mm. If after shrinking on, the radial pressure at the common surface is 12 MN/m², determine the final stresses developed in the compound cylinder.*

Solution. Internal pressure in the cylinder,

$$p_1 = 90 \text{ MN/m}^2$$

Internal radius of the cylinder, $r_1 = \dfrac{180}{2} = 90 \text{ mm} = 0.09 \text{ m}$

External radius of the cylinder, $r_3 = \dfrac{300}{2} = 150 \text{ mm} = 0.15 \text{ m}$

Radius at the junction, $r_2 = \dfrac{240}{2} = 120 \text{ mm} = 0.12 \text{ m}$

Radial pressure at the common surface after shrinking on,

$$p = 12 \text{ MN/m}^2$$

Final stresses developed:

Let the *Lame's equations* be:

For inner tube: $\qquad \sigma_r = \dfrac{b}{r^2} - a \qquad \qquad \text{...(1)}$

and, $\qquad \sigma_c = \dfrac{b}{r^2} + a \qquad \qquad \text{...(2)}$

For outer tube: $\qquad \sigma_r = \dfrac{b'}{r^2} - a' \qquad \qquad \text{...(3)}$

and, $\qquad \sigma_c = \dfrac{b'}{r^2} + a' \qquad \qquad \text{...(4)}$

(a) *Before the fluid is admitted:*

Inner tube:

At, $r = r_1 = 0.09$ m, $\sigma_r = 0$

∴ $\dfrac{b}{0.0081} - a = 0$

or, $123.456\, b - a = 0$ ...(i)

At, $r = r_2 = 0.12$ m,

                                          $\sigma_r = 12$ MN/m²

∴ $\dfrac{b}{0.0144} - a = 12$

or, $69.44\, b - a = 12$ ...(ii)

From eqns. (i) and (ii), we get

$$b = -0.222 \text{ and } a = -27.41$$

Hence circumferential stress at any point in the *inner tube* will be given by

$$\sigma_c = -\frac{0.222}{r^2} - 27.41$$

The *minus sign indicates that the stress will be wholly compressive.*

At, $r = r_1 = 0.09$ m,

$$\sigma_{c(0.09)} = -\frac{0.222}{0.09^2} - 27.41 = 54.82 \ \text{MN/m}^2 \text{ (comp.)}$$

At, $r = 0.12$ m,

$$\sigma_{c(0.12)} = -\frac{0.222}{0.12^2} - 27.41 = 42.82 \ \text{MN/m}^2 \text{ (comp.)}$$

Outer tube:

At, $r = 0.15$ m, $\sigma_r = 0$

∴ $\dfrac{b'}{0.15^2} - a' = 0$

or, $44.44\, b' - a' = 0$ ...(iii)

At, $r = 0.12$ m, $\sigma_r = 12$ MN/m²

∴ $\dfrac{b'}{0.12^2} - a' = 0$

or, $69.44\, b' - a' = 12$ ...(iv)

From eqns. (iii) and (iv), we get

$$b' = +0.48, \text{ and } a' = +21.33$$

Hence the circumferential stress at any point in the *outer tube* will be given by

$$\sigma_c = \frac{0.48}{r^2} + 21.33$$

At, $r = 0.12$ m,

$$\sigma_{c(0.12)} = \frac{0.48}{0.12^2} + 21.33 = 54.66 \ \text{MN/m}^2 \text{ (tensile)}$$

At, $r = 0.15$ m,

$$\sigma_{c(0.15)} = \frac{0.48}{0.15^2} + 21.33 = 42.66 \ \text{MN/m}^2 \text{ (tensile)}$$

(b) *After the fluid is admitted:*

Let the *Lame's equation* be:

$$\sigma_r = \frac{b}{r^2} - a$$

At, $r = 0.09$ m, $\sigma_r = 90$ MN/m^2

∴ $90 = \dfrac{b}{0.09^2} - a$

or, $90 = 123.45\, b - a$ ...(v)

At, $r = 0.15$ m, $\sigma_r = 0$

∴ $0 = \dfrac{b}{0.15^2} - a$

or $0 = 44.44\, b - a$ ...(vi)

From eqns. (v) and (vi), we get

$$b = 1.139 \text{ and } a = 50.61$$

Hence, the circumferential stress at any point in the compound tube is given by,

$$\sigma_c = \frac{b}{r^2} + a$$

At, $r = 0.09$ m, $\sigma_{c(0.09)} = \dfrac{1.139}{0.09^2} + 50.61 = 191.23$ MN/m^2 (tensile)

$r = 0.12$ m, $\sigma_{c(0.12)} = \dfrac{1.139}{0.12^2} + 50.61 = 129.71$ MN/m^2 (tensile)

$r = 0.15$ m, $\sigma_{c(0.15)} = \dfrac{1.139}{0.15^2} + 50.61 = 101.23$ MN/m^2 (tensile)

The *final circumferential stresses* at different points are tabulated below:

Tensile stress........ +

Compressive stress..... −

Circumferential (or hoop) stress (MN/m^2)	Inner tube		Outer tube	
	r = 0.09 m	r = 0.12 m	r = 0.12 m	r = 0.15 m
(i) Initially	− 54·82	− 42·82	+ 54·66	+ 42·66
(ii) Due to fluid pressure	+ 191·23	+ 129·71	+ 129·71	+ 101·23
Final	+ 136·41	+ 86·89	+ 184·31	+ 143·89

Hence the *final circumferential stresses* are:

Inner tube: $\sigma_{c\,(0.09)} = $ **136·41 MN/m^2** *(tensile)*

$\sigma_{c\,(0.12)} = $ **86·89 MN/m^2** *(tensile)* **(Ans.)**

Outer tube: $\sigma_{c\,(0.12)} = $ **184·31 MN/m^2** *(tensile)*

$\sigma_{c\,(0.15)} = $ **143·89 MN/m^2** *(tensile)* **(Ans.)**

Example 11·11. *In example 11·10, calculate necessary difference in diameter of the two tubes at the common surface before shrinking on, so as to produce a radial pressure of 12 MN / m^2.*

Also calculate the minimum temperature to which the outer tube should be heated before it can be slipped on.

Take: $E = 205$ GN/m^2, $\alpha = 0.000011$ per °C, and $\dfrac{1}{m} = 0.3$.

Solution. We know that,

$$\frac{\delta r_2}{r_2} = \frac{1}{E}\left[\left(\frac{b'}{r_2^2} + a'\right) - \left(\frac{b}{r_2^2} + a\right)\right]$$

$$= \frac{1}{205 \times 10^9}[54\cdot66 - (-42\cdot82)] \times 10^6 = 4\cdot755 \times 10^{-4}$$

∴ $$\delta r_2 = 4\cdot755 \times 10^{-4} \times 120 = 0\cdot05706 \text{ mm}$$

or, $$\delta d_2 = 2 \times 0\cdot05706 = \textbf{0·114 mm (Ans.)}$$

Now, $$\pi d_2 \cdot \alpha.t = \pi \cdot \delta d_2 \text{ (where } t = \text{rise of temperature in °C)}$$

∴ $$t = \frac{\delta d_2}{\alpha \cdot d_2} = \frac{0\cdot114}{0\cdot000011 \times 240} = \textbf{43·2°C} \quad \textbf{(Ans.)}$$

Example 11·12. *A steel tube of 300 mm external diameter is to be shrunk on to another steel tube of 90 mm internal diameter. After shrinking the diameter at the junction is 180 mm. Before shrinking on the difference of diameter at the junction is 0·12 mm.*

Find: *(i) The radial pressure at the junction;*

(ii) The circumferential stresses developed in the two tubes after shrinking on.

Take: $E = 200 \text{ GN/m}^2$.

Solution. External radius of outer steel tube,

$$r_3 = \frac{300}{2} = 150 \text{ mm} = 0\cdot15 \text{ m}$$

Internal radius of inner steel tube,

$$r_1 = \frac{90}{2} = 45 \text{ mm} = 0\cdot045 \text{ m}$$

Outer radius of the inner tube (or radius at the *junction*),

$$r_2 = \frac{180}{2} = 90 \text{ mm} = 0\cdot09 \text{ m}$$

$$E = 200 \text{ GN/m}^2$$

Difference of radius at the junction,

$$\delta r_2 = 0\cdot12 \text{ mm}$$

(i) The radial pressure at the junction, p:

Inner tube:

Let the *Lame's equations* be given by:

$$\sigma_r = \frac{b}{r^2} - a \qquad \text{...(1)}$$

and, $$\sigma_c = \frac{b}{r^2} + a \qquad \text{...(2)}$$

At, $$r = r_1 = 0\cdot045 \text{ m, } \sigma_r = 0$$

∴ $$\frac{b}{0\cdot045^2} - a = 0$$

or, $$493\cdot8\, b - a = 0 \qquad \text{...(iii)}$$

At, $$r = r_2 = 0\cdot09 \text{ m,} \quad \sigma_r = p$$

∴ $$\frac{b}{(0\cdot09)^2} - a = p$$

or, $$123\cdot457\, b - a = p \qquad \text{...(ii)}$$

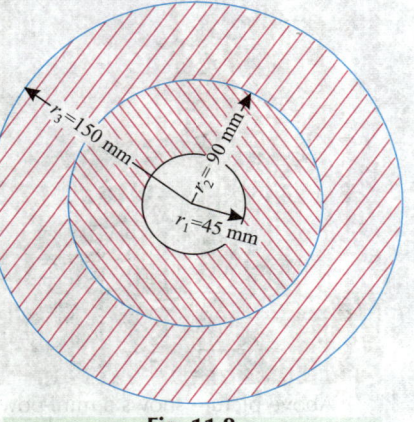

Fig. 11.8

Outer tube:

Let the *Lame's equations* be given by:

$$\sigma_r = \frac{b'}{r^2} - a' \qquad \qquad ...(3)$$

and,

$$\sigma_c = \frac{b'}{r^2} + a' \qquad \qquad ...(4)$$

At, $r = r_3 = 0.15$ m, $\sigma_r = 0$

$$\therefore \qquad = \frac{b'}{0.15^2} - a' = 0$$

or, $\qquad 44.44\, b' - a' = 0 \qquad \qquad ...(iii)$

At, $\qquad r = r_2 = 0.09$ m, $\sigma_r = p$

$$\therefore \qquad \frac{b'}{0.09^2} - a' = p$$

or, $\qquad 123.457\, b' - a' = p \qquad \qquad ...(iv)$

From eqns. (*ii*) and (*iv*), we have

$$123.457\, b - a = 123.457\, b' - a' = p \qquad \qquad ...(v)$$

Finally,

$$\frac{\delta r_2}{r_2} = \frac{0.12}{90} = \frac{1}{E}\left[\left(\frac{b'}{r_2^2} + a'\right) - \left(\frac{b}{r_2^2} + a\right)\right] \qquad ...(vi)$$

From eqns. (*i*) and (*iii*), we have

$$b = 0.002025\, a \text{ and } b' = 0.0225\, a' \qquad \qquad ...(vii)$$

Substituting for b and b' in eqns. (*v*) and (*vi*), we get

$$123.457 \times 0.002025\, a - a = 123.547 \times 0.0225\, a' - a'$$

$$- 0.75\, a = 1.78\, a' \quad \therefore \quad a = -2.73\, a'$$

and, $\left[\left\{\dfrac{0.0225\, a'}{(0.09)^2} + a'\right\} - \left\{\dfrac{0.002025\, a}{(0.09)^2} + a\right\}\right] \times 10^6 = \dfrac{0.12}{90} \times 200 \times 10^9$

$$3.78\, a' - 1.25\, a = 266.66 \qquad \qquad ...(viii)$$

Above picture shows a mini-power plant. Thin and thick shells and tanks are common things in industries.

Solving eqns. (vii) and (viii), we get

$$a' = 39.55, \quad a = -93.7$$
$$b' = 0.89, \quad b = -0.189$$

Hence the *Lame's equations* for the two tubes are as follows:

Inner tube:

$$\sigma_r = -\frac{0.189}{r^2} + 93.7$$

and,

$$\sigma_c = -\frac{0.89}{r^2} - 93.7$$

Outer tube:

$$\sigma_r = \frac{0.89}{r^2} - 39.55$$

$$\sigma_c = \frac{0.89}{r^2} + 39.55$$

At the junction, *i.e.* at $r = 0.09$ m, the radial pressure

$$= p = -\frac{0.189}{0.09^2} + 93.7 = \textbf{70.3 MN/m}^2 \quad \textbf{(Ans)}$$

$$\left[\text{Alternatively;} \quad p = \frac{0.89}{0.09^2} - 39.55 = 70.3 \text{ MN/m}^2 \right]$$

(ii) Circumferential stresses in the two tubes:

The circumferential stresses developed in the two tubes after shrinking on can be calculated as follows:

Inner tube:

$$\sigma_{c(0.09)} = -\frac{0.189}{0.09^2} - 93.7 = \textbf{117 MN/m}^2 \text{ (compressive)} \quad \textbf{(Ans.)}$$

$$\sigma_{c(0.45)} = -\frac{0.189}{0.045^2} - 93.7 = \textbf{187 MN/m}^2 \text{ (compressive)} \quad \textbf{(Ans.)}$$

Outer tube:

$$\sigma_{c(0.15)} = \frac{0.89}{0.15^2} + 39.55 = \textbf{79.1 MN/m}^2 \text{ (tensile)} \quad \textbf{(Ans.)}$$

$$\sigma_{c(0.09)} = \frac{0.89}{0.09^2} + 39.55 = \textbf{149.4 MN/m}^2 \text{ (tensile)} \quad \textbf{(Ans.)}$$

Example 11·13. *The external diameter of steel collar is 240 mm and internal diameter decreases by 0·15 mm when shrunk on to a solid steel shaft of 150 mm diameter. Find:*

(i) *Radial pressure between the collar and the shaft;*

(ii) *Circumferential stress at the inner surface of the tube;*

(iii) *Reduction in diameter of the shaft.*

Take: $E = 205$ GN/m², *and Poisson's ratio* = 0·304.

Solution. External radius of steel collar,

$$r_2 = \frac{240}{2} = 120 \text{ mm} = 0.12 \text{ m}$$

Radius of the solid steel shaft,

$$r_1 = \frac{150}{2} = 75 \text{ mm} = 0.075 \text{ m}$$

Decrease in the internal diameter of the collar = 0·15 mm

$$E = 205 \text{ GN/m}^2$$

Poisson's ratio, $\dfrac{1}{m} = 0·304$

(i) Radial pressure, p:

Let suffix 1 stand for the *shaft* and 2 for *collar*. Due to the radial pressure at the junction, the shaft will have compressive stress throughout.

Let, $\quad\quad\quad\quad\quad\quad p$ = Radial pressure at the junction, and

$\quad\quad\quad\quad\quad\quad \sigma_c$ = Circumferential stress at the junction.

We know that, $\quad\quad \sigma_c = p \left[\dfrac{r_2^2 + r_1^2}{r_2^2 - r_1^2} \right]$

or, $\quad\quad\quad\quad \sigma_c = p \left[\dfrac{0·12^2 + 0·075^2}{0·12^2 - 0·075^2} \right] = 2·28\,p$ ...(i)

Also, $\quad\quad\quad\quad e_2 = \dfrac{1}{E} \left[\sigma_c + \dfrac{p}{m} \right]$

But, $\quad\quad\quad\quad e_2 = \dfrac{0·15}{150} = 0·001$

∴ $\quad\quad\quad 0·001 = \dfrac{1}{205 \times 10^9} \left(2·28\,p + \dfrac{p}{m} \right) \times 10^6$

or, $\quad\quad\quad 205 = 2·28p + 0·304p$

or, $\quad\quad\quad p = \mathbf{79·33\ MN/m^2}$ **(Ans.)**

(ii) Circumferential stress, σ_c:

$$\sigma_c = 2·28p \quad\quad\quad \text{[Equation (i)]}$$
$$= 2·28 \times 79·33 = 180·87 \text{ MN/m}^2$$

i.e. $\quad\quad\quad \sigma_c = \mathbf{180·87\ MN/m^2}$ **(Ans.)**

(iii) Reduction in diameter of the shaft, δd_1:

We know that, $\quad e_1 = \dfrac{p}{E} - \dfrac{p}{mE} = \dfrac{p}{E} \left(1 - \dfrac{1}{m} \right)$

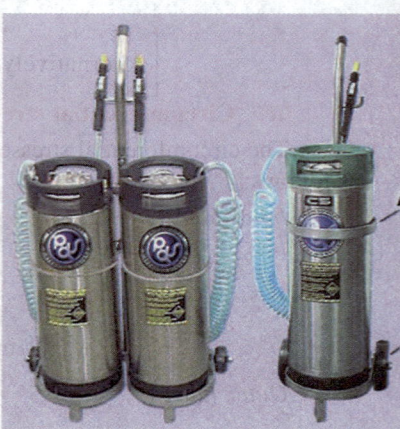

Portable sprayer tanks are widely used spray pesticides and cleaning fluid:

$$= \dfrac{p}{E} (1 - 0·304)$$

$$= \dfrac{79·33 \times 10^6}{205 \times 10^9} (1 - 0·304) = 0·0002693$$

Also, $\quad\quad\quad e_1 = \dfrac{\delta d_1}{d_1}$

or, $\quad\quad 0.0002693 = \dfrac{\delta d_1}{150}$

∴ $\quad\quad\quad \delta d_1 = 0·0002693 \times 150 = 0·04039 \text{ mm}$

i.e. $\quad\quad\quad \delta d_1 = \mathbf{0·04039\ mm}$ **(Ans.)**

11·2. THICK SPHERICAL SHELLS

Fig. 11·9 shows a thick spherical shell.

Let, r_1 = Inner radius,

r_2 = Outer radius,

p = Internal pressure,

σ_r = Radial compressive stress at any radius r,

$\sigma_r + d\sigma_r$ = Radial compressive stress at radius $(r + dr)$, and

σ_c = Circumferential tensile stress (equal in all directions perpendicular to the radius).

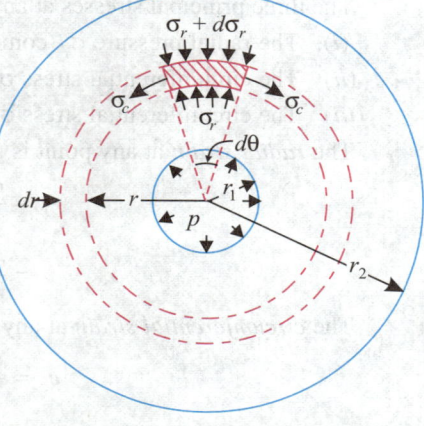

Let us consider the forces on an elementary spherical shell of radius r and thickness dr, to form equation between the principal stresses σ_r and σ_c.

The *bursting force* on any diametral plane on the section of the elemental shell

Fig. 11.9

$$= \sigma_r \cdot \pi r^2 - (\sigma_r + d\sigma_r) \, \pi \, (r + dr)^2$$

Resisting force $= \sigma_c \times 2\pi r \, dr$

Equating the resisting force and the bursting force, we have

$$\sigma_c \times 2\pi \, r \, dr = \sigma_r \cdot \pi r^2 - (\sigma_r + d\sigma_r) \, \pi \, (r + dr)^2$$

Neglecting squares and products of very small quantities, we have

$$2\sigma_c = -2\sigma_r - r \frac{d\sigma_r}{dr} \quad \text{or} \quad \sigma_c = -\sigma_r - \frac{r}{2} \frac{d\sigma_r}{dr} \qquad ...(i)$$

Differentiating the above relation, we get

$$\frac{d\sigma_c}{dr} = -\frac{d\sigma_r}{dr} - \frac{1}{2}\left[r \cdot \frac{d^2\sigma_r}{dr^2} + \frac{d\sigma_r}{dr} \right] \qquad ...(ii)$$

Technicians working at a spherical tank.

The three principal stresses at any point in the elementary spherical shell at a radius r are:

(*i*)   The radial pressure σ_r (compressive);

(*ii*)   The circumferential stress σ_c (tensile);

(*iii*)   The circumferential stress $\sigma'_c = \sigma_c$ (tensile) on a plane perpendicular to that of (*ii*).

The *radial strain* at any point is given by:

$$e_r = \frac{\sigma_r}{E} + \frac{2\sigma_c}{mE} \text{ (compressive)}$$

$$= -\frac{1}{E}\left(\sigma_r + \frac{2\sigma_c}{m}\right)\text{(tensile)} \qquad ...(iii)$$

The *circumferential strain* at any point is given by:

$$e_c = \frac{\sigma_c}{E} - \frac{\sigma_c}{mE} + \frac{\sigma_r}{mE}\text{(tensile)}$$

$$e_c = \frac{1}{E}\left[\frac{m-1}{m}.\sigma_c + \frac{\sigma_r}{m}\right] \quad\text{(tensile)} \qquad ...(iv)$$

Due to internal pressure, let the radius increase from r to $(r + u)$

∴ Radial strain, $$e_r = \frac{d\,(r + u) - dr}{dr} \qquad ...(v)$$

or, $$e_r = \frac{du}{dr} \qquad ...(vi)$$

and circumferential strain, $$e_c = \frac{(r + u)\,d\theta - rd\theta}{rd\theta}$$

or, $$e_c = \frac{u}{r} \qquad ...(vii)$$

$$(\text{or } u = r\cdot e_c)$$

The above strains are tensile if positive.

From eqns. (*v*) and (*vi*), we have

$$e_r = \frac{d}{dr}\,(r \cdot e_c)$$

or, $$e_r = e_c + r \cdot \frac{de_c}{dr} \qquad ...(viii)$$

Substituting the values of e_r and e_c from eqns. (*iii*) and (*iv*), we get

$$-\frac{1}{E}\left[\sigma_r + \frac{2\sigma_c}{m}\right] = \frac{1}{E}\left[\frac{m-1}{m}.\sigma_c + \frac{\sigma_r}{m}\right] + \frac{r}{E}\left[\frac{m-1}{m}\frac{d\sigma_c}{dr} + \frac{1}{m}\frac{d\sigma_r}{dr}\right]$$

Simplifying and rearranging, we get

$$(m + 1)\,(\sigma_r + \sigma_c) + (m - 1)\,r \cdot \frac{d\sigma_c}{dr} + r\frac{d\sigma_r}{dr} = 0$$

Substituting the values of σ_c and $\frac{d\sigma_c}{dr}$ from eqns. (*i*) and (*ii*),

we get

$$(m + 1)\left[\sigma_r - \sigma_r - \frac{r}{2}.\frac{d\sigma_r}{dr}\right] + (m - 1)\,r\left[-\frac{d\sigma_r}{dr} - \frac{1}{2}\left(r.\frac{d^2\sigma_r}{dr^2} + \frac{d\sigma_r}{dr}\right)\right] + r.\frac{d\sigma_r}{dr} = 0$$

On simplification, we get

$$\frac{d^2\sigma_r}{dr^2} + \frac{4}{r}.\frac{d\sigma_r}{dr} = 0$$

Let,
$$\frac{d\sigma_r}{dr} = z$$

then,
$$r\frac{dz}{dr} + 4z = 0$$

$\therefore$
$$\frac{dz}{z} = -4\frac{dr}{r}$$

Integrating both sides, we get
$$\log_e z = -4\log_e r + \log_e C_1$$

(where, $\log_e C_1$ is a constant of integration)

$\therefore$
$$\log_e z = \log_e \left[\frac{C_1}{r^4}\right]$$

$\therefore$
$$z = \frac{C_1}{r^4}$$

$\therefore$
$$\frac{d\sigma_r}{dr} = \frac{C_1}{r^4}$$

or,
$$d\sigma_r = C_1 \frac{dr}{r^4}$$

Integrating both sides, we get
$$\sigma_r = -\frac{C_1}{3r^3} + C_2 \qquad \qquad ...(ix)$$

(where, C_2 is a constant of integration)

Also, we know that
$$\sigma_c = -\sigma_r - \frac{r}{2}\frac{d\sigma_r}{dr}$$

or,
$$\sigma_c = -\left[-\frac{C_1}{3r^3} + C_2\right] - \frac{r}{2}\cdot\frac{d\sigma_r}{dr}$$

$$= \frac{C_1}{3r^3} - C_2 - \frac{r}{2}\times\frac{C_1}{r^4} = \frac{C_1}{3r^3} - C_2 - \frac{C_1}{2r^3}$$

i.e.
$$\sigma_c = -\frac{C_1}{6r^3} - C_2 \qquad \qquad ...(x)$$

Now, let us consider the two expressions for σ_r [eqn. (ix)] and σ_c [eqn. (x)].

$$\sigma_r = -\frac{C_1}{3r^3} + C_2$$

$$\sigma_c = -\frac{C_1}{6r^3} - C_2$$

Putting,
$$C_1 = -6b \text{ and } C_2 = -a, \text{ we get}$$

$$\sigma_r = \frac{2b}{r^3} - a \qquad \qquad ...(11.26)$$

and,
$$\sigma_c = \frac{b}{r^3} + a \qquad \qquad ...(11.27)$$

Applying the given conditions, we have

At,
$$r = r_1, \sigma_r = p$$
$$r = r_2, \sigma_r = 0$$

$$p = \frac{2b}{r_1^3} - a \qquad \qquad ...(xi)$$

and,

$$0 = \frac{2b}{r_2^3} - a \qquad \qquad ...(xii)$$

Solving eqn. (x) and (xi), we get

$$b = \frac{pr_1^3 \, r_2^3}{2 \, (r_2^3 - r_1^3)}, \quad \text{and} \quad a = \frac{pr_1^3}{r_2^3 - r_1^3}$$

Example 11·14. *A thick spherical shell of 180 mm internal diameter is subjected to an internal fluid pressure of 24 MN/m². If the permissible tensile stress is 120 MN/m², find the thickness of the shell.*

Solution. Internal radius of the spherical shell,

$$r_1 = \frac{180}{2} = 90 \text{ mm} = 0.09 \text{ m}$$

Internal fluid pressure, p (or σ_r) = 24 MN/m²
Permissible tensile stress, σ_c = 120 MN/m²

Thickness of the shell:

We know that,

$$\sigma_r = \frac{2b}{r^3} - a \qquad \qquad ...(1)$$

$$\sigma_c = \frac{b}{r^3} + a \qquad \qquad ...(2)$$

At, $\qquad\qquad\qquad r = 0.09 \text{ m}, \ \sigma_r = 24 \text{ MN/m}^2$

∴ $\qquad\qquad \dfrac{2b}{0.09^3} - a = 24$

or, $\qquad\qquad 2743.48 \, b - a = 24 \qquad\qquad ...(i)$

At, $\qquad\qquad\qquad r = 0.09 \text{ m}, \ \sigma_c = 24 \text{ MN/m}^2$

∴ $\qquad\qquad \dfrac{b}{0.09^3} + a = 120$

or, $\qquad\qquad 1371.74 \, b + a = 120 \qquad\qquad ...(ii)$

From eqns. (i) and (ii), we get

$$b = 0.035, \text{ and } a = 72$$

Let, $\qquad\qquad\qquad r_2 = \text{External radius.}$

Then, at $\qquad\qquad\qquad r = r_2, \ \sigma_r = 0$

∴ $\qquad\qquad \dfrac{2 \times 0.035}{r_2^3} - 72 = 0$

or, $\qquad\qquad r_2 = \left(\dfrac{2 \times 0.035}{72} \right)^{1/3} = 0.099 \text{ m} = 99 \text{ mm}$

∴ $\qquad$ *Thickness of the shell* = $r_2 - r_1$ = 99 − 90 = **9 mm** **(Ans.)**

Example 11·15. *A spherical shell of 120 mm internal diameter has to withstand an internal pressure of 30 MN/m². If the permissible tensile stress is 80 MN/m² calculate thickness of the shell.*

Solution. Internal radius of the shell,

$$r_1 = \frac{120}{2} = 60 \text{ mm} = 0.06 \text{ m}$$

Internal fluid pressure, $p = 30$ MN/m^2

Permissible tensile stress, $\sigma_c = 80$ MN/m^2

Thickness of the shell:

Using the relations:

$$\sigma_r = \frac{2b}{r^3} - a \qquad \qquad ...(1)$$

and,

$$\sigma_c = \frac{b}{r^3} + a \qquad \qquad ...(2)$$

At, $r = 0.06$ m, $\sigma_r = 30$ MN/m^2

$\therefore$

$$\frac{2b}{0.06^3} - a = 30$$

or, $9259.26\, b - a = 30 \qquad \qquad ...(i)$

At, $r = 0.06$ m, $\sigma_c = 80$ MN/m^2

$\therefore$

$$\frac{b}{0.06^3} + a = 80$$

or, $4629.63\, b + a = 80 \qquad \qquad ...(ii)$

Solving eqns. (i) and (ii), we get

$$b = 0.00792, \text{ and } a = 43.33$$

Let the external radius be r_2.

Then, at $r = r_2, \sigma_r = 0$

$\therefore$

$$\frac{2 \times 0.00792}{r_2^3} - 43.33 = 0$$

or,

$$r_2 = \left(\frac{2 \times 0.00792}{43.33} \right)^{1/3} = 0.0715 \text{ m} = 71.5 \text{ mm}$$

Hence, thickness of the shell $= (r_2 - r_1) = 71.5 - 60 = $ **11.5 mm** **(Ans.)**

TYPICAL EXAMPLES (For Competitive Examinations)

Example 11·16. *Plot a curve showing the percentage increase in maximum circumferential stress over average circumferential stress for ratios of thickness to inside radius of thick walled cylinder varying from 0 to 3. The cylinder has only internal pressure.*

Solution. *The Lame's equations are:*

$$\sigma_r = \frac{b}{r^2} - a \qquad ...(i)$$

and,

$$\sigma_c = \frac{b}{r^2} + a \qquad ...(ii)$$

Fig. 11·10 shows a portion of a thick cylinder having inside radius r_1 and outside radius r_2. Let t be the thickness of the cylinder. Obviously $t = t_2 - r_1$. It is given in the question that this cylinder is subjected to internal pressure p only.

We know that the conditions are;

when, $r = r_1, \sigma_r = p,$

and, at $r = r_2, \sigma_2 = 0$

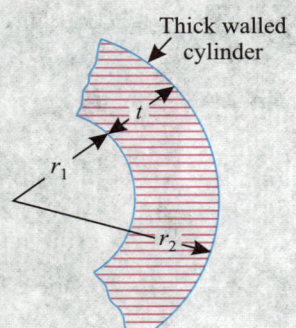

Fig. 11.10

Using the latter relation in eqn. (i), we get

$$0 = \frac{b}{r_2^2} - a$$

or,

$$b = a\,r_2^2$$

Let,

$(\sigma_c)_{max}$ = Maximum circumferential or hoop stress at inside radius r_1, and

$(\sigma_c)_{min}$ = Minimum circumferential or hoop stress at outside radius r_2.

Then,

$$(\sigma_c)_{max} = \frac{b}{r_1^2} + a = \frac{ar_2^2}{r_1^2} + a = a\left(\frac{r_1^2 + r_2^2}{r_1^2}\right) \qquad ...(iii)$$

and,

$$(\sigma_c)_{min} = \frac{b}{r_2^2} + a = \frac{a\,r_2^2}{r_2^2} + a = 2a \qquad ...(iv)$$

Let $(\sigma_c)_{av}$ be the average circumferential or hoop stress then,

$$(\sigma_c)_{av} = \frac{(\sigma_c)_{max} + (\sigma_c)_{min}}{2}$$

$$= \frac{1}{2}\left[a\left(\frac{r_1^2 + r_2^2}{r_1^2}\right) + 2a\right]$$

$$= \frac{a}{2}\left[\frac{r_1^2 + r_2^2}{r_1^2} + 2\right]$$

$$= \frac{a}{2r_1^2}(3r_1^2 + r_2^2) \qquad ...(v)$$

Now,

$$\frac{(\sigma_c)_{max}}{(\sigma_c)_{av}} = \frac{a\left(\dfrac{r_1^2 + r_2^2}{r_1^2}\right)}{\dfrac{a}{2r_1^2}(3r_1^2 + r_2^2)} = \frac{2(r_1^2 + r_2^2)}{(3r_1^2 + r_2^2)} \qquad ...(vi)$$

Steam engine, the leader of industrial revolution, has a boiler which contains steam at high pressures and temperatures. This steam ultimately makes the pistons move.

Next let us determine the percentage increase in maximum circumferential stress over average circumferential stress for ratio of thickness to inside radius of given thick cylinder varying from 0 to 3,

i.e. $\qquad = \dfrac{t}{r_1} = 0, 1, 2, \text{ and } 3.$

Case I: When, $\qquad \dfrac{t}{r_1} = 0$

$\therefore \qquad \dfrac{r_2 - r_1}{r_1} = 0$ $\qquad\qquad (\because t = r_2 - r_1)$

$\therefore \qquad r_1 = r_2$

Thus, $\qquad \dfrac{(\sigma_c)_{max}}{(\sigma_c)_{av}} = \dfrac{2(r_1^2 + r_1^2)}{(3r_1^2 + r_1^2)} = \dfrac{4r_1^2}{4r_1^2} = 1$

$\% \text{ increase} = \dfrac{(\sigma_c)_{max} - (\sigma_c)_{av}}{(\sigma_c)_{av}} \times 100$

$\qquad\qquad = \left[\dfrac{(\sigma_c)_{max}}{(\sigma_c)_{av}} - 1 \right] \times 100$

$\qquad\qquad = (1 - 1) \times 100 = 0\%$

This corresponds to the origin point 0 in Fig. 11.11.

Case II: When $\qquad \dfrac{t}{r_1} = 1$

$\therefore \qquad \dfrac{r_2 - r_1}{r_1} = 1$

or, $\qquad r_2 - r_1 = r_1 \quad \therefore r_2 = 2r_1$

Thus, $\qquad \dfrac{(\sigma_c)_{max}}{(\sigma_c)_{av}} = \dfrac{2(r_1^2 + 4r_1^2)}{(3r_1^2 + 4r_1^2)} = \dfrac{10}{7}$

$\therefore \qquad \% \text{increase} = \left[\dfrac{(\sigma_c)_{max} - (\sigma_c)_{av}}{(\sigma_c)_{av}} \right] \times 100$

$\qquad\qquad = \left[\dfrac{(\sigma_c)_{max}}{(\sigma_c)_{av}} - 1 \right] \times 100$

$\qquad\qquad = \left(\dfrac{10}{7} - 1 \right) \times 100 \approx 43\%$

This corresponds to point A_1 in Fig. 11·11.

Case III: When $\qquad \dfrac{t}{r_1} = 2$

$\therefore \qquad \dfrac{r_2 - r_1}{r_1} = 2$

or, $\qquad r_2 - r_1 = 2r_1$

$\therefore \qquad r_2 = 3r_1$

Thus, $\qquad \dfrac{(\sigma_c)_{max}}{(\sigma_c)_{av}} = \dfrac{2(r_1^2 + 9r_1^2)}{(3r_1^2 + 9r_1^2)} = \dfrac{5}{3}$

$\therefore \qquad \% \text{ increase} = \left[\dfrac{(\sigma_c)_{max} - (\sigma_c)_{av}}{(\sigma_c)_{av}} \right] \times 100$

Fig. 11.11

(graph: Y-axis labelled "Percentage increase" with values 10, 20, 30, 40, 43, 50, 60, 67, 70, 79, 80, 90, 100; X-axis labelled $\dfrac{t}{r_2}$ with values 1, 2; points A_1, A_2, A_3)

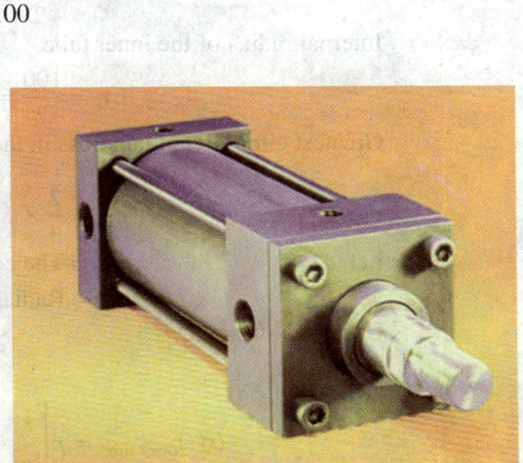

Hydraulic cylinders contain fluids at very high working pressures to move arms.

$$= \left[\frac{(\sigma_c)_{max}}{(\sigma_c)_{av}} - 1 \right] \times 100 = \left(\frac{5}{3} - 1 \right) \times 100 \approx 67\%$$

This corresponds to the point A_2 in Fig. 11.11.

Case IV: When, $\quad \dfrac{t}{r_1} = 3$

∴ $\quad \dfrac{r_2 - r_1}{r_1} = 3$

or, $\quad r_2 - r_1 = 3r_1 \quad \therefore r_2 = 4r_1$

Thus, $\quad \dfrac{(\sigma_c)_{max}}{(\sigma_c)_{av}} = \dfrac{2(r_1^2 + 16r_1^2)}{(3r_1^2 + 16r_1^2)} = \dfrac{34}{19}$

∴ $\quad \%\text{increase} = \left[\dfrac{(\sigma_c)_{max} - (\sigma_c)_{av}}{(\sigma_c)_{av}} \right] \times 100 = \left[\dfrac{(\sigma_c)_{max}}{(\sigma_c)_{av}} - 1 \right] \times 100$

$$= \left(\frac{34}{19} - 1 \right) \times 100 \approx 79\%$$

This corresponds to point A_3 in Fig. 11·11.

Now knowing the values of percentage increase in maximum circumferential stress $(\sigma_t)_{max}$ over average circumferential stress $(\sigma_t)_{av}$ for the ratio of thickness (t) to inside radius (r_1) from 0, 1, 2 and 3 the curve is plotted as shown in Fig. 11·11.

Example 11·17. *A compound cylinder is to be made by shrinking on outer tube of 200 mm external diameter on to an inner tube of 100 mm internal diameter. If the greatest circumferential stress in the inner tube is to be 2/3rd of the greatest circumferential stress in the outer tube, determine the common diameter.*

Solution. External radius of the outer tube,

$$r_3 = \frac{200}{2} = 100 \text{ mm} = 0·1 \text{ m}$$

Internal radius of the inner tube,

$$r_1 = \frac{100}{2} = 50 \text{ mm} = 0·05 \text{ m}$$

Greatest circumferential stress in the inner tube

$$= \frac{2}{3} \times \text{ Greatest circumferential stress in the outer tube.}$$

Let, $\quad r_2 = $ The common radius, and

$\quad p = $ Radial pressure at the junction.

Common diameter:

Outer tube:

$$(\sigma_c)_{\text{outer tube}} = p \left[\frac{r_3^2 + r_2^2}{r_3^2 - r_2^2} \right] = p \left(\frac{K + 1}{K - 1} \right),$$

where, $\quad K = \dfrac{r_3^2}{r_2^2} = \dfrac{0.1^2}{r_2^2}$

Inner tube:

$$(\sigma_c)_{\text{inner tube}} = p\left[\frac{2r_2^2}{r_2^2 - r_1^2}\right] = p\left[\left\{\frac{2}{\left(1 - \frac{r_1^2}{r_2^2}\right)}\right\}\right]$$

$$= p\left[\frac{2}{1 - \frac{(0.05)^2}{r_2^2}}\right] = p\left[\frac{8}{4 - \frac{(0.1)^2}{r_2^2}}\right] = p\left(\frac{8}{4 - K}\right)$$

But, $(\sigma_c)_{\text{inner tube}} = \frac{2}{3}(\sigma_c)_{\text{outer tube}}$ ...Given

$\therefore$

$$p\left(\frac{8}{4 - K}\right) = \frac{2}{3} \cdot p\left[\frac{K + 1}{K - 1}\right]$$

or, $12(K - 1) = (4 - K)(K + 1)$

or, $12K - 12 = 4K + 4 - K^2 - K$ or $K^2 + 9K - 16 = 0$

or, $$K = \frac{-9 \pm \sqrt{81 + 64}}{2} = 1.52 \text{ (neglecting } -ve \text{ value)}$$

or, $$K = 1.52 = \frac{0.1^2}{r_2^2}$$

$\therefore$ $r_2 = 0.081$ m or 81 mm

or, $d_2 = 2r_2 = 2 \times 81 = 162$ mm

Hence, *common diameter* = **162 mm (Ans.)**

Example 11·18. *A bronze liner of 60 mm external diameter is to be shrunk on a steel rod of 48 mm diameter. If the maximum stress in the liner is limited to 144 MN/m², calculate:*

(i) Maximum radial pressure between liner and rod;

(ii) Difference between the liner bore and shaft diameter before shrinking.

Take: $E_s = 200$ GN/m²; $E_b = 90$ GN/m²; Poission's ratio for both steel and bronze = 0·3

Solution. External radius of the bronze liner,

$$r_2 = \frac{60}{2} = 30 \text{ mm} = 0.03 \text{ m}$$

Radius of the steel shaft, $\qquad r_1 = \frac{48}{2} = 24 \text{ mm} = 0.024 \text{ m}$

Maximum stress in the liner, $(\sigma_c)_{r_1} = 144$ MN/m²

Young's modulus for steel, $\qquad E_s = 200$ GN/m²

Young's modulus for bronze, $\qquad E_b = 90$ GN/m²

Poission's ratio, $\qquad \frac{1}{m} = 0.3$

(i) **Maximum radial pressure, p:**

For bronze liner:

$$(\sigma_r)_{r_2} = 0 = \frac{b'}{r_2^2} - a' \text{ or } a' = \frac{b'}{r_2^2}$$

Also,

$$(\sigma_r)_{r_1} = 144 = \frac{b'}{r_1^2} + a' = \frac{b'}{r_1^2} + \frac{b'}{r_2^2} = b'\left[\frac{r_2^2 + r_1^2}{r_1^2 \, r_2^2}\right]$$

$$b' = \frac{144 \times r_1^2\, r_2^2}{\left(r_2^2 + r_1^2\right)} = \frac{144 \times 0.024^2 \times 0.03^3}{[(0.03)^2 + (0.024)^2]} = 0.0506$$

and,

$$a' = \frac{0.0506}{(0.03)^2} = 56.22$$

Also,

$$(\sigma_r)_{r_1} = p = \frac{b}{r_1^2} - a = \frac{0.0506}{(0.024)^2} - 56.22 = 31.63\ \text{MN/m}^2$$

Hence, $p = \mathbf{31.36\ MN/m^2}$ **(Ans.)**

(ii) Difference between the liner bore and shaft diameter before shrinking:

For the steel shaft:

$$\sigma_r = -a = p\ \text{(compressive)}$$

∴

$$a = -p$$

and,

$$\sigma_c = a = -p\ (i.e.\ \text{compressive})$$

∴ The diametral strain in the steel shaft,

$$(e_c)_{\text{steel shaft}} = \frac{1}{E}\left[\sigma_c - \frac{1}{m} \times \sigma_r\right]$$

$$= \frac{1}{200 \times 10^9}[-31.63 + 0.3 \times 31.63] \times 10^6$$

$$= -11.07 \times 10^{-5}$$

∴ Decrease in diameter $= 11.07 \times 10^{-5} \times 48 = 0.005314\ \text{mm}$

For bronze liner at the inner surface:

Diametral strain,

$$(e_c)_{\text{liner}} = \frac{1}{E_b}\left[(\sigma_c)_{r_1} - \frac{1}{m}(\sigma_r)_{r_1}\right]$$

$$= \frac{1}{90 \times 10^9}[144 + 0.3 \times 31.63] \times 10^6$$

$$= 1.705 \times 10^{-3}$$

∴ *Increase in the inner diameter of the liner*

$$= 1.705 \times 10^{-3} \times 48\ \text{mm} = 0.08184\ \text{mm}$$

Hence, *difference in shaft diameter and inner diameter of liner before shrinking*

$$= 0.005314 + 0.08184 = \mathbf{0.087154\ mm}\ \textbf{(Ans.)}$$

Example 11·19. *A solid plug gauge made of steel has a diameter of 30·006 mm and is forced into a ring gauge of the same material. The inside diameter, outside diameter and axial length of the gauge are 30 mm, 48 mm and 24 mm respectively. If $E = 200$ GN/m², Poisson's ratio = 0·286 and co-efficient of friction = 0·3 find:*

(i) *The maximum stress in the ring;*

(ii) *The force required to slide the plug;*

(iii) *Torque required to rotate the plug with respect to the ring.*

Solution. Inside radius of the ring gauge,

$$r_1 = \frac{30}{2} = 15\ \text{mm} = 0.015\ \text{m}$$

Outside radius of the ring gauge,

$$r_2 = \frac{48}{2} = 24 \text{ mm} = 0.024 \text{ m}$$

Length of the ring gauge,

$$24 \text{ mm} = 0.024 \text{ m}$$

Co-efficient of friction, $\mu = 0.3$

Poisson's ratio, $\dfrac{1}{m} = 0.286$; $E = 200 \text{ GN/m}^2$

(i) The maximum stress in the ring $(\sigma_c)_{r_1}$:

Let p (MN/m^2) = Normal radial pressure at the common surface.

For the plug gauge:

$$\sigma_r = -a = p \text{ (compressive)}$$

or, $a = -p$

and, $\sigma_c = a = -p$

∴ Diametral strain in the plug gauge,

$$(\sigma_c)_{\text{plug gauge}} = \frac{1}{E}\left[\sigma_c - \frac{1}{m} \times \sigma_r\right] = \frac{1}{E}\left[-p - \frac{1}{m} \times \sigma_r\right] = -\frac{p}{E}\left(1 - \frac{1}{m}\right)$$

∴ *Decrease* in the diameter of plug

$$= \frac{p}{E}\left(1 - \frac{1}{m}\right) \times 0.03 = \frac{p}{200 \times 10^9}(1 - 0.286) \times 0.03$$

$$= \frac{0.0214\,p}{200 \times 10^9} \text{ m}$$

For the steel ring gauge:

$$(\sigma_r)_{r_2} = 0 = \frac{b'}{r_2^2} - a', \qquad \text{or,} \qquad a' = \frac{b'}{r_2^2}$$

Ship's body partly behaves like a pressure vessel.

and,
$$(\sigma_r)_{r_1} = p = \frac{b'}{r_1^2} - a' \quad \text{or} \quad p = \frac{b'}{r_1^2} - \frac{b'}{r_2^2} = b'\left[\frac{r_2^2 - r_1^2}{r_1^2 \, r_2^2}\right]$$

or,
$$b' = \frac{p \cdot r_1^2 r_2^2}{r_2^2 - r_1^2} = p \cdot \frac{(0.015)^2 \times (0.024)^2}{(0.024)^2 - (0.015)^2} = 0.0002849 \, p$$

and,
$$(\sigma_c)_{r_1} = \frac{b'}{r_1^2} + a'$$

But,
$$a' = \frac{b'}{r_2^2} = \frac{0.0002849 \, p}{(0.024)^2} = 0.495 \, p$$

$$\therefore \quad (\sigma_c)_{r_1} = \frac{b'}{r_1^2} + a' = \frac{0.0002849 \, p}{(0.015)^2} + 0.495 \, p = 1.76 \, p$$

Now, diametral strain at the *inner surface* of the ring,

$$(e_c)_{\text{ring}} = \frac{1}{E}\left[(\sigma_c)_{r_1} - \frac{1}{m}(\sigma_r)_{r_1}\right]$$

$$= \frac{1}{E}(1.76 \, p + 0.286 \times p) = \frac{2.046 \, p}{200 \times 10^9}$$

∴ Increase in the inner diameter of the ring

$$= \frac{2.046 \, p}{200 \times 10^9} \times 0.03 = \frac{0.0614 \, p}{200 \times 10^9}$$

Also,   shrinkage allowance = Decrease in the diameter of the plug

+ increase in the inner diameter of the ring

$$\therefore \quad 0.006 \times 10^{-3} = \frac{0.0214 \, p}{200 \times 10^9} + \frac{0.0614 \, p}{200 \times 10^9}$$

or, $p(0.0214 + 0.0614) = 0.006 \times 10^{-3} \times 200 \times 10^9$

or, $0.0828 p = 0.006 \times 10^{-3} \times 200 \times 10^9$

∴ $p = 14.49 \text{ MN/m}^2$

Thus *maximum stress in the ring*,

$$(\sigma_c)_{r_1} = 17.6 p = 1.76 \times 14.49 = \mathbf{25.5 \text{ MN/m}^2} \quad \textbf{(Ans.)}$$

(ii) Force required to slide the plug:

Force required to slide the plug

= Frictional force

$= 2\pi r_1 \, l p \mu = 2\pi \times 0.015 \times 0.024 \times 14.49 \times 0.3 \times 10^3$ kN

= **9.83 kN (Ans.)**

(iii) Torque required to rotate plug:

Torque required to rotate the plug

= Tangential frictional force × radius

$= 2\pi r_1 \, l p \mu \times r_1 = 2\pi r_1^2 \, l p \mu$

$= 2\pi \times 0.015^2 \times 0.024 \times 14.49 \times 0.3 \times 10^3$ kNm

= **0.1475 kNm (Ans.)**

Example 11·20. *One steel cylinder is shrunk on to another, the compound cylinder having an inside diameter of 120 mm, an outside diameter of 240 mm and a diameter of 180 mm at the surface in contact. If the shrinkage produces a radial pressssure p at the surfaces in contact, after which the compound cylinder is subjected to an internal pressure p_1, find the ratio of p to p_1, so that the maximum circumferential (or hoop) tensions in the two cylinders shall be same.*

Solution. Radius,

$$r_1 = \frac{120}{2} = 60 \text{ mm} = 0·06 \text{ m}$$

Radius,

$$r_2 = \frac{180}{2} = 90 \text{ mm} = 0·09 \text{ m}$$

Radius,

$$r_3 = \frac{240}{2} = 120 \text{ mm} = 0·12 \text{ m}$$

Radial pressure due to shrinkage $= p$

Internal pressure $= p_1$

$$\textbf{Ratio } \frac{\textbf{p}}{\textbf{p}_1} = ?$$

Stresses due to shrinkage only:

For the inner cylinder:

$$\sigma_r = \frac{b}{r^2} - a$$

At,

$$r = r_1 = 0·06 \text{ m}, \ \sigma_r = 0$$

∴

$$0 = \frac{b}{(0·06)^2} - a$$

∴

$$b = 0·0036 \, a$$

At,

$$r = r_2 = 0·09 \text{ m}, \ \sigma_r = p$$

∴

$$p = \frac{b}{(0·09)^2} - a$$

$$p = 123·46b - a = 123·46 \times 0·0036a - a = -0·555a$$

A submarine should be capable of operating at 500 m below the water surface.

$$\therefore \qquad a = -1.8p$$

and, $\qquad b = -0.0036 \times 1.8p = -0.00648p$

Then, $\qquad (\sigma_c)_{max} = \dfrac{b}{r_1^2} + a = \dfrac{-0.00648\,p}{(0.06)^2} - 1.8\,p = -3.6\,p \text{ (comp.)}$

For the outer cylinder:

$$\sigma_r = \dfrac{b'}{r^2} - a'$$

At, $\qquad r = r_3 = 0.12 \text{ m}, \ \sigma_r = 0$

$\therefore \qquad 0 = \dfrac{b'}{(0.12)^2} - a'$

or, $\qquad b' = 0.0144\,a'$

At $\qquad r = r_2 = 0.09 \text{ m}, \ \sigma_r = p$

$\therefore \qquad p = \dfrac{b'}{(0.09)^2} - a'$

or, $\qquad p = \dfrac{0.0144\,a'}{(0.09)^2} - a' = 0.777\,a' \qquad\qquad \text{or } a' = 1.287\,p$

$\therefore \qquad b' = 0.0144 \times 1.287p = 0.0185p$

Then, $\qquad (\sigma_c)_{max} = \dfrac{b'}{(0.09)^2} + a' = \dfrac{0.0185\,p}{(0.09)^2} + 1.287\,p = 3.57\,p$

i.e. $\qquad (\sigma_c)_{max} = 3.57p \text{ (tensile)}$

Stresses due to internal pressure only:

Consider the compound cylinder as one unit.

$$\sigma_r = \dfrac{b''}{r^2} - a''$$

At, $\qquad r = 0.12 \text{ m}, \ \sigma_r = 0$

$\therefore \qquad 0 = \dfrac{b''}{(0.12)^2} - a'' \quad \text{or} \quad b'' = 0.0144\,a''$

At, $\qquad r = 0.06 \text{ m}, \ \sigma_r = p_1$

or, $\qquad p_1 = \dfrac{b''}{(0.06)^2} - a'' = \dfrac{0.0144\,a''}{(0.06)^2} - a'' = 3a''$

$\therefore \qquad a'' = \dfrac{p_1}{3}$

and, $\qquad b'' = 0.0144 \times \dfrac{p_1}{3} = 0.0048\,p_1$

Then, $\qquad (\sigma_c)_{r=0.06} = \dfrac{b''}{(0.06)^2} + a'' = \dfrac{0.0048\,p_1}{(0.06)^2} + \dfrac{p_1}{3} = 1.666\,p_1 \text{ (tensile)}$

and, $\qquad (\sigma_c)_{r=0.09} = \dfrac{b''}{(0.09)^2} + a'' = \dfrac{0.0048\,p_1}{(0.09)^2} + \dfrac{p_1}{3} = 0.926\,p_1 \text{ (tensile)}$

Combined effect of both shrinking and internal pressure:

Maximum resultant tensile circumferential stress in *inner cylinder*

$$= (1.666p_1 - 3.6p) \qquad\qquad\qquad \text{...}(i)$$

Maximum resultant tensile circumferential stress in *outer cylinder*

$$= (0.926p_1 + 3.57p) \qquad\qquad\qquad \text{...}(ii)$$

Equating (*i*) and (*ii*) we get,

$$1.666p_1 - 3.6p = 0.926p_1 + 3.57p$$
$$7.17p = 0.74p_1$$

∴ $$\frac{p}{p_1} = 0.103 \quad \text{(Ans.)}$$

Example 11·21. *A bronze sleeve of outside diameter 120 mm is forced over a steel shaft of diameter 90 mm. The initial inside diameter of the sleeve is less than the diameter of the shaft by 0·09 mm. This compound rod is subjected to external pressure of 37·5 MN / m^2 and the temperature is raised by 120°C. Calculate:*

(*i*) *The radial pressure between the sleeve and the shaft.*

(*ii*) *The maximum hoop stress developed in the sleeve.*

Take: For steel: $\quad E_s = 208 \text{ GN/}m^2; \quad \dfrac{1}{m} = 0.30; \quad \alpha_s = 11 \times 10^{-6} \text{ per } °C.$

For bronze: $\quad E_b = 105 \text{ GN/}m^2; \quad \dfrac{1}{m} = 0.33; \quad \alpha_b = 19 \times 10^{-6} \text{ per } °C.$

Solution. Radius of the shaft, $\quad r_1 = \dfrac{90}{2} = 45 \text{ mm} = 0.045 \text{ m}$

Outside radius of sleeve, $\quad r_2 = \dfrac{120}{2} = 60 \text{ mm} = 0.06 \text{ m}$

External pressure applied = 37·5 MN/m^2

Force fit allowance = 0·09 mm

Let, p = Junction pressure developed due to the force fit of sleeve over shaft.

This compressor cylinder is made of cast iron. Compressor cylinders are thick shells that need to bear high pressures.

(i) The radial pressure between the sleeve and the shaft:

Circumferential stress developed in *sleeve*:

At radius r_1

$$(\sigma_c)_{r_1} = p \times \left[\frac{r_2^2 + r_1^2}{r_2^2 - r_1^2}\right]$$

$$= p \times \left[\frac{(0.06)^2 + (0.045)^2}{(0.06)^2 - (0.045)^2}\right] = 3.57\,p \quad \text{(tensile)}$$

At radius r_2 :

$$(\sigma_c)_{r_2} = p \times \frac{2r_1^2}{r_2^2 - r_1^2}$$

$$= p \times \left[\frac{2 \times (0.045)^2}{(0.06)^2 - (0.045)^2}\right] = 2.57\,p \quad \text{(tensile)}$$

Circumferential stress developed in the *shaft*

$$= p \text{ MN/m}^2 \, (compressive)$$

Diametral strain in the shaft

$$= \frac{p}{E_s} - \frac{1}{m}\frac{p}{E_s}$$

$$= \frac{p}{E_s}\left(1 - \frac{1}{m}\right)$$

$$= \frac{p}{E_s}(1 - 0.3) = \frac{0.7\,p}{E_s} \qquad \dots(contraction)$$

$\therefore$ Decrease in the diameter of the shaft

$$= \frac{0.7\,p}{E_s} \times 0.045 \text{ m}$$

Diametral strain in the sleeve

$$= \frac{(\sigma_c)_{r_1}}{E_b} + \frac{1}{m}\frac{p}{E_b} = \frac{1}{E_b}(3.57\,p + 0.33\,p)$$

$$= \frac{3.9\,p}{E_b} \qquad (expansion)$$

$\therefore$ Increase in the inner diameter of the sleeve

$$= \frac{3.9\,p}{E_b} \times 0.045 \text{ m}$$

Now, force fit allowance = Decrease in the diameter of the shaft
+ increase in the inner diameter of the sleeve

$$0.09 \times 10^{-3} = \frac{0.7\,p}{E_s} \times 0.045 + \frac{3.9\,p}{E_b} \times 0.045$$

$$= \left[\frac{0.7\,p}{208 \times 10^9} \times 0.045 + \frac{3.9\,p}{105 \times 10^9} \times 0.045\right] \times 10^6$$

$$0.09 = 0.0001514\,p + 0.00167\,p$$

$\therefore \qquad p = 49.4 \text{ MN/m}^2$

Circumferential stress in sleeve,

$$(\sigma_c)_{r_1} = 3.57 \times 49.4 = 176.36 \text{ MN/m}^2$$

$$(\sigma_c)_{r_2} = 2.57 \times 49.4 = 126.96 \text{ MN/m}^2$$

Stresses due to external pressure and rise in temperature:

Let, p' (MN/m²) = The pressure at the common radius.

Circumferential stress in the shaft,

$$\sigma_c = -p' \text{ (compressive)}$$

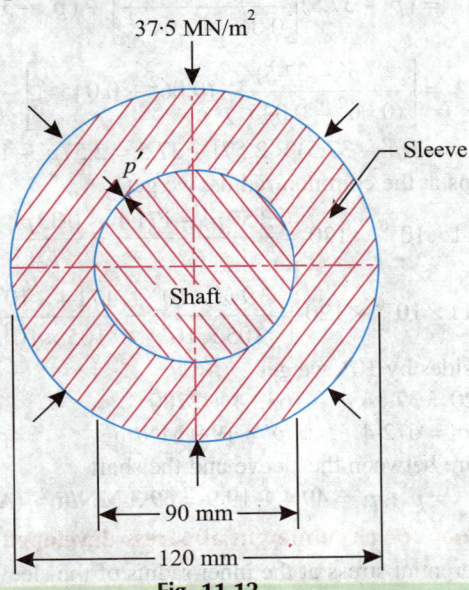

Fig. 11.12

Using *Lame's equations* for the sleeve:

At, $\quad r = r_2 = 0.06$ m; $\quad 37.5 = \dfrac{b}{0.06^2} - a$ $\qquad$ ...(i)

at, $\quad r = r_1 = 0.045$ m; $\quad p' = \dfrac{b}{0.045^2} - a$ $\qquad$ ...(ii)

(where a and b are constants)

or, $\quad 37.5 - p' = \dfrac{b}{0.06^2} - \dfrac{b}{0.045^2} = \dfrac{b\,(0.045^2 - 0.06^2)}{0.045^2 \times 0.06^2}$

[Subtracting (ii) from (i)]

or, $\quad b = -(37.5 - p')\left[\dfrac{0.045^2 \times 0.06^2}{0.06^2 - 0.045^2}\right]$

and, $\quad a = \dfrac{b}{0.06^2} - 37.5 = -(37.5 - p')\left[\dfrac{0.045^2}{0.06^2 - 0.045^2}\right] - 37.5$

[Substituting the values of b in (i)]

Circumferential stress at *inner radius*,

$$[(\sigma_c)_{r_1}]' = \dfrac{b}{0.045^2} + a$$

$$= (p' - 37.5)\left[\dfrac{0.06^2}{0.06^2 - 0.045^2}\right] + (p' - 37.5)\left[\dfrac{0.045^2}{0.06^2 - 0.045^2}\right] - 37.5$$

$$= \left[\dfrac{(p' - 37.5)}{(0.06^2 - 0.045^2)}(0.06^2 + 0.045^2)\right] - 37.5$$

$$= [(p' - 37.5) \times 3.57] - 37.5 = 3.57\,p' - 171.4$$

Circumferential stress at *outer radius*,

$$\left[(\sigma_c)_{r_2}\right]' = \frac{b}{0.06^2} + a$$

$$= (p' - 37.5)\left[\frac{0.045^2}{0.06^2 - 0.045^2}\right] + (p' - 37.5)\left[\frac{0.045^2}{0.06^2 - 0.045^2}\right] - 37.5$$

$$= \left[\frac{(p' - 37.5)}{(0.06^2 - 0.045^2)} (0.06^2 + 0.045^2)\right] - 37.5$$

$$= [(p' - 37.5) \times 2.57] - 37.5 = 2.57p' - 133.87$$

Equating the strains at the common radius, we get

$$\frac{p'}{E_s} - \frac{0.3p'}{E_s} + 11 \times 10^{-6} \times 120 = \frac{3.57p' - 171.4}{E_b} + \frac{0.33p'}{E_b} + 19 \times 10^{-6} \times 120$$

$$\frac{0.7p' \times 10^6}{208 \times 10^9} + 11 \times 10^{-6} \times 120 = \frac{3.9p' \times 10^6}{105 \times 10^9} - \frac{171.4 \times 10^6}{105 \times 10^9} + 19 \times 10^{-6} \times 120$$

Multiplying both sides by 10^6, we get

$$3.36\,p' + 1320 = 37.14\,p' - 1632.4 + 2280$$

or, $\quad\quad 33.78\,p' = 672.4 \quad \therefore \ p' = 19.9 \text{ MN/m}^2$

Final radial pressure between the sleeve and the shaft

$$= p + p' = 49.4 + 19.9 = \textbf{69.3 MN/m}^2 \textbf{ (Ans.)}$$

(ii) **Maximum hoop (or circumferential) stress developed in sleeve:**

Resultant circumferential stress at the inner radius of the sleeve

$$= (\sigma_c)_{r_1} + [(\sigma_c)_{r_1}]' = 176.36 + (3.57p' - 171.4)$$

$$= 176.36 + 3.57 \times 19.9 - 171.87 = 76 \text{ MN/m}^2$$

Resultant circumferential stress at the outer radius of the sleeve

$$= (\sigma_c)_{r_2} + [(\sigma_c)_{r_2}]' = 126.96 + 2.57p' - 133.87$$

$$= 126.96 + 2.57 \times 19.9 - 133.87 = 44.23 \text{ MN/m}^2$$

Hence, maximum hoop stress developed in the sleeve = **76 MN/m²** **(Ans.)**

Example 11·22. *A steel plug 120 mm in diameter is forced into a steel ring 180 mm external diameter and 75 mm wide. From a strain gauge fixed on the outer surface of the ring in the circumferential direction, the strain is found to be 0.615×10^{-4}. If the co-efficient of friction between the mating surfaces = 0·3 and E = 210 GN/m², calculate the axial force required to push the plug out of the ring.*

Solution. Radius of the steel plug,

$$r_1 = \frac{120}{2} = 60 \text{ mm} = 0.06 \text{ m}$$

Inner radius of the steel ring,

$$r_1 = 0.06 \text{ m}$$

Outer radius of the steel ring,

$$r_2 = \frac{180}{2} = 90 \text{ mm} = 0.09 \text{ m}$$

Breadth of the ring

$$b = 75 \text{ mm} = 0.075 \text{ m}$$

Circumferential strain,

$$= 0.615 \times 10^{-4}$$

Co-efficient of friction,

$$\mu = 0.3$$

Axial force required:

Let, p (MN/m^2) = Junction pressure.

Circumferential (or hoop) stress at the outer surface of the ring,

$$(\sigma_c)_{r_2} = + p \times \frac{2r_1^2}{r_2^2 - 2r_1^2}$$

$$= p \times \frac{2 \times 0.06^2}{0.09^2 - 0.06^2} = 1.6\, p$$

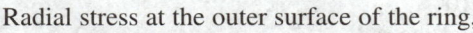

HANSON TANK
800 421 9395
www.hansontank.us

Stainless steel tank.

Radial stress at the outer surface of the ring,

$$(\sigma_c)_{r_2} = 0$$

∴ Circumferential strain on outer surface of the ring

$$= \frac{1.6\, p}{E} = 0.615 \times 10^{-4}$$

∴ $1.6p \times 10^6 = 210 \times 10^9 \times 0.615 \times 10^{-4}$

or, $\qquad p = 8.07$ MN/m^2

Radial force acting on the plug,

$$F_{radial} = p \times 2\pi r_1 \times b = 8.07 \times 2\pi \times 0.06 \times 0.075 = 0.228 \text{ MN}$$

∴ Axial force,

$$F_{axial} = \mu F_{radial} = 0.3 \times 0.228 = 0.0684 \text{ MN or } 68.4 \text{ kN}$$

Hence, *axial force required to push the plug out of the ring* = **68·4 kN   (Ans.)**

Example 11·23. *A steel rod 75 mm diameter is forced into a bronze sleeve 120 mm outside diameter, thereby producing a tension of 60 MN/m^2 at the outer surface of the sleeve. Find:*

(i)   The radial pressure between the bronze sleeve and the steel rod.

(ii) The rise in temperature which would eliminate the force fit.

Given: For steel: $E_s = 210$ GN/m^2; *Poisson's ratio* = 0.28, $\alpha_s = 11.2 \times 10^{-6}$ *per °C.*

For bronze: $E_b = 114$ GN/m^2; *Poisson's ratio* = 0.33, $\alpha_b = 18 \times 10^{-6}$ *per °C.*

Solution. Inside radius of the bronze sleeve

$$(= \text{radius or steel rod}), \quad r_1 = \frac{75}{2} = 37.5 \text{ mm} = 0.0375 \text{ m}$$

Outer radius of the bronze sleeve,

$$r_2 = \frac{120}{2} = 60 \text{ mm} = 0.06 \text{ m}$$

Tension at the outer surface of the sleeve = 60 MN/m^2

(i)   The radial pressure between the bronze sleeve and the steel rod:

Let, $\qquad p$ (MN/m^2) = Radial pressure between the sleeve and the rod.

Circumferential stress at the outer surface of the sleeve,

$$(\sigma_c)_{r_2} = p \cdot \frac{2r_1^2}{r_2^2 - r_1^2}$$

$$60 = p \cdot \frac{2 \times 0.0375^2}{0.06^2 - 0.0375^2}$$

∴ $\qquad p = \frac{60\,(0.06^2 - 0.0375^2)}{2 \times 0.0375^2} = 46.8 \text{ MN/m}^2$

i.e. $\qquad$ **p = 46·8 MN/m^2 (Ans.)**

(ii) **The rise in temperature which would eliminate the force fit:**

Circumferential stress at the inner surface of the sleeve,

$$(\sigma_c)_{r_1} = p \times \left[\frac{r_2^2 + r_1^2}{r_2^2 - r_1^2}\right] = 46.8 \times \left[\frac{0.06^2 + 0.0375^2}{0.06^2 - 0.0375^2}\right] = 106.98 \text{ MN/m}^2$$

Circumferential strain at r_1, in sleeve,

$$[(e_c)_{r_1}]_{\text{sleeve}} = \frac{(\sigma_c)_{r_1}}{E_b} + \frac{p}{m_b \, E_b} = \frac{106.98 \times 10^6}{114 \times 10^9} + \frac{0.33 \times 46.8 \times 10^6}{14 \times 10^9}$$

$$= \frac{106.98 + 0.33 \times 46.8}{114 \times 10^3} = 0.001074$$

Circumferential stress at the outer surface of the rod,

$$(\sigma_c)_{r_1} = -p = -46.8 \text{ MN/m}^2$$

Circumferential strain at r_1, in rod

$$[(e_c)_{r_1}]_{\text{rod}} = -\frac{p}{E_s} + \frac{p}{m_s \, E_s} = -\frac{p}{E_s}\left(1 - \frac{1}{m_s}\right)$$

$$= -\frac{46.8 \times 10^6}{210 \times 10^9}(1 - 0.28) = -0.0001604$$

Force fit allowance $= [(e_c)_{r_1}]_{\text{sleeve}} \times r_1 - \left[(e_c)_{r_1}\right]_{\text{rod}} \times r_1$

$$= (0.001074 + 0.0001604) \times 37.5 = 0.0463 \text{ mm}$$

Let the temperature of rod and sleeve be raised by $t°C$; the sleeve will expand more than the rod as $\alpha_b > \alpha_s$. When the differential expansion at radius r_1 equals the force fit allowance, then force fit will eliminated.

∴   Force fit allowance $= r_1(\alpha_b - \alpha_s)\,t$

or, $0.0463 = 37.5(18 \times 10^{-6} - 11.2 \times 10^{-6}) \times t$

∴ $t = \dfrac{0.0463}{37.5(18 \times 10^{-6} - 11.2 \times 10^{-6})} = 181.5°C$

i.e.   **t = 181·5°C   (Ans.)**

Example 11·24. *A steel sleeve 18 mm thick (radial thickness) is pressed on to a solid steel shaft of 60 mm diameter, the junction pressure being 30 MN/m². An axial tensile force of 120 kN is applied to the shaft. If the Poisson's ratio is 0.285 for steel, determine:*

(i) The change in radial pressure at the common surface.

(ii) The change in circumferential (or hoop) tension in sleeve.

Solution. Radius of the shaft = Inside radius of sleeve,

$$r_1 = 30 \text{ mm} = 0.03 \text{ m}$$

Outer radius of sleeve, $r_2 = 30 + 18 = 48$ mm $= 0.048$ m

Radial pressure at the common surface,

$$p = 30 \text{ MN/m}^2$$

Axial tensile force, $P = 120$ kN

(i) The change in radial pressure at the common surface, p′:

Axial tensile stress, $\sigma_{axial} = \dfrac{P}{\pi r_1^2} = \dfrac{120}{\pi \times 0.03^2} \times 10^{-3}$ MN/m$^2 = 42.43$ MN/m^2

Tensile stress in the shaft produces a lateral strain and its diameter will tend to decrease; consequently the pressure at the common surface decreases.

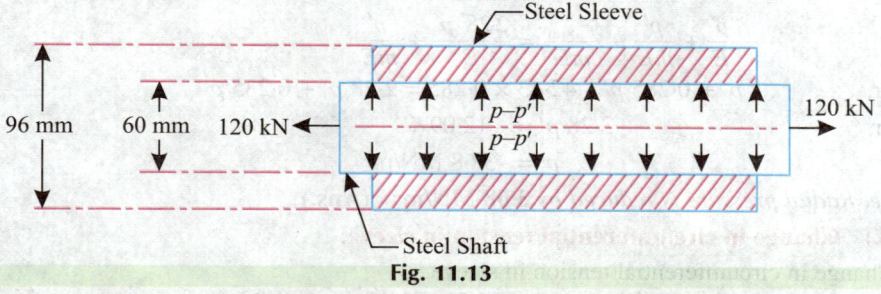

Fig. 11.13

Let, p' = Decrease in radial pressure.

Change in circumferential stress at outer surface of shaft $= -p'$;

Change in circumferential stress at the inner radius of the sleeve

$$= -p' \times \left[\frac{r_2^2 + r_1^2}{r_2^2 - r_1^2} \right]$$

$$= -p' \times \left[\frac{0.048^2 + 0.03^2}{0.048^2 - 0.03^2} \right] = -2.28\,p'$$

Change in circumferential strain in shaft at r_1,

$$[(e_c)_{r_1}]_{\text{shaft}} = -\frac{p'}{E} + \frac{p'}{mE} - \frac{\sigma_{\text{axial}}}{mE}$$

Change in circumferential strain in sleeve at r_1,

$$[(e_c)_{r_2}]_{\text{sleeve}} = -\frac{2.28\,p'}{E} + \frac{p'}{mE}$$

A submarine's doors need to be waterproof and safe.

But for compatibility of strain,

$$[(e_c)_{r_1}]_{\text{shaft}} = [(e_c)_{r_1}]_{\text{sleeve}}$$

$$-\frac{p'}{E} + \frac{p'}{mE} - \frac{\sigma_{\text{axial}}}{mE} = \frac{2 \cdot 28 \; p'}{E} + \frac{p'}{mE}$$

or, $-p' + 0 \cdot 285 \; p' - 42 \cdot 43 \times 0 \cdot 285 = 2 \cdot 28 \; p' + 0 \cdot 285 \; p'$

or, $3 \cdot 28 \; p' = -12 \cdot 09$

∴ $p' = -3 \cdot 68 \; \text{MN/m}^2$

i.e. Radial pressure is reduced by **3·68 MN/m²** **(Ans.)**

(ii) Change in circumferential tension in sleeve:

Change in circumferential tension in sleeve

$$= -2 \cdot 28 \times 3 \cdot 68 = -8 \cdot 39 \; \text{MN/m}^2$$

i.e. Circumferential tension in sleeve is decreased by **8·39 MN/m²** **(Ans.)**

Example 11·25. *A steel sleeve is pressed on to a steel shaft of 60 mm diameter. The radial pressure between the steel shaft and sleeve is 24 MN/m² and the circumferential stress at the inner radius of sleeve is 67·2 MN/m². If an axial compressive force of 60 kN is now applied to the shaft, determine the change in radial pressure.*

Take: $E = 210 \; GN/m^2$, and Poisson's ratio = 0·3.

Solution.   Radius of steel shaft (= inner radius of the sleeve),

$$r_1 = \frac{60}{2} = 30 \; \text{mm} = 0 \cdot 03 \; \text{m}$$

Outer radius of sleeve = r_2

Radial pressure at the junction,

$$p = 24 \; \text{MN/m}^2$$

Circumferential stress at the inner radius of sleeve = 67·2 MN/m²

Compressive force applied to the shaft,

$$P = 60 \; \text{kN}$$

Poisson's ratio, $\dfrac{1}{m} = 0 \cdot 3;$

$$E = 210 \; \text{GN/m}^2$$

Change in radial pressure:

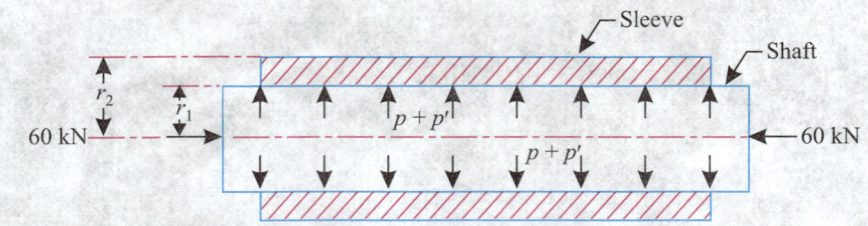

Fig. 11.14

Axial stress, $\sigma_{\text{axial}} = \dfrac{P}{\pi \; r_1^2} = \dfrac{60 \times 10^{-3}}{\pi \times 0 \cdot 03^2} = 21 \cdot 21 \; \text{MN/m}^2$ (compressive)

The compressive stress in shaft introduces a lateral strain and its diameter will tend to increase. Consequently the pressure at the common surface increases.

Let, $p' = $ Increase in radial pressure.

Due to p', increase in circumferential stress (σ_c) at the inner radius of the sleeve

$$= \frac{67 \cdot 2}{24} \times p' = 2 \cdot 8\, p' \qquad\qquad (\because \sigma_c \, \alpha \, p')$$

Additional circumferential strain in sleeve at inner radius

$$[(e_c)_{r_1}]_{\text{sleeve}} = \frac{2 \cdot 8\, p'}{E} + \frac{p'}{mE}$$

Additional circumferential strain in shaft at radius r_1,

$$[(e_c)_{r_1}]_{\text{shaft}} = -\frac{p'}{E} + \frac{p'}{mE} + \frac{\sigma_{\text{axial}}}{mE}$$

For strain compatibility:

$$[(e_c)_{r_1}]_{\text{shaft}} = [(e_c)_{r_1}]_{\text{sleeve}}$$

$$-\frac{p'}{E} + \frac{p'}{mE} + \frac{\sigma_{\text{axial}}}{mE} = \frac{2 \cdot 8\, p'}{E} + \frac{p'}{mE}$$

or,

$$-\frac{p'}{E} + \frac{\sigma_{\text{axial}}}{mE} = \frac{2 \cdot 8\, p'}{E}$$

or, $\quad -p' + 0 \cdot 3 \times 21 \cdot 21 = 2 \cdot 8 p'$

$\therefore \qquad\qquad\qquad p' = 1 \cdot 67 \ \text{MN/m}^2$

i.e. Radial pressure at the common surface is increased by **1·67 MN/m²** **(Ans.)**

A crane's arms are moved by hydraulic cylinders which contain oil at high pressures.
Above, the crane is lifting a heavy component of a tunnel boring machine.

HIGHLIGHTS

1. Thick cylinders are cylindrical vessels containing fluid under pressure and whose wall thickness is not small ($t \geq d / 20$).

2. Lame's equations are given by

$$\sigma_r = \frac{b}{r^2} - a, \qquad \sigma_c = \frac{b}{r^2} + a$$

Special cases:

Case I: External pressure = p_2; internal pressure = p_1

$$\sigma_r = \frac{1}{r_2^2 - r_1^2}\left[p_2 \, r_2^2 - p_1 \, r_1^2 + \frac{r_1^2 \, r_2^2}{r^2} (p_1 - p_2)\right]$$

$$\sigma_c = \frac{1}{r_2^2 - r_1^2}\left[p_1 \, r_1^2 - p_2 \, r_2^2 + \frac{r_1^2 \, r_2^2}{r^2} (p_1 - p_2)\right]$$

Case II: Internal pressure p_1; external pressure = zero

$$\sigma_r = \frac{p_1 \, r_1^2}{r_2^2 - r_1^2}\left[\frac{r_1^2}{r^2} - 1\right] \qquad \left[\text{At, } r = r_1; (\sigma_c)_{r_1} = p_1 \, \frac{r_2^2 + r_1^2}{r_2^2 - r_1^2} \right.$$

$$\sigma_c = \frac{p_1 \, r_1^2}{r_2^2 - r_1^2}\left[\frac{r_2^2}{r^2} + 1\right] \qquad \left. \text{At, } r = r_2; (\sigma_c)_{r_2} = p_1 \, \frac{2r_1^2}{r_2^2 - r_1^2} \right]$$

Case III:

External pressure = p_2; internal pressure = zero

The maximum circumferential (or hoop) stress at radius r_1 is

$$(\sigma_c)_{r_1} = - p_2 \, \frac{2r_2^2}{r_2^2 - r_1^2}$$

Case IV: Solid circular shaft subjected to external pressure p_2:

$$\sigma_r = - a, \, \sigma_c = a$$

3. *Longitudinal and shear stress* for a cylinder with closed ends:

Longitudinal stress, $\sigma_l = \left[\dfrac{r_1^2}{r_2^2 - r_1^2}\right] p$

Maximum shear stress, if no torque acts on the cylinder,

$$\tau_{max} = \frac{b}{r^2}$$

The absolute maximum value of τ_{max} will be on the inner surface, but it will act on the plane which is inclined to the longitudinal axis at 45°.

4. *Compound or shrunk cylinders:*

$$\frac{\delta r' + \delta r''}{r_2} = \frac{\delta r_2}{r_2} \frac{1}{E}\left[\left(\frac{b'}{r_2^2} + a'\right) - \left(\frac{b}{r_2^2} + a\right)\right]$$

where, r_2 = The common radius of the tubes at junction after the shrinkage on,

     $\delta r'$ = The difference between the outer radius of inner tube and r_2,

     $\delta r''$ = The difference between the inner radius of the outer tube and r_2, and

     δr_2 = The difference in the radii before shrinking on.

The original difference of radii at the junction radius r_2, before shrinking on divided by r_2 is equal to the algebraic difference of the circumferential stresses for the two tubes at the junction divided by E.

5. *Thick spherical shells:*

$$\sigma_r = \frac{2b}{r^3} - a; \qquad \sigma_c = \frac{b}{r^3} + a; \text{ Also } b = \frac{p_1 \cdot r_1^3 \, r_2^3}{2 \,(r_2^3 - r_1^3)}, \text{ and } a = \frac{p r_1^3}{r_2^3 - r_1^3}.$$

OBJECTIVE TYPE QUESTIONS

Choose the Correct Answer:

1. Cylindrical vessels, containing fluid under pressure and whose wall thickness is not small ($t \, ^3 \, d/20$) are classified as
   (a) thin cylinders
   (b) thick cylinders
   (c) either of the above
   (d) none of the above.

2. In thick cylinders the radial stress in the wall thickness is
   (a) zero   (b) negligibly small
   (c) not negligible   (d) any of the above.

3. In thick cylinders the circumferential stress
   (a) is zero
   (b) varies along the thickness
   (c) does not vary along the thickness
   (d) none of the above

4. In thick cylinders the radial stress in the wall thickness
   (a) is zero
   (b) negligibly small
   (c) not negligible and varies from the inner surface to the outer surface
   (d) any of the above.

5. In a thick cylinder, the radial sress at the inner surface is
   (a) equal to the magnitude of the fluid pressure
   (b) less than the magnitude of the fluid pressure
   (c) more than the magnitude of the fluid pressure
   (d) independent of the magnitude of the fluid pressure.

6. In a thick cylinder the radial stress at the outer surface is
   (a) always more than zero
   (b) always less than zero
   (c) usually equal to zero
   (d) none of the above.

7. In thick cylinders the variation in the radial as well as circumferential stress across the thickness is obtained with the help of

   (a) Clapeyron's theorem
   (b) Castigliano's theorem
   (c) Lame's theory
   (d) none of the above.

8. Which of the following assumptions is made while solving problems on thick cylinders using Lame's theory ?
   (a) The material is homogeneous and isotropic
   (b) Plane sections perpendicular to the longitudinal axis of the cylinder remain plane after the application of internal pressure
   (c) The material is stressed within the elastic limit.
   (d) All the fibres of the material are free to expand or contract independently without being constrained by the adjacent fibres.
   (e) All of the above.

9. Lame's equations are given by

   (a) $\sigma_r = \dfrac{2b}{r^2} + a$

   $\sigma_c = \dfrac{2b}{r^2} - a$

   (b) $\sigma_r = \dfrac{b}{2r^2} - a$

   $\sigma_c = \dfrac{b}{2r^2} + 2a$

   (c) $\sigma_r = \dfrac{b}{r^2} - a$

   $\sigma_c = \dfrac{b}{r^2} + a$

   (d) $\sigma_r = \dfrac{b}{3r^2} - a$

   $\sigma_c = \dfrac{b}{3r^2} + 2a$.

10. In a thick cylinder if external pressure is p_2 and internal pressure is zero, then the maximum circumferential (or hoop) stress at radius r_1 is

(a) $-p_2 \cdot \dfrac{2r_2^2}{r_2^2 - r_1^2}$ (b) $-p_2 \cdot \dfrac{r_2^2}{r_2^2 + r_1^2}$

(c) $P_2 \cdot \dfrac{3r_2^2}{r_2^2 - r_1^2}$ (d) $P_2 \cdot \dfrac{r_1^2}{r_2^2 + r_1^2}$.

11. In case of a closed ends thick cylinder, the longitudinal stress (s_l) is given by

(a) $\sigma_l = \dfrac{p_1 r_1^2}{r_2^2 - r_1^2}$ (b) $\sigma_l = \dfrac{p_1 r_2^2}{r_2^2 + r_1^2}$

(c) $\sigma_l = \dfrac{2p_1 r_2^2}{r_2^2 - r_1^2}$ (d) $\sigma_l = \dfrac{r_2^2 - r_1^2}{p_1 r_2^2}$.

12. In compound or shrunk cylinders the original difference of radii at the junction radius r_2, before shrinking on divided by r_2, is equal to .... of the circumferential stresses for the two tubes at the junction divided by E.

(a) algebraic sum

(b) algebraic difference

(c) either of the above

(d) none of the above.

13. In thick spherical shells which of the following relations holds good?

(a) $\sigma_r = \dfrac{b}{r^3} - a$ (b) $\sigma_r = \dfrac{2b}{r^3} - a$

$\sigma_c = \dfrac{b}{r^3} + 2a$ $\sigma_c = \dfrac{b}{r^3} + a$

(c) $\sigma_r = \dfrac{b}{2r^3} - a$ (d) $\sigma_r = \dfrac{3b}{r^3} - a$

$\sigma_c = \dfrac{b}{2r^3} + a$ $\sigma_c = \dfrac{b}{r^3} + a$.

14. The cylinders are compounded with a purpose

(a) to make the circumferential stress distribution uniform

(b) to increase the strength of the cylinder

(c) to increase the pressure bearing capacity of a single cylinder

(d) all of the above.

15. The variation of circumferential (or hoop) stress across the thickness of a thick cylinder is

(a) uniform

(b) parabolic

(c) linear

(d) none of the above.

ANSWERS

1. (b)	2. (c)	3. (b)	4. (c)	5. (a)	6. (c)	7. (c)
8. (e)	9. (c)	10. (a)	11. (a)	12. (b)	13. (b)	14. (c)
15. (b)						

UNSOLVED EXAMPLES

1. A pipe of 200 mm internal diameter and 50 mm thickness carries a fluid at a pressure of 5 MN/m². Calculate the maximum and minimum intensities of circumferential stress across the section.

[**Ans.** σ_c = 13 MN/m², 8 MN/m²]

2. Calculate the thickness of the metal necessary for a steel cylindrical shell of internal diameter 0·15 m to withstand an internal pressure of 50 MN/m²; the maximum permissible tensile stress is not to exceed 150 MN/m². [**Ans.** 31 mm]

3. Find the ratio of thickness to internal diameter for a tube subjected to an internal pressure when the pressure is 5/8 of the value of the maximum premissible circumferential stress.

Find the increase in internal diameter of such a tube 100 mm internal diameter when the internal pressure is 100 MN/m².

Also find the change in the wall thickness of the tube.

Take: E = 200 GN/m²; Poisson's ratio = 0·29. [**Ans.** 0·54, 0·09001 mm, 0·018425 mm]

4. A steel cylinder is 1 m inside diameter and is to be designed for an internal pressure of 8 MN/m^2. Calculate the thickness if the maximum shearing stress is not to exceed 35 MN/m^2. Calculate also the increase in volume, due to working pressure, if the cylinder is 6 m long with closed ends. Take $E = 200$ GN/m^2 and Poission's ratio = 1/3.
   Neglect any constraint due to ends. [**Ans.** 70 mm, 0·002835 m^3 (increase)]

5. To measure the longitudinal and circumferential strains, strain gauges were fixed on the outer surface of a closed thick cylinder of diameter ratio 2·5. At an internal pressure of 230 MN/m^2 these strains were recorded as $9·18 \times 10^{-5}$ and $36·9 \times 10^{-5}$ respectively. Determine the values of:
   (*i*) Poisson's ratio; (*ii*) Young's modulus; (*iii*) Modulus of rigidity.
   [**Ans.** (*i*) 0·287 (*ii*) 203 GN/m^2 (*iii*) 78·8 GN/m^2]

6. Strain gauges are fixed on the outer surface of a thick cylinder with diameter ratio of 2·5. The cylinder is subjected to an internal pressure of 150 MN/m^2. The recorded strains are: longitudinal strain = $59·87 \times 10^{-6}$ and circumferential strain = $240·65 \times 10^{-6}$.
   Determine the Young's modulus. [**Ans.** 203 GN/m^2]

7. A thick cylinder 120 mm internal diameter and 180 mm external diameter is used for a working pressure of 15 MN/m^2. Because of external corrosion, the outer diameter of the cylinder is machined to 178 mm. Determine by how much the internal pressure is to be reduced so that the maximum hoop stress remains the same as before.
   [**Ans.** 0·373 MN/m^2]

8. A pressure vessel 200 mm internal radius, 250 mm external radius, 1 metre long is tested under a hydraulic pressure of 20 MN/m^2. If $E = 208$ GN/m^2 and Poisson's ratio = 0·3, determine the change in internal and external diameters. [**Ans.** 0·166 mm, 0·145 mm]

9. A compound cylinder, formed by shrinking one tube on to another is subjected to an internal pressure of 160 MN/m^2. Before the fluid is admitted, the internal and external diameters of the compound cylinder are 120 mm and 200 mm and the diameter at the junction is 160 mm. If after shrinking on, the radial pressure at the common surface is 8 MN/m^2, calculate the final stresses developed in the compound cylinder.

   $$\text{Ans. } Inner\ tube : \sigma_{c(0.06)} = 92.92 \text{ MN/m}^2 \text{ (tensile)}$$
   $$\sigma_{c(0.08)} = 57.9 \text{ MN/m}^2 \text{ (tensile)}$$
   $$Outer\ tube : \sigma_{c(0.08)} = 122.92 \text{ MN/m}^2 \text{ (tensile)}$$
   $$\sigma_{c(0.1)} = 95.94 \text{ MN/m}^2 \text{ (tensile)}$$

10. A compound cylinder is formed by shrinking one cylinder on to another, the final dimensions being internal diameter 120 mm, external diameter 240 mm and diameter at junction 200 mm. After shrinking on the radial pressure at the common surface is 10 MN/m^2.
    (*i*) Calculate the necessary difference in diameters of the two cylinders at the common surface.
    (*ii*) What is the minimum temperature through which outer cylinder should be heated before it can be slipped on ?
    Take: $E = 200$ GN/m^2, and $\alpha = 0·000011$ per °C. [**Ans.** (*i*) 0·0767 mm, (*ii*) 34·86 °C]

11. A compound tube is composed of a tube 250 mm internal diameter and 25 mm thick shrunk on a tube of 250 mm external diameter and 25 mm thick. The radial pressure at the junction is 8 MN/m^2. The compound tube is subjected to an internal fluid pressure of 84·5 MN/m^2. Find the variation of the circumferential stress over the wall of the compound tube.

    $$\text{Ans. } Inner\ tube : \sigma_{c(0.125)} = 128.5 \text{ MN/m}^2 \text{ (tensile)}$$
    $$\sigma_{c(0.1)} = 175.26 \text{ MN/m}^2 \text{ (comp)}$$
    $$Outer\ tube : \sigma_{c(0.125)} = 209.3 \text{ MN/m}^2 \text{ (tensile)}$$
    $$\sigma_{c(0.15)} = 171.56 \text{ MN/m}^2 \text{ (tensile)}$$

12. For the compound tube of unsolved example 11 find the original difference in diameters of the two tubes before shrinking on, so that after shrinking on the radial pressure at the junction may be 8 MN/m². Find also the minimum temperature to which the outer tube should be heated in order that it can be slipped on the inner tube.

    Take: E = 200 GN/m², and α = 0·000012 per °C.   [**Ans.** 0·101 mm, 33·34 °C]

13. A compound cylinder is formed by shrinking one cylinder onto another, the final dimensions being internal diameter 100 mm, external diameter 180 and diameter at the common surface 140 mm, the radial pressure developed at the common surcface is 15 MN/m².

    If E = 102 GN/m², and α = 17·6 × 10^{-6} per °C.

    (i) Calculate the necessary difference in diameters of the two cylinders at the common surface.

    (ii) What is the minimum temperature through which the outer cylinder should be heated before it can be slipped on ?   [**Ans.** 0·1471 mm, 59·7°C]

14. A compound cylinder is obtained by shrink fitting of one cylinder of outer diameter 200 mm over another cylinder of inner 140 mm, such that the diameter at the junction of two cylinders is 170 mm. If the radial pressure developed at the junction is 50 MN/m², what are the circumferential (hoop) stresses at the inner and outer radii of both the cylinders?

$$
\begin{bmatrix}
\textit{Ans. Inner cylinder} : \sigma_{c(0.14)} = 152.91 \text{ MN/m}^2 \text{ (comp.)} \\
\sigma_{c(0.17)} = 102.91 \text{ MN/m}^2 \text{ (comp)} \\
\textit{Outer cylinder} : \sigma_{c(0.17)} = 310.36 \text{ MN/m}^2 \text{ (tensile)} \\
\sigma_{c(0.2)} = 260.36 \text{ MN/m}^2 \text{ (tensile)}
\end{bmatrix}
$$

15. The external diameter of steel collar is 200 mm and the internal diameter decreases by 0·125 mm when shrunk on to a solid steel shaft of 125 mm diameter. If E = 205 GN/m² and Poisson's ratio is 0·304 find:

    (i) Radial pressure between the collar and the shaft;

    (ii) Circumferential stress at the inner surface of the tube;

    (iii) Reduction in diameter of the shaft.

    [**Ans.** (i) 79·33 MN/m² (ii) 180·87 MN/m² (iii) 0·0336 mm]

16. A bronze liner of 50 mm external diameter is to be shrunk on a steel rod of 40 mm diameter. If the maximum stress in the liner is limited to 120 MN/m²,

    (i) Calculate the maximum radial pressure between liner and rod;

    (ii) Difference between the liner bore and shaft diameter before shrinking.

    Take: E_s = 200 GN/m², E_b = 90 GN/m² and Poisson's ratio (for both steel and bronze) = 0·3.

    [**Ans.** (i) 26·35 MN/m² , (ii) 0·0604 mm]

17. A thick spherical shell of 80 mm internal diameter is subjected to an internal fluid pressure of 30 MN/m². If the permissible stress is 80 MN/m², find the thickness of the shell.

    [**Ans.** 7·7 mm]

18. A spherical shell of 150 mm internal diameter has to withstand an internal pressure of 20 MN/m². Calculate the thickness of the shell if the maximum permissible tensile strength is 100 MN/m².

    [**Ans.** 7·55 mm]

19. A compound cylinder is to be made by shrinking an outer tube of 150 mm external radius on to an inner tube of 75 mm internal diameter. Determine the common radius at the junction if the greatest circumferential stress in the inner tube is to be two-third of the greatest circumferential stress in the outer tube.   [**Ans.** 121·7 mm]

20. A solid plug gauge of steel has a diameter of 25·0005 mm, and is forced into a ring gauge of the same material, which measures 25 mm inside diameter and 50 mm outside diameter. Its axial length is 20 mm; what is the maximum stress in the ring, and what force is required to slide the plug, assuming co-efficient of friction is 0·3 ?

    Take: E = 200 GN/m² and Poisson's ratio = 0·286.   [**Ans.** 27·8 MN/m², 5·74 kN]

21. One steel cylinder is shrunk on to another, the compound cylinder having an inside diameter of 100 mm, an outside diameter of 200 mm and a diameter of 150 mm at the surface in contact. If the shrinkage produces a radial pressure p at the surfaces in contact, after which the compound cylinder is subjected to an internal pressure p_1, find the ratio of p to p_1, so that the maximum circumferential tensions in the two cylinders shall be same.

[**Ans.** 0·103]

22. A bronze liner of outside diameter 60 mm and inside diameter 39·94 mm is forced over a steel shaft of 40 mm diameter.

Determine:

(*i*) The radial pressure between shaft and liner;

(*ii*) The maximum circumferential stress in liner;

(*iii*) The change in outside diameter of the liner.

[**Ans.** (*i*) 55·86 MN/m^2, (*ii*) 145·23 MN/m^2, (*iii*) 0·043 mm]

23. A steel tube of 80 mm outside diameter is to be shrunk on another steel tube of 40 mm inside diameter and 60 mm outside diameter. Calculate the shrinkage allowance if in the compound tube the final maximum stress in each tube is same, when subjected to an internal pressure of 50 MN/m^2. Also find the value of this stress and draw the stress distribution diagram.

Take: E = 207 GN/m^2.

[**Ans.** 0·0092 mm; 64·7 MN/m^2]

24. A steel plug 80 mm in diameter is forced into a steel ring 120 mm external diameter and 50 mm wide. From a strain gauge fixed on the outer surface of the ring in the circumferential direction, the strain is found to be 0·41 × 10^{-4}. Considering that the co-efficient of friction between the mating surfaces = 0·3, determine the axial force required to push the plug out of the ring.

Take: E = 210 GN/m^2.

[**Ans.** 20·28 kN]

25. A steel rod 50 mm in diameter is forced into a bronze sleeve 80 mm outside diameter, thereby producing a tension of 40 MN/m^2 at the outer surface of sleeve. Calculate:

(*i*) The radial pressure between the bronze sleeve and steel rod,

(*ii*) The rise in temperature which would eliminate the force fit.

Given that, For steel: E = 210 GN/m^2; Poission's ratio = 0·28; α = 11·2 × 10^{-6} per °C.

For bronze: E = 114 GN/m^2; Poisson's ratio = 0·33; α = 18 × 10^{-6} per °C.

[**Ans.** (*i*) 31·2 MN/m^2 (*ii*) 120·6 °C]

26. A steel sleeve 15 mm thick (radial thickness) is pressed on to a solid steel shaft of 50 mm diameter, the junction pressure being 25 MN/m^2. An axial tensile force of 100 kN is applied to the shaft. If the Poisson's ratio for steel is 0·285, find:

(*i*) The change in radial pressure at the common surface;

(*ii*) The change in circumferential tension in sleeve.

[**Ans.** (*i*) 4·425 MN/m^2 (decrease); (*ii*) 10·09 MN/m^2 (decrease)]

27. A steel sleeve is pressed on to a steel shaft of 50 mm diameter. The radial pressure between the steel shaft and sleeve is 20 MN/m^2 and the hoop stress at the inner radius of the sleeve is 56 MN/m^2. An axial compressive force of 50 kN is applied to the shaft. Determine the change in radial pressure.

Given: Poisson's ratio = 0·3, and E = 210 GN/m^2.

[**Ans.** 2·01 MN/m^2]

Chapter

12

Riveted and Welded Joints

12·1. INTRODUCTION

Invariably we need long plates or bars, due to difficulties in their manufacture, handling or transportation they are not made available in very long sizes. The required length can be had by connecting smaller lengths by *riveting or welding*. The rivets are generally made of steel. Before pushing through the plates or members of structure the rivets are heated and hammered to form a head on other side.

This chapter deals with the analysis and design of several types of riveted and welded joints which occur in the structural and mechanical fields.

12·2. RIVETED JOINTS

12·2·1. Types of Riveted Joints:

The riveted joints can mainly be classified in the following two ways:

1. Lap joints   2. Butt joints.

When the ends of the plates overlap each other the joint is known as the **lap joint** *(Fig. 12·1) and if the ends of the plates butt against each other the joint is known as* **butt joint**.

In case of the lap joints if the joint is made by a single row of rivets, the joint is known as the *single riveted lap joint* [Fig. 12·1 (*a*)], a joint with two rows of rivets the *double riveted lap joint* [Fig. 12·1 (*b*)], a joint with three rows of rivets is known as *triple riveted lap joint*.

A butt joint having one cover plate is called *single cover plate butt joint* and with two cover plates the *double cover plate butt joint* (Fig. 12·2 *a*, *b*).

There can be *three arrangements* of rivets (*a*) *chain riveting*, (*b*) *zig zag riveting* and (*c*) *diamond riveting* as shown in Fig. 12·3 (*a*, *b*) and Fig. 12·10 respectively.

12·2·2. Important Terms used in Riveting:

Refer to Fig. 12·4.

Pitch: The distance from centre to centre of rivets lying in a row is termed as *pitch*. It may vary from row to row.

Gauge line. The line of rivets parallel to the direction of stress is called *gauge line*.

Gauge distance. The distance between two consecutive rivets is called *gauge distance*.

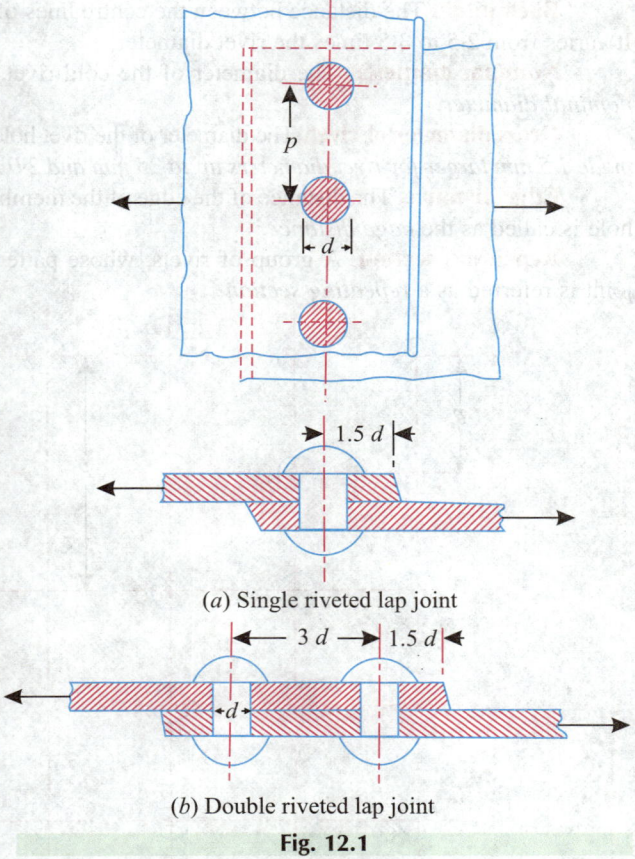

(*a*) Single riveted lap joint

(*b*) Double riveted lap joint

Fig. 12.1

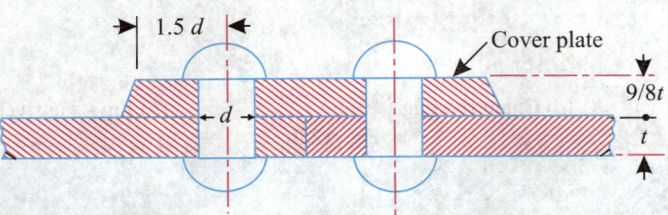

(*a*) Single riveted single cover plate butt joint

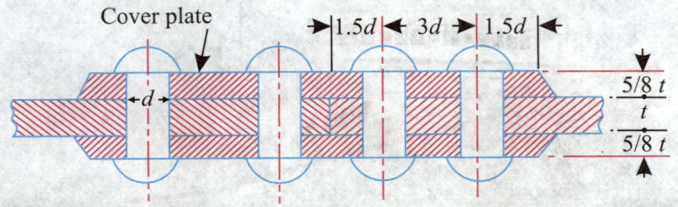

(*b*) Double riveted double cover plate butt joint

Fig. 12.2

Back pitch. The distance between the centre lines of two rows of rivets is called *back pitch*. It varies from 2·5 to 3·5 times the rivet diameter.

Nominal diameter. The diameter of the cold rivet measured before driving is referred as *nominal diameter.*

Gross diameter of rivet. The diameter of the rivet hole is termed as gross diameter of rivet. *It is made 1·5 mm larger for rivet diameters up to 25 mm and 2·0 mm for rivet diameters exceeding 25 mm.*

Edge distance. The distance of the edge of the member or the cover plates from extreme rivet hole is called as the *edge distance.*

Repeating section. A group of rivets, whose pattern repeats itself along the length of the joint is referred as a *repeating section.*

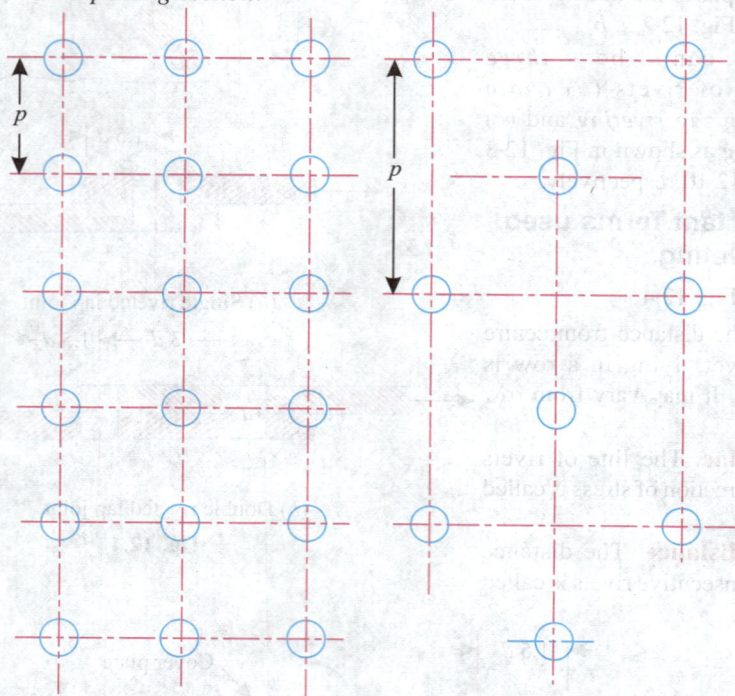

(*a*) Chain riveting   (*b*) Zig - zag riveting

Fig. 12.3

Ship's body and structure have joints of all types.

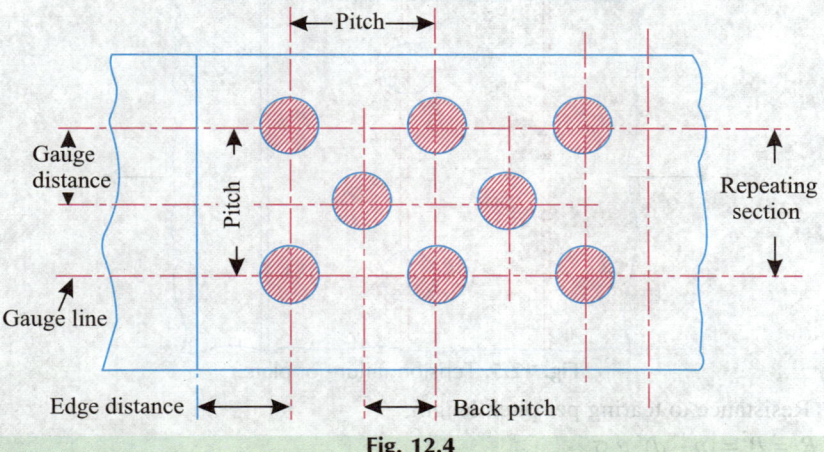

Fig. 12.4

12·2·3. Failure of Riveted Joints

A riveted joint may *fail* in the following ways:

(*i*)   The plate may tear-off along the line of rivets.

(*ii*)   The rivets may shear-off.

(*iii*)   The material may fail to resist the bearing pressure between the rivets and the plate.

(*iv*)   The plate may shear between the rivet hole and the edge of the plate.

(*v*)   The plate may split between rivet holes and the edge of the plate in a line perpendicular to the edge.

Refer to simple riveted lap joint as shown in Fig. 12·1

Let, p = Pitch of the rivets,

d = Gross diameter of the rivet,

t = Thickness of the plate,

σ_t = Safe tensile stress in the plates,

P_t = Tearing strength per pitch length,

σ_{ut} = Ultimate tensile stress of the plates,

P_{ut} = Pull applied per pitch length for tension failure,

τ = Safe shearing stress for the rivet,

P_s = Shearing strength per pitch length,

τ_u = Ultimate shear stress of the rivet,

P_{us} = Pull required per pitch length for shear failure,

σ_c = Allowable crushing/bearing stress of the rivet; or plate whichever is weaker,

P_c = Crushing strength per pitch length of rivet/plate,

σ_{uc} = Ultimate crushing stress of rivet/plate, and

P_{uc} = Pull per pitch length for crushing/bearing failure.

1.   Failure of joint by tearing of the plate along pitch line or along rivet hole:

Fig. 12·5 shows the failure of the joint by tearing of the plate along rivet holes or pitch line.

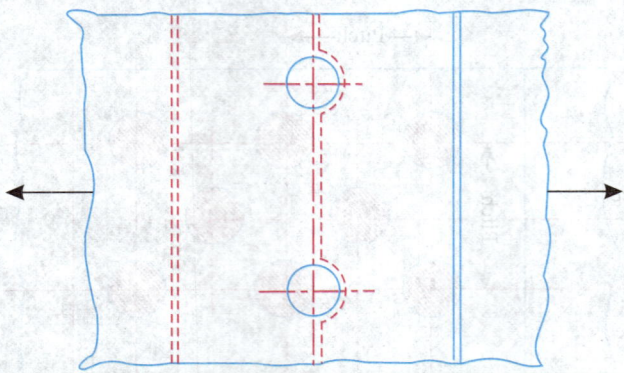

Fig. 12.5. Tension failure of plate.

Resistance to tearing per pitch length,

$$R_t = P_t = (p - d) \cdot t \cdot \sigma_t \qquad \qquad \dots(12 \cdot 1)$$

2. Failure of the joint due to shear of the rivets:

Fig. 12·6 indicates how the failure of the joint occurs when the rivets undergo shear.

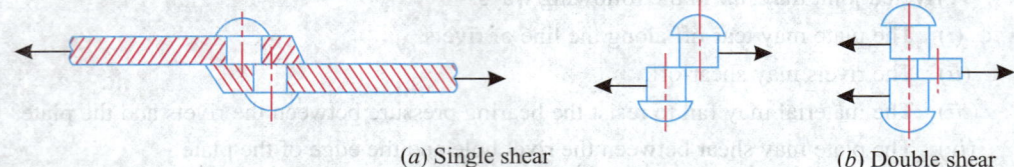

(a) Single shear   (b) Double shear

Fig. 12.6

Resistance to shearing of rivets per pitch length,

$$R_s = P_s = n \times \frac{\pi}{4} \times d^2 \times \tau,$$

where, n = number of rivets per pitch length.   ...(12·2)

In case of *double cover butt joint double shear* of rivets takes place, (Fig. 12·6 b). If number of rivets per pitch length are n; resistance to shearing of rivets will be,

$$R_s = P_s = n \times \frac{\pi}{4} \times d^2 \times \tau \times 2 \qquad \qquad \dots General \ [12 \cdot 3 \ (a)]$$

$$= n \times \frac{\pi}{4} \times d^2 \times \tau \times 1 \cdot 875 \ (or \ 1 \cdot 75)$$

...For design purposes (12·3 (b))

3. Failure of joint due to the failure of the material to resist bearing pressure between rivet and plate.

Refer to Fig. 12·7.

Resistance to bearing or crushing pressure per pitch length,

$$R_c = P_c = n \times d \times t \times \sigma_c \qquad \dots(12 \cdot 4)$$

where, n is the number of rivets per pitch length.

4. Failure of the joint by shearing out the plate between the rivet hole and the edge of the plate (*Fig. 12·8*).

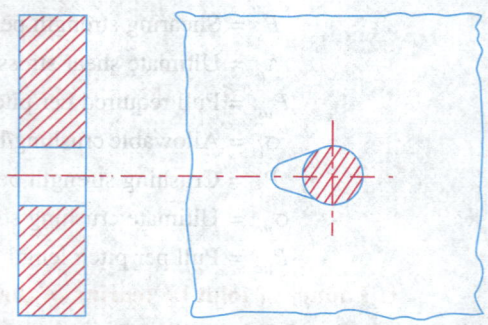

Fig. 12.7

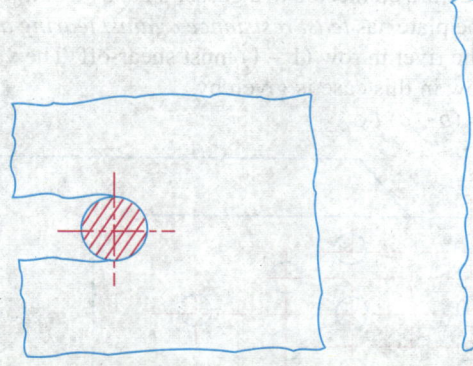

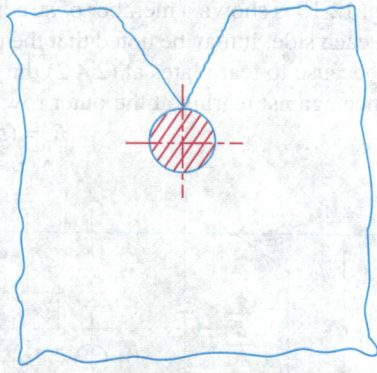

| Fig. 12.8 | Fig. 12.9 |

5. Failure of the joint due to splitting of the plate between rivet holes and edge of the plate (*Fig. 12·9*).

To prevent failure of the joint according to the mode of (4) and (5); generally, distance between the centre of the rivets and edge of the plates is kept as 1·5 times the diameter of the rivet (i.e., 1·5 d).

12·2·4. Efficiency of a Riveted Joint

The efficiency of a riveted joint is the ratio of the *strength of the joint to the strength of the solid plate.*

Thus, the efficiency of the joint,

$$\eta = \frac{\text{Least of } P_t, P_s \text{ and } P_c}{p\, t\, \sigma_t} \qquad (\because \text{Strength of solid plate} = p.\, t.\, \sigma_t)$$

Thus:

Tearing efficiency, $\quad \eta_t = \dfrac{P_t}{p.t.\sigma_t} = \dfrac{(p-d)\, t\, \sigma_t}{p.t.\sigma_t} = \dfrac{p-d}{p}$

Shearing efficiency, $\quad \eta_s = \dfrac{P_s}{p.t.\sigma_t} = \dfrac{n \times \dfrac{\pi}{4} \times d^2\, \tau}{p.t.\sigma_t}$ (for *single shear*)

$$= \frac{n \times \dfrac{\pi}{4} \times d^2\, \tau \times 2}{p.t.\sigma_t} \quad \text{(for rivets in } double\ shear)$$

Crushing efficiency, $\quad \eta_c = \dfrac{P_c}{p.t.\sigma_t} = \dfrac{n\, d\, t\, \sigma_c}{p.t.\sigma_t}$

The lowest of the three efficiencies is taken as the **efficiency of a riveted joint.**

12·2·5. Thickness of Cover Plates

To keep the strength of the cover plates equal to that of the plates to be connected it is obvious that the thickness of the single cover and double cover plates should be *t* and *t*/2 respectively but to be on safer side they are kept as $\dfrac{9}{8}t$ and $\dfrac{5}{8}t$ respectively.

12·2·6. Diamond Riveting

This type of riveting is used in structural units such as roof trusses, bridges, structures, etc.

In Fig. 12·10 is shown a member of structure connected to a gusset plate by a butt joint having six rivets on each side. It may be noted that the plate has *least resistance against tearing at the outer row* (1 – 1) because to tear plates at (2 – 2) the rivet in row (1 – 1) must shear-off. The value of the least resistance against tearing at the outer row in this case is given by,

$$R_t = (b - d) t \, \sigma_t \qquad \qquad ...(12·5)$$

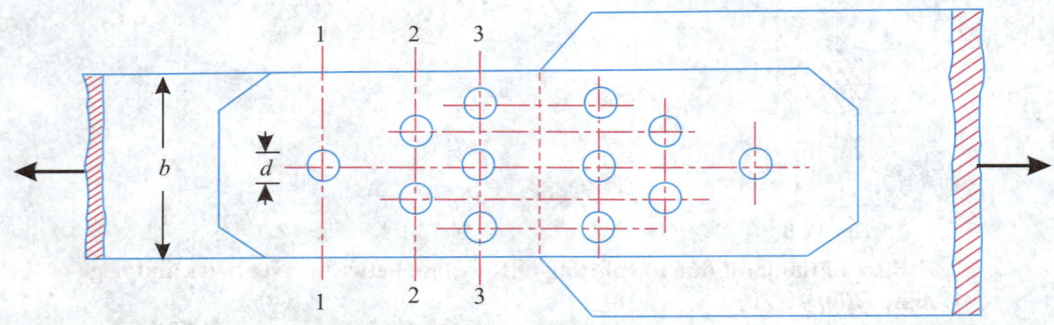

Fig. 12.10

12·2·7. Relation between *d* and *t*

The diameter of the rivet (*d*) and thickness of the plate (*t*) follow the relation: $d = 1.91\sqrt{t}$ when both *d* and *t* are in cm, and $d = 6\sqrt{t}$ if *d* and *t* are in mm, this relation is also known as **Unwin's formula**.

12·2·8. Determination of Pitch (*p*)

The pitch *p* is determined by equating eqn. (12·1) with eqn. (12·2) or (12·3) and (12·1) with (12·4) alternately. Two values of pitch are obtained, the *lesser* value is chosen. The pitch should be found to an accuracy of 3 mm.

While designing a riveted joint, efforts should be made to make it *equally strong against tearing, shearing and crushing* because economically or otherwise there is no point in making it too strong against, say tearing when it is to remain weak against bearing, etc.

Bronze cylinder. Big size cylinders are first made in parts and these parts are later joined by welding.

Example 12·1. *Two plates 15 mm thick are joined by a double riveted lap joint. The pitch in each row of rivets is 60 mm. Rivet diameter = 20 mm.*

Take the following values of permissible stress:

$$\sigma_t = 150 \text{ MN/m}^2; \ \tau = 94·5 \text{ MN/m}^2; \ \sigma_c = 212·5 \text{ MN/m}^2$$

Determine the maximum tensile force permissible on the joint per pitch length. Also determine the efficiency of the joint.

Solution. Thickness of plates, $t = 15$ mm $= 0·015$ m

Pitch in each row of rivets, $p = 60$ mm $= 0·06$ m

Rivet diameter, $d = 20$ mm $= 0·02$ m

Number of rivets per pitch length, $n = 2$ (∵ Joint is *double riveted*)

Maximum tensile force, P_t:

Resistance against tearing of the plate,

$$R_t = P_t = (p - d) \, t \, \sigma_t$$
$$= (0.06 - 0.02) \times 0.015 \times 150 \times 10^6 = 90000 \text{ N}$$

Resistance against shearing of rivets,

$$R_s = P_s = n \times \frac{\pi}{4} \times d^2 \times \tau \times 1 \qquad (\because \text{Rivets are in } single \ shear)$$
$$= 2 \times \frac{\pi}{4} \times (0.02)^2 \times 94.5 \times 10^6 = 59384 \text{ N}$$

Resistance against crushing of rivets,

$$R_c = P_c = n \, d \cdot t \cdot \sigma_c$$
$$= 2 \times 0.02 \times 0.015 \times 212.5 \times 10^6 = 127500 \text{ N}$$

Now, maximum tensile force permissible is equal to least (minimum) of P_t, P_s and P_c i.e., 59384 N.

Hence, *maximum tensile force permissible* = 59384 N or **59·384 kN** (**Ans.**)

Efficiency of the joint:

Efficiency of the joint, $\qquad \eta = \dfrac{\text{Least of } P_t, \, P_s, \, P_c}{\text{Strength of solid plate}} = \dfrac{59.348 \times 10^3}{0.06 \times 0.015 \times 150 \times 10^6}$

$$= \frac{59.348 \times 10^3}{0.06 \times 0.015 \times 150 \times 10^6} = 0.439 \text{ or } \mathbf{43 \cdot 9\%} \quad (\textbf{Ans.})$$

Example 12·2. *A thin cylindrical shell 1·5 m in diameter is made of 1·2 cm thick plates. The circumferential joint is a single riveted lap joint with 2·2 cm diameter rivets at a pitch of 5 cm. If the ultimate tensile stress in the plate is 450 MN/m² and the ultimate shearing and bearing stresses for the rivets are 300 MN/m² and 600 MN/m² respectively, calculate the efficiency of the joint.*

Solution. Thickness of plates, $\quad t = 1.2 \text{ cm} = 0.012 \text{ m}$

Rivet diameter, $\qquad\qquad\qquad d = 2.2 \text{ cm} = 0.022 \text{ m}$

Pitch of rivets, $\qquad\qquad\qquad p = 5 \text{ cm} = 0.05 \text{ m}$

Number of rivets per pitch length, $n = 1$ $\qquad\qquad\qquad (\because \text{Joint is } single \ riveted.)$

Assuming the factor of safety as 3, we find that the permissible tensile stress in plate,

$$\sigma_t = \frac{\sigma_{ut}}{3} = \frac{450}{3} = 150 \text{ MN/m}^2$$

Similarly, $\qquad\qquad\qquad \tau = \dfrac{\tau_u}{3} = \dfrac{300}{3} = 100 \text{ MN/m}^2$

and, $\qquad\qquad\qquad\qquad \sigma_c = \dfrac{\tau_{uc}}{3} = \dfrac{600}{3} = 200 \text{ MN/m}^2$

Now, resistance against tearing of plate,

$$P_t = (p - d) \, t \, \sigma_t$$
$$= (0.05 - 0.022) \times 0.012 \times 150 \times 10^6 = 50400 \text{ N}$$

Resistance against shearing of rivets,

$$P_s = n \times \frac{\pi}{4} \times d^2 \times \tau = 1 \times \frac{\pi}{4} \times (0.022)^2 \times 100 \times 10^6 = 38010 \text{ N}$$

and, resistance against crushing of rivets,

$$P_c = n \, dt \, \sigma_c = 1 \times 0.022 \times 0.012 \times 200 \times 10^6 = 52800 \text{ N}$$

$\therefore$ Efficiency of the joint, $\qquad \eta = \dfrac{\text{Least of } P_t, \, P_s \text{ and } P_c}{\text{Strength of solid plate}} = \dfrac{38010}{p.t.\sigma_t}$

$$= \frac{38010}{0.05 \times 0.012 \times 150 \times 10^6} = 0.422 \text{ or } \mathbf{42 \cdot 2\%} \quad \textbf{(Ans.)}$$

Example 12·3. *A single riveted lap joint for a pair of mild steel plates 1 cm thick has to transmit a load of 250 kN. Determine the diameter, pitch and number of rivets required for the joint. Calculate the width of the plate required. Assume $\sigma_t = 400$ MN/m², $\tau = 320$ MN/m², $\sigma_c = 640$ MN/m². Factor of safety = 4.*

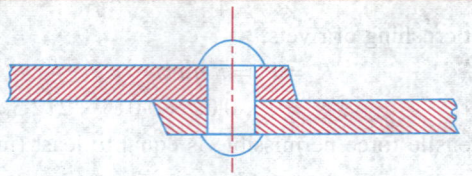

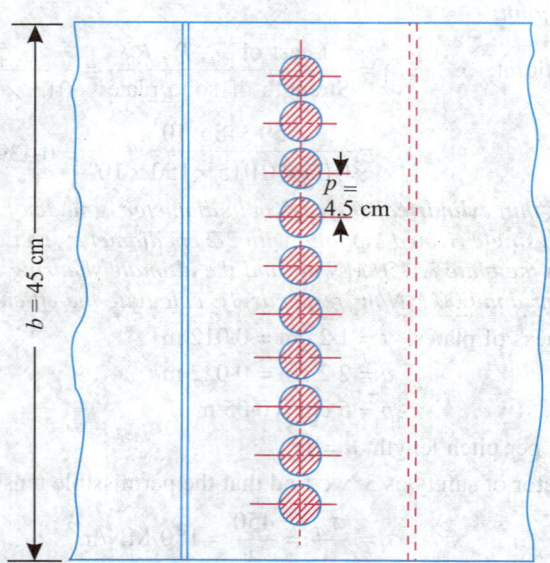

Fig. 12.11

Solution. Using Unwin's formula, we get

$$d = 1.91 \sqrt{t}$$

where, d = Diameter of the rivet in cm, and

t = Thickness of the plate in cm.

∴ $d = 1.91 \sqrt{1} = 1.91$ cm say **2 cm** **(Ans.)**

$$(\because \ t = 1 \text{ cm.} \dots\dots \textit{Given})$$

We know that,

$$\text{Working or permissible strength} = \frac{\text{Ultimate strength}}{\text{Factor of safety}}$$

∴ Permissible $\sigma_t = \dfrac{400}{4} = 100$ MN/m²

Permissible $\tau = \dfrac{320}{4} = 80$ MN/m²

Permissible $\sigma_c = \dfrac{640}{4} = 160$ MN/m²

Now, strength of the rivet (in single shear)

$$P_s = \frac{\pi}{4} \times d^2 \times \tau = \frac{\pi}{4} \times \left(\frac{2}{100}\right)^2 \times 80 \times 10^6 = 25136 \text{ N}$$

Strength of one rivet in crushing $= d\, t\, \sigma_c = \frac{2}{100} \times \frac{1}{100} \times 160 \times 10^6 = 32000 \text{ N}$

Therefore one rivet can take a load of 25136 N which is the *least* of the above two loads.

Hence the number of rivets required $= \dfrac{250 \times 10^3}{25136} = 9.94$ say **10** **(Ans.)**

Let the width of the plate be '*b*' then,

$$(b - 10\, d)\, t\, \sigma_t = 250 \times 10^3$$

or, $\left(b - 10 \times \dfrac{2}{100}\right) \times \dfrac{1}{100} \times 100 \times 10^6 = 250 \times 10^3$

or, $(b - 0.2) \times 10^6 = 250000$

or, $b = 0.45$ m or **45 cm** **(Ans.)**

and, $p = \dfrac{45}{10} = $ **4·5 cm** **(Ans.)**

Fig. 12·11 shows the details of the lap joint.

Example 12·4. *Design a triple riveted lap joint in which the alternate rivets in the outer rows are omitted. The thickness of the plate is 2·5 cm and safe values of σ_t and τ may be assumed as 90 MN/m² and 65 MN/m² respectively.*

Solution. Thickness of the plates $\quad t = 2.5 \text{ cm} = 0.025 \text{ m}$

Diameter of the rivet, $\quad d = 1.91 \sqrt{t} = 1.91 \sqrt{2.5} = 3 \text{ cm} \text{ (or 0·03 m)}$

Refer to Fig. 12·12.

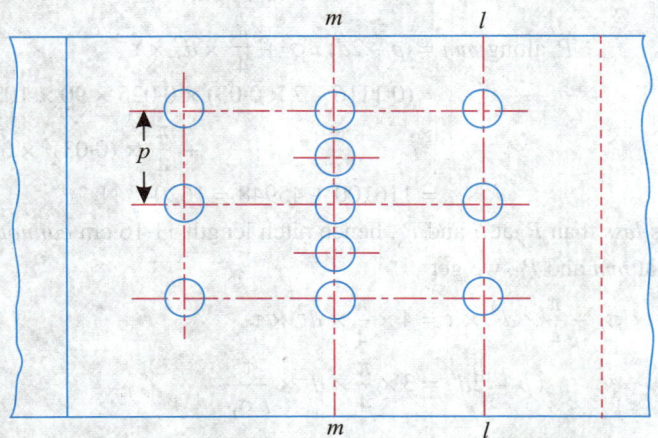

Fig. 12.12

P_t (per pitch length) at section *ll* $\quad = (p - d)\, t\, \sigma_t$

$$= (p - 0.03) \times 0.025 \times 90 \times 10^6 \text{ N}$$

$$P_s = n \times \frac{\pi}{4} \times d^2 \times \tau$$

$$= 4 \times \frac{\pi}{4} \times 0.03^2 \times 65 \times 10^6 = 183807 \text{ N}$$

$$(\because \text{ No. of rivets per pitch length} = 4)$$

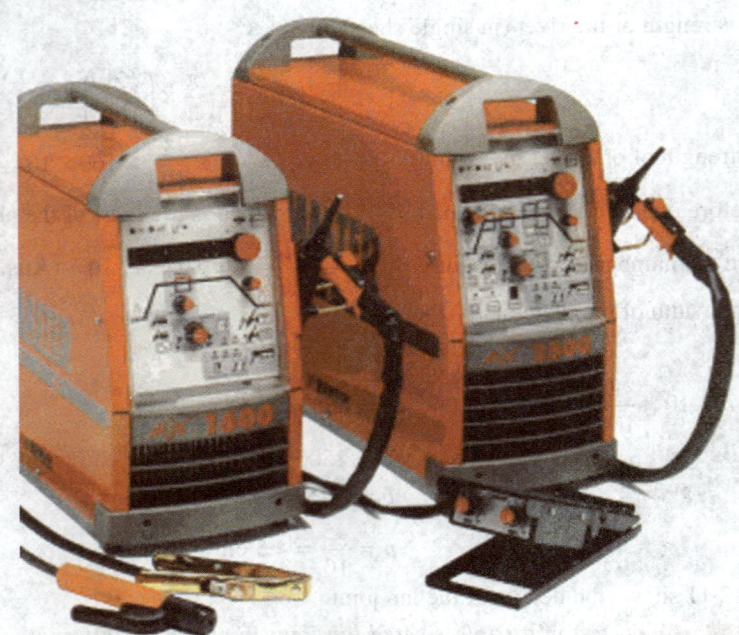

AC and DC welding machines.

Assuming the joint to be equally strong in tearing and shearing,

$$P_t = P_s$$

$$(p - 0.03) \times 0.025 \times 90 \times 10^6 = 183807$$

from which, $p = 0.1116$ m or 11.16 cm

$$P_t \text{ along } mm = (p - 2d)\, t\, \sigma_t + \frac{\pi}{4} \times d^2 \times \tau$$

$$= (0.1116 - 2 \times 0.03) \times 0.025 \times 90 \times 10^6$$

$$+ \frac{\pi}{4} \times (0.03)^2 \times 65 \times 10^6$$

$$= 116100 + 45948 = 162048 \text{ N}$$

This value is *less* than P_t at ll and P_s, hence pitch length 11.16 cm *cannot be accepted.*
Equating P_t at mm and P_s, we get

$$(p - 2d)\, t\, \sigma_t + \frac{\pi}{4} \times d^2 \times \tau = 4 \times \frac{\pi}{4} \times d^2 \times \tau$$

$$(p - 2d) = 3 \times \frac{\pi}{4} \times d^2 \times \frac{\tau}{t\, \sigma_t}$$

$$= 3 \times \frac{\pi}{4} \times (0.03)^2 \times \frac{65 \times 10^6}{0.025 \times 90 \times 10^6} = 0.0612$$

or, $p = 0.0612 + 2d = 0.0612 + 2 \times 0.03$

or, $p = 0.1212$ m or **12·12 cm** (**Ans.**)

Example 12·5. *Design a lap joint to carry a load of 350 kN. The rivets are 2 cm in diameter and placed in a double row. Given* $\sigma_t = 150$ *MN/m²;* $\tau = 100$ *MN/m²;* $\sigma_c = 246$ *MN/m².*

Solution. Load to be carried by the joint = 350 kN

Diameter of the rivet, $d = 2$ cm = 0.02 m

$$\sigma_t = 150 \text{ MN/m}^2, \ \tau = 100 \text{ MN/m}^2, \ \sigma_c = 246 \text{ MN/m}^2$$

Using Unwin's formula, $\quad d = 1.91 \sqrt{t}$, we have

$$2 = 1.91 \sqrt{t}$$

or $\qquad\qquad t = 1.1 \text{ cm} = 0.011 \text{ m}$

Shear value of one rivet in single shear

$$= \frac{\pi}{4} \times (0.02)^2 \times 100 \times 10^6 = 31420 \text{ N}$$

Bearing or crushing value of one rivet

$$= d \, t \, \sigma_c = 0.02 \times 0.011 \times 246 \times 10^6 = 54120 \text{ N}$$

Hence the real strength of rivet is 31420 N (*minimum* of the two values)

Number of rivets required, $\quad n = \dfrac{\text{Load}}{\text{Strength of one rivet}} = \dfrac{350 \times 10^3}{31420} = 11.1 \text{ say } 12$

Adopt pitch $\qquad\qquad p = 3d = 3 \times 2 = 6 \text{ cm}$

Since there are two rows, there will be 6 rivets in each row. Distance of centre of nearest rivet from the edge $\qquad\qquad = 1.5 \, d = 1.5 \times 2 = 3.0 \text{ cm}$

∴ Breadth of the plate, $b \qquad = 5 \times 6 + 2 \times 3 = 36 \text{ cm} = 0.36 \text{ m}$

Back pitch $\qquad\qquad\qquad = 2.5 \, d = 2.5 \times 2 = 5.0 \text{ cm}$

Check the strength of the plate against tearing.

The breadth of the plate is 37 cm and in each row there are six rivet holes,

∴ Plate strength at each row $\quad = (b - 6d) \, t \, \sigma_t$

$$= (0.36 - 6 \times 0.02) \times 0.011 \times 150 \times 10^6$$

$$= 396000 \text{ N or } 396 \text{ kN}$$

which is more than the load on the joint. Hence the plate width is adequate.

The joint is shown in Fig. 12·13.

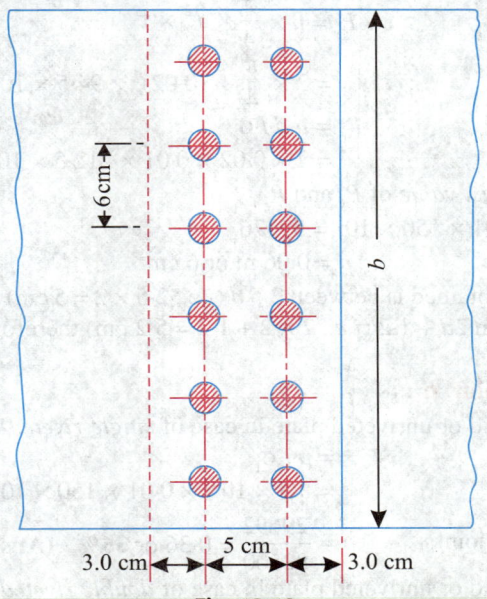

Fig. 12.13

Example 12·6. *Find the suitable pitch for a single riveted lap joint for plates 1 cm thick each, if safe working stress in tension in the plates and crushing and shearing of the rivet material are respectively 150 MN/m², 212·5 MN/m² and 94·5 MN/m² in the following types of joints:*

(i) *Single riveted*

(ii) *Double riveted.*

Find out the efficiency of the joint in the above two cases.

Solution.

(i) Pitch of single riveted lap joint, p:

We know that diameter of the rivet,

$$d = 1·91 \sqrt{t} = 1·91 \sqrt{1} = 1·91 \text{ cm say 2 cm (or 0·02 m)}$$

Now,

$$P_t = (p - d) \, t \, \sigma_t = (p - 0·02) \times 0·01 \times 150 \times 10^6 \text{ N}$$

$$P_s = n \times \frac{\pi}{4} \, d^2 \times \tau \times 1 \qquad (\because \text{ Rivets are in } single \ shear.)$$

$$= 1 \times \frac{\pi}{4} \times (0·02)^2 \times 94·5 \times 10^6 = 29688 \text{ N} \qquad (\because \ n = 1)$$

$$P_c = n.d.t. \, \sigma_c$$

$$= 1 \times 0·02 \times 0·01 \times 212·5 \times 10^6 = 42500 \text{ N}$$

Equating P_s with least value of P_s and P_c

i.e.,

$$P_t = 29688$$

$$(p - 0·02) \times 0·01 \times 150 \times 10^6 = 29688$$

∴

$$p = 0·04 \text{ m or 4 cm}$$

Also pitch should not be *less than 2·5 d (i.e., 2·5 × 2 = 5 cm)* and *2 d + 1·2 (2 × 2 + 1·2 = 5·2 cm)*. So let us provide a pitch of **5·5 cm. (Ans.)**

(ii) Pitch of double riveted lap joint p:

$$P_t = (p - d) \, t·\sigma_t$$

$$= (p - 0·02) \times 0·01 \times 150 \times 10^6 \text{ N}$$

$$P_s = n \times \frac{\pi}{4} \times d^2 \times \tau$$

$$= 2 \times \frac{\pi}{4} \times (0·02)^2 \times 94·5 \times 10^6 = 59376 \text{ N}$$

$$P_c = n·d·t·\sigma_c$$

$$= 2 \times 0·02 \times 0·01 \times 212·5 \times 10^6 = 85000 \text{ N}$$

Equating P_t with least value of P_s and P_c.

i.e., $(p - 0·02) \times 0·01 \times 150 \times 10^6 = 59376$

∴

$$p = 0·06 \text{ m or 6 cm}$$

Since the pitch so obtained is between 2·5 d (i.e., 2·5 × 2 = 5 cm) and 4d (i.e., 4 × 2 = 8 cm), moreover, it is not less than 2d + 1·2 (i.e., 2 × 2 + 1·2 = 5·2 cm) therefore let us provide a pitch of **6 cm. (Ans.)**

Efficiency of the joint, η :

(i) Strength of solid or unriveted plate in case of *single riveted lap joint*

$$= p·t·\sigma_t$$

$$= 5·5 \times 10^{-2} \times 0·01 \times 150 \times 10^6 = 82500 \text{ N}$$

∴ Efficiency of the joint, $\eta = \dfrac{29692}{82500} = 0·36 \text{ or } \textbf{36}\% \textbf{ (Ans.)}$

(ii) Strength of solid or unriveted plate in case of *double riveted lap joint*

$$= p·t·\sigma_t$$

$$= 6 \times 10^{-2} \times 0.01 \times 150 \times 10^6 = 90000 \text{ N}$$

∴ Efficiency of the joint, $\eta = \dfrac{59380}{90000} = 0.66 \text{ or } \mathbf{66\%} \quad \textbf{(Ans.)}$

Example 12·7. *A double riveted double cover butt joint in plates 16 mm thick is made with 25 mm diameter rivets, 100 mm pitch. Permissible stresses are shown below:*

$$\sigma_t = 150 \text{ MN/m}^2, \ \tau = 102.5 \text{ MN/m}^2, \ \sigma_c = 236 \text{ MN/m}^2$$

Compute the pull (per pitch length) which the joint can take and hence work out the efficiency of the joint.

Solution.   Resistance against tearing of the plates,

$$P_t = (p - d) \, t \, \sigma_t$$
$$= (0.1 - 0.025) \times 0.016 \times 150 \times 10^6 = 180000 \text{ N}$$

Resistance against shearing of the rivets,

$$P_s = n \times \frac{\pi}{4} \times d^2 \times \tau \times 2 \qquad (\because \text{ Rivets are in } \textit{double shear.})$$

$$= 2 \times \frac{\pi}{4} \times (0.025)^2 \times 102.5 \times 10^6 \times 2 = 201258 \text{ N}$$

Resistance against crushing of rivets,

$$P_c = n \cdot d \cdot t \cdot \sigma_c = 2 \times 0.025 \times 0.016 \times 236 \times 10^6 = 188800 \text{ N}$$

The safe pull of the joint is the least of the values of P_t, P_s and P_c which is **180000 N (Ans.)**
Strength of solid or unriveted plate per pitch length $= p \cdot t \cdot \sigma_t$

∴ $\eta = \dfrac{\text{Safe pull}}{p.\, t.\, \sigma_t}$

$$= \frac{180000}{0.1 \times 0.016 \times 150 \times 10^6} = 0.75 \text{ or } \mathbf{75\%} \quad \textbf{(Ans.)}$$

A water turbine.

Example 12·8. *Two plates, each 20 mm thick, are to be joined by a double riveted, double strap butt joint. The pitch of the rivets in the outer row is to be twice that of inner row. Zigzag riveting is to be done. Assuming* $\tau = 80 \text{ MN/m}^2$ *and* $\sigma_t = 100 \text{ MN/m}^2$, *calculate the following:*

(i)   *Rivet diameter,*   (ii)   *Rivet pitches in outer and inner rows, and*
(iii)   *Thickness of butt straps.*

Solution.

(i) Rivet diameter, d: Using Unwin's formula, we get

$$d = 6\sqrt{t} = 6 \times \sqrt{20} = 26.8 \text{ mm}$$

From IS code, we adopt **28·5 mm** (or **0·0285 m**)   **(Ans.)**

(ii) Rivet pitches:

Assuming the rivets to be 1·875 times as strong in double shear as in single shear and equating tearing resistance to shearing resistance, we have

$$(p - d)\, t\, \sigma_t = 3 \times \frac{\pi}{4} \times d^2 \times \tau \times 1.875$$

or, $(p - 0.0285) \times 0.02 \times 100 \times 10^6 = 3 \times \dfrac{\pi}{4} \times (0.0285)^2 \times 80 \times 10^6 \times 1.875$

or, $p = 17.2 \text{ cm}$

Rivet pitch in inner row $= \dfrac{1}{2} \times 17.2 = \mathbf{8.6 \text{ cm}}$ **(Ans.)**

(iii) Thickness of butt strap:

Thickness of butt strap, $t_1 = 0.625\, t = 0.625 \times 2 = \mathbf{1.25 \text{ cm}}$ **(Ans.)**

Example 12·9. (a) *Draw neat sketches (plan and section) of a double riveted butt joint with double cover plates.*

(b) *In a double riveted lap joint, the pitch of the rivets is 7·5 cm, thickness of plates = 1·5 cm and rivet diameter = 2·5 cm. What maximum force per pitch length will rupture the joint when ultimate stresses are* $\sigma_{ut} = 400 \text{ MN/m}^2$, $\tau_u = 320 \text{ MN/m}^2$ *and* $\sigma_{uc} = 640 \text{ MN/m}^2$.

Solution. (a) Refer to Fig. 12.14.

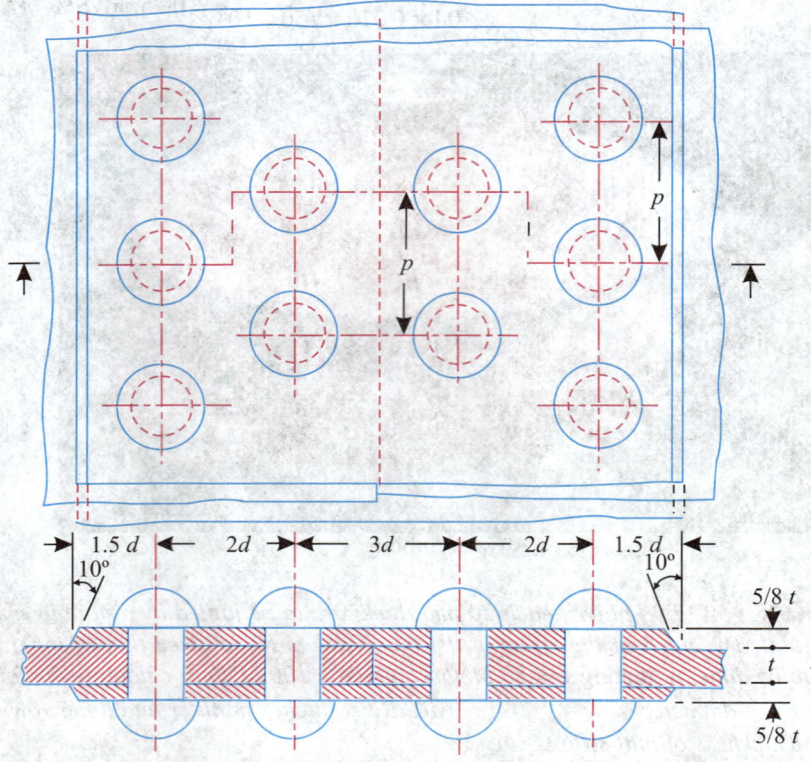

Fig. 12.14

(b) Pitch of the rivets, $p = 7 \cdot 5$ cm or $0 \cdot 075$ m

Thickness of plates, $t = 1 \cdot 5$ cm or $0 \cdot 015$ m

Rivet diameter, $d = 2 \cdot 5$ cm or $0 \cdot 025$ m

Ultimate tensile stress, $\sigma_{ut} = 400$ MN/m^2

Ultimate shear stress, $\tau_u = 320$ MN/m^2

Ultimate crushing stress, $\sigma_{uc} = 640$ MN/m^2

Minimum force per pitch length :

$$P_t = (p - d) \, t \cdot \sigma_{ut} = (0 \cdot 075 - 0 \cdot 025) \times 0 \cdot 015 \times 400 \times 10^6$$
$$= 300000 \text{ N or } 300 \text{ kN}$$

$$P_s = n \times \frac{\pi}{4} \times d^2 \times \tau_u$$

$$= 2 \times \frac{\pi}{4} \times (0 \cdot 025)^2 \times 320 \times 10^6 \qquad [n = \text{no. of rivets}$$
$$\text{per pitch length} = 2]$$

$$= 314159 \text{ N or } 314 \cdot 159 \text{ kN}$$

$$P_c = n \cdot d \cdot t \, \sigma_{uc}$$
$$= 2 \times 0 \cdot 025 \times 0 \cdot 015 \times 640 \times 10^6 = 480000 \text{ or } 480 \text{ kN}$$

∴ The minimum force per pitch length which will rupture the joint is **300 kN** (minimum of the above three values) **(Ans.)**

Example 12·10. *A triple riveted double cover straps butt joint is to be made between 16 mm thick plates having chain riveting with the same rivet pitch in all the rows. Assuming $\sigma_t = 150$ MN / m^2 and $\tau = 75$ MN /m^2 determine the following:*

(i) *Rivet diameter;*   (ii) *Rivet pitch;*

(iii) *Distance between rivet rows;*   (iv) *Strap thickness, and*   (v) *Joint efficiency.*

Solution.

(i) Rivet diameter, d:

Using Unwin's formula, we get $d = 6\sqrt{t} = 6\sqrt{16} = \textbf{24 mm}$ **(Ans.)**

(ii) Rivet pitch p:

Assuming $P_t = P_s$ to get rivet pitch, we have

$$(p - d) \, t \, \sigma_t = n \times \frac{\pi}{4} \times d^2 \times \tau \times 1 \cdot 875$$

$$(p - 0 \cdot 024) \times 0 \cdot 016 \times 150 \times 10^6 = 3 \times \frac{\pi}{4} \times (0 \cdot 024)^2 \times 75 \times 10^6 \times 1.875$$

[Assuming that the rivets in double shear are 1·875 times as strong as those in single shear]

or, $$p = \frac{3 \times \dfrac{\pi}{4} \times (0 \cdot 024)^2 \times 75 \times 10^6 \times 1.875}{150 \times 10^6 \times 0 \cdot 016} + 0 \cdot 024$$

$$\simeq 0 \cdot 104 \text{ m or } \textbf{10·4 cm} \textbf{(Ans.)}$$

(iii) Distance between rivet rows:

Distance between the rows $= 0 \cdot 8 \, p$

$$= 0 \cdot 8 \times 10 \cdot 4 = 8 \cdot 32 \text{ cm say } \textbf{8·4 cm} \textbf{(Ans.)}$$

(iv) Strap thickness:

Strap thickness $= 0 \cdot 625 \, t = 0 \cdot 625 \times 1 \cdot 6 = \textbf{1 cm} \textbf{(Ans.)}$

(v) Efficiency of the joint η:

Efficiency of the joint, $$\eta = \frac{P_t \text{ or } P_s}{p \cdot t \cdot \sigma_t} = \frac{3 \times \pi/4 \times (0.024)^2 \times 75 \times 10^6 \times 1.875}{0 \cdot 104 \times 0 \cdot 016 \times 150 \times 10^6}$$

$$= 0 \cdot 765 \text{ or } \textbf{76·5\%} \textbf{(Ans.)}$$

Example 12·11. *Determine the pitch 'p' and diameter 'd' for the riveted joint shown in the Fig. 12·15 so that efficiencies may be as nearly equal as possible. The thickness of each of the plates to be connected is 1·4 cm. Assume that the shearing resistance of rivets in double shear is 1·75 times that in single shear.*

$$\sigma_t \ for \ plates = 430 \ MN/m^2$$

$$\tau \ for \ rivets = 350 \ MN/m^2$$

Solution. Thickness of plates, $t = 1·4$ cm (or 0·014 m)

Diameter of the rivet, using Unwin's formula

$$d = 1·91 \sqrt{t} = 1·91 \sqrt{1·4} = 2·26 \ cm \ say \ 2·3 \ cm \ (or \ 0·023 \ m)$$

Let the pitch in the outer row be p, then in the inner rows the pitch will be $p/2$ (Fig. 12·15)

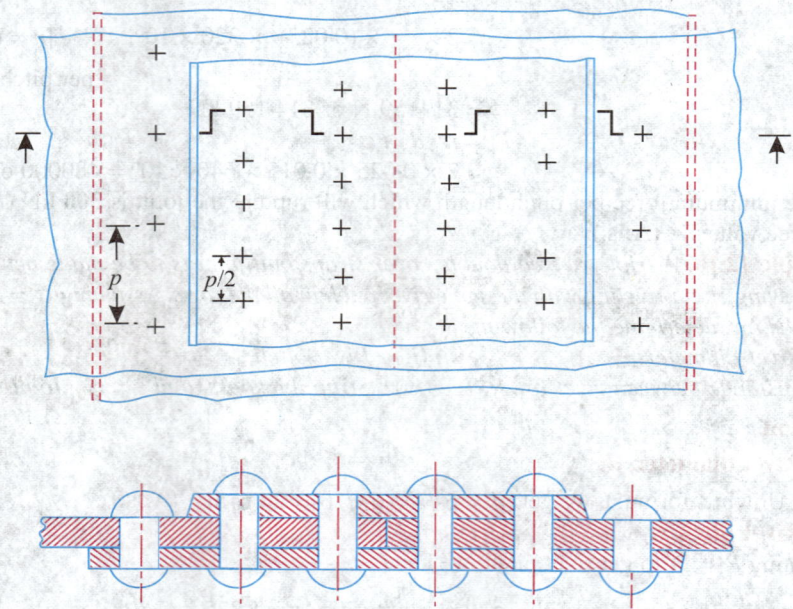

Fig. 12.15

The number of rivets per pitch length are:

(*i*) One rivet in single shear

(*ii*) Four rivets in double shear.

Now,

$$P_s = 1 \times \frac{\pi}{4} \times d^2 \times \tau + 4 \times \pi/4 \times d^2 \times \tau \times 1·75$$

$$= 1 \times \frac{\pi}{4} \times (0·023)^2 \times 350 \times 10^6 + 4 \times \pi/4 \ (0·023)^2$$

$$\times 350 \times 10^6 \times 1·75$$

$$= 1163332 \ N$$

Assuming the section is weakened by one rivet hole only, the resistance against tearing (P_t) is the least in the outermost row.

Now, $P_t = (p - d) \cdot t \cdot \sigma_t$

Also, $P_t \leq P_s$

∴ $(p - d) \cdot t \cdot \sigma_t = 1163332$

or, $(p - 0·023) \times 0·014 \times 430 \times 10^6 \leq 1163332$

or, $p \leq 0·2162$ m

Adopt $p = 0·2162$ m or **21·62 cm** (Ans.)

Metallic framework.

Example 12·12. *A 100 mm × 25 mm steel strap is spliced as shown in Fig. 12·16. The rivets are 25 mm in diameter. The allowable working stresses are 112·5 MN/m² in tension, 70 MN/m² in shear and 190 MN/m² in bearing. Calculate the safe load P for the spliced strap and the efficiency of the joint.*

Solution.   Width of steel plate, $b = 100$ mm $= 0.1$ m
Thickness of steel plate, $t = 25$ mm $= 0.025$ m
Thickness of cover plate, $t_c = 13$ mm $= 0·013$ m
Rivet diameter, $d = 25$ mm $= 0·025$ m
Allowable tensile stress, $\sigma_t = 112·5$ MN/m²
Allowable shear stress, $\tau = 70$ MN/m²
Allowable crushing stress, $\sigma_c = 190$ MN/m²
Refer to Fig. 12·16.

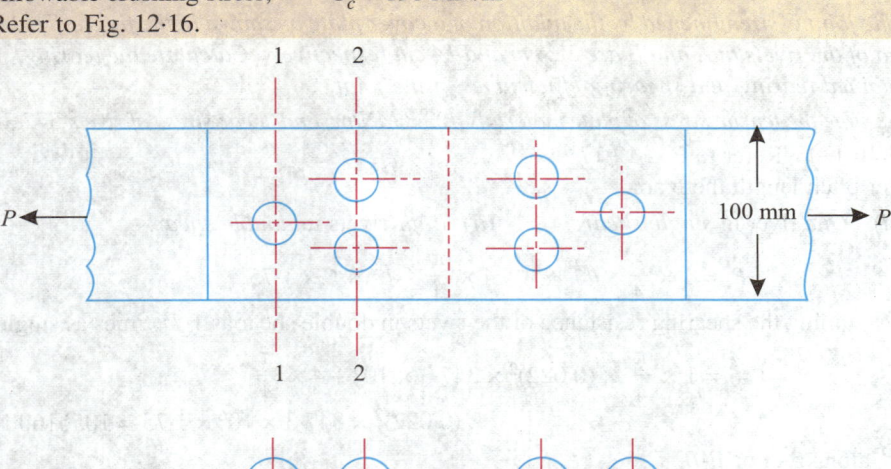

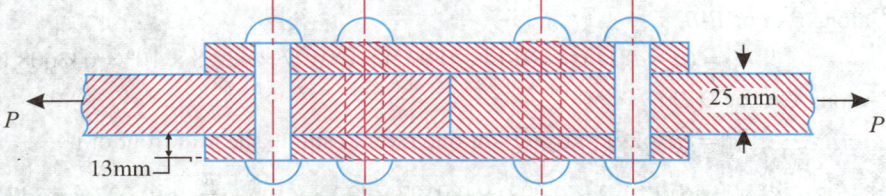

Fig. 12.16

$$P_s = n \times \frac{\pi}{4} \times d^2 \times \tau \times 2 \text{ (Since the rivets are in double shear.)}$$

$$= 3 \times \frac{\pi}{4} \times (0 \cdot 025)^2 \times 70 \times 10^6 \times 2 = 206167 \text{ N} \qquad \dots(i)$$

$$P_c = n \cdot d \cdot t \, \sigma_c$$

$$= 3 \times 0 \cdot 025 \times 0 \cdot 025 \times 190 \times 10^6 = 356250 \text{ N} \qquad \dots(ii)$$

P_t (across one rivet)

$$P_{t_1} = (b - d) \times t \cdot \sigma_t = (0 \cdot 1 - 0 \cdot 025) \times 0 \cdot 025 \times 112 \cdot 5 \times 10^6$$

$$= 210937 \cdot 5 \text{ N} \qquad \dots(iii)$$

and P_t (across two rivets),

i.e., $$P_{t_2} = (b - 2d) \cdot t \cdot \sigma_t + 1 \times \frac{\pi}{4} \times d^2 \times \tau \times 2$$

$$= (0 \cdot 1 - 2 \times 0 \cdot 025) \times 0 \cdot 025 \times 112 \cdot 5 \times 10^6 + \frac{\pi}{4} \times (0 \cdot 025)^2 \times 70 \times 10^6 \times 2$$

$$= 140625 + 68722 = 209347 \text{ N} \qquad \dots(iv)$$

Now *strength of cover plate in tearing,*

$$P_t \text{ (cover plates)} = 2 \, (b - 2d) \cdot t_c \cdot \sigma_t$$

$$= 2 \, (0 \cdot 1 - 2 \times 0 \cdot 025) \times 0 \cdot 013 \times 112 \cdot 5 \times 10^6$$

$$= 146250 \text{ N} \qquad \dots(v)$$

Now, the safe load is the least load of (*i*), (*ii*), (*iii*), (*iv*) and (*v*)

∴ Safe load = 146250 N

∴ Efficiency of the joint,

$$\eta = \frac{\text{Least strength}}{\text{Strength of solid plate}} = \frac{146250}{b \cdot t \cdot \sigma_t} = \frac{146250}{0 \cdot 1 \times 0 \cdot 025 \times 112 \cdot 5 \times 10^6}$$

or, $$\eta = 0 \cdot 52 \text{ or } \mathbf{52\%} \quad \textbf{(Ans.)}$$

Example 12·13. *In a double riveted lap joint shown in the Fig. 12·17 for 1·27 cm plate the diameter and pitch of the rivets are 2·2 cm and 7 cm respectively.*

The joint is strengthened by the addition of a cover plate as shown in the figure. The diameter and pitch of the rivets at A and B are 2·2 cm and 14 cm respectively. Calculate the tearing, shearing and combined tearing and shearing efficiencies of the joint.

Assume the tensile stress of plate materials 422·8 MN/m² and shear stress of rivets 347·3 MN/m².

Solution. Refer to Fig. 12·17

Per pitch length there are,

(*i*) One rivet in *single shear;* (*ii*) *Four* rivets in *double shear.*

$$P_s = 1 \times \frac{\pi}{4} \times d^2 \times \tau + 4 \times \frac{\pi}{4} \times d^2 \times \tau \times 1 \cdot 75$$

[Assuming the shearing resistance of the rivets in double shear as 1·75 times as single shear.]

$$= 1 \times \frac{\pi}{4} \times (0 \cdot 022)^2 \times 347 \cdot 3 \times 10^6 + 4 \times \frac{\pi}{4}$$

$$\times (0 \cdot 022)^2 \times 347 \cdot 3 \times 10^6 \times 1 \cdot 75 = 1056160 \text{ N}$$

P_{t_1} along *A-A* or *B-B*,

$$P_{t_1} = (p - d) \, t \cdot \sigma_t = (0 \cdot 14 - 0 \cdot 022) \times 0 \cdot 0127 \times 422 \cdot 8 \times 10^6 = 633608 \text{ N}$$

P_{t_2} along *C-C* or *D-D*

$$P_{t_2} = (p - 2d) \, t \cdot \sigma_t + P_s \text{ of one rivet in single shear in front of it.}$$

$$= (0 \cdot 14 - 2 \times 0 \cdot 022) \times 0 \cdot 0127 \times 422 \cdot 8 \times 10^6 + \frac{\pi}{4} \times (0 \cdot 022)^2 \times 347 \cdot 3 \times 10^6$$

$$= 647498 \text{ N}$$

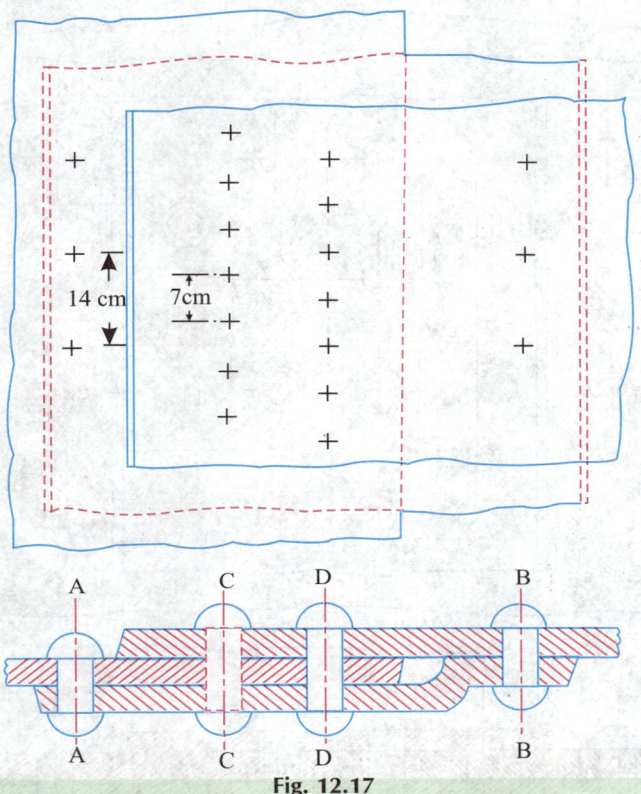

14 cm   7cm

A   C   D   B

A   C   D   B

Fig. 12.17

Strength of solid (undrilled) plate

$$= p.t.\sigma_t = 0.14 \times 0.0127 \times 422.8 \times 10^6$$
$$= 751738 \text{ N}$$

$$\eta_t = \frac{633608}{751738} = 0.8428 \text{ or } \mathbf{84.28\%} \quad \textbf{(Ans.)}$$

$$\eta_s = \frac{1056160}{751738} = 1.405 \text{ or } \mathbf{140.5\%} \quad \textbf{(Ans.)}$$

$$\eta_{\text{combined}} = \frac{647498}{751738} = 0.8613 \text{ or } \mathbf{86.13\%}$$

$$\textbf{(Ans.)}$$

Example 12·14. *Design a triple riveted butt joint with two unequal cover straps for longitudinal seam of an air tank which is 75 cm in diameter and subjected to an internal pressure of 1·5 MN/m² gauge. The plates and rivets are made of the same material having allowable stresses of 80 MN/m² in tension, 60 MN/m² in shear and 135 MN/m² in crushing. The efficiency of the joint must not be less than 80 per cent. Calculate the strength of the joint thus designed for all possible ways of failure and also determine its lowest efficiency. Give a neat dimensioned sketch of the joint.*

Solution. Refer to Fig. 12·18.

Spot welding machine.

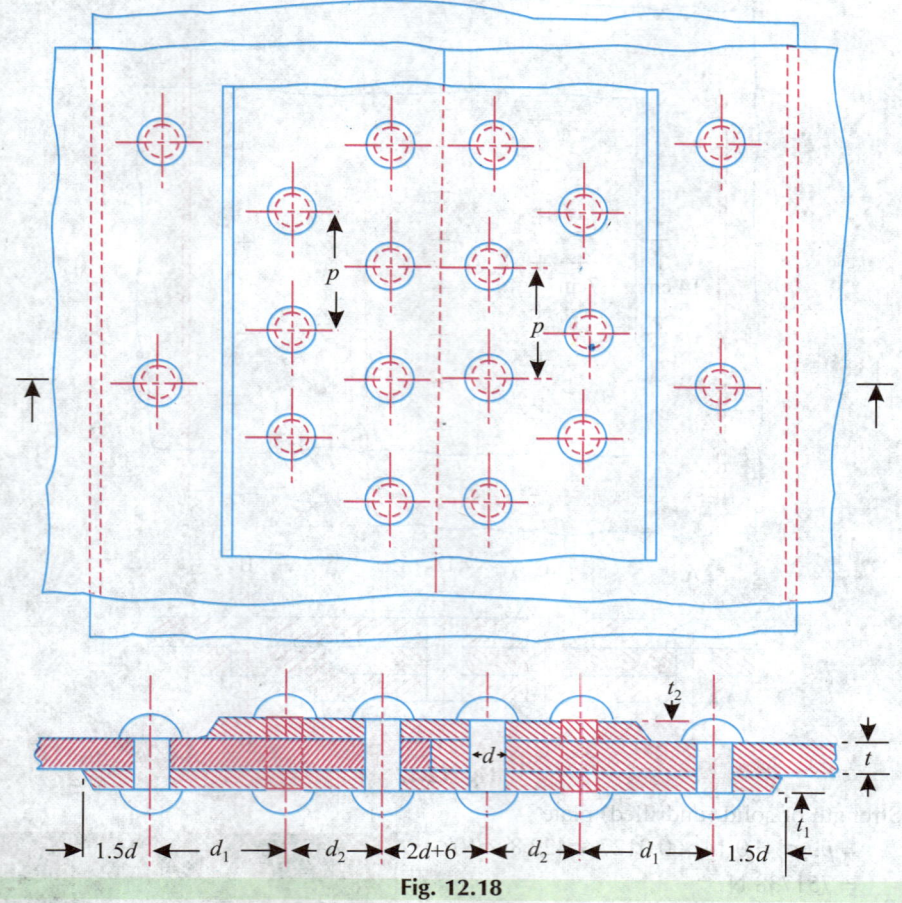

Fig. 12.18

The thickness of the tank plate is given by the formula,

$$t = \frac{pd}{2\,\sigma_t\eta} = \frac{1 \cdot 5 \times 10^6 \times 0 \cdot 75}{2 \times 80 \times 10^6 \times 0 \cdot 8} = 0 \cdot 0088 \text{ m or } 0 \cdot 88 \text{ cm}$$

We adopt $t = 1$ cm

Diameter of the rivet, $d = 6\sqrt{t} = 6\sqrt{10} = 18 \cdot 97$ mm

We adopt $d = 19$ mm (or $0 \cdot 019$ m)

Let us arrange the joint in such a manner that the alternate rivets in outer row are omitted. The rivets in the outer row will be in single shear, while those in inner rows will be in double shear.

Let p be the pitch between the rivets in outer rows. In one length, there are five rivets out of which *four are in double shear* while *one is in single shear*.

$$P_t = (p - d)\, t \cdot \sigma_t$$

$$P_s = 4 \times \frac{\pi}{4} \times d^2 \times \tau \times 1 \cdot 875 + \frac{\pi}{4} \times d^2 \times \tau = 8 \cdot 5\,\frac{\pi}{4} \times d^2 \times \tau$$

To get rivet pitch equating P_t with P_s, we get

$$(p - 0 \cdot 019) \times 0 \cdot 01 \times 80 \times 10^6 = 8 \cdot 5 \times \frac{\pi}{4} \times (0 \cdot 019)^2 \times 60 \times 10^6$$

from which $p = 0 \cdot 20$ m or 20 cm.

Now, $P_t = P_s = 8 \cdot 5 \times \dfrac{\pi}{4} \times d^2 \times \tau$

$$= 8 \cdot 5 \times \frac{\pi}{4} \times (0 \cdot 019)^2 \times 60 \times 10^6 = 144600 \text{ N}$$

and, $\qquad P_c = n\,d \cdot t\,\sigma_c = 5 \times 0.019 \times 0.01 \times 135 \times 10^6 = 128250$ N

∴ Strength of the joint = least of P_t, P_s

and, $\qquad P_c = 128250$ N

∴ Efficiency of the joint, $\quad \eta = \dfrac{128250}{p \cdot t \cdot \sigma_t} = \dfrac{128250}{0.2 \times 0.01 \times 80 \times 10^6}$

$\qquad\qquad\qquad = 0.8015$ or **80·15%** **(Ans.)**

Example 12·15. *Design a double cover butt joint to withstand a load of 250 kN. The plates to be joined are 20 cm wide and 1·25 cm thick, 2 cm diameter rivets are to be used in diamond fashion of rivets rows so as to increase the efficiency of the joint. Permissible stresses are: shear 70 MN/m², bearing 190 MN/m² and tension 110 MN/m². What is the efficiency of the joint?*

Solution. Thickness of plates,

$\qquad\qquad\qquad t = 1.25$ cm (or 0·0125 m)

Diameter of the rivets $\qquad d = 2$ cm (or 0·02 m)

Width of plates, $\qquad\qquad b = 20$ cm (or 0·2 m)

Permissible stresses are:

$\qquad \sigma_t = 110$ MN/m²; $\tau = 70$ MN/m²; $\sigma_c = 190$ MN/m²

$\qquad P_s = n \times \dfrac{\pi}{4} \times d^2 \times \tau \times 2 \qquad\qquad$ (∵ Rivets are in *double shear*.)

$\qquad\quad = n \times \dfrac{\pi}{4} \times (0.020)^2 \times 70 \times 10^6 \times 2$

$\qquad\quad = 43982\,n$ N, *n* being number of rivets.

The strength must be equal to applied load of 250 kN.

∴ $\qquad\qquad 43982\,n = 250 \times 10^3$ or $n = 5.68$

Let us provide 6 rivets

Again, $\qquad\qquad P_c = n.\,d.\,t.\,\sigma_c = n \times 0.02 \times 0.0125 \times 190 \times 10^6 = 47500\,n$

Similarly, $\qquad 47500\,n = 250 \times 10^3 \quad$ ∴ $n = \dfrac{250000}{47500} = 5.26$ say **6**

∴ Efficiency of the joint, $\quad \eta = \dfrac{25000}{\text{Strength of solid plate}}$

$\qquad\qquad\qquad = \dfrac{250000}{b\,t\,\sigma_t} = \dfrac{250000}{0.2 \times 0.0125 \times 110 \times 10^6}$

$\qquad\qquad\qquad = 0.909 = $ **90·9 %** **(Ans.)**

Minimum pitch $\qquad\qquad p = 3\,d = 3 \times 2 = 6$ cm

Distance of centre of rivet nearest to the edge from the edge

$\qquad\qquad \not< 1.75\,d = \not< 1.75 \times 2 = \not< 3.5$ cm

The width of plate is 20 cm.

$\qquad\qquad$ Back pitch $= 2.5\,d = 2.5 \times 2 = 5$ cm

The arrangement of the rivets is shown in Fig. 12·19.

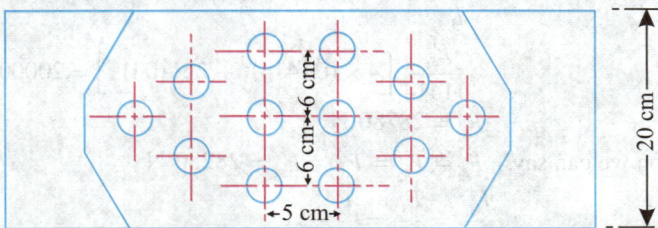

Fig. 12.19

Example 12·16. *An eccentrically loaded lap riveted joint is to be designed for a steel bracket as shown in Fig. 12-20. The bracket plate is 2·5 cm thick. All rivets are to be made of the same size. Load on the bracket P = 50 kN, rivet spacing C = 10 cm, load arm l = 40 cm. Permissible shear stress is 65 MN/m² and crushing stress is 120 MN/m². Determine the size of the rivets to be used for the joint.*

Solution. Refer to Fig. 12·20.

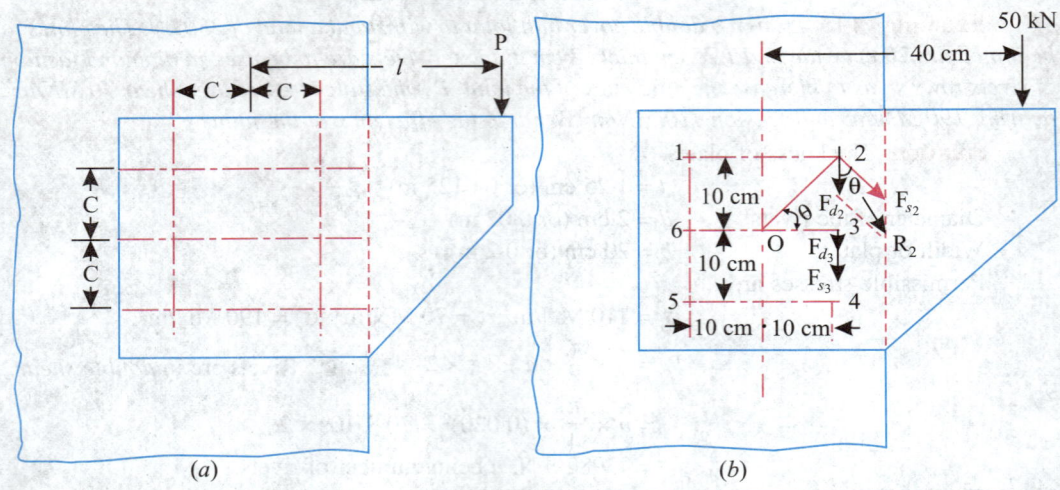

(a)   (b)

Fig. 12.20

Direct shear on each rivet acting vertically downwards

$$F_d = \frac{50 \times 10^3}{6} = 8333 \text{ N}$$

The centroid of the rivet group is at 0. The distance of rivets 3 and 6 from centre of rivets group is 10 cm. The distance of rivets 1, 2, 4 and 5 from centroid

$$= \sqrt{10^2 + 10^2} = \sqrt{200} = 14 \cdot 14 \text{ cm (or } 0 \cdot 1414 \text{ m)}$$

Turning moment due to load P = Load × eccentricity

$$= P \times l = 50 \times 10^3 \times 0 \cdot 4 = 20000 \text{ Nm}$$

This turning moment is resisted by six rivets. Let $F_{s_1}, F_{s_2}, F_{s_3}, F_{s_4}, F_{s_5}, F_{s_6}$ be the secondary shear due to turning moment in the rivets 1, 2, 3, 4, 5 and 6 placed at l_1, l_2, l_3, l_4, l_5 and l_6 respectively from the centroid.

Then, $l_1 = l_2 = l_4 = l_5 = 14 \cdot 14 \text{ cm} = 0 \cdot 1414 \text{ m}$
$$l_3 = l_6 = 10 \text{ cm} = 0 \cdot 1 \text{ m}$$

Now, equating the resisting moment to turning moment, we have

$$\frac{F_{s_1}}{l_1} (l_1^2 + l_2^2 + l_3^2 + l_4^2 + l_5^2 + l_6^2) = P \times l$$

$$\frac{F_{s_1}}{0 \cdot 1414} \left[4 \times (0 \cdot 1414)^2 + 2 \times (0 \cdot 1)^2 \right] = 20000$$

or, $F_{s_1} = 28280 \text{ N}$

By inspection we can say, $F_{s_1} = F_{s_2} = F_{s_4} = F_{s_5} = 28280 \text{ N}$

Since, $\frac{F_{s_1}}{l_1} = \frac{F_{s_3}}{l_3} = \frac{F_{s_6}}{l_6}$

$$F_{s_3} = F_{s_1} \times \frac{l_3}{l_1} = 28280 \times \frac{0.1}{0.1414} = 20000 \text{ N}$$

Also by inspection, $\qquad F_{s_3} = F_{s_6} = 20000 \text{ N}$

The magnitude of the resultant load will not be necessarily same for rivets 1, 2, 4 and 5 because the *resultant load depends upon the inclined angle*,

$$\cos \theta = \frac{0.1}{0.1414}, \text{ angle } \theta \text{ is indicated in the figure. The maximum resultant force will be on rivet 2.}$$

$$R_2 = R = \sqrt{(8333)^2 + (28280)^2 + 2 \times 8333 \times 28280 \times \frac{0.1}{0.1414}}$$

$$\approx 34680 \text{ N}$$

Rivet 5 will also be carrying the same resultant force (R).

Again, $\qquad R = \frac{\pi}{4} \times d^2 \times \tau$ where, d is the diameter of the rivet

$\therefore \qquad 34680 = \frac{\pi}{4} \times d^2 \times 65 \times 10^6$

or, $\qquad$ **d = 0.026 m or 26 mm (Ans.)**

Bearing/crushing stress in the rivet

$$= \frac{R}{d \times t} = \frac{34680}{0.026 \times 0.025} \times 10^{-6} \text{ MN/m}^2$$

$$= 53.35 \text{ MN/m}^2 \text{ which is well within the limit.}$$

12·3. WELDED JOINTS

For many purposes, welded connections are found to be more suitable than riveted or bolted connections. **Welding** *is a process of joining two pieces of metal by fusion.* In the types of welded joints discussed here, the pieces to be joined are heated and molten material is deposited between them. While there are a great variety of welding processes in use, today, the heating is most frequently accomplished by means of an *electric arc*. This section will deal with lap and butt joints, although there are several other types of welded joints of importance.

12·3·1. Advantages and Disadvantages of Welded Joints

Forge welding.

Following are the *advantages and disadvantages of welded joints over riveted joints*:

Advantages:

1. Efficiency of the welded joint is more than that of the riveted joint.
2. The welded structures are usually lighter than the riveted structures.
3. A welded joint has a greater strength.
4. Additions and alterations can be easily made in the existing welded structures.

5. In welded connections the tension members are not weakened as in the case of riveted joints.
6. The welding provides very rigid joints. This is in line with the modern trend of providing rigid frames.
7. The process of welding takes less time than riveting.
8. Layout for punching or drilling of holes is not required in welding.
9. Comparatively smaller sections can be used for the same load.
10. As the welded structure is smooth in appearance, therefore it looks pleasing. Moreover, its painting is easier and economical.

Disadvantages:

1. It requires highly skilled labour and supervision.
2. As there is an uneven heating and cooling during welding, the members may get distorted or additional stress may develop.
3. The inspection of welding work is more difficult than the riveting work.
4. Since no provision is kept for expansion or contraction in the frame, therefore cracks may develop in it.

12·3·2. Types of Welds:

For joining plates in structural practice the following *two types of welded joints* are used:

1. Butt welds
2. Lap welds or fillet welds

Butt welds. Fig. 12·21 shows the various types of butt welds, most of which require advanced preparation of the plate edges at the joint. After the ends of the plates are suitably shaped and butted end-to-end the metal is deposited in place between them as shown. But welds generally act in tension or compression only. These welds are most commonly used for circumferential or girth joints of pipes and tanks for gases, steam, oil etc. The end covers also have butt joints. The allowable stresses for such welds are generally 90 MN/m² in tension and 20 MN/m² in compression. The commercial sizes of welds available vary from 5 mm to 25 mm with variation of 5 mm. The size of the weld chosen should be equal to thickness of the plates to be welded. *If the size of the plates to be welded is not of equal thickness, the size of the weld should be equal to the thickness of thinner plate.*

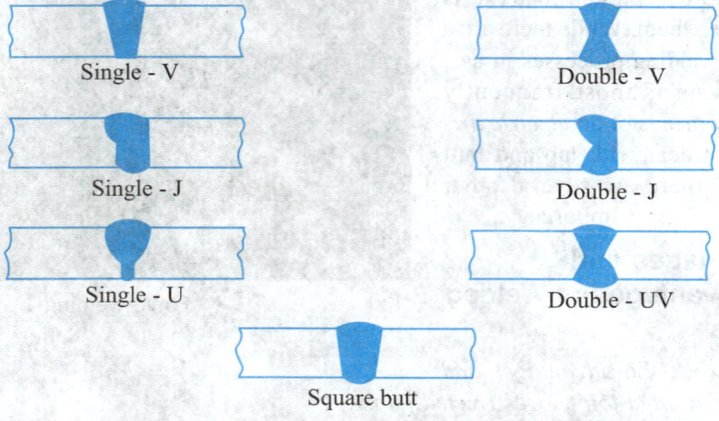

Single - V   Double - V

Single - J   Double - J

Single - U   Double - UV

Square butt

Fig. 12.21

Lap welds or Fillet welds. Fig. 12·22 shows what is known as a *side fillet lap weld* with a corner return. The fused metal is deposited in the angle between the two pieces. Shearing stresses are developed in this weld when either tensile or compressive forces are applied to the long axis of the plates. Fig. 12·23 shows a joint using *end lap welds* with corner returns. This type of weld is usually used in conjunction with side welds.

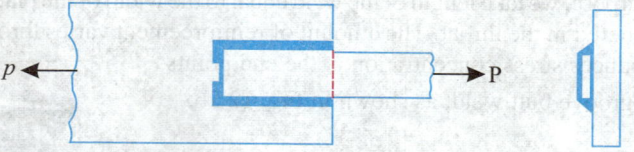

Fig. 12.22. Side fillet welds (with corner returns)

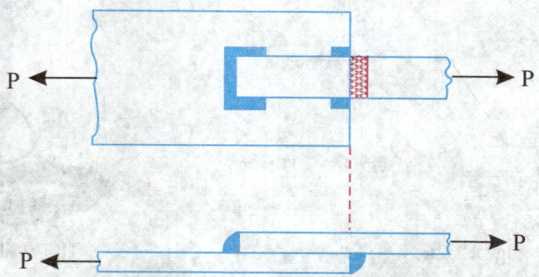

Fig. 12.23. End fillet welds (with corner returns)

The size of the fillet welds is designated by the length of the leg as shown in Fig. 12·24 (*a*). Fillet welds resist loads on throat area, which is the minimum cross-section of the weld. The *throat area is the product of the throat thickness and 'effective' length of the weld.* The throat thickness is determined from Fig. 12·14 (*b*) as Leg sin 45° or 0·707 Leg.

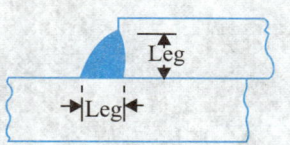

(*a*) Cross section of a fillet weld

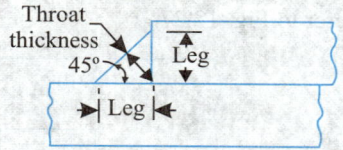

(*b*) Simplified cross-section of a fillet weld

Fig. 12.24

12·3·3. Strength of Butt Welds.

Fig. 12·25 shows a single *V* groove butt weld. It may act only in tension or compression.

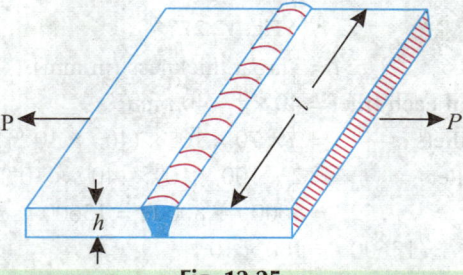

Fig. 12.25

Let,   P = Tensile force,

                                        h = Weld throat,

                                        l = Length of weld, and

                                        σ_t = Permissible stress intensity in the weld material.

Then,   P = Area of weld × permissible stress = $h\,l\,\sigma_t$

For design purpose, we take length of the weld equal to the width of the plate. The reinforcement of weld is not considered in the throat. The amount of reinforcement varies throughout the length of the weld, and it produces stress concentration at the end points.

In double *V*-groove butt weld, as shown in Fig. 12·26.

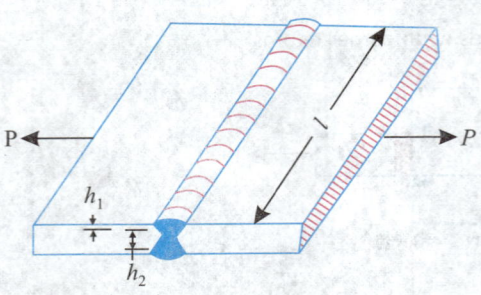

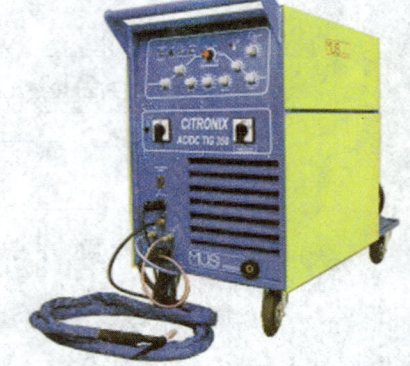

Fig. 12.26	TIG welding machine.

$$P = (h_1 + h_2)\, l\, \sigma_t$$

where, h_1 and h_2 are the amount of the *V*-grooves.

Example 12·17. *A plate 40 mm wide carrying a load of 100 kN is to be welded by four equal fillets to another plate as shown in Fig. 12·27. Find the necessary size of each fillet. Allow working stress of 70 MN/m² in the side fillets and 100 MN/m² in the end fillets.*

Solution. Refer to Fig. 12·27.

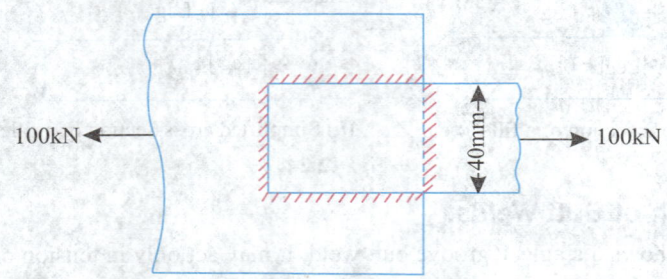

Fig. 12.27

Let, $\quad\quad\quad\quad\quad\quad\quad\quad\quad$ t = Throat thickness (in mm).

Then, throat section of each fillet = $40 \times t = 40\, t$ mm²

Load taken by side fillets $\quad = 2 \times 70 \times 10^6 \times (40\, t \times 10^{-6}) = 5600t$ N

Load taken by end fillets $\quad = 2 \times 100 \times 10^6 \times (40\, t \times 10^{-6}) = 8000t$ N

Total load $\quad\quad\quad\quad\quad = 5600\, t + 8000\, t = 13600\, t$

But, $\quad\quad\quad\quad\quad 13600\, t = 100 \times 10^3$

∴ $\quad\quad\quad\quad\quad\quad\quad t = 7·35$ mm

Hence size of leg of the weld $\quad = 7·35 \times \sqrt{2}$

$$= \mathbf{10·4\ mm}\quad \textbf{(Ans.)}$$

$$\left[\begin{array}{l} \because t = 0·707\ \text{Leg} \\ \\ \therefore \text{Leg} = \dfrac{t}{0·707} = \sqrt{2}\, t \end{array} \right]$$

Example 12·18. *A welded joint is provided to connect two tie bars 150 mm × 10 mm as shown in Fig. 12·28. The working stress in the bar is 120 MN/m². Investigate the design, if the size of the fillet is 12 mm. Take the working stress in the end fillet as 102·5 MN/m² and that in the diagonal fillet as 70 MN/m².*

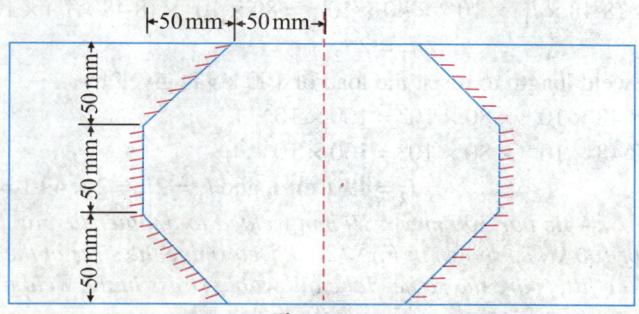

Fig. 12.28

Solution. Area of the bar $= 150 \times 10 = 1500 \text{ mm}^2$ $(= 1500 \times 10^{-6} \text{ m}^2)$

Working stress in the bar $= 120 \text{ MN/m}^2$

Size of the fillet $= 12 \text{ mm (or } 0·012 \text{ m)}$

Working stress in the end fillet $= 70 \text{ MN/m}^2$

Maximum load the tie bar can resist $= 1500 \times 10^{-6} \times 120 \times 10^6 = 180000 \text{ N}$

Effective throat thickness, $t = 0·707 \times 12 = 8·4 \text{ mm } (0·0084 \text{ m})$

Strength of weld, P:

Using the relation $P = l\,t\,\sigma$, we get

$$P_1 = (2 \times 50 \times 10^{-3}) \times 0·0084 \times 102·5 \times 10^6 = 86100 \text{ N}$$

Similarly, $$P_2 = [4\,(\sqrt{50^2 + 50^2}\,) \times 10^{-3}] \times 0·0084 \times 70 \times 10^6$$

$$= 166311 \text{ N} \qquad \text{...For diagonal fillet welds.}$$

∴ Total strength of the weld, $P = P_1 + P_2 = 86100 + 166311 = 252411 \text{ N}$

Since the total strength of the weld is more than the strength of bar i.e. 180000 N, therefore, **joint is safe. (Ans.)**

Example 12·19. *Fig. 12·29 shows 12 mm thick plates loaded by the forces of 100 kN applied eccentrically. Determine the required lengths l_1 and l_2 of the fillet welds so that they will be equally stressed in shear. Take working stress in shear for side fillets to be equal to 80 MN / m².*

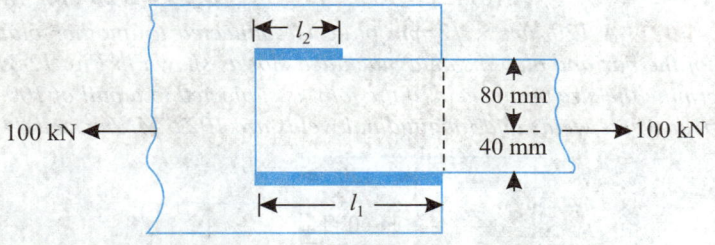

Fig. 12.29

Solution. Refer to Fig. 12·29.

Throat of the weld $= 0·707 \times 12 = 8·48 \text{ mm}$

The effective weld area that resists shear on this throat section

$$= (l_1 + l_2) \times 8·48 \text{ mm}^2$$

The lengths l_1 and l_2 are so proportioned that the welds are equally stressed in shear. This means that the resultant of these two shearing stresses in welds must coincide with the line of action of 100 kN load. In other words, the moment of the two shear forces in the welds about any joint on the line of action must be equal.

$$\therefore \quad 80 \times 10^6 \times (8.48 \times l_1) \times 10^{-6} \times 40 \times 10^{-3} = 80 \times 10^6 \times (8.48 \times l_2) \times 10^{-6} \times 80 \times 10^{-3}$$

or, $$l_1 = 2\,l_2$$

The required weld length to resist the load of 100 kN is given by

$$(l_1 + l_2) \times 8.48 \times 10^{-6} \times 80 \times 10^6 = 100 \times 10^3$$
$$3l_2 \times 8.48 \times 10^{-6} \times 80 \times 10^6 = 100 \times 10^3 \qquad\qquad (\because l_1 = 2l_2)$$

$$\therefore \qquad\qquad l_2 = 49.1 \text{ mm, and } l_1 = 2l_2 = 2 \times 49.1 = 98.2 \text{ mm} \quad \textbf{(Ans.)}$$

Example 12·20. *A tie bar 100 mm × 10 mm welded to another tie bar 100 mm × 15 mm is subjected to a load of 100 kN as shown in Fig. 12·30. Determine the sizes of the end fillets such that the stresses in both the fillets are the same. Take allowable stress in the weld as 102·5 MN/m².*

Solution. Refer to Fig. 12·30.

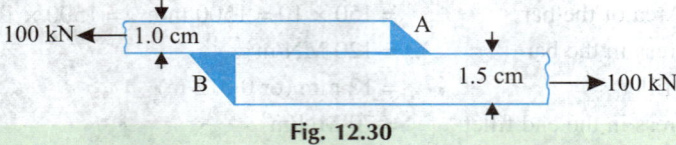

Fig. 12.30

For equal stresses in welds A and B, the load shared by fillet welds will be proportional to their effective throat thickness or size of the weld.

Let, $\qquad\qquad\qquad P$ = Force of resistance of the weld, and

$\qquad\qquad\qquad s$ = Size of the weld (A) in the upper plate.

$\therefore$ Effective throat thickness, $\quad t = 0.707\,s$ mm,

and, throat thickness of the weld (B) in the lower plate

$$= 1.5 \times 0.707\,s = 1.06\,s \text{ mm}$$

Using the relation: $\qquad\qquad P = l\,t\,\sigma$, we have

$$P_1 = 100 \times 0.707\,s \times 10^{-6} \times 102.5 \times 10^6$$
$$= 7247\,s \text{ N} \qquad\qquad\qquad \text{...(for weld A)}$$

Similarly, $\qquad\qquad P_2 = 100 \times 1.06\,s \times 10^{-6} \times 102.5 \times 10^6 = 10865\,s \text{ N}$

$\therefore$ Total force of resistance of the welds,

$$P = P_1 + P_2 = 7247s + 10865s = 18112\,s$$

Equating the total force of the fillets to the load on the plates, we get

$$18112s = 100 \times 10^3 \qquad\qquad \therefore s = 5.25 \text{ mm} \quad \textbf{(Ans.)}$$

Example 12·21. *A 120 mm × 12 mm plate is connected to another plate by fillet welds around the end of the bar and also inside a machined slot as shown in Fig. 12·31. All dimensions are in mm. Determine the size of the weld, if the joint is subjected to a pull of 100 kN. The working stresses for the transverse welds and longitudinal welds are 102·5 MN/m² and 84 MN/m².*

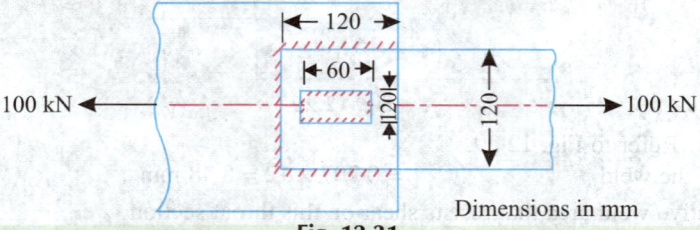

Dimensions in mm

Fig. 12.31

Solution. Length of *longitudinal weld*,

$$l = (2 \times 120) + (2 \times 60) = 360 \text{ mm}$$

Length of *transverse weld*

$$= 120 + 2 \times 20 = 160 \text{ mm}$$

Let, P = Force of resistance of the weld, and s = size of the weld

∴ Effective thickness of weld,

$$t = 0.707 \, s$$

Using the relation:

$$P = l \, t \, \sigma, \text{ we have}$$
$$P_1 = 360 \times 0.707 \, s \times 10^{-6} \times 84 \times 10^6$$
$$= 21379 s \text{ N}$$

Similarly,

$$P_2 = 160 \times 0.707 \, s \times 10^{-6} \times 102.5 \times 10^6$$
$$= 11595 s \text{ N}$$

∴ Total force of resistance of the welds,

$$P = P_1 + P_2$$
$$= 21379 \, s + 11595 \, s = 32974 s \text{ N}$$

A metallic frame is being welded.

Now, equating the total force of resistance to the pull on the joint, we get

$$32974 \, s = 100 \times 10^3$$
$$s = 3.03 \text{ mm say } \textbf{3 mm} \quad \textbf{(Ans.)}$$

Example 12·22. *A circular plate of diameter 150 mm is welded on to another plate by means of 10 mm fillet. Determine the maximum twisting moment which can be applied to the circular plate if the permissible shear stress is 100 MN/m².*

Solution. Refer to Fig. 12·32.

Size of the fillet $= 10$ mm

∴ Thickness of throat $= 0.707 \times 10 = 7.07$ mm

Length of the fillet $=$ Circumference of the circular plate

i.e., $l = \pi d = \pi \times 150 = 471.2$ mm

Total shear force $=$ Throat area × shear stress $= l \, t \, \tau$

$$= (471.2 \times 7.07) \times 10^{-6} \times 100 \times 10^6 = 333138 \text{ N}$$

∴ *Maximum twisting moment* $=$ 333138 × radius of the fillet $= 333138 \times 75 \times 10^{-3}$

$$= \textbf{24985·4 Nm} \quad \textbf{(Ans.)}$$

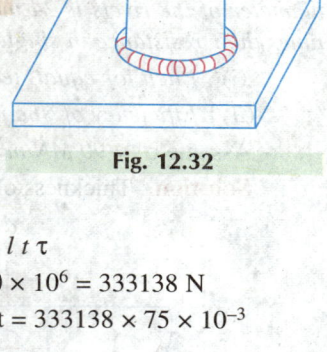

Fig. 12.32

Example 12·23. *Find the size of the fillet weld required to connect the bracket plate to the column as shown in Fig. 12·33. The stress in the weld is not to exceed 100 MPa.* **(N.U.)**

Solution. Refer to Fig. 12·33.

Moment of inertia of the weld length about the horizontal axis passing through its centroid

$$= \frac{1 \times 30^3}{12} = 2250 \text{ cm}^4$$

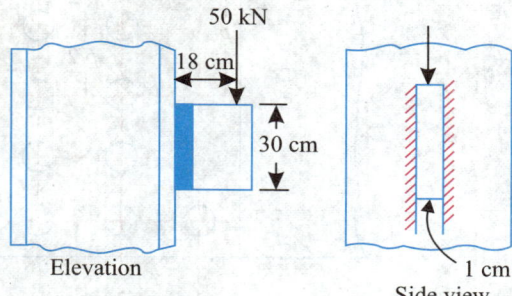

Elevation

Side view

Fig. 12.33

Resistance against translation per cm length,

$$V = \frac{P}{l} = \frac{50 \times 1000}{2 \times 30} = 833 \cdot 33 \text{ N/cm}$$

Maximum resistance against rotation per cm length,

$$H = \frac{M}{I} \, y_{max}$$

$$= \frac{(50 \times 1000) \times 18}{2250} \times \left(\frac{30}{2}\right) = 6000 \text{ N/cm}$$

∴ Resultant resistance per mm length of weld

$$= \sqrt{V^2 + H^2} = \sqrt{(833 \cdot 33)^2 + (6000)^2} = 6057 \cdot 6 \text{ N/cm}$$

Let the size of the weld be s cm.

Equating the strenght of weld per cm length to the maximum resistance per cm length, we have

Thickness (t) × stress in the weld (σ) = 6057·6

$$0 \cdot 707 \, s \times 10^4 = 6057 \cdot 6$$

[∵ Thickness, $t = 0 \cdot 707 \, s$ and $\sigma = 100$ MPa $= 100 \times 10^6$ N/m$^2 = 10^4$ N/cm^2 (given)]

or,

$$s = \frac{6057 \cdot 6}{0 \cdot 707 \times 10^4} = 0 \cdot 86 \text{ cm or } 8 \cdot 6 \text{ mm say 9 mm}$$

Hence, *provide a filled weld of 9 mm size.* **(Ans.)**

TYPICAL EXAMPLES (For Competitive Examinations)

Example 12·24. *Two plates 25 mm thick are connected by a treble riveted butt joint with two cover straps. The pitch of the rivets in the outer row is twice the pitch of those in other rows and the diameter of the rivets is 24 mm. Taking the resistance of the rivets in double shear equal to 1·75 times their resistance in single shear, determine:*

(i) Pitch for equal tearing and shearing resistance;

(ii) Efficiency of the joint.

Assume $\sigma_t = 90$ MN/m^2, and $\tau = 60$ MN/m^2 **(U.P.S.C.)**

Solution. Thickness of each plate $t = 25$ mm ; Gross diameter of the rivets,

$$d = 24 + 1 \cdot 5 = 25 \cdot 5 \text{ mm}$$

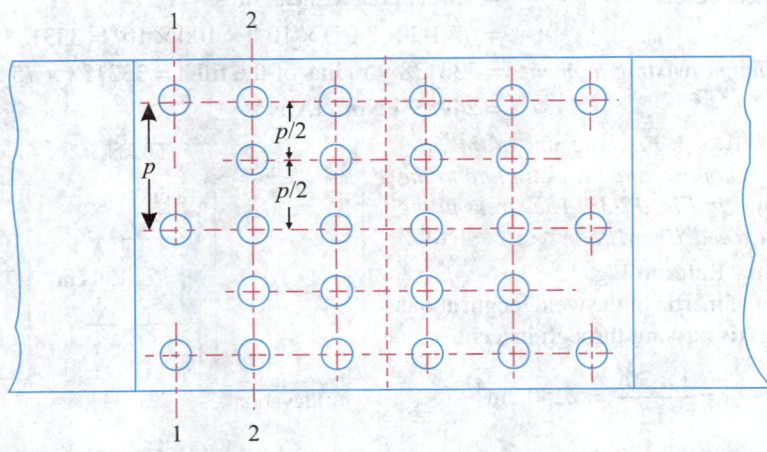

Fig. 12.34

(i) **Pitch:**

Let, p = Pitch (in mm).

Consider *tearing along the section 1–1* :

Permissible load per pitch length

$$= (p - d)\, t\, \sigma_t$$
$$= (p - 25 \cdot 5) \times 10^{-3} \times 25 \times 10^{-3} \times 90 \times 10^{6} \qquad \text{...(i)}$$
$$= 2250\, p - 57375\ N$$

Consider *tearing along the section 2–2* and *shearing of rivets in section 1–1* :

$$= (p - 2d)\, t\, \sigma_t + 1 \cdot 75 \times \frac{\pi}{4}\, d^2 \times \tau$$
$$= (p - 2 \times 25 \cdot 5) \times 10^{-3} \times 25 \times 10^{-3} \times 90 \times 10^{6} + 1 \cdot 75 \times \frac{\pi}{4}$$
$$\times \left(\frac{25 \cdot 5}{1000}\right)^2 \times 60 \times 10^{6}$$
$$= 2250\, p - 114750 + 53624 = 2250\, p - 61126\ N \qquad \text{...(ii)}$$

Eqns. *(i)* and *(ii)* give two modes of failure due to tearing of plates out of which eqn. *(ii)* gives the smaller load.

∴ Permissible load, per pitch length, in tearing

$$= 2250\, p - 61126 \qquad \text{...(1)}$$

Consider mode of failure due to *shearing of rivets*:

There are 5 rivets $(1 + 2 + 2 = 5)$ in shear (double) and hence permissible load, per pitch length

$$= 5 \times \frac{\pi}{4} \times d^2 \times \tau \times 1 \cdot 75$$
$$= 5 \times \frac{\pi}{4} \times \left(\frac{25 \cdot 5}{1000}\right)^2 \times 60 \times 10^{6} \times 1 \cdot 75 = 268120\ N \qquad \text{...(2)}$$

Equating (1) and (2), we get

$$2250\, p - 61126 = 268120$$

∴ $\qquad\qquad\qquad$ **p = 146·3 mm (Ans.)**

(ii) **Efficiency of the joint:**

Permissible load which can be carried by solid plate

$$= p.t.\ \sigma_t = 146 \cdot 3 \times 10^{-3} \times 25 \times 10^{-3} \times 90 \times 10^{6}$$
$$= 329175\ N$$

∴ Efficiency of the joint, $\qquad \eta = \dfrac{268120}{329175} \times 100 = \textbf{81·4\%} \quad \textbf{(Ans.)}$

Example 12·25. *A bridge (truss) diagonal made of 16 mm thick flat has to transmit a pull of 620 kN. The diagonal is to be connected to 16 mm thick gusset plate by a double cover butt joint with 20 mm rivets. Assuming permissible stresses, $\sigma_t = 150\ MN/m^2$, $\tau = 100\ MN/m^2$ and $\sigma_c = 300\ MN/m^2$ calculate:*

(i)   Number of rivets;$\qquad\qquad$ *(ii) Width of the flat required;*

(iii) Efficiency of the joint;$\qquad$ *(iv) Actual stresses induced in the flat and rivets;*

(v) Thickness of cover plates.

$\qquad$ *Sketch the joint.*

Solution. Thickness of the flat = 16 mm

Magnitude of pull,

$$P = 620 \text{ kN}$$

Formed/gross diameter of rivet,

$$d = 20 + 1.5 = 21.5 \text{ mm}$$

Permissible stresses:

$$\sigma_t = 150 \text{ MN/m}^2;$$
$$\tau = 100 \text{ MN/m}^2;$$
$$\sigma_c = 300 \text{ MN/m}^2.$$

(i) Number of rivets, n:

Strength of rivet in shear

$$P_s = \frac{\pi}{4} d^2 \sigma_t \times 2$$

$$= \frac{\pi}{4} \times (21.5 \times 10^{-3})^2 \times 100 \times 10^6 \times 2 = 72610 \text{ N}$$

Strength of rivets in crushing,

$$P_c = d \times t \times \sigma_c$$
$$= (21.5 \times 10^{-3}) \times (16 \times 10^{-3}) \times 300 \times 10^6 = 103200 \text{ N}$$

∴ Strength of rivet = 72610 N (minimum of P_s and P_c)

∴ Number of rivets $= \dfrac{P}{\text{Strength of rivet}} = \dfrac{620 \times 1000}{72610} = 8.5$ say 9

Electric arc welding.

Hence 9 rivets will be provided on each side and arranged in diamond riveting pattern as shown in Fig. 12·35.

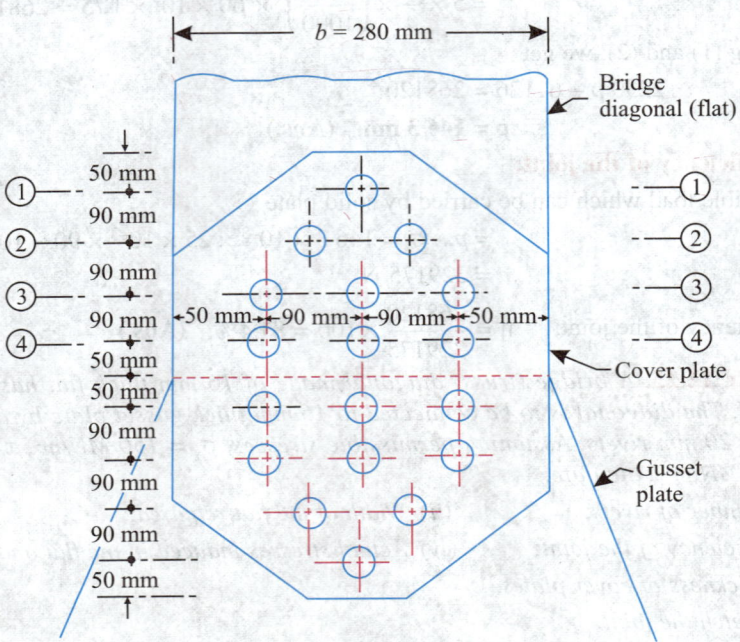

Fig. 12.35

(ii) Width of flat:

Let, b = Width of the flat (in mm).

Assuming the section to be weakened by one rivet hole only,

$$P_t = (b - d) \, t \, \sigma_t$$
$$= (b - 21 \cdot 5) \times 10^{-3} \times 16 \times 10^{-3} \times 150 \times 10^6$$

By equating this to external load, we get

$$(b - 21 \cdot 5) \times 10^{-3} \times 16 \times 10^{-3} \times 150 \times 10^6 = 620 \times 10^3 \text{ or } (b - 21 \cdot 5) \times 16 \times 150 = 620 \times 10^3$$

or, $\quad b = 279 \cdot 8$ mm say **280 mm** **(Ans.)**

(iii) Efficiency of the joint, η:

$$\eta = \frac{b - d}{b} = \frac{280 - 21 \cdot 5}{280} \times 100 = \mathbf{92 \cdot 32\%} \quad \textbf{(Ans.)}$$

Let us provide the following:

Edge distance = 50 mm, and

Pitch and back pitch = 90 mm.

(iv) Actual stresses induced:

Rivets:

We know that, $\quad P = n \times \dfrac{\pi}{4} d^2 \times \tau \times 2 \qquad (\because \text{Rivets are in } double \ sheajr.)$

$$620 \times 10^3 = 9 \times \frac{\pi}{4} \times (21 \cdot 5 \times 10^{-3})^2 \times \tau \times 2$$

$\therefore \qquad \tau = \dfrac{620 \times 10^3}{9 \times \dfrac{\pi}{4} (21 \cdot 5 \times 10^{-3})^2 \times 2} \times 10^{-6} \text{ MN} / \text{m}^2 = \mathbf{94.87 \ MN/m^2} \quad \textbf{(Ans.)}$

Also, $\quad P = n \times d \times t \times \sigma_c$

$$620 \times 10^3 = 9 \times (21 \cdot 5 \times 10^{-3}) \times (16 \times 10^{-3}) \times \sigma_c$$

$\therefore \qquad \sigma_c = \dfrac{620 \times 10^3}{9 \times (21 \cdot 5 \times 10^{-3}) \times (16 \times 10^{-3})} \times 10^{-6} \text{ MN/m}^2$

$$= \mathbf{200.26 \ MN/m^2} \quad \textbf{(Ans.)}$$

Flat:

At section 1–1 : $\quad P = \sigma_1 (b - d) \, t$

or, $\quad 620 \times 10^3 = \sigma_1 (280 - 21 \cdot 5) \times 10^{-3} \times 16 \times 10^{-3}$

or, $\quad \sigma_1 = 150 \times 10^6 \text{ N/m}^2, \text{ or, } \mathbf{150 \ MN/m^2} \quad \textbf{(Ans.)}$

At section 2–2 : $\quad P = (b - 2d) \, t \, \sigma_2 + \dfrac{P}{n}$

$$620 \times 10^3 = (280 - 2 \times 21 \cdot 5) \times 10^{-3} \times 16 \times 10^{-3} \times \sigma_2 + \frac{620 \times 10^3}{9}$$

or, $\quad 620 \times 10^9 = 3792 \, \sigma_2 + 68 \cdot 89 \times 10^9 \qquad \text{(Multiplying both the sides by } 10^6)$

or, $\quad \sigma_2 = \mathbf{145 \cdot 3 \ MN/m^2}$

At section 3–3 : $\quad P = (b - 3d) \, t \, \sigma_3 + \dfrac{3P}{n}$

$$620 \times 10^3 = (280 - 3 \times 21 \cdot 5) \times 10^{-3} \times 16 \times 10^{-3} \times \sigma_3 + \frac{3 \times 620 \times 10^3}{9}$$

or, $\quad 620 \times 10^9 = 3448 \, \sigma_3 + 206 \cdot 67 \times 10^9$ $\qquad \text{(Multiplying both sides by } 10^6)$

or, $\quad \sigma_3 = \mathbf{119 \cdot 87 \ MN/m^2} \quad \textbf{(Ans.)}$

At section 4–4 : $\quad P = (b - 3d) t \sigma_4 + \dfrac{6P}{n}$

$$620 \times 10^3 = (280 - 3 \times 21\cdot5) \times 10^{-3} \times 16 \times 10^{-3} \times \sigma_4 + \frac{6 \times 620 \times 10^3}{9}$$

or, $\qquad 620 \times 10^9 = 3448\ \sigma_4 + 413\cdot33 \times 10^9 \qquad$ (Multiplying both sides by 10^6)

or, $\qquad \boldsymbol{\sigma_4 = 59\cdot94\ MN/m^2}$ **(Ans.)**

(v) Thickness of cover plate, t′:

Let, $\qquad t' =$ Thickness of the cover plate (in mm).

Strength of cover plates at section 4-4, against *tearing*

$$= (b - 3d) \times 2t' \times \sigma_t = (280 - 3 \times 21\cdot5) \times 10^{-3} \times 2 \times t' \times 10^{-3} \times 150 \times 10^6$$
$$= 64650\ t'\ N$$

∴ $\qquad 64650\ t' = 620 \times 10^3$ or $t' = 9\cdot59$ mm $\quad$ (Also $t' = 5/8t = 5/8 \times 16 = 10$ mm)

Hence, assuming **t′ = 10 mm (Ans.)**

Actual stress in cover plate at section 4–4 :

$$= \frac{620 \times 10^3}{(280 - 3 \times 21\cdot5) \times 10^{-3} \times 2 \times 10 \times 10^{-3}} \times 10^{-6}\ MN/m^2$$

$$= 143\cdot8\ MN/m^2 \text{ (Within limit)}$$

Example 12·26. *A load of 150 kN is acting on a bracket as shown in the fig. 12·36. The bracket is welded to a stanchion by means of three lines of weld or three sides as indicated in the figure. Find out the size of the welds so that the load is carried safely. Assume permissible shear stress as 150 N/mm².*

Solution. Load carried by the bracket, $P = 150$ kN

Allowable shear stress = 150 N/mm²

Size of the weld:

Refer to Fig. 12·36.

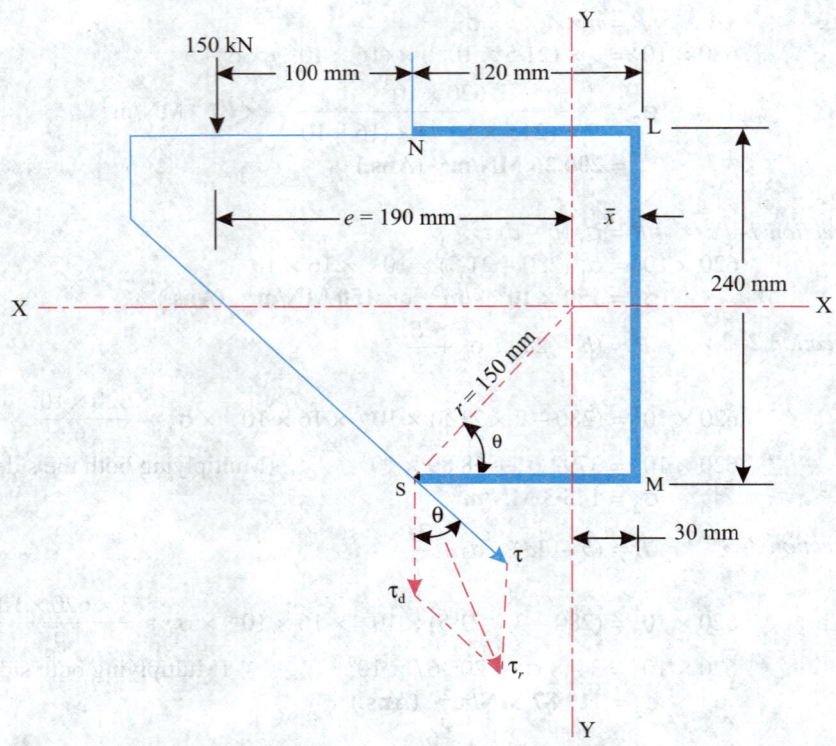

Fig. 12.36

Let, $\qquad \bar{x}$ = Distance of centroid of weld area from LM,

Then, $\qquad \bar{x} = \dfrac{2 \times 120t \times 60}{2 \times 120t + 240t} = \dfrac{14400t}{480t} = 30$ mm

(where, t = thickness of the weld)

∴ Eccentricity of the load, $e = 100 + 120 - 30 = 190$ mm

Moment of Inertia about XX-axis,

$$I_{XX} = \frac{t \times (240)^3}{12} + 2 \times 120t \times (120)^2 = 460 \cdot 8 \times 10^4 \, t \text{ mm}^4$$

Similarly, moment of inertia about YY-axis,

$$I_{YY} = 240t \times 30^2 + 2\left[\frac{t \times 120^3}{12} + 120t \times 30^2\right]$$

$$= 216000 \, t + 2t \times (144000 + 108000) = 72 \times 10^4 \, t \text{ mm}^4$$

∴ Polar moment of inertia,

$$I_p \, (= I_{ZZ}) = I_{XX} + I_{YY}$$
$$= 460 \cdot 8 \times 10^4 \, t + 72 \times 10^4 \, t$$
$$= 532 \cdot 8 \times 10^4 \, t \text{ mm}^4$$

Area, $\qquad A = 2 \times 120 \, t + 240 \, t = 480 \, t \text{ mm}^2$

The maximum stress due to torsion will occur either at N or S.

Length of radius vector for N or S

$$= \sqrt{120^2 + 90^2} = 150 \text{ mm}$$

Using the 'Torsion equation',

$$\frac{T}{I_p} = \frac{\tau}{r}, \text{ we have}$$

Maximum stress, $\qquad \tau = \dfrac{Tr}{I_p} = \dfrac{(150 \times 10^3 \times e) \times 150}{532 \cdot 8 \times 10^4 \, t}$

MIG (Metal Inert Gas) welding machine set.

$$= \frac{150 \times 10^3 \times 190 \times 150}{532 \cdot 8 \times 10^4 \times t} = \frac{802 \cdot 4}{t} \text{ N/mm}^2$$

Direct stress, $\qquad \sigma_d = \dfrac{P}{A} = \dfrac{150 \times 10^3}{480t} = \dfrac{312 \cdot 5}{t} \text{ N/mm}^2$

The angle between the stresses is θ (Fig. 12·36) such that

$$\cos \theta = \frac{90}{150} = 0 \cdot 6$$

∴ Resultant stress, $\qquad \tau_r = \sqrt{\sigma_d^2 + \tau^2 + 2\tau_d \cdot \tau \cos \theta}$

$$= \sqrt{\left(\frac{312 \cdot 5}{t}\right)^2 + \left(\frac{802 \cdot 4}{t}\right)^2 + 2 \times \frac{312 \cdot 5}{t} \times \frac{802 \cdot 4}{t} \times 0 \cdot 6}$$

$$= \frac{1}{t}\sqrt{97656 \cdot 25 + 643845 \cdot 76 + 300900} = \frac{1021}{t} \text{ N/mm}^2$$

But allowable shear stress $\quad = 150 \text{ N/mm}^2$

∴ $\qquad \dfrac{1021}{t} = 150 \quad$ or $\quad t = 6 \cdot 8$ mm

Hence size of the weld $\quad = \sqrt{2} \, t = \sqrt{2} \times 6 \cdot 8 = 9 \cdot 6$ mm say **10 mm** **(Ans.)**

HIGHLIGHTS

1. The riveted joints can mainly be classified as follows:

   (i) Lap joints   (ii) Butt joints

   When the ends of plates overlap each other the joint is known as the *lap joint* and if the ends of the plates butt against each other the joint is known as *butt joint*.

2. A riveted joint may fail in the following ways:

   (i) The plate may tear-off along line of rivets;
   $$R_t = P_t = (p - d)\, t\, \sigma_t \qquad \qquad \text{...(i)}$$

   (ii) The rivets may shear off;

   $$R_s = P_s = n \times \frac{\pi}{4} \times d^2\, \tau \times K \qquad \qquad \text{...(ii)}$$

   where,   K = 1 in case of *single shear* of rivets.

                  K = 2 in case of *double shear* of rivets [= 1·875 (for design purposes)]

   (iii) Failure of joint due to failure of the material to resist bearing pressure between the rivets and plate.

   $$R_c = P_c = n\, d\, t\, \sigma_c \qquad \qquad \text{...(iii)}$$

   where, n is the number of rivets per pitch length.

   (iv) Failure of joint by shearing out plate between the rivet hole and edge of the plate.

   (v) Failure of the joint due to splitting of the plate between the rivet holes and edge of the plates.

3. The *efficiency* of the riveted joint is the ratio of the strength of the joint to the strength of the solid plate.

   Thus, efficiency of the joint,

   $$\eta = \frac{\text{Least of } P_t,\ P_s \text{ and } P_c}{p\, t\, \sigma_t}$$

4. Relation between d and t:

   $$d = 1 \cdot 91 \sqrt{t}\ , \text{ when both } d \text{ and } t \text{ are in cm.}$$
   $$d = 6 \sqrt{t}\ , \text{ when both } d \text{ and } t \text{ are in mm.}$$

   The above relation is known as *Unwin's formula*.

5. *Welding* is a process of joining two pieces of metal by fusion.

   For joining plates in structural practice the following two types of welded joints are used:

   (i) Butt welds   (ii) Lap welds or fillet welds.

6. In lap welds, throat area is the product of throat thickness and 'effective length' of the weld.

   Throat thickness = 0·707 Leg.

7. In V groove butt welds:

   $$P = h \times l \times \sigma_t \qquad \qquad \text{...(i)}$$

   In double V-groove butt welds:

   $$P = (h_1 + h_2)\, t\, \sigma_t \qquad \qquad \text{...(ii)}$$

OBJECTIVE TYPE QUESTIONS

Choose the Correct Answer:

1. The rivets are *generally* made of

   (a) brass   (b) tin

   (c) steel   (d) copper.

2. When the ends of the plates overlap each other the joint is known as

   (a) lap joint

   (b) butt joint

   (c) either of the above

   (d) none of the above.

3. The distance between the consecutive rivets is called
   (a) gauge distance (b) back pitch
   (c) gauge line   (d) pitch.

4. The distance from centre to centre of rivets lying in a row is termed as
   (a) gauge line   (b) gauge distance
   (c) pitch   (d) any of the above.

5. The diameter of the cold rivet measured before driving is referred as
   (a) nominal diameter
   (b) gross diameter
   (c) either of the above
   (d) none of the above.

6. The distance between the centre lines of two rows of rivets is called
   (a) pitch   (b) back pitch
   (c) gauge distance (d) None of the above.

7. .... riveting is used structural units.
   (a) Chain   (b) Zig-zag
   (c) Diamond   (d) None of the above.

8. The diameter of the rivet (d) and thickness of the plate (t) follow the relation.
   (a) $d = 3\sqrt{t}$ (b) $d = 4\sqrt{t}$
   (c) $d = 5\sqrt{t}$ (d) $d = 6\sqrt{t}$.

   Where d and t are in mm.

9. .... is process of joining two pieces of metal by fusion.
   (a) Riveting
   (b) Welding
   (c) Either of the above
   (d) None of the above.

10. As compared to a riveted joint a welded joint has .... strength.

11. Efficiency of the welded joint is........... than that of the riveted joint.
    (a) less
    (b) more
    (c) either of the above
    (d) none of the above.

12. The welding provides........... joints.
    (a) very weak
    (b) weak
    (c) very rigid
    (d) none of the above.

Welding gun that is used to weld plastics.
This works at 600°C.

ANSWERS

1. (c)	2. (a)	3. (a)	4. (c)	5. (a)	6. (b)
7. (c)	8. (d)	9. (b)	10. (b)	11. (b)	12. (c).

UNSOLVED EXAMPLES

1. Two 12·5 mm thick plates are connected by a double riveted double cover butt joint. Using Unwin's formula select a suitable diameter of rivets and calculate the pitch if permissible stresses are 112 MN/m² in tension for the plates; 88 MN/m² and 176 MN/m² in shearing and bearing respectively for rivets. Find the efficiency of the joint.

   [**Ans.** d = 20 mm ; p = 83 mm, η = 75·7%]

2. A double riveted double cover joint is used for connecting plates 15 mm thick. The diameter of the rivet is 22 mm. The permissible stresses are 100 MN/m² in tension, 80 MN/m² in shear and 100 MN/m² in bearing. Calculate the necessary pitch and efficiency of the joint.

   [**Ans.** p = 66 mm ; η = 66·6%]

3. Two 15 mm thick plates are to be joined by a triple riveted double cover butt joint. The rivet pitch is to be the same in all the rows and chain riveting is to be used. Determine the rivet diameter, pitch, distance between rows of rivets, strap thickness and efficiency of the joint. The following stresses are to be used:

$$\sigma_t = 84 \text{ MN/m}^2; \; \tau = 63 \text{ MN/m}^2$$

[**Ans.** $d = 23$ mm; $p = 140$ mm ; 112 mm, $t_c = 10$ mm, 83·5%]

4. The longitudinal seam of a boiler consists of double riveted lap joint. The diameter is 3 m, the plate thickness 20 mm, the rivets are of 25 mm diameter and width of repeating section is 75 mm. The steel has the following ultimate stresses; tension 400 MN/m², shear 300 MN/m², and compression 700 MN/m². Find (i) the efficiency of the lap joint and (ii) the allowable internal pressure in the boiler if a factor of safety of 5 is used in the design.

[**Ans.** 49·2%; 0·52 MN/m²]

5. A boiler shell 1 m diameter has to withstand a steam pressure of 0·14 MN/m². Design the thickness of the shell for a permissible stress in the plate of 112 MN/m², assuming the efficiency of longitudinal joint as 70%.

Calculate the pitch of rivets required for the joint which is to be double cover, double riveted using 25 mm diameter rivets. For the rivets the permissible stresses are $\tau = 80$ MN/m² and $\sigma_c = 160$ MN/m². Check the efficiency of the joint.

[**Ans.** 16 mm; 90 mm; 72·2%]

6. Two plates 16 mm thick are joined by a double riveted lap joint. The pitch of each row of rivets is 20 mm. The rivets are 25 mm in diameter. The permissible stresses are as follows:

Tension = 140 MN/m²; Shear = 110 MN/m²; Bearing = 240 MN/m².

Determine the maximum tensile force permissible on the joint, and calculate the efficiency of the joint.

[**Ans.** 108·03 kN, 53·6%]

7. A triple riveted lap joint is to be made between 6 mm thick plates, having zig-zag riveting. Calculate the diameter, pitch and distance between rows of rivets for the joint. Also find the efficiency of the joint.

Take: $\tau = 75$ MN/m², $\sigma_t = 100$ MN/m², and $\sigma_c = 150$ MN/m²

[**Ans.** $d = 16$ mm, $p = 92$ mm, 55·2 mm, 78·2%]

8. Efficiency of a treble riveted lap joint of 20 mm plates is not to be less than 75%. Find the pitch when 30 mm diameter rivets are used and permissible stresses are:

σ_t in plates = 55 MN/m²; τ in rivets = 47 MN/m²; σ_c in rivets or plates = 80 MN/m².

[**Ans.** $p = 120$ mm]

9. In a triple riveted lap joint, the pitch of the rivets in the middle row is half the pitch of rivets in the outer rows. Calculate the pitch of rivets in the outer rows and resistance of the joint against tearing, shearing and bearing and real efficiency

Semi-automatic overlap welding tool for all flexible thermoplastic materials.

of the joint. Thickness of the plates is 10 mm and rivets diameter 15 mm. The permissible stresses are:

$\sigma_t = 120$ MN/m^2, $\tau = 100$ MN/m^2, and $\sigma_c = 160$ MN/m^2

(**Hint.** $n = 4$)   [**Ans.** $p = 74$ mm; $P_t = P_s = 70\cdot6$ kN, $P_c = 96$ kN, $\eta = 80\%$]

10. A riveted boiler 2.25 m in diameter is to be designed for an internal pressure of 0.56 MN/m^2. Taking the permissible safe tensile stress in the plate to be 60 MN/m^2 and assuming the efficiency of longitudinal joint as 70%, determine the thickness of the shell (using 20 mm diameter rivets) and the pitch for a double cover double riveted butt joint.

         [**Ans.** $t = 15$ mm, $p = 80$ mm]

11. The longitudinal seam of a boiler consists of a double riveted lap joint. The diameter of the boiler is 3 m, the thickness of the plate is 20 mm, the rivets are of 25 mm diameter and the width of a repeating section is 75 mm. The steel has the following ultimate stresses:

Tension = 385 MN/m^2, shear = 308 MN/m^2, and compression 665 MN/m^2.

Find the efficiency of the lap joint and allowable internal pressure in the boiler if factor of safety is 5 in design.   [**Ans.** 52.5%, 53.9 MN/m^2]

12. Two tie bars, each 300 mm wide and 20 mm thick are joined by a double cover-plate butt joint. The diameter of the rivets being 25 mm. Design a suitable joint. The various permissible stresses are:

$\sigma_t = 155$ MN/m^2; $\tau = 90$ MN/m^2; $\sigma_c = 200$ MN/m^2

Also determine the working strength, real efficiency of the joint and the actual stress in the plates and the rivets.

$$\left[\begin{array}{l} \textbf{Ans. } n = 10, \text{ working strength} = 852\cdot5 \text{ kN;} \\ \text{Real efficiency of the joint} = 91\cdot66\%; \\ \text{Actual } \sigma_t = 155 \text{ MN/m}^2, \tau = 86\cdot8 \text{ MN/m}^2 \text{ and } \sigma_c = 170\cdot5 \text{ MN/m}^2 \end{array}\right]$$

13. Two plates of size 180 mm × 10 mm and 250 mm × 12 mm are to be joined. Design a suitable fillet welded joint so that the full tension is developed in the thinner plate. Draw a neat sketch of the joint also. Given safe tensile stress = 150 MN/m^2, and shear stress = 102.5 MN/m^2.

         [**Ans.** 145 mm on sides and 180 mm at end]

14. A circular plate of diameter 180 mm is welded onto another plate by means of 8 mm fillet. Determine the maximum twisting moment which can be applied to the circular plate if permissible shear stress is 102.5 MN/m^2.   [**Ans.** 29.283 Nm]

15. A tie bar 120 mm × 10 mm welded to another tie bar 120 mm × 15 mm is subjected to a load of 170 kN as shown in Fig. 12.37. Determine the sizes of the end fillets, such that the stresses in both the fillets are the same. Take the allowable stress in the weld as 102.5 MN/m^2.

         [**Ans.** 8 mm]

Fig. 12.37

Chapter

13

Torsion of Circular and Non-Circular Shafts

13·1. SHAFTS

The shafts are *usually* cylindrical in section, solid or hollow. They are made of mild steel, alloy steel and copper alloys.

Shafts may be subjected to the following loads:

1. Torsional load
2. Bending load
3. Axial load
4. Combination of above three loads.

The shafts are designed on the *basis of strength and rigidity.*

The following values are usually adopted for the design of shaft:

$\sigma = 112$ MN/m², the maximum permissible tensile or compressive stress.

$\tau = 56$ MN/m², the maximum permissible shear stress.

The ultimate tensile stress for commercial steel shafting may be 315 MN/m² for hot rolled and turned low carbon steel and 490 MN/m² for cold finished low carbon steel, corresponding stresses at the elastic limit would be about 160 MN/m^2 and 315 MN/m^2 respectively. In shafts with key ways the allowable stresses are 75% of the values given.

13·2. TORSION OF SHAFTS

To transmit energy by rotation it is necessary to apply a turning force. In case of a shaft if the force is applied tangentially and in the plane of transverse cross-section the torque or twisting moment may be calculated by multiplying the force with the radius of the shaft. If the shaft is subjected to two opposite turning moments it is said to be in *pure torsion* and it will exhibit the tendency of shearing off at every cross-section which is perpendicular to longitudinal axis.

CIRCULAR SHAFTS

13·3. TORSION EQUATION

The *torsion equation* is based on the following **assumptions:**

1. The material of the shaft is uniform throughout.
2. The shaft circular in section remains circular after loading.
3. A plane section of shaft normal to its axis before loading remains plane after the torques have been applied.
4. The twist along the length of shaft is uniform throughout.
5. The distance between any two normal cross-sections remains the same after the application of torque.
6. Maximum shear stress induced in the shaft due to application of torque does not exceed its elastic limit value.

Let, T = Maximum twisting torque,

D = Diameter of the shaft,

I_p = Polar moment of inertia,

τ = Shear stress,

C = Modulus of rigidity

θ = The angle of twist (radians), and

l = Length of the shaft.

In Fig. 13·1 is shown a shaft fixed at one end and torque being applied at the other end. If a line LM is drawn on the shaft, it will be distorted to LM' on the application of the torque ; thus cross-section will be twisted through angle θ and surface by angle ϕ.

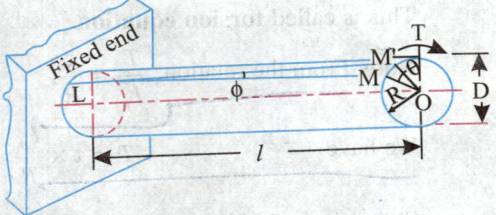

Fig. 13.1

Here, shear strain, $\phi = \dfrac{MM'}{l}$

Also, $\phi = \dfrac{\tau}{C}$

$\therefore$ $\dfrac{MM'}{l} = \dfrac{\tau}{C}$

or, $\dfrac{R\theta}{l} = \dfrac{\tau}{C}$ $\left[\begin{array}{l} \because MM' = R \times \theta, \\ R \text{ being radius of the shaft} \end{array} \right]$

$\therefore$ $\dfrac{\tau}{R} = \dfrac{C\theta}{l}$ ...(13·1)

Refer to Fig. 13·2. Consider an elementary ring of thickness dx at a radius x and let the shear stress at this radius be τ_x.

The turning force on the elementary ring

$$= \tau_x \cdot 2\pi x \cdot dx$$

Turning moment due to this turning force,

$$dT = \tau_x \cdot 2\pi x \cdot dx \times x$$

To get total turning moment integrating both sides, we get

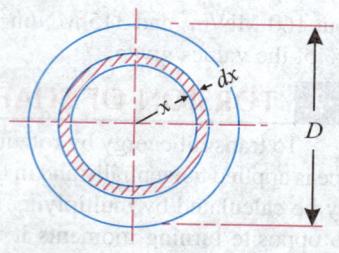

Fig. 13.2

$$\int dT = \int_0^R \tau_x \cdot 2\pi x \, dx \times x$$

or,

$$\int dT = 2\pi \int_0^R \tau_x \cdot x^2 \cdot dx = 2\pi \int_0^R \frac{\tau \cdot x}{R} \cdot x^2 \cdot dx \qquad \left[\because \frac{\tau}{R} = \frac{\tau_x}{x} \right. \\ \left. \text{or } \tau_x = \frac{\tau \cdot x}{R} \right]$$

$$= 2\pi \frac{\tau}{R} \int_0^R x^3 \, dx$$

or,

$$T = 2\pi \frac{\tau}{R} \left| \frac{x^4}{4} \right|_0^R = 2\pi \frac{\tau}{R} \cdot \frac{R^4}{4}$$

$$T = \tau \cdot \frac{\pi R^3}{2} = \tau \cdot \frac{\pi}{16} D^3 \qquad \text{...(Strength of solid shaft)}$$

or,

$$T = \frac{\tau}{R} \cdot \frac{\pi R^4}{2} = \frac{\tau}{R} I_p \qquad \left[\because I_p = \frac{\pi}{32} D^4 = \frac{\pi}{2} R^4 \right]$$

∴

$$\frac{T}{I_p} = \frac{\tau}{R}$$

...(13·2)

From eqns. (13·1) and (13·2), we have

$$\frac{T}{I_p} = \frac{\tau}{R} = \frac{C\theta}{l}$$

...(13·3)

This is called **torsion equation.**

Note. From the relation, $\dfrac{T}{I_p} = \dfrac{\tau}{R}$,

We have

$$T = \tau \times \frac{I_p}{R}$$

For a given shaft I_p and R are constants and $\dfrac{I_p}{R}$ is thus a constant and is known as **polar modulus** of the shaft section.

Thus

$$T = \tau \times Z_p$$

... (13.4)

For a shaft of given material τ, the maximum permissible shear stress is fixed and thus the greatest twisting moment that the shaft can withstand is proportional to the polar modulus of the shaft. *Polar modulus of the section is thus measure of strength of shaft in torsion.*

Motor-generator set. Shafts are essential parts in motors and generators.

13·4. HOLLOW CIRCULAR SHAFTS

Eqn. (13·3) (torsion equation) equally holds good for hollow shafts and can be established in the same way as in Article 13·3.

Consider a hollow circular shaft subject to a torque T.

Refer to Fig. 13·3.

Let, R = Outer radius of the shaft,
                 r = Inner radius of the shaft, and
                 τ = Shear stress at radius R.

Following the same procedure, we have

dT = Turning moment on the elementary ring

$$= \tau_x . 2\pi x . dx . x$$

Integrating both sides, we get

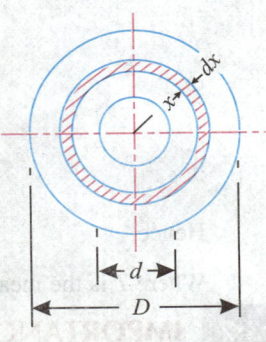

Fig. 13.3

$$\int dT = \int_r^R \tau_x . 2\pi x . dx . x$$

but, $\dfrac{\tau_x}{x} = \dfrac{\tau}{R}$ or $\tau_x = \dfrac{\tau}{R} \cdot x$

$\therefore$ $\displaystyle\int dT = \int_r^R \dfrac{\tau x}{R} . 2\pi x . dx . x = \dfrac{2\pi\tau}{R}\int_r^R x^3 \, dx$

or, $T = \dfrac{2\pi\tau}{R} \left| \dfrac{x^4}{4} \right|_r^R = \dfrac{\pi}{2} \cdot \dfrac{\tau}{R} (R^4 - r^4)$

or, $T = \dfrac{\pi}{16} \tau \left[\dfrac{D^4 - d^4}{D} \right]$ ....(Strength of hollow shaft)

But, $I_p = \dfrac{\pi}{32} (D^4 - d^4) = \dfrac{\pi}{2} (R^4 - r^4)$

$\therefore$ $T = \dfrac{\tau}{R} \cdot I_p$

or, $\dfrac{T}{I_p} = \dfrac{\tau}{R} = \dfrac{C\theta}{l}$ ...(Torsion equation)

13·5. TORSIONAL RIGIDITY

From the relation $\dfrac{T}{I_p} = \dfrac{C\theta}{l}$

we have $\theta = \dfrac{Tl}{CI_p}$

Since C, l and I_p are constants for a given shaft, θ *the angle of twist is directly proportional to the twisting moment*. The quantity $\dfrac{CI_p}{l}$ is known as **torsional rigidity** and is represented by k or μ. From the above relation, we have

$$k = \frac{CI_p}{l} = \frac{T}{\theta} \qquad \qquad ...(13\cdot5)$$

13·6. POWER TRANSMITTED BY THE SHAFT

Consider a force F newtons acting tangentially on the shaft of radius R. If the shaft due to this turning moment ($F \times R$) starts rotating at N r.p.m. then,

Work supplied to the shaft/sec.

$$= F \times \text{distance moved/sec.}$$
$$= F \times 2\pi RN/60 \text{ Nm/s}$$

or, $$P = \frac{F \times 2\pi RN}{60} \text{ watts}$$

$$= \frac{T \times 2\pi N}{60 \times 1000} \text{ kW} \qquad \qquad (\because T = F \times R)$$

Hence, $$P = \frac{2\pi NT}{60 \times 1000} \qquad \qquad ...(13\cdot6)$$

Where T is the mean/average torque in Nm.

13·7. IMPORTANCE OF ANGLE OF TWIST AND VARIOUS STRESSES IN SHAFT

Angle of Twist

In many problems such as torsion of shafts of accurate milling and drilling machines the angle of twist (θ) is required to be restricted. In such applications, the shafts are designed on the basis of limiting angle of twist and checked for shearing stresses. In torsional vibration problems, the angle of twist is also needed. These considerations make it necessary to find suitable expression for angle of twist.

A shaft, for which the angle of twist is significant, should always be designed or checked for angle of twist in addition to the design stresses in shafts.

Stresses in shafts:

In a shaft the following significant stresses occur:

1. A *maximum shear stress* occurs on the cross-section of the shaft at its outermost surface.

2. The *maximum longitudinal shear stress* occurs at the surface of the shaft on the longitudinal planes passing through the longitudinal axis of the shaft.

3. The *maximum tensile stress (i.e., major principal stress)* occurs at planes 45° to the maximum shearing stress planes at the surface of the shaft. This stress is *equal to the maximum shear stress on the cross-section of the shaft.*

4. The *maximum compressive stress* (*i.e., minor principal stress*) occurs on the planes at 45° to the longitudinal and the cross-sectional planes at the surface of the shaft. This stress is *equal to the maximum shear stress on the cross-section.*

These stresses are important/significant because they govern the failure of the shaft. These stresses *develop simultaneously* and therefore they should be *considered simultaneously for design purposes.*

— For most engineering materials, fortunately the shear strength is the smallest as compared to the tensile and compressive stresses and in such cases only the *maximum shear stress on the cross-section of the shaft is the significant stress for design.*

— For materials for which tensile and compressive strengths are lower than the shear strength, the *shaft design* should be carried for the *lowest strength.*

13·8. MODULUS OF RUPTURE

A *modulus of rupture*, corresponding to the modulus of rupture in bending, may be defined as follows:

"*The maximum fictitious shear stress calculated by the torsion formula by using the experimentally found maximum torque (i.e.* ultimate torque*) required to rupture a shaft.*"

Mathematically, $\tau_r = \dfrac{T_u R}{I_p}$

where, τ_r = Modulus of rupture in torsion (also called *computed ultimate twisting strength*),

       T_u = Ultimate torque at failure, and

       R = Outer radius of the shaft.

The above expression for τ_r gives fictitious value of shear stress at the ultimate torque because the torsion formula $\dfrac{T}{I_p} = \dfrac{\tau}{R}$ is *not applicable beyond the limit of proportionality.* The actual shear stress at the ultimate torque is quite different from the shearing modulus of rupture because the shear stress does not vary linearly from zero to maximum but it is uniformly distributed at the ultimate torque.

Although shearing modulus of rupture is a fictitious shear stress, still it can be used as a convenient tool for knowing the maximum torque carrying capacity of shafts by using the normal torsion formula.

Example 13·1. *In a tensile test a test piece of 25 mm diameter, 200 mm gauge length, stretched 0·0975 mm under a pull of 50 kN. In a torsion test, the same rod twisted 0·025 radian over a length of 200 mm when a torque of 0·4 kNm was applied. Evaluate Poisson's ratio and the three elastic moduli for the material.*

Solution. Young's Modulus:

Let, E = Young's modulus for the test piece.

Using the following relation, we have

$$\delta l = \frac{Wl}{AE}, \text{ where } A = \frac{\pi}{4} \times \left(\frac{25}{1000}\right)^2 m^2 = 4\cdot91 \times 10^{-4}\ m^2$$

$$0\cdot0975 \times 10^{-3} = \frac{50 \times 10^3 \times 0\cdot2}{4\cdot91 \times 10^{-4} \times E}$$

$$E = \frac{50 \times 10^3 \times 0\cdot2}{(0\cdot0975 \times 10^{-3}) \times 4\cdot91 \times 10^{-4}} \times 10^{-9}\ GN/m^2$$

or, $E = \textbf{208 GN/m}^2$ **(Ans.)**

Modulus of rigidity:

Let,

C = Modulus of rigidity for the test piece, and

I_p = Polar moment of inertia of a solid circular shaft

$$= \frac{\pi}{32} \times D^4 = \frac{\pi}{32} \times \left(\frac{25}{1000}\right)^4 = 3.835 \times 10^{-8} \text{ m}^4.$$

Using the following relation, we have

$$\frac{T}{I_p} = \frac{C\theta}{l}$$

$$C = \frac{Tl}{\theta I_p} = \frac{(0.4 \times 10^3) \times 0.2}{0.025 \times 3.835 \times 10^{-8}} \times 10^{-9} \text{ GN/m}^2$$

or,

$C = \textbf{83.44 GN/m}^2$ **(Ans.)**

Poisson's ratio:

Let,

$1/m$ = Poisson's ratio for the test piece.

Using the following relation, we have

$$C = \frac{mE}{2(m+1)}$$

$$83.44 \times 10^9 = \frac{m \times 208 \times 10^9}{2(m+1)}$$

$$166.88\ m + 166.88 = 208\ m$$

or,

$1/m = \textbf{0.246}$ **(Ans.)**

Bulk modulus:

Let,

K = Bulk modulus for the test piece.

Then,

$$K = \frac{mE}{3(m-1)} = \frac{\dfrac{1}{0.246} \times 208 \times 10^9}{3\left[\dfrac{1}{0.246} - 2\right]} \times 10^{-9} \text{ GN/m}^2$$

or,

$K = \textbf{136.4 GN/m}^2$ **(Ans.)**

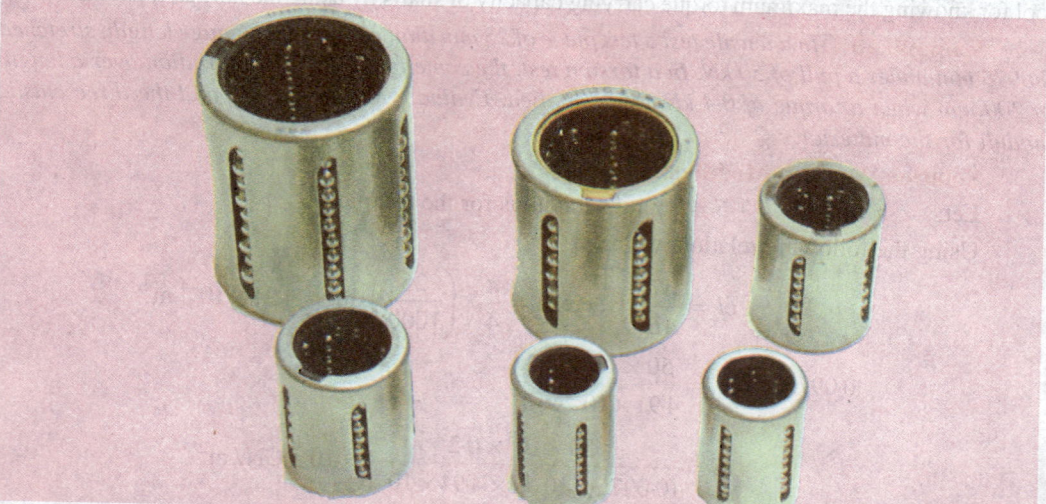

Linear motion bearings and bushings in which shafts can move longitudinally.

Example 13·2. *What must be the length of a 5 mm diameter aluminium wire so that it can be twisted through one complete revolution without exceeding a shearing stress of 42 MN/m² ? Take: C = 27 GN/m².*

Solution. *Given:* $D = 5$ mm $= 0.005$ m;

$\theta = 2\pi$ rad;

$\tau = 42$ MN/m²;

$C = 27$ GN/m².

Let, $l = $ Length of the aluminium wire.

Torque transmitted by the wire,

$$T = \tau \cdot \frac{\pi}{16} D^3 = 42 \times 10^6 \times \frac{\pi}{16} \times (0.005)^3 \text{ Nm} = 1.031 \text{ Nm}$$

Polar moment of inertia of a circular section,

$$I_p = \frac{\pi}{32} \times D^4 = \frac{\pi}{32}(0.005)^4 = 6.136 \times 10^{-11} \text{ m}^4$$

We know that, $\dfrac{T}{I_p} = \dfrac{C\theta}{l}$

$$\frac{1.031}{6.136 \times 10^{-11}} = \frac{27 \times 10^9 \times 2\pi}{l}$$

or, $l = \dfrac{27 \times 10^9 \times 2\pi \times 6.136 \times 10^{-11}}{1.031} = \mathbf{10.096 \text{ m}}$ **(Ans.)**

Example 13·3. *A solid steel shaft has to transmit 75 kW at 200 r.p.m. Taking allowable shear stress as 70 MN/m², find suitable diameter for the shaft, if the maximum torque transmitted on each revolution exceeds the mean by 30%.*

Solution. Power to be transmitted $= 75$ kW

Speed, $N = 200$ r.p.m.

Allowable shear stress, $\tau = 70$ MN / m²

$T_{max} = 1.3 \, T_{mean}$

Using the relation: $P = \dfrac{2\pi N T}{60 \times 1000}$, we get

An electric motor has the advantage where the electromagnetic force turns the shaft.

$$75 = \frac{2\pi \times 200 \times T}{60 \times 1000}$$

$$T = (T_{mean}) = \frac{75 \times 60 \times 1000}{2\pi \times 200} = 3581 \text{ Nm}$$

$$T_{max} = 1.3 \times 3581 = 4655.3 \text{ Nm}$$

Also, $T = \tau \times \dfrac{\pi}{16} \times D^3$

$$4655.3 = 70 \times 10^6 \times \frac{\pi}{16} \times D^3$$

$$D^3 = \frac{4655.3 \times 16}{70 \times 10^6 \times \pi} = 338.7 \times 10^{-6}$$

∴ $D = 0.0697$ m or **69.7 mm**   **(Ans.)**

Example 13·4. *A solid circular shaft transmits 75 kW power at 200 r.p.m. Calculate the shaft diameter, if the twist in the shaft is not to exceed 1° in 2 metres length of shaft, and shear stress is limited to 50 MN/m². Take C = 100 GN/m².*

Solution. *Given:*

$$P = 75 \text{ kW}; \ N = 200 \text{ r.p.m.};$$
$$\theta = 1° = 1 \times \pi/180 \text{ radian }; \ l = 2\text{m};$$
$$\tau = 50 \text{ MN/m}^2; \ C = 100 \text{ GN/m}^2.$$

Using the relation,

$$P = \frac{2\pi \ NT}{60 \times 1000}, \text{ we get}$$

$$75 = \frac{2\pi \times 200 \times T}{60 \times 1000}$$

∴

$$T = \frac{75 \times 60 \times 1000}{2\pi \times 200} = 3581 \text{ Nm}$$

First case: *Allowable shear stress (50 MN/m²),* **D:**

Using the relation :

$$T = \tau \times \frac{\pi}{16} \times D^3, \text{ we get}$$

$$3581 = 50 \times 10^6 \times \frac{\pi}{16} \times D^3$$

$$D^3 = \frac{3581 \times 16}{50 \times 10^6 \times \pi}$$

∴

$$D = 0.0714 \text{ m} \quad \text{or} \quad 71.4 \text{ mm}$$

Second case: *Angle of twist (1°),* **D:**

Using the relation:

$$\frac{T}{I_p} = \frac{C\theta}{l}, \text{ we get,}$$

$$\frac{3581}{\frac{\pi}{32} \times D^4} = \frac{100 \times 10^9 \times 1 \times \pi/180}{2}$$

$$D^4 = \frac{3581 \times 2 \times 180 \times 32}{\pi \times \pi \times 100 \times 10^9}$$

∴

$$D = 0.0804 \text{ m} \quad \text{or} \quad 80.4 \text{ mm}$$

From the above two cases we find that suitable diameter for the shaft is 80·4 mm or say **80 mm** (*i.e. greater of the two values*) **(Ans.)**

Example 13·5. *A hollow shaft is to transmit 300 kW at 80 r.p.m. If the shear stress is not to exceed 60 MN/m² and internal diameter is 0·6 of the external diameter, find the external and internal diameters assuming that the maximum torque is 1·4 times the mean.*

Solution. *Given: P = 300 kW ; N = 80 r.p.m.; τ = 60 MN/m², d = 0·6 D; $T_{max} = 1·4 \ T_{mean}$*

Using the relation:

$$P = \frac{2 \pi \ NT}{60 \times 1000}, \text{ we get}$$

$$300 = \frac{2\pi \times 80 \times T}{60 \times 1000}$$

$$T = \frac{300 \times 60 \times 1000}{2\pi \times 80} = 35809 \text{ Nm} \quad \therefore \ T = T_{mean} = 35809$$

$$T_{max} = 1·4 \ T_{mean} = 1·4 \times 35809 = 50132 \text{ Nm}$$

Also,

$$T_{max} = \tau \cdot \frac{\pi}{16} \left[\frac{D^4 - d^4}{D} \right]$$

$$50132 = 60 \times 10^6 \times \frac{\pi}{16} \left[\frac{D^4 - (0.6\,D)^4}{D} \right]$$

$$\frac{50132 \times 16}{60 \times 10^6 \times \pi} = \frac{D^4 [1 - (0.6)^4]}{D}$$

or, $\qquad 0.004255 = 0.87\,D^3 \qquad \therefore \ D = 0.169 \text{ m}$

Hence, $\qquad\qquad D = 0.169 \text{ m or } 169 \text{ mm say } \textbf{170 mm} \ \textbf{(Ans.)}$

and, $\qquad\qquad d = 0.6\,D = 0.6 \times 170 = \textbf{102 mm} \ \textbf{(Ans.)}$

Example 13·6. *A hollow shaft of diameter ratio 3/8 is required to transmit 600 kW at 110 r.p.m., the maximum torque being 20% greater than the mean. The shear stress is not to exceed 63 MN/m² and the twist in a length of 3 m not to exceed 1.4 degrees. Calculate the maximum external diameter satisfying these conditions.*

Take: $\qquad\qquad C = 84 \text{ GN/m}^2$

Solution. Let, $\qquad d =$ Internal diameter of the hollow shaft, and

$\qquad\qquad\qquad D =$ External diameter of the hollow shaft.

Given: $d = 3/8\ D = 0.375\ D$; $P = 600$ kW ; $N = 110$ r.p.m; $\tau = 63$ MN/m² ; $\theta = 1.4°$; $l = 3$ m

$$T_{max} = 1.2\ T_{mean}$$

We know that, $\qquad P = \dfrac{2\pi N T}{60 \times 1000} \quad \text{or} \quad 600 = \dfrac{2\pi \times 110 \times T}{60 \times 1000}$

$\therefore \qquad\qquad T = \dfrac{600 \times 60 \times 1000}{2\pi \times 110} = 52087 \text{ Nm}$

$\therefore \qquad\qquad T = T_{mean} = 52087 \text{ Nm}$

$\therefore \qquad T_{max} = 1.2\ T_{mean} = 1.2 \times 52087 = 62504 \text{ Nm}$

1st Case: When shear stress is not to exceed 63 MN/m², D:

From torsion equation:

$$\frac{T}{I_p} = \frac{\tau}{R} = \frac{2\tau}{D},$$

A helicopter propeller shaft should bear direct, torsional and bending stresses.

we have,
$$T = I_p \cdot \frac{2\tau}{D} = \frac{\pi}{32}(D^4 - d^4)\frac{2\tau}{D} \quad \text{or} \quad T = \tau \cdot \frac{\pi}{16}\left(\frac{D^4 - d^4}{D}\right)$$

or,
$$62504 = 63 \times 10^6 \times \frac{\pi}{16}\left[\frac{D^4 - (0.375\,D)^4}{D}\right]$$

or,
$$62504 = 63 \times 10^6 \times \frac{\pi}{16} \times 0.9802\,D^3$$

∴
$$D^3 = \frac{62504 \times 16}{63 \times 10^6 \times \pi \times 0.9802} = 5.155 \times 10^{-3} \text{ m}^3$$

or
$$D = 0.1727 \text{ m} \quad \text{or} \quad 172.7 \text{ mm} \qquad \qquad ...(i)$$

2nd Case: *When angle of twist is not to exceed 1.4°, D:*

We know that,
$$\frac{T}{I_p} = \frac{C\theta}{l}$$

$$T = I_p \cdot \frac{C\theta}{l} = \frac{\pi}{32}(D^4 - d^4) \times \frac{84 \times 10^9 \times 1.4 \times \pi}{3 \times 180}$$

$$62504 = \frac{\pi}{32}\left[D^4 - (0.375\,D)^4\right] \times \frac{84 \times 10^9 \times 1.4 \times \pi}{3 \times 180}$$

$$62504 = 6.584 \times 10^7 \, D^4$$

∴
$$D = 0.1755 \text{ m} \quad \text{or} \quad 175.5 \text{ mm} \qquad \qquad ...(ii)$$

From above two values (*i*) and (*ii*), we find that external diameter of the shaft should be **175.5 mm** (*greater of the two values*) **(Ans.)**

Example 13.7. *A circular bar-made of cast-iron is to resist an occasional torque of 2.2 kNm acting in transverse plane. If the allowable stresses in compression, tension and shear are 100 MN/m^2, 35 MN/m^2 and 50 MN/m^2 respectively, find:*

(i)   Diameter of the bar;

(ii) Angle of twist under the applied torque per metre length of bar.

Take: C (for cast-iron) = 40 GN/m^2

Solution. Torque, $T = 2.2$ kNm

Allowable stresses: $\sigma_c = 100$ MN/m^2;

$\sigma_t = 35$ MN/m^2;

$\tau = 50$ MN/m^2;

(i) Diameter of the bar, D:

Since cast-iron is weakest in tension, so it will fail due to tensile principal stress. Due to the tensile stress σ_t, the maximum shear stress is also equal to σ_t.

$$(\tau_{max})_{allowable} = (\sigma_t)_{allowable}$$

$$\frac{T}{I_p} = \frac{\tau_{max}}{R} \quad \text{i.e.} \quad \frac{2.2 \times 1000}{\frac{\pi}{32}D^4} = \frac{35 \times 10^6}{D/2}$$

∴
$$D^3 = \frac{2.2 \times 1000 \times 16}{\pi \times 35 \times 10^6} = 3.2 \times 10^{-4}$$

or,
$$D = 0.0684 \text{ m or } \textbf{68.4 mm} \quad \textbf{(Ans.)}$$

(ii) **Angle of twist, θ:**

$$\frac{\tau_{max}}{R} = \frac{C\theta}{l}$$

$$\therefore \qquad \theta = \frac{\tau_{max} \cdot l}{CR} = \frac{35 \times 10^6 \times (1)}{40 \times 10^9 \times (0.0684/2)} \times \frac{180}{\pi} \text{ degrees} = 1.46°$$

Hence, $\qquad \theta = 1.46°$ **(Ans.)**

Example 13·8. *A hollow circular shaft 20 mm thick transmits 294 kW at 200 r.p.m. Determine the diameters of the shaft if shear strain due to torsion is not to exceed 8·6 × 10⁻⁴.*

Take: Modulus of rigidity as 80 GN/m².

Solution. Let, $\qquad D_H$ = External diameter of the hollow shaft, m,

$\qquad\qquad d_H$ = Internal diameter of the hollow shaft, m, and

$\qquad\qquad t$ = Thickness of the shaft (= 20 mm = 0·02 m).

Then, $\qquad D_H - d_H = 2t = 0.04$ m [or $d_H = (D_H - 0.04)$ m]

Shear strain due to torsion,

$$\left. \begin{array}{l} e_s = 8.6 \times 10^{-4} \\ C = 80 \text{ GN/m}^2 \\ P = 294 \text{ kW} \\ N = 200 \text{ r.p.m.} \end{array} \right\} \quad (Given)$$

Modulus of rigidity, $\qquad C = 80$ GN/m²

Power transmitted, $\qquad P = 294$ kW

Speed, $\qquad N = 200$ r.p.m.

Diameters of the shaft, D_H and d_H:

We know that, $\qquad P = \dfrac{2\pi NT}{60 \times 1000}$ kW

$$\therefore \qquad 294 = \frac{2\pi \times 200 \times T}{60 \times 1000} \quad \text{or} \quad T = \frac{294 \times 60 \times 1000}{2\pi \times 200} = 14037 \text{ Nm}$$

and, $\qquad \dfrac{T}{I_p} = \dfrac{\tau}{R} \quad \text{or} \quad \tau = \dfrac{TR}{I_p} = \dfrac{14037 \times (D_H/2)}{\dfrac{\pi}{32}(D_H^4 - d_H^4)}$

$$= \frac{71489.8 \, D_H}{[D_H^4 - (D_H - 0.04)^4]} \times 10^{-6} \text{ MN/m}^2$$

Also, $\qquad e_s = \dfrac{\tau}{C} \quad \text{or} \quad \tau = e_s \times C$

$$= 8.6 \times 10^{-4} \times 80 \times 10^9 \times 10^{-6} \text{ MN/m}^2 = 68.6 \text{ MN/m}^2$$

$$\therefore \qquad \frac{71489.8 \, D_H}{[D_H^4 - (D_H - 0.04)^4]} \times 10^{-6} = 68.8$$

or, $\qquad \dfrac{D_H^4 - (D_H - 0.04)^4}{D_H} = \dfrac{71489.8}{68.8 \times 10^6} = 1.039 \times 10^{-3}$

By trial and error, $\qquad D_H = 0.108$ m **108 mm** **(Ans.)**

and, $\qquad d_H = 108 - 2 \times 20 = $ **68 mm** **(Ans.)**

Example 13·9. *A solid steel shaft is subjected to a torque of 45 kNm. If the angle of twist is 0·5° per metre length of the shaft and the shear stress is not to be allowed to exceed 90 MN/m² find:*

(i) Suitable diameter for the shaft;

(ii) Final maximum shear stress and angle of twist; and

(iii) Maximum shear strain in the shaft.

Take: $\qquad C = 80 \text{ GN/m}^2$

Solution. Torque $\qquad T = 45$ kNm

Angle of twist, $\qquad \theta = 0.5° = 0.5 \times \pi/180$ rad. $= 0.008727$ rad.

Maximum shear stress, $\tau_{max} = 90$ MN/m^2

(i) Diameter of the shaft, D:

Diameter on the basis of twist:

$$\frac{T}{I_p} = \frac{C\theta}{l}$$

i.e. $\qquad I_p = \frac{Tl}{C\theta} = \frac{45 \times 10^3 \times (1)}{80 \times 10^9 \times 0.008727} = 6.445 \times 10^{-5} \text{ m}^4$

i.e. $\qquad \frac{\pi}{32} D^4 = 6.445 \times 10^{-5}$

or $\qquad D = \left(\frac{6.445 \times 10^{-5} \times 32}{\pi} \right)^{1/4} = 0.16 \text{ m or } 160 \text{ mm}$

Diameter on the basis of shear stress:

$$\frac{T}{I_p} = \frac{\tau}{R}$$

$$T = I_p \times \frac{\tau}{R} = \frac{\pi}{32} D^4 \times \frac{\tau}{D/2} = \frac{\pi}{16} D^3 \tau$$

$\therefore \qquad D = \left(\frac{16\,T}{\pi\,\tau} \right)^{1/3} = \left[\frac{16 \times 45 \times 10^3}{\pi \times 90 \times 10^6} \right]^{1/3} = 0.1365 \text{ m} = 136.5 \text{ mm}$

So the diameter of the shaft based on twist should be used because it *gives higher value for diameter.*

∴ *Diameter of the shaft* = **160 mm** (Ans.)

(ii) Final maximum shear stress and angle of twist:

Since the diameter is given by the angle of twist, the *final angle of twist is 0.5° per metre length.* The maximum shear stress will be less than the given value of 90 MN/m^2. The final maximum shear stress is given by

$$\frac{T}{I_p} = \frac{\tau}{R}$$

Prototype of a shaft is being tested in a laboratory.

$\therefore \qquad \tau = \frac{TR}{I_p} = \frac{45 \times 10^3 \times \dfrac{(0.16)}{2}}{\dfrac{\pi}{32} \times (0.16)^4} \times 10^{-6} \text{ MN/m}^2 = 55.95 \text{ MN/m}^2$

(iii) Maximum shear strain in the shaft $(e_s)_{max}$:

$$(e_s)_{max} = \frac{R\theta}{l} = \frac{(0.16/2) \times 0.008727}{1} = 6.98 \times 10^{-4} \text{ (Ans.)}$$

$$\left[Check : (e_s)_{max} = \frac{\tau_{max}}{C} = \frac{55.95 \times 10^6}{80 \times 10^9} = 6.99 \times 10^{-4} \text{ (Ans.)} \right]$$

13·9. COMPARISON OF SOLID AND HOLLOW SHAFTS

(a) Comparison by strength:

In this case it is assumed that both the shafts have *same length, material, same weight and hence the same maximum shear stress.*

Let,
D_S = Diameter of the solid shaft,
d_H = Internal diameter of the hollow shaft,
D_H = External diameter of the hollow shaft,
A_S = Cross-sectional area of solid shaft,
A_H = Cross-sectional area of hollow shaft,
T_S = Torque transmitted by the solid shaft, and
T_H = Torque transmitted by the hollow shaft.

Now,
$$T_S = \tau \cdot \frac{\pi}{16} \times D_S^3$$

$$T_H = \tau \cdot \frac{\pi}{16} \left[\frac{D_H^4 - d_H^4}{D_H} \right]$$

$$\therefore \quad \frac{\text{Strength of hollow shaft}}{\text{Strength of a solid shaft}} = \frac{T_H}{T_S} = \frac{\tau \cdot \frac{\pi}{16} \left[\frac{D_H^4 - d_H^4}{D_H} \right]}{\tau \cdot \frac{\pi}{16} D_S^3}$$

or,
$$\frac{T_H}{T_S} = \frac{D_H^4 - d_H^4}{D_H \cdot D_S^3} \qquad \qquad ...(13.7)$$

Let,
$$\frac{D_H}{d_H} = n$$

$\therefore \qquad D_H = nd_H$. Substituting it in equation (13.7), we get

$$\frac{T_H}{T_S} = \frac{n^4 d_H^4 - d_H^4}{nd_H D_S^3} = \frac{d_H^4 (n^4 - 1)}{nd_H D_S^3} = \frac{d_H^3 (n^4 - 1)}{n D_S^3} \qquad ...(13.8)$$

As the weight, material and length of both the shafts are same,

$\therefore$ Cross-sectional area of solid shaft = Cross-sectional area of hollow shaft $A_S = A_H$

$$\therefore \quad \frac{\pi}{4} D_S^2 = \frac{\pi}{4} (D_H^2 - d_H^2) \text{ or } D_s = \sqrt{D_H^2 - d_H^2}$$

or,
$$D_S^3 = (D_H^2 - d_H^2) \sqrt{D_H^2 - d_H^2}$$

$$D_S^3 = (n^2 d_H^2 - d_H^2) \sqrt{n^2 d_H^2 - d_H^2}$$

$$D_S^3 = d_H^3 (n^2 - 1) \sqrt{n^2 - 1} \qquad ...(13.9)$$

Substituting the value of D_S^3 in equation (13.8), we get

$$\frac{T_H}{T_S} = \frac{d_H^3 (n^4 - 1)}{n d_H^3 (n^2 - 1) \sqrt{n^2 - 1}}$$

$$= \frac{(n^2 + 1)(n^2 - 1)}{n (n^2 - 1) \sqrt{n^2 - 1}} = \frac{n^2 + 1}{n \sqrt{n^2 - 1}} \qquad ...(13.10)$$

Since $D_H > d_H$ and $= \dfrac{D_H}{d_H} > n$, it is obvious that the value of 'n' is greater than unity.

∴ Suppose, $\qquad\qquad\qquad n = 2$

Then, $\qquad\qquad \dfrac{T_H}{T_S} = \dfrac{2^2 + 1}{2\sqrt{2^2 - 1}} = 1.44$

This shows that the torque transmitted by the hollow shaft is greater than the solid shaft, thereby proving that the hollow shaft is stronger than the solid shaft.

(b) Comparison by weight:

In this case it is assumed that both the shafts have the *same length and material*. Now, if the torque applied to both shafts is same, then, the maximum shear stress will also be same in both the cases.

Now, $\qquad \dfrac{\text{Weight of hollow shaft}}{\text{Weight of solid shaft}} = \dfrac{W_H}{W_S} = \dfrac{A_H}{A_S}$

$$= \dfrac{\dfrac{\pi}{4}(D_H^2 - d_H^2)}{\dfrac{\pi}{4}D_S^2} = \dfrac{D_H^2 - d_H^2}{D_S^2} \qquad\qquad ...(13 \cdot 11)$$

Let, $\qquad\qquad\qquad \dfrac{D_H}{d_H} = n$

∴ $\qquad\qquad D_H = nd_H$ and substituting this value in equation (13·10), we get

$$\dfrac{W_H}{W_S} = \dfrac{n^2 d_H^2 - d_H^2}{D_S^2} = \dfrac{d_H^2 (n^2 - 1)}{D_S^2} \qquad\qquad ...(13 \cdot 12)$$

Torque applied in both the cases is same *i.e.*, $T_S = T_H$

$$\tau \cdot \dfrac{\pi}{16} D_s^3 = \tau \cdot \dfrac{\pi}{16} \left[\dfrac{D_H^4 - d_H^4}{D_H} \right]$$

Electric motor can transfer mechanical energy to other machines via gears that are fitted at the end of the shaft.

$$D_s^3 = \frac{D_H^4 - d_H^4}{D_H} = \frac{n^4 d_H^4 - d_H^4}{n d_H} = \frac{d_H^3 (n^4 - 1)}{n}$$

$$\therefore \qquad D_s = d_H \left[\frac{n^4 - 1}{n} \right]^{1/3}$$

or $$\qquad D_s^2 = d_H^2 \left[\frac{n^4 - 1}{n} \right]^{2/3} \qquad \qquad \qquad \qquad ...(13\cdot13)$$

Substituting the value of D_S^2 in equation (13·12), we have

$$\frac{W_H}{W_S} = \frac{d_H^2 (n^2 - 1)}{d_H^2 \left(\frac{n^4 - 1}{n} \right)^{2/3}} = \frac{(n^2 - 1) n^{2/3}}{(n^4 - 1)^{2/3}} \qquad \qquad ...(13\cdot14)$$

If, $n = 2$ then, $$\qquad \frac{W_H}{W_S} = \frac{(2^2 - 1) \times (2)^{2/3}}{(2^4 - 1)^{2/3}} = 0\cdot7829$$

which shows that for *same material, length and given torque, weight of hollow shaft will be less. So, hollow shafts are economical compared to solid shafts as regards torsion.*

Example 13·10. *Two shafts of the same material and same length are subjected to the same torque. If the first shaft is of a solid circular section, and the second shaft is of a hollow circular section, whose internal diameter is 2/3 of the outside diameter and the maximum shear stress developed in each shaft is the same, compare the weights of the two shafts.*

Solution. Let $\quad D_S$ = Diameter of the solid shaft,

$\qquad D_H$ = External diameter of the hollow shaft,

$\qquad d_H$ = Internal diameter of the hollow shaft, and

$\qquad \tau$ = Maximum shear stress developed.

$$d_H = \frac{2}{3} D_H \qquad (Given)$$

The torque transmitted by the solid shaft,

$$T_s = \tau \cdot \frac{\pi}{16} D_S^3 \qquad \qquad \qquad \qquad ...(i)$$

and, the torque transmitted by the hollow shaft,

$$T_H = \tau \cdot \frac{\pi}{16} \left[\frac{D_H^4 - d_H^4}{D_H} \right] = \tau \cdot \frac{\pi}{16} \left[\frac{D_H^4 - (2/3 D_H)^4}{D_H} \right]$$

$$= \tau \cdot \frac{\pi}{16} \left[\frac{D_H^4 - \left(\frac{16}{81} D_H^4 \right)}{D_H} \right] = \tau \cdot \frac{\pi}{16} \times \frac{65}{81} D_H^3 \qquad ...(ii)$$

Since both the torques are equal, therefore equating (i) and (ii), we get

$$T_S = T_H$$

$$\tau \cdot \frac{\pi}{16} D_S^3 = \tau \cdot \frac{\pi}{16} \cdot \frac{65}{81} D_H^3$$

$$\therefore \qquad D_H^3 = 1\cdot246 D_S^3 , or, \ D_H = 1\cdot08 D_S$$

We know that,

$$\frac{\text{Weight of solid shaft}}{\text{Weight of hollow shaft}} = \frac{W_S}{W_H}$$

$$= \frac{A_S \times l_S \times w_S}{A_H \times l_H \times w_H} = \frac{A_S}{A_H}$$

$$\left[\begin{array}{l} \because \ l_S = l_H \\ w_S = w_H \ \text{where, } w \text{ stands for weight density} \end{array} \right]$$

$$= \frac{\dfrac{\pi}{4} D_S^2}{\dfrac{\pi}{4}(D_H^2 - d_H^2)} = \frac{D_S^2}{[D_H^2 - \left(\dfrac{2}{3} D_H\right)^2]} = \frac{D_S^2}{D_H^2 \left(1 - \dfrac{4}{9}\right)}$$

$$= \frac{D_S^2}{\dfrac{5}{9} \times (1.08 \, D_S)^2} = \frac{1.543}{1} \quad \textbf{(Ans.)}$$

Example 13·11. *A solid cylindrical shaft is to transmit 300 kW at 100 r.p.m.*

(i)   If the shear stress is not to exceed 80 MN / m², find its diameter.

(ii) What percentage saving in weight would be obtained if this shaft is replaced by a hollow one whose internal diameter equals 0.6 of the external diameter, the length, the material and maximum shear stress being the same?

Solution. *Given :* $P = 300$ kW; $N = 100$ r.p.m.; $\tau = 80$ MN/m²; $d_H = 0.6 \, D_H$.

(i)   Diameter of solid shaft, D_S:

We know, $$P = \frac{2\pi NT}{60 \times 1000}$$

$$300 = \frac{2\pi \times 100 \times T}{60 \times 1000} \quad \therefore \ T = \frac{300 \times 60 \times 1000}{2\pi \times 100} = 28648 \text{ Nm}$$

Also, $$T = \tau . \frac{\pi}{16} D_S^3 \qquad\qquad [\text{Assuming } T(T_{\text{mean}}) = T_{\text{max}}]$$

$\therefore$ $$28648 = 80 \times 10^6 \times \frac{\pi}{16} D_S^3$$

$$D_S^3 = \frac{28648 \times 16}{80 \times 10^6 \times \pi} = 1.824 \times 10^{-3}$$

$\therefore$ $$D_S = 0.122 \text{ m or } \textbf{122 mm} \quad \textbf{(Ans.)}$$

(ii) Percentage saving in weight:

$$T_H = T_S$$

$$\tau . \frac{\pi}{16} \left[\frac{D_H^4 - d_H^4}{D_H} \right] = \tau . \frac{\pi}{16} D_S^3 \quad \text{or} \quad \frac{D_H^4 [1 - (0.6 D_H)^4]}{D_H} = D_S^3$$

or, $$\frac{D_H^4 - (0.6 D_H)^4}{D_H} = 1.824 \times 10^{-3} \qquad \text{(calculated above)}$$

$$D_H^3 (1 - 0.1296) = 1.824 \times 10^{-3} \quad \text{or} \quad D_H = 0.128 \text{ m}$$

and, $$d_H = 0.6 \, D_H = 0.6 \times 0.128 = 0.0768 \text{ m}$$

Again, $$\frac{W_H}{W_S} = \frac{A_H \, l_H \, w_H}{A_S \, l_S \, w_S} = \frac{A_H}{A_S} \qquad \left[\begin{array}{l} \because \ l_H = l_S, \\ \text{and, } w_H = w_S \end{array} \right]$$

$$= \frac{\frac{\pi}{4} \times (D_H^2 - d_H^2)}{\frac{\pi}{4} D_s^2} = \frac{0.128^2 - 0.0768^2}{0.122^2} = 0.704$$

Percentage saving in weight

$$= \left(1 - \frac{W_H}{W_s}\right) \times 100 = (1 - 0.704) \times 100 = \mathbf{29.6\ \%} \quad \textbf{(Ans.)}$$

Example 13·12. *A hollow shaft, having an inside diameter 60% of its outer diameter is to replace a solid shaft transmitting the same power at the same speed. Calculate the percentage saving in material, if the material to be used is also the same.*

Solution. Using the relation

$$T = \tau \cdot \pi/16\ D^3, \text{ we have}$$

$$T_S = \tau \cdot \frac{\pi}{16} D_S^3 \qquad \qquad \qquad ...(i)$$

Similarly,

$$T_H = \tau \cdot \frac{\pi}{16}\left[\frac{D_H^4 - d_H^4}{D_H}\right] = \tau \cdot \frac{\pi}{16}\left[\frac{D_H^4 - (0.6\,D_H)^4}{D_H}\right] \quad (\because d_H = 0.6\,D_H)$$

$$= \tau \cdot \frac{\pi}{16}\left[\frac{D_H^4\,(1 - 0.6^4)}{D_H}\right] = \tau \times \pi/16 \times 0.8704\,D_H^3 \qquad ...(ii)$$

Since the torque transmitted and the allowable shear stress is the same, therefore equating (*i*) and (*ii*), we get

$$\tau \times \frac{\pi}{16} \times D_S^3 = \tau \times \frac{\pi}{16} \times 0.8704\,D_H^3 \quad \therefore\ D_S = 0.9548\,D_H$$

Percentage saving in material:

$$= \left[\frac{W_S - W_H}{W_S}\right] \times 100 = \left[1 - \frac{W_H}{W_S}\right] \times 100 = \left[1 - \frac{A_H\,l_H\,w_H}{A_S\,l_S\,w_S}\right] \times 100$$

$$= \left[1 - \frac{A_H}{A_S}\right] \times 100 = \left[1 - \frac{(\pi/4)\left[D_H^2 - (0.6\,D_H)^2\right]}{(\pi/4)\,D_S^2}\right] \times 100$$

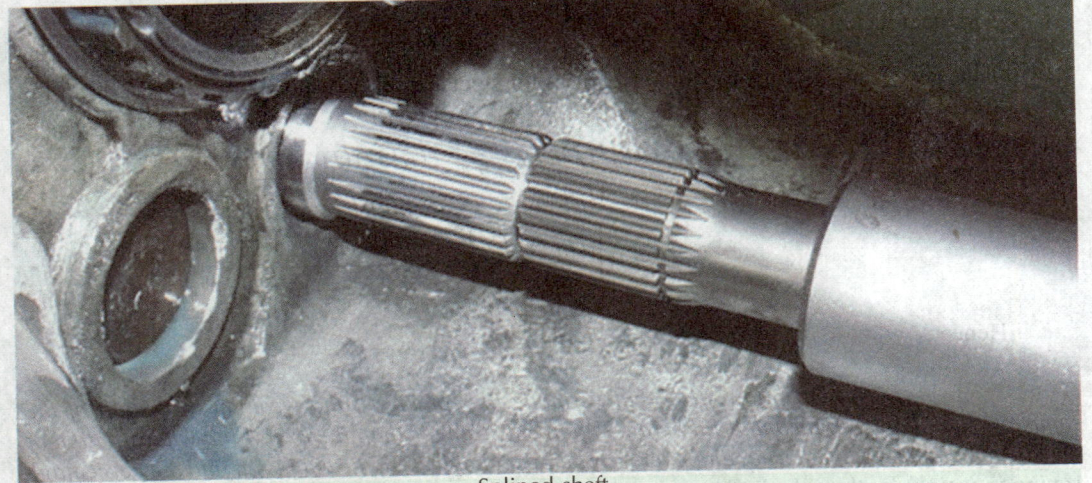

Splined shaft.

$$= \left[1 - \frac{(\pi/4) D_H^2 (1 - 0.36)}{(\pi/4) \times (0.9548 D_H)^2} \right] \times 100$$

$$= \left[1 - \frac{0.64 D_H^2}{0.9116 D_H^2} \right] \times 100 = (1 - 0.702) \times 100 = \mathbf{29 \cdot 8\%} \quad \textbf{(Ans.)}$$

Example 13·13. *Hollow steel shaft of 100 mm internal diameter and 150 mm external diameter is to be replaced by a solid alloy shaft. If the polar modulus has the same value for both, calculate the diameter of the latter and the ratio of their torsional rigidities.*

Given: $\qquad C_{steel} = 2 \cdot 4 \ C_{alloy}$

Solution. Internal diameter of hollow shaft,

$$d_H = 100 \text{ mm} = 0 \cdot 1 \text{ m}$$

External diameter of hollow shaft,

$$D_H = 150 \text{ mm} = 0 \cdot 15 \text{ m}$$

$$C_{steel} = 2 \cdot 4 \ C_{alloy}$$

Diameter of solid shaft, D_S:

Since polar modulus is $\dfrac{I_p}{R}$ and its value is same in both cases

$$\therefore \qquad \frac{(I_p)_{steel}}{\dfrac{D_H}{2}} = \frac{(I_p)_{alloy}}{\dfrac{D_S}{2}}$$

or $\qquad \dfrac{(\pi/32) (D_H^4 - d_H^4)}{\dfrac{D_H}{2}} = \dfrac{(\pi/32) D_S^4}{\dfrac{D_S}{2}}$

or $\qquad (D_H^4 - d_H^4) D_S = D_S^4 \times D_H$

or $\qquad D_S^3 = \dfrac{D_H^4 - d_H^4}{D_H} = \dfrac{(0 \cdot 15)^4 - (0 \cdot 1)^4}{0 \cdot 15} = 0.002708 \text{ m}^3$

$\therefore \qquad D_S = 0 \cdot 1394 \text{ m or } \mathbf{139 \cdot 4 \text{ mm}} \quad \textbf{(Ans.)}$

Ratio of torsional rigidities:

Torsional rigidity, $\qquad k = \dfrac{C I_p}{l}$

Thus if k_{steel} and k_{alloy} be torsional rigidities for steel and alloy respectively and lengths of the two shafts being equal, we have

$$\frac{k_{steel}}{k_{alloy}} = \frac{C_{steel} (I_p)_{steel}}{C_{alloy} (I_p)_{alloy}} = 2 \cdot 4 \times \frac{(\pi/32)(D_H^4 - d_H^4)}{(\pi/32) D_S^4}$$

$$= 2 \cdot 4 \times \left[\frac{D_H^4 - d_H^4}{D_S^4} \right] = 2 \cdot 4 \times \frac{(0 \cdot 15)^4 - (0 \cdot 1)^4}{(0 \cdot 139)^4} = 2 \cdot 61$$

Hence, $\qquad \dfrac{k_{steel}}{k_{alloy}} = \mathbf{2 \cdot 61} \quad \textbf{(Ans.)}$

Example 13·14. *A hollow steel shaft is made to replace a solid wrought iron shaft of the same external diameter, the material being 35 percent stronger than the iron. Find what fraction of the outside diameter the internal diameter may be. Also, neglecting the coupling, find the percentage saving in weight by the substitution, assuming that steel is 2 percent heavier than wrought iron.*

Solution. Let,

D_S = Diameter of solid wrought iron shaft,

D_H = External diameter of hollow steel shaft, and

d_H = Internal diameter of hollow steel shaft.

Ratio of internal diameter to external diameter of hollow shaft, $\dfrac{d_H}{D_H}$:

Since the torque transmitted will be equal

∴
$$T_S = T_H$$

$$\tau \times \frac{\pi}{16} \times D_S^3 = 1\cdot35\ \tau \times \frac{\pi}{16} \times \left[\frac{D_H^4 - d_H^4}{D_H}\right]$$

$$D_H^3 = 1\cdot35 \times \frac{D_H^4 - d_H^4}{D_H} \qquad\qquad (\because D_S = D_H)$$

$$D_H^4 = 1\cdot35\,(D_H^4 - d_H^4)$$

$$1\cdot35\,d_H^4 = 0\cdot35\,D_H^4$$

$$\frac{d_H}{D_H} = \left[\frac{0\cdot35}{1\cdot35}\right]^{1/4} = 0\cdot714$$

Hence the internal diameter should be **0·714 of the external diameter (Ans.)**

Percentage saving in weight:

Area of cross-section of iron shaft $\quad = \dfrac{\pi}{4} \times D_s^2 = \dfrac{\pi}{4} \times D_H^2$

Area of cross-section of the steel shaft

$$= \frac{\pi}{4} \times (D_H^2 - d_H^2) = \frac{\pi}{4} \times D_H^2 \left[1 - \frac{d_H^2}{D_H^2}\right]$$

$$\frac{\pi}{4} \times D_H^2\,(1 - 0\cdot714^2) = \frac{\pi}{4}\,D_H^2 \times 0\cdot49$$

Let w be the weight of unit volume of wrought iron. Then $1\cdot02\,w$ is the weight of unit volume of steel.

∴ Weight of unit length of iron shaft $= (\pi/4) \times D_H^2 \times w$

Weight of unit length of steel shaft $= (\pi/4) \times D_H^2 \times 0\cdot49 \times 1\cdot02\,w$

$$\frac{\text{Weight of steel shaft}}{\text{Weight of iron shaft}} = \frac{(\pi/4) \times D_H^2 \times 0\cdot49 \times 1\cdot02\,w}{(\pi/4) \times D_H^2 \times w} = \frac{0\cdot49 \times 1\cdot02}{1} = 0\cdot5$$

Hence, *saving in weight* $\quad = (1 - 0\cdot5) \times 100 = \mathbf{50\%}$ **(Ans.)**

Example 13·15. *A solid shaft of mild steel 200 mm in diameter is to be replaced by hollow shaft of alloy steel for which the allowable shear stress is 22 percent greater. If the power to be transmitted is to be increased by 20 percent and the speed of rotation increased by 6 percent determine the maximum internal diameter of the hollow shaft. The external diameter of the hollow shaft is to be 200 mm.*

Solution. *For solid shaft:*

Diameter, $\qquad\qquad D_S = 200$ mm $= 0\cdot2$m *(Given)*

Let, $\qquad\qquad \tau_S$ = Maximum shear stress on the surface of solid shaft,

$\qquad\qquad P_S$ = Power transmitted,

$\qquad\qquad N_S$ = Rate of rotation,

$(I_p)_S$ = Polar moment of inertia, and

T_S = Torque transmitted.

For hollow shaft:

Let τ_H, P_H, N_H, $(I_p)_H$ and T_H be the corresponding notations respectively for the hollow shaft.

Given:
$$\tau_H = 1 \cdot 22\tau_S$$
$$P_H = 1 \cdot 2 P_S$$
$$N_H = 1 \cdot 06 N_S$$

Outer diameter of hollow shaft,
$$D_H = 220 \text{ mm}$$

Maximum internal diameter of hollow shaft, d_H:

Now,
$$P = \frac{2\pi NT}{60 \times 1000} \text{ kW}$$

∴
$$P_S = \frac{2\pi N_S T_S}{60 \times 1000} \text{ kW, and } P_H = \frac{2\pi N_H T_H}{60 \times 1000}$$

But,
$$P_H = 1 \cdot 2 P_S$$

∴
$$\frac{2\pi N_H T_H}{60 \times 1000} = 1 \cdot 2 \times \frac{2\pi N_S T_S}{60 \times 1000}$$

or,
$$(1 \cdot 06 N_S) \, T_H = 1 \cdot 2 \, N_S \, T_S$$
$$T_H = 1 \cdot 132 \, T_S$$

From the relation
$$\frac{T}{I_p} = \frac{\tau}{R}, \text{ we have}$$

$$\frac{T_S}{(I_p)_S} = \frac{\tau_S}{\left(\dfrac{0 \cdot 2}{2}\right)} \quad \dots(i)$$

and,
$$\frac{T_H}{(I_p)_H} = \frac{\tau_H}{\left(\dfrac{0 \cdot 2}{2}\right)} \quad \dots(ii)$$

Dividing (*i*) and (*ii*), we get

These gears have key-ways to fit on shafts.

$$\frac{T_S \times (I_p)_H}{T_H \times (I_p)_S} = \frac{\tau_S}{\tau_H}$$

or,
$$\frac{T_S \times \dfrac{\pi}{32}(D_H^4 - d_H^4)}{1 \cdot 132 T_S \times \dfrac{\pi}{32} D_S^4} = \frac{\tau_S}{1 \cdot 22\tau_S} = 0 \cdot 8196 \text{ or } \frac{((0 \cdot 2)^4 - d_H^4)}{1 \cdot 132 \times (0 \cdot 2)^4} = 0 \cdot 8196$$

or,
$$d_H^4 = (0 \cdot 2)^4 - 0 \cdot 8196 \times 1 \cdot 132 \times (0 \cdot 2)^4 = 1 \cdot 1554 \times 10^{-4}$$

∴
$$d_H = 0 \cdot 1037 \text{ m} = \mathbf{103 \cdot 7 \text{ mm}} \quad \textbf{(Ans.)}$$

13·10. SHAFTS IN SERIES

In order to form a composite shaft sometimes two shafts are connected in series. In such cases, *each shaft transmits the same torque. The angle of twist is the sum of the angle of twist of the two shafts connected in series.*

Thus, total angle of twist is given by

$$\theta = \theta_1 + \theta_2 = \frac{Tl_1}{C_1 I_{p_1}} + \frac{Tl_2}{C_2 I_{p_2}} = T\left[\frac{l_1}{C_1 I_{p_1}} + \frac{l_2}{C_2 I_{p_2}}\right] \quad \dots(13 \cdot 15)$$

where,

T = Torque transmitted by each shaft,

l_1, l_2 = Respective lengths of the two shafts,

C_1, C_2 = Respective moduli of rigidity, and

I_{p_1}, I_{p_2} = Respective polar moment of intertias.

When shafts are made of same material,

$$C_1 = C_2 = C \text{ say}$$

Then,

$$\theta = \frac{T}{C}\left[\frac{l_1}{I_{p_1}} + \frac{l_2}{I_{p_2}}\right]$$

Here, the driving torque is applied at one end and the resisting torque at the other.

13·11. SHAFTS IN PARALLEL

The shafts are said to be in *parallel* when the driving torque is applied at the junction of the shafts and the resisting torque is at the other ends of the shafts. Here, the angle of twist is same for each shaft, but the applied torque is divided between the two shafts.

i.e.

$$\theta_1 = \theta_2$$

or,

$$\frac{T_1 l_1}{C_1 I_{p_1}} = \frac{T_2 l_2}{C_2 I_{p_2}} \qquad \qquad ...(13·16)$$

and,

$$T = T_1 + T_2 \qquad \qquad ...(13·17)$$

If the shafts are made of same material

$$C_1 = C_2$$

Then,

$$\frac{T_1 l_1}{I_{p_1}} = \frac{T_2 l_2}{I_{p_2}} \text{ or } \frac{T_1}{T_2} = \frac{I_{p_1} l_2}{I_{p_2} l_1} \qquad \qquad ...[13·18\ (a)]$$

When torque is shared equally by both the shafts

$$T_1 = T_2, \text{ then, } I_{p_1} l_2 = I_{p_2} l_1 \qquad \qquad ...[13·18\ (b)]$$

Example 13·16. *The stepped steel shaft shown in Fig. 13·4 is subjected to a torque T at the free end and a torque of 2T in the opposite direction at the junction of two sizes. What is the total angle of twist at the free end, if the maximum shear stress in the shaft is limited to 70 MN/m^2 ? Assume the modulus of rigidity to be 84 GN/m^2.*

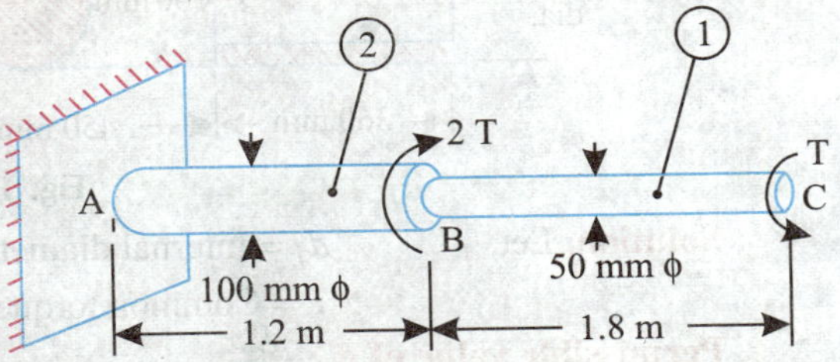

Solution. Refer to Fig. 13·4.

The torque *2T* at *B* is equivalent to two torques each of value *T*. Then *BC* is subject to a torque *T* at *C* and an opposite torque *T* at *B*, while *AB* is also subject to equal and opposite torque *T* at *A* and *B*.

Fig. 13.4

For the length BC;

Torque,

$$T = T, l = 1·8 \text{ m}.$$

$$I_p = (\pi/32) \times (0·05)^4$$

$$= 6·136 \times 10^{-7}\ m^4$$

∴

$$\theta_1 = \text{Angle of twist of } C \text{ relative to } B.$$

$$= \frac{Tl}{C I_p} = \frac{T \times 1·8}{84 \times 10^9 \times 6·136 \times 10^{-7}} \qquad \qquad ...(i)$$

For the length AB :

Torque, $\qquad T = T, l = 1 \cdot 2$ m,

$$I_p = (\pi/32) \times (0 \cdot 1)^4 = 9 \cdot 817 \times 10^{-6} \text{m}^4$$

∴ $\qquad \theta_2$ = Angle of twist of B relative to A.

$$= \frac{T \times 1 \cdot 2}{84 \times 10^9 \times 9 \cdot 817 \times 10^{-6}} \qquad \qquad ...(ii)$$

θ_1 and θ_2 are in opposite directions. Hence θ_C, the total angle of twist at C,

$$\theta_C = \theta_1 - \theta_2$$

The maximum shear stress occurs in *BC*, and its value is 70 MN/m^2 (given)

Also, $\qquad \dfrac{T}{I_p} = \dfrac{\tau}{R}$

∴ $\qquad T = \dfrac{\tau \cdot I_p}{R} = \dfrac{70 \times 10^6 \times 6 \cdot 136 \times 10^{-7}}{0.025} = 1718 \cdot 1 \text{ Nm}$

∴ From eqn. (*i*), we have

$$\theta_1 = \frac{1718 \cdot 1 \times 1 \cdot 8}{84 \times 10^9 \times 6 \cdot 136 \times 10^{-7}} = 0 \cdot 06 \text{ radian}$$

and, from eqn. (*ii*)

$$\theta_2 = \frac{1718 \cdot 1 \times 1 \cdot 2}{84 \times 10^9 \times 9 \cdot 817 \times 10^{-6}} = 0 \cdot 0025 \text{ radian}$$

∴ $\qquad \theta_C = \theta_1 - \theta_2 = 0 \cdot 06 - 0 \cdot 0025 = 0 \cdot 0575 \text{ radian}$

$$= \textbf{3·29 degrees} \quad \textbf{(Ans.)}$$

Example 13·17. *A steel shaft LMNP is made as shown in Fig. 13.5. If equal opposite torques are applied at the end of the shaft find the maximum permissible value of d_1 for the maximum shearing stress in LM not to exceed that in NP. If torque applied is 10 kNm, what is the total angle of twist ?*
Take: $\qquad C = 80GN/m^2.$

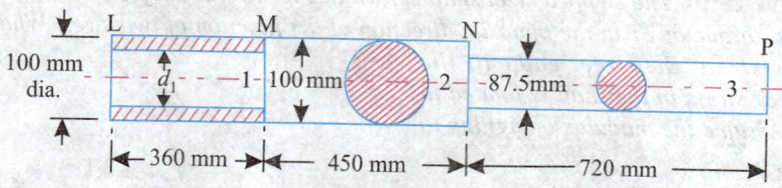

Fig. 13.5

Solution. Let, $\qquad d_1$ = Internal diameter of the shaft *LM*, and

$\qquad T$ = Common torque.

Permissible value of d_1:

For the shaft LM : $\qquad T = \tau_1 \times \dfrac{\pi}{16} \left(\dfrac{D_1^4 - d_1^4}{D_1} \right) = \tau_1 \times \pi/16 \left[\dfrac{0 \cdot 1^4 - d_1^4}{0 \cdot 1} \right]$

∴ $\qquad \tau_1 = \dfrac{T \times 0 \cdot 1 \times 16}{\pi (0 \cdot 1^4 - d_1^4)} = \dfrac{1 \cdot 6T}{\pi (0 \cdot 1^4 - d_1^4)}$

For the shaft NP: $\qquad T = \tau_3 \times (\pi/16) D_3^3 = \tau_3 \times \pi/16 \times (0 \cdot 0875)^3$

∴ $\qquad \tau_3 = \dfrac{16T}{\pi \times (0 \cdot 0875)^3}$

But, $\qquad\qquad \tau_1 = \tau_3 \qquad\qquad$ ...(*Given*)

$\therefore \qquad \dfrac{1 \cdot 6\,T}{\pi\,(0 \cdot 1^4 - d_1^4)} = \dfrac{16\,T}{\pi \times (0 \cdot 0875)^3}$

or $\qquad 0 \cdot 1^4 - d_1^4 = \dfrac{(0 \cdot 0875)^3 \times 1 \cdot 6}{16} = 6 \cdot 7 \times 10^{-5}$

or $\qquad d_1^4 = 0 \cdot 1^4 - 6 \cdot 7 \times 10^{-5} = 3 \cdot 3 \times 10^{-5}$

$\therefore \qquad d_1 = 0 \cdot 0758 \text{ m or } \mathbf{75 \cdot 8\ mm}$ **(Ans.)**

Total angle of twist θ:

This picture shows disk or drum brakes. When brakes are applied they exert a strong torsional stress on the axle shaft.

$$\theta = \sum \dfrac{Tl}{Cl_p} = \dfrac{T}{C} \sum \dfrac{l}{I_p}$$

[∵ In this case T and C are constant]

$$= \dfrac{T}{C}\left[\dfrac{l_1}{I_{P_1}} + \dfrac{l_2}{I_{P_2}} + \dfrac{l_3}{I_{P_3}}\right]$$

$$= \dfrac{10 \times 10^3}{80 \times 10^9}\left[\dfrac{0 \cdot 36}{\dfrac{\pi}{32}(0 \cdot 1^4 - 0 \cdot 0758^4)} + \dfrac{0 \cdot 45}{\dfrac{\pi}{32} \times 0 \cdot 1^4} + \dfrac{0 \cdot 72}{\dfrac{\pi}{32} \times 0 \cdot 0875^4}\right] \text{rad.}$$

$$= \dfrac{1}{8 \times 10^6}[54740 \cdot 4 + 45836 \cdot 6 + 125112 \cdot 4] \times \dfrac{180}{\pi} \text{ degrees}$$

$$= 1 \cdot 616° \text{ or } 1°37'$$

Hence, *total angle of twist,*

$$\theta = 1°37' \quad \textbf{(Ans.)}$$

Example 13·18. *A solid phosphor bronze shaft of 80 mm diameter is coupled to a hollow steel shaft 80 mm outside diameter. The torque applied to the compound shaft develops a maximum shear stress of 40 MN/m² in the bronze shaft and a maximum shear stress of 72 MN/m² in steel shaft. The length of steel shaft is 1 m and of bronze shaft is 1·2 m. Angle of twist for the steel shaft is not to exceed 1°. If C_{steel} = 80 GN/m² and C_{bronze} = 40 GN/m² find:*

(i) Internal diameter of the steel shaft;

(ii) Total angle of twist for whole of the shaft.

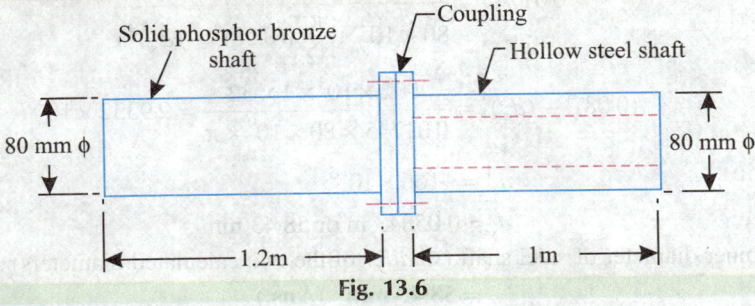

Fig. 13.6

Solution. *Given:* Diameter of solid bronze shaft,

$$D_b = 80 \text{ mm} = 0·08 \text{ m}$$

External diameter of the hollow steel shaft,

$$D_s = 80 \text{ mm} = 0·08 \text{ m}$$

Length of solid bronze shaft, $\quad l_b = 1·2 \text{ m}$

Length of hollow steel shaft, $\quad l_s = 1 \text{ m}$

Maximum shear stress developed in bronze shaft,
$$\tau_b = 40 \text{ MN/m}^2$$

Maximum shear stress developed in steel shaft,
$$\tau_s = 72 \text{ MN/m}^2$$

Angle of twist, $\theta_s = \theta_b = 1° = 1 \times \pi/180 = 0\cdot01745$ rad.

Modulus of rigidity for steel, $C_s = 80 \text{ GN/m}^2$

Modulus of rigidity for phosphor bronze,
$$C_b = 40 \text{ GN/m}^2.$$

(i) Internal diameter of the steel shaft, d_s:

Torque transmitted by the phosphor bronze shaft,

$$T_b = \frac{\pi}{16} D_b^3 \cdot \tau_b$$

$$= \frac{\pi}{16} \times (0\cdot08)^3 \times 40 \times 10^6 \times (10^{-3}) \text{ kNm} = 4\cdot02 \text{ kNm}$$

∴ Torque transmitted, $\qquad T = T_b = T_s = 4\cdot02 \text{ kNm}$

Internal diameter of the steel shaft on the basis of shear stress.

$$T_s = \frac{\pi}{16}\left[\frac{D_s^4 - d_s^4}{D_s}\right] \times \tau_s$$

$$4\cdot02 \times 10^3 = \frac{\pi}{16}\left[\frac{(0\cdot08)^4 - (d_s)^4}{0\cdot08}\right] \times 72 \times 10^6$$

or, $\qquad (0\cdot08)^4 - (d_s)^4 = \dfrac{4\cdot02 \times 10^3 \times 16 \times 0\cdot08}{\pi \times 72 \times 10^6} = 2\cdot2748 \times 10^{-5}$

or, $\qquad (d_s)^4 = (0\cdot08)^4 - 2\cdot2748 \times 10^{-5} = 1\cdot826 \times 10^{-5}$

∴ $\qquad d_s = 0\cdot06537 \text{ or } 65\cdot37 \text{ mm}$

Internal diameter of the steel shaft on the basis of angle of twist:

We know that, $\qquad \theta_s = \dfrac{T_s \, l_s}{C_s \, I_{P_s}}$

∴ $\qquad 0\cdot01745 = \dfrac{4\cdot02 \times 10^3 \times 1}{80 \times 10^9 \times \dfrac{\pi}{32}((0\cdot08)^4 - (d_s)^4)}$

or, $\qquad (0\cdot08)^4 - (d_s)^4 = \dfrac{4\cdot02 \times 10^3 \times 1 \times 32}{0\cdot01745 \times 80 \times 10^9 \times \pi} = 2\cdot9332 \times 10^{-5}$

or, $\qquad d_s^4 = 1\cdot166 \times 10^{-5}$

or, $\qquad d_s = 0\cdot05843 \text{ m or } 58\cdot43 \text{ mm}$

∴ The inner diameter of steel shaft (*smaller of the two calculated diameters*)

$$= \textbf{58·43 mm} \quad \textbf{(Ans.)}$$

(ii) Total angle of twist, θ:

$$\theta = \theta_s + \theta_b$$

$$= \frac{T_s \, l_s}{C_s \, I_{ps}} + \frac{T_b \, l_b}{C_b \, I_{pb}} = T\left[\frac{l_s}{C_s \, I_{ps}} + \frac{l_b}{C_b \, I_{pb}}\right]$$

$$(\because T = T_s = T_b)$$

$$= 4.02 \times 10^3 \left[\frac{1}{80 \times 10^9 \times \dfrac{\pi}{32} \{(0.08)^4 - (0.05843)^4\}} \right.$$

$$\left. + \frac{1.2}{40 \times 10^9 \times (\pi/32) \times (0.08)^4} \right]$$

$$= 4.02 \times 10^3 [4.3449 \times 10^{-6} + 7.46 \times 10^{-6}] = 0.0474 \text{ rad.}$$

$$\therefore \qquad \theta = 0.0474 \times \frac{180}{\pi} = \mathbf{2.719°} \quad \textbf{(Ans.)}$$

Example 13·19. *A solid alloy shaft 50 mm diameter is to be coupled in series with a hollow steel shaft of the same external diameter. If the angle of twist per unit length of the steel shaft is to be 70 percent of that of the alloy shaft find the internal diameter of the steel shaft.*

Also find the speed at which the shafts should be driven to transmit 20 kW if allowable shearing stresses in alloy and steel are 56 MN/m² and 80 MN/m² respectively.

Take: $\qquad C_{steel} = 2.25\ C_{alloy}$

Solution. *Given:* Diameter of the solid alloy shaft,

$$D_a = 50 \text{ mm} = 0.05 \text{ m}$$

External diameter of the hollow steel shaft,

$$D_s = 50 \text{ mm} = 0.05 \text{ m}$$

Allowable shearing stresses:

For alloy shaft, $\qquad \tau_a = 56 \text{ MN/m}^2$

For steel shaft, $\qquad \tau_s = 80 \text{ MN/m}^2$

$$C_s = 2.25\ C_a.$$

Internal diameter of the hollow steel shaft, d_s:

Angle of twist per unit length of a shaft is given by

$$\frac{\theta}{l} = \frac{T}{C I_p}$$

$$\therefore \qquad \frac{T_s}{C_s\ I_{ps}} = 0.7\ \frac{T_a}{C_a\ I_{pa}} \qquad \qquad \text{...(Given)}$$

A large shaft is in process on a big lathe.

or,
$$\frac{I_{pa}}{I_{ps}} = 0.7 \times \frac{C_s}{C_a}$$

$\therefore$
$$\frac{\frac{\pi}{32} \times 0.05^4}{\frac{\pi}{32} \times (0.05^4 - d_s^4)} = 0.7 \times 2.25$$

or,
$$0.05^4 - d_s^4 = \frac{0.05^4}{0.7 \times 2.25} \quad 3.968 \times 10^{-6}$$

or,
$$d_s^4 = 0.05^4 - 3.968 \times 10^{-6} = 2.282 \times 10^{-6}$$

$\therefore$
$$d_s = 0.03887 \text{ m or } \textbf{38.87 mm} \quad \textbf{(Ans.)}$$

Speed, N:

Also for a shaft, we know that
$$\frac{\tau}{R} = \frac{C\theta}{l}$$

$\therefore$ For steel shaft:
$$\frac{\tau_s}{R_s} = \frac{C_s \theta_s}{l_s} \qquad \qquad \qquad ...(i)$$

and, for alloy shaft:
$$\frac{\tau_a}{R_a} = \frac{C_a \theta_a}{l_a} \qquad \qquad \qquad ...(ii)$$

Dividing (i) by (ii), we get
$$\frac{\tau_s}{\tau_a} = \frac{C_s}{C_a} \times \frac{R_s}{R_a} \times \left[\frac{\theta_s}{l_s} \times \frac{l_a}{\theta_a} \right] = 2.25 \times (1) \times 0.7 = 1.575$$

$\therefore$
$$\tau_s = 1.575 \, \tau_a$$

When,
$$\tau_s = 80 \text{ MN/m}^2$$

$$\tau_a = \frac{80}{1.575} = 50.78 \text{ MN/m}^2$$

[This is less than the permissible stress of 56 MN/m²]

Now the torque can be determined by considering any one of the shafts.

Considering solid alloy shaft,

Torque,
$$T_a = \tau_a \times \frac{\pi}{16} \times D_a^3 = 50.78 \times \frac{\pi}{16} \times (0.05)^3 \times 10^6 \text{ Nm} = 1246 \text{ Nm}$$

$\therefore$ Power transmitted,
$$P = \frac{2\pi NT}{60 \times 10000} \text{ kW} \qquad \qquad \text{(where } T \text{ is Nm)}$$

or,
$$20 = \frac{2\pi N \times 1246}{60 \times 1000}$$

$\therefore$
$$N = \frac{20 \times 60 \times 1000}{2\pi \times 1246} = 153.3 \text{ r.p.m.}$$

Hence, speed
$$N = \textbf{153.3 r.p.m.} \quad \textbf{(Ans.)}$$

Example 13·20. *A solid steel shaft 6 m long is securely fixed at each end. A torque of 1250 Nm is applied to the shaft at a section 2.4 m from one end. What are the fixing torques set up at the ends of the shaft? If the diameter of the shaft is 40 mm what are the maximum shear stresses in the two portions? Calculate also the angle of twist for the section where the torque is applied, modulus of rigidity = 84 GN/m².*

Solution. Refer to Fig. 13·7.

Angle of twist θ:

We know that, in this case, $\theta_1 = \theta_2$

$$\frac{T_1 l_1}{C I_p} = \frac{T_2 l_2}{C I_p}$$

or,

$$T_1 l_1 = T_2 l_2$$

(C and I_p being same for portion 1 and 2)

...(i)

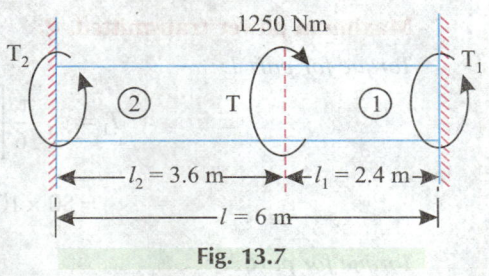

Fig. 13.7

and,

$$T_1 + T_2 = 1250$$ ...(ii)

From (i),

$$T_2 = \frac{T_1 l_1}{l_2} = \frac{T_1 \times 2·4}{3·6}$$

∴

$$T_1 + \frac{T_1 \times 2·4}{3·6} = 1250$$

$$T_1 (1 + 0·667) = 1250$$

∴

$$T_1 = 749·8 \text{ Nm}$$

$$T_2 = 1250 - 749·8 = 500·2 \text{ Nm}$$

$$\theta = \theta_1 = \theta_2 = \frac{749·8 \times 2·4}{84 \times 10^9 \times \pi/32 \times (0·04)^4} = 0·0852 \text{ radian}$$

θ = 4·88 degrees (Ans.)

Maximum shear stresses in the two portions:

$$\tau_1 = \frac{16 T_1}{\pi D^3} = \frac{16 \times 749·8}{\pi \times (0·04)^3} \times 10^{-6}$$

$$= 59·66 \text{ MN/m}^2 \text{ (Ans.)}$$

$$\tau_2 = \frac{16 T_2}{\pi D^3} = \frac{16 \times 500·2}{\pi \times (0·04)^3} \times 10^{-6} = 39·8 \text{ MN/m}^2 \text{ (Ans.)}$$

Example 13·21. *A hollow shaft is 1 m long and has external diameter 50 mm. It has 20 mm internal diameter for a part of length and 30 mm internal diameter for the rest of the length. If the maximum shear stress is not to exceed 80 MN/m², determine the maximum power transmitted by it at a speed of 300 r.p.m. If the twists produced in the two portions of the shaft are equal, find the lengths of the two portions.*

Solution. Refer to fig. 13·8.

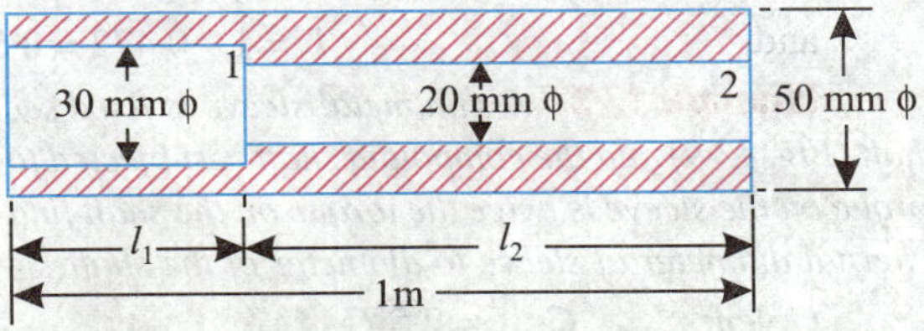

Given: Hollow shaft :

Length of shaft = 1 m;

Diameters of length l_1 are:

50 mm and 30 mm;

Diameters of length l_2 are:

50 mm and 20 mm;

Max. allowable shear stress,

$$\tau = 80 \text{ MN/m}^2;$$

Speed,

$$N = 300 \text{ r.p.m.}$$

Fig. 13.8

Maximum power transmitted, P:

Torque for part 1:

$$T_1 = \tau \cdot \frac{\pi}{16} \left[\frac{(0.05)^4 - (0.03)^4}{0.05} \right]$$

$$= 80 \times 10^6 \times \frac{\pi}{16} \times \left[\frac{(0.05)^4 - (0.03)^4}{0.05} \right] = 1709 \text{ Nm}$$

Torque for part 2:

$$T_2 = 80 \times 10^6 \times \frac{\pi}{16} \times \left[\frac{(0.05)^4 - (0.02)^4}{0.05} \right] = 1913 \text{ Nm}$$

∴ The safe torque which the shaft can transmit

$$= 1709 \text{ Nm}$$

Thus power, transmitted, $P = \dfrac{2\pi NT}{60 \times 1000} = \dfrac{2\pi \times 300 \times 1709}{60 \times 1000}$

or, $\qquad \qquad \mathbf{P = 53.7 \text{ kW} \quad (Ans.)}$

Lengths of two portions, l_1 and l_2 :

From relation: $\qquad \dfrac{T}{I_P} = \dfrac{C\theta}{l}$, we have

$$\theta = \frac{Tl}{CI_p}$$

As per given condition θ is same for both parts

∴ $\qquad \qquad \dfrac{T_1 l_1}{C_1 I_{P_1}} = \dfrac{T_2 l_2}{C_2 I_{P_2}}$

or, $\qquad \qquad l_1 I_{P_2} = l_2 I_{P_1} \qquad \qquad$ (∵ T and C are same for both the parts)

or, $l_1 \times \dfrac{\pi}{32} [(0.05)^4 - (0.02)^4] = l_2 \times \dfrac{\pi}{32} [(0.05)^4 - (0.03)^4]$

or, $\qquad \qquad l_1 \times 6.09 \times 10^{-6} = l_2 \times 5.44 \times 10^{-6}$

Also, $\qquad \qquad l_1 + l_2 = 1$

or, $\qquad \qquad l_2 = 1 - l_1$

∴ $\qquad \qquad 6.09 \times 10^{-6} \, l_1 = 5.44 \times 10^{-6} (1 - l_1)$

or, $\qquad \qquad l_1 = \dfrac{5.44}{11.53} = \mathbf{0.472 \text{ m} \quad (Ans.)}$

and, $\qquad \qquad l_2 = 1 - 0.472 = \mathbf{0.528 \text{ m} \quad (Ans.)}$

Example 13.22. *A gun metal sleeve is fixed securely to a steel shaft (Fig. 13.9) and the compound shaft is subjected to a torque. If the torque on the sleeve is twice the torque on the shaft find the ratio of the external diameter of sleeve to diameter of the shaft.*

Given: $\qquad C_{\text{steel}} = 2.5 \, C_{\text{gun metal}}.$

Solution. Let, D = External diameter of the sleeve,

$\qquad \qquad d$ = Diameter of the shaft,

$\qquad \qquad T_s$ = Torque on steel shaft,

$\qquad \qquad T_g$ = Torque on gun metal sleeve,

$\qquad \qquad = 2T_s$ *(Given)*

Massive steel shafts.

C_s = Modulus of rigidity for steel,

C_g = Modulus of rigidity for gun metal,

I_{p_s} = Polar moment of inertia for steel shaft, and

I_{p_g} = Polar moment of inertia for gun metal sleeve.

Ratio $\dfrac{D}{d}$:

Since the shaft and sleeve are securely fixed together their lengths and the angle of twist at the common surface shall be same *i.e. l and θ for both are the same.*

From the relation,

$$\frac{T}{I_p} = \frac{C\theta}{l} , \text{ we have}$$

$$\frac{\theta}{l} = \frac{T}{CI_p}$$

∴
$$\frac{T_s}{C_s I_{ps}} = \frac{T_g}{C_g I_{pg}}$$

or,
$$\frac{T_s}{T_g} = \frac{C_s I_{ps}}{C_g I_{pg}}$$

But, $T_g = 2T_s$

and, $C_s = 2 \cdot 5 \, C_g$ ...(given)

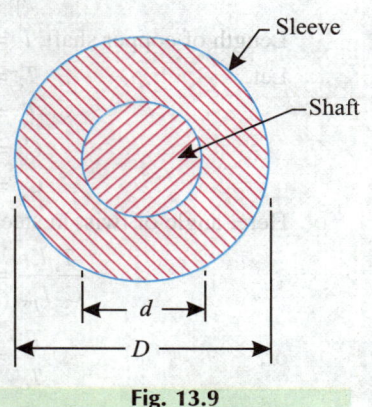

Fig. 13.9

∴
$$\frac{T_s}{2T_s} = 2 \cdot 5 \times \frac{I_{ps}}{I_{pg}} = 2 \cdot 5 \times \frac{(\pi/32) \times d^4}{\dfrac{\pi}{32}(D^4 - d^4)}$$

or, $\dfrac{D^4 - d^4}{d^4} = 5$

or, $D^4 = 6d^4$

or, $\dfrac{D}{d} = (6)^{1/4} = 1 \cdot 565$

Hence, $\dfrac{D}{d} = 1 \cdot 565$ **(Ans.)**

Example 13·23. *Figure 13·10 shows a composite shaft. If 2 kNm torque is applied at the junction determine the maximum shear stress developed in steel and copper shaft. Assume that* $C_{steel} = 2C_{copper}$.

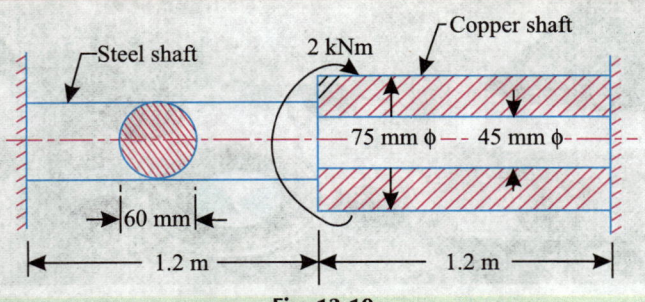

Fig. 13.10

Solution. Diameter of steel shaft,
$$D_s = 60 \text{ mm} = 0.06 \text{ m}$$
∴ Polar moment of inertia,
$$I_{ps} = \frac{\pi}{32} \times 0.06^4 = 1.272 \times 10^{-6} \text{ m}^4$$
Length of steel shaft, $l_s = 1.2$ m
Polar moment of inertia of copper shaft,
$$I_{ps} = \frac{\pi}{32} [(0.75)^4 - (0.045)^4] = 2.704 \times 10^{-6} \text{ m}^4$$
Length of copper shaft, $l_c = 1.2$ m
Let, $\quad\quad\quad T_s$ = Torque shared by steel shaft,
$\quad\quad\quad\quad T_c$ = Torque shared by copper shaft,
$\quad\quad\quad\quad C_s$ = Modulus of rigidity of steel shaft, and
$\quad\quad\quad\quad C_c$ = Modulus of rigidity of copper shaft.
Here, angle of twist in steel shaft (θ_s) = Angle of twist in copper shaft (θ_c)
$$\therefore \quad\quad \frac{T_s l_s}{C_s I_{ps}} = \frac{T_c l_c}{C_c I_{pc}}$$

or, $\quad\quad \dfrac{T_s}{T_c} = \dfrac{C_s}{C_c} \times \dfrac{l_c}{l_s} \times \dfrac{I_{ps}}{I_{pc}} = 2 \times \dfrac{1.2}{1.2} \times \dfrac{1.272 \times 10^{-6}}{2.704 \times 10^{-6}} = 0.94$

i.e. $\quad\quad\quad\quad T_s = 0.94\, T_c$ ...(i)

Also, $\quad\quad\quad T_s + T_c = 2$ ...(ii)

Solving (*i*) and (*ii*), we get
$$T_c = 1.031 \text{ kNm, and } T_s = 0.969 \text{ kNm}$$
Shear stress developed in steel shaft,
$$\tau_s = \frac{T_s \times (D_s/2)}{I_{ps}}$$

A car wheel, although looks simple, needs to be carefully designed.

$$= \frac{0.969 \times (0.06/2)}{1.272 \times 10^{-6}} \times 10^{-3} \text{ MN/m}^2 \qquad \left[\because \frac{T}{I_p} = \frac{\tau}{R} \text{ or } \tau = \frac{TR}{I_p} \right]$$

$$= \textbf{22.85 MN/m}^2 \quad \textbf{(Ans.)}$$

Shear stress developed in copper shaft

$$\tau_c = \frac{T_c \times (D_c/2)}{I_{pc}} = \frac{1.031 \times (0.075/2)}{2.704 \times 10^{-6}} \times 10^{-3} \text{ MN/m}^2$$

$$= \textbf{14.29 MN/m}^2 \quad \textbf{(Ans.)}$$

Example 13.24. *Fig. 13.11 (a) shows a horizontal shaft LM subjected to axial twisting moments.*

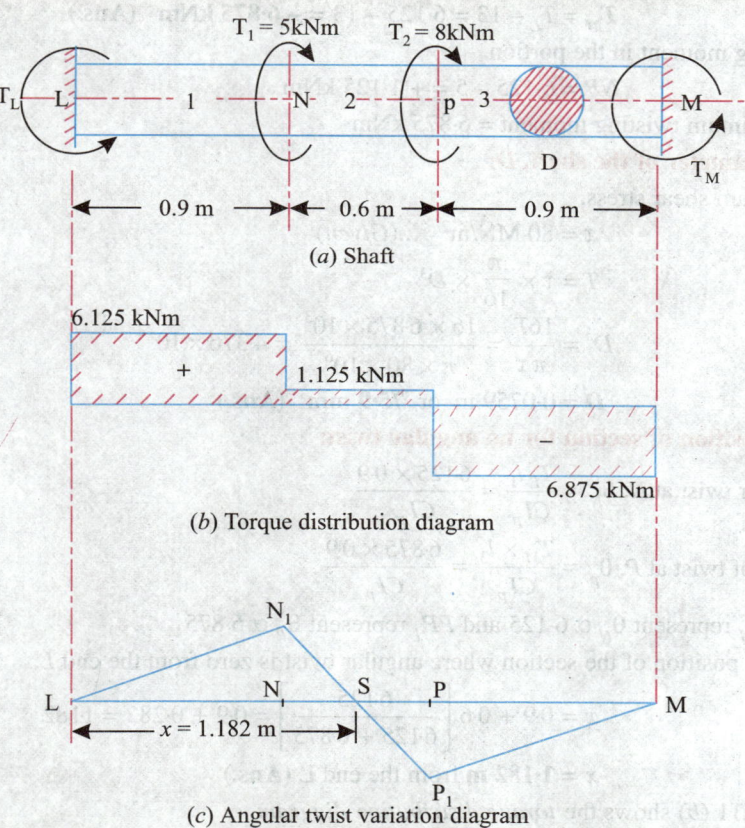

(a) Shaft

(b) Torque distribution diagram

(c) Angular twist variation diagram

Fig. 13.11

Determine :

(i)   *The end fixing couples in magnitude and direction.*

(ii) *The diameter of the shaft if the maximum shearing stress is not to exceed 80 MN/m².*

(iii) *The position of section where the shaft suffers no angular twist.*

Solution. Let, $\qquad T_L$ = End fixing couple at L, and

$\qquad\qquad\qquad T_M$ = End fixing couple at M.

(i)   End fixing couples :

Torque on the portion $\quad NP = T_L - 5$ kNm

Torque on the portion $\quad PM = T_L - 5 - 8 = T_L - 13$ kNm

Total angle of twist between L and M

$$= \theta_1 + \theta_2 + \theta_3 = 0 \qquad\qquad (\because \text{ Both ends are fixed.})$$

$\therefore\qquad \dfrac{T_L \cdot l_1}{CI_p} + \dfrac{(T_L - 5) \, l_2}{CI_p} + \dfrac{(T_L - 13) \, l_3}{CI_p} = 0$

or,$\qquad 0 \cdot 9 \, T_L + 0 \cdot 6 \, (T_L - 5) + 0 \cdot 9 \, (T_L - 13) = 0$

or,$\qquad 0 \cdot 9 \, T_L + 0 \cdot 6 \, T_L - 3 + 0 \cdot 9 \, T_L - 11 \cdot 7 = 0$

or,$\qquad\qquad 2 \cdot 4 \, T_L = 14 \cdot 7$

or,$\qquad\qquad \boldsymbol{T_L = 6 \cdot 125 \text{ kNm} \quad \text{(Ans.)}}$

and,$\qquad\qquad T_M = T_L - 13 = 6 \cdot 125 - 13 = \boldsymbol{-6 \cdot 875 \text{ kNm} \quad \text{(Ans.)}}$

Twisting moment in the portion

$$NP = 6 \cdot 125 - 5 = + 1 \cdot 125 \text{ kNm}$$

$\therefore$ Maximum twisting moment $= 6 \cdot 875$ kNm

(ii) Diameter of the shaft, D:

Maximum shear stress,

$$\tau = 80 \text{ MN/m}^2 \quad ...(Given)$$

$$T = \tau \times \frac{\pi}{16} \times D^3$$

$\therefore\qquad D^3 = \dfrac{16T}{\pi \tau} = \dfrac{16 \times 6 \cdot 875 \times 10^3}{\pi \times 80 \times 10^6} = 4 \cdot 376 \times 10^{-4}$

or $\qquad\qquad D = 0 \cdot 0759 \text{ m or } \boldsymbol{75 \cdot 9 \text{ mm} \quad \text{(Ans.)}}$

(iii) Position of section for no angular twist:

Angular twist at N, $\theta_N = \dfrac{T_L \, l_1}{CI_p} = \dfrac{6 \cdot 125 \times 0 \cdot 9}{CI_p}$

Angle of twist at P, $\theta_p = \dfrac{T_M \times l_3}{CI_p} = \dfrac{6 \cdot 875 \times 0 \cdot 9}{CI_p}$

Let NN_1 represent $\theta_N \, \alpha \, 6 \cdot 125$ and PP_1 represent $\theta_p \, \alpha \, 6 \cdot 875$

$\therefore$ The position of the section where angular twist is zero from the end L,

$$x = 0 \cdot 9 + 0 \cdot 6 \left[\frac{6 \cdot 125}{6 \cdot 125 + 6 \cdot 875} \right] = 0 \cdot 9 + 0 \cdot 282 = 1 \cdot 182$$

Hence,$\qquad\qquad \boldsymbol{x = 1 \cdot 182 \text{ m}}$ from the end L (Ans.)

Fig. 13.11 (b) shows the *torque distribution diagram.*

Fig. 13.11 (c) shows the *angular twist variation diagram.*

13·12. TORSIONAL RESILIENCE

Consider a hollow shaft of external diameter D, internal diameter d, and length l, subjected to a gradually applied torque T. Let θ be the angle of twist. Energy is stored in the shaft due to this angular distortion. This is called torsional energy or the *torsional resilience*, denoted by U.

Torsional energy = Work done by the torque = Average torque × angular twist.

i.e.$\qquad\qquad U = \dfrac{T}{2} \cdot \theta \qquad\qquad\qquad ...(13 \cdot 19)$

From torsion equation:

$$\frac{T}{I_p} = \frac{\tau}{R} = \frac{2\tau}{D}$$

or,
$$T = I_p \frac{2\tau}{D} = \frac{\pi}{32}(D^4 - d^4)\frac{2\tau}{D} = \frac{\pi\tau}{16}\left[\frac{D^4 - d^4}{D}\right]$$

and,
$$\frac{\tau}{R} = \frac{C\theta}{l} \quad \text{or} \quad \frac{2\tau}{D} = \frac{C\theta}{l}$$

$$\therefore \qquad \theta = \frac{2\tau}{D} \cdot \frac{l}{C}$$

Substituting the value of T and θ in eqn. (13.19), we get

$$U = \frac{1}{2} \times \frac{\pi}{16} \cdot \tau \left(\frac{D^4 - d^4}{D}\right) \times \frac{2\tau}{D} \cdot \frac{l}{C} = \frac{\tau^2}{4C}\left[\frac{D^2 + d^2}{D^2}\right]\left[\frac{D^2 - d^2}{4}\right]\pi l$$

$$= \frac{\tau^2}{4C}\left[\frac{D^2 + d^2}{D^2}\right] \times \text{volume of the shaft} \qquad ...(13\cdot20)$$

Average torsional energy/unit volume

$$= \frac{\tau^2}{4C} \cdot \frac{D^2 + d^2}{D^2} \qquad ....(13\cdot21)$$

For solid shaft, $\qquad d = 0$

$$U = \frac{\tau^2}{4C} \times \text{volume of shaft} \qquad ...(13\cdot22)$$

For a very thin hollow shaft D is nearly equal to d

In that case, $\qquad U = \frac{\tau^2}{2C} \times \text{volume of shaft} \qquad ...(13\cdot23)$

Example 13·25. *A hollow shaft subjected to pure torque, attains a maximum shear stress* τ.
If the strain energy per unit volume is $\dfrac{\tau^2}{3C}$ *calculate the ratio of shaft diameters.*

Solution. Let, $\qquad\qquad D = $ External diameter of the shaft, and

$\qquad\qquad\qquad\qquad d = $ Internal diameter of the shaft.

Strain energy per unit volume, $\quad U = \dfrac{\tau^2}{3C} \qquad\qquad ...(Given)$

Ratio of shaft diameters D/d :

For hollow shaft the strain energy is

$$U = \frac{\tau^2}{4C}\left[\frac{D^2 + d^2}{D^2}\right] \times \text{volume of the shaft}$$

or, strain energy per unit volume $= \dfrac{\tau^2}{4C}\left[\dfrac{D^2 + d^2}{D^2}\right] = \dfrac{\tau^2}{3C} \qquad ...(Given)$

i.e. $\qquad\qquad \dfrac{\tau^2}{4C}\left[\dfrac{D^2 + d^2}{D^2}\right] = \dfrac{\tau^2}{3C}$

or, $\qquad\qquad \dfrac{D^2 + d^2}{D^2} = \dfrac{4}{3} \quad \text{or} \quad D^2 = 3d^2$

$\therefore \qquad\qquad \dfrac{D}{d} = \sqrt{3} = \mathbf{1\cdot732} \quad \textbf{(Ans.)}$

Example 13·26. *Two shafts are of length l and outside diameter D, the first one is solid while the second one is hollow with inside diameter D/2. What is the ratio of strain energies that two steel shafts can absorb without exceeding the allowable shear stress?*

Solution. Diameter of the solid shaft, $D_S = D$

Outside diameter of hollow shaft $= D_H = D$

Inside diameter of hollow shaft, $d_H = D/2$

Length of each shaft $= l$

Torque in each shaft $= T$

Maximum shear stress on surface of each shaft $= \tau$

Modulus of rigidity of each shaft $= C$

Volume of solid shaft, $V_S = \dfrac{\pi}{4} \times D^2 \times l$

Circular shaft.

Volume of hollow shaft, $V_H = \dfrac{\pi}{4}\,[D^2 - (D/2)^2] \times l = \dfrac{3\pi}{16}\,D^2\,l$

For *solid shaft* :

$$U_S = \frac{\tau^2}{4C} \times V_S$$

For *hollow shaft* :

$$U_H = \frac{\tau^2}{4C} \times \left[\frac{D_H^2 + d_H^2}{D_H^2}\right] \times V_H$$

$$= \frac{\tau^2}{4C} \left[\frac{D^2 + (D/2)^2}{D^2}\right] \times V_H = \frac{5\,\tau^2}{16C} \times V_H$$

$$\therefore \quad \frac{U_S}{U_H} = \frac{\dfrac{\tau^2}{4C} \times V_S}{\dfrac{5\,\tau^2}{16C} \times V_H} = \frac{4}{5} \times \frac{V_S}{V_H} = \frac{4}{5} \times \frac{\dfrac{\pi}{4} \times D^2 \times l}{\dfrac{3\pi}{16}\,D^2\,l} = \frac{16}{15}$$

Hence, $\qquad \dfrac{U_S}{U_H} = \dfrac{16}{15}$ **(Ans.)**

13·13. SHAFT COUPLINGS

A number of devices are used to connect sections of co-axial shafting. The simplest and most widely used of these devices is the bolted flanged couplings. This type of coupling provides a clear application of torsional shearing stress induced by power transmission. Fig. 13·12 represents such a coupling. The torque is transmitted from one shaft to the other as follows:

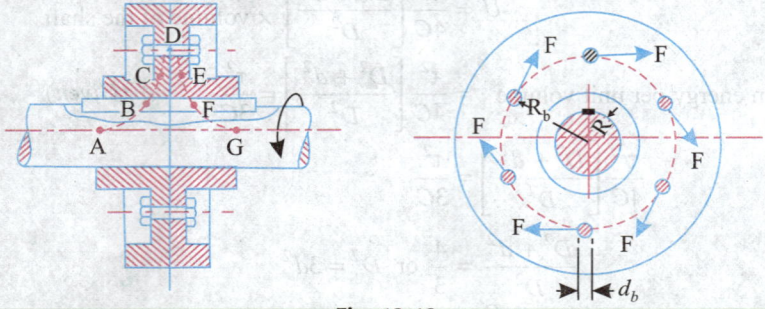

Fig. 13.12

— From shaft A to key B,
— From key B to the left side of the coupling C,
— From the left side of the coupling C to six bolts D,
— From the bolts D to the right side of the coupling E,
— From the right side of the coupling E to the key F,
— From key F to the other shaft G.

Bolts. The transmitted torque will develop a resisting torque in the bolts equal and opposite to it.

$$T = F \times R_b, \text{ where } F \text{ is the resisting force in each bolt.}$$

$$T = n_b \times \frac{\pi}{4} \times d_b^2 \times \tau_b \times R_b \qquad ...(13\cdot24)$$

where,

n_b = Number of bolts,

d_b = Diameter of bolt,

τ_b = Shear stress in bolt, and

R_b = Radius of bolt–circle.

Keys. There is tendency for the key to shear on the cross hatched rectangular area [Fig. 13·13 (a)] at the surface of the shaft. The force F which produces shear at section AA [Fig. 13·13 (b)] is equal to T/R (R being radius of the shaft).

Thus, $$T = l_k \times b_k \times \tau_k \times R \qquad ...(13\cdot25)$$

where l_k, b_k and τ_k are the length, breadth and maximum permissible shear stress in the key respectively.

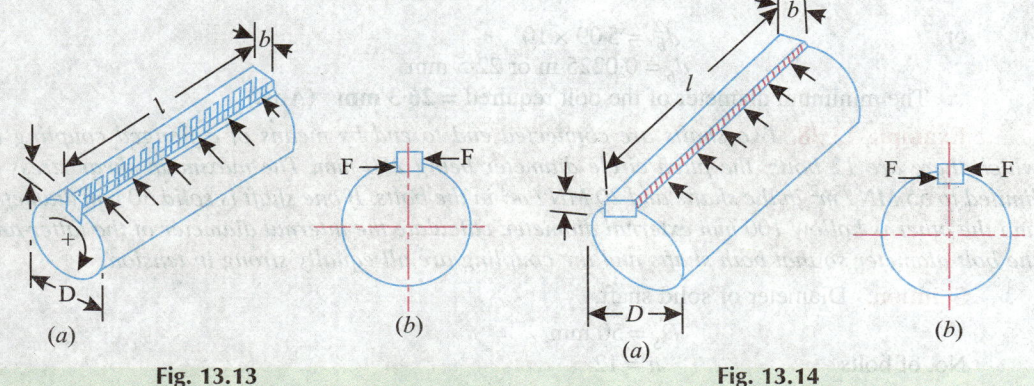

Fig. 13.13   **Fig. 13.14**

The analysis of the key may be carried further to determine the bearing/crushing stress. The force F will tend to crush the key at the cross-hatched area [Fig. 13·14 (a)]. This force may be taken as the same force in shear analysis of the key. Therefore, the bearing stress developed in the key is

$$\sigma_{ck} = \frac{F}{\text{Area}} = \frac{F}{l_k \cdot \dfrac{t_k}{2}} = \frac{T}{\dfrac{t_k l_k R}{2}}$$

$$\therefore \quad T = l_k \cdot \frac{t_k}{2} \, \sigma_{ck} \times R \qquad ...(13\cdot26)$$

Example 13·27. *A shaft is to be fitted with a flanged coupling having 8 bolts on a circle of diameter 150 mm. The shaft may be subjected to either a direct tensile load of 400 kN or a twisting moment of 18 kNm. If the maximum direct and shearing stresses permissible in the bolt material are 125 MN/m² and 55 MN/m² respectively find the minimum diameter of the bolt required. Assume that each bolt takes an equal share of the load or torque.*

Solution. Number of bolts,

$$n = 8$$

Radius of the bolt circle,

$$R_b = \frac{150}{2} = 75 \text{ mm} = 0.075 \text{ m}$$

Twisting moment, $T = 18$ kNm

Direct tensile load $= 400$ kN

$$\sigma_t = 125 \text{ MN/m}^2$$

$$\tau_b = 55 \text{ MN/m}^2$$

Diameter of bolt, d_b :

Strength due to shearing effect :

$$T = n \times \frac{\pi}{4} \times d_b^2 \times \tau_b \times R_b$$

$$18 \times 1000 = 8 \times \frac{\pi}{4} \times d_b^2 \times 55 \times 10^6 \times 0.075$$

∴ $d_b = 0.0263$ m or 26·3 mm

Strength due to tensile load :

$$400 \times 1000 = n \times \frac{\pi}{4} d_b^2 \times \sigma_t$$

or $$= 8 \times \frac{\pi}{4} \times d_b^2 \times 125 \times 10^6$$

or $$d_b^2 = 5.09 \times 10^{-4}$$

∴ $d_b = 0.0225$ m or 22·5 mm

∴ The minimum diameter of the bolt required = **26·3 mm   (Ans.)**

Gear shaft.

Example 13·28. *Two shafts are connected end to end by means of a flanged coupling in which there are 12 bolts, the pitch circle diameter being 250 mm. The maximum shear stress is limited to 55 MN / m² in the shafts and 20 MN / m² in the bolts. If one shaft is solid 50 mm diameter and the other is hollow 100 mm external diameter, calculate the internal diameter of the latter and the bolt diameter so that both shafts and the coupling are all equally strong in tension.*

Solution.   Diameter of solid shaft,

$$D_S = 50 \text{ mm}$$

No. of bolts $n = 12$

Bolt circle radius, $$R_b = \frac{250}{2} = 125 \text{ mm or } 0.125 \text{ m}$$

Maximum shear stress in the shafts,

$$\tau = 55 \text{ MN/m}^2$$

Maximum shear stress in the bolts,

$$\tau_b = 20 \text{ MN/m}^2$$

Diameter of the bolt, d_b :

We know that, $$T = \tau \times \pi/16 \times D_s^3 = 55 \times 10^6 \times \pi/16 \times (0.05)^3 = 1350 \text{ Nm}$$

Also, $$T = n \times \frac{\pi}{4} \times d_b^2 \times \tau_b \times R_b$$

$$1350 = 12 \times \frac{\pi}{4} \times d_b^2 \times 20 \times 10^6 \times 0.125$$

∴   **d_b = 0.0075 m or 7·5 mm   (Ans.)**

Internal diameter of the hollow shaft, d_H :

Using the relation, $T = \tau \cdot \dfrac{\pi}{16} \left[\dfrac{D_H^4 - d_H^4}{D_H} \right]$, we get

$$1350 = 55 \times 10^6 \times \frac{\pi}{16} \left[\frac{0 \cdot 1^4 - d_H^4}{0 \cdot 1} \right]$$

$$1 \cdot 25 \times 10^{-5} = (0 \cdot 1)^4 - d_H^4$$

$$d_H^4 = 8 \cdot 75 \times 10^{-5}$$

∴ $\qquad \mathbf{d_H} = 0 \cdot 0967 \text{ m or } \mathbf{96 \cdot 7 \text{ mm}}$ **(Ans.)**

13·14. COMBINED BENDING AND TORSION

Generally we assume that shaft is subjected to torsion only but in actual practice due to weight of the pulley, couplings, pull in belts or ropes etc. the shaft is subjected to bending too. Thus, actually, in the shaft, both the '*shear stress*' due to torsion and '*direct stress*' due to bending are induced.

As per torsion equation, $\qquad \dfrac{T}{I_p} = \dfrac{\tau}{R}$

where, $\qquad T = \tau \times \dfrac{\pi}{16} \times D^3$ (in case of solid shaft)

Also the *bending stress* is given by the relation,

$$\frac{M}{I} = \frac{\sigma_b}{y} \quad \text{where, } \sigma_b \text{ is the bending stress}$$

or, $\qquad M = \dfrac{\sigma_b \cdot I}{y} = \dfrac{\sigma_b \times \pi D^4}{64 \times D/2}$

or, $\qquad M = \dfrac{\sigma_b \cdot \pi D^3}{32}$ (in case of solid shaft) ...(13·27)

If a certain material is loaded in such a way that at a point a direct stress σ_d (due to bending), and a shear stress τ in the plane of σ_d is induced then the maximum principal stress σ_{max} and maximum τ_{max} in the material are given by

$$\sigma_{max} = \frac{\sigma_d}{2} + \sqrt{\left(\frac{\sigma_d}{2}\right)^2 + \tau^2} \qquad \qquad ...(13 \cdot 28)$$

and, $$\tau_{max} = \sqrt{\left(\frac{\sigma_d}{2}\right)^2 + \tau^2} \qquad \qquad ...(13 \cdot 29)$$

Again, $$\sigma_{max} = \frac{\sigma_d}{2} + \sqrt{\left(\frac{\sigma_d}{2}\right)^2 + \tau^2}$$

Multiplying both sides by $\dfrac{\pi D^3}{32}$, we get

$$\sigma_{max} \cdot \frac{\pi D^3}{32} = \frac{\sigma_d}{2} \cdot \frac{\pi D^3}{32} + \sqrt{\left(\frac{\sigma_d}{2} \cdot \frac{\pi D^3}{32}\right)^2 + \left(\tau \cdot \frac{\pi D^3}{32}\right)^2}$$

$$M_e = \frac{M}{2} + \sqrt{\left(\frac{M}{2}\right)^2 + \left(\frac{T}{2}\right)^2} \qquad \qquad \left(\because T = \tau \times \frac{\pi}{16} \times D^3\right)$$

or,
$$M_e = \frac{M + \sqrt{M^2 + T^2}}{2} \qquad \qquad ...(13\cdot30)$$

where, M_e is the *equivalent bending moment*.

Also,
$$\tau_{max} = \sqrt{\left(\frac{\sigma_d}{2}\right)^2 + (\tau)^2}$$

Multiplying both sides by $\dfrac{\pi D^3}{16}$, we get

$$\tau_{max} \times \frac{\pi D^3}{16} = \sqrt{\left(\frac{\sigma_d}{2} \cdot \frac{\pi D^3}{16}\right)^2 + \left(\tau \frac{\pi D^3}{16}\right)^2}$$

$$T_e = \sqrt{M^2 + T^2} \qquad \qquad ...(13\cdot31)$$

where, T_e is the *equivalent torque*.

T_e is also used to evaluate the magnitude of the maximum shear stress. It does not give the location of the planes of maximum shear stress which are at angles of 45° to the principal planes.

Example 13·29. *A hollow shaft is subjected to a torque of 40 kNm and a bending moment of 30 kNm. The internal diameter of the shaft is one-half the external diameter. If the maximum shear stress is not to exceed 80 MN/m², find the diameter of the shaft.*

Solution. Let, D = External diameter of the shaft, and

d = Internal diameter of the shaft.

Given: $d = 0.5D$;

$T = 40$ kNm;

$M = 30$ kNm.

We know that,

$$T_e = \sqrt{M^2 + T^2} = \sqrt{30^2 + 40^2} = 50 \text{ kNm}$$

Also
$$T_e = \tau \cdot \frac{\pi}{16} \left[\frac{D^4 - d^4}{D}\right]$$

$$50 \times 10^3 = 80 \times 10^6 \times \frac{\pi}{16}\left[\frac{D^4 - (0\cdot5\,D)^4}{D}\right]$$

or,
$$\frac{D^4 - (0\cdot5\,D)^4}{D} = \frac{50 \times 10^3 \times 16}{80 \times 10^6 \times \pi} = 3\cdot183 \times 10^{-3}$$

or,
$$0\cdot9375\,D^3 = 3\cdot183 \times 10^{-3}$$

$$D^3 = 3\cdot3952 \times 10^{-3} \text{ m}^3$$

∴ $\mathbf{D = 0\cdot15 \text{ m} = 150 \text{ mm}}$ **(Ans.)**

and, $\mathbf{d = 150/2 = 75 \text{ mm}}$ **(Ans.)**

A light-weight cam gear made to fit on shafts of IC engines.

Example 13·30. *A solid shaft is subjected to a bending moment of 2·3 kNm and a twisting moment of 3·45 kNm. Find the diameter of the shaft if the permissible tensile and shear stresses for the material of the shaft are limited to 703 and 421·8 MN/m² respectively.*

Solution. *Given*: $M = 2\cdot3$ kNm;

$T = 3\cdot45$ kNm.

$$T_e = \sqrt{M^2 + T^2}$$

$$= \sqrt{2 \cdot 3^2 + 3 \cdot 45^2} = 4 \cdot 15 \text{ kNm}$$

Also, $\qquad M_e = \dfrac{M + \sqrt{M^2 + T^2}}{2} = \dfrac{2 \cdot 3 + \sqrt{2 \cdot 3^2 + 3 \cdot 45^2}}{2} = 3 \cdot 22 \text{ kNm}$

When permissible shear stress, $\tau = 421 \cdot 8$ MN/m^2:

We know that, $\qquad T_e = \tau \cdot \dfrac{\pi}{16} D^3$

$$4 \cdot 15 \times 10^3 = 421 \cdot 8 \times 10^6 \times \dfrac{\pi}{16} D^3$$

∴ $\qquad\qquad D = 0 \cdot 037$ m or 37 mm

When permissible tensile stress, σ_t, or, $\qquad \sigma_b = 703$ MN/m^2 :

We know, $\qquad \dfrac{M_e}{I} = \dfrac{\sigma_t}{y} \quad \dfrac{M_e}{\dfrac{\pi D^4}{64}} = \dfrac{\sigma_t}{\dfrac{D}{2}}$ or $M_e = \dfrac{\pi}{32} \sigma_t \cdot D^3$

$$3 \cdot 22 \times 10^3 = \dfrac{\pi}{32} \times 703 \times 10^6 \times D^3$$

$$D = 0 \cdot 036 \text{ m or 36 mm}$$

Hence, *the diameter of the shaft will be taken as* **37 mm,** *being the greater value* **(Ans.)**

Example 13·31. *The maximum normal stress and the maximum shear stress analysed for a shaft of 150 mm diameter under combined bending and torsion, were found to be 120 MN/m^2 and 80 MN/m^2 respectively. Find the bending moment and torque to which the shaft is subjected.*

If the maximum shear stress be limited to 100 MN/m^2, find by how much the torque can be increased if the bending moment is kept constant. **(AMIE Summer, 2002)**

Solution. *Given:* $\sigma_{max} = 120$ MN/m^2; $\tau_{max} = 80$ MN/m^2; $d = 150$ mm $= 0 \cdot 15$ m; $\tau_{max} = 100$ MN/m^2

Part I: **M : ; T :**

We know that for combined bending and torsion, we have the following expressions:

$$\sigma_{max} = \dfrac{16}{\pi d^3} \left[M + \sqrt{M^2 + T^2} \right] \qquad \qquad ...(i)$$

and, $\qquad \tau_{max} = \dfrac{16}{\pi d^3} \left[\sqrt{M^2 + T^2} \right] \qquad \qquad ...(ii)$

Substituting the given values in the above equations, we have

$$120 = \dfrac{16}{\pi \times (0 \cdot 15)^3} \left[M + \sqrt{M^2 + T^2} \right] \qquad \qquad ...(iii)$$

$$80 = \dfrac{16}{\pi \times (0 \cdot 15)^3} \left[\sqrt{M^2 + T^2} \right] \qquad \qquad ...(iv)$$

or, $\qquad \sqrt{M^2 + T^2} = \dfrac{80 \times \pi \times (0 \cdot 15)^3}{16} = 0 \cdot 053 \qquad \qquad ...(v)$

Substituting this value in eqn. (*iii*), we get

$$120 = \dfrac{16}{\pi \times (0 \cdot 150^3)} [M + 0 \cdot 053]$$

∴ $\qquad M = \dfrac{120 \times \pi \times (0 \cdot 15)^3}{16} - 0 \cdot 053 = \mathbf{0 \cdot 0265} \text{ MNm} \textbf{ (Ans.)}$

Substituting for M in eqn. (v), we have

$$\sqrt{(0.0265)^2 + T^2} = 0.053$$

or, $\qquad (0.0265)^2 + T^2 = 0.002809$

or, $\qquad T = [0.002809 - (0.0265)^2]^{1/2} = \mathbf{0.0459\ MNm}$ **(Ans.)**

Part II: Increase in torque:

Bending moment (M) to be kept constant = 0.0265 MNm (already calculated)

Using the relation as given in eqn. (ii), we get

$$100 = \frac{16}{\pi \times (0.15)^3} \left[\sqrt{(0.0265)^2 + T^2} \right]$$

or, $\qquad (0.0265)^2 + T^2 = \left[\dfrac{100 \times \pi \times (0.15)^3}{16} \right]^2 = 0.004391$

∴ $\qquad T = [0.004391 - (0.0265)^2]^{1/2} = 0.0607$ MNm

∴ *The increased torque* $\quad = 0.0607 - 0.0459 = \mathbf{0.0148\ MNm}$ **(Ans.)**

Example 13·32. *An electric generator rotates at 200 r.p.m. and receives 260 kW from the driving engine. The armature is 610 mm long and is located between two bearings 1·22 m apart (centre to centre of bearing surface). Owing to the combined weight of armature and magnetic pull of the poles, the shaft is subjected to a force of 82 kN acting at right angles to the shaft. The shaft is forged from steel that has ultimate tensile and shear strengths of 435 MN / m^2 and 380 MN / m^2 respectively. Determine the diameter of the shaft for a factor of safety 6.*

Solution. Refer to Fig. 13·15.

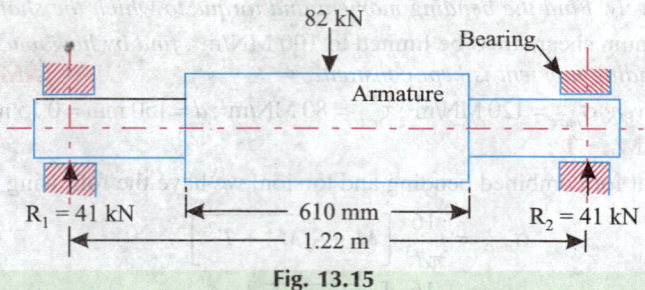

Fig. 13.15

Shaft for reduction gear.

$$R_1 = R_2 = \frac{82}{2} = 41 \text{ kN}$$

The maximum bending moment will occur at the centre and its value is given by,

$$M = 41 \times \frac{1 \cdot 22}{2} = 25 \cdot 01 \text{ kNm}$$

$$P = \frac{2\pi NT}{60} \text{ kW (where } T \text{ is in kNm)}$$

or,

$$260 = \frac{2\pi \times 200 \times T}{60}$$

∴

$$T = \frac{260 \times 60}{2\pi \times 200} = 12 \cdot 41 \text{ kNm}$$

Now,

$$T_e = \sqrt{M^2 + T^2} = \sqrt{(25 \cdot 01)^2 + (12 \cdot 41)^2} = 27 \cdot 92 \text{ kNm}$$

Also,

$$T_e = \tau \cdot \frac{\pi}{16} \times D^3 \text{, where } \tau \text{ is the permissible shear stress}$$

$$27 \cdot 92 \times 10^3 = \left(\frac{380}{6} \times 10^6\right) \times \frac{\pi}{16} D^3$$

from which,

$$D = 0 \cdot 131 \text{ m or } \textbf{131 mm} \quad \textbf{(Ans.)}$$

We know that,

$$M_e = \frac{M + \sqrt{M^2 + T^2}}{2}$$

$$= \frac{25 \cdot 01 + \sqrt{(25 \cdot 01)^2 + (12 \cdot 41)^2}}{2} = 26 \cdot 46 \text{ kNm}$$

Also,

$$M_e = \frac{\pi}{32} D^3 \cdot \sigma_t$$

$$26 \cdot 46 \times 10^3 = \frac{\pi}{32} D^3 \times \left(\frac{435}{6}\right) \times 10^6$$

from which,

$$D = 0 \cdot 155 \text{ m or } 155 \text{ mm}$$

The diameter of the shaft will be taken as **155 mm,** *being higher value.* **(Ans.)**

Example 13·33. *A solid phosphor bronze shaft 60 mm in diameter is rotating at 800 r.p.m. It is subjected to torsion only. An electrical resistance strain gauge mounted on the surface of the shaft with its axis at 45° to the shaft axis, gives the strain reading as $3·98 \times 10^{-4}$. If the modulus of elasticity for bronze is 105 GN/m² and Poisson's ratio is 0·3, find the power being transmitted.*

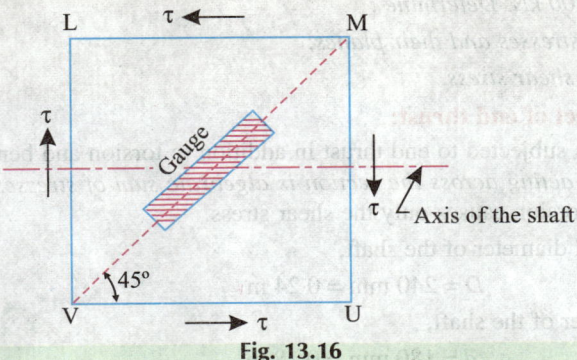

Fig. 13.16

Solution. Diameter of the shaft,
$$D = 60 \text{ mm} = 0.06 \text{ m}$$

Speed of the shaft, $\quad N = 800 \text{ r.p.m.}$

Strain gauge reading $\quad = 3.98 \times 10^{-4}$

Modulus of elasticity for bronze,
$$E_b = 105 \text{ GN/m}^2$$

Poisson's ratio, $\quad \dfrac{1}{m} = 0.3$

Power transmitted, P:

If maximum shear stress on the cross-sectional plane *MU* is τ, then (refering to chapter 2)

Pulley shaft.

Principal stress along $\quad VM = \sigma_1 = -\dfrac{1}{2}\sqrt{4\tau^2} = -\tau$ (*i.e.* comp.)

Principal stress along $\quad LU = \sigma_2 = \dfrac{1}{2}\sqrt{4\tau^2} = \tau$ (*i.e.* tensile)

Thus magnitude of the compressive strain along *VM* is
$$= \dfrac{\tau}{E}\left(1 + \dfrac{1}{m}\right) = 3.98 \times 10^{-4}$$

i.e.
$$\tau = \dfrac{3.98 \times 10^{-4} \times E}{\left(1 + \dfrac{1}{m}\right)}$$

$$= \dfrac{3.98 \times 10^{-4} \times 105 \times 10^9}{(1 + 0.3)} \times 10^{-6} \text{ MN/m}^2 = 32.15 \text{ MN/m}^2$$

∴ Torque being transmitted,
$$T = \tau \times \dfrac{\pi}{16} D^3 = 32.15 \times 10^6 \times \dfrac{\pi}{16} \times 0.06^3 = 1363.5 \text{ Nm}$$

∴ Power being transmitted,
$$P = \dfrac{2\pi NT}{60 \times 1000} \text{ kW} = \dfrac{2\pi \times 800 \times 1363.5}{60 \times 1000} = 114.23 \text{ kW}$$

i.e. $\quad$ **P = 114·23 kW (Ans.)**

Example 13·34. *A propeller shaft of 240 mm external diameter and 180 mm internal diameter has to transmit 1100 kW at 100 r.p.m. It is additionally subjected to a bending moment of 10 kNm and an end thrust of 200 kN. Determine :*

(i) Principal stresses and their planes;

(ii) Maximum shear stress.

Solution. Effect of end thrust:

When a shaft is subjected to end thrust in addition to torsion and bending moment then the *resultant direct stress acting across the section is algebraic sum of stresses caused by end thrust and due to bending*. Torsion causes only the shear stress.

Given: External diameter of the shaft,
$$D = 240 \text{ mm} = 0.24 \text{ m}$$

Internal diameter of the shaft,
$$d = 180 \text{ mm} = 0.18 \text{ m}$$

Power to be transmitted, $P = 1100 \text{ kW}$

Speed of the shaft, $\qquad N = 100$ r.p.m.

Magnitude of the bending moment,

$$M = 10 \text{ kNm}$$

Magnitude of end thrust $\qquad = 200$ kN

(i) Principal stresses and their planes:

Direct stress due to end thrust $= \dfrac{200}{\dfrac{\pi}{4}(0\cdot24^2 - 0\cdot18^2)} \times 10^{-3} \text{ MN/m}^2$

$$= 10\cdot1 \text{ MN/m}^2 \text{ (comp.)}$$

Bending stress, $\qquad \sigma_b = \dfrac{M \times y}{I} = \dfrac{10 \times (0\cdot24/2)}{\dfrac{\pi}{64}(0\cdot24^4 - 0\cdot18^4)} \times 10^{-3} \text{ MN/m}^2$ (comp.)

$$= 10\cdot78 \text{ MN/m}^2 \text{ (comp.)}$$

∴ Maximum direct stress, $\quad \sigma = 10\cdot1 + 10\cdot78 = 20\cdot88 \text{ MN/m}^2$ (comp.)

Now, $\qquad\qquad\qquad P = \dfrac{2\pi NT}{60} \text{ kW}$ (where T is in kNm)

$$1100 = \dfrac{2\pi \times 100 \times T}{60} \text{ or } T = \dfrac{1100 \times 60}{2\pi \times 100} = 105 \text{ kNm}$$

∴ Maximum shear stress due to torsion,

$$\tau = \dfrac{TR}{I_p} \qquad\qquad \left(\because \dfrac{T}{I_p} = \dfrac{\tau}{R}, \text{or, } \tau = \dfrac{TR}{I_p}\right)$$

$$= \dfrac{105 \times (0.24/2)}{\dfrac{\pi}{32}[(10\cdot24)^4 - (0\cdot18)^4]} \times 10^{-3} \text{ MN/m}^2 = 56\cdot59 \text{ MN/m}^2$$

Principal stresses are :

$$\sigma = \dfrac{\sigma}{2} \pm \sqrt{\left(\dfrac{\sigma}{2}\right)^2 + \tau^2} = \dfrac{20\cdot88}{2} \pm \sqrt{\left(\dfrac{20\cdot88}{2}\right)^2 + (56\cdot59)^2}$$

$$= 10\cdot44 \pm 57\cdot54$$

∴ $\qquad\qquad \sigma_{max} = 10\cdot44 + 57\cdot54 = \mathbf{67\cdot98 \ MN/m^2}$ *(comp.)* **(Ans.).**

and, $\qquad\qquad \sigma_{min} = 10\cdot44 - 57\cdot54 = -47\cdot1 \text{ MN/m}^2$

$$= \mathbf{47\cdot1 \ MN/m^2} \text{ *(tensile)* (Ans.)}$$

If θ be the angle that the principle planes make with the axis of the shaft then,

$$\tan 2\theta = \dfrac{2\tau}{\sigma} = \dfrac{2 \times 56\cdot59}{20\cdot88} = 5\cdot42$$

or, $\qquad\qquad \theta = \dfrac{1}{2}\tan^{-1}5\cdot42$ **(Ans.)**

(ii) Maximum shear stress, τ_{max}:

$$\tau_{max} = \dfrac{\sigma_{max} - \sigma_{min}}{2} = \dfrac{67\cdot98 - (-47\cdot1)}{2} = \mathbf{57\cdot54 \ MN/m^2} \text{ (Ans.)}$$

13·15. TORSION OF A TAPERING SHAFT

Refer to Fig. 13·17.

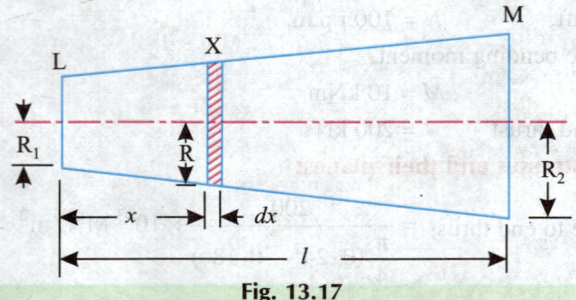

Fig. 13.17

Let, T = Twisting moment applied to a tapering shaft,

           l = Length of the shaft,

    R_1 and R_2 = Respective radii at the two ends of the shaft,

          τ_1 = Maximum shear stress at the left end,

          τ_2 = Maximum shear stress at the right end, and

          τ = Maximum shear stress at a section X distant x from the left end.

Let the radius at the section X be R.

$$\therefore \qquad T = \tau_1 \times \frac{\pi}{2} \times R_1^3 = \tau_2 \times \frac{\pi}{2} \times R_2^3 = \tau \times \frac{\pi}{2} \times R^3$$

or, $\tau_1 R_1^3 = \tau_2 R_2^3 = \tau R^3$

Consider a short length dx of the shaft. For this short length the shaft may be considered as having a uniform radius R.

$\therefore$ Angle of twist of the small length dx of the shaft,

$$= d\theta = \frac{T\,dx}{C I_p} = \frac{T\,dx}{C \times \frac{\pi}{2} R^4} = \frac{2T\,dx}{C\pi R^4}$$

But, $R = R_1 + \left[\dfrac{R_2 - R_1}{l} \right] \times x = R_1 + kx$ ...(13·32)

where, $k = \dfrac{R_2 - R_1}{l}$ (a constant for a given shaft)   ...(13·33)

$$d\theta = \frac{2T}{C\pi} \cdot \frac{dx}{(R_1 + kx)^4}$$

$\therefore$ Total angle of twist for the whole length l of the shaft is $= \theta = \int d\theta$

$$= \int_0^l \frac{2T}{C\pi} \cdot \frac{dx}{(R_1 + kx)^4} = -\frac{2T}{C\pi} \cdot \frac{1}{3k} \left[\frac{1}{(R_1 + kx)^3} \right]_0^l$$

$$= -\frac{2T}{C\pi} \cdot \frac{1}{3k} \left[\frac{1}{(R_1 + kl)^3} - \frac{1}{R_1^3} \right]$$

But, $k = \dfrac{R_2 - R_1}{l}$

$\therefore$ $kl = R_2 - R_1$ (or $R_2 = R_1 + kl$)

$\therefore$

$$\theta = -\frac{2}{3k} \cdot \frac{T}{C\pi} \left[\frac{1}{R_2^3} - \frac{1}{R_1^3} \right] = \frac{2}{3k} \cdot \frac{T}{C\pi} \left[\frac{1}{R_1^3} - \frac{1}{R_2^3} \right]$$

$$\therefore \qquad \theta = \frac{2T}{3C\pi}\left(\frac{1}{R_2 - R_1}\right)\left(\frac{R_2^3 - R_1^3}{R_1^3 \ R_2^3}\right)$$

$$\therefore \qquad \theta = \frac{2}{3}\frac{Tl}{C\pi}\left[\frac{R_1^2 + R_1 R_2 + R_2^2}{R_1^3 \ R_2^3}\right] \qquad\qquad ...(13\cdot34)$$

For the particular case of shaft of uniform radius,

$$R_1 = R_2 = R$$

and, $$\theta = \frac{2}{3}\frac{Tl}{C\pi}\cdot\frac{3R^2}{R^6} = \frac{2Tl}{C\pi R^4} = \frac{Tl}{C\times\left(\dfrac{\pi R^4}{2}\right)} = \frac{Tl}{Cl_p}$$

Example 13·35. *A shaft 1 m long tapers uniformly from a radius of 40 mm to a radius of 60 mm. If the shaft transmits a torque of 12 kNm, find:*

(i) Angle of twist;

(ii) Maximum shear stress developed.

Take: $\qquad C = 80 \ GN/m^2$

Solution. Radius of the smaller end,

$$R_1 = 40 \text{ mm} = 0.04 \text{ m}$$

Radius at the larger end,

$$R_2 = 60 \text{ mm} = 0.06 \text{ m}$$

Torque transmitted, $\quad T = 12$ kNm

Modulus of rigidity, $\quad C = 80 \ GN/m^2$.

(i) Angle of twist, θ:

We know that in a tapering shaft, angle of twist,

$$\theta = \frac{2}{3}\frac{Tl}{C\pi}\left[\frac{R_1^2 + R_1 R_2 + R_2^2}{R_1^3 \ R_2^3}\right]$$

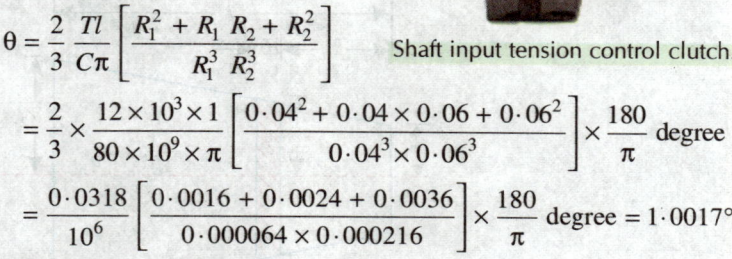

Shaft input tension control clutch.

$$= \frac{2}{3}\times\frac{12\times10^3\times1}{80\times10^9\times\pi}\left[\frac{0\cdot04^2 + 0\cdot04\times0\cdot06 + 0\cdot06^2}{0\cdot04^3\times0\cdot06^3}\right]\times\frac{180}{\pi}\text{ degree}$$

$$= \frac{0\cdot0318}{10^6}\left[\frac{0\cdot0016 + 0\cdot0024 + 0\cdot0036}{0\cdot000064\times0\cdot000216}\right]\times\frac{180}{\pi}\text{ degree} = 1\cdot0017°$$

Hence, $\qquad \theta = 1\cdot0017°$ **(Ans.)**

(ii) Maximum shear stress developed, T_{max}:

Using the relation, $\quad T = \tau_{max}\times\dfrac{\pi}{16}\times D_1^3$, we get

(maximum shear stress occurs at the surface of the smallest diameter of the shaft)

$$12\times10^3 = \tau_{max}\times\frac{\pi}{16}\times(0\cdot08)^3$$

$$\tau_{max} = \frac{12\times10^3\times16}{(0\cdot08)^3}\times10^{-6}\ MN/m^2 = 119\cdot36\ MN/m^2$$

Hence, $\qquad \tau_{max} = 119\cdot36 \ MN/m^2$ **(Ans.)**

Example 13·36. *A shaft tapers uniformly from a radius (R + a) at one end to (R − a) at the other. If it is under the action of an axial torque T and a = 0·1 R find the percentage error in the angle of twist for a given length when calculated on the assumption of a constant radius R.*

Solution. Radius at the smaller end,

$$R_1 = R - a$$

Radius at the larger end, $R_2 = R + a = R + 0 \cdot 1\,R = 1 \cdot 1R$

For shafts of tapering sections, we have

$$\theta = \frac{2}{3}\,\frac{Tl}{C\pi}\left[\frac{R_1^2 + R_1\,R_2 + R_2^2}{R_1^3\,R_2^3}\right]$$

$$= \frac{2}{3}\,\frac{Tl}{C\pi}\left[\frac{(0 \cdot 9R)^2 + (0 \cdot 9R)\times(1 \cdot 1R) + (1 \cdot 1R)^2}{(0 \cdot 9R)^3 \times (1 \cdot 1R)^3}\right]$$

$$= \frac{2}{3}\,\frac{Tl}{C\pi}\left[\frac{0 \cdot 81R^2 + 0 \cdot 99R^2 + 1 \cdot 21R^2}{0 \cdot 729R^3 \times 1 \cdot 331R^3}\right] = \frac{2 \cdot 068\,Tl}{C\pi R^4}$$

For a shaft of uniform radius R,

$$\theta = \frac{Tl}{CI_p} = \frac{2Tl}{C\pi R^4} \qquad \left[\because I_p = \frac{\pi D^4}{32} = \frac{\pi \times (2R)^4}{32} = \frac{\pi R^4}{2}\right]$$

∴ Percentage error

$$= \frac{2 \cdot 068 - 2}{2} \times 100 = 3 \cdot 4 \text{ percent} \quad \textbf{(Ans.)}$$

Example 13·37. *A tapered shaft having radii r_1 and r_2 at the two ends and length l is subjected to equal and opposite torque T at the ends. Determine the angle of twist of one end relative to other.*

Solution. Fig. 13·18 shows a tapered shaft having radii r_1 and r_2 at the two ends of length l. It is subjected to equal and opposite torque T at the ends. Let us set up a coordinate system with variable x denoting the distance from the small end of the shaft. The radius at a section at the distance x from the small end is $r = r_1 + \dfrac{(r_2 - r_1)\,x}{l}$.

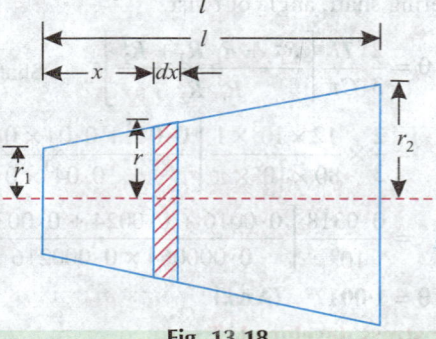

Fig. 13.18

Provided the angle of taper is small it is sufficient to consider the angle $d\theta$ through which the shaded element of length dx is twisted. This is obtained by applying the expression $\theta = \dfrac{Tl}{CI_p}$, where T represents torque, C represents the modulus of rigidity and I_p represents the polar moment of inertia of the section. Note that the above expression is applied to the element of length dx and radius $r = r_1 + \dfrac{(r_2 - r_1)\,x}{l}$. For such an element the polar moment of inertia I_p is given by:

$$I_p = \frac{\pi}{32}\,d^4 = \frac{\pi}{2}\,r^4 = \left[r_1 + \frac{(r_2 - r_1)\,x}{l}\right]^4$$

Thus, $$d\theta = \frac{T \cdot dx}{C \dfrac{\pi}{2}\left[r_1 + \dfrac{(r_2 - r_1)\,x}{l}\right]^4} \qquad \qquad ...(i)$$

The angle of twist in the length l is found by integrating eqn. (i), thus

$$\theta = \frac{2T}{C\pi}\int_0^l \frac{dx}{\left[r_1 + \dfrac{(r_2 - r_1)\,x}{l}\right]^4}$$

$$= \frac{2T}{C\pi}\left(-\frac{1}{3}\right)\left(\frac{l}{r_2 - r_1}\right)\left|\frac{1}{\left\{r_1 + \dfrac{(r_2 - r_1)\,x}{l}\right\}^3}\right|_0^l$$

or, $$\theta = \frac{2T\,l}{3C\pi\,(r_2 - r_1)}\left[\frac{1}{r_1^3} - \frac{1}{r_2^3}\right] \qquad \qquad ...\textbf{Required expression (Ans.)}$$

13·16. THIN CIRCULAR TUBE SUBJECTED TO TORSION

Refer Fig. 13·19. Consider a thin circular tube of external diameter D and thickness t; t being very small as compared with D.

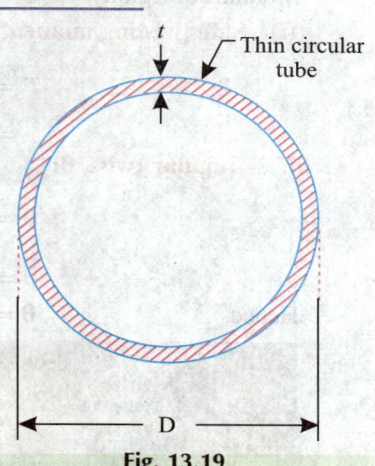

Thin circular tube

Fig. 13.19

Polar moment of inertia of the tube section

$$= I_p = \text{area of the section} \times \text{square of the radius}$$

$$= \pi D t \times \left(\frac{D}{2}\right)^2 = \frac{\pi D^3 t}{4}$$

Torsional resistance (T):

$$T = \tau \times \frac{I_p}{R} = \tau \times \frac{\pi D^3 t}{4R}$$

$$= \tau \times \frac{\pi D^3 t}{4 \times D/2} = \frac{\tau\pi D^2 t}{2}$$

i.e. $$T = \frac{\tau\pi D^2 t}{2} \qquad \qquad ...(13\cdot35)$$

Twist of the tube in a length l, θ:

$$\theta = \frac{Tl}{CI_p} = \frac{Tl}{C \times \dfrac{\pi D^3 t}{4}} = \frac{4Tl}{\pi D^3 t C}$$

i.e. $$\theta = \frac{4Tl}{\pi D^3 t C} \qquad \qquad ...(13\cdot36)$$

Strength weight ratio $= \dfrac{T}{W}$:

If w is the weight density of the material, the weight of the tube of length l,

$$W = \pi\,Dt l\,w$$

Also, torque, $T = \dfrac{\tau \pi D^2 t}{2}$

∴ Strength-weight ratio,

$$T/W = \frac{\tau \pi D^2 t}{2 \times \pi D t l w} = \frac{\tau D}{2lw}$$

i.e. $\dfrac{T}{W} = \dfrac{\tau D}{2lw}$ ...(13·37)

This ratio (i.e. strength-weight ratio) is very important in 'aircraft design'.

Example 13·38. *A thin steel tube 75 mm in diameter is 3 mm thick. If the allowable shear stress is 80 MN/m² and modulus of rigidity is 80 GN/m² find:*

(*i*) *Safe twisting moment that can be applied to the tube;*

(*ii*) *The twist in a length of 600 mm.*

Solution. Diameter of the thin steel tube,

$$D = 75 \text{ mm} = 0.075 \text{ m}$$

Thickness of the tube, $t = 3 \text{ mm} = 3 \times 10^{-3} \text{ m}$

Allowable shear stress, $\tau = 80 \text{ MN/m}^2$

Modulus of rigidity, $C = 80 \text{ GN/m}^2$.

(*i*) **Safe twisting moment, *T*:**

$$T = \frac{\tau \pi D^2 t}{2} = \frac{80 \times 10^6 \times \pi \times 0.075^2 \times 3 \times 10^{-3}}{2}$$

$$= 2120.6 \text{ Nm } i.e. \ T = \textbf{2120.6 Nm} \quad \textbf{(Ans.)}$$

(*ii*) **Angular twist, θ:**

$$\theta = \frac{4Tl}{\pi D^3 t C} = \frac{4 \times 2120.6 \times (600 \times 10^{-3})}{\pi \times (0.075)^3 \times (3 \times 10^{-3}) \times 80 \times 10^9} \times \frac{180}{\pi} \text{ degree}$$

$$= 0.917°$$

Hence, $\theta = \textbf{0.917°} \quad \textbf{(Ans.)}$

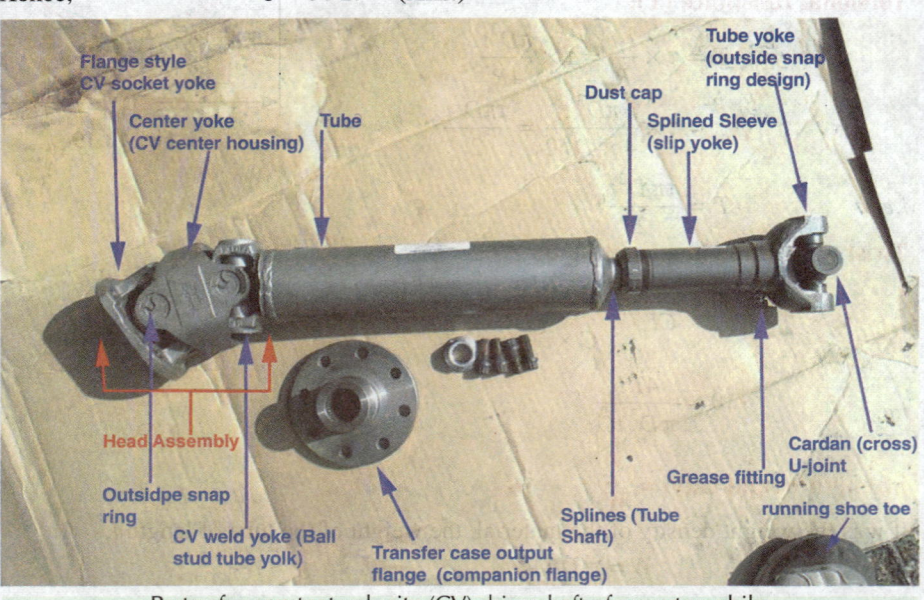

Parts of a constant velocity (CV) drive shaft of an automobile.

13·17. TORSION OF NON-CIRCULAR SOLID SECTIONS

While dealing with the theory of torsion of circular shafts it was assumed that the plane transverse sections remain plane and that the shear strain varies linearly but this assumption does not hold good in case of torsion of non-circular solid sections. Warping of non-circular cross-sections during application of torque makes the analysis of non-circular shafts difficult.

For analysis of non-circular shafts one of the important methods (based on experimental results) is called the *'Membrane analogy'* or *'Soap film analogy method'* (mathematical analysis for non-circular sections based on theory of elasticity is rather difficult).

Soap film analogy method:

— A soap film is stretched over a hole geometrically similar to the cross-section of the shaft and a light differential pressure is applied to it.

— The shape of the film after the application of slight pressure is given by a partial differential equation which is similar to the partial differential equation under torsion. This similarity or analogy makes the job of finding the stresses very simple.

Under this method the *following points are worth noting:*

(*i*) *The slope of a tangenial line* at any point *A* on the surface of inflated soap film is proportional to the shearing stress, τ, at a point vertically below the point *A*.

(*ii*) The direction of the shear stress, τ, vertically below *A* is *in the direction of the tangent to the curve* obtained by joining all the points on the surface of the soap film having heights equal to the height of point *A*.

(*a*) Cross-section

(*iii*) The twisting torque is proportional to *twice* the volume enclosed by soap film.

In case of hollow sections soap film analogy is difficult to use because of the difficulty of maintaining a suitable surface for the film.

Mathematical analysis gives the following useful results for the simple sections.

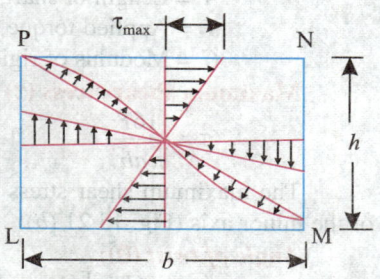

(*b*) Shear stress distribution

Fig. 13.20

1. Rectangular sections:

Fig. 13·20 (*a*) shows the cross-section of the rectangular shaft.

Let,

b = Larger side of the rectangular shaft,

h = Smaller side of the rectangular shaft,

l = Length of the shaft,

C = Modulus of rigidity,

α = Slope of the bubble film at a particular point

$$= 3 + 1 \cdot 8 \left(\frac{h}{b} \right) \quad \text{...(approximate value), and}$$

T = Applied torque,

(For *square section*, put $h = b$, then $\alpha = 4 \cdot 8$).

Maximum shear stress (τ):

The maximum shear stress is given by:

$$\tau = \alpha \frac{T}{bh^2} \qquad \dots(13\cdot38)$$

$$\left[\text{For square section, } \tau = 4\cdot8 \frac{T}{b^3}, \text{ putting } h = b \right]$$

Fig. 13·20 (b) shows the shear stress distribution for a rectangular section. It may be noted that:

(i) The maximum shear stress occurs *at the centres of the larger sides.*

(ii) The maximum shear stress occurs on a surface nearest to the shaft axis and not at a point of maximum distance from the axis.

(iii) The points (L, M, N, P) farthest *from the centre* have *zero shear stresses.*

Angle of twist (θ) :

Angle of twist is given by, $\theta = \dfrac{\beta}{bh^3} \cdot \dfrac{Tl}{C}$ $\qquad \dots(13\cdot39)$

where, $\beta = 3\cdot5 \left[1 + \left(\dfrac{h}{b} \right)^2 \right]$ .... approximately

$$\beta = \frac{3}{1 - 0\cdot63 \dfrac{h}{b} \left(1 - \dfrac{h^4}{12b^4} \right)} \quad \dots \text{ more accurately}$$

2.   Elliptical section:

Refer to Fig. 13·21 (a).

Let, a = Major axis,
   b = Minor axis,
   l = Length of shaft,
   T = Applied torque, and
   C = Modulus of rigidity.

Maximum shear stress (τ) :

$$\tau = \frac{16T}{\pi ab^2} \qquad \dots(13\cdot40)$$

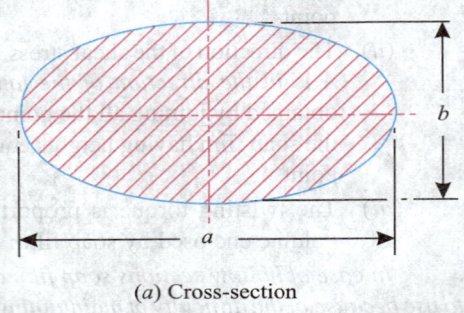

(a) Cross-section

The maximum shear stress occurs at the ends of the minor axis (Fig. 13·21 (b))

Angle of twist (θ):

$$\theta = \frac{16lT}{\pi abC} \left[\frac{1}{a^2} + \frac{1}{b^2} \right] \qquad \dots(13\cdot41)$$

Torsional stiffness (k):

$$k = \frac{C\pi a^3 b^3}{16 (a^2 + b^2)} \qquad \dots(13\cdot42)$$

3.   Equilateral triangle:

Refer to Fig. 13·22 (a).

Let, a = Side of the triangle,
   l = Length of the shaft,
   T = Applied torque, and
and,   C = Modulus of rigidity.

Maximum shear stress (τ):

$$\tau = \frac{20T}{a^3} \qquad \dots(13\cdot43)$$

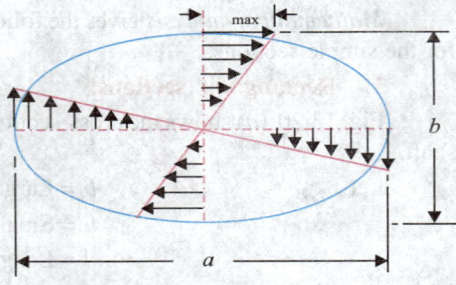

(b) Shear stress distribution

Fig. 13.21

Railway coach wheels and axle shaft set.

Maximum shear stress occurs at the *centre of each side* while the *shear stress at each corner is equal to zero* (Fig. 13·22 (*b*)).

Angle of twist (θ) :

$$\theta = \frac{80}{a^4 \sqrt{3}} \cdot \frac{Tl}{C} \qquad ...(13·44)$$

Torsional stiffness (*k*):

$$k = \frac{\sqrt{3}}{80} a^4 C \qquad ...(13·45)$$

(*a*) Cross-section

4.   Regular hexagon:

Let,   *A* = Area of cross–section,

   d = Diameter of inscribed circle,

   T = Applied torque, and

   C = Modulus of rigidity.

Maximum shear stress (τ):

$$\tau = \frac{T}{0·217\,Ad} \qquad ...(13·46)$$

Angle of twist (θ) :

$$\theta = \frac{T}{0·133\,Ad^2 C} \qquad ...(13·47)$$

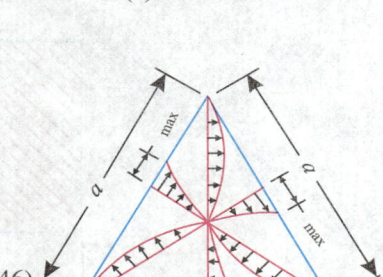

(*b*) Shear stress distribution

Fig. 13.22

5.   Regular octagon:

Let,   *A* = Area of cross–section, and

   d = Diameter of inscribed circle.

Maximum shear stress (τ):

$$\tau = \frac{T}{0·223\,Ad} \qquad ...(13·48)$$

Angle of twist (θ):

$$\theta = \frac{T}{0 \cdot 130 \, Ad^2 C} \qquad \qquad ...(13 \cdot 49)$$

Following points are worth noting:

— In the above cases, the *maximum shear stress occurs at a point where a largest inscribed circle touches.*

— In general the greatest shear stress occurs at or near one of the points of contact of the largest inscribed circle.

— If the surface is concave, the maximum shear stress may occur at the *point of greater concavity.*

Example 13·39. *A rectangular steel shaft 50 mm × 25 mm is subjected to a torque of 2 kNm. Find:*

(*i*) *Maximum shear stress developed in the shaft;*

(*ii*) *Angular twist per metre length.*

Take: $C = 80 \, GN/m^2$

Solution. Larger side of the rectangular shaft,

$$b = 50 \text{ mm} = 0 \cdot 05 \text{ m}$$

Smaller side of the rectangular shaft,

$$h = 25 \text{ mm} = 0 \cdot 025 \text{ m}$$

Applied torque, $T = 2 \cdot 0$ kNm

Modulus of rigidity, $C = 80 \, GN/m^2$

(*i*) **Maximum shear stress developed, τ:**

In case of rectangular shaft,

$$\alpha = 3 + 1 \cdot 8 \left(\frac{h}{b} \right) \quad ... \text{ Approximate value}$$

$$= 3 + 1 \cdot 8 \times \frac{0 \cdot 025}{0 \cdot 05} = 3 \cdot 9$$

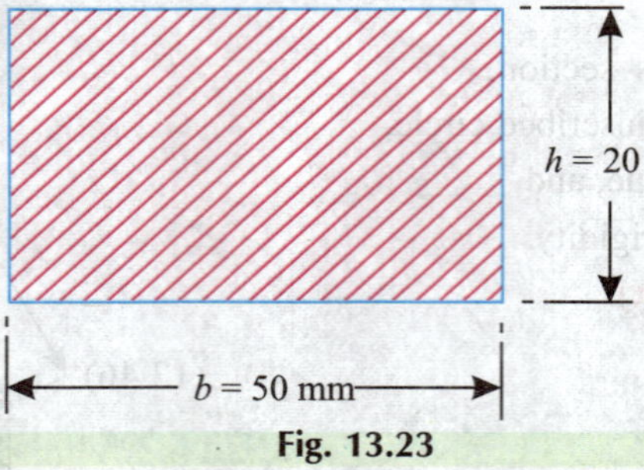

$$h = 20 \text{ mm}$$

$$b = 50 \text{ mm}$$

Fig. 13.23

and, $$\tau = \alpha \cdot \frac{T}{bh^2} = 3 \cdot 9 \times \frac{2 \times 10^3}{0 \cdot 05 \times (0 \cdot 025)^2} \times 10^{-6} \, MN/m^2$$

$$= 249 \cdot 6 \, MN/m^2$$

i.e. $\tau = 249 \cdot 6 \, MN/m^2$ **(Ans.)**

(*ii*) **Angular twist per metre length, $\dfrac{\theta}{l}$:**

Angle of twist is given by

$$\theta = \frac{\beta}{bh^3} \cdot \frac{Tl}{C}$$

or, $$\frac{\theta}{l} = \frac{\beta T}{bh^3 C}$$ ...(i)

where, $$\beta = 3 \cdot 5 \left[1 + \left(\frac{h}{b} \right)^2 \right] \quad \text{...approximately}$$

$$= 3 \cdot 5 \left[1 + \left(\frac{0 \cdot 025}{0 \cdot 05} \right)^2 \right] = 4 \cdot 375$$

Substituting the value of β in eqn. (i), we get

$$\frac{\theta}{l} = \frac{4 \cdot 375 \times 2 \cdot 0 \times 10^3}{0 \cdot 05 \times (0 \cdot 025)^3 \times 80 \times 10^9} \times \frac{180}{\pi} \text{ degree} = 8 \cdot 02°$$

Hence, *angular twist per metre length* = **8·02°** **(Ans.)**

Example 13·40. *A rectangular steel shaft is transmitting power at 300 r.p.m., lifting a load of 40 kN at a speed of 10 m/min. If the maximum permissible shear stress in the shaft is 45 MN/m² and efficiency of the crane gearing is 60 percent, determine:*

(i) Size of the shaft;

(ii) Angle of twist per metre length.

Take: $\qquad$ *C = 78·4 MN/m²; Breadth to depth ratio = 1·5.*

Solution. Load lifted = 40 kN

Speed at which load is lifted = 10 m/min

Speed of the shaft, $\qquad N = 300$ r.p.m.

Maximum shear stress (permissible),

$\qquad \tau = 45$ MN/m²

Efficiency of the crane gearing,

$\qquad \eta_g = 60\%$

Modulus of rigidity, $\qquad C = 78 \cdot 4$ MN/m²

Breadth (b) to depth ratio (h) = 1·5

i.e. $\qquad b = 1 \cdot 5\, h.$

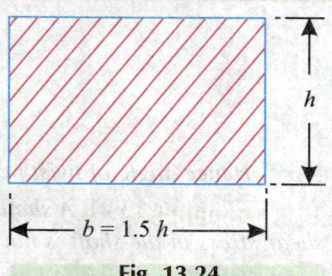

Fig. 13.24

(i) **Size of the shaft; b, h:**

Work done per minute $\quad$ = load lifted × speed = 40 × 10 = 400 kNm/min

Input work per minute $\quad = \dfrac{400}{\eta_g} = \dfrac{400}{0 \cdot 6} = 666 \cdot 67$ kNm/ min

Also work done $\qquad = T \times 2\pi N$

where, $\qquad T = $ Torque

∴ $\qquad T \times 2\pi N = 666 \cdot 67$

or, $\qquad T = \dfrac{666 \cdot 67}{2\pi \times 300}$ kN = 0·3536 kNm

In the case of a rectangular shaft,

$$\tau = \alpha \cdot \frac{T}{bh^2}$$

where, $\qquad \alpha = 3 + 1 \cdot 8 \left(\dfrac{h}{b} \right) \qquad$ ...approximately

$$= 3 + 1 \cdot 8 \times \frac{h}{1 \cdot 5\, h} = 4 \cdot 2$$

$$\therefore \qquad 45 \times 10^6 = 4 \cdot 2 \times \frac{(0 \cdot 3536 \times 10^3)}{(1 \cdot 5h \times h^2)}$$

or, $\qquad h^3 = \dfrac{4 \cdot 2 \times 0 \cdot 3536 \times 10^3}{45 \times 10^6 \times 1 \cdot 5} = 0 \cdot 000022$

$\therefore \qquad h = 0 \cdot 028$ m or 28 mm

and, larger side, $\qquad b = 1 \cdot 5 \times 28 = 42$ mm

(ii) Angle of twist per metre length $\dfrac{\theta}{l}$:

Angle of twist is given by,

$$\theta = \frac{\beta}{bh^3} \times \frac{Tl}{C}$$

$$\therefore \qquad \frac{\theta}{l} = \frac{\beta}{bh^3} \times \frac{T}{C} \qquad\qquad\qquad ...(i)$$

where, $\qquad \beta = 3 \cdot 5 \left[1 + \left(\dfrac{h}{b} \right)^2 \right] = 3 \cdot 5 \left[1 + \left(\dfrac{h}{1 \cdot 5h} \right)^2 \right] = 5 \cdot 055 \quad$ ...approximately

Substituting the value of β in equation (i), we get

$$\frac{\theta}{l} = \frac{5 \cdot 055}{0 \cdot 042 \times (0 \cdot 028)^3} \times \frac{0 \cdot 3536 \times 10^3}{78 \cdot 4 \times 10^9} = 0 \cdot 0247 \text{ rad.}$$

$$= 0 \cdot 0247 \times \frac{180}{\pi} = 1 \cdot 415°$$

Hence angle of twist per metre length = **1·415°** (Ans.)

Example 13·41. *A shaft of elliptical section is subjected to torque of 2·5 kNm. If the maximum shear stress in the shaft is not to exceed 80 MN/m² determine:*

(i)   The major and minor axes, if major axis = 1·5 minor axis.

(ii) The angular twist per metre length.

Take: $\qquad\qquad C = 80 \ GN/m^2.$

A submarine propulsion shaft.

Solution. Applied torque,

$$T = 2.5 \text{ kNm}$$

Maximum shear stress (permissible),

$$\tau = 80 \text{ MN/m}^2$$

Major axis $(a) = 1.5 \times$ minor axis (b)

i.e.

$$a = 1.5 \, b$$

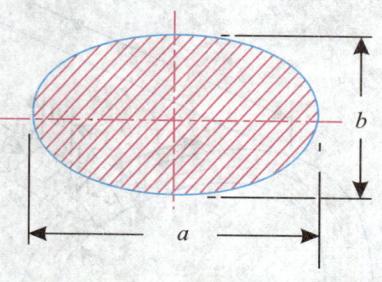

Fig. 13.25

(i) Major axis (a) and minor axis (b):

We know,

$$\tau = \frac{16T}{\pi a b^2}$$

or,

$$80 \times 10^6 = \frac{16 \times 2.5 \times 10^3}{\pi \times 1.5 b \times b^2}$$

∴

$$b^3 = \frac{16 \times 2.5 \times 10^3}{80 \times 10^6 \times \pi \times 1.5} = 1.061 \times 10^{-4}$$

or,

$$b = 0.0473 \text{ m or } \mathbf{47.3 \text{ mm}} \quad \textbf{(Ans.)}$$

∴

$$a = 1.5 b = 1.5 \times 47.3 = \mathbf{70.95 \text{ mm}} \quad \textbf{(Ans.)}$$

(ii) Angular twist per metre length, $\dfrac{\theta}{l}$:

Angular twist is given by:

$$\theta = \frac{16lT}{\pi a b C}\left(\frac{1}{a^2} + \frac{1}{b^2}\right)$$

∴

$$\frac{\theta}{l} = \frac{16T}{\pi a b C}\left(\frac{1}{a^2} + \frac{1}{b^2}\right)$$

$$= \frac{16 \times 2.5 \times 10^3}{\pi \times (70.95 \times 10^{-3}) \times (47.3 \times 10^{-3}) \times 80 \times 10^9}\left[\frac{1}{(70.95 \times 10^{-3})^2} + \frac{1}{(47.3 \times 10^{-3})^2}\right]$$

$$= 4.742 \times 10^{-5} \, (198.65 + 446.97) \text{ rad.}$$

$$= 0.0306 \text{ rad.} = 0.0306 \times \frac{180}{\pi} \text{ degree} = 1.75°$$

Hence, *angular twist per metre length* = **1.75°**   **(Ans.)**

13·18. TORSION OF NON-CIRCULAR THIN TUBULAR SECTIONS

In some applications such as *aeroplane structure* the torsion of a non-circular thin section is quite important. The method used to find shear stresses and angle of twist is simplified by assuming uniform shear stress distribution in the cross-section of the tube; for thin sections this assumption is fairly accurate.

Fig. 13·26 shows a tube of non-circular thin section with varying thickness. Consider an element *LMNSS'N'M'L'* of the tube; this element is shown separately in Fig. 13·27 with shearing stresses acting on various faces. The element is in equilibrium under the action of the shearing forces due to the shearing stresses. Since the longitudinal forces must be equal and opposite for equilibrium, therefore,

Shear force on face *LL'S'S* = Shear force on face *MM'N'N*

∴

$$\tau_1 \times dl \times t_1 = \tau_2 \times dl \times t_2$$

i.e.

$$\tau_1 \cdot t_1 = \tau_2 \cdot t_2 = q, \text{ a constant} = \text{shear force per unit length}$$

The quantity $\tau \cdot t$, a constant is called **shear flow** q.

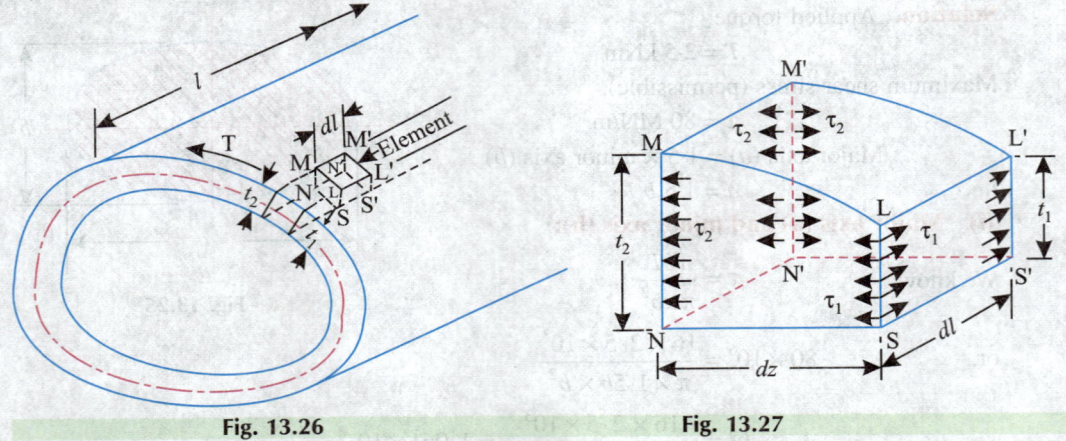

Fig. 13.26   Fig. 13.27

Now consider a across-section of the given tube (Fig. 13·28). Take a small length δs along the centre line. Consider a point P on the small length δs and let t be the thickness of the tube at point P.

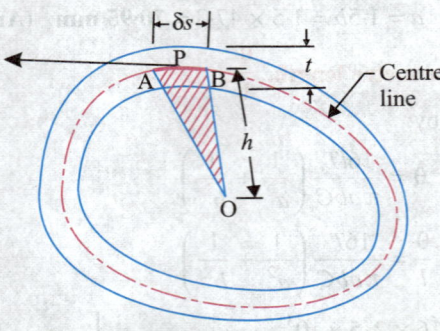

Fig. 13.28

Torque taken by the area of length δs and of thickness t,

$$\delta T = (t \times \delta s) \times \tau \times h$$

$$\left[\begin{array}{l} \text{where, } \tau = \text{shear stress, and} \\ \quad\quad h = \text{lever length from any point} \\ \quad\quad\quad 0 \text{ within the section.} \end{array}\right]$$

∴ $\delta T = (\tau \cdot t) \times h \, \delta s = q \cdot h \, \delta s$ $(\because \ \tau \cdot t = q); \ dT = qh \cdot ds.$

∴ Total torque taken by the section,

$$T = \int dT = \int hq \, ds \qquad \qquad \text{...(13·50)}$$

or $T = q \int h ds$

Now shaded area OAB $= \dfrac{1}{2} \cdot h \cdot \delta s$

∴ $h \, \delta s = 2 \times$ shaded area OAB

Hence twice the area within the centre line (*i.e.* central contour line) $= \int h \, ds$

Let, $A = $ Area enclosed by the centre line of the tube.

Then, $\int h ds = 2A,$

and, $T = 2 \cdot q \cdot A.$ ...(13·51)

This equation is generally known as **Bredt-Batho formula.**

From equation (13·50), we get

$$q = \frac{T}{2A}$$

and,

$$\tau = \frac{T}{2At}$$

...(13·52) $(\because q = \tau \cdot t)$

This equation gives the shear stress in a thin section at any point where thickness is t.

Angle of twist (θ). To obtain angle of twist we equate the strain energy to the external work done by the torque.

The strain energy of a body of small volume δV subjected to shear stress τ is given by:

$$\delta V = \frac{\tau^2}{2C} \times \delta V$$

Universal joint.

Work done due to torque $T = \frac{1}{2} T\theta$

(where, θ = angle of twist)

$\therefore$

$$\frac{1}{2} T\theta = \int_V \frac{\tau^2}{2C} dV$$

i.e.,

$$\theta = \frac{1}{CT} \int_V \tau^2 \, dV = \frac{1}{CT} \oint \tau^2 \, l \cdot t \cdot ds.$$

$(\because dV = l \cdot t \cdot ds)$

$$= \frac{l}{CT} \oint \tau^2 \cdot t \cdot ds = \frac{l}{CT} \oint \left(\frac{q}{t}\right)^2 \cdot t \cdot ds = \frac{q^2 l}{CT} \oint \frac{ds}{t}$$

$$= \left(\frac{T}{2A}\right)^2 \frac{l}{CT} \oint \frac{ds}{t}$$

$\therefore$

$$\theta = \frac{Tl}{4A^2 C} \oint \frac{ds}{t}$$

...(13·53)

13·19. TORSION OF THIN RECTANGULAR SECTIONS

Consider a thin rectangular section shown in Fig. 13·29.

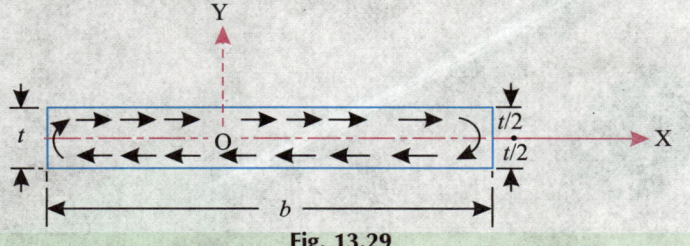

Fig. 13.29

Let, b = Width of the thin section,

t = Thickness of the section (small as compared to b),

T = Applied torque,

θ = Angular twist per unit length, and

C = Modulus of rigidity.

This section consists of only one boundary.

In this case following relations hold good:

Torque on the section:

$$T = \frac{1}{3} b t^3 C \theta \qquad \qquad ...(13\cdot54)$$

Angular twist per unit length:

$$\theta = \frac{1}{C} \cdot \frac{3T}{bt^3} \qquad \qquad ...(13\cdot55)$$

Maximum shear stress:

$$\tau = \pm \frac{3T}{bt^2} \qquad \qquad ...(13\cdot56)$$

$$\left(\text{Maximum shear stress occurs at } y = \pm \frac{t}{2} \right)$$

The above results can be applied to sections built up of rectangular strips and having only one boundary such as angle section, T section etc. as shown in Fig. 13·30.

The following relations hold good for the sections shown in Fig. 13·30

Sections	Angle of twist per unit length	Maximum shear stress (τ_{max})
Angle section and T-section	$$\theta = \frac{3T}{C \sum\limits_{i=1}^{2} b_i t_i^3} \quad ...(13\cdot57)$$ Torque, $T = \frac{1}{3} C\theta (b_1 t_1^3 + b_2 t_2^3)$	$$\tau = \frac{3T}{\sum\limits_{i=1}^{2} b_i t_i^2} \quad ...(13\cdot58)$$
Channel section and I-section,	$$\theta = \frac{3T}{C \sum\limits_{i=1}^{3} b_i t_i^3} \quad ...(13\cdot59)$$ Torque, $T = \frac{1}{3} C\theta (b_1 t_1^3 + b_2 t_2^3 + b_3 t_3^3)$	$$\tau = \frac{3T}{\sum\limits_{i=1}^{3} b_i t_i^2} \quad ...(13\cdot60)$$

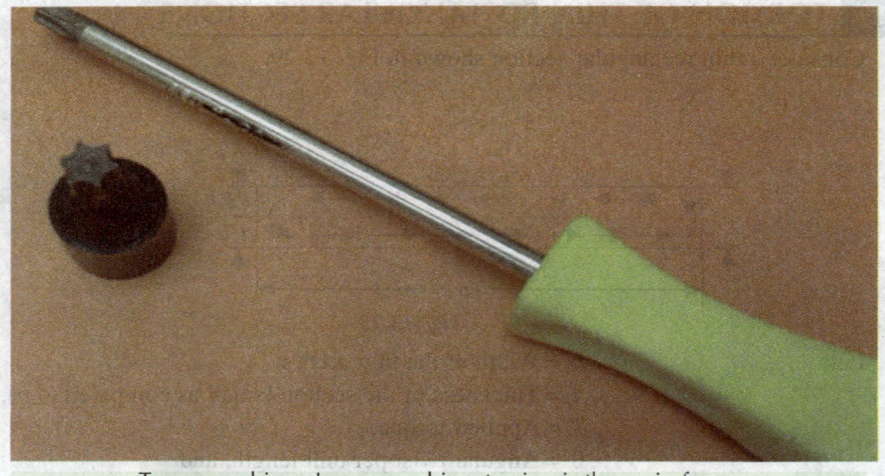

Tox screwdriver. In a screwdriver torsion is the main force.

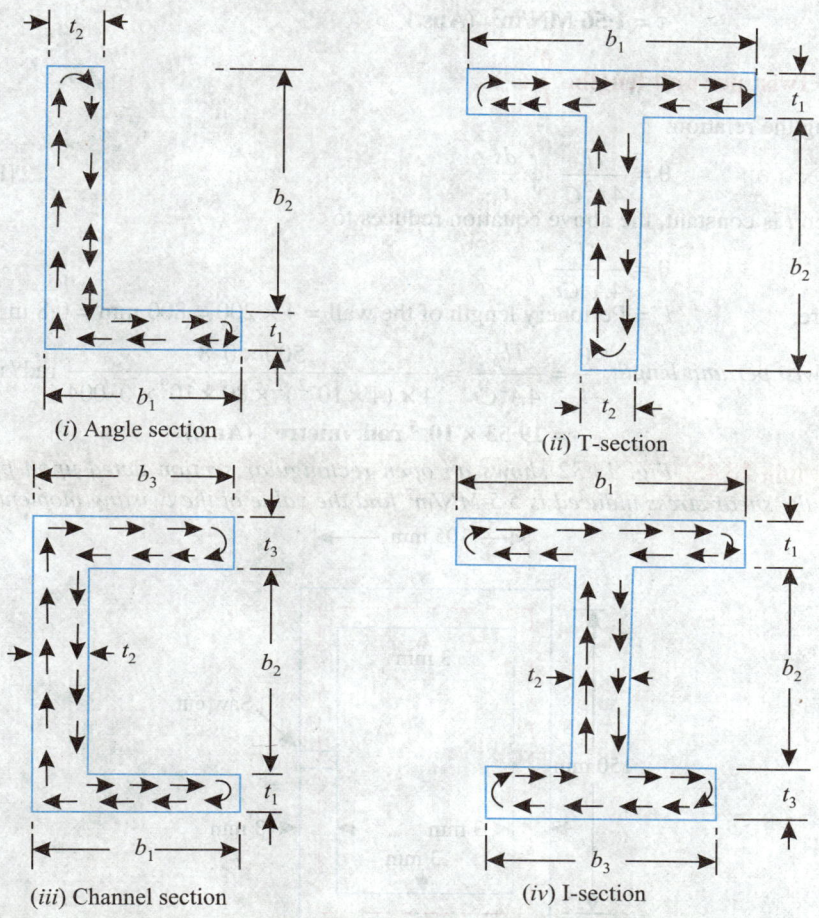

(i) Angle section

(ii) T-section

(iii) Channel section

(iv) I-section

Fig. 13.30

Example 13·42. *A closed cellular square section, shown in Fig. 13·31, is subjected to a torque of 500 Nm. Neglecting stress concentration, determine:*

(i)   *Maximum shear stress;*

(ii) *Twist per unit length.*

Take: $C = 80 \text{ GN/m}^2$.

Solution. Area, $A = 200 \times 200 = 4 \times 10^4 \text{ mm}^2$

$= 4 \times 10^{-2} \text{ m}^2$

Thickness, $t = 4 \text{ mm} = 0.004 \text{ m}$

Applied torque, $T = 500 \text{ Nm}$.

(i) Maximum shear stress, τ:

Using the relation:

$$\tau = \frac{T}{2At} \quad [\text{Eq. 13·51}]$$

or, $\tau = \dfrac{500}{2 \times 4 \times 10^{-2} \times 0.004} \times 10^{-6} \text{ MN/m}^2$

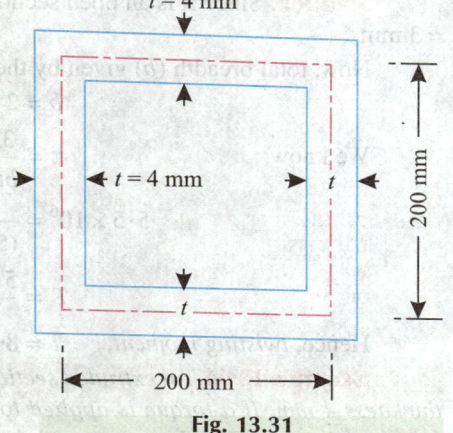

$t = 4 \text{ mm}$

$t = 4 \text{ mm}$

t

200 mm

200 mm

Fig. 13.31

or, $\tau = 1.56 \text{ MN/m}^2$ **(Ans.)**

(ii) **Twist per unit length,** $\dfrac{\theta}{l}$ **:**

Using the relation:

$$\theta = \frac{Tl}{4A^2C} \oint \frac{ds}{t} \qquad\qquad \text{[Eqn. 13·53]}$$

When t is constant, the above equation reduces to

$$\theta = \frac{Tl}{4A^2Ct} l_p$$

where, l_p = Periphery length of the wall = 4 × 200 = 800 mm = 0·8 m

∴ *Twist per unit length,* $\dfrac{\theta}{l} = \dfrac{Tl_p}{4A^2Ct} = \dfrac{500 \times 0.8}{4 \times (4 \times 10^{-2})^2 \times 80 \times 10^9 \times 0.004}$ rad / metre

$$= 19.53 \times 10^{-5} \text{ rad. /metre} \quad \textbf{(Ans.)}$$

Example 13·43. *Fig. 13·32 shows an open rectangular section acted upon by a twisted moment. If the shear stress induced is 5·5 MN/m² find the value of the twisting moment.*

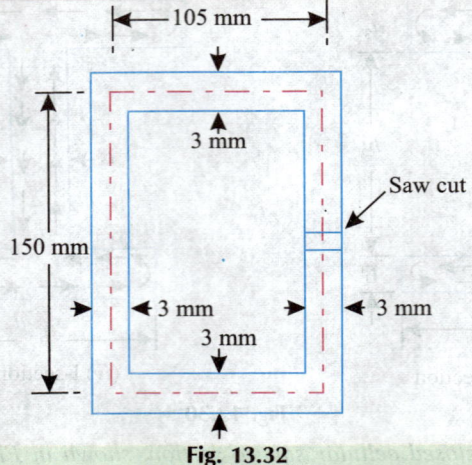

— 105 mm —

3 mm

Saw cut

150 mm

◄ 3 mm   ◄ 3 mm

3 mm

Fig. 13.32

Solution. Since it is an open section, it can be replaced by a rectangular section of thickness = 3 mm

Now, total breadth (b) given by the total length,

$$b = 2 \,(150 + 105) = 510 \text{ mm}$$

We know, $\tau = \dfrac{3T}{bt^2}$ ...(Eqn. 13·56)

∴ $5.5 \times 10^6 = \dfrac{3T}{(510 \times 10^{-3}) \times (3 \times 10^{-3})^2}$

or, $T = \dfrac{5.5 \times 10^6\,(510 \times 10^{-3}) \times (3 \times 10^{-3})^2}{3} \text{ Nm} = 8.415 \text{ Nm}$

Hence, *twisting moment,* $T = \textbf{8·415 Nm}$ **(Ans.)**

Example 13·44. *An extruded section in light alloy is in the form of a semicircle of 90 mm and thickness 4 mm. If a torque is applied to the section and the angle of twist is limited to 3·5° in a length of 1 metre, find the torque and the maximum shear stress.*

Take: $C = 26 \text{ GN/m}^2.$

Solution. Mean radius of the semicircular section,

$$R = 45 \text{ mm} = 0.045 \text{ m}$$

Thickness of the section,

$$t = 4 \text{ mm} = 0.004 \text{ m}$$

Angle of twist, $\theta = 3.5°$ per metre length

$$= 3.5 \times \frac{\pi}{180} = 0.06108 \text{ radian}$$

$$C = 26 \text{ GN/m}^2$$

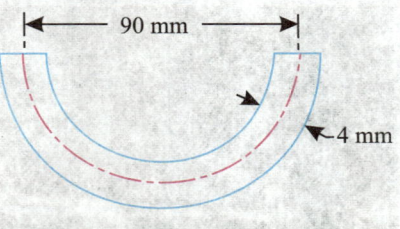

Fig. 13.33

The semicircular section having only one boundary can be treated as thin rectangular section of width (*b*) πR and thickness *t*.

Width, $\qquad b = \pi R = \pi \times 0.045 = 0.1414 \text{ m}$

Torque, T:

We know (from eqn. 13·54),

$$T = \frac{1}{3} b t^3 C \theta = \frac{1}{3} \times 0.1414 \times (0.004)^3$$

$$\times 26 \times 10^9 \times 0.06108 = 4.79 \text{ Nm}$$

i.e. $\qquad T = 4.79 \text{ Nm}$ **(Ans.)**

Maximum shear stress, τ :

We know (from eqn. 13·56)

$$\tau = \frac{3T}{bt^2} = \frac{3 \times 4.79}{0.1414 \times (0.004)^2} \times 10^{-6} \text{ MN/m}^2 = 6.53 \text{ MN/m}^2$$

i.e. $\qquad \tau = 6.35 \text{ MN/m}^2$ **(Ans.)**

Example 13·45. *A shaft of hollow square section is of uniform wall thickness of 4 mm and centre line of the wall forms a square of 200 mm side. It is to be replaced by a solid circular shaft of the same material and having the same torsional stiffness. If the stress concentration factor K at the inner corners of the hollow square section is 1·7 and the twisting moment applied is 800 Nm find :*

(*i*) *Diameter of the solid shaft;*

(*ii*) *Maximum shear stresses in both the shafts.*

Solution. Area, $\qquad A = 200 \times 200 = 4 \times 10^4 \text{ mm}^2 = 0.04 \text{ m}^2$

Thickness, $\qquad t = 4 \text{ mm} = 0.004 \text{ m}$

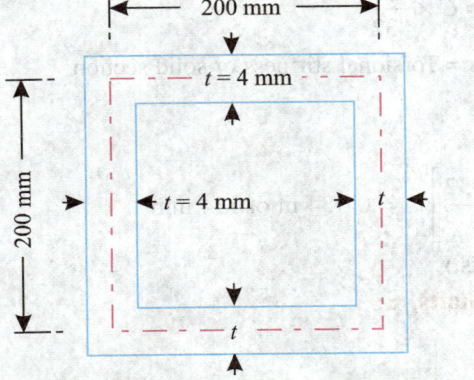

Fig. 13.34

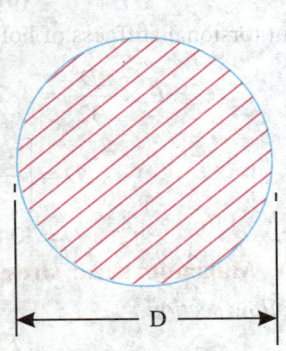

Fig. 13.35

Shaft is being machined on a lathe.

Diameter of the solid shaft, D :

We know,
$$\theta = \frac{Tl}{4A^2C} \oint \frac{ds}{t}$$

When t is constant, the above equation reduces to

$$\theta = \frac{Tl}{4A^2Ct} \cdot l_p$$

where,
l_p = Peripheral length of wall
$$= 4 \times 200 = 800 \text{ mm} = 0.8 \text{ m}$$

∴ Torsional stiffness $= \dfrac{T}{\theta/l} = \dfrac{4A^2Ct}{l_p}$

$$= \frac{4 \times (0.04)^2 \times C \times 0.004}{0.8} = 3.2 \times 10^{-5}\,C \qquad \qquad ...(i)$$

For the solid circular shaft of diameter D, we have

$$\frac{T}{I_p} = \frac{C\theta}{l}$$

∴ Torsional stiffness $= \dfrac{T}{(\theta/l)} = CI_p = C \times \dfrac{\pi D^4}{32}$ $\qquad \qquad ...(ii)$

But torsional stiffness of hollow section = Torsional stiffness of solid section $\qquad ... (given)$

∴
$$C \times \frac{\pi D^4}{32} = 3.2 \times 10^{-5}\,C$$

or
$$D = \left[\frac{3.2 \times 10^{-5} \times 32}{\pi}\right]^{1/4} = 0.134 \text{ m or } 134 \text{ mm}$$

i.e.
$$D = \textbf{134 mm} \quad \textbf{(Ans.)}$$

(ii) Maximum shear stresses in the shafts, τ :

Hollow section :

$$\tau = \frac{T}{2At} \times (\text{stress concentration factor } K)$$

$$= \frac{800}{2 \times 0 \cdot 04 \times 0 \cdot 004} \times 1 \cdot 7 \times 10^{-6} \ MN/m^2 = 4 \cdot 25 \ MN/m^2$$

[where, $T = 800 \ Nm \ ...(Given)$]

i.e. $\qquad \tau = 4 \cdot 25 \ MN/m^2 \ $ **(Ans.)**

Solid circular section:

$$\tau = \frac{16T}{\pi D^3} = \frac{16 \times 800}{\pi \times (0 \cdot 134)^3} \times 10^{-6} \ MN/m^2 = 1 \cdot 693 \ MN/m^2$$

i.e. $\qquad \tau = 1 \cdot 693 \ MN/m^2 \ $ **(Ans.)**

Example 13·46. *A 250 mm × 250 mm I-section with flanges and web 10 mm thick is subjected to a torque of 700 Nm. Neglecting the stress concentration find:*

(i) *Maximum shear stress developed in the section;*

(ii) *Twist per unit length.*

Take: $\qquad C = 80 \ GN/m^2$

Solution. Applied torque, $T = 700 \ Nm$

Modulus of rigidity, $\qquad C = 80 \ GN/m^2$

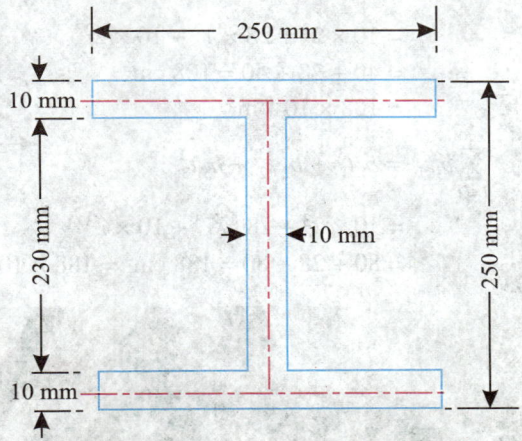

Fig. 13.36

(i) Maximum shear stress, τ:

Total length of the equivalent rectangular section,

$$b = 250 + 250 + 230 = 730 \ mm = 0 \cdot 73 \ m$$

Thickness, $\qquad t = 10 \ mm = 0 \cdot 01 \ m$

Using the relation, $\quad \tau = \dfrac{3T}{bt^2}$, we get

$$\tau = \frac{3 \times 700}{0 \cdot 73 \times (0 \cdot 01)^2} \times 10^{-6} \ MN/m^2 = \textbf{28·76 MN/m}^2 \ \textbf{(Ans.)}$$

(ii) Twist per unit length:

Using the relation, $\quad T = \dfrac{1}{3} bt^3 C\theta$, we have

∴ $\qquad 700 = \dfrac{1}{3} \times 0 \cdot 73 \times (0 \cdot 01)^3 \times 80 \times 10^6 \times \theta$

or $\qquad \theta = \dfrac{700 \times 3}{0 \cdot 73 \times (0 \cdot 01)^3 \times 80 \times 10^9}$ rad./m

$$= 0 \cdot 03595 \text{ rad./m} = 2 \cdot 06°$$

Hence, $\qquad \boldsymbol{\theta = 2 \cdot 06°}$ **(Ans.).**

Example 13·47. *An I-section with flanges 10 cm × 2 cm and web 28 cm × 1 cm is subjected to torque 6 kNm. Find:*

  (*i*)   *Maximum shear stress;*

 (*ii*)   *Angle of twist per unit length.*

Take: $\qquad C = 80 \text{ GN/m}^2$

Solution. Refer to Fig. 13·37

Fig. 13.37

$$b_1 = 10 \text{ cm}, \; t_1 = 2 \text{ cm}$$
$$b_2 = 28 \text{ cm}, \; t_2 = 1 \text{ cm}$$
$$b_3 = 10 \text{ cm}, \; t_3 = 2 \text{ cm}$$

$$\sum_{i=1}^{3} b_i t_i^2 = b_1 t_1^2 + b_2 t_2^2 + b_3 t_3^2$$

$$= 10 \times 2^2 + 28 \times 1^2 + 10 \times 2^2$$
$$= 40 + 28 + 40 = 108 \text{ cm}^3$$
$$= 108 \times 10^{-6} \text{ m}^3$$

$$\sum_{i=1}^{3} b_i t_i^3 = b_1 t_1^3 + b_2 t_2^3 + b_3 t_3^3$$

$$= 10 \times 2^3 + 28 \times 1^3 + 10 \times 2^3$$
$$= 80 + 28 + 80 = 188 \text{ cm}^4 = 188 \times 10^{-8} \text{m}^4$$

Crankshaft in a crankcase.

(*i*) **Maximum shear stress, τ:**

$$\tau = \frac{3T}{\sum\limits_{i=1}^{3} b_i t_i^2} = \frac{3 \times 6}{108 \times 10^{-6}} \times 10^{-3}\ \text{MN}/\text{m}^2 = 166 \cdot 67\ \text{MN}/\text{m}^2$$

[where, T (torque applied) = 6 kNm ...(*Given*)]

i.e. **τ = 166·67 MN/m²   (Ans.)**

(*ii*) **Angle of twist per unit length:**

$$\theta = \frac{3T}{C \sum\limits_{i=1}^{3} b_i t_i^3} = \frac{3 \times 6 \times 10^3}{80 \times 10^9 \times 188 \times 10^{-8}} = 0 \cdot 1197\ \text{rad.}/\text{m}$$

$$= 0 \cdot 1197 \times \frac{180}{\pi} = 6 \cdot 858°$$

i.e. **θ = 6·858°   (Ans.)**

Example 13·48. *A thin-walled bar section shown in Fig. 13·38 has a constant wall thickness t, one compartment is slit open; it is subjected to a twisting moment T. If C is the shear modulus of the section find:*

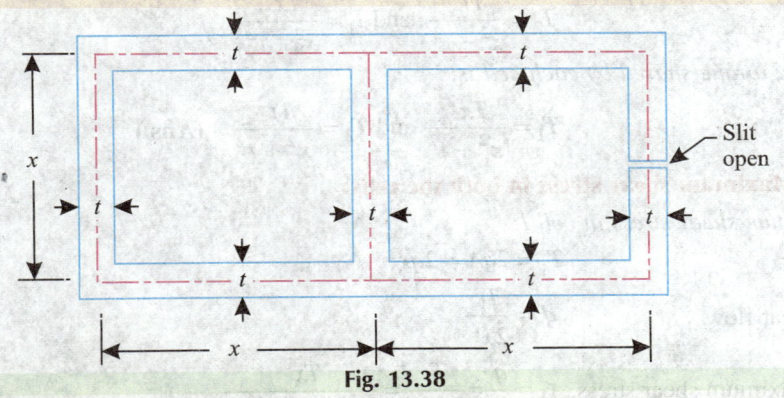

Fig. 13.38

(*i*) *Torque shared by each cell;*

(*ii*) *Maximum shear stress in both the cells;*

(*iii*) *Angular twist per unit length.*

Solution. (*i*) **Torque shared by each cell:**

For the closed cell portion (Cell I): Area, $A = x^2$

$$\oint \frac{ds}{t} = \frac{4x}{t}$$

Let, q = Shear flow in cell *I*, and

T_1 = Torque shared by cell I.

Angular twist per unit length,

$$\theta_1 = \frac{T_1}{4A^2 C} \oint \frac{ds}{t} = \frac{T_1}{4x^4 C} \times \frac{4x}{t} = \frac{T_1}{Cx^3 t} \qquad \text{...(i)}$$

For the slit-up cell (Cell II):

Breadth, $b = x + x + \dfrac{x}{2} + \dfrac{x}{2} = 3x$

Thickness = t

Also, $$\sum bt^3 = 3xt^3; \quad \sum bt^2 = 3xt^2$$

Let, T_2 = Torque shared by cell II.

Angular twist per unit length,

$$\theta_2 = \frac{3T_2}{C\sum bt^3} = \frac{3T_2}{C \times 3xt^3} = \frac{T_2}{Cxt^3} \qquad \qquad ...(ii)$$

But both the cells I and II are integral, for continuity.

Now, $$\theta_1 = \theta_2$$

$\therefore$ $$\frac{T_1}{Cx^3t} = \frac{T_2}{Cxt^3}$$

or, $$T_1 Cxt^3 = T_2 \cdot Cx^3t \quad \text{or} \quad \frac{T_1}{T_2} = \frac{x^2}{t^2} \quad \text{or} \quad T_1 = T_2 \times \frac{x^2}{t^2}$$

But, $$T_1 + T_2 = T$$

$\therefore$ $$T_2 \times \frac{x^2}{t^2} + T_2 = T \qquad \text{or} \qquad T_2\left(1 + \frac{x^2}{t^2}\right) = T$$

or, $$T_2 = \frac{Tt^2}{x^2 + t^2}, \text{ and } T_1 = \frac{Tx^2}{x^2 + t^2} \qquad \qquad ...(iii)$$

Hence, torque shared by each cell is:

$$T_1 = \frac{Tx^2}{x^2 + t^2} \text{ and } T_2 = \frac{Tt^2}{x^2 + t^2} \quad \textbf{(Ans.)}$$

(ii) Maximum shear stress in both the cells:

Maximum shear stress in cell I:

$$T_1 = 2qA = 2qx^2$$

or, shear flow, $$q = \frac{T_1}{2x^2}$$

$\therefore$ Maximum shear stress, $$\tau_1 = \frac{q}{t} = \frac{T_1}{2x^2t} = \frac{Tx^2}{2x^2t\,(x^2 + t^2)}$$

$$= \frac{T}{2t\,(x^2 + t^2)} \quad \textbf{(Ans.)}$$

Maximum shear stress in cell II:

$$\tau_2 = \frac{3T_2}{\sum bt^2} = \frac{3 \times Tt^2}{(x^2 + t^2)\,(3xt^2)} = \frac{T}{x\,(x^2 + t^2)}$$

i.e. $$\tau_2 = \frac{T}{2t\,(x^2 + t^2)} \quad \textbf{(Ans.)}$$

(iii) Angular twist per unit length:

$$\theta_1 = \theta_2 = \frac{3T_2}{C\sum bt^3} = \frac{3 \times Tt^2}{(x^2 + t^2)\,C\,(3xt^3)}$$

$$= \frac{T}{Cxt\,(x^2 + t^2)} \quad \textbf{(Ans.)}$$

13·20. TORSION OF THIN-WALLED MULTI-CELL SECTIONS

Fig. 13·39 (a) shows a two cell section.

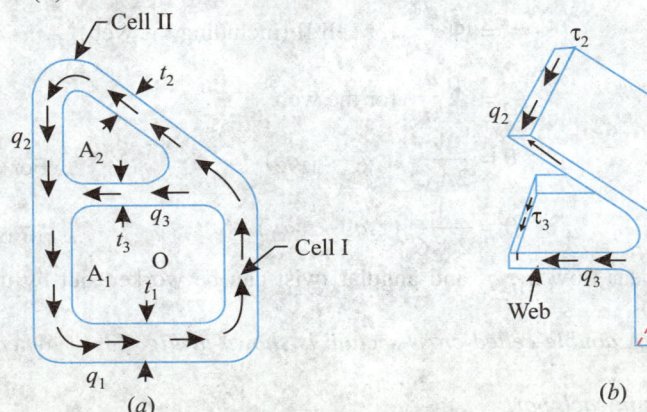

Fig. 13.39

Let,
$\qquad q_1 =$ Shear flow in cell I,

$\qquad q_2 =$ Shear flow in cell II,

$\qquad q_3 =$ Shear flow in the web,

$\qquad A_1 =$ Area of cell I, and

$\qquad A_2 =$ Area of cell II.

Taking a small length δl [Fig. 13·39 (b)] consider the equilibrium of shear forces at the junction of the two cells.

$$\tau_1 \cdot t_1 \cdot \delta l - \tau_2 \, t_2 \, \delta l - \tau_3 \, t_3 \, \delta l = 0 \;\ldots\text{(Equilibrium in the direction of the axis of the tube)}$$

(where τ_1, τ_2, τ_3 are the complementary shear stresses on the longitudinal sections of length δl)

or, $\qquad\qquad \tau_1 \cdot t_1 = \tau_2 \cdot t_2 + \tau_3 \cdot t_3$

or, $\qquad\qquad q_1 = q_2 + q_3 \qquad\qquad\qquad\qquad\qquad$ ...(13·61) $\quad (\because q = \tau \cdot t)$

or, shear flow in web, $\quad q_3 = q_1 - q_2$

Twisting moments:

Twisting moment due to q_1 in cell I, about O,

$$T_1 = 2 \, q_1 \, A_1 \qquad\qquad\qquad\qquad\qquad\text{...(13·62)}$$

Twisting moment due to q_2 in cell II,

$$T_2 = 2 \, q_2 \, (A_2 + A'_1) - 2 \, q_2 \, A'_1$$

or, $\qquad\qquad T_2 = 2 q_2 \, A_2 \qquad\qquad\qquad\qquad\qquad\text{...(13·63)}$

(where $2 \, q_2 \, A'_1$ is the moment due to the shear flow q_2 in the middle web).

∴ Total twisting moment, $\quad T = T_1 + T_2$

or, $\qquad\qquad T = 2 q_1 \, A_1 + 2 q_2 \, A_2 \qquad\qquad\qquad\text{...(13·64)}$

Angular twist per unit length:

The angular twist per unit length in each cell, *for continuity*, will be *same*.

For closed thin sections, $\qquad \theta = \dfrac{q}{2AC} \displaystyle\oint \dfrac{ds}{t}$

But in this case since shear flow is changing,

$$\therefore \qquad\qquad \theta = \dfrac{1}{2AC} \int \dfrac{q \, ds}{t}$$

Let,
$$z_1 = \oint \frac{ds}{t} \text{ for cell I (including the web)}$$

$$z_2 = \oint \frac{ds}{t} \text{ for cell II (including the web)}$$

$$z_{12} = \oint \frac{ds}{t} \text{ for the web}$$

Now,
$$\theta = \frac{1}{2A_1 C} (z_1 q_1 - z_{12} q_2) \qquad \text{... For cell I ...(13·65)}$$

and,
$$\theta = \frac{1}{2A_2 C} (z_2 q_2 - z_{12} q_1) \qquad \text{... For cell II ...(13·66)}$$

The values of shear flow q_1, q_2 and angular twist can be worked out by using equations (13·64), (13·65) (13·66).

Example 13·49. *A double-celled cross-section is shown in Fig. 13·40. When a torque of 5 kNm is applied find:*

(i) Shear stress in each part;

(ii) Angular twist per metre length.

Take: $\qquad\qquad C = 80 \text{ GN/m}^2.$

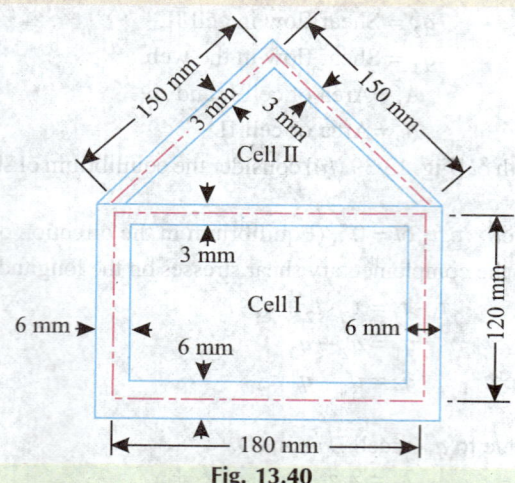

6 mm   Cell I   6 mm

Fig. 13.40

An anti-submarine warfare helicopter which is specially built for sea warfare.

Solution. Area, $\qquad A_1 = 180 \times 120 = 21600 \text{ mm}^2 = 0.0216 \text{ m}^2$

Area, $\qquad A_2 = \dfrac{1}{2} \times 180 \times \sqrt{150^2 - 90^2}$

$$= 10800 \text{ mm}^2 = 0.0108 \text{ m}^2$$

Applied torque, $\qquad T = 5 \text{ kNm}$

Line integrals $\qquad z_1 = \dfrac{180}{6} + \dfrac{120}{6} + \dfrac{120}{6} + \dfrac{180}{3} = 130$

$$z_2 = \dfrac{180}{3} + \dfrac{150}{3} + \dfrac{150}{3} = 160$$

$$z_{12} = \dfrac{180}{3} = 60$$

Let, $\qquad q_1 = $ Shear flow in cell I, and

$\qquad q_2 = $ Shear flow in cell II.

Now total torque, $\qquad T = T_1 + T_2 = 2\, q_1\, A_1 + 2\, q_2\, A_2$

or, $\qquad 5 \times 1000 = 2 \times q_1 \times 0.0216 + 2 \times q_2 \times 0.0108$

or, $\qquad 5000 = 0.0432\, q_1 + 0.0216\, q_2$

or, $\qquad 2q_1 + q_2 = 231481$ $\qquad\qquad$ ...(i)

Also $\qquad (\theta)_{cell–I} = (\theta)_{cell–II}$

$\therefore \qquad \dfrac{1}{2A_1 C}\,(z_1\, q_1 - z_{12}\, q_2) = \dfrac{1}{2A_2 C}\,(z_2\, q_2 - z_{12}\, q_1)$

or, $\qquad A_2\,(z_1\, q_1 - z_{12}\, q_2) = A_1\,(z_2\, q_2 - z_{12}\, q_1)$

or, $\qquad 0.0108\,(130q_1 - 60q_2) = 0.0216\,(160q_2 - 60q_1)$

or, $\qquad 130q_1 - 60q_2 = 2\,(160q_2 - 60q_1)$

or, $\qquad 250q_1 = 380q_2$

$\therefore \qquad q_1 = 1.52\, q_2$ $\qquad\qquad$ ...(ii)

Substituting the value of q_1 in eqn. (i), we get

$$2 \times 1.52\, q_2 + q_2 = 231481$$

$\therefore \qquad q_2 = 57297 \text{ N/m and } q_1 = 87091 \text{ N/m}$

(i) Shear stress in each part:

Shear stress in *rectangular part*

$$= \dfrac{87091}{0.006} \times 10^{-6} \text{ MN/m}^2 = \mathbf{14.5 \text{ MN/m}^2} \quad \textbf{(Ans.)}$$

Shear stress in *triangular part*

$$= \dfrac{57297}{0.003} \times 10^{-6} \text{ MN/m}^2 = \mathbf{19.1 \text{ MN/m}^2} \quad \textbf{(Ans.)}$$

Shear stress in *web*

$$= \dfrac{87091 - 57297}{0.003} \times 10^{-6} \text{ MN/m}^2 = \mathbf{9.93 \text{ MN/m}^2} \quad \textbf{(Ans.)}$$

(ii) Angular twist per metre length, θ:

$$\theta = \dfrac{1}{2A_1 C}\,(z_1\, q_1 - z_{12}\, q_2)$$

$$= \frac{1}{2 \times 0 \cdot 0216 \times 80 \times 10^9} [130 \times 87091 - 60 \times 57297]$$

$$\times \frac{180}{\pi} \text{ degree} = 0 \cdot 1307° \text{ per metre length}$$

i.e. $\qquad \theta = \mathbf{0 \cdot 1307°}$ **per metre length** **(Ans.)**

TYPICAL EXAMPLES (For Competitive Examinations)

Example 13·50. *The shaft shown in fig. 13·41 rotates at 200 r.p.m. with 30 kW and 15 kW taken off at A and B respectively and 45 kW applied at C. Find :*

(*i*)   *Maximum shear stress developed in the shaft;*

(*ii*)   *Angle of twist.*

Take: $\qquad C = 85 \ GN/m^2.$

Solution. Refer to Fig. 13·41

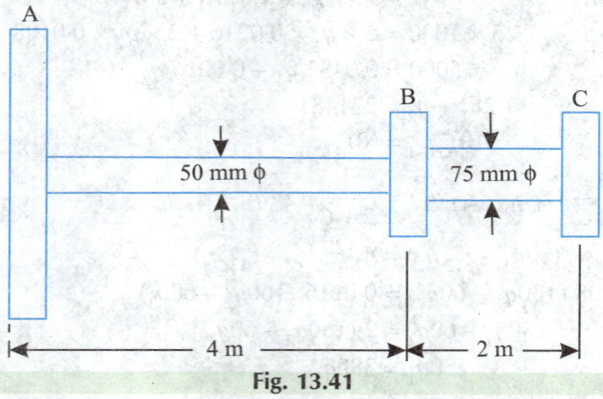

Fig. 13.41

Power transmitted across CB = Power applied at C = 45 kW

Power taken off at B = 15 kW

∴ Power transmitted across BA = 45 – 15 = 30 kW

(*i*)   **Maximum shear stress developed:**

For BC:

$$P = 45 \text{ kW}, N = 200 \text{ r.p.m}, D = 75 \text{ mm} = 0 \cdot 075 \text{ m}, l = 2 \text{ m}$$

The torque transmitted across BC is given by,

$$P = \frac{2 \pi NT}{60 \times 1000}$$

∴ $\qquad T = \dfrac{P \times 60 \times 1000}{2 \pi N} = \dfrac{45 \times 60 \times 1000}{2 \pi \times 200} = 2148 \cdot 6 \text{ Nm}$

The maximum shear stress in BC is given by,

$$T = \tau \times \frac{\pi}{16} \times D^3$$

$$\tau = \frac{16T}{\pi D^3} = \frac{16 \times 2148 \cdot 6}{\pi \times (0 \cdot 075)^3} \times 10^{-6} \text{ MN/m}^2 = 25 \cdot 94 \text{ MN/m}^2$$

The maximum angle of twist θ_1 in BC of B relative to C is given by

$$\frac{T}{I_p} = \frac{C \theta}{l}$$

$$\theta_1 = \frac{Tl}{CI_p} = \frac{2148 \cdot 6 \times 2}{85 \times 10^9 \times \frac{\pi}{32} \times (0 \cdot 075)^4} = 0 \cdot 01628 \text{ radian}$$

For AB:

$$P = 30 \text{ kW}; \ N = 200 \text{ r.p.m}; \ D = 50 \text{ mm}$$
$$= 0 \cdot 05 \text{ m}; \ l = 4 \text{ m}.$$

The torque transmitted across AB is given by,

$$T = \frac{30 \times 60 \times 1000}{2\pi \times 200} = 1432 \cdot 4 \text{ Nm}$$

The maximum shear stress in AB is given by,

$$\tau = \frac{16T}{\pi D^3} = \frac{16 \times 1432 \cdot 4}{\pi \times (0 \cdot 05)^3} \times 10^{-6} \text{ MN/m}^2$$
$$= 58 \cdot 36 \text{ MN/m}^2$$

The maximum angle of twist θ_2 in AB of A relative to B is given by,

$$\theta_2 = \frac{T \times l}{CI_p} = \frac{1432 \cdot 4 \times 4}{85 \times 10^9 \times \frac{\pi}{32} \times (0 \cdot 05)^4} = 0 \cdot 1098 \text{ radian}.$$

∴ The *maximum shear stress* in the shaft is developed in AB, being equal to **58·36 MN/m²(Ans.)**

(ii) **Angle of twist:**

Angle of twist of A relative to $C = \theta_1 + \theta_2 = 0 \cdot 01628 + 0 \cdot 1098$
$$= 0 \cdot 1261 \text{ rad.} = \textbf{7·22 degrees} \quad \textbf{(Ans.)}$$

Example 13·51. *Fig. 13·42 shows a circular steel shaft 45 mm in diameter provided with enlarged portions L and M. On to this enlarged portion a steel tube 3 mm thick is shrunk. While the shrinking process is going on the 45 mm diameter shaft is held twisted by a couple of magnitude 120 Nm. When the tube is firmly set on the shaft, this twisting couple is removed. Calculate what twisting couple is left on the shaft.*

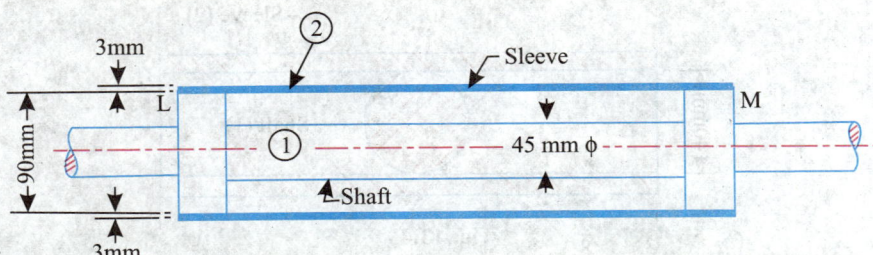

Fig. 13.42

Assume that the shaft and the tube are made of the same material.

Solution. *For shaft*: Diameter of the shaft = 45 mm = 0·045 m

∴ Polar moment of inertia, $I_{P_1} = \frac{\pi}{32} \times 0 \cdot 045^4 = 4 \cdot 026 \times 10^{-7} \text{ m}^4$

∴ $$\theta_1 = \frac{T_1 l}{C I_{P_1}} = \frac{120}{4 \cdot 026 \times 10^{-7}} \cdot \frac{l}{C} = 29 \cdot 81 \times 10^7 \times \frac{l}{C}$$

(where, T_1 = torque or couple applied on the shaft = 120 Nm)

For sleve: Inner diameter $\quad$ = 90 mm = 0·09 m

Outer diameter $\quad$ = 90 + 2 × 3 = 96 mm = 0·096 m

Polar moment of inertia, $\quad I_{p_2} = \dfrac{\pi}{32}(0·096^4 - 0·09^4) = 1·897 \times 10^{-6}\,\text{m}^4$

Let, $\quad T_s$ = Couple left on the shaft.

∴ Corresponding twist of shaft $= \theta'_1 = \dfrac{T_s l}{C I_{p_1}} = \dfrac{T_s l}{C \times 4·026 \times 10^{-7}}$

Corresponding twist in sleeve $= \theta_2 = \dfrac{T_s l}{C I_{p_2}} = \dfrac{T_s l}{C \times 1·897 \times 10^{-6}}$

But, $\quad\quad \theta'_1 + \theta_2 = \theta_1$

$$\dfrac{T_s l}{C \times 4·026 \times 10^{-7}} + \dfrac{T_s l}{C \times 1·897 \times 10^{-6}} = 29·8 \times 10^7 \times \dfrac{l}{C}$$

or, $\quad T_s (0·248 \times 10^7 + 0·0527 \times 10^7) = 29·81 \times 10^7$

or, $\quad\quad\quad 0·3007\,T_s = 29·8$

or, $\quad\quad\quad\quad T_s = \textbf{99 Nm} \quad \textbf{(Ans.)}$

Example 13·52. *A compound shaft is made by rigidly securing a sleeve of phosphor bronze on a solid steel shaft 40 mm diameter. If the maximum shear stresses to be attained in the two materials of the shafts are 40 MN/m² and 64 MN/m², for bronze and steel respectively, determine:*

(*i*) *Outer diameter of the sleeve;*

(*ii*) *Torque transmitted by the compound shaft;*

(*iii*) *Power transmitted by the shaft at 200 r.p.m.;*

(*iv*) *Angle of twist of the shaft per metre length.*

Show the shear stress and strain variations on the shaft cross-section.

Take: $\quad\quad\quad C_{steel} = 80\ \text{GN/m}^2,\ and\ C_{bronze} = 40\ \text{GN/m}^2.$

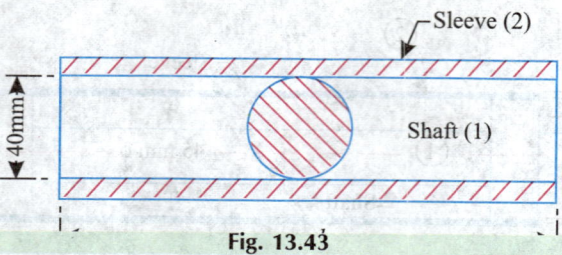

Fig. 13.43

Solution. Let steel and phosphor bronze be represented by the subscripts 1 and 2 respectively.

Given: Diameter of the shaft,

$$D_1 = 40\ \text{mm} = 0·04\ \text{m}$$

Maximum shear stress in steel,

$$\tau_1 = 64\ \text{MN/m}^2$$

Maximum shear stress in phosphor bronze,

$$\tau_2 = 40\ \text{MN/m}^2$$
$$C_1 = 80\ \text{GN/m}^2;\ C_2 = 40\ \text{GN/m}^2.$$

(i) Outer diameter of the sleeve, D_2:

Using the relation: $\dfrac{\tau}{R} = \dfrac{C\theta}{l}$; we have $\dfrac{\theta}{l} = \dfrac{\tau}{CR}$

$\therefore \qquad \dfrac{\tau_1}{C_1 R_1} = \dfrac{\tau_2}{C_2 R_2}$ $\qquad \left[\because \dfrac{\theta}{l} \text{ is constant in this case} \right]$

or, $\qquad R_2 = \dfrac{\tau_2 C_1 R_1}{\tau_1 C_2}$

Substituting the various values, we get

$$R_2 = \dfrac{(40 \times 10^6) \times (180 \times 10^9) \times (0.04/2)}{(64 \times 10^6) \times (40 \times 10^9)} = 0.025 \text{ m}$$

$\therefore \qquad D_2 = 0.025 \times 2 = 0.05 \text{ m or } \textbf{50 mm} \quad \textbf{(Ans.)}$

(ii) Torque transmitted by the compound shaft, T:

Using the relation,

$$\dfrac{T}{I_p} = \dfrac{C\theta}{l}, \qquad \text{we have } \dfrac{\theta}{l} = \dfrac{T}{C I_p}$$

$\therefore \qquad \dfrac{T_1}{C_1 I_{p_1}} = \dfrac{T_2}{C_2 I_{p_2}}$ $\qquad \left(\because \dfrac{\theta}{l} \text{ is constant} \right)$

or, $\qquad \dfrac{T_1}{T_2} = \dfrac{C_1 I_{p_1}}{C_2 I_{p_2}} = \dfrac{80}{40} \times \dfrac{\dfrac{\pi}{32} \times 0.04^4}{\dfrac{\pi}{32} [(0.05)^4 - (0.04)^4]} = 1.387$

But, $\qquad T_1 = \tau_1 \times \dfrac{\pi}{16} \times D_1^3 = 64 \times 10^6 \times \dfrac{\pi}{16} \times (0.04)^3 = 804 \text{ Nm}$

$\therefore \qquad T_2 = \dfrac{804}{1.387} = 580 \text{ Nm}$

Total torque transmitted,

$$T = T_1 + T_2 = 804 + 580 = 1384 \text{ Nm or } 1.384 \text{ kNm}$$

Hence, $\qquad T = \textbf{1.384 kNm} \quad \textbf{(Ans.)}$

(iii) Power transmitted, P:

$$P = \dfrac{2\pi NT}{60} \text{ kW} \qquad \qquad (\text{where } T \text{ is in kN/m})$$

$$= \dfrac{2\pi \times 2000 \times 1.384}{60} = 289.8 \text{ kW}$$

Hence, $\qquad P = \textbf{289.8 kW} \quad \textbf{(Ans.)}$

(iv) Angle of twist per metre length, $\dfrac{\theta}{l}$:

We know, $\qquad \dfrac{\theta}{l} = \dfrac{T}{C I_p}$

As $\dfrac{\theta}{l}$ is constant,

$\therefore \qquad \dfrac{\theta}{l} = \dfrac{T_1}{C_1 I_{p_1}} \left(= \dfrac{T_2}{C_2 I_{p_2}} \right) = \dfrac{804}{80 \times 10^9 \times \dfrac{\pi}{32} \times 0.04^4} = 0.03998 \text{ rad.}$

$$= 0.03998 \times \dfrac{180}{\pi} \text{ degrees} = 2.29°$$

Hence, *angle of twist per metre length* = **2·29°** (Ans.)

$$\left[\text{Check:} \ \frac{\theta}{l} = \frac{T_2}{C_2 I_{P2}} = \frac{580}{40 \times 10^9 \times \frac{\pi}{32}(0.05^4 - 0.04^4)} \times \frac{180}{\pi} \ \text{degrees} = 2.29° \right]$$

Shear stress and strain variations:

Shear stress variations :

Shear stresses at the junction of two materials are given by,

$$\tau_1 = 64 \ \text{MN/m}^2$$
$$\tau'_2 = \text{Shear stress in the sleeve at the junction}$$
$$= \frac{T_2}{I_{P2}} \times R' = \frac{580 \times (0.04/2)}{\frac{\pi}{32}[0.05^4 - 0.04^4]} \times 10^{-6} \ \text{MN/m}^2 = 32 \ \text{MN/m}^2$$

$$\left[\text{Otherwise:} \ \tau'_2 = \tau_2 \times \frac{R'}{R_2} = 40 \times \frac{0.02}{0.025} = 32 \ \text{MN/m}^2 \right]$$

Shear strain variations:

Shear strain at the outer surface is

$$= \frac{\tau_2}{C_2} = \frac{40 \times 10^6}{40 \times 10^9} = 1 \times 10^{-3}$$

Shear strain at the common surface is

$$= \frac{\tau_1}{C_1} = \frac{64 \times 10^6}{80 \times 10^9} = 0.8 \times 10^{-3}$$

Shear stress and shear strain variations are shown in Fig. 13·44.

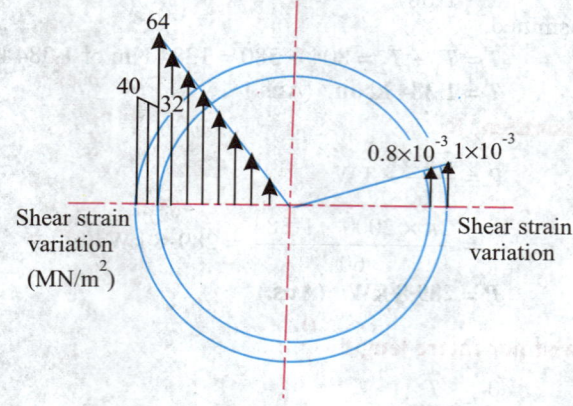

Fig. 13.44

Example 13·53. *Fig 13·45 (a) shows a stepped circular bar subjected to torques as shown:*

(i)   Find the twist of the free end and show the variation of twist for the bar.

(ii) Show the variation of shear stress at the surface of the bar for whole length of the bar. What is the maximum shear stress?

(iii) Where is the twist equal to zero?

Take: 80 GN/m².

Solution. Fig 13·45 (*c*) shows the variation of torque. When seen from the right end in the direction of the axis *UL*, a clockwise torque is taken positive otherwise it is taken negative.

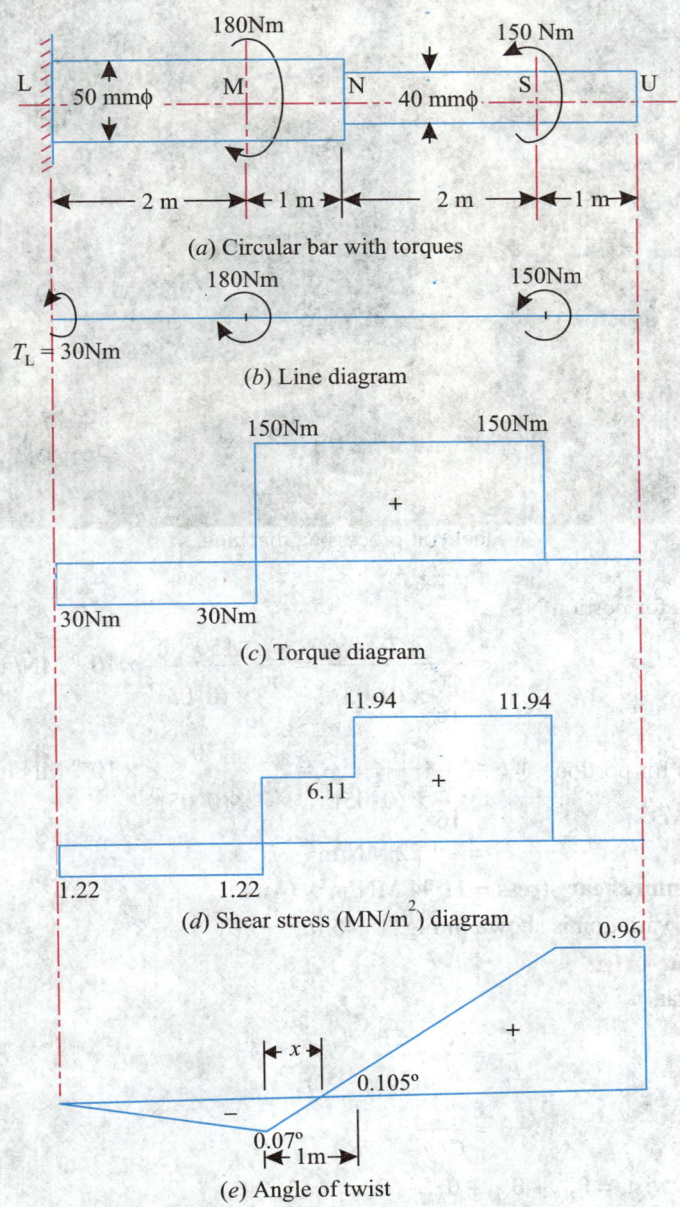

(*a*) Circular bar with torques

(*b*) Line diagram

(*c*) Torque diagram

(*d*) Shear stress (MN/m^2) diagram

(*e*) Angle of twist

Fig. 13.45

(*i*) Variation of shear stresses:

Shear stress for the portion US = 0

Shear stress for portion SN

$$= \frac{T}{\dfrac{\pi}{16} \times (0 \cdot 04)^3} = \frac{150}{\dfrac{\pi}{16} \times (0 \cdot 04)^3} \times 10^{-6} \ \text{MN/m}^2 = 11 \cdot 94 \ \text{MN/m}^2$$

Steel coil processing machine.

Shear stress for portion NM

$$= \frac{T}{\frac{\pi}{16} \times (0 \cdot 05)^3} = \frac{150}{\frac{\pi}{16} \times (0 \cdot 05)^3} \times 10^{-6} \text{ MN/m}^2 = 6 \cdot 11 \text{ MN/m}^2$$

Shear stress for portion $ML = \dfrac{T}{\dfrac{\pi}{16} \times (0 \cdot 05)^3} = \dfrac{-30}{\dfrac{\pi}{16} \times (0 \cdot 05)^3} \times 10^{-6} \text{ MN/m}^2$

$$= - 1 \cdot 22 \text{ MN/m}^2$$

∴   **Maximum shear stress = 11·94 MN/m² (Ans.)**

Shear stress variation is shown in Fig. 13·45 (*d*)

(*ii*) Angular twist:

We know that,

$$\frac{T}{I_p} = \frac{\tau}{R} = \frac{C\theta}{l}$$

∴

$$\theta = \frac{Tl}{CI_p} = \frac{\tau l}{CR}$$

∴ Angle of twist $= \theta_{US} + \theta_{SN} + \theta_{NM} + \theta_{ML}$

$$= 0 + \frac{T_{SN}\, l_{SN}}{C_{SN}\, I_{p_{SN}}} + \frac{T_{NM}\, l_{NM}}{C_{NM}\, I_{p_{NM}}} + \frac{T_{ML}\, l_{ML}}{C_{ML}\, I_{p_{ML}}}$$

$$= \frac{I}{C} \left[\frac{150 \times 2}{\frac{\pi}{32} \times (0 \cdot 04)^4} + \frac{150 \times 1}{\frac{\pi}{32} \times (0 \cdot 05)^4} - \frac{30 \times 2}{\frac{\pi}{32} \times (0 \cdot 05)^4} \right]$$

$$= \frac{10^9}{80 \times 10^9} (1 \cdot 194 + 0 \cdot 244 - 0 \cdot 0978) \times \frac{180}{\pi} \text{ degree}$$

$$= 0 \cdot 716 (1 \cdot 194 + 0 \cdot 244 - 0 \cdot 0978) \text{ degree}$$

$$= 0 \cdot 855° + 0 \cdot 175° - 0 \cdot 07°$$

Twist at $M = -0 \cdot 07°$

Twist at $N = 0 \cdot 175 - 0 \cdot 07 = 0 \cdot 105°$

Twist at $S = 0 \cdot 855 + 0 \cdot 175 - 0 \cdot 07 = 0 \cdot 96°$

Twist at $U = 0 \cdot 96 + 0 = 0 \cdot 96°$

Hence twist at the free end = 0·96°   **(Ans.)**

Note. [The same results will be obtained by using $\theta = \sum \dfrac{\tau l}{CR}$ with due regard to positive and negative signs of shear stresses.]

The variation of angle of twist with respect to the fixed end is shown in Fig. 13·45 (*e*).

(*iii*) Position of zero twist:

The point of zero twist lies between L and N. Let it be at a distance of x metre from point M towards N.

$$0 = 0 \cdot 716 \left(-0 \cdot 0978 + \frac{0 \cdot 244x}{1} \right)$$

∴ $0 \cdot 244x = 0 \cdot 0978$ or   **$x = 0 \cdot 4$ m**   **(Ans.)**

It may be noted that the results of above problem apply if a shaft is rotating at constant speed under the given condition of torques and dimensions.

Example 13·54. *Fig. 13·46 shows a circular tube with 8 similar longitudinal fins attached to it. It is subjected to a twisting moment T. Find :*

(i) Percentage of T taken by the fins;

(ii) Maximum shear stresses induced in the fins and the tube if T = 400 Nm.

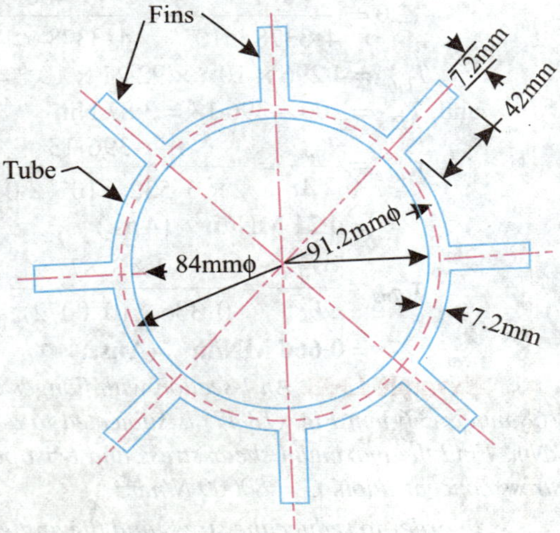

Solution.

(*i*) Percentage of *T* taken by fins:

Let, θ = Angular twist per unit length of whole section.

For tube:

$$\theta = \frac{T}{4A^2C} \oint \frac{ds}{t}$$

$$= \frac{T}{4A^2C} \cdot \frac{l_p}{t}$$

or, $T = \dfrac{4A^2 Ct\theta}{l_p}$

Fig. 13.46

where, l_p = Peripherial length = πD_{mean} = $\pi \times 91 \cdot 2 \times 10^{-3}$ = 0·286 m.

$A = \pi/4 \times (D_{mean})^2 = \pi/4 \times (91 \cdot 2 \times 10^{-3})^2 = 6 \cdot 532 \times 10^{-3} \text{ m}^2$

∴ $T_{tube} = \dfrac{4 \times (6 \cdot 532 \times 10^{-3})^2 \times C \times (7 \cdot 2 \times 10^{-3}) \, \theta}{0 \cdot 286} = 4 \cdot 296 \times 10^{-6} \, C.\theta$

For fins:

$$b = 8 \times 42 = 336 \text{ mm or } 0.336 \text{ m}$$
$$t = 7.2 \text{ mm} = 0.0072 \text{ m}$$
$$\theta = \frac{3T}{Cbt^3}$$

$$\therefore \quad T_{fins} = \frac{Cbt^3 \theta}{3} = \frac{C \times 0.336 \times (0.0072)^3 \theta}{3}$$

$$= 4.18 \times 10^{-8} \, C.\theta$$

$$T = T_{tube} + T_{fins}$$
$$= 4.296 \times 10^{-6} \, C.\theta + 4.18 \times 10^{-8} \, C.\theta$$

i.e. $\quad T = 4.3378 \times 10^{-6} \, C.\theta$

$\therefore$ Percentage of

$$T_{fins} = \frac{T_{fins}}{T} \times 100$$
$$= \frac{4.18 \times 10^{-8} \, C.\theta}{4.3378 \times 10^{-6} \, C.\theta} \times 100 = \textbf{96.4\%} \textbf{ (Ans.)}$$

Another view of crankshaft and crankcase.

(ii) **Maximum shear stress induced:**

$$T = 400 \text{ Nm}$$

$\therefore$ From eqn. (*i*), we have

$$C.\theta = \frac{T}{4.3378 \times 10^{-6}} = \frac{400}{4.3378 \times 10^{-6}} = 92.21 \times 10^6$$

$$\therefore \quad T_{tube} = 4.296 \times 10^{-6} \times 92.21 \times 10^{-6} = 396.13 \text{ Nm}$$

and, $T_{fins} = 400 - 396.13 = 3.87$ Nm

$$\tau_{tube} = \frac{T_{tube}}{2 \, At} = \frac{396.13}{2 \times 6.532 \times 10^{-3} \times 0.0072} \times 10^{-6} \text{ MN/m}^2$$

$$= \textbf{4.21 MN/m}^2 \textbf{ (Ans.)}$$

$$\tau_{fins} = \frac{3T_{fins}}{bt^2} = \frac{3 \times 3.87}{0.336 \times (0.0072)^2} \times 10^{-6} \text{ MN/m}^2$$

$$= \textbf{0.666 MN/m}^2 \textbf{ (Ans.)}$$

Example 13·55. *An I-section with flanges 60 mm × 6 mm and web 168 mm × 3·6 mm Fig. (13·47) is subjected to a twisting moment of 0.24 kNm. Find the maximum shear stress and twist per unit length neglecting stress concentration. C = 80000 N/mm².*

In order to reduce the stress and the angle of twist per unit length, the I-section is reinforced by welding steel plates 168 mm × 6 mm as shown in Fig. 13·47. Find the maximum shear stress due to same twisting moment and also the value of angle of twist per unit length.

Solution.

Case I: *I-section.* Refer to Fig 13.47.

Flanges (2 No.) : 60 mm × 6 mm

Web (1 No.) : 168 mm × 3·6 mm

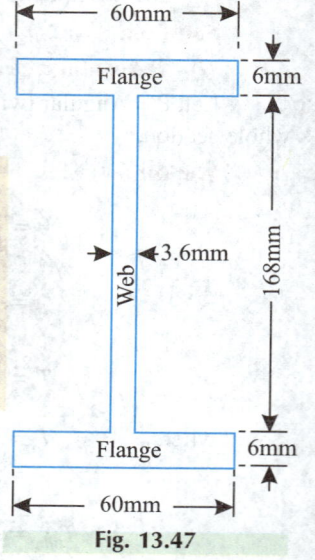

Fig. 13.47

$$\Sigma bt^2 = 2 \times 60 \times 6^2 + 168 \times 3 \cdot 6^2 = 6497 \text{ mm}^3$$
$$\Sigma bt^3 = 2 \times 60 \times 6^3 + 168 \times 3 \cdot 6^3 = 33758 \text{ mm}^4$$

Maximum shear stress,

$$\tau = \frac{3T}{\Sigma bt^2} = \frac{3 \times 0 \cdot 24 \times 10^6}{6497} \text{ N/mm}^2$$
$$= 110 \cdot 8 \text{ N/mm}^2 \text{ or } 110 \cdot 8 \text{ MN/m}^2 \quad \textbf{(Ans.)}$$

Angular twist per unit length,

$$\theta = \frac{3T}{C\,\Sigma bt^3} = \frac{3 \times 0 \cdot 24 \times 10^6}{80000 \times 33758}$$
$$= 2 \cdot 66 \times 10^{-4} \text{ rad./mm}$$

Case II. *Reinforced I-section:* Refer to Fig. 13.48.

There are two cells of area,

$$A_1 = A_2 = 27 \times 174 = 4698 \text{ mm}^2$$

Line integrals, $\quad z_1 = z_2 = \dfrac{27}{6} + \dfrac{27}{6} + \dfrac{174}{6} + \dfrac{174}{3 \cdot 6} = 86 \cdot 33$

$$z_{12} = \frac{174}{3 \cdot 6} = 48 \cdot 33$$

Let, $\qquad q_1$ = Shear flow in cell I, and

$\qquad\qquad q_2$ = Shear flow in cell II.

Then, torque $\quad T = 2q_1 A_1 + 2q_2 A_2$

or, $\qquad T = (q_1 + q_2)\,(2 \times 4698) = 9396\,(q_1 + q_2)$
$$\dots(i)$$

Also, $\quad \dfrac{1}{2A_1C}\,(z_1 q_1 - z_{12} q_2) = \dfrac{1}{2A_2C}\,(z_2 q_2 - z_{12} q_1)$

But, $\qquad A_1 = A_2$

$\therefore \quad 86 \cdot 33\, q_1 - 48 \cdot 33\, q_2 = 86 \cdot 33\, q_2 - 48 \cdot 33\, q_1$

or, $\qquad q_1 = q_2 \quad \dots(ii)$

From (i) and (ii), we get

$$q_1 = q_2 = \frac{T}{2 \times 9396}$$
$$= \frac{0 \cdot 24 \times 10^6}{2 \times 9396} = 12 \cdot 77 \text{ N/mm}$$

and, $\quad q_3 = 0$ (in the web)

Maximum shear stress,

$$\tau = \frac{q_1}{t} = \frac{12 \cdot 77}{6}$$
$$= 2 \cdot 13 \text{ N/mm}^2 \text{ (or } 2 \cdot 13 \text{ MN/m}^2) \quad \textbf{(Ans.)}$$

Angular twist per metre length;

$$\theta = \frac{1}{2A_1\,C}\,(z_1 q_1 - z_{12} q_2)$$
$$= \frac{12 \cdot 77\,(86 \cdot 33 - 48 \cdot 33)}{2 \times 4698 \times 80000} = 6 \cdot 456 \times 10^{-7} \text{ rad./mm}$$
$$= 0 \cdot 0396 \times 10^{-3} \text{ degree/mm}$$

or, $\qquad \theta = 0 \cdot 0369°/\text{metre length} \quad \textbf{(Ans.)}$

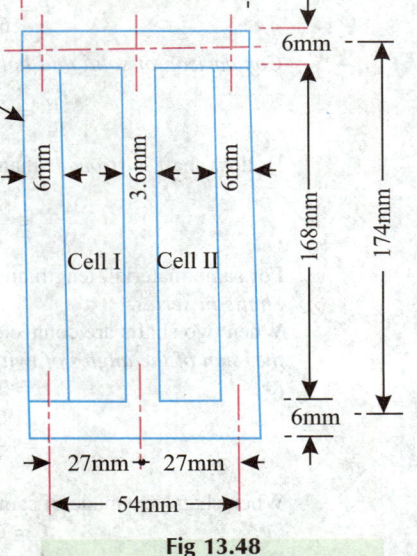

Fig 13.48

Shaft of an asynchronous electric motor.

HIGHLIGHTS

1. *Torsion equation:*

$$\frac{T}{I_p} = \frac{\tau}{R} = \frac{C\theta}{l}$$

2. *Polar moment of intertia of the shafts:*

$$I_p = \frac{\pi}{32} D^4 \quad \text{... for a } solid \text{ shaft}$$

and,

$$I_p = \frac{\pi}{32} (D^4 - d^4) \quad \text{.... for a } hollow \text{ shaft}$$

3. *Torque transmitted by the shafts:*

$$T = \tau \cdot \frac{\pi}{16} D^3 \quad \text{... for a } solid \text{ shaft}$$

$$T = \tau \cdot \frac{\pi}{16} \left(\frac{D^4 - d^4}{D} \right) \quad \text{... for a } hollow \text{ shaft}$$

4. *Power transmitted by the shaft:*

$$P = \frac{2\pi NT}{60 \times 1000} \text{ kW}, \text{ where } T \text{ is the mean/average torque in Nm.}$$

5. *Comparison of solid and hollow shafts:*

$$\frac{T_H}{T_S} = \frac{n^2 + 1}{n\sqrt{n^2 - 1}}, \text{ where } n = \frac{D_H}{d_H}$$

   Hollow shaft is *stronger* than a solid shaft,

$$\frac{W_H}{W_S} = \frac{(n^2 - 1) \, n^{2/3}}{(n^4 - 1)^{2/3}}$$

   For same material, length and torque, *weight of hollow shaft is less than that of a solid shaft.*

6. *Shafts in series:*
   When two shafts are connected in series *each shaft transmits the same torque; the angle of twist is the sum of the angles of twist of the two shafts.*
   i.e. $\quad \theta = \theta_1 + \theta_2$

$$= \frac{T l_1}{C_1 I_{p1}} + \frac{T l_2}{C_2 I_{p2}} = T \left[\frac{l_1}{C_1 I_{p1}} + \frac{l_2}{C_2 I_{p2}} \right]$$

   When shafts are made of same material
   i.e. $\quad C = C_1 = C_2$

   Then, $\qquad \theta = \frac{T}{C} \left[\frac{l_1}{I_{p1}} + \frac{l_2}{I_{p2}} \right]$

   Here the driving torque is applied at one end and the resisting torque at the other.

7. *Shafts in parallel:*
   The shafts are said to be in parallel when the driving torque is applied at the junction of the shafts and resisting torque is at the other ends of the shafts. Here the *angle of twist is same for each shaft, but the applied torque is divided between the two shafts.*
   i.e. $\quad \theta_1 = \theta_2$

   or $\qquad \dfrac{T_1 l_1}{C_1 I_{p1}} = \dfrac{T_2 l_2}{C_2 I_{p2}}, \qquad$ and $\qquad T = T_1 + T_2$

   If the shafts are of the same material
   i.e. $\qquad C_1 = C_2$

Then, $\dfrac{T_1 l_1}{I_{p1}} = \dfrac{T_2 l_2}{I_{p2}}$ or $\dfrac{T_1}{T_2} = \dfrac{I_{p1} l_2}{I_{p2} l_1}$

When torque is shared equally by both the shafts

i.e. $T_1 = T_2$,   then $I_{p1} l_2 = I_{p2} l_1$

8. *Torsional resilience:*

$$U = \frac{\tau^2}{4C} \cdot \frac{D^2 + d^2}{D^2} \times \text{volume of shaft ...... for a \textit{hollow} shaft}$$

$$U = \frac{\tau^2}{4C} \times \text{volume of shaft ...... for a \textit{solid} shaft } (d = 0)$$

$$U = \frac{\tau^2}{2C} \times \text{volume of shaft ..... for a very thin \textit{hollow} shaft}$$

9. *Shaft couplings:*

$$T = n_b \times \frac{\pi}{4} \times d_b^2 \times \tau_b \times R_b$$

where, T = Torque transmitted,

n_b = Number of bolts,

τ_b = Shear strees in bolts, and

R_b = Radius of bolt circle.

Also, $T = l_k \times b_k \times \tau_k \times R$

where, l_k, b_k and τ_k are the length, breadth and maximum permissible shear stress in the key.

and,

$$T = l_k \times \frac{t_k}{2} \times \sigma_{ck} \times R.$$

$(R$ = radius of the shaft).

Parallel shaft helical geared motor.

10. *Combined bending and torsion:*

Equivalent bending moment,

$$M_e = \frac{M + \sqrt{M^2 + T^2}}{2}$$

Equivalent torque, $T_e = \sqrt{M^2 + T^2}$

Also, $\sigma_{max} = \dfrac{\sigma_d}{2} + \sqrt{\left(\dfrac{\sigma_d}{2}\right)^2 + \tau^2}$,   and $\tau_{max} = \sqrt{\left(\dfrac{\sigma_d}{2}\right)^2 + \tau^2}$

$[\sigma_d$ is caused by bending$]$

11. *Torsion of a tapering shaft:*

$$\theta = \frac{2}{3} \frac{Tl}{C\pi} \left[\frac{R_1^2 + R_1 R_2 + R_2^2}{R_1^3 R_2^3} \right]$$

12. *Thin circular tube subjected to torsion:*

$$T = \frac{\tau \pi D^2 t}{2} \qquad\qquad\qquad ...(i)$$

$$\theta = \frac{4Tl}{\pi D^3 tC} \qquad\qquad\qquad ...(ii)$$

where, D = External diameter of the tube.

Strength-weight ratio, $\dfrac{T}{W} = \dfrac{\tau D}{2lw}$

(where, w = Weight density of the material)

13. *Torsion of Non-circular solid sections*:

    (*i*) *Rectangular section*:

        Maximum shear stress, $\tau = \alpha\, \dfrac{T}{bh^2}$ ...(*i*)

        Angular twist, $\theta = \dfrac{\beta}{bh^3} \cdot \dfrac{Tl}{C}$ ...(*ii*)

        where, b = Larger side of the rectangular shaft,

                              h = Smaller side of the rectangular shaft,

                              l = Length of the shaft,

                              T = Applied torque, and

                              C = Modulus of rigidity.

$$\alpha = 3 + 1\cdot 8 \left(\dfrac{h}{b}\right) \ \text{... approximate value, and}$$

$$\beta = 3\cdot 5 \left[1 + \left(\dfrac{h}{b}\right)^2\right] \ \text{... approximate value.}$$

    (*ii*) *Elliptical section* :

        Maximum shear stress, $\tau = \dfrac{16T}{\pi ab^2}$ ...(*i*)

        Angular twist, $\theta = \dfrac{16lT}{\pi abC} \left(\dfrac{1}{a^2} + \dfrac{1}{b^2}\right)$ ...(*ii*)

        Torsional stiffness, $k = \dfrac{C\pi a^3 b^3}{16\,(a^2 + b^2)}$ ...(*iii*)

        where, a = major axis; b = minor axis.

    (*iii*) *Equivalent triangle*:

        Maximum shear stress, $\tau = \dfrac{20T}{a^3}$ ...(*i*)

        Angular twist, $\theta = \dfrac{80}{a^4 \sqrt{3}} \cdot \dfrac{Tl}{C}$ ...(*ii*)

        Torsional stiffness, $k = \dfrac{\sqrt{3}}{80}\, a^4 C$ ...(*iii*)

14. *Torsion of non-circular thin tubular sections*:

    $T = 2\,q\,A$, this equation is generally known as *Bredt-Batho formula*

    where , q = Shear flow.

    Shear stress, $\tau = \dfrac{T}{2At}$ $(\because q = \tau . t)$

    Angular twist, $\theta = \dfrac{Tl}{4A^2 C} \oint \dfrac{ds}{t}$

    where, $\oint \dfrac{ds}{t}$ is the line integral of the ratio of length divided by thickness.

15. For a section built up of thin rectangular sections such as *T*, Channel, *I* etc., subjected to twisting moment *T*:

    Maximum shear stress, $\tau = \dfrac{3\,T}{\Sigma bt^2}$;

    Angular twist per unit length, $\theta = \dfrac{3T}{C\Sigma bt^3}$

16. *Torsion of thin-walled two-cell section:*

Shear flow $\qquad q_1 = q_2 + q_3$

Twisting moment, $\qquad T = T_1 + T_2$, $T_1 = q_1 A_1$ and $T_2 = q_2 A_2$

where, A_1, A_2 = Areas enclosed by the centre lines of the cells I and II respectively

$\qquad q_1$ = Shear flow in cell I;

$\qquad q_2$ = Shear flow in cell II;

$\qquad q_3$ = Shear flow in middle web.

Angle of twist per unit length:

$$Q = \frac{1}{2A_1 C}(z_1 q_1 - z_{12} q_2) = \frac{1}{2A_2 C}(z_2 q_2 - z_{12} q_1)$$

where, $\qquad z_1 = \oint \dfrac{ds}{t}$ for cell I (including the web)

$\qquad z_2 = \oint \dfrac{ds}{t}$ for cell II (including the web)

$\qquad z_{12} = \oint \dfrac{ds}{t}$ for the middle web.

OBJECTIVE TYPE QUESTIONS

Choose the Correct Answer:

1. The shafts are made of

(*a*) mild steel $\qquad$ (*b*) alloy steel

(*c*) copper alloys $\quad$ (*d*) any of the above.

2. The shafts are designed on the basis of

(*a*) strength

(*b*) rigidity

(*c*) either of the above

(*d*) both (*a*) and (*b*).

3. In shafts with keyways the allowable stresses are usually....of the value given.

(*a*) 25 percent $\qquad$ (*b*) 50 percent

(*c*) 75 percent $\qquad$ (*d*) 95 percent.

4. The angle of twist is .... proportional to the twisting moment.

(*a*) directly

(*b*) inversely

(*c*) either (*a*) or (*b*)

(*d*) none of the above.

5. For the same material, length and given torque a hollow shaft weighs ....a solid shaft.

(*a*) less than

(*b*) more than

(*c*) equal to

(*d*) none of the above.

6. The strength of a hollow shaft for the same length, material and weight is ....a solid shaft.

(*a*) less than

(*b*) more than

(*c*) equal to

(*d*) none of the above.

7. In case of a hollow shaft the average torsional energy/unit volume is given by

(*a*) $\dfrac{\tau^2}{4C} \times \dfrac{D^2 + d^2}{D^2}$

(*b*) $\dfrac{\tau^2}{C} \times \dfrac{D^2 + d^2}{D^2}$

(*c*) $\dfrac{\tau^2}{4C} \times \dfrac{D + d}{D^2}$

(*d*) $\dfrac{\tau}{C} \times \dfrac{D^2 + d^2}{D^2}$

ANSWERS

1. (*d*) $\qquad$ **2.** (*d*) $\qquad$ **3.** (*c*) $\qquad$ **4.** (*a*) $\qquad$ **5.** (*a*) $\qquad$ **6.** (*b*) $\qquad$ **7.** (*a*).

UNSOLVED EXAMPLES

1. A solid shaft is to transmit a torque of 80 kNm. If the shearing stress is not to exceed 45 MN/m², find the maximum diameter of the shaft. **[Ans.** 208 mm]

2. A solid circular shaft of 100 mm diameter is transmitting 110 kW at 150 r.p.m. Find the intensity of induced shear stress in the shaft. **[Ans.** 35·66 MN/m²]

3. A solid shaft of 60 mm diameter is running at 160 r.p.m. Find power in kW which the shaft can transmit, if the permissible shear stress is 80 MN/m² and maximum torque is likely to exceed the mean by 20%. **[Ans.** 47·37 kW]

4. A solid shaft of 100 mm diameter is to transmit 117·6 kW at 100 r.p.m. Find the maximum intensity of shear stress induced and the angle of twist for a length of 6 metres.

   Take $C = 80$ GN/m². **[Ans.** 57·19 MN/m²; 4·91 degrees]

5. Design a suitable diameter for a circular shaft required to transmit 80·2 kW (120 H.P.) at 180 r.p.m. The shear stress in the shaft is not to exceed 70 MN/m² and the maximum torque exceeds the mean by 40%. Also calculate the angle of twist in a length of 2 metres.

   Take $C = 90$ GN/m². **[Ans.** 75·67 mm; 2·356 degrees]

6. A hollow shaft of diameter ratio 3/8 is required to transmit 588 kW at 110 r.p.m., the maximum torque being 20% greater than the mean. The shear stress is not to exceed 63 MN/m², and the twist in a length of 3 m not to exceed 1·4 degrees. Calculate the external diameter of the shaft which would satisfy these conditions.

   Take: $C = 84$ GN/m². **[Ans.** 175·4 mm say 180 mm]

7. A hollow shaft of internal diameter 400 mm and of metal thickness 30 mm is required to transmit power at 180 r.p.m. Determine the power it can transmit, if the shear stress in the shaft is not to exceed 60 MN/m² and the maximum torque exceeds the mean by 30%.

   [Ans. 5225·6 kW]

8. A hollow shaft of diameter ratio 3/5 transmits 600 kW at 110 r.p.m. Maximum torque is 12% greater than the mean torque, shear stress should not exceed 60 MN/m². The twist in 3 m length should not exceed 1°. Calculate the maximum size of the shaft. $C = 80$ GN/m².

   [Ans. $d = 108$ mm, $D = 180$ mm]

9. A solid aluminium shaft 1 m long and of 50 mm diameter is to be replaced by a tubular steel shaft of the same length and the same outside diameter (*i.e.* 50 mm) such that each of the two shafts could have the same angle of twist per unit torsional moment over the total length. What must the inner diameter of the tubular steel shaft be? Modulus of rigidity of steel is three times that of aluminium.

   [Ans. 45·2 mm]

Due to the bigger diameter of steering wheel, a small force applied on the steering wheel, can exert a greater torque on the steering shaft which has a smaller diameter.

10. A hollow steel shaft of 150 mm external diameter and 100 mm bore twists through an angle of 0·8 degree in a length of 2 m when subjected to an axial torque. What are the values of shearing stresses at the inner and outer surfaces of the shaft?
Calculate the power that is transmitted by the shaft at a speed of 250 r.p.m. when the above torque is applied. C = 90 GN/m^2. **[Ans.** 31·41MN/m^2; 47·12 MN/m^2; 656 kW**]**

11. Calculate the maximum torque and the mean power being transmitted in the case of a hollow shaft of which outer diameter is 200 mm and the inner 100 mm, if the shear stress is not to exceed 60 MN/m^2 and the shaft speed is 300 r.p.m. Assume maximum torque to be 25 percent more than the mean. **[Ans.** 88·4 kNm, 2221·7 kW**]**

12. A hollow shaft has an external diameter of 300 mm and internal diameter of 150 mm. Compare its strength with that of a solid shaft of the same weight per unit length.

$$\left[\text{Ans.} \frac{T_H}{T_s} = 1\cdot 44\right]$$

13. A solid shaft of 200 mm diameter is to be replaced by a hollow steel shaft with internal diameter equal to 0·5 D where D is the external diameter. Design the hollow shaft and find out the saving in material. The value of maximum shear stress may be assumed as same for both the shafts.

[Ans. Saving in material = 21·68%**]**

14. Two shafts of the same material and same length are subjected to the same torque. If the first shaft is of a solid circular section and the second shaft is of hollow circular section whose internal diameter is 2/3 of the outside and in each shaft, maximum shear is same, then, compare the weight of two shafts.

$$\left[\text{Ans.} \frac{W_H}{W_s} = 0\cdot 648\right]$$

15. What power can be transmitted at 300 r.p.m. by a hollow steel shaft of 7·5 cm external and 5 cm internal diameters, when permissible shear stress for the steel is 70 MN/m^2 and maximum torque is 1·3 times the mean?
Compare the strength of this hollow shaft with that of a solid shaft of the same material, weight and length.

$$\left[\text{Ans.} \ 110 \text{ kW}, \frac{T_H}{T_S} = 1\cdot 935\right]$$

16. A maximum shear stress of 500 MN / m^2 is induced in a hollow shaft of 100 mm and 60 mm external and internal diameters respectively. What maximum shear stress will be developed in a solid shaft of the same weight, material and length subjected to the same torque.

[Ans. 850 MN/m^2**]**

17. A compound shaft 1 m long is fixed at one end and is subjected to a twisting moment of 150 kNm at the free end and of 200 kNm at a distance of 750 mm from the fixed end. The shaft has a diameter of 120 mm for 750 mm from the fixed end and 90 mm for the remaining portion. Determine:
(*i*) Maximum shearing stress in each portion of the shaft; (*ii*) Angle of twist at a distance of 750 mm from the fixed end and at the free end. C = 82 GN/m^2.
[Ans. 14·6 MN/m^2; 104·9 MN/m^2, 0·00225 radian and 0·01956 radian**]**

18. A steel shaft 1 m long, 30 mm diameter is rigidly fixed at the ends. A torque of 600 Nm is applied at a distance of 250 mm from one end. Calculate: (*i*) Fixing couples at the ends, (*ii*) maximum shearing stress, and (*iii*) angle of twist at the point of application of torque. C = 82 GN/m^2.
[Ans. 150 Nm, 450 Nm. 84·8 MN/m^2; 0·01725 radian**]**

19. A solid shaft 3·6 m long and 75 mm in diameter is fixed at both its ends. A twisting moment of 22·8 kNm is applied at a distance of 15 mm from one end. Calculate (*i*) twisting moment shared by each portion. (*ii*) the angle of twist on both sides of the place of application of the twisting moment, and (*iii*) the maximum shear stress in each side. C = 90 GN/m^2.
[Ans. 13·3 kN m, 9·5 kN m, 0·0715 radian, 114·7 MN/m^2, 160·6 MN/m^2**]**

20. A solid steel shaft 6 m long is securely fixed at each end. A torque of 1250 Nm is applied to the shaft at a section 2·4 metres from one end. What are fixing torques set up at the ends of the shaft? If the diameter of the shaft is 40 mm what are the maximum shear stresses in the two portions? Calculate also the angle of twist for the section where the torque is applied. Modulus of rigidity. $C = 84$ GN/m². [**Ans.** $\tau = 32\cdot8$ MN/m², $59\cdot7$ MN/m²; $\theta = 4\cdot885°$]

21. A hollow steel shaft 200 mm external diameter and 150 mm internal diameter transmits 735 kW at a speed of 100 r.p.m. Calculate the shearing stresses at the inner and outer surfaces of the shaft and the strain energy stored in the material per metre length. $C = 80$ GN/m². [**Ans.** $49\cdot02$ MN/m²; $65\cdot36$ MN/m², 295 mN]

22. Two lengths of a 100 mm diameter shaft are joined by a flange coupling with 6 bolts placed symmetrically on a bolt-circle of 250 mm diameter. If the permissible shear stress in the shaft is 50 MN/m² and that in the bolts 25% excess of this, calculate:
    (i) The tangential force acting on the shaft at its outer surface,
    (ii) The size of the bolts,
    (iii) The shear stress in the bolts if they were placed at 150 mm bolt circle radius, and
    (iv) Shear in key if it is 150 mm long and 25 mm wide. [**Ans.** 196·4 kN, $d_b = 16\cdot3$ mm, $\tau_b = 52\cdot3$ MN/m², $\tau_k = 52\cdot4$ MN/m²]

23. Two shafts, one of aluminium alloy and other of steel, are coupled end to end. The steel shaft is 2 m long and is hollow with external diameter 100 mm and internal diameter 50 mm while the aluminium shaft is solid with 100 mm diameter throughout its length of 1 m. The compound shaft is fixed at ends and a torque of 15 kNm is applied to the coupling. Neglecting the effect of coupling find:
    (i) The maximum shear stresses in the two portions of the shaft;
    (ii) The angle of twist at the coupling.
    Given: $C_{steel} = 80$ GN/m²; $C_{aluminium} = 25$ GN/m² [**Ans.** (i) $48\cdot8$ MN/m², $30\cdot4$ MN/m² (ii) $1\cdot4°$]

24. A solid phosphor bronze shaft 50 mm diameter is rotating at 740 r.p.m. It is subjected to torsion only. An electrical resistance strain gauge mounted on the surface of the shaft with its axis at 45° to the shaft axis, gives the strain reading as $4\cdot17 \times 10^{-4}$. Find the power being transmitted if the modulus of elasticity for bronze is 105 GN/m². Take: Poisson's ratio = 0·3. [**Ans.** 64 kW]

25. A propeller shaft of a ship is 450 mm in diameter and it supports a propeller of mass 15 tonnes. The propeller can be considered as a load concentrated at the end of a cantilever of length 2m. When the speed of the ship is 32 km/h the propeller is revolving at 100 r.p.m. If the engine develops 15 MW, calculate the principal stresses and maximum shear stress in the shaft. It may be assumed that the propulsive efficiency of propeller is 85 percent. [**Ans.** $\sigma_{max} = 61\cdot8$ MN/m² (comp.); $\sigma_{min} = 103\cdot7$ MN/m² (tensile); $\tau_{max} = 82\cdot75$ MN/m².]

26. A shaft 1 m long tapers uniformly from a diameter of 80 mm to a diameter of 120 mm. If the shaft transmits a torque of 8 kNm, find the angle of twist and maximum shear stress developed. Assume $C = 80$ GN/m². [**Ans.** $0\cdot67°$, $79\cdot6$ MN/m²]

27. A solid circular shaft has a radius R_1 at one end and R_2 at the outer end. Derive an expression for the angle of twist in a length l. Calculate the angle if $R_1 = 40$ mm, $R_2 = 45$ mm and $l = 1$ m. Also calculate the percentage error committed if θ is calculated on the basis of a mean radius of 42·5 mm.

$$\left[\textbf{Ans. } \theta = 0\cdot0197 \frac{T}{C} \text{ rad; } 0\cdot508\%\right]$$

28. A thin steel tube 50 mm in diameter is 2 mm thick. Find the safe twisting moment that can be applied to the tube, if the allowable shear stress is 80 MN/m². Find also the twist in a length of 400 mm. Take: $C = 80$ GN/m². [**Ans.** $628\cdot32$ Nm, $0\cdot917°$]

29. A closed cellular square section is of uniform wall thickness of 6 mm and centre line of the walls form a square of 300 mm side. It is subjected to a torque of 600 Nm. Neglecting stress concentration, find maximum shear stress and the twist per unit length.
    Take: $C = 80$ GN/m² [**Ans.** $0\cdot56$ MN/m², $4\cdot63 \times 10^{-5}$ rad./m]

Eccentric shaft.

30. A shaft of hollow square section of outer side 48 mm and inner side 40 mm is subjected to a twisting moment such that the maximum shear stress developed in 200 MN/m². Find:

(*i*) Torque acting on the shaft; (*ii*) Angular twist if the shaft is 1·6 m long.

Take: C = 80 GN/m². **[Ans.** (*i*) 3·0976 kNm (*ii*) 2·6°]

31. A shaft of hollow square section is of uniform wall thickness of 6 mm and centre line of the walls form a square of 300 mm side. It is to be replaced by a solid circular shaft of the same material and having the same torsional stiffness. Find the diameter of the solid shaft and the maximum shear stresses in both the shafts for a twisting moment of 980 Nm. Assume the stress concentration factor at the inner corners of the hollow square section as 1·75.

[Ans. 201 mm; 1·59 MN/m², 0·615 MN/m²]

32. A 300 mm × 300 mm I-section with flanges and web 12 mm thick is subjected to a torque of 615 Nm. Neglecting the stress concentration find: (*i*) Maximum shear stress developed; (*ii*) Twist per unit length. Take: C = 80 GN/m² **[Ans.** (*i*) 14·63 MN/m², (*ii*) 0·859°]

33. An aluminium box section has been designed (Fig. 13·49) for the maximum shear stress of 33 MN/m². Neglecting stress concentration find: (*i*) The twisting moment which can be taken by the section; (*ii*) The twist per metre length. Take: C = 27 GN/m²

[Ans. (*i*) 990 Nm (*ii*) 0·27 rad/m]

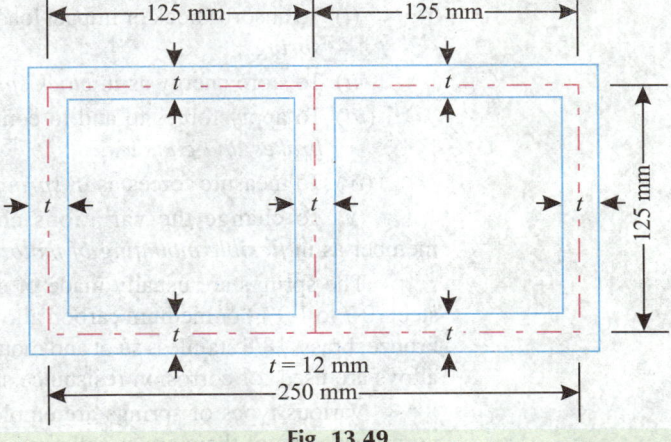

Fig. 13.49

Chapter

14

Springs

14·1. INTRODUCTION

Springs are elastic members which distort under load and regain their original shape when load is removed. They are used in railway carriages, motor cars, scooters, motorcycles, rickshaws, governors etc. According to their uses, the springs perform the following **functions:**

(*i*) To absorb shock or impact loading as in *carriage springs*.

(*ii*) To store energy as in *clock springs*.

(*iii*) To apply forces to and to control motions as in *brakes and clutches*.

(*iv*) To measure forces as in *spring balances*.

(*v*) To change the variations characteristic of a member as in *flexible mounting of motors*.

The springs are usually made of either high carbon steel (0·7 to 1·0%) or medium carbon alloy steels. Phosphor bronze, brass, 18/8 stainless steel and monel and other metal alloys are used for corrosion resistance springs.

Various types of springs are employed for different purposes, some of them are as follows:

1. **Helical springs:**
   - (*i*) Close-coiled helical springs;
   - (*ii*) Open-coiled helical springs;
   - (*iii*) Tension helical springs;
   - (*iv*) Compression helical springs.

2. **Leaf springs:**
   - (*i*) Full-elliptic; (*ii*) Semi-elliptic; (*iii*) Cantilever.

3. Torsion springs

4. Circular springs

5. Belleville springs

6. Flat springs.

14·2. HELICAL SPRINGS

A helical spring is a length of wire or bar wound into a helix. There are mainly two types of helical springs : (*i*) *Close-coiled*, and (*ii*) *open-coiled*.

14·3. THE CLOSE-COILED HELICAL SPRINGS

14·3·1. Close-coiled helical spring with 'Axial load'

A. Circular section wire springs:

In Fig. 14·1 is shown a close-coiled helical spring loaded with an axial load W.

Let, R = Radius of the coil,

   d = Diameter of the wire of the coil,

   δ = Deflection of coil under the load W,

   C = Modulus of rigidity,

   n = Number of coils or turns,

   θ = Angle of twist,

   l = Length of wire = $2\pi Rn$,

   τ = Shear stress, and

   I_p = Polar moment of inertia
   $= \dfrac{\pi}{32} d^4$.

It may be noted that each section of the coil is under torsion but there are small bending and shearing stresses which being small are usually neglected.

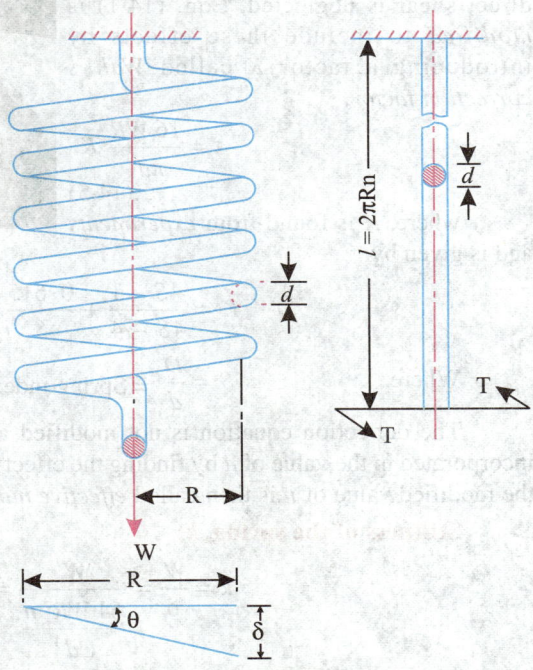

Fig. 14.1. Close-coiled helical spring.

Shear stress τ :

From torsion equation,

$$\frac{T}{I_p} = \frac{C\theta}{l} = \frac{\tau}{r}; \quad \frac{T}{I_p} = \frac{\tau}{r}$$

or, $T = \dfrac{\tau I_p}{r} = \dfrac{\tau \times \pi d^4}{32} \times \dfrac{2}{d} = \tau \cdot \dfrac{\pi}{16} d^3$

or, $\qquad \tau = \dfrac{16T}{\pi d^3}$

or, $\qquad \tau = \dfrac{16\,WR}{\pi d^3} \qquad (\because T = WR)$ ...(14·1)

Deflection, δ:

Again, $\qquad \dfrac{T}{I_p} = \dfrac{C\theta}{l}$

$$\theta = \dfrac{Tl}{CI_p} = \dfrac{WR \times 2\pi\,Rn \times 32}{C \times \pi d^4} = \dfrac{64\,WR^2\,n}{Cd^4} \qquad ...(14\cdot2)$$

but, $\qquad \delta = R \times \theta$ ...(14·3)

$\therefore \qquad \delta = \dfrac{64\,WR^3\,n}{Cd^4}$ ...(14·4)

Wahl's correction factor:

While deriving eqns. (14·1) and (14·2) the effect of curvature of spring and direct shear is neglected. Eqn. (14·1) is *modified* to include these effects by introducing a factor K called *Wahl's correction factor.*

$\therefore \qquad \tau = \dfrac{16\,WR}{\pi d^3}\,K$

...(14·5)

where, K is found from *experiments* and is given by

Shock absorbers of automobiles have springs.

$$K = \dfrac{4S - 1}{4S - 4} + \dfrac{0\cdot615}{S} \qquad ...(14\cdot6)$$

Where, $\qquad S = \dfrac{D}{d}$ = Spring index (where, D = mean diameter of the coil)

The deflection equation is not modified as the effect, if any, is considered to have been incorporated in the value of n by finding the effect on deflection due to end coils experimentally and the modified value of n is then called *effective number* of coils.

Stiffness of the spring, k:

$$k = \dfrac{W}{\delta} = \dfrac{W}{\dfrac{64\,WR^3\,n}{Cd^4}} = \dfrac{Cd^4}{64R^3 n}$$

i.e. $\qquad k = \dfrac{Cd^4}{64\,R^3 n}$ ...(14·7)

Energy stored, U:

Energy stored, $\qquad U = \dfrac{1}{2} \times T \times \theta = \dfrac{1}{2}\,W \cdot R \times \dfrac{64\,WR^2\,n}{Cd^4}$

$$= \dfrac{1}{2} \cdot \dfrac{1}{2}\,\dfrac{16\,WR}{Cd^3} \cdot \dfrac{8\,WR^2\,n}{d} = \dfrac{1}{4C} \cdot \dfrac{16\,WR}{\pi d^3} \cdot \dfrac{16\,WR}{\pi d^3}\left[2\pi R\,nd^2 \times \dfrac{\pi}{4}\right]$$

$$= \dfrac{1}{4C} \cdot \tau^2 \times \text{volume of wire}$$

i.e. $$U = \frac{\tau^2}{4C} \times \text{volume of wire} \qquad \qquad \qquad ...(14\cdot8)$$

Again, energy stored,

$$U = \frac{1}{2} \cdot T \cdot \theta = \frac{1}{2} \cdot W.R. \frac{\delta}{R} = \frac{1}{2} \cdot W \cdot \delta \qquad (\because \delta = R\theta)$$

i.e. $$U = \frac{1}{2} W\delta \qquad \qquad \qquad ...(14\cdot9)$$

B. Rectangular and square section wire springs:

Rectangular and square section wire springs are also used in many applications.

Here, $$\tau = \alpha \frac{T}{bh^2} \qquad \qquad ...(14\cdot10) \text{ (Refer to eqn. 13.38)}$$

Where, $$\alpha = 3 + 1\cdot8 \frac{h}{b}$$

[b = longer side, h = smaller side (For rectangular section)

$b = h$ and $\alpha = 4\cdot8$ ...(For square section)]

$\therefore$ $$\tau = \alpha \frac{WR}{bh^2} \cdot K \text{ with correction factor } K \qquad \qquad ...(14\cdot11)$$

where, $$K = \frac{4S - 1}{4S - 4} + \frac{0\cdot615}{S} \quad (S = \text{spring index})$$

where, $$S = \frac{2R}{\text{Side of the section perpendicular to spring axis}}$$

Also, $$\delta = R\theta, \text{ where } \theta = \frac{\beta}{bh^3} \cdot \frac{Tl}{C} \qquad [\text{Refer to eqn. 13.39}]$$

$\therefore$ $$\delta = \frac{R\beta}{bh^3} \cdot \frac{Tl}{C} \quad \text{or} \quad \delta = \beta \cdot \frac{TlR}{bh^3 C}$$

where, $$\beta = \frac{3\cdot5 \,(b^2 + h^2)}{b^2}$$

$$l = 2\pi Rn \; ; \; T = WR$$

$\therefore$ $$\delta = \frac{3\cdot5 \,(b^2 + h^2)}{b^2} \cdot \frac{WR \times 2\pi Rn \times R}{bh^3 \cdot C}$$

i.e. $$\delta = 7\pi \times \frac{WR^3 n}{C} \left[\frac{b^2 + h^2}{b^3 h^3} \right] \qquad \qquad ...(14\cdot12)$$

By using more accurate value of α and β from Article 13·17 more accurate values of τ and δ can be calculated.

CIRCULAR-SECTION WIRE SPRINGS

Example 14·1. *A closely coiled helical spring is to carry a load of 500 N. Its mean coil diameter is to be 10 times that of the wire diameter. Calculate these diameters if the maximum shear stress in the material of the spring is to be 80 MN/m².*

Solution. Load to be carried, $W = 500$ N

Mean coil diameter, $\qquad D = 10\, d$ (wire diameter),

Shear stress, $\qquad \tau = 80$ MN/m²

Diameters, D and d:

Using the relation: $\tau = \dfrac{16\,WR}{\pi d^3}$, we have

$$80 \times 10^6 = \frac{16 \times 500 \times 5d}{\pi d^3}$$

∴ $\qquad d^2 = \dfrac{16 \times 500 \times 5}{80 \times 10^6} = 1{\cdot}591 \times 10^{-4}\ \text{m}^2$

∴ $\qquad d = 0{\cdot}0126$ m or **12·6 mm** (Ans.)

and, $\qquad D = 10\,d = 10 \times 12{\cdot}6 = 126$ mm

i.e. $\qquad$ **D = 126 mm** (Ans.)

Example 14·2. *A helical spring is made of 12 mm diameter steel wire wound on a 120 mm diameter mandrel. If there are 10 active coils, what is spring constant ? Take: C = 82 GN/m². What force must be applied to the spring to elongate it by 40 mm?*

Solution. Diameter of steel wire,

$$d = 12\ \text{mm} = 0{\cdot}012\ \text{m}$$

Diameter of mandrel, $D = 120$ mm $= 0{\cdot}12$ m

Number of active coils,

$$n = 10$$

Modulus of rigidity, $\quad C = 82$ GN/m²

Elongation of the spring,

$$\delta = 40\ \text{mm} = 0{\cdot}04\ \text{m}$$

Picture shows valve guides and spring mechanisms in an aircraft engine.

Spring constant:

We know that,

Spring constant = stiffness of spring (k),

$$k = \frac{W}{\delta} = \frac{Cd^4}{64R^3 n} = \frac{82 \times 10^9 \times (0{\cdot}012)^4}{64 \times \left[\dfrac{0{\cdot}12}{2}\right]^3 \times 10}$$

or, $\qquad\qquad$ **k = 12300 N/m** (Ans.)

Force to be applied to the spring, W:

Again, $\qquad \dfrac{W}{\delta} = 12300$ or $\dfrac{W}{0{\cdot}04} = 12300$

or, $\qquad$ **W = 492 N** (Ans.)

Example 14·3. *(a) Draw neat illustrative sketches to bring about the difference between a helical coil tension spring and helical coil compression spring.*

(b) A helical coil spring is made of round steel wire 6·35 mm in diameter. The mean radius of helix is 31·75 mm, number of complete turns, 12; the spring is close-coiled. If C = 84·36 GN/m², find:

(i) The pull required to extend the spring by 25·4 mm, and

(ii) The stress in the wire.

Solution. *(a)* The helical coil springs consist of a rod or wire wound in the form of helix. Fig. 14·2 *(a, b)* clearly indicates the difference between the helical coil tension and compression springs.

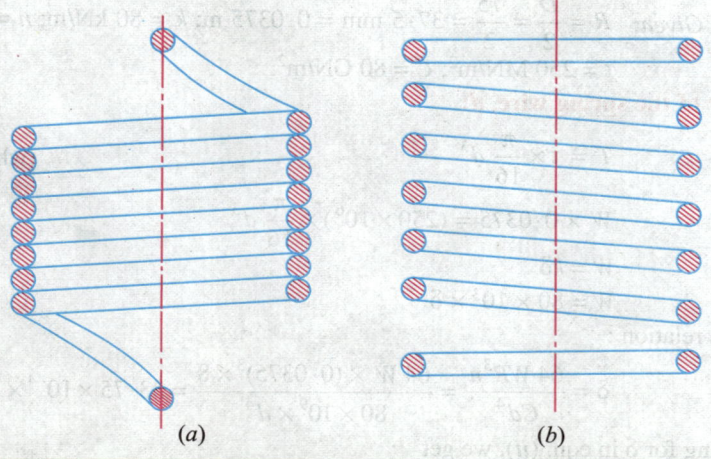

Fig. 14.2

It may be noted that in case of a helical tension spring it is not imperative to provide spacing between coils because helix angle is small while in case of helical compression spring, the helix angle being comparatively more, spacing is provided between the coils.

(b)   Radius of the coil, $R = 31 \cdot 75$ mm $= 0 \cdot 03175$ m

Diameter of the wire, $d = 6 \cdot 35$ mm $= 0 \cdot 00635$ m

Number of coils, $n = 12$

Deflection of the spring, $= 25 \cdot 4$ mm $= 0 \cdot 0254$ m

(i)   Pull required to extend the spring, W:

Using the relation,

$$\delta = \frac{64\,WR^3 n}{Cd^4}, \text{ we have}$$

$$0 \cdot 0254 = \frac{64W \times (0 \cdot 03175)^3 \times 12}{84 \cdot 36 \times 10^9 \times (0 \cdot 00635)^4}$$

$$W = \frac{0 \cdot 0254 \times 84 \cdot 36 \times 10^9 \times (0 \cdot 00635)^4}{64 \times (0 \cdot 03175)^3 \times 12}$$

or, $W = \mathbf{141 \cdot 7 \, N}$ **(Ans.)**

(ii)   Shear stress in the wire, τ :

Using the relation,

$$\tau = \frac{16\,WR}{\pi d^3}, \text{ we have}$$

$$\tau = \frac{16 \times 141 \cdot 7 \times 0 \cdot 03175}{\pi \times (0 \cdot 00635)^3}$$

$$= 89 \cdot 5 \times 10^6 \text{ N/m}^2 = \mathbf{89 \cdot 5 \, MN/m^2} \qquad \textbf{(Ans.)}$$

Example 14·4. *A close-coiled helical spring has mean diameter of 75 mm and spring constant of 80 kN/m. It has 8 coils. What is the suitable diameter of the spring wire if maximum shear stress is not to exceed 250 MN/m² ? Modulus of rigidity of the spring wire material is 80 GN/m².*

What is the maximum axial load the spring can carry? **(AMIE Summer, 2000)**

Solution. *Given:* $R = \dfrac{D}{2} = \dfrac{75}{2} = 37 \cdot 5$ mm $= 0 \cdot 0375$ m; $k = 80$ kN/m; $n = 8$

$\tau = 250$ MN/m^2; $C = 80$ GN/m^2.

Diameter of the spring wire, d:

We know,

$$T = \tau \times \frac{\pi}{16} d^3 \qquad \text{(where, } T = W \times R)$$

$$W \times 0 \cdot 0375 = (250 \times 10^6) \times \frac{\pi}{16} d^3 \qquad \text{...(i)}$$

Also,

$$W = k\delta$$

or,

$$W = 80 \times 10^3 \times \delta \qquad \text{...(ii)}$$

Using the relation :

$$\delta = \frac{64 \, WR^3 n}{Cd^4} = \frac{64 \, W \times (0 \cdot 0375)^3 \times 8}{80 \times 10^9 \times d^4} = 33 \cdot 75 \times 10^{-4} \times \frac{W}{d^4}$$

Substituting for δ in eqn. (ii), we get

$$W = 80 \times 10^3 \times 33 \cdot 75 \times 10^{-14} \times \frac{W}{d^4}$$

or,

$$d^4 = 80 \times 10^3 \times 33 \cdot 75 \times 10^{-14}$$

∴

$$d = 0 \cdot 0128 \text{ m or } \textbf{12·8 mm} \quad \textbf{(Ans.)}$$

Maximum axial load the spring can carry W :

From eqn. (i), we have

$$W \times 0 \cdot 0375 = (250 \times 10^6) \times \frac{\pi}{16} \times (0 \cdot 0128)^3$$

∴

$$W = \textbf{2745·2 N} \quad \textbf{(Ans.)}$$

Example 14·5. *A close-coiled helical spring is to have a stiffness of 900 N/m in compression, with a maximum load of 45N and a maximum shearing stress of 120 N/mm^2. The solid length of the spring (i.e., coils touching) is 45 mm. Find:*

(i) The wire diameter,

(ii) The mean coil radius, and

(iii) The number of coils.

Take modulus of rigidity of material of the spring = $0 \cdot 4 \times 10^5$ N/mm^2.

Solution. *Given:* $k = 900$ N/m $= 0 \cdot 9$ N/mm; $W = 45$ N; $\tau = 120$ N/mm^2; $C = 0 \cdot 4 \times 10^5$ N/mm^2

(i) **The wire diameter, d:**

$$\delta = \frac{64 \, WR^3 n}{Cd^4}$$

or,

$$k = \frac{W}{\delta} = \frac{Cd^4}{64 \, R^3 n}$$

or,

$$0 \cdot 9 = \frac{0 \cdot 4 \times 10^5 \times d^4}{64 \, R^3 n}$$

or,

$$d^4 = \left(\frac{0 \cdot 9 \times 64}{0 \cdot 4 \times 10^5} \right) R^3 n \qquad \text{...(1)}$$

Triggering mechanisms of guns are operated by springs besides other components.

Also,
$$\tau = \frac{16\,WR}{\pi\,d^3}$$

∴
$$120 = \frac{16 \times 45 \times R}{\pi\,d^3} \quad \text{or} \quad R = \frac{120 \times \pi\,d^3}{16 \times 45 \times R}$$

or,
$$R = 0.52\,d^3 \qquad \qquad ...(2)$$

Solid length of the spring when the coils are touching $= nd = 45$

∴
$$n = \frac{45}{d} \qquad \qquad ...(3)$$

Substituting the values of R and n in eqn. (1), we get

$$d^4 = \left(\frac{0.9 \times 64}{0.4 \times 10^5}\right) \times (0.52\,d^3)^3 \times \frac{45}{d}$$

$$= \frac{0.9 \times 64}{0.4 \times 10^5} \times (0.52)^3 \times 45 \times d^8$$

or,
$$d^4 = \frac{0.4 \times 10^5}{0.9 \times 64 \times (0.52)^3 \times 45} = 109.75$$

∴
$$d = (109.75)^{1/4} \simeq \textbf{3.24 mm} \quad \textbf{(Ans.)}$$

(ii) **The mean coil radius, R:**

$$R = 0.52\,d^3 \qquad \qquad ... [\text{Eqn. (2)}]$$

$$= 0.52 \times (3.24)^3 = \textbf{17.68 mm} \quad \textbf{(Ans.)}$$

(iii) **The number of coils, n:**

$$n = \frac{45}{d} \qquad \qquad ... [\text{Eqn. (3)}]$$

$$= \frac{45}{3.24} = \textbf{13.88} \quad \textbf{(Ans.)}$$

Example 14·6. *A close-coiled helical spring of 100 mm mean diameter is made of 10 mm diameter rod and has 20 turns. The spring carries an axial load of 200 N. Determine the shearing stress. Taking the value of modulus of rigidity = 84 GN/m² determine the deflection when carrying this load. Also calculate the stiffness of the spring and frequency of free vibrations for a mass hanging from it.*

Solution. Mean diameter of the spring,
$$D = 100 \text{ mm} = 0.1 \text{ m}$$

or,
$$R = 0.1/2 = 0.05 \text{ m}$$

Diameter of the rod, $\quad d = 10 \text{ mm} = 0.01 \text{ m}$

Axial load, $\quad W = 200 \text{ N};$

No. of turns, $\quad n = 20$

$$C = 84 \text{ GN/m}^2.$$

Shear stress, τ :

Using the relation: $\quad \tau = \dfrac{16\,WR}{\pi d^3}$, we have

$$\tau = \frac{16 \times 200 \times 0.05}{\pi \times (0.01)^3}$$

$$= 50.93 \times 10^6 \text{ N/m}^2 \ \textbf{50.93 MN/m}^2 \quad \textbf{(Ans.)}$$

Deflection of the spring, δ :

Using the relation: $\quad \delta = \dfrac{64\,WR^3 n}{Cd^4}$, we have $\quad \delta = \dfrac{64 \times 200 \times (0.05)^3 \times 20}{84 \times 10^9 \times (0.01)^4}$

or,
$$\delta = 0.03809 \text{ m or } \textbf{38.09 mm} \quad \textbf{(Ans.)}$$

Stiffness, *k*:

$$k = \frac{W}{\delta} = \frac{200}{38 \cdot 09} = 5 \cdot 25 \text{ N/mm} \quad \textbf{(Ans.)}$$

Frequency of free vibration, *f*:

Using the relation: $\quad f = \frac{1}{2\pi}\sqrt{\frac{g}{\delta}}$, we get $\quad f = \frac{1}{2\pi}\sqrt{\frac{9 \cdot 81}{0 \cdot 03809}}$

$$= \textbf{2·55 vibrations/s} \quad \textbf{(Ans.)}$$

Example 14·7. *A close - coiled helical spring is made out of 10 mm diameter steel rod. The coil consists of 10 complete turns with a mean diameter of 120 mm. The spring carries an axial pull of 200 N. Find the maximum shear stress induced in the section of the rod. If C = 80 GN/m², find the deflection in the spring, the stiffness and strain energy stored in the spring.*

Solution. Diameter of steel rod,

$$d = 10 \text{ mm or } 0 \cdot 01 \text{ m}$$

Number of turns, $\quad n = 10$

Mean radius of each turn, $R = \dfrac{120}{2} = 60 \text{ mm or } 0 \cdot 06 \text{ m}$

Axial pull, $\quad W = 200 \text{ N}; \, C = 80 \text{ GN/m}^2$

Deflection of the spring, δ :

Using the relation: $\quad \delta = \dfrac{64\,WR^3\,n}{Cd^4}$, we have $\quad \delta = \dfrac{64 \times 200 \times (0 \cdot 06)^3 \times 10}{80 \times 10^9 \times (0 \cdot 01)^4}$

$$= 0 \cdot 03456 \text{ m or } = \textbf{34·56 mm} \quad \textbf{(Ans.)}$$

Stiffness of the spring, *k*:

Using the relation: $\quad k = \dfrac{W}{\delta}$, we have $\quad k = \dfrac{200}{34 \cdot 56} = 5 \cdot 79 \text{ N/mm} \quad \textbf{(Ans.)}$

Maximum shear stress, τ :

We know, $\quad \tau = \dfrac{16\,WR}{\pi\,d^3}$ or $\quad \tau = \dfrac{16 \times 200 \times 0 \cdot 06}{\pi \times (0 \cdot 01)^3}$

$$= 61 \cdot 11 \times 10^6 \text{ N/m}^2 = \textbf{61·11 MN/m}^2 \quad \textbf{(Ans.)}$$

Strain energy stored, *U*:

Using the relation: $\quad U = \dfrac{1}{2} \cdot W \cdot \delta$, we have $\quad U = \dfrac{1}{2} \times 200 \times 0 \cdot 03456$

$$= \textbf{3·456 Nm} \quad \textbf{(Ans.)}$$

Example 14·8. *For a close-coiled helical spring subjected to an axial load of 300 N having 12 coils of wire diameter of 16 mm, and made with coil diameter of 250 mm, find:*

(i) *Axial deflection;*

(ii) *Strain energy stored;*

(iii) *Maximum torsional shear stress in the wire;*

(iv) *Maximum shear stress using Wahl's correction factor.*

Take: $\quad C = 80 \text{ GN/m}^2$

Solution. Number of coils, $n = 12$ coils

Wire diameter, $\quad d = 16 \text{ mm} = 0 \cdot 016 \text{ m}$

Coil diameter, $\quad D = 250 \text{ mm} = 0 \cdot 25 \text{ m}$

Modulus of rigidity, $\quad C = 80 \text{ GN/m}^2$

Axial load, $\quad W = 300 \text{ N}$

$$0 \cdot 15 = \frac{64 \times 33493 \times (0 \cdot 04)^3 \times n}{84 \times 10^9 \times (0 \cdot 014)^4}$$

$$\therefore \qquad n = \frac{0 \cdot 15 \times 84 \times 10^9 \times (0 \cdot 014)^4}{64 \times 33493 \times (0 \cdot 04)^3} = \mathbf{3 \cdot 53} \quad \textbf{(Ans.)}$$

Example 14·11. *A weight of 200 N is dropped on to a helical spring made of 15 mm wire closely coiled to a mean diameter of 120 mm with 20 coils. Determine the height of drop if the instantaneous compression is 80 mm.*

Assume: $\qquad\qquad C = 84 \text{ GN/m}^2$.

Solution. Magnitude of falling weight, $W = 200$ N

Diameter of wire, $\qquad\qquad d = 15$ mm or 0·015 m

Mean diameter of coils, $\qquad D = 120$ mm or 0·12 m

Number of coils, $\qquad\qquad n = 20$.

Instantaneous compression, $\delta = 80$ mm or 0·08 m

Height of drop, h:

Using the relation: $\qquad\qquad \delta = \dfrac{64\,WR^3\,n}{Cd^4}$, we have

$$0 \cdot 08 = \frac{64\,W \times (0 \cdot 06)^3 \times 20}{84 \times 10^9 \times (0 \cdot 015)^4}$$

$$W = \frac{0 \cdot 08 \times 84 \times 10^9 \times (0 \cdot 015)^4}{64 \times (0 \cdot 06)^3 \times 20} = 1230 \text{ N}$$

(where, W = gradually applied load)

Also, energy supplied by the impact load = Energy stored

$$P\,(h + \delta) = \frac{1}{2}\,W\delta$$

$$200\,(h + 0 \cdot 08) = \frac{1}{2} \times 1230 \times 0 \cdot 08$$

$$h + 0 \cdot 08 = 0 \cdot 246$$

$$h = 0 \cdot 166 \text{ m or } \mathbf{166 \text{ mm}} \quad \textbf{(Ans.)}$$

Example 14·12. *Determine amount of compression and maximum shear stress produced when a load of 2100 N is dropped axially on a close-coiled helical spring from a height of 240 mm. The spring has 22 coils each of mean diameter 180 mm and wire diameter is 25 mm, $C = 84000 \text{ N/mm}^2$.*

Solution. Diameter of wire, $d = 25$ mm

Mean diameter of coil, $\qquad D = 180$ mm

Number of coils, $\qquad\qquad n = 22$

Height of fall, $\qquad\qquad h = 240$ mm

Falling load, $\qquad\qquad P = 2100$ N

Modulus of rigidity, $\qquad C = 84000 \text{ N/mm}^2$.

Amount of compression, δ (mm) :

Let, W_e = Equivalent gradually applied load which shall produce the same effect as produced by the given falling load of 2100 N

Now, work done by the falling load = Work done by W_e

$$P(h + \delta) = \frac{1}{2} \cdot W_e \cdot \delta$$

$$\left[\because \delta = \frac{64\,W_e\,R^3 n}{Cd^4}\right.$$

$$2100(240 + \delta) = \frac{1}{2}\frac{\delta\,Cd^4}{64\,R^3 n}\cdot\delta$$

$$\left.\therefore W_e = \frac{\delta\,Cd^4}{64\,R^3 n}\right]$$

or,

$$2100(240 + \delta) = \frac{1}{2}\cdot\frac{84000 \times (25)^4}{64 \times (180/2)^3 \times 22}\cdot\delta^2 = 15\cdot98\,\delta^2$$

or,

$$\delta^2 - 131\cdot41\delta - 31539 = 0$$

or,

$$\delta = \frac{131\cdot41 \pm \sqrt{131\cdot41^2 + 4 \times 31539}}{2} = 255 \text{ mm}$$

i.e.

$$\delta = 255 \text{ mm} \quad \text{(Ans.)}$$

Maximum shear stress, τ :

Now substituting the value of δ in the following relation, we get

$$W_e = \frac{\delta\,Cd^4}{64\,R^3 n} = \frac{255 \times 84000 \times (25)^4}{64 \times 90^3 \times 22} = 8151 \text{ N}$$

$\therefore$

$$\tau = \frac{16\,W_e\,R}{\pi\,d^3} = \frac{16 \times 8151 \times 90}{\pi \times (25)^3} = 239 \text{ N/mm}^2$$

Hence,

$$\tau = 239 \text{ N/mm}^2 \text{ (Ans.)}$$

14·3·2. Subjected to 'Axial twist'

When a twisting couple is applied to the spring parallel to the axis of the spring wire it produces a bending effect on it. Depending upon the direction of application of the twisting couple or turning moment the spring coils will close or open out. In both cases the radius of coils will close or open out. In both cases the radius of coils changes and bending stresses will be induced.

Let, n_1 = Number of coils before application of twisting moment,

       n_2 = Number of coils after application of twisting moment,

       ϕ = Angle of rotation,

       I = Moment of inertia of coil section,

       R_1 = Mean radius of coil,

       R_2 = Changed radius of coil,

       σ_b = Bending stress, and

       E = Young's modulus of elasticity.

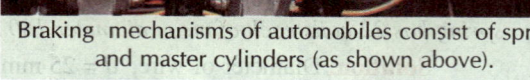

Braking mechanisms of automobiles consist of spri and master cylinders (as shown above).

Initial curvature $= \dfrac{1}{R_1}$

Final curvature $= \dfrac{1}{R_2}$

Change in curvature $= \dfrac{1}{R_2} - \dfrac{1}{R_1}$

Also, as per bending equation,

$$\frac{M}{I} = \frac{E}{R} \quad \text{or} \quad \frac{1}{R} = \frac{M}{EI}$$

$$\frac{M}{EI} = \frac{1}{R_2} - \frac{1}{R_1}$$

Since length of the wire remains unchanged before and after applying the twisting couple,

∴ $\quad l = 2\pi R_1 \, n_1 = 2\pi R_2 \, n_2$

But, $\quad \phi = $ Final helix angle – initial helix angle

$$= (2\pi n_2 - 2\pi n_1) \qquad \qquad ...(14\cdot13)$$

$$\frac{M}{EI} = \frac{1}{R_2} - \frac{1}{R_1} = \frac{1}{\dfrac{l}{2\pi n_2}} - \frac{1}{\dfrac{l}{2\pi n_1}} = \frac{2\pi}{l}(n_2 - n_1) = \frac{\phi}{l}$$

∴ $\quad \phi = \dfrac{Ml}{EI} \qquad \qquad ...(14\cdot14)$

$$= \frac{M \times 2\pi Rn}{E \times \dfrac{\pi}{64} d^4} = \frac{128 \, MRn}{Ed^4}$$

Also, $\quad \sigma_b = \dfrac{M}{Z} = \dfrac{My}{I} = \dfrac{Md/2}{\dfrac{\pi d^4}{64}} = \dfrac{32 \, M}{\pi \, d^3} \qquad \qquad ...(14\cdot15)$

Now, energy stored, $\quad U = \dfrac{1}{2} M \phi = \dfrac{1}{2} M \cdot \dfrac{Ml}{EI} = \dfrac{1}{2}\dfrac{M^2 l}{EI}$

$$= \frac{1}{2} \cdot \frac{\sigma_b \cdot \pi \, d^3}{32} \times \frac{\sigma_b \cdot \pi d^3}{32} \times \frac{l \times 64}{E \times \pi d^4}$$

$$= \frac{\pi^2 \, d^6 \, \sigma_b^2 \, l \times 64}{2 \times 32 \times 32 \times 6 \times E \times \pi d^4} = \frac{\sigma_b^2 \times \pi d^2 l}{4 \times 8E} = \frac{\sigma_b^2}{8E} \times \frac{\pi d^2 l}{4}$$

$$= \frac{\sigma_b^2}{8E} \times \text{volume of spring wire}$$

Hence, energy stored, $\quad U = \dfrac{\sigma_b^2}{8E} \times \text{volume of spring} \qquad \qquad ...(14\cdot16)$

Example 14·13. *A closely coiled helical spring made of wire 5 mm in diameter and having an inside diameter of 40 mm joins two shafts. The effective number of coils between the shafts is 15 and 0·735 kW is transmitted through the spring at 1000 r.p.m. Calculate the relative axial twist in degrees between the ends of spring and also the intensity of bearing stress in the material. $E = 200$ GN/m^2.*

Solution. $\quad d = 5 \text{ mm} = 0\cdot005 \text{ m}$

Mean diameter of the coil, $D = 0\cdot04 + 0\cdot005 = 0\cdot045 \text{ m}$

or, $\quad R = \dfrac{0\cdot045}{2} = 0\cdot0225 \text{ m}$

We know that, $\quad P = \dfrac{2\pi NT}{60 \times 1000}$

or, $\qquad 0 \cdot 735 = \dfrac{2\pi MN}{60 \times 1000} \qquad (\because M = T)$

or, $\qquad M = \dfrac{0 \cdot 735 \times 60 \times 1000}{2\pi \times 1000} = 7 \text{ Nm}$

Now, $\qquad \phi = \dfrac{128 \, MRn}{Ed^4} = \dfrac{128 \times 7 \times 0 \cdot 0225 \times 15}{200 \times 10^9 \times (0 \cdot 005)^4}$

or, $\qquad \phi = 2 \cdot 4 \text{ radians} = \mathbf{137 \cdot 5°} \text{ (Ans.)}$

$\qquad \sigma_b = \dfrac{32M}{\pi d^3} = \dfrac{32 \times 7}{\pi \, (0 \cdot 005)^3} = 570 \cdot 4 \text{ MN/m}^2 \text{ (Ans.)}$

Example 14·14. *A close-coiled helical spring made of round steel wire 6 mm diameter, having 10 complete turns is subjected to an axial couple M. The mean coil radius is 42 mm. If the maximum bending stress in spring wire is not to exceed 240 MN/m², determine:*

(i) The magnitude of axial couple M;

(ii) The angle through which one end of spring is turned relative to the other end.

Take: $\qquad E_{steel} = 200 \text{ GN/m}^2.$

Solution. Diameter of steel wire,

$\qquad d = 6 \text{ mm} = 0·006 \text{ m}$

Number of complete turns, $n = 10$

Mean coil radius, $\qquad R = 42 \text{ mm} = 0·042 \text{ m}$

Maximum bending stress, $\sigma_b = 240 \text{ MN/m}^2$

$\qquad E_{steel} = 200 \text{ GN/m}^2.$

(i) Axial couple, M:

Using the relation: $\qquad \sigma_b = \dfrac{32M}{\pi d^3}, \text{ we have } 240 \times 10^6 = \dfrac{32M}{\pi \times (0·006)^3}$

$\therefore \qquad M = \dfrac{240 \times 10^6 \times \pi \times (0·006)^3}{32} \text{ Nm} = \mathbf{5 \cdot 089 \text{ Nm} \text{ (Ans.)}}$

(ii) Angle of rotation ϕ :

Using the relation: $\qquad \phi = \dfrac{128 \, MRn}{Ed^4}, \text{ we have}$

$\qquad \phi = \dfrac{128 \times 5 \cdot 089 \times 0 \cdot 042 \times 10}{200 \times 10^9 \times (0·006)^4} \times \dfrac{180}{\pi} \text{ degrees} = 60 \cdot 47°$

Hence, $\qquad \phi = \mathbf{60 \cdot 47°} \text{ (Ans.)}$

14·4. OPEN-COILED HELICAL SPRINGS

14·4·1. With Axial Load

Employing the same symbols as used in previous articles, the *slope of coils α is introduced additionally.*

The wire length is now $l = 2\pi R \, sec \, \alpha \times n$, where R is the radius of coil.

The couple applied to the material under the applied load W will be WR, and at each point along the centre line of wire this couple may be resolved into two components, one of **torsion** and one of **bending**.

The couple producing *torsion,* $T = WR \cos \alpha$

The couple producing *bending,* $M = WR \sin \alpha$.

Refer to Fig. 14·3. The couple WR will act in a plane passing through the axes OY and OX, the centre line of the wire being at angle α to OX.

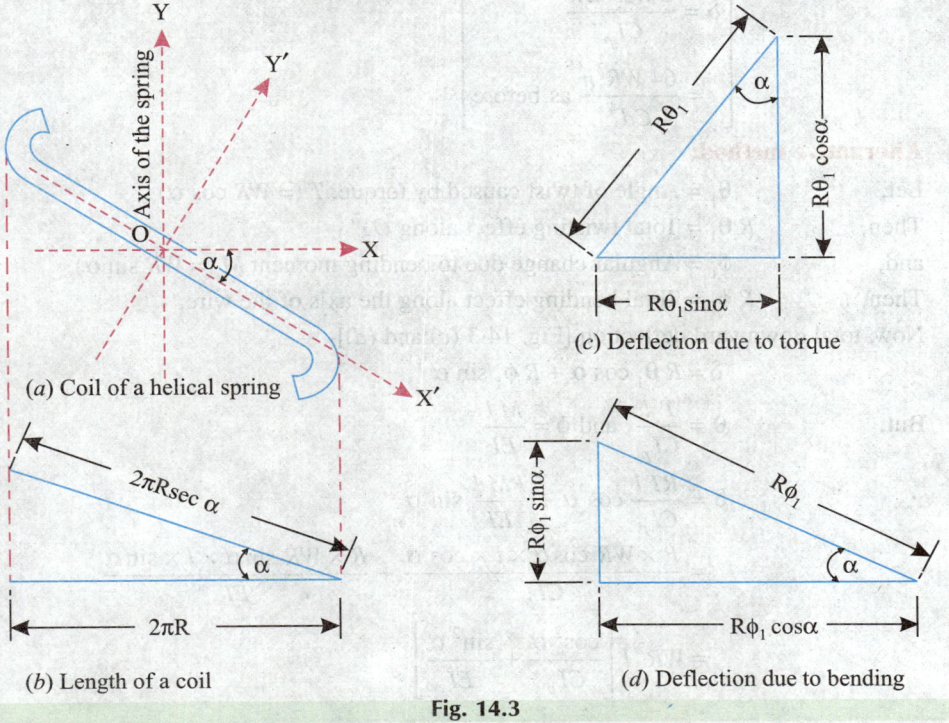

(a) Coil of a helical spring

(b) Length of a coil

(c) Deflection due to torque

(d) Deflection due to bending

Fig. 14.3

The bending moment will tend to wind the coils of the spring more tightly, and to a smaller radius of curvature.

The axial extension of the spring may be most easily calculated by *equating the work done by the load to the internal strain energy of the material.*

The *strain energy due to bending* $= \dfrac{M^2 l}{2EI}$

The strain energy for twisting $= \dfrac{T^2 l}{2\, CI_p}$

∴ Where the deflection is δ,

$$\frac{1}{2}\, W \cdot \delta = \frac{T^2 l}{2\, CI_p} + \frac{M^2 l}{2\, EI} = \frac{W^2 R^2 \cos^2 \alpha \cdot l}{2\, CI_p} + \frac{W^2 R^2 \sin^2 \alpha\, l}{2\, EI}$$

$$= \frac{W^2 R^2 l}{2} \left[\frac{\cos^2 \alpha}{CI_p} + \frac{\sin^2 \alpha}{EI} \right]$$

$$= \frac{W^2 R^2\, 2\pi\, R n \sec \alpha}{2} \left[\frac{\cos^2 \alpha}{CI_p} + \frac{\sin^2 \alpha}{EI} \right]$$

∴ Deflection, $\qquad \delta = 2\, WR^3\, n\, \pi\, \sec \alpha \left[\dfrac{\cos^2 \alpha}{CI_p} + \dfrac{\sin^2 \alpha}{EI} \right] \qquad$ ...(14·17)

$$\left[\begin{array}{l} \text{If } \alpha \text{ is taken as zero,} \\[2mm] \delta = \dfrac{2\,WR^3\,n\pi}{CI_p} \\[4mm] = \dfrac{64\,WR^3 n}{Cd^4} \text{ as before} \end{array}\right]$$

Alternative method:

Let, θ_1 = Angle of twist caused by torque, $T (= WR \cos \alpha)$

Then, $R\,\theta_1$ = Total twisting effect along OY'

and, ϕ_1 = Angular change due to bending moment $M (= WR \sin \alpha)$,

Then, $R\,\phi_1$ = Total bending effect along the axis of the wire.

Now, total downward deflection [Fig. 14·3 (c) and (d)]

$$\delta = R\,\theta_1 \cos \alpha + R\,\phi_1 \sin \alpha$$

But, $\theta = \dfrac{T\,l}{CI_p}$, and $\phi = \dfrac{M\,l}{EI}$

$\therefore$

$$\delta = \frac{RT\,l}{CI_p} \cos \alpha + \frac{RM\,l}{EI} \sin \alpha$$

$$= \frac{R \times WR \cos \alpha \times l \times \cos \alpha}{CI_p} + \frac{R \times WR \sin \alpha \times l \times \sin \alpha}{EI}$$

$$= WR^2\,l \left[\frac{\cos^2 \alpha}{CI_p} + \frac{\sin^2 \alpha}{EI}\right]$$

Shock-absorber of a motorcycle.

$$= WR^2 \times 2\pi R \sec \alpha \times n \left[\frac{\cos^2 \alpha}{CI_p} + \frac{\sin^2 \alpha}{EI} \right] \qquad (\because l = 2\pi R \sec \alpha \times n)$$

or, $$\delta = 2WR^3 \, n \, \pi \sec \alpha \left[\frac{\cos^2 \alpha}{CI_p} + \frac{\sin^2 \alpha}{EI} \right]$$

The result is the same as in eqn. (14·17)

Let, $\psi =$ The resultant of the rotations θ_1 and α_1.

$\therefore \qquad \psi = \theta_1 \sin \alpha - \phi_1 \cos \alpha$

$$= \frac{Tl}{CI_p} \sin \alpha - \frac{Ml}{EI} \cos \alpha$$

$$= \frac{WR \cos \alpha \times 2\pi R \sec \alpha \times n}{CI_p} \sin \alpha - \frac{WR \sin \alpha \times 2\pi R \sec \alpha \times n}{EI} \cos \alpha$$

$$= 2 \, WR^2 \, n \, \pi \sin \alpha \left(\frac{1}{CI_p} - \frac{1}{EI} \right) \qquad \qquad ...(14\cdot18)$$

14·4·2. With Axial Thrust

Let the torque T applied about axis of spring OY' [Fig. 14·3 (a)] be resolved about OX' and OY' ; Component about OX',

$T' = T \sin \alpha$ ....causes *torsion of spring*

Component about OY' ,

$M' = T \cos \alpha$ ....causes *bending of coil.*

If ϕ' be the angular twist due to T', θ' due to M' and ϕ due to T then, by principle of conservation of energy, we have

$$\frac{1}{2} T \phi = \frac{1}{2} T' \phi' + \frac{1}{2} M' \theta' = \frac{1}{2} T' \times \frac{T' l}{CI_p} + \frac{1}{2} M' \times \frac{M' l}{EI}$$

$$= \frac{1}{2} T'^2 \frac{l}{CI_p} + \frac{1}{2} M'^2 \frac{l}{EI} = \frac{1}{2} \frac{T^2 \sin^2 \alpha \cdot l}{CI_p} + \frac{1}{2} \frac{T^2 \cos^2 \alpha \times l}{EI}$$

$\therefore \qquad \phi = \frac{T \sin^2 \alpha \cdot l}{CI_p} + \frac{T \cos^2 \alpha \cdot l}{EI}$

But, $\qquad l = 2\pi R \sec \alpha \times n$

$\therefore \qquad \phi = 2 \, TR \, n \, \pi \sec \alpha \left[\frac{\sin^2 \alpha}{CI_p} + \frac{\cos^2 \alpha}{EI} \right] \qquad ...(14\cdot19)$

For axial deflection/extension resolve rotations as before:

$$\delta = TRl \sin \alpha \cos \alpha \left(\frac{1}{CI_p} - \frac{1}{EI} \right)$$

$$= TR \times 2\pi \, Rn \sec \alpha \times \sin \alpha \cos \alpha \left(\frac{1}{CI_p} - \frac{1}{EI} \right)$$

$$= 2 \, TR^2 \, n \, \pi \sin \alpha \left(\frac{1}{CI_p} - \frac{1}{EI} \right) \qquad \qquad ...(14\cdot20)$$

14·4·3. Stresses in Circular Wire of Open Coil Spring

Consider on open coil helical spring of a mean coil radius R, angle of helix α and wire diameter d. The stresses in the circular wire are calculated as follows:

Case I. Subjected to an axial load W:

On any section of the spring wire,

Twisting moment, $\qquad T = WR \cos \alpha$

Bending moment, $\qquad M = WR \sin \alpha$

Maximum torsional stress on any section,

$$\tau_1 = \frac{16\,T}{\pi\,d^3} = \frac{16\,WR \cos \alpha}{\pi\,d^3} \qquad \qquad ...(i)$$

Direct shear stress due to axial load,

$$\tau_2 = \frac{W}{\dfrac{\pi}{4} \times d^2} = \frac{4\,W}{\pi\,d^2} \qquad \qquad ...(ii)$$

Maximum shear stress at inner coil radius,

$$\tau = \tau_1 + \tau_2 = \frac{16\,W \cos \alpha}{\pi\,d^3} + \frac{4\,W}{\pi\,d^2} \qquad \qquad ...(14·21)$$

Minimum shear stress at outer coil radius,

$$\tau = \tau_1 - \tau_2 = \frac{16\,WR \cos \alpha}{\pi\,d^3} - \frac{4\,W}{\pi\,d^2} \qquad \qquad ...(14·22)$$

Maximum stress due to bending,

$$\sigma_b = \frac{M}{Z} = \frac{WR \sin \alpha}{\dfrac{\pi\,d^3}{32}} = \frac{32\,WR \sin \alpha}{\pi\,d^3} \qquad \qquad ...(14·23)$$

Maximum principal stress occurs at the inner coil radius.

Principal stresses: $\sigma_{max},\ \sigma_{min} = \dfrac{\sigma_b}{2} \pm \sqrt{\left(\dfrac{\sigma_b}{2}\right)^2 + \tau^2} \qquad \qquad ...(14·24)$

Case II. Subjected to axial torque T :

Twisting moment, $\qquad T' = T \sin \alpha$

Bending moment, $\qquad M' = T \cos \alpha$

Maximum torsional shear stress due to T',

$$\tau = \frac{16\,T'}{\pi\,d^3} = \frac{16\,T \sin \alpha}{\pi\,d^3} \qquad \qquad ...(i)$$

Maximum stress due to bending,

$$\sigma_b = \frac{32\,M'}{\pi\,d^3} = \frac{32\,T \cos \alpha}{\pi\,d^3} \qquad \qquad ...(ii)$$

Principal stresses at the extreme radii (inner and outer radii of coil),

$$\sigma_{max},\ \sigma_{min} = \frac{\sigma_b}{2} \pm \sqrt{\left(\frac{\sigma_b}{2}\right)^2 + \tau^2} \qquad \qquad ...(14·25)$$

Example 14·15. *An open-coiled helical spring made from wire of circular cross-section is required to carry a load of 120 N. The wire diameter is 8 mm and mean coil radius is 48 mm. If the helix angle of the spring is 30° and the number of turns is 12, calculate:*

(i)   *Axial deflection;*

(ii) *Angular rotation of free end with respect to the fixed end of the spring.*

Take: $C_{steel} = 80 \text{ GN/m}^2; E_{steel} = 200 \text{ GN/m}^2.$

Solution. Diameter of wire, $d = 8 \text{ mm} = 0·008 \text{ m}$

Mean radius of coil, $R = 48 \text{ mm} = 0·048 \text{ m}$

Helix angle, $\alpha = 30°$

Number of turns, $n = 12$

Axial load, $W = 120 \text{ N}$

Modulus of rigidity, $C = 80 \text{ GN/m}^2$

Young's Modulus, $E = 200 \text{ GN/m}^2.$

(i) Axial deflection, δ:

Axial deflection is given by :

$$\delta = 2 \, WR^3 \, n \, \pi \sec \alpha \left[\frac{\cos^2 \alpha}{CI_p} + \frac{\sin^2 \alpha}{EI} \right] \qquad \text{(Eq. 14·17)}$$

$$= 2 \times 120 \times (0·048)^3 \times 12 \times \pi \times \sec 30°$$

$$\left[\frac{(\cos 30°)^2}{80 \times 10^9 \times \dfrac{\pi}{32} \times (0·008)^4} + \frac{(\sin 30°)^2}{200 \times 10^9 \times \dfrac{\pi}{64} \times (0·008)^4} \right]$$

$$= 1·1554 \, (0·0233 + 0·006217) = 0·0341 \text{ m} = 34·1 \text{ mm}$$

Hence, $\delta = \mathbf{34·1 \text{ mm}}$ **(Ans.)**

(ii) Angular rotation, ψ:

Angular rotation, $\psi = 2WR^2 \, n \, \pi \sin \alpha \left(\dfrac{1}{CI_p} - \dfrac{1}{EI} \right)$

$$= 2 \times 120 \times (0·048)^2 \times 12 \times \pi \times \sin 30° \times$$

$$\left[\frac{1}{80 \times 10^9 \times \dfrac{\pi}{32} \times (0·008)^4} - \frac{1}{200 \times 10^9 \times \dfrac{\pi}{64} (0·008)^4} \right]$$

$$= 10·423 \, (0·03108 - 0·02486) = 0·0648 \text{ rad.} = 3·71°$$

Hence, $\psi = \mathbf{3·71°}$ **(Ans.)**

Example 14·16. *An open-coiled helical spring consists of 12 coils, each of mean diameter 60 mm, the wire forming the coil being 6 mm in diameter. Each coil makes an angle of 30° with the plane perpendicular to the axis of the spring.*

(i)   *Determine the load required to elongate the spring by 25 mm and the bending and shear stresses caused by that load;*

(ii) *Calculate the axial twist that would cause a bending stress of 50 MN/m² in the coils.*

Take: $E = 200 \text{ GN/m}^2, \text{ and } C = 82 \text{ GN/m}^2,$

Solution. Wire diameter, $d = 6 \text{ mm} = 0·006 \text{ m}$

Coil diameter (mean),

$$D = 60 \text{ mm} = 0.06 \text{ m}$$

Number of turns, $\quad n = 12$

Helix angle, $\quad\quad \alpha = 30°$

Deflection $\quad\quad\quad \delta = 25 \text{ mm} = 0.025 \text{ m}$

$$E = 200 \text{ GN/m}^2$$
$$C = 82 \text{ GN/m}^2.$$

(i) Load W, σ_b, τ :

Using the relation, $\delta = 2 WR^3 n\pi \sec \alpha \left(\dfrac{\cos^2 \alpha}{CI_p} + \dfrac{\sin^2 \alpha}{EI} \right)$ with usual notations (Eq. 14·17),

we have $\quad\quad 0.025 = 2W \times (0.06/2)3 \times 12 \times \pi \sec 30° \times$

$$\left[\frac{(\cos 30°)^2}{80 \times 10^9 \times \dfrac{\pi}{32} \times (0.006)^4} + \frac{(\sin 30°)^2}{200 \times 10^9 \times \dfrac{\pi}{64} \times (0.006)^4} \right]$$

$$= 0.00235 \ W (0.07188 + 0.01965)$$

$$\therefore \quad\quad\quad W = 116 \text{ N} \quad \textbf{(Ans.)}$$

Bending moment, $M = WR \sin \alpha = 116 \times (0.06/2) \times \sin 30° = 1.74 \text{ Nm}$

Now bending stress,

$$\sigma_b = \frac{M}{Z} = \frac{M}{I/y} = \frac{My}{I} = \frac{M \times d/2}{\dfrac{\pi}{64} \times d^4} = \frac{32 \ M}{\pi d^3} \quad\quad\quad ...(i)$$

$$= \frac{32 \times 1.74}{\pi \times (0.006)^3} \times 10^{-6} \text{ MN/m}^2 = 82.05 \text{ MN/m}^2$$

i.e. $\quad\quad\quad \sigma_b = \textbf{82·05 MN/m}^2 \ \textbf{(Ans.)}$

Shock-absorbers on a special purpose motor vehicle.

Twisting moments about the axis of the spring,
$$T = WR \cos \alpha = 116 \times (0.06/2) \times \cos 30° = 3.013 \text{ Nm}$$

But,
$$\frac{T}{I_p} = \frac{\tau}{r} \qquad \therefore \ \tau = \frac{Tr}{I_p} = \frac{T \times r}{\dfrac{\pi}{32} \times d^4} = \frac{T \times d}{2 \times \dfrac{\pi d^4}{32}} = \frac{16\,T}{\pi\,d^3}$$

$$= \frac{16 \times 3.013}{\pi \times (0.006)^3} \times 10^{-6} = 71.04 \text{ MN/m}^2$$

i.e. $\tau = \mathbf{71.04 \ MN/m^2}$ **(Ans.)**

(ii) Axial twist, T':

Let, $T' = $ Axial torque required to cause bending stress of 50 MN/m².

Component of axial torque causing bending $= T' \cos \alpha$

From eqn. (*i*), we have :
$$50 \times 10^6 = \frac{32\,T' \cos \alpha}{\pi\,d^3}$$

$$\therefore \qquad T' = \frac{50 \times 10^6 \times \pi \times (0.006)^3}{32 \times \cos 30°}$$

Hence, $T' = \mathbf{1.22 \ Nm}$ **(Ans.)**

Example 14·17. *An open-coiled helical spring of wire diameter 12 mm, mean coil radius 84 mm, helix angle 20° carries an axial load of 480 N. Determine the shear stress and direct stress developed at inner radius of the coil.*

Solution. Diameter of wire,
$$d = 12 \text{ mm} = 0.012 \text{ m}$$

Mean coil radius, $R = 84 \text{ mm} = 0.084 \text{ m}$

Helix angle, $\alpha = 20°$

Axial load, $W = 480 \text{ N.}$

Shear stress; direct stress:

Twisting moment, $T = WR \cos \alpha$

Bending moment, $M = WR \sin \alpha$

Torsional shear stress, $\tau_1 = \dfrac{16\,T}{\pi\,d^3} = \dfrac{16 \times WR \cos \alpha}{\pi\,d^3}$

$$= \frac{16 \times 480 \times 0.084 \times \cos 20°}{\pi \times (0.012)^3} \times 10^{-6} \text{ MN/m}^2 = 111.66 \text{ MN/m}^2$$

Direct shear stress, $\tau_2 = \dfrac{W}{\dfrac{\pi}{4}\,d^2} = \dfrac{4\,W}{\pi\,d^2} = \dfrac{4 \times 480}{\pi \times (0.012)^2} \times 10^{-6} \text{ MN/m}^2 = 4.24 \text{ MN/m}^2$

∴ Total shear stress at the inner coil radius,
$$\tau = \tau_1 + \tau_2 = 111.66 + 4.24 = 115.9 \text{ MN/m}^2$$

Hence, $\tau = \mathbf{115.9 \ MN/m^2}$ **(Ans.)**

Direct stress due to bending,
$$\sigma_b = \frac{32\,M}{\pi\,d^3} = \frac{32 \times WR \sin \alpha}{\pi\,d^3}$$

$$= \frac{32 \times 480 \times 0.084 \times \sin 20°}{\pi \times (0.012)^3} \times 10^{-6} \text{ MN/m}^2 = 81.28 \text{ MN/m}^2$$

Hence, $\sigma_b = 81.28 \text{ MN/m}^2$ **(Ans.)**

Example 14·18. *A open-coiled helical spring made of steel wire 6 mm diameter and 30 mm mean coil radius, with 65° inclination of the coils with the spring axis, is subjected to an axial torque T. If number of turns in the spring increases by 1/8 and the original number of turns is 12 calculate:*

(i) *Magnitude of axial torque T;*

(ii) *Change in axial length of the spring.*

Take: $C_{steel} = 84 \text{ GN/m}^2$, *and* $E_{steel} = 210 \text{ GN/mm}^2$.

Solution. Diameter of steel wire,

$$d = 6 \text{ mm} = 0.006 \text{ m}$$

∴ Polar moment of inertia,

$$I_p = \frac{\pi}{32} \times d^4 = \frac{\pi}{32} \times (0.006)^4 = 1.272 \times 10^{-10} \text{ m}^4$$

Moment of inertia, $\quad I = \dfrac{I_p}{2} = 6.36 \times 10^{-11} \text{ m}^4$

Mean radius of coil, $\quad R = 30 \text{ mm} = 0.03 \text{ m}$

Angle of helix, $\quad \alpha = 90 - 65 = 25°$

Number of turns, $\quad n = 12$

Angular rotation, $\quad \phi = \dfrac{1}{8} \text{ turn} = \dfrac{1}{8} \times 360° = 45° = 0.7854 \text{ radian}$

(i) Magnitude of axial torque, T:

Angular rotation, $\quad \phi = 2\, TRn\, \pi \sec \alpha \left[\dfrac{\sin^2 \alpha}{CI_p} + \dfrac{\cos^2 \alpha}{EI} \right]$ ...(Eqn. 14·19)

$$\therefore 0.7854 = 2T \times 0.03 \times 12\, \pi \sec 25° \left[\frac{(\sin 25°)^2}{84 \times 10^9 \times 1.272 \times 10^{-10}} + \frac{(\cos 25°)^2}{210 \times 10^9 \times 6.36 \times 10^{-11}} \right]$$

$$= 2.495\, T\ (0.0167 + 0.0615) \quad \therefore T = 4.33 \text{ Nm}$$

(ii) Change in axial length of spring, δ:

Change in axial length of spring is given by:

$$\delta = 2\, TR^2\, n\, \pi \sin \alpha \left(\frac{1}{CI_p} - \frac{1}{EI} \right)$$

$$= 2 \times 4.033 \times (0.03)^2 \times 12\, \pi \sin 25° \times$$

$$\left[\frac{1}{84 \times 10^9 \times 1.272 \times 10^{-10}} - \frac{1}{210 \times 10^9 \times 6.36 \times 10^{-11}} \right]$$

$$= 0.1156\ (0.0936 - 0.0748) = 0.002173 \text{ m} = 2.173 \text{ mm}$$

Hence, $\quad \delta = 2.173 \text{ mm}$ **(Ans.)**

14·5. SPRINGS IN SERIES

Fig. 14·4 shows two springs connected in series.

Let, $\quad W = $ Load applied, $\qquad\qquad k_1 = $ Stiffness of spring 1,

$k_2 = $ Stiffness of spring 2, $\qquad \delta_1 = $ Extension of spring 1,

$\delta_2 = $ Extension of spring 2, and $\quad k = $ Stiffness of composite spring.

It can be easily imagined that each spring will be subjected to load W and the total extension produced will be the sum of extensions of two springs.

∴ Total extension, $\delta = \delta_1 + \delta_2$

∴ $$\frac{W}{k} = \frac{W}{k_1} + \frac{W}{k_2} \qquad \left(\because \delta = \frac{W}{k} \right)$$

or, $$\frac{1}{k} = \frac{1}{k_1} + \frac{1}{k_2} \qquad \qquad ...(14 \cdot 26)$$

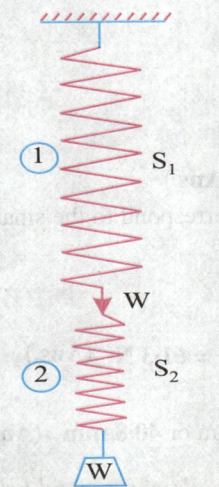

Fig. 14.4. Springs in series.　　　　**Fig. 14.5.** Springs in parallel.

14·6. SPRINGS IN PARALLEL

Fig. 14·5 shows two springs, connected in parallel. When subjected to load, W, they will extend equally say by an amount δ. The load will be shared such that,

$$W = W_1 + W_2$$

i.e., $$\delta \cdot k = \delta \cdot k_1 + \delta \cdot k_2$$

or, $$k = k_1 + k_2$$
$$...(14 \cdot 27)$$

Example 14·19. *A composite spring has two close-coiled springs connected in series; one spring has 12 coils of a mean diameter of 25 mm and wire diameter 2·5 mm. Find the wire diameter of the other spring, if it has 15 coils of mean diameter 40 mm. The stiffness of the composite spring is 1·5 kN/m.*

Suspension spring near the front wheel of an automobile.

Determine the greatest load that can be carried by the composite spring and the corresponding extension if maximum stress is 250 MN/m². C = 80 GN/m².

Solution. In case of spring connected in series,

$$\delta = \delta_1 + \delta_2$$

or, $$\frac{W}{k} = \frac{W}{k_1} + \frac{W}{k_2}$$

or, $$\frac{1}{k} = \frac{1}{k_1} + \frac{1}{k_2} \qquad \qquad ...(i)$$

but, $$k_1 = \frac{Cd_1^4}{64R_1^3\, n_1} = \frac{80 \times 10^9 \times (0 \cdot 0025)^4}{64 \times (0 \cdot 0125)^3 \times 12} = 2083 \text{ N/m}$$

$$k_2 = \frac{Cd_2^4}{64R_2^3\, n_2} = \frac{80 \times 10^9 \times d_2^4}{64 \times (0 \cdot 02)^3 \times 15} = 10416 \times 10^9\, d_2^4$$

Putting these values in eqn. (i), we get

$$\frac{1}{1500} = \frac{1}{2083} + \frac{1}{10416 \times 10^9\, d_2^4}$$

$$0 \cdot 00067 = 0 \cdot 00048 + \frac{1}{10416 \times 10^9\, d_2^4}$$

From which, $d_2 = 0.00474$ m or **4·74 mm** (Ans.)

The greatest load that can be carried by the spring will correspond to the smaller diameter.

$$\therefore \qquad \tau = \frac{16\, WR}{\pi\, d^3}, \quad \text{or} \quad W = \frac{\tau \cdot \pi\, d^3}{16\, R}$$

$$W = \frac{250 \times 10^6 \times \pi \times (0 \cdot 0025)^3}{16 \times 0 \cdot 0125} = \textbf{61·3 N} \text{ (Ans.)}$$

Total extension, $\qquad \delta = \dfrac{W}{k} = \dfrac{61 \cdot 3}{1 \cdot 5 \times 1000} = 0.0408$ m or **40·8 mm** (Ans.)

Example 14·20. *A helical spring B is placed inside the coils of a second helical spring A, having the same number of coils and free axial length and of same material. The two springs are compressed by an axial load of 210 N which is shared between them. The mean coil diameters of A and B are 90 mm and 60 mm and the wire diameters are 12 mm and 7 mm respectively. Calculate the load taken and the maximum stress in each spring.* **(I.E.S.)**

Solution. Let, $\qquad W_A =$ Load shared by spring A,

$\qquad\qquad\qquad W_B =$ Load shared by spring B,

$\qquad\qquad\qquad \delta_A =$ Deflection of spring A, and

$\qquad\qquad\qquad \delta_B =$ Deflection of spring B.

Now, $\qquad\qquad W_A + W_B = 210$ N .... (Given) $\qquad\qquad\qquad ...(i)$

Also, $\qquad\qquad \delta_A = \delta_B$

$\qquad$ (∵ Springs are connected in *parallel*)

$$\therefore \qquad \frac{64\, W_A\, R_A^3\, n_A}{C_A\, d_A^4} = \frac{64\, W_B\, R_B^3\, n_B}{C_B\, d_B^4}$$

or, $$\frac{W_A\, R_A^3}{d_A^4} = \frac{W_B\, R_B^3}{d_B^4} \qquad \left[\begin{array}{l} \because\ n_A = n_B; \\ C_A = C_B \end{array} \right]$$

or, $$\frac{W_A}{W_B} = \left[\frac{R_B}{R_A}\right]^3 \cdot \left[\frac{d_A}{d_B}\right]^4, \quad \text{or,} \quad \frac{W_A}{W_B} = \left[\frac{30}{45}\right]^3 \times \left[\frac{12}{7}\right]^4$$

or, $$\frac{W_A}{W_B} = 2 \cdot 559 \quad \text{or,} \quad W_A = 2 \cdot 559\, W_B \qquad\qquad ...(ii)$$

Substituting this value of W_A in (i) we get

$2 \cdot 559 \, W_B + W_B = 210$

$W_B = \mathbf{59 \, N}$ (**Ans.**)

and, $W_A = 210 - 59 = \mathbf{151 \, N}$ (**Ans.**)

Now, $\tau_A = \dfrac{16 \, W_A \, R_A}{\pi \, d_A^3} = \dfrac{16 \times 151 \times 0 \cdot 045}{\pi \times (0 \cdot 012)^3} = 20 \times 10^6 \, \text{N/m}^2$

or, $\tau_A = \mathbf{20 \, MN/m^2}$ (**Ans.**)

Similarly, $\tau_B = \dfrac{16 \, W_B \, R_B}{\pi \, d_B^3} = \dfrac{16 \times 59 \times 0 \cdot 03}{\pi \times (0 \cdot 007)^3} = 26 \cdot 28 \times 10^6 \, \text{N/m}^2$

or, $\tau_B = \mathbf{26 \cdot 28 \, MN/m^2}$ (**Ans.**)

14·7. FLAT SPIRAL SPRING

Fig. 14·6 shows a flat spiral spring which consists of a uniform thin rectangular metallic strip (of thickness t and width b) wound into a spiral in one plane, the outer end being anchored to a pin D and the inner end being attached to the winding spindle C for winding the spring. Let H and V be respectively the reactions at D along and perpendicular to the line joining the axis of the spindle to the centre of the pin D. Let AB be a small length dl with co-ordinates x and y as shown in Fig. 14·6.

Bending moment M at the element (taken positive if the number of turns increase) is given by

$$M = V \cdot x - H \cdot y \qquad \qquad \qquad ...(14 \cdot 28)$$

Let, $d\phi$ = Rotation of the end B with respect to end A of the element dl due to the bending moment M,

Then, $d\phi = \dfrac{M \, dl}{EI}$

or, $d\phi = \dfrac{(V \cdot x - H \cdot y)}{EI} \cdot dl$ (from eqn. 14·28)

Integrating both sides, we get

$$\phi = \frac{V}{EI} \int x \, dl - \frac{H}{EI} \int y \, dl \qquad \qquad ...(14 \cdot 29)$$

If the centroid of the spring is assumed to be at the centre of the spindle C, the first moment of the length l about the line CD will be zero.

i.e. $\int y \, dl = 0$ ...[14·30 (a)]

and, moment of the length about YY line gives

$\int x \, dl = l \, R$ ...[14·30 (b)]

where, l = Length of spring strip, and

R = The distance between points C and D.

Also, the spring strip will be in equilibrium, when

$T = V \cdot R$ ...(14·31)

(where, T = winding torque)

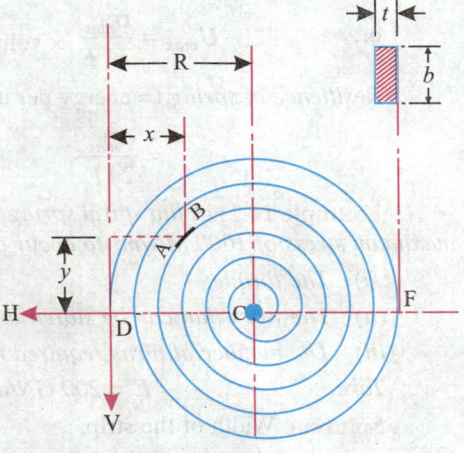

Fig. 14.6

Substituting eqns. (14·30) and (14·31) in eqn. (14·29), we have

$$\phi = \frac{(T/R)}{EI} \cdot (l \cdot R) = \frac{Tl}{EI} \qquad ...(14.32)$$

From eqn. (14.28) we find that the maximum bending moment will be obtained on such a section of strip where *x is maximum and y is minimum*. Obviously the point *F* is such a point where *y* is zero and *x* is *maximum*.

Let, $\qquad D F \simeq 2 R$

Then from eqn. (14·31), we get

$$M_{max} = M_F = \frac{T}{R} \cdot (2 R) = 2 T \qquad ...(14.33)$$

The corresponding maximum bending stress on the cross-section of the strip,

$$\sigma_{max} = \frac{2T}{I} \times \frac{t}{2} = \frac{12 T t}{bt^3} = \frac{12 T}{bt^2} \qquad ...[14.34 \ (a)]$$

or, $\qquad T = \frac{bt^2 \cdot \sigma_{max}}{12} \qquad ...[14.34 \ (b)]$

[If σ_{max} is known, the value of *T* can be calculated]

Energy stored (U):

The energy stored (corresponding to the torque *T*) can be calculated as follows:

We know, $\qquad U = \frac{1}{2} T \phi = \frac{1}{2} T \cdot \frac{Tl}{EI} = \frac{T^2 l}{2 EI} = \frac{T^2 l}{2E \times \frac{bt^3}{12}} = \frac{6 T^2 l}{E \ bt^3}$

i.e. $\qquad U = \frac{6 T^2 l}{E \ b \ t^3} \qquad ...(14.35)$

To find maximum energy which can be stored (when the stress reaches its maximum value) will be obtained by substituting the value of *T* from eqn. [14·34 (*b*)] in eqn. (14·35).

$\therefore \qquad U_{max} = \frac{6 l}{E \ bt^3} \left[\frac{bt^2 \sigma_{max}}{12} \right]^2 = \frac{\sigma_{max}^2}{24 \ E} (btl)$

or, $\qquad U_{max} = \frac{\sigma_{max}^2}{24 \ E} \times \text{volume of spring} \qquad ...(14.36)$

Resilience of spring (= energy per unit volume)

$$= \frac{\sigma_{max}^2}{24 \ E} \qquad ...(14.37)$$

Example 14·21. *A flat spiral spring is 5 mm wide, 0·25 mm thick and 3 metres long. Assuming maximum stress of 1000 MN/m² to occur at the point of greatest bending moment, calculate:*

(*i*) *The torque;*

(*ii*) *The work that can be stored in the spring; and*

(*iii*) *The number of turns required to wind up the spring.*

Take: $\qquad E = 200 \ GN/m^2.$ **(Madras University)**

Solution. Width of the strip,

$\qquad b = 5 \ mm$

Thickness of the strip, $t = 0·25 \ mm$

Length of the strip, $l = 3$ m

Maximum stress, $\sigma_{max} = 1000$ MN/m^2 (or 1000 N/mm^2).

(i) The torque T:

Now torque, $\qquad T = \dfrac{bt^2 \cdot \sigma_{max}}{12}$ with usual notations $\qquad$ ...[Eqn. 14·34 (b)]

$$= \dfrac{5 \times 0 \cdot 25^2 \times 1000}{12} = 26 \text{ Nmm} = 0 \cdot 026 \text{ Nm}$$

i.e. $\qquad T = \mathbf{0 \cdot 026 \ Nm}$ **(Ans.)**

(ii) Work that can be stored, U:

$$U = \dfrac{\sigma_{max}^2}{24\,E} \times \text{volume of spring} \qquad \text{...(Eqn. 14·36)}$$

$$= \dfrac{(1000 \times 10^6)^2}{24 \times 200 \times 10^9} \times [(5 \times 10^{-3}) \times (0 \cdot 25 \times 10^{-3} \times 3] = 0 \cdot 781 \text{ Nm}$$

i.e. $\qquad U = \mathbf{0 \cdot 781 \ Nm \ (or \ J)}$ **(Ans.)**

(iii) Number of turns:

We know, $\qquad \phi = \dfrac{Tl}{EI} \qquad \text{...(Eqn. 14·32)}$

$$= \dfrac{0 \cdot 026 \times 3}{200 \times 10^9 \times \dfrac{bt^3}{12}}$$

$$= \dfrac{0 \cdot 026 \times 3 \times 12}{200 \times 10^9 \times 0 \cdot 005 \times (0 \cdot 00025)^3} = 59 \cdot 9 \text{ radians}$$

Shock-absorber system.

Since, 1 turn = 2π radians,

$\therefore$ *Number of turns* $= \dfrac{59\cdot9}{2\pi} = 9\cdot533$ **turns** **(Ans.)**

Example 14·22. *A flat spiral spring is made of a strip 6 mm wide, 0·25 mm thick and 12 m long. The torque is applied at the winding spindle and 9 complete turns are given. Calculate:*

(*i*)   *The torque;*

(*ii*) *Maximum stress developed at the point of greatest bending moment;*

(*iii*) *The energy stored.*

Take: $E = 210 \ GN/m^2.$

Solution. Width of the strip,

$b = 6$ mm (= 0·006 m)

Thickness of the strip, $t = 0\cdot25$ mm (= 0·00025 m)

Length of the strip, $l = 12$ m

Number of complete turns, $n = 9$

Young's modulus, $E = 210 \ GN/m^2.$

(*i*)   The torque, *T*:

Angular rotation,

$\phi = 2\pi \, n = 2\pi \times 9 = 56\cdot55$ radians

Also, $\phi = \dfrac{Tl}{EI}$ $\therefore T = \dfrac{\phi \, EI}{l} = \dfrac{56\cdot55 \times 210 \times 10^9 \times (bt^3/12)}{l}$

$= \dfrac{56\cdot55 \times 210 \times 10^9 \times (0\cdot006) \times (0\cdot00025)^3}{12 \times 12}$ Nm

$= 0\cdot00773$ Nm $= 7\cdot73$ Nmm.

i.e. $T = \mathbf{7\cdot73}$ **Nmm**   **(Ans.)**

(*ii*) Maximum stress, σ_{max} :

$\sigma_{max} = \dfrac{12 \, T}{bt^2} = \dfrac{12 \times 0\cdot00773}{0\cdot006 \times (0\cdot00025)^2} \times 10^{-6}$ MN/m^2 $= 247\cdot4$ MN/m^2

[Eqn. 14·34 (*a*)]

i.e. $\sigma_{max} = \mathbf{247\cdot4}$ **MN/m**^{2} **(Ans.)**

(*iii*) The energy stored, *U*:

$U = \dfrac{1}{2} \, T \, \phi = \dfrac{1}{2} \times 7\cdot73 \times 56\cdot55$

$= 218.56$ Nmm

i.e. $U = \mathbf{218\cdot56}$ **Nmm**   **(Ans.)**

14·8. LAMINATED SPRINGS

These springs are called *semielliptical, leaf* or *carriage springs* and find their use in *trucks, trains, trolleys* etc. They consist of a number of leaves of *spring steel* held together at the centre with clamps. The plates are provided with curvature initially and the ends of the top plate are pin jointed to chassis of the vehicle. (Fig. 14·7). The load at which the plates become straight is called *"Proof load"*.

14·8·1. Semi-elliptical Spring:

Refer to Fig. 14·7.

Let, $b =$ Width of each plate,

$t =$ Thickness of each plate,

a = Overlap at each end,

N = Number of plates in the spring,

l = The spring span length,

σ_b = Bending stress, and

E = Young's modulus of elasticity.

The maximum bending moment $\dfrac{Wl}{4}$ occurs at the centre and is resisted by all N plates equally.

∴ Resisting moment of each plate,

$$M = \frac{W\,l}{4\,N}$$

Now, $\sigma_b = \dfrac{My}{I}$

or, $\sigma_b = \dfrac{W\,l}{4\,N} \cdot \dfrac{t}{2} \cdot \dfrac{1}{\dfrac{bt^3}{12}}$

or, $\sigma_b = \dfrac{3\,W\,l}{2\,Nbt^2}$

...(14·38)

Let, δ = Initial deflection of the plates, and

R_c = Radius of curvature of the plates.

Refer to Fig. 14·7.

$$R_c{}^2 = (l/2)^2 + (R_c - \delta)^2$$
$$R_c{}^2 = (l/2)^2 + R_c{}^2 + \delta^2 - 2\,R_c\delta$$

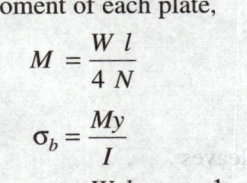

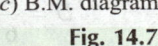

(c) B.M. diagram

Fig. 14.7

A tunnel making machine is being transported. Heavy vehicles use leaf springs to bear loads and shocks.

Neglecting δ^2, we get

$$2 R_c \cdot \delta = (l/2)^2 \text{ or } \delta = \frac{l^2}{8 R_c}$$

Also,

$$\frac{M}{I} = \frac{E}{R_c}$$

or,

$$R_c = \frac{EI}{M} = \frac{E\,bt^3}{12 \times \dfrac{W\,l}{4\,N}} = \frac{E\,N\,bt^3}{3\,Wl}$$

∴

$$\delta = \frac{l^2}{8 \times \dfrac{EN\,bt^3}{3\,W\,l}}$$

or,

$$\delta = \frac{3\,W\,l^3}{8\,EN\,bt^3} \qquad \qquad \qquad \qquad ...(14\cdot39)$$

Strain energy,

$$U = \frac{M^2}{2\,EI} \times \text{total length of leaves}$$

$$= \frac{\left[\dfrac{Wl}{4\,N}\right]^2 \times 12}{2\,Ebt^3} \times \text{total length of leaves}$$

$$= \frac{3\,W^2\,l^2}{8\,EN^2\,bt^3} \times \text{total length of leaves}$$

$$= \left[\frac{3\,Wl}{2\,N\,bt^2}\right]^2 \times \frac{bt}{2 \times 3\,E} \times \text{total length of leaves}$$

$$= \frac{\sigma_b^2}{6\,E} \times bt \times \text{total length of leaves} = \frac{\sigma_b^2}{6\,E} \times \text{volume of spring}$$

i.e.

$$U = \frac{\sigma_b^2}{6\,E} \times \text{volume of spring} \qquad \qquad ...(14\cdot40)$$

14·8·2. Quarter Elliptical Spring

These springs are called *cantilever laminated springs*. In Fig. 14·8 is shown a spring of this type.

It has an effective span length l and carries a load W. If observed carefully it will be noted that the spring is equivalent to a laminated semi-elliptical type spring of length $2l$ carrying a load $2W$ at the centre. Hence by replacing l by $2l$ and W by $2W$ the deflection and stress can be obtained as follows:

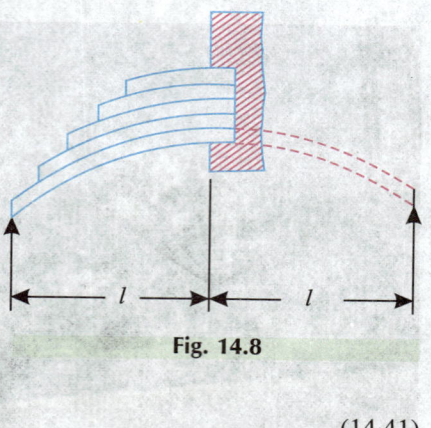

Fig. 14.8

$$\delta = \frac{3\,(2W) \cdot (2l)^3}{8\,EN\,bt^3}$$

or,

$$\delta = \frac{6\,Wl^3}{EN\,bt^3} \qquad \qquad \qquad ...(14\cdot41)$$

Similarly,

$$\sigma_b = \frac{3\,(2W)\,(2l)}{2\,N\,bt^2}$$

$$\therefore \qquad \sigma_b = \frac{6\,Wl}{N\,bt^2} \qquad\qquad\qquad ...(14.42)$$

Example 14·23. *A carriage spring is to be 600 mm long and made of 9·5 mm thick steel plates and 50 mm broad. How many plates are required to carry a load of 4·5 kN, without the stress exceeding 230 MN/m².*

What would be central deflection and the initial radius of curvature, if plates straighten under the load? E = 200 GN/m²

Solution. Span length, $\qquad l = 600$ mm $= 0.6$ m

Thickness of each plate, $\qquad t = 9.5$ mm $= 0.0095$ m

Width of each plate, $\qquad b = 50$ mm $= 0.05$ m

Load, $\qquad\qquad\qquad W = 4.5$ kN

$\qquad\qquad\qquad\qquad E = 200$ GN/m²

$\qquad\qquad\qquad\qquad \sigma_b = 230$ MN/m².

Number of plates, N:

We know, $\qquad\qquad M = \dfrac{W\,l}{4\,N}$ (where, M = bending moment per plate)

but, $\qquad\qquad M = \sigma_b \times Z$ (where, Z = section modulus of one plate $= \dfrac{1}{6}\,bt^2$)

$\therefore \qquad\qquad \dfrac{W\,l}{4\,N} = \sigma_b \times \dfrac{bt^2}{6}$

or, $\qquad \dfrac{4.5 \times 10^3 \times 0.6}{4 \times N} = 230 \times 10^6 \times \dfrac{0.05 \times (0.0095)^2}{6}$

$\therefore \qquad\qquad N = \dfrac{4.5 \times 10^3 \times 0.6 \times 6}{4 \times 230 \times 10^6 \times 0.05 \times (0.0095)^2}$

or, $\qquad\qquad N = \textbf{3·9 say 4}$ **(Ans.)**

Initial radius of carvature, R_c:

Using the relation $\qquad \dfrac{\sigma_b}{y} = \dfrac{E}{R_c}$, we get

$$\dfrac{230 \times 10^6}{t/2} = \dfrac{200 \times 10^9}{R_c}$$

$\therefore \qquad\qquad R_c = \dfrac{200 \times 10^9 \times 0.0095}{230 \times 10^6 \times 2} = \textbf{4·13 m}$ **(Ans.)**

Central deflection, δ:

Using the relation: $\qquad \delta = \dfrac{3\,Wl^3}{8\,EN\,bt^3}$, we have

$$\delta = \dfrac{3 \times 4.5 \times 10^3 \times (0.6)^3}{8 \times 200 \times 10^9 \times 4 \times 0.05 \times (0.0095)^3} = \textbf{10·6 mm} \text{ (Ans.)}$$

Example 14·24. *A laminated steel spring 1 m long is to support central load of 5·8 kN. If the maximum deflection of spring is not to exceed 45 mm and maximum stress should not exceed 300 MN/m², calculate:*

(i) The thickness of the leaves;

(ii) Their number if each plate is to be 80 mm wide.

Take: $\qquad\qquad E = 200$ GN/m².

Solution. Span length, $l = 1$ m

Width of each plate, $b = 80$ mm $= 0.08$ m

Load, $W = 5.8$ kN

Deflection, $\delta = 45$ mm $= 0.045$ m

Stress, $\sigma_b = 300$ MN/m^2

(i)   **Thickness, t:**

(ii) **Number of plates, N:**

We know that, $\sigma_b = \dfrac{3}{2} \times \dfrac{W\,l}{N\,bt^2}$

$$300 \times 10^6 = \frac{3 \times 5.8 \times 10^3 \times 1}{2 \times N \times 0.08 \; t^2}$$

or, $Nt^2 = \dfrac{3 \times 5.8 \times 10^3 \times 1}{300 \times 10^6 \times 2 \times 0.08} = 3.625 \times 10^{-4}$ ...(i)

Also, $\delta = \dfrac{3\,Wl^3}{8\,EN\,bt^3}$

$$0.045 = \frac{3 \times 5.8 \times 10^3 \times 1^3}{8 \times 200 \times 10^9 \times N \times 0.08 \times t^3}$$

$$Nt^3 = \frac{3 \times 5.8 \times 10^3 \times 1^3}{0.045 \times 8 \times 200 \times 10^9 \times 0.08} = 3.02 \times 10^{-6}$$ ...(ii)

Dividing (ii) by (i), we get

$$\frac{Nt^3}{Nt^2} = \frac{3.02 \times 10^{-6}}{3.625 \times 10^{-4}}$$

∴ $t = 0.00833$ m $= \mathbf{8.33}$ **mm**   **(Ans.)**

and, $Nt^3 = 3.02 \times 10^{-6}$

or, $N = \dfrac{3.02}{t^3} = \dfrac{3.02 \times 10^{-6}}{(0.00833)^3}$

or, $N = 5.22$ say **5**   **(Ans.)**

Example 14·25. *A leaf spring of semi-elliptic type has 11 plates each 9 cm wide and 1·5 cm thick. The length of spring is 1·5 m. The plates are made of steel having a proof stress (bending) of 650 MN/m². To what radius should the plates be bent initially?*

From what height can a load of 600 N fall on to centre of the spring, if maximum stress is to be one-half of the proof stress?

$$E = 200 \text{ GN/m}^2.$$ **(Bombay University)**

Solution. The bending equation is given by,

$$\frac{M}{I} = \frac{\sigma_b}{y} = \frac{E}{R_c}$$

$$R_c = \frac{E \times y}{\sigma_b}$$ (where, R_c = radius of curvature)

$$= \frac{E \times t/2}{\sigma_b} = \frac{200 \times 10^9 \times (0.75/100)}{650 \times 10^6}$$

or, $R_c = \mathbf{2.31}$ **m**   **(Ans.)**

The stress in the second case is *half* the proof stress

i.e. $\dfrac{650}{2} = 325 \text{ MN/m}^2$

Let, $\qquad W$ = The equivalent static load which will produce this stress.

Then, $\qquad \sigma_b = \dfrac{3\,W\,l}{2\,N\,bt^2}$

$$W = \frac{2\,N\,bt^2\,\sigma_b}{3\,l} = \frac{2 \times 11 \times (9/100) \times (1 \cdot 5/100)^2 \times 325 \times 10^6}{3 \times 1 \cdot 5}$$

$$= 32175 \text{ N or } 32 \cdot 175 \text{ kN}$$

Deflection under this load, $\delta = \dfrac{3\,W\,l^3}{8\,EN\,bt^3} = \dfrac{3 \times 32175 \times (1 \cdot 5)^3}{8 \times 200 \times 10^9 \times 11 \times (9/100) \times (1 \cdot 5/100)^3}$

$$= 0 \cdot 061 \text{ m} = 61 \text{ mm}$$

Let, $\qquad P$ = Falling weight

Then, work done by the falling weight

$$= P\,(h + \delta)$$

where, $\qquad h$ = Height through which the weight falls.

But, $\qquad P\,(h + \delta)$ = Energy stored in the spring due to static load,

∴ $\qquad P\,(h + \delta) = \dfrac{1}{2}\,W\delta$

$$600\,(h + 0 \cdot 061) = \frac{1}{2} \times 32175 \times 0 \cdot 061$$

$$h = 1 \cdot 574 \text{ m} \quad \textbf{(Ans.)}$$

Leaf spring.

Example 14·26. *A quarter elliptical spring has a length of 50 cm and consists of plates each 6 cm wide and 0·6 cm thick. Find the least number of plates which can be used if deflection under gradually applied load of 3 kN is not to exceed 8 cm. E = 200 GN/m².*

Solution. Span length, $\qquad l = 50 \text{ cm} = 0 \cdot 5 \text{ m}$

Width of each plate, $\qquad b = 6 \text{ cm} = 0 \cdot 06 \text{ m}$

Thickness of each plate, $\qquad t = 0 \cdot 6 \text{ cm} = 0 \cdot 006 \text{ m}$

Deflection, $\qquad \delta = 8 \text{ cm} = 0 \cdot 08 \text{ m}$

Load, $\qquad W = 3 \text{ kN}$

Least number of plates, N:

Using the relation: $\qquad \delta = \dfrac{6\,W\,l^3}{EN\,bt^3}$, we have

$$0 \cdot 08 = \frac{6 \times 3 \times 10^3 \times (0 \cdot 5)^3}{200 \times 10^9 \times N \times 0 \cdot 06 \times (0 \cdot 006)^3}$$

$$\therefore \quad N = \frac{6 \times 3 \times 10^3 \times (0 \cdot 5)^3}{200 \times 10^9 \times 0 \cdot 08 \times 0 \cdot 06 \times (0 \cdot 006)^3}$$

$$N = 10 \cdot 85 \text{ say } 11 \quad \textbf{(Ans.)}$$

Example 14·27. *A quarter elliptical leaf spring has a length of 600 mm and consists of plates each 50 mm wide and 6 mm thick.*

(i) Determine the least number of plates which can be used if the deflection under a gradually applied load of 1·8 kN is not to exceed 80 mm.

(ii) If the applied load of 1·8 kN, instead of being gradually applied, falls a distance of 6 mm on to the undeflected spring, find the maximum deflection and stress produced.

Take: $\qquad E = 200 \text{ GN/m}^2.$

Solution. Length of the spring,

$$l = 600 \text{ mm} = 0 \cdot 6 \text{ m}$$

Width of each plate, $\qquad b = 50 \text{ mm} = 0 \cdot 05 \text{ m}$

Thickness of each plate, $\quad t = 6 \text{ mm} = 0 \cdot 006 \text{ m}$

Gradually applied load, $\quad W = 1 \cdot 8 \text{ kN}$

Deflection under the above load,

$$\delta = 80 \text{ mm} = 0 \cdot 08 \text{ m}$$

(i) Number of plates, N:

We know, $\qquad \delta = \dfrac{6\, W\, l^3}{E\, N\, bt^3}$

Leaf spring fatigue testing system.

or, $\qquad 0 \cdot 08 = \dfrac{6 \times 1 \cdot 8 \times 10^3 \times 0 \cdot 6^3}{200 \times 10^9\, N \times 0 \cdot 05 \times (0 \cdot 006)^3}$

or, $\qquad N = \dfrac{6 \times 1 \cdot 8 \times 10^3 \times 0 \cdot 6^3}{0 \cdot 08 \times 200 \times 10^9 \times 0 \cdot 05 \times (0 \cdot 006)^3} = 13 \cdot 5 \text{ say } 14$

Hence, number of plates $\quad = \textbf{14} \quad \textbf{(Ans.).}$

(ii) δ' ; σ_{max}:

Let, W_e = Equivalent gradually applied load which would produce the same deflection as is caused by the impact load.

Then, $\qquad \delta' = \dfrac{6\, W_e\, l^3}{EN\, bt^3} = \dfrac{6\, W_e \times 0 \cdot 6^3}{200 \times 10^9 \times 14 \times 0 \cdot 05 \times (0 \cdot 006)^3}$

or, $\qquad W_e = 23 \cdot 33 \times 10^3\, \delta'\, N$

Loss of potential energy $\quad = 1 \cdot 8 \times 10^3 \,(6 \times 10^{-3} + \delta')$

Strain energy absorbed by the spring

$$= \frac{1}{2} \cdot W_e \cdot \delta' = \frac{1}{2} \times 23 \cdot 33 \times 10^3 \times \delta'^2$$

$$\therefore \quad 1 \cdot 8 \times 10^3 \,(0 \cdot 006 + \delta') = \frac{1}{2} \times 23 \cdot 33 \times 10^3 \times \delta'^2$$

or, $\qquad 0 \cdot 006 + \delta' = 6 \cdot 48\, \delta'^2 \quad \text{or} \quad \delta'^2 - 0 \cdot 154\, \delta' - 0 \cdot 000926 = 0$

$$\therefore \qquad \delta' = \frac{0 \cdot 154 \pm \sqrt{(0 \cdot 154)^2 + 4 \times 0 \cdot 000926}}{2}$$

$$= \frac{0 \cdot 154 \pm 0 \cdot 1656}{2} = 0 \cdot 1598 \text{ m} = 159 \cdot 8 \text{ mm}$$

Hence, deflection $= \mathbf{159 \cdot 8 \text{ mm}}$ (Ans.)

$\therefore \qquad W_e = 23 \cdot 33 \times 10^3 \times 0 \cdot 0598 \times 10^{-3} \text{ kN} = 3 \cdot 728 \text{ kN}$

Maximum stress, $\sigma_{max} = \dfrac{6 \, W_e \, l}{N \, bt^2} = \dfrac{6 \times 3 \cdot 728 \times 10^3 \times 0 \cdot 6}{14 \times 0 \cdot 05 \times (0 \cdot 006)^2} \times 10^{-6} \text{ MN/m}^2$

$$= 532 \cdot 57 \text{ MN/m}^2$$

Hence, $\qquad \sigma_{max} = \mathbf{532 \cdot 57 \text{ MN/m}^2}$ (Ans.)

TYPICAL EXAMPLES (For Competitive Examinations)

Example 14·28. *A close-coiled helical spring has stiffness of 10 N/mm. Its length when fully compressed with adjacent coils touching each other is 400 mm. The modulus of rigidity of material of the spring = 80000 N/mm².*

(i) *Determine the wire diameter and mean coil diameter, if their ratio is 1/10.*

(ii) *If the gap between any two adjacent coils is 2 mm, what maximum load can be applied before the spring becomes solid i.e. adjacent coils touch.*

(iii) *What is the corresponding maximum shear stress in the spring?*

Solution. Let, $\qquad d$ = Wire diameter, mm,

$\qquad\qquad\qquad D$ = Mean coil diameter, mm, and

$\qquad\qquad\qquad n$ = No. of turns of the coil.

$$\frac{d}{D} = \frac{1}{10} \quad ..... \text{(Given)}$$

i.e. $\qquad\qquad D = 10 \, d$, and $R = 5 \, d$

(i) Wire diameter, d:

When the coils are touching, length of spring,

$$400 = nd \qquad\qquad \therefore \qquad n = \frac{400}{d}$$

Using the relation: $\quad \delta = \dfrac{64 \, WR^3 n}{Cd^4}$, we get

$$\frac{W}{\delta} = \frac{Cd^4}{64 \, WR^3 n}$$

$$10 = \frac{80000 \times d^4}{64 \times (5d)^3 \times \dfrac{400}{d}}, \qquad \text{or,} \qquad d^2 = \frac{10 \times 64 \times 125 \times 400}{80000}$$

$\therefore \qquad\qquad \mathbf{d = 20 \text{ mm}}$ (Ans.)

Mean coil diameter, D:

$$D = 10 \, d = 10 \times 20 = \mathbf{200 \text{ mm}} \text{ (Ans.)}$$

(ii) Load that can make the spring solid, W:

Difference between any two adjacent coils = 2 mm

No. of coils, $\qquad n = \dfrac{400}{d} = \dfrac{400}{20} = 20$

$\therefore$ Total deflection of the load

$$= 20 \times 2 = 40 \text{ mm}$$

and, load that can make the spring solid,

$$W = k \cdot \delta$$
(where, k = stiffness)

$$= 10 \times 40 \quad \text{or} \quad W = \textbf{400 N} \quad \textbf{(Ans.)}$$

(iii) Maximum shear stress in the spring, τ :

Using the relation: $\tau = \dfrac{16\,WR}{\pi d^3}$, we get

$$\tau = \frac{16 \times 400 \times 100}{\pi \times 20^3} = \textbf{25·46 N/mm}^2 \quad \textbf{(Ans.)}$$

Example 14·29. *A close-coiled helical spring is made of a round wire having n turns and the mean coil radius R is 5 times the wire diameter. Show that the stiffness of such a spring is $\dfrac{R}{n} \times$ constant. Determine the constant when the modulus of rigidity C of the spring wire is 82000 N/mm^2.*

If the above spring is to support a load of 1·2 kN with 120 mm compression and the maximum shear stress 250 N/mm^2, calculate:

(i) Mean radius of the coil;

(ii) Number of turns;

(iii) Weight of the spring.

Assume density of material as 76·5 kN/m^3.

Solution. Let, $\qquad d$ = Diameter of the spring wire.

Then, mean radius of the coil,

$$R = 5d$$

$\therefore \qquad\qquad d = 0\cdot2\,R$

Number of turns = n

Stiffness of the spring, $k = \dfrac{W}{\delta} = \dfrac{Cd^4}{64\,nR^3}$

$$= \frac{C \times (0\cdot2\,R)^4}{64\,nR^3} = \frac{C \times 1\cdot6 \times 10^{-3} \times R^4}{64\,nR^3} = \frac{R}{n} \times \text{constant}$$

Hence, *stiffness* $\qquad = \dfrac{R}{n} \times$ constant $\qquad\qquad\qquad$ ......**(Proved)**

where, constant $\qquad = \dfrac{C \times 1\cdot6 \times 10^{-3}}{64} = \dfrac{82000 \times 1\cdot6 \times 10^{-3}}{64} = 2\cdot05$

$\therefore$ Stiffness, $\qquad k = 2\cdot05 \times \dfrac{R}{n}$

Another view of leaf spring.

Axial load on the spring,

$$W = 1.2 \times 10^3 = 1200 \text{ N}$$

Compression,
$$\delta = 120 \text{ mm}$$

(i) Mean radius of the coil, R:

Stiffness,
$$k = 2.05 \times \frac{R}{n}$$

$\therefore$
$$\frac{W}{\delta} = 2.05 \times \frac{R}{n} \quad \text{or} \quad \frac{1200}{120} = 2.05 \times \frac{R}{n}$$

or,
$$\frac{R}{n} = \frac{10}{2.05}$$

Also, shear stress in the wire,
$$\tau = \frac{16 \, WR}{\pi d^3}$$

But,
$$\tau = 250 \text{ N/mm}^2 \quad \text{...(Given)}$$

$\therefore$
$$250 = \frac{16 \times 1200 \times 5d}{\pi \, d^3} \quad \text{or} \quad d^2 = \frac{16 \times 1200 \times 5}{250 \times \pi} = 122.23$$

$\therefore$
$$d = 11 \text{ mm}$$

and,
$$R = 11 \times 5 = \textbf{55 mm} \quad \textbf{(Ans.)}$$

(ii) Number of turns, n:

Now,
$$\frac{R}{n} = \frac{10}{2.05} \quad \text{or} \quad \frac{55}{n} = \frac{10}{2.05}$$

$\therefore$
$$n = \textbf{11.275} \quad \textbf{(Ans.)}$$

(iii) Weight of spring:

Weight of spring
$$= \text{Volume of wire} \times \text{density of the wire material}$$

$$= \left(\frac{\pi}{4} \, d^2 \right) \times (2\pi \, Rn) \times \text{density}$$

$$= \frac{\pi}{4} \times (11 \times 10^{-3})^2 \times (2\pi \times 55 \times 10^{-3} \times 11.275)$$

$$\times (76.5 \times 10^3) \, N = 28.32 \text{ N}$$

i.e., *Weight of spring* = **28·32 N** **(Ans.)**

Example 14·30. *A close-coiled helical spring of 18 mm mean coil diameter and 10 turns is arranged within and concentric with an outer spring. The free length of the inner spring is 4 mm more than that of the outer. The outer spring has 12 coils of mean diameter 30 mm and wire diameter 3·5 mm. The spring load against which a valve is opened is provided by the inner spring. The initial compression in outer spring is 6 mm when the valve is closed. Calculate:*

(i) *The stiffness of the inner spring if the greatest force required to open the valve by 9 mm is 150 N.*

(ii) *The diameter of the wire of the inner spring.*

Take: $\qquad\qquad C = 80 \times 10^3 \text{ N/mm}^2.$

Solution. Let suffix '1' represent inner spring and suffix '2' represent outer spring. Then, mean coil diameter of the *inner spring,*

$$D_1 = 18 \text{ mm}$$

Number of turns,
$$n_1 = 10$$

Free length of inner spring $\qquad = 4$ mm more than the outer one

Wire diameter of *outer spring,* $d_2 = 3.5$ mm

Mean diameter,
$$D_2 = 30 \text{ mm}$$

Number of turns,
$$n_2 = 12$$

Initial compression $\qquad\qquad = 6$ mm

∴ Initial compression in the inner spring

$$= 4 + 6 = 10 \text{ mm}$$

Stiffness of inner spring:

Let, k_1 = Stiffness of inner spring in N/mm, and

k_2 = Stiffness of outer spring in N/mm.

Initial load on the valve, $F_1 = 10 k_1 + 6 k_2$

Stiffness of the outer spring, $k_2 = \dfrac{C d_2^4}{64 R_2^3 n_2} = \dfrac{80 \times 10^3 \times 3 \cdot 5^4}{64 \times 15^3 \times 12} = 4 \cdot 63 \text{ N/mm}$

The valve is to open by 9 mm, additional force required to open the valve,

$$F_2 = 9 k_1 + 9 k_2$$

Total load to lift the valve by 9 mm

$$F = F_1 + F_2 = 19 k_1 + 15 \, k_2 = 150$$

or, $19 k_1 + 15 \times 4 \cdot 63 = 150$

∴ $k_1 = 4 \cdot 24 \text{ N/mm}$ (Ans.)

(ii) Wire diameter of the inner spring, d_1:

$$k_1 = \frac{C d_1^4}{64 R_1^3 n_1} \quad \text{or} \quad 4 \cdot 24 = \frac{80 \times 10^3 \times d_1^4}{64 \times 9^3 \times 10}$$

or, $d_1^4 = \dfrac{4 \cdot 24 \times 64 \times 9^3 \times 10}{80 \times 10^3} = 24 \cdot 727$

∴ $d_1 = 2 \cdot 23 \text{ mm}$ (Ans.)

Example 14·31. *In a compound helical spring, the inner spring is arranged within and concentric with the outer one, but is 7 mm shorter in length. The outer spring has 12 coils of mean diameter 30 mm and the wire diameter is 3·5 mm.*

(i) Find the stiffness of the inner spring if an axial load of 150 N causes the outer spring to compress by 20 mm.

(ii) If the radial clearance between the springs is 1·5 mm, find the wire diameter of the inner spring, if it has 10 coils.

Take: $C = 77000 \text{ N/mm}^2.$

Solution. Outer Spring:

Wire diameter, $d = 3 \cdot 5 \text{ mm}$

Mean coil diameter, $D = 30 \text{ mm}$

Number of coils, $n = 12$

Compression, $\delta = 20 \text{ mm}$

Load required, $W_{outer} = \dfrac{C d^4 \delta}{64 R^3 n} = \dfrac{77000 \times (3 \cdot 5)^4 \times 20}{64 \times 15^3 \times 12} = 89 \cdot 16 \text{ N}$

Inner Spring:

Load shared by the inner spring,

$$W_{inner} = 150 - 89 \cdot 16 = 60 \cdot 84 \text{ N}$$

Compression, $\delta = 20 - 7 = 13 \text{ mm}$

Number of coils $= 10$

(*i*) **Stiffness of inner spring k_{inner}:**

$$k_{inner} = \frac{W_{inner}}{\delta} = \frac{60 \cdot 84}{13} = 4 \cdot 68 \text{ N/mm} \quad \textbf{(Ans.)}$$

(*ii*) **Wire diameter of the inner spring:**

Let, d = Wire diameter of the inner spring.

Now, mean coil radius of outer spring

$$= 15 \text{ mm}$$

Wire diameter of outer spring   = 3·5 mm

Inner coil radius of outer spring $= 15 - \dfrac{3 \cdot 5}{2} = 13 \cdot 25$ mm

Radial clearance between the two springs = 1·5 mm

∴ Outer radius of inner spring   = 13·25 − 1·5 = 11·75 mm

Mean coil radius, $R_{inner} = \left(11 \cdot 75 - \dfrac{d}{2}\right)$ mm

Now, $W_{inner} = \dfrac{77000 \times d^4 \times 13}{64 \times (11 \cdot 75 - d/2)^3 \times 10}$

or, $60 \cdot 84 = \dfrac{1564 \, d^4}{(11 \cdot 75 - d/2)^3}$

$$(11 \cdot 75 - d/2)^3 = 25 \cdot 7 \, d^4$$
$$(23 \cdot 5 - d)^3 = 205 \cdot 6 \, d^4$$

From which, $d = 2 \cdot 58$ mm say **2·6 mm** **(Ans.)**

Example 14·32. *A close-coiled helical spring has 30 turns, the mean radius of the coils is 75 mm while the wire diameter of the wire is 15 mm. Find the work done in rotating one end of the spring by 80° relative to the other end (fixed end), by a couple whose axis coincides with the axis of the spring.*

Take: $E = 210000 \text{ N/mm}^2.$

Solution. Wire diameter of the spring,

$$d = 15 \text{ mm}$$

Mean coil radius, $R = 75$ mm

Number of turns, $n = 30$

Young's modulus, $E = 210000 \text{ N/mm}^2$

Moment of inertia, $I = \dfrac{\pi}{64} d^4 = \dfrac{\pi}{64} \times 15^4 = 2485 \text{ mm}^4$

Length of the wire, $l = 2\pi Rn = 2\pi \times 75 \times 30 = 14137$ mm

Angular rotation, $\theta = 80° = 80 \times \pi/180 = 1 \cdot 396$ radians

Another view of an automobile wheel and spring suspension.

We know, $\phi = \dfrac{Ml}{EI}$

∴ $M = \dfrac{\phi \, EI}{l} = \dfrac{1 \cdot 396 \times 210000 \times 2485}{14137} \times 10^{-3} \text{ Nm} = 51 \cdot 53 \text{ Nm}$

∴ Work done on the spring,

$$U = \frac{1}{2} M \, \theta = \frac{1}{2} \times 51 \cdot 53 \times 1 \cdot 396 = \textbf{35·97 Nm} \quad \textbf{(Ans.)}$$

Example 14·33. *A stiff bar of negligible weight transmits a load P to combination of three springs as shown in the Fig. 14·9. The springs are made up of the same material and out of rods of equal diameters. They are of the same length before loading. The number of coils in the three springs are 10, 12 and 15 respectively, while the mean radii are in ratio of 1 : 1·2 : 1·4 respectively. Find the distance x such that the bar remains horizontal after applying the load.*

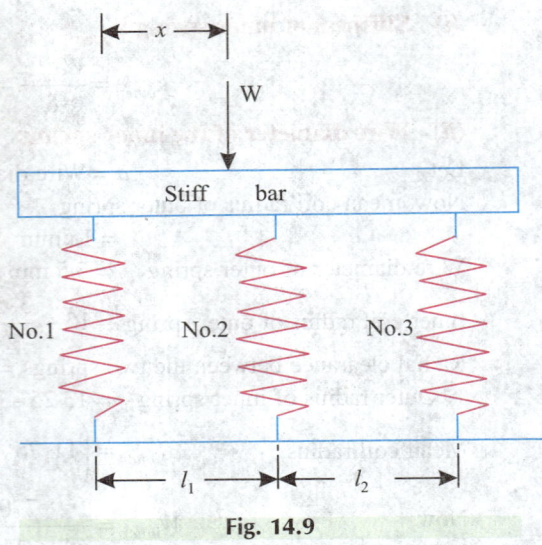

Fig. 14.9

Solution. Refer to Fig. 14·9.

The stiff bar will only remain horizontal if the springs get compressed by the same amount, say δ. Let W_1, W_2 and W_3 be the loads carried by the three springs respectively. Let the mean radii of the coils of three springs be R, $1·2 R$ and $1·4 R$ respectively as per the proportion of three coils.

For first spring:

$$\delta = \frac{64\,W_1 R_1^3\, n_1}{Cd^4} = \frac{64\,W_1 R_1^3 \times 10}{Cd^4}$$

For second spring:

$$\delta = \frac{64\,W_2 R_2^3 \times n_2}{Cd^4} = \frac{64\,W_2 R_2^3 \times 12}{Cd^4}$$

For third spring:

$$\delta = \frac{64\,W_3 R_3^3 n_3}{Cd^4} = \frac{64\,W_3 R_3^3 \times 15}{Cd^4}$$

Since δ, C and d are same for all springs and

$$R_1 = R,\ R_2 = 1·2\,R \text{ and } R_3 = 1·4\,R$$

∴ $$10\,W_1 = (1·2)^3 \times 12\,W_2 = (1·4)^3 \times 15\,W_3$$

or, $$W_1 = \frac{(1·4)^3 \times 15\ W_3}{10} \quad \text{and,} \quad W_2 = \frac{(1·4)^3 \times 15\ W_3}{(1·2)^3 \times 12}$$

Now let the load W act at a distance x from the left end as shown. Taking moments about the point where the load is acting, we have

$$W_1 x = W_2\,(l - x) + W_3\,(2l - x)$$

where, $$l_1 = l_2 = l$$

or, $$\frac{(1·4)^3 \times 15\ W_3 \cdot x}{10} = \frac{(1·4)^3 \times 15\ W_3}{(1·2)^3 \times 12}\,(l - x) + W_3\,(2l - x)$$

$$4·116x = 1·985\,(l - x) + (2l - x)$$

$$4·116x = 1·985\,l - 1·985x + 2l - x$$

$$4·116x = 3·985\,l - 2·985x$$

$$7·101x = 3·985l$$

$$x = 0·561l \quad \textbf{(Ans.)}$$

Example 14·34. *Find the weight of a close-coiled helical spring which would absorb the energy of truck weighing 95 kN and moving with a velocity of 1·2 m / s if:*

    (*i*)   *spring is compressed by the impact,*

    (*ii*) *spring is wound up by the impact.*

Working stress: 290 MN/m² (bending); 240 MN/m² (torsion)

E = 200 GN/m², C = 80 GN/m²; specific gravity of material = 7·9. **(Banglore University)**

Solution. Kinetic energy to be used,

$$K.E. = \frac{1}{2}\frac{W}{g}v^2$$

where, W = Weight of truck = 95 kN,

                         v = Velocity of truck = 1·2 m/s, and

                         g = Acceleration due to gravity = 9·81 m/s².

i.e., $K.E. = \frac{1}{2} \times \frac{95}{9\cdot 81} \times 1\cdot 2^2 = 6\cdot 97$ kNm

(*i*) Spring is compressed by impact:

If the spring absorbs the energy in direct compression, then the wire of the spring is in torsion. Hence strain energy,

$$U = \frac{\tau^2}{4C} \times \text{volume of spring}$$

Equating the strain energy to *K.E.* and substituting values, we get

$$\frac{\tau^2}{4C} \times \text{volume of spring} = K.E.$$

∴ Volume of spring $= \dfrac{K.E. \times 4\,C}{\tau^2} = \dfrac{(6\cdot 97 \times 10^3) \times 4 \times 80 \times 10^9}{(240 \times 10^6)^2} = 0\cdot 03872 \text{ m}^3$

Weight of spring   = Volume of spring × specific gravity × density of water

                      = 0·03872 × 7·9 × 9·81 kN   (∵ Density of water = 9·81 kN/m³)

                      = **3 kN** **(Ans.)**

Drawbars behind a tractor are meant to pull the loads.

(ii) **Spring is wound up by impact:**

Strain energy, $\qquad U = \dfrac{\sigma_b^2}{8E} \times$ volume (where, σ_b = maximum bending stress in the wire)

Equating strain energy to *K.E.*, we get

$$\dfrac{\sigma_b^2}{8E} \times \text{volume of spring} = K.E.$$

Volume of spring $= \dfrac{K.E. \times 8E}{\sigma_b^2} = \dfrac{6 \cdot 97 \times 10^3 \times 8 \times 200 \times 10^9}{(290 \times 10^6)^2} = 0 \cdot 1326 \text{ m}^3$

and, $\qquad$ weight of spring = Volume of spring × density

$\qquad\qquad\qquad\qquad = 0 \cdot 1326 \times (7 \cdot 9 \times 9 \cdot 81) \text{ kN}$

$\qquad\qquad\qquad\qquad = \textbf{10} \cdot \textbf{276 kN} \quad \textbf{(Ans.)}$

HIGHLIGHTS

1. Types of springs: (*i*) Helical springs, (*ii*) Leaf springs, (*iii*) Torsion springs, (*iv*) Circular springs, (*v*) Belleville springs and (*vi*) Flat springs.

2. *Close-coiled helical springs* subjected to '*axial load*':

   Shear stress, $\qquad\qquad \tau = \dfrac{16\, W\, R}{\pi d^3}$ $\qquad\qquad\qquad\qquad$ ...(*i*)

   Angle of twist, $\qquad\qquad \theta = \dfrac{64\, W\, R^2 n}{C d^4}$ $\qquad\qquad\qquad$ ...(*ii*)

   Deflection, $\qquad\qquad \delta = R\theta$ $\qquad\qquad\qquad\qquad\qquad$ ...(*iii*)

   or, $\qquad\qquad\qquad \delta = \dfrac{64\, W\, R^3 n}{C d^4}$ $\qquad\qquad\qquad$ ...(*iv*)

   Stiffness, $\qquad\qquad k = \dfrac{W}{\delta} = \dfrac{C d^4}{64\, R^3 n}$ $\qquad\qquad\qquad$ ...(*v*)

   Energy stored, $\qquad\qquad U = \dfrac{1}{2} \times T \times \theta$ $\qquad\qquad\qquad$ ...(*vi*)

   $\qquad\qquad\qquad\qquad U = \dfrac{\tau^2}{4C} \times \text{volume of spring}$ $\qquad\qquad$ ...(*vii*)

   $\qquad\qquad\qquad\qquad U = \dfrac{1}{2} W\delta$

3. Helical springs subjected to '*axial twist*':

   Angle of twist, $\qquad\qquad \theta = \dfrac{Ml}{EI}$ $\qquad\qquad\qquad\qquad\qquad$ ...(*i*)

   $\qquad\qquad\qquad\qquad \theta = \dfrac{128\, MRn}{E d^4}$ $\qquad\qquad\qquad\qquad$ ...(*ii*)

   Bending stress, $\qquad\qquad \sigma_b = \dfrac{32M}{\pi d^3}$ $\qquad\qquad\qquad\qquad$ ...(*iii*)

   Energy stored, $\qquad\qquad U = \dfrac{\sigma_b^2}{8E} \times \text{volume of spring wire}$ $\qquad$ ...(*iv*)

4. *Open-coiled helical spring*

(i) *With axial load:*

The couple producing *torsion*,

$$T = W R \cos \alpha$$

The couple producing *bending*,

$$M = W R \sin \alpha$$

Deflection,

$$\delta = 2 W R^3 n \pi \sec \alpha \left[\frac{\cos^2 \alpha}{CI_p} + \frac{\sin^2 \alpha}{EI} \right]$$

Angular rotation,

$$\psi = 2 W R^2 n \pi \sin \alpha \left[\frac{1}{CI_p} - \frac{1}{EI} \right]$$

where, W = Applied load, R = Mean coil radius,

             n = Number of coils, α = Helix angle,

             C = Modulus of rigidity, and E = Young's modulus.

(ii) *With axial thrust:*

Angular rotation,

$$\phi = 2 \, TRn \pi \sec \alpha \left[\frac{\sin^2 \alpha}{CI_p} + \frac{\cos^2 \alpha}{EI} \right]$$

Deflection,

$$\delta = 2 \, TR^2 n \pi \sin \alpha \left[\frac{1}{CI_p} - \frac{1}{EI} \right] \qquad (\text{where, } T = \text{Applied torque})$$

5. *Springs in series:*

$$\frac{1}{k} = \frac{1}{k_1} + \frac{1}{k_2}$$

Springs in parallel: $k = k_1 + k_2$

where, k_1 = Stiffness of spring 1,

            k_2 = Stiffness of spring 2, and

            k = Stiffness of composite spring.

6. *Flat spiral spring:*

Angular rotation, $\phi = \dfrac{Tl}{EI}$ ...(i)

Maximum bending moment,

$$M_{max} = 2T \qquad \qquad \qquad \qquad \text{...(ii)}$$

Maximum stress, $\sigma_{max} = \dfrac{12T}{bt^2}$ ...(iii)

Winding torque, $T = \dfrac{bt^2 \cdot \sigma_{max}}{12}$ ...(iv)

Energy stored, $U = \dfrac{6T^2 l}{Ebt^3}$ ...(v)

Maximum energy stored,

$$U_{max} = \frac{\sigma_{max}^2}{24 \, E} \times \text{volume of spring} \qquad \qquad \text{...(vi)}$$

Resilience of spring (= energy per unit volume) $= \dfrac{\sigma_{max}^2}{24E}$ ...(vii)

7. *Laminated spring:*

A. *Semi-elliptical spring:*

Bending stress, $\sigma_b = \dfrac{3\,W\,l}{2\,Nbt^2}$ ...(i)

Deflection, $\delta = \dfrac{3\,Wl^3}{8\,EN\,bt^3}$ ...(ii)

Strain energy, $U = \dfrac{\sigma_b^2}{6E} \times$ volume of spring ...(iii)

Where, $W =$ Applied load,

$b =$ Width of each plate,

$t =$ Thickness of each plate,

$N =$ Number of plates in the spring,

$l =$ The spring span length, and

$E =$ Young' modulus of elasticity.

B. *Quarter elliptical spring:*

Bending stress, $\sigma_b = \dfrac{6\,Wl}{Nbt^2}$ ...(i)

Deflection, $\delta = \dfrac{6\,Wl^3}{ENbt^3}$ ...(ii)

OBJECTIVE TYPE QUESTIONS

Choose the Correct Answer:

1. If a close-coiled helical spring is subjected to load W and the deflection produced is δ, then stiffness of the spring is given by

(a) W/δ (b) $W\cdot\delta$

(c) δ/W (d) $W^2\cdot\delta$.

2. Wahl's connection factor (K) is given by the relation

(a) $K = \dfrac{3S-1}{3S-4} + \dfrac{0\cdot 615}{S}$

(b) $K = \dfrac{4S-1}{4S-4} + \dfrac{0\cdot 615}{S}$

(c) $K = \dfrac{5S-1}{5S-4} + \dfrac{0\cdot 615}{S}$

(d) $K = \dfrac{6S-1}{6S-4} + \dfrac{0\cdot 615}{S}$.

(where $S =$ spring index)

3. The energy stored in a close-coiled helical spring, when subjected to an 'axial twist', is given by

(a) $\dfrac{\sigma_b^2}{6E} \times$ volume of spring

(b) $\dfrac{\sigma_b^2}{8E} \times$ volume of spring

(c) $\dfrac{\sigma_b^2}{4E} \times$ volume of spring

(d) $\dfrac{\sigma_b^2}{2E} \times$ volume of spring.

(where s$_b$ = bending stress)

4. Two springs of stiffness k_1 and k_2 respectively are connected in series, the stiffness of the composite spring (k) will be given by

(a) $k = k_1 + k_2$ (b) $k = k_1 \times k_2$

(c) $k = \dfrac{k_1 k_2}{k_1 + k_2}$ (d) $k = \dfrac{k_1 + k_2}{k_1 k_2}$.

5. The resilience of a flat spiral spring is given by

(a) $\dfrac{\sigma_{max}}{24E}$ (b) $\dfrac{\sigma_{max}^2}{24E}$

(c) $\dfrac{\sigma_{max}^2}{12E}$ (d) $\dfrac{\sigma_{max}^2}{8E}$.

(where, σ_b = bending stress)

6. In case of a laminated spring, the load at which the plates become straight is called

(a) working load   (b) safe load

(c) proof load   (c) none of the above.

7. .... are called cantilever laminated springs.

(a) Semi-elliptical springs

(b) Quarter elliptical springs

(c) Both (a) and (b)

(d) None of the above.

Even good seats need springs to cushion shocks. The above picture shows a tractor driver seat.

ANSWERS

1. (a)   2. (b)   3. (b)   4. (c)   5. (b)   6. (c)   7. (b)

UNSOLVED EXAMPLES

Close-coiled helical springs :

1. A close-coiled helical spring, with the coil diameter as 100 mm and wire diameter as 12 mm consists of 16 coils. If it is subjected to an axial tension of 400 N, find the maximum stress induced in the coil, the extension suffered by the spring and the energy stored in it. Modulus of rigidity, $C = 84$ GN/m^2. [**Ans.** 59 MN/m^2; 29·4 mm; 5·88 Nm]

2. A close-coiled helical spring having 100 mm mean diameter is made of 20 turns of 10 mm diameter steel rod. The spring carries an axial load of 100 N. Find the shearing stress developed in the spring and the deflection of the load. $C = 84$ GN/m^2.

   [**Ans.** 25·5 MN/m^2; 19·1 mm]

3. A close-coiled helical spring, made of 6 mm diameter steel wire has 20 coils, each of 100 mm mean diameter. When subjected to axial load of 70 N, calculate:

   (i) The maximum shear stress produced,   (ii) The deflection,

   (iii) Stiffness, and   (iv) The energy stored.

   Take: C = 84 GN/m^2.   [**Ans.** 82·6 MN/m^2; 103 mm; 0·68 N/mm; 3·6 Nm]

4. A close-coiled helical spring is to carry a load of 100 N and the mean coil diameter is to be eight times the wire diameter. Calculate these diameters if the maximum shear stress is to be 75 MN/m^2.

   [**Ans.** 5·21 mm; 41·68 mm]

5. A close-coiled spring has a radius of 40 mm and length 320 mm. It is required to extend 21 mm under a pull of 185 N. If $C = 84$ GN/m^2 determine the diameter of the wire.

   [**Ans.** 4·84 mm]

6. The mean diameter of a spring is six times the diameter of the wire. It extends by 60 mm under a pull of 550 N. If the maximum allowable shear stress is 350 MN/m^2, find the size of the wire and the number of coils. Take $C = 84·4$ MN/m^2.   [**Ans.** $d = 4·9$ mm; $n = 26·1$]

7. A close-coiled spring is to have a stiffness of 1 N/mm of compression under a maximum load of 45 N and a maximum shearing stress 126 MN/m². The solid length of spring (*i.e.* when the coils are touching) is to be 45 mm. Find the diameter of the wire, the mean diameter of the coils and number of coils required. $C = 42$ GN/m².

[**Ans.** $d = 3.06$ mm, $D = 31.5$ mm; $n = 14.7$]

8. A safety valve of 84 mm diameter is to blow off at a pressure of 1.2 MN / m² gauge. It is held by a closely coiled compression spring of a circular steel. The mean diameter is 200 mm and initial compression is 20 mm. Find the diameter of steel bar and the number of turns necessary. $\tau = 70$ MN/ m² and $C = 80$ GN/m². [**Ans.** 34.5 mm; 5.9]

9. Design a close-coiled helical spring which when put to a load of 40 kN may deflect 80 mm. The diameter of each coil is to be 10 times that of wire of the spring and maximum shear stress is not to exceed 55 MN/m². $C = 85$ GN/m². Also determine, what suddenly applied load will elongate spring by 30 mm. [**Ans.** $d = 28$ mm; $D = 280$ mm; $n = 29.75$]

10. Calculate the modulus of rigidity of a spring of 10 coils, 64 mm mean diameter and wire of 6.4 mm diameter. The spring extends 100 mm under axial force of 700 N.

[**Ans.** $C = 87.5$ GN/m²]

11. A close-coiled helical spring has the stiffness of 40 N/mm. Determine its number of turns when diameter of the wire of the spring is 10 mm and mean diameter of the coils is 80 mm. $C = 80$ GN/m².

[**Ans.** $n = 4.88$]

12. A helical spring, in which the mean diameter of the coils is 12 times the wire diameter, is to be designed to absorb 300 Nm of energy with an extension of 150 mm. The maximum shear stress is not to exceed 140 MN / m². Determine the mean diameter of the helix, diameter of the wire, and the number of turns. Also, find the load with which an extension of 50 mm could be produced in the spring. $C = 80$ GN/m².

[**Ans.** $d = 30$ mm; $D = 360$ mm, $n = 6.52$; $W = 1331$ N]

13. A closely wound helical spring made of round wire is required to absorb 35 Nm of energy with an axial deflection 100 mm. The mean diameter of the coils is to be 10 times the diameter of the wire and the shear stress produced is not to exceed 160 MN/m². Find:

    (*i*) The diameter of wire,   (*ii*) The maximum diameter of the coils, and

    (*iii*) Free number of coils.

    Take $C = 80$ GN/m². [**Ans.** 11 mm; 110 mm; 15.7]

14. A helical spring is made of 4 mm diameter wire and the mean diameter of each coil being 20 mm. Find what axial load may be applied and the corresponding deflection of spring, if the maximum shear stress is not to exceed 300 MN/m². $C = 80$ MN/ m². There are 8 coils in the spring.

[**Ans.** 37.7 N; 9.42 mm]

15. A truck weighing 20 kN and moving at 2.5 km / h has to be brought to rest by a buffer. Find how many springs each of 15 coils will be required to store the energy of motion during the compression of 150 mm. The diameter of the wire is 20 mm and the mean radius of curvature for the coil is 100 mm. $C = 80$ GN/m².

[**Ans.** 4]

Carburettor mixes air and fuel for combustion.

16. A 200 N weight falls freely from the height of 500 mm on a closely wound spring which is compressed by a maximum amount of 100 mm under the impact. The spring is made of steel rod of 25 mm diameter and of mean diameter of coils 200 mm. Determine the instantaneous stress produced by the impact and number of coils in the spring.

[**Ans.** 72·8 MN/m^2; 20·35]

17. A spring of valve 70 mm diameter is to be designed such that the valve may blow off at a pressure of 0·8 MN/m^2 gauge, with an initial spring tension of 20 mm. Determine the diameter of the spring wire and the number of coils required, when the mean diameter of the coils of the spring is to be 125 mm and the maximum shear stress is not to exceed 95 MN/m^2.
Take: $C = 80$ GN/m^2. [**Ans.** 22 mm; 7·8]

18. A weight of 150 N is dropped on a compression spring made of 1·2 mm steel wire closely coiled to a mean diameter of 150 mm with 20 coils. If the instantaneous compression is 14 mm, calculate the height of drop. $C = 80$ GN/m^2. [**Ans.** 61 mm]

19. A railway wagon weighing 50 kN and moving at the rate of 3 m/s has to be brought to rest by a buffer, which is made up of springs each containing 10 coils and having a wire diameter of 60 mm and mean coil diameter of 240 mm. Find the number of springs required if the compression of each spring is to be 200 mm and $C = 84$ GN/m^2. [**Ans.** 7]

20. Two springs are joined in series to form a composite spring of stiffness 1N/mm. Each spring has a diameter 6 times the wire diameter. One of the springs has 32 coils and wire diameter 2 mm. Find the diameter of the wire in the other spring which has 48 coils. Find the greatest tensile load that can be applied to the composite spring and the corresponding extension, the maximum permissible shear stress being 250 MN/m^2 and $C = 84$ GN m^2.

[**Ans.** $d = 2$ mm; 35·45N; 35·45 mm]

21. A helical spring B is placed inside the coils of a second helical spring A having the same number of coils and free axial length and of the same material. The two springs are compressed by an axial load of 2·3 kN which is shared between them. The mean coil diameters of A and B are 10 cm and 7 cm and the wire diameters are 1·3 cm and 0·8 cm respectively. Calculate the load taken and the maximum stress in each spring.

[**Ans.** 1·62 kN; 0·68 kN; 187·9 MN/m^2; 236 MN/m^2]

22. Determine the maximum shear stress and the amount of compression produced when a load of 2 kN is dropped axially on a close-coiled helical spring, from a height of 250 mm. The spring has 20 coils each of mean diameter 200 mm and the wire diameter is 25 mm. $C = 84$ MN/m^2.

[**Ans.** 242 MN/m^2; 290 mm]

23. A closely coiled helical spring having 12 coils each of 75 mm mean diameter and wire of 7 mm diameter is held at one end and a twisting moment of 6 Nm applied axially about the other end. Determine:
   (*i*) The maximum bending stress produced,
   (*ii*) The angle of twist, and
   (*iii*) Resilience.
Take: $E = 200$ GN/m^2. [**Ans.** (*i*) 181 MN/m^2; (*ii*) 0·72 radian, (*iii*) 0·26 Nm]

Open-coiled helical springs:

24. An open-coiled helical spring made from wire of circular cross-section is required to carry a load of 100 N. The wire diameter is 8 mm and the mean coil radius is 40 mm. If the helix angle of the spring is 30° and number of turns is 12, Calculate:
   (*i*) Axial deflection;
   (*ii*) Angular rotation of free end with respect to the fixed end of the spring.

[**Ans.** (*i*) 16·45 mm; (*ii*) 0·0376 radian]

25. An open-coiled spring consists of 10 coils, each of mean diameter 50 mm, the wire forming the coil being 6 mm in diameter. Each coil makes an angle of 30° with the plane perpendicular to the axis of the spring.
    (*i*) Determine the load required to elongate the spring by 20 mm and the bending and shear stresses caused by that load.
    (*ii*) Calculate the axial twist that would cause a bending stress of 60 MN/m².
    Take: E = 200 GN/m², and C = 82 GN/m².
    [**Ans.** (*i*) W = 192·7 N, 113·6 MN/m², 98·37 MN/m² (*ii*) 1·469 Nm]

26. An open-coiled helical spring of wire diameter 10 mm, mean coil radius 70 mm, helix angle 20° carries an axial load of 400 N. Determine the shear stress and direct stress developed at the inner radius of the coil. [**Ans.** τ = 139·09 MN/m², σ_b = 97·54 MN/m²]

27. An open-coiled helical spring has the following data:
    Wire diameter = 6 mm
    Mean coil radius = 36 mm
    Inclination of the coils = 65°.
    Modulus of rigidity of steel = 84 GN/m²
    Young's modulus = 210 GN/m²
    Original number of turns = 10
    The above spring is subjected to axial torque T and the number of turns in the spring increases by 1/8. Calculate:
    (*i*) Magnitude of axial torque T;   (*ii*) Change in axial length of the spring.
    [**Ans.** (*i*) 4·029 Nm; (*ii*) 2·585 mm]

Spiral springs :

28. A flat spiral spring is 6 mm wide and 0·25 mm thick, the length being 2·5 m. Assuming the maximum stress of 800 *MN / m²* to occur at the point of greatest bending moment, calculate:
    (*i*) The torque;   (*ii*) The work that can be stored in the spring;
    (*iii*) The number of turns to wind up the springs.
    Take: E = 200 GN/m².   [**Ans.** (*i*) 0·025 Nm, (*ii*) 0·5 Nm (*iii*) 6·37 turns]

29. A flat spring is made of a strip 5 mm wide and 0·25 mm thick, 10 m long. The torque is applied at the winding spindle and 8 complete turns are given. Calculate:
    (*i*) The torque;
    (*ii*) Maximum stress developed at the point of greatest bending moment;
    (*iii*) The energy stored.
    Take: E = 210 GN/m².   [**Ans.** (*i*) 6·872 Nmm, (*ii*) 263·9 MN/m². (*iii*) 172·7 mm]

Laminated springs :

30. A carriage spring made of laminations each 100 mm wide and 8 mm thick, is 0·8 m long. Determine the number of plates required when the spring is designed for a maximum central load of 6 kN and maximum allowable stress in the material is not to exceed 200 MN/m². What would be the deflection under the load when E = 200 GN/m² ?
    [**Ans.** 5·625 say 6; 18·75 mm]

31. A laminated spring 1 m long is made up of plates each 50 mm wide and 10 mm thick. If the bending stress in the plates is limited to 100 MN/m², how many plates would be required to enable the spring to carry a central point load of 2 kN ? If E = 200 GN/m² what is the deflection under the given load of 2 kN ?   [**Ans.** N = 6, δ = 11·9 mm]

32. A leaf of semi-elliptical type is built-up with 9 leaves of 0·5 cm × 5 cm cross-section. The length of spring is 0·75 m and the leaves are of steel having proof stress of 400 MN/m². Determine to what radius should the leaves be initially bent ? Also calculate the value of proof load. From what maximum height should a load of 200 *N* fall without exceeding the proof stress in the leaves of the spring? E = 200 GN/m².   [**Ans.** 1·25 m, 4 kN, 50·63 cm]

Strain Energy and Deflection due to Shear and Bending

Chapter 15

15.1   STRAIN ENERGY OR RESILIENCE

When an elastic body is loaded it undergoes deformation *i.e.* its dimensions change and when it is relieved of the load it regains its original shape. For the time loaded energy is stored in it, the same is given up or released by the loading when the load is removed. This energy is called **strain energy.** *The strain energy stored by the body 'within' elastic limit, when loaded externally is called* **'Resilience',** *and the maximum energy which a body stores 'upto' elastic limit is called* **'Proof resilience'.**

Proof resilience is the mechanical property of materials and it indicates their *capacity to bear shocks. Proof resilience per unit volume of piece is called* **'Modulus of resilience'.**

15.2   STRAIN ENERGY IN SIMPLE TENSION AND COMPRESSION

Let us take the case of a bar of cross-sectional area A and length l and subjected to a load W. Suppose this load extends the bar by an amount δl and produces a maximum stress σ.

The work done by W and hence the strain energy (U) stored in the material is equal to the area (shaded in Fig. 15.1, under the force-extension curve.

Strain energy stored in the bar = Work done by the load.

$$U = \frac{1}{2} \cdot W \cdot \delta l$$

$$= \frac{W}{2} \cdot \frac{\sigma l}{E} \qquad \left[\because \delta l = \frac{\sigma l}{E} \right]$$

$$= \frac{\sigma A}{2} \cdot \frac{\sigma l}{E} \qquad \left[\begin{array}{l} \because \text{ Load = stress} \times \text{area} \\ \text{or} \quad W = \sigma \times A \end{array} \right]$$

or $\qquad U = \frac{\sigma^2 Al}{2E} = \frac{\sigma^2 V}{2E} \qquad \qquad$ ...(15.1)

$$\left[\begin{array}{l} \because \text{ Volume = area} \times \text{length} \\ \text{or,} \quad V = A \times l \end{array} \right]$$

Fig. 15.1

If σ_p be the proof stress or maximum stress to which the bar is stressed upto the elastic limit, then

Proof resilience,

$$U_p = \frac{\sigma_p^2}{2E} \times V \qquad \qquad \qquad \text{...(15.2)}$$

and, *modulus of resilience* $= \dfrac{\sigma_p^2}{2E}.$ $\qquad \qquad \qquad$ ... (15.3)

15.3. STRESSES DUE TO DIFFERENT TYPES OF LOADS

A body may be subjected to following types of loads :

1. Gradually applied loads
2. Suddenly applied loads
3. Falling or impact loads.

1. Gradually applied loads:

A body is said to be acted upon by a gradually applied load if the load increases from zero and reaches its final value stepwise. Let W be the load applied gradually on a body and let δl and σ be the corresponding change in length and maximum stress induced in it.

An old model of 3-cylinder IC engine. Most moving parts in engines absorb and release strain energies.

Energy due to external load $= \dfrac{1}{2} \sigma \cdot A \cdot \delta l$

Also, Work done on the body $= \dfrac{1}{2} W \cdot \delta l$

But strain energy stored = Work done on the body

$$\frac{1}{2} \sigma A \cdot \delta l = \frac{1}{2} W \cdot \delta l, \quad \text{or,} \quad \sigma = \frac{W}{A}$$

It may be remembered that *unless specifically mentioned, the load is always gradually applied.*

2. Suddenly applied loads:

When the load is applied *all of a sudden* and not stepwise is called *suddenly applied load.*

Now let the load W is applied all of a sudden and maximum stress thus produced be σ_{su}; the extension being the same as δl.

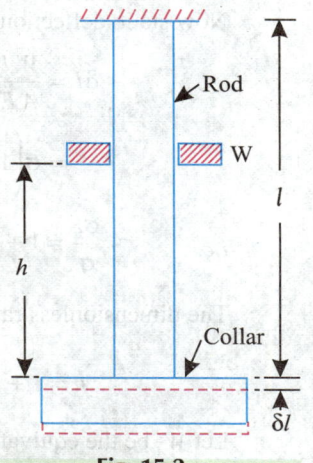

Then,   Energy stored = External work done

$$W \times \delta l = \frac{1}{2}\sigma_{su} . A \times \delta l,$$

or, $$\sigma_{su} = \frac{2W}{A}$$

This shows that stress (σ_{su}) *due to suddenly applied load is double that of gradually applied load.* Evidently the *instantaneous strains will also be in the same ratio.*

3. Falling or impact loads:

The load which falls from a height or strike the body with certain momentum is called *falling or impact load.*

Refer to Fig. 15.2. Consider a weight W falling through a height 'h' on a collar fitted on the rod which is of length 'l' and has a cross-sectional area 'A'. Let the extension and maximum stress thus produced be δl_i and σ_i respectively.

Fig. 15.2

Now,   External work done on the bar = Energy stored in the bar

i.e. $$W (h + \delta l_i) = \frac{1}{2}\sigma_i A \times \delta l_i$$

or, $$W\left[h + \frac{\sigma_i l}{E}\right] = \frac{1}{2}\sigma_i A \times \frac{\sigma_i l}{E} \left[\because \delta l_i = \frac{\sigma_i l}{E}\right]$$

or, $$W\left[h + \frac{\sigma_i l}{E}\right] = \frac{\sigma_i^2 A l}{2E}$$

or, $$\frac{\sigma_i^2 A l}{2E} - \frac{\sigma_i W l}{E} - Wh = 0$$

or, $$\sigma_i = \frac{\dfrac{Wl}{E} \pm \sqrt{\dfrac{W^2 l^2}{E^2} + \dfrac{2WhAl}{E}}}{\dfrac{Al}{E}}$$

or, $$\sigma_i = \frac{\dfrac{Wl}{E} \pm \dfrac{Wl}{E}\sqrt{1 + \dfrac{2WhAl}{E} \times \dfrac{E^2}{W^2 l^2}}}{\dfrac{Al}{E}} \quad \text{or} \quad \sigma_i = \frac{W + W\sqrt{1 + \dfrac{2hAE}{Wl}}}{A}$$

[Taking +ve sign only (max. value)]

$$\sigma_i = \frac{W}{A}\left[1 + \sqrt{1 + \frac{2hAE}{Wl}}\right] \qquad \qquad ...(15.4)$$

If δl_i is negligible as compared to h,

then, $$Wh = \frac{\sigma_i^2 A l}{2E}$$

and, $$\sigma_i = \sqrt{\frac{2WhE}{Al}} \qquad \qquad ...15.4\ (a)$$

and, if $$h = 0 \quad \sigma_i = \frac{2W}{A} \qquad \qquad ...15.4\ (b)$$

Now static deflection δl due to load W is given by

$$\delta l = \frac{W\,l}{A\,E} \qquad \therefore\ \sigma_i = \frac{W}{A}\left(1 + \sqrt{1 + \frac{2h}{\delta l}}\right)$$

$$= \sigma\left(1 + \sqrt{1 + \frac{2h}{\delta l}}\right) \qquad \left(\because\ \sigma = \frac{W}{A}\right)$$

i.e.
$$\frac{\sigma_i}{\sigma} = 1 + \sqrt{1 + \frac{2h}{\delta l}} \qquad\qquad\qquad ...(15.5)$$

The dimensionless ratio $\dfrac{\sigma_i}{\sigma}$ is usually called **load factor** and is denoted by n.

$$\therefore \qquad n = 1 + \sqrt{1 + \frac{2h}{\delta l}}$$

Let W_e be the equivalent static load that produces the same stress σ_i as produced by the load W applied with impact.

$$\therefore \qquad \sigma_i = \frac{W_e}{A}, \qquad \text{and,} \qquad \sigma = \frac{W}{A}$$

$$\therefore \qquad \frac{\sigma_i}{\sigma} = \frac{W_e}{W}, \qquad \text{or,} \qquad n = \frac{W_e}{W} = \frac{\sigma_i}{\sigma}$$

or $\qquad\qquad W_e = n\,W \qquad\qquad\qquad ...(15.6)$

The load factor is important because it gives the factor by which the load producing impact should be multiplied to give an equivalent static load for design. Once W_e is known, the solution of a design problem can be carried out as usual by taking W_e instead of the load W.

Following relations are worth noting :

$$n \cdot \delta l = \delta l_i \qquad\qquad\qquad ...(i)$$
$$n\,\sigma = \sigma_i \qquad\qquad\qquad ...(ii)$$
$$n\,W = W_e \qquad\qquad\qquad ...(iii)$$

Sometimes a different factor known as **impact factor** n' is defined as follows:

$$n' = \sqrt{1 + \frac{2h}{\delta l}} \qquad \therefore\ n = 1 + n'$$

Thus the *load factor is equal to the impact factor plus one. The impact factor gives the factor responsible for producing the impact load over and above the static load.* It can be found that

$$n' = n - 1 = \frac{W_e - W}{W}$$

Impact factors and load factors in many cases are usually determined from tests and experience for use in design.

If δl is neglected as compared to h, then

$$n = \sqrt{\frac{2h}{\delta l}}$$

Example 15.1. *A steel bar 4 cm by 4 cm in section, 3 m long is subjected to an axial pull of 128 kN. Taking E = 200 GN/m² calculate the alteration in the length of the bar. Calculate also the amount of energy stored in the bar during the extension.*

Solution. Refer to Fig. 15.3.

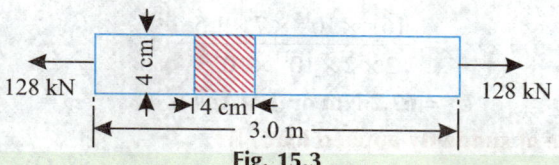

Fig. 15.3

Cross-sectional area of the bar,

$$A = 4 \text{ cm} \times 4 \text{ cm}$$
$$= 16 \text{ cm}^2 = 16 \times 10^{-4} \text{ m}^2$$

Axial pull applied, $W = 128$ kN

Length of the bar, $l = 3$

Modulus of elasticity, $E = 200$ GN/m².

Elongation of the bar, δl :

Using the following relation, we have

$$\delta l = \frac{W\,l}{A\,E} = \frac{(128 \times 1000) \times 3}{16 \times 10^{-4} \times 200 \times 10^9}$$

$$= \frac{(128 \times 1000) \times 3 \times 10^4}{16 \times 200 \times 10^9}$$

or $\delta l = 0.0012$ m or **1.2 mm** **(Ans.)**

Energy stored in the bar during elongation, U:

We know that, $U = \dfrac{\sigma^2}{2E} \times Al = \dfrac{(8 \times 10^7)^2 \times 16 \times 10^{-4} \times 3}{2 \times 200 \times 10^9}$

$$\left[\because \quad \sigma = \frac{W}{A} = \frac{(128 \times 1000)}{16 \times 10^{-4}} = \frac{128 \times 10^7}{16} \right.$$
$$\left. = 8 \times 10^7 \text{ N/m}^2 \right]$$

$$= \frac{64 \times 10^{14} \times 16 \times 3}{2 \times 2 \times 10^{11} \times 10^4}$$

or, $U = $ **76.8 Nm or J** **(Ans.)**

Example 15.2. *A uniform metal bar has a cross-sectional area of 7 cm² and a length of 1.5 m. With an elastic limit of 160 MN/m², what will be its proof resilience ? Determine also the maximum value of an applied load which may be suddenly applied without exceeding the elastic limit. Calculate the value of gradually applied load which will produce the same extension as that produced by the suddenly applied load above.*

Take: $E = 200$ GN/m².

Solution. Cross-sectional area of the bar,

$$A = 7 \text{ cm}^2 = 7 \times 10^{-4} \text{ m}^2$$

Length of the bar, $l = 1.5$ m

Elastic limit, $\sigma = 160$ MN/m².

Proof resilience, U_p :

Using the relation : $U_p = \dfrac{\sigma_p^2 \, Al}{2E} = \dfrac{(160 \times 10^6)^2 \times 7 \times 10^{-4} \times 1.5}{2 \times (200 \times 10^9)}$

$$= \frac{(16)^2 \times 10^{14} \times 7 \times 1.5}{2 \times 2 \times 10^{11} \times 10^4}$$

or, $\qquad U_p = \textbf{67.2 Nm or J} \quad \textbf{(Ans.)}$

Maximum value of suddenly applied load, W:

We knw that $\qquad W \times \delta l = \frac{1}{2} \sigma \times A \times \sigma l$

or, $\qquad W = \frac{\sigma \cdot A}{2} = \frac{(160 \times 10^6) \times 7 \times 10^{-4}}{2} = \frac{160 \times 10^6 \times 7}{2 \times 10^4} = 56000 \, \text{N}$

or, $\qquad W = \textbf{56 kN} \quad \textbf{(Ans.)}$

Equivalent gradually applied load, W_e :

We know that, $\qquad \frac{W_e}{2} \times \delta l = \frac{1}{2} \sigma \times A \times \delta l$

or, $\qquad W_e = \sigma \times A = 160 \times 10^6 \times 7 \times 10^{-4} = 112000 \, \text{N}$

or, $\qquad W_e = \textbf{112 kN} \quad \textbf{(Ans.)}$

Special purpose automobile with shock absorbers. Springs, beams, levers, etc. can store and release strain energy.

Example 15.3. *A steel specimen 1.5 cm² in cross-section stretches 0.05 mm over 5 cm gauge length under an axial load of 30 kN. Calculate the strain energy stored in the specimen at this point. If the load at the elastic limit for specimen is 50 kN, calculate the elongation at the elastic limit and the resilience.*

Solution. Cross-sectional area of specimen,

$$A = 1.5 \text{ cm}^2 = 1.5 \times 10^{-4} \text{ m}^2$$

Increase in length over 5 cm gauge length,

$$\delta l = 0.05 \text{ mm} = 0.05 \times 10^{-3} \text{ m}$$

Axial load, $\qquad W = 30 \text{ kN}$

Load at elastic limit $\qquad = 50 \text{ kN}$

Strain energy stored in the specimen, U:

$$U = \frac{\sigma^2 \, Al}{2\,E} = \frac{1}{2}W \cdot \delta l = \frac{1}{2} \times (30 \times 1000) \times 0.05 \times 10^{-3}$$

or, $\qquad U = \textbf{0.75 Nm or J} \quad \textbf{(Ans.)}$

Also, $\qquad E = \dfrac{W \cdot l}{A \cdot \delta l}$ $\qquad \left[\because E = \dfrac{\text{Stress}}{\text{Strain}} = \dfrac{W/A}{\delta l / l} = \dfrac{W \cdot l}{A \cdot \delta l} \right]$

$$= \frac{(30 \times 1000) \times (5/100)}{(1.5 \times 10^{-4}) \times 0.05 \times 10^{-3}} = \frac{30 \times 1000 \times 5 \times 10^4 \times 10^3}{1.5 \times 100 \times 0.05}$$

$$= 200 \times 10^9 = 200 \text{ GN/m}^2$$

Elongation at elastic limit, δl:

$$\delta l = \frac{W \, l}{A \, E} = \frac{(50 \times 1000) \times (5/100)}{(1.5 \times 10^{-4}) \times (200 \times 10^9)} = \frac{50 \times 1000 \times 5 \times 10^4}{1.5 \times 100 \times 2 \times 10^{11}} = \frac{25 \times 10^8}{3 \times 10^{13}}$$

or, $\qquad \delta l = 0.000083.3 \text{ m or } \textbf{0.0833 mm} \quad \textbf{(Ans.)}$

Example 15.4. *A bar 100 cm in length is subjected to an axial pull, such that the maximum stress is equal to 150 MN/m². Its area of cross-section is 2 cm² over a length of 95 cm and for the middle 5 cm length it is only 1 cm². If E = 200 GN/m², calculate the strain energy stored in bar.*

Solution. Refer to Fig. 15.4.

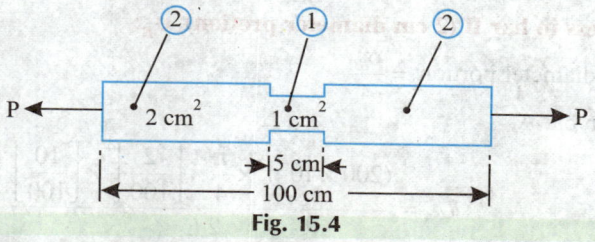

Fig. 15.4

Maximum sterss in portion of 1 cm² cross-section,

$\qquad \sigma_1 = 150 \text{ MN/m}^2$

Maximum stress in portion of 2 cm² cross-section,

$\qquad \sigma_2 = 75 \text{ MN/m}^2$

$$\left[\begin{array}{c} \because \qquad \sigma_1 A_1 = \sigma_2 A_2 \\ 150 \times 1 \times 10^{-4} = \sigma_2 \times 2 \times 10^{-4} \\ \therefore \qquad \sigma_2 = \dfrac{150}{2} = 75 \, \text{MN/m}^2 \end{array} \right]$$

Strain energy stored in the bar,

$$U = \frac{\sigma_1^2 \, A_1 \, l_1}{2\,E} + \frac{\sigma_2^2 \, A_2 \, l_2}{2\,E}$$

$$= \frac{(150 \times 10^6)^2 \times (1 \times 10^{-4}) \times (5/100)}{2 \times (200 \times 10^9)} + \frac{(75 \times 10^6)^2 \times (2 \times 10^{-4}) \times (95/100)}{2 \times (200 \times 10^9)}$$

$$= \frac{225 \times 10^{14} \times 5}{10^4 \times 100 \times 4 \times 10^{11}} + \frac{5625 \times 10^{12} \times 2 \times 95}{10^4 \times 2 \times 2 \times 10^{11} \times 100}$$

$$= \frac{1125}{4 \times 1000} + \frac{5625 \times 2 \times 95}{4 \times 10^5} = 0.28125 + 2.6718$$

or, $U = \textbf{2.953 Nm or J}$ **(Ans.)**

Example 15.5. *Two similar bars A and B are each 30 cm long as shown in Fig. 15.5. The bar A receives an axial blow, which produces a maximum stress of 200 MN/m². Find maximum stress produced by the same blow on the bar B. If the bar is stressed to 200 MN/m² determine the ratio of energy stored by the bars and A and B.*

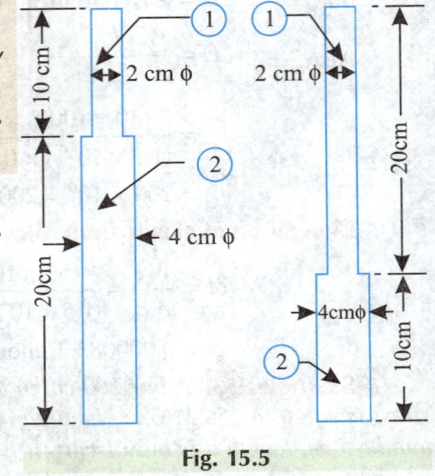

Fig. 15.5

Solution. Refer to Fig. 15.5.

Maximum stress in the bar A (2 cm diameter portion),

$$\sigma_A = 200 \text{ MN/m}^2$$

∴ Stress in the 4 cm diameter portion = 50 MN/m²

$$\left[\begin{array}{l} \because \qquad \sigma_1 A_1 = \sigma_2 A_2 \\[2mm] 200 \times \dfrac{\pi}{4} \times \left[\dfrac{2}{100}\right]^2 = \sigma_2 \times \dfrac{\pi}{4} \times \left[\dfrac{4}{100}\right]^2 \\[4mm] \text{or} \qquad \sigma_2 = \dfrac{200}{4} = 50 \text{ MN/m}^2 \end{array}\right]$$

Maximum stress in bar B (2 cm diameter protion), σ_B:

Stress in 4 cm diameter portion $= \dfrac{\sigma_B}{4}$

Energy stress in bar A,

$$U_A = \frac{(200 \times 10^6)^2 \times \dfrac{\pi}{4} \times \left[\dfrac{2}{100}\right]^2 \times \left[\dfrac{10}{100}\right]}{2E}$$

$$+ \frac{(50 \times 10^6)^2 \times \dfrac{\pi}{4} \times \left[\dfrac{4}{100}\right]^2 \times \dfrac{20}{100}}{2E}$$

$$= \frac{4 \times 10^{16} \times \dfrac{\pi}{4} \times 4}{10^5 \times 2E} + \frac{25 \times 10^{14} \times \dfrac{\pi}{4} \times 32}{10^5 \times 2E}$$

$$= \frac{2\pi \times 10^{11}}{E} + \frac{\pi \times 10^{11}}{E} = \frac{3\pi \times 10^{11}}{E} \qquad \qquad ...(i)$$

Similarly energy stored by the bar B,

$$U_B = \frac{\sigma_B^2 \times \dfrac{\pi}{4} \times \left[\dfrac{2}{100}\right]^2 \times \left[\dfrac{20}{100}\right]}{2E} + \frac{\left[\dfrac{\sigma_B}{4}\right]^2 \times \dfrac{\pi}{4} \times \left[\dfrac{4}{100}\right]^2 \times \dfrac{10}{100}}{2E}$$

$$= \frac{\sigma_B^2 \times \pi \times 2}{10^5 \times 2E} + \frac{\sigma_B^2 \times \pi}{4 \times 10^5 \times 2E} = \frac{\sigma_B^2}{10^5 \times 2E}\left[2\pi + \dfrac{\pi}{4}\right]$$

$$= \frac{2.25 \pi \sigma_B^2}{10^5 \times 2E} \qquad \qquad ...(ii)$$

Since the blow on the bars A and B is the same, therefore the two energies are equal.

∴ Equating (*i*) and (*ii*), we get

$$\frac{3\pi \times 10^{11}}{E} = \frac{2.25\,\pi \times \sigma_B^2}{10^5 \times 2E}$$

or,

$$\sigma_B^2 = \frac{3 \times 10^{11} \times 10^5 \times 2}{2.25}, \quad \text{or,} \quad \sigma_B = \sqrt{\frac{6 \times 10^{16}}{2.25}}$$

$$= 1.633 \times 10^8 \text{ N/m}^2$$

or,

$$\sigma_B = \textbf{163.3 MN/m}^2 \quad \textbf{(Ans.)}$$

Ratio of energy stored by bars A and B, $\dfrac{U_A}{U_B}$:

Energy stored in bar B when it is also stressed to 200 MN/m²,

$$U_B = \frac{2.25\,\pi\sigma_B^2}{10^5 \times 2E} = \frac{2.25\,\pi \times (200 \times 10^6)^2}{10^5 \times 2E}$$

$$= \frac{2.25\,\pi \times 4 \times 10^{16}}{10^5 \times 2E} = \frac{4.5\,\pi \times 10^{11}}{E}$$

∴ Ratio of energy stored by A and B,

$$\frac{U_A}{U_B} = \frac{\dfrac{3\pi \times 10^{11}}{E}}{\dfrac{4.5\,\pi \times 10^{11}}{E}} = \frac{2}{3} \quad \textbf{(Ans.)}$$

Example 15.6 *A wagon weighing 35 kN is attached to a wire-rope and moving down an incline at speed of 3.6 km/hour when the rope jams and the wagon is suddenly brought to rest. If the length of the rope is 60 metres at the time of sudden stoppage, calculate the maximum instantaneous stress and maximum instantaneous elongation produced. Diameter of rope = 30 mm. E = 200 GN/m².*

Solution. Weight of the wagon,

$$W = 35 \text{ kN}$$

Speed of the wagon,

$$v = 3.6 \text{ km/hour} = \frac{3.6 \times 1000}{60 \times 60} = 1 \text{ m/s}$$

Diameter of the rope, $d = 30 \text{ mm} = 30 \times 10^{-3} = 0.03 \text{ m}$

Length of the rope at the time of sudden stoppage,

$$l = 60 \text{ m}$$

Components of a clutch.

Maximum instantaneous stress, σ_i:

The kinetic energy of the wagon

$$= \frac{1}{2}mv^2 = \frac{1}{2} \times \frac{35 \times 1000}{9.81} \times 1^2 = 1783 \text{ Nm or J} \qquad \text{...(i)}$$

This energy is to be absorbed by the rope at a stress σ_i.

Now, Strain energy stored $= \dfrac{\sigma_i^2\, Al}{2E}$

$$= \frac{\sigma_i^2 \times \frac{\pi}{4} \times (0.03)^2 \times 60}{2 \times 200 \times 10^9} = \frac{0.0106\sigma_i^2}{10^{11}} \qquad ...(ii)$$

Equating (i) and (ii), we get

$$\frac{0.0106\sigma_i^2}{10^{11}} = 1783$$

or, $\qquad\qquad \sigma_i^2 = \dfrac{1783}{0.0106} \times 10^{11} = 16820.75 \times 10^{12}$

or, $\qquad\qquad \sigma_i = 129.69 \times 10^6$ N/m^2 or **129.69 MN/m^2** **(Ans.)**

Maximum instantaneous elongation of the rope, δl :

Using the relation, $\qquad \delta l_i = \dfrac{\sigma_i\, l}{E} = \dfrac{129.69 \times 10^6 \times 60}{200 \times 10^9}$

or, $\qquad\qquad \delta l_i = 389.07 \times 10^{-4}$ m or **38.9 mm** **(Ans.)**

Example 15.7. *An object of 100 N weight falls by gravity a vertical distance of 5 m, when it is suddenly stopped by a collar at the end of a vertical rod of length 10 m and diameter 20 mm. The top of the bar is rigidly fixed. Calculate the maximum stress and strain induced in the bar due to impact. Take E = 200 GN/m^2 for the material of the rod.*

Solution. Weight of the object,

$$W = 100 \text{ N}$$

Height of fall, $\qquad h = 5$ m

Length of vertical rod, $\qquad l = 10$ m

Diameter of the rod, $\qquad d = 20$ mm $= 0.02$ m

Modulus of elasticity $\qquad E = 200$ GN/m^2

Maximum stress induced in the bar, σ$_i$:

Using the following relation, we have:

$$\sigma_i = \frac{W}{A}\left[1 + \sqrt{1 + \frac{2\,h\,A\,E}{W\,l}}\right]$$

$$= \frac{100}{\frac{\pi}{4} \times (0.02)^2} \times \left[1 + \sqrt{1 + \frac{2 \times 5 \times \frac{\pi}{4}(0.02)^2 \times 200 \times 10^9}{100 \times 10}}\right]$$

$$= 318268.6[1 + \sqrt{1 + 628400}]$$

or $\qquad\qquad \sigma_i = 252.6 \times 10^6$ N/m^2 or **252.6 MN/m^2** **(Ans.)**

Strain induced in the bar due to impact, e$_i$:

We know that, $\qquad e_i = \dfrac{\sigma_i}{E} = \dfrac{252.6 \times 10^6}{200 \times 10^9} = \textbf{0.001263}$ **(Ans.)**

Example 15.8 *A bar 1.2 cm diameter gets stretched by 0.3 cm under a steady load of 8 kN. What stress would be produced in the same bar by a weight of 0.8 kN which falls freely vertically through a distance of 8 cm to a rigid collar attached at its end?*

Take: $\qquad\qquad E = 200 \text{ GN/m}^2.$

Solution. Cross-sectional area of the bar,

$$A = \frac{\pi}{4} \times d^2 = \frac{\pi}{4} \times \left[\frac{1.2}{100}\right]^2 = 0.0001131 \text{ m}^2$$

Steady load $= 8 \text{ kN}$

Elongation under steady load,

$$\delta l = 0.3 \text{ cm} = 0.003 \text{ m}$$

Falling load $= 0.8 \text{ kN}$

Distance through which the weight falls,

$$h = 8 \text{ cm} = 0.08 \text{ m}$$

Modulus of elasticity, $E = 200 \text{ GN/m}^2$

Instantaneous stress produced due to the falling load, σ_i:

In order to first find length of the bar, using the following relation, we have

$$\delta l = \frac{W l}{A E}, \text{ or, } l = \frac{\delta l \times A E}{W}$$

$$\therefore \quad l = \frac{0.003 \times 0.0001131 \times 200 \times 10^9}{8 \times 1000} = 8.48 \text{ m}$$

Now to calculate instantaneous stress σ_i due to falling load 0.8 kN using the following relation, we have:

$$\sigma_i = \frac{W}{A}\left[1 + \sqrt{1 + \frac{2 h A E}{W l}}\right]$$

$$= \frac{0.8 \times 1000}{0.0001131} \times \left[1 + \sqrt{1 + \frac{2 \times 0.08 \times 0.0001131 \times 200 \times 10^9}{0.8 \times 1000 \times 8.48}}\right]$$

$$= 7073386.3 \ (1 + 23.097)$$

or, $\qquad \sigma_i = 170.4 \times 10^6 \text{ N/m}^2 \text{ or } \textbf{170.4 MN/m}^2 \quad \textbf{(Ans.)}$

Example 15.9. *A steel wire 2.5 mm diameter is firmly held in clamp from which it hangs vertically. An anvil the weight of which may be neglected, is secured to the wire 1.8 m below clamp. The wire is to be tested allowing a weight bored to slide over the wire to drop freely from 1 m above the anvil. Calculate the weight required to stress the wire to 1000 MN/m² assuming the wire to be elastic upto this stress.*

Take: $\qquad E = 210 \text{ GN/m}^2.$

Solution. Diameter of the steel wire,

$$d = 2.5 \text{ mm} = 2.5 \times 10^{-3} \text{ m}$$

Height of fall, $\qquad h = 1 \text{ m}$

Length of the wire, $\qquad l = 1.8 \text{ m}$

Instantaneous stress produced

$$\sigma_i = 1000 \text{ MN/m}^2$$

$$E = 210 \text{ GN/m}^2.$$

Clutch of a car.

Weight required to stress the wire, W:

Instantaneous extension

$$\delta l = \frac{\sigma_i \, l}{E} = \frac{1000 \times 10^6 \times 1.8}{210 \times 10^9} = 0.00857 \text{ m}$$

Equating the loss of potential energy to strain energy stored by the wire, we have

$$W(h + \delta l) = \frac{\sigma_i^2 \, Al}{2E}$$

$$W(1 + 0.00857) = \frac{(1000 \times 10^6) \times \frac{\pi}{4} \times (2.5 \times 10^{-3})^2 \times 1.8}{2 \times 210 \times 10^9}$$

or $\qquad\qquad\qquad W = \textbf{20.85 N} \quad \textbf{(Ans.)}$

Example 15.10 *An unknown weight falls through a height of 10 mm on a collar rigidly attached to the lower end of a vertical bar 5 m long and 600 mm^2 in section. If the maximum extension of the rod is to be 2 mm, what is the corresponding stress and magnitude of the unknown weight?*

Take: $\qquad\qquad\qquad E = 200 \, GN/m^2.$

Solution. Height of fall, $\qquad\qquad h = 10 \text{ mm} = 10 \times 10^{-3} \text{ m}$

Length of the bar, $\qquad\qquad\qquad l = 5 \text{ m}$

Area of cross-section of the bar, $\quad A = 600 \text{ mm}^2 = 600 \times 10^{-6} \text{ m}^2$

Maximum extension of the rod, $\quad \delta l_i = 2 \text{ mm} = 2 \times 10^{-3} \text{ m}$

Instantaneous stress, σ_i:

Using the following relation, we have:

$$E = \frac{\text{Stress}}{\text{Strain}} = \frac{\sigma_i}{\delta l_i / l}$$

or $\qquad\qquad\qquad \sigma_i = E \times \frac{\delta l_i}{l} = \frac{200 \times 10^9 \times 2 \times 10^{-3}}{5}$

or $\qquad\qquad\qquad \sigma_i = 80 \times 10^6 \text{ N/m}^2 \text{ or } \textbf{80 MN/m}^2 \quad \textbf{(Ans.)}$

Unknown weight, W:

Using the relation:

$$\sigma_i = \frac{W}{A}\left[1 + \sqrt{1 + \frac{2hAE}{Wl}}\right], \text{ we get}$$

$$80 \times 10^6 = \frac{W}{600 \times 10^{-6}}$$

$$\times \left[1 + \sqrt{1 + \frac{2 \times 10 \times 10^{-3} \times 600 \times 10^{-6} \times 200 \times 10^9}{W \times 5}}\right]$$

$$\frac{80 \times 10^6 \times 600 \times 10^{-6}}{W} = \left[1 + \sqrt{1 + \frac{480000}{W}}\right]$$

$$\frac{48000}{W} = \left[1 + \sqrt{1 + \frac{480000}{W}}\right]$$

$$\frac{48000}{W} - 1 = \sqrt{1 + \frac{480000}{W}}$$

$$\frac{48000 - W}{W} = \sqrt{1 + \frac{480000}{W}}$$

Squaring both sides, we get

$$\left[\frac{48000 - W}{W}\right]^2 = 1 + \frac{480000}{W}$$

$$\frac{(48000)^2 + W^2 - 2 \times 48000 \times W}{W^2} = 1 + \frac{480000}{W}$$

$$(48000)^2 + W^2 - 96000W = W^2 + 480000W$$
$$576000W = (48000)^2$$
$$W = \frac{(48000)^2}{576000}$$

or
$$W = \textbf{4000 N} \quad \textbf{(Ans.)}$$

Example 15.11. *A vertical round steel rod 1.82 m long is securely held at its upper end. A weight can slide freely on the rod and its fall is arrested by a stop provided at the lower end of the rod. When the weight falls from a height of 3 cm above the stop, the maximum stress reached in the rod is estimated to be 157 MN/m². Determine the stress in the rod if the load had been applied gradually and also the maximum stress if the load had fallen from a height of 4.75 cm.*

Take: $E = 210 \ GN/m^2.$

Solution. Length of vertical round steel rod,
$$l = 1.82 \text{ m}$$
Modulus of elasticity, $E = 210$ GN/m².

(*i*)   **Stress in the rod if the load had fallen gradually:**

We know that, $\delta l_i = \dfrac{\sigma_i l}{E} = \dfrac{157 \times 10^6 \times 1.82}{210 \times 10^9} = 0.00136$ m

Also, $W (h + \delta l_i) = \dfrac{\sigma_i^2 \, Al}{2E}$

$$W\left[\frac{3}{100} + 0.00136\right] = \frac{(157 \times 10^6)^2 \times A \times 1.82}{2 \times 210 \times 10^9}$$

∴ $\dfrac{W}{A} = \dfrac{(157 \times 10^6)^2 \times 1.82}{(0.03 + 0.00136) \times 2 \times 210 \times 10^9} = \textbf{3.406 MN/m}^2$ **(Ans.)**

(*ii*)   **Maximum stress when the load had fallen from a height of 4.75 cm, σ_{max} :**

Using the following relation, we have :

$$\sigma_{max} = \frac{W}{A}\left[1 + \sqrt{1 + \frac{2\,hAE}{Wl}}\right] = \frac{W}{A} + \sqrt{\frac{W^2}{A^2} + \frac{2WhE}{Al}}$$

Biker. The moving vehicle possesses kinetic energy. The parts that are subjected to various pressures and forces possess strain energy.

$$= 3.406 \times 10^6$$

$$+ \sqrt{(3.406 \times 10^6)^2 + \frac{2 \times 3.406 \times 10^6 \times (4.75 \times 10^{-2}) \times 210 \times 10^9}{1.82}}$$

$$\left(\because \frac{W}{A} = 3.406 \times 10^6 \, \text{N}/\text{m}^2 \right)$$

$$= 3.406 \times 10^6 + \sqrt{(3.406 \times 10^6)^2 + 37335 \times 10^{12}}$$

$$= 340.6 \times 10^6 + \sqrt{11.6 \times 10^{12} + 37335 \times 10^{12}}$$

$$= 10^6 + \left[3.406 + \sqrt{11.6 \times 37335} \right] = 10^6 \, (3.406 + 193.25)$$

or, $\qquad \sigma_{max} = 196.66 \times 10^6 \, \text{N}/\text{m}^2$ or **196.66 MN/m²** **(Ans.)**

Example 15.12. *A vertical compound member fixed rigidly at its upper end consists of a steel rod 2.5 m long and 20 mm diameter placed within an equally long brass tube 21 mm internal diameter and 30 mm external diameter. The rod and the tube are fixed together at the ends. The compound member is then suddenly loaded in the tension by a weight of 10 kN falling through a height of 3 mm on to a flange fixed to its lower end. Calculate the maximum stress in steel and brass.*

Assume: $\qquad E_s = 200 \, GN/m^2;$

$\qquad\qquad\qquad E_b = 100 \, GN/m^2.$

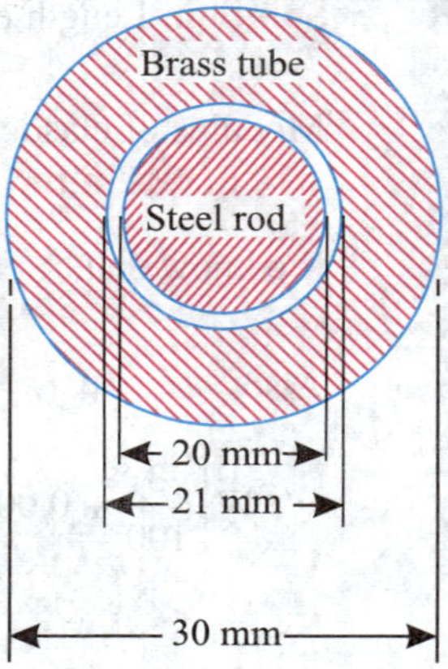

Fig. 15.6

Solution. Refer to Fig. 15.6

Area of steal rod, $\qquad A_s = \dfrac{\pi}{4} \times (20 \times 10^{-3})^2 = 0.0003142 \, \text{m}^2$

Area of brass tube, $\qquad A_b = \dfrac{\pi}{4} \left[(30 \times 10^{-3})^2 - (21 \times 10^{-3})^2 \right]$

$$= \frac{\pi}{4} \, (0.0009 - 0.000441) = 0.0003605 \, \text{m}^2$$

Strain in steel = Strain in brass

$$\frac{(\sigma_i)_s}{E_s} = \frac{(\sigma_i)_b}{E_b} \, , \quad \text{or,} \quad \frac{(\sigma_i)_s}{(\sigma_i)_b} = \frac{E_s}{E_b} = \frac{200}{100} = 2$$

$\therefore \qquad\qquad (\sigma_i)_s = 2(\sigma_i)_b$ $\qquad\qquad\qquad\qquad\qquad\qquad\qquad\qquad$ ...(i)

Elongation of each material,

$$\delta l_i = \frac{(\sigma_i)_s}{E_s} \times l_s = \frac{(\sigma_i)_b}{E_b} \times l_b \quad (l_s = l_b = l = 2.5 \, \text{m})$$

We know that,

Loss of potential energy = Total strain energy stored by both the materials

i.e. $\qquad W \, (h + \delta l_i) = \dfrac{(\sigma_i)_s^2 \, A_s \, l_s}{2 \, E_s} + \dfrac{(\sigma_i)_b^2 \, A_b \, l_b}{2 \, E_b}$

$$= 10000 \left[3 \times 10^{-3} + \frac{(\sigma_i)_b \times 2.5}{100 \times 10^9} \right]$$

$$= \frac{\left[2(\sigma_i)_b \right]^2 \times 0.0003142 \times 2.5}{2 \times 200 \times 10^9} + \frac{(\sigma_i)_b^2 \times 0.0003605 \times 2.5}{2 \times 100 \times 10^9}$$

or, $30 + 0.00025(\sigma_i)_b = 0.0007855 \times 10^{-11}(\sigma_i)_b^2 + 0.0004506 \times 10^{-11}(\sigma_i)_b^2$

or, $30 + 0.00000025(\sigma_i)_b = 0.0012361 \times 10^{-11}(\sigma_i)_b^2$

or, $(\sigma_i)_b^2 - 2.02 \times 10^7 (\sigma_i)_b - 2.427 \times 10^{15} = 0$

or,

$$(\sigma_i)_b = \frac{2.02 \times 10^7 + \sqrt{(2.02 \times 10^7)^2 \times 4 \times 2.427 \times 10^{15}}}{2}$$

$$= \frac{2.02 \times 10^7 + 10^7\sqrt{4.0804 + 97.080}}{2} = \frac{10^7(2.02 + 10.06)}{2}$$

i.e. $(\sigma_i)_b = 6.04 \times 10^7 = \mathbf{60.4\ MN/m^2}$ **(Ans.)**

and, $(\sigma_i)_s = 2(\sigma_i)_b = 2 \times 60.4$

or, $(\sigma_i)_s = \mathbf{120.8\ MN/m^2}$ **(Ans.)**

Example 15.13. *A bar 50 cm long has 1.5 cm² cross-sectional area for 30 cm of its length and 1cm² for the remaining length. If a load of 50 N falls on the collar which is provided at one end of the rod, the other end being fixed, from a height of 3 cm, find the maximum stress induced in the bar. E = 200 GN/m².*

Solution. Refer to Fig. 15.7.

Let, W = Falling load, and

W_e = Equivalent gradually applied load which produces the same maximum stress and extension as is caused by the falling load W.

Now, the total extension,

$$\delta l_i = (\delta l_i)_1 + (\delta l_i)_2$$

$$= \frac{W_e l_1}{A_1 E} + \frac{W_e l_2}{A_2 E} \quad \left[\because \delta l = \frac{Wl}{AE}\right]$$

$$= \frac{W_e}{E}\left(\frac{l_1}{A_1} + \frac{l_2}{A_2}\right)$$

$$= \frac{W_e}{200 \times 10^9}\left(\frac{0.3}{1.5 \times 10^{-4}} + \frac{0.2}{1 \times 10^{-4}}\right)$$

$$= \frac{W_e}{200 \times 10^9}(2000 + 2000), \text{ or, } \delta l_i = \frac{2W_e}{10^8}$$

Also, $W(h + \delta l_i) = \dfrac{1}{2} \times W_e \times \delta l_i$

or, $50\left(\dfrac{3}{100} + \dfrac{2W_e}{10^8}\right) = \dfrac{1}{2} \times W_e \times \dfrac{2W_e}{10^8},$ or, $1.5 + \dfrac{W_e}{10^6} = \dfrac{W_e^2}{10^8}$

or, $1.5 \times 10^8 + 100 W_e = W_e^2,$ or, $W_e^2 - 100 W_e - 1.5 \times 10^8 = 0$

or,

$$W_e = \frac{100 \pm \sqrt{(100)^2 + 4 \times 1.5 \times 10^8}}{2}$$

$$= \frac{100 + 100\sqrt{1 + 60000}}{2} \quad \left[\begin{array}{l}\text{Considering only +ve sign}\\ \text{for maximum stress}\end{array}\right]$$

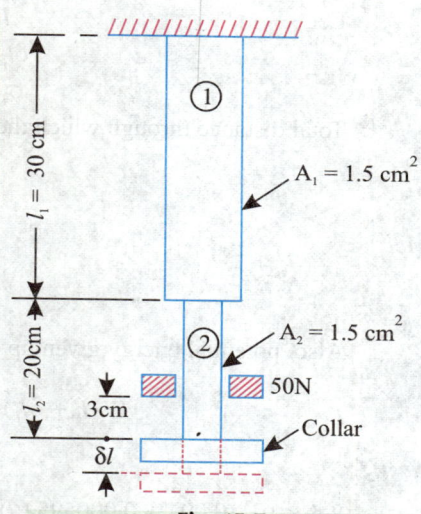

$A_1 = 1.5\ \text{cm}^2$

$A_2 = 1.5\ \text{cm}^2$

50N

3cm

Collar

δl

Fig. 15.7

$l_1 = 30$ cm

$l_2 = 20$cm

$$= \frac{100 + 24495}{2} = 12297.5 \, \text{N}$$

Maximum stress (in the smaller section),

$$(\sigma_i)_2 = \frac{W_e}{A_2} = \frac{12297.5}{1 \times 10^{-4}} = 122975000 \, \text{N/m}^2.$$

or $\qquad\qquad (\sigma_i)_2 = \textbf{122.975 MN/m}^2 \quad \textbf{(Ans.)}$

Example 15.14. *A vertical steel rod of 25 mm diameter checks the fall on its end of weight of 2.5 kN which drops through a distance of 4 mm before it strikes the rod. Find the shortest length of rod which will bear the impact if the stress is not to exceed 125 MN/m².*

Take: $\qquad\qquad E = 210 \, GN/m^2.$

Solution.

Let, $\qquad\qquad l = \text{The least possible length of rod, m,}$

$\qquad\qquad \delta l_i = \text{Maximum instantaneous elongation, m, and}$

$\qquad\qquad \sigma_i = \text{Maximum stress, } MN/m^2.$

Total distance through which the weight falls

$$= \text{Height of drop } (h) + \text{elongation of the bar } (\delta l)$$
$$= 0.004 + \delta l$$
$$= 0.004 + \frac{125 \times 10^6 \times l}{210 \times 10^9} \qquad \left(\because \delta l_i = \frac{\sigma_i \, l}{E}\right)$$
$$= 0.004 + 0.0005952 \, l$$

Also, potential energy given up by the weight = Strain energy stored in rod.

i.e. $\qquad\qquad W(h + \delta l_i) = \dfrac{\sigma_i^2 \, Al}{2E}$

$$2.5 \times 10^3 \, (0.004 + 0.0005952 \, l) = \frac{(125 \times 10^6)^2 \times \dfrac{\pi}{4} \times 0.025^2 \times l}{2 \times 210 \times 10^9}$$

$$10 + 1.488 \, l = 18.264 \, l$$

∴ $\qquad\qquad l = \textbf{0.6 m approx.} \quad \textbf{(Ans.)}$

15.4. STRAIN ENERGY IN PURE SHEARING

Consider a rectangular block of material subjected to shearing forces S acting across two of its opposite faces (Fig.15.8). The face LM will move, relative to face NP, a distance $MM' = MN \times \phi$, where ϕ is the angle of shear produced.

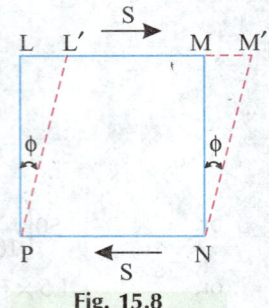

Fig. 15.8

The work done $\qquad = \dfrac{1}{2} S \times MM'$

$$= \frac{1}{2} S \times MN \times \phi$$

$$\left[\because \frac{MM'}{MN} = \tan\phi = \phi, \text{because } \phi \text{ is very small}\right]$$

Now, $\qquad\qquad S = \tau \times LM, \text{where } \tau \text{ is the shearing stress, and } \phi = \dfrac{\tau}{C}$

$\qquad\qquad (\phi = e_s = \text{shear strain})$

Taking *unit depth* normal to diagram, we have

Strain energy $= \text{Work done} = \dfrac{1}{2}\tau \times LM \times MN \times \dfrac{\tau}{C} = \dfrac{1}{2}\dfrac{\tau^2}{C}LM \times MN$

Now $(LM \times MN)$ is the volume of the rectangular block, since it has unit depth normal to $LMNP$.

$\therefore$ *Strain energy,* $U = \dfrac{\tau^2}{2C} \times$ volume of block ...(15.7)

This is the shearing strain energy for a block of material subjected to a constant shearing stress throughout.

15.5. STRAIN ENERGY IN TORSION

Consider a solid circular shaft of length l and radius R, subjected to a torque T producing a twist θ in the length of the shaft (Fig.15.9).

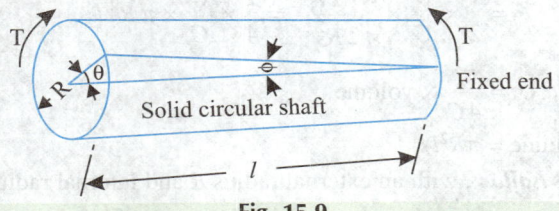

Fig. 15.9

The work done $= \dfrac{1}{2}T\theta$, which is stored in the shaft as strain energy.

But, $\dfrac{T}{I_p} = \dfrac{C\theta}{l} = \dfrac{\tau}{R}$

Wheel for different automobiles.

where, T = Torque applied,

I_P = Polar moment of inertia,

C = Modulus of rigidity,

l = Length of the shaft, and

τ = Maximum shear stress on the surface of the shaft.

$\therefore$ $T = \dfrac{\tau \times I_p}{R}$, and, $\theta = \dfrac{\tau l}{CR}$

$\therefore$ Work done $= \dfrac{1}{2} \dfrac{\tau \times I_p}{R} \times \dfrac{\tau \times l}{C \times R} = \dfrac{1}{2} \times \dfrac{\tau^2}{C} \times \dfrac{I_p \times l}{R^2}$

Now, $I_p = \dfrac{\pi R^4}{2}$

$\therefore$ Workdone $= \dfrac{1}{2} \times \dfrac{\tau^2}{C} \times \dfrac{\pi R^4 \times l}{2R^2} = \dfrac{1}{4} \times \dfrac{\tau^2}{C} \times \pi R^2 l$

or, strain energy, $U = \dfrac{\tau^2}{4C} \times \text{volume}$ ...(15.8)

$(\because \text{Volume} = \pi R^2 l)$

When the shaft is *hollow*, with an external radius R and internal radius r :

Again, Work done $= \dfrac{1}{2} T\theta$, and, $\theta = \dfrac{\tau l}{CR}$, and, $T = \dfrac{\tau I_p}{R}$

Work done $= \dfrac{\tau^2}{2C} \times \dfrac{I_p l}{R^2}$

But, $I_p = \dfrac{\pi}{2}(R^4 - r^4)$

$\therefore$ Work done $= \dfrac{\tau^2}{2C} \times \dfrac{\pi l (R^4 - r^4)}{2R^2} = \dfrac{\tau^2}{4C} \times \dfrac{\pi l (R^2 + r^2)(R^2 - r^2)}{R^2}$

$= \dfrac{\tau^2}{4C} \times \dfrac{(R^2 + r^2)}{R^2} \pi (R^2 - r^2) l$

$\therefore$ *Strain energy,* $U = \dfrac{\tau^2}{4C} \times \dfrac{R^2 + r^2}{R^2} \times \text{volume}$ ...(15.9)

Example 15.15. *The external diameter of a hollow shaft is twice the internal diameter. It is subjected to pure torque and it attains a maximum shear stress τ. Show that the strain energy stored per unit volume of the shaft is* $\dfrac{5\tau^2}{16C}$. *Such a shaft is required to transmit 5400 kW at 110 r.p.m. with uniform torque, the maximum stress not exceeding 84 MN/m^2. Determine:*

(i) The shaft diameters;   *(ii) The energy stored per m^3.*

Take: $C = 90$ GN/m^2. **(Panjab University)**

Solution. Let, R = External radius of the hollow shaft, and

r = Internal radius of the hollow shaft = $R/2$ (*Given*).

Then strain energy is given by:

$$U / \text{volume} = \dfrac{\tau^2}{4C} \times \left(\dfrac{R^2 + r^2}{R^2} \right) = \dfrac{\tau^2}{4CR^2} \left(R^2 + \dfrac{R^2}{4} \right) = \dfrac{5\tau^2}{16C}$$

i.e. $\qquad U / \text{volume} = \dfrac{5\tau^2}{16C}$ **(Ans.)**

Power required to be transmitted,

$$P = 5400 \text{ kW}$$

Speed, $\qquad N = 110 \text{ r.p.m.}$

Maximum shear stress,

$$\tau = 84 \text{ MN/m}^2$$

(i) The shaft diameter, D :

Now, $\qquad P = 2\pi \, NT, \text{ or, } 5400 \times 1000 = 2\pi \times \dfrac{110}{60} \times T$

or, $\qquad T = \dfrac{5400 \times 1000 \times 60}{2\pi \times 110} = 468783 \text{ Nm}$

Also, $\qquad \dfrac{T}{I_p} = \dfrac{\tau}{R}$

or, $\qquad T = I_p \cdot \dfrac{\tau}{R} = \dfrac{\pi}{32}(D^4 - d^4) \times \dfrac{\tau}{D/2}$

$$468783 = \dfrac{\pi}{16}\left(\dfrac{D^4 - d^4}{D}\right)\tau = \dfrac{\pi}{16}\left[\dfrac{D^4 - (D/2)^4}{D}\right]\tau$$

$$= \dfrac{\pi}{16} \times \dfrac{15 D^4}{16} \times \dfrac{1}{D} \times \tau = \dfrac{15\pi}{256} \times D^3 \times 84 \times 10^6 = 15.46 \times 10^6 \, D^3$$

$\therefore \qquad D^3 = \dfrac{468783}{15.46 \times 10^6} = 30322.3 \times 10^{-6}$

or, $\qquad D = 0.312 \text{ m} = 312 \text{ mm}$

$\therefore \qquad d = \dfrac{312}{2} = \textbf{156 mm (Ans.)}$

(ii) Energy stored per m³:

$$U / \text{volume} = \dfrac{5}{16} \cdot \dfrac{\tau^2}{C} = \dfrac{5}{16} \times \dfrac{(84 \times 10^6)^2}{90 \times 10^9} = 24500 \text{ J/m}^3 = 24.5 \text{ kJ/m}^3$$

Hence, energy stored per m³

$$= 24.5 \text{ kJ/m}^3 \quad \textbf{(Ans.)}$$

Example 15.16. *Compare the strain energies of the following two shafts subjected to the same maximum shear stress in torsion:*

(i) A hollow shaft having outer diameter n times the inner diameter.

(ii) A solid shaft.

Masses, lengths and materials of the two shafts are the same.

Solution. Let, $\quad D_H$ = Outer diameter of the hollow shaft,

$\qquad\qquad d_H$ = Inner diameter of the hollow shaft, and

$\qquad\qquad D_S$ = Diameter of the solid shaft.

For the same mass :

$$\dfrac{\pi}{4}(D_H^2 - d_H^2)l \times \rho = \dfrac{\pi}{4}D_S^2 \times l \times \rho \left[\begin{array}{l} \text{where, } \rho = \text{Mass density of the materials of the shaft;} \\ \qquad\qquad l = \text{length of each shaft.} \end{array}\right.$$

$$\therefore \qquad D_H^2 - d_H^2 = D_S^2$$

or, $\qquad n^2 d_H^2 - d_H^2 = D_S^2 \qquad (\because D_H = n\, d_H \,...\, Given)$

$$\therefore \qquad D_S^2 = (n^2 - 1)\, d_H^2$$

$$U_{hollow} = \frac{\tau^2}{4C} \times \left(\frac{D_H^2 + d_H^2}{D_H^2} \right) \times volume$$

$$= \frac{\tau^2}{4C} \times \frac{D_H^2 + d_H^2}{D_H^2} \times \frac{\pi}{4}(D_H^2 - d_H^2) \times l = \frac{\tau^2}{4C} \times \frac{\pi}{4} \times l \times \frac{D_H^4 - d_H^4}{D_H^2}$$

$$U_{solid} = \frac{\tau^2}{4C} \cdot \frac{\pi}{4} D_S^2 \, l$$

$$\therefore \qquad \frac{U_{hollow}}{U_{solid}} = \frac{D_H^4 - d_H^4}{D_H^2 \cdot D_S^2} = \frac{(n^4 - 1)\, d_H^4}{n^2 d_H^2 \times (n^2 - 1)\, d_H^2}$$

$$= \frac{n^2 + 1}{n^2} = 1 + \frac{1}{n^2} \quad \textbf{(Ans.)}$$

Therefore, *hollow shaft is able to absorb more strain energy as compared to a solid shaft.* $\dfrac{U_{hollow}}{U_{solid}} = 1$ to 2 for all shafts under the conditions of the problem.

Example 15.17. *The external diameter of hollow shaft is n times its internal diameter. It transmits a torque T_H and develops a strain energy U_H. Another solid shaft has the same external diameter as the hollow shaft and transmits torque T_S and develops a strain energy U_S.*

Find the ratios $\dfrac{U_H}{U_S}$ *and* $\dfrac{T_H}{T_S}$ *if the two shafts are to be subjected to the same maximum shear stress. Assume the shafts to be of the same material and of the same length. Hence show that*

$$\frac{T_H}{T_S} = \frac{U_H}{U_S}.$$

Garden foot bridge. Even small bridges such as this, store and release strain energies.

Solution. Let, $\quad D_H$ = External diameter of the hollow shaft, and

$\qquad\quad d_H$ = Internal diameter of the hollow shaft.

$\qquad\quad (D_H = n\, d_H \quad ... Given)$

$\qquad\quad D_s$ = diameter of the solid shaft

$\qquad\quad D_s = D_H \qquad ...(Given).$

Now, $\qquad T_H = \tau \times \dfrac{\pi}{16}\left(\dfrac{D_H^4 - d_H^4}{D_H}\right) ...$ for hollow shaft

$\qquad\quad T_S = \tau \times \dfrac{\pi}{16} D_s^3 \;...$for solid shaft

∴ $\qquad \dfrac{T_H}{T_S} = \dfrac{D_H^4 - d_H^4}{D_H \times D_s^3} = \dfrac{D_H^4 - d_H^4}{D_H^4} \qquad (\because D_s = D_H)$

$\qquad\qquad = \dfrac{(n^4 - 1)d_H^4}{n^4\, d_H^4} = \dfrac{n^4 - 1}{n^4}$

Hence, $\qquad \dfrac{T_H}{T_S} = \dfrac{n^4 - 1}{n^4} \quad$ **(Ans.)** $\qquad\qquad\qquad ...(1)$

$\qquad U_H = \dfrac{\tau^2}{4C}\left(\dfrac{D_H^2 + d_H^2}{D_H^2}\right) \times \text{volume} = \dfrac{\tau^2}{4C}\,\dfrac{D_H^2 + d_H^2}{D_H^2} \times \dfrac{\pi}{4}(D_H^2 - d_H^2)l$

$\qquad\qquad = \dfrac{\tau^2}{4C} \times \dfrac{\pi}{4} \times \dfrac{D_H^4 - d_H^4}{D_H^2}\, l$

$\qquad U_S = \dfrac{\tau^2}{4C} \times \dfrac{\pi}{4} D_H^2\, l \qquad (\because D_s = D_H)$

∴ $\qquad \dfrac{U_H}{U_S} = \dfrac{D_H^4 - d_H^4}{D_H^4} = \dfrac{(n^4 - 1)d_H^4}{n^4\, d_H^4} = \dfrac{n^4 - 1}{n^4} \qquad\qquad ...(2)$

Hence, $\qquad \dfrac{U_H}{U_S} = \dfrac{n^4 - 1}{n^4} \quad$ **(Ans.)** $\qquad\qquad\qquad ...(2)$

From eqns. (1) and (2), we have $\dfrac{T_H}{T_S} = \dfrac{U_H}{U_S} \qquad\qquad$ **...Proved**...(2)

15.6. STRAIN ENERGY DUE TO BENDING

15.6.1 Bending Under Gradually Applied Loads

Fig.15.10 shows a beam of uniform cross-section with certain end conditions such that the bending moment varies along its length. Consider a small length dx of a beam where the bending moment is M. Consider further a small strip $EFGH$ of thickness dy at a distance y from the neutral axis. Let b be the width of the strip. Volume of the strip is $dx \cdot dy \cdot b$.

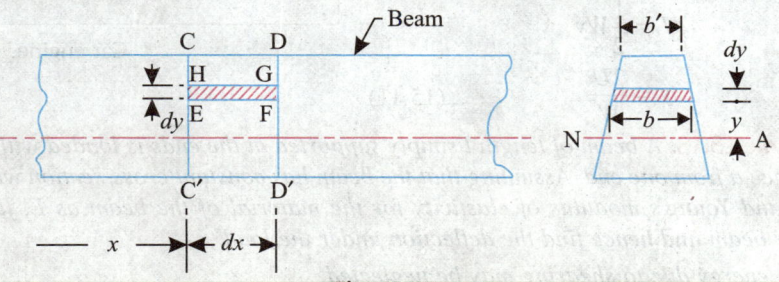

Fig. 15.10

∴ Strain energy of the small volume $(dx \cdot dy \cdot b)$

$$= \frac{(\text{Stress on } EFGH)^2}{2E} \times \text{volume} (dx \cdot dy \cdot b)$$

$$= \frac{\sigma^2}{2E} \times b \cdot dx \cdot dy$$

$$= \left(\frac{My}{I}\right)^2 \frac{b}{2E} \cdot dx \cdot dy \qquad \left[\because \frac{M}{I} = \frac{\sigma}{y} \text{ or } \sigma = \frac{My}{I}\right]$$

$$= \frac{M^2}{2EI^2} by^2 \cdot dx \cdot dy$$

∴ dU = Strain energy of the volume within $CC'D'D$

$$= \int \left(\frac{M^2 \, dx}{2EI^2}\right)(y^2 \cdot b \cdot dy) = \frac{M^2 \, dx}{2EI^2} \int y^2 \cdot b \cdot dy$$

Now, $\int by^2 \, dy$ = Sum of second moments of areas $b \cdot dy$

= Moment of inertia of beam cross-section = I

∴ $$dU = \frac{M^2 \, dx}{2EI^2} \cdot I = \frac{M^2}{2EI} \cdot dx$$

The above expression gives the strain energy of the beam of length dx

∴ Strain energy of the whole of the beam is given by

$$U = \int \frac{M^2}{2EI} \, dx \qquad \qquad ...(15.10)$$

For any given load and end conditions, M can be expressed in terms of x and then the total strain energy can be evaluated with the help of equation (15.10).

In case M is constant over the length l

$$U = \frac{M^2 l}{2EI} \qquad ...[15.10 \, (a)]$$

Strain energy and deflection due to bending:

In order to calculate the deflection under the load in the cases of beams under the action of a *single point load*, after calculating the strain energy of beam, it is equated to the work done by that load for its gradual movement equal to the deflection. If y is he deflection under the load W then,

$$U = \frac{1}{2}Wy$$

Car engine.

or, $$y = \frac{2U}{W} \qquad ...(15.11)$$

Example 15.18. *A beam of length l simply supported at the ends is loaded with a point load W at a distance a from one end. Assuming that the beam has constant cross-section with moment of inertia as I and Young's modulus of elasticity for the material of the beam as E, find the strain energy of the beam and hence find the deflection under the load.*

Strain energy due to shearing may be neglected.

Solution. Fig. 15.11 shows the given beam with the loading.

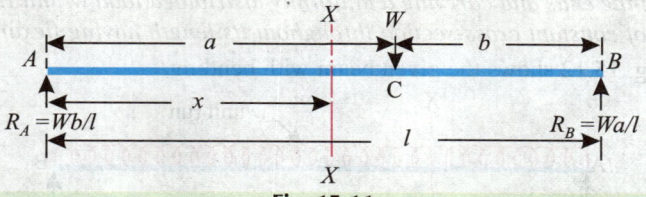

Fig. 15.11

To find reaction R_B, take moments about A.

i.e. $\qquad R_B \times l = W \times a$

$\therefore \qquad R_B = \dfrac{Wa}{l}$

$\therefore \qquad R_A = W - \dfrac{Wa}{l} = W\left(\dfrac{l-a}{l}\right) = \dfrac{Wb}{l} \qquad [\because b = (l-a)]$

For any section XX lying between A and C *i.e.* $0 < x < a$

$$M_x = R_A \cdot x = \frac{Wb}{l} \cdot x \quad \text{and for } a < x < l$$

$$M_x = R_B \times (l - x) = \frac{Wa}{l}(l - x)$$

Strain energy, $U = U_{AC} + U_{BC}$

$$= \int_0^a \frac{1}{2EI} \frac{W^2 b^2}{l^2} \cdot x^2 \, dx + \int_a^l \frac{1}{2EI} \frac{W^2 a^2}{l^2}(l - x)^2 \, dx$$

$$\left[\because \text{Strain energy} = \int \frac{M^2}{2EI} dx \right]$$

$$= \frac{W^2 b^2}{2EIl^2}\left(\frac{a^3}{3}\right) + \frac{W^2 a^2}{2EIl^2}\left[-\frac{(l-x)^3}{3}\right]_a^l = \frac{W^2 b^2 a^3}{6EIl^2} + \frac{W^2 a^2}{2EIl^2}\left[\frac{(l-a)^3}{3}\right]$$

$$= \frac{W^2 b^2 a^3}{6EIl^2} + \frac{W^2 a^2}{6EIl^2}\left[(l-a)^2(l-a)\right]$$

$$= \frac{W^2 b^2 a^3}{6EIl^2} + \frac{W^2 a^2 b^2}{6EIl^2}(l-a) \qquad [\because (l-a) = b]$$

$$= \frac{W^2 a^2 b^2}{6EIl^2}[a + (l-a)] = \frac{W^2 a^2 b^2}{6EIl}$$

i.e. $\qquad \mathbf{U = \dfrac{W^2 a^2 b^2}{6EIl}} \quad \textbf{(Ans.)}$

Let, $\qquad y_C = $ Deflection under the load.

$\therefore$ Work done by the load W on the beam $= \dfrac{1}{2}W \cdot y_C$.

Since work done = Strain energy

$\therefore \qquad \dfrac{1}{2}W \cdot y_C = \dfrac{Wa^2 b^2}{6EIl}$

$\therefore \qquad y_C = \dfrac{Wa^2 b^2}{3EIl} \quad \textbf{(Ans.)}$

Example 15.19. *Find an expression for the strain energy due to bending for a beam of length l simply supported at the ends and carrying a uniformly distributed load w/unit run over whole of its span. The beam is of constant cross-section throughout its length having flexural rigidity as EI.*

Solution. Fig. 15.12 shows the given beam with bending.

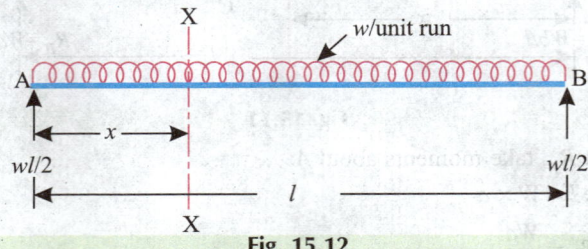

Fig. 15.12

Consider any section *XX* at a distance x from the end *A*. The bending moment at the section is given as,

$$M_x = \frac{wl}{2} \cdot x - \frac{wx^2}{2}$$

∴ Strain energy

$$U = \int_0^l \frac{M^2}{2EI} dx$$

$$= \frac{1}{2EI} \int_0^l \left(\frac{wl}{2} \cdot x - \frac{wx^2}{2} \right)^2 dx$$

$$= \frac{w^2}{8EI} \int_0^l (l^2 x^2 + x^4 - 2lx^3) dx = \frac{w^2}{8EI} \left[\frac{l^2 x^3}{3} + \frac{x^5}{5} - \frac{2l x^4}{4} \right]_0^l$$

$$= \frac{w^2}{8EI} \left[\frac{l^5}{3} + \frac{l^5}{5} - \frac{l^5}{2} \right] = \frac{w^2 l^5}{240 EI}$$

Hence,

$$U = \frac{w^2 l^5}{240 EI} \quad \textbf{(Ans.)}$$

Example 15.20. *Show that the strain energy U due to bending of a cantilever of length l with a concentrated load W at the free end is given by $U = \dfrac{W^2 l^3}{6EI}$ where EI is the flexural rigidity of the beam. Hence show that the deflection of the free end of the beam is $\dfrac{W l^3}{3EI}$. Show further that for rectangular beam cross-section the strain energy can be expressed as $U = \dfrac{\sigma^2}{18E} \times volume of the beam$ where σ is the maximum bending stress in the beam.*

Solution. Fig. 15.13 shows a cantilever of length *l* carrying a point load *W* at the free end.

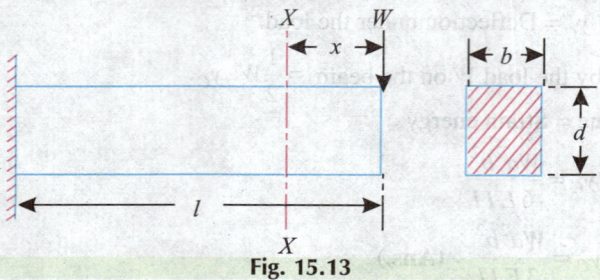

Fig. 15.13

Consider a section XX at a distance x from the end A. The bending moment at the section is given as :

$$M_x = -Wx$$

Strain energy,

$$U = \int_0^l \frac{M^2}{2EI}\,dx = \int_0^l \frac{W^2 x^2}{2EI}\,dx = \frac{W^2}{2EI}\int_0^l x^2\,dx$$

$$= \frac{W^2}{2EI}\left[\frac{x^3}{3}\right]_0^l = \frac{W^2 l^3}{6EI}$$

Hence,

$$U = \frac{W^2 l^3}{6EI} \quad \textbf{...Proved}$$

Let, y_{max} = Deflection at the end.

∴ Work done on the beam by the load W

$$= \frac{1}{2} W\, y_{max}$$

But work done = Strain energy

∴

$$\frac{1}{2} W\, y_{max} = \frac{W^2 l^3}{6EI}$$

Hence,

$$y_{max} = \frac{W l^3}{3EI} \quad \textbf{.....Proved.}$$

Now,

$$\sigma = \text{Maximum bending stress} = \frac{M_{max}}{Z} = \frac{Wl}{\left(\dfrac{bd^2}{6}\right)} = \frac{6Wl}{bd^2}$$

[where, b = width of the beam, and, d = depth of the beam.]

∴

$$W = \frac{\sigma b d^2}{6l}$$

Steam engine mechanisms. In countries like China where coal is abundant and cheap, steam engines are still widely used.

Putting this value of W in the expression for U, we get

$$U = \left(\frac{\sigma b d^2}{6l}\right)^2 \times \frac{l^3}{6EI} = \frac{\sigma^2 b^2 d^4 l^3}{36 l^2 \times 6E \times \dfrac{bd^3}{12}}$$

$$= \frac{\sigma^2}{18E} \times (bdl) = \frac{\sigma^2}{18E} \times \text{volume of the beam}$$

Hence, $\qquad U = \dfrac{\sigma^2}{18E} \times \text{volume of the beam} \qquad$ ...**Proved.**

Example 15.21. *Show that the strain energy U due to bending of a beam of rectangular section simply supported at the ends with a concentrated load W at the centre can be expressed as*

$$U = \frac{\sigma^2}{18E} \times \text{volume of beam where } \sigma \text{ is the bending stress in the beam and } E \text{ is the Young's modulus.}$$

Compare the strain energy when the beam is loaded axially by load W, if the ratio of length of beam to depth is 6.

Solution. Fig. 15.14 shows the given beam with a concentrated load W at the centre.

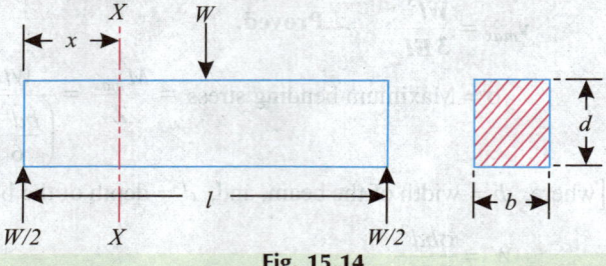

Fig. 15.14

Consider a section XX at a distance x from the left end for left half of the beam. The B.M. at the section is given as

$$M_x = +\frac{W}{2}x$$

∴ Strain energy for left half of the beam

$$= \int_0^{1/2} \frac{W^2 x^2}{4} \times \frac{1}{2EI} \, dx = \frac{W^2}{8EI} \int_0^{1/2} x^2 \, dx = \frac{W^2}{8EI}\left[\frac{x^3}{3}\right]_0^{l/2} = \frac{W^2 l^3}{192 EI}$$

∴ $\qquad U = $ Strain energy for whole of the beam

$$= 2 \times \frac{W^2 l^3}{192 EI} = \frac{W^2 l^3}{96 EI}$$

∴ $\qquad U = \dfrac{W^2 l^3}{96 EI} \qquad\qquad$ ...(i)

Now, $\qquad \sigma = \dfrac{M_{max}}{Z} = \dfrac{\dfrac{Wl}{4}}{\dfrac{bd^2}{6}} = \dfrac{3Wl}{2bd^2} \qquad \left[\begin{array}{l}\text{where, } b = \text{width of the beam;} \\ d = \text{depth of the beam.}\end{array}\right]$

∴ $\qquad W = \dfrac{2\sigma bd^2}{3l}$

Putting this value of W in (i), we get

$$U = \left(\frac{2\sigma bd^2}{3l}\right)^2 \times \frac{l^3}{96\,EI} = \frac{4\sigma^2 b^2 d^4}{9l^2} \times \frac{l^3}{96\,E \times \frac{bd^3}{12}}$$

$$= \frac{\sigma^2}{18\,E}(lbd) = \frac{\sigma^2}{18\,E} \times \text{volume of beam}$$

Hence, $\qquad U = \dfrac{\sigma^2}{18\,E} \times \text{volume of beam} \quad \textbf{...Proved}$

Comparison of strain energies:

$\qquad U_a = \text{Strain energy of beam under axial load}$

$$= \frac{\sigma^2}{2E} \times \text{volume} = \frac{1}{2E}\left(\frac{W}{A}\right)^2 \times lbd \qquad \left[\because \ \sigma = \frac{W}{A}\right]$$

$$= \frac{1}{2E} \times \frac{W^2}{b^2 d^2} \times lbd = \frac{W^2 l}{2Ebd} \qquad \text{...(ii)}$$

Dividing (ii) by (i), we get

$$\frac{U_a}{U} = \frac{W^2 l}{2Ebd} \times \frac{96EI}{W^2 l^3} = \frac{48I}{bdl^2} = \frac{48 \times \dfrac{bd^3}{12}}{bdl^2} = 4\left(\frac{d}{l}\right)^2 = 4 \times \left(\frac{1}{6}\right)^2 = \frac{1}{9}$$

Hence, $\qquad \dfrac{U_a}{U} = \dfrac{1}{9} \quad \textbf{(Ans.)}$

Example 15.22. *A beam 4m in length is simply supported at the ends and carries a uniformly distributed load of 6kN/m length. Determine the strain energy stored in the beam.*

Take: $\qquad E = 200\ GN/m^2,\ and\ I = 1440\ cm^4.$

Solution. *Given:* $l = 4\text{m};\ w = 6\text{kN/m length};\ E = 200\ \text{GN/m}^2;\ I = 1440\ \text{cm}^4.$

Strain energy stored in the beam:

Refer to Fig. 15.15.

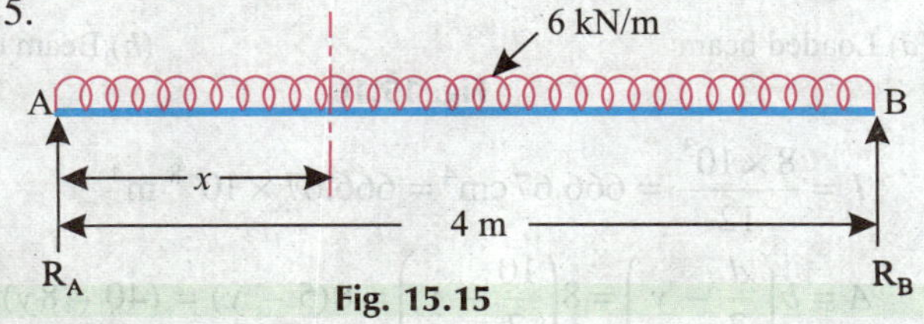

Fig. 15.15

Taking moments about A, we get

$$R_B \times 4 = 6 \times 4 \times 2$$

$$R_B = 12\ \text{kN};\quad R_A = 6 \times 4 - 12 = 12\ \text{kN}$$

Strain energy in bending is given by,

$$U = \int_0^l \frac{M^2\,dx}{2EI}$$

$$M_A = R_A \times x - (w \times x) \times \frac{x}{2} = 12 \times x - (6 \times x) \times \frac{x}{2} = (12x - 3x^2)\,\text{kNm}$$

$$\therefore \qquad U = \int_0^4 \frac{(12x - 3x^2)^2\,dx}{2EI}$$

$$= \frac{10^6}{2EI} \int_0^4 \left[(12)^2\,x^2 + (3)^2\,x^4 - (2 \times 12 \times 3)\,x^3\right]dx$$

$$= \frac{10^6}{2EI} \left[\left| 144 \times \frac{x^3}{3} + 9 \times \frac{x^5}{5} - 72 \times \frac{x^4}{4} \right|_0^4 \right]$$

$$= \frac{10^6}{2EI} \left[144 \times \frac{4^3}{3} + 9 \times \frac{4^5}{5} - 72 \times \frac{4^4}{4} \right]$$

$$= \frac{10^6}{2 \times 200 \times 10^9 \times 1440 \times 10^{-8}} (3072 + 1843.2 - 4608) = \textbf{53.33 Nm} \quad \textbf{Ans.}$$

Example 15.23. *A simply supported beam of span 3m is carrying a concentrated load of 18 kN at the mid-span. Determine the strain energy stored in the beam due to horizontal shear. The beam is 8 cm wide and 10 cm deep. E = 200 GN/m²; Poisson's ratio = 0.32* **(Panjab University)**

Solution. *Given:* $l = 3$ m; $W = 18$ kN; $P = 8$ cm; $d = 10$ cm; $E = 200$ GN/m², $\frac{1}{m} = 0.32$.

Strain energy due to horizontal shear :

Strain energy due to horizontal shear is given by,

$$U = \left(\frac{1 + \frac{1}{m}}{E I^2} \right) \int_0^l F^2 \, dx \iint_A \frac{(A \bar{y})^2}{b^2} \, dA$$

Here, $F = \frac{18}{2} = 9$ kN

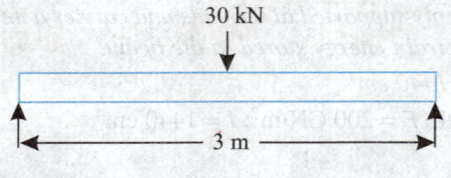

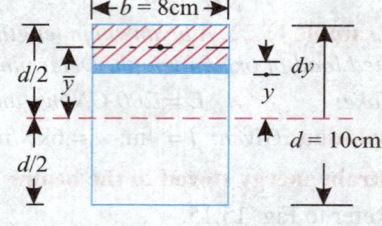

(a) Loaded beam (b) Beam cross-section

Fig. 15.16

$$I = \frac{8 \times 10^3}{12} = 666.67 \text{ cm}^4 = 666.67 \times 10^{-8} \text{ m}^4$$

$$A = b \left(\frac{d}{2} - y \right) = 8 \left(\frac{10}{2} - y \right) = 8(5 - y) = (40 - 8y) \text{ cm}^2$$

$$\bar{y} = y + \frac{1}{2} \left(\frac{d}{2} - y \right) = y + \frac{1}{2} \left(\frac{10}{2} - y \right)$$

$$= y + 2.5 - \frac{y}{2} = (0.5y + 2.5) \text{ cm}$$

$$dA = b \times dy = 8 \, dy$$

Substitutions the values in the above equation, we have

$$U = \frac{(1 + 0.32)}{200 \times 10^9 \times (666.67 \times 10^{-8})^2}$$

$$\int_0^3 (9 \times 10^3)^2 \, dx \int_{-5}^{+5} \frac{\{(40 - 8y)(0.5y + 2.5)\}^2 \times 10^{-12}}{(8)^2} \times 8 \, dy$$

$$= \frac{1.32 \times 10^{16}}{200 \times 10^9 - (666.67)^2} \int\limits_0^3 (9 \times 10^3)^2 \, dx \int\limits_{-5}^{+5} \frac{16(y^2 - 25)^2 \times 10^{-12}}{8} \, dy$$

$$= \frac{1.32 \times 10^{16} \times 10^6}{200 \times 10^9 \times (666.67)^2} \int\limits_0^3 \left| (9)^2 \, x \right|_0^3 \times \int\limits_{-5}^{+5} 2(y^2 - 25)^2 \, dy \times 10^{-12}$$

$$= \frac{1.32}{20 \times (666.67)^2} \left[81 \times 3 \times 2 \int\limits_{-5}^{+5} (y^4 + 625 - 50 y^2) \, dy \right]$$

$$= \frac{1.32}{20 \times (666.67)^2} \left[81 \times 6 \left| 0.2 \, y^5 + 625 \, y - 16.667 \, y^3 \right|_{-5}^{+5} \right]$$

$$= \frac{1.32}{20 \times (666.67)^2} \left[81 \times 6 \left\{ 0.2 \left(5^5 + 5^5 \right) + 625(5 + 5) - 16.667(125 + 125) \right\} \right]$$

$$= \frac{1.32}{20 \times (666.67)^2} \left[81 \times 6 \times 2 \{ 625 + 3125 - 2083.4 \} \right] = 0.24 \text{ Nm}$$

i.e., $U = 0.24$ **Nm (Ans.)**

15.6.2. Bending Under Impact Loads

Fig. 15.17 shows a simply supported beam AB. Let weight W drop from a height h at the point C on the beam. To find the instantanous deflection under the load the following procedure is adopted:

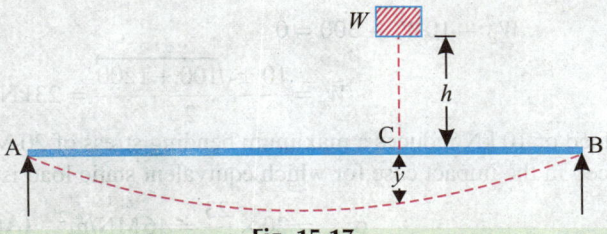

Fig. 15.17

Let W_e be the equivalent weight which when applied gradually at the point C will produce the same deflection y.

Since in the two cases, the deflection is same, the strain energies in the two cases will be equal. Further since the strain energy of the beam is equal to the work done by the individual loads in the two cases, we have

$$W(h + y) = \frac{1}{2} W_e \, y \qquad \qquad ...(15.12)$$

Depending on the end conditions the deflection y can be found in terms of W_e (chapter 8) and in general

$$y \, \alpha \, W_e$$

or $$y = k W_e \qquad \qquad ...(15.13)$$

(where, k = constant which depends upon the beam and its end conditions)

From eqns. (15.12) and (15.13))

$$W(h + k W_e) = \frac{1}{2} k W_e^2 \qquad \qquad ...(15.14)$$

The solution of the quadratic equation (15.14) gives the value of W_e. Then the value of y can be evaluated from (15.13).

Example 15.24. *A point load of 10 kN applied to a simply supported beam at midspan, produces a deflection of 6 mm and a maximum bending stress of 20 MN/m². Calculate the maximum value of the momentary stress produced when a weight of 5 kN is allowed to fall through a height of 18 mm on the beam at middle of the span.* **(Panjab University)**

Solution. Point load, $\qquad\qquad\qquad W_s = 10$ kN

Deflection produced $\qquad\qquad\qquad = 6$ mm

Maximum bending stress produced, $\sigma_{max} = 20$ MN/m²

Magnitude of the falling weight $\qquad W = 5$ kN

Height of fall, $\qquad\qquad\qquad\qquad h = 18$ mm $= 0.018$ m

Let, W_e = The static load equivalent to the given impact/falling load, kN. Since 10 kN static load produces a deflection of 6 mm, then W_e will produce a deflection y such that

$$y = \frac{6}{10} \times W_e = 0.6\, W_e \text{ mm} = 0.0006\, W_e \text{ m}$$

Now by equating the work done by the equivalent static load and the given falling load, we get

$$W(h + y) = \frac{1}{2} W_e \cdot y$$

$$5(0.018 + 0.0006\, W_e) = \frac{1}{2} W_e \times 0.006\, W_e$$

$$0.09 + 0.003\, W_e = 0.0003\, W_e^2$$

$$300 + 10\, W_e = W_e^2$$

or, $\qquad\qquad W_e^2 - 10\, W_e - 300 = 0$

$$W_e = \frac{10 \pm \sqrt{100 + 1200}}{2} = 23\,\text{kN}$$

Since a static load of 10 kN induces a maximum bending stress of 20 MN/m², the maximum bending stress produced in the impact case for which equivalent static load is 23 kN, will be

$$\sigma_{max} = 20 \times \frac{23}{10} = 46\,\text{MN/m}^2 \quad \textbf{(Ans.)}$$

Hence, $\qquad\qquad\qquad\qquad \sigma_{max} = \textbf{46 MN/m}^2 \quad \textbf{(Ans.)}$

Automobile brake mechanism with springs.

Example 15.25. *A 1 m long beam rectangular in section 30 mm wide × 40 mm deep is supported on rigid supports at its ends. If it is struck at the centre by a 12 kg mass falling through a height of 60 mm find:*

(i)   *The instantaneous stress developed;*

(ii) *The instantaneous strain energy stored in the beam.*

Take: $E = 200 \ GN/m^2$.

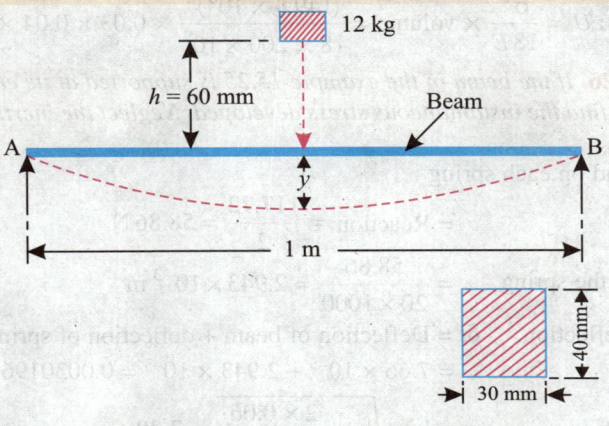

Fig. 15.18

Solution. Magnitude of falling load,

$$W = 12 \times 9.81 \ N = 117.72 \ N$$

Height of fall, $h = 60 \ mm = 0.06 \ m$

Width of the beam, $b = 30 \ mm = 0.03 \ m$

Depth of the beam, $d = 40 \ mm = 0.04 \ m$

∴ $I = \dfrac{bd^3}{12} = \dfrac{0.03 \times (0.04)^3}{12} = 1.6 \times 10^{-7} \ m^4$

Static deflection due to load W

$$\delta l = \frac{Wl^3}{48 \, EI}$$

$$= \frac{117.72 \times 1^3}{48 \times 200 \times 10^9 \times 1.6 \times 10^{-7}} \ m = 7.66 \times 10^{-5} \ m$$

∴ $n = 1 + \sqrt{1 + \dfrac{2h}{\delta l}} = 1 + \sqrt{1 + \dfrac{2 \times 0.06}{7.66 \times 10^{-5}}} = 40.6$

i.e. $n = 40.6$ (where, n = load factor)

(i)   Instantaneous stress developed, σ_i :

Static stress, $\sigma = \dfrac{M}{Z} = \dfrac{Wl}{4Z} = \dfrac{117.72 \times 1}{4 \times \dfrac{1}{6} \times 0.03 \times 0.04^2} = 3.68 \times 10^6 \ N/m^2$

or $\sigma = 3.68 \ MN/m^2$

∴ $\sigma_i = n\sigma = 40.6 \times 3.68 = 149.4 \ MN/m^2$

Hence $\sigma_i = \textbf{149.4 MN/m}^2$ **(Ans.)**

(ii) Strain energy stored:

Strain energy, $U = \dfrac{1}{2} W_e \cdot \delta l_i$

$$= \frac{1}{2}(nW) \times (n\delta l)$$

$$[\because W_e = nW, \delta l_i = n\delta l]$$

$$= \frac{n^2}{2} \cdot W \cdot \delta l = \frac{40.6^2}{2} \times 117.72 \times 7.66 \times 10^{-5} = 7.4 \, \text{J}$$

Hence, $\qquad U = 7.4 \, \text{J (Ans.)}$

$$\left[\text{Alternately}: U = \frac{\sigma^2}{18E} \times \text{volume} = \frac{(149.4 \times 10^6)^2}{18 \times 200 \times 10^9} \times 0.03 \times 0.04 \times 1 = 7.4 \, \text{J (Ans.)} \right]$$

Example 15.26. *If the beam in the example 15.25 is supported at its ends on springs having stiffness of 20 kN/m, find the instantaneous stress developed. Neglect the inertia of the beam and the springs.*

Solution. Load on each spring

$$= \text{Reaction} = \frac{117.72}{2} = 58.86 \, \text{N}$$

Deflection of the spring $\qquad = \dfrac{58.86}{20 \times 1000} = 2.943 \times 10^{-3} \, \text{m}$

Total static deflection, $\quad \delta l = $ Deflection of beam + deflection of spring.

$$= 7.66 \times 10^{-5} + 2.943 \times 10^{-3} = 0.0030196 \, \text{m}$$

$\therefore$ Load factor, $\qquad n = 1 + \sqrt{1 + \dfrac{2 \times 0.06}{0.0030196}} = 7.38$

$\therefore \qquad \sigma_i = n\sigma = 7.38 \times 3.68 = 27.16 \, \text{MN/m}^2$

Hence, $\qquad \sigma_i = \textbf{27.16 MN/m}^2 \ \textbf{(Ans.)}$

15.7. STRAIN ENERGY DUE TO PRINCIPAL STRESSES—STRAIN ENERGY IN A GENERAL CASE

Any compound stress condition can always be reduced to the condition of three principal stresses; thus general strain energy equation deduced in terms of three principal stresses can be used for all cases of compound loading on a body.

Let a triaxial system, σ_1, σ_2 and σ_3 as the principal stresses, be acting on a small block of length δl, width δb and height δh.

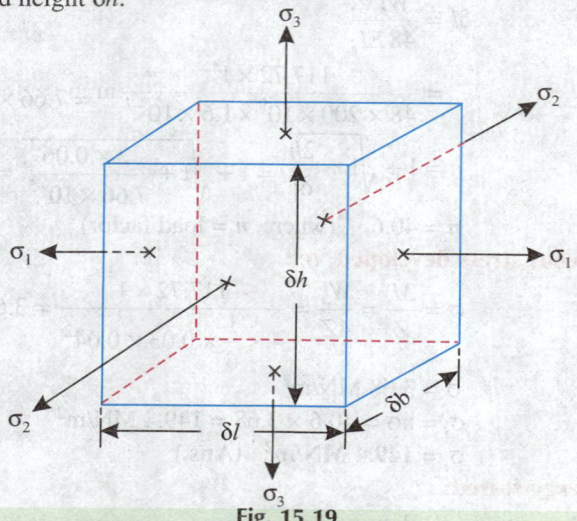

Fig. 15.19

The strains in the respective three directions are:

$$e_1 = \frac{1}{E}[\sigma_1 - \mu(\sigma_2 + \sigma_3)] \qquad \qquad ...[15.15\ (a)]$$

$$e_2 = \frac{1}{E}[\sigma_2 - \mu(\sigma_3 + \sigma_1)] \qquad \qquad ...[15.15\ (b)]$$

$$e_3 = \frac{1}{E}[\sigma_3 - \mu(\sigma_1 + \sigma_2)] \qquad \qquad ...[15.15\ (c)]$$

$$\left[\text{where, } \mu\left(=\frac{1}{m}\right) = \text{Poisson's ratio for the material;} \right.$$
$$\left. E = \text{Young's modulus.} \right]$$

The extensions in three directions are:

$$x_1 = e_1 \cdot \delta l; \quad x_2 = e_2 \cdot \delta b, \quad \text{and,} \quad x_3 = e_3 \cdot \delta h$$

∴ Strain energy due to

$$\sigma_1 = \frac{1}{2} \ (\text{load due to } \sigma_1 \text{ in the direction of } \sigma_1) \times \text{extension in the direction of } \sigma_1$$

$$= \frac{1}{2}(\sigma_1 \cdot \delta b \cdot \delta h) \times \frac{\delta l}{E}[\sigma_1 - \mu(\sigma_2 + \sigma_3)]$$

$$= \frac{1}{2E}\left[\sigma_1^2 - \mu(\sigma_1 \sigma_2 + \sigma_1 \sigma_3)\right](\delta l \cdot \delta b \cdot \delta h)$$

$$= \frac{1}{2E}\left[\sigma_1^2 - \mu(\sigma_1 \sigma_2 + \sigma_1 \sigma_3)\right]\delta V$$

(where, δV = volume of block = $\delta l \cdot \delta b \cdot \delta h$)

Similarly, strain energy due to

$$\sigma_2 = \frac{1}{2E}[\sigma_2^2 - \mu(\sigma_2 \sigma_1 + \sigma_2 \sigma_3)]\delta V$$

and, strain energy due to

$$\sigma_3 = \frac{1}{2E}[\sigma_3^2 - \mu(\sigma_3 \sigma_1 + \sigma_3 \sigma_2)]\delta V$$

∴ δU = Total strain energy for volume δV = sum of strain energies due to σ_1, σ_2 and σ_3

$$= \frac{1}{2E}\left[\sigma_1^2 - \mu(\sigma_1 \sigma_2 + \sigma_3 \sigma_1)\right]\delta V + \frac{1}{2E}[\sigma_2^2 - \mu(\sigma_2 \sigma_3 + \sigma_1 \sigma_2)]\delta V$$

$$+ \frac{1}{2E}\left[\sigma_3^2 - \mu(\sigma_3 \sigma_1 + \sigma_2 \sigma_3)\right]\delta V$$

∴
$$\delta U = \frac{1}{2E}\left[(\sigma_1^2 + \sigma_2^2 + \sigma_3^2) - 2\mu(\sigma_1\sigma_2 + \sigma_2\sigma_3 + \sigma_3\sigma_1)\right]\delta V$$

For a body of volume V subjected to the principal stress σ_1, σ_2 and σ_3 throughout its volume

$$U = \frac{1}{2E}\left[(\sigma_1{}^2 + \sigma_2{}^2 + \sigma_3{}^2) - 2\mu(\sigma_1 \sigma_2 + \sigma_2 \sigma_3 + \sigma_3 \sigma_1)\right]V \qquad ...(15.16)$$

∴ Energy per unit volume,

$$\textbf{U/unit volume} = \frac{1}{2E}\left[(\sigma_1^2 + \sigma_2^2 + \sigma_3^2) - 2\mu(\sigma_1 \sigma_2 + \sigma_2 \sigma_3 + \sigma_3 \sigma_1)\right] \qquad ...[15.16\ (a)]$$

Note. (i) Principal stresses if *compressive* will be considered *negative* for use in eqn. (15.16)

(iii) The order of the principal stresses is immaterial.

Deductions for simple cases of loading :

For simple cases of stresses the expression for strain energy can be easily deduced from the eqn. (15.16).

(a) Direct stress:

If σ is the direct stress under gradually applied load, then

$$\sigma_1 = \sigma, \sigma_2 = 0 \text{ and } \sigma_3 = 0$$

Putting the above values of σ_1, σ_2 and σ_3 in eqn. (15.16),

we get, $$U = \frac{\sigma^2}{2E} \times V$$

(b) Simple shear:

Let τ be the simple shear stress in volume V. Then the principal stresses will be τ and $-\tau$, third principal stress being zero.

$\therefore$ $$\sigma_1 = \tau, \sigma_2 = -\tau, \sigma_3 = 0$$

Putting these values of σ_1, σ_2 and σ_3 in eqn. (15.16), we get

$$U = \frac{1}{2E}[\tau^2 + (-\tau)^2 + 0 - 2\mu(\tau)(-\tau)]V$$

$$= \frac{1}{2E}(2\tau^2 + 2\mu\tau^2)V = \frac{1+\mu}{E}\tau^2 V$$

But, $$E = 2C(1+\mu) \text{ i.e. } \frac{1+\mu}{E} = \frac{1}{2C}$$

$\therefore$ $$U = \frac{\tau^2}{2C}V$$

(c) Strain energy due to hydrostatic pressure or equal tensions applied to a volume:

Let p be the hydrostatic pressure or hydrostatic tension.

$\therefore$ $$\begin{bmatrix} \sigma_1 = -p \\ \sigma_2 = -p \\ \sigma_3 = -p \end{bmatrix} \text{ or } \begin{bmatrix} \sigma_1 = p \\ \sigma_2 = p \\ \sigma_3 = p \end{bmatrix}$$

Bridges undergo deflection due to shear and bending.

Putting any one of these sets in eqn (15.16), we get

$$U = \frac{1}{2E}\left[(p^2 + p^2 + p^2) - 2\mu(p^2 + p^2 + p^2)\right]V$$

or, $\qquad U = \frac{3p^2}{2E}(1 - 2\mu)V$ ...(15.17)

But, $\qquad E = 3K(1 - 2\mu)$

$\therefore \qquad U = \frac{3p^2(1 - 2\mu)}{2 \times 3k(1 - 2\mu)}V = \frac{p^2}{2K}V$

i.e. $\qquad U = \frac{p^2}{2K} \cdot V$ ...(15.18)

15.8. ENERGY OF DISTORTION (SHEAR STRAIN ENERGY)

The strain energy given by eqn. (15.16) can be split up into the following *two strain energies*:

(i) *Strain energy of distortion (shear strain energy)*

(ii) *Strain energy of uniform compression or tension (volumetric strain energy or energy of dilatation or energy of dilation)*

Let e_1, e_2 and e_3 be the principal strains in the directions of principal stresses σ_1, σ_2 and σ_3 respectively.

Then, $\qquad e_1 = \frac{1}{E}[\sigma_1 - \mu(\sigma_2 + \sigma_3)]$

$$e_2 = \frac{1}{E}[\sigma_2 - \mu(\sigma_3 + \sigma_1)]$$

$$e_3 = \frac{1}{E}[\sigma_3 - \mu(\sigma_1 + \sigma_2)]$$

Adding the above equations we get,

$$e_1 + e_2 + e_3 = \frac{1}{E}[\sigma_1 + \sigma_2 + \sigma_3) - 2\mu(\sigma_1 + \sigma_2 + \sigma_3)] = \frac{\sigma_1 + \sigma_2 + \sigma_3}{E}(1 - 2\mu)$$

But, $\qquad e_1 + e_2 + e_3 = e_v$ (volumetric strain)

$\therefore \qquad e_v = \frac{1 - 2\mu}{E}(\sigma_1 + \sigma_2 + \sigma_3)$

If $\sigma_1 + \sigma_2 + \sigma_3 = 0$, $e_v = 0$. This means that if sum of the three principal stress is zero there is no volumetric change, but only the distortion occurs.

If three principal stress σ_1, σ_2 and σ_3 are equal, $e_1 = e_2 = e_3$. This means that there is *no distortion (i.e. no shearing stresses and shearing strains will be present anywhere in the block)* but only the volumetric change occurs.

From the above discussion we arrive at the following conclusions:

1. *When the sum of three principal stresses is zero, there is no volumetric change but only the distortion occurs.*

2. *When the three principal stresses are equal to one another there is no distortion but only volumetric change occurs.*

Energy of Distortion and Energy of Dilation

The given three principal stresses can be broken (by using the above two conclusions) into two sets of principal stresses such that one set produces *distortion only*, while other produces *volumetric change only*.

Refer to Fig 15.19.

Let,
$$\sigma_1 = \sigma'_1 + \sigma \qquad \qquad ...(i)$$
$$\sigma_2 = \sigma'_2 + \sigma \qquad \qquad ...(ii)$$
$$\sigma_3 = \sigma'_3 + \sigma \qquad \qquad ...(iii)$$

The set of principal stresses σ'_1, σ'_2 and σ'_3 produces *distortion only*.

The set σ, σ, σ produces *volumetric change only*.

$$\therefore \qquad \sigma'_1 + \sigma'_2 + \sigma'_3 = 0$$

Adding the equations (i), (ii) and (iii), we get

$$\sigma_1 + \sigma_2 + \sigma_3 = \sigma'_1 + \sigma'_2 + \sigma'_3 + 3\sigma$$

But, $\qquad \sigma'_1 + \sigma'_2 + \sigma'_3 = 0$

$$\therefore \qquad 3\sigma = \sigma_1 + \sigma_2 + \sigma_3$$

or, $\qquad\qquad\qquad \sigma = \dfrac{\sigma_1 + \sigma_2 + \sigma_3}{3}$

When σ is known, the values of σ'_1, σ'_2 and σ'_3 can be found out from eqns. (i), (ii) and (iii).

Strain energy of dilation $\quad$ = Strain energy due to principal stresses σ, σ and σ

$$= \frac{1}{2E}[\sigma^2 + \sigma^2 + \sigma^2 - 2\mu(\sigma^2 + \sigma^2 + \sigma^2)] V$$

$$= \frac{3\sigma^2(1 - 2\mu)}{2E} V = \frac{3(1 - 2\mu)}{2E}\left(\frac{\sigma_1 + \sigma_2 + \sigma_3}{3}\right)^2$$

U_v = Energy of dilation

or, $\qquad\qquad\qquad U_v = \dfrac{3(1 - 2\mu)}{2E}\left(\dfrac{\sigma_1 + \sigma_2 + \sigma_3}{3}\right)^2 V \qquad\qquad ...(15.19)$

Energy of distortion = Total strain energy – energy of dilation

$$= \frac{1}{2E}[\sigma_1^2 + \sigma_2^2 + \sigma_3^2 - 2\mu(\sigma_1\sigma_2 + \sigma_2\sigma_3 + \sigma_3\sigma_1)]V - \frac{3(1 - 2\mu)}{2E}\left(\frac{\sigma_1 + \sigma_2 + \sigma_3}{3}\right)^2 V$$

$$= \frac{V}{6E}\left[3\{\sigma_1^2 + \sigma_2^2 + \sigma_3^2 - 2\mu(\sigma_1\sigma_2 + \sigma_2\sigma_3 + \sigma_3\sigma_1)\} - (1 - 2\mu)(\sigma_1 + \sigma_2 + \sigma_3)^2\right]$$

On simplification, we get $U_v = \dfrac{1 + \mu}{6E}[2\sigma_1^2 + 2\sigma_2^2 + 2\sigma_3^2 - 2\sigma_1\sigma_2 - 2\sigma_2\sigma_3 - 2\sigma_3\sigma_1]V$

$$= \frac{1 + \mu}{6E}[(\sigma_1 - \sigma_2)^2 + (\sigma_2 - \sigma_3)^2 + (\sigma_3 - \sigma_1)^2]V$$

But, $\qquad\qquad E = 2C(1 + \mu)$

$$U_d = \frac{V}{12C}[(\sigma_1 - \sigma_2)^2 + (\sigma_2 - \sigma_3)^2 + (\sigma_3 - \sigma_1)^2] \qquad\qquad ...(15.20)$$

Putting, $\qquad\qquad E = 3(1 - 2\mu) K$, we get

$$U_v = \frac{1}{2K}\left(\frac{\sigma_1 + \sigma_2 + \sigma_3}{3}\right)^3 V = \frac{\sigma_m^2}{2K} V \qquad\qquad ...(15.21)$$

where, $\qquad\qquad \sigma_m = \dfrac{\sigma_1 + \sigma_2 + \sigma_3}{3}$

15.9. STRAIN ENERGY AND DEFLECTION DUE TO SHEAR

In the chapter on 'Deflection of beams' the deflection of the beam was calculated by taking into account bending moments only. Actually due to change of bending moment there will be shearing force (= dM/dx) on each and every section of the beam. As a result of this the transverse sections will slip with respect to the adjacent sections and will give rise what is called the *deflection due to shear*.

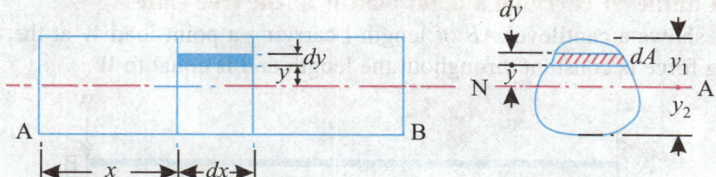

Fig. 15.20

The deflection due to shear can be calculated by the use of *strain energy method* as follows:

The strain energy due to shear $= \dfrac{\tau^2}{2C} \times$ volume

The value of τ varies over the section of the beam.

Consider an elemental area dA of length dx (of the beam) and at a distance y from the neutral axis. For this elemental area shear stress can be assumed to be constant. The shear strain energy for the whole beam AB is given by

$$U_s = \int_0^l \int_{-y_2}^{y_1} \frac{\tau^2}{2C}\, dA\, dx$$

...(15.22)

The above integration can be done for particular shape of the section over which the variation of τ is known.

In case of *rectangular section* the distribution of τ in terms of 'b' (width) and 'd' (depth) of the section is given as

Shock absorbers absorb strain energy and release it gradually.

$$\tau = \frac{6S}{bd^3}\left(\frac{d^2}{4} - y^2\right)$$

where,　　　　　　　S = Shearing force over the section.

Therefore the shear strain energy in case of rectangular section is given by:

$$U_s = \int_0^l \int_{-d/2}^{d/2} \frac{1}{2C} \cdot \frac{36S^2}{b^2 d^6}\left[\frac{d^4}{16} + y^4 - \frac{d^2 y^2}{6}\right](b\, dy)\, dx$$

$$= \frac{18}{Cb^2 d^6} \int_0^l S^2 \left[\frac{d^4 y}{16} + \frac{y^5}{5} - \frac{d^2 y^3}{6}\right]_{-d/2}^{d/2} dx$$

or,

$$= \frac{18}{Cbd} \int_0^l S^2 \left(\frac{1}{16} + \frac{1}{80} - \frac{1}{24}\right) dx$$

or,

$$U_s = \frac{3}{5Cbd} \int_0^l S^2\, dx$$

...(15.23)

Depending upon the *type of loading and end conditions* of the beam the value of S as function of x can be substituted in eqn. (15.23) and integration done.

Let us consider different cases :

Case I. Cantilever carrying a point load W at the free end:

Fig.15.21 shows a cantilever AB of length l carrying a point load W at the free end. In this case the shearing force is constant throughout the length and is equal to W.

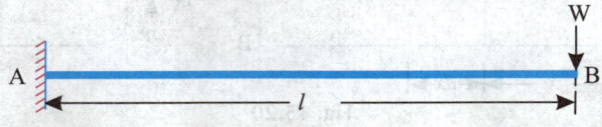

Fig. 15.21

i.e.

$$S = W$$

$$U_s = \frac{3}{5Cbd} \int_0^l W^2 \, dx \qquad \text{...[from eqn. (15.23)]}$$

$$= \frac{3W^2 l}{5\,Cbd} \qquad \text{...(15.24)}$$

Let,

$$\delta_s = \text{Deflection due to shear load.}$$

Then, work done due to $W = \frac{1}{2} W \delta_s$

$$\therefore \qquad \frac{1}{2} W \delta_s = \frac{3W^2 l}{5\,Cbd} \quad \text{or} \quad \delta_s = \frac{6Wl}{5\,Cbd} \qquad \text{...(15.25)}$$

Case II. Cantilever carrying a uniformly distributed load:

Fig. 15.22 shows a cantilever AB of length l carrying a uniformly distributed load w / unit run. Consider a section XX at a distance x from A. The load $(w.dx)$ on a small length dx can be considered to be a point load which will produce a deflection due to shear

$$= \frac{6(w\,dx)\,x}{5\,Cbd} \qquad \text{...(from 15.25)}$$

at x as well as at any point towards the right of section XX.

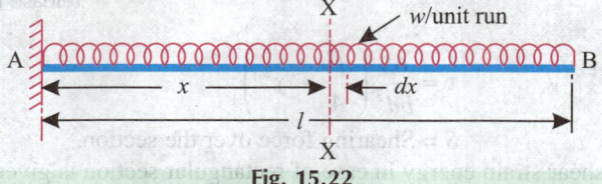

Fig. 15.22

$\therefore$ Total deflection at the free end

$$(\delta_s)_B = \int_0^l \frac{6(w\,dx)\,x}{5\,Cbd} = \frac{3\,wl^2}{5\,Cbd} \qquad \text{...(15.26)}$$

Case III. Simply supported beam carrying a cental load W:

Fig. 15.23 shows a simply supported beam carrying a central load W. The magnitude of shearing force over AC and

CB is $W/2$, therefore for the whole length

$$S^2 = \frac{W^2}{4}$$

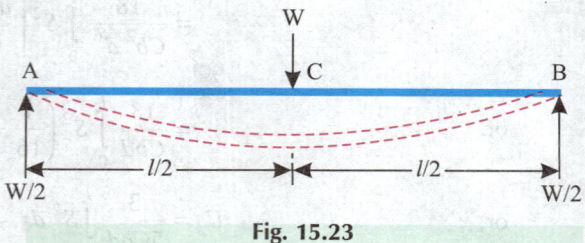

Fig. 15.23

∴ $$U_s = \frac{3}{5Cbd} \int_0^l \frac{W^2}{4} \, dx$$

...[from eqn. 15.23]

$$= \frac{3}{5Cbd} \cdot \frac{W^2 l}{4} = \frac{3W^2 l}{20 Cbd}$$ ...(15.27)

Also, work done $= \frac{1}{2} \cdot W \delta_s$

∴ $$\frac{1}{2} W \delta_s = \frac{3W^2 l}{20 Cbd}$$

or, $$\delta_s = \frac{3Wl}{10 Cbd}$$ ...(15.28)

Case IV. Simply supported beam with load not at the centre:

Fig. 15.24 shows the beam with the loading.

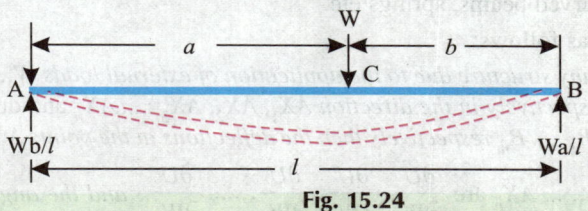

This drilling tool has strong tips that are capable of drilling into rock.

Fig. 15.24

Magnitude of shearing force over $AC = \dfrac{Wb}{l}$

Magnitude of shearing force over $CB = \dfrac{Wa}{l}$

Treating either of the portions AC or CB as cantilever, the deflection due to shear at C will be obtained from eqn. (15.25) by replacing W with the end reaction and l by the length of the corresponding portion.

i.e. $$(\delta_s)_C = \frac{6 \times \left(\dfrac{Wb}{l}\right) a}{5 \, Cbd} = \frac{6Wab}{5 \, Cbdl}$$ ...(15.29)

Case V. Simply supported beam carrying uniformly distributed load:

Fig. 15.25 shows a simply supported beam carrying a uniformly distributed load. Consider a section XX at a distance x from the end A. The deflection at $x < l/2$ due to a point load (wdx) spread over a small length dx

$$= \frac{6(wdx)x(l-x)}{5 \, Cbdl}$$ ...(from eqn. 15.29)

Fig. 15.25

The deflection at the centre C due to (wdx), by proportion, will be

$$= \frac{6(wdx)\, x\, (l - x)}{5\,Cbdl} \cdot \frac{(l/2)}{(l - x)} = \frac{3\,wx\,dx}{5\,Cbd}$$

$\therefore$ Total deflection at C due to shear

$$(\delta_s)_C = 2 \int_0^{l/2} \frac{3\,wx\,dx}{5\,Cbd} = \frac{3\,wl^2}{20\,Cbd}$$

i.e.
$$(\delta_s)_C = \frac{3\,wl^2}{20\,Cbd} \qquad\qquad ...(15.30)$$

15.10. CASTIGLIANO'S THEOREM

Castigliano's theorem can be used in the following *cases*:

1. To determine the displacements of complicated structures.
2. To find the deflection of beams due to shearing or bending if the total strain energy due to shearing forces or bending moments (as the case may be) is known.
3. To find the deflections of curved beams, springs etc.

Castigliano's theorem is stated as follows:

"If U is the total strain energy of any structure due to the application of external loads W_1, W_2, W_3, ... W_n at points A_1, A_2, A_3 ...... A_n respectively in the direction AX_1, AX_2, AX_3 ......AX_n and due to couples M_1, M_2, ...M_m, at points B_1, B_2, B_3, ... B_m respectively then the deflections at the points A_1, A_2, A_3, ... A_n in the directions AX_1, AX_2, AX_3, ... AX_n are $\dfrac{\partial U}{\partial W_1}$, $\dfrac{\partial U}{\partial W_2}$, $\dfrac{\partial U}{\partial W_3}$, ... $\dfrac{\partial U}{\partial W_n}$, and the angular positions of the couples are $\dfrac{\partial U}{\partial M_1}$, $\dfrac{\partial U}{\partial M_2}$, $\dfrac{\partial U}{\partial M_3}$, ... $\dfrac{\partial U}{\partial M_n}$ at the respective points of application".

The *proof* for this theorem for point or concentrated load is given below :

Let, $\quad x_1, x_2, x_3 \ldots x_n$ = Displacements or deflections at the points A_1, A_2, A_3 ...... A_n in the direction of respective gradually applied loads,

and, $\quad \phi_1, \phi_2, \phi_3 \ldots\ldots \phi_m$ = Rotations due to respective gradually applied couples M_1, M_2, M_3 ... M_m at their points of application.

The total strain energy, $U = \left[\dfrac{1}{2}W_1\, x_1 + \dfrac{1}{2}W_2\, x_2 + \dfrac{1}{2}W_3\, x_3 + ... + \dfrac{1}{2}W_n\, x_n\right]$

$$+ \left[\dfrac{1}{2}M_1\, \phi_1 + \dfrac{1}{2}M_2\, \phi_2 + \dfrac{1}{2}M_3\, \phi_3 + ... + \dfrac{1}{2}M_m\, x_m\right] \qquad ...(15.29)$$

Now let the load W_1 alone be increased by δW_1 gradually and then let,

$\delta x_1, \delta x_3, \delta x_3 \ldots\ldots \delta x_n$ = Further deflections of the points A_1, A_2, A_3 ...... A_4 in respective direction of loads,

and, $\delta \phi_1, \delta \phi_2, \delta \phi_3 \ldots\ldots \delta \phi_n$ = Rotations of the points of application of respective couples.

Obviously as the loads W_2, W_3 ..... W_n and couples M_1, M_2, ... M_m, do not change and continue to act at their respective points of applications during the time this change in W_1 occurs, the increase in external work will then be

$$\delta U = \left[\left(W_1 + \dfrac{1}{2}\delta W_1\right)\delta x_1 + W_2\, \delta x_2 + W_3\, \delta x_3 + ...W_n\, \delta x_n\right]$$

$$+ \left[M_1\, \delta\phi_1 + M_2\, \delta\phi_2 + M_3\, \delta\phi_3 + ...M_m\, \delta\phi_m\right]$$

If the product of two small quantities is neglected the above equation reduces to

$$\delta U = [W_1\,\delta x_1 + W_2\,\delta x_2 + W_3\,\delta x_3 + \ldots\ldots W_n\,\delta x_n]$$
$$+[M_1\,\delta\phi_1 + M_2\,\delta\phi_2 + M_3\,\delta\phi_3 + \ldots\ldots M_m\,\delta\phi_m] \qquad \ldots(15.30)$$

Further if the loads $(W_1 + \delta W_1)$, W_2, W_3, ... W_n and couples M_1, M_2, M_3 ...... M_m were all applied gradually from zero, the total strain energy would have been $U + \delta U$ and the corresponding deflections and rotations would be respectively,

$$(x_1 + \delta x_1),\ (x_2 + \delta x_2),\ (x_3 + \delta x_3),\ \ldots\ (x_n + \delta x_n),\ \text{and}$$
$$(\phi_1 + \delta\phi_1),\ (\phi_2 + \delta\phi_2),\ (\phi_3 + \delta\phi_3),\ \ldots\ (\phi_m + \delta\phi_m)$$

The total strain energy $(U + \delta U)$ is given by,

$$(U + \delta U) = \left[\frac{1}{2}(W_1 + \delta W_1)(x_1 + \delta x_1) + \frac{1}{2}W_2(x_2 + \delta x_2) + \frac{1}{2}W_3(x_3 + \delta x_3) + \ldots\ldots\right.$$
$$\left. + \frac{1}{2}W_n(x_n + \delta x_n)\right]$$
$$+\left[\frac{1}{2}M_1(\phi_1 + \delta\phi_1) + \frac{1}{2}M_2(\phi_2 + \delta\phi_2) + \frac{1}{2}M_3(\phi_3 + \delta\phi_3) + \ldots\ldots\right.$$
$$\left. + \frac{1}{2}M_m(\phi_m + \delta\phi_m)\right] \qquad \ldots(15.31)$$

Subtracting eqn. (15.29) from (15.31), we get

$$\delta U = \frac{1}{2}x_1\,\delta W_1 + \left[\frac{1}{2}W_1\,\delta x_1 + \frac{1}{2}W_2\,\delta x_2 + \frac{1}{2}W_3\,\delta x_3 \ldots\ldots \frac{1}{2}W_n\,\delta x_n\right]$$
$$+\left[\frac{1}{2}M_1\,\delta\phi_1 + \frac{1}{2}M_2\,\delta\phi_2 + \frac{1}{2}M_3\,\delta\phi_3 + \ldots\ldots \frac{1}{2}W_m\,\delta\phi_m\right]$$

or

$$2\delta U = x_1\,\delta W_1 + [W_1\,\delta x_1 + W_2\,dx_2 + W_3\,\delta x_3 + \ldots\ldots W_n\,\delta x_n]$$
$$+[M_1\,\delta\phi_1 + M_2\,\delta\phi_2 + M_3\,\delta\phi_3 + \ldots\ldots M_m\,\delta\phi_m] \qquad \ldots(15.32)$$

Subtracting eqn. (15.30) from eqn. (15.32), we get

$$\delta U = x_1\,\delta W_1 \qquad \text{or} \qquad \frac{\delta U}{\delta W_1} = x_1$$

When $\delta W_1 \to 0$ (in the limiting case), we get

$$\frac{\partial U}{\partial W_1} = x_1 \qquad \ldots(15.33)$$

Similarly

$$\frac{\partial U}{\partial W_2} = x_2,\ \frac{\partial U}{\partial W_3} = x_3,\ \ldots\ldots\ \frac{\partial U}{\partial W_n} = x_n \qquad \ldots(15.34)$$

If we proceed in the same way, for the couples we shall get

$$\frac{\partial U}{\partial M_1} = \phi_1,\ \frac{\partial U}{\partial M_2} = \phi_2,\ \frac{\partial U}{\partial M_m} = \phi_m \qquad \ldots(15.35)$$

Points to remember while applying Castigliano's theorem:

1. (a) Treat all the loads and couples/moments as variables and carry out the partial differentiation.

   (b) Substitute the numerical values of different loads and couples in the above equation.

2. To find out the deflection or rotation at a point of the structure where there is no load or couple acting, then it may be assumed that a dummy load W or dummy moment/couple is acting at that point and give a value zero at the end.

i.e.
$$x = \left(\frac{\partial U}{\partial W}\right)_{W=0}, \quad \text{and,} \quad \phi = \left(\frac{\partial U}{\partial M}\right)_{M=0}$$

(i) Deflection under axial load :

Strain energy under axial load W,

$$U = \int \frac{1}{2}\frac{W^2\,dx}{AE}$$

$\therefore$
$$\delta = \frac{\partial U}{\partial W} = \int \frac{W\,dx}{AE} \qquad\qquad ...(15.36)$$

(ii) Deflection under bending:

Strain energy under bending moment M,

$$U = \int \frac{M^2\,dx}{2EI}$$

$\therefore$
$$\delta = \frac{\partial U}{\partial W} = \frac{\partial U}{\partial M} \times \frac{\partial M}{\partial W} = \int \frac{M}{EI}\,dx \times \frac{\partial M}{\partial W} = \int \frac{M}{EI}\frac{\partial M}{\partial W}\,dx \qquad ...(15.37)$$

(iii) Deflection under forsion:

Strain energy under torque T,

$$U = \int \frac{T^2\,dx}{2CI_p}$$

$\therefore$
$$\delta = \frac{\partial U}{\partial W} = \frac{\partial U}{\partial T} \times \frac{\partial T}{\partial W} = \int \frac{T\,dx}{CI_p}\frac{\partial T}{\partial W} = \int \frac{T}{CI_p}\frac{\partial T}{\partial W}\,dx \qquad ...(15.38)$$

(iv) Deflection under shear :

Strain energy under shear free F,

$$U = \int \frac{F^2\,dx}{2AC}$$

$\therefore$
$$\delta = \frac{\partial U}{\partial F} = \int \frac{F\,dx}{AC}$$

Aircraft wheels are fitted with landing gears specially made to absorb and gradually release shock and strain energies.

(v) Deflection under horizontal shear :

$$U = \left(\frac{1 + \dfrac{1}{m}}{E I^2}\right) \int F^2 \, dx \iint \frac{(A\bar{y})^2}{b^2} \, dA$$

where, F is a function of W.

$$\therefore \qquad \delta = \frac{\partial U}{\partial W} \qquad \qquad \qquad ...(15.39)$$

(vi) Rotation under bending :

$$U = \int \frac{M^2}{2 E I}$$

$$\therefore \qquad \phi = \frac{\partial U}{\partial M} = \int \frac{M \, dx}{E I} \qquad \qquad \qquad ...(15.40)$$

(vii) Rotation under torsion :

$$U = \int \frac{T^2 \, dx}{2 C I_p}$$

$$\therefore \qquad \theta = \frac{\partial U}{\partial T} = \int \frac{T \, dx}{C I_p} \qquad \qquad \qquad ...(15.41)$$

Example 15.27. *Using castigliano's theorem, obtain the deflection under a single concentrated load applied to a simply supported beam shown in Fig. 15.26. EI = 2.2 MNm².*

Solution. Refer to Fig. 15.26. Let the load at C be denoted by W. Taking moments about B, we get

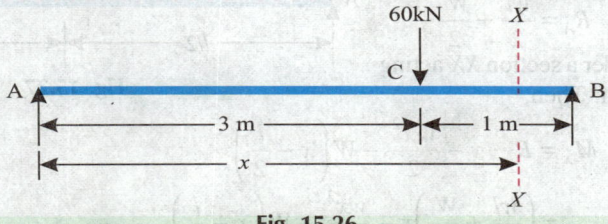

Fig. 15.26

$$R_A \times 4 = W \times 1$$

or, $$R_A = \frac{W}{4}$$

Consider a section XX at a distance x from A.

$$M_x = R_A \times x - W \times (x - 3) = \frac{Wx}{4} - W(x - 3)$$

$$\frac{\partial M}{\partial W} = \frac{x}{4} - (x - 3)$$

Now, $$\delta = \int \frac{M}{EI} \cdot \frac{\partial M}{\partial W} \, dx \qquad \qquad ...[Eqn. 15.37]$$

$$= \frac{1}{EI} \int_0^3 \frac{Wx}{4} \times \frac{x}{4} \, dx + \frac{1}{EI} \int_3^4 \left\{\frac{Wx}{4} - W(x - 3)\right\} \times \left\{\frac{x}{4} - (x - 3)\right\} dx$$

$$= \frac{W}{16 EI} \int_0^3 x^2 \, dx + \frac{W}{EI} \int_3^4 \left\{\left(\frac{x}{4} - (x - 3)\right)\right\}^2 dx$$

$$= \frac{W}{16\,EI} \int_0^3 x^2\, dx + \frac{W}{EI} \int_3^4 \left(\frac{x - 4x + 12}{4} \right)^2 dx$$

$$= \frac{W}{16\,EI} \int_0^3 x^2\, dx + \frac{9W}{16\,EI} \int_3^4 (x^2 - 8x + 16)\, dx$$

$$= \frac{W}{16\,EI} \left| \frac{x^3}{3} \right|_0^3 + \frac{9W}{16\,EI} \left| \frac{x^3}{3} - \frac{8x^2}{2} + 16x \right|_3^4$$

$$= \frac{9W}{16\,EI} + \frac{9W}{16\,EI} (12.33 - 28 + 16)$$

$$= \frac{9W}{16\,EI} (1 + 0.33) = \frac{0.75W}{EI}$$

$$\therefore \qquad \delta = \frac{0.75 \times 60 \times 10^3}{2.2 \times 10^6} = 0.02045\,\text{m} \text{ or } \mathbf{20.45\ mm\ (Ans)}$$

Example 15.28. *Using Castigliano's theorem, calculate the vertical deflection at the middle of a simply supported beam which carries a uniformly distributed load of intensity w over the full span. The flexural rigidity EI of the beam is constant and only strain energy of bending is to be considered.* **(IAS)**

Solution. Refer to fig. 15.27. Consider a dummy load W acting at the centre of the beam.

$$R_A = \frac{wl}{2} + \frac{W}{2}$$

Further, consider a section XX acting at a distance x from A. Then,

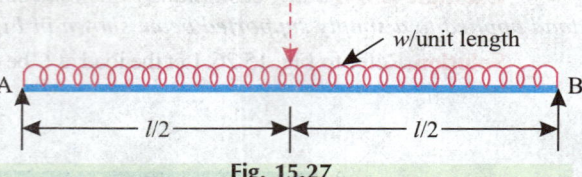

Fig. 15.27

$$M_x = R_A \cdot x - \frac{wx^2}{2} - W \left(x - \frac{l}{2} \right)$$

$$= \left(\frac{wl}{2} + \frac{W}{2} \right) x - \frac{wx^2}{2} - W \left(x - \frac{l}{2} \right)$$

$$= \frac{wlx}{2} + \frac{Wx}{2} - \frac{wx^2}{2} - W \left(x - \frac{l}{2} \right)$$

For, $\qquad x \le l/2$

$$M_x = \frac{wlx}{2} + \frac{Wx}{2} - \frac{wx^2}{2}$$

$$\frac{\partial M_x}{\partial W} = \frac{x}{2}$$

For, $\qquad \dfrac{l}{2} \le x \le l$

$$\frac{\partial M_x}{\partial W} = \frac{x}{2} - x + \frac{l}{2} = \frac{l}{2} - \frac{x}{2} = \frac{l - x}{2}$$

$$\delta = \frac{1}{EI} \int_0^{l/2} \left(\frac{wlx}{2} + \frac{Wx}{2} - \frac{wx^2}{2} \right) \frac{x}{2}\, dx$$

$$+ \frac{1}{EI} \int_{l/2}^{l} \left[\frac{wlx}{2} + \frac{Wx}{2} - \frac{wx^2}{2} - W \left(x - \frac{l}{2} \right) \right] \left[\left(\frac{l - x}{2} \right) \right] dx$$

Putting $W = 0$, we get

$$\delta = \frac{1}{EI} \int_0^{l/2} \left(\frac{wlx^2}{4} - \frac{wx^3}{4} \right) dx + \frac{1}{EI} \int_{l/2}^l \left(\frac{wlx}{2} - \frac{wx^2}{2} \right) \left(\frac{l-x}{2} \right) dx$$

$$= \frac{w}{4EI} \int_0^{l/2} (lx^2 - x^3)\, dx + \frac{w}{4EI} \int_{l/2}^l (l^2x - 2lx^2 + x^3)\, dx$$

$$= \frac{w}{4EI} \left| \frac{lx^3}{3} - \frac{x^4}{4} \right|_0^{l/2} + \frac{w}{4EI} \left| \frac{l^2x^2}{2} - \frac{2}{3}lx^3 + \frac{x^4}{4} \right|_{l/2}^l$$

$$= \frac{w}{4EI} \left[\left(\frac{l^4}{24} - \frac{l^4}{64} \right) + \frac{l^2}{2} \left(l^2 - \frac{l^2}{4} \right) - \frac{2l}{3} \left\{ l^3 - \left(\frac{l}{2} \right)^3 \right\} + \frac{1}{4} \left\{ l^4 - \left(\frac{l}{2} \right)^4 \right\} \right]$$

$$= \frac{w}{4EI} \left[\frac{l^4}{24} - \frac{l^4}{64} + \frac{3l^4}{8} - \frac{7l^4}{12} + \frac{15l^4}{64} \right]$$

$$= \frac{wl^4}{4EI} (8 - 3 + 72 - 112 + 45)$$

$$= \frac{wl^4}{4EI} \times \frac{10}{192} = \frac{5wl^4}{384EI}$$

i.e. $\delta = \dfrac{5wl^4}{384EI}$ **(Ans.)**

Example 15.29. *A beam simply supported over a span of 3m carries a uniformly distributed load of 20 kN/m over the entire span. Taking EI = 2.25 MNm² and using Castiglian's theorem determine the deflection at the centre of the beam.*

Solution. *Given*: Span length, $l = 3$m; $w = 20$ kN/m; $EI = 2.25$ MNm².

Deflection at the centre, δ:

Refer Fig. 15.28

Let W (kN) be the dummy load at the mid-span.

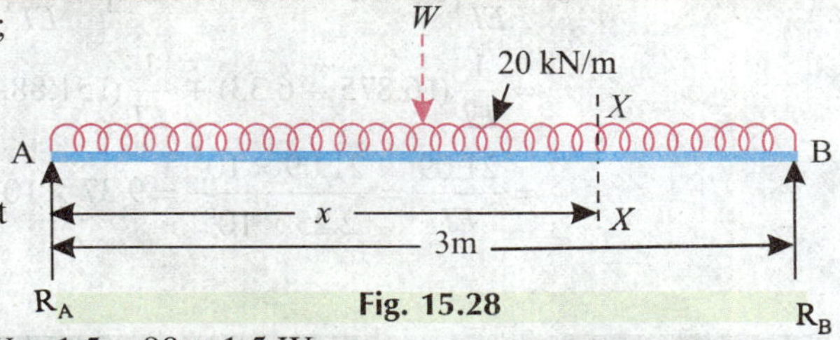

Taking moments about B, we get

$$R_A \times 3 = 20 \times 3 \times 1.5 + W \times 1.5 = 90 + 1.5\,W$$

$$R_A = (30 + 0.5\,W) \text{ kN}$$

Fig. 15.28

Consider a section XX at a distance x from A. Then,

$$M_x = (30 + 0.5\,W)x - 20 \times x \times \frac{x}{2} - W(x - 1.5)$$

$$\frac{\partial M_x}{\partial W} = 0.5x - (x - 1.5)$$

Deflection, $\delta = \displaystyle\int \frac{M}{EI} \frac{\partial M}{\partial W}\, dx = \frac{1}{EI} \int_0^{1.5} \left\{ (30 + 0.5\,W)x - 10x^2 \right\} \times 0.5x\, dx$

$$+ \frac{1}{EI} \int_{1.5}^3 \left\{ (30 + 0.5\,W)x - 10x^2 - W(x - 1.5) \right\} \times \left\{ 0.5x - (x - 1.5) \right\} dx$$

$$= \frac{1}{EI} \int_0^{1.5} \left\{ (15 + 0.25\,W)x^2 - 5x^3 \right\} dx$$

$$+ \frac{1}{EI} \int_{1.5}^3 \left\{ (30x + 0.5\,Wx - 10x^2 - Wx + 1.5\,W)(0.5x - x + 1.5)\, dx \right\}$$

$$= \frac{1}{EI} \int_0^{1.5} \left\{ (15 + 0.25\,W)\,x^2 - 5x^3 \right\} dx$$

$$+ \frac{1}{EI} \int_{1.5}^{3} \left\{ (30 - 0.5\,W)\,x - 10x^2 + 1.5\,W \right\} \left\{ 1.5 - 0.5\,x \right\} dx$$

$$= \frac{1}{EI} \int_0^{1.5} \left\{ (15 + 0.25\,W)\,x^2 - 5x^3 \right\} dx + \frac{1}{EI} \int_{1.5}^{3} \left\{ (45 - 0.75\,W)\,x - 15x^2 \right.$$

$$\left. + 2.25\,W - (15 - 0.25\,W)x^2 + 5x^3 - 0.75\,Wx \right\} dx$$

$$= \frac{1}{EI} \int_0^{1.5} \left\{ (15 + 0.25\,W)\,x^2 - 5x^3 \right\} dx$$

$$+ \frac{1}{EI} \int_{1.5}^{3} \left\{ (45 - 1.5\,W)\,x + (0.25\,W - 30)\,x^2 + 5x^3 + 2.25\,W \right\} dx$$

$$= \frac{1}{EI} \left[\left| (15 + 0.25\,W) \frac{x^3}{3} - \frac{5}{4} x^4 \right|_0^{1.5} \right.$$

$$\left. + \frac{1}{EI} \left| (45 - 1.5\,W) \frac{x^2}{2} + (0.25\,W - 30) \frac{x^3}{3} + \frac{5}{4} x^4 + 2.25\,Wx \right|_{1.5}^{3} \right.$$

Putting $W = 0$, we get

$$\delta = \frac{1}{EI} \left[5 \times 1.5^3 - 1.25 \times 1.5^4 \right] + \frac{1}{EI} \left[22.5 \times 6.75 - 10 \times 23.625 + 1.25 \times 75.94 \right]$$

$$= \frac{1}{EI} (16.875 - 6.33) + \frac{1}{EI} (151.88 - 236.25 + 94.92)$$

$$= \frac{21.09}{EI} = \frac{21.09 \times 10^3}{2.25 \times 10^6} = 9.37 \times 10^{-3} = \textbf{9.37 mm} \quad \textbf{(Ans.)}$$

A rail track maintenance machine. Even slight distortions in the rail tracks, due to stresses and strains, need to be rectified immediate.

Example 15.30. *Using Castigliano's theorem, determine the deflection of the free end of the cantileve beam shown in Fg. 15.29. Take EI = 4.9 MNm².*

Solution. Refer to Fig. 15.29 Apply dummy load W at B.

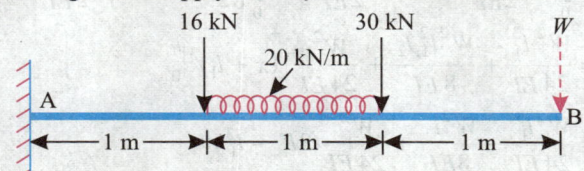

Fig. 15.29

Consider a section XX at a distance x from B. Then

$$M_x = Wx + 30(x-1) + 20 \times 1 \times (x-1.5) + 16(x-2)$$

$$\frac{\partial M_x}{\partial W} = x$$

$$\delta = \int \frac{M}{EI} \frac{\partial M}{\partial W} \, dx$$

$$= \frac{1}{EI} \left[\int_0^1 Wx \times x \, dx + \int_1^2 \left\{ Wx \times x + 30(x-1) \times x + 20(x-1) \times \frac{(x-1)}{2} \times x \right\} dx \right.$$

$$+ \int_2^3 \left\{ Wx \times x + 30(x-1) \times x + 20 \times 1(x-1.5) \times x + 16(x-2) \times x \right\} dx \bigg]$$

$$= \frac{1}{EI} \left[W \left| \frac{x^3}{3} \right|_0^1 + \left| \frac{Wx^3}{3} + 30 \left(\frac{x^3}{3} - \frac{x^2}{2} \right) + 10 \left(\frac{x^4}{4} - \frac{2x^3}{3} + \frac{x^2}{2} \right) \right|_1^2 \right.$$

$$+ \left| \frac{Wx^3}{3} + 30 \left(\frac{x^3}{3} - \frac{x^2}{2} \right) + 20 \left(\frac{x^3}{3} - 0.75 x^2 \right) + 16 \left(\frac{x^3}{3} - x^2 \right) \right|_2^3 \bigg]$$

Putting $W = 0$, we get

$$\delta = \frac{1}{EI} \left[30 \left(\frac{7}{3} - \frac{3}{2} \right) + 10 \left(\frac{15}{4} - \frac{14}{3} + \frac{3}{2} \right) + 30 \left(\frac{19}{3} - \frac{5}{2} \right) \right.$$

$$+ 20 \left(\frac{19}{3} - 3.75 \right) + 16 \left(\frac{19}{3} - 5 \right) \bigg]$$

$$= \frac{1}{EI} \left(30 \times \frac{5}{6} + 10 \times \frac{7}{12} + 30 \times \frac{23}{6} + 20 \times 2.58 + 16 \times \frac{4}{3} \right)$$

$$= \frac{1 \times 10^3}{4.9 \times 10^6} (25 + 5.83 + 115 + 51.6 + 21.33) = 0.446 \, \text{m or } 44.64 \, \text{mm}$$

i.e. $\delta = \textbf{44.64 mm (Ans.)}$

Example 15.31. *Using Castigliano's theorem find the central deflection of the uniform bend shown in Fig. 15.30.*

Solution. Refer to Fig. 15.30.
Strain energy for the portion $ABCG$,

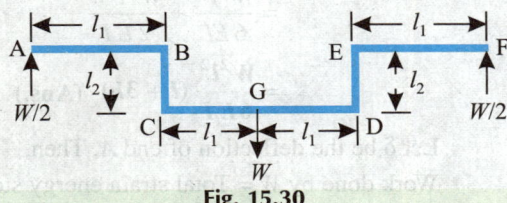

Fig. 15.30

$$U_{ABCG} = \int_0^{l_1} \frac{\left(\dfrac{W}{2}x\right)^2 dx}{2EI} + \int_0^{l_2} \frac{\left(\dfrac{W}{2}l_1\right)^2}{2EI} + \int_0^{l_1} \frac{\left\{\dfrac{W}{2}(x+l_1)\right\}^2 dx}{2EI}$$

$$= \frac{W^2 l_1^3}{24\,EI} + \frac{W^2 l_1^2 l_2}{8\,EI} + \frac{W^2}{24\,EI}\Big|(x+l_1)^3\Big|_0^{l_1}$$

$$= \frac{W^2 l_1^3}{24\,EI} + \frac{W^2 l_2}{8EI} + \frac{W^2}{24\,EI}(8l_1^3 - l_1^3)$$

$$= \frac{W^2 l_1^3}{24\,EI} + \frac{W^2 l_1^2 l_2}{8\,EI} + \frac{7W^2 l_1^3}{24\,EI}$$

$$= \frac{W^2 l_1^2}{24\,EI}(l_1 + 3l_2 + 7l_1)$$

$$= \frac{W^2 l_1^2}{24\,EI}(8l_1 + 3l_2)$$

Total strain energy

$$= 2 \times U_{ABCG} = 2 \times \frac{W^2 l_1^2}{24\,EI}(8l_1 + 3l_2) = \frac{W^2 l_1^2}{12\,EI}(8l_1 + 3l_2)$$

Vertical deflection of G $= \dfrac{\partial U}{\partial W}$

$$= \frac{Wa^2}{6\,EI}(8l_1 + 3l_2) \quad \textbf{(Ans.)}$$

Example 15.32. *In the Fig. 15.31. is shown a structure. Assuming the member to be of uniform cross-section throughout find the strain energy stored by the structure and hence determine the vertical deflection of end A.*

Solution. Refer to Fig. 15.31.

Section AB :

$$M_x = W \cdot x$$

$$U_{AB} = \int_0^l \frac{M_x^2\, dx}{2\,EI} = \int_0^l \frac{W^2 x^2 dx}{2\,EI} = \frac{W^2 l^3}{6\,EI}$$

Section BC :

$$M_y = W \cdot l$$

$$U_{BC} = \int_0^h \frac{M_y^2\, dy}{2\,EI} = \int_0^h \frac{W^2 l^2\, dy}{2\,EI} = \frac{W^2 l^2 h}{2\,EI}$$

Total strain energy,

$$U = U_{AB} + U_{BC}$$

$$= \frac{W^2 l^3}{6\,EI} + \frac{W^2 l^2 h}{2\,EI}$$

$$= \frac{W^2 l^2}{6\,EI}(l + 3h) \quad \textbf{(Ans.)}$$

Let δ be the deflection of end A. Then,

Work done by W = Total strain energy stored

$$\frac{1}{2}W\delta = \frac{W^2 l^2}{6\,EI}(l + 3h)$$

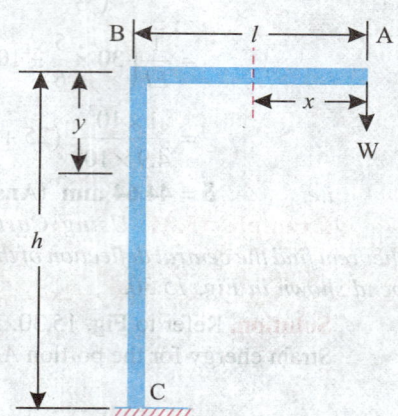

Fig. 15.31

or, $$\delta = \frac{Wl^2}{3EI}(l + 3h) \quad \text{(Ans.)}$$

Example 15.33. *A shaft is supported by two anti-friction beanings with loads of 140 N each actiong at B and D as shown in Fig. 15.32. The portion of shaft between B and C has a diameter of 2d compared to a diameter d for the portion of the shaft between A and B; and between C and D. Using the Castigliano's theorem, determine the deflection of shaft at points B and D.*

(IAS)

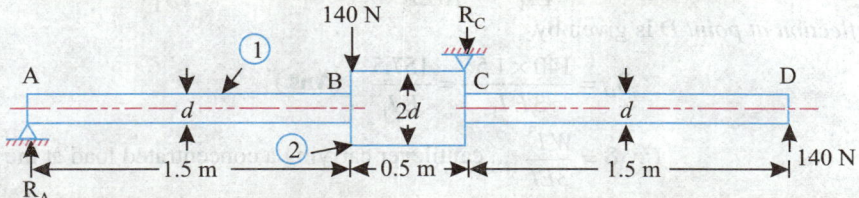

Fig. 15.32

Solution. $$I_1 = \frac{\pi d^4}{64}; \ I_2 = \frac{\pi}{64}(2d)^4 = \frac{\pi d^4}{64} \times 16 = 16 I_1$$

Let W_1 and W_2 be the loads acting at points B and D respectively.

Then, $\qquad R_A \times 2 = W_1 \times 0.5 + W_2 \times 1.5$

or $\qquad R_A = 0.25 \ W_1 + 0.75 \ W_2$

Also, $\qquad R_A + W_2 = W_1 + R_C$

or $\qquad R_A - R_C = W_1 - W_2$

∴ $\qquad R_C = R_A - W_1 + W_2 = 0.25 \ W_1 + 0.75 \ W_2 - W_1 + W_2$

$\qquad\qquad = 1.75 \ W_2 - 0.75 \ W_1$

Consider a section XX at a distance x from A. Then,

$$M_x = (0.25 \ W_1 + 0.75 \ W_2)x - W_1(x - 1.5) - (1.75 \ W_2 - 0.75 \ W_1)(x - 2)$$

For $x < 1.5$ m, we have

$$M_x = (0.25 \ W_1 + 0.75 \ W_2) \ x$$

$$\frac{\partial M_x}{\partial W_1} = 0.25 \ x$$

For $1.5m \le x \le 2m$

$$M_x = (0.25 \ W_1 + 0.75 \ W_2)x - W_1 \ (x - 1.5)$$

$$\frac{\partial M_x}{\partial W_1} = 0.25 \ x - x + 1.5 = 1.5 - 0.75 \ x$$

Now, *deflection at B* is given by,

$$\delta_B = \frac{1}{EI_1} \int_0^{1.5} (0.25W_1 + 0.75W_2) \ x \times 0.25 \ x \ dx$$

$$+ \frac{1}{EI_2} \int_{1.5}^2 \left[(0.25W_1 + 0.75W_2)x - W_1 (x - 1.5) \right] (1.5 - 0.75 x) dx$$

Putting $\qquad W_1 = W_2 = 140$ N, we get

$$\delta_B = \frac{1}{EI_1} \int_0^{1.5} 35 x^2 \ dx + \frac{1}{EI_2} \int_{1.5}^2 210(1.5 - 0.75 x) dx$$

$$= \frac{35}{EI_1} \left| \frac{x^3}{3} \right|_0^{1.5} + \frac{210}{EI_2} \left| 1.5x - 0.375x^2 \right|_{1.5}^2$$

$$= \frac{39.375}{EI_1} + \frac{210}{16EI_1} \left[1.5(2 - 1.5) - 0.375(2^2 - 1.5^2) \right]$$

$$(\because I_2 = 16 I_1 \dots \text{calculated earlier})$$

$$= \frac{39.375}{EI_1} + \frac{210}{16EI_1} [0.75 - 0.656] = \frac{40.61}{EI_1} \quad \textbf{(Ans.)}$$

Deflection at point D is given by,

$$\delta_D = \frac{140 \times 1.5^3}{3EI_1} = \frac{157.5}{EI_1} \quad \textbf{(Ans.)}$$

$$\left(\because \delta = \frac{Wl^3}{3EI} \dots \text{cantilever carrying a concentrated load at the free end} \right)$$

Example 15.34. *A beam of length l is simply supported at its ends and carries a concentrated load W at its centre. It has a rectangular cross-section breadth b, depth d. If C is the modulus of rigidity for the beam find:*

(*i*) *Shear strain energy in the beam;*

(*ii*) *Deflection due to shear.*

Solution. Fig. 15.33 (*a*) shows loaded beam simply supported at *A* and *B*.

Fig. 15.33 (*b*) shows the S.F. diagram for the beam.

Fig. 15.33 (*c*) shows the cross-section of the beam.

Shear force at a distance x from the end B, $S_x = + W/2$

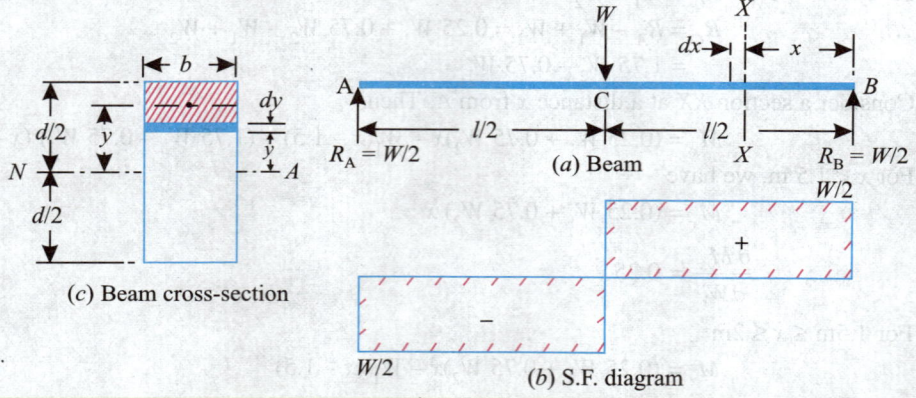

(*c*) Beam cross-section

(*a*) Beam

(*b*) S.F. diagram

Fig. 15.33

(*i*) **Shear strain energy in the beam, U_s:**

Consider a layer of thickness dy at a distance y from the neutral axis [Fig. 15.33 (*c*)].

Shear stress at the layer, $\tau = \dfrac{S A \bar{y}}{Ib}$

where, $S =$ Shear force at the section $= \dfrac{W}{2}$,

$A\bar{y} =$ First moment of area above the layer about neutral axis, and

$I =$ Moment of inertia $= \dfrac{bd^3}{12}$.

∴ Shear stress,

$$\tau = \frac{\frac{W}{2} \times b\left(\frac{d}{2} - y\right) \times \left(y + \frac{d/2 - y}{2}\right)}{\frac{bd^3}{12} \times b}$$

$$= \frac{6W}{bd^3}\left(\frac{d}{2} - y\right)\left(\frac{d/2 + y}{2}\right) = \frac{3W}{bd^3}\left(\frac{d^2}{4} - y^2\right)$$

Volume of the layer = $b \cdot dx \cdot dy$

Shear strain energy in the layer

$$= \frac{\tau^2}{2C} b \cdot dx \cdot dy = \frac{1}{2C} \cdot b\left[\frac{3W}{bd^3}\left(\frac{d^2}{4} - y^2\right)\right]^2 dy \cdot dx$$

$$= \frac{1}{2C} b \times \frac{9W^2}{b^2 d^6}\left(\frac{d^4}{16} + y^4 - \frac{d^2 y^2}{2}\right)dy \cdot dx$$

Total shear strain energy for the portion BC,

$$(U_s)_{BC} = \frac{1}{2C} \int_0^{l/2} \int_{-d/2}^{+d/2} \frac{9W^2}{bd^6}\left(\frac{d^4}{16} + y^4 - \frac{d^2 y^2}{2}\right)dy \cdot dx$$

$$= \frac{1}{2C} \int_0^{l/2} \frac{9W^2}{bd^6}\left[\frac{d^4 y}{16} + \frac{y^5}{5} - \frac{d^2 y^3}{6}\right]_{-d/2}^{+d/2} dx$$

$$= \frac{1}{2C} \int_0^{l/2} \frac{9W^2}{bd^6}\left[\left(\frac{d^5}{32} + \frac{d^5}{32}\right) + \left(\frac{d^5}{160} + \frac{d^5}{160}\right) - \left(\frac{d^5}{48} + \frac{d^5}{48}\right)\right]dx$$

$$= \frac{1}{2C} \int_0^{l/2} \frac{9W^2}{bd^6}\left(\frac{d^5}{16} + \frac{d^5}{80} - \frac{d^5}{24}\right)dx = \frac{1}{2C} \int_0^{l/2}\left(\frac{9W^2}{bd^6} \times \frac{d^5}{30}\right)dx$$

$$= \frac{1}{2C} \int_0^{l/2} \frac{9W^2}{30bd} dx = \frac{1}{2C} \times \frac{9W^2}{30bd} \times l/2 = \frac{3W^2 l}{40Cbd}$$

Similarly, $(U_s)_{CA} = \dfrac{3W^2 l}{40Cbd}$ [∵ Shear force in portion CA is also W/2 (–)]

∴ Shear energy for the beam,

$$U_s = (U_s)_{BC} + (U_s)_{CA} = 2 \times \frac{3W^2 l}{40Cbd} = \frac{3W^2 l}{20Cbd}$$

Hence, $U_s = \dfrac{3W^2 l}{20Cbd}$ **(Ans.)**

(ii) Deflection due to shear, δ_s :

$$\delta_s = \frac{\partial U_s}{\partial W} = \frac{\partial}{\partial W}\left(\frac{3W^2 l}{20Cbd}\right) = \frac{3Wl}{10Cbd}$$

Hence, $\delta_s = \dfrac{3Wl}{10Cbd}$ **(Ans.)**

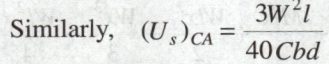
Crankshaft of an aircraft engine.

Example 15.35. *For the cantilever shown in the Fig. 15.34 determine:*

(i) Deflection at the free end;   *(ii) Slope at the free end.*

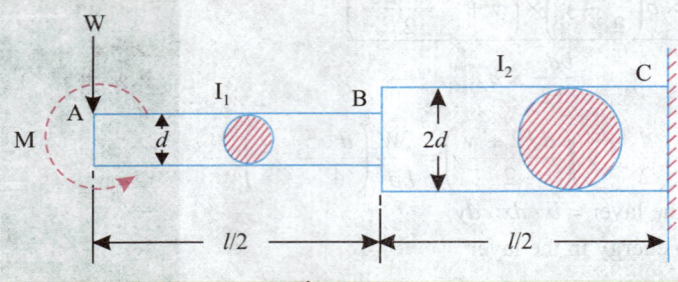

Fig. 15.34

Solution. As we have to find out the slope at the free end A, let us apply a *fictitious moment* $M = 0$ at the free end.

(i) Deflection at the free end, δ_A:

For portion AB: $M_x = M + Wx$

where, $x = 0$ to $l/2$ (with A as origin)

For portion BC: $M_x = M + W(x + l/2)$

where, $x = 0$ to $l/2$ (with B as origin)

Strain energy, $U = \int\limits_0^{l/2} \dfrac{(M + Wx)^2}{2EI_1}\,dx + \int\limits_0^{l/2} \dfrac{[M + W(x + l/2)]^2}{2EI_2}\,dx$

where, $I_1 = \dfrac{\pi d^4}{64},\ I_2 = \dfrac{\pi(2d)^4}{60} = \dfrac{\pi d^4}{4}$

and, $E = $ Young's modulus of elasticity.

Deflection at A, $\delta_A = \left(\dfrac{\partial U}{\partial W}\right)_{M=0}$

$$= \int\limits_0^{l/2} \dfrac{(M + Wx)x}{EI_1}\,dx + \int\limits_0^{l/2} \left(\dfrac{M + W(x + l/2)}{EI_2}\right) \times (x + l/2)\,dx$$

$$= \dfrac{1}{EI_1}\left[\dfrac{Mx^2}{2} + \dfrac{Wx^3}{3}\right]_0^{l/2} + \dfrac{1}{EI_2}\left[\dfrac{Mx^2}{2} + \dfrac{Mlx}{2} + \dfrac{Wlx^2}{2} + \dfrac{Wx^3}{3} + \dfrac{Wl^2x}{4}\right]_0^{l/2}$$

$$= \dfrac{1}{EI_1} \times \dfrac{Wl^3}{24} + \dfrac{1}{EI_2}\left(\dfrac{Wl^3}{8} + \dfrac{Wl^3}{24} + \dfrac{Wl^3}{8}\right) \qquad [\because M = 0]$$

$$= \dfrac{Wl^3}{24EI_1} + \dfrac{7Wl^3}{24EI_2} = \dfrac{Wl^3}{24E}\left(\dfrac{1}{I_1} + \dfrac{7}{I_2}\right)$$

Substituting the values of I_1 and I_2, we get

$$\delta_A = \dfrac{Wl^3}{24E}\left(\dfrac{64}{\pi d^4} + \dfrac{7 \times 4}{\pi d^4}\right) = \dfrac{Wl^3}{24E} \times \dfrac{92}{\pi d^4} = \dfrac{23Wl^3}{6E\pi d^4}$$

Hence, $\delta_A = \dfrac{23Wl^3}{6E\pi d^4}$ **(Ans.)**

(ii) Slope at the free end, φ :

$$\phi = \frac{\partial U}{\partial M} = \int_0^{l/2} \frac{M + Wx}{EI_1} \times (1)\, dx + \int_0^{l/2} \frac{M + W(x + l/2)}{EI_2} \times (1)\, dx$$

$$= \frac{1}{EI_1}\left[Mx + \frac{Wx^2}{2} \right]_0^{l/2} + \frac{1}{EI_2}\left[Mx + \frac{Wx^2}{2} + \frac{Wlx}{2} \right]_0^{l/2}$$

$$= \frac{1}{EI_1} \times \frac{Wl^2}{8} + \frac{1}{EI_2} \times \left(\frac{Wl^2}{8} + \frac{Wl^2}{4} \right) \qquad\qquad [\because M = 0]$$

$$= \frac{Wl^2}{8\,EI_1} + \frac{3Wl^2}{8\,EI_2} = \frac{Wl^2}{8E}\left(\frac{1}{I_1} + \frac{3}{I_2} \right)$$

Substituting the values of I_1 and I_2, we get

$$\phi = \frac{Wl^2}{8E}\left(\frac{64}{\pi d^4} + \frac{3 \times 4}{\pi d^4} \right) = \frac{Wl^2}{8E} \times \frac{76}{\pi d^4} = \frac{19 Wl^2}{2\,E\,\pi d^4}$$

Hence, $$\phi = \frac{\mathbf{19}\,\mathbf{W}\mathbf{l^2}}{\mathbf{2}\,\mathbf{E}\,\mathbf{\pi}\mathbf{d^4}} \quad \textbf{(Ans.)}$$

Example 15.36. *A cantilever of rectangular section breadth b, depth d and of length l carries uniformly distributed load spread from free end to the mid section of the cantilever. Using Castigliano's theorem find :*

(i) *Deflection due to shear at the free end;*

(ii) *Deflection due to bending at the free end.* **(Panjab University)**

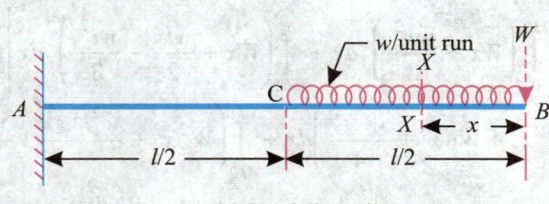

(a) Loaded cantilever $\qquad\qquad$ (b) Cross-section

Fig. 15.35

Solution. (i) **Deflection due to shear at the free end B, $(\delta_s)_B$:**

Let us assume that a point load W acts at B.

The shearing force (S) at any section XX distant x from B is as follows:

$$S = W + wx \qquad \text{For} \quad 0 < x < l/2$$

$$S = W + \frac{wl}{2} \qquad \text{For} \quad l/2 < x < l$$

The strain energy of cantilever due to shearing (using eqn. 15.23) is

$$U_s = \frac{3}{5Cbd}\left[\int_0^{l/2} (W + wx)^2\, dx + \int_{l/2}^{l} \left(W + \frac{wl}{2} \right)^2 dx \right]$$

or, $$\frac{5C\,bd}{3} U_s = \left[\frac{(W + wx)^3}{3w} \right]_0^{l/2} + \left(W + \frac{wl}{2} \right)^2 [x]_{l/2}^{l}$$

$$= \frac{1}{3w}\left[\left(W + \frac{wl}{2} \right)^3 - W^3 \right] + \left(W + \frac{wl}{2} \right)^2 \frac{l}{2}$$

Differentiating, U_s with respect to W partially, we get

$$\frac{5Cbd}{3}\left(\frac{\partial U_s}{\partial W}\right) = \frac{1}{3w}\left[3\left(W + \frac{wl}{2}\right)^2 - 3W^2\right] + 2\left(W + \frac{wl}{2}\right)\frac{l}{2}$$

Equating $W = 0$, we get

$$\frac{5Cbd}{3}\left(\frac{\partial U_s}{\partial W}\right)_{W=0} = \frac{1}{3w}\left(\frac{3w^2l^2}{4}\right) + \frac{wl^2}{2} = \frac{wl^2}{4} + \frac{wl^2}{2} = \frac{3wl^2}{4}$$

i.e.
$$(\delta_s)_B = \left(\frac{\partial U_s}{\partial W}\right)_{W=0} = \frac{3wl^2}{4} \times \frac{3}{5Cbd} = \frac{9wl^2}{20Cbd}$$

Hence,
$$(\delta_s)_B = \frac{9wl^2}{20Cbd} \quad \textbf{(Ans.)}$$

(ii) Deflection due to bending at the free end, δ_B:

Let us assume again that a point load W act at the point where deflection due to bending is to be found.

The bending moment at any section *XX*, distant x from the free end (Fig. 15.35) is given as follows:

$$M = Wx + \frac{wx^2}{2} \quad \text{For } 0 < x < l/2$$

$$M = Wx + \frac{wl}{2}(x - l/4) \quad \text{For } l/2 < x < l$$

$$= Wx + \frac{wl}{2}x - \frac{wl^2}{8}$$

The strain energy of the cantilever due to bending (Eqn. 15.10) is

$$U = \frac{1}{2EI}\left[\int_0^{l/2}\left(Wx + \frac{wx^2}{2}\right)^2 dx + \int_{l/2}^{l}\left(Wx + \frac{wlx}{2} - \frac{wl^2}{8}\right)^2 dx\right]$$

or,
$$2EIU = \int_0^{l/2}\left(W^2x^2 + Wwx^3 + \frac{w^2x^4}{4}\right)dx + \int_{l/2}^{l}\left(W^2x^2 + \frac{w^2l^2x^2}{4} + \frac{w^2l^4}{64}\right.$$
$$\left. + Wwlx^2 - \frac{Wwl^2x}{4} - \frac{w^2l^3x}{8}\right)dx$$

Chinook's propeller shafts and blades are subject to high torsion, bending and direct stresses.

$$= \left[\frac{W^2 x^3}{3} + \frac{Wwx^4}{4} + \frac{w^2 x^5}{20} \right]_0^{l/2}$$

$$+ \left[\frac{W^2 x^3}{3} + \frac{W^2 l^2 x^3}{12} + \frac{w^2 l^4 x}{64} + \frac{Wwlx^3}{3} - \frac{Wwl^2 x^2}{8} - \frac{w^2 l^3 x^2}{16} \right]_{l/2}^{l}$$

$$= \left[\frac{W^2 l^3}{24} + \frac{Wwl^4}{64} + \frac{w^2 l^5}{640} \right] + \left(\frac{W^2}{3} \left(l^3 - \frac{l^3}{8} \right) + \frac{w^2 l^2}{12} \left(l^3 - \frac{l^3}{8} \right) \right.$$

$$+ \frac{w^2 l^4}{64} (l - l/2) \right) + \frac{Wwl}{3} \left(l^3 - \frac{l^3}{8} \right) - \frac{W \, wl^2}{8} \left(l^2 - \frac{l^2}{4} \right)$$

$$- \frac{w^2 l^3}{16} \left(l^2 - \frac{l^2}{4} \right) \Bigg]$$

$$= \left[\frac{W^2 l^3}{24} + \frac{Wwl^4}{64} + \frac{w^2 l^5}{640} \right] + \left[\frac{7}{24} W^2 l^3 + \frac{7}{96} w^2 l^5 + \frac{w^2 l^5}{128} \right.$$

$$+ \frac{7}{24} W \, wl^4 - \frac{3}{32} W \, wl^4 - \frac{3}{64} w^2 \, l^5 \Bigg]$$

Differentiating with respect to *W* partially, we get

$$2EI \frac{\partial U}{\partial W} = \frac{Wl^3}{12} + \frac{wl^4}{64} + \frac{7}{12} Wl^3 + \frac{7}{24} wl^4 - \frac{3}{32} wl^4$$

Equating *W* = 0, we get

$$2EI \left(\frac{\partial U}{\partial W} \right)_{W=0} = \frac{wl^4}{64} + \frac{7}{24} wl^4 - \frac{3}{32} wl^4 = \frac{41}{192} wl^4$$

or,

$$\delta_B = \frac{41 wl^4}{192} \quad \textbf{(Ans.)}$$

Example 15.37. *Fig. 15.36 shows a cantilever, 8m long, carrying a point load 5 kN at the centre and a uniformly distributed load of 2kN/m for a length 4m from the end B. If EI is the flexural rigidity of the cantilever find the reaction at the prop.*

Solution. Refer to Fig. 15.36. **Reaction at the prop, R (in kN) :**

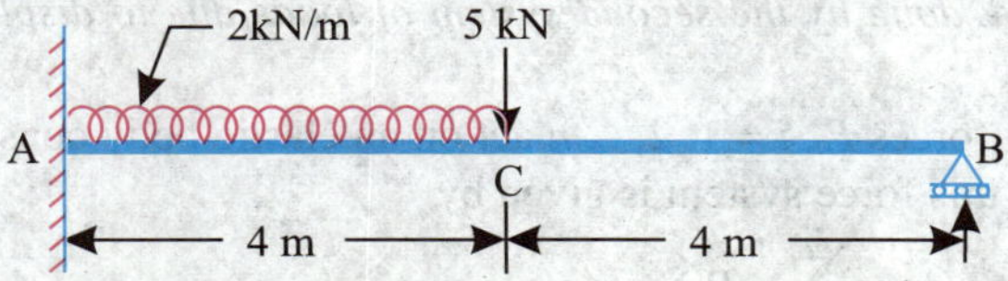

Fig. 15.36

Portion AC (Origin at A):

$$U_1 = \int_0^4 \frac{(Rx)^2 \, dx}{2 EI} = \left[\frac{R^2 x^3}{6 EI} \right]_0^4 = \frac{64 R^2}{6 EI} = \frac{32 R^2}{3 EI}$$

Portion CB (Origin at C) :

Bending moment, $M_x = R(x + 4) - 5x - \dfrac{2 \times x^2}{2}$

$\qquad = R(x + 4) - 5x - x^2$

$\therefore \qquad U_2 = \displaystyle\int_0^4 \dfrac{(M_x)^2\, dx}{2\, EI}$

Total strain energy $= U_1 + U_2$

At the propped end, $\dfrac{\partial U}{\partial R} = 0$

$\therefore$

$$\dfrac{\partial U}{\partial R} = \dfrac{64\,R}{3\,EI} + \int_0^4 \left(\dfrac{M_x}{EI} \cdot \dfrac{dM_x}{dR} \right) dx$$

$$= \dfrac{64\,R}{3\,EI} + \dfrac{1}{EI} \int_0^4 [R(x + 4) - 5x - x^2](x + 4)\, dx$$

$$= \dfrac{64\,R}{3\,EI} + \dfrac{1}{EI} \int_0^4 [R(x + 4)^2 - 5x(x + 4) - x^2(x + 4)]\, dx$$

$$= \dfrac{64\,R}{3\,EI} + \dfrac{1}{EI} \int_0^4 [R(x^2 + 8x + 16) - 5(x^2 + 4x) - (x^3 + 4x^2)]\, dx$$

$$0 = \dfrac{64\,R}{3\,EI} + \dfrac{1}{EI} \left[R\left(\dfrac{x^3}{3} + 4x^2 + 16x \right) - 5\left(\dfrac{x^3}{3} + 2x^2 \right) - \left(\dfrac{x^4}{4} + \dfrac{4x^3}{3} \right) \right]_0^4$$

$$= \dfrac{64\,R}{3} + \left[R\left(\dfrac{64}{3} + 64 + 64 \right) - 5\left(\dfrac{64}{3} + 32 \right) - \left(\dfrac{256}{4} + \dfrac{256}{3} \right) \right]$$

$$= 21.33\,R + [149.33\,R - 266.67 - 149.33]$$

$$= 21.33\,R + (149.33\,R - 416)$$

i.e. $\qquad 21.33R + 149.33R - 416 = 0$

or, $\qquad 170.66\,R = 416$

or, $\qquad R = 2.437\ \text{kN}\ \ \textbf{(Ans.)}$

15.11. MAXWELL'S THEOREM

The **Maxwell's reciprocal theorem** states as under:

"The work done by the first system of loads due to displacements caused by a second system of loads equals the work done by the second system of loads due to displacements caused by the first system of loads."

Proof : Let point forces P_i, $i = 1, 2, ...n$ act on an elastic body constrained in a space. Then the strain energy due to this force system is given by,

$$U_A = \sum_{i=1}^{n} \dfrac{1}{2} P_i\, \delta_i \qquad\qquad ...(i)$$

where, δ_i are the corresponding deflections under the loads P_i.

Let point forces P_j, $j = 1, 2, ..., m$ be the new set of point forces (due to another loading process) which act on the body. Then the strain energy due to the force system is given by

$$U_B = \sum_{j=1}^{m} \dfrac{1}{2} P_j\, \delta_j \qquad\qquad ...(ii)$$

For finding the strain energy of the body under the *combined action* of both the force systems A and B, let us assume that the *principle of superposition holds good* so that the order of application of the loading system does not change the final result.

Now, by applying force/load system A to the body the strain energy U_A is given by

$$U_A = \frac{1}{2} \sum_{i=1}^{n} (P_i)_A (\delta_i)_A \qquad ...(iii)$$

Further, apply force system B to the body. Since the force system A remains constant during the application of force system B, the strain energy $U_{A,B}$ is given by

$$U_{A,B} = \sum_{i=1}^{n} (P_i)_A (\delta_i)_B \qquad ...(iv)$$

where $(\delta_i)_B$ is the displacement due to force system B at the position of the point of application of force $(P_i)_A$ of system A.

The *increase in strain energy U_B* is given by

$$U_B = \frac{1}{2} \sum_{j=1}^{m} (P_j)_B (\delta_j)_B \qquad ...(v)$$

The total strain energy,

$$U = U_A + U_{A,B} + U_B \qquad ...(vi)$$

Now, when the body is *reloaded in reverse,* the strain energy U' becomes:

$$U' = U_B + U_{B,A} + U_A \qquad ...(vii)$$

Accordings to the principle of superposition, we have

$$U = U'$$
$$U_A + U_{A,B} + U_B = U_B + U_{B,A} + U_A$$
$$U_{A,B} = U_{B,A}$$

or, $$\sum_{i=1}^{n} (P_i)_A (\delta_i)_B = \sum_{j=1}^{m} (P_j)_B (\delta_j)_A \qquad ...(15.42)$$

Eample 15.38. *A simply supported beam of span l is carrying a concentrated W at the centre and a uniformly distributed load of intensity of w per unit length. Show that Maxwell's reciprocal theorem holds good at the centre of the beam.*

Solution. Refer to Fig. 15.37

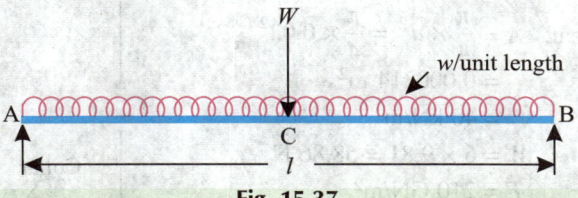

Fig. 15.37

Let the load W is applied first and then the uniformly distributed load w.

Deflection due to load w at the centre of the beam is given by ,

$$\delta_w = \frac{5\,wl^4}{384\,EI}$$

Hence work done by W due to w is given by;

$$U_{A,B} = W \times \frac{5\,wl^4}{384\,EI} = \frac{5}{384} \frac{W\,wl^4}{EI}$$

Deflection at a distance x from the left end due to W is given by

$$\delta_{W(x)} = \frac{W}{48\,E\,I}\left(3l^2x - 4x^3\right) \qquad\qquad \text{...[Refer to eqn. 8.20]}$$

Work done by w per unit length due to W,

$$U_{B,A} = 2\int_0^{l/2} w \times \frac{W}{48\,EI}\left(3l^2x - 4x^3\right)dx$$

$$= \frac{Ww}{24\,EI}\left|\frac{3l^2x^2}{2} - 4\times\frac{x^4}{4}\right|_0^{l/2}$$

$$= \frac{W\,w}{24\,EI}\left[\frac{3l^2}{2}\left(\frac{l}{2}\right)^2 - (l/2)^4\right]$$

$$= \frac{Ww}{24\,E\,I}\left(\frac{3l^4}{8} - \frac{l^4}{16}\right)$$

$$= \frac{5}{384}\frac{Wwl^4}{EI} \qquad \textbf{...Proved.}$$

TYPICAL EXAMPLES (For Competitive Examinations)

Example 15.39. *A 1.25 m long bar and of 20 mm diameter is supported rigidly in the vertical position at the top and is provided with a hollow falling mass and a collar at the bottom which supports a spring 100 mm long (Fig. 15.38). If spring stiffness = 40 kN/m and for the bar material E = 210 GN/m², determine:*

    (i) The stress developed if the falling mass is 6kg and it falls from a height of one metre measured from the collar top;

   (ii) The stress developed if spring is not provided.

Solution. Length of the bar,

$$l = 1.25 \text{ m}$$

Diameter of the bar, $d = 20 \text{ mm} = 0.02 \text{ m}$

$\therefore$ Area of the bar, $\quad A = \dfrac{\pi}{4}\times d^2 = \dfrac{\pi}{4}\times 0.02^2$

$$= 0.000314 \text{ m}^2$$

Spring stiffness $\quad = 40 \text{ kN/m}$

$$W = 6\times 9.81 = 58.86 \text{ N}$$

$$E = 210 \text{ GN/m}^2.$$

(i) Stress developed (with spring) :

We know that $\qquad n = 1 + \sqrt{1 + \dfrac{2h}{\delta l}}$

$$\text{...[Eqn. 15.5 }(a)]$$

(where, n = load factor)

$\qquad\qquad \delta l$ = Static deflection of bar + static deflection of spring = $\delta l_1 + \delta l_2$

Now, $\qquad\qquad \delta l = \dfrac{Wl}{AE} = \dfrac{58.86\times 1.25}{0.000314\times 210\times 10^9} = 1.116\times 10^{-6} \text{ m}$

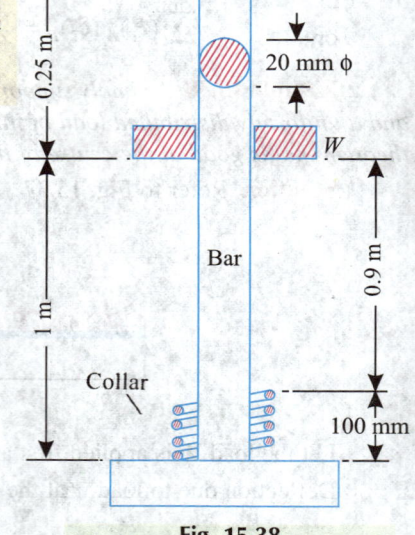

Fig. 15.38

and,
$$\delta l_2 = \frac{W}{\text{Stiffness}} = \frac{58.86}{40 \times 1000} = 1471.5 \times 10^{-6}\, m$$

∴
$$\delta l = \delta l_1 + \delta l_2 = 1.116 \times 10^{-6} + 1471.5 \times 10^{-6}\, m = 1472.616 \times 10^{-6}\, m$$

∴
$$n = 1 + \sqrt{1 + \frac{2 \times 0.9}{1472.616 \times 10^{-6}}} = 35.976 \qquad [(\because h = 0.9\, m\ (Given)]$$

∴
$$\sigma_i = \frac{nW}{A} = \frac{35.976 \times 58.86}{0.000314} = 6.74 \times 10^6\, N/m^2 = 6.74\, MN/m^2$$

Hence,
$$\sigma_i = 6.74\, MN/m^2 \quad (Ans.)$$

(ii) Stress developed (without spring):

$$n = 1 + \sqrt{1 + \frac{2h}{\delta l_1}}$$

$$= 1 + \sqrt{1 + \frac{2 \times 1}{1.116 \times 10^{-6}}} = 1339.7 \qquad (h = 1m)$$

∴
$$\sigma_i = \frac{nW}{A} = \frac{1339.7 \times 58.86}{0.000314} = 251.13 \times 10^6\, N/m^2 = 251.13\, MN/m^2$$

Hence,
$$\sigma_i = 251.13\, MN/m^2 \quad (Ans.)$$

[Note the use of spring in reducing the stress under impact]

Alternative method [for Part (i)]

Let the instantaneous maximum deflections of the spring and rod be $\delta l_i'$ and $\delta l_i''$ respectively and the corresponding maximum stress induced in the rod due to impact be σ_i N/m^2

Let W_e be the equivalent static load when applied on the collar through the spring gradually will deflect the spring and the rod through the same amount viz $\delta l_i'$ and $\delta l_i''$ respectively and also induce the same stress σ_i in the rod as is being done by the falling mass 6 kg.

$$W_e = \sigma_i A = 0.000314\, \sigma_i\, N$$

$$\delta l_i' = \frac{\sigma_i l}{E} = \frac{1.25\sigma_i}{210 \times 10^9} = 5.95 \times 10^{-12}\, \sigma_i m$$

$$\delta l_i'' = \frac{W_e}{\text{Stiffness}} = \frac{0.000314\sigma_i}{40 \times 1000} = 7.85 \times 10^{-9}\, \sigma_i\, m$$

Strain energy of spring
$$= \frac{1}{2} W_e\, \delta l_i'' = \frac{1}{2} \times 0.00314\sigma_i \times 7.85 \times 10^{-9}\, \sigma_i$$
$$= 1.232 \times 10^{-12}\, \sigma_i^2\, Nm$$

Strain energy of rod
$$= \frac{\sigma_i^2}{2E} \times \text{volume}$$

$$= \frac{\sigma_i^2}{2 \times 210 \times 10^9} \times 0.000314 \times 1.25\sigma_i^2$$
$$= 9.34 \times 10^{-16}\, \sigma_i^2$$

Loss of potential energy of faling mass
$$= mg\ (h + \delta l_i' + \delta l_i'')$$
$$= 6 \times 9.81\ (0.9 + 5.95 \times 10^{-12}\, \sigma_i + 7.85 \times 10^{-9}\, \sigma_i)$$

Thus *equating the strain energy to the loss of instantaneous potential energy, we get*
$$1.232 \times 10^{-12}\, \sigma_i^2 + 9.34 \times 10^{-16}\, \sigma_i^2 = 6 \times 9.81\ (0.9 + 5.95 \times 10^{-12}\, \sigma_i + 7.85 \times 10^{-9}\, \sigma_i)$$

or, $\qquad 12329.34 \times 10^{-16} \sigma_i^2 = 58.86 \,(0.9 + 7855.95 \times 10^{-12} \,\sigma_i)$

$$\sigma_i^2 = 42.96 \times 10^{12} + 375041\, \sigma_i$$

or, $\quad \sigma_i^2 - 375041\, \sigma_i - 42.96 \times 10^{12} = 0$

$$\sigma_i = \frac{375041 + \sqrt{(375041)^2 + 4 \times 42.96 \times 10^{12}}}{2}$$

$$= \frac{375041 + 10^6 \sqrt{171.98}}{2}$$

$$= \frac{10^6 \,(0.375 + 13.11)}{2} = 6.74 \times 10^6 \text{ N/m}^2$$

$$= 6.74 \text{ MN/m}^2$$

Hence, $\qquad\qquad \sigma_i = 6.74 \text{ MN/m}^2 \quad \textbf{(Ans.)}$

Example 15.40. *A beam of I-section 240 mm × 160 m (web = 10 mm thick, flanges = 20 mm thick) is simply supported over a span of 5 metres. A concentrated load of 10 kN acts at a distance of 2 metres from one of the ends. Determine the total deflection produced under the concentrated load, assuming that the shearing force is carried by the web only and the shearing stress is uniformly distributed over the web. Take: E = 208 GN/m², and C = 80 GN/m².*

Solution. Refer to Fig. 15.39

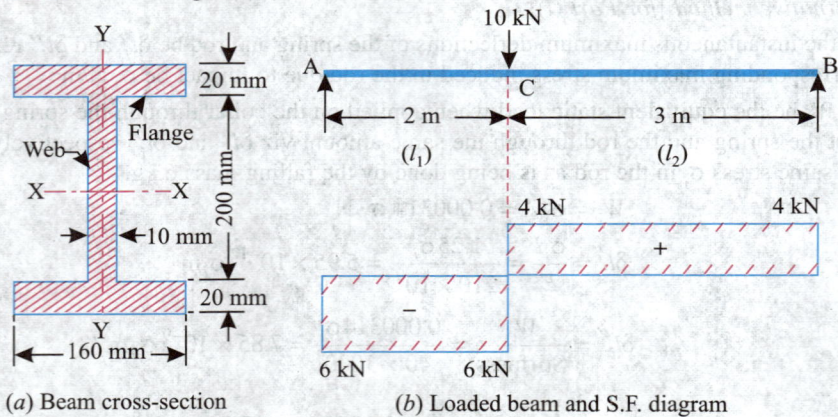

(a) Beam cross-section   (b) Loaded beam and S.F. diagram

Fig. 15.39

Total deflection under the load:

Moment of inertia of I-section about XX axis is

$$I_{XX} = \frac{160 \times 240^3}{12} - \frac{150 \times 200^3}{12}$$

$$= 84.32 \times 10^6 \text{ mm}^4 = 84.32 \times 10^{-6} \text{ m}^4$$

Deflection due to bending:

Deflection due to bending at the point of load,

$$y_C = \frac{W l_1^2 \, l_2^2}{3 \, EIl}$$

where, $\qquad\qquad W = 10 \text{ kN} = 10 \times 10^3 \text{ N}$

$$l_1 = 2 \text{ m}, \, l_2 = 3 \text{ m}$$

$$E = 208 \times 10^9 \text{ N/m}^2 \quad (Given)$$

$$I = 84.32 \times 10^{-6} \text{ m}^4 \quad \text{(Calculated earlier)}$$
$$l = 5\text{m}$$

$$\therefore \qquad y_C = \frac{10 \times 10^3 \times 2^2 \times 3^2}{3 \times 208 \times 10^9 \times 84.32 \times 10^{-6} \times 5}$$

$$= 0.001368 \text{ m or } 1.368 \text{ mm}$$

Deflection due to shear:

Cross-sectional area of web $= 200 \times 10 = 2000 \text{ mm}^2 = 2000 \times 10^{-6} \text{ m}^2$

Shear force in the portion AC,

$$S_1 = 6 \text{ kN}$$

$\therefore$ Shear stress in the portion AC,

$$\tau_{AC} = \frac{6}{2000 \times 10^{-6}} \times 10^{-3} \text{ MN/m}^2 = 3 \text{ MN/m}^2$$

Shear force in the portion $CB = 4$ kN

$\therefore$ Shear stress in the portion CB,

$$\tau_{CB} = \frac{4}{2000 \times 10^{-6}} \times 10^{-3} \text{ MN/m}^2 = 2 \text{ MN/m}^2$$

Shear strain energy in the portion AC,

$$U_{AC} = \int_0^2 \int_{-0.1}^{+0.1} \frac{\tau_{AC}^2}{2C} \cdot b \, dy \cdot dx$$

$$= \int_0^2 \int_{-0.1}^{+0.1} \frac{(3 \times 10^6)^2}{2 \times 80 \times 10^9} \times 0.01 \times dy \cdot dx$$

$$[\because b = 10 \text{ mm} = 0.01 \text{ m}]$$

$$= \int_0^2 \left[0.5625 \, y \right]_{-0.1}^{+0.1} dx = \int_0^2 (0.5625 \times 0.2) \, dx$$

$$= \left[0.1125 \, x \right]_0^2 = 0.225 \text{ Nm}$$

A crane's arm undergoes deflection and bears shear and bending stresses while lifting weights. Above picture shows the inside view of a water tank under construction.

Shear strain energy in the portion CB,

$$U_{CB} = \int_0^3 \int_{-0.1}^{+0.1} \frac{\tau_{CB}^2}{2C} \cdot b\, dy \cdot dx = \int_0^3 \frac{(2 \times 10^6)^2}{2 \times 80 \times 10^9} \times 0.01 \times dy \times dx$$

$$= \int_0^3 [0.25]_{-0.1}^{0.1}\, dx = [0.05x]_0^3 = 0.15\,\text{Nm}$$

Total shear strain energy

$$U_s = U_{AC} + U_{CB} = 0.225 + 0.15 = 0.375\,\text{Nm}$$

If y_s is the deflection due to shear, then

$$\frac{1}{2}Wy_s = 0.375$$

or,

$$y_s = \frac{2 \times 0.375}{10 \times 10^3} \times 10^3\,\text{mm} = 0.075\,\text{mm}$$

$\therefore$ Total deflection under the load,

$$= 1.368 + 0.075 = \textbf{1.443 mm} \quad \textbf{(Ans.)}$$

Example 15.41. *A simply supported beam of rectangular section 100 m × 200 mm is of length l (= 5m) and caries a uniformly distributed load w (= 2kN/m run) over its entire length. Find the deflection due to shear a point at C distant l_1 (= 3 m) and l_2 (= 2m) from the ends respectively. Also find the deflection at the mid section. Take C = 80 GN/m²*

Solution. Width of the section,

$$b = 100\,\text{mm}$$

Depth of the section, $d = 200$ mm

Length of span $l = 5$m

U.D.L., $w = 2$ kN/m

Modulus of rigidity, $C = 80$ GN/m².

(i) Deflection due to shear at point C:

Assume a point load W to be acting at the point C. Then the magnitude of shearing force (S) at a section XX, distant x from A is given as follows:

$$S = \left[\frac{wl}{2} + \frac{wl_2}{l} - wx\right] \quad \text{For } 0 < x < l_1$$

$$S = \left[\frac{wl}{2} + \frac{wl_2}{l} - wx - W\right] \quad \text{For } l_1 < x < l$$

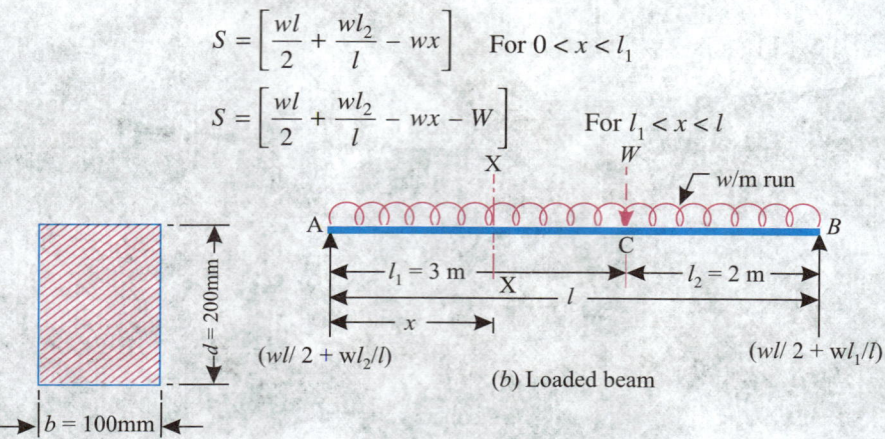

(a) Beam cross - section

(b) Loaded beam

Fig. 15.40

∴ The strain energy of the beam due to shearing is,

$$U_s = \frac{3}{5\,Cbd}\left[\int_0^{l_1}\left(\frac{wl}{2}+\frac{wl_2}{l}-wx\right)^2 dx + \int_{l_1}^{l}\left(\frac{wl}{2}+\frac{wl_2}{l}-wx-W\right)^2 dx\right]$$

$$\frac{5\,Cbd}{3}U_s = \frac{1}{-3w}\left[\left(\frac{wl}{2}+\frac{wl_2}{l}-wl_1\right)^3 - \left(\frac{wl}{2}+\frac{wl_2}{l}\right)^3\right]$$

$$+\frac{1}{-3w}\left[\left(\frac{wl}{2}+\frac{wl_2}{l}-wl-W\right)^3 - \left(\frac{wl}{2}+\frac{wl_2}{l}-wl_1-W\right)^3\right]$$

Differentiating partially with respect to W, we get

$$5wCbd\left(\frac{\partial U_s}{\partial W}\right) = -3\left(\frac{wl}{2}+\frac{wl_2}{l}-wl_1\right)^2\frac{l_2}{l} + 3\left(\frac{wl}{2}+\frac{wl_2}{l}\right)^2\frac{l_2}{l}$$

$$-3\left(\frac{wl}{2}+\frac{wl_2}{l}-wl-W\right)^2\left(\frac{l_2}{l_1}-1\right) + 3\left(\frac{wl}{2}+\frac{wl_2}{l}-wl_1-W\right)\left(\frac{l_2}{l_1}-1\right)$$

Now equating $W = 0$, we get

$$5wCbd\left(\frac{\partial U_s}{\partial W}\right)_{W=0} = -3\left(\frac{wl}{2}-wl_1\right)^2\frac{l_2}{l} + 3\left(\frac{wl}{2}\right)^2\frac{l_2}{l}$$

$$-3\left(\frac{wl}{2}-wl\right)^2\left(\frac{l_2}{l_1}-1\right) + 3\left(\frac{wl}{2}-wl_1\right)^2\left(\frac{l_2}{l_1}-1\right)$$

$$= 3\left(\frac{l_2}{l}\right)\left[\left(\frac{wl}{2}\right)^2 - \left(\frac{wl}{2}-wl_1\right)^2\right]$$

$$+3\left(\frac{l_2}{l}-1\right)\left[\left(\frac{wl}{2}-wl_1\right)^2 - \left(\frac{wl}{2}-wl\right)^2\right]$$

An IC engine is being machined.

$$= 3\frac{l_2}{l}\left[\left(\frac{wl}{2} + \frac{wl}{2} - wl_1\right)\left(\frac{wl}{2} - \frac{wl}{2} + wl_1\right)\right]$$

$$+ 3\left(\frac{l_2}{l} - 1\right)\left[\left(\frac{wl}{2} - wl_1 + \frac{wl}{2} - wl\right)\left(\frac{wl}{2} - wl_1 - \frac{wl}{2} + wl\right)\right]$$

$$= 3\frac{l_2}{l}[(wl - wl_1)(wl_1)] + 3\left(\frac{l_2}{l} - 1\right)[(-wl_1)(wl - wl_1)]$$

$$= \frac{3w^2 l_2}{l}[(l - l_1)l_1] + \frac{3w^2 (l_2 - l)(-l_1)(l - l_1)}{l}$$

$$= \frac{3w^2 l_1 l_2^2}{l} + \frac{3w^2 (l - l_2)l_1 (l - l_1)}{l}$$

$$= \frac{3w^2 l_1 l_2^2}{l} + \frac{3w^2 l_1^2 l_2}{l} \qquad\qquad (\because\ l_1 + l_2 = l)$$

$$= \frac{3w^2 l_1 l_2}{l}(l_2 + l_1) = 3w^2 l_1 l_2$$

or, $\qquad \left(\dfrac{\partial U_s}{\partial W}\right)_{W=0} = \dfrac{3w^2 l_1 l_2}{5wCbd} = \dfrac{3wl_1 l_2}{5Cbd}$

or, $\qquad\qquad (\delta s)_C = \dfrac{3wl_1 l_2}{5Cbd}$ ...(i)

Substituting the values in eqn. (i) we get

$$(\delta s)_C = \frac{3 \times (2 \times 10^3) \times 3 \times 2}{5 \times 80 \times 10^9 \times (100 \times 10^{-3}) \times (200 \times 10^{-3})} \times 10^3 \text{ mm} = 0.0045 \text{ mm}$$

Hence, *deflection at C* = **0.0045 mm** **(Ans.)**

Deflection at mid span:

Putting, $\qquad l_1 = l_2 = l/2$ *i.e.* when C is the mid point, then

$$(\delta s)_C = \frac{3}{5}\frac{w(l/2)^2}{Cbd} = \frac{2wl^2}{20 Cbd}$$

$$= \frac{3 \times (2 \times 10^3) \times 5^2}{20 \times 80 \times 10^9 \times (100 \times 10^{-3}) \times (200 \times 10^{-3})} \times 10^3 \text{ mm} = 0.00468 \text{ mm}$$

Hence, *deflection at mid span* = **0.00468 mm** **(Ans.)**

HIGHLIGHTS

1. The strain energy stored by the body, *'within'* elastic limit when loaded externally is called *'Resilience'* and the maximum energy which a body stores *'upto'* elastic limit is called the *'Proof resilience'*. Proof resilience per unit volume of piece is called *'modulus of resilience'*.

2. Strain energy stored in the bar:

$$U = \frac{\sigma^2 Al}{2E} = \frac{\sigma^2 V}{2E} \qquad\qquad ...(i)$$

Proof resilience,

$$U_p = \frac{\sigma_p^2 V}{2E} \qquad\qquad ...(ii)$$

and, modulus of resilience

$$= \frac{\sigma_p^2}{2E} \qquad \qquad ...(iii)$$

3. Stress (instantaneous) due to suddenly applied load

$$\sigma_{su} = \frac{2W}{A} = 2 \times \frac{W}{A}$$

$$= 2 \times \text{Stress due to gradually applied load}$$

4. Instantaneous stress due to falling load,

$$\sigma_i = \frac{W}{A} \left[1 + \sqrt{1 + \frac{2hAE}{Wl}} \right] \qquad \qquad ...(i)$$

If δl is negligible as compared to h, then

$$\sigma_i = \sqrt{\frac{2WhE}{Al}} \qquad \qquad ...(ii)$$

and, if $\quad h = 0 \quad \sigma i = \dfrac{2W}{A} \qquad \qquad ...(iii)$

which means the load is suddenly applied and not the falling one.

5. Shearing strain energy for a block of material subjected to a constant shearing stress throughout,

$$U = \frac{\tau^2}{2C} \times \text{volume of block (body)}$$

6. Strain energy in torsion:
   (a) For a *solid shaft*,

$$U = \frac{\tau^2}{4C} \times \text{volume} \qquad \qquad ...(i)$$

   (b) For a *hollow shaft*

$$U = \frac{\tau^2}{4C} \times \left[\frac{R^2 + r^2}{R} \right] \times \text{volume} \qquad \qquad ...(ii)$$

7. Strain energy caused by bending:

$$U = \int \frac{M^2}{2EI} \, dx$$

8. Strain energy due to principal stresses:

$$U = \frac{1}{2E} [(\sigma_1^2 + \sigma_2^2 + \sigma_3^2) - 2\mu(\sigma_1\sigma_2 + \sigma_2\sigma_3 + \sigma_3\sigma_1)]V$$

where, $\sigma_1, \sigma_2, \sigma_3$ = Principal stresses

$$\mu \left(= \frac{1}{m} \right) = \text{Poisson's ratio , and}$$

$$V = \text{Volume of the block.}$$

9. Strain energy due to shear:

$$U_s = \frac{3}{5Cbd} \int_0^l S^2 \, dx$$

10. Castigliano's theorem: It states as follows:
    "If U is the total strain energy of any structure due to the application of external loads W_1, W_2, W_3 ...... W_n at points A_1, A_2, A_3 ...... A_n respectively in the directions AX_1, AX_2, AX_3 .... AX_n and due to couples $M_1, M_2, ... M_m$ at points $B_1, B_2, B_3, ...$ B_m respectively then the deflections at the points A_1,

$A_2, A_3, ... A_n$ in the directions $AX_1, AX_2, AX_3, ... AX_n$ are $\dfrac{\partial U}{\partial W_1}, \dfrac{\partial U}{\partial W_2}, \dfrac{\partial U}{\partial W_3}, ... \dfrac{\partial U}{\partial W_n}$ and the angular

positions of the couples are $\dfrac{\partial U}{\partial M_1}, \dfrac{\partial U}{\partial M_2}, \dfrac{\partial U}{\partial M_3}, ... \dfrac{\partial U}{\partial M_m}$ at the respective points of application."

OBJECTIVE TYPE QUESTIONS

Choose the Correct Answer:

1. The strain energy stored by the body within elastic limit when loaded externally is called
   - (a) resilience
   - (b) proof resilience
   - (c) modulus of resilience
   - (d) none of the above.

2. Proof resilience is the mechanical property of materials which indicates their capacity to bear
   - (a) static tensile loads
   - (b) static compressive loads
   - (c) shocks
   - (d) none of the above.

3. If σ_p be the proof stress or maximum stress to which the bar is stressed upto the elastic limit, then modulus of resilience is equal to
   - (a) $\dfrac{\sigma_p}{2E}$
   - (b) $\dfrac{\sigma_p^2}{2E}$
   - (c) $\dfrac{\sigma_p^2}{4E}$
   - (d) $\dfrac{\sigma_p}{4E}$.

4. The stress due to suddenly applied load is ..... times that of gradually applied load.
   - (a) two
   - (b) three
   - (c) four
   - (d) five.

5. The shearing strain energy for a block of material (per unit volume) subjected to a constant shearing stress throughout is given by
   - (a) $\dfrac{\tau^2}{C}$
   - (b) $\dfrac{\tau}{C^2}$
   - (c) $\dfrac{\tau^2}{2C}$
   - (d) $\dfrac{\tau^2}{C^2}$.

6. In case of a solid shaft the strain energy in torsion, per unit volume, is equal to
   - (a) $\dfrac{\tau^2}{2C}$
   - (b) $\dfrac{\tau^2}{4C}$
   - (c) $\dfrac{\tau^2}{6C}$
   - (d) $\dfrac{\tau^2}{8C}$.

7. Strain energy (U) caused by bending is given by the relation
   - (a) $U = \int \dfrac{M}{2EI}\, dx$
   - (b) $U = \int \dfrac{M^2}{2EI}\, dx$
   - (c) $U = \int \dfrac{M^2}{EI}\, dx$
   - (d) $U = \int \dfrac{M^2}{3EI}\, dx$.

ANSWERS

| 1. (a) | 2. (c) | 3. (b) | 4. (a) | 5. (c) | 6. (b) | 7. (b) |

Gauge board on a ship.

UNSOLVED EXAMPLES

1. An axial pull of 50 kN is suddenly applied to a steel rod 2 m long and 10 cm^2 in cross-section. Calculate the strain energy that can be absorbed, if E = 200 GN/m^2

[**Ans.** 50 Nm or J]

2. A wrought iron bar 50 mm in diameter and 2.5 long has to transmit a shock energy of 100 Nm (J). Calculate the maximum instantaneous stress and elongation produced. E = 200 GN/m^2.

[**Ans.** 90.26 MN/m^2; 1.128 mm]

3. An unknown weight falls through 10 mm on a collar rigidly attached to the lower end of a vertical bar, 2 m long and 6 cm^2 in section. If the maximum instantaneous extension is known to be 2 mm, what is the corresponding stress and the value of unknown weight. Take E = 200 GN/m^2.

[**Ans.** 200 MN/m^2; 10 kN]

4. A wagon weighing 40 kN is attached to a wire rope and moving down an incline at a speed of 3.6 km/h when the rope jams and wagon is suddenly brought to rest. If the length of the rope is 50 metres at the time of sudden stoppage, calculate the maximum instantaneous stress and the maximum instantaneous elongation produced. Diameter of the rope = 36 mm. E = 205 GN/m^2.

[**Ans.** 128.3 MN/m^2, 31.2 mm]

5. A vertical steel rod, 1.3 m long, is rigidly secured at its upper end, a weight of 80 N is allowed to slide freely on the rod through a distance a 100 mm on to the stop at the lower end. The upper 700 mm length of rod has a diameter of 20 mm while the lower 600 mm length is 16 mm in diameter. Calculate the maximum stress induced in the bar ignoring the extension of the bar in determining the potential energy given up by the weight. E = 200 GN/m^2

[**Ans.** 123.2 MN/m^2]

6. A vertical tie fixed rigidly at the top consists of a steel rod 3.5 long and 25 mm diameter encased throughout in a brass tube 25 mm internal diameter and 35 mm external diameter. The casing and the rod are fixed together at both ends. The compound rod is suddenly loaded in tension by weight of 15 kN falling through 5 mm before being arrested by the tie. Determine the maximum stresses in steel and brass.

Given: E_s = 200 GN/m^2, and E_b = 100 GN/m^2

[**Ans.** σ_b = 65.66 MN/m^2; σ_s = 131.32 MN/m^2]

7. A hollow shaft having external diameter twice the internal diameter, subjected to pure torque, attains a maximum shear stress τ. Show that the strain energy stored per unit volume of the shaft is $5\,\tau^2/16\,C$.

If such a shaft is required to transmit 4500 kW at 110 r.p.m. with uniform torque, the maximum stress not exceeding 70 MN/m^2, calculate:

(i) Shaft diameters: (ii) Energy stored per m^3.

Take : C = 83 GN/m^2

[**Ans.** 312 mm, 156 mm, 1844 kJ/m^3]

8. A hollow shaft subjected to a pure torque attains a maximum shearing stress τ. Given that the strain energy stored per unit volume is $\tau^2/3\,C$, calculate the ratio of shaft diameters.

If such a shaft is required to transmit 3700 kW at 110 r.p.m. with uniform torque and the energy stored is 20 kJ/m^3 determine the actual diameters. Take: C = 80 GN/m^2.

[**Ans.** $\sqrt{3}$: 1; 298.2 mm; 172 mm]

9. A beam 8 metres in length carries loads of 40 kN each at a distance of 2 metres and 6 metres from one end. The beam is simply supported at the ends. The beam is of rectangular section with breadth b and depth d. If $d = 2b$, and the shear stress is not to exceed , find:

(i) Size of beam;   (ii) Energy stored;

(iii) Deflection due to shear under the load of 40 kN.

Take : C = 80 MN/m^2.

[**Ans.** (i) 20 mm; 40 mm, (ii) 30 Nm (iii) 0.75 mm]

Chapter 16

Columns and Struts

16.1. INTRODUCTION

A member of structure or bar which carries an axial compressive load is called the **strut**. *If the strut is vertical i.e. inclined at 90° to the horizontal is known as* **column, pillar or stanchion.**

Generally, a member in any position *other than vertical*, subjected to a compressive load is called '*strut*', and *vertical member* subjected to a compressive load is called '*column*'.

Another difference between the strut and column is that *strut may have its one or both the ends fixed rigidly or hinged or pin pointed; while the column will have both the ends fixed rigidly.*

The *examples of struts* are: Piston rods, connecting rods, side links in forging machines etc. The failure of such member will occur:

(*i*) By pure compression;

(*ii*) By buckling;

(*iii*) By combination of pure compression and buckling, depending upon a slenderness ratio.

16.2. DEFINITIONS

Column. A column is a long vertical slender bar or vertical member, subjected to an axial compressive load and *fixed rigidly at both ends*.

Strut. A strut is a slender bar or member *in any position other than vertical*, subjected to a compressive load and *fixed rigidly or hinged or pin jointed at one or both the ends*.

Slenderness ratio (k). It is the ratio of unsupported length of the column to the minimum radius of gyration of the cross-sectional ends of the column. It has *no unit* whatsoever.

Buckling factor. It is the ratio between the equivalent length of the column to the *minimum* radius of the gyration.

Buckling load. The maximum limiting load at which the column tends to have lateral displacement or tends to buckle is called *buckling* or *crippling load. The buckling takes place about the axis having minimum radius of gyration, or least moment of inertia.*

Safe load. It is the load to which a column is actually subjected to and is well below the buckling load. It is obtained by dividing the buckling load by a suitable factor of safety (F.O.S.).

i.e.
$$\text{Safe load} = \frac{\text{Buckling load}}{\text{Factor of safety}}$$

16.3. CLASSIFICATION OF COLUMNS

Depending upon the slenderness ratio or length to diameter ratio, columns can be divided into three classes.

1. Short columns:

*Columns which have lengths less than 8 times their respective diameters or slenderness ratio less than 32 are called **short columns or stocky struts**.* When short columns are subjected to compressive loads, their buckling is generally negligible and as such the buckling stresses are very small as compared with direct compressive stress. Therefore it is assumed that short columns are *always subjected to direct compressive stresses only.*

2. Medium size columns:

*The columns which have their lengths varying from 8 times their diameter to 30 times their respective diameters or their slenderness ratio lying between 32 and 120 are called **medium size columns** or **intermediate columns**.* In these columns, both the buckling as well as direct stresses are of significant values. Therefore, in the design of intermediate columns *both these stresses* are taken into account.

3. Long columns:

*The columns having their lengths more than 30 times their respective diameters or slenderness ratio more than 120 are called **long columns**.* They are usually subjected to *buckling stress only.* Direct compressive stress is very small as compared with buckling stress, and hence it is neglected.

Screwjack behaves like a column.

16.4. STRENGTH OF COLUMNS

The strength of a column depends upon the *slenderness ratio*. If the *slenderness ratio is increased the compressive strength of a column decreases as the tendency to buckle is increased. The strength of the column depends upon the end conditions also.*

16.5. END CONDITIONS

The end conditions of a loaded column can be had in *four* ways:

Refer to Fig. 16.1.

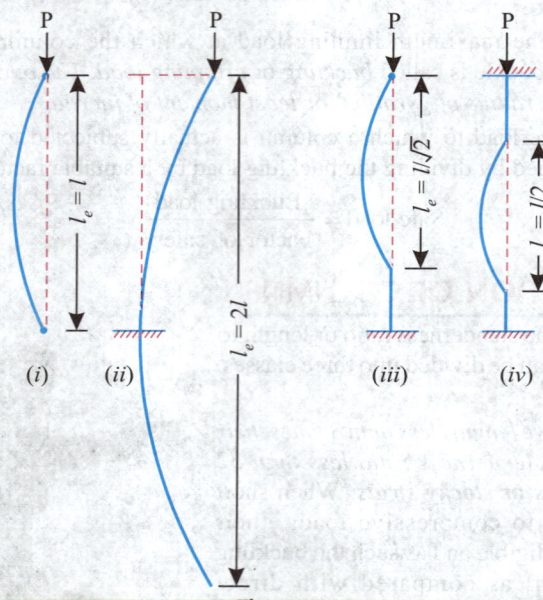

Fig. 16.1

   (*i*)   Both ends pin jointed or hinged or rounded or free.

  (*ii*)   One end fixed and other end free.

 (*iii*)   One end fixed and the other pin jointed.

 (*iv*)   Both ends fixed.

16.6. EQUIVALENT LENGTH (l_e)

The distance between adjacent points of inflexion is called **equivalent length or effective length or simple column length.** A point of inflexion is found at every column end that is free to rotate and at every point where there is a change of the axis.

Hence in case of:

(*i*) Both ends hinged :

Equivalent length = Actual length

$$l_e = l$$

(*ii*) One end fixed and other end free :

$l_e = 2l$, the free end will sway sidewise and the curvature in the length l will be similar to that of the upper half of the simple column.

(*iii*) One end fixed and other pin jointed :

$$l_e = \frac{l}{\sqrt{2}}, \text{ between the top of the column and inflexion point.}$$

(*iv*) **Both ends fixed:**

$l_e = \dfrac{l}{2}$, the distance between the two inflexion points.

16.7. EULER'S THEORY (FOR LONG COLUMNS)

Assumptions :

The following *assumptions* are made while deriving the **Euler's formula:**

1. The column is initially straight and of uniform lateral dimension.
2. The compressive load is exactly axial and it passes through the centroid of the column section.
3. The material of the column is perfectly homogeneous and isotropic.
4. Pin joints are frictionless and fixed ends are perfectly rigid.
5. The weight of the column itself is neglected.
6. The column fails by buckling alone.
7. Limit of proportionality is not exceeded.

16.8. SIGN CONVENTIONS FOR BENDING MOMENTS

A bending moment which bends the column so as to present convexity towards the initial centre line of the member will be regarded as *positive* (Fig. 16.2).

A bending moment which bends the column as to present concavity towards the initial centre line of the member will be regarded as *negative* (Fig. 16.3).

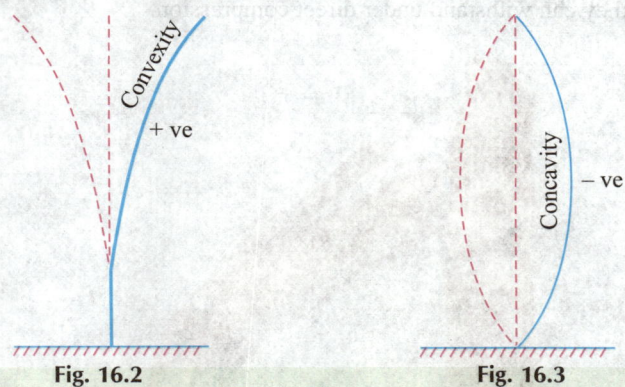

Fig. 16.2   Fig. 16.3

16.9. EULER'S FORMULA

Euler's formula is used for calculating the critical load for a column or strut, and is as follows:

$$P_{Euler} = \frac{\pi^2 EI}{l_e^2} \qquad \qquad ... (16.1)$$

where, P = Critical load,

E = Modulus of elasticity,

I = Least moment of inertia of section of the column, and

l_e = Equivalent length of the strut.

A column of given length, cross-section and material will have different values of buckling loads for different end conditions as given in table 16.1.

TABLE 16.1

Case	End conditions	Equivalent length, l_e	Buckling load, Euler
1	Both ends hinged or pin jointed or rounded or free	l	$\dfrac{\pi^2 EI}{l_e^2} = \dfrac{\pi^2 EI}{l^2}$
2.	One end fixed, other end free	$2l$	$\dfrac{\pi^2 EI}{l_e^2} = \dfrac{\pi^2 EI}{4l^2}$
3.	One end fixed, other end pin jointed	$\dfrac{l}{\sqrt{2}}$	$\dfrac{\pi^2 EI}{l_e^2} = \dfrac{2\pi^2 EI}{l^2}$
4.	Both ends fixed or encastered	$\dfrac{l}{2}$	$\dfrac{\pi^2 EI}{l_e^2} = \dfrac{4\pi^2 EI}{l^2}$

16.10. LIMITATIONS FOR THE USE OF EULER'S FORMULA

The following are the *limitations* due to which Euler's formula is of little practical use.

1. It is applicable to an ideal strut only and in practice, there is always crookedness in the column and the load applied may not be exactly co-axial.

2. It takes no account of direct stress. It means that it may give a buckling load for struts far in excess of load which they can withstand under direct compression.

Mechanical digger's hydraulic arms partly behave like columns.

16.11. APPLICABILITY OF EULER THEORY

It is assumed that only buckling has any effect in developing this theory, that is the strut is assumed to be *slender*.

If the maximum allowable compressive stress in the strut is σ_c, the breaking or crushing load/ strength would be $\sigma_c.A$.

If P exceeds $\sigma_c.A$ then strut will break in crushing.

It is found that for values of slenderness ratio greater than 80, the Euler formula may be applied, but it will not conform closely with the actual buckling load until the slenderness ratio becomes *very large*.

In an attempt to obtain an expression which more truly applies to real struts, a number of empirical formulae have been developed.

16.12. DERIVATIONS OF EULER'S FORMULA (FOR DIFFERENT END CONDITIONS)

Case I. When both ends of the column are hinged or pinned:

Fig. 16.4 shows a column AB of length l and uniform sectional area A, hinged at both the ends A and B. Let P be the crippling load at which the column has just buckled.

Consider any section XX at a distance x from the end B. Let y be the deflection (lateral displacement) at the section.

The bending moment at the section is given by

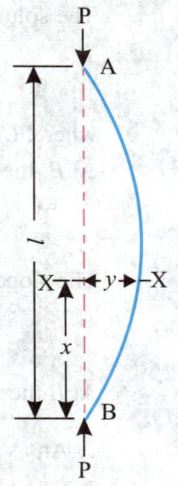

$$EI\frac{d^2y}{dx^2} = -Py$$

or,

$$EI\frac{d^2y}{dx^2} + Py = 0$$

or,

$$\frac{d^2y}{dx^2} + \frac{P}{EI}y = 0$$

The solution to the above differential equation is

$$y = C_1 \cos\left(x\sqrt{\frac{P}{EI}}\right) + C_2 \sin\left(x\sqrt{\frac{P}{EI}}\right)$$

where, C_1 and C_2 are constants of integration.

At B, the deflection is zero.

∴   At $x = 0$, $y = 0$

Hence, $C_1 = 0$

At A also, the deflection is zero.

∴   At $x = l$, $y = 0$

∴

$$0 = C_2 \sin\left(l.\sqrt{\frac{P}{EI}}\right)$$

Since $C_1 = 0$, we conclude that C_2 cannot be zero.

This is because if both C_1 and C_2 are zero the column *will not bend at all*.

Hence $\sin\left(l.\sqrt{\frac{P}{EI}}\right) = 0$

∴

$$l.\sqrt{\frac{P}{EI}} = 0,\ \pi,\ 2\pi,\ 3\pi,\ 4\pi \ ....$$

Fig. 16.4

Considering the first practical value, we have

$$l.\sqrt{\frac{P}{EI}} = \pi$$

$$P = \frac{\pi^2 EI}{l^2} \qquad \qquad ...(16.2)$$

Case II. When one end is fixed and other is free:

Fig. 16.5 shows a column AB of length l whose lower end B is fixed, the upper end A being free. Let due to crippling load P the column just buckle. Let δ be the deflection at the top end.

At any section XX distant x from the fixed end B, the bending moment is given by

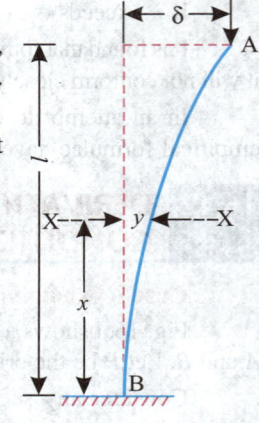

$$EI\frac{d^2 y}{dx^2} = + P\,(\delta - y)$$

$$\therefore \qquad EI\frac{d^2 y}{dx^2} + Py = P\delta$$

$$\therefore \qquad \frac{d^2 y}{dx^2} + \frac{P}{EI}\,y = \frac{P\delta}{EI}$$

The solution to the above differential equation is

Fig. 16.5

$$y = C_1 \cos\left(x\sqrt{\frac{P}{EI}}\right) + C_2 \sin\left(x\sqrt{\frac{P}{EI}}\right) + \delta$$

where, C_1 and C_2 are constants of integration
At B, the deflection is zero.

$\therefore$ At, $\qquad\qquad x = 0,\ y = 0$

$\therefore \qquad\qquad 0 = C_1 + \delta,\ \text{or}\ C_1 = -\delta$

The slope at any section is given by

$$\frac{dy}{dx} = -C_1\sqrt{\frac{P}{EI}}\sin\left(x\sqrt{\frac{P}{EI}}\right) + C_2\sqrt{\frac{P}{EI}}\cos\left(x\sqrt{\frac{P}{EI}}\right)$$

At B the slope is zero.

$\therefore$ At, $\qquad\qquad x = 0,\ \dfrac{dy}{dx} = 0 \qquad \therefore\ 0 = C_2\sqrt{\dfrac{P}{EI}}\ \text{or}\ C_2 = 0$

At A the deflection is δ.

$\therefore$ At, $\qquad\qquad x = l,\ y = \delta$

$$\therefore \qquad\qquad \delta = -\delta\cos\left(l\sqrt{\frac{P}{EI}}\right) + \delta,\ \text{or,}\ \cos\left(l\sqrt{\frac{P}{EI}}\right) = 0$$

$$\therefore \qquad\qquad l\sqrt{\frac{P}{EI}} = \frac{\pi}{2},\ \frac{3\pi}{2},\ \frac{5\pi}{2}$$

Considering the first practical value,

$$l\sqrt{\frac{P}{EI}} = \frac{\pi}{2}$$

$$\therefore \qquad\qquad P = \frac{\pi^2 EI}{4l^2} \qquad\qquad ...(16.3)$$

Case III. When one end of the column is fixed and the other end pinned or hinged:

Fig. 16.6 shows a column AB of length l, whose upper end A is hinged while its lower end is fixed.

Let P be the crippling load. Studying the nature of bending we realize that there will be a restraint moment M_B at the lower fixed end. The existence of restraint moment therefore justifies the need for a horizontal force also at the top end A without which no bending moment can occur at B.

Hence the hinge at A must exert a horizontal force H at A.

Consider any section XX at a distance x from the lower fixed end B.

The bending moment at the section is given by

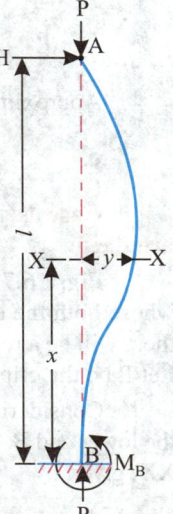

Fig. 16.6

$$EI\,\frac{d^2y}{dx^2} = -Py + H\,(l-x)$$

$\therefore \quad EI\,\frac{d^2y}{dx^2} + Py = H\,(l-x)$

The solution to above differential equation is,

$$y = C_1 \cos\left(x\,\sqrt{\frac{P}{EI}}\right) + C_2 \sin\left(x\,\sqrt{\frac{P}{EI}}\right) + \frac{H}{P}(l-x)$$

where, C_1 and C_2 are constants of integration.

The slope at any section is given by,

$$\frac{dy}{dx} = -C_1\,\sqrt{\frac{P}{EI}}\,\sin\left(x\,\sqrt{\frac{P}{EI}}\right) + C_2\,\sqrt{\frac{P}{EI}}\,\cos\left(x\,\sqrt{\frac{P}{EI}}\right) - \frac{H}{P}$$

At B, the deflection is zero.

$\therefore \qquad$ At $x = 0,\ y = 0$

$\therefore \qquad 0 = C_1 + \dfrac{H}{P}\,l$

$\therefore \qquad C_1 = -\dfrac{H}{P}\,l$

At B, the slope is zero.

$\therefore \qquad$ At $x = 0,\ \dfrac{dy}{dx} = 0$

$\therefore \qquad 0 = C_2\,\sqrt{\dfrac{P}{EI}} - \dfrac{H}{P},\ \text{ or, }\ C_2 = \dfrac{H}{P}\,\sqrt{\dfrac{EI}{P}}$

At A, the deflection is zero

$\therefore \qquad$ At $x = l,\ y = 0$

$\therefore \qquad 0 = -\dfrac{H}{P}\,l\,\cos\left(l\cdot\sqrt{\dfrac{P}{EI}}\right) + \dfrac{H}{P}\,\sqrt{\dfrac{EI}{P}}\,\sin\left(l\,\sqrt{\dfrac{P}{EI}}\right)$

Simplifying, we get

$$\tan\left(l\cdot\sqrt{\frac{P}{EI}}\right) = \left(l\cdot\sqrt{\frac{P}{EI}}\right)$$

The solution to this equation is

$$l\,\sqrt{\frac{P}{EI}} = 4.5 \text{ radians}$$

$$\therefore \qquad \frac{l^2 P}{EI} = (4.5)^2 = 20.25$$

$$\therefore \qquad P = \frac{20.25\, EI}{l^2}$$

Approximately $20.25 = 2\pi^2$

$$\therefore \qquad P = \frac{2\pi^2 EI}{l^2} \qquad \text{... (16.4)}$$

Case IV. When both ends of the column are fixed:

Fig. 16.7 shows a column AB of length l whose both the ends A and B are fixed. Obviously there will be a restraint moment say M_0 at each end. Let P be the crippling load.

Considering any section XX distant x from the lower end B. The bending moment at the section XX, given by

$$EI\, \frac{d^2 y}{dx^2} = M_o - Py$$

$$\therefore \qquad EI\, \frac{d^2 y}{dx^2} + Py = M_o$$

$$\therefore \qquad \frac{d^2 y}{dx^2} + \frac{Py}{EI} = \frac{M_o}{EI}$$

The solution to the above differential equation

Train brake master cylinders and mechanisms. Any machine part can be treated as column if it satisfies the end conditions.

$$y = C_1 \cos\left(x\,\sqrt{\frac{P}{EI}}\right) + C_2 \sin\left(x\,\sqrt{\frac{P}{EI}}\right) + \frac{M_o}{P}$$

where, C_1 and C_2 are constants of integration.
The slope at any section is given by:

$$\frac{dy}{dx} = -\,C_1 \sqrt{\frac{P}{EI}}\, \sin\left(x\,\sqrt{\frac{P}{EI}}\right) + C_2 \sqrt{\frac{P}{EI}}\, \cos\left(x\,\sqrt{\frac{P}{EI}}\right)$$

At B, the deflection is zero.

$$\therefore \qquad \text{At } x = 0,\ y = 0$$

$$\therefore \qquad 0 = C_1 + \frac{M_o}{P}\ \text{ or } C_1 = -\frac{M_o}{P}$$

At B, the slope is zero.

$$\therefore \qquad \text{At } x = 0,\ \frac{dy}{dx} = 0$$

$$\therefore \qquad 0 = C_2 \sqrt{\frac{P}{EI}},\ \text{or},\ C_2 = 0$$

At A, the deflection is zero

$$\therefore \qquad \text{At } x = l,\ y = 0$$

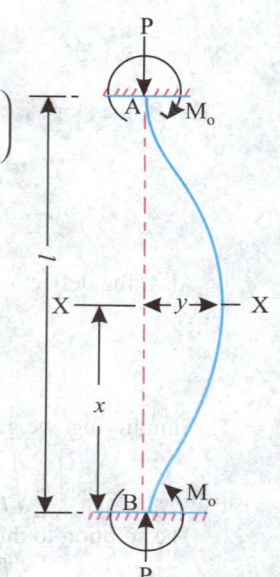

Fig. 16.7

$\therefore \qquad 0 = -\dfrac{M_o}{P} \cos\left(l\sqrt{\dfrac{P}{EI}}\right) + \dfrac{M_o}{P}, \text{ or, } \dfrac{M_o}{P}\left[1 - \cos\left(l\cdot\sqrt{\dfrac{P}{EI}}\right)\right] = 0$

$\therefore \qquad \cos\left(l\cdot\sqrt{\dfrac{P}{EI}}\right) = 1$

$\therefore \qquad l\times\sqrt{\dfrac{P}{EI}} = 0,\ 2\pi,\ 4\pi,\ 6\pi \dots$

Considering the first practical value,

$$l\sqrt{\dfrac{P}{EI}} = 2\pi$$

$$P = \dfrac{4\pi^2 EI}{l^2} \qquad\qquad\qquad \dots(16.5)$$

Example 16.1. *A solid round bar 60 mm in diameter and 2.5 m long is used as a strut. One end of the strut is fixed, while its other end is hinged. Find the safe compressive load for this strut, using Euler's formula. Assume E = 200 GN/m² and factor of safety = 3.*

Solution. Diameter of solid round bar,

$$D = 60 \text{ mm} = 0.06 \text{ m}$$

Modulus of elasticity, $\quad E = 200 \text{ GN/m}^2$

Factor of safety, $\quad$ F.O.S. = 3

Length of round bar (strut),

$$l = 2.5 \text{ m}$$

End conditions : *One end hinged, other fixed*

$\therefore \qquad l_e = \dfrac{l}{\sqrt{2}} = \dfrac{2.5}{\sqrt{2}} = 1.768 \text{ m}$

Safe compressive load :

Euler's crippling load is given by the relation,

$$P_{\text{Euler}} = \dfrac{\pi^2 EI}{l_e^2} = \dfrac{\pi^2 \times 200 \times 10^9 \times \dfrac{\pi}{64} \times (0.06)^4}{(1.768)^2} \times 10^{-3} \text{ kN} = 401.7 \text{ kN}$$

$\therefore$ *Safe compressive load* $= \dfrac{P_{\text{Euler}}}{\text{F.O.S.}} = \dfrac{401.7}{3} =$ **133.9 kN** **(Ans.)**

Example 16.2. *In an experimental determination of the buckling load for 1.2 cm diameter mild steel pin-ended struts of various lengths, two of the values obtained were:*

(i) *When length = 50 cm the load = 10 kN, and*

(ii) *When length = 20 cm, the load = 30 kN.*

Make the necessary calculations and then state whether either of the above values of loads conforms with the Euler's formula for the critical load. Take E = 200 GN/m².

Solution. Moment of inertia, $I = \pi/64 \times 1.2^4 = 0.102 \text{ cm}^4 = 0.102 \times 10^{-8} \text{ m}^4$

When length $(l_1) = 50$ cm; load = 10 kN

and, when length $(l_2) = 20$ cm; load 30 kN

Young's modulus of elasticity,

$$E = 200 \text{ GN/m}^2$$

End conditions: *Both ends pin jointed.*

$\therefore$

$$l_e = l$$

Using the relation,

$$P_{Euler} = \frac{\pi^2 EI}{l_e^2}, \text{ we have}$$

$\therefore$

$$P_1 = \frac{\pi^2 \times (200 \times 10^9) \times 0.102 \times 10^{-8}}{(0.5)^2} \times 10^{-3} \text{ kN} = 8.05 \text{ kN} \quad ...(i)$$

Similarly,

$$P_2 = \frac{\pi^2 \times (200 \times 10^9) \times 0.102 \times 10^{-8}}{(0.2)^2} \times 10^{-3} \text{ kN} = 50.34 \text{ kN} \quad ...(ii)$$

From the above two calculations, we find that the load in case (i) conforms approximately. **(Ans.)**

Example 16.3. *Calculate the safe compressive load on a hollow cast iron column (one end rigidly fixed and the other hinged) of 150 mm external diameter, 100 mm internal diameter and 10 m length. Use Euler's formula with a factor of safety of 5, and E = 95 GN/m².*

Sol. External diameter, $D = 150$ mm $= 0.15$ m

Internal diameter, $\quad d = 100$ mm $= 0.1$ m

Length of the column, $\quad l = 10$ m

Factor of safety, $\quad$ F.O.S. $= 5$

$$E = 95 \text{ GN/m}^2$$

End conditions: *One end rigidly fixed and the other hinged*

$\therefore$

$$l_e = \frac{l}{\sqrt{2}} = \frac{10}{\sqrt{2}} = 7.07 \text{ m}$$

Safe compressive load:

Using the relation, $P_{Euler} = \dfrac{\pi^2 EI}{l_e^2}$, we get

$$P_{Euler} = \frac{\pi^2 \times 95 \times 10^9 \times \pi/64 \, (0.15^4 - 0.1^4)}{(7.07)^2} \times 10^{-3} \text{ kN} = 374 \text{ kN}$$

$\therefore \quad$ Safe load $= \dfrac{P_{Euler}}{\text{F.O.S.}} = \dfrac{374}{5} = \mathbf{74.8 \text{ kN}}$ **(Ans.)**

Example 16·4. *A slender pin ended aluminium column 1.8 m long and of circular cross-section is to have an outside diameter of 50 mm. Calculate the necessary internal diameter to prevent failure by buckling if the actual load applied is 13·6 kN and the critical load applied is twice the actual load. Take E for aluminium as 70 GN/m².*

Solution. Outside diameter of the column,

$$D = 50 \text{ mm} = 0.05 \text{ m}$$

Inside diameter of the column, d:

$$A = \frac{\pi}{4} \, (D^2 - d^2) = \frac{\pi}{4} \, (0.05^2 - d^2)$$

$$I = \frac{\pi}{64} \, (D^4 - d^4) = \frac{\pi}{64} \, (0.05^4 - d^4)$$

Also, critical load $\quad = 2 \times$ safe load (*Given*)

$$= 2 \times 13.6 = 27.2 \text{ kN}$$

End conditions: *Pin-ended*

$\therefore \quad l_e = l = 1.8 \text{ m}$

Using the relation, $P_{Euler} = \dfrac{\pi^2 EI}{l_e^2}$, we get

$$27.2 \times 10^3 = \dfrac{\pi^2 \times 70 \times 10^9 \times \dfrac{\pi}{64}(0.05^4 - d^4)}{1.8^2}$$

or, $\quad (0.05^4 - d^4) = \dfrac{27.2 \times 10^3 \times 1.8^2 \times 64}{\pi^2 \times 70 \times 10^9 \times \pi} = 2.6 \times 10^{-6}$

or, $\qquad d^4 = 6.25 \times 10^{-6} - 2.6 \times 10^{-6} = 3.65 \times 10^{-6}$

$$d = 0.0437 \text{ m} = \textbf{43.7 mm} \quad \textbf{(Ans.)}$$

Example 16·5. *A built-up beam shown in Fig. 16·8 is simply supported at its ends. Compute its length, given that when it is subjected to a load of 40 kN per metre length, it deflects by 1 cm.*

Find out the safe load, if this beam is used as a column with both ends fixed. Assume a factor of safety of 4. Use Euler's formula. E = 210 GN/m².

Solution. Load, $\qquad w = 40$ kN/m

Length of beam :

Let, $\qquad\qquad l =$ Length of beam.

Moment of inertia of section about *X-X* axis.

$$I_{XX} = \dfrac{1}{12}(30 \times 110^3 - 28 \times 100^3)$$

$$= 994166 \text{ cm}^4 = 99.41 \times 10^{-4} \text{ m}^4$$

Using the relation: $\delta = \dfrac{5wl^4}{384\ EI}$, we get

$$0.01 = \dfrac{5 \times (40 \times 10^3) \times l^4}{384 \times 210 \times 10^9 \times 99.41 \times 10^{-4}}$$

∴ $\qquad l^4 = \dfrac{0.01 \times 384 \times 210 \times 10^9 \times 99.41 \times 10^{-4}}{5 \times 40 \times 10^3}$

or, $\qquad\qquad l = \textbf{14.15 m} \quad \textbf{(Ans.)}$

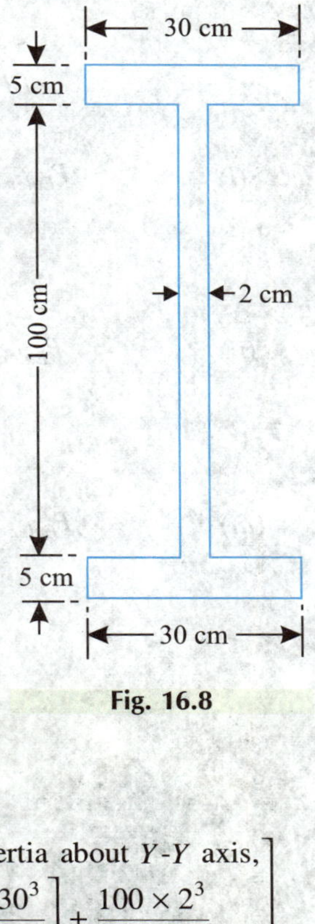

Safe load, the beam can carry as a column:

End condition: *Both ends fixed*

i.e. $\qquad l_e = \dfrac{l}{2} = \dfrac{14.15}{2} = 7.07$ m

$$P_{Euler} = \dfrac{\pi^2 EI}{l_e^2}$$

$$= \dfrac{\pi^2 \times 210 \times 10^9 \times 2.25 \times 10^{-4}}{7.07^2} \times 10^{-3} \text{ kN}$$

$$= 9330 \text{ kN}$$

Safe load $\qquad = \dfrac{P_{Euler}}{\text{F.O.S.}} = \dfrac{9330}{4} = \textbf{2332.5 kN} \quad \textbf{(Ans.)}$

$$\left[\begin{array}{l}\text{Moment of intertia about } Y\text{-}Y \text{ axis,}\\[2mm] I_{YY} = 2 \times \left[\dfrac{5 \times 30^3}{12}\right] + \dfrac{100 \times 2^3}{12}\\[2mm] = 22567 \text{ cm}^4 = 2.25 \times 10^{-4} \text{ m}^4\end{array}\right]$$

Fig. 16.8

Example 16·6. *A bar of length 4 m when used as a simply supported beam and subjected to a u.d.l. of 30 kN/m over the whole span, deflects 15 mm at the centre. Determine the crippling loads when it is used as a column with following end conditions:*

(i) *Both ends pin-jointed;*

(ii) *One end fixed and other end hinged;*

(iii) *Both ends fixed.*

Solution. Length of the bar,

$$l = 4 \text{ m}$$

Uniformly distributed load,

$$w = 30 \text{ kN/m}$$

Deflection,

$$\delta = 15 \text{ mm} = 0 \cdot 015 \text{ m}$$

We know that,

$$\delta = \frac{5 \, wl^4}{384 \, EI}$$

$$0.015 = \frac{5 \times (30 \times 10^3) \times 4^4}{384 \, EI}$$

$$EI = \frac{5 \times (30 \times 10^3) \times 4^4}{0.015 \times 384} = 6.66 \times 10^6 \text{ Nm}^2$$

(i)

$$P_{Euler} = \frac{\pi^2 EI}{l_e^2} \qquad (l_e = l = 4 \text{ m})$$

$$= \frac{\pi^2 \times 6.66 \times 10^6}{4^2} \times 10^{-3} \text{ kN} = \textbf{4108 kN} \quad \textbf{(Ans.)}$$

(ii)

$$P_{Euler} = \frac{\pi^2 EI}{l_e^2} \qquad \left[l_e = \frac{l}{\sqrt{2}} = \frac{4}{\sqrt{2}} = 2.83 \text{ m} \right]$$

$$= \frac{\pi^2 \times 6.66 \times 10^6}{2.83^2} \times 10^{-3} \text{ kN} = \textbf{8207 kN} \quad \textbf{(Ans.)}$$

(iii)

$$P_{Euler} = \frac{\pi^2 EI}{l_e^2} \qquad \left[l_e = \frac{l}{2} = \frac{4}{2} = 2 \text{ m} \right]$$

$$= \frac{\pi^2 \times 6.66 \times 10^6}{2^2} \times 10^{-3} \text{ kN} = \textbf{16432 kN} \quad \textbf{(Ans.)}$$

Prototype of a steam engine in a laboratory. The connecting rods of steam engine behave like columns.

Example 16·7. *Calculate the critical load of a strut which is made of a bar circular in section and 5 m long and which is pin-jointed at both ends. The same bar when freely supported gives mid-span deflection of 10 mm with a load of 80 N at the centre.*

Solution. We know that,

$$P_{Euler} = \frac{\pi^2 EI}{l_e^2} = \frac{\pi^2 EI}{l^2} \qquad\qquad [\because l_e = l \text{ in this case}]$$

and, deflection,

$$\delta = \frac{Wl^3}{48\,EI}$$

or,

$$EI = \frac{Wl^3}{48\,\delta} \qquad \text{or,} \qquad \frac{EI}{l^2} = \frac{Wl}{48\,\delta}$$

∴

$$P_{Euler} = \frac{\pi^2 Wl}{48\,\delta} = \frac{\pi^2 \times 80 \times 5}{48 \times (10 \times 10^{-3})} \times 10^{-3}\ \text{kN} = \textbf{8.22 kN} \quad \textbf{(Ans.)}$$

Example 16·8. *Calculate the maximum value of the slenderness ratio of a mild steel column for which Euler's formula is valid.*

Take:

$$\sigma_c = 330\ MN/m^2$$
$$E = 210\ GN/m^2.$$

Solution. The buckling or crippling load P_{cr} for a column with both ends hinged is given by:

$$P_{cr} = \frac{\pi^2 EI}{l_e^2} = \frac{\pi^2 EI}{l^2} \qquad (\text{where, } l_e = l)$$

Let,
          A = Area of cross-section of the column,

          k = Least radius of gyration, and

          σ_{cr} = Buckling stress.

So that,
          $I = Ak^2$ and $\sigma_{cr} = \dfrac{P_{cr}}{A}$

The maximum value of slenderness ratio (l/k) for which Euler's formula can be applied is found as follows:

Since a long column buckles at the stress lower than the compressive yield stress σ_c of the material of the column, it follows that the Euler's formula will be valid so long as

$$\sigma_{cr} \le \sigma_c$$

For steel,
$$\sigma_c = 330\ MN/m^2$$
$$E = 210\ GN/m^2$$

We know that,
$$P_{cr} = \frac{\pi^2 EI}{l^2} \qquad \text{... Euler's formula}$$

or,
$$P_{cr} = \frac{\pi^2 EAk^2}{l^2} = \frac{\pi^2 EA}{l^2/k^2}$$

or,
$$\frac{P_{cr}}{A} = \frac{\pi^2 E}{l^2/k^2} = \frac{\pi^2 E}{(l/k)^2}$$

or,
$$\sigma_{cr} = \frac{\pi^2 E}{(l/k)^2}$$

Euler's formula for *hinged* ends will apply if

$$\frac{\pi^2 E}{(l/k)^2} \le 330 \times 10^6$$

i.e. if
$$(l/k)^2 \geq \frac{\pi^2 E}{330 \times 10^6}$$

i.e. if
$$(l/k)^2 \geq \frac{\pi^2 \times 210 \times 10^9}{330 \times 10^6}$$

i.e. if
$$l/k \geq 79.25 \text{ say } 80$$

or,
$$l/k \geq 80$$

Hence for a steel column with *hinged ends*, Euler's formula is *valid for* $(l/k) \geq 80$ (**Ans.**)

Example 16·9. *Determine the ratio of the buckling strengths of two columns of circular cross-section one hollow and other solid when both are made of the same material, have the same length, cross-sectional area and end-conditions. The internal diameter of the hollow column is half of its external diameter.*

Solution. Let D_S = Diameter of the solid column,

D_H = External diameter of the hollow column,

d_H = Internal diameter of the hollow column,

A_S = Area of cross-section of the solid column,

A_H = Area of cross-section of the hollow column,

P_S = Buckling load of the solid column, and

P_H = Buckling load of hollow column.

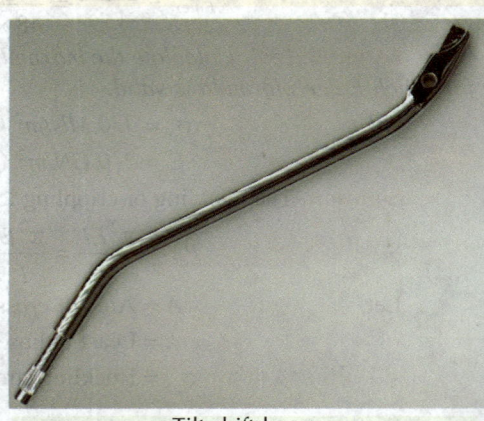

Tilt shift lever.

$$A_S = A_H \qquad (Given)$$

$$\frac{\pi}{4} D_s^2 = \frac{\pi}{4}(D_H^2 - d_H^2)$$

or,
$$D_S^2 = \left[D_H^2 - (0.5\, D_H)^2\right] \qquad [\because d_H = 0.5\, D_H \quad ...(Given)]$$

$$= D_H^2 (1 - 0.25) = 0.75\, D_H^2$$

or,
$$D_S = 0.866\, D_H$$

Solid column:

$$I = \frac{\pi}{64} D_S^4 = \frac{\pi}{64} \times (0.866\, D_H)^4 = 0.0276\, D_H^4$$

$$P_S = \frac{\pi^2 EI}{l_e^2} = \frac{\pi^2 \times E \times 0.0276\, D_H^4}{l_e^2}$$

Hollow column:

$$I = \frac{\pi}{64}\left[D_H^4 - (0.5\, D_H)^4\right] = \frac{\pi}{64} D_H^4 \left[1 - (0.5)^4\right]$$

$$= \frac{\pi}{64} D_H^4 \times 0.9375 = 0.046\, D_H^4$$

$$P_H = \frac{\pi^2 EI}{l_e^2} = \frac{\pi^2 E \times 0.046\, D_H^4}{l_e^2}$$

Since end conditions are same in both cases, therefore, l_e will be same.

∴ Ratio of buckling strengths of the two columns,

$$\frac{P_H}{P_S} = \frac{\pi^2 E \times 0.046 \, D_H^4 / l_e^2}{\pi^2 E \times 0.0276 \, D_H^4 / l_e^2} = 1.66 \quad \text{(Ans.)}$$

16·13. RANKINE'S HYPOTHESIS FOR STRUTS/COLUMNS

Acceptable theoretical expressions are available for both very short and very long struts. The Rankine hypothesis is designed to *link* these two known results to obtain an expression *applicable to struts/columns of all dimensions*.

Now, for a *very short strut*, collapse will result from direct crushing, and crippling load is

$$P_c = \sigma_c \cdot A$$

where, σ_c is the *maximum possible compressive stress* and A is the sectional area.

For a long *strut* the Euler formula applies,

$$P_{Euler} = \frac{\pi^2 EI}{l_e^2} = \frac{\pi^2 EAk^2}{l_e^2} = \pi^2 \, EA \left(\frac{k}{l_e} \right)^2$$

The *Rankine hypothesis* is

$$\frac{1}{P} = \left(\frac{1}{P_{Rankine}} \right) = \frac{1}{P_c} + \frac{1}{P_{Euler}}$$

where, $P_{Rankine}$ is the actual crippling load for a strut.

If the strut is very short, P_{Euler} becomes very large and $\dfrac{1}{P_{Euler}} = 0$, so that $P = P_c$ which is in fact true.

Similarly, for a very long strut, $\dfrac{1}{P_{Euler}}$ becomes very large so that, $P = P_{Euler}$, which is also true.

Then it may be assumed that if the Rankine hypothesis is true for both very long and very short struts, it will also be true for struts of other dimensions.

Substituting, we have

$$\frac{1}{P} = \frac{1}{\sigma_c \cdot A} + \frac{1}{\pi^2 EA \cdot \left(\dfrac{k}{l_e} \right)^2} \quad \text{or} \quad \frac{A}{P} = \frac{1}{\sigma_c} + \frac{1}{\pi^2 E \left(\dfrac{k}{l_e} \right)^2}$$

∴

$$\frac{P}{A} = \frac{1}{\dfrac{1}{\sigma_c} + \dfrac{l_e^2}{\pi^2 E k^2}}$$

∴

$$\frac{P}{A} = \frac{\sigma_c}{1 + \dfrac{\sigma_c}{\pi^2 E} \cdot \left(\dfrac{l_e}{k} \right)^2}$$

∴

$$P_{Rankine} = \frac{\sigma_c \cdot A}{1 + a \left(\dfrac{l_e}{k} \right)^2} \qquad \qquad \dots (16.7)$$

This is Rankine formula for the mean breaking stress of a strut/column, where $a = \dfrac{\sigma_c}{\pi^2 E}$. The expression '$a$' must in practice be obtained empirically.

In Rankine's formula both σ_c and a are constants for given column material.

The table 16·2 shows the values of σ_c and a for different strut/column materials.

TABLE 16·2

Material	σ_c (MN/m²)	$a = \dfrac{\sigma_c}{\pi^2 E}$
Mild steel	320	$\dfrac{1}{7500}$
Cast-iron	550	$\dfrac{1}{1600}$
Wrought iron	250	$\dfrac{1}{9000}$
Strong timber	50	$\dfrac{1}{750}$

Studying Rankine's formula,

$$P_{Rankine} = \frac{\sigma_c \cdot A}{1 + a \cdot \left(\dfrac{l_e}{k}\right)^2}$$

We find,

$$P_{Rankine} = \frac{\text{Crushing load}}{1 + a \left(\dfrac{l_e}{k}\right)^2}$$

The factor $1 + a \left(\dfrac{l_e}{k}\right)^2$ has thus been introduced to *take into account the buckling effect.*

Sometimes it is required to find out the length of a column which shall give the same value of buckling load by Euler and Rankine formulae. This is obtained as follows:

Equating the two formulae, we get

$$P_{Euler} = P_{Rankine}$$

$$\frac{\pi^2 EI}{l_e^2} = \frac{\sigma_c \cdot A}{1 + a \cdot \left(\dfrac{l_e}{k}\right)^2}$$

or, $\pi^2 EI \times \left(1 + a \cdot \dfrac{l_e^2}{k^2}\right) = \sigma_c \cdot A \cdot l_e^2$ or $\sigma_c A l_e^2 - \dfrac{\pi^2 EI \, a l_e^2}{k^2} = \pi^2 EI$

or, $l_e^2 \left(\sigma_c \cdot A - \dfrac{\pi^2 EA \, ak^2}{k^2}\right) = \pi^2 EAk^2$ $\qquad (\because I = Ak^2)$

or, $l_e^2 = \dfrac{\pi^2 Ek^2}{\sigma_c - \pi^2 Ea}$

$\therefore$ $l_e = \left(\dfrac{\pi^2 Ek^2}{\sigma_c - \pi^2 Ea}\right)^{1/2}$ $\qquad ...(16·8)$

It may be noted that the value of 'a' in this equation should be substituted for hinged ends only and the length so obtained will be for hinged ends only. If the problem pertains to end conditions *other than the hinged ends, and 'a' substituted is for the hinged ends, the value of l_e so obtained will be the equivalent length for the given case, and it can be converted into the actual length of the column, the relation between two having been given in the article 16·6.*

16·14. JOHNSON'S PARABOLIC FORMULA

A simple expansion of the Rankine formula will give a parabolic expression which is convenient in use,

$$\frac{P}{A} = \sigma_c \left[1 - a \left(\frac{l_e}{k} \right)^2 \right] \text{ approximately}$$

or,

$$\frac{P}{A} = \sigma_c - b \left(\frac{l}{k} \right)^2 \qquad \qquad ..(16.9)$$

b is approximately $\left(\dfrac{\sigma_c^2}{4\pi^2 E} \right)$ and Johnson accepted the value $\dfrac{\sigma_c^2}{64\,E}$ for pinned ends.

16·15. STRAIGHT-LINE FORMULA

Often in practice it is useful to adopt an empirical expression for a short range of values of l/k. An approximately linear expression can be taken of the form

$$\frac{P}{A} = \sigma_c - z \left(\frac{l}{k} \right) \qquad \qquad ...(16.10)$$

where, z is an empirical constant, and where the range of application is strictly defined.

Some of the approximate empirical formulae used in practical designing are given below:

(i)   Stress at critical load for *cast iron*,

$$\frac{P}{A} = 23.8 - 0.6 \left(\frac{l}{k} \right) \text{N/mm}^2 \qquad \qquad ...(16·11)$$

(ii)   Stress at critical load for *structural steel*,

$$\frac{P}{A} = 367.5 - 2 \left(\frac{l}{k} \right) \text{N/mm}^2 \qquad \qquad ...(16·12)$$

Parts of automobile G-charger which is a modern version of supercharger. G-charger supplies compressed air to the IC engine intake manifold.

(*iii*) Safe working stress for *mild steel*,

$$= 150 \left[1 - 0.0038 \, \frac{1}{k} \right] \text{N/mm}^2 \qquad \qquad ...(16 \cdot 13)$$

In general, strut formulae should be used cautiously, carefully checking the conditions under which the expressions were developed, and limitations on the data employed to calculate the constants. Extrapolation of straight-line formula beyond the limits of applicability will usually produce results seriously in error. Further, it will not be unusual for the strut to have imperfect end fixings, neither pinned nor encastre, which will further introduce error. Frequently the loads will be applied non-axially and additional lateral loads will introduce further complexities. It will perhaps be worthwhile to strive for the perfect strut, truly fixed and loaded, but it will prove to be a rare exception.

Example 16·10. *A hollow C.I. column whose outside diameter is 200 mm has a thickness of 20 mm. It is 4·5 m long and is fixed at both ends. Calculate the safe load by Rankine-Gordon formula using a factor of safety of 4.*

Take: $\sigma_c = 550 \text{ MN/m}^2; a = 1/1600.$

Solution. Outside diameter of the column,

$$D = 200 \text{ mm} = 0 \cdot 2 \text{ m}$$

Thickness of the column, $\qquad t = 20 \text{ mm}$

Inside diameter of the column, $\quad d = D - 2t = 200 - 2 \times 20 = 160 \text{ mm} = 0 \cdot 16 \text{ m}$

Length of the column, $\qquad l = 4 \cdot 5 \text{ m}$

Factor of safety, (F.O.S) = 4

Area of the column, $\qquad A = \dfrac{\pi}{4} (0 \cdot 2^2 - 0 \cdot 16^2) = 0 \cdot 0113 \text{ m}^2$

Moment of inertia, $\qquad I = \dfrac{\pi}{64} (0 \cdot 2^4 - 0 \cdot 16^4) = 4 \cdot 637 \times 10^{-5} \text{ m}^4$

$$k^2 = \frac{I}{A} = \frac{4.637 \times 10^{-5}}{0.0113} = 0.0041 \text{ m}^2$$

End conditions: *Both ends fixed*

i.e. $\qquad l_e = \dfrac{l}{2} = \dfrac{4.5}{2} = 2.25 \text{ m}$

Using the relation, $\qquad P_{Rankine} = \dfrac{\sigma_c \cdot A}{1 + a \cdot \left(\dfrac{l_e^2}{k^2} \right)}$, we get

$$P_{Rankine} = \frac{550 \times 0 \cdot 0113}{1 + \dfrac{1}{1600} \times \dfrac{2 \cdot 25^2}{0 \cdot 0041}} = 3 \cdot 51 \text{ MN}$$

∴ Safe load $= \dfrac{3 \cdot 51}{4} = \mathbf{0 \cdot 877 \text{ MN}}$ **(Ans.)**

Example 16·11. *Compare the crippling loads given by Rankine's and Euler's formulae for tubular strut 2·25 m long having outer and inner diameters of 37·5 mm and 32·5 mm loaded through pin-joint at both ends.*

Take: Yield stress as 315 MN/m²; $a = \dfrac{1}{7500}$, and E = 200 GN/m².

If elastic limit for the material is taken as 200 MN/m², then for what length of the strut does the Euler formula cease to apply?

Solution. Outer diameter of strut, $D = 37 \cdot 5 \text{ mm} = 0 \cdot 0375 \text{ m}$

Inner diameter of the strut,

$$d = 32 \cdot 5 \text{ mm} = 0 \cdot 0325 \text{ m}$$

Length of the strut, $l = 2 \cdot 25$ m

End conditions = *Both ends pin-jointed*

i.e. $l_e = l = 2 \cdot 25$ m

Yield stress $\sigma_c = 315 \text{ MN/m}^2$

Rankine constant, $a = \dfrac{1}{7500}$

$$E = 200 \text{ GN/m}^2.$$

Comparison of loads:

• *Euler's crippling load, P_{Euler}:*

Using the relation: $P_{Euler} = \dfrac{\pi^2 EI}{l_e^2}$, we get

$$P_{Euler} = \frac{\pi^2 \times 200 \times 10^9 \times 4 \cdot 23 \times 10^{-8}}{2 \cdot 25^2} \times 10^{-3} \text{ kN}$$

$$\left[\begin{array}{l} I = \dfrac{\pi}{64}(D^4 - d^4) \\[2mm] = \dfrac{\pi}{64}(0.0375^4 - 0.0325^4) \\[2mm] = 4.23 \times 10^{-8} \text{ m}^4 \end{array}\right]$$

$$= 16 \cdot 493 \text{ kN}$$

• *Rankine's crippling load, $P_{Rankine}$:*

$$P_{Rankine} = \frac{\sigma_c \cdot A}{1 + a\left(\dfrac{l_e^2}{k^2}\right)}$$

$$\left[k^2 = \frac{I}{A} = \frac{\dfrac{\pi}{64}(0.0375^4 - 0.0325^4)}{\dfrac{\pi}{4}(0.0375^2 - 0.0325^2)} = \frac{0.0375^2 + 0.0325^2}{16} = 1.539 \times 10^{-4} \text{ m}^2 \right]$$

$$\therefore \quad P_{Rankine} = \frac{315 \times 10^6 \times \dfrac{\pi}{4}(0.0375^2 - 0.0325^2)}{1 + \dfrac{1}{7500} \times \dfrac{2.25^2}{1 \cdot 539 \times 10^{-4}}} \times 10^{-3} \text{ kN} = 16.078 \text{ kN}$$

$$\frac{P_{Euler}}{P_{Rankine}} = \frac{16 \cdot 493}{16 \cdot 078} = \mathbf{1.026} \quad \textbf{(Ans.)}$$

Length of strut:

Now, Euler's stress = 200 MN/m^2

$$\frac{\text{Euler load}}{\text{Area}} = 200 \times 10^6$$

$$\frac{\pi^2 EI}{A \, l_e^2} = 200 \times 10^6$$

$$\frac{\pi^2 EAk^2}{A \, l_e^2} = 200 \times 10^6 \quad (\because I = Ak^2)$$

or, $$\frac{\pi^2 Ek^2}{l_e^2} = 200 \times 10^6$$

or, $\qquad l_e^2 = \dfrac{\pi^2 E k^2}{200 \times 10^6} = \dfrac{\pi^2 \times 200 \times 10^9 \times 1.539 \times 10^{-4}}{200 \times 10^6} = 1.5189 \text{ m}^2$

or, $\qquad\qquad l_e = 1.232 \text{ m} \quad \therefore \qquad\qquad l_e = l = \textbf{1.232 m} \quad \textbf{(Ans.)}$

Example 16·12. *A 1·5 m long C.I. column has a circular cross-section of 5 cm diameter. One end of the column is fixed in direction and position and the other is free. Taking factor of safety as 3, calculate the safe load, using:*

(i) *Rankine-Gordon formula; take yield stress 560 MN/m², and* $a = \dfrac{1}{1600}$ *for pinned ends.*

(ii) *Euler's formula.*
   Young's modulus for C.I. = 120 GN/m².

Solution. Area, $\qquad A = \dfrac{\pi}{4} \times 5^2 = 19.64 \text{ cm}^2 = 19.64 \times 10^{-4} \text{ m}^2$

and, moment of inertia, $I = \dfrac{\pi}{64} \times 5^4 = 30.7 \text{ cm}^4 = 30.7 \times 10^{-8} \text{ m}^4$

Factor of safety, F.O.S. = 3

(i) Safe load by Rankine-Gordon formula:

Yield stress, $\qquad\qquad \sigma_c = 560 \text{ MN/m}^2$

Rankine's constant, $\quad a = \dfrac{1}{1600}$

End conditions: *One end fixed in direction and position and the other free.*

i.e. $\qquad\qquad l_e = 2l = 2 \times 1.5 = 3.0 \text{ m}$

$$k^2 = \dfrac{I}{A} = \dfrac{(\pi/64) \times 5^4}{(\pi/4) \times 5^2} = 1.5625 \text{ cm}^2 = 1.5625 \times 10^{-4} \text{ m}^2$$

Using the relation,

$$P_{Rankine} = \dfrac{\sigma_c \cdot A}{1 + a \cdot \dfrac{l_e^2}{k^2}}, \text{ we have}$$

$$P_{Rankine} = \dfrac{560 \times 19.64 \times 10^{-4}}{1 + \dfrac{1}{1600} \times \dfrac{3^2}{1.5625 \times 10^{-4}}} = \dfrac{1.09984}{37} \times 10^3 \text{ kN} = 29.72 \text{ kN}$$

$\therefore \qquad$ Safe load $= \dfrac{29.72}{3} = \textbf{9.9 kN} \quad \textbf{(Ans.)}$

(ii) Safe load by Euler's formula:

Young's modulus, $\quad E = 120 \text{ GN/m}^2$
Using the relation,

$$P_{Euler} = \dfrac{\pi^2 EI}{l_e^2}, \text{ we have}$$

$$P_{Euler} = \dfrac{\pi^2 \times 120 \times 10^9 \times 30.7 \times 10^{-8}}{3^2} \times 10^{-3} \text{ kN} = 40.4 \text{ kN}$$

$\therefore \qquad$ Safe load $= \dfrac{40.4}{3} = \textbf{13.47 kN} \quad \textbf{(Ans.)}$

Example 16·13. *A hollow cylindrical cast iron column is 4 m long with both ends fixed. Determine the minimum diameter of the column, if it has to carry a safe load of 250 kN with a factor of safety of 5. Take the internal diameter as 0·8 times the external diameter.*

Take: $\qquad\qquad a = \dfrac{1}{1600}$ *in Rankine's formula and* $\sigma_c = 550 \text{ MN/m}^2$

Solution. Length of the column,

$$l = 4 \text{ m}$$

End conditions: *Both ends fixed,*

i.e., $l_e = \dfrac{l}{2} = \dfrac{4}{2} = 2 \text{ m}$

Safe load $= 250 \text{ kN}$

Factor of safety,

F.O.S. $= 5$

Internal diameter of the column,

$$d = 0.8 \, D$$

$$\sigma_c = 550 \text{ MN/m}^2, \ a = \dfrac{1}{1600}$$

Minimum diameter of the column:

Crippling load,

$$P_{Rankine} = \text{Safe load} \times \text{F.O.S.}$$
$$= 250 \times 5 = 1250 \text{ kN}$$

Using the relation,

$$P_{Rankine} = \dfrac{\sigma_c \cdot A}{1 + a \cdot \left(\dfrac{l_e}{k}\right)^2}, \text{ we get}$$

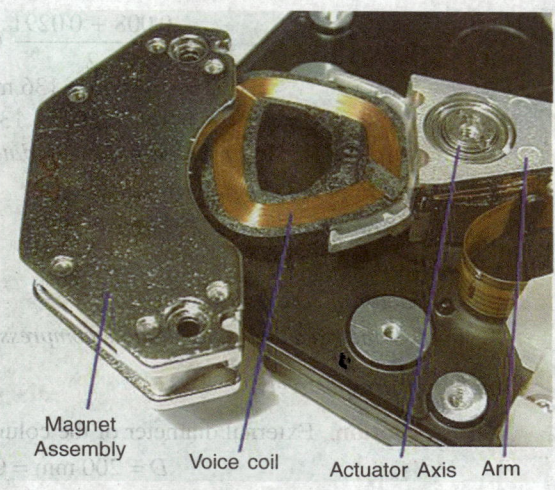

Magnet Assembly   Voice coil   Actuator Axis   Arm

Mechanisms of a hard disk in a computer.

$$1250 \times 10^3 = \dfrac{550 \times 10^6 \times (\pi/4)\,(D^2 - d^2)}{1 + \dfrac{1}{1600}\left(\dfrac{2}{k}\right)^2} \qquad \left[\begin{array}{l} I = (\pi/64)\,(D^4 - d^4) \\[4pt] A = (\pi/4)\,(D^2 - d^2) \\[4pt] \therefore \ k^2 = \dfrac{I}{A} = \dfrac{(\pi/64)\,(D^4 - d^4)}{(\pi/4)\,(D^2 - d^2)} \\[8pt] \therefore \ k^2 = \dfrac{D^2 + d^2}{16} \end{array}\right]$$

or, $$1250 \times 10^3 = \dfrac{550 \times 10^6 \times (\pi/4)\,(D^2 - d^2)}{1 + \dfrac{1}{1600} \times \dfrac{2 \times 2 \times 16}{(D^2 + d^2)}} = \dfrac{550 \times 10^6 \times (\pi/4)\,(D^2 - d^2)}{1 + \dfrac{0.04}{D^2 + d^2}}$$

or, $$\dfrac{1250 \times 10^3}{550 \times 10^6} = (\pi/4)\left[\dfrac{(D^2 - d^2)\,(D^2 + d^2)}{D^2 + d^2 + 0.04}\right]$$

or, $$0.00227 = (\pi/4)\left[\dfrac{D^4 - d^4}{D^2 + d^2 + 0.04}\right]$$

or, $$0.00289 = \dfrac{D^4 - d^4}{D^2 + d^2 + 0.04}$$

or, $$0.00289\,[D^2 + (0.8\,D)^2 + 0.04] = [D^4 - (0.8\,D)^4]$$

or, $$0.00289\,(1.64\,D^2 + 0.04) = 0.59\,D^4$$

$$1.64\,D^2 + 0.04 = 204.1\,D^4$$

or, $$D^4 - 0.008\,D^2 - 0.000196 = 0.$$

$$D^2 = \dfrac{0.008 \pm \sqrt{(0.008)^2 + 4 \times 0.000196}}{2}$$

$$= \frac{0.008 + 0.0291}{2} \text{ (neglecting } -\text{ve sign)} = 0.01855$$

∴ $\qquad D = 0.136 \text{ m} = 136 \text{ mm}$

and, $\qquad d = 0.8 D = 0.8 \times 136 = \mathbf{108.8 \text{ mm}}$ **(Ans.)**

Example 16·14. *From the following data, determine thickness of cast-iron column:*

Length of column = 6 metres

External diameter = 200 mm

Load = 500 kN

Factor of safety = 6

Assume fixed ends and ultimate compressive stress and constant for hinged ends as 570 MN/m²

and $\dfrac{1}{1600}$ *respectively.*

Solution. External diameter of the column,

$$D = 200 \text{ mm} = 0.2 \text{ m}$$

Internal diameter of the column, $d =$?

Thickness of the column, t:

Now, $\qquad\qquad A = \dfrac{\pi}{4} (D^2 - d^2) = \dfrac{\pi}{4} (0.2^2 - d^2)$

$$I = \frac{\pi}{4} (D^4 - d^4)$$

∴ $\qquad k^2 = \dfrac{I}{A} = \dfrac{(\pi/64)(D^4 - d^4)}{(\pi/4)(D^2 - d^2)} = \dfrac{D^2 + d^2}{16} = \dfrac{0.2^2 + d^2}{16}$

Safe load = 500 kN

Factor of safety = 6

∴ $\qquad$ Crippling load = 500 × 6 = 3000 kN

Using Rankine's formula, we get

$$P_{Rankine} = \frac{\sigma_c \cdot A}{1 + a \cdot \left(\dfrac{l_e}{k}\right)^2}$$

$$3000 \times 10^3 = \frac{570 \times 10^6 \times (\pi/4)(0.2^2 - d^2)}{1 + \dfrac{1}{1600} \times \dfrac{3 \times 3 \times 16}{(0.2^2 + d^2)}} \qquad \left[l_e = \frac{l}{2} = \frac{6}{2} = 3 \text{ m}\right]$$

or, $\qquad \dfrac{3000 \times 10^3}{570 \times 10^6} = \dfrac{(\pi/4)(0.04 - d^2)}{1 + \dfrac{0.09}{0.04 + d^2}}$

$$0.0067 \left[1 + \frac{0.09}{0.04 + d^2}\right] = (0.04 - d^2)$$

or, $\qquad 0.0067 (0.04 + d^2 + 0.09) = (0.04 - d^2)(0.04 + d^2)$

$$0.0067 (0.13 + d^2) = (0.0016 - d^4)$$

or, $\qquad d^4 + 0.0067 d^2 - 0.000729 = 0$

$$d^2 = \frac{-0.0067 \pm \sqrt{(0.0067)^2 + 4 \times 0.000729}}{2}$$

$$= \frac{-0.0067 \pm 0.0544}{2} = 0.02385 \text{ m}^2 \qquad \text{(neglecting –ve sign)}$$

∴ $d = 0.1544 \text{ m} = 154.4 \text{ mm}$

∴ $t = \dfrac{D-d}{2} = \dfrac{200 - 154.4}{2} = 22.8 \text{ mm}$

Hence, $t = \textbf{22.8 mm}$ **(Ans.)**

Example 16·15. *Find the Euler crushing load for a hollow cylindrical cast-iron column 200 mm external diameter, 25 mm thick 6, m long and hinged at both ends. E = 120 GN/m².*

Compare the load with the crushing load as given by the Rankine formula taking σ$_c$ = 550 MN/m², and a = $\dfrac{1}{1600}$.

For what length of column would these two formulae give the same crushing load?

Gear teeth are considered as columns with one end fixed and the other end free, while designing.

Solution. External diameter of the column,

$D = 200 \text{ mm} = 0.2 \text{ m}$

Thickness, $t = 25 \text{ mm} = 0.025 \text{ m}$

Internal diameter of the column,

$d = D - 2t = 200 - 2 \times 25 = 150 \text{ mm} = 0.15 \text{ m}$

Length of the column,

$l = 6 \text{ m}$

End conditions: *Both ends hinged*

i.e. $l_e = l = 6 \text{ m}$

$E = 120 \text{ GN/m}^2$

$\sigma_c = 550 \text{ MN/m}^2$

$a = 1/1600.$

Euler crushing load, P_{Euler} :

Using the relation,

$$P_{Euler} = \frac{\pi^2 EI}{l_e^2}, \text{ we get}$$

$$P_{Euler} = \frac{\pi^2 \times 120 \times 10^9 \times (\pi/64)(0.2^4 - 0.15^4)}{6^2} \times 10^{-3} \text{ kN} = \textbf{1766·3 kN (Ans.)}$$

Rankine crushing load, $P_{Rankine}$:

We know, $P_{Rankine} = \dfrac{\sigma_c \cdot A}{1 + a \cdot \left(\dfrac{l_e}{k}\right)^2}$

$$k^2 = \frac{I}{A} = \frac{(\pi/64)(D^4 - d^4)}{(\pi/4)(D^2 - d^2)} = \frac{D^2 + d^2}{16} = \frac{0.2^2 + 0.15^2}{16} = 0.0039 \text{ m}^2$$

$$P_{Rankine} = \frac{550 \times 10^6 \times (\pi/4)(0.2^2 - 0.15^2)}{1 + \frac{1}{1600} \times \frac{6^2}{0.0039}} \times 10^{-3} = 1116.7 \text{ kN}$$

$$\therefore \quad \frac{\text{Euler load}}{\text{Rankine load}} = \frac{1766.3}{1116.7} = 1.58$$

i.e. Euler load is **1·58 times** *Rankine load.* **(Ans.)**

For the same crushing load with both formulae let l_e be the length.

Then,

$$\frac{\pi^2 EI}{l_e^2} = \frac{\sigma_c \cdot A}{1 + a\left(\dfrac{l_e^2}{k^2}\right)}$$

$$\pi^2 EI\left[1 + a \cdot \frac{l_e^2}{k^2}\right] = \sigma_c \cdot A \cdot l_e^2$$

$$\pi^2 EI + \pi^2 EI \cdot a \frac{l_e^2}{k^2} = \sigma_c \cdot A \cdot l_e^2$$

$$\pi^2 EAk^2 + \pi^2 EAk^2 \cdot a \cdot \frac{l_e^2}{k^2} = \sigma_c A \, l_e^2 \quad (\because I = Ak^2)$$

$$\pi^2 EAk^2 + \pi^2 EAal_e^2 = \sigma_c A l_e^2$$

$$l_e^2 \, (\sigma_c - \pi^2 Ea) = \pi^2 Ek^2$$

$$l_e^2 = \frac{\pi^2 Ek^2}{\sigma_c - \pi^2 Ea} = \frac{\pi^2 \times 120 \times 10^9 \times 0.0039}{550 \times 10^6 - \pi^2 \times 120 \times 10^9 \times \dfrac{1}{1600}}$$

$$= -\frac{4.62 \times 10^9}{1.9 \times 10^8} = -24.3$$

$\therefore \; l_e$ is imaginary quantity

i.e. under given conditions there will be no column to give same crushing load with both formulae. **(Ans.)**

Example 16·16. *A short length of tube 3 cm internal and 5 cm external diameter, failed in compression at a load of 240 kN. When a 2 m length of the same tube was tested as a strut with fixed ends, the load at the failure was 158 kN. Assuming that σ_c in Rankine's formula is given by the first test and the value of constant 'a' in the same formula what will be the crippling load of this tube if it is used as a strut 3 metres long with one end fixed and the other end hinged?*

Solution.

$$\sigma_c = \frac{W}{A} = \frac{240}{(\pi/4)(0.05^2 - 0.03^2)} \times 10^{-3} \text{ MN/m}^2 = 191 \text{ MN/m}^2$$

Load at failure, $\quad P = 158 \text{ kN}$

Length of the tube $\quad l = 2 \text{ m}$

$$k^2 = \frac{I}{A} = \frac{(\pi/64)(0.05^4 - 0.03^4)}{(\pi/4)(0.05^2 - 0.03^2)}$$

$$= \frac{0.05^2 + 0.03^2}{16} = \frac{0.0034}{16} = 0.0002125 \text{ m}^2$$

Rankine's constant (a) when both ends are fixed:

$$l_e = \frac{l}{2} = \frac{2}{2} = 1 \text{ m}$$

$$P_{Rankine} = \frac{\sigma_c \cdot A}{1 + a\left(\dfrac{l_e^2}{k^2}\right)}$$

$$158 \times 10^3 = \frac{191 \times 10^6 \times (\pi/4)\,(0.05^2 - 0.03^2)}{1 + a \times \dfrac{1^2}{0.0002125}}$$

$$158 \times 10^3 \left[1 + a \times \frac{1}{0.0002125}\right] = 191 \times 10^6 \times (\pi/4) \times 0.0016$$

$$\frac{a}{0.0002125} = \frac{191 \times 10^6 \times (\pi/4) \times 0.0016}{158 \times 10^3} - 1 = 0.519$$

∴ $$a = 0.519 \times 0.0002125 = 0.0001103 \left(= \frac{1}{9066}\right)$$

Crippling load (P) when one end is fixed and other hinged:

$$l_e = \frac{l}{\sqrt{2}} = \frac{3}{\sqrt{2}} = 2.12 \text{ m}$$

$$P_{Rankine} = \frac{\sigma_c \cdot A}{1 + a\left(\dfrac{l_e^2}{k^2}\right)}$$

$$= \frac{191 \times 10^6 \times (\pi/4)\,(0.05^2 - 0.03^2)}{1 + 0.0001103\left(\dfrac{(2.12)^2}{0.0002125}\right)} \times 10^{-3} \text{ kN} = \textbf{72 kN} \quad \textbf{(Ans.)}$$

Example 16·17. *Find the greatest length for which a mild steel strut of T-shaped cross-section, the area of which is 30 cm^2 and the least moment of inertia of which is 240 cm^4, may be used with one end fixed and other entirely free in order to carry a working load of 70 MN/m^2 of section, the working load being one-fourth the crippling load. Rankine constants for mild steel are:*

$$a = \frac{1}{7500}, \text{ and } \sigma_c = 330 \text{ MN/m}^2.$$

Solution. Least moment of inertia,

$$I = 240 \text{ cm}^4 = 240 \times 10^{-8} \text{ m}^4$$

Area of cross-section, $A = 30 \text{ cm}^2 = 30 \times 10^{-4} \text{ m}^2$

$$k^2 = \frac{I}{A} = \frac{240 \times 10^{-8}}{30 \times 10^{-4}} = 0.0008 \text{ m}^2$$

Crippling load, $\quad P = $ Working stress × area × factor of safety

$$= (70 \times 10^6) \times 30 \times 10^{-4} \times 4 \times 10^{-3} \text{ kN} = 840 \text{ kN}$$

Let, $\qquad l_e = $ Equivalent length of strut.

Now, using Rankine's formula,

$$P = \frac{\sigma_c \cdot A}{1 + a\left(\dfrac{l_e}{k}\right)^2}, \text{ we have}$$

$$840 \times 10^3 = \frac{330 \times 10^6 \times 30 \times 10^{-4}}{1 + \dfrac{1}{7500} \cdot \dfrac{l_e^2}{0 \cdot 0008}} = \frac{990000}{1 + \dfrac{l_e^2}{6}}$$

or, $\qquad 1 + \dfrac{l_e^2}{6} = \dfrac{990000}{840 \times 10^3} \qquad$ or $\quad \dfrac{l_e^2}{6} = \dfrac{99}{84} - 1 = \dfrac{15}{84}$

or, $\qquad l_e^2 = 1 \cdot 0714 \text{ m}^2 \qquad$ or $\quad l_e = 1 \cdot 035 \text{ m}$

We know that for one end fixed the other free case,

$$l_e = 2l$$

∴ $\qquad l = \dfrac{l_e}{2} = \dfrac{1 \cdot 035}{2} = 0 \cdot 5175 \text{ m}$

i.e. $\qquad$ **l = 0·5175 m (Ans.)**

Example 16·18. *From the following data, determine the diameter of the piston rod:*

Diameter of the engine cylinder = 0·3 m

Maximum effective steam pressure in the cylinder = 800 kN/m²

Distance from piston to cross-head centre = 1·5 m.

Factor of safety = 4

Assume : $\qquad \sigma_c = 330 \text{ MN/m}^2; \ a = \dfrac{1}{30000} \ \text{for both ends fixed.}$

Solution. Let, $\qquad d$ = Diameter of the piston rod.

Since the piston rod is firmly fixed in piston on one side and in the cross-head on the other side, *it should always be assumed that for piston rod both ends are fixed.*

Maximum load on the piston = Pressure × area = 800 × (π/4) × 0·3² = 56·55 kN

Crippling load = 56·55 × 4 = 226·2 kN

Using the relation,

$$P_{Rankine} = \frac{\sigma_c \cdot A}{1 + a \cdot \left(\dfrac{l_e}{k}\right)^2}, \text{ we have}$$

Outlet port of a G-charger of an automobile.

$$226 \cdot 2 \times 10^3 = \frac{330 \times 10^6 \times (\pi/4)\, d^2}{1 + \dfrac{1}{30000} \times \dfrac{1 \cdot 5^2 \times 16}{d^2}}$$

$$\left[\because k^2 = \frac{d^2}{16},\ \text{and} = l_e = l = 1\cdot 5\ m \right.$$
because the value of 'a' is for
the given condition i.e. both
$\left. \text{ends fixed.} \right]$

$$226 \cdot 2 \times 10^3 = \frac{259 \cdot 2 \times 10^6\, d^2}{1 + \dfrac{0 \cdot 0012}{d^2}}$$

$$226 \cdot 2 \times 10^3 \left(1 + \frac{0 \cdot 0012}{d^2} \right) = 259 \cdot 2 \times 10^6\, d^2$$

$$1 + \frac{0 \cdot 0012}{d^2} = 1146\, d^2$$

or, $1146\, d^4 - d^2 - 0 \cdot 0012 = 0$

or, $d^4 - 0 \cdot 000873\, d^2 - 1 \cdot 047 \times 10^{-6} = 0$

$$d^2 = \frac{0 \cdot 000873 \pm \sqrt{(0 \cdot 000873)^2 + 4 \times 1 \cdot 047 \times 10^{-6}}}{2}$$

$$= \frac{0 \cdot 000873 \pm 0 \cdot 00222}{2} = 0 \cdot 001546 \ (\text{neglecting } -\text{ve value})$$

∴ $d = 0 \cdot 0393$ m say **40 mm** **(Ans.)**

Example 16·19. *Fig. 16·9 shows built up column consisting of 15 cm × 10 cm RSJ with 12 cm wide plate riveted to each flange. Calculate the safe load, the column can carry, if it is 4 m long having one end fixed and other hinged with a factor of safety 3·5. Take the properties of the joist as A = 21·67 cm², I_{XX} = 839·1 cm⁴, and I_{YY} = 94·8 cm⁴. Assume Rankine's constant as 315 MN/m², and*
$a = \dfrac{1}{7500}.$

(Banglore University)

Solution. Refer to Fig. 16·9.

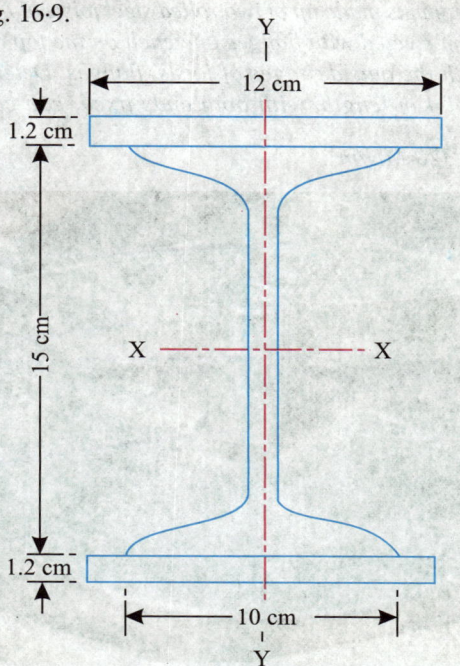

Fig. 16.9

Area of the column,

$$A = 21.67 + (2 \times 12 \times 1.2) = 50.47 \text{ cm}^2 = 50.47 \times 10^{-4} \text{m}^2$$

Moment of inertia of the column about XX axis,

$$I_{XX} = 839.1 + 2 \left[\frac{12 \times 1.2^3}{12} + 12 \times 1.2 \times (7.5 + 0.6)^2 \right] \text{cm}^4$$

$$= 2732.1 \text{ cm}^4 = 2732.1 \times 10^{-8} \text{ m}^4$$

Similarly,

$$I_{YY} = 94.8 + 2 \left[\frac{1.2 \times 12^3}{12} \right] = 440.4 \text{ cm}^4 = 440.4 \times 10^{-8} \text{ m}^4$$

Since I_{YY} is less than I_{XX}, therefore column will tend to buckle in YY direction. Thus we shall take the value of I as $I_{YY} = 440.4 \times 10^{-8} \text{ m}^4$

Also, end conditions are: *One end fixed, other hinged.*

i.e.

$$l_e = \frac{l}{\sqrt{2}} = \frac{4}{\sqrt{2}} = 2.828 \text{ m}$$

and,

$$k^2 = \frac{I}{A} = \frac{440.4}{50.47} = 8.726 \text{ cm}^2 = 8.726 \times 10^{-4} \text{ m}^2$$

Using the relation,

$$P_{Rankine} = \frac{\sigma_c \cdot A}{1 + a \left(\dfrac{l_e^2}{k^2} \right)}, \text{ we get}$$

$$P_{Rankine} = \frac{315 \times 10^6 \times 50.47 \times 10^{-4}}{1 + \dfrac{1}{7500} \times \dfrac{2.828^2}{8.726 \times 10^{-4}}} = \frac{1589805}{2.222} = 715483 \text{ N} = 715.483 \text{ kN}$$

∴ **Safe load**

$$= \frac{715.483}{3.5} = 204.4 \text{ kN} \quad \textbf{(Ans.)}$$

Example 16·20. *A column is made up of two rolled steel joists of I-section, 16 cm × 8 cm × 1 cm thick with plate 20 cm × 1 cm riveted with flanges one each on the top and on the bottom. The edges of the plates being flush with the outside edges of joists' flanges. Determine, by Rankine's formula, the safe load the column of 4 m length, with both ends fixed, can carry with factor of safety 3.*

Take: $a = \dfrac{1}{7500}$, *and* $\sigma_c = 320 \text{ MN/m}^2$.

(Rajasthan University)

Mechanisms of a computer hard disk.

— Actuator Pivot

— Split Band

— Motor Shaft

— Stepper Motor

Solution. Refer to Fig. 16·10.

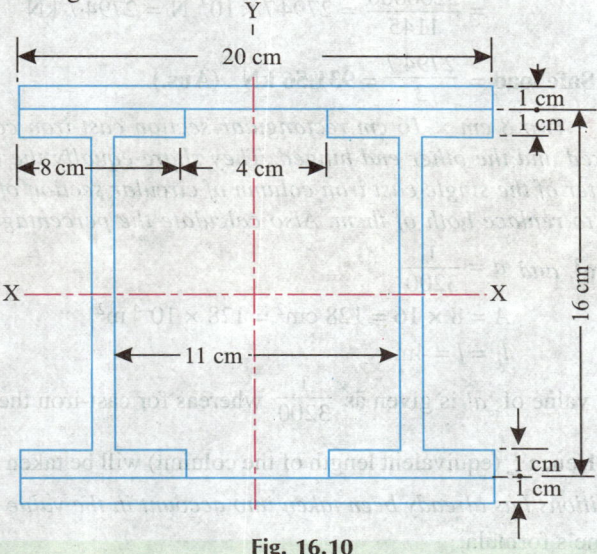

Fig. 16.10

Area of cross-section of the column,

$$A = 2 [8 \times 1 \times 2 + 14 \times 1] + 2 \times 20 \times 1$$
$$= 2 \times 30 + 2 \times 20 = 100 \text{ cm}^2 = 100 \times 10^{-4} \text{ m}^2$$

As the section is symmetrical, the e.g. will lie at the point of intersection of two axes of symmetry.

$$I_{XX} = 4 \left[\frac{8 \times 1^3}{12} + 8 \times 1 \times (8 - 0.5)^2 \right] + 2 \left[\frac{1 \times 14^3}{12} \right]$$

$$+ 2 \left[\frac{20 \times 1^3}{12} + 20 \times 1 \times (8 + 0.5)^2 \right]$$

$$= 1802.6 + 457.3 + 2893.3 = 5153.2 \text{ cm}^4 = 5153.2 \times 10^{-8} \text{ m}^4$$

I_{YY} for one I section only $= 2 \left[\frac{1 \times 8^3}{12} \right] + \frac{14 \times 1^3}{12} = 85.33 + 1.16$

$$= 86.49 \text{ cm}^4 = 86.49 \times 10^{-8} \text{ m}^4$$

Moment of inertia for the whole section about YY axis,

$$I_{YY} = 2 [86.49 + 30 \times 6^2] + 2 \times \frac{1 \times 20^3}{12} = 2332.98 + 1333.33$$

$$= 3666.31 \text{ cm}^4 = 3666.31 \times 10^{-4} \text{ m}^4$$

As I_{YY} is *less* than I_{XX} so column will tend to buckle in YY direction.

End conditions: *Both ends fixed*

i.e.

$$l_e = l / 2 = \frac{4}{2} = 2 \text{ m}$$

$$k^2 = \frac{I}{A} = \frac{3666.31}{100} = 36.66 \text{ cm}^2 = 36.66 \times 10^{-4} \text{ m}^2$$

Now,

$$P_{Rankine} = \frac{\sigma_c \times A}{1 + a \dfrac{l_e^2}{k^2}} = \frac{320 \times 10^6 \times 100 \times 10^{-4}}{1 + \dfrac{1}{7500} \times \dfrac{2^2}{36.66 \times 10^{-4}}}$$

$$= \frac{3200000}{1\cdot145} = 2794\cdot7 \times 10^3 \text{ N} = 2794\cdot7 \text{ kN}$$

Safe load $= \dfrac{2794\cdot7}{3} = \textbf{931.56 kN}$ **(Ans.)**

Example 16.21. *Two 8 cm × 16 cm rectangular section cast-iron columns are each 4 m long with one end fixed and the other end hinged. They share equally the total load carried by them. Find the diameter of the single cast iron column of circular section of the same length and same end conditions to replace both of them. Also calculate the percentage saving in material.*

Take: $\sigma_c = 500 \text{ MN/m}^2,$ *and* $a = \dfrac{1}{3200}.$

Solution. $A = 8 \times 16 = 128 \text{ cm}^2 = 128 \times 10^{-4} \text{ m}^2$

$l_e = l = 4\text{m}$

In this case the value of 'a' is given as $\dfrac{1}{3200}$ whereas for cast-iron the value of 'a' for both

ends hinged is $\dfrac{1}{1600}$. Hence l_e (equivalent length of the column) will be taken as l instead of $\dfrac{l}{\sqrt{2}}$ as the *effect of end conditions has already been taken into account in the value of 'a'.*

Using Rankine's formula,

$$P_{Rankine} = \frac{\sigma_c \cdot A}{1 + a \cdot \dfrac{l_e^2}{k^2}}, \text{ we get} \qquad \left[\begin{array}{l} k^2 = \dfrac{I}{A} \\ \\ = \dfrac{16 \times 8^3}{12 \times 128} \\ \\ = 5\cdot33 \text{ cm}^2 = 5\cdot33 \times 10^{-4} \text{ m}^2 \end{array} \right]$$

$$P_{Rankine} = \frac{500 \times 10^6 \times 128 \times 10^{-4}}{1 + \dfrac{1}{3200} \times \dfrac{4^2}{5\cdot33 \times 10^{-4}}} \times 10^{-3} \text{ kN} = 616.5 \text{ kN}$$

Load taken by two columns $= 2 \times 616\cdot5 = 1233 \text{ kN}$

Let, $d =$ Diameter of the equivalent column with circular section.

$$P_{Rankine} = \frac{\sigma_c \cdot A}{1 + a \cdot \dfrac{l_e^2}{k^2}}$$

$$1233 \times 10^3 = \frac{500 \times 10^6 \times (\pi/4)d^2}{1 + \dfrac{1}{3200} \times \dfrac{4^2}{\dfrac{d^2}{16}}} = \frac{392\cdot7 \times 10^6 \times d^4}{d^2 + 0\cdot08}$$

or $1233 \times 10^3 (d^2 + 0\cdot08) = 392\cdot7 \times 10^6 \times d^4$

$d^4 - 0\cdot00314\, d^2 - 0\cdot0002512 = 0$

$$d^2 = \frac{0\cdot00314 \pm \sqrt{0\cdot00314^2 + 4 \times 0\cdot0002512}}{2}$$

$$= \frac{0\cdot00314 + 0\cdot03185}{2} \quad \text{(neglecting –ve sign),}$$

$= 0\cdot017495$

∴ $d = 0\cdot1322 \text{ m} = \textbf{132·2 mm}$ **(Ans.)**

Let w be the density of the material

Weight of rectangular column = w × volume

$$= w \times (16 \times 8 \times 10^{-4}) \times 4 \times 2 = 0.1024\ w$$

Weight of circular column $\quad = w \times (\pi/4) \times (10.1322)^2 \times 4 = 0.0549\ w$

Percentage saving in weight $\quad = \dfrac{0.1024\ w - 0.0549\ w}{0.1024\ w} \times 100 = \mathbf{46.38\ \%}$ **(Ans.)**

Example 16·22. *A column 8 meters long has a cross-section shown in Fig. 16·11. The column is pinned at both ends. If the column is subject to an axial load equal in value $\dfrac{1}{4}$ of the Euler critical load for the column, determine the factor of safety on Rankine ultimate stress value.*
Take: $\qquad\qquad\qquad \sigma_c = 326\ MN/m^2,$

Rankine's constant $\qquad a = \dfrac{1}{7500},\ and\ E = 200\ GN/m^2$

Properties of one R.S.J.:

$\qquad\qquad$ *Area = 52·05 cm^2; I_{XX} = 5943·1 cm^4;*

$\qquad\qquad I_{YY}$ *= 857·5 cm^4*

Thickness of web $\qquad\qquad$ *= 6·7 mm*

$\qquad\qquad\qquad\qquad\qquad\qquad\qquad\qquad\qquad$ **(Banglore University)**

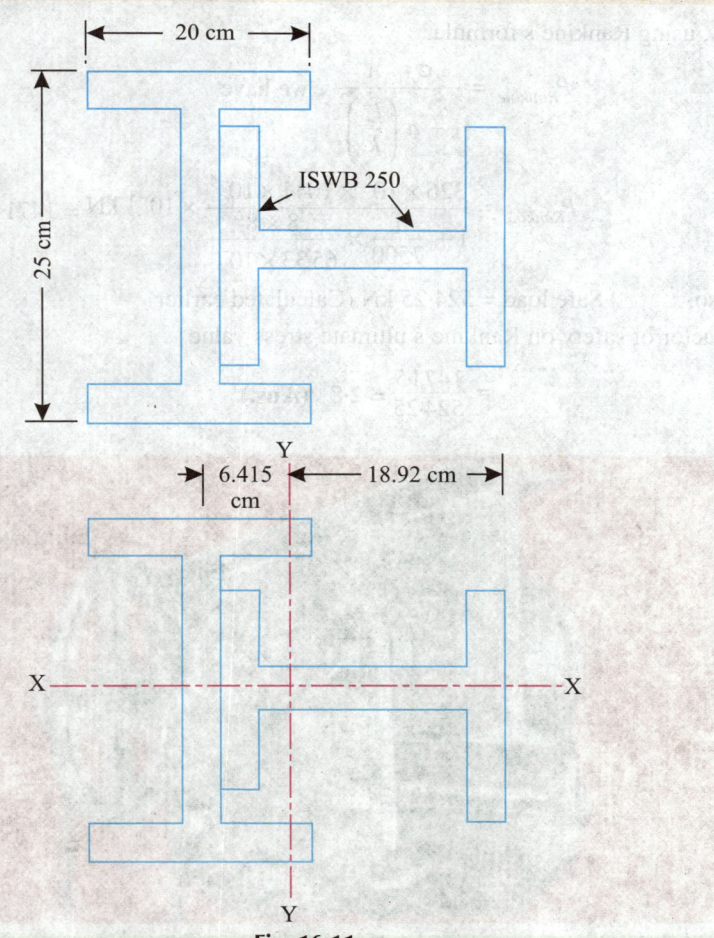

Fig. 16.11

Solution. Area of the combined section,

$$A = 52.05 \times 2 = 104.1 \text{ cm}^2 = 104.1 \times 10^{-4} \text{m}^2$$

Moment of inertia of the combined section about XX-axis,

$$I_{XX} = 5943.1 + 857.5 = 6800.6 \text{ cm}^4$$
$$= 6800.6 \times 10^{-8} \text{ m}^4$$

Since $I_{YY} > I_{XX}$ (evident from the Fig. 16.11)

$$\therefore \quad k^2 = \frac{I_{least}}{A} = \frac{I_{XX}}{A} = \frac{6800.6 \times 10^{-8}}{104.1 \times 10^{-4}}$$
$$= 65.33 \times 10^{-4} \text{ m}^2$$

Using the relation, $p_{Euler} = \dfrac{\pi^2 EI}{l_e^2}$, we have

$$p_{Euler} = \frac{\pi^2 \times 200 \times 10^9 \times 6800.6 \times 10^{-8}}{8^2} \times 10^{-3} \text{ kN} = 2097 \text{ kN} \ (\because l_e = l)$$

Safe load $= \dfrac{1}{4} \times 2097 = 524.25 \text{ kN}$

Now, using Rankine's formula,

$$P_{Rankine} = \frac{\sigma_c \cdot A}{1 + a\left(\dfrac{l_e}{k}\right)^2}, \text{ we have}$$

$$P_{Rankine} = \frac{326 \times 10^6 \times 104.1 \times 10^{-4}}{1 + \dfrac{1}{7500} \times \dfrac{8 \times 8}{65.33 \times 10^{-4}}} \times 10^{-3} \text{ kN} = 1471.5 \text{ kN}$$

Also,　　　　Safe load $= 524.25 \text{ kN}$ (Calculated earlier)

$\therefore$ Factor of safety on Rankine's ultimate stress value

$$= \frac{1471.5}{524.25} = \mathbf{2.8} \quad \textbf{(Ans.)}$$

I-section inside a machine.

Example 16·23. *Fig. 16·12 shows a compound stanchion made up of two channels ISJC 200 weighing 139 N per channel and two 25 cm × 1 cm plates riveted one to each flange. Calculate the safe load that can be carried by the column. The column is 5·5 m long and both its ends are fixed. Allow a factor of safety of 4. Properties of one channel are: A = 17·77 cm², I_{XX} = 1161·2 cm⁴, I_{YY} =*

84·2 cm⁴; distance of centroid from back of web = 1·97 cm. Take σ_c = 320 MN / m² and $a = \dfrac{1}{7500}$.

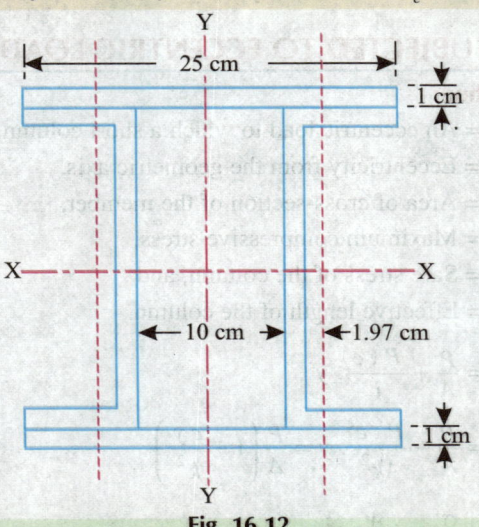

Fig. 16.12

Solution. Area of composite section,

$$A = 2\,(17·77 + 25 \times 1) = 85·54 \text{ cm}^2 = 85·54 \times 10^{-4} \text{ m}^2$$

Moment of inertia of the composite section about *XX*-axis,

$$I_{XX} = 2 \times 1161·2 + 2 \left[\frac{25 \times 1^3}{12} + 25 \times 1 \times (10 + 0·5^2) \right] \text{cm}^4$$

$$= 2322·4 + 5516·6$$
$$= 7839 \text{ cm}^4 = 7839 \times 10^{-8} \text{ m}^4$$

Moment of inertia of the composite section about *YY*-axis,

$$I_{YY} = 2\,[84·2 + 17·77 \times 6·97^2] + 2 \times \frac{1 \times 25^3}{12}$$

$$= 1888·2 + 2604·1$$
$$= 4492.3 \text{ cm}^4 = 4492.3 \times 10^{-8} \text{ m}^4$$

∴ $$k^2 \text{ (least)} = \frac{I_{\text{least}}}{A}$$

$$= \frac{I_{YY}}{A} = \frac{4492·3 \times 10^{-8}}{85·54 \times 10^{-4}} = 0·00525 \text{ m}^2$$

Since both ends are fixed,

∴ $$l_e = \frac{l}{2} = \frac{5·5}{2} = 2·75 \text{ m}$$

Using Rankine's formula, we get

$$P_{Rankine} = \frac{\sigma_c \cdot A}{1 + a \cdot \left(\dfrac{l_e}{k}\right)^2}$$

$$= \frac{320 \times 10^6 \times 85 \cdot 54 \times 10^{-4}}{1 + \dfrac{1}{7500} \times \dfrac{2 \cdot 75^2}{0 \cdot 00525}} \times 10^{-3} \ kN = 2296 \ kN$$

Safe axial load $= \dfrac{2296 \ (\text{crippling load})}{4 \ (\text{factor of safety})} = \mathbf{574 \ kN}$ **(Ans.)**

16·16. COLUMNS SUBJECTED TO ECCENTRIC LOADING

(*a*) Rankine's Method:

Let, P = An eccentric load to which a short column is subjected,

e = Eccentricity from the geometric axis,

A = Area of cross-section of the member,

σ_{max} = Maximum compressive stress,

σ = Safe stress of the column, and

l_e = Effective length of the column.

Now, $\sigma_{max} = \dfrac{P}{A} + \dfrac{P \cdot e}{I} \cdot y$

$= \dfrac{P}{A} + \dfrac{P \cdot e}{Ak^2} \cdot y = \dfrac{P}{A}\left(1 + \dfrac{e\,y}{k^2}\right)$

∴ $P = \dfrac{\sigma_{max} \cdot A}{1 + \dfrac{e\,y}{k^2}}$ ...(16·14)

Safe load for the column at the eccentricity e is given by,

$P = \dfrac{\sigma_c \cdot A}{1 + \dfrac{e\,y}{k^2}}$

I-section bar being lifted by a crane.

and, $P = \dfrac{\sigma_c \cdot A}{\left(1 + \dfrac{e\,y}{k^2}\right)\left(1 + a\,\dfrac{l_e^2}{k^2}\right)}$, when the *effect of buckling is also included* ...(16·15)

(b) Euler's method:

Refer to Fig. 16·13. *AB* is a column of length *l* subjected to an eccentric load *P* at eccentricity *e*. Let us assume that top of the column is free and the bottom of the column fixed.

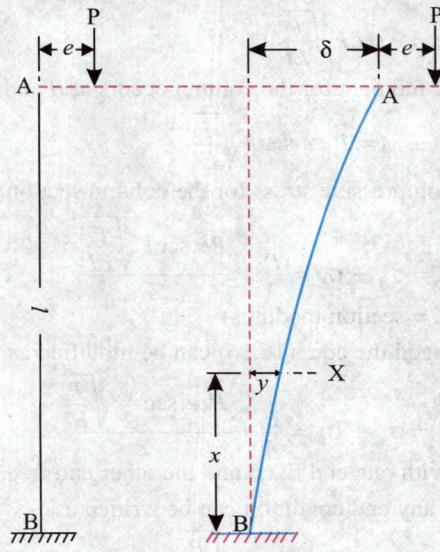

Fig. 16.13

Let, y = Deflection at any section *X* distant *x* from the fixed end *B*, and
               δ = Deflection at *A*.

The bending moment at the section is given by,

$$EI\,\frac{d^2y}{dx^2} = P\,(\delta + e - y)$$

∴ $\dfrac{d^2y}{dx^2} + \dfrac{P}{EI}\,y = \dfrac{P\,(\delta + e)}{EI}$

The solution to the above differential equation is given by,

$$y = C_1 \cos x \sqrt{\frac{P}{EI}} + C_2 \sin x \sqrt{\frac{P}{EI}} + (\delta + e)$$

The slope at any section is given by,

$$\frac{dy}{dx} = -C \sqrt{\frac{P}{EI}} \sin x \sqrt{\frac{P}{EI}} + C_2 \sqrt{\frac{P}{EI}} \cos x \sqrt{\frac{P}{EI}}$$

At *B*, $x = 0$ and $y = 0$, and $\dfrac{dy}{dx} = 0$

∴ $0 = C_1 + (\delta + e)$

and, $0 = C_2 \sqrt{\dfrac{P}{EI}}$

∴ $C_2 = 0$, and, $C_1 = -(\delta + e)$

At *A*, $x = l,\ y = \delta$

$$\therefore \qquad \delta = -(\delta + e)\cos l\sqrt{\frac{P}{EI}} + (\delta + e)$$

$$\therefore \qquad \delta = (\delta + e)\left[1 - \cos l\sqrt{\frac{P}{EI}}\right]$$

$$\therefore \qquad (\delta + e)\cos l\sqrt{\frac{P}{EI}} = e$$

$$\delta + e = e\sec l\sqrt{\frac{P}{EI}}$$

The maximum bending moment for the column occurs at B and is equal to $P(\delta + e)$

$$\therefore \qquad \text{Max. B.M} = M = P\,.\,e\sec l\sqrt{\frac{P}{EI}}$$

Hence the maximum compressive stress for the column section at B,

$$\sigma_{max} = \sigma_d + \sigma_b = \frac{P}{A} + \frac{Pe\,.\sec l\sqrt{\dfrac{P}{EI}}}{Z} \qquad\qquad ...(16\cdot16)$$

(where, Z = section modulus)

If both the ends are hinged the eqn. (16.16) can be modified as

$$\sigma_{max} = \sigma_d + \sigma_b = \frac{P}{A} + \frac{P\,.e\,.\sec\dfrac{l_e}{2}\sqrt{\dfrac{P}{EI}}}{Z} \qquad\qquad ...(16\cdot17)$$

because for a column with one end fixed and the other end free, equivalent length $l_e = 2\,l$

Formula in general for any end condition can be written as

$$\sigma_{max} = \frac{P}{A} + \frac{P\,.e\,.\sec\dfrac{l_e}{2}\sqrt{\dfrac{P}{EI}}}{Z}$$

where, l_e = Equivalent length depending upon the end conditions.

From above it may be noted that in the case of short columns (with no buckling) maximum bending moment is $P.\,e.$ which is increased to $P.e.\sec\dfrac{l_e}{2}\sqrt{\dfrac{P}{EI}}$ in the case of long columns.

Example 16·24. *From the following data of a column of circular section calculate the extreme stresses on the column section. Also find the maximum eccentricity in order that there may be no tension anywhere on the section.*

External diameter = 20 cm

Internal diameter = 16 cm

Length of the column = 4 m

Load carried by the column = 200 kN

Eccentricity of the load = 2.5 cm (from the axis of the column)

End conditions = Both ends fixed

Young's modulus = 94 GN/m².

Solution. Area of the column,

$$A = \frac{\pi}{4}(20^2 - 16^2) = 113\cdot1 \text{ cm}^2 = 113\cdot1 \times 10^{-4}\text{ m}^2$$

Moment of inertia of the section (about a diameter),

$$I = \frac{\pi}{64}(20^4 - 16^4) = 4637 \text{ cm}^4 = 4637 \times 10^{-8}\text{ m}^4$$

Equivalent or effective length of the column,

$$l_e = \frac{l}{2} = \frac{4}{2} = 2 \text{ m}$$

Maximum bending moment,

$$M = P.e \sec \frac{l_e}{2} \sqrt{\frac{P}{EI}}$$

Let us calculate the angle $\frac{l_e}{2} \sqrt{\frac{P}{EI}}$;

$$\frac{l_e}{2} \sqrt{\frac{P}{EI}} = \frac{2}{2} \sqrt{\frac{200 \times 10^3}{94 \times 10^9 \times 4637 \times 10^{-8}}}$$

$$= 0.2142 \text{ radian} = 12 \cdot 27° = 12° \ 16'$$

Sec. $12° \ 16' = 1 \cdot 02$

∴ Maximum bending moment,

$$M_{max} = 200 \times (2 \cdot 5 \times 10^{-2}) \times 1 \cdot 02 = 5 \cdot 1 \text{ kNm}$$

Maximum compressive stress

$$\sigma_{max} = \frac{P}{A} + \frac{M}{Z} = \frac{200}{113 \cdot 1 \times 10^{-4}} + \frac{5 \cdot 1}{463.7 \times 10^{-6}}$$

$$= 17683 + 10998$$

$$= 28681 \text{ kN/m}^2$$

$$\simeq 28.7 \text{ MN/m}^2$$

$$\left[\because Z = \frac{I}{y} = \frac{4637 \times 10^{-8}}{10 \times 10^{-2}} \right.$$
$$\left. = 463 \cdot 7 \times 10^{-6} \text{ m}^3 \right]$$

For *no tension* (corresponding to the maximum eccentricity):

$$\frac{P}{A} = \frac{M}{Z}$$

$$\frac{P}{A} = \frac{P \cdot e \cdot \sec \dfrac{l_e}{2} \sqrt{\dfrac{P}{EI}}}{Z}$$

$$\frac{200}{113 \cdot 1 \times 10^{-4}} = \frac{200 \times e \times 1 \cdot 02}{463 \cdot 7 \times 10^{-6}}$$

∴

$$e = \frac{463 \cdot 7 \times 10^{-6}}{113 \cdot 1 \times 10^{-4} \times 1 \cdot 02}$$

or

$$e = 0.0402 \text{ m}$$

$$= \textbf{40·2 mm} \quad \textbf{(Ans.)}$$

Example 16·25. *Fig. 16·14 shows a compound stanchion made up of two channels ISJC 200 weighing 139 N per metre per channel and two 25 cm × 1 cm plates riveted one to each flange. If the maximum permissible stress is 70 MN/m², find the maximum eccentricity of a 300 kN load from the YY-axis of the column. The load line lies in the vertical plane through the XX-axis. Take: E = 200 GN/m², the effective length of the column being 3 metres.*

Solution. For properties of column section (see example 16·23)

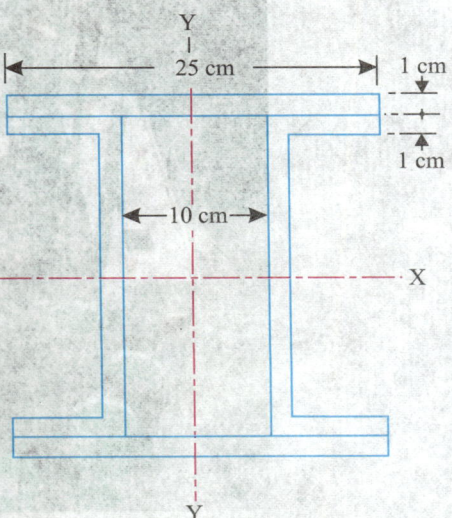

Fig. 16.14

Area of the section

$$= 85{\cdot}54 \text{ cm}^2 = 85{\cdot}54 \times 10^{-4} \text{ m}^2$$
$$I_{YY} = 4492{\cdot}3 \text{ cm}^4 = 4492{\cdot}3 \times 10^{-8} \text{ m}^4$$

Stress due to direct load,

$$\sigma_d = \frac{300}{85{\cdot}54 \times 10^{-4}} \times 10^{-3} = 35{\cdot}07 \text{ MN/m}^2$$

Maximum compressive stress,

$$\sigma_{max} = 70 \text{ MN/m}^2 \qquad \text{...(Given)}$$

∴ Maximum bending stress,

$$\sigma_b = \sigma_{max} - \sigma_d = 70 - 35{\cdot}07 = 34{\cdot}93 \text{ MN/m}^2$$

Section modulus about the YY-axis,

$$Z_{YY} = \frac{4492{\cdot}3}{12{\cdot}5} = 359{\cdot}4 \text{ cm}^3 = 359{\cdot}4 \times 10^{-6} \text{ m}^3$$

∴ Maximum bending moment,

$$M_{max} = 34{\cdot}93 \times 10^6 \times 359{\cdot}4 \times 10^{-6} \times 10^{-3} \text{ kNm} = 12{\cdot}55 \text{ kNm}$$

∴

$$M_{max} = P \, e \, \sec \frac{l_e}{2} \sqrt{\frac{P}{EI}}$$

Now,

$$\frac{l_e}{2} \sqrt{\frac{P}{EI}} = \frac{3}{2} \sqrt{\frac{300 \times 10^3}{200 \times 10^9 \times 4492{\cdot}3 \times 10^{-8}}} = 0{\cdot}274 \text{ radian} = 15{\cdot}7°$$

∴

$$\sec \frac{l_e}{2} \sqrt{\frac{P}{EI}} = \sec 15{\cdot}7° = 1{\cdot}038$$

∴

$$M_{max} = P \cdot e \times 1{\cdot}038 = 12{\cdot}55$$

or,

$$e = \frac{12{\cdot}55}{300 \times 1{\cdot}038} \times 1000 \text{ mm} = \textbf{40·3 mm} \quad \textbf{(Ans.)}$$

Assembling machine.

16·17. PROF. PERRY'S FORMULA

In cases where we have to determine the safe load that can be applied on a column at a given eccentricity Prof. Perry's formula proves quite useful.

Let, σ_d = Stress due to direct load (= P/A),

σ_{max} = Maximum permissible stress,

l_e = Effective length of the column, and

σ_b = Maximum compressive stress due to bending moment

$$= \frac{M}{Z} = \frac{M y_c}{Ak^2} .$$

(where, y_c = distance from the neutral axis of the extreme layer in compression)

$$= \frac{P \cdot e \sec \dfrac{l_e}{2} \sqrt{\dfrac{P}{EI}}}{Ak^2} \cdot y_c = \frac{P \cdot e y_c}{Ak^2} \sec \frac{\pi}{2} \sqrt{\frac{P}{P_{Euler}}}$$

where,

$$P_{Euler} = \frac{\pi^2 EI}{l_e^2}$$

∴

$$\sigma_{max} = \frac{P}{A} + \frac{P \cdot e \cdot y_c}{Ak^2} \sec \cdot \frac{\pi}{2} \sqrt{\frac{P}{P_{Euler}}}$$

But,

$$\frac{P}{A} = \sigma_d$$

∴

$$\sigma_{max} = \sigma_d \left[1 + \frac{e y_c}{k^2} \sec \cdot \frac{\pi}{2} \sqrt{\frac{P}{P_{Euler}}} \right]$$

According to Prof. Perry,

$$\sec \cdot \frac{\pi}{2} \sqrt{\frac{P}{P_{Euler}}} = \frac{1 \cdot 2 \, P_{Euler}}{P_{Euler} - P} \text{ (approximately)}$$

Let,

$$\sigma_{Euler} = \frac{P_{Euler}}{A}$$

∴

$$\sec \frac{\pi}{2} \sqrt{\frac{P}{P_{Euler}}} = \frac{1 \cdot 2 \, P_{Euler}}{P_{Euler} - P} = \frac{1 \cdot 2 \, \sigma_{Euler}}{\sigma_{Euler} - \sigma_d}$$

∴

$$\sigma_{max} = \sigma_d \left[1 + \frac{e y_c}{k^2} \cdot \frac{1 \cdot 2 \, P_{Eular}}{P_{Eular} - P} \right]$$

$$\sigma_{max} = \sigma_d \left[1 + \frac{e y_c}{k^2} \cdot \frac{1 \cdot 2 \, \sigma_{Euler}}{\sigma_{Euler} - \sigma_d} \right]$$

or,

$$\left[\frac{\sigma_{max}}{\sigma_d} - 1 \right] = \frac{e y_c}{k^2} \cdot \frac{1 \cdot 2 \, \sigma_{Euler}}{\sigma_{Euler} - \sigma_d}$$

∴

$$\left(\frac{\sigma_{max}}{\sigma_d} - 1 \right) \left(\frac{\sigma_{Euler} - \sigma_d}{\sigma_{Euler}} \right) = \frac{1 \cdot 2 \, e \cdot y_c}{k^2}$$

$$\left(\frac{\sigma_{max}}{\sigma_d} - 1 \right) \left(1 - \frac{\sigma_d}{\sigma_{Euler}} \right) = \frac{1 \cdot 2 \, e \cdot y_c}{k^2} \qquad \text{...(Prof. Perry's formula)...(16·18)}$$

Example 16·26. *For the column in example 16·25, find the maximum load that can be applied at an eccentricity of 2 cm from the axis YY. The maximum permissible compressive stress is limited to 70 MN/m². Take: E = 200 GN/m².*

Solution.

$$k^2 = \frac{I}{A} = \frac{4492 \cdot 3}{85 \cdot 54} = 52 \cdot 5 \text{ cm}^2 = 0.00525 \text{ m}^2$$

$$P_{Euler} = \frac{\pi^2 EI}{l_e^2} = \frac{\pi^2 \times 200 \times 10^9 \times 4492 \cdot 3 \times 10^{-8}}{3^2} \times 10^{-3} \text{ kN}$$

$$= 9853 \text{ kN}$$

$$\sigma_{Euler} = \frac{P_{Euler}}{A} = \frac{9853}{85 \cdot 54 \times 10^{-4}} \times 10^{-3} \text{ MN/m}^2$$

$$= 1151 \cdot 8 \text{ MN/m}^2$$

$$\sigma_{max} = 70 \text{ MN/m}^2 \qquad \qquad \dots (Given)$$

Let, σ_d = Stress due to direct load.

According to Perry's formula,

$$\left(\frac{\sigma_{max}}{\sigma_d} - 1 \right) \left(1 - \frac{\sigma_d}{\sigma_{Euler}} \right) = \frac{1 \cdot 2 \ e \ y_c}{k^2}$$

$$\left(\frac{70}{\sigma_d} - 1 \right) \left(1 - \frac{\sigma_d}{1151 \cdot 8} \right) = \frac{1 \cdot 2 \times (2 \times 10^{-2}) \times (12 \cdot 5 \times 10^{-2})}{0.0052}$$

$$\left(\frac{70 - \sigma_d}{\sigma_d} \right) \left(\frac{1151 \cdot 8 - \sigma_d}{1151 \cdot 8} \right) = 0 \cdot 5714$$

$$(70 - \sigma_d)(1151 \cdot 8 - \sigma_d) = 0 \cdot 5714 \times 1151 \cdot 8 \ \sigma_d$$

$$\sigma_d^2 - 1222 \ \sigma_d + 80626 = 658 \ \sigma_d$$

∴ $$\sigma_d^2 - 1880 \ \sigma_d + 80626 = 0$$

or, $$\sigma_d = \frac{1880 \pm \sqrt{1880^2 - 4 \times 80626}}{2} = \frac{1880 \pm 1792}{2}$$

or, $$\sigma_d = 44 \text{ MN/m}^2 \text{ (considering } -\text{ve sign)}$$

∴ $$\text{Safe load} = \sigma_d \times A = 44 \times (85 \cdot 54 \times 10^{-4}) \times 10^3 \text{ kN}$$

$$= \mathbf{376 \cdot 4 \ kN} \quad \textbf{(Ans.)}$$

Example 16·27. *Fig. 16·15 shows a stanchion built up of an 250 mm × 125 mm RSJ section with a plate 150 mm × 12 mm plate riveted to each flange. It is 6 metres long and its ends are hinged. If the maximum stress is limited to 88 MN/m², using Perry's formula, calculate the safe load for the stanchion.*

For the joist: Area of cross-section
= 35·53 × 10⁻⁴ m²;
I_{XX} = 3717·8 × 10⁻⁸ m⁴ ;
I_{YY} = 193·4 × 10⁻⁸ m⁴
The eccentricity from the YY-axis is 25 mm.
Young's modulus = 200 GN/m²

Solution. Refer to Fig. 16·15.

Moment of inertia about YY-axis,

$$I_{YY} = 193 \cdot 4 \times 10^{-8} + \left[2 \times \frac{12 \times 150^3}{12} \right] \times 10^{-12}$$

$$= 868 \cdot 4 \times 10^{-8} \text{ m}^4$$

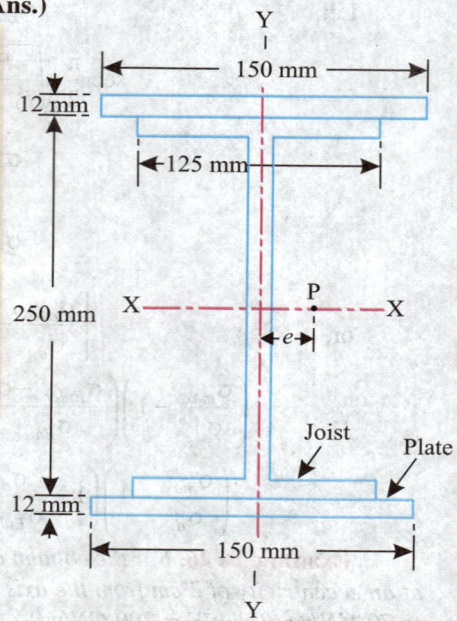

Fig. 16.15

Since the eccentricity is given along *XX*-axis (Fig. 16·15) there is no need to calculate the moment of inertia about *XX*-axis.

Euler buckling load, $\qquad P_{Euler} = \dfrac{\pi^2 \, EI}{l_e^2} = \dfrac{\pi^2 \, EI}{l^2} \qquad\qquad$ (∵ The ends are *hinged*)

Hence, $\qquad\qquad\qquad\qquad l = 6 \text{ m} \quad (Given)$

∴ $\qquad\qquad\qquad P_{Euler} = \dfrac{\pi^2 \times 200 \times 10^9 \times 868\cdot4 \times 10^{-8}}{6^2} \times 10^{-3} \text{ kN} = 476\cdot15 \text{ kN}$

Area of cross-section of the built up section

$$= 35\cdot53 \times 10^{-4} + 2 \times (12 \times 150 \times 10^{-6}) = 71\cdot53 \times 10^{-4} \text{ m}^2$$

Stress, $\qquad\qquad \sigma_{Euler} = \dfrac{P_{Euler}}{A} = \dfrac{476\cdot15}{71\cdot53 \times 10^{-4}} \times 10^{-3} \text{ MN/m}^2 = 66\cdot57 \text{ MN/m}^2$

Maximum stress (permissible),

$$\sigma_{max} = 88 \text{ MN/m}^2$$

Now, $\qquad\qquad\qquad k^2 = \dfrac{I_{YY}}{A} = \dfrac{868\cdot4 \times 10^{-8}}{71\cdot53 \times 10^{-4}} = 12\cdot14 \times 10^{-4} \text{ m}^2$

$$y_c = 150/2 = 75 \text{ mm} = 0\cdot075 \text{ m}$$

(distance of extreme layer in compression from neutral axis *YY*)

Perry's formula is given by:

$$\left(\dfrac{\sigma_{max}}{\sigma_d} - 1 \right)\left(1 - \dfrac{\sigma_d}{\sigma_{Euler}} \right) = \dfrac{1\cdot2 \, e \cdot y_c}{k^2} \qquad\qquad ...[\text{Eqn. (16·18)}]$$

$$\left(\dfrac{88}{\sigma_d} - 1 \right)\left(1 - \dfrac{\sigma_d}{66\cdot57} \right) = \dfrac{1\cdot2 \times (25 \times 10^{-3}) \times (0\cdot075)}{12\cdot14 \times 10^{-4}} = 1\cdot853$$

$$(88 - \sigma_d)\,(66\cdot57 - \sigma_d) = \sigma_d \times 66\cdot57 \times 1\cdot853 = 123\cdot35 \, \sigma_d$$

$$5858\cdot16 - 154\cdot57 \, \sigma_d + \sigma_d^2 = 123\cdot35 \sigma_d$$

or $\qquad \sigma_d^2 - 277\cdot92 \, \sigma_d + 5858\cdot16 = 0$

∴ $\qquad\qquad\qquad \sigma_d = \dfrac{277\cdot92 \pm \sqrt{277\cdot92^2 - 4 \times 5858\cdot16}}{2} = \dfrac{277\cdot92 \pm 231\cdot96}{2}$

$$= 22\cdot98 \text{ MN/m}^2 \qquad (\text{considering } -\text{ve sign only})$$

Clamp of an earthmover.

∴ Safe load on the column

$$= \sigma_d \times A = 22.98 \times 71.53 \times 10^{-4} = 0.1644 \text{ MN or } 164.4 \text{ kN}$$

Hence, *safe load* = **164.4 kN** **(Ans.)**

16·18. COLUMNS WITH INITIAL CURVATURE (AXIAL LOADING)

In the Fig. 16·16 is shown a column *AB* of length *l* with both ends pinned. It has an initial curvature (*AC' B*) having a central deflection δ'. Let *y'* be the initial deflection at a distance *x* from the end *B*.

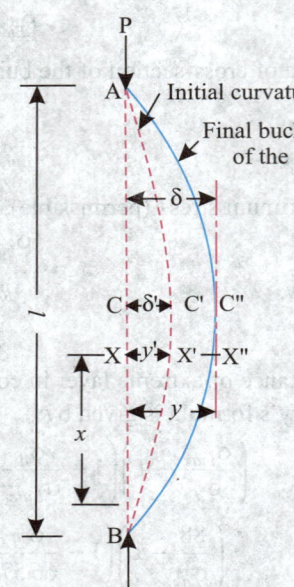

For the purpose of analysis let us assume a sine curve for the initial profile of the centre line of the column, so that,

$$y' = \delta' \sin \frac{\pi x}{l} \qquad \qquad ...(16.19)$$

∴
$$\frac{dy'}{dx} = \frac{\pi \delta'}{l} . \cos \frac{\pi x}{l} \qquad ...(16.20)$$

∴
$$\frac{d^2 y'}{dx^2} = -\frac{\pi^2 \delta'}{l^2} . \sin \frac{\pi x}{l}$$

As the load on the column reaches the critical value *P*, the column will deflect to the form *AC''B*, so that the deflection at *x* changes from *y'* to *y*. This happens due to the bending moment *Py*.

Fig. 16.16

∴
$$EI \frac{d^2(y - y')}{dx^2} = -Py$$

[where, (*y* – *y'*) = change in deflection]

or,
$$\frac{d^2(y - y')}{dx^2} = -\frac{Py}{EI}$$

∴
$$\frac{d^2 y}{dx^2} + \frac{Py}{EI} = \frac{d^2 y'}{dx^2} = -\frac{\pi^2 \delta'}{l^2} . \sin \frac{\pi x}{l} \qquad\qquad ...(16.22)$$

Let the solution to the above differential equation be given by

$$y = C\delta' \sin \frac{\pi x}{l}$$

(where, *C* is a constant of integration)

∴
$$\frac{dy}{dx} = C\delta' \frac{\pi^2}{l^2} \cos \frac{\pi x}{l}$$

and,
$$\frac{d^2 y}{dx^2} = -C \cdot \delta' \frac{\pi^2}{l^2} \sin \frac{\pi x}{l}$$

Inserting the values of *y* and $\dfrac{d^2 y}{dx^2}$ in eqn. 16·22, we get,

$$- C\delta' \frac{\pi^2}{l^2} \sin \frac{\pi x}{l} + \frac{P}{EI} C\delta' \sin \frac{\pi x}{l} = -\frac{\pi^2 \delta'}{l^2} \sin \frac{\pi x}{l}$$

∴
$$C \left[\frac{\pi^2}{l^2} - \frac{P}{EI} \right] = \frac{\pi^2}{l^2}$$

or,

$$C = \frac{\dfrac{\pi^2}{l^2}}{\dfrac{\pi^2}{l^2} - \dfrac{P}{EI}} = \frac{1}{1 - \dfrac{Pl^2}{\pi^2\,EI}} = \frac{1}{1 - \dfrac{P}{P_{Euler}}}$$

or,

$$C = \frac{P_{Euler}}{P_{Euler} - P}$$

Hence the equation to the deflected form of the column is given by :

$$y = \frac{P_{Euler}}{P_{Euler} - P}\,\delta' \sin\frac{\pi x}{l} \qquad\qquad ...(16\cdot23)$$

The deflection will be maximum at the mid-point C and let its (central deflection) value be δ.

∴ At, $\qquad\qquad x = \dfrac{l}{2}, y = \delta$

∴ $\qquad\qquad \delta = \dfrac{P_{Euler}}{P_{Euler} - P} \cdot \delta'$

Maximum bending moment,

$$M = P \times y_{max} = P \cdot \delta \qquad\qquad \text{(bending moment at the mid-section),}$$

$[\because y_{max} = \delta \text{ (say)}]$

$$= \frac{P \cdot P_{Euler}}{P_{Euler} - P} \cdot \delta'$$

Maximum compressive stress, $\sigma_{max} = \sigma_d + \sigma_b = \dfrac{P}{A} + \dfrac{M}{Z} = \dfrac{P}{A} + \dfrac{M y_c}{A k^2}$

(where, y_c = distance of the extreme layer in compression from the neutral axis)

$$= \frac{P}{A} + \frac{P \cdot P_{Euler}}{P_{Euler} - P} \cdot \delta' \cdot \frac{y_c}{A k^2} = \frac{P}{A}\left[1 + \frac{P_{Euler}}{P_{Euler} - P} \cdot \frac{\delta' y_c}{k^2}\right]$$

$$= \sigma_d\left[1 + \frac{P_{Euler}}{P_{Euler} - P} \cdot \frac{\delta' y_c}{k^2}\right]$$

On rearranging, we get

$$\left[\frac{\sigma_{max}}{\sigma_d} - 1\right]\left[1 - \frac{\sigma_d}{\sigma_{Euler}}\right] = \frac{\delta' y_c}{k^2} \qquad\qquad ....(16\cdot24)$$

Example 16·28. *A steel strut has an outside diameter of 180 mm and inside diameter of 120 mm and is 6 m long. It is hinged at both ends and is initially bent. Assuming the centre line of the strut as sinusoidal with maximum deviation of 9 mm, determine the maximum stress developed due to an axial load of 150 kN. Take: E = 208 GN/m².*

Solution. Outside diameter of the strut,

$$D = 180 \text{ mm} = 0\cdot18 \text{ m}$$

Inside diameter of the strut,

$$d = 120 \text{ mm} = 0\cdot12 \text{ m}$$

Length of the strut, $\qquad l = 6 \text{ m}$

Maximum deviation at the centre,

$$\delta' = 9 \text{ mm} = 0\cdot009 \text{ m}$$

Young's modulus, $\qquad E = 208 \text{ GN/m}^2$

Axial load, $\qquad\qquad P = 150 \text{ kN}.$

Maximum stress developed, σ_{max}:

Area of cross-section,

$$A = \frac{\pi}{4}(D^2 - d^2) = \frac{\pi}{4}(0\cdot18^2 - 0\cdot12^2) = 0\cdot01414 \text{ m}^2$$

Moment of inertia,

$$I = \frac{\pi}{64}(D^4 - d^4) = \frac{\pi}{64}(0.18^4 - 0.12^4) = 4.135 \times 10^{-5} \text{ m}^4$$

$$k^2 = \frac{I}{A} = \frac{4.135 \times 10^{-5}}{0.01414} = 2.924 \times 10^{-3} \text{ m}^2$$

(where, k = radius of gyration)

Euler load,

$$P_{Euler} = \frac{\pi^2 EI}{l_e^2}$$

$$= \frac{\pi^2 \times 208 \times 10^9 \times 4.135 \times 10^{-5}}{6^2} \times 10^{-3} \text{ kN} = 2357.9 \text{ kN}$$

$[l_e = l = 6 \text{ m because strut is hinged at both the ends}]$

Stress,

$$\sigma_{Euler} = \frac{P_{Euler}}{A} = \frac{2357.9}{0.01414} \times 10^{-3} \text{ MN/m}^2 = 166.75 \text{ MN/m}^2$$

Direct stress,

$$\sigma_d = \frac{P}{A} = \frac{150}{0.01414} \times 10^{-3} \text{ MN/m}^2 = 10.6 \text{ MN/m}^2$$

Distance of the extreme layer in compression from the neutral axis,

$$y_c = \frac{180}{2} = 90 \text{ mm} = 0.09 \text{ m}$$

We know that,

$$\left(\frac{\sigma_{max}}{\sigma_d} - 1\right)\left(1 - \frac{\sigma_d}{\sigma_{Euler}}\right) = \frac{\delta' y_c}{k^2} \qquad \qquad ...[\text{Eqn. (16.24)}]$$

$$\left(\frac{\sigma_{max}}{10.6} - 1\right)\left(1 - \frac{10.6}{166.75}\right) = \frac{0.009 \times 0.09}{2.924 \times 10^{-3}}$$

$$\left(\frac{\sigma_{max}}{10.6} - 1\right) \times 0.936 = 0.277$$

∴ $\sigma_{max} = \mathbf{13.74 \text{ MN/m}^2}$ **(Ans.)**

16·19. BEAM COLUMNS

Columns having transverse load in addition to the axial compressive load are termed as **beam columns**.

Case I: Strut pinned at both ends and subjected to an axial thrust P and a transverse point load W at the centre.

Refer to Fig. 16·17. Consider any section at a distance x from the end in AC. The bending moment at the section is given by,

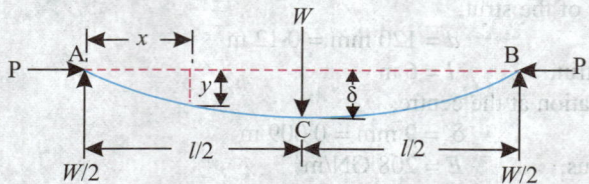

Fig. 16.17

$$EI \frac{d^2 y}{dx^2} = -Py - \frac{W}{2} \cdot x$$

$$\therefore \qquad \frac{d^2y}{dx^2} + \frac{P}{EI}\,y = -\frac{W \cdot x}{2EI} \qquad \qquad \text{...(16·25)}$$

The solution to the above differential equation is,

$$y = C_1 \cos x \sqrt{\frac{P}{EI}} + C_2 \sin x \sqrt{\frac{P}{EI}} - \frac{Wx}{2P} \qquad \qquad \text{...(16·26)}$$

The slope at any section in AC, is given by,

$$\frac{dy}{dx} = -C_1 \sqrt{\frac{P}{EI}} \sin x \sqrt{\frac{P}{EI}} + C_2 \sqrt{\frac{P}{EI}} \cos x \sqrt{\frac{P}{EI}} - \frac{W}{2P} \qquad \text{...(16·27)}$$

At, $\qquad x = 0, \qquad y = 0, \qquad C_1 = 0$

At, $\qquad x = \dfrac{l}{2}, \qquad \dfrac{dy}{dx} = 0$

$$0 = C_2 \sqrt{\frac{P}{EI}} \cos \cdot \frac{l}{2} \sqrt{\frac{P}{EI}} - \frac{W}{2P}$$

$$\therefore \qquad C_2 = \frac{W}{2P} \cdot \sqrt{\frac{EI}{P}} \sec \cdot \frac{l}{2} \sqrt{\frac{P}{EI}}$$

Hence the deflection at any section in AC, is given by,

$$y = \frac{W}{2P} \sqrt{\frac{EI}{P}} \sec \cdot \frac{l}{2} \sqrt{\frac{P}{EI}} \sin x \sqrt{\frac{P}{EI}} - \frac{W \cdot x}{2P}$$

Maximum deflection:

The maximum deflection will occur at the centre.

$$\therefore \text{ At,} \qquad x = \frac{l}{2}, \qquad y = \delta$$

$$\therefore \qquad \delta = \frac{W}{2P} \sqrt{\frac{EI}{P}} \sec \frac{l}{2} \sqrt{\frac{P}{EI}} \sin \frac{l}{2} \sqrt{\frac{P}{EI}} - \frac{Wl}{4P}$$

$$\text{or,} \qquad \delta = \frac{W}{2P} \sqrt{\frac{EI}{P}} \tan \frac{l}{2} \sqrt{\frac{P}{EI}} - \frac{Wl}{4P}$$

Inside view of a car engine.

Maximum bending moment:

$$M_{max} = P\delta + \frac{W}{2} \cdot \frac{l}{2}$$

$$= P\left[\frac{W}{2P} \sqrt{\frac{EI}{P}} \tan \frac{l}{2} \sqrt{\frac{P}{EI}} - \frac{Wl}{4P} \right] + \frac{Wl}{4}$$

$$= \frac{W}{2} \sqrt{\frac{EI}{P}} \tan \frac{l}{2} \sqrt{\frac{P}{EI}} - \frac{Wl}{4} + \frac{Wl}{4}$$

or, $$M_{max} = \frac{W}{2} \sqrt{\frac{EI}{P}} \tan \frac{l}{2} \sqrt{\frac{P}{EI}} \qquad ...(16 \cdot 28)$$

We know that,

$$\tan \theta = \theta + \frac{\theta^3}{3} + ...$$

When θ is small,

$$\tan \theta = \theta + \frac{\theta^3}{3}$$

This drilling tool partly behaves like a column with end fixed and the other end free.

$$\therefore \ M_{max} = -\frac{W}{2} \sqrt{\frac{EI}{P}} \left[\frac{l}{2} \sqrt{\frac{P}{EI}} + \frac{1}{3} \frac{l^3}{8} \frac{P}{EI} \sqrt{\frac{EI}{P}} \right]$$

$$M_{max} = -\left[\frac{Wl}{4} + \frac{Wl^3}{48 EI} \cdot P \right] \qquad ...[16 \cdot 28 \ (a)]$$

Maximum stress:

Stress due to bending, $$\sigma_b = \frac{M_{max}}{Z} = \frac{M_{max} \cdot y_c}{I} = \frac{M_{max} \cdot y_c}{Ak^2}$$

(where, y_c = distance of the extreme layer in compression from the neutral axis)

or, $$\sigma_b = \frac{W}{2} \cdot \frac{y_c}{Ak^2} \sqrt{\frac{EI}{P}} \tan \frac{l}{2} \sqrt{\frac{P}{EI}}$$

Direct stress, $$\sigma_d = \frac{P}{A}$$

∴ Maximum stress, $\sigma_{max} = \sigma_d + \sigma_b$

or, $$\sigma_{max} = \frac{P}{A} + \frac{W}{2} \cdot \frac{y_c}{Ak^2} \sqrt{\frac{EI}{P}} \tan \frac{l}{2} \sqrt{\frac{P}{EI}} \qquad ...(16 \cdot 29)$$

Example 16·29. *A steel strut, 1 m long, is 30 mm in diameter. It is subjected to an axial thrust of 18 kN. In addition, a lateral load W acts at the centre of the strut. If the strut fails at a maximum stress of 350 MN/m². determine the magnitude of W.*

Take: E = 210 GN/m².

Solution. Diameter of the strut,

$$d = 30 \text{ mm} = 0.03 \text{ m}$$

Axial load, $\quad\quad P = 18 \text{ kN}$

Maximum stress, $\quad \sigma_{max} = 350 \text{ MN/m}^2$

Young's modulus, $\quad\quad = 210 \text{ GN/m}^2$

Magnitude of W:

Area of strut,
$$A = \frac{\pi}{4} d^2 = \frac{\pi}{4} \times 0.03^2 = 7.068 \times 10^{-4} \text{ m}^2$$

Moment of inertia,
$$I = \frac{\pi}{64} d^4 = \frac{\pi}{64} \times 0.03^4 = 3.976 \times 10^{-8} \text{ m}^4$$

$$k^2 = \frac{I}{A} = \frac{3.976 \times 10^{-8}}{7.068 \times 10^{-4}} = 5.625 \times 10^{-5} \text{ m}^2$$

(where, k = radius of gyration)

Direct stress,
$$\sigma_d = \frac{P}{A} = \frac{18}{7.068 \times 10^{-4}} \times 10^{-3} \text{ MN/m}^2 = 25.47 \text{ MN/m}^2$$

Also,
$$\sigma_{max} = \sigma_d + \sigma_b$$

or
$$350 = 25.47 + \sigma_b \qquad \therefore \ \sigma_b = 324.53 \text{ MN/m}^2$$

Now
$$\frac{P}{EI} = \frac{18 \times 10^3}{210 \times 10^9 \times 3.976 \times 10^{-8}} = 2.155$$

or
$$\sqrt{\frac{P}{EI}} = 1.468, \quad \text{and} \quad \sqrt{\frac{EI}{P}} = 0.681$$

Moreover,
$$\frac{l}{2} \sqrt{\frac{P}{EI}} = \frac{1}{2} \times 1.468 = 0.734 \text{ rad.}$$

$$\therefore \qquad \tan \frac{l}{2} \sqrt{\frac{P}{EI}} = \tan 40° = 0.9$$

Also,
$$M_{max} = \frac{W}{2} \sqrt{\frac{EI}{P}} \tan \frac{l}{2} \sqrt{\frac{P}{EI}}$$

∴ Bending stress,
$$\sigma_b = \frac{M_{max}}{Z} = \frac{M_{max} \times y}{I}$$

or,
$$\sigma_b = \frac{W \cdot y_c}{2Ak^2} \sqrt{\frac{EI}{P}} \tan \frac{l}{2} \sqrt{\frac{P}{EI}}$$

(where, $y_c = 0.015$ m = distance of the extreme layer in compression from the neutral axis)

or,
$$324.53 \times 10^6 = \frac{W \times 0.015}{2 \times 7.068 \times 10^{-4} \times 5.625 \times 10^{-5}} \times 0.681 \times 0.9$$

$$\therefore \qquad W = \frac{324.53 \times 10^6 \times 2 \times 7.068 \times 10^{-4} \times 5.625 \times 10^{-5}}{0.015 \times 0.681 \times 0.9} \times 10^{-3} \text{ kN}$$

$$= \textbf{2.8 kN} \ \ \textbf{(Ans.)}$$

Case II: Strut pinned at both ends and subjected to an axial thrust P and a lateral uniformly distributed load of intensity w per unit run.

Refer to Fig. 16·18. Consider any section distant x from the end A. The bending moment at any section is given by,

$$EI \frac{d^2 y}{dx^2} = -Py + \frac{wx^2}{2} - \frac{wl}{2} x$$

$$\therefore \qquad \frac{d^2 y}{dx^2} + \frac{P}{EI} y = -\frac{wx}{2} \frac{(l - x)}{EI} \qquad \qquad \dots(16.30)$$

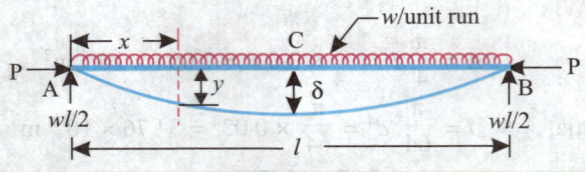

Fig. 16.18

The solution to the above differential equation is,

$$y = C_1 \cos x \sqrt{\frac{P}{EI}} + C_2 \sin x \sqrt{\frac{P}{EI}} - \frac{wx\,(l-x)}{2P} - \frac{w\,EI}{P^2} \qquad ...(16\cdot31)$$

∴
$$\frac{dy}{dx} = - C_1 \sqrt{\frac{P}{EI}} \sin x \sqrt{\frac{P}{EI}} + C_2 \sqrt{\frac{P}{EI}} \cos x \sqrt{\frac{P}{EI}} + \frac{wx}{P} - \frac{wl}{2P}$$

At A, $\qquad x = 0, \qquad y = 0$

∴
$$0 = C_1 - \frac{w\,EI}{P^2}$$

or, $\qquad C_1 = \dfrac{w\,EI}{P^2}$

At, $\qquad x = \dfrac{l}{2}, \qquad \dfrac{dy}{dx} = 0$

∴
$$0 = - \frac{w\,EI}{P^2} \sqrt{\frac{P}{EI}} \sin \frac{l}{2} \sqrt{\frac{P}{EI}} + C_2 \sqrt{\frac{P}{EI}} \cos \frac{l}{2} \sqrt{\frac{P}{EI}}$$

∴
$$C_2 = C_1 \tan \frac{l}{2} \sqrt{\frac{P}{EI}}$$

Inserting the values of C_1 and C_2 in deflection eqn. 16·30, we get

$$y = \frac{w\,EI}{P^2} \left[\cos x \sqrt{\frac{P}{EI}} + \tan \frac{l}{2} \sqrt{\frac{P}{EI}} \sin x \sqrt{\frac{P}{EI}} \right] - \frac{wx\,(l-x)}{2P} - \frac{w\,EI}{P^2}$$

Deflection at the centre:

Let δ be the deflection at the centre.

Milling machine.

At $\qquad x = \dfrac{l}{2}, \qquad y = \delta$

$\therefore \qquad \delta = \dfrac{w\,EI}{P^2}\left[\cos\dfrac{l}{2}\sqrt{\dfrac{P}{EI}} + \tan\dfrac{l}{2}\sqrt{\dfrac{P}{EI}}\,\sin\dfrac{l}{2}\sqrt{\dfrac{P}{EI}}\right] - \dfrac{wl^2}{8P} - \dfrac{w\,EI}{P^2}$

$\therefore \qquad \delta = \dfrac{w\,EI}{P^2}\left[\sec\cdot\dfrac{l}{2}\sqrt{\dfrac{P}{EI}} - 1\right] - \dfrac{wl^2}{8P} \qquad\qquad\qquad …(16 \cdot 32)$

Maximum bending moment which will occur at C is given by

$$M_{max} = -\dfrac{wl^2}{8} - P\delta = -\dfrac{wl^2}{8} - P\left[\dfrac{w\,EI}{P^2}\left(\sec\cdot\dfrac{l}{2}\sqrt{\dfrac{P}{EI}} - 1\right) - \dfrac{wl^2}{8P}\right]$$

$$= -\dfrac{w\,EI}{P}\left[\sec\dfrac{l}{2}\sqrt{\dfrac{P}{EI}} - 1\right]$$

We know that,

$$\sec\theta = 1 + \dfrac{\theta^2}{\lfloor 2} + \dfrac{5\theta^4}{\lfloor 4} + \dfrac{61\theta^6}{\lfloor 6} + …$$

When θ is small,

$$\sec\theta = 1 + \dfrac{\theta^2}{\lfloor 2} + \dfrac{5\theta^4}{\lfloor 4} = 1 + \dfrac{\theta^2}{2} + \dfrac{5\theta^4}{24}$$

$$M_{max} = -\dfrac{w\,EI}{P}\left[\dfrac{1}{2}\cdot\dfrac{l^2}{4}\cdot\dfrac{P}{EI} + \dfrac{5}{24}\cdot\dfrac{l^4}{16}\cdot\dfrac{P^2}{E^2I^2}\right]$$

$$= -\left[\dfrac{wl^2}{8} + \dfrac{5}{384}\cdot\dfrac{wl^4}{EI}\,P\right]$$

i.e. $\qquad M_{max} = -\left[\dfrac{wl^2}{8} + \dfrac{5}{384}\cdot\dfrac{wl^4}{EI}\,P\right] \qquad\qquad\qquad …(16 \cdot 33)$

Maximum stress:

Now maximum stress,

$$\sigma_{max} = \sigma_d + \sigma_b$$

But, $\qquad \sigma_d = \dfrac{P}{A}$ and $\sigma_b = \dfrac{M_{max}}{Z} = \dfrac{M_{max}\cdot y_c}{Ak^2}$

$\therefore \qquad \sigma_{max} = \dfrac{P}{A} + \dfrac{w\,Ey_c}{P}\left[\sec\dfrac{l}{2}\sqrt{\dfrac{P}{EI}} - 1\right] \qquad\qquad …(16 \cdot 34) \quad (\because I = Ak^2)$

(Where, y_c = distance of extreme layer in compression from the neutral axis).

Example 16·30. *A rod, 2 m in length and of rectangular cross-section 88 mm × 44 mm is supported horizontally through pin joints. It carries a vertical load of 3·3 kN/m length and an axial thrust of 110 kN. If E = 208 kN/mm², calculate the maximum stress induced.*

Solution. Length of the rod, $l = 2$ m

Cross-section of the rod, $A = 88$ mm $\times\ 44$ mm $= 3872$ mm²

Moment of inertia, $\qquad I = \dfrac{88 \times 44^3}{12} = 62 \cdot 47 \times 10^4$ mm⁴

Axial thrust, $\qquad P = 110$ kN

Vertical load, $W = 3 \cdot 3$ kN/m $= 3 \cdot 3$ N/mm

Young's modulus, $E = 208$ kN/mm^2

$$\sqrt{\frac{P}{EI}} = \sqrt{\frac{110 \times 10^3}{208 \times 10^3 \times 62 \cdot 47 \times 10^4}} = 9 \cdot 2 \times 10^{-4}$$

$$\frac{l}{2}\sqrt{\frac{P}{EI}} = \frac{2 \times 1000}{2} \times 9 \cdot 2 \times 10^{-4} = 0 \cdot 92 \text{ radian} = 52 \cdot 7°$$

∴ $\sec 52 \cdot 7° = 1 \cdot 65$

Maximum stress, σ_{max} :

Maximum stress is given by (Eqn. 16·34),

$$\sigma_{max} = \frac{P}{A} + \frac{W\,Ey_c}{P}\left[\sec \frac{l}{2}\sqrt{\frac{P}{EI}} - 1\right]$$

(where, $y_c = 44/2 = 22$ mm = distance of extreme layer in compression from the neutral axis).

$$\sigma_{max} = \frac{110 \times 10^3}{3872} + \frac{3 \cdot 3 \times 208 \times 10^3 \times 22}{110 \times 10^3}(1 \cdot 65 - 1)$$

$$= 28 \cdot 41 + 89 \cdot 23 = 117 \cdot 64 \text{ N/mm}^2$$

Hence, $\sigma_{max} = \mathbf{117 \cdot 64}$ **N/mm^2 (or 117·64 MN/m^2)**   **Ans.**

TYPICAL EXAMPLES (For Competitive Examinations)

Example 16·31. *A piston rod of steam engine 80 cm long is subjected to a maximum load of 60 kN. Determine the diameter of the rod using Rankine's formula with permissible compressive stress of 100 N/mm^2. Take constant in Rankine's formula as $\dfrac{1}{7500}$ for hinged ends. The rod may be assumed partially fixed with length coefficient of 0·6.*   **(AMIE Winter, 2002)**

Solution. *Given* : $l = 80$ cm $= 800$ mm; $P = 60$ kN $= 60000$ N, $\sigma_c = 100$ N/mm^2;

$$a = \frac{1}{7500} \text{ for hinged ends; length coefficient} = 0 \cdot 6.$$

Diameter of the rod, d :

Using Rankine's formula,

$$P = \frac{\sigma_c \cdot A}{1 + a\left(\dfrac{l_e}{k}\right)^2}$$

But, $l_e = 0 \cdot 6\, l = 0 \cdot 6 \times 800 = 480$ mm   ..*(Given)*

$$k = \sqrt{\frac{I}{A}} = \sqrt{\frac{\dfrac{\pi}{64}d^4}{\dfrac{\pi}{4}d^2}} = \frac{d}{4} \qquad ∴ \quad 60000 = \frac{100 \times \left(\dfrac{\pi}{4}d^2\right)}{1 + \dfrac{1}{7500}\left[\dfrac{480}{d/4}\right]^2}$$

or, $60000 = \dfrac{78 \cdot 54\, d^2}{1 + \dfrac{491 \cdot 52}{d^2}} = \dfrac{78 \cdot 54\, d^4}{d^2 + 491 \cdot 52}$

or, $78 \cdot 54\, d^4 = 60000\, d^2 + 29491200$

or, $78 \cdot 54\, d^4 - 60000\, d^2 - 29491200 = 0$

or, $d^4 - 763.94\, d^2 - 375492.74 = 0$

or, $d^2 = \dfrac{763.94 + \sqrt{(763.94)^2 + 4 \times 375492.74}}{2} = \dfrac{763.94 \pm 1444.15}{2}$

or, $d^2 = 1104$ mm

or, $d = \mathbf{33.23}$ **mm**   (Ans).

Example 16·32. *A slender column is built-in at one end and an eccentric load is applied at the free end. Working from the first principles find the expression for the maximum length of column such that the deflection of the free end does not exceed the eccentricity of loading.*

(AMIE Winter, 2001)

Solution. Fig. 16·19. shows a slender column of length *l*. The column is built in at one end *B* and eccentric load *P* is applied at the free end *A*.

Let *y* be the deflection at any section *XX* distant *x* from the fixed end *B*. Let δ be the deflection at *A*.

The bending moment at the section *XX* is given by,

$$EI\, \frac{d^2 y}{dx^2} = P\,(\delta + e - y) \qquad ...(i)$$

$$EI\, \frac{d^2 y}{dx^2} + Py = P\,(\delta + e)$$

or, $\dfrac{d^2 y}{dx^2} + \dfrac{P}{EI}\, y = \dfrac{P\,(\delta + e)}{EI}$

The solution to the above differential equation is given by,

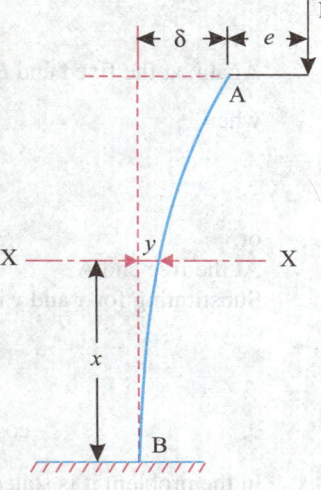

Fig. 16.19

$$y = C_1 \cos\left[x\, \sqrt{\frac{P}{EI}} \right] + C_2 \sin\left[x\, \sqrt{\frac{P}{EI}} \right] + (\delta + e) \qquad ...(ii)$$

Tools of drilling and boring machines act like columns.

where, C_1 and C_2 are the constants of integeration.

At the end B, $x = 0$ and $y = 0$

∴ $0 = C_1 \cos 0 + C_2 \sin 0 + (\delta + e)$

or, $C_1 = - (\delta + e)$

Differentiating eqn. (ii), we have

$$\frac{dy}{dx} = - C_1 \sqrt{\frac{P}{EI}} \sin \left[x \sqrt{\frac{P}{EI}} \right] + C_2 \sqrt{\frac{P}{EI}} \cos \left[x \sqrt{\frac{P}{EI}} \right]$$

$$= (\delta + e) \sqrt{\frac{P}{EI}} \sin \left[x \sqrt{\frac{P}{EI}} \right] + C_2 \sqrt{\frac{P}{EI}} \cos \left[x \sqrt{\frac{P}{EI}} \right]$$

Again, at the fixed end B,

when, $x = 0, \dfrac{dy}{dx} = 0$

∴ $0 = (\delta + e) \sqrt{\frac{P}{EI}} \times 0 + C_2 \sqrt{\frac{P}{EI}} \cos 0$

or, $C_2 = 0$

At the free end A, $x = l, y = \delta$

Substituting for x and y in eqn. (ii), we have

$$\delta = - (\delta + e) \cos \left[l \sqrt{\frac{P}{EI}} \right] = (\delta + e)$$

∴ $$\cos \left[l \sqrt{\frac{P}{EI}} \right] = \frac{e}{\delta + e} \qquad \qquad ...(iii)$$

In the problem it is stated that the deflection of the free end does not exceed the eccentricity of loading.

It follows that $\delta = e$

Substituting this value in eqn. (iii), we have

$$\cos \left[l \sqrt{\frac{P}{EI}} \right] = \frac{e}{\delta + e} = \frac{1}{2}$$

∴ $$l \sqrt{\frac{P}{EI}} = \cos^{-1} \left(\frac{1}{2} \right) = \frac{\pi}{3}$$

∴ $$l = \frac{\pi}{3} \sqrt{\frac{EI}{P}} \quad \textbf{(Ans.)}$$

Example 16·33. *A hollow circular steel pipe 72 mm outside diameter, 60 mm inside diameter, 1·2 m long is fixed at both ends, so as to prevent any expansion in length. The pipe is unstressed at the normal temperature. If the temperature of the pipe rises by 48°C, using Rankine's formula calculate:*

    (i) *Temperature stress in the pipe;*

    (ii) *Factor of safety against failure as a strut.*

    Given: Rankine constants: $\sigma_c = 330 \text{ MN/m}^2$; $a = \dfrac{1}{7500}$ *(for both ends hinged)*

                    $E = 208 \text{ GN/m}^2$, *and* $= 11·1 \times 10^{-6}$ *per* °C.

Solution. Outside diameter of the pipe,

              $D = 72 \text{ mm} = 0·072 \text{ m}$

Inside diameter of the pipe, $d = 60 \text{ mm} = 0·06 \text{ m}$

Length of the pipe $= 1·2 \text{ m}$

Temperature rise of the pipe, $T = 48 \ ^\circ C$

Rankine's constants, $\sigma_c = 330 \text{ MN/m}^2$

$$a = \frac{1}{7500} \text{ (for both the ends hinged)}$$

Young's modulus, $E = 208 \text{ GN/m}^2$

$$\alpha = 11 \cdot 1 \times 10^{-6}/°\text{C}$$

Area of cross-section, $A = \dfrac{\pi}{4}(D^2 - d^2) = \dfrac{\pi}{4}[(0 \cdot 072)^2 - (0 \cdot 06)^2] = 1 \cdot 244 \times 10^{-3} \text{ m}^2$

(Radius of gyration)^2, $k^2 = \dfrac{I}{A} = \dfrac{(\pi/64)(D^4 - d^4)}{(\pi/4)(D^2 - d^2)} = \dfrac{D^2 + d^2}{16}$

$$= \frac{0 \cdot 072^2 + 0 \cdot 06^2}{16} = 5 \cdot 49 \times 10^{-4} \text{ m}^2$$

(i) Temperature stress in the pipe, σ :

Expansion in the pipe is prevented by the fixed ends.

∴ Temperature stress in the pipe,

$$\sigma = \alpha TE = 11 \cdot 1 \times 10^{-6} \times 48 \times (208 \times 10^9) \times 10^{-6} \text{ MN/m}^2$$
$$= 110 \cdot 8 \text{ MN/m}^2 \text{ (comp)}$$

Hence, $\qquad\qquad \sigma = \textbf{110·8 MN/m}^2$ *(comp).* **(Ans.)**

Compressive axial load,

$$P = \sigma_c \cdot A = 110 \cdot 8 \times 1 \cdot 244 \times 10^{-3} \times 10^3 \text{ kN} = 137 \cdot 8 \text{ kN}$$

(ii) Factor of safety:

End conditions: *Both ends fixed.*

∴ Equivalent length, $l_e = \dfrac{l}{2} = \dfrac{1 \cdot 2}{2} = 0 \cdot 6 \text{ m}$

Rankine buckling load is given by,

$$P_{Rankine} = \frac{\sigma_c \cdot A}{1 + a \cdot \dfrac{l_e^2}{k^2}} = \frac{330 \times 10^6 \times 1 \cdot 244 \times 10^{-3}}{1 + \dfrac{1}{7500} \times \dfrac{0 \cdot 6^2}{5 \cdot 49 \times 10^{-4}}} = 377 \cdot 5 \text{ kN}$$

Follower used in cam mechanisms.

$$\therefore \textit{Factory of safety} \quad = \frac{377 \cdot 5}{137 \cdot 8} = 2 \cdot 74 \quad \text{(Ans.)}$$

ECCENTRIC LOADING

Example 16·34. *A steel tube having 88 mm outer diameter, 66 mm inner diameter and 2·8 m long is used as a strut with both the ends hinged. The load is parallel to the axis of the strut but is eccentric. Find the maximum value of eccentricity so that crippling load on strut is 60 percent of the Euler's crippling load.*

Take: E = 210 GN/m², and yield strength = 320 MN/m².

Solution. Outside diameter of steel tube,

$$D = 88 \text{ mm} = 0 \cdot 088 \text{ m}$$

Inside diameter of steel tube,

$$d = 66 \text{ mm} = 0 \cdot 066 \text{ m}$$

∴ Area of cross-section,

$$A = \frac{\pi}{4}(D^2 - d^2) = \frac{\pi}{4}(0 \cdot 088^2 - 0 \cdot 066^2) = 2 \cdot 66 \times 10^{-3} \text{ m}^2$$

and, moment of inertia,

$$I = \frac{\pi}{64}(D^4 - d^4) = \frac{\pi}{64}[0 \cdot 088^4 - 0 \cdot 066^4] = 2 \cdot 012 \times 10^{-5} \text{ m}^4$$

and, (radius of gyration)²,

$$k^2 = \frac{D^2 + d^2}{16} = \frac{0 \cdot 088^2 + 0 \cdot 066^2}{16} = 7 \cdot 562 \times 10^{-4} \text{ m}^2$$

Yield strength $= 320 \text{ MN/m}^2$

Young's modulus, $E = 210 \text{ GN/m}^2$.

Eccentricity, e:

End Conditions: *Both ends hinged.*

∴ $\qquad\qquad l_e = l = 2 \cdot 8 \text{ m}$ $\qquad\qquad\qquad\qquad\qquad\qquad$ ...(Given)

Now, Euler's buckling load,

$$P_{Euler} = \frac{\pi^2 EI}{l_e^2} = \frac{\pi^2 \times 210 \times 10^9 \times 2 \cdot 012 \times 10^{-5}}{2 \cdot 8^2} \times 10^{-3} \text{ kN} = 531 \cdot 9 \text{ kN}$$

Stress due to Euler's crippling load,

$$\sigma_c = \frac{P_{Euler}}{A} = \frac{531 \cdot 9}{2 \cdot 66 \times 10^{-3}} \times 10^{-3} \text{ MN/m}^2 = 199 \cdot 96 \text{ MN/m}^2$$

Crippling load, $\qquad P = 0 \cdot 6 \, P_{Euler} = 0 \cdot 6 \times 531 \cdot 9 = 319 \cdot 14 \text{ kN}$

Direct stress, $\qquad \sigma_d = \frac{P}{A} = \frac{319 \cdot 14}{2 \cdot 66 \times 10^{-3}} \times 10^{-3} \text{ MN/m}^2 = 119 \cdot 97 \text{ MN/m}^2$

Maximum stress, $\sigma_{max} = \sigma_d + \sigma_b = 320$

(where, σ_b is stress due to bending)

∴ $\qquad\qquad \sigma_b = 320 - \sigma_d = 320 - 119 \cdot 97 = 200 \cdot 03 \text{ MN/m}^2$

$$\sec \frac{l_e}{2}\sqrt{\frac{P}{EI}} = \sec \frac{\pi}{2}\sqrt{\frac{P}{P_{Euler}}} = \sec \frac{\pi}{2}\sqrt{0 \cdot 6}$$

$$= \sec (1 \cdot 216 \text{ radian}) = \sec (69 \cdot 7°) = 2 \cdot 882$$

Stress due to bending,

$$\sigma_b = \frac{M}{Z}$$

or,
$$200 \cdot 03 = \frac{P \cdot e \sec \frac{l_e}{2} \sqrt{\frac{P}{EI}}}{I / y} = \frac{P \cdot ey \sec \frac{l_e}{2} \sqrt{\frac{P}{EI}}}{Ak^2}$$

$$= \sigma_d \cdot \frac{ey}{k^2} \cdot \sec \frac{l_e}{2} \sqrt{\frac{P}{EI}} \qquad (\because \sigma_d = P/A)$$

or,
$$200 \cdot 03 = 119 \cdot 97 \times \frac{e \times 44 \times 10^{-3}}{7 \cdot 562 \times 10^{-4}} \times 2 \cdot 882 \qquad \begin{bmatrix} \because \ y = 44 \text{ mm} \\ = 44 \times 10^{-3} \text{ m} \end{bmatrix}$$

or,
$$e = \frac{200 \cdot 03 \times 7 \cdot 562 \times 10^{-4}}{119 \cdot 97 \times 44 \times 10^{-3} \times 2 \cdot 882} = 9 \cdot 94 \times 10^{-3} \text{ m} = 9 \cdot 942 \text{ mm}$$

Hence, **e = 9·942 mm (Ans.)**

Example 16·35. *A rectangular link is 50 cm long and has width twice the thickness. It is subjected to a maximum compressive load of 5 kN. Assuming a factor of safety of 6 find the thickness of the link. The link ends behave as fixed in the direction of weaker axis of the link section (i.e. buckling in plane XX in Fig. 16-20) and hinged in the other direction.*

Rankine constants: $\sigma_c = 325$ MN/m²; $a = \dfrac{1}{7500}$ *(for hinged ends)*

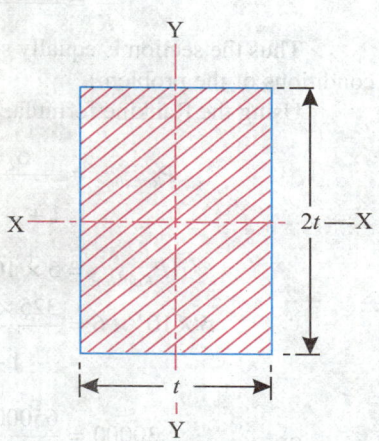

Fig. 16.20

Solution. Let, t = thickness of the link in cm.

∴ Width = $2t$

Area of cross-section,
$$A = t \times 2t = 2t^2$$

Moment of inertia about XX-axis,
$$I_{XX} = \frac{t \times (2t)^3}{12} = \frac{2}{3} t^4$$

$$k_{XX}^2 = \frac{I_{XX}}{A} = \frac{2}{3} \cdot \frac{t^4}{2t^2} = \frac{t^2}{3}$$

Cold rolling machine.

or $\qquad k_{XX} = \dfrac{t}{\sqrt{3}}$ (where, k_{XX} = radius of gyration)

For buckling in plane YY:

$$\frac{l_e}{k} = \frac{50}{t/\sqrt{3}} = \frac{86.7}{t}$$

For buckling in plane XX :

$$I_{YY} = \frac{2t \times t^3}{12} = \frac{t^4}{6}$$

$\therefore \qquad k_{YY}^2 = \dfrac{I_{YY}}{A} = \dfrac{t^4}{6 \times 2t^2} = \dfrac{t^2}{12},$ or, $k_{YY} = \dfrac{t}{\sqrt{12}}$

$\therefore \qquad \dfrac{l_e}{K} = \dfrac{(50/2) \times \sqrt{12}}{t} = \dfrac{86.7}{t}$

Thus the section is equally strong in both directions (and hence most economical for the end conditions of the problem).

Using the Rankine formula,

$$P_{Rankline} = \frac{\sigma_c \cdot A}{1 + a\left(\dfrac{l_e}{k}\right)^2}, \quad \text{we get}$$

$$[P_{Rankline} = 5 \times 10^3 \times \text{factory of safety} = 5 \times 10^3 \times 6 \text{ N}]$$

$\therefore \qquad 5 \times 10^3 \times 6 = \dfrac{325 \times 10^6 \times (2t^2 \times 10^{-4})}{1 + \dfrac{1}{7500} \times \dfrac{86.7^2}{t^2}}$

or, $\qquad 30000 = \dfrac{65000\, t^2}{1 + \dfrac{1}{t^2}},$ or, $\quad 1 + \dfrac{1}{t^2} = 2.167\, t^2$

or, $\quad 2.167\, t^4 - t^2 - 1 = 0$

or, $\qquad t^2 = \dfrac{1 \pm \sqrt{1 + 4 \times 2.167}}{2 \times 2.167} = \dfrac{1 + 3.109}{4.334} = 0.948$

$\therefore \qquad t = 0.974 \text{ cm or } \mathbf{9.74 \text{ mm}} \quad \textbf{(Ans.)}$

Example 16·36. *The forged steel connecting rod of an I.C. engine is 30 cm long and is subjected to an axial load of 24 kN. Using a factor of safety of 7, find the suitable value of t for the cross-section shown in Fig. 16·21. I-section behaves as a column fixed at ends for buckling in plane XX and as a hinged column for buckling in plane YY.*

Rankine constants: σ_c = 325 MN/m², *and for hinged ends* $a = \dfrac{1}{7500}.$

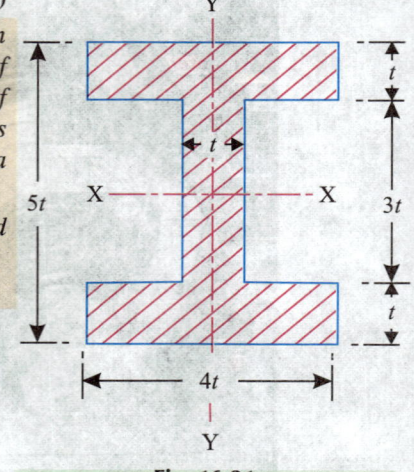

Fig. 16.21

Solution. Length of the connecting rod,

$$l = 30 \text{ cm}$$

Axial load, $\qquad P = 24 \text{ kN}$

Factor of safety, F.O.S. = 7

$$\sigma_c = 325 \text{ MN/m}^2$$

$$a = \frac{1}{7500}$$

Let, t = thickness (in cm)

$\therefore$ Area of cross-section, $A = 2 \times t \times 4t + 3t \times t = 11t^2$

$$I_{XX} = \frac{1}{12} [4 \times 5^3 - 3 \times 3^3] t^4 = \frac{419}{12} t^4$$

$$k_{XX}^2 = \frac{I_{XX}}{A} = = \frac{419 \, t^4}{12 \times 11 \, t^2} = 3 \cdot 174 \, t^2$$

$\therefore$

$$k_{XX} = 1 \cdot 78 \, t$$

$$I_{YY} = \frac{1}{12} [2 \times 4^3 + 3 \times 1^3] t^4 = \frac{131 \, t^4}{12}$$

$$k_{YY}^2 = \frac{I_{YY}}{A} = \frac{131 \, t^4}{12 \times 11 \, t^2} = 0 \cdot 992 \, t^2$$

$\therefore$

$$k_{YY} = 0 \cdot 996 \, t$$

For buckling in plane XX :

Buckling factor (*i.e.* slenderness ratio based on equivalent length)

$$= \frac{l_e}{k_{YY}} = \frac{0 \cdot 5 \times 30}{0 \cdot 996 \, t} = \frac{15}{t}$$

For buckling in plane YY :

Buckling factor

$$= \frac{l_e}{k_{XX}} = \frac{30}{1 \cdot 78 \, t} = \frac{16 \cdot 8}{t}$$

Larger buckling factor governs the design as the smaller critical load corresponds to the larger buckling factor.

We know,

$$P_{Rankline} = \frac{\sigma_c \cdot A}{1 + a \dfrac{l_e^2}{k^2}} = \frac{\sigma_c \cdot A}{1 + a \left(\dfrac{l_e}{k_{XX}} \right)^2}$$

$$(24 \times 10^3) \times 7 = \frac{325 \times 10^6 \times 11 \times (t^2 \times 10^{-4})}{1 + \dfrac{1}{7500} \times \left(\dfrac{16 \cdot 8}{t} \right)^2}$$

or,

$$1 + \frac{1}{7500} \times \frac{16 \cdot 8^2}{t^2} = \frac{325 \times 10^6 \times 11 \, t^2 \times 10^{-4}}{24 \times 10^3 \times 7}$$

or,

$$1 + \frac{1}{26 \cdot 6 \, t^2} = 2 \cdot 128 \, t^2$$

or,

$$26 \cdot 6 \, t^2 + 1 = 56 \cdot 6 \, t^4$$

or,

$$56 \cdot 6 \, t^4 - 26 \cdot 6 \, t^2 - 1 = 0$$

or,

$$t^2 = \frac{26 \cdot 6 \pm \sqrt{26 \cdot 6^2 + 4 \times 56 \cdot 6}}{2 \times 56 \cdot 6}$$

$$= \frac{26 \cdot 6 + 30 \cdot 56}{113 \cdot 2} = 0 \cdot 5 \text{ cm}^2$$

$\therefore$

$$t = 0 \cdot 71 \text{ cm or } \mathbf{7 \cdot 1 \text{ mm}} \quad \textbf{(Ans.)}$$

HIGHLIGHTS

1. A member of structure or bar which carries an axial compressive load is called the *strut*. If the strut is vertical *i.e.* inclined 90° to the horizontal is known as *column, pillar* or *stanchion*.

2. Depending upon the slenderness ratio or length to diameter ratio, columns are divided into three classes:

   (i) *Short columns*: Length < 8 times diameter or slenderness ratio < 32

   (ii) *Medium columns*: Length 8 to 30 times diameter or slenderness ratio between 32 and 120

   (iii) *Long columns*: Length > 30 times diameter or slenderness ratio > 120.

3. *End conditions*:

   (a) Both ends pin jointed or hinged or rounded or free:
   $$l_e = l$$

   (b) One end fixed and other end free:
   $$l_e = 2l$$

   (c) One end fixed and other end pin jointed:
   $$l_e = \frac{l}{\sqrt{2}}$$

   (d) Both ends fixed:
   $$l_e = l/2$$

4. Euler's formula (long columns): $P_{Euler} = \dfrac{\pi^2 EI}{l_e^2}$

5. Rankine's formula: $P_{Rankline} = \dfrac{\sigma_c \cdot A}{1 + a\,(l_e/k)^2}$

6. Johnson's parabolic formula: $\dfrac{P}{A} = \sigma_c - b\left(\dfrac{l}{k}\right)^2$

   b is approximately $= \dfrac{\sigma_c^2}{4\pi^2 E}$ and Johnson accepted the value $= \dfrac{\sigma_c^2}{64E}$ for pinned ends.

7. Straight line formula:

   (i) Stress at critical load for cast-iron,
   $$\frac{P}{A} = 23.8 - 0.6\left(\frac{l}{k}\right) \text{ N/mm}^2$$

   (ii) Stress at critical load for structural steel,
   $$\frac{P}{A} = 367.5 - 2\left(\frac{l}{k}\right) \text{ N/mm}^2$$

   (iii) Safe working stress for mild steel
   $$= 150\left(1 - 0.0038\,\frac{l}{k}\right) \text{ N/mm}^2$$

8. Columns subjected to eccentric loading:

   (a) *Rankine's method*:
   $$P = \frac{\sigma_c \cdot A}{\left(1 + \dfrac{ey}{k^2}\right)\left(1 + a\,\dfrac{l_e^2}{k^2}\right)}, \text{ when the effect of buckling is included.}$$

   (b) *Euler's method*:

   Maximum compressive stress $= \dfrac{P}{A} + \dfrac{p \cdot e \cdot \sec \dfrac{l_e}{2}\sqrt{\dfrac{P}{EI}}}{Z}$

   Bending moment, $M = p \cdot e \cdot \sec \dfrac{l_e}{2}\sqrt{\dfrac{P}{EI}}$

9. *Prof. Perry's formula:*

$$\left(\frac{\sigma_{max}}{\sigma_d} - 1\right)\left(1 - \frac{\sigma_d}{\sigma_{Euler}}\right) = \frac{1\cdot2\, e \times y}{k^2}$$

10. *Laterally loaded struts:*

Case I.   Strut pinned at both ends and subjected to an axial thrust P and a transverse point load W:

Maximum deflection,

$$\delta = \frac{W}{2P}\sqrt{\frac{EI}{P}}\,\tan\frac{l}{2}\sqrt{\frac{P}{EI}} - \frac{Wl}{4P} \qquad ...(i)$$

Maximum bending moment,

$$M_{max} = \frac{W}{2}\sqrt{\frac{EI}{P}}\,\tan\frac{l}{2}\sqrt{\frac{P}{EI}} \qquad ...(ii)$$

Maximum stress,

$$\sigma_{max} = \frac{P}{A} + \frac{W}{2}\cdot\frac{y_c}{Ak^2}\sqrt{\frac{EI}{P}}\,\tan\frac{l}{2}\sqrt{\frac{P}{EI}} \qquad ...(iii)$$

where,
- P = Axial thrust,
- W = Transverse point load,
- y_c = Distance of the extreme layer in compression from the neutral axis,
- A = Area of cross-section of the column,
- I = Moment of inertia,
- k = Radius of gyration, and
- E = Young's modulus.

Case II. Strut pinned at both ends and subjected to an axial thrust P and a lateral uniformly distributed load of intensity w per unit run:

Maximum deflection,

$$\delta = \frac{WEI}{P^2}\left[\sec\frac{l}{2}\sqrt{\frac{P}{EI}} - 1\right] - \frac{wl^2}{8P} \qquad ...(i)$$

Maximum bending moment,

$$M_{max} = -\frac{WEI}{P}\left[\sec\frac{l}{2}\sqrt{\frac{P}{EI}} - 1\right] \qquad ...(ii)$$

Maximum stress,

$$\sigma_{max} = \frac{P}{A} + \frac{wEy_c}{P}\left[\sec\frac{l}{2}\sqrt{\frac{P}{EI}} - 1\right] \qquad ...(iii)$$

OBJECTIVE TYPE QUESTIONS

Choose the Correct Answer:

1. A member of structure or bar which carries an axial compressive load is called
    - (a) strut
    - (b) tie
    - (c) shaft
    - (d) none of the above.

2. When the strut is vertical *i.e.* inclined at 90° to the horizontal is known as
    - (a) column
    - (b) pillar
    - (c) stanchion
    - (d) any of the above.

3. In case of a column
    - (a) one end is hinged and other end fixed
    - (b) one end is fixed and other end free
    - (c) both the ends are hinged
    - (d) both the ends are fixed rigidly.

4. Which of the following is an example of the strut ?
    - (a) Piston rods
    - (b) Connecting rods
    - (c) Side links in forging machines
    - (d) All of the above.

5. The failure of struts may occur
    - (a) by pure compression
    - (b) by buckling
    - (c) by combination of pure compression and buckling, depending upon a slenderness ratio
    - (d) any of the above.

6. The ratio of equivalent length of the column to the minimum radius of gyration is called
    - (a) Poisson's ratio
    - (b) buckling factor
    - (c) factor of safety
    - (d) none of the above.

7. Columns which have length less than 8 times their diameter or slenderness ratio less than 32 are called
   - (a) short columns (b) medium columns
   - (c) long columns (d) any of the above.

8. The ratio between buckling load and safe load is known as
   - (a) slenderness ratio
   - (b) buckling factor
   - (c) factor of safety
   - (d) none of the above.

9. The buckling, in case of a column, takes place about the axis having
   - (a) minimum radius of gyration
   - (b) maximum radius of gyration
   - (c) either of the above
   - (d) none of the above.

10. The strength of a column depends on which of the following factors ?
    - (a) Slenderness ratio
    - (b) End conditions
    - (c) Both (a) and (b)
    - (d) None of the above.

11. Slenderness ratio of a column may be defined as the ratio of its length to the
    - (a) radius of column
    - (b) minimum radius of gyration
    - (c) maximum radius of gyration
    - (d) none of the above.

12. If a slenderness ratio of a column is more than 120 it is termed as
    - (a) short column (b) medium column
    - (c) long column (d) none of the above.

13. Euler's formula is applicable to
    - (a) short columns (b) medium columns
    - (c) long columns (d) none of the above.

14. The radius of gyration of a circular column of diameter d is
    - (a) $d/4$ (b) $d/2$
    - (c) $d^2/4$ (d) $d^2/16$.

15. Euler's critical load for a column of equivalent length l_e, moment of inertia I and modulus of elasticity E is given by
    - (a) $\dfrac{\pi^2\,EI}{l_e}$ (b) $\dfrac{\pi\,EI}{l_e^2}$
    - (c) $\dfrac{\pi^2\,EI}{l_e^2}$ (d) $\dfrac{\pi\,EI}{l_e}$.

16. The ratio of equivalent length of a column, having one end fixed and the other end free, to its length is
    - (a) 2 (b) $\sqrt{2}$
    - (c) 1/2 (d) $1/\sqrt{2}$.

17. The ratio of equivalent length of a column, having one end fixed and the other hinged, to its length is
    - (a) 2 (b) $\sqrt{2}$
    - (c) 1/2 (d) $1/\sqrt{2}$.

18. The ratio of equivalent length of a column, having both ends fixed, to its length is
    - (a) 2 (b) 1/2
    - (c) $\sqrt{2}$ (d) $1/\sqrt{2}$.

19. Euler's buckling formula is applicable for columns
    - (a) subjected to eccentric loads
    - (b) having initial curvature
    - (c) initially straight and subjected to only axial loads
    - (d) none of the above.

20. Rankine formula takes into account which of the following ?
    - (a) The effect of slenderness ratio
    - (b) The initial curvature of the column
    - (c) The eccentricity of loading
    - (d) The effect of direct compressive stress.

21. In the Rankine formula, the material constant for mild steel is
    - (a) $\dfrac{1}{1200}$ (b) $\dfrac{1}{1600}$
    - (c) $\dfrac{1}{7500}$ (d) $\dfrac{1}{5000}$.

Rotary actuator maintained on a ball valve.

22. A beam column is one which carries
  (a) axial load
  (b) transverse loads
  (c) eccentric load
  (d) axial as well as transverse loads.

23. To avoid tension in a short column the load must lie
  (a) at the centre of gravity of the section
  (b) within the middle third of the cross-section
  (c) within the one-third of the cross-section
  (d) outside the middle-third of the cross-section.

24. The secant formula is used for
  (a) long columns under eccentric loading
  (b) long columns under axial loading
  (c) short columns under axial loading
  (d) short columns under eccentric loading.

25. Rankine's constant for the compressive strength of a cast iron column is generally taken as
  (a) 150 MN/m^2
  (b) 320 MN/m^2
  (c) 400 MN/m^2
  (d) 550 MN/m^2.

ANSWERS

1. (a)	2. (d)	3. (d)	4. (d)	5. (d)	6. (b)	7. (a)
8. (c)	9. (a)	10. (c)	11. (b)	12. (c)	13. (c)	14. (a)
15. (c)	16. (a)	17. (d)	18. (b)	19. (c)	20. (d)	21. (c)
22. (d)	23. (b)	24. (a)	25. (d)			

UNSOLVED EXAMPLES

1. A hollow cast iron column rigidly fixed at one end and pin-pointed at the other end, has 150 mm outer diameter and 120 mm inner diameter. Its length is 6 m and E may be taken as 90 GN/m^2. Calculate the critical load of this column by Euler formula. **[Ans. 724 kN]**

2. Calculate the safe compressive load on a hollow cast iron column one end rigidly fixed and other pin-jointed, 150 mm outer and 100 mm inner diameter, 10 metres long. Use Euler's formula with a factor of safety of 5 and take: $E = 95$ GN/m^2. **[Ans. 75 kN]**

3. Determine the ratio of strength of a solid steel column to that of a hollow column of internal diameter equal to 3/4 of its external diameter. Both the columns have the same cross-sectional areas, length and end conditions.

$$\left[\textbf{Ans.}\,\frac{25}{7}\right]$$

4. A cast iron hollow column, having 8 cm external diameter and 6 cm internal diameter, is used as a column of 2 m length. Using Rankine's formula, determine the crippling load, when both the ends are fixed. Take: $\sigma_c = 600$ MN/m^2. **[Ans. 660 kN]**

5. A hollow cast iron column of 30 cm external diameter and 23 cm internal diameter is used as a column 4 m long, with both ends hinged. Determine the Rankine's safe load with a factor of safety of 4. Take: $\sigma_c = 567$ MN/m^2 and $a = \dfrac{1}{1600}$. **[Ans. 2149 kN]**

6. Find the Euler crushing load for a hollow cylindrical cast iron column, 15 cm external diameter and 2 cm thick, if it is 6 m long and hinged at both ends, $E = 80$ GN/m^2. Compare this load with the crushing load given by Rankine's formula, using $\sigma_c = 567$ MN/m^2 and $a = \dfrac{1}{1600}$. For what length of column would these two formulae give the same crushing load ?

[Ans. 387·8 kN, 406·4 kN, 4·82 m]

7. Compare the crippling loads given by Euler's and Rankine formulae for a tubular steel strut 2·5 m long, having outer and inner diameters 4 cm and 3 cm respectively through pin-joints at each end.

Take the yield stress as 330 MN/m^2, the Rankine constant = $\dfrac{1}{7500}$, and E = 200 GN/m^2. For what length of strut of this cross-section does the Euler formula cease to apply ?

$$\left[\textbf{Ans.} \quad \frac{P_{Euler}}{P_{Rankline}} = 0.945,\ 96.675 \text{ cm} \right]$$

8. A hollow cast iron column with fixed ends supports an axial load of 1000 kN. If the column is 5 m long and has an external diameter of 250 mm, find thickness of the metal required. Use Rankine formula taking a constant of $\dfrac{1}{1600}$ and assume a working stress of 80 MN/m^2.

   [Ans. 29·4 mm]

9. Find the greatest length for which a mild strut of T-shaped cross-section the area of which is 30 cm^2 and the least moment of inertia is 240 cm^4 may be used with one end fixed and other entirely free in order to carry a working load of 70 MN/m^2 of section, the working load being one-fourth the crippling load. Rankine constants for mild steel are, $a = \dfrac{1}{7500}$, and $\sigma_c = 330$ MN/m^2. **[Ans.** 51·75 cm]

10. Calculate the diameter of the piston rod from the following data:
    Diameter of engine cylinder = 350 mm
    Maximum effective stress pressure in the cylinder = 700 kN/m^2
    Distance from piston to cross-head centre = 1·5 m
    Factors of safety = 4
    $$\sigma_c = 330 \text{ MN/m}^2$$
    $$a = \frac{1}{30,000}, \text{ for both ends fixed.} \qquad \textbf{[Ans. 41·9 mm]}$$

11. Find the inside diameter of a cast iron column of 200 mm outside diameter which is 5 m long and has to support a safe axial load of 600 kN, one end being rigidly fixed. Use a factor of safety of 5 in conjunction with Rankine formula; $a = \dfrac{1}{1600}$ and $\sigma_c = 567$ MN/m^2.

    [Ans. 179 mm]

12. Find the safe load for a mild steel strut made up of two ties 6 cm × 12 cm × 1 cm riveted back to back to form a cross of 12 cm × 12 cm. Length = 2·27 m; both ends free. $\sigma_c = 472.5$ MN/m^2, $a = \dfrac{1}{7500}$ and factor of safety 5. **[Ans. 126 kN]**

Inside view of a machine shop.

13. A steel stanchion is built of two rolled steel joists of I section 30 cm × 15 cm × 1·25 cm united by plates 2 cm thick and 40 cm wide fastened to flanges. The edges of the plates are flush with the outside edges of the joints. Using Rankine formula for a strut, find the safe load for this stanchion if it is 8 m long, $\sigma_c = 330$ MN/m², and $a = \dfrac{1}{7500}$. F.O.S. = 4. [**Ans.** 1406 kN]

14. Two 8 cm × 16 cm rectangular section cast iron columns are each 4 m long with one end fixed and other hinged. They share equally the total load carried by them. Find by Rankine's formula, the diameter of single solid cast iron column of circular section of the same length and same end conditions to replace both of them. Also calculate the percentage saving in material. Take Rankine's constant as $\dfrac{1}{1600}$ for hinged end conditions and crushing stress for cast-iron as 500 MN/m².

[**Ans.** 13·28 cm, 46·4%]

15. A hollow cylindrical cast iron column is 4 metres long, both ends being fixed. Design the column to carry an axial load of 250 kN. Use Rankine's formula and adopt a factor of safety of 5. The internal diameter may be taken as 0·8 times the external diameter.

Take: $\sigma_c = 350$ MN/m², and $a = \dfrac{1}{1600}$. [**Ans.** $D = 136·8$ mm $d = 109$ mm]

16. A strut 3 metres long with both ends hinged consists of two equal angles 10 cm × 10 cm × 1 cm the spacing between the angles being 1 cm. Find safe compressive load for the strut allowing a factor of 4. Use Rankine's formula. Take: $\sigma_c = 320$ MN/m², and $a = \dfrac{1}{7500}$; . Properties of one angle are :

$A = 19·03$ cm², $I_{XX} = I_{YY} = 177$ cm⁴. [**Ans.** 133 kN]

17. A column of circular section made of cast iron is of 20 cm external diameter, 2 cm thick and 4 m long. Both ends of the column are fixed. The column carries a load of 150 kN at an eccentricity of 2·5 cm from the axis of the column. Find the extreme stresses on the column section. Find also the maximum eccentricity in order there may be no tension anywhere on the section. Take: $E = 94$ GN/m². [**Ans.** 21·5 MN/m², 4·432 cm]

18. A hollow circular steel pipe 60 mm outside diameter, 50 mm inside diameter, 1·2 m long is fixed at both ends, so as to prevent any expansion in length. The pipe is unstressed at the normal temperature. If the temperature of the pipe rises by 40°C, using Rankine's formula calculate:

   (*i*) Temperature stress in the pipe;

   (*ii*) Factor of safety against failure as a strut.

Given: Rankine constants: $\sigma_c = 330$ MN/m², $a = \dfrac{1}{7500}$;

$E = 208$ GN/m², and $\alpha = 11·1 \times 10^{-6}$ per °C [**Ans.** (*i*) 92·35 MN/m², (*ii*) 3·17]

19. A 2·8 m long steel tube having 80 mm outer diameter and 60 mm inner diameter is used as a strut with both ends hinged. The load is parallel to the axis of the strut but is eccentric. Find the maximum value of the eccentricity so that crippling load on strut is equal to 60 percent of the Euler's crippling load. Take: Yield strength = 320 MN/m², and E = 210 GN/m². [**Ans.** 12·07 mm]

20. A steel strut has an outside diameter of 120 mm and inside diameter of 80 mm and is 6 m long. It is hinged at both ends and is initially bent. Assuming the centre line of the strut as sinusoidal with maximum deviation of 6 mm, determine the maximum stress developed due to an axial load of 100 kN. Take: $E = 208$ GN/m². [**Ans.** 20·7 MN/m²]

21. A steel strut, 1 m long is 25 mm in diameter. It is subjected to an axial thrust of 12 kN. In addition, a lateral load W acts at the centre of the strut. If the strut failed at a maximum stress of 320 MN/m², determine the magnitude of W. Take: $E = 210$ GN/m². [**Ans.** 1·343 kN]

22. A rod 2 m in length and of rectangular cross-section 80 mm × 40 mm is supported horizontally through pin joints. It carries a vertical load of 3 N/mm length and an axial thrust of 100 kN. If $E =$ 208 kN/mm², calculate the maximum stress induced. [**Ans.** 161·4 N/mm²]

Chapter

17

Analysis of the Framed Structures

17.1. INTRODUCTION

A framed structure or truss is composed of several bars or rods joined together in a particular fashion; these bars are called the *members* of the structure. Fig. 17.1.

A member under tension is called **'Tie'** *and under compression is known as* **'Strut'**.

The framed structures are of the following three types:

1. *Efficient or Perfect*
2. *Deficient or Imperfect*
3. *Redundant.*

The structure is said to be '*efficient*' if it satisfies the equation :

$$m = 2j - 3 \qquad \qquad ...(17.1)$$

where, m = Number of members of the structure,

and

j = Number of joints of the structure.

Fig. 17.1 (*a*) shows a '*Perfect*' structure in which

$m = 9$, $j = 6$ and satisfies the eqn (17.1).

Fig. 17.1 (*b*) shows a structure in which $j = 6$ and $m = 8$ (*i.e.* one number *less* than the required number as per eqn. 17.1). This type of structure is termed '*deficient*' or '*imperfect*'.

Fig. 17.1 (*c*) shows a framed structure in which $j = 6$ and $m = 10$ (*i.e.* one number *surplus* than the required number) called '*redundant*' frame.

Assumptions:

The following *assumptions* are made while computing the forces in the members of a perfect frame :

1. All members are pin jointed.
2. The frame is loaded only at the joints.
3. The frame is a perfect frame.
4. The self-weight of the members is neglected.

17.2. DETERMINATION OF REACTIONS — GRAPHICAL METHOD

Prior to the discussiion of the method in detail some light willl be thrown on '*Bow's notation*' which plays an important role in the graphical solutions of the problems.

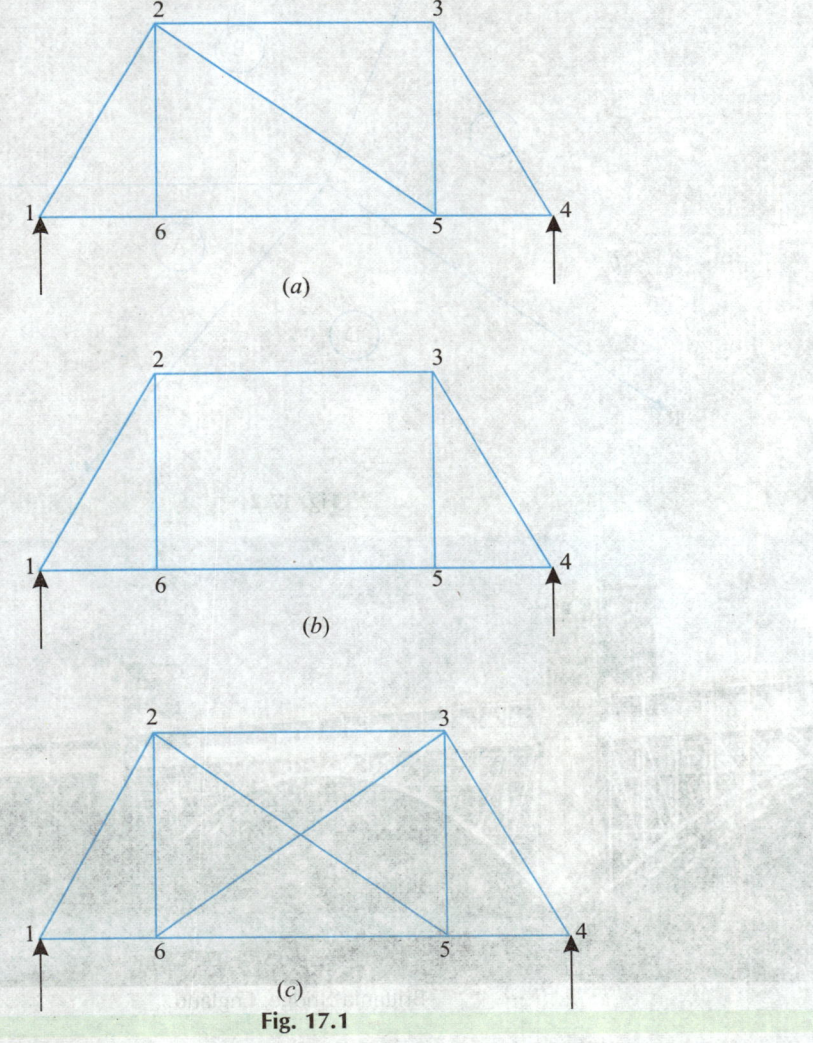

(*a*)

(*b*)

(*c*)

Fig. 17.1

Bow's notation:

It is the method of designating a force by placing capital letters on either side of it.

Refer Fig. 17.2 ; there are four forces, *P*, *Q*, *R* and *S* acting at a point in the same plane. According to Bow's notation the force *P* is represented by *AB*, *Q* by *BC*, *R* by *CD* and *S* by *DA* respectively. The diagram obtained after incorpoarating the letters *A*, *B*, *C* and *D* in the forces is called **Space diagram.**

The *main advantage,* however, of the Bow's notation is that any force in space diagram can be represented in magnitude and direction to a certain scale by drawing a line *parallel to the force.*

Following *types of supports* are commonly used :

1. Simple supports at both ends.
2. One end pin jointed, other supported on roller.
3. One end pin jointed, other supported on smooth surface.
4. Both ends fixed.

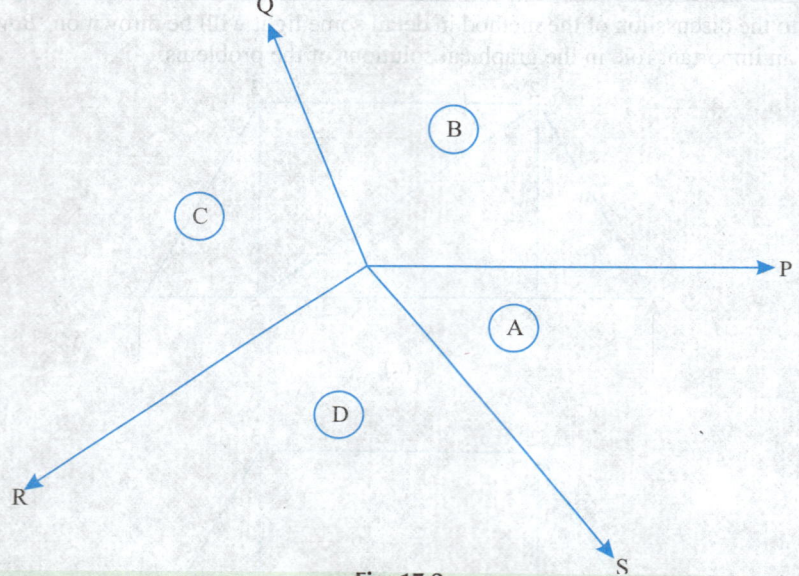

Fig. 17.2

Brittania Bridge, England.

Here follows the details of reactions' determination for above types of supports.

1. Simple supports at both ends :

Refer Fig. 17.3 (*a*). The roof truss carries the vertical loads of magnitudes, say, 100 kN, 200 kN and 100 kN respectively at the joints *P*, *Q*, and *R*. To find the reactions R_L and R_M the following procedure should be adopted :

(*i*) Construct the space diagram to suitable scale.

(*ii*) Draw a load line *ad* to some scale (in this case, say 1 cm = 200 kN) such that *ab*, *bc* and *cd* represent the loads *AB*, *BC* and *CD* respectively in magnitude as well as in direction.

(*iii*) Select any point 0 (called the pole) and join *oa*, *ob*, *oc* and *od* [Fig. 17.3 (*b*)]

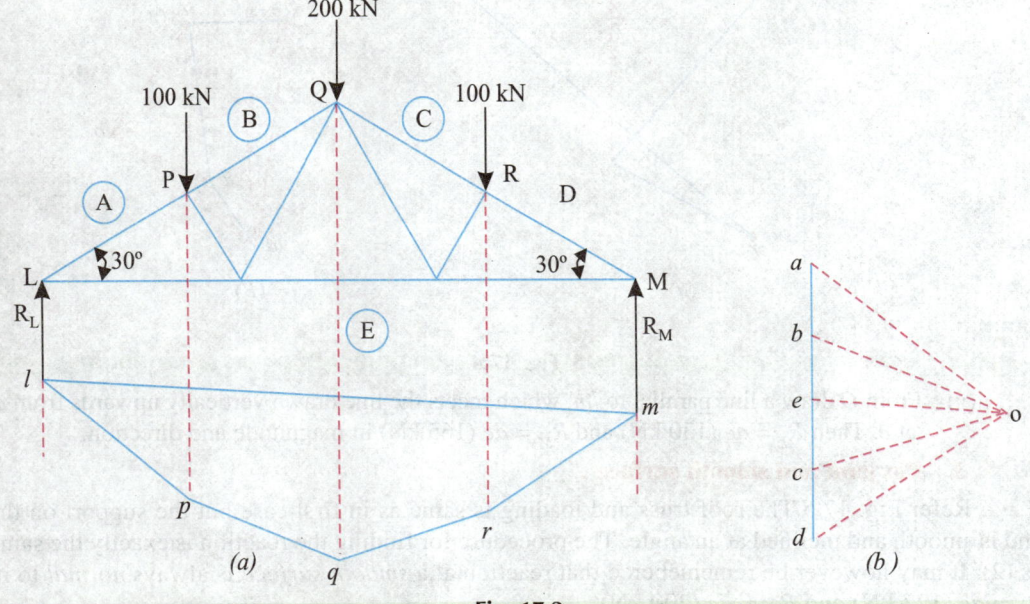

Fig. 17.3

(*iv*) Take any point *l* on the reaction (R_L) line and draw a line *lp* ∥ *oa*, and similarly draw lines *pq* ∥ *ob*, *qr* ∥ *oc* and *rm* ∥*od* respectively. Join *l* and *m*; the polygon *lpqrm* thus obtained is called '*Funicular polygon*'

(*v*) From the pole *O* draw a line *oe* ∥ *lm* (the closing side of the polygon). Then $R_L = de$ (200 kN) and $R_M = ea$ (200 kN) in magnitude as well as in direction.

2. Pin joint and roller support :

In Fig. 17.4 (*a*) is shown the same truss (as in first case) subjected to same loads but with different lines of application.

(*i*) Make the space diagam.

(*ii*) Draw the load line *abcd* to a certain scale (say 1 cm = 100 kN) such that *ab* ∥ *AB*, bc ∥ *BC* and *cd* ∥ *CD*.

(*iii*) Select any pole *O*. Join *oa*, *ab*, *oc* and *od* Fig. [17.4 (*b*)]

(*iv*) Since the reaction R_L must pass through the point *L* therfore starting from this point, draw a line parallel to *oa* meeting at *p* with load (*P*) line produced. From *p* draw line *pq* ∥ *ob* meeting the load *Q*; at *q* draw line *qr* ∥ *oc* intersecting the load (*R*) line at *r* and finally draw line *rm* parallel to *od* meeting the reaction (R_M) line [which is always perpendicular (in this case vertical) to the roller support] at *m*. Join *l* and *m*.

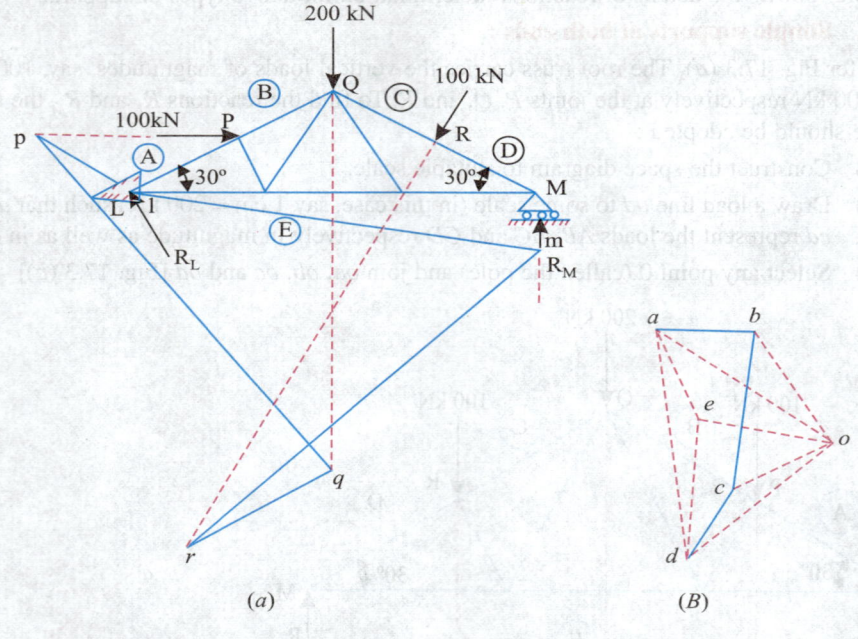

(a)   (B)

Fig. 17.4

(v) From O draw a line parallel to lm, which meets the line drawn vertically upwards from d, at e. Then $R_L = ae$ (130 kN) and $R_M = de$ (165 kN) in magnitude and direction.

3. Pin joint and smooth surface :

Refer Fig. 17.5.The roof truss and loading is same as in first case but the support on the end is smooth and inclined at an angle. The procedure for finding the reaction is exactly the same as (2). It may however be rememebered that reaction at a *smooth surface* is always *normal* to it. $R_L = ae$ (127 kN) and $R_M = de$ (200 kN).

Bridge.

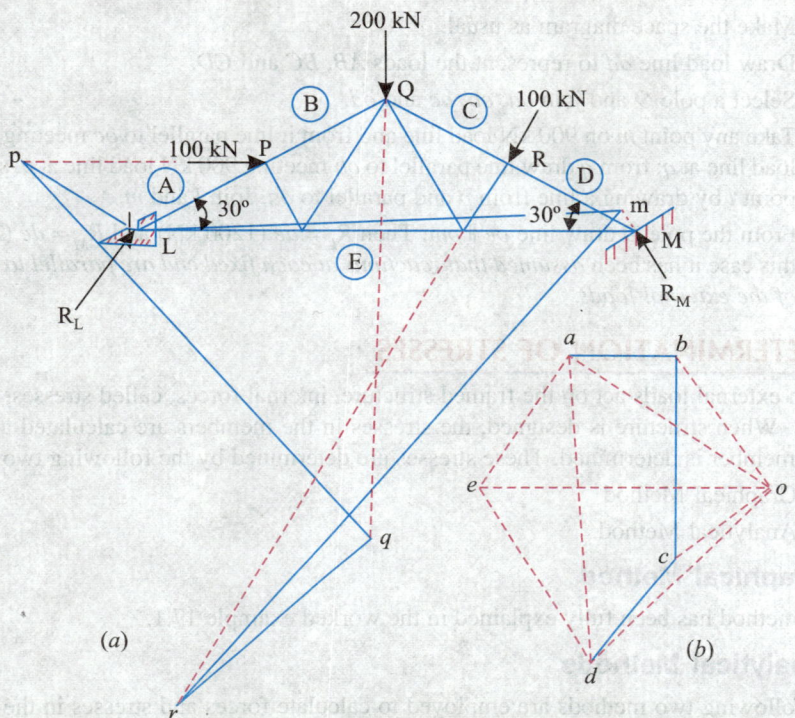

Fig. 17.5

4. Both ends fixed :

A truss fixed at both ends is shown in Fig. 17.6 (a)

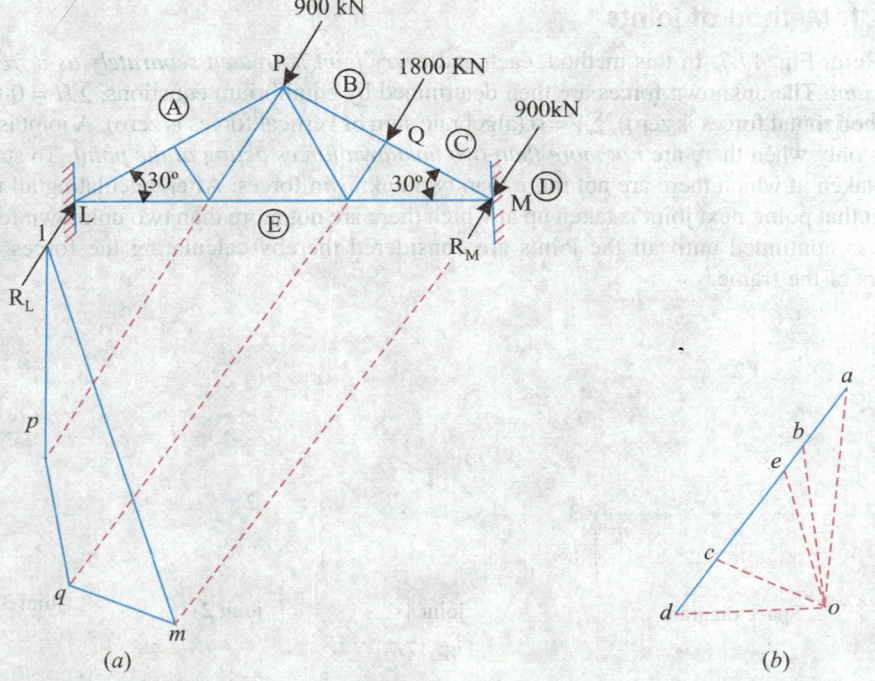

(a)   (b)

Fig. 17.6

    (*i*)   Make the space diagram as usual.

    (*ii*) Draw load line *ad* to represent the loads *AB*, *BC* and *CD*.

    (*iii*) Select a pole *O* and join *oa*, *ob*, *oc* and *od*.

    (*iv*) Take any point *m* on 900 kN load line and from it line parallel to *oc* meeting the 1800 kN load line at *q*; from *q* draw line parallel to *ob* meeting 900 kN load line at *p*, similarly find point *l* by drawing a line from *p* and parallel to *oa*. Join *l* and *m*.

    (*v*) From the pole *O* draw line *oe* ∥ *om*. Then $R_L = ae$ (1200 kN) and $R_M = de$ (2400 kN). In this case it has been *assumed that reactions at each fixed end are parallel to the resultant of the external loads*.

17.3. DETERMINATION OF STRESSES

    When external loads act on the framed structure, internal forces, called stresses, are set up in its members. When structure is designed, the stresses in the members are calculated and from this data size of member is determined. These stresses are determined by the following two methods :

    1. Graphical Method

    2. Analytical Method.

17.3.1. Graphical Method

    This method has been fully explained in the worked example 17.1.

17.3.2. Analytical Methods

    The following two methods are employed to calculate forces and stresses in the members of pin jointed frames.

    1. Method of Joints

    2. Method of Sections.

17.3.2.1. Method of joints

    Refer Fig. 17.7. In this method, each and *every joint is treated separately as a free body in equilibrium*. The unknown forces are then determined by equilibrium equations, $\Sigma H = 0$ (algebraic sum of horizontal forces is zero), $\Sigma V = 0$ (algebraic sum of vertical forces is zero). A joint is taken for analysis only when there are *not more than two unknown forces acting at the point*. To start with, a joint is taken at which there are not more than two unknown forces. After calculating all the forces acting at that point, next joint is taken up at which there are not more than two unknown forces. The process is continued until all the joints are considered thereby calculating the forces in all the members of the frame.

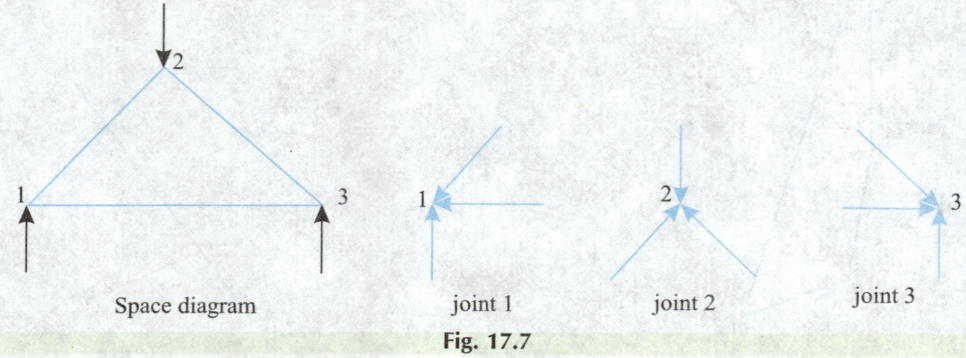

Space diagram   joint 1   joint 2   joint 3

Fig. 17.7

17.3.2. Method of sections (or Method of moments).

This method is *particularly convenient when the forces in few members of a frame are required to be found out*. In this method a *section line* is passed through the members, in which forces are required to be found out as shown in the Fig. 17.8. A part of the structure, on any side of the section line, is then treated as a free body in equilibrium under the action of external forces. The unknown forces are then found by the *application of equilibrium or principle of statics i.e. $\Sigma M = 0$*.

It may be noted that while drawing a *section line*, care should be taken *not to cut more than three members, in which the forces are known*.

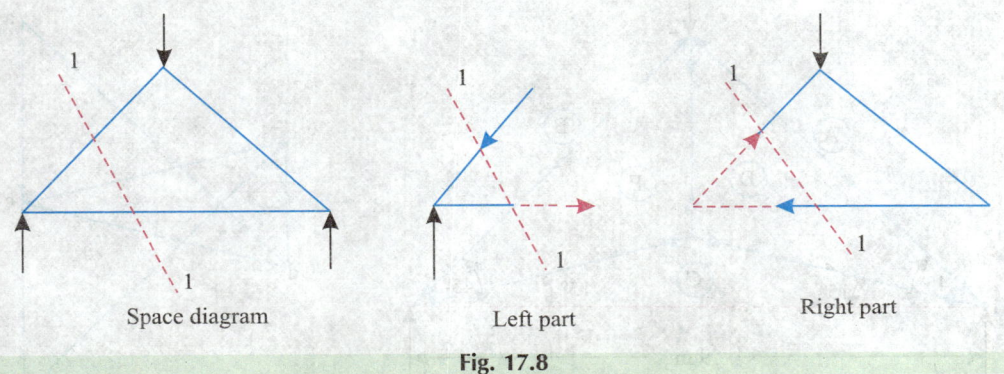

| Space diagram | Left part | Right part |

Fig. 17.8

WORKED EXAMPLES

Solutions by "Graphical method" :

Example 17.1. *Find the forces in different members of the roof truss shown in Fig. 17.9.*

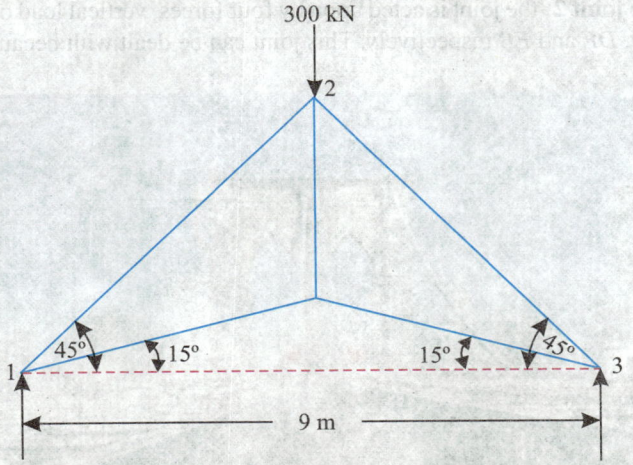

300 kN

9 m

Fig. 17.9

Solution. The procedure for the determination of the forces is as follows :

(*i*)   Draw the *space diagram* (Fig. 17.10) to some scale (say 1 cm = 2 m).

(*ii*) Determine the reactions either graphically or analytically. In this case since frame is symmetrical therefore reaction $R_1 = R_2 = 300/2 = 150$ kN.

(*iii*) Refer Fig. 17.10 and 17.11. After finding the reactions start with the joint in which there are *not more than two unkown forces*, In this case we can start with **joint 1**. This joint is in equilibrium under the reaction CA (R_1) and the forces in the members AD and DC. Commencing from the known force in the space diagram; ca is drawn parallel and equal to the reaction CA (R_1). From a draw a line parallel to the member AD and from c another line parallel to CD, both intersecting at d. Measure ad and cd and multiply with the scale chosen.

Force in member $AD = ad = 280$ kN

Force in member $DC = dc = 205$ kN

Fig. 17.10. Space diagram.　　　　　　　　**Fig. 17.11.** Force/vector diagram.

Incorporate the arrow heads in members AD and DC in the direction ad and dc respectively. The force in the *member AD is compressive and that in DC is tensile.*

Consider the **joint 2;** the joint is acted upon by four forces, vertical load of 300 kN and stresses in the members AD, DE and EB respectively. This joint can be dealt with because only two forces in

Bridge.

the member *DE* and *EB* are unknown. From *b* draw line parallel to member *EB* and from *d* another line parallel to member *DE*, both lines meeting at *e*. Measure *eb* and *de* and multiply with the scale.

Then, force in member *EB* = 280 kN (compressive) and force in member *DE* = 100 kN (tensile) Lastly take the **joint 3**. Join *ce*. The force in the member EC = 205 kN (tensile).

The forces in various members are tabulated in table 17.1.

S.No.	Member	Magnitude of force (kN)	Nature of force
TABLE 17.1			
1.	AD	280	C (compressive)
2.	DC	205	T (tensile)
3.	EB	280	C
4.	EC	205	T
5.	DE	100	T

Example 17.2. *Find the forces and their nature in various members of the truss shown in the Fig. 17.12.*

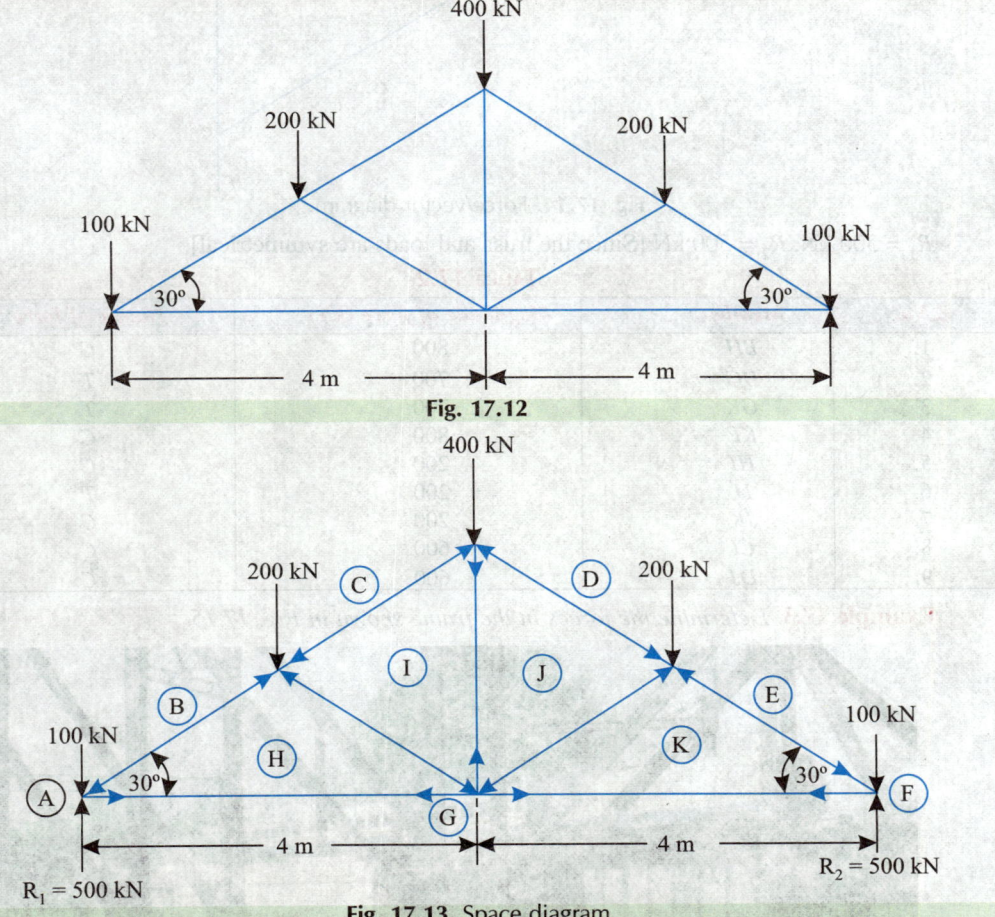

Fig. 17.12

Fig. 17.13. Space diagram.

Solution. Draw the vector/force diagram for the given truss (Fig. 17.14) and tabulate the forces as given below :

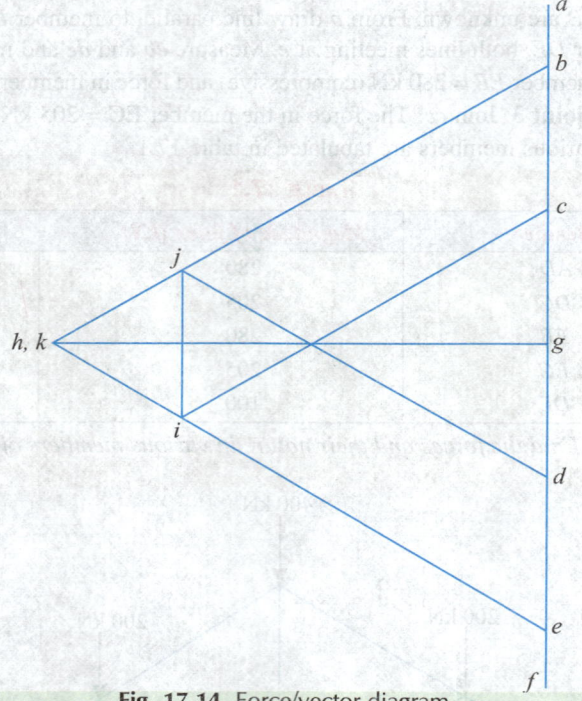

Fig. 17.14. Force/vector diagram.

$R_1 = 500$ kN, $R_2 = 500$ kN [Since the truss and loads are symmetrical]

		TABLE 17.2	
S.No.	*Member*	*Magnitude of force (kN)*	*Nature of force*
1.	BH	800	C
2.	HG	700	T
3.	GK	700	T
4.	KE	800	C
5.	HI	200	C
6.	IJ	200	T
7.	JK	200	C
8.	CI	600	C
9.	DJ	600	C

Example 17.3. *Determine the forces in the frame shown in Fig. 17.15.*

A railway bridge.

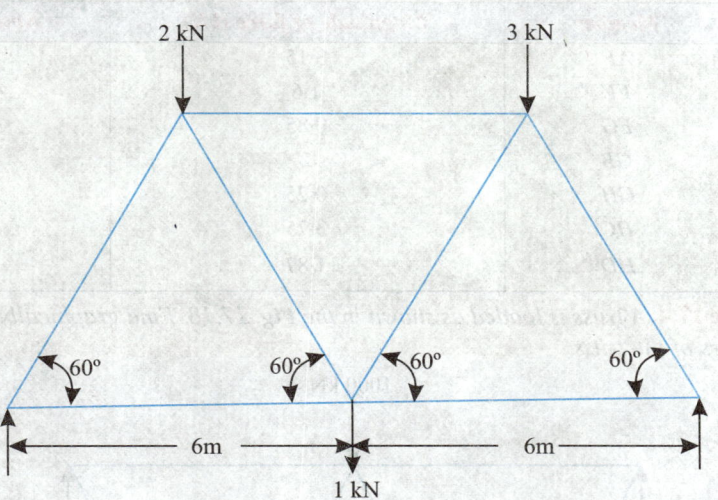

Fig. 17.15

Solution. To find reactions taking moments about L, we get

$$R_M \times 12 = 3 \times 9 + 1 \times 6 + 2 \times 3 = 27 + 6 + 6 = 39$$

∴ $$R_M = \frac{39}{12} = 3 \cdot 25 \text{ kN}$$

Also, $R_L + R_M = 2 + 3 + 1 = 6 \text{ kN}$ ∴ $R_L = 6 - 3 \cdot 25 = 2 \cdot 75 \text{ kN}$

The force/vector diagram for the given frame has been drawn (Fig. 17.17) and the values of forces are given in the table 17.3.

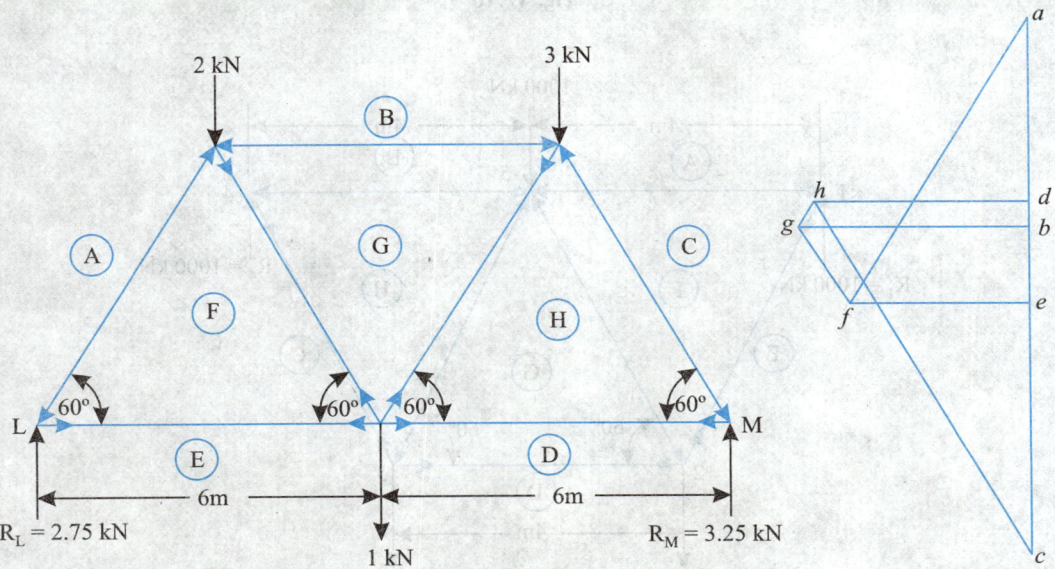

Fig. 17.16. Space diagram.   **Fig. 17.17.** Force/vector diagram.

TABLE 17.3

S.No.	Member	Magnitude of force (kN)	Nature of force
1.	AF	3·15	C
2.	FE	1·6	T
3.	FG	0·85	T
4.	GB	2	C
5.	GH	0·25	T
6.	HC	3·75	C
7.	HD	1·87	T

Example 17.4. *A truss is loaded as shown in the Fig. 17.18. Find graphically the forces in the various members of the truss.*

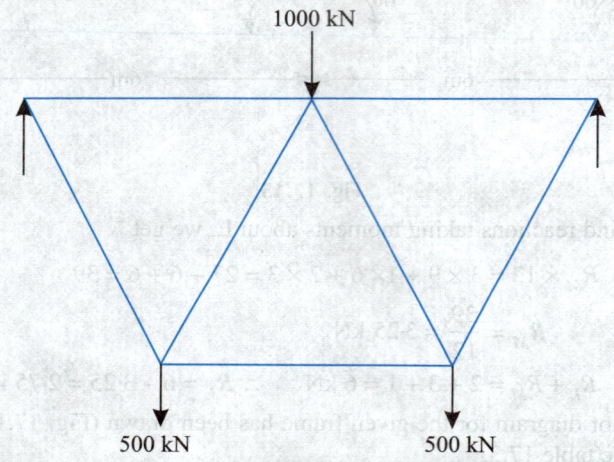

1000 kN

500 kN   500 kN

Fig. 17.18

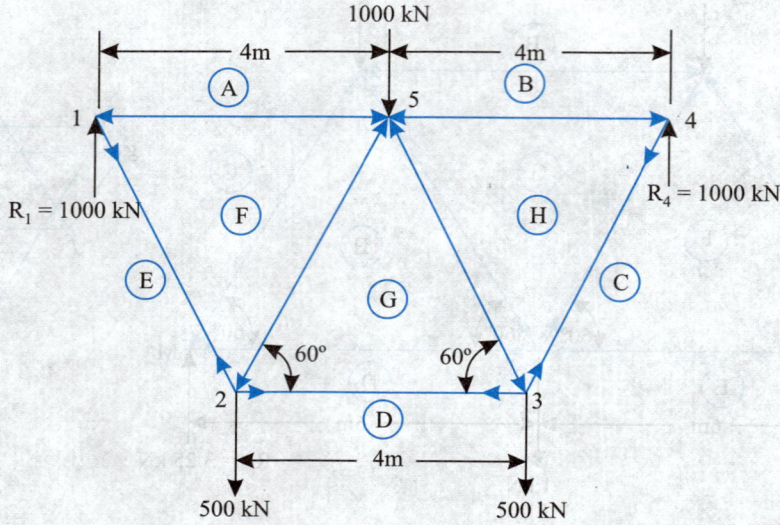

Fig. 17.19. Space diagram.

Solution. $R_1 = R_4 = \dfrac{1000 + 500 + 500}{2} = 1000$ kN (Since the loading and truss are symmetrical)

TABLE 17.4

S.No.	Member	Magnitude of force (kN)	Nature of force
1.	AF	500	C
2.	EF	1120	T
3.	FG	560	C
4.	GH	560	C
5.	HC	1120	T
6.	BH	500	C
7.	GD	760	T

The force/vector diagram is shown in the Fig. 17.20. The magnitudes and nature of the forces are given in the table 17.4.

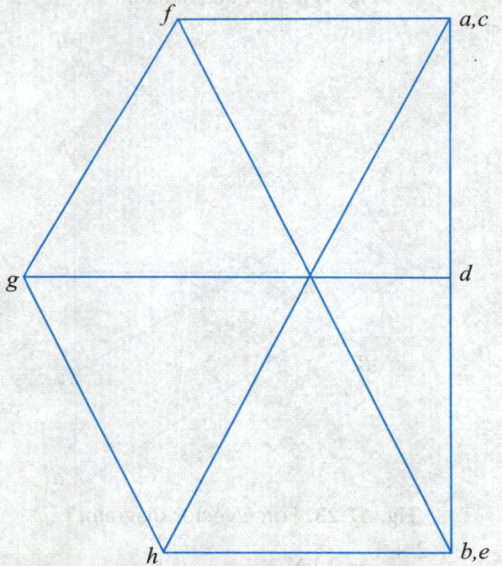

Fig. 17.20. Force/vector diagram

Example 17.5. *A form of roof whose sides have a slope of 34° is shown in Fig. 17.21. The span is 12 metres. Draw the force diagram for the truss and tabulate the forces in the members with their nature.*

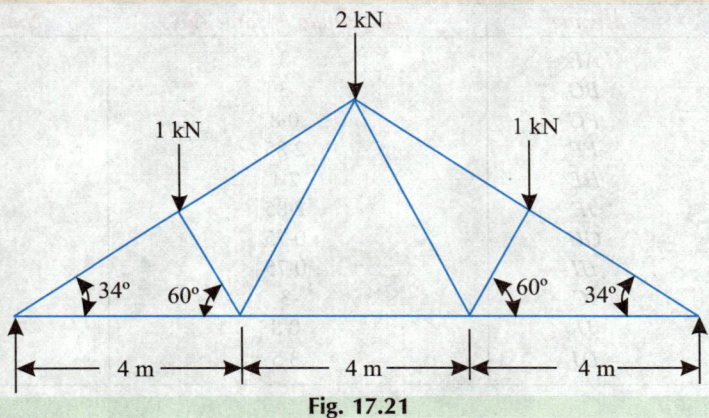

Fig. 17.21

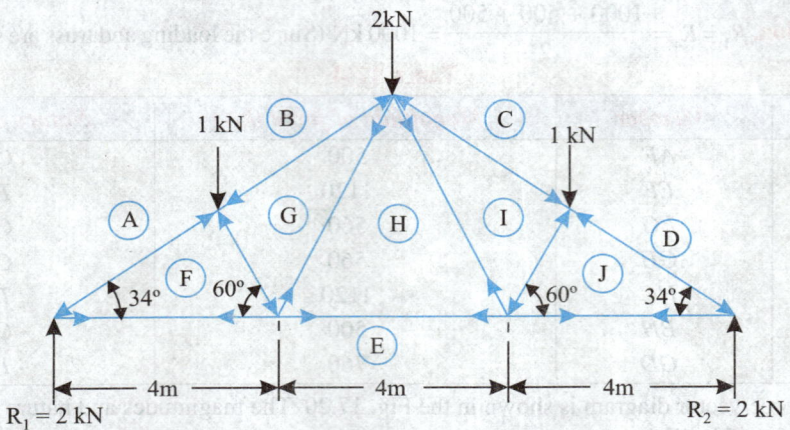

Fig. 17.22. Space diagram

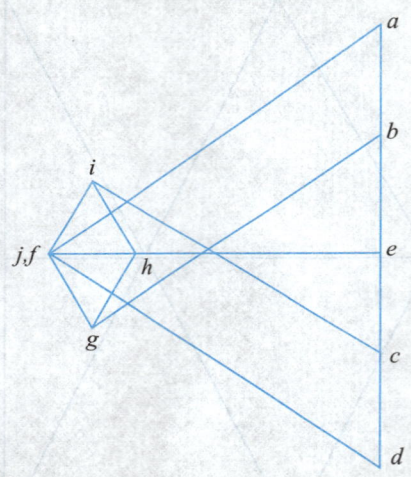

Fig. 17.23. Force/vector diagram.

Solution. $R_1 = R_2 = \dfrac{1 + 2 + 1}{2} = 2$ kN

The magnitude and nature of forces in different members are given in table 17.5.

S.No.	Member	Magnitude of force (kN)	Nature of force
1.	AF	3	C
2.	BG	3	C
3.	FG	0·8	C
4.	FE	2·85	T
5.	HE	2·1	T
6.	JE	2·85	T
7.	GH	0·75	T
8.	HI	0·75	T
9.	IC	3	C
10.	IJ	0·8	C
11.	DJ	3·5	C

TABLE **17.5**

Example 17.6. *Determine the magnitudes and nature of forces in all members of the frame given in Fig. 17.24.*

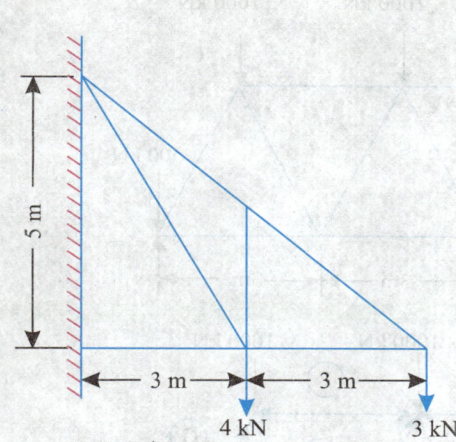

Fig. 17.24

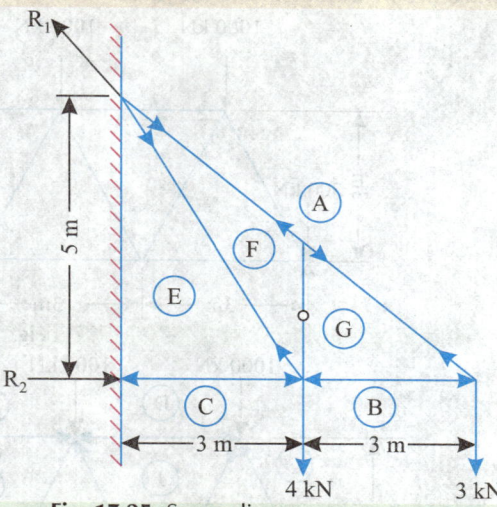

Fig. 17.25. Space diagram.

Solution. The force/vector diagram for the given frame is shown in Fig. 17.26 and the forces in various members in a tabulated form are given in table 17.6.

	TABLE 17.6		
S.No.	*Member*	*Magnitude of force (kN)*	*Nature of force*
1.	AF	4·75	T
2.	AG	4·75	T
3.	GB	3·7	C
4.	EF	4·6	T
5.	EC	6·0	C
6.	FG	Zero	—

Reactions R_1 and R_2 :

To find reactions R_1 and R_2 join, *ae*.

Then, $R_1 = ae = 9·3$ kN, and $R_2 = ec = 6$ kN.

The directions of the reactions R_1 and R_2 are found by drawing reaction lines parallel to *ae* and *ec* respectively as shown in Fig. 17·26.

Fig. 17.26

Small steel bridge.

Example 17.7. *The Fig. 17.27 shows a loaded warren girder for a 12 m span. The loads marked are in kN units. Find the forces in all the members.*

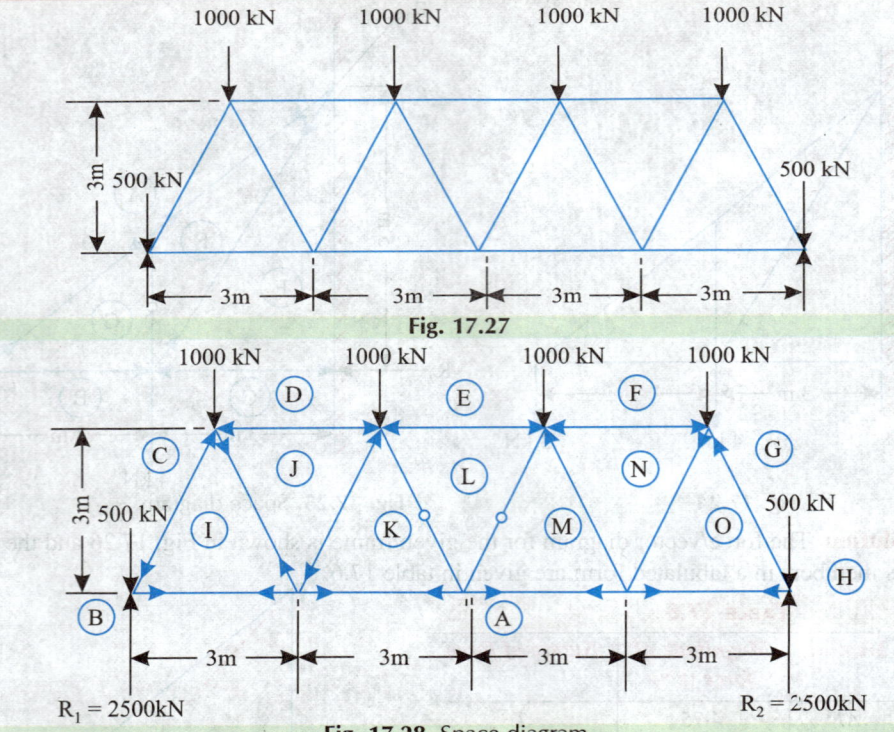

Fig. 17.27

Fig. 17.28. Space diagram.

Solution. Since the frame is symmetrically loaded,

$$R_1 = R_2 = \frac{5000}{2} = 2500 \text{ kN}$$

Complete the force/vector diagram as shown in the Fig. 17.29. By measurement from force diagram the magnitudes and nature of the forces are as given in the table 17.7.

S.No.	Member	Magnitude of force (kN)	Nature of force
		TABLE 17.7	
1.	CI	2250	C
2.	IJ	1125	T
3.	JD	1500	C
4.	JK	1125	C
5.	IA	1000	T
6.	KA	2000	T
7.	LE	2000	C
8.	MA	2000	T
9.	MN	1125	C
10.	NO	1125	T
11.	NF	1500	C
12.	OG	2250	C
13.	OA	1000	T
14.	KL	zero	—
15.	LM	zero	—

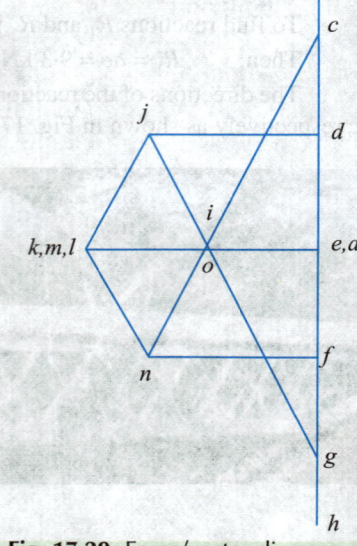

Fig. 17.29. Force/vector diagram.

Example 17·8. *Fig. 17·30 shows a loaded girder of an overhead travelling crane. Determine forces in the members. Tabulate the results and specify whether they are in tension or compression.*

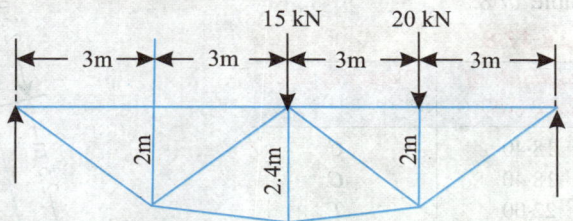

Fig. 17.30

Solution. Draw the space diagram as per scale chosen (Fig. 17·31). To find reactions taking moments about the left hand support, we get

$$R_2 \times 12 = 20 \times 9 + 5 \times 6 = 180 + 90 = 270$$

∴

$$R_2 = \frac{270}{12} = 22.5 \text{ kN}$$

$$R_1 + R_2 = 20 + 15 = 35$$

$$R_1 = 35 - 22.5 = 12.5 \text{ kN}$$

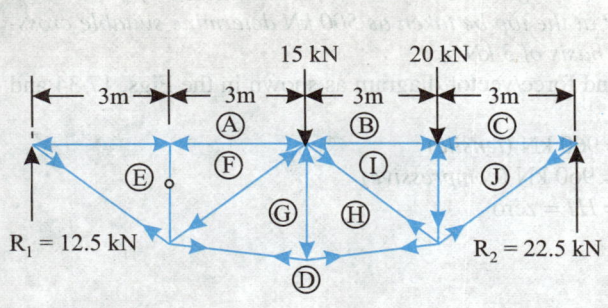

Fig. 17.31. Space diagram.

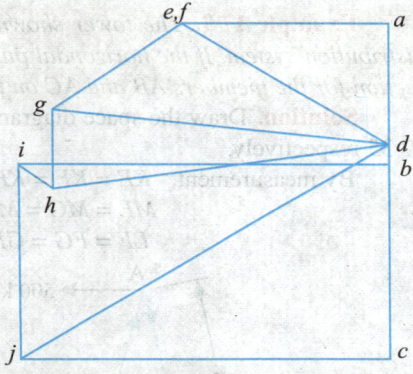

Fig. 17.32. Force/vector diagram.

Monroe Street Bridge, USA.

Complete the force/vector diagram as shown in Fig. 17·32. By measurement from force diagram, the magnitude and nature of the forces are as given in table 17·8.

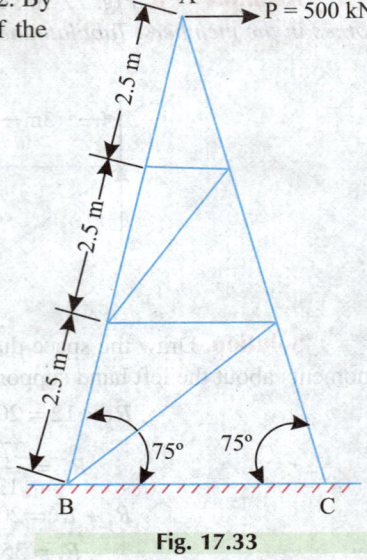

Fig. 17.33

TABLE 17·8			
S. No.	Member	Magnitude of force (kN)	Nature of force
1.	AE	18·40	C
2.	AF	18·40	C
3.	DE	22·00	T
4.	EF	zero	—
5.	FG	16·00	C
6.	GD	32·00	T
7.	GH	8·00	C
8.	HI	3·40	T
9.	BI	34·75	C
10.	IJ	20·00	C
11.	JD	41·4	T
12.	HD	32·20	T
13.	CJ	34·75	C

Example 17·9. *The tower shown in the Fig. 17·33 is used for transmission of power in a distribution system. If the horizondal pull P at the top be taken as 500 kN determine suitable cross-section for the members AB and AC on the basis of 3 kN/m².*

Solution. Draw the space diagram and force/vector diagram as shown in the Figs. 17·34 and 17·35 respectively.

By measurement, $KE = KF = KH = 960$ kN (*tensile*)
$$ME = MG = MJ = 960 \text{ kN } (compressive)$$
$$EF = FG = GH = HJ = \text{zero}$$

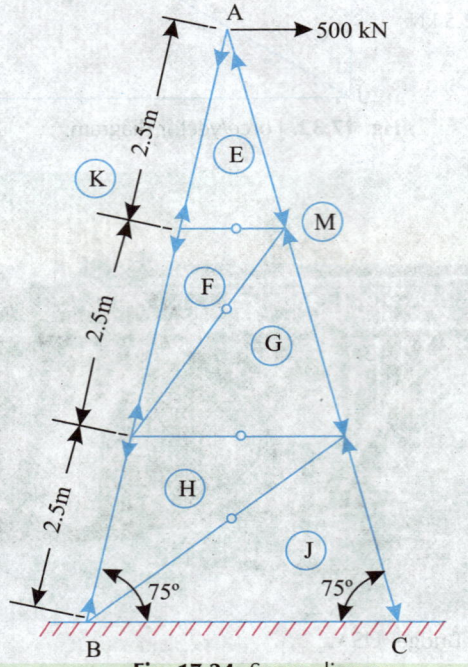

Fig. 17.34. Space diagram.

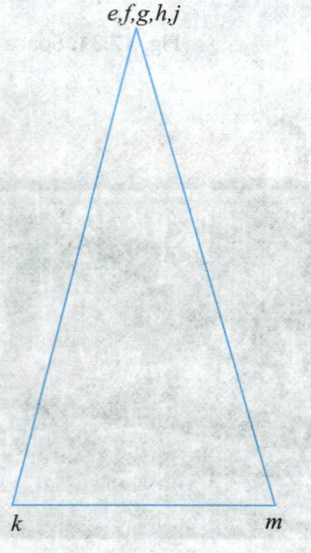

Fig. 17.35. Force/vector diagram.

Therefore the forces in *AB* and *AC* are 960 kN (*tensile*) and 960 kN (*compressive*) respectively. Direct stress (*permissible*) in the material = 3 kN/mm²

∴ Cross-sectional area of the member *AB* or *AC* = $\dfrac{960}{3}$ = **320 mm²** **(Ans.)**

Example 17·10. *The loading and support conditions of a plane truss are shown in the Fig. 17·36. Find graphically or otherwise, the forces in the members AB, AF, BC and FE.*

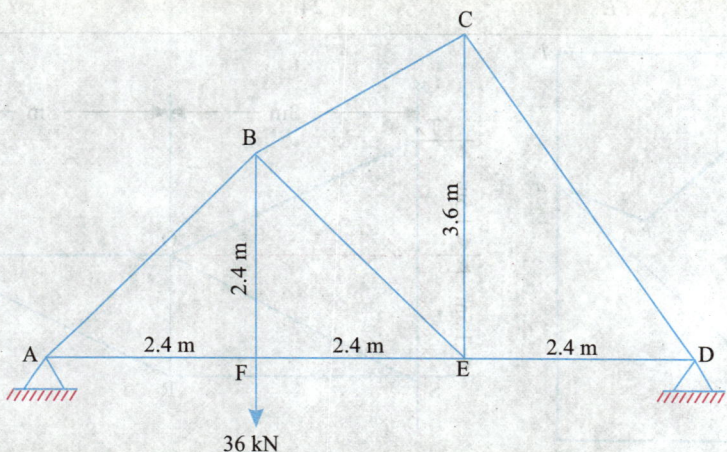

Fig. 17.36

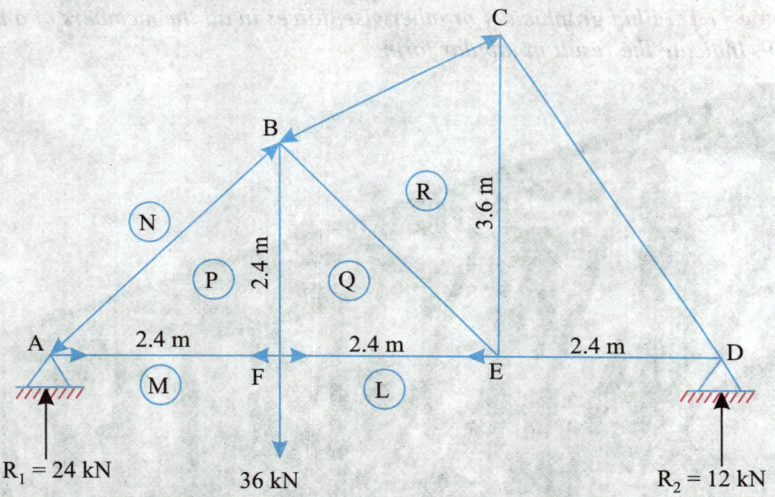

Fig. 17.37. Space diagram.

Solution. Taking moments about *A*, we get

$$R_D \times 7{\cdot}2 = 36 \times 2{\cdot}4$$
$$R_D = 12 \text{ kN}$$

and,
$$R_A = 36 - 12 = 24 \text{ kN}$$

Force/vector diagram is shown in the Fig. 17·38. The magnitudes and nature of the forces in the members *AB*, *AF*, *BC*, in a tabulated form are given in the table. 17·9.

TABLE 17·9

S. No.	Member	Magnitude of force (kN)	Nature of force
1.	AB	34	C
2.	AF	24	T
3.	BC	9	C
4.	FE	24	T

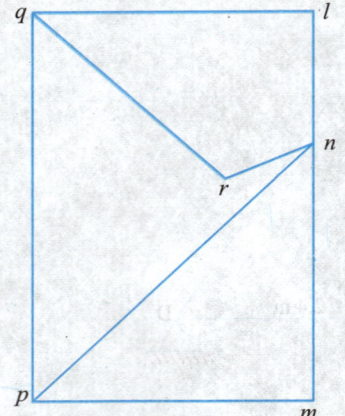

Fig. 17.38. Force/vector diagram.

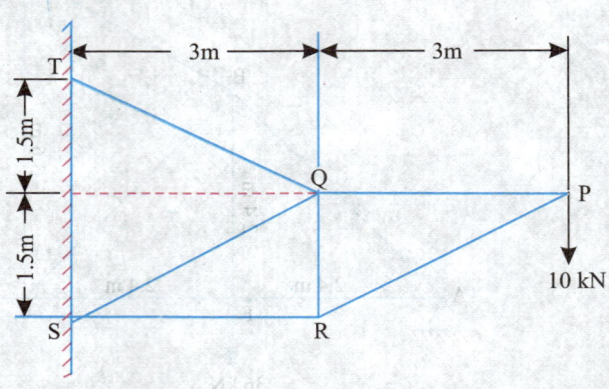

Fig. 17.39

Example 17·11. *Find graphically or otherwise, forces in all the members of a truss shown in the Fig. 17·39. Indicate the result in tabular form.*

Bridge carrying a water pipe across a canal.

Solution. Make the space diagram (Fig. 17·40) to some suitable scale. Draw the force/vector diagram (Fig. 17·41) and tabulate the magnitudes and nature of the forces as given in the table 17·10.

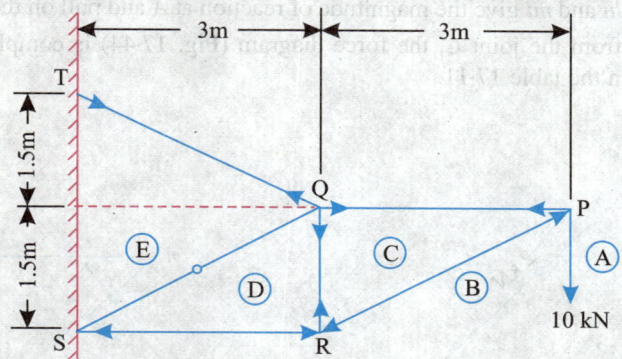

Fig. 17.40. Space diagram.

S. No.	Member	Magnitude of force (kN)	Nature of force
		TABLE 17·10	
1.	PQ	20	T
2.	PR	22·4	C
3.	QR	10	T
4.	QS	zero	–
5.	QT	22·4	T
6.	RS	2·0	C

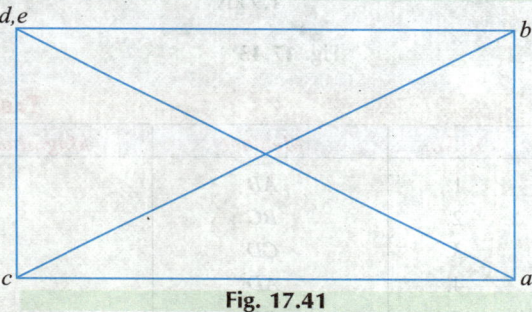

Fig. 17.41

Example 17·12. *A pin jointed frame is as shown in the Fig. 17·42. It is hinged at A and loaded at D. A horizontal chain is attached at C and pulled so that AD is horizontal. Determine the pull on the chain and also the forces in each member, stating whether it is in tension or compression.*

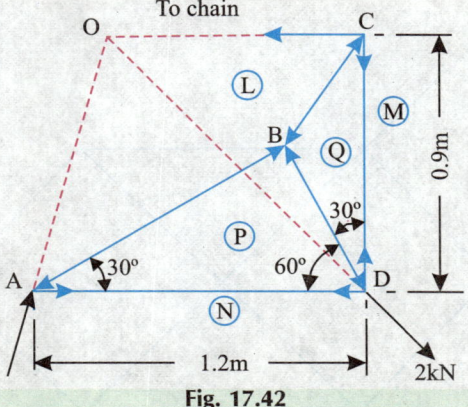

Fig. 17.42

Solution. Draw the space diagram to some suitable scale as shown in Fig. 17·42. Produce the line of 2 kN backward so as to meet the line of action of tension in the chain. As the frame is in equilibrium under the action of the forces *e.g.,* load 2 kN, tension in the chain and the reaction at the joint A, therefore the forces must meet at a point and as such by joining A, with O the direction of the reaction at A could be found out. The magnitudes of the reaction and pull in chain are found out by drawing the force/vector diagram as shown in Fig. 17·43.

To construct this force diagram, to some scale draw *nm* parallel and equal to load 2 kN. From *m* draw a line parallel to the chain and from *n* draw a line parallel to the direction of reaction, both meeting at *l*. Vectors *ln* and *ml* give the magnitude of reaction at *A* and pull on the chain respectively.

Now starting from the joint *C*, the force diagram (Fig. 17·44) is completed as shown. The forces are tabulated in the table 17·11.

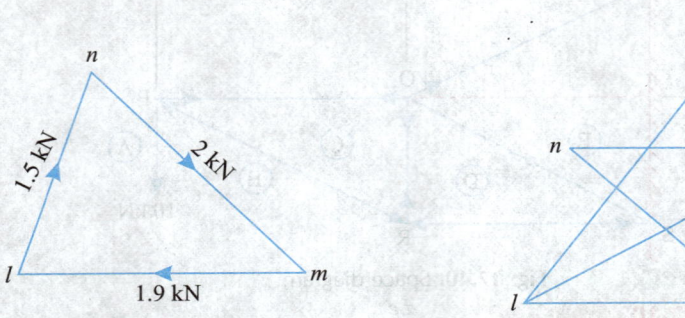

	Fig. 17.43		Fig. 17.44

TABLE 17·11			
S. No.	*Member*	*Magnitude of force (kN)*	*Nature of force*
1.	AB	2·91	C
2.	BC	3·20	C
3.	CD	2·59	T
4.	AD	2·10	T
5.	BD	1·50	C

Example 17·13. *Find the forces in all the members of the truss shown in Fig. 17·45.*

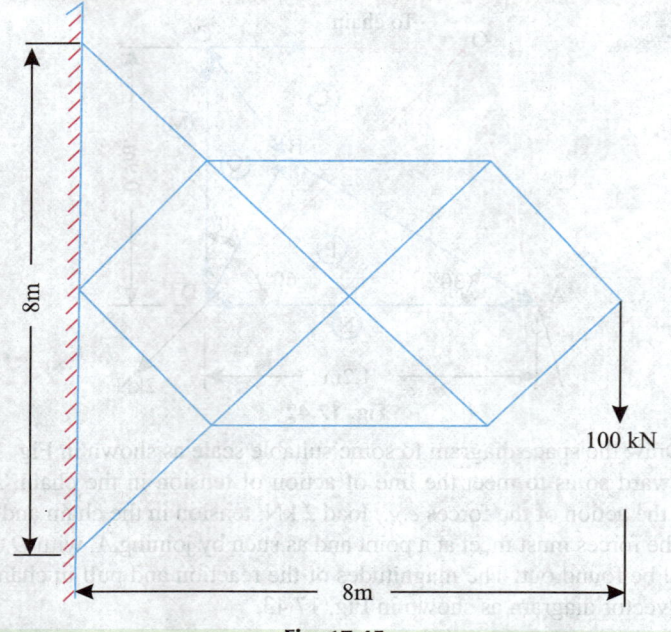

Fig. 17.45

Solution. Draw the space diagram as shown in Fig. 17·46. To draw the force/vector diagram proceed as follows :

(i) Draw *ab* to represent the force *AB* = 100 kN. Draw *bc* and *ac* parallel to *BC* and *AC* and obtain the point *c*.

(ii) Draw *cf* and *af* parallel to *CF* and *AF* to obtain the point *f*.

(iii) Draw *cd* and *bd* parallel to *CD* and *BD* to obtain the point *d*.

(iv) Draw *de* and *fe* parallel to *DE* and *FE* to obtain the point *e*.

(v) Draw *eh* and *ah* parallel to *EH* and *AH* and obtain the point *h*.

(vi) Draw *eg* and *bg* parallel to *EG* and *BG* and obtain point *g*.

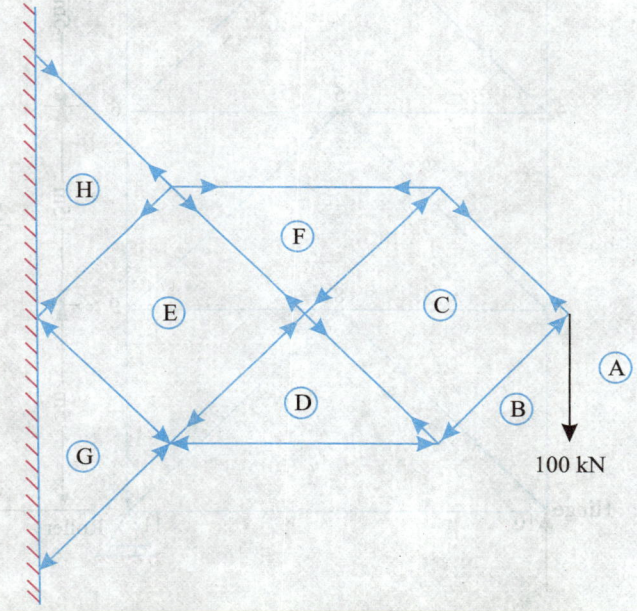

Fig. 17.46. Space diagram

The forces in the various members are now scaled from the force/vector diagram (Fig. 17·47) and tabulated as shown in Table 17·12.

S. No.	Member	Magnitude of force (kN)	Nature of force
TABLE 17·12			
1.	BC	70.7	C
2.	BD	100	C
3.	BG	141.4	C
4.	HA	141·4	T
5.	AF	100	T
6.	AC	70·7	T
7.	HE	70·7	T
8.	GE	70·7	C
9.	EF	70·7	T
10.	DC	70·7	T
11.	FC	70·7	C
12.	ED	70·7	C

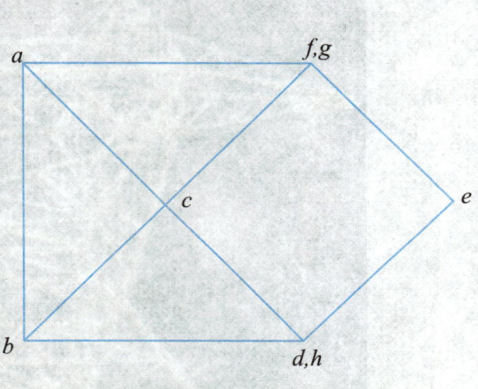

Fig. 17.47. Force/vector diagram

Example 17·14. *Determine the forces in the members of the structure shown in Fig.17·48.*

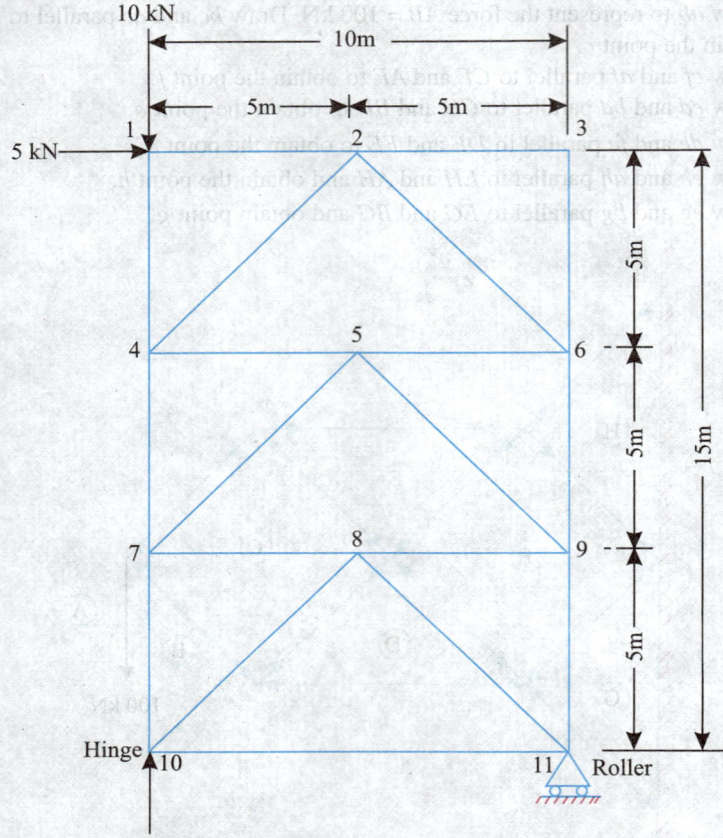

Fig. 17.48

Solution. Draw the space diagram as shown in Fig. 17·49. The procedure for drawing the force/vector diagram (Fig. 17·50) is given below :

Transmission tower built with steel frames.

(i) Select a suitable scale and draw *ab* horizontally to represent force *AB* of 5 kN.

(ii) Draw *bc* vertically to represent the force *BC* of 10 kN.

(iii) Draw *cm* and *am* parallel to the vertical reaction of the roller support and the reaction of hinged support and obtain the point *m*. It will be seen that the points *c* and *f* will coincide.

(iv) Draw *cd* and *ad* horizontally and vertically respectively parallel to the forces *CD* and *AD* and obtain the point *d*.

(v) Draw *de* and *fe* parallel to *DE* and *FE* and obtain the point *e*.

(vi) Draw *ei* and *ai* parallel to *EI* and *AI* and obtain the point *i*.

(vii) Draw *eg* and *cg* parallel to *EG* and *CG* and obtain the point *g*.

(viii) Draw *ih* and *gh* parallel to *IH* and *GH* and obtain the point *h*.

(ix) Draw *hj* and *aj* parallel to *HJ* and *AJ* and obtain the point *j*.

(x) Draw *hl* and *cl* parallel to *HL* and *CL* and obtain the point *l*.

(xi) Draw *lk* and *jk* parallel to *LK* and *JK* and obtain the point *k*.

The forces in the various members may now be scaled from the force/vector diagram (Fig. 17·50). These are tabulated in table 17·13.

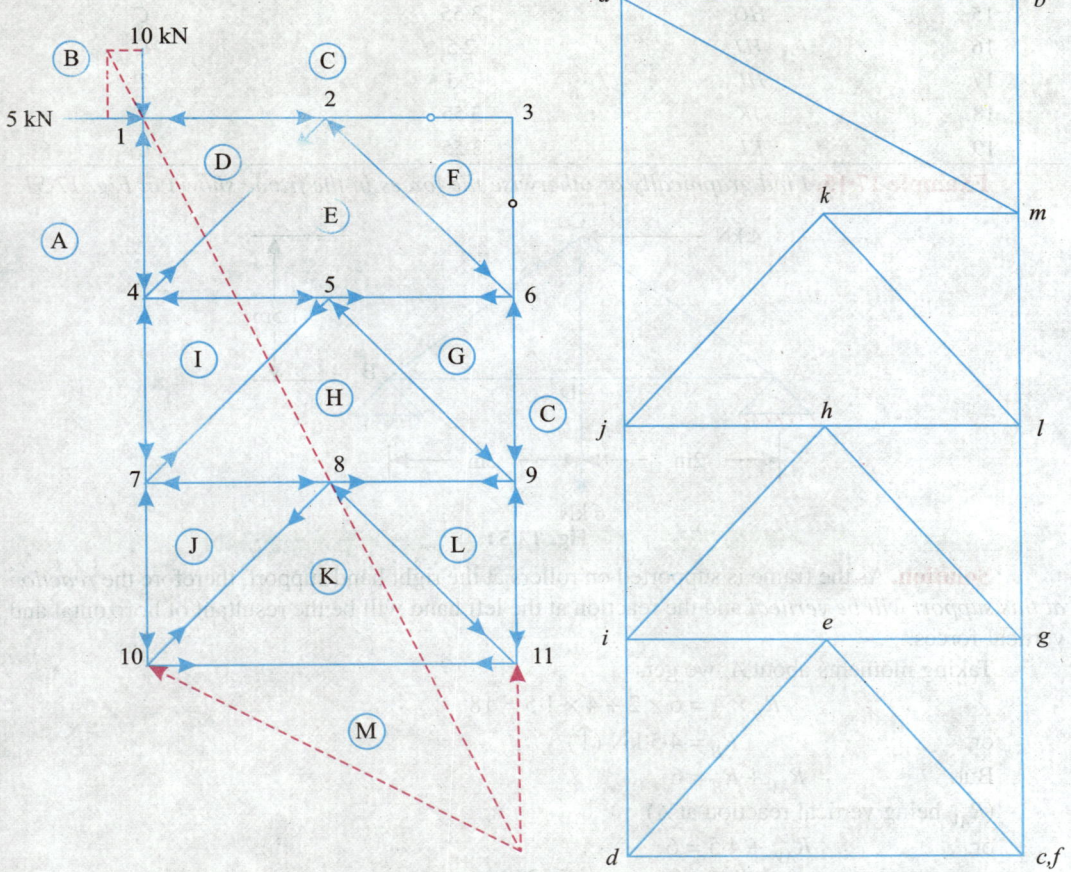

Fig. 17.49. Space diagram   Fig. 17.50. Force/vector diagram.

TABLE 17·13

S. No.	Member	Magnitude of force (kN)	Nature of force
1.	CD	5	C
2.	CF	zero	—
3.	CF	zero	—
4.	CG	2·5	C
5.	CL	5·0	C
6.	MK	2·5	T
7.	AJ	5·0	C
8.	AI	7·5	C
9.	AD	10·0	C
10.	DE	3·55	T
11.	EF	3·55	C
12.	EI	2·5	C
13.	EG	2·5	T
14.	IH	3·55	T
15.	HG	3·55	C
16.	HJ	2·5	C
17.	HL	2·5	T
18.	JK	3·55	T
19.	KL	3·55	C

Example 17·15. *Find graphically or otherwise the forces in the frame shown in Fig. 17·51.*

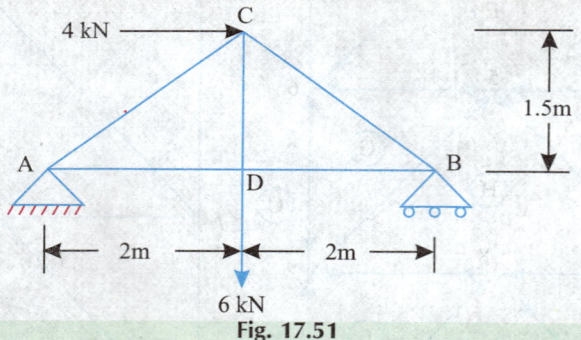

Fig. 17.51

Solution. As the frame is supported on rollers at the right hand support, therefore the *reaction at this support will be vertical* and the reaction at the left hand will be the resultant of horizontal and vertical forces.

Taking moments about A, we get,

$$R_B \times 4 = 6 \times 2 + 4 \times 1·5 = 18$$

or, $$R_B = 4·5 \text{ kN } (\uparrow)$$

But, $$R_{AV} + R_B = 6$$

(R_{AV} being vertical reaction at A)

or, $$R_{AV} + 4·5 = 6$$

$$R_{AV} = 6 - 4·5 = 1·5 \text{ kN}$$

and, horizontal reaction at A at left hand support,

$$R_{AH} = 4 \text{ kN } (\leftarrow)$$

Now draw the space diagram and force/vector diagram as shown in Figs. 17·52 and 17·53 respectively. Measuring the various sides of the vector diagram the results are tabulated as shown in table 17·14.

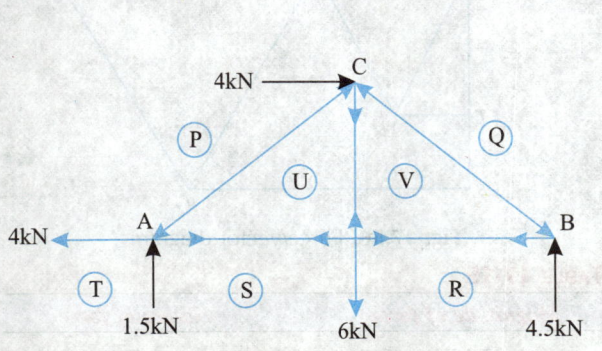

Fig. 17.52. Space diagram

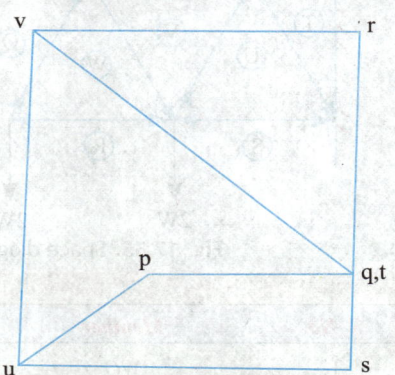

Fig. 17.53. Force/vector diagram

	TABLE 17·14		
S. No.	Member	Magnitude of force (kN)	Nature of force
1.	PU	2·5	C
2.	US	6·0	T
3.	QV	7·5	C
4.	VR	6·0	T
5.	UV	6·0	T

Example 17·16. *A cantilever truss of span 2 l is carrying loads as shown in Fig. 17·54. Determine graphically or otherwise forces in all members of the truss. (Fig. 17·54).*

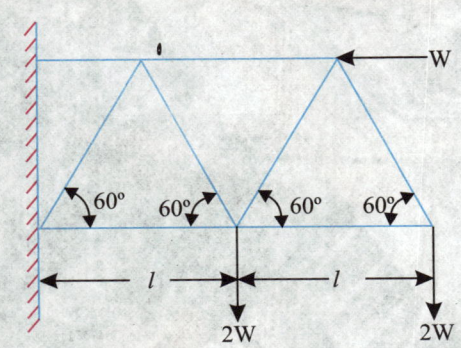

Fig. 17.54

Solution. Draw the space diagram using Bow's notation shown in Fig. 17·55. Now draw force/ vector diagram as shown in Fig. 17·56. Measuring the various sides of the vector diagram, the results are tabulated as shown in Table 17·15.

In tall buildings steel structures are first constructed which are latter covered by concrete.

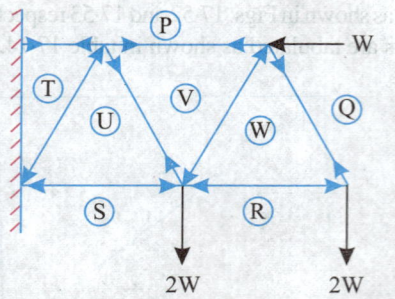

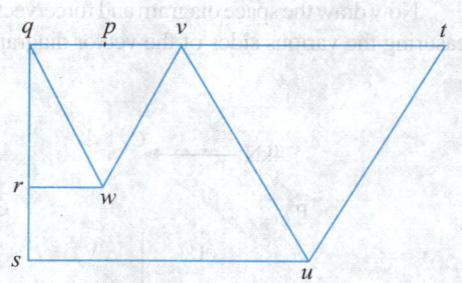

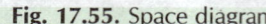

2W   2W

Fig. 17.55. Space diagram.   **Fig. 17.56.** Force/vector diagram

S. No.	Member	Magnitude of force	Nature of force
1.	WQ	2·4 W	T
2.	RW	1·2 W	C
3.	VW	1·2 W	C
4.	PV	1·4 W	T
5.	UV	4·6 W	T
6.	SU	4·6 W	C
7.	UT	4·6 W	C
8.	PT	8·0 W	T

TABLE 17·15

Example 17·17. *Fig. 17·57 represents a north-light roof truss with wind loads acting on it. Find graphically or otherwise the forces in all members of the truss. Give your result in a tabular form.*

Corner gusset block in a construction work.

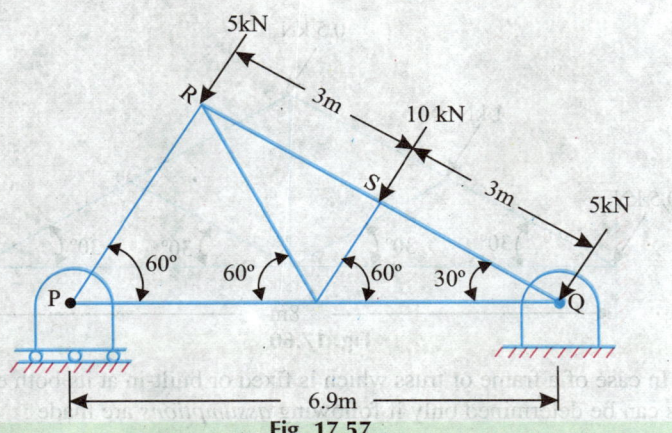

Fig. 17.57

Solution. Draw the space diagram using Bow's notation and find the direction of reactions at *P* and *Q* as shown in Fig. 17·58.

Now draw the force/vector diagram as shown in Fig. 17·59. Measure the various sides of the vector diagram and tabulate the result as shown in table 17·16.

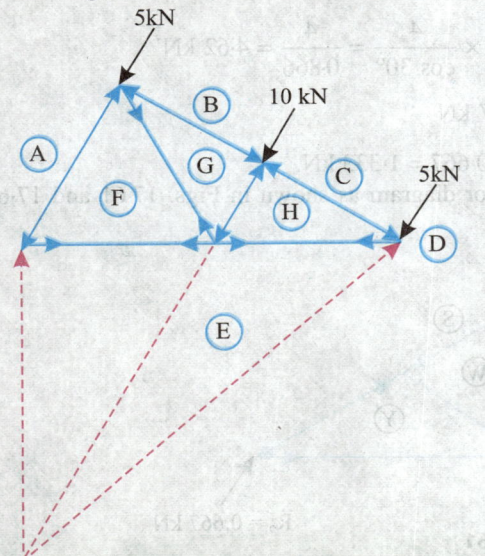

Fig. 17.58. Space diagram.

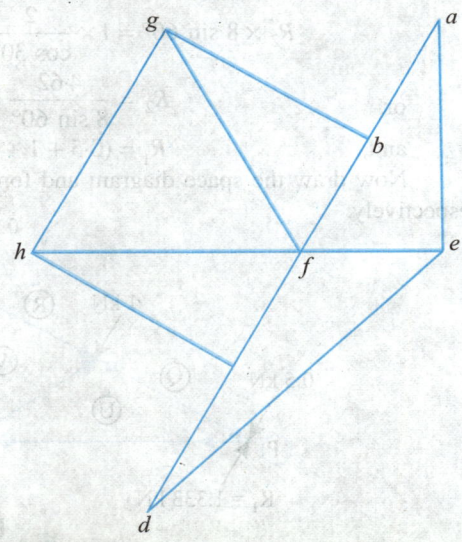

Fig. 17.59. Force/vector diagram.

	TABLE 17·16		
S. No.	**Member**	**Magnitude of force (kN)**	**Nature of force**
1.	AF	10·5	C
2.	FE	5·0	T
3.	FG	10·0	T
4.	BG	8·50	C
5.	GH	10·0	C
6.	HC	8·5	C
7.	HE	15·0	T

Example 17·18. *Fig. 17·60 shows a roof truss with both ends fixed. It is subjected to wind loads normal to main rafter. Determine the forces in various members of the truss.*

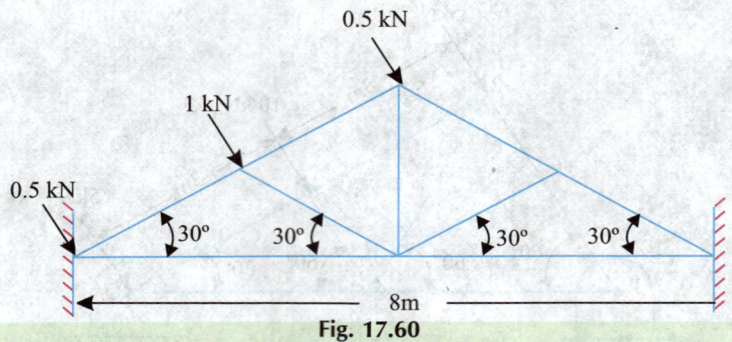

Fig. 17.60

Solution. In case of a frame or truss which is fixed or built-in at its both ends the reactions at both the supports can be determined only if following *assumptions* are made :

1. The reactions are parallel to the direction of the loads.
2. In case of inclined loads, the horizontal thrust is equally shared by the two reactions.

Normally the first assumption is made and the reactions are calculated by taking moments about one of the supports.

Taking moments about left support, we get

$$R_2 \times 8 \sin 60° = 1 \times \frac{2}{\cos 30°} + 0.5 \times \frac{4}{\cos 30°} = \frac{4}{0.866} = 4.62 \text{ kN}$$

or,

$$R_2 = \frac{4.62}{8 \sin 60°} = 0.667 \text{ kN}$$

and,

$$R_1 = (0.5 + 1 + 0.5) - 0.667 = 1.333 \text{ kN}.$$

Now draw the space diagram and force/vector diagram as shown in Figs. 17·61 and 17·62 respectively.

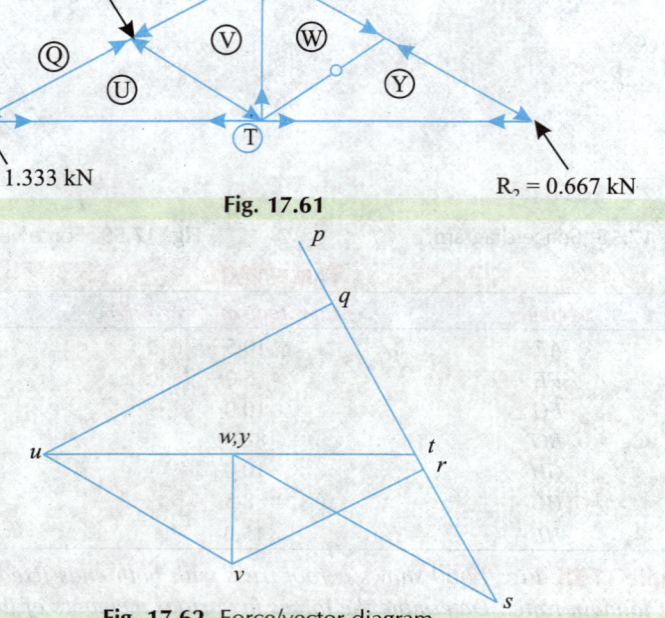

Fig. 17.61

Fig. 17.62. Force/vector diagram.

Measuring the various sides of the vector diagram the results are tabulated as shown in table 17·17.

TABLE 17·17

S. No.	Member	Magnitude of force (kN)	Nature of force
1.	QU	1·5	C
2.	UT	1·725	T
3.	RV	0·95	C
4.	UV	1·2	C
5.	VW	0·6	T
6.	WS	1·2	C
7.	WY	zero	—
8.	YS	1·2	C
9.	YT	0·7	T

Solutions by Analytical Methods :

Example 17·19. *Fig. 17·63 shows a truss with a span of 5 m and carrying a load of 5 kN at its apex. Find the forces in all the members of the truss.*

Solution. To find reactions R_1 and R_2;

Taking moments about A, we have

$$R_2 \times 5 = 5 \times 1·25$$

$$R_2 = \frac{5 \times 1·25}{5} = 1·25 \text{ kN}$$

But, $R_1 + R_2 = 5$

$$R_1 = 5 - R_2$$

$$= 5 - 1·25 = 3·75 \text{ kN}$$

(From the geometry of Fig. 17·63 distance of 5 kN force is 1·25 m from B).

The example may be solved either by :

(1) Method of joints.

       or

(2) Method of sections.

The example may be solved by both the methods as follows :

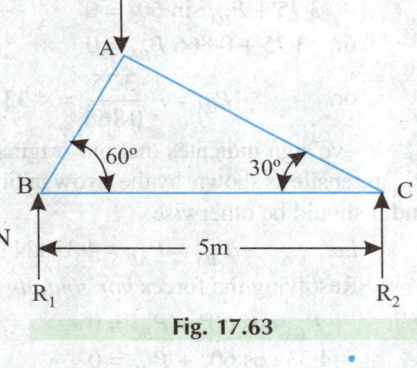

Fig. 17.63

Steel bridge across Spokane river, USA.

1. Method of joints :

While solving problems on trusses by method of joints consider that joint first, at which there *are not more than two unknown forces.* In this case we can consider any joint first since each joint (A, B and C) does not have more than two unknown forces.

— *To start with assume that the forces are 'tensile' in all the members of the joint and arrows are marked accordingly. If we get + ve value for a particular force, our assumption is correct and the force in that member is 'tensile', if we get –ve value it indicates that our assumption is wrong and the force in that member is compressive.*

— Consider $\begin{bmatrix} \text{Upward forces} & \text{+ ve} \\ \text{Downward forces} & \text{– ve} \\ \text{Forces acting towards right} & \text{+ ve} \\ \text{Forces acting towards left} & \text{– ve} \end{bmatrix}$

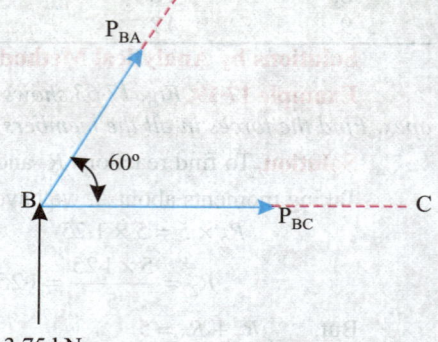

Joint B :

Refer to Fig. 17·64.

Resolving the forces *vertically* and applying the principle of $\Sigma V = 0$, we get

$$3{\cdot}75 + P_{BA} \sin 60° = 0$$

or, $3{\cdot}75 + 0{\cdot}866\, P_{BA} = 0$

or, $$P_{BA} = -\frac{3{\cdot}75}{0.866} = 4{\cdot}33 \text{ kN}$$

–ve sign indicates that our original assumption of P_{BA} as tensile as shown by the arrow in Fig. 17·40 is wrong and it should be otherwise.

Fig. 17.64

i.e., $P_{BA} = P_{AB} = 4{\cdot}33$ kN (*compression*)

Resolving the forces *horizontally* and applying the principle of $\Sigma H = 0$, we get

$$P_{BA} \cos 60° + P_{BC} = 0$$

$$- 4{\cdot}33 \cos 60° + P_{BC} = 0$$

$$P_{BC} = P_{CB} = 43 \cos 60°$$

$$= 2{\cdot}165 \text{ kN (*tension*)}$$

Joint C :

Refer to Fig. 17·65.

Resolving the forces *vertically* and applying the principle of $\Sigma V = 0$, we get

$$P_{CA} \sin 30° + 1{\cdot}25 = 0$$

$$P_{CA} \times 0{\cdot}5 = -1{\cdot}25$$

$$P_{CA} = -\frac{1{\cdot}25}{0{\cdot}5} = -2{\cdot}5 \text{ kN}$$

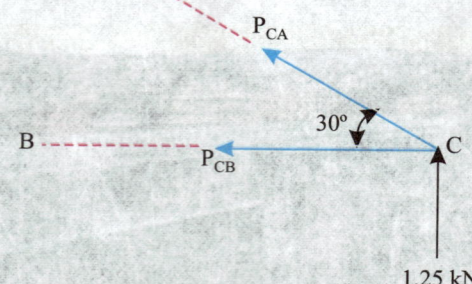

– ve sign indicates that our original assumption of P_{CA} as tensile as shown in Fig. 17·65 is wrong and it should be *otherwise*.

Fig. 17.65

i.e. $P_{CA} = P_{AC} = 2{\cdot}50$ kN (*compression*)

Resolving the forces *horizontally* and using the principle of $\Sigma H = 0$, we get

$$- P_{CA} \cos 30° - P_{CB} = 0$$

or, $- (-2{\cdot}5) \cos 30° = P_{CB}$

or, $P_{CB} = 2.5 \cos 30° = 2.165$ kN

∴ $P_{CB} = P_{BC} = 2.165$ kN (As already obtained)

TABLE 17·18

Tensile +ve, Compressive –ve

Joint	Member	Force (kN)
B	BA	– 4·33
	BC	2·165
A	AB	– 4·33
	AC	– 2·5
C	CA	– 2·5
	CB	2·165

2. Method of sections :

Refer to Fig. 17·66. The reactions R_1 and R_2 are calculated as usual as done in the case of method of Joints.

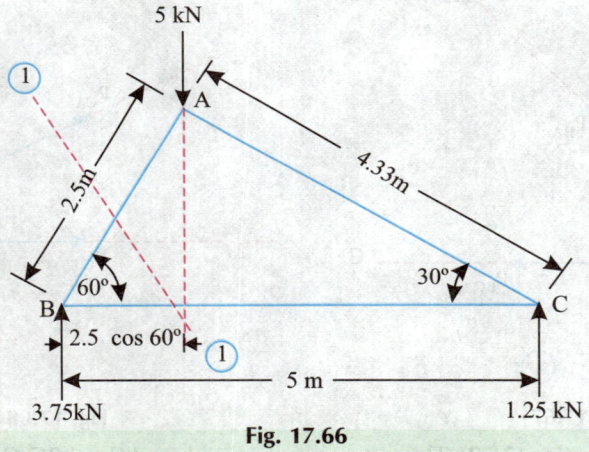

Fig. 17.66

Another bridge.

First of all consider **section (1 – 1)**. Refer to Fig. 17·67. Assume the forces P_{BA} and P_{BC} as tensile. Now consider the equilibrium of the left part of section (1 – 1).

Taking moments of all the forces acting to the left of the section (1 – 1) about C and applying the principle $\Sigma M_C = 0$

$$3·75 \times 5 + P_{BA} \times 5 \sin 60° = 0$$

$$P_{BA} = -\frac{3·75 \times 5}{5 \sin 60°} = -\frac{3·75 \times 5}{5 \times 0·866} = -4·33 \text{ kN}$$

Negative sign indicates that P_{BA} is *compression*.

Next taking moments of the forces (*i.e.* P_{BA} and P_{BC}) acting to the left of section (1 – 1) about A, we have

$$3·75 \times 2·5 \cos 60° = P_{BC} \times 2·5 \sin 60°$$

$$P_{BC} = \frac{3·75 \times 2·5 \cos 60°}{2·5 \sin 60°} = 2·165 \text{ kN (\textit{tension})}$$

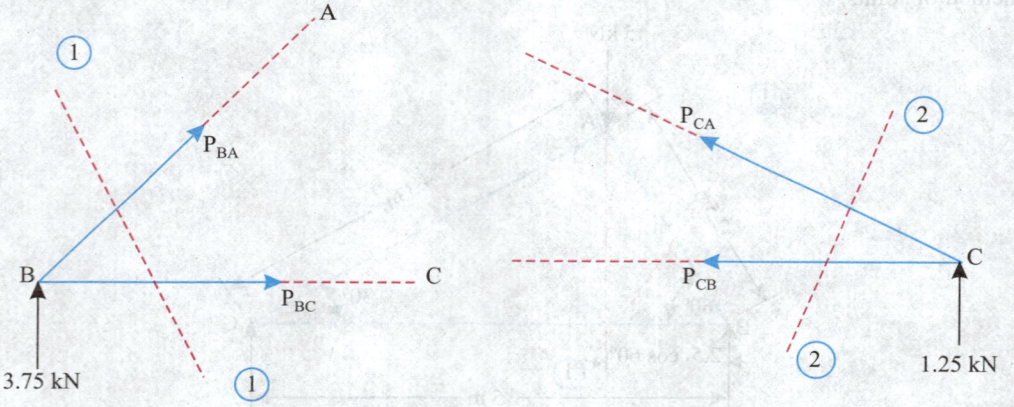

Fig. 17.67   Fig. 17.68

Next consider **section (2 – 2)**. The section cuts the members AC and BC. Considering equilibrium of the right part of the section (2 – 2). Refer to Fig.17·68.

Taking moments about the joint B, we have

$$1·25 \times 5 + P_{CA} \times 5 \sin 30° = 0$$

$$6·25 + P_{CA} \times 5 \times 0·5 = 0$$

$$\therefore \qquad P_{CA} = -\frac{6·25}{2·5} = -2·50 \text{ kN}$$

Negative sign shows that P_{CA} is *compression*. Now taking moments about A (of the right part only), we have

$$P_{CB} \times 2·5 \sin 60° = 1·25 \times 4·33 \cos 30°$$

$$P_{CB} = \frac{1·25 \times 4·33 \cos 30°}{2·5 \sin 60°} = 2·165 \text{ kN (\textit{tension})}$$

(As already obtained)

Example 17·20. *A truss of 12 m span is loaded as shown in Fig. 17·69. Find the forces in the members of the truss by method of joints or method of sections.*

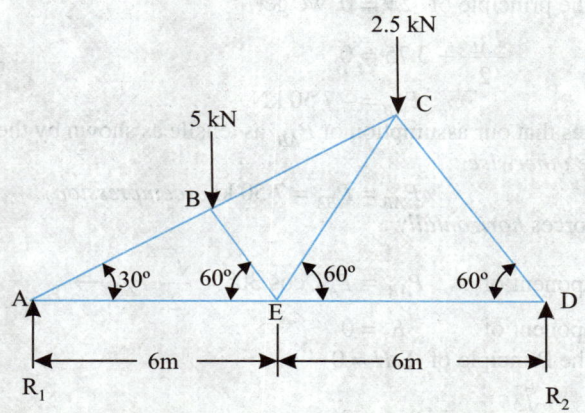

Fig. 17.69

Solution. To calculate reactions R_1 and R_2 taking moments about A, we get

$$R_2 \times 12 = 5 \times 4.5 + 2.5 (6 + 6/2)$$

$$\therefore \qquad R_2 = 3.75 \text{ kN}$$

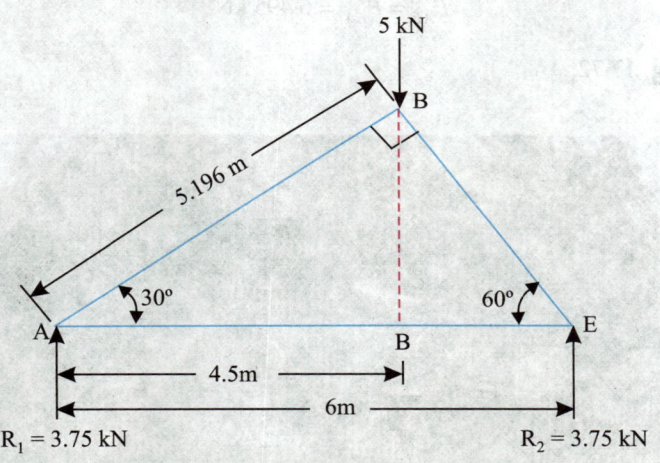

Fig. 17.70

But $\qquad R_1 + R_2 = 5 + 2.5 = 7.5$

$\therefore \qquad R_1 = 7.5 - R_2 = 7.5 - 3.75 = 3.75 \text{ kN}$

$$\left[\begin{array}{l} \text{From the geometry of Fig. 17.70} \\ AB = 6 \sin 60° = 5.196 \text{ m and} \\ AB' = 5.196 \times \cos 30° = 4.5 \text{ m} \end{array} \right]$$

Method of Joints :

Joint A :

Refer to Fig. 17.71.

Resloving the forces *vertically* :

Vertical component of $P_{AB} = P_{AB} \sin 30° = \dfrac{P_{AB}}{2} \uparrow$

Vertical component of $P_{AE} = 0$

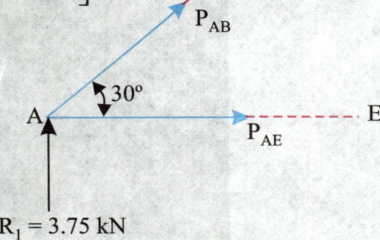

Fig. 17.71

Now applying the principle of $\Sigma V = 0$, we get

$$\frac{P_{AB}}{2} + 3.75 = 0$$

$$\therefore \qquad\qquad P_{AB} = -7.50 \text{ kN}$$

−ve sign indicates that our assumption of P_{AB} as tensile as shown by the arrow in Fig. 17·71 is wrong and it should be *otherwise*.

i.e. $\qquad\qquad P_{AB} = P_{BA} = 7.50 \text{ kN } (compression)$

Resolving the forces *horizontally* :

Horizontal component of $\qquad P_{AB} = P_{AB} \cos 30° = \dfrac{\sqrt{3}}{2} P_{AB} \rightarrow$

Horizontal component of $\qquad R_1 = 0$

Now applying the Principle of $\Sigma H = 0$

$$\frac{\sqrt{3}}{2} P_{AB} + P_{AE} = 0$$

But, $\qquad\qquad P_{AB} = -7.50 \text{ kN}$

$$-7.50 \times \frac{\sqrt{3}}{2} + P_{AE} = 0$$

or, $\qquad\qquad P_{AE} = P_{EA} = 6.495 \text{ kN}$

Joint B :

Refer to Fig. 17·72.

Steel structure under construction.

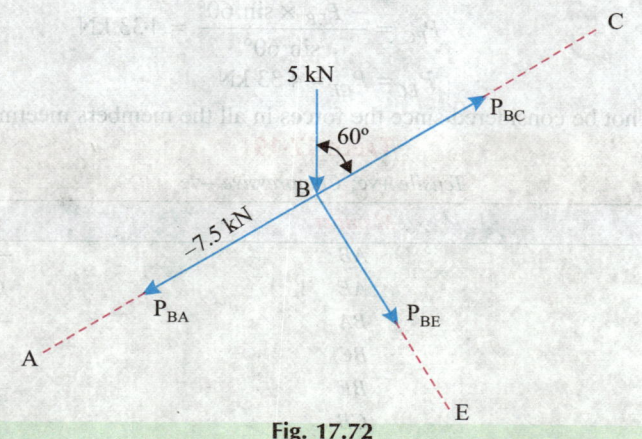

Fig. 17.72

Resolving the forces along AC, we get

$$P_{BC} = P_{BA} + 5 \times \cos 60° = -7.5 + 5 \times 0.5 = -5 \text{ kN}$$
$$P_{BC} = P_{CB} = -5 \text{ kN}$$

Resolving the forces normal to AC, we get

$$P_{BE} + 5 \times \sin 60° = 0$$
$$P_{BE} = -5 \sin 60° = -5 \times 0.866 = -4.33 \text{ kN}$$
$$P_{BE} = P_{EB} = -4.33 \text{ kN}$$

Joint D :

Refer to fig. 17.73.

Resolving the forces *vertically*, we get

$$P_{DC} \times \sin 60° + 3.75 = 0$$
$$P_{DC} = \frac{-3.75}{\sin 60°} = -4.33 \text{ kN}$$

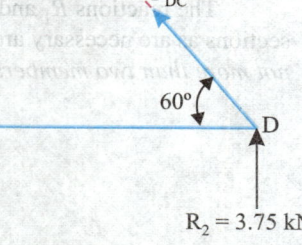

Resolving the forces *horizontally*, we get

$$P_{DC} \cos 60° + P_{DE} = 0$$
$$-4.33 (0.50) + P_{DE} = 0$$
$$\therefore \qquad P_{DE} = P_{ED} = 2.165 \text{ kN}$$

Fig. 17.73

Joint E :

Refer to Fig. 17.74.

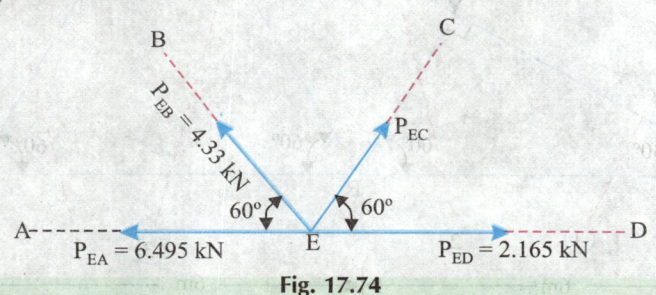

Fig. 17.74

Resolving forces *vertically*, we get

$$P_{EC} \times \sin 60° + P_{EB} \times \sin 60° = 0$$

$$P_{EC} = \frac{-P_{EB} \times \sin 60°}{\sin 60°} = 4.33 \text{ kN}$$

$$\therefore \qquad P_{EC} = P_{CE} = 4.33 \text{ kN}$$

Joint C need not be considered since the forces in all the members meeting at C are known.

<table>
<tr><td colspan="3" align="center">**TABLE 17·19**</td></tr>
<tr><td colspan="3" align="center">*Tensile +ve; Compressive −ve*</td></tr>
</table>

Joint	Member	Force (kN)
A	AB	− 7·50
	AE	6·495
B	BA	− 7·50
	BC	− 5
	BE	− 4·33
	CB	− 5
	CE	4·33
	CD	− 4·33
D	DC	− 4·33
	DE	2·165
E	EA	6·495
	EB	− 4·33
	EC	4·33
	ED	2·165

Method of Sections :

The reactions R_1 and R_2 are calculated as usual as done in case of method of joints. As many sections as are necessary are passed as shown in Fig. 17·75. While considering a section for analysis, *not more than two members should involve in that section in which the forces are unknown.*

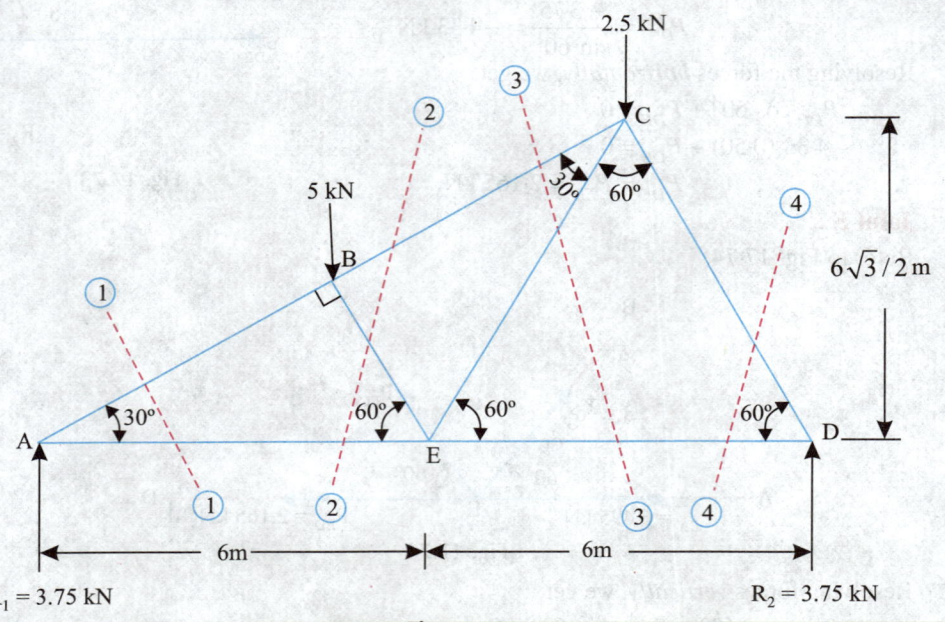

Fig. 17.75

Section (1 –1) :

Refer to Fig. 17·76.

Assume the forces P_{AB} and P_{AE} as tensile.

Now consider the equilibrium of the left part of the section (1 – 1).

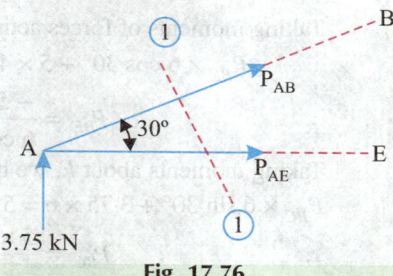

Taking moments of all the forces acting to the left of section (1 – 1) about the joint E and applying the principle $\Sigma M = 0$, we get

Fig. 17.76

$$3{\cdot}75 \times 6 + P_{AB} = (6 \times \sin 30°) = 0$$

$$\therefore \qquad P_{AB} = \frac{-3{\cdot}75 \times 6}{6 \times \sin 30°} = \frac{-3{\cdot}75 \times 6}{6 \times 0{\cdot}5} = -7{\cdot}50 \text{ kN}$$

Negative sign indicates that P_{AB} is compression.

It may be noted that moment of P_{AE} about E is zero since it passes through E. When we take moments about any point, we get only one equation and hence, there should involve only one unknown force in that equation. If there are two unknown forces while considering a section, take moments about any point in the plane of the frame through which one of the two unknown forces passes, thereby eliminating that force in the equation. While considering the equilibrium of the part of section (1 – 1), we can take moments of R_1, P_{AB} and P_{AE} about, B, C, E or D. Conveniently we have taken moment about E eliminating P_{AE} and solving P_{AB}.

Next, taking moments of the forces (*i.e.* R_1, P_{AB} and P_{AE}) acting to the left of the section (1 – 1) about C, we have

$$P_{AE} \times 3\sqrt{3} = 3{\cdot}75 \times 9$$

or, $$P_{AE} = \frac{3{\cdot}75 \times 9}{3\sqrt{3}} = 6{\cdot}495 \text{ kN}$$

Section (2 – 2) :

Refer to Fig. 17·77.

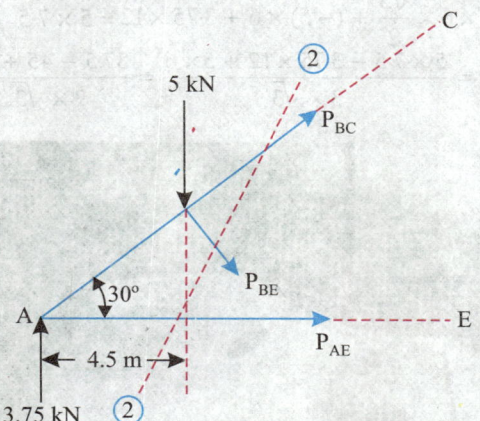

Fig. 17.77

The section line cuts the members *BC, BE* and *AE,* out of which the force P_{AE} is already known. The forces P_{BC} and P_{BE} are unknown (only two unknown forces permitted while considering a section.)

Consider the equilibrium of left part of section (2 – 2).

Taking moments of forces acting to the left of section (2 – 2) about A, we get

$$P_{BE} \times 6 \cos 30° + 5 \times 4.5 = 0$$

∴

$$P_{BE} = \frac{-5 \times 4.5}{6 \cos 30°} = -4.33 \text{ kN}$$

Taking moments about E, we have

$$P_{BC} \times 6 \sin 30° + 3.75 \times 6 = 5 \times 1.5$$

∴

$$P_{BC} = \frac{5 \times 1.5 - 3.75 \times 6}{6 \sin 30°} = \frac{7.5 - 2.25}{3} = -5 \text{ kN}$$

Section (3 – 3) :

Refer to Fig. 17.78.

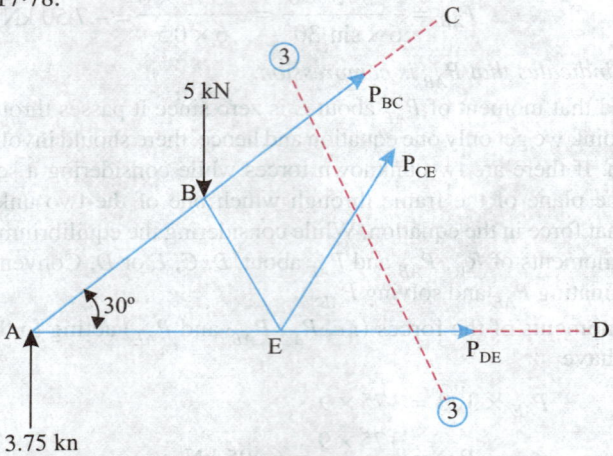

Fig. 17.78

Considering the equilibrium of left part of section (3 – 3) and taking moments about the joint D, we get

$$P_{CE} \times 6 \sin 60° + P_{BC} \times 6 + 3.75 \times 12 = 5 \times 7.5$$

$$P_{CE} \times 6 \times \frac{\sqrt{3}}{2} + (-5) \times 6 + 3.75 \times 12 = 5 \times 7.5$$

∴

$$P_{CE} = \frac{5 \times 7.5 - 3.75 \times 12 + 5 \times 6}{3 \times \sqrt{3}} = \frac{37.5 - 45 + 30}{3 \times \sqrt{3}} = \frac{22.5}{3 \times \sqrt{3}} = 4.33 \text{ kN}$$

Overhead transmission lines for electric trains are often held by steel frames and structures.

Taking moments about C, we get

$$P_{DE} \times 6 \sin 60° + 5 \times 4.5 = 3.75 \times 9$$

$$\therefore \qquad P_{DE} = \frac{3.75 \times 9 - 5 \times 4.5}{6 \times \sin 60°}$$

$$= \frac{33.75 - 22.5}{6 \times \dfrac{\sqrt{3}}{2}} = 2.165 \text{ kN}$$

Section (4–4):

Refer to Fig. 17.79.

Considering equilibrium *right* of section (4 – 4) and taking moments about E, we get

$$P_{CD} \times 6 \sin 60° + 3.75 \times 6 = 0$$

$$P_{CD} = \frac{-3.75 \times 6}{6 \times \sin 60°}$$

$$= -\frac{22.50}{6 \times \dfrac{\sqrt{3}}{2}} = -4.33 \text{ kN}$$

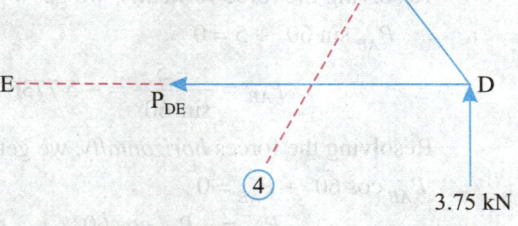

Fig. 17.79

The values are tabulated in the table 17·20

TABLE 17·20			
Tensile +ve; Compressive –ve			
Member	Force (kN)	Member	Force (kN)
AB	– 7·50	AE	6·495
BC	– 5	BE	– 4·33
CD	– 4·33	CE	4·33
DE	2·165		

Example 17·21. *Fig. 17·80 shows a warren girder consisting of seven members each of 4 m length supported at its ends and loaded as shown. Determine the stresses in the members by method of joints.*

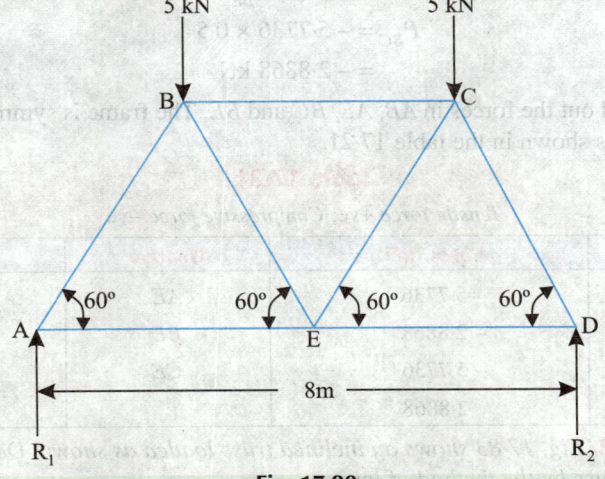

Fig. 17.80

Solution. In the question it has been asked to find the stresses in the members of frame. *Sometimes the force in a member is called total stress.* Hence, here, *stresses mean axial forces.*

Method of joints :

Since the frame is symmetrical, therefore

$$R_1 = R_2 = 5 \text{ kN}$$

Joint A :

Refer to Fig. 17·81.

Resolving the force *vertically*, we get

$$P_{AB} \sin 60° + 5 = 0$$

$$\therefore \qquad P_{AB} = \frac{-5}{\sin 60°} = -5.7736 \text{ kN}$$

Resolving the forces *horizontally*, we get

$$P_{AB} \cos 60° + P_{AE} = 0$$

$$P_{AE} = -P_{AB} \cos 60°$$

$$= -(-5.7736) \times 0.5 = 2.8868 \text{ kN}$$

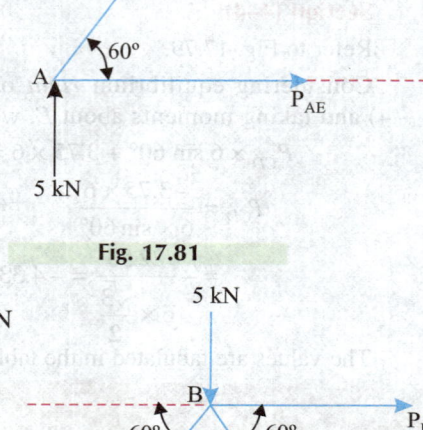

Fig. 17.81

Joint B :

Refer to Fig. 17·82.

Resolving the force *vertically*, we get

$$P_{AB} \sin 60° + P_{BE} \sin 60° + 5 = 0$$

$$-5.7736 \sin 60° + P_{BE} \sin 60° + 5 = 0$$

$$\therefore \ P_{BE} \sin 60° = 5.7736 \sin 60° - 5 = 0$$

$$\therefore \qquad\qquad P_{BE} = 0$$

Resolving force *horizontally*, we get

$$P_{BC} + P_{BE} \cos 60° - P_{AB} \cos 60° = 0$$

$$P_{BC} + 0 - (-5.7736) \cos 60° = 0 \quad (\because P_{BE} = 0)$$

$$\therefore \qquad\qquad P_{BC} = -5.7736 \times 0.5$$

$$= -2.8868 \text{ kN}$$

Fig. 17.82

We have found out the forces in *AB, AE, BC* and *BE*. The frame is symmetrical and hence the forces are tabulated as shown in the table 17·21.

Table 17·21

Tensile force +ve; Compressive force –ve

Member	Force (kN)	Member	Force (kN)
AB	–5·7736	AE	2·8868
BC	– 2·8868	BE	zero
CD	– 5·7736	CE	zero
DE	2·8868		

Example 17·22. *Fig. 17·83 shows an inclined truss loaded as shown. Determine the forces in the members of the truss by the method of joints.*

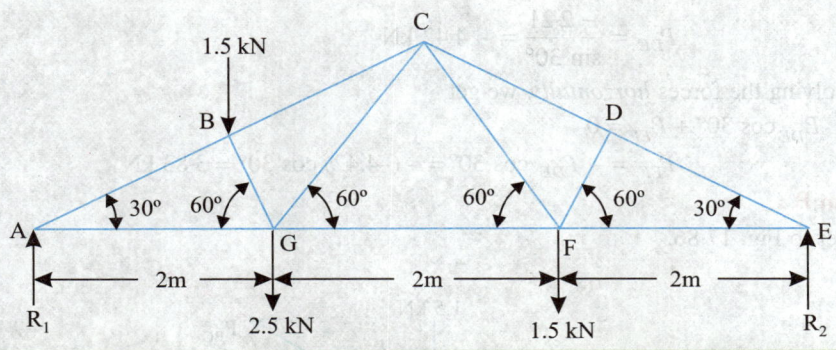

Fig. 17.83

Solution. Let us first calculate the reactions R_1 and R_2.

From the geometry of the figure the load of 1·5 kN at B is acting at a distance of 1·5 m from A.

Taking moments about E, we get

$$R_1 \times 6 = 1.5 \times 4.5 + 2.5 \times 4 + 1.5 \times 2 = 6.75 + 10 + 3 = 19.75$$
$$R_1 = 3.29 \text{ kN}$$
$$R_1 + R_2 = 1.5 + 2.5 + 1.5 = 5.5$$
$$R_2 = 5.5 - R_1 = 5.5 - 3.29 = 2.21 \text{ kN}$$

Joint A :

Refer to Fig. 17·84.

Resolving the forces *vertically,* we get

$$P_{AB} \sin 30° + 3.29 = 0$$
$$P_{AB} = \frac{-3.29}{\sin 30°} = -6.58 \text{ kN}$$

Resolving the forces *horizontally,* we get

$$P_{AB} \cos 30° + P_{AG} = 0$$
$$\therefore \qquad P_{AG} = -P_{AB} \cos 30° = -(-6.58) \cos 30° = 5.70 \text{ kN}$$

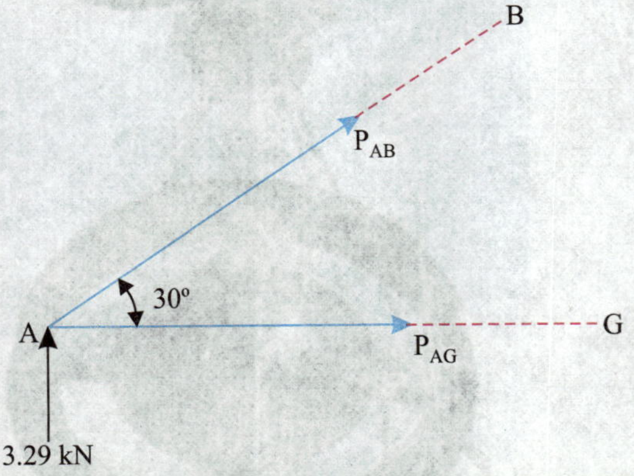

Fig. 17.84

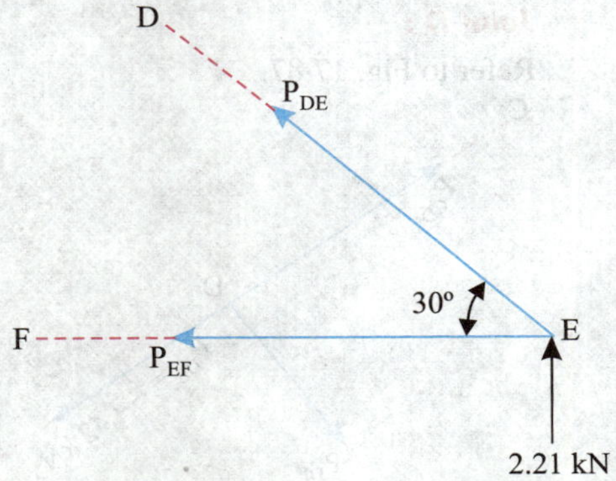

Fig. 17.85

Joint E :

Refer to Fig. 17·85.

Resolving the forces *vertically,* we get

$$P_{DE} \sin 30° + 2.21 = 0$$

$$\therefore \qquad P_{DE} = \frac{-2\cdot21}{\sin 30°} = -4\cdot42 \text{ kN}$$

Resolving the forces *horzontally,* we get

$$P_{DE} \cos 30° + P_{EF} = 0$$

$$\therefore \qquad P_{EF} = -P_{DE} \cos 30° = -(-4.42) \cos 30° = 3\cdot83 \text{ kN}$$

Joint B :

Refer to Fig. 17·86.

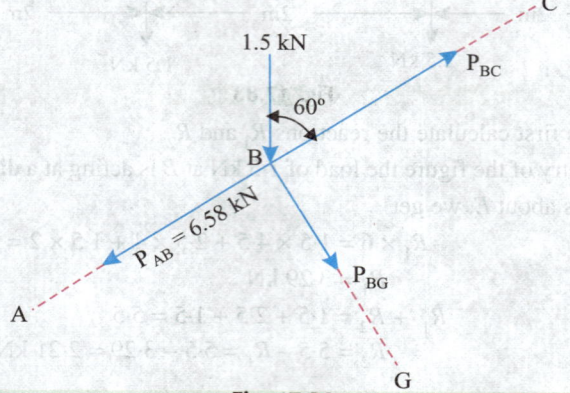

Fig. 17.86

Resolving the forces *parallel* to the member *ABC*

$$P_{BC} = P_{AB} + 1\cdot50 \ \cos 60°$$
$$= -6\cdot58 + 1\cdot50 \times 0\cdot5$$
$$= -5\cdot83 \text{ kN}$$

Resolving the forces *normal* to the member *ABC*

$$P_{BG} + 1\cdot50 \sin 60° = 0$$
$$P_{BG} = -1\cdot50 \sin 60°$$
$$= -1\cdot30 \text{ kN}$$

Joint D :

Refer to Fig. 17·87.

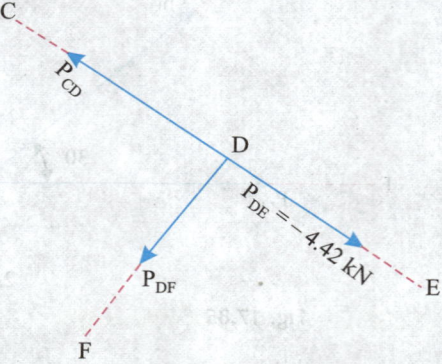

Fig. 17.87

Resolving the forces *parallel* to *CDE*

$$P_{CD} = P_{DC} = -4\cdot42 \text{ kN}$$

Steel mast under construction.

Resolving the forces *normal* to *CDE*

$$P_{DF} = 0$$

Joint G :

Refer to Fig. 17·88.

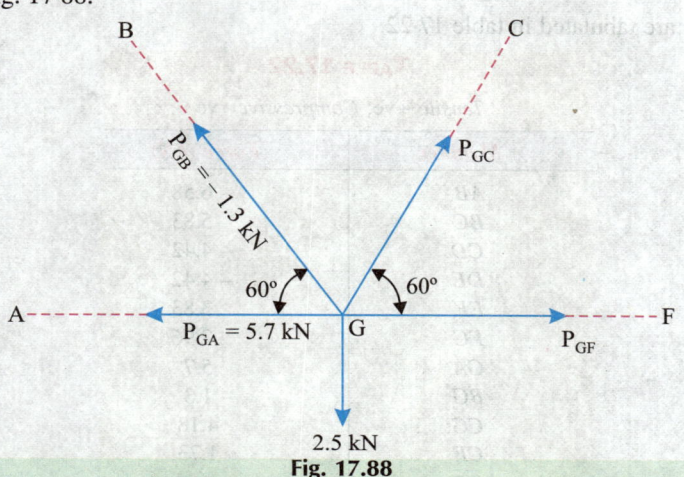

Fig. 17.88

Resolving the forces *vertically*, we get

$$P_{GC} \sin 60° + P_{BG} \sin 60° = 2·50$$

$$P_{GC} = \frac{2·50 - P_{BG} \sin 60°}{\sin 60°}$$

$$= \frac{2·50 - (-1·30) \sin 60°}{\sin 60°} = \frac{2·50 + 1·126}{0·866} = 4·18 \text{ kN}$$

Resolving the forces *horzontally*, we get

$$P_{AG} + P_{BG} \cos 60° = P_{FG} + P_{CG} \cos 60°$$
$$5·70 + (-1·30) \cos 60° = P_{FG} + 4·18 \cos 60°$$
$$5·05 = P_{FG} + 2·09$$
$$\therefore \quad P_{FG} = 2·96 \text{ kN}$$

Joint F :

Refer to Fig. 17·89.

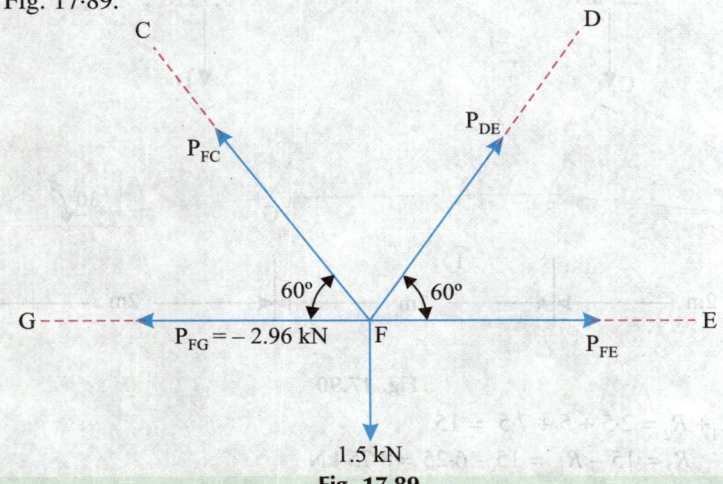

Fig. 17.89

Resolving the forces *vertically,* we get

$$P_{CF} \sin 60° = 1.5$$

$$P_{CF} = \frac{1.5}{\sin 60°} = 1.73 \text{ kN}$$

The forces are tabulated in table 17·22.

TABLE 17.22

Tensile +ve; Compressive –ve

Member	Force (kN)
AB	– 6.58
BC	– 5.83
CD	– 4.42
DE	– 4.42
EF	3.83
FG	2.96
GA	5.7
BG	– 1.3
CG	4.18
CF	1.73
DF	zero

Example 17·23. *Fig. 17·90 shows a roof truss supported and loaded as shown. Determine the forces in the members CE and FG by the method of sections.*

Solution. The truss is supported by an hinge at *A* and on rollers at *B*. Since all the external *loads are vertical* the reactions at *A* (R_1) and *B* (R_2) will be *vertical.* There will be *no horizontal reactions.*

To find R_1 and R_2 taking moments about *B*, we get

$$R_1 \times 6 = 7.5 \times 1.5 + 5 \times 3 + 2.5 \times 4.5 = 11.25 + 15 + 11.25 = 37.5$$

$$\therefore \quad R_1 = \frac{37.5}{6} = 6.25 \text{ kN}$$

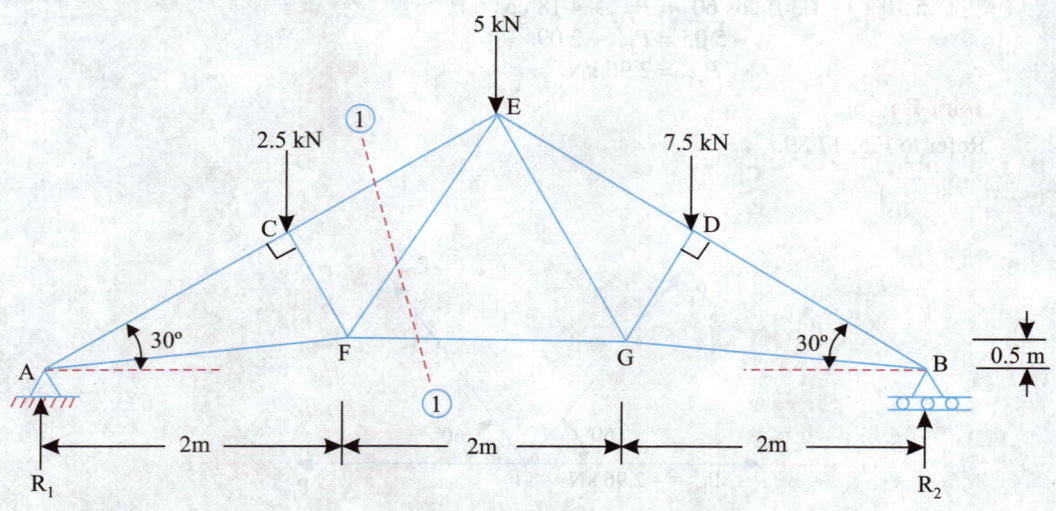

Fig. 17.90

But, $\quad R_1 + R_2 = 2.5 + 5 + 7.5 = 15$

$\therefore \qquad R_2 = 15 - R_1 = 15 - 6.25 = 8.75 \text{ kN}$

From the geometry of figure, the vertical distance of E from
$FG = 3 \tan 30° - 0.5 = 1.23$ m

Length of the member $AF = \sqrt{2^2 + 0.5^2} = 2.06$ m

$\tan \angle FAB = \dfrac{0.5}{2} = 0.25$

$\angle FAB = \tan^{-1} 0.25 = 14.04°$

$\angle CAF = 30 - 14.04° = 15.96°$

Length of the member $CF = AF \sin 15.96° = 0.57$ m

Length of the member $\quad AC = \dfrac{AE}{2} = \dfrac{3}{2 \times \cos 30°} = 1.732$ m

Distance between line of action of load at C and $F = 2 - 1.5 = 0.5$ m

Now consider the equilibrium of left of section (1 – 1). Refer to Fig. 17·91.

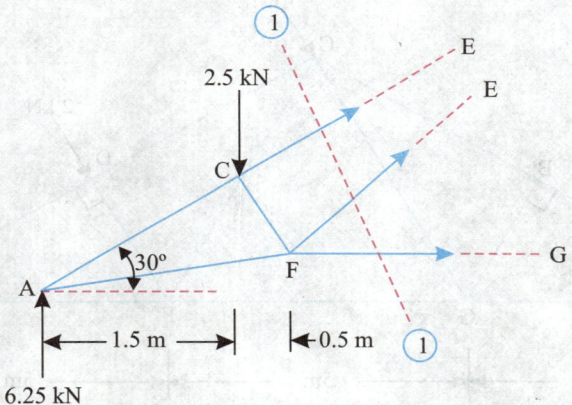

Fig. 17.91

Another bridge

Taking moments of forces about E, we get

$$P_{FG} \times 1 \cdot 23 + 2 \cdot 5 \times 1 \cdot 5 = 6 \cdot 25 \times 3$$

$$P_{FG} = \frac{6 \cdot 25 \times 3 - 2 \cdot 5 \times 1 \cdot 5}{1 \cdot 23} = \frac{18 \cdot 75 - 3 \cdot 75}{1 \cdot 23} = 12 \cdot 19 \text{ kN}$$

Taking moments about F, we get

$$P_{CE} \times 0 \cdot 57 + 6 \cdot 25 \times 2 = 2 \cdot 5 \times 0 \cdot 5$$

$$\therefore \qquad P_{CE} = \frac{2 \cdot 5 \times 0 \cdot 5 - 6 \cdot 25 \times 2}{0 \cdot 57} = \frac{1 \cdot 25 - 12 \cdot 5}{0 \cdot 57} = -19 \cdot 73 \text{ kN}$$

Force in the member CE = **19·73 kN** (*compressive*) (**Ans.**)

and, force in the member FG = **12·19 kN** (*tensile*) (**Ans.**)

Example 17·24. *Fig. 17·92 shows a truss of 15 m span loaded as shown. Find the forces in the members of the truss by the method of Joints.*

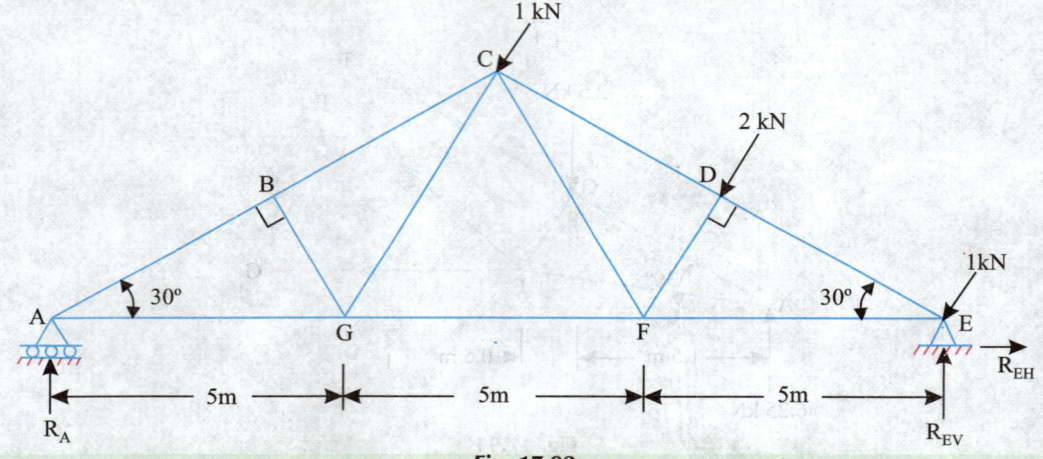

Fig. 17.92

Solution. To start with, reactions at the supports at A and E should be calculated. Since the frame is subjected to inclined loads there will be inclinded reaction at the support E. Let the vertical and horizontal components of the reaction at E be R_{EV} and R_{EH} respectively. Of course there *will only be vertical reaction at the support A since it is supported on rollers.*

Length of inclined members;

$$AC = CE = \frac{7 \cdot 5}{\cos 30°} = 8 \cdot 66 \text{ m}$$

$$CD = DE = \frac{8 \cdot 66}{2} = 4 \cdot 33 \text{ m}$$

Height of the truss

$$= \sqrt{8 \cdot 66^2 - 7 \cdot 5^2} = 4 \cdot 33 \text{ m}$$

$$\angle CGF = 60°$$

Taking moments about E, we get

$$R_A \times 15 = 1 \times 8 \cdot 66 + 2 \times 4 \cdot 33 = 8 \cdot 66 + 8 \cdot 66 = 17 \cdot 32$$

$$\therefore \qquad R_A = \frac{17 \cdot 32}{15} = 1 \cdot 15 \text{ kN (app.)}$$

But,

$$R_A + R_{EV} = 1 \cdot 0 \sin 60° + 2 \sin 60° + 1 \cdot 0 \sin 60°$$

$$= 4 \sin 60° = 2\sqrt{3} = 3 \cdot 46 \text{ kN}$$

$$R_{EV} = 3.46 - 1.15 = 2.31 \text{ kN}$$
$$R_{EH} = 1.0 \cos 60° + 2 \cos 60° + 1.0 \cos 60° = 4 \cos 60° = 2 \text{ kN}$$

Method of Joints :

Joint A :

Refer to Fig. 17·93.

Resolving the force *vertically*, we get

$$P_{AB} \sin 30° + 1.15 = 0$$

∴ $$P_{AB} = \frac{-1.15}{\sin 30°} = -2.30 \text{ kN}$$

Resolving the forces *horizontally*, we get

$$P_{AB} \cos 30° + P_{AG} = 0$$
$$P_{AG} = -P_{AB} \cos 30°$$
$$= -(-2.30) \cos 30°$$
$$= 2.30 \cos 30° = 2 \text{ kN}$$

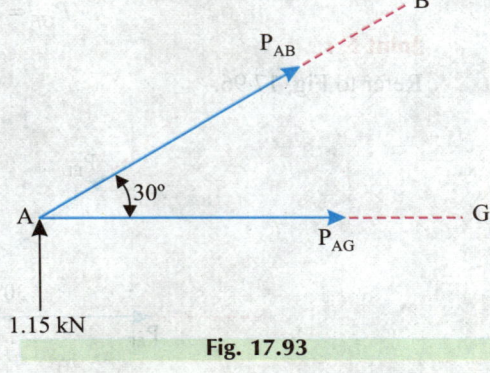

Fig. 17.93

Joint B :

Refer to Fig. 17·94.

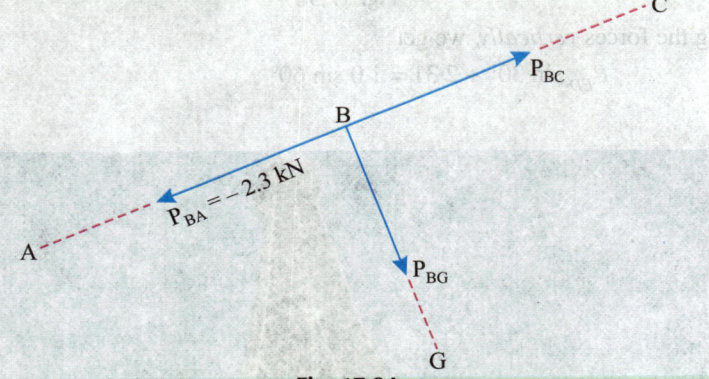

Fig. 17.94

Resolving the forces *parallel* to *ABC*, we get

$$P_{BC} = P_{BA} = -2.3 \text{ kN}$$

Resolving the forces *perpendicular* to *ABC*

$$P_{BG} = 0$$

Joint G :

Refer to Fig. 17·95.

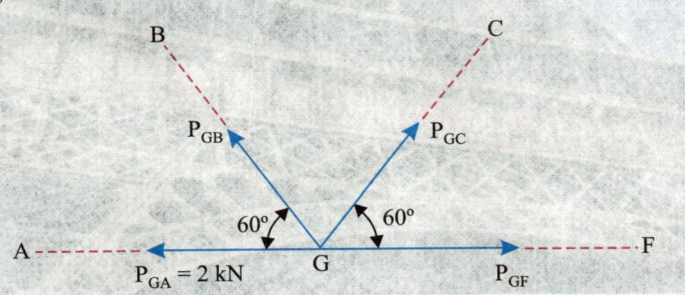

Fig. 17.95

Resolving the forces *vertically*, we get
$$P_{GC} = 0$$
Resolving the forces *horizontally*, we get
$$P_{GF} = P_{GA} = 2 \text{ kN}$$

Joint E :

Refer to Fig. 17·96.

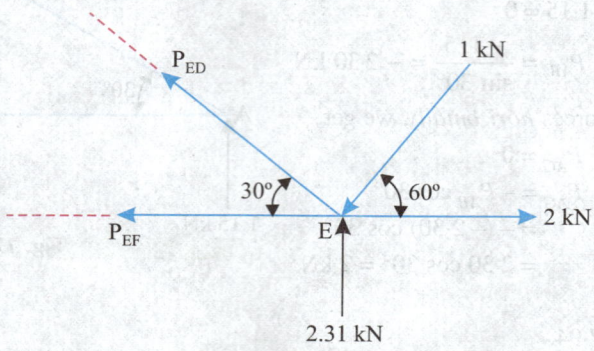

Fig. 17.96

Resolving the forces *vertically*, we get
$$P_{ED} \sin 30° + 2·31 = 1·0 \sin 60°$$

Eiffel Tower, Paris.

$$\therefore \quad P_{ED} = \frac{1 \cdot 0 \sin 60° - 2 \cdot 31}{\sin 30°} = \frac{0 \cdot 866 - 2 \cdot 31}{0 \cdot 5} = -2 \cdot 89 \text{ kN}$$

Resolving the forces *horizontally,* we get

$$P_{EF} + P_{ED} \cos 30° + 1 \cdot 0 \cos 60° = 2$$

$$P_{EF} + (-2 \cdot 89) \times 0 \cdot 866 + 1 \cdot 0 \times 0 \cdot 5 = 2$$

$$P_{EF} - 2 \cdot 5 + 0 \cdot 5 = 2$$

$$\therefore \quad P_{EF} = 4 \text{ kN}$$

Joint D :

Refer to Fig. 17·97.

Resolving the forces *parallel* to the member *CDE*, we get

$$P_{DC} = P_{DE} = -2 \cdot 89 \text{ kN}$$

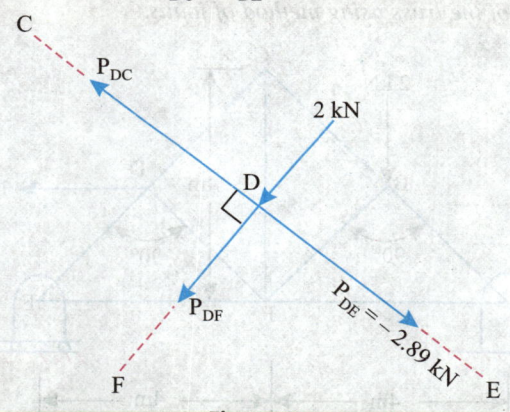

Fig. 17.97

Resolving the forces *perpendicular* to *CDE*, we have

$$P_{DF} + 2 = 0$$

$$\therefore \quad P_{DF} = -2 \text{ kN}$$

Joint F :

Refer to Fig. 17·98.

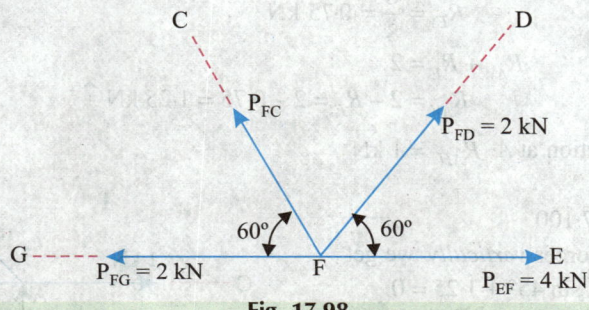

Fig. 17.98

Resolving the forces *vertically,* we get

$$P_{FC} \sin 60° + P_{FD} \sin 60° = 0$$

$$P_{FC} = \frac{-P_{FD} \sin 60°}{\sin 60°} = -(-2) = 2 \text{ kN}$$

The values are tabulated as shown in the Table 17·23.

TABLE 17·23

Tensile +ve; Compressive –ve

Member	Force (kN)	Member	Force (kN)
AB	– 2·31	GA	2·0
BC	– 2·31	BG	zero
CD	– 2·89	CG	zero
DE	– 2·89	CF	2·0
EF	4·0	DF	– 2·0
FG	2·0		

Example 17·25. *Fig. 17·99 shows a truss of 8 metres span and loaded as shown. Find out the forces in all the members of the truss using method of joints.*

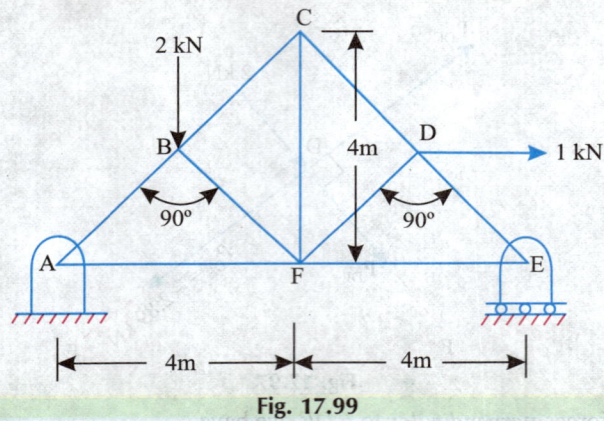

Fig. 17.99

Solution. As the truss is freely supported on rollers at *E*, therefore the reaction at *E* will be *vertical* and the *reaction at A will be the resultant of vertical and horizontal components.*

Taking moments about *A,* we get

$$R_E \times 8 = 2 \times 2 + 1 \times 2 = 6$$

∴
$$R_E = \frac{6}{8} = 0·75 \text{ kN}$$

Also,
$$R_{AV} + R_E = 2$$

∴
$$R_{AV} = 2 - R_E = 2 - 0·75 = 1·25 \text{ kN} \uparrow$$

Horizontal reaction at *A,* R_{AH} = 1 kN ←

Joint A :

Refer to Fig. 17·100.

Resolving the forces *vertically,* we get

$$P_{AB} \sin 45° + 1·25 = 0$$

∴
$$P_{AB} = \frac{-1·25}{\sin 45°} = -1·77 \text{ kN}$$

Resolving the forces *horizontally,* we get

$$P_{AF} + P_{AB} \cos 45° = 1$$

$$P_{AF} - (-1·77) \cos 45° = 1$$

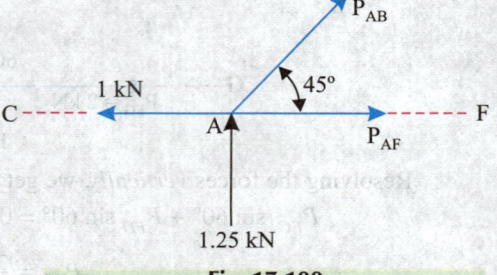

Fig. 17.100

$$P_{AF} - 1 + 1.77 \cos 45° = 2.25 \text{ kN}$$

Joint E :

Refer to Fig. 17·101.

Resolving the forces *vertically*, we get

$$P_{ED} \sin 45° + 0.75 = 0$$

$$\therefore \qquad P_{ED} = \frac{-0.75}{\sin 45°} = -1.06 \text{ kN}$$

Resolving the forces *horizontally*, we get

$$-P_{ED} \cos 45° - P_{EF} = 0$$

$$P_{EF} = -P_{ED} \cos 45° = -(-1.06 \times 0.707) = 0.75 \text{ kN}$$

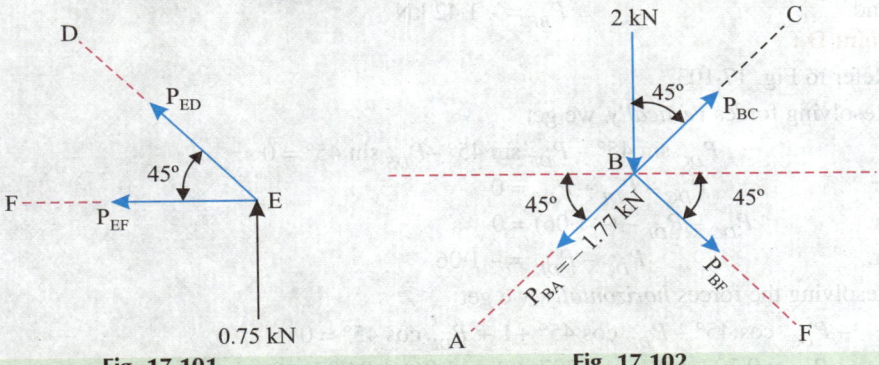

| Fig. 17.101 | Fig. 17.102 |

Joint B :

Refer to Fig. 17·102.

Resolving the forces *vertically*, we get

$$-2 + P_{BC} \cos 45° - P_{BF} \sin 45° - P_{BA} \sin 45° = 0$$

$$-2 + P_{BC} \cos 45° - P_{BF} \sin 45° - (-1.77) \sin 45° = 0$$

Another view of Eiffel Tower, Paris.

or, $P_{BC} \times 0.707 - P_{BF} \times 0.707 = 2 - 1.77 \times 0.707 = 0.75$

∴ $P_{BC} - P_{BF} = \dfrac{0.75}{0.707} = 1.06 \text{ kN}$

Resolving the forces *horizontally*, we get

$$P_{BC} \sin 45° + P_{BF} \cos 45° - P_{BA} \cos 45° = 0$$
$$P_{BC} \times 0.707 + P_{BF} \times 0.707 - (-1.77) \times 0.707 = 0$$

∴ $P_{BC} + P_{BF} = -1.77$

Solving (*i*) and (*ii*), we get,

$$P_{BC} = \dfrac{1.06 - 1.77}{2} = -0.36 \text{ kN (app.)}$$

and, $P_{BF} = -1.42 \text{ kN}$

Joint D :

Refer to Fig. 17.103.

Resolving forces *vertically,* we get

$$P_{DC} \sin 45° - P_{DF} \sin 45° - P_{DE} \sin 45° = 0$$

or, $P_{DC} - P_{DF} - P_{DE} = 0$

or, $P_{DC} - P_{DF} - (-1.06) = 0$

or, $P_{DC} - P_{DF} = -1.06$ ...(*i*)

Resolving the forces *horizontally,* we get

$$- P_{DC} \cos 45° - P_{DF} \cos 45° + 1 + P_{DE} \cos 45° = 0$$
$$- P_{DC} \times 0.707 + P_{DF} \times 0.707 + 1 (-1.06) \times 0.707 = 0$$
$$P_{DC} \times 0.707 + P_{DF} \times 0.707 = 1 - 1.06 \times 0.707$$

∴ $P_{DC} + P_{DF} = \dfrac{0.25}{0.707} = 0.35 \text{ kN}$

Solving (*i*) and (*ii*), we get $P_{DC} = -0.35 \text{ kN (app.)}$

and, $P_{DF} = 0.71 \text{ kN}$

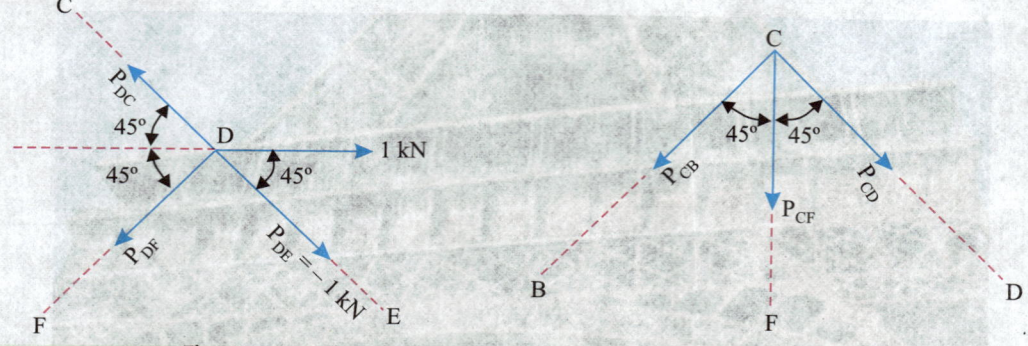

Fig. 17.103   Fig. 17.104

Joint C :

Refer to Fig. 17.104.

Resolving force *vertically,* we get

$$- P_{CF} - P_{CD} \cos 45° - P_{CB} \cos 45° = 0$$
$$- P_{CF} - (-0.35 \cos 45°) - (-0.36 \cos 45°) = 0$$
$$P_{CF} = 0.35 \cos 45° + 0.36 \cos 45° = 0.5 \text{ kN}$$

The values are tabulated as shown in table 17.24

Tensile +ve; Compressive –ve

Member	Force (kN)
AB	– 1·77
BC	– 0·36
CD	– 0·35
DE	– 1·06
EF	0·75
FA	2·25
BF	– 1·42
FD	0·71
CF	0·5

Example 17·26. *A cantilever truss Warren type is loaded as shown in Fig. 17·105. Find the forces in all the members of the truss by method of joints or by method of sections.*

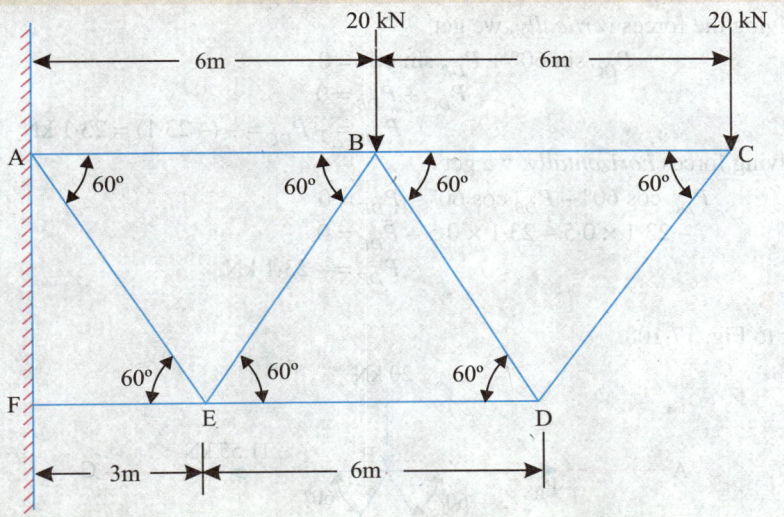

Fig. 17.105

Solution. *In case of cantilever trusses, the reactions need not be calculated.* We can straightway start finding the forces in the members either by method of joints or by method of sections. Note that *AF* is not a member of the truss; it is wall.

Method of joints :

Joint C :

Refer to Fig. 17·106.

Resolving the forces *vertically*, we get

$$- 20 - P_{CD} \sin 60° = 0$$

∴ $$P_{CD} = \frac{-20}{\sin 60°} = -23·1 \text{ kN}$$

Resolving the forces *horizontally*, we get

$$- P_{CB} - P_{CD} \cos 60° = 0$$
$$- P_{CB} - (-23·1) \cos 60° = 0$$

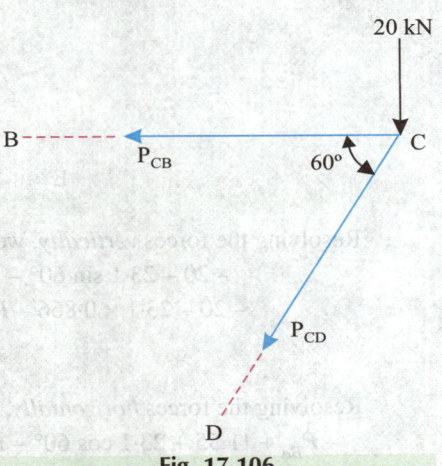

Fig. 17.106

∴ $P_{CB} = 23.1 \times \cos 60° = 11.55$ kN

Joint D :

Refer to Fig. 17.107

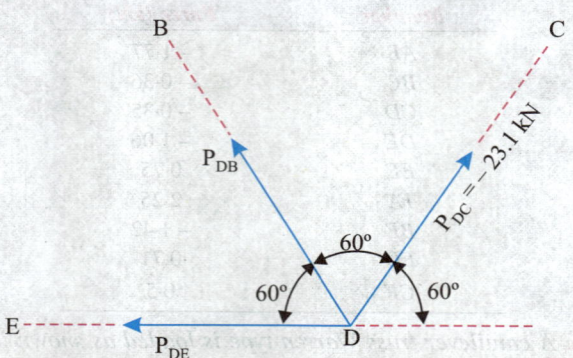

Fig. 17.107

Resolving the forces *vertically*, we get

$$P_{DC} \sin 60° + P_{DB} \sin 60° = 0$$

∴
$$P_{DC} + P_{DB} = 0$$

$$P_{DB} = -P_{DC} = -(-23.1) = 23.1 \text{ kN}$$

Resolving forces *horizontally*, we get

$$P_{DC} \cos 60° - P_{DB} \cos 60° - P_{DE} = 0$$
$$-23.1 \times 0.5 - 23.1 \times 0.5 - P_{DE} = 0$$

∴
$$P_{DE} = -23.1 \text{ kN}$$

Joint B :

Refer to Fig. 17.108.

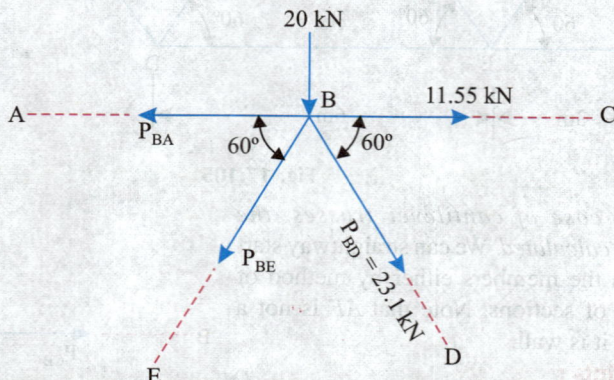

Fig. 17.108

Resolving the forces *vertically*, we get

$$-20 - 23.1 \sin 60° - P_{BE} \sin 60° = 0$$
$$-20 - 23.1 \times 0.866 - P_{BE} \times 0.866 = 0$$

$$P_{BE} = \frac{-40}{0.866} = -46.2 \text{ kN (say)}$$

Resolving the forces *horizontally*, we get

$$-P_{BA} + 11.55 + 23.1 \cos 60° - P_{BE} \cos 60° = 0$$

$$-P_{BA} + 11 \cdot 55 + 23 \cdot 1 \times 0 \cdot 5 - (-46 \cdot 2 \times 0 \cdot 5) = 0$$
$$-P_{BA} = -46 \cdot 2$$
$$P_{BA} = 46 \cdot 2 \text{ kN}$$

Joint E :

Refer to Fig. 17·109.

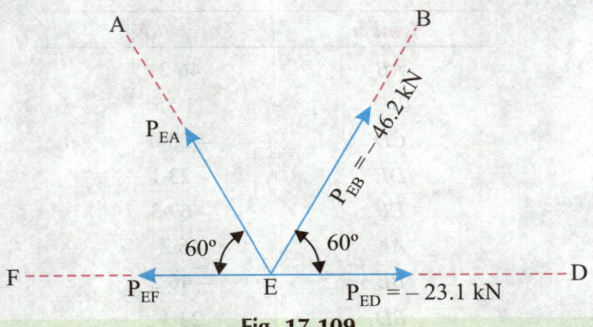

Fig. 17.109

Resolving the forces *vertically*, we get
$$P_{EA} \sin 60° + (-46 \cdot 2) \sin 60° = 0$$
$$P_{EA} = 46 \cdot 2 \text{ kN}$$

Resolving the forces *horizontally*, we get
$$-P_{EF} - P_{EA} \cos 60° + (-46 \cdot 2) \cos 60° + (-23 \cdot 1) = 0$$

Train bridge over a river.

$$-P_{EF} - 46 \cdot 2 \times 0 \cdot 5 - 46 \cdot 2 \times 0 \cdot 5 - 23 \cdot 1 = 0$$

$$P_{EF} = -69 \cdot 3 \text{ kN}$$

The values are tabulated as shown in table 17·25.

TABLE 17·25	
Tension +ve; Compression –ve	
Member	**Force (kN)**
AB	46·2
BC	11·55
CD	– 23·0
DE	– 23·1
EF	– 69·3
AE	46·2
BE	– 46·2
BD	23·1

Method of sections :

Refer to Figs. 17·110 and 17·111.

$$AF = 3 \tan 60° = 5 \cdot 196 \text{ m}$$

Section (1 –1) :

Consider the equilibrium of right part of the section (1 – 1).

Taking moments about *D*, we get

$$P_{CB} \times 5 \cdot 196 = 20 \times 3$$

$$P_{CB} = \frac{20 \times 3}{5 \cdot 196} = 11 \cdot 55 \text{ kN (app.)}$$

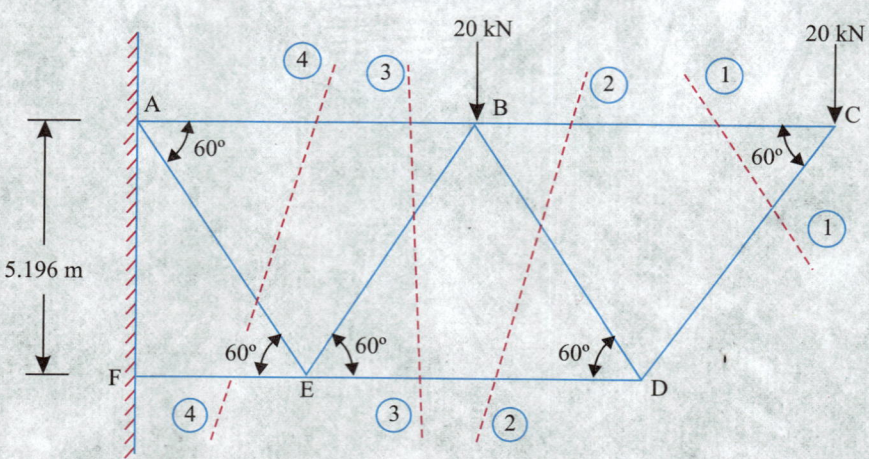

Fig. 17.110

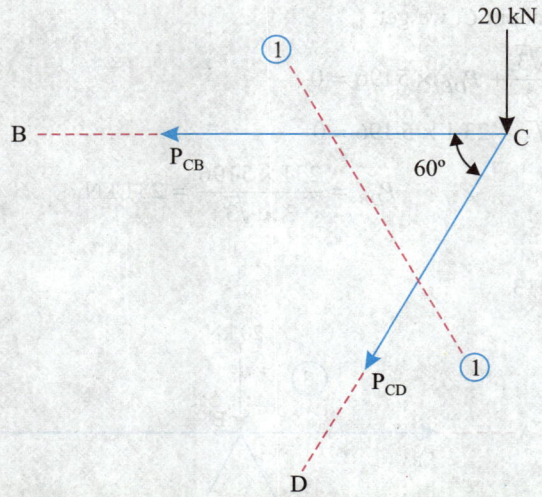

Fig. 17.111

Taking moments about B, we get

$$P_{CD} \times 6 \sin 60° + 20 \times 6 = 0$$

$$P_{CD} \times 6 \times \frac{\sqrt{3}}{2} + 120 = 0$$

$$P_{CD} = \frac{-120}{3\sqrt{3}} = -23.1 \text{ kN}$$

Section (2 – 2) :

Refer to Fig. 17·112.

Consider the equilibrium of the right part of section (2 – 2)

Taking moments about B, we get

$$P_{DE} \times 5·196 + 20 \times 6 = 0$$

$$P_{DE} = \frac{-20 \times 6}{5·196} = -23·1 \text{ kN}$$

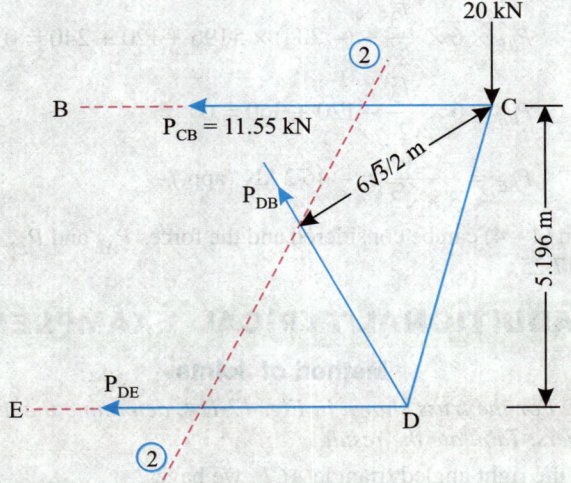

Fig. 17.112

Taking moments about C, we get

$$P_{DB} \times 6 \times \frac{\sqrt{3}}{2} + P_{DE} \times 5 \cdot 196 = 0$$

$$P_{DB} \times 3\sqrt{3} - 23 \cdot 1 \times 5 \cdot 196 = 0$$

$$P_{DB} = \frac{23 \cdot 1 \times 5 \cdot 196}{3 \times \sqrt{3}} = 23 \cdot 1 \text{ kN}$$

Section (3 – 3).

Refer to Fig. 17·113.

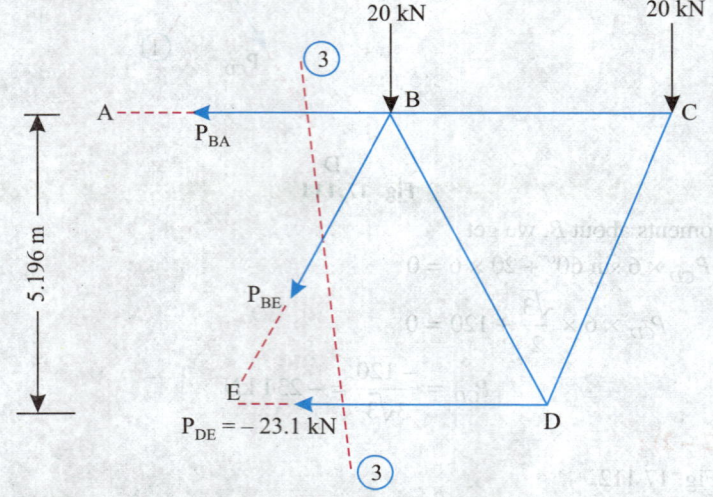

Fig. 17.113

Consider the equilibrium of right part of section (3 – 3).

Taking moments about A, we get

$$P_{BE} \times 6 \times \frac{\sqrt{3}}{2} + P_{DE} \times 5 \cdot 196 + 20 \times 6 + 20 \times 12 = 0$$

$$P_{BE} \times 6 \times \frac{\sqrt{3}}{2} + (- 23 \cdot 1) \times 5 \cdot 196 + 120 + 240 = 0$$

$$P_{BE} \times 6 \times \frac{\sqrt{3}}{2} - 120 + 360 = 0$$

$$P_{BE} = \frac{- 240}{3 \times \sqrt{3}} = - 46 \cdot 2 \text{ kN (app.)}$$

Similarly section (4 – 4) can be considered and the forces P_{AE} and P_{EF} determined. The results are tabulated in table 17·25.

ADDITIONAL/TYPICAL EXAMPLES

Method of Joints

Example 17·27. *For the truss shown in Fig. 17·114, determine the magnitude and nature of the forces in the members. Tabulate the result.* **(SRTMU June, 2000)**

Solution. From the right angled triangle *ACE*, we have

$$\tan 30° = \frac{AE}{AC}$$

or, $\qquad AE = AC \tan 30°$

$$= 8 \tan 30° = 4\cdot619 \text{ m}$$

Similarly, $\qquad \dfrac{BD}{BC} = \tan 30°$

$\therefore \qquad BD = 4 \tan 30° = 2\cdot309 \text{ m}$

In $\triangle ABE$, $\qquad \tan \theta = \dfrac{AE}{AB} = \dfrac{4\cdot619}{4}$

$\therefore \qquad \theta = 49\cdot1°$

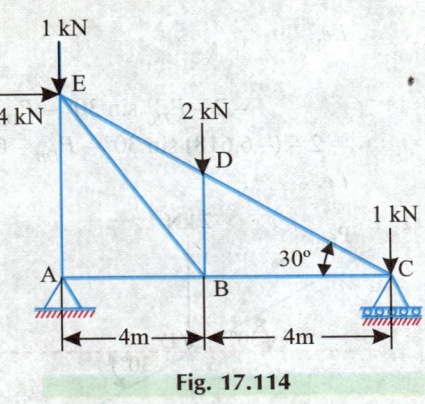

Fig. 17.114

Calculation of reactions at the support :

$$\Sigma M_A = 0$$

$$R_C \times 8 = 4 \times 4\cdot619 + 2 \times 4 + 1 \times 8$$

$\therefore \qquad R_C = 4\cdot309 \text{ kN} \uparrow$

Let R_{AH}, and R_{AV} be the horizontal and vertical components of reaction at 'A' respectively. Then,

$$\Sigma H = 0$$

$$- R_{AH} + 4 = 0$$

$\therefore \qquad R_{AH} = 4 \text{ kN} (\leftarrow)$

$$\Sigma V = 0$$

$$R_{AV} + R_C = 1 + 2 + 1 = 4 \text{ kN}$$

$\therefore \qquad R_{AV} = 4 - 4\cdot309 = - 0\cdot309 \text{ kN}$

Negative sign indicates that the vertical component of reaction at A is acting downward.

Hence, $\qquad R_{AV} = 0\cdot309 \text{ kN} (\downarrow)$

Analysis by method of joints :

Joint C :

Refer to Fig. 17·115.

Applying condition of equilibrium, we get

$$\Sigma H = 0$$

$$- P_{CB} - P_{CD} \cos 30° = 0 \qquad\qquad ...(i)$$

$$\Sigma V = 0$$

$$- 1 + P_{CD} \sin 30° + R_C = 0$$

$$- 1 + 0\cdot5 \, P_{CD} + 4\cdot309 = 0 \qquad \therefore \quad P_{CD} = - 6\cdot618 \text{ kN}$$

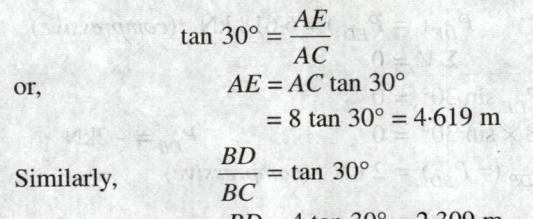

Fig. 17.115

Negative sign indicates that our original assumption of P_{CD} as tensile as shown by the arrow in Fig. is wrong and it should be otherwise.

i.e. $\qquad\qquad P_{CD} (= P_{DC}) = 6\cdot618 \text{ kN} \quad (compressive)$

Substituting the value P_{CD} in (i), we get

$$- P_{CB} - (- 6\cdot618) \cos 30° = 0 \qquad \therefore \quad P_{CB} = 5\cdot731 \text{ kN} \quad (tensile)$$

Joint D :

Refer to Fig. 17·116.

$$\Sigma H = 0$$

$$- P_{DE} \cos 30° + P_{DC} \cos 30° = 0$$

or, $\qquad\qquad P_{DE} - P_{DC} = 0$

or, $\qquad\qquad P_{DE} = P_{DC} = - 6\cdot618 \text{ kN}$

i.e.

$$P_{DE}(= P_{ED}) = 6{\cdot}618 \text{ kN} \quad (compressive)$$
$$\Sigma V = 0$$

$$- 2 - P_{DC} \sin 30° - P_{DB} + P_{DE} \sin 30° = 0$$
$$- 2 - (- 6.618) \sin 30° - P_{DB} - 6.618 \times \sin 30° = 0 \qquad \therefore \quad P_{DB} = - 2 \text{kN}$$

i.e.

$$P_{DB} (= P_{BD}) = 2 \text{ kN} \quad (compressive)$$

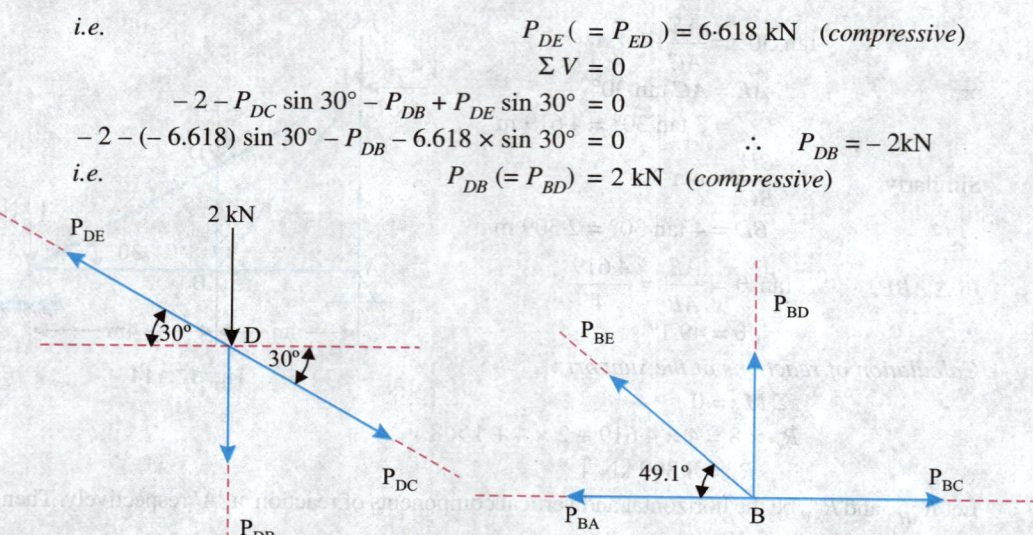

Fig. 17.116   **Fig. 17.117**

Joint B :

Refer to Fig. 17·117.

$$\Sigma V = 0$$
$$P_{BD} + P_{BE} \sin 49{\cdot}1° = 0$$
$$- 2 + P_{BE} \sin 49{\cdot}1° = 0 \qquad \therefore \quad P_{BE} = 2{\cdot}646 \text{ kN} \quad (tensile)$$
$$\Sigma H = 0$$
$$P_{BC} - P_{BE} \cos 49{\cdot}1° - P_{BA} = 0$$
$$5{\cdot}731 - 2{\cdot}646 \cos 49{\cdot}1° - P_{BA} = 0 \qquad \therefore \quad P_{BA} = 4 \text{ kN} \quad (tensile)$$

Joint A :

Refer to Fig. 17·118.

$$\Sigma V = 0$$
$$P_{AE} - R_{AV} = 0$$
$$P_{AE} = 0{\cdot}309 \text{ kN} \quad (tensile)$$

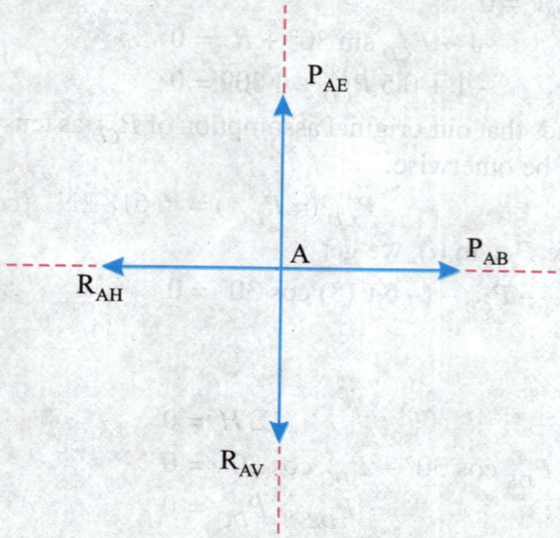

Fig. 17.118

Tabulated result :

TABLE 17·26	
Tensile +ve; Compressive –ve	
Member	**Force (kN)**
AB	4·000
BC	5·731
BD	– 2·000
AE	0·309
CD	– 6·618
DE	– 6·618
EB	2·646

Example 17·28. *A jib crane is shown in Fig. 17·119. Calculate the magnitude and nature of forces in all members of the crane by using method of joints. Also check the force and nature in member CD by method of section.*
(SRTMU Dec., 2001)

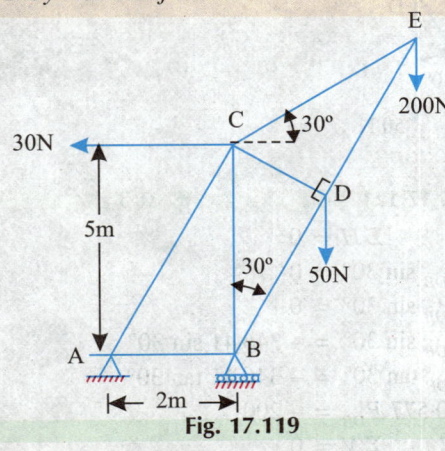

Fig. 17.119

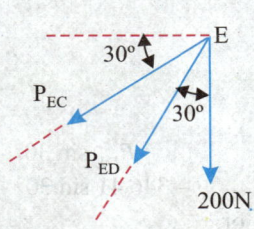

Fig. 17.120

Solution. Refer to Fig. 17·119.

In $\triangle ABC$, $\angle CAB = \theta = \tan^{-1}\left(\dfrac{5}{2}\right) = 68\cdot2°$

Joint E :

Refer to Fig. 17·120.

Applying conditions of equilibrium, we get

$$\Sigma H = 0$$
$$- P_{EC} \cos 30° - P_{ED} \sin 30° = 0$$

or,
$$P_{EC} + P_{ED} \tan 30° = 0 \qquad \qquad ...(i)$$

$$\Sigma V = 0$$
$$-P_{EC} \sin 30° - P_{ED} \cos 30° - 200 = 0$$
$$P_{EC} + P_{ED} \cot 30° = -400 \qquad \qquad ...(ii)$$

Subtracting (*i*) from (*ii*), we get

$$P_{ED}(\cot 30° - \tan 30°) = -400$$

∴
$$P_{ED} = -346\cdot41 \text{ N}$$

–ve sign indicates that our final assumption of P_{EC} as tensile as shown by the arrow in Fig. 17·120 is wrong and it should be otherwise,

i.e. $\qquad\qquad\qquad\qquad\qquad\qquad\qquad P_{ED} (= P_{DE}) = 346·41$ N (*compressive*)

Putting the value of P_{ED} in (*i*), we get

$$P_{EC} - 346·41 \tan 30° = 0$$

or $\qquad\qquad\qquad\qquad\qquad\qquad\qquad\qquad P_{EC} = 200$ N (*tensile*)

Joint D :

Refer to Fig. 17·121.

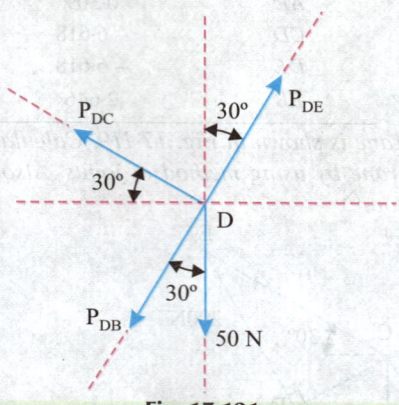

Fig. 17.121

$$\Sigma H = 0$$
$$P_{DE} \sin 30° - P_{DC} \cos 30° - P_{DB} \sin 30° = 0$$
$$- 346·41 \sin 30° - P_{DC} \cos 30° - P_{DB} \sin 30° = 0$$

or, $\qquad\qquad P_{DC} \cos 30° + P_{DB} \sin 30° = - 346·41 \sin 30°$

or, $\qquad\qquad\qquad P_{DC} + P_{DB} \tan 30° = - 346·41 \tan 30°$

or, $\qquad\qquad\qquad P_{DC} + 0·577 P_{DB} = - 200$ $\qquad\qquad\qquad\qquad$ ...(*i*)

$$\Sigma V = 0$$

Railway Station, Antwerp, Germany.

$P_{DE} \cos 30° + P_{DC} \sin 30° - P_{DB} \cos 30° - 50 = 0$

$-346.41 \cos 30° + P_{DC} \sin 30° - P_{DB} \cos 30° - 50 = 0$

$346.41 \cos 30° - P_{DC} \sin 30° + P_{DB} \cos 30° = -50$

$300 - 0.5\, P_{DC} + 0.866\, P_{DB} = -50$

$$P_{DC} - 1.732\, P_{DB} = 700 \qquad \qquad \text{...(ii)}$$

Subtracting (ii) from (i), we get $\qquad 2.309\, P_{DB} = -900$

or $\qquad\qquad\qquad\qquad\qquad\qquad P_{DB} = -389.7\ \text{N}$

i.e. $\qquad\qquad\qquad\qquad\qquad \mathbf{P_{DB}\ (= P_{BD}) = 389.7\ N}$ *(compressive)*

Putting the value of P_{DB} in (ii), we get

$$P_{DC} - 1.732 \times -389.7 = 700$$

or $\qquad\qquad\qquad\qquad\qquad\qquad \mathbf{P_{DC} = 25\ N}$ *(tensile)*

Joint C :

Refer to Fig. 17·122.

$$\Sigma H = 0$$

$P_{CE} \cos 30° + P_{CD} \cos 30° - P_{CA} \cos 68.2° - 30 = 0$

$P_{CE} \cos 30° + 25 \cos 30° - P_{CA} \cos 68.2° - 30 = 0$

$0.866 \times 200 + 21.65 - 0.3714\, P_{CA} - 30 = 0$

or, $\qquad\qquad\qquad\qquad\qquad \mathbf{P_{CA} = 443.9\ N}$ *(tensile)*

$$\Sigma V = 0$$

$P_{CE} \sin 30° - P_{CD} \sin 30° - P_{CB} - P_{CA} \sin 68.2° = 0$

$0.5 \times 200 - 25 \times 0.5 - P_{CB} - 443.9 \times 0.9284 = 0$

or, $\qquad\qquad\qquad\qquad\qquad P_{CB} = -324.6\ \text{N}$

or, $\qquad\qquad\qquad\qquad\qquad \mathbf{P_{CB} = 324.6\ N}$ *(compressive)*

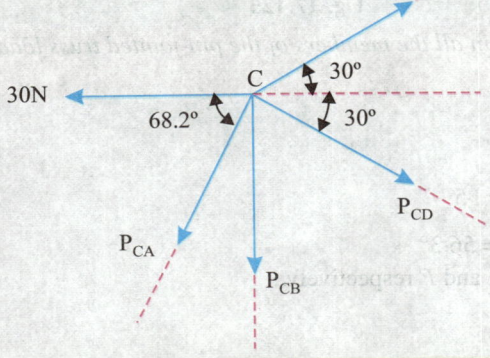

Fig. 17.122

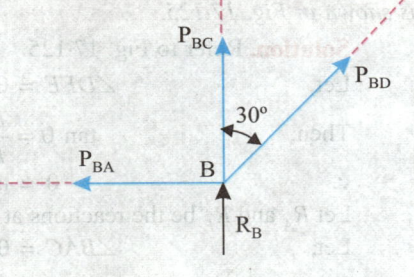

Fig. 17.123

Joint B :

Refer to Fig. 17·123.

$$\Sigma H = 0$$

$$-P_{BA} + P_{BD} \sin 30° = 0$$

$$-P_{BA} - 389.7 \times 0.5 = 0$$

or $\qquad\qquad\qquad\qquad\qquad P_{BA} = -194.85\ \text{N}$

i.e. $\qquad\qquad\qquad\qquad\qquad \mathbf{P_{BA}\ (= P_{AB}) = 194.85\ N}$ *(compressive)*

$$\Sigma V = 0$$
$$R_B + P_{BC} + P_{BD} \cos 30° = 0$$
$$R_B - 324·6 - 389·7 \times 0·866 = 0$$

or
$$R_B = 662 \text{ N} \uparrow$$

Checking of magnitude and nature of force in member 'CD' by using method of sections :
Draw a section line which should cut not more than three unknown member forces of given truss. Hence section line 1 – 1 is drawn as shown in Fig. 17·124.

$$\Sigma M_E = 0$$
$$P_{CD} \times \text{DE} - 50 \times \text{DE} \sin 30° = 0$$

or,
$$P_{CD} = 25 \text{ N} \quad (tensile)$$

Hence magnitude and nature of force in the member '*CD*' by using method of section satisfies the answer.

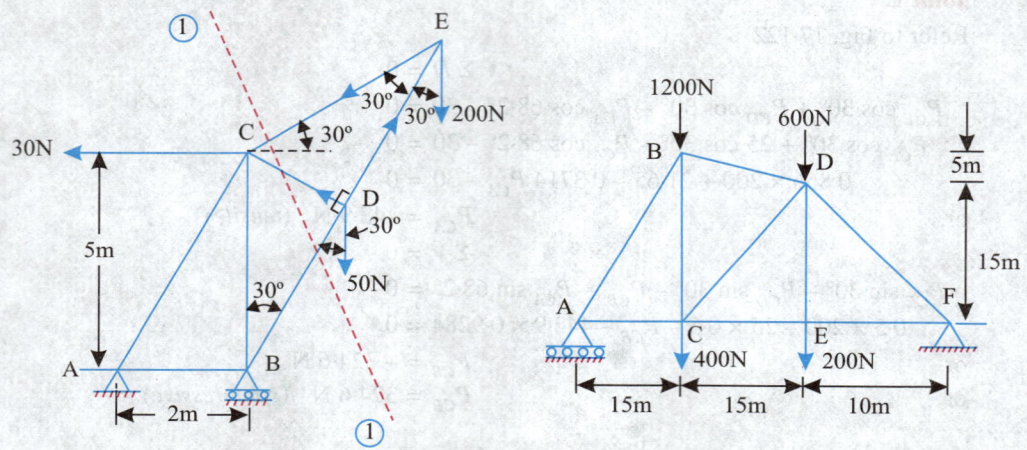

Fig. 17.124   **Fig. 17.125**

Example 17·29. *Determine the axial forces in all the members of the pin-jointed truss loaded as shown in Fig. 17·125.*

Solution. Refer to Fig. 17·125.

Let,   ∠DFE = θ,

Then, $\tan \theta = \dfrac{DE}{EF} = \dfrac{15}{10} = 1·5$,

∴ $\theta = \tan^{-1}(1·5) = 56·3°$

Let R_A and R_F be the reactions at supports A and F respectively.

Let,   ∠BAC = θ₁ $\uparrow$

Then, $\tan \theta_1 = \dfrac{BC}{AC} = \dfrac{20}{15}$

or, $\theta_1 = \tan^{-1}\left(\dfrac{20}{15}\right) = 53·13°$

Taking moments about *F*, we get
$$R_A \times 40 = 1200 \times 25 + 400 \times 25 + 600 \times 10 + 200 \times 10 = 48000$$

∴ $R_A = 1200 \text{ N} \uparrow$

Also, $\Sigma V = 0$
$$R_A + R_F = 1200 + 400 + 600 + 200 = 2400$$

∴ $R_F = 2400 - 1200 = 1200 \text{ N} \uparrow$

and, $\Sigma H = 0$

∴ $R_{FH} = 0$

Joint A :

Refer to Fig. 17·126.

Resolving the forces *vertically* and applying the principle of $\Sigma V = 0$, we get

$$1200 + P_{AB} \sin 53 \cdot 1° = 0$$

∴ $P_{AB} = -\dfrac{1200}{\sin 53 \cdot 13°} = -1500 \text{ N}$

−ve sign indicates that our original assumption of P_{AB} as tensile as shown by the arrow in the Fig. 17·126 is wrong and it should be otherwise.

i.e. $P_{AB} (= P_{BA}) = 1500 \text{ N}$ (*compressive*)

Resolving forces *horizontally* and applying the principle of $\Sigma H = 0$, we get

$$P_{AB} \cos 53 \cdot 13° + P_{AC} = 0$$

$$-1500 \cos 53 \cdot 13° + P_{AC} = 0$$

∴ $P_{AC} (= P_{CA}) = 1500 \cos 53 \cdot 13° = 900 \text{ N}$ (*tensile*)

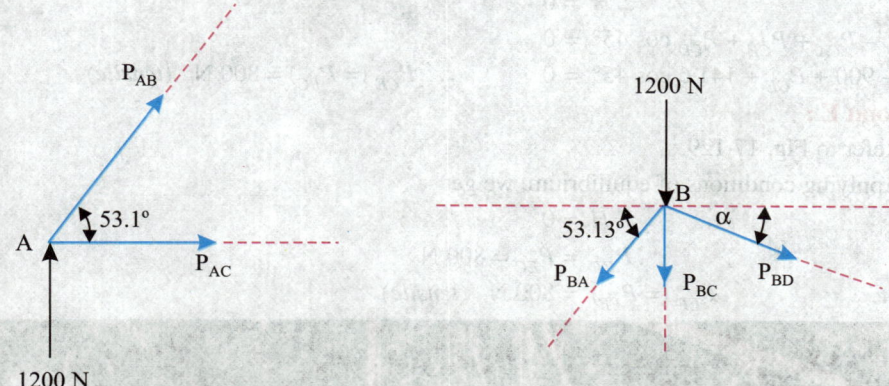

| Fig. 17.126 | Fig. 17.127 |

Joint B :

Refer to Fig. 17·127.

$$\tan \alpha = \frac{5}{15}$$

∴ $\alpha = 18 \cdot 43°$

Applying the conditions of equilibrium, we get

$$\Sigma H = 0$$

or, $P_{BA} \cos 53 \cdot 13° = P_{BD} \cos 18 \cdot 43°$

$$-1500 \cos 53 \cdot 13° = P_{BD} \cos 18 \cdot 43°$$

∴ $P_{BD} = -948 \cdot 66 \text{ N}$

−ve sign indicates that our original assumption of P_{BD} as tensile as shown by the arrow in Fig. 17·127 is wrong and it should be otherwise.

i.e. $P_{BD} = P_{DB} = 948 \cdot 66 \text{ N}$ (*compressive*)

$$\Sigma V = 0$$
$$-1200 - P_{BD} \sin \alpha - P_{BC} - P_{BA} \sin 53{\cdot}15° = 0$$

or
$$-1200 - P_{BD} \sin 18{\cdot}43° - P_{BC} - (-1500) \sin 53{\cdot}13° = 0$$

or,
$$-1200 - (-948{\cdot}66) \sin 18{\cdot}43° - P_{BC} + 1500 \sin 53{\cdot}1° = 0$$

or,
$$-1200 + 299{\cdot}9 - P_{BC} + 1200 = 0$$

or,
$$\boldsymbol{P_{BC}} = \boldsymbol{P_{CB}} = 299{\cdot}9 \text{ N } (tensile)$$

Joint C :

Refer to Fig. 17·128.

$$\tan \theta_2 = \frac{DE}{EC} = \frac{15}{15} = 1,$$

$$\therefore \qquad \theta_2 = 45°$$

Applying conditions of equilibrium, we get

$$\Sigma V = 0$$
$$P_{CB} + P_{CD} \sin 45° - 400 = 0$$
$$299{\cdot}9 + P_{CD} \sin 45° - 400 = 0$$

$$\therefore \qquad P_{CD} (= P_{DC}) = 141{\cdot}6 \text{ N } (tensile)$$

$$\Sigma H = 0$$
$$-P_{CA} + P_{CE} + P_{CD} \cos 45° = 0$$
$$-900 + P_{CE} + 141{\cdot}6 \cos 45° = 0 \qquad \therefore \ \boldsymbol{P_{CE}} \, (= P_{EC}) = 800 \text{ N } (tensile)$$

Fig. 17.128

Joint E :

Refer to Fig. 17·129.

Applying conditions of equilibrium, we get

$$\Sigma H = 0$$
$$\boldsymbol{P_{EF}} = P_{EC} = 800 \text{ N}$$

i.e.
$$\boldsymbol{P_{EF}} \, (= P_{FE}) = 800 \text{ N } (tensile)$$

Steel Structure

$$\Sigma V = 0$$
$$P_{DE} = P_{ED} = 200 \text{ N} \quad (tensile)$$

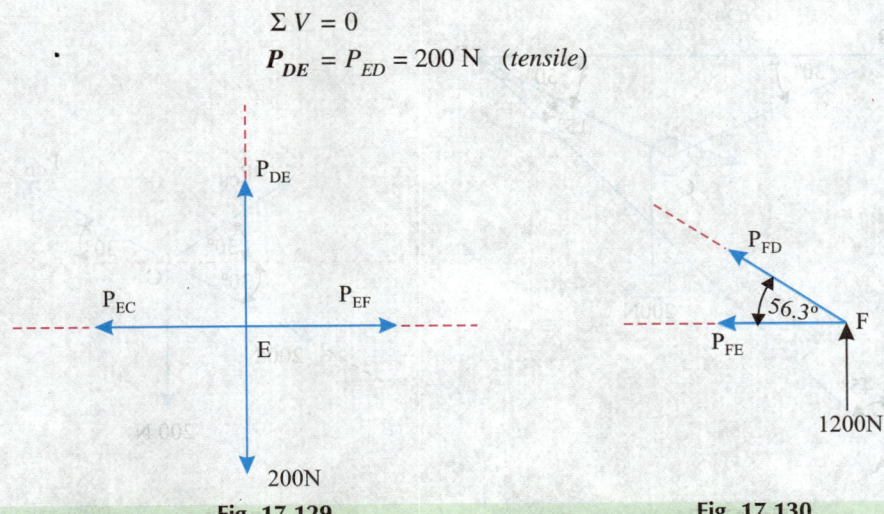

Fig. 17.129   **Fig. 17.130**

Joint F :

Refer to Fig. 17·130.

Applying conditions of equilibrium, we get

$$\Sigma H = 0$$
$$P_{FE} + P_{FD} \cos 56·3° = 0$$
$$800 + P_{FD} \times 0·554 = 0$$
$$\therefore \qquad P_{FD} = -1442 \text{ N}$$

– ve sign indicates that our original assumption of P_{FD} as tensile as shown by the arrow is wrong and it should be otherwise.

i.e. $\qquad P_{FD} (= P_{DF}) = 1442 \text{ N} \quad (compressive)$

TABLE 17·27

Tensile +ve; Compressive –ve

Member	Force (N)
AB	– 1500
AC	900
BC	299·9
BD	– 948·66
CE	800
CD	141·6
DE	200
DF	– 1442
EF	800

Example 17·30. *In the frame shown in Fig. 17·131, there is a small smooth and weightless pulley at the joint C. A weight 200 N is held by a force inclined at 30° as shown in Fig. Find the forces in all the members and reactions at A and B.*

(N.U.)

Solution. Let us analyse this truss by using *method of joints* as follows :

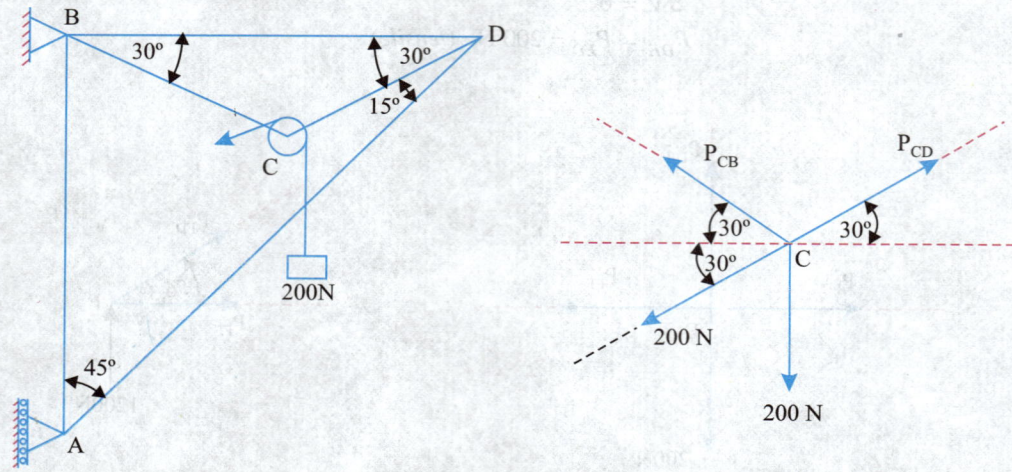

Fig. 17.131   **Fig. 17.132**

Joint C :

Refer to Fig. 17·132.

Applying conditions of equilibrium, we get

$$\Sigma H = 0$$

$$P_{CD} \cos 30° - P_{BC} \cos 30° - 200 \cos 30° = 0$$

$$P_{CD} = P_{BC} + 200 \qquad \qquad ...(i)$$

$$\Sigma V = 0$$

$$P_{CD} \sin 30° + P_{BC} \sin 30° - 200 \sin 30° - 200 = 0$$

$$P_{CD} + P_{BC} = 600 \qquad \qquad ...(ii)$$

Substituting the value of P_{CD} from (i) in (ii), we get

$$(P_{BC} + 200) + P_{BC} = 600$$

∴ $\boldsymbol{P_{BC} = 200 \ N}$ (*tensile*)

Substituting the value of P_{BC} in (i), we get

$$\boldsymbol{P_{CD}} = 200 + 200 = 400 \ N \quad (tensile)$$

Joint D :

Refer to Fig. 17·133.

Applying conditions of equilibrium, we get

$$\Sigma V = 0$$

$$P_{DC} \sin 30° + P_{DA} \sin 45° = 0$$

$$400 \sin 30° + P_{DA} \sin 45° = 0$$

∴ $P_{DA} = -282 \cdot 8 \ N$

– ve sign indicates that our original assumption of P_{DA} as tensile as shown in Fig. 17·133 is wrong and it should be otherwise.

i.e. $\boldsymbol{P_{DA}} \ (= P_{AD}) = 282.8 \ N \ (compressive)$

$$\Sigma H = 0$$

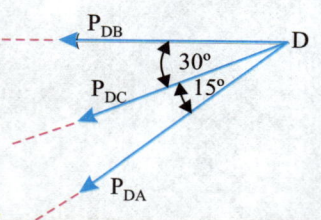

Fig. 17.133

$$-P_{DB} - P_{DC} \cos 30° - P_{DA} \cos 45° = 0$$

$$P_{DB} + P_{DC} \cos 30° + P_{DA} \cos 45° = 0$$

$$P_{DB} + 400 \cos 30° - 282 \cdot 8 \cos 45° = 0$$

∴ $\boldsymbol{P_{DB} = -146 \cdot 4 \ N}$

i.e. $\qquad P_{DB} \, (= P_{BD}) = 146 \cdot 4 \, \text{N} \ (compressive)$

Joint A :

Refer to Fig. 17·134.

Let R_A be the reaction at support A and directed normal to the plane of roller.

Applying conditions of equilibrium, we get

$$\Sigma H = 0$$
$$R_A + P_{AD} \sin 45° = 0$$
$$R_A + (-282 \cdot 8) \sin 45° = 0$$

∴ $\qquad\qquad R_A = 282 \cdot 8 \sin 45°$
$$= 200 \, \text{N} \ (\rightarrow) \ \textbf{(Ans.)}$$
$$\Sigma V = 0$$
$$P_{AB} + P_{AD} \sin 45° = 0$$
$$P_{AB} - 282 \cdot 8 \sin 45° = 0$$

∴ $\qquad\qquad \boldsymbol{P_{AB}} = 200 \, \text{N} \ (tensile)$

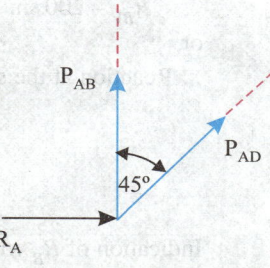

Fig. 17.134

Joint B :

Refer to Fig. 17·135.

Let R_{BH} and R_{BV} are the horizontal and vertical components of reaction at B due to hinge.

Applying conditions of equilibrium at joint B, we have :

$$\Sigma H = 0$$
$$- R_{BH} + P_{BD} + P_{BC} \cos 30° = 0$$
$$- R_{BH} - 146 \cdot 4 + 200 \cos 30° = 0$$

∴ $\qquad\qquad R_{BH} = 26 \cdot 8 \, \text{N} \ (\leftarrow)$
$$\Sigma V = 0$$
$$R_{BV} - P_{BC} \sin 30° - P_{BA} = 0$$

Steel framework on a bridge

$$R_{BV} - 200 \sin 30° - 200 = 0$$

or, $$R_{BV} = 300 \text{ N}$$

∴ Reaction at the support B,

$$R_B = \sqrt{R_{BH}^2 + R_{BV}^2}$$

$$= \sqrt{26 \cdot 8^2 + 300^2}$$

$$= \textbf{301.2 N} \quad \textbf{(Ans.)}$$

Indication of R_B with respect to the horizontal is

$$\tan \theta = \frac{R_{BV}}{R_{BH}} = \frac{300}{26 \cdot 8}$$

∴ $$\theta = \textbf{84·89°} \quad \textbf{(Ans.)}$$

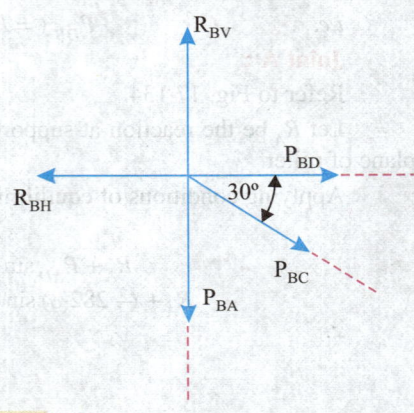

Fig. 17.135

TABLE 17·28

Tensile +ve; Compressive −ve

Member	Force (N)
AB	200
AD	− 282·8
BD	− 146·4
BC	200 N
DC	400

Example 17·31. *Fig 17·136. shows a pin connected truss. The Turnbuckle is adjusted to induce a tensile force of 10 kN in the member containing it. Find the forces in all the members. Indicate nature and magnitude of each force on separate figure of truss.* **(N.U.)**

Solution. As given since turnbuckle induces a tensile force of 10 kN in the member CD; therefore

$$P_{CD} = 10 \text{ kN} \quad \text{(tensile)}$$

From the geometry, $\angle DCF = \tan^{-1}\left(\dfrac{2}{2}\right) = 45°$

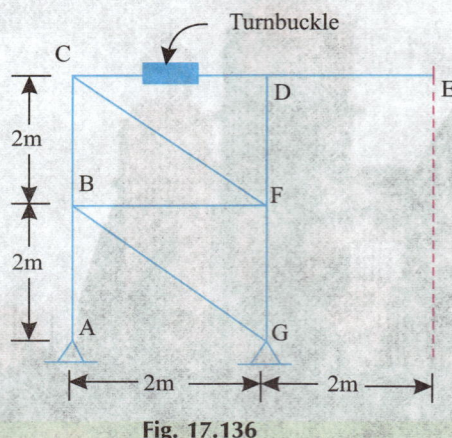

Fig. 17.136

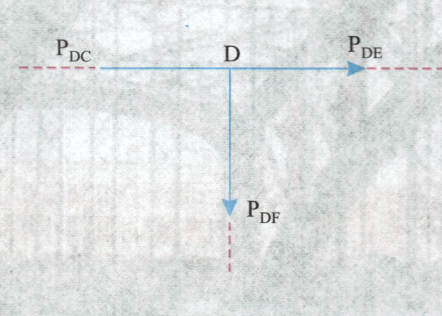

Fig. 17.137

Joint D :

Refer to Fig. 17·137.

Applying conditions of equilibrium, we get

$$- P_{DC} + P_{DE} = 0$$

or, $P_{DE} = 10$ kN *(tensile)*

$\Sigma V = 0$

i.e. $P_{DF} = 0$

Joint C :

Refer to Fig. 17.138.

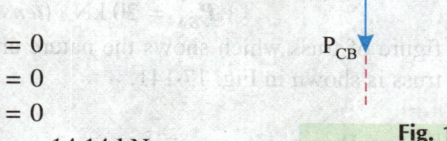

Fig. 17.138

$\Sigma H = 0$

$P_{CD} + P_{CF} \cos 45° = 0$

$10 + P_{CF} \times 0.707 = 0$

or $P_{CF} = -14.14$ kN

–ve sign indicates that our original assumption of force P_{CF} as tensile as shown by the arrow in Fig. 17.138 is wrong and it should otherwise,

i.e. $\boldsymbol{P_{CF}} \, (= \boldsymbol{P_{FC}}) = 14.14$ kN   *(compressive)*

$\Sigma V = 0$

$- P_{CF} \sin 45° - P_{CB} = 0$

$- (- 14.14) \sin 45° - P_{CB} = 0$

∴ $\boldsymbol{P_{CB}} = 10$ kN   *(tensile)*

Joint F :

Refer to Fig. 17.139.

$\Sigma H = 0$

$- P_{FB} - P_{FC} \cos 45° = 0$

$- P_{FB} - (- 14.14) \cos 45° = 0$

∴ $\boldsymbol{P_{FB}} = 10$ kN   *(tensile)*

$\Sigma V = 0$

$P_{FD} + P_{FC} \sin 45° - P_{FG} = 0$

$0 - 14.14 \sin 45° - P_{FG} = 0$ ∴ $P_{FG} = -10$ kN

i.e. $\boldsymbol{P_{FG}} \, (= \boldsymbol{P_{GF}}) = 10$ kN   *(compressive)*

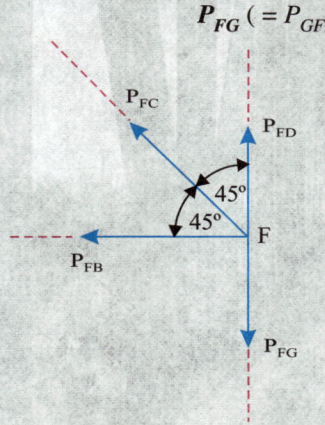

Fig. 17.139

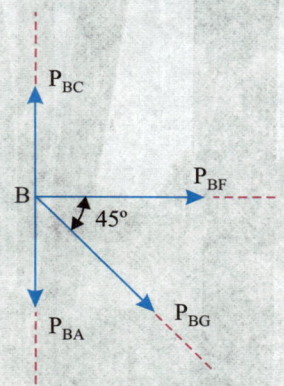

Fig. 17.140

Joint B :

Refer to Fig. 17.140.

$\Sigma H = 0$

$P_{BF} + P_{BG} \cos 45° = 0$

$10 + P_{BG} \cos 45° = 0$

∴ $P_{BG} = -14.14$ kN

i.e. $$P_{BG} \; (= P_{GB}) = 14 \cdot 14 \text{ kN} \quad (compressive)$$
$$\Sigma V = 0$$
$$P_{BC} - P_{BG} \sin 45° - P_{BA} = 0$$
$$10 - (-14 \cdot 14) \sin 45° - P_{BA} = 0$$
$$\therefore \qquad\qquad P_{BA} = 20 \text{ kN} \quad (tensile)$$

Separate figure of truss which shows the nature and magnitude of each force in the various members of the truss is shown in Fig. 17·141.

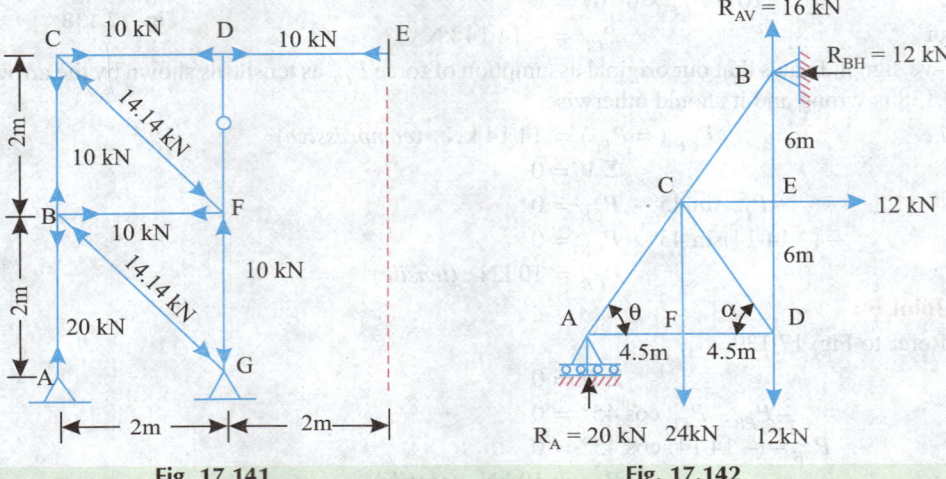

Fig. 17.141   **Fig. 17.142**

Example 17·32. *For the truss shown in Fig. 17·142, determine the forces in the members.*

(Panjab University)

Railway bridge, Chernobyl, Ukraine.

Solution. *Calculations for reactions at supports :*

Applying conditions of equilibrium, we get

$$\Sigma H = 0$$
$$R_{BH} = 12 \text{ kN } (\leftarrow)$$
$$\Sigma V = 0$$
$$R_A + R_{BV} = 24 + 12 = 36$$
$$\Sigma M_B = 0$$
$$R_A \times 9 - 24 \times 4{\cdot}5 - 12 \times 6 = 0$$

∴
$$R_A = 20 \text{ kN,}$$

and,
$$20 + R_{BV} = 36$$

or
$$R_{BV} = 16 \text{ kN } (\uparrow)$$

From geometry,
$$\tan \theta = \frac{12}{9}$$

∴
$$\theta = 53{\cdot}13°,$$

and
$$\tan \alpha = \frac{6}{4{\cdot}5} \text{ or } \alpha = 53{\cdot}13°$$

Joint A :

Refer to Fig. 17·143.

Applying condition of equilibrium, we get

$$\Sigma H = 0$$
$$P_{AF} + P_{AC} \cos 53{\cdot}13° = 0 \qquad \qquad ...(i)$$
$$\Sigma V = 0$$
$$P_{AC} \sin 53{\cdot}13° + 20 = 0$$

or,
$$P_{AC} = -25 \text{ kN}$$

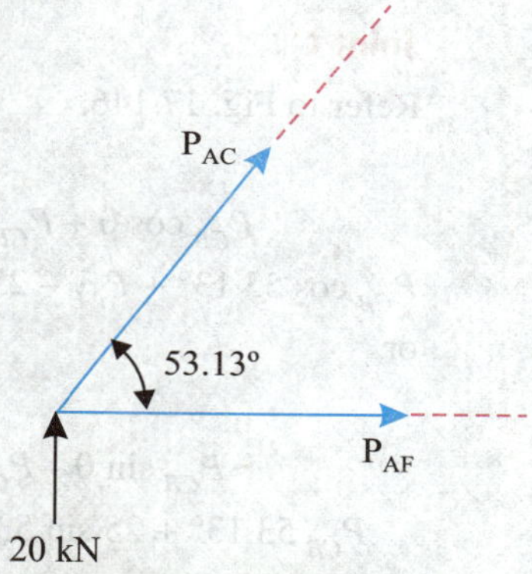

–ve sign indicates that over original assumption of P_{AC} as tensile as shown by the arrow is wrong and it should be otherwise,

i.e.
$$\boldsymbol{P_{AC}} (= P_{CA}) = 25 \text{ kN } \text{ (compressive)}$$

Substituting the value of P_{AC} in eqn. (*i*), we get

$$P_{AF} - 25 \cos 53{\cdot}13° = 0$$

∴
$$\boldsymbol{P_{AF}} = 25 \cos 53{\cdot}13° = 15 \text{ kN } \text{ (tensile)}$$

Joint F :

Refer to Fig. 17.144.

$$\Sigma H = 0$$
$$-P_{FA} + P_{FD} = 0$$

or,
$$\boldsymbol{P_{FD}} = P_{FA}$$
$$= 15 \text{ kN } \text{ (tensile)}$$
$$\Sigma V = 0$$
$$P_{FC} - 24 = 0$$

or,
$$\boldsymbol{P_{FC}} = 24 \text{ kN } \text{ (tensile)}$$

Fig. 17.143

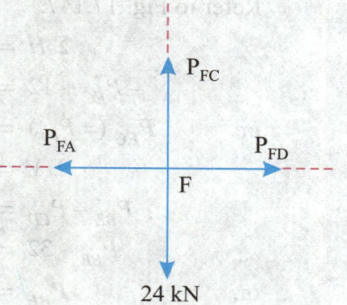

Fig. 17.144

Joint D :

Refer to Fig. 17·145.

$$\Sigma H = 0$$
$$- P_{DF} - P_{DC} \cos \alpha = 0$$
$$- 15 - P_{DC} \cos 53 \cdot 13° = 0$$

or, $$P_{DC} = - 25 \text{ kN}$$

i.e. $$\mathbf{P_{DC}} \, (= P_{CD})$$
$$= 25 \text{ kN} \quad (compressive)$$

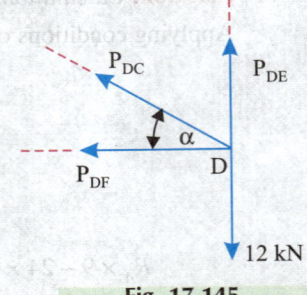

Fig. 17.145

$$\Sigma V = 0$$
$$P_{DC} \sin \alpha + P_{DE} - 12 = 0$$
$$- 25 \sin 53 \cdot 13° + P_{DE} - 12 = 0$$

or $$\mathbf{P_{DE}} = 32 \text{ kN}$$

Joint C :

Refer to Fig. 17·146.

$$\Sigma H = 0$$
$$P_{CB} \cos \theta + P_{CE} + P_{CD} \cos \alpha - P_{CA} \cos \theta = 0$$
$$P_{CB} \cos 53 \cdot 13° + P_{CE} - 25 \cos 53 \cdot 13° + 25 \cos 53 \cdot 13° = 0$$

or, $$P_{CB} \cos 53 \cdot 13° + P_{CE} = 0 \quad ...(i)$$
$$\Sigma V = 0$$
$$P_{CB} \sin \theta - P_{CD} \sin \alpha - P_{CF} - P_{CA} \sin \theta = 0$$
$$P_{CB} \, 53 \cdot 13° + 25 \sin 53 \cdot 13° - 24 + 25 \sin 53 \cdot 13° = 0$$
$$0 \cdot 8 \, P_{CB} + 2 \times 25 \sin 53 \cdot 13° - 24 = 0$$

or, $$P_{CB} = - 20 \text{ kN}$$

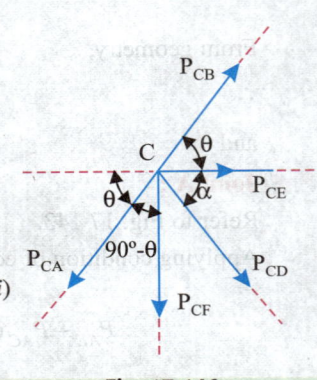

Fig. 17.146

i.e. $$\mathbf{P_{CB}} \, (= P_{BC}) = 20 \text{ kN} \quad (compressive)$$

Substituting the value of P_{CB} in (i), we get

$$- 20 \cos 53 \cdot 13° + P_{CE} = 0$$

or $$P_{CE} = 20 \cos 53 \cdot 13° = 12 \text{ kN}$$

Joint E :

Refer to Fig. 17.147.

$$\Sigma H = 0$$
$$- P_{EC} + 12 = 0 \quad \therefore \quad P_{EC} = 12 \text{ kN}$$

i.e. $$\mathbf{P_{EC}} \, (= P_{CE}) = 12 \text{ kN} \quad (tensile)$$
$$\Sigma V = 0$$
$$P_{BE} - P_{ED} = 0$$
$$P_{BE} - 32 = 0$$

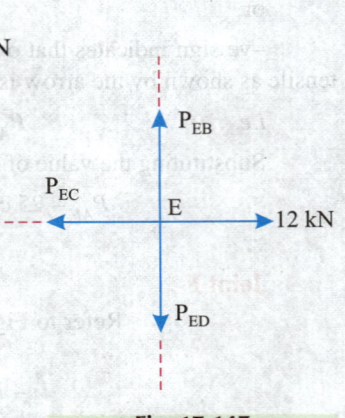

Fig. 17.147

or, $$\mathbf{P_{BE}} = 32 \text{ kN} \quad (tensile)$$

TABLE 17·29

Tensile +ve; Compressive –ve

Member	Force (kN)
AF	15
AC	– 25
BC	– 20
BE	32
CE	12
CD	– 25
CF	24
DF	15
DE	32

Example 17·33. *Determine the magnitude and nature of forces in different members of the truss shown in Fig. 17·148. Tabulate the result.*   **(SRTMU May, 2002)**

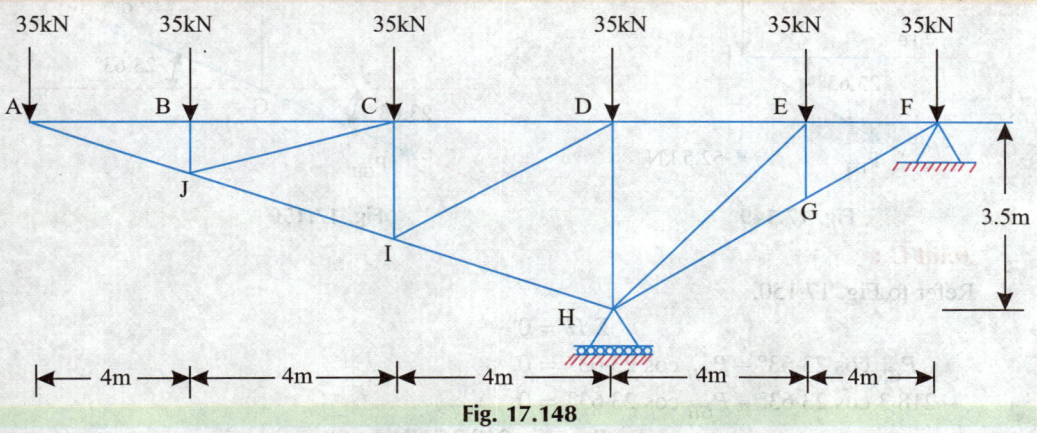

Fig. 17.148

Solution. Consider Δ HDF. $\angle HFD = \theta = \tan^{-1}\left(\dfrac{3 \cdot 5}{8}\right) = 23 \cdot 63°$

Again, from Δ DEH, $\angle DEH = \theta_1 = \tan^{-1}\left(\dfrac{3 \cdot 5}{4}\right) = 41 \cdot 18°$

and, from Δ HAD, $\angle HAD = \theta_2 = \tan^{-1}\left(\dfrac{3 \cdot 5}{12}\right) = 16 \cdot 26°$

Calculations for reactions :-

$$\Sigma M_F = 0$$
$$R_H \times 8 = 35 \times 20 + 35 \times 16 + 35 \times 12 + 35 \times 8 + 35 \times 4$$

$\therefore$
$$R_H = 262 \cdot 5 \text{ kN } (\uparrow)$$
$$\Sigma V = 0$$
$$R_H + R_F = 35 \times 6 = 210$$

$\therefore$
$$R_F = 210 - 262 \cdot 5 = -52 \cdot 5 \text{ kN}   \quad i.e.\ \ 52 \cdot 5 \text{ kN } (\downarrow)$$

Analysis by method of joints :

Joint F :

Refer to Fig. 17·149.

Applying conditions of equilibrium, we get

$$\Sigma V = 0$$
$$-35 - P_{FG} \sin 23 \cdot 63° - 52 \cdot 5 = 0$$
$$\therefore \qquad P_{FG} = -218 \cdot 3 \text{ N}$$

–ve sign shows that our original assumption of P_{FG} as tensile as shown by the arrow in Fig. 17·149 is wrong and it should be otherwise,

i.e. $\qquad P_{FG} (= P_{GF}) = 218 \cdot 3 \text{ N} \quad (compressive)$
$$\Sigma H = 0$$
$$-P_{FE} - P_{FG} \cos 23 \cdot 63° = 0$$
$$-P_{FE} - (-218 \cdot 3) \cos 23 \cdot 63° = 0$$
$$\therefore \qquad P_{FE} = 200 \text{ kN} \quad (tensile)$$

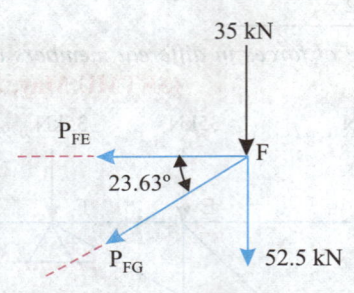

Fig. 17.149   **Fig. 17.150**

Joint G :

Refer to Fig. 17·150.

$$\Sigma H = 0$$
$$P_{GF} \cos 23 \cdot 63° - P_{GH} \cos 23 \cdot 63° = 0$$
$$-218 \cdot 3 \cos 23 \cdot 63° - P_{GH} \cos 23 \cdot 63° = 0$$
$$\therefore \qquad P_{GH} = -218 \cdot 3 \text{ kN}$$

i.e. $\qquad P_{GH} (= P_{HG}) = 218 \cdot 3 \text{ kN} \quad (compressive)$

Yoshida Bridge, USA.

$$\Sigma V = 0$$
$$P_{GE} + P_{GF} \sin 23 \cdot 63° - P_{GH} \sin 23 \cdot 63° = 0$$
$$P_{GE} + (-218 \cdot 3) \sin 23 \cdot 63° - (-218 \cdot 3) \sin 23 \cdot 63° = 0$$
$$\therefore \qquad\qquad P_{GE} = 0$$

Joint E :

Refer to Fig. 17·151.

$$\Sigma V = 0$$
$$- 35 - P_{EH} \sin 41 \cdot 18° = 0$$
$$\therefore \qquad\qquad P_{EH} = -53 \cdot 16 \text{ N}$$

i.e. $\qquad\qquad P_{EH} = P_{HE} = 53 \cdot 16 \text{ N} \quad (compressive)$

$$\Sigma H = 0$$
$$P_{EF} - P_{ED} - P_{EH} \cos 41 \cdot 18° = 0$$
$$200 - P_{ED} - (-53 \cdot 16) \cos 41 \cdot 18° = 0$$
$$\therefore \qquad\qquad P_{ED} = 240 \text{ kN} \quad (tensile)$$

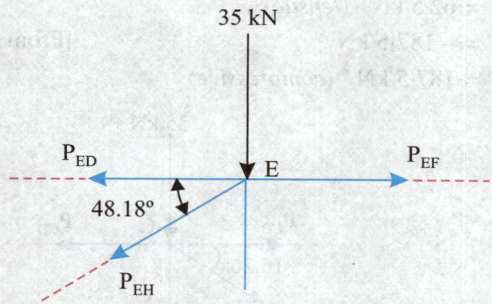

Fig. 17.151

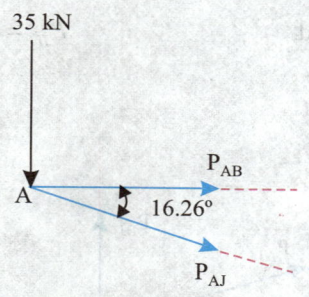

Fig. 17.152

Joint A :

Refer to Fig. 17·152.

$$\Sigma V = 0$$
$$- 35 - P_{AJ} \sin 16 \cdot 26° = 0$$
$$\therefore \qquad\qquad P_{AJ} = -125 \text{ kN}$$

i.e. $\qquad\qquad P_{AJ} (= P_{JA}) = 125 \text{ kN} \quad (compresive)$

$$\Sigma H = 0$$
$$P_{AB} + P_{AJ} \cos 16 \cdot 26° = 0$$
$$P_{AB} + (-125) \cos 16 \cdot 26° = 0$$
$$\therefore \qquad\qquad P_{AB} = 120 \text{ kN} \quad (tensile)$$

Joint B :

Refer to Fig. 17·153.

$$\Sigma H = 0$$
$$- P_{AB} + P_{BC} = 0$$
$$\therefore \qquad\qquad P_{BC} = P_{AB} = 120 \text{ kN} \quad (tensile)$$
$$\Sigma V = 0$$
$$- 35 - P_{BJ} = 0$$

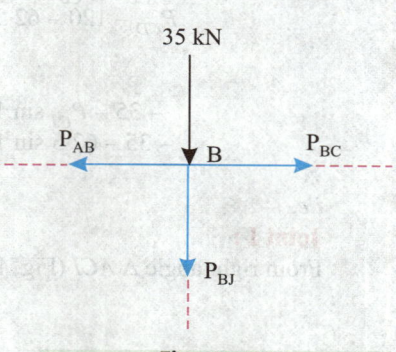

Fig. 17.153

∴ $\qquad\qquad P_{BJ} = -35 \text{ kN}$

i.e. $\qquad\qquad \boldsymbol{P_{BJ}} (= P_{JB}) = 35 \text{ kN}$ (*compressive*)

Joint J :

Refer to Fig. 17·154.

$$\Sigma H = 0$$

$$P_{JC} \cos 16·26° + P_{JI} \cos 16·26° - P_{AJ} \cos 16·26° = 0$$

or, $\qquad\qquad P_{JC} + P_{JI} - P_{AJ} = 0$

or, $\qquad\qquad P_{JC} + P_{JI} - (-125) = 0$

or, $\qquad\qquad P_{JC} + P_{JI} = -125 \qquad\qquad\qquad$ ...(*i*)

$$\Sigma V = 0$$

$$P_{JC} \sin 16·26° - P_{JI} \sin 16·26° + P_{AJ} \sin 16·26° + P_{BJ} = 0$$

$$P_{JC} - P_{JI} + (-125) = \frac{-(-35)}{\sin 16·26°} = 125$$

or, $\qquad\qquad P_{JC} - P_{JI} = 250 \qquad\qquad\qquad$ ...(*ii*)

Adding (*i*) and (*ii*), we get

$$2 P_{JC} = 125$$

or, $\qquad\qquad \boldsymbol{P_{JC}} = 62·5 \text{ kN}$ (*tensile*)

and, $\qquad\qquad P_{JI} = 62·5 - 250 = -187·5 \text{ kN} \qquad$ [From (*ii*)]

i.e. $\qquad\qquad \boldsymbol{P_{JI}} (= P_{IJ}) = 187·5 \text{ kN}$ (*compressive*)

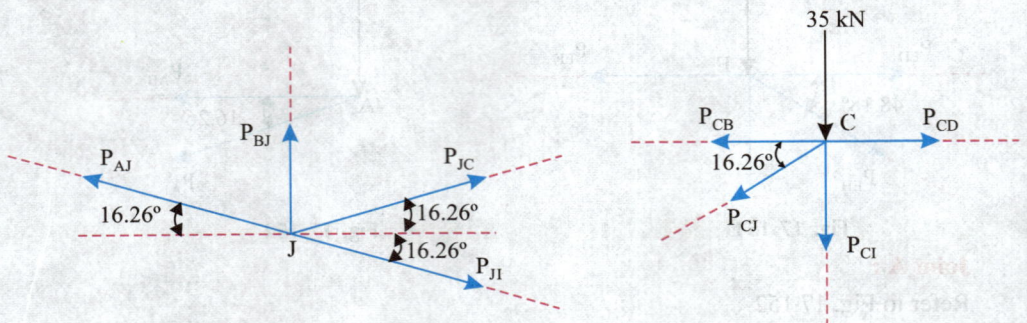

Fig. 17.154	Fig. 17.155

Joint C :

Refer to Fig. 17·155.

$$\Sigma H = 0$$

$$P_{CD} - P_{CB} - P_{CJ} \cos 16·26° = 0$$

$$P_{CD} - 120 - 62·5 \cos 16·26° = 0$$

∴ $\qquad\qquad \boldsymbol{P_{CD}} = 180 \text{ kN}$ (*tensile*)

$$\Sigma V = 0$$

$$-35 - P_{CJ} \sin 16·26° - P_{CI} = 0$$

$$-35 - 62·5 \sin 16·26° - P_{CI} = 0$$

∴ $\qquad\qquad P_{CI} = -52·5 \text{ kN}$

i.e. $\qquad\qquad \boldsymbol{P_{CI}} (= P_{IC}) = 52·5 \text{ kN}$ (*compressive*)

Joint I :

From right angle △ *ACI* (Fig. 17·156)

$$\tan 16·26° = \frac{CI}{8}$$

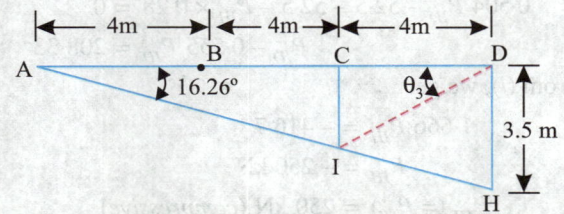

Fig. 17.156

∴ $\qquad\qquad CI = 2\cdot333$ m

$$\tan \theta_3 = \frac{CI}{4} = \frac{2\cdot333}{4}$$

∴ $\qquad\qquad \theta_3 = 30\cdot25°$

Refer to Fig. 17·157.

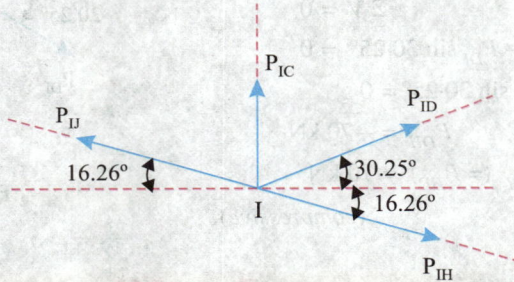

Fig. 17.157

Applying conditions of equilibrium, we get

$$\Sigma H = 0$$

$$P_{ID} \cos 30\cdot25° - P_{IJ} \cos 16\cdot26° + P_{IH} \cos 16\cdot26° = 0$$

$$P_{ID} \cos 30\cdot25° - (- 187\cdot5) \cos 16\cdot26° + P_{IH} \cos 16\cdot26° = 0$$

$$P_{ID} + 187\cdot5 \times \frac{\cos 16\cdot26°}{\cos 30\cdot25°} + 1\cdot111\, P_{IH} = 0$$

or $\qquad\qquad\qquad P_{ID} + 1\cdot111\, P_{IH} = - 208\cdot37$ $\qquad\qquad$ ...(i)

$$\Sigma V = 0$$

$$P_{ID} \sin 30\cdot25° + P_{IC} + P_{IJ} \sin 16\cdot26° - P_{IH} \sin 16\cdot26° = 0$$

$$P_{ID} \sin 30\cdot25° - 52\cdot5 - 18\cdot75 \sin 16\cdot26° - P_{IH} \sin 16\cdot26° = 0$$

Bridge.

$$0.504\, P_{ID} - 52.5 - 52.5 - P_{IH} \times 0.28 = 0$$

$$P_{ID} - 0.555\, P_{IH} = 208.33 \qquad \qquad ...(ii)$$

Subtracting (ii) from (i), we get

$$1.666\, P_{IH} = -416.7$$

$$\therefore \qquad P_{IH} = -250 \text{ kN}$$

i.e. $\qquad \boldsymbol{P_{IH}} (= P_{HI}) = 250 \text{ kN } (\textit{compressive})$

$$P_{ID} - 0.555 \times (-250) = 208.33$$

$$\therefore \qquad \boldsymbol{P_{ID}} = 69.58 \text{ kN}$$

(tensile) $\qquad\qquad$ [From (ii)]

Joint D :

Refer to Fig. 17.158.

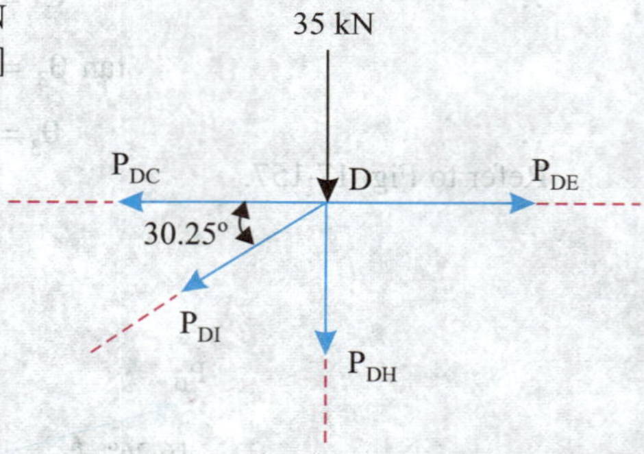

$$\Sigma V = 0$$

$$-35 - P_{DH} - P_{DI} \sin 30.25° = 0$$

$$-35 - P_{DH} - 69.58 \sin 30.25° = 0$$

$$\therefore \qquad P_{DH} = -70 \text{ kN}$$

i.e. $\qquad \boldsymbol{P_{DH}} (= P_{HD}) = 70 \text{ kN}$

(compressive)

Fig. 17.158

Tabular result :

Member	Force (kN)
GF	− 218·30
EF	200·00
GH	− 218·30
GE	00·00
EH	− 53· 16
ED	240
AJ	− 125
AB	120
BC	120
BJ	− 35
CJ	62·50
IJ	− 187·50
CD	180·00
CI	− 52·50
ID	69·56
HI	− 250
HD	− 70

TABLE 17·30

Tensile +ve; Compressive −ve

Example 17·34. *Determine the forces in the members of the truss as shown in Fig. 17·159. Tabulate the result.*

(SRTMU Dec., 2002)

Solution. Refer Fig. 17.159 Geometrical constants of the given truss are as follows :

Angle at A, θ : $\tan \theta = \dfrac{10}{15}$ $\therefore \quad \theta = 63 \cdot 43°$

$$\angle BAF = 90° - \theta = 90° - 63 \cdot 43° = 26 \cdot 57°$$

$$\angle CFG = 63 \cdot 43°; \quad \angle \alpha : \tan \alpha = \dfrac{2}{5} \qquad \therefore \quad \alpha = 21 \cdot 8°$$

$$\angle \beta : \tan \beta = \dfrac{3}{2 \cdot 5} \qquad\qquad \therefore \quad \beta = 50 \cdot 19°$$

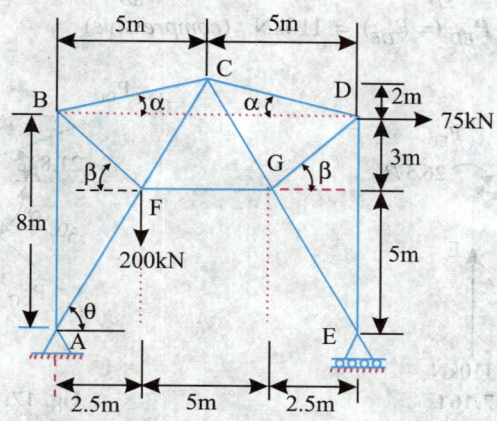

Fig. 17.159

Calculation of reactions :

$$\Sigma M_A = 0$$
$$R_E \times 10 = 200 \times 2 \cdot 5 + 75 \times 8 \qquad \therefore \quad R_E = 110 \text{ kN } (\uparrow)$$
$$\Sigma V = 0$$
$$R_{AV} + R_E = 200 \qquad\qquad \therefore \quad R_{AV} = 200 - 110 = 90 \text{ kN } (\uparrow)$$
$$\Sigma H = 0$$
$$R_{AH} = 75 \text{ kN } (\leftarrow)$$

Analysis of truss by method of joints :

Joint A :

Refer to Fig. 17·160.

Applying conditions of equilibrium, we get

$$\Sigma H = 0$$
$$- 75 + P_{AF} \sin 26 \cdot 57° = 0$$
$$\therefore \qquad P_{AF} = 167 \cdot 67 \text{ kN } \quad (tensile)$$
$$\Sigma V = 0$$
$$P_{AB} + 90 + P_{AF} \cos 26 \cdot 57° = 0$$
$$P_{AB} + 90 + 167 \cdot 67 \cos 26 \cdot 57° = 0$$
$$\therefore \qquad P_{AB} = - 239 \cdot 96 \text{ kN}$$

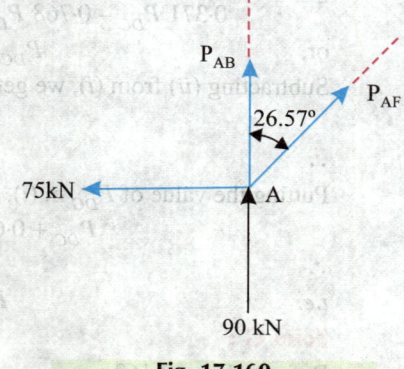

Fig. 17.160

–ve sign indicates that our original assumption of force P_{AB} as tensile as shown by the arrow in Fig.17.160 is wrong and it should be otherwise,

i.e. $\qquad P_{AB} (= P_{BA}) = 239 \cdot 96 \text{ kN } (compressive)$

Joint E :

Refer to Fig. 17·161.

$$\Sigma H = 0$$

$$P_{EG} \sin 26·57° = 0 \qquad \therefore \ P_{EG} = 0$$

$$\Sigma V = 0$$

$$P_{EG} \cos 26·57° + P_{ED} + 110 = 0$$

$$0 + P_{ED} + 110 = 0 \qquad \therefore \ P_{ED} = -110 \text{ kN}$$

i.e. $\qquad P_{ED} \ (= P_{DE}) = 110 \text{kN} \quad (compressive)$

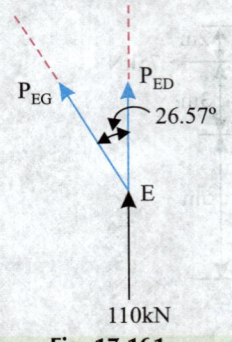

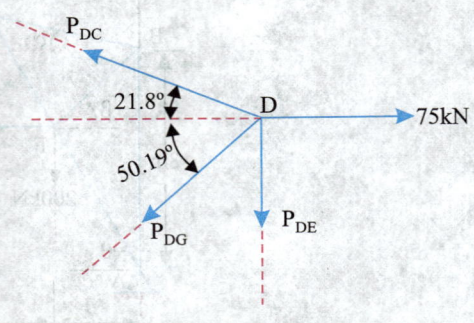

Fig. 17.161   Fig. 17.162

Joint D :

Refer to Fig. 17·162.

$$\Sigma H = 0$$

$$- P_{DC} \cos 21·8° - P_{DG} \cos 50·19° + 75 = 0$$

$$- 0·928 \, P_{DC} - 0·64 \, P_{DG} = -75$$

or, $\qquad P_{DC} + 0·69 \, P_{DG} = 80·82 \qquad\qquad\qquad ...(i)$

$$\Sigma V = 0$$

$$P_{DC} \sin 21·8° - P_{DG} \sin 50·19° - P_{DE} = 0$$

$$0·371 \, P_{DC} - 0·768 \, P_{DG} - (-110) = 0$$

or, $\qquad P_{DC} - 2·07 \, P_{DG} = -296·5 \qquad\qquad\qquad ...(ii)$

Subtracting (ii) from (i), we get

$$2·76 \, P_{DG} = 377·32$$

$\therefore \qquad\qquad\qquad P_{DG} = 136·71 \text{ kN} \quad (tensile)$

Putting the value of P_{DG} in (i), we get

$$P_{DC} + 0·69 \times 136·71 = 80·82$$

$\therefore \qquad\qquad\qquad P_{DC} = -13·5 \text{ kN}$

i.e. $\qquad\qquad P_{DC} \ (= P_{CD}) = 13·5 \text{ kN} \quad (compressive)$

Joint G :

Refer to Fig. 17·163.

$$\Sigma V = 0$$

$$P_{GD} \sin 50·19° + P_{GC} \sin 63·43° = 0$$

$$0·768 \, P_{GD} + 0·894 \, P_{GC} = 0$$

$$0·768 \times 136·71 + 0·894 \, P_{GC} = 0$$

∴ $$P_{GC} = -117{\cdot}44 \text{ kN}$$

i.e. $$P_{GC} (= P_{CG}) = 117{\cdot}44 \text{ kN } (compressive)$$

$$\Sigma H = 0$$

$$P_{GD} \cos 50{\cdot}19° - P_{GC} \cos 63{\cdot}43° - P_{GF} = 0$$

$$136{\cdot}71 \times 0{\cdot}64 - (-117{\cdot}44) \times 0{\cdot}4473 - P_{GF} = 0$$

$$P_{GF} = 140 \text{ kN } (tensile)$$

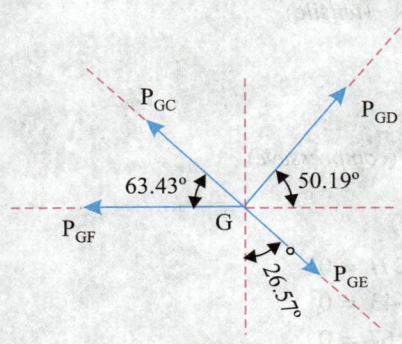

Fig. 17.163

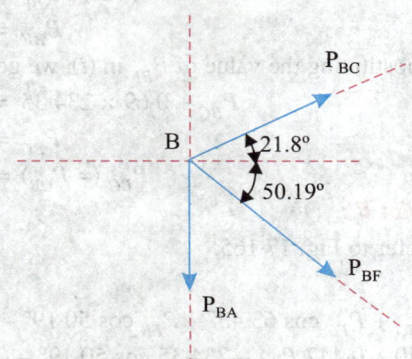

Fig. 17.164

Joint B :

Refer to Fig. 17·164.

$$\Sigma H = 0$$

$$P_{BC} \cos 21{\cdot}8° + P_{BF} \cos 50{\cdot}19° = 0$$

$$P_{BC} + P_{BF} \times \frac{\cos 50{\cdot}19°}{\cos 21{\cdot}8°} = 0$$

USA-Canada bridge over Niagara river.

$$P_{BC} + 0.69 \, P_{BF} = 0 \qquad \ldots(i)$$
$$\Sigma V = 0$$
$$P_{BC} \sin 21.8° - P_{BF} \sin 50.19° - P_{BA} = 0$$
$$0.371 \, P_{BC} - 0.768 \, P_{BF} - (-239.96) = 0$$
$$P_{BC} - 2.07 \, P_{BF} = -646.8 \qquad \ldots(ii)$$

Subtracting (ii) from (i), we get

$$2.76 \, P_{BF} = 646.8$$
∴
$$P_{BF} = 234.35 \text{ kN} \quad (tensile)$$

Substituting the value of P_{BF} in (i), we get

$$P_{BC} + 0.69 \times 234.35 = 0$$
∴
$$P_{BC} = -161.7 \text{ kN}$$
i.e.
$$\boldsymbol{P_{BC}} (= P_{CB}) = 161.7 \text{ kN } (compresseve)$$

Joint F :

Refer to Fig. 17·165.

$$\Sigma H = 0$$
$$P_{FG} + P_{FC} \cos 63.43° - P_{FB} \cos 50.19° - P_{FA} \cos 63.43° = 0$$
$$140 + 0.447 \, P_{FC} - 234.35 \cos 50.19° - 0.447 \times 167.67 = 0$$
$$140 + 0.447 \, P_{FC} - 150.04 - 74.95 = 0$$
∴
$$\boldsymbol{P_{FC}} = 190.13 \text{ kN } (tensile)$$

Tabulated result :

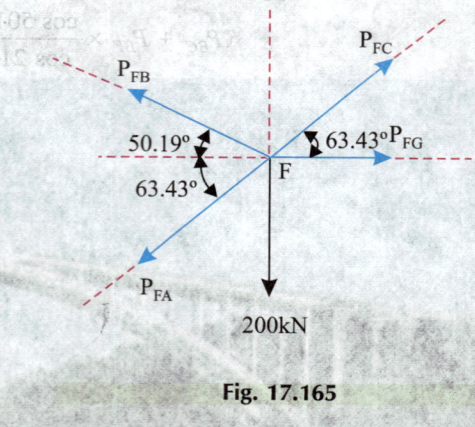

Fig. 17.165

TABLE 17·31	
Tensile + ve; Compressive –ve	
Member	**Force (kN)**
AB	– 239·96
AF	167·67
EG	zero
ED	– 110
GD	136·71
GF	140
CG	– 117·44
CD	– 13·5
CF	190·13
BC	– 161·7
BF	234·35

Method of Sections

Example 17·35. *Determine the forces in the members GC, BC, CD and FE of a truss shown in Fig. 17·166 by the method of sections and tabulate the result.*

Solution. Refer to Fig. 17·166.

From the right angled triangle ABG,

$$\cos 30° = \frac{AB}{AG}$$

i.e.
$$AB = AG \cos 30° = 2 \cos 30° = 1.732 \text{ m}$$

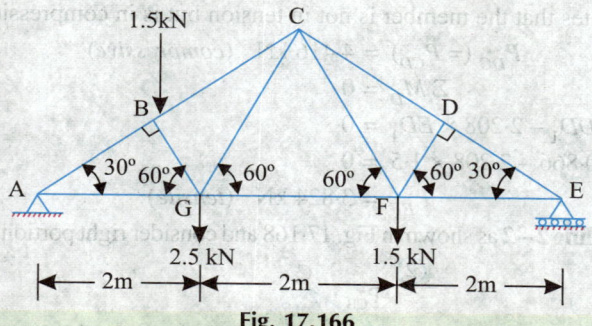

Fig. 17.166

Similarly, $\qquad ED = 1.732$ m

Draw $\perp$ from 'B' on AG at B_1. Then, from right angled triangle ABB_1,

$$\cos 30° = \frac{AB_1}{AB}$$

$\therefore \qquad AB_1 = AB \cos 30° = 1.732 \times 0.866 = 1.5$ m

and, $\qquad \sin 30° = \frac{BB_1}{AB}$

$\therefore \qquad BB_1 = AB \sin 30° = 1.732 \times 0.5 = 0.866$ m

Similarly, $\qquad DD_1 = 0.866$ mm, and, $ED_1 = 1.5$ m,

Reaction at support E, R_E :

$$\Sigma M_A = 0$$

$$R_E \times 6 = 1.5 \times 1.5 + 2.5 \times 2 + 1.5 \times 4 = 0$$

$\therefore \qquad R_E = 2.208$ kN ($\uparrow$)

Analysis of forces by method of sections :

Draw a **section line 1 – 1** as shown in Fig. 17·167 and consider right portion of it under equilibrium.

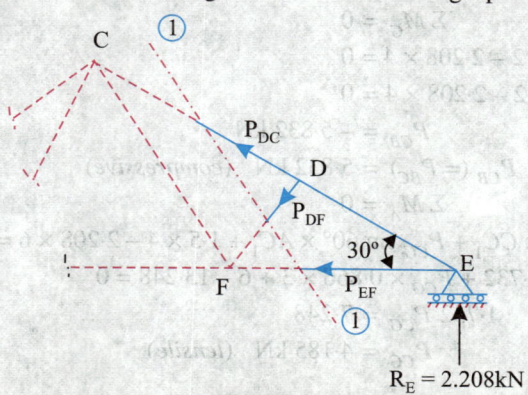

Fig. 17.167

From geometry, $\qquad FD = \sqrt{2^2 - 1.732^2} = 1$ m

Similarly, $\qquad GB = 1$ m

Applying the conditions of equilibrium, for the portion under consideration, we have

$$\Sigma M_F = 0$$

$$- P_{DC} \times 1 - 2.208 \times 2 = 0$$

or $\qquad P_{DC} = -4.416$ kN

−ve sign indicates that the member is not in tension but is in compression.

i.e. $\qquad P_{DC} (= P_{CD}) = 4\cdot416$ kN (*compressive*)

Again, $\qquad \Sigma M_D = 0$

$$P_{EF} \times DD_1 - 2\cdot208 \times ED_1 = 0$$
$$P_{EF} \times 0\cdot866 - 2\cdot208 \times 1\cdot5 = 0$$

∴ $\qquad P_{EF} = 3\cdot824$ kN (*tensile*)

Draw a **section line 2 − 2** as shown in Fig. 17·168 and consider right portion of it under equilibrium.

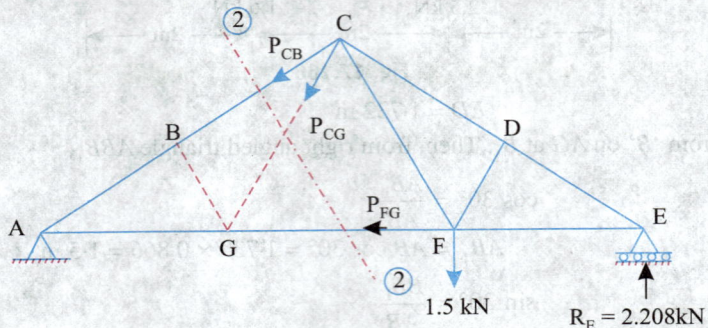

Fig. 17.168

Draw a perpendicular from C to C_1 on GF then,

$$\frac{CC_1}{AC_1} = \tan 30°,$$

or $\qquad \dfrac{CC_1}{3} = \tan 30°$

∴ $\qquad CC_1 = 1\cdot732$ m

and $\qquad C_1F = 1$ m

$$\Sigma M_G = 0$$

$$P_{CB} \times BG + 1\cdot5 \times 2 - 2\cdot208 \times 4 = 0$$
$$- P_{CB} \times 1 + 1\cdot5 \times 2 - 2\cdot208 \times 4 = 0$$

∴ $\qquad P_{CB} = -5\cdot832$ kN

i.e. $\qquad P_{CB} (= P_{BC}) = 5\cdot832$ kN (*compressive*)

$$\Sigma M_A = 0$$

$$- P_{CG} \cos 60° \times CC_1 + P_{CG} \sin 60° \times AC_1 + 1\cdot5 \times 4 - 2\cdot208 \times 6 = 0$$
$$- P_{CG} \times 0\cdot5 \times 1\cdot732 + P_{CG} \times 0\cdot866 \times 3 + 6 - 13\cdot248 = 0$$
$$1\cdot732 \, P_{CG} = 7\cdot248$$

∴ $\qquad P_{CG} = 4\cdot185$ kN (*tensile*)

Tabulated result:

TABLE 17·32

Tensile +ve; Compressive −ve

Member	Force (kN)
BC	− 5·832
CG	4·185
CD	− 4·416
FE	3·824

Example 17·36. *Determine the magnitude and nature of the forces in the members BD, CD and ED of the truss shown in Fig.17·169.*

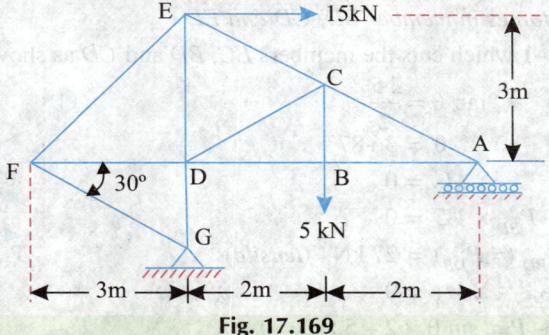

Fig. 17.169

Solution. Consider the entire truss in equilibrium. Its F.B.D. is shown in Fig. 17·170.

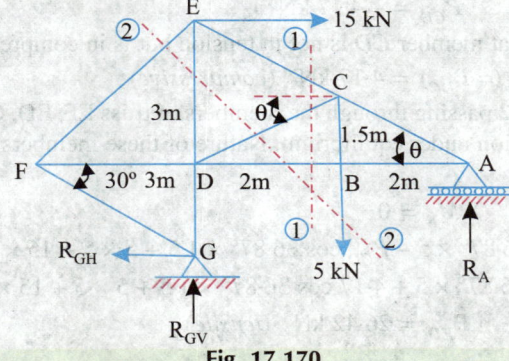

Fig. 17.170

From geometry, $\dfrac{DG}{FD} = \tan 30°$ or $DG = 3 \tan 30° = 1·732$ m

Applying conditions of equilibrium, we get

$$\Sigma M_G = 0$$

Framed structure over railway station.

$$5 \times 2 + 15 \times 4.732 - R_A \times 4 = 0$$

$$\therefore \qquad R_A = 20.25 \text{ kN}$$

Determination of forces in member BD, CD and ED :

● Draw **section 1–1** which cuts the members *EC*, *BD* and *CD* as shown in Fig. 17·171.

From geometry, $\qquad \tan \theta = \dfrac{1.5}{2}$

or, $\qquad\qquad\qquad \theta = 36.87°$

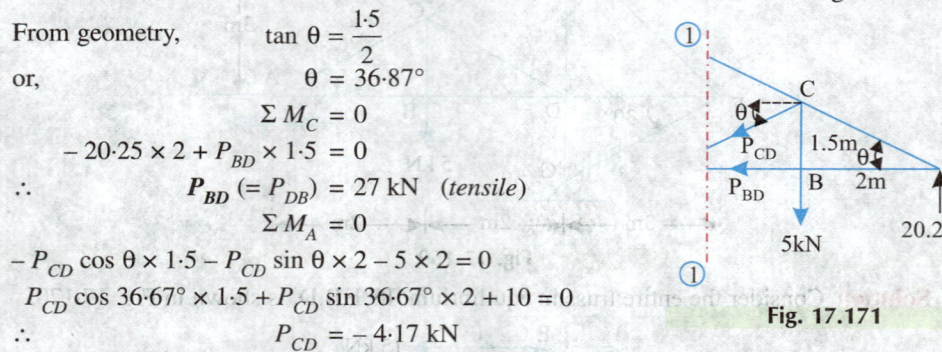

Fig. 17.171

$$\Sigma M_C = 0$$

$$-20.25 \times 2 + P_{BD} \times 1.5 = 0$$

$$\therefore \qquad P_{BD} (= P_{DB}) = 27 \text{ kN} \quad (tensile)$$

$$\Sigma M_A = 0$$

$$-P_{CD} \cos \theta \times 1.5 - P_{CD} \sin \theta \times 2 - 5 \times 2 = 0$$

$$P_{CD} \cos 36.67° \times 1.5 + P_{CD} \sin 36.67° \times 2 + 10 = 0$$

$$\therefore \qquad\qquad P_{CD} = -4.17 \text{ kN}$$

–ve sign indicates that member *CD* is not in tension but is in compression.

i.e. $\qquad\qquad P_{CD} (= P_{DC}) = 4.17 \text{ kN} \quad (compressive)$

● Draw **section 2 – 2** passing through the members of truss EF, ED, CD and BD and consider the right portion of the section under equilibrium. Nature of these members are assumed (*tensile*) as shown in the Fig. 17·172.

$$\Sigma M_F = 0$$

$$P_{ED} \times 3 + P_{CD} \sin 36.87° \times 5 - P_{CD} \cos 36.87° \times 1.5 + 5 \times 5 + 15 \times 3 - 20.25 \times 7 = 0$$

$$P_{ED} \times 3 - 4.17 \sin 36.87° \times 5 + 4.17 \cos 36.87° \times 1.5 + 5 \times 5 + 15 \times 3 - 20.25 \times 7 = 0$$

$$\therefore \qquad\qquad P_{ED} = 26.42 \text{ kN} \quad (tensile)$$

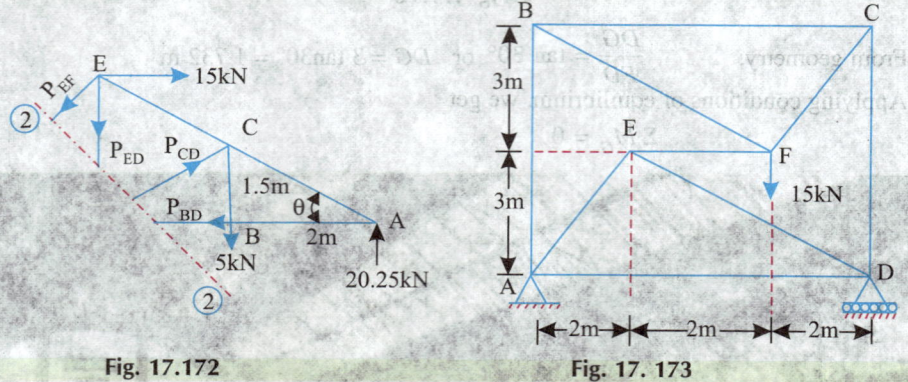

Fig. 17.172   **Fig. 17. 173**

Example 17·37. *Find the forces in the members AB, AD, EF and CD of truss loaded as shown in the Fig. 17·173.*

(P.U.)

Solution. This problem *cannot be solved by the method of* joints because at each joint there are *three unknown members.*

Hence it has to be analysed by using method of sections or combined method of joints and sections wherever it will be applicable.

Consider the entire truss is under equilibrium and assume R_A and R_D are the vertical upward reactions acting normal to the frame of support. As there is no horizontal or inclined load on truss, hence horizontal component at hinge *A* is zero.

Applying conditions of equilibrium, we get :

$$\Sigma M_A = 0$$
$$R_D \times 6 = 15 \times 4 \quad \therefore \ R_D = 10 \text{ kN } (\uparrow)$$
$$\Sigma V = 0$$
$$R_A + R_D = 15$$
$$\therefore \qquad R_A = 15 - R_D$$
$$= 15 - 10 = 5 \text{ kN } (\uparrow)$$

● Draw a **section 1 – 1** through the truss cutting the members AB, EF and CD (Fig. 17·174). Considering the equilibrium of right portion of the section with assumed direction of forces in these members as shown in Fig. 17·174., we have :

$$\Sigma M_A = 0$$
$$- P_{FE} \times 3 + 15 \times 4 + P_{CD} \times 6 = 0$$

or, $\qquad P_{FE} - 2 P_{CD} = 20 \qquad ...(i)$

$$\Sigma M_B = 0$$
$$P_{FE} \times 3 + 15 \times 4 + P_{CD} \times 6 = 0$$

or, $\qquad P_{FE} + 2 P_{CD} = -20 \qquad ...(ii)$

Adding eqns. (*i*) and (*ii*), we get

$$\boldsymbol{P_{FE} = 0}$$

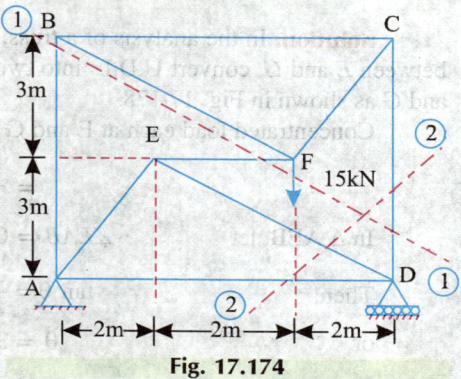

Fig. 17.174

Putting $P_{EF} = 0$ in (*i*), we get

$$\boldsymbol{P_{CD} = -10 \text{ kN}}$$

–ve sign indicates that member is not in tension but in compression.

i.e. $\qquad \boldsymbol{P_{CD} (= P_{DC}) = 10 \text{ kN}}$

$\qquad\qquad$ (*compressive*)

$$\Sigma M_D = 0$$
$$- P_{BA} \times 6 - 15 \times 2 - P_{FE} \times 3 = 0$$
$$P_{BA} + 5 + 0 = 0$$

or $\qquad P_{BA} = -5 \text{ kN}$

i.e. $\qquad \boldsymbol{P_{BA} (= P_{AB}) = 5 \text{kN}}$

$\qquad\qquad$ (*compressive*)

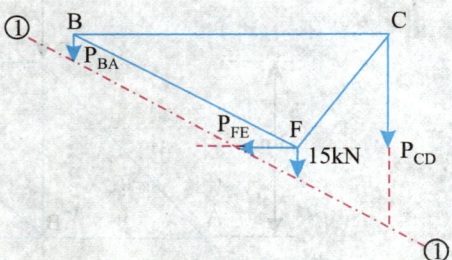

Fig. 17.175

Draw a **section 2 – 2** through the truss cutting members *AD*, *ED* and *CD* (Fig.17·176) Let us assume that lower portion of section is in equilibrium with assumed direction of forces in those members as shown in Fig. 17.176.

$$\Sigma M_E = 0$$
$$P_{DA} \times 3 - P_{DC} \times 4 - 10 \times 4 = 0$$

or $\qquad 3 P_{DA} - (-10) \times 4 - 40 = 0$

$\therefore \qquad\qquad \boldsymbol{P_{DA} = 0}$

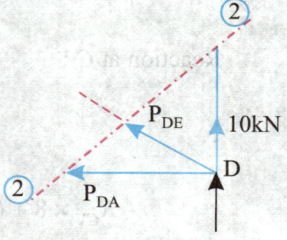

Fig. 17.176

Example 17·38. *Determine the forces in the members BC, BG, EG and GD of the truss shown in Fig. 17·177.* **(SRTMU June, 2001)**

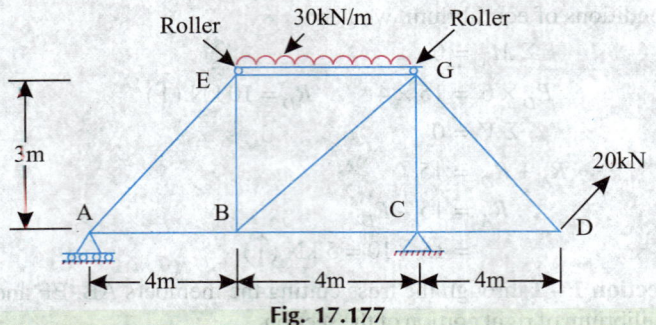

Fig. 17.177

Solution. In the analysis of a truss, it is assumed that loads must be at the joints only. Hence between E and G, convert U.D.L. into two concentrated loads which are acting through rollers at E and G as shown in Fig. 17·178.

Concentrated load each at E and G acting downward

$$= \frac{wl}{2} = \frac{30 \times 4}{2} = 60 \text{ kN}$$

In $\triangle AEB$, let $\qquad \angle EAB = \theta$

Then, $\qquad \tan \theta = \dfrac{3}{4}$

or, $\qquad \theta = 36 \cdot 87°$

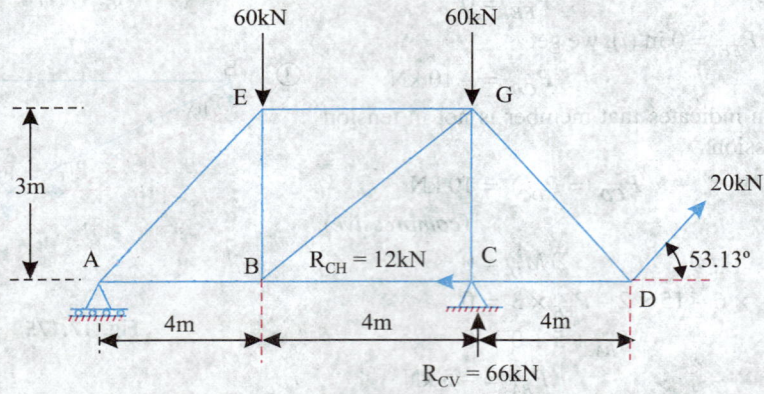

Fig. 17.178

Reaction at C :

$$\Sigma H = 0$$

∴ $\qquad R_{CH} = 20 \cos 53 \cdot 13° = 12 \text{ kN} (\leftarrow)$

$$\Sigma M_A = 0$$

$$- R_{CV} \times 8 + 60 \times 4 + 60 \times 8 - 20 \sin 53 \cdot 13° \times 12 = 0$$

∴ $\qquad R_{CV} = 66 \text{ kN} (\uparrow)$

Determination of forces in the members BC, BG, EG and GD :

Since the forces are required to be determined in the limited members of the truss, let us use the method of sections (which is simple and convenient to use).

● Draw **section 1 – 1** which cuts the members BC, GC and GD, then separate the truss into two parts. Consider the equilibruium of right hand side of section 1 – 1 (prefer the part in which number of forces are less) of the truss shown in Fig. 17·179.

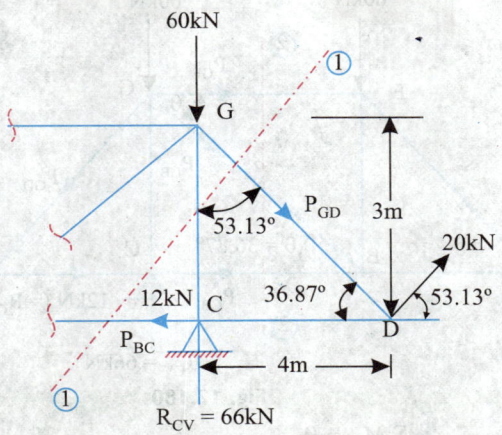

Fig. 17.179

Applying conditions of equilibrium, we get :

$$\Sigma M_C = 0$$
$$P_{GD} \sin 53 \cdot 13° \times 3 - 20 \sin 53 \cdot 13° \times 4 = 0$$

∴
$$P_{GD} = 26 \cdot 67 \text{ kN} \quad (tensile)$$
$$\Sigma M_G = 0$$
$$12 \times 3 + P_{BC} \times 3 - 20 \sin 53 \cdot 13° \times 4 - 20 \cos 53 \cdot 15° \times 3 = 0$$
$$36 + 3 P_{BC} - 64 - 36 = 0$$

∴
$$P_{BC} = 21 \cdot 33 \text{ kN} \quad (tensile)$$

● Draw **section 2 – 2** and consider the equilibrium of right hand side of section 2 – 2 part of truss, which is divided into two parts as shown in Fig. 17·180.

Offshore oilwell.

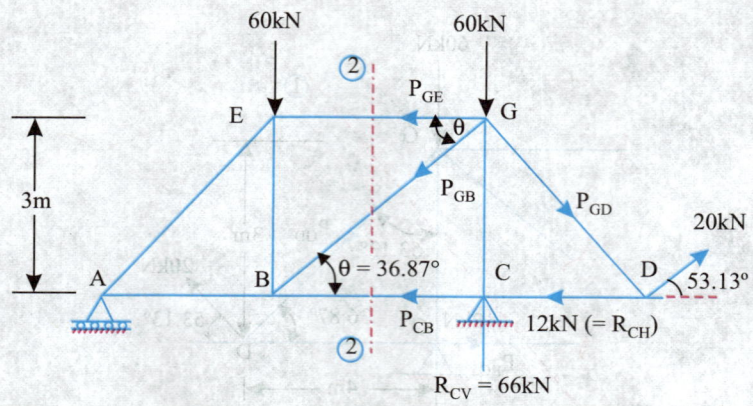

Fig. 17.180

$$\Sigma M_B = 0$$
$$60 \times 4 - P_{GE} \times 3 - 20 \sin 53 \cdot 13° \times 8 - 66 \times 4 = 0$$
$$\therefore \qquad\qquad P_{GE} = -50 \cdot 66 \text{ kN}$$

–ve sign indicates that member *EG* is not in tension but is in compression.

$$\therefore \qquad P_{GE} (= P_{EG}) = 50 \cdot 66 \text{ kN} \quad (compressive)$$
$$\Sigma M_C = 0$$
$$- P_{EG} \times 3 - P_{GB} \cos 36 \cdot 87° \times 3 - 20 \sin 53 \cdot 13° \times 4 = 0$$
$$- (-50 \cdot 66) \times 3 - P_{GB} \cos 36 \cdot 87° \times 3 - 20 \sin 53 \cdot 13° \times 4 = 0$$
$$\therefore \qquad\qquad P_{GB} (= P_{BG}) = 36 \cdot 66 \text{ kN} \quad (tensile)$$

Tabulated result:

TABLE 17·33

Tensile +ve; Compressive –ve

Member	Force (kN)
GD	26·67
BC	21·33
EG	– 50·66
BG	36·66

Example 17·39. *Find the axial force in each of the bar 1, 2 and 3 of the plane truss shown in Fig. 17·181. What will be the change in these forces if horizontal force of 20 kN is removed.*

(Roorkee University)

Solution. Draw a **section 1 – 1** cutting the members 1, 2, and 3 in which forces are to be determined. Considering the upper portion of section with their assumed nature as shown in Fig. 17·182 and applying conditions of equilibrium we have :

$$\Sigma M_F = 0$$
$$- P_1 \times a + 20 \times a - 30 \times a = 0$$
$$\therefore \qquad\qquad P_1 = -10 \text{ kN} \quad i.e. \quad 10 \text{ kN} \text{ (compressive)}$$
$$\Sigma M_C = 0$$
$$P_3 \times a + 20 \times 2a = 0$$
$$P_3 = -40 \text{ kN} \quad i.e. \quad 40 \text{ kN} \text{ (compressive)}$$

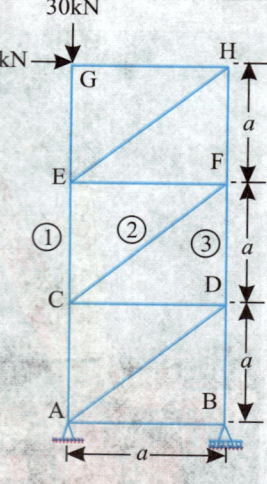

Fig. 17.181

$$\Sigma M_E = 0$$

$$P_3 \times a + P_2 \sin 45° \times a + 20 \times a = 0$$

$$- 40\,a + P_2 \sin 45° \times a + 20\,a = 0$$

$$\therefore \qquad\qquad P_2 = 28\cdot28 \text{ kN} \quad (tensile)$$

If *horizontal force of 20 kN is removed forces in members 2 and 3 will be zero.*

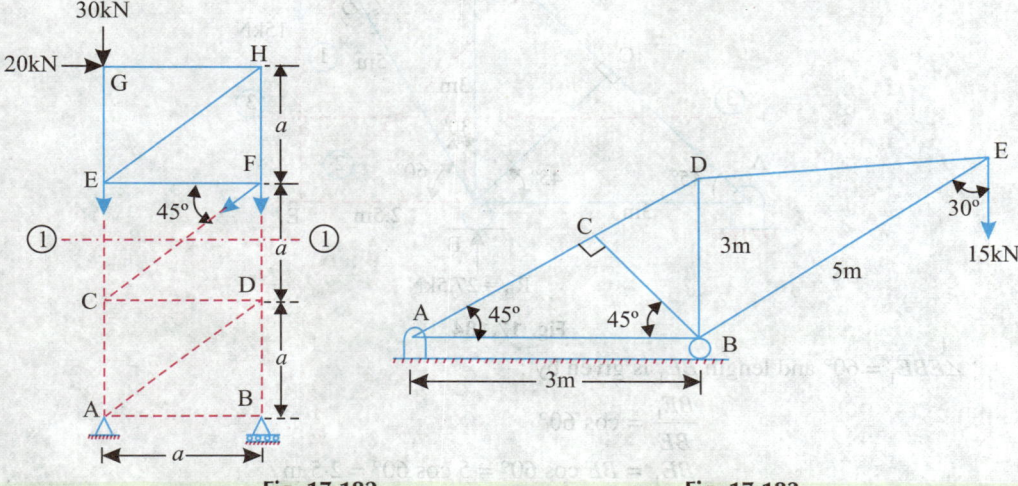

Fig. 17.182	**Fig. 17.183**

Example. 17·40. *Determine the axial forces in all the members of the truss shown in the Fig. 17.183 by using method of sections or method of moments.* **(Banglore University)**

Solution. Consider the entire truss is under equilibrium.

Let R_B be the vertical upward reaction acting normal to the plane of roller.

Geometrical calculations : Refer Fig. 17·184.

Howrah bridge, Kolkata.

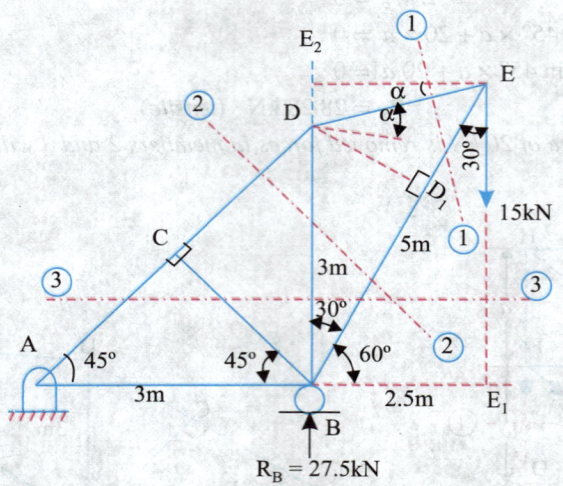

Fig. 17.184

$\angle EBE_1 = 60°$ and length BE_1 is given by :

$$\frac{BE_1}{BE} = \cos 60°$$

$\therefore$ $\quad\quad\quad BE_1 = BE \cos 60° = 5 \cos 60° = 2\cdot5$ m

Calculation for finding R_B :

$$\Sigma M_A = 0$$
$$R_B \times 3 = 15 \times 5\cdot5$$

$\therefore$ $\quad\quad\quad R_B = 27\cdot5$ kN ($\uparrow$)

• Draw a **section line 1 – 1,** cutting member ED and EB in which forces are to be determined. Considering only right portion of section $1 - 1$ under equilibrium with the assumed nature of forces (*i.e.* tensile) as shown in Fig. 17·185.

Refer to Fig. 17·184.

$$\angle DEE_2 = \alpha$$

and, $\quad\quad\quad \angle BEE_2 = 60°$

i.e. $\quad\quad\quad \dfrac{BE_2}{BE} = \sin 60°$

i.e. $\quad\quad\quad BE_2 = BE \times \sin 60°$

or $\quad\quad\quad BE_2 = 5 \times \sin 60° = 4\cdot33$ m

Hence, $\quad\quad\quad DE_2 = BE_2 - BD$

$$= 4\cdot33 - 3 = 1\cdot33 \text{ m}$$

$\therefore$ $\quad\quad\quad \tan \alpha = \dfrac{DE_2}{EE_2} = \dfrac{1\cdot33}{2\cdot5}$

$\therefore$ $\quad\quad\quad \alpha = 28°$

$$\Sigma M_B = 0$$
$$15 \times 2\cdot5 - P_{ED} \cos 28° \times 3 = 0$$

$\therefore$ $\quad\quad\quad P_{ED} = 14\cdot16$ kN (*tensile*)

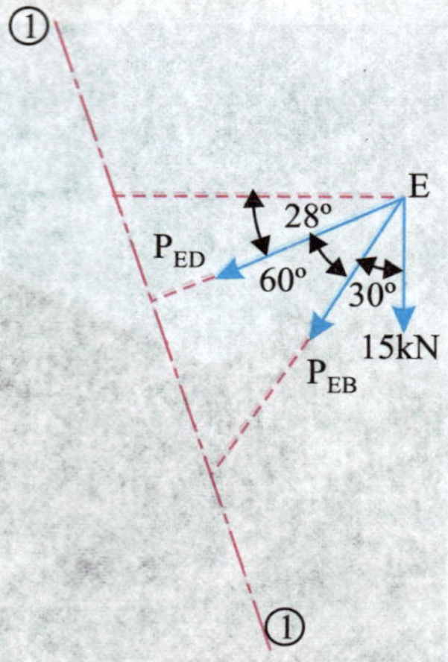

Again using geometry, draw a perpendicular from D on to the BE at D_1.

Now, $\quad\quad\quad \dfrac{DD_1}{DB} = \sin 30°$

Fig. 17.185

$$\therefore \qquad DD_1 = DB \times \sin 30° = 3 \times 0.5 = 1.5 \text{ m}$$
$$\Sigma M_D = 0$$
$$15 \times 2.5 + P_{EB} \times DD_1 = 0$$
$$37.5 + P_{EB} \times 1.5 = 0$$
$$\therefore \qquad P_{EB} = -25 \text{ kN}$$

−ve sign indicates the member is not in tension but is in compression.

i.e. $\qquad P_{EB} (= P_{BE}) = 25\text{kN} \quad$ (*compressive*)

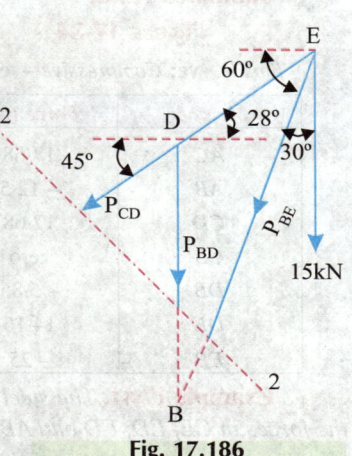

Fig. 17.186

● Draw a **section 2 – 2** which cuts the members CD, BD and BE and consider right portion of section $2 – 2$ is in equilibrium having assumed directions (tensile) as shown in the Fig. 17·186.

$$\Sigma M_B = 0$$
$$- P_{CD} \cos 45° \times 3 + 15 \times 2.5 = 0$$
$$\therefore \qquad P_{CD} = 17.68 \text{ kN} \quad (\textit{tensile})$$
$$\Sigma M_A = 0$$
$$P_{BD} \times 3 + P_{BE} \sin 60° \times 3 + 15 \times 5.5 = 0$$
$$3 \, P_{BD} + (-25) \sin 60° \times 3 + 15 \times 5.5 = 0$$
$$\therefore \qquad P_{BD} = -5.85 \text{ kN}$$

i.e $\qquad P_{BD} (= P_{DB}) = 5.85 \text{ kN} \quad$ (*compressive*)

● Draw a horizontal **section 3 – 3** which cut members AC, BC, BD and BE and consider that upper portion of section $3 – 3$ is in equilibrium having assumed directions (tensile) as shown in Fig. 17·187.

From geometry, $\qquad\qquad \angle ABC = 45°$

Then, $\qquad\qquad\qquad \cos 45° = \dfrac{BC}{AB} = \dfrac{BC}{3}$

$$\therefore \qquad\qquad BC = 3 \cos 45° = 2.121 \text{ m}$$
$$\Sigma M_B = 0$$
$$- P_{CA} \times 2.121 + 15 \times 2.5 = 0$$
$$P_{CA} = 17.68 \text{ kN} \quad (\textit{tensile})$$
$$\Sigma M_D = 0$$
$$P_{BC} \times \cos 45° \times 3 - P_{BE} \times DD_1 + 15 \times 2.5 = 0$$
$$P_{BC} \times \cos 45° \times 3 + (-25) \times 1.5 + 37.5 = 0 \qquad \therefore P_{BC} = 0$$

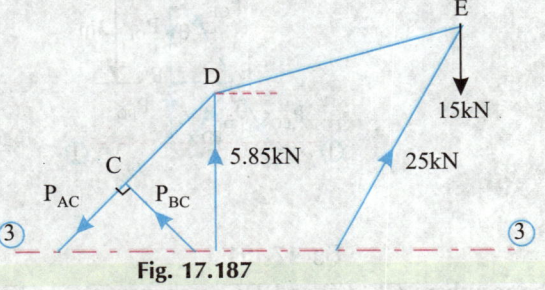

Fig. 17.187

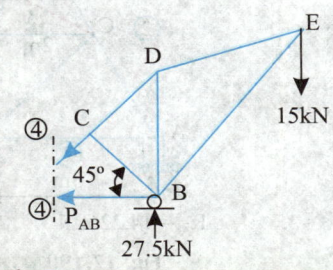

Fig. 17.188

● Draw a **section 4 – 4** in vertical which cuts members AC and AB and consider that right portion of section $4 – 4$ is in equilibrium with assumed directions as shown in Fig. 17·188.

$$\Sigma M_D = 0$$
$$P_{AB} \times 3 + 15 \times 2.5 = 0 \qquad \therefore P_{AB} = -12.5 \text{ kN}$$

i.e.

$$P_{AB} (= P_{BA}) = 12 \cdot 5 \text{ kN} \quad (compressive)$$

Tabulated result :

TABLE 17·34

Tensile +ve; Compressive −ve

Member	Force (kN)
AC	17·68
AB	− 12·5
CD	17·68
CB	zero
DB	− 5·85
DE	14·16
EB	− 25

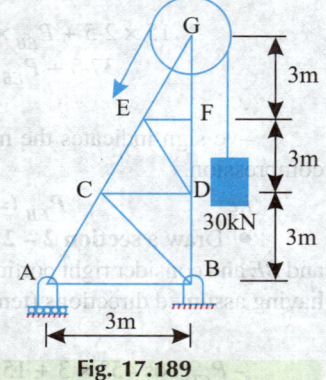

Fig. 17.189

Example 17·41. *The steel frame shown in Fig. 17·189 is hoisting a 30 kN weight. Determine the forces in CE, ED, FD and AB assuming the cable is parallel to the sides as shown in Fig. 17·189.*

(N.U.)

Solution. Assume the pulley weightless and smooth and consider the whole truss in equilibrium, the coresponding F.B.D. is drawn as shown in Fig. 17·190.

$$\tan \theta = \frac{AB}{BG} = \frac{3}{9}$$

or,

$$\theta = 18 \cdot 43°$$

$$EF = 1 \text{ m, and, } CD = 2 \text{ m}$$

Let R_A be the reactions at support A acting normal to the plane of roller.

$$\Sigma M_B = 0$$

$$R_A \times 3 - 30 \sin 18 \cdot 43° \times 9 = 0$$

∴

$$R_A = 28 \cdot 45 \text{ kN } (\uparrow)$$

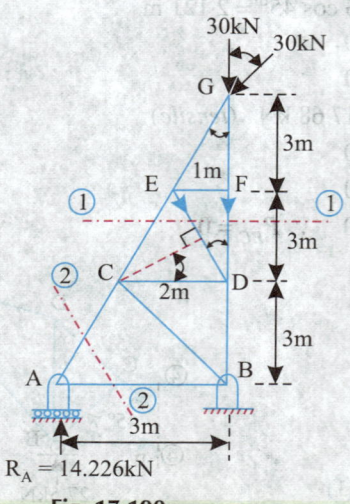

$$R_A = 14.226 \text{kN}$$

Fig. 17.190

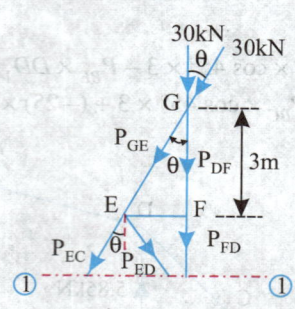

Fig. 17.191

● Draw **section 1 − 1** which cuts the members *CE, ED* and *DF* as shown in Fig. 17.191. Assuming upper portion is in equilibrium with assumed nature of forces in each member of truss, we have :

$$\Sigma M_D = 0$$

$$-P_{EC} \sin 18.43° \times 3 - P_{EC} \cos 18.43° \times 1 - 30 \sin 18.43° \times 6 = 0$$
$$P_{EC} \sin 18.43° \times 3 + P_{EC} \cos 18.43° \times 1 = -30 \sin 18.43° \times 6$$
$$1.897 \, P_{EC} = -56.9$$
$$P_{EC} = -30 \text{ kN}$$

–ve sign indicates that member EC is not in tension but is in compression,

i.e.
$$P_{EC} \, (= P_{CE}) = 30 \text{ kN} \quad (compressive)$$
$$\Sigma \, M_E = 0$$
$$P_{DF} \times 1 + 30 \times 1 = 0$$
∴
$$P_{DF} = -30 \text{ kN}$$
i.e.
$$P_{DF} \, (= P_{FD}) = 30 \text{ kN} \quad (compressive)$$
$$\Sigma \, M_C = 0$$
$$P_{ED} \times 2 \cos \theta + 30 \times 2 + P_{FD} \times 2 = 0$$
$$P_{ED} \times 2 \cos \theta + 60 - 30 \times 2 = 0$$
∴
$$P_{ED} = 0$$

● Draw **section 2 – 2** which cuts the members AC and AB as shown in Fig. 17·192. Considering the left portion of section in equilibrium, we have :
$$\Sigma \, M_C = 0$$
$$-P_{AB} \times 3 + 28.45 \times 1 = 0$$
∴
$$P_{AB} = 9.48 \text{ kN} \quad (tensile)$$

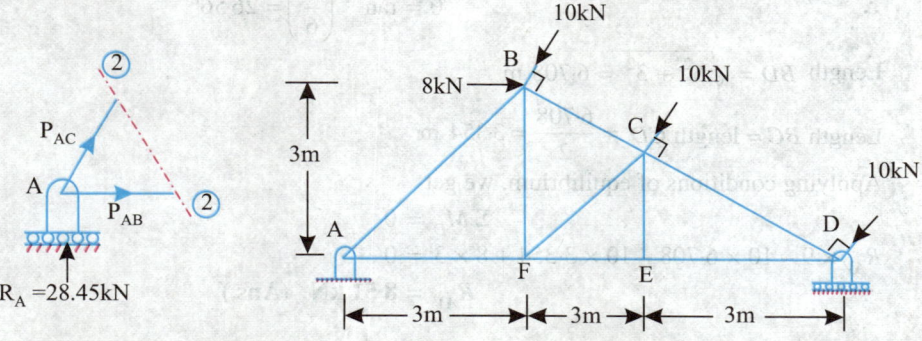

Fig. 17.192   Fig. 17.193

Bridge.

Example 17·42. *Find the support reactions and forces in the members BC, CA, EF and CE of the truss as shown in Fig. 17·193.*

Solution. Consider the entire truss is under equilibrium.

Calculations for reactions : Refer to Fig. 17·194.

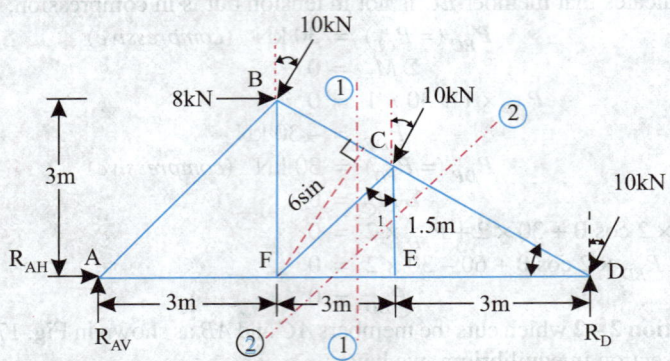

Fig. 17.194

$$\tan \theta = \frac{3}{6}$$

∴

$$\theta = \tan^{-1}\left(\frac{3}{6}\right) = 26.56°$$

Length $BD = \sqrt{6^2 + 3^2} = 6.708$ m

Length BC = length $CD = \dfrac{6.708}{2} = 3.354$ m

Applying conditions of equilibrium, we get

$$\Sigma M_D = 0$$

$$R_{AV} \times 9 - 10 \times 6.708 - 10 \times 3.354 + 8 \times 3 = 0$$

∴ $\qquad\qquad\qquad\qquad R_{AV} = \mathbf{8.51\ kN}$ **(Ans.)**

Framed structure over a railway.

$$\Sigma V = 0$$
$$R_{AV} + R_D - 3 \times 10 \cos 26 \cdot 56° = 0$$
$$8 \cdot 51 + R_D - 3 \times 10 \cos 26 \cdot 56° = 0$$
∴ $$R_D = 18 \cdot 32 \text{ kN } (\uparrow) \quad \textbf{(Ans.)}$$
$$\Sigma H = 0$$
$$R_{AH} + 8 - 3 \times 10 \sin 26 \cdot 56° = 0$$
∴ $$R_{AH} = 5 \cdot 414 \text{ kN } (\rightarrow) \quad \textbf{(Ans.)}$$

Determination of forces in the members BC, CA, EF and CE :

- Draw a **section 1 – 1** which cuts the members *BC, CF* and *EF* as shown in Fig. 17·195.

$$\Sigma M_C = 0$$
$$P_{EF} \times 1 \cdot 5 + 10 \times 3 \cdot 354 - 18 \cdot 32 \times 3 = 0$$
∴ $$P_{EF} (= P_{FE}) = 14 \cdot 28 \text{ kN} \quad (tensile)$$
$$\Sigma M_F = 0$$
$$- P_{CB} \times 6 \sin 26 \cdot 56° + 10 \cos 26 \cdot 56° \times 3 - 10 \sin 26 \cdot 56° \times 1 \cdot 5$$
$$+ 10 \cos 26 \cdot 56° \times 6 - 18 \cdot 32 \times 6 = 0$$
$$- 2 \cdot 683 \, P_{CB} + 26 \cdot 83 - 6 \cdot 707 + 53 \cdot 67 - 109 \cdot 92 = 0$$
∴ $$P_{CB} = - 13 \cdot 46 \text{ kN}$$

–ve sign indicates member BC is not in tension but is in compression.

i.e. $$P_{CB} (= P_{BC}) = 13 \cdot 46 \text{ kN} \quad (compressive)$$
$$\Sigma M_D = 0$$

From geometry, $$\tan \theta_1 = \frac{3}{1 \cdot 5} = 63 \cdot 43°$$

$$- P_{CF} \cos 63 \cdot 43° \times 3 - P_{CF} \sin 63 \cdot 43° \times 1 \cdot 5 - 10 \times 3 \cdot 354 = 0$$
$$- 1 \cdot 342 \, P_{CF} - 1 \cdot 342 \, P_{CF} - 33 \cdot 54 = 0$$
∴ $$P_{CF} = - 12 \cdot 5 \text{ kN}$$
i.e. $$P_{CF} (= P_{FC}) = 12 \cdot 5 \text{ kN} \quad (compressive)$$

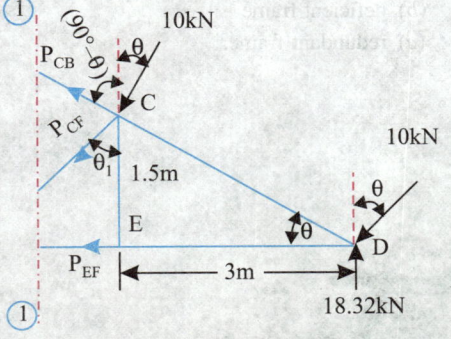

Fig. 17.195

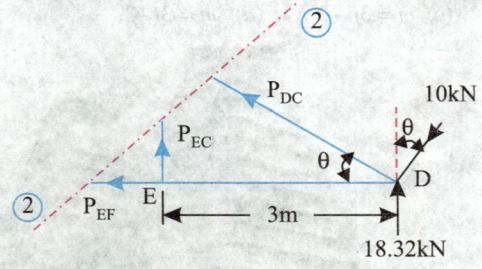

Fig. 17.196

- Draw a **section 2 – 2** which is cutting the members *EF, CE* and *CD* as shown in Fig. 17·196.

$$\Sigma M_D = 0$$
$$P_{EC} \times 3 = 0$$
∴ $$P_{EC} (= P_{CE}) = 0$$

Tabulated result :

TABLE 17·35

Tensile +ve; Compressive –ve

Member	Force (kN)
EF	14·28
BC	– 13·46
CF	– 12·5
EC	zero

HIGHLIGHTS

1. A member under tension is called *Tie* and under compression in known as *strut*.
2. A perfect frame should satisfy the relation, $m = (2j - 3)$
   where, m = Number of members in the frame, and
   j = Number of joints in the frame.
3. A frame, in which the number of members is less than $(2j - 3)$, is known as a *deficient* frame.
4. A frame, in which the number of members is more than $(2j - 3)$, is known as a *redundant* frame.
5. The forces in various members of a perfect frame may be found out by the following methods :
   (*a*) Graphical methods.
   (*b*) Method of joints.
   (*c*) Method of sections.

OBJECTIVE TYPE QUESTIONS

Choose the Correct Answer :

1. A member under tension is called
   (*a*) strut   (*b*) tie
   (*c*) strut-tie.
2. A member under compression is called
   (*a*) tie   (*b*) strut
   (*c*) strut-tie.
3. A perfect frame should satisfy the relation
   (*a*) $m = 2j - 3$ (*b*) $m = 2j - 4$
   (*c*) $m = 3j - 2$ (*d*) $m = 3j - 3$.

4. A frame in which the number of members is less than $(2j - 3)$ is known as
   (*a*) redundant frame
   (*b*) deficient frame
   (*c*) perfect frame.
5. A frame in which the number of members is more than $(2j - 3)$ is known as
   (*a*) perfect frame
   (*b*) deficient frame
   (*c*) redundant frame.

Coal scuttles

ANSWERS

1. (b)   **2.** (b)   **3.** (a)   **4.** (b)   **5.** (c)

THEORETICAL QUESTIONS

1. What is a frame ? How are frames classified ?
2. What is a perfect frame ? How does it differ from an imperfect frame ?
3. Enumerate the assumptions made while finding out the forces in a frame.
4. What is the difference between a deficient and a redundant frame ?
5. What is the difference between a cantilever and a simply supported frame ? How will you find reactions in both the cases ?

UNSOLVED PROBLEMS

1. Warren girder freely supported at ends is loaded as shown in Fig. 17·197. Find the forces in all the members.

> **Ans.** : $AE = 29$ kN comp; $BF = 17$ kN comp;
> $CG = 40$ kN comp; $DE = 15$ kN tensile;
> $DG = 20$ kN tensile; $EF = 5·5$ kN tensile;
> $FG = 5·5$ kN comp.

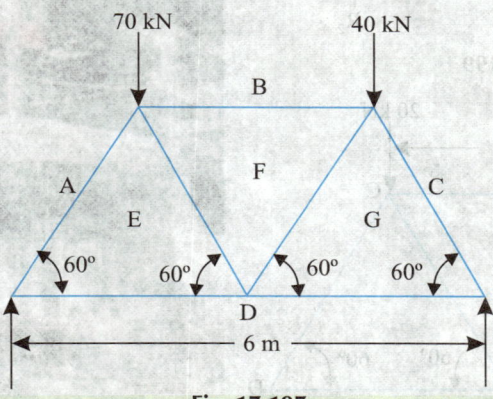

Fig. 17.197

2. A king post truss of 8 metres span is loaded as shown in Fig. 17·198. Find the forces in the members.

> **Ans.** : BH, $EK = 57$ kN comp.
> HG, $KG = 48·5$ kN tensile
> HI, $KJ = 19$ kN comp.
> CI, $DJ = 38$ kN comp.
> $IJ = 20$ kN tensile

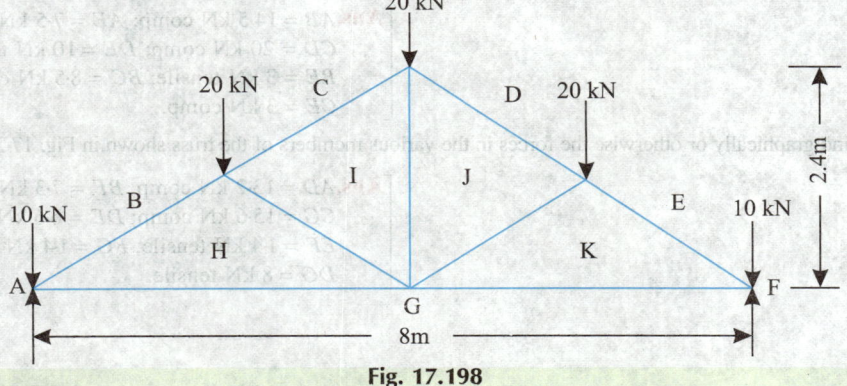

Fig. 17.198

3. A truss of 10 m span is loaded as shown in Fig. 17·199. Find the forces in the members of the truss.

Ans. $AE = 34.5$ kN comp; $ED = 17.5$ kN tensile;
$GC = 50$ kN comp; $GD = 43.5$ kN tenisle;
$FB = 35$ kN comp; $GF = 26$ kN comp;
$EF = 26$ kN tensile.

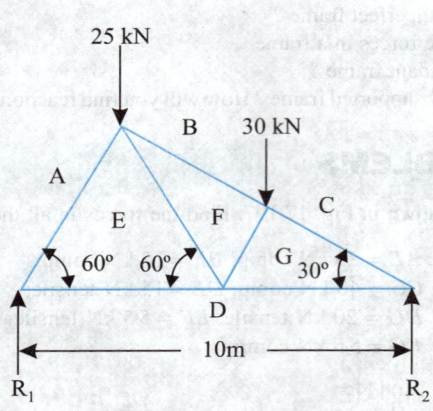

Fig. 17.199

Metallic Staircase.

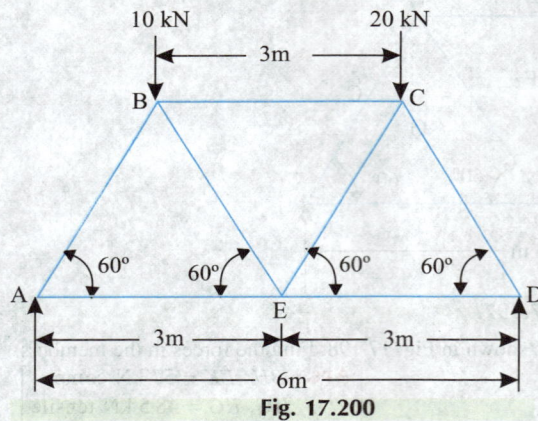

Fig. 17.200

4. Fig. 17·200 shows a Warren girder consisting of seven members each of 3 m length, freely supported at its end points. The girder is loaded at B and C as shown. Find the forces in all the members of the girder.

Ans. $AB = 14.5$ kN comp; $AE = 7.5$ kN tensile;
$CD = 20$ kN comp; $DE = 10$ kN tensile;
$BE = 3$ kN tensile; $BC = 8.5$ kN comp;
$CE = 3$ kN comp.

5. Find graphically or otherwise the forces in the various members of the truss shown in Fig. 17·201.

Ans. $AD = 13.2$ kN comp; $BF = 7.3$ kN comp;
$CG = 15.6$ kN comp; $DE = 6.6$ kN tenisle;
$EF = 1.4$ kN tensile; $FG = 1.4$ kN tensile;
$DG = 8$ kN tensile.

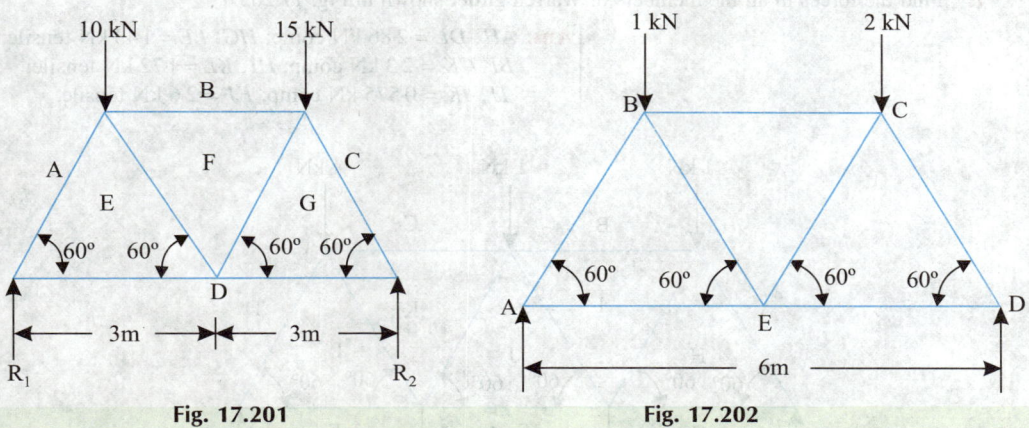

Fig. 17.201   Fig. 17.202

6. Find the forces in the members of the truss shown in Fig. 17·202

Ans. AB = 1·45 kN comp; BC = 0·85 kN comp;
CD = 2 kN comp; DE = 1 kN tenisle;
EA = 0·75 kN tensile; BE = 0·3 kN tensile;
EC = 0·3 kN comp.

Road bridge over a river.

7. Find the forces in all the members of Warren girder shown in Fig. 17·203.

Ans. AH, DL = 2·86 kN comp; HG, LE = 1·43 kN tensile; BI, CK = 2·3 kN comp; HI, KL = 1·72 kN tensile; IJ, JK = 0·575 kN comp; FJ = 2·6 kN tensile.

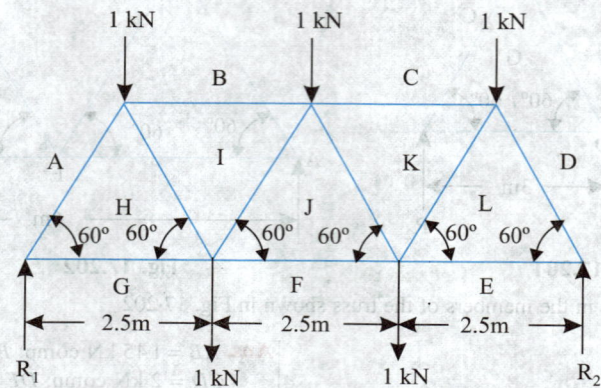

Fig. 17.203

8. Find the forces in the given truss carrying loads as shown in Fig. 17·204

Ans. AB = 10·7 kN comp; AF = 7·5 kN tensile; BF = 0 ; BC = 7·1 kN comp; BE = 3·535 kN; FE = 7·5 kN tensile; EC = 2·475 kN tensile; CD = 8·925 kN comp; ED = 4·825 kN tensile.

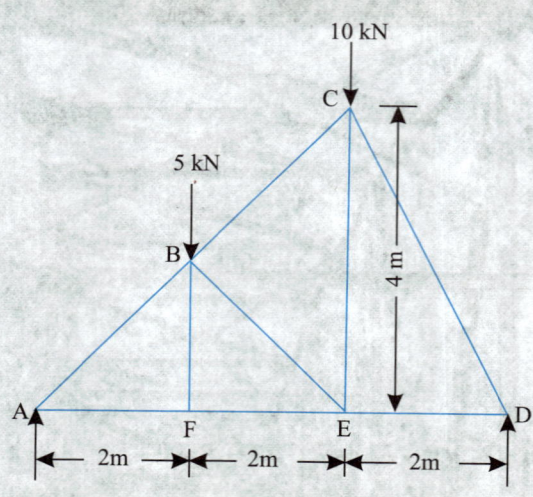

Fig. 17.204

9. A warren girder is shown in Fig. 17·205. Find graphically, or otherwise, forces in members AB, BC and FG.

Ans. AB = 3 kN comp; BC = 2·6 kN comp; FG = 2·6 kN tensile.

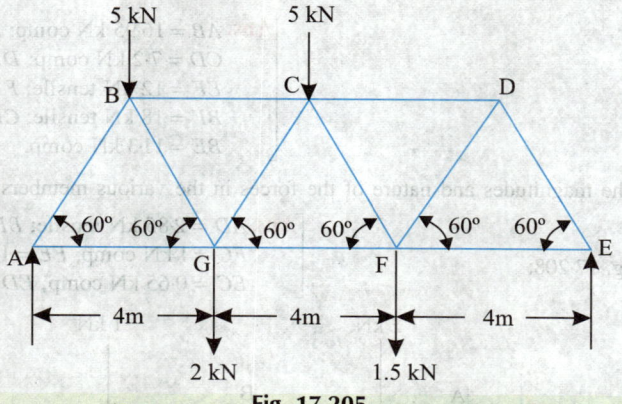

Fig. 17.205

10. Find the forces in all the members of the truss shown in Fig. 17·206.

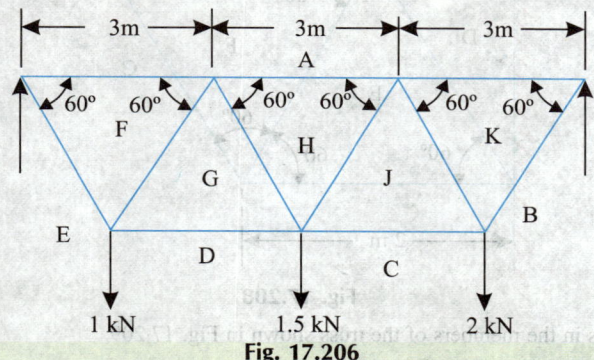

Fig. 17.206

> **Ans.** $AF = 1·25$ kN comp; $EF = 2·15$ kN tensile;
> $FG = 1$ kN comp; $DG = 1·625$ kN tensile;
> $GH = 1$ kN tensile; $AH = 2·125$ kN comp;
> $HJ = 0·7$ kN tensile; $JC = 1·75$ kN tensile;
> $JK = 0·7$ kN comp; $AK = 1·45$ kN comp;
> $KB = 2·875$ kN Tensile.

11. Find the forces in the members of the structure shown in Fig. 17·207.

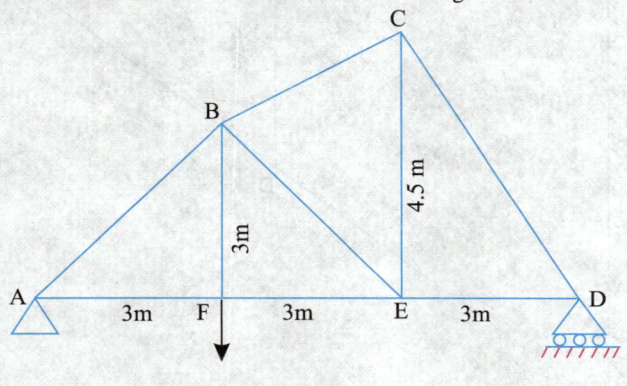

Fig. 17.207

Ans. $AB = 16.95$ kN comp; $BC = 4.45$ kN comp;
$CD = 7.2$ kN comp; $DE = 4$ kN tensile;
$EF = 12$ kN tensile; $FA = 12$ kN tensile;
$BF = 18$ kN tensile; $CE = 8$ kN tensile;
$BE = 11.3$ kN comp.

12. Determine the magnitudes and nature of the forces in the various members of the cantilever truss

Ans. $AD = 2.85$ kN tensile; $BF = 0.575$ kN tensile;
$FC = 1$ kN comp; $FE = 1.15$ kN tensile;
$EC = 0.65$ kN comp; $ED = 3.5$ kN comp.

shown in Fig. 17·208.

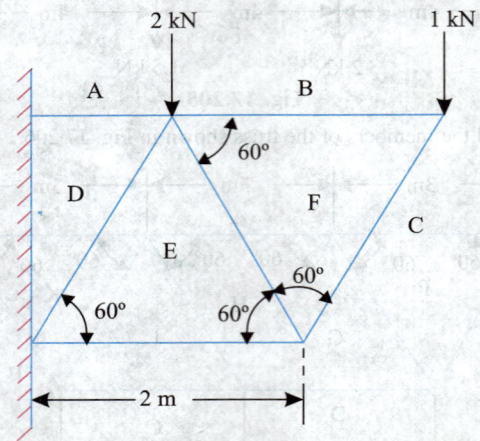

Fig. 17.208

13. Find the forces in the members of the truss shown in Fig. 17·209.

Ans. $AB = BC = 2.666$ kN tensile; $AD = 1.666$ kN tensile;
$CD = 3.334$ kN comp; $DE = 5$ kN comp; $BD = 2$ *kN* comp;

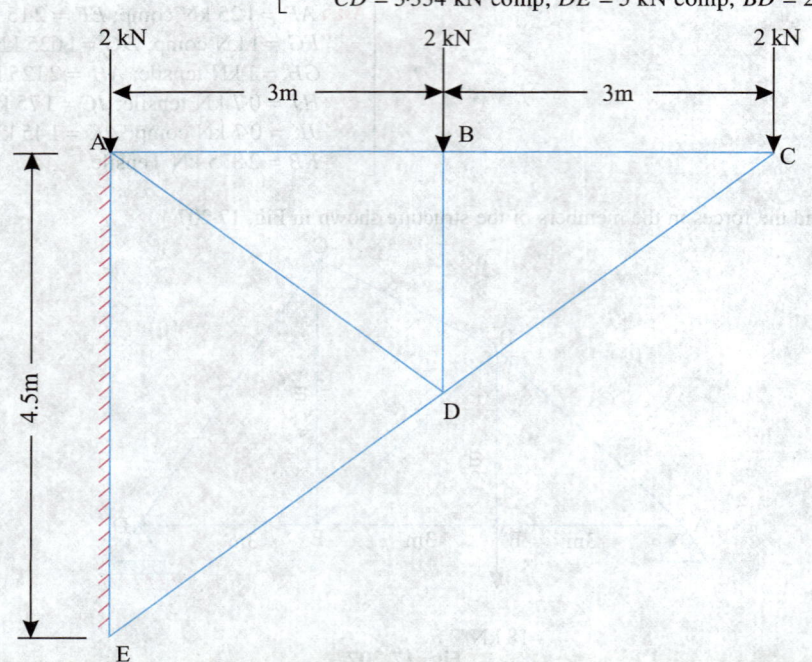

Fig. 17.209

14. Determine the forces and their nature in all the members of the pin-jointed truss shown in Fig. 17·210. The truss is subjected to a vertical load of 16 kN and a horizontal load of 8 kN simultaneously at joint *A*.

$$\left[\text{Ans.} \begin{array}{l} AC = 10 \text{ kN comp; } AD = 16{\cdot}66 \text{ kN comp;} \\ BD = BC = 12{\cdot}02 \text{ kN tensile; } AB = 6{\cdot}66 \text{ kN tensile;} \end{array} \right]$$

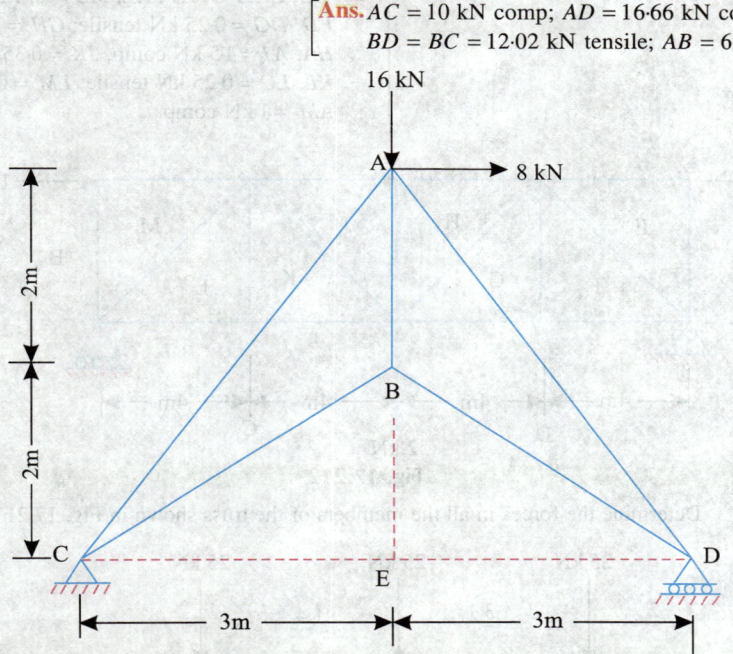

Fig. 17.210

15. Find the forces in the members of the truss shown in Fig. 17·211.

$$\left[\text{Ans.} \begin{array}{l} AD = 30{\cdot}75 \text{ kN comp; } AF = 30{\cdot}23 \text{ kN tensile;} \\ DF = 12 \text{ kN comp; } DC = 30{\cdot}75 \text{ kN comp;} \\ CE = 18{\cdot}73 \text{ kN comp; } CF = 20{\cdot}98 \text{ kN tensile;} \\ EF = 5{\cdot}65 \text{ kN comp; } FB = 17{\cdot}24 \text{ kN tensile;} \\ EB = 24{\cdot}38 \text{ kN comp.} \end{array} \right]$$

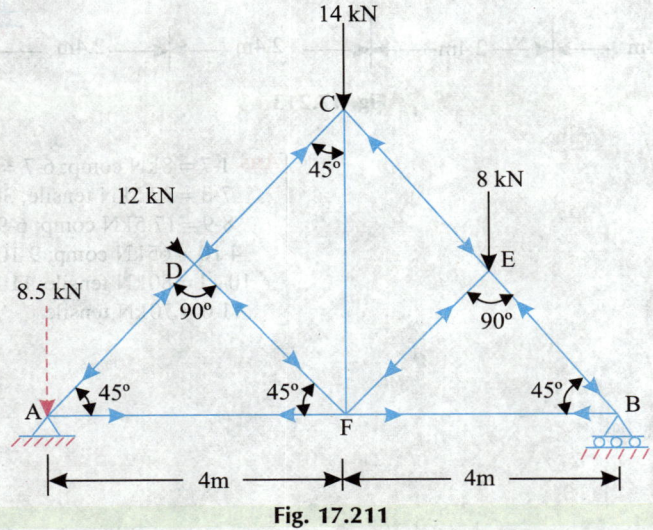

Fig. 17.211

16. Determine graphically or otherwise the forces and their nature in all the members of the truss shown in Fig. 17·212.

Ans. *AE*, *EA*, *FG*, *HJ*, *KL*, *MB* = 0; *FE* = 1 kN comp;
FD, *DG* = 0·25 kN tensile; *GH* = 1 kN tensile;
HA, *AJ* = 1·5 kN comp; *JK* = 0·35 kN tensile;
KL, *LC* = 0·25 kN tensile; *LM* = 0·35 kN comp;
AM = 1 kN comp.

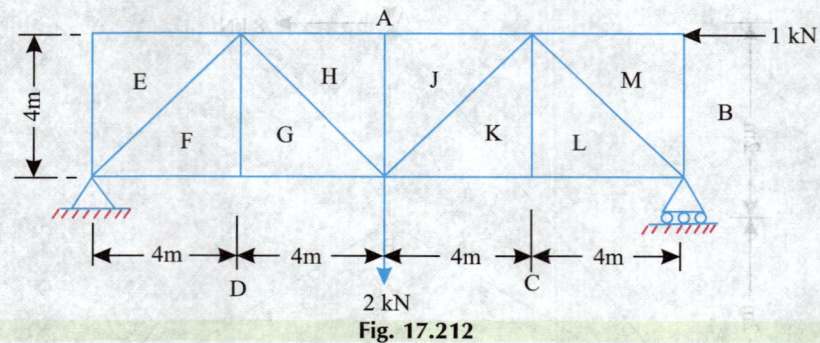

Fig. 17.212

17. Determine the forces in all the members of the truss shown in Fig. 17.213.

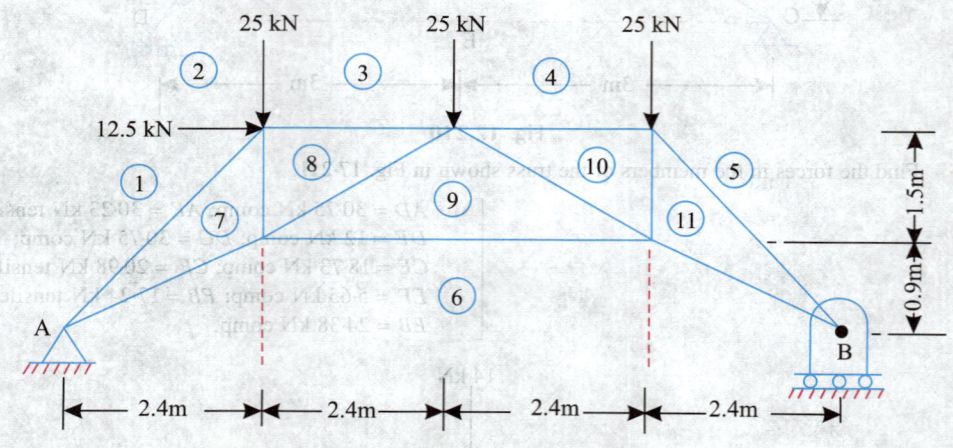

Fig. 17.213

Ans. 1-7 = 8 kN comp; 6-7 = 80 kN tensile;
7-8 = 37.5 kN tensile; 3-8 = 75 kN comp;
8-9 = 17.5 kN comp; 6-9 = 90 kN tensile;
4-10 = 65 kN comp; 9-10 = 35 kN comp;
10-11 = 80 kN tensile; 11-5 = 92.5 kN comp;
11-6 = 70 kN tensile.

Theories of Failure

Chapter 18

18·1. INTRODUCTION

Due to the large number of examples of compound stresses met with in engineering practice, the cause of "failure" or permanent set under such conditions has attracted considerable attention. Certain theories have been advanced to explain the cause of failure and many of theories have received considerable experimental investigation. No great uniformity of opinion has been reached, and there is still room for a great deal of further experimental investigation.

The principal theories are:

1. Maximum principal stress theory
2. Maximum shear stress or stress difference theory
3. Strain energy theory
4. Shear strain energy theory
5. Maximum principal strain theory
6. Mohr's theory.

In all above theories:

σ_{et}, σ_{ec} = Tensile stress at the elastic limit in simple tension and compression respectively.

$\sigma_1, \sigma_2, \sigma_3$ = Principal stresses in any complex system (such that $\sigma_1 > \sigma_2 > \sigma_3$).

It may be assumed that the loading is *gradual* or *static* (and there is no cyclic or impact loading).

18·2. MAXIMUM PRINCIPAL STRESS THEORY

— This theory is usually associated with Rankine, but also received considerable support from other writers.

— This is the simplest and the oldest theory of failure.

— According to this theory *failure will occur when the maximum* principal tensile stress (σ_1) in the complex system reaches the value of the maximum stress at the elastic limit (σ_{et}) in simple tension or the minimum principal stress (that is, the maximum principal compressive stress) reaches the elastic limit stress (σ_{ec}) in simple compression.

i.e., $\qquad \sigma_1 = \sigma_{et}$ (in simple *tension*) $\qquad\qquad$ ...(18·1)

$\qquad | \sigma_3 | = \sigma_{ec}$ (in simple *compression*)

$\qquad | \sigma_3 |$ means numerical value of σ_3.

— If the maximum principal stress is the design criterion, the maximum principal stress must not exceed the working σ for the material. Hence,

$\qquad\qquad \sigma_1 \leqslant \sigma$ $\qquad\qquad\qquad\qquad\qquad\qquad$ ...(18·2)

— This theory disregards the effect of other principal stresses and of the shearing stresses on other planes through the element. For brittle materials which do not fail by yielding *but fail by brittle fracture, the maximum principal stress theory is considered to be reasonably satisfactory.*

This theory appears to be *approximately correct for* **ordinary cast-irons** and **brittle metals**.

The maximum principal stress theory is *contradicted* in the following cases:

1. On a mild steel specimen when simple tension test is carried out sliding occurs approximately 45° to the axis of the specimen; this shows that the failure in this case is due to maximum shear stress rather than the direct tensile stress.

2. It has been found that a material which is even though weak in simple compression yet can sustain hydrostatic pressure far in excess of the elastic limit in simple compression.

Splines and other machine shafts.

Example 18·1. *In a metallic body the principal stresses are +35MN/m² and – 95 MN/m², the third principal stress being zero. The elastic limit stress in simple tension as well as in simple compression is equal and is 220MN/m². Find the factor of safety based on the elastic limit if the criterion of failure for the material is the maximum principal stress theory.*

Solution. *Given:* $\sigma_1 = +35$ MN/m²

$\hspace{3.5cm} \sigma_2 = 0$, and $\hspace{1.5cm}$ ⎤ Principal stresses

$\hspace{3.5cm} \sigma_3 = -95$ MN/m² $\hspace{0.5cm}$ ⎦

$\hspace{3.5cm} \sigma_e = \sigma_{et} = \sigma_{ec} = 220$ MN/m²

where, $\hspace{2.5cm} \sigma_{et} =$ Elastic limit stress (*tension*), and

$\hspace{3.5cm} \sigma_{ec} =$ Elastic limit stress (*compression*).

Now, $\hspace{2.8cm} \sigma_1 = \sigma_t$ (*working stress in tension*)

or, $\hspace{3cm} \sigma_1 = \dfrac{\sigma_{et}}{\text{F.O.S.}} \hspace{1cm}$ (F.O.S. means factor of safety)

F.O.S. $\hspace{2.2cm} = \dfrac{\sigma_e}{\sigma_1} = \dfrac{220}{35} = 6 \cdot 28$

Also, $\hspace{2.5cm} |\sigma_3| = \sigma_c$ (*working stress in compression*)

or, $\hspace{2.9cm} |\sigma_3| = \dfrac{\sigma_{ec}}{\text{F.O.S.}}$

$\hspace{3.2cm} |-95| = \dfrac{\sigma_{ec}}{\text{F.O.S.}}$

∴ $\hspace{2.5cm}$ F.O.S. $= \dfrac{220}{95} = 2 \cdot 3$

So, the material according to the maximum principal stress theory will fail due to the compressive principal stress.

∴ $\hspace{3cm}$ **F.O.S. = 2·3** $\hspace{0.3cm}$ **(Ans.)**

Example 18·2. *In a cast-iron body the principal stresses are + 40 MN/m² and – 100 MN/m² the third principal stress being zero. The elastic limit stresses in simple tension and in simple compression are 80 MN/m² and 400 MN/m² respectively. Find the factor of safety based on the elastic limit if the criterion of failure is the maximum principal stress theory.*

Solution. *Given:* $\sigma_1 = 40$ MN/m²,

$\hspace{3.5cm} \sigma_2 = 0$ $\hspace{1.2cm}$ ⎫ Principal stresses

$\hspace{3.5cm} \sigma_3 = -100$ MN/m² ⎭

$\hspace{3.5cm} \sigma_{et} = 80$ MN/m² (elastic limit stress in *tension*)

$\hspace{3.5cm} \sigma_{ec} = 400$ MN/m² (elastic limit stress in *compression*)

Now, $\hspace{2.8cm} \sigma_1 = \sigma_t$ (*working stress in tension*)

or $\hspace{3cm} \sigma_1 = \dfrac{\sigma_{et}}{\text{F.O.S.}}$, or, $\hspace{0.5cm} 40 = \dfrac{80}{\text{F.O.S.}}$

∴ $\hspace{2.5cm}$ F.O.S. $= \dfrac{80}{40} = 2$

Also, $\hspace{2.5cm} |\sigma_3| = \sigma_c$ (*working stress in compression*)

$\hspace{3.5cm} = \dfrac{\sigma_{ec}}{\text{F.O.S.}}$

$$\therefore \qquad \text{F.O.S.} = \frac{\sigma_{ec}}{|\sigma_3|} = \frac{400}{100} = 4$$

$\therefore$ The material will fail due to *tensile* principal stress.

F.O.S. = 2 (Ans.)

18·3. MAXIMUM SHEAR STRESS OR STRESS DIFFERENCE THEORY

— This theory is also called *Guest's* or *Tresca's theory.*

— This theory implies that *failure will occur when the maximum shear stress* τ_{max} *in the complex system reaches the value of the maximum shear stress in simple tension at the elastic limit i.e.,*

$$\tau_{max} = \frac{\sigma_1 - \sigma_3}{2} = \frac{\sigma_{et}}{2} \text{ in simple } tension$$

or $\qquad\qquad \sigma_1 - \sigma_3 = \sigma_{et}$ ...(18·3)

In actual design σ_{et} in the above equation is replaced by the safe stress.

— This theory gives good correlation with results of experiments on ductile materials. In case if any of the three principal stresses is *compressive* then it must be taken as σ_3 and its proper sign taken in eqn. (18·3). That is, the maximum stress difference is to be equated to σ_{et}. In the case of two dimensional tensile stress system, third stress must be taken as zero and then the maximum stress difference calculated to equate it to σ_{et}.

— This theory has been found to give *quite satisfactory results for* **ductile materials.**

Following points are worth noting:

(*i*) The theory does not give accurate results for the state of stress of pure shear in which the maximum amount of shear is developed (*i.e.*, torsion test).

(*ii*) The theory is not applicable in the case where the state of stress consists of triaxial tensile stresses of nearly equal magnitude reducing the shearing stress to a small magnitude, so that failure would be by brittle fracture rather than by yielding.

(*iii*) The theory does not give as close results as found by experiments on ductile materials. However, it gives *safe results.*

Example 18·3. *A mild steel shaft 120 mm diameter is subjected to a maximum torque of 20 kNm and a maximum bending moment of 12 kNm at a particular section. Find the factor of safety according to the maximum shear stress theory if the elastic limit in simple tension is 220 MN/m².*

Solution. Diameter of the mild steel shaft, $d = 120$ mm $= 0.12$ m

Maximum torque, $T = 20$ kNm

Maximum bending moment, $M = 12$ kNm

Elastic limit in simple tension, $\sigma_{et} = 220$ MN/m².

Factor of safety (F.O.S.):

We know that, $M = \dfrac{\pi}{32} d^3 \cdot \sigma_b$, where, $\sigma_b =$ maximum bending stress

$$\therefore \qquad \sigma_b = \frac{32\,M}{\pi d^3} = \frac{32 \times 20 \times 10^3}{\pi \times (0.12)^3} \times 10^{-6} \text{ MN/m}^2 = 70.74 \text{ MN/m}^2$$

Also, $\qquad\qquad T = \tau \times \dfrac{\pi}{16} d^3$

or, $\qquad\qquad \tau = \dfrac{16\,T}{\pi d^3} = \dfrac{16 \times 20 \times 10^3}{\pi \times (0.12)^3} \times 10^{-6} \text{ MN/m}^2 = 58.95 \text{ MN/m}^2$

Principal stresses are given by

$$\sigma = \frac{\sigma_b}{2} \pm \sqrt{\left(\frac{\sigma_b}{2}\right)^2 + \tau^2} = \frac{70 \cdot 74}{2} \pm \sqrt{\left(\frac{70 \cdot 74}{2}\right)^2 + (58 \cdot 95)^2}$$

$$= 35 \cdot 57 \pm 68 \cdot 75 = 104 \cdot 12 \text{ MN/m}^2, \text{ or } -33 \cdot 38 \text{ MN/m}^2$$

According to the maximum shear stress theory,

$$\sigma_1 - \sigma_3 = \sigma_t$$

Here, $\sigma_1 = 104 \cdot 12 \text{ MN/m}^2$

$\sigma_2 = 0$

$\sigma_3 = -33 \cdot 38 \text{ MN/m}^2$

$[104 \cdot 12 - (-33 \cdot 38)] = \sigma_t$

i.e., $\sigma_t = 137 \cdot 5 \text{ MN/m}^2$

$\therefore$ $\text{F.O.S.} = \dfrac{\sigma_{et}}{\sigma_t} = \dfrac{220}{137 \cdot 5} = \mathbf{1 \cdot 6}$ **(Ans.)**

Example 18·4. *A shaft is subjected to a maximum torque of 10 kNm and a maximum bending moment of 7·5 kNm at a particular section. If the allowable equivalent stress in simple tension is 160 MN/m² find the diameter of the shaft according to the maximum shear stress theory.*

Solution. Maximum torque,

$$T = 10 \text{ kNm}$$

Maximum bending moment,

$$M = 7 \cdot 5 \text{ kNm}$$

Allowable equivalent stress in simple tension,

$$\sigma_t = 160 \text{ MN/m}^2$$

Diameter of the shaft, *d*:

We know that, $M = \dfrac{\pi}{32} d^3 \cdot \sigma_b$ (where, σ_b = maximum bending stress)

Shaft and bearings. Good bearings align the movement of the shaft and minimize deflections and shocks.

$$\therefore \qquad \sigma_b = \frac{32\,M}{\pi d^3}, \quad \text{and,} \; T = \tau \times \frac{\pi}{16} d^3$$

$$\therefore \qquad \tau = \frac{16\,T}{\pi d^3}$$

Principal stresses are given by:

$$\sigma_1, \sigma_2 = \frac{\sigma_b}{2} \pm \sqrt{\left(\frac{\sigma_b}{2}\right)^2 + \tau^2} = \frac{1}{2}\left[\sigma_b \pm \sqrt{\sigma_b^2 + 4\tau^2}\right]$$

$$= \frac{1}{2}\left[\frac{32\,M}{\pi d^3} \pm \sqrt{\left(\frac{32\,M}{\pi d^3}\right)^2 + \left(\frac{32\,T}{\pi d^3}\right)^2}\right]$$

$$= \frac{16}{\pi d^3}\left[M \pm \sqrt{M^2 + T^2}\right]$$

$$\therefore \qquad \sigma_1 = \frac{16}{\pi d^3}\left[M + \sqrt{M^2 + T^2}\right]$$

$$\sigma_2 = 0$$

$$\text{and,} \qquad \sigma_3 = \frac{16}{\pi d^3}\left[M - \sqrt{M^2 + T^2}\right]$$

According to the maximum shear stress theory,

$$\sigma_t = \sigma_1 - \sigma_3 = \frac{16}{\pi d^3}\left[M + \sqrt{M^2 + T^2}\right] - \frac{16}{\pi d^3}\left[M - \sqrt{M^2 + T^2}\right]$$

$$= \frac{32}{\pi d^3}\sqrt{M^2 + T^2}$$

$$\therefore \qquad d^3 = \frac{32}{\pi \sigma_t}\sqrt{M^2 + T^2} = \frac{32 \times 10^3}{\pi \times 160 \times 10^6} \times \sqrt{7 \cdot 5^2 + 10^2} = 7 \cdot 957 \times 10^{-4}$$

$$\therefore \qquad d = 0 \cdot 0926 \text{ m} \quad \text{or} \quad 92 \cdot 6 \text{ mm}$$

Hence, $\qquad \boldsymbol{d = 92 \cdot 6 \text{ mm}}$ **(Ans.)**

18·4. STRAIN ENERGY THEORY

— This theory which has a *thermodynamic analogy* and a *logical basis* is due to Haigh.

— This theory states that *the failure of a material occurs when the total strain energy in the material reaches the total strain energy of the material at the elastic limit in simple tension.*

In a **three dimensional stress system**, the strain energy per unit volume is given by:

$$U = \frac{1}{2E}\left[\sigma_1^2 + \sigma_2^2 + \sigma_3^2 - \frac{2}{m}\,(\sigma_1\sigma_2 + \sigma_2\sigma_3 + \sigma_3\sigma_1)\right]$$

(where, σ_1, σ_2 and σ_3 are of the same sign.)

Hence at the point of failure

$$\frac{1}{2E}\left[\sigma_1^2 + \sigma_2^2 + \sigma_3^2 - \frac{2}{m}\,(\sigma_1\sigma_2 + \sigma_2\sigma_3 + \sigma_3\sigma_1)\right] = \frac{\sigma_e^2}{2E}$$

$$\therefore \qquad \sigma_1^2 + \sigma_2^2 + \sigma_3^2 - \frac{2}{m}\,(\sigma_1\sigma_2 + \sigma_2\sigma_3 + \sigma_3\sigma_1) = \sigma_e^2 \qquad \qquad ...\,(18\cdot4)$$

In actual design σ_e in the above eqn. is replaced by the allowable stress obtained by dividing σ_e by F.O.S.

— Taking two dimensional case ($\sigma_3 = 0$) the eqn. reduces to

$$\sigma_1^2 + \sigma_2^2 - \frac{2}{m} \cdot \sigma_1 \sigma_2 = \sigma_e^2 \qquad \qquad ...(18\cdot5)$$

If σ is the working stress in the material, the design criterion may be stated as follows:

$$\left(\sigma_1^2 + \sigma_2^2 - \frac{2}{m} \sigma_1 \sigma_2\right) \leqslant \sigma^2 \qquad \qquad ...(18\cdot6)$$

The following points are worth noting:

(*i*) The results of this theory are similar to the experimental results for ductile materials (*i.e.*, the materials which fail by general yielding) for which $\sigma_{et} = \sigma_{ec}$ approximately. (It may be noted that order of σ_1, σ_2 and σ_3 is immaterial.)

(*ii*) The theory does not apply to materials for which σ_{et} is quite different from σ_{ec}.

(*iii*) The theory does not give results exactly equal to the experimental results even for ductile materials, even though the results are close to the experimental.

Example 18·5. *Solve example 18·4 using the strain energy theory. Take Poisson's ratio,*

$$\frac{1}{m} = 0\cdot24.$$

Solution.

From example, 18·4

$$\sigma_1 = \frac{16}{\pi d^3} (M + \sqrt{M^2 + T^2})$$

$$\sigma_2 = 0$$

$$\sigma_3 = \frac{16}{\pi d^3} (M - \sqrt{M^2 + T^2})$$

Now according to strain energy theory,

$$\sigma_t^2 = \sigma_1^2 + \sigma_2^2 + \sigma_3^2 - \frac{2}{m} (\sigma_1 \sigma_2 + \sigma_2 \sigma_3 + \sigma_3 \sigma_1)$$

$$= \sigma_1^2 + \sigma_3^2 - \frac{2}{m} \cdot \sigma_3 \sigma_1 \qquad (\because \sigma_2 = 0)$$

$$= \left(\frac{16}{\pi d^3}\right)^2 \left[2\left(M^2 + M^2 + T^2\right) - \frac{2}{m}\left(M^2 - M^2 - T^2\right)\right]$$

$$= \left(\frac{16}{\pi d^3}\right)^2 \left[4M^2 + 2T^2\left(1 + \frac{1}{m}\right)\right]$$

$\therefore \qquad \sigma_t = \frac{16}{\pi d^3} \sqrt{4M^2 + 2\left(1 + \frac{1}{m}\right)T^2} = \frac{32}{\pi d^3} \sqrt{M^2 + \left(\frac{1 + 1/m}{2}\right)T^2}$

$$= \frac{32}{\pi d^3} \sqrt{M^2 + 0\cdot62\,T^2} \qquad \left(\because \frac{1}{m} = 0\cdot24\right)$$

$\therefore \qquad d^3 = \frac{32}{\pi \sigma_t} \sqrt{M^2 + 0\cdot62\,T^2} = \frac{32 \times 10^3}{\pi \times 160 \times 10^6} \sqrt{7\cdot5^2 + 0\cdot62 \times 10^2}$

$$= 6\cdot922 \times 10^{-4}$$

$\therefore \qquad d = 0\cdot0885 \text{ m} = 88\cdot5 \text{ mm}$

Hence, $\qquad$ **$d = 88\cdot5$ mm (Ans.)**

18·5. SHEAR STRAIN ENERGY THEORY

This theory is also called **"Distortion energy theory"** or **"Mises-Henky theory"**.

— According to this theory the *elastic failure occurs where the shear strain energy per unit volume in the stressed material reaches a value equal to the shear strain energy per unit volume at the elastic limit point in the simple tension test.*

Shear strain energy due to the principal stresses σ_1, σ_2 and σ_3 per unit volume of the stress material,

$$U_s = \frac{1}{12\,C}\left[(\sigma_1 - \sigma_2)^2 + (\sigma_2 - \sigma_3)^2 + (\sigma_3 - \sigma_1)^2\right]$$

But for the simple tension test at the elastic limit point where there is only one principal stress *i.e.*, σ_{et} we have the shear strain energy per unit volume which is given by:

$$U_s' = \frac{1}{12\,C}\left[(\sigma_{et} - 0)^2 + (0 - 0)^2 + (0 - \sigma_{et})^2\right] \qquad \begin{bmatrix} \because \sigma_1 = \sigma_{et} \\ \sigma_2 = 0 \\ \sigma_3 = 0 \end{bmatrix}$$

Equating the two energies, we get

$$(\sigma_1 - \sigma_2)^2 + (\sigma_2 - \sigma_3)^2 + (\sigma_3 - \sigma_1)^2 = 2\sigma_{et}^2 \qquad \qquad ...(18\cdot7)$$

In actual design σ_{et} in eqn. (18·7) is replaced by safe equivalent stress σ_t in simple tension.

— The above theory has been found to give *best results for ductile material for which*
$\sigma_{et} = \sigma_{ec}$ *approximately:*

(It may be noted that order of σ_1, σ_2 and σ_3 is immaterial).

The following points are worth noting:

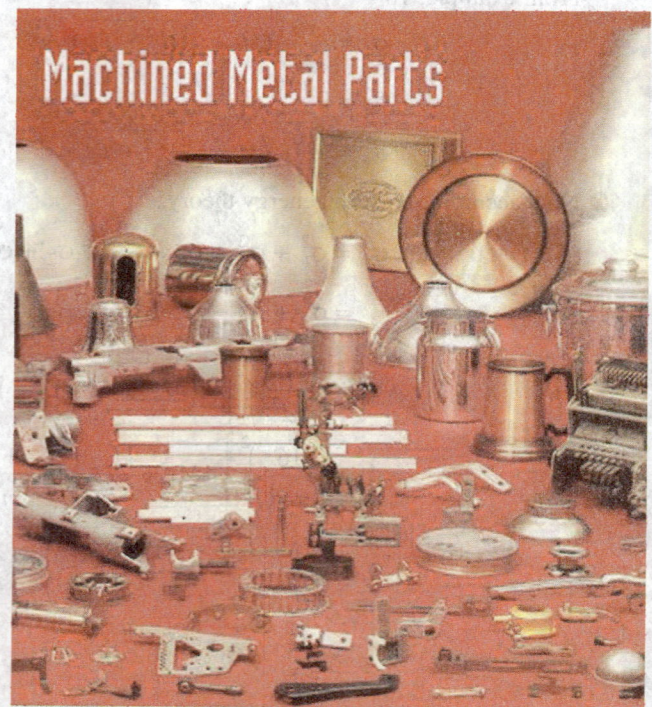

(*i*) The theory does not agree with the experimental results for the material for which σ_{et} is quite different from σ_{ec}.

(*ii*) The theory gives $\sigma_{et} = 0$ for hydrostatic pressure or tension, which means that the material will never fail under any hydrostatic pressure or tension and this is obviously not correct. Actually when three equal tensions are applied in three principal directions, brittle fracture occurs and as such maximum principal stress theory will give reliable results in this case.

(*iii*) This theory is regarded as one to *which conform most of the ductile materials under the action of various types of loading.*

Example 18·6. *Solve example 18·4, using shear strain energy theory.*

Solution. Maximum torque, $T = 10$ kNm

Maximum bending moment, $M = 7·5$ kNm

Allowable equivalent stress in simple tension, $\sigma_t = 160$ MN/m²

Diameter of the shaft, d:

$$\sigma_1 = \frac{16}{\pi d^3}\left(M + \sqrt{M^2 + T^2}\right)$$

$$\sigma_2 = 0$$

$$\sigma_3 = \frac{16}{\pi d^3}\left(M - \sqrt{M^2 + T^2}\right)$$

Now, according to shear strain energy theory

$$2\sigma_t^2 = (\sigma_1 - \sigma_2)^2 + (\sigma_2 - \sigma_3)^2 + (\sigma_3 - \sigma_1)^2 \qquad \text{...(Eqn. 18·7)}$$

$$= \sigma_1^2 - 2\sigma_1\sigma_2 + \sigma_2^2 + \sigma_2^2 - 2\sigma_2\sigma_3 + \sigma_3^2 + \sigma_3^2 - 2\sigma_3\sigma_1 + \sigma_1^2$$

$$= 2(\sigma_1^2 + \sigma_2^2 + \sigma_3^2) - 2(\sigma_1\sigma_2 + \sigma_2\sigma_3 + \sigma_3\sigma_1)$$

∴ $$\sigma_t^2 = \sigma_1^2 + \sigma_2^2 + \sigma_3^2 - (\sigma_1\sigma_2 + \sigma_2\sigma_3 + \sigma_3\sigma_1)$$

$$= \sigma_1^2 + \sigma_3^2 - \sigma_3\sigma_1 \qquad (\because \sigma_2 = 0)$$

$$= \left(\frac{16}{\pi d^3}\right)^2\left[2M^2 + 2M^2 + 2T^2 - \left(M^2 - M^2 - T^2\right)\right]$$

$$= \left(\frac{16}{\pi d^3}\right)^2(4M^2 + 3T^2) = \left[\frac{32}{\pi d^3}\right]^2\left[M^2 + \frac{3}{4}T^4\right]$$

or, $$\sigma_t = \frac{32}{\pi d^3}\sqrt{M^2 + \frac{3}{4}T^2}$$

$$d^3 = \frac{32}{\pi\sigma_t}\sqrt{M^2 + \frac{3}{4}T^2}$$

$$= \frac{32 \times 10^3}{\pi \times 160 \times 10^6}\sqrt{7·5^2 + \frac{3}{4} \times 10^2} = 7·29 \times 10^{-4}$$

∴ $$d = 0·09 \text{ m} = 90 \text{ mm} \quad \textbf{(Ans.)}$$

Note: Out of all the results obtained in Examples 18·4, 18·5, 18·6 the result;

$d = 90$ mm...may be regarded as *most accurate*

$d = 92·6$ mm...is on the *safer side*, and

$d = 88·5$ mm...a *little different from the correct result.*

Example 18·7. *A steel shaft is subjected to an end thrust producing a stress of 90 MPa and the minimum shearing stress on the surface arising from torsion is 60 MPa. The yield point of the material in simple tension was found to be 300 MPa. Calculate the factor of safety of the shaft according to the following theories :*

(i) Maximum shear stress theory;

(ii) Maximum distortion energy theory. **(AMIE Summer, 2000)**

Solution. *Given* $\sigma_1 = ?$, $\sigma_2 = 0$, $\sigma_3 = -90$ MN/m², $\tau_{max} = 60$ MN/m²; $\sigma_{et} = 300$ MN/m²

Factor of safety :

(i) Maximum shear stress theory :

$$\tau_{max} = \frac{\sigma_1 - \sigma_3}{2} = 60$$

or, $\qquad \sigma_1 - (-90) = 120,$ or, $\sigma_1 = 30 \text{ MN/m}^2$

Also, $\qquad \sigma_1 - \sigma_3 = \sigma_t$

$$30 - (-90) = \sigma_t$$

or, $\qquad \sigma_t = 120 \text{ MN/m}^2$

$\therefore \qquad \text{F.O.S.} = \dfrac{\sigma_{et}}{\sigma_t} = \dfrac{300}{120} = 2 \cdot 5 \quad \textbf{(Ans.)}$

(ii) *Maximum distortion energy theory :*

$$\sigma_t^2 = \sigma_1^2 + \sigma_3^2 - \sigma_3 \sigma_1 \qquad \qquad \text{...Eqn.(18·7)}$$
$$= 30^2 + (-90)^2 - (-90)(30) = 11700$$

$\therefore \qquad \sigma_t = 108 \cdot 17 \text{ MN/m}^2$

$\therefore \qquad \text{F.O.S.} = \dfrac{\sigma_{et}}{\sigma_t} = \dfrac{300}{108 \cdot 17} = 2 \cdot 77 \quad \textbf{(Ans.)}$

18·6. MAXIMUM PRINCIPAL STRAIN THEORY

— This theory is associated with St. Venant.

— The theory states that the *failure of a material occurs when the principal tensile strain in the material reaches the strain at the elastic limit in simple tension or when the minimum principal strain (i.e., maximum principal compressive strain) reaches the elastic limit strain in simple compression.*

Principal strain in the direction of principal stress σ_1,

$$e_1 = \frac{1}{E} \left[\sigma_1 - \frac{1}{m}(\sigma_2 + \sigma_3) \right]$$

Principal strain in the direction of the principal stress σ_3,

$$e_3 = \frac{1}{E} \left[\sigma_3 - \frac{1}{m}(\sigma_1 + \sigma_2) \right]$$

The conditions to cause failure according to the maximum principal strain theory are:

$$e_1 > \frac{\sigma_{at}}{E} \qquad (e_1 \text{ must be +ve})$$

The cracks such as the one on the crank rod may appear because of the faulty design, overload and or using it beyond the fatigue limit.

and,
$$|e_3| > \frac{\sigma_{ec}}{E} \qquad (e_3 \text{ must be } -ve)$$

$$\frac{1}{E}\left[\sigma_2 - \frac{1}{m}(\sigma_2 + \sigma_3)\right] > \frac{\sigma_{et}}{E}$$

and,
$$\frac{1}{E}\left[\sigma_3 - \frac{1}{m}(\sigma_1 + \sigma_2)\right] > \frac{\sigma_{ec}}{E}$$

or,
$$\sigma_1 - \frac{1}{m}[\sigma_2 + \sigma_3] > \sigma_{et}$$

$$\sigma_3 - \frac{1}{m}[\sigma_1 + \sigma_2] > \sigma_{ec}$$

To prevent failure:

$$\sigma_1 - \frac{1}{m}(\sigma_2 + \sigma_3) < \sigma_{et}$$

$$\sigma_3 - \frac{1}{m}(\sigma_1 + \sigma_2) < \sigma_{ec}$$

At the point of elastic failure:

$$\sigma_1 - \frac{1}{m}(\sigma_2 + \sigma_3) = \sigma_{et} \qquad \text{...(18·8)}$$

and,
$$\left|\sigma_3 - \frac{1}{m}(\sigma_1 + \sigma_2)\right| = \sigma_{ec} \qquad \text{...(18·9)}$$

For design purposes:

$$\sigma_3 - \frac{1}{m}(\sigma_1 + \sigma_2) = \sigma_t$$

$$\left|\sigma_3 - \frac{1}{m}(\sigma_1 + \sigma_2)\right| = \sigma_c$$

(where, σ_t and σ_c are the safe stresses)

The following points are worth noting:

(*i*) The theory *overestimates the behaviour of ductile materials.*

(*ii*) The theory does not fit well with the experimental results except for brittle materials for biaxial tension—compression state of stress (for which it is sometimes recommended) and is *not much used in practice.*

18·7. GRAPHICAL REPRESENTATION OF THEORIES FOR TWO DIMENSIONAL STRESS SYSTEM

The theories of failure may be represented graphically as follows:

1. **Maximum principal stress theory :**

According to this theory failure will occur when

$$\sigma_1 \text{ or } \sigma_2 = \sigma_{et} \text{ or } \sigma_{ec}$$

If $\sigma_{et} = \sigma_{ec} = \sigma_e$, then these conditions are represented graphically on σ_1, σ_2 coordinates as shown in fig. 18·1.

Failure will take place if any point having co-ordinates (σ_1, σ_2) falls outside the square.

2. Maximum shear stress theory :

— When σ_1 and σ_2 are of opposite sign, the greater shearing stress is given by $\dfrac{\sigma_1 - \sigma_2}{2}$ and the failure is represented by

$$\sigma_1 - \sigma_2 = \sigma_e \qquad ...(i)$$

or,
$$\sigma_2 - \sigma_1 = \sigma_e \qquad ...(ii)$$

The boundaries of these equations are represented by parallel lines EF and CB (Fig. 18.2).

— When σ_1 and σ_2 are like (*i.e.*, I or III quadrant), the maximum shear stress is equal to $\sigma_1/2$ or $\sigma_2/2$ whichever is more. Hence the failure is represented by

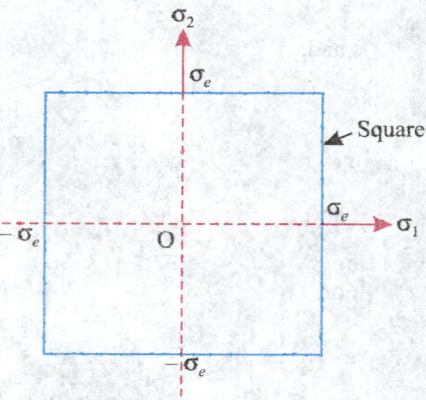

Fig. 18.1. Representation of maximum principal stress theory.

$$\sigma_1 = \sigma_e \qquad ...(iii)$$
or,
$$\sigma_2 = \sigma_e \qquad ...(iv)$$

The conditions are represented by lines FA and AB in the I quadrant and ED and DC in the III quadrant.

Hence according to maximum shear stress theory, material will reach its elastic limit when the point (σ_1, σ_2) falls outside the figure $A\,B\,C\,D\,E\,F\,A$.

3. Maximum principal strain theory:

According to this theory:

For *tensile* yielding : $\qquad \sigma_1 - \dfrac{\sigma_2}{m} = \sigma_e$

For *compressive* yielding : $\qquad \sigma_2 - \dfrac{\sigma_1}{m} = \sigma_e$

(With σ_2 compressive)

By plotting these equations a rhomboid failure envelope is obtained as shown in Fig. 18·3.

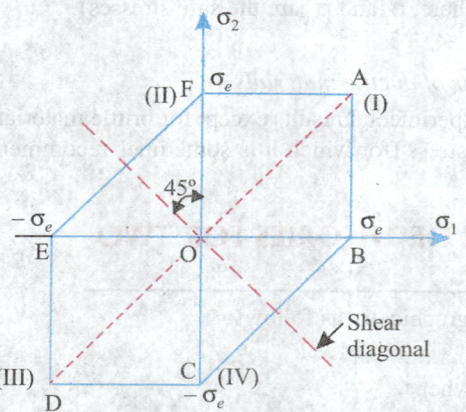

Fig. 18.2. Representation of maximum shear stress theory.

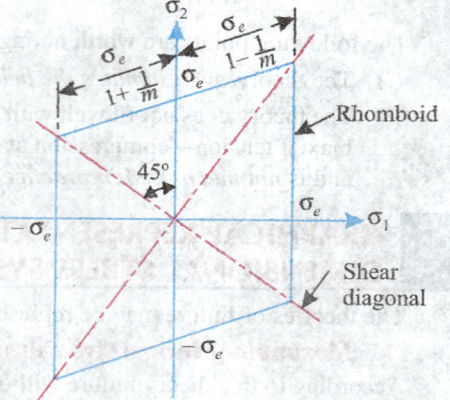

Fig. 18.3. Representation of maximum principal strain theory.

4. Maximum strain energy theory :

For two dimensional case (when $\sigma_3 = 0$), the maximum strain energy theory is represented by :

$$\sigma_1^2 + \sigma_2^2 - \frac{2}{m} \cdot \sigma_1 \sigma_2 = \sigma_e^2$$

$$\left(\frac{\sigma_1}{\sigma_e}\right)^2 + \left(\frac{\sigma_2}{\sigma_e}\right)^2 - \frac{2}{m}\left(\frac{\sigma_1}{\sigma_e}\right)\left(\frac{\sigma_2}{\sigma_e}\right) = 1$$

This is the equation of an ellipse with semi-major and semi-minor axes $\dfrac{\sigma_e}{\sqrt{1 - \dfrac{1}{m}}}$, $\dfrac{\sigma_e}{\sqrt{1 + \dfrac{1}{m}}}$

respectively, each at 45° to the co-ordinate axes as shown in Fig. 18·4.

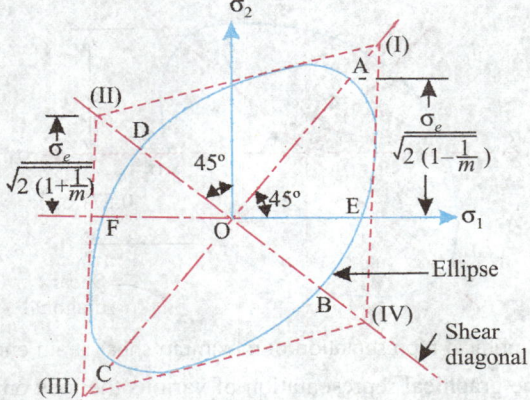

Fig. 18.4. Representation of maximum strain energy theory.

The ellipse is inscribed by the parallelogram given by maximum strain theory. The material will reach its elastic limit when the point (σ_1, σ_2) passes outside the ellipse.

Shaft bearings.

5. Maximum shear strain energy theory :

For two dimensional case (when $\sigma_3 = 0$) this theory is represented by :

$$\sigma_1^2 + \sigma_2^2 - \sigma_1\sigma_2 = \sigma_e^2$$

or

$$\left(\frac{\sigma_1}{\sigma_e}\right)^2 + \left(\frac{\sigma_2}{\sigma_e}\right)^2 - \left(\frac{\sigma_1}{\sigma_e}\right)\left(\frac{\sigma_2}{\sigma_e}\right) = 0$$

This is the equation of an ellipse with semi-major and semi-minor axes $\sqrt{2}\,\sigma_e$ and $\sqrt{\dfrac{2}{3}}\,\sigma_e$ respectively at 45° to the co-ordinate axes (Fig. 18·5).

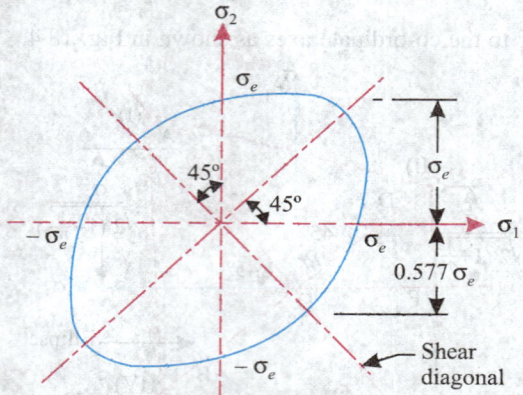

Fig. 18.5. Graphical representation of maximum shear strain energy theory.

Fig. 18·6 shows the graphical representation of various theories on the same diagram when $\sigma_t = \sigma_c$ and $e_t = e_c$.

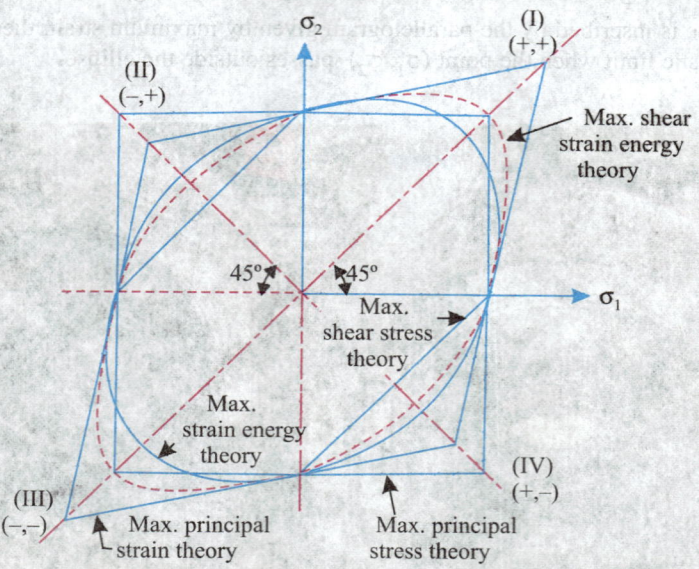

Fig. 18.6. Graphical representation of theories of failure.

Example 18·8. *In a steel member, at a point the major principal stress is 180 MN/m², and the minor principal stress is compressive. If the tensile yield point of the steel is 225 MN/m², find the value of the minor principal stress at which yielding will commence, according to each of the following criteria of failure:*

(i) Maximum shearing stress,

(ii) Maximum total strain energy, and

(iii) Maximum shear strain energy.

Take Poisson's ratio = 0·26.

Solution. Major principal stress,

$$\sigma_1 = 180 \text{ MN/m}^2$$

Yield point stress, $\sigma_e = 225 \text{ MN/m}^2$

Minor principal stress σ_2:

(i) Maximum shearing stress criterion:

$$\sigma_1 - \sigma_2 = \sigma_e$$

$$\sigma_2 = \sigma_1 - \sigma_e = 180 - 225 = -45 \text{ MN/m}^2$$

or   **$\sigma_2 = 45 \text{ MN/m}^2$ (comp.) (Ans.)**

(ii) Maximum total strain energy criterion:

$$\sigma_1^2 + \sigma_2^2 - \frac{2}{m}\sigma_1\sigma_2 = \sigma_e^2 \qquad \text{...(Eqn. 18·5)}$$

$$(180)^2 + \sigma_2^2 - 2 \times 0·26 \times 180 \times \sigma_2 = (225)^2$$

$$32400 + \sigma_2^2 - 93·6\,\sigma_2 = 50625$$

or, $\sigma_2^2 - 93·6\,\sigma_2 - 18225 = 0$

$$\sigma_2 = \frac{93·6 - \sqrt{93·6^2 + 4 \times 18225}}{2}$$

$$= \frac{93·6 - 285·76}{2} = -96·08 \text{ MN/m}^2$$

(Only –ve sign is taken as σ_2 is compressive).

or,   **$\sigma_2 = 96·08 \text{ MN/m}^2$ (comp.)**
(Ans.)

(iii) Maximum shear strain energy criterion:

$$(\sigma_1 - \sigma_2)^2 + (\sigma_2 - \sigma_3)^2 + (\sigma_3 - \sigma_1)^2 = 2\sigma_e^2$$
$$\text{...(Eqn. 18·7)}$$

Putting $\sigma_3 = 0$ in the above eqn., we get

$$(\sigma_1 - \sigma_2)^2 + \sigma_2^2 + \sigma_1^2 = 2\sigma_e^2$$

$$\sigma_1^2 + \sigma_2^2 - 2\sigma_1\sigma_2 + \sigma_2^2 + \sigma_1^2 = 2\sigma_e^2$$

$$\sigma_1^2 + \sigma_2^2 - \sigma_1\sigma_2 = \sigma_e^2$$

$$(180)^2 + \sigma_2^2 - 180\,\sigma_2 = (225)^2$$

$$\sigma_2^2 - 180\,\sigma_2 - 18225 = 0$$

Turbocharger valve that can handle and regulate hot gasses.

∴ $\sigma_2 = \dfrac{180 - \sqrt{180^2 + 4 \times 18225}}{2}$

$$= \frac{180 - 324·5}{2} = -72·25 \text{ MN/m}^2$$

i.e.,   **$\sigma_2 = 72·25 \text{ MN/m}^2$ (comp.) (Ans.)**

Example 18·9. *In a material the principal stresses are 60 MN/m², 48 MN/m² and – 36 MN/m². Calculate :*

(i) *Total strain energy.*

(ii) *Volumetric strain energy.*

(iii) *Shear strain energy.*

(iv) *Factor of safety on the total strain energy criterion if the material yields at 120 MN/m². Take: E = 200 GN/m², and 1/m = 0·3.*

Solution. Principal stresses :

$$\left.\begin{array}{l} \sigma_1 = + 60 \text{ MN/m}^2 \\ \sigma_2 = + 48 \text{ MN/m}^2 \\ \sigma_3 = - 36 \text{ MN/m}^2 \end{array}\right\} \quad Given$$

Yield stress, $\sigma_e = 120$ MN/m²

(i) Total strain energy per unit volume :

Total strain energy per unit volume is

$$= \frac{1}{2E}\left[\sigma_1^2 + \sigma_2^2 + \sigma_3^2 - \frac{2}{m}(\sigma_1\sigma_2 + \sigma_2\sigma_3 + \sigma_3\sigma_1)\right]$$

$$= \frac{10^{12}}{2 \times 200 \times 10^9}\left[60^2 + 48^2 + 36^2 - 2 \times 0\cdot3(60 \times 48 - 48 \times 36 - 36 \times 60)\right]$$

$$= 2\cdot5\,[\,3600 + 2304 + 1296 - 0\cdot6\,(2880 - 1728 - 2160)]\times 10^{-3} \text{ kN/m}^2$$

$$= \textbf{19·51 kNm/m}^3 \quad \textbf{(Ans.)}$$

(ii) Volumetric strain energy per unit volume :

Volumetric strain energy per unit volume is

$$= \frac{1}{3}(\sigma_1 + \sigma_2 + \sigma_3)^2\left[\frac{1 - \dfrac{2}{m}}{2E}\right]$$

$$= \frac{1}{3}(60 + 48 - 36)^2 \times 10^{12}\left[\frac{1 - 2 \times 0\cdot3}{2 \times 200 \times 10^9}\right]\times 10^{-3} \text{ kNm}$$

$$= \textbf{1·728 kNm/m}^3 \quad \textbf{(Ans.)}$$

(iii) Shear strain energy per unit volume :

Shear strain energy per unit volume is

$$= \frac{1}{12C}\left[(\sigma_1 - \sigma_2)^2 + (\sigma_2 - \sigma_3)^2 + (\sigma_3 - \sigma_1)^2\right]$$

$$\left[\text{where, } C = \frac{E}{2\left(1 + \dfrac{1}{m}\right)} = \frac{200}{2\,(1 + 0\cdot3)} = 76\cdot923 \text{ GN/m}^2\right]$$

$$= \frac{1}{12 \times 76\cdot923 \times 10^9}\times 10^{12}\left[(60 - 48)^2 + (48 + 36)^2 + (- 36 - 60)^2\right]$$

$$= 1\cdot083\,(144 + 7056 + 9216)\times 10^{-3} \text{ kNm/m}^3 = \textbf{17·78 kNm/m}^3 \quad \textbf{(Ans.)}$$

(iv) Factor of safety (F.O.S.) :

Strain energy per unit volume under uniaxial loading is

$$= \frac{\sigma_e^2}{2E} = \frac{(120 \times 10^6)^2}{2 \times 200 \times 10^9}\times 10^{-3} \text{ kNm/m}^3 = 36 \text{ kNm/m}^3$$

$$\textbf{F.O.S.} = \frac{36}{19\cdot51} = \textbf{1·845} \quad \textbf{(Ans.)}$$

Example 18·10. *A bolt is under an axial thrust of 9·6 kN together with a transverse force of 4·8 kN. Calculate its diameter according to:*

(i) *Maximum principal stress theory,*

(ii) *Maximum shear stress theory, and*

(iii) *Strain energy theory.*

Given : Factor of safety = 3, yield strength of material of bolt = 270 N/mm², and Poisson's ratio = 0·3.

Solution. Let, d = Diameter of the bolt, mm

Area of cross-section, $A = \dfrac{\pi}{4} \times d^2$, mm²

Axial thrust = 9·6 kN = 9600 N

∴ Axial compressive stress on bolt,

$$\sigma = \frac{9600}{A} \text{ mm}^2$$

Transverse shear force   = 4·8 kN = 4800 N

Shear Stress on the bolt, $\tau = \dfrac{4800}{A} \text{ mm}^2$

Principal stresses :

$$\sigma_{1,2} = \frac{\sigma}{2} \pm \sqrt{\left(\frac{\sigma}{2}\right)^2 + \tau^2}$$

$$= \frac{4800}{A} \pm \sqrt{\left(\frac{4800}{A}\right)^2 + \left(\frac{4800}{A}\right)^2} = \frac{4800}{A} \pm \frac{6788 \cdot 2}{A}$$

∴ Maximum principal stress, $\sigma_1 = \dfrac{11588 \cdot 2}{A} \text{ N/mm}^2$

Minimum principal stress, $\sigma_2 = -\dfrac{1988 \cdot 2}{A} \text{ N/mm}^2$

Yield strength = 270 N/mm²

F.O.S. = 3

∴ Allowable stress $= \dfrac{270}{3} = 90 \text{ N/mm}^2$

Brass machine parts.

(i) *Maximum principal stress theory :*

$$\frac{11588 \cdot 2}{A} \leqslant 90$$

$$A = 128 \cdot 76 = \frac{\pi}{4} \times d^2$$

∴ $$d = 12 \cdot 8 \text{ mm} \quad \text{(Ans.)}$$

(ii) *Maximum shear stress theory :*

$$\sigma_1 - \sigma_2 = \sigma_{et}$$

or, $$\frac{11588 \cdot 2}{A} - \left(-\frac{1988 \cdot 2}{A}\right) = 90$$

or, $$\frac{13576 \cdot 4}{A} = 90$$

or, $$A = 150 \cdot 8 = \frac{\pi}{4} \times d^2$$

∴ $$d = 13 \cdot 8 \text{ mm} \quad \text{(Ans.)}$$

Example 18·11. *The inside and outside diameters of a cast-iron cylinder are 240 mm and 150 mm respectively. If the ultimate strength of cast-iron is 180 MN/m², find, according to each of the following theories the internal pressure which would cause rupture:*

(i) *Maximum principal stress theory,*

(ii) *Maximum strain theory, and*

(iii) *Maximum strain energy theory.*

Poisson's ratio = 0·25. Assume no longitudinal stress in the cylinder.

Solution. Inside radius of the cylinder,

$$r_1 = \frac{150}{2} = 75 \text{ mm} = 0 \cdot 075 \text{ m}$$

Outside radius of the cylinder,

$$r_2 = 240/2 = 120 \text{ mm} = 0 \cdot 12 \text{ m}$$

Ultimate strength of cast iron = 180 MN/m²

Cast-iron cylinder used in air brakes.

Internal pressure, p_1 :

The maximum radial and circumferential stresses at radius r_1 are given as :

$$\sigma_r = p_1 \text{ (comp.)}$$

i.e., $$\sigma_2 = \sigma_r = -p$$

and, $$\sigma_c = p_1 \left[\frac{r_2^2 + r_1^2}{r_2^2 - r_1^2}\right] \text{ tensile}$$

i.e., $$\sigma_1 = \sigma_c = p_1 \left[\frac{0 \cdot 12^2 + 0 \cdot 075^2}{0 \cdot 12^2 - 0 \cdot 075^2}\right] = 2 \cdot 28 \, p_1 \text{ (tensile)}$$

(i) *Maximum principal stress theory :*

$$\sigma_1 = 2 \cdot 28 \, p_1 = 180$$

or, $$p_1 = \frac{180}{2 \cdot 28} = 78 \cdot 95 \text{ MN/m}^2 \quad \text{(Ans.)}$$

(ii) *Maximum strain theory :*

$$\sigma_1 - \frac{\sigma_2}{m} = \sigma_e$$

$$2.28\, p_1 - 0.25\,(-p_1) = 180$$

or
$$p_1 = 71.15 \text{ MN/m}^2 \quad \text{(Ans.)}$$

(iii) *Maximum strain energy theory :*

$$\sigma_1^2 + \sigma_2^2 - \frac{2}{m} \cdot \sigma_1 \sigma_2 = \sigma_e^2$$

$$(2.28\, p_1)^2 + (-p_1)^2 - 2 \times 0.25 \times (2.28\, p_1)(-p_1) = (180)^2$$

$$7.338\, p_1^2 = (180)^2$$

∴
$$p_1 = 66.45 \text{ MN/m}^2$$

TYPICAL EXAMPLES (For Competitive Examinations)

Example 18·12. *A solid shaft transmits 1000 kW at 300 r.p.m. Maximum torque is 2 times the mean. The shaft is subjected to a bending moment which is 1·5 times the mean torque. The shaft is made of a ductile material for which the permissible tensile and shear stresses are 120 MPa and 60 MPa respectively. Determine the shaft diameter using a suitable theory of failure. Give justification for theory used.* **(Magadh University)**

Solution. *Given :* $P = 1000$ kW; $N = 300$ r.p.m., $T_{max} = 2T_{mean}$

$$M = 1.5\, T_{mean};\ \sigma_t = 120 \text{ MPa} = 120 \times 10^6 \text{ N/m}^2;$$

$$\tau = 60 \text{ MPa} = 60 \times 10^6 \text{ N/m}^2.$$

Shaft diameter, d :

In the given case, maximum principal stress theory (Rankine's theory) is used.

Using the relation :

$$\sigma_1 = \frac{16}{\pi d^3} \left[M + \sqrt{M^2 + \tau^2} \right] \qquad \text{...[Eqn. (2·15)]}$$

where,
$\sigma_1 = $ Maximum principal stress

$\sigma_t = 120 \times 10^6 \text{ N/m}^2$ (... Given)

We know,
$$P = \frac{2\,\pi\, NT}{60 \times 1000} \text{ kW}$$

or,
$$1000 = \frac{2\,\pi \times 300 \times T_{mean}}{60 \times 1000}$$

∴
$$T_{mean} = \frac{1000 \times 60 \times 1000}{2\,\pi \times 300} = 31831 \text{ Nm}$$

$$T_{max} = 2 \times T_{mean} = 2 \times 31831 = 63662 \text{ Nm}$$

$$M = 1.5 \times T_{mean} = 1.5 \times 31831 = 47746 \text{ Nm}$$

Substituting the various values in the above relation, we get

$$120 \times 10^6 = \frac{16}{\pi d^3} \left[47746 + \sqrt{(47746)^2 + (63662)^2} \right]$$

or,
$$d^3 = \frac{16}{120 \times 10^6\, \pi} [47746 + 79577] = 0.005404$$

∴
$$d = 0.175 \text{ m} \quad \text{(Ans.)}$$

Justification for theory used :

In case of ductile material the maximum principal stress theory (Rankine's theory) is generally used. In a simple tensile test we can easily determine the stress at which yielding has occurred in material but in the case of machine members subjected to various combination of loads, practically it is impossible to know where and at what stage yielding has started rendering the material useless, but certainly the principal stresses at a critical point can be known. In various theories of failure, the principal stresses have been expressed in terms of yeild stress in the simple tension or compression test and assuming that the stresses developed in the material are proportional to the applied loads, a limit is worked out such that, if the limit is exceeded, yielding is assumed to begin in the material. The materials generally fail by fracture or by excessive deformation at yielding. In ductile materials, failures by yielding is the usual basis. This explains the justification of using maximum principal stress theory in this case.

Example 18·13. *A hollow mild steel shaft having 100 mm external diameter and 50 mm internal diameter is subjected to a twisting moment of 8 kNm and a bending moment of 2·5 kNm. Calculate the principal stresses and find direct stress which, acting alone, would produce the same (i) maximum elastic strain energy, (ii) maximum elastic shear strain energy, as that produced by the principal stresses acting together.*

Take: Poisson's ratio = 0·25

(Engineering Services)

Solution. External diameter of the shaft,

$$D = 100 \text{ mm} = 0·1 \text{ m}$$

Internal diameter of the shaft,

$$d = 50 \text{ mm} = 0·05 \text{ m}$$

Twisting moment, $\quad T = 8$ kNm

Bending moment, $\quad M = 2·5$ kNm

Bending stress, $\quad \sigma_b = \dfrac{M}{Z} = \dfrac{M}{I/y} = \dfrac{My}{I} = \dfrac{M \times D}{(\pi/64)(D^4 - d^4) \times 2}$

$$= \dfrac{32\, MD}{\pi (D^4 - d^4)} = \dfrac{32 \times 2·5 \times 0·1}{\pi (0·1^4 - 0·05^4)} \times 10^{-3} \text{ MN/m}^2$$

$$= 27·16 \text{ MN/m}^2$$

Torsional stress, $\quad \tau = \dfrac{16\, TD}{\pi (D^4 - d^4)}$

$$= \dfrac{16 \times 8 \times 0·1}{\pi (0·1^4 - 0·05^4)} \times 10^{-3} \text{ MN/m}^2$$

$$= 43·46 \text{ MN/m}^2$$

$$\left[\begin{array}{l} \dfrac{T}{I_p} = \dfrac{\tau}{R} = \dfrac{2\tau}{D} \\[2mm] \therefore \tau = \dfrac{TD}{\dfrac{\pi}{32}(D^4 - d^4) \times 2} \\[3mm] = \dfrac{16\, TD}{\pi (D^4 - d^4)} \end{array} \right]$$

Principal stresses are:

$$\sigma_{1,2} = \dfrac{\sigma_b}{2} \pm \sqrt{\left(\dfrac{\sigma_b}{2}\right)^2 + \tau^2} = \dfrac{27·16}{2} \pm \sqrt{\left(\dfrac{27·16}{2}\right)^2 + (43·46)^2}$$

$$= 13·58 \pm \sqrt{(13·58)^2 + (43·46)^2} = 13·58 \pm 45·53$$

∴ $\qquad \sigma_1 = \mathbf{59·11} \text{ MN/m}^2$

and, $\qquad \sigma_2 = \mathbf{31·95} \text{ MN/m}^2$

(i) According to maximum elastic strain theory :

$$\sigma_1^2 + \sigma_2^2 - \frac{2}{m}\sigma_1\sigma_2 = \sigma^2$$

or $(59 \cdot 11)^2 + (31 \cdot 95)^2 - 2 \times 0 \cdot 25 \times 59 \cdot 11 \times (-31 \cdot 95) = \sigma^2$

$3494 + 1020 \cdot 8 + 944 \cdot 28 = \sigma^2$

or $\sigma^2 = 5459 \cdot 08$

or $\boldsymbol{\sigma = 73 \cdot 88 \text{ MN/m}^2}$ **(Ans.)**

(ii) According to maximum elastic shear strain energy theory :

$$\sigma_1^2 + \sigma_2^2 - \sigma_1\sigma_2 = \sigma^2$$

$(59 \cdot 11)^2 + (31 \cdot 95)^2 - 59 \cdot 1 (-31 \cdot 95) = \sigma^2$

$3494 + 1020 \cdot 8 + 1888 \cdot 56 = \sigma^2$

or $\sigma^2 = 6403 \cdot 4$

∴ $\boldsymbol{\sigma = 80 \cdot 02 \text{ MN/m}^2}$ **(Ans.)**

Example 18·14. *The direct stresses on two mutually perpendicular planes, in a two-dimensional stress system, are σ and 144 MN/m². In addition these planes carry a shear stress of 48 MN/m². Assuming the factor of safety on elastic limit as 3,*

(i) *Find the value of σ at which the shear strain energy is least;*

(ii) *If the failure occurs at this value of the shear strain energy, estimate the elastic limit of the material in simple tension.*

Solution. The two principal stresses are given by

$$\sigma_{1,2} = \frac{144 + \sigma}{2} \pm \sqrt{\left(\frac{144 - \sigma}{2}\right)^2 + 48^2}$$

i.e.,
$$\sigma_1 = \left(72 + \frac{\sigma}{2}\right) + \tau_{max}$$

and,
$$\sigma_2 = \left(72 + \frac{\sigma}{2}\right) - \tau_{max}$$

where,
$$\tau_{max} = \sqrt{\left(72 - \frac{\sigma}{2}\right)^2 + 48^2}$$

(i) Value of σ :

Shear strain energy (U) is given by :

$$U = \frac{1}{12C}\left[(\sigma_1 - \sigma_2)^2 + (\sigma_2 - \sigma_3)^2 + (\sigma_3 - \sigma_1)^2\right]$$

$$= \frac{1}{12C}\left[(\sigma_1 - \sigma_2)^2 + (\sigma_2 - 0)^2 + (0 - \sigma_1)^2\right] \qquad (\because \sigma_3 = 0)$$

$$= \frac{1}{12C}\left[(2\tau_{max})^2 + \sigma_1^2 + \sigma_2^2\right] \qquad \left[\because \frac{\sigma_1 - \sigma_2}{2} = \tau_{max}\right]$$

$$= \frac{1}{12C}\left[4\tau_{max}^2 + 2\left\{\left(72 + \frac{\sigma}{2}\right)^2 + \tau_{max}^2\right\}\right]$$

$$= \frac{1}{6C}\left[3\tau_{max}^2 + \left(72 + \frac{\sigma}{2}\right)^2\right]$$

$$= \frac{1}{6C}\left[3\left\{\left(72 - \frac{\sigma}{2}\right)^2 + 48^2\right\} + \left(72 + \frac{\sigma}{2}\right)^2\right]$$

$$= \frac{1}{6C}\left[3\left(5184 + \frac{\sigma^2}{4} - 72\sigma + 2304\right) + 5184 + \frac{\sigma^2}{4} + 72\sigma\right]$$

$$= \frac{1}{6C}\left[15552 + \frac{3\sigma^2}{4} - 216\sigma + 6912 + 5184 + \frac{\sigma^2}{4} + 72\sigma\right]$$

i.e.,
$$U = \frac{1}{6C}\left[\sigma^2 - 144\sigma + 27648\right]$$

For minimum $\quad U, \dfrac{dU}{d\sigma} = 0$

i.e., $\qquad 2\,\sigma - 144 = 0$

∴ $\qquad\qquad \sigma = 72 \text{ MN/m}^2 \quad \text{(Ans.)}$

(*ii*) Elastic limit of the material in tension, σ_e :

For failure at this level of shear strain energy, we have

$$\sigma_1^2 + \sigma_2^2 - \sigma_1\sigma_2 = \left(\frac{1}{3}\sigma_e\right)^2$$

$$(144)^2 + (72)^2 - 144 \times 72 = \frac{\sigma_e^2}{9}$$

$$20736 + 5184 - 10368 = \frac{\sigma_e^2}{9}$$

or $\qquad\qquad 15552 = \dfrac{\sigma_e^2}{9}$

∴ $\qquad\qquad \sigma_e = 374 \text{ MN/m}^2 \quad \text{(Ans.)}$

Example 18·15. *A cylindrical shell made of mild steel plate and 1·2 m in diameter is to be subjected to an internal pressure of 1·5 MN/m². If the material yields at 200 MN/m², calculate the thickness of the plate on the basis of the following three theories, assuming a factor of safety 3 in each case :*

(*i*) *Maximum principal stress theory,*

(*ii*) *Maximum shear stress theory, and*

(*iii*) *Maximum shear strain energy theory.*

(Engineering Services)

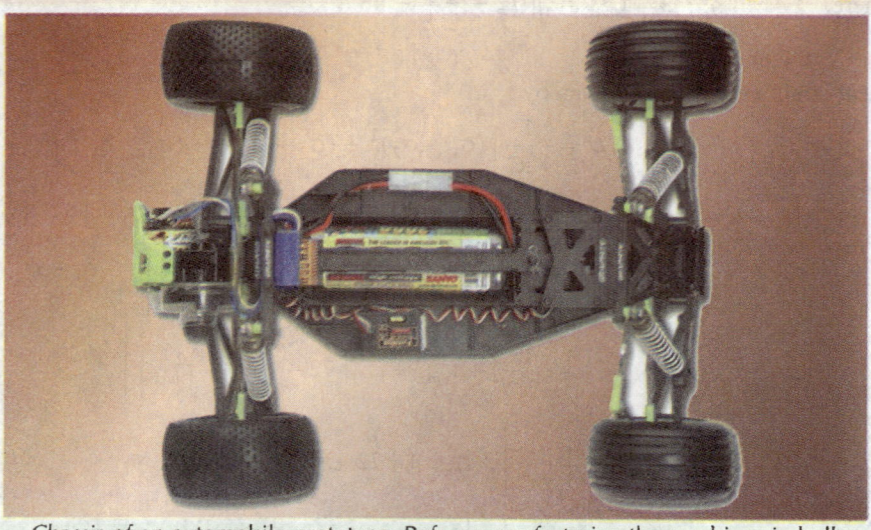

Chassis of an automobile prototype. Before manufacturing the machines in bulk, prototypes are tested in the laboratories for various stresses.

Solution. Diameter of the shell,

$$d = 1 \cdot 2 \text{ m}$$

Internal pressure, $\quad p = 1 \cdot 5 \text{ MN/m}^2$

Yield stress, $\qquad = 200 \text{ MN/m}^2$

Factor of safety $\qquad = 3$

∴ Allowable stress in simple tension,

$$\sigma = \frac{200}{3} \text{ MN/m}^2$$

Thickness, t :

Circumferential (or hoop) stress,

$$\sigma_c = \frac{pd}{2t} = \frac{1 \cdot 5 \times 1 \cdot 2}{2t} = \frac{0 \cdot 9}{t} \text{ MN/m}^2$$

Longitudinal stress, $\sigma_l = \dfrac{pd}{4t} = \dfrac{1 \cdot 5 \times 1 \cdot 2}{4t} = \dfrac{0 \cdot 45}{t} \text{ MN/m}^2$

(i) *According to maximum principal stress theory :*

$$\frac{0 \cdot 9}{t} = \frac{200}{3}$$

$$t = \textbf{0·0135 m or 13·5 mm} \textbf{(Ans.)}$$

(ii) *According to maximum shear stress theory :*

$$\frac{\sigma_c - \sigma_l}{2} = \frac{\sigma}{2}$$

$$\frac{0 \cdot 9 - 0 \cdot 45}{2t} = \frac{200}{6}$$

$$t = \frac{1}{2}\left[\frac{(0 \cdot 9 - 0 \cdot 45)\, 6}{200} \right] = \textbf{0·00675 m or 6·75 mm} \textbf{(Ans.)}$$

(iii) *According to maximum shear strain energy theory :*

$$\sigma_1^2 + \sigma_2^2 - \sigma_1 \sigma_2 = \sigma^2$$

Gear teeth can fail if the design is faulty or loads are too much.

$$\frac{1}{t^2}\left[(0.9)^2 + (0.45)^2 - 0.9 \times 0.45\right] = \left(\frac{200}{3}\right)^2$$

$$\frac{1}{t^2}\left[0.81 + 0.2025 - 0.405\right] = 4444.4$$

or, $$t^2 = \frac{0.6075}{4444.4} = 1.367 \times 10^{-4}$$

or, $$t = 0.0117, \text{ m} = \textbf{11.7 mm} \quad \textbf{(Ans.)}$$

HIGHLIGHTS

* If at a point in a strained body σ_1, σ_2 and σ_3 are the principal stresses such that $\sigma_1 > \sigma_2 > \sigma_3$ and σ_e is the yield point stress of material when tested in simple tension or compression test, then according to :

(i) *Maximum principal stress theory* :

$$\left.\begin{array}{l} \sigma_1 = \sigma_{et} \text{ (in simple tension)} \\ |\,\sigma_3\,| = \sigma_{ec} \text{ (in simple compression)} \end{array}\right] \quad \text{At the point of failure}$$

$$[\,|\,\sigma_3\,| = \text{means numerical value of } \sigma_3\,]$$

$$\begin{array}{l} \sigma_1 = \sigma_t \\ |\,\sigma_3\,| = \sigma_c \end{array} \quad \text{For design purposes}$$

[where, σ_t and σ_c are safe stresses in tension and compression respectively.]

(ii) *Maximum shear stress theory* :

$$\sigma_1 - \sigma_3 = \sigma_{et}$$

In actual design σ_{et} in the above equation is replaced by the safe stress.

(iii) *Strain energy theory* :

$$\sigma_1^2 + \sigma_2^2 + \sigma_3^2 - \frac{2}{m}(\sigma_1\sigma_2 + \sigma_2\sigma_3 + \sigma_3\sigma_1) = \sigma_e^2$$

In actual design σ_e in the above eqn. is replaced by allowable stress obtained by dividing σ_e by F.O.S. For a two-dimensional case, where $\sigma_3 = 0$, we have

$$\sigma_1^2 + \sigma_2^2 - \frac{2}{m}\sigma_1\sigma_2 = \sigma_e^2$$

For design criterion :

$$\sigma_1^2 + \sigma_2^2 - \frac{2}{m}\sigma_1\sigma_2 \leqslant \sigma^2 \quad \text{(where, } \sigma = \text{working/safe stress)}$$

(iv) *Shear strain energy theory* :

$$(\sigma_1 - \sigma_2)^2 + (\sigma_2 - \sigma_3)^2 + (\sigma_3 - \sigma_1)^2 = 2\,\sigma_{et}^2$$

In actual design σ_{et} is replaced by safe equivalent stress σ_t in simple tension.

(v) *Maximum principal strain energy theory* :

At the point of elastic failure :

$$\sigma_1 - \frac{1}{m}(\sigma_2 + \sigma_3) = \sigma_{et} \quad \text{and} \quad |\sigma_3 - \frac{1}{m}(\sigma_1 + \sigma_2)| = \sigma_{ec}$$

For design purpose:

$$\sigma_1 - \frac{1}{m}(\sigma_2 + \sigma_3) = \sigma_t$$

$$|\sigma_3 - \frac{1}{m}(\sigma_1 + \sigma_2)| = \sigma_c \quad \quad \text{(where, } \sigma_t \text{ and } \sigma_c \text{ are the safe stresses.)}$$

OBJECTIVE TYPE QUESTIONS

1. ....theory is suitable for brittle materials.
   - (a) Maximum strain energy
   - (b) Maximum shear stress theory
   - (c) Maximum principal stress theory
   - (d) Distortion energy theory.

2. Which of the following theories is suitable for ductile materials ?
   - (a) Maximum principal stress theory
   - (b) Maximum principal strain theory
   - (c) Maximum shear stress theory
   - (d) Distortion energy theory.

3. Maximum shear stress theory was postulated by
   - (a) St. Venant
   - (b) Mohr
   - (c) Rankine
   - (d) Tresca.

4. Maximum principal stress theory was postulated by
   - (a) St. Venant
   - (b) Rankine
   - (c) Mohr
   - (d) Tresca.

5. Strain energy theory was postulated by
   - (a) Rankine
   - (b) Mohr
   - (c) Tresca
   - (d) Haigh.

ANSWERS

1. (c)   2. (c)   3. (d)   4. (b)   5. (d)

UNSOLVED EXAMPLES

1. A mild steel shaft 100 mm diameter is subjected to a maximum torque of 15 kNm and a maximum bending moment of 10 kNm at a particular section. Find the factor of safety according to the maximum shear stress theory if the elastic limit in simple tension is 240 MN/m². **[Ans.** 1·27]

2. A shaft is subjected to a maximum torque of 12 kNm and a maximum bending moment of 9 kNm at a particular section. If the allowable equivalent stress in simple tension is 180 MN/m² find the diameter of the shaft according to the maximum shear stress theory.

   [**Ans.** 94·7 mm]

3. In a steel member at a point the major principal stress is 200 N/mm², and the minor principal stress is compressive. If the tensile yield point of the steel is 250 N/mm², find the value of the minor principal stress at which yielding will commence, according to each of the following criteria of failure:
   - (i) Maximum shearing stress,
   - (ii) Maximum total strain energy, and
   - (iii) Maximum shear strain energy.

   Take Poisson ratio as 0·28.

   $$\begin{bmatrix} \textbf{Ans.} & (i) \ 50 \ \text{N/mm}^2 \ (comp.) \\ & (ii) \ 104{\cdot}5 \ \text{N/mm}^2 \ (comp.) \\ & (iii) \ 80{\cdot}1 \ \text{N/mm}^2 \ (comp.) \end{bmatrix}$$

Steam turbines and gear reducers used in sugar industry. For safe operation, machine parts need to be highly reliable.

4. In a material the principal stresses are 50 N/mm^2, 40 N/mm^2 and – 30 N/mm^2, calculate :
    (i) Total strain energy,
    (ii) Volumetric strain energy,
    (iii) Shear strain energy, and
    (iv) Factor of safety on the total strain energy criterion if the material yields at 100 N/mm^2.
    Take $E = 200 \times 10^3$ N/mm^2 and Poisson's ratio = 0.3.

    $$\left[\textbf{Ans.} \; (i) \; 13.55 \; \text{kNm/m}^3 \; (ii) \; 1.2 \; \text{kNm/m}^3 \; (iii) \; 12.35 \; \text{kNm/m}^3 \; (iv) \; 1.845 \right]$$

5. A shaft of 100 mm diameter is subjected to a bending moment of 5 kNm. Determine the value of maximum torque which can be applied to the shaft for each of the following conditions :
    (i) Maximum direct stress not to exceed 120 N/mm^2.
    (ii) Maximum shearing stress not to exceed 60 N/mm^2.
    (iii) Maximum shear strain energy per unit volume not to exceed that induced by simple shear stress of 80 N/mm^2. **[Ans.** (i) 17.85 kNm (ii) 10.65 kNm (iii) 14.6 kNm]

6. The inside and outside diameters of a cast-iron cylinder are 200 mm and 125 mm respectively. If the ultimate strength of cast iron is 150 N/mm^2, find according to each of the following theories the internal pressure which would cause rupture:
    (i) Maximum principal stress theory;   (ii) Maximum strain theory;
    (iii) Maximum strain energy theory.
    Take Poisson's ratio as 0.25. Assume no longitudinal stress in the cylinder.
    [Ans. (i) 65.7 N/mm^2 (ii) 59.2 N/mm^2 (iii) 55.2 N/mm^2]

7. The direct stresses, in a two-dimensional stress system, on two mutually perpendicular planes are σ and 120 MN/m^2. Find the value of σ at which the shear strain energy is least. If the failure occurs at this value of the shear strain energy, estimate the elastic limit of the material in simple tension. Take the factor of safety of elastic limit as 3.
    [Ans. 60 MN/m^2, 312 MN/m^2]

In a power screw rotary motion is translated into linear motion.

8. A solid shaft circular in section, is 100 mm in diameter and subjected to combined bending and twisting moments, the bending moment being three times the twisting moments. If the direct tensile yield point of the material is 350 N/mm^2 and factor of safety on the yield is to be 4, calculate the allowable twisting moments by the following three theories of elastic failure :
    (i) Maximum principal stress theory,
    (ii) Maximum shearing stress theory, and
    (iii) Shear strain energy theory. **[Ans.** (i) 2790 Nm, 2715 Nm, 2755 Nm]

9. A circular steel shaft is subjected to combined bending and torsion, the bending moment being 20 kNm and torque 10 kNm. If safe equivalent stress in simple tension is 200 N/mm^2 and Poisson's ratio is 0.25, find suitable diameter of the shaft based on the following theories :
    (i) Maximum principal stress theory,
    (ii) Maximum shear stress theory, and
    (iii) Shear strain energy theory. **[Ans.** (i) 102.56 mm (ii) 104 mm, (iii) 103.5 mm]

10. The principal stresses in a steam boiler are 80, 40 and 0 N/mm^2. Find factor of safety based on elastic limit according to each of the following theories :
    (i) Maximum direct stress theory;   (ii) Maximum shear stress theory;
    (iii) Total strain energy theory;   (iv) Shear strain energy theory.
    Take: Poisson's ratio = 0.3, and elastic limit = 320 N/mm^2

    [Ans. (i) 4, (ii) 4, (iii) 4.1 and (iv) 4.61]

Stresses due to Rotation

Hydel generator

19·1. INTRODUCTION

The bodies like rings, circular discs, cylinders, flywheels, etc. invariably rotate at high speeds and due to rotation they are subjected to large magnitudes of centrifugal forces. The stresses caused by these forces are distributed symmetrically about their axes of rotation. In this chapter we shall study these stresses for machine members of the rotating type; assuming the density of material to be uniform throughout.

19·2. ROTATING RING

Fig. 19·1 (*a*) shows the thin ring rotating about its centre of gravity at 0.

Let, r = Mean radius of the ring, m,

t = Thickness of the ring, m,

ρ = Density of the material of the ring, kg/m^3,

ω = Angular speed of the ring, radian/sec.,

F_c = Centrifugal force, and

σ_c = Circumferential or hoop stress.

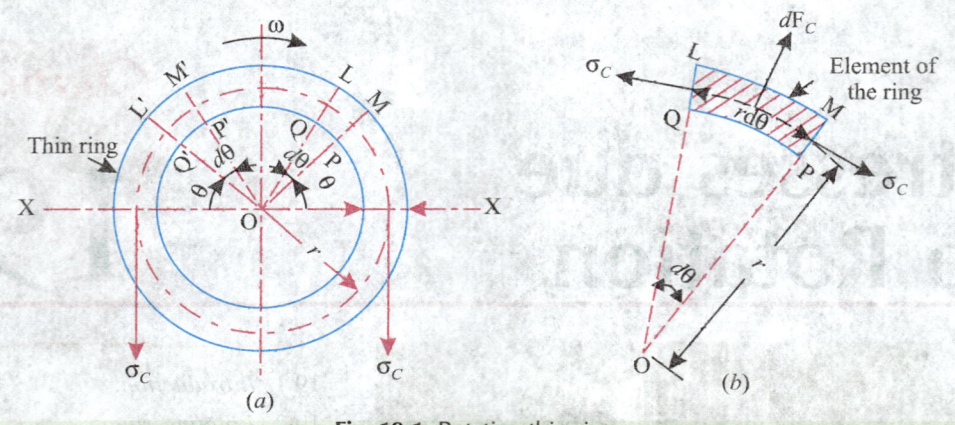

Fig. 19.1. Rotating thin ring.

As a result of rotation each and every element of the ring like *LMPQ* will experience centrifugal (or inertia) force dF_c *which will tend to expand the ring radially outwards. This will in turn induce the circumferential (or hoop stress)* σ_c *in the ring which will be tensile in nature. For evaluating this stress following* **assumptions** *are made :*

   (*i*)   The circumferential stress on the area of cross-section of the ring is uniform.

   (*ii*)   The dimensions of the cross-section of the ring are small as compared to its mean radius.

   (*iii*)   The constraining effect of spokes is negligible.

Now, volume of small element (Fig. 19·1(*b*) *LMPQ* per unit *length* = $r.d\theta.t$

Hence centrifugal force acting on the element,

$$dF_c = \rho\, r.d\theta.t.\,\omega^2\, r \qquad (\because F_c = m\,\omega^2\, r)$$

Vertical component of $\qquad dF_c = dF_c \sin\theta$

$$= \rho.\, r.d\theta.t\,\omega^2\, r \sin\theta$$

Horizontal component of dF_c *will be cancelled when we consider another small element* *L'M'P'Q' in II quadrant at an angle* θ, *but the vertical component of* dF_c *will be added.*

Total vertical component or bursting force across the horizontal diameter *XX*

$$= \int_0^\pi dF_c \sin\theta \;=\; \int_0^\pi \rho\cdot r\cdot d\theta\cdot t\,\omega^2 r \sin\theta$$

$$= \rho\,\omega^2 r^2 t \int_0^\pi \sin\theta\, d\theta$$

$$= \rho\,\omega^2 r^2 t\, |{-\cos\theta}\,|_0^\pi = 2\rho\,\omega^2 r^2 t$$

Now, total resisting force $\qquad = 2\,\sigma_c.t.1$

The ring will be in equilibrium, when :

Total bursting force = Total resisting force

$$2\rho\omega^2 r^2 t = 2\,\sigma_c.t.1$$

∴ $\qquad\qquad\qquad\qquad \sigma_c = \rho\,\omega^2\, r^2 \qquad\qquad\qquad\qquad\qquad\qquad$ ...(19·1)

But, $\qquad\qquad\qquad\qquad v = \omega r$

(where, v = linear velocity of the ring)

∴ $\qquad\qquad\qquad\qquad \sigma_c = \rho\, v^2 \qquad\qquad\qquad\qquad\qquad\qquad\quad$ ...(19·2)

Example 19·1. *A wheel 800 mm in diameter has a thin rim. If density is 7700 kg/m³ and E = 200 GN/m², calculate:*

(i) *How many revolutions per minute may it make without the hoop stress exceeding 130 MN/m² ?*

(ii) *Change in diameter.*

Neglect the effect of spokes.

Solution. Radius of the wheel, $r = \dfrac{800}{2} = 400$ mm $= 0.4$ m

Density of material, $\rho = 7700$ kg/m³

Young's modulus, $E = 200$ GN/m²

Hoop stress, $\sigma_c = 130$ MN/m².

(i) Number of revolutions, N :

We know that $\sigma_c = \rho\, \omega^2\, r^2$

$$130 \times 10^6 = 7700 \times \omega^2 \times 0.4^2$$

∴ $\omega = 324.8$ rad/sec

But, $\omega = \dfrac{2\,\pi\,N}{60} = 324.8$

∴ $N = \dfrac{324.8 \times 60}{2\pi} = \textbf{3101·6 r.p.m}$ **(Ans.)**

(ii) Change in diameter, δd :

Circumferential strain, $e_c = \dfrac{\sigma_c}{E}$

or, $\dfrac{\delta d}{d} = \dfrac{130 \times 10^6}{200 \times 10^9}$

∴ $\delta d = 0.8 \times \dfrac{130 \times 10^6}{200 \times 10^9} = 5.2 \times 10^{-4}$ m $= 0.52$ mm

Hence, $\delta d = \textbf{0·52 mm}$ **(Ans.)**

Example 19·2. *Fig. 19·2 shows a built-up ring. If the ring rotates at 2000 r.p.m. find the stresses set up in steel and copper rings.*

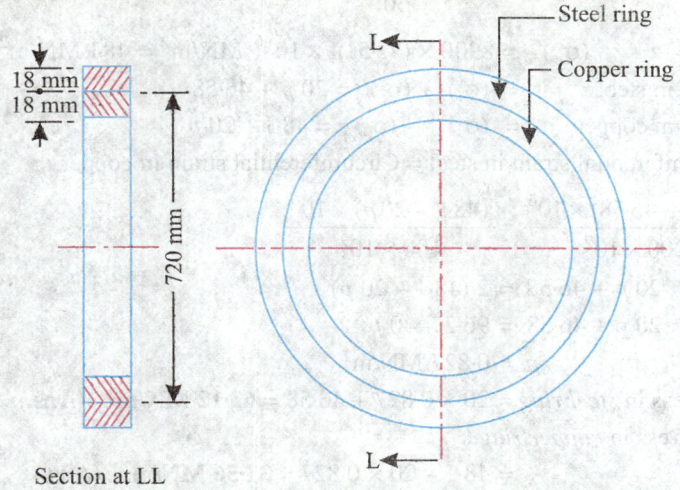

Section at LL

Fig. 19.2

Assume: For steel, $E = 200$ GN/m²; $\rho = 7800$ kg/m³
For copper, $E = 100$ GN/m²; $\rho = 8900$ kg/m³

Solution. *Given:* $E_s = 200$ GN/m², $\rho_s = 7800$ kg/m³
$E_{cu} = 100$ GN/m², $\rho_{cu} = 8900$ kg/m³
Speed, $N = 2000$ *r.p.m.*

Stresses in steel and copper rings :

Let, p = Contact pressure between the steel and copper rings, MN/m²

Circumferential stress in the steel ring,

$$(\sigma_c)_s = \frac{pd}{2t} = \frac{p \times 720}{2 \times 18} = 20\, p \text{ MN/m}^2 \text{ (tensile)}$$

Circumferential stress in the copper ring,

$$(\sigma_c)_{cu} = \frac{pd}{2t} = \frac{p \times 720}{2 \times 18} = 20\, p \text{ MN/m}^2 \text{ (comp.)}$$

Circumferential stresses *due to rotation* in the rings :

Steel ring :

$$(\sigma_c)'_s = \rho_s v^2$$

where,

$$v = \omega r = 2\pi N$$

$$r_s = 2\pi \times \frac{2000}{60} \times [(360 + 9) \times 10^{-3}]$$

$$= 77 \cdot 28 \text{ m/s}$$

$$\therefore \quad (\sigma_c)'_s = 7800 \times (77 \cdot 28)^2 \times 10^{-6} \text{ MN/m}^2$$

$$= 46 \cdot 58 \text{ MN/m}^2 \text{ (tensile)}$$

Copper ring :

$$(\sigma_c)'_{cu} = \rho_{cu}\, v^2$$

where, $v = \omega r = 2\, \pi N r_{cu}$

Wind turbines.

$$= 2\pi \times \frac{2000}{60} \times [(360 - 9) \times 10^{-3}] = 73 \cdot 51 \text{ m/s}$$

$$\therefore \qquad (\sigma_c)'_{cu} = 8900 \times (73 \cdot 51)^2 \times 10^{-6} \text{ MN/m}^2 = 48 \cdot 1 \text{ MN/m}^2 \text{ (tensile)}$$

Total stress in steel $= (\sigma_c)_s + (\sigma_c)'_s = 20\, p + 46 \cdot 58$

Total stress in copper $= (\sigma_c)'_{cu} - (\sigma_c)_{cu} = 48 \cdot 1 - 20\, p$

Now, circumferential strain in steel = Circumferential strain in copper

$$\frac{(20p + 46 \cdot 58) \times 10^6}{200 \times 10^9} = \frac{(48 \cdot 1 - 20p) \times 10^6}{100 \times 10^9}$$

$$20\, p + 46 \cdot 58 = 2\,(48 \cdot 1 - 20\, p)$$

or, $\qquad 20 \text{ p} + 46 \cdot 58 = 96 \cdot 2 - 40\, p$

$$\therefore \qquad\qquad p = 0 \cdot 827 \text{ MN/m}^2$$

∴ Total stress in *steel ring* $= 20 \times 0 \cdot 827 + 46 \cdot 58 = \textbf{63·12 MN/m}^2$ **(Ans.)**

and, total stress in *copper ring*

$$= 48 \cdot 1 - 20 \times 0 \cdot 827 = \textbf{31·56 MN/m}^2 \text{ (Ans.)}$$

19·3. ROTATING THIN DISC

Fig. 19·3 (*a*) shows a circular disc of inner radius r_1 and outer radius r_2 rotating about its axis.

Let us *assume* that the disc is of uniform thickness and that the thickness is so small compared with its diameter that there is *no variation of stress along the thickness*. At the free flat surfaces there can be no stress normal to these faces and there can be no shear stress on or perpendicular to these faces. Thus the direction of axis is the direction of zero principal stress. The displacement of any point due to strain must be radial. The *radial and circumferential stresses*, therefore, represent the *principal stresses*.

Consider an element *ABCD* of the disc, at a radius r, subtending an angle $d\theta$ at the centre, and of radial width dr.

Let ,

$$\sigma_r = \text{Stress on the face } CD,$$
$$\sigma_r + d\sigma_r = \text{Stress on the face } AB,$$
$$\sigma_c = \text{Stress on face } BC,$$
$$\sigma_c = \text{Stress on face } AD.$$

On the flat faces of the disc, there is no normal stress, and hence there is free strain in the direction of the axis.

Volume of the element

$$ABCD = rd\theta.dr.t \quad (\text{where, } t = \text{thickness of the disc})$$

Radial force on the element *ABCD* due to rotation

$$= \rho \, r.d\theta.dr. \, t. \, \omega^2 \, r = \rho \, d\theta. \, dr. \, t \, \omega^2 \, r^2$$

Force on face *AB* (*outward*)

$$= (r + dr) \, d\theta. \, t \, (\sigma_r + d\sigma_r)$$

Force on *CD* (*inward*) $\quad = r. \, d\theta. \, t. \, \sigma_r$

Force on faces *BC* and $AD = \sigma_c. \, t. \, dr$

[Forces acting on the element are shown in Fig. 19·3 (*b*)]

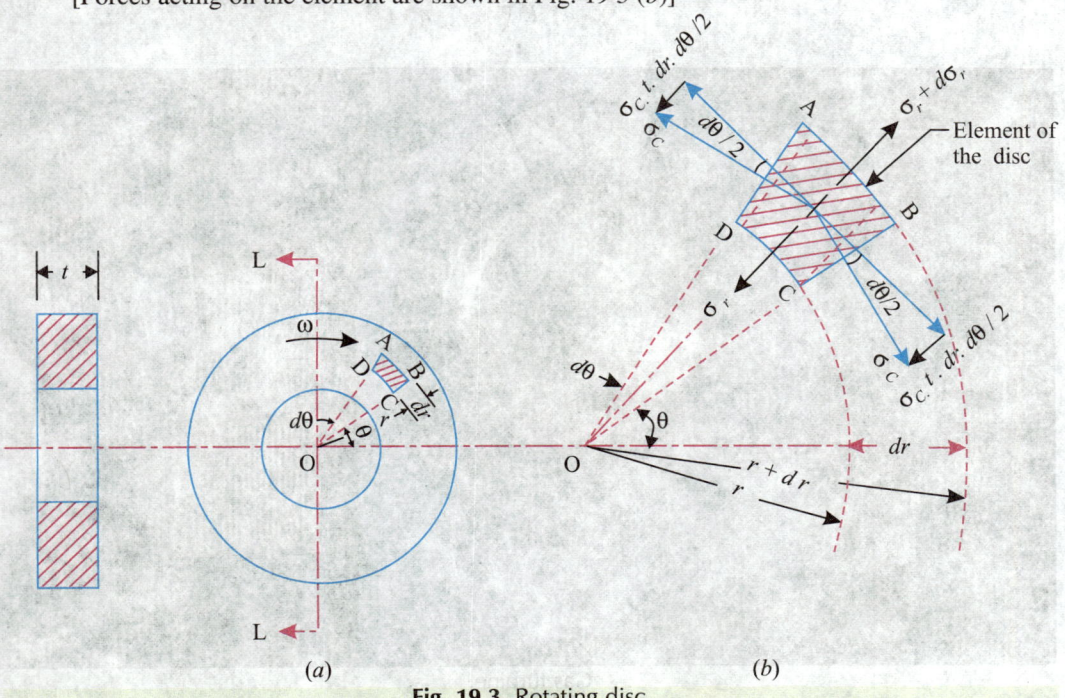

Fig. 19.3. Rotating disc.

Resolving the forces in the radial direction and considering the equilibrium of the forces, we get

$$rd\theta.t.\sigma_r + 2\,\sigma_c\,t.dr\,\sin\left(\frac{d\theta}{2}\right) = (\sigma_r + d\sigma_r)\,(r + dr)\,d\theta.t + \rho\,d\theta.dr.t\,\omega^2 r^2$$

Since $d\theta$ is very small, $\quad \sin\dfrac{d\theta}{2}$ is $\simeq \dfrac{d\theta}{2}$

Now, $t.d\theta$ is common on both the sides, the above expression on simplification becomes

$$\sigma_r.r + \sigma_c.dr = \sigma_r.r + \sigma_r.dr + r.d\sigma_r + d\sigma_r.dr + \rho\omega^2 r^2 dr$$

Neglecting the term $d\sigma_r.dr$ and on further simplification the expression becomes

$$\sigma_c.dr = \sigma_r.dr + r.d\sigma_r + \rho\omega^2 r^2 dr$$

Dividing both sides by dr, we get

$$\sigma_c = \sigma_r + r.\frac{d\sigma_r}{dr} + \rho\omega^2 r^2$$

or,
$$\sigma_c - \sigma_r = r.\frac{d\sigma_r}{dr} + \rho\omega^2 r^2 \qquad \qquad ...(i)$$

Let us now consider circumferential and radial *strains*.

When the disc is rotating at high speed let r become $(r + u)$ and $(r + dr)$ *become* $(r + dr + du)$

Circumferential strain, $\quad e_c = \dfrac{2\pi\,(r + u) - 2\pi r}{2\pi r} = \dfrac{u}{r}$

Radial strain, $\quad e_r = \dfrac{(r + dr + du) - (r + dr)}{dr} = \dfrac{du}{dr}$

Also, $\quad e_c = \dfrac{1}{E}\left(\sigma_c - \dfrac{\sigma_r}{m}\right) = \dfrac{u}{r}$

$\therefore \quad u = \dfrac{r}{E}\left(\sigma_c - \dfrac{\sigma_r}{m}\right) \qquad \qquad ...(ii)$

Gas turbine.

and,
$$\sigma_r = \frac{1}{E}\left(\sigma_r - \frac{\sigma_c}{m}\right) = \frac{du}{dr} \qquad \qquad ...(iii)$$

$$\left(\text{where, } \frac{1}{m} = \text{Poisson's ratio}\right)$$

Differentiating equation (ii) w.r.t r, we get

$$\frac{du}{dr} = \frac{r}{E}\left[\frac{d\sigma_c}{dr} - \frac{1}{m}\frac{d\sigma_r}{dr}\right] + \frac{1}{E}\left(\sigma_c - \frac{\sigma_r}{m}\right) \qquad ...(iv)$$

Comparing equations (iii) and (iv), we get

$$\frac{1}{E}\left(\sigma_r - \frac{\sigma_c}{m}\right) = \frac{r}{E}\left[\frac{d\sigma_c}{dr} - \frac{1}{m}\frac{d\sigma_r}{dr}\right] + \frac{1}{E}\left(\sigma_c - \frac{\sigma_r}{m}\right)$$

$$\frac{1}{E}\left(\sigma_r - \frac{\sigma_c}{m} - \sigma_c + \frac{\sigma_r}{m}\right) = \frac{r}{E}\left(\frac{d\sigma_c}{dr} - \frac{1}{m}\frac{d\sigma_r}{dr}\right)$$

$$(\sigma_r - \sigma_c)\left(1 + \frac{1}{m}\right) = r\left(\frac{d\sigma_c}{dr} - \frac{1}{m}\frac{d\sigma_r}{dr}\right) \qquad ...(v)$$

Substituting eqn. (i) in eqn. (v), we get

$$\left(1 + \frac{1}{m}\right)\left[-r.\frac{d\sigma_r}{dr} - \rho\omega^2 r^2\right] = r\left(\frac{d\sigma_c}{dr} - \frac{1}{m}\frac{d\sigma_r}{dr}\right)$$

On simplification, we get

$$\frac{d}{dr}(\sigma_r + \sigma_c) + \left(1 + \frac{1}{m}\right)\rho\omega^2 r = 0 \qquad ...(vi)$$

Integrating equation (vi), we get

$$\sigma_r + \sigma_c + \left(1 + \frac{1}{m}\right)\frac{\rho\omega^2 r^2}{2} = C_1 \qquad ...(vii)$$

(where, C_1 = constant of integration)

$$\sigma_c = C_1 - \sigma_r - \left(1 + \frac{1}{m}\right)\frac{\rho\omega^2 r^2}{2}$$

Substituting in eqn. (i), we get

$$C_1 - \sigma_r - \left(1 + \frac{1}{m}\right)\frac{\rho\omega^2 r^2}{2} - \sigma_r = r \cdot \frac{d\sigma_r}{dr} + \rho\omega^2 r^2$$

$$2\sigma_r + r\frac{d\sigma_r}{dr} = C_1 - \rho\omega^2 r^2\left[1 + \frac{\left(1 + \frac{1}{m}\right)}{2}\right]$$

$$2\sigma_r + r\frac{d\sigma_r}{dr} = C_1 - \left(\frac{3 + \frac{1}{m}}{2}\right)\rho\omega^2 r^2$$

Multiplying both sides by r, we get

$$2r\sigma_r + r^2\frac{d\sigma_r}{dr} = C_1 r - \left(\frac{3 + \frac{1}{m}}{2}\right)\rho\omega^2 r^3$$

or,
$$\frac{d}{dr}(r^2\sigma_r) = C_1 r - \left(\frac{3 + \frac{1}{m}}{2}\right)\rho\omega^2 r^3$$

Integrating both sides, we get

$$r^2\sigma_r = C_1 \cdot \frac{r^2}{2} - \left(\frac{3 + \frac{1}{m}}{2}\right)\rho\omega^2 \frac{r^4}{4} + C_2$$

or,

$$\sigma_r = \frac{C_1}{2} - \left(\frac{3 + \frac{1}{m}}{8}\right)\rho\omega^2 r^2 + \frac{C_2}{r^2}$$

or,

$$\sigma_r = \frac{C_1}{2} + \frac{C_2}{r^2} - \left(\frac{3 + \frac{1}{m}}{8}\right)\rho\omega^2 r^2 \qquad \qquad ...(19.3)$$

(where, C_2 = constant of integration)

Substituting eqn. (19.3) in equation (vii), we get

$$\left[\frac{C_1}{2} - \left(\frac{3 + \frac{1}{m}}{8}\right)\rho \cdot \omega^2 r^2 + \frac{C_2}{r^2}\right] + \sigma_c + \left(1 + \frac{1}{m}\right)\frac{\rho \cdot \omega^2 r^2}{2} = C_1$$

$$\sigma_c = \frac{C_1}{2} - \frac{C_2}{r^2} + \rho \cdot \omega^2 r^2 \left[\left(\frac{3 + \frac{1}{m}}{8}\right) - \left(\frac{1 + \frac{1}{m}}{2}\right)\right]$$

$$\sigma_c = \frac{C_1}{2} - \frac{C_2}{r^2} - \left(\frac{1 + \frac{3}{m}}{8}\right)\rho\omega^2 r^2 \qquad \qquad ...(19.4)$$

The constants can be evaluated by using the boundary conditions.

Hence eqns. (19.3) and (19.4) give the expressions for radial and circumferential stresses in the disc.

Case I. Solid disc :

At the centre $r = 0$, and stresses cannot be infinite at the centre of the disc, therefore, $C_2 = 0$. Expressions for stresses will now be:

$$\sigma_r = \frac{C_1}{2} - \left(\frac{3 + \frac{1}{m}}{8}\right)\rho\omega^2 r^2$$

$$\sigma_c = \frac{C_1}{2} - \left(\frac{1 + \frac{3}{m}}{8}\right)\rho\omega^2 r^2$$

At the outer radius, $r = r_2$, $\sigma_r = 0$

∴

$$C_1 = \left(\frac{3 + \frac{1}{m}}{4}\right)\rho\omega^2 r_2^2$$

∴

$$\sigma_r = \left(\frac{3 + \frac{1}{m}}{8}\right)\rho\omega^2 (r_2^2 - r^2) \qquad \qquad ...(19.5)$$

and,

$$\sigma_c = \left(\frac{3 + \frac{1}{m}}{8}\right)\rho\omega^2 r_2^2 - \left(\frac{1 + \frac{3}{m}}{8}\right)\rho\omega^2 r^2$$

Racing car wheel.

or, $\qquad \sigma_c = \dfrac{\rho \omega^2}{8}\left[\left(3 + \dfrac{1}{m}\right)r_2^2 - \left(1 + \dfrac{3}{m}\right)r^2\right]$ $\qquad$ ...(19·6)

At $r = r_2$, $\qquad \sigma_c = \left(\dfrac{1 - \dfrac{1}{m}}{4}\right)\rho\omega^2 r_2^2$ $\qquad$ ...(19·7)

At $r = 0$ the values of σ_r and σ_c are maximum; these are :

$$(\sigma_r)_{max} = \left(\dfrac{3 + \dfrac{1}{m}}{8}\right)\rho\omega^2 r_2^2 \qquad \text{...(19·8)}$$

$$(\sigma_c)_{max} = \left(\dfrac{3 + \dfrac{1}{m}}{8}\right)\rho\omega^2 r_2^2 \qquad \text{...(19·9)}$$

Fig. 19·4 shows variations of σ_c and σ_r in a solid disc.

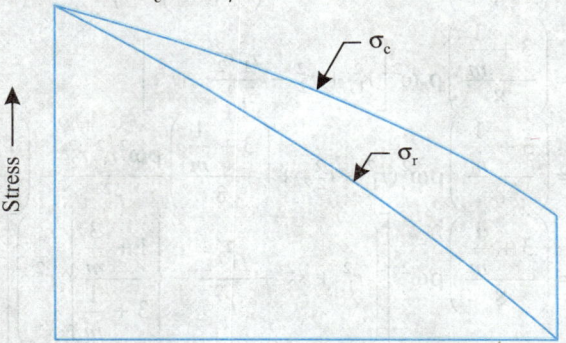

Fig. 19.4. Variations of σ_c and σ_r in a solid disc.

Case II. Hollow disc :

From eqns. (19·3) and (19·4), we have

$$\sigma_r = \dfrac{C_1}{2} + \dfrac{C_2}{r^2} - \left(\dfrac{3 + \dfrac{1}{m}}{8}\right)\rho\omega^2 r^2$$

$$\sigma_c = \dfrac{C_1}{2} - \dfrac{C_2}{r^2} - \left(\dfrac{1 + \dfrac{3}{m}}{8}\right)\rho\omega^2 r^2$$

Boundary conditions are :

At, $r = r_1$, $\sigma_r = 0$ and, at $r = r_2$, $\sigma_r = 0$

$\therefore \qquad 0 = \dfrac{C_1}{2} + \dfrac{C_2}{r_1^2} - \left(\dfrac{3 + \dfrac{1}{m}}{8}\right)\rho\omega^2 r_1^2$ $\qquad$ ...(i)

and, $\qquad 0 = \dfrac{C_1}{2} + \dfrac{C_2}{r_2^2} - \left(\dfrac{3 + \dfrac{1}{m}}{8}\right)\rho\omega^2 r_2^2$ $\qquad$ ...(ii)

From these equations, we get

$$\dfrac{C_2}{r_1^2} - \dfrac{C_2}{r_2^2} = \left(\dfrac{3 + \dfrac{1}{m}}{8}\right)\rho\omega^2 r_1^2 - \left(\dfrac{3 + \dfrac{1}{m}}{8}\right)\rho\omega^2 r_2^2$$

$$C_2 \left[\frac{r_2^2 - r_1^2}{r_1^2 r_2^2} \right] = \left(\frac{3 + \dfrac{1}{m}}{8} \right) \rho \omega^2 (r_1^2 - r_2^2) \qquad \therefore \qquad C_2 = -\left(\frac{3 + \dfrac{1}{m}}{8} \right) \rho \omega^2 r_1^2 \, r_2^2$$

Substituting the value of C_2 in eqn (i), we get

$$0 = \frac{C_1}{2} - \left(\frac{3 + \dfrac{1}{m}}{8} \right) \rho \omega^2 r_2^2 - \left(\frac{3 + \dfrac{1}{m}}{8} \right) \rho \omega^2 r_1^2$$

$$\therefore \qquad C_1 = \left(\frac{3 + \dfrac{1}{m}}{4} \right) \rho \omega^2 (r_1^2 + r_2^2)$$

$$\therefore \qquad \sigma_r = \left(\frac{3 + \dfrac{1}{m}}{8} \right) \rho \omega^2 (r_1^2 + r_2^2) - \left(\frac{3 + \dfrac{1}{m}}{8} \right) \frac{\rho \omega^2 r_1^2 r_2^2}{r^2} - \left(\frac{3 + \dfrac{1}{m}}{8} \right) \rho \omega^2 r^2$$

or,
$$\sigma_r = \left(\frac{3 + \dfrac{1}{m}}{8} \right) \rho \, \omega^2 \left[r_1^2 + r_2^2 - \frac{r_1^2 r_2^2}{r^2} - r^2 \right] \qquad \qquad ...(19 \cdot 10)$$

and,
$$\sigma_c = \left(\frac{3 + \dfrac{1}{m}}{8} \right) \rho \omega^2 (r_1^2 + r_2^2) + \left(\frac{3 + \dfrac{1}{m}}{8} \right) \frac{\rho \omega^2 r_1^2 \, r_2^2}{r^2} - \left(\frac{1 + \dfrac{3}{m}}{8} \right) \omega^2 r^2$$

$$\sigma_c = \left(\frac{3 + \dfrac{1}{m}}{8} \right) \rho \omega^2 \left[\left(r_1^2 + r_2^2 + \frac{r_1^2 r_2^2}{r^2} - \left\{ \frac{1 + \dfrac{3}{m}}{3 + \dfrac{1}{m}} \right\} r^2 \right) \right] \qquad \qquad ...(19 \cdot 11)$$

Inspection of eqn. (19·11) shows that σ_c goes on increasing as r decreases and hence σ_c is maximum when $r = r_1$.

$$\therefore \qquad (\sigma_c)_{max} = \left(\frac{3 + \dfrac{1}{m}}{8} \right) \rho \omega^2 \left[r_1^2 + r_2^2 + r_2^2 - \left\{ \frac{1 + \dfrac{3}{m}}{3 + \dfrac{1}{m}} \right\} r_1^2 \right]$$

Car wheel with tyre. When brakes are applied they exert torque on car wheels.

or, $(\sigma_c)_{max} = \left(\dfrac{3 + \dfrac{1}{m}}{4}\right)\rho\omega^2\left[r_2^2 + \left\{\dfrac{1 - \dfrac{1}{m}}{3 + \dfrac{1}{m}}\right\}r_1^2\right]$ ...(19·12)

For σ_r to be maximum, $\dfrac{d\sigma_r}{dr} = 0$

$\therefore \quad \dfrac{d}{dr}\left[\left(\dfrac{3 + \dfrac{1}{m}}{8}\right)\rho\omega^2\left\{r_1^2 + r_2^2 - \dfrac{r_1^2 r_2^2}{r^2} - r^2\right\}\right] = 0$

$\dfrac{2 r_1^2 r_2^2}{r^3} - 2r = 0$

$\therefore \quad r^4 = r_1^2 r_2^2, \quad \text{or,} \quad r = \sqrt{r_1\, r_2}$ ...(19·13)

$\therefore \quad (\sigma_r)_{max} = \left(\dfrac{3 + \dfrac{1}{m}}{8}\right)\rho\omega^2[(r_2 - r_1)^2]$ ...(19·14)

...substituting the value of r in equation (19·10)

The variations of σ_c and σ_r with r are shown in Fig. 19·5.

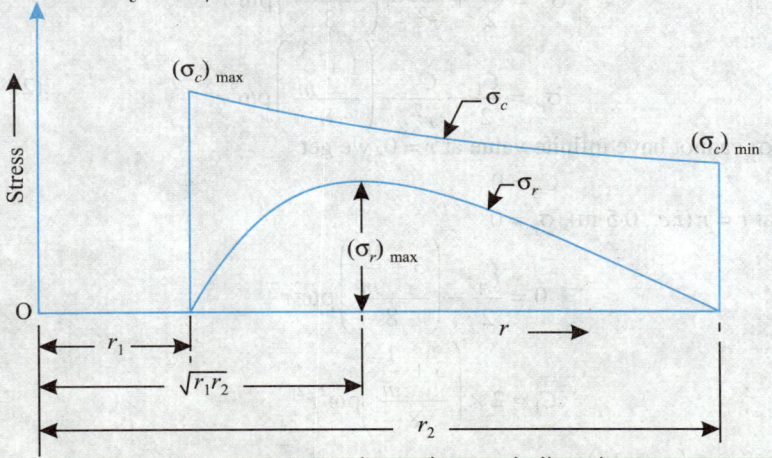

Fig. 19.5. Variations of σ_c and σ_r in a hollow disc.

Hollow disc with a pin hole at the centre :

Here $r_1 \to 0$, then from eqns. (19·12) and (19·13)

$(\sigma_c)_{max} = \left(\dfrac{3 + \dfrac{1}{m}}{4}\right)\rho\omega^2 r_2^2$ ...(19·15)

and, $(\sigma_r)_{max} = \left(\dfrac{3 + \dfrac{1}{m}}{8}\right)\rho\omega^2 r_2^2$ ...(19·16)

Comparing the eqn. (19·15) with eqn. (19·9) it can be concluded that the maximum circumferential stress in a rotating disc is twice as large, when there is a small hole at its axis of rotation, as that when the disc is solid.

Also, when r_1 approaches r_2, such that $r_1 \simeq r_2 \simeq r$, we get

$$(\sigma_c)_{max} = \rho\omega^2 r^2 \qquad \qquad ...(19·17)$$

which is the same as equation (19·1) obtained for the case of a ring.

Example 19·3. *Determine the intensities of principal stresses in flat steel disc of uniform thickness having a diameter of 1 m and rotating at 2400 r.p.m.*

What will be the stresses if the disc has a central hole of 0·2m diameter?

Take Poisson's ratio = 1/3, and $\rho = 7850$ kg/m³.

Solution. Density of material, $\rho = 7850$ kg/m³

Poisson's ratio, $\dfrac{1}{m} = \dfrac{1}{3}$

Speed, $N = 2400$ r.p.m.

∴ Angular speed, $\omega = \dfrac{2\pi N}{60} = \dfrac{2\pi \times 2400}{60} = 80\,\pi$ rad./s

Case I. Solid disc :

Radius of the disc, $r = 0·5$ m

The intensities of radial and circumferential (or hoop) stresses are given by :

$$\sigma_r = \frac{C_1}{2} + \frac{C_2}{r^2} - \left(\frac{3 + \dfrac{1}{m}}{8}\right)\rho\omega^2 r^2 \qquad ...(1)\ [\text{Eqn. (19·3)}]$$

$$\sigma_c = \frac{C_1}{2} - \frac{C_2}{r^2} - \left(\frac{1 + \dfrac{3}{m}}{8}\right)\rho\omega^2 r^2 \qquad ...(2)\ [\text{Eqn. (19·4)}]$$

Since σ_r cannot have infinite value at $r = 0$, we get

$$C_2 = 0 \qquad\qquad ...(i)$$

Also, at $r = r$ (*i.e.*, 0·5 m), $\sigma_r = 0$

∴

$$0 = \frac{C_1}{2} - \left(\frac{3 + \dfrac{1}{m}}{8}\right)\rho\omega^2 r^2$$

∴

$$C_1 = 2 \times \left(\frac{3 + \dfrac{1}{m}}{8}\right)\rho\omega^2 r^2 \qquad\qquad ...(ii)$$

Now,

$$\left(\frac{3 + \dfrac{1}{m}}{8}\right)\rho\omega^2 = \left(\frac{3 + \dfrac{1}{m}}{8}\right) \times 7850 \times (80\pi)^2 = 206·6 \times 10^6 \ \text{N/m}^4$$

∴

$$C_1 = 2 \times 206·6 \times 10^6 \times (0·5)^2 \times 10^{-6} \ \text{MN/m}^2 = 103·3 \ \text{MN/m}^2$$

Also,

$$\left(\frac{1 + \dfrac{3}{m}}{8}\right)\rho\omega^2 = \left(\frac{1 + 3 \times 1/3}{8}\right) \times 7850 \times (80\pi)^2 = 123·9 \times 10^6 \ \text{N/m}^4$$

Substituting the values in (1) and (2), we get

$$\sigma_r = 51·65 - 206·6 \times 10^6\ r^2 \times 10^{-6}\ \text{MN/m}^2 \qquad ...(3)$$

and,

$$\sigma_c = 51·65 - 123·9 \times 10^6\ r^2 \times 10^{-6}\ \text{MN/m}^2 \qquad ...(4)$$

At $r = 0$, $\sigma_r = \sigma_c = \mathbf{51·65\ MN/m^2}$ **(Ans.)**

At $r = 0·5$ m, $\sigma_r = 51·65 - 206·6 \times 10^6 \times (0·5)^2 \times 10^{-6} = 0$

and, $\sigma_c = 51·65 - 123·9 \times 10^6 \times (0·5)^2 \times 10^{-6} = \mathbf{20·67\ MN/m^2}$ **(Ans.)**

It may be noted that the above values of σ_r and σ_c are also the principal stresses.

Case II. Disc with a central hole :

Here, $r_1 = 0.1$ m, $r_2 = 0.5$ m

At $r = 0.1$ m, $\sigma_r = 0$ $\therefore$ $0 = \dfrac{C_1}{2} + \dfrac{C_2}{0.1^2} - \left(\dfrac{3 + \dfrac{1}{m}}{8}\right)\rho\omega^2 \times 0.1^2$

$$0 = 0.5\,C_1 + 100\,C_2 - 206.6 \times 10^6 \times 0.1^2 \times 10^{-6}$$

$\therefore$ $\qquad\qquad C_1 + 200\,C_2 = 4.132$ \qquad\qquad ...(iii)

At $r = 0.5$ m, $\sigma_r = 0$ $\therefore$ $0 = \dfrac{C_1}{2} + \dfrac{C_2}{0.5^2} - \left(\dfrac{3 + \dfrac{1}{m}}{8}\right)\rho\omega^2 \times 0.5^2$

$$0 = 0.5\,C_1 + 4\,C_2 - 206.6 \times 10^6 \times 0.5^2 \times 10^{-6}$$

$\therefore$ $\qquad\qquad C_1 + 8\,C_2 = 103.3$ \qquad\qquad ...(iv)

From (iii) and (iv), we get

$$C_1 = 107.43,\ C_2 = -0.516$$

Hence the stresses are given by :

$$\sigma_r = 53.7 - \dfrac{0.516}{r^2} - 206.6 \times 10^6\,r^2 \times 10^{-6} \qquad\qquad ...(5)$$

and

$$\sigma_c = 53.7 + \dfrac{0.516}{r^2} - 123.9 \times 10^6\,r^2 \times 10^{-6} \qquad\qquad ...(6)$$

At $\qquad\qquad r = 0.1$ m,

$$\sigma_c = (\sigma_c)_{max} = 53.7 + \dfrac{0.516}{0.1^2} - 123.9 \times 10^6 \times 0.1^2 \times 10^{-6}$$

$$= 53.7 + 51.6 - 1.239 = \mathbf{104.06\ MN/m^2}\ \textbf{(Ans.)}$$

At $r = 0.5$ m, $\qquad\qquad \sigma_c = 53.7 + \dfrac{0.516}{0.5^2} - 123.9 \times 10^6 \times 0.5^2 \times 10^{-6}$

$$= 53.7 + 2.064 - 30.97 = \mathbf{24.79\ MN/m^2}\ \textbf{(Ans.)}$$

In hydel generators, the water force turns the turbines which are connected via shafts to generators. These generators convert the torque into electric energy.

σ_r is maximum at $\qquad r = \sqrt{r_1\,r_2} = \sqrt{0.1 \times 0.5} = 0.2236$ m

$$\therefore \qquad (\sigma_r)_{max} = 53.7 - \frac{0.516}{(0.2236)^2} - 206.6 \times 10^6 \times (0.2236)^2 \times 10^{-6}$$

$$= 53.7 - 10.32 - 10.32 = \mathbf{33.06\ MN/m^2} \quad \textbf{(Ans.)}$$

Example 19·4. *A steel disc of uniform thickness and of diameter 400 mm is rotating about its axis at 2000 r.p.m. The density of the material is 7700 kg/m³ and Poisson's ratio is 0·3. Determine the variations of circumferential and radial stresses.*

Solution. Radius of the disc, $r_2 = 200$ mm = 0·2 m

Speed, $\qquad\qquad\qquad N = 2000$ *r.p.m.*

Density of material, $\quad \rho = 7700$ kg/m³

Poisson's ratio, $\qquad 1/m = 0.3$

$$\sigma_r = \left(\frac{3 + \dfrac{1}{m}}{8}\right)\rho\omega^2(r_2^2 - r^2) \qquad\qquad ...(i)\ [\text{Eqn. (19.5)}]$$

$$\sigma_c = \frac{\rho\omega^2}{8}\left[\left(3 + \frac{1}{m}\ r_2^2\right) - 1 + \frac{3}{m}\ r^2\right] \qquad\qquad ...[\text{Eqn. (19·6)}]$$

$$= \frac{\rho\omega^2}{8}\left(3 + \frac{1}{m}\right)r_2^2 - \frac{\rho\omega^2}{8}\left(1 + \frac{3}{m}\right)r^2 \qquad\qquad ...(ii)$$

Now, $\qquad \dfrac{\rho\omega^2}{8}\left(3 + \dfrac{1}{m}\right) = \dfrac{7700 \times (209.44)^2}{8}(3 + 0.3) = 139.3 \times 10^6\ \text{N/m}^4$

$$\left[\text{where, } \omega = \frac{2\pi N}{60} = \frac{2\pi \times 2000}{60} = 209.4\ \text{rad./s}\right]$$

and, $\qquad \dfrac{\rho\omega^2}{8}\left(1 + \dfrac{3}{m}\right) = \dfrac{7700 \times (209.44)^2}{8}(1 + 3 \times 0.3) = 80.22 \times 10^6\ \text{N/m}^4$

Radial stress (σ_r) :

$$\sigma_r = 139.3 \times 10^6\ (0.2^2 - r^2) \qquad\qquad [\text{From eqn. (i)}]$$

At $r = 0$, $\qquad\quad \sigma_r = 139.3 \times 10^6\ (0.2^2 - 0^2) \times 10^{-6} = 5.57\ \text{MN/m}^2 \quad (i.e.,\ \text{at the centre})$

At $r = 0.05$ m, $\quad \sigma_r = 139.3 \times 10^6\ (0.2^2 - 0.05^2) \times 10^{-6} = 5.22\ \text{MN/m}^2$

At $r = 0.1$ m, $\quad\ \ \sigma_r = 139.3 \times 10^6\ (0.2^2 - 0.1^2) \times 10^{-6} = 4.18\ \text{MN/m}^2$

At $r = 0.15$ m, $\quad \sigma_r = 139.3 \times 10^6\ (0.2^2 - 0.15^2) \times 10^{-6} = 2.44\ \text{MN/m}^2$

At $r = 0.2$ m, $\quad\ \ \sigma_r = 139.3 \times 10^6\ (0.2^2 - 0.2^2) \times 10^{-6} = 0$

Circumferential stress (σ_c) :

$$\sigma_c = \frac{\rho\omega^2}{8}\left(3 + \frac{1}{m}\right)r_2^2 - \frac{\rho\omega^2}{8}\left(1 + \frac{3}{m}\right)r^2$$

$$= 139.3 \times 10^6 \times 0.2^2 \times 10^{-6} - 80.22 \times 10^6 r^2 \times 10^{-6}\ \text{MN/m}^2$$

or, $\qquad\qquad\qquad \sigma_c = 5.57 - 80.22\ r^2\ \text{MN/m}^2$

At $r = 0$, $\qquad\quad \sigma_c = 5.57 - 0 = 5.57\ \text{MN/m}^2 \quad (i.e.,\ \text{at the centre})$

At $r = 0.05$ m, $\quad \sigma_c = 5.57 - 80.22 \times 0.05^2 = 5.37\ \text{MN/m}^2$

At $r = 0.1$ m, $\quad\ \ \sigma_c = 5.57 - 80.22 \times 0.1^2 = 4.77\ \text{MN/m}^2$

At $r = 0.15$ m, $\quad \sigma_c = 5.57 - 80.22 \times 0.15^2 = 3.76\ \text{MN/m}^2$

At $r = 0.2$ m, $\quad\ \ \sigma_c = 5.57 - 80.22 \times 0.2^2 = 2.36\ \text{MN/m}^2$

Stress variations:

The variations of circumferential and radial stresses along the radius of the disc is shown in Fig. 19·6.

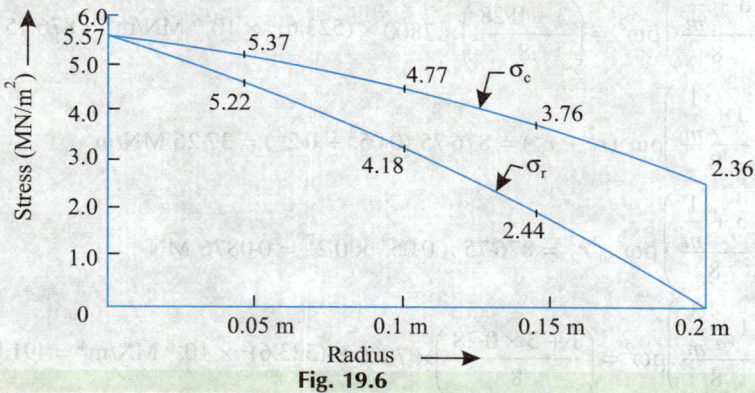

Fig. 19.6

Example 19·5. *A disc of uniform thickness having inner and outer diameters 100 mm and 400 mm respectively is rotating at 5000 r.p.m. about its axis. The density of the material of the disc is 7800 kg/m³ and Poisson's ratio is 0·28.*

Determine the stress variations along the radius of the disc.

Solution. Inner radius of the disc, r_1

$$= 100/2 = 50 \text{ mm} = 0.05 \text{ m}$$

Outer radius of the disc, r_2

$$= 400/2 = 200 \text{ mm} = 0.2 \text{ m}$$

Speed of the disc, $N = 5000$ *r.p.m.*

∴ Angular speed, $\omega = \dfrac{2\pi \times 5000}{60} = 523.6 \text{ rad/s}$

(ii) Principal stresses, maximum shear stress :

$$\sigma_r = \left(\frac{3 + \dfrac{1}{m}}{8} \right) \rho \omega^2 \left[r_1^2 + r_2^2 - \frac{r_1^2 r_2^2}{r^2} - r^2 \right] \qquad \text{[Eqn. ...(19·10)]}$$

or, $$\sigma_r = \left(\frac{3 + \dfrac{1}{m}}{8} \right) \rho \omega^2 [r_1^2 + r_2^2]$$

$$- \left(\frac{3 + \dfrac{1}{m}}{8} \right) \rho \omega^2 \, \frac{r_1^2 r_2^2}{r^2} - \left(\frac{3 + \dfrac{1}{m}}{8} \right) \rho \omega^2 r^2 \qquad \text{...(1)}$$

and, $$\sigma_c = \left(\frac{3 + \dfrac{1}{m}}{8} \right) \rho \omega^2 \left[r_1^2 + r_2^2 + \frac{r_1^2 r_2^2}{r^2} - \left\{ \frac{1 + \dfrac{3}{m}}{3 + \dfrac{1}{m}} \right\} r^2 \right] \qquad \text{...[Eqn. (19·11)]}$$

or, $$\sigma_c = \left(\frac{3 + \dfrac{1}{m}}{8} \right) \rho \omega^2 \, [r_1^2 + r_2^2] + \left(\frac{3 + \dfrac{1}{m}}{8} \right) \rho \omega^2 \, \frac{r_1^2 r_2^2}{r^2} - \left(\frac{1 + \dfrac{3}{m}}{8} \right) \rho \omega^2 r^2$$

$$\text{...(2)}$$

We have :

$$\left(\frac{3 + \frac{1}{m}}{8}\right)\rho\omega^2 = \left(\frac{3 + 0.28}{8}\right) \times 7800 \times (523.6)^2 \times 10^{-6} \text{ MN/m}^4 = 876.75 \text{ MN/m}^4$$

$$\left(\frac{3 + \frac{1}{m}}{8}\right)\rho\omega^2 (r_1^2 + r_2^2) = 876.75 (0.05^2 + 0.2^2) = 37.26 \text{ MN/m}^2$$

$$\left(\frac{3 + \frac{1}{m}}{8}\right)\rho\omega^2 r_1^2 r_2^2 = 876.75 \times 0.05^2 \times 0.2^2 = 0.0876 \text{ MN}$$

$$\left(\frac{1 + \frac{3}{m}}{8}\right)\rho\omega^2 = \left(\frac{1 + 3 \times 0.28}{8}\right) \times 7800 \times (523.6)^2 \times 10^{-6} \text{ MN/m}^4 = 491.8 \text{ MN/m}^4$$

Radial stress, σ_r :

$$\sigma_r = \left(\frac{3 + \frac{1}{m}}{8}\right)\rho\omega^2 (r_1^2 + r_2^2) - \left(\frac{3 + \frac{1}{m}}{8}\right)\rho\omega^2 \frac{r_1^2 r_2^2}{r^2} - \left(\frac{3 + \frac{1}{m}}{8}\right)\rho\omega^2 r^2 \quad \text{[Eqn. (1)]}$$

or, $$\sigma_r = 37.26 - \frac{0.0876}{r^2} - 876.75 \ r^2 \qquad \qquad ...(3)$$

At $r = 0.05$ m,

$$\sigma_r = 37.26 - \frac{0.0876}{0.05^2} - 876.5 \times 0.05^2 = 37.26 - 35.06 - 2.20 = 0$$

At $r = 0.1$ m,

$$\sigma_r = 37.26 - \frac{0.0876}{(0.1)^2} - 876.75 \times 0.1^2 = 37.26 - 8.76 - 8.76 = 19.74 \text{ MN/m}^2 \quad ...(3)$$

At $r = 0.15$ m,

$$\sigma_r = 37.26 - \frac{0.0876}{0.15^2} - 876.75 \times 0.15^2 = 37.26 - 3.89 - 19.73 = 13.64 \text{ MN/m}^2$$

At $r = 0.2$ m,

$$\sigma_r = 37.26 - \frac{0.0876}{0.2^2} - 876.75 \times 0.2^2 = 37.26 - 2.19 - 35.07 = 0$$

Maximum σ_r occurs at $r = \sqrt{r_1 \ r_2} = \sqrt{0.05^2 \times 0.2^2} = 0.1$ m

i.e., $(\sigma_r)_{max} = 19.74 \text{ MN/m}^2$

Circumferential stress, σ_c :

$$\sigma_c = \left(\frac{3 + \frac{1}{m}}{8}\right)\rho\omega^2 (r_1^2 + r_2^2) + \left(\frac{3 + \frac{1}{m}}{8}\right)\rho\omega^2 \frac{r_1^2 r_2^2}{r^2} - \left(\frac{1 + \frac{3}{m}}{8}\right)\rho\omega^2 r^2 \quad \text{[Eqn. (2)]}$$

$$= 37.26 - \frac{0.0876}{r^2} - 491.8 \ r^2$$

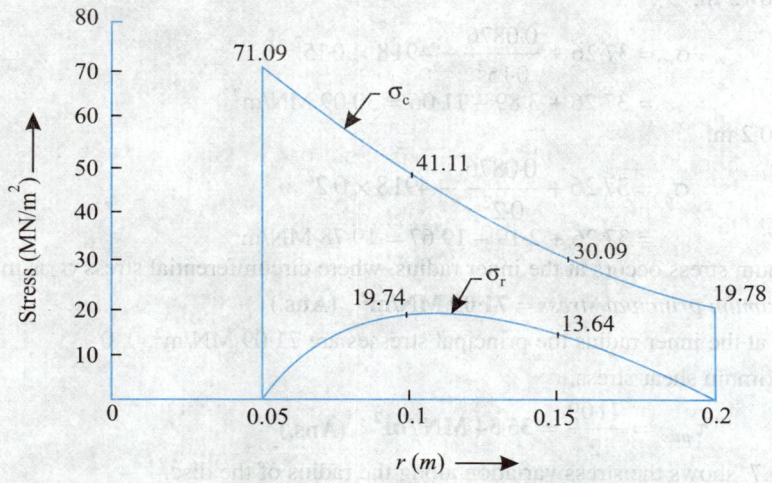

Fig. 19.7

At $r = 0.05$ m,

$$\sigma_c = 37.26 + \frac{0.0876}{(0.05)^2} - 491.8 \times 0.05^2$$

$$= 37.26 + 35.06 - 1.23 = 71.09 \text{ MN/m}^2$$

At $r = 0.1$ m,

$$\sigma_c = 37.26 + \frac{0.0876}{0.1^2} - 491.8 \times 0.1^2$$

$$= 37.26 + 8.76 - 4.92 = 41.11 \text{ MN/m}^2$$

Proper fit of tyre and wheels is essential to safely distribute the torque and shear force.

At $r = 0.15$ m,

$$\sigma_c = 37.26 + \frac{0.0876}{0.15^2} - 491.8 \times 0.15^2$$
$$= 37.26 + 3.89 - 11.06 = 30.09 \text{ MN/m}^2$$

At $r = 0.2$ m,

$$\sigma_c = 37.26 + \frac{0.0876}{0.2^2} - 491.8 \times 0.2^2$$
$$= 37.26 + 2.19 - 19.67 = 19.78 \text{ MN/m}^2$$

Maximum stress occurs at the inner radius, where circumferential stress σ_c is maximum.

∴ *Maximum principal stress =* **71·09 MN/m²** **(Ans.)**

Again, at the inner radius the principal stresses are 71·09 MN/m², 0, 0

∴ Maximum shear stress,

$$\tau_{max} = \frac{71.09}{2} = \textbf{35·54 MN/m}^2 \quad \textbf{(Ans.)}$$

Fig. 19.7. shows the stress variation along the radius of the disc.

19·4. DISC OF UNIFORM STRENGTH

A disc of uniform strength is the one in which the values of radial and circumferential stresses are equal in magnitude for all values of radius r. This means that the *disc of uniform strength must have a varying thickness.*

In industry there are several components such as rotor of a steam turbine which have constant strength throughout the radius and are designed by varying their thickness.

Fig. 19·8 (*a*) shows the elevation of a disc of uniform strength and Fig. 19·8 (*b*) shows the free body diagram of an element *ABCD* of the disc which subtends an angle $d\theta$ at the centre 0. Further the element is bounded by surfaces *DC* and *AB* whose radii are r and $(r + dr)$ respectively and corresponding axial thicknesses as t and $(t + dt)$.

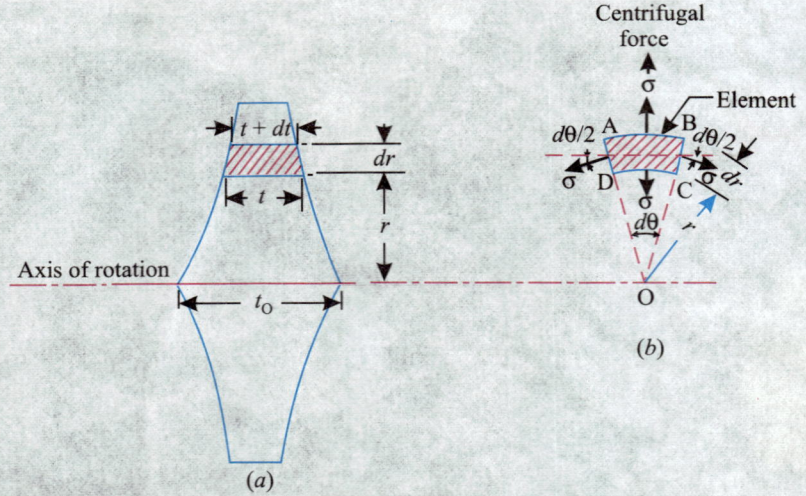

Fig. 19.8. Disc of uniform strength.

Let, σ = Uniform stress in the radial and circumferential directions.

Volume of the element = $r\,d\theta \cdot t\,dr$

Centrifugal force acting on the element *ABCD* due to rotation

$$= \rho \cdot r\,d\theta \cdot t \cdot dr \cdot \omega^2 r = \rho\,d\theta \cdot t \cdot dr \cdot \omega^2 r^2$$

Radial force on the face $DC = r\,d\theta.\,t.\,\sigma$

Radial force on the face $AB = (r + dr)\,d\theta.\,(t + dt)\,\sigma$

Circumferential force on faces BC and $DA = t.\,dr.\,\sigma$

(inclined at an angle $\dfrac{d\theta}{2}$ to the radial direction)

Resolving all the forces along the radial direction and considering equilibrium, we get

$$\rho.d\theta.\,t.\,dr.\,\omega^2\,r^2 + (r + dr)\,d\theta\,(t + dt)\,\sigma = r\,d\theta.\,t.\,\sigma + 2\,t.\,dr.\,\sin\dfrac{d\theta}{2}.\,\sigma \qquad \text{...(1)}$$

$$\left(\sin\dfrac{d\theta}{2} = \dfrac{\theta}{2}, \text{ because } d\theta \text{ is very small.} \right)$$

Cancelling $d\theta$ on both sides, eqn. (1) is simplified as follows:

$$\rho.t.dr.\,\omega^2 r^2 + rt\,\sigma + r\,dt\,\sigma + dr.\,t.\,\sigma + dr.\,dt.\,\sigma = r.t.\sigma + t.dr.\sigma$$

Neglecting the product of small quantities, we get

$$\rho.t.dr.\omega^2\,r^2 + r.dt.\,\sigma = 0$$

$$\sigma\,\dfrac{dt}{t} = -\,\rho\omega^2 r\,dr$$

or,

$$\dfrac{dt}{t} = -\,\dfrac{\rho\omega^2 r}{\sigma}\,dr$$

Integrating both sides, we get

$$\log_e t = \dfrac{\rho\omega^2 r^2}{2\sigma} + \log_e A$$

(where, $\log_e A$ is a constant of integration)

$$\log_e \dfrac{t}{A} = -\,\dfrac{\rho\omega^2 r^2}{2\sigma}$$

$$\dfrac{t}{A} = e^{-\frac{\rho\omega^2 r^2}{2\sigma}}$$

At,

$$r = 0,\, t = t_0$$

$$t_0 = A$$

Torque limiter.

∴ Thickness at any radius, $t = t_0\, e^{-\frac{\rho\omega^2 r^2}{2\sigma}}$ ...(19·18)

Example 19·6. *A steam turbine rotor is to be designed so that the radial and circumferential stresses are to be the same and constant throughout and equal to 90 MN/m², when running at 4000 r.p.m. If the axial thickness at the centre is 20 mm, what is the thickness at a radius of 400 mm?. Assume density of material of the rotor is 7800 kg/m³.*

Sol. Allowable stress, $\sigma = 90$ MN/m²

Speed of the rotor, $N = 4000$ r.p.m.

∴ Angular speed, $\omega = \dfrac{2\pi N}{60} = \dfrac{2\pi \times 4000}{60} = 418.88$ rad./s.

Axial thickness at the centre, $t_0 = 20$ mm

Density of material, $\rho = 7800$ kg/m³

Thickness at a radius of 600 mm, t :

Thickness at any radius is given by

$$t = t_0 \, e^{-\frac{\rho\omega^2 r^2}{2\sigma}}$$

Now,

$$\frac{\rho\omega^2 r^2}{2\sigma} = \frac{7800 \times (418.88)^2 \times (0.4)^2}{2 \times 90 \times 10^6} = 1.216$$

$\therefore$

$$t = 20 \times e^{-1.216} = 5.93 \text{ mm} \quad \text{(Ans.)}$$

19·5. ROTATING LONG CYLINDERS

The analysis of a rotating long cylinder is similar to that of a thin disc, the only difference being that the length of the cylinder along the axis is large as compared to the radius and *axial stress is considered along the length of the cylinder.*

While developing theory for rotating long cylinders following **assumptions** are made :

1. Even at high speeds of rotation plane cross-sections remain plane. Strictly speaking this is true for sections far away from the ends.

2. At the central cross-sectional plane of the cylinder, due to symmetry, shear stress is zero and thus the radial, circumferential and longitudinal stresses will be principal stresses. It will be assumed that it is nearly true at all sections except near the ends.

Let,

$\quad \sigma_r$ = Radial stress,

$\quad \sigma_c$ = Circumferential (or hoop) stress, and

$\quad \sigma_l$ = Longitudinal (or axial) stress.

Let these stresses (σ_r, σ_c, σ_l) act on any element of a section of the cylinder of radius r [Fig. 19·3 (b)]

Then, radial strain,

$$e_r = \frac{1}{E}\left[\sigma_r - \frac{1}{m}(\sigma_c + \sigma_l)\right] = \frac{du}{dr} \qquad \qquad ...(i)$$

Circumferential strain,

$$e_c = \frac{1}{E}\left[\sigma_c - \frac{1}{m}(\sigma_r + \sigma_l)\right] = \frac{u}{r} \qquad \qquad ...(ii)$$

Longitudinal strain,

$$e_l = \frac{1}{E}\left[\sigma_l - \frac{1}{m}(\sigma_r + \sigma_c)\right] \qquad \qquad ...(iii)$$

(where, E = Young's modulus)

From eqn. (*ii*), we have

$$Eu = r\left[\sigma_c - \frac{1}{m}(\sigma_r + \sigma_l)\right]$$

Differentiating w.r.t. r, we get

$$E\frac{du}{dr} = \left[\sigma_c - \frac{1}{m}(\sigma_r + \sigma_l)\right] + r\left[\frac{d\sigma_c}{dr} - \frac{1}{m}\left(\frac{d\sigma_r}{dr} + \frac{d\sigma_l}{dr}\right)\right]$$

$$= \sigma_r - \frac{1}{m}(\sigma_c + \sigma_l) \qquad \qquad \text{[From eqn. (i)]}$$

$\therefore \quad (\sigma_r - \sigma_c)\left(1 + \frac{1}{m}\right) = r\left[\frac{d\sigma_c}{dr} - \frac{1}{m}\left(\frac{d\sigma_c}{dr} + \frac{d\sigma_l}{dr}\right)\right] \qquad ...(iv)$

From eqn. (*iii*), we have

$$Ee_l = \sigma_l - \frac{1}{m}(\sigma_r + \sigma_c) = \text{constant} = C_1$$

∴

$$\sigma_l = C_1 + \frac{1}{m}(\sigma_r + \sigma_c)$$

Differentiating w.r.t. *r*, we get

$$\frac{d\sigma_l}{dr} = \frac{1}{m}\left[\frac{d\sigma_r}{dr} + \frac{d\sigma_c}{dr}\right]$$

Substituting in eqn. (*iv*), we get

$$(\sigma_r - \sigma_c)\left(1 + \frac{1}{m}\right) = r\left[\frac{d\sigma_c}{dr} - \frac{1}{m}\left\{\frac{d\sigma_r}{dr} + \frac{1}{m}\left(\frac{d\sigma_r}{dr} + \frac{d\sigma_c}{dr}\right)\right\}\right]$$

$$= r\left[\left\{1 - \left(\frac{1}{m}\right)^2\right\}\frac{d\sigma_c}{dr} - \frac{1}{m}\left(1 + \frac{1}{m}\right)\frac{d\sigma_r}{dr}\right]$$

or,

$$\sigma_r - \sigma_c = r\left[\left(1 - \frac{1}{m}\right)\frac{d\sigma_c}{dr} - \frac{1}{m}\frac{d\sigma_r}{dr}\right] \qquad \text{...}(v)$$

Also considering the equilibrium of an element of the cylinder between angular position θ and θ + *d*θ and radii *r* and *r* + *dr*, we can get as in the case of a rotating disc

$$\sigma_r - \sigma_c = -\left(r\frac{d\sigma_r}{dr} + \rho\omega^2 r^2\right) \qquad \text{...}(vi)$$

Comparing eqns. (*v*) and (*vi*), we get

$$r\left[\left(1 - \frac{1}{m}\right)\frac{d\sigma_c}{dr} - \frac{1}{m}\frac{d\sigma_r}{dr}\right] = -r\left(\frac{d\sigma_r}{dr} + \rho\omega^2 r\right)$$

$$\left(1 - \frac{1}{m}\right)\frac{d\sigma_c}{dr} - \frac{1}{m}\frac{d\sigma_r}{dr} = -\left(\frac{d\sigma_r}{dr} + \rho\omega^2 r\right)$$

$$\left(1 - \frac{1}{m}\right)\frac{d\sigma_r}{dr} + \left(1 - \frac{1}{m}\right)\frac{d\sigma_c}{dr} = -\rho\omega^2 r$$

$$\left(1 - \frac{1}{m}\right)\left[\frac{d\sigma_r}{dr} + \frac{d\sigma_c}{dr}\right] = -\rho\omega^2 r$$

$$\frac{d\sigma_r}{dr} + \frac{d\sigma_c}{dr} = -\frac{\rho}{\left(1 - \frac{1}{m}\right)}\omega^2 r$$

$$\frac{d}{dr}(\sigma_r + \sigma_c) = -\frac{\rho}{\left(1 - \frac{1}{m}\right)} \cdot \omega^2 r$$

Integrating both sides, we get

$$\sigma_r + \sigma_c = -\frac{\rho}{\left(1 - \frac{1}{m}\right)} \cdot \omega^2 \cdot \frac{r^2}{2} + C_2 \qquad \text{...}(vii)$$

(where, C_2 = constant of integration)

Adding eqns. (vi) and (vii), we get

$$2\sigma_r = -\left[r\frac{d\sigma_r}{dr} + \rho\,\omega^2\,r^2\right] - \frac{\rho}{\left(1 - \dfrac{1}{m}\right)} \cdot \frac{\omega^2 r^2}{2} + C_2$$

$$2\sigma_r + r\frac{d\sigma_r}{dr} = -\rho \cdot \frac{\omega^2 r^2}{2}\left[\frac{3 - \dfrac{2}{m}}{2\left(1 - \dfrac{1}{m}\right)}\right] + C_2$$

Multiplying both sides by r, we get

$$2r\sigma_r + r^2\frac{d\sigma_r}{dr} = -\rho \cdot \frac{\omega^2 r^3}{2}\left[\frac{3 - \dfrac{2}{m}}{1 - \dfrac{1}{m}}\right] + r\,C_2$$

$$\frac{d}{dr}(r^2\sigma_r) = -\rho \cdot \frac{\omega^2 r^3}{2}\left[\frac{3 - \dfrac{2}{m}}{1 - \dfrac{1}{m}}\right] + r\,C_2$$

Integrating both sides, we get

$$r^2\sigma_r = -\rho \cdot \frac{\omega^2 r^4}{8}\left[\frac{3 - \dfrac{2}{m}}{1 - \dfrac{1}{m}}\right] + \frac{r^2}{2}\,C_2 + C_3$$

(where, C_3 = constant of integration)

$$\sigma_r = -\rho \cdot \frac{\omega^2 r^2}{8}\left[\frac{3 - \dfrac{2}{m}}{1 - \dfrac{1}{m}}\right] + \frac{C_2}{2} + \frac{C_3}{r^2}$$

or,

$$\sigma_r = \frac{C_2}{2} + \frac{C_3}{r^2} - \rho \cdot \frac{\omega^2 r^2}{8}\left[\frac{3 - \dfrac{2}{m}}{1 - \dfrac{1}{m}}\right] \qquad \qquad ...(19\cdot19)$$

Substituting in equation (vii), we get

$$\sigma_c = -\frac{\rho}{1 - \dfrac{1}{m}} \cdot \frac{\omega^2 r^2}{2} + C_2 - \left\{-\rho \cdot \frac{\omega^2 r^2}{8}\left(\frac{3 - \dfrac{2}{m}}{1 - \dfrac{1}{m}}\right) + \frac{C_2}{2} + \frac{C_3}{r^2}\right\}$$

$$= -\frac{\rho}{1 - \dfrac{1}{m}} \cdot \frac{\omega^2 r^2}{2} + C_2 + \rho \cdot \frac{\omega^2 r^2}{8}\left(\frac{3 - \dfrac{2}{m}}{1 - \dfrac{1}{m}}\right) - \frac{C_2}{2} - \frac{C_3}{r^2}$$

$$= -\rho \cdot \frac{\omega^2 r^2}{8}\left[\frac{4}{1-\dfrac{1}{m}} - \frac{3-\dfrac{2}{m}}{1-\dfrac{1}{m}}\right] + \frac{C_2}{2} - \frac{C_3}{r^2}$$

$$= -\rho \cdot \frac{\omega^2 r^2}{8}\left[\frac{1+\dfrac{2}{m}}{1-\dfrac{1}{m}}\right] + \frac{C_2}{2} - \frac{C_3}{r^2}$$

or, $$\sigma_c = \frac{C_2}{2} - \frac{C_3}{r^2} - \rho \cdot \frac{\omega^2 r^2}{8}\left[\frac{1+\dfrac{2}{m}}{1-\dfrac{1}{m}}\right] \qquad \qquad ...(19·20)$$

Equations (19·19) and (19·20) are the governing equations for a rotating cylinder in which C_2 and C_3 (constants of integration) are evaluated with the help of end conditions.

19·5·1. Solid Cylinder

Since the stresses cannot be infinite at the centre, therefore $C_3 = 0$. The expressions for stresses will now be as follows :

$$\sigma_r = \frac{C_2}{2} - \rho \cdot \frac{\omega^2 r^2}{8}\left[\frac{3-\dfrac{2}{m}}{1-\dfrac{1}{m}}\right]$$

$$\sigma_c = \frac{C_2}{2} - \rho \cdot \frac{\omega^2 r^2}{8}\left[\frac{1+\dfrac{2}{m}}{1-\dfrac{1}{m}}\right]$$

For a solid cylinder with a free surface, at $r = r_2$, $\sigma_r = 0$

Cylinder boring on a lathe. Most machining processes are performed with rotary motion.

$$\therefore \qquad \frac{C_2}{2} = \rho \cdot \frac{\omega^2 r_2^2}{8} \left[\frac{3 - \dfrac{2}{m}}{1 - \dfrac{1}{m}} \right]$$

$$\therefore \qquad \sigma_r = \rho \cdot \frac{\omega^2 (r_2^2 - r^2)}{8} \left[\frac{3 - \dfrac{2}{m}}{1 - \dfrac{1}{m}} \right] \qquad \qquad \ldots(19\cdot21)$$

$$\sigma_c = \frac{\rho \omega^2}{8} \left[\frac{3 - \dfrac{2}{m}}{1 - \dfrac{1}{m}} \right] \left[r_2^2 - \left(\frac{1 + \dfrac{2}{m}}{3 - \dfrac{2}{m}} \right) r^2 \right] \qquad \ldots(19\cdot22)$$

The *maximum stress occurs at the centre of the cylinder*, where $r = 0$

$$\therefore \qquad (\sigma_r)_{max} = (\sigma_c)_{max} = \rho \cdot \frac{\omega^2 r_2^2}{8} \left[\frac{3 - \dfrac{2}{m}}{1 - \dfrac{1}{m}} \right] \qquad \ldots(19\cdot23)$$

19·5·2. Hollow cylinder

$$\sigma_r = \frac{C_2}{2} + \frac{C_3}{r^2} - \rho \cdot \frac{\omega^2 r^2}{8} \left(\frac{3 - \dfrac{2}{m}}{1 - \dfrac{1}{m}} \right) \qquad \qquad [\text{Eqn. }(19\cdot19)]$$

At $\qquad r = r_1, \sigma_r = 0$

and, $\qquad r = r_2, \sigma_r = 0$

$$\therefore \qquad 0 = \frac{C_2}{2} + \frac{C_3}{r_1^2} - \rho \cdot \frac{\omega^2 r_1^2}{8} \left[\frac{3 - \dfrac{2}{m}}{1 - \dfrac{1}{m}} \right]$$

and, $$\qquad 0 = \frac{C_2}{2} + \frac{C_3}{r_2^2} - \rho \cdot \frac{\omega^2 r_2^2}{8} \left[\frac{3 - \dfrac{2}{m}}{1 - \dfrac{1}{m}} \right]$$

From these boundary conditions

$$C_3 \left(\frac{1}{r_1^2} - \frac{1}{r_2^2} \right) - \frac{\rho \omega^2}{8} \left(\frac{3 - \dfrac{2}{m}}{1 - \dfrac{1}{m}} \right) (r_1^2 - r_2^2) = 0$$

$$C_3 \left(\frac{r_2^2 - r_1^2}{r_1^2 r_2^2} \right) = -\frac{\rho \omega^2}{8} \left(\frac{3 - \dfrac{2}{m}}{1 - \dfrac{1}{m}} \right) (r_2^2 - r_1^2)$$

$$C_3 = -\frac{\rho\omega^2}{8}\left(\frac{3 - \dfrac{2}{m}}{1 - \dfrac{1}{m}}\right)r_1^2 r_2^2$$

Again,

$$\frac{C_2}{2} = -\frac{C_3}{r_1^2} + \frac{\rho\cdot\omega^2 r_1^2}{8}\left(\frac{3 - \dfrac{2}{m}}{1 - \dfrac{1}{m}}\right)$$

$$= \frac{\rho\,\omega^2}{8}\left(\frac{3 - \dfrac{2}{m}}{1 - \dfrac{1}{m}}\right)r_2^2 + \frac{\rho\cdot\omega^2 r_1^2}{8}\left(\frac{3 - \dfrac{2}{m}}{1 - \dfrac{1}{m}}\right)$$

or,

$$\frac{C_2}{2} = \frac{\rho\omega^2}{8}\left(\frac{3 - \dfrac{2}{m}}{1 - \dfrac{1}{m}}\right)(r_1^2 + r_2^2)$$

The expressions for stresses will now be :

Radial stress :

$$\sigma_r = \frac{\rho\omega^2}{8}\left(\frac{3 - \dfrac{2}{m}}{1 - \dfrac{1}{m}}\right)\left[(r_1^2 + r_2^2) - \frac{r_1^2 r_2^2}{r^2} - r^2\right] \qquad ...(19\cdot24)$$

[substituting the values of C_2 and C_3 in eqn. (19·19)]

For σ_r to be *maximum*, $\dfrac{d\sigma_r}{dr} = 0$

$$\frac{2r_1^2 r_2^2}{r^3} - 2r = 0$$

$$r^4 = r_1^2\, r_2^2$$

∴

$$r = \sqrt{r_1\, r_2}$$

∴

$$(\sigma_r)_{max} = \frac{\rho\omega^2}{8}\left(\frac{3 - \dfrac{2}{m}}{1 - \dfrac{1}{m}}\right)\left[(r_1^2 + r_2^2) - \frac{r_1^2 r_2^2}{r_1 r_2} - r_1 r_2\right]$$

$$= \frac{\rho\omega^2}{8}\left(\frac{3 - \dfrac{2}{m}}{1 - \dfrac{1}{m}}\right)[r_1^2 + r_2^2 - 2r_1 r_2]$$

$$(\sigma_r)_{max} = \frac{\rho\omega^2}{8}\left(\frac{3 - \dfrac{2}{m}}{1 - \dfrac{1}{m}}\right)(r_1 - r_2)^2 \qquad ...(19\cdot25)$$

Circumferential stress :

The expression for the circumferential or hoop stress will be

$$\sigma_c = \frac{C_2}{2} - \frac{C_3}{r^2} - \frac{\rho\omega^2 r^2}{8}\left(\frac{1 + \dfrac{2}{m}}{1 - \dfrac{1}{m}}\right)$$

[Eqn. (19·20)]

Substituting the values of C_2 and C_3, we get

$$\sigma_c = \frac{\rho\,\omega^2}{8}\left[\frac{3 - \dfrac{2}{m}}{1 - \dfrac{1}{m}}\right](r_1^2 + r_2^2) + \frac{\rho\omega^2}{8}\left[\frac{3 - \dfrac{2}{m}}{1 - \dfrac{1}{m}}\right]\frac{r_1^2 r_2^2}{r^2} - \frac{\rho\omega^2 r^2}{8}\left[\frac{1 + \dfrac{2}{m}}{1 - \dfrac{1}{m}}\right]$$

$$= \frac{\rho\omega^2}{8}\left[\left(\frac{3 - \dfrac{2}{m}}{1 - \dfrac{1}{m}}\right)(r_1^2 + r_2^2) + \left(\frac{3 - \dfrac{2}{m}}{1 - \dfrac{1}{m}}\right)\frac{r_1^2 r_2^2}{r^2} - \left(\frac{1 + \dfrac{2}{m}}{1 - \dfrac{1}{m}}\right)r^2\right]$$

or,

$$\sigma_c = \frac{\rho\omega^2}{8}\left(\frac{3 - \dfrac{2}{m}}{1 - \dfrac{1}{m}}\right)\left[r_1^2 + r_2^2 + \frac{r_1^2 r_2^2}{r^2} - \left(\frac{1 + \dfrac{2}{m}}{3 - \dfrac{2}{m}}\right)r^2\right] \qquad \ldots(19\cdot26)$$

σ_c is maximum at $r = r_1$

$$\therefore \qquad (\sigma_c)_{max} = \frac{\rho\cdot\omega^2}{8}\left(\frac{3 - \dfrac{2}{m}}{1 - \dfrac{1}{m}}\right)\left[(2r_2^2 + r_1^2) - \left(\frac{1 + \dfrac{2}{m}}{3 - \dfrac{2}{m}}\right)r_1^2\right] \qquad \ldots(19\cdot27)$$

Example 19·7. *A long cylinder of 300 mm radius is rotating at 4500 r.p.m. The density of material is 7800 kg/m³ and Poisson's ratio is 0·3.*

(i)   *Calculate the maximum stress in the cylinder;*

(ii) *Draw the variations of radial and circumferential stresses along the radius.*

Solution. Radius of the cylinder, $r_2 = 300$ mm $= 0\cdot3$ m

Speed of the rotor, $N = 4500$ r.p.m.

$\therefore$ Angular speed, $\omega = \dfrac{2\pi N}{60}$ rad./s $= \dfrac{2\pi \times 4500}{60} = 471\cdot2$ rad/s

(i)   **Maximum stress in the cylinder :**

$$\frac{3 - \dfrac{2}{m}}{1 - \dfrac{1}{m}} = \frac{3 - 2 \times 0\cdot3}{1 - 0\cdot3} = \frac{2\cdot4}{0\cdot7} = 3\cdot428$$

$$\frac{1 + \dfrac{2}{m}}{3 - \dfrac{2}{m}} = \frac{1 + 2 \times 0\cdot3}{3 - 2 \times 0\cdot3} = \frac{1\cdot6}{2\cdot4} = 0\cdot667$$

Maximum radial and circumferential stresses occur at the centre, where $r = 0$

$$(\sigma_r)_{max} = (\sigma_c)_{max} = \frac{\rho\omega^2 r_2^2}{8}\left(\frac{3-\dfrac{2}{m}}{1-\dfrac{1}{m}}\right) \qquad \text{...[Eqn. (19·23)]}$$

$$= \frac{7800\times(471\cdot2)^2\times(0\cdot3)^2}{8}\times3\cdot428\times10^{-6}\ \text{MN/m}^2$$

$$= \mathbf{66\cdot79\ MN/m^2}\quad\textbf{(Ans.)}$$

(ii) Variations of stresses :

Radial stress (σ_r) :

$$\sigma_r = \frac{\rho\ \omega^2\ (r_2^2 - r^2)}{8}\left[\frac{3-\dfrac{2}{m}}{1-\dfrac{1}{m}}\right] \qquad \text{...[Eqn. (19·21)]}$$

$$= \frac{7800\times(471\cdot2)^2\times(0\cdot3^2 - r^2)}{8}\times3\cdot428\times10^{-6}\ \text{MN/m}^2$$

or, $\sigma_r = 742\cdot09\ (0\cdot09 - r^2)$

At $r = 0$, $\sigma_r = 66\cdot79$ MN/m^2

At $r = 0\cdot06$ m $\sigma_r = 64\cdot11$ MN/m^2

At $r = 0\cdot12$ m $\sigma_r = 56\cdot1$ MN/m^2

At $r = 0\cdot18$ m $\sigma_r = 42\cdot74$ MN/m^2

At $r = 0\cdot24$ m $\sigma_r = 24\cdot04$ MN/m^2

At $r = 0\cdot3$ m $\sigma_r = 0$

Circumferential stress, σ_c :

$$\sigma_c = \frac{\rho\omega^2}{8}\left[\frac{3-\dfrac{2}{m}}{1-\dfrac{1}{m}}\right]\left[r_2^2 - \left(\frac{1+\dfrac{2}{m}}{3-\dfrac{2}{m}}\right)r^2\right] \qquad \text{...[Eqn. (19·22)]}$$

$$= \left[\frac{7800\times(471\cdot2)^2}{8}\times3\cdot428\right][(0\cdot3)^2 - 0\cdot667\,r^2]\times10^{-6}\ \text{MN/m}^2$$

or, $\sigma_c = 742\cdot09\ (0\cdot09 - 0\cdot667\,r^2)$ MN/m^2

At $r = 0$, $\sigma_c = 66\cdot79$ MN/m^2

At $r = 0\cdot06$ m, $\sigma_c = 65$ MN/m^2

At $r = 0\cdot12$ m, $\sigma_c = 59\cdot66$ MN/m^2

At $r = 0\cdot18$ m $\sigma_c = 50\cdot75$ MN/m^2

At $r = 0\cdot24$ m, $\sigma_c = 38\cdot28$ MN/m^2

At $r = 0\cdot3$ m, $\sigma_c = 22\cdot24$ MN/m^2

Fig. 19·9 shows the variations of radial and circumferential stresses along the radius of the long cylinder.

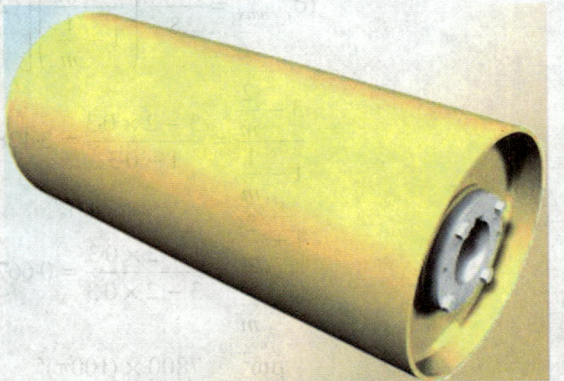

Conveyor belt used in material handling is turned by drum conveyor pulley.

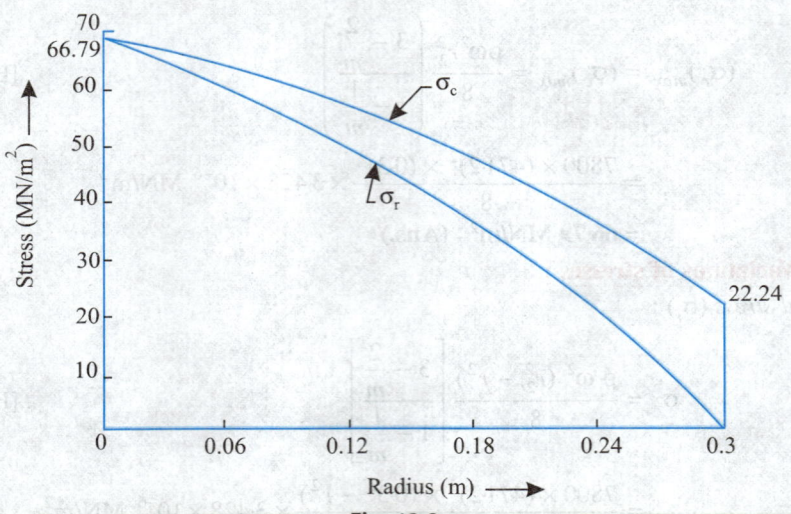

Fig. 19.9

Example 19·8. *A hollow cylinder, 200 mm external radius and 100 mm internal radius is rotating at 3000 r.p.m. The density of material is 7800 kg/m³ and Poisson's ratio is 0·3.*

(i)   Calculate the maximum stress in the cylinder;

(ii)   Draw the variations of radial and hoop stresses in the cylinder.

Solution. Inner radius of the cylinder, $r_1 = 100$ mm $= 0·1$ m

External radius of the cylinder, $r_2 = 200$ mm $= 0·2$ m

Speed of the cylinder, $N = 3000$ r.p.m.

∴ Angular speed, $\omega = \dfrac{2\pi N}{60} = \dfrac{2\pi \times 3000}{60} = 100\pi$ rad/s

Density of material, $\rho = 7800$ kg/m³

Poisson's ratio, $\dfrac{1}{m} = 0·3$

(i)   Maximum stress in the cylinder :

Maximum stress occurs at the inner radius of the cylinder.

$$(\sigma_c)_{max} = \frac{\rho\,\omega^2}{8}\left(\frac{3-\dfrac{2}{m}}{1-\dfrac{1}{m}}\right)\left[(2r_2^2 + r_1^2) - \left(\frac{1+\dfrac{2}{m}}{3-\dfrac{2}{m}}\right)r_1^2\right] \quad \text{...[Eqn. (19·27)]}$$

$$\frac{3-\dfrac{2}{m}}{1-\dfrac{1}{m}} = \frac{3 - 2\times 0·3}{1 - 0·3} = 3·428$$

$$\frac{1+\dfrac{2}{m}}{3-\dfrac{2}{m}} = \frac{1 + 2\times 0·3}{3 - 2\times 0·3} = 0·667$$

$$\frac{\rho\omega^2}{8} = \frac{7800\times(100\pi)^2}{8} = 96·23\times 10^6$$

$\therefore \qquad (\sigma_c)_{max} = 96 \cdot 23 \times 10^6 \times 3 \cdot 428 \ [(2 \times 0 \cdot 2^2 + 0 \cdot 1^2) - 0 \cdot 667 \times 0 \cdot 1^2] \times 10^{-6} \ \text{MN/m}^2$

$\qquad\qquad\qquad = 27 \cdot 5 \ \text{MN/m}^2 \quad \textbf{(Ans.)}$

Radial stress (σ_r) :

$$\sigma_r = \frac{\rho \, \omega^2}{8} \left[\frac{3 - \dfrac{2}{m}}{1 - \dfrac{1}{m}} \right] \left[(r_1^2 + r_2^2) - \frac{r_1^2 \, r_2^2}{r^2} - r^2 \right] \qquad \text{...[Eqn. 19·24]}$$

$$= 96 \cdot 23 \times 10^6 \ (3 \cdot 428) \left[(0 \cdot 1)^2 + (0 \cdot 2)^2 - \frac{(0 \cdot 1)^2 \times (0 \cdot 2)^2}{r^2} - r^2 \right] \times 10^{-6} \ \text{MN/m}^2$$

or, $\qquad\qquad = 329 \cdot 87 \left[0 \cdot 01 + 0 \cdot 04 - \dfrac{0 \cdot 0004}{r^2} - r^2 \right] \text{MN/m}^2$

or, $\qquad \sigma_r = 329 \cdot 87 \left[0 \cdot 05 - \dfrac{0 \cdot 0004}{r^2} - r^2 \right] \text{MN/m}^2$

At $r = 0 \cdot 1$ m,

$$\sigma_r = 329 \cdot 87 \left(0 \cdot 05 - \frac{0 \cdot 0004}{0 \cdot 1^2} - 0 \cdot 1^2 \right) = 0$$

At $r = 0 \cdot 15$ m,

$$\sigma_r = 329 \cdot 87 \left(0 \cdot 05 - \frac{0 \cdot 0004}{0 \cdot 15^2} - 0 \cdot 15^2 \right) = 3 \cdot 21 \ \text{MN/m}^2$$

At $r = 0 \cdot 2$ m,

$$\sigma_r = 329 \cdot 87 \left(0 \cdot 05 - \frac{0 \cdot 0004}{0 \cdot 2^2} - 0 \cdot 2^2 \right) = 0$$

$$(\sigma_r)_{max} \ \text{occurs at} \ r = \sqrt{r_1 \, r_2} = \sqrt{0 \cdot 1 \times 0 \cdot 2} = 0 \cdot 1414 \ \text{m}$$

$\therefore \qquad (\sigma_r)_{max} = 329 \cdot 87 \left(0 \cdot 05 - \dfrac{0 \cdot 0004}{0 \cdot 1414^2} - 0 \cdot 1414^2 \right) = 3 \cdot 3 \ \text{MN/m}^2$

Circumferential stress, σ_c :

$$\sigma_c = \frac{\rho \, \omega^2}{8} \left(\frac{3 - \dfrac{2}{m}}{1 - \dfrac{1}{m}} \right) \left[r_1^2 + r_2^2 + \frac{r_1^2 \, r_2^2}{r^2} - \left(\frac{1 + \dfrac{2}{m}}{3 - \dfrac{2}{m}} \right) r^2 \right] \qquad \text{...[Eqn. (19·26)]}$$

$$= 96 \cdot 23 \times 10^6 \times 3 \cdot 428 \left(0 \cdot 1^2 + 0 \cdot 2^2 + \frac{0 \cdot 1^2 \times 0 \cdot 2^2}{r^2} - 0 \cdot 667 \, r^2 \right) \times 10^{-6} \ \text{MN/m}^2$$

or, $\qquad \sigma_r = 329 \cdot 87 \left(0 \cdot 05 + \dfrac{0 \cdot 0004}{r^2} - 0 \cdot 667 \, r^2 \right) \text{MN/m}^2$

At $r = 0 \cdot 1$ m, $\sigma_r = 329 \cdot 87 \left(0 \cdot 05 + \dfrac{0 \cdot 0004}{0 \cdot 1^2} - 0 \cdot 667 \times 0 \cdot 1^2 \right) = 27 \cdot 5 \ \text{MN/m}^2$

At $r = 0 \cdot 15$ m, $\sigma_r = 329 \cdot 87 \left(0 \cdot 05 + \dfrac{0 \cdot 0004}{0 \cdot 15^2} - 0 \cdot 667 \times 0 \cdot 15^2 \right) = 17 \cdot 41 \ \text{MN/m}^2$

At $r = 0 \cdot 2$ m, $\sigma_r = 329 \cdot 87 \left(0 \cdot 05 + \dfrac{0 \cdot 0004}{0 \cdot 2^2} - 0 \cdot 667 \times 0 \cdot 2^2 \right) = 11 \ \text{MN/m}^2$

The variations of radial and circumferential stresses are shown in figure 19·10.

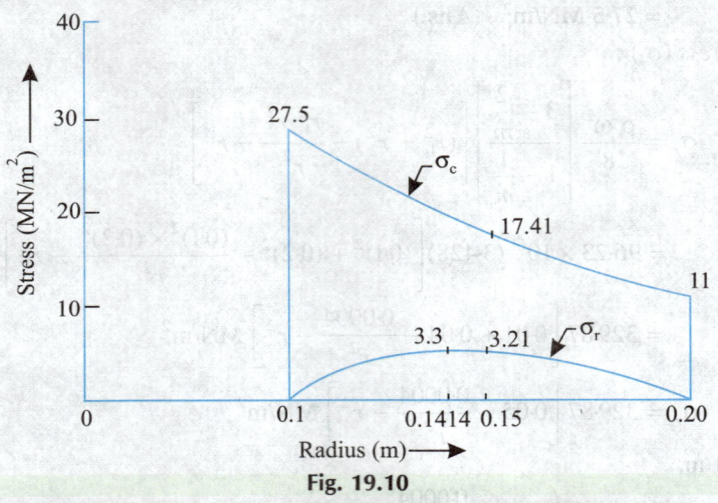

Fig. 19.10

High torque shunt wound drilling motor.

HIGHLIGHTS

1. For a thin ring of mean radius r rotating at an angular speed ω, circumferential stress, $\sigma_c = \rho\, v^2$
   where, ρ = Density of material, kg/m³; v = Linear velocity of the ring, m/s.

2. **Rotating disc :**
   (i) *Solid disc.*

$$\sigma_r = \left(\frac{3 + \dfrac{1}{m}}{8}\right)\rho\omega^2\,(r_2^2 - r_1^2) \qquad \ldots(i)$$

$$(\sigma_r)_{max} = \left(\frac{3 + \dfrac{1}{m}}{8}\right) \rho\omega^2 r_2^2 \qquad \text{...(ii)}$$

$$\sigma_c = \frac{\rho\omega^2}{8}\left[\left(3 + \frac{1}{m}\right) r_2^2 - \left(1 + \frac{3}{m}\right) r^2\right] \qquad \text{...(iii)}$$

$$(\sigma_c)_{max} = \left(\frac{3 + \dfrac{1}{m}}{8}\right) \rho\omega^2 r_2^2 \qquad \text{...(iv)}$$

(ii) *Hollow disc :*

$$\sigma_r = \left(\frac{3 + \dfrac{1}{m}}{8}\right) \rho\omega^2 \left[r_1^2 + r_2^2 - \frac{r_1^2 r_2^2}{r^2} - r^2\right] \qquad \text{...(v)}$$

$$(\sigma_r)_{max} = \left(\frac{3 + \dfrac{1}{m}}{8}\right) \rho\omega^2 (r_2 - r_1)^2 \qquad \text{...(vi)}$$

$$(\text{at } r = \sqrt{r_1 r_2})$$

$$\sigma_c = \left(\frac{3 + \dfrac{1}{m}}{8}\right) \rho\omega^2 \left[r_1^2 + r_2^2 + \frac{r_1^2 r_2^2}{r^2} - \left(\frac{1 + \dfrac{3}{m}}{3 + \dfrac{1}{m}}\right) r^2\right] \qquad \text{...(vii)}$$

$$(\sigma_c)_{max} = \left(\frac{3 + \dfrac{1}{m}}{4}\right) \rho\omega^2 \left[r_2^2 + \left(\frac{1 - \dfrac{1}{m}}{3 + \dfrac{1}{m}}\right) r_1^2\right] \qquad \text{...(viii)}$$

where, σ_r = Radial stress,

$(\sigma_r)_{max}$ = Maximum radial stress,

σ_c = Circumferential stress,

$(\sigma_c)_{max}$ = Maximum circumferential stress,

r_1 = Inner radius,

r_2 = Outer radius, and

ω = Angular velocity.

(iii) *Hollow disc with a pinhole at the centre :*

$$(\sigma_r)_{max} = \left(\frac{3 + \dfrac{1}{m}}{8}\right) \rho\omega^2 r_2^2 \qquad \text{...(ix)}$$

$$(\sigma_c)_{max} = \left(\frac{3 + \dfrac{1}{m}}{4}\right) \rho\omega^2 r_2^2 \qquad \text{...(x)}$$

3. **Disc of uniform strength :**

Thickness at any radius,

$$t = t_0 \, e^{-\frac{\rho \, \omega^2 r^2}{2\sigma}}$$

where, t_0 = Thickness at the centre, and

σ = Uniform stress in the radial and circumferential directions.

4. **Rotating long cylinders :**

(i) *Solid cylinder.*

$$\sigma_r = \rho \cdot \frac{\omega^2 (r_2^2 - r^2)}{8} \left[\frac{3 - \dfrac{2}{m}}{1 - \dfrac{1}{m}} \right] \qquad \dots(i)$$

$$\sigma_r = \frac{\rho \omega^2}{8} \left[\frac{3 - \dfrac{2}{m}}{1 - \dfrac{1}{m}} \right] \left[r_2^2 - \left(\frac{1 + \dfrac{2}{m}}{3 - \dfrac{2}{m}} \right) r^2 \right] \qquad \dots(ii)$$

$$(\sigma_r)_{max} = (\sigma_c)_{max} = \rho \cdot \frac{\omega^2 r_2^2}{8} \left(\frac{3 - \dfrac{2}{m}}{1 - \dfrac{1}{m}} \right) \qquad \dots(iii)$$

(ii) *Hollow cylinder.*

$$\sigma_r = \frac{\rho \cdot \omega^2}{8} \left(\frac{3 - \dfrac{2}{m}}{1 - \dfrac{1}{m}} \right) \left[(r_1^2 + r_2^2) - \frac{r_1^2 r_2^2}{r^2} - r^2 \right] \qquad \dots(iv)$$

$$(\sigma_r)_{max} = \frac{\rho \cdot \omega^2}{8} \left(\frac{3 - \dfrac{2}{m}}{1 - \dfrac{1}{m}} \right) (r_1 - r_2)^2 \qquad \dots(v)$$

$$\sigma_c = \frac{\rho \cdot \omega^2}{8} \left(\frac{3 - \dfrac{2}{m}}{1 - \dfrac{1}{m}} \right) \left[r_1^2 + r_2^2 + \frac{r_1^2 r_2^2}{r^2} - \left(\frac{1 + \dfrac{2}{m}}{3 - \dfrac{2}{m}} \right) r^2 \right] \qquad \dots(vi)$$

$$(\sigma_c)_{max} = \frac{\rho \cdot \omega^2}{8} \left(\frac{3 - \dfrac{2}{m}}{1 - \dfrac{1}{m}} \right) \left[(2r_2^2 + r_1^2) - \left(\frac{1 + \dfrac{2}{m}}{3 - \dfrac{2}{m}} \right) r_1^2 \right] \qquad \dots(vii)$$

OBJECTIVE TYPE QUESTIONS

Choose the Correct Answer :

1. A thin flat ring is rotating at a speed v. The circumferential stress induced is given by

(a) rv (b) rv^2

(c) $\dfrac{1}{2} \rho v^2$ (d) $\dfrac{1}{2} \rho v^3$.

(where, ρ = density of material)

2. In case of a solid rotating circular disc the radial stress is maximum at

(a) the mean radius

(b) the outer radius

(c) square root of the radius

(d) the centre.

3. The circumferential stress in a solid rotating disc is maximum at

(a) the mean radius

(b) the inner radius

(c) the centre

(d) the geometric mean radius.

4. The radial stress in a hollow circular rotating disc is at

(a) the inner radius

(b) the outer radius

(c) the mean radius

(d) the geometric mean radius.

5. The circumferential stress in a hollow circular rotating disc is at

(a) the inner radius

(b) the outer radius

(c) the mean radius

(d) the geometric mean radius.

6. The ratio of the maximum circumferential stress in a circular rotating disc having a very small hole at the centre to the maximum circumferential stress in a solid circular rotating disc is

(a) 1·2   (b) 1·5

(c) 2   (d) 3.

7. In case of rotating disc of uniform strength which of the following statements is *correct*?

(a) Circumferential stress is constant.

(b) Radial stress is constant.

(c) Circumferential and radial stresses are equal to each other and are constant

(d) None of the above.

8. The circumferential stress in a rotating solid circular cylinder is maximum at

(a) the mean radius

(b) the centre

(c) the outer radius

(d) the square root of the outer radius.

9. The radial stress in a rotating hollow circular cylinder is maximum at

(a) the inner radius

(b) the outer radius

(c) the mean radius

(d) the geometric mean radius.

ANSWERS

1. (b)	2. (d)	3. (c)	4. (d)	5. (a)
6. (c)	7. (c)	8. (b)	9. (d).	

Electric motor with shaft and bearings.

UNSOLVED EXAMPLES

1. Calculate the stress in the rim of a pulley when linear velocity of the rim is 80 m/s. Assume density of material of the pulley as 7800 kg/m^3.

   If the speed of the pulley is increased by 20 per cent what will be the stress ?

   [**Ans.** 50·8 MN/m^2, 73·3 MN/m^2]

2. Find the limiting peripheral speed of a cast-iron wheel if allowable stress in cast-iron is 6·6 N/mm^2.

   Take density of material as 7212 kg/m^3. [**Ans.** 30·25 m/s]

3. A composite ring is made of an inner copper ring and outer steel ring. The diameter of the surface of contact of the two rings is 600 mm. If the composite ring rotates at 2500 r.p.m. determine the stresses set up in the steel and copper rings. Both the rings are of rectangular cross-section 15 mm in the radial direction and 20 mm in the direction perpendicular to the plane of the ring.

   Take: E_s = 200 GN/m^2, E_{cu} = 100 GN/m^2; ρ_s = 7800 kg/m^3, ρ_{cu} = 8900 kg/m^3.

   [**Ans.** 68·49 MN/m^2, 32·25 MN/m^2]

4. A thin uniform disc of inner radius 25 mm and outer diameter 125 mm is rotating at 10000 r.p.m. Calculate the maximum and minimum values of circumferential and radial stresses.

   Take : Density of material = 8830 kg/m^3 and Poisson's ratio = 0·33.

   [**Ans.** 129·2 MN/m^2, 30·9 MN/m^2, 41·08 MN/m^2, 0]

5. A steam turbine rotor (to run at a speed of 3500 r.p.m.) is to be designed so that the radial and circumferential stresses are to be same and constant throughout and equal to 80 MN/m^2. If the axial thickness at the centre is 15 mm what is the thickness at a radius of 500 mm ?

   Assume density of material as 7800 kg/m^3. [**Ans.** 2·92 mm]

6. A disc having inner and outer diameters 150 mm and 300 mm respectively is rotating at an angular speed of 150 rad./sec. Calculate the greatest values of radial and circumferential stresses.

   Assume : Density of material = 7700 kg/m^3, and Poisson's ratio = 0·304.

   [**Ans.** 16 MN/m^2, 13·6 MN/m^2]

7. A long steel cylinder of outer radius 375 mm and inner radius 125 mm is rotating about its axis at 4000 r.p.m.

   (*i*) What are the maximum and minimum values of circumferential stress ?

   (*ii*) What is the maximum radial stress and where it occurs ?

   Take : Density of material = 7800 kg/m^3, Poisson's Ratio = 0·3.

   [**Ans.** (*i*) 171·5 MN/m^2, 46·78 MN/m^2 (*ii*) 37·4 MN/m^2 at r = 216·5 mm]

8. A rotor of a turbine having inner and outer radii 100 mm and 200 mm respectively is rotating at 1000 r.p.m. Find the maximum circumferential and radial stresses assuming :

   (*i*) Rotor to be a thin disc;   (*ii*) Rotor to be a long cylinder.

   Assume : Density of material = 7700 kg/m^3 and Poisson's ratio = 0·3.

   [**Ans.** (*i*) 2·93 MN/m^2 , 0·3475 MN/m^2 (*ii*) 3·0157 MN/m^2, 0·3619 MN/m^2]

Bending of Curved Bars

20·1.   INTRODUCTION

In chapter 5 the bending equation $\left(\dfrac{M}{I} = \dfrac{\sigma}{y} = \dfrac{E}{R} \right)$ was derived assuming the beam to be initially straight (besides other fundamental assumptions) before the application of a bending moment. However, machine members and structures subjected to bending are not always straight, as in the case of crane hooks, chain links etc., before a bending moment is applied to them. The simple flexure formula may be used for curved beams for which the radius of curvature is more than five times the depth of the beam. The simple bending formula, however, is not applicable for deeply curved beams where the neutral and centroidal axes do not coincide. To deal with such cases *Winkler-Bach Theory* is used.

20·2.   STRESSES IN CURVED BARS (Winkler-Bach Theory)

Fig. 20·1 shows a bar *ABCD* initially in its unstrained state. Let *AB'C'D* be the strained position of the bar.

Let,   *R* = Radius of curvature of the centroidal axis *HG*,

y = Distance of the fibre EF from the centroidal layer HG,

R' = Radius of curvature of HG',

y' = Distance between the EF' and HG' after straining,

M = Uniform bending moment applied to the beam (*assumed positive when tending to increase the curvature*),

θ = Original angle subtended by the centroidal axis HG at its centre of curvature O, and

θ' = Angle subtended by HG' (after bending) at the centre of curvature O'.

The following **assumptions** are made in this analysis :

1. Plane sections (transverse) remain plane during bending.
2. The material obeys Hooke's law (limit of proportionality is not exceeded).
3. Radial strain is negligible.
4. The fibres are free to expand or contract without any constraining effect from the adjacent fibres.

For finding the strain and stress normal to the section, consider the fibre EF at a distance y from the centroidal axis.

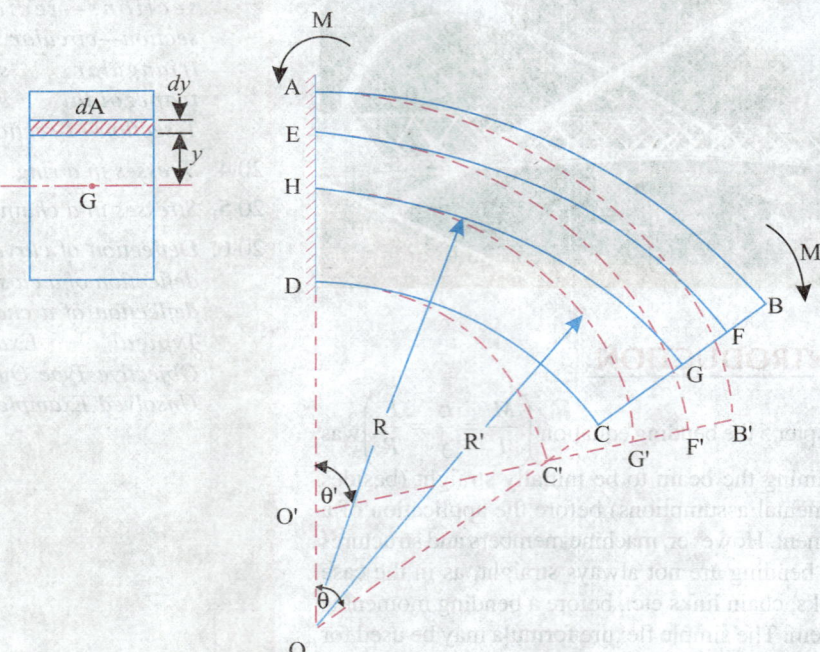

Fig. 20.1. Bending of a curved bar.

Let σ be the stress in the strained layer EF' under the bending moment M and e is strain in the same layer.

Strain, $\quad e = \dfrac{EF' - EF}{EF} = \dfrac{(R' + y')\,\theta' - (R + y)\,\theta}{(R + y)\,\theta}$, or, $e = \dfrac{R' + y'}{R + y} \cdot \dfrac{\theta'}{\theta} - 1$

Also, $\quad e_0$ = Strain in the centroidal layer *i.e.* when $y = 0$

$$= \frac{R'}{R} \cdot \frac{\theta'}{\theta} - 1, \quad \text{or,} \quad 1 + e = \frac{R' + y'}{R + y} \cdot \frac{\theta'}{\theta} \qquad \text{...(20·1)}$$

and, $\quad 1 + e_0 = \dfrac{R'}{R} \cdot \dfrac{\theta'}{\theta}$ $\qquad\qquad\qquad\qquad\qquad\qquad$ ...(20·2)

Dividing eqns. (20·1) and (20·2), we get

$$\frac{1+e}{1+e_0} = \frac{R'+y'}{R+y} \cdot \frac{R}{R'} = \frac{1+\dfrac{y'}{R'}}{1+\dfrac{y}{R}}$$

or,

$$e = (1+e_0)\left[\frac{1+\dfrac{y'}{R'}}{1+\dfrac{y}{R}}\right] - 1$$

or,

$$e = \frac{e_0 \cdot \dfrac{y'}{R'} + \dfrac{y'}{R'} + e_0 - \dfrac{y}{R}}{1+\dfrac{y}{R}}$$

According to assumption (3), radial strain is zero

∴

$$y = y'$$

Strain,

$$e = \frac{e_0 \cdot \dfrac{y}{R'} + \dfrac{y}{R'} + e_0 - \dfrac{y}{R}}{1+\dfrac{y}{R}}$$

Adding and subtracting the term $e_0 \cdot \dfrac{y}{R}$, we get

$$e = \frac{e_0 \cdot \dfrac{y}{R'} + \dfrac{y}{R'} + e_0 - \dfrac{y}{R} + e_0 \cdot \dfrac{y}{R} - e_0 \cdot \dfrac{y}{R}}{1+\dfrac{y}{R}}$$

or,

$$e = e_0 + \frac{(1+e_0)\left(\dfrac{1}{R'} - \dfrac{1}{R}\right)y}{1+\dfrac{y}{R}} \qquad \qquad ...(20·3)$$

In bridges, some members are often curved.

From the Fig. 20·1 it is obvious that, for the given bending moment the layers above the centroidal layer are in tension and layers below the centroidal layer are in compression.

Stress,
$$\sigma = Ee = E\left[e_0 + \frac{(1+e_0)\left(\dfrac{1}{R'} - \dfrac{1}{R}\right)y}{1+\dfrac{y}{R}}\right] \qquad ...(20·4)$$

(where, E = Young's modulus of the material)

Total force on the section, $F = \int \sigma \cdot dA$

Considering a small strip of elementary area dA, at a distance of y from the centroidal layer HG, we have

$$F = E\int e_0 \cdot dA + E\int \frac{(1+e_0)\left(\dfrac{1}{R'} - \dfrac{1}{R}\right)y \cdot dA}{1+\dfrac{y}{R}}$$

$$= E\int e_0 \cdot dA + E(1+e_0)\left(\frac{1}{R'} - \frac{1}{R}\right)\int \frac{y}{1+\dfrac{y}{R}} dA$$

or,
$$F = E\, e_0 \cdot A + E(1+e_0)\left(\frac{1}{R'} - \frac{1}{R}\right)\int \frac{y}{1+\dfrac{y}{R}} dA \qquad ...(20·5)$$

(where, A = area of cross-section of the bar)

The total resisting moment is given by

$$M = \int \sigma \cdot y \cdot dA = E\int e_0 \cdot y\, dA + E\int \frac{(1+e_0)\left(\dfrac{1}{R'} - \dfrac{1}{R}\right)}{1+\dfrac{y}{R}} \cdot y^2\ dA$$

$$= E \cdot e_0 \times 0 + E(1+e_0)\left(\frac{1}{R'} - \frac{1}{R}\right)\int \frac{y^2}{1+\dfrac{y}{R}} dA \quad \left[\because \int y\, dA = 0\right]$$

$\therefore$
$$M = E(1+e_0)\left(\frac{1}{R'} - \frac{1}{R}\right)\int \frac{y^2}{1+\dfrac{y}{R}} \cdot dA$$

Let,
$$\int \frac{y^2}{1+\dfrac{y}{R}} \cdot dA = Ah^2 \qquad ...(20·6)$$

where, h^2 = a constant for the cross-section of the bar

$\therefore$
$$M = E(1+e_0)\left(\frac{1}{R'} - \frac{1}{R}\right)Ah^2 \qquad ...(20·7)$$

Now,
$$\int \frac{y}{1+\dfrac{y}{R}} \cdot dA = \int \frac{Ry}{R+y} \cdot dA = \int \left(y - \frac{y^2}{R+y}\right)dA = \int y\, dA - \int \frac{y^2}{R+y} \cdot dA$$

$$= 0 - \frac{1}{R}\int \frac{y^2}{1+\dfrac{y}{R}} \cdot dA$$

$\therefore$ $$\int \frac{y}{1 + \dfrac{y}{R}} \cdot dA = -\frac{1}{R} \int \frac{y^2}{1 + \dfrac{y}{R}} \cdot dA = -\frac{1}{R} Ah^2 \qquad \text{...(20·8)}$$

Hence eqn. (20·5) becomes

$$F = E\, e_0 \cdot A - E\, (1 + e_0) \left(\frac{1}{R'} - \frac{1}{R} \right) \frac{Ah^2}{R} \qquad \text{...(20·9)}$$

Since transverse plane sections remain plane during bending

$\therefore$ $$F = 0$$

or, $$0 = E\, e_0 \cdot A - E\, (1 + e_0) \left(\frac{1}{R'} - \frac{1}{R} \right) \frac{Ah^2}{R}$$

or, $$E\, e_0 \cdot A = E\, (1 + e_0) \left(\frac{1}{R'} - \frac{1}{R} \right) \frac{Ah^2}{R}$$

or, $$e_0 = E\, (1 + e_0) \left(\frac{1}{R'} - \frac{1}{R} \right) \frac{h^2}{R}$$

or, $$\frac{e_0 R}{h^2} = (1 + e_0) \left(\frac{1}{R'} - \frac{1}{R} \right) \qquad \text{...(20·10)}$$

Substituting the value of $(1 + e_0) \left(\dfrac{1}{R'} - \dfrac{1}{R} \right)$ in eqn. (20·7), we get

Crane-cum-Forklift.

$$M = E \cdot \frac{e_0 R}{h^2} \cdot Ah^2 = e_0 \, EAR$$

or, $$e_0 = \frac{M}{EAR}$$ ...(20·11)

Substituting the value of e_0 in eqn. (20·4), we get

$$\sigma = \frac{M}{AR} + E \times \left(\frac{y}{1 + \dfrac{y}{R}} \right) \times \frac{e_0 R}{h^2}$$

$$= \frac{M}{AR} + E \times \frac{y}{1 + \dfrac{y}{R}} \times \frac{R}{h^2} \times \frac{M}{EAR} \qquad \left(\text{Substituting } e_0 = \frac{M}{EAR} \text{ again} \right)$$

$$= \frac{M}{AR} + \frac{M}{AR} \times \left[\frac{Ry}{1 + \dfrac{y}{R}} \right] \times \frac{1}{h^2}$$

or, $$\sigma = \frac{M}{AR} \left[1 + \frac{R^2}{h^2} \times \left(\frac{y}{R + y} \right) \right] \qquad \qquad \text{...(tensile) ...(20·12)}$$

On the other side of *HG*, *y* will be *negative*, and stress will be *compressive*

∴ $$\sigma = \frac{M}{AR} \left[1 - \frac{R^2}{h^2} \left(\frac{y}{R - y} \right) \right] \qquad \qquad \text{...(20·13)}$$

When the bending moment is applied in such a manner that it tends to *decrease the curvature*, then eqn. (20·12) will give compressive stress and eqn. (20·13), tensile stress.

Position of neutral axis :

At the neutral axis, $\sigma = 0$

∴ $$\frac{M}{AR} \left[1 + \frac{R^2}{h^2} \left(\frac{y}{R + y} \right) \right] = 0$$

or, $$\frac{R^2}{h^2} \left(\frac{y}{R + y} \right) = -1$$

or, $$R^2 y = -h^2 (R + y) = -Rh^2 - h^2 y$$

or, $$y (R^2 + h^2) = -Rh^2$$

$$y = -\left(\frac{Rh^2}{R^2 + h^2} \right) \qquad \qquad \text{...[20·12 (a)]}$$

Hence the neutral axis is located *below the centroidal axis.*

20·3. VALUES OF h² FOR VARIOUS SECTIONS

We know, $$h^2 = \frac{1}{A} \int \frac{y^2}{1 + \dfrac{y}{R}} \cdot dA \qquad \qquad \text{[Eqn. (20·6)]}$$

$$= \frac{R}{A} \int \frac{y^2}{R + y} \cdot dA$$

$$= \frac{R}{A}\left[\int y\,dA - \int R\,dA + \int \frac{R^2\,dA}{R+y}\right]$$

$$= \frac{R}{A}\left[0 - RA + \int \frac{R^2\,dA}{R+y}\right] \qquad \left[\begin{array}{l} \because \int y\,dA = 0 \\ \text{and } \int dA = A \end{array}\right]$$

$$h^2 = \frac{R^3}{A}\int \frac{dA}{R+y} - R^2 \qquad\qquad \ldots(20\cdot14)$$

20·3·1. Rectangular Section

Fig. 20·2 shows the rectangular section with centre of curvature 0 lying on YY-axis. XX-axis is the centroidal bending axis. Consider an elementary strip of with B and depth dy at a distance y from the centroidal layer.

Area of the strip, $\qquad dA = B \cdot dy$

Area of the section, $\qquad A = B \times D$

$$\therefore \qquad h^2 = \frac{R^3}{B \times D}\int_{-D/2}^{+D/2} \frac{B \cdot dy}{R+y} - R^2$$

$$= \frac{R^3}{D}\left|\log_e (R+y)\right|_{-D/2}^{+D/2} - R^2$$

or, $\qquad h^2 = \frac{R^3}{D}\log_e\left(\frac{2R+D}{2R-D}\right) - R^2 \qquad\qquad \ldots(20\cdot15)$

Garden moon bridge.

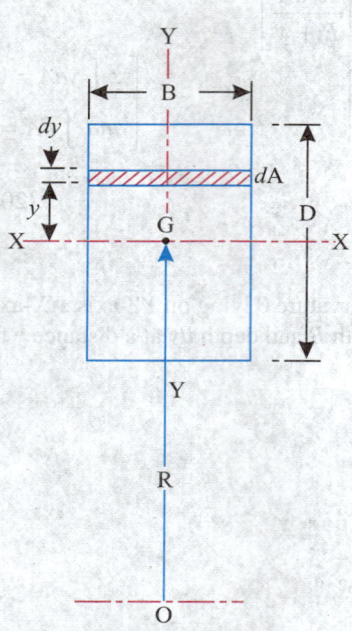

Fig. 20.2. Rectangular section.   **Fig. 20.3.** Circular section.

20·3·2. Circular Section

Fig. 20·3 shows the circular section of diameter d of a curved bar of radius of curvature of R, from the centre of curvature 0 up to the centroid G of the section.

Area of cross-section, $A = \dfrac{\pi}{4} d^2$

Consider a strip of width b and a depth dy at a distance y from the centroidal layer as shown.

$$b = 2 \sqrt{\left(\frac{d}{2}\right)^2 - y^2}$$

Area of strip, $dA = b.dy$

$$= 2 \times \sqrt{\left(\frac{d}{2}\right)^2 - y^2} \cdot dy$$

$$h^2 = \frac{R^3}{A} \int_{-d/2}^{+d/2} \frac{2 \times \sqrt{\left(\frac{d}{2}\right)^2 - y^2}}{R + y} \cdot dy - R^2$$

$$= \frac{8R^3}{\pi d^2} \int_{-d/2}^{+d/2} \frac{\sqrt{\frac{d^2}{4} - y^2}}{R + y} \cdot dy - R^2$$

Expanding the integral by binomial expression and then integrating, we get

$$h^2 = \frac{d^2}{16} + \frac{1}{128} \cdot \frac{d^4}{R^2} + \dots \qquad \dots(20\cdot16)$$

20·3·3. Triangular Section

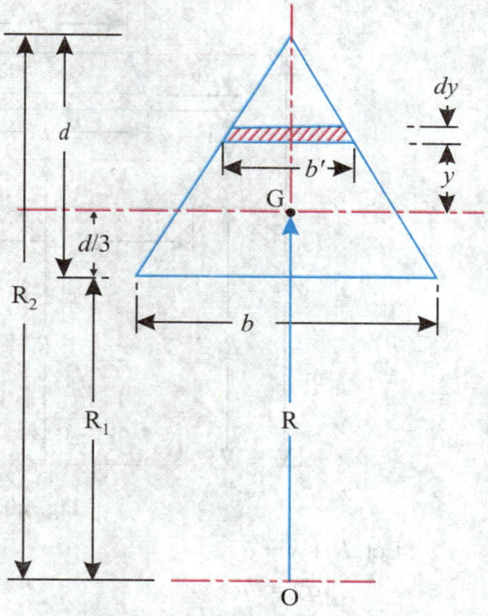

Refer to Fig. 20·4.

Let, $\qquad R + y = a$

$\qquad\qquad dy = da$

Width of elementary strip, $b' = \left(\dfrac{R_2 - a}{d}\right) b$

Area of the elementary strip,

$$dA = b' \cdot dy = b' \cdot da$$

Now, $\qquad h^2 = \dfrac{R^3}{A} \displaystyle\int_{R_1}^{R_2} \dfrac{dA}{R + y} - R^2$

$$= \dfrac{R^3}{A} \int_{R_1}^{R_2} \dfrac{b' da}{a} - R^2$$

$$= \dfrac{R^3}{A} \int_{R_1}^{R_2} \left(\dfrac{R_2 - a}{d}\right) b \cdot \dfrac{da}{a} - R^2$$

$$= \dfrac{R^3}{A} \left[\dfrac{R_2 b}{d} \int_{R_1}^{R_2} \dfrac{da}{a} - \dfrac{b}{d} \int_{R_1}^{R_2} da \right] - R^2$$

$$= \dfrac{R^3}{A} \left[\dfrac{R_2 b}{d} \log_e \dfrac{R_2}{R_1} - \dfrac{b}{d}(R_2 - R_1) \right] - R^2$$

$$= \dfrac{R^3}{A} \left[\left(d + \dfrac{2h}{3}\right) \dfrac{b}{d} \cdot \log_e \left(\dfrac{3R + 2d}{3R - d}\right) - b \right] - R^2$$

Fig. 20.4. Triangular section.

$$\left[\because R_2 = R + \dfrac{d}{3}, \text{ and} \right.$$
$$R_1 = R - \dfrac{d}{3}, \text{ and}$$
$$\left. R_2 - R_1 = d \right]$$

$$= \dfrac{2R^3}{bd} \left[(3R + 2d) \dfrac{b}{3d} \cdot \log_e \left(\dfrac{3R + 2d}{3R - d}\right) - b \right] - R^2 \qquad \left[\because A = \dfrac{1}{2} bd \right]$$

or, $\qquad \boldsymbol{h^2} = \dfrac{2R^3}{d} \left[\left(\dfrac{3R + 2d}{3d}\right) \log_e \left(\dfrac{3R + 2d}{3R - d}\right) - 1 \right] - R^2 \qquad \text{...(20·17)}$

20·3·4. Trapezodial Section

Fig. 20·5 shows a trapezoidal section. Consider an elementary strip of width b' and depth dy at distance y from the centroidal axis.

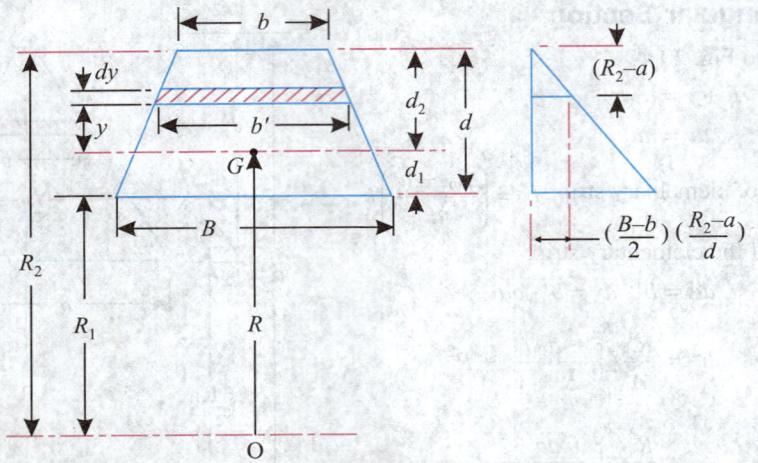

Fig. 20.5. Trapezoidal section.

Let, $R + y = a$

$dy = da$

$$b' = b + \left(\frac{B - b}{d_1 + d_2}\right)(R_2 - a)$$

Area of the strip,

$$dA = b' \, dy = \left[b + \left(\frac{B - b}{d_1 + d_2}\right)(R_2 - a)\right] da$$

$$h^2 = \frac{R^3}{A} \int_{R_1}^{R_2} \left[b + \left(\frac{B - b}{d_1 + d_2}\right)(R_2 - a)\right] \frac{da}{a} - R^2$$

$$= \frac{R^3}{A} \left[\int_{R_1}^{R_2} \frac{b \, da}{a} + \left(\frac{B - b}{d_1 + d_2}\right)\int_{R_1}^{R_2} \left(\frac{R_2 - a}{a}\right) da\right] - R^2$$

$$= \frac{R^3}{A} \left[b \cdot \left|\log_e a\right|_{R_1}^{R_2} + \left(\frac{B - b}{d_1 + d_2}\right)\left| R_2 \log_e a - a \right|_{R_1}^{R_2}\right] - R^2$$

$$= \frac{R^3}{A} \left[b \cdot \log_e \frac{R_2}{R_1} + \left(\frac{B - b}{d_1 + d_2}\right)\left\{R_2 \log_e \frac{R_2}{R_1} - (R_2 - R_1)\right\}\right] - R^2$$

$$= \frac{R^3}{A} \left[b \cdot \log_e \left(\frac{R + d_2}{R - d_1}\right) + \left(\frac{B - b}{d_1 + d_2}\right)(R + d_2) \log_e \left(\frac{R + d_2}{R - d_1}\right) - (B - b)\right] - R^2$$

$$= \frac{R^3}{A} \left[b \cdot \log_e \left(\frac{R + d_2}{R - d_1}\right) + \left(\frac{B - b}{d}\right)(R + d_2)\right.$$

$$\left. \log_e \left(\frac{R + d_2}{R - d_1}\right) - (B - b)\right] - R^2 \qquad ...(20\cdot18)$$

where, $$A = \left(\frac{B + b}{2}\right)d$$

$$d_1 = \frac{d}{3}\left(\frac{B + 2b}{B + b}\right)$$

$$d_2 = d - d_1$$

20·3·5. T-Section

Fig. 20·6 shows a T-section.

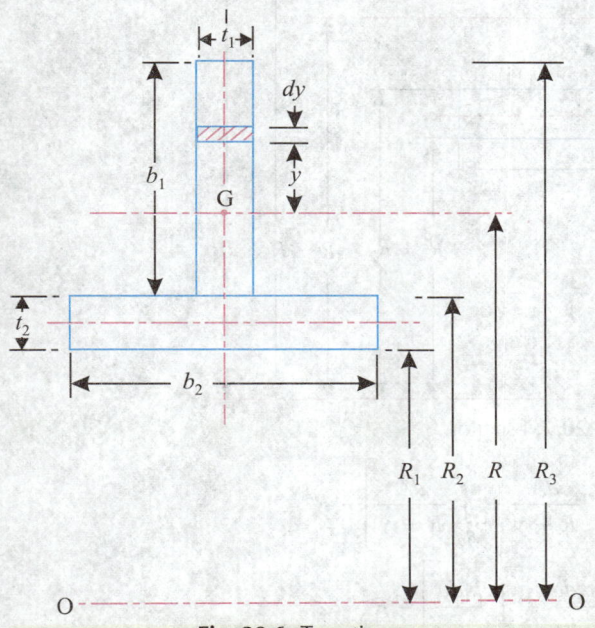

Fig. 20.6. T-section.

Communication tower.

Let, $R + y = a$

∴ $dy = da$

$$h^2 = \frac{R^3}{A}\left[\int_{R_1}^{R_2}\frac{dA}{R + y} + \int_{R_2}^{R_3}\frac{dA}{R + y}\right] - R^2$$

$$= \frac{R^3}{A}\left[\int_{R_1}^{R_2}\frac{b_2 da}{a} + \int_{R_2}^{R_3}\frac{t_1 da}{a}\right] - R^2$$

$$= \frac{R^3}{A}\left[b_2 \log_e \frac{R_2}{R_1} + t_1 \log_e \frac{R_3}{R_2}\right] - R^2$$

i.e. $\mathbf{h^2 = \dfrac{R^3}{A}\left[b_2 \log_e \dfrac{R_2}{R_1} + t_1 \log_e \dfrac{R_3}{R_2}\right] - R^2}$ ...(20·19)

where, $A = b_1 t_1 + b_2 t_2$

20·3·6. I-Section

Consider the I-Section shown in Fig. 20·7

Let, $R + y = a$

∴ $dy = da$

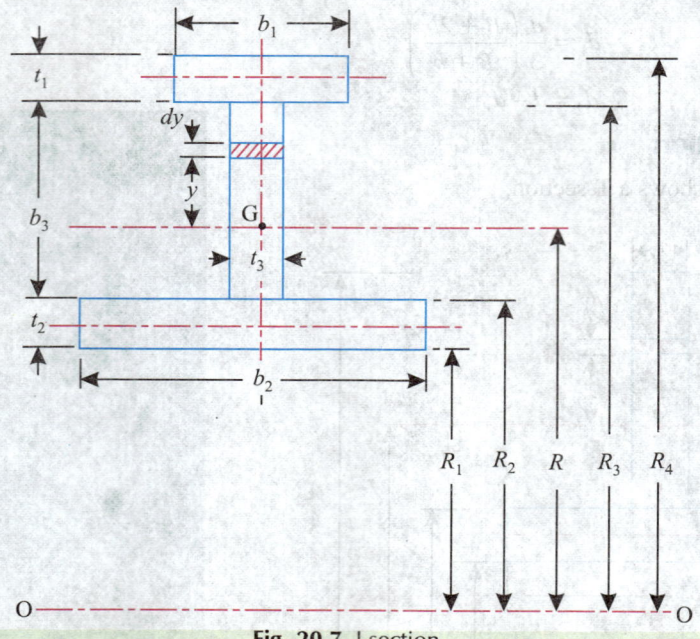

Fig. 20.7. I-section.

$$h^2 = \frac{R^3}{A} \left[\int_{R_1}^{R_2} \frac{dA}{R+y} + \int_{R_2}^{R_3} \frac{dA}{R+y} + \int_{R_3}^{R_4} \frac{dA}{R+y} \right] - R^2$$

$$= \frac{R^3}{A} \left[\int_{R_1}^{R_2} \frac{b_2 \, da}{a} + \int_{R_2}^{R_3} \frac{t_3 \, da}{a} + \int_{R_3}^{R_4} \frac{b_1 \, da}{a} \right] - R^2$$

or, $$h^2 = \frac{R^3}{A} \left[b_2 \, \log_e \frac{R_2}{R_1} + t_3 \, \log_e \frac{R_3}{R_2} + b_1 \, \log_e \frac{R_4}{R_3} \right] - R^2 \qquad ...(20\cdot20)$$

where, $A = b_1 t_1 + b_2 t_2 + b_3 t_3$

Example 20·1. *Fig. 20·8 shows a frame subjected to a load of 2·4 kN.*

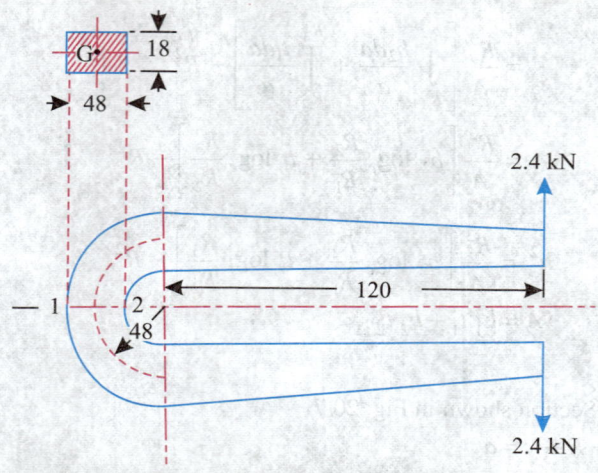

Fig. 20.8. Dimensions in mm.

Find :   (i) *The resultant stresses at points 1 and 2;*

      (ii) *Position of the neutral axis.*

Solution. Area of section at 1–2,

$$A = 48 \times 18 \times 10^{-6} = 8 \cdot 64 \times 10^{-4} \text{ m}^2$$

Bending Moment,

$$M = -2 \cdot 4 \times 10^3 \times (120 + 48) \times 10^{-3} = -403 \cdot 2 \text{ Nm}$$

M is taken -ve because it tends to *decrease the curvature.*

(i) Resultant stresses at points 1 and 2 :

Direct stress, $\sigma_d = \dfrac{2 \cdot 4 \times 10^3}{8 \cdot 64 \times 10^{-4}} \times 10^{-6} = 2 \cdot 77 \text{ MN/m}^2$ (tensile)

$$h^2 = \frac{R^3}{D} \log_e \left(\frac{2R + D}{2R - D} \right) - R^2$$

Here, $R = 48 \text{ mm} = 0 \cdot 048 \text{ m}, D = 48 \text{ mm} = 0 \cdot 048 \text{ m}$

∴ $h^2 = \dfrac{0 \cdot 048^3}{0 \cdot 048} \log_e \left(\dfrac{2 \times 0 \cdot 048 + 0 \cdot 048}{2 \times 0 \cdot 048 - 0 \cdot 048} \right) - 0 \cdot 048^2$

$$= 0 \cdot 048^2 (\log_e^3 - 1) = 2 \cdot 27 \times 10^{-4} \text{ m}^2$$

Bending stress due to *M* at point 2,

$$\sigma_{b2} = \frac{M}{AR} \left[1 - \frac{R^2}{h^2} \left(\frac{y}{R - y} \right) \right]$$

$$= \frac{-403 \cdot 2}{8 \cdot 64 \times 10^{-4} \times 0 \cdot 048} \left[1 - \frac{0 \cdot 048^2}{2 \cdot 27 \times 10^{-4}} \left(\frac{0 \cdot 024}{0 \cdot 048 - 0 \cdot 024} \right) \right] \times 10^{-6} \text{ MN/m}^2$$

$$= -9 \cdot 722 \, (1 - 10 \cdot 149) = +88 \cdot 95 \text{ MN/m}^2 \text{ (tensile)}$$

Bending moment due to *M* at point 1,

$$\sigma_{b1} = \frac{M}{AR} \left[1 + \frac{R^2}{h^2} \left(\frac{y}{R + y} \right) \right]$$

$$= \frac{-403 \cdot 2}{8 \cdot 64 \times 10^{-4} \times 0 \cdot 048} \left[1 + \frac{0 \cdot 048^2}{2 \cdot 27 \times 10^{-4}} \left(\frac{0 \cdot 024}{0 \cdot 048 + 0 \cdot 024} \right) \right] \times 10^{-6} \text{ MN/m}^2$$

$$(\because y = D/2 = 0 \cdot 024 \text{ m})$$

$$= -42 \cdot 61 \text{ MN/m}^2 = 42 \cdot 61 \text{ MN/m}^2 \text{ (comp.)}$$

∴ Resultant stress at **point 2,**

$$\sigma_2 = \sigma_d + \sigma_{b2} = 2 \cdot 77 + 88 \cdot 95 = \textbf{91·72 MN/m}^2 \textbf{ (tensile) (Ans.)}$$

Resultant stress at **point 1,**

$$\sigma_1 = \sigma_d + \sigma_{b1} = 2 \cdot 77 - 42 \cdot 61 = \textbf{39·84 MN/m}^2 \textbf{ (comp.) (Ans.)}$$

(ii) Position of the neutral axis :

We know, $y = - \left(\dfrac{Rh^2}{R^2 + h^2} \right)$ ...[Eqn. 20·12 (*a*)]

$$= - \left(\frac{0 \cdot 048 \times 2 \cdot 27 \times 10^{-4}}{0 \cdot 048^2 + 2 \cdot 27 \times 10^{-4}} \right) = -0 \cdot 00435 \text{ m} = -4 \cdot 35 \text{ mm}$$

Hence, neutral axis is at a radius of **4·35 mm (Ans.)**

Example 20·2. *The curved member shown in Fig. 20·9 has a solid circular cross-section 0·10 m in diameter. If the maximum tensile and compressive stresses in the member are not to exceed 150 MPa and 200 MPa respectively, determine the value of load P that can safely be carried by the member.* **(AMIE Summer, 2001)**

Solution. *Given :* $d = 0{\cdot}10$ m; $R = 0{\cdot}10$ m; $\sigma_1 = 150$ MPa $= 150$ MN/m^2 (tensile)

$\sigma_2 = 200$ MPa $= 200$ MN/m^2 (compressive).

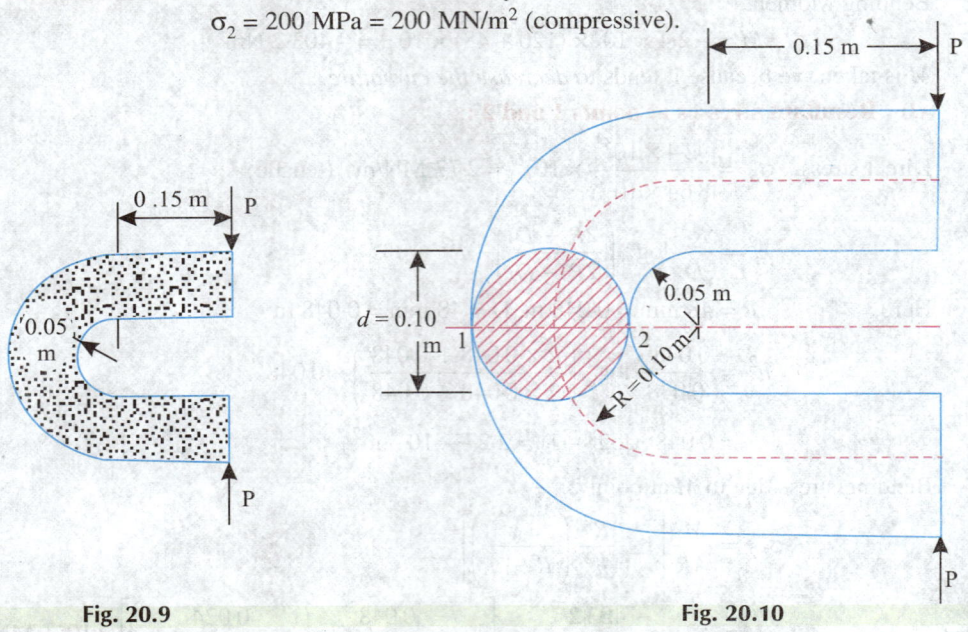

Fig. 20.9   **Fig. 20.10**

Load P:

Refer to Fig. 20·10. Area of cross section,

$$A = \frac{\pi}{4}\, d^2 = \frac{\pi}{4} \times 0{\cdot}10^2 = 7{\cdot}854 \times 10^{-3} \text{ m}^2$$

Bending moment, $M = P\,(0{\cdot}15 + 0{\cdot}10) = 0{\cdot}25\,P$

$$h^2 = \frac{d^2}{16} + \frac{1}{128} \cdot \frac{d^4}{R^2} \qquad\qquad \text{[Eqn. (20·16)]}$$

Substituting the values, we have

$$h^2 = \frac{(0{\cdot}10)^2}{16} + \frac{1}{128} \cdot \frac{(0{\cdot}10)^4}{(0{\cdot}10)^2} = 7{\cdot}031 \times 10^{-4} \text{ m}^2$$

Direct stress, $\quad \sigma_d = \dfrac{P}{A} \quad$ *(comp.)*

Bending stress at point 1 due to M,

$$\sigma_{b1} = \frac{M}{AR}\left[1 + \frac{R^2}{h^2} \times \frac{y}{R + y}\right] \text{(tensile)} \qquad\qquad \text{[Eqn. (20·12)]}$$

Total stress at point 1,

$$\sigma_1 = \sigma_d + \sigma_{b1}$$

$$150 = -\frac{P}{A} + \frac{M}{AR}\left[1 + \frac{R^2}{h^2} \times \frac{y}{R + y}\right] \text{(tensile)}$$

$$150 = -\frac{P}{7.854 \times 10^{-3}} + \frac{0.25\,P}{7.854 \times 10^{-3} \times 0.10}\left[1 + \frac{0.10^2}{7.031 \times 10^{-4}} \times \frac{0.05}{0.10 + 0.05}\right]$$

$$= -127.32\,P + 318.31\,P \times 5.74 = 1699.78\,P$$

$$\therefore \quad P = \frac{150}{1699.78}\,\text{MN} = \frac{150 \times 10^3}{1699.78}\,\text{kN} = 88.25\,\text{kN} \qquad \ldots(i)$$

Bending stress at point 2 due to M,

$$\sigma_{b2} = \frac{M}{AR}\left[\frac{R^2}{h^2} \times \frac{y}{R - y} - 1\right] \qquad (comp.)$$

Total stress at point 2,

$$\sigma_2 = \sigma_d + \sigma_{b2}$$

$$200 = \frac{P}{A} + \frac{M}{AR}\left[\frac{R^2}{h^2} \times \frac{y}{R - y} - 1\right] \qquad (comp.)$$

$$= \frac{P}{7.854 \times 10^{-3}} + \frac{0.25\,P}{7.854 \times 10^{-3} \times 0.10}\left[\frac{0.10^2}{7.031 \times 10^{-4}} \times \frac{0.05}{0.10 - 0.05} - 1\right]$$

$$= 127.32\,P + 318.31\,P \times 13.22 = 4335.38\,P$$

$$\therefore \quad P = \frac{200}{4335.38}\,\text{MN} = \frac{200}{4335.38} \times 10^3\,\text{kN} = 46.13\,\text{kN} \qquad \ldots(ii)$$

Comparing eqn. (i) and (ii) the safe load P will be lesser of these.

Hence, $\qquad P = \mathbf{46.13\ kN}$ **(Ans.)**

Example 20.3. *Fig. 20.11 shows a circular ring of rectangular section, with a slit and subjected to load P.*

  (i) *Calculate the magnitude of the force P if the maximum stress along the section 1-2 is not to exceed 225 MN/m².*

  (ii) *Draw the stress distribution along 1-2.*

Solution. Area of section at 1-2,

$$A = 9 \times 6 = 54\ \text{cm}^2 = 0.0054\ \text{m}^2$$

Permissible stress,

$$\sigma = 225\ \text{MN/m}^2$$

Bending moment,

$$M = P \times (16.5 \times 10^{-2} = +0.165\,P\ \text{Nm}$$

(*M* is taken +ve because it *tends to increase the curvature*)

  (i) **Magnitude of the force P:**

Direct stress,

$$\sigma_d = \frac{P}{A} = \frac{P}{0.0054}$$

$$= 185.18\,P\ \text{N/m}^2\ (comp.)$$

$$h^2 = \frac{R^3}{D}\log_e\left(\frac{2R + D}{2R - D}\right) - R^2$$

Here, $R = 16.5 \times 10^{-2}$ m $= 0.165$ m

$\qquad D = 9$ cm $= 0.09$ m

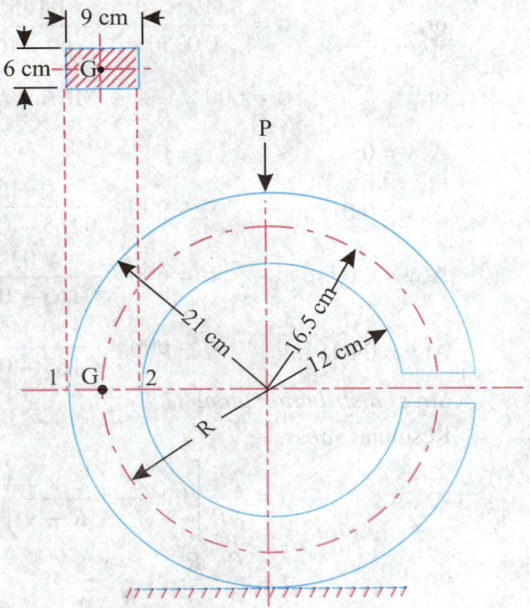

Fig. 20.11

$$h^2 = \frac{0.165^3}{0.09} \log_e \left(\frac{2 \times 0.165 + 0.09}{2 \times 0.165 - 0.09} \right) - 0.165^2$$

$$= 0.0499 \times 0.5596 - 0.165^2 = 6.99 \times 10^{-4}$$

Resultant stress at point 2,

$$\sigma_{max} = \frac{M}{AR} \left[\frac{R^2}{h^2} \left(\frac{y}{R-y} \right) - 1 \right] + \frac{P}{A} \quad \text{(comp.)}$$

$$= \frac{M}{AR} \left(\frac{R^2}{h^2} \times \frac{y}{R-y} \right) - \frac{M}{AR} + \frac{P}{A}$$

$$= \frac{P}{A} \left(\frac{R^2}{h^2} \times \frac{y}{R-y} \right) - \frac{P}{A} + \frac{P}{A}$$

$$\left[\because \frac{M}{AR} = \frac{P \times R}{AR} = \frac{P}{A} \right]$$

$$= \frac{P}{A} \left(\frac{R^2}{h^2} \times \frac{y}{R-y} \right)$$

Elbow part.

or, $\quad 225 \times 10^6 = 185.18\, P \left(\dfrac{0.165^2}{6.99 \times 10^{-4}} \times \dfrac{0.045}{0.165 - 0.045} \right)$

$$(\because y = D/2 = 0.045 \text{ m}) = 2704.68\, P$$

∴ $\qquad P = 83189 \text{ N or } \mathbf{83.19 \text{ kN}} \quad \textbf{(Ans.)}$

(ii) Stress distribution along the section 1-2 :

Stress distribution along G 2:

$$\sigma = \frac{P}{A} \cdot \frac{R^2}{h^2} \times \frac{y}{R-y} \quad (y \text{ varies from 0 to 0.045 m})$$

or, $\qquad \sigma = \dfrac{83.19 \times 10^3}{0.0054} \times \dfrac{0.165^2}{6.99 \times 10^{-4}} \times \dfrac{y}{R-y} \times 10^{-6} \text{ MN/m}^2$

or, $\qquad \sigma = 600 \times \dfrac{y}{R-y} \text{ MN/m}^2$

At $y = 0$: $\qquad \sigma = 0$

At $y = 0.015$ m : $\quad \sigma = 600 \times \dfrac{0.015}{0.165 - 0.015} = 60 \text{ MN/m}^2 \ (comp.)$

At $y = 0.03$ m : $\quad \sigma = 600 \times \dfrac{0.03}{0.165 - 0.03} = 133.3 \text{ MN/m}^2 \ (comp.)$

At $y = 0.045$ m : $\quad \sigma = 600 \times \dfrac{0.045}{0.165 - 0.045} = 225 \text{ MN/m}^2 \ (comp.)$

Stress distribution along G 1:

Resultant stress,

$$\sigma = \frac{M}{AR} \left[1 + \frac{R^2}{h^2} \left(\frac{y}{R+y} \right) \right] - \frac{P}{A} \qquad [\because \sigma_d \,(= P/A) \text{ is compressive}]$$

or, $\qquad \sigma = \dfrac{P}{A} \cdot \dfrac{R^2}{h^2} \times \dfrac{y}{R+y}$

or, $\qquad \sigma = 600 \times \dfrac{y}{R+y} = \text{ MN/m}^2$

At $y = 0$: $\sigma = 0$

At $y = 0.015$ m : $\sigma = 600 \times \dfrac{0.015}{0.165 + 0.015} = 50$ MN/m^2 (tensile)

At $y = 0.03$ m : $\sigma = 600 \times \dfrac{0.03}{0.165 + 0.03} = 92.3$ MN/m^2 (tensile)

At $y = 0.045$ m : $\sigma = 600 \times \dfrac{0.045}{0.165 + 0.045} = 128.6$ MN/m^2 (tensile)

Fig. 20.12 shows the distribution of the stresses along 1-2.

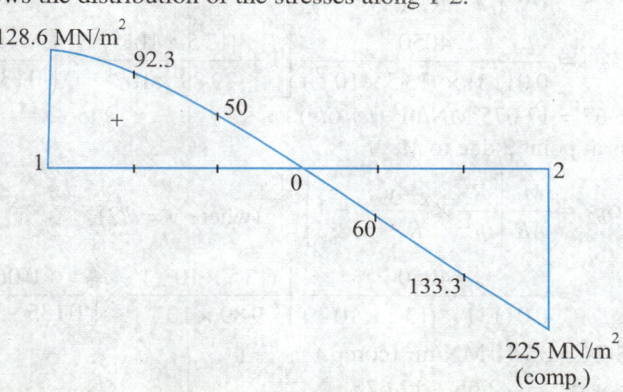

Fig. 20.12

Example 20·4. *Fig. 20·13. shows a ring carrying a load of 30 kN. Calculate the stresses at 1 and 2.*

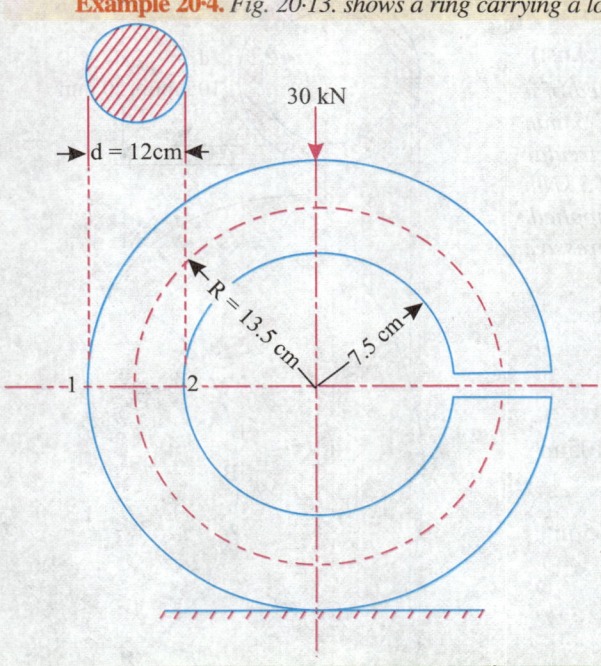

Hand dynamometer.

Fig. 20.13

Solution. Area of cross-section $= \dfrac{\pi}{4} \times 12^2$ cm$^2 = 113.1$ cm$^2 = 0.01131$ m^2

Bending moment, $M = 30 \times 10^3 \times (13.5 \times 10^{-2})$ Nm $= 4050$ Nm

$$h^2 = \frac{d^2}{16} + \frac{1}{128} \cdot \frac{d^4}{R^2} + \dots \qquad \text{...(Eqn. 20·16)}$$

Here, $d = 12$ cm, $R = 7.5 + 6 = 13.5$ cm

$$h^2 = \frac{(12)^2}{16} + \frac{(12)^4}{128 \times 13.5^2} = 9.89 \text{ cm}^2 = 9.89 \times 10^{-4} \text{ m}^2$$

Direct stress, $\sigma_d = \dfrac{30 \times 10^3}{0.01131} \times 10^{-6} = 2.65$ MN/m² (comp.)

Bending stress at point 1 due to M,

$$\sigma_{b1} = \frac{M}{AR}\left[1 + \frac{R^2}{h^2}\left(\frac{y}{R+y}\right)\right]$$

$$= \frac{4050}{0.01131 \times (13.5 \times 10^{-2})}\left[1 + \frac{(13.5 \times 10^{-2})^2}{9.89 \times 10^{-4}} \times \left(\frac{0.06}{0.135 - 0.06}\right)\right] \times 10^{-6}$$

$2.65 \times 6.67 = 17.675$ MN/m² (tensile)

Bending stress at point 2 due to M,

$$\sigma_{b2} = \frac{M}{AR}\left[\frac{R^2}{h^2} \times \frac{y}{R-y} - 1\right] \quad \text{(where } y = d/2)$$

$$= \frac{4050}{0.01131 \times (13.5 \times 10^{-2})}\left[\frac{(13.5 \times 10^{-2})^2}{9.89 \times 10^{-4}} \times \left(\frac{0.06}{0.135 - 0.06}\right) - 1\right] \times 10^{-6}$$

$2.65 \times 13.74 = 36.41$ MN/m² (comp.)

Hence $\sigma_1 = \sigma_d + \sigma_{b1} = -2.65 + 17.675$

$\qquad = \mathbf{15.025}$ **MN/m² (tensile) (Ans.)**

and, $\sigma_2 = \sigma_d + \sigma_{b2} = 2.65 + 36.41$

$\qquad = \mathbf{39.06}$ **MN/m² (comp.) (Ans.)**

Example 20·5. *A curved bar is formed of a tube of 120 mm outside diameter and 7·5 mm thickness. The centre line of this beam is a circular arc of radius 225 mm. A bending moment of 3 kNm tending to increase curvature of the bar is applied. Calculate the maximum tensile and compressive stresses set up in the bar.*

Solution. Outside diameter of the tube,

$d_2 = 120$ mm $= 0.12$ m

Thickness of the tube $= 7.5$ mm

Inside diameter of the tube,

$d_1 = 120 - 2 \times 7.5 = 105$ mm $= 0.105$ m

Area of cross-section,

$$A = \frac{\pi}{4}[0.12^2 - 0.105^2] = 0.00265 \text{ m}^2$$

Bending moment,

$M = 3$ kNm

Area of inner circle,

$$A_1 = \frac{\pi}{4} \times 0.105^2 = 0.00866 \text{ m}^2$$

Area of outer circle,

$$A_2 = \frac{\pi}{4} \times 0.12^2 = 0.01131 \text{ m}^2$$

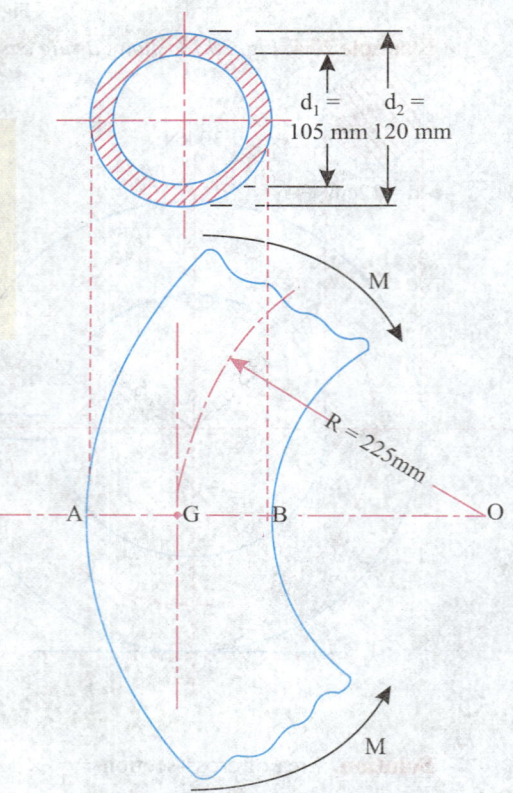

$d_1 = $ 105 mm $\quad$ $d_2 = $ 120 mm

Fig. 20.14

Bending moment, $M = + 3$ kNm (tending to *increase* the curvature)

We know, for a circular section,

$$h^2 = \frac{d^2}{16} + \frac{1}{128} \cdot \frac{d^4}{R^2} + \ldots$$

For inner circle :
$$h_1^2 = \frac{d_1^2}{16} + \frac{1}{128} \cdot \frac{d^4}{R^2}$$

$$= \frac{0{\cdot}105^2}{16} + \frac{1}{128} \times \frac{0{\cdot}105^4}{0{\cdot}225^2} = 7{\cdot}078 \times 10^{-4} \text{ m}^2$$

For outer circle :

$$h_2^2 = \frac{d_2^2}{16} + \frac{1}{128} \cdot \frac{d_2^4}{R^2}$$

$$= \frac{0{\cdot}12^2}{16} + \frac{0{\cdot}12^4}{128 \times 0{\cdot}225^2} = 9{\cdot}32 \times 10^{-4} \text{ m}^2$$

Also,
$$Ah^2 = A_2 h_2^2 - A_1 h_1^2$$

$$0{\cdot}00265\, h^2 = 0{\cdot}01131 \times 9{\cdot}32 \times 10^{-4} - 0{\cdot}00866 \times 7{\cdot}078 \times 10^{-4}$$

$$\therefore \qquad h^2 = 0{\cdot}00166 \text{ m}^2, \text{ and, } \frac{R^2}{h^2} = \frac{0{\cdot}225^2}{0{\cdot}00166} = 30{\cdot}49$$

Maximum stress at A,

$$\sigma_A = \frac{M}{AR}\left[1 + \frac{R^2}{h^2} \times \frac{y}{R + y}\right] \qquad \text{(where, } y = 60 \text{ mm} = 0{\cdot}06 \text{ m)}$$

$$= \frac{3 \times 10^3}{0{\cdot}00265 \times 0{\cdot}225}\left[1 + 30{\cdot}49 \times \frac{0{\cdot}06}{0{\cdot}225 + 0{\cdot}06}\right] \times 10^{-6} \text{ MN/m}^2$$

or, $\qquad \sigma_A = \textbf{37·32 MN/m}^2 \textbf{ (tensile)} \textbf{ (Ans.)}$

Maximum stress at B,

$$\sigma_B = \frac{M}{AR}\left[\frac{R^2}{h^2} \times \frac{y}{R - y} - 1\right]$$

$$= \frac{3 \times 10^3}{0{\cdot}00265 \times 0{\cdot}225}\left[30{\cdot}49 \times \frac{0{\cdot}06}{0{\cdot}225 - 0{\cdot}06} - 1\right] \times 10^{-6} \text{ MN/m}^2$$

or, $\qquad \sigma_B = \textbf{50·75 MN/m}^2 \textbf{ (comp.)} \textbf{ (Ans.)}$

Hook and pulley.

Example 20·6. *Fig. 20·15 shows a crane hook lifting a load of 150 kN. Determine the maximum compressive and tensile stresses in the critical section of the crane hook.*

Solution. Refer Fig. 20·15

Here,

$B = 135$ mm $= 0·135$ m

$b = 45$ mm $= 0·045$ m

$d = 180$ mm $= 0·18$ m

Load, $P = 150$ kN

Area of cross-section, $A = \dfrac{(0·135 + 0·45)}{2} \times 0·18 = 0·0162$ m^2

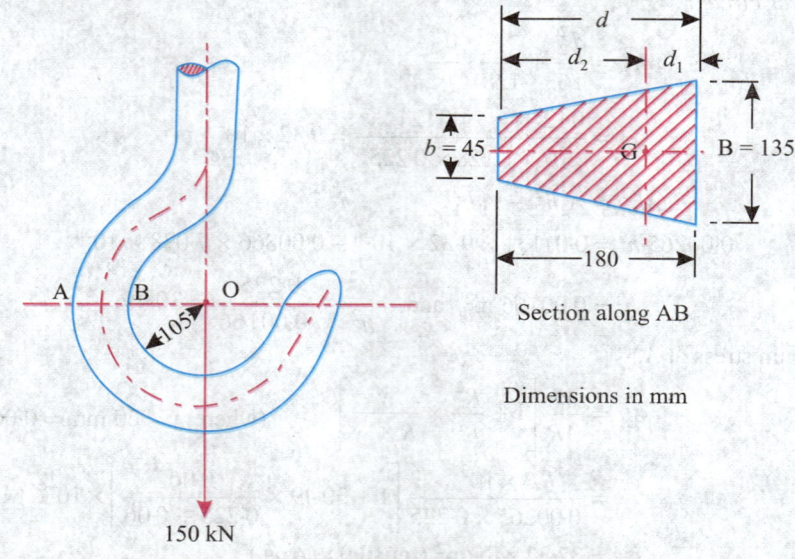

Dimensions in mm

150 kN

Fig. 20.15

$$d_1 = \frac{d}{3}\left[\frac{B + 2b}{B + b}\right] = \frac{0·18}{3}\left[\frac{0·135 + 2 \times 0·045}{0·135 + 0·045}\right] = 0·075 \text{ m}$$

$d_2 = 0·18 - 0·075 = 0·105$ m

$R = 0·105 + 0·075 = 0·18$ m

$$h^2 = \frac{R^3}{A}\left[b \log_e\left(\frac{R + d_2}{R - d_1}\right) + \left(\frac{B - b}{d}\right)(R + d_2) \times \right.$$

$$\left. \log_e\left(\frac{R + d_2}{R - d_1}\right) - (B - b) - R^2\right] \qquad \text{...Eqn. (20·18)}$$

$$= \frac{0·18^3}{0·0162}\left[0·045 \log_e\left(\frac{0·18 + 0·105}{0·18 - 0·075}\right) + \left(\frac{0·135 + 0·045}{0·18}\right)(0·18 + 0·105) \times \right.$$

$$\left. \log_e\left(\frac{0·18 + 0·105}{0·18 - 0·075}\right) - (0·135 - 0·045)\right] - 0·18^2$$

$= 0·36\,(0·0449 + 0·1425 \times 0·998 - 0·09) - 0·0324 = 0·00256$ m^2

Bending moment, $M = -150 \times 10^3 \times 0·18 = -27000$ Nm

[–ve sign is taken because bending moment is tending to *decrease the curvature*]

Direct stress, $\sigma_d = \dfrac{P}{A} = \dfrac{150 \times 10^3}{0·0162} \times 10^{-6} = -9·26$ MN/m^2 (*tensile*)

Bending stress at A,

$$(\sigma_b)_A = \frac{M}{AR}\left[1 + \frac{R^2}{h^2} \times \frac{d_2}{R + d_2}\right]$$

$$= \frac{-27000}{0.0162 \times 0.18}\left[1 + \frac{0.18^2}{0.00256} \times \frac{0.105}{0.18 + 0.105}\right] \times 10^{-6} \text{ MN/m}^2$$

$$= -9.26\,(1 + 4.663) = -52.43 \text{ MN/m}^2$$

$$= 52.43 \text{ MN/m}^2 \text{ (comp.)}$$

Bending stress at B,

$$(\sigma_b)_B = \frac{M}{AR}\left[1 - \frac{R^2}{h^2} \times \frac{d_1}{R - d_1}\right]$$

$$= \frac{-27000}{0.0162 \times 0.18}\left[1 - \frac{0.18^2}{0.00256}\left(\frac{0.075}{0.18 - 0.075}\right)\right]$$

$$= -9.26\,(1 - 9.04) = +74.45 \text{ MN/m}^2 \text{ (tensile)}$$

Stress at A, $\qquad \sigma_A = \sigma_d + (\sigma_b)_A = 9.26 - 52.43 = -43.17 \text{ MN/m}^2$

i.e. $\qquad \boldsymbol{\sigma_A = 43.17 \text{ MN/m}^2 \text{ (comp.) \quad (Ans.)}}$

Stress at B, $\qquad \sigma_B = \sigma_d + (\sigma_b)_B = 9.26 + 74.45 = 83.71 \text{ MN/m}^2$

i.e $\qquad \boldsymbol{\sigma_B = 83.71 \text{ MN/m}^2 \text{ (tensile) \quad (Ans.)}}$

Example 20·7. *A central horizontal section of hook is a symmetrical trapezium 60 mm deep, the inner width being 60 mm and the outer being 30 mm. Estimate the extreme intensities of stress when the hook carries a load of 30 kN, the load line passing 40 mm from the inside edge of the section and the centre of carvature being in the load line. Also plot the stress distribution across the section.*

Solution. Refer to Fig. 20·16.

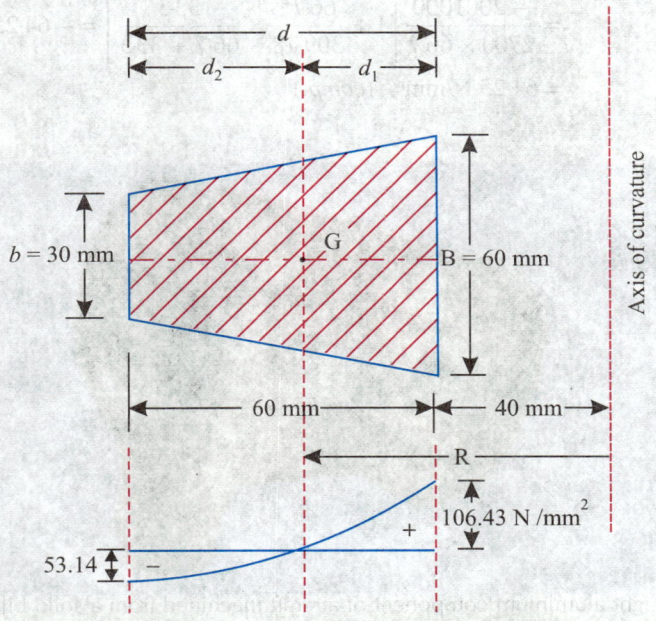

Fig. 20.16

Here, $B = 60$ mm, $b = 30$ mm, $d = 60$ mm, P(load) $= 30$ kN

Area of cross-section, $\quad A = \left(\dfrac{60 + 30}{2}\right) \times 60 = 2700$ mm^2

$$d_1 = \frac{d}{3}\left[\frac{B + 2b}{B + b}\right] = \frac{60}{3}\left[\frac{60 + 2 \times 30}{60 + 30}\right] = 26.7 \text{ mm}$$

$$d_2 = d - d_1 = 60 - 26\cdot7 = 33\cdot3 \text{ mm}$$

$$R = 40 + d_1 = 40 + 26\cdot7 = 66\cdot7 \text{ mm}$$

$$h^2 = \frac{R^3}{A}\left[b\,\log_e\left(\frac{R + d_2}{R - d}\right) + \left(\frac{B - b}{d}\right)(R + d_2)\log_e\left(\frac{R + d_2}{R - d}\right) - (B - b)\right] - R^2$$

$$= \frac{(66\cdot7)^3}{2700}\left[30 \times \log_e\left(\frac{66\cdot7 + 33\cdot3}{66\cdot7 - 26\cdot7}\right) + \left(\frac{60 - 30}{60}\right)(66\cdot7 + 33\cdot3) \times \right.$$

$$\left. \log_e\left(\frac{66\cdot7 + 33\cdot3}{66\cdot7 - 26\cdot7}\right) - (60 - 30)\right] - 66\cdot7^2$$

$$= 109\cdot90\,[27\cdot49 + 45\cdot81 - 30] - 4448\cdot89 = 309\cdot78 \text{ mm}^2$$

Bending moment $M = -\,(30 \times 1000) \times 66\cdot7 = -\,2001000$ Nmm

(–ve sign is taken because bending moment is tending to *decrease the curvature*).

Direct stresses, $\quad \sigma_d = \dfrac{P}{A} = \dfrac{20 \times 1000}{2700} = 11\cdot11$ N/mm^2 (*tensile*)

Bending stress at the *outside edge* of the section,

$$(\sigma_b)_0 = \frac{M}{AR}\left[1 + \frac{R^2}{h^2} \times \frac{d_2}{R + d_2}\right]$$

$$= \frac{-\,2001000}{2700 \times 66\cdot7}\left[1 + \frac{66\cdot7^2}{309\cdot78} \times \frac{33\cdot3}{66\cdot7 + 33\cdot3}\right] = -\,64\cdot25 \text{ N/mm}^2$$

$$= 64\cdot25 \text{ N/mm}^2 \quad (\textit{comp.})$$

Light aluminium component of aircraft machined from a solid billet.

∴ Total stress at the outside edge of the section,

$$\sigma_0 = \sigma_d + (\sigma_b)_0 = 11 \cdot 11 - 64 - 25 = -53 \cdot 14 \text{ N/mm}^2$$

∴ $\qquad\qquad \sigma_0 = \mathbf{53 \cdot 14 \text{ N/m}^2}$ *(comp.)*

Bending stresses at the inside edge of the section,

$$(\sigma_b)_i = \frac{M}{AR}\left[1 - \frac{R^2}{h^2} \times \frac{d_1}{R - d_1}\right]$$

$$= \frac{-2001000}{2700 \times 66 \cdot 7}\left[1 + \frac{66 \cdot 7^2}{309 \cdot 78} \times \frac{26 \cdot 7}{66 \cdot 7 - 26 \cdot 7}\right]$$

$$= -11 \cdot 11 \times -8 \cdot 58 = 95 \cdot 32 \text{ N/mm}^2 \text{ *(tensile)*}$$

Total stress at the inside edge of the section,

$$\sigma_i = \sigma_d + (\sigma_b)_i = 11 \cdot 11 + 95 \cdot 32 = \mathbf{106 \cdot 43 \text{ N/mm}^2} \text{ (tensile)}$$

At any point distant y from the centroidal axis, the total (σ) stress is given by

$$\sigma = \sigma_d + \sigma_b = \sigma_d + \frac{M}{AR}\left[1 + \frac{R^2}{h^2} \times \left(\frac{y}{R + y}\right)\right] \qquad \text{...[Eqn. (20·12)]}$$

or, $\qquad \sigma = +11 \cdot 11 + \frac{-2001000}{2700 \times 66 \cdot 7}\left[1 + \frac{66 \cdot 7^2}{309 \cdot 78} \times \left(\frac{y}{66 \cdot 7 + y}\right)\right]$

or, $\qquad \sigma = 11 \cdot 11 - 11 \cdot 11\left[1 + \frac{14 \cdot 36\, y}{66 \cdot 7 + y}\right]$

At $y = -10$ mm : $\sigma = 11 \cdot 11 - 11 \cdot 11\left[1 + \frac{14 \cdot 36\,(-10)}{66 \cdot 7 - 10}\right] = 28 \cdot 14 \text{ N/mm}^2$

At $y = -20$ mm : $\sigma = 11 \cdot 11 - 11 \cdot 11\left[1 + \frac{14 \cdot 36\,(-20)}{66 \cdot 7 - 20}\right] = 68 \cdot 32 \text{ N/mm}^2$

At $y = 0$: $\qquad \sigma = 11 \cdot 11 - 11 \cdot 11\left[1 + \frac{14 \cdot 36\,(0)}{66 \cdot 7 - 0}\right] = 0$

At $y = +10$ mm : $\sigma = 11 \cdot 11 - 11 \cdot 11\left[1 + \frac{14 \cdot 36 \times 10}{66 \cdot 7 + 10}\right] = -20 \cdot 8 \text{ N/mm}^2$

At $y = +20$ mm : $\sigma = 11 \cdot 11 - 11 \cdot 11\left[1 + \frac{14 \cdot 36 \times 20}{66 \cdot 7 + 20}\right] = -36 \cdot 8 \text{ N/mm}^2$

At $y = +30$ mm : $\sigma = 11 \cdot 11 - 11 \cdot 11\left[1 + \frac{14 \cdot 36 \times 30}{66 \cdot 7 + 30}\right] = -49 \cdot 49 \text{ N/mm}^2$

The stress distribution is shown in Fig. 20·16.

Example 20·8. *A central horizontal section of a hook is an I-section with dimensions shown in Fig. 20·17. The hook carries a load P, the load line passing 60 mm from the inside edge of the section, and the centre of curvature being in the load line.*

Determine the magnitude of the load P if the maximum stress in the hook is not to exceed the permissible stress of 108 N/mm². What will be the maximum compressive stress in hook for that value of load ?

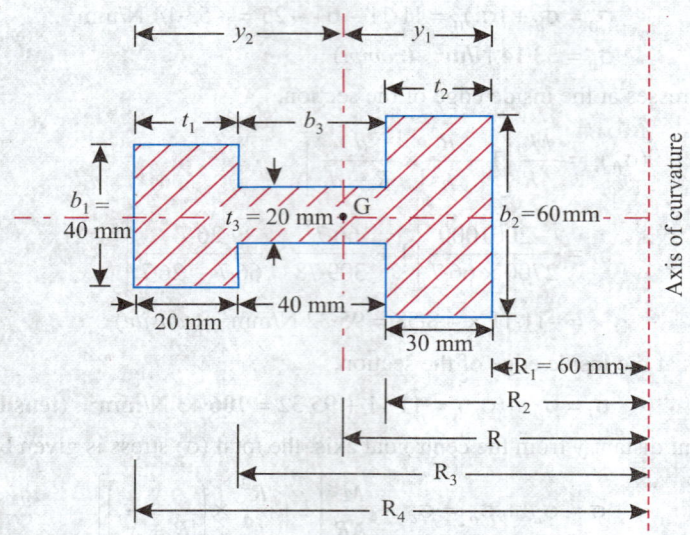

Fig. 20.17

Solution. Permissible stress,

$$\sigma = 105 \text{ N/mm}^2$$

Load P:

Maximum compressive stress :

Area of cross section $= b_1 t_1 + b_2 t_2 + b_3 t_3$

$$= 40 \times 20 + 60 \times 30 + 40 \times 20 = 3400 \text{ mm}^2$$

To find c.g. of the section, taking moments about the edge LL, we get

$$= 40 \times 20 \times 80 + 40 \times 20 \times 50 + 60 \times 30 \times 15 = 3400 \, y_1$$

$\therefore \quad y_1 = 38 \cdot 5 \text{ mm}$

and, $\quad y_2 = 90 - 38 \cdot 5 = 51 \cdot 5 \text{ mm}$

$\quad R_1 = 60 \text{ mm}$

$\quad R_2 = 60 + 30 = 90 \text{ mm}$

$\quad R = 60 + 38 \cdot 5 = 98 \cdot 5 \text{ mm}$

$\quad R_3 = 60 + 30 + 40 = 130 \text{ mm}$

$\quad R_4 = 60 + 30 + 40 + 20 = 150 \text{ mm}$

$$h^2 = \frac{R^3}{A}\left[b_2 \, \log_e\left(\frac{R_2}{R_1}\right) + t_3 \, \log_e\left(\frac{R_3}{R_2}\right) + b_1 \, \log_e\left(\frac{R_4}{R_3}\right) \right] - R^2$$

$$= \frac{(98 \cdot 5)^3}{3400}\left[60 \log_e\left(\frac{90}{60}\right) + 20 \log_e\left(\frac{130}{90}\right) + 40 \log_e\left(\frac{150}{130}\right) \right] - (98 \cdot 5)^2$$

$$= 281 \cdot 08 \, [24 \cdot 33 + 7 \cdot 35 + 5 \cdot 72] = 810 \cdot 14 \text{ mm}^2$$

Direct stress, $\quad \sigma_d = \dfrac{P}{A} = \dfrac{P}{3400} \text{ N/mm}^2$

Beding moment;

$$M = -P \times R = -P \times 98 \cdot 5 = -98 \cdot 5\, P \text{ Nmm}$$

The bending stress at any point is given by

$$\sigma_b = \frac{M}{AR}\left[1 + \frac{R^2}{h^2}\left(\frac{y}{R+y}\right)\right]$$

$$= \frac{-98 \cdot 5\, P}{3400 \times 98 \cdot 5}\left[1 + \frac{98 \cdot 5^2}{810 \cdot 14}\left(\frac{y}{98 \cdot 5 + y}\right)\right]$$

$$= -\frac{P}{3400}\left[1 + \frac{11 \cdot 97\, y}{98 \cdot 5\, y}\right]$$

Maximum bending stress (*tensile*) occurs at $y = -38 \cdot 5$ mm

$$\therefore \qquad (\sigma_b)_{max} = -\frac{P}{3400}\left[1 + \frac{11 \cdot 97\,(-38 \cdot 5)}{98 \cdot 5 - 38 \cdot 5}\right] = +\frac{P}{508 \cdot 9}\ \text{N/mm}^2$$

∴ Maximum *tensile stress*,

$$(\sigma_{max})t = \sigma_d + (\sigma_b)_{max}$$

$$= \frac{P}{3400} + \frac{P}{508 \cdot 9} = \frac{P}{442 \cdot 65}$$

But this is not to exceed the permissible stress of 108 N/mm²;

$$\therefore \qquad \frac{P}{442 \cdot 65} = 108$$

or, $\qquad P = 442 \cdot 65 \times 108 = 47806 \cdot 2 \simeq$ **47·8 kN** **(Ans.)**

Maximum compressive stress occurs at $y = y_2 = 51 \cdot 5$ mm

$$\therefore \qquad (\sigma_{max})_c = \frac{P}{3400} - \frac{P}{3400}\left[1 + \frac{11 \cdot 97 \times 51 \cdot 5)}{98 \cdot 5 + 51 \cdot 5}\right]$$

$$= \frac{P}{3400}\left[1 - \left\{1 + \frac{11 \cdot 97 \times 51 \cdot 5)}{98 \cdot 5 + 51 \cdot 5}\right\}\right]$$

$$= \frac{P}{3400} \times -4 \cdot 11$$

$$= \frac{47806 \cdot 2}{3400} \times -4 \cdot 11 = \textbf{57.79 N/mm}^2 \quad \textbf{(comp.) (Ans.)}$$

Example 20·9. *A curved beam has a T-section (Fig. 20·18). The inner radius is 300 mm. What is the eccentricity of the section?* **(AMIE Winter, 2002)**

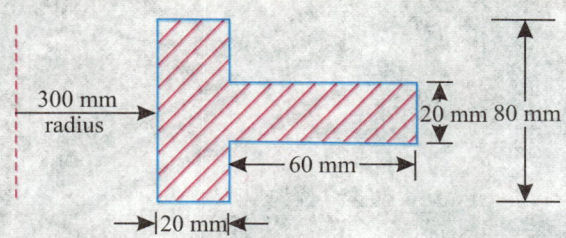

Fig. 20.18

Solution. Refer to Fig. 20·19.

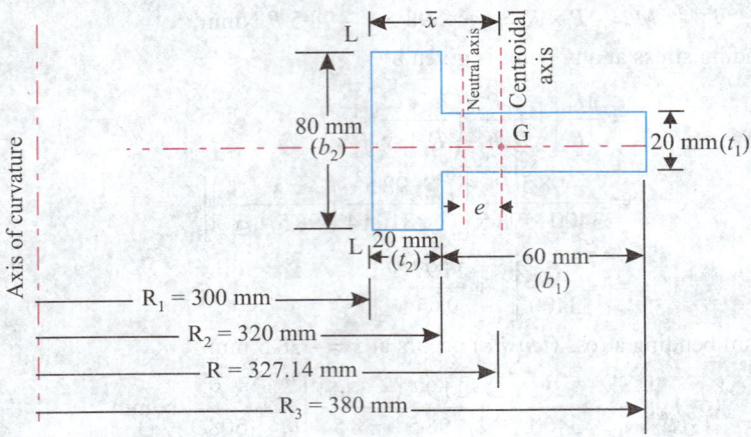

Fig. 20.19

Area of T-section, $= b_1 t_1 + b_2 t_2$
$$= 60 \times 20 + 80 \times 20 = 2800 \text{ mm}^2$$

To find c.g. of T-section, taking moments about the edge LL, we get

$$80 \times 20 \times \frac{20}{2} + 60 \times 20 \times \left(\frac{60}{2} + 20 \right) = [(80 \times 20) + (60 \times 20)] \, \bar{x}$$

or, $16000 + 60000 = 2800 \, \bar{x}$

∴ $\bar{x} = 27 \cdot 14$ mm

Now $R_1 = 300$ mm; $R_2 = 320$ mm; $R_3 = 327 \cdot 14$ mm; $R_3 = 380$ mm.

Using the relation :

$$h^2 = \frac{R^3}{A} \left[b_2 \cdot \log_e \frac{R_2}{R_1} + t_1 \cdot \log_e \frac{R_3}{R_2} \right] - R^2 \qquad \text{...[Eqn. (20·19)]}$$

Brake mechanisms of an automobile.

$$= \frac{(327 \cdot 14)^3}{2800} \left[80 \times \log_e \left(\frac{320}{300} \right) + 20 \log_e \left(\frac{380}{320} \right) \right] - (327 \cdot 14)^2$$

$$= 12503 \cdot 8 \, (5 \cdot 16 + 3 \cdot 44) - 107020 \cdot 6 = 512 \cdot 08$$

We know that
$$y = - \left[\frac{Rh^2}{R^2 + h^2} \right] \qquad \qquad ....[\text{Eqn. } (20 \cdot 12(a))]$$

[where y (= eccentricity e) = distance of the neutral axis from the centroidal axis].

$$\therefore \qquad e \, (= y) = - \frac{Rh^2}{R^2 + h^2} = - \frac{327 \cdot 14 \times 512 \cdot 08}{(327 \cdot 14)^2 + 512 \cdot 08} = \textbf{1·56 mm} \; (-)$$

Negative sign indicates that neutral axes is located *below* the centrodal axis.

Example 20·10. *Fig. 20·20 shows an open ring having T-section. Determine the stresses at the points A and B if the ring is subjected to a load of 15 kN.*

Solution. Refer to Fig. 20·20.

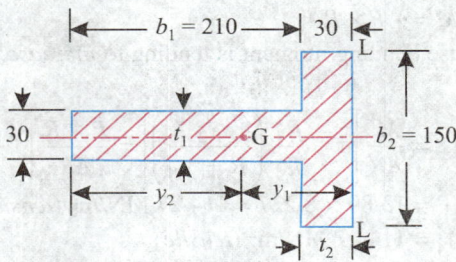

Section along AB

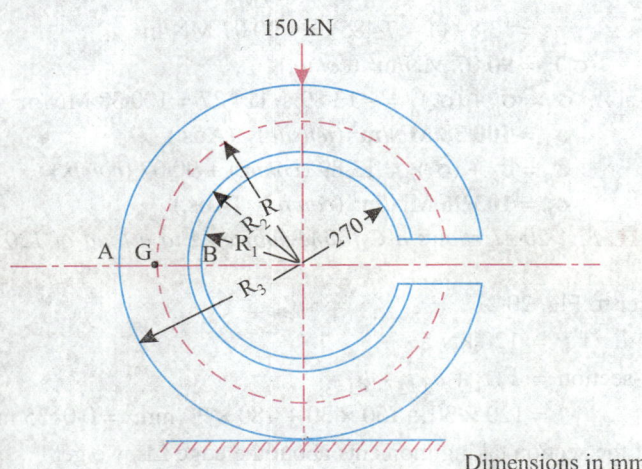

Dimensions in mm

Fig. 20.20

Area of T-section = $b_1 t_1 + b_2 t_2 = 210 \times 30 + 150 \times 30 = 10800 \times 10^{-6}$ m^2 = 0·0108 m^2

Load, $P = 150$ kN

To find c.g. of the T-Section taking moments about the edge LL, we get

$$y_1 = \frac{150 \times 30 \times 15 + 210 \times 30 \times 135}{150 \times 30 + 210 \times 30} = 85 \text{ mm } (= 0 \cdot 085 \text{ m})$$

and, $\qquad y_2 = 240 - 85 = 155 \text{ mm } (= 0.155 \text{ m})$

Now, $\qquad R_1 = 270 \text{ mm} = 0.27 \text{ m}$

$\qquad R_2 = 270 + 30 = 300 \text{ mm} = 0.3 \text{ m}$

$\qquad R = 270 + y_1 = 270 + 85 = 355 \text{ mm} = 0.355 \text{ m}$

$\qquad R_3 = 270 + (30 + 210) = 510 \text{ mm} = 0.51 \text{ m}$

$$h^2 = \frac{R^3}{A}\left[b_2 \cdot \log_e \frac{R_2}{R_1} + t_1 \log_e \frac{R_3}{R_2}\right] - R^2 \qquad \text{...[Eqn. 20·19]}$$

$$= \frac{0.355^3}{0.0108}\left[0.15 \log_e \frac{0.3}{0.27} + 0.03 \log_e \frac{0.51}{0.3}\right] - (0.355)^2$$

$$= 4.142\,[0.0158 + 0.0159] - 0.126$$

or, $\qquad h^2 = 0.0053 \text{ m}^2$

Direct stress, $\qquad \sigma_d = \dfrac{P}{A} = \dfrac{150 \times 10^3}{0.0108} \times 10^{-6} = 13.89 \text{ MN/m}^2 \text{ (comp.)}$

Bending moment $\quad M = +P \times R$

(+sign is taken because bending moment is tending to *increase the curvature*)

Bending stress at A,

$$(\sigma_b)_A = \frac{M}{AR}\left[1 + \frac{R^2}{h^2}\left(\frac{y_2}{R + y_2}\right)\right] = \frac{P \times R}{AR}\left[1 + \frac{0.355^2}{0.0053}\left(\frac{0.155}{0.355 + 0.155}\right)\right]$$

$$= 13.89 \times 8.227 = 114.27 \text{ MN/m}^2 \text{ (tensile)}$$

i.e. $\qquad (\sigma_b)_A = 114.27 \text{ MN/m}^2 \text{ (tensile)}$

$$(\sigma_b)_B = \frac{M}{AR}\left[1 - \frac{R^2}{h^2}\left(\frac{y_1}{R - y_1}\right)\right] = \frac{P \times R}{AR}\left[1 - \frac{0.355^2}{0.0053}\left(\frac{0.085}{0.355 - 0.085}\right)\right]$$

$$= 13.89\,(1 - 7.485) = -90.07 \text{ MN/m}^2$$

i.e. $\qquad (\sigma_b)_B = 90.07 \text{ MN/m}^2 \text{ (comp.)}$

Hence, stress at A, $\;\; \sigma_A = \sigma_d + (\sigma_b)_A = -13.89 + 114.27 = 100.38 \text{ MN/m}^2$

i.e. $\qquad \boldsymbol{\sigma_A = 100.38 \text{ MN/m}^2} \text{ (tensile)} \;\; \textbf{(Ans.)}$

Stress at B, $\qquad \sigma_B = \sigma_d + (\sigma_b)_B = 13.89 \text{ (comp.)} + 90.07 \text{ (comp.)}$

or, $\qquad \boldsymbol{\sigma_B = 103.96 \text{ MN/m}^2} \text{ (comp.)} \;\; \textbf{(Ans.)}$

Example 20·11. *Fig. 20·21 shows a C-frame subjected to a load of 120 kN. Determine the stresses at A and B.*

Solution. Refer to Fig. 20·21

Load, $\qquad P = 120 \text{ kN}$

Area of cross-section $= b_1 t_1 + b_2 t_2 + b_3 t_3$

$$= 120 \times 30 + 150 \times 30 + 180 \times 30 \text{ mm}^2 = 0.0135 \text{ m}^2$$

To find c.g. of the section taking moments about the edge LL, we get

$$y_1 = \frac{120 \times 30 \times 225 + 150 \times 30 \times 15 + 180 \times 30 \times 120}{120 \times 30 + 150 \times 30 + 180 \times 30}$$

$$= 113 \text{ mm} = 0.113 \text{ m}$$

∴ $\qquad y_2 = 240 - 113 = 127 \text{ mm} = 0.127 \text{ m}$

$\qquad R_1 = 225 \text{ mm} = 0.225 \text{ m}$

$\qquad R_2 = 225 + 30 = 255 \text{ mm} = 0.255 \text{ m}$

$\qquad R = 225 + 113 = 338 \text{ mm} = 0.338 \text{ m}$

R_3 = 225 + 210 = 435 mm = 0·435 m

R_4 = 225 + 240 = 465 mm = 0·465 m

$$h^2 = \frac{R^3}{A}\left[b_2 \log_e\left(\frac{R_2}{R_1}\right) + t_3 \log_e\left(\frac{R_3}{R_2}\right) + b_1 \log_e\left(\frac{R_4}{R_3}\right)\right] - R^2$$

$$= \frac{(0·338)^3}{0·0135}\left[0·15 \log_e \frac{0·255}{0·225} + 0·03 \log_e \frac{0·435}{0·255}\right.$$

$$\left. + 0·12 \log_e \frac{0·465}{0·435}\right] - 0·338^2$$

$$= 2·86\,[\,0·01877 + 0·016 + 0·008\,] - 0·1142 = 0·008122 \text{ m}^2$$

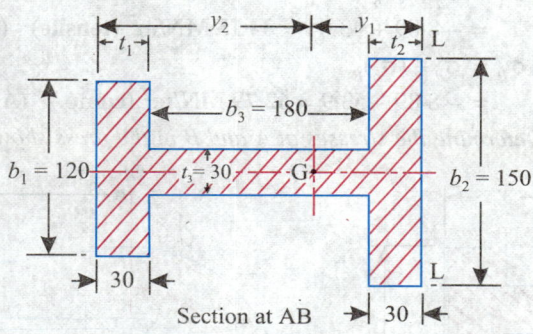

Section at AB

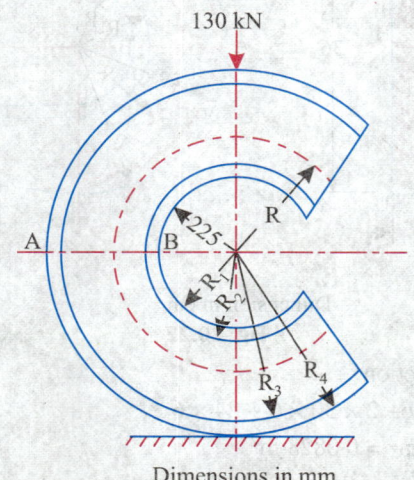

Dimensions in mm

Fig. 20.21

Direct stress, $\quad \sigma_d = \dfrac{P}{A} = \dfrac{120 \times 10^3}{0·0135} \times 10^{-6} = 8·89 \text{ MN/m}^2 \ (comp.)$

Bending moment, $M = P \times R$

Bending stress at A due to the bending moment,

$$(\sigma_b)_A = \frac{M}{AR}\left[1 + \frac{R^2}{h^2} \times \frac{y_2}{R + y_2}\right]$$

$$= \frac{P \times R}{AR}\left[1 + \frac{0·338^2}{0·008122} \times \frac{0·127}{0·338 + 0·127}\right]$$

$$= 8\cdot89\,(1 + 3\cdot842) = 43\cdot04 \text{ MN/m}^2 \text{ (tensile)}$$

Bending stress at B due to the bending moment;

$$(\sigma_b)_B = \frac{M}{AR}\left[1 - \frac{R^2}{h^2}\left(\frac{y_1}{R - y_1}\right)\right]$$

$$= \frac{P \times R}{AR}\left[1 - \frac{0\cdot338^2}{0\cdot008122} \times \frac{0\cdot113}{0\cdot338 - 0\cdot113}\right]$$

$$= 8\cdot89\,(1 - 7\cdot064)$$

$$= -53\cdot9 \text{ MN/m}^2 = 53\cdot9 \text{ MN/m}^2 \text{ (comp.)}$$

Stress at A, $\quad \sigma_A = \sigma_d + (\sigma_b)_A$

$$= -8\cdot89 + 43\cdot04 = \mathbf{34\cdot15\ MN/m^2\ (tensile)}\quad \textbf{(Ans.)}$$

Stress at B, $\quad \sigma_B = \sigma_d + (\sigma_b)_B$

$$= -8\cdot89 - 53\cdot9 = \mathbf{62\cdot79\ MN/m^2\ (comp.)}\quad \textbf{(Ans.)}$$

Example 20·12. *Determine the stresses at A and B in the press shown in Fig. 20·22.*

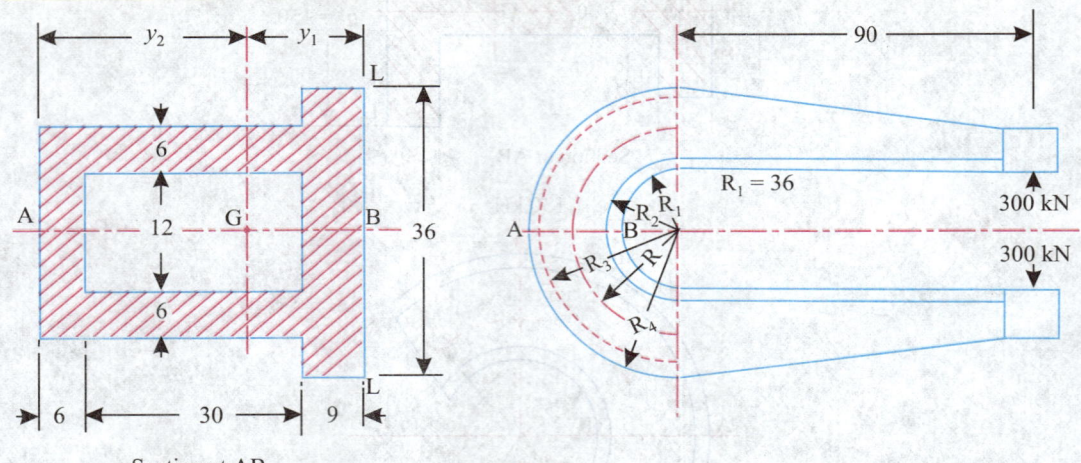

Section at AB

Dimension in cm

Fig. 20.22

Solution. Area of the section

$$= 36 \times 9 + 2 \times 30 \times 6 + 24 \times 6$$

$$= 828 \text{ cm}^2 = 0\cdot0828 \text{ m}^2$$

To find the position of c.g. of the section taking moments about the edge LL, we get

$$y_1 = \frac{36 \times 9 \times 4\cdot5 + 2 \times 30 \times 6 \times 24 + 24 \times 6 \times 42}{36 \times 9 + 2 \times 30 \times 6 + 24 \times 6} = 19\cdot5 \text{ cm}$$

and, $\quad y_2 = 45 - 19\cdot5 = 25\cdot5 \text{ cm}$

Now, $\quad R_1 = 36 \text{ cm} = 0\cdot36 \text{ m (given)}$

$\quad R_2 = 36 + 9 = 45 \text{ cm} = 0\cdot45 \text{ m}$

$\quad R = 36 + 19\cdot5 = 55\cdot5 \text{ cm} = 0\cdot555 \text{ m}$

$\quad R_3 = 36 + 39 = 75 \text{ cm} = 0\cdot75 \text{ m}$

$\quad R_4 = 36 + 45 = 81 \text{ cm} = 0\cdot81 \text{ m}$

$$h^2 = \frac{R^3}{A}\left[0\cdot36 \log_e \frac{R_2}{R_1} + 2 \times 0\cdot06 \times \log_e \frac{R_3}{R_2} + 0\cdot24 \times \log_e \frac{R_4}{R_3}\right] - R^2$$

$$= \frac{0.555^3}{0.0828}\left[0.36\log_e\frac{0.45}{0.36}+2\times0.06\times\log_e\frac{0.75}{0.45}+0.24\times\log_e\frac{0.81}{0.75}\right]-0.555^2$$

$$= 2.065\,(0.0803+0.0613+0.0185)-0.308 = 0.0226\text{ m}^2$$

Direct stress,

$$\sigma_d = \frac{300\times10^3}{0.0828}\times10^{-6} = 3.623\text{ MN/m}^2\ (tensile)$$

Bending moment,

$$M = -\,300\times10^3\,(0.90 + R)$$
$$= -\,300\times10^3\,(0.90+0.555)$$
$$= -\,436.5\times10^3\text{ Nm}$$

(–ve sign indicates that M tends to *decrease the curvature*)

Bending stress at A due to the bending moment,

$$(\sigma_b)_A = \frac{M}{AR}\left[1+\frac{R^2}{h^2}\times\frac{y_2}{R+y_2}\right]$$

$$= \frac{-\,436.5\times10^3}{0.0828\times0.555}\left[1+\frac{0.555^2}{2.0226}\times\frac{0.255}{0.555+0.255}\right]\times10^{-6} = -50.25\text{ MN/m}^2$$

$$= 50.25\text{ MN/m}^2\ (comp.)$$

Bending stress at B due to bending moment,

$$(\sigma_b)_B = \frac{M}{AR}\left[1-\frac{R^2}{h^2}\left(\frac{y_1}{R-y_1}\right)\right]$$

$$= \frac{-\,436.5\times10^3}{0.0828\times0.555}\left[1-\frac{0.555^2}{0.0226}\times\left(\frac{0.195}{0.555-0.195}\right)\right]\times10^{-6} = 60.62\text{ MN/m}^2\ (tensile.)$$

Stress at point A,

$$\sigma_A = \sigma_d + (\sigma_b)_A = 3.623 - 50.25 = -\,46.63\text{ MN/m}^2$$

i.e. $\sigma_A = \mathbf{46.63\ MN/m^2}$ **(comp.)** **(Ans.)**

Stress at point B,

$$\sigma_B = \sigma_d + (\sigma_b)_B = 3.623 + 60.62 = \mathbf{64.24\ MN/m^2}\ \textbf{(tensile)}\ \textbf{(Ans.)}$$

Arms of earthmovers often have curved members.

20·4. STRESSES IN A RING

Fig. 20·23 shows a circular ring under the action of an axial pull P. Consider a section L_1L_2 at an angle θ from the line of the action of the applied pull P. The portion L_2CDL_1 is in equilibrium under the action of the following :

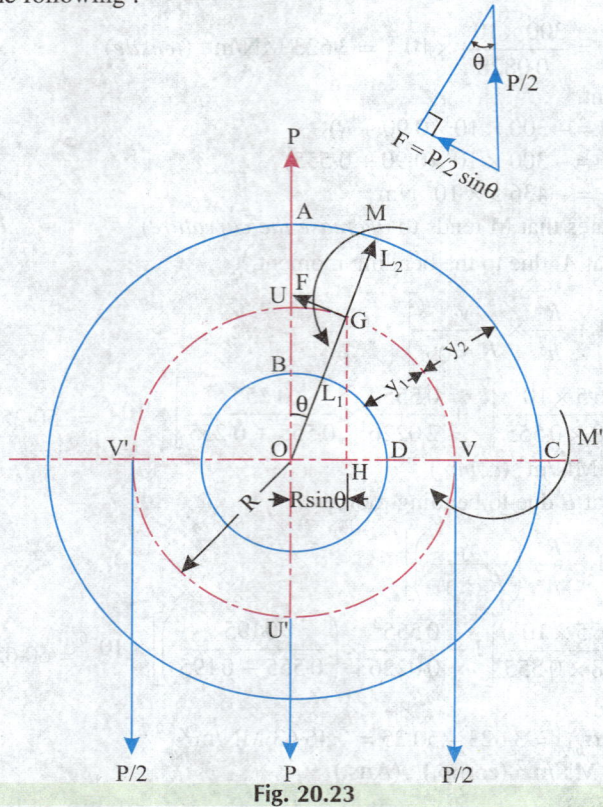

Fig. 20.23

(*i*) Bending moment M' at CD,

(*ii*) Pull $\dfrac{P}{2}$ at CD,

(*iii*) Bending moment M at L_1L_2, and

(*iv*) Pull F at L_1L_2.

Taking moments about L_1L_2 , we get

$$M = M' + \frac{P}{2}(R - R\sin\theta) \qquad\qquad ...(1)$$

Also,

$$M = E(1 + e_0)\left(\frac{1}{R'} - \frac{1}{R}\right)Ah^2 \qquad ...(2)\ [\text{Eqn. (20·7)}]$$

where,

E = Young's modulus,

e_o = Strain in centroidal layer,

R = Initial radius of curvature,

R' = Radius of curvature after bending,

and,

$$Ah^2 = \int \frac{y^2}{1 + \dfrac{y}{R}} \cdot dA \qquad\qquad ...[\text{Eqn. (20·6)}]$$

Comparing eqns. (1) and (2), we get

$$E(1 + e_0)\left(\frac{1}{R'} - \frac{1}{R}\right)Ah^2 = M' + \frac{P}{2}(R - R\sin\theta)$$

Multiplying both sides by $Rd\theta$ and integrating from 0 to $\pi/2$, we get

$$\int_0^{\pi/2} E(1+e_0)\left(\frac{1}{R'}-\frac{1}{R}\right)Ah^2 Rd\theta = \int_0^{\pi/2} M'Rd\theta + \int_0^{\pi/2} \frac{P}{2}R^2(1-\sin\theta)\,d\theta$$

or,

$$E\int_0^{\pi/2}\frac{R(1+e_0)}{R'}Ah^2\,d\theta - E\int_0^{\pi/2}(1+e_0)Ah^2\,d\theta$$

$$= \int_0^{\pi/2} M'Rd\theta + \int_0^{\pi/2}\frac{P}{2}R^2(1-\sin\theta)\,d\theta \qquad ...(3)$$

Now, $$(1+e_0) = \frac{R'}{R}\cdot\frac{\theta'}{\theta} \qquad ...[\text{Eqn. }(20\cdot2)]$$

Also $\theta = 90° = \pi/2$ and final angle $\theta' = 90° = \pi/2$ due to symmetry *i.e.* $\angle UOV$ remains 90° after the application of the pull P.

$$\therefore \qquad (1+e_0)\frac{R}{R'} = 1$$

Substituting this in eqn. (3), we get

$$E\int_0^{\pi/2} Ah^2\cdot d\theta - E(1+e_0)Ah^2\cdot\frac{\pi}{2} = M'R\cdot\frac{\pi}{2} + \frac{PR^2}{2}\left(\frac{\pi}{2}-1\right)$$

$$E\,Ah^2\cdot\frac{\pi}{2} - E(1+e_0)Ah^2\cdot\frac{\pi}{2} = M'R\cdot\frac{\pi}{2} + \frac{PR^2}{2}\left(\frac{\pi}{2}-1\right)$$

$$- E\cdot e_0\,Ah^2\frac{\pi}{2} = M'R\cdot\frac{\pi}{2} + \frac{PR^2}{2}\left(\frac{\pi}{2}-1\right) \qquad ...(4)$$

Also, normal force $F = E\,e_0\,A + E(1+e_0)\left[\dfrac{1}{R'}-\dfrac{1}{R}\right]\displaystyle\int\dfrac{y}{1+\dfrac{y}{R}}\,dA \qquad ...[\text{Eqn. }(20\cdot5)]$

where, $$\int\frac{y}{1+\dfrac{y}{R}}\cdot dA = -\frac{Ah^2}{R} \qquad ...[\text{Eqn. }(20\cdot8)]$$

$$\therefore \qquad F = E\,e_0\,A - E(1+e_0)\left(\frac{1}{R'}-\frac{1}{R}\right)\frac{Ah^2}{R}$$

Normal force on the section $L_1 L_2$,

$$F = \frac{P}{2}\sin\theta$$

$$\therefore \qquad \frac{P}{2}\sin\theta = E\,e_0\,A - EA(1+e_0)\left(\frac{1}{R'}-\frac{1}{R}\right)\frac{h^2}{R}$$

Also, $$M = E(1+e_0)\left(\frac{1}{R'}-\frac{1}{R}\right)Ah^2 \qquad ...[\text{Eqn. }(20\cdot7)]$$

$$\therefore \qquad \frac{P}{2}\sin\theta = E\,e_0\,A - \frac{M}{R}$$

Putting the value of M [from eqn. (1)] in the above eqn. we get

$$\frac{P}{2}\sin\theta = EA\,e_0 - \frac{1}{R}\left[M' + \frac{P}{2}(R - R\sin\theta)\right]$$

or $$\frac{P}{2}\sin\theta = EA\,e_0 - \frac{M'}{R} - \frac{P}{2} + \frac{P}{2}\sin\theta$$

$$\therefore \qquad e_0 = \frac{M'}{EAR} + \frac{P}{2EA} \qquad \qquad ...(5)$$

Substituting the value of e_o in eqn. (4), we get

$$- EAh^2 \times \frac{\pi}{2} \left[\frac{M'}{EAR} + \frac{P}{2EA} \right] = M'R \cdot \frac{\pi}{2} + \frac{PR^2}{2} \left(\frac{\pi}{2} - 1 \right)$$

$$- \frac{M'h^2}{R} \times \frac{\pi}{2} - \frac{Ph^2}{4} \cdot \pi = M' \cdot R \frac{\pi}{2} + \frac{PR^2}{2} \left(\frac{\pi}{2} - 1 \right)$$

$$- \frac{P}{4} \cdot h^2 \cdot \pi - \frac{PR^2}{4} \cdot \pi + \frac{PR^2}{2} = M'R \cdot \frac{\pi}{2} + \frac{M'h^2}{R} \cdot \frac{\pi}{2}$$

$$\frac{PR^2}{\pi} - \frac{PR^2}{2} - \frac{Ph^2}{2} = \frac{M'}{R} (R^2 + h^2)$$

$$M' = \frac{PR^3}{\pi (R^2 + h^2)} - \frac{PR}{2} = \frac{PR}{2} \left[\frac{R^2}{R^2 + h^2} \times \frac{2}{\pi} - 1 \right] \qquad ...(6)$$

Now, $$M = M' + \frac{PR}{2} (1 - \sin \theta) \qquad \qquad ...[\text{Eqn. (1)}]$$

$$= \frac{PR}{2} \left[\frac{R^2}{R^2 + h^2} \times \frac{2}{\pi} - 1 \right] + \frac{PR}{2} (1 - \sin \theta)$$

or, $$M = \frac{PR}{2} \left[\frac{R^2}{R^2 + h^2} \times \frac{2}{\pi} - \sin \theta \right] \qquad \qquad ...[6(a)]$$

The value of M will be maximum when $\theta = 0°$

$$M_{max} = \frac{PR^3}{\pi (R^2 + h^2)} \qquad \qquad ...(7)$$

The value of M will be *zero* when

$$\frac{R^2}{R^2 + h^2} \times \frac{2}{\pi} - \sin \theta = 0$$

or, $$\sin \theta = \frac{R^2}{R^2 + h^2} \times \frac{2}{\pi}$$

or, $$\theta = \sin^{-1} \left[\frac{R^2}{R^2 + h^2} \times \frac{2}{\pi} \right]$$

$$...(8)$$

Thus there will be 4 sections, one in each quadrant where the bending moment M will be zero and consequently the stress due to bending will be zero.

Now substituting the value of M' in eqn. (5) from eqn. (6) we get

$$e_0 = \frac{1}{EAR} \times \frac{PR}{2} \left[\frac{R^2}{R^2 + h^2} \times \frac{2}{\pi} - 1 \right] + \frac{P}{2EA}$$

or, $$e_0 = \frac{P}{AE} \left[\frac{R^2}{\pi (R^2 + h^2)} \right] \qquad \qquad ...(9)$$

This pulley's arms are curved members.

Also, $$e = e_0 + (1 + e_0)\left(\frac{1}{R'} - \frac{1}{R}\right)\left(\frac{y}{1 + y/R}\right) \qquad ...(10) \text{ (Eqn. (20·3)]}$$

Again, from eqns. (1) and (2), we get

$$E\,(1 + e_0)\left(\frac{1}{R'} - \frac{1}{R}\right)Ah^2 = M' + \frac{P}{2}\,R\,(1 - \sin\theta)$$

or, $$(1 + e_0)\left(\frac{1}{R'} - \frac{1}{R}\right) = \frac{M'}{EAh^2} + \frac{PR}{2Ah^2E}\,(1 - \sin\theta)$$

Substituting the value of $(1 + e_0)\left(\frac{1}{R'} - \frac{1}{R}\right)$ in the eqn. (10), we get

$$e = e_o + \left[\frac{M'}{EAh^2} + \frac{PR}{2Ah^2E}\,(1 - \sin\theta)\right]\left(\frac{Ry}{R + y}\right) \qquad ...(11)$$

or, $$e = \frac{P}{AE}\left[\frac{R^2}{\pi\,(R^2 + h^2)}\right] + \left[\frac{PR}{2}\left\{\frac{R^2}{R^2 + h^2} \times \frac{2}{\pi} - 1\right\}\right.$$

$$\left. \times \frac{1}{EAh^2} + \frac{PR}{2Ah^2E}\,(1 - \sin\theta)\right]\left(\frac{Ry}{R + y}\right)$$

$\therefore$ Stress, $$\sigma = Ee$$

$$= \frac{P}{A}\left\{\frac{R^2}{\pi\,(R^2 + h^2)}\right\} + \left[\frac{PR}{2Ah^2}\left\{\frac{2R^2}{\pi\,(R^2 + h^2)} - 1\right\} + \frac{PR}{2Ah^2}\,(1 - \sin\theta)\right]\left(\frac{Ry}{R + y}\right)$$

or, $$\sigma = \frac{P}{A}\left[\frac{R^2}{\pi\,(R^2 + h^2)} + \frac{R^2}{2h^2}\left\{\frac{2R^2}{\pi\,(R^2 + h^2)} - \sin\theta)\right\} \times \left(\frac{y}{R + y}\right)\right] \qquad ...(20·21)$$

Direct stress at any section,

$$\sigma_d = \frac{\dfrac{P}{2}\sin\theta}{A} = \frac{P}{2A}\cdot\sin\theta$$

Resultant stress, $$\sigma_r = \sigma_d \pm \sigma \qquad ...(20·22)$$

(a) Stresses on a section taken along the line of action of *P*, where θ = 0° :

(i)   At *outside of ring* :

i.e. At the point *A*, $$(\sigma_r)_A = \frac{P}{\pi A}\left(\frac{R^2}{R^2 + h^2}\right)\left[1 + \frac{R^2}{h^2}\cdot\left(\frac{y_2}{R + y_2}\right)\right] \text{ tensile} \qquad ...(20·23)$$

(ii)   At *inside of the ring* :
i.e. At the point *B*,

$$(\sigma_r)_B = \frac{P}{\pi A}\left(\frac{R^2}{R^2 + h^2}\right)\left[\frac{R^2}{h^2}\left(\frac{y_1}{R - y_1}\right) - 1\right] \text{ compressive}$$

$$...(20·24)$$

(b) Stresses on a section perpendicular to the line of action of P, where θ = 90° :

(i)   At *outside of the ring* :
i.e. At the point *C*,

$$(\sigma_r)_C = \frac{P}{A}\left[\frac{R^2}{\pi\,(R^2 + h^2)} + \frac{R^2}{2h^2}\left\{\frac{2R^2}{\pi(R^2 + h^2)} - 1\right\}\right.$$

$$\left. \times \left(\frac{y_2}{R + y_2}\right)\right] + \frac{P}{2A} \text{ compressive} \qquad ...(20·25)$$

(ii) *At inside of the ring :*
i.e. At the point D,

$$(\sigma_r)_D = \frac{P}{A}\left[\frac{R^2}{2h^2}\left\{\frac{2R^2}{\pi(R^2+h^2)}-1\right\}\times\left(\frac{-y_1}{R-y_1}\right)\right]$$

$$+ \frac{R^2}{\pi(R^2+h^2)}\right] + \frac{P}{2A} \text{ tensile } ...(20{\cdot}26)$$

Example 20·13. *A ring is made of round steel bar 30 mm diameter and the mean radius of the ring is 180 mm. Calculate the maximum tensile and compressive stresses in the material of the ring if it is subjected to a pull of 12 kN.*

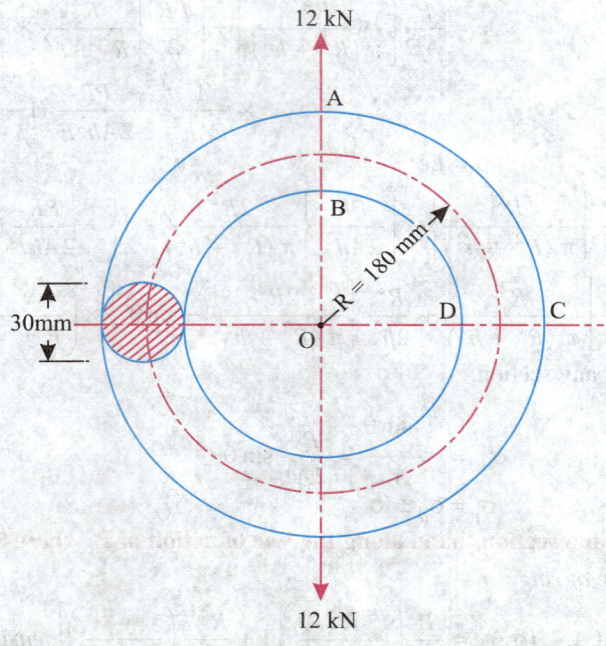

12 kN

A

B

R = 180 mm

D   C

30mm

O

12 kN

Fig. 20.24

Solution. Diameter of round steel bar,

$$d = 30 \text{ mm} = 0{\cdot}03 \text{ m}$$

Area $\qquad A = \frac{\pi}{4}\times 0{\cdot}03^2 = 7{\cdot}068\times 10^{-4} \text{ m}^2$

Mean radius of the ring,

$$R = 180 \text{ mm} = 0{\cdot}18 \text{ m}$$

Pull, $\qquad P = 12 \text{ kN}$

Now, $\qquad h^2 = \dfrac{d^2}{16} + \dfrac{1}{128}\times\dfrac{d^4}{R^2} = \dfrac{(0{\cdot}03)^2}{16} + \dfrac{(0{\cdot}03)^4}{128\times 0{\cdot}18^2} = 5{\cdot}64\times 10^{-5} \text{ m}^2$

$$(\sigma_r)_A = \frac{P}{\pi A}\left(\frac{R^2}{R^2 + h^2}\right)\left[1 + \frac{R^2}{h^2}\left(\frac{y_2}{R + y_2}\right)\right]$$

$$= \frac{12 \times 10^3}{\pi \times 7.068 \times 10^{-4}}\left(\frac{0.18^2}{0.18^2 + 5.64 \times 10^{-5}}\right)$$

$$\left[1 + \frac{0.18^2}{5.64 \times 10^{-5}}\left(\frac{0.015}{0.18 + 0.015}\right)\right] \times 10^{-6} \text{ MN/m}^2$$

$$(\because y_2 = d/2 = 0.015 \text{ m})$$

$$= 243.7 \text{ MN/m}^2 \text{ (tensile)}$$

$$(\sigma_r)_B = \frac{P}{\pi A}\left(\frac{R^2}{R^2 + h^2}\right)\left[\frac{R^2}{h^2}\left(\frac{y_1}{R - y_1}\right) - 1\right] \qquad \text{(where, } y_1 = d/2\text{)}$$

$$= \frac{12 \times 10^3}{\pi \times 7.068 \times 10^{-4}}\left(\frac{0.18^2}{0.18^2 + 5.64 \times 10^{-5}}\right)\left[\frac{0.18^2}{5.64 \times 10^{-5}}\left(\frac{0.015}{0.18 - 0.015}\right) - 1\right] \times 10^{-6} \text{ MN/m}^2$$

$$= 276.3 \text{ MN/m}^2 \text{ (compressive)}$$

$$(\sigma_r)_C = \frac{P}{A}\left[\frac{R^2}{\pi(R^2 + h^2)} + \frac{R^2}{2h^2}\left\{\frac{2R^2}{\pi(R^2 + h^2)} - 1\right\} \times \left(\frac{y_2}{R + y_2}\right)\right] + \frac{P}{2A}$$

$$= \frac{12 \times 10^3 \times 10^{-6}}{7.68 \times 10^{-4}}\left[\frac{0.18^2}{\pi(0.18^2 + 5.64 \times 10^{-5})} + \frac{0.18^2}{2 \times 5.64 \times 10^{-5}}\right.$$

$$\left.\left\{\frac{2 \times 0.18^2}{\pi(0.18^2 + 5.64 \times 10^{-5})} - 1\right\} \times \left(\frac{0.015}{0.18 + 0.015}\right)\right] + \frac{12 \times 10^3}{2 \times 7.068 \times 10^{-4}} \times 10^{-6}$$

$$= 16.98\,[0.3177 + 287.23 \times (-0.3645) \times 0.0769] + 8.49$$

$$= -131.3 + 8.49 = -122.8 \text{ MN/m}^2$$

i.e. $(\sigma_r)_c = 122.8$ MN/m^2 (compressive)

$$(\sigma_r)_D = \frac{P}{A}\left[\frac{R^2}{2h^2}\left\{\frac{2R^2}{\pi(R^2 + h^2)} - 1\right\} \times \left(\frac{-y_1}{R - y_1}\right) + \frac{R^2}{\pi(R^2 - h^2)}\right] + \frac{P}{2A} \text{ (tensile)}$$

$$= \frac{12 \times 10^3 \times 10^{-6}}{7.068 \times 10^{-4}}\left[\frac{0.18^2}{2 \times 5.64 \times 10^{-5}}\left\{\frac{2 \times 0.18^2}{\pi(0.18^2 + 5.64 \times 10^{-5})} - 1\right\}\left(\frac{-0.015}{0.18 - 0.015}\right)\right.$$

Clamp that holds a shaft.

$$+ \frac{0.18^2}{\pi \, (0.18^2 + 5.64 \times 10^{-5})} \Bigg] + \frac{12 \times 10^3}{2 \times 7.068 \times 10^{-4}} \times 10^{-6} \text{ MN/m}^2$$

$$= 16.98 \, [287.23 \, (0.6355 - 1) \, (-0.0909) + 0.3177] + 8.49$$

$$= 166.99 + 8.49 = 175.48 \text{ MN/m}^2 \, (\textit{tensile})$$

∴ **Maximum tensile stress = 243·7 MN/m²   (Ans.)**

Maximum compressive stress = 276·3 MN/m²   (Ans.)

Example 20·14. *A steel ring has a rectangular cross-section, 75 mm in the radial direction and 45 mm perpendicular to the radial direction. If the mean radius of the ring is 150 mm and maximum tensile stress is limited to 180 MN/m² calculate the tensile load the ring can carry.*

Solution. Refer Fig. 20·25

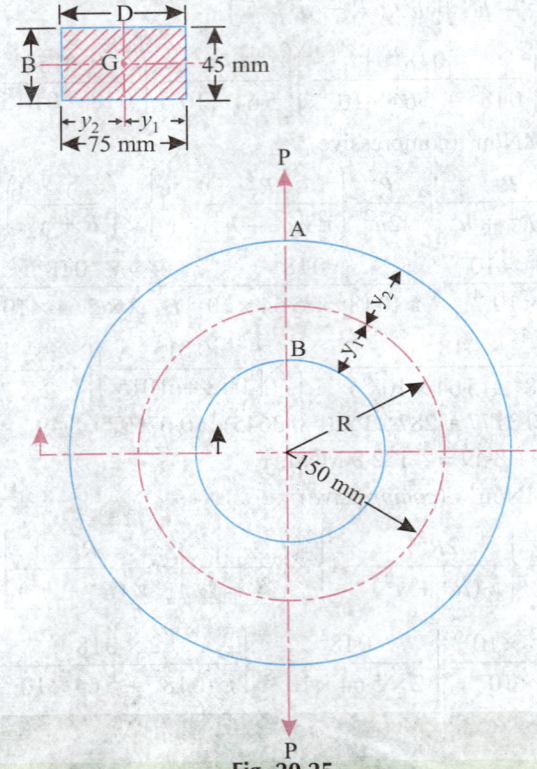

Fig. 20.25

Area of cross-section = $D \times B = 0.075 \times 0.045 = 0.003375 \text{ m}^2$

$$h^2 = \frac{R^3}{D} \log_e \left(\frac{2R + D}{2R - D} \right) - R^2 \qquad \qquad \text{...[Eqn. (20·15)]}$$

$$= \frac{0.15^3}{0.075} \log_e \left(\frac{2 \times 0.15 + 0.075}{2 \times 0.15 - 0.075} \right) - 0.15^2 = 4.87 \times 10^{-4} \text{ m}^2$$

Load, P :

The maximum tensile stress will occur at A (*i.e.* at $\theta = 0°$)

Now, $$\sigma_A = \frac{P}{\pi A} \left(\frac{R^2}{R^2 + h^2} \right) \left[1 + \frac{R^2}{h^2} \left(\frac{y_2}{R + y_2} \right) \right] \qquad \text{...[Eqn. (20·23)]}$$

$$180 \times 10^6 = \frac{P}{\pi \times 0.003375} \left(\frac{0.15^2}{0.15^2 + 4.87 \times 10^{-4}} \right)$$

$$\left[1 + \frac{0.15^2}{4.87 \times 10^{-4}} \left(\frac{0.0375}{0.15 + 0.0375} \right) \right]$$

$$180 \times 10^6 = 92.3 \, P \times 10.24$$

$$\therefore \qquad P = \frac{180 \times 10^6}{92.3 \times 10.24} \times 10^{-3} = 190.44 \text{ kN}$$

Hence, $\qquad$ **P = 19·44 kN (Ans.)**

20·5. STRESS IN A CHAIN LINK

Consider a chain link shown in Fig. 20·26

Let, $\quad P$ = Pull on the chain link,

$\quad R$ = Mean radius of the semi-circular ends,

$\quad I$ = Length of the straight portion of the link,

$\quad M$ = Bending moment on the section $L_1 L_2$, and

$\quad M'$ = Bending moment on the section CD.

Taking moments about the section L_1L_2, we get

$$M = M' + \frac{P}{2} (R - R \sin \theta) \qquad ...(1)$$

Also, $M = E (1 + e_0) \left(\frac{1}{R'} - \frac{1}{R} \right) Ah^2$

$$...[\text{Eqn. (20·7)}]...(2)$$

where, R = Initial radius of curvature,

$\quad R'$ = Final radius of curvature,

$\quad e_0$ = Strain in the centroidal layer,

$\quad E$ = Young's modulus, and

$$Ah^2 = \int \frac{y}{1 + \dfrac{y}{R}} \cdot dA$$

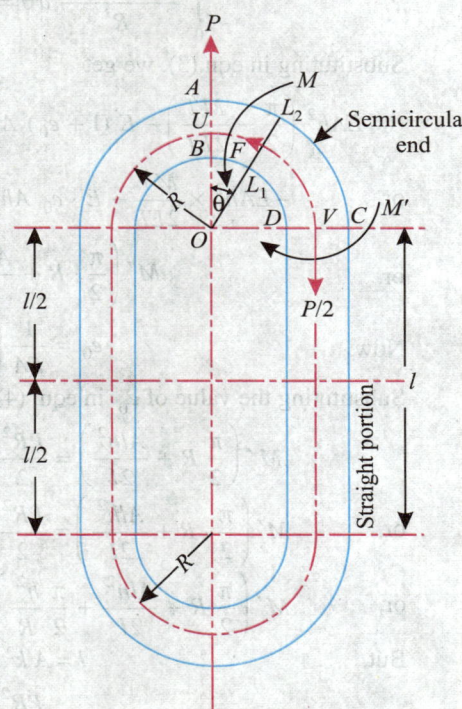

Fig. 20.26

Comparing eqns. (1) and (2), we have

$$E (1 + e_0) \left(\frac{1}{R'} - \frac{1}{R} \right) Ah^2 = M' + \frac{P}{2} (R - R \sin \theta)$$

Multiplying both sides by $Rd\theta$ and integrating from 0 to $\pi/2$, we get

$$\int_0^{\pi/2} E (1 + e_0) \left(\frac{1}{R'} - \frac{1}{R} \right) Ah^2 \, Rd\theta = \int_0^{\pi/2} M' \, Rd\theta + \int_0^{\pi/2} \frac{P}{2} R^2 (1 - \sin \theta) \, d\theta$$

$$\therefore E \int_0^{\pi/2} \frac{R(1 + e_0)}{R'} \, Ah^2 \, d\theta - E \int_0^{\pi/2} (1 + e_0) Ah^2 \, d\theta = \int_0^{\pi/2} M' \, Rd\theta$$

$$+ \int_0^{\pi/2} \frac{P}{2} R^2 (1 - \sin \theta) \, d\theta \qquad ...(3)$$

Now, $$(1 + e_0) = \frac{R'}{R} \cdot \frac{\theta'}{\theta} \qquad \text{...[Eqn. (20·2)]}$$

In this case initial angle $\theta = \angle VOU = 90°$, however, the final angle θ' *will not* be $90°$ and there will be slight change from $90°$.

Slope of the tangent at $V = \dfrac{M' \, l/2}{EI} = \dfrac{M \, l}{2EI}$

(where, I = moment of inertia of the section)

$$\therefore \qquad \int_0^{\pi/2} \frac{R \, (1 + e_0)}{R'} \, d\theta = \frac{\pi}{2} - \frac{M \, l}{2EI}$$

Substituting in eqn.(3), we get

$$EAh^2 \left(\frac{\pi}{2} - \frac{M \, l}{2EI} \right) - E \, (1 + e_0) \, Ah \cdot \frac{\pi}{2} = M'R \times \frac{\pi}{2} + \frac{PR^2}{2} \left(\frac{\pi}{2} - 1 \right)$$

or, $$- EAh^2 \times \frac{M \, l}{2EI} - E \cdot e_0 \, Ah^2 \cdot \frac{\pi}{2} = M' \, R \cdot \frac{\pi}{2} + \frac{PR^2}{2} \left(\frac{\pi}{2} - 1 \right)$$

or, $$M' \left(\frac{\pi}{2} \cdot R + \frac{Alh^2}{2I} \right) = \frac{PR^2}{2} \left(1 - \frac{\pi}{2} \right) - \frac{\pi}{2} EAh^2 \, e_0 \qquad \text{...(4)}$$

Now, $$e_0 = \frac{1}{EA} \left(\frac{P}{2} + \frac{M'}{R} \right) \qquad \text{...(Eqn. 20·27) [Eqn. (5) of article 20·4]}$$

Substituting the value of e_0 in eqn. (4), we get

$$M' \left(\frac{\pi}{2} R + \frac{Alh^2}{2I} \right) = \frac{PR^2}{2} \left(1 - \frac{\pi}{2} \right) - \frac{\pi}{2} EAh^2 \times \frac{1}{EA} \left(\frac{P}{2} + \frac{M'}{R} \right)$$

or, $$M' \left(\frac{\pi}{2} \cdot R + \frac{Alh^2}{2I} \right) = \frac{PR^2}{2} \left(1 - \frac{\pi}{2} \right) - \frac{\pi}{4} h^2 \, P - \frac{\pi}{2} h^2 \cdot \frac{M'}{R}$$

or, $$M' \left(\frac{\pi}{2} R + \frac{Alh^2}{2I} + \frac{\pi}{2} \frac{h^2}{R} \right) = \frac{PR^2}{2} \left(1 - \frac{\pi}{2} \right) - \frac{\pi}{4} h^2 \cdot P$$

But, $$I = A \, k^2 \qquad \text{(where, } k = \text{radius of gyration)}$$

$$\therefore \qquad M' = \frac{\dfrac{PR^2}{2} \left(1 - \dfrac{\pi}{2} \right) - \dfrac{\pi}{4} h^2 \cdot P}{\dfrac{\pi}{2} R + \dfrac{lh^2}{2k^2} + \dfrac{\pi}{2} \dfrac{h^2}{R}}$$

Multiplying numerator and denominator by $\dfrac{2}{\pi}$, we get

$$M' = \frac{P \left(\dfrac{R^2}{\pi} - \dfrac{R^2}{2} - \dfrac{h^2}{2} \right)}{R + \dfrac{lh^2}{\pi k^2} + \dfrac{h^2}{R}} \qquad \text{...(20·28)}$$

and, $$M = \frac{P \left(\dfrac{R^2}{\pi} - \dfrac{R^2}{2} - \dfrac{h^2}{2} \right)}{R + \dfrac{lh^2}{\pi k^2} + \dfrac{h^2}{R}} + \frac{PR}{2} \, (l - \sin \theta) \qquad \text{...(20·29)}$$

[(By substituting the value of M' in eqn. (1)]

Substituting eqn. (20·28) in eqn. (20·27), we get

$$e_0 = \frac{1}{EAR}\left[\frac{P\left(\dfrac{R^2}{\pi} - \dfrac{R^2}{2} - \dfrac{h^2}{2}\right)}{R + \dfrac{lh^2}{\pi k^2} + \dfrac{h^2}{R}}\right] + \frac{P}{2EA} \qquad ...(20\cdot30)$$

Stress at any layer at distance y from the neutral layer is

$$\sigma = Ee_0 + E(1 + e_0)\left(\frac{1}{R'} - \frac{1}{R}\right)\frac{y}{1 + y/R} \qquad\qquad [\text{Eqn. (20·4)}]$$

Also, $\quad E(1 + e_0)\left(\dfrac{1}{R'} - \dfrac{1}{R}\right) = \dfrac{M}{Ah^2}$ $\qquad\qquad\qquad\qquad$ [Eqn. (20·7)]

$\therefore \qquad\qquad\qquad \sigma = Ee_0 + \dfrac{M}{Ah^2} \times \left(\dfrac{Ry}{R + y}\right)$

Substituting the values of M and e_0, stress due to bending moment,

$$\sigma_b = \frac{1}{AR}\left[\frac{P\left(\dfrac{R^2}{\pi} - \dfrac{R^2}{2} - \dfrac{h^2}{2}\right)}{R + \dfrac{lh^2}{\pi k^2} + \dfrac{h^2}{R}}\right] + \frac{P}{2A} + \left(\frac{1}{Ah^2} \times \frac{Ry}{R + y}\right)$$

$$\times \left[\frac{P\left(\dfrac{R^2}{\pi} - \dfrac{R^2}{2} - \dfrac{h^2}{2}\right)}{R + \dfrac{lh^2}{\pi k^2} + \dfrac{h^2}{R}}\right] + \left(\frac{1}{Ah^2} \times \frac{Ry}{R + y}\right) \times \frac{PR}{2}(1 - \sin\theta)$$

Direct stress due to $\qquad F = \dfrac{P\sin\theta}{2A}$

Resultant stress, $\qquad \sigma_r = \sigma_b + \sigma_d$

Leaves of the leaf spring, U-clips, etc. partly behave like curved members.

$$= \frac{P}{AR} \left[\frac{\pi \dfrac{R^2}{2} - \dfrac{R^2}{2} - \dfrac{h^2}{2}}{R + \dfrac{lh^2}{\pi k^2} + \dfrac{h^2}{R}} \right] + \frac{PR}{Ah^2} \times \left(\frac{y}{R+y} \right) \left[\frac{\pi \dfrac{R^2}{2} - \dfrac{R^2}{2} - \dfrac{h^2}{2}}{R + \dfrac{lh^2}{\pi k^2} + \dfrac{h^2}{R}} \right]$$

$$+ \frac{1}{Ah^2} \times \left(\frac{Ry}{R+y} \right) \times \frac{PR}{2} (1 - \sin\theta) + \frac{P}{2A} + \frac{P \sin\theta}{2A}$$

or
$$\sigma_r = \frac{P}{AR} \left[1 + \frac{R^2}{h^2} \times \left(\frac{y}{R+y} \right) \right] \left[\frac{\pi \dfrac{R^2}{2} - \dfrac{R^2}{2} - \dfrac{h^2}{2}}{R + \dfrac{lh^2}{\pi k^2} + \dfrac{h^2}{R}} \right]$$

$$+ \frac{PR^2}{2Ah^2} (1 - \sin\theta) \times \left(\frac{y}{R+y} \right) + \frac{P}{2A} (1 + \sin\theta) \qquad ...(20\cdot31)$$

It may be noted that :

1. Eqn. (20·31) gives the resultant stress in any section along the curved portion of the chain link.
2. On the straight portion of the chain link the bending moment M' will remain constant. The bending stress in the straight portion will be found by using the general flexural formula and to get the resultant stress, direct tensile stress $\dfrac{P}{2A}$ will be added to bending stress.

Stresses on a section taken along the line of action of P, where θ = 0° :

(i) *At outside of ring :*

i.e., At the point A, $(\sigma_r)_A = \dfrac{P}{AR} \left[1 + \dfrac{R^2}{h^2} \times \left(\dfrac{y}{R+y} \right) \right] \left[\dfrac{\pi \dfrac{R^2}{2} - \dfrac{R^2}{2} - \dfrac{h^2}{2}}{R + \dfrac{lh^2}{\pi k^2} + \dfrac{h^2}{R}} \right]$

$$+ \frac{PR^2}{2Ah^2} \times \left(\frac{y_2}{R + y_2} \right) + \frac{P}{2A} \qquad ...(20\cdot32)$$

(ii) *At inside of the ring :*
i.e. At point B,

$$(\sigma_r)_B = \frac{P}{AR} \left[1 + \frac{R^2}{h^2} \times \left(\frac{-y_1}{R - y_1} \right) \right] \left[\frac{\pi \dfrac{R^2}{2} - \dfrac{R^2}{2} - \dfrac{h^2}{2}}{R + \dfrac{lh^2}{\pi k^2} + \dfrac{h^2}{R}} \right]$$

$$+ \frac{PR^2}{2Ah^2} \left(\frac{-y_1}{R - y_1} \right) + \frac{P}{2A} \qquad ...(20\cdot33)$$

Stresses on a section perpendicular to the line of action of P, where θ = 90° :

(i) *At outside of the ring;*

i.e. At the point C, $\quad (\sigma_r)_C = \dfrac{P}{AR} \left[1 + \dfrac{R^2}{h^2} \times \left(\dfrac{y_2}{R + y_2} \right) \right] \left[\dfrac{\pi \dfrac{R^2}{2} - \dfrac{R^2}{2} - \dfrac{h^2}{2}}{R + \dfrac{lh^2}{\pi k^2} + \dfrac{h^2}{R}} \right] + \dfrac{P}{A} \quad ...(20\cdot34)$

(ii)   At inside of the ring :

i.e. At the point D, $(\sigma_r)_D = \dfrac{P}{AR}\left[1 + \dfrac{R^2}{h^2} \times \left(\dfrac{-y_1}{R - y_1}\right)\right]$

$$\left[\dfrac{\dfrac{R^2}{\pi} - \dfrac{R^2}{2} - \dfrac{h^2}{2}}{R + \dfrac{lh^2}{\pi k^2} + \dfrac{h^2}{R}}\right] + \dfrac{P}{A} \qquad \qquad ...(20.35)$$

Maximum stress in straight portion :

Bending moment, $M' = \dfrac{P\left(\dfrac{R^2}{\pi} - \dfrac{R^2}{2} - \dfrac{h^2}{2}\right)}{R + \dfrac{lh^2}{\pi k^2} + \dfrac{h^2}{R}}$ ...[Eqn. (20.28)]

Bending stress, $\sigma_b = \dfrac{My}{I}$

Direct stress, $\sigma_d = \dfrac{P}{2A}$

Resultant stress $\sigma_r = \sigma_d \pm \sigma_b$

Example 20.15. *A chain link (Fig. 20.27) is made of round steel rod of 15 mm diameter. If R = 45 mm, l = 75 mm and load applied is 1.5 kN determine the maximum compressive stress in the link and tensile stress at the same section.*

Solution. $A = \dfrac{\pi}{4} \times (0.015)^2 = 1.767 \times 10^{-4}\, \text{m}^2$

$h^2 = \dfrac{d^2}{16} + \dfrac{d^4}{128\, R^2} = \dfrac{0.015^2}{16} + \dfrac{(0.015)^4}{128 \times 0.045^2} = 1.425 \times 10^{-5}\, \text{m}^2$

Clamp or holder in the above car engine is a curved member.

Radius of gyration, $\quad k = \sqrt{\dfrac{1}{A}} = \left(\dfrac{\dfrac{\pi}{64} d^4}{\dfrac{\pi}{4} d^2} \right)^{\frac{1}{2}} = \dfrac{d}{4} = \dfrac{0.015}{4} = 0.00375 \text{ m}$

$\therefore \qquad\qquad k^2 = 1.406 \times 10^{-5} \text{ m}^2$

$\dfrac{R^2}{\pi} - \dfrac{R^2}{2} - \dfrac{h^2}{2} = \dfrac{0.045^2}{\pi} - \dfrac{0.045^2}{2} - \dfrac{1.425^2 \times 10^{-5}}{2}$

$\qquad = 6.445 \times 10^{-4} - 1.0125 \times 10^{-3} - 0.7125 \times 10^{-5} = -0.375 \times 10^{-3}$

$R + \dfrac{lh^2}{\pi k^2} + \dfrac{h^2}{R} = 0.045 + \dfrac{0.075 \times 1.425 \times 10^{-5}}{\pi \times 1.406 \times 10^{-5}} + \dfrac{1.425 \times 10^{-5}}{0.045}$

$\qquad = 0.045 + 0.0242 + 0.000316 = 0.0695$

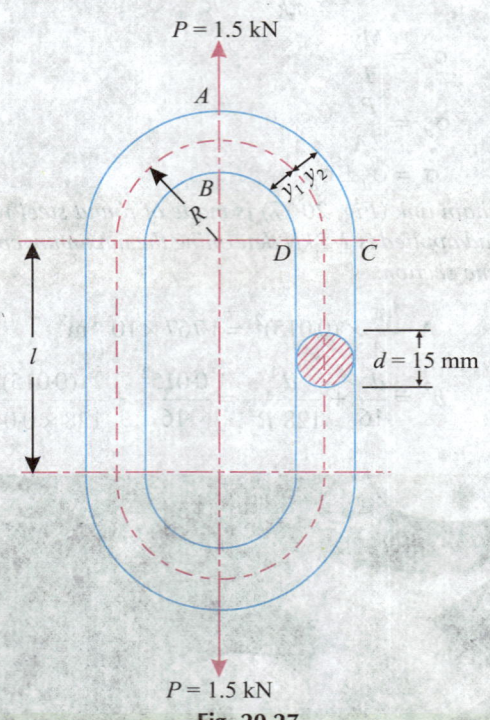

$P = 1.5$ kN

$P = 1.5$ kN

Fig. 20.27

Compressive stress is maximum at $\theta = 0°$ *on inside of the link i.e. at point B.*

$(\sigma_r)_B = \dfrac{P}{AR} \left[1 + \dfrac{R^2}{h^2} \times \left(\dfrac{-y_1}{R - y_1} \right) \right]$

$\qquad \left[\dfrac{\dfrac{R^2}{\pi} - \dfrac{R^2}{2} - \dfrac{h^2}{2}}{R + \dfrac{lh^2}{\pi k^2} + \dfrac{h^2}{R}} \right] + \dfrac{PR^2}{2Ah^2} \left(\dfrac{-y_1}{R - y_1} \right) + \dfrac{P}{2A}$ $\qquad$ [Eqn. (20.33)]

$= \dfrac{1.5 \times 10^3 \times 10^{-6}}{1.767 \times 10^{-4} \times 0.045} \left[1 + \dfrac{0.045^2}{1.425 \times 10^{-5}} \times \left(\dfrac{-0.0075}{0.045 - 0.0075} \right) \right] \left[\dfrac{-0.375 \times 10^{-3}}{0.0695} \right]$

$$+ \frac{1.5 \times 10^3 \times 0.045^2 \times 10^{-6}}{2 \times 1.767 \times 10^{-4} \times 1.425 \times 10^{-5}} \left\{ \frac{-0.0075}{0.045 - 0.0075} \right\} + \frac{1.5 \times 10^3}{2 \times 1.767 \times 10^{-4}} \times 10^{-6}$$

$$= 188.66 \times (-27.42) \times (-0.00539) - 120.63 + 4.244$$

$$= 27.88 - 120.63 + 4.244 = -88.5 \text{ MN/m}^2$$

i.e $(\sigma_r)_B = \textbf{88.5 MN/m}^2$ *(compressive)* **(Ans.)**

Tensile stress at this location on the outside surface of the link *i.e.* at point A,

$$(\sigma_r)_A = \frac{P}{AR} \left[1 + \frac{R^2}{h^2} \times \left(\frac{y_2}{R + y_2} \right) \frac{\dfrac{R^2}{\pi} - \dfrac{R^2}{2} - \dfrac{h^2}{2}}{R + \dfrac{lh^2}{\pi k^2} + \dfrac{h^2}{R}} \right]$$

$$+ \frac{PR^2}{2Ah^2} \times \left(\frac{y_2}{R + y_2} \right) + \frac{P}{2A} \qquad \dots[\text{(Eqn. 20·32)}]$$

$$= \frac{1.5 \times 10^3 \times 10^{-6}}{1.767 \times 10^{-4} \times 0.045} \left[1 + \frac{0.045^2}{1.425 \times 10^{-5}} \times \left(\frac{0.0075}{0.045 + 0.0075} \right) \right] \left[\frac{-0.375 \times 10^3}{0.0695} \right]$$

$$+ \frac{1.5 \times 10^3 \times 0.045^2 \times 10^{-6}}{2 \times 1.767 \times 10^{-4} \times 1.425 \times 10^{-5}} \left(\frac{0.0075}{0.045 + 0.0075} \right) + \frac{1.5 \times 10^3}{2 \times 1.767 \times 10^{-4}} \times 10^{-6}$$

$$= 188.66 \times (21.3) \times (-0.00539) + 86.16 + 4.244$$

$$= -21.66 + 86.16 + 4.244 = 68.74 \text{ MN/m}^2$$

i.e. $(\sigma_r)_A = \textbf{68.74 MN/m}^2$ **(tensile) (Ans.)**

20·6. DEFLECTION OF CURVED BARS

Under situations where it is required to design a curved beam from stiffness or rigidity point of view it becomes necessary to determine the deflection of the curved beam. In case of curved beams with small initial curvature the deflection may be estimated by straight beam theory; but in case of curved beams having large initial curvature the influence of the initial curvature of the beam must be considered while calculating deflection.

Fig. 20·28 shows the centre line *AEB* of curved bar under the action of variable bending moment.

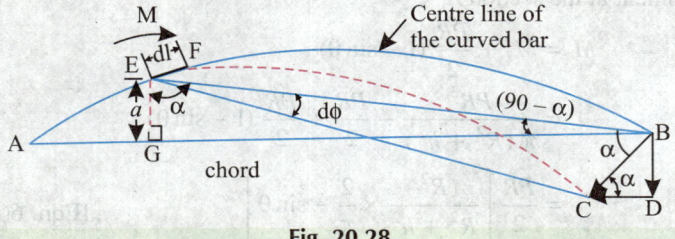

Fig. 20.28

Let, M = Bending moment at any point E (reckoned positive if it increases the curvature)

dl = Length of the small portion *EF*, and

a = Perpendicular distance of the point E from the chord *AB* (*i.e.* *EG*)

Let a small length *EF* (= *dl*) situated about the point E be considered to find the rotation of F with respect to E due to bending moment M and *rest of the curved bar is assumed rigid*. Due to the rotation $d\phi$ of F with respect to E let the portion *EB* shifts to position *BC*. Components of displacement *BC* are :

(*i*) *BD* perpendicular to the chord *AB*, and

(*ii*) *DC* parallel to the chord *AB*, *i.e.* the line joining the ends of the centre line of the curved beam condidered.

DC shows *negative* displacement *towards* the point *A*.

Deflection of the point *B* with respect to *A* is given as

$$\Delta \delta_{BA} = - DC = - BC \cos \alpha$$

or,

$$\Delta \delta_{BA} = - (EB\, d\phi) \cos \alpha \qquad \qquad ...(1)$$

where,

$$d\phi = \angle CEB$$

(The portions *AE* and *FB* being considered rigid, $d\phi$ will also represent the rotation of the end *F* w.r.t. *E* of the element *EF*)

$$\angle GBC = \angle BCD = \alpha \qquad\qquad [\because CD \parallel GB]$$

$$\angle EBG = \left(\frac{\pi}{2} - \alpha\right) \qquad\qquad [\because CB \perp EB \text{ (for small angle of roration } d\phi)]$$

In the right angled triangle *EBG*, $\angle GEB = \alpha$

$$\therefore \qquad\qquad \cos \alpha = \frac{EG}{EB}$$

Eqn. (1) reduces to

$$\Delta \delta_{BA} = - EB\,(d\phi) \times \frac{EG}{EB} = - EG\, d\phi = -a\,(d\phi) \qquad (\because EG = a) ...(2)$$

Also,

$$d\phi = \frac{M.dl}{EI} \qquad\qquad ...(3)$$

$$\delta_{BA} = - \frac{a.M.dl}{EI} \qquad\qquad ...(4)$$

Total deflection of *B* w.r.t. *A* along the chord *AB*,

$$\delta_{BA} = - \int_B^A a \cdot \frac{M.dl}{EI} \qquad\qquad ...(20.36)$$

20·6·1. Deflection of a Closed Ring

(*i*) *Deflection along the load line :*

Consider a small length *dl* at *A* at an angular displacement θ.

$$AA_1 = \text{Perpendicular distance on chord from the point } A$$
$$= a = R \sin \theta$$

Bending moment at the section *A*,

$$M = M' + \frac{PR}{2}\,(1 - \sin \theta)$$

$$= \frac{PR^3}{\pi\,(R^2 + h^2)} - \frac{PR}{2} + \frac{PR}{2}\,(1 - \sin \theta)$$

$$= \frac{PR}{2}\left[\frac{R^2}{R^2 + h^2} \times \frac{2}{\pi} - \sin \theta\right] \qquad ...[\text{Eqn. } 6(a) \text{ of article } 20·4]$$

Deflection along the load line,

$$\delta_{UO} = \text{Deflection along the load}$$

$$= - \int_O^U \frac{R \sin \theta}{EI}\left[\frac{PR}{2} \times \left(\frac{R^2}{R^2 + h^2} \times \frac{2}{\pi} - \sin \theta\right)\right] R d\theta$$

[Substituting the values of *h*, *M*, and *dl* in eqn. (20·36)]

Total deflection, $\delta_{UU} = - 2 \int_O^{\pi/2} \frac{PR^3}{2EI}\left[\frac{R^2}{R^2 + h^2} \times \frac{2}{\pi} \times \sin \theta - \sin^2 \theta\right] d\theta$

$$= -\frac{PR^3}{EI}\left\{\left(\frac{R^2}{R^2+h^2}\times\frac{2}{\pi}\right)\Big|-\cos\theta\Big|_{O}^{\pi/2}\right\}+\frac{PR^3}{EI}\Big|\frac{\theta}{2}-\frac{\sin 2\theta}{4}\Big|_{O}^{\pi/2}$$

$$= -\frac{PR^3}{EI}\left(\frac{R^2}{R^2+h^2}\times\frac{2}{\pi}\right)+\frac{PR^3}{EI}\times\frac{\pi}{2}\times\frac{1}{2}$$

$$= -\frac{PR^3}{EI}\times\frac{2}{\pi}\left(\frac{R^2}{R^2+h^2}\right)+\frac{PR^3}{EI}\times\frac{\pi}{4}$$

or, $$\delta_{UU}=\frac{PR^3}{EI}\left[\frac{\pi}{4}-\frac{2}{\pi}\left(\frac{R^2}{R^2+h^2}\right)\right] \qquad \qquad ...(20\cdot37)$$

(ii) *Deflection perpendicular to the load line :*

Refer to Fig. 20·29. $AA_2 = a = R\cos\theta$

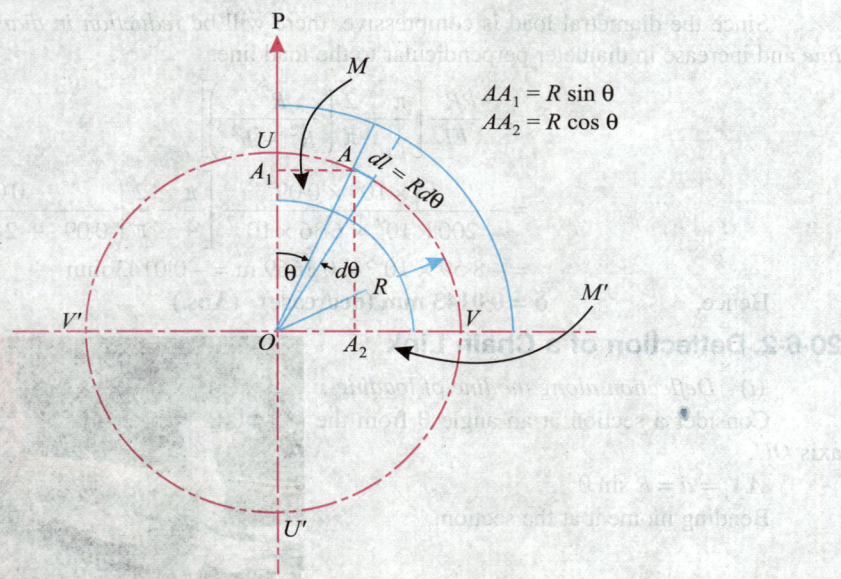

$$AA_1 = R\sin\theta$$
$$AA_2 = R\cos\theta$$

$$dl = Rd\theta$$

Fig. 20.29

$$\delta_{VO} = -\int_{O}^{V}\frac{R\cos\theta}{EI}\left[\frac{PR}{2}\left(\frac{R^2}{R^2+h^2}\times\frac{2}{\pi}-\sin\theta\right)\right]R\,d\theta$$

Total deflection, $$\delta_{VV'}=-2\int_{O}^{\pi/2}\frac{PR^3}{2EI}\left[\frac{2}{\pi}\left(\frac{R^2}{R^2+h^2}\right)\cos\theta-\sin\theta\cos\theta\right]d\theta$$

$$= -\frac{PR^3}{EI}\Big|\frac{2}{\pi}\left(\frac{R^2}{R^2+h^2}\right)\sin\theta+\frac{\cos 2\theta}{4}\Big|_{0}^{\pi/2}$$

$$= -\frac{PR^3}{EI}\left[\frac{2}{\pi}\left(\frac{R^2}{R^2+h^2}\right)\times 1+\frac{1}{4}(-1-1)\right]$$

or, $$\delta_{VV'}=\frac{PR^3}{EI}\left[\frac{1}{2}-\frac{2}{\pi}\left(\frac{R^2}{R^2+h^2}\right)\right] \qquad \qquad ...(20\cdot38)$$

Example 20·16. *A ring with a circular cross-section of 60 mm in diameter and a mean radius of 90 mm is subjected to a compressive load of 15 kN. Calculate the deflection of the ring along the load line. Take : E = 200 GN/m².*

Solution. *Given*:

Diameter of the circular cross-section, d = 60 mm = 0·06 m

Mean radius of the ring, R = 90 mm = 0·09 m

Compressive load, P = 15 kN

Young's modulus, E = 200 GN/m².

Deflection of the ring along the load line, δ :

Moment of inertia, $I = \dfrac{\pi}{64} d^4 = \dfrac{\pi}{64} \times (0·06)^4 = 6·36 \times 10^{-7} \text{ m}^4$

$$h^2 = \frac{d^2}{16} + \frac{d^4}{128\,R^2} = \frac{(0·06)^2}{16} + \frac{(0·06)^4}{128 \times 0·09^2} = 2·375 \times 10^{-4} \text{ m}^2$$

Since the diametral load is compressive, there will be *reduction in diameter along the load line* and increase in diameter perpendicular to the load line.

$$\delta = -\frac{PR^3}{EI}\left[\frac{\pi}{4} - \frac{2}{\pi}\left(\frac{R^2}{R^2 + h^2}\right)\right]$$

$$= -\frac{15 \times 10^3 \times 0·09^3}{200 \times 10^9 \times 6·36 \times 10^{-7}}\left[\frac{\pi}{4} - \frac{2}{\pi}\left(\frac{0·09^2}{0·09^2 + 2·375 \times 10^{-4}}\right)\right]$$

$$= -8·59 \times 10^{-5} \times 0·1669 \text{ m} = -0·0143 \text{ mm}$$

Hence, $\qquad\qquad$ **δ = 0·0143 mm (decrease) (Ans.)**

20·6·2. Deflection of a Chain Link

(*i*) *Deflection along the line of loading :*

Consider a section at an angle θ from the axis *OU*.

$AA_1 = a = R \sin \theta$

Bending moment at the section,

$$M = \frac{P\left(\dfrac{R^2}{\pi} - \dfrac{R^2}{2} - \dfrac{h^2}{2}\right)}{R + \dfrac{lh^2}{\pi k^2} + \dfrac{h^2}{R}} + \frac{PR}{2}(1 - \sin\theta)$$

...[Eqn. (20·29)]

$$= PC' + \frac{PR}{2}(1 - \sin\theta)$$

where,

$$C' = \frac{\dfrac{R^2}{\pi} - \dfrac{R^2}{2} - \dfrac{h^2}{2}}{R + \dfrac{lh^2}{\pi k^2} + \dfrac{h^2}{R}}$$

Protective colour coatings are added to make components corrosion resistant. Corrosion if not taken care can magnify other stresses.

The bending moment (M') over the length $l/2$ is constant and is given by :

$$M' = PC'$$

Deflection along the line of loading,

$$\delta_{UU_1} = -2 \int_0^{\pi/2} \frac{R \sin\theta}{EI} \left\{ PC' + \frac{PR}{2}(1 - \sin\theta) \right\} R\,d\theta - \frac{2R}{EI} \times PC' \times l/2$$

(using theory of simple bending)

$$= -\frac{2R^2}{EI} \int_0^{\pi/2} \left\{ PC' \sin\theta + \frac{PR}{2}\sin\theta - \frac{PR}{2}\sin^2\theta \right\} d\theta - \frac{RPC'l}{EI}$$

$$= -\frac{2R^2}{EI} - \left| PC' \cos\theta \right|_0^{\pi/2} - \frac{2R^2}{EI} \left| - \frac{PR}{2}\cos\theta \right|_0^{\pi/2}$$

$$+ \frac{2R^2}{EI} \left| \frac{PR}{2}\left(\frac{\theta}{2} - \frac{\sin 2\theta}{2}\right) \right|_0^{\pi/2} - \frac{RPC'\,l}{EI}$$

$$= \frac{2R^2}{EI} \times (-PC') - \frac{2R^2}{EI}\left(\frac{PR}{2} \times 1\right) + \frac{2R^2}{EI}\left(\frac{PR\pi}{8}\right) - \frac{RPC'\,l}{EI}$$

$$= -\frac{2R^2}{EI}\left(PC' + \frac{PR}{2} - \frac{PR\pi}{8}\right) - \frac{RPC'\,l}{EI}$$

or,

$$\delta_{UU_1} = \frac{PR^2}{EI}\left(\frac{\pi R}{4} - 2C' - R\right) - \frac{RPC'l}{EI}$$

Deflection due to direct load $P/2$ is given by

$$\delta' = \frac{Pl}{2\,AE}$$

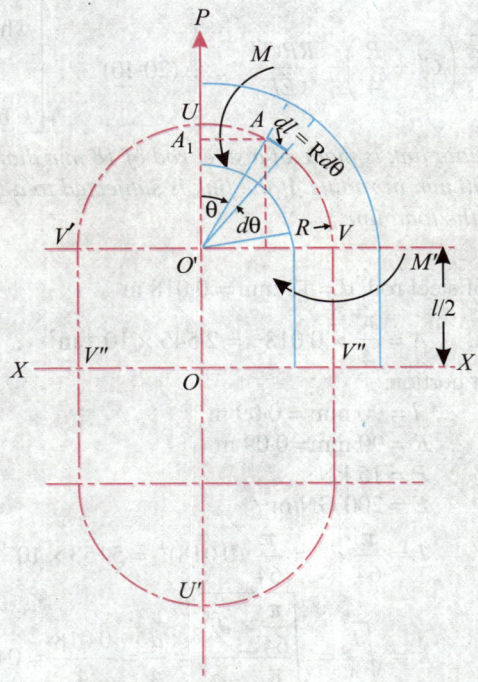

Fig. 20.30

Total deflection, $\delta_{UU'} = \delta_{UU_1} + \delta'$

$$= \frac{PR^2}{EI}\left(\frac{\pi R}{4} - 2C' - R\right) - \frac{RPC'l}{EI} + \frac{Pl}{2AE}$$

i.e. $\qquad \delta_{UU'} = \frac{PR^2}{EI}\left(\frac{\pi R}{4} - 2C' - R\right) - \frac{RPC'l}{EI} + \frac{Pl}{2AE}$ ...(20·39)

$$\left[\text{where, } C' = \frac{\pi \dfrac{R^2}{2} - \dfrac{R^2}{2} - \dfrac{h^2}{2}}{R + \dfrac{lh^2}{\pi k^2} + \dfrac{h^2}{R}} \right]$$

(ii) *Deflection perpendicular to line of loading :*

Here, $\qquad AA_2 = a = R\cos\theta + l/2$

$$\delta_{V''V''} = -2\int_0^{\pi/2} \frac{R\cos\theta}{EI}\left\{PC' + \frac{PR}{2}(1 - \sin\theta)\right\}Rd\theta - \frac{RPC'l}{EI}$$

$$= -\frac{2R^2}{EI}\left[\int_0^{\pi/2} PC'\cos\theta\, d\theta + \int_0^{\pi/2} \frac{PR}{2}\cos\theta\, d\theta - \int_0^{\pi/2} \frac{PR}{4}\sin 2\theta\, d\theta\right] - \frac{RPC'l}{EI}$$

$$= -\frac{2PR^2}{EI}\left[C'\left|\sin\theta\right|_0^{\pi/2} + \frac{R}{2}\left|\sin\theta\right|_0^{\pi/2} + \frac{R}{8}\left|\cos 2\theta\right|_0^{\pi/2}\right] - \frac{RPC'l}{EI}$$

$$= -\frac{2PR^2}{EI}\left[C' + \frac{R}{2} + \frac{R}{8}(-2)\right] - \frac{RPC'l}{EI}$$

or, $\qquad \delta_{V''V''} = -\frac{2PR^2}{EI}\left(C' + \frac{R}{4}\right) - \frac{RPC'l}{EI}$ ...(20·40)

$$\left[\text{where, } C' = \frac{\pi \dfrac{R^2}{2} - \dfrac{R^2}{2} - \dfrac{h^2}{2}}{R + \dfrac{lh^2}{\pi k^2} + \dfrac{h^2}{R}} \right]$$

k being radius of gyration

Example 20·17. *A chain link is made of a steel rod of 18 mm diameter with straight portion 90 mm in length and ends 90 mm in radius. If the link is subjected to a load of 15 kN calculate the deflection of the link along the load line.*

Take : E = 200 GN/m².

Solution. Diameter of steel rod, $d = 18$ mm $= 0·018$ m

Area of cross-section, $\quad A = \dfrac{\pi}{4} \times 0·018^2 = 2·545 \times 10^{-4}$ m²

Length of the straight portion,

$\qquad\qquad\qquad l = 90$ mm $= 0·09$ m

Radius of curvature, $\quad R = 90$ mm$= 0·09$ m

Load, $\qquad\qquad\qquad P = 15$ kN

Young's modulus, $\qquad E = 200$ GN/m²

Moment of inertia, $\qquad I = \dfrac{\pi}{64} d^4 = \dfrac{\pi}{64}(0·018)^4 = 5·153 \times 10^{-9}$ m⁴

Radius of gyration, $\qquad k = \sqrt{\dfrac{I}{A}} = \sqrt{\dfrac{\dfrac{\pi}{64} \times d^4}{\dfrac{\pi}{4} \times d^2}} = \dfrac{d}{4} = \dfrac{0·018}{4} = 0·0045$ m

$$h^2 = \frac{d^2}{16} + \frac{d^4}{128R^2} = \frac{0.018^2}{16} + \frac{(0.018)^2}{128 \times 0.09^2} = 2.035 \times 10^{-5}$$

$$C' = \frac{\dfrac{R^2}{\pi} - \dfrac{R^2}{2} - \dfrac{h^2}{2}}{R + \dfrac{lh^2}{\pi k^2} + \dfrac{h^2}{R}} = \frac{\dfrac{0.09^2}{\pi} - \dfrac{0.09^2}{2} - \dfrac{2.035 \times 10^{-5}}{2}}{0.09 + \dfrac{0.09 \times 2.035 \times 10^{-5}}{\pi \times (0.0045)^2} + \dfrac{2.035 \times 10^{-5}}{0.09}}$$

$$= \frac{0.002578 - 0.00405 - 1.0175 \times 10^{-5}}{0.09 + 0.02879 + 0.000226} = -0.01245$$

Deflection along the load line,

$$\delta_{UU'} = \frac{PR^2}{EI} \left[\frac{\pi R}{4} - 2C' - R \right] - \frac{RPC'l}{EI} + \frac{Pl}{2\,AE}$$

$$= \frac{15 \times 10^3 \times 0.09^2}{200 \times 10^9 \times 5.153 \times 10^{-9}} \left[\frac{\pi \times 0.09}{4} - 2 \times (-0.01245) - 0.09 \right]$$

$$= \frac{0.09 \times 15 \times 10^3 \times (-0.01245) \times 0.09}{200 \times 10^9 \times 5.153 \times 10^{-9}} + \frac{15 \times 10^3 \times 0.09}{2 \times 2.545 \times 10^{-4} \times 200 \times 10^9}$$

$$= 0.1179 \times 0.00558 + 0.001468 + 1.326 \times 10^{-5} = 0.00214 \text{ m} = 2.14 \text{ mm}$$

Hence, *deflection along the load line* = **2·14 mm**   (Ans.)

TYPICAL EXAMPLES   (For Competitive Examinations)

Example 20·18. *Fig. 20·31 shows the frame of a punching machine. If the vertical force P = 120 kN find the circumferential stresses at U and V on a section inclined at angle 45° to vertical force P.*

Solution. Refer Fig. 20·31. Force P = 120 kN

Support frames, holders and parts of chassis are partly curved members.

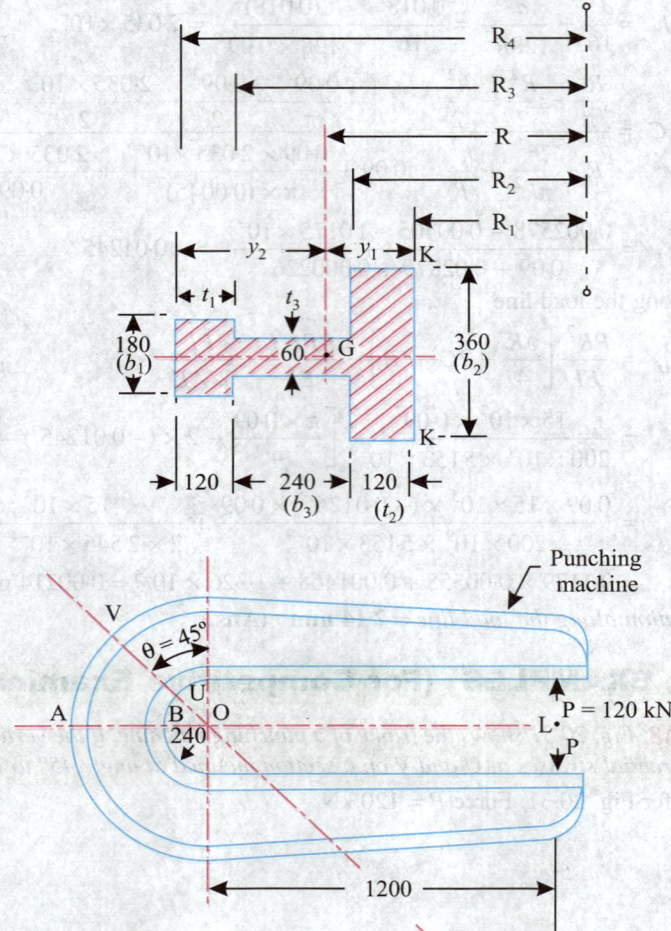

Fig. 20.31

Perpendicular force on the section $UV = P \sin 45° = 120 \times 0.707 = 84.84$ kN

Tangential force on the section $UV = P \cos 45° = 120 \times 0.707 = 84.84$ kN

Cross-sectional area, $\qquad A = 0.36 \times 0.12 + 0.24 \times 0.06 + 0.18 \times 0.12 = 0.0792$ m^2

To find the position of G taking moments about line KK, we get

$$y_1 = \frac{0.36 \times 0.12 \times 0.06 + 0.24 \times 0.06 \times 0.24 + 0.18 \times 0.12 \times 0.42}{0.0792} = 0.191 \text{ m}$$

$\therefore \qquad y_2 = 0.48 - 0.191 = 0.289$ m

Now, $\quad R_1 = 240$ mm $= 0.24$ m

$\qquad R_2 = 240 + 120 = 360$ mm $= 0.36$ m

$\qquad R = R_1 + y_1 = 0.24$ m $+ 0.191$ m $= 0.431$ m

$\qquad R_3 = 0.36 + 0.24 = 0.6$ m

$$R_4 = 0.6 + 0.12 = 0.72 \text{ m}$$

Bending moment on the section UV,

$$M = -84.84 \times (OL' + OU + y_1)$$

$$= -84.84 \times (1.2 \times \sqrt{2} + 0.24 + 0.191) = -180.54 \text{ kNm}$$

(–ve sign is taken because bending moment tends to *decrease the curvature*)

Direct force on the section UV,

$$P' = 0.707 P = 0.707 \times 120 = 84.84 \text{ kN}$$

$$h^2 = \frac{R^3}{A} \left[b_2 \log_e \frac{R_2}{R_1} + t_3 \log_e \frac{R_3}{R_2} + b_1 \log_e \frac{R_4}{R_3} \right] - R^2$$

$$= \frac{(0.431)^3}{0.0792} \left[0.36 \log_e \frac{0.36}{0.24} + 0.06 \times \log_e \frac{0.6}{0.36} + 0.18 \log_e \frac{0.72}{0.6} \right] - 0.431^2$$

$$= 1.0109 (0.1459 + 0.0306 + 0.0328) - 0.1857 = 0.0259$$

Circumferential stress at point U,

$$\sigma_U = \frac{M}{AR} \left[1 - \frac{R^2}{h^2} \times \left(\frac{y_1}{R - y_1} \right) \right] + \frac{P'}{A}$$

$$= -\frac{180.54 \times 10^3 \times 10^{-6}}{0.0792 \times 0.431} \left[1 - \frac{0.431^2}{0.0259} \left(\frac{0.191}{0.431 - 0.191} \right) \right] + \frac{84.84 \times 10^3}{0.0792} \times 10^{-6}$$

$$= -5.289 (1 - 5.708) + 1.071 = 25.97 \text{ MN/m}^2 \text{ (tensile)}$$

i.e. $\sigma_U = \textbf{25.97 MN/m}^2$ **(tensile)   (Ans.)**

Circumferential stress at point V,

$$\sigma_V = \frac{M}{AR} \left[1 + \frac{R^2}{h^2} \times \left(\frac{y_2}{R + y_2} \right) \right] + \frac{P'}{A}$$

$$= -\frac{180.54 \times 10^3 \times 10^{-6}}{0.0792 \times 0.431} \left[1 + \frac{0.431^2}{0.0259} \times \left(\frac{0.289}{0.431 + 0.289} \right) \right] + \frac{84.84 \times 10^3}{0.0792} \times 10^{-6}$$

$$= -5.289 (1 + 2.879) + 1.071 = -19.44 \text{ MN/m}^2$$

i.e. $\sigma_V = \textbf{19.44 MN/m}^2$ **(comp.)   (Ans.)**

Example 20·19. *A curved bar of rectangular section 60 mm wide by 75 mm deep in the plane of bending initially unstressed, is subjected to bending moment of 2·25 kNm which tends to straighten the bar. The mean radius of curvature is 150 mm. Find:*

(i)   The position of the neutral axis;

(ii) The greatest bending stresses.

Draw a diagram to show approximately how the stress varies across the section.

Solution. Depth of the section,

$$D = 75 \text{ mm} = 0.075 \text{ m}$$

Radius of curvature, $R = 150 \text{ mm} = 0.15 \text{ m}$

Bending moment, $M = -2.25 \text{ kNm}$

(–ve sign is taken because bending moment tends to straighten the bar)

Area of the section, $A = 0.06 \times 0.075 = 0.0045 \text{ m}^2$

We know, $h^2 = \dfrac{R^3}{D} \log_e \left(\dfrac{2R + D}{2R - D} \right) - R^2$ ...[Eqn. (20·15)]

$$= \frac{0.15^2}{0.075} \log_e \left(\frac{2 \times 0.15 + 0.075}{2 \times 0.15 - 0.075} \right) - 0.15^2$$

$$= 0.045 \log_e (1.6667) - 0.0225 = 4.88 \times 10^{-4} \text{ m}^2$$

(i) Position of the neutral axis :

We know,
$$y = - \frac{Rh^2}{R^2 + h^2} = - \frac{0.15 \times 4.88 \times 10^{-4}}{0.15^2 + 4.88 \times 10^{-4}} \qquad \text{...[Eqn. (20·12 (a))]}$$

$$= - 0.00318 \text{ m} = - 3.18 \text{ mm}$$

i.e.
$$y = - \mathbf{3.18 \text{ mm}} \quad \textbf{(Ans.)}$$

(ii) The greatest bending stress :

Bending stress at the inside face,

$$(\sigma)_{\text{inside face}} = \frac{M}{AR} \left[1 - \frac{R^2}{h^2} \left(\frac{y}{R - y} \right) \right]$$

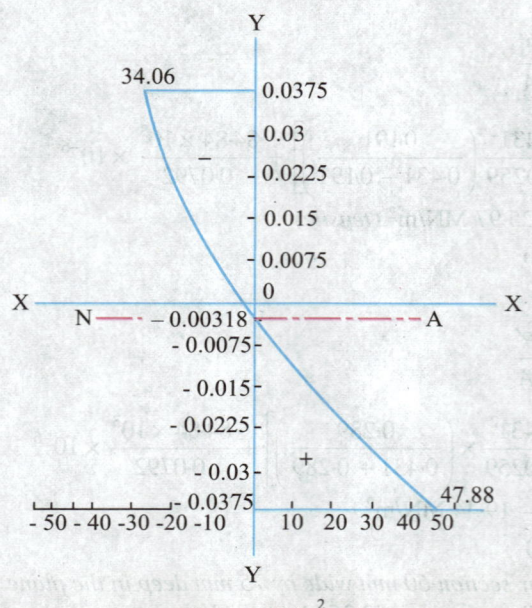

Aircraft component machined from a single billet.

Fig. 20.32

$$= - \frac{2.25 \times 10^3 \times 10^{-6}}{0.0045 \times 0.15} \left[1 - \frac{0.15^2}{4.88 \times 10^{-4}} \left(\frac{0.0375}{0.15 - 0.0375} \right) \right]$$

$$= - 3.333 \times (1 - 15.368) \qquad \left[\because y = \frac{D}{2} = 0.0375 \text{ m} \right]$$

$$= + \mathbf{47.88 \text{ MN/m}^2} \ (\textit{tensile}) \quad \textbf{(Ans.)}$$

Bending stress at the outside face,

$$(\sigma)_{\text{outside face}} = \frac{M}{AR} \left[1 + \frac{R^2}{h^2} \left(\frac{y}{R + y} \right) \right]$$

$$= - \frac{2.25 \times 10^3 \times 10^{-6}}{0.0045 \times 0.15} \left[1 + \frac{0.15^2}{4.88 \times 10^{-4}} \left(\frac{0.0375}{0.15 + 0.0375} \right) \right]$$

$$= -3.333 \ (1 + 9.2213) = -34.06 \ \text{MN/m}^2$$

$$= \mathbf{34.06 \ MN/m^2} \ (comp.) \ \textbf{(Ans.)}$$

Stress at any point,

$$(\sigma) = \frac{M}{AR}\left[1 + \frac{R^2}{h^2}\left(\frac{\pm y}{R \pm y}\right)\right]$$

or,

$$\sigma = -3.333\left[1 + 46.1\left(\frac{\pm y}{0.15 \pm y}\right)\right]$$

y, m	0·0375	0·03	0·015	0	− 0·00318	− 0·015	− 0·03	− 0·0375
σ,MN/m²	− 34·06	− 28·94	− 17·30	− 3·333	0	13·74	35·08	47·88

The variation of stress across the section is shown in Fig. 20·32.

HIGHLIGHTS

1. If a curved bar of radius of curvature R and area of cross-section A is subjected to a

   (i) bending moment (M) *tending to increase the curvature (M is taken as 'positive')*

   Stress in any layer, $\quad \sigma = \dfrac{M}{AR}\left[1 + \dfrac{R^2}{h^2}\times\left(\dfrac{y}{R + y}\right)\right]$ *tensile* when y is +ve

   and, $\quad \sigma = \dfrac{M}{AR}\left[1 - \dfrac{R^2}{h^2}\left(\dfrac{y}{R - y}\right)\right]$ *compressive* when y is −ve

   (ii) bending moment (M) *tending to decrease the curvature (M is taken as 'negative')*

   Stress in any layer, $\quad \sigma = \dfrac{M}{AR}\left[1 + \dfrac{R^2}{h^2}\times\left(\dfrac{y}{R + y}\right)\right]$ *compressive* when y is +ve

   and, $\quad \sigma = \dfrac{M}{AR}\left[1 - \dfrac{R^2}{h^2}\left(\dfrac{y}{R - y}\right)\right]$ *tensile* when y is −ve

2. *Position of neutral axis* $\quad y = -\dfrac{Rh^2}{R^2 + h^2}$

   −ve sign indicates that neutral axis is located below the centroidal axis.

3. *Values of h^2 for various sections;*

   $\quad\quad\quad\quad\quad\quad\quad\quad\quad$ (h = a constant for the cross-section of the bar)

   (i) *For a rectangular section :*

   $$h^2 = \frac{R^3}{D}\log_e\left(\frac{2R + D}{2R - D}\right) - R^2$$

   where, $\quad\quad\quad\quad D$ = Depth of the section, and

   $\quad\quad\quad\quad\quad\quad R$ = Radius of curvature of the curved bar.

   (ii) *For a circular section :*

   $$h^2 = \frac{d^2}{16} + \frac{1}{128}\frac{d^4}{R^2} + ...$$

   where, $\quad\quad\quad\quad d$ = Diameter of the circular section, and

   $\quad\quad\quad\quad\quad\quad R$ = Radius of curvature of the curved bar.

   (iii) *For a triangular section :*

   $$h^2 = \frac{2R^3}{d}\left[\left(\frac{3R + 2d}{3d}\right)\log_e\left(\frac{3R + 2d}{3R - d}\right) - 1\right] - R^2$$

   where, $\quad\quad\quad\quad d$ = Height/depth of the triangle, and

   $\quad\quad\quad\quad\quad\quad R$ = Radius of curvature of the curved bar.

(iv) *For a trapezoidal section :*

$$h^2 = \frac{R^3}{A}\left[b\log_e\left(\frac{R+d_2}{R-d_1}\right) + \left(\frac{B-b}{d}\right)(R+d_2)\log_e\left(\left(\frac{R+d_2}{R-d_1}\right)\right) - (B-b)\right] - R^2$$

where,

$$A = \left(\frac{B+b}{2}\right)d,$$

$$d_1 = \frac{d}{3}\left(\frac{B+2b}{B+b}\right),$$

$$d_2 = d - d_1,$$

and B, b = width of the inner and outer surfaces;

d = depth of the section.

4. A circular ring of mean radius R is subjected to a diametral load P, then resultant stress on any section at an angle θ from the load line is

$$\sigma_r = \frac{P}{A}\left[\frac{R^2}{\pi(R^2+h^2)} + \frac{R^2}{2h^2}\left\{\frac{2R^2}{\pi(R^2+h^2)} - \sin\theta\right\} \times \left(\frac{y}{R+y}\right)\right] + \frac{P}{2A}\sin\theta$$

where, y = distance of the layer under consideration from the centroidal layer.

5. A chain link of radius at end R, length of the straight portion l, and subjected to an axial tensile load P :

Resultant stress on any section inclined at an angle θ from the load line,

$$\sigma_r = \frac{P}{AR}\left[1 + \frac{R^2}{h^2} \times \left(\frac{y}{R+y}\right)\right]C' + \frac{PR^2}{2Ah^2}(1-\sin\theta) \times \left(\frac{y}{R+y}\right) + \frac{P}{2A}(1+\sin\theta)$$

where, $C' = \dfrac{\dfrac{R^2}{\pi} - \dfrac{R^2}{2} - \dfrac{h^2}{2}}{R + \dfrac{lh^2}{\pi k^2} + \dfrac{h^2}{R}}$

and, y = Distance of the layer under consideration from the centroidal layer;

k = Radius of gyration of the section.

6. Deflection in a curved bar along a chord,

$$\delta_{chord} = \int a \cdot \frac{Mdl}{EI}$$

where, M = Bending moment on a section,

a = Perpendicular distance from the point on the centre line of the section to the chord,

E = Young's modulus of elasticity, and

I = Moment of the inertia of the section.

7. A circular ring of mean radius R subjected to a diametral load P :

Deflection *along the load line* $= \dfrac{PR^3}{EI}\left[\dfrac{\pi}{4} - \dfrac{2}{\pi}\left(\dfrac{R^2}{R^2+h^2}\right)\right]$

Deflection *perpendicular to the load line* $= \dfrac{PR^3}{EI}\left[\dfrac{1}{2} - \dfrac{2}{\pi}\left(\dfrac{R^2}{R^2+h^2}\right)\right]$

8. A chain link of mean radius at ends R and length of the straight portion l, subjected to axial load P :

Declection *along the load line*

$$= \frac{PR^2}{EI}\left(\frac{\pi R}{4} - 2C' - R\right) - \frac{RPC'l}{EI} + \frac{Pl}{2AE}$$

Deflection *perpendicular to the load line*

$$= -\frac{2PR^2}{EI}\left(C' + \frac{R}{4}\right) - \frac{RPC'l}{EI}$$

(The value of C' is as above in 5)

OBJECTIVE TYPE QUESTIONS

Choose the Correct Answer :

1. The theory of curved beam was postulated by
 (*a*) Rankline   (*b*) Mohr
 (*c*) Castigliano   (*d*) Winkler-Bach

2. In curved beams the distribution of bending stress is
 (*a*) linear   (*b*) parabolic
 (*c*) uniform   (*d*) hyperbolic.

3. The neutral axis in curved beams
 (*a*) lies at the top of the beam
 (*b*) lies at the bottom of the beam
 (*c*) does not coincide with the geometric axis
 (*d*) coincides with the geometric axis.

4. For a crane hook the most suitable section is
 (*a*) triangular   (*b*) trapezodial
 (*c*) circular   (*d*) rectangular.

5. The nature of stress at the inside surface of a crane hook is
 (*a*) shear
 (*b*) tensile
 (*c*) compressive
 (*d*) none of the above.

6. Which of the following statements is *correct* with reference to the curved beam theory ?
 (*a*) Shear stress is zero
 (*b*) Hoop stress is zero.
 (*c*) Radial stress is zero
 (*d*) Bending stress is zero.

7. The maximum stress in a ring under tension occurs
 (*a*) along the line of action of load
 (*b*) perpendicular to the line of action of load
 (*c*) at 45° with the line of action of the load
 (*d*) none of the above.

8. When fabricating a chain link the joint should come at which of the following locations ?
 (*a*) Parallel to the line of action of the load
 (*b*) 45° with the line of action of the load
 (*c*) 60° with the line of action of the load
 (*d*) Perpendicular to the line of action of the load

9. In a closed ring when a small cut is made at the horizontal diameter the maximum stress will
 (*a*) decrease   (*b*) increase
 (*c*) remain same   (*d*) become infinite.

10. Which of the following *assumptions* is made in the analysis of curved beams ?
 (*a*) Limit of proportionality is not exceeded
 (*b*) Radial strain is negligible
 (*c*) Plane to transverse sections remain plane after bending
 (*d*) The material considered is isotropic and obeys Hooke's law
 (*e*) All of the above.

Aircraft component machined from a single billet.

ANSWERS

1. (*d*)   **2.** (*d*)   **3.** (*c*)   **4.** (*b*)   **5.** (*b*)   **6.** (*c*)   **7.** (*a*)   **8.** (*d*)   **9.** (*b*)   **10.** (*e*).

UNSOLVED EXAMPLES

1. A bar of rectangular section 40 mm × 60 mm is subjected to a bending moment of 2 kNm, its centre line is curved to a radius of 200 mm. If the bending moment tends to increase the curvature determine:

   (*i*) The maximum tensile and compressive stresses in beam;

   (*ii*) Stress at the c.g. of the section.

$$\left[\begin{array}{l} \textbf{Ans.}\ \ 64{\cdot}2\ \text{MN/m}^2\ (tensile) \\ \quad\ 88{\cdot}4\ \text{MN/m}^2\ (comp.) \\ \quad\ 4{\cdot}16\ \text{MN/m}^2\ (tensile) \end{array}\right]$$

2. A steel bar 38 mm in diameter is bent into a curve of mean radius 31·7 mm. If a bending moment of 4·6 Nm tending to increase the curvature, acts on the bar find the intensities of maximum tensile and compressive stresses. [**Ans.** 0·56 MN/m² (*tensile*), 1·6 MN/m² (*comp.*)]

3. A curved bar of rectangular section 60 mm (width) × 40 mm (thickness), is bent in the shape of a horse shoe having a mean radius of 70 mm. Two equal and opposite forces of 10 kN each are applied at a distance of 120 mm from the centre line of the middle section so that they tend to straighten the rod.

   (*i*) Calculate the maximum tensile and compressive stresses;

   (*ii*) Construct a diagram showing the variation of normal stresses over the central section.

   [**Ans.** 75 MN/m² (*tensile*), 34 MN/m² (*comp.*)]

4. A bar of circular cross-section has a radius of curvature of 25 mm at the inner fibres. The bending moment acting on the bar causes a tensile stress of 20 MN/m² in the inner fibres. Find the stress in the outer fibres. [**Ans.** 5·9 MN/m²]

5. A bar of circular cross-section is bent in the shape of a horse shoe. The radius of the section is 40 mm and the mean radius is 80 mm. Two equal and opposite forces of 15 kN each are applied so as to straighten the bar.

   (*i*) Find the maximum tensile and compressive stresses.

   (*ii*) Construct a diagram showing the variation of normal stresses along the central section.

$$\left[\textbf{Ans. } \begin{matrix} 99\text{·}8 \text{ MN/m}^2 \ (tensile) \\ 39\text{·}2 \text{ MN/m}^2 \ (comp.) \end{matrix}\right]$$

6. A curved bar is formed of a tube 40 mm outside radius and 5 mm thickness. The centre line of this beam is a circular arc of radius 150 mm. A bending moment of 2 kNm tending to increase curvature of the bar is appplied. Calculate the maximum tensile and compressive stresses set up in the bar.

$$\left[\textbf{Ans. } \begin{matrix} 83\text{·}61 \text{ MN/m}^2 \ (tensile) \\ 113\text{·}56 \text{ MN/m}^2 \ (comp.) \end{matrix}\right]$$

7. At the critical section of a crane hook, trapezium in section, the inner and outer sides are 4 cm and 2·5 cm respectively and depth is 7·5 cm. The centre of curvature of the section is a distance of 6 cm from the inner fibres and the load line is 5 cm from the inner fibres. If the maximum stress is not to exceed 120 MN/m² what maximum load the hook can carry ?

[**Ans.** 30·56 kN]

8. A curved bar of rectangular section is 40 mm (width) by 50 mm (depth) in the plane of bending and has a mean radius of curvature 100 mm. It is subjected to a bending moment of 1·5 kNm which tends to straighten the bar.

   (*i*) Find the position of the neutral axis;

   (*ii*) Magnitudes of greatest bending stresses;

   (*iii*) Construct a diagram to show approximately the stress variation across the section.

   [**Ans.** (*i*) – 2·12 mm; (*ii*) 107·9 MN/m² (*tensile*); (*iii*) 76·78 MN/m² (*comp.*)]

9. An open ring having channel section as shown in the Fig. 20·33 is subjected to compressive load of 75 kN. Determine the stresses at *A* and *B*.

$$\left[\textbf{Ans. } \begin{matrix} \sigma_A = 153\text{·}5 \text{ MN/m}^2 \ (tensile) \\ \sigma_B = 164\text{·}46 \text{ MN/m}^2 \ (comp.) \end{matrix}\right]$$

Aircraft component machined from a single billet.

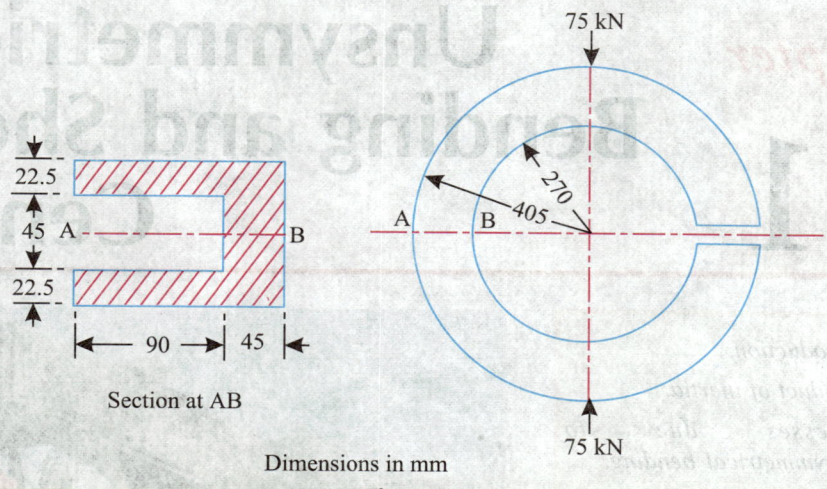

Section at AB

Dimensions in mm

Fig. 20.33

10. A ring is made of round steel bar 25 mm diameter and the mean radius of the ring is 150 mm. Calculae the maximum tensile and compressive stresses in the material of the ring if it is subjected to a pull of 10 kN.

$$\left[\begin{array}{l}\textbf{Ans.}\ 292.2\ \text{MN/m}^2\ (\textit{tensile})\\ 331.2\ \text{MN/m}^2\ (\textit{comp.})\end{array}\right]$$

11. A ring is made of round steel bar 24 mm diameter and the mean diameter of the ring is 144 mm. Determine the greatest intensities of tensile and compressive stresses along a diameter XX if the ring is subjected to a pull of 12 kN along diameter YY.

$$\left[\begin{array}{l}\textbf{Ans.}\ 160.2\ \text{MN/m}^2\ (\textit{tensile})\\ 77.36\ \text{MN/m}^2\ (\textit{comp.})\end{array}\right]$$

12. A steel ring 24 cm mean diameter has a rectangular cross-section 6 cm in the radial direction and 3.6 cm perpendicular to the radial direction. If the maximum tensile stress is limited to 144 MN/m², determine the tensile load that the ring can carry. [**Ans.** 97.5 kN]

13. A chain link is made of round steel rod of 12 mm diameter, $R = 36$ mm, $l = 60$ mm and load applied is 1.2 kN. Determine the maximum compressive stress in the link and tensile stress at the same section.

$$\left[\begin{array}{l}\textbf{Ans.}\ 110.5\ \text{MN/m}^2\ (\textit{compressive})\\ 85.85\ \text{MN/m}^2\ (\textit{tensile})\end{array}\right]$$

14. A ring with a circular cross-section of 45 mm diameter and a mean diameter 225 mm is subjected to a diametral tensile load of 6 kN. Calculate the deflection of the ring along the direction perpendicular to load line.

Take : $E = 200$ GN/m² [**Ans.** 0.02758 mm]

15. A chain link is made of a steel rod of 24 mm diameter. The straight portion is 120 mm in length and the ends are 120 mm in radius. Determine the deflection of the link along the load line when subjected to a load of 20 kN.

Take : $E = 200$ GN/m². [**Ans.** 2.14 mm]

16. A curved bar of rectangular section 80 mm wide by 100 mm deep in the plane of bending, initially unstressed, is subjected to a bending moment of 3 kNm which tends to straighten the bar. The mean radius of curvature is 200 mm.

Find : (*i*) The position of the neutral axis; (*ii*) The greatest bending stresses.

Draw a diagram to show approximately how the stress varies across the section.

[**Ans.** 26.93 MN/m² (*tensile*); 19.16 MN/m² (*comp.*)]

Chapter 21

Unsymmetrical Bending and Shear Centre

21·1. INTRODUCTION

In chapter 5 (Bending stresses), while using the well known bending equation $\dfrac{M}{I} = \dfrac{\sigma}{y}$, it is assumed that the neutral axis of the cross-section of the beam is perpendicular to the plane of loading. This condition implies that the plane of loading or plane of bending, is coincident with, or parallel to, a plane containing a principal centroidal axis of inertia of the cross-section of the beam. *If, however, the plane of loading or that of bending, does not lie in (or parallel to) a plane that contains the principal centroidal axis of the cross-section, the bending is called* **unsymmetrical bending.**

In the case of unsymmetrical bending, the direction of neutral axis is *not* perpendicular to the plane of bending.

Following are the two *reasons* of unsymmetrical bending :

    (*i*)   The section is symmetrical (*viz.* rectangular, circular, I sections) but the load line is inclined to both the principal axes.

    (*ii*) The section itself is unsymmetrical (*viz.* angle section or channel section vertical web) and the load line is along any centroidal axis.

21·2. PRODUCT OF INERTIA

21·2·1. Parallel Axes Theorem for Product of Inertia

21·2·2. Principal Axes and Principal Moments of Inertia

Refer Articles 3·16·1 and 3·16·2 (chapter 3) and Examples 3·17, 3·18 and 3·19.

21·3. STRESSES DUE TO UNSYMMETRICAL BENDING

Fig. 21·1 shows the cross-section of a beam under the action of a bending moment M acting in plane YY.

Also, G = Centroid of the section,

        XX, YY = Co-ordinate axes passing through G, and

        UU, VV = Principal axes inclined at an angle θ to XX and YY axes respectively.

Let us determine the stress distribution over the section.

The moment M in the plane YY can be resolved into its components in the planes UU and VV as follows :

Moment in the plane $UU, M' = M \sin \theta$ (21·1)

Moment in the plane $VV, M'' = M \cos \theta$ (21·2)

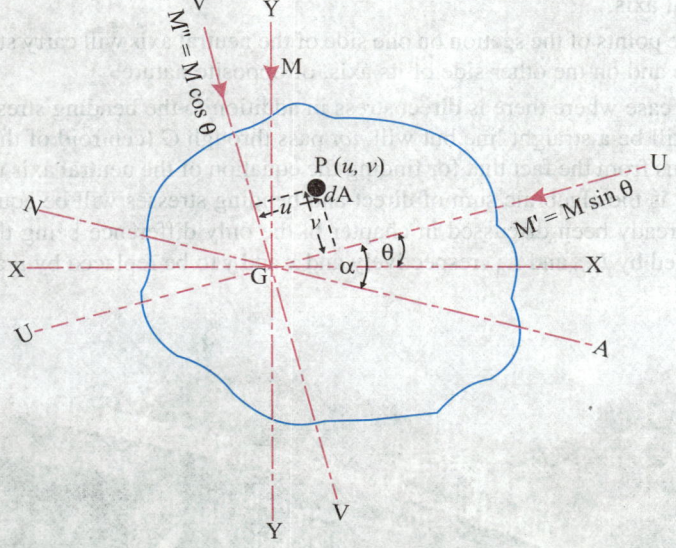

Fig. 21.1

The components M' and M'' have their axes along VV and UU respectively.

The resultant bending stress at the point $P(u, v)$ is given by,

$$\sigma_b = \frac{M' u}{I_{VV}} + \frac{M'' v}{I_{UU}} = \frac{M \sin \theta . u}{I_{VV}} + \frac{M \cos \theta . v}{I_{UU}}$$

or, $$\sigma_b = M \left[\frac{v \cos \theta}{I_{UU}} + \frac{u \sin \theta}{I_{VV}} \right]$$ ...(21·3)

At any point the nature of σ_b will depend upon the quadrant in which it lies. In other words the signs of u and v will have to be taken into account while determining the resultant bending stress.

The equation of the neutral axis (N.A.) can be found by finding the *locus of the points on which the resultant stress is zero*. Thus the points lying on neutral axis will satisfy the condition that $\sigma_b = 0$,

i.e.
$$M \left[\frac{v \cos \theta}{I_{UU}} + \frac{u \sin \theta}{I_{VV}} \right] = 0 \qquad \text{...(From eqn. 21·3)}$$

or,
$$\frac{v \cdot \cos \theta}{I_{UU}} + \frac{u \sin \theta}{I_{VV}} = 0$$

or,
$$v = - \left[\frac{I_{UU}}{I_{VV}} \times \frac{\sin \theta}{\cos \theta} \right] u$$

or,
$$v = - \left[\frac{I_{UU}}{I_{VV}} \tan \theta \right] u \qquad \text{...(21·4)}$$

This is an equation of a straight line passing through the centroid G of the section and inclined at an angle α with UU where,

$$\tan \alpha = - \left[\frac{I_{UU}}{I_{VV}} \tan \theta \right] \qquad \text{...(21·5)}$$

Following points are worth noting :

(*i*)   The maximum stress will occur at a point which is at the greatest distance from the neutral axis.

(*ii*) All the points of the section on one side of the neutral axis will carry stresses of the same nature and on the other side of its axis, of opposite nature.

(*iii*) In the case where there is direct stress in addition to the bending stress, the neutral axis will still be a straight line but will *not* pass through G (centroid of the section). This is obvious from the fact that for finding the equation of the neutral axis the resultant stress which is the algebraic sum of direct and bending stresses will be equated to zero. This has already been discussed in chapter 6, the only difference being that I_{XX} and I_{YY} be replaced by I_{UU} and I_{VV} respectively and x and y to be replaced by u and v respectively.

Bridges have members having both symmetrical and unsymmetrical cross-sections.

21·4. DEFLECTION OF BEAMS DUE TO UNSYMMETRICAL BENDING

Fig. 21·2. shows the transverse section of the beam with centroid *G*. *XX* and *YY* are two rectangular co-ordinate axes and *UU* and *VV* are the principal axes inclined at an angle θ to the *XY* set of co-ordinate axes. *W* is the load acting along line *YY* on the section of the beam. The load *W* can be resolved into the following two components :

(*i*) *W* sin θ.... along *UG*

(*ii*) *W* cos θ.... along *VG*.

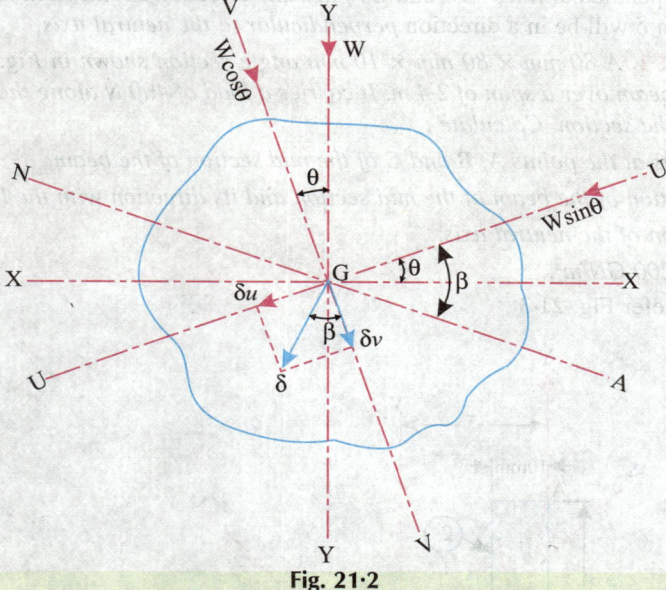

Fig. 21·2

Let, δ*u* = Deflection caused by the component *W* sin θ along the line *GU* for its bending about *VV* axis, and

δ*v* = Deflection caused by the component *W* cos θ along the line *GV* due to bending about *UU* axis.

Then, depending upon the *end conditions* of the beam, the values of δ*u* and δ*v* are given by:

$$\delta u = \frac{K\,(W\,\sin\theta)\,l^3}{EI_{VV}} \qquad\qquad ...(21\cdot5)$$

and, $$\delta v = \frac{K\,(W\,\cos\theta)\,l^3}{EI_{VV}} \qquad\qquad ...(21\cdot6)$$

where, *K* = A constant depending on the end conditions of the beam and position of the load along the beam, and

l = Length of the beam.

The total or resultant deflection δ can then be found as follows :

$$\delta = \sqrt{(\delta u)^2 + (\delta v)^2}$$

$$= \frac{Kl^3}{E}\sqrt{\left(\frac{(W\,\sin\theta)}{I_{VV}}\right)^2 + \left(\frac{(W\,\cos\theta)}{I_{UU}}\right)^2}$$

or,
$$\delta = \frac{KWl^3}{E} \sqrt{\frac{\sin^2 \theta}{I_{VV}^2} + \frac{\cos^2 \theta}{I_{UU}^2}} \qquad \qquad ...(21{\cdot}7)$$

The inclination β of the deflection δ, with the line GV is given by :

$$\tan \beta = \frac{\delta u}{\delta v} = \frac{I_{UU}}{I_{VV}} \tan \theta \qquad \qquad ...(21{\cdot}8)$$

From eqns. (21·5) and (21·8) it is evident that the magnitudes of α and β are the same and are measured from perpendicular lines (GU and GV) in same direction as shown in Figs. 21·1 and 21·2. Thus the deflection δ will be in a direction *perpendicular to the neutral axis*.

Example 21·1. *A 80 mm × 80 mm × 10 mm angle section shown in Fig. 21·3 is used as a simply supported beam over a span of 2·4 m. It carries a load of 400 N along the line YG, where G is the centroid of the section. Calculate :*

(i)   Stresses at the points A, B and C of the mid section of the beam;

(ii)   Deflection of the beam at the mid section and its direction with the load line;

(iii) Position of the neutral axis.

Take : E = 200 GN/m².

Solution. Refer Fig. 21·3

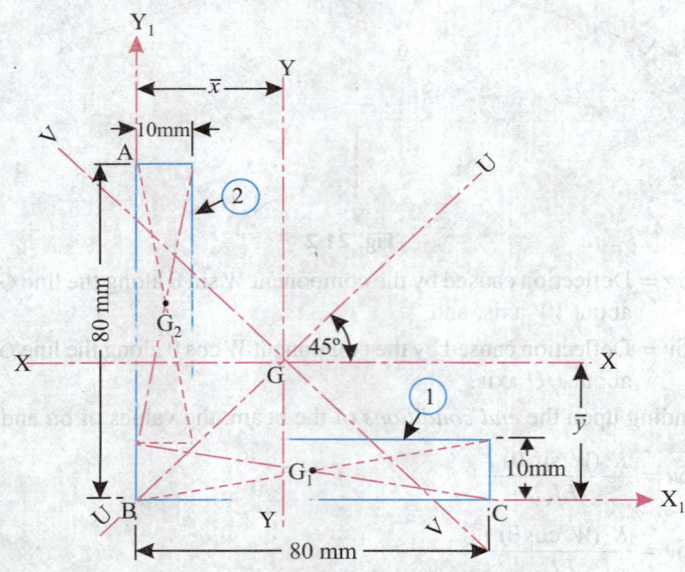

Fig. 21.3

Let $(\bar{x}, \bar{y})$ be the co-ordinates of centroid G, with respect to the rectangular axes BX_1 and BY_1.

Now
$$\bar{x} \,(= \bar{y}) = \frac{80 \times 10 \times 40 + 70 \times 10 \times 5}{80 \times 10 + 70 \times 10} = \frac{32000 + 3500}{800 + 700} = 23{\cdot}66 \text{ mm}$$

Moment of inertia about XX axis.

$$I_{XX} = \left[\frac{80 \times 10^3}{12} + 80 \times 10 \times (23{\cdot}66 - 5)^2\right] + \left[\frac{10 \times 70^3}{12} + 70 \times 10 \times (45 - 23{\cdot}66)^2\right]$$

$$= [6666{\cdot}66 + 278556] + [285833{\cdot}33 + 318777] = 889833 \text{ mm}^4$$

$$= 8{\cdot}898 \times 10^5 \text{ mm}^4 = I_{YY} \text{ (since it is an equal angle section)}$$

Co-ordinates of $\quad G_1 = +(40 - 23 \cdot 66), -(23 \cdot 66 - 5) = (16 \cdot 34, -18 \cdot 66)$

Co-ordinates of $\quad G_2 = -(23 \cdot 66 - 5), +(45 - 23 \cdot 66) = (-18 \cdot 66, +21 \cdot 34)$

Product of inertia, $I_{XY} = 80 \times 10 \, (16 \cdot 34) \, (-18 \cdot 66) + 70 \times 10 \, (-18 \cdot 66) \, (21 \cdot 34)$

$$= -243923 \cdot 5 - 278743 = -522666 \text{ mm}^4$$

$$= -5 \cdot 2266 \times 10^5 \text{ mm}^4$$

[Product of inertia about the centroid axes is zero because portions 1 and 2 are rectangular strips.]

If θ is the inclination of principal axes with *GX*, passing through *G* then,

$$\tan 2\theta = \frac{2I_{XY}}{I_{YY} - I_{XX}} = \infty = \tan 90°, \qquad\qquad (\because I_{XX} = I_{YY})$$

$\therefore \qquad\qquad 2\theta = 90°$

i.e. $\theta_1 = 45°$ and $\theta_2 = 90° + 45° = 135°$ are the inclinations of the principal axes *GU* and *GV* respectively.

Principal moment of inertia :

$$I_{UU} = \frac{1}{2}(I_{XX} + I_{YY}) + \frac{1}{2}(I_{XX} - I_{YY}) \cos 90° - I_{XY} \sin 90° \quad \text{(At } \theta_1 = 45°)$$

$$= \frac{1}{2}(8 \cdot 898 + 8 \cdot 898) \times 10^5 + \frac{1}{2} \times 0 \times \cos 90° - (-5 \cdot 2266 \times 10^5)$$

$$= (8 \cdot 898 + 5 \cdot 2266) \times 10^5 = 14 \cdot 1246 \times 10^5 \text{ mm}^4$$

Also, $\qquad\qquad I_{UU} + I_{VV} = I_{XX} + I_{YY}$

or, $\qquad\qquad I_{VV} = I_{XX} + I_{YY} - I_{UU}$

$$= 2 \times 8 \cdot 898 \times 10^5 - 14 \cdot 1246 \times 10^5 = 3 \cdot 67 \times 10^5 \text{ mm}^4$$

(*i*) Stresses at the points *A*, *B* and *C* :

Bending moment at the mid-section,

$$M = \frac{Wl}{4} = \frac{400 \times 2 \cdot 4 \times 10^3}{4} = 2 \cdot 4 \times 10^5 \text{ Nmm}$$

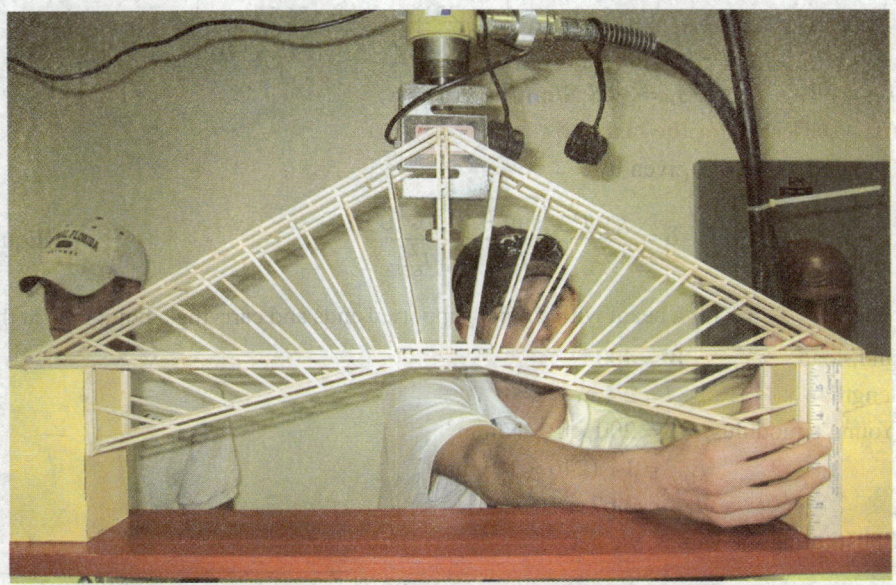

Complex frameworks are often tested in the laboratories by making prototypes,
before they are actually implemented.

The components of the bending moments are :

$$M' = M \sin \theta = 2 \cdot 4 \times 10^5 \sin 45° = 1 \cdot 697 \times 10^5 \text{ Nmm}$$
$$M'' = M \cos \theta = 2 \cdot 4 \times 10^5 \cos 45° = 1 \cdot 697 \times 10^5 \text{ Nmm}$$

u, v co-ordinates :

Point A. $x = -23 \cdot 66, y = 80 - 23 \cdot 66 = 56 \cdot 34 \text{ mm}$

$$u = x \cos \theta + y \sin \theta$$
$$= -23 \cdot 66 \times \cos 45° + 56 \cdot 34 \times \sin 45° = 23 \cdot 1 \text{ mm}$$
$$v = y \cos \theta - x \sin \theta$$
$$= 56 \cdot 34 \cos 45° - (-23 \cdot 66 \times \sin 45°) = 56 \cdot 56 \text{ mm}$$

Point B. $x = -23 \cdot 66, y = -23 \cdot 66 \text{ mm}$

$$u = -23 \cdot 66 \times \cos 45° + (-23 \cdot 66 \sin 45°) = -33 \cdot 45 \text{ mm}$$
$$v = -23 \cdot 66 \cos 45° - (-23 \cdot 66 \sin 45°) = 0$$

Point C. $x = 80 - 23 \cdot 66 = 56 \cdot 34, y = -23 \cdot 66 \text{ mm}$

$$u = 56 \cdot 34 \times \cos 45° - 23 \cdot 66 \sin 45° = 23 \cdot 1 \text{ mm}$$
$$v = -23 \cdot 66 \cos 45° - 56 \cdot 34 \sin 45° = -56 \cdot 56 \text{ mm}$$

$$\sigma_A = \frac{M'u}{I_{VV}} + \frac{M''v}{I_{UU}}$$

$$= 1 \cdot 697 \times 10^5 \left[\frac{23 \cdot 1}{3 \cdot 67 \times 10^5} + \frac{56 \cdot 56}{14 \cdot 1246 \times 10^5} \right] = 17 \cdot 47 \text{ N/mm}^2$$

i.e. $\sigma_A = \mathbf{17 \cdot 47} \text{ N/mm}^2$ **(Ans.)**

$$\sigma_B = 1 \cdot 697 \times 10^5 \left[\frac{-33 \cdot 45}{3 \cdot 67 \times 10^5} + \frac{0}{14 \cdot 1246 \times 10^5} \right] = -15 \cdot 47 \text{ N/mm}^2$$

i.e. $\sigma_B = \mathbf{-15 \cdot 47} \text{ N/mm}^2$ **(Ans.)**

$$\sigma_C = 1 \cdot 697 \times 10^5 \left[\frac{23 \cdot 1}{3 \cdot 67 \times 10^5} - \frac{56 \cdot 56}{14 \cdot 1246 \times 10^5} \right] = 3 \cdot 88 \text{ N/mm}^2$$

i.e. $\sigma_C = \mathbf{3 \cdot 88} \text{ N/mm}^2$ **(Ans.)**

(ii) Deflection of the beam, δ :

The deflection δ is given by :

$$\delta = \frac{KWl^3}{E} \sqrt{\frac{\sin^2 \theta}{I_{VV}^2} + \frac{\cos^2 \theta}{I_{UU}^2}} \qquad \text{[Equ. (21·7)]}$$

where, $K = \dfrac{1}{48}$ for a beam with simply supported ends and carrying a point load at the centre

and, load, $W = 400 \text{ N}$

length, $l = 2 \cdot 4 \text{ m}$

Young's modulus, $E = 200 \times 10^3 \text{ N/mm}^2$

$$I_{UU} = 14 \cdot 1246 \times 10^5 \text{ mm}^4$$
$$I_{VV} = 3 \cdot 67 \times 10^5 \text{ mm}^4$$

Substituting the values, we get

$$\delta = \frac{1}{48} \times \frac{400 \times (2 \cdot 4 \times 10^3)^3}{200 \times 10^3} \sqrt{\frac{(\sin 45°)^2}{(3 \cdot 67 \times 10^5)^2} + \frac{(\cos 45°)^2}{(14 \cdot 1246 \times 10^5)^2}}$$

$$= 5 \cdot 76 \times 10^5 \times \frac{1}{10^5} \sqrt{\frac{1}{2 \times (3 \cdot 67)^2} + \frac{1}{2 \times (14 \cdot 1246)^2}} = 1 \cdot 1466 \text{ mm}$$

i.e. $\qquad$ $\delta = 1 \cdot 1466 \text{ mm}$ (Ans.)

The deflection δ will be inclined at an angle β clockwise with the line GV, given by

$$\tan \beta = \frac{I_{UU}}{I_{VV}} \tan \theta = \frac{14 \cdot 1246 \times 10^5}{3 \cdot 67 \times 10^5} \tan 45° = 3 \cdot 848$$

$\therefore$ $\qquad$ $\beta = 75 \cdot 43°$

Thus the deflection is at $75 \cdot 43° - 45° = 30 \cdot 43°$ clockwise with the load line GY'.

(*iii*) Position of the neutral axis :

The neutral axis will be at $90° - 30 \cdot 43° = 59 \cdot 57°$ *anti-clockwise with the load line*, because the neutral axis is perpendicular to the line of deflection.

Example 21·2. *A cantilever, of I-section, 2·4 metres long is subjected to a load of 200N at the free end as shown in Fig. 21·4. Determine the resulting bending stresses at corners A and B, on the fixed section of the cantilever.*

Solution. Length of the cantilever, $l = 1 \cdot 8$ m

Load, $W = 600$ N

Refer to Fig. 21·4.

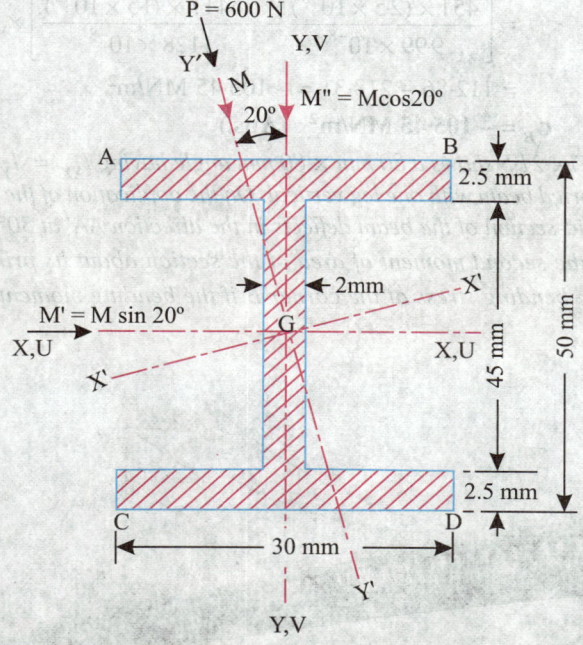

Fig. 21.4

Since I-section is symmetrical about XX and YY axes, therefore XX and YY are the principal axes UU and VV.

Moment of inertia, $I_{UU} = I_{XX} = \dfrac{30 \times 50^3}{12} - \dfrac{28 \times 45^3}{12} = 99875 \text{ mm}^4 = 9 \cdot 99 \times 10^{-8} \text{ m}^4$

$$I_{VV} = I_{YY} = 2 \times \frac{2 \cdot 5 \times 30^3}{12} + \frac{45 \times 2^3}{12} = 1 \cdot 128 \times 10^{-8} \text{ m}^4$$

Maximum bending moment,

$$M = W \times l = 200 \times 2.4 = 480 \text{ Nm}$$

Components of M are :

$$M' = M \sin 20° = 480 \times \sin 20° = 164.17 \text{ Nm}$$

$$M'' = M \cos 20° = 480 \times \cos 20° = 451 \text{ Nm}$$

M' will cause tensile stresses at points A and C and compressive stresses at points B and D

M'' will cause tensile stresses at points A and B and compressive stresses at points C and D

Now, *resultant bending stresses* on A and B are as follows :

$$\sigma_A = \frac{M'' \times (25 \times 10^{-3})}{I_{XX}} + \frac{M' \times (15 \times 10^{-3})}{I_{YY}}$$

$$= \left[\frac{451 \times (25 \times 10^{-3})}{9.99 \times 10^{-8}} + \frac{164.17 \times (15 \times 10^{-3})}{1.128 \times 10^{-8}} \right] \times 10^{-6} \text{ MN/m}^2$$

$$= 112.86 + 218.31 = 331.17 \text{ MN/m}^2$$

i.e. $\qquad \sigma_A = \textbf{331.17 MN/mm}^2$ **(Ans.)**

$$\sigma_B = \frac{M'' \times (2.5 \times 10^{-3})}{I_{XX}} - \frac{M' \times (1.5 \times 10^{-3})}{I_{YY}}$$

$$= \left[\frac{451 \times (2.5 \times 10^{-3})}{9.99 \times 10^{-8}} - \frac{164.17 \times (1.5 \times 10^{-3})}{1.128 \times 10^{-8}} \right] \times 10^{-6} \text{ MN/m}^2$$

$$= 112.86 - 218.31 = - 105.45 \text{ MN/m}^2$$

i.e. $\qquad \sigma_B = \textbf{– 105.45 MN/m}^2$ **(Ans.)**

Example 21·3. *Fig. 21·5 shows a 80 mm × 80 mm angle having* $I_{XX} = I_{YY} = 87.36 \times 10^{-8} \text{ m}^4$ *It is used as a freely supported beam with one leg vertical. On the application of the bending moment in the vertical plane YY the mid-section of the beam deflects in the direction AA' at 30° 15' to the vertical.*

  (i)   Calculate the second moment of area of the section about its principal axis.

 (ii) What is the bending stress at the corner B if the bending moment is 1·5 kNm?

(Panjab University)

Bridge.

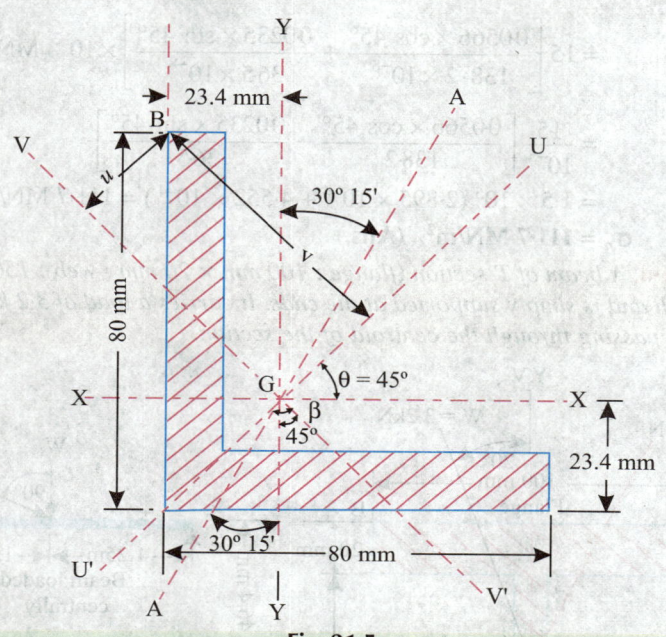

Fig. 21·5

Solution. *Given* : $I_{XX} = I_{YY} = 87·36 \times 10^{-8}$ m^4

Bending moment, $M = 1·5$ kNm

(i) Second moment of area about the principal axes, I_{UU}, I_{YY} :

We know, $I_{UU} + I_{VV} = I_{XX} + I_{YY}$

$$= 2 \times 87·36 \times 10^{-8} = 174·72 \times 10^{-8} \text{ m}^4 \qquad ...(i)$$

Also, $\qquad \tan \beta = \dfrac{I_{UU}}{I_{VV}} \times \tan \theta$

Here, $\qquad \theta = 45°$ (legs of the section being equal)

and, $\qquad \beta$ = Inclination of GA' with $GV' = 45° + 30°\ 15' = 75°\ 15'$

$\therefore \qquad \tan (75°\ 15') = \dfrac{I_{UU}}{I_{VV}} \tan 45°$

or $\qquad \dfrac{I_{UU}}{I_{VV}} = 3·79 \qquad\qquad\qquad ...(ii)$

Solving eqns. (*i*) and (*ii*), we get

$$I_{VV} = \mathbf{36·5 \times 10^{-8}\,m^4} \quad \textbf{(Ans.)}$$

and, $\qquad I_{UU} = \mathbf{138·2 \times 10^{-8}\,m^4} \quad \textbf{(Ans.)}$

(ii) Bending stress at B :

Co-ordinates of B (*u, v*) :

$$u = 80 \cos 45° - \frac{23·4}{\cos 45°} = 23·47 \text{ mm} = 0·0235 \text{ m}$$

$$v = 80 \sin 45° = 56·57 \text{ mm} = 0·0566 \text{ m}$$

Now bending stress at B,

$$\sigma_B = M \left[\frac{v \cos \theta}{I_{UU}} + \frac{u \sin \theta}{I_{VV}} \right] \qquad\qquad \text{Eqn. (21·3)}$$

$$= 1 \cdot 5 \left[\frac{0 \cdot 0566 \times \cos 45°}{138 \cdot 2 \times 10^{-8}} + \frac{0 \cdot 0235 \times \sin 45°}{365 \times 10^{-8}} \right] \times 10^{-3} \text{ MN/m}^2$$

$$= \frac{1 \cdot 5}{10^{-5}} \left[\frac{0 \cdot 0566 \times \cos 45°}{138 \cdot 2} + \frac{0 \cdot 0235 \times \sin 45°}{365} \right]$$

$$= 1 \cdot 5 \times 10^5 \, (2 \cdot 895 \times 10^{-4} + 4 \cdot 552 \times 10^{-4}) = 111 \cdot 7 \text{ MN/m}^2$$

i.e. $\sigma_B = 111 \cdot 7 \text{ MN/m}^2$ **(Ans.)**

Example 21·4. *A beam of T-section (flange : 100 mm × 20 mm ; web : 150 mm × 10 mm) is 2·5 metres in length and is simply supported at the ends. It carries a load of 3·2 kN inclined at 20° to the vertical and passing through the centroid of the section.*

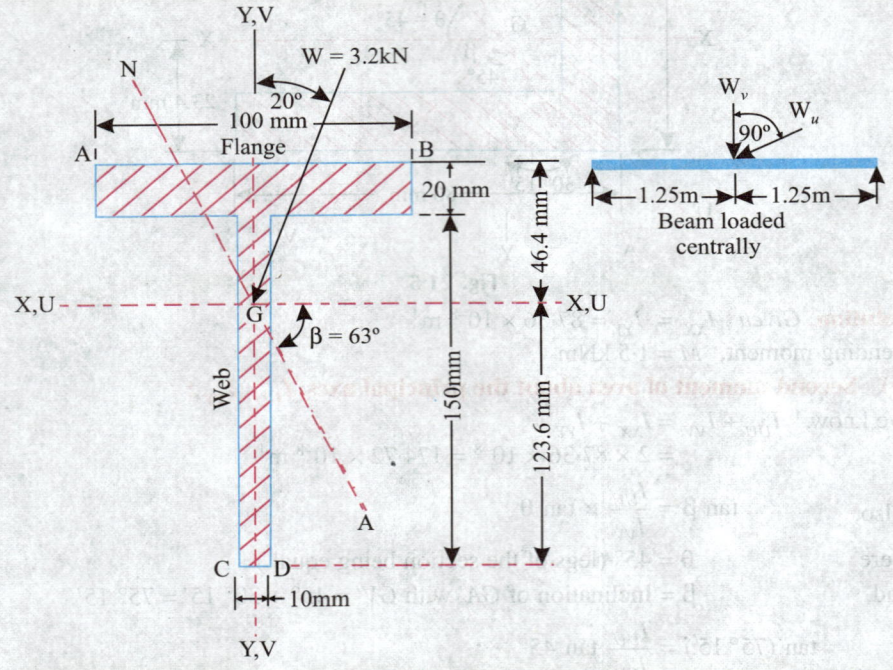

Fig. 21.6

If E = 200 GN/m², calculate :

 (i) Maximum tensile stress;   *(ii) Maximum compressive stress;*

(iii) Deflection due to the load; and   *(iv) Position of the neutral axis.*

Solution. Length of the beam,

$$l = 2 \cdot 5 \text{ m}$$

Load, $W = 3 \cdot 2 \text{ kN}$

Inclination of the load with the vertical = 20°

Young's modulus, $E = 200 \text{ GN/m}^2$

To find centroid of the T-section taking moments about the top of the flange, we get

$$\bar{y} = \frac{100 \times 20 \times 10 + 150 \times 10 \times \left(\dfrac{150}{2} + 20\right)}{100 \times 20 + 150 \times 10}$$

$$= \frac{20000 + 142500}{2000 + 1500} = 46.4 \text{ mm}$$

Since the section is symmetrical about the vertical axis, therefore, the principal axes pass through the centroid G and are along UU and VV axes shown.

$$I_{XX} = I_{UU} = \left[\frac{100 \times 20^3}{12} + 100 \times 20 \times (46.4 - 10)^2 \right]$$

$$+ \left[\frac{10 \times 150^3}{12} + 150 \times 10 \times (123.6 - 75)^2 \right]$$

$$= [66666.67 + 2649920] + [2812500 + 3542940]$$

$$= 9.07 \times 10^6 \text{ mm}^4 = 9.07 \times 10^{-6} \text{ m}^4$$

$$I_{YY} = I_{VV} = \left[\frac{12 \times 100^3}{12} + \frac{150 \times 10^3}{12} \right]$$

$$= 1.679 \times 10^6 \text{ mm}^4 = 1.679 \times 10^{-6} \text{ m}^4$$

Components of W :

$$W_u = W \sin 20° = 3.2 \times \sin 20° = 1.094 \text{ kN}$$
$$W_v = W \cos 20° = 3.2 \times \cos 20° = 3.007 \text{ kN}$$

Bending moments :

$$M_u = \frac{W_u \times l}{4} = \frac{1.094 \times 2.5}{4} = 0.684 \text{ kNm}$$

$$M_v = \frac{W_v \times l}{4} = \frac{3.007 \times 2.5}{4} = 1.879 \text{ kNm}$$

M_u will cause maximum compressive stresses at B and D and maximum tensile stresses at A and C.
M_v will cause maximum compressive stresses at A and B and maximum tensile stresses at C and D.

(i) Maximum tensile stress :

Maximum tensile stress at C,

$$\sigma_c = \frac{M_u \times (5 \times 10^{-3})}{I_{VV}} + \frac{M_v \times (123.6 \times 10^{-3})}{I_{UU}}$$

$$= \left[\frac{0.684 \times 5 \times 10^{-3}}{1.679 \times 10^{-6}} \times 10^{-3} + \frac{1.879 \times 123.6 \times 10^{-3}}{9.07 \times 10^{-6}} \times 10^{-3} \right] \text{MN/m}^2$$

$$= 2.04 + 25.6 = 27.64 \text{ MN/m}^2$$

i.e. $\sigma_C = 27.64 \text{ MN/m}^2$ **(Ans.)**

(ii) Maximum compressive stress :

Maximum compressive stress at B,

$$\sigma_B = \frac{M_u \times (50 \times 10^{-3})}{I_{VV}} + \frac{M_v \times (46.4 \times 10^{-3})}{I_{UU}}$$

$$= \left[\frac{0.684 \times (50 \times 10^{-3})}{1.679 \times 10^{-6}} \times 10^{-3} + \frac{1.879 \times 46.4 \times 10^{-3}}{9.07 \times 10^{-6}} \times 10^{-3} \right] \text{MN/m}^2$$

$$= 20.37 + 9.61 = 29.98 \text{ MN/m}^2$$

i.e. $\sigma_B = 29.98 \text{ MN/m}^2$ **(Ans.)**

(iii) Deflection due to the load, δ :

We know that;

$$\delta = \frac{KWl^3}{E} \sqrt{\frac{\sin^2 \theta}{I_{VV}^2} + \frac{\cos^2 \theta}{I_{UU}^2}}$$

where, $K = \frac{1}{48}$ for a beam with simply supported ends, and carrying a point load at its centre.

or,

$$\delta = \frac{KWl^3}{EI_{UU}} \sqrt{\sin^2 \theta \times \left(\frac{I_{UU}}{I_{VV}}\right)^2 + \cos^2 \theta}$$

$$= \frac{1}{48} \times \frac{3.2 \times 10^3 \times (2.5)^3}{200 \times 10^9 \times 9.07 \times 10^{-6}} \sqrt{(\sin 20°)^2 \times \left(\frac{9.07 \times 10^{-6}}{1.679 \times 10^{-6}}\right) + (\cos 20°)^2}$$

$$= 5.742 \times 10^{-4} \sqrt{3.414 + 0.883} = 11.9 \times 10^{-4} \text{ m} = 1.19 \text{ mm}$$

i.e.   **δ = 1·19 mm (Ans.)**

(iv) Position of the neutral axis :

We know that,

$$\tan \beta = \frac{I_{UU}}{I_{VV}} \tan \theta = \frac{9.07 \times 10^{-6}}{1.679 \times 10^{-6}} \times \tan 20° = 1.966 \qquad \text{[Eqn. 21.8]}$$

∴   **β ≃ 63° (Ans.)**

21·5. SHEAR CENTRE

— The **shear centre** (*for any transverse section of the beam*) *is the point of intersection of the bending axis and the plane of the transverse section.*

— Shear centre of a section can be defined *as a point about which the applied force is balanced by the set of shear forces obtained by summing the shear stresses over the section* (for unsymmetrical sections such as angle section and channel section, summation of shear stresses in each leg gives a set of forces which should be in equilibrium with the applied shear force).

Shear centre is also known as "*centre of twist.*"

— In case of a beam having *two axes of symmetry, the shear centre coincides with the centroid.*

Eiffel Tower, Paris.

— In case of sections having one axis of symmetry, the shear centre *does not coincide with the centroid but lies on the axis of symmetry.*

— When the load passes through the shear centre then there will be only bending in the cross-section and no twisting.

The principle involved in locating the shear centre for a cross-section of a beam is that the loads acting on the beam must lie in a plane which contains the resultant shear force on each cross-section of the beam as computed from the shearing stresses produced in the beam when it is loaded so that it does not twist at its ends.

21·5·1. Shear Centre for Channel Section

Fig. 21·7 shows a channel section (flanges : $b \times t_1$; web : $h \times t_2$) with *XX* as the horizontal symmetric axis.

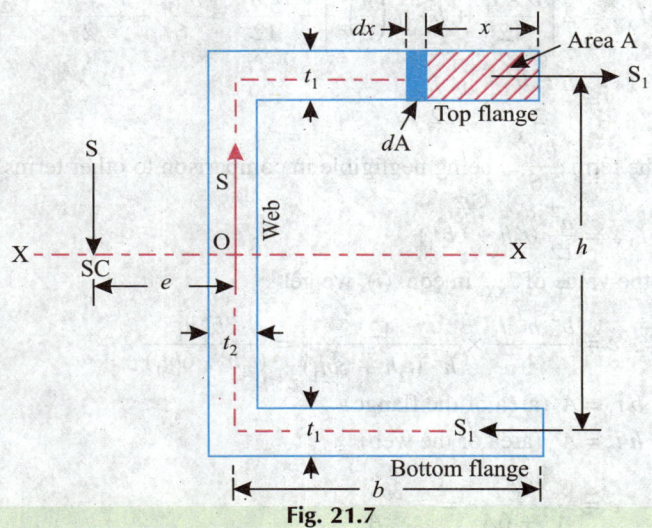

Fig. 21.7

Let, S = Applied shear force (vertical downwards), and
(then S is the shear force in the web in the upward direction).

S_1 = Shear force in the top flange (there will be *equal and opposite shear force in the bottom flange as shown*).

Now, shear stress (τ) in the flange at a distance of x from the right hand edge (of the top flange),

$$\tau = \frac{S A \bar{y}}{I_{XX} t}$$

$$A \bar{y} = (t_1 \cdot x) \frac{h}{2} \qquad \text{(where, } t = t_1 \text{, \textit{thickness of the flange})}$$

$$\therefore \quad \tau = \frac{S t_1 \cdot x}{I_{XX} \cdot t_1} \cdot \frac{h}{2} = \frac{S \, x \, h}{2 I_{XX}}$$

Shear force in elementary area

$$(d A = t_1 \cdot dx) = \tau \cdot dA = \tau \cdot t_1 \cdot dx$$

Total shear force in top flange

$$= \int_0^b \tau \cdot t_1 \cdot dx \qquad \text{(where, } b = \text{breadth of the flange)}$$

$$S_1 = \int_0^b \frac{S \, x \, h}{2 I_{XX}} \cdot t_1 \cdot dx = \frac{S h t_1}{2 I_{XX}} \int_0^b x \, dx$$

or, $\qquad S_1 = \dfrac{Sh\,t_1}{I_{XX}} \cdot \dfrac{b^2}{4}$

Let, $\;e$ = Distance of the shear centre (SC) from the web along the symmetric axis XX.

Taking moments of shear forces about the centre O of the web, we get

$$S \cdot e = S_1 \cdot h$$

$$= \frac{Sh\,t_1}{I_{XX}} \cdot \frac{b^2}{4} \cdot h = \frac{S \cdot t_1\,h^2\,b^2}{4\,I_{XX}}$$

$\therefore \qquad e = \dfrac{b^2\,h^2\,t_1}{4\,I_{XX}}$ $\qquad\qquad$ ...(i)

Now, $\qquad I_{XX} = 2\left[\dfrac{b \times t_1^3}{12} + b \cdot t_1 \left(\dfrac{h}{2}\right)^2\right] + \dfrac{t_2\,h^3}{12} = \dfrac{b\,t_1^3}{6} + \dfrac{b\,t_1\,h^2}{2} + \dfrac{t_2\,h^3}{12}$

$$= \frac{b\,t_1\,h^2}{2} + \frac{t_2\,h^3}{12}$$

(neglecting the term $\dfrac{b\,t_1^3}{6}$, being negligible in comparison to other terms)

or, $\qquad I_{XX} = \dfrac{h^2}{12}\,(t_2 h + 6bt_1)$

Substituting the value of I_{XX} in eqn. (i), we get

$$e = \frac{b^2\,h^2\,t_1}{4} \times \frac{12}{h^2\,(t_2 h + 6bt_1)} = \frac{3b^2\,t_1}{(t_2 h + 6bt_1)}$$

Let, $\qquad b\,t_1 = A_f$ (area of the flange)

and, $\qquad h\,t_2 = A_w$ (area of the web)

Then, $\qquad e = \dfrac{3b\,A_f}{A_w + 6A_f} = \dfrac{3b}{6 + \dfrac{A_w}{A_f}}$

i.e. $\qquad e = \dfrac{3b}{6 + \dfrac{A_w}{A_f}}$ $\qquad$ ...(21.9)

Example 21·5. *A channel section has flanges 12 cm × 2 cm and web 16 cm × 1 cm. Determine the shear centre of the channel.*

Solution. Here, $b = 12 - 0.5 = 11.5$ cm

$\qquad\qquad t_1 = 2$ cm, $t_2 = 1$ cm, $h = 18$ cm

$\qquad\qquad A_w = h\,t_2 = 18 \times 1 = 18$ cm^2

and, $\qquad A_f = bt_1 = 11.5 \times 2 = 23$ cm^2

We know that, $e = \dfrac{3b}{6 + \dfrac{A_w}{A_f}}$ (Eqn. 21·9)

$$= \frac{3 \times 11.5}{6 + \dfrac{18}{23}} = 5.086 \text{ cm}$$

*Hence, position of the shear centre = **5·086 cm** (Ans.)*

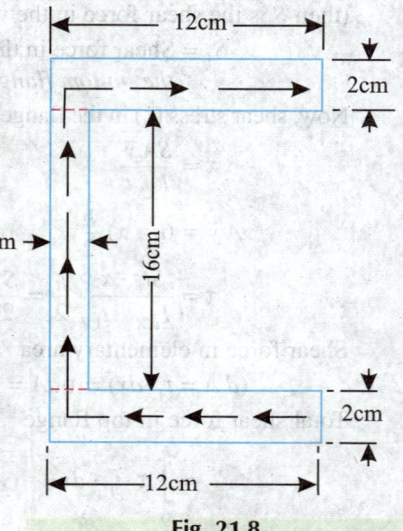

Fig. 21.8

21·5·2. Shear Centre for Unequal I-section

Fig. 21·9 shows an unequal I-section which is symmetrical about XX axis.

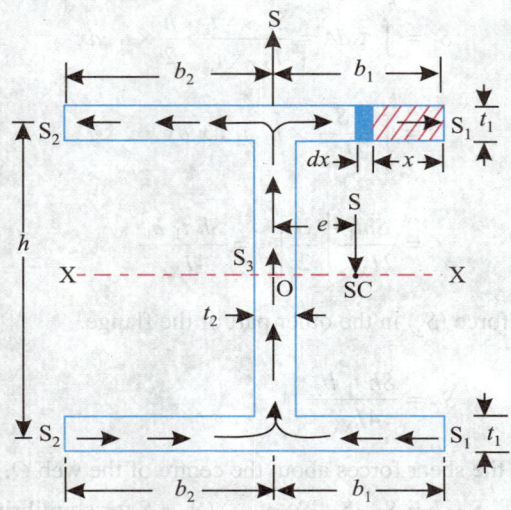

Fig. 21.9

Shear stress in any layer, $\tau = \dfrac{SA\,\bar{y}}{It}$

where, $$I = I_{XX} = 2\left[(b_1 + b_2)\frac{t_1^3}{12} + (b_1 + b_2)\,t_1 \times \frac{h^2}{4}\right] + t_2 \cdot \frac{h^3}{12}$$

Cross-section of the earthmover's bucket is not symmetical althrough. But due to the nature of work this asymmetry to some extent, is unavoidable.

Shear force S_1 :

$$dA = t_1 \cdot dx, \quad A\bar{y} = t_1 \cdot x \cdot \frac{h}{2}$$

$$S_1 = \int_0^{b_1} \tau \, dA = \frac{S \cdot x \cdot t_1}{I_{XX} \cdot t_1} \cdot \frac{h}{2} \times t_1 \cdot dx$$

$$= \int_0^{b_1} \frac{S}{2I_{XX}} \times h \cdot t_1 \cdot x \, dx$$

$$= \frac{Sht_1}{2I_{XX}} \left. \frac{x^2}{2} \right|_0^{b_1} = \frac{Sh \, t_1 \, b_1^2}{4I_{XX}}$$

Similarly the shear force (S_2) in the other part of the flange,

$$S_2 = \frac{Sh \, t_1 \, b_2^2}{4I_{XX}}$$

Taking moments of the shear forces about the centre of the web O, we get

$$S_2 \cdot h = S_1 \cdot h + S \cdot e \qquad (S_3 = S \text{ for equilibrium})$$

(where, e = distance of shear centre from the centre of the web)

or, $\qquad (S_2 - S_1) \, h = S \cdot e$

$$\frac{Sh^2 \, t_1}{4I_{XX}} (b_2^2 - b_1^2) = S \cdot e$$

$$\therefore \qquad e = \frac{t_1 \, h^2 \, (b_2^2 - b_1^2)}{4I_{XX}} \qquad \qquad ...(21 \cdot 10)$$

Example 21·6. *Determine the position of the shear centre of the section of a beam shown in Fig. 21·10.*

Solution. Here, $\qquad t_1 = 4$ cm.

$$b_1 = 6 \text{ cm}$$
$$b_2 = 8 \text{ cm}$$
$$h = 30 - 4 = 26 \text{ cm}$$

$$I_{XX} = 2 \left[\frac{14 \times 4^3}{12} + 14 \times 4 \times 13^2 \right] + \frac{2 \times 22^3}{12}$$

$$= 2 \, (74 \cdot 67 + 9464) + 1774 \cdot 67 = 20852 \text{ cm}^4$$

We know, $\qquad e = \dfrac{t_1 \, h^2 \, (b_2^2 - b_1^2)}{4I_{XX}}$

(where, e = distance of the shear centre from the centre of the web)

$$\therefore \qquad e = \frac{4 \times 26^2 \, (8^2 - 6^2)}{4 \times 20852} = 0 \cdot 9077 \text{ cm} \quad \textbf{(Ans.)}$$

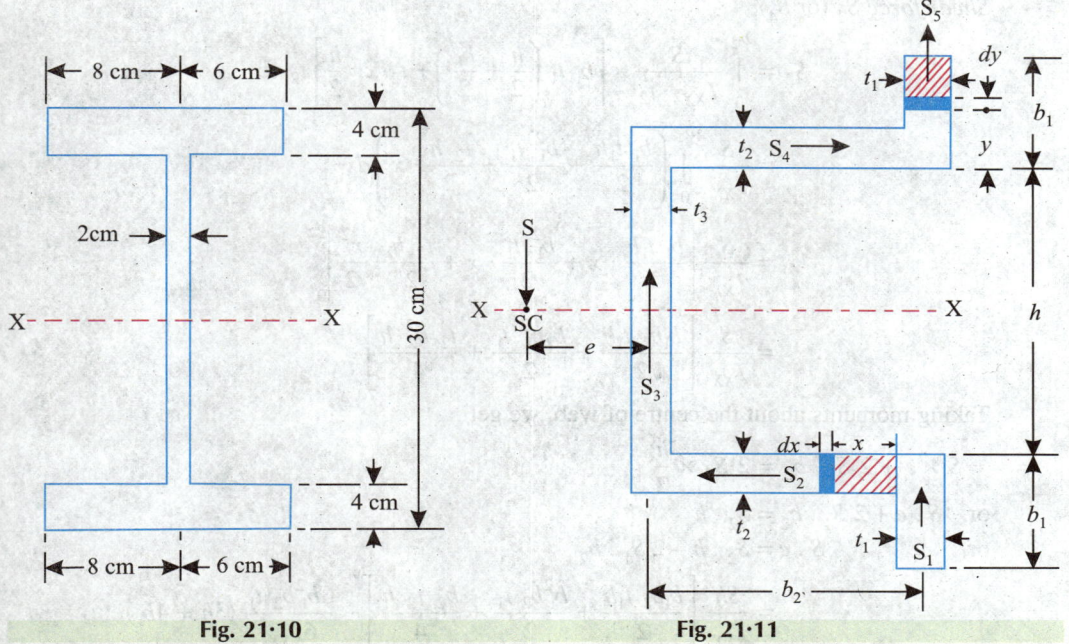

Fig. 21·10   Fig. 21·11

Example 21·7. *Locate the shear centre of the section shown in the Fig. 21·11*

Solution. Refer to Fig. 21·11.

Since the section is symmetrical about XX axis the shear centre will lie on this axis.

Also by symmetry: Shear forces $S_1 = S_5$, $S_2 = S_4$

Shear force S_1 (or S_5).

$$\tau = \frac{SA\bar{y}}{I_{XX} \cdot t_1} = \frac{S(b_1 - y)t_1}{I_{XX} \cdot t_1} \times \left[\frac{h}{2} + y + \frac{b_1 - y}{2}\right]$$

$$= \frac{S(b_1 - y)}{I_{XX}} \left[\frac{h + b_1 + y}{2}\right]$$

(where, S = applied shear force on the section)

Now, $dA = t_1 \cdot dy$

Shear force

$$S_1 = \int_0^{b_1} \frac{S(b_1 - y)}{2I_{XX}} (h + b_1 + y) \, t_1 \, dy$$

$$= \frac{S t_1}{2I_{XX}} \int_0^{b_1} (hb_1 - hy + b_1^2 - b_1 y + b_1 y - y^2) \, dy$$

$$= \frac{S t_1}{2I_{XX}} \left| (hb_1 y - \frac{hy^2}{2} + b_1^2 y - \frac{y^3}{3}) \right|_0^{b_1}$$

$$= \frac{S t_1}{2I_{XX}} \left[b_1^2 h - \frac{h}{2} b_1^2 + b_1^3 - \frac{b_1^3}{3} \right]$$

$$= \frac{S t_1}{2I_{XX}} \left[\frac{b_1^2 h}{2} + \frac{2b_1^3}{3} \right] = \frac{S b_1^2 t_1}{12 \, I_{XX}} (3h + 4b_1)$$

Shear force S_2 (or S_4).

$$S_2 = \int_0^{b_2} \frac{S}{I_{XX} \cdot t_2} \times \left[b_1 \, t_1 \left(\frac{h}{2} + \frac{b_1}{2} \right) + t_2 \, x \cdot \frac{h}{2} \right] t_2 \, dx$$

$$= \frac{S}{I_{XX}} \int_0^{b_2} \left(\frac{b_1 \, t_1 h}{2} + \frac{b_1^2 \, t_1}{2} + \frac{t_2 \, h}{2} \cdot x \right) dx$$

$$= \frac{S}{I_{XX}} \left| \frac{b_1 \, t_1 h}{2} \cdot x + \frac{b_1^2 \, t_1}{2} \cdot x + \frac{t_2 \, h}{2} \cdot \frac{x^2}{2} \right|_0^{b_2}$$

$$= \frac{S}{I_{XX}} \left[\frac{b_1 b_2 \, t_1 h}{2} + \frac{b_2 b_1^2 \, t_1}{2} + \frac{t_2 \, b_2^2 \, h}{4} \right]$$

Taking moments about the centre of web, we get

$$S \times e + 2 \, S_1 \cdot b_2 = 2 \, S_2 \times \frac{h}{2}$$

or, $\quad S \cdot e + 2 \, S_1 \cdot b_2 = S_2 \cdot h$

or, $\qquad\qquad S \cdot e = S_2 \cdot h - 2 \, S_1 \cdot b_2$

$$= \frac{Sh}{I_{XX}} \left[\frac{b_1 b_2 \, t_1 h}{2} + \frac{b_1^2 b_2 \, t_1}{2} + \frac{b_2^2 \, t_2 \, h}{4} \right] - \frac{S b_1^2 b_2 \, t_1}{6 I_{XX}} (3h + 4b_1)$$

$\therefore \qquad e = \dfrac{h^2}{I_{XX}} \left(\dfrac{b_1 b_2 t_1}{2} + \dfrac{t_2 b_2^2}{4} \right) + \dfrac{h \, b_1^2 b_2 t_1}{2 I_{XX}} - \dfrac{b_1^2 b_2 t_1 h}{2 I_{XX}} - \dfrac{2}{3} \dfrac{b_1^3 b_2 t_1}{I_{XX}}$

$$= \frac{h^2}{I_{XX}} \times \frac{b_1 b_2 t_1}{2} + \frac{h^2 t_2 b_2^2}{4 I_{XX}} - \frac{2}{3} \frac{b_1^3 b_2 t_1}{I_{XX}}$$

or, $\qquad e = \dfrac{b_1 b_2 t_1}{I_{XX}} \left[\dfrac{h^2}{2} - \dfrac{2}{3} b_1^2 \right] + \dfrac{h^2 b_2^2 t_2}{4 I_{XX}}$

where, $\qquad I_{XX} = 2 \left[\dfrac{t_1 b_1^3}{12} + b_1 \, t_1 \left(\dfrac{b_1}{2} + \dfrac{h}{2} \right)^2 \right] + 2 \left[\dfrac{b_2 t_2^3}{12} + b_2 \, t_2 \left(\dfrac{h}{2} \right)^2 \right] + \dfrac{t_3 \, h^3}{12}$

Wheels, suspensions and chassis of a train freight car. The centre transverse member has U-cross-section.

Example 21·8. *Find the shear centre of the section shown in Fig. 21·12.*

Solution. Refer to Fig. 21·12

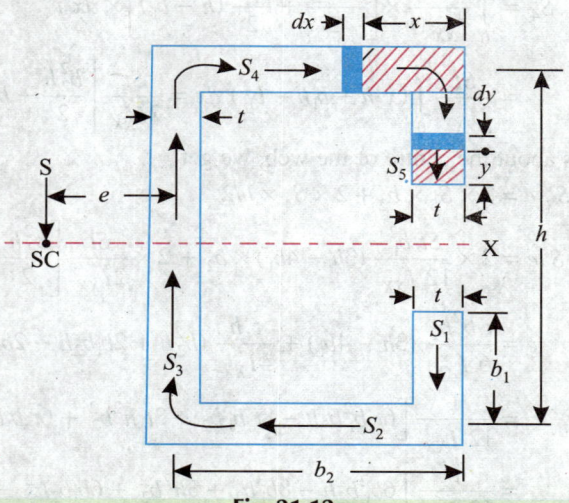

Fig. 21.12

Flanges : $b_2 \times t$

Web : $h \times t$

Projections : $b_1 \times t$

Let, $\qquad\qquad$ S = Applied force, and

$\qquad$ S_1, S_2, S_3, S_4, S_5 = Shear forces in different portions.

Also, $\qquad\qquad\qquad$ $S_1 = S_5$, and $S_2 = S_4$ (By symmetry)

Shear stress in any layer, $\tau = \dfrac{SA\bar{y}}{I_{XX} \cdot t}$

Shear force S_1 ($= S_5$).... vertical projection

Area, $\qquad\qquad$ $A = y \cdot t$

$$\bar{y} = \left(\frac{h}{2} - b_1\right) + \frac{y}{2} = \left(\frac{h - 2b_1 + y}{2}\right)$$

Area, $\qquad\qquad$ $dA = t \cdot dy$

∴ Shear force $\qquad$ $S_5 = \displaystyle\int_0^{b_1} \tau \cdot dA = \int_0^{b_1} \frac{Syt}{I_{XX} \cdot t} \times \left(\frac{h - 2b_1 + y}{2}\right) t \cdot dy$

$$= \frac{St}{2 I_{XX}} \int_0^{b_1} (hy - 2b_1 y + y^2)\, dy = \frac{St}{2 I_{XX}} \left| \frac{hy^2}{2} - b_1 y^2 + \frac{y^3}{3} \right|_0^{b_1}$$

$$= \frac{St}{2 I_{XX}} \left[\frac{hb_1^2}{2} - b_1^3 + \frac{b_1^3}{3} \right] = \frac{St}{2 I_{XX}} \left[\frac{hb_1^2}{2} - \frac{2}{3} b_1^3 \right]$$

or, $\qquad\qquad\qquad$ $S_5 = \dfrac{St\, b_1^2}{12\, I_{XX}} (3h - 4b_1)$

Shear force, $\qquad$ S_4 ($= S_2$) $\qquad$ ...Flange

$$A\bar{y} = (x \cdot t)\frac{h}{2} + b_1 \cdot t \left(\frac{h}{2} - b_1 + \frac{b_1}{2}\right) = x \cdot t \cdot \frac{h}{2} + b_1 \cdot t \left(\frac{h}{2} - \frac{b_1}{2}\right)$$

and,

$$dA = dx \cdot t$$

$$\therefore \qquad S_4 = \int_0^{b_2} \frac{S}{I_{XX} \cdot t} \times \left[\frac{x \cdot t \cdot h}{2} + \frac{b_1 t}{2}(h - b_1) \right] \times dx \cdot t$$

$$= \frac{St}{2I_{XX}} \int_0^{b_2} (x \cdot h + b_1 h - b_1^2) \, dx = \frac{St}{2I_{XX}} \left[\frac{b_2^2 h}{2} + b_1 b_2 h - b_1^2 b_2 \right]$$

Taking moments about the centre of the web, we get

$$S \cdot e = 2 \times S_5 \times b_2 + 2 \times S_4 \times h/2$$

or,

$$S \cdot e = 2 \times \frac{St b_1^2}{12 I_{XX}} (3h - 4b_1) \times b_2 + 2 \times \frac{St}{2I_{XX}} \left[\frac{b_2^2 h}{2} + b_1 b_2 h - b_1^2 b_2 \right] \times \frac{h}{2}$$

or,

$$e = \frac{t \, b_1^2 b_2}{6 \, I_{XX}} (3h - 4b_1) + \frac{t \cdot h}{4 I_{XX}} (b_2^2 h + 2 b_1 b_2 h - 2 b_1^2 b_2)$$

$$= \frac{1}{12 \, I_{XX}} \left[6t \, b_1^2 b_2 h - 8t \, b_1^3 b_2 + 3t \, h^2 b_2^2 + 6t \, b_1 b_2 h^2 - 6t \, h b_1^2 b_2 \right]$$

$$= \frac{t}{12 \, I_{XX}} \left[6 b_1^2 b_2 h - 8 b_1^3 b_2 + 3 h^2 b_2^2 + 6 b_1 b_2 h^2 - 6 h b_1^2 b_2 \right]$$

or,

$$e = \frac{t}{12 \, I_{XX}} \left[- 8 b_1^3 b_2 + 3 h^2 b_2^2 + 6 b_1 b_2 h^2 \right]$$

where,

$$I_{XX} = \frac{t \times h^3}{12} + 2 \left[\frac{b_2 \times t^3}{12} + b_2 t \times \left(\frac{h}{2} \right)^2 \right] + \left[\frac{t \times b_1^3}{12} + b_1 t \times \left(\frac{h}{2} - \frac{b_1}{2} \right)^2 \right]$$

or,

$$I_{XX} = \frac{t \cdot h^3}{12} + \frac{b_2 \cdot t^3}{6} + \frac{b_2 \cdot t \cdot h^2}{2} + \frac{t \cdot b_1^3}{6} + \frac{b_1 \cdot t}{2} (h - b_1)^2$$

Example 21·9. *Locate the shear centre of the section shown in Fig. 21·13.*

Solution. Refer to Fig. 21·13.

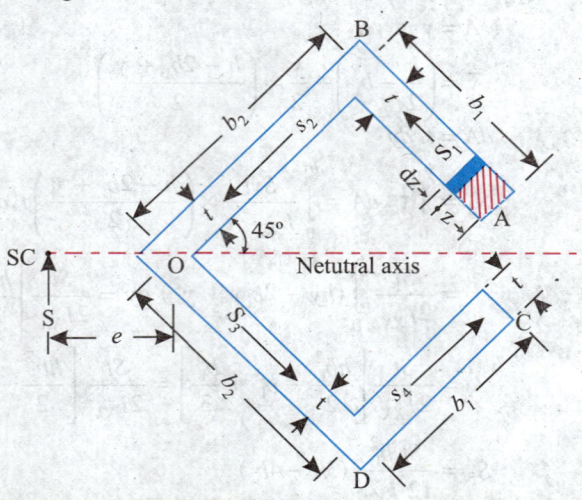

Fig. 21.13

$$S_1 = S_4, \text{ and, } S_2 = S_3 \qquad\qquad \dots \textit{Due to symmetry}$$

Shear stress in any layer, $\tau = \dfrac{S A \bar{y}}{I_{NA}}$

where, S = Applied force, and I_{NA} = Moment of inertia about neutral axis.

Shear force S_1 (= S_4).

$$S_1 = \int_0^{b_1} \tau . dA$$

where, $A = z . t$

∴ $dA = t \, dz$

$$\bar{y} = (b_2 \sin 45° - b_1 \sin 45°) + \frac{z}{2} \sin 45°$$

$$= \left(b_2 - b_1 + \frac{z}{2} \right) \sin 45° = \frac{2b_2 - 2b_1 + z}{2\sqrt{2}}$$

∴ $$S_1 = \int_0^{b_1} \frac{Sz.t}{I_{NA} .t} \left(\frac{2b_2 - 2b_1 + z}{2\sqrt{2}} \right) t.dz = \frac{St}{2\sqrt{2}\, I_{NA}} \int_0^{b_1} (2b_2\, z - 2b_1\, z + z^2)\, dz$$

$$= \frac{St}{2\sqrt{2}\, I_{NA}} \left[b_2 b_1^2 - b_1^3 + \frac{b_1^3}{3} \right] = \frac{St\, b_1^2\, (3b_2 - 2b_1)}{6\sqrt{2}\, I_{NA}}$$

Total moment of inertia of the section :

For rectangle I :

$$I_{U_1 U_1} = \frac{b_1 t^3}{12}, \; I_{V_1 V_1} = \frac{t . b_1^3}{12}$$

Now, $$I_{X_1 X_1} = I_{U_1 U_1} \cos^2 \theta + I_{V_1 V_1} \sin^2 \theta = \frac{I_{UU} + I_{VV}}{2} \qquad\qquad (\because \theta = 45°)$$

or, $$I_{X_1 X_1} = \frac{1}{2} \left(\frac{b_1 t^3}{12} + \frac{t b_1^3}{12} \right) = \frac{b_1 t}{24} (b_1^2 + t^2)$$

$$I_{NA_1} = I_{X_1 X_1} + t.b_1 \left[\left(b_2 - \frac{b_1}{2} \right) \sin 45° \right]^2 = I_{X_1 X_1} + \frac{t b_1 (2b_2 - b_1)^2}{8}$$

Machine part with acme threads.

$$= \frac{b_1 t}{24} (b_1^2 + t^2) + b_1 t \left[\frac{b_2^2}{2} + \frac{b_1^2}{8} - \frac{b_1 b_2}{2} \right] = \frac{b_1 t}{24} [b_1^2 + t^2 + 12 b_2^2 + 3 b_1^2 - 12 b_1 b_2]$$

or, $\quad I_{NA_1} = \frac{b_1 t}{24} [t^2 + 4 b_1^2 + 12 b_2^2 - 12 b_1 b_2]$

For rectangle II :

$$I_{U_2 U_2} = \frac{b_2 t^3}{12}, \quad I_{V_2 V_2} = \frac{t b_2^3}{12}$$

Now, $\quad I_{X_2 X_2} = I_{U_2 U_2} \cos^2 \theta + I_{V_2 V_2} \sin^2 \theta = \frac{I_{U_2 U_2} + I_{V_2 V_2}}{2} \qquad (\because \theta = 45°)$

$$= \frac{1}{2} \left(\frac{b_2 t^3}{12} + \frac{t b_2^3}{12} \right) = \frac{b_2 t}{24} (b_2^2 + t^2)$$

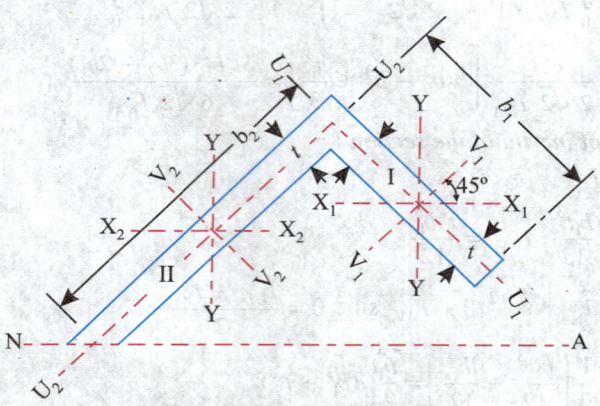

Fig. 21.14

$$I_{NA_2} = I_{X_2 X_2} + b_2 \cdot t \left(\frac{b_2}{2} \sin 45° \right)^2 = \frac{b_2 t}{24} (b_2^2 + t^2) + b_2 t \left(\frac{b_2^2}{8} \right)$$

$$= \frac{b_2 t}{24} (4 b_2^2 + t^2)$$

Total moment of inertia of the section,

$$I_{NA} = 2 \times I_{NA_1} + 2 I_{NA_2}$$

or, $\quad I_{NA} = \frac{b_1 t}{12} (t^2 + 4 b_1^2 + 12 b_2^2 - 12 b_1 b_2) + \frac{b_2 t}{12} (4 b_2^2 + t^2)$

Taking moments of the shear forces about O, we get

$$S \times e = S_1 \times b_2 + S_1 \times b_2 = 2 S_1 b_2$$

$$= \frac{St \, b_1^2 b_2}{3 \sqrt{2} \, I_{NA}} (3 b_2 - 2 b_1)$$

∴ $\quad e = \frac{t \, b_1^2 b_2}{3 \sqrt{2} \, I_{NA}} (3 b_2 - 2 b_1)$

HIGHLIGHTS

1. Reasons of unsymmetrical bending are :
   (i) The section is symmetrical but the load line is inclined to both the principal axes.
   (ii) The section itself is unsymmetrical and the load line is along any centroidal axis.

2. Product of inertia $I_{XY} = \int xy\, dA$. Product of inertia of a section about its principal axis is zero.

3. If the section is symmetrical, the principal axes are along the axes of symmetry.

4. Stresses due to unsymmetrical bending :
   Resultant bending stress at a point,

   $$\sigma_b = M\left[\frac{v\cos\theta}{I_{UU}} + \frac{u\sin\theta}{I_{VV}}\right]$$

   where, M = Bending moment,
   u, v = Co-ordinates of the point, and
   θ = Angle of inclination of axis of M with respect to the principal axis VV.

5. Angle of inclination of neutral axis with respect to principal axis UU,

   $$\alpha = \tan^{-1}\left(\frac{I_{UU}}{I_{VV}}\tan\theta\right)$$

6. Deflection of a beam under load W causing unsymmetrical bending,

   $$\delta = \frac{KWl^3}{E}\sqrt{\frac{\sin^2\theta}{I_{VV}^2} + \frac{\cos^2\theta}{I_{UU}^2}}$$

   where, K = A constant depending on the end conditions of the beam and position of the load along the beam,
   l = Length of the beam, and
   θ = Angle of inclination of load W with respect to VV principal axis.

7. The shear centre for any transverse section of the beam is the point of intersection of the bending axis and the plane of the transverse section. Shear centre is also known as *centre of twist.*

8. In case of a beam having two axes of symmetry, the shear centre coincides with the centroid.

OBJECTIVE TYPE QUESTIONS

Choose the Correct Answer :

1. In the case of unsymmetrical bending, the direction of neutral axis is
   (a) perpendicular to the plane of bending
   (b) not perpendicular to the plane of bending
   (c) either (a) or (b)
   (d) none of the above.

2. Which of the following is the reason of unsymmetrical bending ?
   (a) The section is symmetrical but the load line is inclined to both the principal axes
   (b) The section itself is unsymmetrical and the load line is along centroidal axis
   (c) either (a) or (b)
   (d) both (a) and (b).

3. Unsymmetrical bending is the bending caused by loads that
   (a) lie in a vertical plane
   (b) lie in a horizontal plane
   (c) lie in or parallel to a plane containing the principal centroidal axis of inertia of the cross-section.
   (d) do not lie in or parallel to a plane containing the principal centroidal axis of inertia of the cross-section.

4. When the loads pass through the bending axis of a beam then there shall be
   (a) pure bending of the beam
   (b) twisting of the beam
   (c) bending shall be accompanied by twisting
   (d) non-bending of beam.

5. Under unsymmetrical bending the resultant deflection of a beam is
   (a) parallel to the axis of symmetry
   (b) perpendicular to the axis of symmetry
   (c) parallel to the neutral axis
   (d) perpendicular to the neutral axis.

6. If the load passes through the shear centre of the section of the beam, then there will be

(a) no bending of the beam

(b) only bending in the beam

(c) bending accompanied by twisting

(d) only twisting in the beam.

7. In a channel section symmetrical about XX-axis, shear centre lies at

(a) the centre of the vertical web

(b) the centre of the top flange

(c) the centroid of the section

(d) none of the above.

8. In an I-section, symmetrical about XX and YY axes, shear centre lies at

(a) centroid of the top flange

(b) centroid of the web

(c) at the centroid of the bottom flange

(d) none of the above.

ANSWERS

1. (b) **2.** (d) **3.** (d) **4.** (a) **5.** (d) **6.** (b) **7.** (d) **8.** (b).

UNSOLVED EXAMPLES

1. A 40 mm × 40 mm × 5 mm angle is used as a simply supported beam over a span of 2·4 metres. It carries a load of 200 N along the vertical axis passing through the centroid of the section. Determine the resulting bending stresses on the outer corners of the section, along the middle section of the beam. [**Ans.** 69·88 MN/m^2, − 42·98 MN/m^2, 15·51 MN/m^2]

2. A cantilever of rectangular section 40 mm (width) × 60 mm (depth) is subjected to an inclined load P at the free end. The inclination of the load is 25° to the vertical. If the length of the cantilever is 2 metres and maximum stress due to bending is not to exceed 200 MN/m^2, determine the value of P. [**Ans.** 1·558 kN]

3. A timber beam 250 mm wide by 300 mm deep is used as simply supported beam on a span of 5 m. It is subjected to a concentrated load of 30 N at the mid-section of the span. If the plane of the load makes an angle of 45° with the vertical plane of symmetry find the direction of neutral axis and the maximum stress in the beam. (**Punjab University**) [**Ans.** 55·2° anticlockwise with the vertical; 15·51 MN/m^2]

4. A cantilever of T-section (Flange : 120 mm × 20 mm ; Web : 130 mm × 20 mm) is 2·8 m long and carries a concentrated load W at its free end but inclined at an angle of 45° to the vertical. If $E = 200$ GN/m^2 and the deflection at the free end is not to exceed 2 mm, determine:

(i) The maximum value of W;

(ii) Direction of neutral axis with respect to vertical axis [**Ans.** 221·2 N; 15° 24′]

5. A beam of angle section 150 mm × 100 mm × 10 mm is simply supported over a span of 1·6 m with 150 mm leg vertical. A uniformly distributed vertical load of 10 kN/m is applied throughout the span. Determine :

(i) Maximum bending stress,   (ii) Direction of neutral axis, and   (iii) Deflection at the centre.

Take: $E = 210$ GN/m^2. [**Ans.** (i) 68·3 MN/m^2; (ii) 68° 9′; (iii) 1·55 mm]

Cross-sections of the cement columns are preferably made symmetrical, whereas roof frames are usually made of metallic members having U and L shaped cross-sections.

Competitive Examinations (UPSC, GATE, etc.) Questions with Solutions

Chapter 22

SIMPLE STRESSES AND STRAINS

Example 22.1. *The diameters of the brass and steel segments of the axially loaded bar shown in Fig. 22.1 are 30 mm and 12 mm respectively. The diameter of the hollow section of the brass segment is 20 mm. Determine :*

(i) The maximum normal stress in the steel and brass;
(ii) The displacement of the free end.

Take, $E_s = 210 \ GN/m^2$, *and* $E_b = 105 \ GN/m^2$.

Solution. Refer to Fig. 22.1

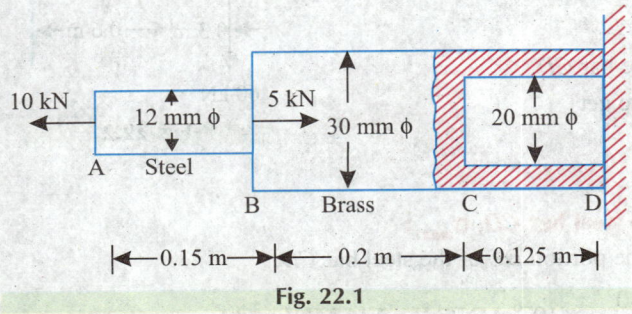

Fig. 22.1

$$A_s = \frac{\pi}{4} \times (12)^2 = 36\pi \text{ mm}^2 = 36\pi \times 10^{-6} \text{ m}^2$$

$$(A_b)_{BC} = \frac{\pi}{4} \times (30)^2 = 225\pi \text{ mm}^2 = 225\pi \times 10^{-6} \text{ m}^2$$

$$(A_b)_{CD} = \frac{\pi}{4} \left[(30)^2 - (20)^2 \right] = 125\pi \text{ mm}^2 = 125\pi \times 10^{-6} \text{ m}^2$$

(i) The maximum normal stress in steel and brass:

$$\sigma_s = \frac{10 \times 10^3}{36\pi \times 10^{-6}} \times 10^{-6} \text{ MN/m}^2 = \textbf{88.42 MN/m}^2 \textbf{ (Ans.)}$$

$$(\sigma_b)_{BC} = \frac{5 \times 10^3}{225\pi \times 10^{-6}} \times 10^{-6} \text{ MN/m}^2 = \textbf{707 MN/m}^2 \textbf{ (Ans.)}$$

$$(\sigma_b)_{BC} = \frac{5 \times 10^3}{125\pi \times 10^{-6}} \times 10^{-6} \text{ MN/m}^2 = \textbf{12.73 MN/m}^2 \textbf{ (Ans.)}$$

(ii) The displacement of the free end:

The displacement of the free end

$$\delta l = (\delta l_s)_{AB} + (\delta l_b)_{BC} + (\delta l_b)_{CD}$$

$$= \frac{88.42 \times 0.15}{210 \times 10^9 \times 10^{-6}} + \frac{7.07 \times 0.2}{105 \times 10^9 \times 10^{-6}} + \frac{12.73 \times 0.125}{105 \times 10^9 \times 10^{-6}} \quad \left(\because \delta l = \frac{\sigma l}{E} \right)$$

$$= 6.316 \times 10^{-5} + 1.347 \times 10^{-5} + 1.515 \times 10^{-5}$$

$$= 9.178 \times 10^{-5} \text{ m or } \textbf{0.09178 mm (Ans.)}$$

Example 22.2. *A beam AB hinged at A is loaded at B as shown in Fig. 22.2. It is supported from the roof by a 2.4 m long vertical steel bar CD which is 3.5 cm square for the first 1.8 m length and 2.5 cm square for the remaining length. Before the load is applied the beam hangs horizontally. Determine :*

(i) The maximum stress in the steel bar CD;

(ii) The total elongation of the bar.

Take $E_s = 210 \text{ GN/m}^2$. **(Panjab University)**

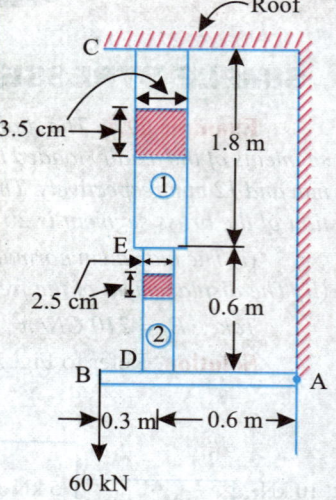

Fig. 22.2

Solution. *Given:*

$$A_2 = 2.5 \times 2.5 = 6.25 \text{ cm}^2 = 6.25 \times \times 10^{-4} \text{ m}^2;$$

$$A_1 = 3.5 \times 3.5 = 12.25 \times 10^{-4} \text{ m}^2; \, l_1 = 1.8 \text{ m}$$

$$l_2 = 0.6 \text{ m}$$

$$E_s = 210 \text{ GN/m}^2.$$

Let, P = The pull in the bar *CD*.

Then, taking moments about *A*, we get

$$P \times 0.6 = 60 \times 0.9$$

∴ $P = 90 \text{ kN}$

(i) The maximum stress in the steel bar CD, σ_{max} :

The stress shall be maximum in the portion *DE* of the steel bar *CD*.

∴ $$\sigma_{max} = \frac{P}{A_2} = \frac{90 \times 10^3}{6.25 \times 10^{-4}} \times 10^{-6} \text{ MN/m}^2 = \textbf{144 MN/m}^2 \textbf{ (Ans.)}$$

(*ii*) **The total elongation of the bar *CD*, δ :**

$$\delta = \delta l_1 + \delta l_2 = \frac{Pl_1}{A_1 E} + \frac{Pl_2}{A_2 E} = \frac{P}{E}\left(\frac{l_1}{A_1} + \frac{l_2}{A_2}\right)$$

$$= \frac{90 \times 10^3}{210 \times 10^9}\left(\frac{1.8}{12.25 \times 10^{-4}} + \frac{0.6}{6.25 \times 10^{-4}}\right)$$

$$= 0.428 \times 10^{-6}(1469.38 + 960) = 0.001039\,\text{m} \simeq \textbf{1.04 mm} \quad \textbf{(Ans.)}$$

Example 22.3. *A flat bar of section 50 mm × 10 mm is subjected to an axial pull of 130 kN. One side of the bar is polished and lines are ruled on it to form a square of 30 mm side, one diagonal of the square being along the middle line of the polished side. If E = 200 GN/m² and Poisson's ratio is 0.25, calculate the change in the angles and sides of the square.* **(GATE)**

Solution. *Given :* Area of the bar = 50 × 10 = 500 mm²

Axial pull, $P = 130\,\text{kN} = 130 \times 10^3\,\text{N}$

Side of the square = 30 mm

Modulus of elasticity,

$$E = 200\,\text{GN/m}^2 = 200 \times 10^9\,\text{N/m}^2 = 200 \times 10^3\,\text{N/mm}^2$$

Poisson's ratio, $\mu = \left(\dfrac{1}{m}\right) = 0.25$

Changes in the angles and sides of the square :

Refer to Fig. 22.3

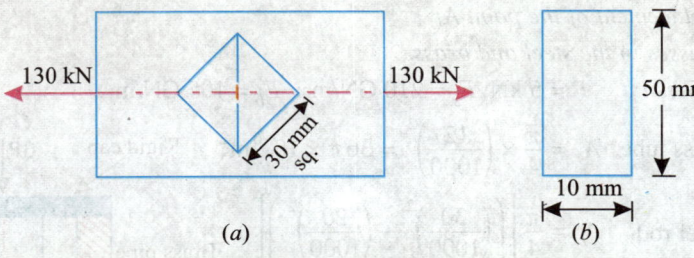

(*a*)   (*b*)

Fig. 22.3

The length of the diagonal of square,

$$l = \sqrt{30^2 + 30^2} = 30\sqrt{2}\,\text{mm} = 42.43\,\text{mm}$$

Change in length (δl) in the *longitudinal direction* due to the force is given as,

$$e_L = \frac{(\delta l)_L}{l} = \frac{P/A}{E} = \frac{130 \times 10^3}{500 \times 200 \times 10^3} = 0.0013$$

∴ $\delta l = 42.43 \times 0.0013 = 0.05516\,\text{mm}$

Now, change in length in *transverse direction, i.e.* $(\delta l)_T$ is given as

$$e_L = \frac{(\delta l)_T}{l} = -\mu e_L = -(0.25)(0.0013) = -0.000325$$

∴ $(\delta l)_T = (-0.000325)(42.43) = -0.01379\,\text{mm}$

After deformation, length of the diagonal along the axis,

$$(l)_L = l + (\delta l)_L = 42.43 + 0.05516 = 42.4852\,\text{mm}$$

Length of the diagonal perpendicular to the axis,

$$(l)_T = l - (\delta l)_T = 42.43 - 0.01379 = 42.4162 \text{ mm}$$

Before deformation, the angle between diagonal and side, $\alpha = 45°$

After deformation the angle, $\alpha' = \tan^{-1}\left[\dfrac{(42.4162/2)}{(42.4852/2)}\right] = $ **44.95° (Ans.)**

Length of the side of the square $= \dfrac{(42.4162/2)}{\sin 44.95°} = $ **30.019 mm (Ans.)**

This is shown in Fig. 22.4.

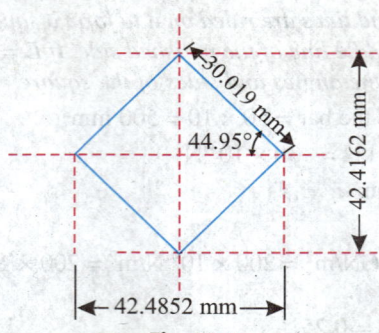

Fig. 22.4

Example 22.4. *A steel rod, 12 mm in diameter, is welded to a rigid plate that is supported by a brass pipe whose outside diameter is 30 mm and inside diameter is 20 mm, as shown in Fig. 22.5. If P = 5 kN, $E_s = 210$ GN/m² and $E_b = 105$ GN/m², determine:* **(GATE)**

(i) The displacement of the point A;

(ii) The stresses in the steel and brass.

Solution. *Given:* $\quad P = 5 \text{ kN}; E_s = 210 \text{ GN/m}^2; E_b = 105 \text{ GN/m}^2.$

Area of brass tube, $A_s = \dfrac{\pi}{4} \times \left(\dfrac{12}{1000}\right)^2 = 36\pi \times 10^{-6} \text{m}^2$

Area of steel rod, $A_b = \dfrac{\pi}{4}\left[\left(\left(\dfrac{30}{1000}\right)^2 - \left(\dfrac{20}{1000}\right)^2\right)\right]$

$\qquad = 125\pi \times 10^{-6} \text{ m}^2$

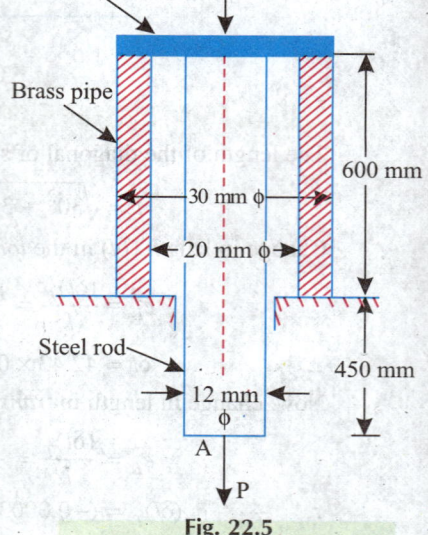

Fig. 22.5

(*i*) The displacement of point A :

Compression of the brass pipe,

$$(\delta l)_b = \frac{4P \times (600/1000)}{125\pi \times 10^{-6} \times 105 \times 10^9}$$

$$= \frac{4 \times 5 \times 10^3 \times (600/1000)}{125\pi \times 10^{-6} \times (105 \times 10^9)}$$

$$= 0.2907 \times 10^{-3} \text{ m or } 0.2907 \text{ mm}$$

Extension of steel rod,

$$(\delta l)_s = \frac{P \times (450/1000)}{36\pi \times 10^{-6} \times 210 \times 10^9}$$

$$= \frac{5 \times 10^3 \times (1050 \times 1000)}{36\pi \times 10^{-6} \times (210 \times 10^9)} = 0.221 \times 10^{-3} \text{m} \text{ or } 0.221 \text{ mm}$$

The displacement of point
$$A = 0.2907 + 0.221 = \textbf{0.5117 mm (Ans.)}$$

(ii) The stresses in steel and brass :

$$\sigma_s = \frac{5 \times 10^3}{36\pi \times 10^{-6}} \times 10^{-6} \text{ MN/m}^2 = \textbf{44.21 MN/m}^2 \text{ (tensile)}$$

$$\sigma_b = \frac{4 \times (5 \times 10^3)}{125\pi \times 10^{-6}} \times 10^{-6} \text{ MN/m}^2 = \textbf{50.3 MN/m}^2 \text{ (comp).}$$

Example 22.5. *The steel bolt shown in Fig. 22.6 has a thread pitch of 1.6 mm. If the nut is initially tightened up by hand so as to cause no stress in the copper spacing tube, calculate the stresses induced in the tube and in the bolt if a spanner is then used to turn the nut through 90°. Take E_c and E_s as 100 GPa and 209 GPa respectively.* **(IAS)**

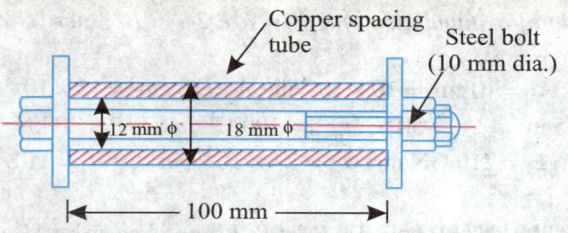

Fig. 22.6

Solution. *Given :* $\quad p = 1.6$ mm, $E_c = 100$ GP_a; $E_s = 209$ GPa.

Stresses induced in the tube and the bolt, σ_c, σ_s :

$$A_s = \frac{\pi}{4} \times \left(\frac{10}{1000}\right)^2 = 7.854 \times 10^{-5} \text{m}^2$$

$$A_c = \frac{\pi}{4} \times \left[\left(\frac{18}{1000}\right)^2 - \left(\frac{12}{1000}\right)^2\right] = 14.14 \times 10^{-5} \text{m}^2$$

Tensile force on steel bolt, P_s = Compressive force in copper tube, $P_c = P$

Also, increase in length of bolt + decrease length of tube = Axial displacement of nut

i.e., $\quad (\delta l)_s + (\delta l)_c = 1.6 \times \dfrac{90}{360} = 0.4 \text{ mm} = 0.4 \times 10^{-3} \text{ m}$

or, $\quad \dfrac{Pl}{A_s E_s} + \dfrac{Pl}{A_c E_c} = 0.4 \times 10^{-3} \quad (\because l_s = l_c = l)$

or, $\quad Pl \left[\dfrac{1}{A_s E_s} + \dfrac{1}{A_c E_c}\right] = 0.4 \times 10^{-3}$

or, $\quad P \times \left(\dfrac{100}{1000}\right)\left[\dfrac{1}{7.854 \times 10^{-5} \times 209 \times 10^9} + \dfrac{1}{14.14 \times 10^{-5} \times 100 \times 10^9}\right] = 0.4 \times 10^{-3}$

or, $\quad 0.1 \, P(6.092 \times 10^{-8} + 7.072 \times 10^{-8}) = 0.4 \times 10^{-3}$

or, $\quad P = \dfrac{0.4 \times 10^{-3}}{0.1(6.092 \times 10^{-8} + 7.072 \times 10^{-8})} = 30386 \text{ N}$

$\therefore \quad \sigma_s = \dfrac{P}{A_s} = \dfrac{30386}{7.854 \times 10^{-5}} \times 10^{-6} \text{ MPa} = \textbf{386.88 MPa (Ans.)}$

and, $\quad \sigma_c = \dfrac{P}{A_c} = \dfrac{30386}{14.14 \times 10^{-5}} \times 10^{-6} \text{ MPa} = \textbf{214.89 MPa (Ans.)}$

Example 22.6. *A composite bar made up of a aluminium bar and steel bar is firmly held between two unyielding supports as shown in Fig. 22.7.*

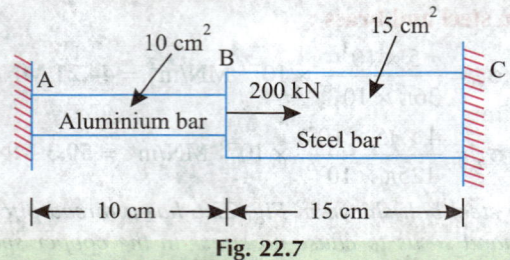

Fig. 22.7

An axial load of 20 kN is applied at B at 50° C. Find the stresses in each material when the temperature is 100° C. Take E for aluminium and steel as 70 GN/m² and 210 GN/m² respectively. Coefficients of expansion for aluminium and steel are 24 × 10⁻⁶ per °C and 11.8 × 10⁻⁶ per °C respectively.

(Panjab University)

Solution. *Given*: $A_{al} = 10 \text{ cm}^2 = 10 \times 10^{-4} \text{ m}^2$; $A_s = 15 \text{ cm}^2 = 15 \times 10^{-4} \text{ m}^2$; $l_{al} = 10 \text{ cm} = 0.1 \text{ m}$; $l_s = 15 \text{ cm} = 0.15 \text{ m}$; $t_{initial} = 50 \,°C$, $t_{final} = 100°C$; Load, $P = 20 \text{ kN}$
$E_{al} = 70 \text{ GN/m}^2$; $E_s = 210 \text{ GN/m}^2$; $\alpha_{al} = 24 \times 10^{-6}$ per °C, $\alpha_s = 11.8 \times 10^{-6}$ per °C.

Stresses σ_{al}; σ_s :

Out of the load of 200 kN (P) applied at B, let P_{al} kN be taken up by AB and $(20 - P_{AB})$ kN by BC.

Since the supports are rigid,

Elongation of AB = Contraction of BC.

$$\frac{P_{al} \times l_{al}}{A_{al} E_{al}} = \frac{P_s l_s}{A_s E_s} = \frac{(P - P_{al}) l_s}{A_s E_s}$$

$$\frac{(P_{al} \times 10^3) \times 0.1}{10 \times 10^{-4} \times 70 \times 10^9} = \frac{(200 - P_{al}) \times 10^3 \times 0.15}{15 \times 10^{-4} \times 210 \times 10^9}$$

$$= \frac{P_{al}}{(200 - P_{al})} = \frac{10^3 \times 0.15 \times 10 \times 10^{-4} \times 70 \times 10^9}{10^3 \times 0.1 \times 15 \times 10^{-4} \times 210 \times 10^9} = 0.333$$

$$P_{al} = 66.67 - 0.333 \, P_{al}$$

∴ $$P_{al} = 50 \text{ kN}$$

Stress in aluminium, $(\sigma_{al})_1 = \dfrac{P_{al}}{A_{al}} = \dfrac{50 \times 10^3}{10 \times 10^{-4}} = 50 \text{ MN/m}^2$ *(tensile)*

Stress in steel, $(\sigma_s)_1 = \dfrac{P_s}{A_s} = \dfrac{P - P_{al}}{A_s}$

$$= \frac{(20 - 5) \times 10^3}{15 \times 10^{-4}} = 100 \text{ MN/m}^2 \ (compressive)$$

These are the stresses in the two materials (aluminium and steel) at 50° C.

Now, let the temperature be raised to 100°C. In order to determine the stresses due to rise of temperature, assume that the support at C is removed and expansion is allowed free.

Rise of temperature $= t_{final} - t_{initial} = 100 - 50 = 50° \text{ C.}$

Expansion $AB = l_{al} \cdot \sigma_{al} \cdot t_{al} = 0.1 \times 24 \times 10^{-6} \times 50 = 120 \times 10^{-6} \text{ m}$ ...(*i*)

Expansion of $BC = l_s \cdot \alpha_s \cdot t_s = 0.15 \times 11.8 \times 10^{-6} \times 50 = 88.5 \times 10^{-6} \text{ m.}$

Let a load be applied at C which causes a total contraction equal to the total expansion and let C be attached to rigid supports.

If this load causes stress σ N/m² in BC, its value must be $15 \times 10^{-4} \sigma$ and hence stress in AB must be

$$\frac{15 \times 10^{-4} \sigma}{10 \times 14^{-4}} = 1.5\sigma \text{ N/m}^2$$

Total contraction caused by the load

$$= \frac{\sigma \times 0.15}{210 \times 10^9} + \frac{150 \times 0.1}{70 \times 10^9} \qquad \qquad ...(ii)$$

From eqns. (i) and (ii), we have

$$\frac{\sigma \times 0.15}{210 \times 10^9} + \frac{1.5\sigma \times 0.1}{70 \times 10^9} = 120 \times 10^{-6} + 88.5 \times 10^{-6} = 208.5 \times 10^{-6}$$

or, $$\frac{0.15\sigma}{210} + \frac{0.15\sigma}{70} = 10^9 \times 208.5 \times 10^{-6} = 208500$$

$$0.6\,\sigma = 210 \times 208500 = 43785000$$

∴ $$\sigma = 72.97 \times 10^6 \text{ N/m}^2 \text{ or } 72.97 \text{ MN/m}^2 \text{ (compressive)}$$

∴ At 100°C,

Stress in aluminium, $\sigma_{al} = -(\sigma_{al})_1 + 1.5 \times 72.97$

$$= -50 + 1.5 \times 72.97 = \textbf{59.45 MN/m}^2 \textbf{ (comp.) (Ans.)}$$

Stress in steel, $\sigma_s = (\sigma_s)_1 + 72.97$

$$= 100 + 72.97 = \textbf{172.97 MN/m}^2 \textbf{ (comp.)(Ans.)}$$

Example 22.7. *A composite bar shown in Fig. 22.8 is subjected to a load of 100 kN. Calculate the forces in both materials if the assembly is heated through 40°C. $E_{steel} = 210$ GPa, $\alpha_{steel} = 12 \times 10^{-6}/°C$, $E_{copper} = 105$ GPa, and $\alpha_{copper} = 18 \times 10^{-6}/°C$.*

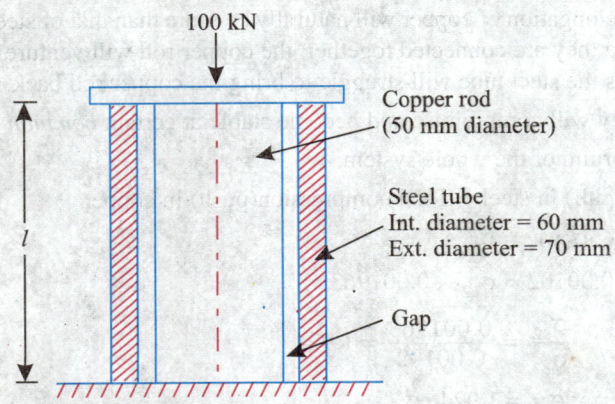

100 kN

Copper rod
(50 mm diameter)

Steel tube
Int. diameter = 60 mm
Ext. diameter = 70 mm

Gap

l

Fig. 22.8

Solution. *Given* : $P = 100$ kN, $D_s = 70$ mm $= 0.07$ m; $d_s = 60$ mm $= 0.06$ m; $D_c = 50$ mm $= 0.05$ m; $E_s = 210$ GPa $= 210 \times 10^9$ N/m²; $E_c = 105$ GPa$= 105 \times 10^9$ N/m²; $t = 40°C$ $\alpha_s = 12 \times 10^{-6}/°C$; $\alpha_c = 18 \times 10^{-6}/°C$

Forces in copper and steel; P_c and P_s :

Cross-sectional area of steel, $A_s = \frac{\pi}{4}(0.07^2 - 0.06^2) = 0.00102 \text{ m}^2$

Cross-sectional area of copper, $A_c = \frac{\pi}{4} \times 0.05^2 = 0.001963 \text{ m}^2$

Let us analyse this problem in two parts:

I. Considering first the total applied load 100 kN on the composite bar, ignoring the rise of temperature.

II. Considering the composite bar, subjected to rise in temperature only, ignoring the stresses caused due to the 100 kN load.

Case I. Stresses due to load 100 kN:

Let σ_{s1} and σ_{c1} be the stresses produced (in N/m²) in steel and copper respectively.

Then, $\qquad P = P_{s1} + P_{c1}$

$$100 \times 10^3 = \sigma_{s1} \times A_s + \sigma_{c1} \times A_c$$

or, $\qquad 100000 = 0.00102\,\sigma_s + 0.01963\,\sigma_c$ ...(i)

Also, Strain in steel = Strain in copper

$$\frac{\sigma_{s1}}{E_s} = \frac{\sigma_{c1}}{E_c} \quad \text{or} \quad \frac{\sigma_{s1}}{\sigma_{c1}} = \frac{E_s}{E_c} = \frac{210 \times 10^9}{105 \times 10^9} = 2$$

or, $\qquad \sigma_{s1} = 2\sigma_{c1}$ ...(ii)

Substituting for σ_s in eqn. (i), we get

$$100000 = 0.00102 \times 2\sigma'_c + 0.001963\sigma'_c$$

or, $\qquad \sigma_{c1} = \dfrac{100000}{(0.00102 \times 2 + 0.001963)} \times 10^{-6} \text{ MN/m}^2 = 24.98 \text{ MN/m}^2 \text{ (compressive)}$

and, $\qquad \sigma_{s1} = 2 \times 24.98 = 49.96 \text{ MN/m}^2 \text{ (compressive)}$

Case II. Stresses due to rise in temperature, σ_{s2}, σ_{c2} :

Let the stresses due to rise in temperature in steel and copper be σ_{s2} and σ_{c2} respectively.

Since $\sigma_c > \sigma_s$, elongation of copper will naturally be more than that of steel for the same rise of temperature, but since they are connected together, the copper rod will venture to pull the steel tube along with it, whereas the steel tube will struggle to bring the copper rod back.

Ultimately, they will compromise and become stable at certain *common position*.

For the equilibrium of the whole system,

Total tension (pull) in steel = Total compression(push) in copper

$$\sigma_{s2}A_s = \sigma_{c2}A_c$$

$$\sigma_{s2} \times 0.00102 = \sigma_{c2} \times 0.001963$$

or, $\qquad \dfrac{\sigma_{s2}}{\sigma_{c2}} = \dfrac{0.001963}{0.00102} = 1.924$

or, $\qquad \sigma_{s2} = 1.924\ \sigma_{c2}$ ...(iii)

Also we know that,

Final increase in length of steel = Final decrease in length of copper

$$\alpha_s t l_s + \frac{\sigma_{s2}}{E_s} l_s = \alpha_c t l_c - \frac{\sigma_{c2}}{E} l_c \qquad \text{(where, } t = \text{rise in temp.} = 40°\text{C ...Given)}$$

Dividing both sides by l, we get

$$\alpha_s t + \frac{\sigma_{s2}}{E_s} = \alpha_c t - \frac{\sigma_{c2}}{E_c} \qquad (\because l_s = l_c = l)$$

or,

$$\frac{\sigma_{c2}}{E_c} + \frac{\sigma_{s2}}{E_s} = \alpha_c t - \alpha_s t = t(\alpha_c - \alpha_s)$$

or,

$$\frac{\sigma_{c2}}{105 \times 10^9} + \frac{\sigma_{s2}}{210 \times 10^9} = 40(18 \times 10^{-6} - 12 \times 10^{-6}) = 240 \times 10^{-6}$$

or,

$$\frac{\sigma_{c2}}{105 \times 10^9} + \frac{1.924 \sigma_{c2}}{210 \times 10^9} = 240 \times 10^6$$

or,

$$3.924 \, \sigma_{c2} = 210 \times 10^9 \times 240 \times 10^{-6}$$

∴

$$\sigma_{c2} = 12.84 \times 10^6 \text{ N/m}^2 \text{ or } 12.84 \text{ MN/m}^2 \text{ (compressive)}$$

and,

$$\sigma_{s2} = 1.924 \times 12.84 = 24.70 \text{ MN/m}^2 \text{ (tensile)}$$

Now, combining case I and case II, we can write down the values of final stresses :

Stress in copper, $\quad \sigma_c = \sigma_{c1} + \sigma_{c2} = 24.98 + 12.84 = 37.82$ MN/m² (Compressive)

Stress in steel, $\quad \sigma_s = \sigma_{s1} + \sigma_{s2} = 49.96 + (-24.70) = 25.26$ MN/m² (Compressive)

Force shared by copper rod,

$$P_c = \sigma_c \times A_c = 37.824 \times 0.001963 = 0.07424 \text{ MN} = \textbf{74.24 kN (Ans.)}$$

Force shared by steel tube,

$$P_s = P - P_c = 100 - 74.24 = \textbf{25.76 kN (Ans.)}$$

Example 22.8. *A component used in the Mars pathfinder can be idealized as a circular bar clamped at its ends. The bar should withstand a torque of 1000 Nm. The component is assembled on earth when the temperature is 30°C. Temperature on Mars at the site of landing is – 70°C. The material of the bar has an allowable shear stress of 300 MPa and its young's modulus is 200 GPa. Design the diameter of the bar taking a factor of safety of 1.5 and assuming a coefficient of thermal expansion for the material of the bar as $12 \times 10^{-6}/°C$.* **(GATE)**

Solution. *Given :* $T_{max} = 1000$ Nm; $t_E = 30°C$; $t_M = -70°C$; $\tau_{allwable} = 300$ MP_a; $E = 200$ *GPa*; F.O.S. $= 1.5$; $\alpha = 12 \times 10^{-6}/°C$

Diameter of the bar, D :

Change in length, $\quad \delta l = l \propto \Delta t$, where, l = original length, m.

Change in length at Mars $= l \times 12 \times 10^{-6} \times [30 - (-70)] = 12 \times 10^{-4} l$ metres

$$\text{Linear strain} = \frac{\text{Change in length}}{\text{Original length}} = \frac{12 \times 10^{-4} l}{l} = 12 \times 10^{-4}$$

$$\sigma_a = \text{Axial stress} = E \times \text{Linear strain}$$

$$= 200 \times 10^9 \times 12 \times 10^{-4} = 2.4 \times 10^8 \text{ N/m}^2$$

From maximum shear stress equation, we have

$$\tau_{max} = \left[\left\{ \frac{K_s \times 16T}{\pi D^3} \right\}^2 + \left(\frac{\sigma_a}{2} \right)^2 \right]$$

where, $\qquad \tau_{max} = \dfrac{\tau_{allowable}}{F.O.S.} = \dfrac{300}{1.5} = 200$ MPa

$$K_s = 1, \text{ for steady loads.}$$

Substituting the values, we get

$$200 \times 10^6 = \left[\left(\frac{16 \times 1000}{\pi \times D^3} \right)^2 + \left(\frac{2.4 \times 10^8}{2} \right)^2 \right]^{1/2}$$

Squaring both sides, we get

$$4 \times 10^{16} = \left(\frac{16 \times 1000}{\pi D^3} \right)^2 + (1.2 \times 10^8)^2$$

or, $$\left(\frac{16 \times 1000}{\pi D^3} \right)^2 = 4 \times 10^{16} - 1.44 \times 10^{16} = 2.56 \times 10^{16}$$

or, $$\frac{16 \times 1000}{\pi D^3} = 1.6 \times 10^8$$

or, $$D = \left(\frac{16 \times 1000}{\pi \times 1.6 \times 10^8} \right)^{1/3} = 0.03169 \text{ m or } \mathbf{31.69 \text{ mm} \text{ (Ans.)}}$$

PRINCIPAL STRESSES AND STRAINS

Example 22.9. *Draw Mohr's circle for a two-dimensional stress field subjected to (a) pure shear (b) pure biaxial tension, (c) pure uniaxial tension, and (d) pure uniaxial compression.* **(IAS)**

Solution. Moh'r circles for two-dimensional stress field subjected to pure shear, pure biaxial tension, pure uniaxial compression and pure uniaxial tension are shown in Fig. 22.9 (*a*) to (*d*).

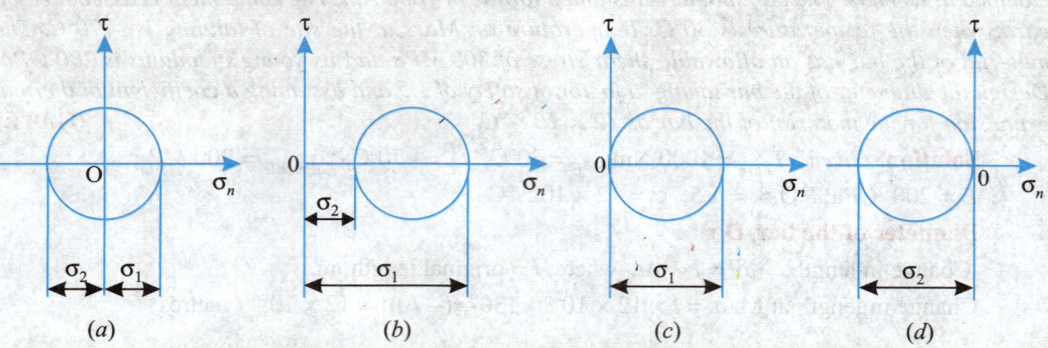

Fig. 22.9

Example 22.10. *A thin cylinder of 100 mm internal diameter and 5 mm thickness is subjected to an internal pressure of 10 MPa and a torque of 2000 Nm. Calculate the magnitudes of the principal stresses.* **(GATE)**

Solution. Given : $d = 100$ mm $= 0.1$ m; $t = 5$ mm $= 0.005$ m; $D = d + 2t = 0.1 + 2 \times 0.005 = 0.11$ m; $p = 10$ MPa $= 10 \times 10^6$ N/m²; $T = 2000$ Nm.

Principal stresses, σ_1, σ_2 :

Longitudinal stress, $$\sigma_l = \sigma_x = \frac{pd}{4t} = \frac{10 \times 10^6 \times 0.1}{4 \times 0.005} = 50 \times 10^6 \text{ N/m}^2 = 50 \text{ MN/m}^2$$

Circumferential stress, $\sigma_c = \sigma_y = \dfrac{pd}{2t} = \dfrac{10 \times 10^6 \times 0.1}{4 \times 0.005} = 100 \text{ NM/m}^2$

To find the stress, using the relation,

$$\frac{T}{I_p} = \frac{\tau}{R}, \text{ we have}$$

$$\tau = \tau_{xy} = \frac{TR}{I_p} = \frac{T \times R}{\dfrac{\pi}{32}(D^4 - d^4)}$$

$$= \frac{2000 \times (0.05 + 0.005)}{\dfrac{\pi}{32}(0.11^4 - 0.1^4)} \times 10^{-6} = 24.14 \text{ MN/m}^2$$

Principal stresses are calculated as follows :

$$\sigma_1, \sigma_2 = \frac{\sigma_x + \sigma_y}{2} \pm \sqrt{\left(\frac{\sigma_x - \sigma_y}{2}\right)^2 + (\tau_{xy})^2}$$

$$= \frac{50 + 100}{2} \pm \sqrt{\left(\frac{50 - 100}{2}\right)^2 + (24.14)^2}$$

$$= 75 \pm 34.75 = 109.75 \text{ and } 40.25 \text{ MN/m}^2$$

Hence, σ_1 (*Major principal stress*) = **109.75 MN/m²** (**Ans.**)

σ_2 (*Minor principal stress*) = **40.25 MN/m²** (**Ans.**)

Example 22.11. *A solid shaft of diameter 30 mm is fixed at one end. It is subjected to a tensile force of 10 kN and a torque of 60 Nm. At a point on the surface of the shaft, determine the principle stresses and the maximum shear stress.* (**UPSC**)

Fig. 22.10

Solution. *Given :* $D = 30 \text{ mm} = 0.03 \text{ m}; \ P = 10 \text{ kN}; \ T = 60 \text{ Nm}$

Principal stresses (σ_1, σ_2) and maximum shear stress (τ_{max}) :

Tensile stress, $\sigma_t = \sigma_x = \dfrac{10 \times 10^3}{\dfrac{\pi}{4} \times 0.03^2} = 14.15 \times 10^6 \text{ N/m}^2 \text{ or } 14.15 \text{ MN/m}^2$

As per torsion equation, $\dfrac{T}{I_p} = \dfrac{\tau}{R}$

∴ Shear stress, $\tau = \dfrac{TR}{I_p} = \dfrac{TR}{\dfrac{\pi}{32}D^4} = \dfrac{60 \times 0.015}{\dfrac{\pi}{32} \times (0.03)^4}$

$$= 11.32 \times 10^6 \text{ N/M}^2 \text{ or } 11.32 \text{ MN/m}^2$$

The principal stresses are calculated by using the relations:

$$\sigma_1, \sigma_2 = \left(\frac{\sigma_x + \sigma_y}{2}\right) \pm \sqrt{\left(\frac{\sigma_x - \sigma_y}{2}\right)^2 + \tau_{xy}^2}$$

Here, $\sigma_x = 14.15 \text{ MN/m}^2, \ \sigma_y = 0; \ \tau_{xy} = \tau = 11.32 \text{ MN/m}^2$

∴ $\sigma_1, \sigma_2 = \dfrac{14.15}{2} \pm \sqrt{\left(\dfrac{14.15}{2}\right)^2 + (11.32)^2}$

$$= 7.075 \pm 13.35 = 20.425 \text{ MN/m}^2, -6.275 \text{ MN/m}^2$$

Hence, major principal stress,

$$\sigma_1 = \mathbf{20.425 \ MN/m^2} \ (tensile) \ \textbf{(Ans.)}$$

Minor principal stress, $\sigma_2 = \mathbf{6.275 \ MN/m^2} \ (compressive) \ \textbf{(Ans.)}$

Maximum sheat stress, $\tau_{max} = \dfrac{\sigma_1 - \sigma_2}{2} = \dfrac{20.425 - (-6.275)}{2} = \mathbf{13.35 \ MN/m^2} \ \textbf{(Ans.)}$

Example 22.12. *A thin cylinder with closed ends has an internal diameter of 50 mm and a wall thickness of 2.5 mm. It is subjected to an axial pull of 10 kN and a torque of 500 Nm while under an internal pressure of 6 MN/m².*

(*i*) *Determine the principal stresses in the tube and the maximum shear stress.*

(*ii*) *Represent the stress configuration on a square element taken in the load direction with direction and magnitude indicated (schematic).*

(UPSC)

Solution. *Given* : $d = 50 \text{ mm} = 0.05 \text{ m}; \ D = d + 2t = 50 + 2 \times 2.5 = 55 \text{ mm} = 0.055 \text{ m};$
Axial pull, $P = 10 \text{ kN}; \ T = 500 \text{ Nm}; \ p = 6 \text{ MN/m}^2.$

(*i*) Principal stresses (σ_1, σ_2) in the tube and the maximum shear stress (τ_{max}) :

$$\sigma_x = \frac{pd}{4t} + \frac{P}{\pi dt} = \frac{6 \times 10^6 \times 0.05}{4 \times 2.5 \times 10^{-3}} + \frac{10 \times 10^3}{\pi \times 0.05 \times 2.5 \times 10^{-3}}$$

$$= 30 \times 10^6 + 25.5 \times 10^6 = 55.5 \times 10^6 \text{ N/m}^2$$

$$\sigma_y = \frac{pd}{2t} = \frac{6 \times 10^6 \times 0.05}{2 \times 2.5 \times 10^{-3}} = 60 \times 10^6$$

Principal stresses are given by the realations :

$$\sigma_1, \sigma_2 = \left(\frac{\sigma_x + \sigma_y}{2}\right) \pm \sqrt{\left(\frac{\sigma_x - \sigma_y}{2}\right)^2 + \tau_{xy}^2} \qquad \qquad \dots(1)$$

We know that, $\dfrac{T}{I_p} = \dfrac{\tau}{R}$

where, $I_p = \dfrac{\pi}{32}(D^4 - d^4) = \dfrac{\pi}{32}\left[(0.055)^4 - (0.05)^4\right] = 2.848 \times 10^{-7} \text{ m}^4$

$(I_p = \text{polar moment of inertia})$

Substituting the values in (i), we get

$$\frac{500}{2.848 \times 10^{-7}} = \frac{\tau}{(0.055/2)}$$

or,

$$\tau = \frac{500 \times (0.055/2)}{2.848 \times 10^{-7}} = 48.28 \times 10^6 \text{ N/m}^2$$

Now, substituting the various values in eqn. (i), we have

$$\sigma_1, \sigma_2 = \left(\frac{55.5 \times 10^6 + 60 \times 10^6}{2}\right)$$

$$\pm \sqrt{\left(\frac{55.5 \times 10^6 - 60 \times 10^6}{2}\right)^2 + (48.28 \times 10^6)^2}$$

$$= \frac{(55.5 + 60) \times 10^6}{2} \pm \sqrt{4.84 \times 10^{12} + 2330.96 \times 10^{12}}$$

$$= 57.75 \times 10^6 \pm 48.33 \times 10^6 = 106.08 \text{ MN/m}^2, 9.42 \text{ MN/m}^2$$

Hence, principal stresses are:

$$\sigma_1 = \textbf{106.08 MN/m}^2; \quad \sigma_2 = \textbf{9.42 MN/m}^2 \textbf{ (Ans.)}$$

Maximum shear stress, $\quad \tau_{max} = \dfrac{\sigma_1 - \sigma_2}{2} = \dfrac{106.08 - 9.42}{2} = \textbf{48.33 MN/m}^2 \textbf{ (Ans.)}$

(ii) Stress configuration on a square element :

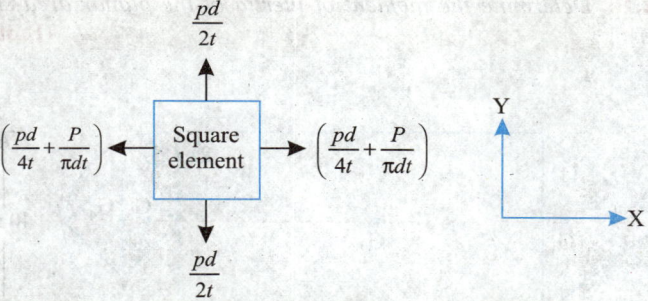

Fig. 22.11

Example 22.13. *At a point in a stressed body the state of stress on two planes 45° apart is as shown in Fig 22.12. Determine the two principal stresses.* **(GATE)**

Solution. Refer to Fig. 22.12

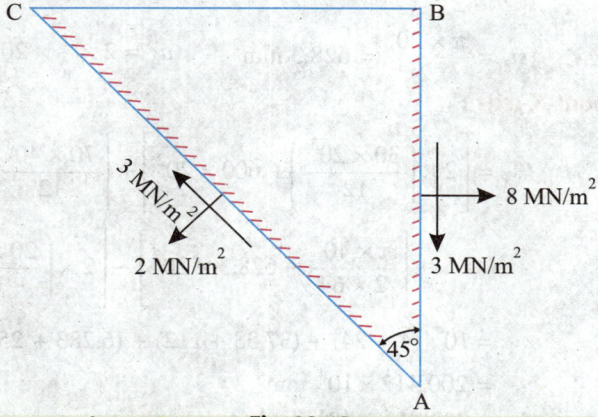

Fig. 22.12

Principal stresses σ_1, σ_2 :

Resolving the forces vertically, we have

$$\sigma_y \times BC + 3 \times AC \times \cos 45° = 3 \times AB + 2 \times AC \times \cos 45°$$

Dividing both sides by BC, we get

$$\sigma_y + 3 \times \frac{AC}{BC} \cos 45° = 3 \times \frac{AB}{BC} + 2 \times \frac{AC}{BC} \times \cos 45°$$

$$\sigma_y + 3 \times \frac{\cos 45°}{\sin 45°} = 3 \times \cot 45° + 2 \times \frac{\cos 45°}{\sin 45°}$$

$$\sigma_y + 3 = 3 + 2 \quad \text{or} \quad \sigma_y = 2\,MN/m^2$$

Now, $\qquad \sigma_x = 8\,MN/m^2; \quad \sigma_y = 2\,MN/m^2; \quad \tau_{xy} = 3\,MN/m^2$

$$\sigma_1, \sigma_2 = \left(\frac{\sigma_x + \sigma_y}{2}\right) \pm \sqrt{\left(\frac{\sigma_x - \sigma_y}{2}\right)^2 + \tau_{xy}^2}$$

$$= \left(\frac{8+2}{2}\right) \pm \sqrt{\left(\frac{8-2}{2}\right)^2 + 3^2} = 5 \pm 4.24$$

i.e., $\qquad \sigma_1 = \textbf{9.24 MN/m}^2; \quad \sigma_2 = \textbf{0.76 MN/m}^2 \textbf{ (Ans.)}$

MOMENT OF INERTIA

Example 22.14. *Determine the moment of inertia of the planar area shown in Fig. 22.13 above X and Y axes.*

(Bangalore University)

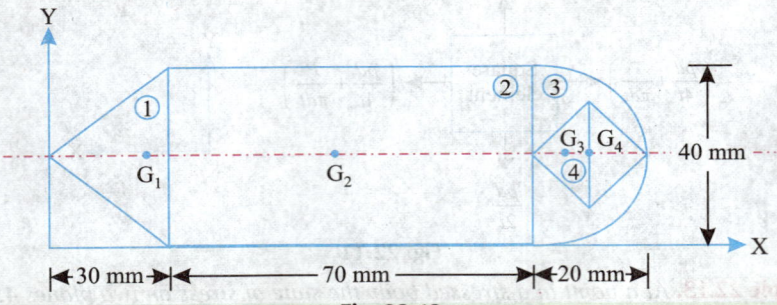

Fig. 22.13

Solution. $\qquad a_1 = \frac{1}{2} \times 40 \times 30 = 600\ mm^2 \qquad a_2 = 70 \times 40 = 2800\ mm^2$

$$a_3 = \frac{\pi \times 20^2}{2} = 628.3\ mm^2 \qquad a_4 = 2 \times \frac{1}{2} \times 20 \times 10 = 200\ mm^2$$

(*i*) M.O.I. about X-axis :

$$I_{XX} = \left[2 \times \left(\frac{30 \times 20^3}{12}\right) + 600 \times 20^2\right] + \left[\frac{70 \times 40^3}{12} + 2800 \times 20^2\right]$$

$$+ \left[\frac{\pi \times 40^4}{2 \times 64} + 628.3 \times 20^2\right] - \left[2 \times \left(\frac{20 \times 10^3}{12}\right) + 200 \times 20^2\right]$$

$$= 10^4\,[(4 + 24) + (37.33 + 112) + (6.283 + 25.132) - (0.383 + 8)]$$

$$= 200.412 \times 10^4\ mm^4$$

Hence, $\qquad I_{XX} = \textbf{200.412} \times \textbf{10}^4\ \textbf{mm}^4 \textbf{ (Ans.)}$

(ii) **M.O.I. about Y-axis:**

$$I_{YY} = \left[\frac{40 \times 30^3}{36} + 600 \times \left(\frac{2}{3} \times 30\right)^2\right] + \left[\frac{40 \times 70^3}{12} + 2800 \times \left(30 + \frac{70}{2}\right)^2\right]$$

$$+ \left[0.11 \times 20^4 + 628.3 \times \left(30 + 70 + \frac{4 \times 20}{3\pi}\right)^2\right] - \left[2 \times \left(\frac{20 \times 10^3}{12}\right) + 200 \times (30 + 70 + 10)^2\right]$$

$$= 10^4[(3 + 24)] + [(114.33 + 1183)] + [(1.76 + 739.49)] - [0.333 + 242]$$

$$= 1823.25 \times 10^4 \text{ mm}^2$$

Hence, $I_{YY} = \mathbf{1823.25 \times 10^4 \text{ mm}^2}$ **(Ans.)**

Example 22.15. *Find the centroid and the moment of inertia of the planar area shown in Fig. 22.14, about its centroidal axis.* **(Roorkee University)**

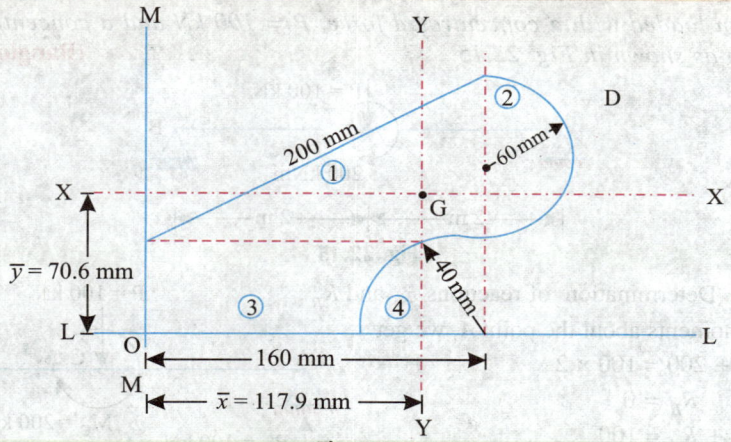

Fig. 22.14

Solution. To determine the location of centroid of the area we have the following table :

Components	Area a (mm²)	Centroidal distance 'x' from MM (mm)	Centroidal distance 'y' from LL (mm)	ax (mm³)	ay (mm³)
Triangle (1)	$\frac{1}{2} \times 160 \times 120$ = 9600	$\frac{2}{3} \times 160 = 106.67$	$40 + \frac{1}{3} \times 120 = 80$	1024032	768000
Semicircle (2)	$\frac{\pi \times 60^2}{2}$ = 5655	$160 + \frac{4 \times 60}{3\pi}$ = 185.5	$40 + 60 = 100$	1049002	565500
Rectangle (3)	160×40 = 6400	$\frac{160}{2} = 80$	$\frac{40}{2} = 20$	512000	128000
Quadrant (4)	$\frac{\pi \times 40^2}{4}$ = 1256.6(–)	$160 - \frac{4 \times 40}{3\pi}$ = 143	$\frac{4 \times 40}{3\pi} = 16.97$	179694 (–)	21325 (–)

$$\Sigma a = 20398.4 \qquad \Sigma ax = 2405340 \qquad \Sigma ay = 1440175$$

$$\bar{x} = \frac{\Sigma ax}{\Sigma a} = \frac{2405340}{20398.4} \approx 117.9 \text{ mm} \quad \bar{y} = \frac{\Sigma ay}{\Sigma a} = \frac{1440175}{20398.4} = 70.6 \text{ mm}$$

M.O.I. about XX-axis :

$$I_{XX} = I_{XX_1} + I_{XX_2} + I_{XX_3} - I_{XX_4}$$

$$= \left[\frac{160 \times 120^3}{36} + 9600 \times (80 - 70.6)^2 \right] + \left[\frac{\pi \times 120^4}{2 \times 64} + 5655 \times (100 - 70.6)^2 \right]$$

$$+ \left[\frac{160 \times 40^3}{12} + 6400 \times (70.6 - 20)^2 \right] - [0.055 \times 40^4 + 1256.6(70.6 - 16.97)^3]$$

$$= 10^6 [(7.68 + 0.848) + (5.089 + 4.868) + (0.853 + 16.386) - (0.141 + 3.614)]$$

$$= 31.969 \times 10^6 \text{ mm}^4$$

Hence, $I_{XX} = \mathbf{31.969 \times 10^6 \text{ mm}^2}$ **(Ans.)**

BENDING MOMENTS AND SHEAR FORCES

Example 22.16. *Draw the shear force and the bending moment diagrams for the simply supported beam loaded with a concentrated force, P = 100 kN and a concentrated moment, M_C = 200 kNm as shown in Fig. 22.15.* **(Banglore University)**

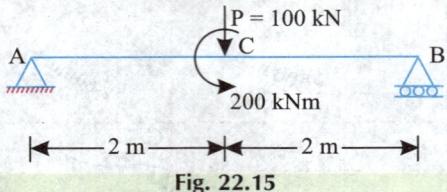

Fig. 22.15

Solution. Determination of reactions R_A and R_B :

Taking moments about the point A, we get

$R_B \times 4 + 200 = 100 \times 2$

∴ $R_B = 0$

Also, $R_A + R_B = 100$

∴ $R_A = 100$ kN

S.F. calculations :

     $S_{B-C} = 0$

     $S_{C-A} = -100$ kN

      $S_A = -100$ kN

S.F. diagram is shown in Fig. 22.15 (*b*)

B.M. calculations :

     $M_B = 0$

     $M_C = +200$ kNm

     $M_A = +200 - 100 \times 2 = 0$

B.M. diagram is shown in Fig. 22.16. (*c*).

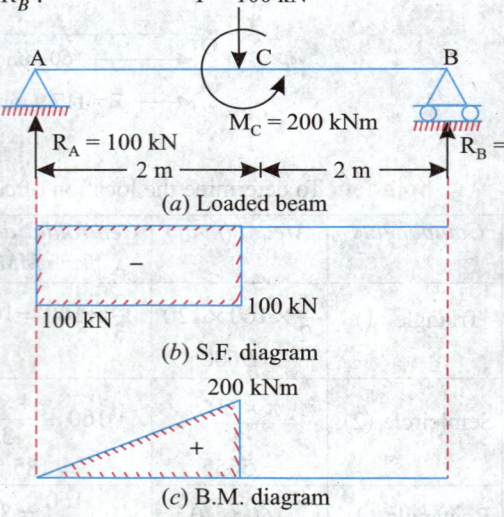

(*a*) Loaded beam

(*b*) S.F. diagram

(*c*) B.M. diagram

Fig. 22.16

Example 22.17. *A bracket CDE is welded to a beam AB of span 10 m at C. The beam is hinged at A and supported on rollers at B. Vertical and horizontal loads are 10 kN each act at E (Fig. 22.17).*

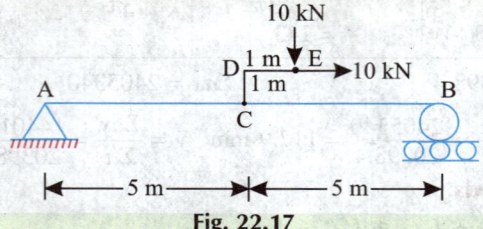

Fig. 22.17

Draw (i) bending moment, (ii) shear force, and (iii) axial force diagrams for the beam indicating values at salient points.

Solution. Fig. 22.18 shows a beam hinged at A and supported on roller at B. Vertical and horizontal loads 10 kN each act at E, through a welded bracket CDE.

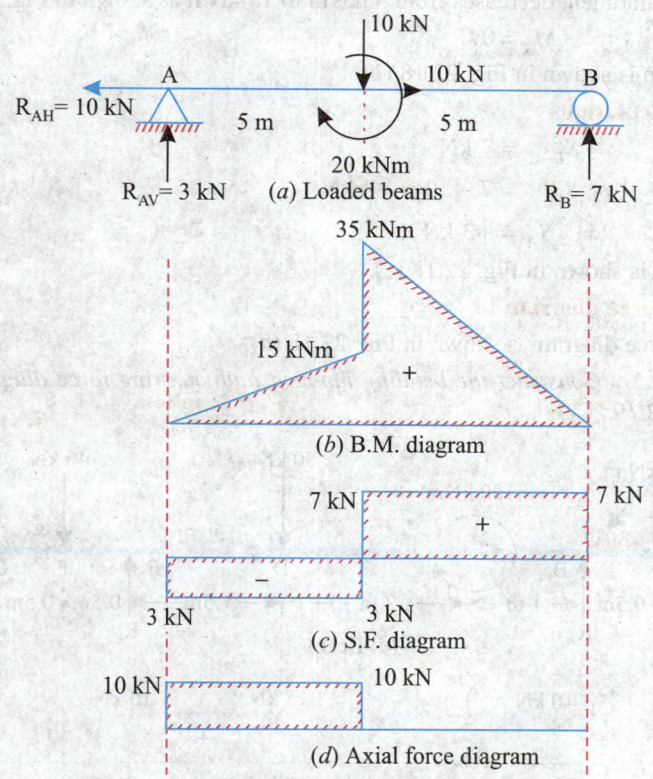

Fig. 22.18

Vertical load of 10 kN acting at E is equivalent to a clockwise moment of the value $(10 \times 1) =$ 10 kNm acting at C together with a vertical downward load of 10 kN at C.

Horizontal load 10 kN acting at E is equal to a clockwise moment of the value of $(10 \times 1) =$ 10 kNm at C together with an axial load of 10 kN at C.

Fig. 22.18 (a) shows the beam loaded with a vertical load of 10 kN acting at C together with total clockwise moment $(10 + 10) = 20$ kNm at C. Also horizontal load (axial load) of 10 kN is shown acting at C.

Calculation of reactions at A and B :

Let, R_{AH} = Horizontal component of the reaction at the hinged support A,

           R_{AV} = Vertical component of the reaction at the hinged support A, and

           R_B = Reaction at the roller support B.

Obviously, $R_{AH} = 10$ kN $(\leftarrow)$

In order to determine R_B, apply $\Sigma M_A = 0$.

$$R_B \times 10 = 10 \times 5 + 20 = 70$$

$\therefore$ $R_B = 7$ kN $(\uparrow)$

Also, $R_{AV} + R_B = 10$

$\therefore$ $R_{AV} = 10 - R_B = 10 - 7 = 3$ kN $(\uparrow)$

(i) B.M. calculations :

$$M_B = 0$$
$$M_C = 7 \times 5 - 20 = 15 \text{ kNm}$$

The bending moment decreases from 35kNm to 15 kNm as shown in Fig. 22.18 (b)]

$$M_A = 0$$

B.M. diagram is shown in Fig. 22.18 (b).

(ii) S.F. calculations :

$$S_{B-C} = 7 \text{ kN}$$
$$S_{C-A} = 7 - 10 = -3 \text{ kN}$$
$$S_A = -3 \text{ kN}$$

S.F. diagram is shown in Fig. 22.18 (c).

(iii) Axial force diagram :

The axial force diagram is shown in Fig. 22.18 (d)

Example 22.18. *Construct the bending moment and shearing force diagrams for the beam shown in the fig. 22.19.* **(GATE)**

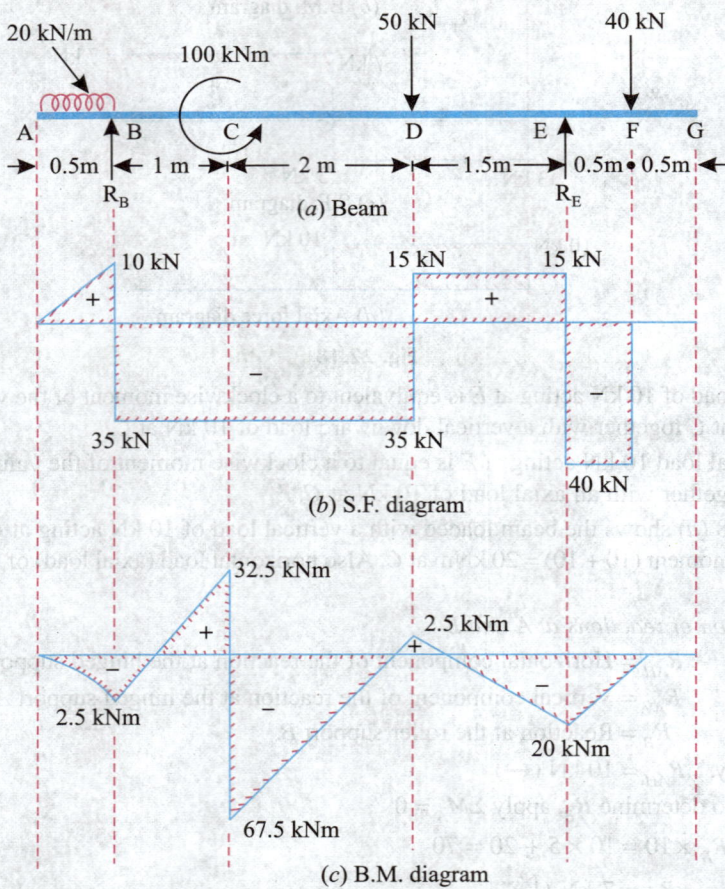

(a) Beam

(b) S.F. diagram

(c) B.M. diagram

Fig. 22.19

Solution. *Reactions at B and E :*

Taking moments, about *B*, we get

$$R_E \times 4.5 + 20 \times 0.5 \times \frac{0.5}{2} + 100 = 50 \times 3 + 40 \times 5$$

or, $\qquad 4.5\, R_E + 2.5 + 100 = 150 + 200$

$$R_E = 55 \text{ kN}$$

Also, $\qquad\qquad R_B + R_E = 20 \times 0.5 + 50 + 40$

or, $\qquad\qquad R_B + 55 = 10 + 50 + 40$

∴ $\qquad\qquad\qquad R_B = 45 \text{ kN}$

S.F. calculations : $\qquad S_F = -40 \text{ kN}$

$$S_E = -40 + 55 = 15 \text{ kN}$$

$$S_D = 15 - 50 = -35 \text{ kN}$$

$$S_B = -35 + 45 = 10 \text{ kN}$$

$$S_A = 10 - (20 \times 0.5) = 0$$

S.F. diagram is shown in Fig. 22.19 (b)

B.M. calculations :

$$M_G = 0$$

$$M_F = 0$$

$$M_E = -40 \times 0.5 = -20 \text{ kNm}$$

$$M_D = -40 \times 2 + 55 \times 1.5 = 2.5 \text{ kNm}$$

$$M_C = -40 \times 4 + 55 \times 3.5 - 50 \times 2 = -67.5 \text{ kNm}$$

The bending moment increases from – 67.5 kNm to 100.

$$M_B = -20 \times 0.5 \times \frac{0.5}{2} = -2.5 \text{ kNm}$$

$$M_A = 0$$

B.M. diagram is shown in Fig. 22.19 (c).

Example 22.19. *Sketch the shear force and bending moment diagrams for shown in Fig. 22.20. Indicate all the important features.*

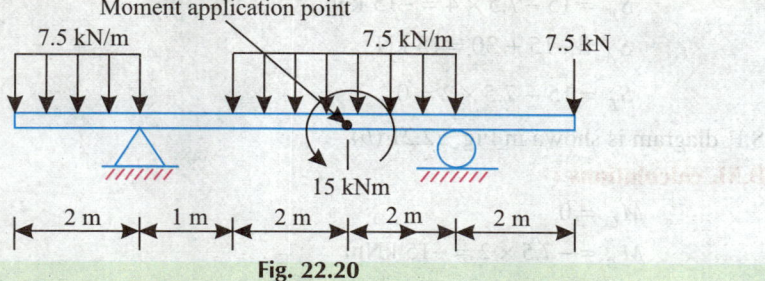

Fig. 22.20

Solution. Fig. 22.21 shows the overhanging beam with the given loads.

Calculations for reactions :

Taking moments about *A*, we get

$$R_B \times 5 + 7.5 \times 2 \times \frac{2}{5} + 15 = 7.5 \times 7 + 7.5 \times 4 \times \left(\frac{4}{2} + 1\right)$$

or, $\qquad\qquad 5R_B + 15 + 15 = 52.5 + 90$

∴ $\qquad\qquad\qquad R_B = 22.5 \text{ kN } (\uparrow)$

Also, $\qquad\qquad R_A + R_B = 7.5 \times 2 + 7.5 \times 4 + 7.5 = 52.5 \text{ kN}$

∴ $R_A = 52.5 - 22.5 = 30$ kN ($\uparrow$)

Fig. 22.21 (a) shows the given loaded beam in the calculated values of R_A and R_B.

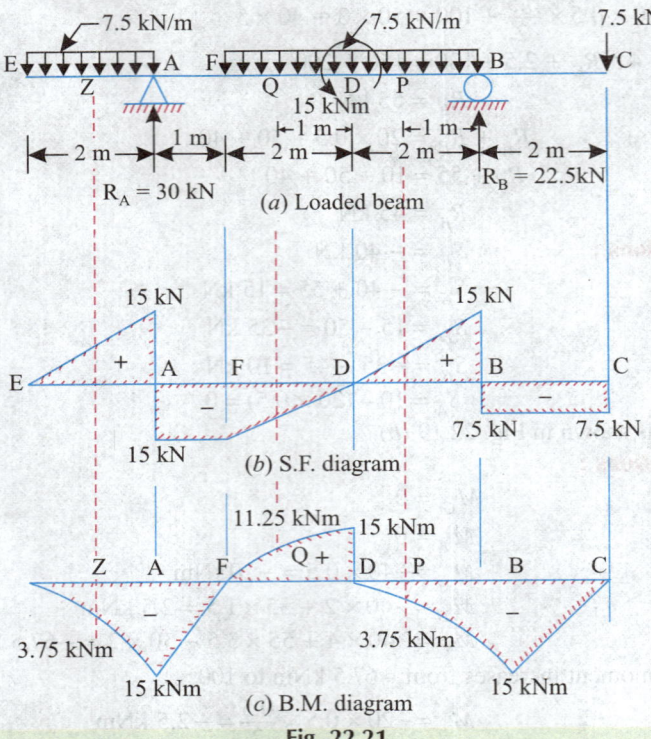

(a) Loaded beam

(b) S.F. diagram

(c) B.M. diagram

Fig. 22.21

S.F. calculations :

$$S_C = -7.5 \text{ kN}$$

$$S_B = -7.5 + 22.5 = 15 \text{ kN}$$

$$S_F = 15 - 7.5 \times 4 = -15 \text{ kN}$$

$$S_A = -15 + 30 = 15 \text{ kN}$$

$$S_E = 15 - 7.5 \times 2 = 0$$

S.F. diagram is shown in Fig. 22.21 (b).

B.M. calculations :

$$M_C = 0$$

$$M_B = -7.5 \times 2 = -15 \text{ kNm}$$

$$M_P = -7.5 \times (2 + 1) + 22.5 \times 1 - 7.5 \times 1 \times \frac{1}{2} = -3.75 \text{ kNm}$$

$$M_D = -7.5 \times 4 + 22.5 \times 2 - 7.5 \times 2 \times \frac{2}{2} + 15$$

$$= -30 + 45 - 15 + 15 = 15 \text{ kNm}$$

              (B.M. changes from 0 to 15 kNm at D)

$$M_Q = -7.5 \times 5 + 22.5 \times 3 - 7.5 \times 3 \times \frac{3}{2} + 15$$

$$= -37.5 + 67.5 - 33.75 + 15 = 11.25 \text{ kNm}$$

$$M_F = -7.5 \times 6 + 22.5 \times 4 - 7.5 \times 4 \times \frac{4}{2} + 15 = -45 + 90 - 60 + 15 = 0$$

$$M_Z = -7.5 \times 1 \times \frac{1}{2} = -3.75 \text{ kNm} \qquad \text{(From L.H.S.)}$$

$$M_E = 0$$

B.M. diagram is shown in Fig. 22.21 (c)

Example 22.20. *Construct shear force and bending moment diagrams for the beam loaded as shown in Fig. 22.22. Also find the location of the inflexion for the beam.* **(GATE)**

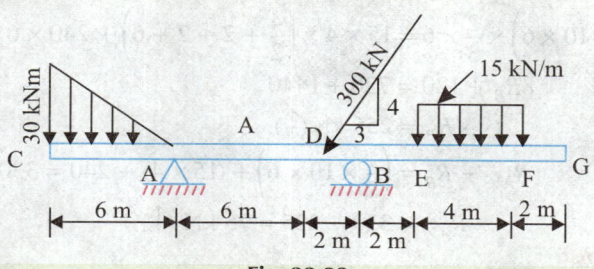

Fig. 22.22

Solution. Fig. 22.23 shows the loaded beam. From this figure we see that an inclined load of 300 kN acts at D. Resolving this force horizontally and vertically, we have:

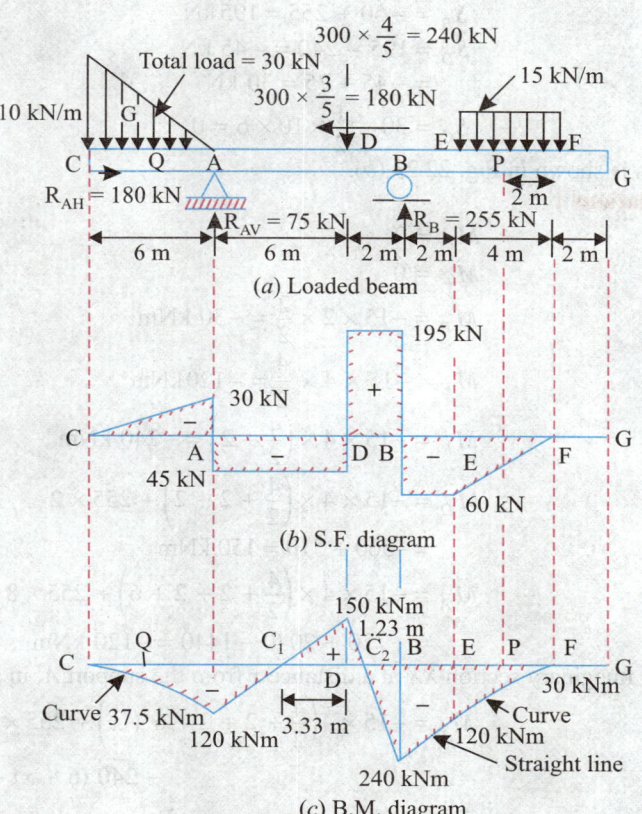

(a) Loaded beam

(b) S.F. diagram

(c) B.M. diagram

Fig. 22.23

Vertical component $= 300 \times \dfrac{4}{5} = 240$ kN ($\downarrow$)

Horizontal component $= 300 \times \dfrac{3}{5} = 180$ kN ($\leftarrow$)

Since the support at B is on rollers it cannot take up any horizontal reaction.
The hinged reaction at A will take up the horizontal reaction.

$\therefore$ $\qquad R_{AH} = 180$ kN ($\rightarrow$)

Reactions:

In order to determine R_B, apply $\Sigma M_A = 0$:

$$R_B \times 8 + \left(\dfrac{1}{2} \times 10 \times 6\right) \times \dfrac{2}{3} \times 6 = 15 \times 4 \times \left(\dfrac{4}{2} + 2 + 2 + 6\right) + 240 \times 6$$

or, $\qquad 8R_B + 120 = 720 + 1440$

$\therefore \qquad R_B = 255$ kN ($\uparrow$)

But, $\qquad R_{AV} + R_B = \left(\dfrac{1}{2} \times 10 \times 6\right) + (15 \times 4) + 240 = 330$

$\therefore \qquad R_{AV} = 330 - 225 = 75$ kN ($\uparrow$)

S.F. calculations:

$$S_{G-F} = 0$$
$$S_F = -15 \times 4 = -60 \text{ kN}$$

(The shearing force increases from 0 to 60 kN).

$$S_B = -60 + 255 = 195 \text{ kN}$$
$$S_D = 195 - 240 = -45 \text{ kN}$$
$$= -45 + 75 = 30 \text{ kN}$$
$$S_C = 30 - \dfrac{1}{2} \times 10 \times 6 = 0$$

S.F. diagram is shown in Fig. 22.23 (b).

B.M. calculations :

$$M_C = 0$$
$$M_F = 0$$
$$M_P = -15 \times 2 \times \dfrac{2}{2} = -30 \text{ kNm}$$
$$M_E = -15 \times 4 \times \dfrac{4}{2} = -120 \text{ kNm}$$
$$M_B = -15 \times 4 \times \left(\dfrac{4}{2} + 2\right) = -240 \text{ kNm}$$
$$M_D = -15 \times 4 \times \left(\dfrac{4}{2} + 2 + 2\right) + 255 \times 2$$
$$= -360 + 510 = 150 \text{ kNm}$$
$$M_A = -15 \times 4 \times \left(\dfrac{4}{2} + 2 + 2 + 6\right) + 255 \times 8 - 240 \times 6$$
$$= -720 + 2040 - 1440 = -120 \text{ kNm}$$

Consider an imaginary section XX at a distance x from the support A, in segment CA.

$$M_x = -15 \times 4\left(\dfrac{4}{2} + 2 + 2 + 6 + x\right) + 255 \times (2 + 6 + x)$$

$$- 240 (6 + x) + 75x - \dfrac{1}{2} x . \dfrac{5}{3} x . \dfrac{1}{3} x$$

$$\left[\begin{array}{l} \dfrac{y}{x} = \dfrac{10}{6} \text{ or } y = \dfrac{10}{6} x = \dfrac{5}{3} x \\ \text{Refer to Fig. 22.23 } (a) \end{array}\right]$$

$$= -60(12 + x) + 255(8 + x) - 240(6 + x) + 75x - \frac{5}{18}x^3$$

$$M_{Q(x=3)} = -60(12 + 3) + 255(8 + 3) - 240(6 + 3) + 75 \times 3 - \frac{5}{18} \times 3^3$$

$$= -900 + 2805 - 2160 + 225 - 7.5 = -37.5 \text{ kNm}$$

$$M_A = -60(12 + 6) + 225(8 + 6) - 240(6 + 6) + 75 \times 6 - \frac{5}{18} \times 6^3$$

$$= -1080 + 3570 - 2880 + 450 - 60 = 0$$

B.M. diagram is shown in the Fig. 22.23 (c).

Points of contraflexure or inflextion points:

Let C_1 and C_2 are the points of contraflexure or the points of inflextion on the beam.

Let point C_2 is located at distance x from point B. Then,

$$M_x = -15 \times 4\left(\frac{4}{2} + 2 + x\right) + 255x = 0$$

or, $-60(4 + x) + 255x = 0$

∴ **$x = 1.23$ m (from B) (Ans.)**

Let point C_1 is located at distance x from point D. Then,

$$M_x = -15 \times 4\left(\frac{4}{2} + 2 + 2 + x\right) + 255(2 + x) - 240x = 0$$

or, $-60(6 + x) + 255(2 + x) - 240x = -360 - 60x + 510 + 255x - 240x = 0$

or, **$x = 3.33$ m (from D) (Ans.)**

The point of contraflexure C_1 and C_2 are shown in Fig. 22.23 (c)

Example 22.21. *Two bars AB and BC are connected by a frictionless hinge at B. The assembly is supported and loaded as shown in Fig. 22.24. Draw the shear force and bending moment diagrams for the combined beam AC, clearly labelling the important values. Also indicate your sign convention.*

(GATE)

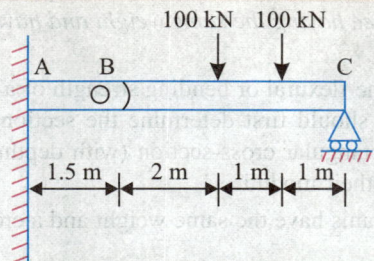

Fig. 22.24

Solution. There small be a vertical reaction at hinge B and we can split the problem in the two parts.

Referring to Fig. 22.24, we get,

$$F_y = 0,$$

or, $R_1 + R_2 = 200$ kN

From, $\Sigma M_B = 0, \quad 100 \times 2 + 100 \times 3 - R_2 \times 4 = 0$

or, $R_2 = \dfrac{500}{4} = 125$ kN

∴ $R_1 = 200 - 125 = 75$ kN

From Fig. (22.25), $R_3 = R_1 = 75$ kN

The important values are labelled. (Refer Fig. 22.25)

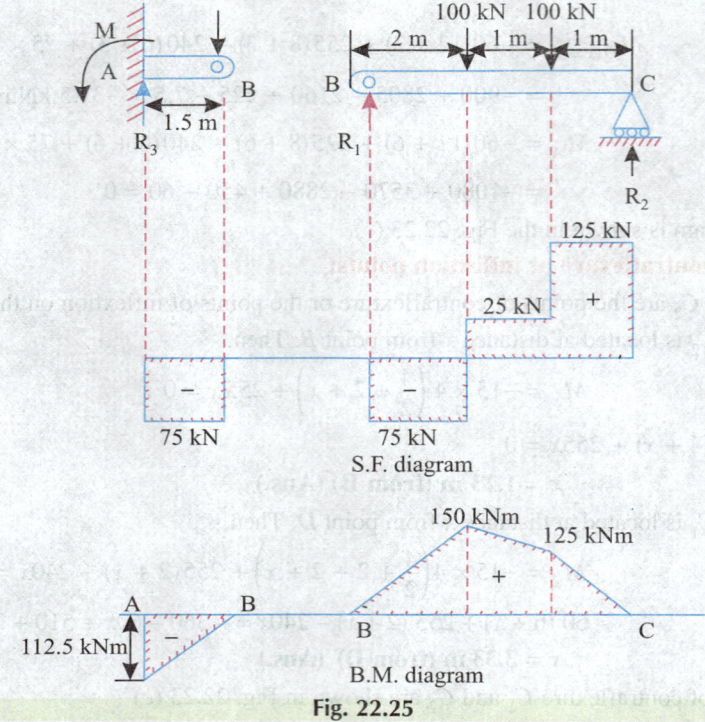

Fig. 22.25

BENDING STRESSES

Example 22.22. *Compare the bending strength of three beams one having a square cross-section, the second one of a rectangular cross-section (with depth twice breadth) and the third of a circular cross-section, all the three having the same weight and having a cross-section of 1000 mm² each.*

(Bangalore University)

Solution. We know that the flexural or bending strength of a beam is directly proportional to its *section modulus*. Hence, we should first determine the section modulus in each of the cases, namely square cross-section, rectangular cross-section (with depth twice the breadth) and circular cross-section, and then compare their moduli.

Given that all the three beams have the same weight and a cross-section of 1000 mm² each.

(i) Square section (1) :

Let, l = The side of the square beam.

Then, area, $A_1 = l^2 = 1000$ or $l = \sqrt{1000} = 31.623$ mm

Section modulus of the square section,

$$Z_1 = \frac{I}{y} = \frac{l \times (l^3/12)}{(l/2)} = \frac{l^4}{12} \times \frac{2}{l} = \frac{l^3}{6}$$

$$= \frac{(31.623)^3}{6} = 5270.6 \text{ mm}^3 \qquad \qquad ...(i)$$

(ii) Rectangular section (2) :

Let, b = Breadth (width of the section).

Then, depth $d = 2b$ (*Given*)

∴ Area, $A_2 = 2b \times b = 2b^2 = 1000$

or, $$b = \left(\frac{1000}{2}\right)^{1/2} = 22.36 \text{ mm}$$

∴ $$d = 2b = 2 \times 22.36 = 44.72$$

Section modulus of the rectangular section,

$$Z_2 = \frac{I}{y} = \frac{[b \times (2b)^3 / 12]}{(2b/2]} = \frac{b \times 8b^3}{12 \times b} = \frac{2b^3}{3}$$

$$= \frac{2 \times (22.36)^3}{3} = 7453 \text{ mm}^3 \qquad \qquad ...(ii)$$

(iii) Circular section (3) :

Let, d = Diameter of the section.

Area, $$A_3 = \frac{\pi}{4} d^2 = 1000 \quad \text{or} \quad d = \left(\frac{1000 \times 4}{\pi}\right)^{1/2} = 35.68 \text{ mm}$$

Section modulus of the circular section,

$$Z_3 = \frac{I}{y} = \frac{\frac{\pi}{64} d^4}{\frac{d}{2}} = \frac{\pi}{32} d^3 = \frac{\pi}{32} \times (35.68)^3 = 4459.4 \text{ mm}^3 \qquad ... (iii)$$

Comparing the values of Z in eqns. (i), (ii) and (iii) we find that the greatest value of $Z = 7453$ mm³, in case of rectangular cross-section.

Comparison :

Hence we conclude that for the same weight and the same cross-sectional area, a beam of the rectangular cross-section has the greatest flexural bending strength whose

$$Z (= Z_1) = 7453 \text{ mm}^3 \text{ (the greatest value)}.$$

The next one corresponding to the flexural or bending strength is the beam of circular section whose $Z (= Z_2) = 5270.6$ mm³.

The last one corresponding to the flexural or bending strength is the beam of circular section whose $Z (= Z_3) = 4459.4$ mm³ (the least value) **(Ans.)**

Example 22.23. *A flat steel bar measuring 50 mm × 12.5 mm was placed in a testing machine and subject to a 50 kN load acting as shown in Fig. 22.26. An extensometer placed in line with the load recorded an extension of 0.17 mm on a gauge length of 210 mm extensometer. Calculate the maximum and minimum stresses set up, and the value of Young's modulus.* **(Panjab University)**

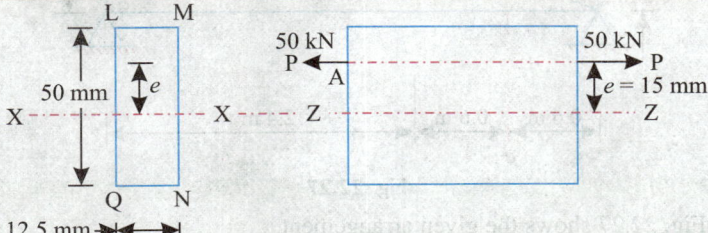

Fig. 22.26

Solution. Eccentricity about XX,

$$e = 15 \text{ mm}$$

Load, $$P = 50 \text{ kN}$$

Moment about $XX = P \times e = 50 \times 10^3 \times 15 = 75 \times 10^4$ N mm

Direct stress, $$\sigma_d = \frac{50 \times 1000}{50 \times 12.5} = 80 \text{ N/mm}^2, \text{ or, } 80 \text{ MN/m}^2 \text{ (tensile)}$$

Bending stress at $$A = (\sigma_b)_A = \frac{M}{I_{XX}} \times y_A = \frac{75 \times 10^4 \times 15}{(12.5 \times 50^3)/12} = \frac{75 \times 10^4 \times 15 \times 12}{12.5 \times 50^3}$$

= 86.4 N/mm² or 86.4 MN/m² (tensile)

Therefore, resultant stress along the load line

$$\sigma_r = 80 + 86.4 \text{ N/mm}^2 \text{ or } 166.4 \text{ MN/m}^2.$$

Therefore, strain along the load line,

$$= \frac{\sigma_r}{E} = \frac{166.4}{E}$$

Equating to the given value, we have

$$\frac{166.4}{E} = \frac{0.17}{210}$$

$$E = \frac{166.4 \times 210}{0.17} = 205553 \text{ MN/m}^2 = \textbf{205.533 GN/m}^2 \textbf{ (Ans.)}$$

The bending stresses at *LM* and *QN* will be equal in magnitude and are given by

$$(\sigma_b)_{LM} = (\sigma_b)_{QN} = \frac{M}{I_{xx}} \times y$$

$$= \frac{75 \times 10^4 \times (50/2)}{(12.5 \times 50^3)/12} = \frac{75 \times 10^4 \times 25 \times 12}{12.5 \times 50^3}$$

$$= 144 \text{ N/m m}^2 \text{ or } 144 \text{ MN/m}^2.$$

Thus, the maximum resultant stress at LM will be

$$= 80 + 144 = \textbf{224 N/mm}^2 \textbf{ or 224 MN/m}^2 \textit{ (tensile)} \textbf{ (Ans.)}$$

The minimum stress at QN will be

$$= 144 - 80 = \textbf{64 N/mm}^2 \textbf{ or 64 MN/m}^2 \textit{ (compressive)} \textbf{ (Ans.)}$$

Example 22.24. *A 100 mm by 500 mm rectangular wooden beam supports a 40 kN load. (see Fig. 22.27). At section a – a, the grain of the wood makes an angle of 20° with the axis of the beam. Find the shearing stress along the grain of the wood at the point A caused by applied concentrated load.*

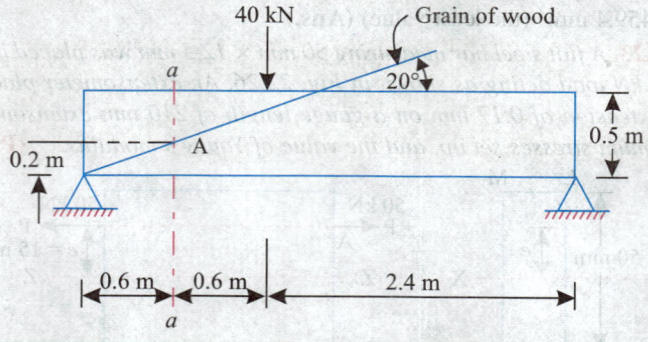

40 kN   Grain of wood

20°

0.5 m

0.2 m

A

0.6 m   0.6 m   2.4 m

Fig. 22.27

Solution. Fig. 22.27 shows the given arrangement.

Fig. 22.28 (*a*) shows the reactions at the left hand support (*R*₁) and at the right hand support (*R*₂); also an end view is drawn showing the section of the beam with the neutral axis (N. A.).

Let us first determine the reactions R_1 and R_2.

Taking moments about the left hand support, we get

$$R_2 \times 3.6 = 40 \times 1.2$$

∴ $R_2 = 13.33 \text{ kN}$

Also, $R_1 + R_2 = 40$

∴ $R_1 = 40 - R_2 = 40 - 13.33 = 26.67 \text{ kN}$

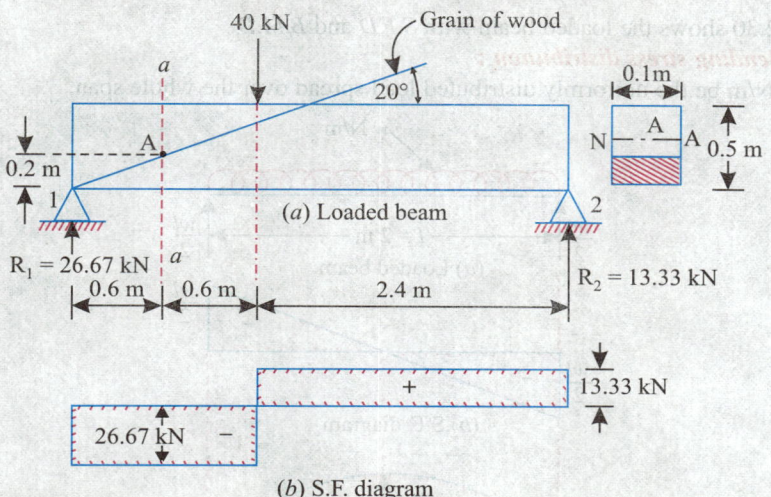

(a) Loaded beam

40 kN

Grain of wood

$R_1 = 26.67$ kN

$R_2 = 13.33$ kN

13.33 kN

26.67 kN

(b) S.F. diagram

Fig. 22.28

From Fig. 22.28 (*b*) we find that shearing force at the section $a - a$.

$$S = 26.67 \text{ kN}$$

Using the relation, $\qquad \tau = \dfrac{SA\bar{y}}{Ib}$ $\qquad\qquad$ ...(*i*)

In this case, $\qquad A\bar{y} = (0.2 \times 0.1)\left(0.25 - \dfrac{0.2}{2}\right) = 0.003 \, \text{m}^3$

$$S = 26.67 \text{ kN}$$

$$I = \frac{0.1 \times 0.5^3}{12} = 0.00104 \, \text{m}^4 \, ; \, b = 0.1 \, \text{m}$$

Substituting the values in eqn. (*i*), we get

$$\tau = \frac{26.67 \times 0.003}{0.00104 \times 0.1} = 769.3 \, \text{kN/m}^2$$

Shearing stress along the grain of wood at the point A

$$= \tau \sin 20° = 769.3 \times \sin 20° = \textbf{263.1 kN/m}^2 \quad \textbf{(Ans.)}$$

Example 22.25. *A timber beam 10 cm wide by 15 cm deep carries a uniformly distributed load over a span of 2 m. If the permissible stresses are 28 N/mm² longitudinally and 2 N/mm² in transverse shear calculate the maximum load per metre run which can be safely carried.* **(Delhi University)**

Solution. *Given* : A timber beam of rectangular section, $b = 10$ cm $= 0.1$ m, $d = 15$ cm $= 0.15$ m.
Span length, $l = 2$ m; $\sigma = 28$ N/mm²; $\tau = 2$ N/mm²

Safe maxmimum load per metre-run, *w* :

Fig. 22.29 shows the given rectangular section.

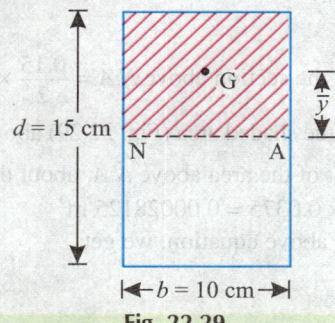

$d = 15$ cm

$b = 10$ cm

Fig. 22.29

Fig. 22.30 shows the loaded beam with *S.F.D* and *B.M.D.*

(i) Bending stress distribution :

Let *w* N/m be the uniformly distributed load spread over the whole span.

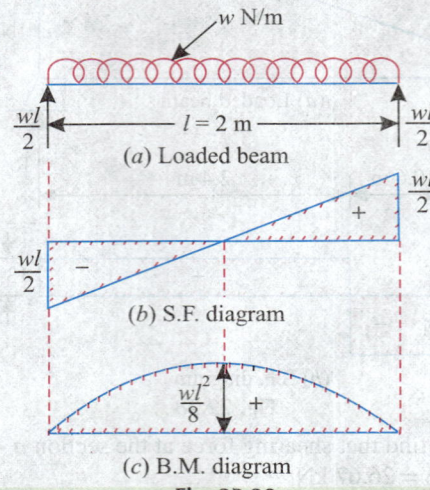

(a) Loaded beam

(b) S.F. diagram

(c) B.M. diagram

Fig. 22.30

From Fig. 22.30 (c) we find the maximum bending moment on the beam

$$M = \frac{wl^2}{8} = \frac{w \times 2^2}{8} \text{ Nm} \quad \text{or} \quad \frac{w}{2} \text{ Nm} \qquad (\because l = 2\text{m})$$

$$I = \frac{bd^3}{12} = \frac{0.1 \times 0.15^3}{12} = 2.8125 \times 10^{-5} \text{ m}^4.$$

Using the bending equation,

$$\frac{M}{I} = \frac{\sigma}{y}, \text{ we get}$$

$$\frac{\frac{w}{2}}{2.8125 \times 10^{-5}} = \frac{28 \times 10^6}{\left(\frac{0.15}{2}\right)} \quad \text{or} \quad \frac{w}{2 \times 2.8125 \times 10^{-5}} = \frac{28 \times 10^6 \times 2}{0.15}$$

$$\therefore \qquad w = \frac{28 \times 10^6 \times 2 \times 2 \times 2.8125 \times 10^{-5}}{0.15} = 21000 \text{ N/m (or 21 kN/m)}$$

(ii) Shear stress distribution :

Next let us consider the effect of transverse shear stress on the beam.

For a simply supported beam loaded with uniformly load *w* N/m, shearing force diagram is shown in Fig. 22.30 (b).

Shear stress at N.A. $\tau = \dfrac{SA\bar{y}}{Ib}$

where, $\qquad$ A = Area of the section above $N.A = \dfrac{0.15}{2} \times 0.1 = 0.0075 \text{ m}^2$

$\bar{y}$ = C.G. of the shaded area (Fig. 22.29) above N.A. = $\dfrac{0.075}{2} = 0.0375$ m, and

$A\bar{y}$ = Moment of the area above *N.A.* about the *N.A.*

$\qquad = 0.0075 \times 0.0375 = 0.00028125 \text{ m}^2$

Substituting the value in the above equation, we get

$$\tau = \frac{wA\bar{y}}{Ib}$$

$$2 \times 10^6 = \frac{w \times 0.00028125}{2.8125 \times 10^{-5} \times 0.1}$$

$$\therefore \qquad w = \frac{2 \times 10^6 \times 2.8125 \times 10^{-5} \times 0.1}{0.00028125} = 20000 \text{ N/m or } 20 \text{ kN/m}$$

Hence, $\qquad w = \textbf{20 kN/m}$ *(smaller of the two values)* **(Ans.)**.

Example 22.26. *A steel beam of rectangular hollow cross-section with a span l = 4m is simply supported and loaded by two equal forces P applied at equal distances, a = 0.5 m from the respective supports. The cross-section dimensions are shown in Fig. 22.31. Find the magnitude of the forces P from the limiting condition dependent on the normal stresses if* $\sigma_{allowable}$ *is 160 Nmm^2. Plot the variation of shear stresses along the depth of the beam.* **(Roorkee University)**

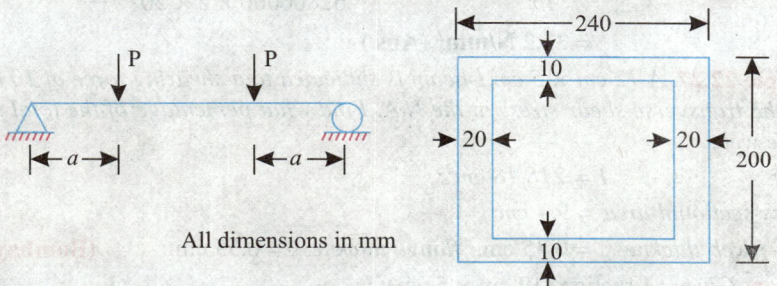

All dimensions in mm

Fig. 22.31

Solution. Moment of inertia of beam section about neutral axis (N.A.)

$$I_{NA} = \frac{1}{12}\left[240 \times 200^3 - 200 \times 180^3\right] = 62800000 \text{ mm}^4$$

Now, bending moment, $M = P \times a$ (Refer to Fig. 22.31)

$$= P \times 0.5 \times 1000 = 500 \, P \text{ Nmm}$$

Using the relation: $\quad \dfrac{M}{I} = \dfrac{\sigma}{y}$, we have

$$\frac{500 \, P}{62800000} = \frac{160}{100} \qquad\qquad \left(\because \ y = \frac{200}{2} = 100 \text{ mm}\right)$$

$$\therefore \qquad P = \frac{160 \times 62800000}{500 \times 100} = \textbf{200960 N (Ans.)}$$

Variation of shear stresses along the depth of the beam :

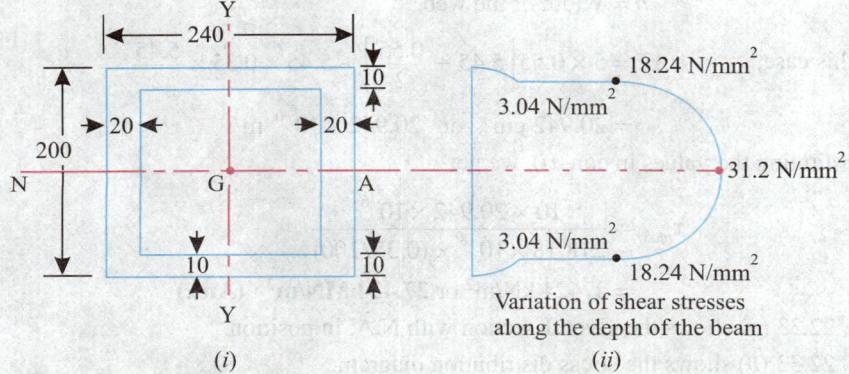

Variation of shear stresses along the depth of the beam

(i)   *(ii)*

Fig. 22.32

Shear stress at 100 mm from the N.A. = 0

Shear stress in the flange at 90 mm from N.A.

$$= \frac{SA\overline{y}}{Ib} = \frac{200960 \times (240 \times 10) \times 95}{62800000 \times 240} = \textbf{3.04 N/mm}^2 \textbf{ (Ans.)}$$

Shear stress in the web at 90 mm from N.A.

$$= 3.04 \times \frac{240}{40} = \textbf{18.24 N/mm}^2 \textbf{ (Ans.)}$$

Shear stress at the N.A.

$$\tau_{NA} = \frac{SA\overline{y}}{Ib} = \frac{200960 \times [240 \times 10 \times 95 + 2 \times 20 \times 90 \times 45]}{62800000 \times 2 \times 20}$$

$$= \textbf{31.2 N/mm}^2 \textbf{ (Ans.)}$$

Example 22.27. *A 12 cm × 5 cm I-beam is subjected to a shearing force of 10 kN. Calculate the value of the transverse shear stress at the N.A. Find what percentage of the total shear force is carried by the web.*

Given: $I = 218.18 \text{ cm}^4;$

cross-sectional area = 9.4 cm²;

web thickness. = 0.35 cm; flange thickness = 0.55 cm.   **(Bombay University)**

Solution. *Given* : I-section (12 cm × 5 cm);

Shearing force, $S = 10$ kN;

Moment of inertia, $I = 218.18$ cm⁴

Cross-sectional area, $A = 9.4$ cm²

Web thickness, $b = 0.35$ cm

Shear stress at the N.A. τ_{max} :

Let τ_{max} be the transverse shear stress at the N.A.

Using the relation:

$$\tau_{max} = \frac{SA\overline{y}}{Ib}$$

where, $S =$ Maximum shearing force,

$A\overline{y} =$ Moment of the area above N.A.,

$I =$ Moment of inertia of the whole section about N.A., and

$b =$ Width of the web.

In this case, $A\overline{y} = 5 \times 0.55 \left[5.45 + \frac{0.55}{2} \right] + 5.45 \times 0.35 \times \frac{5.45}{2}$

$$= 20.942 \text{ cm}^3 \text{ or } 20.942 \times 10^{-6} \text{ m}^3$$

Substituting the values in eqn. (*i*), we get

$$\tau_{max} = \frac{10 \times 20.942 \times 10^{-6}}{218.18 \times 10^{-8} \times (0.35/100)}$$

$$= 27424 \text{ kN/m}^2 \text{ or } \textbf{27.424 MN/m}^2 \textbf{ (Ans.)}$$

Fig. 22.33 (*a*) shows the given I-section with N.A. in position.

Fig. 22.33 (*b*) shows the stress distribution diagram.

Percentage of the total shear force carried by the web :

Refer to Fig. 22.33 (*a*)

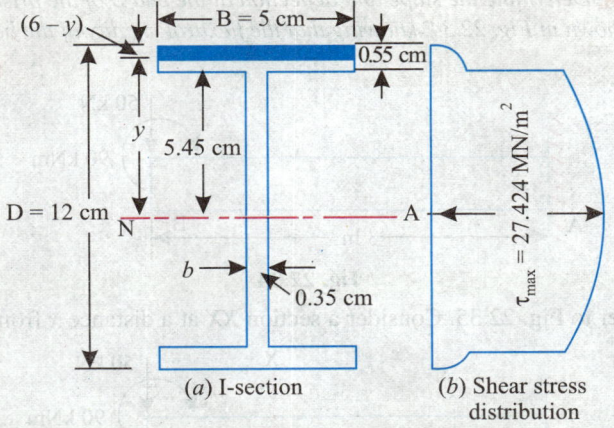

(*a*) I-section

(*b*) Shear stress distribution

Fig. 22.33

Shear stress in the flange at a distance y from the neutral axis (N.A.)

$$\tau = \frac{SA\bar{y}}{IB}$$

$$A\bar{y} = B(6-y)\left(\frac{6+y}{2}\right) = \frac{B}{2}(36-y^2)$$

$\therefore \qquad \tau = \frac{S}{IB} \times \frac{B}{2}(36-y^2) = \frac{S}{2I}(36-y^2)$

$\therefore$ Shear resistance offered by an elementary strip of flange 5 cm wide and dy cm deep is

$$= \tau \, dA = \tau \times 5 \times dy = \frac{S}{2I}(36-y^2) \times 5 \times dy$$

$$= \frac{10}{2 \times 218.18}(36-y^2) \times 5 \times dy = 0.1146(36-y^2)\,dy$$

$\therefore$ Shear resistance of one flange

$$= 0.1146 \int_{5.45}^{6}(36-y^2)\,dy$$

$$= 0.1146\left[\int_{5.45}^{6} 36\,dy - \int_{5.45}^{6} y^2\,dy\right]$$

$$= 0.1146\left[[36y]_{5.45}^{6} - \left\{\frac{y^3}{3}\right\}_{5.45}^{6}\right]$$

$$= 0.1146\left[36(6-5.45) - \frac{1}{3}\left\{6^3 - 5.45^3\right\}\right]$$

$$= 0.1146(19.8 - 18.04) = 0.2 \text{ kN}$$

Total shear resistance of two flanges = $2 \times 0.2 = 0.4$ kN

$\therefore$ Total shear resistance of the web $= 10 - 0.4 = 9.6$ kN

$\therefore$ *Percentage of the total shear force carried by the web*

$$= \frac{9.6}{10} \times 100 = \mathbf{96\%} \quad \textbf{(Ans.)}$$

DEFLECTION OF BEAMS

Example 22.28. *Determine the slope and deflection at the end B of the prismatic cantilever beam when it is loaded as shown in Fig. 22.34, knowing that the flexural rigidity of the beam is EI = 10^4 kNm².*

(Panjab University)

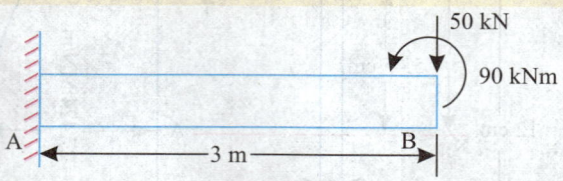

Fig. 22.34

Solution. Refer to Fig. 22.35. Consider a section *XX* at a distance *x* from the end *B*.

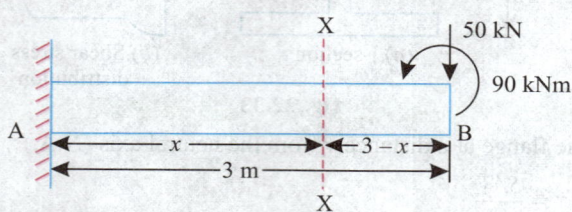

Fig. 22.35

Bending moment at *XX*,

$$M_x = 90 - 50\,(3 - x)$$

$\therefore$ $EI \dfrac{d^2 y}{dx^2} = M_x = 90 - 150 + 50\,x$

Integrating, we get

$$EI \frac{dy}{dx} = 50 \times \frac{x^2}{2} - 60\,x + C_1 \qquad \text{(where, } C_1 = \text{constant of integration)}$$

when $x = 0,\ \dfrac{dy}{dx} = 0$

$\therefore$ $C_1 = 0$

Hence slope equation becomes,

$$EI \frac{dy}{dx} = \left(50 \times \frac{x^2}{2} - 60\,x\right) \qquad \text{...Slope equation ...(i)}$$

$\therefore$ **Slope at B** (at *x* = 3m), θ_B

$$\theta_B = \frac{1}{EI}\left(50 \times \frac{3^2}{2} - 60 \times 3\right) = \frac{1}{10^4} \times 45 = \textbf{0.0045 rad. (Ans.)}$$

(Since *EI* = 10^4 kNm²)

Deflection at B, y_B :

Integrating eqn. (*i*), we get

$$EIy = \left(50 \times \frac{x^3}{6} - 60 \times \frac{x^2}{2}\right) + C_2 \qquad \text{(where, } C_2 = \text{constant of in-}$$

tegration)

When $x = 0,\quad y = 0$ $\therefore\ C_2 = 0$

Hence $EIy = 50 \times \dfrac{x^3}{6} - 60 \times \dfrac{x^2}{2}$

$\therefore \qquad y_B = (\text{at } x = 3m) = \dfrac{1}{EI}\left(50 \times \dfrac{3^3}{6} - 60 \times \dfrac{3^2}{2}\right)$

$\qquad = \dfrac{1}{10^4}\left(50 \times \dfrac{3^3}{6} - 60 \times \dfrac{3^2}{2}\right) = -0.0045 \text{ m or } -4.5 \text{ mm}$ **(Ans.)**

(Negative sign indicates downward deflection)

Example 22.29. *A vertical column 9 m high is fixed at the base and a counter-clockwise moment of 2.1 kNm is applied at the top of the column. A horizontal force P is applied to the column at a height of 4.5 m above the base so as to give a counter-clockwise moment.*

Determine the value P so that the horizontal deflections at the top of the column and at the point of application of P shall be equal,

(i)   *when the deflections are on the same side,*

(ii)   *when the deflections are on opposite sides of the vertical line through the foot of the column.*   **(IAS)**

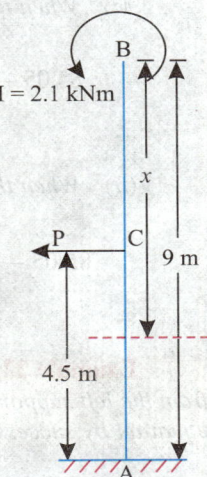

Fig. 22.36

Sol. Choosing the origin at B and using Macaulay's method we have,

$$EI\dfrac{d^2y}{dx^2} = M - P(x - 4.5)$$

When $x < 4.5$ m the effect of the term in bracket is to be neglected.

Integrating, w.r.t. x, we get

$$EI\dfrac{dy}{dx} = Mx - \dfrac{P}{2}(x - 4.5)^2 + C_1 \qquad \text{(where, } C_1 = \text{constant of integration)}$$

When, $x = 9$ m, $\dfrac{dy}{dx} = 0$

$\therefore \qquad 0 = 9M - \dfrac{P}{2}(9 - 4.5)^2 + C_1$

$\therefore \qquad C_1 = 10.125\,P - 9M$

$\therefore \qquad EI\dfrac{dy}{dx} = Mx - \dfrac{P}{2}(x - 4.5)^2 + 10.125P - 9M$

Integrating once again, we get

$$EIy = \dfrac{Mx^2}{x} = \dfrac{P}{6}(x - 4.5)^3 + (10.125P - 9M)x + C_2$$

When, $x = 9$ m, $y = 0$

$\therefore \qquad 0 = \dfrac{M}{2} \times 81 - \dfrac{P}{6}(9 - 4.5)^3 + (10.125P - 9M) \times 9 + C_2$

$\qquad = 40.5M - 15.19P + 91.12P - 81M + C_2$

$\therefore \qquad C_2 = 40.5M - 75.93P$

Thus, the deflection equation reduces to

$$EIy = \dfrac{Mx^2}{2} - \dfrac{P}{6}(x - 4.5)^3 + (10.125P - 9M)x + (40.5M - 75.93P)$$

Therefore, at $x = 0$, neglecting the term within the brackets,

$\qquad EIy_B = 40.5M - 75.93P$

$\qquad = 40.5 \times 2.1 - 75.93P \qquad [\because M = 2.1 \text{ kNm... } Given]$

$\qquad = 85.05 - 75.93P$

and, at $x = 4.5$ m

$$EIy_C = \frac{M \times (4.5)^2}{2} - \frac{P}{6}(45.45)^3 + (10.125P - 9M) \times 4.5 + (40.5M - 75.93P)$$

$$= 10.125\ M - 0 + 45.56\ P - 40.5\ M + 40.5\ M - 75.93\ P$$

$$= 10.125\ M - 30.37\ P = 10.125 \times 2.1 - 30.37\ P = 21.26 - 30.37\ P$$

(i) *When the deflections are on the same side :*

$$y_B = y_C$$

$$85.05 - 75.93\ P = 21.26 - 30.37\ P$$

∴ $$P = \frac{(85.05 - 21.26)}{(75.93 - 30.37)} = 1.4\ \text{kN}\ \ \textbf{(Ans.)}$$

(ii) *When the deflections are on the opposite sides of the vertical line :*

$$y_B = -y_C$$

$$85.05 - 75.93\ P = (21.26 - 30.37\ P) = -21.26 + 30.37\ P$$

∴ $$P = \frac{(85.05 + 21.26)}{(75.93 + 30.37)} = 1.0\ \text{kN}\ \ \textbf{(Ans.)}$$

Example 22.30. *A simple beam supports a concentrated downward load of P at a distance from the left support as shown in Fig. 22.37. The flexural rigidity EI is constant. Find the deflection equation by successive integration method.* **(GATE)**

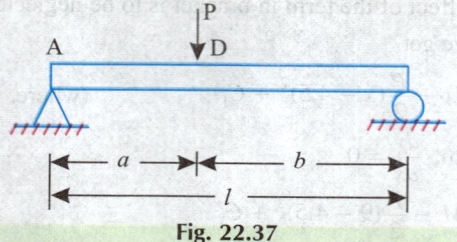

Fig. 22.37

Solution. Refer to Figs. 22.37 and 22.38

Fig. 22.37 shows the given arrangement.

Fig. 22.38 shows the same arrangement with deflection curve; the reactions at the ends *A* and *B* are also indicated.

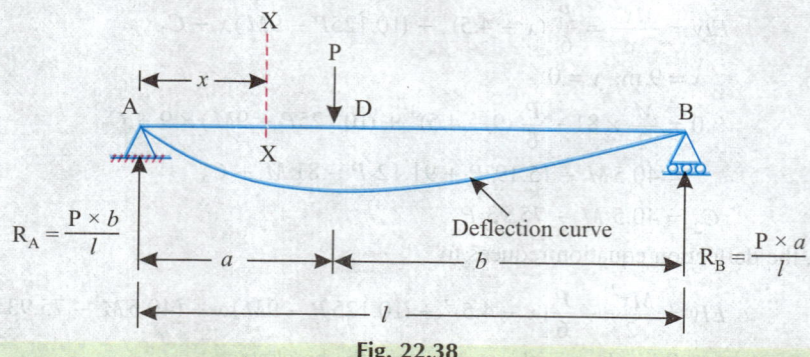

Fig. 22.38

In order to determine R_A, apply $\Sigma M_B = 0$

$$R_A \times l = P \times b$$

∴ $$R_A = \frac{P \times b}{l}\ (\uparrow)$$

Similarly, $R_B = \dfrac{P \times a}{l}(\uparrow)$

When we examine Fig. 22.38 we see clearly there are two segments AD and DB of the beam AB. We have, therefore, to consider two different equations, describing the bending moment in the beam.

One equation is valid to the field of the load P, and the other holds to the right of this load. The integration of each equation gives rise to two constants of integration to be determined.

Consider an imaginary section XX at a distance x from A in the segment AD.

B.M. at XX, $\quad M_x = \dfrac{P \times b}{l} x \quad$ for $\;0 < x < a$

The differential equation of the bent beam thus becomes

$$EI\,\frac{d^2 y}{dx^2} = \frac{P \times b}{l} x \text{ for } 0 < x < a \qquad \ldots(i)$$

The first integration yields

$$EI\,\frac{dy}{dx} = \frac{P \times b}{l} \times \frac{x^2}{2} + C_1 \qquad \ldots(ii)$$

where, C_1 is a constant of integration.

Now, that no numerical information is available about the slope $\dfrac{dy}{dx}$ at any point in this region. Since the load is not applied at the centre of the beam there is no reason to believe the slope is zero at $x = \dfrac{l}{2}$.

However, for the slope of the beam under the point of application of force, P, we can write from eqn. (ii)

$$EI\left(\frac{dy}{dx}\right)_{x=a} = \frac{Pba^2}{2l} + C_1 \qquad \ldots(iii)$$

The next integration of eqn. (ii) yields

$$EIy = \frac{Pb}{2l} \times \frac{x^3}{3} + C_1 x + C_2 \qquad \ldots(iv)$$

where, C_2 is another constant of integration.

At the left support when $x = 0$, $y = 0$. By substituting these values in eqn. (iv), we get $C_2 = 0$.

It is to be noted that it is not possible to use the condition $y = 0$ at $x = l$ in eqn. (iv) since eqn. (i) *is not valid in that region.*

We have for the deflection under the point of application of the force P,

$$EIy_{(x=a)} = \frac{Pba^3}{6l} + C_1 a \qquad \ldots(v)$$

In the region to the right of force P, let us take another imaginary section XX at a distance x from the end A.

B.M. at $\quad XX = M_x = \dfrac{Pb}{l} x - P(x - a)$ for $a < x < l$.

Then, $\quad EI\,\dfrac{d^2 y}{dx^2} = \dfrac{Pb}{l} x - P(x - a) \quad$ for $\;a < x < l \qquad \ldots(vi)$

The first integration of this equation yields

$$EI\,\frac{dy}{dx} = \frac{Pb}{l} \times \frac{x^2}{2} - P\left[\frac{(x-a)^2}{2}\right] + C_3 \qquad \ldots(vii)$$

Although nothing definite may be said about the slope in this portion of the beam, we have for slope under point of application of the force P,

$$EI\left(\frac{dy}{dx}\right)_{x=a} = \frac{Pba^2}{2l} + C_3 \qquad \qquad ...(viii)$$

Under the concentrated load P, the slope as given by eqn. (iii) must be equal to the slope given by eqn. (viii).

Consequently the right sides of these two equations must be equal. Thus, we have

$$\frac{Pba^2}{2l} + C_1 = \frac{Pba^2}{2l} + C_3$$

$$\therefore \qquad \qquad C_1 = C_3$$

Eqn. (vii) may now be integrated to give

$$EIy = \frac{Pb}{2l} \times \frac{x^3}{3} - \frac{P(x-a)^3}{6} + C_3 x + C_4 \qquad \qquad ...(ix)$$

We may write for deflection under the concentrated load P,

$$EIy_{(x=a)} = \frac{Pba^3}{6l} + C_3 a + C_4 \qquad \qquad ...(x)$$

The deflection at $x = a$ by eqn. (v) must be equal to that given by eqn. (x).

Thus, the right sides of these two equations are equal. Hence, we have,

$$\frac{Pba^3}{6l} + C_1 a = \frac{Pba^3}{6l} + C_3 a + C_4$$

Since $C_1 = C_3$, we have $C_4 = 0$

The condition that $y = 0$, when $x = l$ may be substituted in eqn. (ix) yielding

$$0 = \frac{Pbl^3}{6l} - \frac{Pb^3}{6} + C_3 l$$

$$\therefore \qquad \qquad C_3 = \frac{Pb}{6l}(b^2 - l^2)$$

In this manner all four constants of integration are determined.

These values may now be substituted in eqns. (iv) and (ix) to give the required deflection equations,

$$EIy = \frac{Pb}{2l} \times \frac{x^3}{3} + \frac{Pb}{6l}(b^2 - l^2)x + 0$$

or,

$$EIy = \frac{Pb}{6l}\left[x^3 - (l^2 - b^2)x\right] \qquad \text{for } 0 < x < l \qquad ...(1)$$

$$EIy = \frac{Pb}{2l} \times \frac{x^3}{3} - \frac{P(x-a)^3}{6} + \frac{Pb}{6l}(b^2 - l^2)x + 0$$

or,

$$EIy = \frac{Pb}{6l}\left[x^3 - \frac{l}{b}(x-a)^3 - (l^2 - b^2)x\right] \qquad \text{for} \qquad a < x < 1 \qquad ...(2)$$

These two equations (1) and (2) are the required deflection curve of the bent beam. Note that *each equation is valid only in the region indicated.* **(Ans.)**

Example 22.31. *A tube 40 mm outside diameter, 5 mm thick and 1.5 m long simply supported at 125 mm from each end carries a concentrated load of 1 kN at each extreme end.*

(i) Neglecting the weight of the tube, sketch the shearing force and bending moment diagrams;

(ii) Calculate the radius of curvature and deflection at mid-span. Take the modulus of elasticity of the material as 208 GN/m². **(UPSC)**

Solution. *Given :* $d_0 = 40$ mm, $= 0.04$ m; $d_i = d_0 - 2t = 40 - 2 \times 5 = 30$ mm $= 0.03$ m; $W = 1$ kN; $E = 208$ GN/m² $= 208 \times 10^9$ N/m²; $l = 1.5$ m; $a = 125$ mm $= 0.125$ m.

(i) S.F. diagram and B.M. diagrams are shown respectively in Fig. 22.39 (b, c)

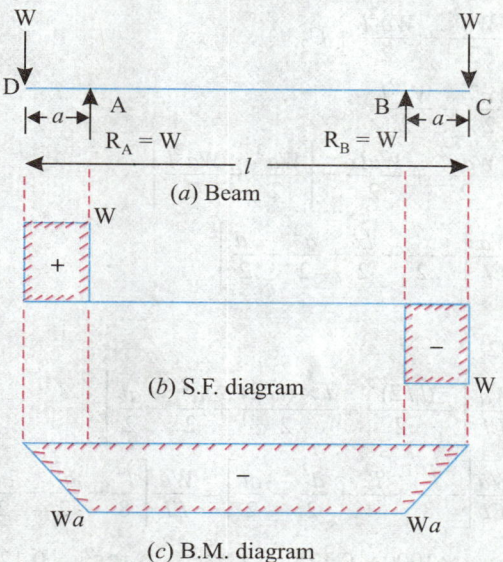

(a) Beam

(b) S.F. diagram

(c) B.M. diagram

Fig. 22.39

(ii) Radius of curvature, R:

As per bending equation:
$$\frac{M}{I} = \frac{\sigma}{y} = \frac{E}{R}$$

or,
$$R = \frac{EI}{M} \qquad \qquad ...(i)$$

Here,
$$M = W \times a = 1 \times 10^3 \times 0.125 = 125 \text{ Nm}$$

$$I = \frac{\pi}{64}(d_0^4 - d_i^4)$$

$$= \frac{\pi}{64}\left[(0.04)^4 - (0.03)^4\right] = 8.59 \times 10^{-8} \text{ m}^4$$

Substituting the values in eqn. (i), we get

$$R = \frac{208 \times 10^8 \times 8.59 \times 10^{-8}}{125} = \textbf{142.9 m (Ans.)}$$

Deflection at mid-span:

$$EI \frac{d^2 y}{dx^2} = M_x = -Wx + W(x - a) = -Wx + Wx - Wa = -Wa$$

Integrating, we get
$$EI \frac{dy}{dx} = -Wax + C_1$$

When,
$$x = \frac{l}{2}, \frac{dy}{dx} = 0$$

∴
$$0 = -Wa\frac{l}{2} + C_1 \quad \text{or} \quad C_1 = \frac{Wal}{2}$$

∴
$$EI \frac{dy}{dx} = -Wax + \frac{Wal}{2}$$

Integrating again, we get

$$EIy = -Wa\frac{x^2}{2} + \frac{Wal}{2}x + C_2$$

When $\qquad x = a, \quad y = 0$

$\therefore \qquad 0 = -\dfrac{Wa^3}{2} + \dfrac{Wa^2l}{2} + C_2$

or, $\qquad C_2 = \dfrac{Wa^3}{2} - \dfrac{Wa^2l}{2}$

$\therefore \qquad EIy = -\dfrac{Wax^2}{2} + \dfrac{Walx}{2} + \left[\dfrac{Wa^3}{2} - \dfrac{Wa^2l}{2} \right]$.

or, $\qquad y = \dfrac{Wa}{EI}\left[-\dfrac{x^2}{2} + \dfrac{lx}{2} + \dfrac{a^2}{2} - \dfrac{al}{2} \right]$

At mid-span, i.e.,

$\qquad x = l/2$

$$y = \dfrac{Wa}{EI}\left[-\dfrac{(l/2)^2}{2} + \dfrac{l \times (l/2)}{2} + \dfrac{a^2}{2} - \dfrac{al}{2} \right]$$

$$= \dfrac{Wa}{EI}\left[-\dfrac{l^2}{8} + \dfrac{l^2}{4} + \dfrac{a^2}{2} - \dfrac{al}{2} \right] = \dfrac{Wa}{EI}\left[\dfrac{l^2}{8} + \dfrac{a^2}{2} - \dfrac{al}{2} \right]$$

$$= \dfrac{1 \times 1000 \times 0.125}{208 \times 10^9 \times 8.59 \times 10^{-8}}\left[\dfrac{1.5^2}{8} + \dfrac{0.125^2}{2} - \dfrac{0.125 \times 1.5}{2} \right]$$

$$= 0.006996\,(0.28125 + 0.007812 - 0.09375)$$

$$= 0.001366 \text{ m} = 1.366 \text{ mm}$$

Hence, *deflection at mid-span* = **1.366 mm** (*upwards*) **(Ans.)**

Example 22.32. *A beam 4m long is freely supported at the ends. It carries concentrated loads of 20 kN each at points 1m from the ends. Calculate the maximum slope and deflection of the beam and slope and deflection under each load. Take EI = 13000 kNm².* **(GATE)**

Solution. Refer to Fig. 22.40. Let R_A and R_B be the reactions at the left and right hand supports respectively.

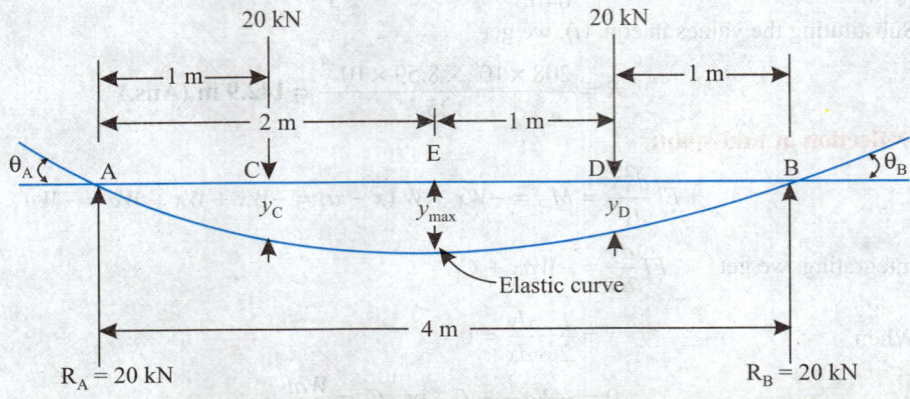

Fig. 22.40

Since the loads are placed symmetrically on the beam,

$$R_A = R_B = \dfrac{20 + 20}{2} = 20 \text{ kN } (\uparrow)$$

Apply Macaulay's method in this problem.

Consider an imaginary section XX at a distance x from A in the segment DB.

$$M_x = EI \frac{d^2y}{dx^2} = 20\,x\lvert -20(x-1) - 20(x-3)$$

Integrating the above expression, we have

$$EI \frac{dy}{dx} = 20 \times \frac{x^2}{2} + C_1 \Big\lvert -\frac{20}{2}(x-1)^2 \Big\rvert -\frac{20}{2}(x-3)^2$$

$$= 10x^2 + C_1 \Big\lvert -10(x-1)^2 - 10(x-3)^2 \qquad \text{...(i)}$$

Integrating again, we get

$$EIy = \frac{10}{3}x^3 + C_1 x + C_2 - \frac{10}{3}(x-1)^3 - \frac{10}{3}(x-3)^3 \qquad \text{...(ii)}$$

When $\qquad x = 0, \ y = 0 \quad \therefore \ C_2 = 0$

When $\qquad x = 4\,\text{m}, \ y = 0$

From eqn. (ii), we have

$$0 = \frac{10}{3} \times 4^3 + C_1 \times 4 - \frac{10}{3}(4-1)^3 - \frac{10}{3}(4-3)^3$$

$$0 = \frac{640}{3} + 4C_1 - 90 - \frac{10}{3}$$

$$\therefore \qquad C_1 = \Big(90 + \frac{10}{3} - \frac{640}{3}\Big)\Big/4 = -30$$

Substituting for constant C_1 and C_2, the slope and deflection equations reduce to

$$EI \frac{dy}{dx} = 10x^2 - 30 - 10(x-1)^2 - 10(x-3)^2 \qquad \text{... Slope equation}$$

$$EIy = \frac{10}{3}x^3 - 30x - \frac{10}{3}(x-1)^3 - \frac{10}{3}(x-3)^3 \qquad \text{... Deflection equation}$$

Maximum slope:

Maximum slope will occur at the supporting ends *A* and *B*. Consider the slope equation,

$$EI \frac{dy}{dx} = 10 \times 0 - 30 = -30$$

$$\therefore \qquad \theta_A \Big(= \frac{dy}{dx}\Big) = -\frac{30}{EI} = -\frac{30}{13000} = -\mathbf{0.002308}\ \textbf{rad. (Ans.)}$$

Similarly slope at

$$\theta_B = +\mathbf{0.002308}\ \textbf{rad. (Ans.)}$$

Maximum deflection, y_{max} :

For obtaining maximum deflection, we should first determine the position where slope is zero. Considering the slope equation we have:

$$0 = 10x^2 - 30 - 10(x-1)^2 \ \text{[Since the term } 10(x-3)^2 \text{ will be negative for the segment } CD \text{ and}$$
hence neglected.]

or, $\qquad 10x^2 - 30 - 10(x^2 - 2x + 1)$

$$10x^2 - 30 - 10x^2 + 20x - 10 = 0$$

$$20x = 40 \ \text{ or } \ x = 2\text{m from } A.$$

Now, in order to determine maximum deflection, putting $x = 2$m in the deflection equation, we have

$$EIy_{max} = \frac{10}{3} \times 2^3 - 30 \times 2 - \frac{10}{3}(2-1)^3 = 26.67 - 60 - 3.33 = -36.66$$

$$\therefore \qquad y_{max} = -\frac{36.66}{13000} = 0.00282\,\text{m or } -2.82\ \text{mm}$$

Hence, $\qquad y_{max} = \mathbf{2.82}\ \textbf{mm} \ (downward)$

Slope under the load 20 kN at C, θ_C :

Substituting $x = 1m$ in the slope equation, we get

$$EI \frac{dy}{dx} = 10 \times 1^2 - 30 = -20$$

$$\therefore \quad \theta_C \left(= \frac{dy}{dx} \right) = \frac{-20}{EI} = \frac{-20}{13000} = -0.001538 \text{ rad. (Ans.)}$$

Slope under the load 20 kN at D, θ_D:

Substituting $x = 3$ m in the slope equation, we get

$$EI \frac{dy}{dx} = 10 \times 3^2 - 30 - 10(3-1)^2 = 90 - 30 - 40 = 20$$

$$\therefore \quad \theta_D \left(= \frac{dy}{dx} \right) = \frac{20}{EI} = \frac{20}{13000} = +0.001538 \text{ rad. (Ans.)}$$

Deflection under the load 20 kN at C, y_c :

Substituting $x = 1$, m in the deflection equation, we get

$$EI \, y_c = \frac{10}{3} \times 1^3 - 30 \times 1 = -\frac{80}{3}$$

$$\therefore \quad y_c = -\frac{80}{3} \times \frac{1000}{13000} \text{ mm} = -2.05 \text{ mm}$$

Hence, $\quad y_c = \textbf{2.05 mm} \textit{ (downwards)} \textbf{(Ans.)}$

Deflection under the load 20 kN at D, y_D:

$$EI y_D = \frac{10}{3} \times 3^3 - 30 \times 3 - \frac{10}{3}(3-1)^3 = 90 - 90 - \frac{80}{3} = -\frac{80}{3}$$

$$\therefore \quad y_D = -\frac{80}{3} \times \frac{1000}{13000} = -2.05 \text{ mm}$$

Hence, $\quad y_D = \textbf{2.05 mm} \textit{ (downward)} \textbf{(Ans.)}$

Example 22.33. *Consider the signboard mounting shown in Fig. 22.41. The wind load acting perpendicular to the plane of the figure is F = 100 N. We wish to limit the deflection, due to bending, at point A of the hollow cylindrical pole of outer diameter 150 mm to 5 mm. Find the wall thickness for the pole. Assume E = 2.0 × 10¹¹ N/m².* **(GATE)**

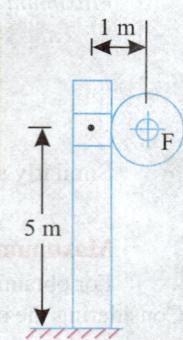

Fig. 22.41

Solution. *Given* : $F = 100$ N; $d_0 = 150$ mm $= 0.15$ m; $y = 5$ mm;
$$E = 2.0 \times 10^{11} \text{ N/m}^2.$$

Thickness of pole, *t*:

The system of signboard mounting can be considered as a cantilever loaded at A *i.e.* $W = 100$ N and also having anticlockwise moment of $M = 100 \times 1 = 100$ Nm at the free end.

Deflection of cantilever having concentrated load at the free end,

$$y = \frac{Wl^3}{3EI} + \frac{Ml^2}{2EI}$$

$$5 \times 10^{-3} = \frac{100 \times 5^3}{3 \times 2.0 \times 10^{11} \times I} + \frac{100 \times 5^2}{2 \times 2.0 \times 10^{11} \times I}$$

or, $$I = \frac{1}{5 \times 10^{-3}} \left[\frac{100 \times 5^3}{3 \times 2.0 \times 10^{11}} + \frac{100 \times 5^2}{2 \times 2.0 \times 10^{11}} \right]$$

or, $$I = \frac{200}{10^{11}} (2083.33 + 625) = 5.417 \times 10^{-6}$$

But,

$$I = \frac{\pi}{64}(d_0^4 - d_i^4)$$

$$\therefore \quad 5.417 \times 10^{-6} = \frac{\pi}{64}(0.15^4 - d_i^4)$$

or, $\quad 0.15^4 - d_i^4 = \dfrac{5.417 \times 10^{-6} \times 64}{\pi} = 110.35 \times 10^{-6}$

or, $\quad d_i^4 = 0.15^4 - 110.35 \times 10^{-6} = 0.003959$

or, $\quad d_i = 0.141 \text{ m or } 141 \text{ mm}$

$$\therefore \quad t = \frac{d_0 - d_i}{2} = \frac{150 - 141}{2} = \textbf{4.5 mm (Ans.)}$$

Example 22.34. *A cantilever beam of 5m length is loaded as shown in Fig. 22.42. Using moment area method or otherwise calculate the slope and deflection at the free end. The beam has uniform flexural rigidity as EI.* **(Roorkee University)**

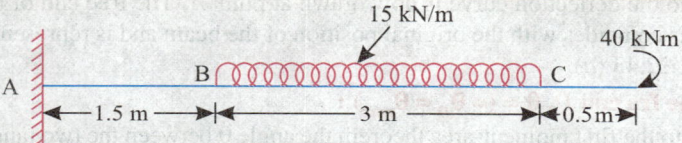

Fig. 22.42

Solution. Refer to Fig. 22.43 (*a*), (*b*) and (*c*).

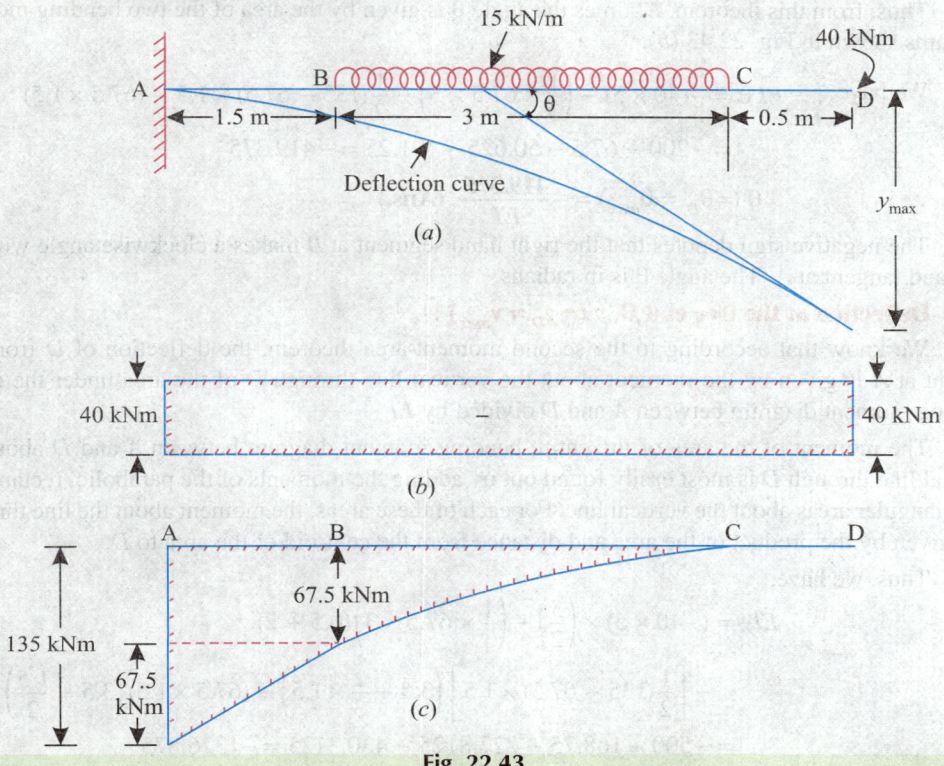

Fig. 22.43

The reaction at the left end of the beam need not be calculated. The heavy curved line represents the approximate shape of the deflection curve of the beam. The curve is an agreement with the conditions of zero slope and zero deflection at the left end of the beam.

In this case, since there are two loads acting on the beam, it is perhaps simplest to construct the bending moment diagram by parts shown in Fig. 22.43 (*b*) and (*c*). One bending moment diagram will be constructed to represent the bending moment due to moment load of 40 kNm alone without regard to the uniform load, and a second moment diagram will be drawn for the uniform load alone without regard to the moment load of 40 kNm. In the construction of each of these diagrams [Fig. 22.43. (*b*) and (*c*)] it is most convenient to work from the right end of the beam towards the left end.

The first bending moment diagram [Fig. 22.43 (*b*)] due to moment load of 40 kNm alone, is evidently a rectangle, since the bending moment due to 40 kNm is the same at all points along the beam. The moment of 40 kNm causes the beam to bend into a configuration that is concave downwards, which according to sign convention constitutes a negative bending moment.

The second bending moment diagram [Fig. 22.42 (*c*)] due to uniform load of 15 kN/m is parabolic except the parabola here corresponds only to the portionof the beam that is subject to the uniform load i.e., the 3m portion of the beam as shown.

A tangent to the deflection curve is now drawn at point *A*. The free end of the beam is designated point *D*. This coincides with the original position of the beam and is represented by the straight line shown in Fig. 22.43 (*a*).

Slope at the fre end *D*, θ = (= θ$_D$ = θ$_{max}$) :

According to the first moment-area theorem the angle θ between the two tangents drawn to the deflection curve at the clamped end *A* and the free end *D*, as shown in Fig. 22.43 (*a*), is equal to the area under the bendign moment diagram between *A* and *D*, divided by *EI*.

Thus, from this theorem, *EI* times the angle θ is given by the area of the two bending moment diagrams shown in Fig. 22.43 (*b*).

We have,
$$EI\,\theta = -(40 \times 5) - \left(\frac{1}{3} \times 67.5 \times 3\right) - \frac{1}{2}(135 - 67.5) \times 1.5 - (67.5 \times 1.5)$$

$$= -200 - 67.5 - 50.625 - 101.25 = -419.375$$

∴
$$\theta\,(=\theta_D = \theta_{max}) = -\frac{419.375}{EI} \quad \text{(Ans.)}$$

The negative sign denotes that the right hand segment at *B* makes a clockwise angle with the left hand tangent at *A*. The angle θ is in radians.

Deflection at the free end *D*, *y* (= *y*$_D$ = *y*$_{max}$) :

We know that according to the second moment-area theorem, the deflection of *D* from the tangent at *A* is given by the moment about the vertical line through *D* of the area under the entire bending moment diagram between *A* and *D* divided by *EI*.

The moment of the area of the entire bending moment diagram between *A* and *D* about the vertical line through *D* is most easily found out by adding the moments of the parabolic, rectangular and triangular areas about the vertical line. For each of these areas, the moment about the line through *D* is given by the product of the area and distance from the centroid of the area to *D*.

Thus, we have,
$$EIy = (-40 \times 5) \times \left(\frac{5}{2}\right) - \left(\frac{1}{3} \times 67.5 \times 3\right)(0.5 + 2)$$

$$- \left[\frac{1}{2}(135 - 67.5) \times 1.5\right]\left(3.5 + \frac{2}{3} \times 1.5\right) - (67.5 \times 1.5)\left(3.5 + \frac{1.5}{2}\right)$$

$$= -500 - 168.75 - 227.8125 - 430.3125 = -1326.875$$

∴
$$y = (= y_D = y_{max}) = -\frac{1326.875}{EI} \quad \text{(Ans.)}$$

The negative sign indicates that the final position of *D* lies below the tangent drawn at *A*.

Example 22.35. *Use either Macaulay's method or the area-moment method to show that the deflection of a simply supported beam at point B with an off-centre load at point A, as shown in Fig.22.44 is given by*

$$y_B = \cdot \frac{PL_2 x (L^2 - x^2 - L_2^2)}{6\,EIL}$$

(IAS)

Solution. Refer to Fig. 22.44.

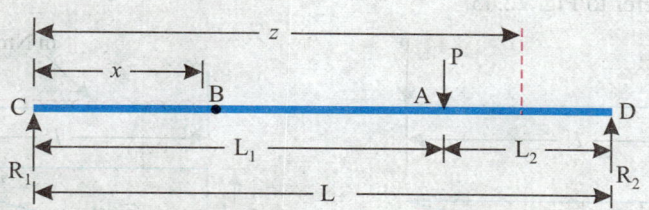

Fig. 22.44

Taking moments about D, we get

$$R_1 \times L = P \times L_2$$

$$\therefore \qquad R_1 = \frac{PL_2}{L}$$

$$M_z = R_1 z - P(z - L_1)$$

or,

$$EI \frac{d^2 y}{dz^2} = R_1 z - P(z - L_1)$$

$$EI \frac{dy}{dz} = \frac{R_1 z^2}{2} - \frac{P(z - L_1)^2}{2} + C_1$$

$$EIy = \frac{R_1 z^3}{6} - \frac{P(z - L_1)^3}{6} + C_1 z + C_2$$

At, $z = 0$, $\qquad y = 0$ gives $C_2 = 0$

At, $z = L$, $\qquad y = 0$, gives

$$0 = \frac{R_1 L^3}{6} - \frac{P(L - L_1)^3}{6} + C_1 L$$

$$= \frac{PL_2 L^2}{6} - \frac{PL_2^3}{6} + C_1 L$$

$$\therefore \qquad C_1 = \frac{PL_2^3}{6L} - \frac{PL_2 L}{6}$$

$$EI z = \frac{PL_2}{6L} z^3 - \frac{P(z - L_1)^3}{6} + z \left(\frac{PL_2^3}{6L} - \frac{PL_2 L}{6} \right)$$

At $z = x < L_1$

$$EIy = \frac{PL_2}{6L} x^3 + x \left(\frac{PL_2^3}{6L} - \frac{PL_2 L}{6} \right)$$

$$= -\frac{PL_2 x}{6L} \left[L^2 - x^2 - L_2^2 \right]$$

$$y_B = \frac{PL_2 x}{6\,EIL} \left(L^2 - x^2 - L_2^2 \right) \qquad (downward)$$

(– ve sign indicates downward deflection) .... **Proved.**

Example 22.36. *A cantilever beam of length l and uniform flexural rigidity EI is subjected to continuously dstributed externally applied moment of intensity m Nm per m of length of the beam.*

Using the area-moment method, show that the deflections of the free end of the beam is $\dfrac{ml^3}{3EI}$, *and explain why this is the same deflection as that obtained for the case of concentrated froce P = m applied at the end of the beam.* **(IAS)**

Solution. Refer to Fig. 22.45.

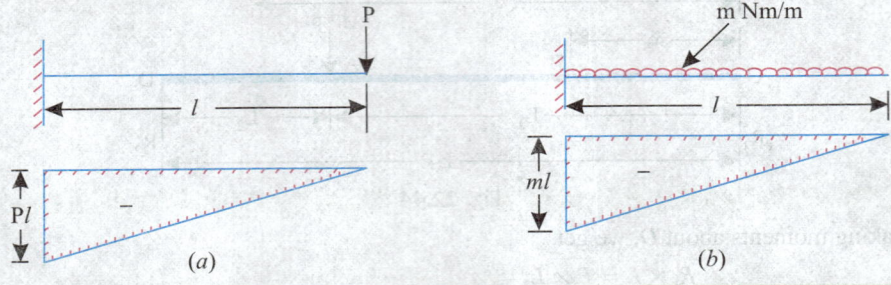

Fig. 22.45

Concentrated load *P* :

$$\delta_1 = \frac{1}{EI}\left[\left(\frac{1}{2} \times Pl \times l\right) \times \frac{2}{3}l\right] = \frac{Pl^3}{3EI}$$

Uniformly distributed moment m Nm/m :

$$\delta_2 = \frac{1}{EI}\left[\left(\frac{1}{2} \times ml \times l\right) \times \frac{2}{3}l\right] = \frac{ml^3}{3EI}$$

Now, $\delta_1 = \delta_2$ for $P = m$

This is so because total moment due to m Nm/m over length *l* is *ml*, which is equal to the maximum bending moment *Pl* due to concentrated end load.

Example 22.37. *A simply supported beam of length l carries a load W at a distance 'a' from one end, 'b' from the other (a > b). Find the position and magnitude of maximum deflection.* **(IES)**

Solution. Refer to Fig. 22.46

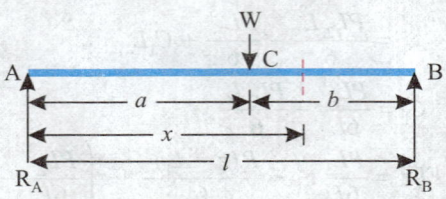

Fig. 22.46

Taking moments about *B*, we get

$$R_A \times l = W \times b$$

∴ $$R_A = \frac{Wb}{l}$$

Now, $$EI\frac{d^2y}{dx^2} = M_x = R_A \cdot x - W(x - a) = \frac{Wb}{l} \cdot x - W(x - a)$$

$$EI\frac{dy}{dx} = \frac{Wb}{l} \cdot \frac{x^2}{2} - \frac{W(x - a)^2}{2} + C_1$$

$$EIy = \frac{Wb}{l} \cdot \frac{x^3}{6} - \frac{W(x-a)^3}{6} + C_1 x + C_2$$

At $x = 0$ and $x = l$, $y = 0$ gives $C_2 = 0$

$$0 = \frac{Wbl^2}{6} - \frac{Wb^3}{6} + C_1 l$$

or,

$$C_1 = -\frac{Wbl}{6} + \frac{Wb^3}{6l}$$

∴

$$EIy = \frac{Wb}{6l} x^3 - \frac{W(x-a)^3}{6} + \left(\frac{Wb^3}{6l} - \frac{Wbl}{6} \right) x$$

For $x < a$, second term on R.H.S. becomes negative and is therefore neglected.

Similarly, $EI \dfrac{dy}{dx} = \dfrac{Wb}{2l} x^2 + \left(\dfrac{Wb^3}{6l} - \dfrac{Wbl}{6} \right)$

For y_{max}, $\dfrac{dy}{dx} = 0$

∴

$$0 = \frac{Wb}{2l} x^2 + \left(\frac{Wb^3}{6l} - \frac{Wbl}{6} \right)$$

$$x^2 = \frac{l^2}{3} - \frac{b^2}{3}$$

or,

$$x = \sqrt{\frac{l^2 - b^2}{3}} = \sqrt{\frac{(a+b)^2 - b^2}{3}} = \sqrt{\frac{a^2 + 2ab}{3}}$$

$$EIy_{max} = \frac{Wb}{6l} \left(\frac{a^2 + 2ab}{3} \right)^{3/2} + \left(\frac{Wb^3}{6l} - \frac{Wbl}{6} \right) \left(\sqrt{\frac{a^2 + 2ab}{3}} \right)$$

$$= \sqrt{\frac{a^2 + 2ab}{3}} \left[\frac{Wb}{6l} \left(\frac{a^2 + 2ab}{3} \right) + \frac{Wb^3}{6l} - \frac{Wbl}{6} \right]$$

$$= \sqrt{\frac{a^2 + 2ab}{3}} \left[\frac{Wb}{6l} \left(\frac{a^2 + 2ab}{3} + b^2 \right) - \frac{Wbl}{6} \right]$$

$$= \sqrt{\frac{a^2 + 2ab}{3}} \times \frac{Wb}{6l} \left[\left(\frac{a^2 + 2ab}{3} + b^2 \right) - l^2 \right]$$

$$= \sqrt{\frac{a^2 + 2ab}{3}} \times \frac{Wb}{6l} \left[\frac{a^2 + 2ab + 3b^2 - 3(a^2 + b^2 + 2ab)}{3} \right]$$

$$= -\sqrt{\frac{a^2 + 2ab}{3}} \times \frac{Wb}{6l} \left[\frac{2a^2 + 4ab}{3} \right]$$

$$= -\sqrt{\frac{a^2 + 2ab}{3}} \times \frac{Wb}{9l} (a^2 + 2ab)$$

$$= -\frac{Wb}{9l \times \sqrt{3}} (a^2 + 2ab)^{3/2}$$

∴

$$y_{max} = \frac{Wb}{9\sqrt{3}\, EIl} (a^2 + 2ab)^{3/2} \quad (downward) \quad \textbf{(Ans.)}$$

Example 22.38. *A horizontal cantilever of uniform section of length l carries two concentrated loads W at the free end and 2W at a distance 'a' from the free end. Starting from the first priciples, determine the maximum deflection for the loading.*

If the cantilever is made from a steel tube of circular section of 100 mm external diameter and 6 mm thickness and l = 1.5 m, a = 0.6 m, determine the value of W so that the maximum bending stress is 140 MN/m². Calculate the maximum deflection for the loading. Take E = 200 GN/m². **(IES)**

Solution. Refer to Fig. 22.47.

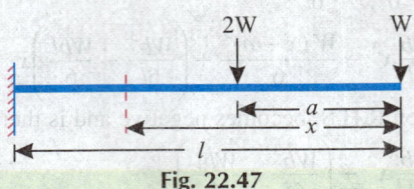

Fig. 22.47

$$EI \frac{d^2y}{dx^2} = M_x = -\left[Wx + 2W(x-a)\right]$$

$$EI \frac{dy}{dx} = -\left[\frac{Wx^2}{2} + W(x-a)^2\right] + C_1$$

$$EIy = -\left[\frac{Wx^3}{6} + \frac{W(x-a)^3}{3}\right] + C_1 x + C_2$$

At $x = l$, $y = 0$ and $\frac{dy}{dx} = 0$, we have

$$0 = -\left[\frac{Wl^2}{2} + W(l-a)^2\right] + C_1$$

and,

$$0 = -\left[\frac{Wl^3}{6} + \frac{W(l-a)^3}{3}\right] + C_1 l + C_2$$

∴

$$C_1 = \frac{Wl^2}{2} + W(l-a)^2 = \frac{Wl^2}{2} + Wl^2 + Wa^2 - 2Wla$$

$$= \frac{3}{2}Wl^2 + Wa^2 - 2Wal$$

and,

$$C_2 = \frac{Wl^3}{6} + \frac{W(l-a)^3}{3} - \frac{3Wl^3}{2} - Wa^2l + 2Wal^2$$

$$= -\frac{4}{3}Wl^3 + \frac{W(l-a)^3}{3} - W\left(a^2l - 2al^2\right)$$

$$= \frac{W}{3}\left[-4l^3 + (l-a)^3 - 3a^2l + 6al^2\right]$$

$$= -\frac{W}{3}\left[4l^3 - \left\{l^3 - a^3 - 3l^2a + 3la^2\right\} + 3a^2l - 6al^2\right]$$

$$= -\frac{W}{3}\left[4l^3 - l^3 + a^3 + 3l^2a - 3la^2 + 3a^2l - 3al^2 - 6al^2\right]$$

$$= -\frac{W}{3}\left(3l^3 - 3al^2 + a^3\right)$$

∴

$$EIy = -\left[\frac{Wx^3}{6} + \frac{W(x-a)^3}{3}\right] + C_1 x - \frac{W}{3}\left(3l^3 - 3al^2 + a^3\right)$$

At, $x = 0$

$$EIy_{max} = -\frac{W}{3}\left(3l^3 - 3al^2 + a^3\right) \qquad ...(i)$$

$$M_{max} = Wl + 2W(l - a) = 3Wl - 2Wa$$
$$= W(3l - 2a)$$
$$= W(3 \times 1.5 - 2 \times 0.6) = 3.3\ W$$

Refer to Fig. 22.48.

$$d_i = 100 - 2 \times 6 = 88\ \text{mm}$$

Section modulus,

$$Z = \frac{I}{y} = \frac{\frac{\pi}{64}\left(d_0^4 - d_i^4\right)}{(d_0 / 2)}$$

$$= \frac{\pi}{32} d_0^3 \left[1 - \left(\frac{d_i}{d_0}\right)^4\right]$$

$$= \frac{\pi}{32} \times 100^3 \left[1 - \left(\frac{88}{100}\right)^4\right] \times 10^{-9}\ \text{m}^3$$

$$= 39.3 \times 10^{-6}\ \text{m}^3$$

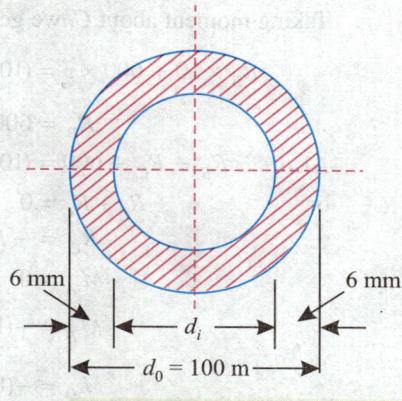

Fig. 22.48. Cross-section of the cantilever.

Now,

$$M_{max} = 3.3\ W = \sigma_{max} Z = 140 \times 10^6 \times 39.3 \times 10^{-6}$$

$$(\text{where, } \sigma_{max} = 140\ \text{MN/m}^2 \dots \text{Given})$$

or,

$$W = 1667.3\ \text{N}$$

Also,

$$I = \frac{\pi}{64}\left(d_0^4 - d_i^4\right) = \frac{\pi}{64}(100^4 - 88^4) = 1964991\ \text{mm}^4$$

From eqn. (i), we have

$$y_{max} = -\frac{1}{EI} \times \frac{W}{3}(3l^3 - 3al^2 + a^3)$$

$$= -\frac{1}{200 \times 10^9 \times 1964991 \times 10^{-12}} \times \frac{1667.3}{3}\left(3 \times 1.5^3 - 3 \times 0.6 \times 1.5^2 + 0.6^3\right)$$

$$= -2.544 \times 10^{-6} \times 3496.3\ \text{m} \quad \text{or} \quad -8.89\ \text{mm}$$

$$\therefore \quad y_{max} = \textbf{8.89 mm} \ (downward) \ \textbf{(Ans.)}$$

Example 22.39. *For the composite beam shown in Fig. 22.49 flexural rigidities EI of AB and DC are equal to 10^5 N cm², and 2×10^5 N cm² respectively. Using moment-area theorem, determine the location and magnitude of maximum deflection between B and C.* **(GATE)**

Solution. Refer to Fig. 22.49.

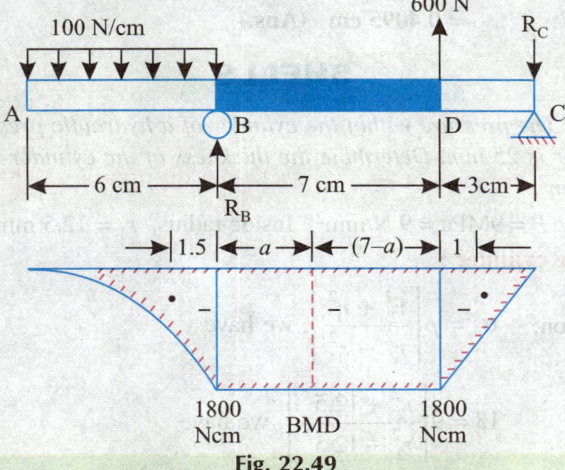

Fig. 22.49

Taking moment about C, we get

$$R_B \times 10 + 600 \times 3 = (100 \times 6) \times \left(\frac{6}{2} + 7 + 3\right) = 7800$$

$\therefore$ $\qquad\qquad R_B = 600$ N

Also, $\qquad R_B + R_C + 600 = (100 \times 6)$

or, $\qquad\qquad R_B + R_C = 0$

$\therefore$ $\qquad\qquad R_C = -R_B = -600$ N

$$M_A = 0$$

$$M_B = -(100 \times 6) \times \frac{6}{2} = -1800 \text{ N cm}$$

$$M_D = -(100 \times 6) \times \left(\frac{6}{2} + 7\right) + 600 \times 7 = -1800 \text{ N cm}$$

$$M_C = 0$$

Let the deflection be maximum at a distance 'a' from B. Then,

$$\frac{1}{10^5}\left[\frac{1}{3} \times 1800 \times 6 \times \left(a + \frac{6}{4}\right)\right] + \frac{1}{2 \times 10^5}\left[1800 \times a \times \frac{a}{2}\right]$$

$$= \frac{1}{10^5}\left[\frac{1}{2} \times 1800 \times 3 \times \left(\frac{3}{3} + 7 - a\right)\right] + \frac{1}{2 \times 10^5}\left[\frac{1800 \times (7-a)^2}{2}\right]$$

$3600\,(a + 1.5) + 450\,a^2 = 2700\,(8 - a) + 450\,(7 - a)^2$

$3600\,a + 5400 + 450\,a^2 = 21600 - 2700\,a + (22050 + 450\,a^2 - 6300\,a)$

$$= 450\,a^2 - 9000\,a + 43650$$

$$12600\,a = 328250$$

or, $\qquad\qquad a = \textbf{3.036 cm}$ **(Ans.)**

Maximum deflection,

$$y_{max} = \frac{1}{10^5}\left[(3600\,a + 5400 + 450\,a^2) + (450\,a^2 - 9000\,a + 43650)\right]$$

$$= \frac{1}{10^5}\left(900\,a^2 - 5400\,a + 49050\right)$$

$$= \frac{1}{10^5}\left(900 \times 3.036^2 - 5400 \times 3.036 + 49050\right)$$

$$= \textbf{0.4095 cm} \quad \textbf{(Ans.)}$$

SHELLS

Example 22.40. *The pressure within the cylinder of a hydraulic press is 9 MPa. The inside diameter of the cylinder is 25 mm. Determine the thickness of the cylinder wall, if the permissible tensile stress is 18 N/mm².* **(UPSC)**

Solution. *Given :* $P = 9$MPa $= 9$ N/mm²; Inside radius, $r_1 = 12.5$ mm; $\sigma_t = 18$ N/mm².

Thickness of the cylinder :

Using the equation; $\qquad \sigma_t = p\left[\dfrac{r_2^2 + r_1^2}{r_2^2 - r_1^2}\right]$, we have

$$18 = 9\left[\frac{r_2^2 + 12.5^2}{r_2^2 - 12.5^2}\right], \text{ we have}$$

or,
$$2 = \frac{r_2^2 + 156.25}{r_2^2 - 156.25}$$

or,
$$2r_2^2 - 312.5 = r_2^2 + 156.25$$

or,
$$r_2^2 = 468.75$$

or,
$$r_2 = 21.65 \text{ mm}$$

∴ Thickness of the cylinder $= r_2 - r_1 = 21.65 - 12.5 = $ **9.15 mm (Ans.)**

Example 22.41. *A thick spherical vessel of inner radius 150 mm is subjected to an internal pressure of 80 MPa. Calculate its wall thickness based upon the*

(i) *Maximum principal stress theory, and*

(ii) *Total strain energy theory.*

Poisson's ratio = 0.30; yield strength = 300 MPa. **(UPSC)**

Solution. *Given :* $\quad r_1 = 150 \text{ mm}; p (\sigma_r) = 80 \text{ MPa} = 80 \times 10^6 \text{ N/m}^2;$

$$\mu = \frac{1}{m} = 0.30;$$

$$\sigma = 300 \text{ MPa} = 300 \times 10^6 \text{ N/m}^2$$

Wall thickness t :

(i) **Maximum principal stress theory :**

We know that, $\quad \sigma_r \left(\dfrac{K^2 + 1}{K^2 - 1} \right) \le \sigma$ [Eqn. (11.17)] $\qquad$ (where, $K = \dfrac{r_2}{r_1}$)

or,
$$80 \times 10^6 \left(\frac{K^2 + 1}{K^2 - 1} \right) \le 300 \times 10^6$$

or,
$$\frac{K^2 + 1}{K^2 - 1} \le 3.75$$

or,
$$(K^2 + 1) \le (3.75 \, K^2 - 3.75)$$

or,
$$K^2 \ge 1.727 \quad \text{or} \quad K \ge 1.314$$

or,
$$K = 1.314$$

i.e.,
$$\frac{r_2}{r_1} = 1.314 \quad \text{or} \quad r_2 = r_1 \times 1.314 = 150 \times 3.14 = 197.1 \text{ mm}$$

∴ *Metal thickness,* $\quad t = r_2 - r_1 = 197.1 - 150 = $ **47.1 mm (Ans.)**

(ii) **Total strain energy theory :**

$$\sigma^2 \ge \frac{2\sigma_r^2 \left[K^4 \left(1 + \dfrac{1}{m}\right) + \left(1 - \dfrac{1}{m}\right) \right]}{(K^2 - 1)^2}$$

∴
$$(300 \times 10^6)^2 \ge \frac{2 \times (80 \times 10^6)^2 \, [K^4 (1 + 0.3) + (1 - 0.3)]}{(K^2 - 1)^2}$$

or,
$$300^2 (K^2 - 1)^2 = 2 \times 80^2 (1.3 \, K^4 + 0.7)$$

$$\frac{300}{2 \times 80^2} (K^4 - 2K^2 + 1) = 1.3 \, K^4 + 0.7$$

$$7.03 \, K^4 - 14.06 \, K^2 + 7.03 = 1.3 \, K^4 + 0.7$$

or,
$$5.73 \, K^4 - 14.06 \, K^2 + 6.33 = 0$$

$$K^2 = \frac{14.06 \pm \sqrt{(14.06)^2 - 4 \times 5.73 \times 6.33}}{2 \times 5.73} = \frac{14.06 \pm 7.25}{11.46} = 1.86 \text{ or } 0.59$$

or,
$$K = 1.364 \quad \text{(Consider + ve sign since } K > 1\text{)}$$

or,
$$\frac{r_2}{r_1} = 1.364 \text{ or } r_2 = 150 \times 1.364 = 204.6 \text{ m}$$

∴
$$t = r_2 - r_1 = 204.6 - 150 = \textbf{54.6 mm (Ans.)}$$

Example 22.42. *The longitudinal joint of a thin cylindrical pressure vessel, 6m internal diameter and 16 mm plate thickness, is double riveted lap point with no staggering between the rows. The rivets are of 20 mm nominal diameter with a pitch of 72 mm.*

What is the efficiency of the joint and what would be the safe pressure inside the vessel?

Allowable stresses for the plate and rivet materials are; 145 MN/m² in tension, 120 MN/m² in shear, and 230 MN/m² in bearing. Take rivet hole diameter as 1.5 mm more than the rivet diameter.

(GATE)

Solution. *Given* : Diameter of rivet = 20 mm

Diameter of hole = 20 + 1.5 = 21.5 mm

Diameter of the pressure vessel, $d = 6$ m

Thickness of the plate, $t = 16$ mm

Type of the joint: Double riveted lap joint

Allwable stresses :

$$\sigma_t = 145 \text{ MN/m}^2; \quad \tau = 120 \text{ MN/m}^2; \quad \sigma_c = 230 \text{ MN/m}^2$$

Strength of plate in tearing/pitch,

$$R_t = \left[\frac{72 - (2 \times 21.5)}{1000} \right] \times \frac{16}{1000} \times 145 = 0.06728 \text{ MN}$$

Strength of the rivet in shear/pitch,

$$R_s = 2 \times \frac{\pi}{4} \times \left(\frac{20}{100} \right)^2 \times 120 = 0.0754 \text{ MN}$$

Strength of plate in crushing/pitch,

$$R_s = 2 \times \left(\frac{20}{1000} \times \frac{16}{1000} \right) \times 230 = 0.1472 \text{ MN}$$

From the above three modes of failure it can be seen that the weakest element is the plate as it will have tear failure at 0.06728 MN/pitch load itself.

Stresses acting on the plate for an inside pressure of pN/m² is shown in Fig. 22.51.

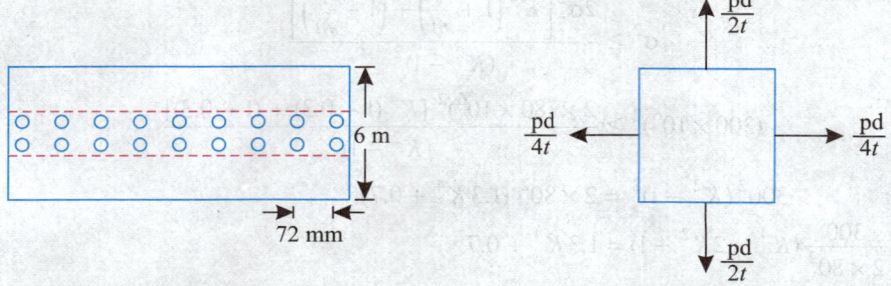

Fig. 22.50. Arrangement of rivets.   **Fig. 22.51.** Stress distribution.

$$\text{Hoop stress} = \frac{pd}{2t} = \frac{p \times 6}{2 \times \left(\frac{16}{1000}\right)} = 187.5\ p$$

$$\text{Longitudinal stress} = \frac{pd}{4t} = \frac{p \times 6}{4 \times \left(\frac{16}{1000}\right)} = 93.75\ p$$

Maximum principal stress acting on the plate $= \dfrac{pd}{2t}$ only (i.e., 187.5 p) as there is no shear stress.

or, $\quad\quad 187.5\ p \le \dfrac{0.06728}{\dfrac{16}{1000} \times \left[\dfrac{72 - (2 \times 21.5)}{1000}\right]} \le 145$

or, $\quad\quad p \le 0.7733\ \text{MN/m}^2$ or **0.7733 MPa (Ans.)**

$$\eta_{\text{joint}} = \frac{R_t}{p.t.\sigma_t} = \frac{0.06728}{\dfrac{72}{1000} \times \dfrac{16}{1000} \times 145} = 0.4028 \text{ or } \textbf{40.28\% (Ans.)}$$

Example 22.43. *A thick walled cylindrical vessel with internal radius $r_1 = 0.25$ m and external radius $r_2 = 0.35$ m, has rigid end plates welded to its two ends (Fig.22.52). It is subjected to a tensile load $F = 1000$ kN through the end plates, a twisting moment $T = 100$ kNm, and an internal pressure $p = 60$ MN/m². Determine the principal stresses and the absolute maximum shear stress on the inside surface of the vessel. Neglect the end effects.* **(IES)**

Solution. *Given :* $r_1 = 0.25$ m; $r_2 = 0.35$ m; $F = 1000$ kN; $T = 100$ kNm; $p = 60$ MN/m²

$\sigma_1;\ \sigma_2;\ \tau_{max}$:

Refer to Fig. 22.52.

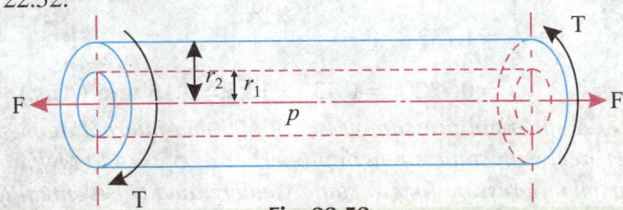

Fig. 22.52

Direct stress, $\quad\quad \sigma_d = \dfrac{F}{A} = \dfrac{1000 \times 10^3}{\pi (0.35^2 - 0.25^2)} = 5.305\ \text{MN/m}^2$

Torque, $\quad\quad T = \tau \times \dfrac{\pi}{16}\left(\dfrac{d_2^{\,4} - d_1^{\,4}}{d_2}\right)$

or, $\quad\quad 100 \times 1000 = \tau \times \dfrac{\pi}{16}\left(\dfrac{0.7^4 - 0.5^4}{0.7}\right) = 0.0498\ \tau$

$\therefore \quad\quad \tau = 2.01\ \text{MN/m}^2$

Hoop stress, $\quad \sigma_c = p\left[\dfrac{r_2^2 + r_1^2}{r_2^2 - r_1^2}\right] = 60\left[\dfrac{0.35^2 + 0.25^2}{0.35^2 - 0.25^2}\right] = 185\ \text{MN/m}^2$

Principal stress, $\quad \sigma_{1,2} = \left(\dfrac{\sigma_c + \sigma_d}{2}\right) \pm \sqrt{\left(\dfrac{\sigma_c - \sigma_d}{2}\right)^2 + \tau^2}$

$$= \left(\dfrac{185 + 5.035}{2}\right) \pm \sqrt{\left(\dfrac{185 - 5.305}{2}\right)^2 + 2.01^2} = 95.15 \pm 89.87$$

i.e., $\quad\quad \sigma_1 = \textbf{185.02 MN/m}^2;\ \sigma_2 = \textbf{5.28 MN/m}^2 \textbf{ (Ans.)}$

$$\tau_{max} = \frac{\sigma_1 - \sigma_2}{2} = \frac{185.02 - 5.28}{2} = 89.87 \text{ MN/m}^2 \text{ (Ans.)}$$

Example 22.44. *The pressure within the cylinder of a hydraulic press is 9 MN/m². The inside diameter of the cylinder is 25 mm. Determine the thickness of the cylinder wall, if the permissible tensile stress is 18 N/mm².* **(IES)**

Solution. *Given :* $p = 9 \text{ MN/m}^2$; $r_1 = \dfrac{25}{2} = 12.5 \text{ m}$; $\sigma_c = 18 \text{ N/mm}^2$ or 18 MN/m^2

Thickness of cylinder wall, t :

$$\sigma_c = p\left(\frac{r_2^2 + r_1^2}{r_2^2 - r_1^2}\right), \qquad \text{where, } r_2 = r_1 + t$$

or,
$$\sigma_c = p\left(\frac{K^2 + 1}{K^2 - 1}\right), \quad \text{where } K = \frac{r_2}{r_1}$$

$$18 = 9\left(\frac{K^2 + 1}{K^2 - 1}\right)$$

or,
$$2(K^2 - 1) = K^2 + 1$$

or,
$$2K^2 - 2 = K^2 + 1$$

or,
$$K = \sqrt{3} = 1.732$$

or,
$$\frac{r_2}{r_1} = \frac{r_1 + t}{r_1} = 1.732$$

or,
$$1 + \frac{t}{r_1} = 1.732$$

or,
$$t = 0.732 \, r_1 = 0.732 \times 12.5 = \textbf{9.15 mm (Ans.)}$$

Example 22.45. *A spherical pressure vessel of 600 mm internal diameter is made of 3 mm thick cold drawn sheet steel with static strength properties of $\sigma_{ut} = 440 \text{ MN/m}^2$ and $\sigma_{yp} = 370 \text{ MN/m}^2$. Determine the maximum pressure for (a) static yielding, and (b) eventual fatigue failure, when the pressure fluctuates between 0 and p_{max} when yielding is also not permitted. (Do not apply any factor of safety.)* **(GATE)**

Solution. *Given :* $d = 600 \text{ mm}$; $t = 3 \text{ mm}$; $\sigma_{ut} = 440 \text{ MN/m}^2$; $\sigma_{yp} = 370 \text{ MN/m}^2$

(*a*) $\sigma_l = \dfrac{pd}{4t}$ or $370 = \dfrac{p \times 600}{4 \times 3}$

∴
$$p = \textbf{7.4 MN/m}^2 \textbf{ (Ans.)}$$

(*b*) When pressure fluctuates between 0 and p_{max}, then $\sigma_m = \sigma_a$. Using Soderberg line of failure,

$$\frac{\sigma_m}{\sigma_{yp}} + \frac{\sigma_a}{\sigma_e} = 1$$

Endurance limit $\sigma_e = 0.5 \, \sigma_{ut} = 0.5 \times 440 = 220 \text{ MN/m}^2$

∴
$$\frac{p \times 600}{2 \times 4 \times 3 \times 370} + \frac{p \times 600}{2 \times 4 \times 3 \times 220} = 1$$

or,
$$600 \, p\left(\frac{1}{2 \times 4 \times 3 \times 370} + \frac{1}{2 \times 4 \times 3 \times 220}\right) = 1$$

or,
$$600 \, p\left(\frac{1}{8880} + \frac{1}{5280}\right) = 1$$

∴
$$p = \textbf{5.52 MN/m}^2 \textbf{ (Ans.)}$$

RIVETED AND WELDED JOINTS

Example 22.46. *Find the maximum permissible value of load P for the riveted joint shown in Fig.22.53 (a), if the allowable yield shear strength of the rivet material is 100 MPa. Rivets are 20 mm in diameter.* **(GATE)**

Solution. *Given* : Yield shear strength of rivet material, $\tau_{yield} = 100$ MPa $= 100$ N/mm^2; diameter of each rivet, $d = 20$ mm.

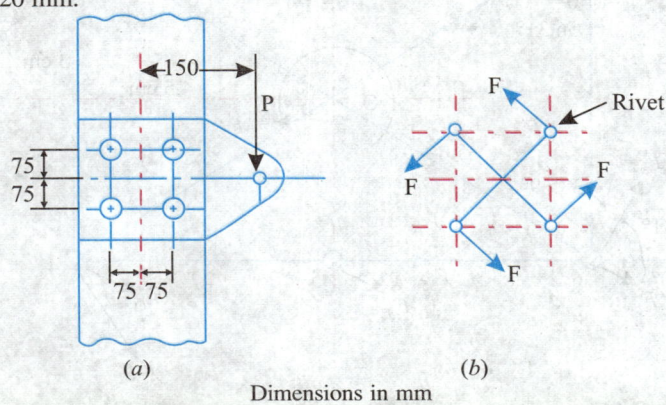

(a)   (b)

Dimensions in mm

Fig. 22.53

Maximum value of load P :

Let, F = Force in a rivet.

Taking the rivets in single shear, we have

$$P \times 150 = 4 \times F \times 75\sqrt{2} = 4 \times (\tau_{yield} \times \frac{\pi}{4}d^2) \times 75\sqrt{2}$$

or,

$$= 4 \times \left[100\left(\frac{\pi}{4} \times 20^2\right)\right] \times 75\sqrt{2}$$

∴ $P = 88858$ N say **88.86 kN (Ans.)**

Example 22.47. *Calculate the forces in all the rivets as shown in Fig. 22.54*

(Roorkee University)

Solution. Refer to Figs. 22.54 and 22.55.

Direct shear force on each rivet; $F_d = \dfrac{180}{6} = 30$ kN

i.e., $F_{d1} = F_{d2} = F_{d3} = F_{d4} = F_{d5} = F_{d6} = F_{d7} = 30$ kN

The centroid of the rivet group is at 0. The distance of the rivets 2 and 5 from the centre of rivets group is 3 cm. The distance of rivets 1, 3, 4 and 6 from centroid $= \sqrt{4.5^2 + 3^2} = 5.4$ cm

Turning moment due to the load 180 kN

$$= 180 \times \text{eccentricity} = 180 \times 30 \sin 30° \times 10^{-2} = 27 \text{ kNm}$$

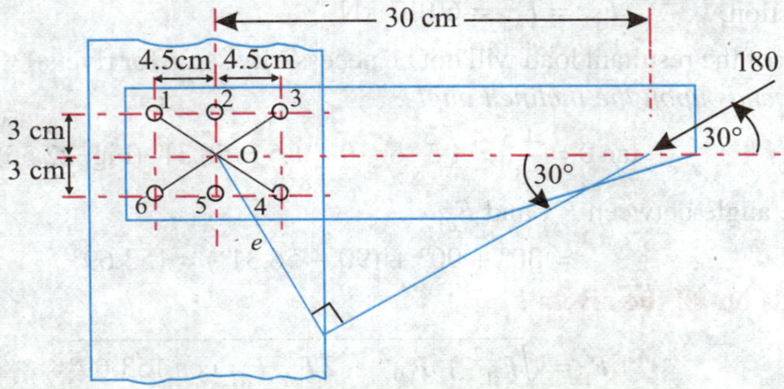

Fig. 22.54

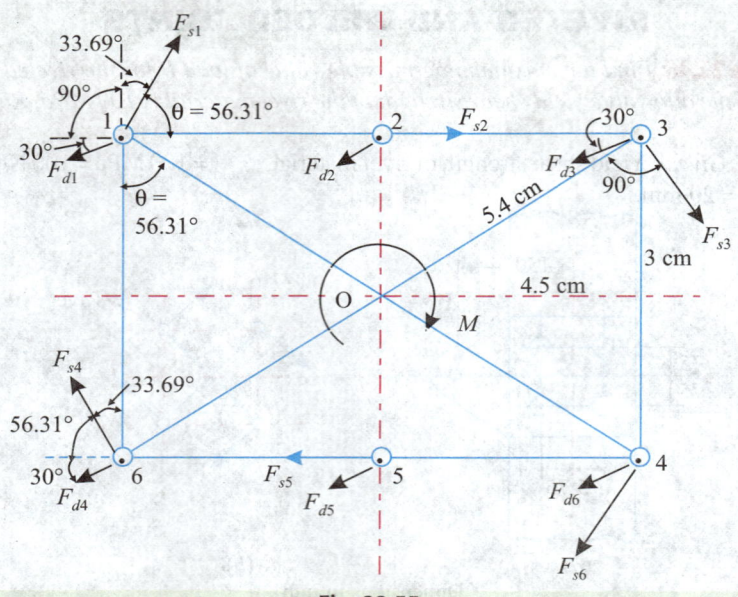

Fig. 22.55

The turning moment is resisted by six rivets.

Let F_{s1}, F_{s2}, F_{s3}, F_{s4}, F_{s5}, F_{s6} be the secondary shear due to turning moment in the rivets 1, 2, 3, 4, 5 and 6 placed at l_1, l_2, l_3, l_4, l_5 and l_6 respectively from the centroid.

Then,
$$l_1 = l_3 = l_4 = l_6 = 5.4 \text{ cm} = 0.054 \text{ m}$$
$$l_2 = l_5 = 3 \text{ cm} = 0.03 = 0.03 \text{ m}$$

Now, equating the resisting moment to turning moment, we have

$$\frac{F_{s1}}{l_1} = (l_1^2 + l_2^2 + l_3^2 + l_4^2 + l_5^2 + l_6^2) = 27$$

$$\frac{F_{s1}}{0.054} = \left[4 \times (0.054)^2 + 2 \times (0.03)^2 \right] = 27$$

∴ $$F_{s1} = 108.3 \text{ kN}$$

By inspection we can say, $F_{s1} = F_{s3} = F_{s4} = F_{s6} = 108.3 \text{ kN}$

Since,
$$\frac{F_{s1}}{l_1} = \frac{F_{s2}}{l_2} = \frac{F_{s5}}{l_5}$$

∴ $$F_{s2} = F_{s1} \times \frac{l_2}{l_1} = 108.3 \times \frac{0.03}{0.054} = 60.17 \text{ kN}$$

Also by inspection, $F_{s2} = F_{s5} = 60.17 \text{ kN}$

The magnitude of the resultant load will not be necessarily same for rivets 1, 3, 4 and 6 because the *resultant load depends upon the inclined angle,*

$$\tan \theta = \frac{4.5}{3} = 1.5 \text{ or } \theta = 1.5 = 56.31° \text{ (Fig. 22.55)}$$

For rivet 1, the angle between F_{s1} and F_{d1}

$$= 30° + 90° + (90 - 56.31°) = 153.69°$$

Resultant loads on all the rivets :

$$R_1 = \sqrt{F_{s1}^2 + F_{d1}^2 + 2 F_{s1} F_{d1} \cos 153.69°}$$

$$= \sqrt{108.3^2 + 30^2 + 2 \times 108.3 \times 30 \cos 153.69°} = \textbf{82.49 kN (Ans.)}$$

$$R_2 = \sqrt{F_{s2}^2 + F_{d2}^2 + 2F_{s2} F_{d2} \cos 150°}$$

$$= \sqrt{60.17^2 + 30^2 + 2 \times 60.17 \times 30 \cos 150°} = \textbf{37.33 kN (Ans.)}$$

$$R_3 = \sqrt{F_{s3}^2 + F_{d3}^2 + 2 F_{s3} F_{d3} \cos [90° + (90° - 56.31° - 30°)]}$$

$$= \sqrt{108.3^2 + 30^2 + 2 \times 108.3 \times 30 \cos 93.69°} = \textbf{110.5 kN (Ans.)}$$

$$R_4 = \sqrt{F_{s4}^2 + F_{d4}^2 + 2 F_{s4} F_{d4} \cos (56.31° - 30°)}$$

$$= \sqrt{108.3^2 + 30^2 + 2 \times 108.3 \times 30 \cos 26.31°} = \textbf{135.84 kN (Ans.)}$$

$$R_5 = \sqrt{F_{s5}^2 + F_{d5}^2 + 2 F_{s5} F_{d5} \cos 30°}$$

$$= \sqrt{60.17^2 + 30^2 + 2 \times 60.17 \times 30 \cos 30°} = \textbf{87.45 kN (Ans.)}$$

$$R_6 = \sqrt{F_{s6}^2 + F_{d6}^2 + 2 F_{s6} F_{d6} \cos (56.31° + 30°)}$$

$$= \sqrt{108.3^2 + 30^2 + 2 \times 108.3 \times 30 \cos 86.31°} = \textbf{114.22 kN (Ans.)}$$

Example 22.48. *A bracket plate 15 mm thick is welded to the flange of a column by fillet welds on either side of the plate as shown in Fig. 22.56. If the size of the weld is 6 mm, determine the maximum stress in the weld.* **(GATE)**

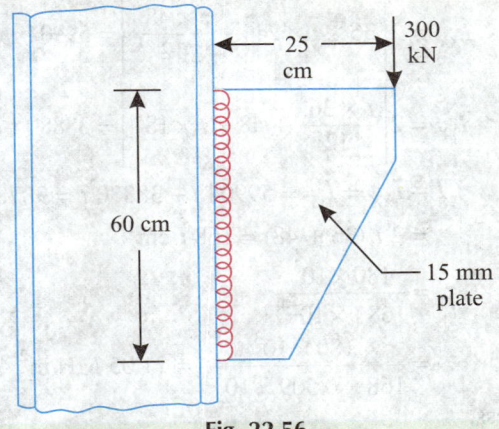

Fig. 22.56

Solution. Thickness of bracket plate = 15 mm

Size of the weld, $\quad h = 6$ mm or 0.6 cm

Maximum stress in the weld :

Moment, $\qquad M = 300 \times 0.25 = 75$ kNm

$$t = 0.707 \times h = 0.707 \times 0.6 = 0.4242 \text{ cm}$$

Direct stress, $\quad \sigma_d = \dfrac{300}{2t \times l} = \dfrac{300 \times 10^3}{2 \times (0.4242 \times 10^{-2}) \times (60 \times 10^{-2})} = 58.93 \text{ MN/m}^2$

Now, $\qquad I = \dfrac{2tl^3}{12} = \dfrac{2 \times 0.4242 \times 60^3}{12} = 15271.2 \text{ cm}^4$ or $15271.2 \times 10^{-8} \text{ m}^4$

∴ Binding stress, $\sigma_b = \dfrac{M \, y}{I} = \dfrac{75 \times 10^3 \times 0.3}{15271.2 \times 10^{-8}} = 147.34 \text{ MN/m}^2$

∴ *Resultant (or maximum) stress,*

$$\sigma_r = \sqrt{\sigma_d^2 + \sigma_b^2} = \sqrt{(58.93)^2 + (147.34)^2} = \textbf{158.69 MN/m}^2 \quad \textbf{Ans.}$$

Example 22.49. *A bracket shown in Fig 22.57 is welded to a column by fillet welds on four sides as shown. If the size of the weld is 7.2 mm, determine the maximum shear stress in the weld.* **(IES)**

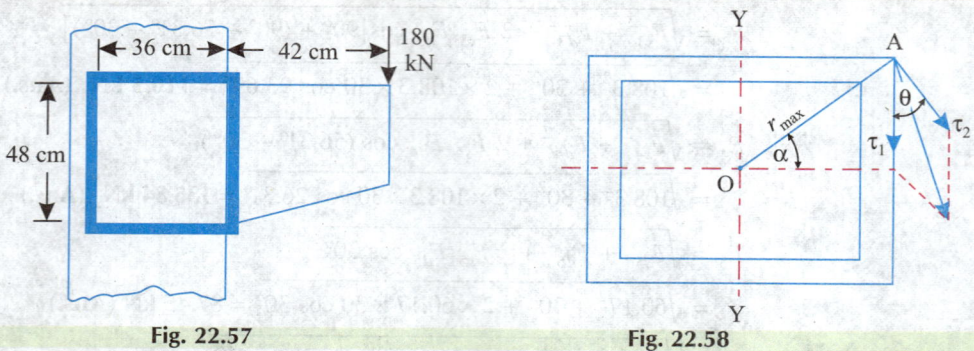

Fig. 22.57   Fig. 22.58

Solution. Maximum shear stress in the weld :

$$M = 180(42 + 18) \times 10^{-2} = 108 \text{ kNm}$$

$$h = 7.2 \text{ mm} \quad \text{or} \quad 0.72 \text{ cm}$$

$$t = 0.707 \, h = 0.707 \times 0.72 = 0.509 \text{ cm}$$

$$I_{XX} = 2\left[36 \times t \times 24^2 + \frac{t \times 48^3}{12} \right] = 59904 \, t \text{ cm}^4$$

$$I_{YY} = 2\left[\frac{t \times 36^3}{12} + 48 \times t \times 18^2 \right] = 38880 \, t \text{ cm}^4$$

$$I = I_{XX} + I_{YY} = 59904 \, t + 38880 \, t = 98784 \, t \text{ cm}^4$$

Total weld area
$$= 2 \, t \, (36 + 48) = 168 \, t \text{ cm}^2$$

Direct shear stress, $\tau_1 = \dfrac{180 \times 10^3}{168 \, t \times 10^{-4}}$

$$= \frac{180 \times 10^3}{168 \times 0.509 \times 10^{-4}} = 21.05 \text{ MN/m}^2$$

Secondary shear stress,

$$\tau_2 = \frac{M \cdot r_{max}}{I} = \frac{108 \times 10^3 \, (\sqrt{18^2 + 24^2}) \times 10^{-2}}{98784 \times 0.509 \times 10^{-8}} = 64.44 \text{ MN/m}^2$$

$$\tan \alpha = \frac{24}{18} = 1.33 \quad \text{or} \quad \alpha = 53.13°$$

Resultant/maximum shear stress

$$= \sqrt{\tau_1^2 + \tau_2^2 + 2 \tau_1 \tau_2 \cos \alpha}$$

$$= \sqrt{21.05^2 + 64.44^2 + 2 \times 21.05 \times 64.44 \times \cos 53.13°}$$

$$= \textbf{78.89 MN/m}^2 \quad \textbf{(Ans.)}$$

Example 22.50. *A 18 mm bracket plate is connected to a column flange as shown in Fig 22.59 and transmits a load of 300 kN to it. Find the suitable size of the weld if allowable shear stress in the weld is 110 MN/m².* **(IAS)**

Solution. Refer to Figs. 22.59 and 22.60

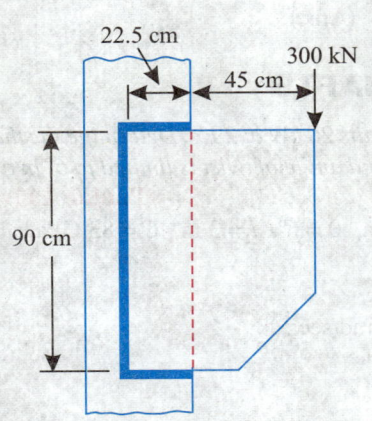

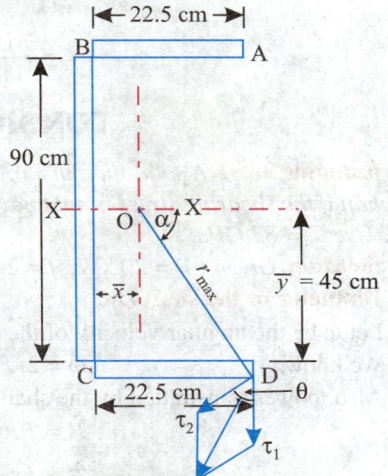

Fig. 22.59	Fig. 22.60

Let t be the throat thickness of weld.

Taking moments about BC, we get

$$t(22.5 + 90 + 22.5)\,\bar{x} = 22.5\,t \times 11.25 + 22.5t \times 11.25$$

$$135\,\bar{x} = 506.25\,t$$

∴ $\qquad\qquad \bar{x} = 3.75$ cm; $\bar{y} = 45$ cm

∴ Eccentricity, $\qquad e = 45 + (22.5 - 3.75) = 63.75$ cm

Moment, $\qquad\qquad M = P \times e = 300 \times 63.75 \times 10^{-2} = 191.25$ kNm

$$I_{XX} = 2 \times 22.5 \times t \times 45^2 + \frac{t \times 90^3}{12} = 151875\,t \text{ cm}^4$$

$$I_{YY} = 2\left(\frac{t \times 22.5^3}{12} + t \times 22.5 \times 7.5^2\right) + 90 \times t\,(3.75)^2$$

$$= 2 \times 2214.84t + 1265.62 = 5695.3t \text{ cm}^4$$

$$I = I_{XX} + I_{YY} = 151875t + 5695.3t = 157570.3t \text{ cm}^4$$

Primary shear stress,

$$\tau_1 = \frac{P}{A} = \frac{300 \times 10^3}{t\,(2 \times 22.5 + 90) \times 10^{-4}} = \frac{22.22}{t} \text{ MN/m}^2$$

Secondary shear stress,

$$\tau_2 = \frac{M\,r_{max}}{I} = \frac{191.25 \times 10^3\,(\sqrt{45^2 + 18.75^2}) \times 10^{-2}}{157570.3t \times 10^{-8}} = \frac{59.17}{t} \text{ MN/m}^2$$

$$\tan \alpha = \frac{45}{18.75} = 2.4$$

∴ $\qquad\qquad \alpha = \tan^{-1}(2.4) = 67.38°$

Resultant stress, $\qquad \sigma_r = \sqrt{\tau_1^2 + \tau_2^2 + 2\tau_1\tau_2 \cos \alpha} = 110$ (*Given*)

$$\sqrt{\left(\frac{22.22}{t}\right)^2 + \left(\frac{59.17}{t}\right)^2 + 2 \times \frac{22.22}{t} \times \frac{59.17}{t} \times \cos 67.38°} = 110$$

$$\frac{70.75}{t} = 110$$

∴ $$t = \frac{70.75}{110} = 0.643 \text{ cm} \quad \textbf{(Ans.)}$$

TORSION OF SHAFTS

Example 22.51. *A solid shaft in a rolling mill transmits 20 kW at 2 Hz. Determine the diameter of the shaft if the shearing stress is not to exceed 40 MPa and angle of twist is limited to 6° in a length of 3 m. Use C = 83 GPa.* **(Panjab University)**

Solution. *Given :* $P = 20$ kW; $f = 2$Hz, $\tau = 40$ MPa, $\theta = 6°$, $l = 3$ m, $C = 83$ GPa.

Diameter of the shaft D :

Let ω be the angular velocity of the shaft is rad/sec.

We know, $\omega = 2\pi f = 2\pi \times 2 = 4\pi$ rad/sec.

Also, power transmitted by the shaft,

$$P = T.\omega$$

∴ $$T = \frac{P}{\omega} = \frac{20 \times 1000}{4\pi} = 1591.55 \text{ Nm}$$

Consider shear stress :

$$T = \tau \times \frac{\pi}{16} D^3$$

∴ $$D = \left(\frac{16T}{\tau \times \pi}\right)^{1/3} = \left(\frac{16 \times 1591.55}{40 \times 10^6 \times \pi}\right)^{1/3} = 0.0587 \text{ m or } 58.7 \text{ mm}$$

Consider angle of twist :

$$\frac{T}{I_p} = \frac{C\theta}{l}$$

$$\frac{1591.55}{\frac{\pi}{32} \times D^4} = \frac{83 \times 10^9 \times 6 \times \frac{\pi}{180}}{3}$$

$$\frac{1591.55 \times 32}{\pi D^4} = 2897246558$$

$$D = \left(\frac{1591.55 \times 32}{\pi \times 2897246558}\right)^{1/4} = 0.0486 \text{ m or } 48.6 \text{ mm}$$

Hence, $D = \textbf{58.7 mm}$ *(larger of the two values)* **(Ans.)**

Example 22.52. *While transmitting power of a steam turbine, the angle of twist of a rotating shaft was measured and was found to be 1.2° over a length of 6 metres. The external and internal diameters of the shaft are 250 mm and 170 mm respectively. The rotational speed of the shaft is 250 revolutions per minute. The shear modulus, C is 80 GPa. Determine the power being transmitted by the shaft and the maximum shear stress developed in it.* **(Bangalore University)**

Solution. *Given :* $\theta = 1.2° = 1.2 \times \frac{\pi}{180} = 0.02094$ rad; $l = 6$m;

$$D = 250 \text{ mm}; d = 170 \text{ mm}; N = 250 \text{ r.p.m.};$$

$$C = 80 \text{ } GPa = 80 \times 10^3 \text{ N/mm}^2.$$

(i) **Maximum shear stress developed, τ_{max} :**

Using the relation,

$$\frac{C\theta}{l} = \frac{\tau_{max}}{R}$$

or,
$$\tau_{max} = \frac{C\theta}{l} \times R = \frac{80 \times 10^3 \times 0.02094}{6 \times 10^3} \times \left(\frac{250}{2}\right) = \textbf{34.9 N/mm}^2 \textbf{ (Ans.)}$$

(ii) Power transmitted by the shaft; P :

Let T be the torque transmitted by the shaft.

We know that,
$$T = \tau_{max} \times \frac{\pi}{16}\left[\frac{D^4 - d^4}{D}\right] = 34.9 \times \frac{\pi}{16}\left[\frac{(250)^4 - (170)^4}{250}\right]$$
$$= 84178422 \text{ N mm} \quad \text{or} \quad 84178.4 \text{ Nm}$$

Now, power transmitted by the shaft,
$$P = \frac{2\pi NT}{60 \times 1000} = \frac{2\pi \times 250 \times 84178.4}{60 \times 1000} = \textbf{2203.78 kW (Ans.)}$$

Example 22.53. *A hollow shaft is 50 mm outside diameter and 30 mm internal diameter. An applied torque of 1.5 Nm is found to produce an angular twist of 0.4°, measured on a length of 0.2 m of the shaft. Calculate the value of the modulus of rigidity.*

Calculate also the maximum power which could be transmitted by the shaft at 2000 r.p.m. If the maximum allowable shearing stress is 70 MN/m². **(Banglore University)**

Solution. *Given :* $D = 50$ mm $= 0.05$ m; $d = 30$ mm $= 0.03$ m, $T_{applied} = 1.5$ kNm,

$$\theta = 0.4° = 0.4 \times \frac{\pi}{180} = 0.00698 \text{ rad.} \; ; \; l = 0.2 \text{ m} \; ; \; \tau_{max} = 70 \text{ MN/m}^2; \; N = 2000 \text{ r.p.m.}$$

Modulus of rigidity, C :

We know that, $\quad \dfrac{T_{applied}}{I_p} = \dfrac{C\theta}{l}$

or,
$$\frac{1.5}{\dfrac{\pi}{32}(0.05^4 - 0.03^4)} = \frac{C \times 0.00698}{0.2}$$

or,
$$C = \frac{1.5 \times 0.2}{\dfrac{\pi}{32}(0.05^4 - 0.03^4) \times 0.00698} = \textbf{80.476 kN/m}^2 \textbf{ (Ans.)}$$

Maximum power transmitted P :

Now,
$$\frac{T}{I_p} = \frac{\tau_{max}}{R}$$

$$T = \frac{\tau_{max} \times I_p}{(D/2)} = \frac{70 \times 10^6 \times \dfrac{\pi}{32}(0.05^4 - 0.03^4)}{(0.05/2)} = 1495.4 \text{ Nm}$$

Hence, maximum power transmitted under the given conditions,

∴
$$P = \frac{2\pi NT}{60 \times 1000} \text{ kW} = \frac{2\pi \times 2000 \times 1495.4}{60 \times 1000} = \textbf{313.2 kW (Ans.)}$$

Example 22.54. *A hollow steel shaft 80 mm internal diamter and 120 mm external diameter is to be replaced by a solid alloy shaft. If the polar modulus has the same value for both, calculate the diameter of the latter and ratio of torsional rigidities. The shear modulus of elasticity for steel is 2.4 times that of the alloy.* **(Osmania University)**

Solution. Let, $(Z_p)_{steel\ shaft}$ = Polar modulus of the steel shaft section,

$(Z_p)_{alloy}$ = Polar modulus of alloy shaft section,

D_s = Outer diameter of steel shaft,

d_s = Inner diameter of steel shaft, and

D_a = Diameter of the alloy shaft.

Diameter of the alloy shaft D_a :

Now, $(Z_p)_{steel\,shaft} = (Z_p)_{alloy\,shaft}$ ...(Given)

$$\frac{\pi}{16}\left[\frac{D_s^4 - d_s^4}{D_s}\right] = \frac{\pi}{16} \times D_a^3$$

Substituting the values, we get

$$\frac{\pi}{16}\left[\frac{0.12^4 - 0.08^4}{0.12}\right] = \frac{\pi}{16} \times D_a^3$$

or, $D_a = \left[\dfrac{0.12^4 - 0.08^4}{0.12}\right]^{1/3} = 0.1115$ m or **111.5 mm (Ans.)**

Ratio of the torsional rigidities :

$$\frac{\text{Torsional rigidity of steel shaft}}{\text{Torsional rigidity of alloy shaft}} = \frac{(CI_p)_{steel\,shaft}}{(CI_p)_{alloy\,shaft}}$$

$$= 2.4\left[\frac{(I_p)_{steel}}{(I_p)_{alloy}}\right] \qquad (\because C_{steel} = 2.4\,C_{alloy} \ ...Given)$$

$$= 2.4\left[\frac{\dfrac{\pi}{32}(D_s^4 - d_s^4)}{\dfrac{\pi}{32}D_a^4}\right] = 2.4\left[\frac{0.12^4 - 0.08^4}{0.1115^4}\right] = \mathbf{2.58} \ \ \textbf{(Ans.)}$$

Example 22.55. *Determine the shaft diameter and bolt size for a marine flange-coupling transmitting 3.75 MW at 150 r.p.m. The allowable shear stress in the shaft and bolts may be taken as 50 MPa. The number of bolts may be taken as 10 and bolt pitch circle diameter as 1.6 times the shaft diameter.* **(GATE)**

Solution. *Given :* $P = 3.75$ MW; $N = 150$ r.p.m.; $\tau_s = \tau_b = 50$ MPa; $n = 10$, $D_b = 1.6\,D$

Shaft diameter, D :

$$P = \frac{2\pi NT}{60}$$

or, $3.75 \times 10^6 = \dfrac{2\pi \times 150 \times T}{60}$

or, $T = \dfrac{3.75 \times 10^6 \times 60}{2\pi \times 150} = 238732$ Nm

Also, $T = \tau_s \times \dfrac{\pi}{16} \times D^3$

or, $238732 = 50 \times 10^6 \times \dfrac{\pi}{16} D^3$

$\therefore \quad D = \left(\dfrac{238732 \times 16}{50 \times 10^6 \times \pi}\right)^{1/3} = 0.289$ m or **290 mm (Ans.)**

Bolt size, d_b :

Bolt pitch circle diameter,

$$D_b = 1.6\,D = 1.6 \times 0.29 = 0.464 \text{ m}$$

Now, $T = n \times \dfrac{\pi}{4} \times d_b^2 \times \tau_b \times \left(\dfrac{D_b}{2}\right)$

or, $\qquad 238732 = 10 \times \dfrac{\pi}{4} \times d_b^2 \times 50 \times 10^6 \times \left(\dfrac{0.464}{2}\right)$

or, $\qquad d_b = \left(\dfrac{238732 \times 4 \times 2}{10 \times \pi \times 50 \times 10^6 \times 0.464}\right) = 0.0512$ m   or   **51.2 m (Ans.)**

Example 22.56. *A 450 mm diameter propeller shaft supports a propeller of mass 15 tonnes. The propeller can be considered as a load concentrated at the end of a cantilever of length 2.4 m. When the speed of ship is 32 km/hr the propeller is running at 100 r.p.m. If the engine develops 18 MW, calculate the principal stresses and maximum shear stress in the shaft. It may be assumed that the propulsive efficiency of the propeller is 85%.* **(UPSC)**

Solution. *Given :* $d = 450$ mm $= 0.45$ m; $m = 15$ tonnes $= 15 \times 1000 = 15000$ kg; $v = 32$ km/hr; $N = 100$ r.pm.

Power developed = 18 MW; $\eta = 85\%$

Principal stresses (σ_1, σ_2) and maximum shear stress (τ_{max}) :

Refer to Fig. 22.61 (*a*).

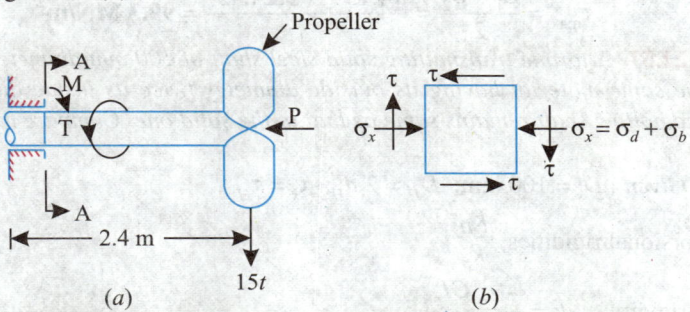

(*a*)   (*b*)

Fig. 22.61

Bending moment, $M = (15000 \times 9.81) \times 2.4$ Nm $= 0.3532$ MNm

Torque $\qquad\qquad T = \dfrac{\text{Power} \times 60}{2\pi N} = \dfrac{18 \times 60}{2\pi \times 100} = 1.719$ MNm

If P is propulsive force and v in the linear speed of the ship, then

Engine power $\qquad = \dfrac{P \times v}{\text{Efficiency}}$

or, $\qquad 18 = \dfrac{P \times \left(\dfrac{32 \times 1000}{60 \times 60}\right)}{0.85}$

∴ $\qquad P = \dfrac{18 \times 0.85}{\left(\dfrac{32 \times 1000}{60 \times 60}\right)} = 1.721$ MN

Therefore, direct stress on the cross-section *AA*,

$$\sigma_d = \dfrac{P}{\text{Area}} = \dfrac{1.721}{\dfrac{\pi}{4} \times 0.45^2} = 10.82 \text{ MN/m}^2 \text{ (comp.)}$$

Maximum bending stress due to M, on the cross-section *AA*,

$$\sigma_b = \dfrac{32M}{\pi d^3} = \dfrac{32 \times 0.3532}{\pi \times (0.45)^3} = 39.48 \text{ MN/m}^2 \text{ (comp. at the lower most point)}$$

∴ $\qquad \sigma_x = \sigma_d + \sigma_b = 10.82 + 39.48 = 50.3 \text{ MN/m}^2 \text{ (comp.)}$

Shear stress due to torque T,

$$\tau = \frac{16T}{\pi d^3} = \frac{16 \times 1.719}{\pi \times (0.45)^3} = 96.07 \text{ MN/m}^2$$

The point diagram for the bottom most point of section AA is given in Fig. 22.61 (b). Thus the principal stresses are given by

$$\sigma_{1,2} = \frac{\sigma_x}{2} \pm \sqrt{\left(\frac{\sigma_x}{2}\right)^2 + \tau^2} \quad \text{[Eqn. (2.12)]}$$

$$= \frac{1}{2}\left[\sigma_x \pm \sqrt{\sigma_x^2 + 4\tau^2}\right]$$

$$= \frac{1}{2}\left[-50.3 \pm \sqrt{(50.3)^2 + 4 \times (96.07)^2}\right] = \frac{1}{2}(-50.3 \pm 198.6)$$

i.e. $\quad \sigma_1 = 74.15 \text{ MN/m}^2$ (*tensile*), **and**

$\quad\quad\quad \sigma_2 = -124.45 \text{ MN/m}^2$ (*i.e. comp.*) **(Ans.)**

$$\tau_{max} = \frac{\sigma_1 - \sigma_2}{2} = \frac{74.15 - (-124.45)}{2} = 99.3 \text{ MN/m}^2$$

Example 22.57. *A torque transmitting solid steel shaft of 100 mm diameter is replaced by a hollow one of the same material having its outside diameter twice its inside diameter. Maximum shear stress in the hollow shaft remains same as that in the solid one. Compare torsional rigidity of the two shafts.* **(IES)**

Solution. *Given* : $D_s = 100$ mm; $D_H = 2 d_H$; $\tau_s = \tau_H$

Ratio of torsional rigidities, $\dfrac{k_H}{k_S}$

Torsional rigidity, $\quad k = \dfrac{\tau}{\theta} = \dfrac{CI_p}{l}$

Since C and l are same for solid and hollow shafts, therfore,

$$k \propto I_p$$

For solid shaft, $\quad (I_p)_S = \dfrac{\pi}{32} D_S^4 \text{ mm}^4$

For hollow shaft, $(I_p)_H = \dfrac{\pi}{32}(D_H^4 - d_H^4) = \dfrac{\pi}{32} D_H^4 \left[1 - \left(\dfrac{d_H}{D_H}\right)^4\right]$

$$= \dfrac{\pi}{32} D_H^4 \left[1 - \left(\dfrac{1}{2}\right)^4\right] = \dfrac{\pi}{32} \times \dfrac{15}{16} D_H^4 \text{ mm}^4$$

Maximum shear stresses :

$$\tau_S = \frac{16T}{\pi D_S^3}$$

$$\tau_H = \frac{16T}{\pi D_H^3 \times \dfrac{15}{16}}$$

Now, $\quad\quad\quad \tau_S = \tau_H$ $\quad\quad\quad\quad\quad\quad\quad\quad\quad\quad\quad$ ...(*Given*)

$$\frac{16T}{\pi D_S^3} = \frac{16T}{\pi D_H^3 \times \dfrac{5}{16}}$$

or, $\quad\quad\quad D_H^3 = \dfrac{16}{15} D_S^3$

or,
$$D_H = D_S \left(\frac{16}{16}\right)^{1/3}$$

or,
$$\frac{D_H}{D_S} = \left(\frac{16}{15}\right)^{1/3} = 1.021746$$

$$\therefore \quad \frac{(k)_H}{(k)_S} = \frac{(I_p)_H}{(I_p)_S} = \frac{\frac{\pi}{32} \times \frac{15}{16} D_H^4}{\frac{\pi}{32} D_S^4} = \frac{15}{16} \times \left(\frac{D_H}{D_S}\right)^4 = \frac{15}{16} (1.021746)^4$$

$$= 1.0217 \text{ (Ans.)}$$

Example 22.58. *A hollow mild steel shaft is supported between bearings 1000 mm apart and has 100 mm external diameter and 50 mm internal diameter. It carries a concentrated load of 10 kN at the mid-span and is subjected to a twisting moment of 8 kNm. Find the equivalent direct stress which acting alone on the shaft at the mid-section would produce the same maximum elastic strain energy as caused by the above loading. Assume Poisson's ratio as 0.25.* **(IAS)**

Solution. *Given :* $l = 1000$ mm $= 1$ m; $D_H = 100$ mm $= 0.1$ m; $D_H = 50$ mm $= 0.05$ m; $W = 10$ kN; $T = 8$ kNm; $\frac{1}{m} = 0.25$.

Equivalent direct stress, σ :

$$I = \frac{\pi}{64} (D_H^4 - d_H^4) = \frac{\pi}{64}(0.1^4 - 0.05^4) = 4.6 \times 10^{-6} \text{ m}^4$$

$$I_p = \frac{\pi}{32} (D_H^4 - d_H^4) = \frac{\pi}{32}(0.1^4 - 0.05^4) = 9.2 \times 10^{-6} \text{ m}^4$$

Bending moment at the mid-span,

$$M = \frac{Wl}{4} = \frac{10 \times 1}{4} = 2.5 \text{ kNm or } 2500 \text{ Nm.}$$

$$U_1 = \frac{M^2 l}{2EI} + \frac{T^2 l}{2CI_p}$$

Now,
$$C = \frac{E}{2\left(1 + \frac{1}{m}\right)} = \frac{E}{2(1 + 0.25)} = \frac{E}{2.5}$$

$$\therefore \quad U_1 = \frac{M^2 l}{2EI} + \frac{T^2 l}{2 \times \frac{E}{2.5} \times 2I} = \frac{M^2 l}{2EI} + \frac{2.5 T^2 l}{4EI} = \frac{l}{2EI}\left(M^2 + \frac{2.5 T^2}{2}\right)$$

$$= \frac{1}{2 \times E \times 4.6 \times 10^{-6}}\left(2500^2 + \frac{2.5 \times 8000^2}{2}\right) = \frac{9375 \times 10^9}{E} \text{ Nm}$$

$$U_2 = \frac{\sigma^2 V}{2E} = \frac{1}{2} \times \frac{\sigma^2}{E} \times \frac{\pi}{4}(0.1^2 - 0.05^2) \times 1$$

$$= \frac{2.945 \sigma^2 \times 10^{-3}}{E} \text{ Nm}$$

$$U_2 = U_1 \qquad \qquad ...Given$$

$$\frac{2.945 \sigma^2 \times 10^{-3}}{E} = \frac{9375 \times 10^9}{E}$$

$$\therefore \quad \sigma = \sqrt{\frac{9375 \times 10^9}{2.945 \times 10^{-3}}} = 56.42 \times 10^6 \text{ N/m}^2 = \textbf{56.42 MN/m}^2 \text{ (Ans.)}$$

Example 22.59. *In order to reduce the weight of the control and power plant equipments used in an aircraft, it is planned to use all hollow shafting for power transmission purposes. The shafts are* *subjected to combined bending and torsion. Develop an expression to determine the weight saving in percentage through the use of hollow shafting in place of equal strength solid shafting for this application.* **(IES)**

Solution. Let, D_S = Diameter of the solid shaft,

D_H = External diameter of the hollow shaft, and

d_H = Internal diameter of the hollow shaft.

$$T_e = \sqrt{M^2 + T^2}$$

Polar modulus for the solid shaft,

$$(Z_p)_S = \frac{\pi}{16} D^3$$

Polar modulus for the hollow shaft,

$$(Z_p)_H = \frac{\pi}{16} D_H^3 \left[1 - \left(\frac{d_H}{D_H}\right)^4\right]$$

For equal shear stress to transmit the same torque,

$$(Z_p)_S = (Z_p)_H$$

$$\frac{\pi}{16} D_S^3 = \frac{\pi}{16} D_H^3 \left[1 - \left(\frac{d_H}{D_H}\right)^4\right]$$

or

$$D_S^3 = D_H^3 \left[1 - \left(\frac{d_H}{D_H}\right)^4\right]$$

or

$$\frac{D_S}{D_H} = \left[1 - \left(\frac{d_H}{D_H}\right)^4\right]^{1/3}$$

Weight of the solid shaft,

$$W_S = \frac{\pi}{4} D_S^2 \, l \, w$$

Weight of the hollow shaft,

$$W_H = \frac{\pi}{4} (D_H^2 - d_H^2) \, l \times w = \frac{\pi}{4} D_H^2 \left[1 - \left(\frac{d_H}{D_H}\right)^2\right] l \times w$$

Percentage saving in weight $= \dfrac{W_S - W_H}{W_S}$

$$= \frac{\frac{\pi}{4} D_S^2 \, l \times w - \frac{\pi}{4} D_H^2 \left[1 - \left(\frac{d_H}{D_H}\right)^2\right] l \times w}{\frac{\pi}{4} D_S^2 \, l \times w} \times 100$$

$$= \frac{D_S^2 - D_H^2 \left[1 - \left(\frac{d_H}{D_H}\right)^2\right]}{D_S^2} \times 100$$

$$= \left[1 - \left(\frac{D_H}{D_S}\right)^2 \left\{1 - \left(\frac{d_H}{D_H}\right)^2\right\}\right] \times 100$$

$$= \left[1 - \frac{\left\{ 1 - \left(\dfrac{d_H}{D_H} \right) \right\}}{\left\{ 1 - \left(\dfrac{d_H}{D_H} \right)^4 \right\}^{2/3}} \right] \times 100$$

$$= \left[1 - \left\{ 1 - \left(\frac{d_H}{D_H} \right)^2 \right\} \left\{ 1 - \left(\frac{d_H}{D_H} \right)^{-2/3} \right\} \right] \times 100$$

$$= \left[1 - \left\{ 1 - \left(\frac{d_H}{D_H} \right)^2 \right\} \left\{ 1 + \frac{2}{3} \left(\frac{d_H}{D_H} \right)^4 \right\} \right] \times 100$$

$$= \left[1 - \left\{ 1 - \left(\frac{d_H}{D_H} \right)^2 + \frac{2}{3} \left(\frac{d_H}{D_H} \right)^4 - \frac{2}{3} \left(\frac{d_H}{D_H} \right)^6 \right\} \right] \times 100$$

$$= \left[\left(\frac{d_H}{D_H} \right)^2 - \frac{2}{3} \left(\frac{d_H}{D_H} \right)^4 + \frac{2}{3} \left(\frac{d_H}{D_H} \right)^6 \right] \times 100 \quad \textbf{(Ans.)}$$

Example 22.60. *A shaft shown in Fig.22.62 is subjected to bending load of 3 kN, pure torque of 1 kNm and an axial pulling force of 15 kN. Calculate the stresses at A and B.* **(IES)**

Solution. Refer to Fig. 22.62. *Given :* $D = 50$ mm $= 0.05$ m; Bending load, $W = 3$kN; Pure torque, $T = 1$ kNm; Axial pulling force $P = 15$ kN.

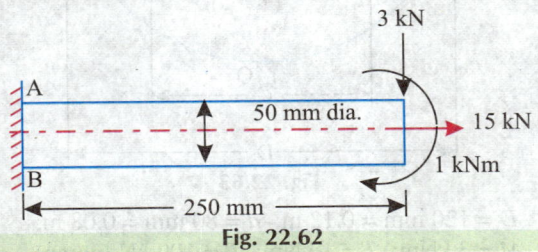

Fig. 22.62

Stresses at A and B :

Direct stress, $\quad \sigma_d = \dfrac{P}{A} = \dfrac{15 \times 10^3}{\dfrac{\pi}{4} \times 0.05^2} \times 10^{-6} = 7.64$ MN/m^2

Bending stress, $\sigma_b = \dfrac{32 M}{\pi D^3} = \dfrac{32 \times (W \times 0.25)}{\pi \times 0.05^3}$

$$= \frac{32 \times (3 \times 1000 \times 0.25)}{\pi \times 0.05^3} \times 10^{-6} = 61.11 \text{ MN/m}^2$$

Shear stress, $\quad \tau = \dfrac{16 T}{\pi D^3} = \dfrac{16 \times (1 \times 1000)}{\pi \times 0.05^3} \times 10^{-6} = 40.74$ MN/m^2

Point A :

$$\sigma_x = \sigma_d + \sigma_b = 7.64 + 61.11 = 68.75 \text{ MN/m}^2$$

$$\sigma_1, \sigma_2 = \frac{\sigma_x}{2} \pm \sqrt{\left(\frac{\sigma_x}{2} \right)^2 + \tau^2}$$

$$= \frac{68.75}{2} \pm \sqrt{\left(\frac{68.75}{2}\right)^2 + (40.74)^2} = 34.37 \pm 53.3$$

i.e., $\quad\sigma_1 = 87.67 \text{ MN/m}^2\,; \quad \sigma_2 = -18.93 \text{ MN/m}^2 \quad$ **(Ans.)**

Point B :

$$\sigma_x = \sigma_d - \sigma_b = 7.64 - 61.11 = -53.47 \text{ MN/m}^2$$

$$\sigma_1, \sigma_2 = \frac{-53.47}{2} \pm \sqrt{\left(\frac{-53.47}{2}\right)^2 + 40.74^2} = -26.73 \pm 48.73$$

$\therefore \quad\quad \sigma_1 = -75.46 \text{ MN/m}^2\,; \quad \sigma_2 = 22 \text{ MN/m}^2 \quad$ **(Ans.)**

Example 22.61. *A shaft 120 mm external diameter and 80 mm internal diameter is subjected to a bending moment of 3 kNm, twisting moment of 1 kNm and a direct thrust of 100 kN. Determine the maximum principal stress and direction in which it acts with reference to the axis of the shaft at the end points P and Q of diameter PQ as shown in Fig. 22.63.* **(IAS)**

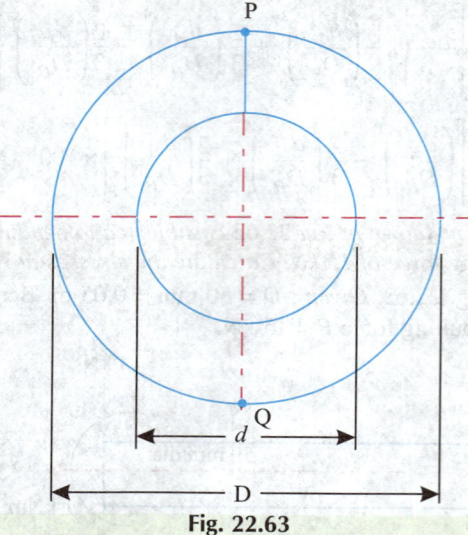

Fig. 22.63

Solution. *Given :* $\quad D = 120 \text{ mm} = 0.12 \text{ m}; \quad d = 80 \text{ mm} = 0.08 \text{ m};$
$\quad\quad\quad\quad\quad\quad\quad M = 3 \text{ kNm}; \quad T = 1 \text{ kNm}; \quad P = 100 \text{ kN (comp).}$

Direct stress, $\quad \sigma_d = -\dfrac{P}{A} = -\dfrac{100 \times 10^3}{\dfrac{\pi}{4}\left(0.12^2 - 0.08^2\right)} \times 10^{-6} = -15.91 \text{ MN/m}^2$

Bending stress, $\quad \sigma_b = \dfrac{32 M}{\pi D^3\left[1 - \left(\dfrac{d}{D}\right)^4\right]}$

$$= \frac{32 \times 3 \times 10^3}{\pi \times 0.12^3\left[1 - \left(\dfrac{0.08}{0.12}\right)^4\right]} \times 10^{-6} = 22.04 \text{ MN/m}^2$$

Shear stress, $\quad \tau = \dfrac{16 T}{\pi D^3\left[1 - \left(\dfrac{d}{D}\right)^4\right]} = \dfrac{16 \times 1 \times 10^3}{\pi \times 0.12^3\left[1 - \left(\dfrac{0.08}{0.12}\right)^4\right]} \times 10^{-6} = 3.67 \text{ MN/m}^2$

Point P :

$$\sigma_x = \sigma_b \text{ (comp.)} + \sigma_d \text{ (comp.)} = -22.04 - 15.91 = -37.95 \text{ MN/m}^2$$

$$\sigma_1, \sigma_2 = \frac{\sigma_x}{2} \pm \sqrt{\left(\frac{\sigma_x}{2}\right)^2 + \tau^2}$$

$$= \frac{-37.95}{2} \pm \sqrt{\left(\frac{-37.95}{2}\right)^2 + 3.67^2} = -18.97 \pm 19.33$$

i.e., $$\sigma_1 = 0.36 \text{ MN/m}^2; \quad \sigma_2 = -38.3 \text{ MN/m}^2 \text{ (Ans.)}$$

$$\tan 2\theta = \frac{2\tau}{\sigma_x} = \frac{2 \times 3.67}{-37.95} = -0.1934 \quad \text{or} \quad 2\theta = -10.946°$$

∴ $$\theta_1 = -5.47°, \quad \theta_2 = 84.53° \text{ Ans.}$$

Point Q :

$$\sigma_x = \sigma_b \text{ (tensile)} + \sigma_d \text{ (comp.)} = 22.04 - 15.91 = 6.13 \text{ MN/m}^2$$

$$\sigma_1, \sigma_2 = \frac{6.13}{2} \pm \sqrt{\left(\frac{6.13}{2}\right)^2 + 3.67^2} = 3.06 \pm 4.78$$

i.e. $$\sigma_1 = 7.84 \text{ MN/m}^2; \quad \sigma_2 = -1.72 \text{ MN/m}^2 \text{ (Ans.)}$$

$$\tan 2\theta = \frac{2 \times 3.67}{6.13} = 11.97 \quad \text{or} \quad 2\theta = 50.12°$$

∴ $$\theta_1 = 25.06°, \quad \theta_2 = 115.06° \text{ (Ans.)}$$

Example 22.62. *A hollow shaft with diameter ratio 0.7 is required to transmit 500 kW at 300 r.p.m. with a uniform twisting moment. Allowable shear stress in the material is 65 N/mm² and twist in a length of 2.4 m is not exceed one degree. Calculate the maximum external diameter of the shaft satisfying these conditions. Modulus of rigidity = 8.2 × 10⁴ N/mm².* **(IES)**

Solution. *Given :* $\dfrac{d_i}{d_0} = 0.7$; $\quad P = 500 \text{ kW}; \quad N = 300 \text{ r.p.m}; \quad l = 2.4 \text{ m}, \quad \tau = 65 \text{ N/mm}^2;$

$\theta = 1° \times \dfrac{\pi}{180} = 0.0175 \text{ rad.}; \quad C = 8.2 \times 10^4 \text{ N/mm}^2.$

Maximum external diameter, d_0 :

Power, $$P = \frac{2\pi N T}{60 \times 1000}$$

$$500 = \frac{2\pi \times 300 \times T}{60 \times 1000}$$

or, $$\tau = \frac{500 \times 60 \times 1000}{2\pi \times 300} = 15915.5 \text{ Nm}$$

Polar moment of inertia,

$$I_p = \frac{\pi}{32}\left(d_0^4 - d_i^4\right) = \frac{\pi}{32} d_0^4\left[1 - \left(\frac{d_i}{d_0}\right)^4\right]$$

$$= \frac{\pi}{32} d_0^4\left[1 - (0.7)^4\right] = 0.0746 \, d_0^4$$

Using torsion equation, we get

$$\frac{\tau}{I_p} = \frac{\tau}{r} = \frac{C\theta}{l}$$

Case 1. $$\tau_{max} = 65 \text{ N/mm}^2$$

$$\therefore \quad \frac{15915.5}{0.0746 d_0^4} = \frac{65 \times 10^6}{(d_0/2)} = \frac{130 \times 10^6}{d_0}$$

or, $\quad d_0 = \left(\dfrac{15915.5}{0.0746 \times 130 \times 10^6}\right)^{\frac{1}{3}} = 0.1179$ m or 117.9 mm

Case 2. $\quad \theta_{max} = 0.01745$ rad.

$$\therefore \quad \frac{15915.5}{0.0746 d_0^4} = \frac{8.2 \times 10^4 \times 10^6 \times 0.01745}{2.4}$$

or, $\quad d_0 = \left(\dfrac{15915.5 \times 2.4}{0.0746 \times 8.2 \times 10^4 \times 10^6 \times 0.01745}\right)^{\frac{1}{4}} = 0.1375$ m or 137.5 mm

Hence, *maximum external diameter satisfying the conditions*,

$$d_0 = \mathbf{137.5 \ mm} \ (larger \ of \ the \ two \ values) \quad \textbf{(Ans.)}$$

Example 22.63. *For the composite shaft shown in Fig. 22.64, determine the maximum shear stress in aluminium and steel.* $C_{al} = 28$ GN/m², $C_s = 84$ GN/m². **(Bangalore University)**

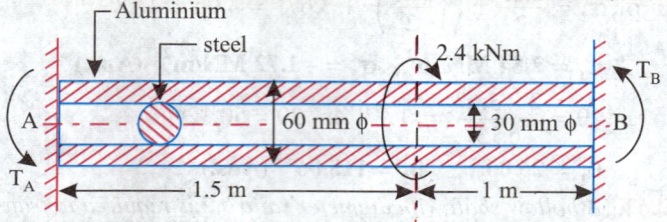

Fig. 22.64

Solution. Refer to Fig. 22.64. *Given* : $D_s = 30$ mm; $d_{al} = 30$ mm; $D_{al} = 60$ mm; $T = 2.4$ kNm; $C_{al} = 28$ GN/m²; $C_s = 84$ GN/m²; $l_A = 1.5$ m; $l_B = 1$ m.

$$\left(I_p\right)_s = \frac{\pi}{32} d_s^4 = \frac{\pi}{32} \times 30^4 = 79522 \ \text{mm}^4$$

$$\left(I_p\right)_{al} = \frac{\pi}{32}(D_{al}^4 - d_{al}^4) = \frac{\pi}{32}[(60)^4 - (30)^4] = 1192823 \ \text{mm}^4$$

We know that, in this case,

$$\theta_A = \theta_B$$

$$T_A \, l_A \left[\frac{1}{C_s\left(I_p\right)_s} + \frac{1}{C(I_p)_{al}}\right] = T_B \, l_B \left[\frac{1}{C_s\left(I_p\right)_s} + \frac{1}{C_{al}(I_p)_{al}}\right]$$

$$1.5 \, T_A = T_B \quad \text{...(i)}$$

$$T_A + T_B = 2.4 \quad \text{...(ii)}$$

From (*i*) and (*ii*), we have

$$T_A = 0.96 \ \text{kNm}; \quad T_B = 1.44 \ \text{kNm}$$

Also, $\quad \dfrac{T_{al}}{T_s} = \dfrac{C_{al}\,(I_p)_{al}}{C_s\,(I_p)_s} = \dfrac{28}{84} \times \dfrac{1192823}{79522} = 5 \quad \text{...(iii)}$

$$T_{al} + T_s = T_B = 1.44 \quad \text{...(iv)}$$

From (*iii*) and (*iv*), we have

$$T_s = 0.24 \ \text{kNm}; \quad T_{al} = 1.2 \ \text{kNm}$$

$$\therefore \qquad \tau_s = \frac{T_s \times (D_s/2)}{(I_p)_s} = \frac{(0.24 \times 1000) \times (15 \times 10^{-3})}{79522 \times 10^{-12}} = \textbf{45.27 MN/m}^2 \quad \textbf{(Ans.)}$$

$$\tau_{al} = \frac{T_{al} \times (D_{al}/2)}{(I_p)_{al}} = \frac{(1.2 \times 1000) \times (30 \times 10^{-3})}{1192823 \times 10^{-12}} = \textbf{30.18 MN/m}^2 \quad \textbf{(Ans.)}$$

Example 22.64. *A solid circular uniformly tapered shaft of length l, with a small angle of taper is subjected to a torque T. The diameters at the two ends of the shaft are D and 1.2 D. Determine the error introduced if its angular twist for a given length is determined on the uniform mean diameter of the shaft.* **(IAS)**

Solution. For shaft of tapering section, we have

$$\theta = \frac{2Tl}{3C\pi}\left[\frac{R_1^2 + R_1 R_2 + R_2^2}{R_1^3 R_2^3}\right]$$

$$= \frac{32Tl}{3C\pi}\left[\frac{D_1^2 + D_1 D_2 + D_2^2}{D_1^3 D_2^3}\right]$$

$$= \frac{32Tl}{3C\pi D^4}\left[\frac{(1.2)^2 + 1.2 \times 1 + (1)^2}{(1.2)^3 \times (1)^3}\right] \qquad \left[\because D_1 = D \text{ and } D_2 = 1.2\,D\right]$$

$$= \frac{32Tl}{3C\pi D^4} \times 2.1065$$

Now, $\qquad D_{avg} = \dfrac{1.2\,D + D}{2} = 1.1\,D$

$$\therefore \qquad \theta' = \frac{32Tl}{3C\pi} \times \left[\frac{3(1.1D)^2}{(1.1D)^6}\right] = \frac{32Tl}{3C\pi} \times \frac{3}{(1.1)^4 . D^4} = \frac{32Tl}{3C\pi D^4} \times 2.049$$

$$\text{Error} = \frac{\theta - \theta'}{\theta} = \frac{2.1065 - 2.049}{2.1065} = 0.0273 \text{ or } \textbf{2.73\%} \quad \textbf{(Ans.)}$$

Example 22.65. *A 300 mm × 300 mm I-section with flange thickness 15 mm and web thickness 12 mm is subjected to a twisting moment of 850 Nm. Neglecting the stress concentration, find the maximum shear stress developed and the twist per unit length. Take C = 80 GN/m².*

(Bombay University)

Solution. Given : $b_1 = b_3 = 300$ mm $= 0.3$ m; $b_2 = 270$ mm $= 0.27$ m; $t_1 = t_3 = 15$ mm $= 0.015$ m; $t_2 = 12$ mm $= 0.012$ m, $T = 850$ Nm; $C = 80$ GN/m².

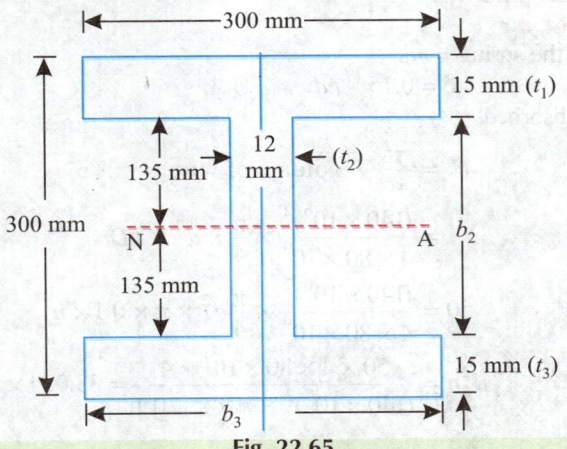

Fig. 22.65

Maximum shear stress, τ :

Maximum shear stress is given by,

$$\tau = \frac{3T}{\sum\limits_{i=1}^{3} b_i t_i^2}$$
[Eqn. (13.60)]

$$= \frac{3T}{b_1 t_1^2 + b_2 t_2^2 + b_3 t_3^2}$$

$$= \frac{3 \times 850}{0.3 \times 0.015^2 + 0.27 \times 0.012^2 + 0.3 \times 0.015^2}$$

$$= 14.66 \times 10^6 \text{ N/m}^2 = \textbf{14.66 MN/m}^2 \textbf{ (Ans.)}$$

Twist per unit length, θ :

$$\theta = \frac{1}{C} \frac{3T}{\sum\limits_{i=1}^{3} b_i t_i^3} = \frac{1}{C} \cdot \frac{3T}{b_1 t_1^3 + b_2 t_2^3 + b_3 t_3^3}$$

$$= \frac{1}{80 \times 10^9} \times \frac{3 \times 850}{(0.3 \times 0.015^3 + 0.27 \times 0.012^3 + 0.3 \times 0.015^3)}$$

$$= \frac{1}{80 \times 10^9} \times \frac{2550}{2.491 \times 10^{-6}} \text{ rad/m.}$$

$$= 0.0128 \text{ rad/m} = \textbf{0.733}°\textbf{/m (Ans.)}$$

SPRINGS

Example 22.66. *A close-coiled helical spring whose free length when not compressed is 15 cm, is required to absorb strain energy equal to 50 Nm when fully compressed with the coils in contact. The maximum shearing stress is limited to 140 MPa. Assuming a mean coil diameter of 10 cm, find the diameter of the steel wire required and the number of coils. C = 80 GPa.*

(Engineering Services)

Solution. *Given* : Free length of the spring when not compressed = 15 cm or 0.15 m

Strain energy to be absorbed by the spring = 50 Nm

Maximum shear stress, τ = 140 MPa or 140 MN/m²

Mean coil diameter, D = 10 cm = 0.1 m

Modulus of rigidity, C = 80 GPa = 80 × 10⁹ N/m²

Diameter of steel wire, d :

Number of coils, n :

Solid length of the spring = nd

∴ Deflection, δ = 0.15 – nd

Strain energy absorbed,

$$U = \frac{\tau^2}{4C} \times \text{volume of wire}$$
... (Eqn. 14.8)

$$= \frac{(140 \times 10^6)^2}{4 \times 80 \times 10^9} \times \frac{\pi}{4} \times d^2 \times \pi D n$$

or,

$$50 = \frac{(140 \times 10^6)^2}{4 \times 80 \times 10^9} \times \frac{\pi}{4} d^2 \times \pi \times 0.1 \times n$$

or,

$$d^2 n = \frac{50 \times 4 \times 80 \times 10^9 \times 4}{(140 \times 10^6)^2 \times \pi \times \pi \times 0.1} = 33.084 \times 10^{-4}$$

or, $$n = \frac{33.084 \times 10^{-4}}{d^2}$$

Now, $$\tau = \frac{16 WR}{\pi d^3}$$

or, $$140 \times 10^6 = \frac{16 \times W \times 0.05}{\pi d^3}$$

or, $$W = \frac{140 \times 10^6 \times \pi d^3}{16 \times 0.05} = 54.978 \times 10^7 \ d^3$$

Also, $$\delta = \frac{64 WR^3 n}{C d^4}$$

or, $$0.15 - nd = \frac{64 \times 54.978 \times 10^7 \ d^3 \times (0.05)^3 \times n}{80 \times 10^9 \ d^4} = \frac{5.498 \times 10^{-5} n}{d}$$

or, $$0.15 - \frac{33.084 \times 10^{-4}}{d} = \frac{5.498 \times 10^{-5} \times 33.084 \times 10^{-4}}{d^3}$$

or, $$0.15 d^3 - 33.084 \times 10^{-4} \ d^2 = 181.896 \times 10^{-9}$$

or, $$d^3 - 0.022 \ d^2 - 1.212 \times 10^{-6} = 0$$

Solving, we get $$d = \mathbf{0.0241 \ m} \quad \text{or} \quad \mathbf{2.41 \ cm} \ \textbf{(Ans.)}$$

∴ $$n = \frac{33.084 \times 10^{-4}}{(0.0241)^2} = \mathbf{5.7} \ \textbf{(Ans.)}$$

Example 22.67. *Inside a helical spring of circular cross-section 20 mm in diameter another spring is placed having the same cross-section. The mean radius of the outer spring is 80 mm whereas that of the inner spring is 50 mm. Both the springs are of the same height and have 10 coils each. With a load being transmitted to both the springs they deflect by 50 mm. In which spring is the maximum shear stress greater, and what are their magnitudes? What is the corresponding load being transmitted? The shear modulus C = 80 GPa for both the spring materials.*

(Roorkee University)

Solution.

Outer spring	Inner spring
$R_0 = 80$ mm	$R_i = 50$ mm
$d_0 = 20$ mm	$d_i = 20$ mm
$n_0 = 10$	$n_i = 10$
$\delta_0 = 50$ mm	$\delta_i = 50$ mm
$C_0 = 80$ GPa $= 80 \times 10^3$ N/mm²	$G_i = 80$ GPa $= 80 \times 10^3$ N/mm²

Let, P_i = Load transmitted to inner spring, and

P_0 = Load transmitted to outer spring.

Then, $$P_i + P_0 = P \qquad \qquad ...(i)$$

Using the relation,

$$\delta_0 = \frac{64 P_0 R_0^3 n_0}{C_0 d_0^4}$$

or, $$50 = \frac{64 \times P_0 \times 80^3 \times 10}{80 \times 10^3 \times 20^4}$$

∴ $$P_0 = \frac{50 \times 80 \times 10^3 \times 20^4}{64 \times 80^3 \times 10} = 1953.125 \ \text{N}$$

Similarly, $\delta_i = \dfrac{64\,P_i\,R_i^3\,n_i}{C_i\,d_i^4}$

or, $50 = \dfrac{64\,P_i \times 50^3 \times 10}{80 \times 10^3 \times 20^4}$

or, $P_i = \dfrac{50 \times 80 \times 10^3 \times 20^4}{64 \times 50^3 \times 10} = 8000 \text{ N}$

Maximum shear stress in the outer spring, τ_0:

$$\tau_0 = \frac{16\,P_0\,R_0}{\pi d_0^3} = \frac{16 \times 1953.125 \times 80}{\pi \times 20^3} = \textbf{99.47 N/mm}^2 \text{ (Ans.)}$$

Maximum stress in the inner spring τ_i:

$$\tau_i = \frac{16\,P_i\,R_i}{\pi d_i^3} = \frac{16 \times 80000 \times 50}{\pi \times 20^3} = \textbf{254.65 N/mm}^2 \text{ (Ans.)}$$

*Hence we conclude that in the inner spring there is maximum shear produced, its magnitude is equal to 254.65 N/mm² (**Ans.**)*

Corresponding load being transmitted, P:

$$P = P_o + P_i = 1953.125 + 8000 = 9953.125 \text{ or } \textbf{9.953 kN} \text{ (Ans.)}$$

Example 22.68. *A truck weighting 25 kN and moving at 2.5 m/s has to be brought to rest by a buffer. Find how many springs each of 25 coils will be required to store energy of motion during compression of 0.2 m. The spring is made of 25 mm diameter steel rod coiled to a mean diameter of 0.2 m. C = 100 G/m².* **(IES)**

Solution. *Given :* $W = 25$ kN; $v = 2.5$ m/s; $n = 25$; $\delta = 0.2$ m; $d = 25$ mm $= 0.025$ m; $D = 0.2$ m; $G = 100$ GN/m².

No. of springs required :

K.E. of the truck $= \dfrac{1}{2} \cdot \dfrac{W}{g}\,v^2 = \dfrac{1}{2} \times \dfrac{25 \times 10^3}{9.81} \times (2.5)^2 = 7963.8$ Nm

Stiffness, $k = \dfrac{C\,d^4}{64\,R^3\,n} = \dfrac{100 \times 10^9 \times (0.025)^4}{64 \times (0.1)^3 \times 25} = 24414$ N/m

Energy stored by the spring $= \dfrac{1}{2}\,k\delta^2 = \dfrac{1}{2} \times 24414 \times (0.2)^2 = 488.28$ Nm

Number of springs required $= \dfrac{7963.8}{488.28} = \textbf{16.3 say 17 (Ans.)}$

Example 22.69. *A close-coiled helical spring is of 80 mm mean coil diameter. The spring extends by 37.75 mm when loaded axially by a weight of 500 N. There is an angular rotation of 45° when this spring is subjected to an axial couple of magnitude 20 Nm. Determine the Poisson's ratio for the material of the spring.* **(IAS)**

Solution. *Given:* $R = \dfrac{D}{2} = \dfrac{80}{2} = 40$ mm or 0.04 m; $d = 37.75$ mm; $W = 500$ N; $\phi = 45°$ $= \dfrac{\pi}{4}$ rad.; $M = 20$ Nm.

Poisson's ratio; $\dfrac{1}{m}$ **:**

Stiffness, $k = \dfrac{W}{\delta} = \dfrac{500}{37.75} = 13.24$ N/mm $= 13240$ N/m

Also, $k = \dfrac{C\,d^4}{64\,R^3\,n}$ ...(Eqn. 14.7)

$\therefore$ $$13240 = \frac{Cd^4}{64R^3n}$$

or, $$\frac{Cd^4}{n} = 13240 \times 64 R^3 = 13240 \times 64 \times (0.04)^3 = 54.23 \qquad ...(i)$$

Angle of rotation ϕ is given by the relation,

$$\phi = \frac{128MRn}{Ed^4} \qquad ...[\text{Eqn. (14.14)}]$$

or, $$\frac{Ed^4}{n} = \frac{128MR}{\phi} = \frac{128 \times 20 \times 0.04}{\pi/4} = 130.38 \qquad ...(ii)$$

Dividing (i) by (ii), we get

$$\frac{C}{E} = \frac{54.23}{130.38} = 0.416$$

Now, $$E = 2C\left(1 + \frac{1}{m}\right) \qquad ...[\text{Eqn. (1.17)}]$$

or, $$\frac{C}{E} = \frac{1}{2\left(1 + \frac{1}{m}\right)} = 0.416$$

or, $$\left(1 + \frac{1}{m}\right) = \frac{1}{2 \times 0.416} = \textbf{1.202 (Ans.)}$$

$\therefore$ Poisson's ratio, $$\frac{1}{m} = \textbf{2.02 (Ans.)}$$

Example 22.70. *A leaf spring 1 m long has 8 plates 75 mm wide and 8 mm thick. Assuming $E = 2 \times 10^5$ N/mm^2, determine the maximum safe load that can be applied on the spring. The maximum deflection must not exceed 25 mm and the maximum bending stress is limited to 157.5 N/mm^2.*

Solution. *Given* : Span length $l = 1$m;

Width of each plate, $b = 75$ mm = 0.075 m;

Thickness of each plate, $t = 8$ mm = 0.008 m;

Young's modulus, $E = 2 \times 10^5$ N/mm^2 = 2×10^5 MN/m^2,

Maximum deflection, $\delta = 25$ mm = 0.025 m;

Maximum bending stress, $\sigma_b = 157.5$ N/mm^2 = 157.5 MN/m^2.

The maximum safe load, W :

First case : *Maximum deflection permissible*, $\delta = 0.025$ m:

Using the relation :

$$\delta = \frac{3Wl^3}{8ENbt^3} \qquad ...[\text{Eqn. 14.39}]$$

or, $$0.025 = \frac{3W \times l^3}{8 \times (2 \times 10^5 \times 10^6) \times 8 \times 0.08 \times (0.008)^3}$$

or, $$W = \frac{0.025 \times 8 \times (2 \times 10^5 \times 10^6) \times 8 \times 0.075 \times (0.08)^3}{3} = 4096 \text{ N}$$

Second case : *Maximum bending stress permissible*, $\sigma_b = 157.5$ MN/m^2 :

Using the relation :

$$\sigma_b = \frac{3Wl}{2Nbt^2} \qquad ...[\text{Eqn. (14.38)}]$$

$$157.5 \times 10^6 = \frac{3W \times 1}{2 \times 8 \times 0.075 \times 0.008}$$

or,
$$W = \frac{157.5 \times 10^6 \times 2 \times 8 \times 0.075 \times 0.008}{3} = 4032 \text{ N}$$

The maximum safe load is equal to **4032 N** (*minimum of the above two values*) **(Ans.)**

Example 22.71. *A laminated steel spring, simply supported at the ends and centrally loaded, with a span of 75 cm is required to carry a proof load of 7.5 kN and the central deflection is not to exceed 5 cm. The bending stress must not be greater than 400 MN/m². Plates are available in multiples of 1 mm for thickness and in multiples of 4 mm for width.*

Assuming width to be twelve times the thickness and E = 200 GN/m², determine suitable values for the thickness, width and number of plates and the radius to which the plates should be formed.

Solution. *Given :* $l = 75$ cm $= 0.75$ m; $W = 7.5$ kN $= 7500$ N; $\delta = 6$ cm $= 0.06$ m; $\sigma_b = 400$ MN/m²; $b = 12\ t$; $E = 200$ GN/m².

t, b, N, R :

Deflection,
$$\delta = \frac{3Wl^3}{8\,ENbt^3} \qquad \qquad ...[\text{Eqn. (14.39)}]$$

$$0.05 = \frac{3 \times 7500 \times 0.75^3}{8 \times 200 \times 10^9 \times N \times 12t \times t^3}$$

or,
$$Nt^4 = 9.888 \times 10^{-9} \qquad \qquad ...(i)$$

Bending stress,
$$\sigma_b = \frac{3Wl}{2\,Nbt^2}$$

$$400 \times 10^6 = \frac{3 \times 7500 \times 0.75}{2N \times 12t \times t^2}$$

or,
$$Nt^3 = 1757.8 \times 10^{-9} \qquad \qquad ...(ii)$$

Dividing (*i*) by (*ii*), we get $t = 5.625$ m say **6 mm** **(Ans.)**

$$b = 12 = 12 \times 6 = \textbf{72 mm} \ \ \textbf{(Ans.)}$$

$$N = \frac{9.888 \times 10^{-9}}{(6 \times 10^{-3})^4} = 7.63 \qquad \qquad ...\text{From (}i\text{)}$$

Also,
$$N = \frac{1757.8 \times 10^{-9}}{(6 \times 10^{-3})^3} = 8.14 \qquad \qquad ...\text{Fig. (}ii\text{)}$$

Adopt **9 plates** **(Ans.)**

$$\therefore \text{ Actual deflection, } \delta_{actual} = \frac{3 \times 7500 \times 0.75^3}{8 \times 200 \times 10^9 \times 9 \times 0.072 \times (0.006)^3}$$

$$= 0.04238 \text{ m} \ \text{ or } \ 42.38 \text{ mm}$$

Radius of curvature
$$R = \frac{l^2}{8\delta_{actual}} = \frac{0.75^2}{8 \times 0.04238} = \textbf{1.66} \ \ \textbf{(Ans.)}$$

Example 22.72. *A quarter-elliptic leaf spring has a length of 60 cm and consists of plates each 5 cm wide and 7.2 mm thick. Find the least number of plates which can be used if the deflection under a graduallly applied load of 2 kN is not to exceed 7 cm.*

If the load of 2 kN, instead of being gradually applied, falls from a distance of 6 mm on to the deflected spring, find the maximum deflection and stress produced. E = 200 GN/m². **(GATE)**

Solution. *Given :* $l = 60$ cm $= 0.6$ m; $b = 5$ cm $= 0.05$ m; $t = 7.2$ mm $= 0.0072$ m; $W = 2$kN; $\delta = 7$ cm $= 0.07$ m; $E = 200$ GN/m²; $h = 6$mm $= 0.006$ m.

Deflection,
$$\delta = \frac{6Wl^3}{ENbt^3} \qquad \qquad ...[\text{Eqn. 14.41}]$$

$$0.07 = \frac{6 \times 2 \times 10^3 \times 0.6^3}{200 \times 10^9 \times N \times 0.05 \times (0.0072)^3}$$

∴ $N = 9.92 \simeq \mathbf{10} \ (say) \ \mathbf{(Ans.)}$

Let W_e be the equivalent gradually applied load which would produce the same deflection as is caused by the impact load.

$$\delta_1 = \frac{6W_e l^3}{ENbt^3} = \frac{6W_e \times (0.6)^3}{200 \times 10^9 \times 10 \times 0.05 \times (0.0072)^3}$$

or, $W_e = 28800 \ \delta_1 \ N$

Loss of P.E. $= 2 \times 10^3 \ (0.006 + \delta_1)$

Strain energy absorbed by the spring $= \frac{1}{2} W_e \delta_1$

∴ $2 \times 10^3 (0.006 + \delta_1) = \frac{1}{2} \times 28800 \ \delta_1 \times \delta_1 = 14400 \ \delta_1^2$

$$12 + 2000\delta_1 = 14400 \ \delta_1^2$$

or, $\delta_1^2 - 0.1389\delta_1 - 0.00083 = 0$

or, $\delta_1 = \dfrac{0.1389 \pm \sqrt{(0.1389)^2 + 4 \times 0.00083}}{2} = 0.1446 \ m$

∴ $W_e = 28800 \times 0.1446 = 4164 \ N$

Maximum stress, $\sigma_{max} = \dfrac{6W_e l}{Nbt^2}$ ...[Eqn. (14.42)]

$$= \frac{6 \times 4164 \times 0.6}{10 \times 0.05 \times (0.0072)^2} \times 10^{-6} = \mathbf{578.3 \ MN/m^2} \ \mathbf{(Ans.)}$$

CONICAL SPRING

Example 22.73. (*a*) *Derive expressions for axial deflection and maximum shear stress for a conical spring.*

    (*b*) *A conical spring of minimum diameter 6 cm and maximum diameter 12 cm is made of wire of 18 mm diameter. If the shear stress is not to exceed 200 MN/m², modulus of rigidity is 84 MN/m² and the spring has 8 coils, determine :*

    (*i*)   *The load which the spring can carry.*

  (*ii*)   *The deflection of the spring.*   **(Roorkee University)**

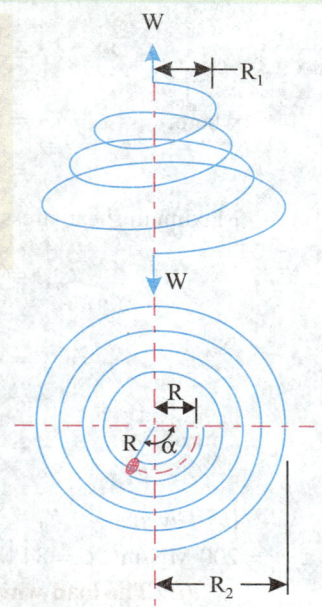

Solution: Fig. 22.66 shows a conical spring.

Let, R_1 = Minimum radius,

          R_2 = Maximum radius,

          d = Diameter of the wire

          n = Number of coils of the spring, and

          W = Load applied.

For any point on the centre line of the spring determined by the magnitude of the angle α, the distance R from the axis of the spring is,

$$R = R_1 + \frac{(R_2 - R_1)\alpha}{2\pi n} \qquad ...(i)$$

Consider a small length ds of the spring at a radius R.

Fig. 22.66. Conical spring.

The angle of twist produced by W in length ds,

$$d\theta = \frac{(WR)\,ds}{CI_p} \qquad\qquad\qquad (\text{where, } WR = T, \text{ torque})$$

Strain energy stored by the length ds of the spring,

$$dU = \frac{1}{2}(WR)\,d\theta = \frac{W^2R^2\,ds}{2CI_p} = \frac{W^2R^3\,d\alpha}{2CI_p} \qquad ...(ii)\ (\because\ ds = R\,d\alpha)$$

Also, $\qquad\qquad dR = \left(\dfrac{R_2 - R_1}{2\pi n}\right)d\alpha$

or, $\qquad\qquad d\alpha = \left(\dfrac{2\pi n}{R_2 - R_1}\right)dR$

Substituting this value in (ii), we get

$\therefore \qquad\qquad dU = \dfrac{W^2R^3}{2CI_p}\left(\dfrac{2\pi}{R_2 - R_1}\right)dR$

or, Total energy, $U = \dfrac{\pi n W^2}{CI_p\,(R_2 - R_1)}\cdot\displaystyle\int_{R_1}^{R_2} R^3\,dR$

or, $\qquad\qquad U = \dfrac{\pi n W^2}{CI_p\,(R_2 - R_1)}\cdot\left(\dfrac{R_2^4 - R_1^4}{4}\right) \qquad\qquad\qquad ...(1)$

Axial deflection, $\delta = \dfrac{\partial U}{\partial W} = \dfrac{\partial}{\partial W}\left[\dfrac{\pi n W^2}{CI_p\,(R_2 - R_1)}\left(\dfrac{R_2^4 - R_1^4}{4}\right)\right]$

$\qquad\qquad = \dfrac{\pi n W}{2CI_p\,(R_2 - R_1)}\left(R_2^4 - R_1^4\right) = \dfrac{n\pi W}{2CI_p\,(R_2 - R_1)}\times(R_2^2 + R_1^2)(R_2^2 - R_1^2)$

or, $\qquad\qquad \delta = \dfrac{\pi n W}{2CI_p}\,(R_1^2 + R_2^2)\,(R_1 + R_2) \qquad\qquad\qquad\qquad ...(2)$

$\qquad\qquad = \dfrac{Wl\,(R_1^2 + R_2^2)}{2CI_p}$

where, $\qquad\qquad l = \pi n\,(R_1 + R_2),$ and

$\qquad\qquad I_p = \dfrac{\pi}{32}d^4$

Maximum shear stress,

$$\tau_{max} = \frac{WR_2 d}{2\,Ip} + \frac{4}{3} \qquad (\text{average shear stress due to } W)$$

$$= \frac{WR_2 d}{2\,I_p} + \frac{4}{3}\times\frac{W}{(\pi/4)d^2}$$

$$= \frac{WR_2 d}{2I_p} + \frac{Wd^2}{6I_p} = \frac{Wd}{6I_p}\,(3R_2 + d) \qquad\qquad\qquad ...(3)$$

(b) Given : $\qquad R_1 = \dfrac{6}{2} = 3\,\text{cm} = 0.03\,\text{m}; \quad R_2 = \dfrac{12}{2} = 6\,\text{cm} = 0.06\text{ m}; \quad d = 18\text{ mm} = 0.018\text{ m};$
$\tau_{max} = 200\ \text{MN/m}^2; \quad C = 84\ \text{MN/m}^2; \quad n = 8.$

(i) The load which the spring can carry W:

$$\tau_{max} = \frac{Wd}{6I_p}\,(3R_2 + d)$$

$$200 \times 10^6 = \frac{W \times 0.018}{6 \times \frac{\pi}{32} \times (0.018)^4} \ (3 \times 0.06 + 0.018)$$

∴ $W = \textbf{3470 N (Ans.)}$

(ii) The deflection of the spring, δ:

$$\delta = \frac{\pi n W}{2 C I_p} \ (R_1^2 + R_2^2)(R_1 + R_1)$$

$$= \frac{\pi \times 8 \times 3470}{2 \times 84 \times 10^9 \times \frac{\pi}{32} \times (0.018)^4} \ (0.03^2 + 0.06^2)(0.03 + 0.06)$$

$$= 0.0204 \ m \ \text{or} \ \textbf{20.4 mm (Ans.)}$$

STRAIN ENERGY

Example 22.74. *The steel and brass bars comprising the stepped shaft in Fig. 22.67 are to suffer the same displacement under tensile force of 40 kN. Determine the diameter of the brass bar, d_b, the stress in each material and the total strain energy stored. Take $E_s = 207$ GPa and $E_b = 82.7$ GPa.*

Solution. *Given :* $d_s = 50 \ mm = 0.05 \ m$; $l_s = 0.5 \ m$; $P = 40 \ kN$, $E_s = 207$ GPa; $E_b = 82.7$ GPa.

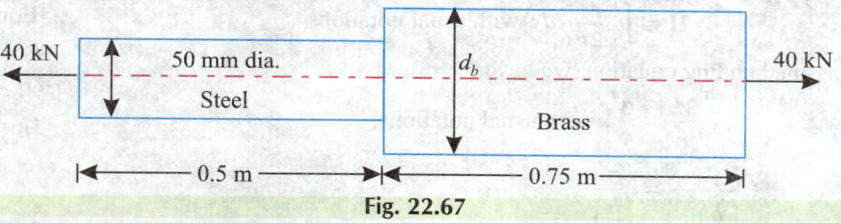

Fig. 22.67

Stress in each material, σ_s, σ_b :

Fig. 22.67. shows a stepped shaft, subjected to an axial tensile force of 40 kN. The total force on each section, steel and brass is the same, *i.e.* 40 kN, but the intensities of stress, σ_s ans σ_b will be different for the two sections.

Hence, $$\sigma_s = \frac{P}{A_s} = \frac{40 \times 10^{-3}}{\frac{\pi}{4} \times (0.05)^2} \ \text{MN/m}^2 = \textbf{20.37 MN/m}^2 \ \textbf{(Ans.)}$$

Also, $$\delta_s = \delta_b \quad ...(Given) \qquad \qquad (\text{where, } \delta \text{ denotes } displacement)$$

$$\frac{\sigma_s l_s}{E_s} = \frac{\sigma_b l_b}{E_b}$$

or, $$\sigma_b = \frac{\sigma_s l_s E_b}{l_b E_s} = \frac{\sigma_s l_s}{l_b} \left(\frac{E_b}{E_s} \right)$$

$$= \frac{20.37 \times 0.5}{0.75} \times \left(\frac{827}{207} \right) \ \text{MN/m}^2 = \textbf{5.42 MN/m}^2 \ \textbf{(Ans.)}$$

Diameter of the brass bar, d_b :

$$P = \sigma_b \times A_b = 5.42 \times \frac{\pi}{4} d_b^2$$

or, $$40 \times 10^{-3} = 5.42 \times \frac{\pi}{4} d_b^2$$

∴ $$d_b = \left(\frac{40 \times 10^{-3} \times 4}{5.42 \times \pi} \right)^{1/2} = 0.09694 \ m \ \text{or} \ \textbf{96.94 mm (Ans.)}$$

Total strain energy stored, U :

$$U = U_s + U_b = \frac{\sigma_s^2 A_s l_s}{2 E_s} + \frac{\sigma_b^2 A_b l_b}{2 E_b}$$

$$= \frac{(20.37)^2 \times \frac{\pi}{4} \times (0.05)^2 \times 0.5}{2 \times (207 \times 10^3)} + \frac{(5.42)^2 \times \frac{\pi}{4} \times (0.09694)^2 \times 0.75}{2 \times (82.7 \times 10^3)}$$

$$= 9.84 \times 10^{-7} + 9.83 \times 10^{-7} \qquad (\because 1\,GPa = 1 \times 10^3 \, MN/m^2)$$

$$= 19.67 \times 10^{-7} \, MNm = \mathbf{1.967 \, Nm \; or \; J} \; \mathbf{(Ans.)}$$

Example 22.75. *Find the elastic strain energy stored in a rectangular cantilever beam subjected to a bending moment M applied at the free end as shown in Fig. 22.68. Express the same is term of maximum bending stress.* **(GATE)**

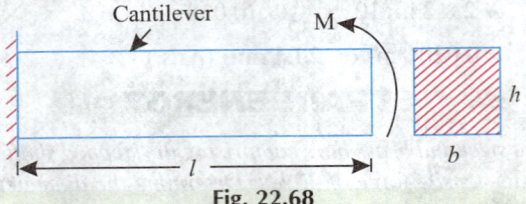

Fig. 22.68

Solution. We know that the strain energy stored in a beam is given by,

$$U = \int_0^l \frac{M^2}{2EI} dx \text{ with usual notations} \qquad \qquad ...[\text{Eqn. (15.10)}]$$

Using the bending equation, we have,

$$\frac{M}{I} = \frac{\sigma}{y}, \text{ with usual notations,}$$

$$\frac{M}{\left(\frac{bh^3}{12}\right)} = \frac{\sigma}{\left(\frac{h}{2}\right)}$$

or, $$M = \sigma \times \left(\frac{bh_3}{12}\right)\left(\frac{2}{h}\right) = \sigma \times \frac{bh^2}{6} \text{ or } \frac{1}{6}\sigma bh^2$$

Substituting for M in the strain energy equation, we get

$$U = \int_0^l \frac{\left(\frac{1}{6}\sigma bh^2\right)^2}{2E \times \left(\frac{bh^3}{12}\right)} dx \qquad \qquad \left(\because I = \frac{bh^3}{12}\right)$$

$$= \int_0^l \left(\frac{1}{36}\sigma^2 b^2 h^4\right) \times \frac{12}{2E \times bh^3} \cdot dx = \left[\frac{\sigma^2 bh}{6E} x\right]_0^l$$

or, $$U = \frac{\sigma^2 bhl}{6E} \; \mathbf{(Ans.)}$$

Example 22.76. *A simply supported beam is subjected to a single force P at a distance b from one of the supports. Obtain the expression for the deflection under the load using Castigliano's theorem. How do you calculate deflection at the mid-point of the beam?* **(GATE)**

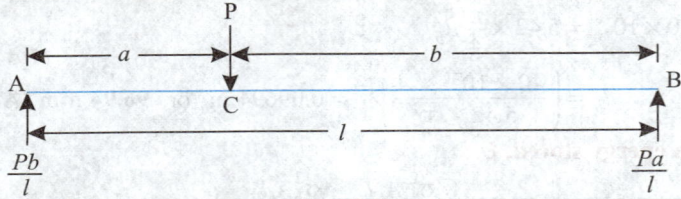

Fig. 22.69

Solution. Let the load P acts at a distance b from the support B, and l be the total length of the beam.

Reaction at A, $\quad R_A = \dfrac{Pb}{l}$, and

Reaction at B, $\quad R_B = \dfrac{Pa}{l}$

Strain energy stored by beam AB,

$$U = \text{Strain energy stored by } AC\ (U_{AC}) + \text{strain energy stored by } BC\ U_{BC})$$

$$= \int_0^a \left(\frac{Pb}{l}.x\right)^2 \frac{dx}{2EI} + \int_0^b \left(\frac{Pa}{l}.x\right)^2 \frac{dx}{2EI}$$

$$= \frac{P^2 b^2 a^3}{6EIl^2} + \frac{P^2 a^2 b^3}{6EIl^2}$$

$$= \frac{P^2 a^2 b^2}{6EIl^2}(a+b) = \frac{P^2(l-b)^2 b^2}{6EIl} \qquad\qquad [\because\ (a+b)=l]$$

Deflection under the load P,

$$\delta = y = \frac{\partial U}{\partial P} = \frac{2P(l-b)^2 b^2}{6EIl} = \frac{P(l-b)^2 b^2}{3EIl} \quad \textbf{(Ans.)}$$

Deflection at the mid-span of the beam can be found by Macaulay's method.
By Macaulay's method, deflection at any section is given by

$$EIy = \frac{Pbx^3}{6l} - \frac{Pb}{6l}(l^2-b^2)x - \frac{P(x-a)^3}{6} \qquad ...[\text{Refer Art. 8.7}]$$

where, y is deflection at any distance x from the support.

At $\qquad x = \dfrac{l}{2}$, *i.e.* at mid-span,

$$EIy = \frac{Pb \times (l/2)^3}{6l} - \frac{Pb}{6l}(l^2-b^2) \times \frac{1}{2} - \frac{P\left(\dfrac{l}{2}-a\right)^3}{6}$$

$$EIy = \frac{Pbl^2}{48} - \frac{Pb(l^2-b^2)}{12} - \frac{P(l-2a)^3}{48}$$

$$y = \frac{P}{48EI}\left[bl^2 - 4b(l^2-b^2) - (l-2a)^3\right] \quad \textbf{(Ans.)}$$

Example 22.77. *A cantilever beam of stepwise constant cross-section as shown in Fig. 22.70 is loaded by a concentrated force P at its tip. Determine the deflection under the point of application of the force using Castigliano's theorem.* **(Bombay University)**

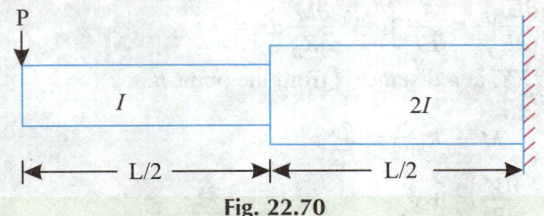

Fig. 22.70

Solution: Refer to Figs. 22.70 and 22.71.

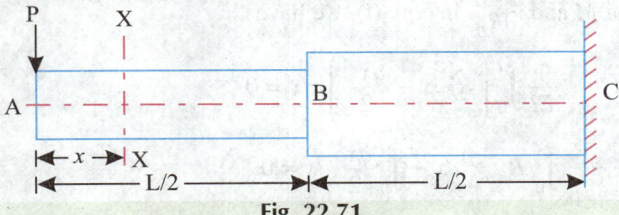

Fig. 22.71

Consider a section XX at a distance x from the free end A of the cantilever beam.

Then, $M_{XX} = Px$

Now the internal strain energy stored in the beam is given by,

$$U = \int_0^L \frac{M^2 dx}{2EI} = \int_0^{L/2} \frac{(Px)^2 dx}{2EI} + \int_{L/2}^L \frac{(Px)^2 dx}{2E(2I)}$$

According to Castigliano's theorem the deflection under the point of application of the force P is given by,

$$\delta_A = \int_0^{L/2} \frac{Px^2 dx}{EI} + \int_{L/2}^L \frac{Px^2 dx}{2EI} = \frac{P}{EI}\int_0^{L/2} x^2 dx + \frac{P}{2EI}\int_{L/2}^L x^2 dx$$

$$= \frac{P}{EI}\left[\frac{x^3}{3}\right]_0^{L/2} + \frac{P}{2EI}\left[\frac{x^3}{3}\right]_{L/2}^L = \frac{P}{3EI} \times \frac{L^3}{8} + \frac{P}{6EI}\left[L^3 - \frac{L^3}{8}\right]$$

$$= \frac{PL^3}{EI}\left[\frac{1}{24} + \frac{7}{48}\right] = \frac{9}{48}\frac{PL^3}{EI} \quad \textbf{(Ans.)}$$

Example 22.78. *A beam AB of span l is fixed at A and simply supported at B. It carries a uniformly distributed load of intensity q per unit length over the entire span. Find the reaction at B, using Castigliano's theorem. EI is constant.* **(Roorkee University)**

Solution. Fig. 22.72 shows a beam AB of span l which is fixed at A and simply supported at B. It carries a uniformly distributed load of intensity q per unit length over the entire span.

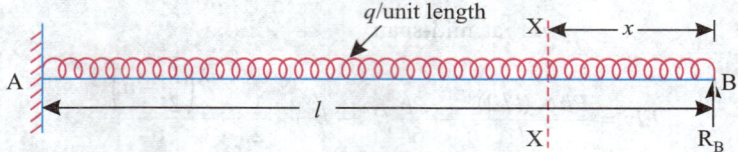

Fig. 22.72

Let R_B be the reaction at B.

Using Castigliano's theorem of minimum strain energy, we have

$$\frac{\partial U}{\partial R} = 0, \text{ where, } U = \text{strain energy stored in an elastic body}$$

Hence we have, $R = $ redundant reaction.

$$\frac{\partial U_{AB}}{\partial R_B} = \frac{1}{EI}\int_A^B M \frac{\partial M}{\partial R_B} dx = 0 \qquad \text{...}(i)$$

Consider a section XX, at a distance x from the point B.

$$M = R_B . x - \frac{qx^2}{2}$$

∴ $$\frac{\partial M}{\partial R_B} = +x$$

Substituting for M and $\dfrac{\partial M}{\partial R_B}$ in eqn. (i), we have

$$\frac{1}{EI}\int_0^l \left(R_B . x - \frac{qx^2}{2}\right) x dx = 0$$

or, $$\int_0^l R_B . x^2 dx - \int_0^l \frac{qx^3}{2} dx = 0$$

or, $R_B \left[\dfrac{x^3}{3} \right]_0^l - \dfrac{q}{2} \left[\dfrac{x^4}{4} \right]_0^l = 0$

or, $R_B \times \dfrac{l^3}{3} - \dfrac{ql^4}{8} = 0$

∴ $R_B = \dfrac{ql^4}{8} \times \dfrac{3}{l^3} = \dfrac{3}{8} ql$ **(Ans.)**

COLUMNS AND STRUTS

Example 22.79. *A column 2.5 m long is pin-connected at both ends. It has 50 mm × 100 mm rectangular cross-section. Young's modulus of material is 2.0×10^5 MPa. Determine.*

(i)   The slenderness ratio.

(ii)   The Euler buckling load, P_{cr}.

(iii)   The axial stress at P_{cr}.

(iv)   The safe load if factors of safety is 2.5.   **(Bombay University)**

Solution. *Given :* $l = 2.5$ m; $l_e = l = 2.5$ m (both ends pin-connected);

$A = 50$ mm × 100 mm = 5000 mm^2 = 0.005 m^2;

$E = 2.0 \times 10^5$ MPa = 2.0×10^5 MN/m^2; F.O.S. = 2.5

Moment of inertia of the section about N.A.,

$$I = \dfrac{bd^3}{12} = \dfrac{50 \times 100^3}{12} \text{ mm}^4 = 4.167 \times 10^{-6} \text{ m}^4$$

$$P_{Euler} = P_{cr} = \dfrac{\pi^2 EI}{l_e^2} = \dfrac{\pi^2 \times 2.0 \times 10^5 \times (4.167 \times 10^{-6})}{2.5^2} = 1.316 \text{ MN}$$

(i)   Slenderness ratio, $\dfrac{l}{k}$:

We know, $P_{cr} = \dfrac{\pi^2 EI}{l_e^2} = \dfrac{\pi^2 EI}{l^2} = \dfrac{\pi^2 EAk^2}{l^2} = \dfrac{\pi^2 EA}{(l^2 / k^2)}$ $(\because I = Ak^2)$

∴ $\dfrac{l^2}{k^2} = \dfrac{\pi^2 EA}{P_{cr}}$

or $\dfrac{l}{k} = \sqrt{\dfrac{\pi^2 EA}{P_{cr}}} = \sqrt{\dfrac{\pi^2 \times 2.0 \times 10^5 \times 0.005}{1.316}} = \textbf{86.6 (say 87)}$ **(Ans.)**

(ii)   The Euler buckling load, P_{cr}:

$P_{cr} = \textbf{1.316 MN}$ **(Ans.)**   ... (calculated above)

(iii)   The axial stress at P_{cr}, σ_a :

$$P_{cr} = \dfrac{\pi^2 EI}{l_e^2} = \dfrac{\pi^2 EI}{l^2} = \dfrac{\pi^2 EAk^2}{l^2}$$

or $\dfrac{P_{cr}}{A} = \sigma_a = \dfrac{\pi^2 Ek^2}{l^2} = \dfrac{\pi^2 E}{(l^2/k^2)} = \dfrac{\pi^2 \times 2.0 \times 10^5}{(86.6)^2} = \textbf{263.2 MN/m}^2$ **(Ans.)**

(iv)   The safe load :

Safe load $= \dfrac{P_{cr}}{F.O.S.} = \dfrac{1.316}{2.5} = \textbf{0.5264 MN}$ **(Ans.)**

Example 22.80. *Find the shortest length of a hinged steel column having a rectangular cross-section 600 mm × 100 mm, for which the elastic Euler formula applies. Take yield strength and modulus of elasticity values for steel as 250 MPa and 200 GPa respectively.* **(GATE)**

Solution. *Given :* Cross-section $(= b \times d)$ = 60 mm × 100 mm = 0.06 m × 0.1 m = 0.006 m²; yield strength $= \dfrac{P}{A} = 250\,\text{MPa} = 250\,\text{MN/m}^2$; $E = 200\,\text{GPa} = 200 \times 10^9\,\text{N/m}^2$

Length of the column, *l* :

$$I = \frac{bd^3}{12} = \frac{0.06 \times 0.1^3}{12} = 5 \times 10^{-6}\ \text{m}^4$$

Also, $I = Ak^2$ (where, A = area of cross-section, k = radius of gyration)

$$k^2 = \frac{I}{A} = \frac{5 \times 10^{-6}}{0.06 \times 0.1} = 0.0008333\ \text{m}^2$$

From Euler's formula for column, we have

Crushing load, $\quad P = \dfrac{\pi^2 EI}{l_e^2} = \dfrac{\pi^2 EI}{l^2}$

For hinged type of column, $l_e = l$

or, $\qquad P = \dfrac{\pi^2 EAk^2}{l^2}$

or, $\qquad \dfrac{P}{A}(= \text{yield stress}) = \dfrac{\pi^2 Ek^2}{l^2}$

or, $\qquad l^2 = \dfrac{\pi^2 Ek^2}{(P/A)}$

Substituting the value, we have

$$l^2 = \frac{\pi^2 \times 200 \times 10^9 \times 0.0008333}{250 \times 10^6} = 6.579$$

∴ $\qquad$ **l = 2.56 m** **(Ans.)**

Example 22.81. *Determine the temperature rise necessary to induce buckling in a 1m long circular rod of diameter 40 mm shown in the figure below. Assume the rod to be pinned at its ends and the coefficient of thermal expansion as $20 \times 10^{-6}/°C$. Assume uniform heating of the bar.* **(GATE)**

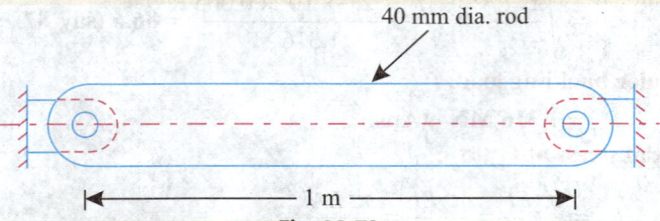

40 mm dia. rod

|← 1 m →|

Fig. 22.73

Solution. Let the buckling load be '*P*'.

$$\delta l = l \cdot \alpha \cdot \Delta t, \text{ where } \Delta t \text{ is the temperature rise.}$$

or, $\qquad \Delta t = \dfrac{\delta l}{l \cdot \alpha}$

Also, $\qquad \delta l = \dfrac{Pl}{AE}$

or,
$$P = \frac{\delta l . AE}{l}$$

$$P_{cr} = \frac{\pi^2 EI}{l_e^2} \qquad \text{(where, } l_e \text{ = equivalent length)}$$

or,
$$\frac{\pi^2 EI}{l^2} = \frac{\delta l . A . E}{l} \qquad [\because l_e = l \text{ (pinned ends)}]$$

or,
$$\delta l = \frac{\pi^2 I l}{l^2 A}$$

$$\Delta t = \frac{\delta l}{l . \alpha} = \frac{\pi^2 I l}{l^2 A . l . \alpha} = \frac{\pi^2 I}{l^2 . A . \alpha}$$

Substituting the values, we get

Temperature rise
$$\Delta t = \frac{\pi^2 \times \dfrac{\pi}{64} \times \left(\dfrac{40}{1000}\right)^4}{(1)^2 \times \dfrac{\pi}{4} \times \left(\dfrac{40}{1000}\right)^2 \times 20 \times 10^{-6}} = 49.35 \text{ °C}$$

Hence, the rod will buckle when the temperature rises beyond **49.35°C.** **(Ans.)**

Example 22.82. *A round steel rod 15 mm diameter and 2m long is subjected to a gradually increasing axial compressive load. Using Euler's formula find the buckling load. Also find the maximum lateral deflection corresponding to buckling condition. Both ends of the column can be taken as hinged. Assume Young's modulus for steel as equal to 200 GPa and the yield stress as 240 N/mm².* **(IAS)**

Solution. Given : $d = 15$ mm; $l = 2$m; $E = 200$ GPa; $\sigma = 240$ N/mm²;

Buckling load, P_{cr}:
Lateral deflection, δ:

Moment of inertia,
$$I = \frac{\pi d^4}{64} = \frac{\pi \times (15)^4}{64} = 2485 \text{ mm}^4 \text{ or } 2485 \times 10^{-12} \text{ m}^4$$

We know that,
$$P_{cr} = \frac{\pi^2 EI}{l_e^2} \qquad ...\text{Euler's formula}$$

$$= \frac{\pi^2 EI}{l^2} \qquad (\because l_e = l \ ... \text{ ends being hinged})$$

$$\frac{\pi^2 \times (200 \times 10^9) \times (2485 \times 10^{-12})}{(2)^2} = \textbf{1226.3 N} \quad \textbf{(Ans.)}$$

Also,
$$\frac{\sigma}{y} = \frac{E}{R}$$

or,
$$R = \frac{Ey}{\sigma} = \frac{(200 \times 10^9) \times (7.5 \times 10^{-3})}{240 \times 10^6} = 6.25 \text{ m}$$

Now,
$$\left(\frac{l}{2}\right)^2 = \delta(2R - \delta)$$

$$\left(\frac{2}{2}\right)^2 = \delta(2 \times 6.25 - \delta)$$

or,
$$I = 12.5\delta - \delta^2$$

or,
$$\delta^2 - 12.5\delta + 1 = 0$$

or,
$$\delta = \frac{12.5 \pm \sqrt{(12.5)^2 - 4 \times 1}}{2}$$

$$= \frac{12.5 \pm 12.34}{2} = 0.08 \,\text{m or } \mathbf{80\ mm} \quad \textbf{(Ans.)}$$

Example 22.83. *A mild steel strut is of single angle section 100 mm × 75 mm × 8 mm. Its length is 3m with both ends fixed. Taking factor of safety 3 and using Rankine's formula, determine the maximum safe load for the column.*

The section has 1.336 m² as area of cross-section. Its radius of gyration about XX, YY, UU and VV axes are 31.4 mm 21.8 mm, 34.8 mm and 15.9 mm respectively. Constants in formula may be taken as $a = \dfrac{1}{7500}$ *and* $f_c = 330 \,N/mm^2$. **(Bangalore University)**

Solution. *Given* : Single angle section : 100 mm × 75 mm × 8 mm;

Length, *l* = 3m = 3000 mm;

End conditions: Both ends fixed;

Factor of safety (F.O.S.) = 3;

Cross-sectional area of the section *A* = 1336 mm²

Radius of gyration about *XX, YY, UU* and *VV* axes = 31.4 mm, 21.8 mm, 34.8 mm and 15.9 mm

Constants : $\qquad a = \dfrac{1}{7500}, \sigma_C = 330 \,\text{N}/\text{mm}^2.$

Maximum safe load, W:

For *both ends fixed* $\quad l_e = \dfrac{l}{2} = \dfrac{3000}{2} = 1500 \,\text{mm}$

Minimum radius of gyration,

$$k_{\text{least}} = 15.9 \,\text{mm}$$

Then, $\qquad \dfrac{l_e}{k} = \dfrac{1500}{15.9} = 94.34$

Using Rankine's formula, we have:

$$P_{\text{Rankine}} = \frac{\sigma_c \times A}{1 + a \left(\dfrac{l_e}{k}\right)^2} = \frac{330 \times 1336}{1 + \dfrac{1}{7500} \times (94.34)^2} = 201621 \,\text{N}$$

$$\therefore \ \textit{Maximum safe load} = \frac{P_{\text{Rankine}}}{F.O.S.} = \frac{201621}{3} = 67207 \,\text{N} = \mathbf{67.2\ kN} \quad \textbf{(Ans.)}$$

Example 22.84. *Calculate the percentage error in finding the bulking load if the equivalent length for the freely hinged strut is assumed as 0.666 l, where l is the actual length of the strut.* **(IAS)**

Solution. *Given* : $\quad l_e = 0.666\, l$

Percentage error in finding the buckling load:

$$l_e = \frac{l}{\sqrt{2}} \text{ (since the strut is fixed at one end and hinged at the other)}$$

$$P_{\text{euler}} = \frac{\pi^2 EI}{l_e^2} = \frac{\pi^2 EI}{(l/\sqrt{2})^2} = \frac{2\pi^2 EI}{l^2} \qquad\qquad ...(i)$$

If $\ l_e = 0.666\, l,$ then

$$P_{\text{euler}} = \frac{\pi^2 EI}{(0.666\,l)^2} = \frac{2.2545\,\pi^2 EI}{l^2} \qquad\qquad ...(iii)$$

From eqns. (*i*) and (*ii*), we have

$$Percentage\ error = \frac{2.2545 - 2}{2} \times 100 = \textbf{12.725\%} \quad \textbf{(Ans.)}$$

THEORIES OF FAILURE

Example 22.85. *Find the maximum principal stress developed in a cylindrical shaft, 8 cm in diameter and subjected to a bending moment of 2.5 kNm and a twisting moment of 4.2 kNm. If the yield stress of the shaft material is 300 MPa, determine the factor of safety of the shaft according to the maximum shearing stress theory of failure.* **(GATE)**

Solution. *Given :* d = 8cm = 0.08 m; M = 2.5 kNm = 2500 Nm; T = 4.2 kNm = 4200 Nm; σ_{yield} (σ_{et}) = 300 MPa = 300 MN/m^2.

Equivalent torque, $T_e = \sqrt{M^2 + T^2} = \sqrt{(2.5)^2 + (4.2)^2} = 4.888$ kNm

Maximum shear stress developed in the shaft.

$$\tau_{max} = \frac{16 T_e}{\pi d^3} = \frac{16 \times 4.888 \times 10^3}{\pi \times (0.08)^3} \times 10^{-6} \text{ MN/m}^2 = 48.62 \text{ MN/m}^2$$

Permissible shear stress = $\dfrac{300}{2}$ = 150 MN/m^2

$\therefore$ *Factor of safety* $\quad = \dfrac{150}{48.62} = \textbf{3.085} \quad$ **(Ans.)**

Example 22.86. *Three exactly similar specimens of mild steel tube are 37.5 mm external diameter and 31.25 mm internal diameter. One of these tubes was tested in tension and the limit of proportionality was achieved at a load of 70 kN. The second was tested in torsion whereas the third was tested in torsion with a superimposed bending moment of 350 Nm. If the failure criteria is maximum shear stress, estimate the torque at which the two specimens would fail.* **(IAS)**

Solution. *Given :* $\quad d_0$ = 37.5 mm; d_i = 31.25 mm; P = 70 kN; M = 350 Nm.

Area of each specimen,

$$A = \frac{\pi}{4}(d_0^2 - d_i^2) = \frac{\pi}{4}(37.5^2 - 31.25^2)$$
$$= 337.476 \text{ mm}^2 = 337.476 \times 10^{-6} \text{ m}^2$$

Proportional limit, $\quad \sigma_p = \dfrac{P}{A} = \dfrac{70 \times 10^3}{337.476 \times 10^{-6}} = 207.42$ MN/m^2

Polar moment of inertia,

$$I_p = \frac{\pi}{32}\left[d_0^4 - d_i^4\right] = \frac{\pi}{32}\left[(37.5)^4 - (31.25)^4\right] \text{ mm}^4$$
$$= 100517.7 \text{ mm}^4 = 100517.7 \times 10^{-12} \text{ m}^4$$

According to maximum shear stress criteria,

$$\tau_{max} = \frac{\sigma_p}{2} = \frac{207.42}{2} = 103.71 \text{ MN/m}^2$$

Let, $\quad T$ = Torque applied in Nm.

Then, equivalent torque,

$$T_e = \sqrt{M^2 + T^2} \qquad \qquad \qquad ...(i)$$

Also, $\quad \dfrac{T_e}{I_p} = \dfrac{\tau_{max}}{(d_0/2)} = \dfrac{2\tau_{max}}{d_0}$

or, $\quad T_e = I_p \times \dfrac{2\tau_{max}}{d_0} = 100517.7 \times 10^{-12} \times \dfrac{2 \times 103.71 \times 10^6}{37.5 \times 10^{-3}} = 555.98$ Nm

Substituting the values in eqn. (*i*), we get

$$555.98 = \sqrt{(350)^2 + T^2}$$

∴ $$T = \sqrt{(555.98)^2 - (350)^2} = \textbf{432 Nm} \quad \textbf{(Ans.)}$$

Example 22.87. *A cube of 5mm side is loaded as shown in Fig. 22.74*

(*i*) *Determine the principal stresses* σ_1, σ_2 *and* σ_3.

(*ii*) *Will the cube yield if the yield strength of the material is 70 MPa? Use Von-Mises theory.* **(GATE)**

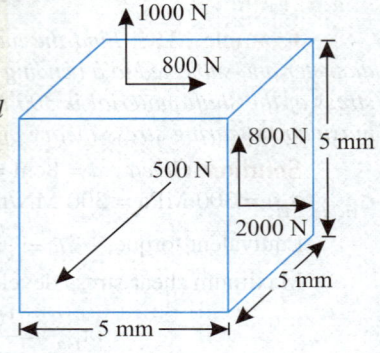

Fig. 22.74

Solution. Yield strength of the material,

$$\sigma_{et} = 70 \text{ MPa} = 70 \text{ MN/m}^2 \text{ or } 70 \text{ N/mm}^2$$

(*i*) **Principal stresses** σ_1, σ_2 **and** σ_3 **:**

$$\sigma_x = \frac{2000}{5 \times 5} = 80 \text{ N/mm}^2$$

$$\sigma_y = \frac{1000}{5 \times 5} = 40 \text{ N/mm}^2$$

$$\sigma_z = \frac{500}{5 \times 5} = 20 \text{ N/mm}^2$$

$$\tau_{xy} = \frac{800}{5 \times 5} = 32 \text{ N/mm}^2$$

$$\sigma = \frac{\sigma_x + \sigma_y}{2} \pm \sqrt{\left(\frac{\sigma_x - \sigma_y}{2}\right)^2 + \tau_{xy}^2}$$

$$= \frac{80 + 40}{2} \pm \sqrt{\left(\frac{80 - 40}{2}\right)^2 + (32)^2}$$

$$= 60 \pm \sqrt{(20)^2 + (32)^2} = 97.74, \ 22.26$$

∴ $$\sigma_1 = 97.74 \text{ N/mm}^2, \text{ or } \textbf{97.74 MPa} \quad \textbf{(Ans.)}$$

and, $$\sigma_2 = 22.26 \text{ N/mm}^2 \text{ or } \textbf{22.26 MPa} \quad \textbf{(Ans.)}$$

$$\sigma_3 = \sigma_z = 20 \text{ N/mm}^2 \text{ or } \textbf{20 MPa} \quad \textbf{(Ans.)}$$

(*ii*) **Will the cube yield ?**

According to Von-Mises yield criteria, yielding will occur if

$$(\sigma_1 - \sigma_2)^2 + (\sigma_2 - \sigma_3)^2 + (\sigma_3 - \sigma_1)^2 \geqslant 2\sigma_{et}^2$$

Now, $(\sigma_1 - \sigma_2)^2 + (\sigma_2 - \sigma_3)^2 + (\sigma_3 - \sigma_1)^2 = (97.74 - 22.26)^2 + (22.26 - 20)^2 + (20 - 97.74)^2$

$$5697.2 + 5.1 + 6043.5 = 11745.8 \qquad \qquad ...(i)$$

and, $$2\sigma_{et}^2 = 2 \times (70)^2 = 9800 \qquad \qquad ...(ii)$$

Since 11745.8 (*i*) > 9800, (*ii*) hence hence **yielding will occur** **(Ans.)**

Example 22.88. *A thin-walled circular tube of wall thickness t and mean radius r is subjected to an axial load P and torque T in a combined tension-torsion experiment.*

(*i*) *Determine the state of stress existing in the tube in terms of P and T.*

(*ii*) *Using Von-Mises–Henky failure criteria show that failure takes place when* $\sqrt{\sigma^2 + 3\tau^2} = \sigma_0$, *where* σ_0 *is the yield stress in uniaxial tension.* **(GATE)**

Solution. Mean radius of the tube = r,

Wall thickness of the tube = t,

Axial load = P, and Torque = T.

(i)   The state of stress in the tube :

Due to axial load, the axial stress in the tube

$$\sigma_x = \frac{P}{2\pi rt}$$

Due to torque, shear stress,

$$\tau_{xy} = \frac{T r}{I_p} = \frac{T r}{2\pi r^3 t} = \frac{T}{2\pi r^2 t}$$

$$\left[I_p = \frac{\pi}{2}\left\{(r+t)^4 - r^4\right\} = 2\pi r^3 t \ \ ...... \ \text{neglecting } t^2 \text{ and higher powers of } t \right]$$

∴ The state of stress in the tube is,

$$\sigma_x = \frac{P}{2\pi rt}, \sigma_y = 0, \tau_{xy} = \frac{T}{2\pi r^2 t}$$

(ii)   Von Mises-Henky failure in tension for 2-dimensional stress is given by,

$$\sigma_0^2 = \sigma_1^2 + \sigma_2^2 - \sigma_1 \sigma_2$$

$$\sigma_1 = \frac{\sigma_x + \sigma_y}{2} + \sqrt{\left(\frac{\sigma_x - \sigma_y}{2}\right)^2 + \tau_{xy}^2}$$

$$\sigma_2 = \frac{\sigma_x + \sigma_y}{2} - \sqrt{\left(\frac{\sigma_x - \sigma_y}{2}\right)^2 + \tau_{xy}^2}$$

In this case, $\sigma_1 = \dfrac{\sigma_x}{2} + \sqrt{\dfrac{\sigma_x^2}{4} + \tau_{xy}^2}$, and $\sigma_2 = \dfrac{\sigma_x}{2} - \sqrt{\dfrac{\sigma_x^2}{4} + \tau_{xy}^2}$ $(\because \sigma_y = 0)$

∴

$$\sigma_0^2 = \left[\frac{\sigma_x}{2} + \sqrt{\frac{\sigma_x^2}{4} + \tau_{xy}^2}\right]^2 + \left[\frac{\sigma_x}{2} - \sqrt{\frac{\sigma_x^2}{4} + \tau_{xy}^2}\right]^2$$

$$- \left[\frac{\sigma_x}{2} + \sqrt{\frac{\sigma_x^2}{4} + \tau_{xy}^2}\right]\left[\frac{\sigma_x}{2} - \sqrt{\frac{\sigma_x^2}{4} + \tau_{xy}}\right]$$

$$= \left[\frac{\sigma_x^2}{4} + \frac{\sigma_x^2}{4} + \tau_{xy}^2 + 2 \cdot \frac{\sigma_x}{2} \cdot \sqrt{\frac{\sigma_x^2}{4} + \tau_{xy}^2}\right]$$

$$+ \left[\frac{\sigma_x^2}{4} + \frac{\sigma_x^2}{4} + \tau_{xy}^2 - 2\frac{\sigma_x}{2}\sqrt{\frac{\sigma_x^2}{4} + \tau_{xy}^2}\right] - \left[\frac{\sigma_x^2}{4} - \frac{\sigma_x^2}{4} - \tau_{xy}^2\right]$$

$$= \sigma_x^2 + 3\tau_{xy}^2$$

$$\sigma_0 = \sqrt{\sigma_x^2 + 3\tau_{xy}^2} \quad \textbf{Proved   (Ans.)}$$

Example 22.89. *The state of plane stress shown in Fig. 22.75 occurs at a critical point of a steel machine component for which the tensile yield strength, $\sigma_{yp} = 250$ MPa. Determine the factor of safety with respect to yield, using (i) the maximum shearstress criterion; (ii) the maximum distortion energy criterion.*

(Panjab University)

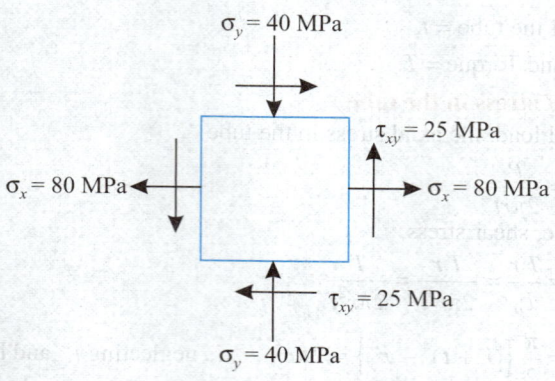

Fig. 22.75

Solution. *Given:* $\sigma_x = 80$ MPa (tensile); $\sigma_y = 40$ MPa (compressive);

$\tau_{xy} = 40$ MPa; $\sigma_{yp} = 250$ MPa.

Principal stresses, σ_1 and σ_2 :

$$\sigma = \frac{\sigma_x + \sigma_y}{2} \pm \sqrt{\left(\frac{\sigma_x - \sigma_y}{2}\right)^2 + \tau_{xy}^2}$$

$$= \frac{80 + (-40)}{2} \pm \sqrt{\left(\frac{80 - (-40)}{2}\right)^2 + 25^2} = 20 \pm 65$$

∴ $\sigma_1 = 85$ MPa (tensile), and $\sigma_2 = 45$ MPa (compressive)

Factr of safety (F.O.S.) :

(i) The maximum shear stress criterion :

Using the expression,

$$\sigma_1 - \sigma_2 = \sigma_{et} \qquad \qquad ...(Eqn.\ 18.3)$$

$$85 - (-45) = \left(\frac{\sigma_{yp}}{F.O.S.}\right) = \left(\frac{250}{F.O.S.}\right)$$

∴ $$\text{F.O.S.} = \frac{250}{[85 - (-45)]} = 1.92 \quad \text{(Ans.)}$$

(ii) The maximum distribution energy criterion :

Using the expression,

$$\sigma_1^2 + \sigma_2^2 - \sigma_1 \sigma_2 = \sigma_{yp}^2 \qquad \qquad [Eqn.\ (18.7)]$$

$$(85)^2 + (-45)^2 - 85 \times (-45) = \left(\frac{250}{F.O.S.}\right)^2$$

$$7225 + 2025 + 3825 = \left(\frac{250}{F.O.S.}\right)^2$$

or $$13075 = \left(\frac{250}{F.O.S.}\right)^2 \quad \text{or} \quad \frac{250}{F.O.S.} = (13075)^{1/2}$$

$\therefore$ $$\text{F.O.S.} = \frac{250}{(13075)^{1/2}} = 2.186 \quad \text{(Ans.)}$$

Example 22.90. *A cast-iron cylinder has outside and inside diameters of 240 mm and 150 mm respectively. If the ultimate strength of cast-iron is 150 MN/m² and Poisson's ratio is 0.25, find the internal pressure, according to each of the following theories, which would cause rupture.*

(i)   Maximum principal stress theory;

(ii)   Maximum strain theory;

(iii)   Maximum strain energy theory.

Assume no longitudinal stress in the cylinder. **(Panjab University)**

Solution. *Given:* $\quad d_1 = 150 \text{ mm}; d_2 = 240 \text{ mm}; \dfrac{1}{m} = 0.25$

Internal pressure, p_1:

If internal pressure is p_1, then

$$\sigma_r = p_1 \text{ (comp.)}$$

i.e., $$\sigma_2 = \sigma_r = -p_1$$

and, $$\sigma_c = \frac{p_1(r_2^2 + r_1^2)}{(r_2^2 - r_1^2)} \text{ tensile} \qquad \text{...(Refer to chapter 11)}$$

i.e., $$\sigma_1 = \sigma_c = p_1\left[\frac{d_2^2 + d_1^2}{d_2^2 - d_1^2}\right] = \frac{p_1(240^2 + 150^2)}{(240^2 - 150^2)} = \frac{p_1 \times 80100}{35100}$$

$$= 2.282 \, p_1 \text{ (tensile)}$$

(i)   Maximum principal stress theory :

$$\sigma_1 = 2.282 \, p_1 = 150$$

or, $$p_1 = \frac{150}{2.282} = \mathbf{65.73 \ MN/m^2} \quad \textbf{(Ans.)}$$

(ii)   Maximum strain theory :

$$\sigma_1 - \frac{1}{m}\sigma_2 = \sigma$$

$$2.282 \, p_1 - 0.25 \, (-p_1) = 150 \quad \text{or} \quad 2.532 \, p_1 = 150$$

or, $$p_1 = \frac{150}{2.532} = \mathbf{59.24 \ MN/m^2} \quad \textbf{(Ans.)}$$

(iii)   Maximum strain energy theory :

$$\sigma_1^2 + \sigma_2^2 - \frac{2}{m}\sigma_1\sigma_2 = \sigma^2$$

$$(2.282 \, p_1)^2 + (-p_1)^2 - 2 \times 0.25 \, (2.282 \, p_1)(-p_1) = (150)^2$$

$$7.348 \, p_1^2 = (150)^2$$

or, $$p_1 = \left[\frac{(150)^2}{7.348}\right]^{1/2} = \mathbf{55.33 \ MN/m^2} \quad \textbf{(Ans.)}$$

Example 22.91. *A shaft of 100 mm diameter is subjected to a bending moment of 1.5 kNm. Find the value of the maximum torque which can be applied to the shaft for each of the following conditions :*

    (i)   Maximum direct stress not to exceed 120 MN/m²;

    (ii)   Maximum shearing stress not to exceed 60 MN/m²;

   (iii)   Maximum shear strain energy per unit volume not to exceed that induced by simple shear stress of 80 MN/m².   **(UPSC)**

Solution. *Given:* $d = 100$ mm $= 0.1$ m; M $= 5$ kNm $= 5000$ Nm.

Let σ_b be the maximum bending stress induced due to bending moment, then

$$\sigma_b = \frac{32M}{\pi d^3} = \frac{32 \times 5000}{\pi \times 0.1^3} = 5.1 \times 10^7 \text{ N/m}^2$$

Let the torque be T Nm corresponding to which the maximum shear stress be τ, then

$$\tau = \frac{16T}{\pi d^3} = \frac{16T}{\pi \times (0.1)^3} \quad 5.1 \times 10^3 \text{ T N/m}^2$$

The two principal stresses are,

$$\sigma_{1,2} = \frac{\sigma_b}{2} \pm \sqrt{\left(\frac{\sigma_b}{2}\right)^2 + \tau^2}$$

$$= \frac{5.1 \times 10^7}{2} \pm \sqrt{\left(\frac{5.1 \times 10^7}{2}\right)^2 + (5.1 \times 10^3)^2 T^2}$$

$$= 2.55 \times 10^7 \pm \sqrt{(2.55 \times 10^7)^2 + 2.6 \times 10^7 T^2} \text{ N/m}^2$$

i.e., $\sigma_1 = 2.55 \times 10^7 + \sqrt{6.5 \times 10^{14} + 2.6 \times 10^7 T^2} \text{ N/m}^2$

and, $\sigma_2 = 2.55 \times 10^7 - \sqrt{6.5 \times 10^{14} + 2.6 \times 10^7 T^2} \text{ N/m}^2$

  (i)   Maximum direct stress,

$$\sigma_1 = 120 \text{ MN/m}^2 = 120 \times 10^6 \text{ N/m}^2$$

$\therefore$ $120 \times 10^6 = 2.55 \times 10^7 + \sqrt{6.5 \times 10^{14} + 2.6 \times 10^7 T^2}$

or, $(120 \times 10^6 - 25.5 \times 10^6)^2 = 6.5 \times 10^{14} \times 2.6 \times 10^7 T^2$

or, $(94.5 \times 10^6)^2 = 6.5 \times 10^{14} + 2.6 \times 10^7 T^2$

or, $T^2 = \dfrac{(9.45 \times 10^7)^2 - 6.5 \times 10^{14}}{2.6 \times 10^7} = \dfrac{(9.45^2 - 6.5) \times 10^{14}}{2.6 \times 10^7}$

$$= \frac{82.8 \times 10^{14}}{2.6 \times 10^7} = 31.85 \times 10^7$$

$\therefore$ $T = 1.785 \times 10^4$ Nm or **17.85 kNm** **(Ans.)**

 (ii)   Maximum shearing stress,

$$\tau_{max} = 60 \text{ MN/m}^2 = 60 \times 10^6 \text{ N/m}^2$$

$\therefore$ $60 \times 10^6 = \sqrt{\left(\frac{\sigma_b}{2}\right)^2 + \tau^2} = \sqrt{6.5 \times 10^{14} + 2.6 \times 10^7 T^2}$

or, $(60 \times 10^6)^2 = 6.5 \times 10^{14} + 2.6 \times 10^7 \, T^2$

or, $36 \times 10^{14} - 6.5 \times 10^{14} = 2.6 \times 10^7 \, T^2$

or, $T^2 = \dfrac{(36 - 6.5)10^{14}}{2.6 \times 10^7} = 11.35 \times 10^7$

or, $T = 1.065 \times 10^4$ Nm or **10.65 kNm**   (**Ans.**)

(iii)   In the present case shear strain energy per unit volume is

$$= \frac{1}{12} C \left[(\sigma_1 - \sigma_2)^2 + (\sigma_2 - 0)^2 + (0 - \sigma_1)^2 \right]$$

$$= \frac{1}{12} C \left[(\sigma_1 - \sigma_2)^2 + \left\{ (\sigma_1 + \sigma_2)^2 - 2\sigma_1 \sigma_2 \right\} \right]$$

$$\frac{1}{12} C \left[\left\{ 2\sqrt{\left(\frac{\sigma_b}{2}\right)^2 + \tau^2} \right\}^2 + (\sigma_1 + \sigma_2)^2 - 2\sigma_1 \sigma_2 \right]$$

$$= \frac{1}{12} C \left[4 \left\{ \left(\frac{\sigma_b}{b}\right)^2 + \tau^2 \right\} + \sigma_b^2 - 2 \left\{ \left(\frac{\sigma_b^2}{2^2}\right) - \left(\frac{\sigma_b^2}{2^2} + \tau^2\right) \right\} \right]$$

$$= \frac{1}{12} C \left[\sigma_b^2 + 4\tau^2 + \sigma_b^2 + 2\tau^2 \right]$$

$$= \frac{1}{12} C \left[2\sigma_b^2 + 6\tau^2 \right]$$

Also the shear strain energy per unit volume for pure shear case

$$= \frac{1}{2C} (80 \times 10^6)^2$$

Equating these two and substituting values of σ_b and τ, we have

$$\frac{1}{12C} \left[2 \times (5.1 \times 10^7)^2 + 6(5.1 \times 10^3 \, T)^2 \right] = \frac{1}{2C} (80 \times 10^6)^2$$

or, $52 \times 10^{14} + 15.6 \times 10^7 \, T^2 = 384 \times 10^{14}$

$$T^2 = \frac{(384 - 52)10^{14}}{15.6 \times 10^7} = 21.28 \times 10^7$$

or, $T = 1.46 \times 10^4$ Nm = **14.6 kNm**   (**Ans.**)

Example 22.92. *A bending moment M applied to a solid shaft carries a maximum direct stress σ at elastic failure. Determine the numerical relationships between M and a twisting moment T which, acting alone on the shaft, will produce elastic failure, according to each of the following theories of failure :*

(i)   *Maximum principal stress;*   (ii)   *Maximum principal strain;*

(iii)   *Maximum strain energy;*   (iv)   *Maximum shear stress;*

(v)   *Shear strain energy.*

Poisson's ratio = 0.30.

                                                            (**Bangalore University**)

Solution. The elastic limit stress related to M is given as,

$$\sigma = \frac{32\,M}{\pi d^3}, \text{ where } d \text{ is the diameter of the shaft.}$$

When the torque T is acting alone on the shaft, then the shear stress on the cross-sectional plane is

$$\tau = \frac{16\,T}{\pi d^3}$$

Further in this pure shear case the two principal stresses σ_1 and σ_2,

$$\sigma_1 = -\sigma_2 = \frac{16\,T}{\pi d^3}$$

Thus, for this case the three principal stresses in descending order are :

$$\frac{16\,T}{\pi d^3},\ 0,\ -\frac{16\,T}{\pi d^3}$$

Numerical relation between M and T:

(i) *Maximum principal stress theory* :

For the maximum principal stress theory,

$$\sigma_1 = \sigma$$

i.e.,

$$\frac{16\,T}{\pi d^3} = \frac{32\,M}{\pi d^3}$$

∴

$$T = 2M \quad \textbf{(Ans.)}$$

(ii) *Maximum principal strain theory* :

$$\sigma_1 - \frac{1}{m}\sigma_2 = \sigma$$

$$\frac{16\,T}{\pi d^3} - 0.3\left(-\frac{16\,T}{\pi d^3}\right) = \frac{32\,M}{\pi d^3}$$

$$1.3 \times \frac{16\,T}{\pi d^3} = \frac{32\,M}{\pi d^3}$$

or,

$$T = 1.538\,M \quad \textbf{(Ans.)}$$

(iii) *Maximum strain energy theory* :

$$\sigma_1^2 + \sigma_2^2 - \frac{2}{m}\sigma_1\sigma_2 = \sigma^2$$

$$2\sigma_1^2 + \frac{2}{m}\sigma_1^2 = \sigma^2 \qquad\qquad [\because \sigma_2 = -\sigma_1)]$$

$$2 \times \left(\frac{16\,T}{\pi d^3}\right)^2\left(1 + \frac{2}{m}\right) = \left(\frac{32\,M}{\pi d^3}\right)^2$$

$$2\,T^2\,(1 + 0.3) = 4M^2$$

or,

$$T = \sqrt{\frac{4}{2.6}}\,M = 1.24\,M \quad \textbf{(Ans.)}$$

(iv) *Maximum shear stress theory* :

Since the principal stresses are unlike in nature, we have

$$\sigma_1 + \sigma_2 = \sigma$$

or,

$$\frac{16\,T}{\pi d^3} + \frac{16\,T}{\pi d^3} = \frac{32\,M}{\pi d^3}$$

$$T = M \quad \text{(Ans.)}$$

(*v*) **Shear strain energy theory :**

$$\sigma_1^2 + \sigma_2^2 - \sigma_1 \sigma_2 = \sigma^2$$

$$2\sigma_1^2 + \sigma_1^2 = \sigma^2 \qquad\qquad (\because \sigma_2 = -\sigma_1)$$

$$3\sigma_1^2 = \sigma^2$$

$$3\left(\frac{16 T}{\pi d^3}\right)^2 = \left(\frac{32 M}{\pi d^3}\right)^2$$

$$T = \sqrt{\frac{4}{3}} M = 1.155 \, M \quad \text{(Ans.)}$$

UNSYMMETRICAL BENDING AND SHEAR CENTRE

Example 22.93. *Fig. 22.76 shows the section of a tensile member of a structure. A tensile load P of magnitude 25 kN passing through 0 acts normal to the section and parallel to the longitudinal axis through the centroid G. Calculate the stress at the point A.* **(Allahabad University)**

Solution. Refer to Fig. 22.76

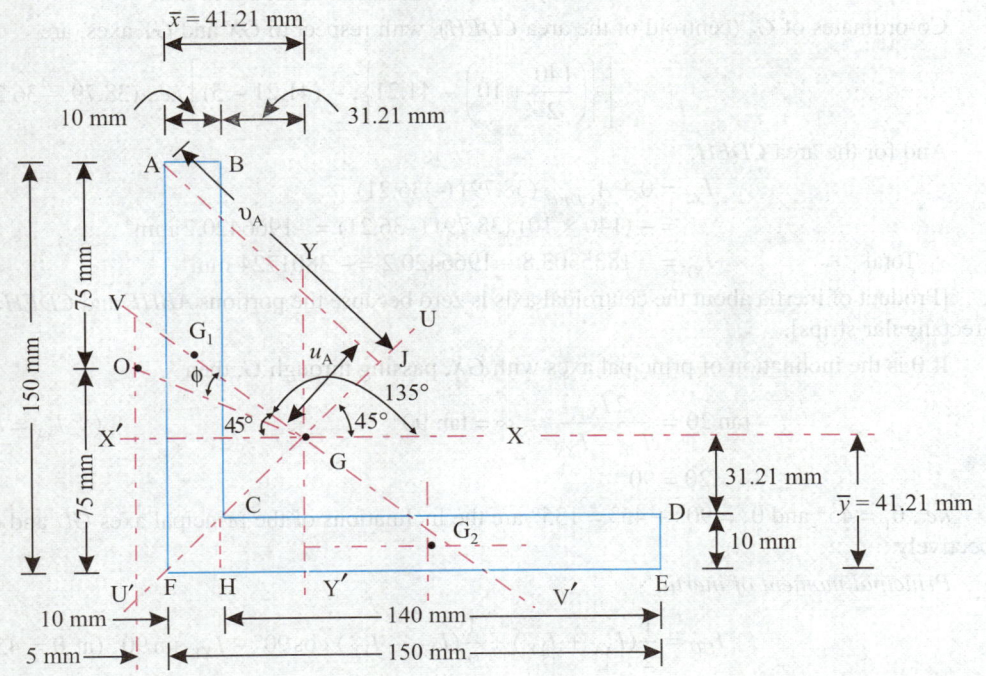

Fig. 22.76

Through the centroid *G* and parallel to the given load apply two equal and opposite loads. The section will be subjected to the following :

(*i*) Direct tensile stress of $\dfrac{P}{A}$, and

(*ii*) Bending moment equal to $P \times OG$

Let us first locate the centroid G of the section, Taking moments about the line EF, we get

$$(150 \times 10 + 140 \times 10)\,\bar{y} = 150 \times 10 \times 75 + 140 \times 10 \times 5$$

$$\therefore \qquad \bar{y} = \frac{150 \times 10 \times 75 + 140 \times 10 \times 5}{150 \times 10 + 140 \times 10} = \frac{112500 + 7000}{1500 + 1400} = 41.21 \text{ mm}$$

Also, $\bar{x} = \bar{y} = 41.21$ mm, since the section is symmetrical.

Now, $$I_{XX} = I_{YY} = \left[\frac{10 \times 150^3}{12} + (10 \times 150) \times (75 - 41.21)^2 \right]$$

$$+ \left[\frac{140 \times 10^3}{12} + (140 \times 10) \times (41.21 - 5)^2 \right]$$

$$= [2812500 + 1712646] + [11666.7 + 1835629.7]$$

$$= 6372442 \text{ mm}^4$$

Co-ordinates of G_1 (centroid of area $ABHF$), with respect to GX and GY axes, are

$$[- (41.21 - 5), (75 - 41.21)] \text{ i.e., } (-36.21, 33.79)$$

Therefore, for the area $ABHF$

$$I_{XY} = 0 + A_{ABHF}\,(-36.21)\,(33.79)$$

$$= (150 \times 10)\,(-36.21)\,(33.79) = -1835303.8 \text{ mm}^4$$

Co-ordinates of G_2 (centroid of the area $CDEH$), with respect to GX and GY axes, are

$$\left[\left\{ \left(\frac{140}{2} + 10 \right) - 41.21 \right\}, \, - (41.21 - 5) \right] \text{ i.e., } (38.79, -36.21)$$

And for the area $CDEH$,

$$I_{XY} = 0 + A_{CDEH}\,(38.79)\,(- 36.21)$$

$$= - (140 \times 10)\,(38.79)\,(-36.21) = -1966420.2 \text{ mm}^4$$

$\therefore$ Total $\qquad I_{XY} = - 1835303.8 - 1966420.2 = - 3801724 \text{ mm}^4$

[Product of inertia about the centroidal axis is zero because the portions $ABHF$ and $CDEH$ are the rectangular strips].

If θ is the inclination of principal axes with GX, passing through G, then

$$\tan 2\theta = \frac{2\,I_{XY}}{I_{YY} - I_{XX}} = \infty = \tan 90° \qquad\qquad (\because I_{XX} = I_{YY})$$

$\therefore \qquad 2\theta = 90°$

i.e., $\theta_1 = 45°$ and $\theta_2 = 90° + 45° = 135°$ are the inclinations of the principal axes GU and GV respectively.

Principal moment of inertia :

$$I_{UU} = \frac{1}{2}(I_{XX} + I_{YY}) + \frac{1}{2}(I_{XX} - I_{YY})\cos 90° - I_{XY} \sin 90° \text{ (at } \theta = 45°)$$

$$= \frac{1}{2}(6372442 + 6372442) + \frac{1}{2} \times 0 \times \cos 90° - (-3801724)$$

$$= 10174166 \text{ mm}^4$$

Also, $\qquad I_{UU} + I_{VV} = I_{XX} + I_{YY} \text{ or } I_{VV} = I_{XX} + I_{YY} - I_{UU}$

$\therefore \qquad I_{VV} = 2 \times 6372442 - 10174166 = 2570718 \text{ mm}^4$

Now, the bending moment M in the plane GO $= P \times (GO)$

Also, $\qquad$ GO $= [(41.21 + 5)^2 + (75 - 41.21)^2]^{1/2} = 57.246$ mm

$$M = (25 \times 10^3) \times \frac{57.246}{1000} = 1431.15 \text{ Nm}$$

$$\angle OGX' = \tan^{-1} \frac{(75 - 41.21)}{(41.21 + 5)} = 36.175°$$

∴ $\qquad$ $\angle OGV = \phi = 45° - 36.175° = 8.825°$

Thus, bending moment in the plane GV,

$$M' = M \cos \phi = 1431.15 \cos (8.825°) = 1414.2 \text{ Nm}$$

The bending moment, M' is anticlockwise if viewed in the direction UG, *i.e.* its effect will be

to give rise to tensile stresses on all points lying on the right of $\overrightarrow{UO}$; particularly the point A will be

experiencing the tensile stress due to M'.

Similarly, the other component M in the plane GU is

$$M'' = M \sin \phi = 1431.15 \sin (8.825°) = 219.56 \text{ Nm}$$

The bending moment M will be anticlockwise as viewed in the direction YG, *i.e.* its effect will

also be to induce tensile stress at A. Further there will also be direct tensile stress.

$$(\sigma_d) = \frac{P}{A} = \frac{25 \times 10^3}{(150 \times 10 + 140 \times 10)} = 8.62 \text{ N/mm}^2 \text{ or MN/m}^2$$

Total bending stress at A,

$$(\sigma_b)_A = \frac{M'}{I_{UU}} v_A + \frac{M''}{I_{VV}} u_A$$

Now, $\qquad$ $u_A = GJ = FJ - FG = 150 \cos 45° - \frac{41.21}{\cos 45°} = 47.79$ mm

and, $\qquad$ $v_A = AJ = 150 \sin 45° = 106.07$ mm

∴ $\qquad$ $(\sigma_b)_A = \frac{1414.2 \times 1000}{10174166} \times 106.07 + \frac{219.56 \times 1000}{2570718} \times 47.79 \text{ N/mm}^2$

$$= 14.74 + 4.08 = 18.82 \text{ N/mm}^2$$

Hence, *resultant stress at A*,

$$(\sigma_r)_A = (\sigma_d)_s + (\sigma_b)_A \text{ (tensile)}$$
$$= 8.62 + 18.82 = \textbf{27.44 MN/m}^2 \text{ (Ans.)}$$

Universities Questions (Recent) with Solutions

Example 23.1. *Fig. 23.1 shows an axially loaded bar.*
Taking $E_s = 210 GN/m^2$ and $E_b = 105 GN/m^2$, determine :
  (i) *The maximum normal stress in steel and brass.*
  (ii) *The displacement of the free end.*

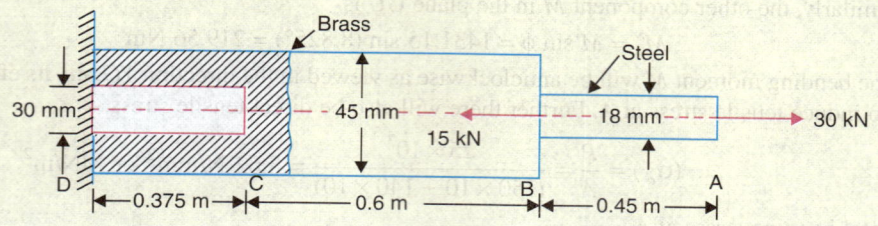

Fig. 23.1

Solution : Given: $E_s = 210$ GN/m^2; $E_b = 105$ GN/m^2.
(*i*) **The maximum normal stress in steel and brass:**

$$A_s = \frac{\pi}{4} \times (18)^2 = 81\,\pi \text{ mm}^2$$

$$(A_b)_{BC} = \frac{\pi}{4} \times 45^2 = 506.25\,\pi \text{ mm}^2$$

$$(A_b)_{CD} = \frac{\pi}{4} \times (45^2 - 30^2) = 281.25\,\pi \text{ mm}^2$$

$$\sigma_s = \frac{30 \times 10^3}{81\pi \times 10^{-6}} \times 10^{-6} \text{ MN/m}^2 = \textbf{117.89 MN/m}^2 \textbf{ (Ans.)}$$

$$(\sigma_b)_{BC} = \frac{15 \times 10^3}{506.25 \times \pi \times 10^{-6}} \times 10^{-6} = \textbf{9.43 MN/m}^2 \textbf{ (Ans.)}$$

$$(\sigma_b)_{CD} = \frac{15 \times 10^3}{281.25 \times \pi \times 10^{-6}} \times 10^{-6} = \textbf{16.97 MN/m}^2 \textbf{ (Ans.)}$$

(*ii*) **The displacement of the free end, δ :**

$$\delta = \delta_s + (\delta_b)_{BC} + (\delta_b)_{CD}$$

$$= \frac{30 \times 10^3 \times 0.45 \times 10^3}{(81\pi \times 10^{-6}) \times 210 \times 10^9} + \frac{15 \times 10^3 \times 0.6 \times 10^3}{(506\,\pi \times 10^{-6}) \times 105 \times 10^9}$$

$$+ \frac{15 \times 10^3 \times 0.375 \times 10^3}{281.25\pi \times 10^{-6} \times 105 \times 10^9} \text{ mm} \qquad \left(\because \delta = \frac{Pl}{AE}\right)$$

$$= 0.2526 + 0.05392 + 0.06062 = 0.367 \text{ mm}$$

i.e. $\qquad\qquad \delta = \textbf{0.367 mm. (Ans.)}$

1342

Example 23.2. *Fig. 23.2 shows a tapered rod carrying a tensile load of 15 kN at its free end. Calculate the total extension of the rod. E = 205 GN/m².*

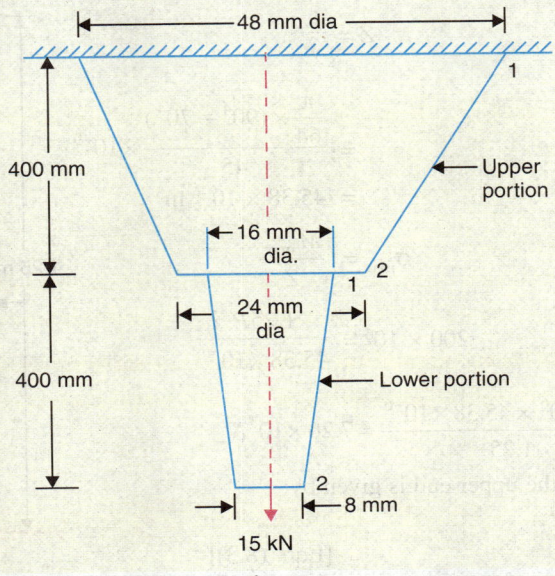

Fig. 23.2

Solution : *Lower portion: $d_1 = 16$ mm $= 0.016$ cm; $d_2 = 8$ mm $= 0.008$ m; $l = 400$ mm $= 0.4$ m;*

Upper portion: $d_1 = 48$ mm $= 0.048$ m; $d_2 = 24$ mm $= 0.024$ m; $l = 400$ mm $= 0.4$ m; $E = 205$ GN/m².

Total extension of the rod, $\delta l = (\delta l_U + \delta l_L)$

The extension of a tapered rod under load ($P = 15$ kN) is given by:

$$\delta l = \frac{4\,Pl}{\pi\,E d_1\,d_2} \qquad \text{....[Equ. (1.15)]}$$

Now,

$$\delta l_L = \frac{4 \times (15 \times 10^3) \times 0.4}{\pi \times (205 \times 10^9) \times 0.016 \times 0.008} \times 1000 \text{ mm}$$

$$= 0.291 \text{ mm}$$

and,

$$\delta l_U = \frac{4 \times (15 \times 10^3) \times 0.4}{\pi \times (205 \times 10^9) \times 0.048 \times 0.024} \times 1000 \text{ mm}$$

$$= 0.0323 \text{ mm}$$

∴ **Total extension, $\delta_l = 0.291 + 0.0323 =$ 0.3233 mm   (Ans.)**

Example 23.3. *Fig.23.3 shows a steel pipe fixed vertically at its bottom end. A block of mass 8 kg strikes the upper end of the pipe with a velocity 'v'. Determine the value of 'v' which produces in the pipe, a maximum stress of 200 MN/m².*

Take E for the pipe material as 200 GN/m².

Solution : Given : *$m = 8$ kg; $\sigma_{max} = 200$ MN/m²; $E = 200$ GN/m².*

Velocity, v :

Let P newtons be the equivalent static load.

Then,
$$M_{max} = P \times 1.25, \text{ at the fixed end of the pipe}$$
$$= 1.25 \, P \, \text{Nm}$$

Section modulus
$$Z = \frac{I}{y}$$

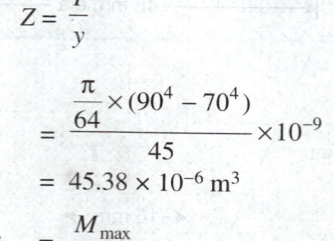

8 kg block

$$= \frac{\dfrac{\pi}{64} \times (90^4 - 70^4)}{45} \times 10^{-9}$$

$$= 45.38 \times 10^{-6} \, \text{m}^3$$

Now,
$$\sigma_{max} = \frac{M_{max}}{Z}$$

or,
$$200 \times 10^6 = \frac{1.25P}{45.38 \times 10^{-6}}$$

or,
$$P = \frac{200 \times 10^6 \times 45.38 \times 10^{-6}}{1.25} = 7.26 \times 10^3 \, N$$

The deflection of the upper end is given by:

$$\delta = \frac{Pl^3}{3EI} \qquad \text{... [Eqn. (8.3)]}$$

Fig. 23.3

Here, $I = \dfrac{\pi}{64}(90^4 - 70^4) \times 10^{-12} = 2.04 \times 10^{-6} m^4$

$$\therefore \quad \delta = \frac{7.26 \times 10^3 \times (1.25)^3}{3 \times 200 \times 10^9 \times 2.04 \times 10^{-6}} = 0.0116 \, m$$

Now, strain energy stored in the pipe,

$$U = \frac{1}{2} P\delta = \frac{1}{2} mv^2$$

or,
$$\frac{1}{2} \times 7.26 \times 10^3 \times 0.0116 = \frac{1}{2} \times 8 \times v^2$$

or,
$$v = 3.24 \, \text{m/s (Ans.)}$$

Example 23.4. *Fig. 23.4 shows a close-coiled helical spring built in at both the ends. A load W is applied at an immediate point on it. If the number of coils above and below this point are 6 and 4 respectively, determine how the load W = 800N will be shared between the top and bottom.*

Solution : Let W_1 and W_2 be this shared load between the top and bottom.

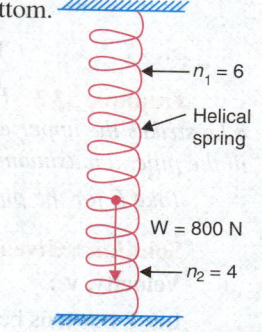

Deflection,
$$\delta = \frac{64 \, WR^3 n}{Cd^4}$$

Now,
$$\delta_1 = \frac{64 \, W_1 R^3 n_1}{Cd^4}$$

$n_1 = 6$

Helical spring

and,
$$\delta_2 = \frac{64 \, W_2 R^3 n_2}{d^4}$$

W = 800 N

Since R, d and C are the same,

$n_2 = 4$

$$\therefore \quad W_1 n_1 = W_2 n_2$$

Fig. 23.4

or, $$W_1 \times 6 = W_2 \times 4, \text{ or, } W_1 = \frac{2}{3}W_2$$

and, $$W_1 + W_2 = 800$$

or, $$\frac{2}{3}W_2 + W_2 = 800$$

or, $$\mathbf{W_2 = 480 \text{ N}; \ W_1 = 320 \text{ N} \ (Ans.)}$$

Example 23.5. *(a) List the applications of close-coiled helical springs.*

(b) A close-coiled helical spring whose free length when not compressed is 15 cm, is required to absorb strain energy equal to 50 Nm when fully compressed with the coils in contact. The maximum shearing stress is limited to 140 MPa. Assuming a mean coil diameter of 10 cm, find the diameter of the steel wire required and number of coils, C = 80 GPa.

Solution : (*a*) Applications of a close-coiled helical spring are:

1. Absorbs shook and vibrations, *e.g.*, vehicle suspension spring.
2. Acts as a reservoir of energy, *e.g.*, clock springs, toys, etc.
3. Measures force as a transducer, *e.g.*, spring balance.
4. Acts as actuating device, *e.g.*, valve spring, clutches, etc.

(*b*) **Given :** Free length (when not compressed), $l_f = 15$ cm, $= 0.15$ m; strain energy, $U = 50$ Nm; maximum shearing stress, $\tau = 140$ *MPa*; mean coil diameter, $D = 10$ cm $= 0.1$ m; $C = 80$ GPa.

Diameter of steel wire, *d* :

We know, $$U = \frac{\tau^2}{4C} \times \text{volume of wire} \qquad \qquad \dots \text{[Eqn. 14.8]}$$

or, $$U = \frac{\tau^2}{4C} \times \frac{\pi}{4}d^2 \times \pi \, D \times n \quad (\text{where, } n = \text{number of turns})$$

or, $$50 = \frac{(140 \times 10^6)^2}{4 \times 80 \times 10^9} \times \frac{\pi}{4}d^2 \times \pi \times 0.1 \times n$$

or, $$d^2 n = \frac{50 \times 4 \times 80 \times 10^9 \times 4}{(140 \times 10^6)^2 \times \pi^2 \times 0.1} = 3.3084 \times 10^{-3} \qquad \dots(i)$$

Also, $$\tau = \frac{16 \, WR}{\pi d^3}$$

or, $$W = \frac{\tau \times \pi d^3}{16R} = \frac{(140 \times 10^6) \times \pi}{16 \times 0.05} = 549.78 \times 10^6 d^3 \qquad \dots(ii)$$

Again, Solid length, $l_s = nd$ $\qquad \qquad \dots(iii)$

Deflection, $\delta = l_f - l_s = 0.15 - nd$ $\qquad \qquad \dots(iv)$

We know, $$\delta = \frac{64 \, WR^2 n}{Cd^4} \qquad \qquad \dots(\text{Eqn. 14.4})$$

$$(0.15 - nd) = \frac{64 \times 549.78 \times 10^6 d^3 \times (0.05)^3 \times n}{80 \times 10^9 \times d^4} \qquad \dots \text{using eqn. } (ii)$$

$$= 0.055 \times 10^{-3} \times \frac{n}{d}$$

or, $$0.15d - nd^2 = 0.055 \times 10^{-3} \times n$$

or, $\qquad 0.15d - 3.3084 \times 10^{-3} = 0.055 \times 10^{-3} \times n$ $\qquad\qquad$ using eqn. (*i*)

Dividing both sides by 10^{-3}, we get

or, $\qquad\qquad 150\,d - 3.3084 = 0.055\,n$

Substituting the value of '*n*' from eqn. (*i*), we have

$$150\,d\; - 3.3084 = \;0.055 \times \frac{3.3084 \times 10^{-3}}{d^2}$$

or, $\qquad\qquad 150\,d^3 - 3.3084\,d^2 - 0.1819 \times 10^{-3} = 0$

Dividing both sides by 150, we get

$$d^3 - 0.022\,d^2 - 1.2127 \times 10^{-6} = 0$$

From which $d \simeq 0.022$ m or **22 mm (Ans.)**

Number of coils, *n* :

Using eqn (*i*), we get,

$$n = \frac{3.3084 \times 10^{-3}}{(0.022)^2} \simeq 6.8 \simeq 7$$

i.e. $\qquad\qquad n \simeq$ **7 (Ans.)**

Example 23.6. (*a*) *Differentiate between the close-coiled and open-coiled helical springs.*

(*b*) *A flat spiral spring is 12mm wide, 0.3 mm thick and 2.5 m long. Assuming the maximum stress of 900 MN/m² to occur at the greatest bending moment; calculate:*

(*i*) *The torque that can be stored in the spring,*

(*ii*) *The work that can be stored in the spring, and*

(*iii*) *The number of turns to wind up the spring.*

$\qquad$ *Take,* $E = 200$ GN/m²

Solution : (*a*) The differences between the close-coiled and open-coiled springs are:

In a close-coiled helical spring the helix angle is small, generally *less than 5-10 degrees*. Only *tortional stress is induced.*

In an open-coiled helical spring the helix angle is *generally more than 10°*. It is subjected to *both bending and shear stresses.*

(*b*) Width of the strip, $\qquad b = 12$ mm

Thickness of the spring, $\qquad t = 0.3$ mm

Length of the spring $\qquad l = 2.5$ m

Maximum stress, $\qquad \sigma_{max} = 900$ MN/m² (or 900 N/mm²)

(*i*) The torque, T :

Now, torque, $\qquad\qquad T = \dfrac{bt^2 \cdot \sigma_{max}}{12}$ with usual notations $\qquad$ [Eqn. 14.34 (*b*)]

$$= \frac{12 \times (0.3)^2 \times 900}{12} = 81\; Nmm = 0.081\; Nm.$$

i.e., $\qquad\qquad T =$ **0.081 Nm (Ans.)**

(*ii*) The work that can be stored, U :

$$U = \frac{\sigma_{max}^2}{24E} \times \text{volume of spring}$$

$$= \frac{(900 \times 10^6)^2}{24 \times (200 \times 10^9)} \times \left[(12 \times 10^{-3}) \times (0.3 \times 10^{-3}) \times 2.5 \right]$$

$$= \textbf{1.52 Nm or J \; (Ans.)}$$

(iii) Number of turns:

We know,

$$\phi = \frac{Tl}{EI} \qquad \text{...[Eqn. (14.32)]}$$

$$= \frac{0.081 \times 2.5}{200 \times 10^9 \times \dfrac{bt^3}{12}}$$

$$= \frac{0.081 \times 2.5 \times 12}{200 \times 10^9 \times (12 \times 10^{-3}) (0.3 \times 10^{-3})^3} = 37.5 \text{ radians}$$

Since

$$1 \text{ turn} = 2\pi \text{ radians}$$

∴

$$Number\ of\ turns = \frac{37.5}{2\pi} = 5.968 \ \textbf{(Ans.)}$$

Example 23.7. *Derive the relations for axial deflection and maximum shear stress in case of conical spring.*

Fig. 23.4 shows a conical spring.

$$R_1 = Minimum\ radius,$$
$$R_2 = Maximum\ radius,\ and$$
$$n = Number\ of\ turns.$$

Solution : The distance R from the axis of the spring is determined by the angle α, given by:

$$R = R_1 + \frac{(R_2 - R_1)\,\alpha}{2\pi n} \qquad ... (i)$$

Let us consider a small length of the spring at a radius R, then the angle of twist $(d\theta)$ produced by W in length ds is given by:

$$d\theta = \frac{WR\,ds}{C\,I_p} \qquad ... (ii)$$

$$\left(\because \frac{T}{I_p} = \frac{c\theta}{l} \text{ or } \theta = \frac{Tl}{C_p} \text{ where } T = WR \text{ and } t = ds \right)$$

Now, strain energy (dU) stored by the length ds (of the spring) may be written as :

$$dU = \frac{1}{2} WR\, d\theta$$

or,

$$dU = \frac{1}{2} WR \times \frac{WR\,ds}{CI_p} = \frac{W^2 R^3\, d\alpha}{2CI_p} \qquad ... (iii)$$

$$(\because ds = R.d\alpha)$$

Also,

$$dR = \left(\frac{R_2 - R_1}{2\pi n} \right) d\alpha$$

or,

$$d\alpha = \left(\frac{2\pi n}{R_2 - R_1} \right) dR \qquad ... (iv)$$

Now,

$$dU = \frac{W^2 R^3}{2CI_p} \left(\frac{2\pi n}{R_2 - R_1} \right) dR$$

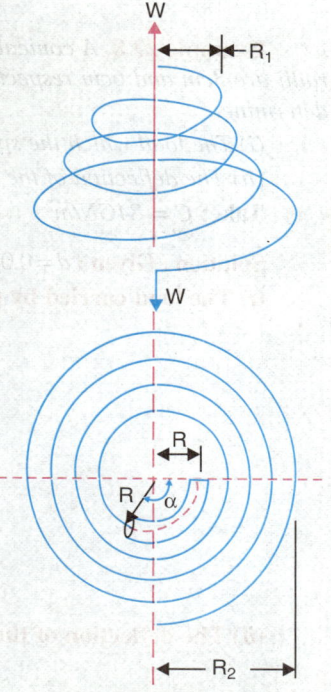

W

R₁

W

R

R

α

R₂

Fig. 23.4 Conical spring

$\therefore$ Total strain energy,

$$U = \frac{\pi n W^2}{CI_p \, (R_2 - R_1)} \int_{R_1}^{R_2} R^3 \, dR$$

or,

$$U = \frac{\pi n W^2}{CI_p \, (R_2 - R_1)} \times \frac{1}{4} (R_2^4 - R_1^4) \qquad \text{... (Eqn. 14.38)}$$

Axial deflection,

$$\delta = \frac{\partial U}{\partial W} = \frac{\pi n W}{2CI_p \, (R_2 - R_1)} (R_2^4 - R_1^4)$$

$$= \frac{\pi n W \, (R_1^2 + R_2^2) \, (R_1 + R_2)}{2CI_p}$$

or,

$$\delta = \frac{Wl \, (R_1^2 + R_2^2)}{2 \, CI_p} \qquad \text{... (Eqn. 14.39)}$$

[where $l = \pi n \, (R_1 + R_2)$ and $I_p = \dfrac{\pi}{32} d^4$, '$d$' being wire diameter]

Maximum shear stress (τ_{max}) can be written as:

$$\tau_{max} = \frac{Wd}{6C_p} (3R_2 + d) \qquad \text{... (Eqn. 14.40)}$$

Example 23.8. *A conical spring is made of 12mm diameter wire. Its minimum and maximum radii are 3cm and 6cm respectively. It has 8 coils. If the shear stress is not to exceed 180 MN/m², determine:*

(i) The load which the spring can carry.

(ii) The deflection of the spring.

Take: $C = 84GN/m^2$

Solution : Given : $d = 0.012$ m; $R_1 = 3$ cm $= 0.03$ m; $R_2 = 6$ cm $= 0.06$ m; $n = 8$; $I_{max} = 180$ MN/m².

(*i*) The load carried by the spring, W :

$$I_{max} = \frac{Wd}{6 \, Ip} (3R_2 + d)$$

$$180 \times 10^6 = \frac{W \times 0.012}{6 \times \dfrac{\pi}{32} \times (0.012)^4} (3 \times 0.06 + 0.012)$$

$$= 188628 \, W$$

$\therefore$

$$W = \frac{180 \times 10^6}{188628} = \textbf{954.26N} \quad \textbf{(Ans.)}$$

(*ii*) The deflection of the spring δ :

$$\delta = \frac{\pi n W}{2CI_p} (R_1 + R_2) \, (R_1^2 + R_2^2)$$

$$= \frac{\pi \times 8 \times 954.26}{2 \times 84 \times 10^9 \times \dfrac{\pi}{32} \times (0.012)^4}$$

$$(0.03 + 0.06) \, (0.03^2 + 0.06^2) \times 1000 \text{ mm}$$

$$\delta = \textbf{28.4mm (Ans.)}$$

Example 23.9. *When spring of an open coil helical spring having 10 coils is axially loaded the stresses due to bending and twisting are 100 MN/m² and 110 MN/m² respectively : Assuming the mean diameter of the coils to be eight times the diameter of the wire, find the maximum permissible load and the diameter of the wire for a maximum extension of 18 mm.*

Take: E = 210 GN/m²; C = 82 GN/m².

Solution : Given : $n = 10$; $\sigma_b = 100$ MN/m²; $\tau = 110$ MN/m²;

$D = 8d$ *i.e.*, $R = 4d$; $\delta = 18$ mm $= 0.018$ m;

$E = 210$ GN/m²; $C = 82$ GN/m².

$$\mathbf{d : ; W :}$$

Axial deflection is given by :

$$\delta = 2WR^3\, n\pi \sec \alpha \left[\frac{\cos^2 \alpha}{CI_p} + \frac{\sin^2 \alpha}{EI} \right] \qquad \dots(i)$$

Also, $\qquad\qquad\qquad T = WR \cos \alpha$

and, $\qquad\qquad\qquad \tau = \dfrac{16T}{\pi d^3} = \dfrac{16WR \cos \alpha}{\pi d^3}$

or, $\qquad\qquad 110 \times 10^6 = \dfrac{16W \times 4d \times \cos \alpha}{\pi d^3}$

or, $\qquad\qquad 110 \times 10^6 = \dfrac{64W \cos \alpha}{\pi d^2} \qquad \dots(ii)$

Again, $\qquad\qquad\qquad M = WR \sin \alpha$

and, $\qquad\qquad\qquad \sigma_b = \dfrac{32M}{\pi d^3} = \dfrac{32WR \sin \alpha}{\pi d^3}$

or, $\qquad\qquad 100 \times 10^6 = \dfrac{32W \times 4d \times \sin \alpha}{\pi d^3}$

or, $\qquad\qquad 100 \times 10^6 = \dfrac{128W \sin \alpha}{\pi d^2} \qquad \dots(iii)$

Dividing eqn. (*ii*) by eqn. (*i*) we get:

$$\frac{100}{110} = \frac{128W \sin \alpha}{64W \cos \alpha} = 2 \tan \alpha$$

or, $\qquad\qquad \tan \alpha = \dfrac{100}{110 \times 2} = 0.4545$

or, $\qquad\qquad \alpha = \tan^{-1} 0.4545 = 24.44°$

Inserting the value of α in eqn. (*ii*), we have

$$110 \times 10^6 = \frac{64\,W \times \cos 24.44°}{\pi d^2}$$

or, $\qquad\qquad W = \dfrac{110 \times 10^6 \times \pi d^2}{64 \times \cos 24.44°} = 5.93 \times 10^6 d^2$

Now, substituting the value of W in eqn. (*i*), we get:

$$0.02 = 2 \times 5.93 \times 10^6\, d^2 \times (4d)^3 \times 10 \times \pi \times \sec 24.44°$$

$$\left[\frac{\cos^2 24.44° \times 32}{82 \times 10^9 \times \pi d^4} + \frac{\sin^2 24.44° \times 64}{210 \times 10^9 \times \pi d^4} \right]$$

$$= \frac{26192.96 \, d^5 \times 10^6}{\pi d^4 \times 10^9}(0.323 + 0.0521)$$

or, $0.02 = 8.34d \times 0.375 = 3.127d$

∴ $d = 6.394 \times 10^{-3} \text{ m} = 6.394 \text{ mm say } \mathbf{6.4 \text{ mm}} \quad \textbf{(Ans.)}$

and, $W = 5.93 \times 10^6 \times (6.4 \times 10^{-3})^2 = \mathbf{242.9 \text{ N}} \quad \textbf{(Ans.)}$

Example 23.10. *The following data relate to an open-coiled helical spring:*

Number of coils = 10; Mean diameter of each coil = 5 cm; Diameter of wire forming the coils = 6 mm, making constant angle of 30° with the planes perpendicular to the axis of the spring. E = 210 GN/m² and G =84 GN/m².

(i) What load will cause the spring to elongate 1.55 cm and what will be the bending and shearing stresses due to this load ?

(ii) What will be the value of axial twist which would cause a bending stress of 60 MN/m² in the coils ?

Soution : Given : $n = 10$; $R = \dfrac{5}{2} = 2.5$ cm, 0.025 m; $d = 6$ mm = 0.006 m; $\alpha = 30°$; $\delta = 1.55$ cm = 0.0155 m; $E = 210 \times 10^9$ N/m²; $C = 84 \times 10^9$ N/m².

(i) Load, W : ; σ_b : ; τ :

Axial deflection is given by :

$$\delta = 2WR^3 \, n \, \pi \, \sec \alpha \left[\frac{\cos^2 \alpha}{CI_p} + \frac{\sin^2 \alpha}{EI} \right] \qquad \text{... (Eqn. 14.17)}$$

$$0.0155 = 2 \, W \times 0.025^3 \times 10 \times \pi \times \sec 30°$$

$$\left[\frac{\cos^2 30°}{84 \times 10^9 \times \left(\dfrac{\pi}{32} \times 0.006^4 \right)} + \frac{\sin^2 30°}{210 \times 10^9 \times \left(\dfrac{\pi}{64} \times 0.006^4 \right)} \right]$$

or, $0.0155 = 1.134 \times 10^{-3} W \left[\dfrac{0.75}{10.815} + \dfrac{0.25}{13.36} \right]$

$$= 0.1 \times 10^{-3} \, W$$

∴ $W = \dfrac{0.0155}{0.1 \times 10^{-3}} = \mathbf{155 \text{ N}} \quad \textbf{(Ans.)}$

Now, $T = WR \cos \alpha = 155 \times 0.025 \times \cos 30° = 3.36$ Nm.

∴   *Shear stress,* $\tau = \dfrac{16T}{\pi d^3} = \dfrac{16 \times 3.36}{\pi \times 0.006^3} \times 10^{-6} = \mathbf{79.2 \text{ MN/m}^2} \quad \textbf{(Ans.)}$

Bending moment, $M = WR \sin \alpha = 155 \times 0.025 \times \sin 30° = 1.937$ Nm.

∴   *Bending stress,* $\sigma_b = \dfrac{32M}{\pi d^3} = \dfrac{32 \times 1.937}{\pi \times (0.006)^3} \times 10^{-6} = \mathbf{91.34 \text{ MN/m}^2} \quad \textbf{(Ans.)}$

(ii)   Axial torque, M_o :

Bending moment, $M = M_o \cos \alpha$

Bending stress, $\sigma_b = \dfrac{32 \, M_o \cos \alpha}{\pi d^3}$

Here $\sigma_b = 60$ MN/m²   ... (Given)

$$\therefore \qquad 60 \times 10^6 = \frac{32 \times M_o \times \cos 30°}{\pi \times (0.006)^3}$$

or, $$M_o = \frac{60 \times 10^6 \times \pi \times (0.006)^3}{32 \times \cos 30°} = \textbf{1.47 Nm} \ \ \textbf{(Ans.)}$$

Example 23.11. *An open coil helical spring is made of wire 10 mm diameter. It has 15 coils of mean coil diameter 6 cm. What is the greatest axial compressive load that can be applied to the ends of the spring if the principal stress is not to exceed 120 MPa, and maximum shear stress is not to exceed 80 MPa. Calculate for this load the axial and angular deflection of the spring. $E = 2 \times 10^5$ MPa, $G = 0.8 \times 10^5$ MPa.*

Solution : Given : $d = 10$ mm, $n = 15$; $D = 6$ cm or $R = 3$ cm $= 30$ mm; $\sigma_1 = 120$ MPa $= 120$ N/mm², $\tau_{max} = 80$ MPa $= 80$ N/mm²; $E = 2 \times 10^5$ MPa $= 2 \times 10^5$ N/mm²;

G or $C = 0.8 \times 10^5$ MPa $= 0.8 \times 10^5$ N/mm².

We know, for axial load in an open coil helical coil spring,

Torque, $\qquad \tau = WR \cos \alpha$

Bending moment, $\qquad M = MR \sin \alpha$

Shear stress, $\qquad \tau = \dfrac{16\,T}{\pi d^3} = \dfrac{16\,WR \cos \alpha}{\pi d^3}$

Bending stress $\qquad \sigma = \dfrac{32M}{\pi d^3} = \dfrac{32WR \sin \alpha}{\pi d^3}$

Principal stresses are:

$$\sigma_{1,\,2} = \frac{\sigma}{2} \pm \frac{1}{2}\sqrt{\sigma^2 + 4\tau^2}$$

$$= \frac{16\,WR \sin \alpha}{\pi d^2} \pm \frac{1}{2}$$

$$\sqrt{\left(\frac{32\,WR \sin \alpha}{\pi d^3}\right)^2 + 4\left(\frac{16\,WR \cos \alpha}{\pi d^3}\right)^2}$$

$$= \frac{16\,WR \sin \alpha}{\pi d^3} \pm \frac{16\,WR}{\pi d^3}\sqrt{\sin^2 \alpha + \cos^2 \alpha}$$

$$= \frac{16\,WR}{\pi d^3}(\sin \alpha \pm 1)$$

i.e., $\qquad \sigma_1 = \dfrac{16\,WR}{\pi d^3}(\sin \alpha + 1)$

$\qquad 120 = 80\,(\sin \alpha + 1)$

or, $\qquad \sin \alpha = 0.5 \ i.e., \ \alpha = 30°$

Also, $\qquad \tau_{max} = \dfrac{16\,WR}{\pi d^3}$

or, $\qquad 80 = \dfrac{16\,W \times 30}{\pi \times (10)^3} \ \ \text{or} \ W = \textbf{523.6 N}$

Axial deflection of the spring, δ :

We know,
$$\delta = 2WR^3 \, n\pi \, \sec\alpha \left[\frac{\cos^2\alpha}{CI_p} + \frac{\sin^2\alpha}{EI}\right] \qquad ...(\text{Eqn. } 14.17)$$

$$= 2 \times 523.6 \times 30^3 \times 15 \times \pi \sec 30°$$

$$\left[\frac{\cos^2 30°}{0.8 \times 10^5 \times \left(\dfrac{\pi}{32} \times 10^4\right)} + \frac{\sin^2 30°}{2 \times 10^5 \times \left(\dfrac{\pi}{64} \times 10^4\right)}\right]$$

$$= 1.539 \times 10^9 \, (9.55 \times 10^{-9} + 2.546 \times 10^{-9}) = 18.61 \text{ mm}$$

i.e., $\qquad\qquad \delta = \textbf{18.61 mm} \textbf{ (Ans.)}$

Angular deflection of the spring, ψ ;

We know,
$$\psi = 2WR^2 n\pi \sin\alpha \left(\frac{1}{CI_p} - \frac{1}{EI}\right) \qquad ... (\text{Eqn. } 14.18)$$

$$= 2 \times 523.6 \times 30^2 \times 15 \times \pi \sin 30°$$

$$\left(\frac{1}{0.8 \times 10^5 \times \left(\dfrac{\pi}{32} \times 10^4\right)} - \frac{1}{2 \times 10^5 \left(\dfrac{\pi}{64} \times 10^4\right)}\right)$$

$$= 0.0222 \times 10^9 \, (12.732 \times 10^{-9} - 10.186 \times 10^{-9})$$

$$= 0.05652 \text{ rad. or } 0.05652 \times \frac{180}{\pi} = 3.24°$$

i.e., $\qquad\qquad \psi = \textbf{3.24°} \textbf{ (Ans.)}$

Example 23.12. *Two springs, one close coiled and the other open coiled are made of the same material and have the same coil radius, the wire diameter and the number of coils. The angle of helix for the open coiled spring is 30°. Compare the stiffness of the two springs when subjected to an axial load. E = 2.5 C.*

Solution : Let '*o*' and '*c*' be the suffices for the open coiled and close coiled springs respectively. The deflection for an open coiled spring is given by:

$$\delta_o = 2WR^3 \, n\pi \sec\alpha \left[\frac{\cos^2\alpha}{CI_p} + \frac{\sin^2\alpha}{EI}\right]$$

$$= 2WR^3 n \, \pi \sec 30° \left[\frac{\cos^2 30°}{C \times 2I} + \frac{\sin^2\alpha}{2.5 \, C \times I}\right] \qquad (\because I_p = 2I)$$

$$= 2WR^3 n \, \pi \, \frac{(1.1547)}{CI}\left[\frac{0.75}{2} + \frac{0.25}{2.5}\right]$$

i.e., $\qquad\qquad \delta_o = 2WR^3 n \, \pi \times \dfrac{0.548}{CI} \qquad\qquad ...(i)$

The deflection for a close coiled springs is given by:

$$\delta_c = \frac{64WR^3n}{Cd^4} = \frac{64WR^3n}{CI_p} \times \frac{\pi}{32} = \frac{2WR^3\pi n}{CI_p}$$

$$= 2WR^3 n\,\pi \times \frac{1}{CI_p}$$

$$= 2WR^3 n\,\pi \times \frac{1}{C \times 2I}$$

i.e.,
$$\delta_c = \frac{2WR^3 n\pi}{CI} \times (0.5) \qquad\qquad ...(ii)$$

Let k_o and k_c be the stiffnesses for the open coiled and close coiled spring respectively.

Then,
$$\frac{k_o}{k_c} = \frac{\delta_c}{\delta_o} = \frac{0.5}{0.548} = 0.912 \text{ (Ans.)}$$

Example 23.13. *Design a laminated spring, simply supported at the ends, and centrally loaded with a span of 800 mm, given the following :*

Proof load = 8.5 kN; Maximum central deflection = 50 mm;

Ratio of width to thickness = 10; $E = 2 \times 10^5$ MPa;

Permissible bending stress = 370 MPa

The plates are available in the multiples of 1 mm for thickness and in the multiple of 3 mm for width.

Solution : Given : $l = 800$ mm; $W = 8.5$ kN; $\delta = 50$ mm; $\dfrac{b}{t} = 10$; $E = 2 \times 10^5$ MPa $= 2 \times 10^5$ N/mm²; $\sigma_b = 370$ MPa $= 370$ N/mm².

We know,
$$\delta = \frac{3\,Wl^3}{8\,ENbt^3} \qquad\qquad ... \text{(Eqn. 14.39)}$$

or,
$$50 = \frac{3 \times (8.5 \times 1000) \times 800^3}{8 \times (2 \times 10^5) \times N \times 10t \times t^3}$$

or,
$$Nt^4 = \frac{3 \times (8.5 \times 1000) \times 800^3}{50 \times 8 \times (2 \times 10^5) \times 10}$$

or,
$$Nt^4 = 16320 \qquad\qquad ...(i)$$

Again, Bending stress,
$$\sigma_b = \frac{3}{2} \times \frac{Wl}{Nbt^2} \qquad\qquad ...\text{(Eqn. 14.38)}$$

$$370 = \frac{3}{2} \times \frac{(8.5 \times 1000) \times 800}{N \times 10t \times t^2}$$

or,
$$Nt^3 = \frac{3 \times (8.5 \times 1000) \times 800}{370 \times 2 \times 10}$$

or,
$$Nt^3 = 2756.76 \qquad\qquad ... (ii)$$

Dividing eqn. (*i*) by (*ii*), we get:

$$t = \frac{16320}{2756.76} = 5.92 \approx \textbf{6 mm (Ans.)}$$

From equ. (*i*), we have

$$N = \frac{16320}{t^4} = \frac{16320}{(6)^4} = 12.59 \simeq \textbf{13 (Ans.)}$$

Also,

$$b = 10\,t = 10 \times 6 = \textbf{60 mm} \text{ in multiples of 3mm } \textbf{(Ans.)}$$

Overlap:

$$a = \frac{l}{2N} = \frac{800}{2 \times 13} = \textbf{30.77 mm } \textbf{(Ans.)}$$

Actual deflection

$$\delta_a = \frac{3}{8} \times \frac{(8.5 \times 1000) \times 800^3}{2 \times 10^5 \times 13 \times 60 \times 6^3}$$

$$= \textbf{48.43 mm } \textbf{(Ans.)}$$

Radius of curvature,

$$R_c = \frac{l^2}{8\delta_a} = \frac{800^2}{8 \times 48.43} = 1656.3 \text{ mm or } 1.656\text{m}$$

i.e.,

$$R_c = \textbf{1.656 m } \textbf{(Ans.)}$$

Example 23.14. *A laminated steel spring, simply supported at the ends is centrally loaded. Its span is 80 cm and is required to carry a proof load of 8000 N, with central deflection not to exceed 6.5 cm. The maximum bending stress not to exceed 420 MN/m². The width is twelve times the thickness. Plates are available in multiples of 1mm for thickness and in multiples of 4 mm for width.*

Determine the suitable values for :

Thickness; Width, Number of plates, and the radius to which the plates should be formed.

Take : E = 200 GN /m².

Solution : Given : $l = 80$ cm = 0.8 m; $W = 8000$ N; $\delta = 6.5$ cm = 0.065 m; $\sigma_b = 420$ MN/m²; $b = 12\,t$, $E = 200$ GN/m²

t : ; b : ; N : ; R$_c$:

Deflection,

$$\delta = \frac{3\,Wl^3}{8\,ENbt^3} \qquad \qquad \text{...[Eqn. (14.39)]}$$

$$0.065 = \frac{3 \times 8000 \times 0.8^3}{8 \times 200 \times 10^9 \times N\,(12 \times t) \times t^3} \qquad \qquad \text{... ($\because b = 12t$)}$$

∴

$$Nt^4 = \frac{3 \times 8000 \times 0.8^3}{0.065 \times 8 \times 200 \times 10^9 \times 12} = 9.846 \times 10^{-9} \qquad \qquad \text{...(i)}$$

Bending stress,

$$\sigma_b = \frac{3\,Wl}{2\,Nbt^2}$$

$$420 \times 10^6 = \frac{3 \times 8000 \times 0.8}{2 \times N \times 12t \times t^2}$$

∴

$$Nt^3 = \frac{3 \times 8000 \times 0.8}{420 \times 10^6 \times 2 \times 12} = 1.905 \times 10^{-6} \qquad \qquad \text{...(ii)}$$

Dividing (*i*) by (*ii*), we get

$$t = \frac{9.846 \times 10^{-9}}{1.905 \times 10^{-6}} = 5.17 \times 10^{-3}\,m$$

say 6 mm (to be on safe side) **(Ans.)**

$$b = 12t = 12 \times 6 = \textbf{72 mm} \text{ (which is multiple of 4 mm as per the requirement) } \textbf{(Ans.)}$$

$$N = \frac{9.846 \times 10^{-9}}{(6 \times 10^{-3})^4} = 7.6$$

Also,

$$N = \frac{1.905 \times 10^{-6}}{(6 \times 10^{-3})^3} = 8.82$$

Let us adop **N = 9 (Ans.)**

∴   *Actual deflection,*

$$\delta = \frac{3\,Wl^3}{8\,ENbt^3}$$

$$= \frac{3 \times 8000 \times 0.8^3}{8 \times 200 \times 10^9 \times 9 \times 72 \times 10^{-3} \times (6 \times 10^{-3})^3}$$

$$= 0.0548 \text{ m } 54.8 \text{ mm}$$

Radius of curvature,

$$R_c = \frac{l^2}{8\delta} = \frac{0.8^2}{8 \times 0.0548} = 1.46 \text{ m } \textbf{(Ans.)}$$

Example 23.15. *(a) Discuss what is 'Fictitious Load Method'.*

(b) Using Castigliano's theorem, determine the reaction X at the end, A of the cantilever shown in the Fig. 23.5 below.

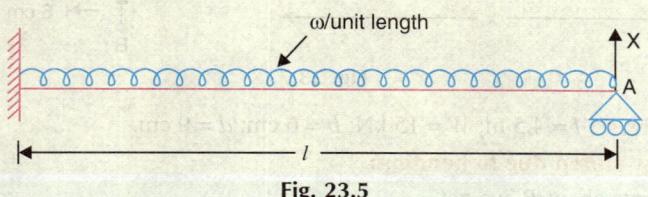

ω/unit length

X

A

l

Fig. 23.5

Solution: *(a)* In order to determine the deflection at a point in a loaded member by Castigliano's theorem, a fictitious load *P* in the direction of the desired deflection is applied. Then the total strain energy for the system is differentiated partially w.r.t. the load *P*. The desired deflection is then obtained by *P* = 0.

(b) Consider a section at a distance *x* from *A* as shown in Fig. 23.6.

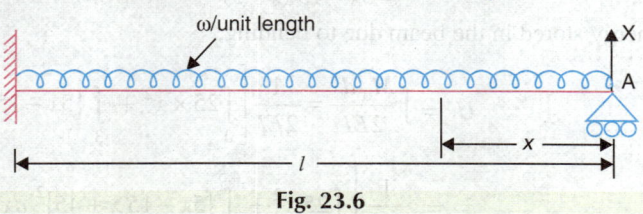

ω/unit length

X

A

x

l

Fig. 23.6

Then,

$$M_x = X.x - \frac{\omega x^2}{2}$$

$$\frac{\partial M_x}{\partial X} = x$$

Deflection,

$$\delta = \int_0^l \frac{M_x}{EI} \cdot \frac{\partial M_x}{\partial X}\, dx$$

$$= \frac{1}{EI}\left[\int_0^l \left(X.x - \frac{\omega x^2}{2} \right) x\, dx \right] = 0$$

$$= \frac{1}{EI}\left[\int_0^l X \cdot x^2 \cdot dx - \frac{\omega x^3}{2}\, dx\right] = 0$$

or,

$$= \left[X \cdot \frac{x^3}{3}\right]_0^l - \left[\frac{W}{2} \cdot \frac{x^4}{4}\right]_0^l = 0$$

or,

$$X \cdot \frac{l^3}{3} - \frac{\omega l^4}{8} = 0$$

or,

$$X = \frac{3}{8}\omega l \quad \text{(Ans.)}$$

Example 23.16. *A simply supported beam is loaded as shown in the Fig. 23.7. Determine the strain energy stored in it due to bending. E = 210 GN/m².*

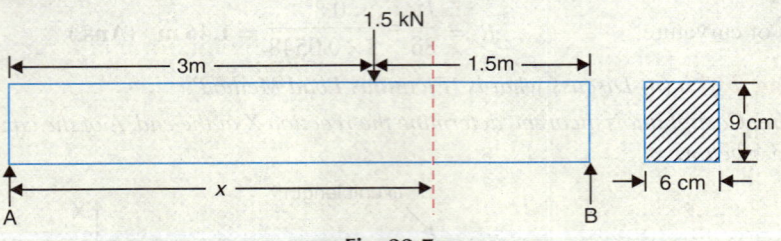

Fig. 23.7

Solution : Given : $l = 4.5$ m; $W = 15$ kN; $b = 6$ cm; $d = 9$ cm.

Stored energy stored due to bending:

Taking moments about B, we get

$$R_A \times 4.5 = 15 \times 1.5$$

or,

$$R_A = \frac{15 \times 1.5}{4.5} = 5\, kN$$

Now,

$$M_x = R_A \times x - 15\,(x - 3)$$
$$= 5x - 15\,(x - 3)$$

and, Strain energy stored in the beam due to bending,

$$U = \int \frac{M_x^2 dx}{2EI} = \frac{1}{2EI}\left[\int_0^3 25 \times x^2 + \int_3^{4.5} \{5x - 15\,(x - 3)\}^2\, dx\right]$$

$$= \frac{1}{2EI}\left[\int_0^3 25\,x^2 + \int_3^{4.5} \{5x - 15x + 45\}^2\, dx\right]$$

$$= \frac{1}{2EI}\left[\int_0^3 25\,x^2 + \int_3^{4.5} \{-10x + 45\}^2\, dx\right]$$

$$= \frac{1}{2EI}\left[\int_0^3 25\,x^2 + \int_3^{4.5} 100\,(x^2 - 9x + 20.25)\, dx\right]$$

$$= \frac{1}{2EI}\left[\frac{25}{3}\,|x^3\,|_0^3 + 100\left\{\left|\frac{x^3}{x}\right|_3^{4.5} - \left|\frac{9}{2}x^2\right|_3^{4.5} + |\,20.25x\,|_3^{4.5}\right\}\right]$$

$$= \frac{1}{2EI}\left[225 + 2137.5 - 5062.5 + 3037.5\right]$$

$$= \frac{337.5 \times 10^6}{2 \times 21.0 \times 10^9 \times \left(\frac{6 \times 9^3}{12} \times 10^{-8}\right)} = 220.46 \text{ Nm}$$

i.e., $\qquad\qquad U = \textbf{220.46 Nm (Ans.)}$

Example 23.17. *(a) What is the relationship between the radial and circumferencial stresses ?*

(b) Show that the tensile hoop stress set up in a thin rotating ring or cylinder is given by $\sigma_c = \rho\omega^2 r^2.$

Hence determine the maximum angular velocity at which the steel disc can be rotated if the hoop stress is limited to 20 MN/m². The ring has a mean diameter of 260 mm.

Take, $\rho = 7800$ kg/m³ for steel.

Solution : (*a*) The radial and circumferential stresses are *mutually perpendicular to each other.*

(*b*) Refer to art. 19.2.

Maximum angular velocity, ω:

Maximum hoop stress, $\sigma_c = 20$ MN/m².

Mean diameter of the ring = 260 mm or $r = 130$ mm = 0.13 m

Density of steel disc, $\qquad\qquad \rho = 7800$ kg/m³

We know that, $\qquad\qquad \sigma_c = \rho\omega^2 r^2 \qquad\qquad$ ...[Eqn. 19.1]

or, $\qquad\qquad 20 \times 10^6 = 7800 \times \omega^2 \times (0.13)^2$

∴ $\qquad\qquad \omega = \sqrt{\dfrac{20 \times 10^6}{7800 \times (0.13)^2}} = \textbf{389.16 rad. /s. (Ans.)}$

Example 23.18. *(a) I-section beams are generally preferred for lateral loading, why ?*

(b) At a point on the surface of an alloy steel ($E = 210$ GPa and γ or $\mu = \dfrac{1}{m} = 0.30$) machine part subjected to a biaxal state of stress, the measured strains were $e_x = + 1394$ μm/m, $\rho_y = -660$ μm/m, $\gamma_{xy} = 2054$ μ rad. Determine the following.

(i) The stress components σ_x, σ_y and τ_{xy} at the point.

(ii) The principal stresses and the maximum shear stress at the point. Locate the planes on which these stresses act and show the stresses on the complete stretch.

Solution : (*a*) The I-section beams are generally preferred for *lateral loading because they provide higher section modulus about an axis perpendicular to the lateral load, resulting into lower bending stresses and higher load carrying capacity.*

(*b*) (*i*) **Given :** e_x (or e_1) = + 1394 μm/m; e_y (or e_2) = – 660 μm/m,

$\qquad\qquad \gamma_{xy}$ (e_s) = 2054 μ rad.; $E = 210$ Gρ_a = 210 × 10⁹ N/m²;

$\qquad\qquad \gamma\left(=\dfrac{1}{m}\right) = 0.30.$

(*i*) **The stress components σ_x, σ_y and τ_{xy} at the point.**

We know, $\qquad\qquad e_x = \dfrac{\sigma_x}{E} - \dfrac{\gamma\sigma_y}{E} = \dfrac{1}{E}(\sigma_x - \gamma\sigma_y)$

or, $\qquad\qquad Ee_x = (\sigma_x - \gamma\sigma_y) \qquad\qquad$ ...(*i*)

and,

$$e_y = \frac{\sigma_y}{E} - \frac{\gamma\sigma_x}{E} = \frac{1}{E}(\sigma_y - \gamma\sigma_x)$$

or,

$$E\sigma_y = (\sigma_y - \gamma\sigma_x) \qquad \qquad ...(ii)$$

Multiplying (ii) by γ we have,

$$\gamma E e_y = (\gamma\sigma_y - \gamma^2\sigma_x) \qquad \qquad ...(iii)$$

Adding eqn. (i) and (iii), we get

$$E(e_x + \gamma e_y) = \sigma_x(1 - \gamma^2) \qquad \qquad ...(iv)$$

or,

$$\sigma_x = \frac{E}{1 - \gamma^2}(e_x + \gamma e_y) \qquad \qquad ...(v)$$

Similarly,

$$\sigma_y = \frac{E}{1 - \gamma^2}(e_y + \gamma e_x) \qquad \qquad ...(vi)$$

Substituting the given values in (v) and (vi) we have,

$$\sigma_x = \frac{210 \times 10^9}{1 - (0.3)^2}[1394 + 0.3 \times (-0.660)] \times 10^{-6}$$

or, $\boldsymbol{\sigma_x} = 2.308 \times 10^4 \times 1.3938 \times 10^{-3} = 3.217 \times 10^8$ or **321.7 MPa (Ans.)**

$$\sigma_y = \frac{210 \times 10^9}{(1 - 0.3^2)}[-660 + 0.3 \times 1394] \times 10^{-6}$$

$$= 2.308 \times 10^{11} \times -0.2418 \times 10^{-3} = -55.6 \text{ MPa}$$

i.e.,

$$\boldsymbol{\sigma_y} = \textbf{--55.8 MPa (Ans.)}$$

Also,

$$E = 2C(1 + \gamma) \qquad \qquad ...(\text{Eqn. 1.17})$$

[where C (or c_x) is modulus of rigidity]

∴

$$C = \frac{E}{2(1 + \gamma)} = \frac{210}{2(1 + 0.3)} = 80.77 \text{ GPa}$$

∴

$$\tau_{xy} = C\gamma_{xy} = (80.77 \times 10^9) \times (2054 \times 10^{-6}) = 165.9 \text{ MPa}$$

i.e.,

$$\boldsymbol{\tau_{xy}} = \textbf{165.9 MPa (Ans.)}$$

(b) The principal stresses (σ_1, σ_2) and maximum shear stress (τ_{xy}) at the point.

We know,

$$\sigma_{1,2} = \frac{(\sigma_x + \sigma_y)}{2} \pm \sqrt{\left(\frac{\sigma_x - \sigma_y}{2}\right)^2 + \tau_{xy}^2} \qquad \qquad ...(\text{Eqn. 2.12})$$

$$= \frac{1}{2}(\sigma_x + \sigma_y) \pm \frac{1}{2}\sqrt{(\sigma_x - \sigma_y)^2 + 4\tau_{xy}^2}$$

$$= \frac{1}{2}(321.7 - 55.8) \pm \frac{1}{2}\sqrt{[321.7 - (-55.8)]^2 + 4 \times 165.9^2}$$

$$= 132.95 \pm 251.29$$

∴ *Max. principal stress,* $\sigma_1 = 132.95 + 251.29 = $ **384.24 MPa (Ans.)**

Minor. principal stress, $\sigma_2 = 132.95 - 251.29 = $ **--118.34 MPa (Ans.)**

Maximum shear stress: $\tau_{max} = \dfrac{\sigma_1 - \sigma_2}{2} = \dfrac{384.24 - (-118.34)}{2} = $ **251.29 MPa (Ans.)**

Stresses on the complete sketch:

Principal planes are given by :

$$\tan 2\theta = \frac{2\,\tau_{xy}}{\sigma_x - \sigma_y} \qquad\qquad ...[Eqn.\ 2.10]$$

$$= \frac{2 \times 251.29}{321.7 - (-55.8)} = 1.33$$

∴ $$2\theta = \tan^{-1} 1.33 = 53°$$

∴ $$\theta_1 = \frac{53}{2} = 26.5° \text{ and } \theta_2 = 116.5°$$

The stresses on complete stresses are shown in Fig. 23.8.

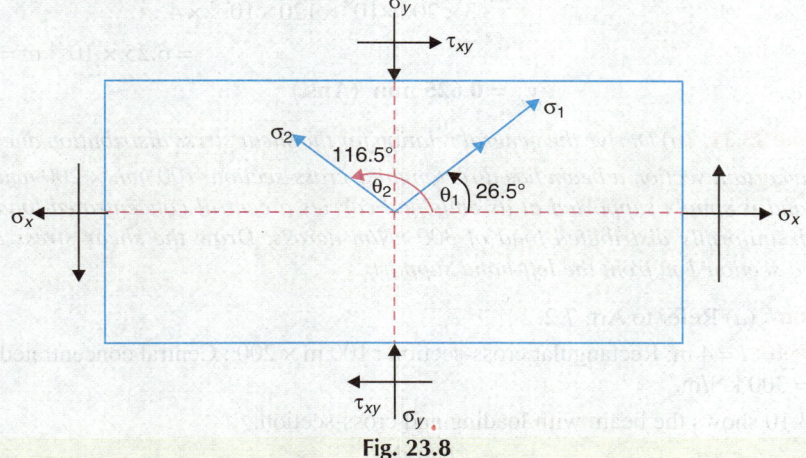

Fig. 23.8

Example 23.19. *(a) What is strain energy ? How is it related between the suddenly applied loading and gradually applied loading ?*

(b) Differentiate between the distortional and dilational stresses.

(c) A beam of square section subject to a shear force S is so placed that one of its diagonals is horizontal. Sketch shear stress distribution for the section.

Solution : *(a)* When an elastic material is subjected to external loads, it is deformed and work is done in deforming it. This work is stored in material as energy. The work stored in the material within elastic limits is call "*strain energy.*"

(b) Refer to Art. 15.8.

(c) Refer to Example 7.36.

Example 23.20. *(a) Derive the relation for the strain energy resulting from the bending of a beam (neglecting shear).*

(b) A beam simply supported at its ends, is of 4 m span and carries, at 3 m from the left-hand support, a load of 20 kN. If I is 120 × 10⁻⁶ m⁴ and E = 200 GN/m², find the deflection under the load using the relation derived in part (a).

Solution : *(a)* Refer to Art. 15.6.1.

(b) Fig. 23.9 shows the given beam with the loading.

Given : $a = 3$ m; $b = 1$ m, $l = 4$ m, $I = 120 \times 10^{-6}$ m⁴;

$W = 20$ kN $= 20 \times 10^3$ N; $E = 200$ GN/m² $= 200 \times 10^9$ N/m².

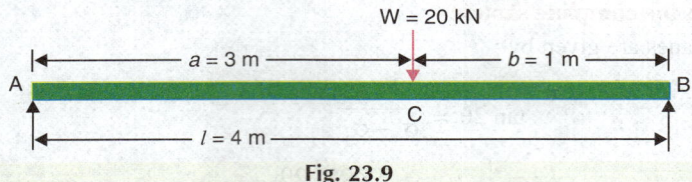

Fig. 23.9

As per Art : 15.6.1, *deflection under the load,*

$$y_c = \frac{wa^2 b^2}{3EI\, l}$$

$$= \frac{(20 \times 10^3) \times 3^2 \times 1^2}{3 \times 200 \times 10^9 \times 120 \times 10^{-6} \times 4}$$

$$= 6.25 \times 10^{-4} \text{ m} = 0.625 \text{ mm}$$

i.e., $\qquad\qquad y_c = \mathbf{0.625\ mm\ (Ans.)}$

Example 23.21. *(a) Derive the general relation for the shear stress distribution due to bending.*

(b) At a certain section a beam has a rectangular cross-section, 100 mm × 200 mm. The beam is 4 m long and is simply supported at its ends and carries a central concentrated load of 500 kN together with uniformly distributed load of 300 kN/m across. Draw the shear stress distribution diagram for a section 1 m from the left-hand support.

Solution : (*a*) Refer to Art. 7.2.

(*b*) **Given :** $l = 4$ m; Rectangular cross-section : 100 m × 200 ; Central concentrated load = 500 kN; U.D.L. = 300 kN/m.

Fig. 23.10 shows the beam with loading and cross-section.

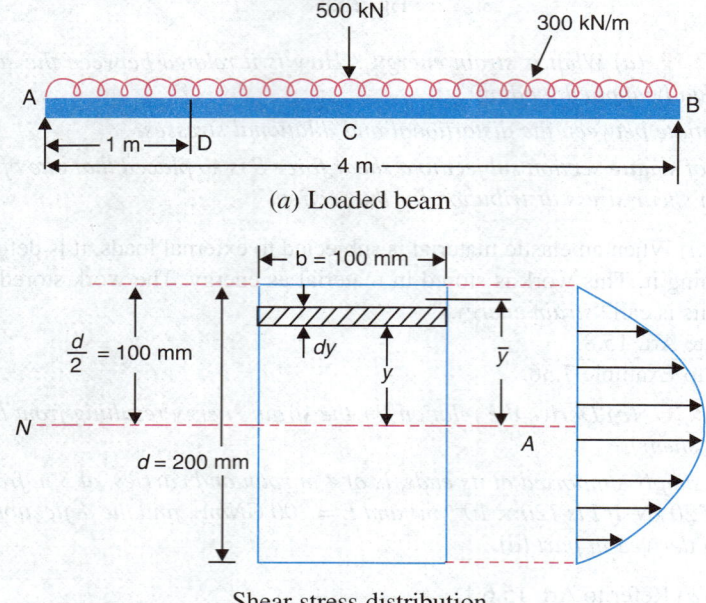

(*a*) Loaded beam

Shear stress distribution
(*b*) Beam cross-section

Fig. 23.10

Shear stress distribution diagram:

Taking moments about B, we get

$$R_A \times 4 = 500 \times 2 + 300 \times 4 \times \frac{4}{2}$$

$$= 1000 + 2400 = 3400$$

$\therefore$ $\qquad\qquad\qquad R_A = 850 \text{ kN}$

Shear force (S) at D, 1m from left hand support A,

$$S = R_A - 300 \times 1$$

$$= 850 - 300 = 550 \text{ kN}$$

$$I = \frac{bd^3}{12} = \frac{100 \times (200)^3 \times 10^{-12}}{12} = 66.67 \times 10^{-6} \ m^4 \ \textbf{...Art 7.2}$$

Shear stress, $\qquad\qquad \tau = \dfrac{SA\bar{y}}{Ib}$ $\qquad\qquad$ ... [Eqn. 7.1]

where, $\qquad\qquad S = $ Shear force,

$A\bar{y}\ = $ Moment of the sectional area above that point about the neutral axis,

$b = $ Width of the beam at the point where the shearing force is considered, and

$I = $ M.O.I.

Consider a section at a distance y from N.A. of depth dy.

$$A = b\left(\frac{d}{2} - y\right)$$

$$\bar{y} = y + \frac{1}{2}\left(\frac{d}{2} - y\right) = \frac{1}{2}\left(\frac{d}{2} + y\right)$$

$$A\bar{y} = b\left(\frac{d}{2} - y\right) \times \frac{1}{2}\left(\frac{d}{2} + y\right) = \frac{b}{2}\left(\frac{d^2}{4} - y^2\right)$$

Now, shear stress, $\qquad\qquad \tau = \dfrac{S \times \dfrac{b}{2}\left(\dfrac{d^2}{4} - y^2\right)}{\dfrac{bd^3}{12} \times b} = \dfrac{6S}{bd^3}\left(\dfrac{d^2}{4} - y^2\right)$

$$= \frac{6 \times 550 \times 1000}{100 \times (200)^3}\left(\frac{200^2}{4} - y^2\right)$$

$$= 4.125 \times 10^{-3}\,(10000 - y^2)$$

The Shear stress distribution is **parabolic** in nature, as shown in Fig. 23.10

y, mm	0	25	50	75	100
τ,N/mm^2	41.25	38.67	30.93	18.05	0

Example 23.22. *A smaller light piston 100 sq. mm in area compresses oil in a rigid container of 1000 c.c. capacity. When a weight of 70 N is gradually applied to the piston, its movement is observed to be 2.5 mm. If a weight of 20 N falls from a height of 40 mm on to the piston, determine the maximum pressure developed in the oil container neglecting the effect of friction and less of energy.*

Solution : Refer to Fig. 23.11.

Given : Area of the piston = 100 mm²;

Capacity of container = 1000 c.c.;

$W_g = 70$ N; $\delta_g = 2.5$ mm;

Falling weight, $W_f = 20$ N ; Height of fall; $h = 40$ mm.

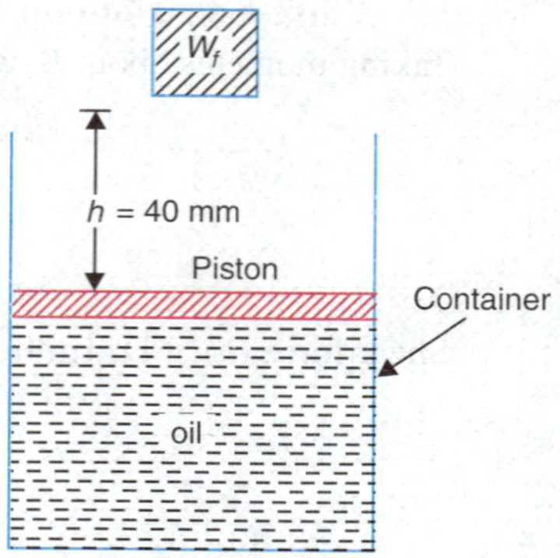

Maximum pressure developed p :

$$\text{Bulk modulus of oil, K} = \frac{p}{\dfrac{\delta V}{V}} = \frac{pV}{\delta V}$$

$$= \frac{W_g}{A} \times \frac{V}{A \times \delta_g}$$

$$= \frac{70}{100^2} \times \frac{(1000 \times 10^3)}{2.5} = 2800 \text{ N/mm}^2$$

Fig. 23.11

Let p = pressure developed in the oil container due to falling weight; and

δ = compression of oil under falling weight.

Now, work done by the falling weight = Strain energy stored in oil.

$$W_f(h + \delta) = \frac{1}{2} p \times (A\delta)$$

$$= \frac{1}{2} p \times \delta V = \frac{1}{2} p \times \frac{pV}{K} = \frac{1}{2} p^2 \frac{V}{K}$$

Also,

$$\delta = \frac{\delta V}{A} = \frac{1}{A} \times \frac{pV}{k} = \frac{pV}{AK}$$

∴

$$W_f\left(h + \frac{pV}{AK}\right) = \frac{1}{2} \times \frac{p^2V}{K}$$

or,

$$20\left[40 + \frac{p \times (1000 \times 10^3)}{100 \times 2800}\right] = \frac{1}{2} \times \frac{p^2 \times (1000 \times 10^3)}{2800}$$

or,

$$800 + 71.43\, p = 178.6\, p^2.$$

or,

$$178.6\, p^2 - 71.43\, p + 800 = 0$$

or,

$$p^2 - 0.4\, p + 4.48 = 0$$

or,

$$p = \frac{0.4 \pm \sqrt{(0.4)^2 + 4 \times 4.48}}{2} = 2.326 \text{ N/mm}^2$$

i.e.,

$$P = \textbf{2.326 N/mm}^2 \quad \textbf{(Ans.)}$$

Example 23.23. *A solid circular shaft is subjected to a B.M. of 80 kN/m and a torque of 120 kNm. In a uniaxial tensile stress the shaft material gave the following results :*

E = 2 × 10⁵ N/mm²; Stress at yield point = 300 N/mm²;

Poisson's ratio = 0.3; Factor of safety = 3

Estimate the least diameter of the shaft using

(i) Maximum principal stress theory;

(ii) Maximum principal strain theory.

Solution : Given : $M = 80$ kNm; $T = 120$ kNm; $E = 2 \times 10^5$ N/mm^2; stress at yield point = 300 N/mm^2; Poisson's ratio = 0.3; Factor of safety = 3.

Least diameter of the shaft, d:

$$\text{Bending stress, } \sigma_b = \frac{32 \, M}{\pi d^3}$$

$$\text{Torsional stress, } \tau = \frac{16 \, T}{\pi d^3}$$

Principal stresses are given by :

$$\sigma_1, \sigma_3, (\sigma_2 = 0) = \frac{\sigma_b}{2} \pm \sqrt{\left(\frac{\sigma_b}{2}\right)^2 + \tau^2} = \frac{1}{2}\left[\sigma_b \pm \sqrt{\sigma_b^2 + 4\tau^2}\right]$$

$$= \frac{1}{2}\left[\frac{32\pi}{\pi d^3} \pm \sqrt{\left(\frac{32M}{\pi d^3}\right)^2 + \left(\frac{32\,T}{\pi d^3}\right)^2}\right]$$

$$= \frac{16}{\pi d^3}\left[M \pm \sqrt{M^2 + T^2}\right]$$

$$\therefore \qquad \sigma_1 = \frac{16}{\pi d^3}\left[M \pm \sqrt{M^2 + T^2}\right]$$

$$= \frac{16 \times 10^3}{\pi d^3}\left[80 \pm \sqrt{80^2 + 120^2}\right]$$

$$= \frac{16 \times 10^3}{\pi d^3}(80 \pm 144.22)$$

$$\sigma_1 = \frac{16 \times 10^3}{\pi d^3} \times 224.22 \text{ N/m}^2$$

$$\sigma_3 = -\frac{16 \times 10^3}{\pi d^3} \times 64.22 \text{ N/m}^2$$

(*i*) Maximum pricipal theory :

$$\sigma_1 = \frac{\sigma_y}{\text{F.O.S.}}$$

$$\frac{16 \times 10^3}{\pi d^3} \times 224.22 = \frac{300}{3} = 100\,\text{N/mm}^2 \text{ or } 100 \times 10^6 \,\text{N/m}^2$$

or, $$d^3 = \frac{16 \times 10^3 \times 224.22}{\pi \times 100 \times 10^6} = 0.01142$$

or, $$d = (0.01142)^{1/3} = 0.225 \text{ m or } 225 \text{ mm}$$

Hence, load diameter, d = **225 mm** **(Ans.)**

(*ii*) Maximum principal strain theory :

$$\sigma_1 - \frac{1}{m} \text{ (or } \mu) \, \sigma_3 = \frac{\sigma_y}{F.O.S.}$$

$$\frac{16 \times 10^3}{\pi d^3}(224.22 + 0.3 \times 64.22) = 100 \times 10^6$$

$$\frac{16 \times 10^3}{\pi d^3} \times 243.48 = 100 \times 10^6$$

$$d^3 = \frac{16 \times 10^3 \times 243.48}{\pi \times 100 \times 10^6} = 0.0124$$

or, $d = (0.0124)^{1/3} = 0.2315$ m

Hence, least diameter, $d = 0.2315$ m or **231.5 mm. (Ans.)**

Example 23.24. *The maximum stress permitted in a thick cylinder of internal and external radii 200 mm and 300 mm respectively is 16 MPa. If the external pressure is 4 MPa, find the internal pressure that can be applied. What will be the change in outer diameter if cylinder has closed ends ?*

Solution : Refer to Example 11.8.

Example 23.25. *A vertical tie rod rigidly fixed at the top consists of steel rod 2 m long and 20 mm diameter encased throughout in a brass tube 20 mm internal diameter and 30 mm external diameter. The rod and the casing are fixed together at the ends. The compound bar is suddenly loaded in tension by a mass of 1530 kg falling freely through 3 mm before being arrested by the lie. Compare the maximum stress in steel and brass.*

Solution : Refer to Example 15.12.

Example 23.26. *A steel rod is to be bent in the form of a hook to lift a load of 8kN such that the maximum stress does not exceed 140 MPa. The ratio of the radius of curvature of centroidal plane to the radius of the rod is to be 4 and load acts through the centre of curvature. Determine the diameter of the rod.*

Solution : Refer to Fig. 23.12.

Given : P = 8kN; Maximum stress, σ_{max} = 140 MPa; 140×10^6 N/m²; the ratio of the radius of curvature of the

centroidal plane to the rod radius = $\dfrac{R}{r} = 4$

Diameter of the rod, *d* :

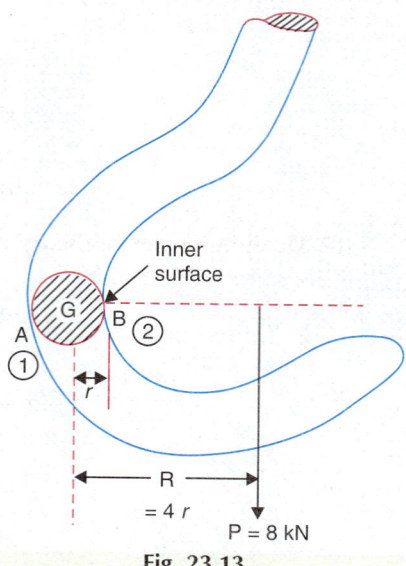

$$h^2 = \frac{d^2}{16} + \frac{1}{128} \cdot \frac{d^4}{R^2} \quad \text{[Eqn. (20.16)]}$$

$$= \frac{(2r)^2}{16} + \frac{1}{128} \times \frac{(2r)^4}{(4r)^2}$$

$$= \frac{r^2}{4} + \frac{1}{128} \times \frac{16r^4}{16r^2}$$

$$= \frac{r^2}{4} + \frac{r^2}{128} = \frac{33r^2}{128} = 0.2578\, r^2$$

Bending moment; $M = -(8 \times 1000) \times 4r = -32000\, r$

(–ve sign is taken because tending moment is bending to *decrease the curvature*)

Direct stress, $\sigma_d = \dfrac{P}{A} = \dfrac{8 \times 1000}{\pi r^2} = \dfrac{8000}{\pi r^2}$ *(tensile)*

Bending stress at '2'

$$(\sigma_b)_2 = \frac{M}{AR} \left[1 - \frac{R^2}{h^2} \times \frac{y_1}{R - y_1} \right] \qquad \qquad ...\text{[Eqn. (20.13)]}$$

Fig. 23.13

Inner surface

P = 8 kN

R = 4 r

$$= \frac{-32000\,r}{\pi r^2 \times 4r}\left[1 - \frac{(4r)^2}{0.2578r^2} \times \frac{r}{(4r-r)}\right]$$

$$= \frac{-8000}{\pi r^2}[1 - 20.68] = \frac{50114.7}{r^2}\ (tensile)$$

Maximum stress occurs in the *inner surfaces* '2'.

∴   Stress at '2',

$$\sigma_2 = \sigma_d + (\sigma_b)_2$$

$$140 \times 10^6 = \frac{8000}{\pi r^2} + \frac{50114.7}{r^2}$$

$$= \frac{52661.2}{r^2}$$

or, $$r = \left(\frac{52661.2}{140 \times 10^6}\right)^{\frac{1}{2}} = 0.0194 \text{ m or } 19.4 \text{ mm}$$

i.e., $r = \textbf{19.4 mm (or } d = \textbf{38.8 mm) (Ans.)}$

Example 23.27. (a) *What are the assumptions of solving the problems on thick cylinder shells?*

(b) *A thick cylinder of 20 mm inside diameter and 30 mm outside diameter has a fluid flowing through it which exerts an internal pressure of 700 N/mm². Evaluate the circumferential stresses at the inside and outside surfaces of the cylinder.*

Solution : (*a*) The assumptions of solving the problems on thick cylindrical shells are :

1. The material of the cylinder is linear, homogeneous and isotropic.

2. Plane transverse sections remain plane under the pressure.

3. The longitudinal strain is constant at all points in the cylinder wall, *i.e.,* It is independent of radius.

(*b*) Internal radius, $r_1 = 10$ mm; Outside radius = 15 mm; Internal pressure, $P_r = 700$ N/mm².

Circumferential stresses at inside and outside surfaces $(\sigma_c)r_1$ and $(\sigma_c)r_2$:

In this case, σ_c is given by :

$$\sigma_c = \frac{p_1 r_1^2}{r_2^2 - r_1^2}\left[\frac{r_2^2}{r_1^2} + 1\right]$$ ... (Eqn. 11.11*a*)

At, $$r = r_1, (\sigma_c)_{r_1} = \frac{p_1 r_1^2}{r_2^2 - r_1^2}\left[\frac{r_2^2}{r_1^2} + 1\right]$$

$$= \frac{p_1 r_1^2}{r_2^2 - r_1^2}\left(\frac{r_2^2 + r_1^2}{r_1^2}\right) = p_1 \frac{r_2^2 + r_1^2}{r_2^2 - r_1^2}$$ ...(*i*)

At, $$r = r_2, (\sigma_c)_{r_2} = \frac{p_1 r_1^2}{r_2^2 - r_1^2}\left[\frac{r_2^2}{r_2^2} + 1\right]$$

$$= \frac{p_1 r_1^2}{r_2^2 - r_1^2}\left(\frac{2r_2^2}{r_2^2}\right) = p_1 \frac{2r_1^2}{r_2^2 - r_1^2}$$ ... (*ii*)

Substituting the values in eqn. (*i*) and (*ii*), we get,

$$(\sigma_c)_{r_1} = 700 \times \frac{(15^2 + 10^2)}{(15^2 - 10^2)} = 1820 \text{ N/mm}^2 \quad \textbf{(Ans.)}$$

$$(\sigma_c)_{r_2} = 700 \times \frac{2 \times 10^2}{(15^2 - 10^2)} = 1120 \text{ N/mm}^2 \quad \textbf{(Ans.)}$$

Example 23.28. *For the section shown below, find the position of the shear centre from the shear centre of the web.*

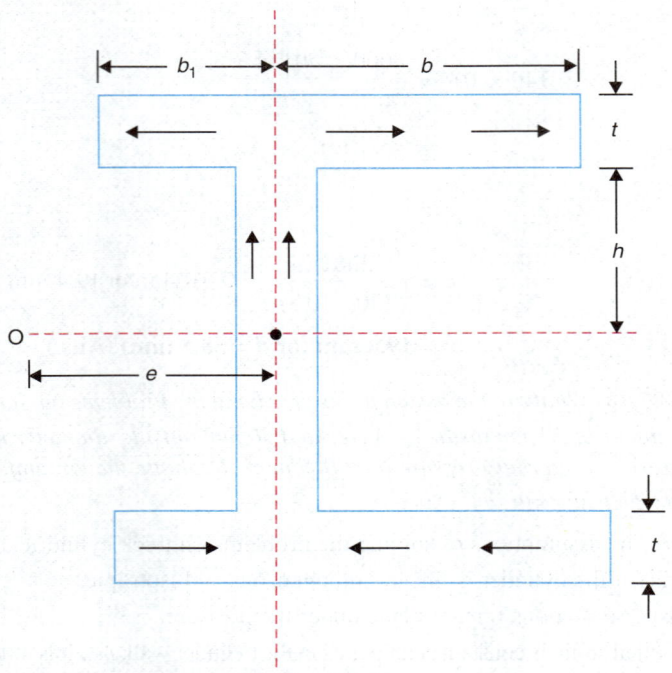

Fig. 23.13

Solution : Refer to Art. 23.5.2.

Example 23.29. *(Springs). An axial load of 20 N is supported on an open helical spring of 10 number of turns with coil mean diameter 200 mm made from 10 mm rod. If the helix angle of twist is 30°, calculate: (i) Axial deflection; (ii) Angular rotation of free end with respect to the fixed end of the spring; (iii) Twist about the horizontal axis.*

Solution : Refers to Example 14.15

(Hint. Twist about the horizontal axis = WR cos α)

ADDITIONAL OBJECTIVE TYPE QUESTIONS

(With Answers and Solutions – Comments)

[Including questions selected from various Universities and Competitive Examinations (ESE, CSE, GATE, etc.)]

ADDITIONAL OBJECTIVE TYPE QUESTIONS

I. UNIVERSITIES' QUESTIONS

A. Choose the Correct Answer :

1. Limit within which Hooke's law holds good is known as
(a) elastic limit
(b) plastic limit
(c) yield point
(d) proportional limit.

2. The magnitude of shear stress induced in a shaft due to applied torque varies
(a) from maximum at the centre to zero at the circumference
(b) from zero at the centre to maximum at the circumference
(c) from maximum at the centre to minimum at the circumference and not zero
(d) from minimum at the centre and not zero to maximum at the circumference.

3. Polar moment of inertia of the cross-section of a shaft of diameter d is

(a) $\dfrac{\pi}{32}d^3$ (b) $\dfrac{\pi}{32}d^4$

(c) $\dfrac{\pi}{16}d^3$ (d) $\dfrac{\pi}{16}d^4$.

4. A hollow shaft of the same cross-sectional area as that of a solid shaft can
(a) resists less torque
(b) resist more torque
(c) resist equal torque.

5. The flexural rigidity of a beam is
(a) EI (b) E/I
(c) I/E (d) E/I^2.

6. A thick curved bar has an original curvature which
(a) is large
(b) is small
(c) is zero
(d) could be any one of the above.

7. A thin curved bar has an original curvature which
(a) is large
(b) is small
(c) is zero
(d) could be any one of the above.

8. A beam is said to be loaded in pure bending when
(a) bending moment and shear force are constant but not zero
(b) bending moment is changing linearly
(c) bending moment and shear force both are changing linearly
(d) bending moment is constant.

9. A high carbon steel specimen as compared with mild steel specimen
(a) is more ductile
(b) is stronger
(c) has a more defined yeild point
(d) has more Poisson's ratio.

10. One kgf / cm^2 when converted into SI unit is
(a) 0.0981 MPa (b) 0.981 Pa
(c) 10^4 Pa (d) 1 Pa.

11. To express the stress-strain relations for a homogeneous isotropic, linearly elastic material, minimum number of material constant needed is
(a) two (b) three
(c) six (d) nine.

12. Deflection of the free end of a cantilever beam, subjected to a concentrated load at the middle of its span is given by

(a) $\dfrac{Pl^3}{3\,EI}$ (b) $\dfrac{Pl^3}{8\,EI}$

(c) $\dfrac{Pl^3}{24\,EI}$ (d) $\dfrac{5\,Pl^3}{48\,EI}$.

13. Shear centre
(a) always coincides with the centroid of the beam cross-section
(b) always lies within the boundaries of the cross-section
(c) is a point in the plane of the beam cross-section through which the resultant of shear forces must pass
(d) is a point in the beam cross-section representing zero shear stress.

14. The maximum eccentricity of a compressive load acting on a short strut of diameter d

without producing tension at the base section is

(a) $d/2$ (b) $d/4$

(c) $d/6$ (d) $d/8$.

15. Pure torsion of a circular shaft produces
(a) longitudinal normal stress in the shaft
(b) only direct shear stress in the transverse section of the shaft
(c) circumferential shear stress on a surface element of the shaft
(d) a longitudinal shear stress and a circumferential shear stress on a surface element of the shaft.

16. The rigidity modulus G, in terms of modulus of elasticity E and Poisson's ratio v is given by

(a) $\dfrac{2E}{(1+v)}$ (b) $\dfrac{E}{2(1-v)}$

(c) $\dfrac{E}{2(1+v)}$ (d) $\dfrac{E}{(1-v)}$.

17. A bar is subjected to an uniaxial load P, having length $= l$, area of cross-section A, and E is modulus of elasticity. The strain energy stored in the bar is given by

(a) $\dfrac{Pl}{AE}$ (b) $\dfrac{Pl^2}{AE}$

(c) $\dfrac{P^2l}{AE}$ (d) $\dfrac{P^2l}{2AE}$.

18. If σ_1 and σ_2 are principal stresses, the shear stress on the principal planes is given by

(a) $\dfrac{\sigma_1 - \sigma_2}{2}$ (b) zero

(c) $\dfrac{\sigma_1 + \sigma}{2}$ (d) $\sigma_1 - \sigma_2$.

19. A circular shaft of length l subjected to a torque T, G is rigidity modulus and J is polar moment of inertia, then the total angle of twist is given by

(a) $\dfrac{Tl}{GJ}$ (b) $\dfrac{TJ}{Gl}$

(c) $\dfrac{TG}{Jl}$ (d) $\dfrac{GJ}{Tl}$.

20. Point of contraflexure where
(a) bending moment is maximum
(b) shear force is maximum
(c) shear force is zero
(d) bending moment is zero.

21. The deflection of a cantilever beam at the free end due to concentrated load P at the free end is given by

(a) $\dfrac{Pl^3}{2EI}$ (b) $\dfrac{Pl^2}{2EI}$

(c) $\dfrac{Pl^3}{3EI}$ (d) $\dfrac{Pl^3}{4EI}$.

22. The maximum stress used in the analysis of bending beyond elastic limit is
(a) ultimate stress
(b) allowable stress
(c) yield stress
(d) none of the above.

23. If two springs with stiffness k_1 and k_2 are connected in series, then stiffness of the composite spring is given by

(a) $k_1 + k_2$ (b) $\dfrac{1}{k_1} + \dfrac{1}{k_2}$

(c) $\dfrac{1}{k_1} - \dfrac{1}{k_2}$ (d) $k_1 - k_2$.

24. A compressive member always tends to buckle in the direction of
(a) axis of load
(b) minimum cross-section
(c) least radius of gyration
(d) perpendicular to the axis of load.

25. If the normal cross-section a of a member is subjected to a tensile force P, the resulting normal stress on an oblique plane inclined at θ to the transverse plane will be

(a) $\dfrac{P}{A} \sin^2 \theta$ (b) $\dfrac{P}{A} \cos^2 \theta$

(c) $\dfrac{P}{2A} \sin 2\theta$ (d) $\dfrac{P}{2A} \cos 2\theta$.

26. The maximum bending moment due to a moving load on a simply supported beam occurs
(a) at the midspan
(b) at the supports
(c) under the load
(d) anywhere on the beam.

27. A simply supported beam carries two equal concentrated load W at distance $L/3$ from either support. The maximum BM is

(a) $\dfrac{WL}{3}$ (b) $\dfrac{WL}{4}$

(c) $\dfrac{5WL}{8}$ (d) $\dfrac{2}{5}WL$.

28. The equivalent length of a column fixed at both ends is
(a) 0.7 l
(b) 0.5 l
(c) l
(d) 2 l.

29. The maximum deflection of a simply supported beam of length L with central concentrated load W is

(a) $\dfrac{WL^2}{48\,EI}$
(b) $\dfrac{WL^2}{24\,EI}$

(c) $\dfrac{WL^3}{48\,EI}$
(d) $\dfrac{WL^2}{8\,EI}$.

30. If a solid shaft is subjected to a torque T at its end such that maximum shear stress does not exceed τ_a, the diameter of shaft will be

(a) $\dfrac{16T}{\pi\tau_a}$

(b) $\sqrt{\dfrac{16T}{\pi\tau_a}}$

(c) $\left(\dfrac{16T}{\pi\tau_a}\right)^{1/3}$

(d) None of the above.

31. If a shaft is simultaneously subjected to a torque T and bending moment M, the ratio of the maximum bending stress to maximum shearing stress is

(a) $\dfrac{M}{T}$
(b) $\dfrac{T}{M}$

(c) $\dfrac{2M}{T}$
(d) $\dfrac{2T}{M}$.

32. Euler's formula states that the buckling load P for a column of length L, both ends hinged and whose least moment of inertia and modulus of elasticity of the material of column are I and E respectively is given by

(a) $P = \dfrac{\pi EI}{L^2}$
(b) $P = \dfrac{\pi^2\,EI}{L^2}$

(c) $P = \dfrac{\pi L^2}{EI}$
(d) $P = \dfrac{\pi EI}{L^2}$.

33. A closely coiled helical spring of radius R contains W. The radius of the coil wire is r and the modulus of rigidity of the coil material is G. The deflection of the coil is

(a) $\dfrac{WR^3N}{Gr^4}$
(b) $\dfrac{2WR^3}{Gr^4}$

(c) $\dfrac{3WR^3N}{Gr^4}$
(d) $\dfrac{4WR^3N}{Gr^4}$.

34. Strain energy of a member may be equated to
(a) average resistance × displacement
(b) $\dfrac{1}{2}$ stress × strain
    × area of its cross-section
(c) $\dfrac{1}{2}$ stress × strain
    × volume of the member
(d) $\dfrac{1}{2}\dfrac{(\text{stress})^2 \times \text{volume of member}}{\text{Young's modulus } E}$.

35. The work done on a unit volume of material as a simple tensile force is gradually increased from zero to the value causing ruputre is called
(a) modulus of resilience
(b) modulus of toughness
(c) tangent modulus
(d) none of the above.

36. The ratio of maximum shear stress to average shear is 4/3 in a beam of
(a) circular cross-section
(b) rectangular cross-section
(c) triangular cross-section
(d) none of the above.

37. The Young's modulus E, the shear modulus G and the Poisson's ratio μ are related by
(a) E = 2G (1 − μ);
(b) E = 2G (1 + 2μ);
(c) E = 2G (1 + μ)
(d) none of these.

38. Poisson's ratio, μ is defined as
(a) ratio of transverse strain to axial strain;
(b) modulus of ratio of transverse strain to axial strain;
(c) modulus of ratio of axial strain to transverse strain
(d) none of the above.

39. The point of contraflexure in a loaded beam is the point where
(a) the bending moment is maximum;
(b) the bending moment changes sign;
(c) the shear force changes sign
(d) none of the above.

40. A thin cylindrical tube of inner diameter d, and thickness t is closed at both ends and subjected to an internal pressure p. The tube also carries a torque T. The stresses at any point (σ_s, σ_q and τ_{xq}) are

(a) $\dfrac{pd}{2t}, \dfrac{pd}{4t}, \dfrac{2T}{\pi d^2 t}$

(b) $\dfrac{pd}{4t}, \dfrac{pd}{2t}, \dfrac{T}{\pi d^2 t}$

(c) $\dfrac{pd}{4t}, \dfrac{pd}{2t}, \dfrac{T}{2\pi d^2 t}$

(d) None of the above.

41. A rod of length L and area of cross-section A, whose material has a modulus of elasticity E and coefficient of thermal expansion α is subjected to a change of temperature ΔT. The change in length is

(a) $L \alpha \Delta T$; (b) $L \alpha \Delta T/E$;

(c) $L \alpha \Delta T/AE$; (d) none of these.

42. An axial core of 100 mm is bored throughout the length of a 200 mm diameter solid circular shaft. For the same maximum shear stress, the percentage torque carrying capacity lost by this operation is

(a) $6\dfrac{1}{4}$; (b) $12\dfrac{1}{2}$;

(c) 25 (d) 30.

43. A hole is to be punched in a mild steel plate of 10 mm thickness with the help of a hardened punch. The allowable crushing strength of the punch is 4 times the shearing strength of the plate. The diameter of the smallest hole that can be punched in the plate is

(a) 20 mm; (b) 10 mm;

(c) 5 mm. (d) 2 mm.

44. A close-coiled helical spring with wire diameter d, mean coil radius R and active number of coils n is subjected to an axial compressive load P. The shear modulus of the spring material is G. The deflection of the spring is given by

(a) $8\,PR^3n/Gd^4$;

(b) $64\,PR^3n/Gd^4$;

(c) $32PR^3n/Gd^4$

(d) none of the above.

45. In a two-dimensional problem the principal stress at a point are σ_1, and σ_2. The normal stress associated with the plane of maximum shear is

(a) $(\sigma_1 + \sigma_2)/2$; (b) $(\sigma_1 - \sigma_2)/2$;

(c) zero (d) none of these.

46. When mild steel is subjected to axial tension, the ratio of engineering stress to actual stress is

(a) < 1; (b) $= 1$;

(c) > 1 (d) none of above.

47. A uniform rod of length L, cross-sectional area A and modulus of elasticity E is held rigidly by fixed supports at the ends. An axial load P is applied at mid-length of the rod. The elastic strain energy stored is

(a) $P^2L / 8AE$; (b) $P^2L/4\,AE$;

(c) $P^2L/16\,AE$ (d) none of these.

48. Maximum normal stress theory is used for

(a) ductile materials

(b) brittle materials

(c) visco-elastic materials.

(d) none of the above.

49. A circular shaft is subjected to a torque T. The maximum shear stress develped is τ. The maximum tensile stress developed in the shaft is

(a) $\tau/2$ (b) 2τ;

(c) τ (d) 3τ.

50. The moment of inertia of a rectangular lamina of sides b and d about its centroidal axis parallel to the side d is given by

(a) $bd^3/12$

(b) $db^3/12$

(c) $bd^3/36$

(d) none of the these.

51. Number of elastic constants sufficient for homogeneous isotropic materials is

(a) 3 (b) 4

(c) 2 (d) 1.

52. Number of equilibrium equations for a two-dimensional system is

(a) 6; (b) 3;

(c) 2 (d) 1

53. For an element under a biaxial state of stress $\sigma_x = -\sigma$ in the $x - y$ plane, the radius of the Mohr's circle is

(a) $2\sigma_x$; (b) $2\sigma_y$;

(c) σ_x (d) $3\sigma_x$.

54. The buckling load for a fixed-fixed column is

(a) π^2EI/L^2 (b) $\pi^2EI/4L^2$

(c) $\pi^2\,EI/2L^2$ (d) none of these.

55. Stress developed in an elastic body due to external force
(a) does not depend on elastic constants
(b) depends on elastic constants
(c) depends partly on elastic constants
(d) none of the above.

56. Modulus of resilience is defined by
(a) strain energy stored in an elastic body
(b) strain energy per unit volume of the elastic body
(c) percentage of elongation of a ductile metal
(d) any of the above.

57. In the simple bending formula relating the B.M, bending stress and curvature, plane of external loading should pass through
(a) the shear centre at any angle
(b) the shear centre vertically
(c) the principal axis of the section
(d) none of the above.

58. Relation amongst Young's modulus, Poisson's ratio and Bulk modulus is given by
(a) $E = 3K(1 - 2\mu)$ (b) $E = 3K/(1 - 2\mu)$
(c) $E = 2K(1 + \mu)$ (d) none of these

59. In a beam of I-Section the maximum shear force is carried by
(a) the upper flange   (b) the web
(c) the lower flange   (d) any of these.

60. In a short column with eccentric loading, the neutral axis
(a) passes through the centroid of the section
(b) passes through the point of application of load
(c) does not pass through the centroid of the section
(d) none of the above.

61. In a long column with hinged ends, the critical stress is
(a) more than the yield stress
(b) equal to the yield stress
(c) less than the yield stress
(d) any of these.

62. In a thick walled cylinder subject to internal pressure, the maximum hoop stress occurs at the
(a) inner wall
(b) outer wall
(c) middle point of thickness
(d) none of these.

63. Castigliano's theorem is valid for
(a) any structure
(b) non-linear structure
(c) linear structure
(d) any of these.

64. In a curved beam subjected to pure bending the neutral axis
(a) passes through the centroid of the section
(b) is shifted towards the centre of curvature
(c) is shifted away from the centre of curvature
(d) none of these.

65. A beam with hinged ends is subjected to a moment at one end. The slope at the other end is given by
(a) $\dfrac{Ml}{2EI}$ (b) $\dfrac{Ml}{6EI}$
(c) $\dfrac{Ml}{3EI}$ (d) $\dfrac{Ml}{12EI}$.

66. A hollow prismatic beam of circular section is subjected to a torsional moment. The maximum shear stress occurs at
(a) the inner wall of the cross-section
(b) the middle of the thickness
(c) the outer surface of the shaft
(d) none of the above.

67. A beam of length l and thermal coefficient α is fixed at two ends without stress. Then temperature drops by $T°C$. Axial force developed is
(a) $AE \propto T$ tensile   (b) $\dfrac{AE}{\alpha T}$ tensile
(c) $A \propto T$ compressive (d) none of these.

68. The reaction at the prop in a propped cantilever beam subjected to u.d.l. is
(a) $\dfrac{wl}{4}$ (b) $\dfrac{3wl}{8}$
(c) $\dfrac{5wl}{8}$ (d) $\dfrac{6wl}{7}$

69. In a long column, with one end fixed and the other free, if slenderness ratio increases, critical stress
(a) remains constant   (b) increases
(c) decreases   (d) any of these.

70. The ratio of maximum shear stress to average shear stress is 1.5 in a beam of
(a) circular cross-section
(b) rectangular cross-section
(c) triangular cross-section
(d) any cross-section

71. A material having identical properties in all directions, is called :
    (a) elastic
    (b) homogeneous
    (c) isotropic
    (d) any of these.

72. The ratio of the deformation of a bar due to its own weight, to the deformation due to axial load equal to its weight is:
    (a) 1
    (b) $\dfrac{1}{2}$
    (c) 2
    (d) 4.

73. If the Poisson's ratio of a material is 0.4, then the ratio of the shear modulus of elasticity to modulus of elasticity is:
    (a) $\dfrac{5}{7}$
    (b) $\dfrac{7}{5}$
    (c) $\dfrac{5}{14}$
    (d) $\dfrac{5}{12}$.

74. At a point in a strained material there acts a tensile stress of 100 N/mm^2 on one plane and a tensile stress of 50 N/mm^2 on a plane at right angles to it. These planes also carry a shear stress of 50 N/mm^2. The angle of inclination of principal stresses with the 100 N/mm^2 stress is :
    (a) $\tan^{-1}(2)$;
    (b) $\dfrac{1}{2}\tan^{-1}(2)$;
    (c) $\dfrac{1}{2}\tan^{-1}\left(\dfrac{2}{3}\right)$
    (d) none of these.

75. A cantilever beam is loaded uniformly throughout its length. The shape of the shear force diagram will be
    (a) a right angle triangle
    (b) an isosceles triangle
    (c) a rectangle
    (d) none of these.

76. Two prismaic beams A and B have same length. The one having larger .... will be stronger in flexure:
    (a) moment of inertia
    (b) section modulus
    (c) area of cross-section
    (d) none of these.

77. The strength of a beam of square cross-section placed with its diagonal horizontal is ......times the strength when it is placed with its sides horizontal;
    (a) $\dfrac{1}{2}$
    (b) $\sqrt{2}$
    (c) $\dfrac{1}{\sqrt{2}}$
    (d) $\dfrac{1}{3}$.

78. Leaf springs are subjected to :
    (a) shear stress
    (b) direct stress
    (c) bending stress
    (d) none of these.

79. Load required to produce unit deflection is known as :
    (a) toughness
    (b) stiffness
    (c) flexibility
    (d) none of these.

80. The maximum energy stored at ...... is called proof resilience
    (a) elastic limit
    (b) plastic limit
    (c) limit of proportionality
    (d) any of these.

81. A cylinder is said to be thick if :
    (a) it is made of thick plates
    (b) the internal pressure is large
    (c) the ratio of its wall thickness to its diameter is more than $\dfrac{1}{10}$
    (d) none of these.

82. The maximum stress produces in a thin cylinder is ...... times that in a thin spherical shell, having same diameter, thickness and internal pressure:
    (a) 1
    (b) 2
    (c) $\dfrac{1}{2}$
    (d) $\dfrac{1}{3}$.

83. The strength of a column with both ends fixed is .... times the strength when its one end is fixed and other free, other parameters remaining same.
    (a) two
    (b) four
    (c) sixteen
    (d) eighteen.

84. Euler's formula is applicable to :
    (a) only long columns
    (b) only short columns
    (c) long as well as short columns
    (d) any of these.

85. The variation of the shear stress on the transverse plane of a normal beam section caused by a transverse shear force is usually:
    (a) linear
    (b) parabolic
    (c) non-linear
    (d) none of these.

86. In all sections having symmetry about the load axis, the shear centre is at
    (a) centroid of the section
    (b) neutral axis of the section
    (c) geometric centre of the section
    (d) any of these.

87. The Poisson's ratio is given by :

(a) $\dfrac{-\text{Lateral strain}}{\text{Longitudinal strain}}$

(b) $\dfrac{+\text{Lateral strain}}{\text{Longitudinal strain}}$

(c) $\dfrac{\text{Longitudinal strain}}{\text{Lateral strain}}$

(d) $\dfrac{\text{Longitudinal strain}}{-\text{Lateral strain}}$.

88. The work done per unit volume in elongating a body by a uniaxial force is :

(a) $\dfrac{\text{Stress}}{\text{Strain}}$

(b) Stress × strain

(c) $\dfrac{1}{2}$ Stress × strain

(d) none of these.

89. 4 wires of same material are stretched by same load. Their dimensions are given below. In which wire will the extension be a max.?

(a) length 2m, dia. 1 mm

(b) $l = 1$m, $d = 0.5$ mm

(c) $l = 4$m, $d = 2$mm

(d) $l = 6$ m, $d = 3$mm.

90. The stresses in thick cylinder subjected to uniform pressure vary proportional to

(a) r

(b) $\dfrac{1}{r}$

(c) r^2

(d) $\dfrac{1}{r^2}$.

91. Max. deflection for a cantilever of span L, loaded at the free end by P is given by:

(a) $\dfrac{PL^2}{3EI}$

(b) $\dfrac{PL^2}{6EI}$

(c) $\dfrac{PL^3}{8EI}$

(d) $\dfrac{PL^3}{3EI}$.

92. A helical spring is subjected to an axial tensile force. The predominant effect of this force on the spring is:

(a) twisting

(b) bending

(c) tension

(d) compression.

93. Polar moment of inertia of a circular area is

(a) $\dfrac{\pi}{32}d^4$

(b) $\dfrac{\pi}{64}d^4$

(c) $\dfrac{\pi}{4}d^4$

(d) $\dfrac{\pi}{4}d^3$.

94. A bar of square cross-section of side a is subjected to a tensile load P. On a plane inclined at 45° to the axis of the bar, the normal stress will be

(a) $\dfrac{2P}{a^2}$

(b) $\dfrac{P}{a^2}$

(c) $\dfrac{P}{2a^2}$

(d) $\dfrac{P}{4a^2}$.

95. The ratio of lateral strain to linear strain is known as

(a) Modulus of elasticity

(b) Modulus of rigidity

(c) Poisson's ratio

(d) Elastic limit.

96. In a cantilever carrying a load whose intensity varies uniformly from zero at the free end to w per unit run at the fixed end, the S.F. changes following a

(a) linear law

(b) parabolic law

(c) cubic law

(d) none of the above.

97. At the point of application of a concentrated load on a beam there is

(a) sudden change of slope of B.M. diagram

(b) maximum B.M.

(c) point of contraflexure

(d) maximum deflection.

98. The diameter of kernel of a circular section of diameter d is

(a) $\dfrac{d}{2}$

(b) $\dfrac{d}{3}$

(c) $\dfrac{d}{\sqrt{2}}$

(d) $\dfrac{d}{4}$.

99. A simply supported beam of span 4m carries a u.d.l. of 30 kN/m throughout. If $EI = 25000$ kN-m^2, the maximum deflection of the beam is

(a) 0.5 mm

(b) 1.5 mm

(c) 3.5 mm

(d) 4.0 mm.

100. Maximum shear stress theory of failure was postulated by

(a) St. Venant

(b) Rankine

(c) Castigliano

(d) Tresca.

101. The secant formula is used for

(a) long column under eccentric loading

(b) long column under axial loading

(c) short column under eccentric loading

(d) short column under axial loading.

102. The stress due to suddenly applied load is ..... times that of gradually applied load.
(a) 1.5   (b) 2
(c) 3   (d) 4.

103. Castigliano's theorems are valid for
(a) elastic structure
(b) truss
(c) beam
(d) linear structure.

104. Over a part of the beam, carrying transverse loads, where S.F. is zero, the B.M. is
(a) maximum   (b) indeterminate
(c) constant   (d) zero.

105. In I-section of a beam subjected to transverse S.F. the maximum shear stress occurs at the
(a) centre of the web
(b) top edge of the top flange
(c) bottom edge of the top flange;
(d) none of the above.

106. A strut of length *l* is fixed at one end and free at the other end. The Euler's buckling load for this strut is 20 kN. If both ends of the strut are fixed, the buckling load will be
(a) 80 kN   (b) 160 kN
(c) 240 kN   (d) 320 kN.

107. A close-coiled helical spring absorbs 80 N mm of energy while extending by 4 mm. The stiffness of the spring is
(a) 5 N/mm;   (b) 10 N/mm;
(c) 16 N/mm;   (d) 20 N/mm.

108. A solid circular shaft under pure torsion develops a maximum shear stress of 50 N/m². The maximum principal stress on the surface of the shaft is
(a) 100 N/mm²   (b) 50 N/mm²
(c) 25 N/mm²   (d) none of these.

109. A mild steel beam develops a bending stress of 80 N/mm² at a distance of 8 cm from the neutral layer. If $E = 200$ GPa, the radius of curvature is
(a) 400 m
(b) 200 m
(c) 100 m
(d) none of these.

110. If the load passes through the shear centre of the section of a beam, then there will be
(a) no bending of the beam
(b) only bending
(c) bending with twisting
(d) only twisting.

II. COMPETITIVE EXAMINATION QUESTIONS

(ESE, CSE, etc. from 1996 onwards)
* (With Solutions – Comments)

111. If the principle stresses corresponding to a two-dimensional state of stress are σ_1 and σ_2 (σ_2 is greater than σ_1) and both are tensile, then which one of the following would be the correct for failure by yielding, according to the maximum shear stress criterion?

(a) $\dfrac{(\sigma_1 - \sigma_2)}{2} = \pm \dfrac{\sigma_{yp}}{2}$

(b) $\dfrac{\sigma_1}{2} = \pm \dfrac{\sigma_{yp}}{2}$

(c) $\dfrac{\sigma_2}{2} = \pm \dfrac{\sigma_{yp}}{2}$

(d) $\sigma_1 = \pm 2\, \sigma_{yp}$.

*112. A length of 10 mm diameter steel wire is coiled to a close helical spring having 8 coils of 75 mm mean diameter, and the spring has a stiffness *k*. If the same length of wire is coiled to 10 coils of 60 mm mean diameter, then the spring stiffness will be.

(a) *k*   (b) 1.25 *k*
(c) 1.56 *k*   (d) 1.95 *k*.

113. The unit of elastic modulus is the same as those of
(a) stress, shear modulus and presssure
(b) strain, shear modulus and force
(c) shear modulus, stress and force
(d) stress, strain and pressure.

114. In the case of an engineering material under unidirectional stress in the *x*-direction, the Poisson's ratio is equal to (symbols have their usual meanings)
(a) $\varepsilon_y / \varepsilon_x$ (b) ε_y / σ_x
(c) ε_y / σ_y (d) σ_y / ε_x.

*115. A rod of length '*l*' and cross-section area 'A' rotates about an axis passing through one end of the rod. The extension production in the rod due to centrifugal forces is (*w* is the weight of the rod per unit length and ω is angular velocity of rotation of the rod)

(a) $\omega wl^2/gE$ (b) $\omega^2 wl^3/3gE$

(c) $\omega^2 \, wl^3/gE$ (d) $3gE/\omega wl^3$.

116. The buckling load will be maximum for a column, if

(a) one end of the column is clamped and the other end is free

(b) both ends of the column are clamped

(c) both ends of the column are hinged

(d) one end of the column is hinged and the other end is free.

117. For the loaded beam shown in Fig. 1 the correct shear force diagram is

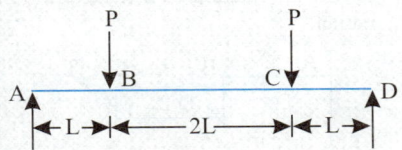

Fig. 1

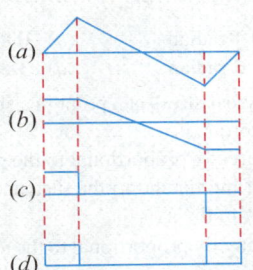

Fig. 2

*118. Which one of the following materials is highly elastic?

(a) Rubber   (b) Brass

(c) Steel   (d) Glass.

*119. The state of stress at a point in a loaded member is shown in the Fig. 3. The magnitude of maximum shear stress is:

(a) 10 MPa   (b) 30 MPa

(c) 50 MPa   (d) 100 MPa.

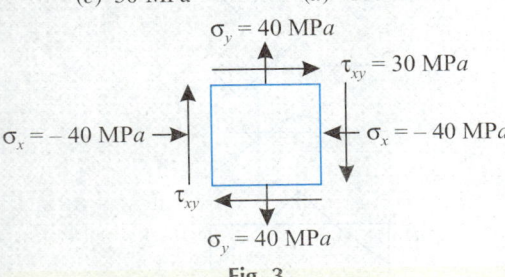

Fig. 3

*120. For a linearly elastic, isotropic and homogeneous material, the number of elastic constants required to relate stress and strain are:

(a) two   (b) three

(c) four   (d) six.

*121. A horizontal beam with square cross-section is simply supported with sides of the square horizontal and vertical and carries a distributed loading that produces maximum bending stress s in the beam. When the beam is placed with one of the diagonals horizontal the maximum bending stress will be

(a) $\dfrac{1}{\sqrt{2}}\sigma$ (b) 2s

(c) $\sqrt{2}\sigma$ (d) 3s.

*122. Shear stress distribution diagram of a beam of rectangular cross-section, subjected to transverse loading will be

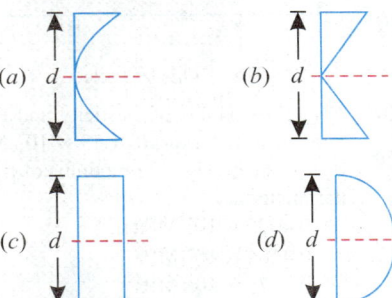

Fig. 4

*123. A shaft was initially subjected to bending moment and then was subjected to torsion. If the magnitude of bending moment is found to be the same as that of the torque, then the ratio of maximum bending stress to shear stress would be

(a) 0.25   (b) 0.50

(c) 2.0   (d) 4.0.

124. In a homogenous, isotropic elastic material, the modulus of elasticity E in terms of G and K is equal to

(a) $\dfrac{G+3K}{9KG}$ (b) $\dfrac{3G+K}{9KG}$

(c) $\dfrac{9KG}{G+3K}$ (d) $\dfrac{9KG}{G+3K}$.

*125. The Fig. 5 shows a bending moment diagram for the beam *CABD*:

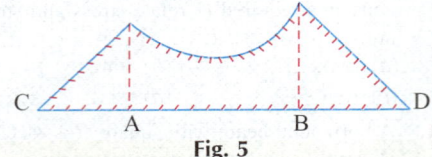

Fig. 5

Load diagram for the above ebam will be

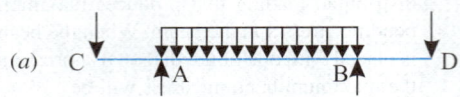

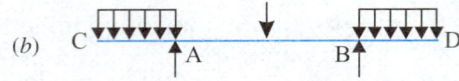

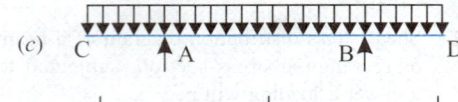

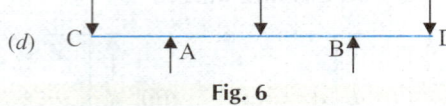

Fig. 6

*126. Young's modulus of elasticity and Poisson's ratio of a material are 1.25×10^5 MPa and 0.34 respectively. The modulus of rigidity of the matierial is:
(a) 0.4025×10^5 MPa
(b) 0.4664×10^5 MPa
(c) 0.8375×10^5 MPa
(d) 0.9469×10^5 MPa.

127. The ends of the leaves of a semi-elliptical leaf spring are made triangular in plane in order to
(a) obtain variable *l* in each leaf
(b) Permit each leaf to act as a overhanging beam
(c) have variable bending moment in each leaf
(d) make M/I constant throughout the length of the leaf.

128. The ratio of circumferential stress to longitudinal stress in a thin cylinder subjected to internal hydrostatic pressure is:
(a) 1/2   (b) 1
(c) 2   (d) 4.

129. The independent elastic constants for a homogeneous and isotropic material are:
(a) *E, G, K, v*   (b) *E, G, K*
(c) *E, G, v*   (d) *E, G*.

*130. A simply supported beam of rectangular section 4 cm by 6 cm carries a mid-span concentrated load such that the 6 cm side lies parallel to line of action of loading; deflection under the load is δ. If the beam is now suppported with the 4 cm side parallel to line of action of loading, the deflection under the load will be
(a) 0.44 δ   (b) 0.67 δ
(c) 1.5 δ   (d) 2.25 δ.

*131. A circular shaft fixed at *A* has diameter *D* for half of its length and diameter *D/2* over the other half. What is the rotation of *C* relative to *B* if the rotation of *B* relative to A is 0.1 radian ?

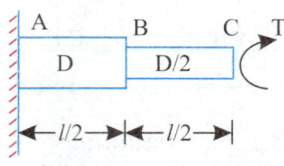

Fig. 7

(a) 0.4 radian   (b) 0.8 radian
(c) 1.6 radian   (d) 3.2 radian

*132. The shear stress at a point in a shaft subjected to a torque is
(a) directly proportional to the polar moment of inertia and to the distance of the point from the axis
(b) directly proportional to the applied torque and inversely proportional to the polar moment of inertia
(c) directly proportional to the applied torque and polar moment of inertia
(d) inversely proportional to the applied torque and the polar moment of inertia.

*133. A beam carries a uniformly distributed load and is supported with two equal overhangs. Which one of the following correctly shows the bending moment diagram for the beam ?

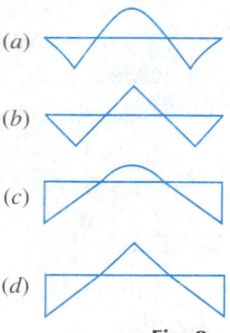

Fig. 8

*134. A beam *AB* is hinge-supported at its ends and is loaded by couple P.c. as shown in the Fig. 9. The magnitude of shearing force at a section *X* of the beam is

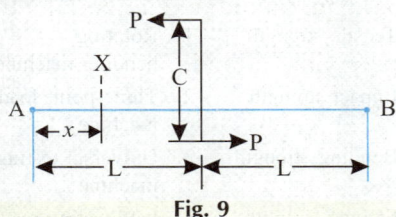

Fig. 9

(*a*) 0   (*b*) *P*
(*c*) *P/2L*   (*d*) P.c./2L

*135. A 0.2 mm thick tape goes over a frictionless pulley of 25 mm diameter. If *E* of material is 100 GPa, then the maximum stress induced in the tape is

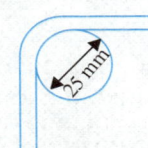

Fig. 10

(*a*) 100 MPa   (*b*) 200 MPa
(*c*) 400 MPa   (*d*) 800 MPa.

136. A solid circular shaft is subjected to a maximum shearing stress of 140 MPa. The magnitude of the maximum normal stress developed in the shaft is
(*a*) 140 MPa   (*b*) 80 MPa
(*c*) 70 MPa   (*d*) 60 MPa.

* 137. A metal pipe of 1 m diameter contains a fluid having a pressure of 1 N/mm². If the permissible tensile stress in the metal is 20 N/mm², then the thickness of the metal required for making the pipe would be
(*a*) 5 mm   (*b*) 10 mm
(*c*) 20 mm   (*d*) 25 mm.

138. Total strain energy stored in simply supported beam of span '*L*' and flexural rigidity '*EI*' subjected to a concentrated load '*W*' at the centre is equal to

(*a*) $\dfrac{W^2 L^3}{40\,EI}$ (*b*) $\dfrac{W^2 L^3}{60\,EI}$

(*c*) $\dfrac{W^2 L^3}{96\,EI}$ (*d*) $\dfrac{W^2 L^3}{240\,EI}$.

*139. A solid thick cylinder is subjected to an external hydrostatic pressure *p*. The state of stress in the material of the cylinder is represented as:

(*a*)

(*b*)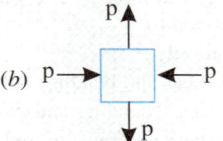

(*c*)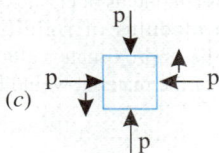

(*d*)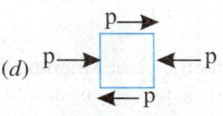

Fig. 11

*140. Circumferential and longitudinal strains in a cylindrical boiler under steam pressure are ε_1 and ε_2 respectively. Change in volume of the boiler cylinder per unit volume will be

(*a*) $\varepsilon_1 + 2\,\varepsilon_2$ (*b*) $\varepsilon_1 \varepsilon_2^2$

(*c*) $2\varepsilon_1 + \varepsilon_2$ (*d*) $\varepsilon_1^2 \varepsilon_2$.

*141. If two shafts of the same length, one of which is hollow, transmit equal torque and have equal maximum stress, then they should have equal
(*a*) polar moment of inertia
(*b*) polar modulus of section
(*c*) diameter
(*d*) angle of twist.

*142. In the assembly of pulley, key and shaft
(*a*) pulley is made the weakest
(*b*) key is made the weakest
(*c*) key is made the strongest
(*d*) all the three are designed for equal strength.

143. A column of length '*l*' is fixed at its both ends. The equivalent length of the column is:
(*a*) 2 *l*   (*b*) 0.5 *l*
(*c*) 4 *l*   (*d*) *l*.

*144. The shafts of same length and material are joined in series. If the ratio of their diameters is 2, then the ratio of their angles of twist will be

(a) 2   (b) 4

(c) 8   (d) 16.

145. An open-coiled helical spring of mean diameter D, number of coils N and wire diameter d is subjected to an axial force P. The wire of the spring is subjected to

(a) direct shear only

(b) combined shear and bending only

(c) combined shear, bending and twisting

(d) combined shear and twisting only.

*146. If a material has a modulus of elasticity of 210 GN/m^2 and a modulus of rigidity of 80 GN/m^2, then the approximate value of the Poisson s ratio of the material would be

(a) 0.26

(b) 0.31

(c) 0.47

(d) 0.5

147. In an axi-symmetric plane strain problem, let u be the radial displacement at r. Then the strain components, ε_r, ε_θ, $\gamma_{r\theta}$ are given by:

(a) $\varepsilon_r = \dfrac{u}{r}, \varepsilon_\theta = \dfrac{\partial u}{\partial r}, \gamma_{r\theta} = \dfrac{\partial^2 u}{\partial r \, \partial \theta}$

(b) $\varepsilon_r = \dfrac{\partial u}{\partial r}, \varepsilon_\theta = \dfrac{u}{r}, \gamma_{r\theta} = 0$

(c) $\varepsilon_r = \dfrac{u}{r}, \varepsilon_\theta = \dfrac{\partial u}{\partial r}, \gamma_{r\theta} = 0$

(d) $\varepsilon_r = \dfrac{\partial u}{\partial r}, \varepsilon_\theta = \dfrac{\partial u}{\partial \theta} = \gamma_{r\theta} = \dfrac{\partial^2 u}{\partial r \, \partial \theta}.$

148. A beam AB of length $2L$ having a concentrated load P at its mid-span is hinge-supported at its two ends A and B or two identical cantilevers as shown in the Fig. 12. The correct value of bending moment at A is:

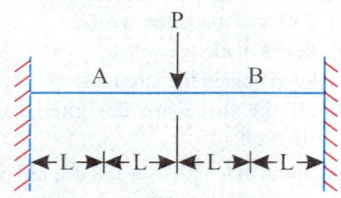

Fig. 12

(a) zero

(b) $PL/2$

(c) PL

(d) $2PL$.

*149. Match List I with List II and select the correct answer from the codes given below :

List I	List II
(*Property*)	(*Testing machines*)
A. Tensile strength	1. Rotating bending machine
B. Impact strength	2. Three-point loading machine
C. Bending strength	3. Universal testing machine
D. Fatigue strength	4. Izod testing machine

Codes:

	A	B	C	D
(a)	4	3	2	1
(b)	3	2	1	4
(c)	2	1	4	3
(d)	3	4	2	1

150.

List I	List II
(*Condition of beam*)	(*Bending moment diagram*)
A. Subject to bending moment at the end of cantilever	1. Triangle
B. Cantilever carrying uniformly distributed load over the whole length	2. Cubic parabola
C. Cantilever carrying linearly varying load from zero at the fixed end to maximum at the support	3. Parabola
D. A beam having load at the centre and supported at the ends	4. Rectangle

Codes :

	A	B	C	D
(a)	4	1	2	3
(b)	4	3	2	1
(c)	3	4	2	1
(d)	3	4	1	2

151.

List I	List II

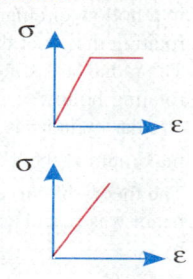

A. Rigid-perfectly

B. Elastic-perfectly

C. Rigid-strain hardening

D. Linearly elastic

Codes:

	A	B	C	D
(a)	3	1	4	2
(b)	1	3	2	4
(c)	3	1	2	4
(d)	1	3	4	2

152.

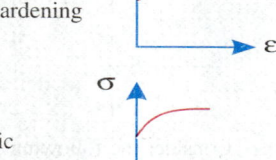

(Material properties)	(Tests to determine material properties)
A. Ductility	1. Impact test
B. Toughness	2. Fatigue test
C. Endurance limit	3. Tension test
D. Resistance to penetration	4. Hardness test

Codes :

	A	B	C	D
(a)	3	2	1	4
(b)	4	2	1	3
(c)	3	1	2	4
(d)	4	1	2	3

153.

List I	List II
A. The ... is the deformation produced by the stress	1. Cantilever
B. ...... is the property that enables the formation of a permanent deformation in a material.	2. Section modulus
C. A .... is a beam whose one end is fixed and the other end free.	3. Plasticity
D. The strength of the beam mainly depends on	4. Strain

Codes:

	A	B	C	D
(a)	4	3	1	2
(b)	4	3	2	1
(c)	1	2	3	4
(d)	2	3	4	1

154.

List I	List II
A. ..... loading induces direct and bending stresses at the section.	1. Middle third
B. The brick chimney is stable if the resultant thrust lies within the	2. Eccentric
C. If the bending moment is consistent there will be no	3. Zero
D. If a beam is fixed at both ends, then slope and deflection at both ends are	4. Shearing stress

Code:

	A	B	C	D
(a)	2	1	3	4
(b)	2	1	4	3
(c)	3	4	1	2
(d)	4	1	2	3

155.

List I	List II
A. Longitudinal stresses act ..... to the longitudinal axis of the shell.	1. Welding
B. In thick cylinders the ..... stress varies along the thickness.	2. Gauge distance
C. The ......provides very rigid joints.	3. Circumferential
D. The distance between the consecutive rivets is called........	4. Parallel

Codes :

	A	B	C	D
(a)	3	4	1	2
(b)	2	3	4	1
(c)	4	3	1	2
(d)	4	2	3	1

156.

	List I		List II
A..	The angle of twiset is directly proportional to the ...	1.	Quarter ellptical
B.	.....springs are called cantilever laminated springs.	2.	Twisting moment
C.	In case of a laminated spring the load at which the plates become straight is called .....load.	3.	Resilience
D.	The strain energy stored by the body within elastic limit when loaded externally is called......	4.	Proof

Codes:

	A	B	C	D
(a)	1	2	3	4
(b)	2	3	4	1
(c)	2	4	1	3
(d)	2	1	4	3

157.

	List I		List II
A.	Stress due to suddenly applied load is two times that of .... load.	1.	Secant formula
B.	A member of structure or bar which carries an axial compressive load is called....	2.	Maximum principal stress
C.	The .... is used for long columns under eccentric loading	3.	Strut
D.	.... theory is suitable for brittle materials	4.	Gradually applied

Codes :

	A	B	C	D
(a)	4	3	1	2
(b)	4	2	3	1
(c)	2	3	4	1
(d)	3	4	1	2

158.

	List I		List II
A.	In case of a solid rotating circular disc the radial stress is maximum at the	1.	Inner radius
B.	The circumferential stress in a hollow circular rotating disc is at the	2.	Centre
C.	The radial stress in a rotating hollow circular cylinder is maximum at the	3.	Winkler-Bach
D.	The theory of curved beam was postulated by.......	4.	Geometric mean radius

Codes:

	A	B	C	D
(a)	2	1	3	4
(b)	2	1	4	3
(c)	3	4	1	2
(d)	4	1	2	3

159. Consider the following statemnts:

State of stress at a point when completely specified, enables one to determine the
1. principal stresses at the point
2. maximum shearing stress at the point
3. stress components on any plane containing the point
Of these statements:
(a) 1, 2 and 3 are correct
(b) 1 and 3 are correct
(c) 2 and 3 are correct
(d) 1 and 2 are correct

*160. State of stress at a point in a strained body is shown in Fig 13. Which one of the figures given below represents correctly the Mohr's circle for the state of stress ?

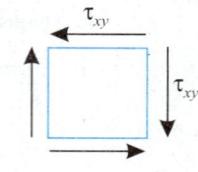

Fig. 13

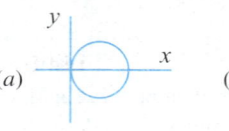 (a)

 (b)

 (c)

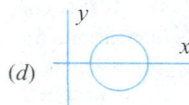 (d)

*161. The given figure shows the shear force diagram for the beam

ABCD. Bending moment in the portion *BC* of the beam

(*a*) is a non-zero constant

(*b*) is zero

(*c*) varies linearly from *B* to *C*

(*d*) varies parabolically from *B* to *C*.

162. The maximum bending moment in a simply supported beam of length *L* loaded by a concentrated load *W* at the mid-point is given by

(*a*) *WL* (*b*) $\dfrac{WL}{2}$

(*c*) $\dfrac{WL}{4}$ (*d*) $\dfrac{WL}{8}$.

*163. A beam, built-in at both ends, carries a uniformly distributed load over its entire span as shown in figure. 14. Which one of the diagrams given below, represents bending moment distribution along the length of the beam ?

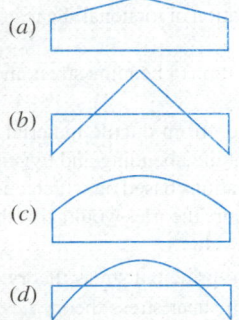

Fig. 14

(*a*)

(*b*)

(*c*)

(*d*)

*164. If the shear force acting at every section of a beam is of the same magnitude and of the same direction then it represents a

(*a*) simply supported beam with a concentrated load at the centre

(*b*) overhang beam having equal overhang at both supports and carrying equal concentrated loads acting in the same direction at the free ends

(*c*) cantilever subjected to concentrated load at the free end

(*d*) simply supported beam having concentrated loads of equal magnitude and in the same direction acting at equal distances from the supports.

*165. A cantilever beam carries a load *W* uniformly distributed over its entire length. If the same load is placed at the free end of the same cantilever, then the ratio of maximum deflection in the first case to that in the second case will be:

(*a*) $\dfrac{3}{8}$ (*b*) $\dfrac{8}{3}$

(*c*) $\dfrac{5}{8}$ (*d*) $\dfrac{8}{5}$.

166. The Fig. 15 shows a cantilever of span '*L*' subjected to a concentrated load '*P*' and a moment '*M*' at the free end. Deflection at the free end is given by:

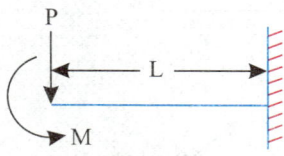

Fig.15

(*a*) $\dfrac{PL^2}{2\,EI} + \dfrac{ML^2}{3\,EI}$

(*b*) $\dfrac{ML^2}{2\,EI} + \dfrac{PL^3}{3\,EI}$

(*c*) $\dfrac{ML^2}{3\,EI} + \dfrac{PL^3}{2\,EI}$

(*d*) $\dfrac{ML^2}{2\,EI} + \dfrac{PL^3}{48\,EI}$

167. For a cantilever beam of length '*l*', flexural rigidity *EI* and loaded at its free end by a concentrated load *W*, match List I with List II and select the correct answer using the codes given below the lists :

List I	List II
A. Maximum bending moment	1. *Wl*
B. Strain energy	2. $\dfrac{Wl^2}{2\,EI}$
C. Maximum slope	3. $\dfrac{Wl^3}{3\,EI}$
D. Maximum deflection	4. $\dfrac{W^2 l^3}{6\,EI}$

Codes :

	A	B	C	D
(a)	1	4	3	2
(b)	1	4	2	3
(c)	4	2	1	3
(d)	4	3	1	2

*168. A hollow shaft is subjected to torsion. The shear stress variation in the shaft along the radius is given by:

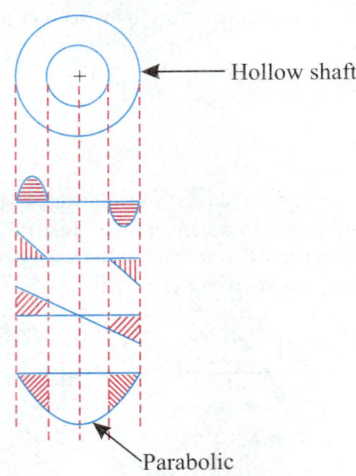

Hollow shaft

Parabolic

Fig. 16

169. The equivalent bending moment under combined action of bending moment M and torque T is:

(a) $\sqrt{M^2 + T^2}$

(b) $\dfrac{1}{2}\sqrt{M^2 + T^2}$

(c) $M + \sqrt{M^2 + T^2}$

(d) $\dfrac{1}{2}\left[M + \sqrt{M^2 + T^2}\right]$

170. The bending moment (M) is constant over a length segment (l) of a beam. The shearing force will also be constant over this length and is given by:

(a) $\dfrac{M}{l}$

(b) $\dfrac{M}{2l}$

(c) $\dfrac{M}{4l}$

(d) none of these.

*171. In a thick cylinder, subjected to internal and external presssures, let r_1 and r_2 be the inter-

nal and external radii respectively. Let u be the radial displacement of a material element at radius r, $r_2 \geqslant r \geqslant r_1$. Identifying the cylinder axis as z-axis, the radial strain component is

(a) $\dfrac{u}{r}$

(b) $\dfrac{u}{\theta}$

(c) $\dfrac{du}{dr}$

(d) $\dfrac{du}{d\theta}$.

172. Auto frettage is the method of

(a) joining thick cylinders

(b) calculating stresses in thick cylinders

(c) prestressing thick cylinders

(d) increasing the life of thick cylinders.

173. Given that, d = diameter of spring, R = mean radius of coils, n = number of coils and G = modulus of rigidity, the stiffness of the close-coiled helical spring subject to an axial load W is equal to:

(a) $\dfrac{Gd^4}{64R^3n}$

(b) $\dfrac{Gd^3}{64R^3n}$

(c) $\dfrac{Gd^4}{32R^3n}$

(d) $\dfrac{Gd^4}{64R^2n}$.

174. When a close-coiled helical spring is subjected to a couple about its axis, the stress induced in the wire material of the spring is

(a) bending stress only

(b) direct shear stress only

(c) a combination of torsional shear stress and bending

(d) a combination of bending stress and direct shear stress.

*175. If a shaft made from ductile material is subjected to combined bending and twisting moments, calculations based on which one of the following failure theories would give the most conservative value?

(a) Maximum principal stress theory

(b) Maximum shear stress theory

(c) Maximum strain energy theory

(d) Maximum distortion energy theory.

176. During tensile-testing of a specimen using a Universal testing machine, the parameters actually measured include:

(a) true stress and true strain

(b) Poisson s ratio and Young s modulus

(c) engineering stress and engineering strain

(d) load and elongation.

177. A bar of uniform cross-section of one sq. cm is subjected to a set of five forces as shown in Fig. 17, resulting in its equilibrium. The maximum tensile stress (in kgf/cm²) produced in the bar is

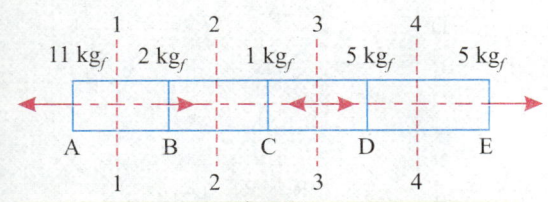

Fig. 17

(a) 1   (b) 2

(c) 10   (d) 11

178. A plane stressed element is subjected to the state of stress given by $\sigma_x = \tau_{xy} = 100$ kgf/cm² and $\sigma = 0$. Maximum shear stress in the element is equal to

(a) $50\sqrt{3}$ kgf/cm²   (b) 100 kgf/cm²

(c) $50\sqrt{5}$ kgf/cm²   (d) 150 kgf/cm².

179. Match List-I (Elastic properties of an isotropic elastic material) with List-II (Nature of strain produced) and select the correct answer using the codes given below the Lists :

List I

A. Young s modulus

B. Modulus of rigidity

C. Bulk modulus of rigidity

D. Poisson s ratio

List II

1. Shear strain

2. Normal strain

3. Transverse strain

4. Volumetric strain

Codes:

	A	B	C	D
(a)	1	2	3	4
(b)	2	1	3	4
(c)	2	1	4	3
(d)	1	2	4	3

180. The relationship between the Lame s constant λ, Young s modulus E and the Poisson s ratio $\pm$ is:

(a) $\lambda = \dfrac{E\mu}{(1+\mu)(1-2\mu)}$

(b) $\lambda = \dfrac{E\mu}{(1+2\mu)(1-\mu)}$

(c) $\lambda = \dfrac{E\mu}{(1+\mu)}$

(d) $\lambda = \dfrac{E\mu}{(1-\mu)}$.

181. A 10 cm long and 5 cm diameter steel rod fits snugly between two rigid walls 10 cm apart at room temperature. Young s modulus of elasticity and coefficient of linear expansion of steel are 2×10^6 kgf/cm² and $12 \times 10^{-6}/°C$ respectively. The stress developed in the rod due to a 100°C rise in temperature will be

(a) 6×10^{-10} kgf/cm²

(b) 6×10^{-9} kgf/cm²

(c) 2.4×10^3 kgf/cm²

(d) 2.4×10^4 kgf/cm².

182. A beam subjected to a load P is shown in the given Fig. 18. The bending moment at the support AA of the beam will be

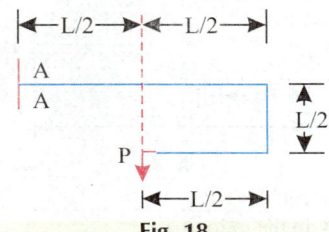

Fig. 18

(a) PL (b) $PL/2$

(c) $2PL$ (d) Zero.

183. A 2m long beam BC carries a single concentrated load at its mid-span and is simply supported at its ends by two cantilevers AB 1 m long and CD 2m long as shown in the Fig. 19.

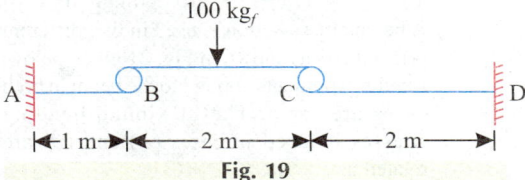

Fig. 19

The shear force at end A of the cantilever AB will be:

(a) Zero   (b) 40 kgf

(c) 50 kgf   (d) 60 kgf.

184. If a beam is subjected to a constant bending moment along its length then the shear force will

(a) also have a constant value everywhere along its length

(b) be zero at all section along the beam

(c) be maximum at the centre and zero at the ends

(d) zero at the centre and maximum at the ends.

185. A simply supported beam with width b and depth d carries a central load W and under -

goes deflection δ at the centre. If the width and depth are interchanged, the deflection at the centre of the beam would attain the value

(a) $\dfrac{d}{b}\delta$

(b) $\left(\dfrac{d}{b}\right)^2 \delta$

(c) $\left(\dfrac{d}{b}\right)^3 \delta$

(d) $\left(\dfrac{d}{b}\right)^4 \delta.$

186. A round shaft of diameter 'd' and length 'l' fixed at both ends 'A' and 'B', is subjected to a twisting moment T at C at a distance of $l/4$ from A (see Fig 20). The torsional stresses in the parts AC and CB will be:

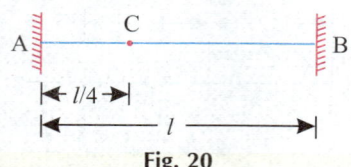

Fig. 20

(a) equal
(b) in the ratio of 1 : 3
(c) in the ratio of 3 : 1
(d) indeterminate.

187. A wooden beam of rectangular cross-section 10 cm deep by 5 cm wide carries maximum shear force of 2000 kgf. Shear stress at neutral axis of the beam section is

(a) zero   (b) 49 kgf/cm²
(c) 260 kgf/cm²   (d) 80 kgf/cm².

188. A beam cross-section is used in two different orientations as shown in Fig. 21 given below. Bending moments applied to the beam in both cases are same. The maximum bending stresses induced in cases (A) and (B) are related as

(a) $s_A = s_B$

(b) $s_A = 2s_B$

(c) $\sigma_A = \dfrac{\sigma_B}{2}$

(d) $\sigma_A = \dfrac{\sigma_B}{4}.$

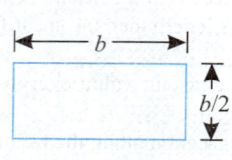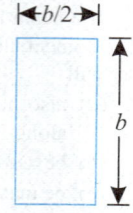

Fig. 21

189. The curve ABC is the Euler's curve for stability of column. The horizontal line DEF is

the strength limit. With reference to Fig. 22 Match List-I with List-II and select the correct answer using the codes given below the

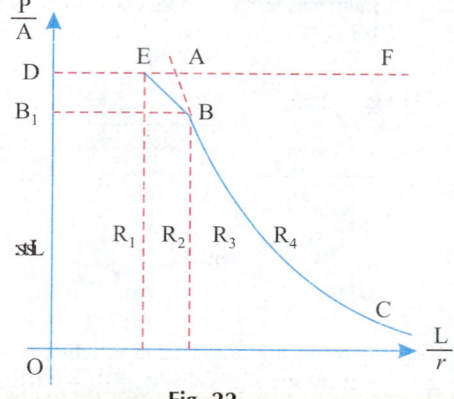

Fig. 22

List I	List II
(Regions)	(Column specifications)
A. R_1	1, Long, stable
B. R_2	2. Short
C. R_3	3. Medium
D. R_4	4. Long unstable

Codes:

	A	B	C	D
(a)	2	4	3	1
(b)	2	3	1	4
(c)	1	2	4	3
(d)	2	1	3	4

190. The ratio of the compressive critical load for a long column fixed at both the ends and a column with one end fixed and the other end free is:

(a) 1 : 2   (b) 1 : 4
(c) 1 : 8   (d) 1 : 16

*191. Maximum shear stress in a solid shaft of diameter D and length L twisted through an angle θ is τ. A hollow shaft of same material and length having outside and inside diameters of D and $D/2$ respectively is also twisted through the same angle of twist θ. The value of maximum shear stress in the hollow shaft will be:

(a) $\dfrac{16}{15}\tau$

(b) $\dfrac{8}{7}\tau$

(c) $\dfrac{4}{3}\tau$

(d) $\tau.$

192. From design point of view, spherical pressure vessels are preferred over cylindrical pressure vessels because they:
    (a) are cost effective in fabrication
    (b) have uniform higher circumferential stress
    (c) uniform lower circumferential stress
    (d) have a larger volume for the same quantity of material used.

*193. If two identical helical springs are connected in parallel and to these two, another identical spring is connected in series and the system is loaded by a weight W, as shown in the Fig. 23, then the resulting deflection will be given by (δ = deflection, S = stiffness, W = load)

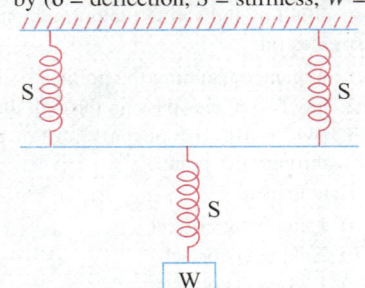

Fig. 23

(a) $\delta = \dfrac{3W}{2S}$ (b) $\delta = \dfrac{W}{2S}$

(c) $\delta = \dfrac{2W}{3S}$ (d) $\delta = \dfrac{W}{3S}$.

194. Which one of the following gives the correct expression for strain energy stored in a beam of length L and of uniform cross-section having moment of inertia I and subjected to constant bending moment M:

(a) $\dfrac{ML}{EI}$ (b) $\dfrac{ML}{2EI}$

(c) $\dfrac{M^2L}{EI}$ (d) $\dfrac{M^2L}{2EI}$.

195. Match List (Failure theories) with List-II (Fig. 24 representing boundaries of these theories) and select the correct answer using the codes given below the Lists:

List-I   **List II**

A. Max. principal stress theory

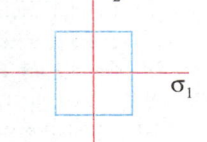

B. Max. shear stress theory

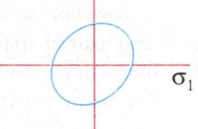

C. Max-octahedral shear stress theory

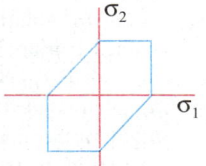

D. Max. shear strain energy theory

Fig. 24

Codes:

	A	B	C	D
(a)	2	1	3	4
(b)	2	4	3	1
(c)	4	2	3	1
(d)	2	4	1	3

196. Two metal plates of thickness 't' and width 'w' are jointed by a fillet weld of 45° as shown in the Fig. 25. When subjected to a pulling force 'F', the stress induced in the weld will be

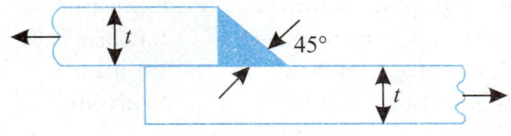

Fig. 25

(a) $\dfrac{F}{wt\sin 45°}$ (b) $\dfrac{F}{wt}$

(c) $\dfrac{F\sin 45°}{wt}$ (d) $\dfrac{2F}{wt}$.

197. Consider the following statements:
    A splined shaft is used for
    1. transmitting power
    2. holding a flywheel rigidly in position
    3. moving axially the gear wheels mounted on it
    4. mounting V-belt pulleys on it

(a) 2 and 3 are correct
(b) 1 and 4 are correct
(c) 2 and 4 are correct
(d) 1 and 3 are correct.

198. When two shafts are neither parallel nor intersecting, power can be transmitted by using

(a) a pair of spur gears
(b) a pair of helical gears
(c) an Oldham's coupling
(d) a pair of spiral gears.

199. In the assembly desing of shaft, pulley and key, the weakest member is:

(a) pulley   (b) key
(c) shaft   (d) none.

200. In the formulation of Lewis equation for toothed gearing, it is assumed that tangential tooth load, F_1 acts on the

(a) pitch point
(b) tip of the tooth
(c) root of the tooth
(d) whole face of the tooth.

201. For a two-dimensional state stress ($\sigma_1 > \sigma_2$, $\sigma_1 > 0$, $\sigma_2 < 0$) the designed values are most conservative if which one of the following failure theories were used:

(a) Maximum principal strain theory
(b) Maximum distortion energy theory
(c) Maximum shear stress theory
(d) Maximum principal stress theory.

202. Match List I with List II and select the correct answer using the codes given below the lists:

List I	List II
A. Single-plate friction clutch	1. Scooters
B. Multi-plate friction clutch	2. Rolling mills
C. Centrifugal clutch	3. Trucks
D. Jaw clutch	4. Mopeds

Codes:

	A	B	C	D
(a)	1	3	4	2
(b)	1	3	2	4
(c)	3	1	2	4
(d)	3	1	4	2

203. Which of the following stresses are associated with the design of pins in bushed pin type flexible coupling?

1. Bearing stress
2. Bending stress,
3. Axial tensile stress
4. Transverse shear stress.

Codes :

(a) 1, 3 and 4   (b) 2, 3 and 4
(c) 1, 2 and 3   (d) 1, 2 and 4

204. The state of plane stress at a point is described by $\sigma_x = \sigma_y = \sigma$ and $\tau_{xy} = 0$. The normal stress on the plane inclined at 45° to the x-plane will be:

(a) σ (b) $\sqrt{2}\sigma$
(c) $\sqrt{3}\sigma$ (d) 2σ.

205. Consider the following statements:
State of stress in two-dimensions at a point in a loaded component can be completely specified by indicating the normal and shear stresses on

1. a plane containing the point
2. any two planes passing through the point
3. two mutually perpendicular planes through the point.

Of these statements:

(a) 1 and 3 are correct
(b) 2 alone is correct
(c) 1 alone is correct
(d) 3 alone is correct.

206. For a composite bar consisting of a bar enclosed inside a tube of another material and when compressed under a load 'W' as a whole through rigid collars at the end of the bar, the equation of compability is given by (suffices 1 and 2 refer to bar and tube respectively)

(a) $W_1 + W_2 = W$
(b) $W_1 + W_2 = $ Constant
(c) $\dfrac{W_1}{A_1 E_1} = \dfrac{W_2}{A_2 E_2}$
(d) $\dfrac{W_1}{A_1 E_2} = \dfrac{W_2}{A_2 E_1}$.

207. A tapering bar (diameters of end sections being d_1 and d_2) and a bar of uniform cross-section 'd' have the same length and are subjected to the same axial pull. Both the bars will have the same extension if 'd' is equal to:

(a) $\dfrac{d_1 + d_2}{2}$ (b) $\sqrt{d_1 d_2}$

(c) $\sqrt{\dfrac{d_1 d_2}{2}}$ (d) $\dfrac{\sqrt{d_1 + d_2}}{2}$.

208. The number of independent elastic constants required to express the stress-strain relationship for a linearly elastic isotropic material is:

(a) one   (b) two
(c) three   (d) four.

209. The deformation of bar under its own weight as compared to that when subjected to a direct axial load equal to its own weight will be:

(a) the same   (b) one-fourth
(c) half   (d) double.

210. A slender bar of 100 mm² cross-section is subjected to loading as shown in the Fig. 26. If the modulus of elasticity is taken as 200 × 10⁹ Pa, then the elongation produced in the bar will be

(a) 10 mm   (b) 5 mm
(c) 1 mm   (d) nil.

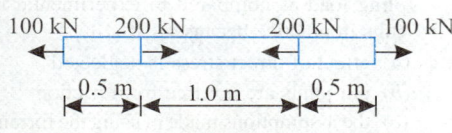

Fig. 26

211. For the beam shown in the figure 27, the elastic curve between the supports *B* and *C* will be

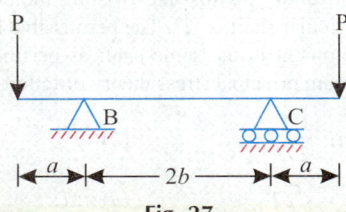

Fig. 27

(a) circular   (b) parabolic
(c) elliptic   (d) a straight line.

*212. A simply supported beam is loaded as shown in the Fig. 28. The maximum shear force in the beam will be:

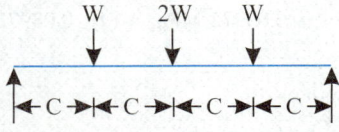

Fig. 28

(a) zero   (b) *W*
(c) 2W   (d) 4W.

*213. A lever is supported on two hinges at *A* and *C*. It carries a force of 3 kN as shown in the

Fig. 29. The bending moment at *B* will be:

(a) 3 kN-m   (b) 2 kN-m
(c) 1 kN-m   (d) zero.

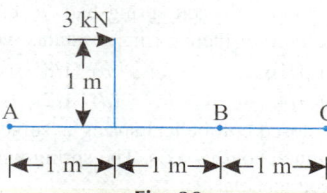

Fig. 29

214. Two hollow shafts of the same material have the same length and outside diameter. Shaft 1 has internal diameter equal to one-third of the outer diameter and shaft 2 has internal diameter equal to half of the outer diameter. If both the shafts are subjected to the same torque, the ratio of their twists $\dfrac{\theta_1}{\theta_2}$ will be equal to:

(a) 16/81   (b) 8/27
(c) 19/27   (d) 243/256.

215. A simply supported beam of constant flexural rigidity and length 2*L* carries a concentrated load '*P*' at its mid-span and the deflection under the load is δ. If a cantilever beam of the same flexural rigidity and length '*L*' is subjected to a load '*P*' at its free end, then the deflection at the free end will be:

(a) $\dfrac{1}{2}\delta$ (b) δ

(c) 2δ   (d) 4δ.

216. A solid shaft of diameter 100 mm, length 1000 mm is subjected to a twisting moment *T*. The maximum shear stress developed in the shaft is 60 N/mm². A hole of 50 mm diameter is now drilled throughout the length of the shaft. To develop a maximum shear stress of 60 N/mm² in the hollow shaft, the torque '*T*' must be reduced by

(a) T/4   (b) T/8
(c) T/12   (d) T/16.

217. A circular shaft is subjected to the combined action of bending, twisting and direct axial loading. The maximum bending stress σ, maximum shearing stress $\sqrt{3}\sigma$ and a uniform axial stress σ (compressive) are produced. The maximum compressive normal stress produced in the shaft will be:

(a) 3σ   (b) 2σ
(c) σ   (d) zero.

218. Two close-coiled springs are made from the small diameter wire, one wound on 2.5 cm diameter core and the other on 1.25 cm diameter core. If each spring had 'n' coils, then the ratio of their spring constants would be:

(a) 1/16   (b) 1/8

(c) 1/4   (d) 1/2.

219. A closed coil helical spring is subjected to a torque about its axis. The spring wire would experience a

(a) bending stress

(b) direct tensile stress of uniform intensity at its cross-section

(c) direct shear stress

(d) torsional shearing stress.

220. When a thin cylinder of diameter 'd' and thickness 't' is pressurized with an internal pressure of 'p' (1/m is the Poissin's ratio and E is the modulus of elasticity), then

(a) the circumferential strain will be equal to

$$\frac{pd}{2tE}\left(\frac{1}{2} - \frac{1}{m}\right)$$

(b) the longitudinal strain will be equal to

$$\frac{pd}{2tE}\left(1 - \frac{1}{2m}\right)$$

(c) the longitudinal stress will be equal to

$$\frac{pd}{2t}$$

(d) the ratio of the longitudinal strain to circumferential strain will be equal to

$$\frac{m - 2}{2m - 1}.$$

221. A thick-walled hollow cylinder having outside and inside radii of 90 mm and 40 mm respectively is subjected to an external pressure of 800 MN/m². The maximum circumferential stress in the cylinder will occur at a radius of

(a) 40 mm

(b) 60 mm

(c) 65 mm

(d) 90 mm.

222. In a thick cylinder pressurized from inside, the hoop stress is maximum at

(a) the centre of the wall thickness

(b) the outer radius

(c) the inner radius

(d) both the inner and the outer radii.

223. The Euler's crippling load for a 2 m long slender steel rod of uniform cross-section hinged at both the ends is 1 kN. The Euler's crippling load for a 1 m long steel rod of the same cross-section and hinged at both ends will be:

(a) 0.25 kN   (b) 0.5 kN

(c) 2kN   (d) 4kN.

224. For the state of stress of pure shear τ, the strain energy stored per unit volume in the elastic, homogeneous isotropic material having elastic constants E and v will be

(a) $\dfrac{\tau^2}{E}(1 + v)$ (b) $\dfrac{\tau^2}{2E}(1 + v)$

(c) $\dfrac{2\tau^2}{E}(1 + v)$ (d) $\dfrac{\tau^2}{2E}(2 + v)$.

225. Euler's formula gives 5 to10% error in crippling load as compared to experimental results in practice because

(a) effect of direct stress is neglected

(b) pin joints are not free from friction

(c) the assumptions made in using the formula are not met in practice

(d) the material does not behave in an ideal elastic way in tension and compression.

226. According to the maximum shear stress theory of failure permissible twisting moment in a circular shaft is 'T'. The permissible twisting moment in the same shaft as per the maximum principal stress theory of failure will be

(a) $\dfrac{1}{2}T$ (b) T

(c) $\sqrt{2}\,T$ (d) $2T$.

*227. $\alpha = 12.5 \times 10^{-6}/°C$, $E = 200$ GPa

If the rod fitted snugly between the supports as shown in the Fig. 30, is heated, the stress induced in it due to 20°C rise in temperature will be

(a) 0.07945 MPa   (b) – 0.07945 MPa

(c) – 0.03972 MPa   (d) 0.03972 MPa.

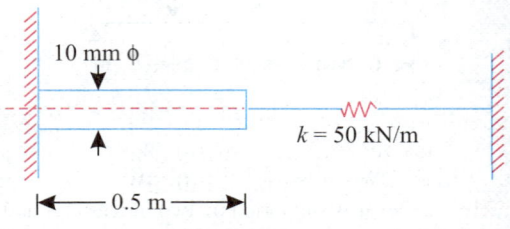

Fig. 30

*228. If permissible stress in plates of joint through a pin as shown in Fig. 31 is 200 MPa, then the width w will be:

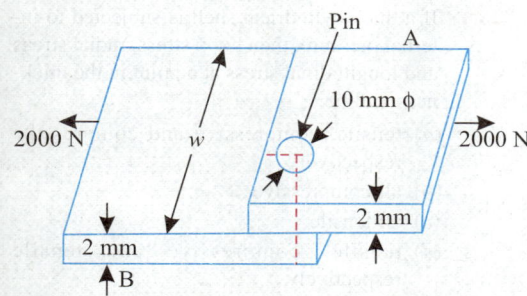

Fig. 31

    (a) 15 mm   (b) 18 mm
    (c) 20 mm   (d) 25 mm.

*229. Circumferential stress in a cylindrical steel boiler shell under internal pressure is 80 MPa, Young's modulus of elasticity and Poisson's ratio are respectively 2×10^5 MPa and 0.28. The magnitude of circumferential strain in the boiler will be

    (a) 3.44×10^{-4} (b) 3.84×10^{-4}
    (c) 4×10^{-4} (d) 4.56×10^{-4}

*230. A thin cylinder with closed lids is subjected to internal pressure and supported at the ends as shown in the Fig. 32.

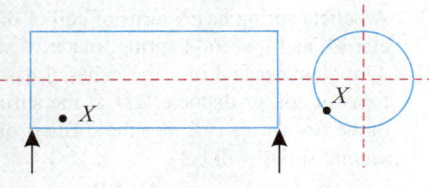

Fig. 32

The state of stress at point X is as represented in

(a)

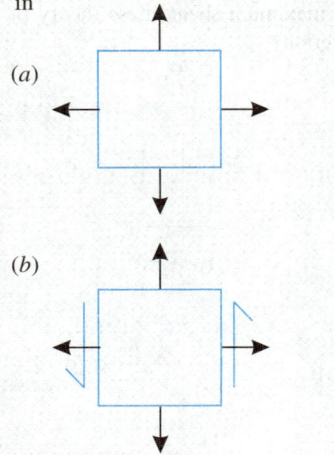

(b)

(c)

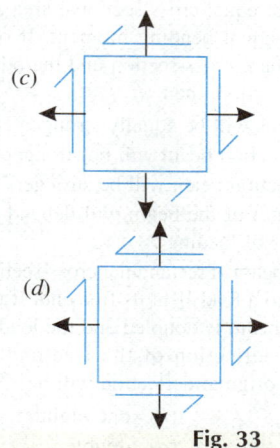

(d)

Fig. 33

*231. The number of elastic constants for a completely anisotropic elastic material which follows Hooke's law is:

    (a) 3   (b) 4
    (c) 21   (d) 25.

*232. The bending moment equation, as a function of distance x measured from the left end, for a simply supported beam of span Lm carrying a uniformly distributed load of intensity w N/m will be given by:

    (a) $M = \dfrac{wL}{2}(L - x) - \dfrac{w}{2}(L - x)^2$ N-m

    (b) $M = \dfrac{wL}{2}x - \dfrac{wx^2}{2}$ N-m

    (c) $M = \dfrac{wL}{2}(L - x)^2 - \dfrac{w}{2}(L - x)^3$ N-m

    (d) $M = \dfrac{wx^2}{2} - \dfrac{wLx}{2}$ N-m.

*233. The bending moment diagram shown in Fig. 34 corresponds to the shear force diagram in

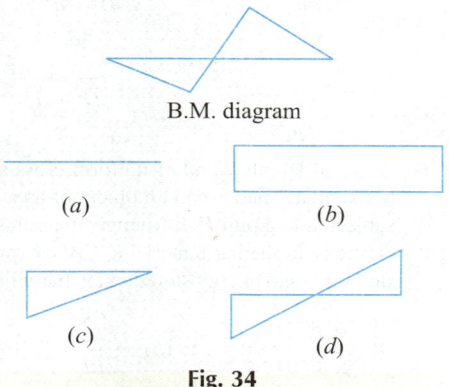

B.M. diagram

(a)

(b)

(c)

(d)

Fig. 34

*234. Two beams of equal cross-sectional area are subjected to equal bending moment. If one beam has square cross-section and the other has circular section, then

(*a*) both beams will be equally strong

(*b*) circular section beam will be stronger

(*c*) square section beam will be stronger

(*d*) the strength of the beam will depend on the nature of loading .

*235. A cantilever beam of rectangular cross-section is subjected to a load *W* at its free end. If the depth of the beam is doubled and the load is halved, the deflection of the free end as compared to original deflection will be:

(*a*) half　　　　　(*b*) one-eighth

(*c*) one-sixteenth　(*d*) double.

*236. Which one of the following portions of the loaded beam shown in the given figure is subjected to pure bending ?

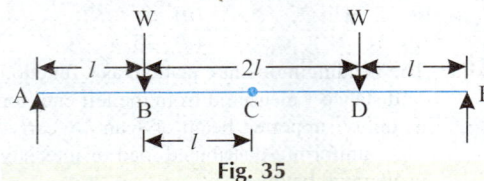

Fig. 35

(*a*) *AB*　　　　(*b*) *DE*

(*c*) *AE*　　　　(*d*) *BD*

*237. A solid circular shaft is subjected to pure torsion. The ratio of maximum shear to maximum normal stress at any point would be

(*a*) 1 : 1　　　(*b*) 1 : 2

(*c*) 2 : 1　　　(*d*) 2 : 3.

238. A short column of external diameter *D* and internal diameter *d* carries an eccentric load *W*. The greatest eccentricity which the load can have without producing tension on the cross-section of the column would be:

(*a*) $\dfrac{D+d}{8}$　　　(*b*) $\dfrac{D^2+d^2}{8d}$

(*c*) $\dfrac{D^2+d^2}{8D}$　　(*d*) $\sqrt{\dfrac{D^2+d^2}{8}}$.

*239. A bar of length *L* and of uniform cross-sectional area *A* and second moment of area *I* is subjected to a pull *P*. If Young's modulus of elasticity of the bar material is *E*, the expression for strain energy stored in the bar will be

(*a*) $\dfrac{P^2L}{2AE}$　　　(*b*) $\dfrac{PL^2}{2EI}$

(*c*) $\dfrac{PL^2}{AE}$　　　(*d*) $\dfrac{P^2L}{AE}$.

*240. If a thick cylindrical shell is subjected to internal pressure, then hoop stress, radial streas and longitudinal stress at a point in the thickness will be

(*a*) tensile, compressive and compressive respectively

(*b*) all compressive

(*c*) all tensile

(*d*) tensile, compressive and tensile respectively.

*241. A thin cylinder with both ends closed is subjected to internal pressure *p*. The longitudinal stress at the surface has been calculated as σ_0. Maximum shear stress at the surface will be equal to

(*a*) $2\sigma_0$　　　(*b*) $1.5\,\sigma_0$

(*c*) σ_0　　　(*d*) $0.5\,\sigma_0$.

242. The maximum shear stress occurs on the outermost fibres of a circular shaft under torsion. In a close coiled helical spring, the maximum shear stress occurs on the

(*a*) outermost fibres

(*b*) fibres at mean diameter

(*c*) innermost fibres

(*d*) end coils.

*243. A helical spring has *N* turns of coil of diameter *D*, and a second spring, made of same wire diameter and of same material has *N*/2 turns of coil of diameter 2*D*. If the stiffness of the first spring is *k*, then the stiffness of the second spring will be

(*a*) *k*/4　　　(*b*) *k*/2

(*c*) 2*k*　　　(*d*) 4*k*.

244. Which one of the following figures represents the maximum shear stress theory of Tresca criterion?

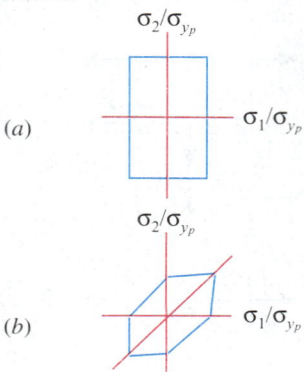

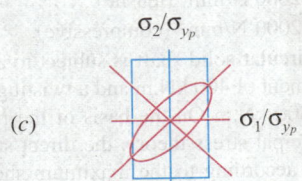

(c)

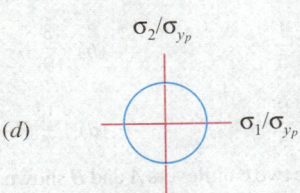

(d)

Fig. 36

245. Match List I (*End conditions of columns*) with List II (*Equivalent length in terms of length of hinged column*) and select the correct answer using the codes given below the Lists:

List I	List II
A. Both ends hinged	1. L
B. One end fixed and other end free	2. $\sqrt{2}$ L
C. One end fixed and the other pin-joined	3. L/2
D. Both ends fixed	4. 2L

Codes :

	A	B	C	D
(a)	1	3	4	2
(b)	1	3	2	4
(c)	3	1	2	4
(d)	3	1	4	2

246. Match List I with List II and select the correct answer using the codes given below the Lists :

List I	List II
A. Bending moment is constant	1. Point of contraflexure
B. Bending moment is maximum or minimum	2. Shear force changes sign
C. Bending moment is zero	3. Slope of shear force diagram is zero over the portion of the beam
D. Loading is constant	4. Shear force is zero over the portion of the beam

Codes:

	A	B	C	D
(a)	4	1	2	3
(b)	3	2	1	4
(c)	4	2	1	3
(d)	3	1	2	4

247. A load beam is shown in the Fig. 37. The bending moment diagram of the beam is best represented as

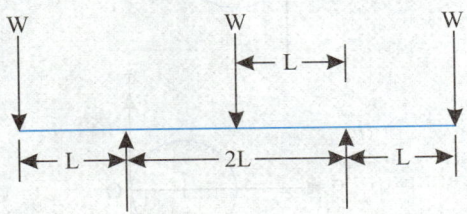

Fig. 37

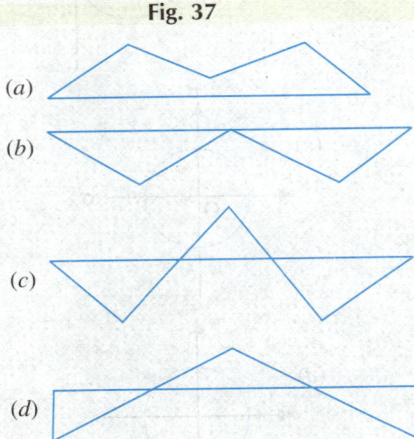

Fig. 38

248. At a certain section at a distance 'x' from one of the supports of a simply supported beam the intensity of loading, bending moment and shear force are W_x, M_x, and V_x respectively. If the intensity of loading is varying continuously along the length of the beam, then the *invalid* relation is:

(a) Slope $Q_x = \dfrac{M_x}{V_x}$

(b) $V_x = \dfrac{dM_x}{dx}$

(c) $W_x = \dfrac{d^2 M_x}{dx^2}$

(d) $W_x = \dfrac{dV_x}{dx}$.

249. Plane stress at a point in a body is defined by principal stresses 3σ and σ. The ratio of the normal stress to the maximum shear stress on the plane of maximum shear stress is

(a) 1   (b) 2

(c) 3   (d) 4.

250. Which one of the following Mohr's circles Fig. 39 represents the state of pure shear ?

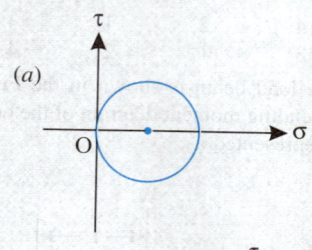

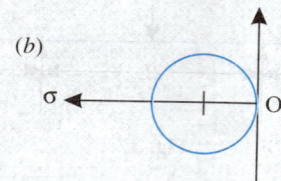

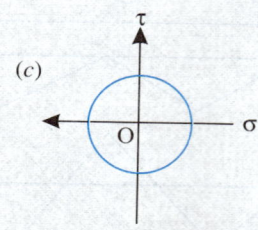

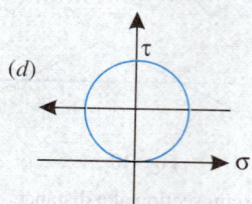

Fig. 39

251. The state of plane stress in a plate of 100 mm thickness is given as:

$\sigma_{xx} = 100$ N/mm^2; $\sigma_{yy} = 200$ N/mm^2; Young's modulus = 300 N/mm^2; Poisson's ratio = 0.3. The stress developed in the direction of thickness is

(a) zero   (b) 90 N/mm^2

(c) 100 N/mm^2 (d) 200 N/mm^2

252. A rod of material with $E = 200 \times 10^3$ MPa and $\alpha = 10^{-3}$ mm/mm°C is fixed at both the ends. It is uniformly heated such that the increase in temperature is 30°C. The stress developed in the rod is

(a) 6000 N/mm^2 (tensile)

(b) 6000 N/mm^2 (compressive)

(c) 2000 N/mm^2 (tensile)

(d) 2000 N/mm^2 (compressive).

253. A circular solid shaft is subject to a bending moment of 400 kN.m and a twisting moment of 300 kN.m On the basis of the maximum principal stress theory, the direct stress is σ and according to the maximum shear stress theory, the shear stress is τ. The ratio σ/τ is

(a) $\dfrac{1}{5}$ (b) $\dfrac{3}{9}$

(c) $\dfrac{9}{5}$ (d) $\dfrac{11}{6}$.

254. The two cantilevers A and B shown in the Fig. 40 have the same uniform cross-section and the same material. Free end deflection of cantilever 'A' is δ. The value of mid-span deflection of the cantilever 'B' is

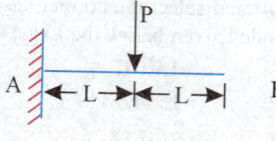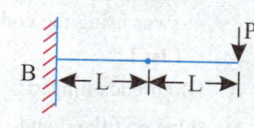

Fig. 40

(a) $\dfrac{1}{2}\delta$ (b) $\dfrac{2}{3}\delta$

(c) δ   (d) 2δ.

255. A link is under a pull which lies on one of the faces as shown in the figure 40. The magnitude of maximum compressive stress in the link would be:

(a) 21.3 N/mm^2 (b) 16.0 N/mm^2

(c) 10.7 N/mm^2 (d) zero.

256. Two coiled springs, each having stiffness K, are placed in parallel. The stiffness of the combination will be:

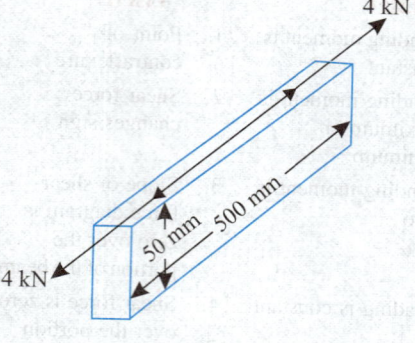

Fig. 41

(a) $4K$ (b) $2K$

(c) $\dfrac{K}{2}$ (d) $\dfrac{K}{4}$.

257. A long slender bar having uniform rectangular cross-section '$B \times H$' is acted upon by an axial compressive force. The sides B and H are parallel to x-and y-axes respectively. The ends of the bar are fixed such that they behave as pin-joined when the bar buckles in a plane normal to x- and they behave as built-in when the bar buckles in a plane normal to y-axis. If load capacity in either mode of buckling is same, then the value of H/B will be:

(a) 2   (b) 4

(c) 8   (d) 16.

258. The property by which an amount of energy is absorbed by material without plastic determation, is called

(a) toughness   (b) impact strength

(c) ductility   (d) resilience.

259. When a weight of 100 N falls on a spring of stiffness 1 kN/m from a height of 2m, the deflection caused in the first fall is:

(a) equal to 0.1 m

(b) between 0.1 and 0.2 m

(c) equal to 0.2 m

(d) more than 0.2 m.

260. Which one of the following features improves the fatigue strength of a metallic material?

(a) Increasing the temperature

(b) Scratching the surface

(c) Overstressing

(d) Understressing.

261. Cermets are

(a) metals for high temperature use with ceramic like properties

(b) ceramics with metallic strength and lustre

(c) coated tool materials

(d) metal-cermaic composites.

262. Percentage of various alloying elements present in different steel materials are given below:

1. 18% W; 4% Cr; 1% V; 5% Co; 0.7% C

2. 8% Mo; 4% Cr; 2% V; 6% W; 0.7% C

3. 27% Cr; 3% Ni; 5% Mo; 0.25% C

4. 18% Cr; 8% Ni; 0.15% C

Which of these relate to that of high speed steel ?

(a) 1 and 3   (b) 1 and 2

(c) 2 and 3   (d) 2 and 4.

263. A thin cylinder contains fluid at a pressure of 500 N/m², the internal diameter of the shell is 0.6 m and the tensile stress in the material is to be limited to 9000 N/m². The shell must have a minimum wall thickness of nearly

(a) 9 mm   (b) 11 mm

(c) 17 mm   (d) 21 mm.

264. From a tension test, the yield strength of steel is found to be 200 N/mm². Using a factory of safety of 2 and applying maximum principal stress theory of failure the permissible stress in the steel shaft subjected to torque will be

(a) 50 N/mm²   (b) 57.7 N/mm²

(c) 86.6 N/mm²   (d) 100 N/mm².

265. Which one of the following properties is more sensitive to increase in strain rate?

(a) Yield strength

(b) Proportional limit

(c) Elastic limit

(d) Tensile strength.

266. Two identical spring labelled as 1 and 2 are arranged in series and subjected to force F as shown in Fig. 42.

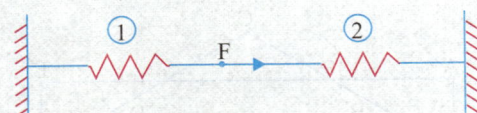

Fig. 42

Assume that each spring constant is K. The strain energy stored in spring 1 is:

(a) $\dfrac{F^2}{2K}$ (b) $\dfrac{F^2}{4K}$

(c) $\dfrac{F^2}{8K}$ (d) $\dfrac{F^2}{16K}$.

267. A rod having cross-sectional area $100 \times 10^{-6}\,\text{m}^2$ is subjected to a tensile load. Based on the Tresca failure criterion, if the uniaxial yield stress of the material is 200 MPa, the failure load is

(a) 10 kN   (b) 20 kN

(c) 100 kN   (d) 200 kN.

268. If diameter of a long column is reduced by 20%, the percentage of reduction in Euler buckling load is:

(a) 4   (b) 36

(c) 49   (d) 59.

269. With one fixed end and other free end, a column of length L buckles at load P_1. Another column of same length and same cross-section fixed at both ends buckles at load P_2. The ratio of P_2/P_1 is

(a) 1   (b) 2

(c) 4   (d) 16.

270. In a two-dimensional problem. the state of pure shear at a point is characterized by

(a) $\varepsilon_x = \varepsilon_y$ and $\gamma_{xy} = 0$

(b) $\varepsilon_x = -\varepsilon_y$ and $\gamma_{xy} \neq 0$

(c) $\varepsilon_x = 2\varepsilon_y$ and $\gamma_{xy} \neq 0$

(d) $\varepsilon_x = 0.5\,\varepsilon_y$ and $\gamma_{xy} \neq 0$.

271. The principal stresses σ_1, σ_2 and σ_3 at a point respectively are 80 MPa, 30 MPa and −40 MPa. The maximum shear stress is

(a) 25 MPa   (b) 35 MPa

(c) 55 MPa   (d) 60 MPa.

272. The Poisson's ratio of a material which has Young's modulus of 120 GPa and shear modulus of 50 GPa is

(a) 0.1   (b) 0.2

(c) 0.3   (d) 0.4.

273. Bending moment distribution in a built beam is shown in the given figure.

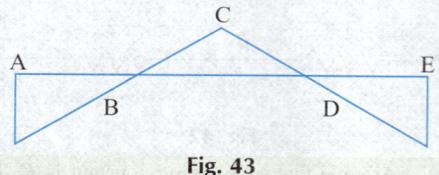

Fig. 43

The shear force distribution in the beam is represented by:

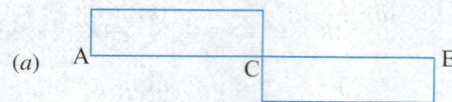

(a)

(b)

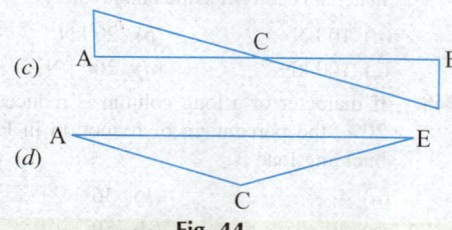

(c)

(d)

Fig. 44

274. A thick cylinder is subjected to internal pressure of 100 N/mm². If hoop stress developed at the outer radius of the cylinder is 100 N/mm², the hoop stress developed at the inner radius is

(a) 100 N/mm²

(b) 200 N/mm²

(c) 300 N/mm²

(d) 400 N/mm².

275. The outside diameter of a hollow shaft is twice that of its inside diameter. The torque-carrying capacity of this shaft is M_{t_1}. A solid shaft of the same material has the diameter equal to the outside diameter of the hollow shaft. The solid shaft can carry a torque of M_{t_2}. The ratio M_{t_1}/M_{t_2} is

(a) $\dfrac{15}{16}$ (b) $\dfrac{3}{4}$

(c) $\dfrac{1}{2}$ (d) $\dfrac{1}{16}$.

276. A body having weight of 1000 N is dropped from a height of 10 cm over a close-coiled helical spring of stiffness 200 N/cm. The resulting deflection of spring is nearly.

(a) 5 cm

(b) 16 cm

(c) 35 cm

(d) 100 cm.

277. The diameter of shaft A is twice the diameter of shaft B and both are made of the same material. Assuming both the shafts to rotate at he same speed. the maximum power transmitted by B is:

(a) the same as that of A

(b) half of A

(c) 1/8th of A

(d) 1/4th of A.

278. The given figure (all dimensions are in mm) shows an I-section of the beam.

The shear stress at point P (very close to the bottom of the flange) is 12 MPa. The stress at point Q in the web (very close to the flange) is

(a) indeterminable due to incomplete data

(b) 60 MPa

(c) 18 MPa

(d) 12 MPa.

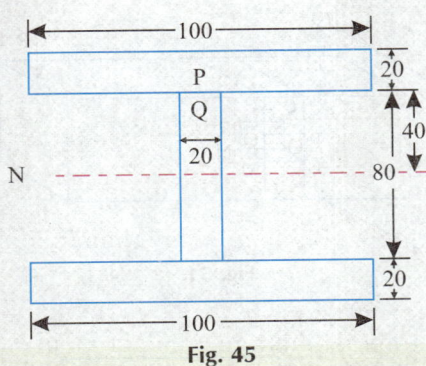

Fig. 45

279. A close-coiled helical spring is made of 5 mm diameter wire coiled to 50 mm mean diameter. Maximum shear stress in the spring under the action of an axial force is 20 N/mm². The maximum shear stress in a spring made of 3 mm diameter wire coiled to 30 mm mean diameter, under the action of the same force will be nearly

(a) 20 N/mm² (b) 33.3 N/mm²
(c) 55.6 N/mm² (d) 92.6 N/mm².

280. A horizontal beam carrying uniformly distributed load is supported with equal overhangs as shown in the Fig. 46.

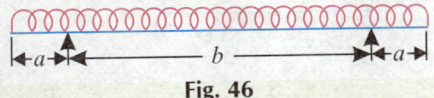

Fig. 46

The resultant bending moment at the mid-span shell be zero if a/b is

(a) 3/4 (b) 2/3
(c) 1/2 (d) 1/3.

281. A short column of symmetric cross-section made of a brittle material is subjected to an eccentric vertical load P at an eccentricity e. To avoid tensile stress in the short column, the eccentricity e should be less than or equal to:

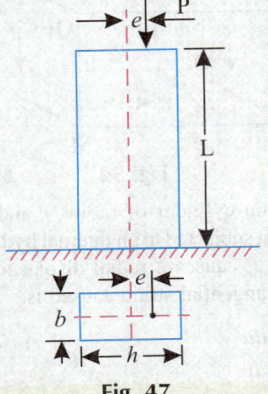

Fig. 47

(a) $h/12$ (b) $h/6$
(c) $h/3$ (d) $h/2$.

282. A thin cylindrical shell is subjected to internal pressure p. The Poisson's ratio of the material of the shell is 0.3. Due to internal pressure, the shell is subjected to circumferential strain and axial strain. The ratio of circumferential strain to axial strain is:

(a) 0.425 (b) 2.25
(c) 0.225 (d) 4.25.

283. A cantilever of length L, moment of inertia I, Young's modulus E carries a concentrated load W at the middle of its length. The slope of cantilever at the free end is

(a) $\dfrac{WL^2}{2EI}$ (b) $\dfrac{WL^2}{4EI}$

(c) $\dfrac{WL^2}{8EI}$ (d) $\dfrac{WL^2}{16EI}$.

284. For a given material, the modulus of rigidity is 100 GPa an Poisson's ratio is 0.25. The value of modulus of elasticity in GPa is

(a) 125 (b) 150
(c) 200 (d) 250.

285. A rigid beam of negligible weight is supported in a horizontal position by two rods of steel and aluminium, 2 m and 1 m long having values of cross-sctional areas 1 cm² and 2 cm² and E of 200 GPa and 100 GPa respectively. A load P is applied as shown in the figure.

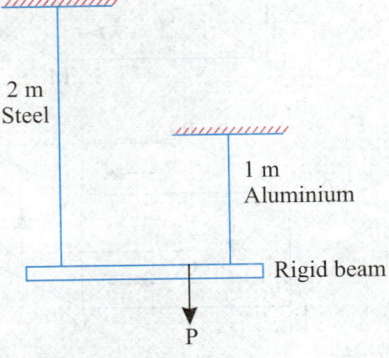

Fig. 48

If the rigid beam is to remain horizontal, then
(a) the force on both rods should be equal
(b) the force on aluminium rod should be twice the force on steel

    (c) the force on the steel rod should be twice the force on aluminium

    (d) the force P must be applied at the centre of the beam.

286.   A shaft is subjected to torsion as shown.

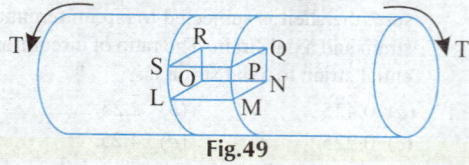

Fig.49

Which of the following figures represents the shear stress on the element LMNOPQRS?

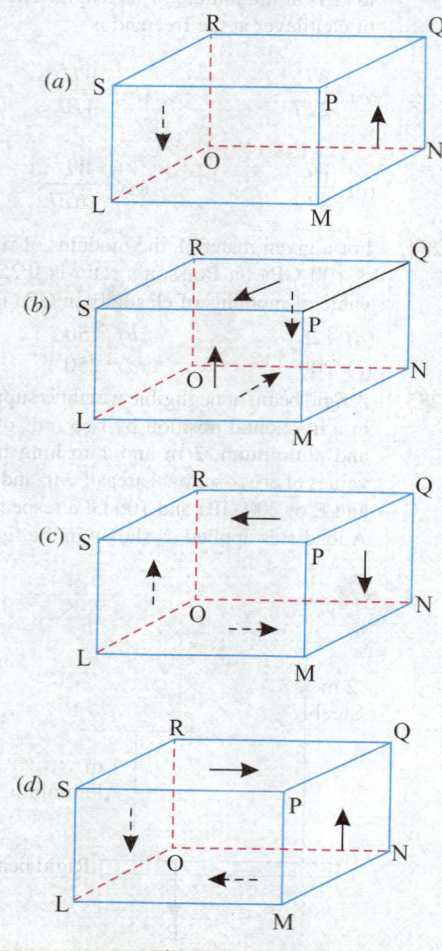

Fig. 50

287.   A cantilever is loaded by a concentrated load P at the free end as shown (Fig.51). The shear stress in the element LMNOPQRS is under consideration. Which of the following figures represents the shear stress directions in the cantilever?

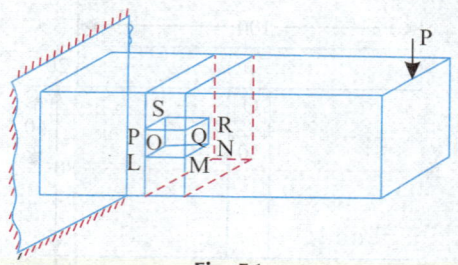

Fig. 51

(a)

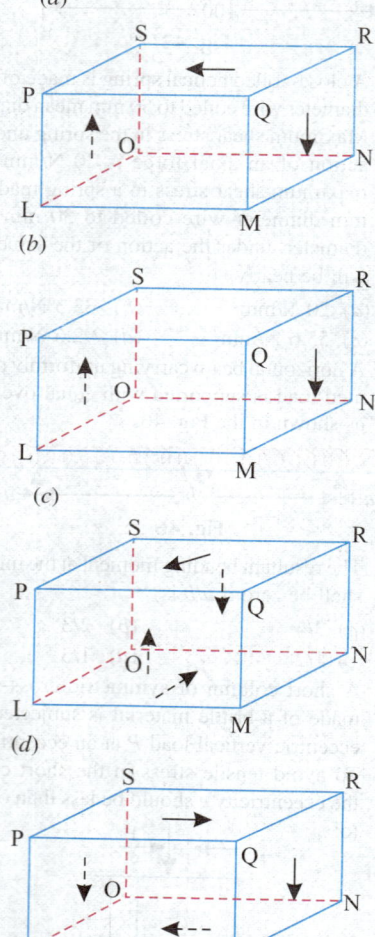

Fig. 52

288.   A thin cylinder of radius r and thickness t when subjected to an internal hydrostatic pressure P causes a radial displacement u, then the tangential strain caused is

(a) $\dfrac{du}{dr}$ (b) $\dfrac{1}{r} \cdot \dfrac{du}{dt}$

(c) $\dfrac{u}{r}$ (d) $\dfrac{2u}{r}$.

*289.

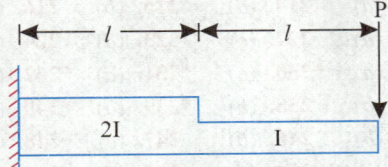

Fig. 53

$I = 375 \times 10^{-6}$ m^4

$l = 0.5$ m

$E = 200$ GPa

Determine the stiffness of the beam shown in the above Fig.53.

(a) 12×10^{10} N/m   (b) 10×10^{10} N/m

(c) 4×10^{10} N/m   (d) 8×10^{10} N/m.

290. Strain energy stored in a body of volume V subjected to uniform stress s is

(a) $\dfrac{sE}{V}$ (b) $\dfrac{sE^2}{V}$

(c) $\dfrac{sV^2}{E}$ (d) $\dfrac{s^2V}{2E}$

*291. A thick open ended cylinder as shown in the Fig. 54, is made of a material with permissible normal and shear stresses 200 MPa and 100 MPa respectively. The ratio of permissible pressure based on the normal and shear stress is:

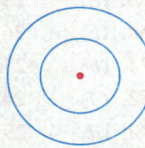

$d_i = 10$ cm

$d_o = 20$ cm

Fig. 54

(a) 9/5   (b) 8/5

(c) 7/5   (d) 4/5

ANSWERS

1. (d)	**2.** (b)	**3.** (b)	**4.** (b)	**5.** (a)	**6.** (a)	**7.** (b)	**8.** (b)
9. (a)	**10.** (a)	**11.** (a)	**12.** (d)	**13.** (c)	**14.** (d)	**15.** (b)	**16.** (c)
17. (d)	**18.** (a)	**19.** (a)	**20.** (d)	**21.** (c)	**22.** (c)	**23.** (b)	**24.** (c)
25. (b)	**26.** (c)	**27.** (a)	**28.** (b)	**29.** (c)	**30.** (c)	**31.** (c)	**32.** (b)
33. (d)	**34.** (d)	**35.** (b)	**36.** (a)	**37.** (c)	**38.** (a)	**39.** (b)	**40.** (b)
41. (a)	**42.** (a)	**43.** (b)	**44.** (b)	**45.** (b)	**46.** (c)	**47.** (a)	**48.** (b)
49. (c)	**50.** (b)	**51.** (b)	**52.** (b)	**53.** (c)	**54.** (d)	**55.** (a)	**56.** (b)
57. (a)	**58.** (a)	**59.** (b)	**60.** (a)	**61.** (c)	**62.** (a)	**63.** (a)	**64.** (a)
65. (c)	**66.** (c)	**67.** (a)	**68.** (b)	**69.** (c)	**70.** (b)	**71.** (c)	**72.** (b)
73. (c)	**74.** (b)	**75.** (a)	**76.** (b)	**77.** (b)	**78.** (c)	**79.** (b)	**80.** (a)
81. (c)	**82.** (b)	**83.** (c)	**84.** (a)	**85.** (b)	**86.** (a)	**87.** (b)	**88.** (c)
89. (b)	**90.** (d)	**91.** (d)	**92.** (a)	**93.** (a)	**94.** (c)	**95.** (c)	**96.** (b)
97. (a)	**98.** (d)	**99.** (d)	**100.** (d)	**101.** (a)	**102.** (b)	**103.** (a)	**104.** (a)
105. (a)	**106.** (d)	**107.** (b)	**108.** (a)	**109.** (b)	**110.** (b)	**111.** (a)	***112.** (c)
113. (a)	***114.** (a)	***115.** (c)	**116.** (b)	**117.** (c)	***118.** (c)	***119.** (c)	***120.** (c)
*121. (a)	***122.** (d)	***123.** (c)	**124.** (c)	***125.** (a)	***126.** (b)	**127.** (c)	**128.** (c)
*129. (a)	***130.** (d)	**131.** (c)	***132.** (b)	**133.** (a)	***134.** (d)	***135.** (d)	**136.** (a)
137. (d)	**138.** (c)	***139.** (b)	***140.** (c)	***141.** (b)	***142.** (b)	**143.** (b)	***144.** (d)
145. (d)	***146.** (b)	**147.** (a)	**148.** (a)	***149.** (d)	**150.** (b)	**151.** (a)	**152.** (c)
153. (a)	**154.** (b)	**155.** (c)	**156.** (d)	**157.** (a)	**158.** (b)	**159.** (a)	***160.** (c)
161. (a)	**162.** (c)	***163.** (d)	***164.** (c)	***165.** (a)	**166.** (b)	**167.** (b)	***168.** (c)
169. (d)	**170.** (d)	**171.** (c)	**172.** (c)	**173.** (a)	**174.** (a)	***175.** (d)	**176.** (d)
177. (d)	**178.** (c)	**179.** (c)	**180.** (c)	**181.** (c)	**182.** (c)	**183.** (c)	**184.** (b)

185. (a)	186. (c)	187. (a)	188. (d)	189. (b)	190. (d)	*191. (a)	192. (d)
*193. (a)	194. (d)	195. (d)	196. (a)	197. (a)	198. (d)	199. (b)	200. (a)
201. (d)	202. (b)	203. (d)	204. (a)	205. (d)	206. (c)	207. (b)	208. (a)
209. (c)	210. (d)	211. (b)	*212. (c)	*213. (d)	214. (d)	215. (b)	216. (d)
217. (d)	218. (b)	219. (a)	220. (b)	221. (d)	222. (c)	223. (d)	224. (c)
225. (b)	226. (b)	*227. (a)	*228. (a)	*229. (a)	*230. (a)	*231. (a)	*232. (b)
*233. (b)	*234. (b)	*235. (c)	*236. (d)	*237. (b)	238. (b)	*239. (a)	*240. (d)
241. (d)	242. (c)	*243. (a)	244. (b)	245. (b)	246. (b)	247. (c)	248. (d)
249. (b)	250. (c)	251. (a)	252. (b)	253. (c)	254. (d)	255. (d)	256. (b)
257. (a)	258. (d)	259. (b)	260. (d)	261. (c)	262. (b)	263. (c)	264. (d)
265. (b)	266. (b)	267. (b)	268. (d)	269. (d)	270. (b)	271. (a)	272. (b)
273. (a)	274. (b)	275. (a)	276. (b)	277. (c)	278. (d)	279. (c)	280. (c)
281. (b)	282. (d)	283. (a)	284. (d)	285. (b)	286. (b)	287. (b)	288. (a)
289. (a)	290. (d)	291. (b)					

SOLUTIONS-COMMENTS

*112. Stiffness of spring $k = \dfrac{Cd^4}{64R^3n}$

where,

C = Modulus of rigidity (same in both cases),

d = Diameter of wire (same in both cases),

n = Number of coils (8 and 10 respectively), and

R = Mean radius of coil $\left(\dfrac{75}{2}\text{mm and }\dfrac{60}{2}\text{mm respectively}\right)$.

∴ $k_1 \, \alpha \, \dfrac{1}{\left(\dfrac{75}{2}\right)^3 \times 8}$ and $k_2 \propto \dfrac{1}{\left(\dfrac{60}{2}\right)^3 \times 10}$

∴ $\dfrac{k_2}{k_1} = \left(\dfrac{75}{60}\right)^3 \times \dfrac{8}{10} = 1.56$ or $k_2 = \textbf{1.56 k1}$ **(Ans.)**

Thus, (c) is the correct choice.

*115. Centrifugal force $= \dfrac{\text{Weight}}{g} \times \omega^2 \times \text{radius} = \dfrac{w/A}{g} \times \omega^2 \times l = \dfrac{w\omega^2 l^2 A}{g}$

∴ Stress due to this force $= \dfrac{w\omega^2 l^2 A}{g \times A} = \dfrac{w\omega^2 l^2}{g}$

Also, $E = \dfrac{\text{Stress}}{\text{Strain}}$ or $\text{Strain} = \dfrac{\text{Stress}}{E}$ or $\dfrac{\delta l}{l} = \dfrac{\text{Stress}}{E}$

or, δl (extension) $= l \times \dfrac{\text{Stress}}{E} = l \times \dfrac{w\omega^2 l^2}{gE} = \dfrac{w\omega^2 l^3}{gE}$ **(Ans.)**

*118. Steel is highly elastic because it undergoes least deformation when loaded, and it regains its original shape/form when the load is removed.

*119. The maximum shear stress

$$= \sqrt{\left(\dfrac{\sigma_y - \sigma_x}{2}\right)^2 + \tau_{xy}^2} = \sqrt{\left[\dfrac{40 - (-40)}{2}\right]^2 + 30^2} = 50\,\text{MPa}$$ **(Ans.)**

Thus, (c) is the corrrect choice.

*120. The number of elastic constants required to relate stress and strain, for a linearly elastic, isotropic and homogeneous material, is four (viz. E, C, K and v).

*121. Bending stress, $\sigma = \dfrac{M}{Z}$, where M = bending moment, and Z = section modulus.

For rectangular beam with sides horizontal and vertical, $Z = \dfrac{a^3}{6}$

For the same section with horizontal diagonal, $Z = \dfrac{a^3 \sqrt{2}}{6}$

∴ Ratio of stress = $\dfrac{1}{\sqrt{2}}$. Hence maximum bending stress in the second case $= \dfrac{1}{\sqrt{2}} \sigma$.

Thus, (a) is the correct choice.

*123. In case of shaft, subjected to bending:

$$\frac{M}{I} = \frac{\sigma}{y} \text{ or } \sigma \text{ (bending stress)} = \frac{M}{I} \times r = \frac{M}{\frac{\pi}{4} r^4} \times r = \frac{4M}{\pi r^3}$$

In case of shaft, subjected to torsion; $\dfrac{T}{I_p} = \dfrac{\tau}{r} \text{ or } \dfrac{T(=M)}{\frac{\pi r^4}{2}} = \dfrac{\tau}{r}$ or τ (shear stress) $= \dfrac{2M}{\pi r^3}$

∴ Ratio of bending stress and shear stress, $\dfrac{\sigma}{\tau} = \dfrac{4M}{\pi r^3} \times \dfrac{\pi r^3}{2M} = 2$

Thus correct choice is (c).

*125. The correct choice is (a) since B.M. diagram between A and B is parabola which is possible with U.D.L. in ths region.

*126. $E = 1.25 \times 10^5$ MPa, $\dfrac{1}{m} = 0.34$

$$E = 2C\left(1 + \frac{1}{m}\right) \text{ or } 1.25 \times 10^5 = 2C\,(1 + 0.34)$$

or, $\qquad C = \dfrac{1.25 \times 10^5}{2 \times 1.34} = \mathbf{0.4664 \times 10^5\ MPa}$ **(Ans.)**

Thus, (b) is the correct choice.

*130. In case of a simply supported beam, deflection at centre with concentrated load in centre,

$$\delta = \frac{wl^3}{48\,EI} \quad i.e., \quad \delta \times \frac{1}{l}\left(\text{where, } I = \frac{bd^3}{12}\right)$$

∴ New deflection $= \delta \times \dfrac{6^3 \times 4 \times 12}{4^3 \times 6 \times 12} = 2.25\delta$

Thus, (d) correct choice.

*131. $\dfrac{T}{I_p} = \dfrac{C\theta}{l}$ or $\theta \propto \dfrac{1}{I_p}$ or $\theta \propto \dfrac{1}{D^4}$

∴ $\qquad \dfrac{0.1}{\theta} = \dfrac{(D/2)^4}{D^4}$ or $\theta = \mathbf{1.6\ radian}$ **(Ans.)**

Thus, correct choice is (c).

*132. For shaft, $\dfrac{T}{I_p} = \dfrac{\tau}{r}$ or $\tau \propto \dfrac{T}{I_p}$ (where, T = torque and I_p = polar moment of inertia).

Hence (b) is the correct choice.

*134. Let F = Shearing force at section x, then

$$F \times 2L = P \times c \text{ or } F = \frac{P \times c}{2L}$$

Thus, (d) is the correct choice.

*135. $E = 100$ GPa or 100×10^3 MPa, $R = \frac{25}{2} = 12.5$ mm $= 0.0125$ m, $t = 0.2$ mm

Now, $\frac{\sigma}{y} = \frac{E}{R}$ where $y = \frac{t}{2} = \frac{0.2}{2} = 0.1$ mm $= 0.0001$ m

$\therefore \qquad \sigma = \frac{E \times y}{R} = \frac{100 \times 10^3 \times 0.0001}{0.0125} = \mathbf{800 \ MPa}$ **(Ans.)**

Hence (d) is the correct choice.

*137. Circumferential or hoop stress, $\sigma_c = \frac{pd}{2t}$

or, $\qquad 20 = \frac{1 \times 1000}{2 \times t}$

or, $\qquad t = \frac{1 \times 1000}{20 \times 2} = 25$ mm

*138. Total strain energy stored in a simply supported beam of span, 'L' and flexural rigidity 'EI' subjected to a concentrated load 'W' at the centre.

$$= 2 \int_0^{L/2} \frac{M^2 dx}{2EI} = 2 \times \frac{1}{EI} \int_0^{L/2} \left(\frac{W}{2} \times x\right)^2 dx$$

$$\left[M = \text{Bending moment up to middle of the shaft} = \frac{W}{2} \times x \right]$$

$$= 2 \times \frac{1}{2EI} \int_0^{L/2} \left(\frac{W^2 \times x^2}{4}\right) dx = \frac{W^2}{4EI} \left[\frac{x^3}{3}\right]_0^{L/2} = \frac{W^2 L^2}{90 EI}$$

Thus, (c) is the correct choice.

*139. (c) is the correct choice since whenever a solid thick cylinder is subjected to an external hydrostatic pressure p, it is compressed equally from all sides.

*140. For a cylindrical boiler under internal pressure,

Volumetric strain $\left(\frac{\delta V}{V}\right) = 2 \times$ circumferential strain (ε_1) + longitudinal strain (ε_2)

$\therefore \qquad \delta V = (2\varepsilon_1 + \varepsilon_2)V = 2\varepsilon_1 + \varepsilon_2 \qquad (\because V = \text{volume} = 1, \text{ in this case})$

*141. For a shaft, under torsion, $\frac{T}{I_p} = \frac{\tau}{r}$

Since T (torque) and τ (shear stress) are same for hollow and solid shaft, so $\frac{I_p}{r}$ (polar modulus of section) should also be the same. Thus, (b) is the correct choice.

*142. In the assembly of pulley, key and shaft, key is made the weakest because it is easy to replace (when failure occurs) and is cheap also.

*144. For a shaft, $\frac{T}{I_p} = \frac{C\theta}{l}$ or $\theta \propto \frac{1}{I_p}$ (where, I_p = polar moment of inertia) or $\theta \propto \frac{1}{d^4}$. Therefore, the ratio of angles of twist of the two shafts, having diameters in the ratio of 2, will be 16. Thus, (d) is the correct choice.

*146. E 210 GN/m^2, $C = 80$ GN/m^2, Poisson's ratio, $\frac{1}{m} = ?$

$E = 2C\left(1 + \frac{1}{m}\right)$ or $210 = 2 \times 80\left(1 + \frac{1}{m}\right)$ or $\frac{1}{m} = \mathbf{0.31}$ **(Ans.)**

Thus, (b) is the correct choice.

* **148.** The correct value of bending moment at A (or B) is zero since due to hinge support between AB and cantilevers, the bending moment cannot be transmitted to cantilever.
   Thus, (a) is the correct choice.

* **149.** The correct choice is (d) because tensile, impact, bending and fatigue strengths are measured on universal testing, izod testing, three-point loading and rotating bending machines respectively.

* **160.** As no tensile or compressive stress exists, the Mohr's diagram is simply a circle of radius = xy and centre at the intersection of the axes.

161. Whenever shear force is zero, bending moment is constant, as depicted in Fig. 55.

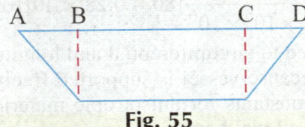

Fig. 55

163. For fixed beams, resultant B.M. diagram is the algebraic addition of fixed B.M. and free B.M. diagram.

165. When W is uniformly distributed over entire length, $\delta_1 = \dfrac{WL^3}{8EIZ}$ and W is concentrated at free end,

$$\delta_2 = \frac{WL^3}{3EI} \qquad \therefore \quad \frac{\delta_1}{\delta_2} = \frac{3}{8}$$

168. As $\dfrac{\tau}{r} = \dfrac{T}{I_p}$, or $\tau = \dfrac{T \cdot r}{I_p}$

which is a linear function and at internal radius there is a certain value of τ (shear stress).

171. Radial strain $= \dfrac{d(r+u) - dr}{dr} = \dfrac{du}{dr}$

175. Most conservative value means safest design, i.e. largest diameter, in case of maximum principal stress theory, maximum shear stress = maximum tensile stress for the material, giving it the highest value.

191. $\tau_{hollow} = \dfrac{16T}{\pi d^3 \left[1 - \left(\dfrac{d/2}{d} \right)^4 \right]} = \dfrac{16}{15} \times \dfrac{16T}{\pi d^3}$

$\tau_{solid} = \dfrac{16T}{\pi d^3} \qquad \therefore \quad \tau_{hollow} = \dfrac{16}{15} \tau_{solid}$

193. $\delta = \delta_1 + \delta_2 = \dfrac{W}{2S} + \dfrac{W}{S} = \dfrac{3W}{2S}$

Refer to Fig, 56, the maximum shear force is 2W.

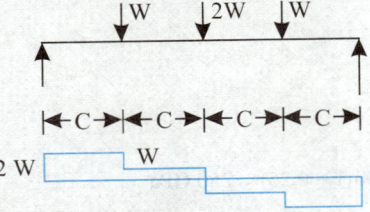

Shear force diagram
Fig. 56

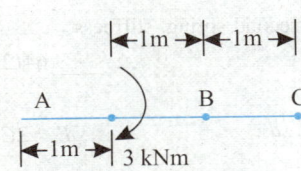

Fig. 57

213. Refer Fig. 57, which shows the equivalent diagram.

227. Expansion of rod $= l \, \alpha \, \Delta t = 0.5 \times 12.5 \times 10^{-6} \times 20 = 0.125 \times 10^{-3}$ m
   Force will be induced due to spring and same $= 0.125 \times 10^{-3} \times 50 \times 10^{3}$ N $= 6.25$ N

$$\text{Stress} = \frac{F}{A} = \frac{6.25}{(\pi/4) \times 0.01^2} = 0.07945 \ 0.07945 \text{ MPa}$$

228. **Ans.** (*a*). $(w - 10) \ 2 \times 10^6 \times 200 \times 10^6 = 2000$ N; or $w - 10 = 5$, and $w = 15$ mm.

229. **Ans.** (*a*). Circumferential strain $= \dfrac{1}{E}\left(\sigma_c - \dfrac{1}{m}\sigma_l\right)$

Since circumferential stress $\sigma_c = 80$ MPa, longitudinal stress $\sigma_l = 20$MPa

$$\therefore \text{ Circumferential strain } = \frac{1}{2 \times 10^5 \times 10^6}[80 - 0.28 \times 40]10^6 = \frac{68.8}{20} \times 10^{-4} = 3.44 \times 10^{-4}$$

230. **Ans.** (*a*). Point '*X*' is subjected to circumferential and longitudinal stress, *i.e.* tension on a all faces, but there is no shear stress because vessel is supported freely outside.

231. **Ans.** (*a*). The three elastic constants for anisotropic material following Hooke's law are Young's modulus, elastic limit stress and yield stress.

232. **Ans.** (*b*). $M = \dfrac{wL}{2}\ x - \dfrac{wx^2}{2}$ Nm

233. **Ans.** (*b*). If shear force is zero, B.M. will also be zero. If shear force varies linearly with length, B.M. diagram will be curved line.

234. **Ans.** (*b*). If D is diameter of circle and '*a*' the side of square section, $\dfrac{\pi}{4}d^2 = a^2$ or $d = \sqrt{\dfrac{4}{\pi}}\ a$

Z for circular section $= \dfrac{\pi d^3}{32} = \dfrac{\pi}{32}\dfrac{4}{\pi}a^2 = \dfrac{a^2}{8}$; and Z for square section $= \dfrac{a^3}{6}$.

Since Z for square section is more, it is stronger.

235. **Ans.** (*c*). Deflection in cantilever $= \dfrac{Wl^3}{3EI} = \dfrac{Wl^3 \times 12}{3Eah^3} = \dfrac{4Wl^3}{Eah^3}$

If *h* is doubled, and *W* is halved, new deflection $= \dfrac{4Wl^3}{2Ea(2h)^3} = \dfrac{1}{16} \times \dfrac{4Wl^3}{Eah^3}$

236. **Ans.** (*d*). Pure bending takes place in the section between two weights W.

237. **Ans.** (*b*). Shear stress $= \dfrac{16T}{\pi d^3}$ and normal stress $= \dfrac{32T}{\pi d^3}$

$\therefore$ Ratio of shear stress and normal stress $= 1 : 2$

239. **Ans.** (*a*) Strain energy $= \dfrac{1}{2} \times \text{load} \times \text{increase in length} = \dfrac{1}{2} \times P \times \dfrac{P}{A} \cdot \dfrac{L}{E} = \dfrac{P^2L}{2AE}$

240. **Ans.** (*d*) Hoop stress-tensile, radial stress-compressive, and longitudinal stress-tensile

241. **Ans.** (*d*) Longitudinal stress $= \sigma_0$; hoop stress $= 2\ \sigma_0$

Max. shear stress $= \dfrac{2\sigma_0 - \sigma_0}{2} = \dfrac{\sigma_0}{2}$

243. **Ans.** (*a*).   Stiffness $= \dfrac{Cd^4}{64R^3N} = k$

For second spring, stiffness $= \dfrac{Cd^4}{64(2R)^3 \times \dfrac{N}{2}} = \dfrac{k}{4}$

284. **Ans.** (*d*). $E = 2C\left(1 + \dfrac{1}{m}\right) = 2 \times 100(1 + 0.25) = 250$ GPa

285. **Ans.** (*b*). Elongation in both rods should be same to keep beam horizontal

$$\frac{P_S l_S}{A_S E_S} = \frac{P_A l_A}{A_A E_A}$$

$$\frac{P_S}{P_A} = \frac{1}{2} \times \frac{1}{2} \times \frac{200}{100} = \frac{1}{2}$$

289. **Ans.** (b). Deflection $\qquad \delta = \dfrac{1}{EI}\left[\dfrac{Pl^3}{2}\right]$

$\therefore$ Stiffness $\qquad = \dfrac{P}{\delta} = \dfrac{2EI}{l^3} = \dfrac{2 \times 200 \times 10^9 \times 375 \times 10^{-6}}{0.5^2} = 12 \times 10^{10}$ N/m

B. Fill in the blanks:

1. The maximum shear stres in a thin cylinder subjected to internal pressure is ....

2. The relation between *E. K* and *v* is ....

3. The point of contraflexure in a beam is a point where bending moment ....

4. The most important assumption in the theory of bending of beams is ....

5. In a concrete beam with rectangular section subjected to lateral loads the reinforcing with steel rods should be done near .... surface.

6. On a principal plane the shear stress is ....

7. The equivalent length of a Euler's column with one end fixed and the other end hinged is .....

8. In a thick cylinder subjected to external pressure the maximum hoop stress occurs at a point ....

9. The best theory of failure for ductile materials is ....

10. The stress concentration due to a small circular hole in a plate subjected to uniaxial tensile force is approximately ....

11. Shear deflection should be taken into account when the length of the beam is ....

12. In a curved beam of rectangular section the maximum stress due to bending occurs at a point on ....

13. The shear centre for the *L*-section is located at ....

14. When a helical spring is cut into two halves the stiffness of the new springs is ....

15. The degree of statical determinacy of a beam with both ends fixed is ....

16. The maximum shear stress in a rectangular bar subjected to torsion occurs at the centre of the ....

17. The fully plastic moment for rectangular beam is .... times the maximum elastic bending moment.

18. For infinite life of machine member made of steel the fatigue stress should be less than......

19. The strain energy stored in a circular shaft subject to a constant twisting moment is ....

20. The shear stress in a cross-section of a beam subjected to lateral load is maximum at ....

21. Hooke's law is valid only for.......... materials.

22. The modulus of resilence of the material is given by.............

23. The maximum shear stress on the surface of thin cylinder subjected to internal pressure is....

24. The maximum possible compressive strain in a cylindrical beam subjected to axial force is ....

25. The shear force at a point in the beam is given by the slope ....

26. The maximum bending moment in a simply supported beam subjected to uniform loading of unit length is ....

27. The point through which the resultant of shear stress of a cross-section passes is known as ....

28. The maximum shear stress in a beam with T-section occurs at ....

29. The angle between the principal plane and plane of maximum shear stress is ....

30. For a column with one end fixed and the other end free the effective length is...........

31. In a thick cylinder subjected to external pressure the maximum hoop stress occurs at ....

32. When two equal springs are joined in series the stiffness of the new spring is .... that of original spring.

33. The best theory of failure for the brittle material is ....

34. The strain energy for unit length due to transverse shear force *S* in a beam with rectangular cross-section is............

35. In a curved beam subjected to bending moment the maximum stress always occurs at ....

36. The fatigue strength of mild steel for infinite number of cycles is known as............

37. If the two principal strains at a point are 1000×10^{-6} and -400×10^{6}, then the maximum shear strain is............

38. In a built in-beam (at both the ends) carrying uniformly distributed load the maximum bending moment occurs at ....

39. Stress concentration occurs due to....... in cross-section.

40. For rectangular shafts subjected to torsion the maximum shear stress occurs at ....

41. In a thin-walled cylindrical pressure vessel subjected to internal pressure, the hoop stress is .... the longitudinal stress.

42. The bending moment is ...... at a section where the shear force changes sign.

43. In a rectangular cross-section of a beam, the ratio of the maximum shear stress to average shear stress is ....

44. In a curved beam subejcted to pure moment, the nature of variation of bending stress across the section is ....

45. An axially loaded bar is subjected to a longitudinal strain of 0.0007. The lateral strain in the rod for a Poisson's ratio of 0.35 will be ....

46. The ratio of maximum deflection of a simply supported beam with (a) a central concentrated load P and (b) a U.D.L. of P over the entire length is given by ....

47. The critical load for a long column with both ends fixed is ..... than that for a column with one end fixed and the other end hinged.

48. If the bending in a beam does not occur about a principal axis, the bending is called ....

49. If the depth of a simply supported beam of rectangular section carrying a concentrated load at the centre is halved, the deflection of the beam at the centre will increase by ....

50. Endurance limit is the maximum stress at which a material, undergoing cyclic stress will not fail in .... number of cylces.

51. In a spring the load to cause unit deflection is called .... of the spring.

52. The ductility of cast-iron is ..... than that of mild steel.

53. For a circular shaft subjected to torsional moment, the value of shear stress at the centre of cross-section is ....

54. Modulus of resilience is the maximum ....stored per unit volume of the material.

55. The central deflection of a simply supported beam when a load is suddenly applied at its centre is .... times the deflection produced when the load is gradually applied.

56. The zone in a short column section within which a compressive load applied parallel to axis will produce no tension is called the .... of the section.

57. The deflection of a beam at a point is given by the bending moment of ....

58. For a beam subjected to pure bending, the horizontal shear stress is ....

59. In a two-dimensional element, the angle between the planes where the maximum shear stress occurs is ....

60. In a thick-walled cylinder with internal pressure, the maximum hoop stress occurs at the ....

61. Percentage elongation is a measure of ....

62. Modulus of rigidity is related to modulus of elasticity through ....

63. A chalk piece, when twisted, breaks along a 45° helix because the fracture is due to ...... stress.

64. Cast-iron is more .... than mild steel.

65. Fatigue life is .... if the amplitude of cyclic loading is less than the endurance limit.

66. Modulus of rupture is equal to the area under the tensile test stress-strain diagram up to ....

67. .... stresses do not produce any distortion of a body.

68. Flange of an I-beam takes most of the .... load.

69. As per the Tresca criterion failure is defined in terms of ....stresses.

70. A steel sheet is heated uniformly. Temperatuer rise is T. The sheet is free to expand in all directions. Stress produced in the sheet is ....

71. If a moment acts at a point in a beam the bending moment diagram will be .... at the location.

72. For simply supported beam with uniformly distributed load the maximum bending stress occurs at the .... section of beam.

73. The ratio of buckling loads of fixed and pin-ended beam is ....

74. For a curved beam the maximum bending stress occurs at the ...... radius.

75. Shear stress at the corner of a rectangular shaft subjected to torsion is ....

76. The ratio of fully plastic and limit moments of a rectangular section is ....

77. Angle between the maximum principal stress and shear stress direction is ....

78. For a welded joint the design area is the area at the ....

79. Mohr's circle diagram for an element under equibiaxial tension is a ....

80. The principal of superposition is applicable when the material is .... elastic.

81. The unit of power is ....

82. The modulus of rigidity of rolled aluminium may be .... GN/m^2.

83. The torsional rigidity of a bar is the product of its .... and modulus of rigidity.

84. Elastic strain energy stored per unit volume up to elastic limit is called ....

85. Shear resistance developed on a cross-section of a beam per unit length along the section is called ....

86. The maximum shear stress theory of failure was propounded by ....

87. The range of stress in cyclic loading is given by the algebraic .... of σ_{max} and σ_{min} of the cycle.

88. Endurance ratio is the ratio of .... and endurance limit.

89. Deflection due to shear in a beam is negligible if the ratio of span and .... is large.

90. The slope of bending moment diagram for a loaded beam at a section gives the .... at that section.

91. A close-coiled helical spring, subjected to axial couple, produces a .... stress in the spring wire material.

92. A theory of failure is necessary when a component is subjected to ....

93. Lame's relations are used for evaluation of ....stress and ..... stress along the wall thickness of a thick cylinder subjected to external and internal and internal fluid pressure.

94. Deflection of a beam for a given load and span is inversely proportional to its ....

95. Maximum bending moment occurs in a beam where the shear force is ..... zero or ....

96. In a lap riveted joint the rivets are subjected to .....stress.

97. The limit up to which the stress is linearly proportional to strain is called ..... limit.

98. Principal planes and planes of maximum shear stress are inclined at .... degree to each other.

99. Bending stresses are maximum at the .... of the cross-section.

100. The shear stress in a beam of rectangular cross-section varies .... along the depth.

101. Two simply supported beams A and B having the same span, depth are subjected to same central point load, but the breadth of beam B is twice that of beam A. The central deflection of the beam B will be .... as compared to beam A.

102. The ratio between equivalent length and actual length of a column for both ends fixed is .....

103. For an angle section the shear centre is at the point of intersection of the ....

104. For a curved beam the linear stress distribution formula can be applied when the radius of curvature of the centre line of the beam to the depth of the beam ratio is greater than ....

105. The stress at which extension of a material takes place more quickly as compared to the increase in load is called ....

106. A steel rod of 15 mm diameter and 5 m long is subjected to an axial pull of 30 kN. If $E = 200 \times 10^9$ Pa, the elongation of the rod will be .... mm.

107. A cylinder is said to be thick if the ratio of its thickness and diameter is less than ....

108. The stress in the wall of a cylinder normal to its longitudinal axis due to force acting along the circumference is known as ...... stress.

109. Shear deflection of a cantilever of length L, cross-sectional area A, and shear modulus C under a concentrated load W at its free end is............

110. The energy stored in a beam of length L (EI constant) subjected to a constant bending moment M is expressed as ....

111. A shaft turning at 150 rpm is transmitting a torque at 1500 Nm. Horse power transmitted by the shaft is ....

112. Maximum shear theory at the failure of a material at elastic limit is known as ....

113. If *p* is the internal pressure in a thin-walled cylinder of diameter *d* and thickness of plates *t*, the hoop stress developed is ....

114. Slenderness ratio of a long column is the length of the column divided by ....

115. The point of contraflexure is the point where ..... changes sign.

116. The maximum stress in a cylinder subjected to internal pressure occurs at .... surface.

117. The maximum possible value of Poisson's ratio is..........

118. The normal stress assumes the maximum value on the plane on which the .... is zero.

119. A timber and a steel beam, having identical dimensions are subjected to identical loads. Then the stress in timber beam will be....... the stress in steel beam.

120. The expression for deflection of a closed coiled helical spring with mean radius *R*, diameter of wire *d* and number of coils *n* is ....

121. If the tensile stress in a specimen under uniaxial tension is 0, the value of maximum shear stress in the specimen is ....

122. The angle of twist in a shaft of length *L* and torsional rigidity *GJ*, subjected to a torque *T* is equal to ....

123. The conjugate beam method is used to obtain .... and .... in a beam.

ANSWERS

B. Fill in the Blanks :

1. $\frac{pd}{8t}$
2. $E = 3K\left(1 - \frac{2}{m}\right)$
3. is zero or changes sign
4. elastic limit is not exceeded
5. lateral
6. zero
7. $l/\sqrt{2}$
8. on the inner radius
9. maximum shear stress theory
10. 3
11. large
12. the outside surface
13. its c.g.
14. halved
15. 3
16. longer side
17. 2
18. maximum allowable stress
19. $\frac{T^2 l}{2Cl_p}$
20. the top of the surface
21. elastic
22. $\frac{\sigma^2}{2E}$
23. $\frac{pd}{8t}$
24. $\frac{\sigma}{E}$
25. M diagram
26. $\frac{wl^2}{2}$
27. c.g.
28. N.A.
29. 45°
30. 2*l*
31. inner radius
32. half
33. principal stress theory
34. $\frac{S^2}{2C}d$
35. outer surface
36. limiting fatigue range
37. 700×10^{-6}
38. centre
39. rapid change
40. $y = \pm\frac{t}{2}$
41. double
42. maximum
43. 1.5
44. hyperbola
45. 0.000245
46. 1.6
47. double
48. pure
49. eight times
50. infinite
51. stiffness
52. less
53. maximum
54. energy
55. two
56. core
57. conjugate beam

58. $\dfrac{SA\bar{y}}{Ib}$

59. 90°

60. inner radius

61. ductility of the material

62. Poisson's ratio

63. shear

64. brittle

65. infinite

66. yield point

67. Direct

68. tensile

69. maximum and minimum principal

70. zero

71. vertical line

72. outermost

73. 4

74. outer

75. zero

76. 1.5

77. 45°

78. throat of filled weld

79. point

80. perfectly

81. watt

82. 1.0

83. polar moment of inertia

84. resilience

85. shear centre

86. Tresca

87. difference

88. fatigue strength (stress)

89. depth

90. shearing force

91. shear

92. combined loading

93. circumferential, radial

94. EI

95. zero, changes sign

96. shear

97. proportional

98. 90°

99. top layer or bottom layer

100. parabolically

101. half

102. $\dfrac{1}{2}$

103. principal axes of the section and plane of the loads

104. 1

105. yield

106. 4.244

107. 20

108. hoop

109. $\dfrac{3}{10}\dfrac{Wl}{CA}$

110. $\dfrac{M^2 L}{2EI}$

111. 23.5 kW

112. Tresca

113. $\dfrac{pd}{2t}$

114. radius of gyration

115. bending moment

116. inner

117. four

118. shear stress

119. $\dfrac{E_{timber}}{E_{steel}}$

120. $\dfrac{64W R^3}{Cd^4}$

121. zero

122. $\dfrac{TL}{GJ}$

123. slope, deflection

MATERIAL TESTING
"EXPERIMENTS"

MATERIAL TESTING — GENERAL ASPECTS

A. CLASSIFICATION OF MATERIALS

The engineering materials may be classified as follows:

1. **Metals** (*e.g.*, Iron, aluminium, copper, zinc, lead, etc.)
2. **Non-metals** (Leather, rubber, asbestos, plastics, carbon, sulphur, phosphorus, timber, concrete, etc.)

Metals are further subdivided as:

(i) *Ferrous metals* (*e.g.*, cast iron, wrought iron and steel) and alloys (*e.g.*, silicon steel, high speed steel, spring steel, etc.)

(ii) *Non-ferrous metals* (copper, aluminium, zinc, lead, etc.) and alloys (brass, bronze, duralumin, etc.)

The iron group which includes all irons and steels are called *ferrous metals* whilst others are specified as *non-ferrous*.

Non-metals. The commonly adopted non-metallic materials are *leather, rubber, asbestos* and *plastics*. Leather is used for belt drives and as a packing or as washers. It is very flexible and will stand considerable wear under suitable conditions. The modulus of elasticity varies according to load. Rubber is used as a packing, belt drive and as an electric insulator. It has a high bulk modulus and must have lateral freedom if used as a packing ring. Asbestos is used for lagging round steam pipes and steam boilers.

Support roller.

Plastic is term applied to a large class of mouldable organic compounds which are sold under different trades names and are being discovered constantly.

They are used for bushing, steering wheels, tubes for oil and water, automobile tyres, etc. Plastics are divided roughly into two classes, called *thermoplastic* and *thermosetting plastics*. Materials in the former group become soft and pliable when heated to moderate temperatures and then hardened when cooled. They will soften every time when heat is applied and reworked as often as desired. *Thermosetting plastics* soften the first time they are heated, hardened when cooled and *cannot be softened by reheating*. Plastics can be moulded, cast, folded into sheets and extruded.

B. MECHANICAL PROPERTIES OF METALS

Mechanical properties of metals are discussed below:

1. Strength

The strength of metal is its ability to withstand various forces to which it is subjected during a test or in service. It is usually defined as tensile strength, compressive strength, proof stress, shear strength, etc. Strength of materials is a general expression for the measure of capacity of resistance possessed by solid masses or pieces of various kinds to any cause tending to produce in them a permanent and disabling change of form or positive fracture. Materials of all kinds owe their strength to the action of the forces residing in and about the molecules of the bodies (the molecular forces) but mainly to those ones' of these known as cohesion; certain modified results of cohesion as toughness or tenacity, hardness, stiffness and elasticity are also important elements, and strength is in relation of the toughness and stiffness combined.

S. No.	Property	Metals	Non-metals
	Differences between Metals and Non-metals		
1.	Structure	All solid metals have crystal line structure	They exist in amorphic or mesomorphic forms
2.	Excitation of valency electrons by E.M.F. (electromotive force)	Easy	Difficult
3.	State	Generally solids at room temperature (except mercury)	Gases and solids at ordinary temperature
4.	Lustre	Possess metallic lustre	Do not possess metallic lustre (except iodine and graphite)
5.	Conductivity	Good conductors of heat and electricity	Bad conductors of heat and electricity (except graphite)
6.	Malleability	Malleable	Not malleable
7.	Ductility	Ductile	Not ductile
8.	Hardness	Generally hard	Hardness varies
9.	Electrolysis	Form cations	Form anions
10.	Density	High density	Low density

2. Elasticity

A material is said to be perfectly elastic *if the whole strain produced by a load disappears completely on the removal of the load*. The modulus of elasticity or Young's modulus (E) is the proportionality constant between stress and strain for elastic materials. Young's modulus is indicative of the property called *stiffness*; small values of E indicate flexible materials and large values of E reflect stiffness and rigidity. The property of spring back is a function of modulus of elasticity and refers to the extent to which metal springs back, when an elastic deforming load is removed. In metal cutting, modulus of elasticity of the workpiece affects its rigidity, and modulus of elasticity of the cutting tool and the tool holder affect their rigidity.

Values of modulus of elasticity for some important materials are given below:

Material	Young's modulus of elasticity, E (GN/m^2)
Cast-iron	98
Wrought iron	197
Mild steel	210
Aluminium	72
Copper	120
Zinc	100
Tungsten	430
Molybdenum	350
Tin	42
Lead	18

3. Plasticity

"Plasticity" is the property that enables the formation of a permanent deformation in a material. It is the *reverse of elasticity*. A plastic material will remain exactly of the shape it takes under load, even after the load is removed. Gold and lead are highly plastic materials. Plasticity is made use of in stamping images on coins and ornamental work.

During plastic deformation there is the displacement of atoms within the metallic grains and consequently the shapes of materials components change.

It is because of this peoperty that certain synthetic materials are given the name 'Plastics'. These materials can be changed into required shape easily.

4. Ductility

It is the ability of a metal to withstand elongation or bending. Due to this property, wires are made by drawing out through a hole. The material shows a considerable amount of plasticity during the ductile extension. This is a valuable property in chains, rope, etc. because they do not snapoff, while in service, without giving sufficient warning by elongation.

Note. The elongation of a test piece of metal which occurs when it is subjected to a sufficiently high-tensile stress is a measure of the ductility of the material. The elongation is measured accurately by an extensometer, etc. The value of this test as a measure of ductility has often been questioned however. The actual property measured is the ability of the metal to distort or 'flow' without breaking, this is the same as ductility, but there is no universal agreement regarding the value of the test to the engineer. For sheet strip and wire material, a bend test is used as a substitute for elongation test. A simple bend test consists of bending the materials through 180°; the radius of the bend depending on the gauge and composition of the material under test. The sample must withstand this test without cracking. In the reverse-bend test as its name suggests, the test piece is bent thought 90° and the back again; if necessary, this reversal is repeated a specified number of times, or until the sample breaks. This test is considerably more searching than the elongation test and reveals defects which the elongation test may not show up. For tube material, a bend test is carried out on a test strip cut from the wall of the tube, while a length of the tube itself is usually flattend between two plates until its internal walls are at a specified distance apart.

5. Malleability

This is the property by virtue of which a material may be hammered or rolled into thin sheets without rupture. This property generally increases with increase of temperature.

The common metals in order of their ductility and malleability (at room temperature) are given below:

Ductility	Malleability
Gold	Gold
Silver	Silver
Platinum	Copper
Iron	Aluminium
Nickel	Tin
Copper	Platinum
Aluminium	Lead
Zinc	Zinc
Tin	Iron
Lead	Nickel

6. Tenacity or Toughness

Tenacity or toughness is the strength with which the material opposes rupture. It is due to the attraction which the molecules have for each other; giving them power to resist tearing apart.

The area under the stress-strain curve indicates the toughness (*i.e.*, energy which can be absorbed by the material up to the point of rupture.)

Although the engineering stress-strain curve is often used for this computation, a more realistic result is obtained from the *true stress-true strain curve*. Toughness is expressed as energy absorbed per unit volume of material participating in absorption. This result is obtained by multiplying the ordinate by the abscissa (in appropriate units) of the stress-strain plot.

Tractor components need different consideration from an ordinary automobile, due to the nature of work.

7. Brittleness

Lack of ductility is brittleness. When a body breaks easily when subjected to shock it is said to be brittle.

8. Hardness

Hardness is usually defined as *resistance of a material to penetration.* Hard materials resist scratches or being worn out by friction with another body.

Hardness is primarily a function of the elastic limit (*i.e.,* yield strength) of the material and to a lesser extent a function of the work hardening coefficient. The modulus of elasticity also exerts a slight effect on hardness.

In the most generally accepted tests, an indenter is pressed into the surface of the material by a slowly applied known load, and the extent of the resulting impression is measured mechanically or optically. A large impression for a given load and indentor indicates a soft material, and the opposite is true for a small impression.

The *converse of hardness is known as "softness."*

C. TESTING OF MATERIALS

Introduction

Materials are tested for one or more of the following *purposes*:

- (*i*) To assess numerically the fundamental mechanical properties of ductility, malleability, toughness, etc.
- (*ii*) To check chemical composition.
- (*iii*) To determine suitability of a material for a particular application.
- (*iv*) To determine data, *i.e.,* force deformation (or stress) values to draw up sets of specifications upon which the engineer can base this design.
- (*v*) To determine the surface or surface defects in raw materials or processed parts.

C-1. Classification of Tests

Tests on materials may be classified as

1. Non-destructive tests.
2. Destructive tests.

In *non-destructive testing* a component does not break and even after being tested, so it can be used for the purpose for which it was made.

Example: Radiography, ultrasonic inspection, etc.

In *destructive testing* the component or specimen either breaks or remains no longer useful for further use.

Example: Tensile test, impact test, torsion test, etc.

C-2. Non-destructive Test

"Non-destructive" *tests may be defined as those which in a specific context would not damage the material being examined to an extent such that it is rendered useless for future for which it was originally meant.*

Although non-destructive tests do not provide direct measurement of mechanical properties, yet they are extremely useful in revealing defects in components that could impair their performance when put in service. These tests make components more reliable, safe and economical.

The various methods used for non-destructive testings are as follows:

1. X-ray radiography
2. Gama radiography
3. Magnetic particle inspection
4. Ultrasonic testing
5. Electrical methods
6. Damping test.

C-3. Destructive Tests (Mechanical tests)

The component or specimen, after being destructively tested, either breaks or remains no longer useful for further use. Examples of destructive or mechanical tests are: tensile test, impact test, torsion test, bend test, fatigue test, etc.

C-4. Importance of Mechanical Tests

Structures, machines and products of various kinds are usually subjected to load and deformation. Therefore, the properties of material under the action of load and deformation so produced under various environments become an important engineering consideration. The microscopic properties of materials under applied forces or loads are broadly classed as *'mechanical properties'*. They are a measure of the strength and lasting characteristic of a material in service and are of great importance particular to the design engineer. Unfortunately, these properties cannot be desired from the structural or bonding considerations alone since most of them are structure-sensitive, are much more affected by crystal imperfections and other factors such as composition, grain size, heat treatment, etc. Therefore, mechanical properties do not depend on them in all situations. A great number of mechanical properties are, therefore, best evaluated by mechanical testing of the materials like metals and alloys.

The following important mechanical tests give valuable information about metals and alloys as given below:

S.No.	Name of test	Information supplied about
1.	Tensile test	Tensile strength, yield point, elastic limit, Young's modulus, ductility, toughness, etc.
2.	Impact test	Toughness of a material under shock loading conditions.
3.	Hardness test	Wear resistance, indentation resistance, scratch resistance or cutting ability of a material.
4.	Fatigue test	Behaviour of a material under repeatedly applid stress and its endurance limit.
5.	Creep test	Behaviour of a material under a steady load over a long period of time and creep limit of a material.

MATERIAL TESTING-EXPERIMENTS

Expt. No. 1. _Tensile test on a metal._

 Object: To conduct a tensile test on a mild steel specimen and determine the following:

 (_i_) Limit of proportionality   (_ii_) Elastic limit

 (_iii_) Yield strength   (_iv_) Ultimate strength

 (_v_) Young's modulus of elasticity   (_vi_) Percentage elongation

 (_vii_) Percentage reduction in area.

Expt. No. 2. _Tensile test on a wire._

 Object: To determine Young's modulus of elasticity for steel wire with Searle's apparatus.

Expt. No. 3. _Hardness test._

 Object: To conduct hardness test on mild steel, carbon steel, brass and aluminium pieces.

Expt. No. 4. _Torsion test on mild steel rod._

 Object: To conduct torsion test on mild steel or cast iron specimens to find out modulus of rigidity.

Expt. No. 5. _Impact strength of steel._

 Object: To determine the impact strength of steel by

 (_a_) Izod test   (_b_) Charpy test

Expt. No. 6. _Young's modulus of elasticity of material of a beam simply supported at ends._

 Object: To find the values of bending stresses and Young's modulus of elasticity of the material of a beam (say a wooden or steel) simply supported at the ends and carrying a concentrated load at the centre.

Expt. No. 7. _Modulus of rupture._

 Object: To determine modulus of rupture of a timber beam.

Expt. No. 8. _Shear test._

 Object: To determine the stiffness of the spring and the modulus of rigidity of the spring wire.

Expt. No. 10. _Fatigue testing._

 Object: To find the endurance limit of a metal specimen.

Expt. No. 11. _Verification of forces in a framed structure._

 Object: To verify the forces in the members of a simple roof truss.

EXPERIMENT NO. 1

Tensile Test on a Metal

Object *To conduct a tensile test on a mild steel specimen and determine the following:*

   (*i*) *Limit of proportionality*   (*ii*) *Elastic limit*

  (*iii*) *Yield strength*   (*iv*) *Ultimate strength*

   (*v*) *Young's modulus of elasticity*   (*vi*) *Percentage elongation*

  (*vii*) *Percentage reduction in area.*

Material and equipment:

   (*i*) Tensile testing machine (Fig. 1).   (*ii*) Mild steel specimens (Refer to Fig. 2).

  (*iii*) Graph paper.   (*iv*) Dividers.

   (*v*) Ruler.

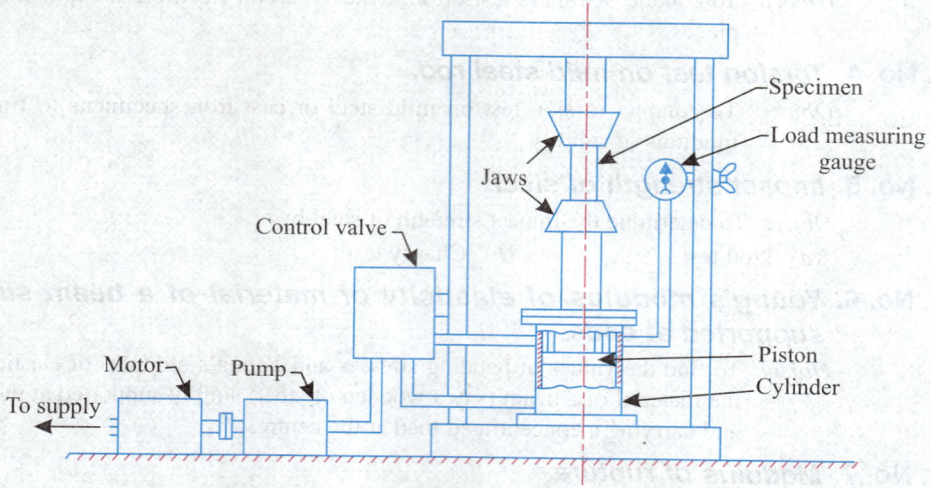

Fig. 1. Tensile testing machine.

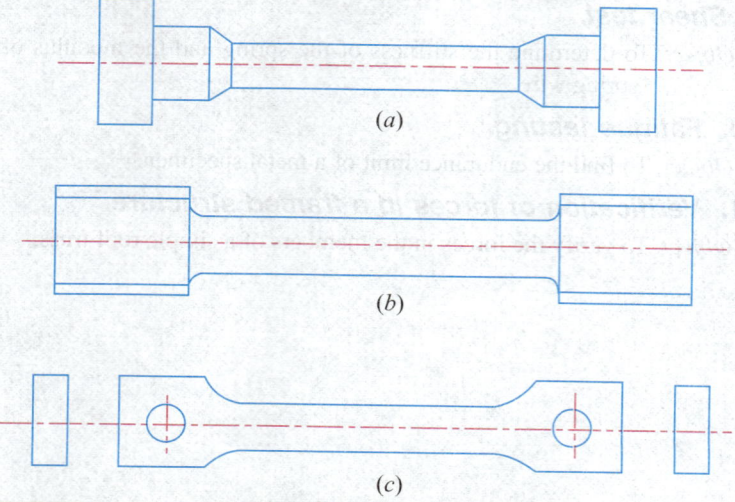

Fig. 2. Mild steel specimens.

Theory:

The tensile test is most applied one, of all mechanical tests. In this test ends of a test piece are fixed into grips connected to a straining device and to a load measuring device. If the applied load is small enough, the deformation of any solid body is entirely elastic. An elastically deformed solid will return to its original form as soon as load is removed. However, if the load is too large, the material can be deformed permanently. The initial part of the tension curve (Fig. 3), which is recoverable immediately after unloading, is termed as *elastic* and the rest of the curve, which represents the manner in which solid undergoes plastic deformation is termed *plastic*. The stress below which the deformation is essentially entirely elastic is known as the *yield strength of material*. In some materials (like mild steel) the onset of plastic deformation is denoted by a sudden drop in load indicating both an upper and a lower yield point. However, some materials do not exhibit a sharp yield point. During plastic deformation, at larger extensions strain hardening

Automated tensile testing machine.

cannot compensate for the decrease in section and thus the load passes through a maximum and then begins to decrease. As this stage the "ultimate strength", which is defined as the ratio of the load on the speciment to original cross-sectional area, reaches a maximum value. Further loading will eventually cause 'neck' formation and rupture.

Usually a tension test is conducted at room temperature and the tensile load is applid slowly. During this test either round or flat specimens (Fig. 2) may be used. The round specimens may have smooth, shouldered or threaded ends. The load on the specimen is applied mechanically or hydraulically depending on the type of testing machine. Fig. 1 shows a hydraulically operated tensile testing machine.

Procedure:

1. Measure the original length and diameter of the specimen. The length may either be length of gauge section which is marked in the specimen with a preset punch or the total length of the specimen.
2. Insert the specimen into grips of the test machine and attach strain-measuring device to it.

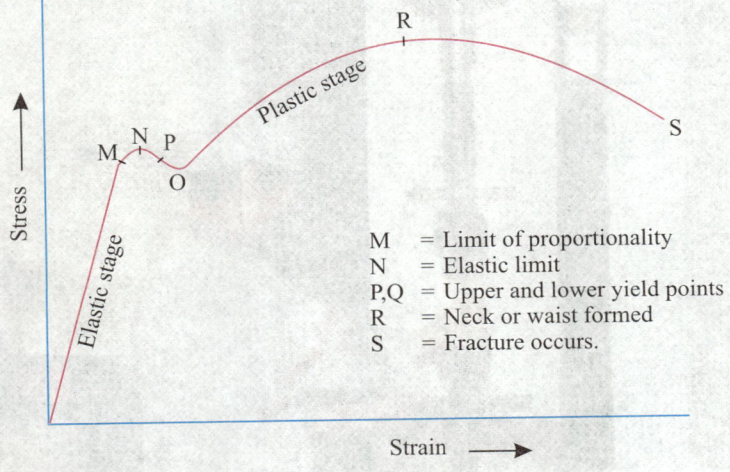

M   = Limit of proportionality
N   = Elastic limit
P,Q = Upper and lower yield points
R   = Neck or waist formed
S   = Fracture occurs.

Fig. 3. Stress-strain curve.

3. Begin the load application and record load versus elongation data.

4. Take readings more frequently as yield point is approached.

5. Measure elongation values with the help of dividers and a ruler.

6. Continue the test till fracture occurs.

7. By joining the two broken halves of the specimen together, measure the final length and diameter of specimen.

Observations:

1. Record the data in the following table:

Material:

Original dimensions: Length = ..., Diameter = ..., Area = ...

Final dimensions: Length = ..., Diameter = ..., Area = ...

S. No.	Load (N)	Extension (mm)	Original gauge length	$Stress = \dfrac{Load}{Area}$ (N/mm^2)	$Strain = \dfrac{Increase\ in\ length}{Original\ length}$
1.					
2.					
3.					
4.					
5.					
6.					
7.					

2. Plot the stress strain curve (Refer to Fig. 3) and determine the following:

(i) Limit of proportionality

$$= \frac{Load\ at\ limit\ of\ proportionality}{Original\ area\ of\ cross\text{-}section} = ...N/mm^2$$

(ii) Elastic limit:

$$= \frac{Load\ at\ elastic\ limit}{Original\ area\ of\ cross\text{-}section} = ...N/mm^2$$

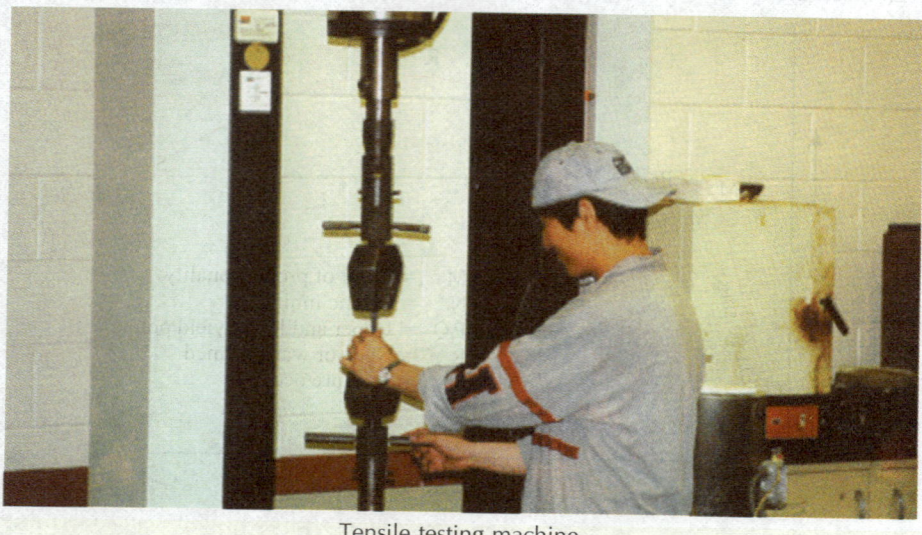

Tensile testing machine.

(*iii*) Yield strength:

$$= \frac{\text{Yield load}}{\text{Original area of cross-section}} = \text{.....N} / \text{mm}^2$$

(*iv*) Ultimate strength:

$$= \frac{\text{Maximum tensile load}}{\text{Original area of cross-section}} = \text{....N/mm}^2$$

(*v*) Young's modulus, *E*:

$$= \frac{\text{Stress below proportionality limit}}{\text{Corresponding strain}} = \text{....N/mm}^2$$

(*vi*) Percentage elongation:

$$= \frac{\text{Final length (at fracture)} - \text{original length}}{\text{Original length}} = \text{....\%}$$

(*vii*) Percentage elongation:

$$= \frac{\text{Original area} - \text{area at fracture}}{\text{Original area}} = \text{....\%.}$$

Precautions:

If the strain mesuring device is an extensometer it should be *removed* before necking begins.

EXPERIMENT NO. 2

Tensile Test on a Wire

Object. *To determine Young's modulus of elasticity for steel wire with Searle's apparatus.*

Material and equipment:

(*i*) Searle's apparatus
(*ii*) Two exactly similar wires nearly 150 cm long and about 1 mm in diameter.
(*iii*) A hanger with weights.
(*iv*) A constant weight iron slab with a hook.
(*v*) A screw gauge.
(*vi*) A metre rod.

Theory:

Fig. 4. shows Searle's aparatus generally employed for finding Young's moduls of a wire. It consists of two metal frames *M* and *M'* having two torsion heads N and N'. The frames are held together by a cross-piece *P* and are suspended from the torsion heads T_2 and T_1 by means of the wires '*y*' and '*x*' of the *same material*, *length* and *area of cross-section*. To keep the wire '*x*' taut a constant weight *W* is suspended from the hook *A* at the lower end of the frame *M'*. The frame *M* is also provided with a hook *B* at its lower end. A hanger is suspended from this hook. To cause an extension of the experimental wire '*y*' slotted weights are supplied on to the hanger *H* when desired.

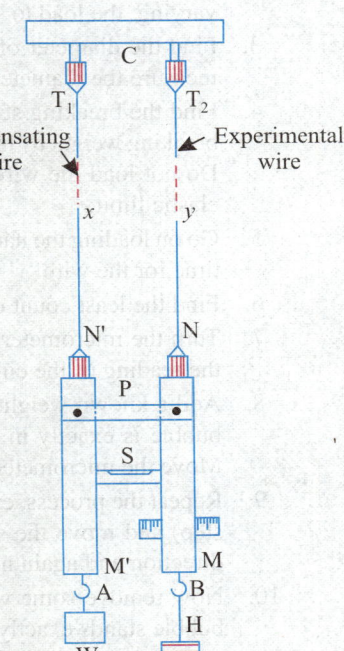

Fig. 4. Searle's apparatus.

One end of a spirit level 'S' is pivoted to the frame M' and the other rests on the tip of micrometer screw which can be worked in the frame M along a vertical scale marked in millimetres. The micrometer screw is adjusted so that the spirit level is in the horizontal position. This is so when the bubble of the spirit level stands exactly in the centre. On loading the hanger H the wire 'y' is elongated and the frame M is lowered. The micrometer screw is raised till the bubble again stands in the centre. The distance by which it is moved measures the increase in length produced in the wire 'y' due to the load added in the hanger.

If l is the length of a wire and an increase in length δl is produced by stretching force P, then

Young's modulus, $E = \dfrac{\text{Stress}}{\text{Strain}} = \dfrac{P/A}{\delta l / l} = \dfrac{P.l}{A.\delta l}$

where 'A' is the area of cross-section of wire.

Procedure:

1. To set up Searle's apparatus:

    (i) Turn the graduated disc of the micrometer screw till it stands against the zero of millimeter scales.

    (ii) Take two exactly similar steel wires 'x' and 'y' each about 150 cm long and nearly 1 mm in diameter, free from kinks. Suspend the apparatus by means of those two wires from a fixed support 'C'. Pass the wires through the torsion heads N and N' and adjust their positions till the bubble stands nearby in the centre of the spirit level S. The wire 'y' is experimental wire and 'x' is the *compensation wire, which eliminates the error due to yielding of the supports and the change in length due to changes in temperature.*

2. Suspend the constant weight slab W from the hook A to keep the wire 'x' taut and the load hanger H from the hook B. The hanger in addition to keeping the wire 'y' taut is used for varying the load to study the extensions.

3. Find the diameter of the wire with a screw gauge at six different points. At each point measure the diameter in two mutually perpendicular directions.

4. Find the breaking stress for the material of wire (from appropriate table). Calculate the breaking weight by the formula: Breaking weight = breaking stress × area of cross-section. Do not load the wire to more than 1/3 of the breaking weight to keep the wire within elastic limit.

5. Go on loading the hanger gradually with weights. After adding each weight, wait to allow time for the wire 'y' to straighten out. Load the hanger upto the calculated value.

6. Find the least count of the micrometer.

7. Turn the micrometer so that bubble of the spirit level stands exactly in the centre. Take the reading of the circular scale of the micrometer against its millimetre scale.

8. Add a known weight. Wait for about two minutes and then adjust the screw so that the bubble is exactly in the centre of the spirit level. Note the reading of the micrometer. Move the micrometer screw always in the same direction to avoid error due to backlash.

9. Repeat the process, every time increasing the load uniformly up to the load (calculated in step) and move the screw forward by about one revolution. Rotate it in the backward direction and again note the reading when the bubble is in the centre.

10. Now remove some weight. Wait for two minutes and then adjust the screw so that the bubble stands exactly in the centre of spirit level. Take the reading of micrometer screw. Similarly, unload gradually and report the observation till the load with which the observation was started is left on the hanger. The readings for the same load while loading and unloading should closely agree. If they do not, it means that the wire is not free from kinks. Similarly, load and unload again and repeat the observation.

11. Measure the length of the wire 'y' from the point of suspension to the point of attachment N is the Searle's apparatus.

Observations:

Least count of the screw gauge = ... mm.

Zero correction = ... mm.

Observed screw gauge reading

Diameter	1	2	3	4	5	6
AB						
CD						

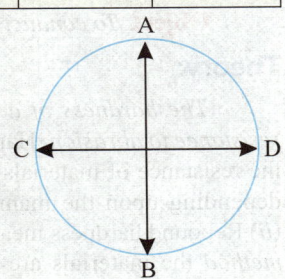

Fig. 5

Observed mean diameter = ... mm.

Correct mean diameter (d) = ... mm.

Area of cross-section = $\frac{\pi}{4} d^2$ = ... N/mm²

Breaking stress = ... N/mm²

Breaking weight = ... N

Maximum load to be applied = ... N

Maximum load actually applied = ... N

Pitch of the micrometer screw = ... mm.

Least count of Searle's apparatus = ... mm.

S. No.	Load P (N)	Micrometer reading on circular scale	Extension for 20N (say)
1	0		
2	5		
3	10		
4	15		
5	20		
6	25		
7	30		
8	35		

Note: To find the extension for 20 N (say) subtract reading against No. 1 from reading against No. 5,2 from 6,3 from 7 and so on.

Mean extension for 20N, δl = ... scale division

= .... mm.

Length of the wire, = ... mm.

Calculation : $E = \dfrac{\text{Stress}}{\text{Strain}} = \dfrac{\dfrac{P}{(\pi/4)d^2}}{\dfrac{\delta l}{l}}$ = ... N/mm²

Actual value from tables = ...

Percentage error = ...

Precautions:

1. There should be no kinks in wire.
2. The wire should be loaded and unloaded alternately.
3. The screw should always be rotated in the same direction in order to avoid error due to backlash.
4. The wire should not be loaded beyond the elastic limit.

EXPERIMENT NO. 3

Hardness Test

Object. *To conduct hardness test on mild steel, carbon steel, brass and aluminium specimens.*

Theory:

The hardness of a material is its resistance to penetration under a localised pressure or resistance to abrasion. Hardness tests provide an accurate, rapid, and economical way of determining the resistance of materials to deformation. There are three general types of hardness measurements depending upon the manner in which the test is conducted: (*i*) Scratch hardness measurement, (*ii*) Rebound hardness measurement and (*iii*) Indentation hardness measurement. In *scratch hardness method* the materials are rated on their ability to scratch one another and it is usually used by mineralogists only. In *rebound hardness measurement,* a standard body is usually dropped on to the material surface and the hardness is measured in terms of the height of its rebound. *The general means of judging the hardness is measuring the resistance of a material to indentation.* The indenter is usually a ball, cone or pyramid of a material much harder than that being used. Hardened steel, sintered tungsten carbide or diamond indenters are generally used. In Indentation tests a load is aplied by pressing the indenter at right angles to the surface being tested. The hardness of the material depends on the resistance which it exerts during a small amount of yielding or plastic straining. The resistance depends on friction, elasticity, viscosity and the intensity and distribution of plastic strain produced by a given tool during indentation. Description of *Rockwell hardness test* and *Brinell hardness test* (indentation test) is given below:

Rockwell hardness test:

The test consists in forcing an indenter of standard cone or ball into the surface of a test piece in two operations and measuring the permanent increase of depth of indentation of this indenter under specified condition. From it Rockwell hardness is deduced. The ball (*B*) is used for soft materials (*e.g.,* mild steel, cast iron, aluminium, brass, etc.) and the cone (*C*) for hard ones (High carbon steel, High speed steel, etc.)

S.No.	Rockwell 'B'	Rockwell 'C'
1.	Diameter of ball = 1.5875 mm (1/16″) (Hardened and tempered steel 65.5 HRC)	Angle of tip of the diamond cone = 120°
2.		Radius of curvature at the tip of the cone = 0.20 mm
3.	Preliminary load = 100N ± 2N	Preliminary load 100N ± 2N
4.	Additional load = 900N	Additional load = 1400N
5.	Total load = 1000N ± 6.5 N	Total load = 1500N ± 9N

HRB means Rockwell hardness measured on *B* scale
HRC means Rockwell hardness measured on C scale.

Test blocks:

(*i*) Standardized metal block shall be of a tihckness not less than 6 mm. Block, if made of steel be demagnetised.

(*ii*) The upper and lower surfaces of the blocks shall be flat with 0.005 mm and parallel in thickness such that it should not vary more than 0.010 mm per 50 mm.

(*iii*) The surfaces should be ground and polished.

Test requirements:

(*i*) The test should be carried out at an ambient temperature of $20 \pm 2°C$ in temperate climate and $27° \pm 2°C$ in tropical climates.

(*ii*) The testing machine shall be protected throughout the test from shock and vibration.

(*iii*) The test piece shall be placed on a rigid support. The contact surfaces shall be clean and free from foreign matter (such as scale, oil and dust).

(*iv*) The thickness of the test piece shall be at least 8 times the permanent indentation of depth. No deformation shall be visible at the back of the test piece after the test.

Rockwell hardness testing machine.

(*v*) The distance between the centres of the two adjacent indentations shall be at least 4 times the diameter of the indentation and the distance from the centre of any indentation to the edge of test piece shall be at least 2.5 times the diameter of the indentation unless agreed otherwise.

Procedure:

Refer to Fig. 6.

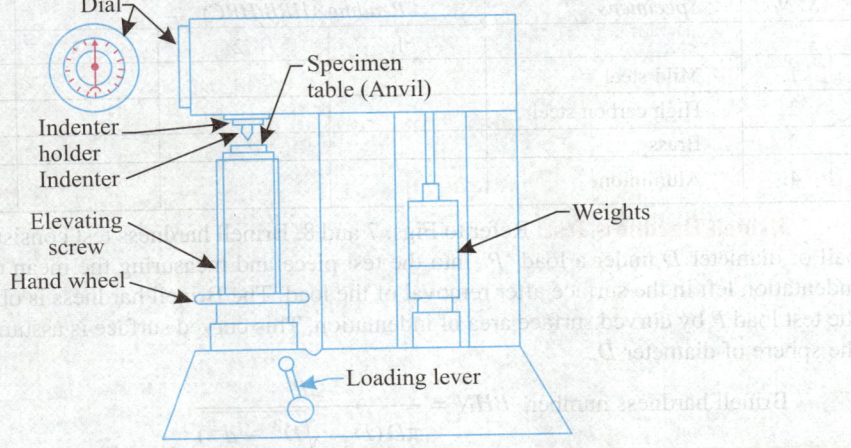

Fig. 6. Rockwell hardness tester.

1. Place the specimen securely upon the anvil.

2. Elevate the specimen so that it comes into contact with the penetrator and put the specimen under a preliminary or minor load of 100 ± 2N without shock (by moving handwheel).

3. Apply the major load 900N by loading lever.

4. Watch the pointer until it comes to rest.

5. Remove the major load.

6. Read the Rockwell hardness number or hardness scale.

Precautions:

(*i*) Successive impressions should not be superimposed on another nor be made too close together when making hardness determinations.

(*ii*) Nor should a measurement be made too close to the edge, or on a specimen so thin that impression comes through the other side.

(*iii*) Small irregularities, dirt, and scale should be avoided because of the greater sensibility of the Rockwell test.

Brinell hardness testing machine.

Observations:

S. No.	Specimens	Reading (HRB/HRC)			
		1	2	3	Mean
1.	Mild steel				HRB =
2.	High carbon steel				HRC =
3.	Brass				HRB =
4.	Aluminium				HRB =

Brinell Hardness Test: Refer to Figs. 7 and 8. Brinell hardness test consists in forcing a steel ball of diameter D under a load 'P' into the test piece and measuring the mean diameter 'd' of the indentation left in the surface after removal of the load. The Brinell hardness is obtained by dividing the test load P by curved surface area of indentation. This curved surface is assumed to be portion of the sphere of diameter D.

Brinell hardness number, $BHN = \dfrac{2P}{\pi D\,(D - \sqrt{D^2 - d^2}\,)}$.

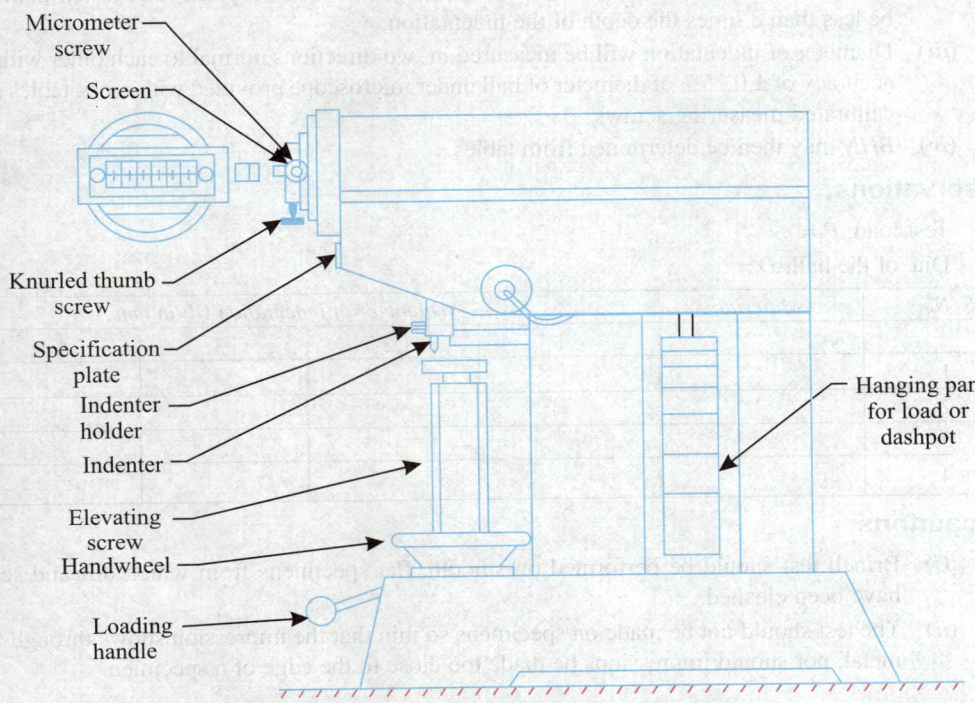

Fig. 7. Brinell hardness tester.

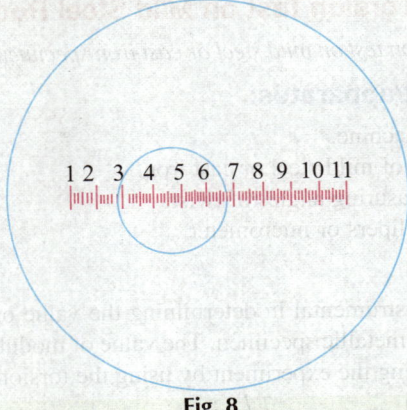

Fig. 8

Test requirements/Procedure:

(i) **Ball.** Usual ball size is 10 mm ±0.0045 mm. It shall be of hardened and tempered steel with a hardness of at least 850 VPN (Vicker's pyramid number). It shall be polished and free from surface defects.

For soft metal, load = 5000 N

For hard metal, load = 30000 N

The load shall be applied for a standard time, usually 30 seconds.

(ii) **Test piece.** Smooth and free from oxide film. Thickness of the piece to be tested shall not be less than 8 times the depth of the indentation.

(iii) Diameter of indentation will be measured in two directions normal to each other with an accuracy of ± 0.25% of diameter of ball under microscope provided with cross-tables and calibrated measuring screws.

(iv) *BHN* may then be determined from tables.

Observations:

Test load, *P* =

Dia. of the ball, *D* =

S. No.	Material	Diameter of indentation (d) in mm			
		1	2	3	4
1					
2					
3					
4					

Precautions:

(i) Brinell test should be performed on smooth, flat specimens from which dirt and scale have been cleaned.

(ii) The test should not be made on specimens so thin that the impression shows through the metal, nor should impressions be made too close to the edge of a specimen.

EXPERIMENT NO. 4

Torsion Test on Mild Steel Rod

Object: *To conduct torsion test on mild steel or cast iron specimens to find out modulus of rigidity.*

Material and equipment/apparatus:

1. A torsion testing machine.
2. Standard specimen of mild steel or cast iron.
3. Twist meter for measuring angles of twist.
4. A steel rule and callipers or micrometer.

Theory:

A torsion test is quite instrumental in determining the value of modulus of rigidity (ratio of shear stress to shear strain) of a metallic specimen. The value of modulus of rigidity can be found out through observations made during the experiment by using the torsion equation:

$$\frac{T}{I_p} = \frac{C\theta}{l}, \text{ or, } C = \frac{Tl}{I_p\theta}$$

where,
T = Torque applied,
I_p = Polar moment of inertia,
C = Modulus of rigidity,
θ = Angle of twist (radians), and
l = Gauge length.

Description:

Refer to Fig. 9. In the testing machine the ends of the specimen are held in suitable grips through one of which the torque is aplied; other is connected to the torque arm, by means of which the torque on the specimen is measured. The torque arm is attached to the indicating unit through intermediate levers housed in the cabinet. The levers are so arranged that the load indicator moves in clockwise direction, irrespective of the direction of torsion in the specimen. The chart range is selected by means of a hand lever at the front of the cabinet, indentical face plates, provided with attachment holes and a tenon slot are fitted to the straining spindle and torque arm spindle. The angular movement of the straining spindle and holder is indicated on a large diameter protractor and vernier which record deflection down to 0.1 degrees.

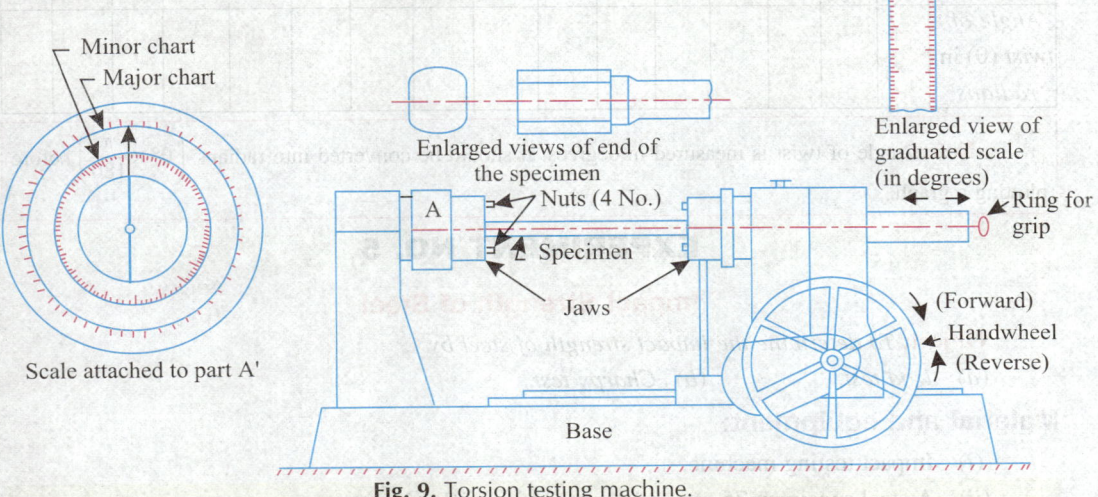

Fig. 9. Torsion testing machine.

Procedure:

1. Select the driving dogs to suit the size of the specimen and clamp it in the machine by adjusting the length of the specimen by means of a sliding spindle.
2. Measure the diameter at about three places and take the average value.
3. Choose the appropriate range by capacity. Change lever.
4. Set the maximum load pointer to zero.
5. Set the protector to zero for convenience and clamp it by means of knurled screw.
6. Carry out straining by rotating the handwheel in either direction.
7. Load the machine in suitable increments, observing and recording strain readings.
8. Then load out to failure as to cause equal increments of strain reading.
9. Plot a torque-twist ($T - \theta$) graph.
10. Read off co-ordinates of a convenient point from the straight line portion of the torque-twist ($T - \theta$) graph and calculate the value of C by using the relation:

$$C = \frac{Tl}{\theta I_p}$$

Torsion testing machine.

Observations:

Gauge length of the specimen, $l = \ldots\ldots$

Diameter of the specimen, $\quad d = \ldots$

Polar moment of inertia, $\qquad I_p = \dfrac{\pi}{32} d^4 = \ldots$

Torque (T)	1	2	3	4	5	6	7	8	9	10	11	12	13	14	15	16
Angle of twist (θ) in 'radians'																

Note. Angle of twist is measured in degrees. It should be converted into radians $\left(\theta° \times \dfrac{\pi}{180}\right)$ before plotting a graph.

EXPERIMENT NO. 5

Impact Strength of Steel

Object: *To determine the impact strength of steel by*

(*a*) *Izod test;*   (*b*) *Charpy test.*

Material and equipment:

(*i*)   Impact testing machine.

(*ii*) A steel specimen 75 mm × 10 mm × 10 mm (Izod test).

(*iii*) A steel specimen 55 mm × 10 mm × 10 mm (Charpy test).

Theory:

An impact test signifies toughness of material that is ability of material to absorb energy during plastic deformation. Static tension tests of unnotched specimens do not always reveal the susceptibility of a metal to brittle fracture. This important factor is determined by impact test. Toughness takes into account both the strength and ductility of the material. Several engineering materials have to withstand impact or suddenly applied loads while in service. *Impact strengths are generally lower as compared to strengths achieved under slowly applied loads.* Of all types of impact tests, the notched bar tests are most extensively used. Therefore, the impact test measures the energy necessary to fracture a standard notch bar by applying an impulse load. The test measures the notch toughness of material under shock loading. Values obtained from these tests are *not of much utility* to design problems directly and are highly arbitrary. Still it is important to note that it provides a good way of

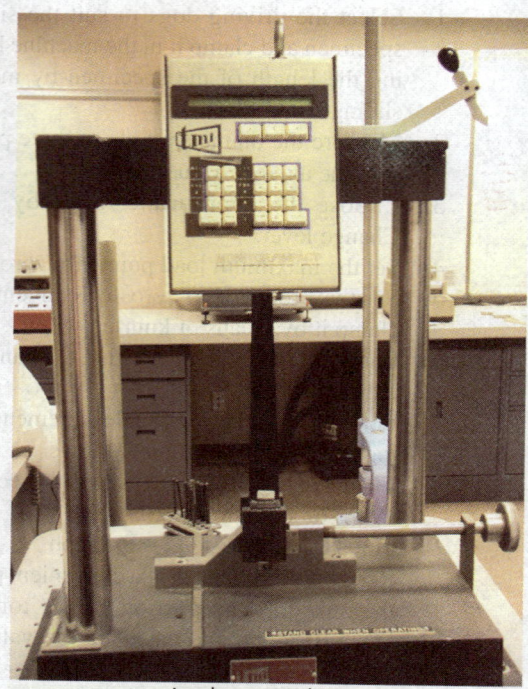

Izod test machine.

comparing toughness of various materials or toughness of the same material under different conditions. This test can also be used to *assess the ductile brittle transition temperature of the material occurring due to lowering of temperature.*

Procedure:

(*a*) Izod test:

Refer to Fig. 10.

1. With the striking hammer (pendulum) in safe rest position, firmly hold the steel specimen in impact testing machine's vice in such a way that the notch faces the hammer and is half inside and half above the top surface of the vice.

2. Bring the striking hammer to its top most striking position unless it is already there, and lock it at that position.

3. Bring indicator of the machine to zero, or follow the instructions of the operating manual supplied with the machine.

4. Release the hammer. It will fall due to gravity and break the specimen through its momentum. The total energy is not absorbed by the specimen. Then it continues to swing. At its topmost height after breaking the specimen, the indicator stops moving, while the pendulum falls back. Note the indicator reading at that topmost final position (Fig. 11)

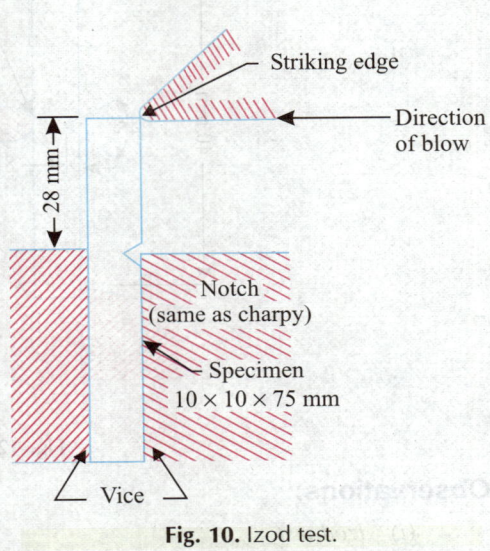

Fig. 10. Izod test.

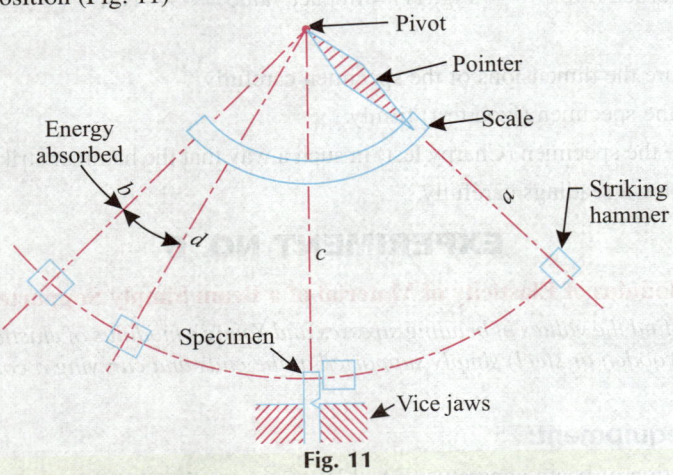

Fig. 11

5. Again bring back the hammer to its idle position and lock.

(*b*) Charpy test:

— This test is more common than Izod test and it uses *simply supported specimen* (Fig. 12) of 10 mm × 10 mm section.

— The specimen is placed on supports or anvil so that the blow of hammer is opposite to the notch.

The procedure of conducting the test is same as that given under Izod test.

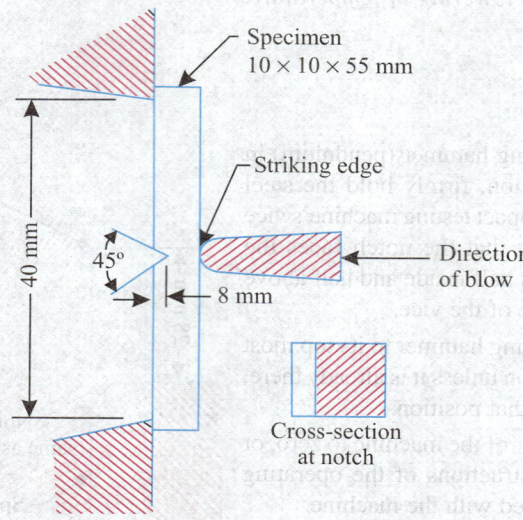

Fig. 12. Charpy test.

Observations:

    (i) *Izod test.*

Energy absorbed (bd ----- Fig. 11) = Impact value = -------- Nm

    (ii) *Charpy test.*

Energy absorbed (bd -------- Fig. 11) = Impact value = ------ Nm

Precautions:

    1. Measure the dimensions of the specimen carefully.

    2. Hold the specimen (Izod test) firmly.

    3. Locate the specimen (Charpy test) in such a way that the hammer strikes it at the middle.

    4. Note down readings carefully.

EXPERIMENT NO. 6

Young's Modulus of Elasticity of Material of a Beam Simply Supported at Ends

    Object: *To find the values of bending stresses and Young's modulus of elasticity of the material of a beam (say a wooden or steel) simply supported at the ends and carrying a concentrated load at the centre.*

Material and equipment:

    (i) Deflection of beam apparatus (Fig. 13).

    (ii) Pan.

    (iii) Weight.

    (iv) Beam of different cross-sections and materials (say wooden and steel beams).

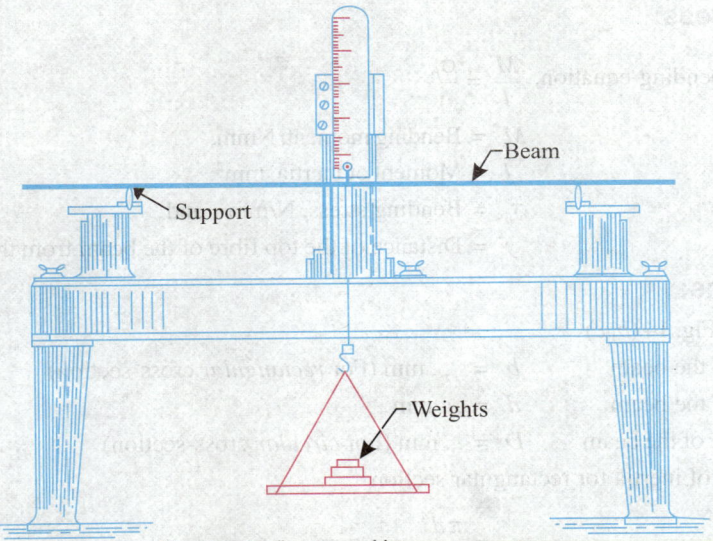

Fig. 13. Deflection of beam apparatus.

Theory: *Young's modulus of elasticity:*

If a beam is simply supported at the ends and carries a concentrated load at its centre, the beam bends concave upwards. The distance between the original position of the beam and its position after bending is different at different points [Fig. 14 (*a*)] along the length of the beam, being maximum at the centre in this case. This difference is known is '*deflection*'.

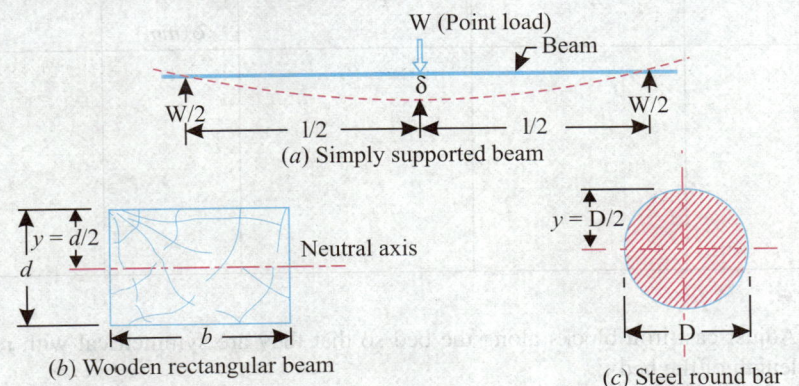

Fig. 14

In this particular type of loading the maximum amount of deflection (δ) is given by the relation,

$$\delta = \frac{Wl^3}{48\ EI} \qquad\qquad ...(i)$$

or,

$$E = \frac{Wl^3}{48\ \delta I} \qquad\qquad ...(ii)$$

where, W = Load acting at the centre, N,

l = Length of the beam between the supports, mm

E = Young's modulus of material of the beam, N/mm^2, and

I = Second moment of area of the cross –section (*i.e.*, moment of inertia) of the beam, about the neutral axis, mm^4.

Bending stress:

As per bending equation, $\dfrac{M}{I} = \dfrac{\sigma_b}{y}$

where,

M = Bending moment, Nmm,

I = Moment of inertia, mm^4,

σ_b = Bending stress, N/mm^2, and

y = Distance of the top fibre of the beam from the neutral axis.

Observations:

Refer to Fig. 14 (b, c)

Width of the beam b = ... mm (For *rectangular* cross-section)

Depth of the beam d = ... mm

Diameter of the beam D = ... mm (For *circular* cross-section)

Moment of inertia for rectangular section

$$= \frac{\pi d^4}{64} = ...\,mm^4$$

Initial reading of the vernier   = ... mm

(It should be subtracted from the reading taken after putting the load)

S.No.	Load W (N)	Bending moment $M = \dfrac{Wl}{4}$ (Nmm)	$\sigma_b = \dfrac{My}{I}$ (N/mm^2)	Deflection, δ (mm)	Young's modulus of elasticity, $E = \dfrac{Wl^3}{48\,\delta I}$
1.					
2.					
3.					
4.					
5.					
6.					

Procedure:

1. Adjust cast-iron blocks along the bed so that they are symmetrical with respect to the length of the bed.

2. Place the beam on the knife edges on the blocks so as to project equally beyond each knife edge. See that the load is applied at the centre of the beam.

3. Note the initial reading of vernier scale.

4. Add a weight of 20N (say) and again note the reading of the vernier scale.

5. Go on taking readings adding 20N (say) each time till you have minimum six readings.

6. Find the deflection (δ) in each case by subtracting the initial reading of vernier scale.

7. Draw a graph between load (W) and deflection (δ). On the graph choose any two convenient points and between these points find the corresponding values of W and δ. Putting these values in the relation, $E = \dfrac{Wl^3}{48\,\delta I}$ calculate the value of E.

8. Calculate the bending stresses for different loads using relation $\left(\sigma_b = \dfrac{My}{I} \right)$ as given in the observation table.

Repeat the experiment for different beams.

Precautions:

1. Make sure that the beam and load are placed a the proper positions.
2. The cross-section of the beam should be large.
3. Note down the readings of the vernier scale carefully.

EXPERIMENT NO. 7

Modulus of Rupture

Object: *To determine modulus of rupture of a timber beam.*

Material and equipment:

(*i*) Universal testing machine.
(*ii*) Wooden beam specimen.

Observations:

Width of the beam $b = \ldots \ldots$ mm

Depth of the beam $d = \ldots$ mm

Moment of inertia (M.O.I.) $I = \dfrac{bd^3}{12} = \ldots$ mm^4

Total length of the beam L $= \ldots$ mm

Effective length of the beam $l = \ldots \ldots$ mm

(between supports)

Load at breaking point $W = \ldots$ N

Stress at failure (modulus of rupture) $\sigma = \dfrac{M}{I} \times y \ldots$ N / mm^2

where, M = bending moment, and $y = d/2 = \ldots$ mm.

Lathe. While cutting materials, in addition to the properties of the cutting machine and tools, mechanical properties of the metal component that is being machined play important role.

Procedure:

1. Measure the dimensions of the wooden beam (*depth of the beam should be greater than width of the beam*).
2. *Fix the beam (depthwise) symmetrically on rollers.*
3. Measure the distance (centre to centre) between he roller supports.
4. Set the load indicator at zero.
5. Increase the load gradually till the wooden beam ruptures/fails.
6. Note down the load indicated by the pointer.

Precautions:

1. Measure the dimensions of the beam accurately.
2. Before starting the experiment, set the drag and load indicator at zero.
3. Apply the load gradually.

   Result: Modulus of rupture = . . . N/mm^2

EXPERIMENT NO. 8

Shear Test

Object: *To conduct shear test on specimens under double shear.*

Material and equipment:

(*i*) Universal testing machine.

(*ii*) Shear test attachment.

(*iii*) Given specimens.

Observations:

Diameter of the pin, d = . . . mm

Cross-sectional area of the pin (in double shear) = $2 \times (\pi/4) \times d^2$ = . . . mm^2

Load taken by the specimen at the time of failure, W = . . . (N)

Strength of pin against shearing = $\tau \times 2 \times (\pi/4) \times d^2$, where τ is the shear stress/strength

∴ $W = \tau \times 2 \times (\pi/4) \times d^2$

or $\tau = \dfrac{W}{2 \times (\pi/4)\,d^4}$ = . . . N/mm^2

Procedure:

1. Insert the specimen in position and grip one end of the attachment in the upper portion and one end in the lower portion.
2. Switch on the main switch of universal testing machine.
3. Bring the drag indicator in contact with the main indicator.
4. Select the suitable range of loads and place the corresponding weight in the pendulum and balance it if necessary with the help of small balancing weights.
5. Operate (push) the buttons for driving the motor to drive the pump.
6. Gradually move the head control lever in left-hand direction till the specimen shears.
7. Note down the load at which the specimen shears.
8. Stop the machine and remove the specimen.

Repeat the experiment with other specimens.

Precautions:

1. The measuring range should not be changed at any stage during the test.
2. The inner diameter of the hole in the shear stress attachment should be slightly greater than that of the specimen.
3. Measure the diameter of the specimen accurately.

Result: *Shear strength of specimen* = ... N/mm^2

EXPERIMENT NO. 9

Spring Testing

Object: *To determine the stiffness of the spring and modulus of rigidity of the spring wire.*

Material and equipment:

(*i*) Spring testing machine.

(*ii*) A spring.

(*iii*) Micrometer.

(*iv*) Vernier calliper.

Procedure:

1. By using the micrometer measure the diameter of the wire of the spring.
2. By using the vernier calliperse measure the diameter of spring coil.
3. Count the number of turns.
4. Insert the spring in the spring testing machine and load the spring by a suitable weight and note the corresponding axial deflection in tension or compression.
5. Increase the load and take the corresponding axial deflection readings.
6. Plot a curve between load and deflection. The slope of the curve gives the stiffness of the spring.

Observations:

Least count of micrometer = . . . mm

Diameter of the spring wire, d = ... mm
(mean of three readings)

Least count of vernier calliper = . . . mm

Diameter of the spring coil, D = ... mm
(mean of three readings)

Mean coil diameter, $D_m = D - d = $ . . . mm

Number of turns, n =

S.No.	Load, W (N)	Deflection, δ (mm)	Stiffness, $k = \dfrac{W}{\delta}$ (N / mm)
1			
2			
3			
4			
5			

Mean $k =$ . . .

Modulus of rigidity,

$$C = \frac{8W D_m^3 n}{\delta d^4}$$

[Spring index = $\dfrac{D_m}{d}$]

• Plot of graph between W and δ.

EXPERIMENT NO. 10

Fatigue Testing

Object: *To find the endurance limit of a metal specimen.*

Material and equipment:

(*i*)   Test pieces

(*ii*)   Rotary bending fatigue machine

(*iii*)   Dial gauge.

Theory:

Fatigue can be defined as *the failure of a material under varying loads, well below the ulti-mate static load, after a finite number of cycles of loading and unloading.* This is very frequent cause of failure of working parts of machines, and load bearing parts of aircraft structures, rockets and missiles, etc. subject to repetitive loading.

A variety of fatigue testing apparatus is available in the market. Basically it should consist of some way to produce alternating loads on the specimen, some counting arrangement for the number of load cycles, and some load measuring device. Control devices like stopping the motor once the specimen breaks and keeping the load amplitude constant, etc. may also be incorporated.

Fig. 15. shows a rotary bending fatigue machine.

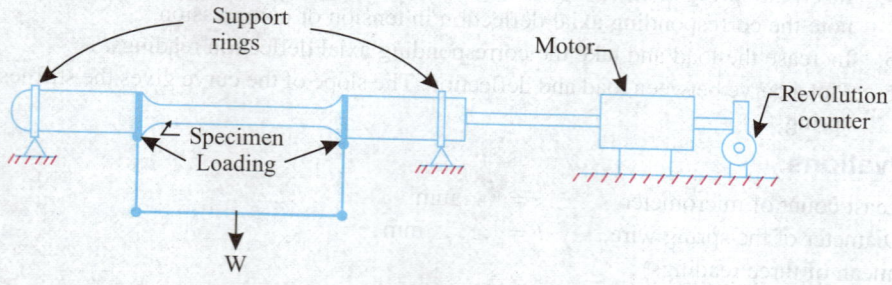

Fig. 15. Rotary bending fatigue machine.

Procedure:

1.   Insert the test piece in the bearing housing of the machine and measure its diameter.

2.   By using a dial gauge and rotating the test piece check the eccentricity which should not be generally more than 0.03 mm.

3.   Apply suitable load by adjusting the jockey weight.

4.   Set the revolution counter to zero.

5.   Start the motor of the machine and record the number of revolutions after which the specimen fails.

6.   Increase the load and test other specimen in a similar way.

7.   In each test calculate the stress (σ) applied.

8.   Plot a curve between σ and log N (where N denotes cycles).

Observations:

Diameter of the test piece, d (mm)	Load, W	Number of cycles (N)	Stress (N/mm^2)

Endurance limit (10^6 cycles) = . . .

EXPERIMENT NO. 11

Verification of Forces in a Framed Structure

Object: *To verify the forces in the members of a simple roof truss.*

Material and equipment:

(*i*)   A simple roof truss apparatus (Refer to Fig. 16).

(*ii*)   Weights (to be added).

(*iii*)   Ruler, etc.

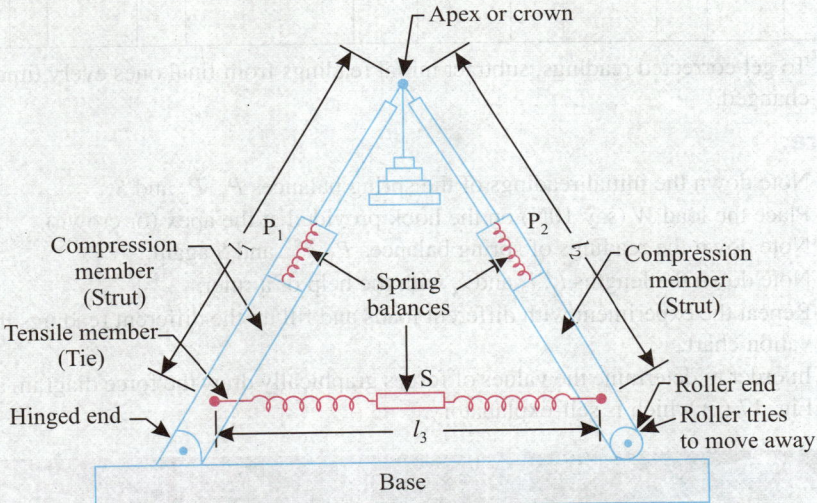

Fig. 16. Simple roof truss apparatus.

Theory:

A framed structure or truss is composed of seveal bars or rods joined together in particular fashion;these bars are called the *members* of the structure. A member under tension is called '*Tie*'and under compression is known as a '*Strut*'. The framed structures are of three types: (*i*) *Efficient or perfect*, (*ii*) *Deficient or imperfect*, and (*iii*) *Redundant*. The structure is said to be efficient if it satisfies the equation $m = (2j - 3)$ where, m and j are the number of members and number of joints of the structure respctively.

Observations:

Initial reading of spring balance $P_1 = . . . N$

Initial reading of spring balance $P_2 = . . . N$

Initial reading of spring balance $S = . . . N$

S. No.	Load at the apex	Length of members (mm)				Readings of spring balances (N)				*Corrected readings (N)		Determined graphically	
	W (N)	l_1	l_2	$\dfrac{l_1 + l_2}{2}$	l_3	P_1	P_2	$P = \dfrac{P_1 + P_2}{2}$	S	P	S	P	S

*　To get corrected readings, subtract initial readings from final ones every time the load is changed.

Procedure:

1. Note down the initial readings of the spring balances P_1, P_2 and S.
2. Place the load W (say 10N) on the hook provided at the apex (or crown).
3. Note down the readings of spring balances P_1, P_2, and S again.
4. Note down the lengths l_1, l_2 and l_3 with the help of a ruler.
5. Repeat the experiment with different loads and fill up the different readings in the observation chart.
6. In order to determine the values of forces graphically draw the force diagram as shown in Fig. 17 (v) which is self-explanatory.

An aircraft part is being tested for fatigue.

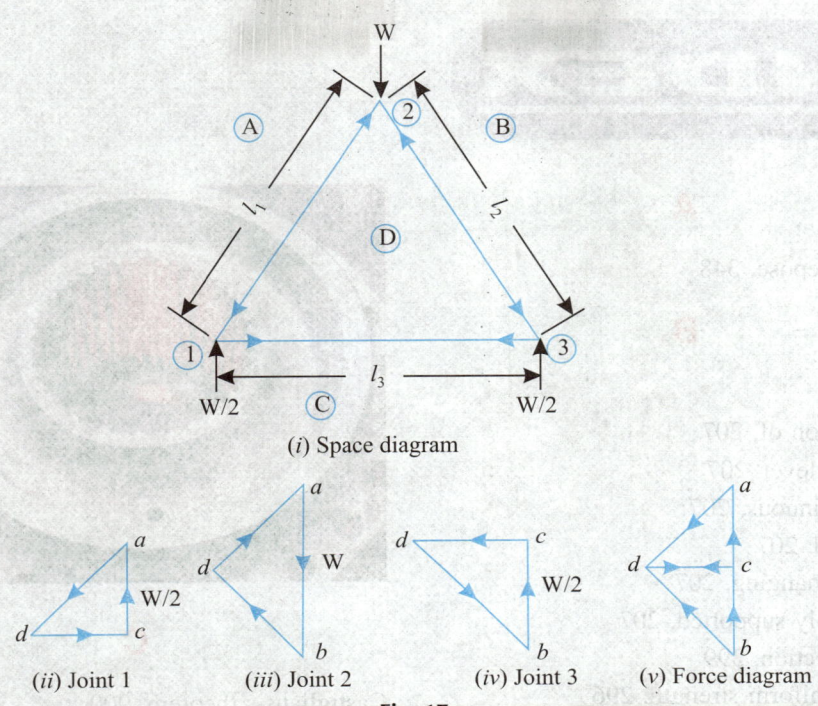

(*i*) Space diagram

(*ii*) Joint 1   (*iii*) Joint 2   (*iv*) Joint 3   (*v*) Force diagram

Fig. 17

Precautions:

1. Lubricate the parts (if not done earlier) before starting the experiment.
2. Measure the lengths accurately.
3. Note down the 'spring balances' readings carefully.

INDEX

E

F

D

M

NOTES

NOTES

NOTES

NOTES

NOTES

NOTES

NOTES

NOTES